CHILTON®

CHRYSLER
MECHANICAL SERVICE
2005 Edition

THOMSON

DELMAR LEARNING

Australia • Canada • Mexico • Singapore • Spain • United Kingdom • United States

Chilton®

Chrysler Mechanical Service

2005 Edition

Vice President, Technology and Trades SBU:

Alar Elken

Executive Director, Professional Business Unit:

Gregory L. Clayton

Publisher, Professional Business Unit:

David Koontz

Marketing Director:

Beth A. Lutz

Marketing Specialist:

Brian McGrath

Production Director:

Mary Ellen Black

Production Manager:

Larry Main

Production Editor:

Elizabeth Hough

Editorial Assistant:

Kristen Shenfield

Editors:

Terry Blomquist

Timothy A. Crain

Thomas A. Mellon

Richard J. Rivele

Christine L. Sheeky

Cover Design:

Melinda Possinger

ISBN: 1-4018-6718-9

ISSN: 1548-0887

NOTICE TO THE READER

Table of Contents

Model Index

EDITORIAL POLICY

Manufacturer and Model Coverage

This manual does not cover every DaimlerChrysler model that is currently available on the market. Rather, the Chilton editorial staff makes judicious decisions as to which models warrant coverage, based on which vehicles are serviced by most technicians.

Model Year Information

This manual is published toward the end of the year prior to the edition year. Every effort is made to gather current data from the Original Vehicle Manufacturers (OEMs) when they publish it. Different OEMs choose to release their new model information at different times of the year. Indeed, the same OEM can publish information early one season and late the next season. As a result, not all models are equally current when each edition of this manual is published.

Although information in this manual is based on industry sources and is as complete as possible at the time of publication, some vehicle manufacturers may make changes which cannot be included here. Information on very late models may not be available in some circumstances. While striving for total accuracy, the publisher cannot assume responsibility for any errors, changes, or omissions that may occur in the compilation of this data.

Safety Notice

Proper service and repair procedures are vital to the safe, reliable operation of all motor vehicles, as well as the personal safety of those performing the repairs. This manual outlines procedures for servicing and repairing vehicles using safe, effective methods. The procedures contain many NOTES, WARNINGS and CAUTIONS which should be followed along with standard safety procedures to reduce the possibility of personal injury or improper service which could damage the vehicle or compromise its safety.

Repair procedures, tools, parts, and technician skill and experience vary widely. It is not possible to anticipate all conceivable ways or conditions under which vehicles may be serviced, or to provide cautions for all possible hazards that may result. Standard and accepted safety precautions and equipment should be used when handling toxic or flammable fluids, and safety goggles or other protection should be used during cutting, grinding, chiseling, prying, or any other process that can cause material removal or projectiles.

Some procedures require the use of tools specially designed for a specific purpose. Before substituting another tool or procedure, you must be completely satisfied that neither your personal safety, nor the performance of the vehicle will be endangered.

LOCATING AND USING THE INFORMATION

Organization

To find where a particular model section or procedure is located, look in the Table of Contents. Main topics are listed with the page number on which they may be found. Following the main topics is an alphabetical listing of all of the procedures within the section and their page numbers.

Part Numbers

Part numbers listed in this book are not recommendations by the publisher for any product by brand name. They are references that can be used with interchanges manuals and aftermarket supplier catalogs to locate each brand supplier's discrete part number.

Special Tools

Special tools are recommended by the vehicle manufacturer to perform specific jobs. When necessary, special tools are referred to in the text by the part number of the tool manufacturer. These tools may be purchased, under the appropriate part number, from your local dealer or regional distributor, or an equivalent tool can be purchased locally from a tool supplier or parts outlet. Before substituting any tool for the one recommended, read the previous Safety Notice.

ACKNOWLEDGEMENT

The publisher would like to express appreciation to DaimlerChrysler for its assistance in producing this publication. No further reproduction or distribution of the material in this manual is allowed without the expressed written permission of DaimlerChrysler and the publisher.

CHRYSLER AND DODGE

300M • Concorde • LHS • Intrepid

1

SPECIFICATION CHARTS

ENGINE AND VEHICLE IDENTIFICATION

Engine							Model Year	
Code ①	Liters (cc)	Cu. In.	Cyl.	Fuel Sys.	Engine Type	Eng. Mfg.	Code ②	Year
G	3.5 (3518)	215	V6	MFI	SOHC	Chrysler	1	2001
J	3.2 (3231)	195	V6	MFI	SOHC	Chrysler	2	2002
R	2.7 (2736)	167	V6	MFI	DOHC	Chrysler	3	2003
U	2.7 (2736) ③	167	V6	MFI	DOHC	Chrysler	4	2004
V	3.5 (3518) ④	215	V6	MFI	SOHC	Chrysler	5	2005
K	3.5 (3518)	215	V6	MFI	SOHC	Chrysler		
M	3.5 (3518)	215	V6	MFI	SOHC	Chrysler		

MFI: Multi-point Fuel Injection

DOHC: Double Overhead Camshafts

SOHC: Single Overhead Camshaft

① 8th position of the Vehicle Identification Number (VIN)

② 10th position of VIN

③ 120 amp alternator

④ Magnum

67189-300M-C01

GENERAL ENGINE SPECIFICATIONS

All measurements are given in inches.

Year	Model	Engine Displacement Liters	Engine Series (ID/VIN)	Net Horsepower @ rpm	Net Torque @ rpm (ft. lbs.)	Bore x Stroke (in.)	Com-pression Ratio	Oil Pressure @ rpm
2001	300M	3.5	G	253@6400	255@3950	3.78x3.19	9.6:1	25-105@3000
	Concorde	2.7	R	200@5800	190@4850	3.39x3.09	9.6:1	25-105@3000
		2.7	U	200@5800	190@4850	3.39x3.09	9.6:1	25-105@3000
		3.5	V	253@6400	255@3950	3.78x3.19	9.6:1	25-105@3000
		3.2	J	225@6300	225@3800	3.62x3.19	9.5:1	25-105@3000
	Intrepid	2.7	R	200@5800	190@4850	3.39x3.09	9.6:1	25-105@3000
		2.7	U	200@5800	190@4850	3.39x3.09	9.6:1	25-105@3000
		3.5	V	253@6400	255@3950	3.78x3.19	9.6:1	25-105@3000
		3.2	J	225@6300	225@3800	3.62x3.19	9.5:1	25-105@3000
	LHS	3.5	G	253@6400	255@3950	3.78x3.19	9.6:1	25-105@3000
2002	300M	3.5	G	253@6400	255@3950	3.78x3.19	9.6:1	25-105@3000
	Concorde	2.7	R	200@5800	190@4850	3.39x3.09	9.6:1	25-105@3000
		2.7	U	200@5800	190@4850	3.39x3.09	9.6:1	25-105@3000
		3.5	K	232@6200	241@4000	3.78X3.19	9.6:1	25-105@3000
		3.5	M	232@6200	241@4000	3.78X3.19	9.6:1	25-105@3000
		3.5	V	253@6400	255@3950	3.78x3.19	9.6:1	25-105@3000
	Intrepid	2.7	R	200@5800	190@4850	3.39x3.09	9.6:1	25-105@3000
		2.7	U	200@5800	190@4850	3.39x3.09	9.6:1	25-105@3000
		3.5	K	232@6200	241@4000	3.78X3.19	9.6:1	25-105@3000
		3.5	M	232@6200	241@4000	3.78X3.19	9.6:1	25-105@3000
		3.5	V	253@6400	255@3950	3.78x3.19	9.6:1	25-105@3000
	Concord Limited	3.5	G	253@6400	255@3950	3.78x3.19	9.6:1	25-105@3000
2003	300M	3.5	G	253@6400	255@3950	3.78x3.19	9.6:1	25-105@3000
	Concorde	2.7	R	200@5800	190@4850	3.39x3.09	9.6:1	25-105@3000
		2.7	U	200@5800	190@4850	3.39x3.09	9.6:1	25-105@3000
		3.5	K	232@6200	241@4000	3.78X3.19	9.6:1	25-105@3000
		3.5	M	232@6200	241@4000	3.78X3.19	9.6:1	25-105@3000
		3.5	V	253@6400	255@3950	3.78x3.19	9.6:1	25-105@3000
	Intrepid	2.7	R	200@5800	190@4850	3.39x3.09	9.6:1	25-105@3000
		2.7	U	200@5800	190@4850	3.39x3.09	9.6:1	25-105@3000
		3.5	K	232@6200	241@4000	3.78X3.19	9.6:1	25-105@3000
		3.5	M	232@6200	241@4000	3.78X3.19	9.6:1	25-105@3000
		3.5	V	253@6400	255@3950	3.78x3.19	9.6:1	25-105@3000
	Concord Limited	3.5	G	253@6400	255@3950	3.78x3.19	9.6:1	25-105@3000
2004	300M	3.5	G	253@6400	255@3950	3.78x3.19	9.6:1	25-105@3000
	Concorde	2.7	R	200@5800	190@4850	3.39x3.09	9.6:1	25-105@3000
		3.5	K	232@6200	241@4000	3.78X3.19	9.6:1	25-105@3000
		3.5	M	232@6200	241@4000	3.78X3.19	9.6:1	25-105@3000
		3.5	V	253@6400	255@3950	3.78x3.19	9.6:1	25-105@3000
	Intrepid	2.7	R	200@5800	190@4850	3.39x3.09	9.6:1	25-105@3000
		3.5	K	232@6200	241@4000	3.78X3.19	9.6:1	25-105@3000
		3.5	M	232@6200	241@4000	3.78X3.19	9.6:1	25-105@3000
		3.5	V	253@6400	255@3950	3.78x3.19	9.6:1	25-105@3000
	Concord Limited	3.5	G	253@6400	255@3950	3.78x3.19	9.6:1	25-105@3000

67189-300M-C02

ENGINE TUNE-UP SPECIFICATIONS

Year	Engine Displacement Liters	Engine ID/VIN	Spark Plug Gap (in.)	Ignition Timing (deg.)	Fuel Pump (psi)	Idle Speed (rpm)	Valve Clearance	
							Intake	Exhaust
2001	2.7	R	0.048-0.053	①	58	①	HYD	HYD
	2.7	U	0.048-0.053	①	58	①	HYD	HYD
	3.5	V	0.048-0.053	①	58	①	HYD	HYD
	3.2	J	0.048-0.053	①	58	①	HYD	HYD
	3.5	G	0.048-0.053	①	58	①	HYD	HYD
2002	2.7	R	0.048-0.053	①	58	①	HYD	HYD
	2.7	U	0.048-0.053	①	58	①	HYD	HYD
	3.5	K	0.048-0.053	①	58	①	HYD	HYD
	3.5	M	0.048-0.053	①	58	①	HYD	HYD
	3.5	V	0.048-0.053	①	58	①	HYD	HYD
	3.5	G	0.048-0.053	①	58	①	HYD	HYD
2003	2.7	R	0.048-0.053	①	58	①	HYD	HYD
	2.7	U	0.048-0.053	①	58	①	HYD	HYD
	3.5	K	0.048-0.053	①	58	①	HYD	HYD
	3.5	M	0.048-0.053	①	58	①	HYD	HYD
	3.5	V	0.048-0.053	①	58	①	HYD	HYD
	3.5	G	0.048-0.053	①	58	①	HYD	HYD
2004	2.7	R	0.048-0.053	①	58	①	HYD	HYD
	3.5	K	0.048-0.053	①	58	①	HYD	HYD
	3.5	M	0.048-0.053	①	58	①	HYD	HYD
	3.5	V	0.048-0.053	①	58	①	HYD	HYD
	3.5	G	0.048-0.053	①	58	①	HYD	HYD

NOTE: The Vehicle Emission Control Information label often reflects specification changes made during production. The label figures must be used if they differ from those in this chart.

HYD: Hydraulic

① Controlled by the Powertrain Control Module (PCM)

67189-300M-C03

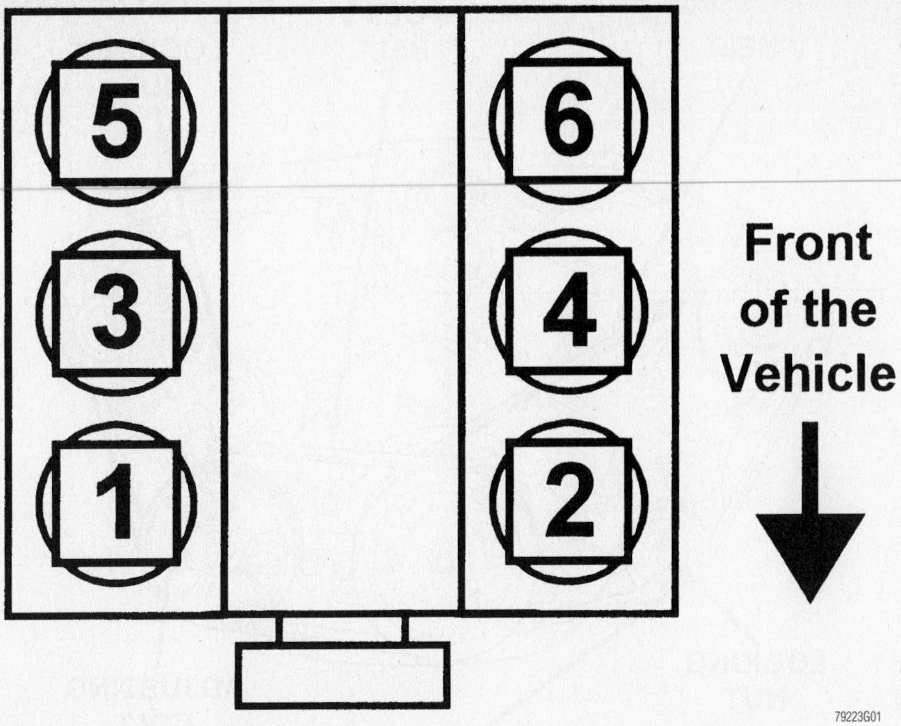

```
5        6
3        4
1        2
```

Front of the Vehicle

79223G01

3.2L and 3.5L (VIN G) Engines
Firing Order: 1–2–3–4–5–6
Distributorless ignition system (One coil per cylinder)

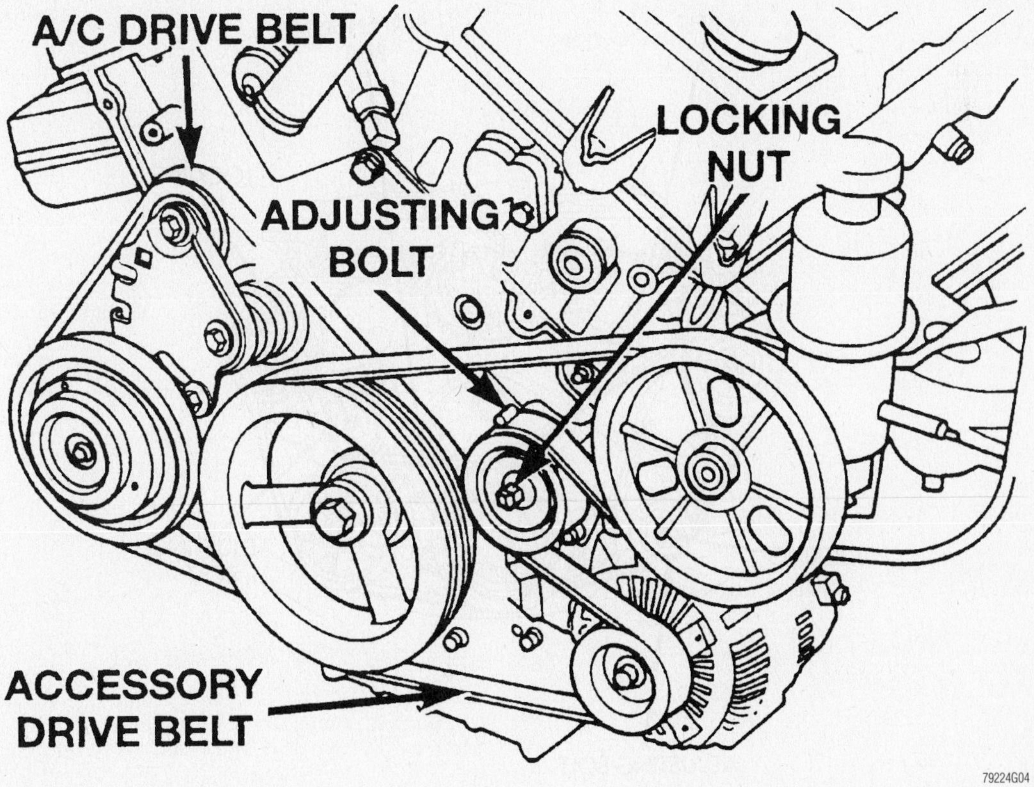

A/C DRIVE BELT

LOCKING NUT

ADJUSTING BOLT

ACCESSORY DRIVE BELT

79224G04

Accessory drive belt routing—2.7L engine

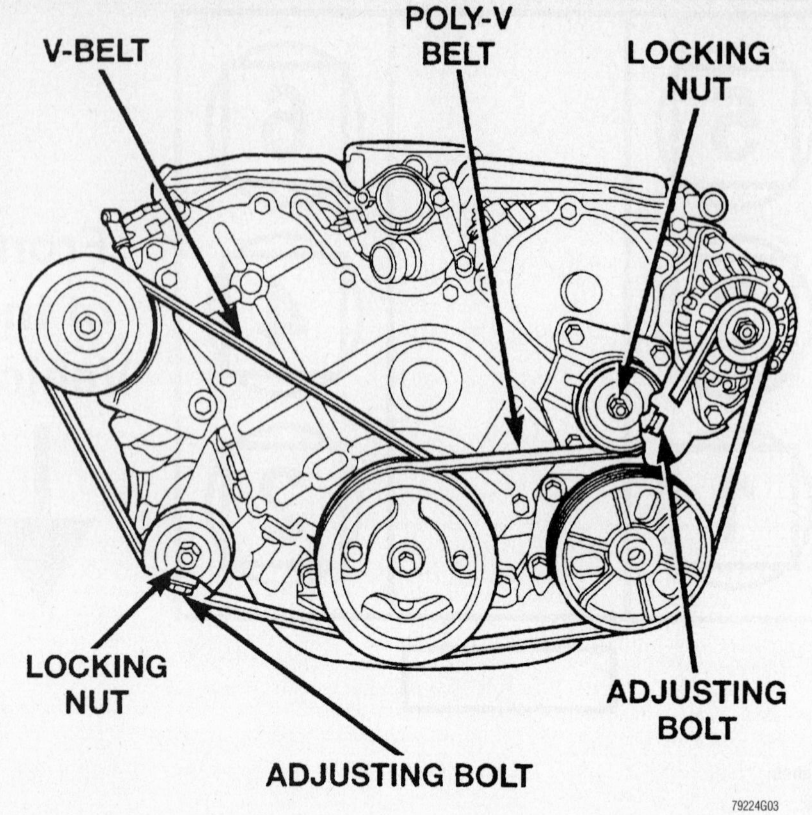

V-BELT

POLY-V BELT

LOCKING NUT

LOCKING NUT

ADJUSTING BOLT

ADJUSTING BOLT

79224G03

Accessory drive belt routing—3.2L engine

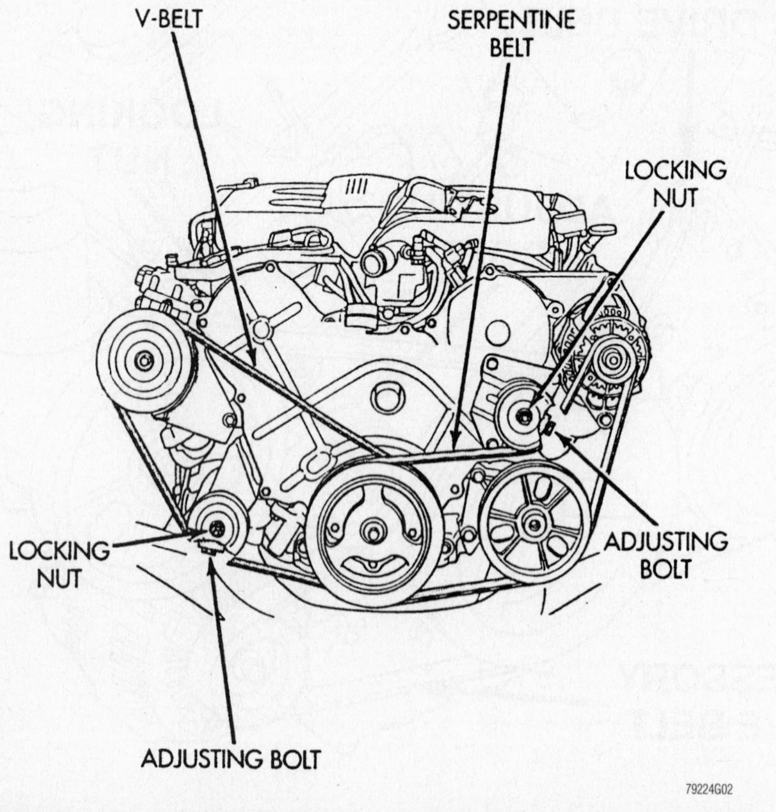

V-BELT

SERPENTINE BELT

LOCKING NUT

LOCKING NUT

ADJUSTING BOLT

ADJUSTING BOLT

79224G02

Accessory drive belt routing—3.5L engine

CAPACITIES

Year	Model	Engine Displacement Liters	Engine ID/VIN	Engine Oil with Filter (qts.)	Auto. Transmission (pts.) ①	Front Drive Axle (pts.)	Fuel Tank (gal.)	Cooling System (qts.)
2001	300M	3.5	G	5.0	19.8	2.0	18.0	9.4
	Concorde	2.7	R	5.0	19.8	2.0	18.0	9.4
		2.7	U	5.0	19.8	2.0	18.0	9.4
		3.5	V	5.0	19.8	2.0	18.0	9.4
		3.2	J	5.0	19.8	2.0	18.0	9.4
	Intrepid	2.7	R	5.0	19.8	2.0	18.0	9.4
		2.7	U	5.0	19.8	2.0	18.0	9.4
		3.5	V	5.0	19.8	2.0	18.0	9.4
		3.2	J	5.0	19.8	2.0	18.0	9.4
	LHS	3.5	G	5.0	19.8	2.0	18.0	9.4
2002	300M	3.5	G	5.0	19.8	2.0	18.0	9.4
	Concorde	2.7	R	5.0	19.8	2.0	18.0	9.4
		2.7	U	5.0	19.8	2.0	18.0	9.4
		3.5	K	5.0	19.8	2.0	18.0	9.4
		3.5	M	5.0	19.8	2.0	18.0	9.4
		3.5	V	5.0	19.8	2.0	18.0	9.4
	Intrepid	2.7	R	5.0	19.8	2.0	18.0	9.4
		2.7	U	5.0	19.8	2.0	18.0	9.4
		3.5	K	5.0	19.8	2.0	18.0	9.4
		3.5	M	5.0	19.8	2.0	18.0	9.4
		3.5	V	5.0	19.8	2.0	18.0	9.4
	Concorde Limited	3.5	G	5.0	19.8	2.0	18.0	9.4
2003	300M	3.5	G	5.0	19.8	2.0	18.0	9.4
	Concorde	2.7	R	5.0	19.8	2.0	18.0	9.4
		2.7	U	5.0	19.8	2.0	18.0	9.4
		3.5	K	5.0	19.8	2.0	18.0	9.4
		3.5	M	5.0	19.8	2.0	18.0	9.4
		3.5	V	5.0	19.8	2.0	18.0	9.4
	Intrepid	2.7	R	5.0	19.8	2.0	18.0	9.4
		2.7	U	5.0	19.8	2.0	18.0	9.4
		3.5	K	5.0	19.8	2.0	18.0	9.4
		3.5	M	5.0	19.8	2.0	18.0	9.4
		3.5	V	5.0	19.8	2.0	18.0	9.4
	Concorde Limited	3.5	G	5.0	19.8	2.0	18.0	9.4
2004	300M	3.5	G	5.0	19.8	2.0	18.0	9.4
	Concorde	2.7	R	5.0	19.8	2.0	18.0	9.4
		3.5	K	5.0	19.8	2.0	18.0	9.4
		3.5	M	5.0	19.8	2.0	18.0	9.4
		3.5	V	5.0	19.8	2.0	18.0	9.4
	Intrepid	2.7	R	5.0	19.8	2.0	18.0	9.4
		3.5	K	5.0	19.8	2.0	18.0	9.4
		3.5	M	5.0	19.8	2.0	18.0	9.4
		3.5	V	5.0	19.8	2.0	18.0	9.4
	Concorde Limited	3.5	G	5.0	19.8	2.0	18.0	9.4

NOTE: All capacities are approximate. Add fluid gradually and ensure a proper fluid level is obtained.

① Overhaul fill capacity with torque converter empty

Estimated service fill: 9 pts.

VALVE SPECIFICATIONS

Year	Engine Displacement Liters	Engine ID/VIN	Seat Angle (deg.)	Face Angle (deg.)	Spring Test Pressure (lbs. @ in.)	Spring Installed Height (in.)	Stem-to-Guide Clearance (in.)		Stem Diameter (in.)	
							Intake	Exhaust	Intake	Exhaust
2001	3.5	G	45-45.5	44.5-45	①	1.496	0.0009-0.0026	0.0020-0.0037	0.2730-0.2737	0.2719-0.2726
	3.2	J	45-45.5	44.5-45	①	1.496	0.0009-0.0026	0.0020-0.0037	0.2730-0.2737	0.2719-0.2726
	2.7	R	44.5-45	45-45.5	②	1.496	0.0009-0.0026	0.0020-0.0033	0.2337-0.2344	0.2326-0.2333
	2.7	U	44.5-45	45-45.5	②	1.496	0.0009-0.0026	0.0020-0.0033	0.2337-0.2344	0.2326-0.2333
	3.5	V	45-45.5	44.5-45	①	1.496	0.0009-0.0026	0.0020-0.0037	0.2730-0.2737	0.2719-0.2726
2002	3.5	G	45-45.5	44.5-45	①	1.496	0.0009-0.0026	0.0020-0.0037	0.2730-0.2737	0.2719-0.2726
	2.7	R	44.5-45	45-45.5	②	1.496	0.0009-0.0026	0.0020-0.0033	0.2337-0.2344	0.2326-0.2333
	2.7	U	44.5-45	45-45.5	②	1.496	0.0009-0.0026	0.0020-0.0033	0.2337-0.2344	0.2326-0.2333
	3.5	K	45-45.5	44.5-45	①	1.496	0.0009-0.0026	0.0020-0.0037	0.2730-0.2737	0.2719-0.2726
	3.5	M	45-45.5	44.5-45	①	1.496	0.0009-0.0026	0.0020-0.0037	0.2730-0.2737	0.2719-0.2726
	3.5	V	45-45.5	44.5-45	①	1.496	0.0009-0.0026	0.0020-0.0037	0.2730-0.2737	0.2719-0.2726
2003	3.5	G	45-45.5	44.5-45	①	1.496	0.0009-0.0026	0.0020-0.0037	0.2730-0.2737	0.2719-0.2726
	2.7	R	44.5-45	45-45.5	②	1.496	0.0009-0.0026	0.0020-0.0033	0.2337-0.2344	0.2326-0.2333
	2.7	U	44.5-45	45-45.5	②	1.496	0.0009-0.0026	0.0020-0.0033	0.2337-0.2344	0.2326-0.2333
	3.5	K	45-45.5	44.5-45	①	1.496	0.0009-0.0026	0.0020-0.0037	0.2730-0.2737	0.2719-0.2726
	3.5	M	45-45.5	44.5-45	①	1.496	0.0009-0.0026	0.0020-0.0037	0.2730-0.2737	0.2719-0.2726
	3.5	V	45-45.5	44.5-45	①	1.496	0.0009-0.0026	0.0020-0.0037	0.2730-0.2737	0.2719-0.2726
2004	3.5	G	45-45.5	44.5-45	①	1.496	0.0009-0.0026	0.0020-0.0037	0.2730-0.2737	0.2719-0.2726
	2.7	R	44.5-45	45-45.5	②	1.496	0.0009-0.0026	0.0020-0.0033	0.2337-0.2344	0.2326-0.2333
	3.5	K	45-45.5	44.5-45	①	1.496	0.0009-0.0026	0.0020-0.0037	0.2730-0.2737	0.2719-0.2726
	3.5	M	45-45.5	44.5-45	①	1.496	0.0009-0.0026	0.0020-0.0037	0.2730-0.2737	0.2719-0.2726
	3.5	V	45-45.5	44.5-45	①	1.496	0.0009-0.0026	0.0020-0.0037	0.2730-0.2737	0.2719-0.2726

① Intake: 69.5-80.5 lbs. @ 1.496 in. valve closed
Intake: 188.0-204.0 lbs. @ 1.1594 in. valve opened
Exhaust: 71-79 lbs. @ 1.4961 in. valve closed
Exhaust: 130-144 lbs. @ 1.239 in. valve opened

② 56-64 lbs. @ 1.496 in. valve closed
Intake: 147.9-162.1 lbs. @ 1.1417 in. valve opened
Exhaust: 138.0-150.8 lbs. @ 1.811 in. valve opened

CRANKSHAFT AND CONNECTING ROD SPECIFICATIONS
All measurements are given in inches.

Year	Engine Displacement Liters	Engine ID/VIN	Crankshaft Main Brg. Journal Dia.	Main Brg. Oil Clearance	Shaft End-play	Thrust on No.	Connecting Rod Journal Diameter	Oil Clearance	Side Clearance
2001	2.7	R	2.4997-2.5004	0.0014-0.0021	0.017 max.	3	2.1067-2.1060	0.0010-0.0026	0.0052-0.0150
	2.7	U	2.4997-2.5004	0.0014-0.0021	0.017 max.	3	2.1067-2.1060	0.0010-0.0026	0.0052-0.0150
	3.5	.V	2.5190-2.5200	0.0004-0.0022	0.0040-0.0120	2	2.2830-2.2840	0.0008-0.0034	0.0050-0.0150
	3.2	J	2.5190-2.5200	0.0007-0.0034	0.0040-0.0120	2	2.2830-2.2840	0.0008-0.0034	0.0050-0.0150
	3.5	G	2.5190-2.5200	0.0004-0.0022	0.0040-0.0120	2	2.2830-2.2840	0.0008-0.0034	0.0050-0.0150
2002	2.7	R	2.4997-2.5004	0.0014-0.0021	0.017 max.	3	2.1067-2.1060	0.0010-0.0026	0.0052-0.0150
	2.7	U	2.4997-2.5004	0.0014-0.0021	0.017 max.	3	2.1067-2.1060	0.0010-0.0026	0.0052-0.0150
	3.5	K	2.5190-2.5200	0.0004-0.0022	0.0040-0.0120	2	2.2830-2.2840	0.0008-0.0034	0.0050-0.0150
	3.5	M	2.5190-2.5200	0.0004-0.0022	0.0040-0.0120	2	2.2830-2.2840	0.0008-0.0034	0.0050-0.0150
	3.5	V	2.5190-2.5200	0.0004-0.0022	0.0040-0.0120	2	2.2830-2.2840	0.0008-0.0034	0.0050-0.0150
	3.5	G	2.5190-2.5200	0.0004-0.0022	0.0040-0.0120	2	2.2830-2.2840	0.0008-0.0034	0.0050-0.0150
2003	2.7	R	2.4997-2.5004	0.0014-0.0021	0.017 max.	3	2.1067-2.1060	0.0010-0.0026	0.0052-0.0150
	2.7	U	2.4997-2.5004	0.0014-0.0021	0.017 max.	3	2.1067-2.1060	0.0010-0.0026	0.0052-0.0150
	3.5	K	2.5190-2.5200	0.0004-0.0022	0.0040-0.0120	2	2.2830-2.2840	0.0008-0.0034	0.0050-0.0150
	3.5	M	2.5190-2.5200	0.0004-0.0022	0.0040-0.0120	2	2.2830-2.2840	0.0008-0.0034	0.0050-0.0150
	3.5	V	2.5190-2.5200	0.0004-0.0022	0.0040-0.0120	2	2.2830-2.2840	0.0008-0.0034	0.0050-0.0150
	3.5	G	2.5190-2.5200	0.0004-0.0022	0.0040-0.0120	2	2.2830-2.2840	0.0008-0.0034	0.0050-0.0150
2004	2.7	R	2.4997-2.5004	0.0014-0.0021	0.017 max.	3	2.1067-2.1060	0.0010-0.0026	0.0052-0.0150
	3.5	K	2.5190-2.5200	0.0004-0.0022	0.0040-0.0120	2	2.2830-2.2840	0.0008-0.0034	0.0050-0.0150
	3.5	M	2.5190-2.5200	0.0004-0.0022	0.0040-0.0120	2	2.2830-2.2840	0.0008-0.0034	0.0050-0.0150
	3.5	V	2.5190-2.5200	0.0004-0.0022	0.0040-0.0120	2	2.2830-2.2840	0.0008-0.0034	0.0050-0.0150
	3.5	G	2.5190-2.5200	0.0004-0.0022	0.0040-0.0120	2	2.2830-2.2840	0.0008-0.0034	0.0050-0.0150

Max: Maximum

PISTON AND RING SPECIFICATIONS
All measurements are given in inches.

Year	Engine Displacement Liters	Engine ID/VIN	Piston Clearance	Ring Gap			Ring Side Clearance		
				Top Compression	Bottom Compression	Oil Control	Top Compression	Bottom Compression	Oil Control
2001	2.7	R	0.0001-0.0016	0.008-0.014	0.0146-0.0249	0.010-0.030	0.0016-0.0031	0.0016-0.0031	0.0025-0.0082
	2.7	U	0.0003-0.0018	0.008-0.014	0.0146-0.0249	0.010-0.030	0.0016-0.0031	0.0016-0.0031	0.0025-0.0082
	3.5	V	0.0003-0.0018	0.008-0.014	0.0087-0.0193	0.010-0.030	0.0016-0.0031	0.0016-0.0031	0.0015-0.0073
	3.2	J	0.0003-0.0018	0.008-0.014	0.0087-0.0193	0.010-0.030	0.0016-0.0031	0.0016-0.0031	0.0015-0.0073
	3.5	G	0.0003-0.0018	0.008-0.014	0.0087-0.0193	0.010-0.030	0.0016-0.0031	0.0016-0.0031	0.0015-0.0073
2002	2.7	R	0.0001-0.0016	0.008-0.014	0.0146-0.0249	0.010-0.030	0.0016-0.0031	0.0016-0.0031	0.0025-0.0082
	2.7	U	0.0003-0.0018	0.008-0.014	0.0146-0.0249	0.010-0.030	0.0016-0.0031	0.0016-0.0031	0.0025-0.0082
	3.5	K	0.0003-0.0018	0.008-0.014	0.0087-0.0193	0.010-0.030	0.0016-0.0031	0.0016-0.0031	0.0015-0.0073
	3.5	M	0.0003-0.0018	0.008-0.014	0.0087-0.0193	0.010-0.030	0.0016-0.0031	0.0016-0.0031	0.0015-0.0073
	3.5	V	0.0003-0.0018	0.008-0.014	0.0087-0.0193	0.010-0.030	0.0016-0.0031	0.0016-0.0031	0.0015-0.0073
	3.5	G	0.0003-0.0018	0.008-0.014	0.0087-0.0193	0.010-0.030	0.0016-0.0031	0.0016-0.0031	0.0015-0.0073
2003	2.7	R	0.0001-0.0016	0.008-0.014	0.0146-0.0249	0.010-0.030	0.0016-0.0031	0.0016-0.0031	0.0025-0.0082
	2.7	U	0.0003-0.0018	0.008-0.014	0.0146-0.0249	0.010-0.030	0.0016-0.0031	0.0016-0.0031	0.0025-0.0082
	3.5	K	0.0003-0.0018	0.008-0.014	0.0087-0.0193	0.010-0.030	0.0016-0.0031	0.0016-0.0031	0.0015-0.0073
	3.5	M	0.0003-0.0018	0.008-0.014	0.0087-0.0193	0.010-0.030	0.0016-0.0031	0.0016-0.0031	0.0015-0.0073
	3.5	V	0.0003-0.0018	0.008-0.014	0.0087-0.0193	0.010-0.030	0.0016-0.0031	0.0016-0.0031	0.0015-0.0073
	3.5	G	0.0003-0.0018	0.008-0.014	0.0087-0.0193	0.010-0.030	0.0016-0.0031	0.0016-0.0031	0.0015-0.0073
2004	2.7	R	0.0001-0.0016	0.008-0.014	0.0146-0.0249	0.010-0.030	0.0016-0.0031	0.0016-0.0031	0.0025-0.0082
	3.5	K	0.0003-0.0018	0.008-0.014	0.0087-0.0193	0.010-0.030	0.0016-0.0031	0.0016-0.0031	0.0015-0.0073
	3.5	M	0.0003-0.0018	0.008-0.014	0.0087-0.0193	0.010-0.030	0.0016-0.0031	0.0016-0.0031	0.0015-0.0073
	3.5	V	0.0003-0.0018	0.008-0.014	0.0087-0.0193	0.010-0.030	0.0016-0.0031	0.0016-0.0031	0.0015-0.0073
	3.5	G	0.0003-0.0018	0.008-0.014	0.0087-0.0193	0.010-0.030	0.0016-0.0031	0.0016-0.0031	0.0015-0.0073

TORQUE SPECIFICATIONS
All readings in ft. lbs.

Year	Engine Displacement Liters	Engine ID/VIN	Cylinder Head Bolts	Main Bearing Bolts	Rod Bearing Bolts	Crankshaft Damper Bolts	Flywheel Bolts	Manifold Intake	Manifold Exhaust	Spark Plugs	Oil Pan Drain Plug
2001	3.5	G	①	②	③	70	70	109④	200④	20	20
	3.2	J	①	②	③	70	70	110④	200④	20	20
	2.7	R	⑤	②	⑥	125	70	111④	200④	15	20
	2.7	U	⑤	②	⑥	125	70	112④	200④	15	20
	3.5	V	①	②	③	70	70	113④	200④	20	20
2002	3.5	G	①	②	③	70	70	109④	200④	20	20
	3.5	K	①	②	③	70	70	109④	200④	20	20
	3.5	M	①	②	③	70	70	109④	200④	20	20
	2.7	R	⑤	②	⑥	125	70	111④	200④	15	20
	2.7	U	⑤	②	⑥	125	70	112④	200④	15	20
	3.5	V	①	②	③	70	70	113④	200④	20	20
2003	3.5	G	①	②	③	70	70	109④	200④	20	20
	3.5	K	①	②	③	70	70	109④	200④	20	20
	3.5	M	①	②	③	70	70	109④	200④	20	20
	2.7	R	⑤	②	⑥	125	70	111④	200④	15	20
	2.7	U	⑤	②	⑥	125	70	112④	200④	15	20
	3.5	V	①	②	③	70	70	113④	200④	20	20
2004	3.5	G	①	②	③	70	70	109④	200④	20	20
	3.5	K	①	②	③	70	70	109④	200④	20	20
	3.5	M	①	②	③	70	70	109④	200④	20	20
	2.7	R	⑤	②	⑥	125	70	111④	200④	15	20
	3.5	V	①	②	③	70	70	113④	200④	20	20

① Step 1: 45 ft. lbs.
Step 2: 65 ft. lbs.
Step 3: 65 ft. lbs.
Step 4: Plus 1/4 turn
Final torque should be over 90 ft. lbs.

② Main cap inside bolts: 15 ft. lbs. plus 1/4 turn
Main cap outside bolts: 20 ft. lbs. plus 1/4 turn
Main cap tie bolts: 250 inch lbs.

③ Step 1: 40 ft. lbs.
Step 2: Plus 1/4 turn

④ Specification is in Inch lbs.

⑤ Step 1: 35 ft. lbs.
Step 2: 55 ft. lbs.
Step 3: 55 ft. lbs.
Step 4: Plus 90 degrees
Step 5: M8 bolts (3 front) 250 inch lbs.

⑥ Step 1: 20 ft. lbs.
Step 2: Plus 1/4 turn

67189-300M-C08

WHEEL ALIGNMENT

Year	Model		Caster Range (+/-Deg.)	Caster Preferred Setting (Deg.)	Camber Range (+/-Deg.)	Camber Preferred Setting (Deg.)	Toe-in (in.)
2001	300M	F	+1.00	+3.00	+0.60	0	0.10 +/- 0.20
		R	—	—	+0.50	-0.20	0.10 +/- 0.30
	Concorde	F	+1.00	+3.00	+0.56	0	0.09 +/- 0.10
		R	—	—	+0.50	-0.20	0.09 +/- 0.15
	LHS	F	+1.00	+3.00	+0.60	0	0.09 +/- 0.10
		R	—	—	+0.50	-0.20	0.05 +/- 0.15
	Intrepid	F	+1.00	+3.00	+0.56	0	0.05 +/- 0.10
		R	—	—	+0.50	-0.20	0.05 +/- 0.15
2002	300M	F	+1.00	+3.00	+0.60	0	0.10 +/- 0.20
		R	—	—	+0.50	-0.20	0.10 +/- 0.30
	Concorde	F	+1.00	+3.00	+0.56	0	0.09 +/- 0.10
		R	—	—	+0.50	-0.20	0.09 +/- 0.15
	Concorde Limited	F	+1.00	+3.00	+0.60	0	0.09 +/- 0.10
		R	—	—	+0.50	-0.20	0.05 +/- 0.15
	Intrepid	F	+1.00	+3.00	+0.56	0	0.05 +/- 0.10
		R	—	—	+0.50	-0.20	0.05 +/- 0.15
2003	300M	F	+1.00	+3.00	+0.60	0	0.10 +/- 0.20
		R	—	—	+0.50	-0.20	0.10 +/- 0.30
	Concorde	F	+1.00	+3.00	+0.56	0	0.09 +/- 0.10
		R	—	—	+0.50	-0.20	0.09 +/- 0.15
	Concorde Limited	F	+1.00	+3.00	+0.60	0	0.09 +/- 0.10
		R	—	—	+0.50	-0.20	0.05 +/- 0.15
	Intrepid	F	+1.00	+3.00	+0.56	0	0.05 +/- 0.10
		R	—	—	+0.50	-0.20	0.05 +/- 0.15
2004	300M	F	+1.00	+3.00	+0.60	0	0.10 +/- 0.20
		R	—	—	+0.50	-0.20	0.10 +/- 0.30
	Concorde	F	+1.00	+3.00	+0.56	0	0.09 +/- 0.10
		R	—	—	+0.50	-0.20	0.09 +/- 0.15
	Concorde Limited	F	+1.00	+3.00	+0.60	0	0.09 +/- 0.10
		R	—	—	+0.50	-0.20	0.05 +/- 0.15
	Intrepid	F	+1.00	+3.00	+0.56	0	0.05 +/- 0.10
		R	—	—	+0.50	-0.20	0.05 +/- 0.15

67189-300M-C09

TIRE, WHEEL AND BALL JOINT SPECIFICATIONS

Year	Model	OEM Tires		Tire Pressures (psi)		Wheel Size	Ball Joint Inspection	Lug Nuts (ft. lbs.)
		Standard	Optional	Front	Rear			
2001	300M	P225/55/R17	None	30	30	7-JJ	①	100
	Concorde	P205/70R15	P225/60R16	32	32	6-JJ	①	100
	Intrepid	P225/55/R17	None	30	30	7-JJ	①	100
	LHS	P225/55/R17	None	30	30	7-JJ	①	100
2002	300M	P225/55/R17	None	30	30	7-JJ	①	100
	Concorde	P205/70R15	P225/60R16	32	32	6-JJ	①	100
	Concorde Limited	P225/55/R17	None	30	30	7-JJ	①	100
	Intrepid	P225/55/R17	None	30	30	7-JJ	①	100
2003	300M	P225/55/R17	None	30	30	7-JJ	①	100
	Concorde	P205/70R15	P225/60R16	32	32	6-JJ	①	100
	Concorde Limited	P225/55/R17	None	30	30	7-JJ	①	100
	Intrepid	P225/55/R17	None	30	30	7-JJ	①	100
2004	300M	P225/55/R17	None	30	30	7-JJ	①	100
	Concorde	P205/70R15	P225/60R16	32	32	6-JJ	①	100
	Concorde Limited	P225/55/R17	None	30	30	7-JJ	①	100
	Intrepid	P225/55/R17	None	30	30	7-JJ	①	100

OEM: Original Equipment Manufacturer

PSI: Pounds Per Square Inch

① Do not lift car. Grasp the grease fitting and attempt to move or rotate. Replace if any movement is found.

67189-300M-C10

BRAKE SPECIFICATIONS
All measurements in inches unless noted

Year	Model		Brake Disc Original Thickness	Brake Disc Minimum Thickness	Brake Disc Maximum Runout	Brake Drum Diameter Original Inside Diameter	Brake Drum Diameter Max. Wear Limit	Brake Drum Diameter Maximum Machine Diameter	Minimum Lining Thickness	Brake Caliper Mounting Bolts (ft. lbs.)
2001	300M	F	1.024	0.960	0.0035	—	—	—	0.250	17
		R	0.468	0.409	0.0035	—	—	—	0.280	17
	Concorde	F	1.024	0.960	0.0035	—	—	—	0.250	17
		R	0.468	0.409	0.0035	—	—	—	0.280	17
	Intrepid	F	1.024	0.960	0.0035	—	—	—	0.250	17
		R	0.468	0.409	0.0035	—	—	—	0.280	17
	LHS	F	1.024	0.960	0.0035	—	—	—	0.250	17
		R	0.468	0.409	0.0035	—	—	—	0.280	17
2002	300M	F	1.024	0.960	0.0035	—	—	—	0.250	17
		R	0.468	0.409	0.0035	—	—	—	0.280	17
	Concorde	F	1.024	0.960	0.0035	—	—	—	0.250	17
		R	0.468	0.409	0.0035	—	—	—	0.280	17
	Intrepid	F	1.024	0.960	0.0035	—	—	—	0.250	17
		R	0.468	0.409	0.0035	—	—	—	0.280	17
	Concord Ltd	F	1.024	0.960	0.0035	—	—	—	0.250	17
		R	0.468	0.409	0.0035	—	—	—	0.280	17
2003	300M	F	1.024	0.960	0.0035	—	—	—	0.250	17
		R	0.468	0.409	0.0035	—	—	—	0.280	17
	Concorde	F	1.024	0.960	0.0035	—	—	—	0.250	17
		R	0.468	0.409	0.0035	—	—	—	0.280	17
	Intrepid	F	1.024	0.960	0.0035	—	—	—	0.250	17
		R	0.468	0.409	0.0035	—	—	—	0.280	17
	Concord Ltd	F	1.024	0.960	0.0035	—	—	—	0.250	17
		R	0.468	0.409	0.0035	—	—	—	0.280	17
2004	300M	F	1.024	0.960	0.0035	—	—	—	0.250	17
		R	0.468	0.409	0.0035	—	—	—	0.280	17
	Concorde	F	1.024	0.960	0.0035	—	—	—	0.250	17
		R	0.468	0.409	0.0035	—	—	—	0.280	17
	Intrepid	F	1.024	0.960	0.0035	—	—	—	0.250	17
		R	0.468	0.409	0.0035	—	—	—	0.280	17
	Concord Ltd	F	1.024	0.960	0.0035	—	—	—	0.250	17
		R	0.468	0.409	0.0035	—	—	—	0.280	17

F: Front

R: Rear

67189-300M-C11

SCHEDULED MAINTENANCE INTERVALS
2001-02 Chrysler—300M, Concorde, LHS & Dodge—Intrepid

TO BE SERVICED	TYPE OF SERVICE	VEHICLE MILEAGE INTERVAL (x1000)												
		7.5	15	22.5	30	37.5	45	52.5	60	67.5	75	82.5	90	97.5
Engine oil & filter	R	✓	✓	✓	✓	✓	✓	✓	✓	✓	✓	✓	✓	✓
Exhaust system	S/I	✓	✓	✓	✓	✓	✓	✓	✓	✓	✓	✓	✓	✓
Brake hoses	S/I	✓	✓	✓	✓	✓	✓	✓	✓	✓	✓	✓	✓	✓
CV joints & front suspension components	S/I	✓	✓	✓	✓	✓	✓	✓	✓	✓	✓	✓	✓	✓
Rotate tires	S/I	✓	✓	✓	✓	✓	✓	✓	✓	✓	✓	✓	✓	✓
Coolant level, hoses & clamps	S/I	✓	✓	✓	✓	✓	✓	✓	✓	✓	✓	✓	✓	✓
Accessory drive belts	S/I					✓			✓		✓		✓	
Brake linings	S/I			✓				✓	✓				✓	
Spark plugs	R				✓				✓				✓	
Air filter element	R					✓							✓	
Lubricate steering linkage & tie rod ends	S/I				✓				✓				✓	
Engine coolant	R						✓				✓			
PCV valve	S/I								✓				✓	
Ignition cables	R								✓					
Camshaft timing belt	R								✓					

R: Replace   S/I: Service or Inspect

FREQUENT OPERATION MAINTENANCE (SEVERE SERVICE)

If a vehicle is operated under any of the following conditions it is considered severe service:

- **Extremely dusty areas**

- **50% or more of the vehicle operation is in 32°C (90°F) or higher temperatures, or constant operation in temperatures below 0°C (32°F)**

Prolonged idling (vehicle operation in stop and go traffic)

- **Frequent short running periods (engine does not warm to normal operating temperatures)**

- **Police, taxi, delivery usage or trailer towing usage**

CV joints & front suspension components: check every 3000 miles

Oil & oil filter change: change every 3000 miles

Rotate tires: every 3000 miles

Brake linings: check every 9000 miles

Air filter element: change every 15,000 miles

Automatic transaxle fluid: change every 15,000 miles

Differential fluid: change every 15,000 miles

Tie rod ends & steering linkage: lubricate every 15,000 miles

PCV valve: check every 30,000 miles

67189-300M-C12

SCHEDULED MAINTENANCE INTERVALS
2003-04 Chrysler—300M, Concorde, LHS & Dodge—Intrepid

TO BE SERVICED	TYPE OF SERVICE	VEHICLE MILEAGE INTERVAL (x1000)												
		6	12	18	24	30	36	42	48	54	60	66	72	78
Engine oil & filter	R	✓	✓	✓	✓	✓	✓	✓	✓	✓	✓	✓	✓	✓
Exhaust system	S/I	✓	✓	✓	✓	✓	✓	✓	✓	✓	✓	✓	✓	✓
Brake hoses	S/I	✓	✓	✓	✓	✓	✓	✓	✓	✓	✓	✓	✓	✓
CV joints & front suspension components	S/I	✓	✓	✓	✓	✓	✓	✓	✓	✓	✓	✓	✓	✓
Rotate tires	S/I	✓	✓	✓	✓	✓	✓	✓	✓	✓	✓	✓	✓	✓
Coolant level, hoses & clamps	S/I	✓	✓	✓	✓	✓	✓	✓	✓	✓	✓	✓	✓	✓
Accessory drive belts	S/I					✓			✓		✓		✓	
Brake linings	S/I			✓				✓	✓				✓	
Spark plugs	R				✓				✓				✓	
Air filter element	R					✓			✓				✓	
Lubricate steering linkage & tie rod ends	S/I				✓				✓				✓	
Engine coolant	R						✓				✓			
PCV valve	S/I								✓				✓	
Ignition cables	R								✓					
Camshaft timing belt	R								✓					

R: Replace   S/I: Service or Inspect

FREQUENT OPERATION MAINTENANCE (SEVERE SERVICE)

If a vehicle is operated under any of the following conditions it is considered severe service:

- Extremely dusty areas

- 50% or more of the vehicle operation is in 32°C (90°F) or higher temperatures, or constant operation in temperatures below 0°C (32°F)

Prolonged idling (vehicle operation in stop and go traffic)

- Frequent short running periods (engine does not warm to normal operating temperatures)

- Police, taxi, delivery usage or trailer towing usage

CV joints & front suspension components: check every 3000 miles

Oil & oil filter change: change every 3000 miles

Rotate tires: every 3000 miles

Brake linings: check every 9000 miles

Air filter element: change every 15,000 miles

Automatic transaxle fluid: change every 15,000 miles

Differential fluid: change every 15,000 miles

Tie rod ends & steering linkage: lubricate every 15,000 miles

PCV valve: check every 30,000 miles

67189-300M-C13

PRECAUTIONS

Before servicing any vehicle, please be sure to read all of the following precautions, which deal with personal safety, prevention of component damage, and important points to take into consideration when servicing a motor vehicle:

• Never open, service or drain the radiator or cooling system when the engine is hot; serious burns can occur from the steam and hot coolant.

• Observe all applicable safety precautions when working around fuel. Whenever servicing the fuel system, always work in a well-ventilated area. Do not allow fuel spray or vapors to come in contact with a spark, open flame or excessive heat (a hot drop light, for example). Keep a dry chemical fire extinguisher near the work area. Always keep fuel in a container specifically designed for fuel storage; also, always properly seal fuel containers to avoid the possibility of fire or explosion. Refer to the additional fuel system precautions later in this section.

• Fuel injection systems often remain pressurized, even after the engine has been turned **OFF**. The fuel system pressure must be relieved before disconnecting any fuel lines. Failure to do so may result in fire and/or personal injury.

• Brake fluid often contains polyglycol ethers and polyglycols. Avoid contact with the eyes and wash your hands thoroughly after handling brake fluid. If you do get brake fluid in your eyes, flush your eyes with clean, running water for 15 minutes. If eye irritation persists, or if you have taken brake fluid internally, IMMEDIATELY seek medical assistance.

• The EPA warns that prolonged contact with used engine oil may cause a number of skin disorders, including cancer! You should make every effort to minimize your exposure to used engine oil. Protective gloves should be worn when changing oil. Wash your hands and any other exposed skin areas as soon as possible after exposure to used engine oil. Soap and water, or waterless hand cleaner should be used.

• All new vehicles are now equipped with an air bag system. The system must be disabled before performing service on or around system components, steering column, instrument panel components, wiring and sensors. Failure to follow safety and disabling procedures could result in accidental air bag deployment, possible personal injury and unnecessary system repairs.

• Always wear safety goggles when working with, or around, the air bag system. When carrying a non-deployed air bag, be sure the bag and trim cover are pointed away from your body. When placing a non-deployed air bag on a work surface, always face the bag and trim cover upward, away from the surface. This will reduce the motion of the module if it is accidentally deployed. Refer to the additional air bag system precautions later in this section.

• Clean, high quality brake fluid from a sealed container is essential to the safe and proper operation of the brake system. You should always buy the correct type of brake fluid for your vehicle. If the brake fluid becomes contaminated, completely flush the system with new fluid. Never reuse any brake fluid. Any brake fluid that is removed from the system should be discarded. Also, do not allow any brake fluid to come in contact with a painted surface; it will damage the paint.

• Never operate the engine without the proper amount and type of engine oil; doing so WILL result in severe engine damage.

• Timing belt maintenance is extremely important! Many models utilize an interference-type, non-freewheeling engine. If the timing belt breaks, the valves in the cylinder head may strike the pistons, causing potentially serious (also time-consuming and expensive) engine damage.

• Disconnecting the negative battery cable on some vehicles may interfere with the functions of the on-board computer system(s) and may require the computer to undergo a relearning process once the negative battery cable is reconnected.

• When servicing drum brakes, only disassemble and assemble one side at a time, leaving the remaining side intact for reference.

• Only an MVAC-trained, EPA-certified automotive technician should service the air conditioning system or its components.

ENGINE REPAIR

Alternator

REMOVAL

2.7L Engine

1. Before servicing the vehicle, refer to the precautions in the beginning of this section.
2. Remove or disconnect the following:
   • Negative battery cable
   • Lower plastic splash shield
   • Transmission cooler and position it aside
   • Lower radiator crossmember support
3. Loosen but do not remove:
   • Adjusting "T" bolt and the pivot bolt
   • Drive belt adjusting bolt
4. Remove or disconnect the following:

• Drive belt
• Alternator field circuit plug
• Alternator B+ terminal nut and wire
• Pivot bolt; be careful not to lose the spacer
• Alternator

3.2L and 3.5L Engines

1. Before servicing the vehicle, refer to the precautions in the beginning of this section.
2. Remove or disconnect the following:
   • Negative battery cable
   • Lower mounting bolt and pivot bolt
   • Drive belt tension bolt
   • Drive belt
   • Bracket, lower mounting bolt and pivot bolt
   • Alternator
   • Alternator field circuit plug
   • Alternator B+ terminal nut and wire

INSTALLATION

2.7L Engine

1. Install or connect the following:
   • Alternator
   • Pivot bolt with the spacer and leave loose
   • Alternator field circuit plug
   • Alternator B+ wire and terminal nut; then, torque the nut to 89 inch lbs. (10 Nm)
   • Drive belt
2. Using a belt tension gauge, torque the adjusting "T" bolt until the tension gauge reads 120 lbs. (534 N) for a used belt or 180–200 lbs. (792–880 N) for a new belt.

   • Lower radiator crossmember support
   • Transmission cooler

- Lower plastic splash shield
- Negative battery cable

3. Torque the bolts to the following specifications:

    a. Pivot bolt to 40 ft. lbs. (54 Nm).

    b. 8mm mounting bolt to 30 ft. lbs. (41 Nm).

    c. 10mm mounting bolt to 40 ft. lbs. (54 Nm).

3.2L and 3.5L Engines

1. Install or connect the following:

- Alternator B+ wire and terminal nut; then, torque the nut to 89 inch lbs. (10 Nm)
- Alternator field circuit plug
- Alternator
- Drive belt
- Pivot bolt and lower mounting bolt

2. Using a belt tension gauge, torque the "V" drive belt until the tension gauge reads 120 lbs. (534 N) for a used belt or 140–160 lbs. (623–711 N) for a new belt.

3. Using a belt tension gauge, torque the Poly "V" belt until the tension gauge reads 120 lbs. (534 N) for a used belt or 180–200 lbs. (792–880 N) for a new belt.

4. Install the bracket.

5. Torque the bolts to the following specifications:

    a. Pivot bolt to 40 ft. lbs. (54 Nm).

    b. 8mm mounting bolt to 30 ft. lbs. (41 Nm).

    c. 10mm mounting bolt to 40 ft. lbs. (54 Nm).

6. Connect the negative battery cable.

Ignition Timing

ADJUSTMENT

All models utilize a Distributorless Ignition System (DIS). It is a fixed ignition timing system, which means that basic ignition timing cannot be adjusted. All spark advance is permanently set by the Powertrain Control Module (PCM).

Engine Assembly

REMOVAL & INSTALLATION

2.7L Engine

1. Before servicing the vehicle, refer to the precautions in the beginning of this section.

2. Relieve the fuel system pressure.

3. Drain the cooling system.

4. Drain the engine oil.

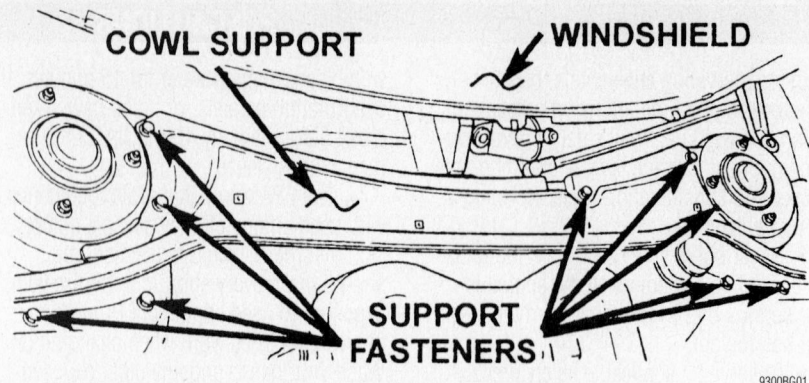

COWL SUPPORT   **WINDSHIELD**

SUPPORT FASTENERS

9300BG01

Remove the cowl support for clearance to remove the engine—2.7L engine

5. Remove or disconnect the following:

- Negative battery cable at the remote terminal near the right strut tower
- Hood
- Wiper arms, cowl covers and cowl support
- Air intake duct and air cleaner assembly
- Upper radiator crossmember
- Hood release cable from the latch
- Cooling fan assembly
- Upper and lower radiator hoses
- A/C condenser-to-radiator fasteners
- Radiator
- Accessory drive belts
- Power steering pump and position it aside without disconnecting the lines

➥Remove the alternator with the engine.

- A/C compressor and position it aside without disconnecting the lines
- V-band clamps at the exhaust manifolds
- Speed control and throttle cables from the throttle body
- Heater hoses and the coolant hoses at the recovery tank
- All vacuum lines, electrical connectors and ground straps from the engine

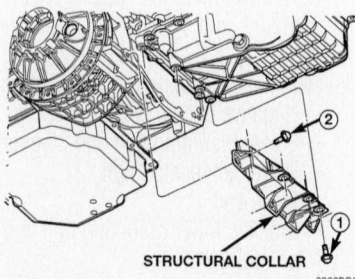

STRUCTURAL COLLAR

9300BG02

Removing the structural collar—2.7L engine

- Catalytic converter down pipes-to-rear mount fasteners
- Structural collar mounting bolts and collar

6. Matchmark the flexplate-to-torque converter and remove the torque converter bolts.

- Both transaxle cooler line-to-engine brackets
- Starter and Crankshaft Position (CKP) sensor
- 2 lower transaxle-to-engine bolts
- Exhaust Gas Recirculation (EGR) valve assembly
- Fuel line and engine harness from the throttle body bracket

7. Remove the following brackets from the double-ended transaxle-to-engine bolts:

- Wiring harness bracket
- Transaxle shift cable bracket
- Throttle body support bracket
- Upper transaxle-to-engine bolts
- Both (right and left) engine mount isolators-to-engine mount bracket fasteners

✳✳ WARNING

Be careful not to damage the intake manifold or cylinder head covers when lifting the engine.

8. Attach an engine lifting fixture to the engine.

9. Support the transaxle with a jack and remove the engine.

To install:

10. Install the engine.

11. Align the engine mounts and install the fasteners, but do not tighten them at this time.

✳✳ WARNING

Do not tighten the transaxle-to-engine bolts until all of the bolts have been hand-started and the engine is flush against the transaxle.

12. Install the transaxle-to-engine bolts. Torque the bolts to 75 ft. lbs. (102 Nm).

13. Remove the engine support fixture.

14. Torque the engine mount isolator-to-bracket fasteners to 45 ft. lbs. (61 Nm).

15. Align the matchmarks and install the flexplate-to-torque converter bolts. Torque the bolts to 55 ft. lbs. (75 Nm).

16. Install the starter and the CKP sensor.

17. Install the structural collar using the following sub-steps:

   a. Torque the vertical collar-to-oil pan bolts to 10 inch lbs. (1.1 Nm).

   b. Torque the collar-to-transaxle bolts to 40 ft. lbs. (55 Nm).

   c. Torque the center vertical bolt to 40 ft. lbs. (55 Nm); then, the remaining vertical bolts to the same torque.

18. Complete the installation by reversing the removal procedure. Keep the following in mind:

- Use new V-band clamps and torque them to 100 inch lbs. (11 Nm)
- Torque the A/C compressor and power steering pump fasteners to 21 ft. lbs. (28 Nm)
- Negative battery cable

19. Fill the cooling system to the proper level.

20. Fill the engine with clean oil.

21. Start the vehicle, check for leaks and repair if necessary.

3.2L and 3.5L Engines

1. Before servicing the vehicle, refer to the precautions in the beginning of this section.

2. Drain the engine oil.

3. Drain the engine coolant.

4. Properly relieve the fuel system pressure.

5. Remove or disconnect the following:

- Negative battery cable
- Hood
- Wiper arms
- Cowl covers and supports
- Air cleaner and air inlet duct
- Upper radiator support and hood release cable
- Fan module
- Accessory drive belts
- Upper and lower radiator hoses
- Engine oil and transmission cooler lines at the radiator
- Alternator, if necessary.
- A/C compressor mounting bolts and position it aside with the lines connected
- Power steering pump mounting bolts and position the pump aside

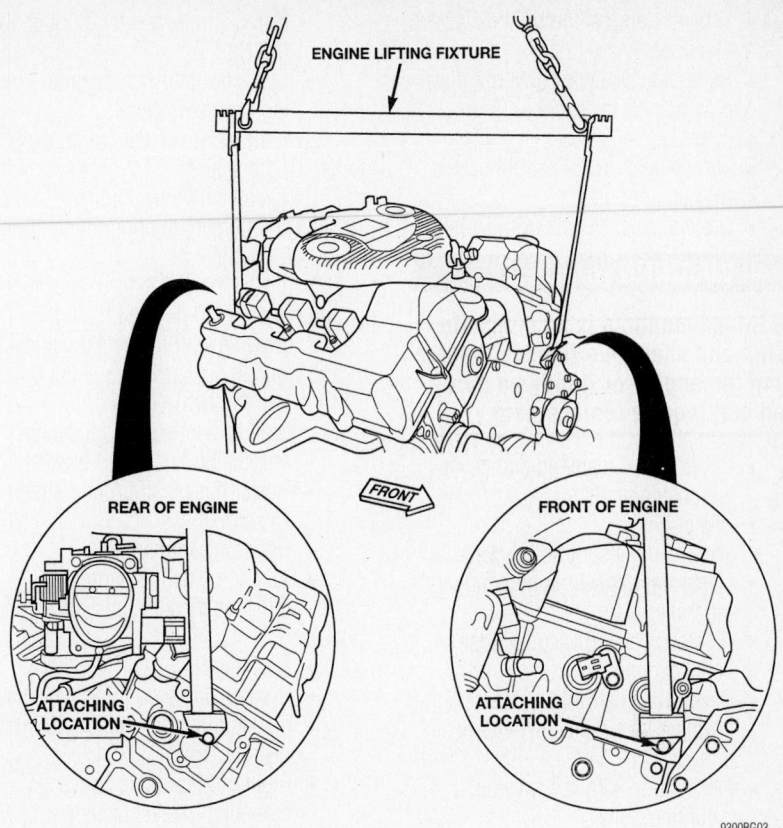

Engine support fixture attaching points—2.7L engine

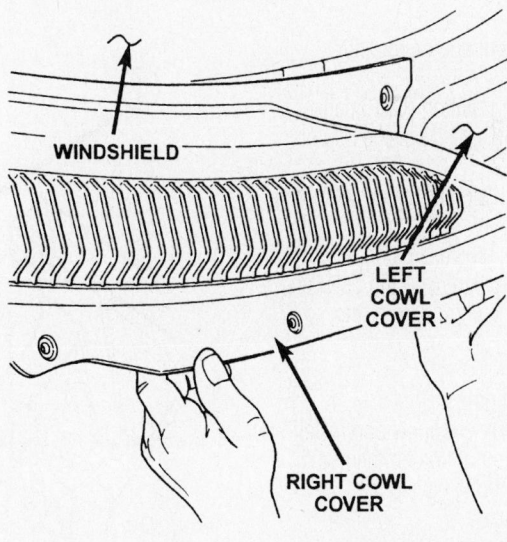

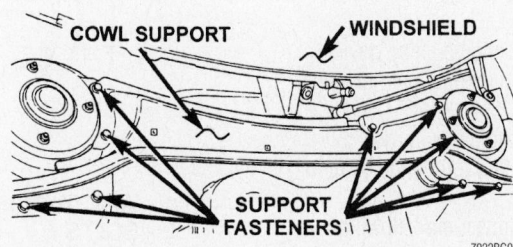

Be careful not to damage the windshield when removing the cowl covers and supports—3.2L and 3.5L models

- V-band clamp at the right exhaust manifold
- Right side catalytic converter down pipe bracket fasteners
- Fuel lines
- Throttle and cruise control cables
- All vacuum hoses
- Ground straps at both cylinder heads

✳✳ WARNING

The intake manifold is a composite design and should be removed before lifting the engine or it may be damaged and require replacement.

- Upper intake manifold and cover the openings
- Heater hoses
- Throttle body support brackets
- Upper transaxle-to-engine mounting bolts
- All electrical and vacuum hose connections
- Structural collar and matchmark the flexplate-to-torque converter position
- Flexplate-to-torque converter mounting bolts
- Left exhaust manifold V-band clamp and left catalytic converter support brackets
- Starter
- Left and right engine mounting bolts
- Crankshaft Position (CKP) sensor
- Lower engine-to-transaxle mounting bolts

6. Attach a suitable lifting device to the engine.

7. Support the transaxle using a floor jack with a small block of wood in between.

8. Slowly lift the engine from the vehicle.

To install:

9. Install the engine.

10. Align the engine mounts and install the fasteners, but do not tighten them until all of the mounting bolts have been installed.

11. Install the engine-to-transaxle mounting bolts and torque to 75 ft. lbs. (102 Nm).

12. Remove the engine lifting device.

13. Torque the engine mount nuts/bolts 45 ft. lbs. (61 Nm).

14. Align the flexplate-to-torque converter matchmarks and torque the bolts to 55 ft. lbs. (75 Nm).

15. Install or connect the following:
- CKP sensor and the starter
- Left-side exhaust manifold V-band

clamp and torque to 100 inch lbs. (11 Nm)
- Left-side catalytic converter mounting bracket fasteners

16. Install the structural collar using the following procedure:

a. Install the vertical collar-to-oil pan bolts and tighten, temporarily to 10 inch lbs. (1.1 Nm).

b. Install the collar-to-transaxle bolts and torque to 40 ft. lbs. (55 Nm).

c. Starting with the center vertical bolt and working outward, torque the bolts to 40 ft. lbs. (55 Nm).

17. Install or connect the following:
- Throttle body support bracket
- Heater hoses, all ground straps and vacuum hoses
- Upper intake manifold
- All engine wiring harnesses

18. Adjust the throttle and cruise control cables.

- Fuel lines
- V-band clamp on the right exhaust manifold and torque to 100 inch lbs. (11 Nm)
- Right-side catalytic converter mounting bracket fasteners
- A/C compressor and torque the bolts to 21 ft. lbs. (28 Nm)

- Alternator and its wiring harness
- Radiator
- Power steering pump
- Engine oil and transaxle cooler lines
- Radiator hoses and accessory drive belt
- Hood release cable and upper radiator support
- Air cleaner and inlet hose
- Cowl covers, supports and wiper arms
- Hood
- Negative battery cable

19. Fill the cooling system to the proper level.

20. Fill the engine with clean oil.

21. Start the vehicle, check for leaks and repair if necessary.

Water Pump

REMOVAL & INSTALLATION

The water pump has a die cast aluminum body and a stamped steel impeller. It bolts directly to the chain case cover using an O-ring for sealing. It is driven by the back side of the serpentine belt.

It is normal for a small amount of coolant

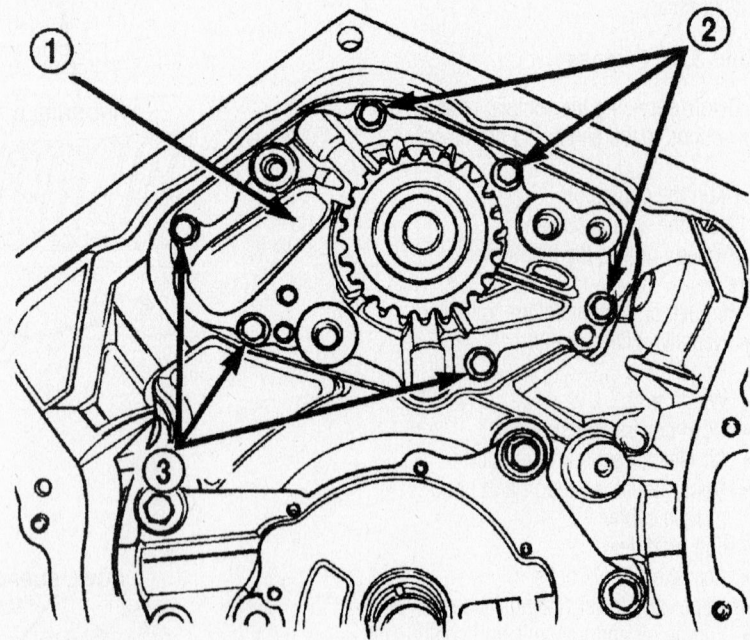

1 - WATER PUMP
2 - BOLTS
3 - BOLTS

Water pump mounting—2.7L engine

to drip from the weep hole located on the water pump body (small black spot). If this condition exists, DO NOT replace the water pump. Only replace the water pump if a heavy deposit or steady flow of brown/green coolant is visible on the water pump body from the weep hole, which would indicate shaft seal failure. Before replacing the water pump, be sure to perform a thorough inspection. A defective pump will not be able to circulate heated coolant through the long heater hose.

2.7L Engine

1. Before servicing the vehicle, refer to the precautions in the beginning of this section.
2. Drain the cooling system.
3. Remove or disconnect the following:
   - Negative battery cable
   - Upper radiator crossmember
   - Fan module
   - Accessory drive belts

➡**The water pump is driven by the primary timing chain.**

   - Crankshaft damper, timing chain cover, timing chain and all guides
   - Water pump mounting bolts
   - Water pump
4. Clean the water pump mounting surfaces.
To install:
5. Install or connect the following:
   - Water pump and gasket and torque the bolts to 105 inch lbs. (12 Nm)
   - Guides, timing chain and timing chain cover
   - Crankshaft damper and torque the center bolt to 125 ft. lbs. (170 Nm)
   - Accessory drive belts
   - Fan module and upper radiator crossmember
   - Negative battery cable
6. Fill the cooling system to the proper level.
7. Start the vehicle, check for leaks and repair if necessary.

3.2L and 3.5L Engines

1. Before servicing the vehicle, refer to the precautions in the beginning of this section.
2. Drain the cooling system.

✳✳ WARNING

Do not use pliers to open the plastic drain.

3. Remove or disconnect the following:
   - Negative battery cable

   - Coolant recovery cap and open the thermostat bleed valve
   - Drive belts
   - Timing belt

➡**It is good practice to turn the crankshaft until the No. 1 cylinder is at Top Dead Center (TDC) of its compression stroke (firing position).**

   - Water pump mounting bolts, noting the position of the longer bolt
   - Water pump. Discard the O-ring seal.
4. Clean the gasket sealing surfaces, being careful not to scratch the aluminum surfaces.
To install:
5. Install or connect the following:
   - New O-ring and wet with clean coolant prior to installation. Be sure to keep the new O-ring free of any oil or grease.
   - Water pump with a new O-ring. Torque the water pump-to-engine bolts to 105 inch lbs. (12 Nm).

✳✳ WARNING

Rotate the pump and check for freedom of movement.

   - Timing belt
   - Drive belts
6. Fill the cooling system by performing the following procedure:
   a. Close the radiator drain.
   b. Open the thermostat bleed valve. Install a ¼ in. (6mm) clear hose about 48 in. (1.2m) long to the end of the bleed valve and the other end into a clean container. The intent is to keep coolant off of the drive belt(s).
   c. Slowly, refill the coolant recovery bottle until a steady stream of coolant flows out of the thermostat bleed valve. Gently squeeze the upper radiator hose

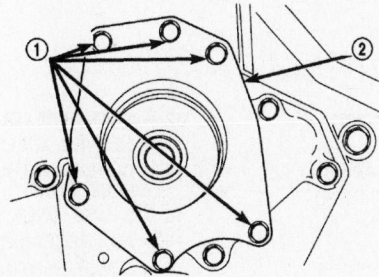

1 - SCREWS
2 - WATER PUMP BODY

67189-300M-G02

Water pump bolt locations, make sure to note the position of the longer bolt—3.5L engine

until all of the air is removed from the system.
   d. Close the bleed valve and continue to fill the coolant recovery bottle to the proper level. Install the cap on the bottle and remove the hose from the bleed valve.
7. Reconnect the negative battery cable. Start the engine and allow it to reach normal operating temperatures.
8. Check the cooling system for leaks and correct coolant level. Be sure that the thermostat bleed valve is closed once the cooling system has been bled of any trapped air.

Heater Core

REMOVAL & INSTALLATION

1. Disconnect the negative battery cable from the remote battery post located near the right strut tower.

✳✳ CAUTION

After disconnecting the negative battery cable, wait 2 minutes for the driver's/passenger's air bag system capacitor to discharge before attempting to do any work around the steering column or instrument

2. Remove the instrument panel by removing or disconnecting the following:
   - Shifter knob Allen screw
   - Both instrument panel end covers

➡**On the LHS or 300M, remove the center bezel prior to the shifter bezel**

3. Using a trim stick tool, gently pry upward on the shifter bezel; then, disconnect both wiring connectors and a light bulb socket connector; then, remove the socket.
4. Remove the 2 lower instrument panel cover screws (outside end) and disconnect the deck lid release switch wiring connector. Pull the lower instrument panel cover rearward releasing the clips. Remove the cable-to-brake release handle. Pull the cover rearward and remove it.
5. If equipped for 6 passengers, remove the lower floor bin.
6. If equipped for 5 passengers, remove or disconnect the following items:
   - Center bezel using the trim stick
   - Heater/air conditioning housing assembly control switch and the traction control switch wiring harnesses
   - Bezel from the vehicle
   - Two left side console side cover

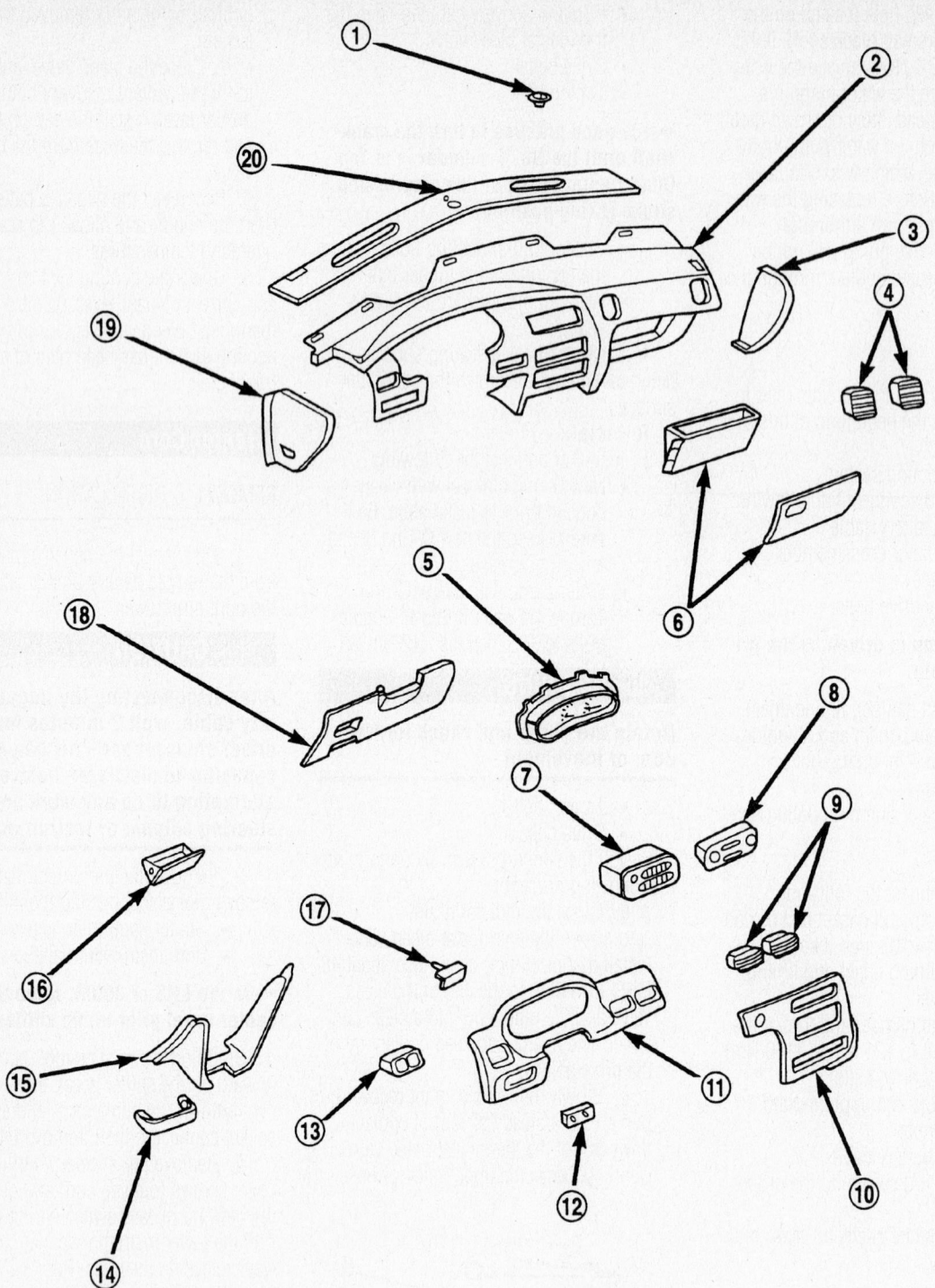

1 – SPEAKER, INSTRUMENT PANEL CENTER
2 – INSTRUMENT PANEL ASSEMBLY
3 – BEZEL, INSTRUMENT PANEL END CAP
4 – LOUVER, AIR OUTLET
5 – HOUSING INSTRUMENT CLUSTER
6 – GLOVE BOX ASSEMBLY
7 – RADIO
8 – CONTROL ASSEMBLY, INSTRUMENT PANEL
9 – LOUVER, AIR OUTLET
10 – BEZEL, INSTRUMENT PANEL TRIM-CENTER
11 – BEZEL INSTRUMENT CLUSTER

12 – SWITCH HEADLAMP
13 – LOUVER, AIR OUTLET
14 – COVER, INSTRUMENT PANEL CENTER SUPPORT/BIN (6 PASS. ONLY)
15 – 6 PASS. ONLY
16 – ASH RECEIVER
17 – LEVER, PARKING BRAKE
18 – COVER, LOWER INSTRUMENT PANEL
19 – BEZEL, INSTRUMENT PANEL END CAP
20 – COVER, UPPER INSTRUMENT PANEL

93111G81

Exploded view of the instrument panel—Intrepid

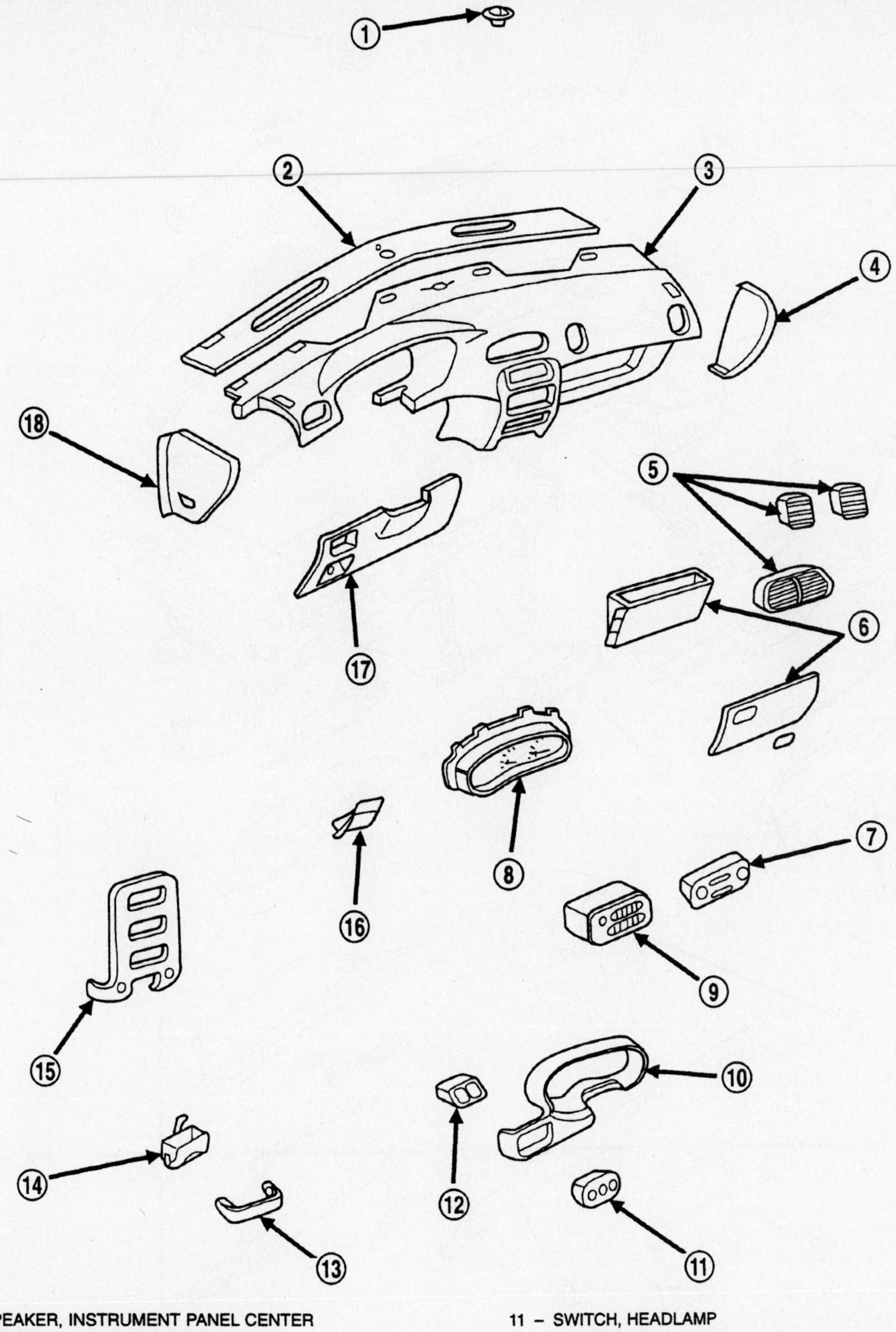

1 – SPEAKER, INSTRUMENT PANEL CENTER
2 – COVER, UPPER INSTRUMENT PANEL
3 – INSTRUMENT PANEL ASSEMBLY
4 – BEZEL, INSTRUMENT PANEL END CAP
5 – LOUVER, AIR OUTLET
6 – GLOVE BOX ASSEMBLY
7 – CONTROL ASSEMBLY, INSTRUMENT PANEL
8 – HOUSING, INSTRUMENT CLUSTER
9 – RADIO
10 – BEZEL, INSTRUMENT CLUSTER

11 – SWITCH, HEADLAMP
12 – LOUVER, AIR OUTLET
13 – COVER, INSTRUMENT PANEL CENTER SUPPORT/BIN (6 PASS. ONLY)
14 – ASH RECEIVER
15 – BEZEL, INSTRUMENT PANEL TRIM-CENTER
16 – LEVER, PARKING BRAKE
17 – COVER, LOWER INSTRUMENT PANEL
18 – BEZEL, INSTRUMENT PANEL END CAP

93111G82

Exploded view of the instrument panel—Concorde

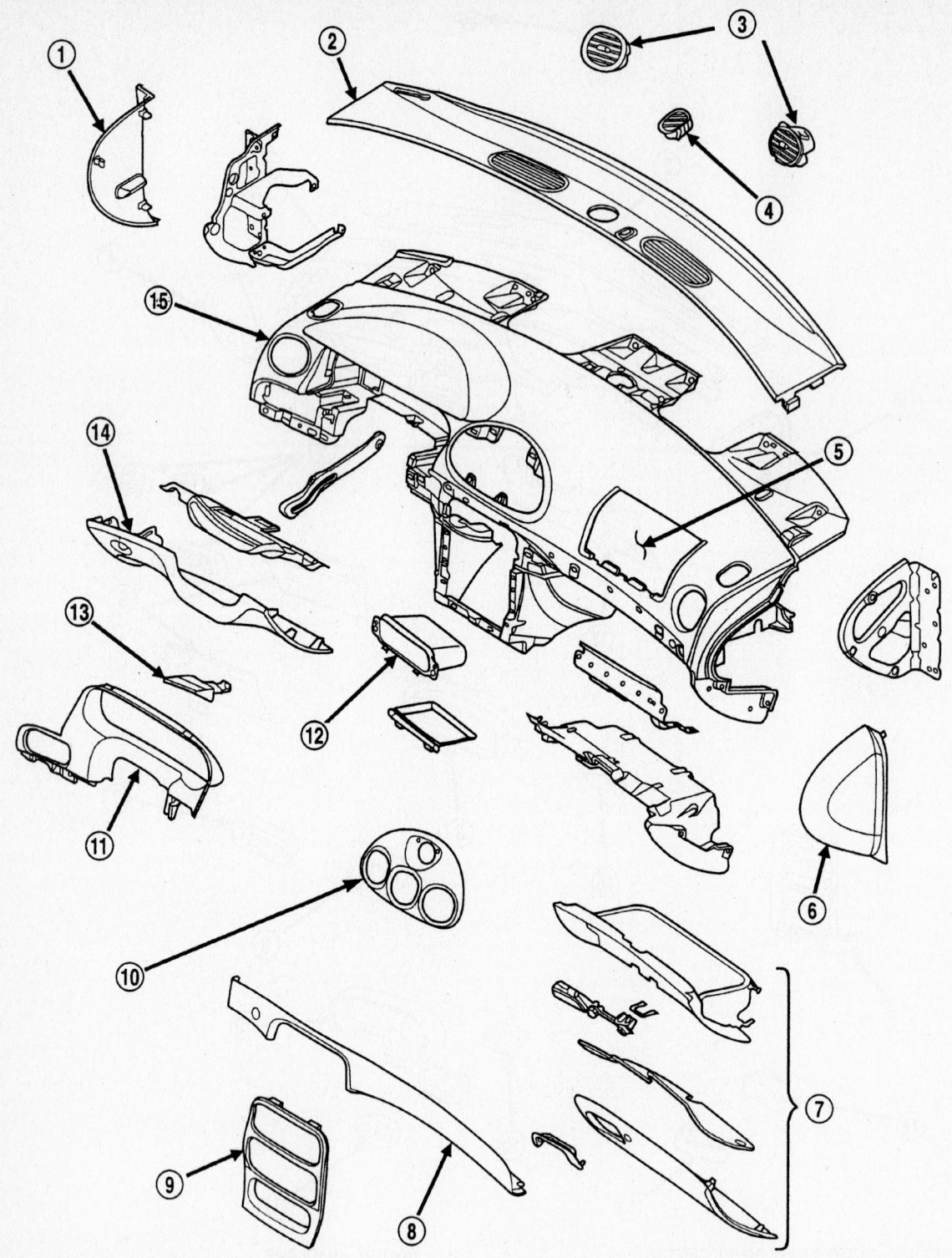

1 – BEZEL, INSTRUMENT PANEL END CAP
2 – COVER, UPPER INSTRUMENT PANEL
3 – LOUVER, AIR OUTLET
4 – LOUVER/DEMISTER SIDE WINDOW
5 – MODULE, PASSENGER SIDE AIRBAG
6 – BEZEL, INSTRUMENT PANEL END CAP
7 – GLOVE BOX ASSEMBLY, INSTRUMENT PANEL
8 – BEZEL, INSTRUMENT PANEL UPPER RIGHT TRIM

9 – BEZEL, INSTRUMENT PANEL TRIM-CENTER
10 – BEZEL, INSTRUMENT PANEL AIR DISTRIBUTION OUTLET
11 – BEZEL, INSTRUMENT CLUSTER
12 – STORAGE COMPARTMENT CUBBY BOX
13 – LEVER, PARKING BRAKE
14 – COVER, LOWER INSTRUMENT PANEL-LEFT SIDE
15 – INSTRUMENT PANEL ASSEMBLY

93111G83

Exploded view of the instrument panel—LHS and 300M

screws, pull the cover outward and remove from the vehicle
- Lower glove box door; the 2 screws are located at the right console side cover, pull outward and remove it
- 2 bracket screws and the 2 inside the console bin; then, remove the console
- 2 center lower instrument panel nuts
- 4 steel reinforcement-to-lower instrument panel cover bottom bolts
- 16-way DLC from the reinforcement
- left floor duct/silencer pad screw and the pad
- Steering column shrouds
- Shift interlock cable at the ignition switch
- Column wiring
- The under column duct section
- Left panel air conditioning outlet duct
- The 4 brake pedal support bracket-to-column mounting bolts
- Lower the steering column
- Heater/air conditioning housing and the Air bag Control Module (ACM) harness connectors then, the 2 ground eyelets from the floor tunnel near the bulkhead

7. Using a trim stick, carefully, pry out on the left and right A-pillar trim moldings and slide them rearward to remove them.

8. Using a trim stick, carefully pry upward on the instrument panel top cover, and remove it.

9. From both sides, remove the scuff plate screw. Using a trim stick, carefully, pry out the scuff plates and remove them.

10. Remove or disconnect the following:
- 3 right side cowl panel screws and the panel
- 3 left side cowl panel screws and the panel
- Remove the right side under dash silencer and pad
- 2 right side radio antenna and amplifier harness connectors
- Left side junction block and Body Control Module (BCM) harness connectors
- 8 instrument panel-to-chassis screws
- Pull the instrument panel rearward
- Instrument panel from the vehicle with the help of an assistant

11. Discharge and recover the air conditioning system refrigerant.

12. Drain the cooling system into a clean container for reuse.

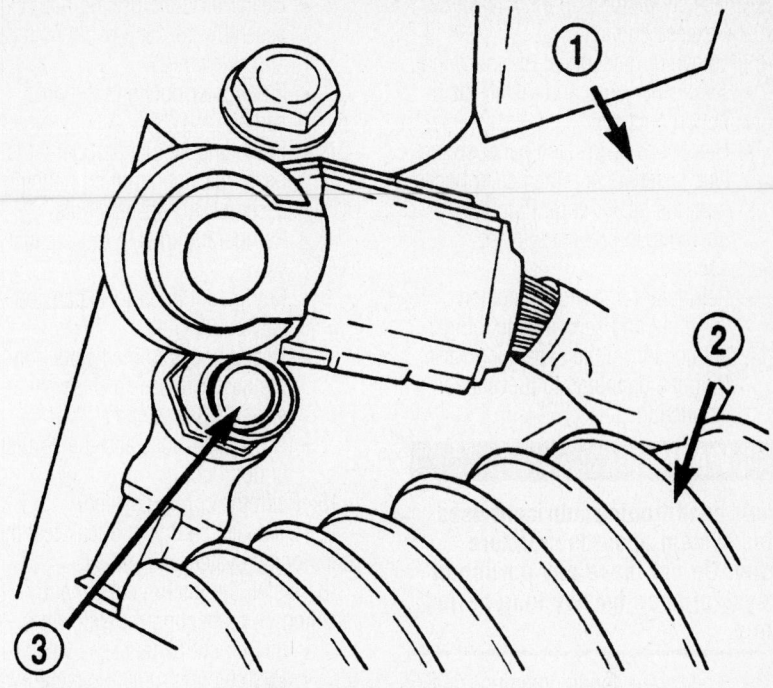

1. Right strut tower
2. Air cleaner inlet tube
3. Remote terminal

93111G84

View of the remote battery cable terminal—LHS, 300M, Concorde and Intrepid

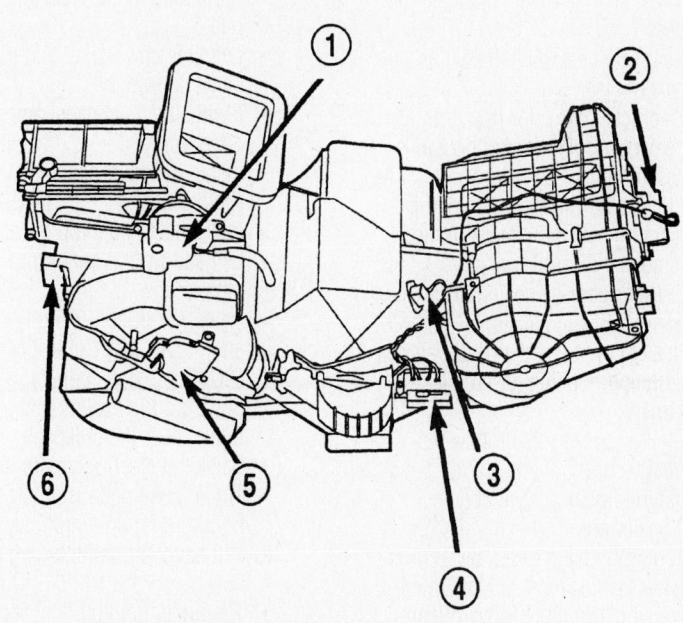

1 – MODE DOOR ACTUATOR
2 – RECIRCULATION DOOR ACTUATOR
3 – EVAPORATOR TEMPERATURE SENSOR
4 – POWER MODULE OR BLOWER RESISTOR
5 – BLEND DOOR ACTUATOR
6 – HVAC PLENUM CONNECTOR

93111G85

View of the heater/air conditioning housing assembly—LHS, 300M, Concorde and Intrepid

13. Place shop cloths in the interior to catch any excess coolant.

14. Remove or disconnect the following:
- Air cleaner hose and the air distribution duct
- Heater hoses-to-dash panel spring type fasteners and the heater hoses from the heater core. Plug the openings to prevent coolant spillage.
- Refrigerant lines-to-expansion valve nut and separate the refrigerant lines from the expansion valve. Plug the openings to prevent contamination.

✳✳ WARNING

The air conditioning lubricant used in this system absorbs moisture readily. Do not leave any portion of the system open for any long period of time.

- 3 heater/air conditioning housing assembly-to-chassis nuts from the engine compartment
- 2 defroster duct screws and the duct
- Heater/air conditioning housing assembly-to-dash panel nuts (2) and screws (2)
- 4 rear seat heat duct nuts and the duct
- Rear seat heat duct elbow push-pin fastener
- Electrical connector from the heater/air conditioning housing assembly
- Pull the heater/air conditioning housing assembly rearward and remove it from the vehicle
- 2 heater core-to-heater/air conditioning housing assembly screws
- Heater core from the heater/air conditioning housing assembly

To install:

15. Install or connect the following:
- Heater core to the heater/air conditioning housing assembly
- 2 heater core-to-heater/air conditioning housing assembly screws
- Push the heater/air conditioning housing assembly forward (carefully) and install it to the vehicle
- Electrical connector to the heater/air conditioning housing assembly
- Rear seat heat duct elbow push pin fastener
- Duct and the 4 rear seat heat duct nuts

- Heater/air conditioning housing assembly-to-dash panel nuts (2) and screws (2)
- 2 defroster duct and the duct screws
- In the engine compartment, the 3 heater/air conditioning housing assembly-to-chassis nuts
- Refrigerant lines to the expansion valve
- Refrigerant lines-to-expansion valve nut
- Heater hoses to the heater core and the heater hoses-to-dash panel spring type fasteners
- Air cleaner hose and the air distribution duct

16. Refill the cooling system.

17. Evacuate, charge and leak-test the air conditioning system refrigerant.

18. Install the instrument panel by installing or connecting the following:
- Instrument panel to the vehicle
- Push the instrument panel forward
- 8 instrument panel-to-chassis screws
- Left side junction block and Body Control Module (BCM) harness connectors
- 2 right side radio antenna and amplifier harness connectors
- Right side under dash silencer and pad
- Left side cowl panel and the 3 panel screws
- Right side cowl panel and the 3 panel screws
- Install the scuff plate and the screw (both sides)
- Instrument panel top cover
- Left and right A-pillar trim moldings
- Heater/air conditioning housing and the Air bag Control Module (ACM) harness connectors then, the 2 ground eyelets to the floor tunnel near the bulkhead
- Raise the steering column
- The 4 brake pedal support bracket-to-column mounting bolts
- Left panel air conditioning outlet duct
- The under column duct section
- Column wiring
- Shift interlock cable at the ignition switch
- Steering column shrouds
- Left floor duct/silencer pad and the pad screw
- 4 steel reinforcement-to-lower instrument panel cover bottom bolts

- 6-way Diagnostic Link Connector (DLC) to the reinforcement
- 2 center lower instrument panel nuts
- The 2 bracket screws and the 2 inside the console bin; then, the console (5 passenger models only)
- 2 screws are located at the right console side cover (5 passenger models only)
- Two left side console side cover screws (5 passenger models only)
- Bezel to the vehicle (5 passenger models only).
- Heater/air conditioning housing assembly control switch and the traction control switch wiring harnesses.
- Center bezel using the trim stick.
- Lower floor bin (6 passenger models only).
- 2 lower instrument panel cover screws (outside end) and connect the deck lid release switch wiring connector. Install the cable-to-brake release handle
- Light bulb socket. Both wiring connectors and a light bulb socket connector
- Shifter bezel
- Center bezel prior to the shifter bezel (LHS or 300M models)
- Both instrument panel end covers
- Shifter knob Allen screw

19. Connect the negative battery cable to the remote battery post located near the right strut tower.

20. Operate the engine to normal operating temperatures; then, check the climate control operation and check for leaks.

Cylinder Head

REMOVAL & INSTALLATION

2.7L Engine

1. Before servicing the vehicle, refer to the precautions in the beginning of this section.

2. Properly relieve the fuel system pressure.

3. Drain the cooling system.

4. Remove or disconnect the following:
- Negative battery cable
- Accessory drive belts
- Crankshaft damper
- Intake plenum, lower intake manifold and exhaust manifold

➡**Place shop rags in the openings to prevent debris from entering the engine.**

- Valve and timing chain covers
- Coolant connections for the cylinder heads

5. Rotate the crankshaft until the crankshaft timing mark aligns with the timing mark on the oil pump.

- Primary timing chain
- Camshaft bearing caps, gradually, in the reverse order of the tightening sequence
- Camshafts

✳✳ WARNING

Be sure the head bolts 9–11 are removed before attempting to remove the cylinder head, the head and/or block may be damaged.

- Cylinder head bolts in reverse order of installation starting with bolts 11–9, then 8–1
- Cylinder head(s)

To install:

6. Thoroughly clean and dry the mating surfaces of the head and block. Check the cylinder head for cracks, damage or engine coolant leakage. Remove scale, sealing compound and carbon. Clean the oil passages thoroughly.

7. Place a new head gasket on the cylinder block over the locating dowels.

8. Inspect the cylinder head bolts for necking (stretching) by holding a straightedge against the threads of each bolt. If all of the threads are not contacting the straightedge, the bolt should be replaced. New head bolts are recommended.

✳✳ WARNING

Due to the cylinder head bolt torque method used, it is imperative that the bolt threads be inspected for necking (stretching) prior to installation. If the threads are necked down, the bolt should be replaced. Failure to do so may result in parts failure or damage.

9. Lubricate the bolt threads with clean engine oil, then install them.

10. Torque the head bolts in the sequence shown in the illustration, utilizing the following steps and tightening values:

    a. Step 1: bolts 1 through 8: 35 ft. lbs. (48 Nm).

    b. Step 2: bolts 1 through 8: 55 ft. lbs. (75 Nm).

    c. Step 3: bolts 1 through 8: 55 ft. lbs. (75 Nm).

    d. Step 4: bolts 1 through 8: plus 90 degree turn using a torque angle meter.

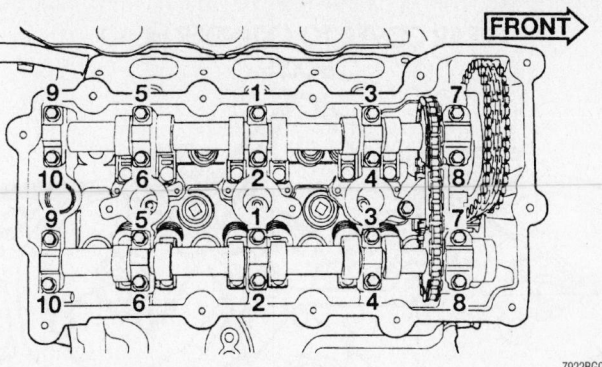

Camshaft bolt tightening sequence—2.7L engine

    e. Step 5: bolts 9 through 11: 21 ft. lbs. (28 Nm).

11. Install or connect the following:

- Camshafts, timing chain and sprockets
- Water connections to the cylinder head
- Valve and timing chain covers
- Crankshaft damper and torque the center bolt to 125 ft. lbs. (170 Nm)
- Lower intake manifold and intake plenum
- Exhaust manifolds
- Accessory drive belts
- Negative battery cable

12. Fill the cooling system to the proper level.

13. Start the vehicle, check for leaks and repair if necessary.

3.2L and 3.5L Engines

1. Before servicing the vehicle, refer to the precautions in the beginning of this section.

This engine uses aluminum alloy cylinder heads. Use care when working with light alloy components. The heads are common to either cylinder bank, but in practice a cylinder head should be returned

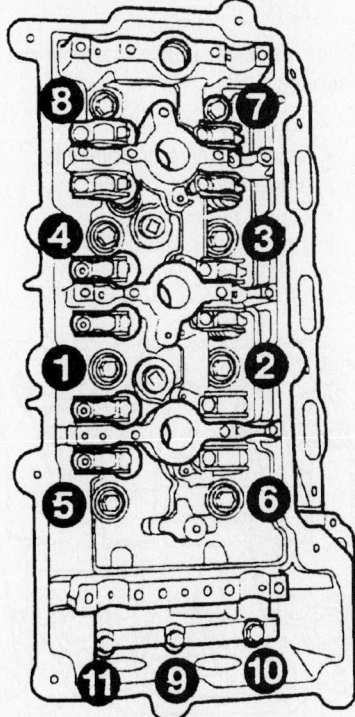
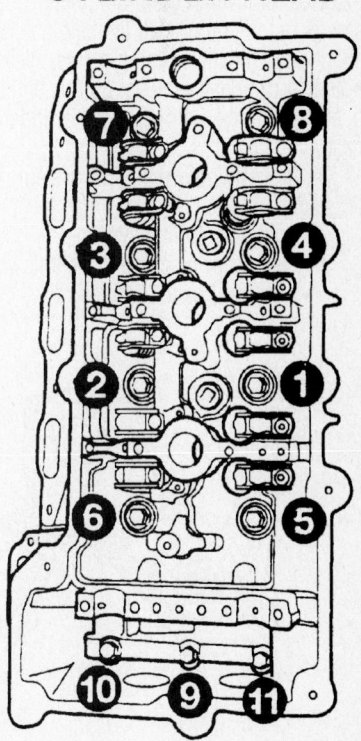

Cylinder head bolt tightening sequence—2.7L engine

REAR COVER TO CYLINDER HEAD BOLTS

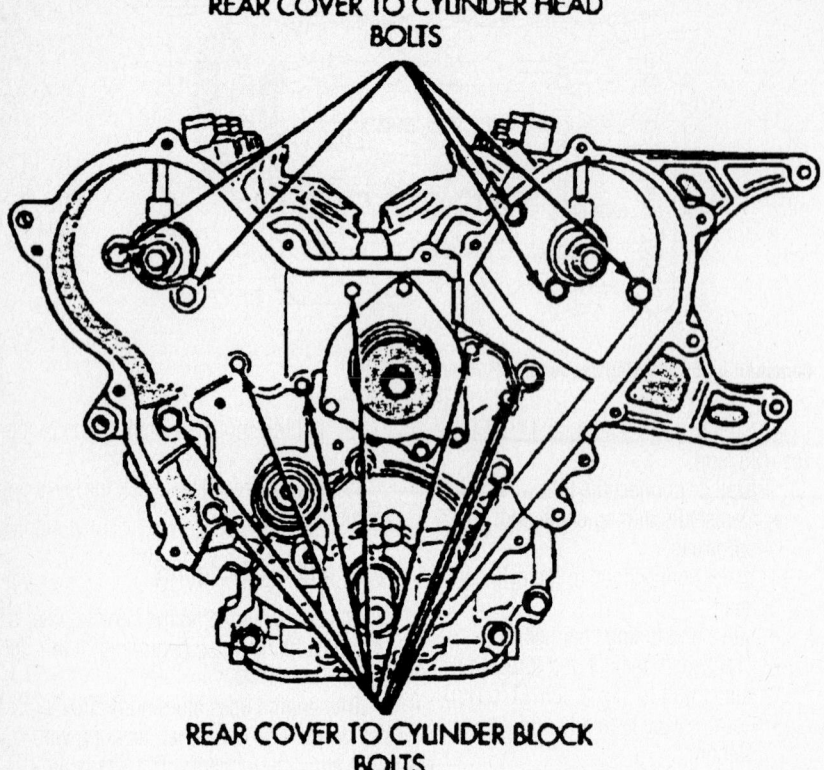

REAR COVER TO CYLINDER BLOCK BOLTS

7922BG09

Remove the rear timing belt cover-to-cylinder head bolts, noting the bolt locations—3.2L and 3.5L engines

to the side of the engine from which it was removed. Removal of a cylinder head involves removal of the timing belt. Great care is required to install the belt, paying attention to all valve timing marks. Please note that camshaft removal on this engine does require the removal of the cylinder head.

2. Properly relieve the fuel system pressure.
3. Drain the cooling system.
4. Remove or disconnect the following:
   • Negative battery cable

   • Radiator and cooling fan assemblies
   • Air cleaner assembly and intake manifold plenum

➡**Cover the lower intake manifold during service.**

   • Accessory drive belts
   • Crankshaft damper using the proper puller
   • Engine valve covers
   • Timing belt covers

➡Mark the timing belt running direction for installation. Align the camshaft sprockets with the marks on the rear covers.

   • Timing belt and tensioner
5. Pre-load the timing belt tensioner as follows:
   a. Place tensioner in a vise the same way it is mounted on the engine.
   b. Slowly compress the plunger into the tensioner body.
   c. Once the plunger is compressed, install a pin through the body and plunger to retain it in place until the tensioner is installed.
6. Hold the camshaft sprocket with a 36mm box wrench, loosen and remove the sprocket retaining bolt and washer.

➡**To remove the camshaft sprocket retainer bolt while the engine is in the vehicle, it may be necessary to raise that side of the engine due to the length of the retainer bolt. The right bolt is 8.370 in. (212.6mm) long, while the left bolt is 10.0 in. (253mm) long. These bolts are not interchangeable and their original location during removal should be noted.**

7. Remove or disconnect the following:
   • Camshaft sprocket from the camshaft

➡**The camshaft sprockets are not interchangeable.**

   • Intake manifold assembly using the recommended procedure
   • Rear timing belt cover-to-cylinder head fasteners

➡**If the right timing belt cover is to be removed, there are O-rings located behind it for the water pump passages.**

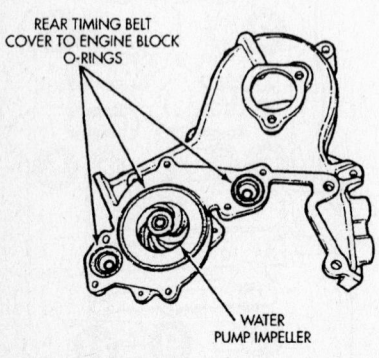

REAR TIMING BELT COVER TO ENGINE BLOCK O-RINGS

WATER PUMP IMPELLER

7922BG10

Right side belt cover, water pump and O-rings—3.2L and 3.5L engines

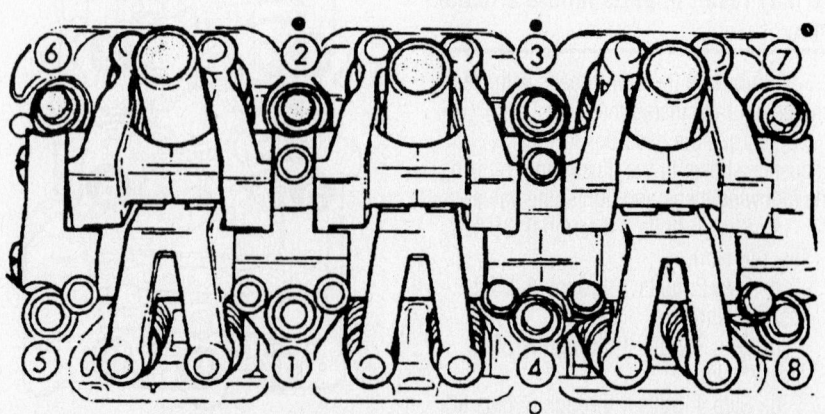

7922BG12

Cylinder head bolt tightening sequence—3.2L and 3.5L engines

- Cylinder head mounting bolts in the reverse order of the tightening sequence
- Cylinder head

To install:

8. Thoroughly clean and dry the mating surfaces of the head and block.

❊❊ WARNING

When cleaning the cylinder head and block mating surfaces, do not use a metal scraper because the soft aluminum surfaces could be cut or damaged. Instead, use a scraper made of wood or plastic.

9. Check the cylinder head for cracks, damage or engine coolant leakage. Check the head for flatness. End-to-end, the head should be within 0.002 in. (0.051mm) normally with 0.008 in. (0.203mm) the maximum allowed out of true. The resurface limit is 0.008 in. (0.203mm) maximum, the combined total dimension of stock removal from the cylinder head, if any, and block top surface.

10. Place a new head gasket on the cylinder block locating dowels, being sure the gasket is on the correct side.

11. Inspect the cylinder head bolts for necking (stretching) by holding a straight-edge against the threads of each bolt. If all of the threads are not contacting the scale, the bolt should be replaced.

❊❊ WARNING

Due to the cylinder head bolt torque method used, it is imperative that the threads of the bolts be inspected for necking prior to installation. If the threads are necked down, the bolt

should be replaced. Failure to do so may result in parts failure or damage. New bolts are always recommended.

12. Install the cylinder head into position on the engine block and over the dowels. Install the cylinder head bolts, lubricating the threads with clean engine oil prior to installation.

13. Torque the cylinder head bolts using the proper sequence as follows:

    a. Step 1: torque in sequence to 45 ft. lbs. (61 Nm).

    b. Step 2: torque in sequence to 65 ft. lbs. (88 Nm).

    c. Step 3: torque in sequence to 65 ft. lbs. (88 Nm).

    d. Step 4: torque in sequence an additional ¼ turn by using a torque angle meter.

➡**Inspect the bolt torque after tightening. The torque should be over 90 ft. lbs. (122 Nm). If not, replace the cylinder head bolt.**

14. Install the rear timing belt cover bolts and torque as follows:

    a. M6 bolts: 105 inch lbs. (12 Nm).

    b. M8 bolts: 21 ft. lbs. (28 Nm).

    c. M10 bolts: 40 ft. lbs. (54 Nm).

15. Install the intake manifold assembly and torque the bolts following the proper sequence to 21 ft. lbs. (28 Nm).

➡**The following procedure can only be used when the camshaft sprockets have been loosened or removed from the shafts.**

16. When the camshaft sprockets are loosened or removed, the camshafts must

be timed to the engine. Install the Camshaft Alignment tools 6642-A, to the rear of the cylinder heads.

17. Install both camshaft sprockets to the appropriate shafts. The left camshaft sprocket has the Distributorless Ignition System (DIS) pickup as part of the sprocket.

18. Apply thread locking compound to the threads of the camshaft sprocket retainer bolts and install to the appropriate shafts. The right bolt is 8.380 in. (21.3cm) long, while the left bolt is 10.0 in. (25.4cm) long. These bolts are not interchangeable. Do not tighten the bolts at this time. The camshaft marks should be positioned between the marks on the cover.

19. Place the crankshaft sprocket to the Top Dead Center (TDC) mark on the oil pump housing. Install the timing belt starting at the crankshaft sprocket and working in a counterclockwise direction.

20. After the belt is installed around the last sprocket keep tension on the belt until it is past the tensioner pulley.

21. Holding the tensioner pulley against the belt, install the tensioner housing and torque to 21 ft. lbs. (28 Nm).

22. When the tensioner is in place pull the retainer pin to allow the tensioner to extend to the pulley bracket.

23. Hold the right camshaft sprocket hex with a 36mm box wrench and torque the right camshaft sprocket bolt to 75 ft. lbs. (102 Nm). Turn the sprocket bolt an additional 90 degrees.

24. Hold the left camshaft sprocket hex with a 36mm box wrench and torque the left camshaft sprocket bolt to 85 ft. lbs. (115 Nm). Turn the sprocket bolt an additional 90 degrees.

25. Install or connect the following:

- Camshaft alignment tools from the back of the cylinder heads
- Cam covers and new O-rings. Torque the fasteners to 20 ft. lbs. (27 Nm). Repeat this procedure on the other camshaft.
- Timing belt covers and crankshaft damper and torque the crankshaft damper bolt to 85 ft. lbs. (115 Nm)
- Valve covers and torque the bolts to 105 inch lbs. (12 Nm)
- Spark plug tube nut and O-ring and torque to 60 inch lbs. (7 Nm)
- Spark plug and torque to 20 ft. lbs. (28 Nm)
- A/C compressor and torque the mounting bracket bolts to 30 ft. lbs. (41 Nm)
- Spark plug wires
- Accessory drive belts and adjust to the proper tension

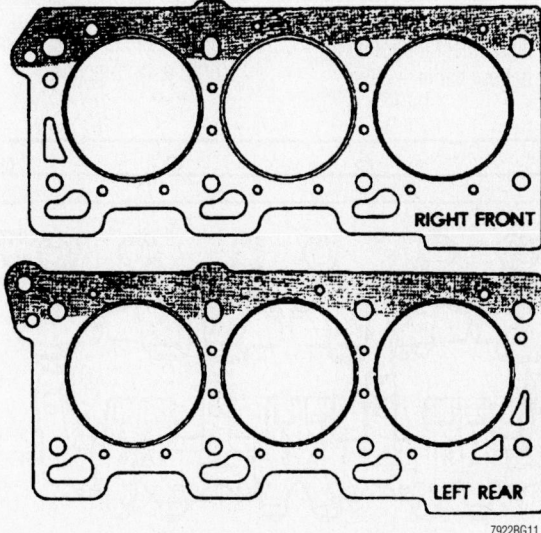

7922BG11

Correct positioning for the head gaskets—3.2L and 3.5L engines

- Intake manifold plenum using the recommended procedure
- Air cleaner assembly
- Radiator and cooling fan assemblies
- Negative battery cable

26. Check to be sure that all hoses, wiring connectors, cables, fluid and vacuum lines are reconnected.

27. Change the engine oil and oil filter.

28. Fill and bleed the cooling system.

29. Run the vehicle with the radiator cap off so coolant can be added as required until the thermostat opens. Watch for leaks and for unusual engine noises. Refill the radiator completely as required.

30. Once the vehicle has cooled, recheck the coolant and oil level.

Rocker Arm/Shafts

REMOVAL & INSTALLATION

2.7L Engine

1. Before servicing the vehicle, refer to the precautions in the beginning of this section.

2. Disconnect the negative battery cable.

3. Remove the valve covers.

4. Position the camshaft so that the base circle (heel) is facing the rocker arm being serviced.

> ❊❊ **WARNING**
>
> **Depress the valve spring only enough to remove the rocker arm or damage to the spring may result.**

5. Using Valve Spring tool 8215 and Adapter 8216, depress the valve spring enough to release the tension on the rocker arm.

➡ **If the rocker arms are to be reused, identify their positions for reassembly in their original positions.**

6. Repeat this procedure for each rocker arm being removed.

To install:

7. Lubricate the rocker arms with clean engine oil, prior to installation.

8. Position the camshaft so that the base circle (heel) is facing the rocker arm being installed.

> ❊❊ **WARNING**
>
> **Depress the valve spring only enough to install the rocker arm or damage to the spring may result.**

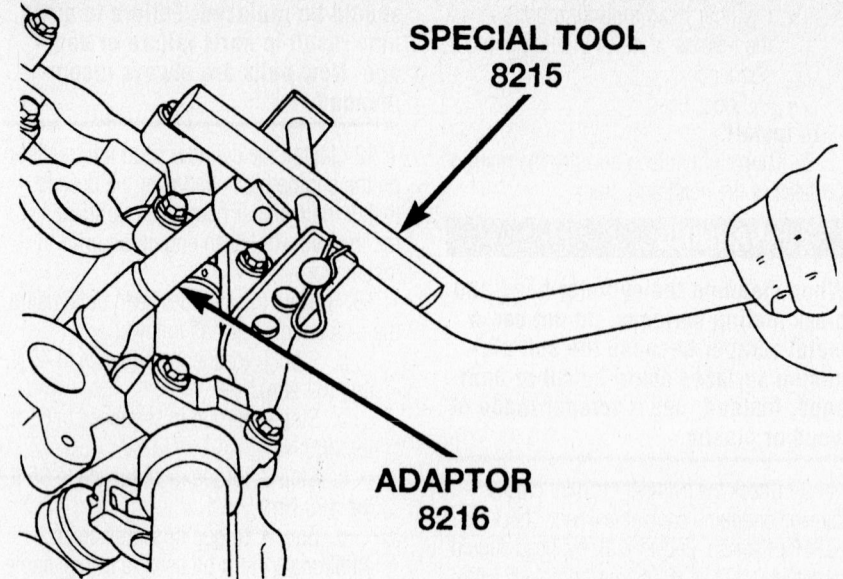

Only depress the valve spring enough to remove the rocker arm—2.7L engine

9. Using Valve Spring tool 8215 and Adapter 8216, depress the valve spring enough to install the rocker arm.

10. Install the rocker arm in the original position (if reused) over the valve and lash adjuster.

➡ **Inspect the rocker arm for proper engagement into the lash adjuster and valve tip.**

11. Release the tension on the valve spring and remove the tools.

12. Install the valve covers and connect the negative battery cable.

3.2L and 3.5L Engines

1. Before servicing the vehicle, refer to the precautions in the beginning of this section.

2. Relieve the fuel system pressure.

3. Remove or disconnect the following:
   - Negative battery cable
   - Air cleaner assembly and the intake manifold plenum

➡ **Cover the lower intake manifold during service.**

   - Cylinder head covers
   - Rocker arm assembly

4. Inspect the rocker arms for wear or damage. Inspect the roller for scuffing or wear. Replace assembly as necessary.

> ❊❊ **WARNING**
>
> **Do not remove the lash adjusters from the rocker arm assembly. The rocker arm and the adjuster are serviced as an assembly.**

5. Identify the rocker arm assemblies and rocker arms and disassemble the shaft as follows:

   a. Thread a nut, washer and spacer onto a 4mm screw.

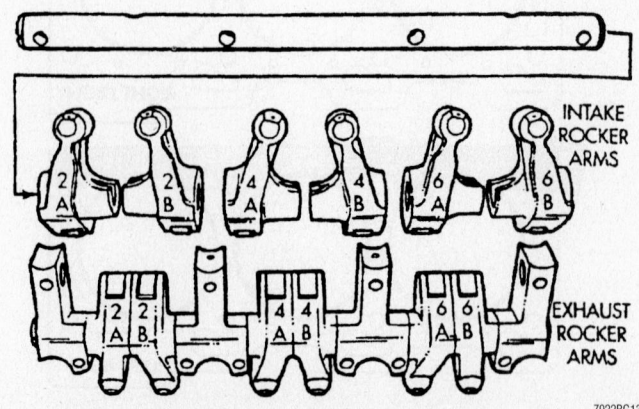

Left bank rocker arm and shaft identification—3.2L and 3.5L engines

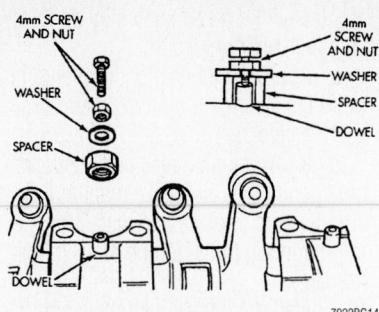

Remove the dowel pin using a 4mm screw, nut, spacer and washer installed into the pin—3.2L and 3.5L engines

b. Insert and tighten the 4mm screw into the dowel pin on the shaft.

c. Loosen the nut on the screw. This will pull the dowel pin from the shaft support.

d. Remove the rocker arms and pedestals, keeping them in order.

e. Check the oil holes for restrictions with a small wire and clean as required.

To install:

6. Assemble the rocker shaft as follows:

a. Install the rocker arms and pedestals onto the shaft, keeping them in the original order.

b. Press the dowel pins into the pedestals until they bottom out in the pedestals.

7. Position the camshaft so that the timing mark on the right camshaft timing belt sprocket aligns with the timing mark on the rear timing belt cover and the timing mark on the left sprocket is 45 degrees from the mark on the rear timing belt cover. There will be no load on the shaft during installation. Install the rocker shafts so the identification marks are facing toward the front of the engine.

8. Install the oil feed bolt in the correct location on the rocker shaft retainer. Torque the bolts in proper sequence to 23 ft. lbs. (31 Nm).

9. Install or connect the following:
- Valve covers and torque the bolts to 105 inch lbs. (12 Nm)
- Intake manifold plenum
- Air cleaner assembly
- Negative battery cable

Intake Manifold

REMOVAL & INSTALLATION

2.7L Engine

1. Before servicing the vehicle, refer to the precautions in the beginning of this section.

2. Disconnect the negative battery cable.

3. Properly relieve the fuel system pressure.

4. Drain the cooling system.

5. Disconnect the air tube from the air cleaner and the throttle body.

6. Hold the throttle lever in the wide-open position and remove the throttle cable and the speed control cable from the lever. Compress the locking tabs on the cables and remove them from the mounting brackets.

7. Unplug the electrical connections from the following:
- Solenoid on the Exhaust Gas Recirculation (EGR) valve transducer
- Manifold Absolute Pressure (MAP) sensor
- Throttle Position (TP) sensor
- Idle Air Control (IAC) motor
- Vacuum hose from the Positive Crankcase Ventilation (PCV) valve as well as the power brake booster at the intake manifold nipple
- Vacuum line at the fuel pressure regulator
- Purge hose from the throttle body
- Electrical connector from the TP sensor and the IAC motor
- EGR tube-to-intake manifold plenum screws
- Intake manifold plenum (upper part of the manifold)

➡**Cover the lower part of the intake manifold to prevent foreign material from entering the engine.**

- Fuel supply and return tubes from the fuel rail at the rear of the intake manifold

8. Disconnect the fuel/return tubes by pushing the quick-connect fitting toward the fuel tube while depressing the built-in disconnect tool with Quick-Connect Fitting tool 6751. To disconnect the fitting from the fuel rail, slightly twist the fitting while maintaining downward pressure on tool 6751. Wrap shop towels around the fuel hoses to absorb any fuel spillage.

9. Plug the fuel line openings to prevent system contamination.

Proper torque sequence for the rocker arm and shaft assemblies—3.2L and 3.5L engines

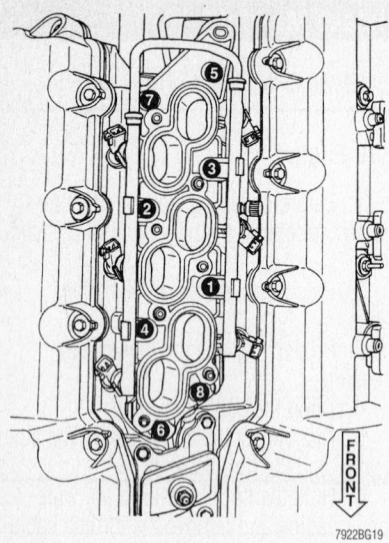

Intake manifold loosening and tightening sequence—2.7L engine

7922BG19

10. Remove or disconnect the following:
- Fuel clamp screw and tubes from the bracket
- Electrical harness from the injectors and turn toward the center of the engine
- Fuel rail mounting bolts and lift the fuel rail with the injectors attached straight up and off the engine. Cover the injector openings.
- Intake manifold bolts and the manifold
- Intake manifold seal retainer screws and intake manifold gasket

11. Clean all mating surfaces.

12. Inspect the manifold for damage, cracks or clogged passages. Repair, clean or replace the manifold as required.

To install:

13. Verify that all intake manifold and cylinder head sealing surfaces are clean. Place a drop of sealant onto each of the 4 corners of the intake manifold gasket, where the cylinder head meets the engine block.

✲✲ CAUTION

The intake manifold gasket is made of very thin metal and can cause cuts if handled carelessly.

14. Install the intake manifold and 8 mounting bolts. Snug down evenly to just 10 inch lbs. (1.1 Nm).

15. Torque the lower intake manifold bolts in the proper sequence to 105 inch lbs. (12 Nm).

16. Install the fuel injectors by performing the following procedure:

a. Apply a light coat of clean engine oil to the O-ring on the nozzle end of each injector.

b. Insert the fuel injector nozzles into the openings in the intake manifold. Seat the injectors in place and install the fuel rail mounting bolts, torque to 16 ft. lbs. (22 Nm).

17. Install or connect the following:
- Electrical connectors to each fuel injector. Rotate the injectors toward the cylinder head covers.
- Fuel supply and return tubes to the fuel rail. Be sure that the black plastic release ring to the quick-connect fitting is in the OUT position. Place special tool 6751 under the largest diameter of the quick-connect fitting.

18. Pull tool 6751 toward the fuel rail until the quick-connect fitting clicks into place. Place the special tool between the shoulder of the built-in disconnect tool and top of the quick-connect fitting, then inspect the security of the fitting by applying a slight downward force against the fitting. It should be locked in place.
- Intake plenum with new gasket onto the intake manifold. Loosely install the mounting bolts.
- EGR tube to the manifold with a new gasket in place. Loosely install the mounting screws.
- Torque the intake manifold plenum mounting bolts to 105 inch lbs. (12 Nm) following the outlined sequence
- Left and right support brackets to the manifold and torque the lower fasteners to 50 inch lbs. (6 Nm) and the upper fasteners to 105 inch lbs. (12 Nm)
- EGR tube mounting bolts
- PCV valve hose and power brake booster hose
- Electrical connectors to the EGR transducer solenoid, IAC motor, Map and TP sensors
- Throttle cable and speed control cable to the mounting bracket and connect to the throttle body lever while holding lever in the wide-open position
- Purge hose to the throttle body
- Reconnect the air tube to the air cleaner and the throttle body
- Negative battery cable

19. Change the engine oil and oil filter.

20. Fill the cooling system. Run the vehicle with the radiator cap removed until

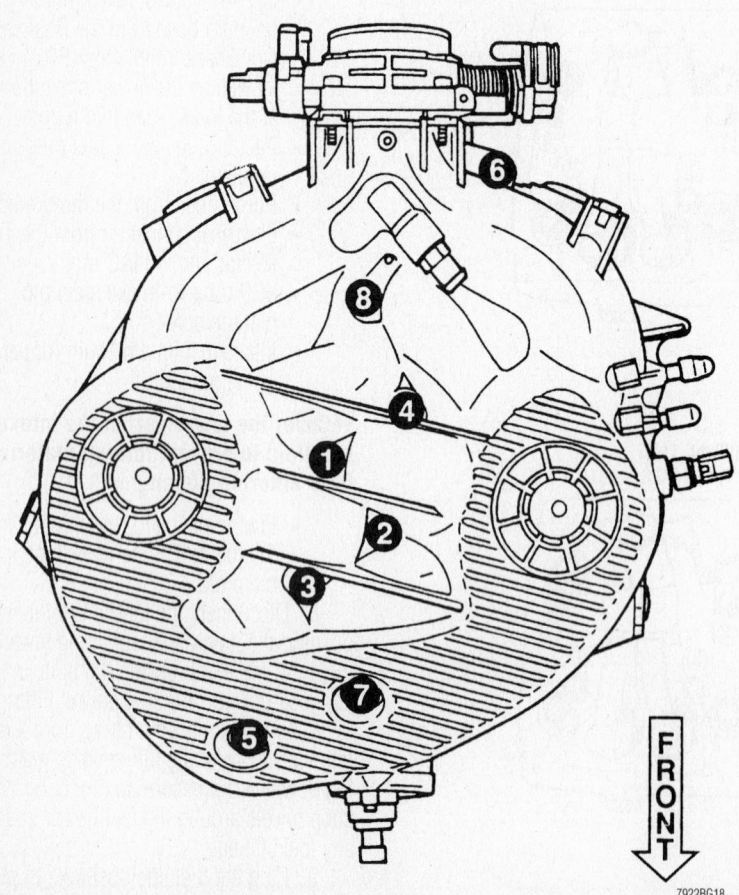

Intake plenum tightening sequence—2.7L engine

7922BG18

the thermostat opens, adding coolant as required. Watch for fuel and coolant leaks and for correct engine operation.

21. Once the vehicle has cooled, recheck the coolant level and add, if necessary.

3.2L and 3.5L Engines

1. Before servicing the vehicle, refer to the precautions in the beginning of this section.

2. Properly relieve the fuel system pressure.

3. Drain the cooling system.

4. Remove or disconnect the following:
- Negative battery cable
- Engine cover from the top of the intake manifold
- Accelerator and the speed control cable from the throttle lever
- Idle Air Control (IAC) motor
- Intake Air Temperature (IAT) sensor
- Manifold Absolute Pressure (MAP) sensor
- Ground screw from the intake manifold
- Electrical connector from the Throttle Position (TP) sensor
- Vacuum hoses from the manifold tuning valve, Positive Crankcase Ventilation (PCV) make-up air hose, IAC motor supply hose and the purge hose from the throttle bodies
- Brake booster hose, PCV hose and the remaining vacuum hoses from the intake manifold
- Exhaust Gas Recirculation (EGR) tube-to-intake manifold plenum bolts
- Plenum support bracket mounting bolts on each side of the plenum
- Intake plenum mounting bolts

➡**The intake manifold plenum (upper half of the intake manifold assembly) uses 2 different length bolts. Take note of their position and be sure they are installed in the same location during installation.**

5. Remove the intake manifold plenum from the intake manifold.

➡**Discard the old gasket. Cover the intake manifold openings with tape to keep debris from entering the engine.**

6. Remove or disconnect the following:
- Upper radiator hose from the thermostat housing
- Heater hose from the rear of the intake manifold
- Lower intake manifold bolts and manifold

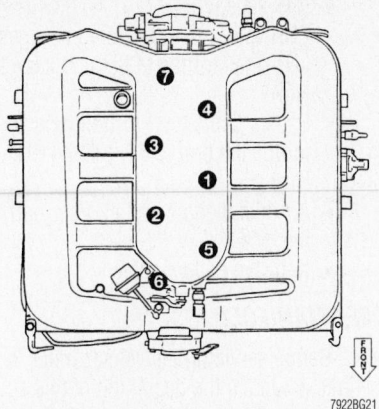

7922BG21

Upper intake manifold loosening and tightening sequence—3.2L and 3.5L engines

➡**Clean all gasket mating surfaces and inspect for distortion with a good straightedge.**

To install:

➡**Verify that all intake manifold and cylinder head sealing surfaces are clean.**

7. Install or connect the following:

- Intake manifold gasket, then the lower manifold. Torque the bolts in the proper sequence to 21 ft. lbs. (28 Nm).
- Upper radiator hose to the thermostat housing
- Heater hose to the rear of the intake manifold

➡**Ensure the ignition cables are routed out of the way of the intake plenum.**

- Intake manifold plenum with a new gasket in place. Torque the mounting bolts, working from the center outward, to 21 ft. lbs. (28 Nm)

➡**Do not overtighten bolts when working with light alloys.**

- Support bracket bolts
- Electrical connectors to the MAP sensor, TP sensor, IAC motor and IAT sensor
- Vacuum hose to the manifold tuning valve
- EGR tube. Torque the EGR tube-to-intake manifold plenum screws to 17 ft. lbs. (22 Nm).

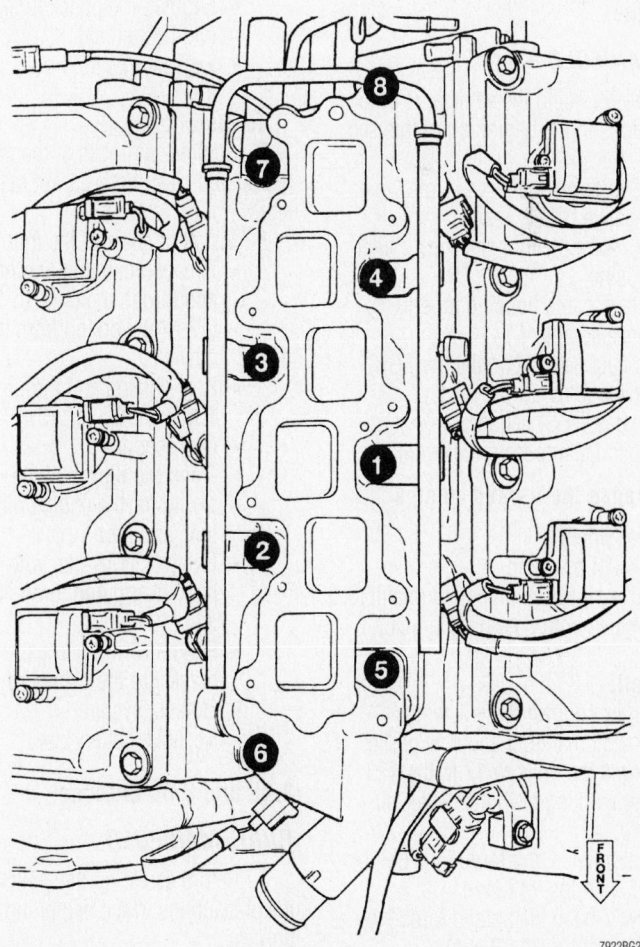

7922BG22

Lower intake manifold loosening and tightening sequence—3.2L and 3.5L engines

➡️**Be sure that the insulation on the EGR tube aligns with and contacts the insulation on the vacuum harness at the rear of the engine. Rotate the throttle lever to the wide-open position and reconnect the speed control and throttle cables.**

- PCV valve hose
- Air cleaner plenum and plenum hose
- Ground wire to the intake manifold plenum
- Brake booster hose to intake manifold plenum fitting
- Throttle body purge tubes
- Intake manifold plenum cover
- Negative battery cable

8. Fill and bleed the cooling system.
9. Change the engine oil and filter.
10. Test run the engine, check for fuel and coolant leaks and verify correct engine operation.

Exhaust Manifold

REMOVAL & INSTALLATION

2.7L Engine

RIGHT MANIFOLD

1. Before servicing the vehicle, refer to the precautions in the beginning of this section.
2. Remove or disconnect the following:
   - Negative battery cable
   - Air intake plenum and the air filter housing
   - Battery cable housing tube-to-transaxle bolt
   - Exhaust Gas Recirculation (EGR) valve and tube
   - Oxygen (O2S) sensor
   - V-band clamp from the manifold

➡️**Do not reuse the V-band clamps.**

- Heat shield
- Exhaust manifold

3. Remove all traces of the old manifold gasket and clean both gasket mating surfaces.

To install:

4. Install or connect the following:
   - Exhaust manifold and new gasket. Torque the bolts to 17 ft. lbs. (23 Nm) working from the center outward.
   - Heat shields and torque the bolts to 105 inch lbs. (12 Nm)
   - New V-band clamp and torque to 100 inch lbs. (11.3 Nm)

- O2S sensor
- EGR valve and tube using new gaskets, then torque to 95 inch lbs. (11 Nm)
- Battery cable tube-to-transaxle and torque the bolt to 75 ft. lbs. (101 Nm)
- Air inlet plenum and air filter housing
- Negative battery cable

LEFT MANIFOLD

1. Before servicing the vehicle, refer to the precautions in the beginning of this section.
2. Remove or disconnect the following:
   - Negative battery cable
   - Exhaust system
   - V-band clamps and left catalytic converter

➡️**Do not reuse the V-band clamps.**

3. Loosen and rotate the transaxle dipstick tube out of the way.
4. Remove or disconnect the following:
   - Engine wiring harness support bracket from the cylinder head
   - Oxygen (O2S) sensor
   - Engine oil dipstick tube and manifold heat shield
   - Exhaust manifold bolts and manifold

To install:
5. Remove all traces of the old manifold gasket and clean both gasket mating surfaces.
6. Install or connect the following:
   - Exhaust manifold and new gasket and torque the bolts to 17 ft. lbs. (23 Nm) working from the center outward
   - Heat shields and torque the bolts to 105 inch lbs. (12 Nm)
   - O2S sensor
   - Transaxle dipstick tube
   - o-ring for the engine oil dipstick tube and tube
   - Catalytic converter with a new V-band clamp and torque to 89 inch lbs. (10 Nm)
   - Engine wiring harness support bracket to the cylinder head
   - Exhaust system
   - Negative battery cable

3.2L and 3.5L Engines

RIGHT MANIFOLD

1. Before servicing the vehicle, refer to the precautions in the beginning of this section.
2. Remove or disconnect the following:

- Negative battery cable
- Exhaust pipes from the exhaust manifold
- Loosen the converter pipe support attaching bolt at the transaxle mount
- A/C belt, loosen only
- Air cleaner housing and air inlet tube
- V-Band clamp from the exhaust manifold connector

➡️**Do not reuse the V-band clamps.**

- A/C compressor bolts and position aside
- Engine oil dipstick
- A/C compressor bracket
- Heated Oxygen Sensor (HO2S) electrical connector and sensor
- Heat shield-to-exhaust manifold screws
- Exhaust manifold bolts and manifold

3. Inspect the manifold for damage or cracks. Check for distortion against a straight-edge or thickness gauge. Replace manifold if required.
4. Remove all traces of the old manifold gasket and clean both gasket mating surfaces.

To install:

5. Install or connect the following:
   - New manifold gasket and exhaust manifold to the cylinder head and torque to 17 ft. lbs. (23 Nm)
   - Heat shield and torque the retaining screws to 11 ft. lbs. (15 Nm)
   - HO2S and electrical connector
   - A/C compressor mounting bracket
   - New V-band clamp and tighten to 100 inch lbs. (11 Nm)
   - Engine oil dipstick tube
   - A/C compressor and drive belt
   - New V-band clamp and tighten to 100 inch lbs. (11 Nm)
   - Exhaust pipe to the exhaust manifold and torque the nuts to 21 ft. lbs. (28 Nm)
   - Nut attaching converter pipe support to the transaxle mount. Torque to 35 ft. lbs. (47 Nm).
   - Air cleaner housing and air inlet tube
   - Negative battery cable

6. Operate the vehicle and inspect for exhaust leaks.

LEFT MANIFOLD

1. Before servicing the vehicle, refer to the precautions in the beginning of this section.
2. Remove or disconnect the following:

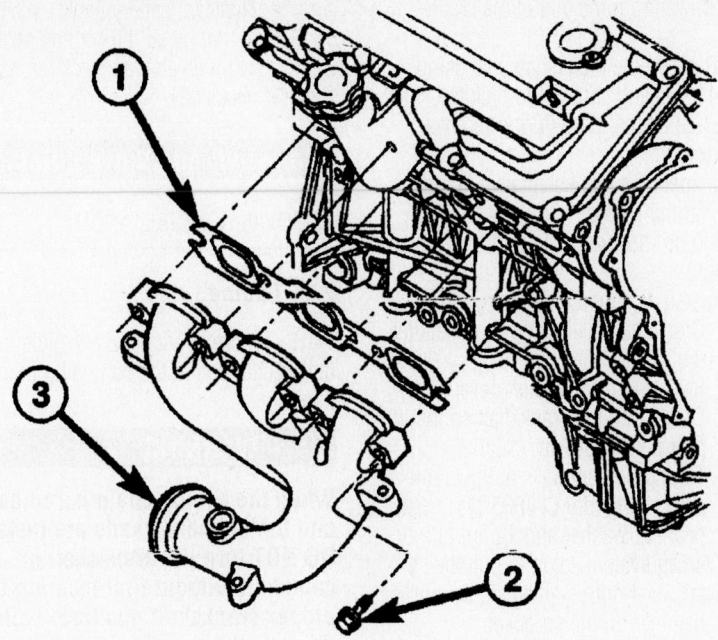

1 - GASKET
2 - BOLT - EXHAUST MANIFOLD
3 - EXHAUST MANIFOLD

67189-300M-G03

Exploded view of the right side exhaust manifold mounting—3.5L engine

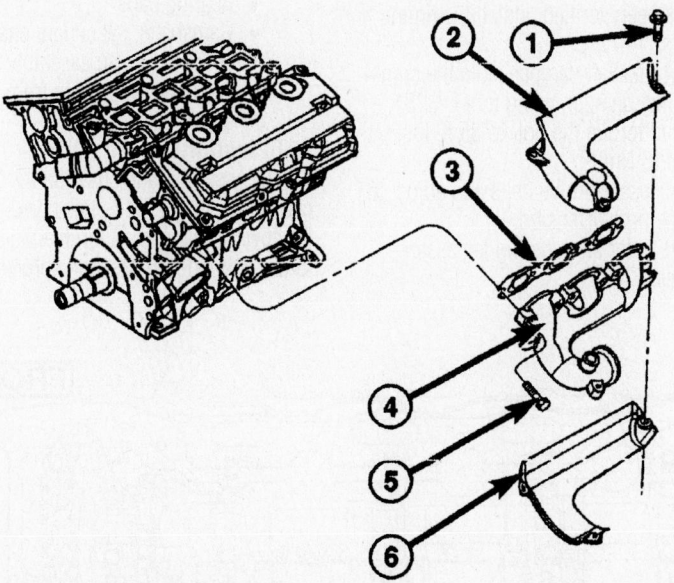

1 - BOLT - HEAT SHIELD
2 - HEAT SHIELD
3 - GASKET
4 - EXHAUST MANIFOLD
5 - BOLT - EXHAUST MANIFOLD
6 - HEAT SHIELD

67189-300M-G04

Exploded view of the left side exhaust manifold mounting—3.5L engine

- Negative battery cable
- Exhaust pipes or system, as necessary from the exhaust manifold
- Loosen the converter pipe support attaching bolt at the transaxle mount
- V-Band clamp from the exhaust manifold connector

➡ **Do not reuse the V-band clamps.**

- Electrical connector harness bracket
- Heated Oxygen Sensor (HO2S) electrical connector and sensor
- Heat shield-to-exhaust manifold screws
- Exhaust manifold bolts and manifold

3. Inspect the manifold for damage or cracks. Check for distortion against a straight-edge or thickness gauge. Replace manifold if required.

4. Remove all traces of the old manifold gasket and clean both gasket mating surfaces.

To install:

5. Install or connect the following:
- New manifold gasket and exhaust manifold to the cylinder head and torque to 17 ft. lbs. (23 Nm)
- Heat shield and torque the retaining screws to 11 ft. lbs. (15 Nm)
- HO2S and electrical connector
- Electrical connector bracket to brace
- New V-band clamp and tighten to 100 inch lbs. (11 Nm)
- Nut attaching converter pipe support to 35 ft. lbs. (47 Nm)
- Exhaust pipes/system to the manifold
- Negative battery cable

6. Operate the vehicle and inspect for exhaust leaks.

Front Crankshaft Seal

➡ **The front crankshaft seal procedures are for timing belt equipped engines only. For engines that utilize timing chains, please refer to the applicable procedure later in this section.**

REMOVAL & INSTALLATION

3.2L and 3.5L Engines

Note that the timing belt must be removed from the vehicle to perform this service. Use care to be sure all valve timing marks are carefully aligned both before removing the belt and after belt installation

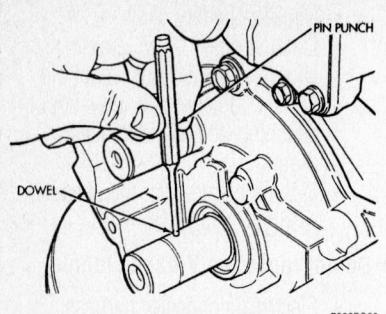

Removing the timing belt sprocket dowel pin from the crankshaft—3.2L and 3.5L engines

and all service has been completed. It may be good practice to set the engine to Top Dead Center (TDC) No. 1 cylinder compression stroke (firing position) and aligning all timing marks before removing the timing belt. This serves as a reference for all work that follows.

1. Before servicing the vehicle, refer to the precautions at the beginning of this section.
2. Properly relieve the fuel system pressure.
3. Drain the cooling system.
4. Remove or disconnect the following:
   - Negative battery cable
   - Radiator and cooling fan module assembly
   - Accessory drive belts
   - Crankshaft damper bolt
   - Timing belt front cover

➡The sealer on the timing belt front cover may be reusable and should not be removed. Use silicone rubber adhesive sealant to replace any missing sealer.

   - Timing belt and tensioner. Refer to the appropriate section for the removal and installation procedure.
   - Crankshaft timing belt sprocket
5. Locate the small dowel pin in the crankshaft. With a small punch, carefully tap

out the dowel from the end of the crankshaft.

6. Remove the crankshaft seal using tool 6341A, taking care not to nick the shaft seal surface or seal bore during removal.

To install:

7. Inspect the crankshaft seal lip surface for varnish and dirt. Polish the area using 400 grit sandpaper to remove varnish as necessary.
8. Install or connect the following:
   - Crankshaft seal using seal installer tool 6342
   - Rear lower timing belt cover
   - Dowel into the crankshaft so that it protrudes 0.047 in. (1.2mm)
   - Timing belt sprocket at the crankshaft using tool C-4685C1, thrust bearing, washer and 12mm bolt or an equivalent setup to pull the sprocket onto crankshaft. Do not hammer on the sprocket.
9. Verify that all valve timing marks are aligned.
   - Timing belt and tensioner using the recommended procedure
10. Rotate the crankshaft 2 complete turns and recheck the timing marks on the camshafts and crankshaft. The marks must align with their respective locations. If the marks do not align, repeat the timing belt installation procedure. When correct valve timing has been verified, install the timing belt covers.
    - Crankshaft damper. Hold the crankshaft damper, using tool L-3281, and torque the bolt to 85 ft. lbs. (115 Nm).
    - Accessory drive belts and adjust to the proper tension
    - Radiator and cooling fan assemblies

   - Negative battery cable
11. Fill and bleed the cooling system.
12. Start the vehicle, check for leaks and repair if necessary.

Camshaft and Valve Lifters

REMOVAL & INSTALLATION

2.7L Engine

1. Before servicing the vehicle, refer to the precautions in the beginning of this section.

❄❄ WARNING

When the timing chain is removed and the cylinder heads are installed, DO NOT turn the crankshaft or camshaft without first locating the proper crankshaft position. Failure to do so will result in piston-to-valve contact.

2. Remove or disconnect the following:
   - Primary timing chain
   - Second chain tensioner mounting bolts
3. Slowly loosen the camshaft bearing cap retaining bolts in the reverse order of the tightening sequence.
   - Bearing caps
   - Camshafts, secondary chain and tensioner as an assembly
   - Tensioner and chain from the crankshaft

To install:

4. Assemble the chain on the camshafts. Ensure the plated links are facing toward the front. Align the plated links to the dots on the camshaft sprockets.

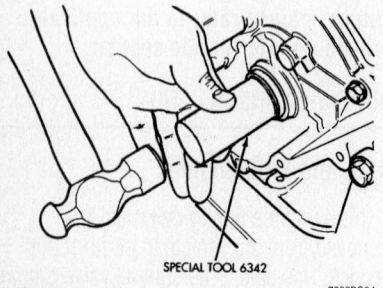

Installing the crankshaft oil seal—3.2L and 3.5L engines

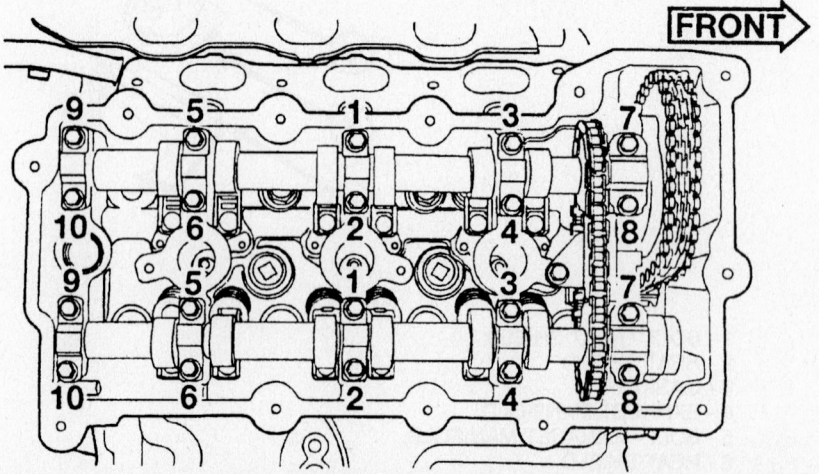

Camshaft bearing cap tightening sequence—2.7L engine

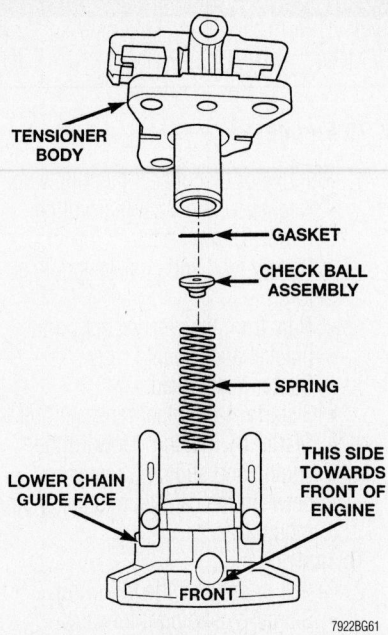

Exploded view of the camshaft (secondary) chain tensioner, early build—2.7L engine

➡There are 2 different styles of camshaft (secondary) chain tensioners. The Early Build tensioners will separate into sub-components, the Later Build tensioners will not. Compress the camshaft (secondary) chain as follows:

5. For early build vehicles:

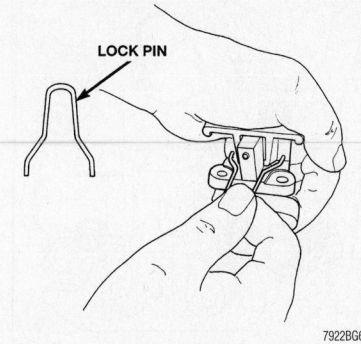

Fabricate a lockpin, as shown, to keep the tensioner compressed—2.7L engine

a. Separate the tensioner cylinder from the tensioner housing.

b. Carefully drain the oil from the housing using care not to remove the internal tensioner components.

c. Assemble the tensioner housing.

d. Using hand pressure, compress and lock the tensioner using a fabricated lockpin.

6. For late build vehicles:

a. Place the tensioner in a soft-jawed vise.

b. Slowly compress the tensioner until the fabricated lockpin can be installed.

c. Remove the compressed and locked tensioner from the vise.

7. Insert the compressed and locked camshaft chain tensioner in between the camshafts and chain.

8. Position the camshafts so that the plated links and dots are facing upward.

9. Install or connect the following:

- Camshafts

❋❋ WARNING

Ensure that the rocker arms are correctly seated and in proper positions.

- Camshaft bearing caps and torque the bolts gradually, in sequence, to 105 inch lbs. (12 Nm)
- Secondary chain tensioner and torque the bolts to 105 inch lbs. (12 Nm)
- Primary timing chain
- Negative battery cable

10. Remove the lockpin from the secondary chain tensioner.

3.2L and 3.5L Engines

Camshafts are serviced from the rear of the cylinder head. Although the engine does not need to be removed for camshaft service, the cylinder head must be removed from the vehicle. Note too, that the camshaft sprockets have a D-shaped hole that allows it to rotate several degrees in each direction on its shaft.

1. Before servicing the vehicle, refer to the precautions in the beginning of this section.

2. Properly relieve the fuel system pressure.

3. Drain the cooling system.

4. Remove or disconnect the following:

- Negative battery cable
- Radiator/cooling fan assemblies and the accessory drive belts
- Crankshaft damper and the timing belt covers

➡Mark the timing belt rotation direction for installation. Align the timing belt sprockets with marks on the rear timing belt covers before removing the timing belt.

- Timing belt tensioner and timing belt. Refer to the appropriate section for the removal and installation procedure.
- Camshaft timing belt sprockets
- Intake manifold assembly using the recommended procedure
- Exhaust manifold

➡Be sure to clean the gasket mating surfaces between the exhaust manifold and the cylinder head.

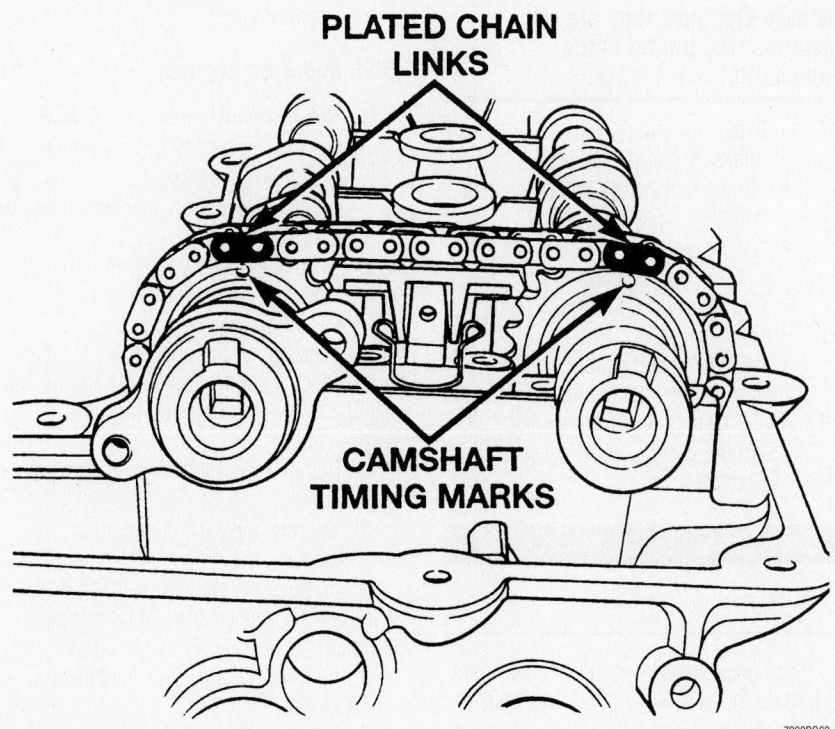

Proper camshaft (secondary) chain alignment—2.7L engine

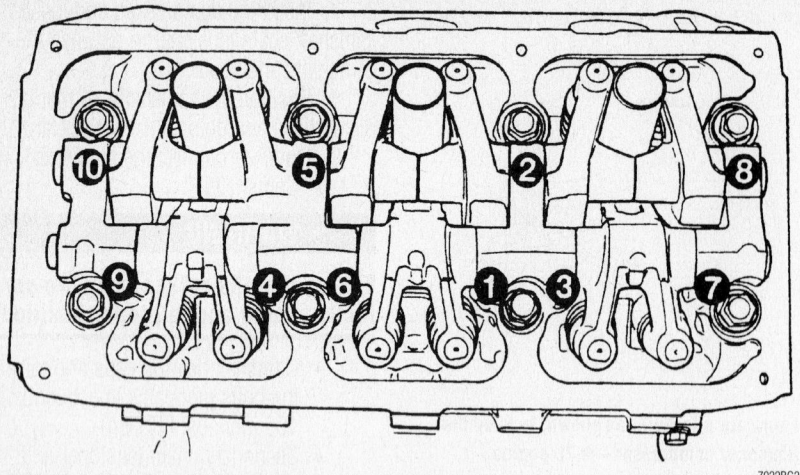

Rocker arm/shaft tightening sequence—3.2L and 3.5L engines

7922BG27

5. The rear timing belt cover must be removed to remove the cylinder heads. Remove the rear timing belt cover-to-cylinder head bolts. Remove the rear timing belt covers.

➡ **The right-hand side timing belt cover has O-rings located behind it for the water pump passages.**

6. Remove the cylinder head bolts and cylinder head.

➡ **Mark the rocker arm assembly to note component locations before disassembly.**

7. Remove the rocker arm and shaft assemblies.

8. Remove the rear camshaft cover and O-ring.

✳✳ WARNING

Carefully, remove the camshaft from the rear of the head taking care not to nick or scratch the journals.

9. Inspect camshaft journals for wear or damage. If wear is present, inspect the cylinder head for damage. Inspect the head oil holes for clogging. Replace the camshaft as required.

10. Measure the height of the cam using a micrometer. Measure in 2 places: the unworn area and in the wear zone. Subtract the figures to get cam wear. The standard specification is 0.001 in. (0.0254mm) with the wear limit being 0.010 in. (0.254mm). Replace the camshaft if it is worn beyond this specification.

To install:

11. Lubricate the camshaft journals and lobes with clean engine oil.

12. Install or connect the following:
   • Camshaft

   • Camshaft cover and O-ring and torque the bolts to 21 ft. lbs. (28 Nm)
   • Rocker arm assemblies
   • Cylinder head assembly

✳✳ WARNING

New head bolts are recommended.

   • Rear timing belt covers
   • Timing belt sprocket, timing belt and timing belt tensioner. Refer to the appropriate section for the removal and installation procedure.

✳✳ WARNING

Be sure that once they are all installed, the timing of the camshaft(s) is accurate.

   • Exhaust manifold with new gasket
   • Intake manifold assembly
   • Timing belt covers and crankshaft damper
   • Accessory drive belts and set them to the proper tension
   • Radiator and cooling fan assembly
   • Negative battery cable

13. Fill and bleed the cooling system. An oil and filter change is recommended.

14. Start the vehicle, check for leaks and repair if necessary.

Valve Lash

ADJUSTMENT

These engines use hydraulic roller lifters to take up the free-play in the valve train system, therefore no lash adjustments are necessary.

Starter Motor

REMOVAL & INSTALLATION

2.7L Engine

1. Remove or disconnect the following:
   • Negative battery cable from the remote ground post
   • Battery feed and posi-lock connectors
   • Nuts from the catalyst support bracket and mount
   • Starter heat shield
   • 3 starter-to-engine/transaxle bolts
   • Starter by rotating it toward the engine and sliding it rearward between the catalyst and the engine mount

To install:

2. Install or connect the following:
   • Starter by sliding it forward between the catalyst and the engine mount and rotating it away from the engine
   • Starter and torque the 3 starter-to-engine/transaxle bolts to 40 ft. lbs. (54 Nm)
   • Starter heat shield
   • Nuts to the catalyst support bracket and mount
   • Positive battery cable and torque the nut to 89 inch lbs. (10 Nm)
   • Posi-lock connectors
   • Negative battery cable to the remote ground post

3.2L and 3.5L Engines

1. Remove or disconnect the following:
   • Negative battery cable from the remote ground post
   • Starter-to-engine/transaxle nut and bolts
   • Positive battery feed wire from the starter
   • Starter and position it to gain access to the posi-lock connector

2. Place a support under the engine and slightly relieve the pressure from the left engine mount.

3. Remove the 3 left engine mount-to-engine bolts.

4. Slightly, raise the engine to provide more room.

5. Remove the starter by sliding it rearward between the catalyst and the engine mount.

6. Remove the posi-lock connector.

To install:

7. Install or connect the following:
   • Posi-lock connector

- Starter by sliding it forward between the catalyst and the engine mount

8. Slightly, lower the engine.
   - 3 left engine mount-to-engine bolts
   - Positive battery feed wire to the starter and torque the nut to 89 inch lbs. (10 Nm)
   - Starter and torque the starter-to-engine/transaxle bolts to 40 ft. lbs. (54 Nm)
   - Negative battery cable to the remote ground post

Oil Pan

REMOVAL & INSTALLATION

1. Before servicing the vehicle, refer to the precautions in the beginning of this section.
2. Drain the engine oil and remove the oil filter.
3. Remove or disconnect the following:
   - Negative battery cable
   - Dipstick and housing
   - Structural collar from the rear of the oil pan and transmission housing
   - Engine oil cooler lines from the oil pan, if equipped
   - Transmission oil cooler line clips, if necessary
   - Oil pan mounting bolts, oil pan and gasket

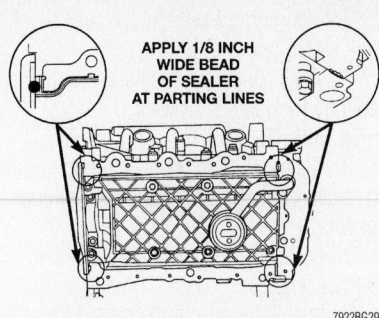

APPLY 1/8 INCH WIDE BEAD OF SEALER AT PARTING LINES

7922BG29

To ensure a proper seal, apply sealer as shown—2.7L and 3.5L engines

4. Clean the oil pan and all gasket surfaces.

To install:

5. Apply a ⅛ in. (3mm) bead of sealer at the parting line of the oil pump body and the rear seal retainer.
6. Install or connect the following:
   - Oil pan and torque the M8 nuts/bolts to 21 ft. lbs. (28 Nm) and the M6 nuts/bolts to 105 inch lbs. (12 Nm)
7. Install the structural collar using the following procedure:
   - a. Install the vertical collar to the oil pan mounting bolts and tighten, temporarily to 10 inch lbs. (1.1 Nm).
   - b. Install the collar-to-transaxle bolts and torque to 40 ft. lbs. (55 Nm).
   - c. Starting with the center vertical bolt and working outward. Torque the bolts to 40 ft. lbs. (55 Nm).

8. Install the dipstick and housing and connect the negative battery cable.
9. Fill the engine with the proper amount of clean SAE 5W-30 or SAE 10W-30 engine oil only. Do not mix the two grades of oil.
10. Start the engine, check for leaks and repair if necessary.

Oil Pump

REMOVAL & INSTALLATION

2.7L Engine

1. Before servicing the vehicle, refer to the precautions in the beginning of this section.
2. Drain the engine oil.
3. Remove or disconnect the following:
   - Crankshaft damper
   - Timing chain cover
   - Timing chain
   - Crankshaft sprocket
   - Oil pan
   - Oil pickup tube and o-ring
   - Oil pump

To install:

4. Fill the oil pump rotor cavity with clean engine oil.
5. Carefully, install the oil pump over the crankshaft and into position.
6. Install the oil pump mounting bolts and torque to 21 ft. lbs. (28 Nm).
7. Lubricate the new pickup tube O-ring with clean engine oil and torque the pickup tube mounting bolts to 21 ft. lbs. (28 Nm).
8. Install or connect the following:
   - Oil pan
   - Crankshaft sprocket
   - Timing chain
   - Timing chain cover
   - Crankshaft damper

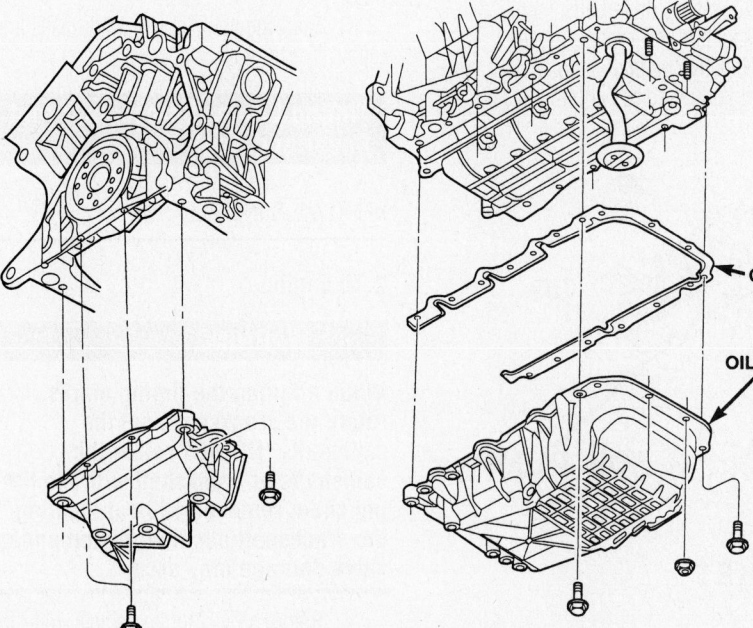

GASKET

OIL PAN

7922BG28

Exploded view of the oil pan removal and installation—2.7L and 3.5L engines

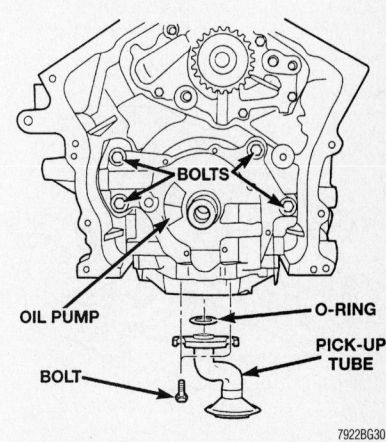

BOLTS

OIL PUMP

BOLT

O-RING

PICK-UP TUBE

7922BG30

Oil pump mounting bolt locations—2.7L engine

9. Fill the engine with clean engine oil.

10. Start the vehicle, check for leaks and repair if necessary.

3.2L and 3.5L Engines

The timing belt must be removed to access the oil pump located behind the crankshaft drive sprocket. It is good practice to turn the crankshaft to Top Dead Center (TDC) No. 1 cylinder compression stroke (firing position) before starting disassembly. This should align all timing marks and be a good point of reference for all work to follow.

1. Before servicing the vehicle, refer to the precautions in the beginning of this section.

2. Drain the engine oil.

3. Drain the cooling system.

4. Remove or disconnect the following:

- Negative battery cable
- Accessory drive belts
- Oil filter
- Oil pan
- Oil pump pickup tube
- Windage tray/oil pan gasket
- Crankshaft damper using a suitable puller tool
- Timing belt covers

5. Place matchmarks on the timing belt to aid installation. Align the matchmarks on the camshaft sprockets to marks on the rear timing belt covers before removing the timing.

6. Remove or disconnect the following:

- Timing belt. Refer to the appropriate section for the removal and installation procedure.
- Crankshaft sprocket using a suitable puller tool
- Oil pump-to-engine screws and pump

- Oil pump cover screws and cover
- Oil pump rotors

7. Wash all parts in solvent and inspect carefully for damage or wear.

To install:

8. Clean all parts well. There should be no traces of old gasket/sealer on any components.

9. Assemble the oil pump with new parts as required.

10. Install the oil pump cover. Torque the fasteners to 108 inch lbs. (12 Nm).

11. Prime the oil pump prior to installation by filling the rotor cavity with clean engine oil.

12. Install the oil pump and tighten the oil pump-to-engine screws as follows:

    a. M8 screws: 21 ft. lbs. (28 Nm).

    b. M10 screws: 40 ft. lbs. (55 Nm).

13. Tighten the oil pan drain plug and install a new oil filter.

14. Install or connect the following:

- Oil pump pickup tube
- Windage tray/oil pan gasket
- Oil pan and torque fasteners to 108 inch lbs. (12 Nm). Pay attention to sealing the oil pan gasket and its integral windage tray.
- Crankshaft sprocket using tool C-4685C1, thrust bearing, washer and 12mm bolt to draw the sprocket onto the crankshaft
- Timing belt. Refer to the appropriate section for the removal and installation procedure.
- Timing belt covers, thrust bearing, washer plate and vibration damper using tool L-4524
- Accessory drive belts
- Radiator and radiator hoses
- Negative battery cable

15. Fill and bleed the cooling system.

16. Fill the engine with the correct amount of clean SAE 5W-30 or SAE 10W-30 engine oil only. Do not mix the two grades of oil.

17. Start the engine, check for leaks and proper oil pressure.

Rear Main Seal

REMOVAL & INSTALLATION

1. Before servicing the vehicle, refer to the precautions in the beginning of this section.

2. Drain the transaxle fluid.

3. Remove the negative battery cable.

4. Remove the transaxle, inspection cover and flywheel/flexplate.

5. Using a small prytool, carefully pry out the rear oil seal. Be careful not to nick or damage the crankshaft flange seal surface or the retainer bore.

To install:

6. Place the Seal Pilot tool C-4681 on the crankshaft.

7. Lightly coat the oil seal outside diameter with Loctite® Stud N' Bearing Mount® or the equivalent.

8. Apply a light coating of engine oil to the entire circumference of the oil seal lip.

9. Place the seal over the special tool and tap the seal in place with a plastic mallet.

10. Install the flexplate/flywheel and transaxle.

11. Connect the negative battery cable.

12. Fill the transaxle with the proper fluid.

13. Start the vehicle, check for leaks and repair if necessary.

Timing Chain, Sprockets, Front Cover and Seal

REMOVAL & INSTALLATION

2.7L Engine

✴✴ WARNING

When aligning the timing marks, rotate the crankshaft, not the camshafts. DO NOT rotate the camshafts or crankshaft with the timing chain removed without locating the crankshaft position, piston and/or valve damage may occur.

1. Before servicing the vehicle, refer to the precautions in the beginning of this section.

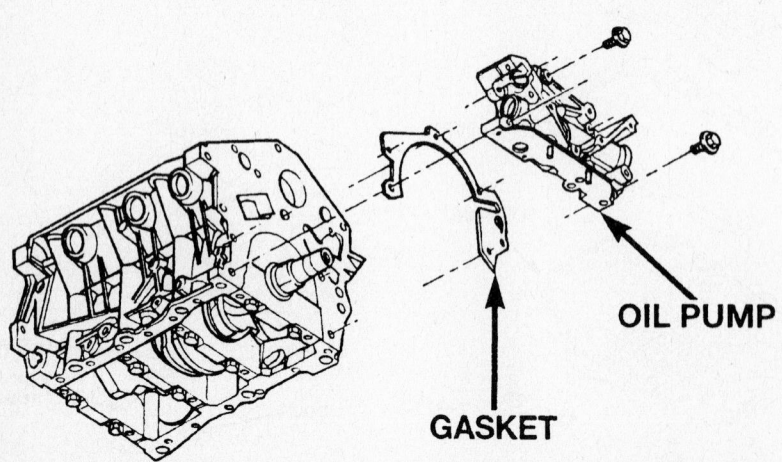

OIL PUMP

GASKET

7922BG31

Prime the oil pump before installation, because a dry pump will wear prematurely and cause low oil pressure—3.2L and 3.5L engines

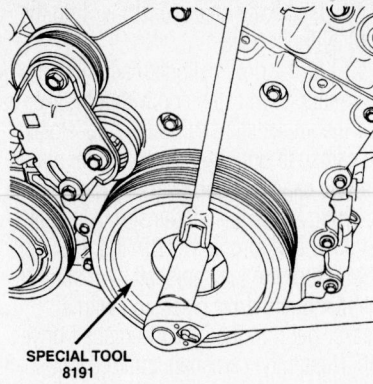

SPECIAL TOOL 8191

7922BG32

Removing the crankshaft center bolt using the Crankshaft Damper Holder tool—2.7L engine

2. Remove or disconnect the following:
- Upper intake manifold and valve covers
- Upper radiator crossmember
- Fan module
- Accessory drive belts

3. Using Crankshaft Damper Holder tool 8191, hold the crankshaft and remove the center bolt.
- Damper using a 3-jaw puller
- Power steering pump and position

it aside without disconnecting the hydraulic lines
- Accessory drive belt tensioner pulley
- Timing chain cover bolts

4. Clean and inspect the sealing surfaces.

5. Align the crankshaft sprocket timing mark with the oil pump housing mark.

➡**The mark on the oil pump housing is 60 degrees at Top Dead Center (TDC).**

6. Remove or disconnect the following:
- Primary timing chain tensioner from the right cylinder head
- Camshaft Position (CKP) sensor and timing chain access plug from the left cylinder head

➡**The camshafts will rotate clockwise, when the camshaft sprocket bolts are removed.**

- Right camshaft sprocket mounting bolts, camshaft damper and sprocket
- Left camshaft sprocket bolts and sprocket
- Lower timing chain guide, tensioner arm and primary timing chain

To install:

➡**Lubricate the timing chain and guides with clean engine oil before installation.**

7. Install the timing chain by performing the following procedure:

a. Verify that the crankshaft sprocket timing mark is aligned with the mark on the oil pump housing.

b. Place the left side primary timing chain sprocket onto the chain, while aligning the timing mark on the sprocket to the 2 plated links on the chain.

c. Lower the chain with the left sprocket through the left cylinder head opening.

d. Loosely position the left camshaft sprocket over the camshaft hub.

e. Align the plated link to the crankshaft sprocket timing mark.

f. Position the timing chain around the water pump drive sprocket.

g. Align the right camshaft sprocket timing mark to the plated link on the timing chain and loosely position the sprocket over the camshaft hub.

h. Verify that all the plated links are aligned to their proper timing marks.

8. Install the left lower timing chain

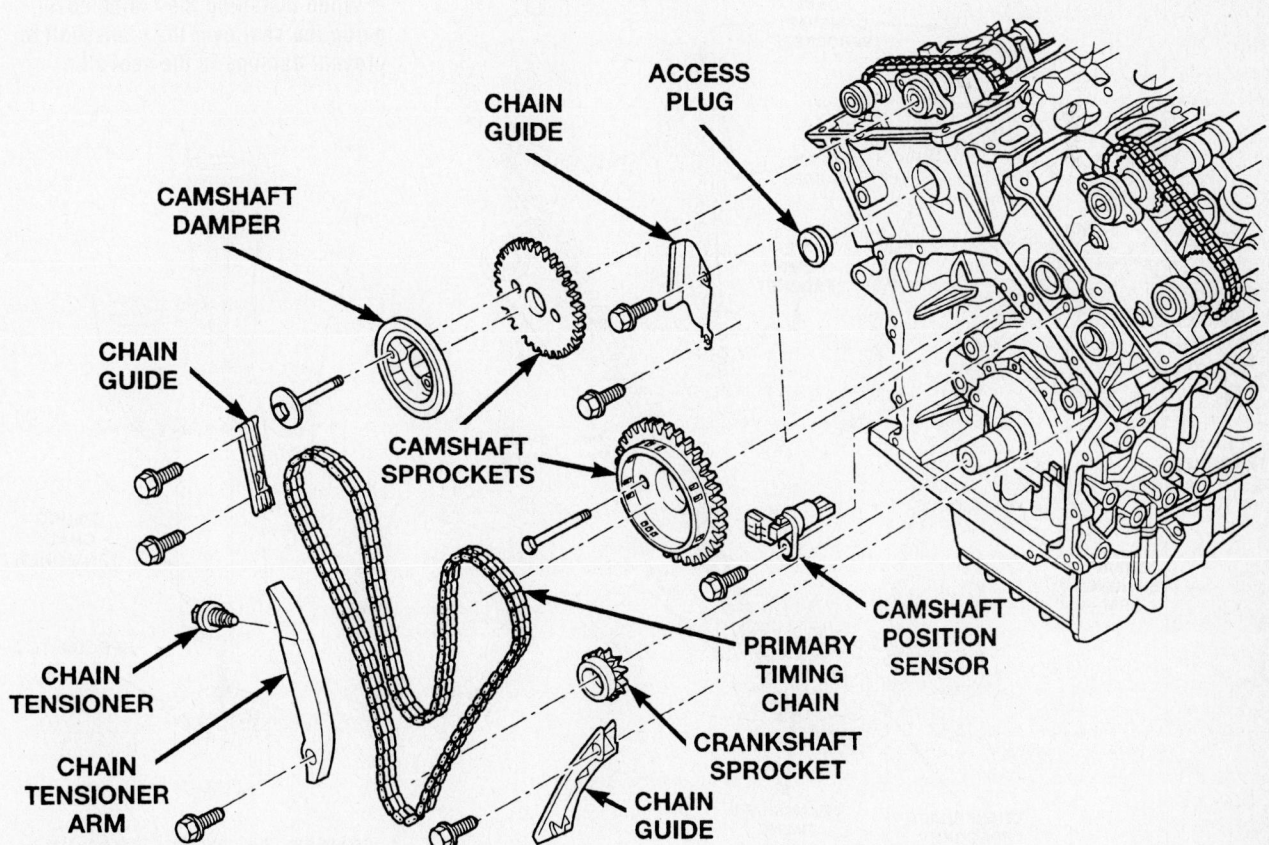

CHAIN GUIDE · **ACCESS PLUG** · **CAMSHAFT DAMPER** · **CHAIN GUIDE** · **CAMSHAFT SPROCKETS** · **CHAIN TENSIONER** · **CHAIN TENSIONER ARM** · **PRIMARY TIMING CHAIN** · **CRANKSHAFT SPROCKET** · **CHAIN GUIDE** · **CAMSHAFT POSITION SENSOR**

7922BG33

Exploded view of the timing chain drive assembly—2.7L engine

guide and tensioner. Torque the bolts to 21 ft. lbs. (28 Nm).

➡**Inspect the timing chain guide access plug O-rings before installing. Replace damaged O-rings as necessary.**

9. Install the timing chain guide access plug to the left cylinder head and torque to 15 ft. lbs. (20 Nm).

➡**To reset the timing chain tensioner, oil will first need to be purged from the tensioner.**

10. Purge oil from the timing chain tensioner using the following procedure:

a. Remove the tensioner from the tensioner housing.

b. Place the check ball end of the tensioner into the shallow end of the Tensioner Resetting Special tool 8186.

c. Using hand pressure, slowly depress the tensioner until oil is purged from the cylinder.

d. Reinstall the tensioner into the tensioner housing.

11. Reset the timing chain tensioner using the following procedure:

a. Position the cylinder plunger into the deeper side of the Tensioner Resetting special tool 8186.

b. Apply a downward force until the tensioner is reset.

✺✺ WARNING

Ensure that the tensioner is properly reset. The tensioner body must be bottomed against the top edge of the Tensioner Resetting Special tool 8186. Failure to properly perform the resetting procedure may cause tensioner jamming.

12. Install the chain tensioner into the right cylinder head.

13. At the right cylinder head, insert a ⅜ in. square drive extension with a breaker bar into the intake camshaft drive hub. Rotate the camshaft until the camshaft hub aligns with the camshaft sprocket and damper attaching holes. Install the sprocket attaching bolts and torque to 21 ft. lbs. (28 Nm).

14. Turn the left camshaft by inserting a ⅜ in. square drive extension with a breaker bar into the intake camshaft drive hub. Rotate the camshaft until the camshaft hub aligns with the camshaft sprocket and damper attaching holes. Install the sprocket attaching bolts and torque to 21 ft. lbs. (28 Nm).

15. If necessary, rotate the engine slightly clockwise to remove any slack in the timing chain.

16. To arm the timing chain tensioner: Use a flat-bladed prytool to gently pry the tensioner arm towards the tensioner slightly. Then, release the tensioner arm. Verify the tensioner extends.

17. Inspect and replace the timing chain cover gasket and oil seal.

18. Apply a ⅛ in. (3mm) bead of sealer at the parting line of the oil pan and engine block.

➡**When installing the timing cover, guide the seal over the crankshaft to prevent damage to the seal's lip.**

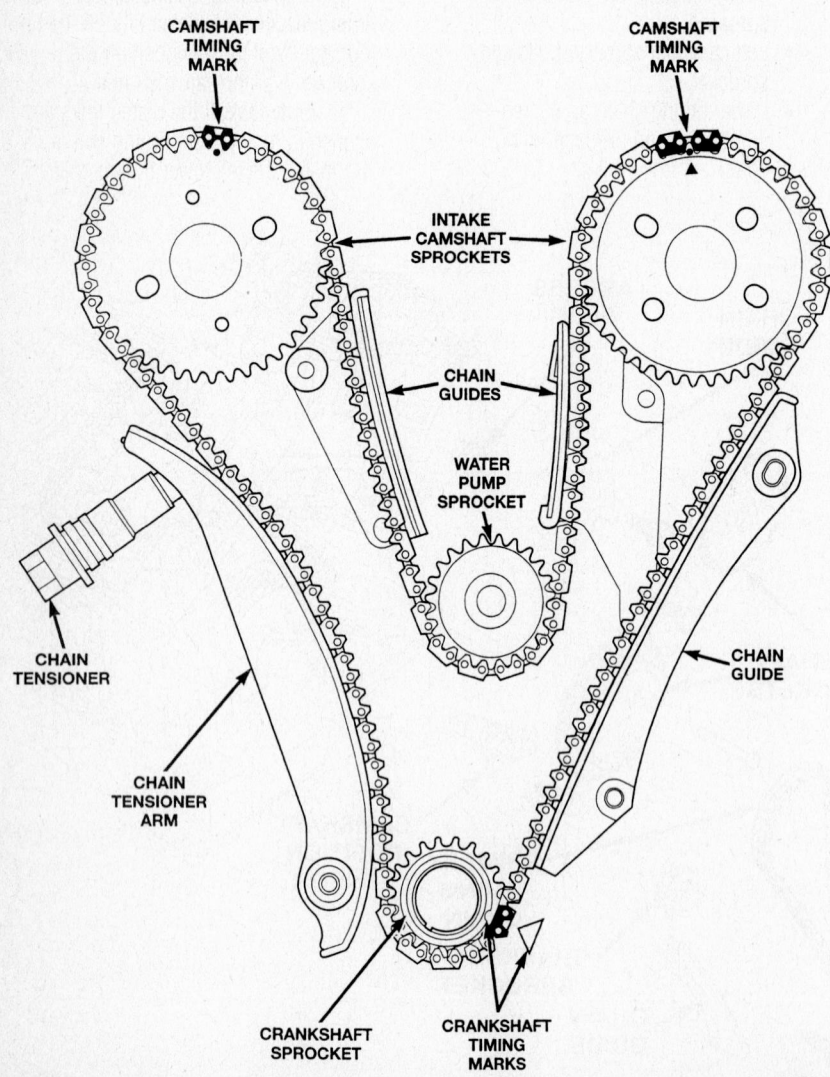

After the timing chain is installed, the timing marks should be aligned—2.7L engine

7922BG34

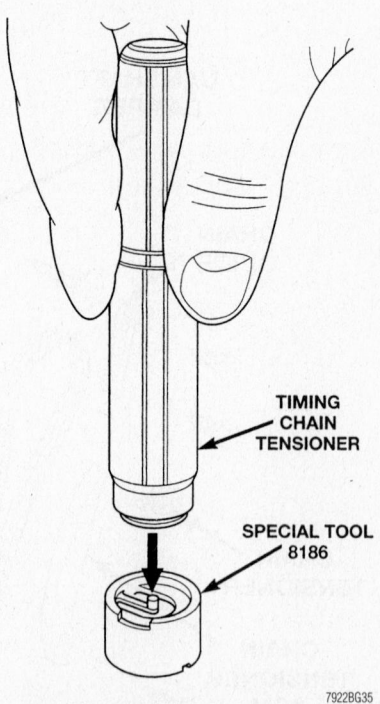

Using the Tensioner Resetting Special tool, to purge the oil from the tensioner—2.7L engine

7922BG35

19. Install or connect the following:
- Timing cover and gasket and torque the M10 bolts to 40 ft. lbs. (54 Nm) and the M6 bolts to 105 inch lbs. (12 Nm)
- Crankshaft damper
- Accessory drive belt tensioner pulley
- Power steering pump
- Crankshaft damper
- Crankshaft center bolt and torque the bolt to 125 ft. lbs. (170 Nm) using tool 8191
- Accessory drive belts
- Fan module and electrical wiring harness
- Upper radiator crossmember
- Negative battery cable

Timing Belt

REMOVAL & INSTALLATION

3.2L and 3.5L Engines

Use care when servicing a timing belt. Valve timing is absolutely critical to engine performance. If the valve timing marks on all drive sprockets are not properly aligned, engine damage will result. If only the belt and tensioner are being serviced, do not loosen the camshaft drive sprockets unless they are to be replaced. The sprockets have oversized openings and can be rotated several degrees in each direction on their shafts. This means the sprockets must be re-timed, requiring some special tools.

❋❋ CAUTION

Fuel injection systems remain under pressure, even after the engine has been turned off. The fuel system pressure must be relieved before disconnecting any fuel lines. Failure to do so may result in fire and/or personal injury.

1. Before servicing the vehicle, refer to the precautions in the beginning of this section.
2. Disconnect the negative battery cable.
3. Rotate the engine to Top Dead Center (TDC) on the compression stroke for cylinder No. 1.
4. Release the fuel system pressure using the recommended procedure.
5. Place a pan under the radiator and drain the coolant.
6. Remove or disconnect the following:

- Radiator and cooling fan assemblies
- Accessory drive belts
- Upper radiator hose
- Crankshaft damper with a quality puller tool gripping the inside of the pulley
- Stamped steel cover. Do not remove the sealer on the cover; it may be reusable.
- Left side cast cover
- Lower belt cover, located behind the crankshaft damper, if necessary

7. If the timing belt is to be reused, mark the timing belt with the running direction for installation.
8. Align the camshaft sprockets with the marks on the rear covers.

- Timing belt and tensioner

9. If it is necessary to service the camshaft sprockets, use the following procedure:

   a. Hold the camshaft sprocket with a 36mm box end wrench, loosen and remove the sprocket retaining bolt and washer.

➡ **To remove the camshaft sprocket retainer bolt while the engine is in the vehicle, it may be necessary to raise that side of the engine due to the length of the retainer bolt. The right bolt is 8⅜ in. (213mm) long, while the left bolt is 10 in. (254mm) long. These bolts are not interchangeable and their original location during removal should be noted.**

b. Remove the camshaft sprocket from the camshaft. The camshaft sprockets are not interchangeable from side-to-side.

c. Remove the crankshaft sprocket using Puller L-4407A.

To install:

10. If it was necessary to remove the camshaft sprockets, use the following procedure:

➡ **This procedure can only be used when the camshaft sprockets have been loosened or removed from the camshafts. Each sprocket has a D shaped hole that allows it to be rotated several degrees in each direction on its shaft. This design must be timed with the engine to ensure proper performance.**

   a. Install the crankshaft sprocket, using tool C-4685-C1, thrust bearing, washer and 12mm bolt.

   b. When the camshaft sprockets are loosened or removed, the camshafts must be timed to the engine. Install the Camshaft Alignment tools 6642-A, to the rear of the cylinder heads. These tools lock the camshafts in the proper position.

11. Preload the belt tensioner as follows:

   a. Place the tensioner in a vise the same way it is mounted on the engine.

   b. Slowly compress the plunger into the tensioner body.

   c. When the plunger is compressed into the tensioner body, install a pin through the body and plunger to retain

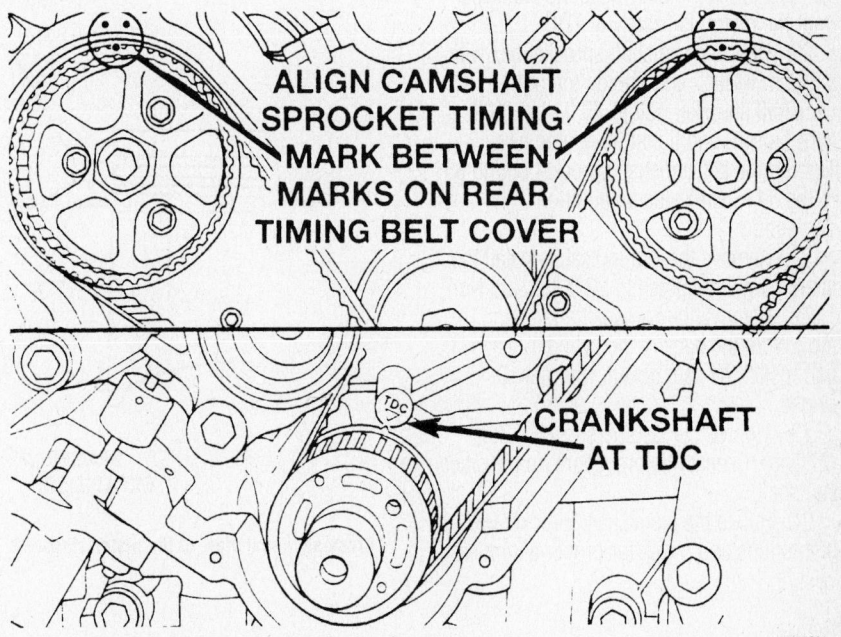

ALIGN CAMSHAFT SPROCKET TIMING MARK BETWEEN MARKS ON REAR TIMING BELT COVER

CRANKSHAFT AT TDC

79225G01

Timing belt alignment marks—3.2L and 3.5L engines

the plunger in place until the tensioner is installed.

12. Install both camshaft sprockets to the appropriate shafts. The left camshaft sprocket has the Distributorless Ignition System (DIS) pick-up as part of the sprocket.

➡**The right bolt is 8⅜ in. (213mm) long, while the left bolt is 10 in. (254mm) long. These bolts are not interchangeable.**

13. Apply Loctite®271, to the threads of the camshaft sprocket retainer bolts and install to the appropriate shafts. Do not tighten the bolts at this time.

14. Align the camshaft sprockets between the marks on the rear belt covers.

15. Align the crankshaft sprocket with the TDC mark on the oil pump cover.

16. Install the timing belt, starting at the crankshaft sprocket and going in a counter-clockwise direction. After the belt is installed on the right sprocket, keep tension on the belt until it is past the tensioner pulley.

17. Holding the tensioner pulley against the belt, install the timing belt tensioner into the housing and tighten to 21 ft. lbs. (28 Nm).

18. When the tensioner is in place, pull the retainer pin out to allow tensioner to extend to the pulley bracket.

➡**Be sure that the timing marks on the cam sprockets are still between the marks on the rear cover.**

19. Remove the spark plug in the No.1 cylinder and install a dial indicator to check for TDC of the piston. Rotate the crankshaft until the piston is exactly at TDC.

20. Hold the camshaft sprocket hex with a 36mm wrench and tighten the right camshaft sprocket bolt to 75 ft. lbs. (102 Nm) plus an additional 90 degree turn. Tighten the left camshaft sprocket bolt to 85 ft. lbs. (115 Nm) plus an additional 90 degree turn.

21. Remove the dial indicator. Install the spark plug and tighten to 20 ft. lbs. (28 Nm).

22. Remove the camshaft alignment tools from the back of the cylinder heads and install the cam covers with new O-rings.

23. Tighten the fasteners to 20 ft. lbs. (27 Nm). Repeat this procedure on the other camshaft.

24. Rotate the crankshaft sprocket two revolutions and check for proper alignment of the timing marks on the camshaft and the crankshaft. If the timing marks do not align, repeat the procedure.

25. Before installing, inspect the sealer on the stamped steel cover. If some sealer is missing, use MOPAR Silicone Rubber Adhesive sealant to replace the missing sealer.

26. Install or connect the following:
- Lower belt cover behind the crankshaft damper, if necessary
- Stamped steel cover and the left side cast cover. Tighten the 6mm bolts to 105 inch lbs. (12 Nm), the 8mm bolts to 250 inch lbs. (28 Nm) and the 10mm bolts to 40 ft. lbs. (54 Nm).
- Crankshaft damper using special tool L-4524, a 5.9 in. long bolt, thrust bearing and washer. Tighten the center bolt to 85 ft. lbs. (115 Nm).
- Upper radiator hose
- Accessory drive belts and adjust them to the proper tension
- Radiator and cooling fan assemblies

27. Refill and bleed the cooling system.

28. Connect the negative battery cable.

29. With the radiator cap off so coolant can be added, run the engine. Watch for leaks and listen for unusual engine noises.

Piston and Rings

POSITIONING

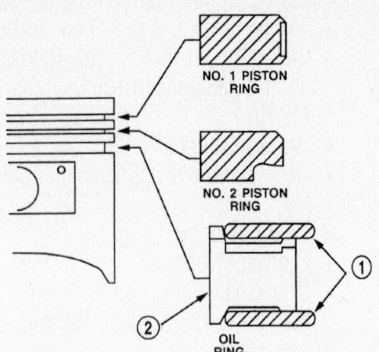

1 – SIDE RAIL
2 – SPACER EXPANDER

9306BG04

Cross-sectional view of the piston rings—3.5L Engine

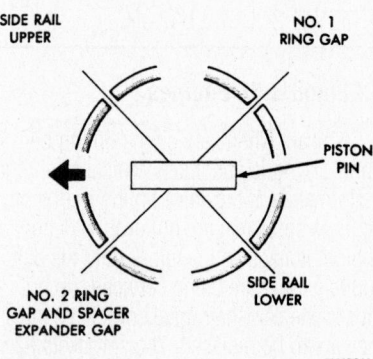

9306BG03

Piston ring gap positions—2.7L, 3.2L and 3.5L Engines

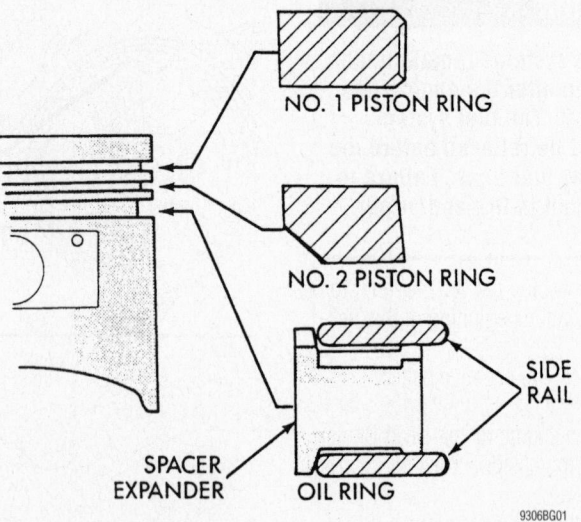

9306BG01

Cross-sectional view of the piston rings—2.7L and 3.2L Engines

FUEL SYSTEM

Fuel System Service Precautions

Safety is the most important factor when performing not only fuel system maintenance but any type of maintenance. Failure to conduct maintenance and repairs in a safe manner may result in serious personal injury or death. Maintenance and testing of the vehicle's fuel system components can be accomplished safely and effectively by adhering to the following rules and guidelines:

• To avoid the possibility of fire and personal injury, always disconnect the negative battery cable unless the repair or test procedure requires that battery voltage be applied.

• Always relieve the fuel system pressure prior to disconnecting any fuel system component (injector, fuel rail, pressure regulator, etc.), fitting or fuel line connection. Exercise extreme caution whenever relieving fuel system pressure, to avoid exposing skin, face and eyes to fuel spray. Please be advised that fuel under pressure may penetrate the skin or any part of the body that it contacts.

• Always place a shop towel or cloth around the fitting or connection prior to loosening to absorb any excess fuel due to spillage. Ensure that all fuel spillage (should it occur) is quickly removed from engine surfaces. Ensure that all fuel soaked cloths or towels are deposited into a suitable waste container.

• Always keep a dry chemical (Class B) fire extinguisher near the work area.

• Do not allow fuel spray or fuel vapors to come into contact with a spark or open flame.

• Always use a back-up wrench when loosening and tightening fuel line connection fittings. This will prevent unnecessary stress and torsion to fuel line piping.

• Always replace worn fuel fitting O-rings with new. Do not substitute fuel hose, where fuel pipe is installed.

Before servicing the vehicle, also make sure to refer to the precautions in the beginning of this section as well.

Fuel System Pressure

RELIEVING

1. Before servicing the vehicle, refer to the precautions in the beginning of this section.
2. At the Power Distribution Center (PDC), remove the Fuel Pump Relay.
3. Operate the engine until the engine

stalls; then, continue restarting the engine until it will no longer run.
4. Turn the ignition switch **OFF**.

✳✳ CAUTION

The previous steps must be performed to relieve the high pressure fuel from the fuel rail. The following steps must be performed to remove excess fuel from the fuel rail. Do not use the following steps to relieve high pressure for excessive fuel will be forced into a cylinder chamber.

5. Disconnect the electrical connector from any injector.
6. Connect 1 end of a jumper wire (with an alligator clip) to either injector terminal and the other end to the positive side of the battery.
7. Connect 1 end of a second jumper wire to the other injector terminal.

✳✳ WARNING

Applying power to an injector for more than a few seconds will damage the injector.

8. Momentarily, touch the other end of the jumper wire to a ground for no more than a few seconds.
9. At the Power Distribution Center (PDC), install the fuel pump relay.

➡ **When the fuel pump relay is removed, 1 or more Diagnostic Trouble Codes (DTC's) may be stored in the Powertrain Control Module (PCM) memory. A DRB scan tool must by used to clear the DTC's.**

Fuel Filter

The fuel filter mounts to the frame rail in front of the fuel tank. The inlet and outlet ends of the filter are marked for installation purposes. Install the fuel filter making certain that it is properly orientated.

REMOVAL & INSTALLATION

➡ **The fuel filter is part of the fuel pressure regulator mounted on the fuel pump module.**

1. Before servicing the vehicle, refer to the precautions in the beginning of this section.

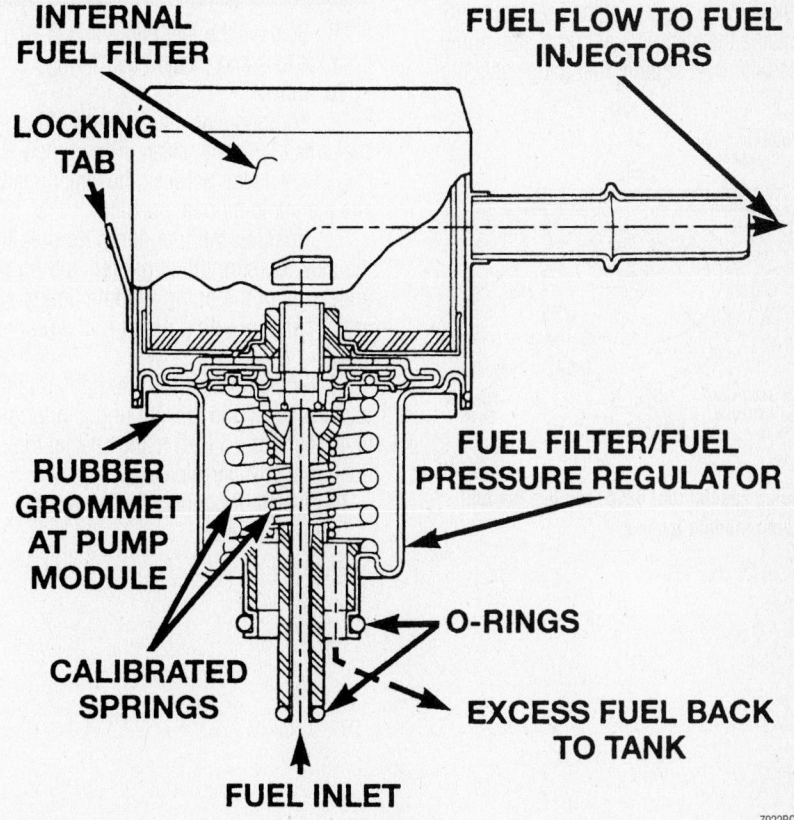

Cut away view of the fuel filter/pressure regulator

7922BG40

2. Properly relieve the fuel system pressure.

3. Lower the fuel tank.

4. Remove or disconnect the following:
- Negative battery cable
- Purge and vent lines
- Fuel line from the pressure regulator
- Filter/regulator by pushing in the locking tab, turning the regulator to unlock it and pulling the it straight up

To install:

5. Push the fuel filter/regulator into the fuel pump module and turn to lock it into position.

6. Connect the fuel lines and install the tank.

7. Start the engine, check for leaks and repair if necessary.

Fuel Pump

REMOVAL & INSTALLATION

The in-tank fuel pump module contains the fuel pump and pressure regulator which adjusts fuel system pressure. Fuel pump voltage is supplied through the fuel pump relay.

The fuel pump is serviced as part of the fuel pump module. The fuel pump module is installed in the top of the fuel tank and contains the electric fuel pump, fuel pump reservoir, inlet strainer fuel gauge sending

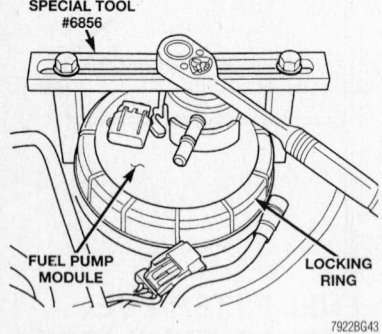

SPECIAL TOOL
#6856

FUEL PUMP
MODULE

LOCKING
RING

7922BG43

Using special tool 6856, remove the fuel pump module locknut

unit, fuel supply and return line connections and the pressure regulator. The inlet strainer, fuel pressure regulator and level sensor are the only serviceable items. If the fuel pump requires service, replace the fuel pump module.

1. Before servicing the vehicle, refer to the precautions in the beginning of this section.

2. Properly relieve the fuel system pressure.

3. Remove or disconnect the following:
- Negative battery cable
- Fuel tank

4. Clean the top of the tank to remove any loose dirt.
- Fuel lines from the fuel pump module by squeezing the quick-connect fitting with thumb and forefinger
- Fuel pump module electrical connector from the top of the fuel pump module

5. Using special tool 6856, remove the fuel pump locknut by turning it counter-clockwise.

✳✳ CAUTION

The fuel reservoir of the fuel pump module does not empty out when the tank is drained. The fuel in the reservoir may spill out when the module is removed.

6. Remove the fuel pump and O-ring from the tank and discard the O-ring.

To install:

7. Thoroughly clean all parts. Wipe the seal area of the tank clean. Place a new O-ring on the ledge between the tank threads and the pump module opening.

8. Position the fuel pump module in the tank. Be sure the alignment tab on the underside of the pump module flange sits in the corresponding notch in the fuel tank.

9. While holding the fuel pump module in place install the locking ring and torque to 40 inch lbs. (5 Nm) using special tool 6856 or a spanner-type tool.

10. Install or connect the following:
- Fuel tank

- Fuel pump module electrical connector
- Negative battery cable

11. Fill the fuel tank with fuel. Install the fuel filler cap. Turn the ignition switch to the **ON** position to pressurize the system. Check the fuel system for leaks.

Fuel Injector

REMOVAL & INSTALLATION

1. Before servicing the vehicle, refer to the precautions in the beginning of this section.

2. Relieve the fuel system pressure.

3. Remove or disconnect the following:
- Negative battery cable
- Intake manifold plenum and cover opening with a clean cloth

4. Place a shop rag under the fuel rail's quick-connect fitting; then, squeeze the quick-connect fitting's retainer tabs together and pull the fitting assembly off of the fuel tube nipple.
- Fuel injector electrical connectors
- Fuel rail-to-engine bolts and fuel rail
- Fuel injector-to-fuel rail retainer clips
- Fuel injectors

To install:

5. Lubricate the injector O-rings with clean engine oil.

6. Install or connect the following:
- Fuel injector and secure with retaining clips
- Fuel rail onto cylinder head and press rail into place. Make sure that the injectors are fully seated.
- Fuel rail-to-cylinder head bolts and torque bolts to 100 inch lbs. (11 Nm) for 2.7L engines, or to 21 ft. lbs. (28 Nm) for 3.2L and 3.5L engine.
- Intake plenum

7. Lubricate the quick-connect fitting's O-rings with clean engine oil; then, push the connector together until the retainer seats and a click is heard.

8. Connect the negative battery cable.

DRIVE TRAIN

Transaxle Assembly

REMOVAL & INSTALLATION

The 42LE four speed transaxle uses fully-adaptive controls. Adaptive controls are those which perform their functions based on real-time feedback sensor information. The transaxle is conventional in the use of hydraulically applied clutches to shift a planetary gear train. However, it uses electronics to control virtually all other functions. The following components are serviceable in the vehicle: valve body assembly, solenoid pack, manual valve lever position sensor, input and output speed sensors, transfer chain and sprockets, short (right side) stub shaft seal and the long (left side) stub shaft and ball bearing. Note that the factory recommends that before attempting any repair on the 42LE four-speed automatic transaxle, always check for proper shift linkage adjustment. Also, check for diagnostic trouble codes with the Chrysler DRB scan tool.

Use MOPAR Type 7176 Automatic Transmission Fluid only. Do not substitute transaxle fluid. If the differential sump requires fluid, use 80W-90 petroleum based Hypoid gear lubricant.

2001 Vehicles

1. Before servicing the vehicle, refer to the precautions in the beginning of this section.
2. Remove or disconnect the following:
   - Negative battery cable
   - Engine air inlet tube
   - Crankshaft Position (CKP) sensor connector
   - CKP sensor from the upper right side of the transaxle bell housing
   - Transaxle wiring connector block located on the right shock tower. To free the connector from the harness, remove the wire ties.
   - Front wheels
   - Strut-to-steering knuckle bolts on both sides of the vehicle and/or tie rod ends, if required
   - Antilock Brake System (ABS) wheel speed sensor, if equipped
   - Halfshafts

➡**Remove the halfshafts by inserting a prybar between the halfshaft and the transaxle case and prying the shafts from the transaxle housing. Swing the shafts out of the way, keeping the** joints straight and suspend using wire. Be careful not to damage the halfshaft seals.

❊❊ WARNING

Do not let the halfshafts or CV-joints hang unsupported. Internal joint damage may result if allowed to hang free.

3. Remove or disconnect the following:
   - Engine-to-transaxle brackets
   - Transaxle bell housing cover
4. Mark the driveplate to the torque converter and remove the torque converter bolts. The driveplate-to-torque converter bolts are not to be reused.
   - Starter assembly from the bell housing and allow the starter motor to sit between the engine and the frame
   - Transaxle oil cooler lines and plug the openings
   - Transaxle dipstick
   - Transaxle gear selector cable
   - Exhaust pipe from the exhaust manifold and position out of the way

➡**If the clearance will not allow for transaxle removal, remove the exhaust system from the vehicle.**

5. Support the transaxle using a transmission jack. Raise the transaxle slightly to relieve the weight off the rear transaxle mount.
   - Engine-to-transaxle brackets and transaxle mount through-bolt
   - Rear crossmember

➡**Pry the transaxle mount rearward to separate the mount from the transaxle.**

6. Lower the rear of the transaxle to gain access to the bell housing bolts. Remove the bell housing bolts.
7. Place a drain pan under the dipstick in the transaxle to catch transaxle fluid that will drain out of the case.
8. Remove or disconnect the following:
   - Transaxle dipstick tube and plug the hole
   - Engine-to-transaxle bolts and transaxle

➡**The driveplate-to-torque converter bolts and the driveplate-to-crankshaft bolts must not be reused. Install new bolts whenever these bolts are removed.**

9. Inspect the driveplate for cracks. If cracks are present, replace the driveplate.
To install:
➡**Apply a light coating of grease to the pilot hole of the crankshaft if the torque converter is being replaced.**

❊❊ WARNING

When installing the transaxle, be careful that the fuel tubes at the rear of the engine do not contact the following:

10. Install or connect the following:
   - Transaxle wiring harness
   - Driveplate and torque the fastener to 75 ft. lbs. (101 Nm)
   - Transaxle and torque the engine-to-transaxle case bolts to 75 ft. lbs. (101 Nm)
   - Rear transaxle case mount and rear crossmember in position and secure all fasteners
   - Transaxle dipstick tube
   - Exhaust pipe-to-engine exhaust manifold
   - Transaxle gear selector cable and oil cooler lines
   - Starter. Torque the bolts to 40 ft. lbs. (54 Nm). Be sure that the starter ground strap is installed correctly.
   - Align torque converter matchmarks and torque new torque converter-to-driveplate bolts to 60 ft. lbs. (81 Nm)
   - Bell housing cover and engine-to-transaxle brackets
11. While pulling the top of the steering knuckle outward, install the inner CV-joint, with new retainer clip in place, into the transaxle.
   - ABS wheel sensor (if removed) and strut-to-steering knuckle bolts
   - Front wheels and torque the lug nuts, in a star pattern to 100 ft. lbs. (135 Nm)
   - Transaxle dipstick
   - Transaxle wiring harness connector on the right shock tower
   - CKP sensor
   - Air inlet tube and negative battery cable
   - Transaxle with fluid
12. Start the engine and allow it to idle for 2 minutes. Apply the parking brake and move the selector through each gear position, ending in **N**. Recheck the fluid level and add if

necessary. Be sure the vehicle is level when refilling the transaxle. Use Mopar Type 7176 Automatic Transmission Fluid (ATF) only. Do not substitute transaxle fluid. If the differential sump requires fluid, use 80W-90 petroleum based Hypoid gear lubricant.

13. Check the transaxle or proper operation. Adjust the shift linkage, if necessary. Be sure the reverse lamps come on when in reverse.

2002–04 Vehicles

1. Before servicing the vehicle, refer to the precautions in the beginning of this section.

2. Remove or disconnect the following:
- Negative battery cable
- Wiper blades
- Right and left wiper module covers
- Steel cowl/strut support

✼✼ CAUTION

Be careful of sharp edges cowl/wiper area.

- Engine air inlet tube
- Transaxle harness connectors at cowl area
- Upper bell housing studs
- Loosen clamps at intersection of rear exhaust system to front catalytic converter pipe
- Rear exhaust from left catalytic converter pipe and right extension pipe
- Rear exhaust system
- Nuts securing the exhaust pipes to transaxle mount
- Loosen clamp at right extension at right catalytic converter
- Right extension
- Crankshaft Position (CKP) sensor connector and remove sensor. The sensor is located on the upper right side of the transaxle bell housing.
- Dipstick tube
- Gear selector cable from the transaxle
- Transmission range sensor connector
- Input and output speed sensor connector
- Transaxle cooler lines from the transaxle and plug them
- Lower control arm pinch bolts
- Lower control arms from steering knuckles by prying them out

3. Use a prybar to separate the inner tripod joints from the transaxle

4. Pull the bottom of the knuckles and halfshafts outward to allow clearance during

transaxle removal. The halfshafts do not have to be completely removed from the vehicle.

✼✼ WARNING

Do not let the halfshafts or CV-joints hang unsupported. Internal joint damage may result if allowed to hang free.

5. Remove or disconnect the following:
- Heated Oxygen Sensor (HO2S) wiring and left catalytic converter pipe
- Starter bolts and position it between the engine and transaxle. Do not disconnect the wiring or completely remove the starter.
- Engine oil pan cooler

➡**Matchmark the installed position of the torque converter to the flex plate before removal.**

- Torque converter bolts and discard them. New bolts must be used during installation.

6. Use a transmission jack to slightly raise the transaxle, removing weight from the rear transaxle mount.

➡**The crossmember bridge bolts are different lengths for each side, so note their locations during removal.**

- Rear crossmember bridge bolts
- Rear mount adapter plate mounting bolts
- Rear crossmember bridge, mount and adapter plate as an assembly

➡**Lower the rear of the transaxle to access to the bell housing bolts.**

- Side bell housing bolts
- Dipstick tube from the transaxle and plug to prevent leakage
- Solenoid/pressure switch assembly connector from the top of the transaxle
- Transaxle by lowering it from the vehicle.

To install:

7. To prevent damaging the structural collar, hand-tighten all fasteners before torquing them to specifications.

8. If the torque converter is being replaced, apply a light coat of oil to the crankshaft pilot hole.

9. Check the drive plate for cracks before reinstalling the transaxle and replace the plate as necessary.

10. Use new torque converter-to-drive plate bolts, as different year models use different bolts.

11. Installation is the reverse of the removal procedure, noting the following torque specifications:

    a. Converter-to-driveplate bolts: 65 ft. lbs. (88 Nm)

    b. Driveplate-to-crankshaft: 70 ft. lbs. (95 Nm)

    c. Engine-to-transaxle case bolts to 75 ft. lbs. (101 Nm)

    d. Input and output speed sensor-to-case bolts: 20 ft. lbs. (27 Nm).

12. Fill the transaxle with fluid. Start the engine and allow it to idle for 2 minutes. Apply the parking brake and move the selector through each gear position, ending in **N**. Recheck the fluid level and add if necessary. Be sure the vehicle is level when refilling the transaxle. Use Mopar Type 9602 Automatic Transmission Fluid (ATF+4) only. Do not substitute transaxle fluid.

13. Check the transaxle or proper operation. Adjust the shift linkage, if necessary. Be sure the reverse lamps come on when in reverse.

Halfshaft

REMOVAL & INSTALLATION

✼✼ WARNING

Allowing the CV-joint assemblies to dangle unsupported, or pulling or pushing the ends, can damage boots or CV-joints. Always support both ends of the halfshaft to prevent damage or disengagement of the Tri-pot joint.

1. Before servicing the vehicle, refer to the precautions in the beginning of this section.

2. Remove or disconnect the following:
- Negative battery cable
- Front wheels
- Front caliper assembly from steering knuckle
- Front brake rotor from the hub
- Speed sensor cable routing bracket from strut assembly
- Hub and bearing-to-stub axle retainer nut

3. Install a puller tool onto the hub and bearing assembly and secure it into place using the wheel lug nuts.

4. Protect wheel stud threads by installing a wheel lug nut onto a wheel stud. Use a flat-bladed prying tool to prevent the hub from turning. Using the puller tool, force the halfshaft outer stub axle from the hub and bearing assembly.

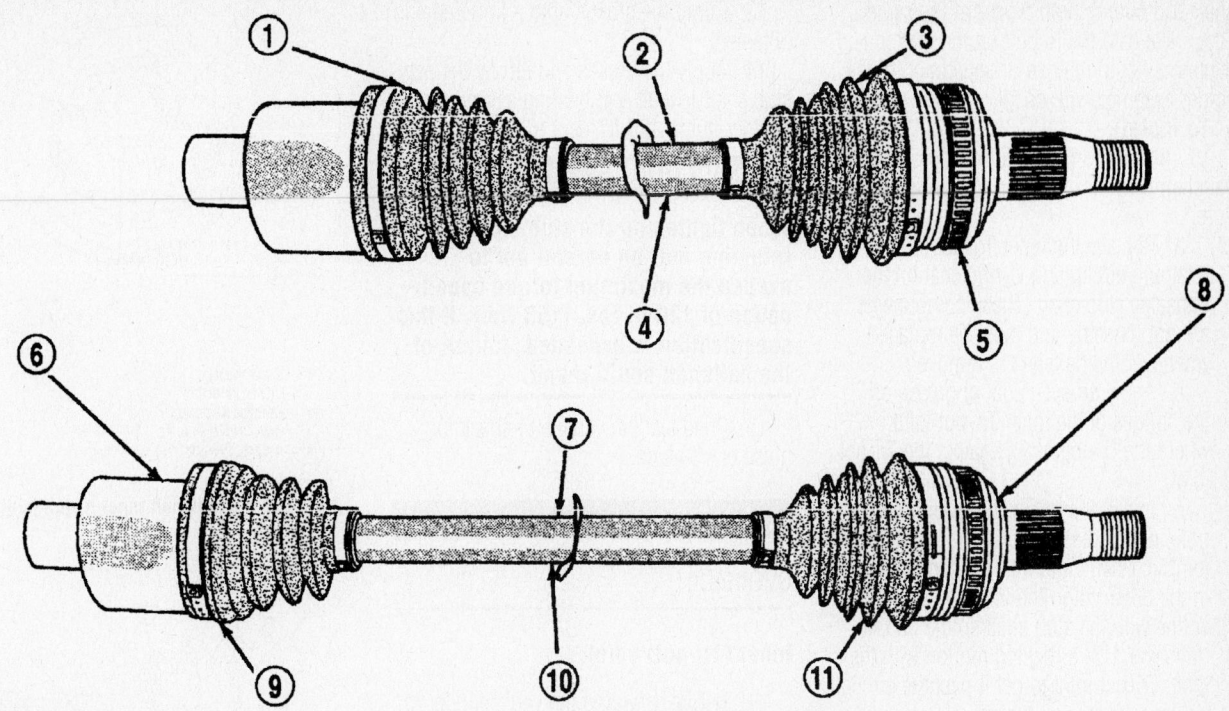

1 - INNER BOOT
2 - INTERCONNECTING SHAFT
3 - OUTER BOOT
4 - RIGHT HALF SHAFT
5 - TONE WHEEL
(WHEN EQUIPPED WITH ABS)
6 - INNER TRIPOD JOINT

7 - INTERCONNECTING SHAFT
8 - OUTER RZEPPA JOINT
9 - INNER BOOT
10 - LEFT HALF SHAFT
11 - OUTER BOOT

67189-300M-G05

Front halfshaft assemblies—2004 shown, others similar

5. Dislodge the inner Tri-pot joint from the stub shaft retaining snapring on the transaxle. To do this, insert a prybar between the transaxle case and the inner Tri-pot joint and pry on Tri-pot joint.

➡**Do not try to remove the inner Tri-pot joint from the transaxle stub shaft at this time. Only disengage the inner Tri-pot joint from the retainer snapring.**

6. Remove the strut assembly-to-steer-

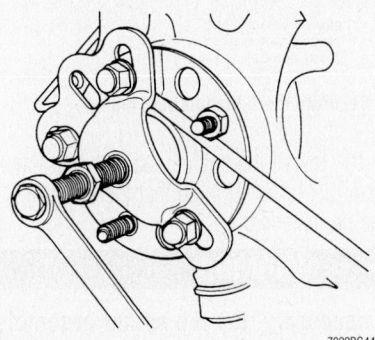

7922BG44

Removing the stub axle from the front hub/bearing assembly—halfshaft service

ing knuckle attaching bolts from the strut assembly.

✱✱ WARNING

The strut assembly-to-steering knuckle bolts are serrated (toothed) where they go through the strut assembly and steering knuckle. When removing the bolts, turn the nuts off the bolt; do not turn the bolts in the steering knuckle or damage to the steering knuckle will result.

7. Separate the top of the steering knuckle from the lower end of the strut.

8. Hold the outer joint assembly with one hand. Grasp the steering knuckle with the other hand and rotate it out and to the rear of the vehicle, until the outer CV-joint clears the hub and bearing assembly.

✱✱ WARNING

When removing the outer CV-joint from the hub and bearing assembly, do not allow the flange disc on the

hub and bearing assembly to become damaged. If this happens, dirt and water can enter the bearing, which will cause premature bearing failure.

9. Remove the halfshaft inner joint from the transaxle stub shaft by grasping the inner Tri-pot joint and the interconnecting

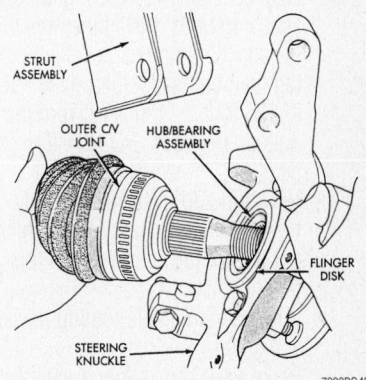

7922BG45

Be careful not to damage the threads for the axle nut when removing the outer CV-joint from the steering knuckle—halfshaft service

shaft and pulling both pieces at the same time. Take care not to pull on the interconnecting shaft to remove or separation of the spider assembly will occur.

To install:

10. Install the Tri-pot joint side of the halfshaft by performing the following procedure:

a. Replace the inner Tri-pot joint retaining circlip and O-ring seal on the transaxle stub shaft. These components are not reusable and must be replaced whenever the halfshaft is removed.

b. Apply an even coat of grease on the splines of the inner Tri-pot joint, where the O-ring seats against the Tri-pot joint.

c. Install the halfshaft through the hole in the splash shield. Grasp the inner joint in 1 hand and interconnecting shaft in the other. Align the inner Tri-pot joint spline with the stub shaft spline on the transaxle. Use a rocking motion with the inner Tri-pot joint to get it past the circlip on the transaxle stub shaft.

d. Continue pushing the Tri-pot joint onto transaxle stub shaft until it stops moving. The O-ring on the stub shaft should not be visible when the inner Tri-pot joint is fully installed. Check that the inner Tri-pot joint is locked in position by grasping the inner joint and pulling. If locked in position, the joint will not move on the stub shaft.

11. Hold the outer CV-joint assembly with one hand. Grasp the steering knuckle with the other and rotate it out and to the rear of the vehicle. Install the outer CV-joint into the hub and bearing assembly.

12. Install or connect the following:

- Top of the steering knuckle into the strut assembly. Align the steering knuckle-to-strut assembly mounting holes.
- Strut assembly-to-steering knuckle attaching bolts. Install the nuts to the attaching bolts and while holding the bolt heads, torque the nuts to 125 ft. lbs. (170 Nm). Turn the nuts on the bolts. DO NOT turn the bolts.
- New hub and bearing assembly-to-stub shaft retainer nut. Tighten but do not torque the nut at this time.
- Speed sensor cable routing bracket and screw
- Brake rotor and caliper assembly; then, tighten the caliper guide pin bolts to 30 ft. lbs. (41 Nm)
- Front wheels and lug nuts
- Negative battery cable

13. Pump the brakes until a firm pedal is obtained.

14. Apply the brakes and torque the new stub shaft-to-hub and bearing assembly retainer nut to 120 ft. lbs. (163 Nm).

✳✳ WARNING

When tightening the stub shaft retaining nut, be careful not to exceed the maximum torque specification of 120 ft. lbs. (163 Nm). If this specification is exceeded, failure of the halfshaft could result.

15. Road test the vehicle to check for noise or vibration.

CV-Joints

OVERHAUL

Inner (Tri-pot) Joint

1. Before servicing the vehicle, refer to the precautions in the beginning of this section.

2. Disconnect the negative battery cable.

3. Remove the halfshaft and retaining clamps.

4. Slide the boot down the shaft away from the tri-pot housing.

➡**When separating the spider joint from the tri-pot joint housing, hold the rollers in place on the trunnions to prevent the rollers and needle bearings from falling away.**

5. Carefully, slide the shaft/spider assembly from the tri-pot housing.

6. Remove the spider assembly-to-shaft snapring; then, slide the spider assembly off the shaft.

✳✳ WARNING

If necessary, tap the spider assembly off the shaft using a brass drift; be careful not to hit the outer bearings.

7. Slide the boot off the shaft.

8. Thoroughly, inspect all parts for signs of excessive wear; if necessary, replace the halfshaft.

➡**Component parts are not serviceable and must be replaced as an assembly.**

To install:

9. Slide the inner tri-pot boot clamp and boot onto the shaft; then, position the boot so that only the thinnest (sight) groove is visible on the shaft.

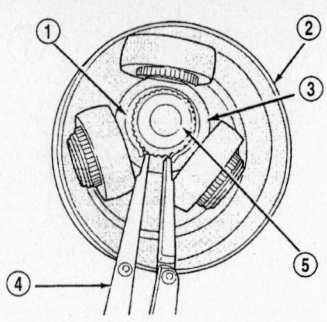

1 – SNAP RING
2 – SEALING BOOT
3 – SPIDER ASSEMBLY
4 – SNAP RING PLIERS
5 – INTERCONNECTING SHAFT

9306BG11

View of the halfshaft inner tri-pot joint and snapring

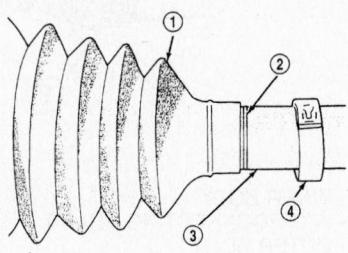

1 – SEALING BOOT
2 – INTERCONNECTING SHAFT THINNEST GROOVE
3 – INTERCONNECTING SHAFT
4 – BOOT CLAMP

9306BG12

View of the halfshaft boot, shaft and boot clamp

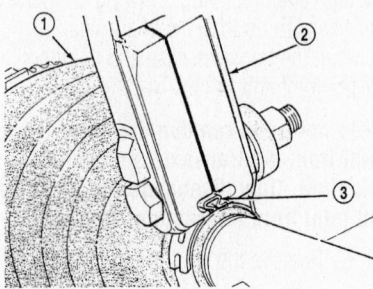

1 – SEALING BOOT
2 – SPECIAL TOOL C-4975
3 – CLAMP BRIDGE

9306BG13

Securing the halfshaft boot clamp

10. Install the spider assembly onto the shaft just far enough so that the snapring can be installed.

✳✳ WARNING

If necessary, tap the spider assembly onto the shaft using a brass drift; be careful not to hit the outer bearings.

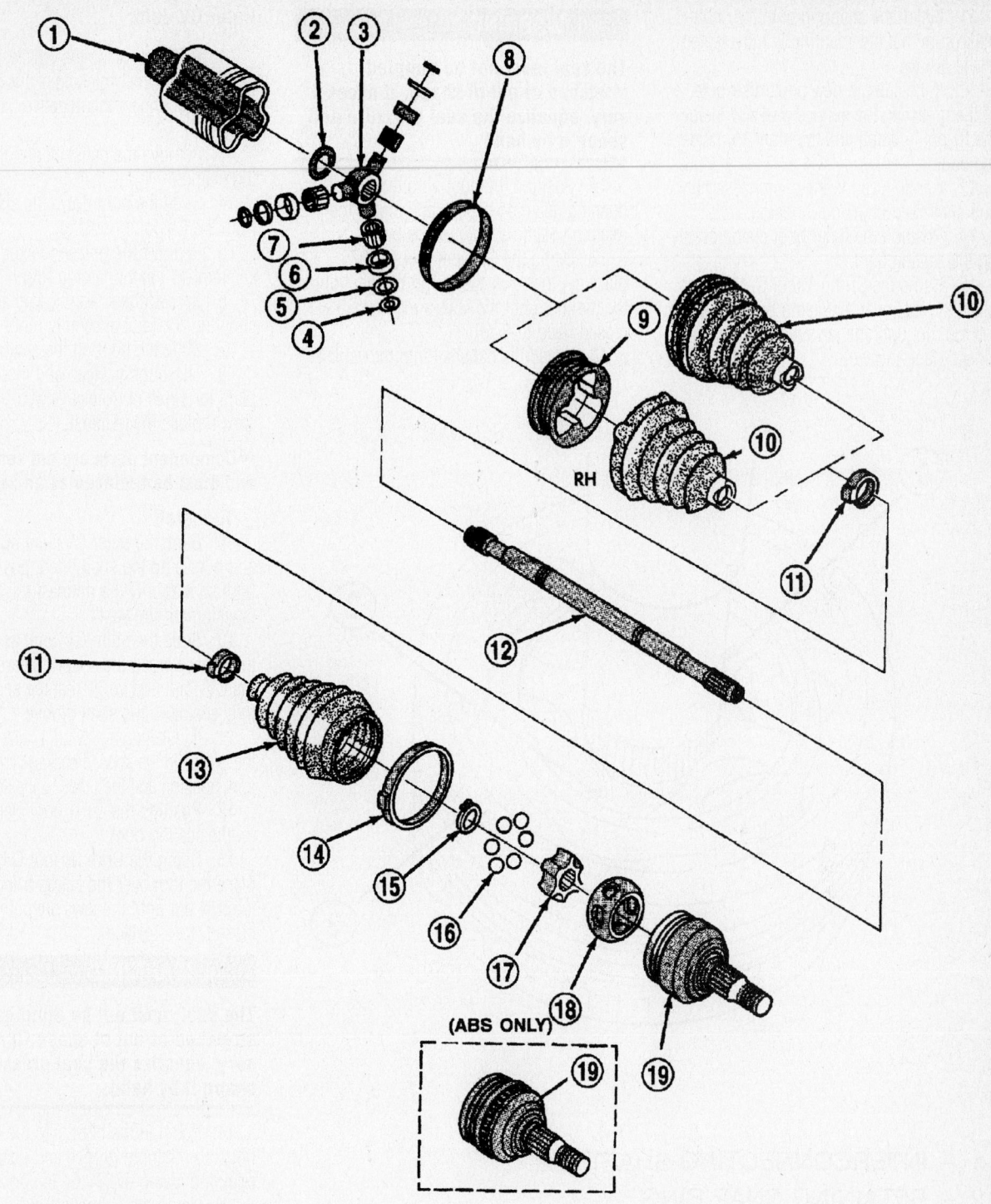

1 - HOUSING ASM, RETAINER
2 - RING, SPACER
3 - SPIDER, TRIPOD JOINT
4 - RING, RETAINING
5 - RETAINER, BALL & ROLLER
6 - BALL, TRIPOD JOINT
7 - ROLLER, NEEDLE
8 - CLAMP, SEAL RETAINING
9 - BUSHING, TRILOBAL TRIPOD
10 - SEAL, DRIVE AXLE INBOARD

11 - CLAMP, SEAL RETAINING
12 - SHAFT, AXLE (RH SHOWN, LH SIMILAR)
13 - SEAL, DRIVE AXLE OUTBOARD
14 - CLAMP, SEAL RETAINING
15 - RING, RACE RETAINING
16 - BALL, CHROME ALLOY
17 - RACE, C/V JOINT INNER
18 - CAGE, C/V JOINT
19 - RACE, C/V JOINT OUTER

67189-300M-G06

Exploded view of the halfshaft assemblies

11. Install the snapring onto the shaft; make sure that the snapring is fully seated in the groove.

12. If installing a new boot, distribute ½ of the grease in the service package inside the tri-pot housing and the other ½ inside the boot.

13. Carefully, slide the spider assembly and shaft into the tri-pot housing.

14. Position the inner boot clamp evenly on the sealing boot.

15. Using the Crimper tool C-4975, place the tool over the clamp bridge, tighten the tool nut until the jaws are completely closed (face-to-face).

✳✳ WARNING

The seal must not be dimpled, stretched or out of shape. If necessary, equalize the seal pressure and shape it by hand.

16. Position the boot onto the tri-pot housing retaining groove and install the retaining clamp evenly on the boot.

17. Using the Crimper tool C-4975, place the tool over the clamp bridge, tighten the tool nut until the jaws are completely closed (face-to-face).

18. Install the halfshaft into the vehicle.

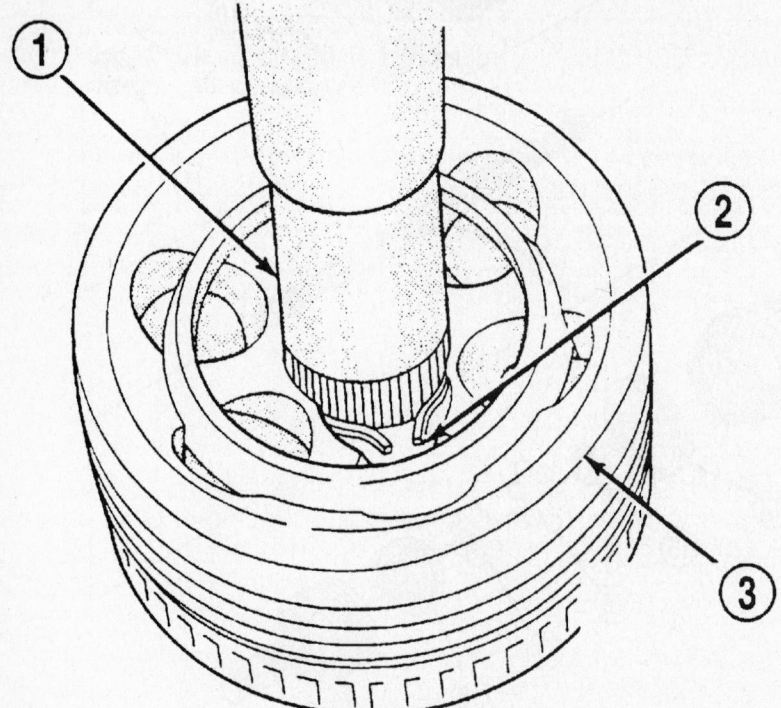

1 – INTERCONNECTING SHAFT
2 – RETAINING SNAP RING
3 – OUTER C/V JOINT ASSEMBLY

9306BG14

View of the halfshaft outer CV-joint and snapring

Outer CV-Joint

1. Before servicing the vehicle, refer to the precautions in the beginning of this section.

2. Disconnect the negative battery cable.

3. Remove the halfshaft and the retaining clamps.

4. Slide the boot down the shaft away from the CV-joint housing.

5. Remove the grease to expose the CV-joint-to-shaft retaining ring.

6. Spread the snapring ears apart and slide the CV-joint assembly off of the shaft.

7. Slide the boot off the shaft.

8. Thoroughly, clean and inspect all parts for signs of excessive wear; if necessary, replace the halfshaft.

➡**Component parts are not serviceable and must be replaced as an assembly.**

To install:

9. Slide the outer CV-joint boot clamp and boot onto the shaft; then, position the boot so that only the thinnest (sight) groove is visible on the shaft.

10. Slide the outer CV-joint assembly on the shaft, spread the snapring ears, position the CV-joint and verify that the snapring is fully seated in the shaft groove.

11. If installing a new boot, distribute ½ of the grease in the service package into the CV-joint housing and the other ½ inside the boot.

12. Position the outer boot clamp evenly on the sealing boot.

13. Using the Crimper tool C-4975, place the tool over the clamp bridge, tighten the tool nut until the jaws are completely closed (face-to-face).

✳✳ WARNING

The seal must not be dimpled, stretched or out of shape. If necessary, equalize the seal pressure and shape it by hand.

14. Position the boot onto the CV-joint housing retaining groove and install the retaining clamp evenly on the boot.

15. Using the Crimper tool C-4975, place the tool over the clamp bridge, tighten the tool nut until the jaws are completely closed (face-to-face).

16. Install the halfshaft into the vehicle.

STEERING AND SUSPENSION

Air Bag

✳✳ CAUTION

Some vehicles are equipped with an air bag system. The system must be disabled before performing service on or around system components, steering column, instrument panel components, wiring and sensors. Failure to follow safety and disabling procedures could result in accidental air bag deployment, possible personal injury and unnecessary system repairs.

PRECAUTIONS

Several precautions must be observed when handling the inflator module to avoid accidental deployment and possible personal injury.

• Never carry the inflator module by the wires or connector on the underside of the module.

• When carrying a live inflator module, hold securely with both hands, and ensure that the bag and trim cover are pointed away.

• Place the inflator module on a bench or other surface with the bag and trim cover facing up.

• With the inflator module on the bench, never place anything on or close to the module which may be thrown in the event of an accidental deployment.

Before servicing the vehicle, also make sure to refer to the precautions in the beginning of this section as well.

DISARMING

✳✳ CAUTION

The Air Bag system must be disarmed before repair and/or removal of any component in its immediate area including the air bag itself. Failure to do so may cause accidental deployment of the air bag, resulting in unnecessary system repairs and/or personal injury.

1. Disconnect the negative battery cable and isolate the cable using an appropriate insulator (wrap with quality electrical tape).

2. Allow the system capacitor to discharge for 2 minutes before starting any repair on any air bag system or related

components. This will disable the air bag system.

✳✳ CAUTION

Always wear safety goggles when working with or around the air bag system. When carrying a live air bag, be sure the bag and trim cover are pointed away from the body. In the unlikely event of an accidental deployment, the bag will, then deploy with minimal chance of injury. When placing a live air bag on a bench or other surface, always face the bag and trim cover up, away from the surface. This will reduce the motion of the module if it is accidentally deployed.

Power Rack and Pinion Steering Gear

REMOVAL & INSTALLATION

1. Before servicing the vehicle, refer to the precautions in the beginning of this section.

2. Turn the front wheels to the straight-ahead position.

3. Remove or disconnect the following:
• Negative battery cable
• Wiper arms
• Wiper module cover and cowl cover
• Reinforcement from the strut towers and wiper module
• Throttle body's intake air duct and resonator

4. Clamp the steering wheel in the center position with some type of holding device.
• Intermediate shaft from the steering column coupler
• Both tie rods from the steering gear and place them on the transaxle bell housing
• Siphon the power steering fluid from the reservoir
• Power steering gear fluid lines

5. If equipped with speed proportional steering, disconnect the wiring from the solenoid control valve on the end of the steering gear under the master cylinder.
• Master cylinder from the booster and position it to the side without disconnecting the fluid lines
• Booster vacuum line

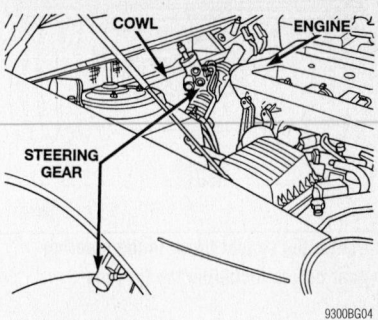

9300BG04

The steering gear is removed or installed through the right side of the vehicle

➡**It may be helpful to loosen the 2 bolts attaching the right mounting bracket to the steering gear in order the clear the air conditioning lines.**

• 4 steering gear mounting bolts (2 on each side)

6. Move the steering gear and intermediate shaft into the engine compartment to gain access to the roll pin.

➡**Special tools are commercially available for removing roll pins.**

• Roll pin and separate the intermediate shaft from the steering gear
• Right front wheel
• Tie rod end from the steering arm on the right strut

➡**If equipped with a 2.7L engine, turn the left front tire towards the left as far as possible to provide clearance.**

7. Slide the right end of the steering gear out through the tie rod hole while raising the left end of the steering gear upward.

To install:

8. Position the tie rod attaching points at the center of the steering gear travel.

9. Install or connect the following:
• Steering gear

➡**If equipped with a 2.7L engine, turn the left front tire straight-ahead.**

• Right tie rod on the steering arm and torque the nut to 27 ft. lbs. (37 Nm)
• Right front wheel. Torque the lug nuts in sequence to half of the specified torque, then tighten the nuts in sequence again to 100 ft. lbs. (135 Nm).
• Intermediate shaft to the steering gear using the roll pin. Be sure the pin is centered in the joint.
• 4 steering gear mounting bolts and torque the bolts to 43 ft. lbs. (58 Nm)

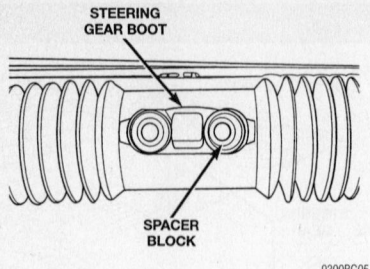

Center the spacer block in the steering gear before installing the tie rods

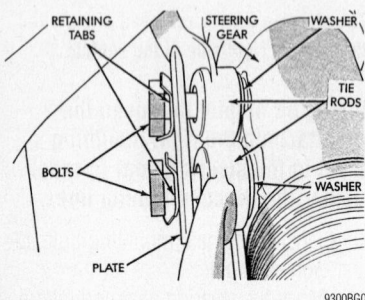

Be sure to position the washers between the tie rods and the steering gear as shown

- Bracket bolts; tighten them (if loosened) to 27 ft. lbs. (37 Nm)
- Steering gear fluid lines; torque the nuts to 35 ft. lbs. (47 Nm)
- Solenoid valve, if equipped with speed proportional steering

✳✳ WARNING

Be sure the spacer block in the rack is centered before connecting the tie rods to the steering gear.

- Tie rods to the steering gear with the washers between the rods and the steering gear. Torque the bolts to 60 ft. lbs. (82 Nm).
- Remaining components in the reverse order of the removal steps
- Negative battery cable

10. Add Mopar® power steering fluid to the reservoir and let it settle for at least 2 minutes. Start the engine for a few seconds and turn it **OFF**. Add fluid as necessary. Repeat this procedure until the fluid level remains constant.

11. Raise the front wheels off the floor and start the engine. Turn the steering from lock-to-lock several times. Do not hold the wheel in the locked position for more than 2 seconds at a time. Refill the reservoir as needed.

12. Lower the vehicle and repeat the procedure. If the fluid is extremely foamy, allow the vehicle to stand with the engine **off** for a few minutes, then repeat the procedure.

Strut

REMOVAL & INSTALLATION

Front

1. Before servicing the vehicle, refer to the precautions in the beginning of this section.

➡**Service of the coil spring requires the use of a coil spring compressor tool. It is required that 5 coils be captured within the jaws of the compressor tool.**

➡**Do not support the vehicle by placing supports under the suspension arms. The suspension arms must hang freely.**

2. Remove or disconnect the following:
- Negative battery cable
- Front wheel(s)
- Stabilizer bar attaching link at the strut assembly

3. Loosen, but do not remove the outer tie rod end-to-strut assembly steering arm attaching nut. Then, remove the outer tie rod end from the steering arm using puller MB-990635/C-3894A, or equivalent.
- Speed sensor wiring harness mounting bracket from the strut
- Brake caliper assembly. Support the caliper assembly from the vehicle frame with a strong piece of wire. Do not allow the assembly to hang by the brake hose.
- Front brake rotor

✳✳ WARNING

The strut assembly-to-steering knuckle bolts are serrated where they go through the strut and steering knuckle. Do not turn the bolts during removal. If the bolts are turned, damage to the steering knuckle will result.

4. The strut assembly-to-steering knuckle bolts must not be turned during strut removal. Hold the bolt head with a wrench and turn the nuts off the bolts.

5. Remove the 3 or 4 strut assembly upper mount-to-shock tower mounting nuts and washers. Remove the strut from the vehicle.

6. Disassemble the strut by performing the following procedure:
   a. Securely mount the strut assembly into a vise. Using paint, mark the strut

unit, lower spring isolator, spring and upper strut mount for indexing of the parts at assembly.

   b. Position the spring compressor tool onto the strut. Compress the coil spring until all load is off the upper strut mount assembly.

   c. Install Strut Rod Socket tool L-4558A on the strut shaft nut and a 10mm socket on the end of the strut shaft to prevent it from turning. Remove the strut shaft nut.

   d. Remove the upper mount assembly, jounce bumper and seat bearing and dust shield as an assembly.

   e. Remove the coil spring and compressor as an assembly from the strut. Remove the lower spring isolator from the strut assembly lower spring seat.

   f. Inspect all components for abnormal wear, oil leakage or failure. Replace parts as required.

To install:

7. Assemble the strut by performing the following procedure:
   a. Inspect the strut assembly for signs of leakage. Actual leakage will be a stream of fluid running down the side and dripping off the lower end of the strut. A slight amount of seepage between the strut rod and strut shaft seal is not unusual and does not affect performance of the strut assembly.

   b. Install the lower spring isolator on the strut unit. Install the compressed coil spring onto the strut assembly aligning the paint marks made during removal.

   c. Install the strut bearing into the bearing seat. The bearing must be installed into the seat with the notches on the bearings facing down.

   d. Lower the seat bearing and dust shield onto the strut and spring assembly. Align the paint marks made during removal.

   e. Install the jounce bumper and upper mount on the strut shaft, aligning the paint marks.

   f. Install the strut mount-to-shaft retainer nut. Inspect all alignment marks made during removal and align as required. While holding the strut shaft from turning with a 10mm socket, torque the strut shaft nut to 70 ft. lbs. (94 Nm).

   g. Equally loosen the spring compressor tool until all tension is released. Remove the spring compressor tool.

8. Install the front strut into the strut tower. Torque the 3 or 4 upper nuts to 25–28 ft. lbs. (33–37 Nm).

9. Position the steering knuckle neck into the strut assembly. Install the strut

assembly-to-steering knuckle bolts. Install the nuts onto the attaching bolts and torque to 125 ft. lbs. (169 Nm) for 2001 vehicles, or to 150 ft. lbs. (203 Nm) for 2002–04 vehicles. Do not turn the serrated bolt heads during installation. Turn only the nuts.

✳✳ WARNING

The strut assembly-to-steering knuckle bolts are serrated (toothed) where they go through the strut and steering knuckle. Do not turn the bolts during removal. If bolts are turned, damage to the steering knuckle will result.

10. Install or connect the following:
  • Brake rotor and caliper assembly to the adapter and torque the caliper bolts to 14 ft. lbs. (19 Nm)
  • Front speed sensor cable routing

bracket onto the front strut, if equipped
  • Outer tie rod on the steering arm and torque the attaching nut to 27 ft. lbs. (37 Nm)
  • Stabilizer link assembly onto the strut assembly and torque the attaching nut to 70 ft. lbs. (95 Nm)
  • Front wheel and lug nuts. Torque the lug nuts, in sequence, to 100 ft. lbs. (135 Nm).
  • Negative battery cable

Rear

1. Before servicing the vehicle, refer to the precautions in the beginning of this section.

The rear strut assemblies support the weight of the vehicle using coil springs positioned around the struts. The coil springs are contained between the upper mount of the strut assembly and a lower

spring seat on the body of the strut assembly. The strut is attached to the spindle by a split collar on the rear spindle with a pinch bolt to hold the spindle to the strut.

2. Remove or disconnect the following:
  • Rear wheel
  • Caliper assembly and rotor from the hub, if equipped with rear disc brakes
  • Brake flex hose from the support bracket and wheel cylinder, if equipped with rear drum brakes. Plug the brake flex hose to prevent system contamination. Do not allow the rear caliper to hang by the brake hose. Support the caliper off of the frame with a strong piece of wire.
  • Speed sensor cable routing bracket and tube, if equipped with Anti-lock Brake System (ABS)
  • Lateral links to the rear spindle assembly bolts

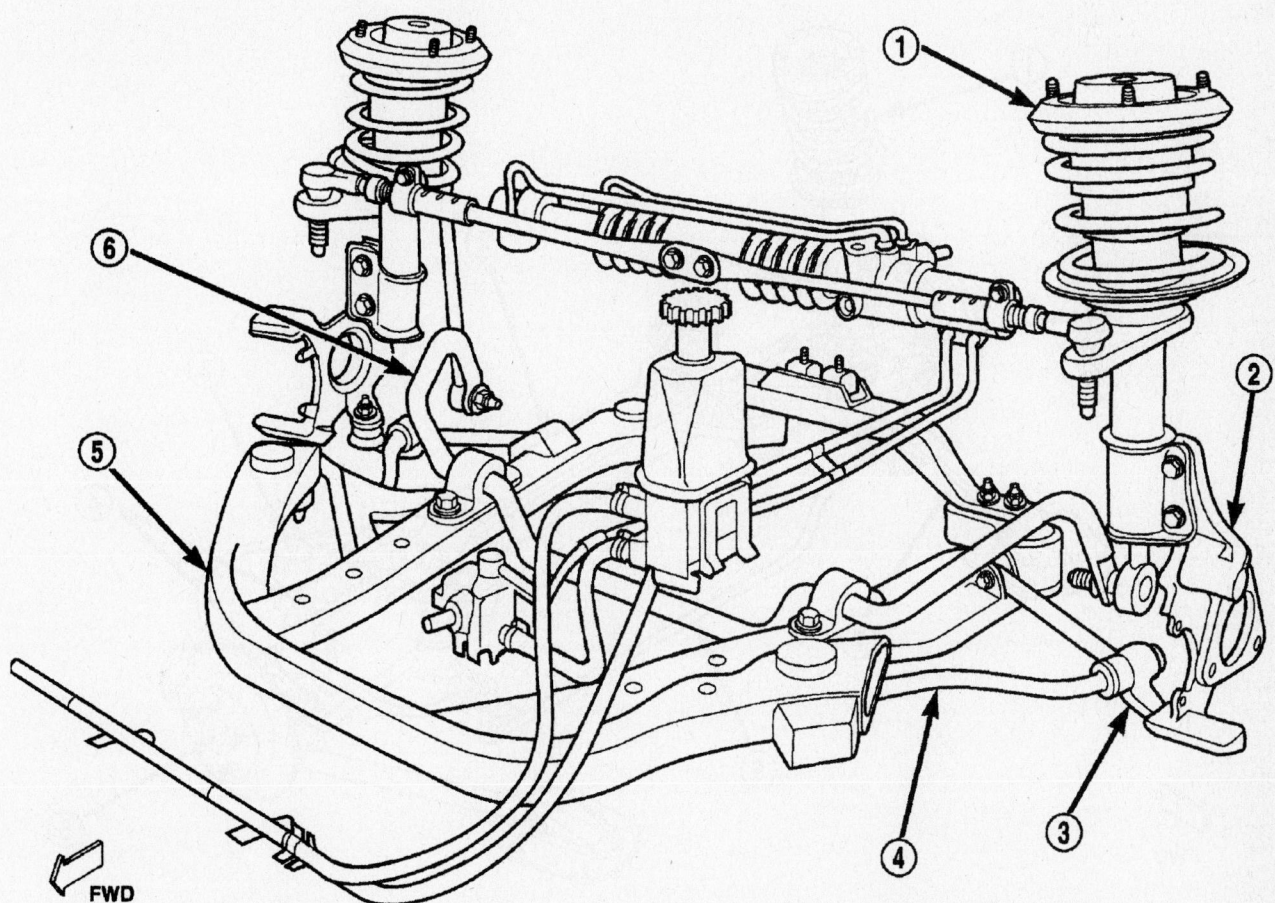

1 - STRUT ASSEMBLY
2 - STEERING KNUCKLE
3 - LOWER CONTROL ARM
4 - TENSION STRUT
5 - ENGINE CRADLE CROSSMEMBER
6 - STABILIZER BAR

67189-300M-G07

View of the front suspension components—2004 vehicle shown, others similar

- Rear strut assembly-to-stabilizer bar attaching link at the stabilizer bar

➡ **Hold the hex on the attaching link stud while breaking the nut loose. The attaching link does not have to be removed from the strut.**

3. Remove the rear spindle-to-strut assembly pinch bolt. Install a center punch in the hole on the spindle and tap the punch into the hole until jammed. This will spread the spindle casting allowing it to be removed from the strut.

4. Using a hammer, tap on the top surface of the spindle, driving the spindle down and off the end of the strut assembly. Let the spindle and assembled components hang from the trailing arm while the strut is being serviced.

5. From inside the trunk of the vehicle, remove the 3 upper strut mounting bolts and remove the strut from the vehicle.

6. Disassemble the strut by performing the following procedure:

➡ **Service of the coil spring requires the use of a coil spring compressor tool. It is required that 5 coils be captured within the jaws of the compressor tool.**

a. Securely mount the strut assembly into a vise. Using paint, mark the strut assembly, lower spring isolator, spring and upper strut mount for indexing of the parts at reassembly.

b. Position a spring compressor tool onto the coil spring. Compress the coil spring until all load is off of the upper strut mount assembly.

c. Install the strut rod socket tool L-4558 on the strut shaft nut and an 8mm Allen wrench on the end of the strut shaft to prevent it from turning. Remove the strut shaft nut.

d. Remove the upper strut mount

assembly off of the strut shaft. Remove the coil spring and compressor tool as an assembly from the strut.

e. Remove the plate, dust shield and jounce bumper off of the strut unit.

f. Inspect all components for abnormal wear, oil leakage or failure. Replace parts as required.

To install:

7. Assemble the strut by performing the following procedure:

a. Install the lower spring isolator on the strut unit. If it is the original isolator, align the paint marks.

b. Install the jounce bumper into the dust shield. Install the plate on top of the dust shield and into the jounce bumper.

c. Install the dust shield, jounce bumper and the top plate onto the strut unit as an assembly.

d. Install the coil spring and compressor tool onto the strut unit and align

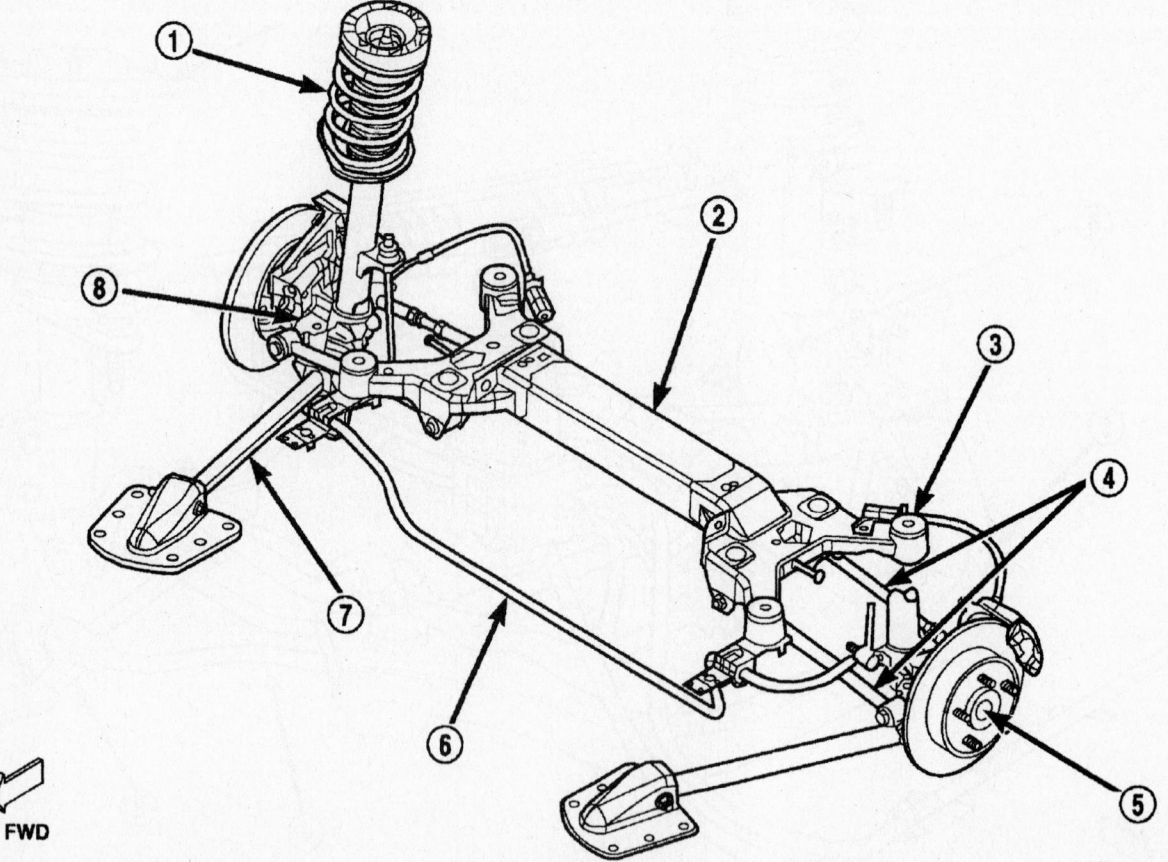

FWD

1 - STRUT ASSEMBLY
2 - REAR SUSPENSION CROSSMEMBER
3 - REAR SUSPENSION CROSSMEMBER BUSHING
4 - LATERAL LINKS
5 - HUB AND BEARING
6 - STABILIZER BAR
7 - TRAILING ARM
8 - SPINDLE

67189-300M-G08

View of the rear suspension components—2004 vehicle shown, others similar

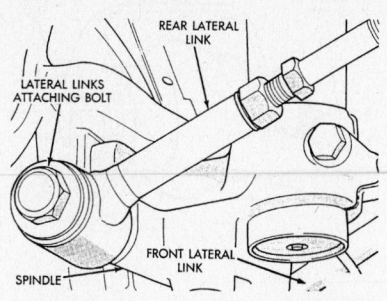

Separate the lateral links from the spindle—rear strut service

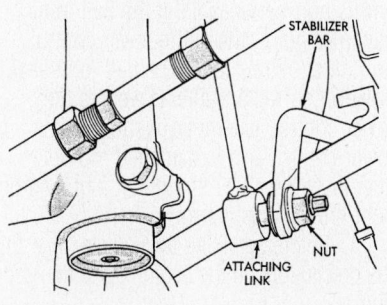

Remove the nut from the stabilizer-to-strut attaching link stud at the bar—rear strut service

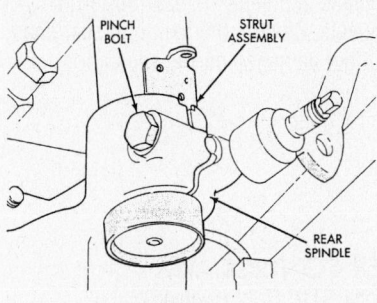

Loosen, then remove the rear spindle-to-strut pinch bolt—rear strut service

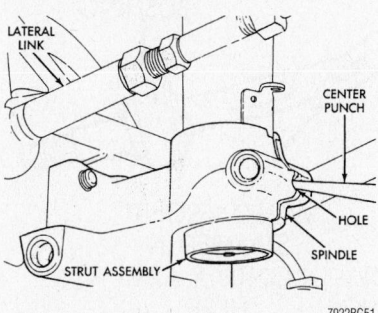

Insert a center punch into the hole on the spindle and tap until the casting is spread—rear strut service

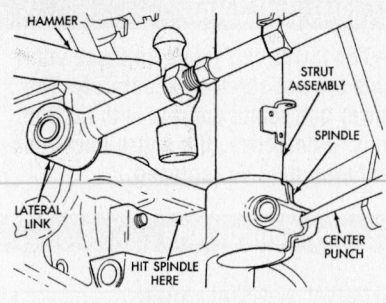

Tap with a hammer on the surface of the spindle driving it down and off the end of the strut—rear strut service

the paint marks on the spring to that of the strut unit.

e. Install the upper strut mount assembly onto the strut shaft. Align the paint marks and install the strut shaft retaining nut.

f. Using the strut rod socket tool L-4558 and the 8mm Allen wrench to prevent the strut shaft from turning, torque the strut shaft nut to 70 ft. lbs. (95 Nm).

g. Equally loosen the spring compressor tool until all tension is released. Remove the spring compressor tool.

8. Position the strut in the vehicle.

Torque the 3 upper mounting nuts to 20 ft. lbs. (28 Nm).

9. Install the spindle assembly onto the bottom of the strut. Push or tap the spindle assembly onto the strut, until the notch in the spindle is tightly seated against the locating tap on the strut assembly. Remove the center punch from the hole in the spindle.

10. Install or connect the following:
- Strut-to-spindle pinch bolt and torque to 40 ft. lbs. (55 Nm)
- Lateral link-to-spindle attaching bolt, hand-tight only

❊❊ WARNING

Do not tighten the lateral link-to-spindle bolt at this time, or the bushings will be contorted and fail. You must tighten the bolt at curb height.

- Stabilizer bar attaching link and torque the stabilizer link-to-stabilizer bar attaching nut to 70 ft. lbs. (95 Nm), while holding the stabilizer link stud at the hex with a wrench
- Rear speed sensor cable routing tube and bracket, if equipped with ABS
- Rotor and caliper assembly (if

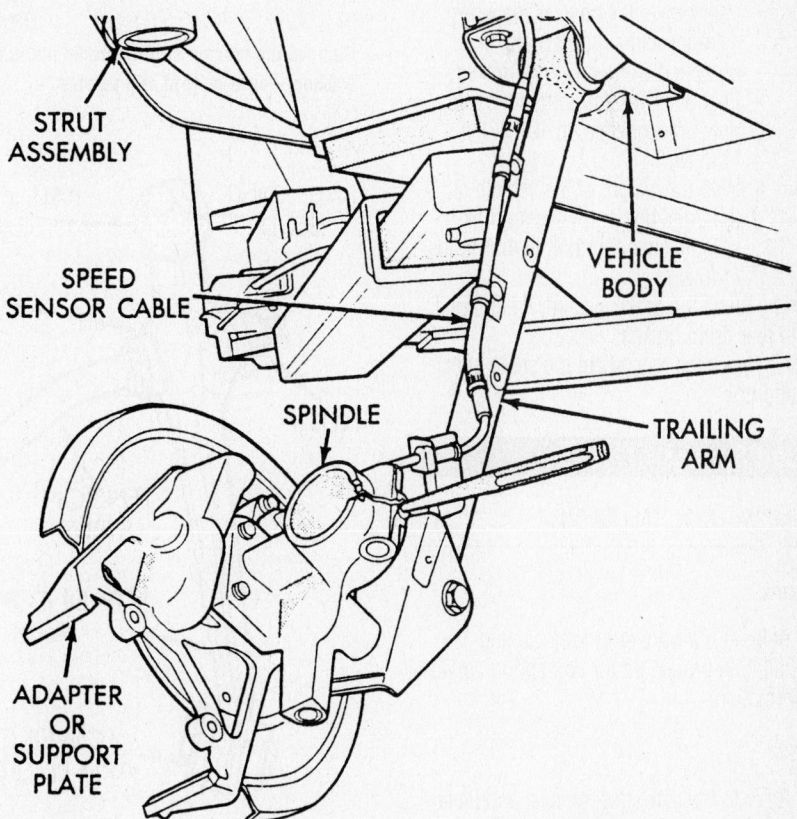

Allow the components to hang from the trailing arm as shown—rear strut service

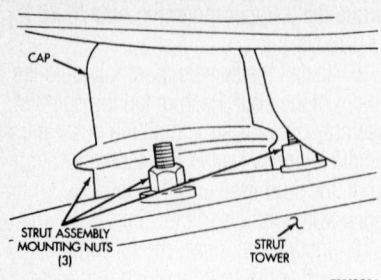

Access the 3 upper strut mounting nuts through the trunk—rear strut service

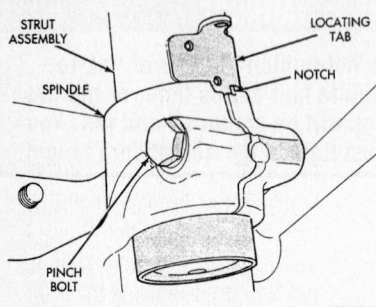

During reassembly, tap the spindle onto the strut until the notch in the spindle is tightly seated against the tab—rear strut service

equipped with rear disc brakes) and torque the bolts to 16 ft. lbs. (22 Nm)
- Rear brake flex hose to the wheel cylinder and support plate, if equipped with rear drum brakes
- Rear wheel(s) and torque the lug nuts, in sequence, to 100 ft. lbs. (135 Nm)
- With the weight of the vehicle on the tires, tighten the lateral link-to-spindle attaching bolt to 105 ft. lbs. (140 Nm)

11. Bleed the brake system, if equipped with rear drum brakes.
12. Have the rear wheel toe set to specifications.

Coil Springs

REMOVAL & INSTALLATION

Front

Refer to the front strut removal and installation procedure for coil spring service information.

Rear

Refer to the rear strut removal and installation procedure for coil spring service information.

Lower Ball Joint

➡The lower ball joints on these vehicles are not serviced separately. The lower ball joints operate with no free-play. If defective, the entire lower control arm must be replaced.

Lower Control Arm

REMOVAL & INSTALLATION

The front lower control arm is a steel forging with 2 rubber bushings isolating the lower control arm from the front cradle assembly. The isolator bushings consist of a metal encased pivot bushing and a solid rubber tension strut bushing. The lower control arm is bolted to the cradle assembly using a pivot bolt through the center of the rubber pivot bushing at the

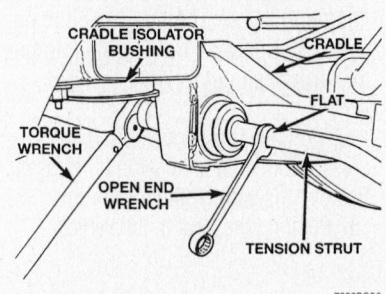

Remove the tension strut-to-cradle nut and washer—lower control arm service

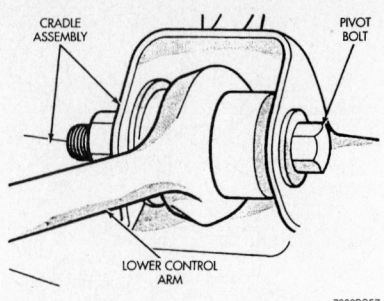

Loosen the pivot bolt and remove it—lower control arm service

tension strut isolator bushing. The ball joint is built into the lower control arm and is non-serviceable. If the ball joint becomes worn, the entire lower control arm must be replaced. The ball joint seal, however, is replaceable as well as the lower control arm inner bushing. If the lower control arm is damaged, do not attempt to repair or straighten a broken or bent lower control arm.

1. Before servicing the vehicle, refer to the precautions in the beginning of this section.
2. Remove the front wheel(s).
3. Remove the ball joint stud-to-steering knuckle clamp nut and bolt.
4. Carefully insert a prybar between the lower control arm and the steering knuckle and separate ball joint from knuckle. Be sure the ball joint seal does not get damaged during separation.

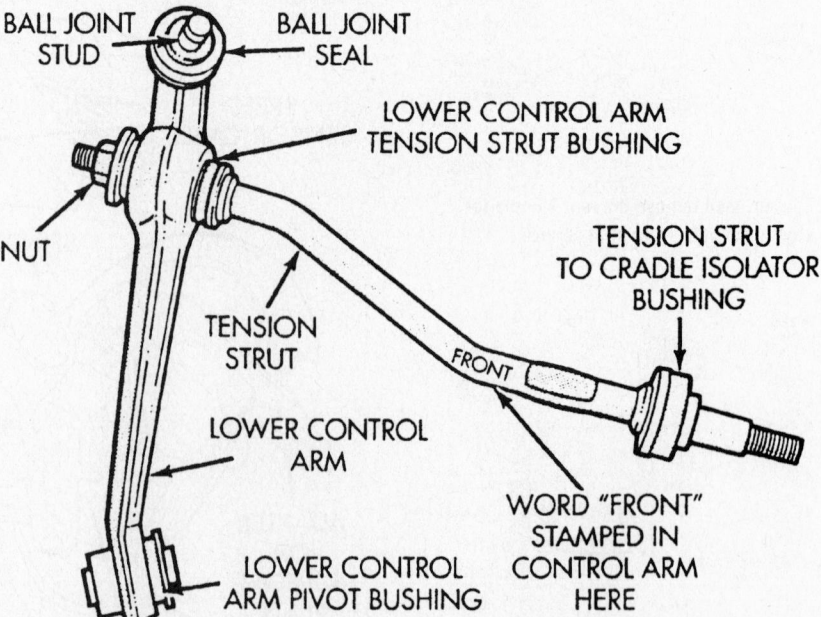

Inspect the control arm and tension strut for distortion—lower control arm service

Pulling the steering knuckle out from the vehicle after releasing from the ball joint can separate the inner CV-joint. Do not separate the inner CV-joint or it can be damaged.

5. Remove the tension strut-to-cradle attaching nut and washer from the end of the tension strut. When removing the nut, keep the strut from turning by holding the tension strut at the flats using an open end wrench. Discard the tension strut-to-cradle retainer nut. A new nut must be used during installation.

➡ **A new tension strut-to-cradle attaching nut must be used when installing the tension strut.**

6. Loosen and remove the lower control arm pivot bushing-to-cradle assembly pivot bolt.

7. Separate the lower control arm and tension strut from the cradle as an assembly by first removing the pivot bushing from the cradle, then sliding the tension strut out of the isolator bushing. Inspect the control arm and tension strut for distortion, check the rubber bushings for excessive wear or deterioration and replace these components, if necessary.

To install:

8. Install or connect the following:
- Tension strut and isolator bushing into the cradle first, then install lower control arm pivot bushing into bracket on the cradle
- Lower control arm-to-cradle bracket attaching bolt. Do not tighten the bolt at this time.
- Washer and new nut on the end of the tension strut. Torque the tension strut-to-cradle bracket retainer nut to 110 ft. lbs. (150 Nm), while holding the tension strut flat with an open end wrench.

9. Inspect the ball joint seal and replace it if damaged. Install the lower ball joint stud into the steering knuckle and install the clamp bolt and nut. Torque the bolt to 40 ft. lbs. (55 Nm).

10. Install the front wheel and lug nuts. Torque the lug nuts, in a star pattern, to 100 ft. lbs. (135 Nm). Lower the vehicle so the suspension is supporting the weight of the vehicle.

11. Torque the lower control arm pivot bushing-to-cradle bracket attaching bolt to 105 ft. lbs. (142 Nm).

CONTROL ARM BUSHING REPLACEMENT

1. Remove the lower control arm/tension strut assembly from the vehicle.

2. Remove the nut and separate the tensions strut from the lower control arm.

3. Using an arbor press, position the lower control arm with the large end of the pivot bushing inside the Receiver tool MB-990799.

4. Position Remover tool 6644-2 on top of the pivot bushing.

5. Using the arbor press, press the pivot bushing out of the lower control arm.

To install:

6. Turn the lower control arm over and reposition it on the Receiver tool MB-990799.

7. Using a new pivot bushing, position it in the lower control arm so that it is square with the bushing hole.

8. Position Installer tool 6644-1 on top of the pivot bushing with the pivot bushing setting in the recessed area of the installer tool.

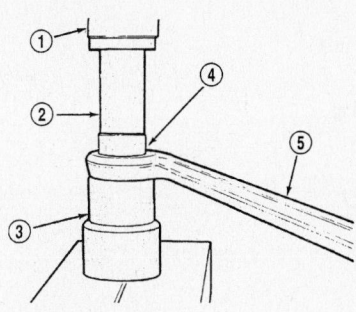

```
1 – ARBOR PRESS RAM
2 – SPECIAL TOOL 6644-2
3 – SPECIAL TOOL MB990799
4 – PIVOT BUSHING
5 – LOWER CONTROL ARM
```
9306BG25

Removing the lower control arm pivot bushing

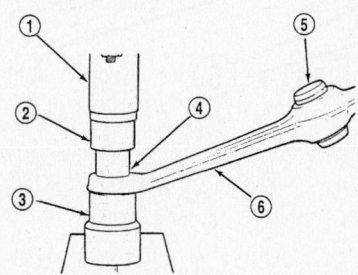

```
1 – ARBOR PRESS RAM
2 – SPECIAL TOOL 6644-1
3 – SPECIAL TOOL MB990799
4 – PIVOT BUSHING
5 – TENSION STRUT BUSHING
6 – LOWER CONTROL ARM
```
9306BG22

Installing the lower control arm pivot bushing

9. Using the arbor press, press the pivot bushing into the lower control arm until the installer tool squarely bottoms against the surface of the lower control arm.

10. Install the tension strut into the lower control arm's tension strut bushing with the word FRONT, stamped on the tension strut, facing away from the control arm.

11. Position an open end wrench on the flat of the tension strut to keep it from turning. Torque the tension strut-to-lower control arm nut to 95 ft. lbs. (130 Nm).

TENSION STRUT BUSHING REPLACEMENT

1. Remove the lower control arm/tension strut assembly from the vehicle.

2. Remove the nut and separate the tensions strut from the lower control arm.

3. Using an arbor press, position the lower control arm with the tension strut bushing inside the Receiver tool MB-990799.

4. Position Remover tool 6644-4 on top of the tension strut bushing.

5. Using the arbor press, press the tension strut bushing out of the lower control arm.

➡ **As the Remover tool is press through the tension strut bushing, it will cut the bushing into 2 pieces.**

6. Remove the lower control arm, both bushing pieces and the Remover tool.

To install:

7. Thoroughly, lubricate the new tension strut bushing, the lower control arm and the Installer tool 6644-3 with Rubber Bushing Installation Lube.

8. Press the new tension strut bushing by hand, into the large end of the Installer tool 6644-3 as far as it will go.

9. Position the lower control arm onto on the arbor press so that the tension strut hole is centered on the Receiver tool MB-990799.

10. Position Installer tool 6644-3, with the bushing installed, inside the tension strut bushing hole in the lower control arm.

11. Position Installer tool 6644-2 on top of the tension strut bushing.

12. Using the arbor press, press the tension strut bushing into the lower control arm.

➡ **As the bushing is being installed, a pop will be heard and Installer tool 6644-3 will move slightly up off the control arm.**

13. Remove the control arm assembly from the arbor press and remove Installer

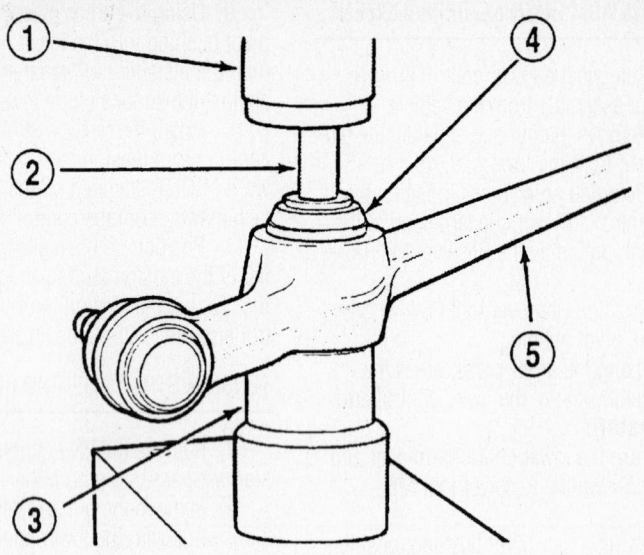

1 – ARBOR PRESS RAM
2 – SPECIAL TOOL 6644-4
3 – SPECIAL TOOL MB990799
4 – TENSION STRUT BUSHING
5 – LOWER CONTROL ARM

9306BG23

Removing the lower control arm's tension strut bushing

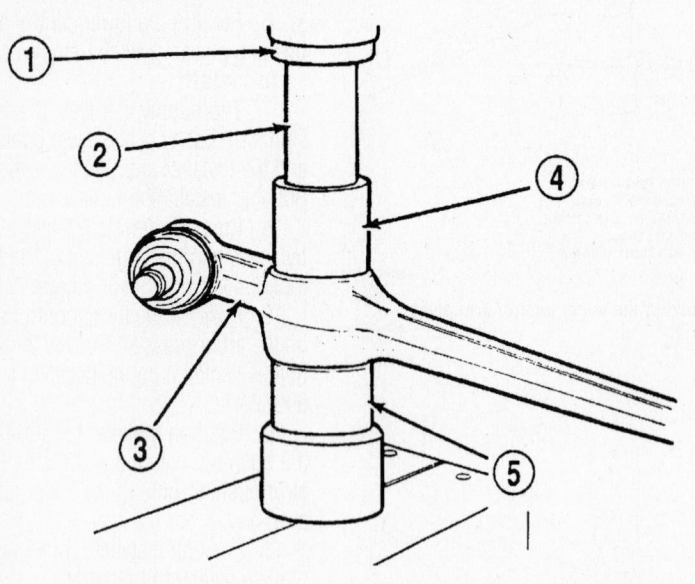

1 – ARBOR PRESS
2 – SPECIAL TOOL 6644-2
3 – LOWER CONTROL ARM
4 – SPECIAL TOOL 6644-3
5 – SPECIAL TOOL MB990799

9306BG24

Installing the lower control arm's tension strut bushing

tool 6644-3 from the tension strut bushing; the tension strut bushing is now installed.

14. Install the tension strut into the lower control arm's tension strut bushing with the word FRONT, stamped on the tension strut, facing away from the control arm.

15. Position an open end wrench on the flat of the tension strut to keep it from turning. Torque the tension strut-to-lower control arm nut to 95 ft. lbs. (130 Nm).

Wheel Bearings

ADJUSTMENT

These front wheel drive vehicles are equipped with permanently sealed front and rear wheel bearings. There is no periodic lubrication or maintenance recommended for these units.

REMOVAL & INSTALLATION

Front

1. Before servicing the vehicle, refer to the precautions in the beginning of this section.

2. Remove the front wheel.

3. Remove the front caliper assembly from the steering knuckle by removing the 2 guide pin bolts, then rotating the top of the caliper away from the knuckle and lifting the caliper off the machined abutment on the steering knuckle.

4. Suspend the caliper out of the way with a piece of wire.

5. Remove the front brake rotor from the hub by pulling it straight off of the wheel mounting stud.

6. Remove the hub and bearing-to-stub axle retainer nut.

➡ **This hub nut is a torque prevailing retaining nut and can not be reused. A NEW retaining nut MUST be used when assembling the hub.**

7. Remove the 3 attaching bolts that mount the hub and bearing assembly to the steering knuckle assembly.

➡ **If the metal seal on the hub and bearing assembly is seized to the steering knuckle and becomes dislodged on the hub and bearing during removal, the hub and bearing must be replaced. If the flinger disc becomes damaged during the removal procedure, the hub and bearing assembly must be replaced.**

8. Remove the hub and bearing assembly from the steering knuckle by sliding it

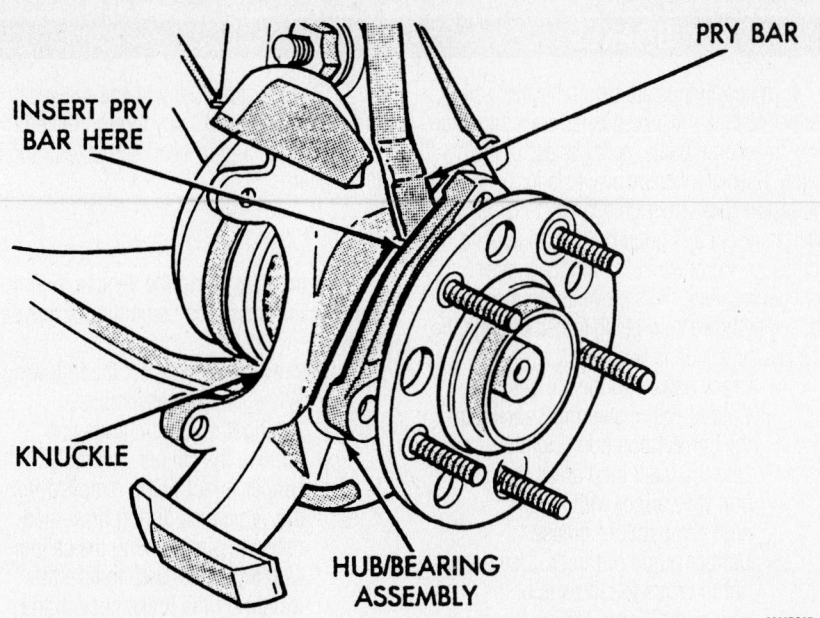

PRY BAR

INSERT PRY
BAR HERE

KNUCKLE

HUB/BEARING
ASSEMBLY

9300BG07

Carefully pry the front bearing/hub assembly away from the steering knuckle

straight out of the knuckle and off the ends of the stub shaft.

9. Gently pry the assembly out with a prybar or tap it out with a soft-faced hammer, if necessary. Be very careful not to damage the hub and bearing assembly.

To install:

10. Clean the hub and bearing mounting surfaces of dirt and be sure there are no nicks present.

11. Install or connect the following:
- Hub and bearing squarely onto the stub shaft and the steering knuckle
- Bearing assembly mounting bolts and tighten equally until the bearing assembly is seated squarely against the front of the steering knuckle. Torque the mounting bolts to 80 ft. lbs. (110 Nm).

- New hub and bearing assembly-to-stub shaft retainer nut. A NEW retaining nut MUST be used when assembling the hub. Tighten but do not torque the nut at this time.
- Brake rotor and the caliper assembly. Torque the brake caliper guide bolts to 16 ft. lbs. (22 Nm).
- Wheel and lug nuts. Torque the lug nuts in a star pattern to 100 ft. lbs. (135 Nm).

12. Pump the brakes until a firm pedal is obtained.

13. With the weight on the vehicle on its wheels, apply the brakes to keep the vehicle from moving. Torque the hub and bearing assembly-to-stub shaft retaining nut to 120 ft. lbs. (163 Nm).

✸✸ WARNING

When tightening the hub and bearing assembly to stub shaft retaining nut, do not exceed the maximum torque of 120 ft. lbs. (163 Nm). If the maximum torque is exceeded this may result in a failure of the halfshaft.

14. Inspect the toe setting on the vehicle and adjust, if necessary.

Rear

1. Before servicing the vehicle, refer to the precautions in the beginning of this section.

2. Remove or disconnect the following:
- Rear wheel
- Brake caliper and rotor, if equipped with rear disc brakes
- Brake drum, if equipped with drum brakes
- Bearing dust cap
- Cotter pin, nut retainer, nut, washer and bearing/hub assembly from the spindle

To install:

3. Install or connect the following:
- Bearing/hub assembly, the bearing/hub assembly washer and torque the nut to 124 ft. lbs. (168 Nm)
- Nut retainer and a new cotter pin
- Dust cap
- Brake drum or rotor and caliper assembly
- Rear wheel and torque the lug nuts, in a star pattern, to 100 ft. lbs. (135 Nm)

4. Road test the vehicle to verify no excessive noise from the rear wheel bearing area.

BRAKES

Caliper

REMOVAL & INSTALLATION

Front

1. Before servicing the vehicle, refer to the precautions in the beginning of this section.

2. Remove or disconnect the following:
   • Wheel and tire assemblies
   • 2 caliper guide pin bolts and the caliper assembly. If the caliper is not being removed from the vehicle as during brake pad renewal, simply hang the caliper with a piece of wire to take the weight off the brake hose. If the caliper is being removed for rebuild or replacement, continue to Step 5.
   • Bolt retaining the brake hose to the caliper. Be sure to plug the end of the brake hose or cover it with a plastic bag to prevent contamination from entering the hydraulic system.
   • Caliper from the vehicle

To install:

3. Install or connect the following:
   • Brake hose to the caliper, if removed, using new sealing washers

4. If new linings are being installed, the caliper pistons must be pushed back into their bore to accommodate the thickness of the new lining. Special tools are available for pushing the piston back although a large C-clamp can often be used. It is good practice to remove some (⅓–½) brake fluid from the master cylinder reservoir. This prevents overflow caused by brake fluid being forced through the lines as the piston is pushed back.
   • Brake pads into the caliper
   • Caliper to the steering knuckle in the correct position. Lightly lubricate the machined areas that support the caliper with high-temperature grease.
   • Caliper guide pin bolts. Use care not to cross the threads of the caliper pin bolts. Torque the guide pin bolts to 15 ft. lbs. (20 Nm).
   • Brake hose fittings to 35 ft. lbs. (48 Nm)

5. Be sure to bleed the brake system.
   • Wheels and lug nuts. Torque the wheel lug nuts in a star pattern sequence to 100 ft. lbs. (135 Nm).

6. Before attempting to move the vehicle, pump the brake pedal to seat the pads against the rotors. Make sure the vehicle has a firm brake pedal. Check the level of the brake fluid and add DOT 3 brake fluid, if necessary.

7. Road test the vehicle and make several stops to wear off any foreign material on the brakes and to seat the brake linings, if replaced.

Rear

1. Before servicing the vehicle, refer to the precautions in the beginning of this section.

2. Remove or disconnect the following:
   • Wheel and tire assemblies
   • 2 caliper guide pin bolts and remove the caliper assembly. If the caliper is not being removed from the vehicle as during brake pad renewal, simply hang the caliper with a piece of wire to take the weight off the brake hose. If the caliper is being removed for rebuild or replacement, continue to Step 5.
   • Bolt retaining the brake hose to the caliper. Be sure to plug the end of the brake hose or cover it with a plastic bag to prevent contamination from entering the hydraulic system.
   • Caliper from the vehicle

To install:

3. Install or connect the following:
   • Brake hose to the caliper, if removed, using new sealing washers

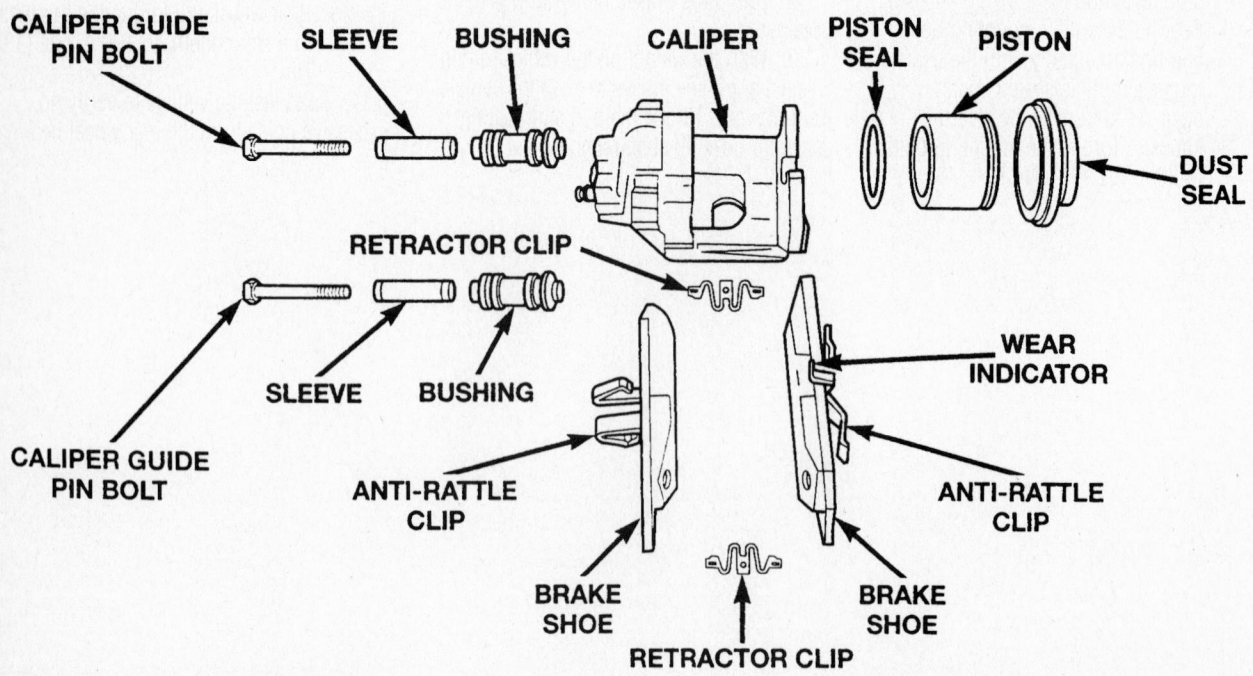

Front brake caliper and related components

93006G01

4. If new linings are being installed, the caliper pistons must be pushed back into their bore to accommodate the thickness of the new lining. Special tools are available for this, although a large C-clamp can often be used. It is good practice to remove some brake fluid from the master cylinder reservoir. This prevents overflow caused by brake fluid being forced through the lines as the piston is pushed back.

- Brake pads into the caliper
- Brake caliper to the caliper adapter. Lightly lubricate machined areas that support the caliper with high-temperature grease.
- Caliper guide pin bolts. Use care not to cross the threads of the caliper pin bolts. Torque the guide pin bolts to 17 ft. lbs. (22 Nm).
- Brake hose fittings to 35 ft. lbs. (48 Nm)

5. Be sure to bleed the brake system.
- Wheels and lug nuts. Torque the wheel lug nuts in a star pattern sequence to 100 ft. lbs. (135 Nm).

6. Before attempting to move the vehicle, pump the brake pedal to seat the pads against the rotors. Make sure the vehicle has a firm brake pedal.

7. Check the level of the brake fluid and add DOT 3 brake fluid, if necessary.

8. Road test the vehicle and make several stops to wear off any foreign material on the brakes and to seat the brake linings, if replaced.

Disc Brake Pads

REMOVAL & INSTALLATION

1. Before servicing the vehicle, refer to the precautions in the beginning of this section.

2. Remove or disconnect the following:
- Some of the fluid from the master cylinder
- Wheels
- 2 caliper guide pin bolts
- Caliper assembly by swinging the top part of the caliper away from the brake rotor edge, then lift the caliper assembly up

➡**Prevent strain or other damage to the brake hose by supporting the caliper assembly with a strong piece of wire hanging from the strut.**

- Outboard brake pad by prying the brake pad retaining clip over raised area on the caliper. Then slide the pad down and off the caliper.

3. Before removing the inboard brake pad, use a large C-clamp to press the piston back into the caliper. This will prevent possible damage to the caliper piston. It is good practice to remove some (⅓–½) of the brake fluid from the reservoir. This is because as the caliper piston is pushed back into the caliper, brake fluid will be pushed back through the lines, back into the master cylinder and fluid reservoir, possibly causing the reservoir to overflow.

- Inboard brake pad by pulling away from piston until the retainer clip is free from the cavity in the piston.

To install:

4. Lubricate both the caliper mating surface and the machined abutment surfaces with multi-purpose lubricant.

5. Before brake pad installation, be sure to lightly coat the outer backing plate surface of the new brake pads with a disc brake pad anti-squeal lubricant, usually a gel-like material that deadens any high-frequency vibration that can be the source of disc brake squeal.

6. Install or connect the following:
- Brake pads into the caliper assembly making sure both pads are seated securely onto the caliper
- Caliper assembly back into position over the brake rotor
- Caliper guide pin bolts and torque to 17 ft. lbs. (23 Nm)
- Wheels and lug nuts. Torque the lug nuts, in a star pattern sequence, to 100 ft. lbs. (135 Nm).

7. Top off the master cylinder to the appropriate level, using DOT 3 type brake fluid only.

8. Before moving the vehicle, pump the brakes until a firm pedal is obtained. Road test the vehicle to make sure the brake operation is normal.

Brake Drums

REMOVAL & INSTALLATION

1. Before servicing the vehicle, refer to the precautions in the beginning of this section.

2. Remove or disconnect the following:
- Rear wheels
- Brake drum from the rear hub and bearing assembly

3. If the drum is difficult to remove, increase the clearance between the brake shoes and the drum as follows:
   a. Remove the rubber plug from the top of the brake support plate.
   b. Rotate the automatic shoe adjuster

screw with an upward motion using a medium size flat tipped tool.
   c. Remove the brake drum from the rear hub and bearing.

4. Inspect the brake shoe linings and drums for wear, contamination and scoring.

To install:

5. Install the rear brake drum onto the rear hub and bearing assembly.

6. Adjust the brake shoes as follows:
   a. Rotate the automatic shoe adjuster in a downward motion using a medium size flat tip tool. Turn the adjuster until there is a slight drag felt while turning the drum.

7. Install or connect the following:
- Rubber plug back into the top part of the brake support plate
- Rear wheels and lug nuts. Torque the lug nuts in a star pattern sequence to 100 ft. lbs. (135 Nm).

Brake Shoes

REMOVAL & INSTALLATION

1. Before servicing the vehicle, refer to the precautions in the beginning of this section.

2. Remove or disconnect the following:
- Rear wheels and brake drums
- Dust cap from the rear hub and bearing assembly
- Cotter pin, nut retainer and wave washer. Discard the old cotter pin.
- Rear hub and bearing assembly retainer nut and washer
- Rear hub and bearing assembly from the spindle
- Automatic adjuster spring from the adjuster lever

3. Rotate the automatic adjuster star-wheel enough so both shoes move out far enough to be free of the wheel cylinder boots.

- Parking brake cable from the actuating lever. Disconnect parking brake cable one side at a time.
- Both lower brake shoe to anchor springs
- 2 brake shoe hold-down springs from the brake shoes
- Brake shoes, upper shoe-to-shoe return spring, automatic adjuster and automatic adjuster lever from the backing plate as an assembly
- Brake shoes from the automatic adjuster mechanism
- Brake shoe automatic adjuster lever from the leading brake shoe

To install:

4. Thoroughly clean and dry the backing plate. To prepare the backing plate, lubricate the bosses, anchor pin and parking brake actuating lever pivot surface lightly with lithium based grease.

5. Remove, clean and dry all parts still on the old shoes. Lubricate the starwheel shaft threads with anti-seize lubricant.

6. Assemble both brake shoes, the top shoe to shoe return spring, automatic adjuster and automatic adjuster lever before mounting on vehicle. Make sure the ends of the automatic adjusters are positioned above the extruded pins in the webbing of the brake shoes prior to installation.

7. Install or connect the following:
- Brake shoe assembly onto the brake support plate and secure with the hold-down springs
- Lower anchor springs and reconnect the parking brake cable to the park brake lever of the trailing brake shoe

8. Rotate the serrated adjuster nut to remove the free-play from the adjuster assembly.
- Automatic adjuster lever spring on the lead brake shoe assembly and the automatic adjuster lever
- Rear hub and bearing assembly
- Washer and retainer nut and torque to 124 ft. lbs. (168 Nm)
- Wave washer, nut retainer and a new cotter pin onto the spindle. Install dust cap.

9. Adjust brake shoes so not to interfere with brake drum installation. Install the rear brake drum.

➡ **After installing the brake drums, pump the brake pedal several times to partially adjust the brake shoes. To verify proper operation of the self-adjusting parking brake, be sure that both rear brakes are not dragging when the parking brake pedal is released.**

- Rear wheels and lug nuts. Torque the lug nuts, in a star pattern sequence, to 100 ft. lbs. (135 Nm).

10. Road test the vehicle. The automatic adjusters will continue brake adjustment during the road test of the vehicle.

CHRYSLER, DODGE AND PLYMOUTH

2

Caravan • Town & Country • Voyager

SPECIFICATION CHARTS

ENGINE AND VEHICLE IDENTIFICATION

		Engine							Model Year	
Code ①	Liters (cc)	Cu. In.	Cyl.	Fuel Sys.	Engine Type	Eng. Mfg.		Code ②		Year
B	2.4 (2429)	148	4	SMFI	DOHC	Chrysler		1		2001
L	3.8 (3785)	231	6	SMFI	OHV	Chrysler		2		2002
R	3.3 (3300)	201	6	SMFI	OHV	Chrysler		3		2003

SMFI: Sequential Multi-port Fuel Injection

DOHC: Double Overhead Camshaft

SOHC: Single Overhead Camshaft

OHV: Overhead Valve

① 8th position of VIN

② 10th position of VIN

	Year
4	2004
5	2005

67189-MINV-C01

GENERAL ENGINE SPECIFICATIONS

Year	Model	Engine Displacement Liters (VIN)	Net Horsepower @ rpm	Net Torque @ rpm (ft. lbs.)	Bore x Stroke (in.)	Com- pression Ratio	Oil Pressure @ rpm
2001	Caravan	2.4 (B)	150@5200	167@4000	3.44x3.98	9.4:1	25-80@3000
		3.3 (R)	158@4800	203@3600	3.66x3.19	8.9:1	30-80@3000
		3.8 (L)	180@4400	240@3300	3.78x3.43	9.0:1	30-80@3000
	Town & Country	3.3 (R)	158@4800	203@3600	3.66x3.19	8.9:1	30-80@3000
		3.8 (L)	180@4400	240@3300	3.78x3.43	9.0:1	30-80@3000
	Voyager	2.4 (B)	150@5200	167@4000	3.44x3.98	9.4:1	25-80@3000
		3.3 (R)	158@4800	203@3600	3.66x3.19	8.9:1	30-80@3000
		3.8 (L)	180@4400	240@3300	3.78x3.43	9.0:1	30-80@3000
2002	Caravan	2.4 (B)	150@5200	167@4000	3.44x3.98	9.4:1	25-80@3000
		3.3 (R)	158@4800	203@3600	3.66x3.19	8.9:1	30-80@3000
		3.8 (L)	180@4400	240@3300	3.78x3.43	9.0:1	30-80@3000
	Town & Country	3.3 (R)	158@4800	203@3600	3.66x3.19	8.9:1	30-80@3000
		3.8 (L)	180@4400	240@3300	3.78x3.43	9.0:1	30-80@3000
	Voyager	2.4 (B)	150@5200	167@4000	3.44x3.98	9.4:1	25-80@3000
		3.3 (R)	158@4800	203@3600	3.66x3.19	8.9:1	30-80@3000
		3.8 (L)	180@4400	240@3300	3.78x3.43	9.0:1	30-80@3000
2003	Caravan	2.4 (B)	150@5200	167@4000	3.44x3.98	9.4:1	25-80@3000
		3.3 (R)	158@4800	203@3600	3.66x3.19	8.9:1	30-80@3000
	Town & Country	3.3 (R)	158@4800	203@3600	3.66x3.19	8.9:1	30-80@3000
		3.8 (L)	180@4400	240@3300	3.78x3.43	9.0:1	30-80@3000
	Voyager	2.4 (B)	150@5200	167@4000	3.44x3.98	9.4:1	25-80@3000
		3.3 (R)	158@4800	203@3600	3.66x3.19	8.9:1	30-80@3000
		3.8 (L)	180@4400	240@3300	3.78x3.43	9.0:1	30-80@3000
2004-05	Caravan	2.4 (B)	150@5200	167@4000	3.44x3.98	9.4:1	25-80@3000
		3.3 (R)	158@4800	203@3600	3.66x3.19	8.9:1	30-80@3000
	Town & Country	3.3 (R)	158@4800	203@3600	3.66x3.19	8.9:1	30-80@3000
		3.8 (L)	180@4400	240@3300	3.78x3.43	9.0:1	30-80@3000

SMFI: Sequential Multi-port Fuel Injection

67189-MINV-C02

ENGINE TUNE-UP SPECIFICATIONS

Year	Engine Displacement Liters (VIN)	Spark Plug Gap (in.)	Ignition Timing (deg.)	Fuel Pump (psi)	Idle Speed (rpm)	Valve Clearance In.	Ex.
2001	2.4 (B)	0.048-0.053	①	49	②	HYD	HYD
	3.3 (R)	0.048-0.053	①	55	②	HYD	HYD
	3.8 (L)	0.048-0.053	①	49	②	HYD	HYD
2002	2.4 (B)	0.048-0.053	①	49	②	HYD	HYD
	3.3 (R)	0.048-0.053	①	55	②	HYD	HYD
	3.8 (L)	0.048-0.053	①	49	②	HYD	HYD
2003	2.4 (B)	0.048-0.053	①	49	②	HYD	HYD
	3.3 (R)	0.048-0.053	①	55	②	HYD	HYD
	3.8 (L)	0.048-0.053	①	49	②	HYD	HYD
2004-05	2.4 (B)	0.048-0.053	①	49	②	HYD	HYD
	3.3 (R)	0.048-0.053	①	55	②	HYD	HYD
	3.8 (L)	0.048-0.053	①	49	②	HYD	HYD

NOTE: The Vehicle Emission Control Information label often reflects specification changes made during production. The label figures must be used if they differ from those in this chart.

HYD: Hydraulic

① Ignition timing is regulated by the Powertrain Control Module (PCM), and cannot be adjusted.

② Idle speed is controled by the Powertrain Control Module (PCM), and cannot be adjusted.

67189-MINV-C03

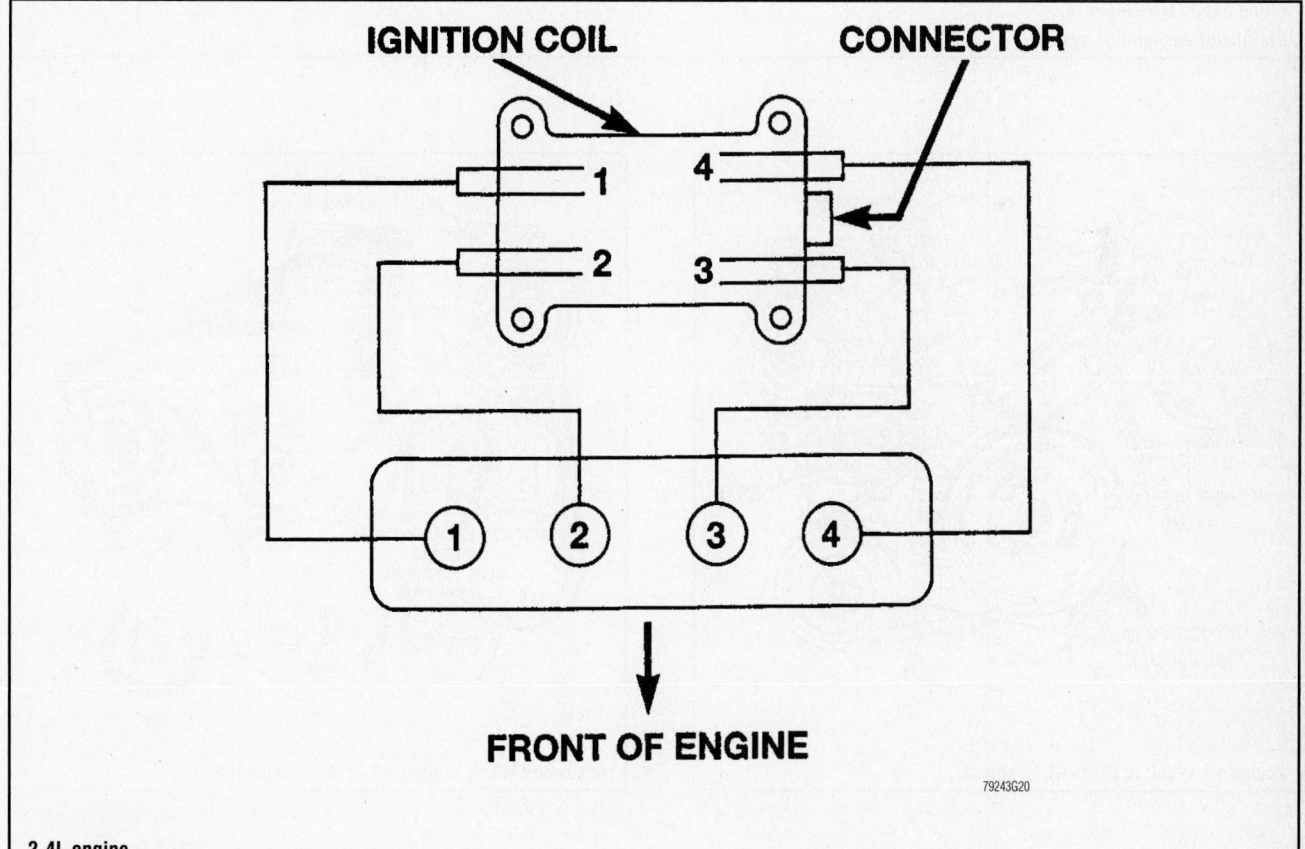

IGNITION COIL   **CONNECTOR**

FRONT OF ENGINE

79243G20

2.4L engine
Firing order: 1–3–4–2
Distributorless ignition system

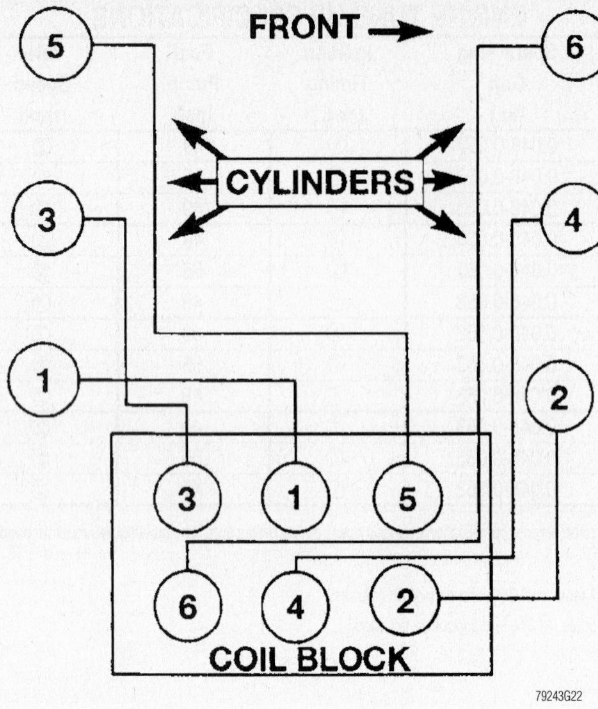

FRONT →

CYLINDERS

COIL BLOCK

79243G22

3.3L and 3.8L engines
Firing order: 1–2–3–4–5–6
Distributorless ignition system

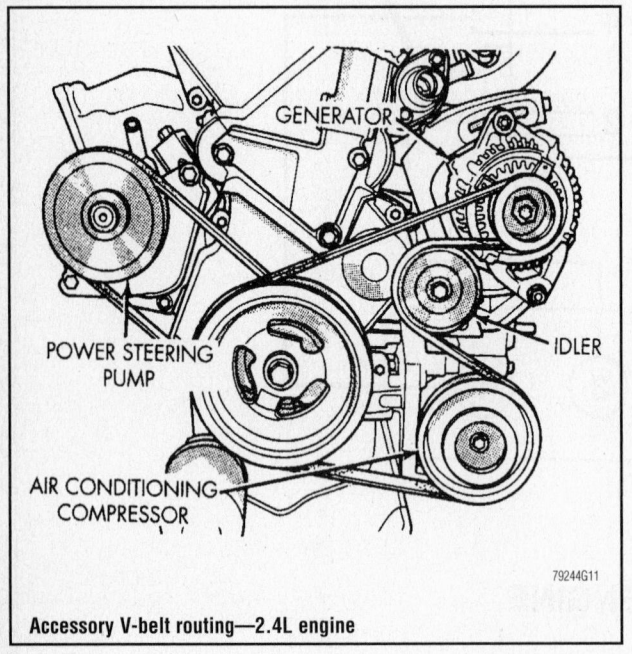

Accessory V-belt routing—2.4L engine

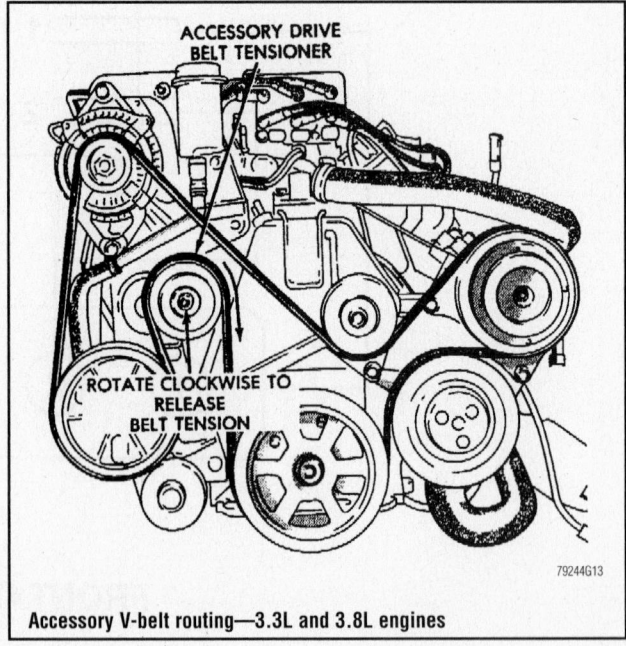

Accessory V-belt routing—3.3L and 3.8L engines

CAPACITIES

Year	Model	Engine Displacement Liters (VIN)	Engine Oil with Filter (qts.)	Automatic Transaxle (qts.)	Transfer Case (qts.)	Rear Drive Axle (pts.)	Fuel Tank (gal.)	Cooling System (qts.)
2001	Caravan	2.4 (B)	4.5	8.5	—	—	20.0	9.5
		3.3 (R)	4.5	9.1	—	—	20.0	10.5
	Town & Country	3.3 (R)	4.5	9.1	—	—	20.0	10.5
		3.8 (L)	4.5	9.1	1.22	4.0	20.0	10.5
	Voyager	2.4 (B)	4.5	8.5	—	—	20.0	9.5
		3.3 (R)	4.5	9.1	—	—	20.0	10.5
		3.8 (L)	4.5	9.1	—	—	20.0	10.5
2002	Caravan	2.4 (B)	4.5	8.5	—	—	20.0	9.5
		3.3 (R)	4.5	9.1	—	—	20.0	10.5
		3.8 (L)	4.5	9.1	1.22	4.0	20.0	10.5
	Town & Country	3.3 (R)	4.5	9.1	—	—	20.0	10.5
		3.8 (L)	4.5	9.1	1.22	4.0	20.0	10.5
	Voyager	2.4 (B)	4.5	8.5	—	—	20.0	9.5
		3.3 (R)	4.5	9.1	—	—	20.0	10.5
		3.8 (L)	4.5	9.1	—	—	20.0	10.5
2003	Caravan	2.4 (B)	4.5	8.5	—	—	20.0	9.5
		3.3 (R)	4.5	9.1	—	—	20.0	10.5
		3.8 (L)	4.5	9.1	1.22	4.0	20.0	10.5
	Town & Country	3.3 (R)	4.5	9.1	—	—	20.0	10.5
		3.8 (L)	4.5	9.1	1.22	4.0	20.0	10.5
	Voyager	2.4 (B)	4.5	8.5	—	—	20.0	9.5
		3.3 (R)	4.5	9.1	—	—	20.0	10.5
		3.8 (L)	4.5	9.1	—	—	20.0	10.5
2004-05	Caravan	2.4 (B)	4.5	8.5	—	—	20.0	9.5
		3.3 (R)	4.5	9.1	—	—	20.0	10.5
	Town & Country	3.3 (R)	4.5	9.1	—	—	20.0	10.5
		3.8 (L)	4.5	9.1	1.22	4.0	20.0	10.5

NOTE: All capacities are approximate. Add fluid gradually and check to be sure a proper fluid level is obtained.

67189-MINV-C04

VALVE SPECIFICATIONS

Year	Engine Displacement Liters (VIN)	Seat Angle (deg.)	Face Angle (deg.)	Spring Test Pressure (lbs. @ in.)	Spring Installed Height (in.)	Stem-to-Guide Clearance (in.)		Stem Diameter (in.)	
						Intake	Exhaust	Intake	Exhaust
2001	2.4 (B)	45	44.5-45.0	129-143@ 1.17	1.50	0.0018-0.0025	0.0029-0.0037	0.2340	0.2330
	3.3 (R)	45.0-45.5	①	207-229@ 1.169	1.62-1.68	0.0010-0.0030	0.0020-0.0060	0.3120-0.3130	0.3112-0.3119
	3.8 (L)	45.0-45.5	①	207-229@ 1.169	1.62-1.68	0.0010-0.0030	0.0020-0.0060	0.3120-0.3130	0.3112-0.3119
2002	2.4 (B)	45	44.5-45.0	129-143@ 1.17	1.50	0.0018-0.0025	0.0029-0.0037	0.2340	0.2330
	3.3 (R)	45.0-45.5	①	207-229@ 1.169	1.62-1.68	0.0010-0.0030	0.0020-0.0060	0.3120-0.3130	0.3112-0.3119
	3.8 (L)	45.0-45.5	①	207-229@ 1.169	1.62-1.68	0.0010-0.0030	0.0020-0.0060	0.3120-0.3130	0.3112-0.3119
2003	2.4 (B)	45	44.5-45.0	129-143@ 1.17	1.50	0.0018-0.0025	0.0029-0.0037	0.2340	0.2330
	3.3 (R)	45.0-45.5	①	207-229@ 1.169	1.62-1.68	0.0010-0.0030	0.0020-0.0060	0.3120-0.3130	0.3112-0.3119
	3.8 (L)	45.0-45.5	①	207-229@ 1.169	1.62-1.68	0.0010-0.0030	0.0020-0.0060	0.3120-0.3130	0.3112-0.3119
2004-05	2.4 (B)	45	44.5-45.0	129-143@ 1.17	1.50	0.0018-0.0025	0.0029-0.0037	0.2340	0.2330
	3.3 (R)	45.0-45.5	①	207-229@ 1.169	1.62-1.68	0.0010-0.0030	0.0020-0.0060	0.3120-0.3130	0.3112-0.3119
	3.8 (L)	45.0-45.5	①	207-229@ 1.169	1.62-1.68	0.0010-0.0030	0.0020-0.0060	0.3120-0.3130	0.3112-0.3119

① Intake valve: 44.5 degrees
Exhaust valve: 45 degrees

67189-MINV-C05

CRANKSHAFT AND CONNECTING ROD SPECIFICATIONS

All measurements are given in inches.

Year	Engine Displacement Liters (VIN)	Crankshaft				Connecting Rod		
		Main Brg. Journal Dia.	Main Brg. Oil Clearance	Shaft End-play	Thrust on No.	Journal Diameter	Oil Clearance	Side Clearance
2001	2.4 (B)	2.3610-2.3625	0.0007-0.0023	0.0035-0.0094	2	1.9670-1.9685	0.0009-0.0027	0.0051-0.0150
	3.3 (R)	2.5202-2.5195	0.0023-0.0043	0.0036-0.0095	2	2.1240-2.1250	0.0008-0.0026	0.0050-0.0150
	3.8 (L)	2.5202-2.5195	0.0023-0.0043	0.0036-0.0095	2	2.1240-2.1250	0.0008-0.0026	0.0050-0.0150
2002	2.4 (B)	2.3610-2.3625	0.0007-0.0023	0.0035-0.0094	2	1.9670-1.9685	0.0009-0.0027	0.0051-0.0150
	3.3 (R)	2.5202-2.5195	0.0023-0.0043	0.0036-0.0095	2	2.1240-2.1250	0.0008-0.0026	0.0050-0.0150
	3.8 (L)	2.5202-2.5195	0.0023-0.0043	0.0036-0.0095	2	2.1240-2.1250	0.0008-0.0026	0.0050-0.0150
2003	2.4 (B)	2.3610-2.3625	0.0007-0.0023	0.0035-0.0094	2	1.9670-1.9685	0.0009-0.0027	0.0051-0.0150
	3.3 (R)	2.5202-2.5195	0.0023-0.0043	0.0036-0.0095	2	2.1240-2.1250	0.0008-0.0026	0.0050-0.0150
	3.8 (L)	2.5202-2.5195	0.0023-0.0043	0.0036-0.0095	2	2.1240-2.1250	0.0008-0.0026	0.0050-0.0150
2004-05	2.4 (B)	2.3610-2.3625	0.0007-0.0023	0.0035-0.0094	2	1.9670-1.9685	0.0009-0.0027	0.0051-0.0150
	3.3 (R)	2.5202-2.5195	0.0023-0.0043	0.0036-0.0095	2	2.1240-2.1250	0.0008-0.0026	0.0050-0.0150
	3.8 (L)	2.5202-2.5195	0.0023-0.0043	0.0036-0.0095	2	2.1240-2.1250	0.0008-0.0026	0.0050-0.0150

67189-MINV-C06

PISTON AND RING SPECIFICATIONS
All measurements are given in inches.

Year	Engine Displacement Liters (VIN)	Piston Clearance	Ring Gap			Ring Side Clearance		
			Top Compression	Bottom Compression	Oil Control	Top Compression	Bottom Compression	Oil Control
2001	2.4 (B)	0.0009-0.0022	0.0098-0.0200	0.0090-0.0180	0.0098-0.0250	0.0011-0.0031	0.0011-0.0031	0.0004-0.0070
	3.3 (R)	0.0010-0.0022	0.0118-0.0217	0.0118-0.0217	0.0098-0.0394	0.0012-0.0037	0.0012-0.0037	0.0005-0.0089
	3.8 (L)	0.0010-0.0022	0.0118-0.0217	0.0118-0.0217	0.0098-0.0394	0.0012-0.0037	0.0012-0.0037	0.0005-0.0089
2002	2.4 (B)	0.0009-0.0022	0.0098-0.0200	0.0090-0.0180	0.0098-0.0250	0.0011-0.0031	0.0011-0.0031	0.0004-0.0070
	3.3 (R)	0.0010-0.0022	0.0118-0.0217	0.0118-0.0217	0.0098-0.0394	0.0012-0.0037	0.0012-0.0037	0.0005-0.0089
	3.8 (L)	0.0010-0.0022	0.0118-0.0217	0.0118-0.0217	0.0098-0.0394	0.0012-0.0037	0.0012-0.0037	0.0005-0.0089
2003	2.4 (B)	0.0009-0.0022	0.0098-0.0200	0.0090-0.0180	0.0098-0.0250	0.0011-0.0031	0.0011-0.0031	0.0004-0.0070
	3.3 (R)	0.0010-0.0022	0.0118-0.0217	0.0118-0.0217	0.0098-0.0394	0.0012-0.0037	0.0012-0.0037	0.0005-0.0089
	3.8 (L)	0.0010-0.0022	0.0118-0.0217	0.0118-0.0217	0.0098-0.0394	0.0012-0.0037	0.0012-0.0037	0.0005-0.0089
2004-05	2.4 (B)	0.0009-0.0022	0.0098-0.0200	0.0090-0.0180	0.0098-0.0250	0.0011-0.0031	0.0011-0.0031	0.0004-0.0070
	3.3 (R)	0.0010-0.0022	0.0118-0.0217	0.0118-0.0217	0.0098-0.0394	0.0012-0.0037	0.0012-0.0037	0.0005-0.0089
	3.8 (L)	0.0010-0.0022	0.0118-0.0217	0.0118-0.0217	0.0098-0.0394	0.0012-0.0037	0.0012-0.0037	0.0005-0.0089

① Oil control ring side rails must be free to rotate after assembly

67189-MINV-C07

TORQUE SPECIFICATIONS
All readings in ft. lbs.

Year	Engine Displacement Liters (cc)	Cylinder Head Bolts	Main Bearing Bolts	Rod Bearing Bolts	Crankshaft Damper Bolts	Flywheel Bolts	Manifold Intake	Manifold Exhaust	Spark Plugs	Oil Pan Drain Plug
2001	2.4 (B)	①	②	③	100	70	④	15	20	20
	3.3 (R)	⑤	⑥	⑦	40	70	⑧	⑨	20	20
	3.8 (L)	⑤	⑥	⑦	40	70	⑧	⑨	20	20
2002	2.4 (B)	①	②	③	100	70	④	15	20	20
	3.3 (R)	⑤	⑥	⑦	40	70	⑧	⑨	20	20
	3.8 (L)	⑤	⑥	⑦	40	70	⑧	⑨	20	20
2003	2.4 (B)	①	②	③	100	70	④	15	20	20
	3.3 (R)	⑤	⑥	⑦	40	70	⑧	⑨	20	20
	3.8 (L)	⑤	⑥	⑦	40	70	⑧	⑨	20	20
2004-05	2.4 (B)	①	②	③	100	70	④	15	20	20
	3.3 (R)	⑤	⑥	⑦	40	70	⑧	⑨	20	20
	3.8 (L)	⑤	⑥	⑦	40	70	⑧	⑨	20	20

① Step 1: 25 ft. lbs.
Step 2: 50 ft. lbs.
Step 3: 50 ft. lbs.
Step 4: Plus 1/4 turn

② M8 bolts: 21 ft. lbs.
M11 bolts: 30 ft. lbs. plus 90 degrees

③ Step 1: 20 ft. lbs.
Step 2: Plus 90 degrees

④ 2.4L: Upper and lower 250 inch lbs

⑤ Step 1: 45 ft. lbs.
Step 2: 65 ft. lbs.
Step 3: 65 ft. lbs.
Step 4: Plus 90 degrees

⑥ Step 1: 30 ft. lbs.
Step 2: Plus 90 degrees

⑦ Step 1: 40 ft. lbs.
Step 2: Plus 90 degrees

⑧ Step 1: 10 inch lbs.
Step 2: 17 ft. lbs.
Step 3: 17 ft. lbs.

⑨ Refer to procedure for torque sequence and specifications

67189-MINV-C08

WHEEL ALIGNMENT

Year	Model		Caster Range (+/-Deg.)	Caster Preferred Setting (Deg.)	Camber Range (+/-Deg.)	Camber Preferred Setting (Deg.)	Toe-in (in.)
2001	All	F	1.00	+1.40	0.30	+0.05	0.10+/-0.20
		R	—	—	0.25	0	0+/-0.30
		F	1.00	+1.40	0.40	+0.15	0.10+/-0.20
		R	—	—	0.25	0	0+/-0.30
2002	All	F	1.00	+1.40	0.30	+0.05	0.10+/-0.20
		R	—	—	0.25	0	0+/-0.30
		F	1.00	+1.40	0.40	+0.15	0.10+/-0.20
		R	—	—	0.25	0	0+/-0.30
2003	All	F	1.00	+1.40	0.30	+0.05	0.10+/-0.20
		R	—	—	0.25	0	0+/-0.30
		F	1.00	+1.40	0.40	+0.15	0.10+/-0.20
		R	—	—	0.25	0	0+/-0.30
2004-05	All	F	1.00	+1.40	0.30	+0.05	0.10+/-0.20
		R	—	—	0.25	0	0+/-0.30
		F	1.00	+1.40	0.40	+0.15	0.10+/-0.20
		R	—	—	0.25	0	0+/-0.30

67189-MINV-C09

TIRE, WHEEL AND BALL JOINT SPECIFICATIONS

| Year | Model | OEM Tires | | Tire Pressures (psi) | | Wheel Size | Ball Joint Inspection | Lug Nuts |
		Standard	Optional	Front	Rear			
2001	Town & Country base, LX	P215/70R15	None	①	①	6-J	0.030 in. ②	100
	Except Town & Country base, LX	P215/65R16	None	①	①	6.5-J	0.030 in. ②	100
	Caravan/Voyager, base	P205/75SR14	None	①	①	6-J	0.030 in. ②	100
	Caravan/Voyager SE, LE	P215/65SR15	P215/65R16	①	①	6.5-J	0.030 in. ②	100
	Caravan/Voyager ES	P215/65R16	None	①	①	6.5-J	0.030 in. ②	100
2002	Town & Country base, LX	P215/70R15	None	①	①	6-J	0.030 in. ②	100
	Except Town & Country base, LX	P215/65R16	None	①	①	6-J	0.030 in. ②	100
	Caravan/Voyager, base	P205/75SR14	None	①	①	6-J	0.030 in. ②	100
	Caravan/Voyager SE, LE	P215/65SR15	P215/65R16	①	①	6-J	0.030 in. ②	100
	Caravan/Voyager ES	P215/65R16	None	①	①	6-J	0.030 in. ②	100
2003	Town & Country base, LX	P215/70R15	None	①	①	6-J	0.030 in. ②	100
	Except Town & Country base, LX	P215/65R16	None	①	①	6.5-J	0.030 in. ②	100
	Caravan/Voyager, base	P205/75SR14	None	①	①	6-J	0.030 in. ②	100
	Caravan/Voyager SE, LE	P215/65SR15	P215/65R16	①	①	6-J	0.030 in. ②	100
	Caravan/Voyager ES	P215/65R16	None	①	①	6-J	0.030 in. ②	100
2004-05	Town & Country base, LX	P215/70R15	None	①	①	6-J	0.030 in. ②	100
	Except Town & Country base, LX	P215/65R16	None	①	①	6.5-J	0.030 in. ②	100
	Caravan base	P205/75SR14	None	①	①	6-J	0.030 in. ②	100
	Caravan SE, LE	P215/65SR15	P215/65R16	①	①	6-J	0.030 in. ②	100
	Caravan ES	P215/65R16	None	①	①	6-J	0.030 in. ②	100

OEM: Original Equipment Manufacturer

PSI: Pounds Per Square Inch

① Refer to label on drivers side door pillar

② Both upper and lower

67189-MINV-C10

BRAKE SPECIFICATIONS
All measurements in inches unless noted

Year	Model		Brake Disc Original Thickness	Brake Disc Minimum Thickness	Brake Disc Maximum Run-out	Brake Drum Diameter Original Inside Diameter	Brake Drum Diameter Max. Wear Limit	Brake Drum Diameter Maximum Machine Diameter	Min. Lining Thickness	Caliper Guide Pin Bolts (ft. lbs.)
2001	Caravan	F	0.939-0.949	0.881	0.005	—	—	—	0.313	35
		R	0.482-0.502	0.443	0.005	9.84	9.93	9.90	①	35
	Town & Country	F	0.939-0.949	0.881	0.005	—	—	—	0.313	35
		R	0.482-0.502	0.443	0.005	9.84	9.93	9.90	①	35
	Voyager	F	0.939-0.949	0.881	0.005	—	—	—	0.313	35
		R	—	—	—	9.84	9.93	9.90	0.031	35
2002	Caravan	F	0.939-0.949	0.881	0.005	—	—	—	0.313	35
		R	0.482-0.502	0.443	0.005	9.84	9.93	9.90	①	35
	Town & Country	F	0.939-0.949	0.881	0.005	—	—	—	0.313	35
		R	0.482-0.502	0.443	0.005	9.84	9.93	9.90	①	35
	Voyager	F	0.939-0.949	0.881	0.005	—	—	—	0.313	35
		R	—	—	—	9.84	9.93	9.90	0.031	35
2003	Caravan	F	0.939-0.949	0.881	0.005	—	—	—	0.313	35
		R	0.482-0.502	0.443	0.005	9.84	9.93	9.90	①	35
	Town & Country	F	0.939-0.949	0.881	0.005	—	—	—	0.313	35
		R	0.482-0.502	0.443	0.005	9.84	9.93	9.90	①	35
2004-05	Caravan	F	0.939-0.949	0.881	0.005	—	—	—	0.313	35
		R	0.482-0.502	0.443	0.005	9.84	9.93	9.90	①	35
	Town & Country	F	0.939-0.949	0.881	0.005	—	—	—	0.313	35
		R	0.482-0.502	0.443	0.005	9.84	9.93	9.90	①	35

F: Front

R: Rear

① Drum brakes: 0.031 in
   Disc brakes: 0.281 in.

67189-MINV-C11

SCHEDULED MAINTENANCE INTERVALS
CHRYSLER—TOWN & COUNTRY, DODGE—CARAVAN & PLYMOUTH—VOYAGER

TO BE SERVICED	TYPE OF SERVICE	7.5	15	22.5	30	37.5	45	52.5	60	67.5	75	82.5	90	97.5
Engine oil & filter	R	✓	✓	✓	✓	✓	✓	✓	✓	✓	✓	✓	✓	✓
Driveshaft boots	S/I	✓	✓	✓	✓	✓	✓	✓	✓	✓	✓	✓	✓	✓
Exhaust system	S/I	✓	✓	✓	✓	✓	✓	✓	✓	✓	✓	✓	✓	✓
Engine coolant level, hoses & clamps	S/I	✓	✓	✓	✓	✓	✓	✓	✓	✓	✓	✓	✓	✓
Rotate tires	S/I	✓	✓	✓	✓	✓	✓	✓	✓	✓	✓	✓	✓	✓
Drive belts	S/I		✓		✓		✓		✓		✓		✓	
Brake hoses & linings	S/I			✓			✓			✓			✓	
Automatic transaxle fluid & filter	R				✓				✓				✓	
Air filter	R				✓				✓				✓	
Spark plugs ①	R				✓				✓				✓	
Serpentine belts (3.0L & 3.3L)	S/I								✓		✓		✓	
Lubricate tie rod ends	S/I				✓				✓				✓	
PCV valve	S/I				✓				✓				✓	
Engine coolant	R								✓			✓		
Timing belt (3.0L)	R								✓					
Distributor cap & rotor	R							✓						
Ignition cables (3.0L)	R								✓					
Ignition timing	S/I								✓					

R: Replace   S/I: Service or Inspect

① Platinum tip spark plugs & ignition cables (3.3L & 3.8L): replace every 100,000 miles.

FREQUENT OPERATION MAINTENANCE (SEVERE SERVICE)

If a vehicle is operated under any of the following conditions it is considered severe service:

- Extremely dusty areas.

- 50% or more of the vehicle operation is in 32°C (90°F) or higher temperatures, or constant operation in temperatures below 0°C (32°F).

- Prolonged idling (vehicle operation in stop and go traffic).

- Frequent short running periods (engine does not warm to normal operating temperatures).

- Police, taxi, delivery usage or trailer towing usage.

Oil & oil filter change: change every 3000 miles.

Automatic transaxle fluid & filter: change every 15,000 miles.

Brake hoses & linings: check every 9000 miles.

CV-joints & front suspension ball joints: check every 3000 miles.

Tie rod ends & steering linkage: check every 15,000 miles.

Air filter: change every 15,000 miles.

67189-MINV-C12

PRECAUTIONS

Before servicing any vehicle, please be sure to read all of the following precautions, which deal with personal safety, prevention of component damage, and important points to take into consideration when servicing a motor vehicle:

• Never open, service or drain the radiator or cooling system when the engine is hot; serious burns can occur from the steam and hot coolant.

• Observe all applicable safety precautions when working around fuel. Whenever servicing the fuel system, always work in a well-ventilated area. Do not allow fuel spray or vapors to come in contact with a spark, open flame, or excessive heat (a hot drop light, for example). Keep a dry chemical fire extinguisher near the work area. Always keep fuel in a container specifically designed for fuel storage; also, always properly seal fuel containers to avoid the possibility of fire or explosion. Refer to the additional fuel system precautions later in this section.

• Fuel injection systems often remain pressurized, even after the engine has been turned **OFF**. The fuel system pressure must be relieved before disconnecting any fuel lines. Failure to do so may result in fire and/or personal injury.

• Brake fluid often contains polyglycol ethers and polyglycols. Avoid contact with the eyes and wash your hands thoroughly after handling brake fluid. If you do get brake fluid in your eyes, flush your eyes with clean, running water for 15 minutes. If eye irritation persists, or if you have taken brake fluid internally, IMMEDIATELY seek medical assistance.

• The EPA warns that prolonged contact with used engine oil may cause a number of skin disorders, including cancer! You should make every effort to minimize your exposure to used engine oil. Protective gloves should be worn when changing oil. Wash your hands and any other exposed skin areas as soon as possible after exposure to used engine oil. Soap and water, or waterless hand cleaner should be used.

• All new vehicles are now equipped with an air bag system, often referred to as a Supplemental Restraint System (SRS) or Supplemental Inflatable Restraint (SIR) system. The system must be disabled before performing service on or around system components, steering column, instrument panel components, wiring and sensors. Failure to follow safety and disabling procedures could result in accidental air bag deployment, possible personal injury and unnecessary system repairs.

• Always wear safety goggles when working with, or around, the air bag system. When carrying a non-deployed air bag, be sure the bag and trim cover are pointed away from your body. When placing a non-deployed air bag on a work surface, always face the bag and trim cover upward, away from the surface. This will reduce the motion of the module if it is accidentally deployed. Refer to the additional air bag system precautions later in this section.

• Clean, high quality brake fluid from a sealed container is essential to the safe and proper operation of the brake system. You should always buy the correct type of brake fluid for your vehicle. If the brake fluid becomes contaminated, completely flush the system with new fluid. Never reuse any brake fluid. Any brake fluid that is removed from the system should be discarded. Also, do not allow any brake fluid to come in contact with a painted surface; it will damage the paint.

• Never operate the engine without the proper amount and type of engine oil; doing so WILL result in severe engine damage.

• Timing belt maintenance is extremely important! Many models utilize an interference-type, non-freewheeling engine. If the timing belt breaks, the valves in the cylinder head may strike the pistons, causing potentially serious (also time-consuming and expensive) engine damage. Refer to the maintenance interval charts in this manual for the recommended replacement interval for the timing belt, and to the timing belt repair procedure in this section for belt replacement and inspection.

• Disconnecting the negative battery cable on some vehicles may interfere with the functions of the on-board computer system(s) and may require the computer to undergo a relearning process once the negative battery cable is reconnected.

• When servicing drum brakes, only disassemble and assemble one side at a time, leaving the remaining side intact for reference.

• Only an MVAC-trained, EPA-certified automotive technician should service the air conditioning system or its components.

ENGINE REPAIR

➡Disconnecting the negative battery cable on some vehicles may interfere with the functions of the on board computer system. The computer may undergo a relearning process once the negative battery cable is reconnected.

Alternator

REMOVAL

2.4L Engine

1. Before servicing the vehicle, refer to the precautions in the beginning of this section.
2. Remove or disconnect the following:
   • Negative battery cable

• Air inlet temperature sensor connector
• Air cleaner assembly
• Evaporative emissions (EVAP) purge solenoid from the bracket and position aside
• Alternator harness connections
• Accessory drive belt
• Alternator

3.3L and 3.8L Engines

1. Before servicing the vehicle, refer to the precautions in the beginning of this section.
2. Remove or disconnect the following:
   • Negative battery cable
   • Alternator harness connectors
   • Right front lower splash shield

• Accessory drive belt
• Lower oil dipstick tube bolt and the wiring harness from the tube
• 3 mounting bolts
• Dipstick tube
• Alternator

INSTALLATION

2.4L Engine

1. Install or connect the following:
   • Alternator
   • Accessory drive belt. Tighten the alternator mounting fasteners to 20 ft. lbs. (28 Nm).
   • Alternator harness connections
   • EVAP purge solenoid

- Air cleaner assembly
- Air inlet temperature sensor connector
- Negative battery cable

3.3L and 3.8L Engines

1. Install or connect the following:
   - Alternator in position and the 3 mounting bolts. Tighten to 40 ft. lbs. (54 Nm).
   - Dipstick tube, lubricate the O–ring with clean engine oil prior to installation, then tighten the upper dipstick tube bolt.
   - Alternator harness connectors
   - Lower oil dipstick tube bolt
   - Accessory drive belt
   - Right front lower splash shield
   - Wiring harness to the dipstick tube
   - Negative battery cable

Ignition Timing

ADJUSTMENT

The base ignition timing cannot be adjusted. The Powertrain Control Module (PCM) regulates the ignition timing automatically.

Engine Assembly

REMOVAL & INSTALLATION

2.4L Engine

1. Before servicing the vehicle, refer to the precautions in the beginning of this section.
2. Drain the cooling system.
3. Drain the engine oil.
4. Relieve the fuel system pressure.
5. Recover the A/C refrigerant.
6. Remove or disconnect the following:
   - Negative battery cable
   - Air cleaner and hoses
   - Fuel line
   - Vacuum hoses
   - Radiator fans
   - Radiator hoses

➡**When the transaxle lines are removed from the fittings at the transaxle damage to the inner wall of the hose will occur. To prevent leakage, cut the cooler hoses off flush at the transaxle fitting and use a service cooler hose splice kit upon installation.**

   - Transaxle cooler lines, cut the lines flush at the transaxle fittings

- Transaxle shift linkage
- Throttle body linkage
- Engine wiring harness
- Heater hoses
- Front wheels
- Right inner splash shield
- Drive belts
- Axle halfshafts
- Crossmember cradle plate
- Exhaust front pipe
- Front motor mount
- Structural collar
- Rear motor mount
- Power steering pump. Pinch off the supply hose at the pump, disconnect the hose but leave the pressure line attached and then set the pump assembly aside.
- A/C compressor lines
- Body ground straps

7. Raise the vehicle enough to position dolly 6135, cradle tool 6710 with posts 6848 under vehicle.

8. Loosen the cradle posts to allow movement for proper positioning. Position the two rear posts (right side of engine) into the engine bedplate holes. Position the two front posts (left side of engine) on the oil pan rails. Lower the vehicle and position the

cradle mounts until the engine is resting on the mounts and then tighten the mounts to the cradle frame to prevent mounts from moving when removing/installing the powertrain assembly.

9. Attach safety straps around the engine-to-cradle assembly. Lower the vehicle so the weight of the powertrain assembly, not the vehicle are on the cradle.

10. Remove the engine and transmission bolts.

11. Raise the vehicle away from the powertrain.

To install:

12. Lower the vehicle over the powertrain.

13. Align the engine and transmission mounts to their attaching points. Install and tighten the right mount-to-rail and engine and left mount-to-frame brackets as shown in the two accompanying illustrations.

14. Remove the safety straps and the dolly.

15. Install the rear mount bracket and through bolt.

16. Install or connect the following:
   - Torque converter
   - Axle halfshafts

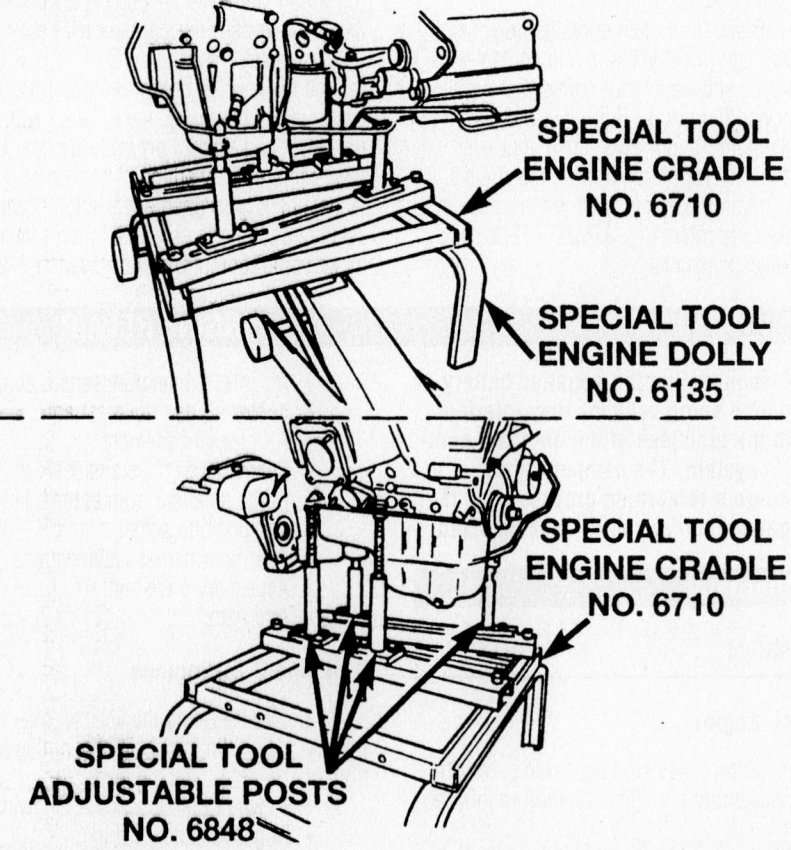

SPECIAL TOOL
ENGINE CRADLE
NO. 6710

SPECIAL TOOL
ENGINE DOLLY
NO. 6135

SPECIAL TOOL
ENGINE CRADLE
NO. 6710

SPECIAL TOOL
ADJUSTABLE POSTS
NO. 6848

Chrysler powertrain support tools

9302CG01

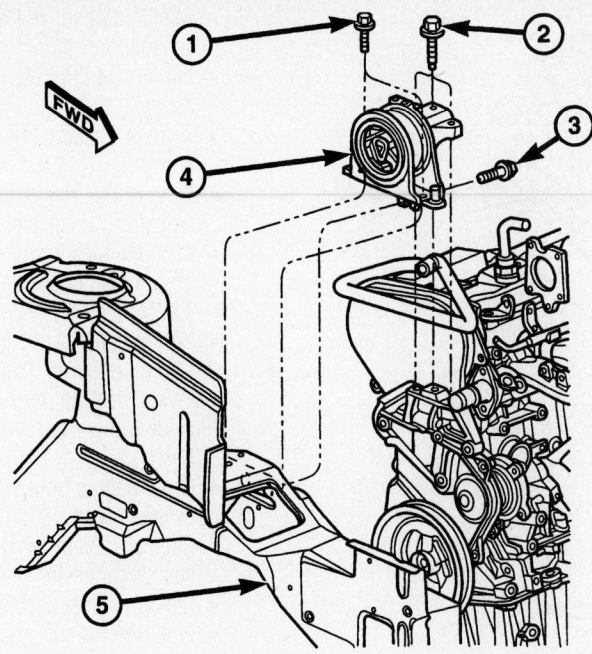

1 - BOLT - MOUNT TO RAIL 68 N·m (50 ft. lbs.)
2 - BOLT - MOUNT TO ENGINE 54 N·m (40 ft. lbs.)
3 - BOLT - MOUNT TO RAIL (HORIZONTAL) 68 N·m (50 ft. lbs.)
4 - RIGHT ENGINE MOUNT
5 - RIGHT FRAME RAIL

67189-MINV-G03

Right mount-to-rail and engine installation—2.4L engine

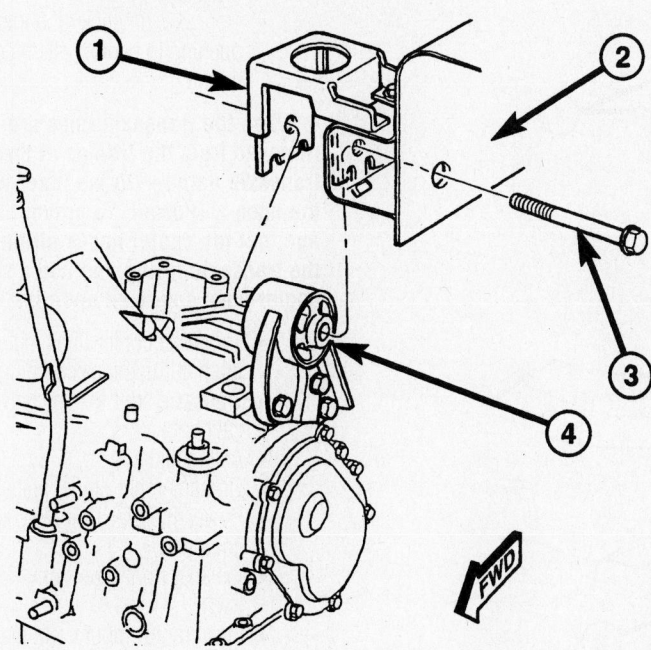

1 - FRAME BRACKET
2 - FRAME RAIL - LEFT
3 - BOLT
4 - TRANSAXLE MOUNT

67189-MINV-G04

Left mount-to-frame bracket installation—2.4L engine

17. Install the bending struts and structural collar, as follows:

a. Step 1: Position the collar between the transaxle and oil pan and install bolts hand tight.

b. Step 2: Position the collar to oil pan bolts and install bolts hand tight.

c. Step 3: Tighten collar to transaxle bolts to 75 ft. lbs. (101 Nm).

d. Step 4: Tighten collar to oil pan bolts to 40 ft. lbs. (54 Nm).

e. Step 5: Install the engine front mount assembly.

18. Install or connect the following:
- Exhaust front pipe
- Crossmember cradle plate
- Power steering pump and lines
- A/C compressor lines
- Drive belts
- Right inner splash shield
- Axle halfshafts
- Front wheels
- Transaxle cooler lines using a splice service kit
- Transaxle shift linkage
- Heater hoses
- Body ground straps
- Engine wiring harness
- Vacuum lines
- Throttle body linkage
- Engine cooling fans
- Fuel line
- Radiator hoses
- Air cleaner and hoses
- Negative battery cable

19. Fill the cooling system.
20. Fill the engine with clean oil.
21. Recharge the A/C system.
22. Start the engine and check for leaks.

3.3L and 3.8L Engines

1. Before servicing the vehicle, refer to the precautions in the beginning of this section.

2. Drain the cooling system.
3. Drain the engine oil.
4. Relieve the fuel system pressure.
5. Recover the A/C refrigerant.
6. Remove or disconnect the following:
- Battery and tray
- Air cleaner and hoses
- Fuel line from the rail
- Wiper module

7. Block off the heater hoses to the rear 6system by using pinch off pliers, if equipped.
- Heater hoses
- Radiator upper support crossmember
- Engine cooling fan
- Throttle body linkage

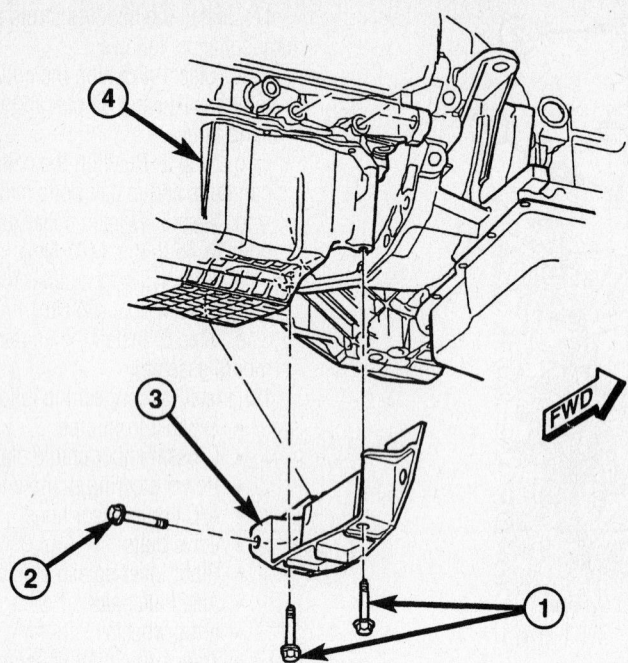

1 - BOLT - COLLAR TO OIL PAN
2 - BOLT - COLLAR TO TRANSAXLE
3 - STRUCTURAL COLLAR
4 - OIL PAN

67189-MINV-G05

Structural collar and bending strut torque sequence—2.4L engine

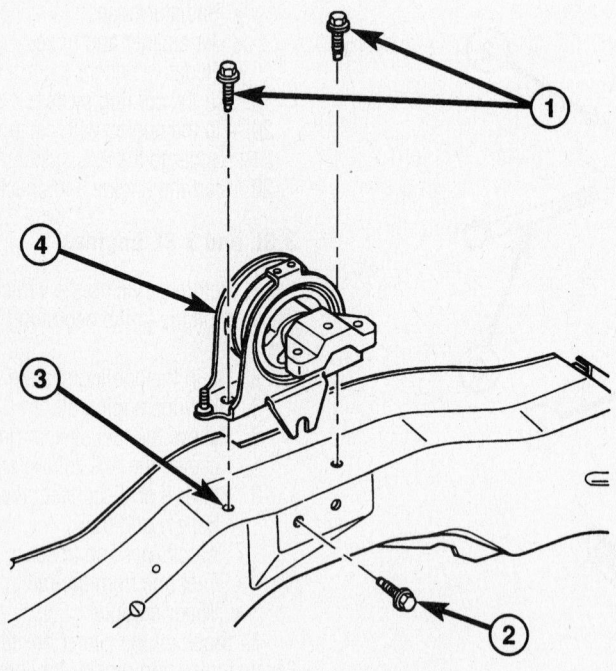

1 - BOLT
2 - BOLT
3 - FRAME RAIL
4 - RIGHT MOUNT - 2.4L ENGINE

67189-MINV-G06

Exploded view of the right side engine hydro–type mount—2.4L engine

- Manifold Absolute Pressure (MAP) sensor connector
- Idle Air Control (IAC) valve connector
- Throttle Position (TPS) sensor connector
- Exhaust Gas Recirculation (EGR) transducer connector
- Vacuum lines from the throttle body, brake booster and speed control
- Wiring harness clip from the right side mount
- Power steering reservoir and set aside without disconnecting the lines
- Ground strap from the rear of the cylinder head
- Engine Coolant Temperature (ECT) sensor connector
- Ignition coil connectors
- Fuel injector connectors and clip from the bracket
- Camshaft (CMP) and Crankshaft (CKP) position sensor connectors
- A/C compressor electrical connectors and hoses. Cap the lines to avoid system contamination.
- Radiator upper hose
- Electrical connector at the transaxle dipstick tube
- Transaxle dipstick tube and seal the opening to avoid system contamination

➡**When the transaxle lines are removed from the fittings at the transaxle damage to the inner wall of the hose will occur. To prevent leakage, cut the cooler hoses off flush at the transaxle fitting and use a service cooler hose splice kit upon installation.**

- Transaxle cooler lines, cut the lines flush at the transaxle fittings
- Transaxle shift linkage and electrical connectors
- Axle halfshafts
- Crossmember cradle plate
- Power Transfer Unit (PTO), if equipped
- Exhaust front pipe from the manifold
- Front motor mount and bracket
- Rear motor mount bracket
- Engine-to-transaxle struts
- Transaxle case cover
- Torque converter
- Power steering pressure hose clip bolt
- Knock sensor connector, if equipped

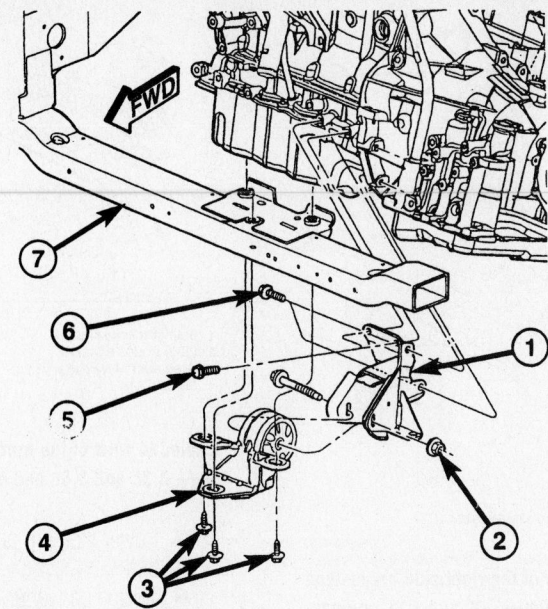

1 - BRACKET - FRONT MOUNT
2 - NUT - 68 N·m (50 ft. lbs.)
3 - BOLT - 54 N·m (40 ft. lbs.)
4 - MOUNT - FRONT INSULATOR
5 - BOLT - 68 N·m (50 ft. lbs.)
6 - BOLT - 68 N·m (50 ft. lbs.)
7 - FRONT CROSSMEMBER

67189-MINV-G07

Exploded view of the front mount and bracket assembly—2.4L engine

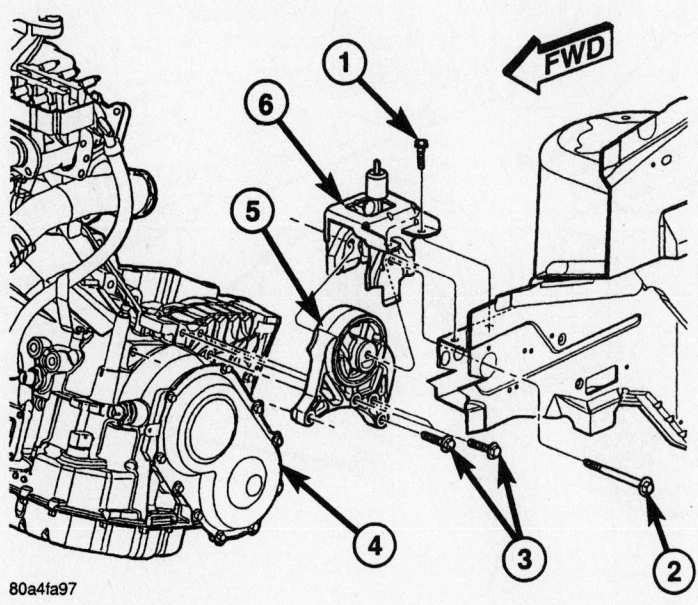

80a4fa97

1 - BOLT - BRACKET TO FRAME RAIL 68 N·m (50 ft. lbs.)
2 - BOLT - MOUNT TO RAIL THRU 75 N·m (55 ft. lbs.)
3 - BOLT - LEFT MOUNT TO TRANSAXLE 54 N·m (40 ft. lbs.)
4 - TRANSAXLE
5 - MOUNT - LEFT
6 - BRACKET - LEFT MOUNT

67189-MINV-G08

Exploded view of the left mount and bracket assembly—2.4L engine

- Engine block heater connector, if equipped
- Drive belt splash shield
- Accessory drive belt
- Lower radiator hose
- A/C compressor
- Alternator
- Water pump pulley bolts and position the pulley between the pump and the housing
- Oil pressure switch connector
- Wiring harness clip from the dipstick tube

8. Install adapter tools on the right side of the engine block as illustrated
- Power steering pump and set aside

9. Raise the vehicle enough to position dolly 6135, cradle tool 6710 with posts 6848 and adapter tool 6909 under vehicle.

10. Loosen the cradle posts to allow movement for proper positioning. Lower the vehicle and position the cradle/post mounts until the engine is resting on the posts and then tighten the mounts to the cradle frame to prevent mounts from moving when removing/installing the powertrain assembly.

11. Attach safety straps around the engine-to-cradle assembly. Lower the vehicle so the weight of the powertrain assembly, not the vehicle are on the cradle.

12. Remove or disconnect the following:
- Engine right side mount-to-engine bolts
- Left mount through bolt

13. Raise the vehicle away from the powertrain.

To install:

14. Lower the vehicle over the powertrain.

15. Align the engine and transmission mounts to their attaching points. Install and tighten the right mount and left transmission mounts as shown in the two accompanying illustrations.

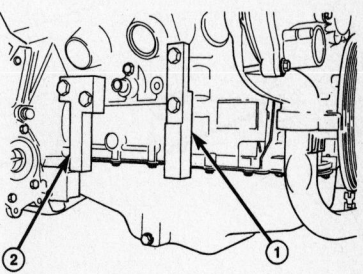

1 - SPECIAL TOOL 6912
2 - SPECIAL TOOL 8444

9302CMVG09

Adapter tools 6912 and 8444 mounted on the block—3.3L and 3.8L engines

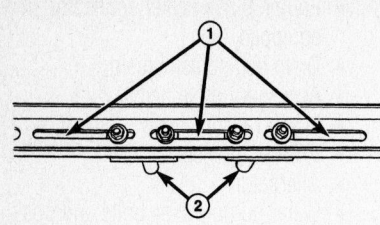

1 - SLOTS
2 - SPECIAL TOOLS 6848

9302CMVG10

Position tool 6848 for use with adapters 8444 and 6912–3.3L and 3.8L engines

16. Remove the safety straps and the dolly and adapters.

17. Install or connect the following:
- Power steering pump and pressure line
- Alternator
- Wiring harness clip to the dipstick tube
- Oil pressure switch connector
- A/C compressor
- Water pump pulley
- Lower radiator hose
- Accessory drive belt

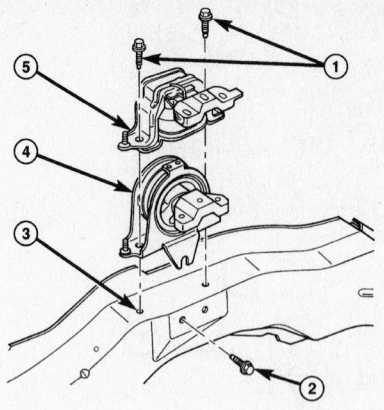

1 - BOLT
2 - BOLT
3 - FRAME RAIL
4 - RIGHT MOUNT - 3.3/3.8L ENGINE

67189-MINV-G21

Exploded view of the right side hrydo–type engine mounting—3.3L and 3.8L engines

- Drive belt splash shield
- Engine block heater connector, if equipped
- Knock sensor connector, if equipped

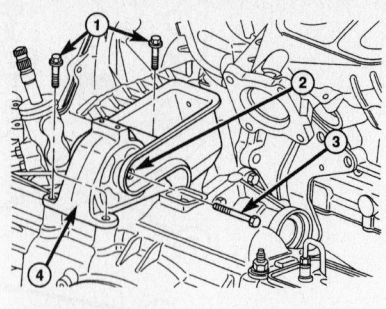

1 - BOLT 54 N·m (40 ft. lbs.)
2 - REAR MOUNT BRACKET
3 - THRU-BOLT 54 N·m (40 ft. lbs.)
4 - REAR MOUNT

67189-MINV-G22

Exploded view of the rear engine mounting—3.3L and 3.8L engines

- Power steering pressure hose clip bolt
- Torque converter
- Transaxle case cover
- Engine-to-transaxle struts
- Rear motor mount bracket
- Front motor mount and bracket
- PTO, if equipped
- Axle halfshafts
- Exhaust front pipe to the manifold

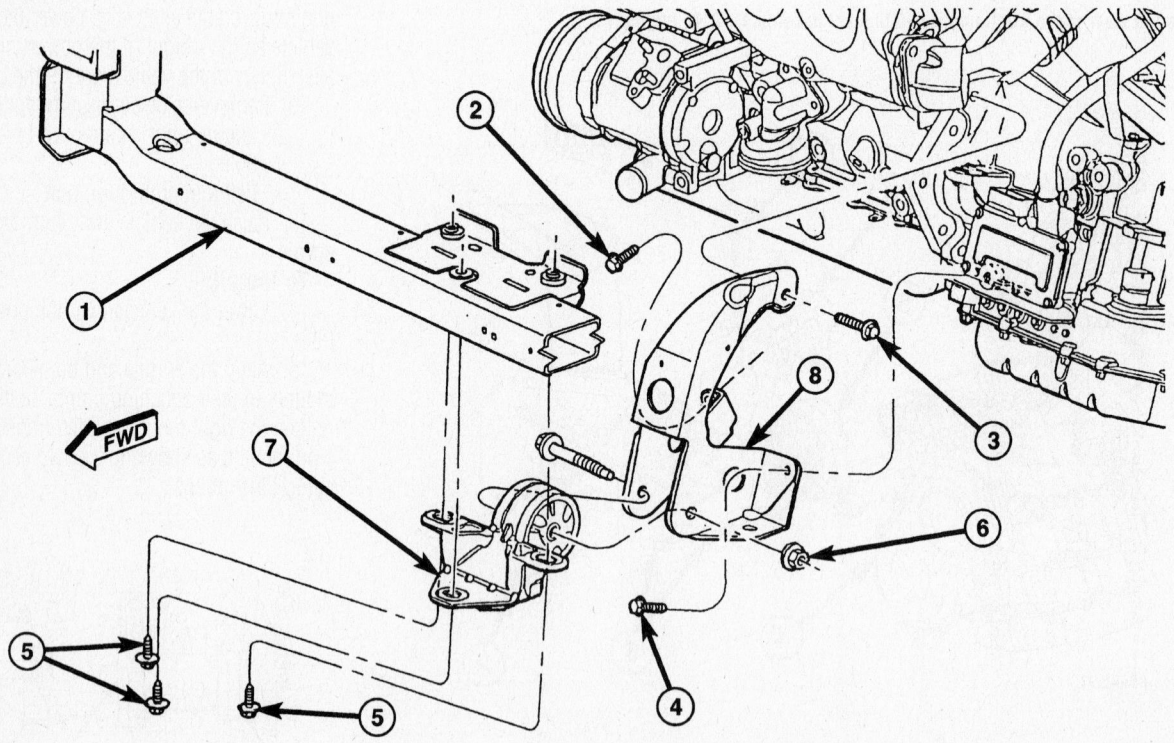

1 - CROSSMEMBER
2 - BOLT - 68 N·m (50 ft. lbs.)
3 - BOLT - 102 N·m (75 ft. lbs.)
4 - BOLT - 68 N·m (50 ft. lbs.)

5 - BOLT 54 N·m (40 ft. lbs.)
6 - NUT - 68 N·m (50 ft. lbs.)
7 - MOUNT - ENGINE FRONT
8 - BRACKET - ENGINE FRONT MOUNT

67189-MINV-G20

Exploded view of the front engine mounting—3.3L and 3.8L engines

- Crossmember cradle plate
- Transaxle shift linkage and electrical connectors
- Transaxle cooler lines using a splice kit
- Transaxle dipstick tube and
- Electrical connector at the transaxle dipstick tube
- A/C compressor electrical connectors and hoses
- CMP and CKP position sensor connectors
- Fuel injector connectors and clip to the bracket
- ECT sensor connector
- Ignition coil connectors
- Ground strap to the rear of the cylinder head
- Power steering reservoir
- Wiring harness clip to the right side mount
- Vacuum lines to the throttle body, brake booster and speed control

- EGR transducer connector
- TPS sensor connector
- IAC valve connector
- MAP sensor connector
- Throttle body linkage
- Engine cooling fan
- Radiator upper hose
- Heater hoses
- Radiator upper support crossmember
- Wiper module
- Fuel line to the rail
- Air cleaner and hoses
- Battery and tray
18. Fill the cooling system.
19. Fill the engine with clean oil.
20. Recharge the A/C system.
21. Start the engine and check for leaks.

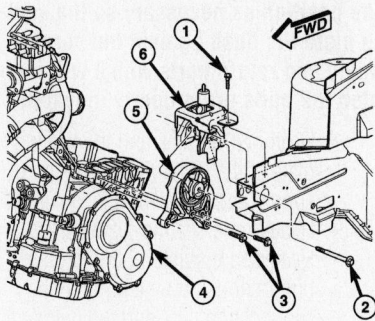

1 - BOLT - BRACKET TO FRAME RAIL 68 N·m (50 ft. lbs.)
2 - BOLT - MOUNT TO RAIL THRU 75 N·m (55 ft. lbs.)
3 - BOLT - LEFT MOUNT TO TRANSAXLE 54 N·m (40 ft. lbs.)
4 - TRANSAXLE
5 - MOUNT - LEFT
6 - BRACKET - LEFT MOUNT

67189-MINV-G23

Exploded view of the left engine mounting—3.3L and 3.8L engines

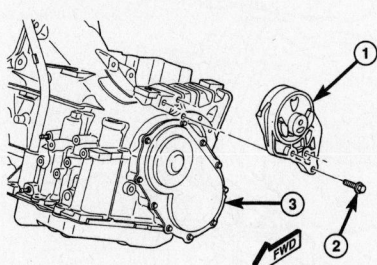

1 - LEFT MOUNT ASEMBLY
2 - BOLT - 54 N·m (40 ft. lbs.)
3 - TRANSAXLE - 31TH

67189-MINV-G24

Exploded view of the left engine mounting on models equipped with a 31TH transaxle—3.3L and 3.8L engines

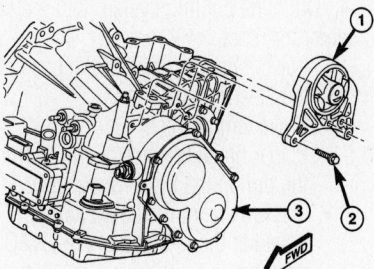

1 - LEFT MOUNT ASSEMBLY
2 - BOLT - 54 N·m (40 ft. lbs.)
3 - TRANSAXLE - 41TE

67189-MINV-G25

Exploded view of the left engine mounting on models equipped with a 41TE transaxle—3.3L and 3.8L engines

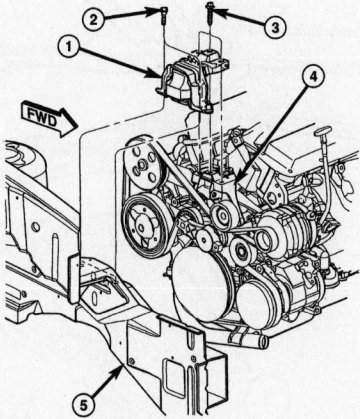

1 - RIGHT ENGINE MOUNT
2 - BOLT - MOUNT TO FRAME RAIL
3 - BOLT - MOUNT TO ENGINE
4 - ENGINE MOUNT BRACKET
5 - RIGHT FRAME RAIL

67189-MINV-G26

Exploded view of the right engine mounting—3.3L and 3.8L engines

Water Pump

REMOVAL & INSTALLATION

2.4L Engine

1. Before servicing the vehicle, refer to the precautions in the beginning of this section.
2. Drain the cooling system.
3. Remove or disconnect the following:
   - Negative battery cable
   - Right inner splash shield
   - Accessory drive belts
   - Right motor mount and bracket, support the engine from below before removing the mount
   - Front cover
   - Timing belt
   - Timing belt idler pulley
   - Camshaft sprockets
   - Timing belt rear cover
   - Alternator and bracket
   - Water pump

To install:

4. Install or connect the following:
   - Water pump with a new O–ring gasket. Tighten the bolts to 105 inch lbs. (12 Nm). Rotate the pump by hand to make sure it moves freely.
   - Timing belt rear cover
   - Camshaft sprockets. Tighten the bolts to 75 ft. lbs. (101 Nm).
   - Timing belt idler pulley. Tighten the bolt to 45 ft. lbs. (61 Nm).
   - Timing belt
   - Front cover
   - Alternator and bracket
   - Right motor mount
   - Accessory drive belts
   - Right inner splash shield
   - Negative battery cable
5. Fill the cooling system.
6. Start the engine and check for leaks.

3.3L and 3.8L Engines

1. Before servicing the vehicle, refer to the precautions in the beginning of this section.
2. Drain the cooling system.
3. Remove or disconnect the following:
   - Negative battery cable
   - Right inner splash shield
   - Accessory drive belt
   - Water pump pulley bolts

➡️**To remove the water pump it must be positioned between the pump housing and the drive hub. Refer to the accompanying illustration for further clarification.**

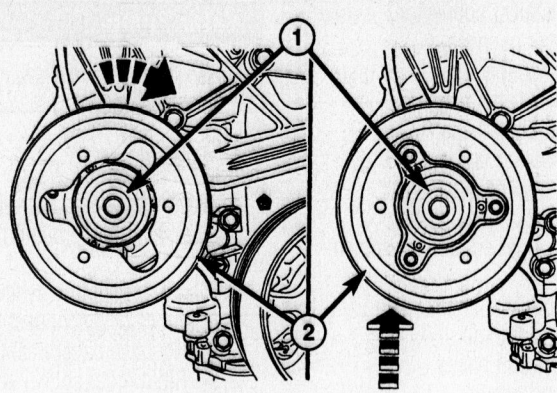

1 - HUB - WATER PUMP
2 - PULLEY - WATER PUMP

67189-MINV-G11

The water pump must be positioned between the pump housing and the drive hub so it can be removed—3.3L and 3.8L engines

- Water pump with the pulley loosely positioned between the hub and pump body

To install:

4. Install or connect the following:
   - New seal into the water pump housing groove
   - Water pump with the pulley loosely positioned between the hub and pump body. Tighten the bolts to 105 inch lbs. (12 Nm).
   - Water pump pulley and tighten the bolts to 250 inch lbs. (28 Nm). Rotate the pump by hand to make sure it moves freely.
   - Accessory drive belt
   - Right inner splash shield
   - Negative battery cable
5. Fill the cooling system.
6. Start the engine and check for leaks.

Heater Core

REMOVAL & INSTALLATION

Front System

1. Before servicing the vehicle, refer to the precautions in the beginning of this section.
2. Align the wheels in the straight-ahead position.
3. Disconnect the negative battery cable.

✷✷ CAUTION

Before working around the steering wheel or the instrument panel all 2 minutes to pass to allow the air bag module to discharge.

4. Drain the cooling system into a clean container for reuse.
   - Lower silencer boot, at the base of the steering shaft
   - Brake lamp switch
   - Power brake booster push rod from the pin on the brake pedal arm
   - 3 screws from the heater core shield at the left side of the HVAC housing
   - Heater core shield rearwards so the 2 tabs that position the front of the shield can be disengaged
   - Heater core shield. Position some towels under the heater core hoses.
   - Heater core plate screw

5. Push both core tubes at the same time towards the dash far enough to disengage the fittings from the heater core supply and return ports.
   - 2 screws that attach the core mounting plate to the distribution housing
6. Lift the accelerator pedal and slide the heater core past it.
7. Depress the brake pedal and remove the heater core from the heater/air conditioning housing assembly.

To install:

8. Depress the brake pedal and install the heater core into the heater/air conditioning housing assembly.
9. Lift the accelerator pedal and slide the heater core past it.
10. Install or connect the following:
    - 2 screws that attach the core plate to the distribution housing

➡**The heater core tubes each have a slot that must be indexed to a location tab within each of the ports. Adjust the tube position as necessary so the sealing plate fits flush against the core supply and return ports which will indicated the ports are properly indexed.**

- Heater core tubes and the sealing plate at the same time to the core supply and return ports.
- Heater core sealing plate screw.
- Heater core shield onto the distribution housing making sure the two location tabs and fully engaged
- 3 screws that attach the core

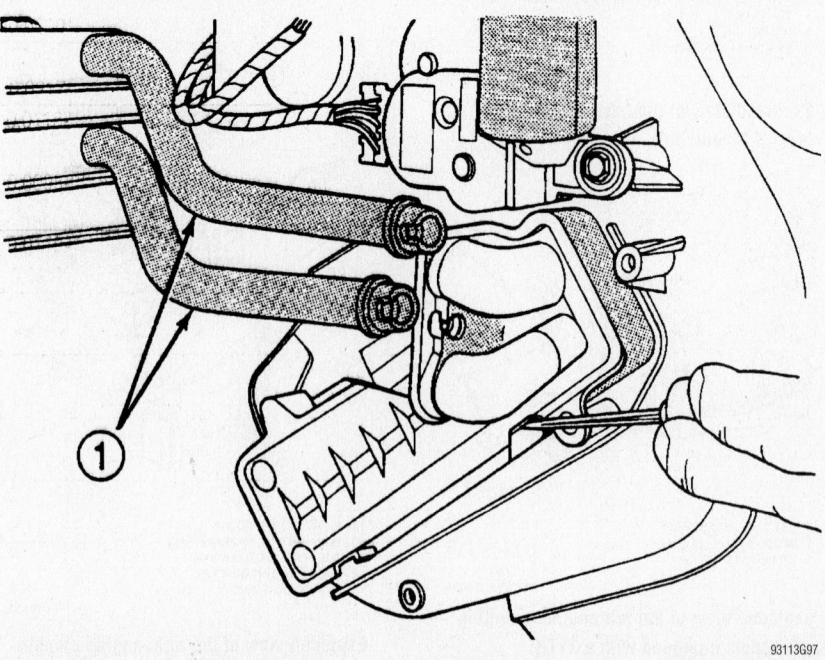

View of the front heater core assembly

93113G97

shield to the left end of the HVAC
housing
- Power brake booster push rod to
  the pin on the brake pedal arm
- Brake lamp switch
- Lower silencer boot, at the base of
  the steering shaft

11. Refill the cooling system.
12. Connect the negative battery cable.
13. Run the engine to normal operating
temperatures; then, check the climate control operation and check for leaks.

Rear Auxiliary System

1. Disconnect the negative battery
cable.
2. Partially drain the cooling system
into a clean container for reuse.
3. Remove the right rear quarter trim
panel by removing or disconnecting the following:
- 1st row seat
- 2nd row seat, if equipped
- Sliding door sill trim panel and the
  quarter trim bolster
- C-pillar and D-pillar trim panels
- 1st row seat belt anchor and the
  2nd row seat anchor, if equipped
- Quarter trim-to-quarter panel
  screws from the bolster area
- Quarter trim's rear edge-to-bracket
  screws
- Quarter trim-to-quarter panel hidden clips, located rearward of the
  sliding door
- Quarter trim from the quarter panel
- Electrical connector from the accessory power outlet, if equipped
- Pass the 2nd row seat belt through
  the slot in the trim panel, if
  equipped
- Pass the 1st row seat belt through
  the slot in the trim panel
- Quarter trim panel from the vehicle

4. Remove the rear heater distribution
duct from the right quarter inner panel.
5. Remove the screw attaching the back
of the rear heater/AC unit to the to the right
D panel.
6. Remove the screw that attaches the
front of the rear heater/AC unit to the right
quarter inner panel.
7. Remove the rear auxiliary heater
hoses clamps from the rear heater core.
8. Remove the rear heater hoses from
the rear heater core.
9. Remove the 4 latch tabs that attach
the core in the heater/AC housing.
10. Carefully lift the rear heater core and
tubes up and out of the rear auxiliary housing.

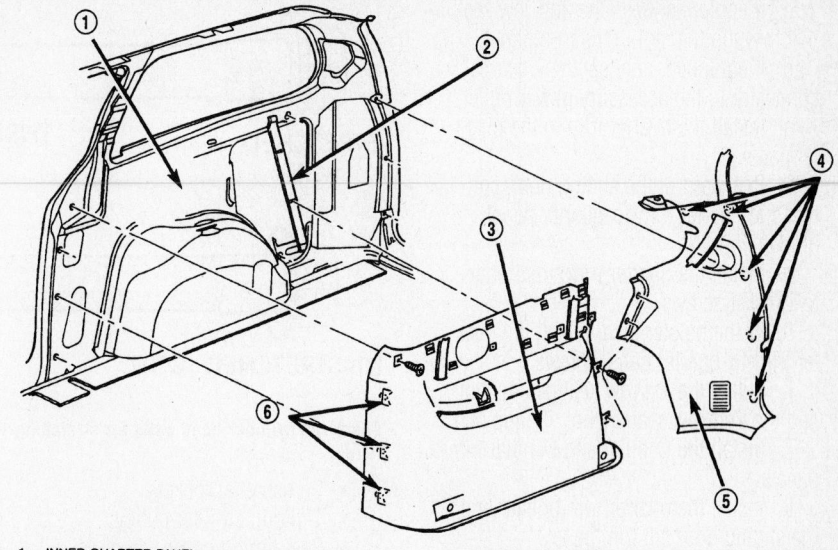

1 – INNER QUARTER PANEL
2 – QUARTER TRIM PANEL ATTACHING BRACKET
3 – QUARTER TRIM PANEL
4 – CLIPS
5 – D-PILLAR TRIM PANEL
6 – CLIPS

93113G98

View of the right quarter and D-trim panels

To install:

11. Carefully, install the rear heater core
into the housing. Press firmly and evenly to
engage the 4 latch tabs.
12. Install the rear heater hoses to the
rear heater core.
13. Install the rear auxiliary heater hoses
clamps to the rear heater core.
14. Install the screw that attaches the
front of the rear heater/AC unit to the right
quarter inner panel.

15. Install the screw attaching the back
of the rear heater/AC unit to the to the right
D panel.
16. Install the rear heater distribution
duct to the right quarter inner panel.
17. Install the right rear quarter trim panel
by performing the following procedure:
   a. Install the quarter trim panel to the
   vehicle.
   b. Pass the 1st row seat belt through
   the slot in the trim panel.

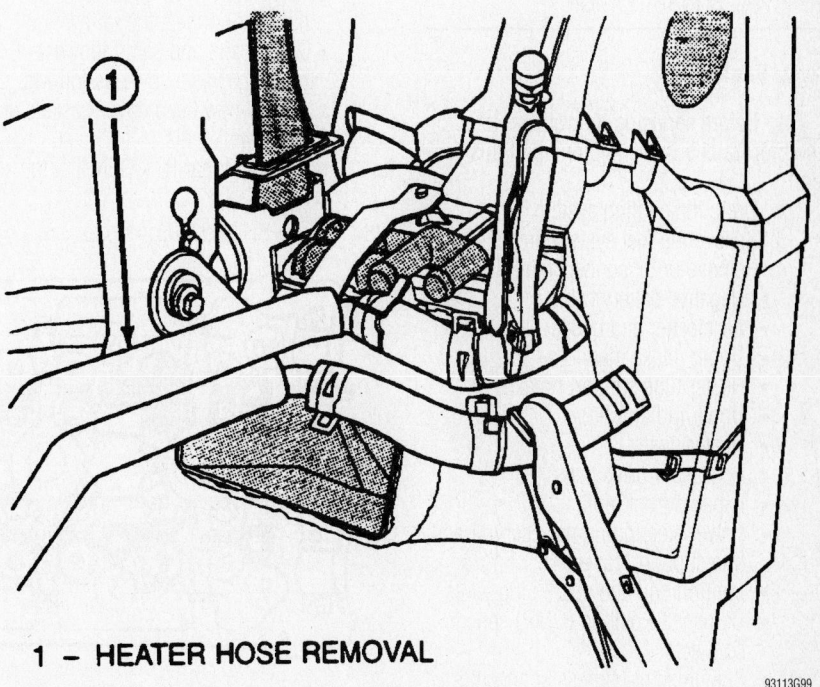

1 – HEATER HOSE REMOVAL

93113G99

View of the rear heater core

c. If equipped, pass the 2nd row seat belt through the slot in the trim panel.

d. If equipped, connect the electrical connector to the accessory power outlet.

e. Install the quarter trim to the quarter panel.

f. Rearward of the sliding door, connect the quarter trim-to-quarter panel hidden clips.

g. Install the quarter trim's rear edge-to-bracket screws.

h. At the bolster area, install the quarter trim-to-quarter panel screws.

i. Install the 1st row seat belt anchor and the 2nd row seat anchor, if equipped.

j. Install the C-pillar and D-pillar trim panels.

k. Install the quarter trim bolster and the sliding door sill trim panel.

l. If equipped, install the 2nd row seat.

m. Install the 1st row seat.

18. Refill the cooling system.

19. Connect the negative battery cable.

20. Run the engine to normal operating temperatures; then, check the climate control operation and check for leaks.

➡**If the rear heater core was emptied and not refilled, it will be necessary to thermal cycle the vehicle twice. Run the engine until the thermostat opens; then, turn the engine OFF and allow it to cool.**

Cylinder Head

REMOVAL & INSTALLATION

2.4L Engine

1. Before servicing the vehicle, refer to the precautions in the beginning of this section.

2. Drain the cooling system.

3. Relieve the fuel system pressure.

4. Remove or disconnect the following:
- Negative battery cable
- Air cleaner and hoses
- Upper intake manifold
- Heater tube support bracket
- Upper radiator hose
- Heater hose
- Accessory drive belts
- Exhaust front pipe
- Power steering pump reservoir and line support bracket
- Ignition coil and spark plug wires
- Camshaft Position (CMP) sensor connector
- Fuel injector harness connectors
- Timing belt

Check the cylinder head bolts for stretching—2.4L engine

- Camshaft sprockets
- Timing belt idler pulley
- Rear timing cover
- Valve cover

➡**Keep all valvetrain components in order for assembly.**

- Camshafts and cam followers
- Cylinder head bolts in the reverse of the tightening sequence using several passes
- Cylinder head and gasket

To install:

5. Examine the cylinder head bolts and replace any that have stretched.

6. Install the cylinder head and tighten the bolts in sequence, as follows:

a. Step 1: 25 ft. lbs. (34 Nm).

b. Step 2: 50 ft. lbs. (68 Nm).

c. Step 3: 50 ft. lbs. (68 Nm).

d. Step 4: Plus 90 degrees.

7. Install or connect the following:
- Camshafts and cam followers

8. Install the valve cover as follows:

a. Install new head cover gaskets and spark plug well seals.

b. Apply Mopar RTV GEN II at the camshaft cap corners and the top edges of the ½ round seal.

c. Valve cover and use new bolts making sure the single stud used to attach the upper intake manifold support bracket is located in the # 8 as shown in the illustration. Tighten the bolts in the sequence illustrated in 3 steps. First step to 40 inch lbs. (4.5 Nm). Second step to 80 inch lbs. (9 Nm) and the third step to 105 inch lbs. (12 Nm).

9. Install or connect the following:
- Rear timing cover
- Timing belt idler pulley. Tighten the bolt to 45 ft. lbs. (61 Nm).
- Camshaft sprockets. Tighten the bolts to 75 ft. lbs. (101 Nm).
- Timing belt
- Front cover
- CMP sensor connector
- Fuel injector harness connectors
- Ignition coil and spark plug wires
- Power steering pump reservoir and line support bracket
- Exhaust front pipe
- Accessory drive belts
- Upper radiator hose

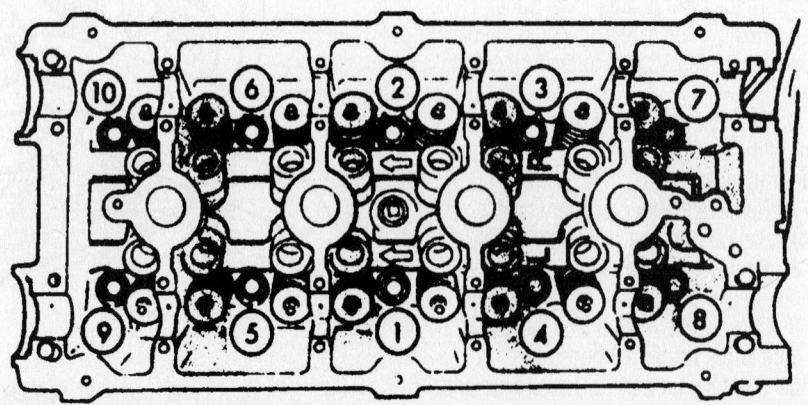

Cylinder head torque sequence—2.4L engine

- Heater hose
- Intake manifold and the EGR tube with new gaskets
- Air cleaner and hoses
- Negative battery cable

10. Fill the cooling system.

11. Start the engine, check for leaks and repair if necessary.

3.3L and 3.8L Engines

1. Before servicing the vehicle, refer to the precautions in the beginning of this section.

2. Drain the cooling system.

3. Relieve the fuel system pressure.

4. Remove or disconnect the following:
- Negative battery cable
- Wiper module if removing the right cylinder head
- Upper and lower intake manifolds

✳✳ CAUTION

The intake manifold gaskets are made of a very thin metal and can cause injury if not properly handled.

- Valve covers.
- Spark plugs
- Dipstick tube
- Exhaust manifolds

➡ **Keep all valvetrain components in order for assembly**

- Rocker arm and shaft assemblies
- Cylinder head bolts, heads and gaskets

5. Clean all gasket mating surfaces.

To install:

6. Examine the cylinder head bolts and replace any that have stretched.

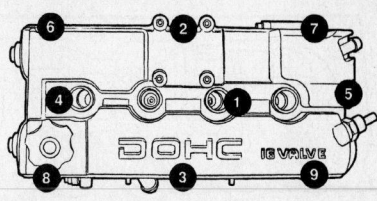

67189-MINV-G13

Valve cover torque sequence. The stud should be located at the # 8 position— 2.4L engines

7. Install the cylinder head with a new gasket. The left bank gasket has an L on it and is located at the front of the engine. The right gasket has an R stamped on it and is located at the rear of the engine. Tighten the bolts in sequence, as follows:

a. Step 1: Tighten bolts 1–8 to 45 ft. lbs. (61 Nm).

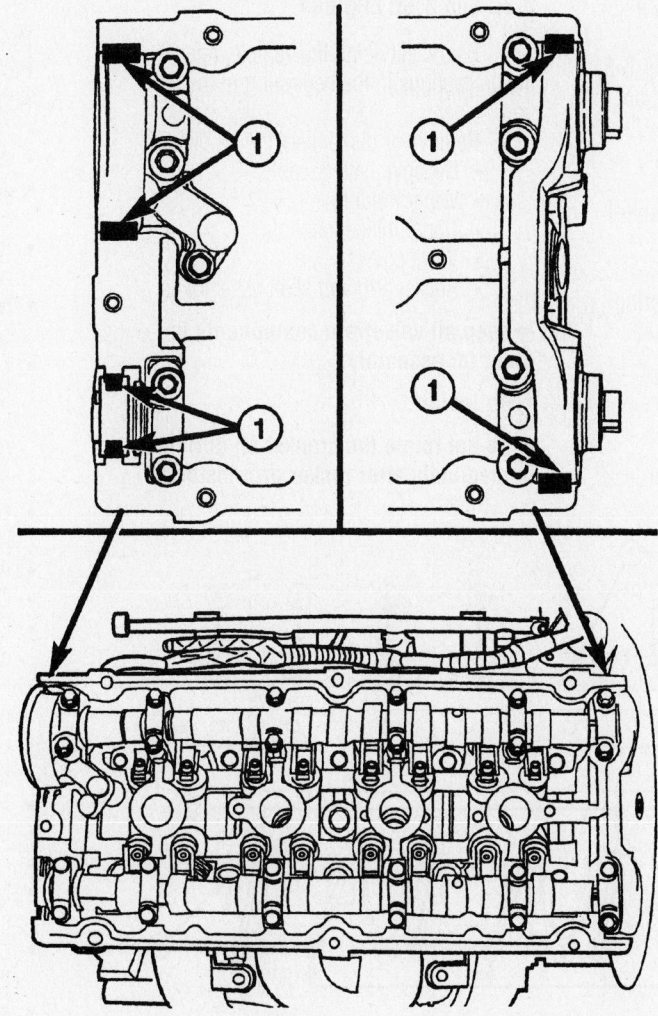

1 – SEALER LOCATION

Apply sealant at these locations when installing the valve cover—2.4L engines

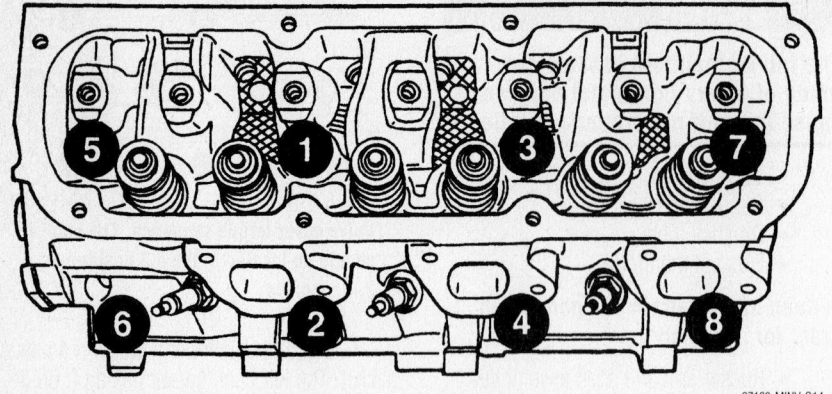

67189-MINV-G14

Cylinder head torque sequence—3.3L and 3.8L engines

b. Step 2: Tighten bolts 1–8 to 65 ft. lbs. (88 Nm).

c. Step 3: Tighten bolts 1–8 to 65 ft. lbs. (88 Nm).

d. Step 4: Bolts 1–8 plus 90 degrees.

e. Step 5: Tighten bolt 9 to 25 ft. lbs. (33 Nm).

8. Check that the torque on bolts 1–8 has exceeded 90 ft. lbs. (122 Nm). If not, replace the bolt.

9. Install or connect the following:
- Pushrods
- Rocker arm and shaft assemblies
- Valve covers
- Exhaust manifolds
- Dipstick tube with a new O–ring
- Spark plugs
- Intake manifolds
- Wiper module, if removed
- Negative battery cable

10. Fill the cooling system.

11. Start the engine and check for leaks.

Rocker Arms/Shafts

REMOVAL & INSTALLATION

3.3L and 3.8L Engines

1. Before servicing the vehicle, refer to the precautions in the beginning of this section.

2. Remove or disconnect the following:
- Negative battery cable
- Wiper module
- Upper intake manifold
- Valve covers
- Rocker arm and shaft assemblies

➡**Keep all valvetrain components in order for assembly.**

To install:

➡**Do not rotate the crankshaft during or immediately after rocker arm installa-**

tion. **Wait 20 minutes for the hydraulic lash adjusters to bleed down.**

3. Install or connect the following:
- Rocker arm and shaft assemblies. Tighten the bolts to several passes to 200 inch lbs. (23 Nm).
- Valve covers
- Upper intake manifold
- Wiper module
- Negative battery cable

4. Start the engine and check for proper operation.

Intake Manifold

REMOVAL & INSTALLATION

2.4L Engine

1. Before servicing the vehicle, refer to the precautions in the beginning of this section.

2. Drain the cooling system.

3. Relieve the fuel system pressure.

4. Remove or disconnect the following:
- Negative battery cable
- Air cleaner and tube
- Throttle Position (TP) sensor connector
- Idle Air Control (IAC) valve connector
- Manifold Absolute Pressure (MAP) sensor connector
- Evaporative Emissions (EVAP) canister purge solenoid vacuum line
- Positive Crankcase Ventilation (PCV) valve and hose
- Brake booster vacuum line
- Cruise control vacuum reservoir line
- Accelerator cable
- Cruise control cable
- Exhaust Gas Recirculation (EGR) tube
- Front and rear intake manifold support brackets
- Engine oil dipstick tube
- Upper intake manifold
- Fuel line
- Upper radiator hose
- Heater hose
- Engine Coolant Temperature (ECT) sensor connector
- Lower manifold support upper bolts and lower bolt
- Fuel injector harness connectors
- Power steering reservoir bolts and position aside without disconnecting the lines
- Lower intake manifold bolts and the manifold

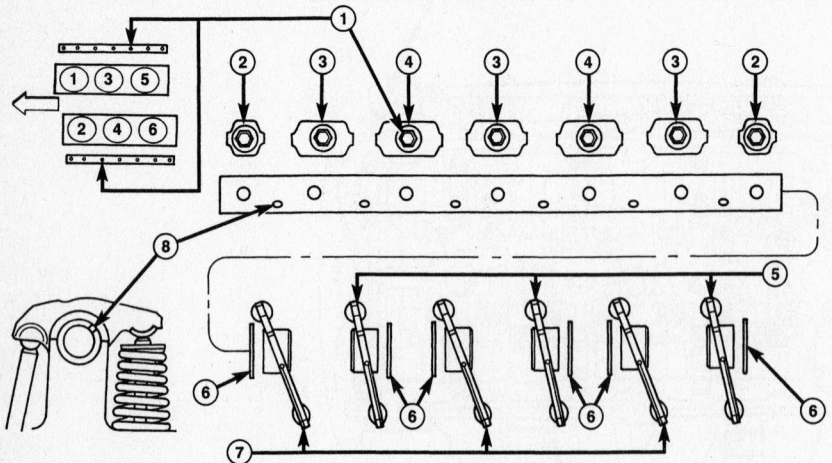

1 - BOLT (ROCKER SHAFT OIL FEED - LONGER LENGTH)
2 - SHAFT RETAINER/SPACER - 21.5 mm (0.84 in.)
3 - SHAFT RETAINER/SPACER - 37.5 mm (1.47 in.)
4 - SHAFT RETAINER/SPACER - 40.9 mm (1.61 in.)

5 - ROCKER ARM - EXHAUST
6 - WASHER
7 - ROCKER ARM - INTAKE (LARGER OFFSET)
8 - ROCKER ARMS LUBRICATION FEED HOLE (POSITION UPWARD & TOWARD VALVE SPRING)

67189-MINV-G19

Rocker arm and shaft assembly—3.3L and 3.8L engines

STUD LOCATIONS

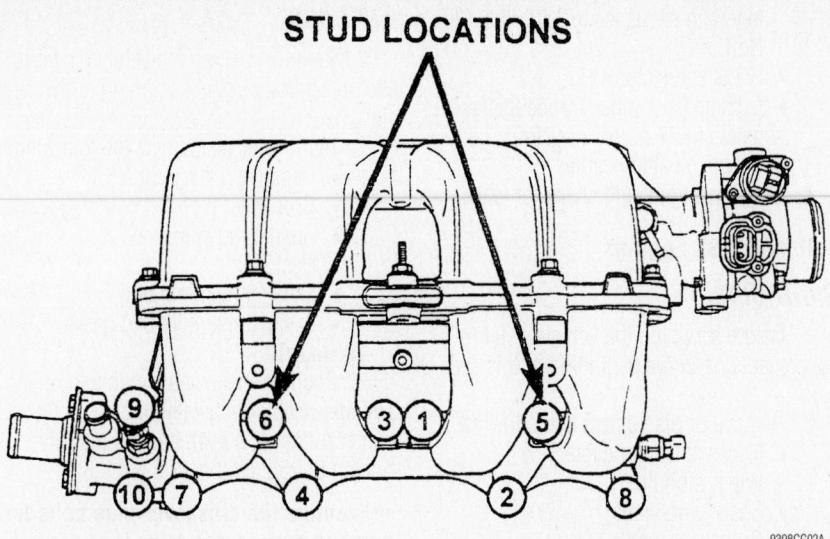

Lower intake manifold torque sequence—2.4L engine

To install:

5. Install or connect the following:
- Lower intake manifold using a new gasket. Tighten the bolts in several steps to 250 inch lbs. (23 Nm).
- Fuel injector harness connectors
- Intake manifold Y-bracket. Tighten the block bolts to 40 ft. lbs. (54 Nm) and the intake manifold bolts to 250 inch (23 Nm).
- Power steering reservoir
- Fuel line
- ECT sensor connector
- Heater hose
- Upper radiator hose

6. Apply a 0.060 inch (1.5mm) bead of gasket maker to the perimeter of the lower intake manifold runner openings.
- Upper intake manifold. Tighten the bolts in several passes to 250 inch lbs. (23 Nm).
- Engine oil dipstick tube
- Upper bolt on the manifold-to-front support bracket to 250 inch lbs. (28 Nm)
- EGR tube
- Cruise control cable
- Accelerator cable
- Cruise control vacuum reservoir line
- Brake booster vacuum line
- PCV valve and hose
- EVAP canister purge solenoid vacuum line
- MAP sensor connector
- IAC valve connector
- TP sensor connector
- Air cleaner and tube
- Negative battery cable

7. Fill the cooling system.

8. Start the engine, check for leaks and repair if necessary.

3.3L and 3.8L Engines

1. Before servicing the vehicle, refer to the precautions in the beginning of this section.

2. Drain the cooling system.

3. Relieve the fuel system pressure.

4. Remove or disconnect the following:
- Negative battery cable
- Air cleaner and tube
- Idle Air Control (IAC) valve connector
- Accelerator cable
- Cruise control cable
- Make up air hose support clip from the throttle cable bracket
- Automatic Idle Speed (IAS) and Throttle Position (TP) sensor connectors
- Manifold Absolute Pressure (MAP) sensor connector
- Exhaust Gas Recirculation (EGR) transducer connector, if equipped
- EGR tube, if equipped
- Vapor purge vacuum hose from the throttle body
- Positive Crankcase Ventilation (PCV) hose
- Power steering reservoir bolts and loosen the nut, lift the reservoir up to disengage the lower mount from the stud and position aside with the lines still attached.
- Brake booster and leak detection pump hoses
- Upper intake manifold
- Fuel line
- Ignition coil and bracket

- Heater supply hose and Engine Coolant temperature (ECT) sensor
- Fuel injector harness connectors
- Fuel supply manifold
- Upper radiator hose
- Lower intake manifold

To install:

5. Install a ¼ inch bead of gasket maker onto each of the 4 manifold-to-cylinder head gasket corners.

6. Install the lower intake manifold gasket and tighten the seal retainer screws to 105 inch lbs. (12 Nm)

7. Install the lower intake manifold and tighten the bolts in sequence, as follows:
a. Step 1: 10 inch lbs. (1 Nm).
b. Step 2: 17 ft. lbs. (22 Nm).
c. Step 3: 17 ft. lbs. (22 Nm).

8. Install or connect the following:
- Fuel supply manifold
- Fuel injector harness connectors
- ECT sensor connector
- Heater supply hose
- Radiator hose
- Fuel line

9. Place a new gasket in the channel and press lightly in place.
- Upper intake manifold. Apply Mopar thread and lock adhesive (medium strength) to each upper intake manifold bolt. Tighten the bolts to 105 inch lbs. (12 Nm) in the sequence illustrated.

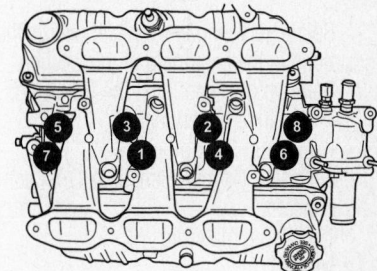

Lower intake manifold torque sequence— 3.3L and 3.8L engines

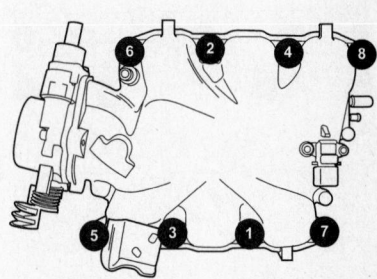

Upper intake manifold torque sequence— 3.3L and 3.8L engines

- MAP sensor connector
- Brake booster and leak detection pump hoses

➡**The special screws used to retain the EGR tube and power steering reservoir must be tightened slowly with hand tools only to avoid stripping the threads.**

- Power steering reservoir
- PCV hose
- Vapor purge vacuum hose from the throttle body
- EGR tube, if equipped
- EGR transducer connector, if equipped
- MAP sensor connector
- IAS and TP sensor connectors
- Make up air hose support clip to the throttle cable bracket
- Cruise control cable
- Accelerator cable
- IAC valve connector
- Air cleaner and tube
- Negative battery cable

10. Fill the cooling system.
11. Start the engine, check for leaks and repair if necessary.

Exhaust Manifold

REMOVAL & INSTALLATION

2.4L Engine

1. Before servicing the vehicle, refer to the precautions in the beginning of this section.
2. Remove or disconnect the following:
- Negative battery cable
- Exhaust front pipe
- Heated Oxygen Sensor (HO2S) connector
- Exhaust manifold

To install:
3. Install or connect the following:
- Exhaust manifold. Torque the fas-

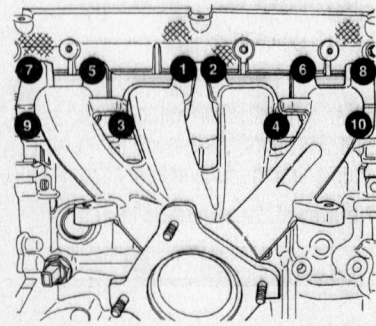

67189-MINV-G30

Exhaust manifold torque sequence—2.4L engine

teners in sequence to 15 ft. lbs. (20 Nm).
- HO2S connector
- Exhaust front pipe. Torque the fasteners to 27 ft. lbs. (37 Nm).
- Negative battery cable
4. Start the engine and check for leaks.

3.3L and 3.8L Engines

RIGHT SIDE

1. Before servicing the vehicle, refer to the precautions in the beginning of this section.
2. Remove or disconnect the following:
- Negative battery cable
- Wiper module
- Spark plug wires
- Crossover pipe
- Upstream Oxygen Sensor (O2S) connectors
- Heat shield
- Accessory drive belt
- Power steering pump support strut lower bolt
- Downstream O2S connectors
- Catalytic converter pipe from the manifold
- Power steering support strut upper bolt and strut
- Exhaust manifold bolt
- Exhaust manifold and gasket

To install:
3. Position the exhaust manifold on the cylinder head and install the bolts to the center runner (cyl # 3), and temporarily tighten to 25 inch lbs. (2.8 Nm).

➡**Examine the crossover pipe bolts for damage caused due to heat or corrosion and replace using new OEM bolts if found to be defective.**

4. Install a new gasket, attach the crossover pipe to the manifold and tighten the bolts to 30 ft. lbs. (41 Nm).
5. Install or connect the following:
- Remaining manifold bolts. Torque the fasteners to 17 ft. lbs. (23 Nm).
- Power steering support strut and upper bolt
- Heat shield
- Upstream O2S connectors
- Catalytic converter pipe to the manifold
- Downstream O2S connectors
- Power steering pump support strut lower bolt
- Accessory drive belt
- Spark plug wires
- Wiper module
- Negative battery cable
6. Start the engine and check for leaks.

LEFT SIDE

1. Before servicing the vehicle, refer to the precautions in the beginning of this section.
2. Remove or disconnect the following:
- Negative battery cable
- Crossover pipe
- Spark plug wires
- Heat shield
- Exhaust manifold bolt
- Exhaust manifold and gasket

To install:
3. Position the exhaust manifold on the cylinder head and install the bolts to the center runner (cyl # 4), and temporarily tighten to 25 inch lbs. (2.8 Nm).

➡**Examine the crossover pipe bolts for damage caused due to heat or corrosion and replace using new OEM bolts if found to be defective.**

4. Install a new gasket, attach the crossover pipe to the manifold and tighten the bolts to 30 ft. lbs. (41 Nm).
5. Install or connect the following:
- Remaining manifold bolts. Torque the fasteners to 17 ft. lbs. (23 Nm).
- Heat shield
- Negative battery cable
6. Start the engine and check for leaks.

Front Crankshaft Seal

REMOVAL & INSTALLATION

2.4L Engine

1. Before servicing the vehicle, refer to the precautions in the beginning of this section.
2. Remove or disconnect the following:
- Negative battery cable
- Accessory drive belts
- Crankshaft pulley
- Front cover
- Timing belt
- Crankshaft timing sprocket using puller 6793 and insert C-4685-C2, being careful not to damage the seal bore
- Front crankshaft seal using tool 6771

To install:
3. Install or connect the following:
- Front crankshaft seal using tool 6780. Place the seal in position, with the seal spring towards the engine and install so that it is flush with the cover.
- Crankshaft timing sprocket using tool 6792

- Timing belt
- Front cover
- Crankshaft pulley. Torque the bolt to 105 ft. lbs. (142 Nm).
- Accessory drive belts
- Negative battery cable

4. Start the engine and check for leaks.

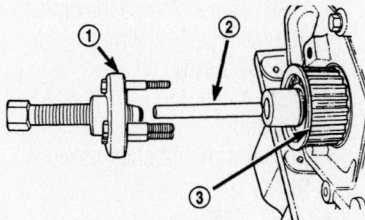

1 - SPECIAL TOOL 6793
2 - SPECIAL TOOL C-4685-C2
3 - CRANKSHAFT SPROCKET

67189-MINV-G31

Remove the crankshaft timing sprocket using puller 6793 and insert C–4685–C2— 2.4L engine

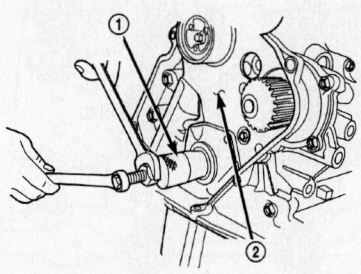

1 - SPECIAL TOOL 6771
2 - REAR TIMING BELT COVER

67189-MINV-G32

Remove the front crankshaft seal using tool 6771—2.4L engine

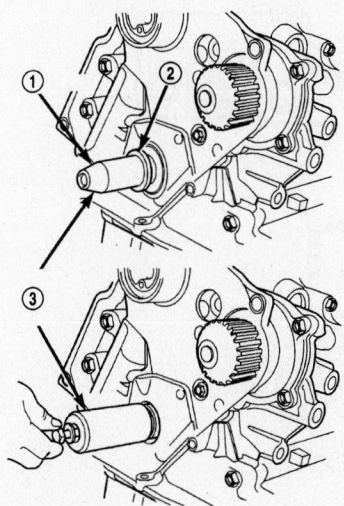

1 - PROTECTOR
2 - SEAL
3 - SPECIAL TOOL 6780

67189-MINV-G33

Install the front crankshaft seal using tool 6780—2.4L engine

3.3 and 3.8L Engines

1. Before servicing the vehicle, refer to the precautions in the beginning of this section.
2. Remove or disconnect the following:
   - Negative battery cable
   - Right wheel and splash shield
   - Accessory drive belts
   - Crankshaft pulley
3. Place tool 6341A on the crankshaft nose and screw the tool into the seal until it

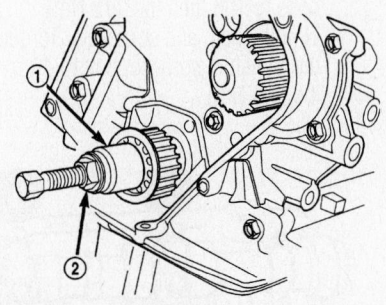

1 - SPECIAL TOOL 6792
2 - TIGHTEN NUT TO INSTALL

67189-MINV-G34

Install the crankshaft timing sprocket using tool 6792—2.4L engine

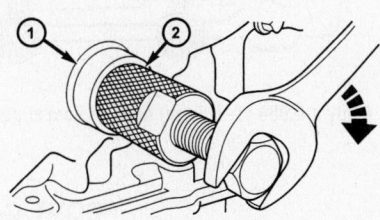

1 - SEAL
2 - SPECIAL TOOL 6341A

67189-MINV-G38

Removing the front seal using tool 6341A—3.3L and 3.8L engines

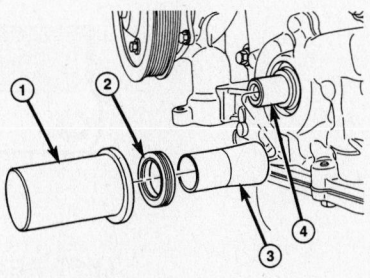

1 - SPECIAL TOOL C-4992-1
2 - SEAL
3 - SPECIAL TOOL C-4992-2
4 - CRANKSHAFT

67189-MINV-G35

Removing the front seal using tools C–4992–2 and C–42992–1—3.3L and 3.8L engines

is engaged being careful not to damage the crankshaft seal surface cover.

4. Turn the forcing screw on the tool to remove the seal

To install:

5. Position guide C–4992–2 on the crankshaft nose.
6. Position a new seal over the guide with the seal spring facing the engine cover.
7. Install the seal using tool C–42992–1 until the seal is flush with the cover and remove the tools.
8. Install or connect the following:
   - Crankshaft pulley and tighten the bolt to 40 ft. lbs. (54 Nm)
   - Accessory drive belts
   - Right wheel and splash shield
   - Negative battery cable

Camshaft and Valve Lifters

REMOVAL & INSTALLATION

2.4L Engine

1. Before servicing the vehicle, refer to the precautions in the beginning of this section.
2. Remove or disconnect the following:
   - Negative battery cable
   - Valve cover
   - Camshaft Position (CMP) sensor and camshaft target magnet

67189-MINV-G15

The camshaft bearing caps are marked for location—2.4L engines

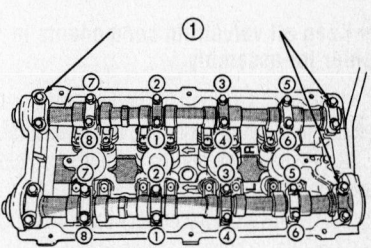

1 - REMOVE OUTSIDE BEARING CAPS FIRST

67189-MINV-G16

Camshaft bearing caps removal sequence—2.4L engines

- Timing belt
- Camshaft sprockets
- Rear timing cover

➡**The bearing caps are marked for location, Remove the outside caps first. The caps are not interchangeable the intake cam # 6 thrust bearing face spacing is wider.**

3. Make sure to identify the caps before removal, then loosen the caps in the sequence illustrated one at a time and remove the camshafts and cam followers. Always make sure to remove the outside bearing caps first.

To install:

➡**Make sure none of the pistons are at Top Dead Center (TDC) when installing the camshafts.**

4. Install or connect the following:
- Lubricate then install the cam followers in their original positions
- Lubricate then install the camshafts

5. Install right and left camshaft bearing caps 2–5 and right # 6. Tighten the M6 fasteners to 105 inch lbs. (12 Nm) in the sequence illustrated.

6. Apply a 0.060 (1–1.5mm) bead of gasket maker to the # 1 and # 6 bearing caps. Install the bearing caps in their original positions and tighten the M8 fasteners to 250 inch lbs. (28 Nm)
- Camshaft oil seals
- Camshaft target magnet and CMP sensor
- Valve cover
- Rear timing cover
- Timing belt idler pulley
- Camshaft sprockets. Tighten the bolts to 75 ft. lbs. (101 Nm).
- Timing belt
- Front cover
- Negative battery cable

7. Start the engine and check for proper operation.

3.3L and 3.8L Engines

➡**Keep all valvetrain components in order for assembly.**

1. Before servicing the vehicle, refer to the precautions in the beginning of this section.

2. Remove the engine from the vehicle and mount it on a stand.

3. Remove or disconnect the following:
- Valve covers
- Rocker arm and shaft assemblies
- Pushrods
- Intake manifold
- Cylinder heads
- Yoke retainer

- Aligning yokes
- Hydraulic lifters
- Oil pan
- Oil pump pickup tube
- Crankshaft pulley
- Front cover
- Timing chain and sprockets
- Camshaft thrust plate
- Camshaft

To install:

4. Install or connect the following:
- Camshaft
- Camshaft thrust plate. Torque the bolts to 105 inch lbs. (12 Nm).
- Timing chain and sprockets. Torque the camshaft sprocket bolt to 40 ft. lbs. (54 Nm).

- Front cover. Torque the bolts to 20 ft. lbs. (27 Nm).
- Crankshaft pulley. Torque the bolt to 40 ft. lbs. (54 Nm).
- Oil pump pickup tube
- Oil pan
- Hydraulic lifters
- Aligning yokes
- Yoke retainer. Torque the bolts to 105 inch lbs. (12 Nm).
- Cylinder heads
- Intake manifold
- Pushrods
- Rocker arm and shaft assemblies
- Valve covers

5. Install the engine to the vehicle.

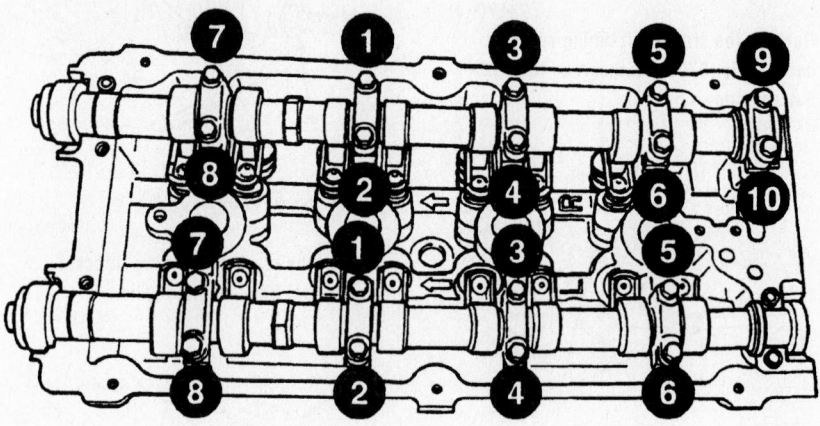

Apply a 0.060 (1–1.5mm) bead of gasket maker to the # 1 and # 6 bearing caps—2.4L engines

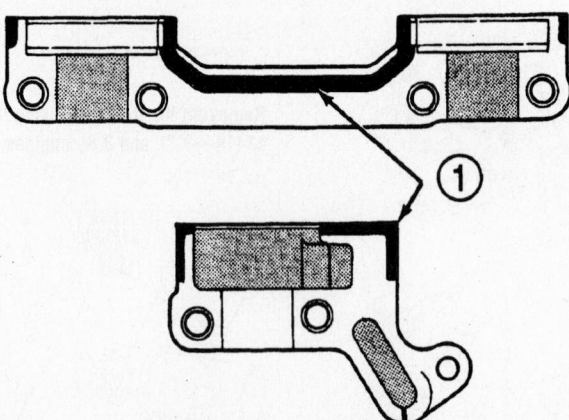

FRONT CAM CAP

LEFT REAR CAM CAP

1 - 1.5 mm (.060 in.) DIAMETER BEAD OF MOPAR GASKET MAKER

Camshaft bearing caps tightening sequence—2.4L engines

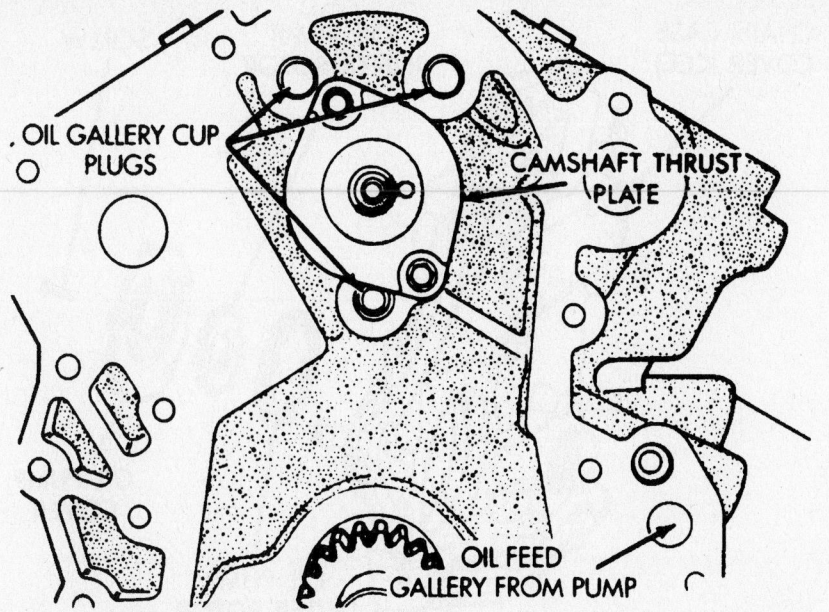

Remove the thrust plate and withdraw the camshaft from the engine—3.3L and 3.8L engines

7924CG45

Valve Lash

ADJUSTMENT

All engines covered in this section are equipped with hydraulic lash adjusters. No adjustment is necessary.

Starter Motor

REMOVAL & INSTALLATION

1. Before servicing the vehicle, refer to the precautions in the beginning of this section.
2. Remove or disconnect the following:
   - Negative battery cable
   - Starter harness connectors
   - Starter motor

To install:
3. Install or connect the following:
   - Starter motor. Torque the bolts to 35 ft. lbs. (47 Nm).
   - Starter harness connectors. Torque the battery cable nut to 100 inch lbs. (11 Nm).
   - Negative battery cable

Oil Pan

REMOVAL & INSTALLATION

2.4L Engine

1. Before servicing the vehicle, refer to the precautions in the beginning of this section.

2. Drain the engine oil.
3. Remove or disconnect the following:
   - Negative battery cable
   - Structural collar
   - A/C compressor bracket to oil pan bolt
   - Oil pan
4. Clean the oil pan and all gasket mating surfaces.

To install:
5. Install or connect the following:
   - Oil pan gasket to the block after applying engine RTV at the oil pump parting line
   - Oil pan. Torque the bolts to 105 inch lbs. (12 Nm).
   - Structural collar
   - Negative battery cable
6. Fill the crankcase to the correct level.
7. Start the engine, check for leaks and repair if necessary.

3.3L and 3.8L Engines

1. Before servicing the vehicle, refer to the precautions in the beginning of this section.
2. Drain the engine oil.
3. Remove or disconnect the following:
   - Negative battery cable
   - Engine oil dipstick
   - Drive belt splash shield
   - Strut to transaxle attaching bolt and loosen the strut to engine attaching bolts
   - Transaxle case cover
   - Oil pan fasteners
   - Oil pan and gasket

To install:
4. Clean the oil pan and all mating surfaces.
5. Apply a ⅛ inch bead of gasket material at the parting line of the chain case cover and the real seal retainer.
6. Install or connect the following:
   - New gasket on the oil pan
   - Oil pan. Torque the bolts to 105 inch lbs. (12 Nm).
   - Transaxle case cover
   - All bending brace bolts
   - Drive belt splash shield
   - Engine oil dipstick
   - Negative battery cable
7. Fill the crankcase to the correct level.
8. Start the engine, check for leaks and repair if necessary.

Oil Pump

REMOVAL & INSTALLATION

2.4L Engine

1. Before servicing the vehicle, refer to the precautions in the beginning of this section.
2. Drain the engine oil.
3. Remove or disconnect the following:
   - Negative battery cable
   - Timing belt
   - Rear timing belt cover
   - Oil pan
   - Crankshaft sprocket using puller 6793 and insert C–4685–C2
   - Crankshaft key
   - Oil pump pickup tube
   - Oil pump

To install:
4. Clean all mating surfaces of any remaining gasket material.

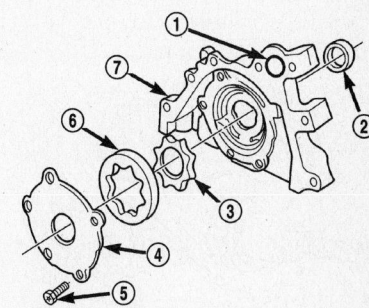

1 - O-RING
2 - SEAL
3 - INNER ROTOR
4 - OIL PUMP COVER
5 - FASTENER
6 - OUTER ROTOR
7 - OIL PUMP BODY

67189-MINV-G36

Exploded view of the oil pump—2.4L engine

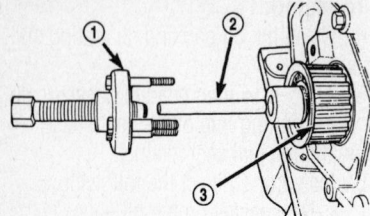

1 - SPECIAL TOOL 6793
2 - SPECIAL TOOL C-4685-C2
3 - CRANKSHAFT SPROCKET

67189-MINV-G31

Remove the crankshaft timing sprocket using puller 6793 and insert C-4685-C2—2.4L engine

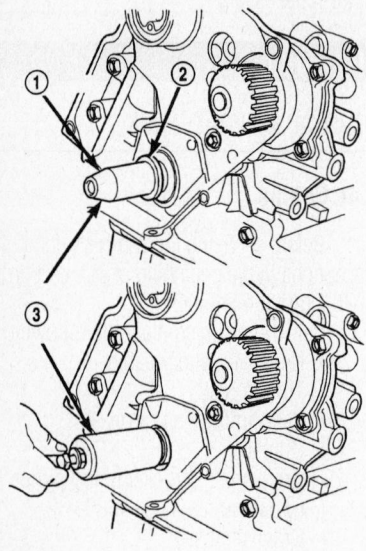

1 - PROTECTOR
2 - SEAL
3 - SPECIAL TOOL 6780

67189-MINV-G33

Install the front crankshaft seal using tool 6780—2.4L engine

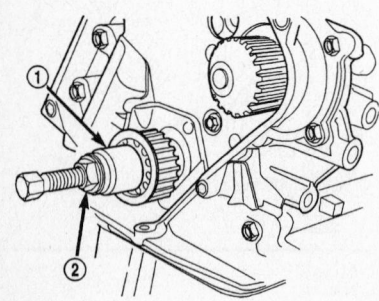

1 - SPECIAL TOOL 6792
2 - TIGHTEN NUT TO INSTALL

67189-MINV-G34

Install the crankshaft timing sprocket using tool 6792—2.4L engine

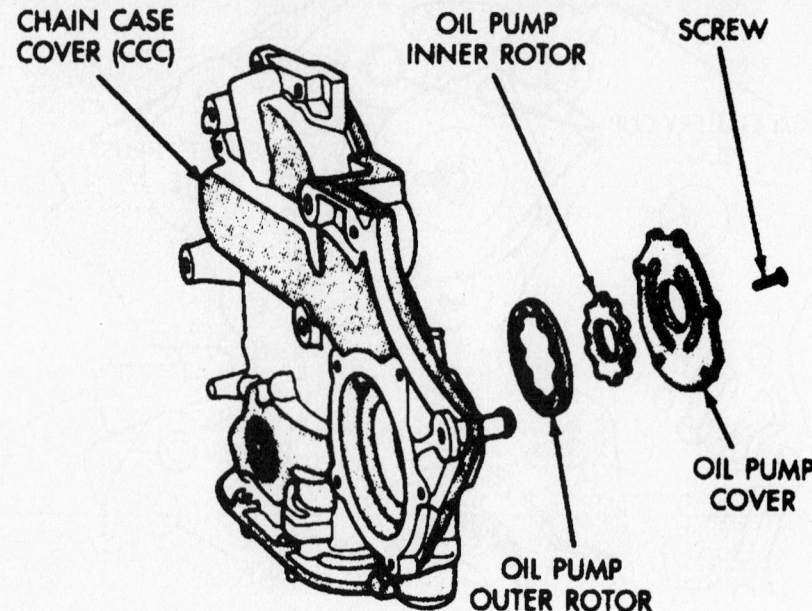

CHAIN CASE COVER (CCC)

OIL PUMP INNER ROTOR

SCREW

OIL PUMP COVER

OIL PUMP OUTER ROTOR

7924CG27

Exploded view of the oil pump assembly—3.3L and 3.8L engines

5. Apply gasket maker material to the oil pump and replace the O-ring in the oil pump discharge passage.

6. Prime the pump with oil prior to installation.

7. Install or connect the following:
- Oil pump, align the flats on the rotor with the flats on the crankshaft. Torque the bolts to 21 ft. lbs. (28 Nm).

➡The front crankshaft seal must be out of the pump to align it or the pump may be damaged.

- Front crankshaft seal using tool 6780
- Crankshaft key
- Crankshaft sprocket using tool 6792
- Oil pump pickup tube. Torque the bolt to 21 ft. lbs. (28 Nm).
- Oil pan
- Rear timing belt cover
- Timing belt
- Negative battery cable

8. Fill the engine with clean oil.

9. Start the engine, check for leaks and repair if necessary.

3.3L and 3.8L Engines

1. Before servicing the vehicle, refer to the precautions in the beginning of this section.

2. Drain the cooling system.

3. Drain the engine oil.

4. Remove or disconnect the following:
- Negative battery cable
- Oil pan

- Timing chain cover
- Oil pump from the case cover

To install:

5. Install or connect the following:
- Oil pump to the front cover. Torque the cover screws to 105 inch lbs. (12 Nm).
- Front cover. Torque the bolts to 20 ft. lbs. (27 Nm).
- Oil pan
- Negative battery cable

6. Fill the engine with clean oil.

7. Fill the cooling system.

8. Start the engine, check for leaks and repair if necessary.

Rear Main Seal

REMOVAL & INSTALLATION

1. Before servicing the vehicle, refer to the precautions at the beginning of this section.

2. Remove or disconnect the following:
- Negative battery cable
- Transaxle
- Flexplate
- Rear main seal

To install:

3. Place seal guide tool 6926-1 on the crankshaft and place the seal over the tool. Make sure the lip of the seal is facing towards the crankcase.

4. Install or connect the following:
- Rear main seal flush with the cylinder block surface using the tools illustrated.

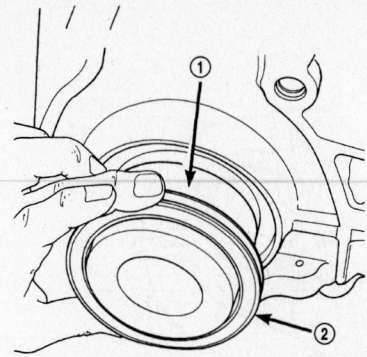

1 - SPECIAL TOOL 6926–1 PILOT
2 - SEAL

67189-MINV-G37

Install the rear main seal flush with the cylinder block surface

- Flexplate
- Transaxle
- Negative battery cable
5. Run the engine and check for leaks.

Timing Chain, Sprockets, Front Cover and Seal

REMOVAL & INSTALLATION

3.3L and 3.8L Engines

1. Before servicing the vehicle, refer to the precautions in the beginning of this section.
2. Drain the cooling system.
3. Drain the engine oil.
4. Remove the timing chain cover as follows:
- Negative battery cable
- Right wheel and splash shield
- Oil pan and pick up tube
- Drive belt
- A/C compressor and set aside with the lines still attached
- Crankshaft damper
- Lower radiator hose
- Camshaft Position (CMP), if necessary
- Heater hose from the timing cover or water pump inlet tub if equipped with an engine oil cooler
- Right side engine mount
- Idler pulley from the bracket
- Engine mount bracket
- Water pump
- Power steering support strut-to-cover bolt
- Timing chain cover bolts and the cover

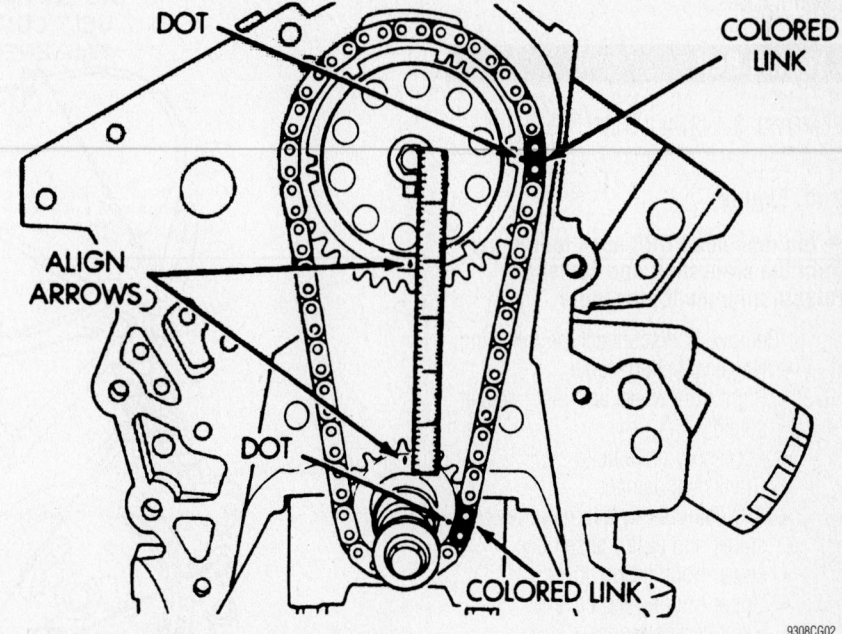

Timing mark alignment—3.3L and 3.8L engines

9308CG02

5. Rotate the engine so that the timing marks are aligned.
6. Remove or disconnect the following:
- Camshaft sprocket attaching bolt
- Timing chain and camshaft sprocket
- Crankshaft sprocket with special tools 8539, 5048–6 and 5048–1

To install:

7. Rotate the engine so the timing arrow is at the 12 o'clock position.
8. Lubricate the chain and sprockets with clean oil.
9. Hold the camshaft sprocket and chain and place the chain around the sprocket aligning the plated link with the dot on the sprocket. Position the timing arrow at the 6 o'clock position.
10. Place the chain around the crankshaft sprocket with the plated link lined up with the dot on the sprocket and install the camshaft sprocket.
11. Use a straight edge to check the timing alignment marks.
12. Align the timing chain colored links with the dots on the timing sprockets.
13. Tighten the camshaft sprocket bolt to 40 ft. lbs. (54 Nm).
14. Rotate the crankshaft 2 revolutions and verify the proper timing chain alignment, if not remove the components and reinstall as described above.
15. Clean the timing cover mating surfaces.
16. Install or the front cover as follows:
- New gasket on the cover making

sure the lower edge of the gasket is flush to 0.020 inch (0.5mm) past the lower edge of the cover. Rotate the crankshaft so the oil pump drive flats are in a vertical position.
- Position the oil pump inner rotor so the mating flats are in the same position as the crankshaft drive flats or damage may occur
- Timing chain cover and bolts. Tighten the M8 bolts to 20 ft. lbs. (27 Nm) and the M10 bolts to 40 ft. lbs. (54 Nm).
- Crankshaft front oil seal
- Water pump
- Power steering support strut-to-cover bolt
- Crankshaft damper
- Engine mount bracket. Tighten the M10 bolts to 40 ft. lbs. (54 Nm) and the M8 bolts to 21 ft. lbs. (28 Nm).
- Idler pulley on the bracket
- Right side engine mount
- CMP sensor, if removed
- Heater hose to the timing cover or water pump inlet tub if equipped with an engine oil cooler
- Lower radiator hose
- A/C compressor
- Drive belt
- Oil pick up tube and pan
- Right splash shield and wheel
- Negative battery cable
17. Fill the engine with clean oil.
18. Fill the cooling system.

19. Start the engine, check for leaks and repair if necessary.

Timing Belt

REMOVAL & INSTALLATION

2.4L Engine

➡**You may need DRB scan tool to perform the crankshaft and camshaft relearn alignment procedure.**

1. Remove or disconnect the following:
   • Negative battery cable
   • Right front wheel and inner splash shield
   • Accessory drive belts
   • Crankshaft damper
   • A/C compressor/alternator tensioner and pulley assembly
   • Lower front timing cover
   • Upper front timing cover
   • Right engine mount
   • Engine mount bracket

➡**This is an interference engine. Do not rotate the crankshaft or the camshafts after the timing belt has been removed. Damage to the valve components may occur. Before removing the timing belt, always align the timing marks.**

2. Rotate the crankshaft until the Top Dead Center (TDC) mark on the oil pump housing aligns with the TDC mark on the camshaft sprocket (located on the trailing edge of the tooth).

3. Loosen the timing belt tensioner lock bolt.

4. Insert a 6mm Allen wrench into the hexagon opening located on the top plate of the belt tensioner pulley. Rotate the top plate clockwise until there is enough slack to remove the belt.

5. If necessary, remove the camshaft timing belt sprockets.

6. If necessary, remove the crankshaft timing belt sprocket using removal tool No. 6793, or equivalent.

7. Place the tensioner into a soft-jawed vise to compress the tensioner.

8. After compressing the tensioner, insert a pin (a 5/64 in. Allen wrench will also work) into the plunger side hole to retain the plunger until installation.

To install:

9. If necessary, use tool No. 6792, or equivalent, to install the crankshaft timing belt sprocket onto the crankshaft.

10. If necessary, install the camshaft sprockets onto the camshafts. Install and

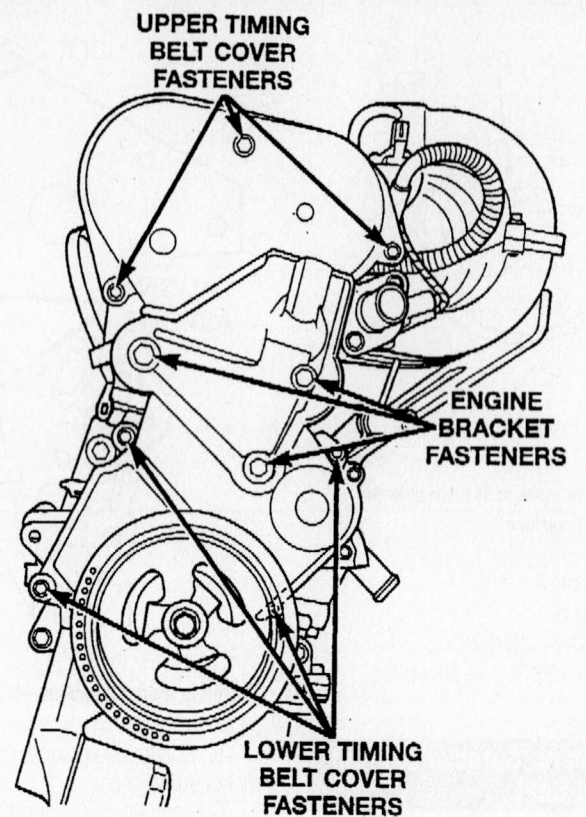

Timing cover and engine mounting bracket bolt locations—2.4L engine

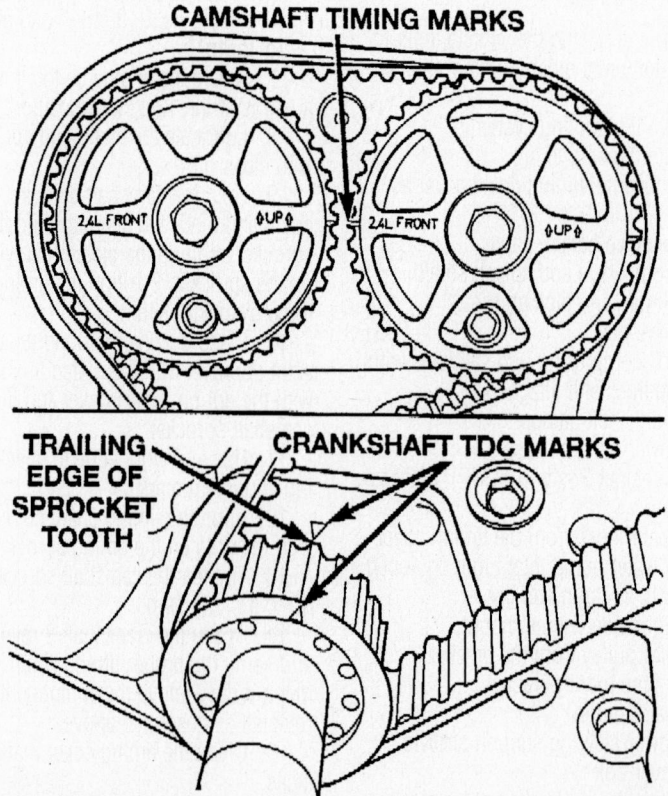

Camshaft and crankshaft alignment marks—2.4L engine

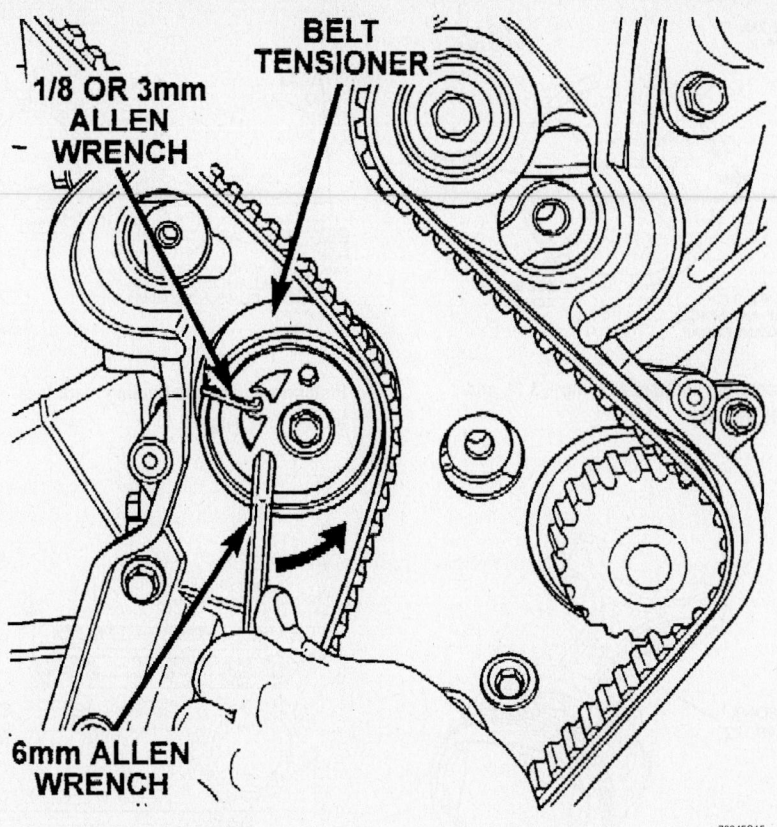

To lock the timing belt tensioner, be sure to fully insert the smaller Allen wrench into the tensioner as shown—2.4L engine

79245G15

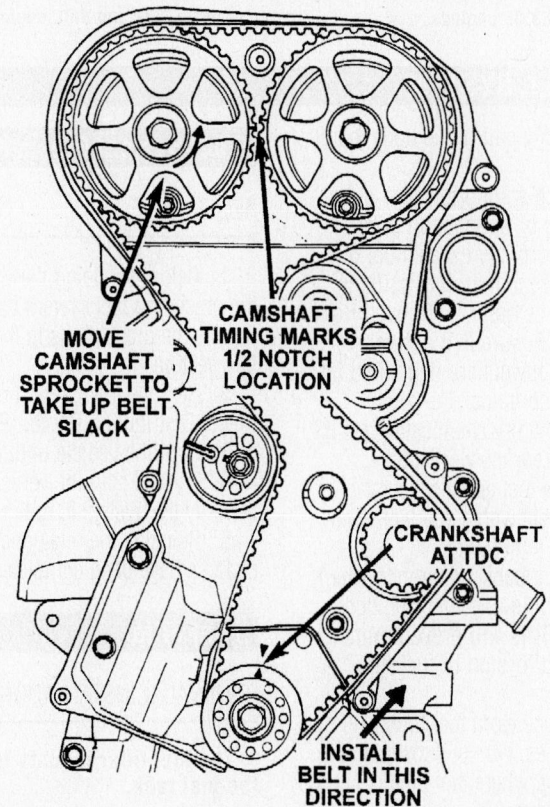

Installation of the timing belt, notice the camshaft alignment— 2.4L (VIN B) engine

79245G16

tighten the camshaft sprocket bolts to 75 ft. lbs. (101 Nm).

11. Set the crankshaft sprocket to Top Dead Center (TDC) by aligning the notch on the sprocket with the arrow on the oil pump housing.

12. Set the camshafts timing marks so the exhaust camshaft is ½ notch below intake camshaft sprocket. Make sure the arrows on both camshaft sprockets are facing up

13. Install the timing belt starting at the crankshaft, then around the water pump sprocket, idler pulley, camshaft sprockets and around the tensioner pulley.

14. Move the exhaust camshaft sprocket counterclockwise to take up the belt slack.

15. Insert a 6mm Allen wrench into the hexagon opening located on the top plate of the belt tensioner pulley. Rotate the top plate counterclockwise until there is no slack on the belt. The tensioner setting notch will start to move clockwise. Watch the notch and continue rotating the top plate counterclockwise until the setting notch is aligned with the spring tang. Using the Allen wrench to prevent the top plate from moving, tighten the tensioner lock bolt to 220 inch lbs. (25 Nm). The setting notch and spring tang should remain aligned after the lock nut is tightened.

16. Rotate the crankshaft 2 revolutions and recheck the timing marks.

17. Check the spring tang is within the tolerance window. If not within the window, reinsert the Allen wrench and into the hexagon opening located on the top plate of the belt tensioner pulley. Rotate the top plate counterclockwise until there is no slack on the belt. The tensioner setting notch will start to move clockwise. Watch the notch and continue rotating the top plate counterclockwise until the setting notch is aligned with the spring tang. Using the Allen wrench to prevent the top plate from moving, tighten the tensioner lock bolt to 220 inch lbs. (25 Nm). The setting notch and spring tang should remain aligned after the lock nut is tightened.

18. Install the engine mount bracket.

19. Install the front timing belt covers.

20. Install the A/C compressor/alternator tensioner and pulley assembly.

21. Install the right engine mount.

22. Install the crankshaft damper

23. Install the accessory drive belts and adjust to the proper tension.

24. Install the right inner splash shield and wheel.

25. Reconnect the negative battery cable.

26. Perform the crankshaft and camshaft relearn alignment procedure using the DRB scan tool, or equivalent.

Piston and Ring

POSITIONING

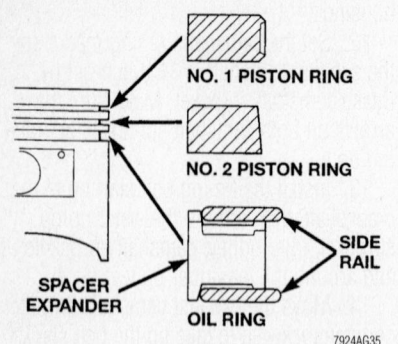

Piston ring positioning—2.4L engine

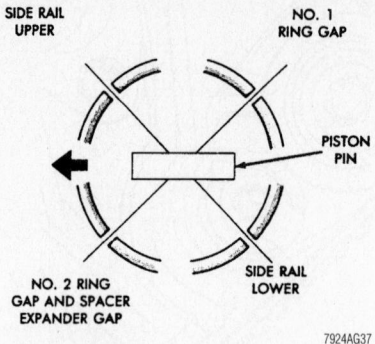

Piston ring end-gap spacing—3.3L and 3.8L engines

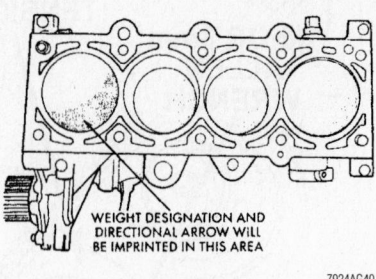

Piston-to-engine positioning mark locations—2.4L engine

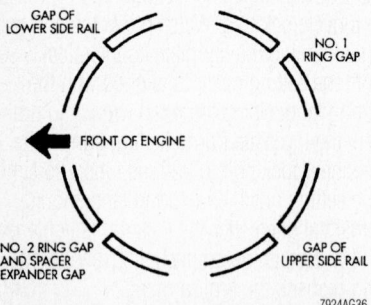

Piston ring end-gap spacing—2.4L engine

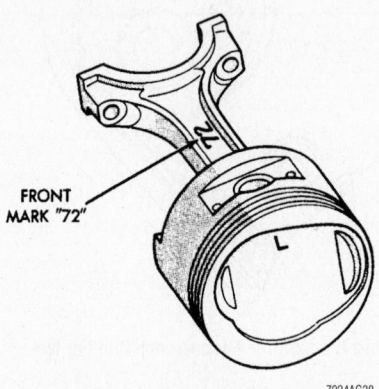

Piston and connecting rod front mark locations—3.3L and 3.8L engines

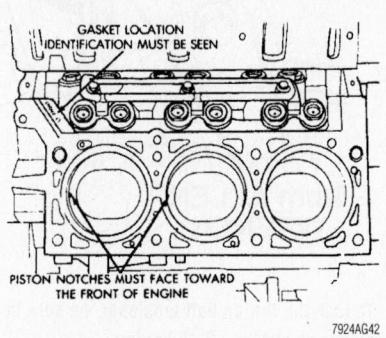

Piston-to-engine positioning mark locations—3.3L and 3.8L engines

FUEL SYSTEM

Fuel System Service Precautions

Safety is the most important factor when performing not only fuel system maintenance but any type of maintenance. Failure to conduct maintenance and repairs in a safe manner may result in serious personal injury or death. Maintenance and testing of the vehicle's fuel system components can be accomplished safely and effectively by adhering to the following rules and guidelines.

• To avoid the possibility of fire and personal injury, always disconnect the negative battery cable unless the repair or test procedure requires that battery voltage be applied.

• Always relieve the fuel system pressure prior to disconnecting any fuel system component (injector, fuel rail, pressure regulator, etc.), fitting or fuel line connection. Exercise extreme caution whenever relieving fuel system pressure to avoid exposing skin, face and eyes to fuel spray. Please be advised that fuel under pressure may penetrate the skin or any part of the body that it contacts.

• Always place a shop towel or cloth around the fitting or connection prior to loosening to absorb any excess fuel due to spillage. Ensure that all fuel spillage (should it occur) is quickly removed from engine surfaces. Ensure that all fuel soaked cloths or towels are deposited into a suitable waste container.

• Always keep a dry chemical (Class B) fire extinguisher near the work area.

• Do not allow fuel spray or fuel vapors to come into contact with a spark or open flame.

• Always use a back-up wrench when loosening and tightening fuel line connection fittings. This will prevent unnecessary stress and torsion to fuel line piping.

• Always replace worn fuel fitting O-rings with new ones. Do not substitute fuel hose or equivalent, where fuel pipe is installed.

Fuel System Pressure

RELIEVING

1. Before servicing the vehicle, refer to the preceding fuel system precautions, as well as the precautions in the beginning of this section.

2. Remove the fuel pump relay from the Power Distribution Center (PDC).

3. Start the vehicle until it stalls.

4. Try and start the vehicle several times to verify the system is fully relieved.

5. Reinstall the relay once all repair procedures have been completed.

Fuel Filter

REMOVAL & INSTALLATION

➡The fuel filter mounts to the top of the fuel tank.

1. Before servicing the vehicle, refer to the precautions in the beginning of this section.

2. Relieve the fuel system pressure.
3. Remove or disconnect the following:
   • Negative battery cable
4. Raise the vehicle and support the fuel tank with a transmission jack.
   • Fuel line from the front of the tank
   • Ground strap
   • Inboard side of the fuel tank straps and front T strap fastener and lower the tank about 6 inches
   • Fuel lines from the fuel pump module
   • Fuel filter

To install:

5. Installation is the reverse of removal. Tighten the fuel filter bolts to 40 inch lbs. (4.5 Nm), the main tank straps to 40 ft. lbs. (54 Nm) and the T strap to 250 inch lbs. (28 Nm).
6. Start the engine and check for leaks.

Fuel Pump

REMOVAL & INSTALLATION

1. Before servicing the vehicle, refer to the precautions in the beginning of this section.
2. Relieve the fuel system pressure.
3. Drain the fuel tank.
4. Remove or disconnect the following:
   • Negative battery cable
   • Fuel tank straps. Support the fuel tank before loosening the strap bolts.
   • Fuel lines
   • Fuel pump module harness connector
5. Lower the tank for access and remove the fuel pump module locking ring and the fuel pump module.

To install:

6. Install or connect the following:
   • Fuel pump module. Tighten the locking ring to 40 ft. lbs. (54 Nm).
   • Fuel pump module harness connector
   • Fuel lines
   • Fuel tank straps. Tighten the main tank straps to 40 ft. lbs. (54 Nm) and the T strap to 250 inch lbs. (28 Nm).
   • Negative battery cable
7. Start the engine and check for leaks.

Fuel Injector

REMOVAL & INSTALLATION

2.4L Engine

1. Before servicing the vehicle, refer to the precautions in the beginning of this section.
2. Relieve the fuel system pressure.
3. Remove or disconnect the following:
   • Negative battery cable
   • Wiring connectors for the injector harness
   • Wiring harness from the brackets
   • Fuel injector harness connectors
   • Harness from the vehicle
   • Quick connect fuel hose fittings from the chassis tube
   • Fuel rail bolts
   • Fuel rail with injectors attached
4. Rotate and pull the injectors to separate them from the fuel supply manifold.

To install:

5. Install or connect the following:
   • Injectors with new O-ring seals

   • Fuel rail with injectors attached. Tighten the bolts to 16 ft. lbs. (22 Nm).
   • Quick connect fuel hose fittings to the chassis tube
   • Harness from the vehicle
   • Fuel injector harness connectors
   • Wiring harness from the brackets
   • Wiring connectors to the injector harness
   • Negative battery cable
6. Start the engine, check for leaks and repair if necessary.

3.3L and 3.8L Engines

1. Before servicing the vehicle, refer to the precautions in the beginning of this section.
2. Relieve the fuel system pressure.
3. Remove or disconnect the following:
   • Negative battery cable
   • Upper intake manifold
   • Quick connect fuel hose fittings from the chassis tube
   • Fuel rail bolts
   • Fuel rail with injectors attached
4. Rotate and pull the injectors to separate them from the fuel supply manifold.

To install:

5. Install or connect the following:
   • Injectors with new O-ring seals
   • Fuel rail with injectors attached. Tighten the bolts to 16 ft. lbs. (22 Nm).
   • Quick connect fuel hose fittings to the chassis tube
   • Upper intake manifold
   • Negative battery cable
6. Start the engine, check for leaks and repair if necessary.

DRIVE TRAIN

Manual Transaxle Assembly

REMOVAL & INSTALLATION

1. Before servicing the vehicle, refer to the precautions in the beginning of this section.
2. Drain the transaxle fluid.
3. Remove or disconnect the following:
   • Negative battery cable
   • Gearshift cables from the shift levers cover assembly
   • Gearshift cable from the mounting bracket
   • 3 right engine mount bracket-to-transaxle bolts
   • Front wheels

   • Axle shafts
   • Front harness retainer and move the harness aside
   • Clutch quick connect fitting from the slave cylinder tube using tool 6638A by depressing towards the case and turning counter-clockwise 60 degrees while lifting the anti-rotation tab out of the case slot using a suitable pry-tool
   • Engine left mount bracket
   • Starter
   • Back-up lamp switch connector
   • Structural collar
   • Modular clutch-to-drive plate bolts
4. Support the engine with a screw jack and wood block.

   • Upper mount through bolt from the left frame rail
5. Lower the engine/transaxle assembly.
   • 4 upper mount-to-transaxle bolts and the mount
6. Attach a transmission jack to the transaxle.
   • Transaxle

To install:

7. Install or connect the following:
   • Transaxle and tighten the bolts to 70 ft. lbs. (95 Nm)
8. Raise the engine/transaxle assembly.
   • Upper mount through bolt to the left frame rail and tighten to 55 ft. lbs. (75 Nm)

- Modular clutch-to-drive plate bolts and tighten to 65 ft. lbs. (88 Nm)
- Structural collar
- Back-up lamp switch connector
- Starter
- Front mount bracket and tighten the bolts and nut to 50 ft. lbs. (68 Nm)
- Slave cylinder making sure to note the location of the different size lugs. While pressing inward, rotate the slave cylinder 60 degrees clockwise until the nylon locating tab rests inside the case cutout and the hydraulic tube is vertical. Attach the quick connect fitting, a click should be heard when it is properly attached.
- Front harness
- Axle shafts
- Front wheels
- 3 right engine mount bracket-to-transaxle bolts
- Gearshift cable
- Gearshift cables to the shift levers cover assembly
- Negative battery cable

9. Fill the transaxle with the correct type and amount of fluid

10. Start the engine and check for proper operation.

Automatic Transaxle Assembly

REMOVAL & INSTALLATION

31TH Transaxle

1. Before servicing the vehicle, refer to the precautions in the beginning of this section.

2. Attach a support fixture to the engine lifting eyes.

3. Remove or disconnect the following:
- Negative battery cable
- Dipstick tube and plug the opening
- Torque converter clutch harness connector
- Kickdown cable
- Shift cable
- Gear position switch connector
- Back-up lamp switch connector
- 2 upper transaxle bolts
- 3 rear mount bracket-to-case bolts
- Front wheels
- Axle halfshafts
- Starter motor
- Transaxle fluid cooler lines. Cut the lines flush with the fittings. A service kit will be used upon installation.

- Structural collar
- Vehicle Speed (VSS) sensor connector
- Rear mount shield
- Rear mount through bolt

4. Support the engine with a screw jack and wood block.
- Cradle plate
- Torque converter bolts
- Left wheel splash shield
- Left upper mount through bolt

5. Lower the engine/transaxle assembly and attach a transmission jack to the transaxle.
- Remaining transaxle bolts
- Transaxle

To install:

6. Install or connect the following:
- Transaxle. Torque the flange bolts to 70 ft. lbs. (95 Nm).

7. Raise the transaxle/engine assembly into position using a screw jack and a wood block.
- Left mount bolt. Torque the through bolt to 55 ft. lbs. (75 Nm).
- Left wheel splash shield
- Torque converter bolts and tighten to 65 ft. lbs. (88 Nm)
- Cradle plate
- Rear mount bracket-to-transaxle and hand tighten the bolts. Align the rear mount bracket to the mount and install the through bolt. Hand tighten the bolt.
- VSS sensor connector
- Structural collar
- Transaxle fluid cooler lines. Follow the instruction with service kit to install.
- Starter motor
- Front mount and bracket. Torque the large bracket bolts to 80 ft. lbs. (108 Nm), the small bracket bolts to 40 ft. lbs. (54 Nm) and the through bolt to 45 ft. lbs. (61 Nm).
- Axle halfshafts
- Rear mount bracket vertical bolts to 75 ft. lbs. (102 Nm)
- Back-up lamp switch connector
- Gear position switch connector
- Shift cable
- Kickdown cable
- Torque converter clutch harness connector
- Dipstick tube
- Tighten the rear mount bract horizontal bolt to 75 ft. lbs. (102 Nm)
- Tighten the rear mount through bolt to 40 ft. lbs. (54 Nm)
- Negative battery cable

8. Fill the transaxle with the correct type and amount of fluid

9. Start the engine and check for proper operation.

41TE Transaxle

1. Before servicing the vehicle, refer to the precautions in the beginning of this section.

2. Attach a support fixture to the engine lifting eyes.

3. Remove or disconnect the following:
- Battery cables
- Battery shield
- Coolant recovery bottle
- Dipstick tube and plug the opening
- Torque converter clutch harness connector
- Transaxle fluid cooler lines. Cut the lines flush with the fittings. A service kit will be used upon installation.
- Input and output shaft sensor connectors
- Transmission Range Sensor (TRS) connector
- Solenoid/pressure switch connector
- Shift cable
- Crankshaft Position (CKP) sensor from the bell housing, if equipped
- Position the leak detection pump harness and hoses aside
- Rear mount bracket-to-case bolts
- Upper transaxle bolts
- Transaxle pan
- Front wheels
- Axle halfshafts
- Power Transfer Unit (PTO), if equipped
- Rear mount bracket-to-case lower bolt
- Front mount/bracket assembly
- Starter motor
- Lateral bending brace
- Inspection cover
- Torque converter bolts

4. Support the engine with a screw jack and wood block.
- Left wheel splash shield
- Left upper mount through bolt

5. Lower the engine/transaxle assembly and attach a transmission jack to the transaxle.
- Upper mount bracket from the transaxle
- Remaining transaxle bolts
- Transaxle

To install:

6. Install or connect the following:
- Transaxle. Torque the flange bolts to 70 ft. lbs. (95 Nm).

- Upper mount assembly and tighten the bolts to 40 ft. lbs. (54 Nm)

7. Raise the transaxle/engine assembly into position using a screw jack and a wood block.

- Left upper mount through bolt and tighten to 55 ft. lbs. (75 Nm)
- Left wheel splash shield
- Torque converter bolts and tighten to 65 ft. lbs. (88 Nm)
- Inspection cover
- Lateral bending brace
- Starter motor
- Front mount/bracket assembly
- Rear mount bracket-to-case lower bolt. Align the assembly, hand tighten the bolt first, then tighten to 75 ft. lbs. (102 Nm) once the assembly is aligned.
- PTO, if equipped
- Axle halfshafts
- Front wheels
- Transaxle pan
- Rear mount bracket-to-case vertical bolts to 75 ft. lbs. (102 Nm)
- Upper transaxle bolts to 70 ft. lbs. (95 Nm)
- CKP sensor to the bell housing, if equipped
- Shift cable
- Attach the leak detection pump harness and hoses
- Solenoid/pressure switch connector
- TRS connector
- Input and output shaft sensor connectors
- Transaxle fluid cooler lines. Attach using the service kit supplied following the kit instructions.
- Torque converter clutch harness connector
- Dipstick tube
- Coolant recovery bottle
- Battery shield
- Battery cables

8. Fill the transaxle with the correct type and amount of fluid

9. Start the engine and check for proper operation.

Modular Clutch Assembly

REMOVAL & INSTALLATION

1. Before servicing the vehicle, refer to the precautions in the beginning of this section.
2. Remove or disconnect the following:
   - Transaxle
   - Clutch assembly from the input shaft
3. Installation is the reverse of removal

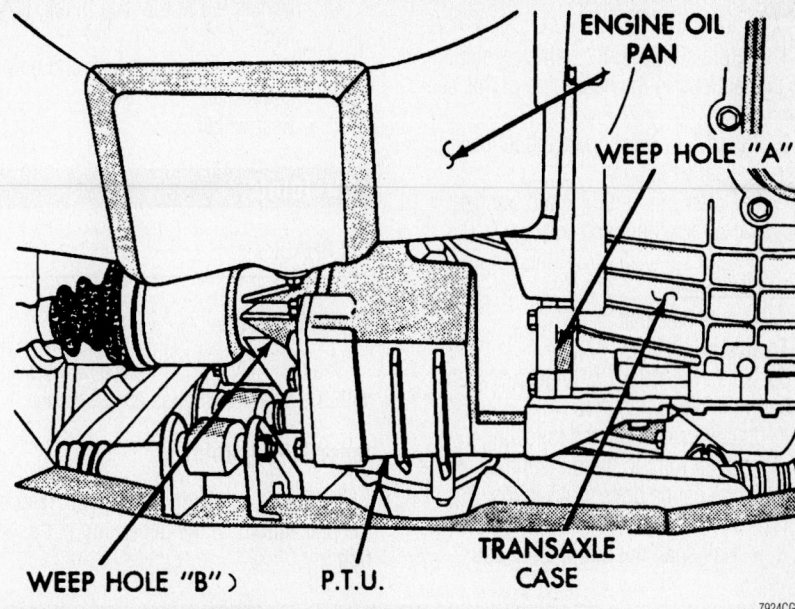

Power transfer unit and related components

Power Transfer Unit

REMOVAL & INSTALLATION

1. Before servicing the vehicle, refer to the precautions in the beginning of this section.
2. Remove or disconnect the following:

- Right front wheel
- Right axle halfshaft
- Propeller shaft
- Cradle plate
- Power transfer unit brackets
- Power transfer unit

To install:
3. Install or connect the following:
- Power transfer unit and brackets. Tighten the bolts to 37 ft. lbs. (50 Nm).
- Rear driveshaft
- Cradle plate. Tighten the bolts to 123 ft. lbs. (166 Nm).
- Right axle halfshaft
- Right front wheel

Halfshaft

REMOVAL & INSTALLATION

Front

1. Before servicing the vehicle, refer to the precautions in the beginning of this section.
2. Remove or disconnect the following:

- Front wheel
- Split pin
- Nut lock
- Spring washer
- Hub nut
- Brake caliper and rotor
- Outer tie rod end
- Wheel speed sensor harness, if equipped
- Lower ball joint

3. Separate the outer CV-joint stub shaft from the steering knuckle.
4. Pry the inner tri-pot joint out of the transaxle and remove the axle halfshaft.

To install:
5. Install the axle halfshaft so that the inner joint circlip seats in the transaxle side gear.
6. Guide the outer CV-joint stub shaft through the steering knuckle hub.
7. Install or connect the following:
- Lower ball joint. Torque the nut to 60 ft. lbs. (81 Nm) plus a 90 degree turn.
- Wheel speed sensor harness, if equipped
- Outer tie rod end. Torque the nut to 55 ft. lbs. (75 Nm).
- Brake caliper and rotor. Torque the caliper bolts to 35 ft. lbs. (47 Nm).
- Hub nut. Torque the nut to 180 ft. lbs. (245 Nm).
- Spring washer
- Nut lock
- Split pin
- Front wheel

8. Check the wheel alignment and adjust as necessary.

Rear

1. Before servicing the vehicle, refer to the precautions in the beginning of this section.
2. Remove or disconnect the following:
   - Rear wheel
   - Cotter pin, nut lock and washer
   - Half shaft nut and washer
   - Rear halfshafts from the output flanges
   - Axle halfshaft

To install:

3. Guide the outer CV-joint stub shaft through the rear wheel hub.
4. Install or connect the following:
   - Inner joint to the differential. Torque the flange bolts to 45 ft. lbs. (61 Nm).
   - Half shaft nut and washer and tighten the nut to 180 ft. lbs. (244 Nm)
   - Waver washer, nut lock and new cotter pin
   - Rear wheel

CV-Joints

OVERHAUL

Outer CV-Joint

The outer CV-joint and boot are serviced with the axle halfshaft as an assembly.

Inner Tri-pot Joint

1. Before servicing the vehicle, refer to the precautions in the beginning of this section.
2. Remove or disconnect the following:
   - Negative battery cable
   - Axle halfshaft from the vehicle
   - Inner tri-pot joint boot clamps
   - Tri-pot joint housing
   - Snapring
   - Tri-pot joint

To install:

➡**Use new snaprings, clips, and boot clamps for assembly.**

3. Install or connect the following:
   - Tri-pot joint
   - Snapring
   - Tri-pot joint housing
4. Fill the tri-pot joint housing and boot with grease and tighten the boot clamps.
5. Install the axle halfshaft.
6. Connect the negative battery cable

STEERING AND SUSPENSION

Air Bag

❊❊ CAUTION

Some vehicles are equipped with an air bag system. The system must be disarmed before performing service on, or around, system components, the steering column, instrument panel components, wiring and sensors. Failure to follow the safety precautions and the disarming procedure could result in accidental air bag deployment, possible injury and unnecessary system repairs.

PRECAUTIONS

Several precautions must be observed when handling the inflator module to avoid accidental deployment and possible personal injury.

- Never carry the inflator module by the wires or connector on the underside of the module.
- When carrying a live inflator module, hold securely with both hands, and ensure that the bag and trim cover are pointed away from your body. In the unlikely event of an accidental deployment, the bag will then deploy with minimal chance of injury.
- Place the inflator module on a bench or other surface with the bag and trim cover facing up. This will reduce the motion of the module if accidentally deployed.
- With the inflator module on the bench, never place anything on or close to the module which may be thrown in the event of an accidental deployment.

DISARMING

1. Disconnect and isolate the negative battery cable from the battery.
2. Allow the SIR system capacitor to discharge for at least two (2) minutes, before performing any repairs.
3. When repairs are completed, connect the negative battery cable.

Power Rack and Pinion Steering Gear

REMOVAL & INSTALLATION

1. Before servicing the vehicle, refer to the precautions in the beginning of this section.
2. Lock the steering wheel to the straight ahead position.
3. Drain the power steering fluid from the reservoir.
4. Remove or disconnect the following:
   - Negative battery cable
   - Front wheels
   - Steering column shaft coupler
   - Front emissions vapor canister, if equipped
   - Single hose at the power steering cooler and let the fluid drain into a suitable container
   - Power steering cooler-to-cradle bolts, if equipped
   - Leak detection pump-to-cradle crossmember bolts, if equipped
   - Outer tie rod ends

➡**The bolts retaining the cradle crossmember are different sizes, makes sure to note the locations of the bolts and their sizes prior to removal.**

   - Lower control arm rear bushing bolts
   - Crossmember reinforcement
   - Power transfer unit, if equipped with AWD
   - Power steering pressure and return lines
   - Steering gear mounting fasteners
   - Intermediate shaft coupler
   - Power steering gear

To install:

5. Install or connect the following:
   - Power steering gear
   - Intermediate shaft coupler. Start the roll pin into the coupler before installing the coupler, start the roll pin and then user a hammer to tap it into the coupler. Install the coupler on the shaft of the gear. Install removal/installer tool 6831A through the center of the roll pin, using the knurled nut to secure it. Hold the threaded rod stationary while turning the nut to pull the pin into the coupler.
   - Steering gear mounting fasteners. Torque all bolts on 2001 models to 135 ft. lbs. (183 Nm). On all other models tighten the 14mm fasteners to 135 ft. lbs. (183 Nm) and the 12mm fastener to 70 ft. lbs. (95 Nm).
   - Power steering pressure and return lines. Torque the line to 25 ft. lbs. (31 Nm).

- Power transfer unit, if equipped with AWD
- Crossmember reinforcement. Tighten the M14 bolts to 113 ft. lbs. (153 Nm) and the M12 bolts to 78 ft. lbs. (106 Nm).
- Lower control arm rear bushing

bolts. Torque the bolts to 45 ft. lbs. (61 Nm).
- Outer tie rod ends to the steering knuckle. Torque the nut to 55 ft. lbs. (75 Nm).
- Leak detection pump-to-cradle crossmember bolts, if equipped

- Power steering cooler-to-cradle bolts, if equipped
- Single hose at the power steering cooler and fasten the clamp
- Front emissions vapor canister, if equipped
- Front wheels

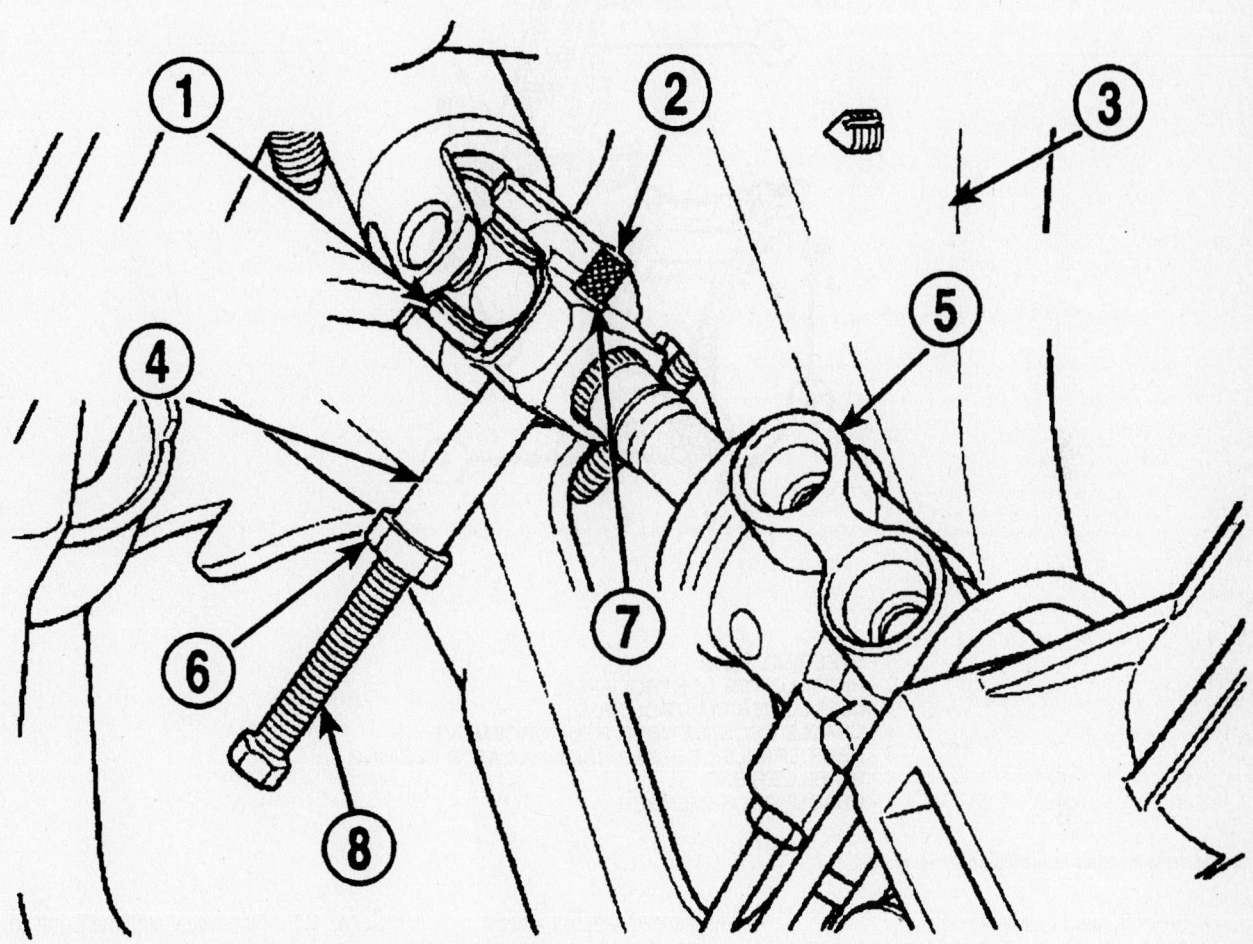

1 - INTERMEDIATE COUPLER

2 - KNURLED NUT

3 - SUSPENSION CRADLE

4 - REMOVER SPECIAL TOOL 6831A

5 - STEERING GEAR

6 - NUT

7 - ROLL PIN

8 - THREADED ROD

Installing the intermediate shaft coupler to the gear

67189-MINV-G02

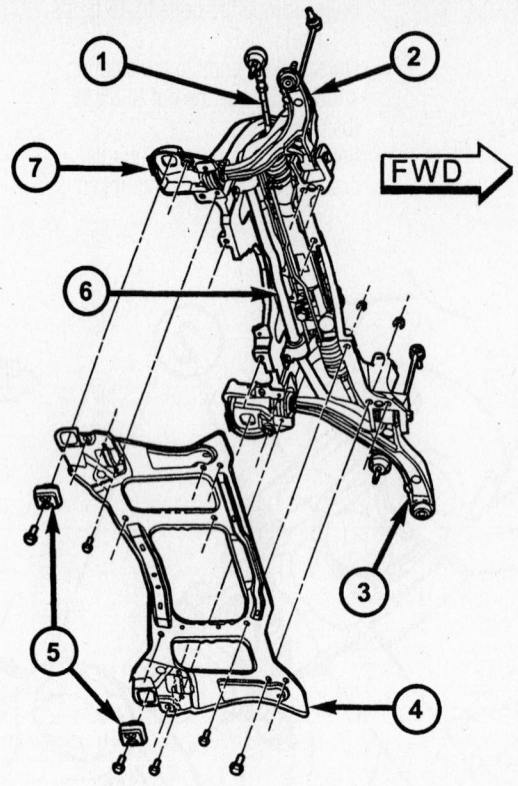

1 - STEERING GEAR
2 - RIGHT LOWER CONTROL ARM
3 - LEFT LOWER CONTROL ARM
4 - CRADLE CROSSMEMBER REINFORCEMENT
5 - REAR CRADLE CROSSMEMBER ISOLATOR BUSHING
6 - STABILIZER BAR
7 - CRADLE CROSSMEMBER

67189-MINV-G01

Cradle crossmember reinforcement mounting

- Steering column shaft coupler. Tighten the pinch bolt to 21 ft. lbs. (28 Nm).
- Negative battery cable

6. Fill and bleed the power steering reservoir

7. Inspect the power steering system for leaks and repair if necessary.

8. Check the wheel alignment and adjust as necessary.

Strut

REMOVAL & INSTALLATION

1. Before servicing the vehicle, refer to the precautions in the beginning of this section.

2. Remove or disconnect the following:
- Front wheel
- Brake hose bracket

- Wheel speed sensor harness bracket, if equipped
- Stabilizer bar link
- Steering knuckle bracket bolts
- Upper strut mount nuts
- Strut assembly

To install:
3. Install or connect the following:
- Strut assembly. Torque the upper strut mount nuts to 21 ft. lbs. (28 Nm).
- Steering knuckle bracket bolts. Torque the bolts to 60 ft. lbs. (81 Nm) plus 90 degrees.
- Stabilizer bar link. Torque the nut to 65 ft. lbs. (88 Nm).
- Wheel speed sensor harness bracket, if equipped. Torque the bolt to 10 ft. lbs. (13 Nm).
- Brake hose bracket. Tighten the bolt to 10 ft. lbs. (13 Nm).
- Front wheel

4. Check the wheel alignment and adjust as necessary.

Shock Absorber

REMOVAL & INSTALLATION

Rear

1. Before servicing the vehicle, refer to the precautions in the beginning of this section.

2. Support the rear axle with a jackstand.

3. Remove the top and bottom shock absorber bolts.

4. Remove the shock absorber.

To install:
5. Install the shock absorber. Torque the mounting bolts to 75 ft. lbs. (101 Nm) on 2001–02 models or 65 ft. lbs. (88 Nm) on 2003–05 models.

6. Remove the jackstand.

Coil Spring

REMOVAL & INSTALLATION

Front

1. Before servicing the vehicle, refer to the precautions in the beginning of this section.
2. Remove the strut assembly from the vehicle.
3. Compress the coil spring and remove the piston rod nut.
4. Remove or disconnect the following:
   • Upper strut mount
   • Pivot bearing
   • Spring upper seat
   • Coil spring
   • Jounce bumper and dust shield
   • Lower spring isolator

To install:
5. Install or connect the following:
   • Lower spring isolator
   • Jounce bumper and dust shield
   • Coil spring
   • Spring upper seat
   • Pivot bearing
   • Upper strut mount. Torque the piston rod nut to 75 ft. lbs. (100 Nm).
6. Remove the spring compressor and install the strut assembly to the vehicle.

Leaf Springs

REMOVAL & INSTALLATION

1. Before servicing the vehicle, refer to the precautions in the beginning of this section.
2. Remove the axle halfshaft, if equipped.
3. Support the vehicle at the frame rail with jackstands.
4. Support the axle with a floor jack.
5. Remove or disconnect the following:
   • Shock absorber
   • Axle plate from the axle and spring
6. Slowly lower the axle so that the leaf spring hangs free.
7. Remove or disconnect the following:
   • Front spring mount
   • Rear spring shackle plate
   • Leaf spring

To install:
8. Install or connect the following:
   • Leaf spring
   • Rear spring shackle plate. Torque the nuts to 45 ft. lbs. (61 Nm).
   • Axle plate. Torque the bolts to 75 ft.

lbs. (101 Nm) on AWD models or 70 ft. lbs. (95 Nm) on FWD models.
   • Front spring mount. Torque the through bolt to 115 ft. lbs. (156 Nm) and the mount bolts to 45 ft. lbs. (61 Nm).
   • Shock absorber
   • Axle halfshaft, if equipped

Lower Ball Joint

REMOVAL & INSTALLATION

1. Before servicing the vehicle, refer to the precautions in the beginning of this section.
2. Remove or disconnect the following:
   • Front wheel
   • Lower control arm
   • Ball joint seal boot
3. Remove the ball joint from the control arm with a press.

To install:
4. Install the ball joint to the control arm with a press so that the ball joint flange contacts the control arm with no visible gaps.
5. Install or connect the following:
   • Ball joint seal boot
   • Lower control arm
   • Front wheel
6. Check the wheel alignment and adjust as necessary.

Lower Control Arm

REMOVAL & INSTALLATION

1. Before servicing the vehicle, refer to the precautions in the beginning of this section.
2. Remove or disconnect the following:
   • Negative battery cable
   • Front wheels
   • Power steering cooler

➡**The bolts retaining the cradle crossmember are different sizes, makes sure to note the locations of the bolts and their sizes prior to removal.**

   • Cradle plate
   • Lower ball joints
   • Rear control arm bushing retainers
3. Matchmark the suspension cradle and the frame rail.
4. Loosen the left suspension cradle bolts and lower the cradle to allow the pivot bolt to be removed.
5. Remove or disconnect the following:
   • Front pivot bolts
   • Lower control arms

To install:
➡**The suspension must be at curb height for final tightening of the control arm pivot bolts and the rear bushing retainer bolts.**

6. Install or connect the following:
   • Lower control arms
   • Front pivot bolts. Torque the M14 suspension cradle bolts to 120 ft. lbs. (163 Nm) and the M12 bolts to 80 ft. lbs. (108 Nm).
   • Rear control arm bushing retainers. Torque the bolts to 45 ft. lbs. (61 Nm).
   • Lower ball joints. Tighten the pinch bolts 105 ft. lbs. (145 Nm).
   • Cradle plate. Tighten the bolts to 120 ft. lbs. (163 Nm).
   • Power steering cooler and tighten the bolts to 100 inch lbs. (11 Nm)
   • Front wheels
   • Negative battery cable
7. Raise the suspension to curb height and tighten the pivot bolts to 135 ft. lbs. (183 Nm).
8. Check the wheel alignment and adjust as necessary.

CONTROL ARM BUSHING REPLACEMENT

1. Before servicing the vehicle, refer to the precautions in the beginning of this section.
2. Remove the control arm from the vehicle.
3. Press the front bushing out of the control arm.
4. Cut the rear bushing lengthwise and remove it from the control arm.

To install:
5. Press the front bushing into the control arm until the bushing flange contacts the control arm.
6. Lubricate the rear bushing with silicone spray lubricant and push it onto the control arm.
7. Install the control arm to the vehicle.

Wheel Bearings

ADJUSTMENT

The front and rear wheel bearings are designed for the life of the vehicle and require no type of adjustment or periodic maintenance. The bearing is a sealed unit with the wheel hub and can only be removed and/or replaced as an assembly.

REMOVAL & INSTALLATION

Front

1. Before servicing the vehicle, refer to the precautions in the beginning of this section.
2. Remove or disconnect the following:
   - Front wheel
   - Brake caliper and rotor
   - Hub retaining bolts
   - Hub and bearing assembly

To install:

3. Install or connect the following:
   - Hub and bearing assembly. Torque the bolts to 45 ft. lbs. (65 Nm).
   - Caliper and adapter to the brake rotor and align the assembly to the steering knuckle. Torque the mounting bolts to 125 ft. lbs. (169 Nm).
   - Hub retaining nut. Tighten the nut to 180 ft. lbs. (244 Nm).
   - Front wheel
4. Check and adjust the front end alignment, if necessary.

Rear

FRONT WHEEL DRIVE

1. Before servicing the vehicle, refer to the precautions in the beginning of this section.
2. Remove or disconnect the following:
   - Rear wheel
   - Brake drum or caliper
   - Wheel speed sensor, if equipped
   - Hub and bearing assembly from the rear axle

To install:

3. Install or connect the following:
   - Hub and bearing assembly. Tighten the bolts to 95 ft. lbs. (129 Nm).
   - Wheel speed sensor, if equipped. Tighten the bolt to 105 inch lbs. (12 Nm).
   - Brake drum or caliper
   - Rear wheel

ALL WHEEL DRIVE

1. Before servicing the vehicle, refer to the precautions in the beginning of this section.
2. Set the parking brake.
3. Remove or disconnect the following:
   - Rear wheel
   - Brake caliper and rotor
   - Mounting bolts from the driveshaft inner joint to the output shaft
   - Wheel speed sensor and release the parking brake
   - Axle halfshaft
   - Hub and bearing assembly

To install:

4. Install or connect the following:
   - Hub and bearing assembly. Torque the bolts to 95 ft. lbs. (129 Nm).
   - Axle halfshaft
   - Wheel speed sensor and set the parking brake. Torque the fastener to 105 inch lbs. (12 Nm).
   - Driveshaft inner joint to output shaft mounting bolts. Torque the bolts to 45 ft. lbs. (61 Nm).
   - Torque the outer CV Joint hub nut to 180 ft. lbs. (244 Nm).
   - Brake caliper and rotor
   - Rear wheel

BRAKES

Brake Caliper

REMOVAL & INSTALLATION

1. Before servicing the vehicle, refer to the precautions in the beginning of this section.
2. Raise and safely support the front of the vehicle. Remove the front wheels.

3. If the caliper is only being removed from the bracket (as for a brake pad change), move to Step 3. If the caliper is being removed from the vehicle (as for replacement or an overhaul and reseal), remove the brake hose attaching bolt from the caliper. Remove the hose from the caliper and discard the washers. New seal washers will be required at assembly. Plug the brake hose to prevent fluid leakage.

4. Remove the caliper guide pin bolts that secure the caliper to the steering knuckle.

5. Remove the caliper by slowly sliding it away from the steering knuckle. Slide the opposite end of the brake caliper out from under the machined abutment on the steering knuckle.

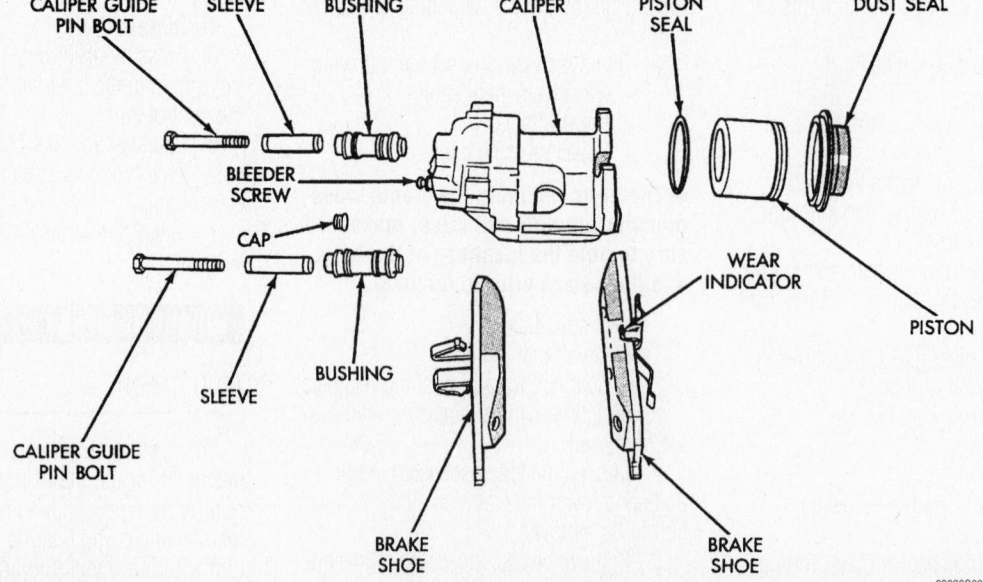

Front brake caliper assembly

93026G03

6. Using a strong piece of wire, support the brake caliper assembly off the strut unit. Do NOT allow the caliper to hang from the brake fluid flex hose or damage to the hose will result.

To install:

7. Clean both steering knuckle abutment surfaces of any dirt, grease or corrosion. Then lubricate the abutment surfaces with a liberal amount of MOPAR® Multipurpose Lubricant or equivalent.

8. Properly position the brake caliper over the brake pads and disc rotor. Be careful not to allow the caliper seals or guide pin bushings to get damaged by the steering knuckle bosses. Install the caliper guide pin bolts and torque to: 26 ft. lbs. (35 Nm). Be careful not to cross thread the guide pin bolts.

9. If removed, attach the brake hose to the caliper using new washers. Tighten the banjo bolt to 35 ft. lbs. (47 Nm).

10. Bleed the brake system.

11. Install the front wheels and lug nuts. Torque the lug nuts, in a star pattern sequence, to ½ torque specifications. Then repeat the tightening sequence to the full torque specification of 100 ft. lbs. (135 Nm). Lower the vehicle.

12. Pump the brake pedal several times to insure that the brake pedal is firm. Road-test the vehicle.

Disc Brake Pads

REMOVAL & INSTALLATION

1. Before servicing the vehicle, refer to the precautions in the beginning of this section.

2. Remove brake fluid from the master cylinder brake fluid reservoir until the reservoir is approximately ½ full. Discard the removed fluid.

3. Raise and safely support the front of the vehicle. Remove the front wheels.

4. Remove or disconnect the following:
   - Front brake caliper guide pin bolts
   - Brake caliper by slowly sliding it up and off the adapter and brake rotor. Support the caliper out of the way with a strong piece of wire. Do not let the caliper hang by the brake hose or damage to the brake hose will result.

5. If necessary, compress the caliper piston into the bore using a C-clamp. Insert a suitable piece of wood between the C-clamp and caliper piston to protect the piston.
   - Outboard disc brake pad from the

caliper by prying the brake pad retaining clip over the raised area on the caliper. Slide the brake pad down and off the caliper.
   - Inboard disc brake pad from the caliper by pulling the brake pad away from the caliper piston until the retaining clip on the pad is free from the caliper piston cavity

To install:

6. Be sure the caliper piston has been completely retracted into the piston bore of the caliper assembly. This is required when installing the brake caliper equipped with new brake pads.

7. If equipped, remove the protective paper from the noise suppression gaskets on the new disc brake pads.

8. Install or connect the following:
   - New inboard disc brake pad into the caliper piston by pressing the pad firmly into the cavity of the caliper piston. Be sure the new inboard brake pad is seated squarely against the face of the brake caliper piston.
   - Outboard disc brake pad by sliding it onto the caliper assembly
   - Brake caliper assembly over the brake rotor and onto the steering knuckle adapter
   - Caliper guide pin bolts and torque to: 35 ft. lbs. (47 Nm)
   - Front. Apply the brake pedal several times until a firm pedal is obtained.

9. Check the fluid level in the master cylinder and add fluid as necessary. Road-test the vehicle.

Brake Drums

REMOVAL & INSTALLATION

1. Before servicing the vehicle, refer to the precautions in the beginning of this section.

2. Raise and safely support the vehicle.

3. Remove or disconnect the following:
   - Rear wheels
   - Brake drum from the hub assembly by pulling the drum straight off the wheel studs

4. Inspect the brake drum for thickness and runout. Replace or machine as necessary.

To install:

5. Install the brake drum to the hub assembly.

6. Adjust the brake shoes.

7. Install the rear wheel.

Brake Shoes

REMOVAL & INSTALLATION

1. Before servicing the vehicle, refer to the precautions in the beginning of this section.

2. Raise and safely support the vehicle.

3. Remove or disconnect the following:
   - Rear wheels and the brake drums

4. Be sure the parking brake pedal is in the released position. Create slack in the rear parking brake cables by grasping an exposed section of the front parking brake cable, pulling it down and rearward. Maintain the slack in the brake cable by clamping a pair of locking pliers onto the parking brake cable just rearward of **the rear** body outrigger bracket.
   - Adjustment lever spring from the automatic adjustment lever and front brake shoe (leading brake shoe)
   - Automatic adjustment lever from the front brake shoe (leading brake shoe)
   - Brake shoe-to-brake shoe lower return spring
   - Tension clip that secures the upper return spring to the automatic adjuster assembly
   - Brake shoe-to-brake shoe upper return spring
   - Rear brake shoe (trailing brake shoe) hold-down clip and pin
   - Trailing brake shoe, parking brake actuating lever and parking brake actuator strut from the brake support plate
   - Automatic adjuster assembly from the leading brake shoe
   - Leading brake shoe hold-down clip and pin
   - Leading brake shoe
   - Parking brake actuator plate from the leading brake shoe and install onto the replacement brake shoe

To install:

5. Thoroughly clean and dry the backing plate. To prepare the backing plate, lubricate the 8 brake shoe contact areas and brake shoe anchor, using suitable grease.

6. Install or connect the following:
   - Leading brake shoe into position on the brake shoe support plate. Secure the leading brake shoe by installing the brake shoe hold-down clip and pin.
   - Parking brake actuating strut onto the leading brake shoe and then

install the parking brake actuating lever onto the strut

7. Lubricate the shaft threads of the automatic adjuster screw assembly with anti-seize lubricant.

- Automatic adjuster screw assembly onto the leading brake shoe
- Trailing brake shoe onto the parking brake actuating lever and parking brake actuating strut
- Trailing brake shoe into position on the brake support plate and install the brake shoe hold-down clip and pin
- Brake shoe-to-brake shoe upper return spring

- Tension clip that secures the upper return spring to the automatic adjuster assembly. Be sure the tension clip is positioned on the threaded area of the adjuster assembly or the function of the automatic adjuster will be affected.
- Brake shoe-to-brake shoe lower return spring
- Automatic adjustment lever onto the leading brake shoe
- Actuating spring onto the automatic adjustment lever and leading brake shoe. Make sure the automatic adjustment lever makes positive contact with the star wheel.

8. Once the brake shoes and all other brake system components are fully and correctly installed, remove the locking pliers from the front parking brake cable. This will remove the slack and correctly adjust the parking brake cables.

9. Make sure there is no grease on the brake shoe linings, then install the brake drums.

10. Adjust the rear brakes, then lower the vehicle and check the brakes for proper operation.

11. Install the wheels.

12. Road-test the vehicle. The automatic adjuster will continue to adjust the brake shoes during the road-test.

JEEP

Cherokee • Grand Cherokee • Liberty • Wrangler

3

SPECIFICATION CHARTS

ENGINE AND VEHICLE IDENTIFICATION

	Engine						Model Year	
Code ①	Liters (cc)	Cu. In.	Cyl.	Fuel Sys.	Engine Type	Eng. Mfg.	Code ②	Year
N	4.7 (4701)	287	8	MFI	SOHC	Chrysler	1	2001
P	2.5 (2458)	150	4	MFI	OHV	Chrysler	2	2002
S	4.0 (3966)	242	6	MFI	OHV	Chrysler	3	2003
K	3.7 (3701)	226	6	MFI	SOHC	Chrysler	4	2004
1	2.4 (2429)	148	4	MFI	DOHC	Chrysler	5	2005

MFI: Multi-port Fuel Injection

OHV: Over Head Valve

SOHC: Single Overhead Camshaft

① 8th position of VIN

② 10th position of VIN

67189-JEEP-C01

GENERAL ENGINE SPECIFICATIONS

Year	Model	Engine Displ. Liters	Engine VIN	Net Horsepower @ rpm	Net Torque @ rpm (ft. lbs.)	Bore x Stroke (in.)	Comp. Ratio	Oil Pressure @ rpm
2001	Cherokee	2.5	P	125@5400	149@3250	3.88x3.19	9.1:1	37@1600
		4.0	S	190@4750	225@4000	3.88x3.44	8.8:1	37@1600
	Grand Cherokee	4.0	S	195@4600	230@3000	3.88x3.44	8.8:1	37@1600
		4.7	N	235@4800	295@3200	3.66x3.40	9.3:1	25@3000
	Wrangler	2.5	P	120@5400	139@3500	3.88x3.19	9.1:1	37@1600
		4.0	S	181@4600	222@2800	3.88x3.44	8.8:1	37@1600
2002	Grand Cherokee	4.0	S	195@4600	230@3000	3.88x3.44	8.8:1	37@1600
		4.7	N	235@4800	295@3200	3.66x3.40	9.3:1	25@3000
	Wrangler	2.5	P	120@5400	139@3500	3.88x3.19	9.1:1	37@1600
		4.0	S	181@4600	222@2800	3.88x3.44	8.8:1	37@1600
	Liberty	2.4	1	150@5200	165@4000	3.44x3.98	9.5:1	25-80@3000
		3.7	K	210@5200	225@4200	3.66x3.40	9.1:1	25-110@3000
2003	Grand Cherokee	4.0	S	195@4600	230@3000	3.88x3.44	8.8:1	37@1600
		4.7	N	235@4800	295@3200	3.66x3.40	9.3:1	25@3000
	Wrangler	2.4	1	150@5200	165@4000	3.44x3.98	9.5:1	25-80@3000
		4.0	S	181@4600	222@2800	3.88x3.44	8.8:1	37@1600
	Liberty	2.4	1	150@5200	165@4000	3.44x3.98	9.5:1	25-80@3000
		3.7	K	210@5200	225@4200	3.66x3.40	9.1:1	25-110@3000
2004	Grand Cherokee	4.0	S	195@4600	230@3000	3.88x3.44	8.8:1	37@1600
		4.7	N	235@4800	295@3200	3.66x3.40	9.3:1	25@3000
	Wrangler	2.4	1	150@5200	165@4000	3.44x3.98	9.5:1	25-80@3000
		4.0	S	181@4600	222@2800	3.88x3.44	8.8:1	37@1600
	Liberty	2.4	1	150@5200	165@4000	3.44x3.98	9.5:1	25-80@3000
		3.7	K	210@5200	225@4200	3.66x3.40	9.1:1	25-110@3000

MFI: Multi Port Fuel Injection

67189-JEEP-C02

ENGINE TUNE-UP SPECIFICATIONS

Year	Engine Displ. Liters	Engine VIN	Spark Plug Gap (in.)	Ignition Timing (deg.)	Fuel Pump (psi)	Idle Speed (rpm)	Valve Clearance	
							Intake	Exhaust
2001	2.5	P	0.035	①	47-51	①	HYD	HYD
	4.0	S	0.035	①	47-51	①	HYD	HYD
	4.7	N	0.040	①	47-51	①	HYD	HYD
2002	2.4	1	0.050	①	44-54	①	HYD	HYD
	2.5	P	0.035	①	47-51	①	HYD	HYD
	3.7	K	0.042	①	44-54	①	HYD	HYD
	4.0	S	0.035	①	47-51	①	HYD	HYD
	4.7	N	0.040	①	47-51	①	HYD	HYD
2003	2.4	1	0.050	①	44-54	①	HYD	HYD
	3.7	K	0.042	①	44-54	①	HYD	HYD
	4.0	S	0.035	①	47-51	①	HYD	HYD
	4.7	N	0.040	①	47-51	①	HYD	HYD
2004	2.4	1	0.050	①	44-54	①	HYD	HYD
	3.7	K	0.042	①	44-54	①	HYD	HYD
	4.0	S	0.040	①	47-51	①	HYD	HYD
	4.7	N	0.040	①	47-51	①	HYD	HYD

Note: The information on the Vehicle Emission Control labe must be used, if different from the figures in this chart.

HYD: Hydraulic

① Ignition timing and idle speed are controlled by the PCM. No adjustment is necessary.

67189-JEEP-C03

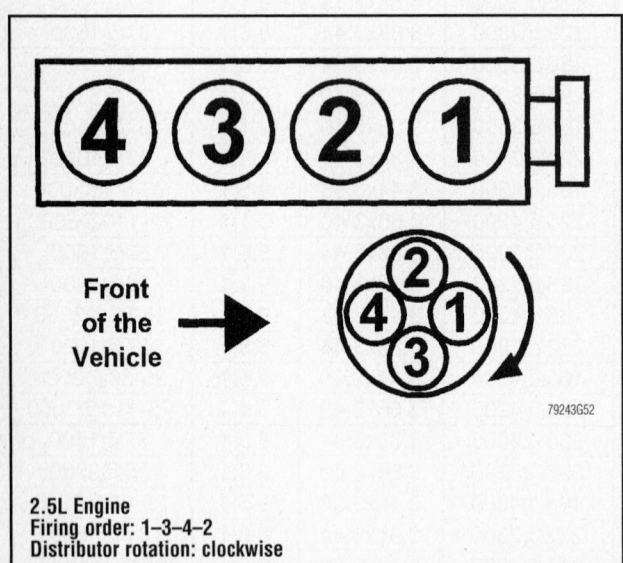

2.5L Engine
Firing order: 1–3–4–2
Distributor rotation: clockwise

79243G52

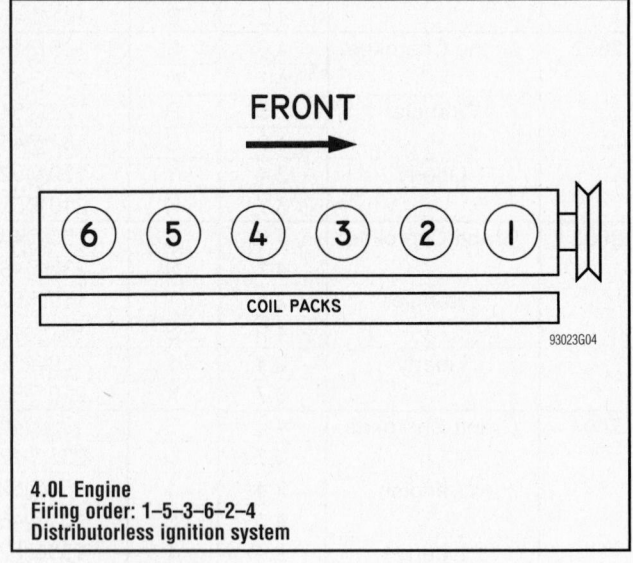

4.0L Engine
Firing order: 1–5–3–6–2–4
Distributorless ignition system

93023G04

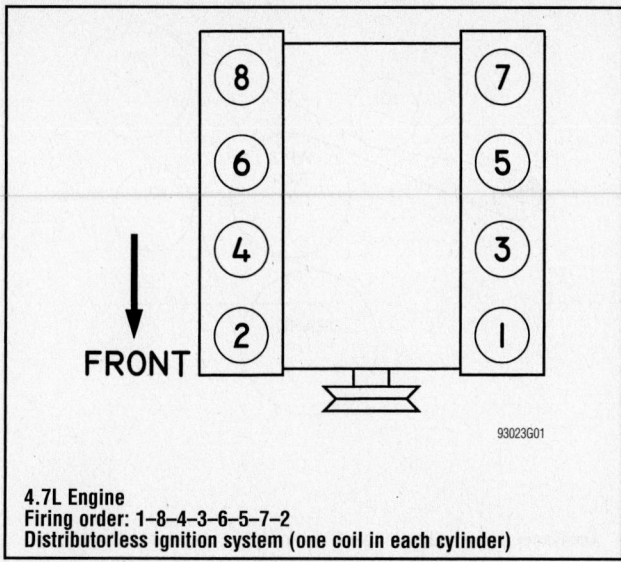

4.7L Engine
Firing order: 1–8–4–3–6–5–7–2
Distributorless ignition system (one coil in each cylinder)

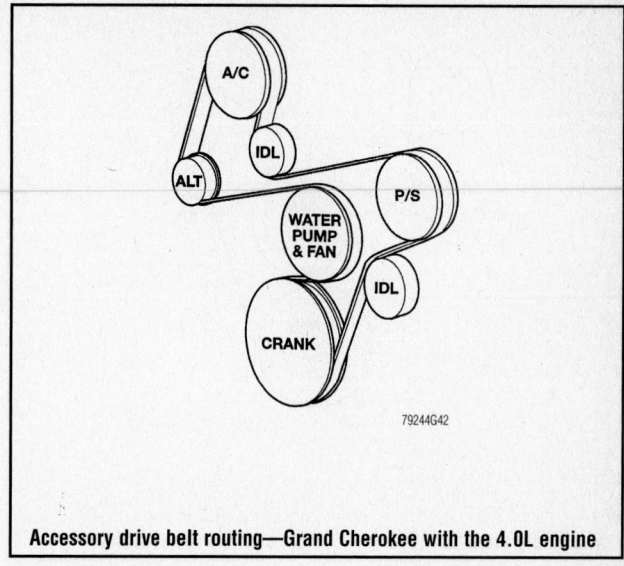

Accessory drive belt routing—Grand Cherokee with the 4.0L engine

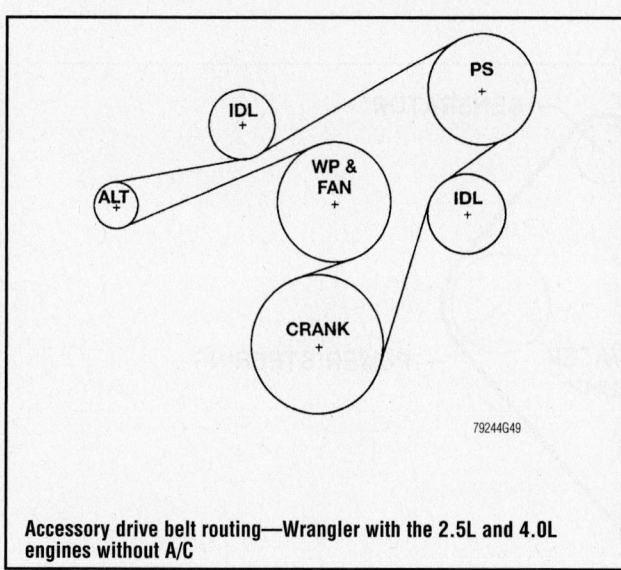

Accessory drive belt routing—Wrangler with the 2.5L and 4.0L engines without A/C

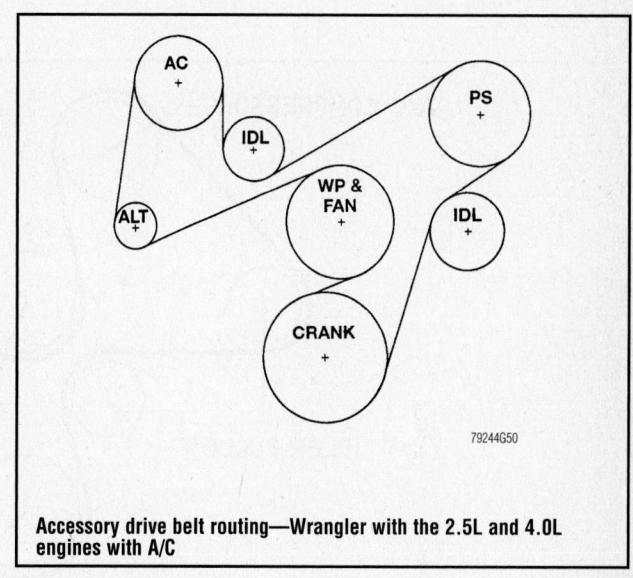

Accessory drive belt routing—Wrangler with the 2.5L and 4.0L engines with A/C

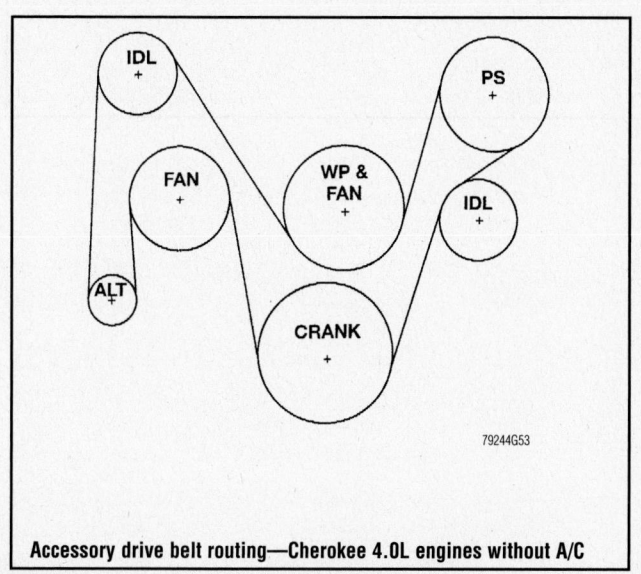

Accessory drive belt routing—Cherokee 4.0L engines without A/C

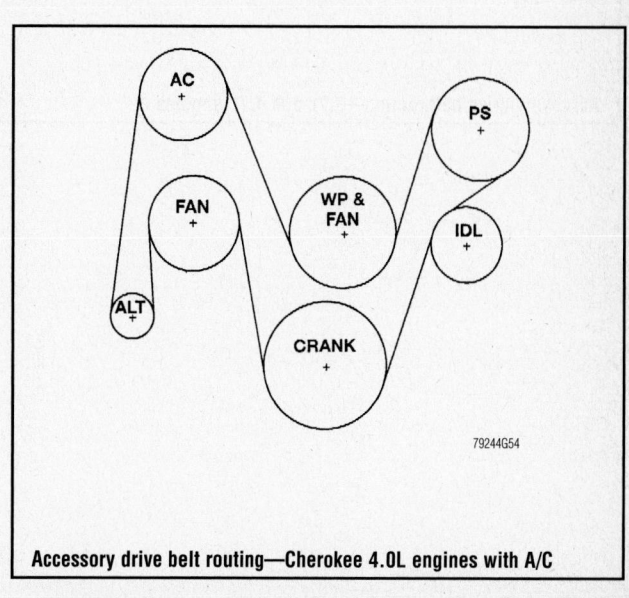

Accessory drive belt routing—Cherokee 4.0L engines with A/C

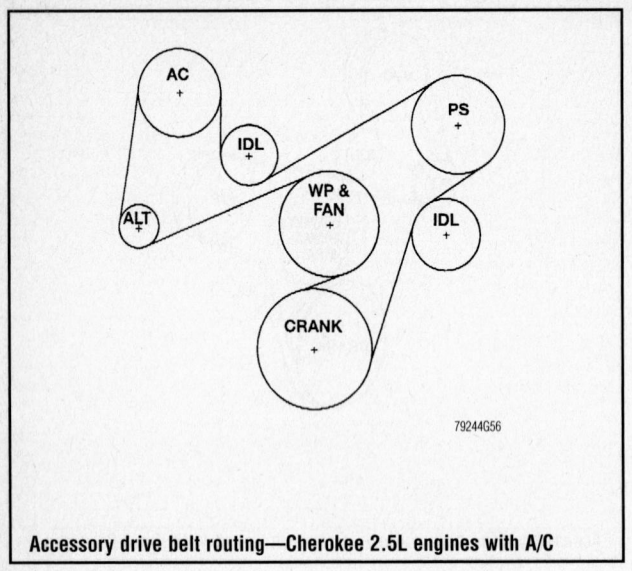

Accessory drive belt routing—Cherokee 2.5L engines with A/C

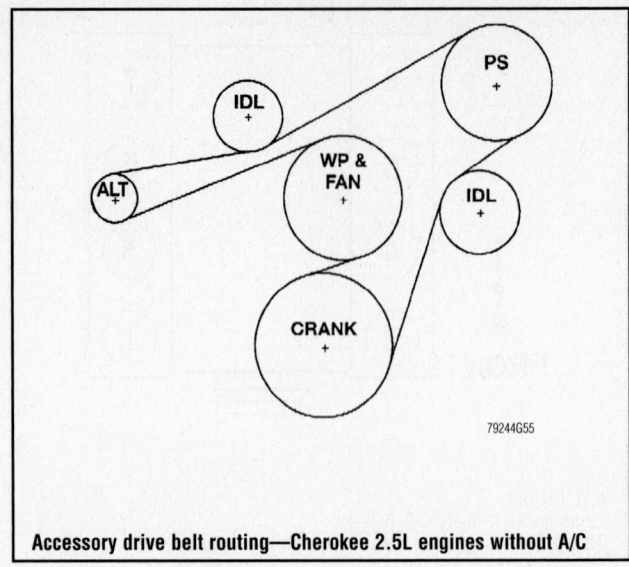

Accessory drive belt routing—Cherokee 2.5L engines without A/C

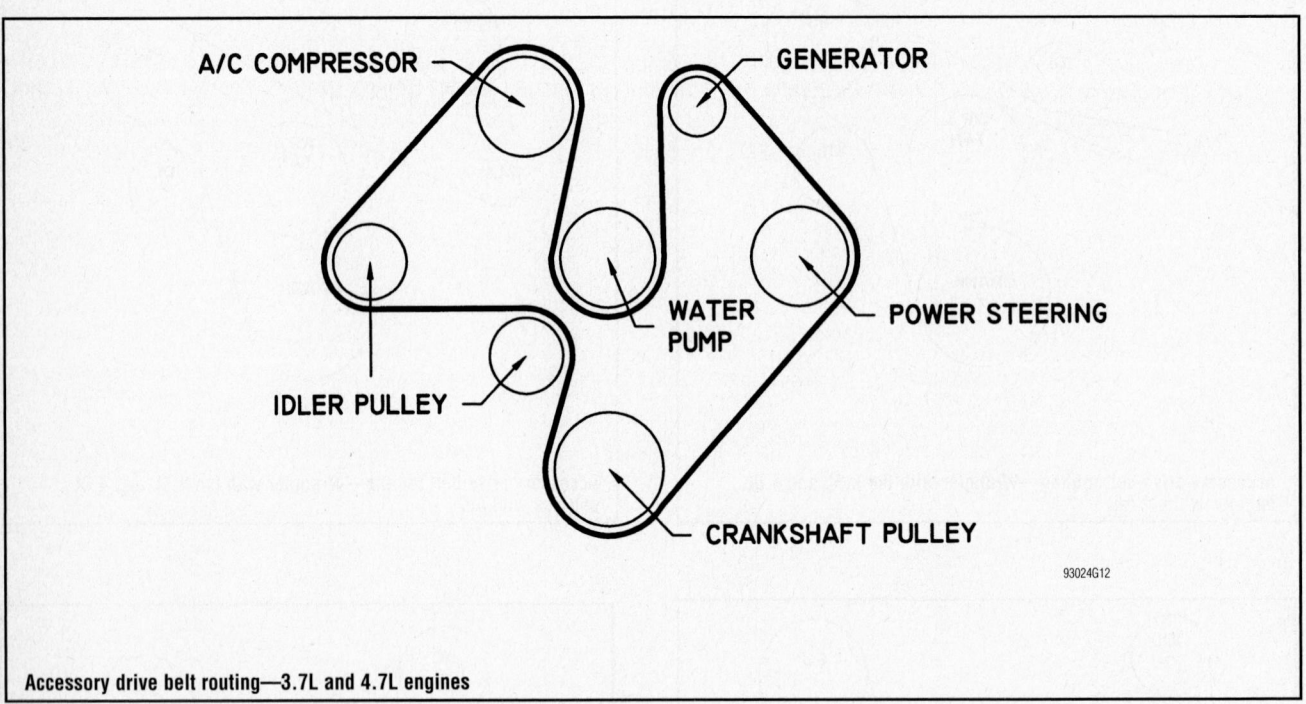

Accessory drive belt routing—3.7L and 4.7L engines

CAPACITIES

Year	Model	Engine Displ. Liters	Engine VIN	Engine Oil with Filter	Transmission (pts.) Man.	Transmission (pts.) Auto.**	Transfer Case (pts.)	Drive Axle Front (pts.)	Drive Axle Rear* (pts.)	Fuel Tank (gal.)	Cooling System (qts.)
2001	Cherokee	2.5	P	4.0	6.6①	17.0	3.0②	3.1	3.5	20.2	10.0
		4.0	S	6.0	6.6	17.0	3.0②	3.1	3.5	20.2	12.0
	Grand Cherokee	4.0	S	6.0	—	③	④	3.1	⑤	20.5	12.0
		4.7	N	6.0	—	③	④	2.5	⑤	20.5	13.0
	Wrangler	2.5	P	4.0	6.6	17.5	⑥	3.7	3.5	⑦	9.0
		4.0	S	6.0	6.6	17.5	⑥	3.7	3.5	⑦	10.5
2002	Grand Cherokee	4.0	S	6.0	—	③	④	3.1	⑤	20.5	12.0
		4.7	N	6.0	—	③	④	2.5	⑤	20.5	13.0
	Wrangler	2.5	P	4.0	6.6	17.5	⑥	3.7	3.5	⑦	9.0
		4.0	S	6.0	6.6	17.5	⑥	3.7	3.5	⑦	10.5
	Liberty	2.4	1	5.0	4.8	—	⑧	2.6	⑨	18.5	10.1
		3.7	K	5.0	4.8	10.0	⑧	2.6	⑨	18.5	13.0
2003	Grand Cherokee	4.0	S	6.0	—	⑩	⑪	3.1	⑤	20.0	15.0
		4.7	N	6.0	—	⑩	⑪	2.5	⑤	20.0	14.5
	Wrangler	2.4	1	4.0	⑭	17.5	⑬	⑭	⑮	19.0	9.0
		4.0	S	6.0	⑭	17.5	⑬	⑭	⑮	19.0	10.5
	Liberty	2.4	1	5.0	4.8	—	⑯	2.6	4.4	18.5	10.1
		3.7	K	5.0	4.8	⑰	⑯	2.6	4.4	18.5	14.0
2004	Grand Cherokee	4.0	S	6.0	—	⑩	⑪	2.5	⑤	20.0	15.0
		4.7	N	6.0	—	⑩	⑪	2.5	⑤	20.0	14.5
	Wrangler	2.4	1	4.0	⑱	17.5	⑬	⑭	⑮	19.0	9.0
		4.0	S	6.0	⑱	17.5	⑬	⑭	⑮	19.0	10.5
	Liberty	2.4	1	5.0	4.8	—	⑯	2.6	4.4	19.5	10.1
		3.7	K	5.0	4.8	⑰	⑯	2.6	4.4	19.5	14.0

* When equipped with Trac Lok, add 4 oz. of limited slip additive

**Overhaul

① 2WD: 7.0 pts.

② Command-Trac - 2.2 pts.

③ 42RE: 19-22 pts.
   44RE: 19-22 pts.
   45RFE: 28.0 pts.
   46RE: 19.5-28.0 pts.

④ 242 NVG: 3.0 pts.
   247 NVG: 2.5 pts.
   249 NVG: 2.5 pts.

⑤ 194 RBI: 3.5 pts.
   198 RBI: 3.75 pts.
   216 RBA: 4.75 pts.
   226 RBA: 4.75 pts.

⑥ Command-Trac:
   Automatic: 2.2 pts.
   Manual: 3.3 pts.

⑦ Standard: 15.0 gals.
   Optional: 19.6 gals.

⑧ NV231: 2.2 pts.
   NV242: 2.85 pts.

⑨ Model 35: 3.7 pts.
   8 1/4 inch: 4.4 pts.

⑩ 42RE: 19-20 pts.
   545RFE: 28.0 pts.
   NAG1: 16.3 pts.

⑪ NV147: 3.4 pts.
   NV242: 2.85 pts.
   NV247: 3.4 pts.

⑫ NV1500: 4.8 pts
   NV3550: 4.2 pts

⑬ NV231: 2.2 pts.
   NV241: 4.2 pts.

⑭ Model 30: 2.5 pts.
   Model 44: 4.0 pts.

⑮ Model 35: 3.5 pts.
   Model 44: 4.0 pts.

⑯ NV231: 2.95 pts.
   NV241: 3.4 pts.

⑰ 45RFE: 28.0
   42RLE: 17.6

⑱ NV231: 2.2 pts.
   NV241: 4.2 pts.

VALVE SPECIFICATIONS

Year	Engine Displ. Liters	Engine VIN	Seat Angle (deg.)	Face Angle (deg.)	Spring Test Pressure (lbs. @ in.)	Spring Installed Height (in.)	Stem-to-Guide Clearance (in.)		Stem Diameter (in.)	
							Intake	Exhaust	Intake	Exhaust
2001	2.5	P	44.5	45	184-196@ 1.216	1.640	0.0010-0.0030	0.0010-0.0030	0.3110-0.3120	0.3110-0.3120
	4.0	S	44.5	45	184-196@ 1.216	1.640	0.0010-0.0030	0.0010-0.0030	0.3110-0.3120	0.3110-0.3120
	4.7	N	44.5-45	45-45.5	176.2-192.4 @1.1532	1.602	0.0011-0.0017	0.0029	0.2728-0.2739	0.2717-0.2728
2002	2.4	1	44.5-45	44.5-45	136@1.172	1.496	0.0018-0.0025	0.0029-0.0037	0.2337-0.2344	0.2326-0.2333
	2.5	P	44.5	45	184-196@ 1.216	1.640	0.0010-0.0030	0.0010-0.0030	0.3110-0.3120	0.3110-0.3120
	3.7	K	44.5-45	45-45.5	221-242@ 1.107	1.619	0.0008-0.0028	0.0019-0.0039	0.2729-0.2739	0.2717-0.2728
	4.0	S	44.5	45	184-196@ 1.216	1.640	0.0010-0.0030	0.0010-0.0030	0.3110-0.3120	0.3110-0.3120
	4.7	N	44.5-45	45-45.5	176.2-192.4 @1.1532	1.602	0.0011-0.0017	0.0029	0.2728-0.2739	0.2717-0.2728
2003	2.4	1	44.5-45	44.5-45	136@1.172	1.496	0.0018-0.0025	0.0029-0.0037	0.2337-0.2344	0.2326-0.2333
	3.7	K	44.5-45	45-45.5	221-242@ 1.107	1.619	0.0008-0.0028	0.0019-0.0039	0.2729-0.2739	0.2717-0.2728
	4.0	S	44.5	45	184-196@ 1.216	1.640	0.0010-0.0030	0.0010-0.0030	0.3110-0.3120	0.3110-0.3120
	4.7	N	44.5-45	45-45.5	176.2-192.4 @1.1532	1.602	0.0011-0.0017	0.0029	0.2728-0.2739	0.2717-0.2728
2004	2.4	1	44.5-45	44.5-45	136@1.172	1.496	0.0018-0.0025	0.0029-0.0037	0.2337-0.2344	0.2326-0.2333
	3.7	K	44.5-45	45-45.5	221-242@ 1.107	1.619	0.0008-0.0028	0.0019-0.0039	0.2729-0.2739	0.2717-0.2728
	4.0	S	44.5	46.5	202-218@ 1.216	1.640	0.0010-0.0030	0.0010-0.0030	0.3110-0.3120	0.3110-0.3120
	4.7	N	44.5-45	45-45.5	176.2-192.4 @1.1532	1.602	0.0011-0.0017	0.0029	0.2728-0.2739	0.2717-0.2728

NA: Information not available

67189-JEEP-C05

CRANKSHAFT AND CONNECTING ROD SPECIFICATIONS

All measurements are given in inches.

Year	Engine Displ. Liters	Engine VIN	Crankshaft				Connecting Rod		
			Main Brg. Journal Dia.	Main Brg. Oil Clearance	Shaft End-play	Thrust on No.	Journal Diameter	Oil Clearance	Side Clearance
2001	2.5	P	2.4996-2.5001	0.0010-0.0025	0.0015-0.0065	2	2.2080-2.2085	0.0015-0.0020	0.0100-0.0190
	4.0	S	2.4996-2.5001 ①	0.0010-0.0025	0.0015-0.0065	2	2.0934-2.0955	0.0015-0.0020	0.0100-0.0190
	4.7	N	2.4996-2.5005	0.0002-0.0013	0.0021-0.0112	2	2.0076-2.0082	0.0004-0.0019	0.0040-0.0138
2002	2.4	1	2.3620-2.3625	0.0007-0.0024	0.0035-0.0094	NA	1.9680-1.9685	0.0009-0.0027	0.0050-0.0150
	2.5	P	2.4996-2.5001	0.0010-0.0025	0.0015-0.0065	2	2.2080-2.2085	0.0015-0.0020	0.0100-0.0190
	3.7	K	2.4996-2.5005	0.0020-0.0034	0.0021-0.0112	2	2.2794-2.2797	0.0004-0.0019	0.0040-0.0138
	4.0	S	2.4996-2.5001 ①	0.0010-0.0025	0.0015-0.0065	2	2.0934-2.0955	0.0015-0.0020	0.0100-0.0190
	4.7	N	2.4996-2.5005	0.0002-0.0013	0.0021-0.0112	2	2.0076-2.0082	0.0004-0.0019	0.0040-0.0138
2003	2.4	1	2.3620-2.3625	0.0007-0.0024	0.0035-0.0094	NA	1.9680-1.9685	0.0009-0.0027	0.0050-0.0150
	3.7	K	2.4996-2.5005	0.0020-0.0034	0.0021-0.0112	2	2.2794-2.2797	0.0004-0.0019	0.0040-0.0138
	4.0	S	2.4996-2.5001 ①	0.0010-0.0025	0.0015-0.0065	2	2.0934-2.0955	0.0015-0.0020	0.0100-0.0190
	4.7	N	2.4996-2.5005	0.0002-0.0013	0.0021-0.0112	2	2.0076-2.0082	0.0004-0.0019	0.0040-0.0138
2004	2.4	1	2.3620-2.3625	0.0007-0.0024	0.0035-0.0094	NA	1.9680-1.9685	0.0009-0.0027	0.0050-0.0150
	3.7	K	2.4996-2.5005	0.0020-0.0034	0.0021-0.0112	2	2.2794-2.2797	0.0004-0.0019	0.0040-0.0138
	4.0	S	2.4996-2.5001 ①	0.0010-0.0025	0.0015-0.0065	2	2.0934-2.0955	0.0015-0.0020	0.0100-0.0190
	4.7	N	2.4996-2.5005	0.0002-0.0013	0.0021-0.0112	2	2.0076-2.0082	0.0004-0.0019	0.0040-0.0138

NA: Information not available

① No 7: 2.4980-2.4995

67189-JEEP-C06

PISTON AND RING SPECIFICATIONS

All measurements are given in inches.

Year	Engine Displ. Liters	Engine VIN	Piston Clearance	Ring Gap			Ring Side Clearance		
				Top Compression	Bottom Compression	Oil Control	Top Compression	Bottom Compression	Oil Control
2001	2.5	P	0.0013-0.0021	0.0090-0.0240	0.0190-0.0380	0.0100-0.0600	0.0017-0.0033	0.0017-0.0033	0.0024-0.0083
	4.0	S	0.0008-0.0015	0.0090-0.0240	0.0190-0.0380	0.0100-0.0600	0.0017-0.0033	0.0017-0.0033	0.0024-0.0083
	4.7	N	0.0008-0.0020	0.0146-0.0249	0.0146-0.0249	0.0100-0.0500	0.0020-0.0041	0.0016-0.0032	0.0007-0.0091
2002	2.4	1	0.0009-0.0022	0.0098-0.0200	0.0090-0.0180	0.0098-0.0250	0.0011-0.0031	0.0011-0.0031	0.0004-0.0070
	2.5	P	0.0013-0.0021	0.0090-0.0240	0.0190-0.0380	0.0100-0.0600	0.0017-0.0033	0.0017-0.0033	0.0024-0.0083
	3.7	K	0.0014	0.0146-0.0249	0.0146-0.0249	0.0100-0.0300	0.0020-0.0037	0.0016-0.0031	0.0007-0.0091
	4.0	S	0.0008-0.0015	0.0090-0.0240	0.0190-0.0380	0.0100-0.0600	0.0017-0.0033	0.0017-0.0033	0.0024-0.0083
	4.7	N	0.0008-0.0020	0.0146-0.0249	0.0146-0.0249	0.0100-0.0500	0.0020-0.0041	0.0016-0.0032	0.0007-0.0091
2003	2.4	1	0.0009-0.0022	0.0098-0.0200	0.0090-0.0180	0.0098-0.0250	0.0011-0.0031	0.0011-0.0031	0.0004-0.0070
	3.7	K	0.0014	0.0146-0.0249	0.0146-0.0249	0.0100-0.0300	0.0020-0.0037	0.0016-0.0031	0.0007-0.0091
	4.0	S	0.0008-0.0015	0.0090-0.0240	0.0190-0.0380	0.0100-0.0600	0.0017-0.0033	0.0017-0.0033	0.0024-0.0083
	4.7	N	0.0008-0.0020	0.0146-0.0249	0.0146-0.0249	0.0100-0.0500	0.0020-0.0041	0.0016-0.0032	0.0007-0.0091
2004	2.4	1	0.0009-0.0022	0.0098-0.0200	0.0090-0.0180	0.0098-0.0250	0.0011-0.0031	0.0011-0.0031	0.0004-0.0070
	3.7	K	0.0014	0.0146-0.0249	0.0146-0.0249	0.0100-0.0300	0.0020-0.0037	0.0016-0.0031	0.0007-0.0091
	4.0	S	0.0008-0.0015	0.0090-0.0240	0.0190-0.0380	0.0100-0.0600	0.0017-0.0033	0.0017-0.0033	0.0024-0.0083
	4.7	N	0.0008-0.0020	0.0146-0.0249	0.0146-0.0249	0.0100-0.0500	0.0020-0.0041	0.0016-0.0032	0.0007-0.0091

NA: Information not available

67189-JEEP-C07

TORQUE SPECIFICATIONS
All readings in ft. lbs.

Year	Engine Displ. Liters	Engine VIN	Cylinder Head Bolts	Main Bearing Bolts	Rod Bearing Bolts	Crankshaft Damper Bolts	Flywheel Bolts	Manifold		Spark Plugs	Oil Pan Drain Plug
								Intake	Exhaust		
2001	2.5	P	①	80	33	80	105	②	②	27	25
	4.0	S	③	80	33	80	105	④	④	27	25
	4.7	N	⑤	⑥	15 ⑦	130	45	9	18	27	25
2002	2.4	1	⑨	⑩	⑪	100	⑫	21	17	20	20
	2.5	P	①	80	33	80	105	②	②	27	25
	3.7	K	⑧	⑬	⑭	130	45	9	18	27	25
	4.0	S	③	80	33	80	105	④	④	27	25
	4.7	N	⑤	⑥	15 ⑦	130	45	9	18	27	25
2003	2.4	1	⑨	⑩	⑪	100	⑫	21	17	20	20
	3.7	K	⑧	⑬	⑭	130	45	9	18	27	25
	4.0	S	③	80	33	80	105	④	④	27	25
	4.7	N	⑤	⑥	15 ⑦	130	45	9	18	27	25
2004	2.4	1	⑨	⑮	⑯	100	⑫	21	17	20	20
	3.7	K	⑧	⑬	⑭	130	45	9	18	27	25
	4.0	S	③	80	33	80	105	⑰	④	27	25
	4.7	N	⑤	⑥	15 ⑦	130	45	9	18	27	25

NA: Information not available

① Step 1: 22 ft. lbs.
Step 2: 45 ft. lbs.
Step 3: Bolts 1-6 to 110 ft. lbs.
Step 4: Bolt 7 to 100 ft. lbs.
Step 5: Bolts 8-10 to 110 ft. lbs.

② Bolt 1: 30 ft. lbs.
Bolts 2-7: 23 ft. lbs.

③ Step 1: 22 ft. lbs.
Step 2: 45 ft. lbs.
Step 3: Bolts 1-10 to 110 ft. lbs.
Step 4: Bolt 11 to 100 ft. lbs.
Step 5: Bolts 12-14 to 110 ft. lbs.

④ Bolts 1-5 and 8-11: 24 ft. lbs.
Bolts 6-7: 23 ft. lbs.

⑤ M11 bolts: 60 ft. lbs.
M8 bolts: 250 inch lbs.

⑥ See text
Step 1: Bolts 1-10 to 25 inch lbs
Step 2: Bolts 1-10 plus 90 degrees
Step 3: Bolts A-K to 40 ft. lbs.
Step 4: Bolts A1-A5 to 20 ft. lbs.

⑦ Plus 110 degrees

⑧ See procedure

⑨ Step 1: 25 ft. lbs.
Step 2: 50 ft. lbs.
Step 3: 50 ft. lbs.
Step 4: Plus 1/4 turn

⑩ M8 bolts: 21 ft. lbs.
M11 bolts: 30 ft. lbs. plus 1/4 turn

⑪ 40 ft. lbs. plus 1/4 turn

⑫ Flexplate: 70 ft. lbs.
Flywheel: 60 ft. lbs.

⑬ Bed plate torque. Refer to procedure

⑭ 20 ft. lbs. + 90 degrees

⑮ M8 bolts: 21 ft. lbs.
M11 bolts: 55 ft. lbs.

⑯ 20 ft. lbs. plus 1/4 turn

⑰ 1-5 and 8-11: 24 ft. lbs.
6 and 7: 11 ft. lbs.

67189-JEEP-C08

WHEEL ALIGNMENT

Year	Model		Caster Range (+/-Deg.)	Caster Preferred Setting (Deg.)	Camber Range (+/-Deg.)	Camber Preferred Setting (Deg.)	Toe-in (deg.)
2001	Cherokee Sport		1.50	+7.00	0.50	-0.25	0.25+/-0.25
	Grand Cherokee	F	0.75	+6.75	0.37	-0.37	0.20+/-0.20
		R	—	—	0.25	-0.25	0.25+/-0.25
	Wrangler	F	1.00	+7.00	0.63	-0.25	0.15+/-0.07
		R	—	—	0.25	-0.25	0.25+/-0.25
2002	Liberty	F	0.60	+3.5	0.375	0	0.20+/-0.125
		R	—	—	0.375	-0.25	0.25+/-0.41
	Grand Cherokee	F	0.75	+6.75	0.37	-0.37	0.20+/-0.20
		R	—	—	0.25	-0.25	0.25+/-0.25
	Wrangler	F	1.00	+7.00	0.63	-0.25	0.15+/-0.07
		R	—	—	0.25	-0.25	0.25+/-0.25
2003	Liberty	F	0.50	+3.9	0.50	0	0.20+/-0.125
		R	—	—	0.375	-0.25	0.25+/-0.41
	Grand Cherokee	F	0.75	+6.75	0.37	-0.37	0.20+/-0.20
		R	—	—	0.25	-0.25	0.25+/-0.25
	Wrangler	F	1.00	+7.00	0.63	-0.25	0.15+/-0.07
		R	—	—	0.25	-0.25	0.25+/-0.25
2004	Liberty	F	0,50	+3.9	0.375	-0.375	0.20+/-0.125
		R	—	—	0.375	-0.25	0.25+/-0.41
	Grand Cherokee Standard	F	0.75	+6.75	0.37	-0.37	0.20+/-0.20
		R	—	—	0.25	-0.25	0.25+/-0.25
	Grand Cherokee Up-Country	F	0.80	+6.50	0.37	-0.75	0.20+/-0.20
		R	—	—	0.25	-0.25	0.25+/-0.25
	Wrangler	F	1.00	+7.00	0.63	-0.25	0.15+/-0.07
		R	—	—	0.25	-0.25	0.25+/-0.25

67189-JEEP-C09

TIRE, WHEEL AND BALL JOINT SPECIFICATIONS

| Year | Model | OEM Tires | | Tire Pressures (psi) | | Wheel Size | Ball Joint Inspection | Lug Nut Torque (ft. lbs.) |
		Standard	Optional	Front	Rear			
2001	Cherokee	P215/75R15	P225/75R15 P225/70R15	33	33	7-J	①	85-115
	Grand Cherokee	P215/75R15	P225/75R15 P235/75R15 P245/70R15 P225/70R16	33	33	7-J	①	85-115
	Wrangler SE	P205/75R15	P215/75R15 P225/75R15	33	33	6JJ/7JJ 7JJ	①	85-115
	Wrangler Sport	P215/75R15	P225/75R15 30x9.5R15LT	33 29	33 29	6JJ/7JJ 8JJ	①	85-115
	Wrangler Sahara	P225/70R15	30x9.5R15LT	Std: 33 Opt: 29	Std: 33 Opt: 29	7JJ/8JJ	①	85-115
2002	Liberty	P215/75R16	P235/70R16 P225/70R15	B	B	B	①	85-115
	Grand Cherokee	P215/75R15	P225/75R15 P235/75R15 P245/70R15 P225/70R16	33	33	7-J	①	85-115
	Wrangler SE	P205/75R15	P215/75R15 P225/75R15	33	33	6JJ/7JJ 7JJ	①	85-115
	Wrangler Sport	P215/75R15	P225/75R15 30x9.5R15LT	33 29	33 29	6JJ/7JJ 8JJ	①	85-115
	Wrangler Sahara	P225/70R15	30x9.5R15LT	Std: 33 Opt: 29	Std: 33 Opt: 29	7JJ/8JJ	①	85-115
2003	Liberty	P215/75R16	P235/70R16 P235/65R17	②	②	②	①	85-115
	Grand Cherokee	P215/75R15	P225/75R15 P235/75R15 P245/70R15 P225/70R16	33	33	7-J	①	85-115
	Wrangler	P215/75R15	P225/75R15 LT245/75R16 30x9.5R15LT	②	②	②	①	85-115

67189-JEEP-C10

TIRE, WHEEL AND BALL JOINT SPECIFICATIONS

| Year | Model | OEM Tires | | Tire Pressures (psi) | | Wheel Size | Ball Joint Inspection | Lug Nut Torque (ft. lbs.) |
		Standard	Optional	Front	Rear			
2004	Liberty	P215/75R16	P235/70R16 P235/65R17	②	②	②	①	85-115
	Grand Cherokee	P225/75R16	P245/70R16 P235/65R17 235/65HR17	②	②	②	①	85-115
	Wrangler SE	P215/75R15	P225/75R15	②	②	std: 6 opt.: 7	①	85-115
	Wrangler Sport	P215/75R15	P225/75R15 30x9.5R15LT	②	②	6JJ/7JJ 8JJ	①	85-115
	Wrangler Sahara	30x9.5R15LT	none	②	②	8	①	85-115
	Wrangler X	P215/75R15	P225/75R15	②	②	7	①	85-115
	Wrangler Rubicon	LT245/75R16	none	②	②	8	①	85-115

NA: Information not available
OEM: Original Equipment Manufacturer
PSI: Pounds Per Square Inch
STD: Standard
OPT: Optional
① Replace if any measurable movement is found.
② See placard on vehicle

67189-JEEP-C11

BRAKE SPECIFICATIONS
All measurements in inches unless noted

| Year | Model | | Brake Disc | | | Brake Drum | | | Minimum Lining Thickness | | Caliper Mounting Bolts (ft. lbs.) |
			Original Thickness	Minimum Thickness	Maximum Run-out	Original Inside Diameter	Max. Wear Limit	Maximum Machine Diameter	Front	Rear	
2001	Cherokee		0.94	0.89	0.005	①	—	②	0.030	③	11
	Grand Cherokee		—	④	0.003	—	—	—	0.030	0.030	26-30
	Wrangler		0.94	0.89	0.005	9.00	—	9.06	0.030	③	11
2002	Liberty		NA	0.8937	0.005	①	—	②	0.030	③	11
	Grand Cherokee		—	④	0.003	—	—	—	0.030	0.030	26-30
	Wrangler		0.94	0.89	0.005	9.00	—	9.06	0.030	③	11
2003	Liberty	F	1.102	1.0236	0.004	—	—	—	NA	NA	11
		R	0.472	0.4331	0.004	—	—	—	NA	NA	18
	Grand Cherokee		—	④	0.003	—	—	—	0.030	0.030	⑤
	Wrangler		⑥	⑦	⑧	9.00	—	9.06	0.030	③	⑨
2004	Liberty	F	1.102	1.0236	0.004	—	—	—	NA	NA	11
		R	0.472	0.4331	0.004	—	—	—	NA	NA	18
	Grand Cherokee		—	④	⑤	—	—	—	0.030	0.030	⑤
	Wrangler	F	0.940	0.0937	0.005	—	—	—	0.030	—	11
		R	0.472	0.4330	0.004	9.00	—	9.06	—	③	18

NA: Information not available
① Standard: 9.00 in.
  Optional 10.00 in.
② Standard: 9.06 in.
  Optional 10.06 in.
③ Riveted brake shoes: 0.030 in.
  Bonded brake shoes: 0.060 in.

④ Front: 0.965 in.
  Rear: 0.335 in.
⑤ Slide pin bolts: Front, 53; Rear, 26 ft. lbs.
  Anchor bolts: 76 ft. lbs.
⑥ F: 0.94 in.
  R: 0.47 in.

⑦ F: 0.8937 in.
  R: 0.4330 in.
⑧ F: 0.005 in.
  R: 0.004 in.
⑨ F: 11 ft. lbs.
  R: 220 inch lbs.

67189-JEEP-C12

SCHEDULED MAINTENANCE INTERVALS
2001-03 Wrangler, Cherokee, Grand Cherokee, Liberty

TO BE SERVICED	TYPE OF SERVICE	VEHICLE MILEAGE INTERVAL (x1000)												
		7.5	15	22.5	30	37.5	45	52.5	60	67.5	75	82.5	90	97.5
Engine oil & filter	R	✓	✓	✓	✓	✓	✓	✓	✓	✓	✓	✓	✓	✓
Brake hoses & linings	S/I	✓	✓	✓	✓	✓	✓	✓	✓	✓	✓	✓	✓	✓
Engine coolant level, hoses & clamps	S/I	✓	✓	✓	✓	✓	✓	✓	✓	✓	✓	✓	✓	✓
Exhaust system	S/I	✓	✓	✓	✓	✓	✓	✓	✓	✓	✓	✓	✓	✓
Lubricate steering linkage (4x2)	S/I	✓	✓	✓	✓	✓	✓	✓	✓	✓	✓	✓	✓	✓
Lubricate steering linkage (4x4)	S/I	✓		✓		✓		✓		✓		✓		✓
Air filter	R				✓				✓				✓	
Automatic transmission fluid & filter	R				✓				✓				✓	
Spark plugs	R				✓				✓				✓	
Transfer case fluid	R				✓				✓				✓	
Drive belts	S/I				✓				✓				✓	
Front & rear axle oil	R				✓				✓				✓	
Prop shaft universal joints	S/I				✓				✓				✓	
Rotate tires	S/I				✓				✓				✓	
Engine coolant	R						✓				✓			
Manual transmission fluid	R					✓					✓			
Distributor cap & rotor	R								✓					
Fuel filter	R								✓					
Ignition cables	R								✓					

R: Replace   S/I: Service or Inspect

FREQUENT OPERATION MAINTENANCE (SEVERE SERVICE)

 If a vehicle is operated under any of the following conditions it is considered severe service:

- **Extremely dusty areas.**
- **50% or more of the vehicle operation is in 32°C (90°F) or higher temperatures, or constant operation in temperatures below 0°C (32°F).**
- **Prolonged idling (vehicle operation in stop and go traffic.)**
- **Frequent short running periods (engine does not warm to normal operating temperatures).**
- **Police, taxi, delivery usage or trailer towing usage.**

Oil & oil filter change: change every 3000 miles.

Automatic transmission fluid, filter & bands: change & adjust every 12,000 miles.

Brake hoses & linings: check every 12,000 miles.

Liubricate steering linkage: check every 3000 miles.

Manual transmission fluid: change every 18,000 miles.

Prop shaft universal joints: lubricate every 3000 miles.

Front & rear axle oil: change every 12,000 miles.

67189-JEEP-C13

SCHEDULED MAINTENANCE INTERVALS
2004 Liberty

TO BE SERVICED	TYPE OF SERVICE	VEHICLE MILEAGE INTERVAL (x1000)												
		6	12	18	24	30	36	42	48	54	60	66	72	78
Engine oil & filter	R	✓	✓	✓	✓	✓	✓	✓	✓	✓	✓	✓	✓	✓
Brake hoses & linings	I			✓			✓			✓			✓	
Engine coolant	R	Replace every 5 years, regardless of mileage												
Air filter	R					✓					✓			
Spark plugs	R				✓				✓				✓	
Transfer case fluid level	I				✓						✓			
Transfer case fluid	R	Drain, flush and refill every 120,000 miles												
Accessory drive belt ①	I/R												✓	
Timing belt (2.4L)	R	Replace every 120,000 miles												
PCV valve	R										✓			
Ignition cables (2.4L)	R										✓			

R: Replace   S/I: Service or Inspect   I/R: Inspect and replace as necessary

① Replace at 72,000 miles if not replaced at 60,000 during inspection

FREQUENT OPERATION MAINTENANCE (SEVERE SERVICE)

If a vehicle is operated under any of the following conditions it is considered severe service:

- Extremely dusty areas.

- 50% or more of the vehicle operation is in 32°C (90°F) or higher temperatures, or constant operation in temperatures below 0°C (32°F).

- Prolonged idling (vehicle operation in stop and go traffic.)

- Frequent short running periods (engine does not warm to normal operating temperatures).

- Police, taxi, delivery usage or trailer towing usage.

Oil & oil filter change: change every 3000 miles.

Air filter element: inspect and replace as necessary every 12,000 miles

PCV valve: replace every 30,000 miles

Automatic transmission fluid, filter: change every 12,000 miles.

Brake hoses & linings: check every 12,000 miles.

Accessory drive belt: inspect and replace as necessary every 45,000 miles

Transfer case fluid: replace every 60,000 miles

Engine coolant: replace every 99,000 miles

Front & rear axle oil: change every 12,000 miles.

67189-JEEP-C14

SCHEDULED MAINTENANCE INTERVALS
2004 Wrangler

TO BE SERVICED	TYPE OF SERVICE	VEHICLE MILEAGE INTERVAL (x1000)												
		6	12	18	24	30	36	42	48	54	60	66	72	78
Engine oil & filter	R	✓	✓	✓	✓	✓	✓	✓	✓	✓	✓	✓	✓	✓
Brake hoses & linings	S/I			✓			✓			✓			✓	
Engine coolant level, hoses & clamps	S/I	✓	✓	✓	✓	✓	✓	✓	✓	✓	✓	✓	✓	✓
Lubricate steering linkage and outer tie rod ends	S/I	✓	✓	✓	✓	✓	✓	✓	✓	✓	✓	✓	✓	✓
Lubricate steering and suspension ball joints	S/I		✓		✓		✓		✓		✓		✓	
Air filter	R					✓					✓			
Spark plugs	R					✓					✓			
Transfer case fluid level	I					✓					✓			
Transfer case fluid	R	Replace every 120,000 miles												
PCV valve	I/R										✓			
Accessory drive belt ①	S/I										✓			
Timing belt	R	Replace every 120,000 miles												
Engine coolant	R						✓				✓			
Ignition cables (2.4L)	R										✓			

R: Replace   S/I: Service or Inspect

① Replace every 72,000 miles

FREQUENT OPERATION MAINTENANCE (SEVERE SERVICE)

If a vehicle is operated under any of the following conditions it is considered severe service:

- Extremely dusty areas.

- 50% or more of the vehicle operation is in 32°C (90°F) or higher temperatures, or constant operation in temperatures below 0°C (32°F).

- Prolonged idling (vehicle operation in stop and go traffic.)

- Frequent short running periods (engine does not warm to normal operating temperatures).

- Police, taxi, delivery usage or trailer towing usage.

Oil & oil filter change: change every 3000 miles.

Air filter element: inspect and replace as necessary every 12,000 miles

PCV valve: replace every 30,000 miles

Automatic transmission fluid, filter: change every 12,000 miles.

Brake hoses & linings: check every 12,000 miles.

Accessory drive belt: inspect and replace as necessary every 45,000 miles

Transfer case fluid: replace every 60,000 miles

Engine coolant: replace every 99,000 miles

Front & rear axle oil: change every 12,000 miles.

67189-JEEP-C15

SCHEDULED MAINTENANCE INTERVALS
2004 Grand Cherokee

TO BE SERVICED	TYPE OF SERVICE	VEHICLE MILEAGE INTERVAL (x1000)												
		6	12	18	24	30	36	42	48	54	60	66	72	78
Accessory drive belt ①	S/I													✓
Engine oil & filter	R	✓	✓	✓	✓	✓	✓	✓	✓	✓	✓	✓	✓	✓
Brake hoses & linings	S/I			✓			✓			✓			✓	
Lubricate steering and suspension ball joints	S/I		✓		✓		✓		✓		✓		✓	
Brake caliper pins	C/L			✓			✓			✓			✓	
Air filter	R					✓					✓			
Spark plugs	R					✓					✓			
Transfer case fluid level ②	I					✓					✓			
Transfer case fluid ③	R					✓					✓			
Transfer case fluid ②	R	Replace every 120,000 miles												
PCV valve ④	I/R										✓			
Automatic transmission	⑤	every 100,000 miles												
Engine coolant	R										✓			

R: Replace   S/I: Service or Inspect   C/L: Clean and lubricate

① Replace as necessary....4.0L only

② Selec-Trac only

③ Quadra-Trac only

④ 4.7L only

⑤ 4.0L only: Drain and refill the fluid, adjust the bands and change the filter

   4.7L Only: Drain and refill the automatic transmission fluid, replace main sump filter, and spin-on cooler return filter (if equipped)

FREQUENT OPERATION MAINTENANCE (SEVERE SERVICE)

 If a vehicle is operated under any of the following conditions it is considered severe service:

- Extremely dusty areas.

- 50% or more of the vehicle operation is in 32°C (90°F) or higher temperatures, or constant operation in temperatures below 0°C (32°F).

- Prolonged idling (vehicle operation in stop and go traffic.)

- Frequent short running periods (engine does not warm to normal operating temperatures).

- Police, taxi, delivery usage or trailer towing usage.

Oil & oil filter change: change every 3000 miles.

Air filter element: inspect and replace as necessary every 12,000 miles.

PCV valve: replace every 30,000 miles.

Automatic transmission fluid, filter: change every 12,000 miles.

Brake hoses & linings: check every 12,000 miles.

Accessory drive belt: inspect and replace as necessary every 45,000 miles

Front & rear axle oil: change every 12,000 miles.

PRECAUTIONS

Before servicing any vehicle, please be sure to read all of the following precautions, which deal with personal safety, prevention of component damage, and important points to take into consideration when servicing a motor vehicle:

• Never open, service or drain the radiator or cooling system when the engine is hot; serious burns can occur from the steam and hot coolant.

• Observe all applicable safety precautions when working around fuel. Whenever servicing the fuel system, always work in a well-ventilated area. Do not allow fuel spray or vapors to come in contact with a spark, open flame, or excessive heat (a hot drop light, for example). Keep a dry chemical fire extinguisher near the work area. Always keep fuel in a container specifically designed for fuel storage; also, always properly seal fuel containers to avoid the possibility of fire or explosion. Refer to the additional fuel system precautions later in this section.

• Fuel injection systems often remain pressurized, even after the engine has been turned **OFF**. The fuel system pressure must be relieved before disconnecting any fuel lines. Failure to do so may result in fire and/or personal injury.

• Brake fluid often contains polyglycol ethers and polyglycols. Avoid contact with the eyes and wash your hands thoroughly after handling brake fluid. If you do get brake fluid in your eyes, flush your eyes with clean, running water for 15 minutes. If eye irritation persists, or if you have taken brake fluid internally, IMMEDIATELY seek medical assistance.

• The EPA warns that prolonged contact with used engine oil may cause a number of skin disorders, including cancer. You should make every effort to minimize your exposure to used engine oil. Protective gloves should be worn when changing oil. Wash your hands and any other exposed skin areas as soon as possible after exposure to used engine oil. Soap and water, or waterless hand cleaner should be used.

• All new vehicles are now equipped with an air bag system, often referred to as a Supplemental Restraint System (SRS) or Supplemental Inflatable Restraint (SIR) system. The system must be disabled before performing service on or around system components, steering column, instrument panel components, wiring and sensors. Failure to follow safety and disabling procedures could result in accidental air bag deployment, possible personal injury and unnecessary system repairs.

• Always wear safety goggles when working with, or around, the air bag system. When carrying a non-deployed air bag, be sure the bag and trim cover are pointed away from your body. When placing a non-deployed air bag on a work surface, always face the bag and trim cover upward, away from the surface. This will reduce the motion of the module if it is accidentally deployed. Refer to the additional air bag system precautions later in this section.

• Clean, high quality brake fluid from a sealed container is essential to the safe and proper operation of the brake system. You should always buy the correct type of brake fluid for your vehicle. If the brake fluid becomes contaminated, completely flush the system with new fluid. Never reuse any brake fluid. Any brake fluid that is removed from the system should be discarded. Also, do not allow any brake fluid to come in contact with a painted surface; it will damage the paint.

• Never operate the engine without the proper amount and type of engine oil; doing so WILL result in severe engine damage.

• Timing belt maintenance is extremely important. Many models utilize an interference-type, non-freewheeling engine. If the timing belt breaks, the valves in the cylinder head may strike the pistons, causing potentially serious (also time-consuming and expensive) engine damage. Refer to the maintenance interval charts for the recommended replacement interval for the timing belt, and to the timing belt section for belt replacement and inspection.

• Disconnecting the negative battery cable on some vehicles may interfere with the functions of the on-board computer system(s) and may require the computer to undergo a relearning process once the negative battery cable is reconnected.

• When servicing drum brakes, only disassemble and assemble one side at a time, leaving the remaining side intact for reference.

• Only an MVAC-trained, EPA-certified automotive technician should service the air conditioning system or its components.

ENGINE REPAIR

➡**Disconnecting the negative battery cable on some vehicles may interfere with the functions of the on-board computer system. The computer may undergo a relearning process once the negative battery cable is reconnected.**

Distributor

REMOVAL

2.5L Engines

1. Before servicing the vehicle, refer to the precautions in the beginning of this section.

2. Remove or disconnect the following:
   • Negative battery cable

   • Distributor cap
   • Camshaft Position (CMP) sensor connector
   • Distributor wiring harness from the main engine harness
   • Cylinder No. 1 spark plug
3. Matchmark the distributor housing and the rotor.
4. Remove the distributor.

INSTALLATION

Timing Not Disturbed

2.5L ENGINES

1. Before servicing the vehicle, refer to the precautions in the beginning of this section.

➡**The rotor will rotate clockwise as the gears engage.**

2. Position the rotor slightly counterclockwise of the matchmark made during removal.
3. Install the distributor. Ensure that the rotor moves into alignment with the matchmark.
4. Align the locating fork with the clamp bolt hole. Install the clamp and bolt. Tighten the bolt to 17 ft. lbs. (23 Nm).
5. Install or connect the following:
   • Cylinder No. 1 spark plug
   • Distributor wiring harness to the main engine harness
   • CMP sensor connector
   • Distributor cap
   • Air cleaner tube
   • Negative battery cable

FRONT ➡

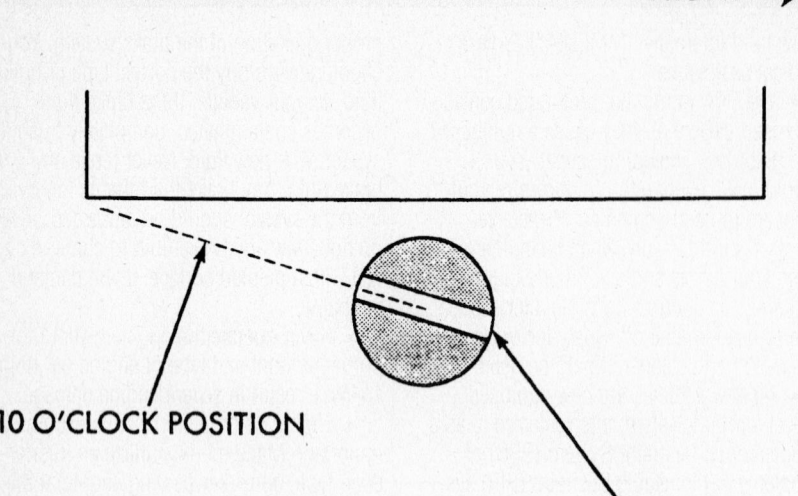

10 O'CLOCK POSITION

OIL
PUMP
SLOT

7924PG30

Slot in the oil pump gear at 10 o'clock position—2.5L engine

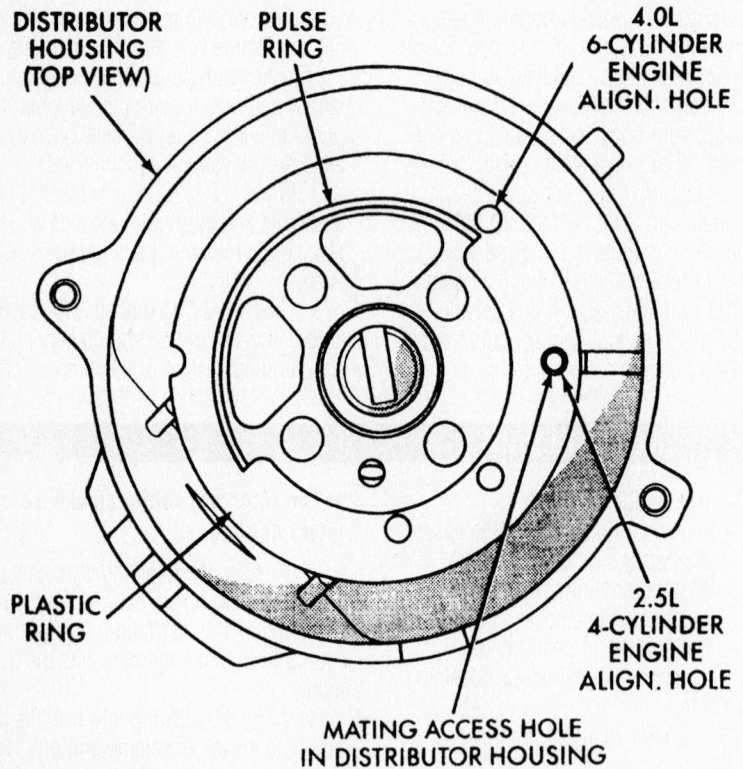

DISTRIBUTOR
HOUSING
(TOP VIEW)

PULSE
RING

4.0L
6-CYLINDER
ENGINE
ALIGN. HOLE

PLASTIC
RING

2.5L
4-CYLINDER
ENGINE
ALIGN. HOLE

MATING ACCESS HOLE
IN DISTRIBUTOR HOUSING

7924PG01

Distributor pin alignment holes—2.5L engines

Timing Disturbed

2.5L ENGINES

1. Before servicing the vehicle, refer to the precautions in the beginning of this section.
2. Set the engine at Top Dead Center (TDC) of the No. 1 cylinder compression stroke.
3. Position the slot in the oil pump drive gear as shown.
4. Locate the alignment holes in the plastic ring and align the correct hole with the mating hole in the distributor housing as shown. Install a locking pin.

➡**The distributor will rotate clockwise as the gears engage.**

5. Position the base mounting slot at the 1 o'clock position and install the distributor.
6. Check that the centerline of the mounting slot aligns with the centerline of the clamp bolt hole.
7. Install the clamp and bolt. Tighten the bolt to 17 ft. lbs. (23 Nm).
8. Remove the locking pin.
9. Install or connect the following:
   - CMP sensor connector
   - Distributor cap
   - Air cleaner tube
   - Negative battery cable

Alternator

REMOVAL

1. Before servicing the vehicle, refer to the precautions in the beginning of this section.
2. Remove or disconnect the following:
   - Negative battery cable
   - Accessory drive belt
   - Alternator harness connectors
   - Mounting bolts and alternator

➡**The 3.7L engine has 1 vertical and 2 horizontal bolts.**

INSTALLATION

1. Before servicing the vehicle, refer to the precautions in the beginning of this section.
2. Install the alternator and tighten the bolts to the following specifications:
   - 2.4L, 2.5L and 4.0L engines: 41 ft. lbs. (55 Nm).
   - 3.7L engine: Tighten the horizontal bolts to 42 ft. lbs. (57 Nm), then the vertical bolt to 29 ft. lbs. (40 Nm)
   - 4.7L engine: Vertical bolt and long horizontal bolt to 41 ft. lbs. (56 Nm), short horizontal bolt to 55 ft. lbs. (74 Nm)
3. Install or connect the following:
   - Alternator harness connectors
   - Accessory drive belt
   - Negative battery cable

Ignition Timing

ADJUSTMENT

Ignition timing is controlled by the Powertrain Control Module (PCM). No adjustment is possible.

Engine Assembly

REMOVAL & INSTALLATION

Liberty

2.4L ENGINE

1. Disconnect the battery negative cable.
2. Remove hood. Mark hood hinge location for reinstallation.
3. Remove air cleaner assembly.
4. Remove radiator core support bracket.
5. Remove fan shroud with electric fan assembly.
6. Remove drive belt.

➡**It is not necessary to discharge the A/C system to remove the engine.**

7. Remove A/C compressor and secure away from engine with lines attached.
8. Remove generator and secure away from engine.

➡**Do not remove the phenolic pulley from the P/S pump. It is not required for P/S pump removal.**

9. Remove power steering pump with lines attached and secure away from engine.
10. Drain cooling system.
11. Remove coolant bottle.
12. Disconnect the heater hoses from the engine.
13. Disconnect heater hoses from heater core and remove hose assembly.
14. Disconnect throttle and speed control cables.
15. Remove upper radiator hose from engine.
16. Remove lower radiator hose from engine.
17. Disconnect the engine to body ground straps at the left side of cowl.
18. Disconnect the engine wiring harness at the following points:
   - Intake air temperature (IAT) sensor
   - Fuel Injectors
   - Throttle Position (TPS) Switch
   - Idle Air Control (IAC) Motor
   - Engine Oil Pressure Switch
   - Engine Coolant Temperature (ECT) Sensor
   - Manifold Absolute Pressure MAP) Sensor
   - Camshaft Position (CMP) Sensor
   - Coil Over Plugs
   - Crankshaft Position Sensor
19. Remove coil over plugs.
20. Release fuel rail pressure.

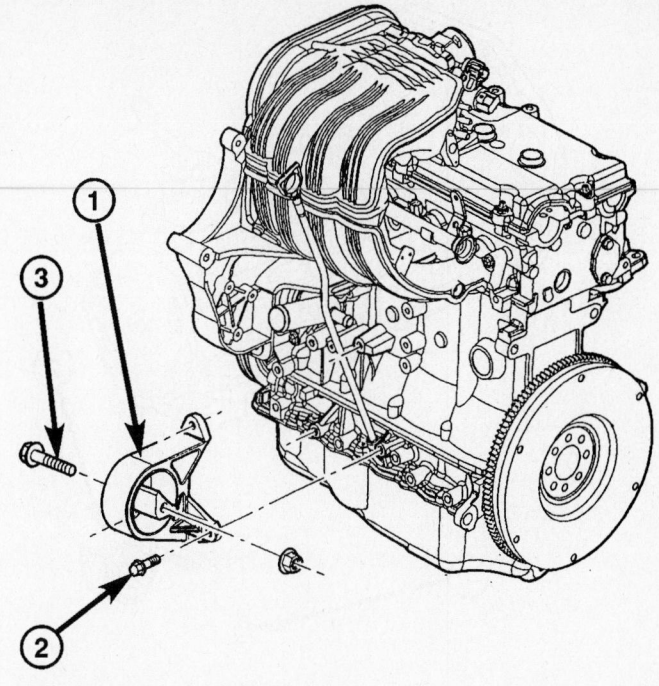

1 - ENGINE MOUNT
2 - ENGINE MOUNT BOLT (3)
3 - ENGINE MOUNT THROUGH BOLT

67189-JEEP-G01

Left side engine mount—2.4L engines

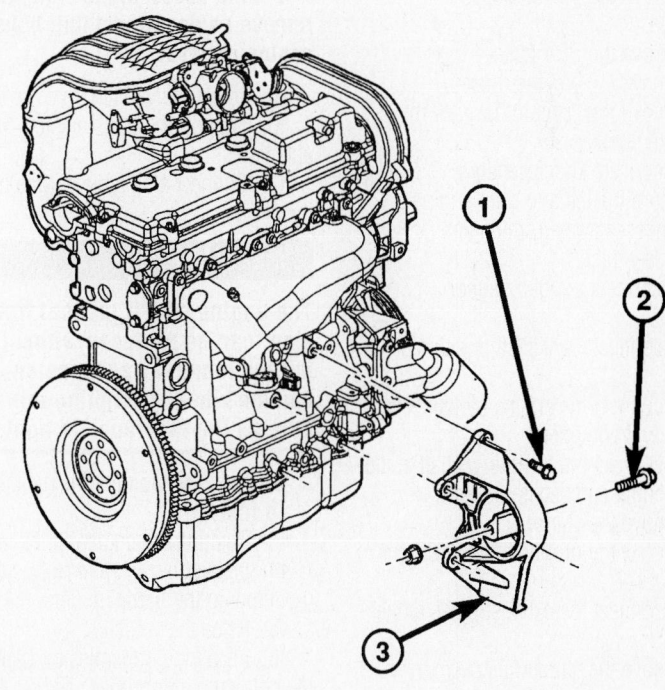

1 - ENGINE MOUNT BOLT (4)
2 - ENGINE MOUNT THROUGH BOLT
3 - ENGINE MOUNT

67189-JEEP-G02

Right side engine mount—2.4L engines

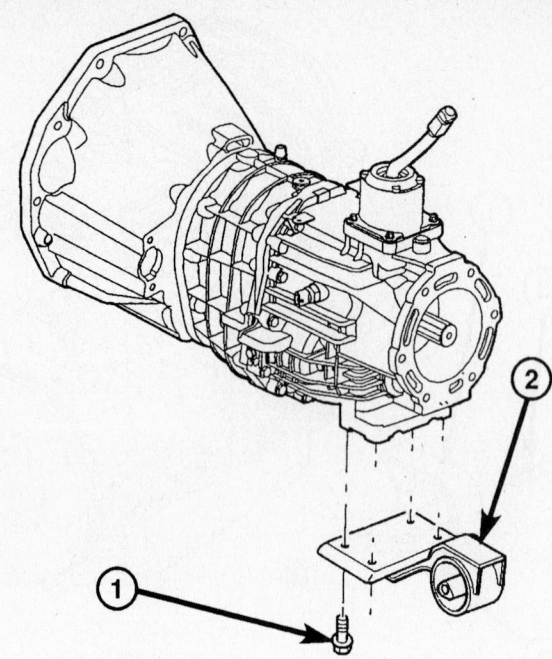

1 - TRANSMISSION MOUNT
2 - MOUNTING BOLT

67189-JEEP-G03

Transmission mount—2.4L engines

21. Remove fuel rail and secure away from engine.
22. Remove the PCV hose.
23. Remove the breather hoses.
24. Remove the vacuum hose for the power brake booster.
25. Disconnect knock sensors.
26. Secure the left and right engine wiring harnesses away from engine.
27. Raise vehicle.
28. Disconnect oxygen sensor wiring.
29. Disconnect crankshaft position sensor.
30. Disconnect the engine block heater power cable, if equipped.
31. Disconnect the front driveshaft at the front differential and secure out of way.
32. Remove the starter.
33. Remove the ground straps from the engine
34. Disconnect the exhaust pipes at the manifold.
35. Remove the structural cover, if equipped.
36. Remove torque converter bolts, and mark location for reassembly.
37. Remove transmission bellhousing to engine bolts.
38. Loosen left and right engine mount thru bolts.

➡ It is not necessary to completely remove engine mount thru bolts, for engine removal.

39. Lower the vehicle.
40. Support the transmission with a suitable jack.
41. Connect a suitable engine hoist to the engine.

✳✳ WARNING

The engine with a manual transmission, can be removed without removing the manual transmission. Use caution when attempting this procedure as the clearance is tight.

42. Remove engine from vehicle.
To install:
43. Position the engine in the vehicle.
44. Install both left and right side engine mounts into the frame mounts.
45. Raise the vehicle.
46. Install the transmission bellhousing to engine mounting bolts. Tighten the bolts to 41 N·m (30 ft. lbs.).
47. Tighten the engine mount thru bolts.
48. Install the torque converter bolts.
49. Connect the ground straps on the left and right side of the engine.
50. Install the starter.

51. Connect the crankshaft position sensor.
52. Install the engine block heater power cable, if equipped.

✳✳ WARNING

The structural collar requires a specific torque sequence. Failure to follow this sequence may cause severe damage to the collar.

53. Install the structural collar.
Perform the following steps for installing structural collar.
   a. Step 1: Position collar between transmission and oil pan. Install collar to transmission bolts, hand start only.
   b. Step 2: Install collar to oil pan bolts, hand snug only.
   c. Step 3: Tighten collar to transmission bolts.
   d. Step 4: Tighten collar to oil pan bolts.
54. Install the exhaust pipe.
55. Connect the oxygen sensors.
56. Lower vehicle.
57. Connect the knock sensors.
58. Connect the engine to body ground straps.
59. Install the power brake booster vacuum hose.
60. Install the breather hoses.
61. Install the PCV hose.
62. Install the fuel rail.
63. Install the coil over plugs.
64. Reconnect the engine wiring harness at the following points:
   • Intake air temperature (IAT) sensor
   • Fuel Injectors
   • Throttle Position (TPS) Switch
   • Idle Air Control (IAC) Motor
   • Engine Oil Pressure Switch
   • Engine Coolant Temperature (ECT) Sensor
   • Manifold Absolute Pressure MAP) Sensor
   • Camshaft Position (CMP) Sensor
   • Coil Over Plugs
   • Crankshaft Position Sensor
65. Connect lower radiator hose.
66. Connect upper radiator hose.
67. Connect throttle and speed control cables.
68. Install the heater hose assembly.
69. Install coolant recovery bottle.
70. Install the power steering pump.
71. Install the generator.
72. Install the A/C compressor.
73. Install the drive belt.
74. Install the fan shroud with the electric fan assembly.

75. Install the radiator core support bracket.
76. Install the air cleaner assembly.
77. Refill the engine cooling system.
78. Install the hood.
79. Check and fill engine oil.
80. Connect the battery negative cable.
81. Start the engine and check for leaks.

3.7L ENGINE

1. Before servicing the vehicle, refer to the precautions in the beginning of this section.
2. Properly relieve the fuel system pressure.
3. Drain the cooling system.
4. Drain the engine oil.
5. Remove or disconnect the following:
- Negative battery cable
- Hood
- Air cleaner assembly
- Radiator
- Electric and mechanical fan assemblies
- A/C compressor, if equipped, and secure it out of the way with the lines attached. DO NOT DISCHARGE!
- Power steering pump, with the lines attached
- Alternator
- Coolant bottle
- Heater hoses
- Accelerator and speed control cables
- Lower and upper radiator hoses
- Engine ground straps
- Intake Air Temperature (IAT) sensor
- Fuel injection wiring connectors
- Throttle Position (TP) sensor
- Idle Air Control (IAC) motor
- Oil pressure sender connector
- Engine Coolant Temperature (ECT) sensor
- Manifold Absolute Pressure (MAP) sensor
- Camshaft position sensor
- Ignition coil wiring connector
- Crankshaft Position (CKP) sensor
- Coil pack
- Fuel rail
- PCV hose
- Vacuum hoses from the intake manifold
- Knock sensor connectors
- Oil dipstick tube
- Intake manifold
- Heated Oxygen (HO2S) sensor connector
- Block heater connector
- Front driveshaft at the differential
- Starter

- Structural cover
- With a manual transmission, remove the transmission
- Torque converter bolts and match-mark the converter
- Automatic transmission-to-engine bolts
- Exhaust front pipes
- Left and right engine mounts

6. Place a support stand under the transmission.
7. Install an engine lift plate
8. Lift the engine out of the vehicle.

To install:

9. If equipped with a manual transmission, install the transmission
10. Lower the engine and install the mounts. Don't tighten the bolts yet.
11. If equipped with an automatic transmission, perform the following steps:
   a. Align the torque converter housing to the engine.
   b. Torque the bolts to 30 ft. lbs. (41 Nm).
   - Install the torque converter to flexplate bolts. Torque the bolts to 50 ft. lbs. (68 Nm).
12. Install or connect the following:
- Torque the through bolts to 45 ft. lbs. (61 Nm).
- Engine ground strap

- Starter motor
- CKP sensor
- Block heater cable
- Structural cover.

✳✳ WARNING

The structural cover must be held tightly against the engine and bellhousing during tightening. The torque for all bolts is 40 ft. lbs. (54 Nm); the bolts must be tightened in the order shown.

- Exhaust pipes. New flange clamps MUST be used!
- HO2S sensor connectors
- KS sensors
- Intake Manifold
- Dipstick tube
- Vacuum hoses to the intake manifold
- PCV and breather hoses
- Fuel rail
- Ignition coil
- IAT sensor
- Fuel injector connectors
- TP sensor
- IAC motor
- Oil pressure sender
- ECT sensor electrical connector
- MAP sensor

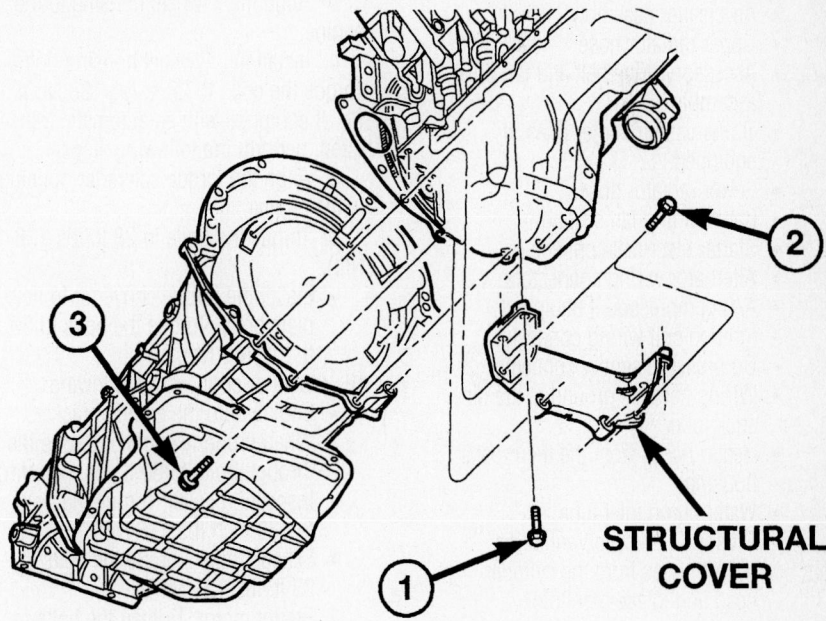

1 - BOLT
2 - BOLT
3 - BOLT

Tighten the structural cover bolts in this order—3.7L

9355PG01

- CMP sensor
- Radiator hoses
- Cruise control cable, if equipped
- Throttle cable
- Heater hoses
- Coolant bottle
- Power steering pump
- Alternator
- A/C compressor
- Radiator
- Fan assemblies
- Air cleaner assembly
- Negative battery cable

13. Fill and bleed the power steering system.

14. Fill the engine with clean oil.

15. Start the engine and check for leaks, repair if necessary.

Wrangler

1. Before servicing the vehicle, refer to the precautions in the beginning of this section.

2. Properly relieve the fuel system pressure.

3. Drain the cooling system.

4. Drain the engine oil.

5. Evacuate the A/C system.

6. Drain the power steering fluid, if equipped.

7. Remove or disconnect the following:
- Negative battery cable
- Air cleaner assembly
- Upper radiator hose
- Accessory drive belt and fan drive assembly
- Transmission cooler lines, if equipped
- Lower radiator hose
- Radiator and fan shroud
- Starter electrical connectors
- Alternator wiring connectors
- A/C compressor, if equipped
- Ignition coil wiring connector
- Oil pressure sender connector
- Wiring harness ground at the dipstick tube
- Heater hoses from the thermostat housing
- Water pump inlet tube
- Closed Crankcase Ventilation (CCV) hoses from the cylinder head and intake manifold
- Accelerator, transmission line pressure and speed control cables, if equipped
- Engine ground strap
- Power steering pressure switch, if equipped
- Engine Coolant Temperature (ECT) sensor

- Fuel injection wiring connectors
- Intake Air Temperature (IAT) sensor
- Idle Air Control (IAC) motor
- Throttle Position (TP) sensor
- Manifold Absolute Pressure (MAP) sensor
- Crankshaft Position (CKP) sensor
- Heated Oxygen (HO2S) sensor connector
- Vacuum hoses from the intake manifold
- Fuel line and bracket
- Power steering hoses from the steering gear, if equipped
- Oil filter
- Starter
- Engine support through bolts
- Exhaust front pipe
- Flywheel cover
- Torque converter, if equipped with automatic transmission
- Transmission flange bolts
- Left and right motor mounts
- Engine shock damper bracket

8. Place a support stand under the transmission.

9. Lift the engine out of the vehicle.

To install:

10. If equipped with a manual transmission, perform the following steps:

a. Insert the transmission shaft into the clutch spline.

b. Align the flywheel housing to the engine.

c. Install the flywheel housing bolts. Torque the bolts to 28 ft. lbs. (38 Nm).

11. If equipped with an automatic transmission, perform the following steps:

a. Align the torque converter housing to the engine.

b. Torque the bolts to 28 ft. lbs. (38 Nm).

- Install the torque converter to flexplate bolts. Torque the bolts to 50 ft. lbs. (68 Nm).

12. Install or connect the following:
- Engine onto the engine brackets. When properly aligned, torque the through bolts to 60 ft. lbs. (81 Nm).
- Inspection cover. Torque the bolts to 138 inch lbs. (16 Nm).
- Exhaust pipe. Torque the bolts to 23 ft. lbs. (31 Nm).
- Starter motor. Tighten the bolts to 33 ft. lbs. (45 Nm).
- Oil filter
- Power steering hoses
- Fuel supply line to the fuel rail
- Fuel supply rail to the intake manifold
- Vacuum hoses to the intake manifold

- Power steering pressure switch electrical connector
- ECT sensor electrical connector
- Fuel injector connectors
- IAT sensor
- IAC motor
- TP sensor
- MAP sensor
- CKP sensor
- HO2S sensor connector
- Engine ground strap
- Heater hoses and water pump inlet tube
- Cruise control cable, if equipped
- Throttle cable
- Fan shroud and radiator. Torque the bolts to 75 inch lbs. (8 Nm).
- Transmission cooler lines, if equipped. Torque the bolts to 10 ft. lbs. (15 Nm).
- Fan drive assembly. Torque the bolts to 20 ft. lbs. (27 Nm).
- Drive belt
- Radiator hoses
- Ignition coil
- Distributor
- A/C compressor
- A/C high pressure switch
- Alternator
- Oil pressure sender
- Harness ground at the oil dipstick tube bracket
- A/C hoses
- Air cleaner assembly
- Negative battery cable

13. Fill and bleed the power steering system.

14. Fill the engine with clean oil.

15. Fill and bleed the cooling system.

16. Recharge the A/C system.

17. Start the engine and check for leaks, repair if necessary.

Cherokee

1. Before servicing the vehicle, refer to the precautions in the beginning of this section.

2. Drain the cooling system.

3. Recover the A/C refrigerant.

4. Drain the engine oil.

5. Drain the power steering system.

6. Properly relieve the fuel system pressure.

7. Remove or disconnect the following:
- Battery
- Hood
- Air cleaner assembly
- Radiator hoses
- Fan shroud
- Electric cooling fan, if equipped
- Radiator

- Cooling fan
- Heater hoses
- Throttle cable
- Cruise control cable
- Transmission cable
- Fuel injector harness
- Camshaft Position (CMP) sensor connector
- Ignition coil wiring connector
- Oil pressure sender connector
- Body ground cable
- Starter solenoid connectors
- Fuel injection wiring harness
- Fuel line and bracket
- A/C compressor suction/discharge hose assembly
- Brake booster vacuum line
- Power steering hoses
- Intake manifold vacuum lines
- Crankshaft Position (CKP) sensor connector
- Engine speed sensor
- Oil filter
- Starter motor
- Heated Oxygen (HO2S) sensor connector
- Exhaust front pipe
- Flywheel housing access cover
- Torque converter, if equipped with automatic transmission
- Transmission flange bolts
- Left and right motor mounts

8. Support the transmission and lift the engine out of the vehicle.

To install:

9. Lower the engine in to the vehicle and position to the transmission.

10. Install or connect the following:
- Left and right motor mounts. Tighten the through bolts to 51 ft. lbs. (69 Nm).
- Transmission flange bolts. Tighten the bolts to 28 ft. lbs. (38 Nm).
- Torque converter, if equipped with automatic transmission. Tighten the bolts to 23 ft. lbs. (31 Nm).
- Flywheel housing access cover
- Exhaust front pipe
- Engine speed sensor electrical connector
- HO2S sensor connector
- Starter motor. Tighten the bolts to 33 ft. lbs. (45 Nm).
- Oil filter
- Vacuum hoses to the intake manifold
- CKP sensor connector
- Power steering hoses
- Brake booster vacuum line
- A/C compressor suction/discharge hose assembly
- Fuel line and bracket

- Fuel injection wiring harness
- Starter solenoid connectors
- Body ground cable
- Oil pressure sender connector
- Ignition coil wiring connector
- CMP sensor connector
- Fuel injector harness
- Transmission cable
- Cruise control cable
- Throttle cable
- Heater hoses
- Cooling fan
- Radiator
- Electric cooling fan, if equipped
- Fan shroud
- Radiator hoses
- Air cleaner assembly
- Hood
- Battery

11. Fill and bleed the cooling system.

12. Fill the engine with clean oil.

13. Fill and bleed the power steering system.

14. Recharge the A/C system.

15. Start the engine, check for leaks and repair if necessary.

Grand Cherokee

4.0L ENGINE

1. Before servicing the vehicle, refer to the precautions in the beginning of this section.

2. Drain the cooling system.

3. Drain the engine oil.

4. Drain the power steering fluid.

5. Properly relieve the fuel system pressure.

6. Recover the A/C refrigerant.

7. Remove or disconnect the following:
- Negative battery cable
- Hood
- Radiator hoses
- Upper radiator support
- Cooling fan and shroud
- Radiator
- A/C condenser
- Heater hoses
- Accelerator cable
- Cruise control cable
- Transmission cable
- Body ground cable
- Power steering pressure switch connector
- Engine Coolant Temperature (ECT) sensor connector
- Intake Air Temperature (IAT) sensor
- Fuel injector connectors
- Throttle Position (TP) sensor connector
- Manifold Absolute Pressure (MAP) sensor connector

- Crankshaft Position (CKP) sensor connector
- Heated Oxygen (HO2S) sensor connector
- Camshaft Position (CMP) sensor connector
- Ignition coil wiring connector
- Oil pressure sender connector
- Fuel line and bracket
- Air cleaner assembly
- Brake booster vacuum line
- Power steering hoses
- Starter motor
- Exhaust front pipe
- Flywheel cover
- Torque converter
- Transmission flange bolts
- Left and right motor mounts

8. Place a support stand under the transmission.

9. Lift the engine out of the vehicle.

To install:

10. Lower the engine in to the vehicle and position to the transmission.

11. Install or connect the following:
- Left and right motor mounts. Tighten the through bolts to 51 ft. lbs. (69 Nm).
- Transmission flange bolts. Tighten the bolts to 30 ft. lbs. (41 Nm).
- Torque converter. Tighten the bolts to 23 ft. lbs. (31 Nm).
- Flywheel cover
- Exhaust front pipe
- Starter motor. Tighten the bolts to 33 ft. lbs. (45 Nm).
- Power steering hoses
- Brake booster vacuum line
- Air cleaner assembly
- Fuel line and bracket
- Oil pressure sender connector
- Ignition coil wiring connector
- CMP sensor connector
- HO2S sensor connector
- CKP sensor connector
- MAP sensor connector
- TP sensor connector
- Fuel injector connectors
- IAT sensor
- ECT sensor connector
- Power steering pressure switch connector
- Body ground cable
- Transmission cable
- Cruise control cable
- Accelerator cable
- Heater hoses
- A/C condenser
- Radiator
- Cooling fan and shroud
- Upper radiator support
- Radiator hoses

- Hood
- Negative battery cable
12. Fill and bleed the cooling system.
13. Fill and bleed the cooing system.
14. Fill the engine with clean oil
15. Recharge the A/C system.
16. Start the engine, check for leaks and repair if necessary.

4.7L ENGINE

1. Before servicing the vehicle, refer to the precautions in the beginning of this section.
2. Drain the cooling system.
3. Drain the engine oil.
4. Drain the power steering system.
5. Properly relieve the fuel system pressure.
6. Recover the A/C refrigerant.
7. Remove or disconnect the following:
- Negative battery cable
- Front fascia
- Exhaust crossover pipe
- Engine ground straps
- Crankshaft Position (CKP) sensor
- Structural collar
- Starter
- Left and right inner fender liners
- Headlamp mounting module
- Air intake resonator

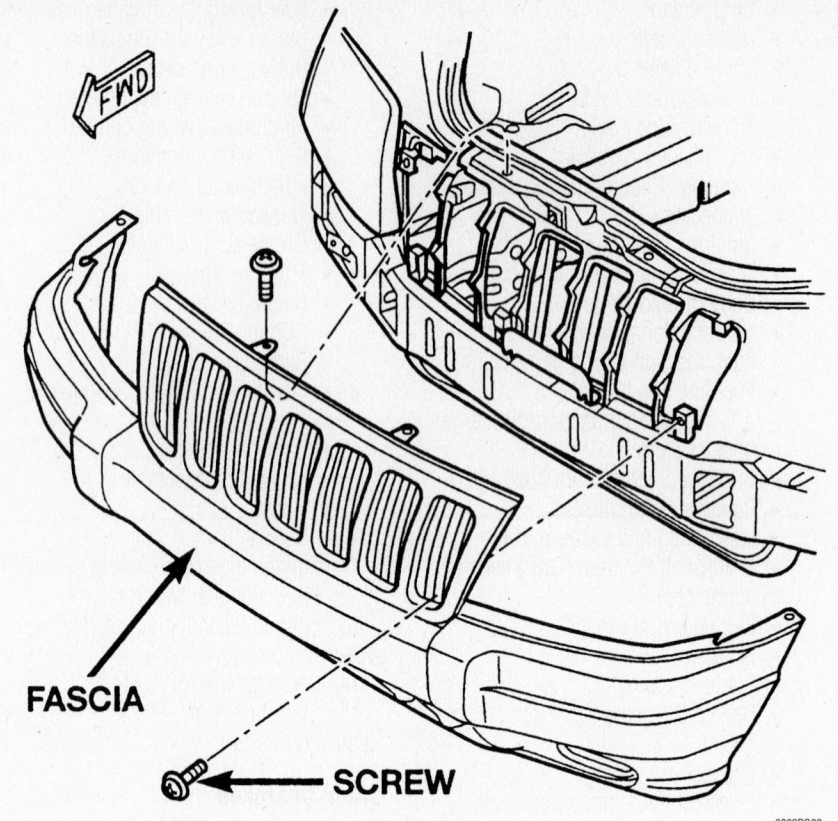

Exploded view of the front fascia panel—2001 Grand Cherokee

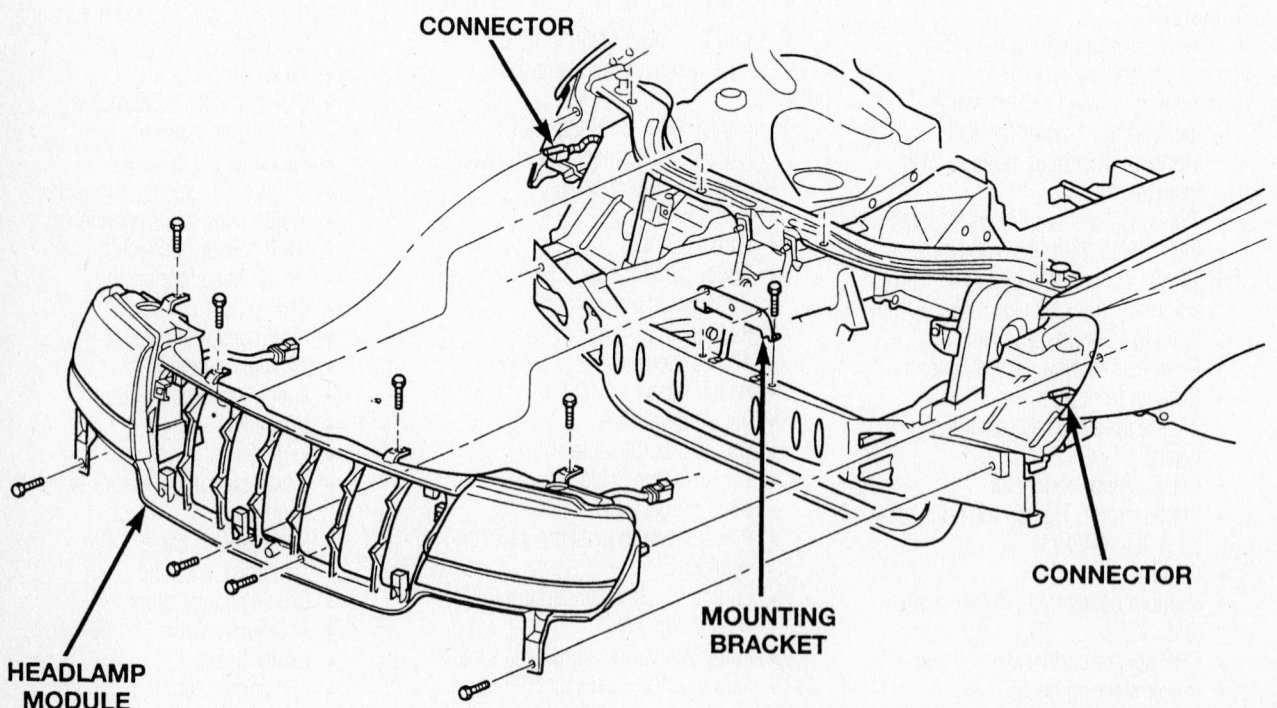

Exploded view of the headlamp module assembly—2001 Grand Cherokee

- Accelerator cable
- Cruise control cable
- Crankcase breather tubes
- Accessory drive belt
- A/C compressor
- Cooling fan assemblies
- Radiator hoses
- Transmission oil cooler lines
- Radiator
- A/C condenser
- Alternator
- Heater hoses
- Throttle Position (TP) sensor connector
- Intake Air Temperature (IAT) sensor connector
- Fuel injector harness connectors
- Engine Coolant Temperature (ECT) sensor connector
- Idle Air Control (IAC) valve connector
- Manifold Absolute Pressure (MAP) sensor connector
- Ignition coils
- Fuel line
- Power steering pump
- Oil fill tube
- Oil dipstick tube
- Heated Oxygen (HO$_2$S) sensor connectors
- Engine oil filter
- Exhaust crossover pipe
- Structural cover
- Rubber splash shield
- Starter motor
- Crankshaft Position (CKP) sensor connector
- Camshaft Position (CMP) sensor connector
- Torque converter
- Engine ground straps
- Left and right motor mounts
- Transmission flange bolts

8. Install Engine Lifting Fixture 8347 as shown.

9. Place a support stand under the transmission.

10. Lift the engine out of the vehicle.

To install:

11. Lower the engine in to the vehicle and position to the transmission.

12. Remove the engine lifting fixture.

13. Install or connect the following:
- Transmission flange bolts. Tighten the bolts to 50 ft. lbs. (68 Nm).
- Left and right motor mounts. Tighten the bolts to 45 ft. lbs. (61 Nm).
- Engine ground straps
- Torque converter. Tighten the bolts to 23 ft. lbs. (31 Nm).
- CMP sensor connector

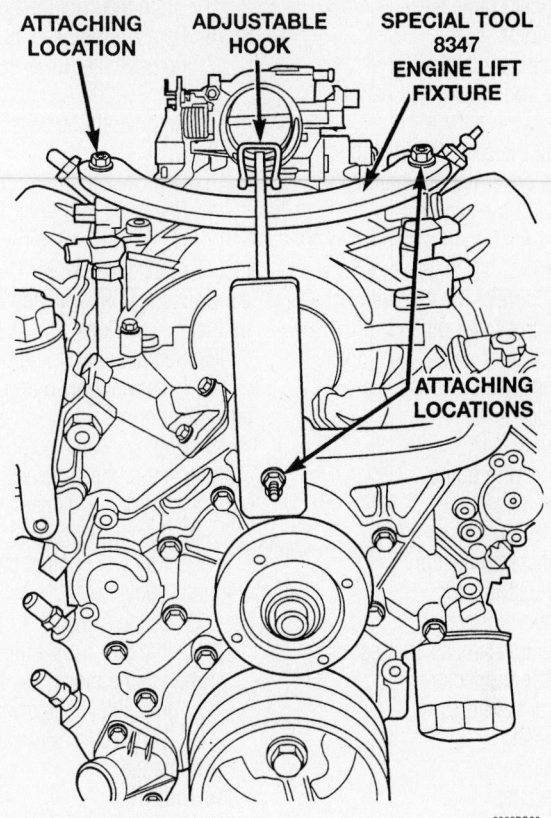

Engine Lifting Fixture—4.7L engine

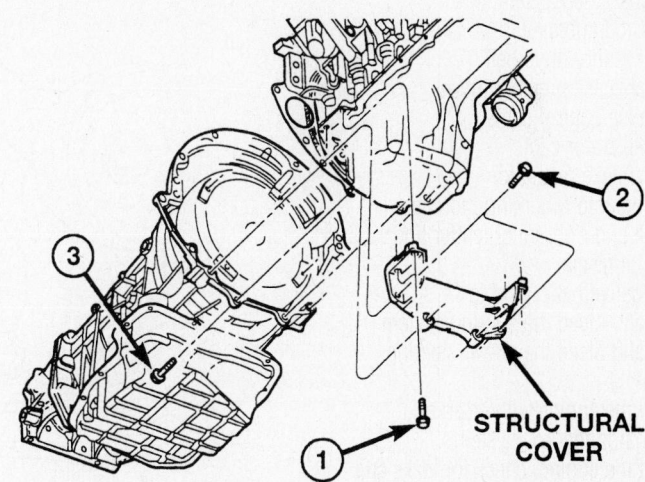

Structural cover torque sequence—4.7L engine

SEQUENCE	ITEM	TORQUE
1	BOLT (Qty 4)	54 N·m (40 ft. lbs.)
2	BOLT (Qty 2)	54 N·m (40 ft. lbs.)
3	BOLT (Qty 2)	54 N·m (40 ft. lbs.)

- CKP sensor connector
- Starter motor
- Rubber splash shield

14. Install the structural cover as follows:

a. Install all the bolts finger-tight.

b. Hold the cover tightly against the transmission and the engine.

c. Tighten the bolts in sequence to 40 ft. lbs. (54 Nm).

15. Install or connect the following:

- Exhaust crossover pipe
- Engine oil filter
- HO2S sensor connectors
- Oil dipstick tube
- Oil fill tube
- Power steering pump
- Fuel line
- Ignition coils
- MAP sensor connector
- IAC valve connector
- ECT sensor connector
- Fuel injector harness connectors
- IAT sensor connector
- TP sensor connector
- Heater hoses
- Alternator
- A/C condenser
- Radiator
- Transmission oil cooler lines
- Radiator hoses
- Cooling fan assemblies
- A/C compressor
- Accessory drive belt
- Crankcase breather tubes
- Cruise control cable
- Accelerator cable
- Air intake resonator
- Headlamp mounting module
- Left and right inner fender liners
- Front fascia
- Negative battery cable

16. Fill and bleed the cooling system.

17. Fill and bleed the power steering system.

- Fill the engine with clean oil.

18. Recharge the A/C system.

19. Start the engine, check for leaks and repair if necessary.

Water Pump

REMOVAL & INSTALLATION

2.4L Engine

1. Disconnect negative cable from battery.

2. Raise vehicle on a hoist.

3. Remove the accessory drive belts.

4. Remove the belt tensioner.

5. Drain the cooling system.

6. Remove the generator.

7. Remove the power steering pump.

8. Remove the A/C compressor.

9. Remove the accessory drive bracket.

10. Remove the timing belt.

11. Remove timing belt idler pulley.

12. Hold camshaft sprocket with Special tool C-4687 and adaptor C-4687-1, or equivalent, while removing bolt. Remove both cam sprockets.

13. Remove the timing belt rear cover.

14. Remove water pump to engine attaching screws.

15. Replace water pump body assembly if it has any of these defects:

a. Cracks or damage on the body.

b. Coolant leaks from the shaft seal, evident by wet coolant traces on the pump body.

c. Loose or rough turning bearing.

d. Impeller rubs either the pump body or the engine block.

e. Impeller loose or damaged.

f. Sprocket or sprocket flange loose or damaged.

To install:

16. Install new O-ring gasket in water pump body O-ring locating groove.

➡**Make sure O-ring is properly seated in water pump groove before tightening screws. An improperly located O-ring may be damaged and cause a coolant leak.**

17. Assemble pump body to block and tighten screws to 12 Nm (105 inch lbs.). Pressurize cooling system to 103.4 Kpa (15 psi) with pressure tester and check water pump shaft seal and O-ring for leaks.

18. Rotate pump by hand to check for freedom of movement.

19. Install the timing belt rear cover.

20. Install camshaft sprockets and target ring. Torque bolts to 101 Nm (75 ft. lbs.) while holding camshaft sprocket.

21. Install timing belt idler pulley and torque mounting bolt to 61 Nm (45 ft. lbs.).

22. Install the timing belt.

23. Install the accessory drive mounting bracket.

24. Install the power steering pump.

25. Install the generator.

26. Install the A/C compressor.

27. Install the belt tensioner.

28. Install the accessory drive belts.

29. Fill the cooling system.

30. Lower vehicle and connect battery cable.

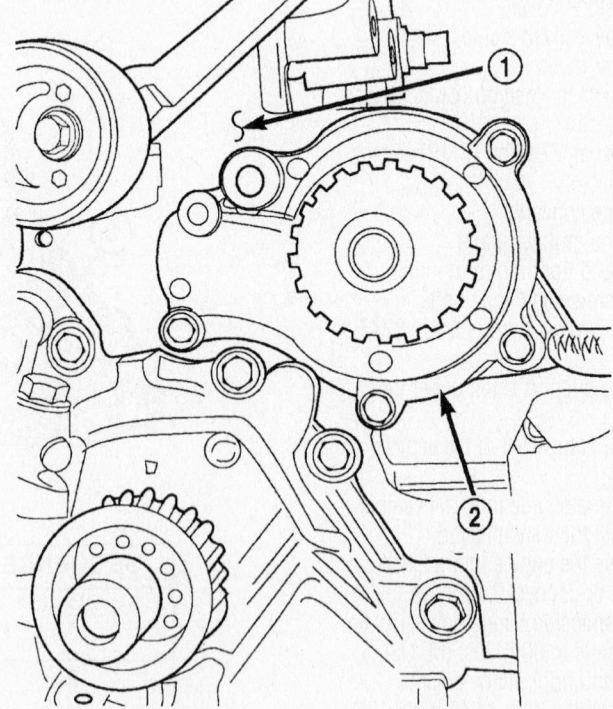

1 - CYLINDER BLOCK
2 - WATER PUMP

Water pump mounting—2.4L engines

67189-JEEP-G04

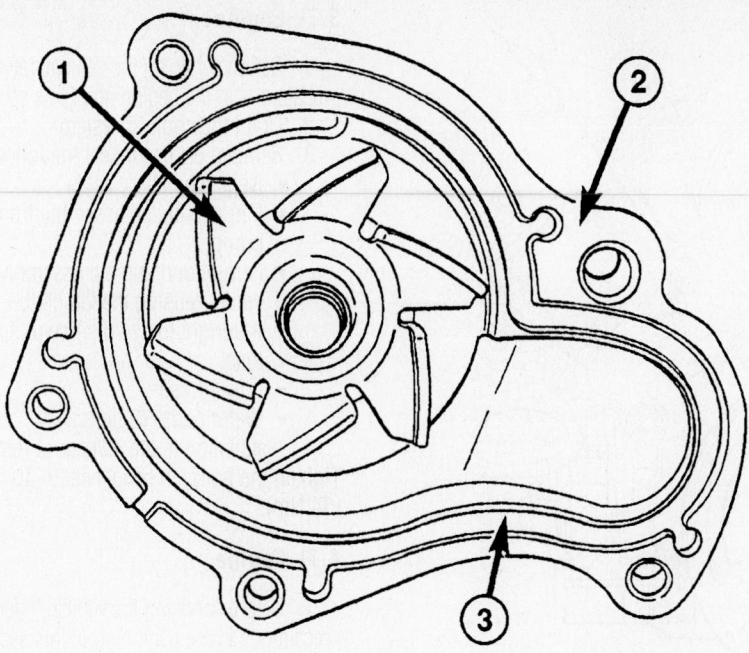

1 - IMPELLER
2 - WATER PUMP BODY
3 - O-RING LOCATING GROOVE

67189-JEEP-G05

Water pump body—2.4L engines

2.5L and 4.0L Engines

➡The 2.5L and 4.0L engines covered use a reverse rotation water pump. The letter R is stamped on the impeller to identify. Engines from previous years may be equipped with forward rotation water pumps. Installation of the wrong water pump will cause engine over heating.

1. Before servicing the vehicle, refer to the precautions in the beginning of this section.
2. Drain the cooling system.
3. Remove or disconnect the following:
- Negative battery cable
- Electric cooling fan connector
- Accessory drive belt
- Engine cooling fan and pulley

➡Some 4.0L engines are equipped with a fan clutch that threads directly on to the water pump shaft. This fan clutch is equipped with right-hand threads.

➡Do not store the fan clutch assembly horizontally, silicone may leak into the bearing grease and cause contamination.

- Water pump pulley
- Power steering pump

- Lower radiator hose from the water pump
- Heater hose
- Water pump and discard the gasket

➡**One of the water pump bolts is longer than the others. Note the location for reassembly.**

To install:

4. Clean the mating surfaces of all gasket material.
5. Install or connect the following:
- Water pump using a new gasket. Torque the bolts to 17 ft. lbs. (23 Nm).
- Heater hose
- Lower radiator hose
- Water pump pulley. Torque the bolts to 20 ft. lbs. (27 Nm).
- Power steering pump
- Engine cooling fan and shroud. Torque the bolts to 31 inch lbs. (3 Nm).
- Accessory drive belt
- Electric cooling fan connector
- Negative battery cable
6. Fill and bleed the cooling system.
7. Start the engine, check for leaks and repair if necessary.

FRONT VIEW

ROTATION DIRECTION
AS VIEWED

BACK VIEW

ROTATION DIRECTION
AS VIEWED

R STAMPED
INTO IMPELLER

7924PG02

Reverse rotation water pump—2.5L and 4.0L engines

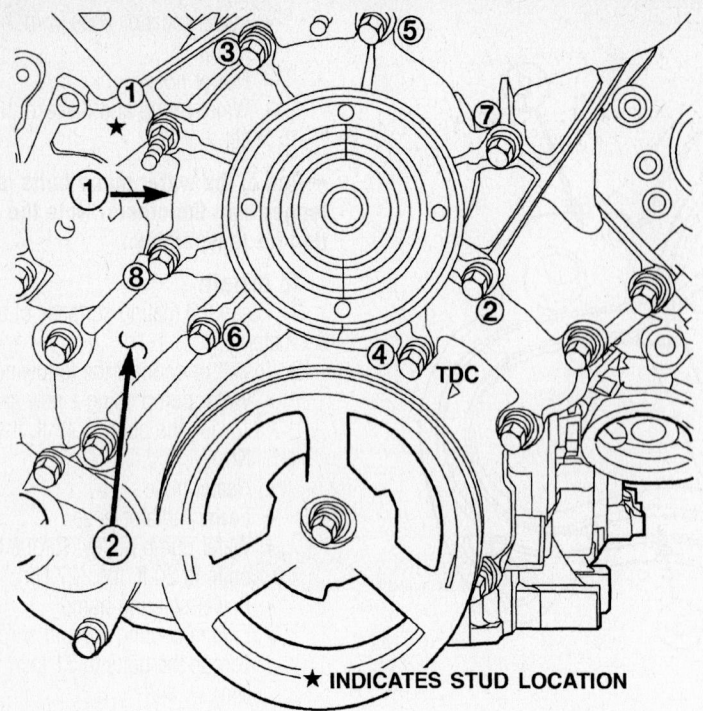

1 - WATER PUMP
2 - TIMING CHAIN COVER

Water pump tightening sequence—3.7L

9355PG02

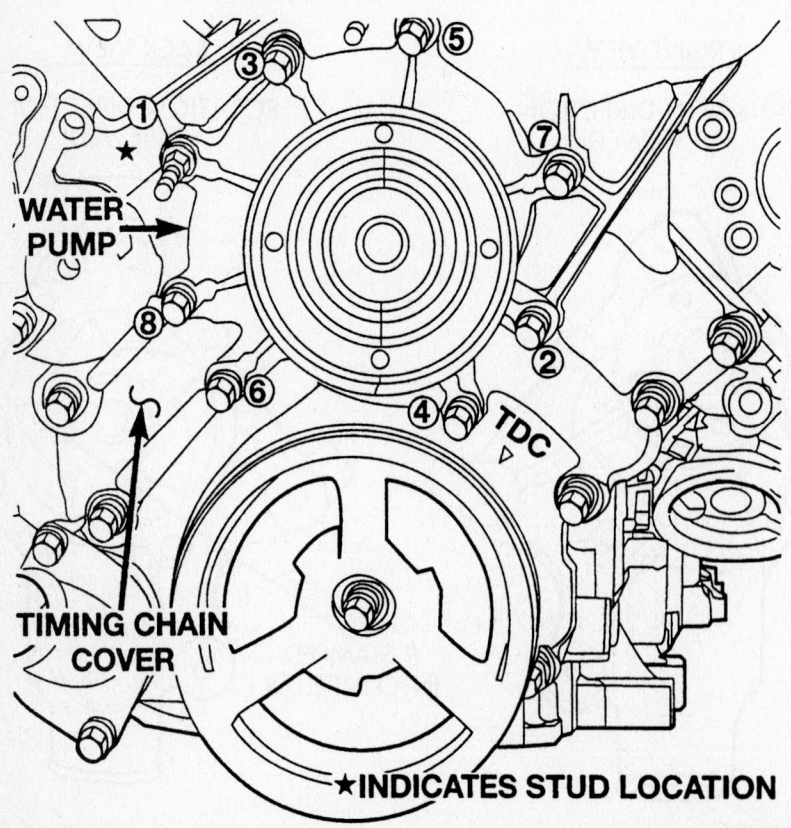

Water pump torque sequence—4.7L engine

9302PG06

3.7L Engine

1. Before servicing the vehicle, refer to the precautions in the beginning of this section.
2. Drain the cooling system.
3. Remove or disconnect the following:
   - Negative battery cable
   - Fan and clutch assembly from the pump
   - Fan shroud and fan assembly. If you're reusing the fan clutch, keep it upright to avoid silicone fluid loss!
   - Lower hose
   - Water pump (8 bolts)
4. Installation is the reverse of removal. Tighten the bolts, in sequence, to 40 ft. lbs. (54 Nm).

4.7L Engine

1. Before servicing the vehicle, refer to the precautions in the beginning of this section.
2. Drain the cooling system.
3. Remove or disconnect the following:
   - Negative battery cable
   - Fan and clutch assembly from the pump
   - Drive belt
   - Lower radiator hose from the water pump
   - Water pump
4. Clean the mating surfaces of all gasket material.

   To install:
2. Install or connect the following:
   - Water pump using a new gasket. Torque the bolts to 40 ft. lbs. (54 Nm).
   - Lower radiator hose
   - Drive belt
   - Negative battery cable
5. Fill and bleed the cooling system.
6. Start the engine, check for leaks and repair if necessary.

Heater Core

REMOVAL & INSTALLATION

Cherokee

1. Disconnect and remove the negative battery.

❊ CAUTION

After disconnecting the negative battery cable, wait 2 minutes for the driver's/passenger's air bag system capacitor to discharge before attempting to do any work around the steering column or instrument.

2. Drain the cooling system into a clean container for reuse.

3. Remove the instrument panel by performing the following procedure:

a. Turn the steering wheel in the straight-ahead position.

b. Remove the knee blocker from the instrument panel.

c. Remove the steering column; do not remove the air bag module, the steering wheel or switches from the steering column.

d. From under the driver's side of the instrument panel, disconnect the following items:

- Instrument panel wiring harness connector from the 100-way wiring harness connector at the left side of the inner panel.
- Side window demister hose at the heater/air conditioning housing demister/defroster duct on the driver's side.

e. Remove the glove box.

f. Reaching through the glove box opening, disconnect the following items:

- Two halves of the heater/air conditioning system vacuum harness connector.
- Instrument panel wiring harness connector from the heater/air conditioning system wiring harness connector.
- Instrument panel wiring harness connector from the passenger's side air bag module wiring harness connector.
- Side window demister hose at the heater/air conditioning housing demister/defroster duct (passenger's side).
- Two halves of the radio antenna coaxial cable connector.
- Two instrument panel wiring harness connectors from the passenger air bag ON/OFF switch wiring harness connector.
- Passenger's side air bag ON/OFF switch wiring harness from the retainer clip on the plenum bracket that supports the heater/air conditioning housing just inboard of the fuse block module.
- Two lower passenger's side air bag module bracket-to-dash panel nuts.

g. Remove the upper cover from the instrument panel.

h. Remove the 3 instrument panel-to-door hinge pillar screws.

i. Remove the 4 upper instrument panel-to-dash nuts.

j. Using an assistant, remove the instrument panel from the vehicle.

4. If equipped with air conditioning, discharge and recover the air conditioning system refrigerant.

5. Disconnect the refrigerant lines from the evaporator. Plug the refrigerant openings to prevent evaporation.

6. Disconnect the heater hoses from the heater core tubes.

7. Disconnect the heater/air conditioning system vacuum supply line connector from the T-fitting near the heater core tubes.

8. In the engine compartment, remove the 5 heater/air conditioning housing-to-chassis nuts. If necessary, loosen the battery hold-downs and reposition the battery for access.

9. Remove the cowl plenum drain tube from the heater/air conditioning housing stud; it's located behind the cylinder head on the cowl.

10. From the bottom of the heater/air conditioning housing, remove the floor duct.

11. On the passenger side, remove the heater/air conditioning housing-to-plenum bracket screw.

12. Pull the heater/air conditioning housing down far enough to clear the defrost/demist and fresh air ducts, then, rearward far enough to clear the mounting studs and the evaporator drain tube to clear the dash panel holes.

13. Remove the heater/air conditioning housing assembly from the vehicle.

14. Remove the heater/air conditioning housing upper case.

15. Lift the heater core from the lower half of the heater/air conditioning housing.

To install:

16. Assemble the heater core into the lower half of the heater/air conditioning housing.

17. Install the heater/air conditioning housing upper case.

18. Install the heater/air conditioning housing assembly to the vehicle.

19. On the passenger's side, install the heater/air conditioning housing-to-plenum bracket screw.

20. At the bottom of the heater/air conditioning housing, install the floor duct.

21. Install the cowl plenum drain tube to the heater/air conditioning housing stud; it's located behind the cylinder head on the cowl.

22. In the engine compartment, install the 5 heater/air conditioning housing-to-chassis nuts.

23. Connect the heater/air conditioning system vacuum supply line connector to the T-fitting near the heater core tubes.

24. Connect the heater hoses to the heater core tubes.

25. Connect the refrigerant lines to the evaporator.

26. If equipped with air conditioning, evacuate and charge the air conditioning system refrigerant.

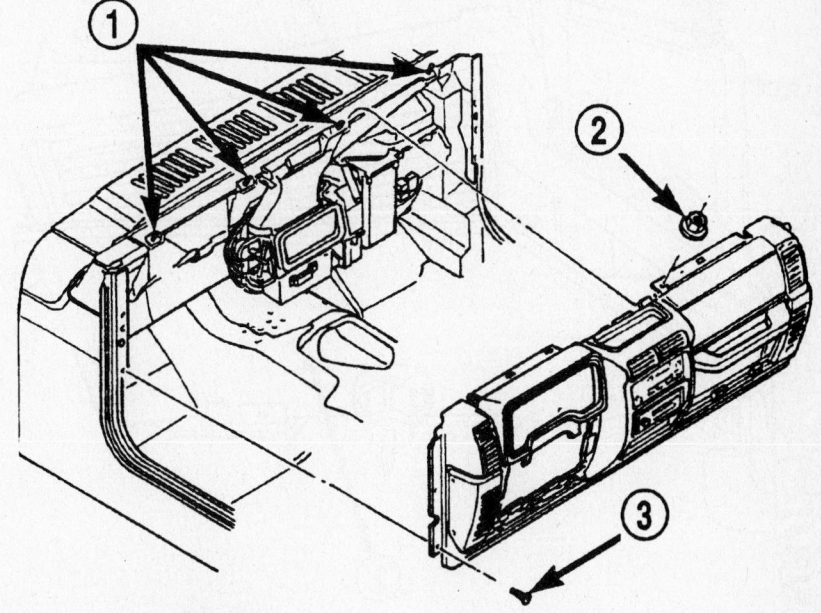

1 – STUDS
2 – NUT
3 – SCREW

93113GA6

View of the instrument panel and fasteners—Jeep Cherokee

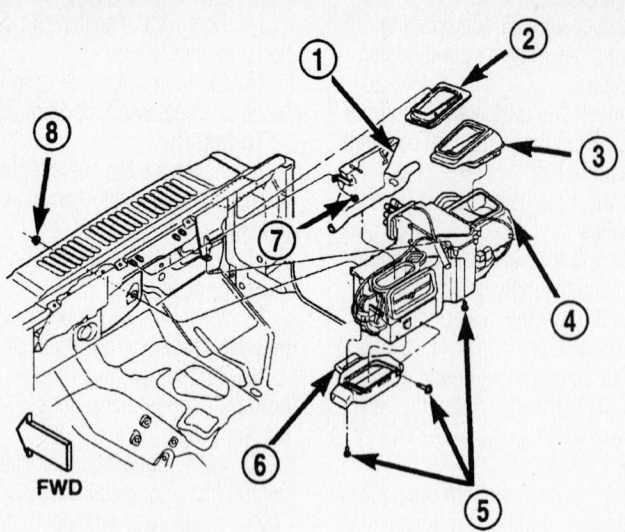

1 – DEFROST/DEMIST DUCT
2 – COLLAR
3 – FRESH AIR DUCT
4 – HEATER-A/C HOUSING
5 – SCREWS
6 – FLOOR DUCT
7 – NUT
8 – NUT

93113GA7

Exploded view of the heater core assembly—Jeep Cherokee

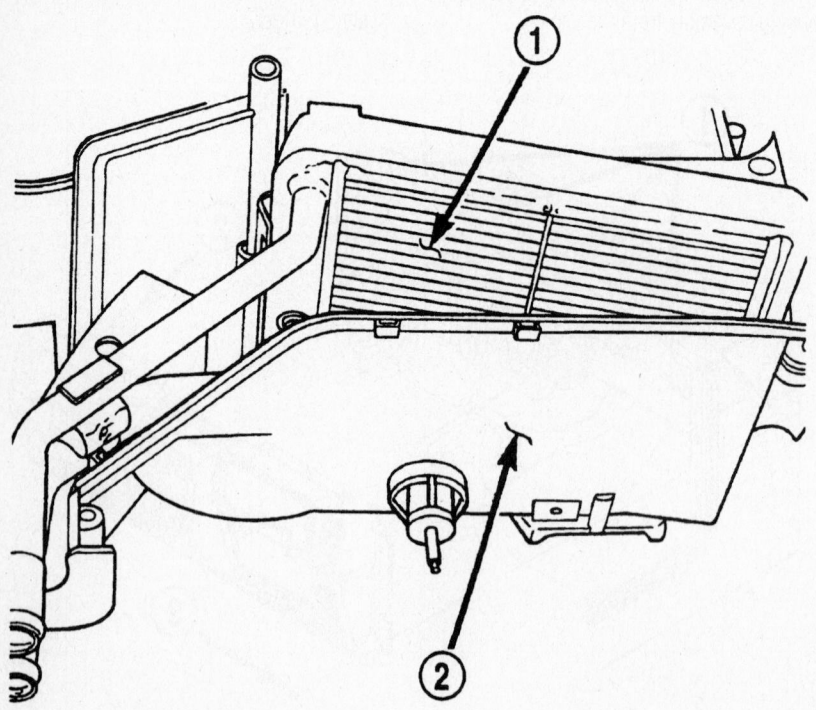

1 – HEATER CORE
2 – LOWER HEATER-A/C HOUSING

93113GA8

View of the heater core—Jeep Cherokee

27. Install the instrument panel by performing the following procedure:

a. Using an assistant, install the instrument panel to the vehicle.

b. Install the 4 upper instrument panel-to-dash nuts.

c. Install the 3 instrument panel-to-door hinge pillar screws.

d. Install the upper cover to the instrument panel.

e. Reaching through the glove box opening, connect the following items.

- Two lower passenger's side air bag module bracket-to-dash panel nuts.
- Passenger's side air bag ON/OFF switch wiring harness to the retainer clip on the plenum bracket that supports the heater/air conditioning housing just inboard of the fuse block module.
- Two instrument panel wiring harness connectors to the passenger air bag ON/OFF switch wiring harness connector.
- Two halves of the radio antenna coaxial cable connector.
- Side window demister hose at the heater/air conditioning housing demister/defroster duct (passenger's side).
- Instrument panel wiring harness connector to the passenger's side air bag module wiring harness connector.
- Instrument panel wiring harness connector to the heater/air conditioning system wiring harness connector.
- Two halves of the heater/air conditioning system vacuum harness connector.

f. Install the glove box.

g. Under the driver's side of the instrument panel, connect the following items:

- Side window demister hose at the heater/air conditioning housing demister/defroster duct on the driver's side.
- Instrument panel wiring harness connector to the 100-way wiring harness connector at the left side of the inner panel.

h. Install the steering column.

i. Install the knee blocker to the instrument panel.

28. Connect and remove the negative battery.

29. Refill the cooling system.

30. Run the engine to normal operating temperatures; then, check the climate control operation and check for leaks.

Wrangler

1. Disconnect and remove the negative battery.

✳✳ CAUTION

After disconnecting the negative battery cable, wait 2 minutes for the driver's/passenger's air bag system capacitor to discharge before attempting to do any work around the steering column or instrument.

2. Drain the cooling system into a clean container for reuse.

3. Remove the instrument panel by performing the following procedure:

 a. Turn the steering wheel in the straight-ahead position.

 b. Remove the knee blocker from the instrument panel.

 c. Remove the steering column; do not remove the air bag module, the steering wheel or switches from the steering column.

 d. From under the driver's side of the instrument panel, disconnect the following items:

 • Instrument panel wiring harness connector from the 100-way wiring harness connector at the left side of the inner panel.

 • Side window demister hose at the heater/air conditioning housing demister/defroster duct on the driver's side.

 e. Remove the glove box.

 f. Reaching through the glove box opening, disconnect the following items:

 • Two halves of the heater/air conditioning system vacuum harness connector.

 • Instrument panel wiring harness connector from the heater/air conditioning system wiring harness connector.

 • Instrument panel wiring harness connector from the passenger's side air bag module wiring harness connector.

 • Side window demister hose at the heater/air conditioning housing demister/defroster duct (passenger's side).

 • Two halves of the radio antenna coaxial cable connector.

 • Two instrument panel wiring harness connectors from the passenger air bag ON/OFF switch wiring harness connector.

 • Passenger's side air bag ON/OFF switch wiring harness from the

retainer clip on the plenum bracket that supports the heater/air conditioning housing just inboard of the fuse block module.

 • Two lower passenger's side air bag module bracket-to-dash panel nuts.

 g. Remove the upper cover from the instrument panel.

 h. Remove the 3 instrument panel-to-door hinge pillar screws.

 i. Remove the 4 upper instrument panel-to-dash nuts.

 j. Using an assistant, remove the instrument panel from the vehicle.

4. If equipped with air conditioning, discharge and recover the air conditioning system refrigerant.

5. Disconnect the refrigerant lines from the evaporator. Plug the refrigerant openings to prevent evaporation.

6. Disconnect the heater hoses from the heater core tubes.

7. Disconnect the heater/air conditioning system vacuum supply line connector from the T-fitting near the heater core tubes.

8. In the engine compartment, remove the 5 heater/air conditioning housing-to-chassis nuts. If necessary, loosen the battery hold-downs and reposition the battery for access.

9. Remove the cowl plenum drain tube from the heater/air conditioning housing

stud; it's located behind the cylinder head on the cowl.

10. From the bottom of the heater/air conditioning housing, remove the floor duct.

11. On the passenger side, remove the heater/air conditioning housing-to-plenum bracket screw.

12. Pull the heater/air conditioning housing down far enough to clear the defrost/demist and fresh air ducts, then, rearward far enough to clear the mounting studs and the evaporator drain tube to clear the dash panel holes.

13. Remove the heater/air conditioning housing assembly from the vehicle.

14. Remove the heater/air conditioning housing upper case.

15. Lift the heater core from the lower half of the heater/air conditioning housing.

To install:

16. Assemble the heater core into the lower half of the heater/air conditioning housing.

17. Install the heater/air conditioning housing upper case.

18. Install the heater/air conditioning housing assembly to the vehicle.

19. On the passenger's side, install the heater/air conditioning housing-to-plenum bracket screw.

20. At the bottom of the heater/air conditioning housing, install the floor duct.

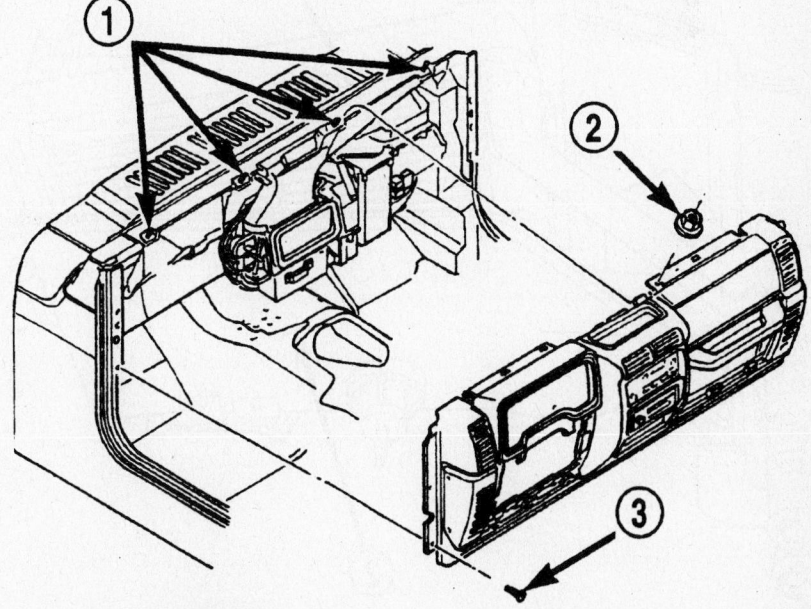

1 – STUDS
2 – NUT
3 – SCREW

93113GA6

View of the instrument panel and fasteners—Jeep Wrangler

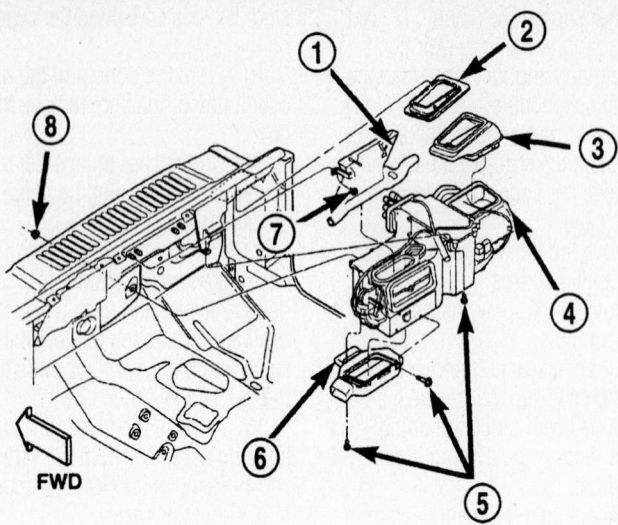

1 – DEFROST/DEMIST DUCT
2 – COLLAR
3 – FRESH AIR DUCT
4 – HEATER-A/C HOUSING
5 – SCREWS
6 – FLOOR DUCT
7 – NUT
8 – NUT

93113GA7

Exploded view of the heater core assembly—Jeep Wrangler

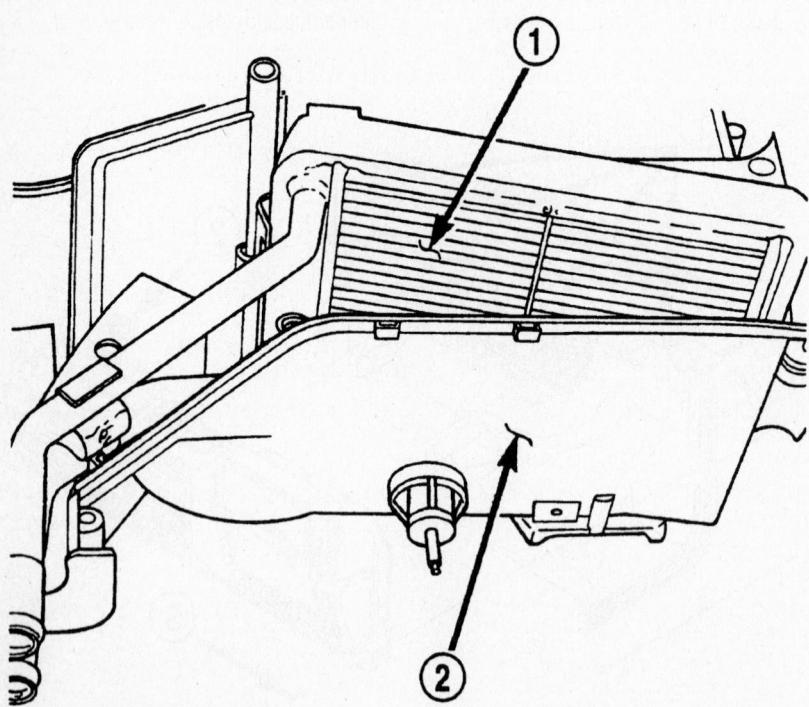

1 – HEATER CORE
2 – LOWER HEATER-A/C HOUSING

93113GA8

View of the heater core—Jeep Cherokee Wrangler

21. Install the cowl plenum drain tube to the heater/air conditioning housing stud; it's located behind the cylinder head on the cowl.

22. In the engine compartment, install the 5 heater/air conditioning housing-to-chassis nuts.

23. Connect the heater/air conditioning system vacuum supply line connector to the T-fitting near the heater core tubes.

24. Connect the heater hoses to the heater core tubes.

25. Connect the refrigerant lines to the evaporator.

26. If equipped with air conditioning, evacuate and charge the air conditioning system refrigerant.

27. Install the instrument panel by performing the following procedure:

 a. Using an assistant, install the instrument panel to the vehicle.

 b. Install the 4 upper instrument panel-to-dash nuts.

 c. Install the 3 instrument panel-to-door hinge pillar screws.

 d. Install the upper cover to the instrument panel.

 e. Reaching through the glove box opening, connect the following items.

 • Two lower passenger's side air bag module bracket-to-dash panel nuts.

 • Passenger's side air bag ON/OFF switch wiring harness to the retainer clip on the plenum bracket that supports the heater/air conditioning housing just inboard of the fuse block module.

 • Two instrument panel wiring harness connectors to the passenger air bag ON/OFF switch wiring harness connector.

 • Two halves of the radio antenna coaxial cable connector.

 • Side window demister hose at the heater/air conditioning housing demister/defroster duct (passenger's side).

 • Instrument panel wiring harness connector to the passenger's side air bag module wiring harness connector.

 • Instrument panel wiring harness connector to the heater/air conditioning system wiring harness connector.

 • Two halves of the heater/air conditioning system vacuum harness connector.

 f. Install the glove box.

 g. Under the driver's side of the instrument panel, connect the following items:

- Side window demister hose at the heater/air conditioning housing demister/defroster duct on the driver's side.
- Instrument panel wiring harness connector to the 100-way wiring harness connector at the left side of the inner panel.

h. Install the steering column.

i. Install the knee blocker to the instrument panel.

28. Connect and remove the negative battery.

29. Refill the cooling system.

30. Run the engine to normal operating temperatures; then, check the climate control operation and check for leaks.

Grand Cherokee

1. Disconnect and remove the negative battery.

❋❋ CAUTION

After disconnecting the negative battery cable, wait 2 minutes for the driver's/passenger's air bag system capacitor to discharge before attempting to do any work around the steering column or instrument.

2. Drain the cooling system into a clean container for reuse.

3. Remove the instrument panel by performing the following procedure:

a. Turn the steering wheel in the straight-ahead position.

b. Remove the A-pillar trim from both sides of the vehicle.

c. Remove the top cover from the instrument panel.

d. Near the windshield line, remove the 4 instrument panel-to-chassis nuts.

e. Remove the scuff plates from both front door sills.

f. Remove the trim panels from both sides of the inner cowl.

g. Remove the floor console.

h. Remove the fuse cover from the junction box.

i. Remove the instrument panel cluster bezel.

j. Remove the steering column opening cover from the instrument panel.

k. Remove the steering column bracket from the instrument panel column support bracket.

l. Remove the lower steering column shroud cover-to-multifunction switch screw; then, unsnap both halves of the shroud cover from the steering column.

m. Disconnect the instrument panel

wiring harness connectors from the following steering column components:

- Both lower clockspring connector receptacles
- Left multifunction switch receptacle
- Right multifunction switch receptacle
- Both ignition switch receptacles
- Shifter interlock solenoid receptacle
- Sentry Key Immobilizer Module (SKIM) receptacle, if equipped

n. Turn the ignition switch to ON position; then, release and remove the shifter interlock cable connector from the ignition lock housing receptacle.

o. Turn the ignition switch to OFF position; this will prevent the steering wheel from turning and the loss of the clockspring centering following steering column removal.

p. Remove the 4 steering column-to-instrument panel steering column bracket nuts.

q. Remove the steering column from the instrument panel.

r. Disconnect both side body wiring harness bulkhead connectors, the Ignition Off Draw (IOD) wiring harness connector and the fused wiring harness connector from the junction block connector receptacles.

s. Disconnect the instrument panel wiring harness-to-floor console component connectors:

- Air bag control module connector receptacle
- Parking brake switch terminal
- Transmission shifter connector receptacle

t. Remove the 2 instrument panel wiring harness-to-floor console ground terminals located behind the air bag control module.

u. Disconnect the instrument panel wiring harness-to-floor console retainers.

v. Remove the instrument panel-to-floor console bracket screws and the bracket.

w. Remove the driver's side floor duct-to-heater/air conditioning housing assembly screw and remove the duct.

x. If equipped with a manual heating-air conditioning system, disconnect the vacuum harness connector from behind the driver's side floor duct.

y. Remove the instrument panel steering column support bracket-to-driver's side of the heater/air conditioning housing assembly screw

z. Remove the instrument panel steering column support bracket-to-intermediate bracket screw.

aa. Remove the instrument panel

steering column support bracket-to-driver's side cowl plenum panel nut.

bb. Remove the 2 instrument panel-to-driver's side cowl side inner panel screws.

cc. Remove the instrument panel end cap.

dd. Remove the lower right center bezel from the instrument panel.

ee. At the passenger's side cowl side inner panel, disconnect the instrument panel wiring harness bulkhead connector from the lower cavity of the inline connector.

ff. Near the right side cowl inner panel, located under the end of the instrument panel, disconnect both halves of the radio antenna coaxial cable connector.

gg. Disconnect the 2 instrument panel-to-heater/air conditioning assembly wiring harness connectors.

hh. At the passenger's side, remove the 2 instrument panel structural duct-to-heater/air conditioning housing assembly screws.

ii. At the passenger's side cowl side inner panel, remove the 2 instrument panel-to-passenger's side cowl side inner panel screws.

jj. With the help of an assistant, lift the instrument panel from the vehicle.

4. Discharge and recover the air conditioning system refrigerant.

5. Disconnect the air conditioning system lines at the evaporator. Plug the openings to prevent contamination.

6. Disconnect the heater hoses from the heater core. Plug the openings to prevent coolant loss.

7. If equipped with a manual temperature control system, unplug the heater/air conditioning system vacuum supply line connector from the T-fitting located near the heater core tubes.

8. From the passenger's side inner fender shield, remove the coolant reservoir/overflow bottle.

9. From the passenger's side in the engine compartment dash panel, remove the PCM; DO NOT unplug it, just move it aside.

10. In the engine compartment, remove the heater/air conditioning housing-to-chassis nuts.

11. At the center of the dashboard, remove the rear floor ducts from the floor heat duct outlets.

12. Disconnect the heater/air conditioning housing wire harness connectors.

13. In the passenger compartment, remove the heater/air conditioning housing-to-chassis nuts.

14. Place covers inside the vehicle to catch any spilt coolant.

15. Remove the heater/air conditioning housing assembly from the vehicle.

16. Remove the foam gasket from around the heater core tubes.

➡ **Note the position of the irregular shaped gasket so that it may be reinstalled in its correct position.**

17. Remove the heater core retainers and screws.

18. If necessary, remove the mode door actuator for clearance to remove the core.

19. Remove the heater core from the heater/air conditioning housing assembly.

To install:

20. Install the heater core tom the heater/air conditioning housing assembly.

21. If removed, install the mode door actuator.

22. Install the heater core retainers and screws.

23. Install the foam gasket around the heater core tubes.

24. Install the heater/air conditioning housing assembly to the vehicle.

25. In the passenger compartment, install the heater/air conditioning housing-to-chassis nuts.

26. Connect the heater/air conditioning housing wire harness connectors.

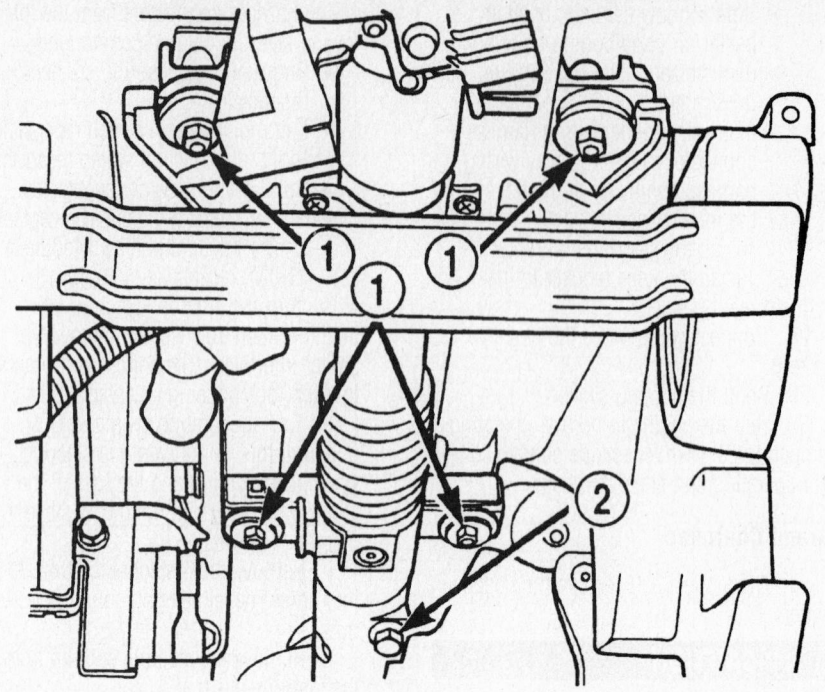

1 – COLUMN MOUNTING NUTS
2 – COUPLER BOLT

93113GA1

View of the steering column mounting nuts—Jeep Grand Cherokee

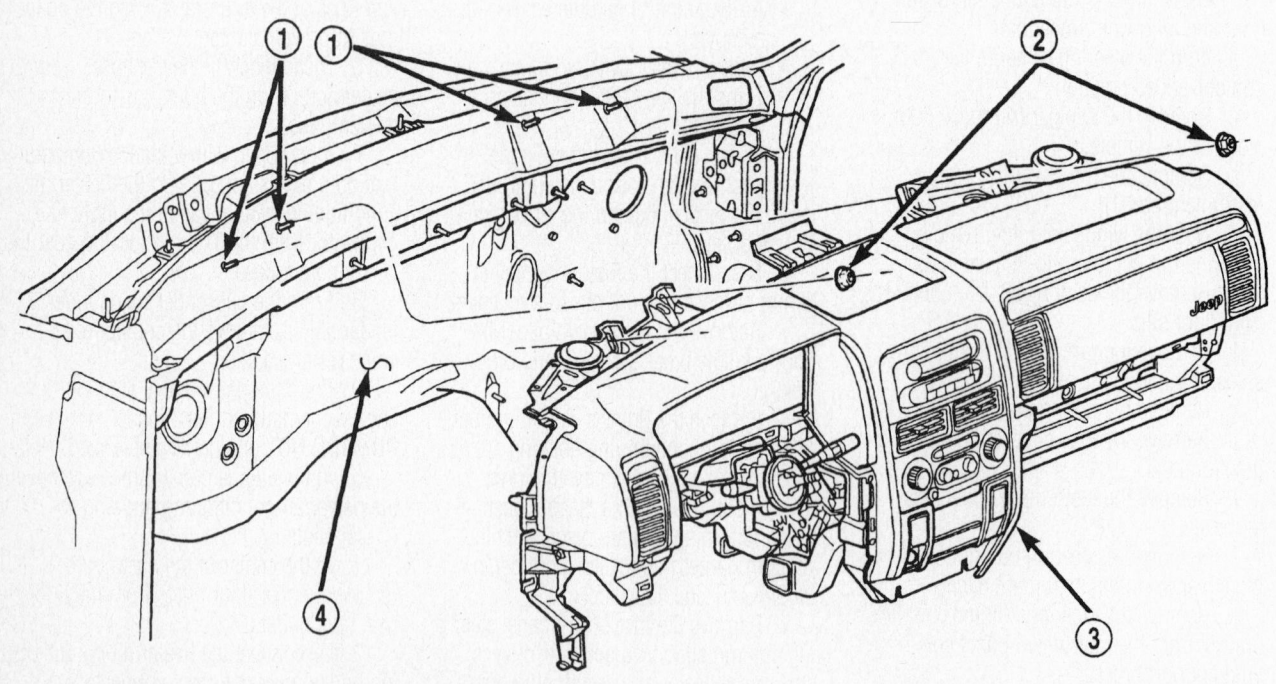

1 – STUD (4)   3 – INSTRUMENT PANEL
2 – NUT (4)   4 – DASH PANEL

93113GA2

View of the instrument panel assembly— Jeep Grand Cherokee

27. At the center of the dashboard, install the rear floor ducts to the floor heat duct outlets.

28. In the engine compartment, install the heater/air conditioning housing-to-chassis nuts.

29. At the passenger's side in the engine compartment dash panel, install the PCM.

30. At the passenger's side inner fender shield, install the coolant reservoir/overflow bottle.

31. If equipped with a manual temperature control system, plug the heater/air conditioning system vacuum supply line connector to the T-fitting located near the heater core tubes.

32. Connect the heater hoses to the heater core.

33. Connect the air conditioning system lines to the evaporator.

34. Evacuate and charge the air conditioning system refrigerant.

35. Remove the instrument panel by performing the following procedure:

a. Turn the steering wheel in the straight-ahead position.

b. Install the A-pillar trim to both sides of the vehicle.

c. Install the top cover to the instrument panel.

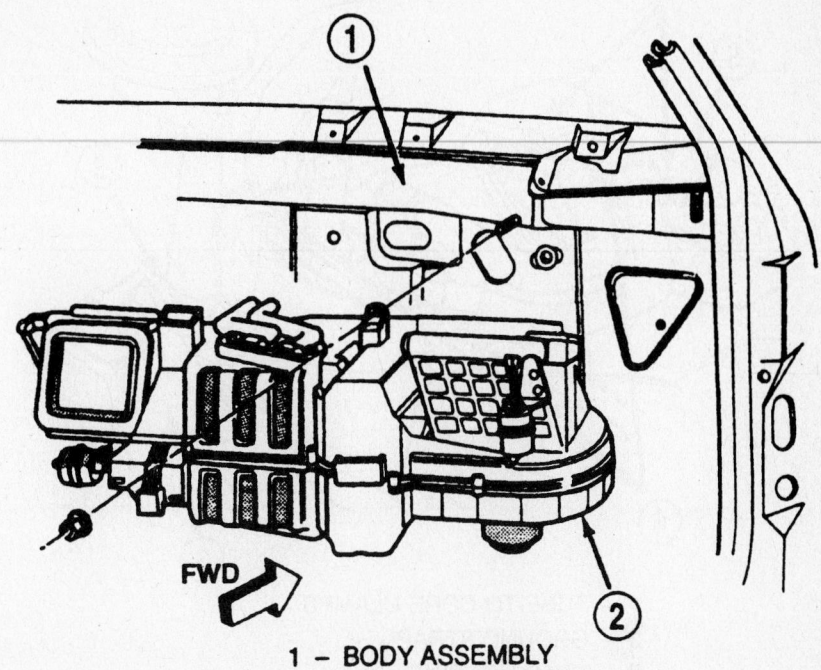

FWD

1 – BODY ASSEMBLY
2 – HEATER A/C UNIT

93113GA3

View of the heater/air conditioning housing assembly—Jeep Grand Cherokee

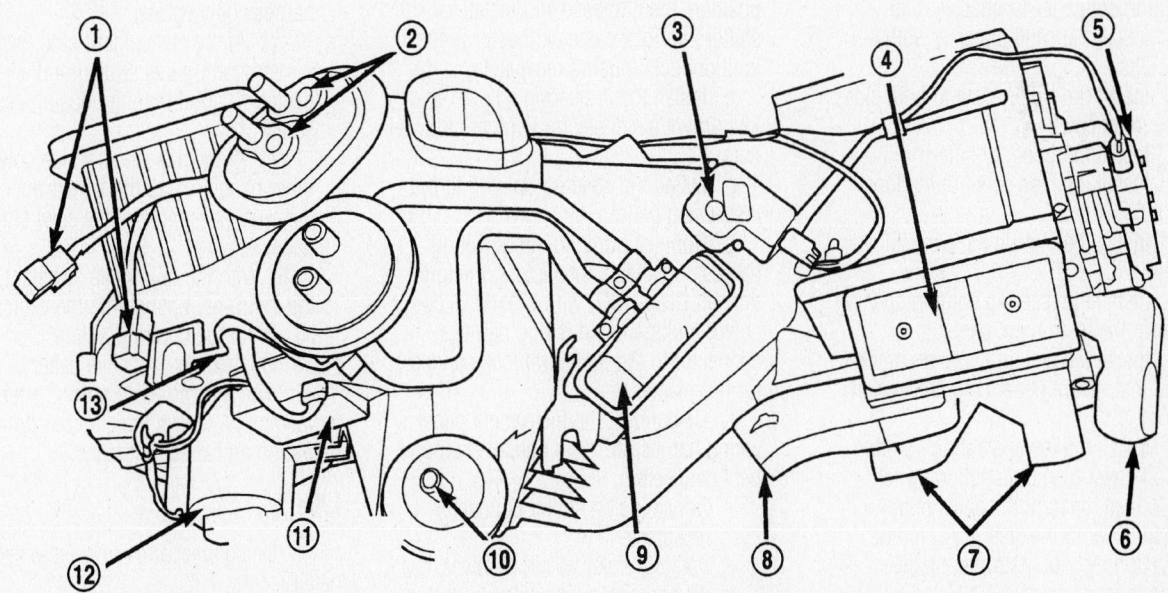

1 – ELECTRICAL CONNECTORS
2 – EVAPORATOR FITTINGS (CAPPED)
3 – ELECTRIC ACTUATOR
4 – OUTLET TO DEFROSTER DUCTS
5 – ELECTRIC ACTUATOR
6 – FLOOR DUCT
7 – TO REAR PASSENGER FLOOR AIR DUCTS

8 – FLOOR DUCT
9 – HEATER CORE AND TUBES
10 – HOUSING DRAIN
11 – BLOWER MOTOR CONTROLLER/POWER MODULE
12 – BLOWER MOTOR
13 – GROUND STRAP

93113GA4

View of the heater core, the heater/air conditioning housing and related components—Jeep Grand Cherokee

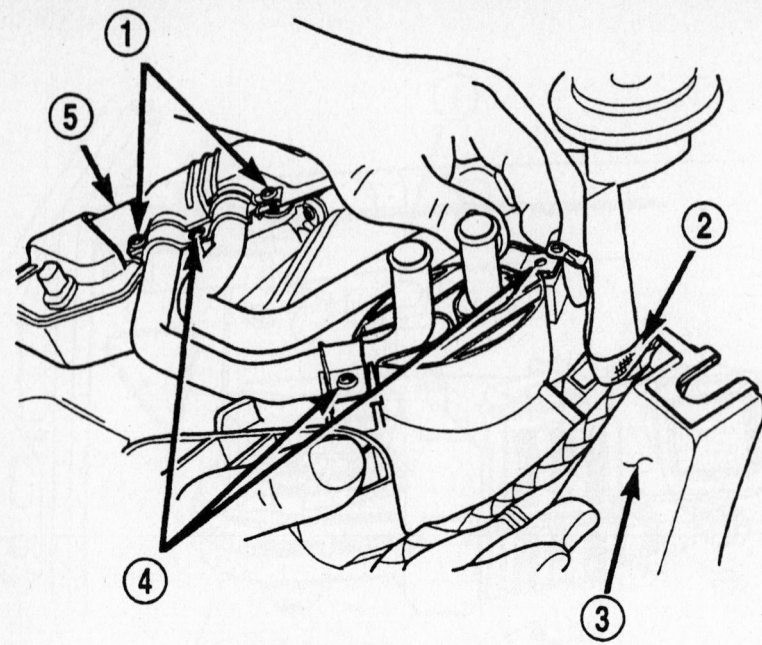

1 – TUBE-TO-CORE CLAMPS
2 – GROUND STRAP
3 – HVAC HOUSING
4 – TUBE RETAINERS AND SCREWS
5 – HEATER CORE

93113GA5

View of the heater core screws, gasket and retainers—Jeep Grand Cherokee

d. Near the windshield line, install the 4 instrument panel-to-chassis nuts.

e. Install the scuff plates to both front door sills.

f. Install the trim panels to both sides of the inner cowl.

g. Install the floor console.

h. Install the fuse cover to the junction box.

i. Install the instrument panel cluster bezel.

j. Install the steering column opening cover to the instrument panel.

k. Install the steering column bracket to the instrument panel column support bracket.

l. Install the lower steering column shroud cover-to-multifunction switch screw; then, snap both halves of the shroud cover to the steering column.

m. Connect the instrument panel wiring harness connectors to the following steering column components:

- Sentry Key Immobilizer Module (SKIM) receptacle, if equipped
- Shifter interlock solenoid receptacle
- Both ignition switch receptacles
- Right multifunction switch receptacle
- Left multifunction switch receptacle
- Both lower clockspring connector receptacles

n. Turn the ignition switch to ON position; then, release and install the shifter interlock cable connector to the ignition lock housing receptacle.

o. Install the 4 steering column-to-instrument panel steering column bracket nuts.

p. Install the steering column to the instrument panel.

q. Connect both side body wiring harness bulkhead connectors, the Ignition Off Draw (IOD) wiring harness connector and the fused wiring harness connector to the junction block connector receptacles.

r. Disconnect the instrument panel wiring harness-to-floor console component connectors:

- Transmission shifter connector receptacle
- Parking brake switch terminal
- Air bag control module connector receptacle

s. Install the 2 instrument panel wiring harness-to-floor console ground terminals located behind the air bag control module.

t. Connect the instrument panel wiring harness-to-floor console retainers.

u. Install the instrument panel-to-floor console bracket and the screws.

v. Install the driver's side floor duct and the duct-to-heater/air conditioning housing assembly screw.

w. If equipped with a manual heating-air conditioning system, connect the vacuum harness connector behind the driver's side floor duct.

x. Install the instrument panel steering column support bracket-to-driver's side of the heater/air conditioning housing assembly screw.

y. Install the instrument panel steering column support bracket-to-intermediate bracket screw.

z. Install the instrument panel steering column support bracket-to-driver's side cowl plenum panel nut.

aa. Install the 2 instrument panel-to-driver's side cowl side inner panel screws.

bb. Install the instrument panel end cap.

cc. Install the lower right center bezel to the instrument panel.

dd. At the passenger's side cowl side inner panel, connect the instrument panel wiring harness bulkhead connector to the lower cavity of the inline connector.

ee. Near the right cowl side inner panel located under the end of the instrument panel, connect both halves of the radio antenna coaxial cable connector.

ff. Connect the 2 instrument panel-to-heater/air conditioning assembly wiring harness connectors.

gg. At the passenger's side, install the 2 instrument panel structural duct-to-heater/air conditioning housing assembly screws.

hh. At the passenger's side cowl side inner panel, install the 2 instrument panel-to-passenger's side cowl side inner panel screws.

ii. With the help of an assistant, lift the instrument panel into the vehicle.

36. Refill the cooling system.

37. Connect the negative battery.

38. Run the engine to normal operating temperatures. Check the climate control operation and check for leaks.

Liberty

1. Disconnect and remove the negative battery.

⁂ CAUTION

After disconnecting the negative battery cable, wait 2 minutes for the driver's/passenger's air bag system capacitor to discharge before attempting to do any work around the steering column or instrument.

2. Drain the cooling system into a clean container for reuse.

3. Remove the instrument panel by performing the following procedure:

a. Turn the steering wheel in the straight-ahead position.

b. Remove the A-pillar trim from both sides of the vehicle.

c. Remove the top cover from the instrument panel.

d. Remove the speakers.

e. Remove the floor console.

f. Remove the radio.

g. Remove the center support bracket.

h. Remove the trim panels from both sides of the inner cowl.

i. Remove the fuse cover from the junction box.

j. Remove the instrument panel cluster bezel.

k. Remove the steering column opening cover from the instrument panel.

l. Remove the steering column bracket from the instrument panel column support bracket.

m. Remove the lower steering column shroud cover-to-multifunction switch screw; then, unsnap both halves of the shroud cover from the steering column.

n. Disconnect the instrument panel wiring harness connectors from the following steering column components:
- Both lower clockspring connector receptacles
- Left multifunction switch receptacle
- Right multifunction switch receptacle
- Both ignition switch receptacles
- Shifter interlock solenoid receptacle
- Sentry Key Immobilizer Module (SKIM) receptacle, if equipped

o. Turn the ignition switch to ON position; then, release and remove the shifter interlock cable connector from the ignition lock housing receptacle.

p. Turn the ignition switch to OFF position; this will prevent the steering wheel from turning and the loss of the clockspring centering following steering column removal.

q. Remove the 4 steering column-to-instrument panel steering column bracket nuts.

r. Remove the steering column from the instrument panel.

4. Remove the driver's side cowl trim cover.

5. Disconnect the green and light blue wire harness bulk connectors at the junction block.

6. Disconnect the electrical connector at the inner side of the pedal support bracket.

7. Remove the 2 bolts from the front and the 2 from the side of the pedal support bracket.

8. Remove the glove box.

9. Remove the 2 HVAC mounting bolts behind the center trim.

10. Remove the passenger trim bezel.

11. Remove the HVAC mount bolt above the glove box.

12. Remove the HVAC bolt at the lower outside glove box opening.

13. Remove the passenger trim cover, disconnect the blower resistor, remove the roll-down brackets at the right cowl side panel.

14. Disconnect the vacuum check valve and the vacuum reservoir.

15. Disconnect the blower connectors.

16. Remove the 4 top bolts connecting the instrument panel to the cowl.

17. Roll the instrument panel rearward and disconnect the wiring.

18. Remove the panel.

19. Discharge and recover the air conditioning system refrigerant.

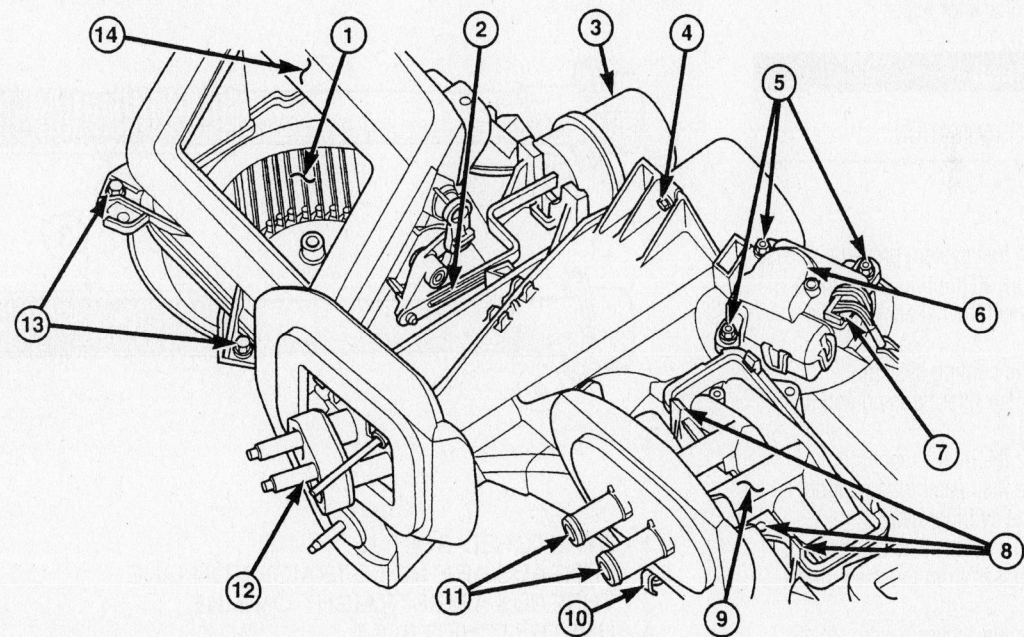

1 - BLOWER MOTOR AND CAGE
2 - RECIRCULATION DOOR ACTUATOR LINKAGE
3 - RECIRCULATION DOOR VACUUM ACTUATOR
4 - CASE RETAINER SCREW
5 - BLEND DOOR ACTUATOR MOUNTING SCREWS
6 - ELECTRIC BLEND DOOR ACTUATOR
7 - ELECTRICAL CONNECTOR FOR BLEND DOOR ACTUATOR
8 - HEATER CORE RETAINER TABS (4) AND SCREWS (2)
9 - HEATER CORE
10 - HVAC CASE RETAINER CLIP
11 - HEATER CORE INPUT AND OUTPUT CONNECTIONS
12 - EVAPORATOR CONNECTION FLANGE
13 - HVAC CASE RETAINER SCREWS
14 - HVAC HOUSING

HVAC case components—Liberty

9355PG99

20. Disconnect the air conditioning system lines at the evaporator. Plug the openings to prevent contamination.

21. Disconnect the heater hoses from the heater core. Plug the openings to prevent coolant loss.

22. If equipped with a manual temperature control system, unplug the heater/air conditioning system vacuum supply line connector from the T-fitting located near the heater core tubes.

23. Remove all remaining fasteners and connections and remove the HVAC unit.

24. Disconnect all remaining hoses and wires.

25. Remove the blower motor.

26. Pop out the grommet on the vacuum supply line and slide hole.

27. Remove the foam gasket from around the heater core tubes.

28. Pry off the 4 snap clips that hold the halves of the unit together and separate the unit halves.

29. Installation is the reverse of removal.

30. Refill the cooling system.

31. Connect the negative battery.

32. Evacuate, charge and leak test the system.

33. Run the engine to normal operating temperatures. Check the climate control operation and check for leaks.

Cylinder Head

REMOVAL & INSTALLATION

2.4L Engine

1. Perform fuel system pressure release procedure before attempting any repairs.

2. Disconnect the battery negative cable.

3. Drain the cooling system.

4. Remove air filter housing and inlet tube.

5. Remove the intake manifold.

6. Remove the heater tube support bracket from the cylinder head.

7. Disconnect the radiator upper and heater supply hoses from the water outlet connections.

8. Remove the accessory drive belts.

9. Raise the vehicle and remove the exhaust pipe from the manifold.

10. Remove the power steering pump and set aside. Do not disconnect lines.

11. Remove the accessory drive bracket.

12. Remove the ignition coil and wires from the engine.

13. Disconnect the cam sensor and fuel injector wiring connectors.

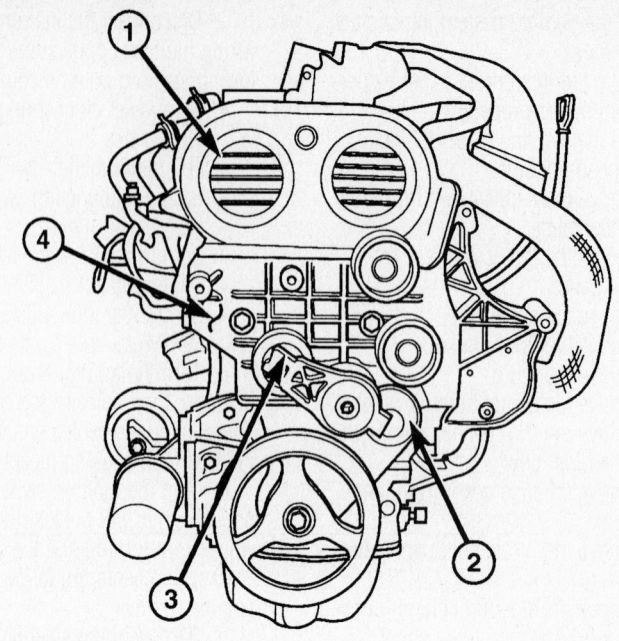

1- UPPER TIMING BELT COVER
2- LOWER TIMING BELT COVER
3- BELT TENSIONER
4- ACCESSORY DRIVE BRACKET

67189-JEEP-G06

Accessory drive belt bracket—2.4L engines

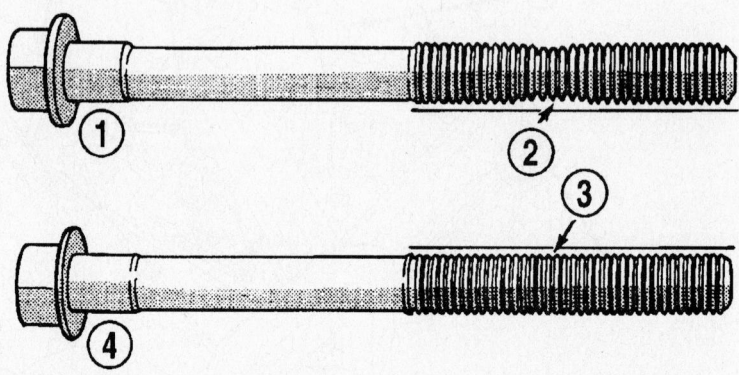

1 - STRETCHED BOLT
2 - THREADS ARE NOT STRAIGHT ON LINE
3 - THREADS ARE STRAIGHT ON LINE
4 - UNSTRETCHED BOLT

67189-JEEP-G07

Checking the head bolts for stretching—2.4L engines

14. Remove the timing belt and camshaft sprockets.

15. Remove the timing belt idler pulley and rear timing belt cover.

16. Remove the cylinder head cover.

17. Remove camshafts.

➥Identify the rocker arm positions to ensure correct installation in original position, if reused.

18. Remove the rocker arms.

19. Remove the cylinder head bolts in reverse of the tightening sequence.

20. Remove the cylinder head from the engine block.

21. Inspect and clean the cylinder head.

To install:

➡ **The cylinder head bolts should be examined before reuse. If the threads are necked down, the bolts must be replaced.**

22. Before installing the bolts, the threads should be coated with engine oil.

23. Position the cylinder head gasket on the engine block.

24. Install the cylinder head on the engine block.

25. Tighten the cylinder head bolts in the sequence shown. Tighten in 4 steps:
- First: 34 Nm (25 ft. lbs.)
- Second: 68 Nm (50 ft. lbs.)
- Third: 68 Nm (50 ft. lbs.)
- Fourth: an additional ¼ turn

26. Install the rocker arms.

27. Install the camshafts.

28. Install the cylinder head cover.

29. Install the timing belt rear cover and timing belt idler pulley.

30. Install the timing belt and camshaft sprockets.

31. Connect the cam sensor and fuel injectors wiring connectors.

32. Install the ignition coil and wires. Connect the ignition coil wiring connector.

33. Install the accessory drive bracket.

34. Install the power steering pump to the cylinder head.

35. Raise the vehicle and install the exhaust pipe to the manifold.

36. Install the accessory drive belts.

37. Install the heater tube support bracket to the cylinder head.

38. Install the intake manifold.

39. Connect all vacuum lines, electrical wiring, ground straps and fuel line.

40. Fill the cooling system.

41. Connect the battery negative cable.

2.5L Engine

1. Before servicing the vehicle, refer to the precautions in the beginning of this section.

2. Properly relieve the fuel system pressure.

3. Drain the cooling system.

4. Remove or disconnect the following:
- Negative battery cable
- Crankcase Ventilation (CCV) hoses
- Air cleaner assembly
- Valve cover

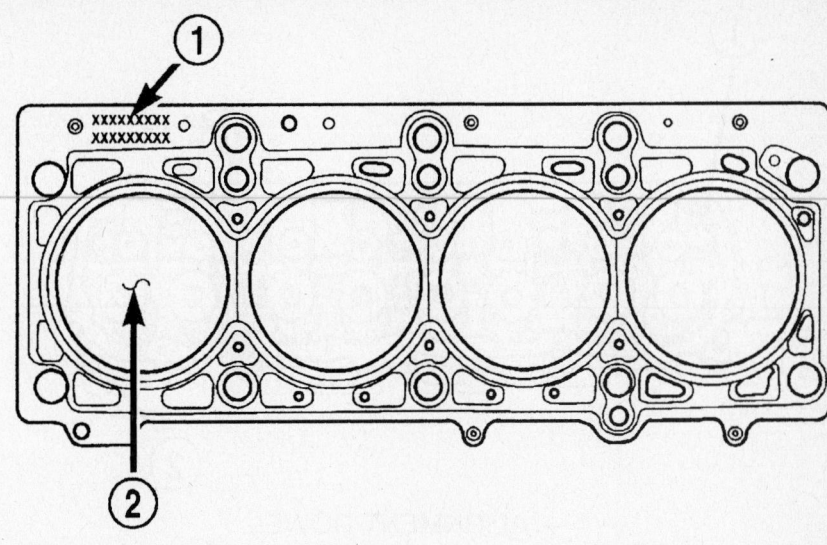

1 - PART NUMBER FACES UP
2 - NO. 1 CYLINDER

67189-JEEP-G08

Cylinder head gasket positioning—2.4L engines

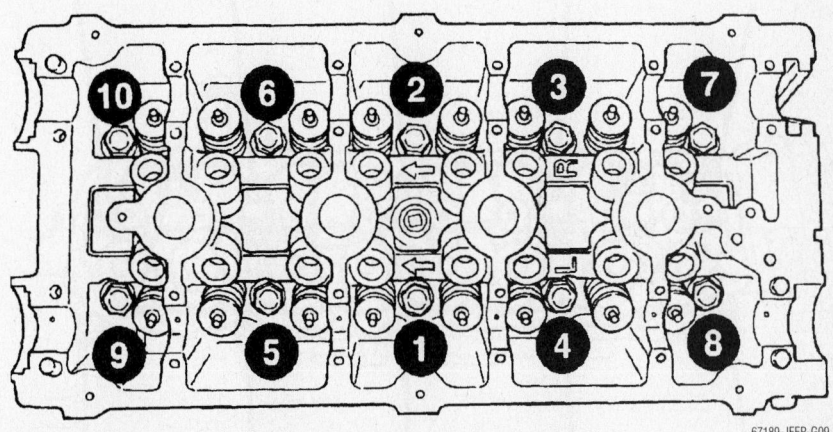

67189-JEEP-G09

Cylinder head tightening sequence—2.4L engines

➡ **Keep valvetrain components in order for reassembly.**

- Rocker arms
- Pushrods
- Accessory drive belt
- A/C compressor and bracket, if equipped
- Power steering pump and bracket, if equipped
- Fuel line
- Combination manifold
- Thermostat housing coolant hoses
- Spark plugs
- Engine Coolant Temperature (ECT) sensor connector
- Cylinder head

To install:

✳✳ WARNING

Cylinder head bolts may only be reused one time. If reusing a cylinder head bolt, place a paint mark on the bolt after installation. If a cylinder head bolt has a paint mark, discard it and use a new bolt.

5. Fabricate two alignment dowels from old cylinder head bolts. Cut the hex head off of the bolts, and cut a slot in each dowel to ease removal.

6. Install or connect the following:
- One dowel in bolt hole No. 8, and one dowel in bolt hole No. 10.

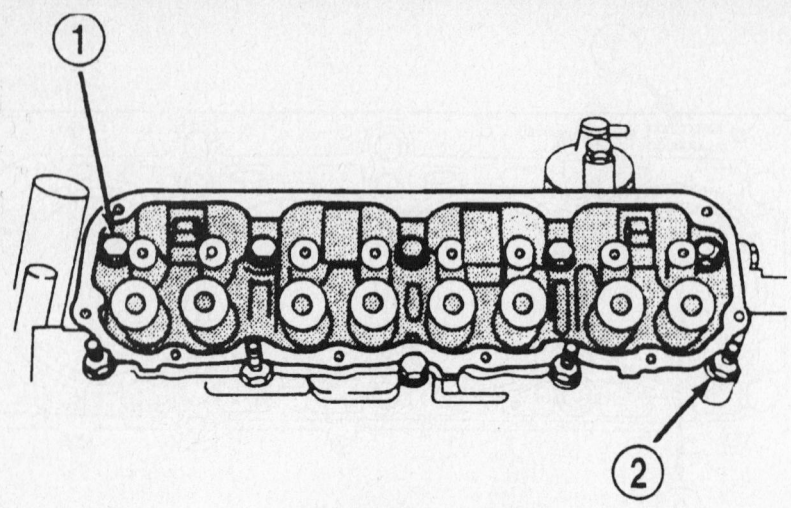

1 – ALIGNMENT DOWEL
2 – ALIGNMENT DOWEL

9308PG03

Alignment dowel locations—2.5L engine

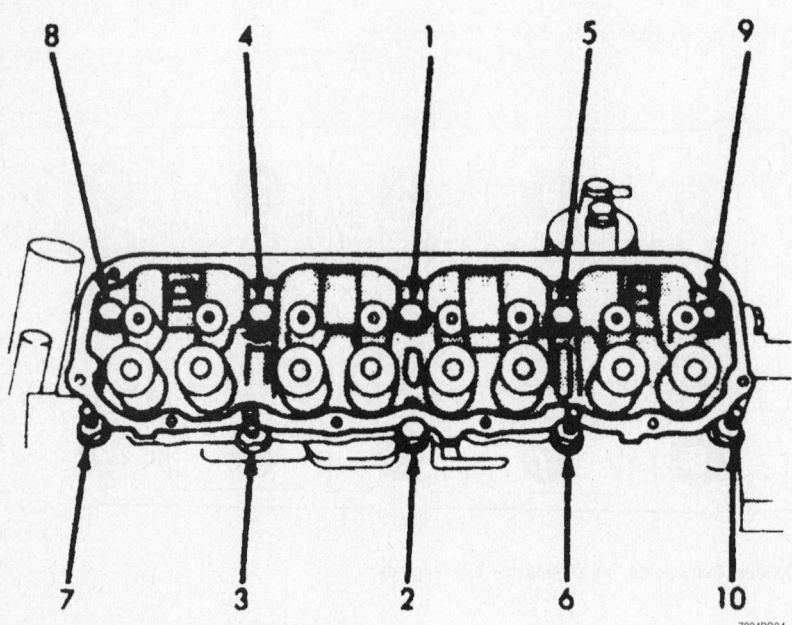

7924PG04

Cylinder head torque sequence—2.5L engine

- Cylinder head and gasket.
- Cylinder head bolts except for No 8 and No 10. Coat the threads of bolt No. 7 with Loctite® 592 sealant.

7. Remove the alignment dowels and install the No. 8 and No. 10 head bolts.

✳✳ WARNING

During the final tightening sequence, bolt No. 7 will be tightened to a lower torque value than the rest of the bolts. Do not overtighten bolt No. 7.

8. Tighten the cylinder head bolts, in sequence, as follows:
  a. Step 1: 22 ft. lbs. (30 Nm)
  b. Step 2: 45 ft. lbs. (61 Nm)
  c. Step 3: 45 ft. lbs. (61 Nm)
  d. Step 4: Bolts 1–6 to 110 ft. lbs. (149 Nm)
  e. Step 5: Bolt 7 to 100 ft. lbs. (136 Nm)
  f. Step 6: Bolts 8–10 to 110 ft. lbs. (149 Nm)
  g. Step 7: Repeat steps 4, 5 and 6
9. Install or connect the following:
- ECT sensor connector

- Spark plugs
- Thermostat housing coolant hoses
- Combination manifold
- Fuel line
- Power steering pump and bracket, if equipped
- A/C compressor and bracket, if equipped
- Accessory drive belt
- Pushrods and rocker arms in their original positions
- Valve cover
- Air cleaner assembly
- CCV hoses
- Negative battery cable

10. Fill and bleed the cooling system.
11. Start the engine, check for leaks and repair if necessary.

4.0L Engines

1. Before servicing the vehicle, refer to the precautions in the beginning of this section.
2. Drain the cooling system.
3. Properly relieve the fuel system pressure.
4. Remove or disconnect the following:
- Negative battery cable
- Crankcase Ventilation (CCV) hoses
- Air cleaner assembly
- Accelerator cable
- Cruise control cable, if equipped
- Transmission cable, if equipped
- Control cable bracket
- Valve cover

➡**Keep valvetrain components in order for reassembly.**

- Rocker arms
- Pushrods
- Accessory drive belt
- A/C compressor and bracket, if equipped
- Power steering pump and bracket, if equipped
- Fuel line
- Combination manifold
- Thermostat housing coolant hoses
- Spark plugs
- Engine Coolant Temperature (ECT) sensor connector
- Cylinder head

To install:

✳✳ WARNING

Cylinder head bolts may only be reused one time. If reusing a cylinder head bolt, place a paint mark on the bolt after installation. If a cylinder head bolt has a paint mark, discard it and use a new bolt.

⬅ FRONT

Cylinder head torque sequence—4.0L engine

7924PG05

5. Install the cylinder head with a new gasket. Coat the threads of bolt No. 11 with Loctite®F 592 sealant.

❋❋ CAUTION

During the final tightening sequence, bolt No. 11 will be tightened to a lower torque value than the rest of the bolts. Do not overtighten bolt No. 11.

6. Tighten the cylinder head bolts, in sequence, as follows:
   a. Step 1: 22 ft. lbs. (30 Nm)

   b. Step 2: 45 ft. lbs. (61 Nm)
   c. Step 3: 45 ft. lbs. (61 Nm)
   d. Step 4: Bolts 1–10 to 110 ft. lbs. (149 Nm)
   e. Step 5: Bolt 11 to 100 ft. lbs. (136 Nm)
   f. Step 6: Bolts 12–14 to 110 ft. lbs. (149 Nm)
   g. Step 7: Repeat steps 4, 5 and 6

7. Install or connect the following:
   • ECT sensor connector
   • Spark plugs
   • Thermostat housing coolant hoses
   • Combination manifold

   • Fuel line
   • Power steering pump and bracket, if equipped
   • A/C compressor and bracket, if equipped. Torque the bolts to 30 ft. lbs. (40 Nm).
   • Accessory drive belt
   • Pushrods and rocker arms in their original positions
   • Valve cover
   • Control cable bracket
   • Transmission cable, if equipped
   • Cruise control cable, if equipped
   • Accelerator cable
   • Air cleaner assembly
   • CCV hoses
   • Negative battery cable

8. Fill and bleed the cooling system.
9. Start the engine, check for leaks and repair if necessary.

3.7L Engine

LEFT SIDE

1. Before servicing the vehicle, refer to the precautions in the beginning of this section.
2. Drain the cooling system.
3. Properly relieve the fuel system pressure.
4. Remove or disconnect the following:
   • Negative battery cable

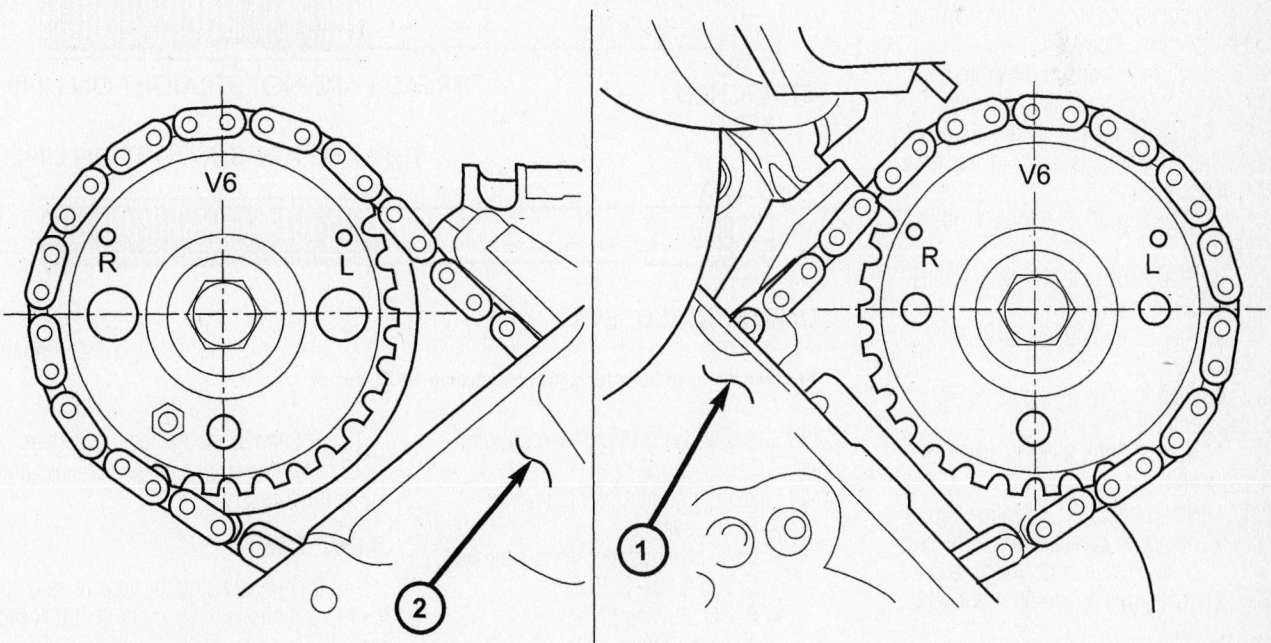

1 - LEFT CYLINDER HEAD
2 - RIGHT CYLINDER HEAD

9355PG04

Camshaft sprocket timing marks—3.7L

- Exhaust Y-pipe
- Intake manifold
- Cylinder head cover
- Engine cooling fan and shroud
- Accessory drive belt
- Power steering pump

5. Rotate the crankshaft so that the crankshaft timing mark aligns with the Top Dead Center (TDC) mark on the front cover, and the **V6** marks on the camshaft sprockets are at 12 o'clock as shown.

- Crankshaft damper
- Front cover

6. Lock the secondary timing chain to the idler sprocket with Timing Chain Locking tool 8429.

7. Matchmark the secondary timing chain one link on each side of the V6 mark to the camshaft sprocket.

- Left secondary timing chain tensioner
- Cylinder head access plug
- Secondary timing chain guide
- Camshaft sprocket
- Cylinder head

➡**The cylinder head is retained by twelve bolts. Four of the bolts are smaller and are at the front of the head.**

To install:

8. Check the cylinder head bolts for signs of stretching and replace as necessary.

9. Lubricate the threads of the 11mm bolts with clean engine oil.

10. Coat the threads of the 8mm bolts with Mopar® Lock and Seal Adhesive.

11. Install the cylinder heads. Use new gaskets and tighten the bolts, in sequence, as follows:

a. Step 1: Bolts 1–8 to 20 ft. lbs. (27 Nm)

b. Step 2: Bolts 1–10 verify torque without loosening

c. Step 3: Bolts 9–12 to 10 ft. lbs. (14 Nm)

d. Step 4: Bolts 1–8 plus ¼ (90 degree) turn

e. Step 5: Bolts 9–12 to 19 ft. lbs. (26 Nm)

12. Install or connect the following:

- Camshaft sprocket. Align the secondary chain matchmarks and tighten the bolt to 90 ft. lbs. (122 Nm).
- Secondary timing chain guide
- Cylinder head access plug
- Secondary timing chain tensioner. Refer to the timing chain procedure in this section.

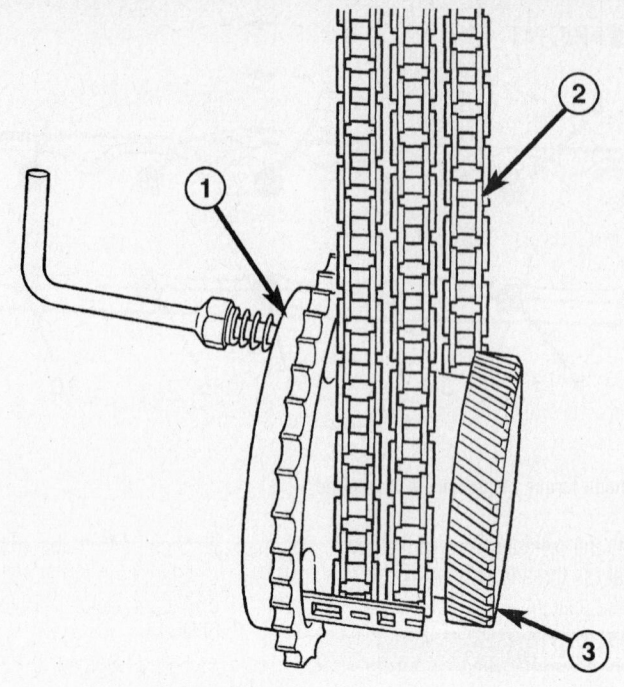

1 - SPECIAL TOOL 8429

2 - CAMSHAFT CHAIN

3 - CRANKSHAFT TIMING GEAR

9355PG05

Camshaft locking tool—3.7L

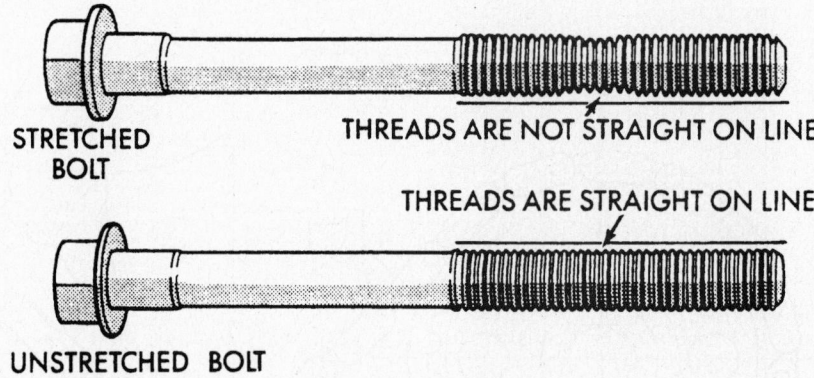

STRETCHED BOLT

THREADS ARE NOT STRAIGHT ON LINE

THREADS ARE STRAIGHT ON LINE

UNSTRETCHED BOLT

9302PG10

Examine the head bolts for signs of stretching—3.7L engine

13. Remove the Timing Chain Locking tool.

14. Install or connect the following:

- Front cover
- Crankshaft damper. Torque the bolt to 130 ft. lbs. (175 Nm).
- Power steering pump
- Accessory drive belt
- Engine cooling fan and shroud
- Cover
- Intake manifold
- Exhaust Y-pipe
- Negative battery cable

15. Fill and bleed the cooling system.

16. Start the engine, check for leaks and repair if necessary.

RIGHT SIDE

1. Before servicing the vehicle, refer to the precautions in the beginning of this section.

2. Drain the cooling system.

3. Properly relieve the fuel system pressure.

4. Remove or disconnect the following:

- Negative battery cable

LEFT BANK

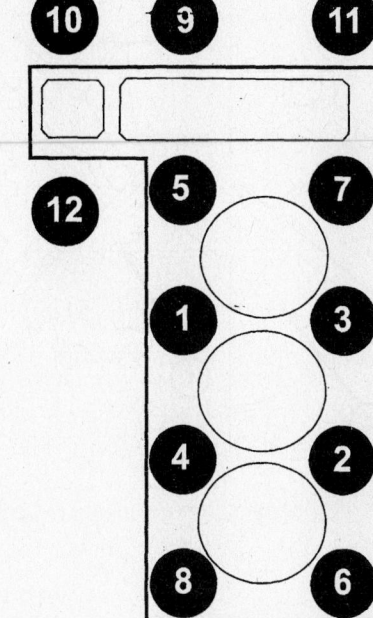

RIGHT BANK

9355PG03

Cylinder head bolt torque sequence—3.7L

- Exhaust Y-pipe
- Intake manifold
- Valve cover
- Engine cooling fan and shroud
- Accessory drive belt
- Oil fill housing
- Power steering pump

5. Rotate the crankshaft so that the crankshaft timing mark aligns with the Top Dead Center (TDC) mark on the front cover, and the **V6** marks on the camshaft sprockets are at 12 o'clock as shown.

6. Remove or disconnect the following:

- Crankshaft damper
- Front cover

7. Lock the secondary timing chains to the idler sprocket with Timing Chain Locking tool 8429.

8. Matchmark the secondary timing chains to the camshaft sprockets.

9. Remove or disconnect the following:

- Secondary timing chain tensioners
- Cylinder head access plugs
- Secondary timing chain guides
- Camshaft sprockets
- Cylinder heads

➡ **Each cylinder head is retained by 8 11mm bolts and four 8mm bolts.**

To install:

10. Check the cylinder head bolts for signs of stretching and replace as necessary.

11. Lubricate the threads of the 11mm bolts with clean engine oil.

12. Coat the threads of the 8mm bolts with Mopar® Lock and Seal Adhesive.

13. Install the cylinder heads. Use new gaskets and tighten the bolts, in sequence, as follows:

   a. Step 1: Bolts 1–8 to 20 ft. lbs. (27 Nm)

   b. Step 2: Bolts 1–10 verify torque without loosening

   c. Step 3: Bolts 9–12 to 10 ft. lbs. (14 Nm)

   d. Step 4: Bolts 1–8 plus ¼ (90 degree) turn

   e. Step 5: Bolts 9–12 to 19 ft. lbs. (26 Nm)

14. Install or connect the following:

- Camshaft sprockets. Align the secondary chain matchmarks and tighten the bolts to 90 ft. lbs. (122 Nm).
- Secondary timing chain guides
- Cylinder head access plugs
- Secondary timing chain tensioners. Refer to the timing chain procedure in this section.

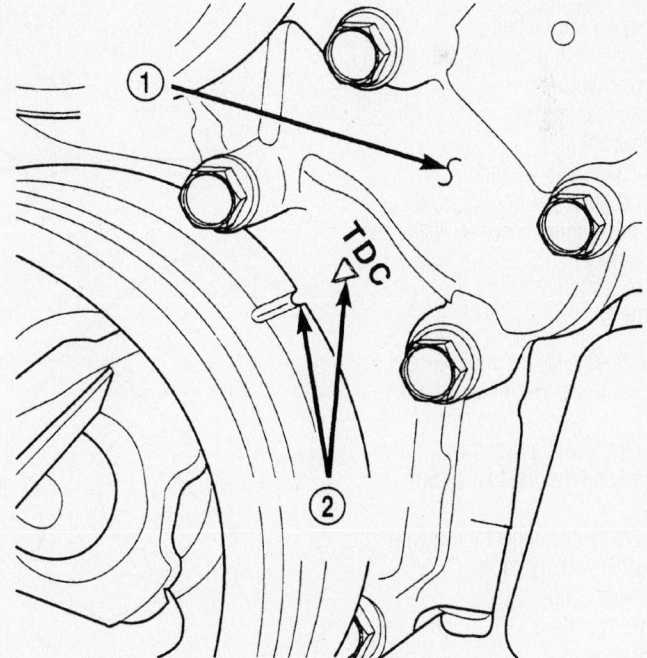

1 – TIMING CHAIN COVER
2 – CRANKSHAFT TIMING MARKS

9308PG04

Crankshaft timing marks—3.7L engine

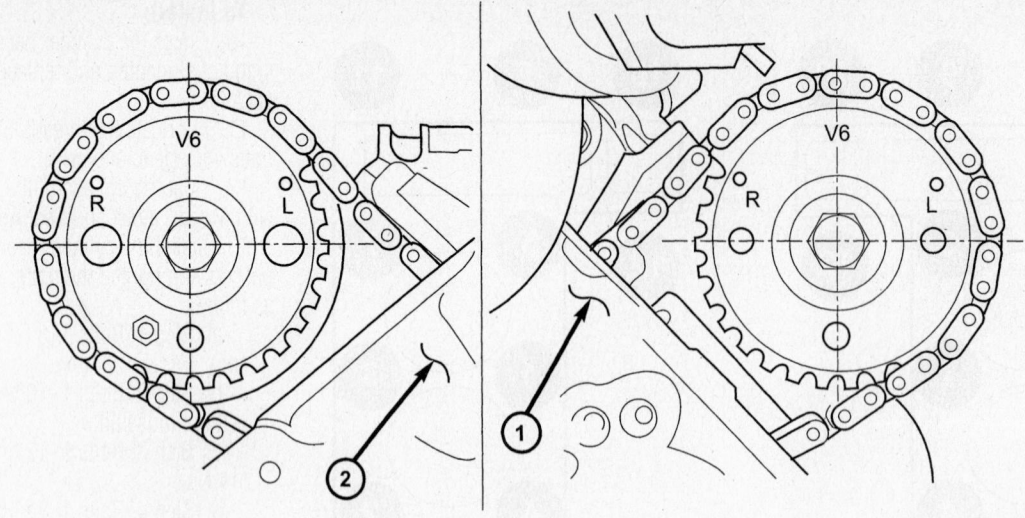

1 - LEFT CYLINDER HEAD
2 - RIGHT CYLINDER HEAD

9355PG04

Camshaft positioning—3.7L engine

15. Remove the Timing Chain Locking tool.

16. Install or connect the following:
- Front cover
- Crankshaft damper. Torque the bolt to 130 ft. lbs. (175 Nm).
- Rocker arms
- Power steering pump
- Oil fill housing
- Accessory drive belt
- Engine cooling fan and shroud
- Valve covers
- Intake manifold
- Exhaust Y-pipe
- Negative battery cable

17. Fill and bleed the cooling system.

18. Start the engine, check for leaks and repair if necessary.

4.7L Engine

1. Before servicing the vehicle, refer to the precautions in the beginning of this section.

2. Drain the cooling system.

3. Properly relieve the fuel system pressure.

4. Remove or disconnect the following:
- Negative battery cable
- Exhaust Y-pipe
- Intake manifold
- Valve covers
- Engine cooling fan and shroud
- Accessory drive belt
- Oil fill housing
- Power steering pump
- Rocker arms

5. Rotate the crankshaft so that the crankshaft timing mark aligns with the Top Dead Center (TDC) mark on the front cover, and the **V8** marks on the camshaft sprockets are at 12 o'clock as shown.

6. Remove or disconnect the following:
- Crankshaft damper
- Front cover

7. Lock the secondary timing chains to the idler sprocket with Timing Chain Locking tool 8515.

8. Matchmark the secondary timing chains to the camshaft sprockets.

9. Remove or disconnect the following:
- Secondary timing chain tensioners
- Cylinder head access plugs

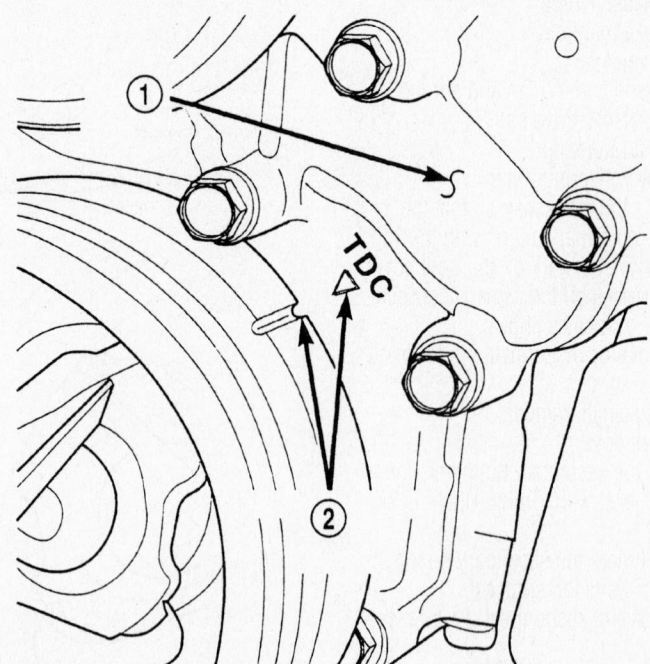

1 – TIMING CHAIN COVER
2 – CRANKSHAFT TIMING MARKS

9308PG04

Crankshaft timing marks—4.7L engine

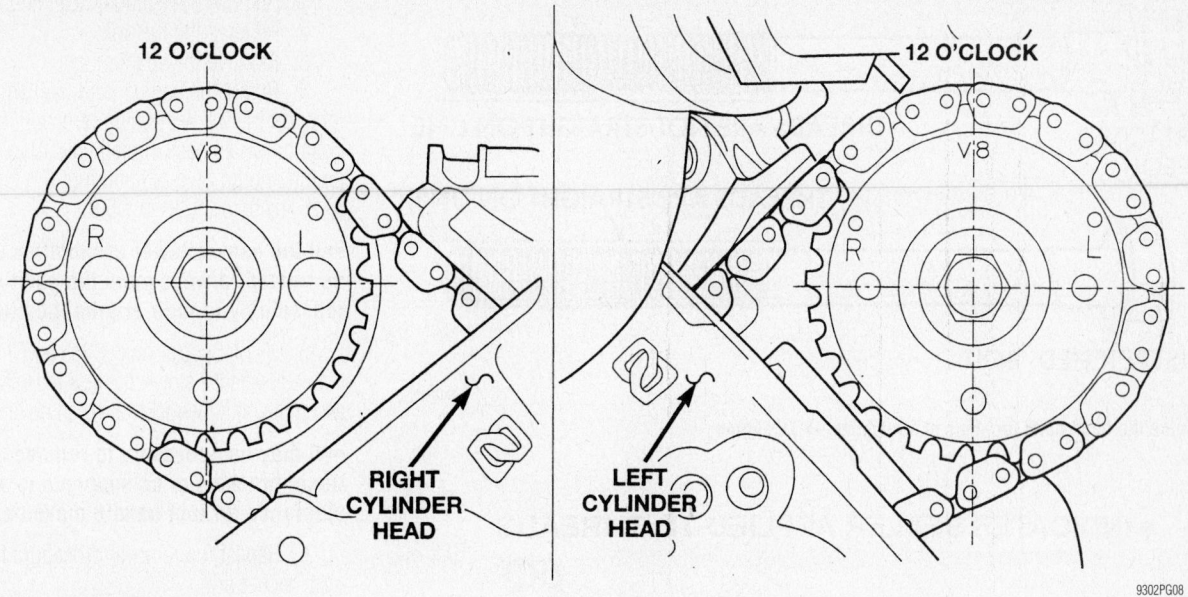

Camshaft positioning—4.7L engine

- Secondary timing chain guides
- Camshaft sprockets
- Cylinder heads

➡**Each cylinder head is retained by ten 11mm bolts and four 8mm bolts.**

To install:

10. Check the cylinder head bolts for signs of stretching and replace as necessary.

11. Lubricate the threads of the 11mm bolts with clean engine oil.

12. Coat the threads of the 8mm bolts with Mopar® Lock and Seal Adhesive.

13. Install the cylinder heads. Use new gaskets and tighten the bolts, in sequence, as follows:

  a. Step 1: Bolts 1–10 to 15 ft. lbs. (20 Nm)

  b. Step 2: Bolts 1–10 to 35 ft. lbs. (47 Nm)

  c. Step 3: Bolts 11–14 to 18 ft. lbs. (25 Nm)

  d. Step 4: Bolts 1–10 plus ¼ (90 degree) turn

  e. Step 5: Bolts 11–14 to 19 ft. lbs. (26 Nm)

14. Install or connect the following:

- Camshaft sprockets. Align the secondary chain matchmarks and tighten the bolts to 90 ft. lbs. (122 Nm).
- Secondary timing chain guides
- Cylinder head access plugs
- Secondary timing chain tensioners. Refer to the timing chain procedure in this section.

15. Remove the Timing Chain Locking tool 8515.

16. Install or connect the following:

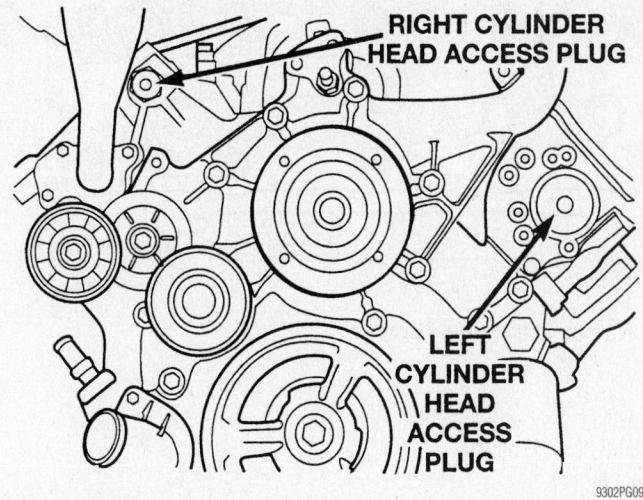

Cylinder head access plug locations—4.7L engine

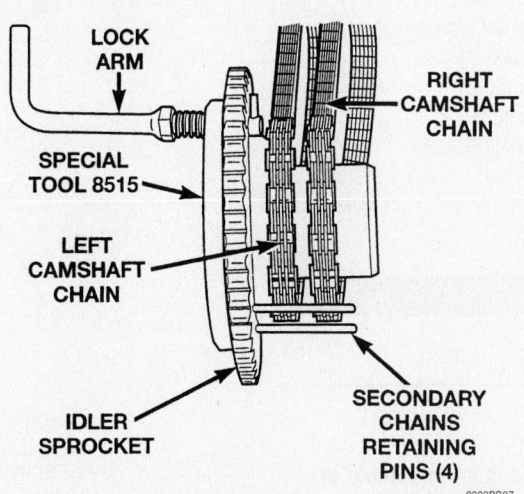

Use the special tool to lock the timing chains on the idler gear—4.7L engine

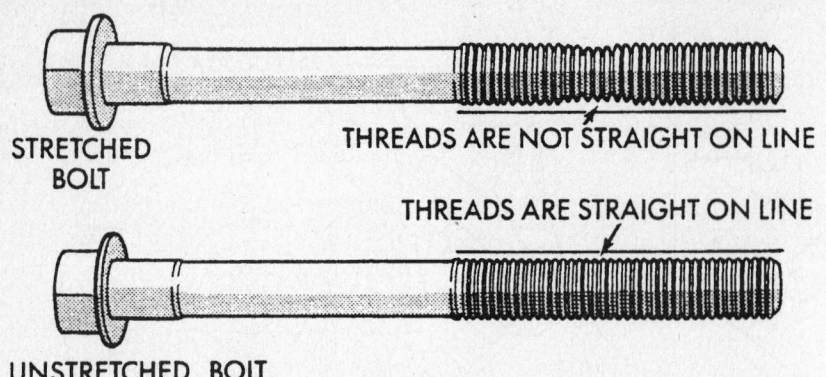

STRETCHED BOLT

THREADS ARE NOT STRAIGHT ON LINE

THREADS ARE STRAIGHT ON LINE

UNSTRETCHED BOLT

9302PG10

Examine the head bolts for signs of stretching—4.7L engine

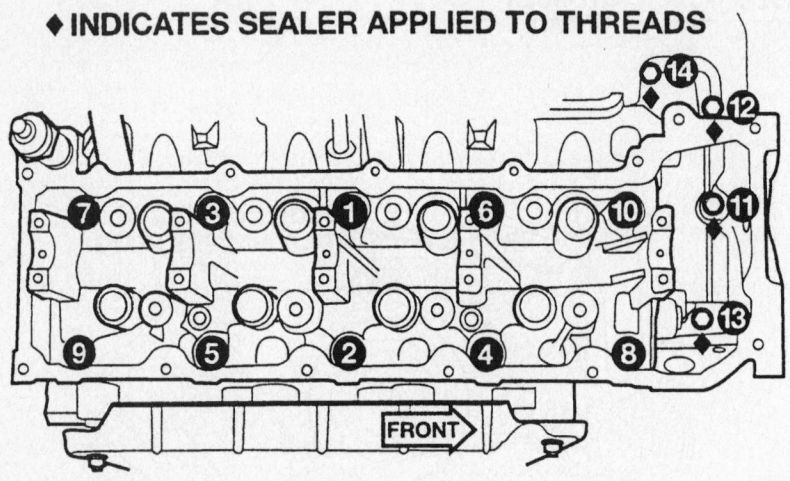

♦ INDICATES SEALER APPLIED TO THREADS

9302PG11

Cylinder head torque sequence—4.7L engine

- Front cover
- Crankshaft damper. Torque the bolt to 130 ft. lbs. (175 Nm).
- Rocker arms
- Power steering pump
- Oil fill housing
- Accessory drive belt
- Engine cooling fan and shroud
- Valve covers
- Intake manifold
- Exhaust Y-pipe
- Negative battery cable

17. Fill and bleed the cooling system.
18. Start the engine, check for leaks and repair if necessary.

Rocker Arms

REMOVAL & INSTALLATION

2.4L Engines

➡This procedure is for in-vehicle service with the camshafts installed.

1. Remove the cylinder head cover.
2. Remove the fuel rail.
3. Remove the spark plugs.
4. Rotate the engine until the camshaft lobe, on the follower being removed, is position on its base circle(heel). Also, the piston should be a minimum of 6.3 mm (0.25 in) below TDC position.

➡If the cam follower assemblies are to be reused, always mark the position for reassembly in their original positions.

5. Using Special Tools 8215 and 8436, or their equivalents, slowly depress the valve assembly until the rocker arm can be removed.

➡It may be necessary to remove additional brackets or components to allow clearance for tool handle movement.

6. Repeat the removal procedure for each rocker arm.
7. Inspect the rocker arm for wear or damage. Replace as necessary.

To install:

8. Lubricate the rocker arm with clean engine oil.
9. Using Special Tools 8215 and 8436, slowly depress the valve assembly until the rocker arm can be installed on the hydraulic lifter and valve stem.
10. Repeat the installation procedure for each rocker arm.
11. Install the spark plugs.
12. Install the fuel rail.
13. Install the cylinder head cover.

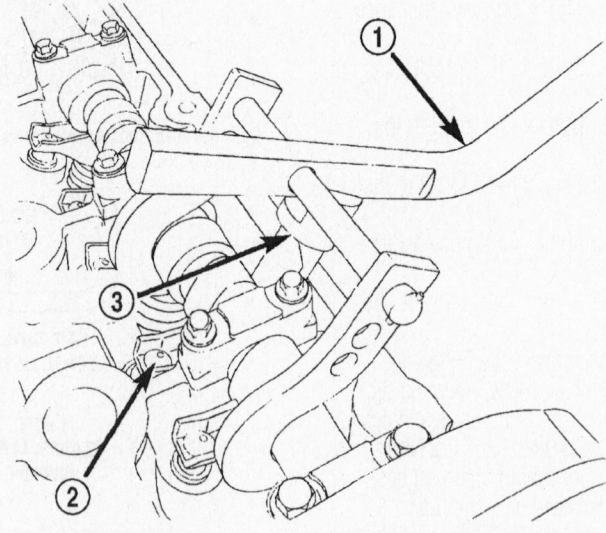

1 - SPECIAL TOOL 8215
2 - ROCKER ARM
3 - SPECIAL TOOL 8436

67189-JEEP-G10

Rocker arm removal—2.4L engines

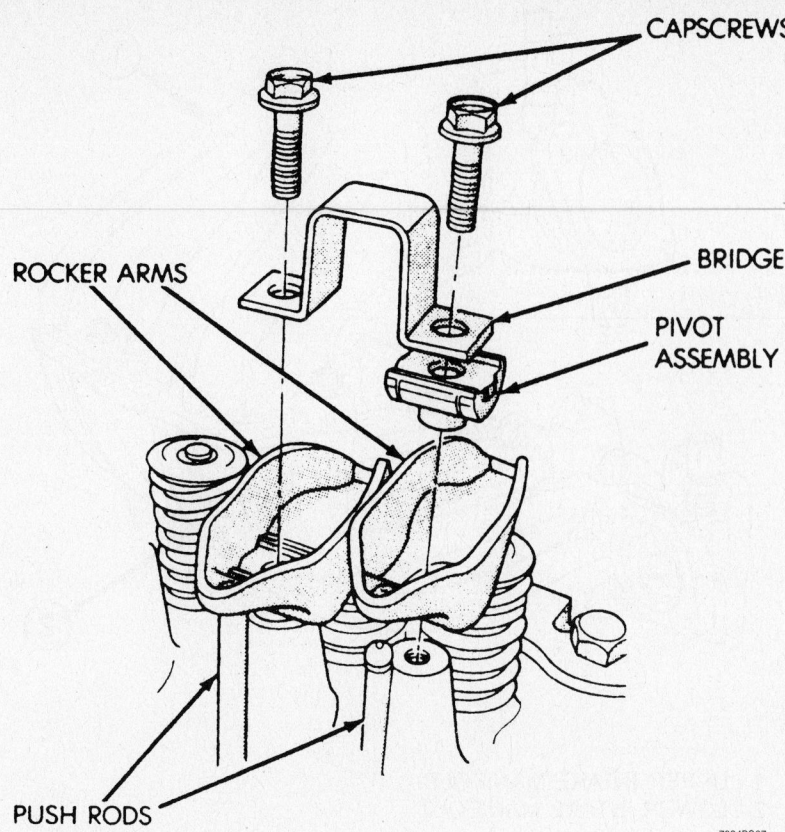

Exploded view of the rocker arm assembly—2.5L and 4.0L engines

CAPSCREWS

ROCKER ARMS

BRIDGE

PIVOT ASSEMBLY

PUSH RODS

7924PG07

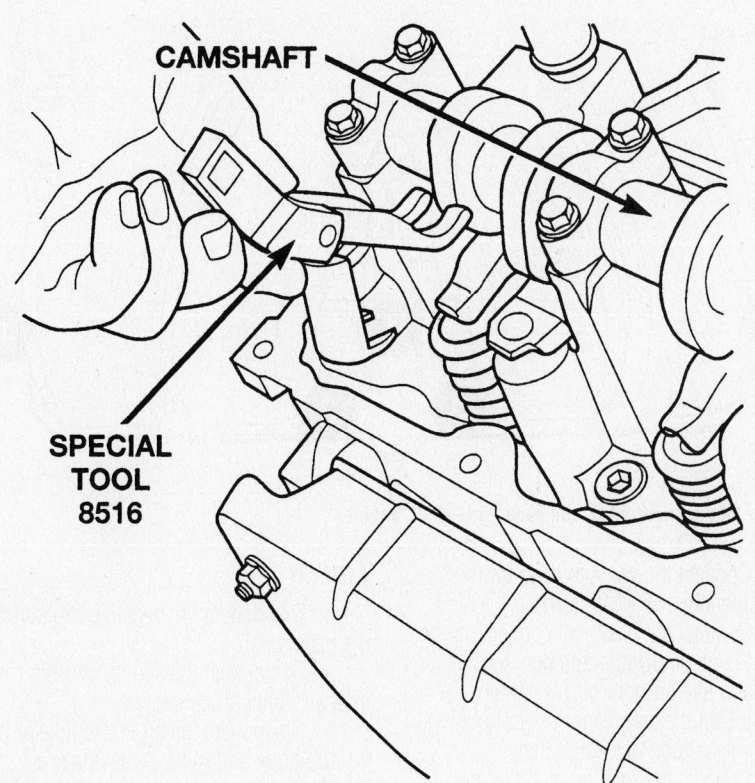

Rocker arm service—3.7L and 4.7L engines

CAMSHAFT

SPECIAL TOOL 8516

9302PG13

2.5L and 4.0L Engines

1. Before servicing the vehicle, refer to the precautions in the beginning of this section.
2. Remove or disconnect the following:
   - Negative battery cable
   - Valve cover
   - Rocker arm bolts, loosen them evenly to avoid damaging the alignment bridges
   - Rocker arms

➡ **Keep valvetrain components in order for reassembly.**

To install:

3. Install or connect the following:
   - Rocker arms, pivots and bridges in their original positions. Torque the bolts for each bridged pair one turn at a time to 21 ft. lbs. (28 Nm).
   - Valve cover
   - Negative battery cable

3.7L and 4.7L Engines

1. Before servicing the vehicle, refer to the precautions in the beginning of this section.
2. Remove or disconnect the following:
   - Negative battery cable
   - Valve covers
3. Rotate the crankshaft so that the piston of the cylinder to be serviced is at Top Dead Center (TDC) and both valves are closed.
4. Use special tool 8516 to depress the valve and remove the rocker arm.
5. Repeat for each rocker arm to be serviced.

➡ **Keep valvetrain components in order for reassembly.**

To install:

6. Rotate the crankshaft so that the piston of the cylinder to be serviced is at BDC.
7. Compress the valve spring and install each rocker arm in its original position.
8. Repeat for each rocker arm to be installed.
9. Install or connect the following:
   - Cylinder head cover
   - Negative battery cable

Intake Manifold

REMOVAL & INSTALLATION

2.4L Engine

UPPER

1. Disconnect the negative cable from the battery.

2. Disconnect the connector from the inlet air temperature sensor.

3. Disconnect the air intake tube at the throttle body and remove the upper air cleaner housing.

4. Disconnect the connector from the throttle position sensor (TPS).

5. Disconnect the connector from the idle air control (IAC) motor.

6. Disconnect the connector from the MAP sensor.

7. Remove the vacuum lines for the purge solenoid and the PCV valve at the intake manifold.

8. Remove the vacuum lines for the power brake booster, LDP, EGR transducer, and speed control vacuum reservoir (if equipped) at the upper intake manifold fittings.

9. Disconnect the throttle, speed control (if equipped), and transaxle control (if equipped) and cables from the throttle lever and bracket.

10. Perform the fuel system pressure release procedure before attempting any repairs.

11. Disconnect the fuel line.

12. Disconnect the coolant temperature sensor.

➥**Cover the intake manifold openings to prevent foreign material from entering the engine.**

13. Disconnect the fuel injector harness.

14. Remove the intake manifold to cylinder head fasteners.

15. Remove the manifold from engine.

To install:

16. Clean the manifold sealing surfaces.

17. Apply a 1.5 mm (0.060 in.) bead Mopar® Gasket Maker, or equivalent, to the perimeter of the lower intake manifold runner openings.

18. Install the upper intake manifold and tighten the fasteners to 28 Nm (250 inch lbs.) in the sequence shown in. Repeat this procedures until all fasteners are at the specified torque.

19. Install the engine oil dipstick.

20. Install the upper bolt in the intake manifold to front support bracket.

21. Install the EGR tube.

22. Install the throttle cables in the bracket.

23. Connect the throttle, speed control, (if equipped), cables to the throttle lever.

24. Connect the vacuum lines for the power brake booster, LDP, EGR transducer, and speed control vacuum reservoir (if equipped) at the upper intake manifold fittings.

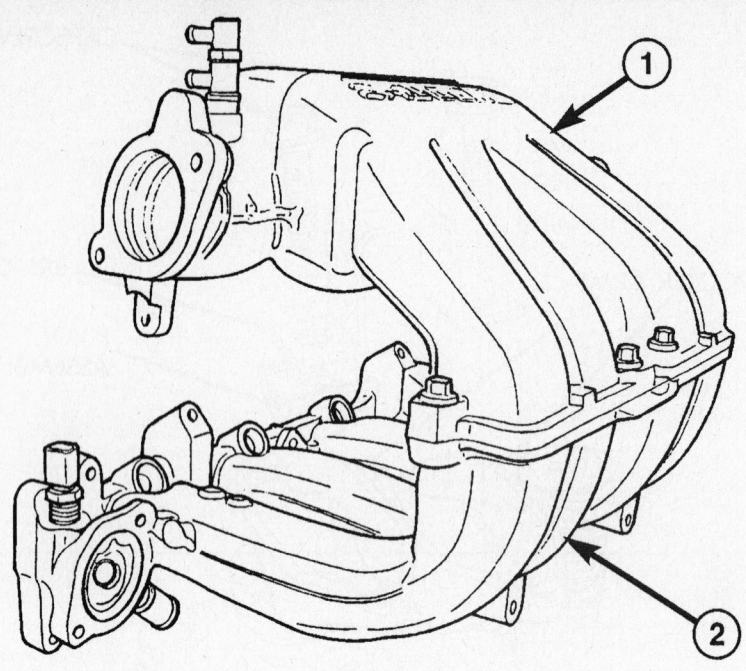

1 - UPPER INTAKE MANIFOLD
2 - LOWER INTAKE MANIFOLD

67189-JEEP-G11

Upper and lower intake manifolds—2.4L engines

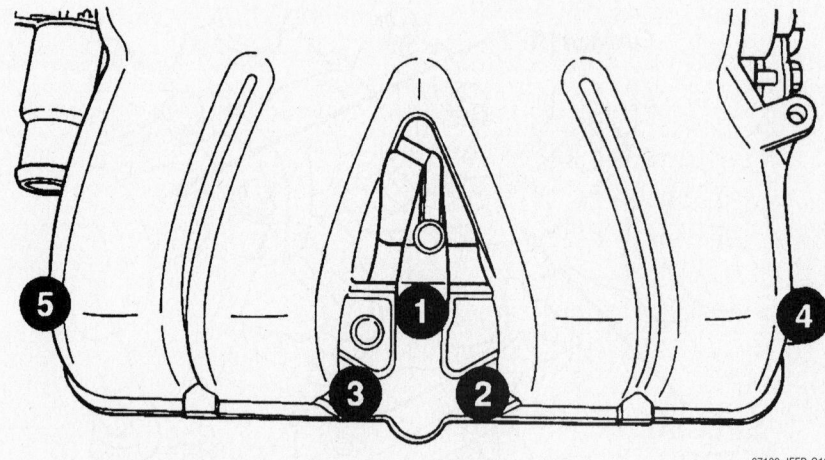

67189-JEEP-G12

Upper intake manifold torque sequence—2.4L engines

25. Connect the vacuum lines for the purge solenoid and PCV valve.

26. Connect the electrical connectors for the MAP sensor, throttle position sensor (TPS), and idle air control (IAC) motor.

27. Install the air cleaner upper housing and air intake tube to throttle body.

28. Connect the inlet air temperature sensor connector.

29. Connect the negative cable to the battery.

LOWER

1. Disconnect the negative cable from the battery.

2. Disconnect the connector from the inlet air temperature sensor.

3. Disconnect the air intake tube at the throttle body and remove the upper air cleaner housing.

4. Disconnect the connector from the throttle position sensor (TPS).

5. Disconnect the connector from the idle air control (IAC) motor.

6. Disconnect the connector from the MAP sensor.

7. Remove the vacuum lines for the purge solenoid and PCV valve at the intake manifold.

8. Remove the vacuum lines for the power brake booster, LDP, EGR transducer, and speed control vacuum reservoir (if equipped) at the intake manifold fittings.

9. Disconnect the throttle, speed control (if equipped), and transaxle control (if equipped) and cables from the throttle lever and bracket.

10. Perform the fuel system pressure release procedure before attempting any repairs.

11. Disconnect the fuel line.

12. Disconnect the coolant temperature sensor/fuel injector wire harness connector.

13. Disconnect the fuel injector harness.

14. Remove the intake manifold to cylinder head fasteners.

15. Remove the manifold from engine.

➡**Cover the intake manifold openings to prevent foreign material from entering engine.**

16. Inspect the manifold.

17. Check the manifold surfaces for flatness with straight edge. Surface must be flat within 0.15 mm per 300 mm (0.006 in. per foot) of manifold length.

18. Inspect the manifold for cracks or distortion. Replace the manifold if necessary.

To install:
If the following items were removed, install and torque to specifications:

- Fuel rail bolts: 22 Nm (200 inch lbs.)
- Coolant outlet connector bolts: 28 Nm (250 inch lbs.)
- Coolant temperature sensor: 7 Nm (60 inch lbs.)

19. Position a new gasket on the cylinder head and install the lower manifold.

20. Install and tighten the intake manifold fasteners to 28 Nm (250 inch lbs.) in the sequence shown in. Repeat the procedure until all bolts are at the specified torque.

21. Install the lower intake manifold support bracket bolts. Torque the bolts to the intake to 28 Nm (250 inch lbs.). Torque the bolts to the engine block to 54 Nm (40 ft. lbs.)

22. Connect the fuel line.

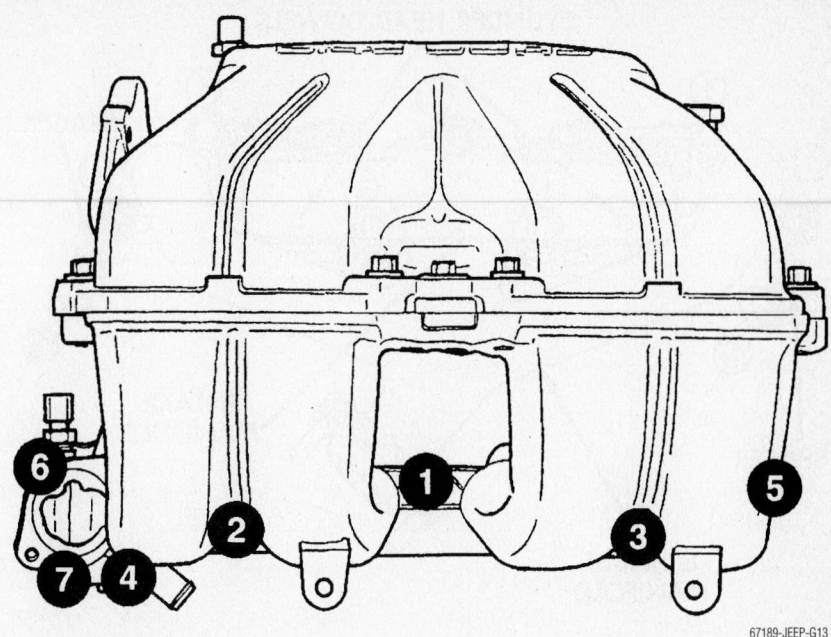

Lower intake manifold torque sequence—2.4L engines

67189-JEEP-G13

23. Connect the coolant temperature sensor/fuel injector wiring harness electrical connector.

24. Install the radiator upper and heater supply hoses.

25. Install the upper intake manifold.

26. Fill the cooling system.

2.5L Engine

1. Before servicing the vehicle, refer to the precautions in the beginning of this section.

2. Properly relieve the fuel system pressure.

3. Drain the cooling system.

4. Remove or disconnect the following:

- Negative battery cable
- Air intake hose
- Accessory drive belt
- Power steering pump and brackets, if equipped
- Fuel line
- Accelerator cable
- Cruise control cable, if equipped
- Transmission cable, if equipped
- Throttle Position (TP) sensor connector
- Idle Air Control (IAC) valve connector
- Engine Coolant Temperature (ECT) sensor connector
- Intake Air Temperature (IAT) sensor connector
- Heated Oxygen (HO2S) sensor connector
- Fuel injector connectors

- Closed Crankcase Ventilation (CCV) system
- Manifold Absolute Pressure (MAP) sensor vacuum line
- Brake booster vacuum line
- Exhaust front pipe
- Intake manifold and discard the gaskets
- Exhaust manifold and discard the gaskets

To install:
5. Install or connect the following:

- New gasket over the locating dowels
- Exhaust manifold to the studs and tighten the nuts finger-tight

6. Install the intake manifold. Torque the fasteners in sequence as follows:

a. Step 1: Bolt 1 to 30 ft. lbs. (41 Nm).

b. Step 2: Fasteners 2–7 to 23 ft. lbs. (31 Nm).

7. Install or connect the following:

- Exhaust front pipe
- Brake booster vacuum line
- MAP sensor vacuum line
- CCV system
- Fuel injector connectors
- HO2S sensor connector
- IAT sensor connector
- ECT sensor connector
- IAC valve connector
- TP sensor connector
- Transmission cable, if equipped
- Cruise control cable, if equipped
- Accelerator cable
- Fuel line

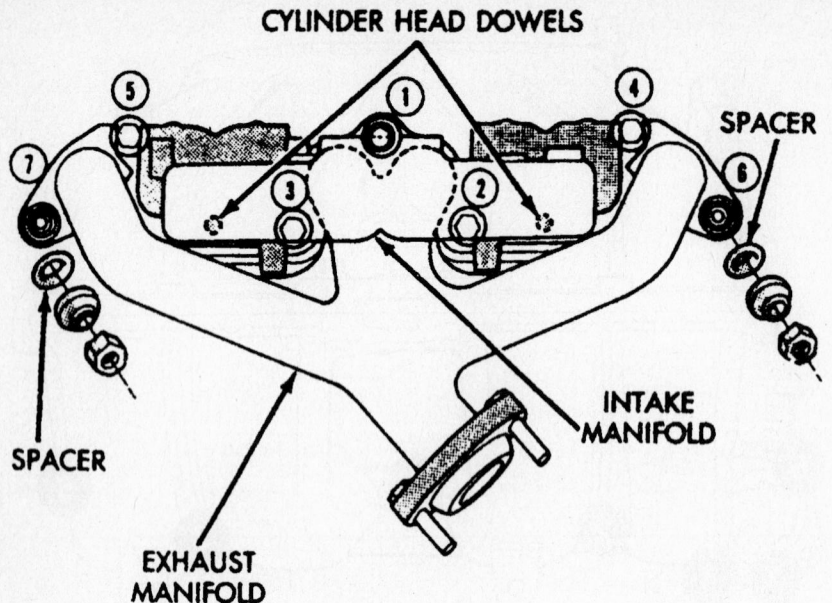

Manifold torque sequence—2.5L engine

7924PG08

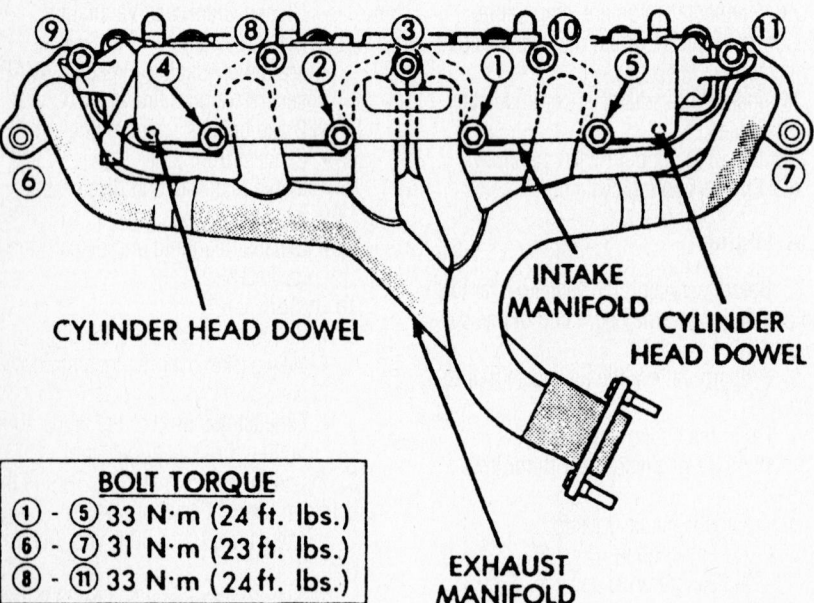

BOLT TORQUE
① - ⑤ 33 N·m (24 ft. lbs.)
⑥ - ⑦ 31 N·m (23 ft. lbs.)
⑧ - ⑪ 33 N·m (24 ft. lbs.)

Manifold torque sequence—4.0L engine

7924PG12

- Power steering pump and brackets, if equipped
- Accessory drive belt
- Air intake hose
- Negative battery cable

8. Fill and bleed the cooling system.
9. Start the engine, check for leaks and repair if necessary..

4.0L Engine

1. Before servicing the vehicle, refer to the precautions in the beginning of this section.
2. Drain the cooling system.

3. Properly relieve the fuel system pressure.
4. Remove or disconnect the following:
- Negative battery cable
- Air cleaner assembly
- Accessory drive belt
- Power steering pump and brackets, if equipped
- Fuel line
- Fuel injector connectors
- Fuel supply manifold and injectors
- Accelerator cable
- Cruise control cable, if equipped

- Transmission cable, if equipped
- Throttle Position (TP) sensor connector
- Idle Air Control (IAC) valve connector
- Engine Coolant Temperature (ECT) sensor connector
- Intake Air Temperature (IAT) sensor connector
- Heated Oxygen (HO2S) sensor connector
- Closed Crankcase Ventilation (CCV) system
- Manifold Absolute Pressure (MAP) sensor vacuum line
- Brake booster vacuum line
- Exhaust front pipe
- Intake and exhaust manifolds and discard the gaskets

To install:
5. Install or connect the following:
- New gasket over the locating dowels
- Exhaust manifold to the studs and tighten the nuts finger-tight
6. Install the intake manifold. Torque the fasteners in sequence as follows:
  a. Step 1: Fasteners 1–5 to 30 ft. lbs. (41 Nm).
  b. Step 2: Fasteners 6 and 7 to 23 ft. lbs. (31 Nm).
  c. Step 3: Fasteners 8–11 to 24 ft. lbs. (33 Nm).
7. Install or connect the following:
- Exhaust front pipe
- Brake booster vacuum line
- MAP sensor vacuum line
- CCV system
- HO2S sensor connector
- IAT sensor connector
- ECT sensor connector
- IAC valve connector
- TP sensor connector
- Transmission cable, if equipped
- Cruise control cable, if equipped
- Accelerator cable
- Fuel supply manifold and injectors
- Fuel injector connectors
- Fuel line
- Power steering pump and brackets, if equipped
- Accessory drive belt
- Air cleaner assembly
- Negative battery cable
8. Fill and bleed the cooling system.
9. Start the engine, check for leaks and repair if necessary.

3.7L and 4.7L Engines

1. Before servicing the vehicle, refer to the precautions in the beginning of this section.

2. Drain the cooling system.
3. Properly relieve the fuel system pressure.
4. Remove or disconnect the following:
   • Negative battery cable
   • Air cleaner assembly
   • Accelerator cable
   • Cruise control cable
   • Manifold Absolute Pressure (MAP) sensor connector
   • Intake Air Temperature (IAT) sensor connector
   • Throttle Position (TP) sensor connector
   • Idle Air Control (IAC) valve connector
   • Engine Coolant Temperature (ECT) sensor
   • Positive Crankcase Ventilation (PCV) valve and hose
   • Canister purge vacuum line
   • Brake booster vacuum line
   • Cruise control servo hose
   • Accessory drive belt
   • Alternator
   • A/C compressor
   • Engine ground straps
   • Ignition coil towers
   • Oil dipstick tube
   • Fuel line
   • Fuel supply manifold
   • Throttle body and mounting bracket
   • Cowl seal
   • Right engine lifting stud
   • Intake manifold. Remove the fasteners in reverse of the tightening sequence.

To install:

5. Install or connect the following:
   • Intake manifold using new gaskets. Torque the bolts, in sequence, to 105 inch lbs. (12 Nm).
   • Right engine lifting stud
   • Cowl seal
   • Throttle body and mounting bracket
   • Fuel supply manifold
   • Fuel line
   • Oil dipstick tube
   • Ignition coil towers
   • Engine ground straps
   • A/C compressor
   • Alternator
   • Accessory drive belt
   • Cruise control servo hose
   • Brake booster vacuum line
   • Canister purge vacuum line
   • PCV valve and hose
   • ECT sensor
   • IAC valve connector
   • TP sensor connector
   • IAT sensor connector
   • MAP sensor connector
   • Cruise control cable

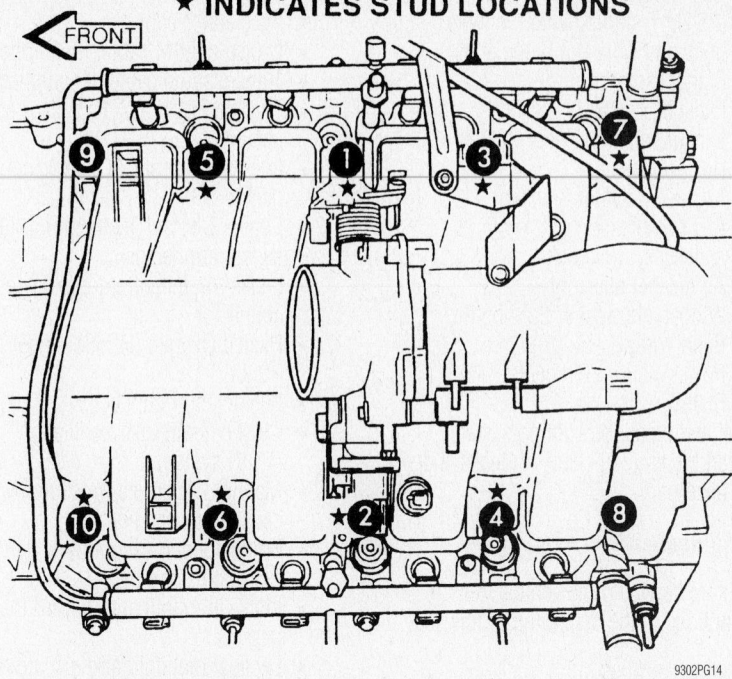

★ **INDICATES STUD LOCATIONS**

Intake manifold torque sequence—3.7 & 4.7L engines

   • Accelerator cable
   • Air cleaner assembly
   • Negative battery cable
6. Fill and bleed the cooling system.
7. Start the engine, check for leaks and repair if necessary.

Exhaust Manifold

REMOVAL & INSTALLATION

2.4L Engine

1. Raise the vehicle and disconnect the exhaust pipe from the exhaust manifold.
2. Lower the vehicle.
3. Disconnect the upstream oxygen sensor connector at the rear of the exhaust manifold.
4. Remove the air cleaner bracket.
5. Remove the heat shield.
6. Remove the bolts attaching the manifold to the cylinder head.
7. Remove the exhaust manifold.
8. Inspect the manifold.
9. Discard the gasket (if equipped) and clean all surfaces of the manifold and cylinder head.
10. Inspect the manifold gasket surfaces for flatness with a straight edge. Surface must be flat within 0.15 mm per 300 mm (0.006 in. per foot) of manifold length.
11. Inspect the manifolds for cracks or distortion. Replace the manifold as necessary.

To install:
12. Clean the manifold mating surfaces.
13. Install the exhaust manifold with a new gasket. Tighten the attaching nuts to 175 inch lbs. (20 Nm).
14. Attach the exhaust pipe to the exhaust manifold and tighten the fasteners to 27 ft. lbs. (37 Nm)
15. Install and connect the oxygen sensor.
16. Install the heat shield.
17. Install the air cleaner bracket.

4.7L Engine

1. Before servicing the vehicle, refer to the precautions in the beginning of this section.
2. Drain the cooling system.
3. Remove or disconnect the following:
   • Battery
   • Power distribution center
   • Battery tray
   • Windshield washer fluid bottle
   • Air cleaner assembly
   • Accessory drive belt
   • A/C compressor
   • A/C accumulator bracket
   • Heater hoses
   • Exhaust manifold heat shields
   • Exhaust Y-pipe
   • Starter motor
   • Exhaust manifolds

To install:
4. Install or connect the following:
   • Exhaust manifolds, using new gas-

kets. Torque the bolts to 18 ft. lbs. (25 Nm), starting with the inner bolts and work out to the ends.
- Starter motor
- Exhaust Y-pipe
- Exhaust manifold heat shields
- Heater hoses
- A/C accumulator bracket
- A/C compressor
- Accessory drive belt
- Air cleaner assembly
- Windshield washer fluid bottle
- Battery tray
- Power distribution center
- Battery

5. Fill and bleed the cooling system.
6. Start the engine, check for leaks and repair if necessary.

3.7L Engines

1. Before servicing the vehicle, refer to the precautions in the beginning of this section.
2. Remove or disconnect the following:
- Negative battery cable
- Exhaust manifold heat shields
- Exhaust Gas Recirculation (EGR) tube
- Exhaust Y-pipe
- Exhaust manifolds

To install:

➥**If the exhaust manifold studs came out with the nuts when removing the exhaust manifolds, replace them with new studs.**

3. Install or connect the following:
- Exhaust manifolds. Torque the fasteners to 20 ft. lbs. (27 Nm), starting with the center nuts and work out to the ends.
- Exhaust Y-pipe
- EGR tube
- Exhaust manifold heat shields
- Negative battery cable
4. Start the engine, check for leaks and repair if necessary.

2.5L Engine

1. Before servicing the vehicle, refer to the precautions in the beginning of this section.
2. Properly relieve the fuel system pressure.
3. Drain the cooling system.
4. Remove or disconnect the following:
- Negative battery cable
- Air intake hose
- Accessory drive belt
- Power steering pump and brackets, if equipped

- Fuel line
- Accelerator cable
- Cruise control cable, if equipped
- Transmission cable, if equipped
- Throttle Position (TP) sensor connector
- Idle Air Control (IAC) valve connector
- Engine Coolant Temperature (ECT) sensor connector
- Intake Air Temperature (IAT) sensor connector
- Heated Oxygen (HO2S) sensor connector
- Fuel injector connectors
- Closed Crankcase Ventilation (CCV) system
- Manifold Absolute Pressure (MAP) sensor vacuum line
- Brake booster vacuum line
- Exhaust front pipe
- Intake manifold and discard the gaskets
- Exhaust manifold and discard the gaskets

To install:

5. Install or connect the following:
- New gasket over the locating dowels
- Exhaust manifold to the studs and tighten the nuts finger-tight
6. Install the intake manifold. Torque the fasteners in sequence as follows:
   a. Step 1: Bolt 1 to 30 ft. lbs. (41 Nm).
   b. Step 2: Fasteners 2–7 to 23 ft. lbs. (31 Nm).

7. Install or connect the following:
- Exhaust front pipe
- Brake booster vacuum line
- MAP sensor vacuum line
- CCV system
- Fuel injector connectors
- HO2S sensor connector
- IAT sensor connector
- ECT sensor connector
- IAC valve connector
- TP sensor connector
- Transmission cable, if equipped
- Cruise control cable, if equipped
- Accelerator cable
- Fuel line
- Power steering pump and brackets, if equipped
- Accessory drive belt
- Air intake hose
- Negative battery cable
8. Fill and bleed the cooling system.
9. Start the engine, check for leaks and repair if necessary..

4.0L Engine

1. Before servicing the vehicle, refer to the precautions in the beginning of this section.
2. Drain the cooling system.
3. Properly relieve the fuel system pressure.
4. Remove or disconnect the following:
- Negative battery cable
- Air cleaner assembly
- Accessory drive belt
- Power steering pump and brackets, if equipped

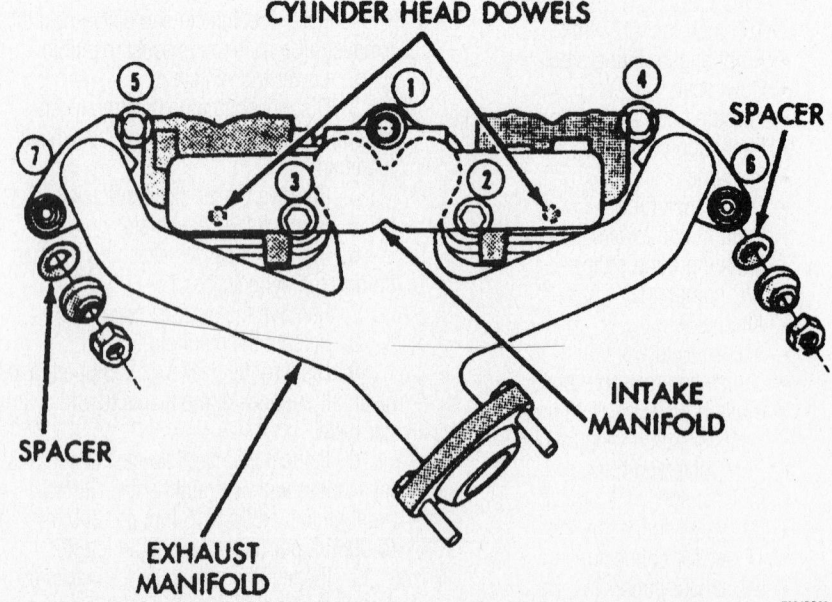

Manifold torque sequence—2.5L engine

7924PG08

- Fuel line
- Fuel injector connectors
- Fuel supply manifold and injectors
- Accelerator cable
- Cruise control cable, if equipped
- Transmission cable, if equipped
- Throttle Position (TP) sensor connector
- Idle Air Control (IAC) valve connector
- Engine Coolant Temperature (ECT) sensor connector
- Intake Air Temperature (IAT) sensor connector
- Heated Oxygen (HO2S) sensor connector
- Closed Crankcase Ventilation (CCV) system
- Manifold Absolute Pressure (MAP) sensor vacuum line
- Brake booster vacuum line
- Exhaust front pipe
- Intake and exhaust manifolds and discard the gaskets

To install:

5. Install or connect the following:
   - New gasket over the locating dowels
   - Exhaust manifold to the studs and tighten the nuts finger-tight
6. Install the intake manifold. Torque the fasteners in sequence as follows:
   a. Step 1: Fasteners 1–5 to 30 ft. lbs. (41 Nm).
   b. Step 2: Fasteners 6 and 7 to 23 ft. lbs. (31 Nm).
   c. Step 3: Fasteners 8–11 to 24 ft. lbs. (33 Nm)
7. Install or connect the following:
   - Exhaust front pipe
   - Brake booster vacuum line
   - MAP sensor vacuum line
   - CCV system
   - HO2S sensor connector
   - IAT sensor connector
   - ECT sensor connector
   - IAC valve connector
   - TP sensor connector
   - Transmission cable, if equipped
   - Cruise control cable, if equipped
   - Accelerator cable
   - Fuel supply manifold and injectors
   - Fuel injector connectors
   - Fuel line
   - Power steering pump and brackets, if equipped
   - Accessory drive belt
   - Air cleaner assembly
   - Negative battery cable
8. Fill and bleed the cooling system.
9. Start the engine, check for leaks and repair if necessary.

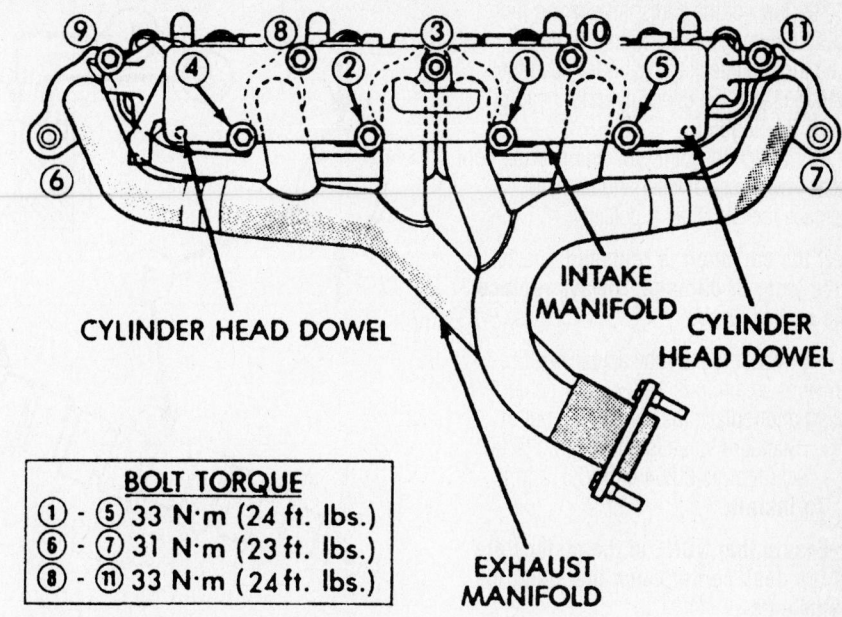

BOLT TORQUE
① - ⑤ 33 N·m (24 ft. lbs.)
⑥ - ⑦ 31 N·m (23 ft. lbs.)
⑧ - ⑪ 33 N·m (24 ft. lbs.)

7924PG12

Manifold torque sequence—4.0L engine

Camshaft and Valve Lifters

REMOVAL & INSTALLATION

2.4L Engine

1. Remove the cylinder head cover.
2. Remove the camshaft position sensor and camshaft target magnet.
3. Remove the timing belt.
4. Remove the camshaft sprockets and timing belt rear cover.
5. Bearing caps are identified for location. Remove the outside bearing caps first.

6. Loosen the camshaft bearing cap attaching fasteners in the sequence shown, one camshaft at a time.

➡**Camshafts are not interchangeable. The intake cam number 6 thrust bearing face spacing is wider.**

7. Identify the camshafts before removing from the head. The camshafts are not interchangeable.
8. Remove the camshafts from the cylinder head.

➡**If removing the rocker arms, identify for reinstallation in the original position.**

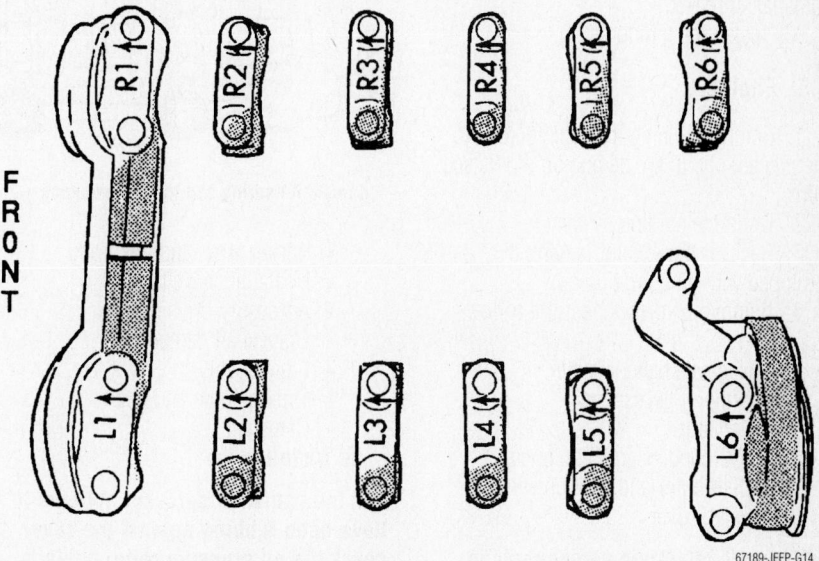

Camshaft bearing cap identification—2.4L engines

67189-JEEP-G14

9. Inspect the camshaft bearing journals for damage and binding. If the journals are binding, check the cylinder head for damage. Also the check cylinder head oil holes for clogging.

10. Check the cam lobe and bearing surfaces for abnormal wear and damage. Replace the camshaft if defective.

➡**If the camshaft is replaced due to lobe wear or damage, always replace the rocker arms.**

11. Measure the lobe actual wear (unworn area minus wear zone = actual wear) and replace the camshaft if out of limit. Standard value is 0.0254 mm (0.001 in.), wear limit is 0.254 mm (0.010 in.).

To install:

➡**Ensure that NONE of the pistons are at top dead center when installing the camshafts.**

12. Lubricate all camshaft bearing journals, rocker arms and camshaft lobes.

13. Install all rocker arms in original positions, if reused.

14. Position the camshafts on cylinder head bearing journals. Install the right and left camshaft bearing caps No. 2—5 and right No. 6. Tighten M6 fasteners to 12 Nm (105 inch lbs.) in the sequence shown.

15. Apply Mopar® Gasket Maker, or equivalent, to the No. 1 and No. 6 bearing caps. Install the bearing caps and tighten the M8 fasteners to 28 Nm (250 inch lbs.).

16. Install the camshaft oil seals.

17. Install the camshaft target magnet and camshaft position sensor.

18. Install the cylinder head cover.

19. Install the timing belt rear cover and camshaft sprocket.

20. Install the timing belt.

2.5L Engine

1. Before servicing the vehicle, refer to the precautions in the beginning of this section.

2. Drain the cooling system.

3. Recover the A/C refrigerant, if equipped with air conditioning.

4. Remove or disconnect the following:
- Negative battery cable
- Grille, if necessary
- Radiator
- A/C condenser, if equipped
- Distributor and ignition wires
- Valve cover

➡**Keep all valvetrain components in order for assembly.**

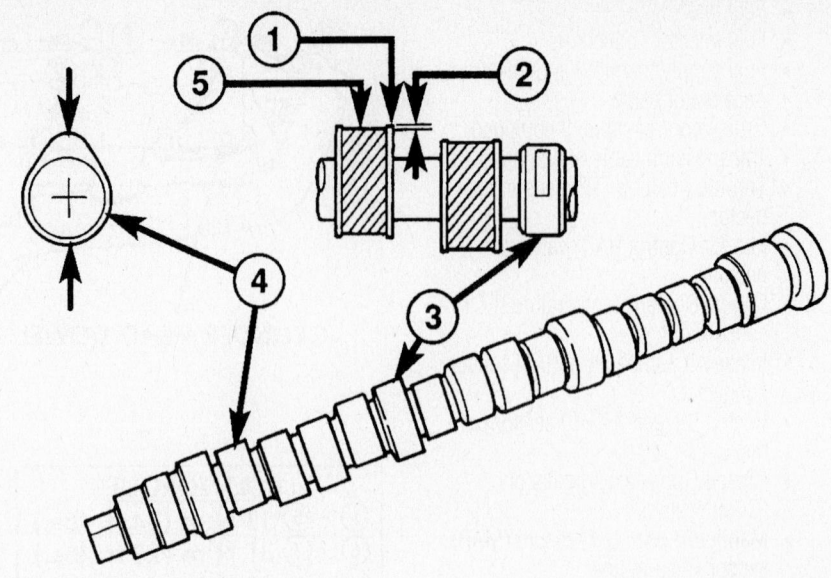

1 - UNWORN AREA
2 - ACTUAL WEAR
3 - BEARING JOURNAL
4 - LOBE
5 - WEAR ZONE

67189-JEEP-G16

Checking camshaft wear—2.4L engines

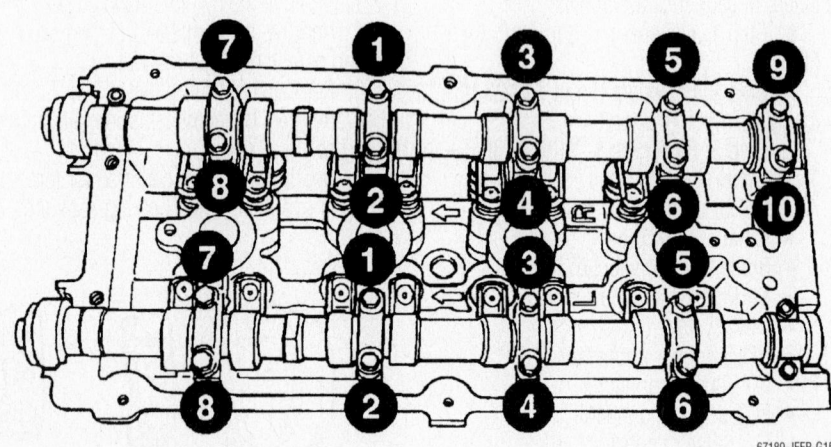

67189-JEEP-G15

Camshaft bearing cap torque sequence—2.4L engines

- Rocker arms and pushrods
- Hydraulic valve tappets
- Accessory drive belt
- Crankshaft damper
- Front cover
- Timing chain and gears
- Camshaft

To install:

➡**If the camshaft sprocket appears to have been rubbing against the cover, check the oil pressure relief holes in the rear cam journal for debris.**

5. Lubricate the camshaft with clean engine oil.

6. Install or connect the following:
- Camshaft
- Timing chain and gears
- Front cover
- Crankshaft damper
- Accessory drive belt
- Hydraulic valve tappets
- Rocker arms and pushrods
- Valve cover
- Distributor

- A/C condenser, if equipped
- Radiator
- Grille, if removed
- Negative battery cable

7. Fill and bleed the cooling system.

8. Recharge the A/C system, if equipped.

9. Start the engine, check for leaks and repair if necessary.

4.0L Engine

1. Before servicing the vehicle, refer to the precautions in the beginning of this section.

2. Drain the cooling system.

3. Recover the A/C refrigerant, if equipped with air conditioning.

4. Remove or disconnect the following:
- Negative battery cable
- Grille, if necessary
- Radiator
- A/C condenser, if equipped
- Distributor or camshaft sensor housing
- Valve cover

➡**Keep all valvetrain components in order for assembly.**

- Rocker arms and pushrods
- Cylinder head
- Hydraulic valve tappets
- Accessory drive belt
- Crankshaft damper
- Front cover
- Timing chain and gears
- Thrust plate
- Camshaft

To install:

5. Lubricate the camshaft with clean engine oil.

6. Install or connect the following:
- Camshaft
- Thrust plate. Torque the bolts to 18 ft. lbs. (24 Nm).
- Timing chain and gears
- Front cover
- Crankshaft damper
- Accessory drive belt
- Hydraulic valve tappets
- Cylinder head
- Rocker arms and pushrods
- Valve cover
- Distributor or camshaft sensor housing
- A/C condenser, if equipped
- Radiator
- Grille, if removed
- Negative battery cable

7. Fill and bleed the cooling system.

8. Recharge the A/C system, if equipped.

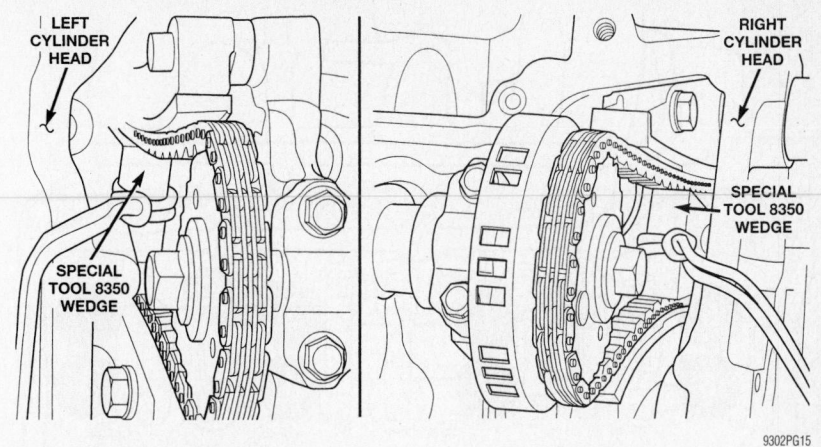

9302PG15

Chain Tensioner Retaining Wedges

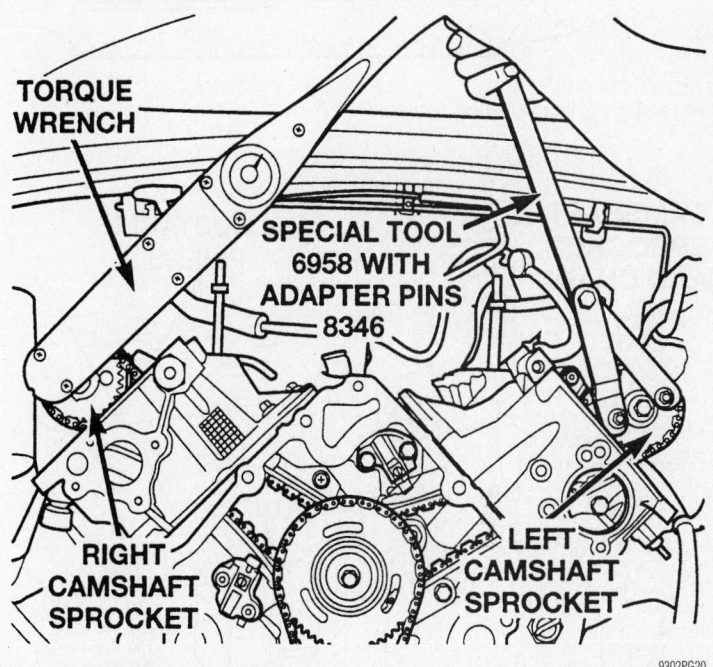

9302PG20

Hold the left camshaft sprocket with a spanner wrench while removing or installing the camshaft sprocket bolts—4.7L engine

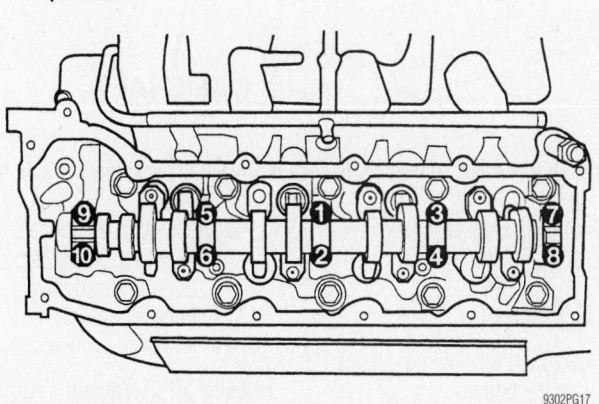

9302PG17

Camshaft bearing cap bolt tightening sequence—4.7L engine

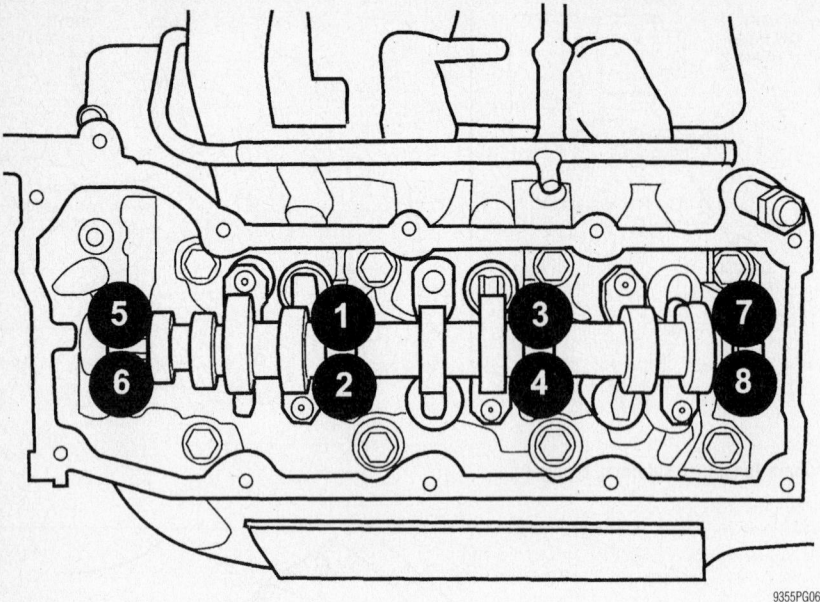

Camshaft bearing cap bolt tightening sequence—3.7L

9355PG06

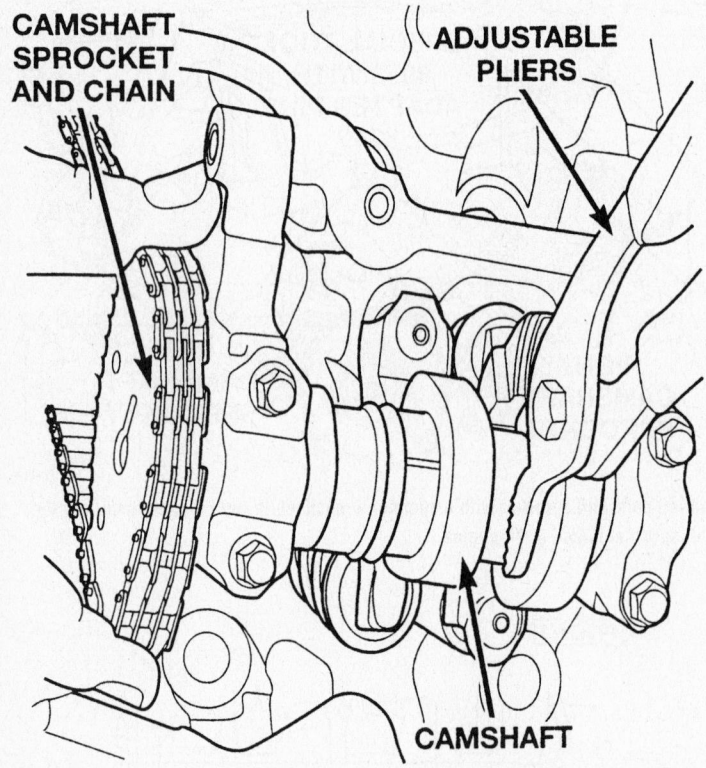

CAMSHAFT SPROCKET AND CHAIN

ADJUSTABLE PLIERS

CAMSHAFT

9302PG16

Turn the camshaft with pliers, if needed, to align the dowel in the sprocket—4.7L engine

9. Start the engine, check for leaks and repair if necessary.

3.7L and 4.7L Engines

1. Before servicing the vehicle, refer to the precautions in the beginning of this section.

2. Remove or disconnect the following:
   • Negative battery cable
   • Cylinder head covers
   • Rocker arms
   • Hydraulic lash adjusters

→Keep all valvetrain components in order for assembly.

3. Set the engine at Top Dead Center (TDC) of the compression stroke for the No. 1 cylinder.

4. Install Timing Chain Wedge (8350 4.7L; 8379 3.7L) to retain the chain tensioners.

5. Matchmark the timing chains to the camshaft sprockets.

6. Install Camshaft Holding Tool (6958 and Adapter Pins 8346 4.7L; 8428 3.7L) to the left camshaft sprocket.

7. Remove or disconnect the following:
   • Right camshaft timing sprocket and target wheel
   • Left camshaft sprocket
   • Camshaft bearing caps, by reversing the tightening sequence
   • Camshafts

To install:

8. Install or connect the following:
   • Camshafts. Torque the bearing cap bolts in ½ turn increments, in sequence, to 100 inch lbs. (11 Nm).
   • Target wheel to the right camshaft
   • Camshaft timing sprockets and chains, by aligning the matchmarks

9. Remove the tensioner wedges and tighten the camshaft sprocket bolts to 90 ft. lbs. (122 Nm).

10. Install or connect the following:
   • Hydraulic lash adjusters in their original locations
   • Rocker arms in their original locations
   • Cylinder head covers
   • Negative battery cable

Valve Lash

ADJUSTMENT

These engines are equipped with hydraulic valve lifters. No valve clearance adjustments are necessary.

Starter Motor

REMOVAL & INSTALLATION

1. Before servicing the vehicle, refer to the precautions in the beginning of this section.

2. Remove or disconnect the following:
   • Negative battery cable
   • Starter mounting bolts

→On the 3.7L engine, the left side exhaust pipe and front driveshaft must be disconnected.

- Starter solenoid harness connections
- Starter

To install:

3. Connect the starter solenoid wiring connectors.

4. Install the starter and torque the bolts to the following specifications:

- 2.5L engine: 33 ft. lbs. (45 Nm).
- 4.0L engine: Upper bolt to 40 ft. lbs. (54 Nm) and lower bolt to 30 ft. lbs. (41 Nm).
- 2.4L, 3.7L and 4.7L engine: 40 ft. lbs. (54 Nm).
- On the 3.7L engine, the left side exhaust pipe and front driveshaft

5. Install the negative battery cable and check for proper operation.

Oil Pan

REMOVAL & INSTALLATION

2.4L Engine

1. Remove air cleaner assembly.
2. Raise the vehicle on a hoist and drain the engine oil.
3. Loosen the engine mount through bolts.
4. Disconnect the exhaust pipe at the manifold.
5. Remove the structural collar, if equipped.
6. Remove the front axle mounting bolts, and lower the axle as far possible, if equipped.
7. Position Special Tool 8534, or equivalent, on the fender lip and align the slots in the brackets with the fender mounting holes.
8. Secure the brackets to the fender using four M6 X 1.0 X 25 mm flanged cap screws.
9. Tighten the thumbscrews to secure the sleeves to the support tube.
10. Secure the support tube in an upright position.
11. Assemble the flat washer, thrust bearing, hook and T handle.
12. Using the M10 X 1.5 X 40 mm cap-screw supplied with the support fixture, secure the chain to the front cover and the hook.
13. Support the engine as needed.
14. Remove the oil pan attaching bolts.
15. Remove the oil pan.
16. Clean the oil pan and all gasket surfaces.

To install:

17. Install the oil pan gasket to the block.

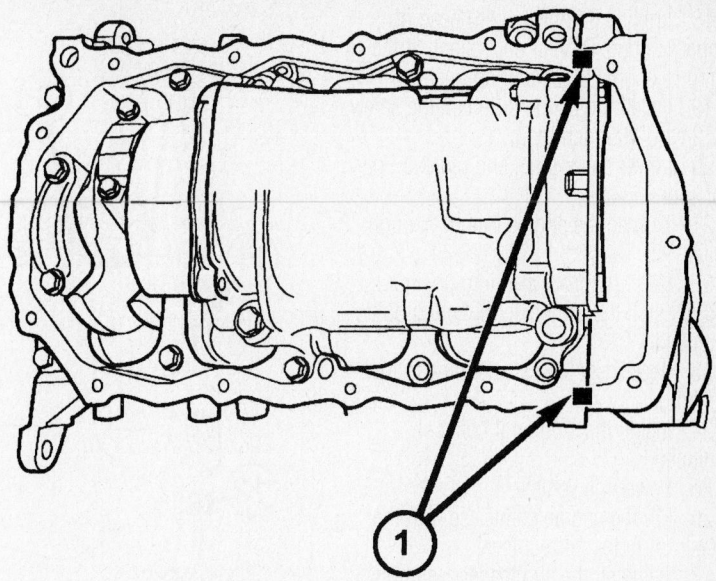

1 - SEALER LOCATION

Oil pan gasket positioning—2.4L engines

67189-JEEP-G17

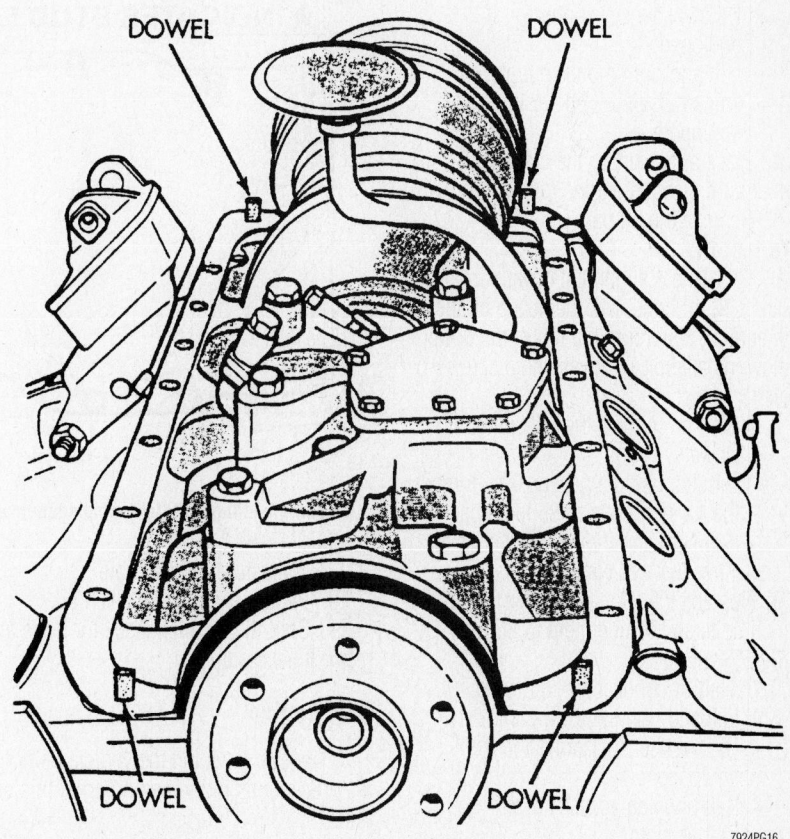

Oil pan alignment dowel placement—2.5L and 4.0L engines

7924PG16

18. Apply a 3mm (⅛ inch) bead of Mopar® Engine RTV, or equivalent, at the oil pump to engine block parting line.

19. Install the pan and tighten the screws to 12 Nm (105 inch lbs.).

20. Lower the engine, and remove Special Tool 8534.

21. Tighten the engine mount through bolts.

22. Raise the front axle into position, and reinstall the front axle mounting bolts, if equipped.

23. Reconnect the exhaust pipe to the manifold.

24. Install the structural collar, if equipped.

25. Lower the vehicle.

26. Fill the engine crankcase with the proper oil to the correct level.

27. Reinstall the air cleaner assembly.

2.5L and 4.0L Engines

1. Before servicing the vehicle, refer to the precautions in the beginning of this section.

2. Drain the engine oil.

3. Remove or disconnect the following:
- Negative battery cable
- Exhaust front pipe
- Starter motor
- Bell housing access cover
- Oil level sensor connector, if equipped
- Left and right motor mounts
- Transmission oil cooler lines, if equipped

4. Place a jack under the crankshaft damper and raise the engine for clearance.

5. Remove the oil pan.

To install:

6. Fabricate 4 alignment dowels from 1½ in. x ¼ in. bolts. Cut the heads off the bolts and cut a slot into the top of the dowel to allow installation/removal with a screwdriver.

7. Install or connect the following:
- Dowels
- Oil pan, using a new gasket. Torque the ¼ inch bolts to 85 inch lbs. (9.5 Nm) and the ⁵⁄₁₆ inch bolts to 11 ft. lbs. (15 Nm).

8. Replace the alignment dowels with ¼ inch bolts and torque them to 85 inch lbs. (9.5 Nm).

9. Install or connect the following:
- Left and right motor mounts
- Oil level sensor connector, if equipped
- Bell housing access cover
- Starter motor
- Exhaust front pipe

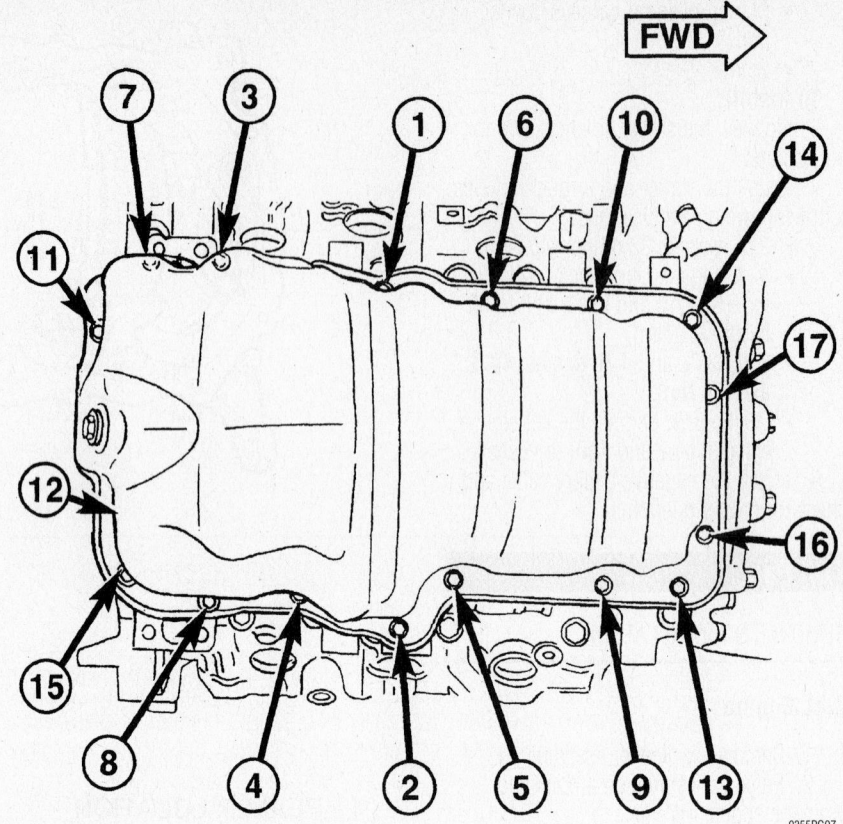

Oil pan bolt torque sequence—3.7L

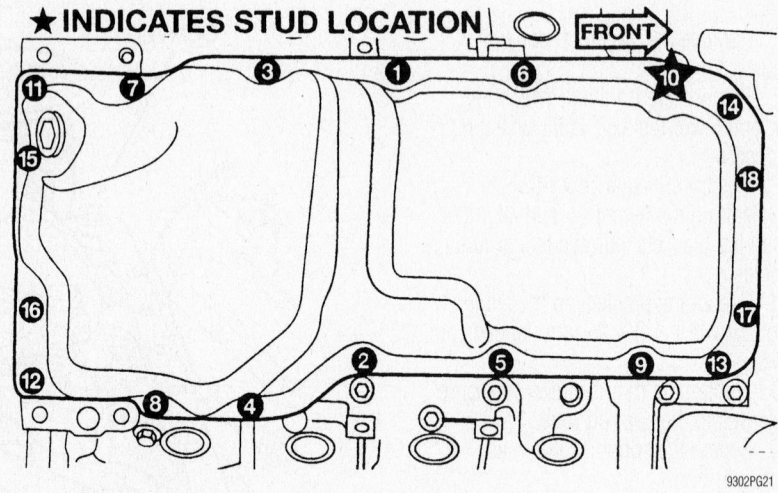

Oil pan mounting bolt tightening sequence—4.7L engine

- Negative battery cable

10. Fill the engine with clean oil.

11. Start the engine, check for leaks and repair if necessary.

3.7L Engine

1. Before servicing the vehicle, refer to the precautions in the beginning of this section.

2. Remove or disconnect the following:
- Engine from the vehicle

- Oil pan
- Oil pump pickup tube
- Oil pan gasket

3. Installation is the reverse of removal. Torque the bolts, in sequence, to 11 ft. lbs. (15 Nm).

4.7L Engine

1. Before servicing the vehicle, refer to the precautions in the beginning of this section.

2. Drain the engine oil.

3. Remove or disconnect the following:
   - Negative battery cable
   - Structural cover
   - Exhaust Y-pipe
   - Starter motor
   - Transmission oil cooler lines
   - Oil pan
   - Oil pump pickup tube
   - Oil pan gasket

To install:

4. Install or connect the following:
   - Oil pan gasket
   - Oil pump pickup tube, using a new O-ring. Torque the tube bolts to 20 ft. lbs. (28 Nm); torque the O-ring end bolt first.
   - Oil pan. Torque the bolts, in sequence, to 11 ft. lbs. (15 Nm).
   - Transmission oil cooler lines
   - Starter motor
   - Exhaust Y-pipe
   - Structural cover
   - Negative battery cable

5. Fill the engine to the proper level with clean oil.

6. Start the engine, check for leaks and repair if necessary.

Oil Pump

REMOVAL & INSTALLATION

2.4L Engine

1. Disconnect the negative cable from battery.
2. Remove the timing belt.
3. Remove the timing belt rear cover.
4. Remove the oil pan.
5. Remove the crankshaft sprocket using Special Tools 6793 and C-4685-C2, or equivalent.
6. Remove the crankshaft key.
7. Remove the oil pick-up tube.
8. Remove the oil pump and front crankshaft seal.

To install:

9. Make sure all surfaces are clean and free of oil and dirt.
10. Apply Mopar® Gasket Maker, or equivalent, to the oil pump as shown. Install the O-ring into the oil pump body discharge passage.
11. Prime the oil pump with engine oil before installation.
12. Align the oil pump rotor flats with the flats on the crankshaft. Install the oil pump to the block.

➡**To align, the front crankshaft seal MUST be out of the pump, or damage may result.**

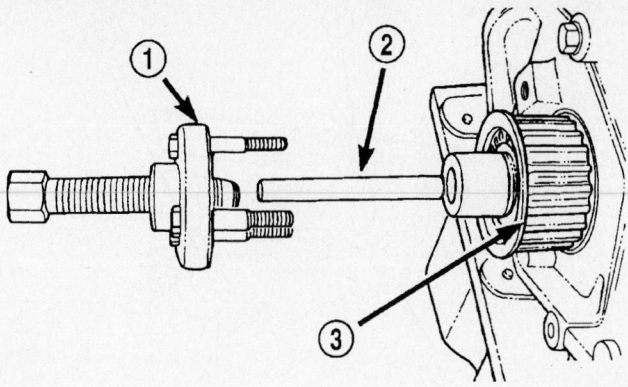

1 - SPECIAL TOOL 6793
2 - SPECIAL TOOL C-4685–C2
3 - CRANKSHAFT SPROCKET

67189-JEEP-G18

Crankshaft sprocket removal—2.4L engines

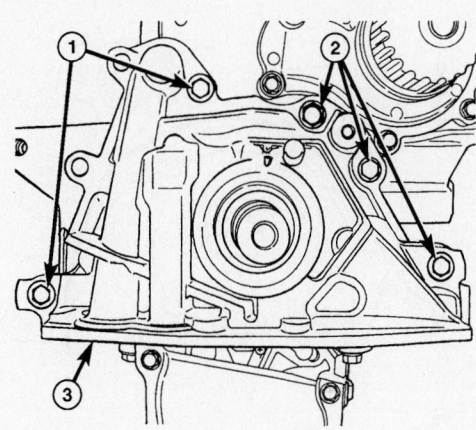

1 - BOLTS
2 - BOLTS
3 - OIL PUMP

67189-JEEP-G19

Oil pump—2.4L engines

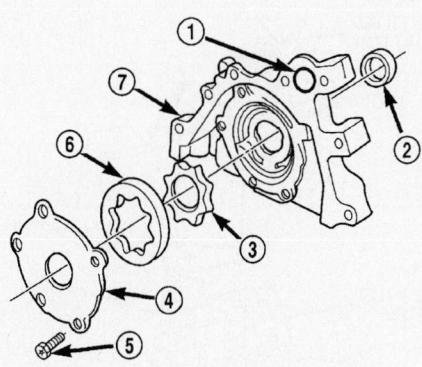

1 - O-RING
2 - SEAL
3 - INNER ROTOR
4 - OIL PUMP COVER
5 - FASTENER
6 - OUTER ROTOR
7 - OIL PUMP BODY

67189-JEEP-G20

Oil pump parts—2.4L engines

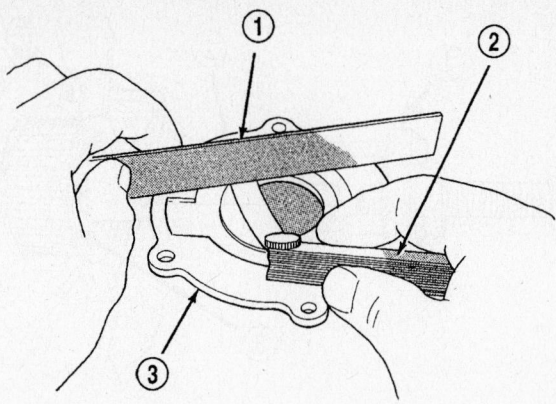

1 - STRAIGHT EDGE
2 - FEELER GAUGE
3 - OIL PUMP COVER

67189-JEEP-G21

Checking the oil pump body for flatness—2.4L engines

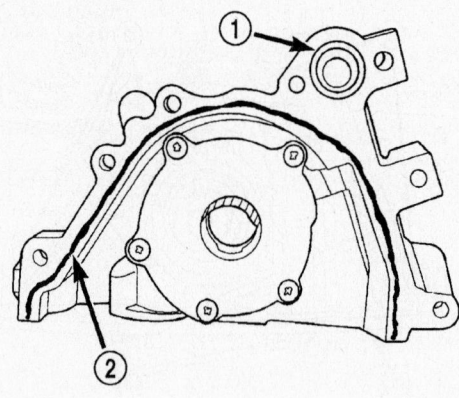

1 - O-RING
2 - SEALER LOCATION

67189-JEEP-G22

Oil pump sealing—2.4L engines

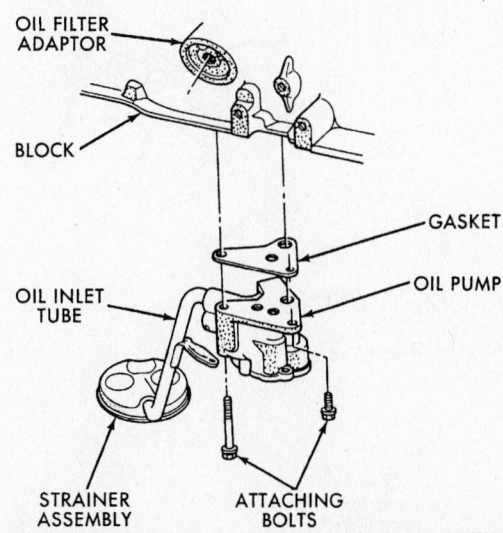

OIL FILTER
ADAPTOR

BLOCK

GASKET

OIL PUMP

OIL INLET
TUBE

STRAINER
ASSEMBLY

ATTACHING
BOLTS

7924PG17

Exploded view of the oil pump assembly—2.5L and 4.0L engine

13. Install a new front crankshaft seal.
14. Install the crankshaft key.

➡ **The crankshaft sprocket is set to a predetermined depth from the factory for correct timing belt tracking. If removed, use of Special Tool 6792, or equivalent, is required to set the sprocket to original installation depth. An incorrectly installed sprocket will result in timing belt and engine damage.**

15. Install the crankshaft sprocket using Special Tool 6792.
16. Install the oil pump pick-up tube.
17. Install the oil pan.
18. Install the timing belt rear cover.
19. Install the timing belt.

2.5L and 4.0L Engines

1. Before servicing the vehicle, refer to the precautions in the beginning of this section.
2. Drain the engine oil.
3. Remove or disconnect the following:
   • Negative battery cable
   • Oil pan
   • Oil pump and pickup tube

➡ **If the oil pump is not to be serviced, do not disturb the position of the oil inlet tube and strainer assembly in the pump body. If the tube is moved within the pump body, a replacement tube and strainer assembly must be installed to assure an airtight seal.**

To install:
4. Install or connect the following:
   • Oil pump. Torque the mounting bolts to 17 ft. lbs. (23 Nm).
   • Oil pan
   • Negative battery cable
5. Fill the engine with the proper type and quantity of oil.
6. Start the engine, check for leaks and repair if necessary.

3.7L Engine

1. Before servicing the vehicle, refer to the precautions in the beginning of this section.
2. Remove or disconnect the following:
   • Oil Pan
   • Timing chain cover
   • Timing chains and tensioners
   • Oil pump
3. Installation is the reverse of removal. Torque the pump bolts, in sequence, to 21 ft. lbs. (28 Nm),

4.7L Engine

1. Before servicing the vehicle, refer to the precautions in the beginning of this section.

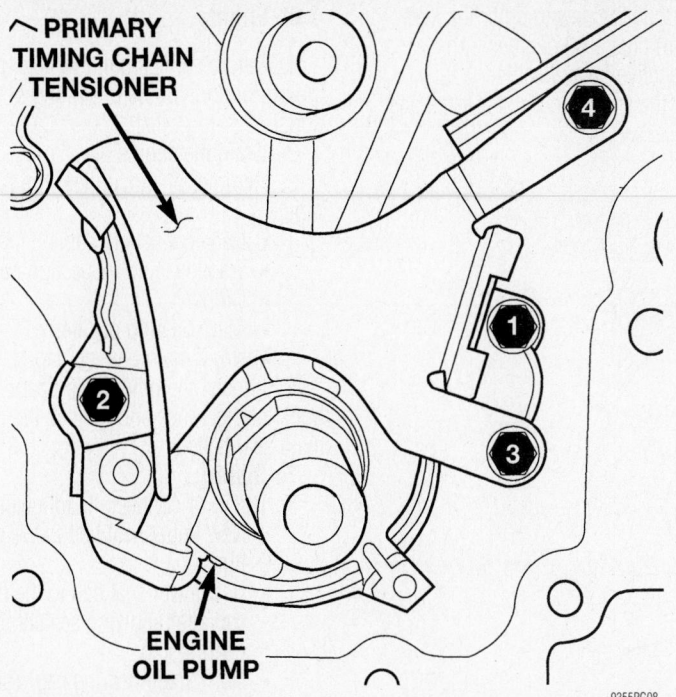

Oil pump bolt torque sequence—3.7L

2. Drain the engine oil.
3. Remove or disconnect the following:
   - Valve covers
   - Front cover
   - Timing chains and sprockets
   - Oil pan and pickup tube
   - Oil pump and primary timing chain tensioner

To install:

4. Install or connect the following:
   - Oil pump and primary timing chain tensioner. Torque the bolts in sequence to 21 ft. lbs. (28 Nm).
   - Oil pan and pickup tube
   - Timing chains and sprockets
   - Front cover

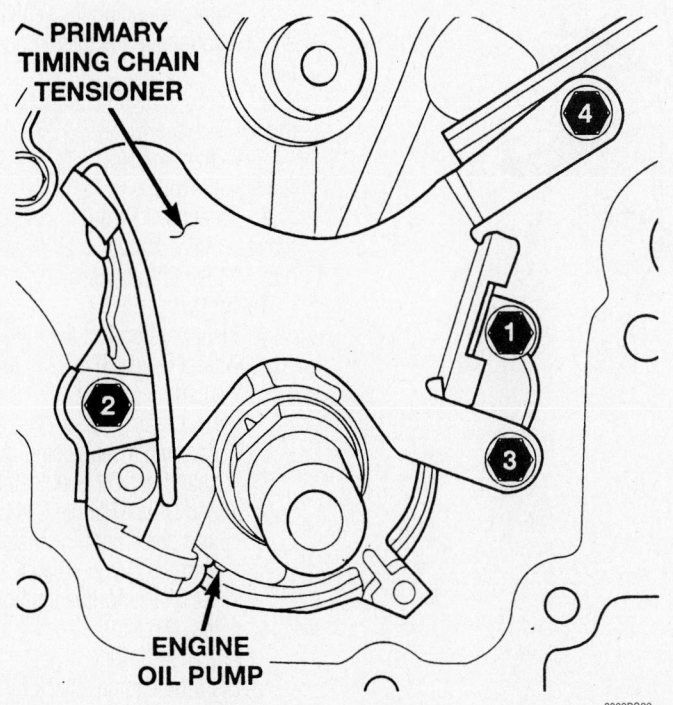

Oil pump and chain tensioner torque sequence—4.7L engine

- Valve covers
5. Fill the engine with clean oil.
6. Start the engine, check for leaks and repair if necessary.

Rear Main Seal

REMOVAL & INSTALLATION

2.4L Engine

1. Remove the transmission.
2. Remove the flex plate.
3. Insert a ³⁄₁₆ flat bladed screwdriver between the dust lip and the metal case of the crankshaft seal. Angle the screwdriver through the dust lip against metal case of the seal. Pry out seal.

➡ **Do not permit the screwdriver blade to contact the crankshaft seal surface. Contact of the screwdriver blade against the crankshaft edge (chamfer) is permitted.**

To install:

➡ **If burrs or scratches are present on the crankshaft edge (chamfer), cleanup with 400 grit sand paper to prevent seal damage during the installation of the new seal.**

4. Lubricate the crankshaft flange with engine oil.
5. Place Special Tool 6926-1 Seal Guide, or equivalent, on the crankshaft.
6. Position the seal over the guide tool. The guide tool should remain on the crankshaft during installation of the seal. Ensure that the lip of the seal is facing towards the crankcase during installation.

➡ **If the seal is driven into the block past flush, this may cause an oil leak.**

7. Drive the seal into the block using Special Tool 6926-2 and handle C-4171, or equivalent, until the tool bottoms out against the block.
8. Install the flex plate. Apply Mopar® Lock & Seal Adhesive to the bolt threads and tighten the bolts to 95 Nm (70 ft. lbs.).
9. Install the transmission.

2.5L Engine

1. Before servicing the vehicle, refer to the precautions in the beginning of this section.
2. Remove or disconnect the following:
   - Negative battery cable
   - Flywheel/converter drive plate and discard the bolt
   - Rear main seal

To install:
3. Install or connect the following:
- Rear main seal so that it is flush with the cylinder block
- Flywheel/ converter drive plate.

Torque the new to 50 ft. lbs. (68 Nm) plus a 60 degrees turn
- Negative battery cable

4. Start the engine, check for leaks and repair if necessary.

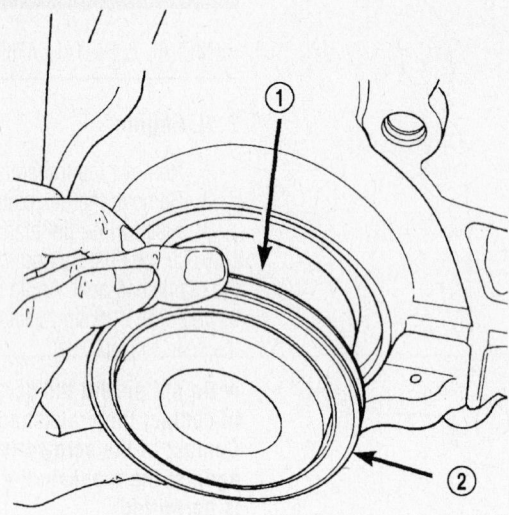

1 - SPECIAL TOOL 6926-1 PILOT
2 - SEAL

67189-JEEP-G23

Rear crankshaft seal—2.4L engines

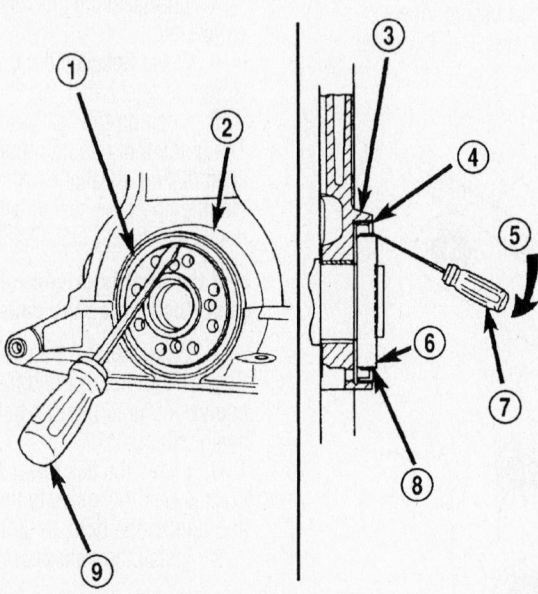

1 - REAR CRANKSHAFT SEAL
2 - ENGINE BLOCK
3 - ENGINE BLOCK
4 - REAR CRANKSHAFT SEAL METAL CASE
5 - PRY IN THIS DIRECTION
6 - CRANKSHAFT
7 - SCREWDRIVER
8 - REAR CRANKSHAFT SEAL DUST LIP
9 - SCREWDRIVER

67189-JEEP-G24

Rear crankshaft seal removal—2.4L engines

4.0L Engine

1. Before servicing the vehicle, refer to the precautions in the beginning of this section.
2. Drain the engine oil.
3. Remove or disconnect the following:

- Negative battery cable
- Transmission inspection cover
- Oil pan
- Main bearing cap brace
- No. 7 main bearing cap

4. Loosen the other main bearing cap bolts for clearance and remove the rear main seal halfs.

To install:
5. Install or connect the following:
- New upper seal half to the cylinder block
- New lower seal half to the bearing cap after applying sealant to the bearing cap
- No. 7 main bearing cap. Torque **all** main bearing cap bolts to 80 ft. lbs. (108 Nm).
- Main bearing cap brace. Tighten the nuts to 35 ft. lbs. (47 Nm).
- Oil pan
- Transmission inspection cover
- Negative battery cable

6. Fill the engine with clean oil.
7. Start the engine, check for leaks and repair if necessary.

3.7L and 4.7L Engines

1. Before servicing the vehicle, refer to the precautions in the beginning of this section.
2. Remove or disconnect the following:

- Transmission
- Flexplate

3. Thread Oil Seal Remover 8506 into the rear main seal as far as possible and remove the rear main seal.

To install:
4. Install or connect the following:
- Seal Guide 8349-2 onto the crankshaft
- Rear main seal on the seal guide
- Rear main seal, using the Crankshaft Rear Oil Seal Installer 8349 and Driver Handle C-4171; tap it into place until the installer is flush with the cylinder block
- Flexplate. Torque the bolts to 45 ft. lbs. (60 Nm).
- Transmission

5. Start the engine, check for leaks and repair if necessary.

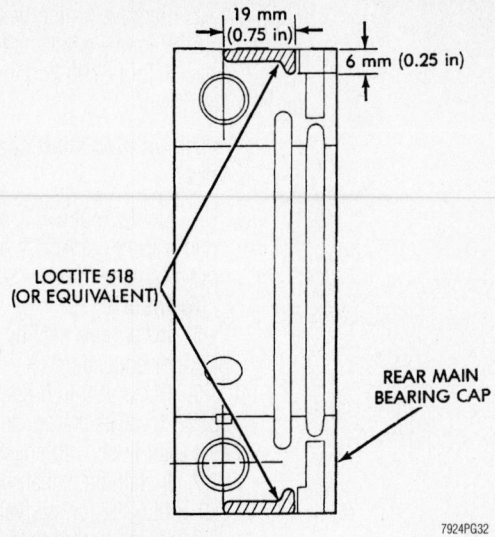

Sealant application locations—4.0L engine

Timing Belt Covers

REMOVAL & INSTALLATION

2.4L Engine

FRONT COVER

1. Remove the crankshaft vibration damper.
2. Remove the generator drive belt tensioner assembly.
3. Remove the timing belt front cover bolts, and remove the covers.

To install:

4. Install the timing belt front covers. Tighten the fasteners to 7 Nm (60 inch lbs.).
5. Install the generator drive belt tensioner.
6. Install the crankshaft vibration damper.

REAR COVER

1. Remove the front covers.
2. Remove the timing belt.
3. Hold the camshaft sprocket with Special Tool 6847, or equivalent, while removing the center bolt.
4. Remove the timing belt idler pulley.
5. Remove the rear cover fasteners and remove the cover from engine.

To install:

6. Install the timing belt rear cover and bolts.

➡**Do not use an impact wrench for tightening the camshaft sprocket bolt. Damage to the timing locating pin can occur. Hand tighten using a wrench ONLY.**

7. Install the camshaft sprockets. Hold the sprockets with Special Tool 6848, or equivalent, and tighten the center bolt to 101 Nm (75 ft. lbs.).
8. Install the timing belt idler pulley and tighten the mounting bolt to 61 Nm (45 ft. lbs.).
9. Install the timing belt.
10. Install the front covers.

Front Crankshaft Seal

The following applies only to the 2.4L engine, which has a timing belt.

REMOVAL & INSTALLATION

2.4L Engines

1. Remove the crankshaft vibration damper.

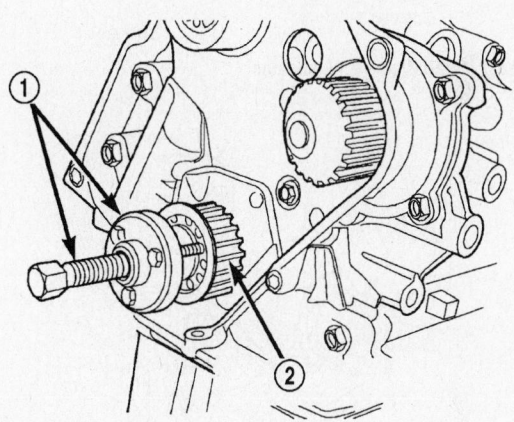

1 - SPECIAL TOOL 6793
2 - CRANKSHAFT SPROCKET

Crankshaft sprocket removal—2.4L engines

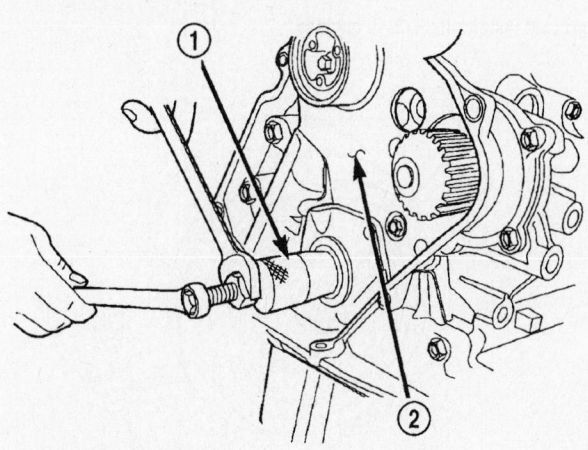

1 - SPECIAL TOOL 6771
2 - REAR TIMING BELT COVER

Front crankshaft seal removal—2.4L engines

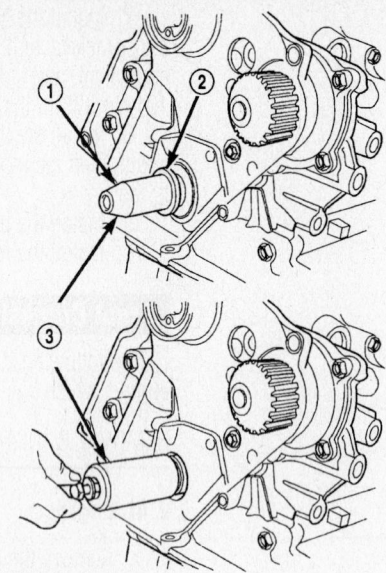

1 - PROTECTOR
2 - SEAL
3 - SPECIAL TOOL 6780

67189-JEEP-G28

Front crankshaft seal installation—2.4L engines

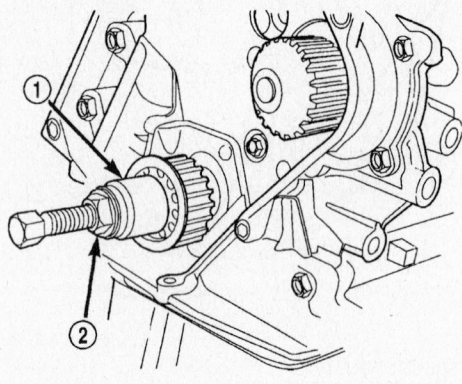

1 - SPECIAL TOOL 6792
2 - TIGHTEN NUT TO INSTALL

67189-JEEP-G27

Crankshaft sprocket installation—2.4L engines

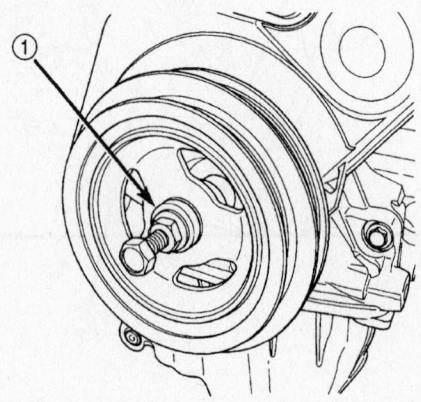

1 - SPECIAL TOOL 6792

67189-JEEP-G29

Crankshaft damper installation—2.4L engines

2. Remove timing belt.
3. Remove crankshaft sprocket using Special Tool 6793 and insert C-4685-C2, or equivalent.

➡**Do not nick shaft seal surface or seal bore.**

4. Using Tool 6771, or equivalent, remove front crankshaft oil seal. Be careful not to damage the seal surface of cover.

To install:

5. Install new seal by using Special Tool 6780, or equivalent.
6. Place seal into opening with seal spring towards the inside of engine. Install seal until flush with cover.
7. Install crankshaft sprocket using Special Tool 6792, or equivalent.
8. Install timing belt.
9. Install crankshaft vibration damper.

Timing Belt and Crankshaft Sprocket

REMOVAL & INSTALLATION

2.4L Engines

1. Remove the air cleaner upper cover, housing, and clean air tube.
2. Raise the vehicle on hoist.
3. Remove the accessory drive belts.
4. Remove the crankshaft vibration damper.
5. Remove the air conditioner/generator belt tensioner and pulley assembly.
6. Remove the timing belt lower front cover bolts and remove cover.
7. Lower the vehicle.
8. Remove the bolts attaching timing belt upper front cover and remove cover.

➡**When aligning the crankshaft and camshaft timing marks always rotate engine by the crankshaft. Camshaft should not be rotated after the timing belt is removed. Damage to the valve components may occur. Always align the timing marks before removing the timing belt.**

9. Before removal of the timing belt, rotate the crankshaft until the TDC mark on the oil pump housing aligns with the TDC mark on the crankshaft sprocket (trailing edge of the sprocket tooth).

➡**The crankshaft sprocket TDC mark is located on the trailing edge of the sprocket tooth. Failure to align the trailing edge of sprocket tooth to TDC mark on oil pump housing will cause the camshaft timing marks to be misaligned.**

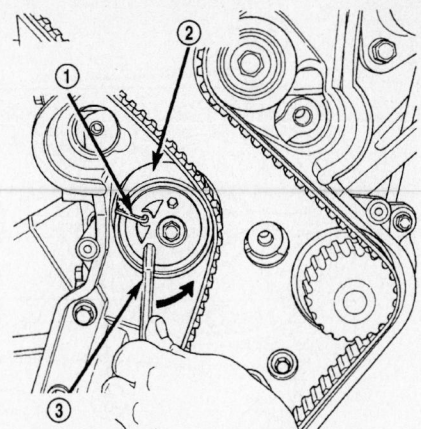

1 - 1/8 OR 3mm ALLEN WRENCH
2 - BELT TENSIONER
3 - 6mm ALLEN WRENCH

67189-JEEP-G30

Locking the tensioner—2.4L engines

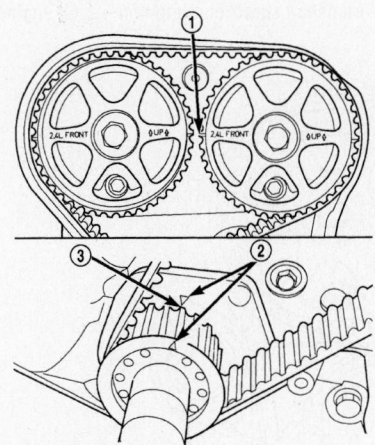

1 - CAMSHAFT TIMING MARKS
2 - CRANKSHAFT TDC MARKS
3 - TRAILING EDGE OF SPROCKET TOOTH

67189-JEEP-G31

Timing mark alignment—2.4L engines

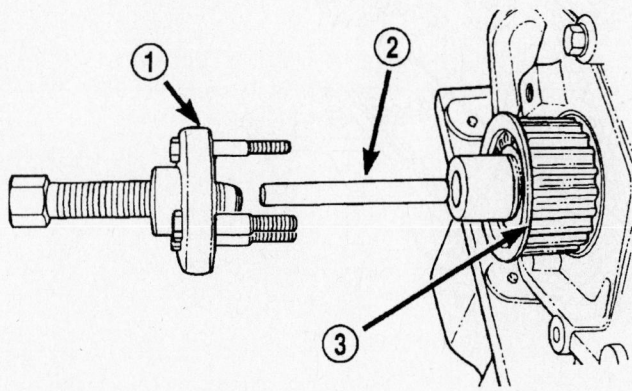

1 - SPECIAL TOOL 6793
2 - SPECIAL TOOL C-4685–C2
3 - CRANKSHAFT SPROCKET

67189-JEEP-G32

Crankshaft sprocket removal—2.4L engines

10. Install 6 mm Allen wrench into belt tensioner. Before rotating the tensioner, insert the long end of a ⅛ inch or 3 mm Allen wrench into the pin hole on the front of the tensioner. While rotating the tensioner counterclockwise, push in lightly on the ⅛ inch or 3 mm Allen wrench, until it slides into the locking hole.

11. Remove timing belt.

12. Remove crankshaft sprocket using Special Tools 6793 and insert C-4685-C2, or equivalent.

To install:

➡**The crankshaft sprocket is set to a predetermined depth from the factory for correct timing belt tracking. If removed, use of Special Tool 6792 is required to set the sprocket to original installation depth. An incorrectly installed sprocket will result in timing belt and engine damage.**

13. Install crankshaft sprocket using Special Tool 6792, or equivalent.

14. Set crankshaft sprocket to TDC by aligning the sprocket with the arrow on the oil pump housing.

15. Set camshafts timing marks so that the exhaust camshaft sprocket is a ½ notch below the intake camshaft sprocket.

❊❊ CAUTION

Ensure that the arrows on both camshaft sprockets are facing up.

16. Install timing belt. Starting at the crankshaft, go around the water pump sprocket, idler pulley, camshaft sprockets and then around the tensioner.

17. Move the exhaust camshaft sprocket counterclockwise to align marks and take up belt slack.

18. Insert a 6 mm Allen wrench into the hexagon opening located on the top plate of the belt tensioner pulley. Rotate the top plate COUNTERCLOCKWISE. The tensioner pulley will move against the belt and the tensioner setting notch will eventually start to move clockwise. Watching the movement of the setting notch, continue rotating the top plate counterclockwise until the setting notch is aligned with the spring tang. Using the allen wrench to prevent the top plate from moving, torque the tensioner lock nut to 30 Nm (22 ft. lbs.). Setting notch and spring tang should remain aligned after lock nut is torqued.

19. Remove allen wrench and torque wrench.

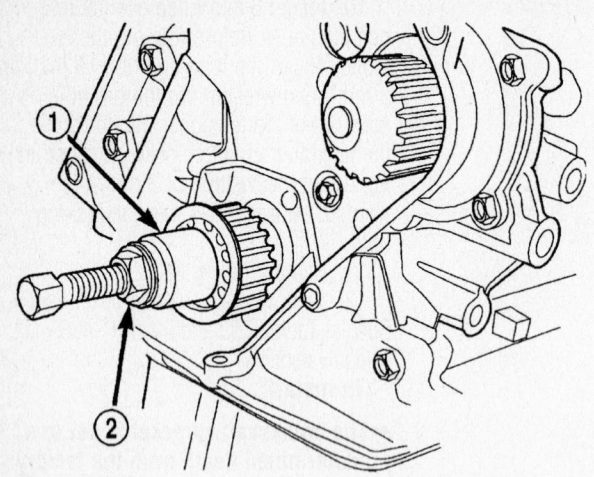

1 - SPECIAL TOOL 6792
2 - TIGHTEN NUT TO INSTALL

67189-JEEP-G33

Crankshaft sprocket installation—2.4L engines

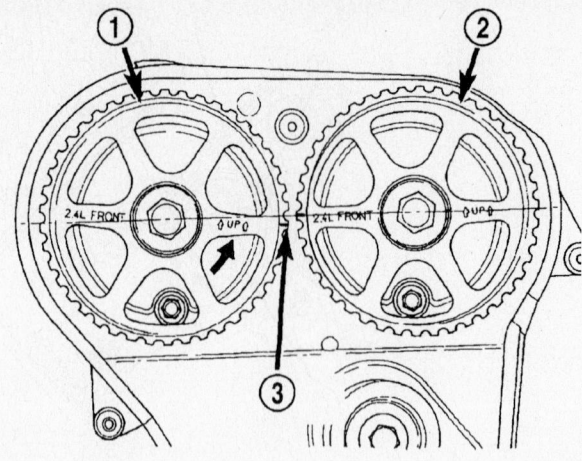

1 - CAMSHAFT SPROCKET-EXHAUST
2 - CAMSHAFT SPROCKET-INTAKE
3 - 1/2 NOTCH LOCATION

67189-JEEP-G34

Camshaft sprocket alignment—2.4L engines

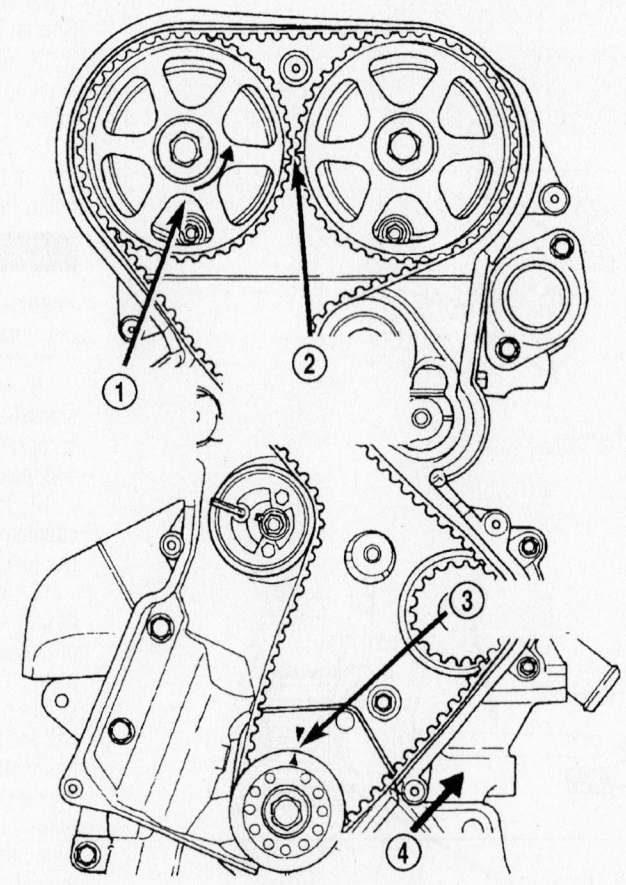

1 - ROTATE CAMSHAFT SPROCKET TO TAKE UP BELT SLACK
2 - CAMSHAFT TIMING MARKS 1/2 NOTCH LOCATION
3 - CRANKSHAFT AT TDC
4 - INSTALL BELT IN THIS DIRECTION

67189-JEEP-G35

Timing belt installation—2.4L engines

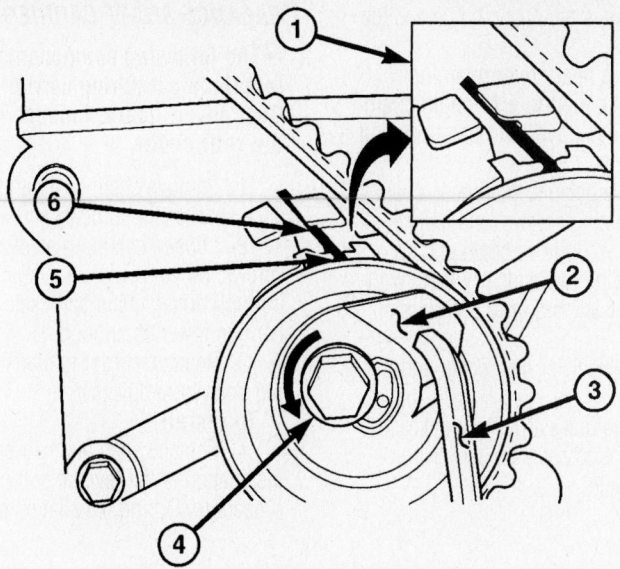

1 - ALIGN SETTING NOTCH WITH SPRING TANG
2 - TOP PLATE
3 - 6mm ALLEN WRENCH
4 - LOCK NUT
5 - SETTING NOTCH
6 - SPRING TANG

67189-JEEP-G36

Timing belt tension adjustment—2.4L engines

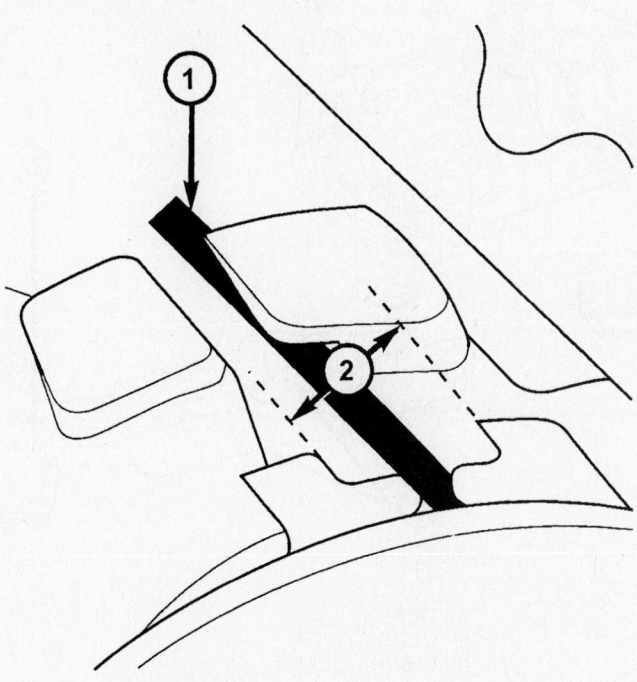

1 - SPRING TANG
2 - TOLERANCE WINDOW

67189-JEEP-G37

Timing belt tension verification—2.4L engines

➡Repositioning the crankshaft to the TDC position must be done only during the CLOCKWISE rotation movement. If TDC is missed, rotate a further two revolutions until TDC is achieved. DO NOT rotate crankshaft counterclockwise as this will make verification of proper tensioner setting impossible.

20. Once the timing belt has been installed and tensioner adjusted, rotate the crankshaft CLOCKWISE two complete revolutions manually for seating of the belt, until the crankshaft is repositioned at the TDC position. Verify that the camshaft and crankshaft timing marks are in proper position.

21. Check if the spring tang is within the tolerance window. If the spring tang is within the tolerance window, the installation process is complete and nothing further is required. If the spring tang is not within the tolerance window, repeat Steps 5 through 7.

22. Install timing belt front covers and bolts.

23. Install air conditioning/generator belt tensioner and pulley.

24. Install crankshaft vibration damper.

25. Install accessory drive belts.

26. Install drive belt splash shield.

27. Install air cleaner housing, upper cover, and clean air tube.

Timing Belt Tensioner and Pulley

REMOVAL & INSTALLATION

2.4L Engine

1. Remove the timing belt.

2. Remove the timing belt idler pulley.

3. Hold the camshaft sprocket with Special Tool 6847, or equivalent, while removing the bolt. Remove both cam sprockets.

4. Remove the rear timing belt cover fasteners and remove the cover from engine.

5. Remove the lower bolt attaching the timing belt tensioner assembly to engine and remove the tensioner as an assembly.

To install:

6. Align the timing belt tensioner assembly to engine and install the lower mounting bolt but do not tighten. To properly align the tensioner assembly, install one of the engine bracket mounting bolts (M10) 5 to 7 turns into the tensioner's upper mounting location.

7. Torque the tensioner's lower mounting bolt to 61 Nm (45 ft. lbs.). Remove the upper bolt used for tensioner alignment.

8. Install the rear timing belt cover and fasteners.

9. Install the timing belt idler pulley and torque the mounting bolt to 61 Nm (45 ft. lbs.).

10. Install the camshaft sprockets. Use Special Tool 6847 to hold the sprockets. Torque the bolts to 101 Nm (75 ft. lbs.).

11. Install the timing belt.

Balance Shafts and Carrier

REMOVAL & INSTALLATION

2.4L Engine

BALANCE SHAFTS

1. Drain engine oil.
2. Remove the oil pan.

3. Remove chain cover, guide and tensioner.

4. Remove gear cover retaining stud (double ended to also retain chain guide). Remove cover and balance shaft gears.

5. Remove balance shaft gear, chain sprocket retaining screws, and crankshaft chain sprocket. Remove chain and sprocket assembly. Using two wide pry bars, work the sprocket back and forth until it is off the shaft.

6. Remove carrier gear cover and balance shafts.

7. Remove four carrier to crankcase attaching bolts to separate carrier from engine bedplate.

BALANCE SHAFT CARRIER

➡ The following components will remain intact during carrier removal. Gear cover, gears, balance shafts and the rear cover.

1. Remove chain cover and driven balance shaft chain sprocket screw.

2. Loosen tensioner pivot and adjusting screws, move driven balance shaft inboard through driven chain sprocket. Sprocket will hang in lower chain loop.

3. Remove carrier to crankcase attaching bolts to remove carrier.

To install:

4. Balance shaft and carrier assembly installation is the reverse of the removal procedure. During installation crankshaft-

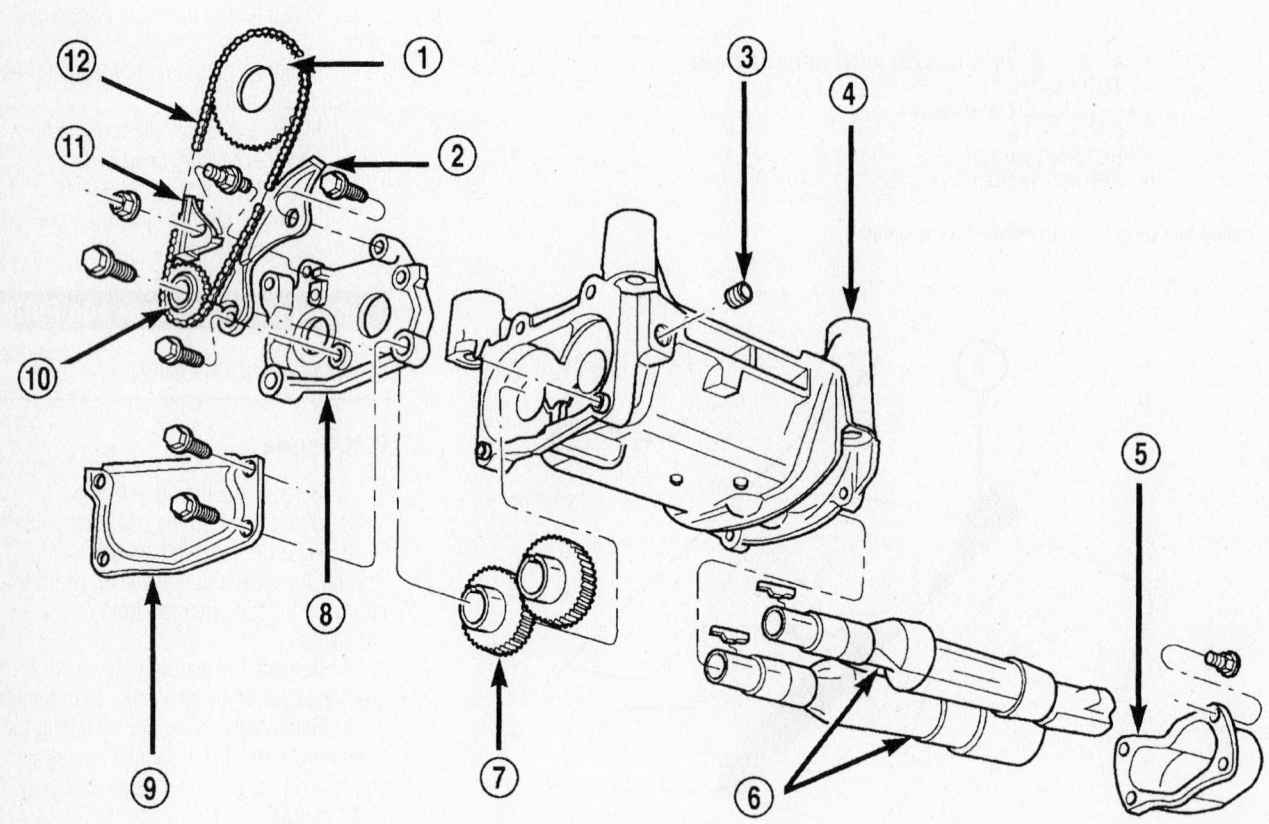

1 - SPROCKET
2 - TENSIONER
3 - PLUG
4 - CARRIER
5 - REAR COVER
6 - BALANCE SHAFTS

7 - GEARS
8 - GEAR COVER
9 - CHAIN COVER
10 - SPROCKET
11 - GUIDE
12 - CHAIN

67189-JEEP-G38

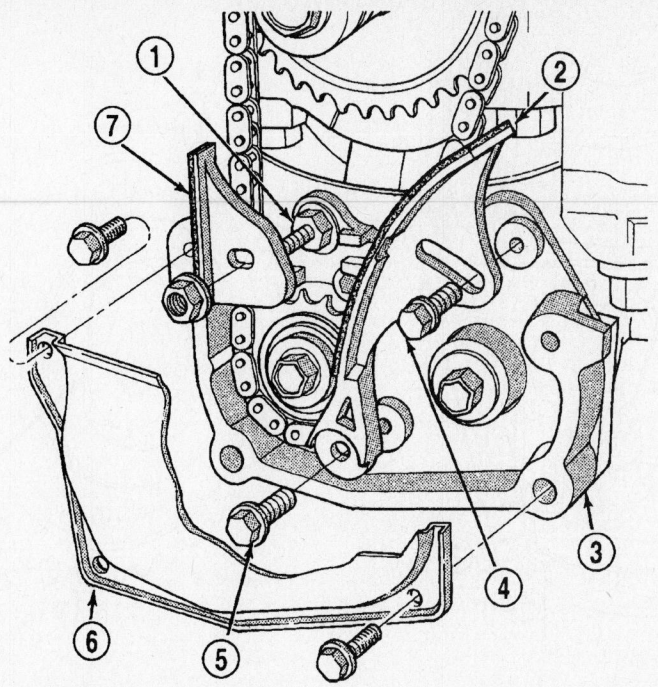

1 - STUD
2 - TENSIONER (ADJUSTER)
3 - GEAR COVER
4 - ADJUST SCREW
5 - PIVOT SCREW
6 - CHAIN COVER (CUTAWAY)
7 - GUIDE

67189-JEEP-G39

Balance shaft chain cover, guide and tensioner—2.4L engines

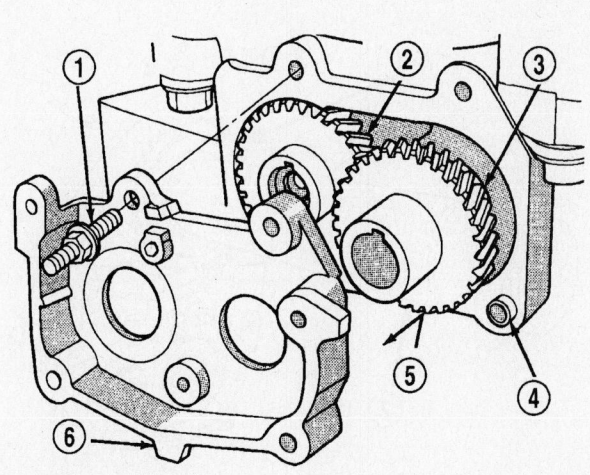

1 - STUD (DOUBLE ENDED)
2 - DRIVE GEAR
3 - DRIVEN GEAR
4 - CARRIER DOWEL
5 - GEAR(S)
6 - GEAR COVER

67189-JEEP-G40

Balance shaft gear cover and gears—2.4L engines

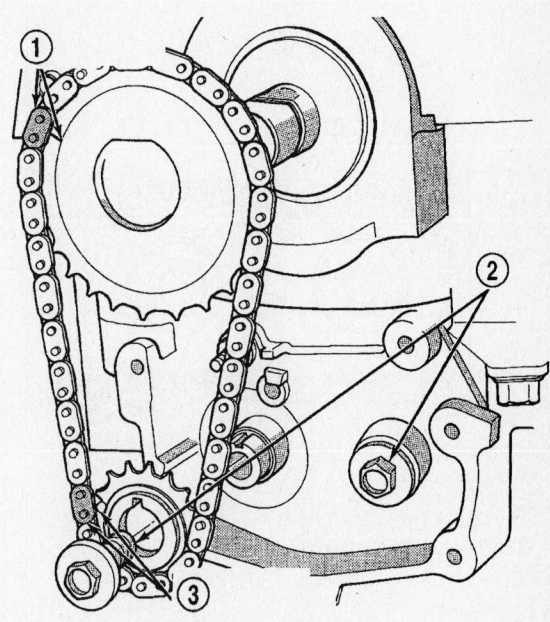

1 - NICKEL PLATED LINK AND MARK
2 - GEAR/SPROCKET SCREWS
3 - NICKEL PLATED LINK AND DOT

67189-JEEP-G41

Balance shaft chain and sprockets—2.4L engines

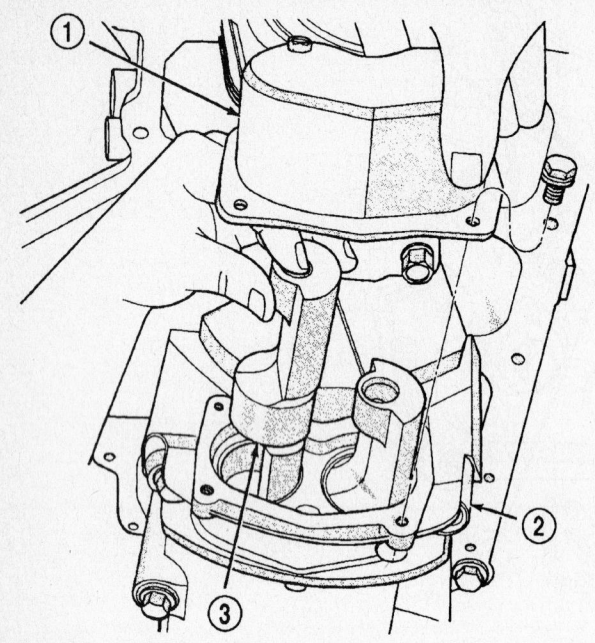

1 - REAR COVER
2 - CARRIER
3 - BALANCE SHAFT

67189-JEEP-G42

Balance shaft removal—2.4L engines

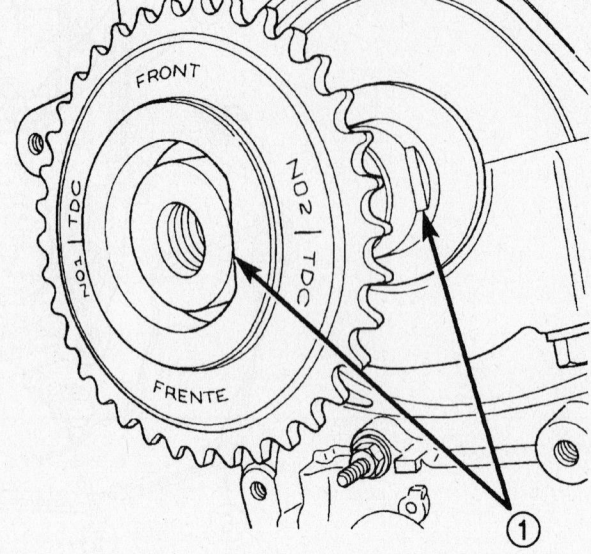

1 - ALIGN FLATS

67189-JEEP-G43

Balance shaft installation—2.4L engines

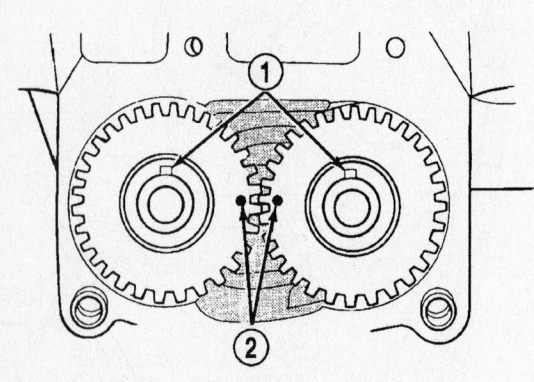

1 - KEY WAYS UP
2 - GEAR ALIGNMENT DOTS

67189-JEEP-G44

Balance shaft gear timing—2.4L engines

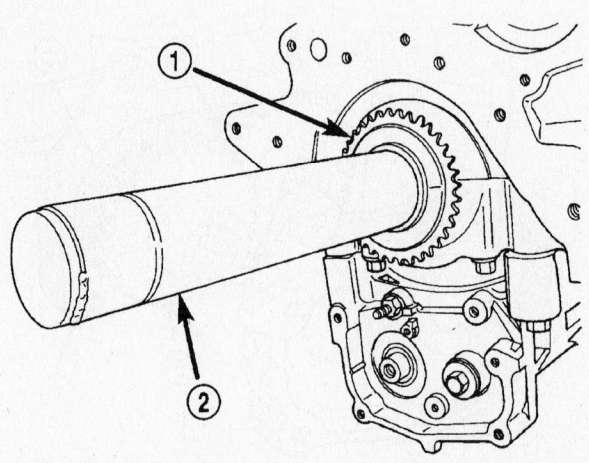

1 - SPROCKET
2 - SPECIAL TOOL 6052

67189-JEEP-G45

Balance shaft drive sprocket installation—2.4L engines

to-balance shaft timing must be established. Timing is performed as follows:

5. With balance shafts installed in carrier position carrier on crankcase and install four attaching bolts and tighten to 54 Nm (40 ft. lbs.).

6. Turn balance shafts until both shaft key ways are up, parallel to vertical centerline of engine. Install short hub drive gear on sprocket driven shaft and long hub gear on gear driven shaft. After installation gear and balance shaft keyways must be up with gear timing marks meshed as shown in.

7. Install gear cover and tighten double ended stud/washer fastener to 12 Nm (105 inch lbs.).

8. Align flat on balance shaft drive sprocket to the flat on crankshaft.

9. Install balance shaft drive sprocket on crankshaft using Special Tool 6052, or equivalent.

10. Turn crankshaft until number 1 cylinder is at top dead center (TDC). The timing marks on the chain sprocket should line up with the parting line on the left side of number one main bearing cap.

11. Place chain over crankshaft sprocket so that the plated link of the chain is over the number 1 cylinder timing mark on the balance shaft crankshaft sprocket.

12. Place balance shaft sprocket into the timing chain and align the timing mark on the sprocket (dot) with the (lower) plated link on the chain.

➡**The lower plated link is 8 links from the upper link.**

13. With balance shaft keyways pointing up (12 o'clock) slide the balance shaft sprocket onto the nose of the balance shaft. The balance shaft may have to be pushed in slightly to allow for clearance.

14. If the sprockets are timed correctly, install the balance shaft bolts and tighten to 28 Nm (250 in. lbs.). A wood block placed between crankcase and crankshaft counterbalance will prevent crankshaft and gear rotation.

CHAIN TENSIONING

1. Install chain tensioner loosely assembled.

2. Position guide on double ended stud making sure tab on the guide fits into slot on the gear cover. Install and tighten nut/washer assembly to 12 Nm (105 inch lbs.).

3. Place a shim 1 mm (0.039 in.) thick x 70 mm (2.75 in.) long or between tensioner and chain. Push tensioner and shim up against the chain. Apply firm pressure 2.5–3 Kg (5.5–6.6 lbs.) directly behind the

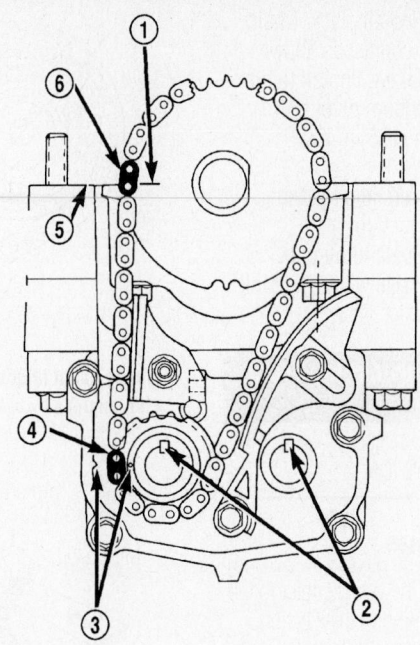

1 - MARK ON SPROCKET
2 - KEYWAYS UP
3 - ALIGN MARKS
4 - PLATED LINK
5 - PARTING LINE (BEDPLATE TO BLOCK)
6 - PLATED LINK

67189-JEEP-G46

Balance shaft timing—2.4L engines

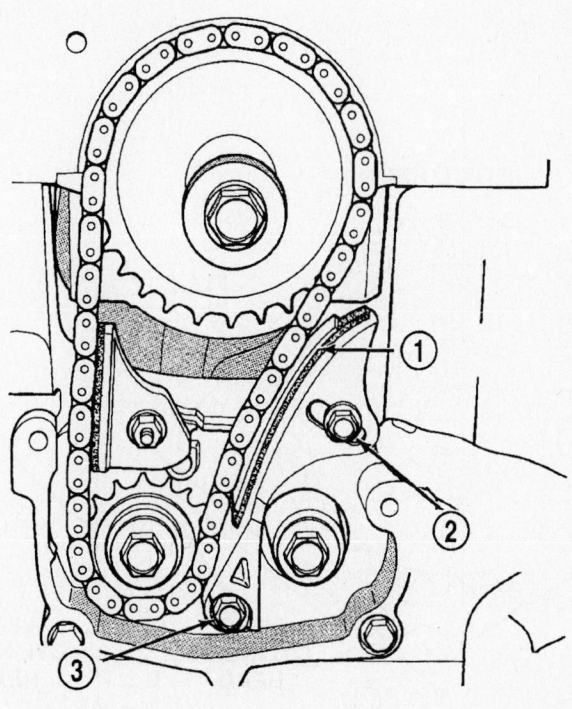

1 - 1MM (0.039 IN.) SHIM
2 - TENSIONER (ADJUSTER) BOLT
3 - PIVOT BOLT

67189-JEEP-G47

Balance shaft chain tension adjustment—2.4L engines

adjustment slot to take up all slack. Chain must have shoe radius contact as shown.

4. With the load applied, tighten top tensioner bolt first, then bottom pivot bolt. Tighten bolts to 12 Nm (105 in. lbs.). Remove shim.

5. Install carrier covers and tighten screws to 12 Nm (105 inch lbs.).

6. Install pick-up tube and oil pan.

7. Fill engine crankcase with proper oil to correct level.

Timing Chain, Sprockets, Front Cover and Seal

REMOVAL & INSTALLATION

2.5L and 4.0L Engines

1. Before servicing the vehicle, refer to the precautions in the beginning of this section.

2. Remove or disconnect the following:
- Negative battery cable
- Accessory drive belt
- Cooling fan and shroud
- Crankshaft damper
- Front crankshaft seal
- Accessory brackets
- Front cover
- Oil slinger

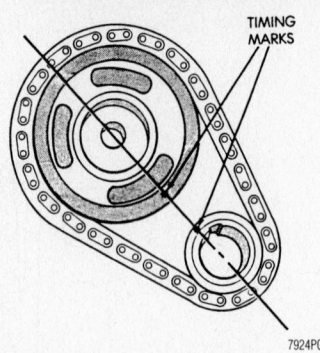

Timing chain alignment marks—2.5L and 4.0L engines

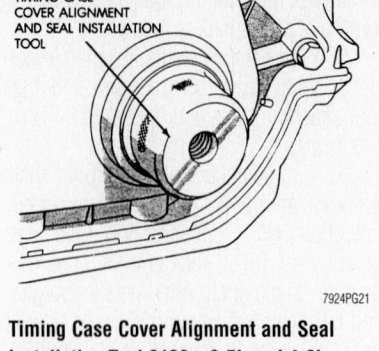

Timing Case Cover Alignment and Seal Installation Tool 6139—2.5L and 4.0L engines

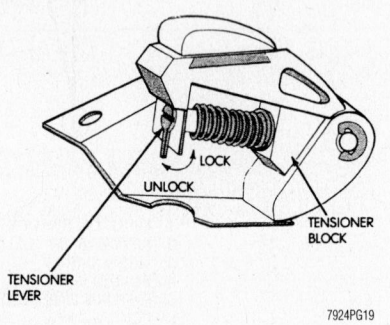

Timing chain tensioner—2.5L engines

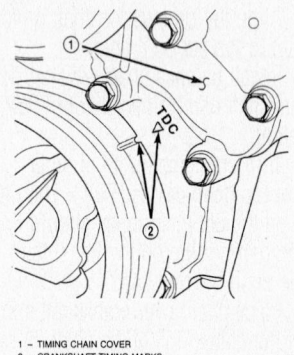

Crankshaft timing marks—4.7L engine

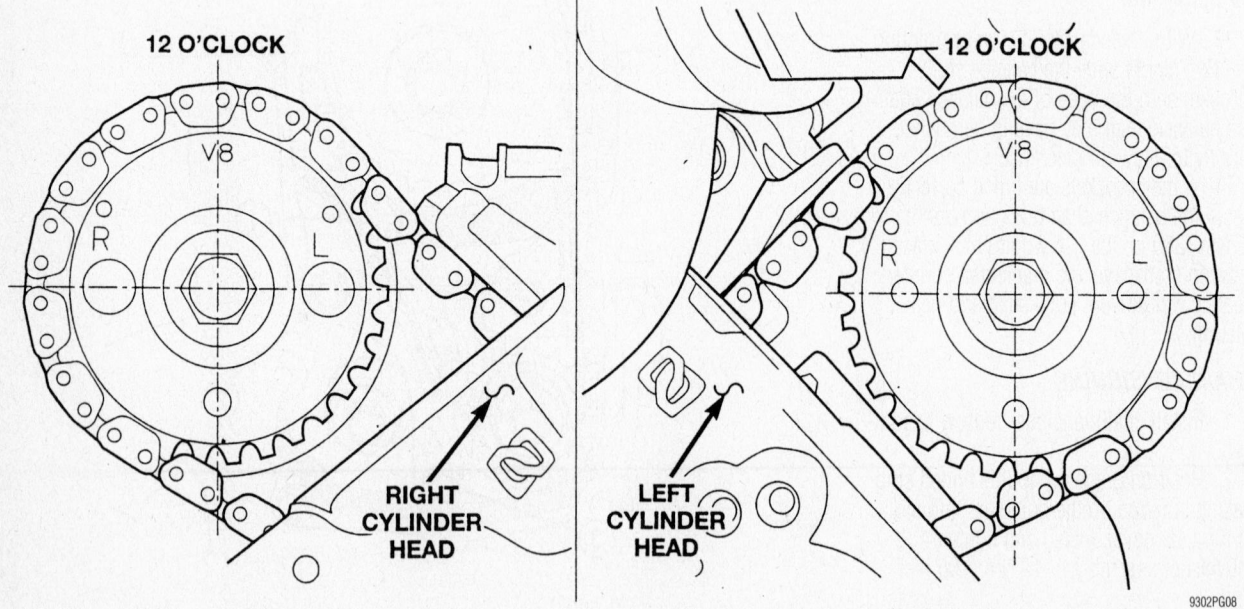

Camshaft positioning—4.7L engine

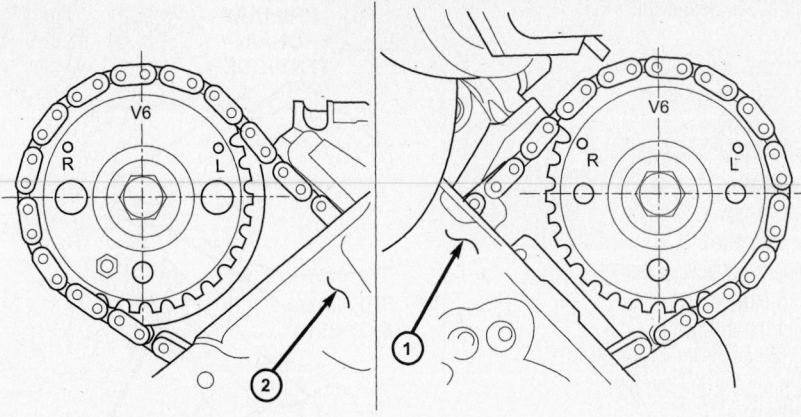

1 - LEFT CYLINDER HEAD
2 - RIGHT CYLINDER HEAD

9355PG09

Camshaft sprocket timing marks—3.7L

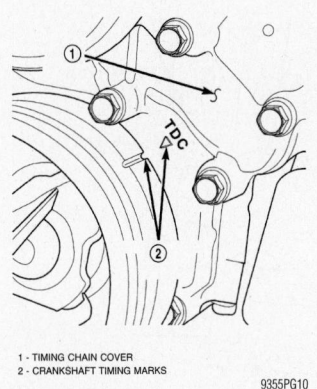

1 - TIMING CHAIN COVER
2 - CRANKSHAFT TIMING MARKS

9355PG10

Crankshaft timing marks—3.7L

3. Rotate the crankshaft so that the timing marks are aligned.

4. Remove the timing chain and sprockets.

To install:

5. For 2.5L engines, turn the timing chain tensioner lever to the unlock (down) position. Pull the tensioner block toward the tensioner lever to compress the spring. Hold the block and turn the tensioner lever to the lock (up) position.

6. Install or connect the following:
- Timing chain and sprockets with the timing marks aligned. Torque the camshaft sprocket bolt to 80 ft. lbs. (108 Nm) for 2.5L engines or to 50 ft. lbs. (68 Nm) for 4.0L engines.

7. For 2.5L engines, release the timing chain tensioner.
- Oil slinger
- New front crankshaft seal to the front cover
- Front cover, using a new gasket
- Timing Case Cover Alignment and Seal Installation Tool 6139 in the

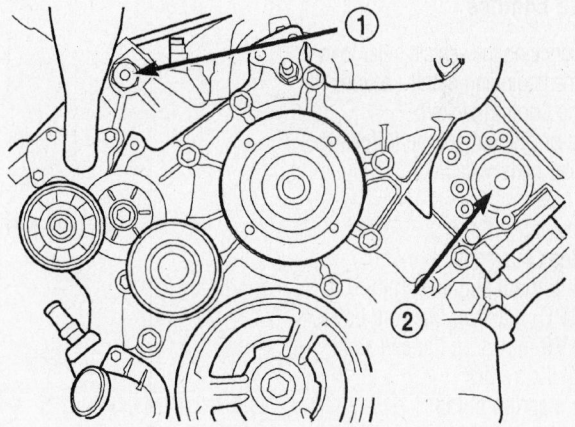

1 - RIGHT CYLINDER HEAD ACCESS PLUG
2 - LEFT CYLINDER HEAD ACCESS PLUG

9355PG11

Cylinder head access plugs—3.7L

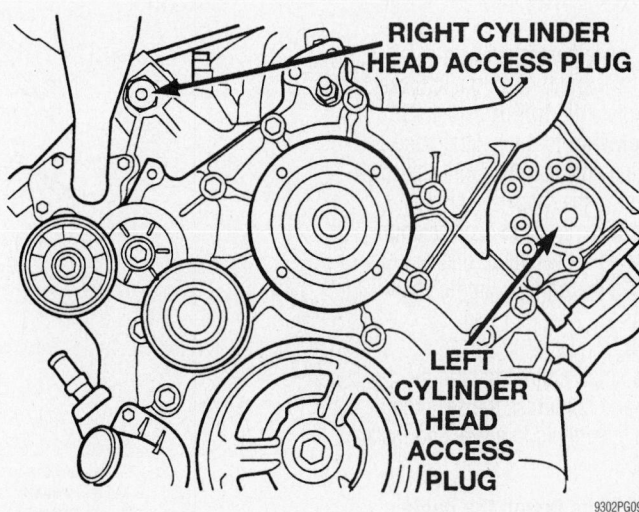

RIGHT CYLINDER
HEAD ACCESS PLUG

LEFT
CYLINDER
HEAD
ACCESS
PLUG

9302PG09

Cylinder head access plug locations—4.7L engine

crankshaft opening to center the front cover

8. Torque the front cover bolts as follows:

a. Step 1: Cover-to-block ¼ inch bolts to 60 inch lbs. (7 Nm).

b. Step 2: Cover-to-block ⁵⁄₁₆ inch bolts to 16 ft. lbs. (22 Nm).

c. Step 3: Oil pan-to-cover ¼ inch bolts to 85 inch lbs. (9.5 Nm).

d. Step 4: Oil pan-to-cover ⁵⁄₁₆ inch bolts to 11 ft. lbs. (15 Nm).

9. Install or connect the following:

• Accessory brackets
• Crankshaft damper. Torque the bolt to 80 ft. lbs. (108 Nm).
• Cooling fan and shroud
• Accessory drive belt
• Negative battery cable

10. Start the engine, check for leaks and repair if necessary.

3.7L and 4.7L Engines

1. Before servicing the vehicle, refer to the precautions in the beginning of this section.

2. Drain the cooling system.

3. Remove or disconnect the following:

• Negative battery cable
• Valve covers
• Radiator fan

4. Rotate the crankshaft so that the crankshaft timing mark aligns with the Top Dead Center (TDC) mark on the front cover, and the **V8 or V6** marks on the camshaft sprockets are at 12 o'clock.

• Power steering pump
• Access plugs from the cylinder heads
• Oil fill housing
• Crankshaft damper

5. Compress the primary timing chain tensioner and install a lockpin.

6. Remove the secondary timing chain tensioners.

7. Hold the left camshaft with adjustable pliers and remove the sprocket and chain. Rotate the **left** camshaft 15 degrees **clockwise** to the neutral position.

8. Hold the right camshaft with adjustable pliers and remove the camshaft sprocket. Rotate the **right** camshaft 45 degrees **counterclockwise** to the neutral position.

9. Remove the primary timing chain and sprockets.

To install:

10. Use a small prytool to hold the ratchet pawl and compress the secondary timing chain tensioners in a vise and install locking pins.

➡The black bolts fasten the guide to the engine block and the silver bolts fasten the guide to the cylinder head.

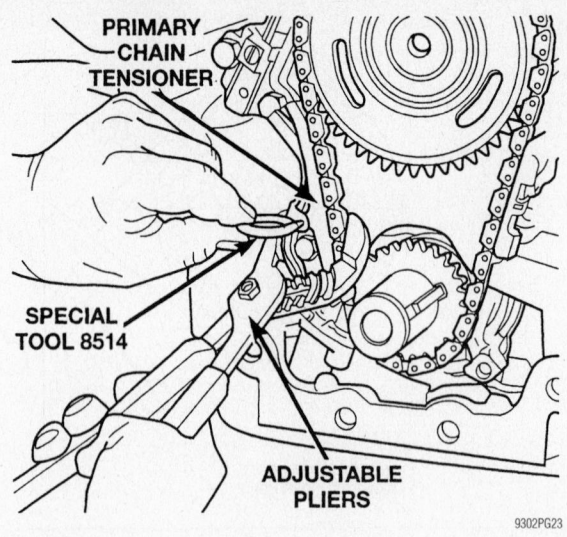

Compress and lock the primary chain tensioner—4.7L engine

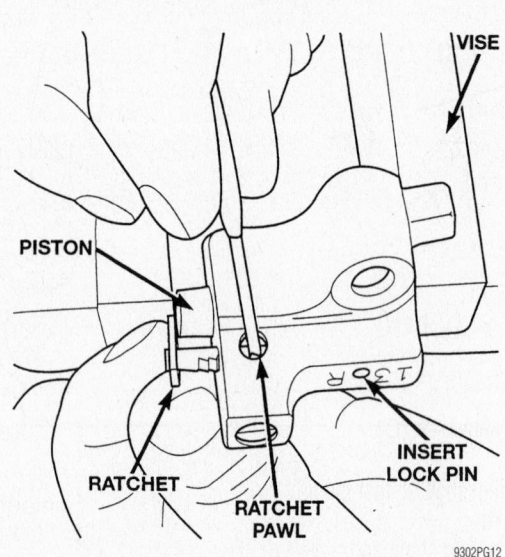

Secondary timing chain tensioner preparation—4.7L engine

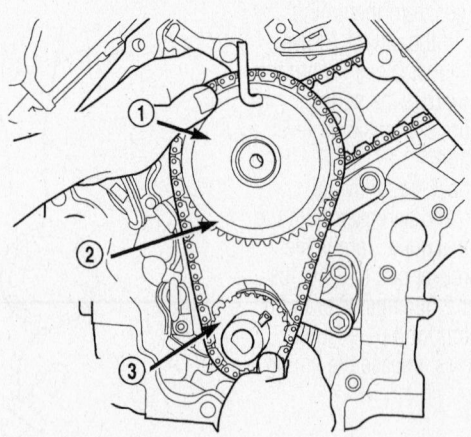

1 - SPECIAL TOOL 8429
2 - PRIMARY CHAIN IDLER SPROCKET
3 - CRANKSHAFT SPROCKET

Installing the idler gear and timing chains—3.7L

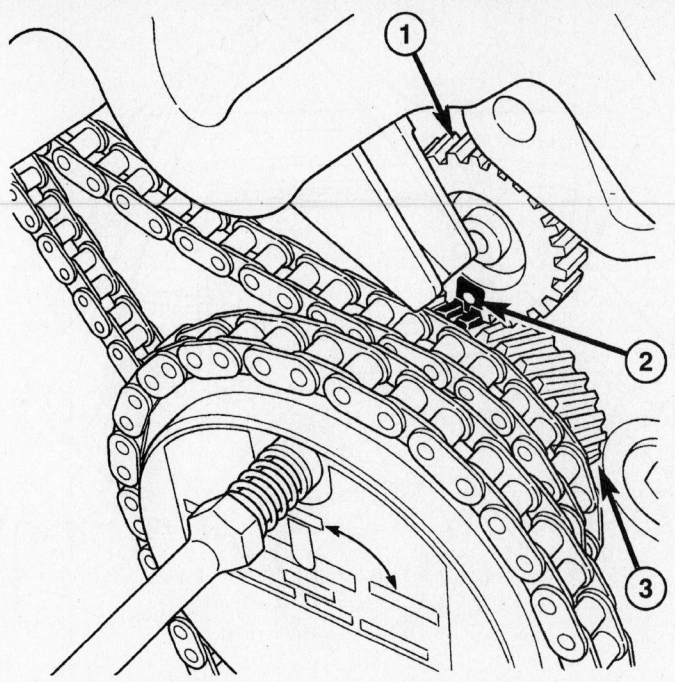

1 - COUNTERBALANCE SHAFT

2 - TIMING MARKS

3 - IDLER SPROCKET

9355PG13

Counterbalance shaft timing marks—3.7L

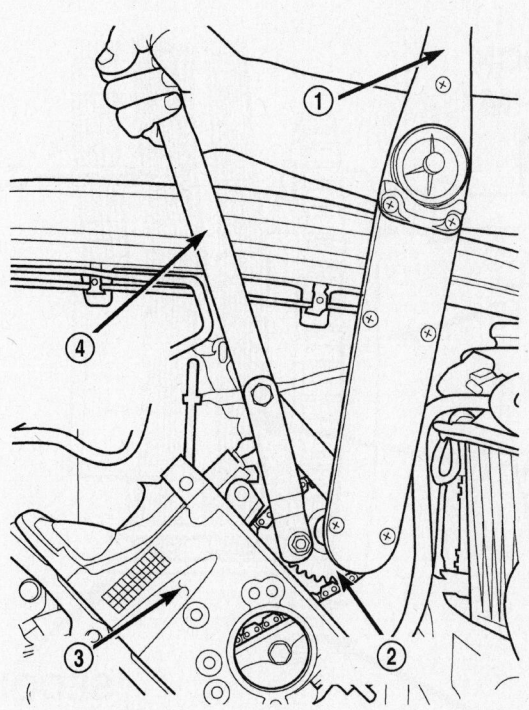

1 - TORQUE WRENCH

2 - CAMSHAFT SPROCKET

3 - LEFT CYLINDER HEAD

4 - SPECIAL TOOL 6958 SPANNER WITH ADAPTER PINS 8346

9355PG14

Tightening the left side camshaft sprocket—3.7L

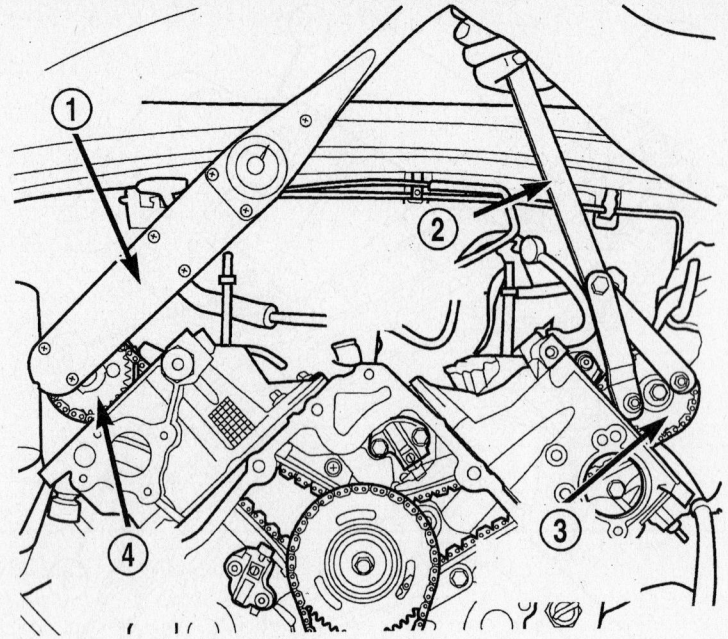

1 - TORQUE WRENCH
2 - SPECIAL TOOL 6958 WITH ADAPTER PINS 8346
3 - LEFT CAMSHAFT SPROCKET
4 - RIGHT CAMSHAFT SPROCKET

9355PG15

Tightening the right side camshaft sprocket—3.7L

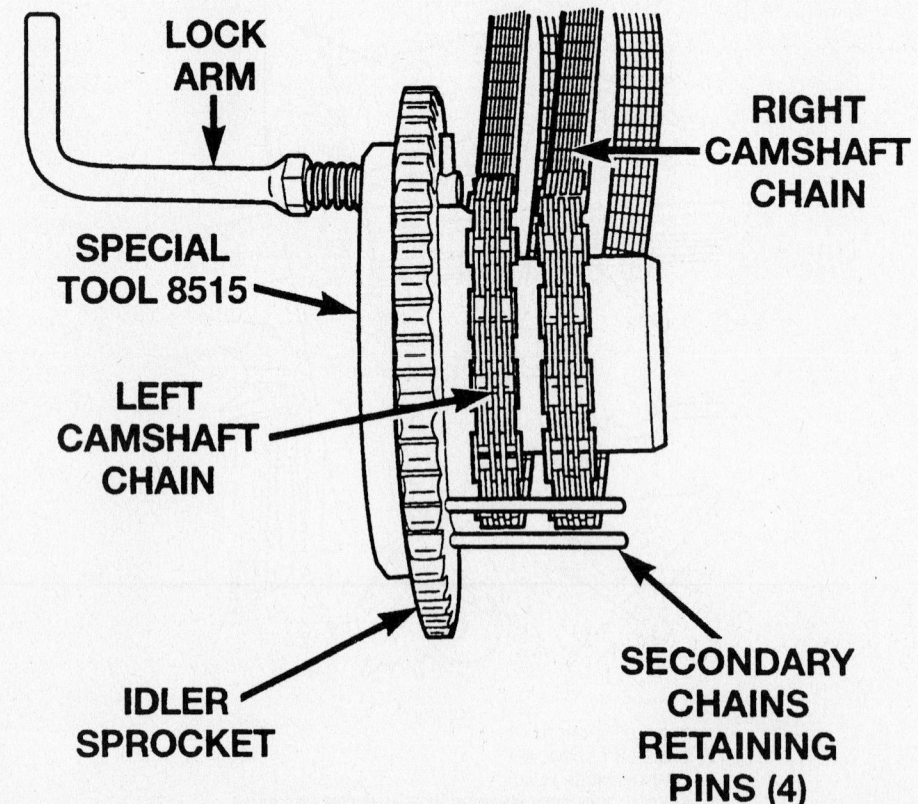

LOCK ARM

RIGHT CAMSHAFT CHAIN

SPECIAL TOOL 8515

LEFT CAMSHAFT CHAIN

IDLER SPROCKET

SECONDARY CHAINS RETAINING PINS (4)

9302PG07

Use the Timing Chain Locking tool to lock the timing chains on the idler gear—4.7L engine

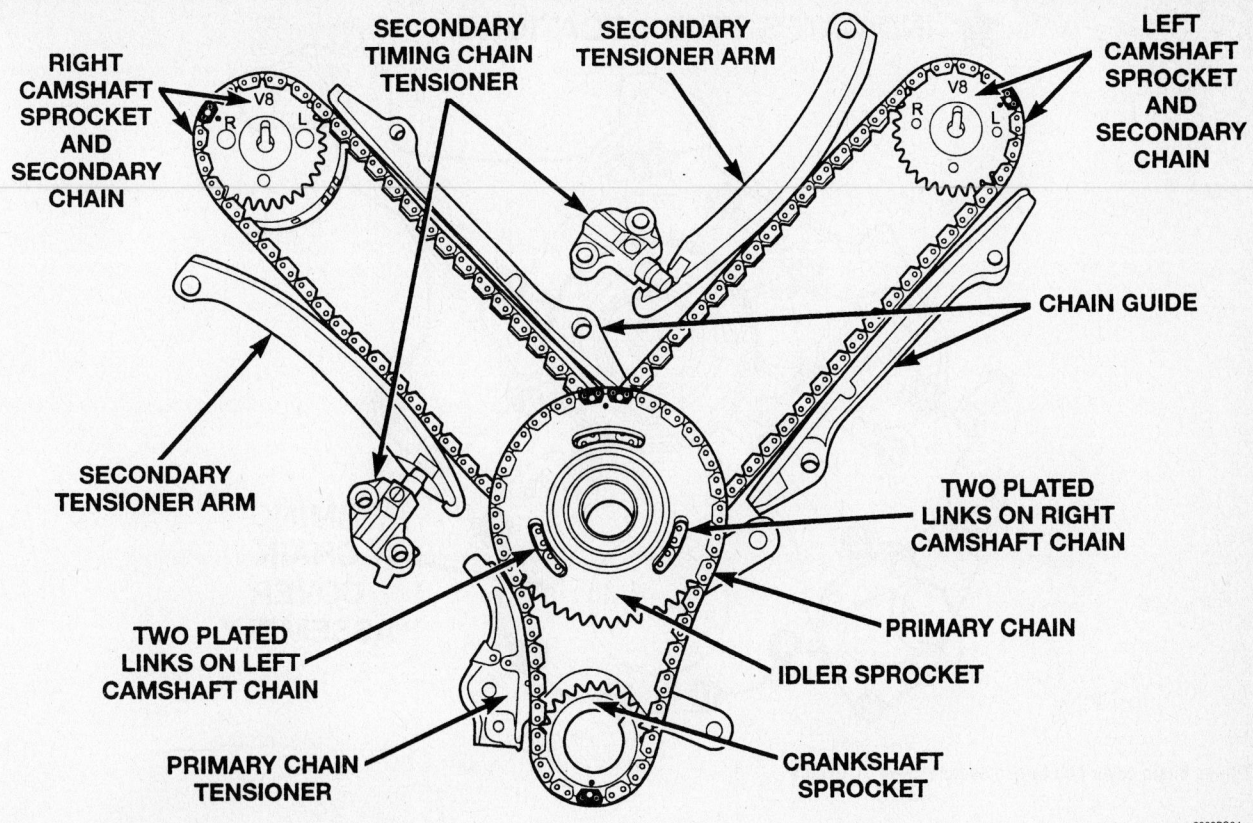

RIGHT CAMSHAFT SPROCKET AND SECONDARY CHAIN

SECONDARY TIMING CHAIN TENSIONER

SECONDARY TENSIONER ARM

LEFT CAMSHAFT SPROCKET AND SECONDARY CHAIN

CHAIN GUIDE

SECONDARY TENSIONER ARM

TWO PLATED LINKS ON RIGHT CAMSHAFT CHAIN

TWO PLATED LINKS ON LEFT CAMSHAFT CHAIN

PRIMARY CHAIN

IDLER SPROCKET

PRIMARY CHAIN TENSIONER

CRANKSHAFT SPROCKET

9302PG24

Timing chain system and alignment marks—4.7L engine

11. Install or connect the following:
  • Secondary timing chain guides. Tighten the bolts to 21 ft. lbs. (28 Nm).
  • Secondary timing chains to the idler sprocket so that the double plated links on each chain are visible through the slots in the primary idler sprocket

12. Lock the secondary timing chains to the idler sprocket with Timing Chain Locking tool as shown.

13. Align the primary chain double plated links with the idler sprocket timing mark and the single plated link with the crankshaft sprocket timing mark.

14. Install the primary chain and sprockets. Tighten the idler sprocket bolt to 25 ft. lbs. (34 Nm).

15. Align the secondary chain single plated links with the timing marks on the secondary sprockets. Align the dot at the **L** mark on the left sprocket with the plated link on the left chain and the dot at the **R** mark on the right sprocket with the plated link on the right chain.

16. Rotate the camshafts back from the neutral position and install the camshaft sprockets.

17. Remove the secondary chain locking tool.

★ **INDICATES STUD LOCATIONS**

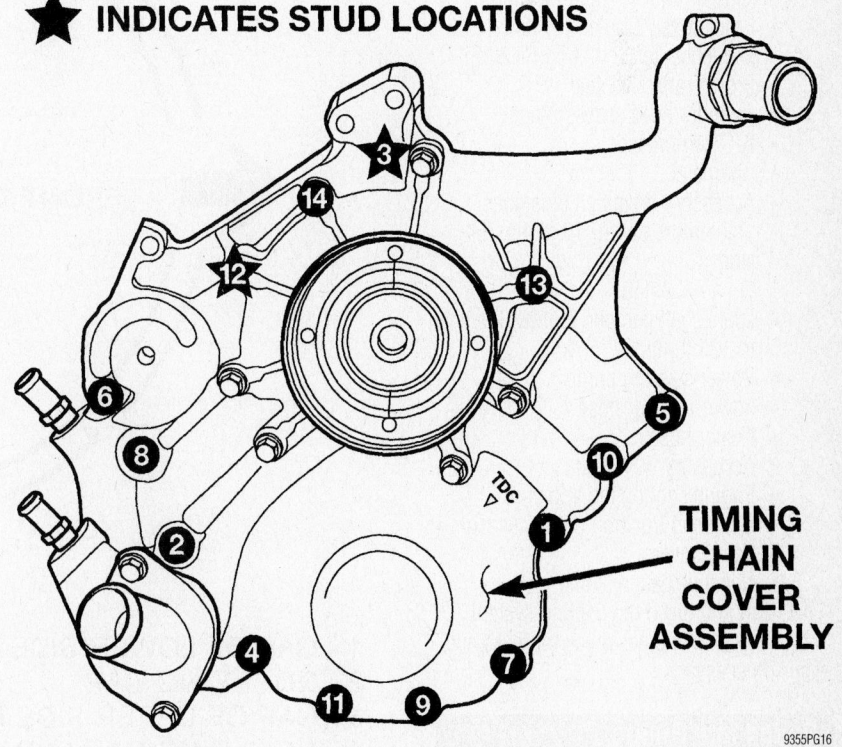

TIMING CHAIN COVER ASSEMBLY

9355PG16

Timing cover bolt torque sequence—3.7L

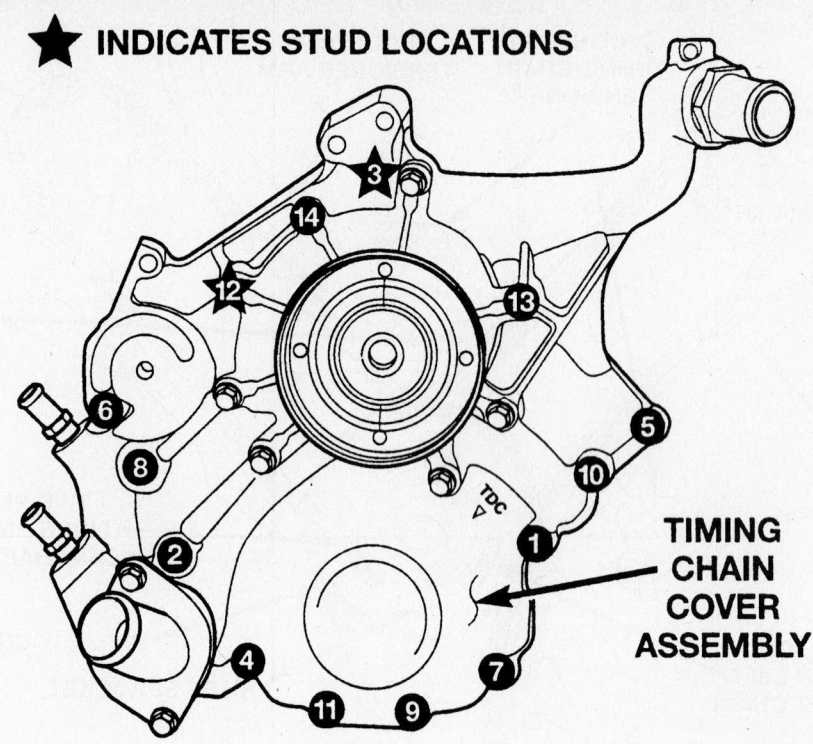

★ INDICATES STUD LOCATIONS

TIMING CHAIN COVER ASSEMBLY

Timing chain cover bolt torque sequence—4.7L engine

18. Remove the primary and secondary timing chain tensioner locking pins.

19. Hold the camshaft sprockets with a spanner wrench and tighten the retaining bolts to 90 ft. lbs. (122 Nm).

20. Install or connect the following:
- Front cover. Tighten the bolts, in sequence, to 40 ft. lbs. (54 Nm).
- Front crankshaft seal
- Cylinder head access plugs
- A/C compressor
- Alternator
- Accessory drive belt tensioner. Tighten the bolt to 40 ft. lbs. (54 Nm).
- Oil fill housing
- Crankshaft damper. Tighten the bolt to 130 ft. lbs. (175 Nm).
- Power steering pump
- Lower radiator hose
- Heater hoses
- Accessory drive belt
- Engine cooling fan and shroud
- Camshaft Position (CMP) sensor
- Valve covers
- Negative battery cable

21. Fill and bleed the cooling system.

22. Start the engine, check for leaks and repair if necessary.

Piston and Ring

POSITIONING

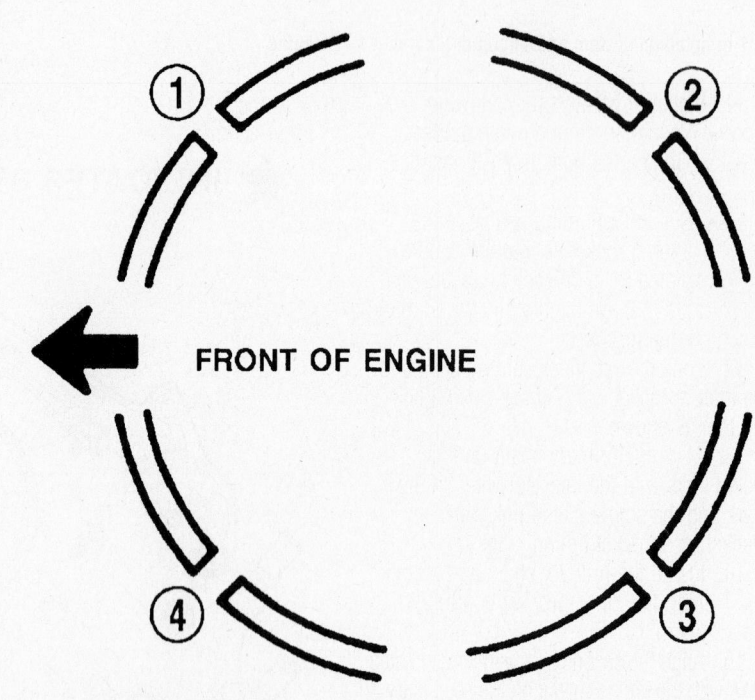

FRONT OF ENGINE

1 - GAP OF LOWER SIDE RAIL
2 - NO. 1 RING GAP
3 - GAP OF UPPER SIDE RAIL
4 - NO. 2 RING GAP AND SPACER EXPANDER GAP

Piston ring end gap spacing—2.4L engines

9302PG26

67189-JEEP-G48

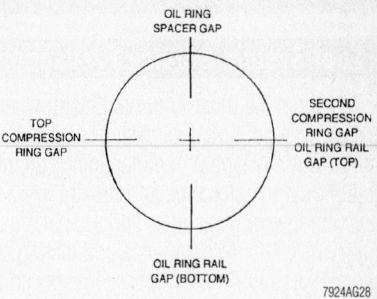

7924AG28

Piston ring end-gap spacing—2.5L engines

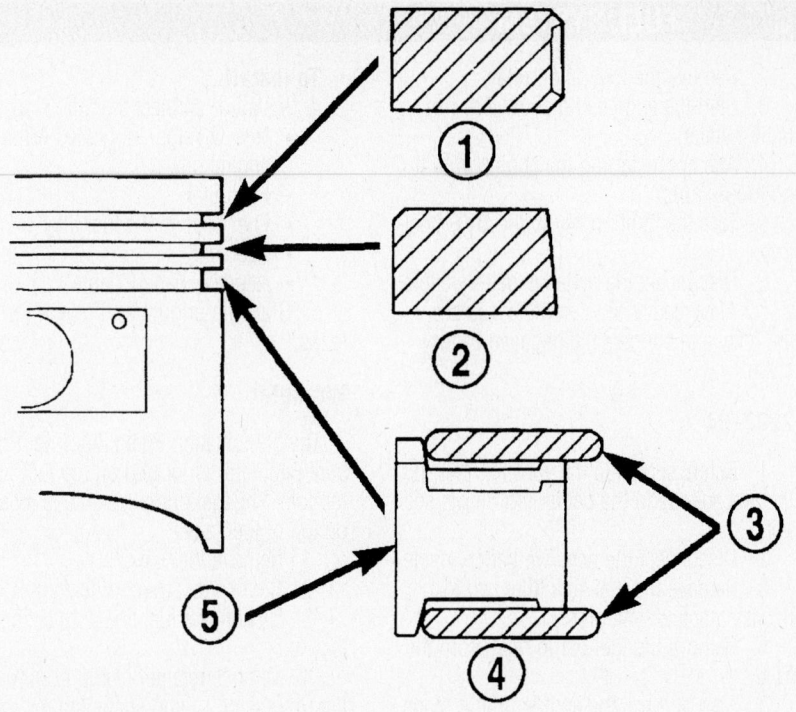

1 - NO. 1 PISTON RING
2 - NO. 2 PISTON RING
3 - SIDE RAIL
4 - OIL RING
5 - SPACER EXPANDER

67189-JEEP-G49

Piston ring installation—2.4L engines

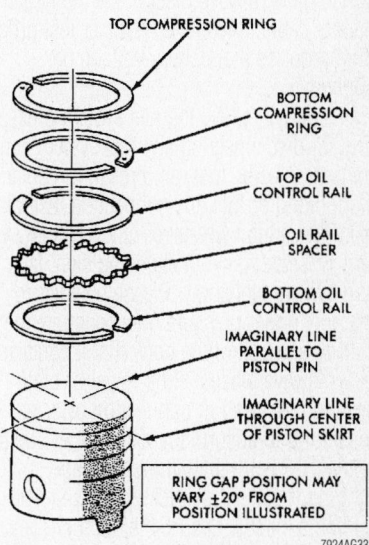

7924AG33

Piston ring end-gap spacing—4.0L engine

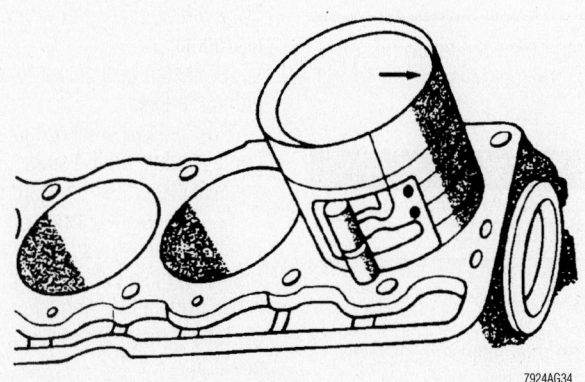

7924AG34

Piston to engine positioning—2.5L, 4.0L engines

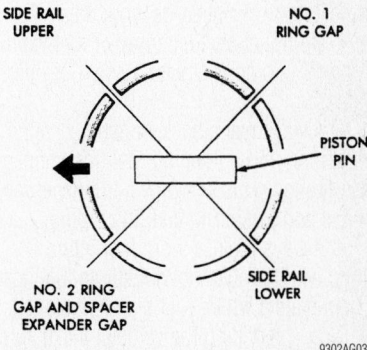

9302AG03

Piston ring end-gap spacing. Position raised "F" on piston towards front of engine—3.7L and 4.7L engines

FUEL SYSTEM

Fuel System Service Precautions

Safety is the most important factor when performing not only fuel system maintenance but any type of maintenance. Failure to conduct maintenance and repairs in a safe manner may result in serious personal injury or death. Maintenance and testing of the vehicle's fuel system components can be accomplished safely and effectively by adhering to the following rules and guidelines.

• To avoid the possibility of fire and personal injury, always disconnect the negative battery cable unless the repair or test procedure requires that battery voltage be applied.

• Always relieve the fuel system pressure prior to disconnecting any fuel system component (injector, fuel rail, pressure regulator, etc.), fitting or fuel line connection. Exercise extreme caution whenever relieving fuel system pressure to avoid exposing skin, face and eyes to fuel spray. Please be advised that fuel under pressure may penetrate the skin or any part of the body that it contacts.

• Always place a shop towel or cloth around the fitting or connection prior to loosening to absorb any excess fuel due to spillage. Ensure that all fuel spillage (should it occur) is quickly removed from engine surfaces. Ensure that all fuel soaked cloths or towels are deposited into a suitable waste container.

• Always keep a dry chemical (Class B) fire extinguisher near the work area.

• Do not allow fuel spray or fuel vapors to come into contact with a spark or open flame.

• Always use a back-up wrench when loosening and tightening fuel line connection fittings. This will prevent unnecessary stress and torsion to fuel line piping.

• Always replace worn fuel fitting O-rings with new. Do not substitute fuel hose or equivalent where fuel pipe is installed.

Before servicing the vehicle, make sure to also refer to the precautions in the beginning of this section as well.

Fuel System Pressure

RELIEVING

2001–02

1. Before servicing the vehicle, refer to the precautions in the beginning of this section.

2. Remove the fuel pump relay.

3. Start the engine and allow it to run until it stalls.

4. Attempt restarting the engine until it no longer runs.

5. Turn the ignition key to the **OFF** position.

6. Disconnect the negative battery cable.

7. After repairs are complete, replace the relay and connect the negative battery cable.

2003–04

1. Before servicing the vehicle, refer to the precautions in the beginning of this section.

2. Disconnect the negative battery cable.

3. Remove the fuel tank filler cap to release any fuel tank pressure.

4. Remove the fuel pump relay from the PDC.

5. Start and run the engine until it stops.

6. Unplug the connector from any injector and connect a jumper wire from either injector terminal to the positive battery terminal. Connect another jumper wire to the other terminal and momentarily touch the other end to the negative battery terminal.

※※ WARNING

Just touch the jumper to the battery. Powering the injector for more than a few seconds will permanently damage it.

7. Place a rag below the quick-disconnect coupling at the fuel rail and disconnect it.

Fuel Filter

REMOVAL & INSTALLATION

Cherokee

The fuel pump inlet filter is located on the bottom of the fuel pump module.

1. Before servicing the vehicle, refer to the precautions in the beginning of this section.

2. Relieve the fuel system pressure.

3. Remove or disconnect the following:
   • Negative battery cable
   • Fuel tank
   • Fuel pump module
   • Filter from the bottom of the fuel pump module

To install:

4. Install or connect the following:
   • New O-rings, lubricated with engine oil
   • Fuel filter
   • Fuel lines to the fuel filter
   • Fuel tank
   • Negative battery cable

5. Start the engine and check for leaks.

Wrangler

The combination Fuel Filter/Fuel Pressure Regulator is located on the fuel pump module. The fuel pump module is located on top of fuel tank.

1. Remove fuel tank.

2. Clean area around filter/regulator.

3. Disconnect fuel line at filter/regulator.

4. Remove retainer clamp from top of filter/regulator. Clamp snaps to tabs on pump module. Discard old clamp.

5. Pry filter/regulator from top of pump module with 2 screwdrivers. Unit is snapped into module.

6. Discard gasket below filter/regulator.

To install:

7. Clean recessed area in pump module where filter/regulator is to be installed.

8. Obtain new filter/regulator (two new O-rings should already be installed).

9. Apply a small amount of clean engine oil to O-rings. Do not install o-rings separately into fuel pump module. They will be damaged when installing filter/regulator.

10. Install new gasket to top of fuel pump module.

11. Press new filter/regulator into top of pump module until it snaps into position (a positive click must be heard or felt).

12. The molded arrow on top of fuel pump module should be pointed towards front of vehicle (12 o'clock position).

13. Rotate filter/regulator until fuel supply tube (fitting) is pointed to 10 o'clock position.

14. Install new retainer clamp (clamp snaps over top of filter/regulator and locks to flanges on pump module).

15. Connect fuel line at filter/regulator.

16. Install fuel tank.

Grand Cherokee

The fuel filter/pressure regulator is mounted to the body above the rear axle and near the front of the fuel tank.

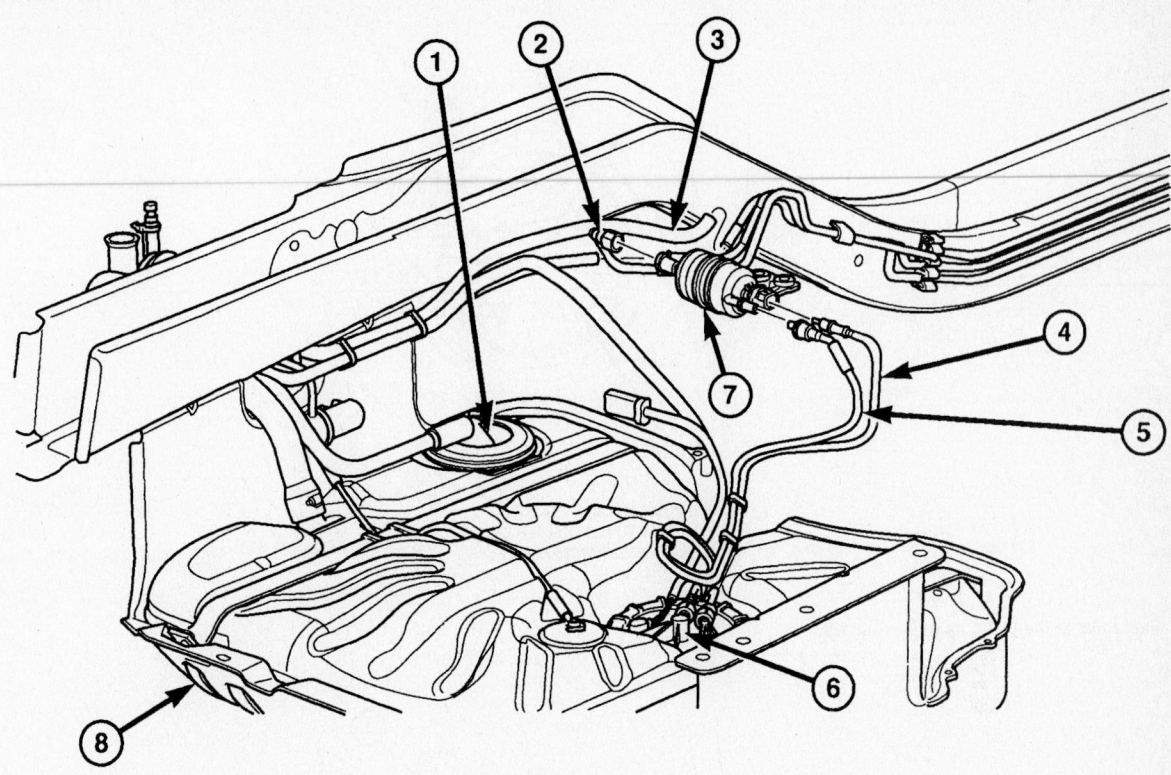

1 - ORVR CONTROL VALVE
2 - FUEL SUPPLY LINE (TO FUEL RAIL)
3 - EVAP LINE
4 - FUEL RETURN LINE (MALE)

5 - FUEL PRESSURE LINE (FEMALE)
6 - FUEL PUMP MODULE ASSEMBLY
7 - FUEL FILTER/PRESSURE REGULATOR
8 - FUEL TANK

67189-JEEP-G55

Fuel filter/pressure regulator—Grand Cherokee

1. Before servicing the vehicle, refer to the precautions in the beginning of this section.
2. Relieve the fuel system pressure.
3. Remove or disconnect the following:
   - Negative battery cable
   - Fuel supply, return and pressure lines
   - Mounting bolts
   - Fuel filter/pressure regulator

To install:

4. Install or connect the following:
   - Fuel filter/pressure regulator to the body. Torque the bolts to 30 inch lbs. (3 Nm).
   - Fuel pressure, return and supply lines
   - Negative battery cable
5. Start the vehicle, check for leaks and repair if necessary.

Liberty

The fuel filter is attached to the front of the fuel tank.

1. Before servicing the vehicle, refer to the precautions in the beginning of this section.
2. Relieve the fuel system pressure.
3. Remove or disconnect the following:
   - Negative battery cable

- 2 rear cargo hold-down clamps by drilling the rivets
- 4 fuel pump module access plate nuts

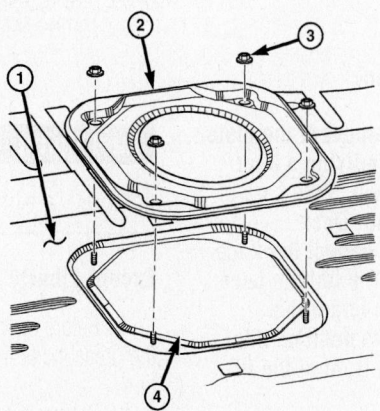

1 - FLOORPAN AT REAR
2 - FUEL PUMP MODULE ACCESS PLATE
3 - NUTS (4)
4 - OPENING TO PUMP MODULE

9355PG17

Access plate—Liberty

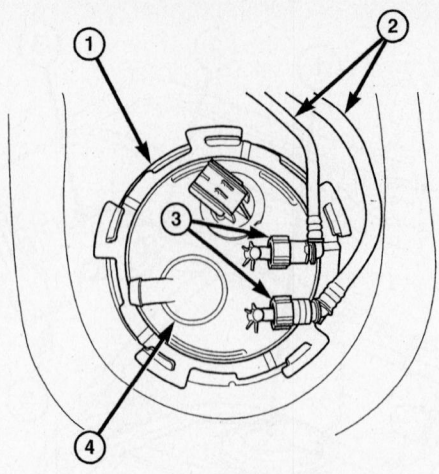

1 - FUEL PUMP MODULE LOCKRING
2 - FUEL LINES TO FUEL FILTER (2)
3 - QUICK-CONNECT FITTINGS (2)
4 - ROLLOVER VALVE

9355PG18

Fuel lines at the pump module—Liberty

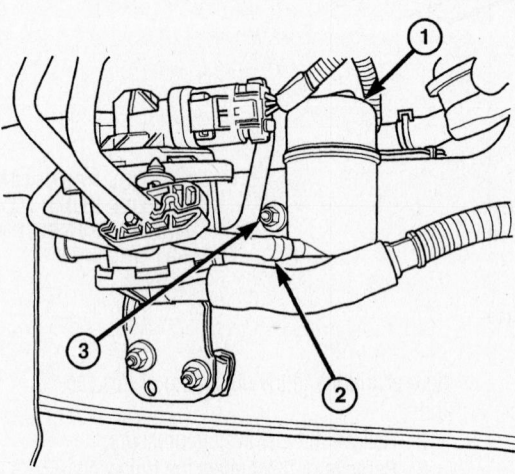

1 - FUEL FILTER
2 - 3RD FUEL LINE TO ENGINE
3 - FILTER MOUNTING NUT

9355PG19

Fuel filter location—Liberty

➡Once the nuts are removed, the plate must be removed by applying a heat gun to melt the adhesive. Take care to avoid bending the plate. Once removed, you can disconnect the 2 top hoses from the filter. The bottom hose must be removed from under the vehicle. The disconnect point on this hose is about a foot in front of the filter.

- Ground strap
- Mounting nut and filter.
4. Installation is the reverse of removal.

Fuel Pump

REMOVAL & INSTALLATION

Except Liberty

1. Before servicing the vehicle, refer to the precautions in the beginning of this section.
2. Relieve the fuel system pressure.
3. Remove or disconnect the following:
   - Negative battery cable
   - Fuel tank
   - Fuel lines

- Fuel pump module locknut
- Fuel pump module
To install:
4. Install or connect the following:
   - Fuel pump module with a new gasket. Rotate the module until the arrow is pointed toward the front of the vehicle, on all models except Grand Cherokee. The arrow must pointed toward the rear of the vehicle.
   - Locknut. Torque the locknut to 55 ft. lbs. (74 Nm).
   - Fuel lines
   - Fuel tank

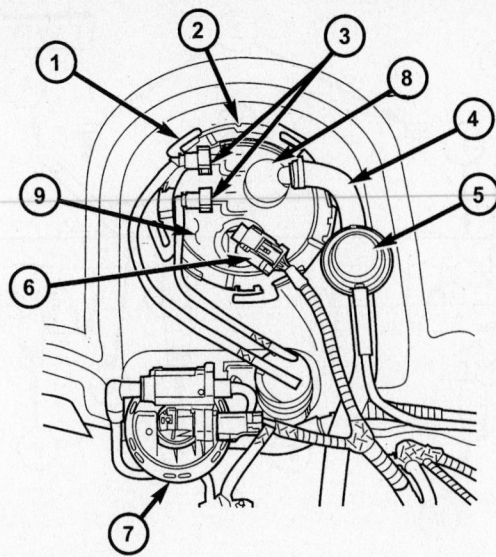

1 - LOCK RING
2 - ALIGNMENT NOTCH
3 - FUEL FILTER FITTINGS (2)
4 - ORVR SYSTEM HOSE AND CLAMP
5 - FLOW MANAGEMENT VALVE
6 - ELECTRICAL CONNECTOR
7 - LEAK DETECTION PUMP
8 - FUEL TANK CHECK (CONTROL) VALVE
9 - FUEL PUMP MODULE (UPPER SECTION)

9355PG20

Top view of the fuel pump module—Liberty

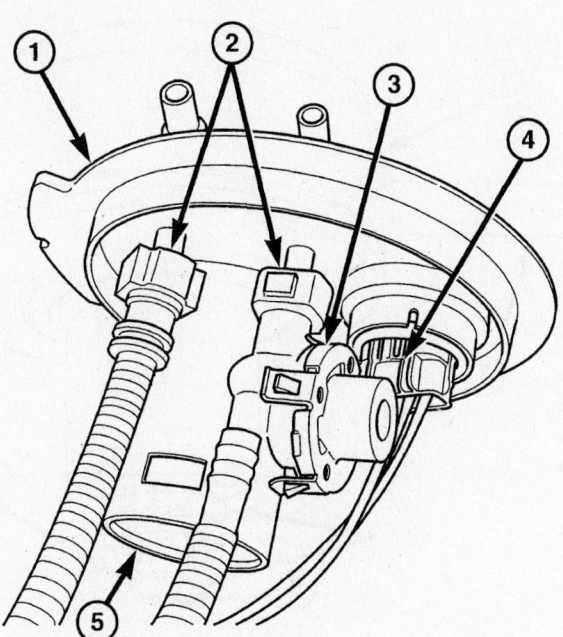

1 - UPPER SECTION OF PUMP MODULE
2 - QUICK-CONNECT FITTINGS
3 - FUEL PRESSURE REGULATOR
4 - 4-WIRE ELECTRICAL CONNECTOR
5 - FUEL TANK CHECK (CONTROL) VALVE

9355PG21

Fuel pressure regulator—Liberty

- Negative battery cable
5. Start the engine, check for leaks and repair if necessary.

Liberty

The fuel filter is attached to the front of the fuel tank.

1. Before servicing the vehicle, refer to the precautions in the beginning of this section.
2. Relieve the fuel system pressure.
3. Remove or disconnect the following:
   - Negative battery cable
   - 2 rear cargo hold-down clamps by drilling the rivets
   - 4 fuel pump module access plate nuts

➡Once the nuts are removed, the plate must be removed by applying a heat gun to melt the adhesive. Take care to avoid bending the plate.

   - Fuel lines at the module
   - Electrical connector by first sliding the red tab, then pushing the gray tab
   - ORVR hose
   - Module lockring
   - Module from the tank

✳✳ WARNING

Lift the module out slowly and carefully until you can secure the rubber gasket. If you're not careful, the gasket will fall into the tank.

   - Electrical connector from the upper module section
   - Fule pressure regulator
   - Fuel return line
   - Upper module section
4. Drain the tank through the module opening.
5. Remove or disconnect the following:
   - Lower module section by pushing gently on the release tab while sliding the lock tab upward
6. Installation is the reverse of removal.

Fuel Rail

REMOVAL & INSTALLATION

2.4L Engine

✳✳ CAUTION

the fuel system is under constant pressure even with engine off. Before servicing fuel rail, fuel system pressure must be released.

The fuel rail can be removed without removing the intake manifold if the following procedures are followed.

1. Remove fuel tank filler tube cap.

2. Perform Fuel System Pressure Release Procedure.

3. Remove negative battery cable at battery.

4. Remove air duct at throttle body.

5. Disconnect fuel line latch clip and fuel line at fuel rail. A special tool will be necessary for fuel line disconnection.

6. Remove necessary vacuum lines at throttle body.

7. Drain engine coolant and remove thermostat and thermostat housing.

8. Remove PCV hose and valve at valve cover.

9. Remove 3 upper intake manifold mounting bolts, but only loosen 2 lower bolts about 2 turns.

10. Disconnect 2 main engine harness connectors at rear of intake manifold.

11. Disconnect 2 injection wiring harness clips at harness mounting bracket.

12. Disconnect electrical connectors at all 4 fuel injectors. Push red colored slider away from injector. While pushing slider, depress tab and remove connector from injector. The factory fuel injection wiring harness is numerically tagged (INJ 1, INJ 2, etc.) for injector position identification. If harness is not tagged, note wiring location before removal.

13. Remove 2 injection rail mounting bolts.

14. Gently rock and pull fuel rail until fuel injectors just start to clear machined holes in intake manifold.

15. Remove fuel rail (with injectors attached) from intake manifold.

16. If fuel injectors are to be removed, refer to Fuel Injector Removal/Installation.

To install:

17. If fuel injectors are to be installed, refer to Fuel Injector Removal/Installation.

18. Clean out fuel injector machined bores in intake manifold.

19. Apply a small amount of engine oil to each fuel injector O-ring. This will help in fuel rail installation.

20. Position fuel rail/fuel injector assembly to machined injector openings in intake manifold.

21. Guide each injector into cylinder head. Be careful not to tear injector O-rings.

22. Push fuel rail down until fuel injectors have bottomed on shoulders.

23. Install 2 fuel rail mounting bolts and tighten to 28 Nm (250 inch lbs.).

24. Connect electrical connectors at all fuel injectors. Push connector onto injector

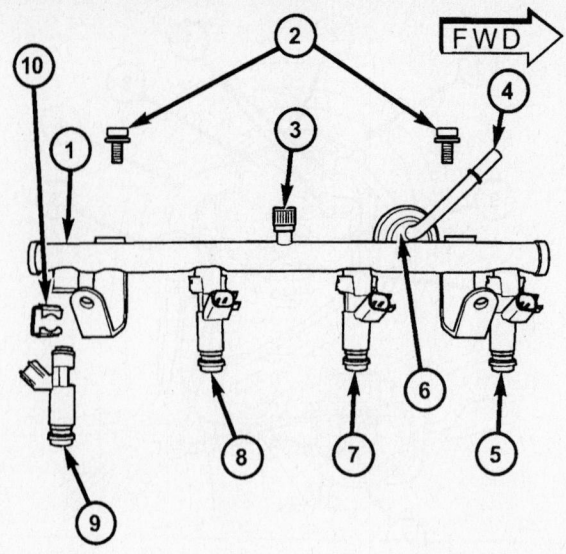

1 - FUEL RAIL
2 - MOUNTING BOLTS
3 - TEST PORT
4 - QUICK-CONNECT FITTING
5 - INJ. #1
6 - DAMPER
7 - INJ #2
8 - INJ #3
9 - INJ #4
10- INJECTOR RETAINING CLIP

67189-JEEP-G51

Fuel rail—2.4L engines

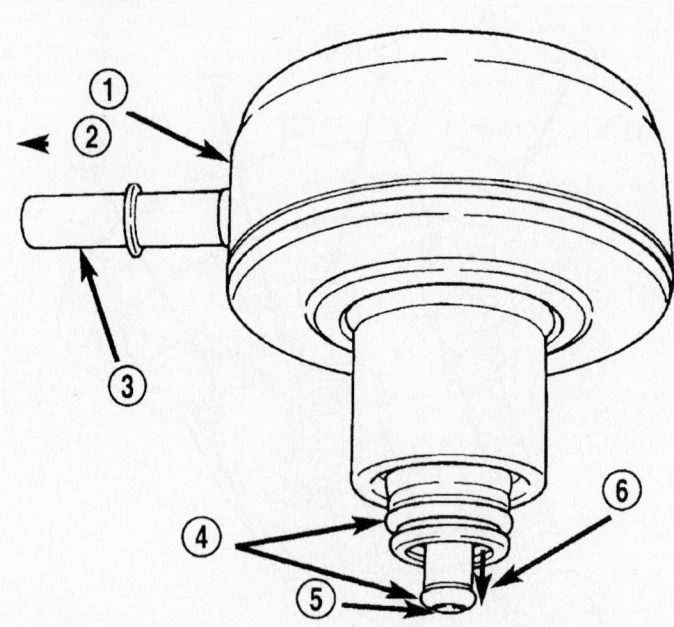

1 - FUEL FILTER/FUEL PRESSURE REGULATOR
2 - TO FUEL INJECTORS
3 - FUEL SUPPLY TUBE
4 - O-RINGS
5 - FUEL INLET FROM PUMP
6 - FUEL RETURN TO TANK

67189-JEEP-G53

Fuel pressure regulator—2.4L engines

and then push and lock red colored slider. Verify connector is locked to injector by lightly tugging on connector.

25. Snap 2 injection wiring harness clips into brackets.

26. Connect 2 main engine harness connectors at rear of intake manifold.

27. Tighten 5 intake manifold mounting bolts.

28. Install PCV valve and hose.

29. Install thermostat and radiator hose. Fill with coolant.

30. Connect necessary vacuum lines to throttle body.

31. Connect fuel line latch clip and fuel line to fuel rail.

32. Install air duct to throttle body.

33. Connect battery cable to battery.

34. Start engine and check for leaks.

3.7L Engine

※※ CAUTION

the fuel system is under constant pressure even with engine off. Before servicing fuel rail, fuel system pressure must be released.

➡**The left and right fuel rails are replaced as an assembly. Do not attempt to separate rail halves at connector tube. Due to design of tube, it does not use any clamps. Never attempt to install a clamping device of any kind to tube. When removing fuel rail assembly for any reason, be careful not to bend or kink tube.**

1. Remove fuel tank filler tube cap.

2. Perform Fuel System Pressure Release Procedure.

3. Remove negative battery cable at battery.

4. Remove air duct at throttle body air box.

5. Remove air box at throttle body.

6. Disconnect fuel line latch clip and fuel line at fuel rail. A special tool will be necessary for fuel line disconnection.

7. Remove necessary vacuum lines at throttle body.

8. Disconnect electrical connectors at all 6 fuel injectors. Push red colored slider away from injector. While pushing slider, depress tab and remove connector from injector. The factory fuel injection wiring harness is numerically tagged (INJ 1, INJ 2, etc.) for injector position identification. If harness is not tagged, note wiring location before removal.

9. Disconnect electrical connectors at throttle body sensors.

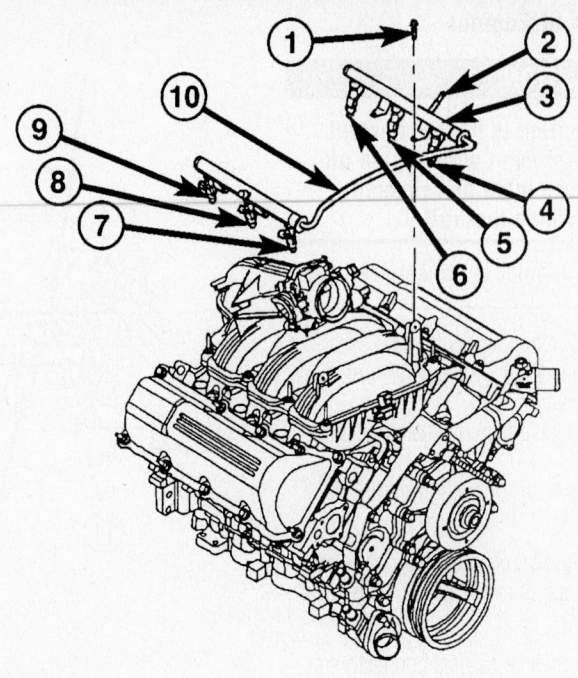

1 - MOUNTING BOLTS (4)
2 - QUICK-CONNECT FITTING
3 - FUEL RAIL
4 - INJ. #1
5 - INJ. #3
6 - INJ. #5
7 - INJ. #2
8 - INJ. #4
9 - INJ. #6
10 - CONNECTOR TUBE

67189-JEEP-G52

Fuel rail—3.7L engines

10. Remove 6 ignition coils.

11. Remove 4 fuel rail mounting bolts.

12. Gently rock and pull left side of fuel rail until fuel injectors just start to clear machined holes in cylinder head. Gently rock and pull right side of rail until injectors just start to clear cylinder head holes. Repeat this procedure (left/right) until all injectors have cleared cylinder head holes.

13. Remove fuel rail (with injectors attached) from engine.

14. If fuel injectors are to be removed, refer to Fuel Injector Removal/Installation.

To install:

15. If fuel injectors are to be installed, refer to Fuel Injector Removal/Installation.

16. Clean out fuel injector machined bores in intake manifold.

17. Apply a small amount of engine oil to each fuel injector O-ring. This will help in fuel rail installation.

18. Position fuel rail/fuel injector assembly to machined injector openings in cylinder head.

19. Guide each injector into cylinder head. Be careful not to tear injector O-rings.

20. Push right side of fuel rail down until fuel injectors have bottomed on cylinder head shoulder.

Push left fuel rail down until injectors have bottomed on cylinder head shoulder.

21. Install 4 fuel rail mounting bolts and tighten to 11 Nm (100 inch lbs.)

22. Install 6 ignition coils.

23. Connect electrical connectors to throttle body.

24. Connect electrical connectors at all fuel injectors. Push connector onto injector and then push and lock red colored slider. Verify connector is locked to injector by lightly tugging on connector.

25. Connect necessary vacuum lines to throttle body.

26. Connect fuel line latch clip and fuel line to fuel rail.

27. Install air box to throttle body.

28. Install air duct to air box.

29. Connect battery cable to battery.

30. Start engine and check for leaks.

2.5L and 4.0L Engines

> **✳✳ CAUTION**
>
> The fuel system is under constant fuel pressure even with engine off. This pressure must be released before servicing fuel rail.

➡ The fuel damper is not serviced separately

1. Remove fuel tank filler tube cap.
2. Perform Fuel System Pressure Release Procedure.
3. Disconnect negative battery cable from battery.
4. Remove air tube at top of throttle body.

➡ Some engine/vehicles may require removal of air cleaner ducts at throttle body.

5. Disconnect electrical connectors at all 6 fuel injectors. Push red colored slider away from injector. While pushing slider, depress tab and remove connector from injector. The factory fuel injection wiring harness is numerically tagged (INJ 1, INJ 2, etc.) for injector position identification. If harness is not tagged, note wiring location before removal.
6. Disconnect fuel supply line latch clip and fuel line at fuel rail.
7. Disconnect throttle cable at throttle body.
8. Disconnect speed control cable at throttle body (if equipped).
9. Disconnect automatic transmission cable at throttle body (if equipped).
10. Remove cable routing bracket at intake manifold.
11. If equipped, remove wiring harnesses at injection rail studs by removing nuts.
12. Clean dirt/debris from each fuel injector at intake manifold.
13. Remove fuel rail mounting nuts/bolts.
14. Remove fuel rail by gently rocking until all the fuel injectors are out of intake manifold.

To install:

15. Clean each injector bore at intake manifold.
16. Apply a small amount of clean engine oil to each injector O-ring. This will aid in installation.
17. Position tips of all fuel injectors into the corresponding injector bore in intake manifold. Seat injectors into manifold.
18. Install and tighten fuel rail mounting bolts to 11 ± 3 Nm (100 ± 25 inch lbs.) torque.

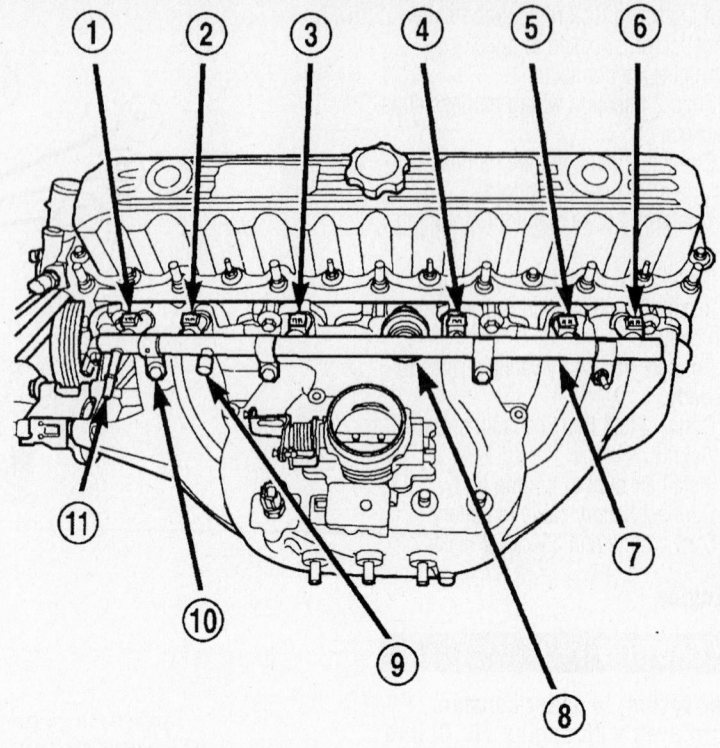

1 - INJ. #1
2 - INJ. #2
3 - INJ. #3
4 - INJ. #4
5 - INJ. #5
6 - INJ. #6
7 - FUEL INJECTOR RAIL
8 - FUEL DAMPER
9 - PRESSURE TEST PORT CAP
10 - MOUNTING BOLTS (4)
11 - QUICK-CONNECT FITTING

67189-JEEP-G54

Fuel rail—4.0L engines

19. If equipped, connect wiring harnesses to injection rail studs.
20. Connect electrical connectors at all fuel injectors. Push connector onto injector and then push and lock red colored slider. Verify connector is locked to injector by lightly tugging on connector.
21. Connect fuel line and fuel line latch clip to fuel rail.
22. Install protective cap to pressure test port fitting (if equipped).
23. Install cable routing bracket to intake manifold.
24. Connect throttle cable at throttle body.
25. Connect speed control cable at throttle body (if equipped).

26. Connect automatic transmission cable at throttle body (if equipped).
27. Install air tube (or duct) at top of throttle body.
28. Install fuel tank cap.
29. Connect negative battery cable to battery.
30. Start engine and check for fuel leaks.

4.7L Engine

1. Remove fuel tank filler tube cap.
2. Perform Fuel System Pressure Release Procedure.
3. Remove negative battery cable at battery.
4. Remove air duct at throttle body air box.

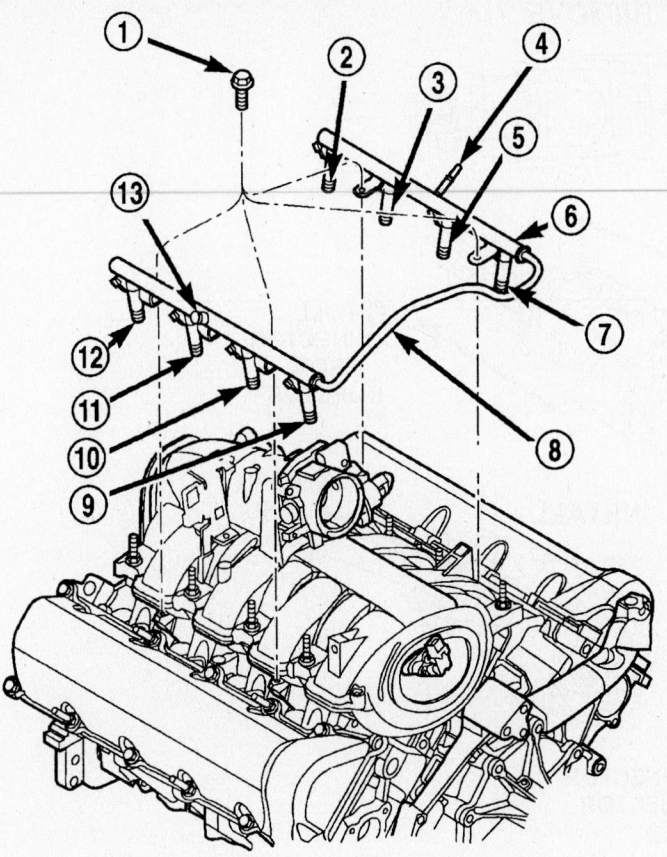

1 - MOUNTING BOLTS (4)
2 - INJ.#7
3 - INJ.#5
4 - QUICK-CONNECT FITTING
5 - INJ.#3
6 - FUEL INJECTOR RAIL
7 - INJ.#1
8 - CONNECTOR TUBE
9 - INJ.#2
10 - INJ.#4
11 - INJ.#6
12 - INJ.#8
13 - PRESSURE TEST PORT CAP

67189-JEEP-G56

Fuel rail—4.7L engines

5. Remove air box at throttle body.

6. Remove wiring at rear of generator.

7. Disconnect fuel line latch clip and fuel line at fuel rail. A special tool will be necessary for fuel line disconnection.

8. Remove vacuum lines at throttle body.

9. Disconnect electrical connectors at all 8 fuel injectors. Push red colored slider away from injector. While pushing slider, depress tab and remove connector from injector. The factory fuel injection wiring harness is numerically tagged (INJ 1, INJ 2, etc.) for injector position identification. If harness is not tagged, note wiring location before removal.

10. Disconnect electrical connectors at throttle body.

11. Disconnect electrical connectors at MAP and IAT sensors.

12. Remove first three ignition coils on each bank (cylinders #1, 3, 5, 2, 4 and 6).

13. Remove 4 fuel rail mounting bolts.

14. Gently rock and pull left side of fuel rail until fuel injectors just start to clear machined holes in cylinder head. Gently rock and pull right side of rail until injectors just start to clear cylinder head holes. Repeat this procedure (left/right) until all injectors have cleared cylinder head holes.

15. Remove fuel rail (with injectors attached) from engine.

16. If fuel injectors are to be removed, refer to Fuel Injector Removal/Installation.

To install:

17. If fuel injectors are to be installed, refer to Fuel Injector Removal/Installation.

18. Apply a small amount of engine oil to each fuel injector o-ring. This will help in fuel rail installation.

19. Position fuel rail/fuel injector assembly to machined injector openings in cylinder head.

20. Guide each injector into cylinder head. Be careful not to tear injector o-rings.

21. Push right side of fuel rail down until fuel injectors have bottomed on cylinder head shoulder. Push left fuel rail down until injectors have bottomed on cylinder head shoulder.

22. Install 4 fuel rail mounting bolts and tighten to 27 Nm (20 ft. lbs.).

23. Install ignition coils.

24. Connect electrical connectors to throttle body.

25. Connect electrical connectors to MAP and IAT sensors.

26. Connect electrical connectors at all fuel injectors. Push connector onto injector and then push and lock red colored slider. Verify connector is locked to injector by lightly tugging on connector.

27. Connect vacuum lines to throttle body.

28. Connect fuel line latch clip and fuel line to fuel rail.

29. Connect wiring to rear of generator.

30. Install air box to throttle body.

31. Install air duct to air box.

32. Connect battery cable to battery.

33. Start engine and check for leaks.

Fuel Injector

REMOVAL & INSTALLATION

1. Before servicing the vehicle, refer to the precautions in the beginning of this section.

2. Relieve fuel system pressure.

3. Remove or disconnect the following:
   • Negative battery cable
   • Fuel rail
   • Fuel injector retaining clips
   • Fuel injectors

To install:

4. Install or connect the following:
   • Fuel injectors, using new O-rings coated with a small amount of engine oil
   • Fuel injector clips to the fuel rail
   • Fuel rail
   • Negative battery cable

5. Start the vehicle and check for leaks, repair if necessary.

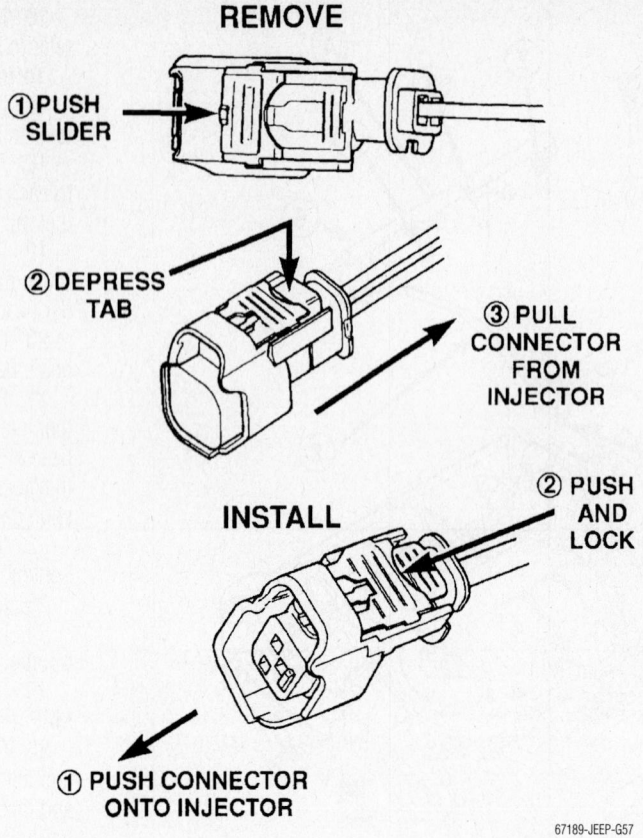

Fuel injector connector

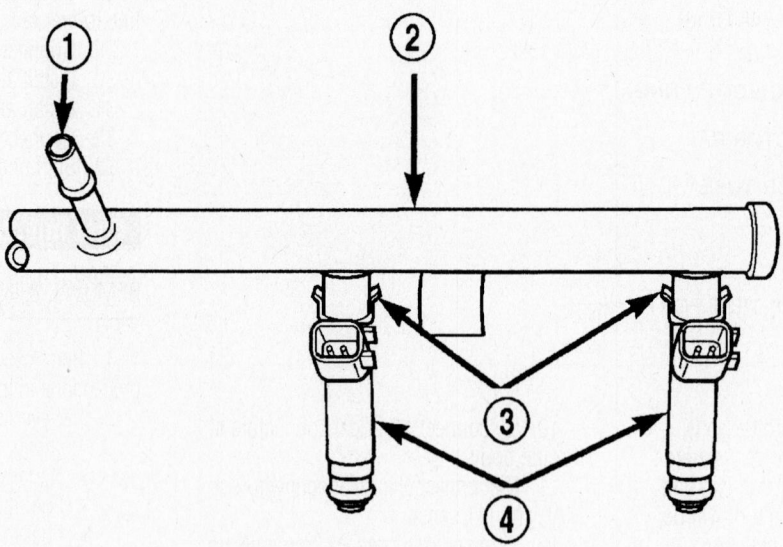

1 - INLET FITTING
2 - FUEL INJECTOR RAIL
3 - CLIP
4 - FUEL INJECTOR

Typical fuel injector mounting—4.7L engine shown

DRIVE TRAIN

Transmission

REMOVAL & INSTALLATION

2001–03 Automatic

1. Before servicing the vehicle, refer to the precautions in the beginning of this section.
2. Drain the transmission fluid.
3. Remove or disconnect the following:
   - Negative battery cable
   - Exhaust front pipe
   - Transmission braces, if equipped
   - Transmission oil cooler lines
   - Starter motor
   - Crankshaft Position (CKP) sensor
   - Torque converter access cover
   - Transmission oil pan
   - Skid plate
   - Transmission oil dipstick tube
   - Propellar shafts after matchmarking them
   - Park/Neutral switch connector
   - Shift cable
   - Throttle valve cable
   - Shift rod from the transfer case shift lever
   - Crossmember
   - Transfer case vent hose
   - Transfer case
4. Slide the transmission/torque converter assembly rearward off the engine dowels
5. Lower and remove the transmission assembly

To install:

6. Install or connect the following:
   - Transmission. Torque the flange bolts to 55 ft. lbs. (75 Nm).
   - Front driveshaft and transfer case, if equipped
   - Transfer case vent hose
   - Transmission mount and crossmember
   - Throttle valve cable
   - Shift cable
   - Park/Neutral switch connector
   - Propeller shafts using the matchmarks made during removal
   - Transmission oil dipstick tube
   - Skid plate
   - Transmission oil pan
   - Torque converter access cover
   - CKP sensor
   - Starter motor
   - Transmission oil cooler lines
   - Transmission braces, if equipped. Torque the bolts to 30 ft. lbs. (41 Nm).

   - Exhaust front pipe
   - Negative battery cable
7. Fill the transmission with the proper fluid.
8. Start the vehicle, check for leaks and repair if necessary.

2001–03 Manual

1. Before servicing the vehicle, refer to the precautions in the beginning of this section.
2. Place the transmission in first or third gear.
3. Drain the fluid from the transmission.
4. Remove or disconnect the following:
   - Negative battery cable
   - Exhaust front pipe
   - Skid plate
   - Clutch slave cylinder
   - Rear propeller shaft
   - Front propeller shaft, if equipped
   - Wire harness from the transmission and transfer case, if equipped
   - Transfer case shift linkage, if equipped
   - Transfer case, if equipped
   - Crankcase Position (CKP) sensor
   - Rear cushion and transmission bracket after supporting the engine and transmission
   - Rear crossmember
   - Transmission shift lever
   - Clutch housing brace rod
   - Transmission

To install:

5. Install or connect the following:
   - Transmission. Tighten the ⅜ inch bolts to 27 ft. lbs. (37 Nm), the ⁷⁄₁₆ inch bolts to 43 ft. lbs. (58 Nm), and the 12mm bolts to 55 ft. lbs. (75 Nm).
   - Shift lever
   - Rear crossmember. Torque the bolts to 31 ft. lbs. (41 Nm).
   - Rear cushion and bracket. Torque the bolts to 33 ft. lbs. (45 Nm).
   - CKP sensor
   - Transfer case, if equipped. Torque the bolts to 26 ft. lbs. (35 Nm).
   - Transfer case shift linkage, if equipped
   - Wire harness to the transmission
   - Front driveshaft and transfer case, if equipped. Torque the bolts to 12 ft. lbs. (19 Nm).
   - Rear driveshaft. Torque the bolts to 12 ft. lbs. (19 Nm).
   - Clutch slave cylinder
   - Skid plate. Torque the bolts to 31 ft. lbs. (42 Nm).

   - Exhaust front pipe
   - Negative battery cable
6. Fill the transmission with the proper fluid.
7. Start the vehicle, check for leaks and repair if necessary.

2004 Automatic

LIBERTY

1. Disconnect the negative battery cable.
2. Raise and support the vehicle
3. Remove any necessary skid plates.
4. Mark propeller shaft and axle companion flanges for assembly alignment.
5. Remove the rear propeller shaft.
6. Remove the front propeller shaft, if necessary.
7. Disconnect wires from the input and output speed sensors.
8. Disconnect wires from the transmission range sensor and the solenoid/pressure switch assembly.
9. Remove the bolts holding the exhaust crossover pipe to the pre-catalytic converter pipe flanges.
10. Remove the bolts holding the exhaust crossover pipe to the catalytic converter flange.
11. Disconnect gearshift cable from transmission manual valve lever.
12. Disengage the shift cable from the cable support bracket.
13. Remove the starter motor.
14. Remove the engine to transmission collar.
15. Rotate crankshaft in clockwise direction until converter bolts are accessible. Then remove bolts one at a time. Rotate crankshaft with socket wrench on dampener bolt.
16. Disconnect the transmission vent hose from the transmission.
17. Remove transfer case.
18. Support rear of engine with safety stand or jack.
19. Raise transmission slightly with service jack to relieve load on crossmember and supports.
20. Remove bolts securing rear support and cushion to transmission and crossmember.
21. Remove bolts attaching crossmember to frame and remove crossmember.
22. Disconnect transmission fluid cooler lines at transmission fittings and clips.
23. Remove all remaining converter housing bolts.

24. Carefully work transmission and torque converter assembly rearward off engine block dowels.

25. Hold torque converter in place during transmission removal.

26. Lower transmission and remove assembly from under the vehicle.

27. To remove torque converter, carefully slide torque converter out of the transmission.

To install:

28. Check torque converter hub and hub drive flats for sharp edges burrs, scratches, or nicks. Polish the hub and flats with 320/400 grit paper and crocus cloth if necessary. Verify that the converter hub O-ring is properly installed and is free of any debris. The hub must be smooth to avoid damaging pump seal at installation.

29. If a replacement transmission is being installed, transfer any components necessary, such as the manual shift lever and shift cable bracket, from the original transmission onto the replacement transmission.

30. Lubricate oil pump seal lip with transmission fluid.

31. Align converter and oil pump.

32. Carefully insert converter in oil pump. Then rotate converter back and forth until fully seated in pump gears.

33. Check converter seating with steel scale and straightedge. Surface of converter lugs should be at least 13mm (½ in.) to rear of straightedge when converter is fully seated.

34. Temporarily secure converter with C-clamp.

35. Position transmission on jack and secure it with chains

36. Check condition of converter driveplate. Replace the plate if cracked, distorted or damaged.

➡**Be sure transmission dowel pins are seated in engine block and protrude far enough to hold transmission in alignment.**

37. Apply a light coating of Mopar® High Temp Grease to the torque converter hub pocket in the rear pocket of the engine's crankshaft.

38. Raise transmission and align the torque converter with the drive plate and transmission converter housing with the engine block.

39. Move transmission forward. Then raise, lower or tilt transmission to align the converter housing with engine block dowels.

40. Carefully work transmission forward and over engine block dowels until converter hub is seated in crankshaft. Verify that no wires, or the transmission vent hose, have become trapped between the engine block and the transmission.

41. Install two bolts to attach the transmission to the engine.

42. Install remaining torque converter housing to engine bolts. Tighten to 68 Nm (50 ft. lbs.).

43. Install transfer case, if equipped. Tighten transfer case nuts to 35 Nm (26 ft. lbs.).

44. Install rear transmission crossmember. Tighten crossmember to frame bolts to 68 Nm (50 ft. lbs.).

45. Install rear support to transmission. Tighten bolts to 47 Nm (35 ft. lbs.).

46. Lower transmission onto crossmember and install bolts attaching transmission mount to crossmember.
Tighten clevis bracket to crossmember bolts to 47 Nm (35 ft. lbs.). Tighten the clevis bracket to rear support bolt to 68 Nm (50 ft. lbs.).

47. Remove engine support fixture.

48. Connect gearshift cable to support bracket and transmission manual lever.

49. Connect input and output speed sensor wires.

50. Connect wires to the transmission range sensor and the solenoid/pressure switch assembly.

51. Install torque converter-to-driveplate bolts. Tighten bolts to 88 Nm (65 inch lbs.).

52. Install starter motor and cooler line bracket.

53. Connect cooler lines to transmission.

54. Install transmission fill tube.

55. Install exhaust components.

56. Align and connect propeller shaft(s).

57. Adjust gearshift cable if necessary.

58. Install any skid plates removed previously.

GRAND CHEROKEE W/42RE

➡**The transmission and torque converter must be removed as an assembly to avoid component damage. The converter driveplate, pump bushing, or oil seal can be damaged if the converter is left attached to the driveplate during removal. Be sure to remove the transmission and converter as an assembly.**

1. Disconnect battery negative cable.

2. Disconnect and lower or remove necessary exhaust components.

3. Disconnect fluid cooler lines at transmission.

4. Remove starter motor.

5. Disconnect and remove crankshaft position sensor. Retain sensor attaching bolts.

➡**The crankshaft position sensor will be damaged if the transmission is removed, or installed, while the sensor is still bolted to the engine block, or transmission (4.0L only). To avoid damage, be sure to remove the sensor before removing the transmission.**

6. Remove the bolts holding the bell housing brace to the transmission.

7. Remove nut holding the bell housing brace to the engine to transmission bending brace.

8. Remove the bell housing brace from the transmission.

9. Remove the bolt holding the torque converter cover to the transmission.

10. Remove the torque converter cover from the transmission.

11. If transmission is being removed for overhaul, remove transmission oil pan, drain fluid and reinstall pan.

12. Remove fill tube bracket bolts and pull tube out of transmission. Retain fill tube seal. On 4 x 4 models, it will also be necessary to remove bolt attaching transfer case vent tube to converter housing.

13. Rotate crankshaft in clockwise direction until converter bolts are accessible. Then remove bolts one at a time. Rotate crankshaft with socket wrench on dampener bolt.

14. Mark propeller shaft and axle yokes for assembly alignment. Then disconnect and remove propeller shaft. On 4 x 4 models, remove both propeller shafts.

15. Disconnect wires from park/neutral position switch and transmission solenoid.

16. Disconnect gearshift cable from transmission manual valve lever.

17. Disconnect throttle valve cable from transmission bracket and throttle valve lever.

18. Disconnect transfer case shift cable from the transfer case shift lever.

19. Remove the clip securing the transfer case shift cable into the cable support bracket.

20. Disconnect transmission fluid cooler lines at transmission fittings and clips.

21. Support rear of engine with safety stand or jack.

22. Raise transmission slightly with service jack to relieve load on crossmember and supports.

23. Remove bolts securing rear support and cushion to transmission and crossmember.

24. Remove bolts attaching crossmember to frame and remove crossmember.
25. Remove transfer case.

➡**To install:**

26. Check torque converter hub and hub drive notches for sharp edges burrs, scratches, or nicks. Polish the hub and notches with 320/400 grit paper and crocus cloth if necessary. The hub must be smooth to avoid damaging pump seal during installation.
27. Lubricate oil pump seal lip with transmission fluid.
28. Align converter and oil pump.
29. Carefully insert converter in oil pump. Then rotate converter back and forth until fully seated in pump gears.
30. Check converter seating with steel scale and straightedge. Surface of converter lugs should be ½ in. to rear of straightedge when converter is fully seated.
31. Temporarily secure converter with C-clamp.
32. Position transmission on jack and secure it with chains.
33. Check condition of converter driveplate. Replace the plate if cracked, distorted or damaged. Also be sure transmission dowel pins are seated in engine block and protrude far enough to hold transmission in alignment.
34. Apply a light coating of Mopar® High Temp grease to the torque converter hub pocket in the rear of the crankshaft.
35. Raise transmission and align converter with drive plate and converter housing with engine block.
36. Move transmission forward. Then raise, lower or tilt transmission to align converter housing with engine block dowels.
37. Carefully work transmission forward and over engine block dowels until converter hub is seated in crankshaft.
38. Install two bolts to attach converter housing to engine.
39. Install the upper transmission bending braces to the torque converter housing and the overdrive unit. Tighten the bolts to 41 Nm (30 ft. lbs.).
40. Install remaining torque converter housing to engine bolts. Tighten to 68 Nm (50 ft. lbs.).
41. Install rear transmission crossmember. Tighten crossmember to frame bolts to 68 Nm (50 ft. lbs.).
42. Install rear support to transmission. Tighten bolts to 47 Nm (35 ft. lbs.).
43. Lower transmission onto crossmember and install bolts attaching transmission mount to crossmember. Tighten clevis

bracket to crossmember bolts to 47 Nm (35 ft. lbs.). Tighten the clevis bracket to rear support bolt to 68 Nm (50 ft. lbs.).
44. Remove engine support fixture.
45. Install crankshaft position sensor.
46. Install new plastic retainer grommet on any shift cable that was disconnected. Grommets should not be reused. Use pry tool to remove rod from grommet and cut away old grommet. Use pliers to snap new grommet into cable and to snap grommet onto lever.
47. Connect gearshift and throttle valve cable to transmission.
48. Connect wires to park/neutral position switch and transmission solenoid connector. Be sure transmission harnesses are properly routed.

➡**It is essential that correct length bolts be used to attach the converter to the driveplate. Bolts that are too long will damage the clutch surface inside the converter.**

49. Install all torque converter-to-driveplate bolts by hand.
50. Verify that the torque converter is pulled flush to the driveplate. Tighten bolts to 31 Nm (270 inch lbs.).
51. Install converter housing access cover. Tighten bolt to 23 Nm (200 inch lbs.).
52. Install the bell housing brace to the torque converter cover and the engine to transmission bending brace. Tighten the bolts and nut to 41 Nm (30 ft. lbs.).
53. Install starter motor and cooler line bracket.
54. Connect cooler lines to transmission.
55. Install transmission fill tube. Install new seal on tube before installation.
56. Install exhaust components.
57. Install transfer case. Tighten transfer case nuts to 35 Nm (26 ft. lbs.).
58. Install the transfer case shift cable to the cable support bracket and the transfer case shift lever.
59. Align and connect propeller shaft(s).
60. Adjust gearshift linkage and throttle valve cable if necessary.
61. Lower vehicle.
62. Fill transmission with Mopar® ATF +4, type 9602, fluid.

GRAND CHEROKEE W/545RFE

➡**The transmission and torque converter must be removed as an assembly to avoid component damage. The converter driveplate, converter hub O-ring, or oil seal can be damaged if the converter is left attached to the drive-**

plate during removal. Be sure to remove the transmission and converter as an assembly.

1. Disconnect the negative battery cable.
2. Raise and support the vehicle
3. Mark propeller shaft and axle yokes for assembly alignment.
4. Remove the rear propeller shaft
5. Remove the front propeller shaft.
6. Remove the engine to transmission collar.
7. Remove the exhaust support bracket from the rear of the transmission.
8. Disconnect and lower or remove any necessary exhaust components.
9. Remove the starter motor.
10. Rotate crankshaft in clockwise direction until converter bolts are accessible. Then remove bolts one at a time. Rotate crankshaft with socket wrench on dampener bolt.
11. Disconnect wires from solenoid and pressure switch assembly, input and output speed sensors, and line pressure sensor.
12. Disconnect gearshift cable from transmission manual valve lever.
13. Disconnect transfer case shift cable from the transfer case shift lever.
14. Remove the clip securing the transfer case shift cable into the cable support bracket.
15. Disconnect transmission fluid cooler lines at transmission fittings and clips.
16. Disconnect the transmission vent hose from the transmission.
17. Support rear of engine with safety stand or jack.
18. Raise transmission slightly with service jack to relieve load on crossmember and supports.
19. Remove bolts securing rear support and cushion to transmission and crossmember.
20. Remove bolts attaching crossmember to frame and remove crossmember.
21. Remove transfer case.
22. Remove all remaining converter housing bolts.
23. Carefully work transmission and torque converter assembly rearward off engine block dowels.
24. Hold torque converter in place during transmission removal.
25. Lower transmission and remove assembly from under the vehicle.
26. To remove torque converter, carefully slide torque converter out of the transmission.

To install:
27. Check torque converter hub and hub drive flats for sharp edges burrs, scratches,

or nicks. Polish the hub and flats with 320/400 grit paper and crocus cloth if necessary. Verify that the converter hub O-ring is properly installed and is free of any debris. The hub must be smooth to avoid damaging pump seal at installation.

28. If a replacement transmission is being installed, transfer any components necessary, such as the manual shift lever and shift cable bracket, from the original transmission onto the replacement transmission.

29. Lubricate oil pump seal lip with transmission fluid.

30. Align converter and oil pump.

31. Carefully insert converter in oil pump. Then rotate converter back and forth until fully seated in pump gears.

32. Check converter seating with steel scale and straightedge. Surface of converter lugs should be at least 13 mm (½ in.) to rear of straightedge when converter is fully seated.

33. Temporarily secure converter with C-clamp.

34. Position transmission on jack and secure it with chains.

35. Check condition of converter driveplate. Replace the plate if cracked, distorted or damaged.

➡ **Be sure transmission dowel pins are seated in engine block and protrude far enough to**

hold transmission in alignment.

36. Apply a light coating of Mopar® High Temp Grease to the torque converter hub pocket in the rear pocket of the engine's crankshaft.

37. Raise transmission and align the torque converter with the drive plate and the transmission converter housing with the engine block.

38. Move transmission forward. Then raise, lower, or tilt transmission to align the converter housing with the engine block dowels.

39. Carefully work transmission forward and over engine block dowels until converter hub is seated in crankshaft. Verify that no wires, or the transmission vent hose, have become trapped between the engine block and the transmission.

40. Install two bolts to attach the transmission to the engine.

41. Install remaining torque converter housing to engine bolts. Tighten to 68 Nm (50 ft. lbs.).

42. Install rear transmission crossmember. Tighten crossmember to frame bolts to 68 Nm (50 ft. lbs.).

43. Install rear support to transmission. Tighten bolts to 47 Nm (35 ft. lbs.).

44. Lower transmission onto crossmember and install bolts attaching transmission mount to crossmember. Tighten clevis bracket to crossmember bolts to 47 Nm (35 ft. lbs.). Tighten the clevis bracket to rear support bolt to 68 Nm (50 ft. lbs.).

45. Remove engine support fixture.

46. Install new plastic retainer grommet on any shift cable that was disconnected. Grommets should not be reused. Use pry tool to remove rod from grommet and cut away old grommet. Use pliers to snap new grommet into cable and to snap grommet onto lever.

47. Connect gearshift cable to transmission.

48. Connect wires to solenoid and pressure switch assembly connector, input and output speed sensors, and line pressure sensor. Be sure transmission harnesses are properly routed.

➡ **It is essential that correct length bolts be used to attach the converter to the driveplate. Bolts that are too long will damage the clutch surface inside the converter.**

49. Install all torque converter-to-driveplate bolts by hand.

50. Verify that the torque converter is pulled flush to the driveplate. Tighten bolts to 31 Nm (270 inch lbs.).

51. Install starter motor and cooler line bracket.

52. Connect cooler lines to transmission.

53. Install transmission fill tube.

54. Install exhaust components.

55. Install transfer case. Tighten transfer case nuts to 35 Nm (26 ft. lbs.).

56. Install the transfer case shift cable to the cable support bracket and the transfer case shift lever.

57. Install the structural dust cover onto the transmission and the engine.

58. Align and connect propeller shaft(s).

59. Adjust gearshift cable if necessary.

60. Lower vehicle.

61. Fill transmission with Mopar® ATF +4, Automatic Transmission fluid.

WRANGLER

1. Disconnect battery negative cable.

2. Raise and support vehicle.

3. Disconnect and lower or remove necessary exhaust components.

4. Remove engine-to-transmission bending braces or engine collar.

5. Remove starter motor.

6. On 4.0L engine equipped vehicles, disconnect and remove crankshaft position sensor. Retain sensor attaching bolt.

➡ **The crankshaft position sensor can be damaged during transmission removal (or installation) if the sensor is left in place. To avoid damage, remove the sensor before removing the transmission.**

7. If transmission is being removed for overhaul, remove transmission oil pan, drain fluid and reinstall pan.

8. Remove torque converter access cover.

9. Rotate crankshaft in clockwise direction until converter bolts are accessible. Then remove bolts one at a time. Rotate crankshaft with socket wrench on dampener bolt.

10. Mark propeller shaft and axle yokes for assembly alignment. Then disconnect and remove propeller shafts.

11. Disconnect wires from the input and output speed sensors.

12. Disconnect wires from the transmission range sensor and the solenoid/pressure switch assembly.

13. Disconnect gearshift cable from transmission manual valve lever.

14. Disconnect shift rod from transfer case shift lever or remove shift lever from transfer case.

15. Support rear of engine with safety stand or jack.

16. Raise transmission slightly with service jack to relieve load on skid plate and transmission support.

17. Remove bolts securing rear support and cushion to transmission and skid plate. Raise transmission slightly, slide exhaust hanger arm from bracket and remove rear support.

18. Remove bolts attaching skid plate to frame and remove skid plate.

19. Disconnect transfer case vent hose.

20. Remove transfer case.

21. Remove fill tube bracket bolts and pull tube out of transmission. Retain fill tube seal. Remove the bolt attaching transfer case vent tube to converter housing.

22. Disconnect fluid cooler lines at transmission.

23. Remove all converter housing bolts.

24. Carefully work transmission and torque converter assembly rearward off engine block dowels.

25. Hold torque converter in place during transmission removal.

26. Lower transmission and remove assembly from under the vehicle.

27. To remove torque converter, carefully slide torque converter out of the transmission.

To install:

28. Check torque converter hub and hub drive notches for sharp edges burrs, scratches, or nicks. Polish the hub and notches with 320/400 grit paper and crocus cloth if necessary. The hub must be smooth to avoid damaging pump seal at installation.

29. Lubricate converter drive hub and oil pump seal lip with transmission fluid.

30. Align converter and oil pump.

31. Carefully insert converter in oil pump. Then rotate converter back and forth until fully seated in pump gears.

32. Check converter seating with steel scale and straightedge. Surface of converter lugs should be ½ in. to rear of straightedge when converter is fully seated.

33. Temporarily secure converter with C-clamp.

34. Lightly grease crankshaft flange hole.

35. Position transmission on jack and secure it with safety chains.

36. Check condition of converter drive-plate. Replace the plate if cracked, distorted or damaged.

➡**Be sure transmission dowel pins are seated in engine block and protrude far enough to hold transmission in alignment.**

37. Raise transmission and align converter with drive plate and converter housing with engine block.

38. Move transmission forward. Then raise, lower or tilt transmission to align converter housing with engine block dowels.

39. Carefully work transmission forward and over engine block dowels until converter hub is seated in crankshaft.

40. Install and tighten bolts that attach transmission converter housing to engine block.

➡**Be sure the converter housing is fully seated on the engine block dowels before tightening any bolts.**

41. Install torque converter attaching bolts. Tighten bolts to 88 Nm (65 ft. lbs.).

42. On 4.0L engine equipped vehicles, install the crankshaft position sensor.

43. Install transmission fill tube and seal. Install new fill tube seal in transmission before installation.

44. Connect transmission cooler lines to transmission.

45. Install transfer case onto transmission.

46. Install skid plate and attach transmission rear support to skid plate.

47. Remove engine support fixture.

48. Remove transmission jack.

49. Connect input and output speed sensor wires.

50. Connect wires to the transmission range sensor and the solenoid/pressure switch assembly.

51. Install converter housing access cover.

52. Install exhaust pipes and support brackets, if removed.

53. Install starter motor and cooler line bracket.

54. Install new plastic retainer grommet on any shift linkage rod or lever that was disconnected. Grommets should not be reused. Use pry tool to remove rod from grommet and cut away old grommet. Use pliers to snap new grommet into lever and to snap rod into grommet at assembly.

55. Connect gearshift cable.

56. Connect transfer case shift linkage.

57. Adjust gearshift linkage, if necessary.

58. Align and connect propeller shaft(s).

59. Fill transfer case to bottom edge of fill plug hole.

60. Lower vehicle and connect battery negative cable.

61. Fill transmission to correct level with Mopar® ATF +4

2004 Manual

LIBERTY W/NV1500

1. Shift transmission into neutral.
2. Raise and support the vehicle.
3. Remove skid plate if equipped.
4. Remove wiring connectors from the transmission.
5. Remove propeller shaft/shafts.
6. Remove transfer case shift cable, vent hose and transfer case, if equipped.
7. Remove slave cylinder from clutch housing.
8. Remove starter.
9. Support engine with jack stand. Position wood block between jack and oil pan to avoid damaging pan.
10. Support transmission with a transmission jack.
11. Remove crossover pipe from manifold extensions.
12. Remove exhaust hanger from the transmission crossmember.
13. Remove transmission dust shield.
14. Remove transmission mount and crossmember.
15. Lower transmission jack enough to remove shift tower bolts.
16. Lower transmission jack and remove transmission from under vehicle.
17. Pull transmission jack rearward until input shaft clears clutch.
18. Remove clutch release bearing, release fork and retainer clip.

19. Remove clutch housing from transmission.

To install:

20. Install clutch housing on transmission and tighten housing bolts to 46 Nm (34 ft. lbs.).

21. Lubricate contact surfaces of release fork pivot ball stud and release fork with high temp grease.

22. Install release bearing, fork and retainer clip.

23. Position and secure transmission on transmission jack.

24. Lightly lubricate pilot bearing and transmission input shaft splines with Mopar high temp grease.

25. Raise transmission and align transmission input shaft and clutch disc splines. Then slide transmission into place.

26. Install clutch housing-to-engine bolts and tighten to 58 Nm (43 ft. lbs.).

➡**Be sure the housing is properly seated on engine block before tightening bolts.**

27. Install shift tower and bolts. Tighten bolts to 7–10 Nm (5–7 ft. lbs.).

28. Install rear crossmember and tighten crossmember bolts to 41 Nm (31 ft. lbs.).

29. Install transmission mount bolts and to 54 Nm (40 ft. lbs.).

30. Install exhaust bracket to crossmember.

31. Install crossover pipe to manifold extensions.

32. Remove support stands from engine and transmission.

33. Install transfer case, shift cable and vent hose if equipped.

34. Install wire connectors to transmission/transfer case.

35. Install propeller shaft/shafts.

36. Install slave cylinder in clutch housing.

37. Install skid plate if equipped.

38. Fill transmission and transfer case if equipped, with recommended lubricants.

LIBERTY W/NV3550

1. Shift transmission into neutral.
2. Raise and support the vehicle.
3. Remove skid plate if equipped.
4. Remove wiring connectors from the transmission.
5. Remove propeller shaft/shafts.
6. Remove transfer case shift cable and vent hose from transfer case, if equipped.
7. Remove slave cylinder from clutch housing.
8. Support engine with jack stand. Position wood block between jack and oil pan to avoid damaging pan.

9. Support transmission with a transmission jack.

10. Remove exhaust hanger from the transmission crossmember.

11. Remove transmission mount and crossmember.

12. Lower transmission jack enough to remove shift tower bolts.

13. Remove clutch housing-to-engine bolts.

14. Pull transmission jack rearward until input shaft clears clutch.

15. Remove clutch release bearing, release fork and retainer clip.

16. Remove clutch housing from transmission.

To install:

17. Install clutch housing on transmission and tighten housing bolts to 46 Nm (34 ft. lbs.).

18. Lubricate contact surfaces of release fork pivot ball stud and release fork with high temp grease.

19. Install release bearing, fork and retainer clip.

20. Position and secure transmission on transmission jack.

21. Lightly lubricate the transmission input shaft splines with Mopar high temp grease.

22. Raise transmission and align transmission input shaft and clutch disc splines. Then slide transmission into place.

23. Install clutch housing-to-engine bolts and tighten to 58 Nm (43 ft. lbs.).

➡**Be sure the housing is properly seated on engine block before tightening bolts.**

24. Install shift tower and bolts. Tighten bolts to 7–10 Nm (5–7 ft. lbs.).

25. Install rear crossmember and tighten crossmember bolts to 41 Nm (31 ft. lbs.).

26. Install transmission mount bolts and to 54 Nm (40 ft. lbs.).

27. Install exhaust bracket to crossmember.

28. Remove support stands from engine and transmission.

29. Install transfer case, shift cable and vent hose if equipped.

30. Install wire connectors to transmission/transfer case.

31. Install propeller shaft(s).

32. Install slave cylinder in clutch housing.

33. Fill transmission and transfer case if equipped, with recommended lubricants.

34. Install skid plate if equipped.

WRANGLER W/NV1500

1. With vehicle in neutral, position vehicle on hoist.

2. Support engine with jack stand. Position wood block between jack and oil pan to avoid damaging pan.

3. Remove skid plate/crossmember.

4. Support transmission with a transmission jack.

5. Remove transmission mount from transmission and exhaust.

6. Remove propeller shafts.

7. Remove transfer case shift linkage and vent hose.

8. Remove wiring connectors from transmission and transfer case.

9. Remove transfer case.

10. Remove slave cylinder from clutch housing.

11. Remove starter.

12. Remove transmission dust shield.

13. Lower transmission jack enough to remove shift tower bolts.

14. Lower transmission jack and remove transmission from under vehicle.

15. Pull transmission jack rearward until input shaft clears clutch.

16. Remove clutch release bearing, release fork and retainer clip.

17. Remove clutch housing from transmission.

To install:

18. Install clutch housing on transmission and tighten housing bolts to 46 Nm (34 ft. lbs.).

19. Lubricate contact surfaces of release fork pivot ball stud and release fork with high temp grease.

20. Install release bearing, fork and retainer clip.

21. Position and secure transmission on transmission jack.

22. Lightly lubricate pilot bearing and transmission input shaft splines with Mopar high temp grease.

23. Raise transmission and align transmission input shaft and clutch disc splines. Then slide transmission into place.

24. Install clutch housing-to-engine bolts and tighten to 75 Nm (55 ft. lbs.).

➡**Be sure the housing is properly seated on engine block before tightening bolts.**

25. Install shift tower and bolts. Tighten bolts to 11 Nm (8 ft. lbs.).

26. Install transmission mount and tighten bolts to 54 Nm (40 ft. lbs.).

27. Install transfer case, shift linkage and vent hose.

28. Install wire connectors to transmission and transfer case.

29. Install skid plate/crossmember and tighten bolts to 41 Nm (31 ft. lbs.).

30. Remove support stands from engine and transmission.

WRANGLER W/NV3550

1. Shift transmission into first or third gear.

2. Remove floor console and shift boot as necessary to access the bottom of the shift lever at the shift tower attachment.

3. Remove shift tower bolts and remove shift tower and shift lever assembly.

4. Raise and support vehicle on suitable safety stands.

5. Support engine with adjustable jack stand. Position wood block between jack and oil pan to avoid damaging pan.

6. Remove skid plate, if equipped.

7. Remove crossmember.

8. Disconnect necessary exhaust system components.

9. Remove slave cylinder from clutch housing.

10. Remove propeller shafts.

11. Unclip wire harnesses from transmission and transfer case.

12. Disconnect transfer case shift linkage at transfer case.

13. Remove nuts attaching transfer case to transmission.

14. Remove transfer case.

15. Remove crankshaft position sensor.

➡**The crankshaft position sensor must be removed prior to transmission removal. Failure to heed caution may result in damage.**

16. Support engine with adjustable jack stand. Position wood block between jack and oil pan to avoid damaging pan.

17. Support transmission with transmission jack.

18. Secure transmission to jack with safety chains.

19. Disconnect rear cushion and bracket from transmission.

20. Remove rear crossmember.

21. Remove clutch housing-to-engine bolts.

22. Pull transmission jack rearward until input shaft clears clutch. Then slide transmission out from under vehicle.

23. Remove clutch release bearing, release fork and retainer clip.

24. Remove clutch housing from transmission.

To install:

25. Install clutch housing on transmission and tighten housing bolts to 46 Nm (34 ft. lbs.).

26. Lubricate contact surfaces of release fork pivot ball stud and release fork with high temp grease.

27. Install release bearing, fork and retainer clip.

28. Position and secure transmission on transmission jack.

29. Lightly lubricate pilot bearing and transmission input shaft splines with Mopar high temp grease.

30. Raise transmission and align transmission input shaft and clutch disc splines. Then slide transmission into place.

➡**Be sure the housing is properly seated on engine block before tightening bolts.**

31. Install and tighten clutch housing-to-engine bolts to:
- ⅜ in. diameter bolts to 37 Nm (27 ft. lbs.)
- 7⁄16 in. diameter bolts to 58 Nm (43 ft. lbs.)
- M12 bolts to 75 Nm (55 ft. lbs.)

32. Be sure transmission is in first or third gear.

33. Install crossmember and tighten crossmember-to-frame bolts to 41 Nm (31 ft. lbs.).

34. Install fasteners to hold rear cushion and bracket to transmission. Then tighten transmission-to- rear support bolts/nuts to 54 Nm (40 ft. lbs.).

35. Remove support stands from engine and transmission.

36. Install and connect crankshaft position sensor.

37. Install transfer case.

38. Install propeller shafts.

39. Install slave cylinder in clutch housing.

40. Install skid plate, if equipped and tighten bolts to 42 Nm (31 ft. lbs.). Tighten stud nuts to 17 Nm (150 inch lbs.).

41. Fill transmission and transfer case if equipped, with recommended lubricants.

42. Install nuts on two M6 x 1.0 bolts and thread the bolts into the threaded holes at the base of the shift lever.

43. Tighten the nuts equally until the shift lever will slide over the shift tower stub shaft.

44. Install the floor console and shift boot.

Clutch

REMOVAL & INSTALLATION

1. Before servicing the vehicle, refer to the precautions in the beginning of this section.

2. Remove or disconnect the following:
- Negative battery cable
- Transfer case, if equipped
- Transmission
- Pressure plate. Loosen the bolts evenly in ½ turn steps.
- Clutch disc

To install:

3. Install or connect the following:
- Clutch disc and pressure plate. Tighten the pressure plate bolts evenly in ½ turns to 23 ft. lbs. (31 Nm) for 2.4L and 2.5L engines, 37 ft. lbs. (50 Nm) for 3.7L engines, or to 40 ft. lbs. (54 Nm) for 4.0L engines.
- Transmission
- Transfer case, if equipped
- Negative battery cable

Hydraulic Clutch System

BLEEDING

➡**The clutch master cylinder, slave cylinder and fluid line are serviced only as an assembly. Bleeding is not possible.**

Transfer Case Assembly

REMOVAL & INSTALLATION

2001–03

1. Before servicing the vehicle, refer to the precautions in the beginning of this section.

2. Shift the transfer case into **N**.

3. Drain the transfer case fluid.

4. Remove or disconnect the following:
- Negative battery cable
- Front and rear driveshafts
- Transmission mount and crossmember. Support the transmission with a jackstand.
- Vehicle Speed (VSS) sensor connector
- Indicator switch connector
- Vacuum hose
- Shift linkage
- Vent hose
- Transfer case attaching nuts
- Transfer case

To install:

5. Install or connect the following:
- Transfer case. Tighten the nuts to 26 ft. lbs. (35 Nm).
- Indicator switch connector
- Vacuum hose
- Vent hose
- Shift linkage
- VSS sensor connector
- Transmission mount and cross-

member. Torque the bolts to 30 ft. lbs. (41 Nm).
- Front and rear driveshafts
- Negative battery cable

6. Fill the transfer case with the proper fluid.

7. Start the vehicle, check for leaks and repair if necessary.

2004

GRAND CHEROKEE W/NV147

1. Raise vehicle.

2. Remove transfer case drain plug and drain transfer case lubricant.

3. Mark front and rear propeller shaft yokes for alignment reference.

4. Support transmission with jack stand.

5. Remove rear crossmember and skid plate, if equipped.

6. Disconnect front propeller shaft from transfer case at companion flange. Remove rear propeller shaft from vehicle.

➡**Do not allow driveshafts to hang at attached end. Damage to joint can result.**

7. Disconnect transfer case vent hose.

8. Support transfer case with transmission jack.

9. Secure transfer case to jack with chains.

10. Remove nuts attaching transfer case to transmission.

11. Pull transfer case and jack rearward to disengage transfer case.

12. Remove transfer case from under vehicle.

To install:

13. Mount transfer case on a transmission jack.

14. Secure transfer case to jack with chains.

15. Position transfer case under vehicle.

16. Align transfer case and transmission shafts and install transfer case on transmission.

17. Install and tighten transfer case attaching nuts to 35 Nm (26 ft. lbs.) torque.

18. Connect front propeller shaft and install rear propeller shaft.

19. Fill transfer case with correct fluid. Check transmission fluid level. Correct as necessary.

20. Install rear crossmember and skid plate, if equipped. Tighten crossmember bolts to 41 Nm (30 ft. lbs.) torque.

21. Remove transmission jack and support stand.

22. Lower vehicle and verify transfer case shift operation.

GRAND CHEROKEE W/NV242

1. Shift transfer case into NEUTRAL.
2. Raise vehicle.
3. Remove transfer case drain plug and drain transfer case lubricant.
4. Mark front and rear propeller shaft yokes for alignment reference.
5. Support transmission with jack stand.
6. Remove rear crossmember and skid plate, if equipped.
7. Disconnect front/rear propeller shafts at transfer case.
8. Disconnect transfer case cable from range lever.
9. Disconnect transfer case vent hose and transfer case position sensor.
10. Support transfer case with transmission jack.
11. Secure transfer case to jack with chains.
12. Remove nuts attaching transfer case to transmission.
13. Pull transfer case and jack rearward to disengage transfer case.
14. Remove transfer case from under vehicle.

To install:

15. Mount transfer case on a transmission jack.
16. Secure transfer case to jack with chains.
17. Position transfer case under vehicle.
18. Align transfer case and transmission shafts and install transfer case on transmission.
19. Install and tighten transfer case attaching nuts to 35 Nm (26 ft. lbs.) torque.
20. Align and connect propeller shafts.
21. Fill transfer case with correct fluid. Check transmission fluid level. Correct as necessary.
22. Install rear crossmember and skid plate, if equipped. Tighten crossmember bolts to 41 Nm (30 ft. lbs.) torque.
23. Remove transmission jack and support stand.
24. Connect shift rod to transfer case range lever.
25. Connect transfer case vent hose and transfer case position sensor.
26. Adjust transfer case shift cable.
27. Lower vehicle and verify transfer case shift operation.

GRAND CHEROKEE W/NV247

1. Shift transfer case into NEUTRAL.
2. Raise vehicle.
3. Remove transfer case drain plug and drain transfer case lubricant.

4. Mark front and rear propeller shaft yokes for alignment reference.
5. Support transmission with jack stand.
6. Remove rear crossmember and skid plate, if equipped.
7. Disconnect front propeller shaft from transfer case at companion flange. Remove rear propeller shaft from vehicle.

➡**Do not allow driveshafts to hang at attached end. Damage to joint can result.**

8. Disconnect transfer case cable from range lever.
9. Disconnect transfer case vent hose.
10. Support transfer case with transmission jack.
11. Secure transfer case to jack with chains.
12. Remove nuts attaching transfer case to transmission.
13. Pull transfer case and jack rearward to disengage transfer case.
14. Remove transfer case from under vehicle.

To install:

15. Mount transfer case on a transmission jack.
16. Secure transfer case to jack with chains.
17. Position transfer case under vehicle.
18. Align transfer case and transmission shafts and install transfer case on transmission.
19. Install and tighten transfer case attaching nuts to 35 Nm (26 ft. lbs.).
20. Connect front propeller shaft and install rear propeller shaft.
21. Fill transfer case with correct fluid. Check transmission fluid level. Correct as necessary.
22. Install rear crossmember and skid plate, if equipped. Tighten crossmember bolts to 41 Nm (30 ft. lbs.).
23. Remove transmission jack and support stand.
24. Verify transfer case is in NEUTRAL. Connect shift cable to transfer case range lever.
25. Lower vehicle and verify transfer case shift operation.
26. Adjust the transfer case shift cable, if necessary.

WRANGLER W/NV231

1. Shift transfer case into NEUTRAL.
2. Raise vehicle.
3. Drain transfer case lubricant.
4. Mark front and rear propeller shaft yokes for alignment reference.

5. Support transmission with jack stand.
6. Remove rear crossmember, or skid plate.
7. Disconnect front/rear propeller shafts at transfer case.
8. Disconnect vehicle speed sensor wires.
9. Disconnect transfer case linkage rod from range lever.
10. Disconnect transfer case vent hose and indicator switch harness, if necessary.
11. Support transfer case with transmission jack.
12. Secure transfer case to jack with chains.
13. Remove nuts attaching transfer case to transmission.
14. Pull transfer case and jack rearward to disengage transfer case.
15. Remove transfer case from under vehicle.

To install:

16. Mount transfer case on a transmission jack.
17. Secure transfer case to jack with chains.
18. Position transfer case under vehicle.
19. Align transfer case and transmission shafts and install transfer case on transmission.
20. Install and tighten transfer case attaching nuts to 35 Nm (26 ft. lbs.) torque.
21. Connect vehicle speed sensor wires, and vent hose.
22. Connect indicator switch harness to transfer case switch, if necessary. Secure wire harness to clips on transfer case.
23. Align and connect propeller shafts.
24. Fill transfer case with correct fluid. Check transmission fluid level. Correct as necessary.
25. Install rear crossmember, or skid plate. Tighten crossmember bolts to 41 Nm (30 ft. lbs.).
26. Remove transmission jack and support stand.
27. Connect shift rod to transfer case range lever.
28. Adjust transfer case shift linkage.
29. Lower vehicle and verify transfer case shift operation.

WRANGLER W/NV242

1. Raise and support vehicle.
2. Remove skid plate, if equipped.
3. Position drain oil container under transfer case.
4. Remove transfer case drain plug and drain lubricant into container.
5. Disconnect vent hose and vacuum harness at transfer case switch.

6. Disconnect shift rod from grommet in transfer case shift lever, or from floor shift arm whichever provides easy access. Use pliers to press rod out of lever grommet.

7. Support transmission with jack stand.

8. Remove rear crossmember.

9. Mark front and rear propeller shafts for assembly reference.

10. Remove front and rear propeller shafts.

11. Support transfer case with suitable jack. Secure transfer case to jack with safety chains.

12. Remove nuts attaching transfer case to transmission.

13. Move transfer case assembly rearward until free of transmission output shaft.

14. Lower jack and move transfer case from under vehicle.

To install:

15. Align and seat transfer case on transmission. Be sure transfer case input gear splines are aligned with transmission output shaft. Align splines by rotating transfer case rear output shaft yoke if necessary.

➡**Do not install any transfer case attaching nuts until the transfer case is completely seated against the transmission.**

16. Install and tighten transfer case attaching nuts. Tighten nuts to 30–41 Nm (20–30 ft. lbs.).

17. Install rear crossmember.

18. Remove jack stand from under transmission.

19. Align and connect propeller shafts.

20. Connect vacuum harness and vent hose.

21. Connect shift rod to transfer case lever or floor shift arm. Use pliers to press rod back into lever grommet.

22. Adjust shift linkage, if necessary.

23. Fill transfer case with recommended transmission fluid and install fill plug.

24. Install skid plate, if equipped.

25. Lower vehicle

LIBERTY W/NV231

1. Shift transfer case into NEUTRAL.

2. Raise vehicle.

3. Remove skid plate.

4. Drain transfer case lubricant.

5. Mark front and rear propeller shaft yokes for alignment reference.

6. Disconnect front/rear propeller shafts at transfer case.

7. Disconnect transfer case position sensor connector.

8. Disconnect transfer case shift cable at the range lever.

9. Disconnect the transfer case shift cable from the shift cable bracket.

10. Disconnect transfer case vent hose.

11. Support transfer case with transmission jack.

12. Secure transfer case to jack with chains.

13. Remove nuts attaching transfer case to transmission.

14. Pull transfer case and jack rearward to disengage transfer case.

15. Remove transfer case from under vehicle.

To install:

16. Mount transfer case on a transmission jack.

17. Secure transfer case to jack with chains.

18. Position transfer case under vehicle.

19. Align transfer case and transmission shafts and install transfer case on transmission.

20. Install and tighten transfer case attaching nuts to 35 Nm (26 ft. lbs.) torque.

21. Connect vent hose.

22. Connect transfer case position sensor connector to sensor.

23. Align and connect propeller shafts.

24. Fill transfer case with correct fluid. Check transmission fluid level. Correct as necessary.

25. Install skid plate.

26. Remove transmission jack and support stand.

27. Connect shift cable to transfer case range lever.

28. Lower vehicle and verify transfer case shift operation.

LIBERTY W/NV242

1. Shift transfer case into NEUTRAL.

2. Raise vehicle.

3. Remove skid plate.

4. Drain transfer case lubricant.

5. Mark front and rear propeller shaft yokes for alignment reference.

6. Disconnect front/rear propeller shafts at transfer case.

7. Disconnect transfer case position sensor connector.

8. Disconnect transfer case shift cable at the range lever.

9. Disconnect the transfer case shift cable from the shift cable bracket.

10. Disconnect transfer case vent hose.

11. Support transfer case with transmission jack.

12. Secure transfer case to jack with chains.

13. Remove nuts attaching transfer case to transmission.

14. Pull transfer case and jack rearward to disengage transfer case.

15. Remove transfer case from under vehicle

To install:

16. Mount transfer case on a transmission jack.

17. Secure transfer case to jack with chains.

18. Position transfer case under vehicle.

19. Align transfer case and transmission shafts and install transfer case on transmission.

20. Install and tighten transfer case attaching nuts to 35 Nm (26 ft. lbs.) torque.

21. Connect vent hose.

22. Connect transfer case position sensor connector to sensor.

23. Align and connect propeller shafts.

24. Fill transfer case with correct fluid. Check transmission fluid level. Correct as necessary.

25. Install skid plate

26. Remove transmission jack and support stand.

27. Connect shift cable to transfer case range lever.

28. Lower vehicle and verify transfer case shift operation.

CV-Joints

Driveshaft and CV-joints are serviced only as an assembly, as are halfshafts and CV-joints.

Halfshafts

REMOVAL & INSTALLATION

Liberty

1. Before servicing the vehicle, refer to the precautions in the beginning of this section.

2. Remove or disconnect the following:
   • Wheel
   • Hub nut
   • Stabilizer link
   • Lower clevis bolt
   • Ball joint from the lower arm

1. Pull on the hub and push the halfshaft from the knuckle

➡**The right side has a splined axle shaft that will stay in the axle.**

To install:

2. Apply a light coating of wheel bearing grease on the splines of the inner joint, and in the hub bearing bore.

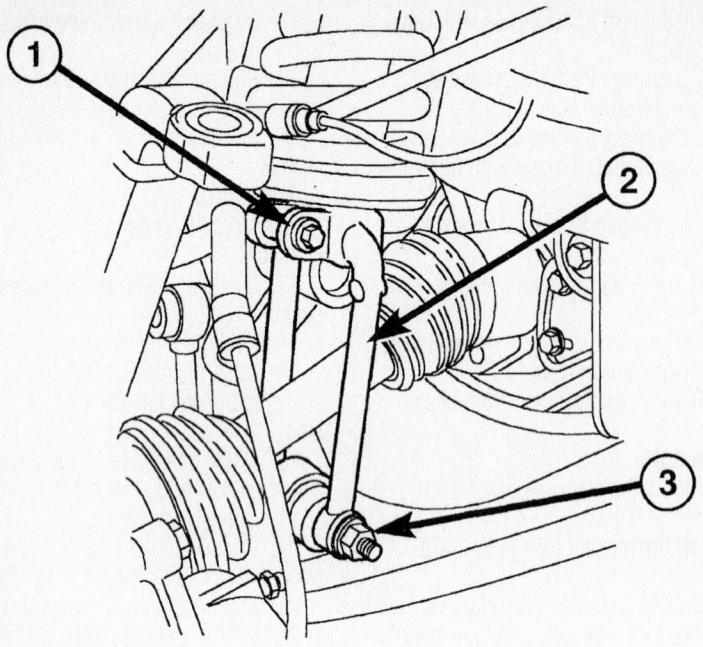

1 - UPPER BOLT
2 - CLEVIS BRACKET
3 - LOWER BOLT

9355PG22

Clevis bracket—Liberty

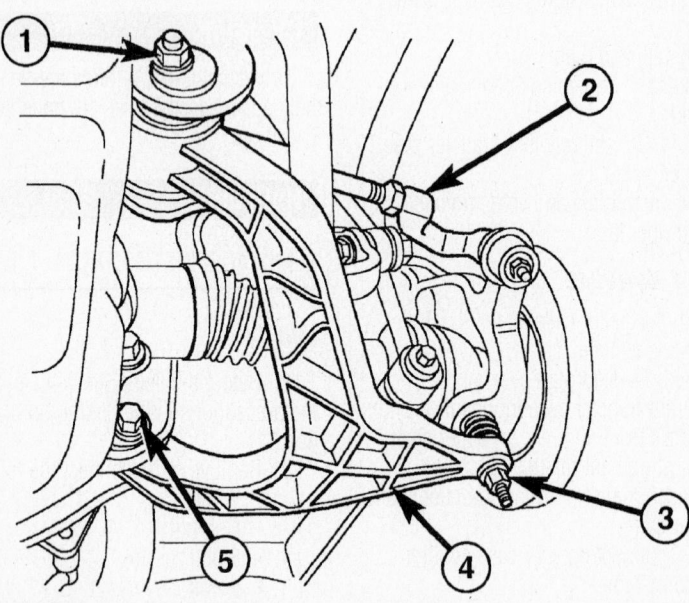

1 - FRONT CAM BOLT
2 - OUTER TIE ROD END
3 - LOWER BALL JOINT NUT
4 - LOWER CONTROL ARM
5 - REAR CAM BOLT

9355PG23

Lower control arm—Liberty

3. Install or connect the following:
   • Halfshaft on the axle shaft splines

➡**Push firmly enough to engage the snapring. Pull on it to verify engagement.**

   • Halfshat into the knuckle
   • Lower ball joint and pinch bolt
   • Lower clevis bolt
   • Stabilizer link
   • Hub nut. Torque to 100 ft. lbs. (136 Nm).
   • Wheel

Front Axle Shaft, Bearing and Seal

REMOVAL & INSTALLATION

Liberty

RIGHT SIDE SHAFT ONLY

1. Before servicing the vehicle, refer to the precautions in the beginning of this section.
2. Remove or disconnect the following:
   • Negative battery cable
   • Right front wheel
   • Halfshaft
   • Snapring from the axle
   • Axle shaft using remover 8420A and a slide hammer

To install:

3. Coat the bearing bore and seal lip with gear oil.
4. Install or connect the following:
   • Axle shaft and snapring
   • Halfshaft
   • Front wheel
   • Negative battery cable

RIGHT SIDE SEAL AND BEARING ONLY

1. Before servicing the vehicle, refer to the precautions in the beginning of this section.
2. Remove or disconnect the following:
   • Negative battery cable
   • Right front wheel
   • Halfshaft
   • Snapring from the axle
   • Axle shaft using remover 8420A and a slide hammer
   • Seal, using Remover 7794A and a slide hammer
   • Bearing, using Remover 7794A and a slide hammer

To install:

3. Coat the bearing bore and seal lip with gear oil.
4. Install or connect the following:
   • New seal, using Installer 8806 and handle C4171, or their equivalents

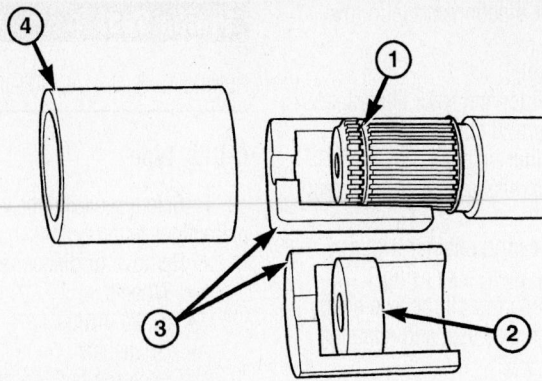

1 - SNAP RING GROVE
2 - SLID HAMMER THREADS
3 - REMOVER BLOCKS
4 - REMOVER COLLAR

9355PG24

Axle shaft remover tool—Liberty

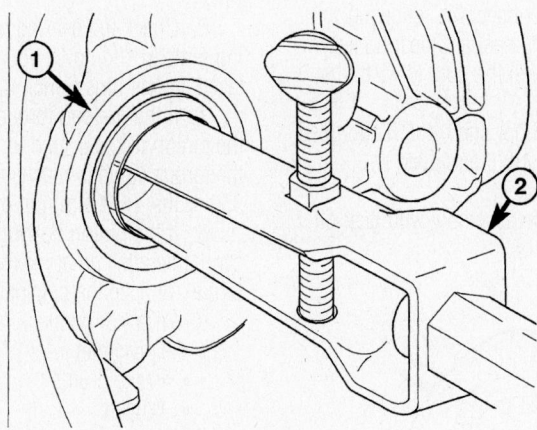

1 - SHAFT SEAL
2 - REMOVER

9355PG25

Seal removal—Liberty

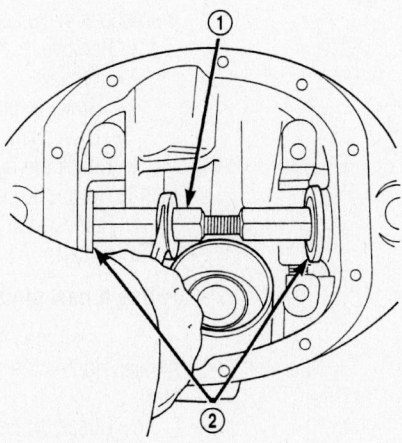

1 – TURNBUCKLE 6797
2 – DISCS 8110

9308PG05

Axle seal installation

- Axle shaft and snapring
- Halfshaft
- Front wheel
- Negative battery cable

Except Liberty

FRONT AXLE SHAFT

1. Before servicing the vehicle, refer to the precautions in the beginning of this section.
2. Remove or disconnect the following:
   - Negative battery cable
   - Front wheel
   - Brake caliper and rotor
   - Wheel speed sensor, if equipped
   - Axle hub nut
   - Wheel bearing and hub assembly
   - Axle shaft

To install:

3. Install or connect the following:
   - Axle shaft
   - Wheel bearing and hub assembly
   - Axle hub nut. Tighten the nut to 175 ft. lbs. (237 Nm).
   - Wheel speed sensor, if equipped
   - Brake caliper and rotor
   - Front wheel
   - Negative battery cable

FRONT SHAFT SEAL

1. Before servicing the vehicle, refer to the precautions in the beginning of this section.
2. Remove or disconnect the following:
   - Front axle shafts
   - Differential cover
   - Differential and ring gear assembly
   - Axle seals

To install:

3. Press the axle seals into the differential housing with Turnbuckle 6797 and Disc set 8110.
4. Install or connect the following:
   - Differential and ring gear assembly. Tighten the bearing cap bolts to 45 ft. lbs. (61 Nm).
   - Differential cover. Tighten the bolts to 30 ft. lbs. (41 Nm).
   - Front axle shafts
5. Fill the axle assembly with gear oil and check for leaks.

Rear Axle Shaft, Bearing and Seal

REMOVAL & INSTALLATION

C-Clip Type Rear Axle Shaft

1. Before servicing the vehicle, refer to the precautions in the beginning of this section.

2. Remove or disconnect the following:
- Negative battery cable
- Rear wheel
- Brake drum
- Differential cover
- Differential gear shaft retainer
- Differential gear shaft
- C-clip
- Axle shaft
- Axle seal
- Axle bearing

To install:

3. Install or connect the following:
- Axle bearing
- Axle seal
- Axle shaft
- C-clip
- Differential gear shaft. Use Loctite® and tighten the retainer to 14 ft. lbs. (19 Nm).
- Differential cover. Tighten the bolts to 30 ft. lbs. (41 Nm).
- Brake drum
- Rear wheel

4. Fill the axle assembly with gear oil and check for leaks.

Non C-Clip Type Rear Axle Shaft

1. Before servicing the vehicle, refer to the precautions in the beginning of this section.

2. Remove or disconnect the following:
- Rear wheel
- Brake caliper and rotor, if equipped
- Brake drum, if equipped
- Axle retainer nuts
- Axle shaft, seal and bearing assembly

3. Split the bearing retainer with a chisel and remove the retainer ring.

4. Press the bearing off the axle shaft.

5. Remove the axle seal and retaining plate.

To install:

6. Install the retaining plate and axle seal onto the axle shaft.

7. Pack the wheel bearing with axle grease and press the bearing on to the axle shaft.

8. Press the retaining ring onto the axle shaft.

9. Install or connect the following:
- Axle shaft, seal and bearing assembly. Tighten the nuts to 45 ft. lbs. (61 Nm).
- Brake caliper and rotor, if equipped
- Brake drum, if equipped
- Rear wheel

10. Fill the axle assembly with gear oil and check for leaks.

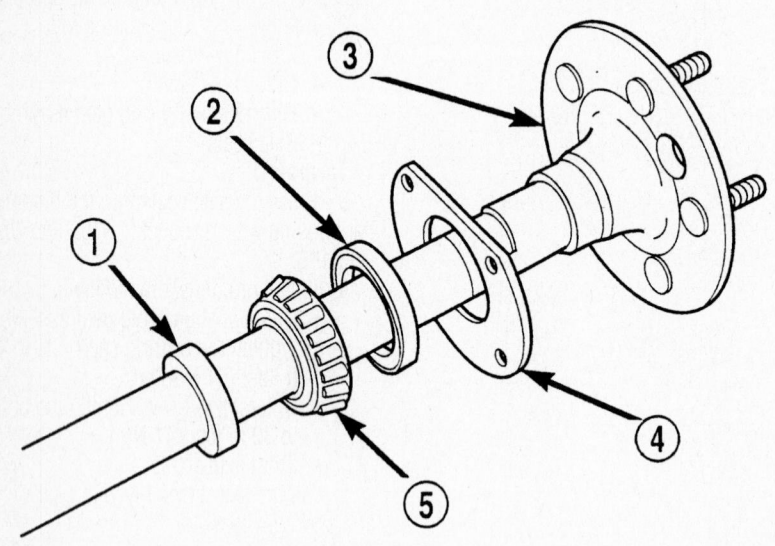

1 – RETAINING RING
2 – SEAL
3 – AXLE
4 – RETAINING PLATE
5 – AXLE BEARING

9308PG06

Rear axle seal and bearing components

Pinion Seal

REMOVAL & INSTALLATION

C-Clip Type

1. Before servicing the vehicle, refer to the precautions in the beginning of this section.

2. Remove or disconnect the following:
- Wheels
- Brake drums
- Driveshaft

3. Check the bearing preload with an inch lb. torque wrench.

4. Remove the pinion flange and seal.

To install:

➡ **Use a new pinion nut for assembly.**

5. Install the new pinion seal and flange. Tighten the nut to 200 ft. lbs. (271 Nm).

6. Check the bearing preload. The bearing preload should be equal to the reading taken earlier, plus 5 inch lbs.

7. If the preload torque is low, tighten the pinion nut in 5 inch lb. increments until the torque value is reached. Do not exceed 350 ft. lbs. (474 Nm) pinion nut torque.

8. If the pinion bearing preload torque cannot be attained at maximum pinion nut torque, replace the collapsible spacer.

9. Install or connect the following:
- Driveshaft
- Brake drums
- Wheels

10. Fill the axle assembly with gear oil and check for leaks.

Non C-Clip Type

FRONT

1. Before servicing the vehicle, refer to the precautions in the beginning of this section.

2. Remove or disconnect the following:
- Wheels
- Brake rotors
- Driveshaft

3. Check the bearing preload with an inch lb. torque wrench.

4. Remove the pinion flange and seal.

To install:

➡ **Use a new pinion nut for assembly.**

5. Install the new pinion seal and flange. Tighten the nut to 160 ft. lbs. (217 Nm).

6. Check the bearing preload. The bearing preload should be equal to the reading taken earlier, plus 5 inch lbs.

7. If the preload torque is low, tighten the pinion nut in 5 inch lb. increments until

the torque value is reached. Do not exceed 260 ft. lbs. (353 Nm) pinion nut torque.

8. If the pinion bearing preload torque can not be attained at maximum pinion nut torque, replace the collapsible spacer.

9. Install or connect the following:
- Driveshaft
- Brake rotors
- Wheels

10. Fill the axle assembly with gear oil and check for leaks.

REAR

1. Before servicing the vehicle, refer to the precautions in the beginning of this section.

2. Remove or disconnect the following:
- Wheels
- Brake rotors or drums
- Driveshaft

3. Check the bearing preload with an inch lb. torque wrench.

4. Remove the pinion flange and seal.

To install:

➡**Use a new pinion nut for assembly.**

5. Install the new pinion seal and flange. Tighten the nut to 160 ft. lbs. (217 Nm).

6. Check the bearing preload. The bearing preload should be equal to the reading taken earlier, plus 5 inch lbs.

7. If the preload torque is low, tighten the pinion nut in 5 inch lb. increments until the torque value is reached. Do not exceed 260 ft. lbs. (353 Nm) pinion nut torque.

8. If the pinion bearing preload torque can not be attained at maximum pinion nut torque, remove one or more pinion preload shims.

9. Install or connect the following:
- Driveshaft
- Brake rotors or drums
- Wheels

10. Fill the axle assembly with gear oil and check for leaks.

STEERING AND SUSPENSION

Air Bag

✳✳ CAUTION

Some vehicles are equipped with an air bag system. The system must be disarmed before performing service on, or around, system components, the steering column, instrument panel components, wiring and sensors. Failure to follow the safety precautions and the disarming procedure could result in accidental air bag deployment, possible injury and unnecessary system repairs.

PRECAUTIONS

Several precautions must be observed when handling the inflator module to avoid accidental deployment and possible personal injury.

- Never carry the inflator module by the wires or connector on the underside of the module.
- When carrying a live inflator module, hold securely with both hands, and ensure that the bag and trim cover are pointed away.
- Place the inflator module on a bench or other surface with the bag and trim cover facing up.
- With the inflator module on the bench, never place anything on or close to the module, which may be thrown in the event of an accidental deployment.

Before servicing the vehicle, also make sure to refer to the precautions in the beginning of this section as well.

DISARMING

Disconnect and isolate the negative battery cable. Wait 2 minutes for the system

capacitor to discharge before performing any service.

ARMING

To arm the system, connect the negative battery cable.

Recirculating Ball Power Steering Gear

REMOVAL & INSTALLATION

1. Before servicing the vehicle, refer to the precautions in the beginning of this section.

2. Place the front wheels in the straight ahead position.

3. Drain the power steering system.

4. Remove or disconnect the following:
- Negative battery cable
- Air cleaner housing
- Power steering pressure and return lines
- Column coupler shaft bolt
- Left front wheel assembly
- Pitman arm
- Windshield washer reservoir
- Steering gear

To install:

5. Install or connect the following:
- Steering gear. Tighten the bolts to 80 ft. lbs. (108 Nm).
- Pitman arm. Tighten the nut to 185 ft. lbs. (251 Nm).
- Windshield washer reservoir
- Power steering pressure and return lines. Torque the fasteners to 14 ft. lbs. (20 Nm).
- Intermediate shaft. Tighten the pinch bolt to 36 ft. lbs. (49 Nm).
- Air cleaner assembly
- Left front wheel
- Negative battery cable

6. Fill and bleed the power steering fluid reservoir.

7. Start the engine, check for leaks and repair if necessary.

Rack and Pinion Power Steering Gear

REMOVAL & INSTALLATION

2-Wheel Drive

1. Siphon the power steering fluid from the power steering reservoir.

➡**The steering column on vehicles with an automatic transmission may not be equipped with an internal locking shaft that allows the ignition key cylinder to be locked with the key. Alternative methods of locking the steering wheel for service will have to be used.**

2. Lock the steering wheel to prevent spinning of the clockspring.

3. Raise and support the vehicle.

4. Remove the skid plate from under the front end to gain access to the gear.

5. Remove the tire and wheel assembly.

➡**Mark the alignment adjusting cams and tie rod end jam nuts on the steering gear for easier installation.**

6. Remove the tie rod end nuts.

7. Separate tie rod ends from the knuckles.

8. Remove the lower intermediate shaft coupler bolt and slide the coupler off the gear.

9. Remove power steering pressure hose bracket.

10. Remove the power steering lines from the gear.

11. Remove the mounting bolts from the gear to the front cradle.

12. Remove the steering gear from the vehicle.

To install:

13. Transfer the outer tie rod ends to the new steering gear (if needed).

14. Install the steering gear to the vehicle.

15. Install the gear mounting bolts to the front cradle. Tighten the gear mounting bolts to 162 Nm (120 ft. lbs.)

16. Install the power steering lines to the gear.

17. Install the power steering pressure hose bracket.

18. Install the lower coupler bolt and slide the coupler on to the gear.

19. Install the tie rod end to the knuckle and tighten the nuts.

20. Install the tire and wheel assembly.

21. Install the skid plate.

22. Lower the vehicle.

23. Unlock the steering wheel.

24. Fill the power steering fluid.

25. Reset the toe and center the steering wheel.

4-Wheel Drive

1. Siphon the power steering fluid from the power steering reservoir.

➡ **The steering column on vehicles with an automatic transmission may not be equipped with an internal locking shaft that allows the ignition key cylinder to be locked with the key. Alternative methods of locking the steering wheel for service will have to be used.**

2. Lock the steering wheel to prevent spinning of the clockspring.

3. Raise and support the vehicle.

4. Remove the skid plate from under the front end to gain access to the gear.

5. Remove the front tire and wheel assemblies.

➡ **Mark the alignment adjusting cams for easier installation.**

6. Remove the lower control arms.

7. Remove the front axle.

8. Remove the tie rod end nuts.

9. Separate tie rod ends from the knuckles.

10. Remove the intermediate shaft lower coupler bolt and slide the coupler off the gear).

11. Remove power steering pressure hose bracket.

12. Remove the power steering lines from the gear.

13. Remove the mounting bolts from the gear to the front cradle.

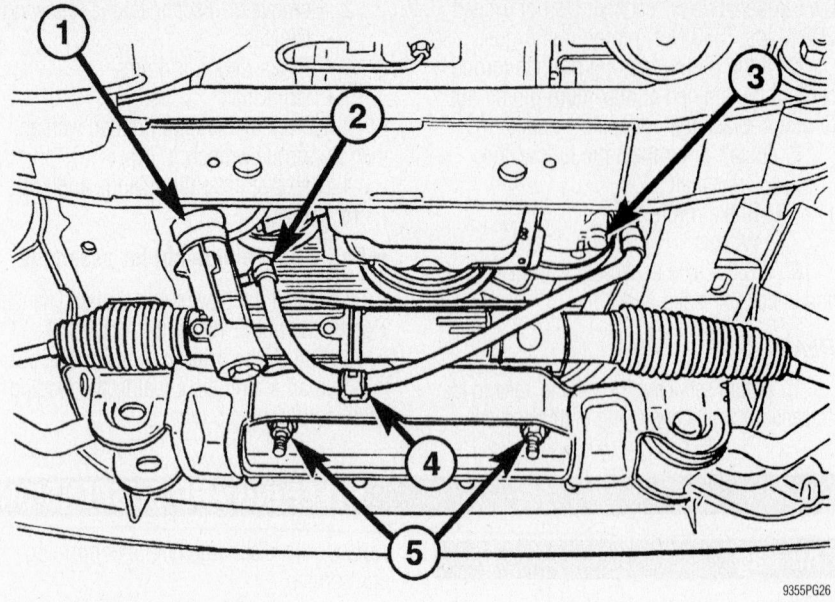

Rack mounting points—Liberty

14. Remove the steering gear from the vehicle.

To install:

15. Transfer the tie rod ends to the new steering gear (if needed).

16. Install the steering gear to the vehicle.

17. Install the gear mounting bolts to the front cradle. Tighten the gear mounting bolts to 162 Nm (120 ft. lbs.)

18. Install the power steering lines to the gear.

19. Install the power steering pressure hose bracket.

20. Install the lower coupler bolt and slide the coupler on to the gear.

21. Install the tie rod end to the knuckle and tighten the nuts.

22. Install the front axle.

23. Install the lower control arms.

24. Install the tire and wheel assembly.

25. Install the skid plate.

26. Lower the vehicle.

27. Unlock the steering wheel.

28. Fill the power steering fluid.

29. Reset the toe and center the steering wheel.

Front Struts

REMOVAL & INSTALLATION

Liberty

LEFT SIDE

1. Before servicing the vehicle, refer to the precautions in the beginning of this section.

2. Remove or disconnect the following:
- Negative battery cable
- Battery
- Power center
- Battery tray
- Battery temperature sensor
- The 4 upper strut nuts
- Left wheel
- Lower clevis bracket bolt
- Stabilizer link
- Lower ball joint from the arm
- Clevis bracket
- Strut

3. Installation is the reverse of removal. Observe the following torques:
- Upper strut nuts: 80 ft. lbs. (108 Nm)
- Clevis bracket-to-strut bolt: 65 ft. lbs. (88 Nm)
- Ball joint nut: 60 ft. lbs. (81 Nm)
- Clevis bracket-to-arm bolt: 110 ft. lbs. (150 Nm)
- Stabilizer link-to-arm bolt: 100 ft. lbs. (136 Nm)

RIGHT SIDE

1. Before servicing the vehicle, refer to the precautions in the beginning of this section.

2. Remove or disconnect the following:
- Negative battery cable
- Air box
- Cruise control servo mounting nuts
- The 4 upper strut nuts
- Right wheel
- Lower clevis bracket bolt
- Stabilizer link
- Lower ball joint from the arm
- Clevis bracket

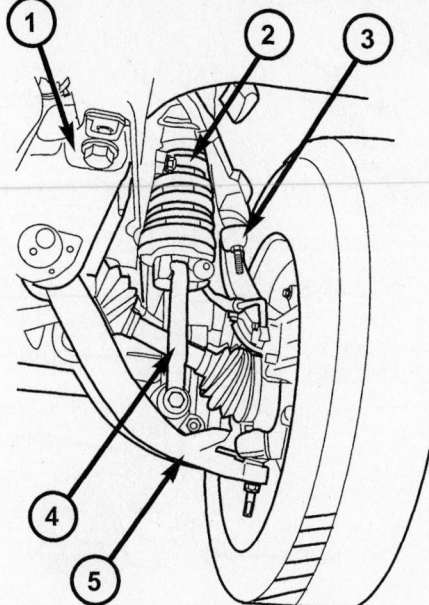

1 - FRONT CRADLE
2 - SPRING & SHOCK ASSEMBLY
3 - STEERING KNUCKLE
4 - CLEVIS BRACKET
5 - LOWER CONTROL ARM

9355PG27

Strut and clevis assembly—Liberty

- Strut

3. Installation is the reverse of removal. Observe the following torques:
- Upper strut nuts: 80 ft. lbs. (108 Nm)
- Clevis bracket-to-strut bolt: 65 ft. lbs. (88 Nm)
- Ball joint nut: 60 ft. lbs. (81 Nm)
- Clevis bracket-to-arm bolt: 110 ft. lbs. (150 Nm)
- Stabilizer link-to-arm bolt: 100 ft. lbs. (136 Nm)

Shock Absorber

REMOVAL & INSTALLATION

Front

1. Before servicing the vehicle, refer to the precautions in the beginning of this section.
2. Remove or disconnect the following:
- Upper nut, washer and grommet from the upper stud
- Lower fasteners
- Shock absorber

To install:
3. Install or connect the following:
- Shock absorber. Torque the lower fasteners to 17 ft. lbs. (23 Nm).

- Upper grommet, washer, and nut to the stud. Torque it to 16 ft. lbs. (22 Nm).

Rear

1. Before servicing the vehicle, refer to the precautions in the beginning of this section.
2. Remove or disconnect the following:
- Upper locknut and washer from the frame bracket stud, on the Grand Cherokee and Liberty
- Upper mounting bolts, on the Wrangler and Cherokee
- Lower bolt, nut and washers from the axle shaft tube bracket
- Shock absorber

To install:
3. Place the shock absorber upper end in position and tighten the fasteners to the following specifications:
- Wrangler: 23 ft. lbs. (31 Nm).
- Cherokee: 17 ft. lbs. (23 Nm).
- Liberty and Grand Cherokee: 80 ft. lbs. (108 Nm).
4. Place the shock absorber lower end in position and tighten the fasteners to the following specifications:
- Wrangler: 74 ft. lbs. (100 Nm).

- Cherokee: 46 ft. lbs. (62 Nm).
- Liberty and Grand Cherokee: 85 ft. lbs. (115 Nm).

Coil Spring

REMOVAL & INSTALLATION

Front

EXCEPT LIBERTY

1. Before servicing the vehicle, refer to the precautions in the beginning of this section.
2. Remove or disconnect the following:
- Front wheels
- Front driveshaft, if equipped
- Lower suspension arm
- Stabilizer bar links
- Track bar
- Drag link
- Brake hose brackets
- Spring retainers
- Coil spring

To install:
3. Install or connect the following:
- Coil spring
- Spring retainers. Torque the bolt to 16 ft. lbs. (21 Nm).
- Stabilizer bar links. Torque the bolts to 70 ft. lbs. (95 Nm).
- Track bar. Torque the bolt to 35 ft. lbs. (47 Nm).
- Front propeller shaft to the axle
- Drag link
- Lower suspension arm. Torque the bolt to 133 ft. lbs. (180 Nm).
- Front driveshaft, if equipped
- Front wheels

LIBERTY

1. Before servicing the vehicle, refer to the precautions in the beginning of this section.
2. Remove the strut and place it in a Pentastar W7200 spring compressor, or equivalent.
3. Compress the spring, remove the strut mount nut and remove the strut from the compressor.
4. Installation is the reverse of removal. Torque the nut to 30 ft. lbs. (41 Nm).

Rear

EXCEPT LIBERTY

1. Before servicing the vehicle, refer to the precautions in the beginning of this section.
2. Remove or disconnect the following:
- Rear wheels

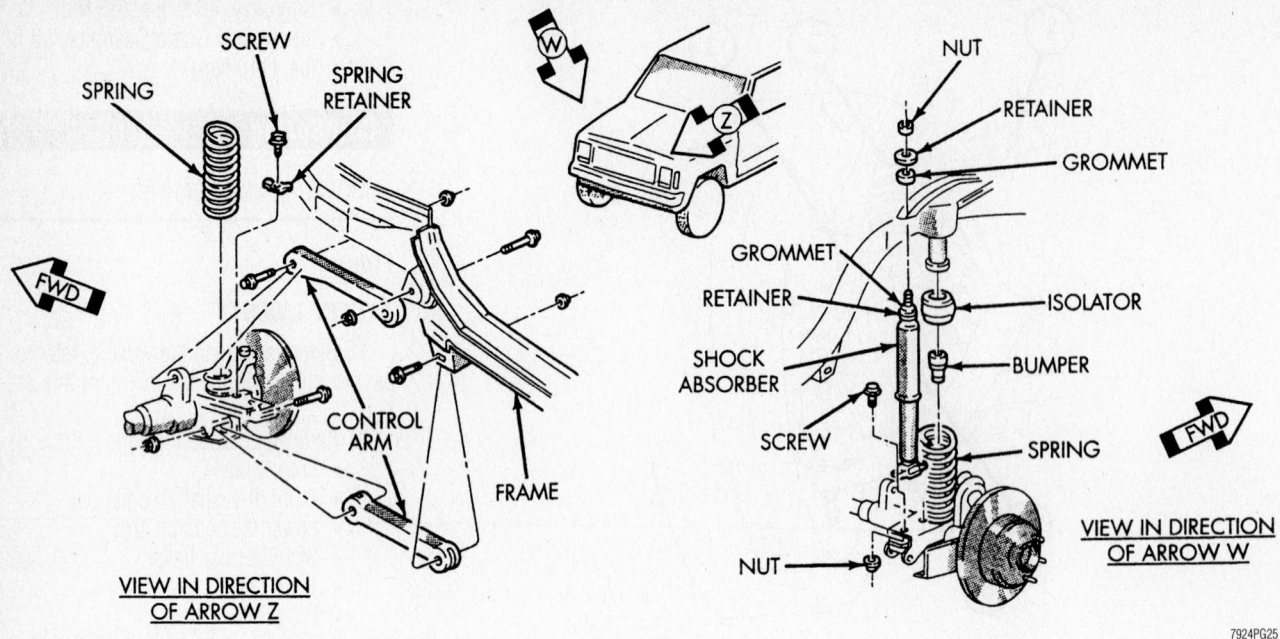

Exploded view of the front suspension—Cherokee

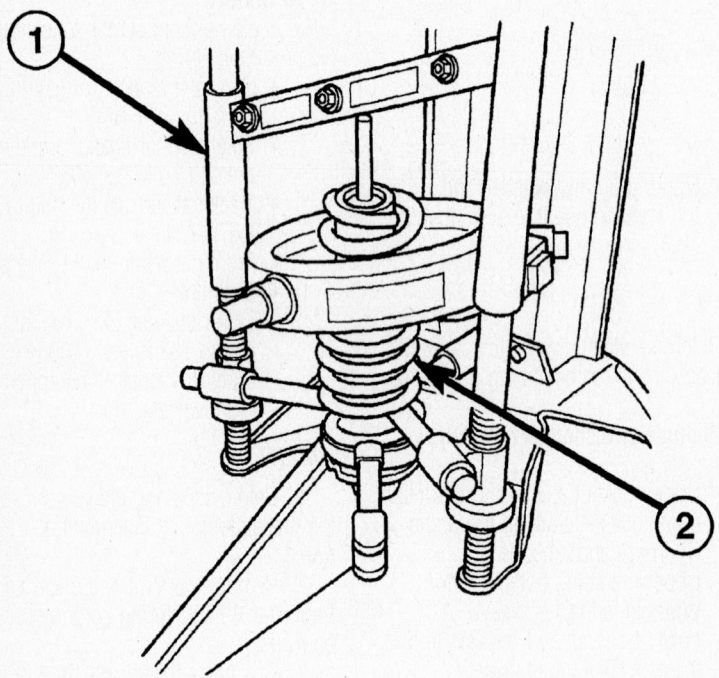

1 - SPRING COMPRESSOR
2 - SPRING

Front coil spring removal—Liberty

- Stabilizer bar links
- Shock absorbers
- Track bar
- Spring retainers
- Coil springs

To install:

3. Install or connect the following:
- Coil springs
- Spring retainers
- Track bar

- Shock absorbers
- Stabilizer bar links
- Rear wheels

LIBERTY

1. Before servicing the vehicle, refer to the precautions in the beginning of this section.
2. Support the axle with a jack.
3. Remove or disconnect the following:
- Rear wheels
- Shock absorbers from the axle
4. Lower the axle and tilt it to remove the springs.
5. Installation is the reverse of removal.

Leaf Springs

REMOVAL & INSTALLATION

Cherokee

1. Before servicing the vehicle, refer to the precautions in the beginning of this section.
2. Support the vehicle at the frame rails.
3. Support the rear axle with a jack.
4. Remove or disconnect the following:
- Rear wheel
- Stabilizer bar link
- Axle U-bolts
- Spring bracket
- Leaf spring

To install:

➡**The weight of the vehicle must be supported by the springs when the spring eye and stabilizer bar fasteners are tightened.**

5. Install or connect the following:
- Leaf spring
- Spring bracket
- Axle U-bolts. Torque the nuts to 52 ft. lbs. (70 Nm).
- Stabilizer bar link
- Rear wheel

6. Torque the front spring eye bolt and nut to 115 ft. lbs. (156 Nm).

7. Torque the rear spring eye bolt and nut to 80 ft. lbs. (108 Nm).

8. Torque the stabilizer bar nuts 55 ft. lbs. (74 Nm).

Upper Ball Joint

REMOVAL & INSTALLATION

Except Liberty

1. Before servicing the vehicle, refer to the precautions in the beginning of this section.
2. Remove or disconnect the following:
- Front wheel
- Disc brake caliper and rotor
- Wheel bearing
- Axle shaft
- Steering knuckle
- Upper ball joint

To install:

3. Install or connect the following:
- Upper ball joint
- Steering knuckle. Torque the nuts to 100 ft. lbs. (135 Nm).
- Axle shaft
- Wheel bearing
- Disc brake caliper and rotor
- Front wheel

Liberty

The upper ball joint is serviced as an assembly with the control arm.

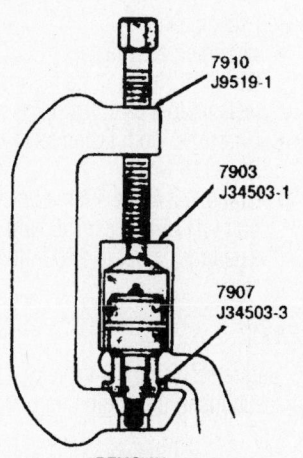

Upper ball joint removal and installation

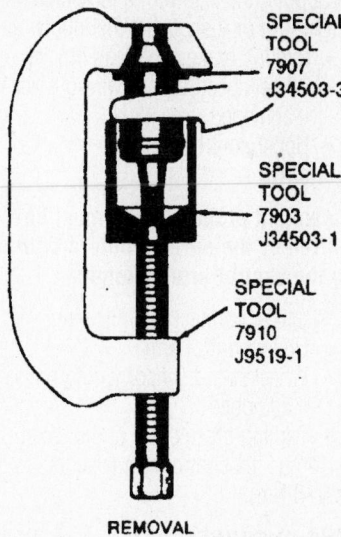
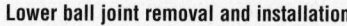

REMOVAL

Lower ball joint removal and installation

Lower Ball Joint

REMOVAL & INSTALLATION

Except Liberty

1. Before servicing the vehicle, refer to the precautions in the beginning of this section.
2. Remove or disconnect the following:
- Front wheel
- Disc brake caliper and rotor
- Wheel bearing
- Axle shaft
- Steering knuckle
- Lower ball joint

To install:

3. Install or connect the following:
- Lower ball joint

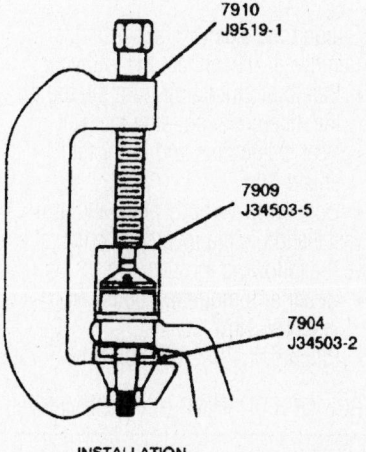

INSTALLATION

7924PG26

INSTALLATION

7924PG27

- Steering knuckle. Torque the nuts to 100 ft. lbs. (135 Nm).
- Axle shaft
- Wheel bearing
- Disc brake caliper and rotor
- Front wheel

Liberty

1. Before servicing the vehicle, refer to the precautions in the beginning of this section.
2. Remove or disconnect the following:
- Front wheel
- Lower clevis bracket bolt from the control arm
- Stabilizer link at the control arm
- Lower ball joint nut
- Control arm from lower ball joint with tool C4150A

3. Installation is the reverse of removal. Torque the ball joint nut to 60 ft. lbs. (81 Nm); the stabilizer link bolt to 100 ft. lbs. (136 Nm); the lower clevis bracket bolt to 110 ft. lbs. (150 Nm).

Upper Control Arm

REMOVAL & INSTALLATION

Front

EXCEPT LIBERTY

1. Before servicing the vehicle, refer to the precautions in the beginning of this section.
2. Support the axle with a jackstand.
3. Unbolt and remove the upper control arm.

To install:

➡The weight of the vehicle must be supported by the springs before tightening the control arm fasteners.

4. Install the control arms.
5. Torque the axle fastener to 55 ft. lbs. (75 Nm) and the frame fastener to 66 ft. lbs. (90 Nm).

LIBERTY-RIGHT SIDE

1. Before servicing the vehicle, refer to the precautions at the beginning of this section.
2. Remove or disconnect the following:
   • Wheel
   • Upper ball joint nut
   • Upper Ball joint from the knuckle
   • Air box
   • Cruise control servo mounting nuts
   • Upper arm rear bolt
   • Upper arm front bolt
   • Upper arm
3. Installation is the reverse of removal. Observe the following torques:
   • Front and rear bolts: 90 ft. lbs. (122 Nm)
   • Ball joint stud nut: 60 ft. lbs. (81 Nm)

LIBERTY-LEFT SIDE

1. Before servicing the vehicle, refer to the precautions at the beginning of this section.
2. Remove or disconnect the following:
   • Wheel
   • Ball joint nut
   • Ball joint from the knuckle
   • Battery
   • Power center
   • Battery tray
   • Battery temperature sensor
   • Control arm rear bolt, by using a ratchet and extension under the steering shaft, positioned by the power steering reservoir
   • Control arm front bolt
   • Control arm
3. Installation is the reverse of removal. Observe the following torques:
   • Front and rear bolts: 90 ft. lbs. (122 Nm)
   • Ball joint stud nut: 60 ft. lbs. (81 Nm)

Rear

EXCEPT LIBERTY AND GRAND CHEROKEE

1. Before servicing the vehicle, refer to the precautions in the beginning of this section.

2. Support the axle with a jackstand.
3. Remove or disconnect the following:
   • Parking brake cable and bracket
   • Wheel speed sensor wiring bracket, if equipped
   • Upper control arm

To install:

➡The weight of the vehicle must be supported by the springs before tightening the control arm fasteners.

4. Install or connect the following:
   • Upper control arm
   • Wheel speed sensor wiring bracket, if equipped
   • Parking brake cable and bracket
5. Torque the control arm fasteners to 55 ft. lbs. (75 Nm).

GRAND CHEROKEE

1. Before servicing the vehicle, refer to the precautions in the beginning of this section.
2. Support the axle with a jackstand.
3. Remove or disconnect the following:
   • Parking brake cable brackets
   • Brake hose brackets
   • Axle ball joint
   • Upper control arm

To install:

➡Use a new axle ball joint nut.

4. Install or connect the following:
   • Upper control arm. Torque the frame bracket bolts to 74 ft. lbs. (100 Nm) and the axle ball joint nut to 105 ft. lbs. (142 Nm).
   • Brake hose brackets
   • Parking brake cable brackets

LIBERTY

1. Before servicing the vehicle, refer to the precautions in the beginning of this section.
2. Support the axle with a jackstand.
3. Remove or disconnect the following:
   • Ball joint pinch bolt from the top of the differential housing bracket.
   • Heat shield nuts and lower the shield
   • Upper arm-to-body bolts and arm
4. Installation is the reverse of removal. Observe the following torques:
   • Upper arm mounting bolts: 74 ft. lbs. (100 Nm)
   • Pinch bolt: 70 ft. lbs. (95 Nm)

CONTROL ARM BUSHING REPLACEMENT

The upper control arm bushings are serviced with the control arms as complete

assemblies, with the exception of the front upper axle bushing, which may be replaced after removing the upper control arm.

Front Upper Control Arm Axle Bushing

1. Before servicing the vehicle, refer to the precautions in the beginning of this section.
2. Remove the upper control arm.
3. Press the old bushing out of the axle housing.
4. Press the new bushing into the axle housing.
5. Install the upper control arm.

Lower Control Arms

REMOVAL & INSTALLATION

Front

EXCEPT LIBERTY

1. Before servicing the vehicle, refer to the precautions in the beginning of this section.
2. Support the axle with a jackstand.
3. Remove or disconnect the following:
   • Wheel speed sensor wiring, if equipped
   • Lower control arm

To install:

➡The weight of the vehicle must be supported by the springs before tightening the control arm fasteners.

4. Install or connect the following:
   • Lower control arm
   • Wheel speed sensor wiring, if equipped
5. Tighten the control arm bolts to the following specifications:
   • Wrangler: Axle fastener to 85 ft. lbs. (115 Nm) and frame bracket fastener to 130 ft. lbs. (176 Nm).
   • Cherokee: Both fasteners to 85 ft. lbs. (115 Nm).
   • Grand Cherokee: Frame bracket bolt to 115 ft. lbs. (156 Nm) and axle bracket nut to 120 ft. lbs. (163 Nm).

LIBERTY

1. Before servicing the vehicle, refer to the precautions in the beginning of this section.
2. Remove or disconnect the following:
   • Front wheel

- Lower clevis bracket bolt from the control arm
- Stabilizer link at the control arm
- Lower ball joint nut
- Control arm from lower ball joint with tool C4150A

➡ **Matchmark the front and rear control arm pivot bolts.**

- Front pivot bolt
- Rear pivot bolt
- Control arm

To install:

3. Install or connect the following:
- Lower control arm
- Rear pivot bolt
- Front pivot bolt
- Ball joint nut. Torque the ball joint nut to 60 ft. lbs. (81 Nm)

4. Align the matchmarks and tighten the pivot bolts to 125 ft. lbs. (170 Nm).

5. The remainder of the installation is the reverse of removal. Torque the stabilizer link bolt to 100 ft. lbs. (136 Nm); the lower clevis bracket bolt to 110 ft. lbs. (150 Nm).

Rear

1. Before servicing the vehicle, refer to the precautions in the beginning of this section.

2. Support the axle with a jackstand.

3. On the Liberty, disconnect the stabilizer bar from the arm.

4. Unbolt and remove the lower control arm.

➡ **On the Liberty's right arm, it will be necessary to pry the exhaust pipe slightly to get to the frame rail-to-arm bolt.**

To install:

➡ **The weight of the vehicle must be supported by the springs before tightening the control arm fasteners.**

5. Install the lower control arm.

6. Tighten the lower control arm fasteners to the following specifications:
- Wrangler: Both fasteners to 130 ft. lbs. (177 Nm)
- Liberty: Both fasteners to 120 ft. lbs. (163 Nm)
- Grand Cherokee: Frame bracket nut to 115 ft. lbs. (156 Nm) and axle bracket nut to 120 ft. lbs. (163 Nm)

CONTROL ARM BUSHING REPLACEMENT

The lower control arm bushings are serviced with the control arms as complete assemblies.

Front Wheel Bearings

ADJUSTMENT

2-Wheel Drive

EXCEPT LIBERTY

1. Before servicing the vehicle, refer to the precautions in the beginning of this section.

2. Remove or disconnect the following:
- Front wheel, grease cap, cotter pin and nut cap
- Wheel bearing nut, loosen it

3. While turning the rotor, tighten the nut to 25 ft. lbs. (34 Nm) to seat the bearings.

4. Back the nut off ½ turn.

5. While turning the rotor, tighten the nut to 19 inch lbs. (2 Nm).

6. Install or connect the following:
- Nut cap, new cotter pin and the grease cap
- Wheel

4-WHEEL DRIVE AND ALL LIBERTY MODELS

The front wheel bearings are not adjustable.

REMOVAL & INSTALLATION

2WD Models

EXCEPT LIBERTY

1. Before servicing the vehicle, refer to the precautions in the beginning of this section.

2. Remove or disconnect the following:
- Front wheel
- Brake caliper
- Grease cap
- Split pin
- Nut retainer
- Nut and washer
- Outer bearing
- Brake rotor and hub assembly
- Grease seal
- Inner bearing

To install:

➡ **Use new grease seals and split pins.**

3. Pack the wheel bearings and the inside of the hub with high temperature wheel bearing grease. Add grease to the hub until it is flush with the inside diameter of the bearing cup.

4. Install or connect the following:
- Inner bearing
- Grease seal
- Brake rotor and hub assembly
- Outer bearing
- Nut and washer
- Nut retainer
- Split pin
- Grease cap
- Brake caliper
- Front wheel

4WD MODELS AND ALL LIBERTY MODELS

1. Before servicing the vehicle, refer to the precautions in the beginning of this section.

2. Remove or disconnect the following:
- Front wheel
- Brake caliper and rotor
- Wheel speed sensor and bracket
- Axle stub shaft nut (except 2-WD Liberty)
- Hub and wheel bearing assembly

To install:

3. Install or connect the following:
- Hub assembly over the axle stub shaft. Tighten the hub bolts to 75 ft. lbs. (102 Nm) except Liberty; 96 ft. lbs. (130 Nm) for Liberty models
- Axle stub shaft nut. Torque the nut to 175 ft. lbs. (237 Nm) except Liberty; 100 ft. lbs. (136 Nm) for 4-WD Liberty models
- Brake caliper and rotor
- Front wheel

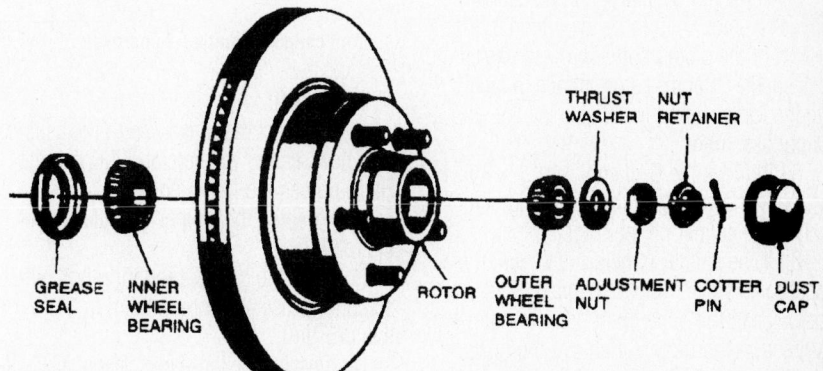

GREASE SEAL   INNER WHEEL BEARING   ROTOR   OUTER WHEEL BEARING   THRUST WASHER   NUT RETAINER   ADJUSTMENT NUT   COTTER PIN   DUST CAP

7924PG28

Exploded view of the front wheel bearings—2WD models

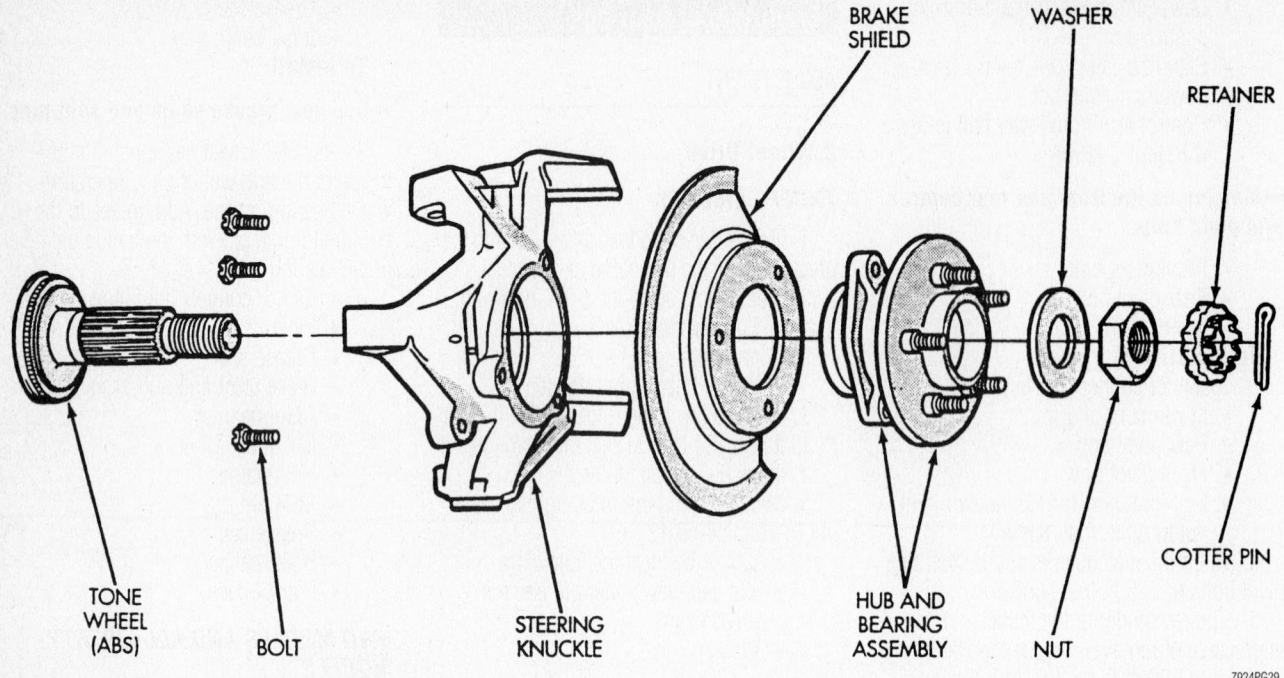

Exploded view of the hub assembly—4WD models

BRAKES

Brake Caliper

REMOVAL & INSTALLATION

Cherokee and Wrangler Front

1. Drain ⅔ of the brake fluid from the front reservoir. Use the bleeder screw at the front outlet port to drain the fluid. If equipped with anti-lock brakes, relieve the system pressure.

2. Raise and safely support the vehicle.

3. Remove the wheels.

4. Place a C-clamp on the caliper so the solid end contacts the back of the caliper and screw end contacts the metal part of the outboard brake pad.

5. Tighten the clamp until the caliper moves far enough to force the piston to the bottom of the piston bore. This will back the brake pads off of the rotor surface to facilitate the removal and installation of the caliper assembly.

6. Remove the C-clamp.

7. Remove both of the mounting bolts and lift the caliper off the rotor.

8. If the caliper is being removed, it is necessary to disconnect the brake fluid hose. Clean the brake fluid hose-to-caliper connection thoroughly. Remove the hose-to-caliper bolt. Cap or tape the open ends to keep dirt out. Discard the copper gaskets.

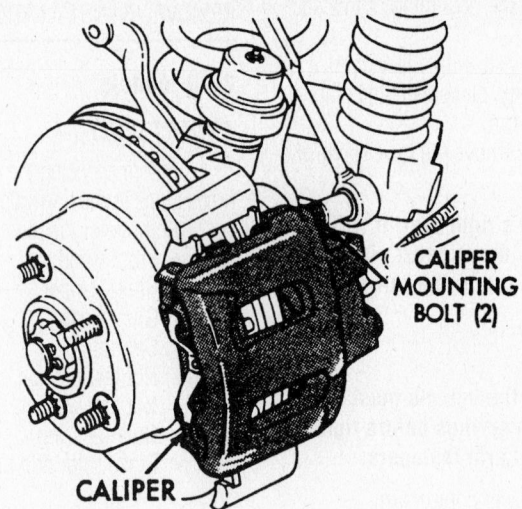

Front caliper mounting—Cherokee

To install:

9. Connect the brake line to the caliper with new sealing washers and fitting bolt. Hand-tighten the fitting bolt.

10. Position the caliper into place over the rotor.

11. Coat the caliper mounting bolt with silicone grease and torque them to 11 ft. lbs. (15 Nm).

12. Position the brake line clear of all chassis components, untwisted and free of kinks. Torque the fitting bolt to 23 ft. lbs. (31 Nm).

13. Install the wheels.

14. Fill the master cylinder with fluid and bleed the brake system.

15. Before driving the vehicle, pump the brakes several times to seat the pads.

Wrangler Rear

1. Drain ⅔ of the brake fluid from the front reservoir. If equipped with anti-lock brakes, relieve the system pressure.

2. Raise and safely support the vehicle.

3. Remove the wheels.

4. Insert a small prybar through the caliper opening and pry the caliper (using the outboard brake pad) to bottom the piston in the caliper bore.

➡️**This will back the brake pads off of the rotor surface to facilitate the removal and installation of the caliper assembly.**

5. Remove the brake hose-to-caliper bolt, hose and washers.

6. Remove both caliper slide pin bushing caps and slide pins.

7. Lift the caliper from the anchor.

To install:

8. Position the caliper into place on the anchor.

9. Coat the caliper slide pins with silicone grease and torque them to 11 ft. lbs. (15 Nm). Install the slide pin bushing caps.

10. Using new gasket washers, install the brake line and torque the fitting bolt to 23 ft. lbs. (31 Nm).

11. Fill the master cylinder with fluid and bleed the brake system.

12. Before driving the vehicle, pump the brakes several times to seat the pads.

13. Install the wheels.

2001–2002 Grand Cherokee

FRONT

1. Drain ⅔ of the brake fluid from the front reservoir. Use the bleeder screw at the front outlet port to drain the fluid. If equipped with anti-lock brakes, relieve the system pressure.

2. Raise and safely support the vehicle.

3. Remove the wheels.

4. Insert a small prybar through the caliper opening and pry the caliper (using the outboard brake pad) to bottom the pistons in the caliper bore.

➡️**This will back the brake pads off of the rotor surface to facilitate the removal and installation of the caliper assembly.**

5. Remove the brake hose-to-caliper bolt, hose and washers.

6. Pry the caliper support spring out of the caliper.

7. Remove both caliper slide pin bushing caps and slide pins.

8. Lift the caliper from the anchor.

To install:

9. Position the caliper into place on the anchor.

10. Coat the caliper slide pins with silicone grease and torque them to 21–30 ft. lbs. (29–41 Nm). Install the slide pin bushing caps.

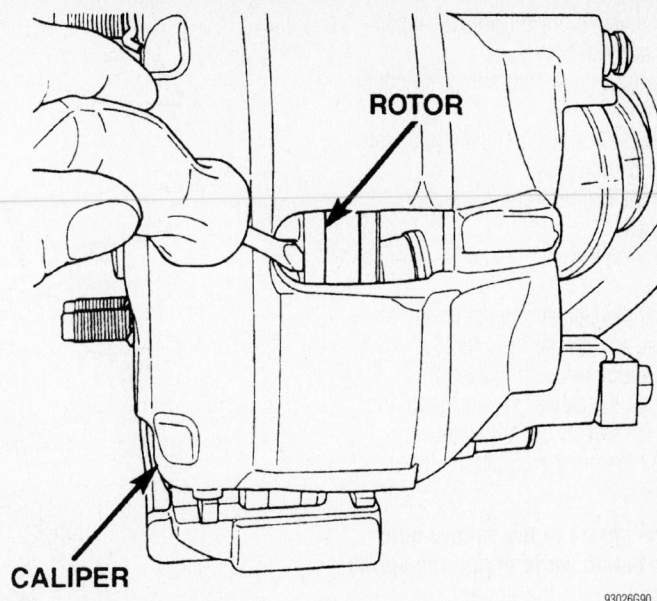

Bottoming the pistons in the front caliper—Grand Cherokee

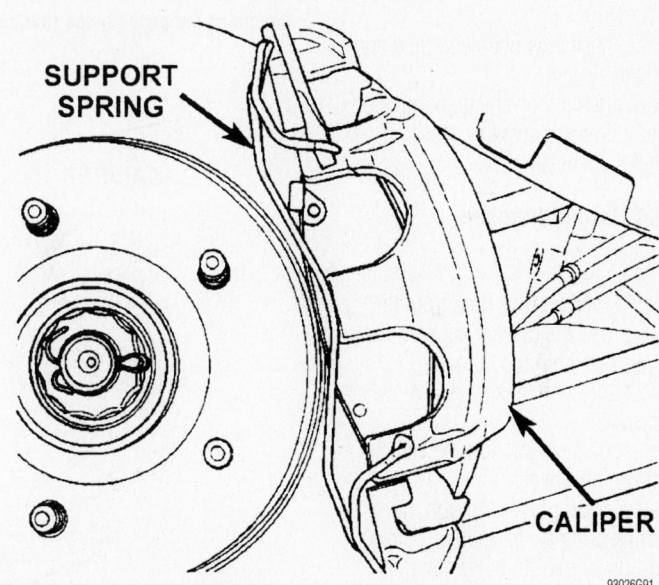

View of the support springs on the front caliper—Grand Cherokee

11. Install the caliper support spring in the top of the caliper under the anchor; then, install the other end into the lower caliper hole.

➡️**Hold the spring in the caliper hole with your thumb while prying the spring end out and under the anchor.**

12. Using new gasket washers, install the brake line and torque the fitting bolt to 23 ft. lbs. (31 Nm).

13. Fill the master cylinder with fluid and bleed the brake system.

14. Before driving the vehicle, pump the brakes several times to seat the pads.

15. Install the wheels.

REAR

1. Drain ⅔ of the brake fluid from the front reservoir. Use the bleeder screw at the front outlet port to drain the fluid. If equipped with anti-lock brakes, relieve the system pressure.

2. Raise and safely support the vehicle.

3. Remove the wheels.

4. Insert a small prybar through the caliper opening and pry the caliper (using the outboard brake pad) to bottom the piston in the caliper bore.

➡️**This will back the brake pads off of the rotor surface to facilitate the removal and installation of the caliper assembly.**

5. Remove the brake hose-to-caliper bolt, hose and washers.

6. Pry the caliper support spring out of the caliper.

7. Remove both caliper slide pin bushing caps and slide pins.

8. Lift the caliper from the anchor.

To install:

9. Position the caliper into place on the anchor.

10. Coat the caliper slide pins with silicone grease and torque them to 26 ft. lbs. Install the slide pin bushing caps.

11. Install the caliper support spring in the top of the caliper under the anchor; then, install the other end into the lower caliper hole.

➡**Hold the spring in the caliper hole with your thumb while prying the spring end out and under the anchor.**

12. Using new gasket washers, install the brake line and torque the fitting bolt to 23 ft. lbs. (31 Nm).

13. Fill the master cylinder with fluid and bleed the brake system.

14. Before driving the vehicle, pump the brakes several times to seat the pads.

15. Install the wheels.

2003–2004 Grand Cherokee

FRONT

1. Drain ⅔ of the brake fluid from the front reservoir. Use the bleeder screw at the front outlet port to drain the fluid. If equipped with anti-lock brakes, relieve the system pressure.

2. Raise and safely support the vehicle.

3. Remove the wheels.

4. Insert a small prybar through the caliper opening and pry the caliper (using the outboard brake pad) to bottom the pistons in the caliper bore.

➡**This will back the brake pads off of the rotor surface to facilitate the removal and installation of the caliper assembly.**

5. Remove the brake hose-to-caliper bolt, hose and washers.

6. Remove both caliper slide pin bushing caps and slide pins.

7. Lift the caliper from the anchor.

To install:

8. Position the caliper into place on the anchor.

9. Coat the caliper slide pins with silicone grease and torque them to 53 ft. lbs. (72 Nm). Install the slide pin bushing caps.

10. Using new gasket washers, install the brake line and torque the fitting bolt to 23 ft. lbs. (31 Nm).

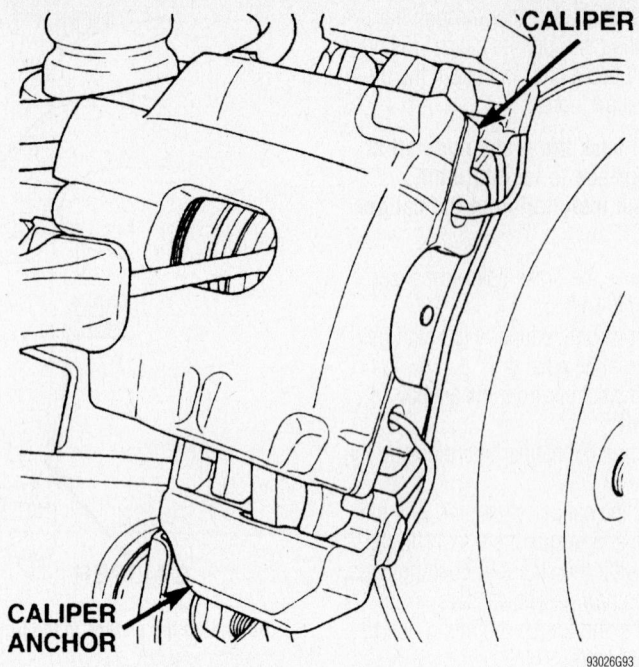

Bottoming the piston in the rear caliper—Grand Cherokee

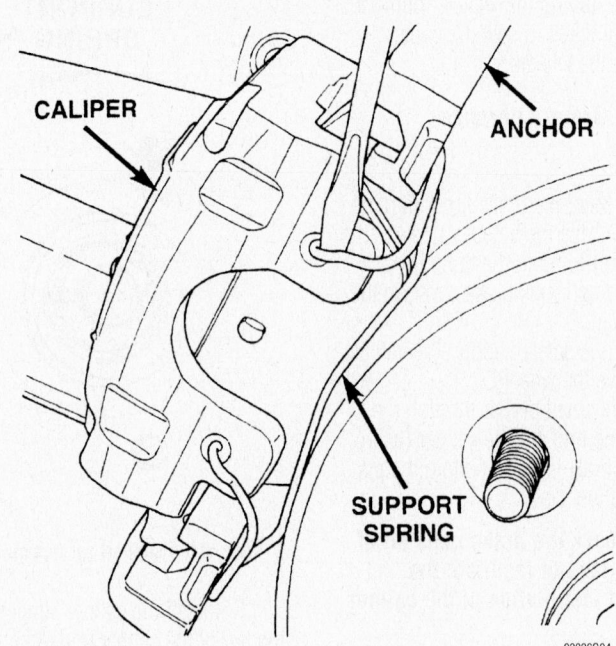

View of the support springs on the rear caliper—Grand Cherokee

11. Fill the master cylinder with fluid and bleed the brake system.

12. Before driving the vehicle, pump the brakes several times to seat the pads.

13. Install the wheels.

REAR

1. Drain ⅔ of the brake fluid from the front reservoir. Use the bleeder screw at the front outlet port to drain the fluid. If equipped with anti-lock brakes, relieve the system pressure.

2. Raise and safely support the vehicle.

3. Remove the wheels.

4. Insert a small prybar through the caliper opening and pry the caliper (using the outboard brake pad) to bottom the piston in the caliper bore.

➡**This will back the brake pads off of the rotor surface to facilitate the removal and installation of the caliper assembly.**

5. Remove the brake hose-to-caliper bolt, hose and washers.

6. Pry the caliper support spring out of the caliper.

7. Remove both caliper slide pin bushing caps and slide pins.

8. Lift the caliper from the anchor.

To install:

9. Position the caliper into place on the anchor.

10. Coat the caliper slide pins with silicone grease and torque them to 26 ft. lbs. (35 Nm). Install the slide pin bushing caps.

11. Install the caliper support spring in the top of the caliper under the anchor; then, install the other end into the lower caliper hole.

➡ **Hold the spring in the caliper hole with your thumb while prying the spring end out and under the anchor.**

12. Using new gasket washers, install the brake line and torque the fitting bolt to 23 ft. lbs. (31 Nm).

13. Fill the master cylinder with fluid and bleed the brake system.

14. Before driving the vehicle, pump the brakes several times to seat the pads.

15. Install the wheels.

Liberty

FRONT

1. Drain ⅔ of the brake fluid from the front reservoir. Use the bleeder screw at the front outlet port to drain the fluid. If equipped with anti-lock brakes, relieve the system pressure.

2. Raise and safely support the vehicle.

3. Remove the wheels.

4. Remove both of the mounting bolts and lift the caliper off the rotor.

5. If the caliper is being removed, it is necessary to disconnect the brake fluid hose. Clean the brake fluid hose-to-caliper connection thoroughly. Remove the hose-to-caliper bolt. Cap or tape the open ends to keep dirt out. Discard the copper gaskets.

To install:

6. Connect the brake line to the caliper with new sealing washers and fitting bolt. Hand-tighten the fitting bolt.

7. Position the caliper into place over the rotor.

8. Coat the caliper mounting bolt with silicone grease and torque them to 11 ft. lbs. (15 Nm).

9. Position the brake line clear of all chassis components, untwisted and free of kinks. Torque the fitting bolt to 23 ft. lbs. (31 Nm).

10. Install the wheels.

11. Fill the master cylinder with fluid and bleed the brake system.

12. Before driving the vehicle, pump the brakes several times to seat the pads.

REAR

1. Drain ⅔ of the brake fluid from the front reservoir. If equipped with anti-lock brakes, relieve the system pressure.

2. Raise and safely support the vehicle.

3. Remove the wheels.

4. Remove the brake hose-to-caliper bolt, hose and washers.

5. Remove both caliper slide pin bushing caps and slide pins.

6. Lift the caliper from the anchor.

To install:

7. Position the caliper into place on the anchor.

8. Coat the caliper slide pins with silicone grease and torque them to 18 ft. lbs. Install the slide pin bushing caps.

9. Using new gasket washers, install the brake line and torque the fitting bolt to 23 ft. lbs. (31 Nm).

10. Fill the master cylinder with fluid and bleed the brake system.

11. Before driving the vehicle, pump the brakes several times to seat the pads.

12. Install the wheels.

Disc Brake Pads

REMOVAL & INSTALLATION

Cherokee

1. Raise and safely support the vehicle.

2. Drain ⅔ of the brake fluid from the front reservoir. Use the bleeder screw at the front outlet port to drain the fluid.

3. Raise and support the vehicle safely.

4. Remove the wheels.

5. Remove the brake caliper. Use a suitable tool to compress the caliper piston into the bore.

6. Hold the anti-rattle clip against the caliper anchor plate and remove the outboard brake pad.

7. Remove the inboard pad and its anti-rattle clip.

To install:

8. Clean all the mounting holes and bushing grooves in the caliper ears. Clean the mounting bolts. Replace the bolts if they are corroded or if the threads are damaged. Wipe the inside of the caliper clean, including the exterior of the dust boot. Inspect the dust boot for cuts or cracks and for proper

seating in the piston bore. If evidence of fluid leakage is noted, the caliper should be rebuilt.

➡ **Do not use abrasives on the bolts. This will destroy their protective plating.**

9. Install the inboard anti-rattle clip on the trailing end of the anchor plate. The split end of the clip must face away from the rotor.

10. Install the inboard pad in the caliper. The pad must lay flat against the piston.

11. Install the outboard pad in the caliper while holding the anti-rattle clip.

12. With the pads installed, position the caliper over the rotor. Line up the mounting holes in the caliper and the support bracket and insert the mounting bolts. Make sure the bolts pass under the retaining ears on the inboard shoes. Push the bolts through until they engage the holes of the outboard pad and caliper ears. Thread the bolts into the support bracket and tighten them to 11 ft. lbs. (15 Nm).

13. Fill the master cylinder with brake fluid and pump the brake pedal to seat the pads.

14. Install the wheel assembly and lower the vehicle. Check the level of the brake fluid in the master cylinder and fill as necessary. Test the operation of the brakes before taking the vehicle onto the road.

2001–02 Grand Cherokee

FRONT

1. Drain ⅔ of the brake fluid from the front reservoir. Use the bleeder screw at the front outlet port to drain the fluid. If equipped with anti-lock brakes, relieve the system pressure.

2. Raise and safely support the vehicle.

3. Remove the wheels.

4. Insert a small prybar through the caliper opening and pry the caliper (using the outboard brake pad) to bottom the pistons in the caliper bore.

➡ **This will back the brake pads off of the rotor surface to facilitate the removal and installation of the caliper assembly.**

5. Pry the caliper support spring out of the caliper.

6. Remove both caliper slide pin bushing caps and slide pins.

7. Lift the caliper from the anchor.

8. Using a piece of mechanic's wire, support the caliper so there is not tension on the brake hose.

9. Remove the brake pads from the caliper.

To install:

10. Position the brake pads onto the caliper.

11. Position the caliper into place on the anchor.

12. Coat the caliper slide pins with silicone grease and torque them to 21–30 ft. lbs. (29–41 Nm). Install the slide pin bushing caps.

13. Install the caliper support spring in the top of the caliper under the anchor; then, install the other end into the lower caliper hole.

➡**Hold the spring in the caliper hole with your thumb while prying the spring end out and under the anchor.**

14. Fill the master cylinder with fluid and bleed the brake system.

15. Before driving the vehicle, pump the brakes several times to seat the pads.

16. Install the wheels.

REAR

1. Drain ⅔ of the brake fluid from the front reservoir. Use the bleeder screw at the front outlet port to drain the fluid. If equipped with anti-lock brakes, relieve the system pressure.

2. Raise and safely support the vehicle.

3. Remove the wheels.

4. Insert a small prybar through the caliper opening and pry the caliper (using the outboard brake pad) to bottom the piston in the caliper bore.

➡**This will back the brake pads off of the rotor surface to facilitate the removal and installation of the caliper assembly.**

5. Pry the caliper support spring out of the caliper.

6. Remove both caliper slide pin bushing caps and slide pins.

7. Lift the caliper from the anchor.

8. Using a piece of mechanics wire, support the caliper so there is not tension on the brake hose.

9. Remove the brake pads from the caliper.

To install:

10. Position the brake pads onto the caliper.

11. Position the caliper into place on the anchor.

12. Coat the caliper slide pins with silicone grease and torque them to 21–30 ft. lbs. (29–41 Nm). Install the slide pin bushing caps.

13. Install the caliper support spring in the top of the caliper under the anchor;

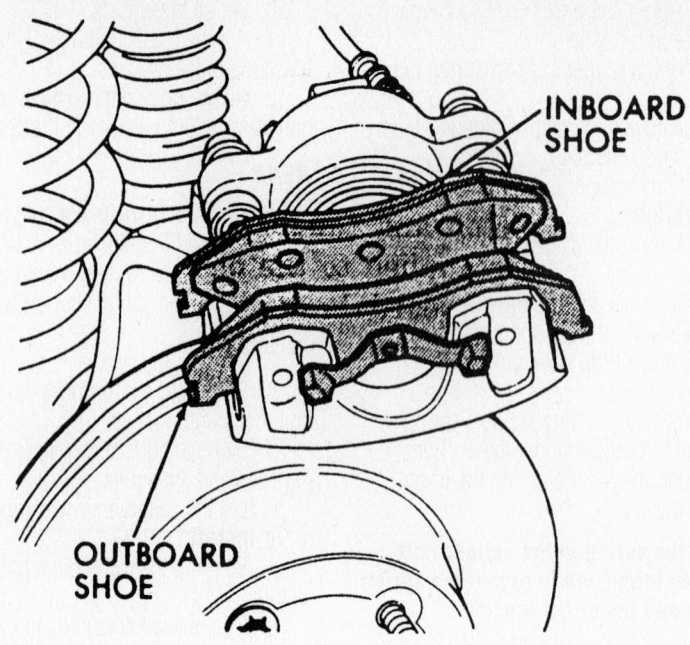

Front disc brake pad installation—Cherokee

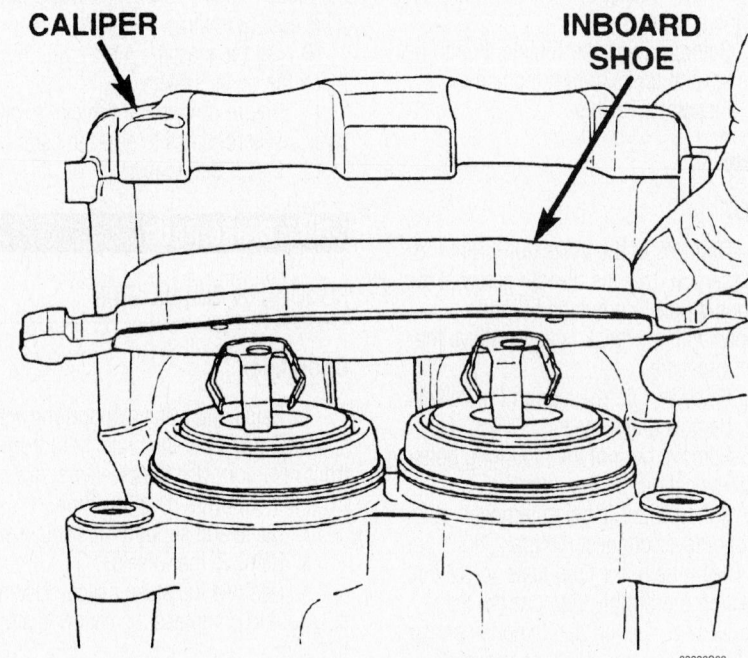

Installing the inward brake pad on the front caliper pistons—Grand Cherokee

then, install the other end into the lower caliper hole.

➡**Hold the spring in the caliper hole with your thumb while prying the spring end out and under the anchor.**

14. Fill the master cylinder with fluid and bleed the brake system.

15. Before driving the vehicle, pump the brakes several times to seat the pads.

16. Install the wheels.

2003–04 Grand Cherokee

FRONT

1. Drain ⅔ of the brake fluid from the front reservoir. Use the bleeder screw at the front outlet port to drain the fluid. If equipped with anti-lock brakes, relieve the system pressure.

2. Raise and safely support the vehicle.

3. Remove the wheels.

4. Insert a small prybar through the

caliper opening and pry the caliper (using the outboard brake pad) to bottom the pistons in the caliper bore.

➡**This will back the brake pads off of the rotor surface to facilitate the removal and installation of the caliper assembly.**

5. Remove both caliper slide pin bushing caps and slide pins.
6. Lift the caliper from the anchor.
7. Using a piece of mechanic's wire, support the caliper so there is not tension on the brake hose.
8. Remove the brake pads from the caliper.

To install:
9. Position the brake pads onto the caliper.
10. Position the caliper into place on the anchor.
11. Coat the caliper slide pins with silicone grease and torque them to 53 ft. lbs. (72 Nm). Install the slide pin bushing caps.

➡**Hold the spring in the caliper hole with your thumb while prying the spring end out and under the anchor.**

12. Fill the master cylinder with fluid and bleed the brake system.
13. Before driving the vehicle, pump the brakes several times to seat the pads.
14. Install the wheels.

REAR

1. Drain ⅔ of the brake fluid from the front reservoir. Use the bleeder screw at the front outlet port to drain the fluid. If equipped with anti-lock brakes, relieve the system pressure.
2. Raise and safely support the vehicle.
3. Remove the wheels.
4. Insert a small prybar through the caliper opening and pry the caliper (using the outboard brake pad) to bottom the piston in the caliper bore.

➡**This will back the brake pads off of the rotor surface to facilitate the removal and installation of the caliper assembly.**

5. Pry the caliper support spring out of the caliper.
6. Remove both caliper slide pin bushing caps and slide pins.
7. Lift the caliper from the anchor.
8. Using a piece of mechanics wire, support the caliper so there is not tension on the brake hose.
9. Remove the brake pads from the caliper.

To install:
10. Position the brake pads onto the caliper.
11. Position the caliper into place on the anchor.
12. Coat the caliper slide pins with silicone grease and torque them to 21–30 ft. lbs. (29–41 Nm). Install the slide pin bushing caps.
13. Install the caliper support spring in the top of the caliper under the anchor; then, install the other end into the lower caliper hole.

➡**Hold the spring in the caliper hole with your thumb while prying the spring end out and under the anchor.**

14. Fill the master cylinder with fluid and bleed the brake system.
15. Before driving the vehicle, pump the brakes several times to seat the pads.
16. Install the wheels.

Liberty

FRONT

1. Drain ⅔ of the brake fluid from the front reservoir. Use the bleeder screw at the front outlet port to drain the fluid. If equipped with anti-lock brakes, relieve the system pressure.
2. Raise and safely support the vehicle.
3. Remove the wheels.
4. Remove both caliper slide pin bushing caps and slide pins.
5. Lift the caliper from the anchor.

6. Using a piece of mechanic's wire, support the caliper so there is not tension on the brake hose.
7. Remove the brake pads from the caliper.

To install:
8. Position the brake pads onto the caliper.
9. Position the caliper into place on the anchor.
10. Coat the caliper slide pins with silicone grease and torque them to 11 ft. lbs. (15 Nm). Install the slide pin bushing caps.

➡**Hold the spring in the caliper hole with your thumb while prying the spring end out and under the anchor.**

11. Fill the master cylinder with fluid and bleed the brake system.
12. Before driving the vehicle, pump the brakes several times to seat the pads.
13. Install the wheels.

REAR

1. Drain ⅔ of the brake fluid from the front reservoir. Use the bleeder screw at the front outlet port to drain the fluid. If equipped with anti-lock brakes, relieve the system pressure.
2. Raise and safely support the vehicle.
3. Remove the wheels.
4. Pry the caliper support spring out of the caliper.
5. Remove both caliper slide pin bushing caps and slide pins.

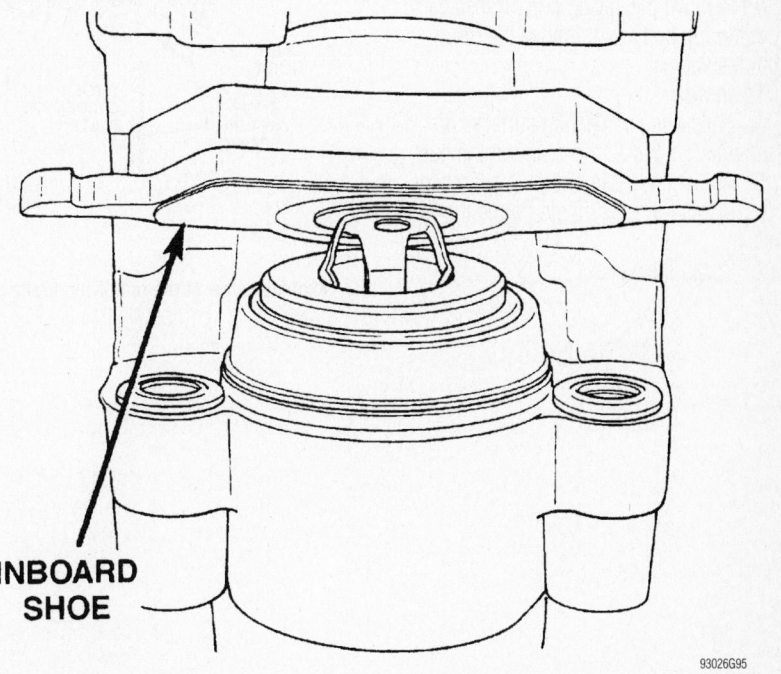

INBOARD SHOE

93026G95

Installing the inward brake pad on the rear caliper piston—Grand Cherokee

6. Lift the caliper from the anchor.

7. Using a piece of mechanics wire, support the caliper so there is not tension on the brake hose.

8. Remove the brake pads from the caliper.

To install:

9. Position the brake pads onto the caliper.

10. Position the caliper into place on the anchor.

11. Coat the caliper slide pins with silicone grease and torque them to 11 ft. lbs. (15 Nm). Install the slide pin bushing caps.

12. Install the caliper support spring in the top of the caliper under the anchor; then, install the other end into the lower caliper hole.

➡ **Hold the spring in the caliper hole with your thumb while prying the spring end out and under the anchor.**

13. Fill the master cylinder with fluid and bleed the brake system.

14. Before driving the vehicle, pump the brakes several times to seat the pads.

15. Install the wheels.

Brake Drum

REMOVAL & INSTALLATION

Cherokee and Wrangler

1. Raise and safely support the vehicle.
2. Remove the wheel.
3. Remove the spring nuts (if installed) from the lug bolts and remove the drum from the vehicle.

To install:

4. Ensure the contacting surfaces are clean and flat. Install the drum on the hub.
5. Adjust the brake shoes, if necessary.
6. Install the spring nuts on the lug bolts.
7. Install the wheel.

Brake Shoes

REMOVAL & INSTALLATION

Cherokee and Wrangler

1. Raise and safely support the vehicle.
2. Remove the wheel and brake drum.
3. Remove the U-clip and washer securing the adjuster cable to the parking brake lever.
4. Remove the primary and secondary return springs from the anchor pin.
5. Remove the hold-down springs, retainers and pins.
6. Install spring clamps on the wheel cylinders to hold the pistons in place.
7. Remove the adjuster lever, adjuster screw and spring.
8. Remove the adjuster cable and cable guide.
9. Remove the brake shoes and parking brake strut.
10. Disconnect the cable from the parking brake lever and remove the lever.

To install:

11. Clean the support plate with brake cleaner.

12. Apply multi-purpose grease to the brake shoe contact surfaces on the backing plate.

13. Lubricate the adjuster screw threads.

14. Attach the parking brake lever to the secondary brake shoe. Use a new washer and U-clip.

15. Remove the wheel cylinder clamps.

16. Attach the parking brake cable to the lever.

17. Install the brake shoes on the support plate. Secure the shoes with new hold-down springs, pins and retainers.

18. Install the parking brake strut and spring.

19. Install the guide plate and adjuster cable to the anchor pin.

20. Install the return springs.

21. Install the adjuster cable guide on the secondary shoe.

22. Install the adjuster screw, spring and lever. Connect to the adjuster cable.

23. Adjust the shoes to the drum. Install the drum.

24. Install the wheel/tire assemblies and lower the vehicle.

25. Verify a firm brake pedal before moving the vehicle.

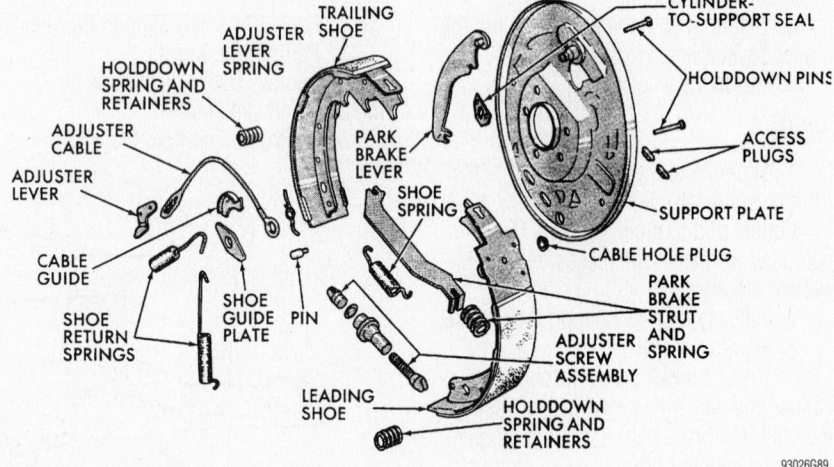

Exploded view of the rear drum brake components—Cherokee

93026G89

CHRYSLER AND DODGE

4

Cirrus • Sebring Convertible & Sedan • Stratus

SPECIFICATION CHARTS

ENGINE AND VEHICLE IDENTIFICATION

Code ①	Liters (cc)	Cu. In.	Cyl.	Fuel Sys.	Engine Type	Eng. Mfg.
X	2.4 (2429)	143	I4	MFI	DOHC	Chrysler
U	2.7 (2736)	167	V6	MFI	DOHC	Chrysler
S	2.4 (2429)	148	I4	MFI	DOHC	Chrysler
R	2.7 (2736)	167	V6	MFI	DOHC	Chrysler

Code ②	Year
1	2001
2	2002
3	2003
4	2004
5	2005

MFI: Multi-point Fuel Injection

SOHC: Single Overhead Camshaft

DOHC: Double Overhead Camshaft

① 8th position of VIN

② 10th position of VIN

67189-CIRR-C01

GENERAL ENGINE SPECIFICATIONS

All measurements are given in inches.

Year	Model	Engine Displacement Liters	Engine Series (ID/VIN)	Net Horsepower @ rpm	Net Torque @ rpm (ft. lbs.)	Bore x Stroke (in.)	Compression Ratio	Oil Pressure @ rpm
2001	Sebring Convertible	2.7	U	200@5900	192@4300	3.39x3.09	9.7:1	45-105@3000
	Sebring Sedan	2.4	X	150@5200	167@4000	3.44x3.98	9.4:1	25-80@3000
	Sebring Sedan	2.7	U	200@5900	192@4300	3.39x3.09	9.7:1	45-105@3000
	Stratus	2.4	X	150@5200	167@4000	3.44x3.98	9.4:1	25-80@3000
	Stratus	2.7	U	200@5900	192@4300	3.39x3.09	9.7:1	45-105@3000
2002	Sebring Convertible	2.4	S	150@5200	167@4000	3.44x3.98	9.4:1	25-80@3000
	Sebring Convertible	2.7	R	200@5800	190@4850	3.38x3.09	9.7:1	45-105@3000
	Sebring Sedan	2.4	S	150@5200	167@4000	3.44x3.98	9.4:1	25-80@3000
	Sebring Sedan	2.7	R	200@5900	192@4300	3.39x3.09	9.7:1	45-105@3000
	Stratus	2.4	S	150@5200	167@4000	3.44x3.98	9.4:1	25-80@3000
	Stratus	2.7	R	200@5900	192@4300	3.39x3.09	9.7:1	45-105@3000
2003	Sebring Convertible	2.4	S	150@5200	167@4000	3.44x3.98	9.4:1	25-80@3000
	Sebring Convertible	2.7	R	200@5800	190@4850	3.38x3.09	9.7:1	45-105@3000
	Sebring Sedan	2.4	S	150@5200	167@4000	3.44x3.98	9.4:1	25-80@3000
	Sebring Sedan	2.7	R	200@5900	192@4300	3.39x3.09	9.7:1	45-105@3000
	Stratus	2.4	S	150@5200	167@4000	3.44x3.98	9.4:1	25-80@3000
	Stratus	2.7	R	200@5900	192@4300	3.39x3.09	9.7:1	45-105@3000
2004	Sebring Convertible	2.4	S	150@5200	167@4000	3.44x3.98	9.4:1	25-80@3000
	Sebring Convertible	2.7	R	200@5800	190@4850	3.38x3.09	9.7:1	45-105@3000
	Sebring Sedan	2.4	S	150@5200	167@4000	3.44x3.98	9.4:1	25-80@3000
	Sebring Sedan	2.7	R	200@5900	192@4300	3.39x3.09	9.7:1	45-105@3000
	Stratus	2.4	S	150@5200	167@4000	3.44x3.98	9.4:1	25-80@3000
	Stratus	2.7	R	200@5900	192@4300	3.39x3.09	9.7:1	45-105@3000

MFI: Multi-point Fuel Injection

① California: 9.0:1

67189-CIRR-C02

GASOLINE ENGINE TUNE-UP SPECIFICATIONS

Year	Engine Displacement Liters	Engine ID/VIN	Spark Plug Gap (in.)	Ignition Timing (deg.) MT	Ignition Timing (deg.) AT	Fuel Pump (psi) ①	Idle Speed (rpm) MT	Idle Speed (rpm) AT	Valve Clearance In.	Valve Clearance Ex.
2001	2.4	X	0.050	—	①	58	②	②	HYD	HYD
	2.7	U	0.048-0.053	①	①	58	②	②	HYD	HYD
2002	2.4	S	0.050	—	①	58	②	②	HYD	HYD
	2.7	R	0.048-0.053	①	①	58	②	②	HYD	HYD
2003	2.4	S	0.050	—	①	58	②	②	HYD	HYD
	2.7	R	0.048-0.053	①	①	58	②	②	HYD	HYD
2004	2.4	S	0.050	—	①	58	②	②	HYD	HYD
	2.7	R	0.048-0.053	①	①	58	②	②	HYD	HYD

NOTE: The Vehicle Emission Control Information label often reflects specification changes made during production. The label figures must be used if they differ from those in this chart.

HYD: Hydraulic

① Ignition timing cannot be adjusted. Base engine timing is set at TDC during assembly.

② Refer to the Vehicle Emission Control Information label for correct specifications.

67189-CIRR-C03

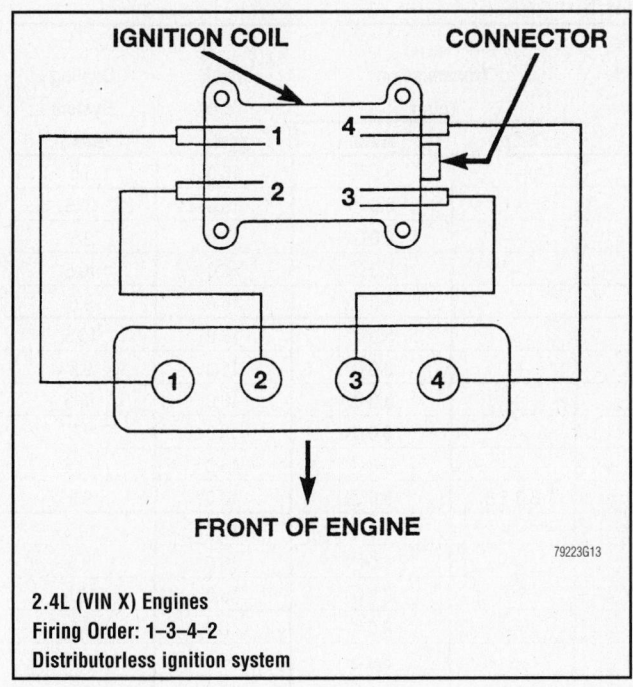

IGNITION COIL   **CONNECTOR**

1   4

2   3

1   2   3   4

FRONT OF ENGINE

79223G13

2.4L (VIN X) Engines
Firing Order: 1–3–4–2
Distributorless ignition system

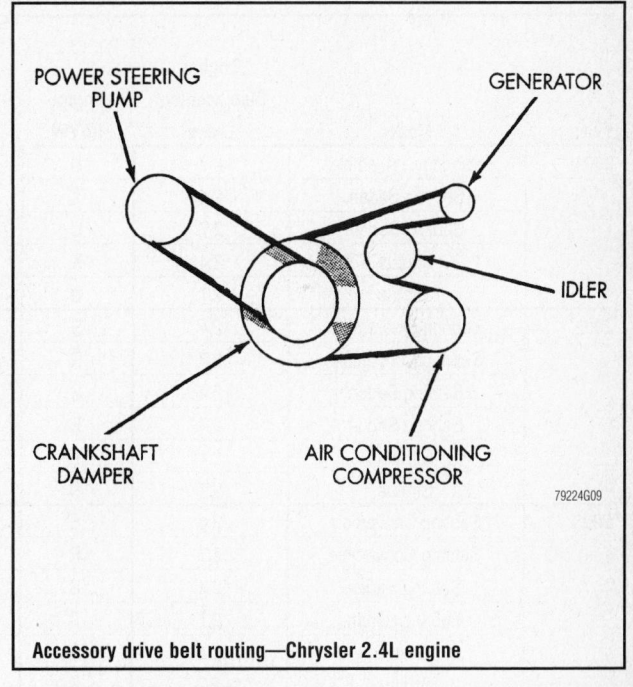

POWER STEERING PUMP   GENERATOR

IDLER

CRANKSHAFT DAMPER   AIR CONDITIONING COMPRESSOR

79224G09

Accessory drive belt routing—Chrysler 2.4L engine

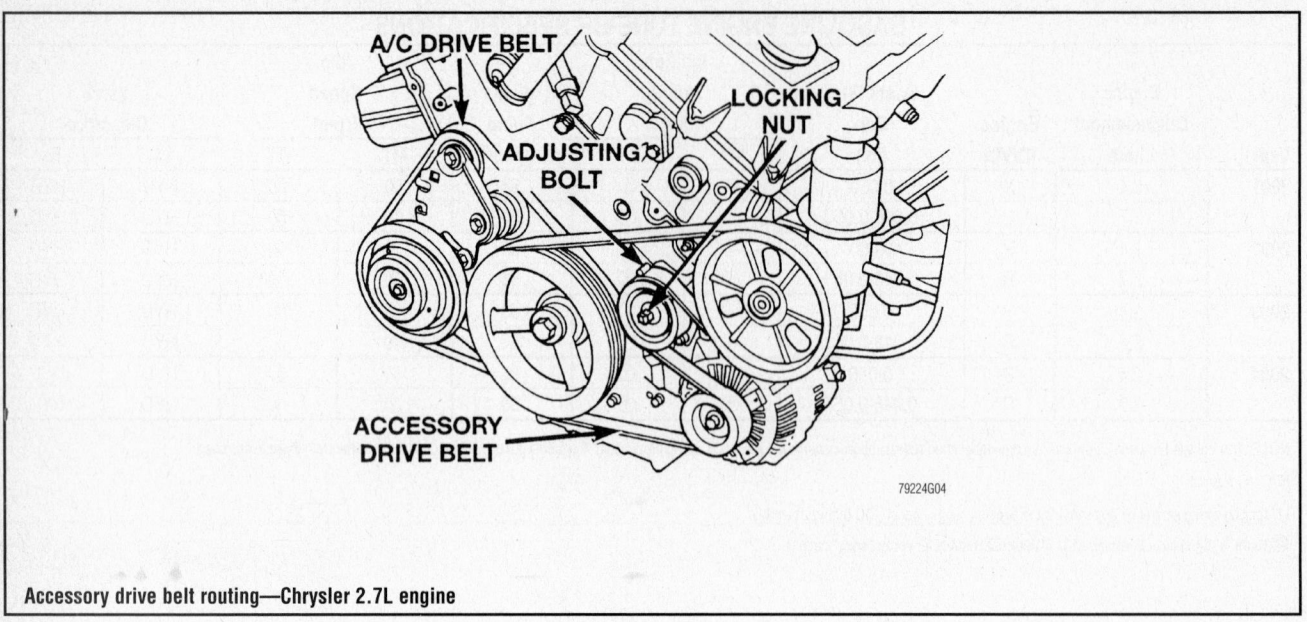

Accessory drive belt routing—Chrysler 2.7L engine

CAPACITIES

Year	Model	Engine Displacement Liters	Engine ID/VIN	Engine Oil with Filter (qts.)	Transmission (pts.) 5-Spd	Transmission (pts.) Auto.	Fuel Tank (gal.)	Cooling System (qts.)
2001	Sebring Convertible	2.7	U	5.0	—	8.0 ①	16.0	9.5
	Sebring Sedan	2.4	X	5.0	—	8.0 ①	16.0	10.5
	Sebring Sedan	2.7	U	5.0	—	8.0 ①	16.0	9.5
	Stratus	2.4	X	5.0	—	8.0 ①	16.0	10.5
	Stratus	2.7	U	5.0	—	8.0 ①	16.0	9.5
2002	Sebring Convertible	2.4	S	5.0	—	8.0 ①	16.0	10.5
	Sebring Convertible	2.7	R	5.0	—	8.0 ①	16.0	9.5
	Sebring Sedan	2.4	S	5.0	—	8.0 ①	16.0	10.5
	Sebring Sedan	2.7	R	5.0	—	8.0 ①	16.0	9.5
	Stratus	2.4	S	5.0	—	8.0 ①	16.0	10.5
	Stratus	2.7	R	5.0	5.0-5.6	8.0 ①	16.0	9.5
2003	Sebring Convertible	2.4	S	5.0	—	8.0 ①	16.0	10.5
	Sebring Convertible	2.7	R	5.0	—	8.0 ①	16.0	9.5
	Sebring Sedan	2.4	S	5.0	—	8.0 ①	16.0	10.5
	Sebring Sedan	2.7	R	5.0	—	8.0 ①	16.0	9.5
	Stratus	2.4	S	5.0	—	8.0 ①	16.0	10.5
	Stratus	2.7	R	5.0	5.0-5.6	8.0 ①	16.0	9.5
2004	Sebring Convertible	2.4	S	5.0	—	8.0 ①	16.0	10.5
	Sebring Convertible	2.7	R	5.0	—	8.0 ①	16.0	9.5
	Sebring Sedan	2.4	S	5.0	—	8.0 ①	16.0	10.5
	Sebring Sedan	2.7	R	5.0	—	8.0 ①	16.0	9.5
	Stratus	2.4	S	5.0	—	8.0 ①	16.0	10.5
	Stratus	2.7	R	5.0	5.0-5.6	8.0 ①	16.0	9.5

NOTE: All capacities are approximate. Add fluid gradually and ensure a proper fluid level is obtained.

① Overhaul fill capacity with torque converter empty: 18.4 pts.

CRANKSHAFT AND CONNECTING ROD SPECIFICATIONS

All measurements are given in inches.

Year	Engine Displacement Liters	Engine ID/VIN	Crankshaft Main Brg. Journal Dia.	Crankshaft Main Brg. Oil Clearance	Crankshaft Shaft End-play	Crankshaft Thrust on No.	Connecting Rod Journal Diameter	Connecting Rod Oil Clearance	Connecting Rod Side Clearance
2001	2.4	X	2.3610-2.3625	0.0007-0.0023	0.0035-0.0094	3	1.9670-1.9685	0.0009-0.0027	0.0050-0.0150
	2.7	U	2.4997-2.5004	0.0014-0.0021	0.0019-0.0108	3	2.1067-2.1060	0.0010-0.0026	0.0052-0.0150
2002	2.4	S	2.3610-2.3625	0.0007-0.0023	0.0035-0.0094	3	1.9670-1.9685	0.0009-0.0027	0.0051-0.0150
	2.7	R	2.4997-2.5004	0.0014-0.0021	0.017 max.	3	2.1067-2.1060	0.0010-0.0026	0.0052-0.0150
2003	2.4	S	2.3610-2.3625	0.0007-0.0023	0.0035-0.0094	3	1.9670-1.9685	0.0009-0.0027	0.0051-0.0150
	2.7	R	2.4997-2.5004	0.0014-0.0021	0.017 max.	3	2.1067-2.1060	0.0010-0.0026	0.0052-0.0150
2004	2.4	S	2.3610-2.3625	0.0007-0.0023	0.0035-0.0094	3	1.9670-1.9685	0.0009-0.0027	0.0051-0.0150
	2.7	R	2.4997-2.5004	0.0014-0.0021	0.017 max.	3	2.1067-2.1060	0.0010-0.0026	0.0052-0.0150

67189-CIRR-C05

VALVE SPECIFICATIONS

Year	Engine Displacement Liters	Engine ID/VIN	Seat Angle (deg.)	Face Angle (deg.)	Spring Test Pressure (lbs. @ in.)	Spring Installed Height (in.)	Stem-to-Guide Clearance (in.) Intake	Stem-to-Guide Clearance (in.) Exhaust	Stem Diameter (in.) Intake	Stem Diameter (in.) Exhaust
2001	2.4	X	45	44.5-45	76@1.50	1.496	0.0018-0.0025	0.0029-0.0037	0.2337-0.2344	0.2326-0.2333
	2.7	U	44.5-45	44.5-45.5	148@1.14	1.496	0.0009-0.0026	0.0020-0.0037	0.2337-0.2344	0.2326-0.2333
2002	2.4	S	45	44.5-45	76@1.50	1.496	0.0018-0.0025	0.0029-0.0037	0.2337-0.2344	0.2326-0.2333
	2.7	R	44.5-45	44.5-45.5	148@1.14	1.496	0.0009-0.0026	0.0020-0.0037	0.2337-0.2344	0.2326-0.2333
2003	2.4	S	45	44.5-45	76@1.50	1.496	0.0018-0.0025	0.0029-0.0037	0.2337-0.2344	0.2326-0.2333
	2.7	R	44.5-45	44.5-45.5	148@1.14	1.496	0.0009-0.0026	0.0020-0.0037	0.2337-0.2344	0.2326-0.2333
2004	2.4	S	45	44.5-45	76@1.50	1.496	0.0018-0.0025	0.0029-0.0037	0.2337-0.2344	0.2326-0.2333
	2.7	R	44.5-45	44.5-45.5	148@1.14	1.496	0.0009-0.0026	0.0020-0.0037	0.2337-0.2344	0.2326-0.2333

67189-CIRR-C06

PISTON AND RING SPECIFICATIONS
All measurements are given in inches.

| Year | Engine Displacement Liters | Engine ID/VIN | Piston Clearance | Ring Gap | | | Ring Side Clearance | | |
				Top Compression	Bottom Compression	Oil Control	Top Compression	Bottom Compression	Oil Control
2001	2.4	X	0.0009- ② 0.0022	0.010- 0.020	0.009- 0.018	0.010- 0.025	0.0011- 0.0031	0.0011- 0.0031	0.0004- 0.0070
	2.7	U	0.0003- 0.0016	0.008- 0.014	0.0146- 0.0249	0.010- 0.030	0.0013- 0.0032	0.0016- 0.0031	0.0022- 0.0080
2002	2.4	S	0.0009- ② 0.0022	0.010- 0.020	0.009- 0.018	0.010- 0.025	0.0011- 0.0031	0.0011- 0.0031	0.0004- 0.0070
	2.7	R	0.0003- 0.0016	0.008- 0.014	0.0146- 0.0249	0.010- 0.030	0.0013- 0.0032	0.0016- 0.0031	0.0022- 0.0080
2003	2.4	S	0.0009- ② 0.0022	0.010- 0.020	0.009- 0.018	0.010- 0.025	0.0011- 0.0031	0.0011- 0.0031	0.0004- 0.0070
	2.7	R	0.0003- 0.0016	0.008- 0.014	0.0146- 0.0249	0.010- 0.030	0.0013- 0.0032	0.0016- 0.0031	0.0022- 0.0080
2004	2.4	S	0.0009- ② 0.0022	0.010- 0.020	0.009- 0.018	0.010- 0.025	0.0011- 0.0031	0.0011- 0.0031	0.0004- 0.0070
	2.7	R	0.0003- 0.0016	0.008- 0.014	0.0146- 0.0249	0.010- 0.030	0.0013- 0.0032	0.0016- 0.0031	0.0022- 0.0080

NA: Not Available

① Clearance at 11/16 inch from bottom of skirt

② Clearance at 9/16 inch from bottom of skirt

67189-CIRR-C07

TORQUE SPECIFICATIONS
All readings in ft. lbs.

| Year | Engine Displacement Liters | Engine ID/VIN | Cylinder Head Bolts | Main Bearing Bolts | Rod Bearing Bolts | Crankshaft Damper Bolts | Flywheel Bolts | Manifold | | Spark Plug | Oil Pan Drain Plug |
								Intake	Exhaust		
2001	2.4	X	①	②	③	100	70	9	17	21	20
	2.7	U	④	⑤	③	125	—	9	17	15	20
2002	2.4	S	①	⑥	③	100	70	21	17	13	20
	2.7	R	④	⑤	③	125	—	9	17	15	20
2003	2.4	S	①	⑥	③	100	70	21	17	13	20
	2.7	R	④	⑤	③	125	—	9	17	15	20
2004	2.4	S	①	⑥	③	100	70	21	17	13	20
	2.7	R	④	⑤	③	125	—	9	17	15	20

① Step 1: 25 ft. lbs.
　Step 2: 50 ft. lbs.
　Step 3: 50 ft. lbs.

② M8 bolts: 21 ft. lbs.
　M11 bolts: 30 ft. lbs. Plus 1/4 turn
　Step 4: Plus 1/4 turn

③ Step 1: 20 ft. lbs.
　Step 2: Plus 1/4 turn

④ Step 1: 35 ft. lbs.
　Step 2: 55 ft. lbs.
　Step 3: 55 ft. lbs.
　Step 4: Plus 1/4 turn
　Step 5: 21 ft. lbs.

⑤ Main cap inside bolts: 15 ft. lbs. Plus 1/4 turn
　Main cap outside bolts: 20 ft. lbs. Plus 1/4 turn
　Main cap tie bolts: 21 ft. lbs.

⑥ M8 bolts: 21 ft. lbs.
　M11 bolts: 55 ft. lbs.

67189-CIRR-C08

WHEEL ALIGNMENT

Year	Model		Caster Range (+/-Deg.)	Caster Preferred Setting (Deg.)	Camber Range (+/-Deg.)	Camber Preferred Setting (Deg.)	Toe-in (in.)
2001	All	F	+1.00	+3.31	+0.59	0	0.06 +/- 0.10
		R	—	—	+0.38	-0.19	0.06 +/- 0.10
2002	All	F	+1.00	+3.31	+0.59	0	0.06 +/- 0.10
		R	—	—	+0.38	-0.19	0.06 +/- 0.10
2003	All	F	+1.00	+3.31	+0.59	0	0.06 +/- 0.10
		R	—	—	+0.38	-0.19	0.06 +/- 0.10
2004	All	F	+1.00	+3.31	+0.59	0	0.06 +/- 0.10
		R	—	—	+0.38	-0.19	0.06 +/- 0.10

67189-CIRR-C09

TIRE, WHEEL AND BALL JOINT SPECIFICATIONS

Year	Model	OEM Tires Standard	OEM Tires Optional	Tire Pressures (psi) Front	Tire Pressures (psi) Rear	Wheel Size	Ball Joint Inspection	Lug Nut
2001	Sebring Conv.	P205/65R15	P215/55R16	32	32	6-JJ	②	100
	Sebring	P205/65R15	P205/60HR16	32	32	6-JJ	①	100
	Stratus	P195/70R14	P195/65HR15	31	31	6-JJ	①	100
2002	Sebring Conv.	P205/65R15	P205/55R16	32	32	6-JJ	②	100
	Sebring	P205/65TR15	P205/60TR16	32	32	6-JJ	①	100
	Stratus	P205/65TR15	P205/60TR16 ③	31	31	6-JJ	①	100
2003	Sebring Conv.	P205/65R15	P205/55R16	32	32	6-JJ	②	100
	Sebring	P205/65TR15	P205/60TR16	32	32	6-JJ	①	100
	Stratus	P205/65TR15	P205/60TR16 ③	31	31	6-JJ	①	100
2004	Sebring Conv.	P205/65R15	P205/55R16	32	32	6-JJ	②	100
	Sebring	P205/65TR15	P205/60TR16	32	32	6-JJ	①	100
	Stratus	P205/65TR15	P205/60TR16 ③	31	31	6-JJ	①	100

OEM: Original Equipment Manufacturer

PSI: Pounds Per Square Inch

STD: Standard

OPT: Optional

L: Lower

U: Upper

① Do not lift car. Grasp the grease fitting and attempt to move or rotate. Replace if any movement is found.

② Replace if any measurable movement is found

③ P215/50VR17 tires are available on the ES models with the Performance Handling Group

67189-CIRR-C10

BRAKE SPECIFICATIONS
All measurements in inches unless noted

Year	Model		Brake Disc Original Thickness	Brake Disc Minimum Thickness	Brake Disc Maximum Run-out	Brake Drum Diameter Original Inside Diameter	Brake Drum Diameter Max. Wear Limit	Brake Drum Diameter Maximum Machine Diameter	Minimum Lining Thickness	Brake Caliper Guide Pin Bolts (ft. lbs.)
2001	Sebring Convertible	F	0.911	0.843	0.005	—	—	—	③	16
		R	0.400	0.330	0.005	9.80	①	①	0.040	31
	Sebring Sedan	F	0.911	0.843	0.005	—	—	—	③	26
		R	0.360	0.285	0.005	NA	①	①	②	26
	Stratus	F	0.911	0.843	0.003	—	—	—	③	16
		R	0.360	0.285	0.005	NA	①	①	②	16
2002	Sebring Convertible	F	0.911	0.843	0.005	—	—	—	③	16
		R	0.400	0.330	0.005	9.80	①	①	0.040	31
	SebringSedan	F	0.911	0.843	0.005	—	—	—	③	26
		R	0.360	0.285	0.005	NA	①	①	②	26
	Stratus	F	0.911	0.843	0.003	—	—	—	③	16
		R	0.360	0.285	0.005	NA	①	①	②	16
2003	Sebring Convertible	F	0.911	0.843	0.005	—	—	—	③	16
		R	0.400	0.330	0.005	9.80	①	①	0.040	31
	SebringSedan	F	0.911	0.843	0.005	—	—	—	③	26
		R	0.360	0.285	0.005	NA	①	①	②	26
	Stratus	F	0.911	0.843	0.003	—	—	—	③	16
		R	0.360	0.285	0.005	NA	①	①	②	16
2004	Sebring Convertible	F	0.911	0.843	0.005	—	—	—	③	16
		R	0.400	0.330	0.005	9.80	①	①	0.040	31
	SebringSedan	F	0.911	0.843	0.005	—	—	—	③	26
		R	0.360	0.285	0.005	NA	①	①	②	26
	Stratus	F	0.911	0.843	0.003	—	—	—	③	16
		R	0.360	0.285	0.005	NA	①	①	②	16

NA: Not Available

① Maximum diameter is stamped on drum

② Disc brake lining and backing total thickness: 9/32 inch

   Drum brake shoe lining and backing total thickness: 1/8 inch

③ Disc brake lining and backing: 3/8 inch

SCHEDULED MAINTENANCE INTERVALS
Chrysler—Cirrus, Sebring & Dodge—Stratus

TO BE SERVICED	TYPE OF SERVICE	VEHICLE MILEAGE INTERVAL (x1000)												
		7.5	15	22.5	30	37.5	45	52.5	60	67.5	75	82.5	90	97.5
Engine oil & filter	R	✓	✓	✓	✓	✓	✓	✓	✓	✓	✓	✓	✓	✓
Brake hoses	S/I	✓	✓	✓	✓	✓	✓	✓	✓	✓	✓	✓	✓	✓
Coolant level, hoses & clamps	S/I	✓	✓	✓	✓	✓	✓	✓	✓	✓	✓	✓	✓	✓
CV joints & front suspension components	S/I	✓	✓	✓	✓	✓	✓	✓	✓	✓	✓	✓	✓	✓
Exhaust system	S/I	✓	✓	✓	✓	✓	✓	✓	✓	✓	✓	✓	✓	✓
Rotate tires	S/I	✓	✓	✓	✓	✓	✓	✓	✓	✓	✓	✓	✓	✓
Accessory drive belts	S/I		✓		✓		✓		✓		✓		✓		
Brake linings	S/I			✓			✓			✓			✓		
Air filter element	R				✓				✓				✓		
Spark plugs ① ②	R														
Lubricate front & rear ball joints	S/I				✓				✓			✓	✓		
Engine coolant	R						✓				✓				
PCV valve	S/I								✓				✓		
Ignition cables ② ③	R														
Camshaft timing belt ④	R														

R: Replace   S/I: Service or Inspect

① 4-cylinder: every 30,000 miles.

② 6-cylinder: 100,000 miles.

③ 4-cylinder: 60,000 miles.

④ Replace at 105,000 miles for normal service; replace at 102,000 miles for severe service

FREQUENT OPERATION MAINTENANCE (SEVERE SERVICE)

If a vehicle is operated under any of the following conditions it is considered severe service:

- Extremely dusty areas.

- 50% or more of the vehicle operation is in 32°C (90°F) or higher temperatures, or constant operation in temperatures below 0°C (32°F).

- Prolonged idling (vehicle operation in stop and go traffic).

- Frequent short running periods (engine does not warm to normal operating temperatures).

- Police, taxi, delivery usage or trailer towing usage.

Oil & oil filter change: change every 3000 miles.

Rotate tires every 6000 miles.

Brake linings: check every 12,000 miles.

Air filter element: change every 15,000 miles.

Automatic transaxle fluid: service or inspect every 15,000 miles.

PCV valve: check every 30,000 miles.

Engine coolant, replace at 36,000, 51,000 & 81,000 miles.

67189-CIRR-C12

PRECAUTIONS

Before servicing any vehicle, please be sure to read all of the following precautions. The following precautions deal with personal safety, preventing of component damage, and important points to take into consideration when servicing a motor vehicle:

• Never open, service or drain the radiator or cooling system when the engine is hot; serious burns can occur from the steam and hot coolant.

• Observe all applicable safety precautions when working around fuel. Whenever servicing the fuel system, always work in a well-ventilated area. Do not allow fuel spray or vapors to come in contact with a spark, open flame or excessive heat (a hot drop light, for example). Keep a dry chemical fire extinguisher near the work area. Always keep fuel in a container specifically designed for fuel storage; also, always properly seal fuel containers to avoid the possibility of fire or explosion. Refer to the additional fuel system precautions later in this section.

• Fuel injection systems often remain pressurized, even after the engine has been turned **OFF**. The fuel system pressure must be relieved before disconnecting any fuel lines. Failure to do so may result in fire and/or personal injury.

• Brake fluid often contains polyglycol ethers and polyglycols. Avoid contact with the eyes and wash your hands thoroughly after handling brake fluid. If you do get brake fluid in your eyes, flush your eyes with clean, running water for 15 minutes. If eye irritation persists, or if you have taken brake fluid internally, IMMEDIATELY seek medical assistance.

• The EPA warns that prolonged contact with used engine oil may cause a number of skin disorders, including cancer! You should make every effort to minimize your exposure to used engine oil. Protective gloves should be worn when changing the oil. Wash your hands and any other exposed skin areas as soon as possible after exposure to used engine oil. Soap and water, or waterless hand cleaner should be used.

• All vehicles are equipped with an air bag system, often referred to as a Supplemental Restraint System (SRS) or as a Supplemental Inflatable Restraint (SIR) system. The system must be disabled before performing service on or around system components, steering column, instrument panel components, wiring and sensors. Failure to follow safety and disabling procedures could result in accidental air bag deployment, possible personal injury and unnecessary system repairs.

• Always wear safety goggles when working with, or around, the air bag system. When carrying a non-deployed air bag, be sure the bag and trim cover are pointed away from your body. When placing a non-deployed air bag on a work surface, always face the bag and trim cover upward, away from the surface. This will reduce the motion of the module if it is accidentally deployed.

• Clean, high quality brake fluid from a sealed container is essential for the safe and proper operation of the brake system.

You should always buy the grade of fluid recommended for your vehicle. If the brake fluid becomes contaminated, drain and flush the system, then refill the master cylinder with new fluid. Never reuse any brake fluid. Any brake fluid that is removed from the system should be discarded. Also, do not allow any brake fluid to come in contact with a painted surface; it will damage the paint.

• Never operate the engine without the proper amount and type of engine oil; doing so WILL result in severe engine damage.

• Timing belt maintenance is extremely important! Many models may utilize an interference-type, non-free-wheeling engine. If the timing belt breaks, the valves in the cylinder head may strike the pistons, causing potentially serious (also time-consuming and expensive) engine damage. Refer to the maintenance interval charts in the front of this section for the recommended replacement interval for the timing belt, and to the timing belt procedure for belt replacement and inspection.

• Disconnecting the negative battery cable on some vehicles may interfere with the functions of the on board computer system(s) and may require the computer to undergo a relearning process once the negative battery cable is reconnected.

• When servicing drum brakes, only disassemble and assemble one side at a time, leaving the remaining side intact for reference.

• Only an MVAC-trained, EPA-certified, automotive technician should service the air conditioning system or its components.

ENGINE REPAIR

Distributor

These vehicles are equipped with distributorless ignition systems.

Alternator

REMOVAL

2.4L Engine

1. Before servicing the vehicle, refer to the precautions in the beginning of this section.

2. Remove or disconnect the following:
   • Negative battery cable from the shock tower

• Accessory belt cover
• Alternator electrical connectors
• Accessory belt splash shield
• Accessory belt
• Manifold Absolute Pressure (MAP) sensor from the intake manifold
• Alternator

2.7L Engine

1. Before servicing the vehicle, refer to the precautions in the beginning of this section.

2. Remove or disconnect the following:
   • Negative battery cable from the shock tower
   • Accessory belt splash shield and loosen the belt

• Alternator electrical connectors
• A/C pressure switch and clutch electrical connectors
• Engine oil dipstick
• Alternator

INSTALLATION

2.4L Engine

1. Install or connect the following:
   • Alternator and torque the bolts to 33 ft. lbs. (44 Nm)
   • MAP sensor
   • Drive belt and splash shield
   • Alternator electrical connectors
   • Accessory belt cover
   • Negative battery cable

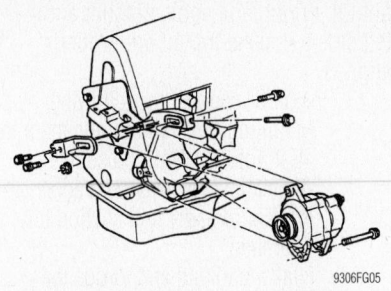

Exploded view of the alternator and related components—2.4L engine

2.7L Engine

1. Install or connect the following:
   - Alternator and torque the bolts to 40 ft. lbs. (54 Nm)
   - Engine oil dipstick tube
   - A/C pressure switch and clutch electrical connectors
   - Alternator electrical connectors
   - Drive belt and splash shield
   - Negative battery cable

Ignition Timing

ADJUSTMENT

These engines use a fixed ignition system. The Powertrain Control Module (PCM) regulates the ignition timing. Basic ignition timing is not adjustable.

Engine Assembly

REMOVAL & INSTALLATION

2.4L Engine

1. Before servicing the vehicle, refer to the precautions in the beginning of this section.

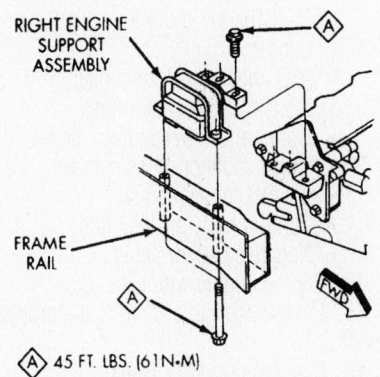

Exploded view of the right side engine mount—all engines

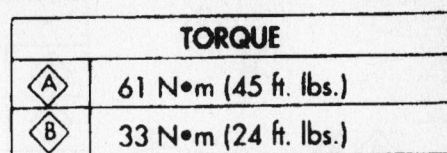

TORQUE	
Ⓐ	61 N•m (45 ft. lbs.)
Ⓑ	33 N•m (24 ft. lbs.)

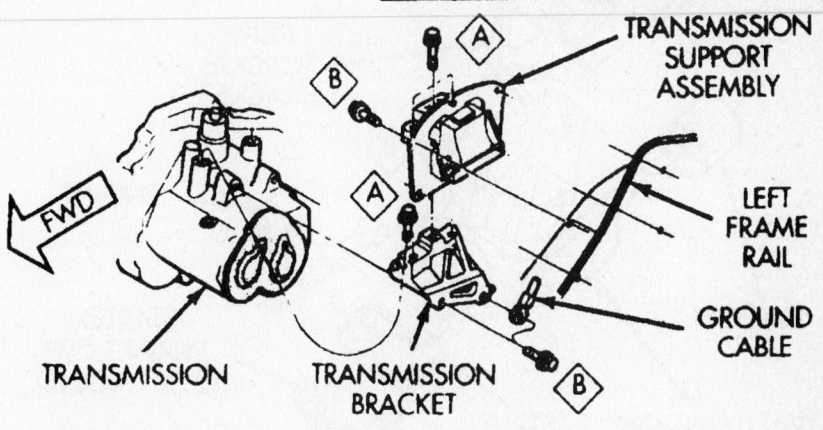

Exploded view of the left side engine mount—Type 1

TORQUE	
Ⓐ	61 N•m (45 ft. lbs.)
Ⓑ	33 N•m (24 ft. lbs.)

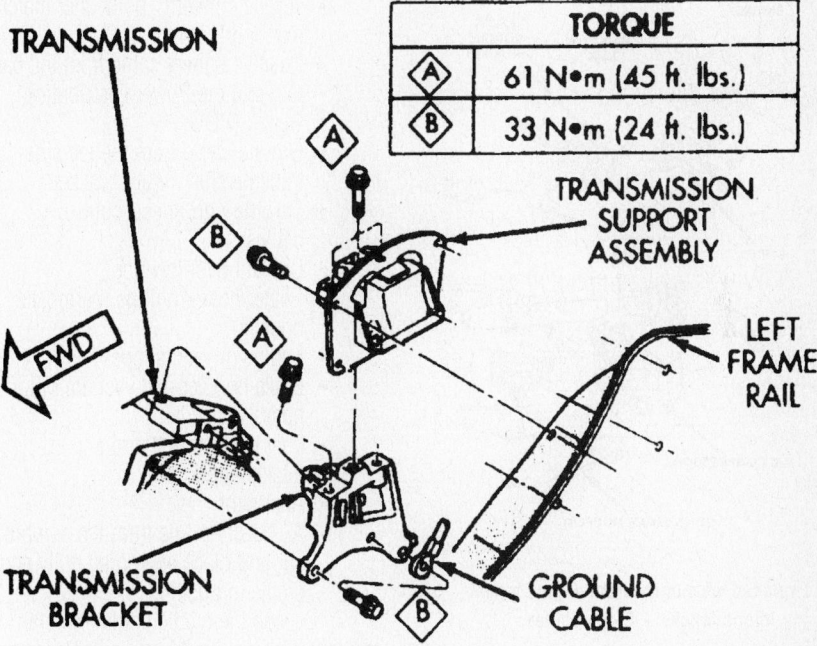

Exploded view of the left side engine mount—Type 2

2. Properly relieve the fuel system pressure.
3. Drain the engine oil.
4. Drain the cooling system.
5. Recover the A/C refrigerant.
6. Remove or disconnect the following:
   - Negative battery cable
   - Throttle body air inlet hose
   - Air cleaner assembly
   - Upper radiator crossmember
   - Upper and lower radiator hoses
   - Transmission cooler lines
   - A/C lines at the condenser
   - Radiator, fan module and A/C condenser
   - Transmission electrical harness connectors
   - Transmission shift cable
   - Engine electrical harness from the Powertrain Control Module (PCM) and bulkhead
   - Both front wheels and splash shields
   - Both halfshafts

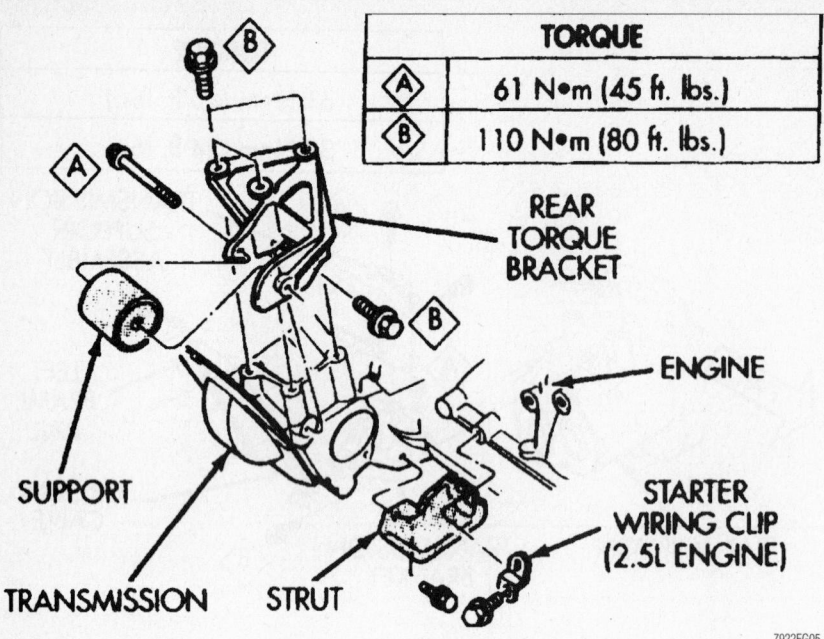

TORQUE	
Ⓐ	61 N•m (45 ft. lbs.)
Ⓑ	110 N•m (80 ft. lbs.)

Exploded view of the rear engine mounting torque bracket—2.4L engines

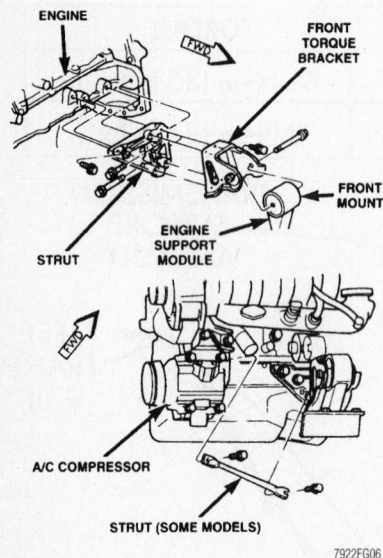

Exploded view of the front engine mounting torque bracket—2.4L engines

- Drive belts
- Power steering pump from the bracket and move it aside
- Heater return hose from the right front frame rail
- A/C compressor electrical connectors
- Exhaust pipe from the manifold
- Through bolts from the front and rear engine mounts
- Rear mount bracket from the transmission
- Structural collar and torque reaction bracket

- Torque converter bolts after match-marking them
- Positive battery cable from the battery and the Power Distribution Center (PDC)
- Ground cable from the left side transmission mount bracket
- Throttle and speed control cables
- Coolant overflow hose
- Heater hose from the thermostat housing
- Engine ground straps
- Brake booster and vacuum purge hoses
- Fuel lines from the rail
- Intake manifold
- Alternator
- A/C suction line from the compressor and place an engine dolly and cradle in position

7. Loosen the cradle engine mounts and lower the engine/transmission assembly onto the dolly.

8. Remove the vertical engine mount bolts

9. Slowly raise the vehicle and make certain that the cradle is positioned properly.

10. Secure the engine/transmission assembly to the cradle and remove the assembly.

To install:

11. Position the engine/transmission assembly under the vehicle and slowly lower the vehicle into position.

12. Continue to the lower the vehicle

until the right side engine mount and left side transaxle mount are properly aligned.

13. Install or connect the following:
- Mounting bolts and torque them to 45 ft. lbs. (61 Nm). Raise the vehicle and remove the cradle
- A/C compressor and suction line
- Alternator
- Intake manifold and torque the bolts to 105 inch lbs. (12 Nm)
- Alternator electrical connectors
- Fuel line to the rail
- Brake booster and vacuum purge hoses
- Engine ground straps
- Heater hose to the thermostat housing
- Coolant overflow hose
- Throttle and speed control cables
- Ground cable to the left side transmission mount bracket
- Positive battery cable to the PDC
- Torque converter bolts
- Structural collar and the torque reaction bracket
- Rear mount bracket to the transmission
- Front and rear engine mount through bolts. Torque the bolts to 45 ft. lbs. (61 Nm)
- Exhaust pipe to the manifold and torque the bolts to 21 ft. lbs. (28 Nm)
- A/C compressor electrical connectors
- Heater return hose to the right front frame rail
- Power steering pump to the bracket
- Drive belts
- Both halfshafts
- Splash shields and front wheels
- Engine electrical harness to the PCM and the bulkhead connectors
- Transmission shift cable
- Transmission electrical connectors
- Radiator, fan module and A/C condenser
- A/C lines to the condenser
- Transmission cooler lines
- Upper and lower radiator hoses
- Upper radiator crossmember
- Air cleaner assembly
- Throttle body air inlet hose
- Negative battery cable

14. Fill the engine with clean oil.

15. Fill the cooling system to the proper level.

16. Recharge the A/C system.

17. Start the vehicle, check for leaks and repair if necessary.

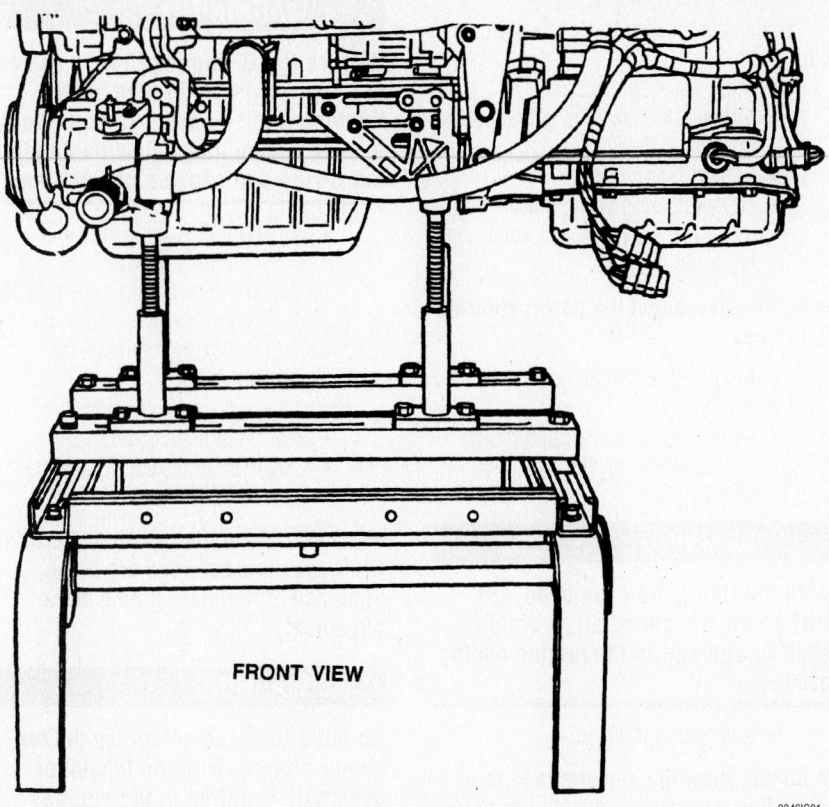

FRONT VIEW

9346IG01

Place the engine/transmission assembly on the cradle as shown

2.7L Engine

1. Before servicing the vehicle, refer to the precautions in the beginning of this section.

2. Properly relieve the fuel system pressure.

3. Drain the engine oil.

4. Drain the cooling system.

5. Recover the A/C refrigerant.

6. Remove or disconnect the following:
- Negative battery cable
- Throttle body air inlet hose
- Air cleaner assembly
- Both front wheels and splash shields
- Lower front fascia from the cross-member
- Lower air shield from the cross-member
- Front fascia
- Upper radiator crossmember
- Upper and lower radiator hoses
- Transmission oil cooler lines
- A/C lines from the condenser
- Radiator, fan and A/C condenser
- Transmission electrical harness connectors
- Transmission shift cable
- Engine electrical harness from the Powertrain Control Module (PCM) and bulkhead connectors

- Antilock Brake System (ABS) brake module from the lower radiator crossmember
- Brake line from the lower radiator crossmember
- Both halfshafts
- Front engine mount through bolt
- Front engine mount from the lower radiator crossmember
- Lower radiator crossmember
- Drive belts
- Power steering pump and bracket. Do not disconnect the lines
- Heater return hose from the right front frame rail
- A/C compressor electrical connectors and reposition the compressor
- Structural collar
- Exhaust system cross under pipe
- Rear engine mount and transaxle bracket
- Torque converter housing cover and match mark the bolts
- Positive battery cable from the Power Distribution Center (PDC)
- Ground cable from the left side transaxle mount bracket
- Throttle and speed control cables
- Coolant overflow hose

- Heater hose
- Ground strap from the right shock tower
- Fuel lines
- Brake booster and vacuum purge hoses
- Engine ground straps

7. Loosen the cradle engine mounts and lower the engine/transmission assembly onto the dolly.

8. Remove the vertical engine mount bolts

9. Slowly raise the vehicle and make certain that the cradle is positioned properly.

10. Secure the engine/transmission assembly to the cradle and remove the assembly.

To install:

11. Position the engine/transmission assembly under the vehicle and slowly lower the vehicle into position.

12. Continue to the lower the vehicle until the right side engine mount and left side transaxle mount are properly aligned.

13. Install or connect the following:
- Mounting bolts and torque them to 45 ft. lbs. (61 Nm). Raise the vehicle and remove the cradle
- Engine ground straps
- Brake booster and vacuum purge hoses
- Fuel lines
- Ground strap to the right shock tower
- Heater hose
- Coolant overflow hose
- Throttle and speed control cables
- Ground cable to the left transaxle mount bracket
- Positive battery cable to the PDC
- Torque converter bolts and cover
- Rear engine mount and transaxle bracket
- Exhaust system cross under pipe
- Structural collar
- A/C compressor to the bracket
- A/C compressor clutch electrical connectors
- Heater return hose to the right side frame rail
- Power steering pump and bracket
- Drive belts
- Lower radiator crossmember
- Front engine mount to the lower radiator crossmember
- Both halfshafts
- Brake line to the lower radiator crossmember
- ABS module to the lower radiator crossmember

- Engine electrical harness to the PCM and bulkhead connectors
- Transmission shift cable
- Transmission electrical harness connectors
- Radiator, fan and A/C condenser
- A/C lines to the condenser
- Transmission oil cooler lines
- Upper and lower radiator hoses
- Upper radiator crossmember
- Front bumper fascia and lower air shield
- Splash shields and both front wheels
- Air cleaner assembly and throttle body air inlet hose
- Negative battery cable

14. Fill the engine with clean oil and replace the filter.

15. Fill the cooling system to the proper level.

16. Recharge the A/C system with the proper refrigerant.

17. Start the vehicle, check for leaks and repair if necessary.

Water Pump

REMOVAL & INSTALLATION

2.4L Engine

This engine uses a die-cast aluminum body water pump with a stamped steel impeller. The water pump bolts directly to the block. Cylinder block-to-water pump sealing is provided by a large rubber O-ring. The water pump is driven by the timing belt, that must be removed to service the water pump.

1. Before servicing the vehicle, refer to the precautions in the beginning of this section.

2. Disconnect the negative battery cable from the left shock tower.

➡The ground cable is equipped with an insulator grommet which should be placed on the stud to prevent the negative battery cable from accidentally grounding.

➡This procedure requires removing the engine timing belt and the auto-tensioner. The factory specifies that the timing marks should always be aligned before removing the timing belt. Set the engine at Top Dead Center (TDC) on No. 1 compression stroke. This should align all timing marks on the crankshaft sprocket and both camshaft sprockets.

3. Drain the cooling system.

4. Remove or disconnect the following:

- Right inner splash shield
- Accessory drive belts and support the engine
- Right motor mount
- Power steering pump bracket bolts and move the pump/bracket assembly aside

➡Do not disconnect the power steering fluid lines.

- Right engine mount bracket
- Timing belt front covers
- Timing belt tensioner and timing belt by loosening the tensioner screws

> ※※ **WARNING**
>
> With the timing belt removed, DO NOT rotate the camshaft or crankshaft or damage to the engine could occur.

- Camshaft sprockets

➡Do not allow the camshafts to turn when the camshaft sprockets are being removed.

- Rear timing belt cover
- Water pump

To install:

5. Thoroughly clean all sealing surfaces. Replace the water pump if there are any cracks, signs of coolant leakage from the shaft seal, loose or rough turning bearings, damaged impeller or sprocket or sprocket flange loose or damaged.

6. Install or connect the following:

- New rubber O-ring into the water pump

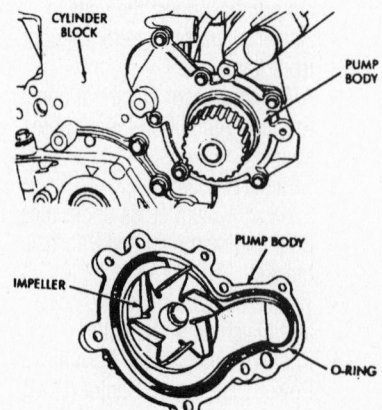

CYLINDER
BLOCK

PUMP
BODY

PUMP BODY

IMPELLER

O-RING

7922FG08

When installing the water pump, properly install the O-ring to ensure a tight seal— 2.4L engines

> ※※ **WARNING**
>
> Be sure the O-ring is properly seated in the water pump groove before tightening the screws. An improperly located O-ring may cause damage to the O-ring and cause a coolant leak.

- Water pump. Torque the bolts to 105 inch lbs. (12 Nm)

7. Pressurize the cooling system to 15 psi (103.4 kPa) and check for leaks. If okay, release the pressure and continue the engine assembly process.

8. Install or connect the following:

- Rear timing belt cover
- Camshaft sprockets. Torque the bolts to 75 ft. lbs. (101 Nm)

➡DO NOT allow the camshafts to turn while the sprocket bolts are being tightened to maintain timing mark alignment.

> ※※ **WARNING**
>
> Do not attempt to compress the tensioner plunger with the tensioner assembly installed in the engine. This will cause damage to the tensioner and other related components. The tensioner MUST be compressed in a vise.

- Timing belt tensioner and timing belt. Properly tension the timing belt.
- Front upper and lower timing belt covers
- Right engine mount bracket and engine mount
- Crankshaft damper. Torque the center bolt to 105 ft. lbs. (142 Nm)
- Right inner splash shield
- Power steering pump bracket and power steering pump. Torque the bracket mounting bolts to 40 ft. lbs. (54 Nm)
- Accessory drive belts. Properly tension the drive belts

9. Refill and bleed the cooling system.

10. Start the engine and check for proper operation.

11. Check and top off cooling system, if necessary.

2.7L Engine

1. Before servicing the vehicle, refer to the precautions in the beginning of this section.

2. Drain the cooling system.

3. Remove or disconnect the following:

- Negative battery cable
- Timing chain cover
- Timing chain and guides
- Water pump and discard the gasket

To install:

4. Clean all mating surfaces of any residual gasket material.

5. Install or connect the following:
- Water pump with a new gasket and torque the bolts to 105 inch lbs. (12 Nm)
- Timing chain guides and chain
- Timing chain cover
- Negative battery cable

6. Fill the cooling system to the proper level.

7. Start the vehicle, check for leaks and repair if necessary.

Heater Core

REMOVAL & INSTALLATION

1. Disconnect the negative battery cable.
2. Drain the cooling system into a clean container for reuse.
3. From the center of the instrument panel, grasp and pull to remove the radio/control module bezel.
4. At the right side of the instrument panel, remove the side trim.
5. Remove or disconnect the following:
- 2 lower right side support beam screws
- Instrument panel support-to-A-pillar bolt

- On left side of the instrument panel, remove the side trim
- Upper instrument panel bezel
- Lower knee bolster
- Console-to-instrument panel screws
- Gearshift knob and the shifter bezel
- Rear console screws and the rear console half
- Front console screws and the front console half
- Right side instrument panel support strut
- Heater hoses from the heater core tubes
- Heater core cover-to-heater/air conditioning housing assembly screws and the cover

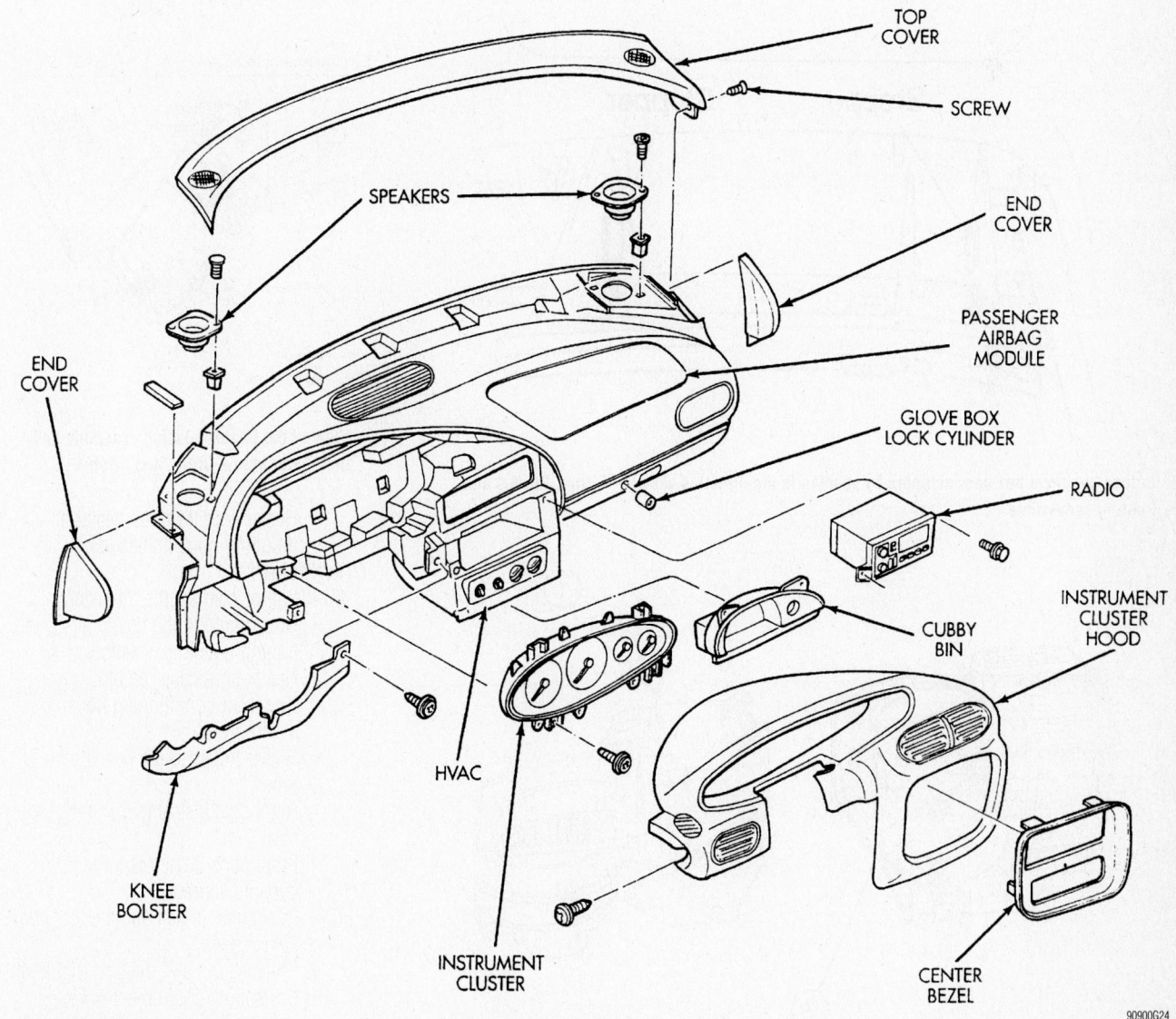

Exploded view of the instrument panel assembly

90900G24

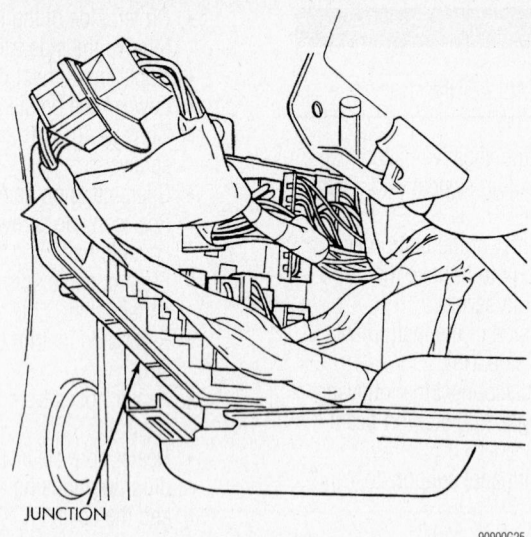

Junction block location—Cirrus, Stratus and Sebring convertible

90900G25

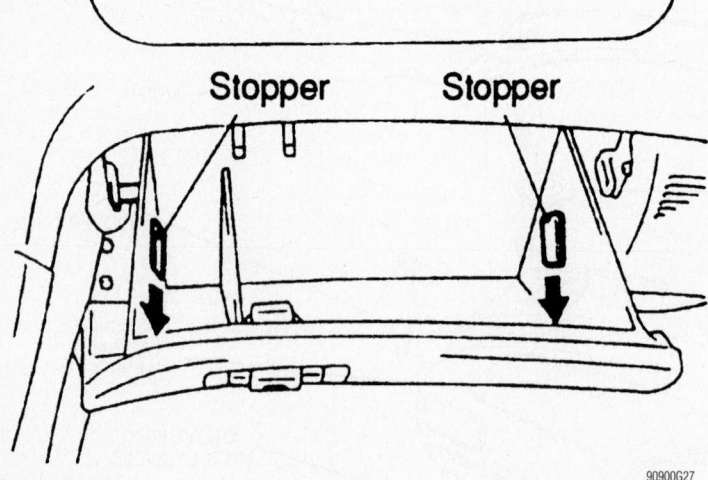

Unlock the glove box door stoppers by pushing in the direction shown—Cirrus, Stratus and Sebring convertible

90900G27

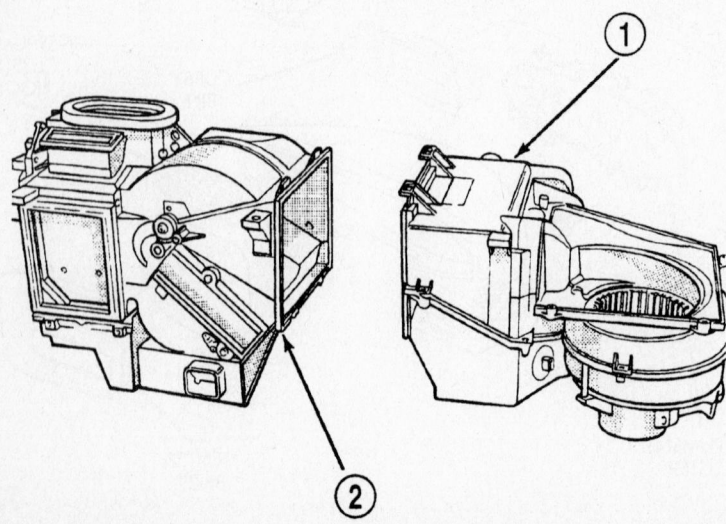

View of the heater/air conditioning housing assembly—Cirrus and Stratus

93111G79

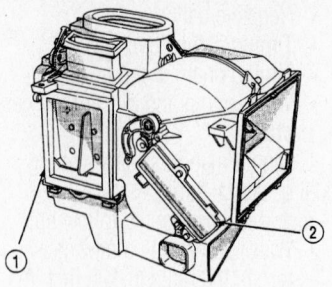

1 – HEATER DISTRIBUTION HOUSING
2 – HEATER CORE COVER

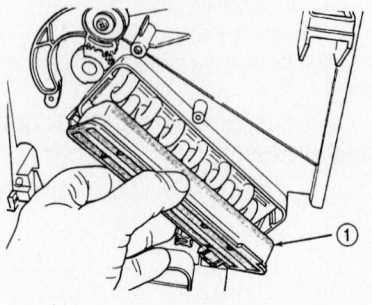

1 – HEATER CORE COVER

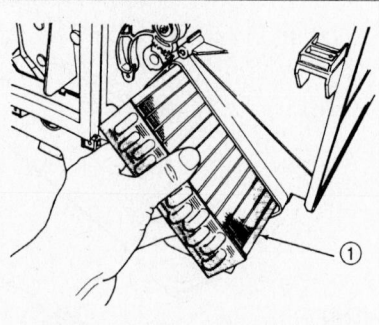

1 – HEATER CORE

93111G80

View of the heater housing assembly and the heater core—Cirrus and Stratus

- Heater core from the heater/air conditioning housing assembly

To install:

6. Install or connect the following:
   - Heater core to the heater/air conditioning housing assembly
   - Heater core cover and the cover-to-heater/air conditioning housing assembly screws
   - Heater hoses to the heater core tubes
   - Right side instrument panel support strut
   - Front console half and the front console screws
   - Rear console half and the rear console screws
   - Shifter bezel and the gearshift knob
   - Console-to-instrument panel screws
   - Lower knee bolster
   - Upper instrument panel bezel

- On left side of the instrument panel, install the side trim
- Instrument panel support-to-A-pillar bolt
- 2 lower right side support beam screws
- Right side of the instrument panel, install the side trim
- On center of the instrument panel, install the radio/control module bezel

7. Refill the cooling system.
8. Connect the negative battery cable.
9. Operate the engine to normal operating temperatures; then, check the climate control operation and check for leaks.

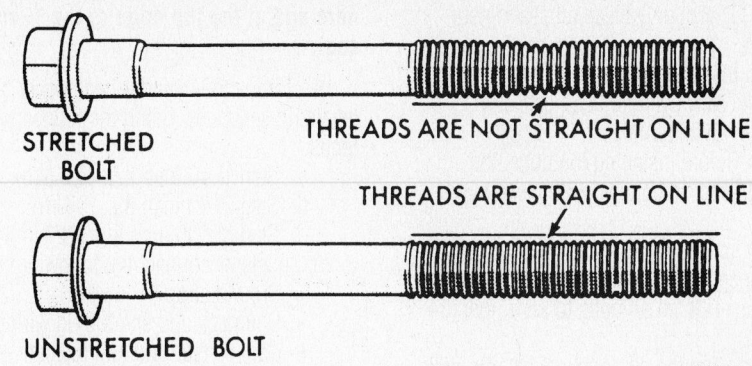

Before reusing old cylinder head bolts, check them for necking (stretching)—2.4L engine

Cylinder Head

REMOVAL & INSTALLATION

➡The cylinder head bolts should be checked for stretching before reuse. If the thread area of the bolt is necked-down the bolts must be replaced with new. New head bolts are recommended.

2.4L Engine

This engine uses a Dual Over Head Camshaft (DOHC) 4-valves per cylinder cross flow aluminum cylinder head. The valves are actuated by roller cam followers which pivot on stationary hydraulic valve adjusters. Care must be taken to be sure all valve timing marks align after cylinder head and valvetrain service.

1. Before servicing the vehicle, refer to the precautions in the beginning of this section.
2. Relieve the fuel system pressure.
3. Drain the cooling system.
4. Disconnect the negative battery cable from the left shock tower.

➡The ground cable is equipped with an insulator grommet which should be placed on the stud to prevent the negative battery cable from accidentally grounding.

5. Remove or disconnect the following:
- Air cleaner assembly
- All vacuum lines, electrical connectors and fuel lines from the throttle body
- Throttle linkage
- Accessory drive belts
- Power brake vacuum hose from the intake manifold
- Exhaust pipe from the exhaust manifold

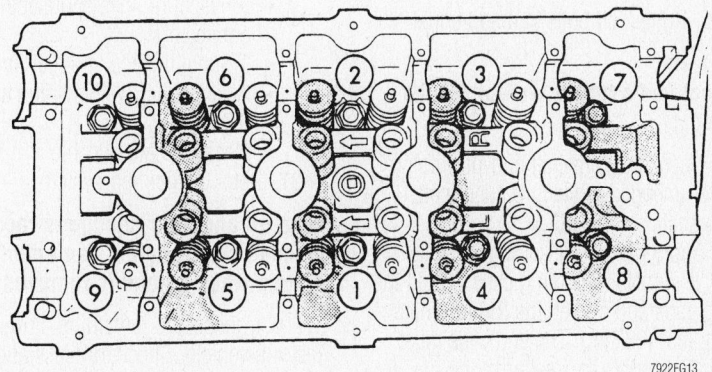

Cylinder head bolt torque sequence—2.4L engine

- Power steering pump and move it aside. Do not disconnect the fluid lines
- Spark plug wires
- Coil pack electrical connector
- Coil pack with the spark plug wires
- Cam sensor and fuel injector electrical connectors
- Timing belt covers
- Timing belt
- Camshaft sprockets
- Timing belt idler pulley
- Rear timing belt cover
- Cylinder head cover
- Ground strap

6. Identify the camshafts if they are to be reused for later installation. The camshafts are not interchangeable.
7. Remove or disconnect the following:
- Camshaft bearing cap bolts in sequence
- Camshafts

➡Refer to the camshaft removal and installation procedure for correct bolt removal/installation sequence.

- Camshaft followers

➡Any components that are to be reused must be installed in their original locations. Use care to identify and mark the positions of any removed valvetrain components so they may be reinstalled correctly.

- Intake and exhaust manifolds
- Cylinder head bolts
- Cylinder head

➡Be careful not to damage the aluminum gasket surfaces.

8. Remove all gasket material from the cylinder head and engine block. Be careful not to gouge or scratch the sealing surface of the aluminum head. The cylinder head should be checked for flatness using a good straightedge and feeler gauges. The cylinder head must be flat within 0.004 in. (0.1mm).
9. Inspect the camshaft bearing oil feed holes in the cylinder head for clogging. Inspect the camshaft bearing journals for wear or scoring. Check the cam surface for abnormal wear and damage. A visible worn groove in the roller path or on the cam lobes is cause for replacement. Valve service may be performed at this time.

To install:
10. Thoroughly clean all parts.

➡The cylinder head bolts are stretch-type. New cylinder head bolts are recommended.

11. Thoroughly clean all sealing surfaces.

12. Install or connect the following:
- New cylinder head gasket
- Cylinder head

13. Before installing the bolts, the threads should be oiled with clean engine oil.

14. Torque the cylinder head bolts in sequence, using the following 4 Steps:

a. Tighten all bolts to 25 ft. lbs. (34 Nm).

b. Tighten all bolts to 50 ft. lbs. (68 Nm).

c. Tighten all bolts again to 50 ft. lbs. (68 Nm).

d. Tighten all bolts and additional ¼ turn.

➡**Do not use a torque wrench for the 4th step.**

15. Check the camshaft end-play using the recommended procedure, then install the camshaft.

16. Apply Mopar Gasket Maker sealer to the No. 1 and No. 6 bearing caps. Install the bearing caps and tighten the M8 fasteners to 21 ft. lbs. (28 Nm). The end caps must be installed before the seals may be installed.

17. Apply a light coating of clean engine oil to the lip of the new camshaft seal. Install the camshaft seal until it fits flush with the cylinder head.

18. Install or connect the following:
- Camshaft sprockets, if removed
- Rear timing belt cover
- Timing belt and properly align the timing marks
- Timing belt cover

✳✳ WARNING

Verify that all timing marks are correct. If the timing belt or sprockets are incorrectly installed, engine damage will occur. Take time to be sure all timing marks are correctly aligned.

- Intake and exhaust manifolds
- New cylinder head cover gasket
- Cylinder head cover

✳✳ WARNING

DO NOT allow oil or solvents to contact the timing belt as they can deteriorate the rubber and cause tooth skipping.

➡**Apply Mopar Silicone Rubber Adhesive Sealant at the camshaft cap cor-**
ners and at the top edge of the ½ round seal.

19. Torque the cylinder head cover fasteners, in sequence, using the following Steps:

a. Step 1: 40 inch lbs. (4.5 Nm).

b. Step 2: 80 inch lbs. (9 Nm).

c. Step 3: 105 inch lbs. (12 Nm).

20. Install or connect the following:
- Ground strap
- Coil pack and spark plug wiring
- Cam sensor and fuel injector wiring
- Power steering pump assembly
- Exhaust pipe to the exhaust manifold
- All vacuum lines and electrical connectors
- Throttle linkage and fuel lines
- Accessory drive belts and adjust them
- Negative battery cable

21. Refill the cooling system.

➡**An oil and filter change is recommended since coolant can enter the oil system when a head is removed.**

22. Connect the remaining air ducting. Test run vehicle. Check for leaks and for proper operation.

2.7L Engine

1. Before servicing the vehicle, refer to the precautions in the beginning of this section.

2. Relieve the fuel system pressure.

3. Drain the cooling system.

4. Remove or disconnect the following:
- Negative battery cable
- Drive belts
- Vibration damper
- Exhaust cross under pipe
- Catalytic converter
- Upper and lower intake manifolds
- Rocker arm covers
- Water outlet connector and rotate the crankshaft until the sprocket timing mark aligns with the timing mark on the oil pump housing
- Primary timing chain
- Camshaft bearing caps gradually
- Camshafts and valve train components
- Cylinder head and discard the gasket

To install:

5. Clean all mating surfaces of any residual gasket material.

6. Lubricate the threads of the cylinder head bolts with clean engine oil.

RIGHT CYLINDER HEAD   **LEFT CYLINDER HEAD**

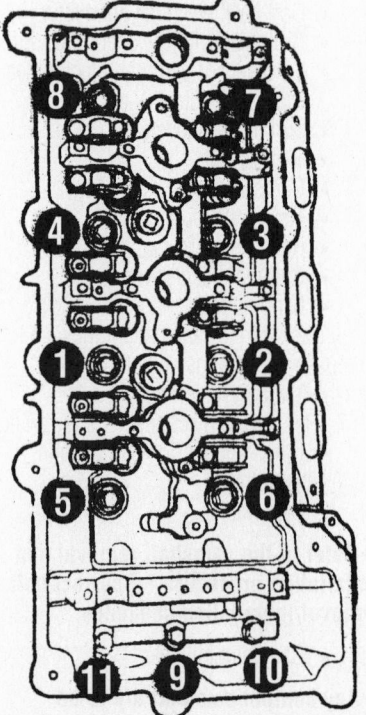
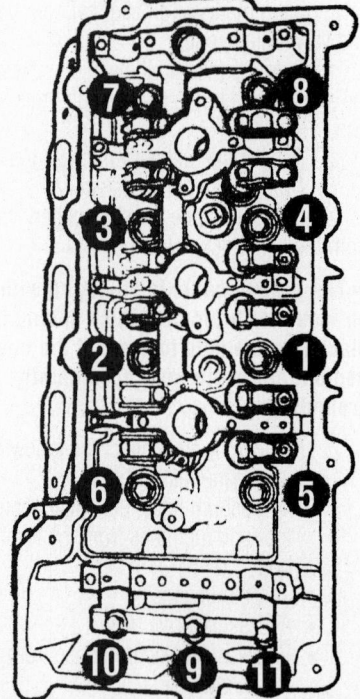

cylinder head bolt torque sequence—2.7L engine

9346IG03

7. Install or connect the following:
- Cylinder head with a new gasket

8. Torque the bolts, in sequence, using the following procedure:

   a. Step 1: Bolts 1–8: 35 ft. lbs. (48 Nm).

   b. Step 2: Bolts 1–8: 55 ft. lbs. (75 Nm).

   c. Step 3: Bolts 1–8: 55 ft. lbs. (75 Nm).

   d. Step 4: Bolts 1–8: An additional 90 degrees. Do not use a torque wrench for this step.

   e. Step 5: Bolts 9–11: 21 ft. lbs. (28 Nm).

- Valve train components and the camshafts
- Timing chain and sprockets
- Water outlet connector
- Rocker arm covers
- Timing chain cover
- Crankshaft vibration damper
- Lower and upper intake manifolds
- Catalytic converter
- Exhaust cross under pipe
- Drive belts
- Negative battery cable

9. Fill the cooling system to the proper level.

10. Start the vehicle, check for leaks and repair if necessary.

Rocker Arm/Shaft

REMOVAL & INSTALLATION

2.7L Engine

1. Before servicing the vehicle, refer to the precautions in the beginning of this section.

2. Relieve the fuel system pressure using the recommended procedure.

3. Remove or disconnect the following:
- Negative battery cable
- Rocker arm cover and rotate the engine until the cam lobe is on its base circle

4. Using special tools 8215—A and 8216—A, depress the valve spring enough to release the tension on the rocker arm.

5. Remove the rocker arm from the cylinder head.

To install:

6. Lubricate the rocker arms with clean engine oil.

7. Depress the valve spring far enough to install the rocker arm.

8. Install or connect the following:
- Rocker arm into its original position over the valve and lash adjuster.

- Rocker arm cover
- Negative battery cable

Intake Manifold

REMOVAL & INSTALLATION

2.4L Engine

The intake manifold is a long branch design made of cast aluminum. It is attached to the cylinder head with 8 fasteners.

1. Before servicing the vehicle, refer to the precautions in the beginning of this section.

2. Disconnect the negative battery cable from the left shock tower.

➡ **The ground cable is equipped with an insulator grommet which should be placed on the stud to prevent the negative battery cable from accidentally grounding.**

3. Relieve the fuel system pressure using the recommended procedure.

4. Remove the air inlet resonator as follows:

- Both air inlet resonator-to-intake manifold bolts
- Resonator-to-throttle body screw, loosen it
- Air inlet resonator-to-air inlet tube clamp, loosen it
- Resonator

5. Remove or disconnect the following:

- Fuel supply line quick-disconnect at the fuel tube assembly

➡ **Squeeze the retainer tabs together and pull the fuel tube/quick-disconnect fitting assembly from the fuel tube nipple. The retainer will remain on the fuel tube. Use shop towels to catch any dripping fuel.**

- Fuel rail

❋❋ WARNING

Use care when handling the fuel injectors. Do not set them on their tips. Cover the fuel injector openings after fuel rail removal.

- Accelerator, kickdown and speed control cables, from the throttle lever and bracket
- Idle Air Control (IAC) motor and Throttle Position (TP) sensor electrical connections
- Throttle body vacuum hoses
- Manifold Absolute Pressure (MAP) sensor and Intake Air Temperature (IAT) sensor electrical connections
- Vapor and brake booster hoses
- Knock sensor electrical connector
- Wiring harness from the intake manifold tab
- Transaxle-to-throttle body support bracket fasteners at the throttle body and loosen the fastener at the transaxle end
- Throttle body

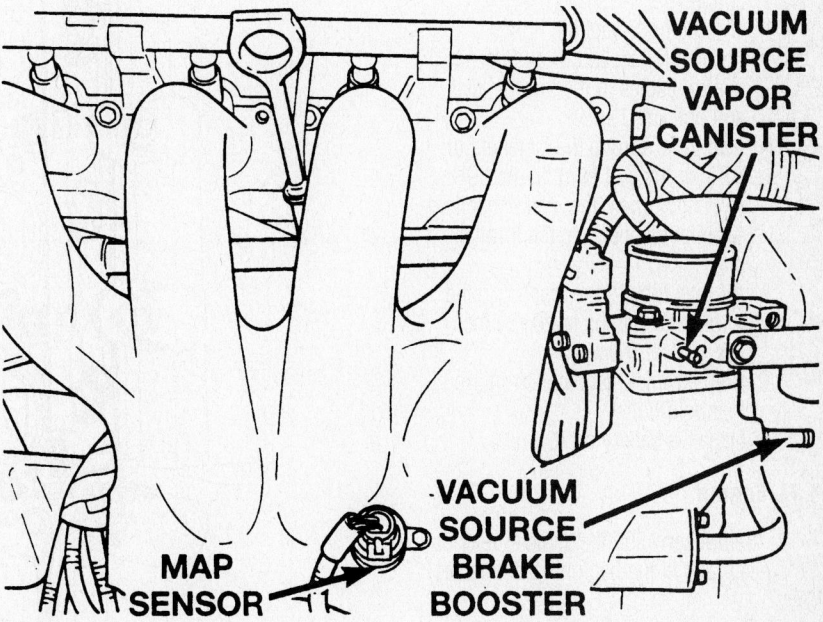

MAP sensor location—2.4L engine

7922FG25

- Exhaust Gas Recirculation (EGR) tube
- Intake manifold support bracket
- Intake manifold

To install:

6. Thoroughly clean all parts. Clean all sealing surfaces

7. Install or connect the following:
   - New gasket
   - Intake manifold. Torque the fasteners to 17 ft. lbs. (23 Nm), in correct sequence, starting at the center and working outward

➡ **Make sure the fuel injector openings clean.**

- Fuel rail assembly to the intake manifold and torque the screws to 17 ft. lbs. (23 Nm)
- Positive Crankcase Ventilation (PCV) and brake booster hoses

8. Lubricate the fuel tube with engine oil.

9. Install or connect the following:
   - Fuel supply line to the fuel rail. Pull on the connector to insure it is locked into position
   - Throttle body and torque the fasteners to 17 ft. lbs. (23 Nm)
   - Transaxle-to-throttle body support bracket. Torque the fasteners to 105 inch lbs. (12 Nm) at the throttle body first; then, at the transaxle
   - MAP and IAT electrical connectors
   - Knock sensor electrical connector
   - Wiring harness to the intake manifold tab
   - IAC motor and TP sensor electrical connectors
   - Throttle body vacuum hoses
   - Accelerator, kickdown and speed control cables to the throttle lever and bracket
   - EGR tube. Torque the fasteners to 95 inch lbs. (11 Nm), at the EGR valve first; then, the intake manifold
   - Air inlet resonator to the throttle body
   - Air inlet tube to the resonator and torque the clamps to 20–30 inch lbs. (2.5–3.5 Nm)
   - Both air inlet resonator-to-intake manifold bolts
   - Negative battery cable

2.7L Engine

1. Before servicing the vehicle, refer to the precautions in the beginning of this section.

2. Properly relieve the fuel system pressure.

3. Remove or disconnect the following:
   - Negative battery cable
   - Throttle body air inlet hose
   - Air cleaner assembly
   - Throttle cable shield
   - Throttle and speed control cables
   - Throttle cable bracket
   - Manifold Absolute Pressure (MAP) sensor connector
   - Throttle Position Sensor (TPS) connector
   - Idle Air Control (IAC) motor connector
   - Manifold Tuning Valve (MTV) connector
   - Vapor purge hose
   - Brake booster vacuum hose
   - Speed Control Servo
   - Positive Crankcase Ventilation (PCV) hose
   - Exhaust Gas Recirculation (EGR) upper tube
   - Upper throttle body support bracket
   - Upper intake manifold and discard the gasket

4. To remove the lower intake manifold proceed as follows:

5. Remove the fuel injector electrical connectors.

6. Remove the fuel supply hose from the fuel rail.

7. Remove the fuel rail support bracket.

8. Remove the fuel rail and injectors as an assembly.

9. Remove the lower intake manifold and discard the gasket.

To install:

10. Clean all mating surfaces of any residual gasket material.

11. Install or connect the following:
   - Lower intake manifold with a new gasket
   - Fuel rail and injectors and torque the lower intake manifold bolts, in sequence, to 105 inch lbs. (12 Nm)
   - Fuel supply hose to the fuel rail
   - Fuel rail support bracket to the throttle body
   - Fuel injector electrical connectors
   - Upper intake manifold with a new gasket on the lower intake manifold and torque the bolts, in sequence, to 105 inch lbs. (12 Nm)
   - EGR upper tube
   - Speed control servo
   - PCV, brake booster and vapor purge hoses
   - MAP sensor electrical connector
   - TPS sensor electrical connector

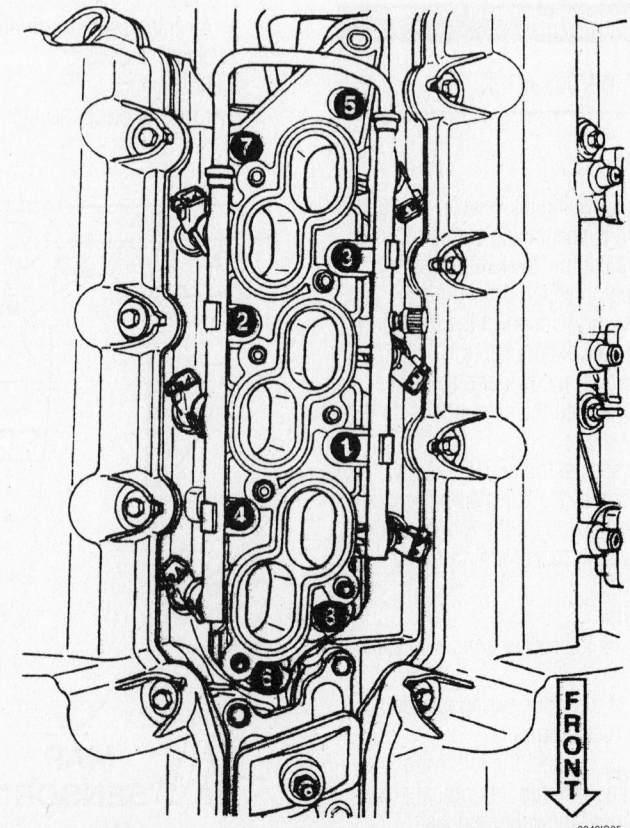

Intake manifold bolt torque sequence 2.7L engine

93461G05

- IAC motor electrical connector
- MTV electrical connector
- Throttle cable bracket
- Throttle and speed control cables
- Throttle cable shield
- Throttle body air inlet hose
- Air cleaner assembly
- Negative battery cable

12. Start the vehicle, check for leaks and repair if necessary.

Exhaust Manifold

REMOVAL & INSTALLATION

2.4L Engine

1. Before servicing the vehicle, refer to the precautions in the beginning of this section.

2. Disconnect the negative battery cable from the left shock tower.

➡**The ground cable is equipped with an insulator grommet which should be placed on the stud to prevent the negative battery cable from accidentally grounding.**

3. Remove or disconnect the following:
- Exhaust pipe from the exhaust manifold
- Exhaust manifold heat shield
- Heated Oxygen Sensor (HO2S), if necessary
- Exhaust manifold and discard the gasket

To install:

4. Thoroughly clean all parts. Clean all sealing surfaces of the manifold and cylinder head. Check the manifold gasket surface for flatness with a straightedge and feeler gauge. The surface must be flat within 0.006 in. (0.15mm) per foot (30cm) of manifold length. Inspect the manifold for cracks or distortion. Replace if necessary.

5. Install or connect the following:

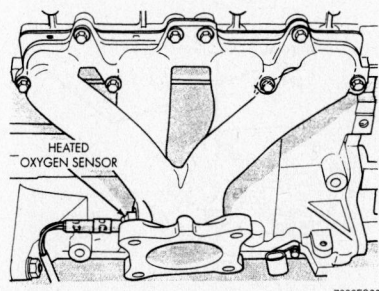

Be careful not to damage the oxygen sensor when servicing the manifold—2.4L engine

- New gasket
- Exhaust manifold. Torque the bolts, starting at the center and working outward, to 17 ft. lbs. (23 Nm)
- HO2S sensor
- Heat shield
- Exhaust pipe and torque the fasteners to 21 ft. lbs. (28 Nm)
- Negative battery cable

6. Start the engine and allow it to idle while inspecting the manifold for exhaust leaks.

2.7L Engine

FRONT

1. Before servicing the vehicle, refer to the precautions in the beginning of this section.

2. Remove or disconnect the following:
- Negative battery cable
- Cross under pipe
- Front catalytic converter
- Exhaust manifold and discard the gasket

To install:

3. Clean all mating surfaces of any residual gasket material.

4. Install or connect the following:
- Exhaust manifold with a new gasket. Torque the bolts, starting in the center and working outward, to 17 ft. lbs. (23 Nm)
- Front catalytic converter and heat shield and torque the bolts to 21 ft. lbs. (28 Nm)
- Cross under pipe and torque the bolts to 21 ft. lbs. (28 Nm)
- Negative battery cable

5. Start the vehicle, check for leaks and repair if necessary.

REAR

1. Before servicing the vehicle, refer to the precautions in the beginning of this section.

2. Remove or disconnect the following:
- Negative battery cable
- Throttle body air inlet hose
- Air cleaner assembly
- Exhaust Gas Recirculation (EGR) tube from the exhaust manifold and EGR valve
- Exhaust system
- Cross under pipe
- Rear catalytic converter
- Rear Oxygen Sensor (O 2S)
- Exhaust manifold and discard the gasket

To install:

3. Clean all mating surfaces of any residual gasket material.

4. Install or connect the following:
- Exhaust manifold with a new gasket. Torque the bolts, starting in the center and working outward, to 17 ft. lbs. (23 Nm)
- Rear catalytic converter and heat shield and torque the bolts to 21 ft. lbs. (28 Nm)
- Rear Oxygen Sensor (O 2S)
- Cross under pipe and torque the bolts to 21 ft. lbs. (28 Nm)
- Exhaust system
- EGR tube with new gaskets
- Air cleaner assembly and air intake hose
- Negative battery cable

5. Start the vehicle, check for leaks and repair if necessary.

Front Crankshaft Seal

REMOVAL & INSTALLATION

2.4L Engine

The timing belt must be removed for this procedure. Use care that all timing marks are aligned after installation or the engine will be damaged.

1. Drain the engine oil.

2. Before servicing the vehicle, refer to the precautions in the beginning of this section.

3. Disconnect the negative battery cable from the left shock tower.

➡**The ground cable is equipped with an insulator grommet which should be placed on the stud to prevent the negative battery cable from accidentally grounding.**

4. Remove or disconnect the following:
- Accessory drive belts
- Crankshaft damper/pulley using a jaw puller tool
- Timing belt
- Crankshaft timing belt sprocket using a gear/sprocket puller

➡**Be careful not to nick the seal surface of the crankshaft or the seal bore.**

- Front crankshaft seal using Seal Removal Tool No. 6771

➡**Be careful not to damage the seal contact area of the crankshaft.**

To install:

5. Lubricate the new oil seal lip with engine oil.

6. Install or connect the following:
- New crankshaft oil seal with the

spring facing inward using Oil Seal Installer Tool 6780–1, until it is flush with the front cover

- Crankshaft timing belt sprocket using Tool No. 6792

➡**Be sure the word "FRONT" on the timing belt sprocket is facing outward.**

- Timing belt and timing belt cover
- Crankshaft damper/pulley using the thrust bearing/washer and 12M-1.75 x 150mm bolt from special Tool No. 6792. Torque the bolt to 105 ft. lbs. (142 Nm)
- Accessory drive belts and adjust the tension
- Negative battery cable

❉❉ WARNING

Operating the engine without the proper amount and type of engine oil will result in severe engine damage.

7. Fill the engine with clean oil.
8. Start the engine and check for leaks.

2.7L Engine

1. Before servicing the vehicle, refer to the precautions in the beginning of this section.

2. Remove or disconnect the following:
- Negative battery cable
- Crankshaft damper
- Front crankshaft seal with special Tool 6771

To install:
3. Install or connect the following:
- New seal with special Tools 6780–2, 8179 and 6780–1
- Crankshaft damper
- Negative battery cable

Camshaft

REMOVAL & INSTALLATION

2.4L Engine

This engine uses a DOHC, 4-valves per cylinder, cross-flow aluminum cylinder head. The valves are actuated by roller cam followers which pivot on stationary hydraulic valve adjusters. Care must be taken to ensure all valve timing marks align after cylinder head and valvetrain service.

1. Before servicing the vehicle, refer to the precautions in the beginning of this section.

2. Disconnect the negative battery cable from the left shock tower.

➡**The ground cable is equipped with an insulator grommet which should be placed on the stud to prevent the negative battery cable from accidentally grounding.**

3. Relieve the fuel system pressure using the recommended procedure.
4. Remove or disconnect the following:
- Spark plugs cables
- Ignition coil pack with the spark plug cables
- Cylinder head cover and discard the gasket
- Ground strap
- Timing belt covers and timing belt
- Camshaft sprockets by holding them with Tool 6847, while removing the bolt

5. Take note that the camshaft bearing caps are numbered for correct location during installation. Remove the outer bearing caps first.

6. Loosen, but do not remove, the camshaft bearing cap retaining fasteners in the correct sequence, inside working outward. Perform this step on one camshaft at a time.

7. Identify the camshafts, if they are to be reused for later installation. The camshafts are not interchangeable.

8. Remove or disconnect the following:
- Camshaft bearing caps
- Camshafts
- Camshaft followers

9. Any components that are to be reused must be installed in their original locations. Use care to identify and mark the positions of any removed valvetrain components so they may be reinstalled correctly.

10. Inspect the camshaft bearing oil feed holes in the cylinder head for clogging. Inspect the camshaft bearing journals for wear or scoring. Check the cam surface for abnormal wear and damage. A visible worn

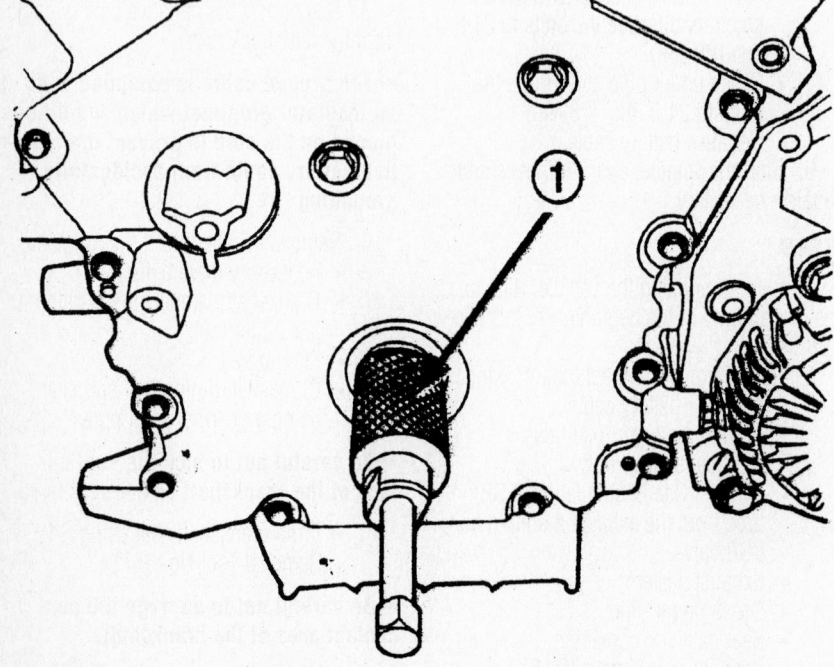

1 - SPECIAL TOOL 6771

9346IG06

Remove the front crankshaft seal with special tool 6771

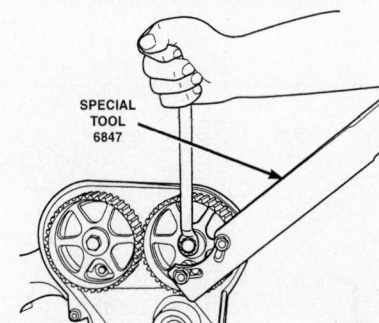

9300FG04

Use special tool 6847 to hold the camshaft sprocket while removing or installing the center bolt—2.4L engine

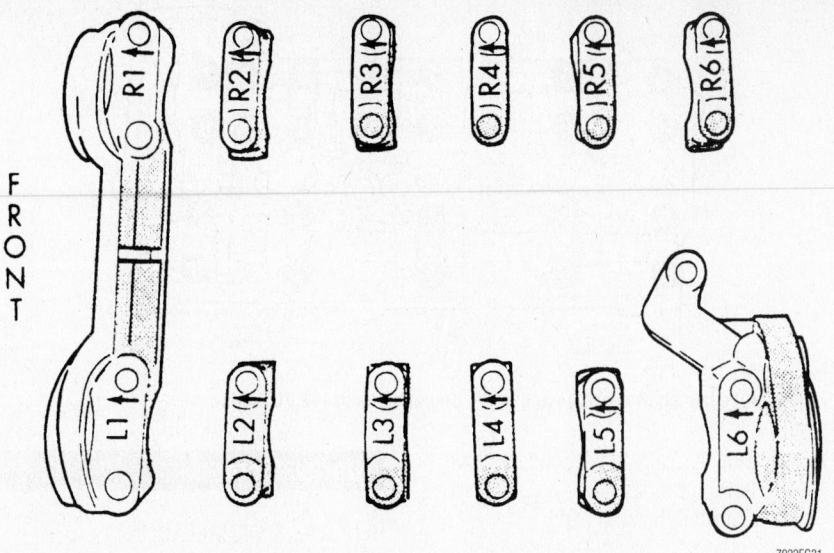

Camshaft bearing cap identification—2.4L engine

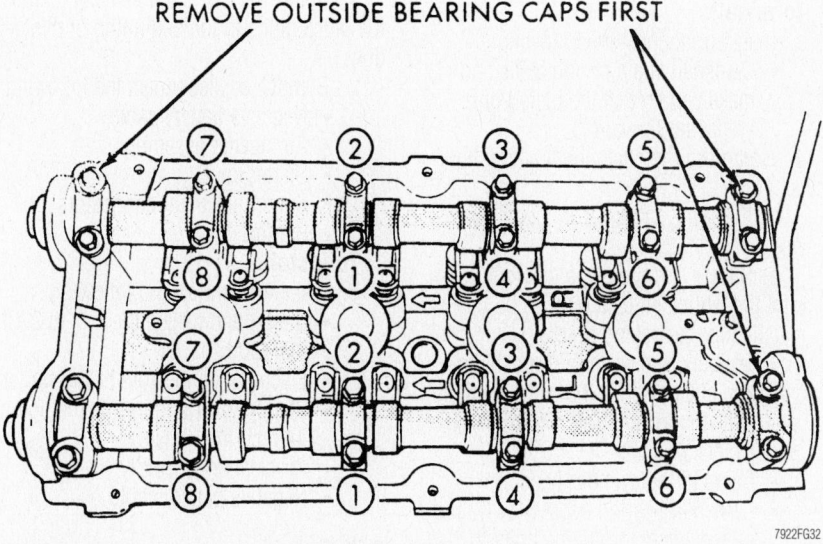

Camshaft bearing cap bolt removal sequence—2.4L engine

as it will go. Read the dial indicator. End-play specification is 0.002–0.010 in. (0.05–0.15mm).

   e. If excessive end-play is present, inspect the cylinder head and camshaft for wear; replace if necessary.

13. If the fit and condition of the camshafts are acceptable, remove the camshafts for installation of the cam followers.

14. The hydraulic valve lash adjusters are inside the roller cam followers. Be sure they are clean, well lubricated with engine oil and properly positioned. Install the cam followers in their original positions on the hydraulic adjuster and valve stem.

✳✳ WARNING

Be sure NONE of the pistons are at Top Dead Center (TDC) when installing the camshafts.

15. Lubricate the camshaft bearing journals and cam followers with clean engine oil

16. Install or connect the following:
   • Camshafts
   • Right/left-side camshaft bearing caps Nos. 2 through 5 and right-side No. 6. Torque the M6 fasteners to 105 inch lbs. (12 Nm) in the correct sequence.

17. Apply Mopar® Gasket Maker sealer to the No. 1 and left-side No. 6 bearing caps.

18. Install or connect the following:
   • Bearing caps and torque the M8 fasteners to 21 ft. lbs. (28 Nm)
   • Camshaft end seals
   • Camshaft sprockets, if removed, and torque the bolts to 75 ft. lbs. (101 Nm)

groove in the roller path or on the cam lobes is cause for replacement.

   To install:

11. Thoroughly clean all camshaft and related parts.

12. Inspect the camshaft end-play using the following procedure:

   a. Lubricate the camshaft journals and install the camshaft **WITHOUT** the cam follower assemblies. Install the rear cam caps and tighten to 21 ft. lbs. (28 Nm).

   b. Push the camshaft rearward as far as it will go.

   c. Adjust a dial indicator to rest against the front of the camshaft (the sprocket end). Zero the indicator.

   d. Move the camshaft forward as far

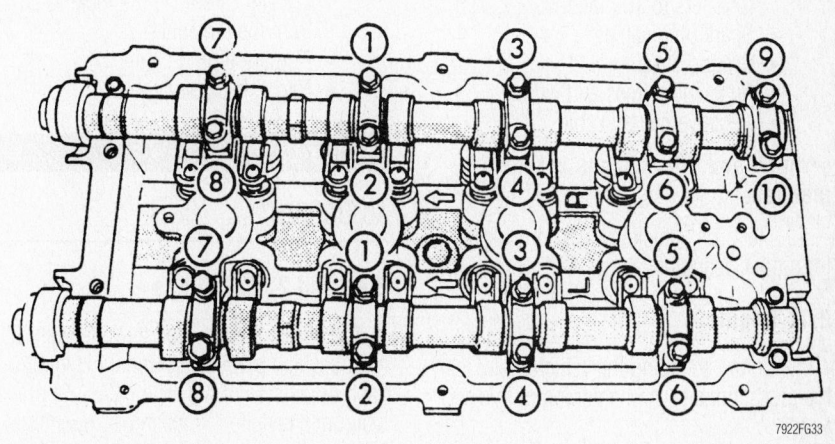

Camshaft bearing cap tightening sequence—2.4L engine

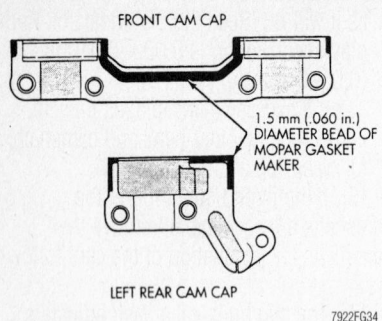

FRONT CAM CAP

1.5 mm (.060 in.)
DIAMETER BEAD OF
MOPAR GASKET
MAKER

LEFT REAR CAM CAP

7922FG34

Apply sealer as shown to prevent oil leakage from the camshaft bearing end caps—2.4L engine

- Timing belt, making sure all timing marks are aligned
- Timing belt covers

✳✳ WARNING

If the timing belt or sprockets are incorrectly installed, engine damage will occur. Take time to be sure all timing marks are correctly aligned.

19. Clean all sealing surfaces. Make certain the rails are flat.

20. Install new cylinder head cover gaskets. Apply Mopar® Silicone Rubber Adhesive Sealant, at the camshaft cap corners and at the top edge of the ½ round seal.

➡Inspect the spark plug well seals for cracking and/or swelling and replace if necessary.

21. Install the cylinder head cover and torque the fasteners, in sequence, using the following:
   a. Step 1: 40 inch lbs. (4.5 Nm).
   b. Step 2: 80 inch lbs. (9 Nm).
   c. Step 3: 105 inch lbs. (12 Nm).

22. Install or connect the following:
   - Ignition coil pack and torque the fasteners to 105 inch lbs. (12 Nm)
   - Spark plug cables
   - Ground strap
   - All vacuum lines and wiring
   - Negative battery cable

➡An oil and filter change is recommended.

23. Test run vehicle. Check for leaks and for proper operation.

2.7L Engine

1. Before servicing the vehicle, refer to the precautions in the beginning of this section.

2. Remove or disconnect the following:
   - Negative battery

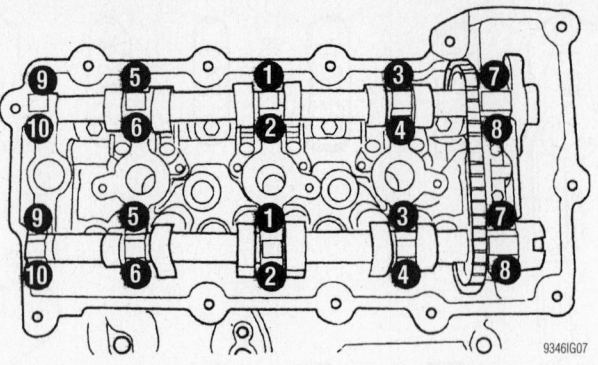

93461G07

Tighten the camshaft bearing caps in the proper sequence–2.7L Engine

- Timing chain and secondary chain tensioner bolts
- Camshaft bearing caps
- Camshafts, secondary chain and tensioner as an assembly
- Tensioner and chain from the camshafts

To install:

3. Install or connect the following:
   - Camshaft chain on the cams and make certain that the plated links are facing forward

4. Compress the chain tensioner as follows:
   a. Place the tensioner in a soft jaw vise.
   b. Slowly compress the tensioner until the fabricated lock pin can be installed in the locking holes.
   - Tensioner between the camshafts and chain. Rotate the cams until the plated links and dots are in the 12 o'clock position
   - Cams to the cylinder head
   - Camshaft bearing caps. Slowly, and in sequence, torque the bolts to 105 inch lbs. (12 Nm)
   - Secondary chain tensioner bolts and torque them to 105 inch lbs. (12 Nm). Remove the locking pin from the tensioner
   - Primary timing chain
   - Negative battery cable

Valve Lash

ADJUSTMENT

2.4L and 2.7L Engines

The valves are actuated by roller cam followers which pivot on stationary Hydraulic Lash Adjusters (HLAs). The HLAs are precision units installed in machined openings of the cam follower. Valve clearance adjustments are not performed.

Starter Motor

REMOVAL & INSTALLATION

2.4L Engine

1. Before servicing the vehicle, refer to the precautions in the beginning of this section.

2. Remove or disconnect the following:
   - Negative battery cable
   - Air cleaner assembly
   - Starter cover
   - Starter electrical connectors
   - Starter

To install:

3. Install or connect the following:
   - Starter and torque the bolt to 23 ft. lbs. (30 Nm)
   - Electrical connectors to the starter
   - Starter cover and torque the bolt to 44 inch lbs. (5 Nm)
   - Air cleaner assembly
   - Negative battery cable

2.7L Engine

1. Before servicing the vehicle, refer to the precautions in the beginning of this section.

2. Remove or disconnect the following:
   - Negative battery cable
   - Oxygen Sensor (O2S) electrical connector
   - Front engine mount through bolt
   - Starter electrical connectors
   - Starter

To install:

3. Install or connect the following:
   - Starter and hand tighten the lower bolt
   - Starter electrical connectors and torque the starter bolts to 40 ft. lbs. (54 Nm)
   - Engine mount through bolt and torque it to 45 ft. lbs. (61 Nm)

- O$_2$ sensor and torque it to 20 ft. lbs. (27 Nm)
- Negative battery cable

Oil Pan

REMOVAL & INSTALLATION

2.4L Engine

1. Before servicing the vehicle, refer to the precautions in the beginning of this section.
2. Drain the engine oil.
3. Remove or disconnect the following:
   - Negative battery cable
   - Oil dipstick
   - Front exhaust pipe
   - Bell housing cover
   - Oil pan and discard the gasket
4. Clean all mating surfaces of any residual gasket material.

To install:

5. Install or connect the following:
   - Oil pan with a new gasket and torque the bolts to 62 inch lbs. (7 Nm)
   - Bell housing cover and torque the upper bolt to 80 inch lbs. (9 Nm) and the lower bolt to 19 ft. lbs. (26 Nm)
   - Front exhaust pipe
   - Oil dipstick
   - Negative battery cable
6. Fill the engine with clean oil.
   - Start the vehicle, check for leaks and repair if necessary.

2.7L Engine

1. Before servicing the vehicle, refer to the precautions in the beginning of this section.
2. Drain the engine oil.
3. Remove the engine support module as follows:
   - Negative battery cable
   - Oil dipstick and tube
   - Structural collar
   - Exhaust cross under pipe
   - Torque converter cover
   - Lower bolt from the A/C compressor
   - Oil pan and discard the gasket

To install:

4. Clean all mating surfaces of any residual gasket material.
   - Oil pan with a new gasket
5. Torque the oil pan bolts as follows:
   a. Timing chain cover to oil pan bolts and torque them to 105 inch lbs. (12 Nm).

   b. Oil pan bolts to 21 ft. lbs. (28 Nm).
   c. Oil pan nuts to 105 inch lbs. (12 Nm).
   d. A/C compressor bolt to 21 ft. lbs. (28 Nm).
   - Torque converter cover
   - Exhaust cross under pipe
   - Structural collar
   - Oil dipstick and tube
   - Negative battery cable
6. Fill the engine with clean oil.
7. Start the vehicle, check for leaks and repair if necessary.

Oil Pump

REMOVAL & INSTALLATION

2.4L Engine

The oil drawn up through the pick-up tube is pressurized by the pump and routed through the full flow filter to the main oil galley running the length of the cylinder block. The oil pick-up, pump and check valve provide oil flow to the main oil gallery. A vertical hole at the No. 5 bulkhead routes pressurized oil through a restrictor up past a cylinder head bolt to an oil galley running the length of the cylinder head. The camshaft journals are slotted to allow pressurized oil to pass into the bearing cap cavities. Small holes in the bearing caps direct oil to the camshaft lobes.

1. Before servicing the vehicle, refer to the precautions in the beginning of this section.
2. Drain the engine oil.
3. Remove the engine support module as follows:
   - Negative battery cable
   - Crankshaft damper
   - Timing belt and tensioner
   - Camshaft sprocket
   - Rear timing belt cover
   - Oil pan
   - Oil pick up tube
   - Oil pump and discard the gasket
4. Clean all mating surfaces of any residual gasket material.

To install:

5. Prime the oil pump before installation.
6. Install or connect the following:
   - Oil pump with the flats aligned with the crankshaft flats and torque the bolts to 21 ft. lbs. (28 Nm)
   - New front crankshaft seal
   - Crankshaft sprocket
   - Oil pick up tube and oil pan
   - Camshaft sprocket

- Rear timing belt cover
- Timing belt and tensioner
- Front timing belt cover
- Crankcase damper
- Negative battery cable
7. Fill the engine with clean oil.
8. Start the vehicle, check for leaks and repair if necessary.

2.7L Engine

1. Before servicing the vehicle, refer to the precautions in the beginning of this section.
2. Drain the engine oil.
3. Remove the engine support module as follows:
   - Negative battery cable
   - Crankshaft damper
   - Timing chain cover
   - Timing chain and sprockets
   - Oil pan
   - Oil pick up tube and O-ring and make certain that the crankshaft is at 60 degrees Above Top Dead Center (ATDC)
   - Oil pump

To install:

4. Prime the pump before installation by filling the rotor cavity with clean engine oil.
5. Install or connect the following:
   - Oil pump over the crankshaft and torque the bolts to 21 ft. lbs. (28 Nm)
   - Oil pick up tube with a new O-ring and torque the bolt to 21 ft. lbs. (28 Nm)
   - Oil pan with a new gasket
   - Timing chain and sprockets
   - Timing chain and cover
   - Crankshaft damper
   - Negative battery cable
6. Fill the engine with clean oil.
7. Start the vehicle, check for leaks and repair if necessary.

Rear Main Seal

REMOVAL & INSTALLATION

1. Before servicing the vehicle, refer to the precautions in the beginning of this section.
2. Remove or disconnect the following:
   - Negative battery cable from the left shock tower.

➡**The ground cable is equipped with an insulator grommet which should be placed on the stud to prevent the negative battery cable from accidentally grounding.**

- Transaxle
- Flexplate/flywheel
- Rear crankshaft oil seal using a flat-bladed prying tool

To install:

➡ **When installing the seal there is no need to lubricate the sealing surface.**

3. Install or connect the following:
   - New seal using a suitable driver or seal installer
   - Flexplate/flywheel and torque the bolts to 70 ft. lbs. (95 Nm)
   - Transaxle
   - Negative battery cable
4. Start the engine and check for leaks.

Balance Shaft Chain, Sprockets And Cover

REMOVAL & INSTALLATION

2.4L Engine

1. Before servicing the vehicle, refer to the precautions in the beginning of this section.
2. Drain the engine oil.
3. Remove or disconnect the following:
   - Oil pan and discard the gasket
   - Chain cover, guide and tensioner
   - Gear cover and balance shaft gears
   - Balance shaft gear and crankshaft chain sprocket
   - Carrier gear cover and balance shafts

Balance Shaft Timing and Installation

➡ **During installation of the balance shafts to crankshaft, the proper timing procedure must be followed to prevent internal engine damage.**

1. Install or connect the following:
2. Install the balance shafts and carrier on the crankshaft. Torque the bolts to 40 ft. lbs. (54 Nm)
3. Turn the balance shafts until both key ways are pointing up. Install the short hub drive gear on the sprocket driven shaft.
4. Make certain that the key ways and the gear timing marks are properly aligned.
5. Install the gear cover and torque the double ended stud and washer to 105 inch lbs. (12 Nm)
6. Align the flat of the balance shaft drive sprocket to the flat on the crankshaft
7. Install the balance shaft drive sprocket to the crankshaft with special tool 6052.
8. Turn the crankshaft until the No. 1 cylinder is at Top Dead Center (TDC).

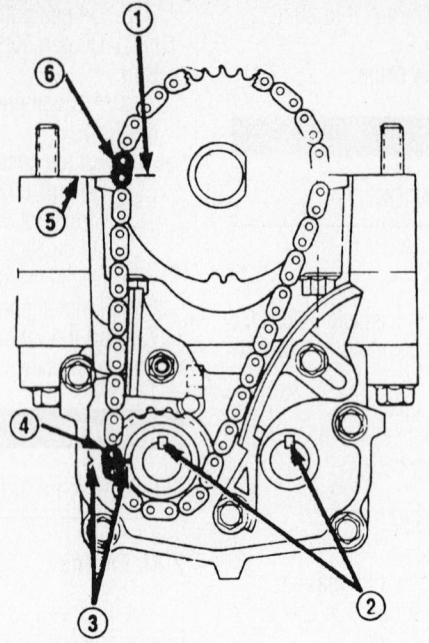

1 - MARK ON SPROCKET
2 - KEYWAYS UP
3 - ALIGN MARKS
4 - PLATED LINK
5 - PARTING LINE (BEDPLATE TO BLOCK)
6 - PLATED LINK

9346IG10

Align the key ways and the gear timing marks

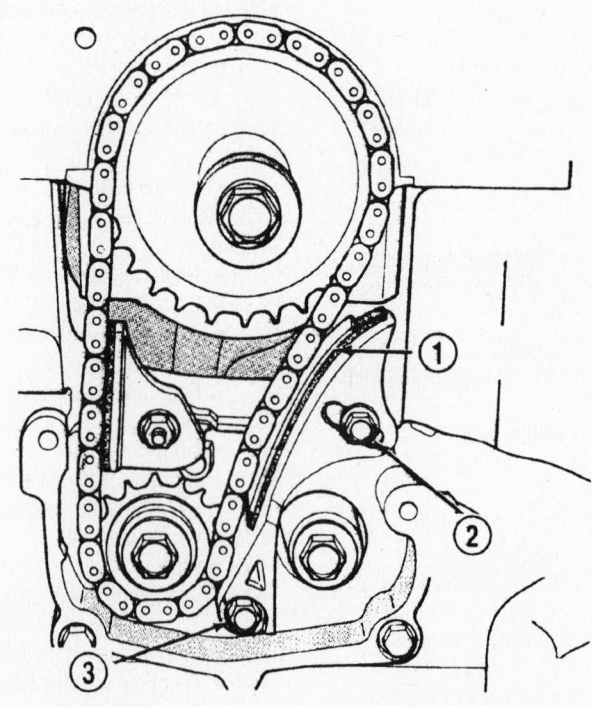

1 - 1MM (0.039 IN.) SHIM
2 - TENSIONER (ADJUSTER) BOLT
3 - PIVOT BOLT

9346JG11

Properly adjust the chain tension

9. Make certain that the timing marks on the chain sprocket are aligned with the parting line on the left side main bearing cap.

10. Install the chain over the crankshaft and be sure the plated link is over the No. 1 cylinder timing mark on the balance shaft crankshaft sprocket.

11. Place the balance shaft sprocket into the timing chain and align the sprocket dot to the lower plate link on the chain.

12. Slide the balance shaft sprocket on the nose of the balance shaft with the key ways pointing at the 12 o'clock position.

13. When properly timed, install the balance shaft bolts and torque to 21 ft. lbs. (28 Nm)

14. To tighten the chain use the following procedure:

    a. Loosely install the chain tensioner.

    b. Position the guide on the double ended stud and make certain that the tab on the guide fits into the slot on the gear cover.

    c. Install and tighten the nut/washer assembly to 105 inch lbs. (12 Nm).

    d. Place a 1 mm thick by 70 mm long shim between the tensioner and chain.

    e. With a load applied, torque the top tensioner bolt to 105 inch lbs. (12 Nm).

    f. Torque the bottom pivot bolt to 105 inch lbs. (12 Nm).

    g. Remove the shim.

15. Install the carrier covers and torque the bolts to 105 inch lbs. (12 Nm).

16. Install the oil pan.

17. Fill the engine with clean oil.

18. Start the vehicle, check for leaks and repair if necessary.

Timing Belt

REMOVAL & INSTALLATION

2.4L (VIN X) Engine

1. Disconnect the negative battery cable from the left strut tower. The ground cable is equipped with a insulator grommet which should be placed on the stud to prevent the negative battery cable from accidentally grounding.

2. Remove the right inner splash-shield.

3. Remove the accessory drive belts.

4. Remove the crankshaft damper.

5. Remove the right engine mount.

6. Place a suitable floor jack under the vehicle to support the engine.

7. Remove the engine mount bracket.

8. Remove the timing belt cover.

➡ **Do not rotate the crankshaft or the camshafts after the timing belt has been removed. Damage to the valve components may occur. Before removing the timing belt, always align the timing marks.**

9. Align the timing marks of the timing belt sprockets to the timing marks on the rear timing belt cover and oil pump cover. Loosen the timing belt tensioner bolts.

10. Remove the timing belt and the tensioner.

11. Remove the camshaft timing belt sprockets.

12. Remove the crankshaft timing belt sprocket using special removal tool No. 6793.

13. Place the tensioner into a soft-jawed vise to compress the tensioner.

14. After compressing the tensioner, place a pin (a ⁵⁄₆₄ in. Allen wrench will work) into the plunger side hole to retain the plunger until installation.

To install:

15. Using special tool No. 6792, install the crankshaft timing belt sprocket onto the crankshaft.

16. Install the camshaft sprockets onto the camshafts. Install and tighten the camshaft sprocket bolts to 75 ft. lbs. (101 Nm).

17. Set the crankshaft sprocket to Top Dead Center (TDC) by aligning the notch on the sprocket with the arrow on the oil pump housing.

18. Set the camshafts to align the timing marks on the sprockets.

19. Move the crankshaft to ½ notch before TDC.

20. Install the timing belt starting at the crankshaft, then around the water pump sprocket, idler pulley, camshaft sprockets and the tensioner pulley.

21. Move the crankshaft sprocket to TDC to take up the belt slack.

22. Reinstall the tensioner to the block but do not tighten it at this time.

23. Using a torque wrench on the tensioner pulley, apply 250 inch lbs. (28 Nm) of torque to the tensioner pulley.

24. With torque being applied to the tensioner pulley, move the tensioner up against the tensioner pulley bracket and tighten the fasteners to 275 inch lbs. (31 Nm).

25. Remove the tensioner plunger pin, the tension is correct when the plunger pin can be removed and replaced easily.

26. Rotate the crankshaft two revolutions and recheck the timing marks. Wait several minutes, then recheck that the plunger pin can easily be removed and installed.

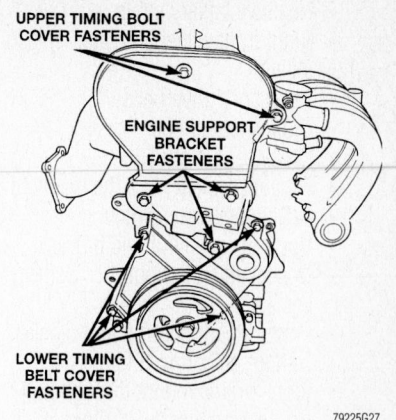

Timing belt cover bolt locations—2.4L (vin x) Engine

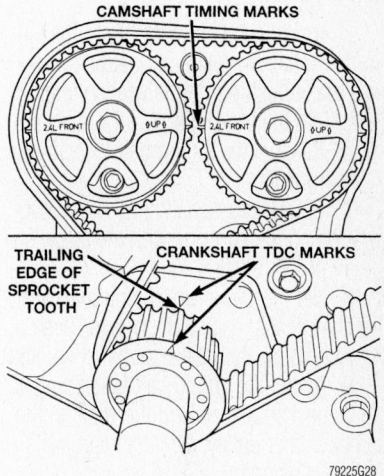

Crankshaft and camshaft alignment marks—2.4L (vin x) Engine

27. Reinstall the front timing belt cover.

28. Reinstall the engine mount bracket.

29. Reinstall the right engine mount.

30. Remove the floor jack from under the vehicle.

31. Install the crankshaft damper and tighten to 105 ft. lbs. (142 Nm).

32. Install and adjust the accessory drive belts.

33. Install the right inner splash-shield.

34. Reconnect the negative battery cable.

35. Perform the crankshaft and camshaft relearn alignment procedure using the DRB scan tool or equivalent.

2.4L (VIN S) Engine

1. Before servicing the vehicle, refer to the precautions in the beginning of this section.

2. Remove the A/C refrigerant using approved equipment.

3. Remove or disconnect the following:

- Negative battery cable
- Right front wheel and inner splash shield
- Accessory drive belts
- Crankshaft damper
- Lower torque strut
- Exhaust system from the manifold
- A/C compressor switch
- Upper torque strut and bracket
- Upper radiator support crossmember
- Power steering pump and bracket without disconnecting the lines
- Right engine mount through bolt after supporting the engine
- Engine support bracket

4. Rotate the crankshaft until the Top Dead Center (TDC) mark on the oil pump housing aligns with the TDC mark on the crankshaft sprocket.

5. Install an allen wrench into the tensioner. Rotate the tensioner counterclockwise while pushing on the wrench until it slides into the locking hole.

6. Remove the timing belt.

To install:

7. Set the crankshaft sprocket at TDC by aligning the sprocket with the arrow on the oil pump housing.

8. Set the camshafts timing marks so that the exhaust camshaft sprocket is a ½notch below the intake camshaft sprocket.

9. Install the timing belt by starting at the crankshaft. Go around the water pump sprocket idler pulley, camshaft sprockets and the tensioner.

10. Move the exhaust camshaft sprocket counterclockwise to align the marks and to remove any slack.

11. Remove the wrench from the belt tensioner.

12. Rotate the crankshaft two full revolutions and verify that the TDC marks are properly aligned.

13. Install or connect the following:
- Lower timing belt cover and torque the bolts to 40 inch (4.5 Nm)
- Upper timing belt cover and torque the bolts to 40 inch lbs. (4.5 Nm)
- Right engine support bracket and reposition the power steering pump. Torque the bracket bolts to 45 ft. lbs. (61 Nm).
- Right engine mount through bolt and torque it to 87 ft. lbs. (118 Nm)
- Upper radiator support crossmember
- Torque strut bracket
- Upper torque strut
- A/C lines and pressure switch
- Exhaust system to the manifold
- Crankshaft damper

- Accessory drive belts
- Lower torque strut
- Right splash shield and wheel
- Negative battery cable

14. Recharge the A/C system.

15. Perform the camshaft/crankshaft synchronization procedure.

Piston and Ring

POSITIONING

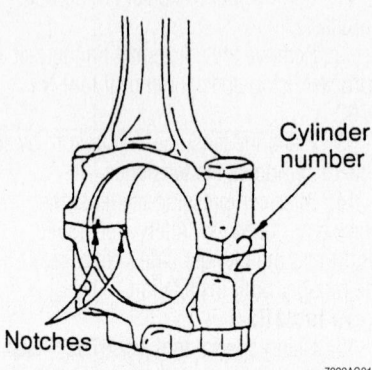

Engine connecting rod and cap installation—ensure to matchmark the cap and rod prior to disassembly

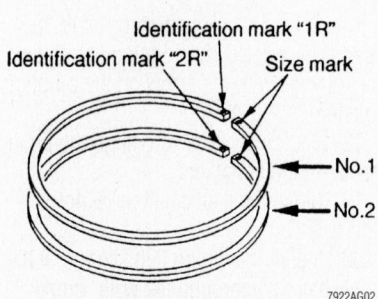

Piston ring identification mark locations

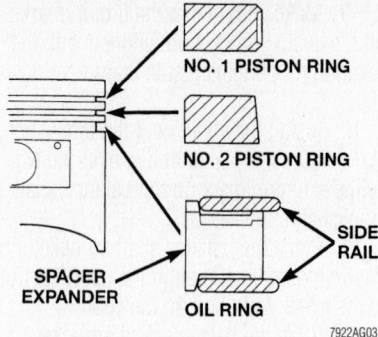

Piston ring orientation—2.4L and 2.7L engines

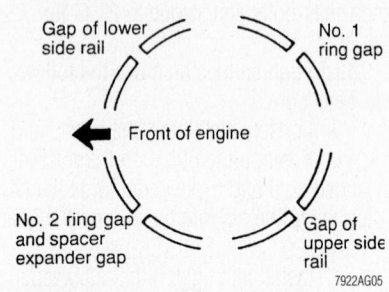

Piston ring end-gap spacing—2.4L engines

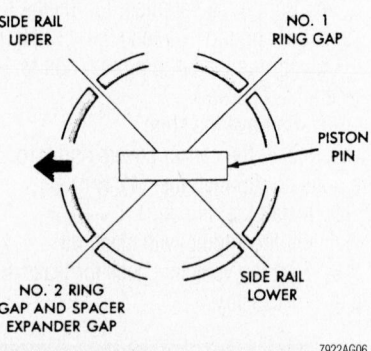

Piston ring end-gap spacing—2.7L engines

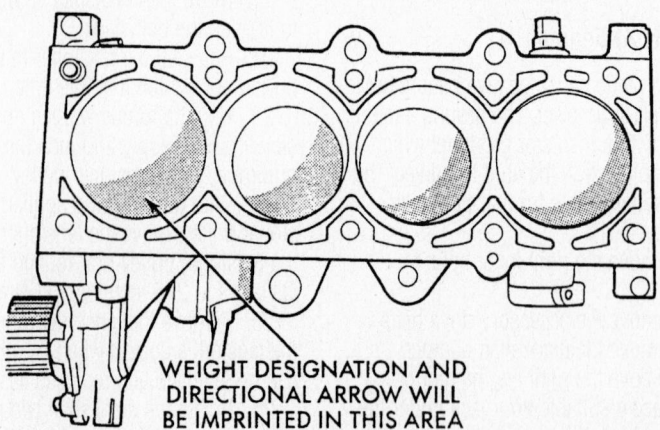

Piston positioning. The arrow or weight marking (L or H) must face toward the timing belt side of the engine—2.4L engines

FUEL SYSTEM

Fuel System Service Precautions

Safety is the most important factor when performing not only fuel system maintenance but any type of maintenance. Failure to conduct maintenance and repairs in a safe manner may result in serious personal injury or death. Maintenance and testing of the vehicle's fuel system components can be accomplished safely and effectively by adhering to the following rules and guidelines.

- To avoid the possibility of fire and personal injury, always disconnect the negative battery cable unless the repair or test procedure requires that battery voltage be applied.
- Always relieve the fuel system pressure prior to disconnecting any fuel system component (injector, fuel rail, pressure regulator, etc.), fitting or fuel line connection. Exercise extreme caution whenever relieving fuel system pressure to avoid exposing skin, face and eyes to fuel spray. Please be advised that fuel under pressure may penetrate the skin or any part of the body that it contacts.
- Always place a shop towel or cloth around the fitting or connection prior to loosening to absorb any excess fuel due to spillage. Ensure that all fuel spillage (should it occur) is quickly removed from engine surfaces. Ensure that all fuel soaked cloths or towels are deposited into a suitable waste container.
- Always keep a dry chemical (Class B) fire extinguisher near the work area.
- Do not allow fuel spray or fuel vapors to come into contact with a spark or open flame.
- Always use a back-up wrench when loosening and tightening fuel line connection fittings. This will prevent unnecessary stress and torsion to fuel line piping.
- Always replace worn fuel fitting O-rings with new. Do not substitute fuel hose where fuel pipe is installed.

Fuel System Pressure

RELIEVING

1. Before servicing the vehicle, refer to the fuel system precautions and to the precautions in the beginning of this section.
2. Remove or disconnect the following:
3. Remove the fuel pump relay from the Power Distribution Center (PDC).
4. Start and run the engine until it stalls.

5. Attempt to restart the engine until it no longer runs.
6. Turn the ignition key to the off position.

Fuel Filter

REMOVAL & INSTALLATION

Frame Mounted

The fuel delivery system contains a replaceable inline filter. The fuel filter mounts to the frame above the rear of the fuel tank. The fuel tank assembly must be loosened and lowered slightly to access the filter. The inlet and outlet tubes are permanently attached to the filter. Please note that the fuel system pressure must be relieved before servicing fuel system components. In addition, quick-disconnect fittings are used on fuel line connections. When the tubes are fully connected, the locking ears and the fuel tube shoulder are visible in the windows of the connector.

1. Before servicing the vehicle, refer to the precautions in the beginning of this section.
2. Disconnect the negative battery cable from the left shock tower.

➡**The ground cable is equipped with an insulator grommet which should be placed on the stud to prevent the negative battery cable from accidentally grounding.**

3. Relieve the fuel system pressure using the recommended procedure.
4. From inside the trunk, disconnect the fuel pump module wiring jumper from the main body harness. The 4-pin connector is located under the trunk mat on the left side of the trunk near the base of the shock tower. Locate the body grommet for the jumper near the base of the rear seat. Push the grommet out and feed the jumper completely through the hole in the body.

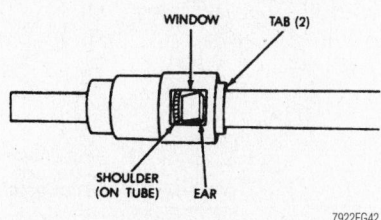

Be sure the shoulder and locking tabs are visible through the window of the connector when attached

5. Remove the fuel cap slowly to release tank pressure.
6. Raise and safely support the vehicle.
7. Locate the drain plug on the bottom left of the fuel tank. Place an approved fuel container with a capacity of at least 16 gallons, under the drain plug. Remove the plug and drain the fuel tank. When finished draining, install the plug since there will be 1–2 gallons of fuel remaining. Tighten the drain plug to 32 inch lbs. (3.6 Nm).
8. Remove or disconnect the following:
   - Driver's side fuel tank strap
   - Passenger's side fuel tank strap loosen it

➡**Allow the tank to lower until the fuel tank neck touches the rear suspension crossmember.**

✳✳ CAUTION

Wrap shop towels around the fuel hoses to catch any gasoline spillage.

- Fuel line quick-disconnect fittings from the fuel pump module. Squeeze fuel filter hose connector releasing tabs and detach the fuel lines
- Fuel filter

To install:

➡**The fuel supply (to filter) tube and the return tube (to fuel pump module) are permanently attached to the fuel filter. The ends of the fuel supply and return tubes have different size quick-disconnect fittings. The large quick-disconnect fitting attaches to the large nipple (supply side) on the fuel pump module. The smaller quick-disconnect fitting attaches the small nipple (return side) on the fuel pump module.**

9. Lubricate the fuel filter nipples with engine oil.

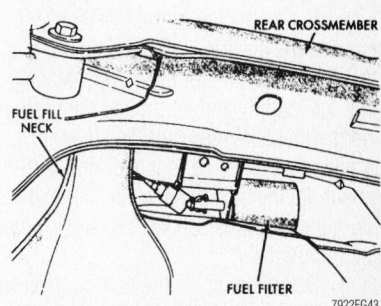

fuel filter mounting location

10. Install or connect the following:
- Fuel tubes
- Fuel tank raise it into position
- Fuel tank straps and torque the bolts to 17 ft. lbs. (23 Nm)

➡Be sure the fuel pump module electrical harness grommet is installed in the body as the tank is raised into position.

- Fuel pump module connector
- Negative battery cable

11. Refill the fuel tank.

Inlet Filter

1. Before servicing the vehicle, refer to the precautions in the beginning of this section.

2. Properly relieve the fuel system pressure.

3. Remove or disconnect the following:
- Negative battery cable
- Fuel pump module
- Locking tabs on the fuel pump
- Strainer and O-ring
- Inlet filter from the assembly

To install:

4. Install or connect the following:
- Inlet filter
- New O-ring and strainer
- Locking tabs on the fuel pump module
- Fuel pump Module
- Negative battery cable

5. Start the vehicle, check for leaks and repair if necessary.

Fuel Pump

REMOVAL & INSTALLATION

The in-tank fuel pump module contains the fuel pump and pressure regulator which adjusts fuel system pressure to approximately 49 psi (337.8 kPa). Voltage to the fuel pump is supplied through the fuel pump relay.

The fuel pump is serviced as part of the fuel pump module. The fuel pump module is installed in the top of the fuel tank and contains the electric fuel pump, fuel pump reservoir, inlet strainer fuel gauge sending unit, fuel supply and return line connections and the pressure regulator. The inlet strainer, fuel pressure regulator and level sensor are the only serviceable items. If the fuel pump requires service, replace the fuel pump module.

1. Before servicing the vehicle, refer to the precautions in the beginning of this section.

2. Disconnect the negative battery cable from the left shock tower.

➡The ground cable is equipped with an insulator grommet which should be placed on the stud to prevent the negative battery cable from accidentally grounding.

3. Remove the fuel filler cap and relieve the fuel system pressure using the recommended procedure.

4. Remove or disconnect the following:
- Fuel pump harness electrical connector on the left side of the trunk near the base of the shock tower
- Fuel pump harness grommet and push the harness through the hole

5. Raise the vehicle and drain the fuel tank.

6. Remove or disconnect the following:
- Fuel tank inlet and from the filler hose
- Fuel tank straps. Support with a jack prior to loosening the straps.

7. Carefully lower the tank.

8. Clean the top of the tank to remove any loose dirt.

9. Remove or disconnect the following:
- Fuel lines from the fuel pump module
- Fuel pump locknut using Spanner wrench tool 6856

✳✳ CAUTION

The fuel reservoir of the fuel pump module does not empty out when the tank is drained. The fuel in the reservoir may spill out when the module is removed.

- Fuel pump
- Fuel tank O-ring and discard it

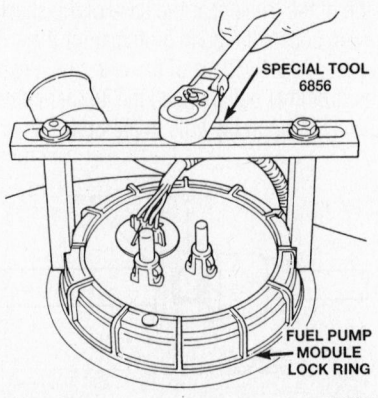

Removing the fuel pump module lock ring using special tool 6856

7922FG44

To install:

10. Thoroughly clean all parts. Wipe the seal area of the tank clean.

11. Install or connect the following:
- New O-ring onto the fuel tank
- Fuel pump module in the tank

➡Be sure the alignment tab on the underside of the pump module flange sits in the corresponding notch in the fuel tank.

- Locking ring and torque it to 40–45 ft. lbs. (54–61 Nm), using Spanner Wrench Tool 6856
- Fuel tank assembly
- Negative battery cable

12. Refill the fuel tank.

13. Turn the ignition switch **ON** to pressurize the system. Check the fuel system for leaks.

Fuel Injector

REMOVAL & INSTALLATION

2.4L Engine

1. Before servicing the vehicle, refer to the precautions in the beginning of this section.

2. Disconnect the negative battery cable from the left shock tower.

➡The ground cable is equipped with an insulator grommet which should be placed on the stud to prevent the negative battery cable from accidentally grounding.

3. Relieve the fuel system pressure using the recommended procedure.

4. Remove or disconnect the following:
- Fuel supply line quick quick-connect fitting from the fuel rail
- Fuel injector electrical connectors
- Fuel rail from the intake manifold

➡Cover the fuel injector holes in the intake manifold.

- Fuel injector clip
- Fuel injector(s) from the fuel rail and discard the O-rings

To install:

5. Install or connect the following:
- Fuel injector to the fuel rail using new O-rings lubricated with engine oil
- Fuel injector clip
- Fuel rail and torque the screws to 14–19 ft. lbs. (19.5–25.5 Nm)
- Fuel injector electrical connectors

- Fuel supply line quick quick-connect fitting to the fuel rail
- Negative battery cable

6. Use a Diagnostic Readout Box (DRB) scan tool Automatic Shutdown (ASD) fuel system test to pressurize the fuel system. Check for leaks.

2.7L Engine

1. Before servicing the vehicle, refer to the precautions in the beginning of this section.

2. Disconnect the negative battery cable from the left shock tower.

➡**The ground cable is equipped with an insulator grommet which should be placed on the stud to prevent the negative battery cable from accidentally grounding.**

3. Relieve the fuel system pressure using the recommended procedure.

4. Remove or disconnect the following:
- Fuel supply line quick quick-connect fitting from the fuel rail
- Upper intake manifold plenum and discard the gasket
- Fuel injector electrical connectors
- Fuel rail from the intake manifold

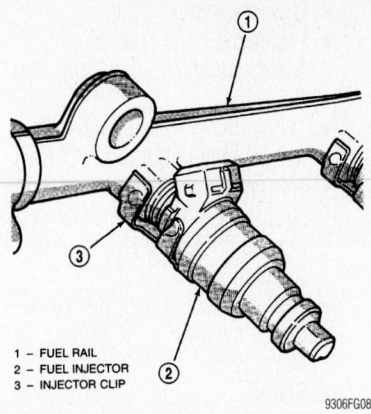

1 – FUEL RAIL
2 – FUEL INJECTOR
3 – INJECTOR CLIP

9306FG08

View of the fuel injector—2.4L engine

➡**Cover the intake manifold openings to keep dirt from entering the system.**

- Fuel injector clip
- Fuel injector(s) from the fuel rail and discard the O-rings

To install:

5. Install or connect the following:
- Fuel injector to the fuel rail using new O-rings lubricated with engine oil
- Fuel injector clip
- Fuel rail and torque the bolts to 8 ft. lbs. (12 Nm)

1 – FUEL RAIL BOLTS

9306FG07

View of the fuel rail assembly—2.7L engine

➡**Be sure the spacers are located under the fuel rail**

- Fuel injector electrical connectors
- Fuel supply line quick quick-connect fitting to the fuel rail
- Upper intake manifold plenum using a new gasket and torque the bolts to 13 ft. lbs. (18 Nm)
- Throttle cables
- Sensor electrical connectors
- Negative battery cable

6. Use a Diagnostic Readout Box (DRB) scan tool Automatic Shutdown (ASD) fuel system test to pressurize the fuel system. Check for leaks.

DRIVE TRAIN

Transaxle Assembly

REMOVAL & INSTALLATION

Manual

✳✳ WARNING

If the vehicle is going to be rolled on its wheels while the transaxle is out of the vehicle, obtain 2 outer CV-joints to install in the hubs. If the vehicle is rolled without the proper torque applied to the front wheel bearings, the bearings will no longer be usable.

1. Before servicing the vehicle, refer to the precautions in the beginning of this section.

2. Remove or disconnect the following:
- Negative battery cable from the left shock tower.

➡**The ground cable is equipped with an insulator grommet which should be placed on the stud to prevent the negative battery cable from accidentally grounding.**

- Air cleaner and intake hoses
- Clutch housing vent cap
- Clutch cable from the bell housing

✳✳ WARNING

Using equal force, pry up on both sides of the shifter cable isolator bushings to avoid damaging the cable isolator bushing.

- Selector lever and crossover cables from the transaxle
- Shift cable mounting bracket from the transaxle
- Accelerator cables from the throttle body
- Accelerator cable bracket from the throttle body
- Upper starter-to-throttle body support bracket bolt
- Upper bell housing-to-throttle body support bracket stud nut
- Throttle body support bracket
- Left transaxle mount upper bolts
- Upper bell housing bolts
- Vehicle Speed Sensor (VSS)

- Back-up light electrical connector from the transaxle

3. Install an engine support fixture tool and support the engine.

4. Remove or disconnect the following:
- Front wheels
- Halfshafts
- Lower splash shield/battery cover from the left side
- Lower bracket bolts, from the left transaxle mount
- Engine-to-lower crossbar bolts
- Front steel engine mount bracket
- 3 front aluminum engine mount bracket bolts
- Starter
- Rear transaxle mount bracket
- Transaxle-to-rear lateral bending strut from engine/transaxle assembly
- Transaxle bell housing cover

5. Using a transaxle jack, support the transaxle.

6. Rotate the engine clockwise to gain access to the driveplate clutch bolts.
- Driveplate clutch bolts
- Lower engine-to-transaxle bolts
- Transaxle

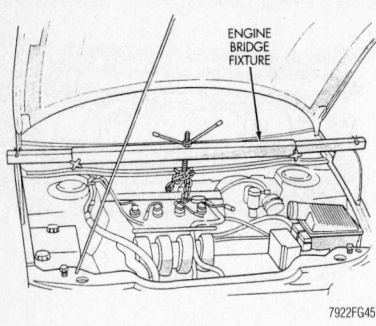

Secure the engine assembly with an appropriate support fixture

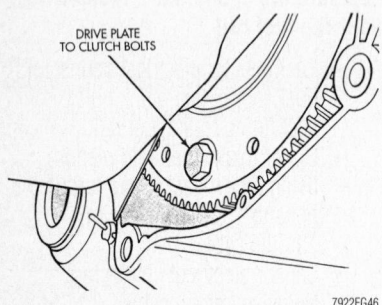

Remove the driveplate clutch bolts—rotate the engine clockwise to advance to the next bolt

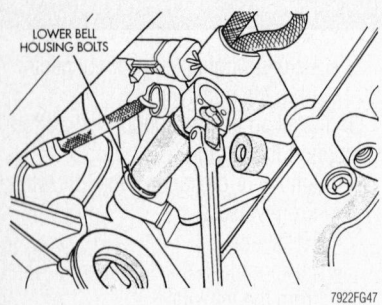

Removing the lower engine-to-transaxle mounting bolts

7. To prepare the vehicle for rolling, support the engine with a suitable support or reinstall the front motor mount to the engine. Then, reinstall the ball joints to the steering knuckle and install the retaining bolt. Install the obtained outer CV-joints to the hubs, install the washers and tighten the axle nuts to 180 ft. lbs. (244 Nm). The vehicle may now be safely rolled.

To install:

8. Install or connect the following:
- Transaxle and torque the transaxle-to-engine bolts to 70 ft. lbs. (95 Nm)
- Driveplate clutch bolts
- Lower engine-to-transaxle bolts and torque the bolts to 70 ft. lbs. (95 Nm)
- Transaxle bell housing cover and torque bolts to 9 ft. lbs. (12 Nm)
- Transaxle-to-rear lateral bending strut to the engine/transaxle assembly and torque the bolt to 40 ft. lbs. (54 Nm)
- Rear transaxle mount bracket and torque the bolts to 40 ft. lbs. (54 Nm)
- Starter
- 3 front aluminum engine mount bracket bolts and torque the bolts to 40 ft. lbs. (54 Nm)
- Front steel engine mount bracket and torque the bolt to 45 ft. lbs. (61 Nm)
- Engine-to-lower crossbar bolts
- Lower bracket bolts to the left transaxle mount and torque the bolts to 40 ft. lbs. (54 Nm)
- Lower splash shield/battery cover to the left side
- Halfshafts
- Front wheels

9. Remove engine support fixture tool.
10. Install or connect the following:
- Back-up light electrical connector
- VSS sensor
- Upper bell housing bolts and torque the bolts to 70 ft. lbs. (95 Nm)
- Left transaxle mount upper bolts and torque the bolts to 40 ft. lbs. (54 Nm)
- Throttle body support bracket
- Upper bell housing-to-throttle body support bracket stud nut and torque the nut to 32 ft. lbs. (43 Nm)

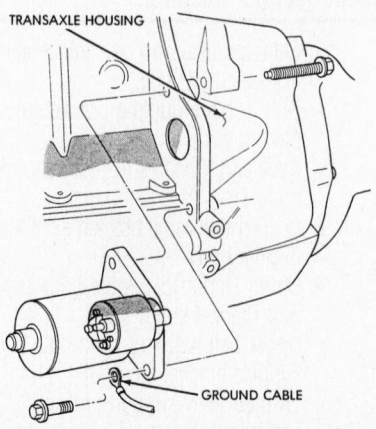

Be sure to connect the ground cable at the starter

- Upper starter-to-throttle body support bracket bolt
- Accelerator cable bracket to the throttle body
- Accelerator cables to the throttle body
- Shift cable mounting bracket to the transaxle
- Selector lever and crossover cables to the transaxle
- Clutch cable to the bell housing
- Clutch housing vent cap
- Air cleaner and intake hoses
- Negative battery cable

11. Check to be sure that all fasteners are tightened and connections made.
12. Refill the transaxle.
13. Check the transaxle for proper operation. Be sure the reverse lights turn **ON** when in reverse.

Automatic

✳✳ WARNING

If the vehicle is going to be rolled on its wheels while the transaxle is out of the vehicle, obtain 2 outer CV-joints to install to the hubs. If the vehicle is rolled without the proper torque applied to the front wheel bearings, the bearings will no longer be usable.

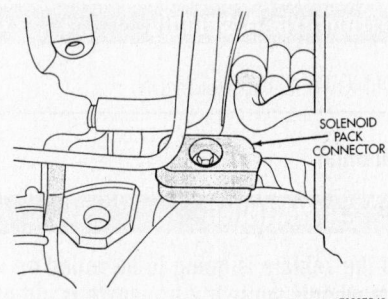

Location of the transaxle solenoid assembly 8-way connector and retaining bolt

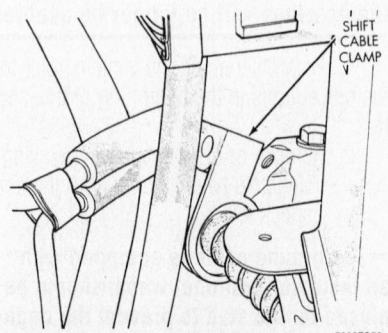

Removing the shift cable and clamp

1. Before servicing the vehicle, refer to the precautions in the beginning of this section.

2. Drain the transaxle.

3. Disconnect the negative battery cable from the left shock tower.

➡ **The ground cable is equipped with an insulator grommet which should be placed on the stud to prevent the negative battery cable from accidentally grounding.**

4. Remove or disconnect the following:
- Air cleaner duct
- Transmission Control Module (TCM)
- Solenoid pack electrical connector
- Dipstick tube from the transaxle
- Transaxle cooler lines
- Shift lever cable from the transaxle lever

5. Using an engine support fixture tool, support the engine assembly.

6. Remove or disconnect the following:
- Upper transaxle mount top bolts from the left side
- Front wheels
- Lower splash shields from both sides
- Exhaust pipe from the exhaust manifold
- Upper transaxle mount remaining bolts from the left side
- Engine oil filter
- Starter
- Front engine mount bracket
- Rear mount bracket through-bolt
- Centermember bolts
- Rear mount bracket
- Radiator lower crossmember
- Both lateral bending strut brackets
- Flexplate cover

7. Matchmark the converter-to-flexplate location. Rotate the crankshaft clockwise to align the converter bolts.

8. Install or connect the following:
- Converter-to-flexplate bolts
- Crankshaft Position (CKP) sensor, if equipped
- Transaxle electrical connectors
- Right side steering gear and K-frame bolts
- Sway bar mounts

9. Using a transmission jack and a safety chain, secure the transaxle and support it.

10. Install or connect the following:
- Bell housing-to-engine bolts
- Transaxle, by moving the K-frame rearward

11. To prepare the vehicle for rolling, support the engine with a suitable support or reinstall the front motor mount to the engine. Then, reinstall the ball joints to the steering knuckle and install the retaining bolt. Install the obtained outer CV-joints to the hubs, install the washers and tighten the axle nuts to 180 ft. lbs. (244 Nm). The vehicle may now be safely rolled.

To install:

12. Install or connect the following:
- Transaxle, by moving the K-frame rearward
- Bell housing-to-engine bolts and torque the bolts to 70 ft. lbs. (95 Nm)
- Sway bar mounts
- Right side steering gear and K-frame bolts
- Transaxle electrical connectors
- CKP sensor, if equipped

13. Align the converter-to-flexplate matchmark. Rotate the crankshaft clockwise to align the converter bolts.

14. Install or connect the following:
- Torque converter and torque the converter-to-flexplate bolts to 55 ft. lbs. (74 Nm)
- Flexplate cover and torque the bolts to 108 inch lbs. (12 Nm)
- Both lateral bending strut brackets and torque the bolts to 45 ft. lbs. (61 Nm)
- Radiator's lower crossmember and torque the bolts to 45 ft. lbs. (61 Nm)
- Rear mount bracket
- Centermember bolts and torque the bolts to 45 ft. lbs. (61 Nm)
- Rear mount bracket through-bolt and torque the through-bolt to 45 ft. lbs. (61 Nm)
- Front engine mount bracket and torque the bolts to 24 ft. lbs. (33 Nm)
- Starter
- New engine oil filter
- Upper transaxle mount remaining bolts to the left side
- Exhaust pipe to the exhaust manifold
- Lower splash shields to both sides
- Front wheels
- Upper transaxle mount top bolts, to the left side
- Shift lever cable to the transaxle lever and torque the nut to 14 ft. lbs. (19 Nm)
- Transaxle cooler lines
- Dipstick tube to the transaxle
- Solenoid pack electrical connector
- TCM
- Air cleaner duct
- Negative battery cable

15. Adjust the gearshift and throttle cables.

16. Refill the transaxle.

17. Check the transaxle for proper operation. Be sure the back-up lights and speedometer are working properly.

Clutch

ADJUSTMENT

Free-Play

The manual transaxle clutch release system has a unique self-adjusting mechanism to compensate for clutch disc wear. This adjuster mechanism is located with the clutch cable assembly. The preload spring maintains tension on the cable. This tension keeps the clutch release bearing continuously loaded against the fingers of the clutch cover assembly. No manual adjustment is necessary.

When servicing this vehicle or if removing and installing the clutch cable, do not pull on the clutch cable housing to remove it from the dash panel. Damage to the cable self-adjuster may occur.

To check the function of the adjuster mechanism, use the following procedure:

1. With slight pressure, pull the clutch release lever end of the cable to draw the cable taut.

2. Push the clutch cable housing toward the dash panel. With less than 25 lbs. (11 kg) of effort, the cable housing should move 1.2–2.0 in. (30–50mm). This indicates proper adjuster mechanism function.

3. If the cable does not adjust, determine if the mechanism is properly seated on the bracket.

REMOVAL & INSTALLATION

➡ **The transaxle assembly must be removed to service the clutch assembly.**

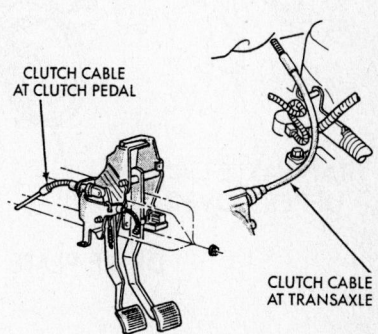

CLUTCH CABLE AT CLUTCH PEDAL

CLUTCH CABLE AT TRANSAXLE

7922FG51

Clutch cable routing

1. Before servicing the vehicle, refer to the precautions in the beginning of this section.

2. Remove or disconnect the following:
   - Negative battery cable from the left shock tower

➡**The ground cable is equipped with an insulator grommet which should be placed on the stud to prevent the negative battery cable from accidentally grounding.**

   - Starter
   - Rear and front transaxle support brackets
   - Clutch inspection cover
   - Modular clutch-to-flywheel bolts
   - Transaxle assembly with the clutch as an assembly
   - Clutch assembly from the transaxle input shaft

To install:

3. Clean all parts well. Inspect for oil leakage through the engine rear crankshaft oil seal and transaxle input shaft seal. If leakage is noted, it should be corrected at this time.

4. Examine the throwout or clutch release bearing. It is pre-lubricated and sealed and should not be washed in solvent. The bearing should turn smoothly when held in the hand with a light thrust load. A light drag caused by the lubricant fill is normal. If the bearing is noisy, rough or dry, replace the complete bearing assembly. In most cases where a clutch is being serviced, the complete clutch assembly and release bearing are usually replaced together.

5. Check the condition of the stud pivot spring clips on the back side of the clutch fork. If the clips are broken or distorted, replace the clutch fork. The pivot ball pocket in the fork is Teflon® coated and should be installed **WITHOUT** any lubricant such as grease which will break down the Teflon® coating. Be sure the ball stud and fork pocket are clean of contamination and dirt. When assembling the fork to the bearing, the small pegs on the bearing must go over the fork arms.

6. Check the flywheel for cracks, glazing or grooves. If any of these conditions exist, machine (reface) or replace the flywheel to prevent clutch chatter and premature clutch wear.

➡**The manual transaxle is equipped with a reverse brake. It functions as a synchronizer, but only if the vehicle is not moving. When the clutch pedal is depressed to the floor and held for 3 seconds, and the transaxle shifts to reverse, no gear clash should be present. If there is, the input shaft should be checked. When the transaxle is removed for clutch service, check the input clutch shaft, clutch disc splines and release bearing for dry rust. If present, clean rust off and apply a light coat of high temperature bearing grease to the input shaft splines. Apply grease on the input shaft splines only where the clutch disc slides. Verify that the clutch disc slides freely along the input shaft splines.**

7. Install or connect the following:
   - Modular clutch assembly onto the transaxle input shaft
   - Transaxle assembly
   - New clutch-to-driveplate (flywheel) bolts. Tighten the bolts, in a criss-cross pattern, a few turns at a time to 55 ft. lbs. (75 Nm)
   - Clutch inspection cover
   - Transaxle lower support brackets
   - Starter
   - Negative battery cable

8. Road test the vehicle to check for proper clutch operation.

Halfshafts

REMOVAL & INSTALLATION

➡**If the vehicle is going to be rolled while the halfshafts are out of the vehicle, obtain 2 outer CV-joints or proper equivalent tools and install to the hubs.**

❋❋ **WARNING**

If the vehicle is rolled without the proper torque applied to the front wheel bearings, the bearings will no longer be usable.

1. Before servicing the vehicle, refer to the precautions in the beginning of this section.

2. Remove or disconnect the following:
   - Negative battery cable
   - Cotter pin, nut lock and spring washer
   - Halfshaft nut, loosen it while the vehicle is on the floor with the brakes applied
   - Wheel
   - Brake caliper assembly and support it on a wire
   - Brake rotor
   - Halfshaft nut and washer
   - Tie rod end from the steering knuckle using joint separation Tool MB991113

❋❋ **WARNING**

Use of improper methods of joint separation can result in damage to the joint, leading to possible failure.

   - Speed sensor cable routing bracket, if equipped with an Anti-lock Brake System (ABS)
   - Sway bar link from the damper fork, if necessary

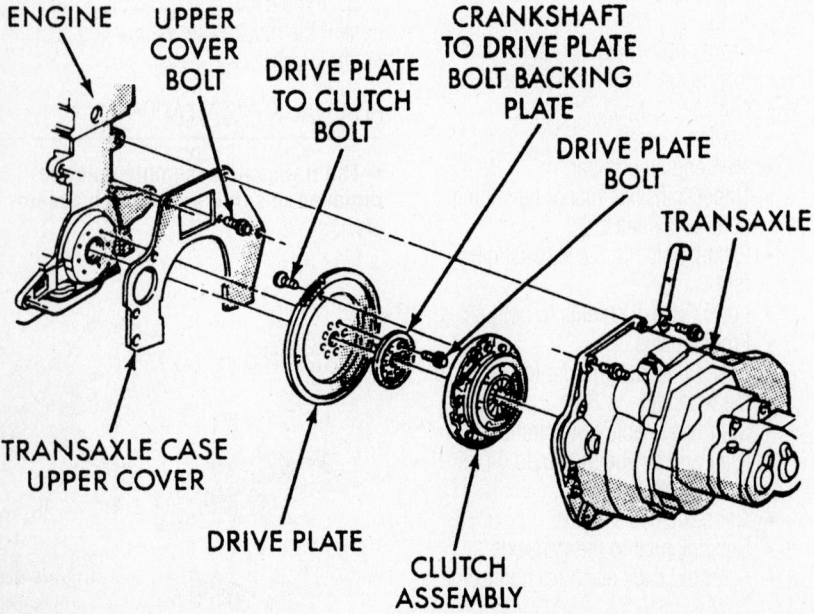

ENGINE   UPPER COVER BOLT   DRIVE PLATE TO CLUTCH BOLT   CRANKSHAFT TO DRIVE PLATE BOLT BACKING PLATE   DRIVE PLATE BOLT   TRANSAXLE   TRANSAXLE CASE UPPER COVER   DRIVE PLATE   CLUTCH ASSEMBLY

7922FG52

Exploded view of the clutch assembly

- Damper fork assembly
- Steering knuckle from the lower control arm
- Halfshaft by pressing it from the hub

➡ **After pressing the outer shaft, insert a prybar between the transaxle case and the halfshaft and pry the shaft from the transaxle.**

❊❊ WARNING

Do not pull on the shaft. Doing so damages the inboard joint. Do not insert the prybar too far or the oil seal in the case may be damaged.

To install:

3. Inspect the halfshaft boot for damage or deterioration. Check the ball joints and splines for wear.

4. Replace the circlips on the ends of the halfshaft(s).

5. Install or connect the following:
   - Halfshaft into the transaxle until it is fully seated
   - Halfshaft into the hub by pulling the knuckle assembly outward
   - Washer. Make sure the chamfered edge faces outward
   - Halfshaft nut and tighten temporarily

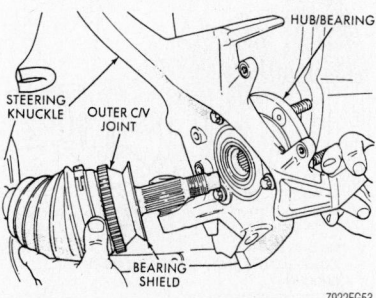

Carefully remove the outer CV-joint from the steering knuckle—be careful NOT to damage the threads or splines on the joint

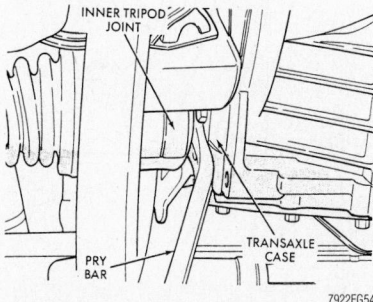

Inserting the prybar too far may damage the transaxle seal

- Control arm to the steering knuckle and torque the nuts to 43–52 ft. lbs. (59–71 Nm)
- Damper fork and torque the lower through-bolt/nut to 65 ft. lbs. (88 Nm) and the upper pinch bolt to 76 ft. lbs. (103 Nm)
- Tie rod end, to the steering knuckle and torque the nut to 17–25 ft. lbs. (24–33 Nm) and install a new cotter pin
- Sway bar link to the damper fork and torque the link nut to 29 ft. lbs. (39 Nm)
- Lockwasher and axle nut and torque the nut to 145–188 ft. lbs. (200–260 Nm)

➡ **Before securely tightening the axle nut, be sure there is no load on the wheel bearings.**

- Brake rotor and caliper assembly
- New cotter pin
- Wheel
- Negative battery cable

6. Refill the transaxle.

7. Test drive the vehicle and check for proper operation.

CV-Joints

OVERHAUL

Inner Tri-Pod Joint

1. Remove or disassemble the following:

   - Negative battery cable
   - Halfshaft
   - Large and small boot retaining clamps

2. Slide the boot down the shaft away from the tri-pod housing.

➡ **When separating the spider joint from the tri-pod joint housing, hold the rollers in place on the trunnions to prevent the rollers and needle bearings from falling away.**

3. Carefully slide the shaft/spider assembly from the tri-pod housing.

4. Remove the spider assembly-to-shaft snapring; then, slide the spider assembly off the shaft.

❊❊ WARNING

If necessary, tap the spider assembly off the shaft using a brass drift; be careful not to hit the outer bearings.

5. Slide the boot off the shaft.

6. Thoroughly inspect all parts for signs of excessive wear. If necessary, replace the halfshaft.

➡ **Component parts are not serviceable and must be replaced as an assembly.**

To assemble:

❊❊ WARNING

The Tri-pod sealing boots are made of 2 different types of material; silicon rubber (high temperature) which is soft and pliable or Hytrel plastic (standard temperature) which is stiff and rigid. Be sure to replace the boot made of the correct material.

7. Slide the inner tri-pod boot clamp and boot onto the shaft; then, position the boot so that only the thinnest (sight) groove is visible on the shaft.

8. Install the spider assembly onto the shaft just far enough so that the snapring can be installed.

❊❊ WARNING

If necessary, tap the spider assembly onto the shaft using a brass drift; be careful not to hit the outer bearings.

9. Install the snapring onto the shaft; make sure that the snapring is fully seated in the groove.

10. If installing a new boot, distribute ½ of the grease in the service package inside the tri-pod housing and the other ½ inside the boot.

11. Carefully, slide the spider assembly and shaft into the tri-pod housing.

12. Position the inner boot clamp evenly on the sealing boot.

❊❊ WARNING

The seal must not be dimpled, stretched or out of shape. If necessary, use a trim stick to seat and equalize the seal pressure and shape it by hand.

➡ **If using a Hytrel (hard plastic) boot, be sure the stick is inserted between the soft rubber insert and the tri-pod housing, not between the hard plastic sealing boot and the soft rubber insert.**

13. Position the boot onto the tri-pod housing retaining groove and install the retaining clamp evenly on the boot.

14. If using a crimp type boot clamp, perform the following procedure:

   a. Using the Crimper Tool C-4975-A,

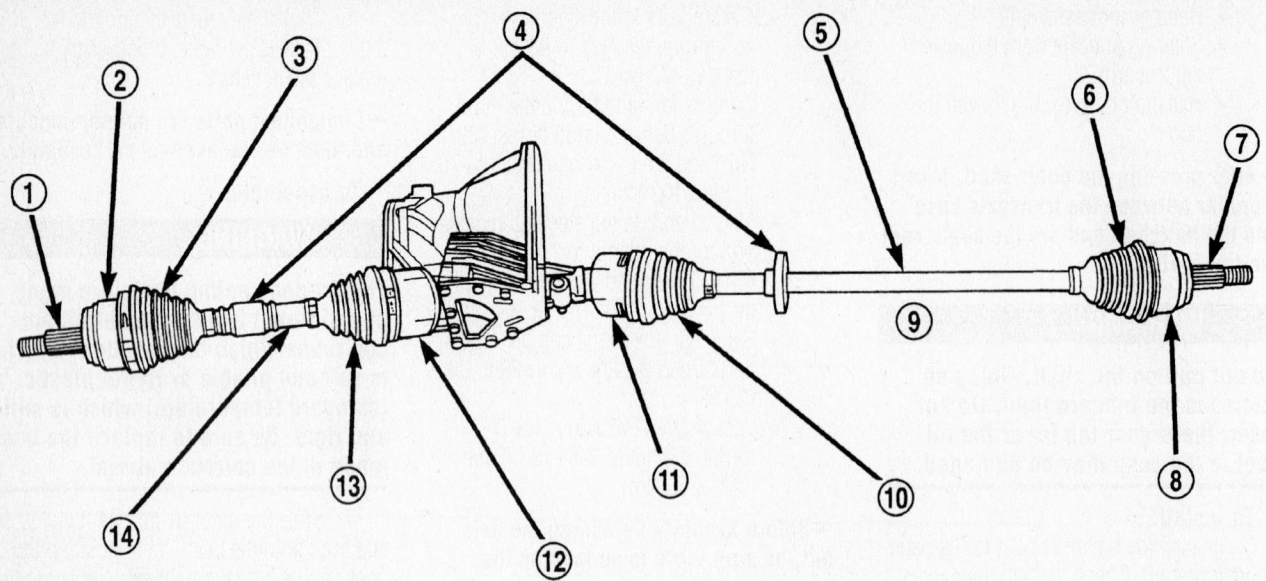

1 – STUB AXLE
2 – OUTER C/V JOINT
3 – OUTER C/V JOINT BOOT
4 – TUNED RUBBER DAMPER WEIGHT
5 – INTERCONNECTING SHAFT
6 – OUTER C/V JOINT BOOT
7 – STUB AXLE

8 – OUTER C/V JOINT
9 – RIGHT DRIVESHAFT
10 – INNER TRIPOD JOINT BOOT
11 – INNER TRIPOD JOINT
12 – INNER TRIPOD JOINT
13 – INNER TRIPOD JOINT BOOT
14 – INTERCONNECTING SHAFT LEFT DRIVESHAFT

9306EG02

View of the halfshaft assemblies—Typical

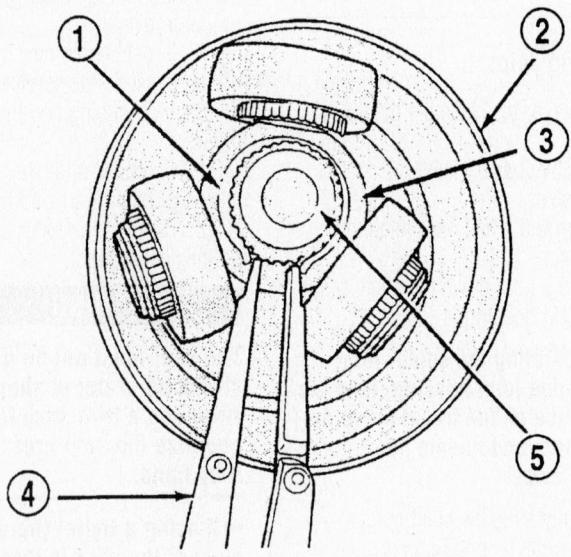

1 – SNAP RING
2 – SEALING BOOT
3 – SPIDER ASSEMBLY
4 – SNAP RING PLIERS
5 – INTERCONNECTING SHAFT

9306BG11

View of the halfshaft inner tri-pod joint and snapring

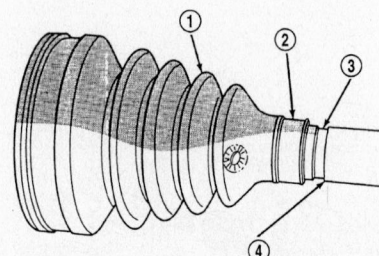

1 – SEALING BOOT
2 – RAISED BEAD IN THIS AREA OF SEALING BOOT
3 – GROOVE
4 – INTERCONNECTING SHAFT

9306EG04

View of the halfshaft boot and shaft

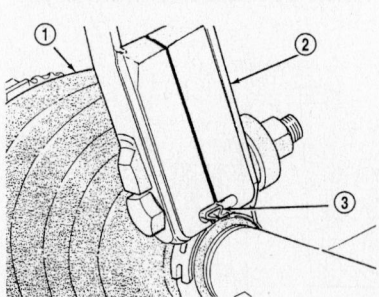

1 – SEALING BOOT
2 – SPECIAL TOOL C-4975
3 – CLAMP BRIDGE

9306BG13

Securing the halfshaft boot clamp

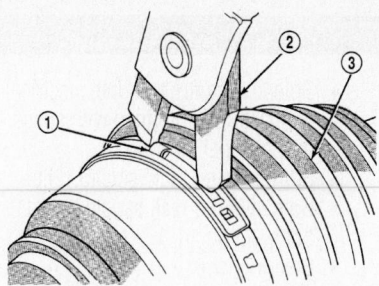

1 – CLAMP
2 – SPECIAL TOOL YA3050
3 – SEALING BOOT

9306EG07

Tightening the low profile boot clamp—Silicone rubber Tri-pod boot

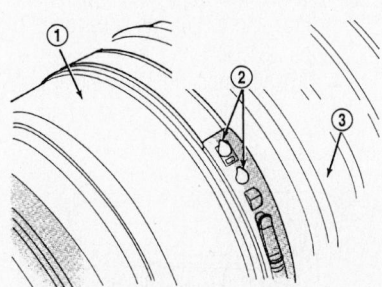

1 – INNER TRIPOD JOINT HOUSING
2 – TOP BANK OF CLAMP MUST BE RETAINED BY TABS AS SHOWN HERE TO CORRECTLY LATCH BOOT CLAMP
3 – SEALING BOOT

9306EG08

Secured the low profile boot clamp—Silicone rubber Tri-pod boot

place the tool over the clamp bridge, tighten the tool nut until the jaws are completely closed (face-to-face).

b. Using the Crimper Tool C-4975-A, place the tool over the clamp bridge, tighten the tool nut until the jaws are completely closed (face-to-face).

15. If using a latching type boot clamp, perform the following procedure:

a. Position Snap-On® Clamp Locking Tool YA3050 prongs in the clamp holes.

b. Squeeze the tool until the upper clamp band is latched behind the 2 tabs on the lower clamp band.

16. Install the halfshaft into the vehicle.

Rzeppa (Outer) Joint

1. Remove or disassemble the following:
   • Halfshaft and place it in a soft-jawed vise
   • Rzeppa joint boot clamps and slide the boot down the shaft
   • Rzeppa joint housing by sharply hitting it with a soft-faced hammer to drive it off the shaft
   • Circlip from the shaft
   • Boot by sliding it off the shaft

※※ WARNING

If any parts show excessive wear, replace the halfshaft assembly; the component parts are not serviceable.

To assemble:

2. Install or assemble the following:
   • New small boot clamp and slide it onto the shaft
   • Boot and slide it onto the shaft
   • Circlip, if removed

3. Position the boot so that the raised bead on the inside the boot seal is in the shaft groove.

4. Install or assemble the following:
   • Halfshaft hub nut onto the Rzeppa joint threaded shaft so it is flush with the end
   • Rzeppa joint and align the shaft splines and tap it onto the shaft with a soft-faced hammer so it locks on the circlip

5. Distribute ½ of the grease in the service package inside the Rzeppa joint housing and the other ½ inside the boot.

6. Install or connect the following:
   • New small boot clamp and position it evenly on the sealing boot

7. Using the Crimper Tool C-4975-A, place the tool over the clamp bridge, tighten the tool nut until the jaws are completely closed (face-to-face).

※※ WARNING

The seal must not be dimpled, stretched or out of shape. If neces-

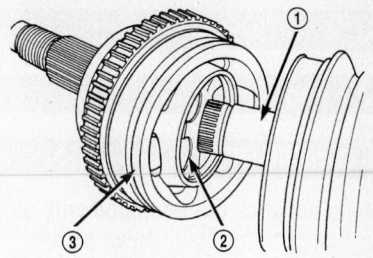

1 – INTERCONNECTING SHAFT
2 – CROSS
3 – OUTER C/V JOINT ASSEMBLY

9306EG09

Aligning the cross splines with the shaft splines—Rzeppa joint

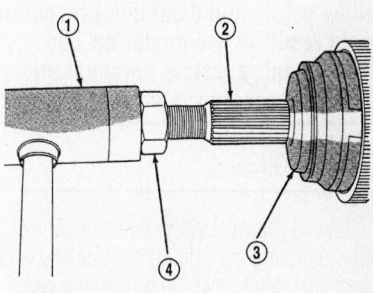

1 – SOFT FACED HAMMER
2 – STUB AXLE
3 – OUTER C/V JOINT
4 – NUT

9306EG10

Driving the Rzeppa joint onto the shaft

sary, equalize the seal pressure and shape it by hand.

8. Position the boot onto the Rzeppa housing retaining groove and install the retaining clamp evenly on the boot.

9. Using the Crimper Tool C-4975-A, place the tool over the clamp bridge, tighten the tool nut until the jaws are completely closed (face-to-face).

※※ WARNING

The seal must not be dimpled, stretched or out of shape. If necessary, equalize the seal pressure and shape it by hand.

10. Install the halfshaft into the vehicle.

STEERING AND SUSPENSION

Air Bag

❋❋ CAUTION

Some vehicles are equipped with an air bag system, also known as the Supplemental Inflatable Restraint (SIR) or Supplemental Restraint System (SRS). The system must be disabled before performing service on or around system components, steering column, instrument panel components, wiring and sensors. Failure to follow safety and disabling procedures could result in accidental air bag deployment, possible personal injury and unnecessary system repairs.

PRECAUTIONS

Several precautions must be observed when handling the inflator module to avoid accidental deployment and possible personal injury.

➡**Before servicing the vehicle, also refer to the precautions in the beginning of this section.**

• Never carry the inflator module by the wires or connector on the underside of the module.

• When carrying a live inflator module, hold securely with both hands, and ensure that the bag and trim cover are pointed away.

• Place the inflator module on a bench or other surface with the bag and trim cover facing up.

• With the inflator module on the bench, never place anything on or close to the module which may be thrown in the event of accidental deployment.

DISARMING

This air bag system is a sensitive, complex, electromechanical unit. Proper SRS (also called Supplemental Inflatable Restraint, or SIR, or air bag system) disarming can be obtained by disconnecting and isolating the negative battery cable (wrapping the battery cable end with electrical tape is a good method for isolating it). Failure to disconnect the battery could result in accidental air bag deployment and possible personal injury. Before beginning service work, allow the system capacitor 2 minutes to discharge after disconnecting and isolating the negative battery cable.

Power Rack and Pinion Steering Gear

REMOVAL & INSTALLATION

1. Before servicing the vehicle, refer to the precautions in the beginning of this section.
2. Remove or disconnect the following:
   • Negative battery cable from the left shock tower

➡**The ground cable is equipped with an insulator grommet which should be placed on the stud to prevent the negative battery cable from accidentally grounding.**

   • Intermediate shaft coupler pin bolt retaining pin
   • Intermediate shaft coupler pin bolt
   • Front wheels
   • Tie rod ends by holding the tie rod end stud with a 11/32 in. socket and loosen the retaining nut
   • Tie rod end from the steering knuckle

➡**Before removing the front suspension crossmember from the vehicle scribe the front suspension crossmember and the vehicle body. This must be done to retain the proper alignment. The caster and camber are not adjustable.**

3. Scribe a line on the body and on the crossmember on all 4 sides.
4. Remove or disconnect the following:
   • Stabilizer
   • 3 anti-lock brake controller-to-crossmember bolts and secure it to the chassis, if equipped with Anti-lock Brake System (ABS)
   • Strut clevis from the lower control arm
   • Both engine support bracket-to-crossmember bolts
   • Engine support bracket-to-transaxle mounting bracket bolt
5. Place a lifting device under the front suspension crossmember.
6. Remove or disconnect the following:
   • 8 crossmember-to-chassis bolts
7. Lower the lifting device enough to gain access to the steering rack.
8. Remove or disconnect the following:
   • Power steering lines and drain the fluid
   • Power steering pressure switch electrical connector
   • Solenoid control module electrical connector, if equipped with speed proportional steering
   • Both steering rack isolator bolts
   • Both steering rack saddle bracket bolts
   • Steering rack

To install:

9. Install or connect the following:
   • Steering rack
   • Isolator and saddle bracket and torque the bolts to 50 ft. lbs. (68 Nm)
   • Power steering pressure and return lines and torque the lines to 23 ft. lbs. (31 Nm)
   • Crossmember, install the rear bolts

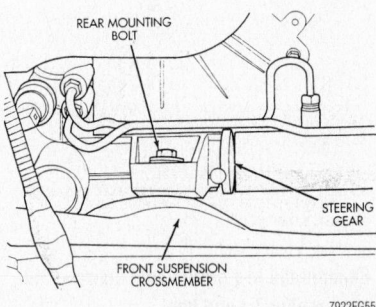

Steering gear rear mounting bolt location

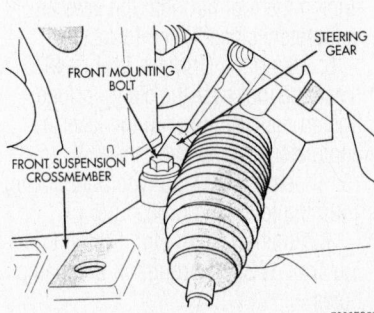

Steering gear front mounting bolt location

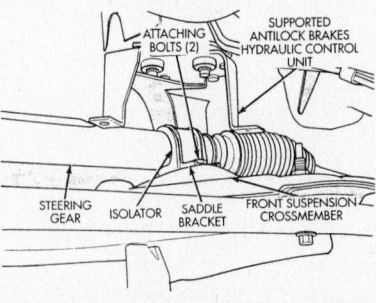

Check the condition of the isolator—if worn or oil soaked, replace

first and torque the bolts to 20 inch lbs. (2 Nm)

10. Using a soft faced hammer tap the crossmember into position.

➡**Be sure to align the scribed marks on the crossmember.**

11. Install or connect the following:
- Crossmember, starting with the rear bolts and torque the bolts to 120 ft. lbs. (163 Nm)
- Both engine support bracket-to-crossmember bolts
- Engine support bracket-to-transaxle bracket bolt and torque the 3 bolts to 55 ft. lbs. (75 Nm)
- Power steering pressure switch
- ABS control unit and torque the bolts to 21 ft. lbs. (28 Nm)
- Heat shield on the tie rod ends
- Shock clevis to the lower control arm
- Tie rod ends and torque the nuts to 45 ft. lbs. (61 Nm)
- Both stabilizer clamps
- Strut clevis bolt and torque the bolt to 68 ft. lbs. (92 Nm)
- Wheels and torque the lug nuts to 95 ft. lbs. (129 Nm)
- Intermediate shaft pin bolt and retaining pin and torque the pin bolt to 20 ft. lbs. (27 Nm)
- Negative battery cable

12. Refill the power steering system.

13. Start the engine and allow it to run for a few minutes.

14. Shut **OFF** the engine and check the power steering fluid.

15. Add power steering fluid if necessary.

16. Raise the front wheels off the ground.

17. Start the engine and turn the wheel from stop-to-stop to bleed any air from the system.

18. Check the fluid level and add, if necessary.

19. Check and adjust the alignment.

Strut

REMOVAL & INSTALLATION

Front

1. Before servicing the vehicle, refer to the precautions in the beginning of this section.

2. Remove or disconnect the following:
- Front wheel
- Steering knuckle

- Strut-to-shock clevis pin bolt
- Clevis-to-lower control arm through-bolt
- Clevis from the strut by tapping it with a brass drift
- 4 strut-to-shock tower bolts
- Strut/upper control arm mounting bracket as an assembly

To install:

3. Install or connect the following:
- Strut into the shock tower
- 4 upper strut mounting bolts and torque the bolts to 23 ft. lbs. (31 Nm)
- Clevis onto the strut, using a brass drift until the clevis is fully seated against the locating tab

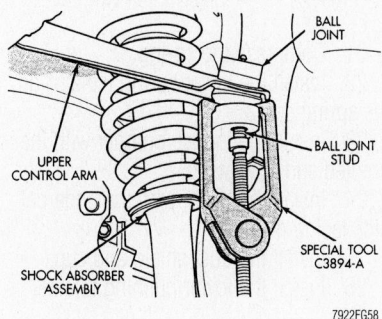

Separating the upper ball joint from the steering knuckle

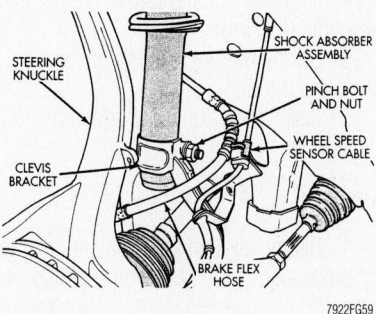

View of the front strut mount and related components

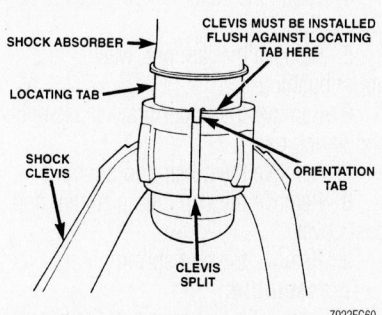

Be sure the orientation tab is situated into the clevis split

- Clevis pin bolt
- Clevis onto the lower control arm
- Clevis through-bolt
- Steering knuckle and torque the clevis-to-strut pin bolt to 65 ft. lbs. (88 Nm)

4. Lower the vehicle to support the lower control arm.

5. Torque the clevis-to-lower control arm mounting bolt 40 ft. lbs. (54 Nm)

6. Install the front wheel

Rear

1. Before servicing the vehicle, refer to the precautions in the beginning of this section.

2. Remove or disconnect the following:
- Carpet, pull it back from the rear strut tower
- Plastic cover from the top of the strut tower
- Both strut assembly-to-chassis nuts
- Rear wheel
- Splash shield from the upper strut mount
- Strut-to-rear knuckle bolt
- Strut by pushing the rear suspension downward and tilting the top of the strut outward

To install:

3. Install or connect the following:
- Strut by pushing the rear suspen-

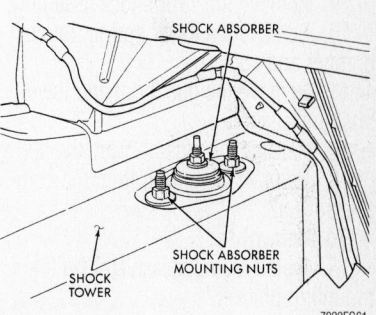

Access the rear strut upper mounting from inside the trunk

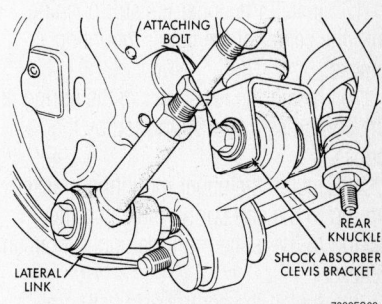

View of the rear strut lower mounting

sion downward and inserting the top of the strut into the vehicle
- Strut-to-rear knuckle bolt and torque the bolt to 70 ft. lbs. (95 Nm)
- Strut upper mounting nuts and torque the nuts to 25 ft. lbs. (34 Nm)
- Strut top cover
- Rear wheel and torque the nuts to 95 ft. lbs. (125 Nm)

Coil Spring

REMOVAL & INSTALLATION

Front

1. Before servicing the vehicle, refer to the precautions in the beginning of this section.
2. Disassemble as follows:
3. Remove the strut/shock assembly and place it in a suitable compressor tool
4. Set the lower hooks and install the clamp on the lower end of the spring.
5. Rotate the spring so the ball joint sits directly below the front upper hook.
6. Compress the spring until all tension is removed.
7. Remove the retaining nut and washer.
8. Remove the clamp from the bottom of the spring.
9. Remove the strut/shock assembly.
10. Remove the lower spring, jounce bumper and cup.
11. Remove the dust boot and lower bushing washer.
12. Release the tension from the spring.
13. Remove the spring from the compressor tool.

To assemble:

14. Install the upper isolator on the mounting bracket.
15. Install the sleeve into the lower isolator bushing.
16. Install the bushing and sleeve into the bottom of the upper mounting bracket.
17. Install the upper isolator bushing into the center of the upper mounting bracket.
18. Install the lower end of the spring in the compressor tool supported by the hooks.
19. Install the upper mounting bracket on top of the coil spring matching the spring to the isolator on the upper mounting bracket.
20. Install the control arm ball joint directly below the front upper hook.

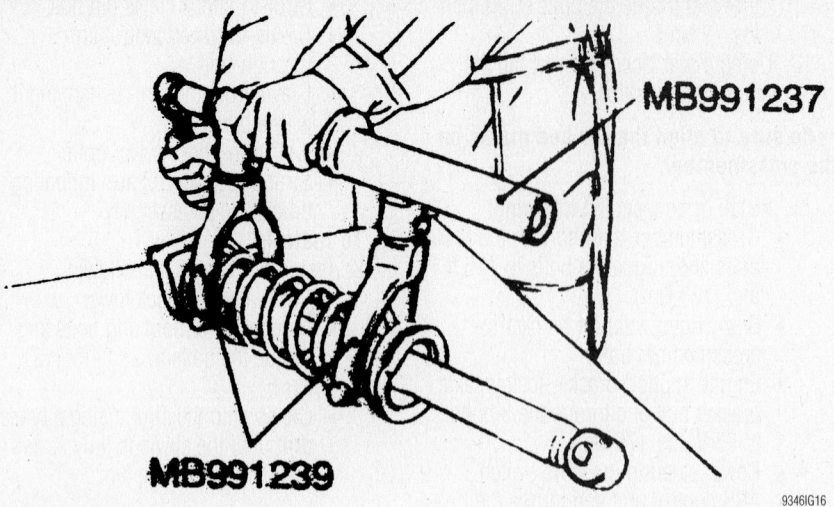

MB991237

MB991239

9346IG16

Compress the coil spring as shown

21. Compress the spring.
22. Install the lower spring isolator on the spring seat.
23. Install the jounce bumper with the pointed end downward.
24. Install the collar with the undercut side facing down.
25. Install the dust shield and cup
26. Install the lower bushing retainer washer
27. Install the strut/shock assembly through the coil spring.
28. When properly aligned, install the upper mounting nut and torque the nut to 40 ft. lbs. (55 Nm).
29. Slowly release the tension from the compressor tool.
30. Install the strut/shock assembly in the vehicle.

Rear

1. Before servicing the vehicle, refer to the precautions in the beginning of this section.
2. Disassemble as follows:
3. Remove the rear shock from the vehicle.
4. Install the assembly in a spring compressor tool.
5. Remove the jam nut, washer and upper bushing.
6. Remove the upper bracket assembly and spring pad.
7. Remove the collar and bushing.
8. Remove the cup, bump rubber and dust cover.
9. Remove the coil spring

To assemble:

10. Install the coil spring to the shock absorber using compressor Tools MB991237 and MB991239.

11. Install the dust cover and bump rubber.
12. Install the cup and bushing
13. Install the collar, supper spring pad and bracket assembly.
14. Install the upper bushing and washer.
15. When properly aligned, install the jam nut and torque to 17 ft. lbs. (23 Nm).
16. Remove the compressor tools and install the shock absorber.

Upper Ball Joint

REMOVAL & INSTALLATION

The upper ball joint is an integrated part of the upper control arm assembly, and cannot be serviced separately. A worn or damaged ball joint requires replacement of upper control arm assembly.

Lower Ball Joint

REMOVAL & INSTALLATION

On all vehicles, the ball joint cannot be serviced separately. If the ball joint is defective it will require replacement of the lower control arm.

Upper Control Arm

REMOVAL & INSTALLATION

Front

1. Before servicing the vehicle, refer to the precautions in the beginning of this section.

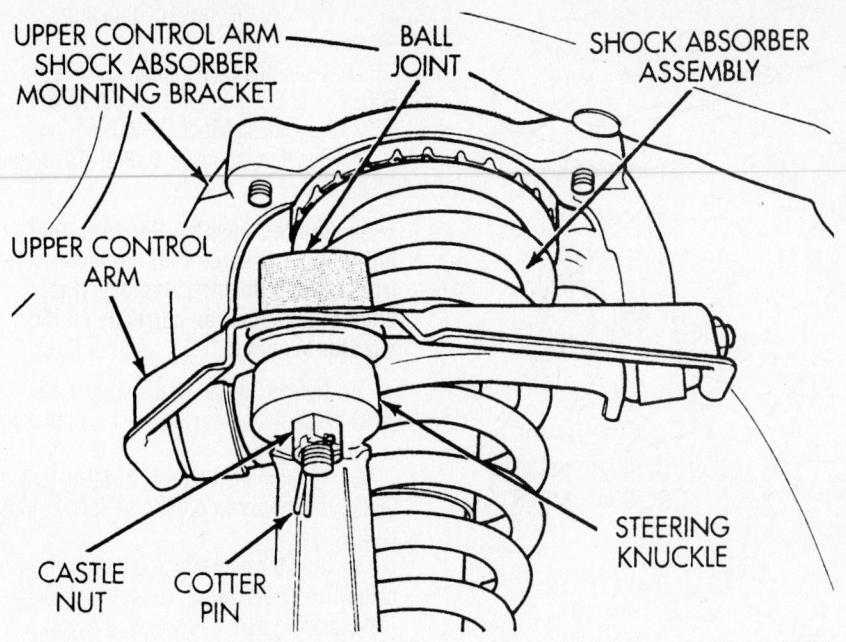

Front upper control arm component identification

2. Remove or disconnect the following:
   - Front wheel
   - Ball joint from the steering knuckle using the joint separation Tool MB991113
   - Strut assembly
   - Upper control arm mounting bracket from the strut assembly
   - Upper control arm shaft-to-bracket nuts
   - Upper control arm, from the bracket using the joint separation tool
   - Upper control arm assembly

To install:

3. Install or connect the following:
   - Upper control arm assembly
   - Upper control arm to the bracket and torque the nuts to 62 ft. lbs. (86 Nm)
   - Upper control arm mounting bracket to the strut assembly
   - Strut assembly
   - Ball joint to the steering knuckle and torque the locking nut to 20 ft. lbs. (28 Nm)
   - Front wheel

4. Check and/or adjust the wheel alignment, if necessary.

Rear

1. Before servicing the vehicle, refer to the precautions in the beginning of this section.

2. Remove or disconnect the following:
   - Both rear wheels
   - Both struts from the knuckles
   - Muffler hanger from the frame rail
   - Rear exhaust pipe hanger from the rear crossmember and allow it to hang
   - Upper control arm ball joint stud from the knuckle

3. Support the center of the rear crossmember with a jack and a block of wood.

4. Remove or disconnect the following:
   - Speed sensor wiring from the upper control arms, if equipped

with an Anti-lock Brake System (ABS)
   - 4 crossmember-to-frame bolts

✲✲ WARNING

Do not damage the rear brake hoses while lowering the crossmember.

5. Lower the crossmember to access the 2 upper control arm pivot bar mounting bolts.

6. Remove or disconnect the following:
   - Both upper control arm-to-crossmember bolts
   - Upper control arm

To install:

7. Install or connect the following:
   - Upper control arm to the crossmember and torque the bolts to 80 ft. lbs. (108 Nm)

➡**Be sure to place the flat washers between the crossmember and the pivot bar.**

   - Crossmember using a drift to align the holes and tighten the bolts to 80 ft. lbs. (108 Nm)

8. Remove the jack

9. Install or connect the following:
   - Speed sensor wiring to the upper control arms, if equipped with ABS
   - Upper control arm ball joint stud to the knuckle and torque the ball joint stud nut to 63 ft. lbs. (85 Nm)
   - Rear exhaust pipe hanger to the rear crossmember
   - Muffler hanger to the frame rail

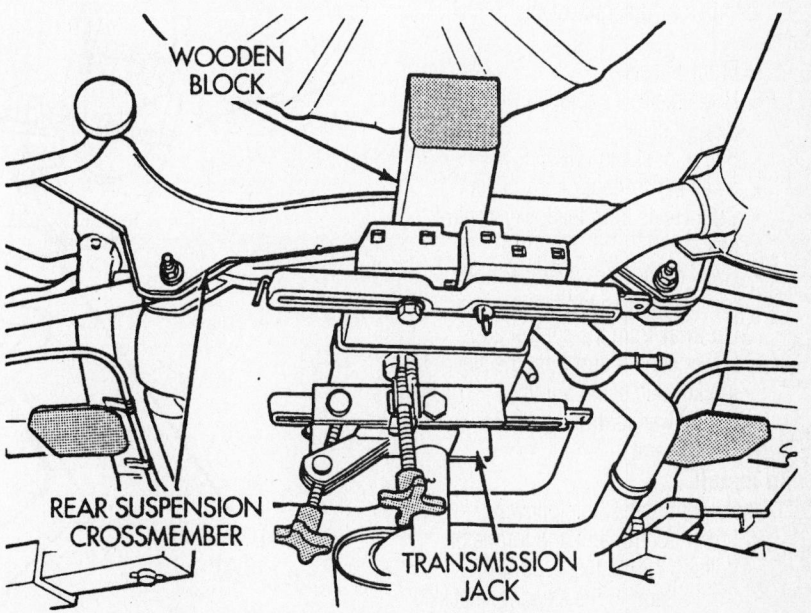

A jack with a block of wood should be used to support the rear crossmember

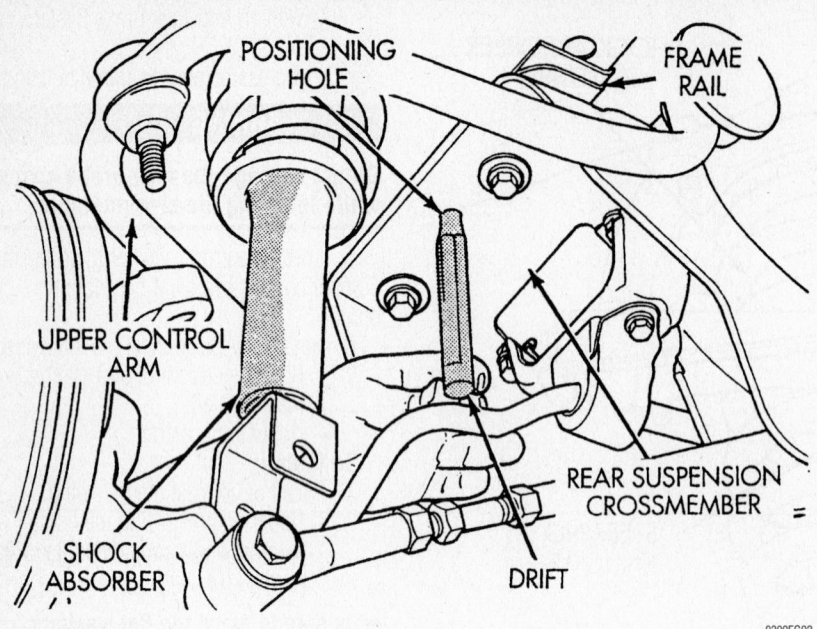

Use a drift to align the rear crossmember to the frame

- Both struts to the knuckles and torque the lower strut bolt to 70 ft. lbs. (95 Nm)
- Both rear wheels

Lower Control Arm

REMOVAL & INSTALLATION

Front

1. Before servicing the vehicle, refer to the precautions in the beginning of this section.
2. Remove or disconnect the following:
   - Front wheels
   - Heat shield, if equipped with 15 inch wheels
   - Ball joint clamp nut
   - Sway bar-to-control arm bolts
   - Strut clevis from the lower control arm
   - Sway bar-to-crossmember bolts, loosen them and rotate the sway bar away from the control arm
   - Lower control arm from the steering knuckle using a prybar
   - Both lower control arm bolts
   - Control arm

To install:
3. Install or connect the following:
   - Lower control arm and torque the bolts to 120 ft. lbs. (163 Nm)
   - Control arm to the steering knuckle and torque the nut to 70 ft. lbs. (95 Nm)

4. Rotate the sway bar up to the control arms.
   - Strut clevis
   - Sway bar and torque the bolts to 21 ft. lbs. (28 Nm)
   - Rear wheels

CONTROL ARM BUSHING REPLACEMENT

Front Isolator Bushing

1. Before servicing the vehicle, refer to the precautions in the beginning of this section.

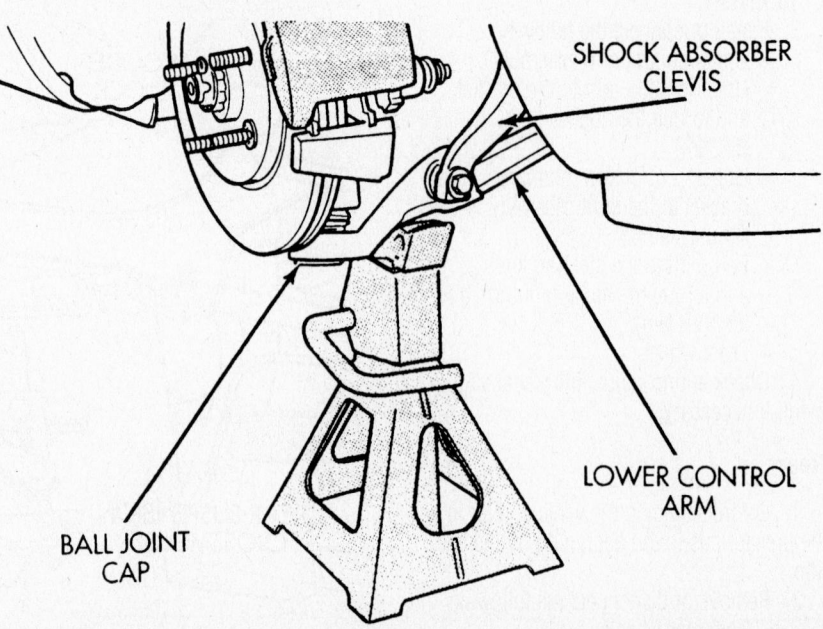

Pre-load the suspension before final tightening the control arm to the steering knuckle

2. Remove the lower control arm assembly
3. Install the Bushing Remover Tool 6602-5 and Bushing Receiver Tool MB-990799 on Special Tool C-4212-F.
4. Position the lower control arm on the assembled removal tools.

➡ **Be sure the Bushing Receiver Tool MB-990799 is square on the lower control arm and Bushing Remover Tool 6602-5 is positioned correctly on the isolator bushing.**

5. Tighten Special Tool C-4212-F to press the bushing from the lower control arm.

To assemble:
6. Position the Bushing Installer Tool 6876 onto the screw portion of Special Tool C-4212-F.
7. Start the new bushing into the lower control arm hole machined surface side by hand, making sure it is square with its mounting hole.
8. Assemble Special Tools 6758, 6876 and C-4212-F; then, position the lower control arm on the assembly, making sure that the tools are aligned.
9. Tighten Special Tool C-4212-F to press the bushing into the lower control arm until it is flush on the machined surface.
10. Install the lower control arm assembly.

Rear Isolator Bushing

1. Before servicing the vehicle, refer to the precautions in the beginning of this section.

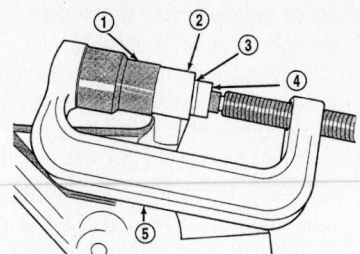

1 – SPECIAL TOOL MB-990799
2 – LOWER CONTROL ARM
3 – FRONT ISOLATOR BUSHING
4 – SPECIAL TOOL 6602-5
5 – SPECIAL TOOL C-4212-F

9306FG09

Removing the front isolator bushing from the lower control arm

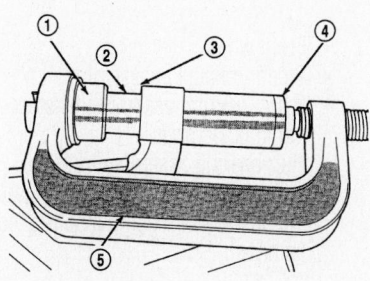

1 – SPECIAL TOOL 6876
2 – ISOLATOR BUSHING
3 – MACHINED SURFACE SIDE OF LOWER CONTROL ARM
4 – SPECIAL TOOL 6758
5 – SPECIAL TOOL C-4212-F

9306FG10

Installing the front isolator bushing to the lower control arm

2. Remove the lower control arm assembly

3. Install the Bushing Remover Tool 6756 and Bushing Receiver Tool C-4366-2 on Special Tool C-4212-F.

4. Position the lower control arm on the assembled removal tools.

➡**Be sure the Bushing Receiver Tool C-4366-2 is square on the lower control**

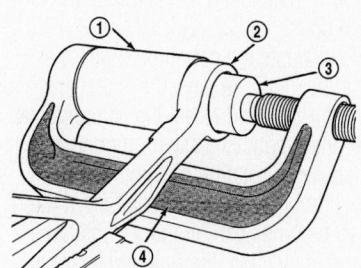

1 – SPECIAL TOOL C-4366-2
2 – LOWER CONTROL ARM
3 – SPECIAL TOOL 6756
4 – SPECIAL TOOL C-4212-F

9306FG11

Removing the rear isolator bushing from the lower control arm

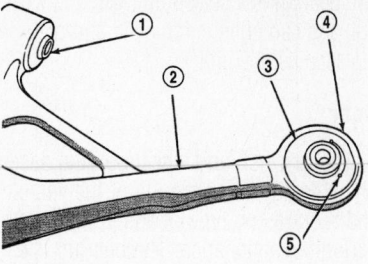

1 – FRONT ISOLATOR BUSHING
2 – LOWER CONTROL ARM
3 – REAR ISOLATOR BUSHING
4 – MACHINED SURFACE
5 – VOID IN BUSHING IN THIS DIRECTION

9306FG12

Positioning the rear isolator bushing to the lower control arm

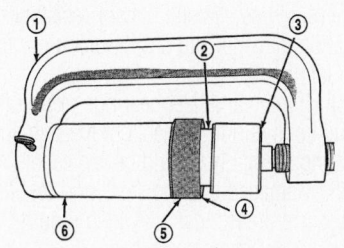

1 – SPECIAL TOOL C-4212-F
2 – REAR BUSHING
3 – SPECIAL TOOL 6760
4 – MACHINED SURFACE ON LOWER CONTROL ARM
5 – LOWER CONTROL ARM
6 – SPECIAL TOOL 6756

9306FG13

Installing the rear isolator bushing to the lower control arm

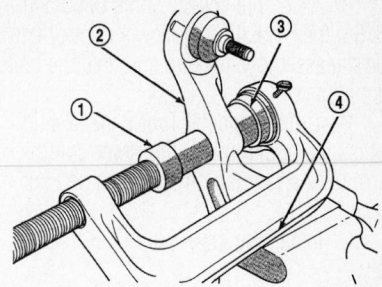

1 – SPECIAL TOOL 6877
2 – LOWER CONTROL ARM
3 – SPECIAL TOOL 6876
4 – SPECIAL TOOL C-4212-F

9306FG14

Removing the clevis bushing from the lower control arm

arm and Bushing Remover Tool 6756 is positioned correctly on the isolator bushing.

5. Tighten Special Tool C-4212-F to press the bushing from the lower control arm.

To assemble:

6. Position the Bushing Installer Tool 6760 onto the screw portion of special Tool C-4212-F.

7. Start the new bushing into the lower control arm hole's machined surface side by hand, making sure it is square with its mounting hole with the void in the rubber portion facing away from the ball joint.

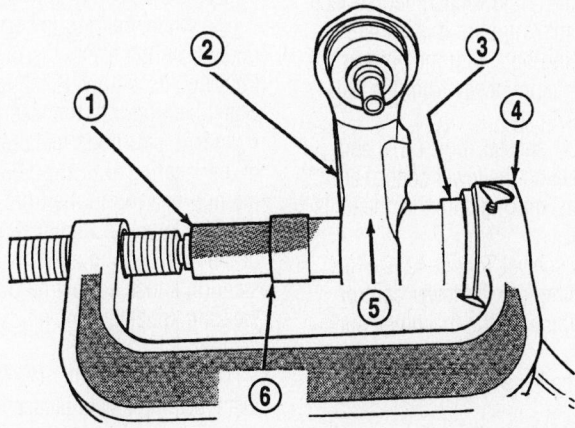

1 – SPECIAL TOOL 6877
2 – MACHINED SURFACE SIDE OF LOWER CONTROL ARM
3 – SPECIAL TOOL 6876
4 – SPECIAL TOOL C-4212-F
5 – LOWER CONTROL ARM
6 – CLEVIS BUSHING

9306FG15

Installing the clevis bushing to the lower control arm

8. Assemble special Tools 6760 and 6756; then, position the lower control arm on the assembly, making sure that the tools are aligned.

9. Tighten special Tool C-4212-F to press the bushing into the lower control arm until it is flush on the machined surface.

10. Install the lower control arm assembly.

Clevis Bushing

1. Before servicing the vehicle, refer to the precautions in the beginning of this section.

2. Remove the lower control arm assembly

3. Install the Bushing Remover Tool 6877 and Bushing Receiver Tool 6876 on special Tool C-4212-F.

4. Position the lower control arm on the assembled removal tools.

➡ **Be sure the Bushing Receiver Tool 6876 is square on the lower control arm and Bushing Remover Tool 6877 is positioned correctly on the clevis bushing.**

5. Tighten special Tool C-4212-F to press the bushing from the lower control arm.

To assemble:

6. Position the Bushing Installer Tool 6877 onto the screw portion of special Tool C-4212-F.

7. Start the new bushing into the lower control arm hole's machined surface side by hand, making sure it is square with its mounting hole with the void in the rubber portion facing away from the ball joint.

8. Assemble special Tools 6876 and 6877; then, position the lower control arm on the assembly, making sure that the tools are aligned.

9. Tighten special Tool C-4212-F to press the bushing into the lower control arm until it is flush on the machined surface.

10. Install the lower control arm assembly.

Wheel Bearings

ADJUSTMENT

Front

The front hub wheel bearing is designed for the life of the vehicle and requires no type of adjustment or periodic maintenance.

The bearing is a sealed unit with the wheel hub and can only be removed and/or replaced as one unit.

Rear

The rear hub and wheel bearing assembly is designed for the life of the vehicle and requires no type of adjustment or periodic maintenance. The bearing is a sealed unit with the wheel hub and can only be removed and/or replaced as one unit.

The following procedure may be used for evaluation of bearing condition:

1. Raise and safely support the vehicle.

2. Remove the rear wheels and brake drums.

3. Turn the hub flange carefully. Excessive roughness, lateral play or resistance to rotation may indicate dirt intrusion or bearing failure.

4. If the rear wheel bearings exhibit the conditions during inspection, the hub and bearing assembly should be replaced.

5. Damaged bearing seals and resulting excessive grease loss may also require bearing replacement. Moderate grease loss from the bearing is considered normal and should not require replacement of the hub and bearing assembly.

REMOVAL & INSTALLATION

Front

The front wheel bearing used on this vehicle is a bolt-in type wheel bearing.

The wheel bearing is serviced separately from the front steering knuckle and front hub assembly. Retention of the front wheel bearing into the steering knuckle is by means of 3 bolts installed from the rear of the steering knuckle. The 3 bolts attach the hub/bearing to the front surface of the steering knuckle. Removal and installation of the hub/bearing assembly from the steering knuckle must be done with the steering knuckle removed from the vehicle.

The face of the outer CV-joint has a metal bearing shield pressed on it. This design deters direct water splash on the bearing seal while allowing any water that gets in to run out the bottom of the steering knuckle. It is important to thoroughly clean the outer CV-joint and the wheel bearing area in the steering knuckle before it is assembled after servicing the front wheel bearing or driveshaft.

At no time when servicing this vehicle, can a sheet metal screw, bolt or other metal fastener be installed in the shock tower to take the place of an original plastic clip. Also, NO holes can be drilled into the front shock tower for the installation of any metal fasteners into the shock tower. Because of the minimum clearance in this area installation of metal fasteners could damage the coil spring's protective coating and lead to corrosion failure of the spring. If a plastic clip is missing, lost or broken during servicing a vehicle, replace only with the equivalent part listed in the Mopar parts catalog.

1. Before servicing the vehicle, refer to the precautions in the beginning of this section.

2. Remove or disconnect the following:

- Front wheel
- Steering knuckle assembly
- 3 hub/bearing assembly-to-steering knuckle bolts
- Hub/bearing assembly from the steering knuckle. If necessary, tap the bearing out, using a soft-faced hammer

➡ **The wheel bearing is transferable to a replacement steering knuckle if the bearing is found in serviceable condition.**

To install:

3. Clean all parts well. Thoroughly, clean all the hub/bearing assembly mounting surfaces on the steering knuckle.

4. Install or connect the following:

- Hub/bearing assembly and torque the bolts to 80 ft. lbs. (110 Nm)
- Steering knuckle
- Front wheel

Rear

All vehicles are equipped with permanently lubricated and sealed for life rear wheel bearings. There is no periodic lubrication or maintenance recommended for these units.

To evaluate the condition of the rear wheel bearings, remove the wheel and brake drum or rotor and rotate the flanged outer ring of the hub. Excessive roughness or resistance to rotation may indicate dirt intrusion or wheel bearing failure. If the rear wheel bearings exhibit these conditions during inspection, the hub and bearing assembly should be replaced. Damaged bearing seals and resulting excessive grease loss may also require bearing replacement. Moderate grease loss from the bearing is considered normal and should not require replacement of the hub and

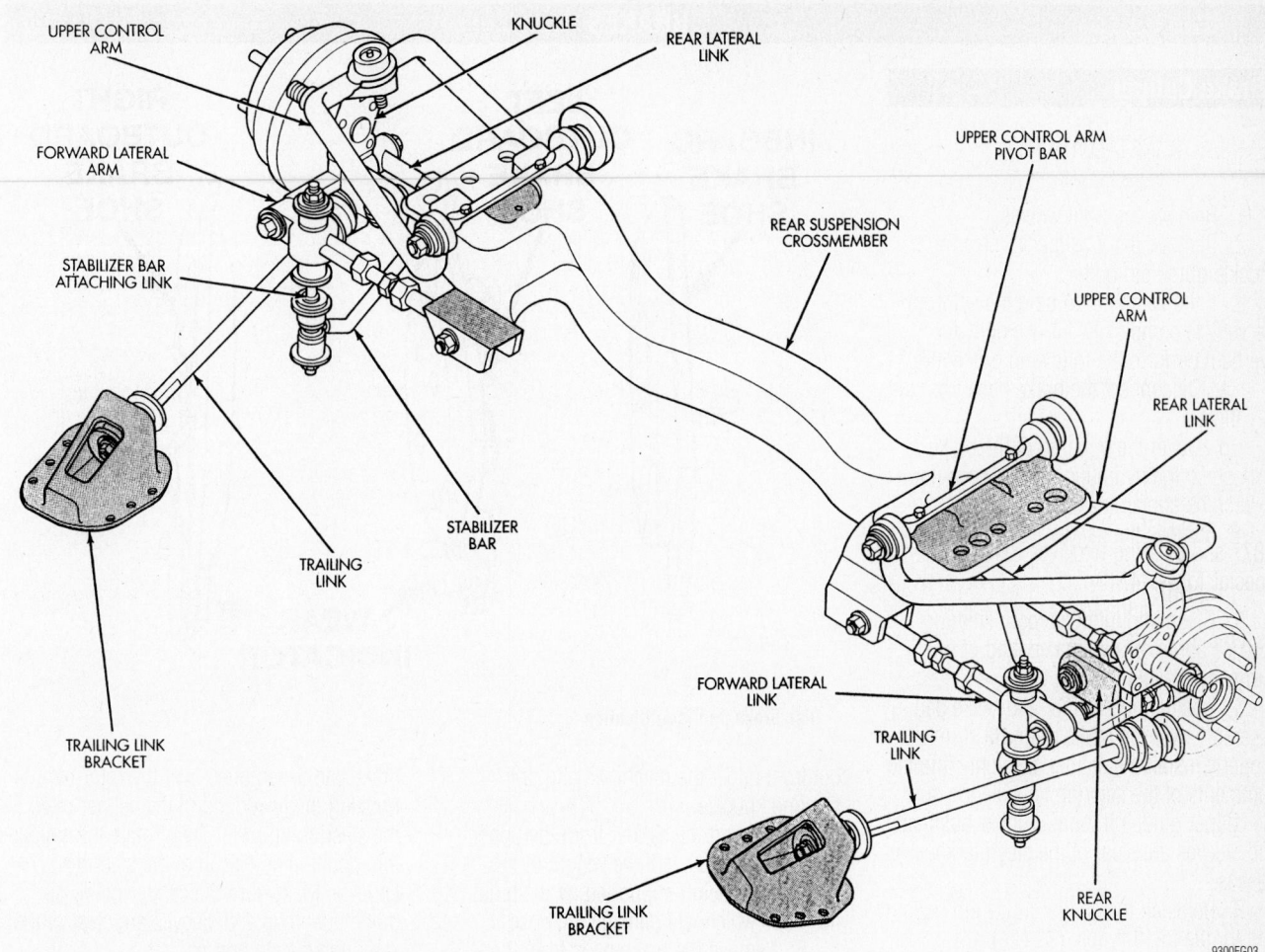

UPPER CONTROL ARM

KNUCKLE

REAR LATERAL LINK

FORWARD LATERAL ARM

UPPER CONTROL ARM PIVOT BAR

REAR SUSPENSION CROSSMEMBER

UPPER CONTROL ARM

STABILIZER BAR ATTACHING LINK

REAR LATERAL LINK

TRAILING LINK

STABILIZER BAR

TRAILING LINK BRACKET

FORWARD LATERAL LINK

TRAILING LINK

TRAILING LINK BRACKET

REAR KNUCKLE

9300FG03

Independent rear suspension component identification

bearing assembly. If service requires removal for inspection or replacement of the rear wheel bearing and hub assembly, use the following procedure.

1. Before servicing the vehicle, refer to the precautions in the beginning of this section.

2. Remove or disconnect the following:

- Rear wheel
- Brake drum, if equipped with drum brakes
- Brake caliper and rotor, if equipped with disc brakes
- Hub dust cap
- Hub nut and discard it
- Hub/bearing assembly by pulling straight off the spindle

To install:

3. Install or connect the following:
- New bearing on the spindle
- Hub using a new nut and torque the nut to 185 ft. lbs. (250 Nm)
- Dust cap
- Brake drum or rotor and caliper
- Rear wheel and torque the lug nuts to 95 ft. lbs. (129 Nm)

BRAKES

Brake Caliper

REMOVAL & INSTALLATION

1. Remove the front wheels.
2. Remove the 2 caliper-to-steering knuckle guide pin bolts.
3. If the caliper is to be removed from the vehicle completely, for example, for overhaul perform the following procedure.
   a. Disconnect the brake hose from the caliper.
   b. Cover the opening of the brake hose so the hydraulic system does not become contaminated.
4. Remove the caliper from the steering knuckle.

To install:

5. Clean and lubricate both steering knuckle abutments with a coating of multi-purpose grease.
6. Position the caliper and brake pad assembly over the brake rotor. Be sure to properly install the caliper assembly into the abutments of the steering knuckle. Be sure the caliper guide pin bolts, rubber bushings and sleeves are clear of the steering knuckle bosses.
7. Reinstall the caliper guide pin bolts and torque to 16 ft. lbs. (22 Nm). On Sebring and Avenger models, torque the caliper guide pin bolts to 54 ft. lbs. (74 Nm).
8. If removed, connect the brake hose to the caliper and torque to 35 ft. lbs. (48 Nm).
9. Properly bleed the brake system.
10. Reinstall the wheel and tire and torque the lug nuts in a star pattern sequence to half specification. Repeat the tightening procedure to full specified torque of 95–100 ft. lbs. (129–135 Nm).
11. Pump the brake pedal to seat the front brake pads before moving the vehicle.
12. Road test the vehicle and check for proper operation.

Disc Brake Pads

REMOVAL & INSTALLATION

1. Remove the front wheels.
2. Remove the 2 caliper-to-steering knuckle guide pin bolts.
3. Lift the caliper away from the steering knuckle by first rotating the free end of the caliper away from the steering knuckle. Then slide the opposite end of the caliper

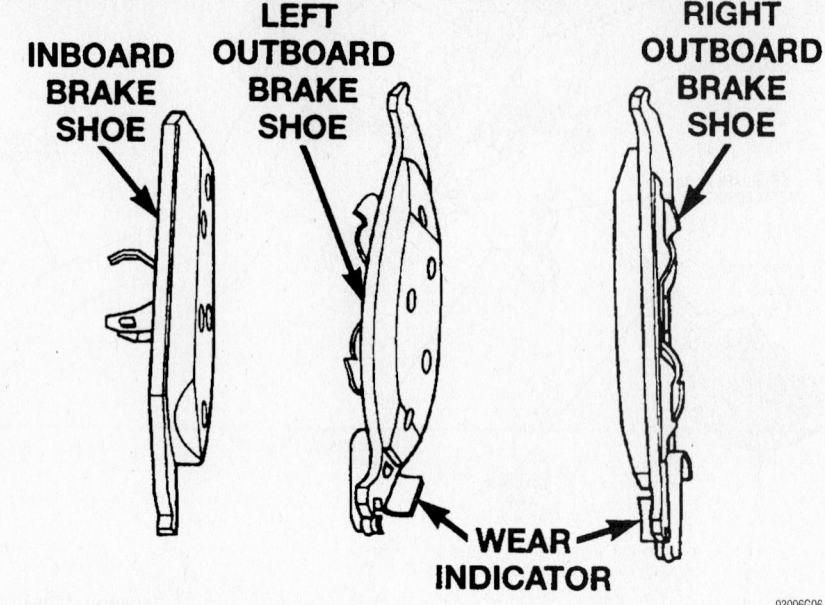

Disc brake pad identification

out from under the machined end of the steering knuckle.
4. Support the caliper from the upper control arm to prevent the weight of the caliper from being supported by the brake flex hose which will damage the hose.
5. Remove the brake pads from the caliper. Remove the outboard brake pad by prying the pad retaining clip over the raised area on the caliper. Then slide the pad down and off the caliper. Pull the inboard brake pad away from the piston until the retaining clip is free from the cavity in the piston.

To install:

6. Completely depress the piston into the caliper using a large C-clamp or other suitable tool.
7. Lubricate the area on the steering knuckle where the caliper slides with high temperature grease.
8. Install the new inboard brake pad into the caliper piston by firmly pressing into the piston bore. Install the brake pads into the caliper. Note that the inboard and outboard pads are different. Make sure the inboard brake pad assembly is positioned squarely against the face of the caliper piston.

➡ **Be sure to remove the noise suppression gasket paper cover if the pads come so equipped.**

9. Install the new outboard brake pad onto the caliper assembly.
10. Carefully position the caliper and

brake pad assemblies over the rotor by hooking the lower end of the caliper over the steering knuckle. Then rotate the caliper into position at the top of the steering knuckle. Make sure the caliper guide pin bolts, bushings and sleeves are clear of the steering knuckle bosses.
11. Reinstall the caliper guide pin bolts and torque to 16 ft. lbs. (22 Nm). On Sebring and Avenger models, torque the caliper guide pin bolts to 54 ft. lbs. (74 Nm).
12. Reinstall the wheel and tire. Torque the lug nuts in 2 steps, in a star pattern sequence to 95–100 ft. lbs. (129–135 Nm).
13. Pump the brake pedal until the brake pads are seated and a firm pedal is achieved before attempting to move the vehicle.
14. Road test the vehicle to check for proper operation.

Brake Drums

REMOVAL & INSTALLATION

All vehicles except Sebring Convertible, are equipped with rear wheel, 2-shoe leading/trailing, internal expanding type of drum brakes with automatic self-adjuster mechanisms. The automatic self-adjuster mechanisms used on these vehicles are new designs and function differently than the screw type adjusters used in the past. These new self-adjusters are still actuated each time the vehicle's service brakes are applied.

The new adjusters are located directly below the wheel cylinders.

The Sebring Convertible's rear wheel drum brake is a 2-shoe leading/trailing internal expanding type with an automatic self-adjuster mechanism. The automatic self-adjuster mechanism used on this vehicle is the screw type adjuster. The self-adjuster mechanism is actuated each time the vehicle service brakes are applied. Generally, drum brakes with a self-adjusting mechanism do not require manual brake shoe adjustment. Although, in the event that the brake shoes are replaced, it is advisable to make the initial adjustment manually to speed up the initial adjustment time. The initial adjustment procedure must be done prior to driving the vehicle.

1. Remove the rear wheel assembly.

2. For all vehicles, except Sebring Convertible, use the following procedure:

    a. Locate and remove the rubber plug from the brake support plate (backing plate).

    b. Insert a brake adjuster tool or similarly shaped prytool through the automatic adjuster access hole and engage the teeth on the adjuster wheel. Rotate the adjuster wheel so it is moved toward the front of the vehicle. Continue moving the adjuster until it stops; this will back off the adjustment of the rear brake shoes.

3. For the Sebring Convertible, use the following procedure for releasing the self-adjusting mechanism:

    a. Locate and remove the rubber plug from the brake support plate (backing plate).

    b. Insert a brake adjuster tool or similarly shaped prytool through the automatic adjuster access hole and carefully push the adjuster actuating lever out of engagement with the adjuster starwheel. While holding the lever away from the starwheel, insert a second prytool through the access hole and engage the teeth on the adjuster wheel. Rotate the adjuster wheel upward away from the ground; this will back off the adjustment of the rear brake shoes.

4. Remove the rear brake drum from the hub assembly.

To install:

5. Inspect the brake drums for cracks or signs of overheating. Measure the drum run-out and diameter. If not to specification, resurface the drum. Run-out should not exceed 0.006 in. (0.15mm). The diameter variation (oval shape) of the drum braking surface must not exceed either 0.0025 in. (0.064mm) in 30 degrees rotation, or

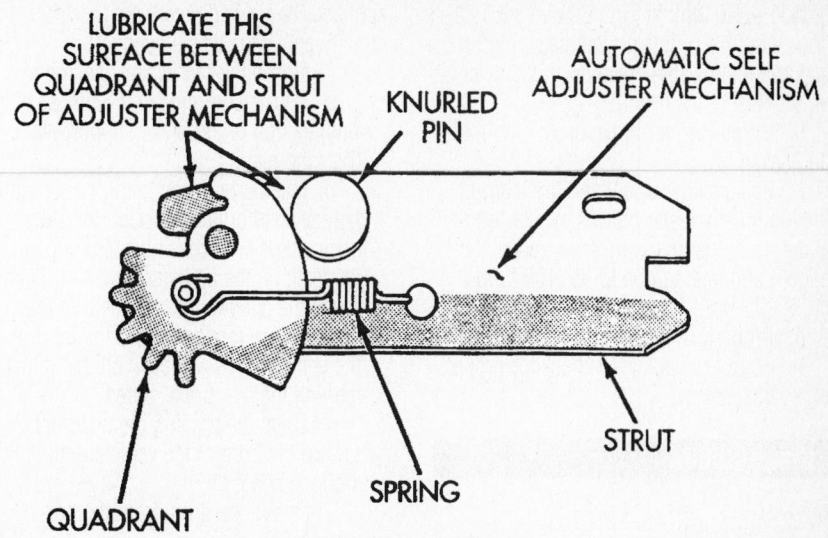

Automatic self-adjuster mechanism—except convertible

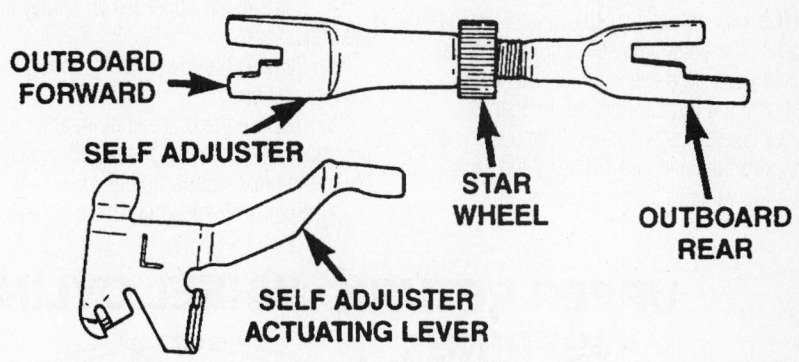

Rear brake shoe automatic self-adjuster mechanism and actuating lever—Sebring Convertible

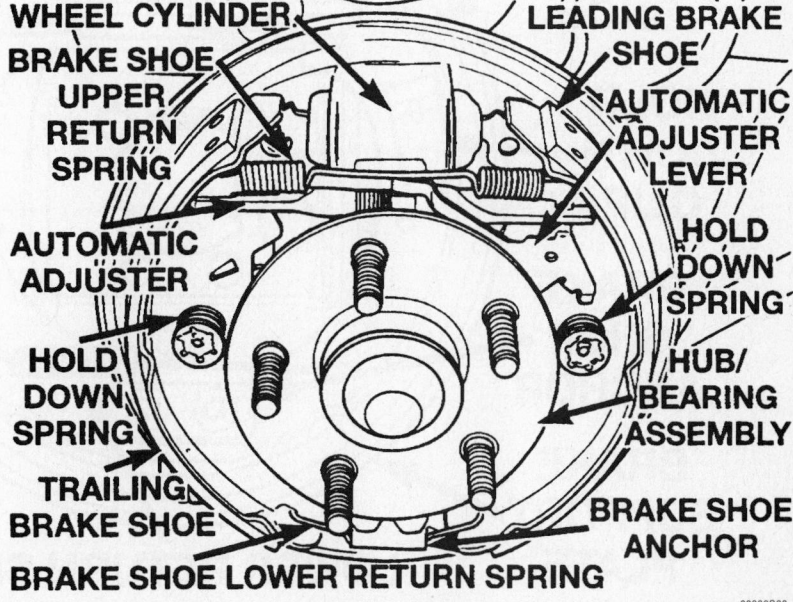

Kelsey Hayes rear brake assembly—Sebring Convertible

0.0035 in. (0.089mm) in 360 degrees rotation. All brake drums are marked with the maximum allowable brake drum diameter on the face of the drum.

6. Install the rear brake drum onto the hub assembly.

7. Reinstall the wheel and tire. Torque the lug nuts in a star pattern sequence to about 45 ft. lbs. (61 Nm); then, repeat the pattern and final torque to 95–100 ft. lbs. (129–135 Nm).

8. Properly adjust the rear brakes.

9. Road test vehicle to check for proper brake operation.

Brake Shoes

REMOVAL & INSTALLATION

All vehicles except Sebring Convertible, are equipped with rear wheel, 2 shoe leading/trailing, internal expanding type of drum brakes with automatic self-adjuster mechanisms. The automatic self-adjuster mechanisms used on these vehicles are new designs and function differently than the screw type adjusters used in the past. These new self-adjusters are still actuated each time the vehicles' service brakes are applied. The new adjusters are located directly below the wheel cylinders.

1. Remove the rear wheel assembly.

   a. Locate and remove the rubber plug from the top of the brake support plate (backing plate).

   b. Insert a brake adjuster tool or similarly shaped prytool through the automatic adjuster access hole and engage the teeth on the adjuster quadrant. Then rotate the quadrant so the teeth of the quadrant are moved toward the front of the vehicle. This will back off the adjustment of the rear brake shoes.

   c. Continue moving the quadrant toward the front of the vehicle until it stops moving.

2. Remove the drum from the hub assembly.

3. Remove the actuating spring from the adjuster mechanism and trailing brake shoe.

4. Remove the upper return spring from the brake shoes.

5. Remove the lower return spring from the brake shoes.

6. Remove the brake shoe retainer and pin attaching the leading brake shoe assembly to the brake support plate.

7. Remove the leading brake shoe and the adjuster mechanism as an assembly from the rear brake support plate. The adjuster mechanism cannot be separated from the leading brake shoe until the brake shoe and the adjuster mechanism is removed from the support plate.

8. Remove the trailing brake shoe retainer and pin attaching the trailing brake shoe assembly to the brake support plate. Remove the trailing brake shoe assembly.

➡On this vehicle, the parking brake actuating lever is permanently attached to the trailing brake shoe assembly. Do not attempt to remove it from the original brake shoe assembly or reuse the original actuating lever on a replacement brake shoe assembly. All replacement brake shoe assemblies for this vehicle must have the actuating lever as part of the trailing brake shoe assembly.

9. Remove the parking brake cable from the parking brake lever. Do not remove the lever from the brake shoe.

10. Remove the automatic adjuster mechanism from the brake shoe by fully extending the adjuster and rotate the adjuster out to release from the brake shoe.

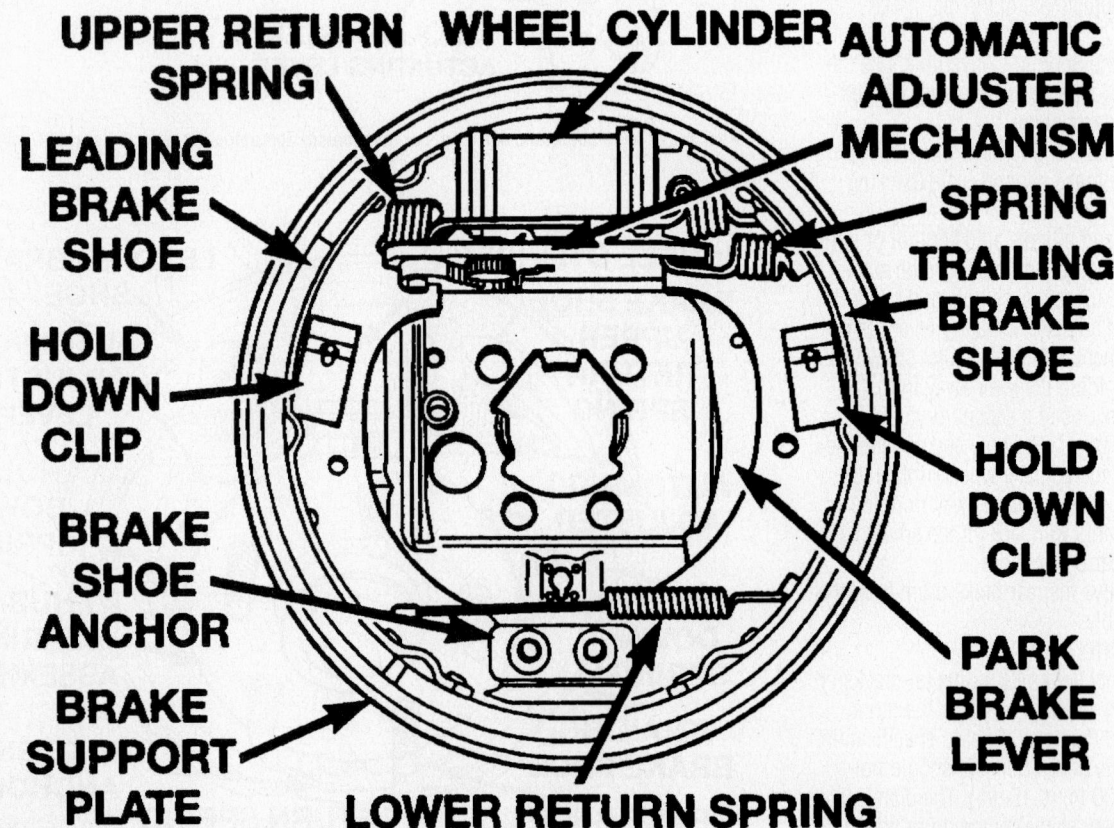

Varga rear wheel brake assembly (left side shown)—except convertible

93006G14

To inspect:

11. Thoroughly clean all parts. The brake lining should show contact across the entire width and from heel to toe; otherwise, replace. Clean and inspect the brake support plate and the automatic adjuster mechanism. Be sure the quadrant (toothed part) of the adjuster is free to rotate throughout its entire tooth contact range and is free to slide the full length of its mounting slot. Check the knurled pin. It should be securely attached to the adjuster mechanism and its teeth should be in good condition. If the adjuster is worn or damaged, replace it. If the adjuster is serviceable, lubricate lightly with high-temperature grease between the strut and the quadrant. Check the brake springs. Overheating indications are paint discoloration or distorted end coils. Replace parts as required.

12. Inspect the brake drums for cracks or signs of overheating. Measure the drum run-out and diameter. If not to specification, reface the drum. Run-out should not exceed 0.006 in. (0.15mm). The diameter variation (oval shape) of the drum braking surface must not exceed either 0.0025 in. (0.064mm) in 30 degrees rotation, or 0.0035 in. (0.089mm) in 360 degrees rotation. All brake drums are marked with the maximum allowable brake drum diameter on the face of the drum.

To install:

13. Lubricate the 8 brake shoe contact points with high-temperature grease.

➡**The trailing brake shoe assemblies used on the rear brakes of this vehicle are unique to the left and right side of the vehicle. Care must be taken to ensure the brake shoes are properly installed in their correct side of the vehicle. When the trailing shoes are properly installed on their correct side of the vehicle, the park brake actuating lever will be positioned under the brake shoe web.**

14. Reinstall the parking brake cable onto the parking brake lever and install the trailing brake shoe and attaching pin.

15. Reinstall the automatic self-adjuster on the leading brake shoe by rotating it inward to attach. Install the leading shoe and adjuster assembly to the brake support plate.

16. Make sure the leading brake shoe is squarely seated on the brake support plate shoe contact areas, and install the brake retainer on the retainer pin.

17. Reinstall the lower return spring.

➡**The upper brake shoe return spring and adjuster mechanism actuating spring are unique to the side of the vehicle they are used on. The springs are colored for identification. The left side springs are green and the right side springs are blue.**

18. Reinstall the upper return spring (blue, right side; green, left side) on the leading brake shoe first, then on the trailing brake shoe.

19. Reinstall the self-adjuster spring on the trailing brake shoe first then attach it to the adjuster.

20. Reinstall the rear brake drums.

21. Reinstall the wheel and tire. Torque the lug nuts in a star pattern sequence to about 45 ft. lbs. (61 Nm), then repeat the pattern and final torque to 95–100 ft. lbs. (129–135 Nm).

22. Adjust the rear brakes by depressing the brake pedal. Brake shoe adjustment will occur the first time the brake pedal is depressed, pushing the rear brake shoes against the braking surface of the rear brake drums. Brake shoes should now be correctly adjusted and will not require any type of manual adjustment.

23. Road test vehicle to check for proper brake operation.

Sebring Convertible

The Sebring Convertible's rear wheel drum brake is a 2-shoe leading/trailing internal expanding type with an automatic self-adjuster mechanism. The automatic self-adjuster mechanism used on this vehicle is the screw type adjuster. The self-adjuster mechanism is actuated each time the vehicle service brakes are applied. Generally, drum brakes with a self-adjusting mechanism do not require manual brake shoe adjustment. Although, in the event that the brake shoes are replaced, it is advisable to make the initial adjustment manually to speed up the initial adjustment time. The initial adjustment procedure must be done prior to driving the vehicle.

➡**When removing the rear brake shoes, replace the brake shoes from only one side of the vehicle at a time. This is due to the automatic adjustment feature of the parking brake system. If the brake shoes are removed from both sides of the vehicle at the same time, the automatic adjuster will remove all slack from the parking brake cables, which will make brake shoe installation extremely difficult.**

1. Remove the rear wheel assembly.

a. Locate and remove the rubber plug from the brake support plate (backing plate).

b. Insert a brake adjuster tool or similarly shaped prytool through the automatic adjuster access hole and carefully push the adjuster actuating lever out of engagement with the adjuster starwheel. While holding the lever away from the starwheel, insert a second prytool through the access hole and engage the teeth on the adjuster wheel. Rotate the adjuster wheel upward away from the ground. This will back off the adjustment of the rear brake shoes.

2. Remove the drum from the hub assembly.

3. Remove the adjusting lever actuating spring from the leading brake shoe. Remove the automatic adjuster actuating lever from the leading brake shoe.

4. Thread the adjuster starwheel all the way into the adjuster, which will remove all tension from the adjuster.

5. Remove the upper and lower return springs from the brake shoes.

6. Remove the brake shoe hold-down spring and pin attaching the leading brake shoe assembly to the brake support plate.

7. Remove the leading brake shoe from the support plate.

8. Remove the automatic adjuster from the parking brake actuating lever and trailing brake shoe.

9. Remove the retaining clip securing the parking brake actuating lever to the trailing brake shoe.

10. Remove the trailing brake shoe hold-down spring and pin attaching the trailing brake shoe assembly to the brake support plate.

11. Remove the trailing brake shoe from the brake support plate and separate the shoe from the parking brake actuating lever.

To inspect:

12. Clean all parts well. The brake lining should show contact across the entire width and from heel to toe; otherwise, replace. Clean and inspect the brake support plate and the automatic adjuster mechanism. Be sure the adjuster is free to rotate throughout its entire range. If the adjuster is worn or damaged, replace it. If the adjuster is serviceable, lightly lubricate the threaded portion with high-temperature grease. Check the brake springs. Overheating indications are paint discoloration or distorted end coils. Replace parts as required.

13. Inspect the brake drums for cracks or signs of overheating. Measure the drum

run-out and diameter. If not to specification, reface the drum. Run-out should not exceed 0.006 in. (0.15mm). The diameter variation (oval shape) of the drum braking surface must not exceed either 0.0025 in. (0.064mm) in 30° rotation, or 0.0035 in. (0.089mm) in 360° rotation. All brake drums are marked with the maximum allowable brake drum diameter on the face of the drum.

To install:

14. Lubricate the 6 brake shoe contact points and the brake shoe anchor points with high-temperature grease.

15. Reinstall the wave washer on the pivot pin of the parking brake actuating lever.

16. Install the trailing brake shoe onto the attaching pin of the parking brake actuating lever.

17. Position the trailing brake shoe onto the brake support plate and be sure the trailing brake shoe is squarely seated on the support plate shoe contact areas and install the brake shoe hold-down spring on the hold-down pin.

18. Reinstall the parking brake actuating lever-to-trailing brake shoe retaining clip.

19. Reinstall the automatic adjuster on the trailing brake shoe and the parking brake actuating lever.

20. Place the leading brake shoe onto the brake support plate in proper position and install the attaching pin and hold-down spring.

21. Reinstall the lower and upper return springs.

22. Reinstall the automatic adjuster actuating lever and spring onto the leading brake shoe.

23. Manually adjust the brake shoes to the furthest adjusted position but not so far as to interfere with the installation of the brake drum.

24. Reinstall the rear brake drums. Check and adjust the brake shoes as necessary.

25. Reinstall the wheel and tire. Torque the lug nuts in a star pattern sequence to about 45 ft. lbs. (61 Nm), then repeat the pattern and final torque to 100 ft. lbs. (135 Nm).

26. Road test vehicle to check for proper brake operation.

CHRYSLER

5

Crossfire

SPECIFICATION CHARTS

ENGINE AND VEHICLE IDENTIFICATION

		Engine						Model Year	
Code	Liters (cc)	Cu. In.	Cyl.	Fuel Sys.	Engine Type	Eng. Mfg.		Code ①	Year
L	3.2 (3198)	195	6	MFI	SOHC	Daimler/Chrysler		5	2005

MFI: Multi-point Fuel Injection

SOHC: Single Overhead Camshaft

① 10th digit of the Vehicle Identification Number (VIN)

67189-CSFR-C01

GENERAL ENGINE SPECIFICATIONS

Year	Model	Engine Displacement Liters (VIN)	Net Horsepower @ rpm	Net Torque @ rpm (ft. lbs.)	Bore x Stroke (in.)	Compression Ratio	Oil Pressure @ rpm
2005	Crossfire	3.2 (L)	215@5700	229@3000	3.54x3.31	10.0:1	45-105@3000

67189-CSFR-C02

ENGINE TUNE-UP SPECIFICATIONS

Year	Engine Displacement Liters (VIN)	Spark Plug Gap (in.)	Ignition Timing (deg.) MT	Ignition Timing (deg.) AT	Fuel Pump psi ②	Idle Speed (rpm) MT	Idle Speed (rpm) AT	Valve Clearance In.	Valve Clearance Ex.
2005	3.2 (L)	0.039	①	①	54-61	700-800	650-750	HYD	HYD

NOTE: The Vehicle Emission Control Information label reflects specification changes made during production and must be used if different from this chart.

HYD: Hydraulic

① The basic setting is controlled by the ECU and is not adjustable

② Fuel pump pressure specifications with the fuel pressure regulator vacuum hose attached.

67189-CSFR-C03

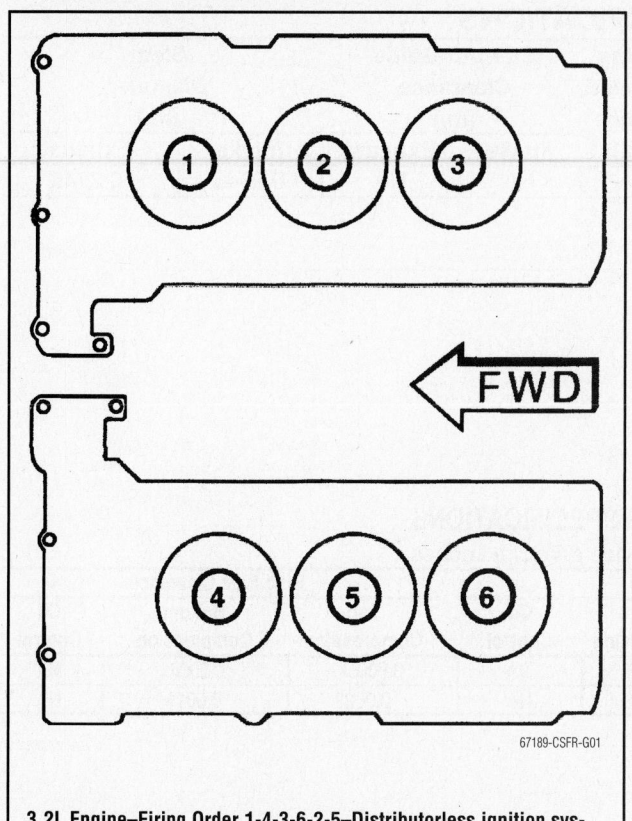

3.2L Engine–Firing Order 1-4-3-6-2-5–Distributorless ignition system (one coil on each cylinder)

67189-CSFR-G01

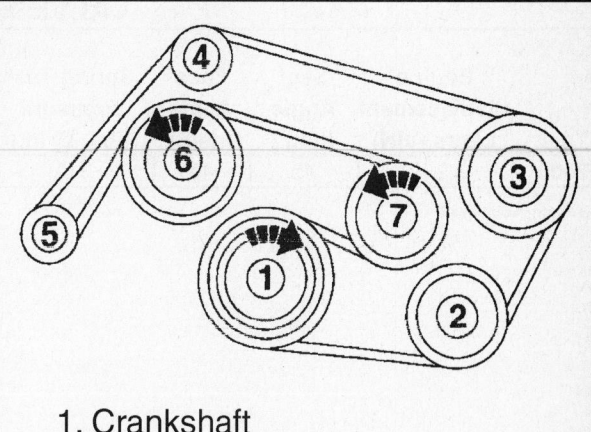

1. Crankshaft
2. A/C compressor
3. Power steering pump
4. Idler pulley
5. Alternator
6. Coolant pump
7. Automatic belt tensioner

67189-CSFR-G02

Accessory drive belt routing–3.2L Engine

CAPACITIES

Year	Model	Engine Displacement Liters (VIN)	Engine Oil with Filter	Transmission (pts.)		Drive Axle		Fuel Tank (gal.)	Cooling System (qts.)
				6-Spd	Auto	Front (pts.)	Rear (pts.)		
2005	Crossfire	3.2 (L)	6.1	3.2	16.7	—	2.8	15.8	11.8

67189-CSFR-C04

CRANKSHAFT AND CONNECTING ROD SPECIFICATIONS

All measurements are given in inches.

Year	Engine Displacement Liters (VIN)	Crankshaft				Connecting Rod		
		Main Brg. Journal Dia.	Main Brg. Oil Clearance	Shaft End-play	Thrust on No.	Journal Diameter	Oil Clearance	Side Clearance
2005	3.2 (L)	NA	0.0011-0.0020	NA	3	NA	0.0010-0.0021	NA

NA: Not Available

67189-CSFR-C05

VALVE SPECIFICATIONS

Year	Engine Displacement Liters (VIN)	Seat Angle (deg.)	Face Angle (deg.)	Spring Test Pressure (lbs. @ in.)	Spring Installed Height (in.)	Stem-to-Guide Clearance (in.)		Stem Diameter (in.)	
						Intake	Exhaust	Intake	Exhaust
2005	3.2 (L)	45	45	NA	NA	NA	NA	0.2746	0.2745

NA: Not Available

67189-CSFR-C06

PISTON AND RING SPECIFICATIONS
All measurements are given in inches

Year	Engine Displacement Liters (VIN)	Piston Clearance	Ring Gap			Ring Side Clearance		
			Top Compression	Bottom Compression	Oil Control	Top Compression	Bottom Compression	Oil Control
2005	3.2 (L)	0.0010-	0.0078-	0.0078-	NA	0.0005-	0.0004-	NA
		0.0020	0.0138	0.0158	NA	0.0023	0.0011	NA

67189-CSFR-C07

TORQUE SPECIFICATIONS
All readings in ft. lbs.

Year	Engine Displacement Liters (VIN)	Cylinder Head Bolts	Main Bearing Bolts	Rod Bearing Bolts	Crankshaft Damper Bolts	Flywheel Bolts	Manifold		Spark Plugs	Oil Pan Drain Plug
							Intake	Exhaust		
2005	3.2 (L)	①	②	③	④	⑤	15	26	21	22

① Bolts 1 - 8
  Step 1: 15 ft. lbs.
  Step 2: 37 ft. lbs.
  Step 3: 65 degrees
  Step 4: 65 degrees
  Bolts 9 and 10: 18 ft. lbs.

② M8 Bolt
  Step 1: 15 ft. lbs.
  Step 2: plus 90 degrees
  M10 bolt
  Step 1: 22 ft. lbs.
  Step 2: plus 90 degres

③ Step 1: 44 inch lbs.
  Step 2: 18 ft. lbs.
  Step 3: 90 degrees
④ Step 1: 148 ft. lbs.
  Step 2: 90 degrees

⑤ Step 1: 33 ft. lbs.
  Step 2: 90 degrees

67189-CSFR-C08

WHEEL ALIGNMENT

Year	Model		Caster Range (+/-Deg.)	Caster Preferred Setting (Deg.)	Camber Range (+/-Deg.)	Camber Preferred Setting (Deg.)	Toe-in (in.)
2005	Crossfire	F	—	5.12	0.33	-0.63	0.16 +/- 0.12
		R	—	—	0.33	-0.86	0.33 +/- 0.16

67189-CSFR-C09

TIRE, WHEEL AND BALL JOINT SPECIFICATIONS

Year	Model		OEM Tires Standard	OEM Tires Optional	Tire Pressure (psi) Front	Tire Pressure (psi) Rear	Wheel Size	Wheel Lug Nut Torque
2005	Crossfire	F	225/40ZR18	—	32	—	—	①
		R	255/35ZR19	—	—	33	—	①

F: Front

R: Rear

OEM: Original Equipment Manufacturer

PSI: Pounds Per Square Inch

① 81 ft. lbs.

67189-CSFR-C10

BRAKE SPECIFICATIONS

All measurements in inches unless noted

Year	Model		Brake Disc Original Thickness	Brake Disc Minimum Thickness	Brake Disc Maximum Runout	Minimum Lining Thickness Front	Minimum Lining Thickness Rear	Brake Caliper Bracket Bolts (ft. lbs.)	Brake Caliper Mounting Bolts (ft. lbs.)
2005	Crossfire	F	1.100	0.899	0.002	0.078	—	85	25
		R	0.350	0.299	0.003	—	0.078	—	41

67189-CSFR-C11

PRECAUTIONS

Before servicing any vehicle, please be sure to read all of the following precautions, which deal with personal safety, prevention of component damage, and important points to take into consideration when servicing a motor vehicle:

• Never open, service or drain the radiator or cooling system when the engine is hot; serious burns can occur from the steam and hot coolant.

• Observe all applicable safety precautions when working around fuel. Whenever servicing the fuel system, always work in a well-ventilated area. Do not allow fuel spray or vapors to come in contact with a spark, open flame, or excessive heat (a hot drop light, for example). Keep a dry chemical fire extinguisher near the work area. Always keep fuel in a container specifically designed for fuel storage; also, always properly seal fuel containers to avoid the possibility of fire or explosion. Refer to the additional fuel system precautions later in this section.

• Fuel injection systems often remain pressurized, even after the engine has been turned **OFF**. The fuel system pressure must be relieved before disconnecting any fuel lines. Failure to do so may result in fire and/or personal injury.

• Brake fluid often contains polyglycol ethers and polyglycols. Avoid contact with the eyes and wash your hands thoroughly after handling brake fluid. If you do get brake fluid in your eyes, flush your eyes with clean, running water for 15 minutes. If eye irritation persists, or if you have taken brake fluid internally, IMMEDIATELY seek medical assistance.

• The EPA warns that prolonged contact with used engine oil may cause a number of skin disorders, including cancer! You should make every effort to minimize your exposure to used engine oil. Protective gloves should be worn when changing oil. Wash your hands and any other exposed skin areas as soon as possible after exposure to used engine oil. Soap and water, or waterless hand cleaner should be used.

• All new vehicles are now equipped with an air bag system, often referred to as a Supplemental Restraint System (SRS) or Supplemental Inflatable Restraint (SIR) system. The system must be disabled before performing service on or around system components, steering column, instrument panel components, wiring and sensors. Failure to follow safety and disabling procedures could result in accidental air bag deployment, possible personal injury and unnecessary system repairs.

• Always wear safety goggles when working with, or around, the air bag system. When carrying a non-deployed air bag, be sure the bag and trim cover are pointed away from your body. When placing a non-deployed air bag on a work surface, always face the bag and trim cover upward, away from the surface. This will reduce the motion of the module if it is accidentally deployed. Refer to the additional air bag system precautions later in this section.

• Clean, high quality brake fluid from a sealed container is essential to the safe and proper operation of the brake system. You should always buy the correct type of brake fluid for your vehicle. If the brake fluid becomes contaminated, completely flush the system with new fluid. Never reuse any brake fluid. Any brake fluid that is removed from the system should be discarded. Also, do not allow any brake fluid to come in contact with a painted surface; it will damage the paint.

• Never operate the engine without the proper amount and type of engine oil; doing so WILL result in severe engine damage.

• Timing belt maintenance is extremely important! Many models utilize an interference-type, non-freewheeling engine. If the timing belt breaks, the valves in the cylinder head may strike the pistons, causing potentially serious (also time-consuming and expensive) engine damage. Refer to the maintenance interval charts in the front of this manual for the recommended replacement interval for the timing belt, and to the timing belt section for belt replacement and inspection.

• Disconnecting the negative battery cable on some vehicles may interfere with the functions of the on-board computer system(s) and may require the computer to undergo a relearning process once the negative battery cable is reconnected.

• When servicing drum brakes, only disassemble and assemble one side at a time, leaving the remaining side intact for reference.

• Only an MVAC-trained, EPA-certified automotive technician should service the air conditioning system or its components.

ENGINE REPAIR

➡ **Disconnecting the negative battery cable on some vehicles may interfere with the functions of the on board computer system. The computer may undergo a relearning process once the negative battery cable is reconnected.**

Distributor

➡ **The 3.2L engine does not use a distributor.**

Alternator

REMOVAL

1. Before servicing the vehicle, refer to the precautions in the beginning of this section.

2. Remove or disconnect the following:
   • Negative battery cable
   • Right side air inlet tube
   • Accessory drive belt
   • Alternator harness connectors
   • Alternator

INSTALLATION

1. Before servicing the vehicle, refer to the precautions in the beginning of this section.

2. Install or connect the following:
   • Alternator and tighten the bolts to 30 ft. lbs. (41 Nm)
   • Alternator harness connectors
   • Accessory drive belt
   • Right side air inlet tube
   • Negative battery cable

Ignition Timing

ADJUSTMENT

The ignition timing is controlled by the Powertrain Control Module (PCM) and is not adjustable.

Engine Assembly

REMOVAL & INSTALLATION

1. Before servicing the vehicle, refer to the precautions in the beginning of this section.
2. Drain the cooling system.
3. Recover the A/C refrigerant, if equipped.
4. Drain the engine oil.

5. Relieve the fuel system pressure.
6. Remove or disconnect the following:
   - Negative battery cable
   - Air cleaner housing
   - Radiator hoses
   - Radiator and fan shroud
   - Mass Air Flow (MAF) sensor
   - Vacuum hoses from brake booster, intake manifold and purge valve
   - Power steering fluid from pump
   - Ground lead at power steering pump
   - Power steering pressure and return lines at pump
   - Fuel supply line
   - Engine wire harness
7. Raise and support vehicle on jack stands.
8. Remove or disconnect the following:
   - Engine undercover
   - Driveshaft
   - Ground cable at transmission
   - Transmission harness connector
   - Pressure line from clutch slave cylinder on manual transmission models
   - Reverse lockout cable
   - Shift rod on manual transmission models
   - Starter
   - Front engine mount bolt
9. Lower the vehicle.
10. Place a transmission jack under the transmission.
11. Remove or disconnect the following:
    - Heater hoses
    - A/C compressor and place aside
    - Rear crossmember/transmission mount

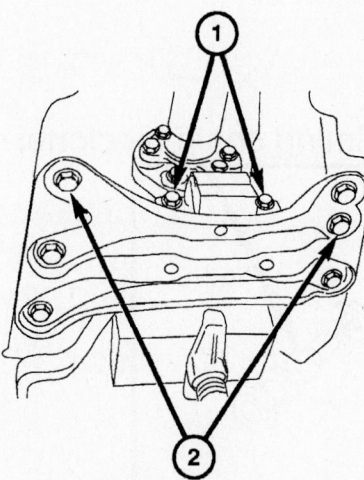

1. Transmission mount bolts
2. Crossmember bolts

67189-CSFR-G03

**Rear crossmember/transmission mount—
3.2L engine**

- Engine and transmission assembly
- Separate engine from transmission

To install:
12. Install or connect the following:
    - Transmission to engine. Tighten the bolts to 28 ft. lbs. (38 Nm) on manual transmission, or 30 ft. lbs. (40 Nm) on automatic transmission
    - Rear crossmember. Tighten the transmission mount bolts to 22 ft. lbs. (30 Nm), and the crossmember bolts to 30 ft. lbs. (40 Nm).
    - Front engine mount. Tighten the bolt to 41 ft. lbs. (55 Nm).
    - A/C compressor. Tighten the bolts to 17 ft. lbs. (23 Nm).
    - Manual transmission shift rod
    - Reverse lockout cable
    - Slave cylinder pressure line
    - Transmission harness connector
    - Transmission ground cable
    - Driveshaft
    - Engine undercover
    - Heater hoses
    - Radiator hoses
    - Engine wiring harness
    - Fuel supply line. Tighten fitting to 28 ft. lbs. (38 Nm).
    - Power steering lines. Tighten fittings to 33 ft. lbs. (45 Nm).
    - Power steering pump ground
    - Vacuum hoses
    - MAF sensor
    - Accessory drive belt
    - Radiator and fan
    - Air cleaner housing
    - Negative battery cable
13. Fill the crankcase to the correct level.
14. Fill the cooling system.
15. Start the engine and check for leaks.

Water Pump

REMOVAL & INSTALLATION

1. Before servicing the vehicle, refer to the precautions in the beginning of this section.
2. Drain the cooling system.
3. Remove or disconnect the following:
   - Negative battery cable
   - Air cleaner housing
   - Radiator fan
   - Accessory drive belt
   - Belt tensioner
   - Radiator hoses
   - Pump pulley and idler pulley
   - 16 water pump bolts
4. To install:
5. Installation is the reverse of removal. Tighten the bolts to 28 ft. lbs. (35 Nm).

Heater Core

REMOVAL & INSTALLATION

1. Disconnect the negative battery cable.

❋❋ CAUTION

After disconnecting the negative battery cable, wait 2 minutes for the driver's/passenger's air bag system capacitor to discharge before attempting to do any work around the steering column or instrument panel.

2. Drain the cooling system.
3. Recover the A/C refrigerant, if equipped.
4. Remove or disconnect the following:
   - Heater hoses
   - A pillar trim
   - Fuse block covers
   - Left and right air vents
   - Center console-to-instrument top panel screws
   - Glove box
   - Defroster vents
   - A pillar sheet metal clips
   - Instrument panel top section
   - Steering column cover
   - Lower instrument panel screws
   - Lower instrument panel cover
   - Steering wheel
   - Instrument cluster cover
   - Instrument cluster
   - Sentry Key Remote Entry Module (SKREM)
   - Wiring connectors at transmission tunnel
   - Steering column bolts-to-instrument panel support
   - Position steering column aside
   - Left and right heater ducts
   - Heater core attaching nuts
   - Instrument panel support bolts
   - Vacuum reservoir lines
   - Instrument panel support
   - HVAC housing electrical connectors
   - HVAC housing
5. Separate the heater core from HVAC housing
To install:
6. Install or connect the following:
   - Heater core to HVAC housing
   - HVAC housing
   - HVAC housing electrical connectors
   - Instrument panel support
   - Instrument panel support bolts
   - Steering column bolts-to-instrument panel support
   - Wiring connectors at transmission tunnel
   - SKREM

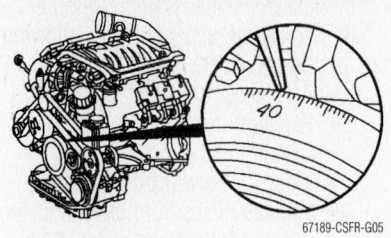

Crankshaft timing mark set to 40 degrees
ATDC—3.2L engine

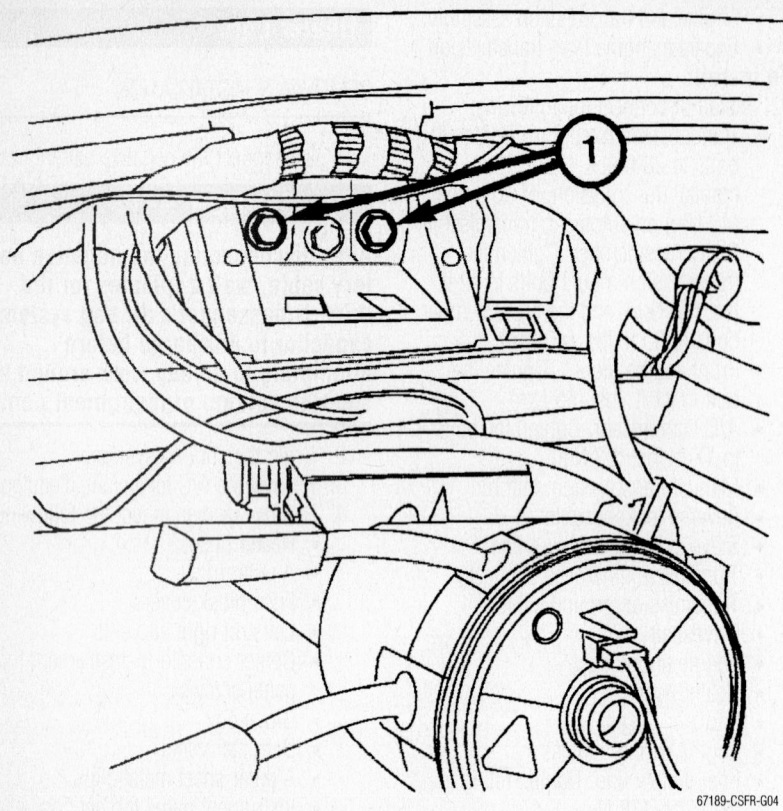

Removing instrument panel support bolts from bulkhead—Crossfire

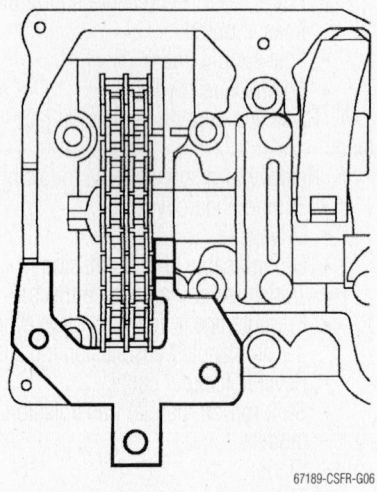

Camshaft locking plate—3.2L engine

- Instrument cluster
- Instrument cluster cover
- Steering wheel
- Left and right heater ducts
- Vacuum reservoir lines
- HVAC housing attaching nuts
- Position steering column into place
- Steering column bolts-to-instrument panel support
- Lower instrument panel cover
- Lower instrument panel screws
- Steering column cover
- Instrument panel top section
- A pillar sheet metal clips
- Defroster vents
- Glove box
- Center console-to-instrument top panel screws
- Left and right air vents
- Fuse block covers
- A pillar trim
- Heater hoses

7. Recharge the A/C refrigerant.
8. Fill the cooling system.

Cylinder Head and Camshafts

REMOVAL & INSTALLATION

1. Before servicing the vehicle, refer to the precautions in the beginning of this section.

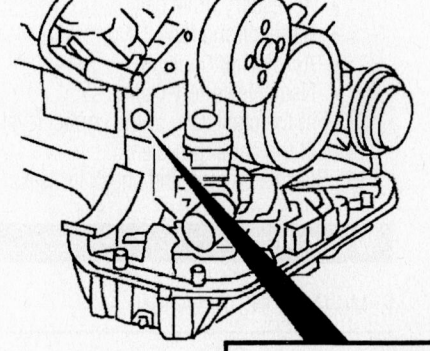

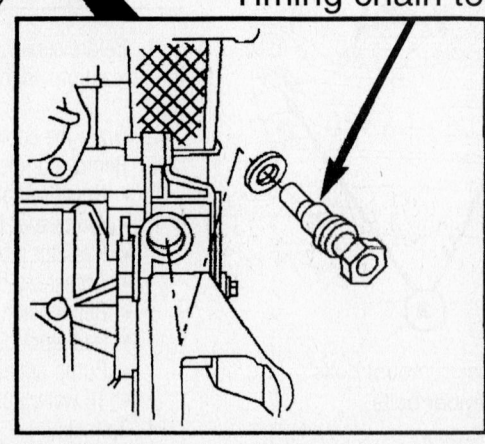

Timing chain tensioner

Removing timing chain tensioner—3.2L engine

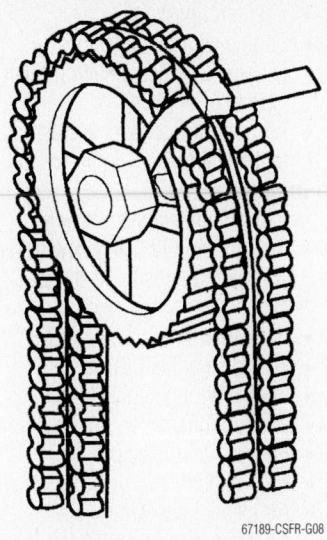

Securing timing chain to camshaft sprocket—3.2L engine

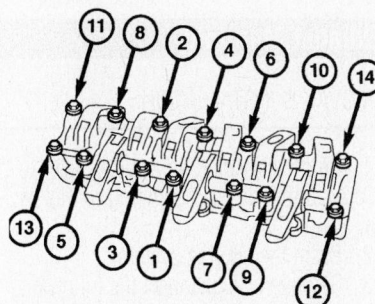

Camshaft bearing cap bolt removal sequence—3.2L engine

2. Drain the cooling system.

3. Drain the engine oil.

4. Properly relieve the fuel system pressure.

5. Remove or disconnect the following:
- Negative battery cable
- Air cleaner housing
- Mass Air Flow (MAF) sensor
- Engine cooling fan and shroud
- Radiator
- Camshaft position sensor
- Accessory drive belt

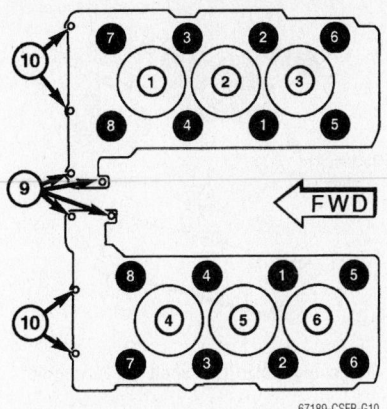

Cylinder head bolt removal sequence—3.2L engine

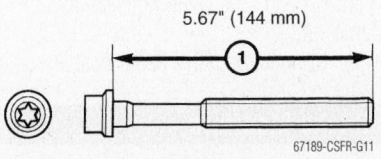

Examine the head bolts for signs of stretching—3.2L engine

- Ignition coil connectors
- Spark plug wires
- Cylinder head covers
- Exhaust pipes from manifold

6. Rotate the crankshaft so that the crankshaft timing mark aligns with the 40 degree After Top Dead Center (ATDC) mark on the front cover as shown. Grooves in the camshafts must be toward the inside of the wedge.

7. Lock the camshafts in place with Camshaft Locking Plates 9104 and 9105.

8. Remove or disconnect the following:
- Timing chain tensioner as shown.

9. Use a cable tie and secure the timing chain the camshaft sprocket.
- Camshaft sprockets.
- Camshaft locking plate tools.
- Camshaft bearing cap bolts in the sequence shown.
- Rocker arms and the camshafts.
- Cylinder head bolts in the sequence shown.
- Cylinder head.

➡The cylinder head is retained by six-teen bolts. Eight of the bolts are smaller and are at the front of the head.

To install:

10. Check the cylinder head bolts for signs of stretching. If length exceeds 5.67

inches (144 mm), replace bolts as necessary.

11. Lubricate the threads of the bolts with clean engine oil.

12. Install the cylinder heads. Use new gaskets and tighten the bolts, in sequence, as follows:
- a. Step 1: Bolts 1–8 to 15 ft. lbs. (20 Nm).
- b. Step 2: Bolts 1–8 to 37 ft. lbs. (50 Nm).
- c. Step 3: Bolts 1–8 plus 65 degrees clockwise.
- d. Step 4: Bolts 1–8 plus 65 degrees clockwise.

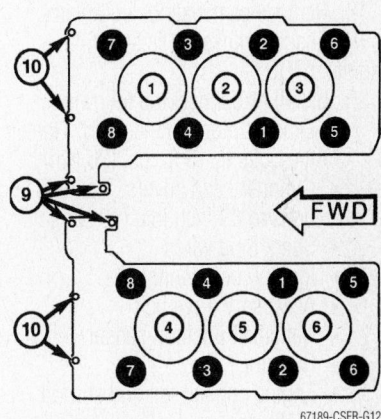

Cylinder head bolt torque sequence—3.2L engine

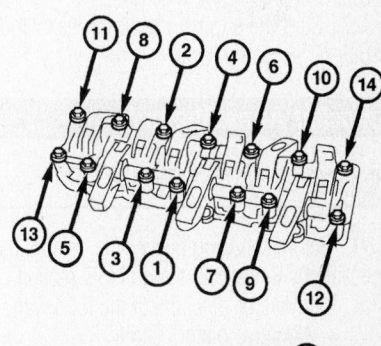

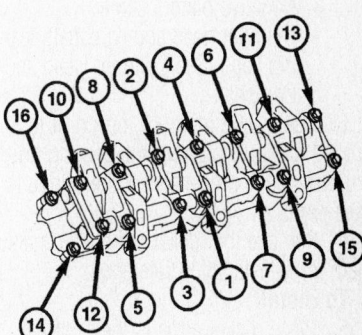

Camshaft bearing cap bolt torque sequence—3.2L engine

e. Step 5: Bolts 9 and 10 to 18 ft. lbs. (25 Nm).

13. Install the rocker arms and the camshafts.

14. Install the camshaft bearing cap bolts in the sequence shown. Tighten the bolts to 11 ft. lbs. (15 Nm), plus an additional 90 degrees.

15. Lock the camshafts in place with Camshaft Locking Plates 9104 and 9105.

16. Install or connect the following:
- Camshaft sprockets. Tighten the bolts to 37 ft. lbs. (50 Nm), plus an additional 90 degrees.
- Timing chain tensioner. Tighten the bolt to 59 ft. lbs. (80 Nm).

17. Remove cable tie from timing chain.

18. Remove camshaft locking plates.

19. Ensure crankshaft is set to 40 degrees ATDC.

20. Install or connect the following:
- Exhaust pipes to manifold. Tighten the bolts to 15 ft. lbs. (20 Nm).
- Cylinder head covers. Tighten the bolts to 89 inch lbs. (10 Nm).
- Spark plug wires
- Ignition coil connectors
- Accessory drive belt
- Camshaft position sensor
- Radiator
- Engine cooling fan and shroud
- MAF sensor
- Air cleaner housing
- Negative battery cable

21. Fill and bleed the cooling system.

22. Fill the engine with oil.

23. Start the engine, check for leaks and repair if necessary.

Rocker Arms/Shafts

REMOVAL & INSTALLATION

1. Before servicing the vehicle, refer to the precautions in the beginning of this section.

2. Remove or disconnect the following:
- Negative battery cable
- Cylinder head and camshaft bearing caps. See Cylinder Head and Camshaft.

3. Use a 0.63 inch (16 mm) drift to drive out the rocker shaft from the bearing bridge. If resistance is present, heat the bridge to ease removal.

4. Remove the rocker arms after marking them for reinstallation reference.

To install:

5. Place rocker arms in their correct locations.

6. Drive the rocker arm shaft into the bearing bridge using a mallet.

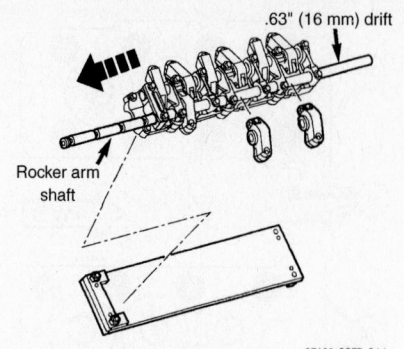

Removing rocker arm shaft—3.2L engine

7. Insert 2 camshaft bearing bridge bolts to secure the shaft. The oil supply holes in the shaft point downward toward the cylinder head.

8. Install the camshaft bearing caps and cylinder head. See Cylinder Head and Camshaft.

9. Connect negative battery cable.

Intake Manifold

REMOVAL & INSTALLATION

1. Before servicing the vehicle, refer to the precautions in the beginning of this section.

2. Drain the cooling system.

3. Properly relieve the fuel system pressure.

4. Remove or disconnect the following:
- Negative battery cable
- Air cleaner assembly
- Mass Air Flow (MAF) sensor
- Fuel rail and injectors

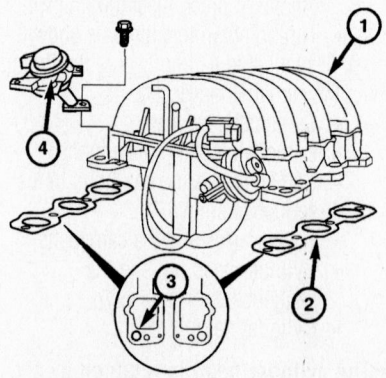

1. Intake manifold
2. Gasket
3. Locating hole
4. Air pump switchover valve

67189-CSFR-G15

Exploded view of intake manifold—3.2L engine

- Vacuum lines
- Engine wiring connectors
- Exhaust Gas Recirculation (EGR) pipe at EGR valve
- Air pump switchover valves
- Intake manifold

To install:

5. Install or connect the following:
- Intake manifold using new gaskets. Torque the bolts to 15 ft. lbs. (20 Nm).
- Air pump switchover valves
- EGR pipe at EGR valve
- Engine wiring connectors
- Vacuum lines
- Fuel rail and injectors
- MAF sensor
- Air cleaner assembly
- Negative battery cable

6. Fill and bleed the cooling system.

7. Start the engine, check for leaks and repair if necessary.

Exhaust Manifold

REMOVAL & INSTALLATION

1. Before servicing the vehicle, refer to the precautions in the beginning of this section.

2. Drain the cooling system.

3. Remove or disconnect the following:
- Air cleaner housing

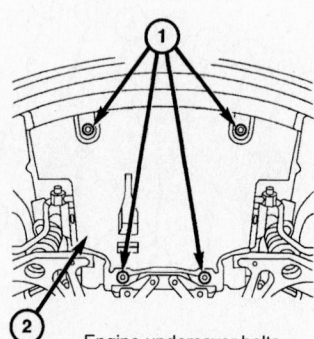

Engine undercover bolts

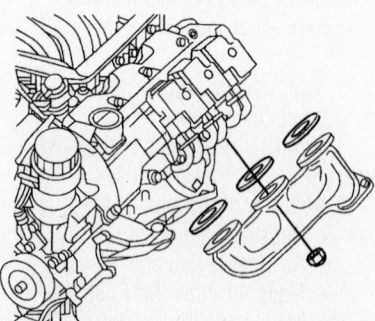

67189-CSFR-G16

Exhaust manifold left side, right side similar—3.2L engine

- Engine undercover
- Catalytic converters
- Exhaust manifolds

To install:

4. Install or connect the following:
   - Exhaust manifolds, using new gaskets. Tighten the bolts to 26 ft. lbs. (35 Nm).
   - Catalytic converters
   - Engine undercover
   - Air cleaner housing
5. Fill the cooling system.
6. Start the engine and check for leaks.

Valve Lash

ADJUSTMENT

The 3.2L engine uses hydraulic lifters. No maintenance or periodic adjustment is required.

Starter Motor

REMOVAL & INSTALLATION

1. Before servicing the vehicle, refer to the precautions in the beginning of this section.
2. Raise and support the vehicle.
3. Remove or disconnect the following:
   - Negative battery cable
   - Engine undercover
   - Right side Oxygen (O2S) sensor connector
   - Right side exhaust pipe
   - Starter harness connections
   - Starter

To install:

4. Connect the starter wiring connectors.
5. Install the starter and torque the bolts to 31 ft. lbs. (42 Nm).
6. Install the right side exhaust pipe.
7. Install the right side Oxygen (O2S) sensor connector.
8. Install the engine undercover.
9. Install the negative battery cable and check for proper operation.

Oil Pan

REMOVAL & INSTALLATION

Lower Oil Pan

1. Before servicing the vehicle, refer to the precautions in the beginning of this section.
2. Drain the engine oil.
3. Remove or disconnect the following:

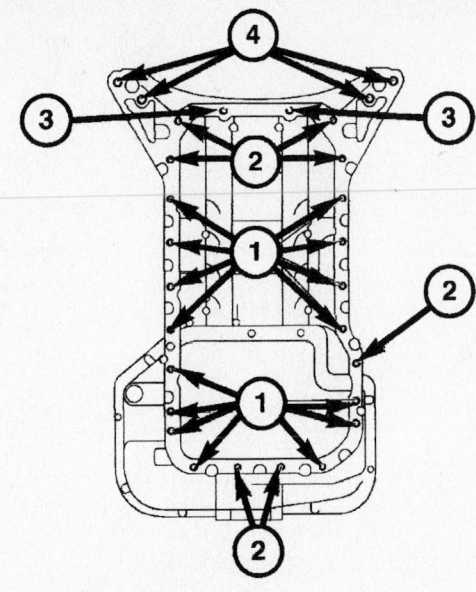

1. M6 x 20 bolts
2. M6 x 40 bolts
3. M6 x 90 bolts
4. M6 x 30 bolts

67189-CSFR-G17

Upper oil pan mounting bolt locations—3.2L engine

- Engine undercover
- Position transmission oil cooler line aside
- Oil pan
- Oil pan gasket
4. To install
5. Installation is the reverse of removal. Install a bead of sealant around oil pan perimeter. Torque the bolts to 10 ft. lbs. (14 Nm).

Upper Oil Pan

1. Before servicing the vehicle, refer to the precautions in the beginning of this section.
2. Remove or disconnect
   - Engine from the vehicle. Separate the transmission and remove the clutch, if equipped.
   - Lower oil pan
   - 24 upper oil pan bolts and remove oil pan.

To install:

3. Install or connect
   - Bead of sealant around oil pan perimeter.
   - Correct bolts in the proper locations as shown. Torque the M6 bolts to 89 inch lbs. (10 Nm), and M8 bolts to 15 ft. lbs. (20 Nm).
4. The remainder if the installation is the reverse of removal.

Oil Pump

REMOVAL & INSTALLATION

1. Before servicing the vehicle, refer to the precautions in the beginning of this section.
2. Remove or disconnect the following:
   - Lower oil pan
   - Oil pump bolts
   - Release the oil pump timing chain tensioner and remove the oil pump

To install:

3. Fill the oil pump with clean engine oil.
4. Position the pump driven sprocket into the drive chain.
5. Install and torque the bolts to 15 ft. lbs. (20 Nm).
6. Install the lower oil pan.
7. Fill the engine with oil and check for proper operation.

Rear Main Seal

REMOVAL & INSTALLATION

1. Before servicing the vehicle, refer to the precautions in the beginning of this section.

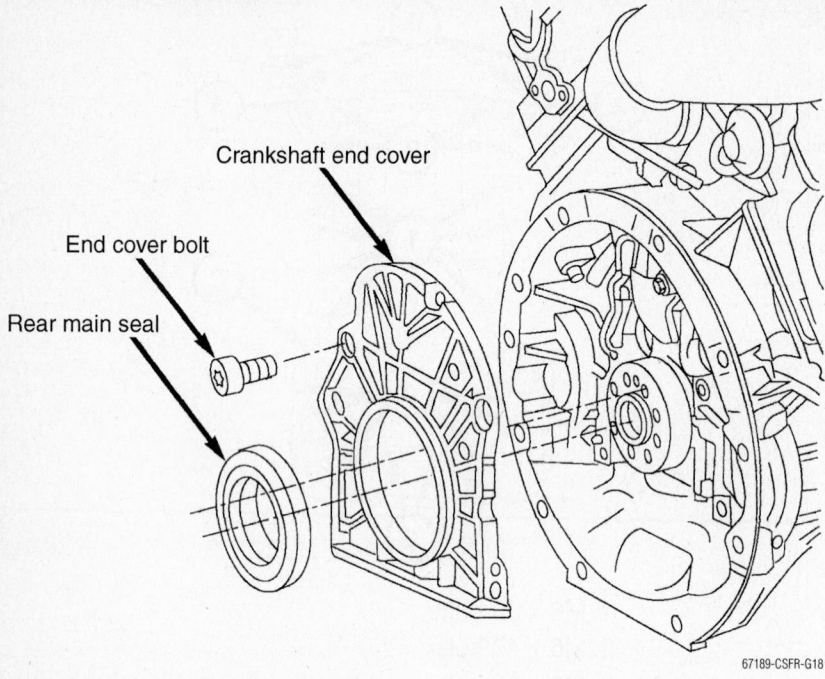

Crankshaft end cover

End cover bolt

Rear main seal

Crankshaft end cover assembly—3.2L engine

67189-CSFR-G18

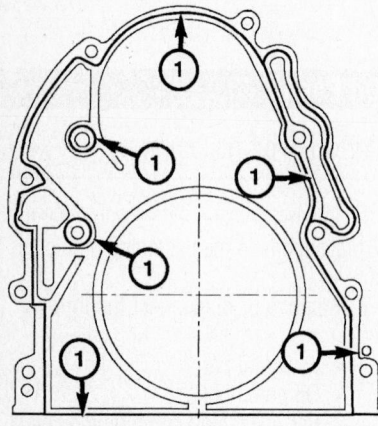

1. Sealant application

67189-CSFR-G19

Crankshaft end cover sealant application—3.2L engine

➡**The rear main seal cannot be replaced separately. The crankshaft end cover and oil seal are replaced as a set.**

2. Remove or disconnect the following:
   • Transmission
   • Flywheel
   • Crankshaft end cover
3. Clean the engine block and oil pan sealing surfaces.
   To install:
4. Install or connect
   • Bead of sealant as shown on the NEW end cover

• End cover. Torque the bolts to 89 inch lbs. (10 Nm).
• Use Seal Guide 9100 and install the real seal into the end cover
• Flywheel
• Transmission

Timing Chain, Balance Shafts, Front Cover and Seal

REMOVAL & INSTALLATION

1. Before servicing the vehicle, refer to the precautions in the beginning of this section.

➡**The timing chain and cover can only be removed when the engine is removed from the vehicle.**

2. Remove the engine and separate the transmission.
3. Remove or disconnect the following:
   • Accessory drive belt

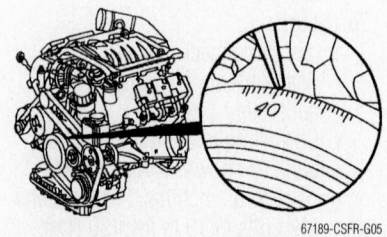

67189-CSFR-G05

Crankshaft timing mark set to 40 degrees ATDC—3.2L engine

• Lower and upper oil pans
• Power steering pump
• Idler pulley
• Crankshaft damper bolt and damper
• Alternator
• Rotate crankshaft to 40 degrees After Top Dead Center (ATDC)
• Starter
• Lock flywheel in position
• Cylinder heads
• Timing chain tensioner
• Timing chain cover
• Oil pump drive chain and tensioner
• Timing chain and camshaft sprockets. Refer to cylinder head and camshafts procedure in this section
• Crankshaft sprocket
• Crankshaft rear main seal/end cover assembly

4. Hold the balance shaft rear counter weight using a drift, and remove the rear retaining bolt.
   • Balance shaft counterweight
   • Balance shaft locking plate at the front of the engine block.
   • Balance shaft
   • Front crankshaft seal using a suitable pry tool.
   To install:
5. Position the front crankshaft seal into Seal Installation tool 9103 as shown.
6. Align the slot of the installation tool into the crankshaft keyway.
7. Tap the tool in until the crankshaft damper bolt can be inserted.
8. Tighten the damper bolt until the seal is installed.
9. Remove the installer tool.
10. Install or connect
    • Balance shaft into the bore in the block from the front of the engine
    • Rear counter weight onto the shaft and torque the retaining bolt to 15 ft. lbs. (20 Nm), plus an additional 90 degrees
    • Locking plate and bolt and torque to 15 ft. lbs. (20 Nm)
    • Rear crankshaft seal/end cover assembly
    • Crankshaft sprocket
11. Ensure that the crankshaft is still positioned at 40 degrees ATDC.
12. Align the balance shaft sprocket with the timing chain (3) and check that the Copper teeth (1) of the timing chain are aligned on the camshaft sprockets (6).
13. Route the timing chain within the timing chain guides as shown.
14. Lock the camshafts in place with Camshaft Locking Plates 9104 and 9105.

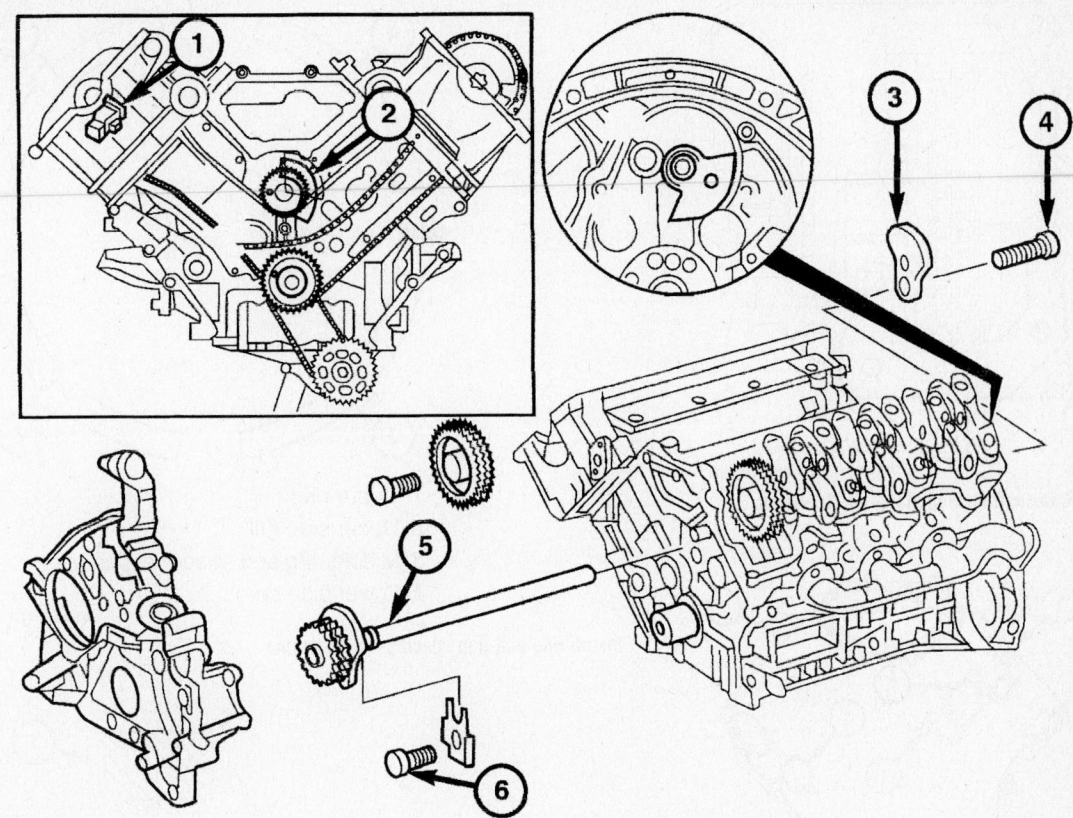

1 - CAMSHAFT POSITION SENSOR
2 - BALANCE SHAFT FRONT COUNTER WEIGHT
3 - BALANCE SHAFT REAR COUNTER WEIGHT
4 - REAR RETAINING BOLT
5 - BALANCE SHAFT
6 - LOCKING PLATE BOLT

67189-CSFR-G20

Exploded view of balance shaft assembly—3.2L engine

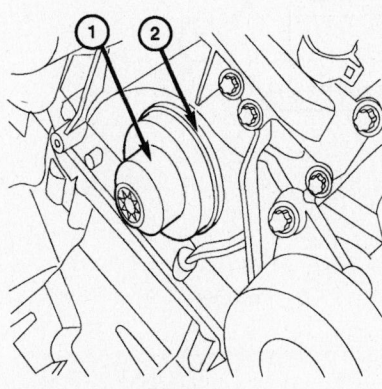

1. Crankshaft front seal
2. Seal installer 9103

67189-CSFR-G21

Positioning crankshaft front seal into Seal Installer—3.2L engine

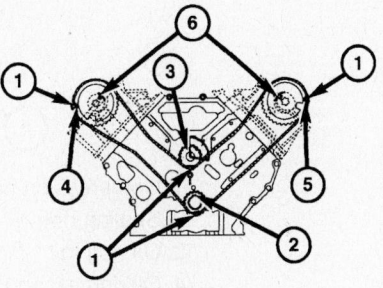

1. Copper teeth
2. Crankshaft sprocket
3. Balance shaft timing mark
4. Camshaft sprocket timing mark
5. Camshaft sprocket timing mark
6. Camshaft sprockets

67189-CSFR-G22

Positioning timing chain on camshaft sprockets—3.2L engine

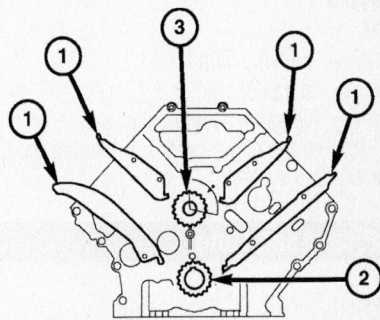

1. Timing chain guides
2. Crankshaft sprocket
3. Balance shaft sprocket

67189-CSFR-G23

Timing chain guide locations—3.2L engine

15. Install or connect the following:
   - Oil pump drive chain and tensioner
   - Cylinder heads
   - Apply sealant to front cover as shown
   - Timing chain cover. Torque the bolts to 15 ft. lbs. (20 Nm).

   - Timing chain tensioner. Torque the bolt to 59 ft. lbs. (80 Nm).
   - Unlock flywheel
   - Starter and torque the bolts to 31 ft. lbs. (42 Nm)
   - Alternator and torque the bolts to 31 ft. lbs. (42 Nm)

   - Vibration damper and torque the bolt to 148 ft. lbs. (200 Nm)
   - Idler pulley
   - Power steering pump
   - Upper and lower oil pans
   - Accessory drive belt
   - Clutch (if equipped)

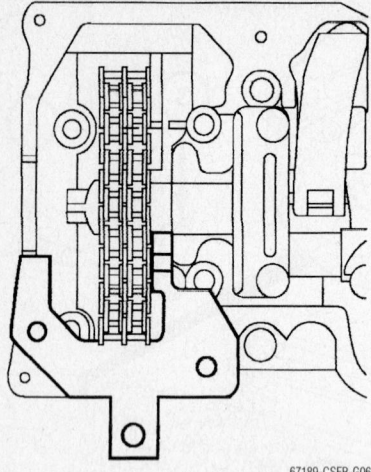

Camshaft locking plate—3.2L engine

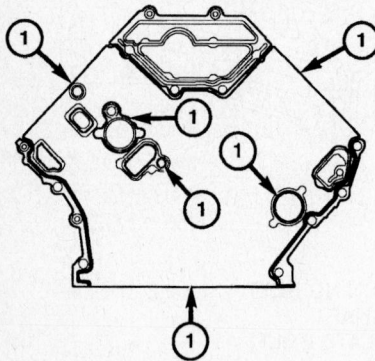

1. Sealant application areas

Applying sealant to front cover—3.2L engine

- Transmission
- Install engine
- Negative battery cable

16. Fill the engine with oil. Fill and bleed the cooling system.

Piston and Ring

POSITIONING

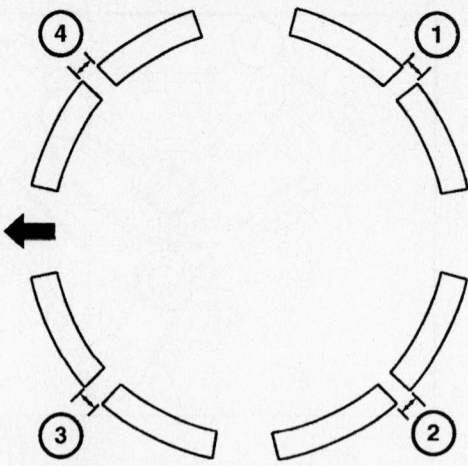

1. Upper ring
2. Upper side rail
3. Middle ring and spacer expander
4. Lower side rail

Piston ring end-gap spacing—3.2L engine

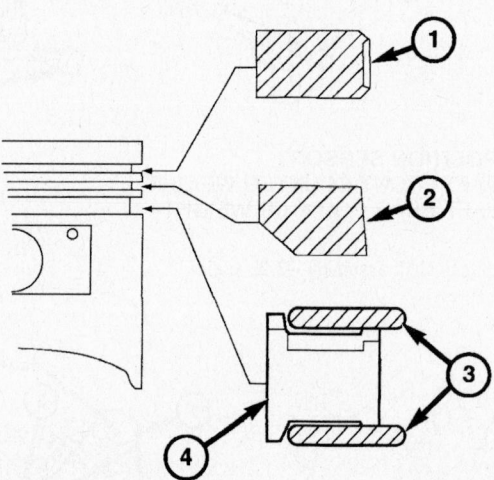

1. Compression ring
2. Sealing ring
3. Oil ring rail
4. Oil ring expander

Ring to piston positioning—3.2L engine

FUEL SYSTEM

Fuel System Service Precautions

Safety is the most important factor when performing not only fuel system maintenance but any type of maintenance. Failure to conduct maintenance and repairs in a safe manner may result in serious personal injury or death. Maintenance and testing of the vehicle's fuel system components can be accomplished safely and effectively by adhering to the following rules and guidelines.

• To avoid the possibility of fire and personal injury, always disconnect the negative battery cable unless the repair or test procedure requires that battery voltage be applied.

• Always relieve the fuel system pressure prior to detaching any fuel system component (injector, fuel rail, pressure regulator, etc.), fitting or fuel line connection. Exercise extreme caution whenever relieving fuel system pressure to avoid exposing skin, face and eyes to fuel spray. Please be advised that fuel under pressure may penetrate the skin or any part of the body that it contacts.

• Always place a shop towel or cloth around the fitting or connection prior to loosening to absorb any excess fuel due to spillage. Ensure that all fuel spillage (should it occur) is quickly removed from engine surfaces. Ensure that all fuel soaked cloths or towels are deposited into a suitable waste container.

• Always keep a dry chemical (Class B) fire extinguisher near the work area.

• Do not allow fuel spray or fuel vapors to come into contact with a spark or open flame.

• Always use a back-up wrench when loosening and tightening fuel line connection fittings. This will prevent unnecessary stress and torsion to fuel line piping.

• Always replace worn fuel fitting O-rings with new. Do not substitute fuel hose or equivalent, where fuel pipe is installed.

Before servicing the vehicle, also make sure to refer to the precautions in the beginning of this section as well.

Fuel System Pressure

RELIEVING

1. Before servicing the vehicle, refer to the precautions in the beginning of this section.

2. Disconnect the negative battery cable.

3. Remove the fuel tank filler cap to release any fuel tank pressure.

4. Remove the fuel pump relay from the Power Distribution Center (PDC).

5. Start and run the engine until it stops.

➡**One or more Diagnostic Trouble Codes (DTC) may be stored in Powertrain Control Module (PCM) memory when the fuel pump fuse is removed. A DRB III scan tool must be used to erase the code.**

Fuel Filter

REMOVAL & INSTALLATION

1. Before servicing the vehicle, refer to the precautions in the beginning of this section.

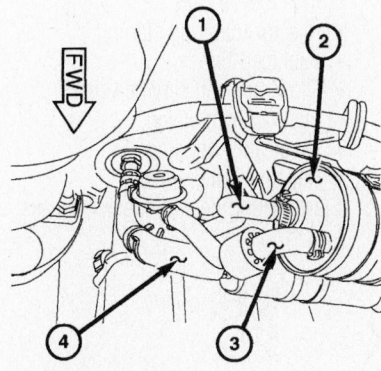

1. Degaussing line
2. Fuel filter/pressure regulator
3. Fuel supply line
4. Fuel return line

67189-CSFR-G27

Fuel filter/pressure regulator assembly—3.2L engine

➡**The fuel filter is integral with the fuel pressure regulator. Both must be replaced as an assembly.**

2. Relieve the fuel system pressure.
3. Remove or disconnect the following:
   • Splash shield
   • Degassing line
   • Fuel supply line
   • Fuel delivery line
   • Fuel return line
   • Filter/regulator out of the mounting clamp

To install:

4. Install or connect the following:
   • Filter/regulator into the mounting clamp
   • Fuel return line
   • Fuel delivery line
   • Fuel supply line
   • Degassing line
   • Splash shield

5. Start the engine and check for leaks.

Fuel Pump

REMOVAL & INSTALLATION

1. Before servicing the vehicle, refer to the precautions in the beginning of this section.

2. Relieve the fuel system pressure.

3. Remove or disconnect the following:

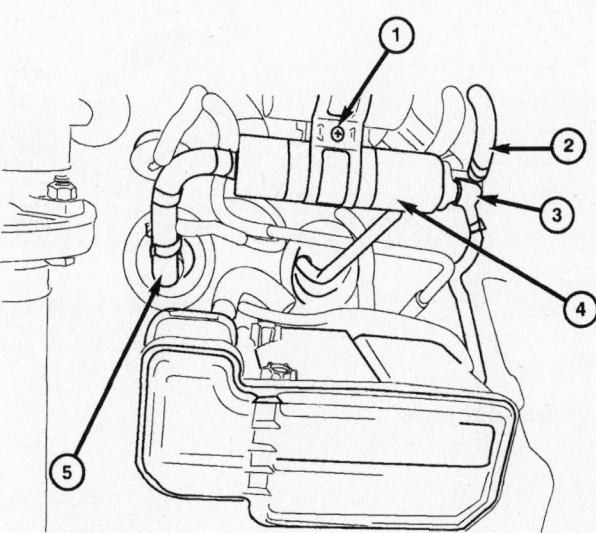

1. Retaining clamp
2. Fuel delivery hose
3. Harness connectors
4. Fuel pump
5. Suction hose

67189-CSFR-G28

Fuel pump removal—3.2L engine

- Disconnect the negative battery cable
- Splash shield
- Fuel suction hose
- Fuel delivery hose
- Fuel pump electrical connectors
- Fuel pump retaining clamp
- Fuel pump

To install:

4. Install or connect the following:
- Fuel pump
- Fuel pump retaining clamp
- Electrical connectors
- Fuel delivery hose
- Fuel suction hose
- Splash shield
- Negative battery cable

5. Start the engine and check for leaks.

Fuel Rail And Injectors

REMOVAL & INSTALLATION

1. Before servicing the vehicle, refer to the precautions in the beginning of this section.

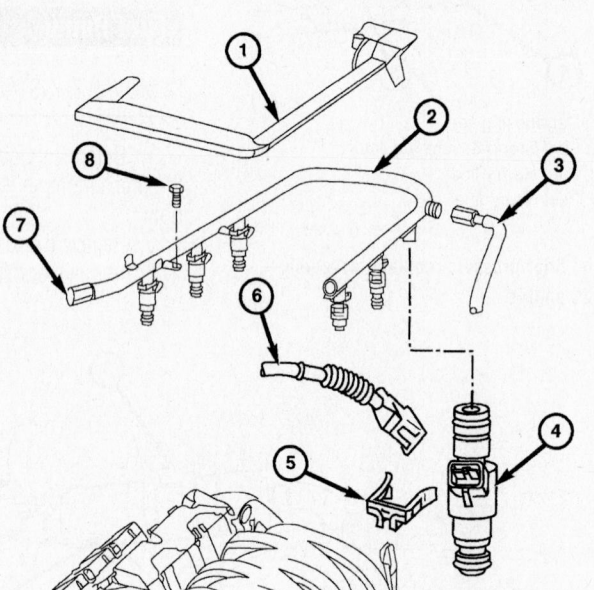

1 - FUEL RAIL TRIM PANEL
2 - FUEL RAIL
3 - FUEL LINE
4 - FUEL INJECTOR

5 - FUEL INJECTOR RETAINING CLIP
6 - FUEL INJECTOR HARNESS CONNECTOR
7 - FUEL RAIL SERVICE VALVE
8 - FUEL RAIL BOLT

67189-CSFR-G29

Fuel rail and injectors—3.2L engine

2. Relieve the fuel system pressure.
3. Remove or disconnect the following:
- Negative battery cable
- Air cleaner housing
- Fuel rail trim cover
- Open fuel rail service valve
- Fuel feed line
- Fuel injector retaining clamps
- Fuel supply manifold with injectors attached
- Fuel injectors

To install:

4. Install or connect the following:
- Fuel injectors
- Fuel supply manifold with injectors attached. Tighten the bolts to 80 inch lbs. (9 Nm).
- Injector retaining clamps
- Fuel feed line
- Close fuel rail service valve
- Fuel rail trim cover
- Air cleaner housing
- Negative battery cable

5. Start the engine and check for leaks.

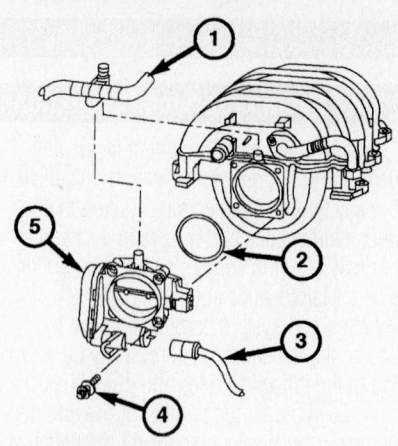

1. PCV hose
2. O ring
3. Harness connector
4. Bolts (4)
5. Throttle body

67189-CSFR-G30

Throttle body—3.2L engine

Throttle Body

REMOVAL & INSTALLATION

1. Before servicing the vehicle, refer to the precautions in the beginning of this section.
2. Relieve the fuel system pressure.
3. Remove or disconnect the following:

- Negative battery cable
- Air cleaner housing
- Mass air flow (MAF) sensor
- Positive Crankcase Valve (PCV) tube
- Throttle body electrical connector
- Throttle body with O-ring

To install:

4. Install or connect the following:
- Throttle body with new O-ring. Tighten bolts to 80 inch lbs. (9 Nm).
- Electrical connector
- PCV tube
- MAF sensor
- Air cleaner housing
- Negative battery cable

DRIVE TRAIN

Transmission Assembly

REMOVAL & INSTALLATION

Automatic

1. Before servicing the vehicle, refer to the precautions in the beginning of this section.
2. Remove or disconnect the following:
   - Negative battery cable
   - Exhaust system
   - Center exhaust heat shield
   - Transmission support bracket
   - Transmission mount and cross-member. Support the transmission with a suitable jack.
   - Driveshaft
   - Front engine undercover
   - Transmission ground bolt
   - Drain transmission fluid
   - Shift cable
   - Transmission oil cooler fittings
   - Torque converter cover
   - Torque converter bolts
   - Oxygen (O_2S) sensor connectors
   - Controller harness connector
   - Heat shield
   - Starter motor
   - Transmission

To install:

3. Install or connect the following:
   - Transmission. Torque the mounting bolts to 28 ft. lbs. (28 Nm).
   - Torque converter. Torque the mounting bolts to 37 ft. lbs. (50 Nm).
   - Torque converter access cover
   - Starter. Torque the mounting bolts to 31 ft. lbs. (42 Nm).
   - Controller harness connector
   - Heat shield. Torque the mounting bolts to 15 ft. lbs. (20 Nm).
   - (O_2S) sensor connector
   - Transmission cooler fittings. Torque the mounting bolts to 25 ft. lbs. (34 Nm).
   - Shift cable
   - Driveshaft. Torque the mounting bolts to 44 ft. lbs. (60 Nm).
   - Transmission support bracket. Torque the mounting bolts to 15 ft. lbs. (20 Nm).
   - Transmission mount and cross-member. Torque the mounting bolts to 33 ft. lbs. (45 Nm).
   - Center heat shield
   - Exhaust system

   - Engine undercover
   - Negative battery cable
4. Fill the transmission with Shell Automatic Transmission fluid no. 3403.
5. Start engine and check for proper transmission operation.

Manual

1. Before servicing the vehicle, refer to the precautions in the beginning of this section.
2. Remove or disconnect the following:
   - Negative battery cable
   - Exhaust system
   - Center exhaust heat shield
   - Transmission support bracket
   - Transmission mount and cross-member. Support the transmission with a suitable jack
   - Driveshaft
   - Reverse light switch connector
   - Clutch slave cylinder
   - Reverse lockout cable
   - Shift rod
   - Ground cable
   - Transmission

To install:

3. Install or connect the following:
   - Transmission. Tighten the bolts to 30 ft. lbs. (40 Nm).
   - Ground cable
   - Transmission mount and cross-member. Torque the mounting bolts to 30 ft. lbs. (40 Nm).
   - Shift rod
   - Reverse lockout cable
   - Clutch slave cylinder
   - Reverse light connector
   - Driveshaft. Torque the mounting bolts to 44 ft. lbs. (60 Nm).
   - Fill transmission with Automatic Transmission Fluid+4.
   - Bleed clutch slave cylinder
   - Negative battery cable
4. Check for proper clutch operation.

Clutch

REMOVAL & INSTALLATION

1. Before servicing the vehicle, refer to the precautions in the beginning of this section.
2. Remove or disconnect the following:
   - Negative battery cable
   - Transmission
   - Pressure plate. Loosen the bolts

evenly in ½ turn steps in a criss-cross pattern.
   - Clutch disc

To install:

3. Install or connect the following:
   - Clutch disc and pressure plate. Tighten the pressure plate bolts evenly in 1 to ½ turns to 18 ft. lbs. (25 Nm) in a criss-cross pattern.
   - Transmission
   - Negative battery cable

Hydraulic Clutch System

BLEEDING

1. Before servicing the vehicle, refer to the precautions in the beginning of this section.
2. Connect a pressure bleeder to the brake master cylinder.
3. Open the bleed screw on the clutch slave cylinder.
4. Allow the fluid to flow until there are no bubbles and the fluid is clear.
5. Close the bleed screw and disconnect the pressure bleeder.
6. Check the clutch for proper operation.

Halfshaft

REMOVAL & INSTALLATION

1. Before servicing the vehicle, refer to the precautions in the beginning of this section.
2. Raise and support the rear of the vehicle.
3. Remove or disconnect the following:
   - Rear wheels
   - Muffler
   - Hub nut
   - Halfshaft outer end through the wheel hub
   - Halfshaft to differential flange bolts
   - Halfshaft

To install:

4. Install or connect
   - Axle halfshaft to the differential
   - NEW flange bolts, lubricate with oil and torque the bolts to 52 ft. lbs. (70 Nm)
   - NEW hub nut and torque to 164 ft. lbs. (220 Nm)
   - Muffler
   - Rear wheels and tighten the lug nuts to 81 ft. lbs. (110 Nm)

CV-Joints

OVERHAUL

Outer CV-Joint

1. Before servicing the vehicle, refer to the precautions in the beginning of this section.
2. Remove or disconnect the following:
   • Axle halfshaft from the vehicle
   • Inner CV-joint boot and clamps
   • Outer CV-joint and boot

To install:

3. Pack the outer CV-joint with grease supplied in repair kit.
4. Install or connect the following:
   • CV-joint into inner seal
5. Crimp the retaining boot clamps.
6. Install the axle halfshaft.

Inner CV Joint

1. Before servicing the vehicle, refer to the precautions in the beginning of this section.
2. Remove the axle halfshaft from the vehicle.
3. Remove the inner joint boot clamps.
4. Clamp the halfshaft in a vise.
5. Pry the joint cover from the bearing retainer.
6. Remove the boot from the inner seal.
7. Pry the inner seal from the bearing retainer.

8. Slide the boot, inner seal and bearing retainer off the halfshaft.
9. Remove the circlip from the axle shaft.
10. Using a press, remove the inner joint bearing from the halfshaft.

To install:

➡**Use new circlips and boot clamps for assembly.**

11. Install or connect the following:
   • Inner boot
   • Inner joint bearing
   • Inner seal
   • Circlip
12. Fill the bearing housing with grease.
13. Apply a bead of sealant to the outer sealing surface of the joint ring.
14. Using a mallet, tap on a NEW joint cover.
15. Install the boot into the sealing groove of the inner seal.
16. Install the retaining clamps and crimp in place.
17. Install the axle halfshaft.

Pinion Seal

REMOVAL & INSTALLATION

1. Before servicing the vehicle, refer to the precautions in the beginning of this section.

2. Drain the differential of fluid.
3. Remove or disconnect the following:
   • Wheels
   • Muffler
   • Driveshaft
   • Halfshafts at differential
4. Check the bearing preload with an inch lb. torque wrench.
5. Hold the pinion flange and remove the collared nut.
6. Pry out the pinion flange seal.

To install:

➡**Use a new pinion nut for assembly.**

7. Install the new pinion seal. Tighten the nut to 133 ft. lbs. (180 Nm).
8. Check the bearing preload. The bearing preload should be equal to the reading taken earlier.
9. If the preload torque is low, tighten the pinion nut in 44 inch lb. increments until the torque value is reached.
10. If the pinion bearing preload torque cannot be attained at maximum pinion nut torque, replace the collapsible spacer.
11. Install or connect the following:
   • Halfshafts
   • Driveshaft
   • Muffler
   • Wheels
12. Fill the axle assembly with 75W-140 gear oil and check for leaks.

STEERING AND SUSPENSION

Air Bag

✳✳ CAUTION

Some vehicles are equipped with an air bag system. The system must be disarmed before performing service on, or around, system components, the steering column, instrument panel components, wiring and sensors. Failure to follow the safety precautions and the disarming procedure could result in accidental air bag deployment, possible injury and unnecessary system repairs.

PRECAUTIONS

Several precautions must be observed when handling the inflator module to avoid accidental deployment and possible personal injury.
   • Never carry the inflator module by the

wires or connector on the underside of the module.
   • When carrying a live inflator module, hold securely with both hands, and ensure that the bag and trim cover are pointed away.
   • Place the inflator module on a bench or other surface with the bag and trim cover facing up.
   • With the inflator module on the bench, never place anything on or close to the module which may be thrown in the event of an accidental deployment.

Before servicing the vehicle, also make sure to refer to the precautions in the beginning of this section as well.

DISARMING

1. Disconnect and isolate the negative battery cable. Wait 2 minutes for the system capacitor to discharge before performing any service.
2. When repairs are completed, connect the negative battery cable.

Recirculating Ball Power Steering Gear

REMOVAL & INSTALLATION

1. Before servicing the vehicle, refer to the precautions in the beginning of this section.
2. Remove ignition key and allow the steering wheel to lock.
3. Remove or disconnect the following:
   • Negative battery cable
   • Drag link and tie rod
   • Left front cross brace
   • Power steering pressure and return lines

✳✳ WARNING

On vehicles with horn/airbag clock spring contact, the steering wheel must not be turned when the steering gear is removed, otherwise the clock spring contact will be damaged beyond repair.

4. Remove the ignition key and allow the steering wheel to lock.
- Pinch bolt from the steering coupling
- Intermediate shaft
- Lower engine mounting bolts on both sides

5. Place a jack under the left side of the engine and raise the engine about 2 inches.

6. Remove 3 bolts and remove the steering gear.

✳✳ CAUTION

DO NOT use force, otherwise the lower steering shaft will be damaged on vehicles with a rigid steering column.

To install:

7. Align the marks on the steering shaft with the housing cover to ensure gear is in the center position.

8. Install or connect the following:
- Gear coupling onto shaft

✳✳ CAUTION

DO NOT use force, otherwise the lower steering shaft (collapsible tubing) will be damaged. Ensure that the connection is assembled correctly.

- Steering gear. Tighten the bolts to 44 ft. lbs. (60 Nm).
- Engine mount bolts. Tighten the bolts to 18 ft. lbs. (25 Nm).
- Intermediate shaft. Tighten the pinch bolt to 22 ft. lbs. (30 Nm).
- Power steering pressure and return lines
- Drag link/tie rod. Tighten the nut to 44 ft. lbs. (60 Nm).
- Left cross brace
- Negative battery cable

9. Fill the power steering fluid reservoir.

10. Start the engine and check for leaks.

Shock Absorber

REMOVAL & INSTALLATION

Front

1. Before servicing the vehicle, refer to the precautions in the beginning of this section.

2. Remove or disconnect the following:
- Front wheel
- Upper mounting bolt
- Lower mounting bolts
- Shock absorber

To install:

3. Install or connect the following:
- Shock absorber. Tighten the upper bolt nut to 13 ft. lbs. (18 Nm) and the lower bolt to 41 ft. lbs. (55 Nm).
- Front wheel

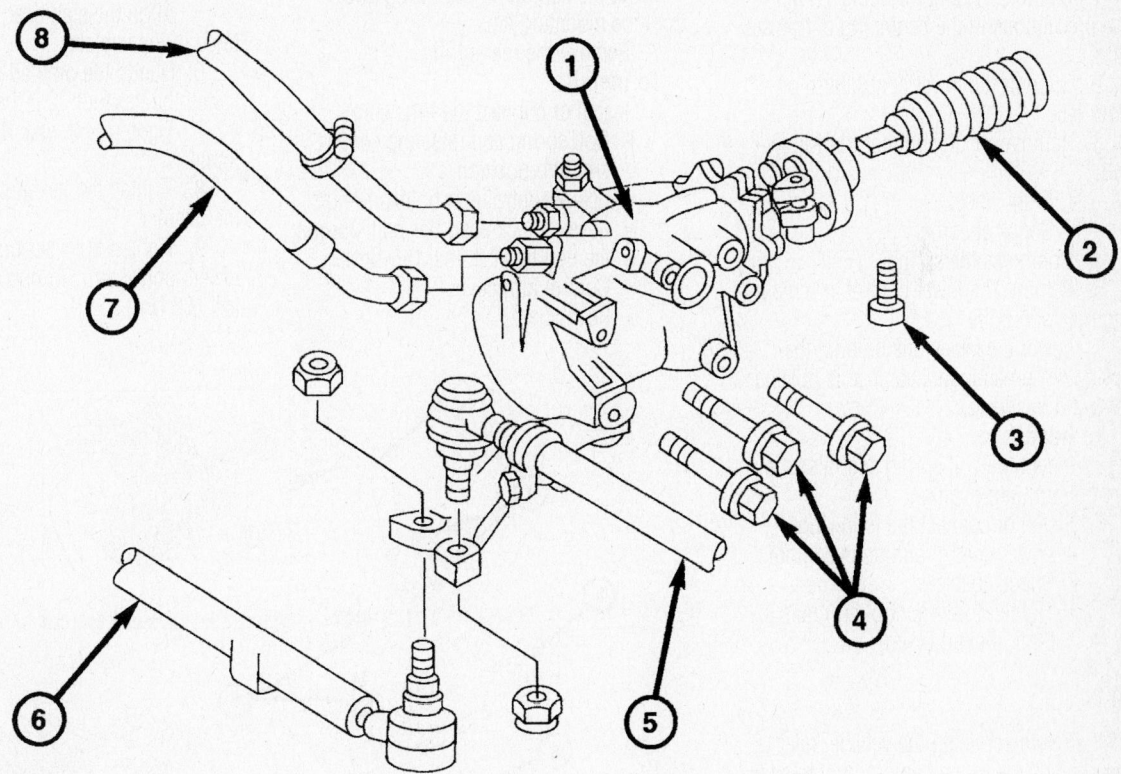

1 - STEERING GEAR
2 - STEERING SHAFT
3 - PINCH BOLT
4 - STEERING GEAR MOUNTING BOLTS

5 - TIE ROD
6 - DRAG LINK
7 - PRESSURE HOSE
8 - RETURN HOSE

67189-CSFR-G31

Exploded view of power steering gear—Crossfire

Rear

1. Before servicing the vehicle, refer to the precautions in the beginning of this section.
2. Remove or disconnect the following:
   - Rear wheel
   - Upper mounting bolt
   - Lower control arm cover
   - Lower mounting bolts
   - Shock absorber

To install:

3. Install or connect the following:
   - Shock absorber. Tighten the upper bolt nut to 13 ft. lbs. (18 Nm) and the lower bolt to 41 ft. lbs. (55 Nm).
   - Lower control arm cover
   - Rear wheel

Coil Spring

REMOVAL & INSTALLATION

Front

1. Before servicing the vehicle, refer to the precautions in the beginning of this section.
2. Support the lower control arm on a floor jack.
3. Remove or disconnect the following:
   - Front wheel
   - Shock absorber
4. Compress the spring.
5. Remove the lower control arm mounting nuts and bolts.
6. Lower the jack while holding the spring. When enough clearance is obtained, remove the spring.

To install:

7. Install the coil spring and raise the control arm into position.
8. Install or connect the following:
   - Lower control arm bolts and nuts
   - Shock absorber
   - Tighten lower control arm nuts to 88 ft. lbs. (120 Nm)

Rear

1. Before servicing the vehicle, refer to the precautions in the beginning of this section.
2. Raise and support the vehicle.
3. Remove or disconnect the following:
   - Rear wheel
   - Lower control arm cover
   - Lower shock absorber bolts
4. Raise the lower control arm with a jack until the halfshaft is horizontal.

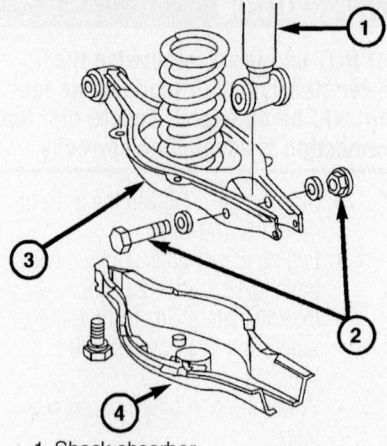

1. Shock absorber
2. Lower shock bolt
3. Lower control arm
4. Control arm cover

67189-CSFR-G32

Rear coil spring mounting—Crossfire

5. Using a spring compressor, compress the rear spring.
6. Remove the rear lower control arm attaching bolts, and then lower the jack allowing the control arm to swing away from the mounting tabs.
7. Remove the rear spring.

To install:

8. Install or connect the following:
   - Coil spring and raise the control arm into position
   - Lower control arm bolts and nuts
   - Lower shock absorber bolt. Torque the bolt to 41 ft. lbs. (55 Nm).
   - Control arm cover
   - Rear wheel

Stabilizer Bar

REMOVAL & INSTALLATION

Front

1. Before servicing the vehicle, refer to the precautions in the beginning of this section.
2. Remove or disconnect the following:
   - Stabilizer bar brackets on both sides
   - Retainer bracket and mounting plate
   - Stabilizer bar
   - Stabilizer bar bushings

To install:

3. Assemble the stabilizer bushings onto the bar.

➡ **The lower control arms may have to be raised to allow the stabilizer bar bushings to be inserted into their mating surfaces.**

4. Install the stabilizer bar by loosely installing the bolts and nuts to all 4 brackets.
5. Align the stabilizer bar so it is centered in the vehicle.
6. Due to the preload of the bracket, install the hexagon bolt first.
7. Tighten both stabilizer bolts to the frame.
8. Tighten the hexagon socket bolts to the retainer.
9. Tighten both bar bracket nuts to the lower control arm. Torque the bolts to 15 ft. lbs. (20 Nm).

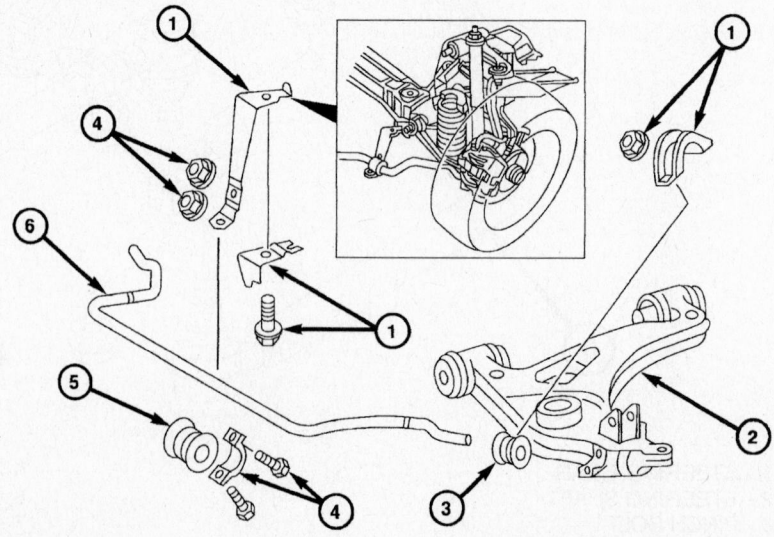

1 - STABILIZER BAR BRACKETS
2 - LOWER CONTROL ARM
3 - RUBBER BUSHING
4 - RETAINER BRACKET BOLTS AND NUTS
5 - RUBBER BUSHINGS
6 - STABILIZER BAR

67189-CSFR-G33

Front stabilizer bar mounting—Crossfire

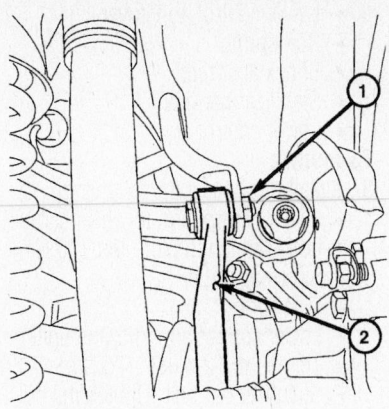

1. Link bolt
2. Stabilizer bar

67189-CSFR-G34

Rear stabilizer bar link bolt—Crossfire

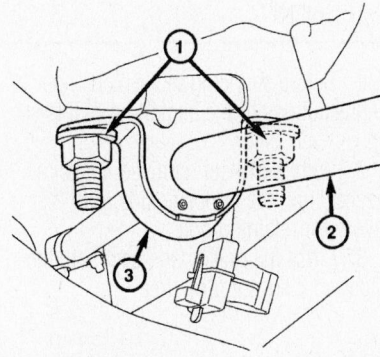

1. Mounting nuts
2. Stabilizer bar
3. Bar clamps

67189-CSFR-G35

Rear stabilizer bar clamp bolts—Crossfire

10. Install the bar retainer and mounting plate.

11. Tighten the retainer bolts to 44 ft. lbs. 60 Nm).

12. Tighten the lower retainer bracket nut to 15 ft. lbs. (20 Nm).

13. Tighten the upper retainer nuts to 29 ft. lbs. (40 Nm).

Rear

1. Before servicing the vehicle, refer to the precautions in the beginning of this section.

2. Raise and support the vehicle.

3. Remove or disconnect the following:
- Rear wheels
- Wheel speed sensors
- Brake hoses from caliper
- Coil springs
- Shock absorbers
- Parking brake cables from equalizer
- Rear driveshaft bolts
- Stabilizer bar links

4. Raise and support the rear axle carrier.

5. Remove 4 bolts attaching the rear axle carrier to the frame.

6. Lower the rear axle carrier enough to allow the stabilizer bar to be removed.

7. Remove the stabilizer bar clamps and remove the stabilizer bar.

To install:

8. Position the stabilizer bar into place.

9. Raise the rear axle carrier and install the 4 carrier to frame bolts. Torque the bolts to 66 ft. lb. (90 Nm).

10. Install or connect the following:
- Rear springs
- Shock absorbers
- Stabilizer bar clamps
- Stabilizer bar links. Torque the bolts to 22 ft. lbs. (30 Nm).
- Tighten the bar clamp nuts 15 ft. lbs. (20 Nm)
- Rear driveshaft bolts.
- Parking brake cables to equalizer
- Brake hoses to calipers
- Wheel seed sensors
- Rear wheels

Upper Ball Joint and Upper Control Arm

REMOVAL & INSTALLATION

➡**If replacing the ball joint on the right side, the air cleaner housing must be removed for access. The shock absorbers must remain installed.**

1. Before servicing the vehicle, refer to the precautions in the beginning of this section.

2. Remove the front wheel.

3. Wire the steering knuckle to the shock absorber.

4. Remove the ball joint nut.

5. Using a puller, press the ball joint out of the steering knuckle.

6. From inside the engine compartment, remove the upper control arm nut and bolt, and remove the control arm.

7. Installation is the reverse of the removal procedure. Tighten the upper control arm nut to 48 ft. lbs. (65 Nm). Tighten the ball joint nut to 33 ft. lbs. (45 Nm).

Lower Ball Joint

REMOVAL & INSTALLATION

1. Before servicing the vehicle, refer to the precautions in the beginning of this section.

2. Raise and support the vehicle.

3. Remove or disconnect the following:
- Front wheels
- Brake caliper
- Brake rotor
- Dust shield
- Lower shock mount bolt
- Coil spring
- Ball joint nuts
- Outer ball joint stud from steering knuckle using a puller
- Upper ball joint stud from lower control arm using a puller

To install:

4. Install or connect the following:
- Ball joint. Tighten ball joint nuts to 77 ft. lbs. (105 Nm).
- Coil spring
- Lower shock bolt
- Dust shield
- Brake rotor and caliper
- Front wheel

Lower Control Arm

REMOVAL & INSTALLATION

Front

1. Before servicing the vehicle, refer to the precautions in the beginning of this section.

2. Remove or disconnect the following:
- Front wheel
- Coil spring
- Lower shock bolt
- Stabilizer bar clamp
- Ball joint nuts
- Press ball joint out of steering knuckle
- Lower control arm

To install:

3. Install or connect the following:
- Lower control arm. Tighten the bolts to 88 ft. lbs. (120 Nm).
- Ball joint to steering knuckle. Tighten the nut to 77 ft. lbs. (105 Nm).
- Stabilizer bar clamp. Tighten the nuts to 15 ft. lbs. (20 Nm).
- Lower shock bolt. Tighten the nut to 41 ft. lbs. (55 Nm).
- Coil spring
- Front wheel

Rear

1. Before servicing the vehicle, refer to the precautions in the beginning of this section.

2. Remove or disconnect the following:
- Rear wheel

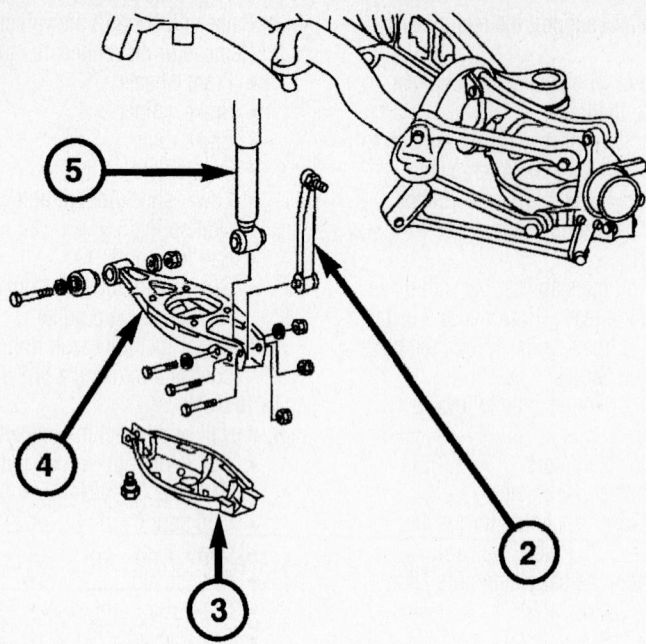

2. Stabilizer bar link
3. Lower control arm cover
4. Lower control arm
5. Shock absorber

67189-CSFR-G36

Exploded view of rear lower control arm—Crossfire

- Lower control arm cover
- Coil spring
- Shock absorber
- Stabilizer bar link
- Lower control arm

To install:

3. Install or connect the following:
   - Lower control arm to wheel carrier bolt. Tighten to 52 ft. lbs. (70 Nm).
   - Shock absorber
   - Coil spring
   - Lower control arm to frame bolt. Tighten to 52 ft. lbs. (70 Nm).
   - Stabilizer bar link. Tighten the nut to 15 ft. lbs. (20 Nm).
   - Lower control arm cover
   - Rear wheel

Front Wheel Bearings

ADJUSTMENT

1. Before servicing the vehicle, refer to the precautions in the beginning of this section.

2. With the wheel removed, install a wheel bolt on the opposite side from the brake rotor retaining bolt

3. Press the brake pads back into the caliper so the rotor turns free.

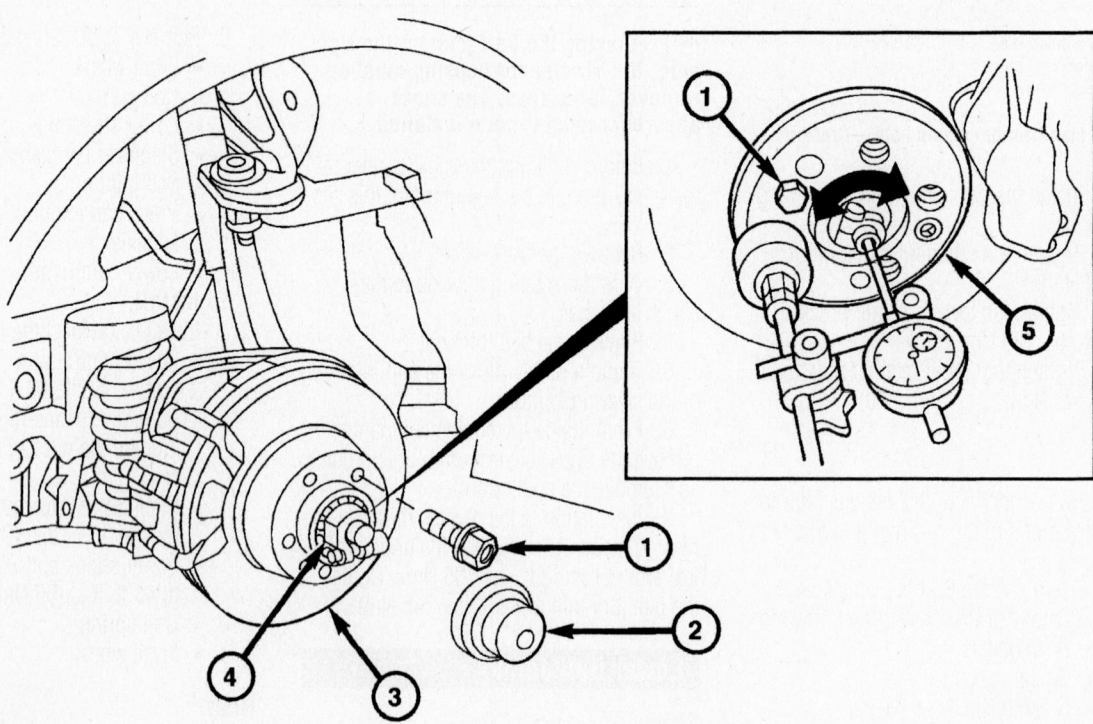

1 - WHEEL BOLT
2 - DUST CAP
3 - ROTOR

4 - HUB NUT
5 - ROTOR LOCK BOLT

67189-CSFR-G37

Adjusting front wheel bearing—Crossfire

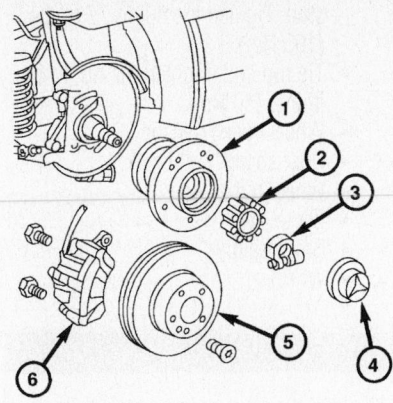

1. Wheel hub
2. Bearing
3. Hub nut
4. Grease cap
5. Rotor
6. Caliper

67189-CSFR-G38

**Exploded view of front wheel bearing
assembly—Crossfire**

4. Remove the bearing grease cap.

5. Loosen the wheel bearing adjusting nut until some end play is present.

6. Attach a dial indicator to the hub.

7. Turn the hub nut in stages while pushing and pulling the rotor firmly back and forth. Adjust the end play to 0.0004-0.0008 inches (0.01-0.02 mm).

8. Tighten the hub nut to 97 inch lbs. (11 Nm).

9. Remove dial indicator and install grease cap.

10. Remove the wheel bolt from rotor and install the wheel.

REMOVAL & INSTALLATION

1. Before servicing the vehicle, refer to the precautions in the beginning of this section.

2. Remove or disconnect the following:
- Front wheel
- Brake caliper
- Brake rotor

- Grease cap
- Adjusting nut
- Bearing
- Wheel hub

To install:

3. Pack the wheel hub with grease and install the hub, bearing and seal ring onto the spindle.

4. Install or connect the following:
- Brake rotor
- Hub nut
- Adjust bearing end play
- Brake caliper
- Grease cap
- Front wheel

Steering Knuckle

REMOVAL & INSTALLATION

1. Before servicing the vehicle, refer to the precautions in the beginning of this section.

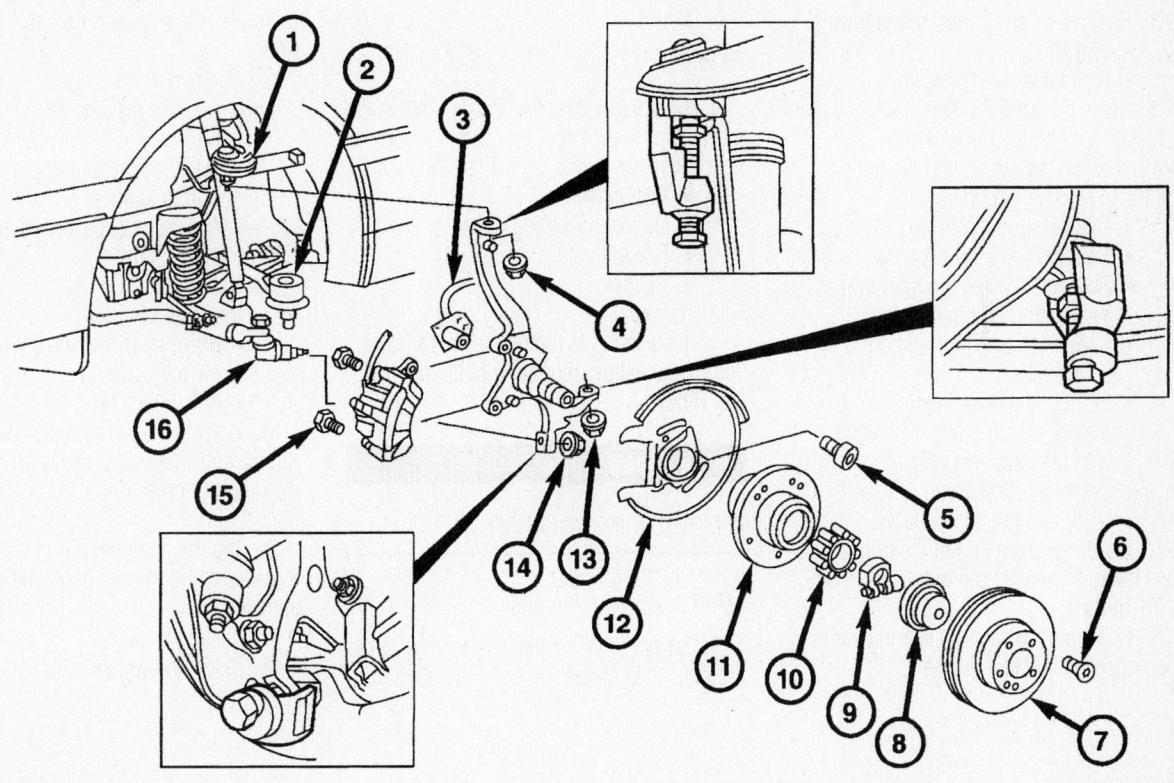

1 - UPPER CONTROL ARM
2 - TIE ROD END
3 - WHEEL SPEED SENSOR
4 - UPPER BALL JOINT NUT
5 - DUST SHIELD BOLT
6 - ROTOR LOCK BOLT
7 - ROTOR
8 - DUST CAP

9 - HUB NUT
10 - BEARING
11 - HUB
12 - DUST SHIELD
13 - TIE ROD END NUT
14 - LOWER BALL JOINT NUT
15 - CALIPER BOLT
16 - LOWER BALL JOINT

67189-CSFR-G39

Exploded view of front steering knuckle assembly—Crossfire

2. Raise and support the vehicle.

3. Remove or disconnect the following:

- Front wheel
- Brake caliper
- Brake rotor
- Wheel hub
- Dust shield
- Wheel speed sensor
- Tie rod end

- Upper and lower ball joints from knuckle
- Steering knuckle

To install:

4. Install or connect the following:

- Steering knuckle to upper ball joint stud. Tighten the nut to 33 ft. lbs. (45 Nm).
- Steering knuckle to lower ball joint

stud. Tighten the nut to 77 ft. lbs. (105 Nm).
- Tie rod end. Tighten the nut to 37 ft. lbs. (50 Nm).
- Wheel speed sensor
- Dust shield
- Wheel hub
- Brake rotor
- Brake caliper
- Wheel

BRAKES

Brake Caliper

REMOVAL & INSTALLATION

Front

1. Before servicing the vehicle, refer to the precautions in the beginning of this section.

2. Raise and support the front end on jackstands.

3. Remove or disconnect the following:
   - Wheels
   - Brake pad wear indicator

4. Press the caliper piston back into the bore with a suitable prytool. Use a large C-clamp to drive the piston into the bore of additional force is required.
   - Caliper support spring.
   - Caliper mounting pins
   - Bolt that secures the front brake hose fitting bolt in the caliper
   - Caliper off the rotor and out from its mount
   - Front brake hose fitting bolt completely, then remove the caliper with the pads installed as an assembly. Take care not to drip fluid onto the pad surfaces.

5. Cover the open end of the front brake hose fitting to prevent dirt entry.

To install:

6. Lubricate the caliper slide pins and bushings with silicone grease.

7. Install or connect
   - Caliper over the rotor and seat it in its original position until flush
   - Slide pins by hand, and then tighten them to 25 ft. lbs. (34 Nm)
   - Brake hose
   - Brake pad wear indicator
   - Caliper support spring
   - Wheels

8. Lower the vehicle.

9. Pump the brakes several times to seat the pads.

Rear

1. Before servicing the vehicle, refer to the precautions in the beginning of this section.

2. Remove or disconnect the following:
   - Rear wheel
   - Caliper mounting bolts
   - Brake hose
   - Caliper

3. To install

4. Installation is the reverse of removal. Torque the caliper mounting bolts to 41 ft. lbs. (55 Nm).

Disc Brake Pads

REMOVAL & INSTALLATION

Front

1. Before servicing the vehicle, refer to the precautions in the beginning of this section.

2. Remove or disconnect the following:
   - Wheels
   - Caliper
   - Brake pads

3. To install

4. Installation is the reverse of the removal procedure.

Rear

1. Before servicing the vehicle, refer to the precautions in the beginning of this section.

2. Remove or disconnect
   - Remove the wheels
   - Caliper
   - Brake pad retaining pin using a drift to knock it out
   - Anti-rattle clips and remove the brake pads

To install:

3. Install or connect
   - Brake pads into the caliper through the top opening
   - Anti-rattle clip

4. Hold the anti-rattle clip with your thumb and install the retaining pin through the hole in the side of the caliper.

5. Tap the retaining pin in until it is threaded through both pads and over the anti-rattle clip.

6. Install the caliper.

7. Install the wheels.

DODGE

Dakota

SPECIFICATION CHARTS

ENGINE AND VEHICLE IDENTIFICATION

Engine							Model Year	
Code ①	Liters (cc)	Cu. In.	Cyl.	Fuel Sys.	Engine Type	Eng. Mfg.	Code ②	Year
P	2.5 (2507)	153	4	SMFI	OHV	Chrysler	1	2001
K	3.7 (3701)	226	6	MFI	SOHC	Chrysler	2	2002
X	3.9 (3916)	238	6	SMFI	OHV	Chrysler	3	2003
N	4.7 (4701)	287	8	SMFI	SOHC	Chrysler	4	2004
Z	5.9 (5899)	360	8	SMFI	OHV	Chrysler	5	2005

OHV: Overhead Valve

SMFI: Sequential Multi-port Fuel Injection

① 8th position of VIN

② 10th position of VIN

67189-DAKO-C01

GENERAL ENGINE SPECIFICATIONS

Year	Model	Engine Displ. Liters	Engine (VIN)	Net Horsepower @ rpm	Net Torque @ rpm (ft. lbs.)	Bore x Stroke (in.)	Compression Ratio	Oil Pressure @ rpm
2001	Dakota	2.5	P	120@5200	145@3400	3.88x3.19	9.2:1	25-80@3000
		3.9	X	175@4800	220@3200	3.91x3.31	9.1:1	30-80@3000
		4.7	N	235@4800	295@3200	3.66x3.40	9.3:1	25@3000
		5.9	Z	230@4000	330@3250	4.00x3.58	9.1:1	30-80@3000
2002	Dakota	2.5	P	120@5200	145@3400	3.88x3.19	9.2:1	25-80@3000
		3.9	X	175@4800	220@3200	3.91x3.31	9.1:1	30-80@3000
		4.7	N	235@4800	295@3200	3.66x3.40	9.3:1	25@3000
		5.9	Z	230@4000	330@3250	4.00x3.58	9.1:1	30-80@3000
2003	Dakota	3.9	X	175@4800	220@3200	3.91x3.31	9.1:1	30-80@3000
		4.7	N	235@4800	295@3200	3.66x3.40	9.3:1	25@3000
		5.9	Z	230@4000	330@3250	4.00x3.58	9.1:1	30-80@3000
2004	Dakota	3.7	K	210@5200	225@4200	3.66x3.40	9.2:1	25-110@3000
		4.7	N	235@4800	295@3200	3.66x3.40	9.0:1	25-110@3000

SMFI: Sequential Multi-port Fuel Injection

67189-DAKO-C02

GASOLINE ENGINE TUNE-UP SPECIFICATIONS

Year	Engine Displ. Liters	Engine VIN	Spark Plug Gap (in.)	Ignition Timing (deg.)	Fuel Pump (psi)	Idle Speed (rpm)	Valve Clearance	
							Intake	Exhaust
2001	2.5	P	0.035	①	44.2-54.2	②	HYD	HYD
	3.9	X	0.040	①	44.2-54.2	②	HYD	HYD
	4.7	N	0.040	①	47-51	②	HYD	HYD
	5.9	Z	0.040	①	44.2-54.2	②	HYD	HYD
2002	2.5	P	0.035	①	44.2-54.2	②	HYD	HYD
	3.9	X	0.040	①	44.2-54.2	②	HYD	HYD
	4.7	N	0.040	①	47-51	②	HYD	HYD
	5.9	Z	0.040	①	44.2-54.2	②	HYD	HYD
2003	3.9	X	0.040	①	44.2-54.2	②	HYD	HYD
	4.7	N	0.040	①	47-51	②	HYD	HYD
	5.9	Z	0.040	①	44.2-54.2	②	HYD	HYD
2004	3.7	K	0.042	①	44-54	②	HYD	HYD
	4.7	N	0.040	①	47-51	②	HYD	HYD

NOTE: The Vehicle Emission Control Information (VECI) label often reflects specification changes made during production.
The label figures must be used if they differ from those in this chart.

HYD: Hydraulic

① Ignition timing is controlled by the PCM and is not adjustable.

② Idle speed is controlled by the PCM and is not adjustable

67189-DAKO-C03

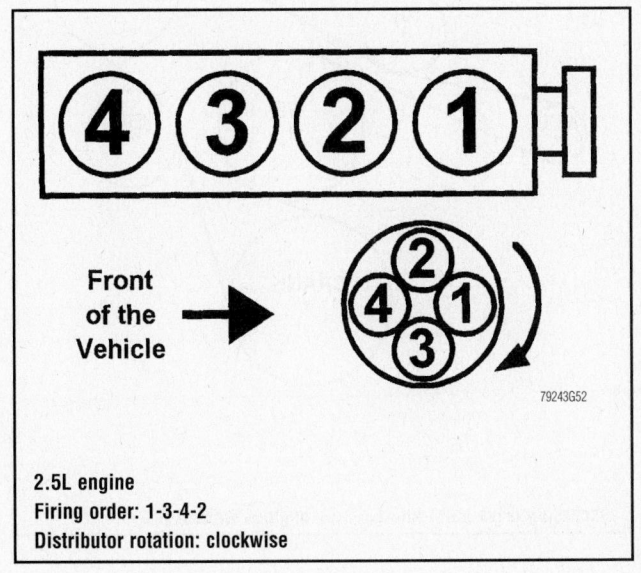

2.5L engine
Firing order: 1-3-4-2
Distributor rotation: clockwise

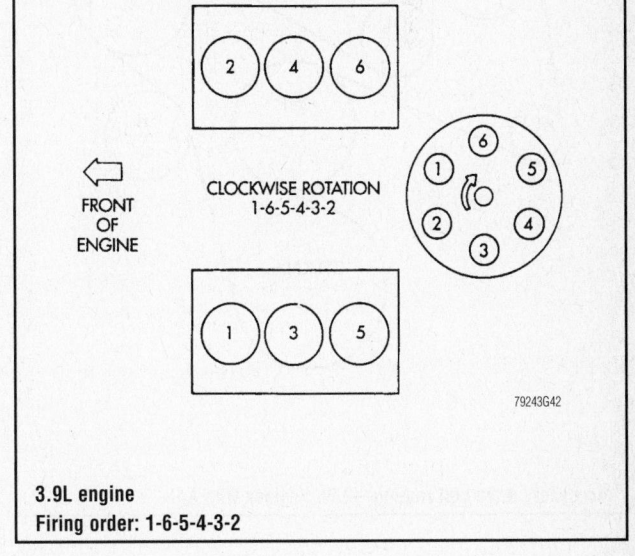

3.9L engine
Firing order: 1-6-5-4-3-2

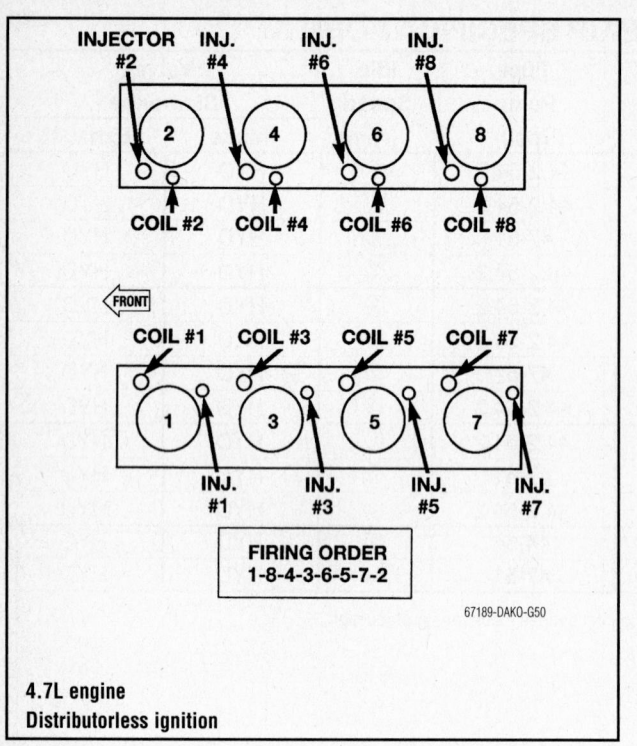

FIRING ORDER
1-8-4-3-6-5-7-2

67189-DAKO-G50

4.7L engine
Distributorless ignition

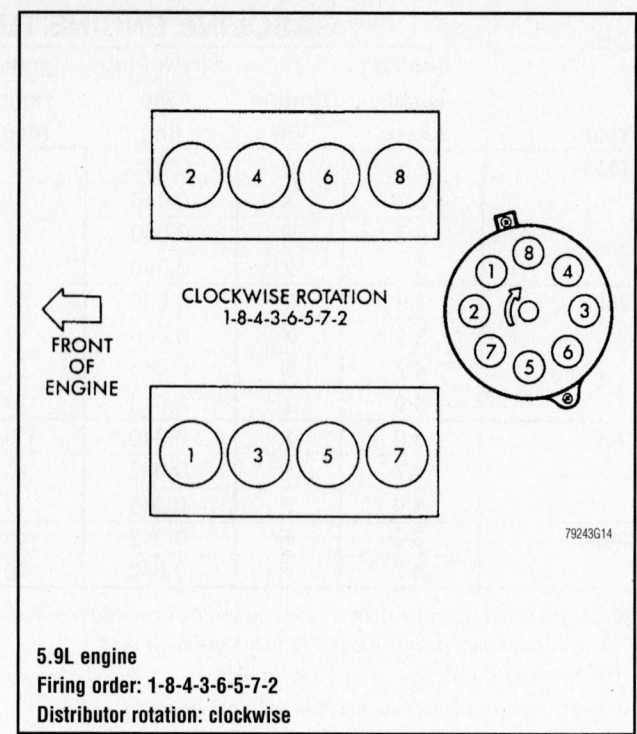

CLOCKWISE ROTATION
1-8-4-3-6-5-7-2

FRONT
OF
ENGINE

79243G14

5.9L engine
Firing order: 1-8-4-3-6-5-7-2
Distributor rotation: clockwise

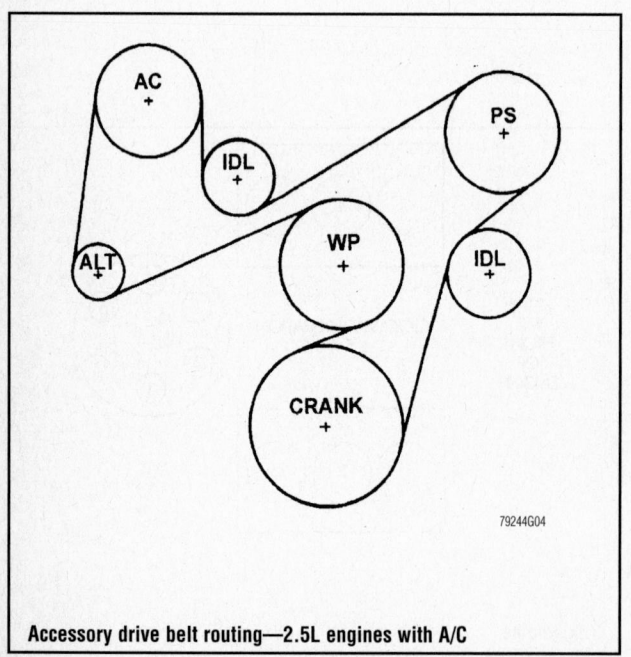

79244G04

Accessory drive belt routing—2.5L engines with A/C

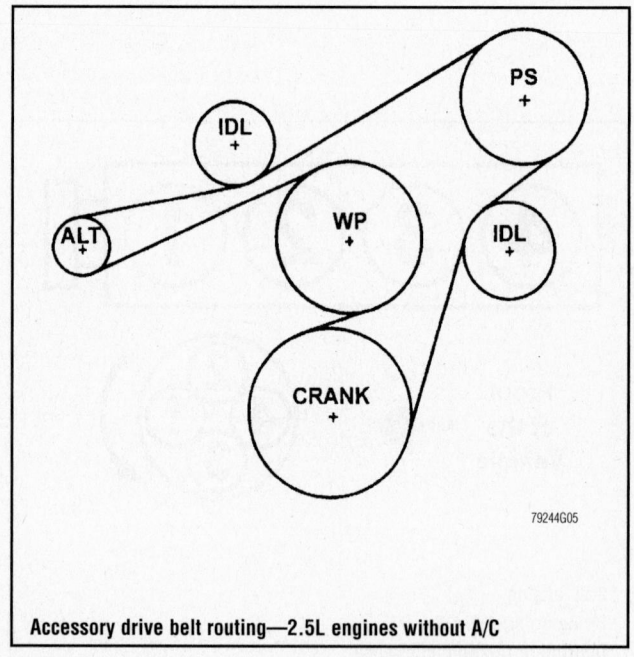

79244G05

Accessory drive belt routing—2.5L engines without A/C

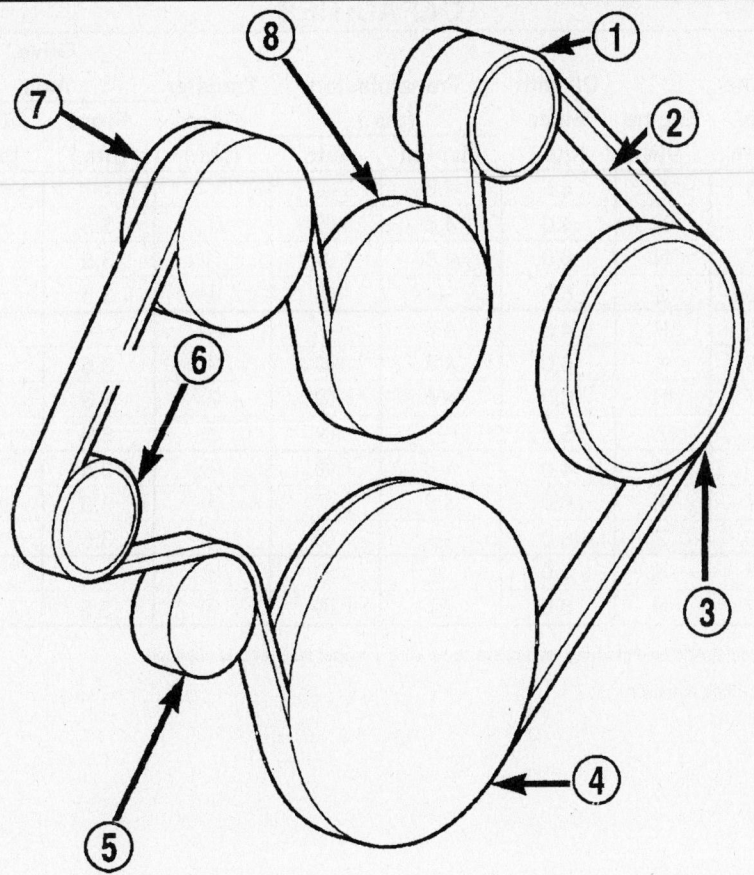

1 - GENERATOR PULLEY
2 - ACCESSORY DRIVE BELT
3 - POWER STEERING PUMP PULLEY
4 - CRANKSHAFT PULLEY
5 - IDLER PULLEY
6 - TENSIONER
7 - A/C COMPRESSOR PULLEY
8 - WATER PUMP PULLEY

67189-DAKO-G51

Accessory drive belt routing—3.7L and 4.7L engines

CAPACITIES

Year	Model	Engine Displ. Liters	Engine VIN	Oil with Filter (qts.)	Transmission (pts.) Manual	Transmission (pts.) Auto.	Transfer Case (pts.)	Drive Axle Front (pts.)	Drive Axle Rear (pts.)	Fuel Tank (gal.)	Cooling System (qts.)
2001	Dakota	2.5	P	4.5	4.8	—	—	—	①	②	9.8
		3.9	X	4.0	4.8	③	④	3.5	①	②	14.0
		4.7	N	6.0	4.8	③	④	3.5	①	②	13.0
		5.9	Z	5.0	—	③	④	3.5	①	②	14.6
2002	Dakota	2.5	P	4.5	4.8	—	—	—	①	②	9.8
		3.9	X	4.0	4.8	③	④	3.5	①	②	14.0
		4.7	N	6.0	4.8	③	④	3.5	①	②	13.0
		5.9	Z	5.0	—	③	④	3.5	①	②	14.6
2003	Dakota	3.9	X	4.0	4.8	③	④	3.5	①	②	14.0
		4.7	N	6.0	4.8	③	④	3.5	①	②	13.0
		5.9	Z	5.0	—	③	④	3.5	①	②	14.6
2004	Dakota	3.7	K	5.0	⑤	⑥	④	3.5	①	②	13.0
		4.7	N	6.0	⑤	⑥	④	3.5	①	②	13.0

NOTE: All capacities are approximate. Add fluid gradually and check to be sure a proper fluid level is obtained.

① The following values include 0.25 pt. of friction
  modifier for LSD axles.
  8.25 axle: 4.4 pts.
  9.25 axle: 4.9 pts.

② 2-door: 22 gal.
  4-door: 24 gal.

③ Drain and refill: 8.0 pts.
  42RE overhaul: 20 pts.
  46RE overhaul: 20 pts.
  45RFE overhaul: 28.0 pts.

④ NV233: 2.5 pts.
  NV244: 2.85 pts.

⑤ 2wd: 4.8 pts.
  4wd: 4.2 pts.

⑥ 42RLE: 8.0 pts.
  545RFE 2wd: 11.0
  545RFE 4wd: 13 pts.

67189-DAKO-C04

VALVE SPECIFICATIONS

Year	Engine Displ. Liters	Engine VIN	Seat Angle (deg.)	Face Angle (deg.)	Spring Test Pressure (lbs. @ in.)	Spring Installed Height (in.)	Stem-to-Guide Clearance (in.)		Stem Diameter (in.)	
							Intake	Exhaust	Intake	Exhaust
2001	2.5	P	44.5	45.0	184-196@ 1.216	1.64	0.001-0.003	0.001-0.003	0.311-0.312	0.311-0.312
	3.9	X	44.25-44.75	43.25-43.75	200@1.21	1.64	0.001-0.003	0.001-0.003	0.311-0.312	0.311-0.312
	4.7	N	44.5-45	45-45.5	176.7-193.3 @1.1670	1.601	0.0008-0.0028	0.0019-0.0039	0.2729-0.2739	0.2717-0.2728
	5.9	Z	44.25-44.75	43.25-43.75	200@1.212	1.640	0.0010-0.0030	0.0020-0.0040	0.3720-0.3730	0.3710-0.3720
2002	2.5	P	44.5	45.0	184-196@ 1.216	1.64	0.001-0.003	0.001-0.003	0.311-0.312	0.311-0.312
	3.9	X	44.25-44.75	43.25-43.75	200@1.21	1.64	0.001-0.003	0.001-0.003	0.311-0.312	0.311-0.312
	4.7	N	44.5-45	45-45.5	176.7-193.3 @1.1670	1.601	0.0008-0.0028	0.0019-0.0039	0.2729-0.2739	0.2717-0.2728
	5.9	Z	44.25-44.75	43.25-43.75	200@1.212	1.640	0.0010-0.0030	0.0020-0.0040	0.3720-0.3730	0.3710-0.3720
2003	3.9	X	44.25-44.75	43.25-43.75	200@1.21	1.64	0.001-0.003	0.001-0.003	0.311-0.312	0.311-0.312
	4.7	N	44.5-45	45-45.5	176.7-193.3 @1.1670	1.601	0.0008-0.0028	0.0019-0.0039	0.2729-0.2739	0.2717-0.2728
	5.9	Z	44.25-44.75	43.25-43.75	200@1.212	1.640	0.0010-0.0030	0.0020-0.0040	0.3720-0.3730	0.3710-0.3720
2004	3.7	K	44.5-45	45-45.5	221-234@ 1.107	1.579	0.0008-0.0028	0.0019-0.0039	0.2729-0.2739	0.2717-0.2728
	4.7	N	44.5-45	45-45.5	176.7-193.3 @1.1670	1.601	0.0008-0.0028	0.0019-0.0039	0.2729-0.2739	0.2717-0.2728

67189-DAKO-C05

CRANKSHAFT AND CONNECTING ROD SPECIFICATIONS

All measurements are given in inches.

Year	Engine Displ. Liters	Engine VIN	Crankshaft				Connecting Rod		
			Main Brg. Journal Dia.	Main Brg. Oil Clearance	Shaft End-play	Thrust on No.	Journal Diameter	Oil Clearance	Side Clearance
2001	2.5	P	2.4996-2.5001	0.0010-0.0025	0.0015-0.0065	2	2.2080-2.2085	0.0015-0.0020	0.010-0.0190
	3.9	X	2.4995-2.5005	0.0005-0.0015	0.0020-0.0070	2	2.1240-2.1250	0.0005-0.0022	0.0060-0.0140
	4.7	N	2.4996-2.5005	0.0008-0.0021	0.0021-0.0112	2	2.0076-2.0082	0.0006-0.0022	0.0040-0.0138
	5.9	Z	2.8095-2.8105	①	0.0020-0.0070	2	2.1240-2.1250	0.0005-0.0022	0.0060-0.0140
2002	2.5	P	2.4996-2.5001	0.0010-0.0025	0.0015-0.0065	2	2.2080-2.2085	0.0015-0.0020	0.010-0.0190
	3.9	X	2.4995-2.5005	0.0005-0.0015	0.0020-0.0070	2	2.1240-2.1250	0.0005-0.0022	0.0060-0.0140
	4.7	N	2.4996-2.5005	0.0008-0.0021	0.0021-0.0112	2	2.0076-2.0082	0.0006-0.0022	0.0040-0.0138
	5.9	Z	2.8095-2.8105	①	0.0020-0.0070	2	2.1240-2.1250	0.0005-0.0022	0.0060-0.0140
2003	3.9	X	2.4995-2.5005	0.0005-0.0015	0.0020-0.0070	2	2.1240-2.1250	0.0005-0.0022	0.0060-0.0140
	4.7	N	2.4996-2.5005	0.0008-0.0021	0.0021-0.0112	2	2.0076-2.0082	0.0006-0.0022	0.0040-0.0138
	5.9	Z	2.8095-2.8105	①	0.0020-0.0070	2	2.1240-2.1250	0.0005-0.0022	0.0060-0.0140
2004	3.7	K	2.4996-2.5005	0.0020-0.0034	0.0021-0.0112	2	2.2794-2.2797	0.0004-0.0019	0.0040-0.0138
	4.7	N	2.4996-2.5005	0.0008-0.0021	0.0021-0.0112	2	2.0076-2.0082	0.0006-0.0022	0.0040-0.0138

① No. 1: 0.0005-0.0015

Nos. 2-5: 0.0005-0.0020

67189-DAKO-C06

PISTON AND RING SPECIFICATIONS

All measurements are given in inches.

Year	Engine Displ. Liters	Engine VIN	Piston Clearance	Ring Gap Top Comp.	Ring Gap Bottom Comp.	Ring Gap Oil Control	Ring Side Clearance Top Comp.	Ring Side Clearance Bottom Comp.	Ring Side Clearance Oil Control
2001	2.5	P	0.0013-0.0021	0.0090-0.0240	0.0190-0.0380	0.0100-0.0600	0.0017-0.0033	0.0017-0.0033	0.0024-0.0083
	3.9	X	0.0005-0.0015	0.0100-0.0200	0.0100-0.0200	0.0020-0.0080	0.0015-0.0030	0.0015-0.0030	0.1515-0.1565
	4.7	N	0.0014	0.0146-0.0249	0.0146-0.0249	0.0099-0.0300	0.0020-0.0037	0.0016-0.0031	0.0175-0.0185
	5.9	Z	0.0005-0.0015	0.0120-0.0220	0.0220-0.0310	0.0150-0.0550	0.0016-0.0033	0.0016-0.0033	0.0020-0.0080
2002	2.5	P	0.0013-0.0021	0.0090-0.0240	0.0190-0.0380	0.0100-0.0600	0.0017-0.0033	0.0017-0.0033	0.0024-0.0083
	3.9	X	0.0005-0.0015	0.0100-0.0200	0.0100-0.0200	0.0020-0.0080	0.0015-0.0030	0.0015-0.0030	0.1515-0.1565
	4.7	N	0.0014	0.0146-0.0249	0.0146-0.0249	0.0099-0.0300	0.0020-0.0037	0.0016-0.0031	0.0175-0.0185
	5.9	Z	0.0005-0.0015	0.0120-0.0220	0.0220-0.0310	0.0150-0.0550	0.0016-0.0033	0.0016-0.0033	0.0020-0.0080
2003	3.9	X	0.0005-0.0015	0.0100-0.0200	0.0100-0.0200	0.0020-0.0080	0.0015-0.0030	0.0015-0.0030	0.1515-0.1565
	4.7	N	0.0014	0.0146-0.0249	0.0146-0.0249	0.0099-0.0300	0.0020-0.0037	0.0016-0.0031	0.0175-0.0185
	5.9	Z	0.0005-0.0015	0.0120-0.0220	0.0220-0.0310	0.0150-0.0550	0.0016-0.0033	0.0016-0.0033	0.0020-0.0080
2004	3.7	K	0.0014	0.0079-0.0142	0.0146-0.0249	0.0100-0.0300	0.0020-0.0037	0.0016-0.0031	0.0007-0.0091
	4.7	N	0.0014	0.0146-0.0249	0.0146-0.0249	0.0099-0.0300	0.0020-0.0037	0.0016-0.0031	0.0175-0.0185

67189-DAKO-C07

TORQUE SPECIFICATIONS

All readings in ft. lbs.

Year	Engine Displ. Liters	Engine VIN	Cylinder Head Bolts	Main Bearing Bolts	Rod Bearing Bolts	Crankshaft Damper Bolts	Flywheel Bolts	Manifold		Spark Plugs	Oil Pan Drain Plug
								Intake	Exhaust		
2001	2.5	P	①	80	33	80	105	②	②	27	25
	3.9	X	③	85	45	18	55	④	25	30	25
	4.7	N	⑤	⑥	⑦	130	45	⑧	18	20	25
	5.9	Z	③	85	45	18	55	④	25	30	25
2002	2.5	P	①	80	33	80	105	②	②	27	25
	3.9	X	③	85	45	18	55	④	25	30	25
	4.7	N	⑤	⑥	⑦	130	45	⑧	18	20	25
	5.9	Z	③	85	45	18	55	④	25	30	25
2003	3.9	X	③	85	45	18	55	④	25	30	25
	4.7	N	⑤	⑥	⑦	130	45	⑧	18	20	25
	5.9	Z	③	85	45	18	55	④	25	30	25
2004	3.7	K	⑤	⑤	⑦	130	70	9	18	27	25
	4.7	N	⑨	⑥	⑦	130	45	④	18	20	25

① See illustration in text section
   Bolts 1-10 and 12-14: 110 ft. lbs
   Bolt 11: 100 ft. lbs.

② Exhaust manifold bolt 1: 30 ft. lbs.
   Intake/exhaust manifold bolts 2-5: 23 ft. lbs.
   Exhaust manifold nuts 6 & 7: 23 ft. lbs.

③ Step 1: 50 ft. lbs.
   Step 2: 105 ft. lbs.

④ See illustration in text section
   Step 1: 1-4 to 72 inch lbs. in 12 inch lb. Increments
   Step 2: bolts 5-12: 72 inch lbs.
   Step 3: Check that all bolts are at 72 inch lbs.
   Step 4: All bolts, in sequence, to 12 ft. lbs.
   Step 5: Check that all bolts are at 12 ft. lbs.

⑤ See text

⑥ Bed plate bolt sequence. Refer to illustration in text section
   Step 1: Bolts A-L to 40 ft. lbs.
   Step 2: Bolts 1-10 25 inch lbs.
   Step 3: Bolts 1-10 plus 90 degrees
   Step 4: Bolts A1-A6 20 ft. lbs.

⑦ 20 ft. lbs. plus 90 degrees

⑧ 105 inch lbs.

⑨ M8 bolts: 19 ft. lbs.
   M11 bolts: 60 ft. lbs.

67189-DAKO-C08

BRAKE SPECIFICATIONS
All measurements in inches unless noted

| Year | Model | Brake Disc | | | Brake Drum | | | Minimum Lining Thickness | | Brake Caliper | |
		Original Thickness	Minimum Thickness	Maximum Run-out	Original Inside Diameter	Max. Wear Limit	Maximum Machine Diameter	Front	Rear	Bracket Bolts (ft. lbs.)	Mounting Bolts (ft. lbs.)
2001	Dakota	0.944	0.890	0.004	①	②	②	③	④	47	22
2002	Dakota	0.944	0.890	0.004	①	②	②	③	④	47	22
2003	Dakota	F 1.000	F 0.965	F 0.004	11.00	②	②	③	④	⑤	⑥
		R 0.600	R 0.585	R 0.0014							
2004	Dakota	F 1.000	F 0.965	F 0.004	11.00	②	②	③	④	⑤	⑥
		R 0.600	R 0.585	R 0.0014							

F: Front

R: Rear

NA: Not Available

① Available with both 9 in. and 11 in. rear brakes

② Maximum allowable drum diameter, either from wear or machining, is stamped on the drum.

③ Riveted brake pads: 0.0625 in.
Bonded brake pads: 0.1875 in.

④ Riveted brake shoes: 0.031 in.
Bonded brake shoes: 0.0625 in.

⑤ Front adapter bolts: 147.5 ft. lbs.
Support plate: 50 ft. lbs.

⑥ Front: 22 ft. lbs.
Rear: 19 ft. lbs.

67189-DAKO-C09

TIRE, WHEEL AND BALL JOINT SPECIFICATIONS

| Year | Model | OEM Tires | | Tire Pressures (psi) | | Wheel Size | Ball Joint Inspection | Lug Nut Torque ft. lbs. |
		Standard	Optional	Front	Rear			
2001	Dakota, 2wd	P215/75R15	P235/75R15	①	①	6.5-JJ	0.060 in. ②	100
	Dakota, 4wd	P235/75R15 XL	P255/65R16	①	①	NA		
			P265/70R16	①	①			
	Dakota RT	P255/55R17	none	①	①	NA		
2002	Dakota, 2wd	P215/75R15	P235/75R15	①	①	6.5-JJ	0.060 in. ②	100
	Dakota, 4wd	P235/75R15 XL	P255/65R16	①	①	NA		
			P265/70R16	①	①			
	Dakota RT	P255/55R17	none	①	①	NA		
2003	Dakota, 2wd	P215/75R15	P235/75R15	①	①	6.5-JJ	0.060 in. ②	100
	Dakota, 4wd	P235/75R15 XL	P255/65R16	①	①	NA		
			P265/70R16	①	①			
	Dakota RT	P255/55R17	none	①	①	NA		
2004	Dakota, 2wd	P245/70R16	P255/65R16	①	①	std: 7	0.060 in. ②	100
	Dakota, 4wd	P245/70R16	P265/70R16	①	①	opt: 8		

OEM: Original Equipment Manufacturer

PSI: Pounds Per Square Inch

STD: Standard

OPT: Optional

① See the tire placard on the vehicle

② Both upper and lower

67189-DAKO-C10

WHEEL ALIGNMENT

Year	Model	Wheel Base (in.)	Caster Range (+/-Deg.)	Caster Preferred Setting (Deg.)	Camber Range (+/-Deg.)	Camber Preferred Setting (Deg.)	Toe-in (in.)
2001	2WD	111.9	0.50	+2.99	0.50	-0.25	0.10+/-0.06
	2WD	130.9	0.50	+3.13	0.50	-0.25	0.10+/-0.06
	4WD	111.9	0.50	+3.16	0.50	-0.25	0.10+/-0.06
	4WD	130.9	0.50	+3.27	0.50	-0.25	0.10+/-0.06
	RT	111.9	0.50	+3.67	0.50	-0.34	0.10+/-0.06
	RT	130.9	0.50	+3.81	0.50	-0.34	0.10+/-0.06
2002	2WD	111.9	0.50	+2.99	0.50	-0.25	0.10+/-0.06
	2WD	130.9	0.50	+3.13	0.50	-0.25	0.10+/-0.06
	4WD	111.9	0.50	+3.16	0.50	-0.25	0.10+/-0.06
	4WD	130.9	0.50	+3.27	0.50	-0.25	0.10+/-0.06
	RT	111.9	0.50	+3.67	0.50	-0.34	0.10+/-0.06
	RT	130.9	0.50	+3.81	0.50	-0.34	0.10+/-0.06
2003	2WD	111.9	0.50	+2.99	0.50	-0.25	0.10+/-0.06
	2WD	130.9	0.50	+3.13	0.50	-0.25	0.10+/-0.06
	4WD	111.9	0.50	+3.16	0.50	-0.25	0.10+/-0.06
	4WD	130.9	0.50	+3.27	0.50	-0.25	0.10+/-0.06
	RT	111.9	0.50	+3.67	0.50	-0.34	0.10+/-0.06
	RT	130.9	0.50	+3.81	0.50	-0.34	0.10+/-0.06
2004	2WD	111.9	0.50	+2.99	0.50	0	0.10+/-0.06
	2WD	130.9	0.50	+3.13	0.50	0	0.10+/-0.06
	4WD	111.9	0.50	+3.16	0.50	0	0.10+/-0.06
	4WD	130.9	0.50	+3.27	0.50	0	0.10+/-0.06
	RT	111.9	0.50	+3.67	0.50	-0.09	0.10+/-0.06
	RT	130.9	0.50	+3.81	0.50	-0.09	0.10+/-0.06

67189-DAKO-C11

SCHEDULED MAINTENANCE INTERVALS
2001-2003 Dodge Dakota

TO BE SERVICED	TYPE OF SERVICE	VEHICLE MILEAGE INTERVAL (x1000)													
		7.5	15	22.5	30	37.5	45	52.5	60	67.5	75	82.5	90	97.5	100
Engine oil & filter	R	✓	✓	✓	✓	✓	✓	✓	✓	✓	✓	✓	✓	✓	
Ball joints	L			✓		✓				✓			✓		
Front wheel bearings	S/I			✓		✓				✓			✓		
Brake linings	S/I			✓		✓				✓			✓		
Air cleaner element	R				✓				✓				✓		
Spark plugs	R				✓				✓				✓		
Transfer case fluid	R					✓					✓				
Automatic transmission fluid	R					✓					✓				
Automatic transmission bands	Adj					✓					✓				
Engine coolant ①	R						✓						✓		
Spark plug cables	R								✓						
PCV valve ②	S/I								✓						
Drive belt tensioner (3.9L & 5.9L) ②	S/I								✓						
Drive belt tension (2.5L)	Adj								✓						

R: Replace   S/I: Service or Inspect   L: Lubricate   Adj: Adjust

① Change every 36 months, regardless of mileage

② Replace if necessary.

FREQUENT OPERATION MAINTENANCE (SEVERE SERVICE)

If a vehicle is operated under any of the following conditions it is considered severe service:

- Extremely dusty areas.
- 50% or more of the vehicle operation is in 32°C (90°F) or higher temperatures, or constant operation in temperatures below 0°C (32°F).
- Prolonged idling (vehicle operation in stop and go traffic.
- Frequent short running periods (engine does not warm to normal operating temperatures).
- Police, taxi, delivery usage or trailer towing usage.

Oil & oil filter change: change every 3000 miles.

Air filter/air pump air filter: change every 24,000 miles.

Engine coolant level, hoses & clamps: check every 6,000 miles.

Exhaust system: check every 6000 miles.

Drive belts: check every 18,000 miles; replace every 24,000 miles.

Crankcase inlet air filter (6 & 8 cyl.): clean every 24,000 miles.

Oxygen sensor: replace every 82,500 miles.

Automatic transmission fluid, filter & bands: change & adjust every 12,000 miles.

Steering linkage: lubricate every 6000 miles.

Rear axle fluid: change every 12,000 miles.

67189-DAKO-C12

SCHEDULED MAINTENANCE INTERVALS
2004 Dodge Dakota

TO BE SERVICED	TYPE OF SERVICE	VEHICLE MILEAGE INTERVAL (x1000)													
		6	12	18	24	30	36	42	48	54	60	66	72	78	84
Engine coolant ①	R						✓						✓		
Accessory drive belt ②	S/I										✓				
Engine oil & filter	R	✓	✓	✓	✓	✓	✓	✓	✓	✓	✓	✓	✓	✓	
PCV valve ②	S/I										✓				
Ball joints	L			✓			✓			✓			✓		
Front wheel bearings	S/I			✓			✓			✓			✓		
Brake linings	S/I			✓			✓			✓			✓		
Air cleaner element	R					✓					✓				
Spark plugs	R					✓					✓				
Auto. Trans. fluid ③	R	every 100,000 miles													
Transfer case fluid ④	I										✓				

R: Replace   S/I: Service or Inspect   L: Lubricate   Adj: Adjust

① Change every 36 months, regardless of mileage

② Replace if necessary.

③ w/3.7L engine: change fluid, filter and adjust bands
   w/4.7L engine: change fluid, main filter and cooler return filter (if equipped)

④ Replace every 120,000 miles

FREQUENT OPERATION MAINTENANCE (SEVERE SERVICE)

If a vehicle is operated under any of the following conditions it is considered severe service:

- **Extremely dusty areas.**
- **50% or more of the vehicle operation is in 32°C (90°F) or higher temperatures, or constant operation in temperatures below 0°C (32°F).**
- **Prolonged idling (vehicle operation in stop and go traffic.**
- **Frequent short running periods (engine does not warm to normal operating temperatures).**
- **Police, taxi, delivery usage or trailer towing usage.**

Oil & oil filter change: change every 3000 miles.

Air filter/air pump air filter: change every 24,000 miles.

Engine coolant level, hoses & clamps: check every 6,000 miles.

Exhaust system: check every 6000 miles.

Drive belts: check every 18,000 miles; replace every 24,000 miles.

Crankcase inlet air filter (6 & 8 cyl.): clean every 24,000 miles.

Oxygen sensor: replace every 82,500 miles.

Automatic transmission fluid, filter & bands: change & adjust every 12,000 miles.

Steering linkage: lubricate every 6000 miles.

Rear axle fluid: change every 12,000 miles.

67189-DAKO-C13

PRECAUTIONS

Before servicing any vehicle, please be sure to read all of the following precautions, which deal with personal safety, prevention of component damage, and important points to take into consideration when servicing a motor vehicle:

• Never open, service or drain the radiator or cooling system when the engine is hot; serious burns can occur from the steam and hot coolant.

• Observe all applicable safety precautions when working around fuel. Whenever servicing the fuel system, always work in a well-ventilated area. Do not allow fuel spray or vapors to come in contact with a spark, open flame, or excessive heat (a hot drop light, for example). Keep a dry chemical fire extinguisher near the work area. Always keep fuel in a container specifically designed for fuel storage; also, always properly seal fuel containers to avoid the possibility of fire or explosion. Refer to the additional fuel system precautions later in this section.

• Fuel injection systems often remain pressurized, even after the engine has been turned **OFF**. The fuel system pressure must be relieved before disconnecting any fuel lines. Failure to do so may result in fire and/or personal injury.

• Brake fluid often contains polyglycol ethers and polyglycols. Avoid contact with the eyes and wash your hands thoroughly after handling brake fluid. If you do get brake fluid in your eyes, flush your eyes with clean, running water for 15 minutes. If eye irritation persists, or if you have taken

brake fluid internally, IMMEDIATELY seek medical assistance.

• The EPA warns that prolonged contact with used engine oil may cause a number of skin disorders, including cancer! You should make every effort to minimize your exposure to used engine oil. Protective gloves should be worn when changing oil. Wash your hands and any other exposed skin areas as soon as possible after exposure to used engine oil. Soap and water, or waterless hand cleaner should be used.

• All new vehicles are now equipped with an air bag system, often referred to as a Supplemental Restraint System (SRS) or Supplemental Inflatable Restraint (SIR) system. The system must be disabled before performing service on or around system components, steering column, instrument panel components, wiring and sensors. Failure to follow safety and disabling procedures could result in accidental air bag deployment, possible personal injury and unnecessary system repairs.

• Always wear safety goggles when working with, or around, the air bag system. When carrying a non-deployed air bag, be sure the bag and trim cover are pointed away from your body. When placing a non-deployed air bag on a work surface, always face the bag and trim cover upward, away from the surface. This will reduce the motion of the module if it is accidentally deployed. Refer to the additional air bag system precautions later in this section.

• Clean, high quality brake fluid from a sealed container is essential to the safe and

proper operation of the brake system. You should always buy the correct type of brake fluid for your vehicle. If the brake fluid becomes contaminated, completely flush the system with new fluid. Never reuse any brake fluid. Any brake fluid that is removed from the system should be discarded. Also, do not allow any brake fluid to come in contact with a painted surface; it will damage the paint.

• Never operate the engine without the proper amount and type of engine oil; doing so WILL result in severe engine damage.

• Timing belt maintenance is extremely important! Many models utilize an interference-type, non-freewheeling engine. If the timing belt breaks, the valves in the cylinder head may strike the pistons, causing potentially serious (also time-consuming and expensive) engine damage. Refer to the maintenance interval charts in the front of this manual for the recommended replacement interval for the timing belt, and to the timing belt section for belt replacement and inspection.

• Disconnecting the negative battery cable on some vehicles may interfere with the functions of the on-board computer system(s) and may require the computer to undergo a relearning process once the negative battery cable is reconnected.

• When servicing drum brakes, only disassemble and assemble one side at a time, leaving the remaining side intact for reference.

• Only an MVAC-trained, EPA-certified automotive technician should service the air conditioning system or its components.

ENGINE REPAIR

➡**Disconnecting the negative battery cable on some vehicles may interfere with the functions of the on board computer system. The computer may undergo a relearning process once the negative battery cable is reconnected.**

Distributor

REMOVAL

2.5L Engine

1. Before servicing the vehicle, refer to the precautions in the beginning of this section.

2. Remove or disconnect the following:
   • Negative battery cable

   • Distributor cap
   • Camshaft Position (CMP) sensor connector

3. Matchmark the distributor housing and the rotor.
   • Distributor

3.9L and 5.9L Engines

1. Before servicing the vehicle, refer to the precautions in the beginning of this section.

2. Remove or disconnect the following:

   • Negative battery cable
   • Air cleaner tube
   • Distributor cap
   • Camshaft Position (CMP) sensor connector

3. Matchmark the distributor housing and the rotor.

4. Matchmark the distributor housing and the intake manifold.

5. Remove the distributor.

INSTALLATION

Timing Not Disturbed

2.5L ENGINE

1. Before servicing the vehicle, refer to the precautions in the beginning of this section.

➡**The rotor will rotate clockwise as the gears engage.**

2. Position the rotor slightly counterclockwise of the matchmark made during removal.

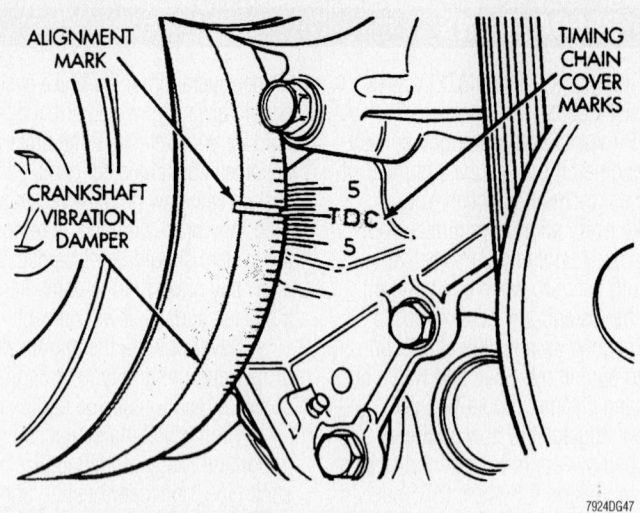

Crankshaft pulley and timing chain cover marks aligned at Top Dead Center (TDC)

3. Install the distributor. Ensure that the rotor moves into alignment with the match-mark.

4. Align the locating fork with the clamp bolt hole. Install the clamp and bolt. Tighten the bolt to 17 ft. lbs. (23 Nm).

5. Install or connect the following:
- CMP sensor connector
- Distributor cap
- Air cleaner tube
- Negative battery cable

3.9L AND 5.9L ENGINES

1. Before servicing the vehicle, refer to the precautions in the beginning of this section.

➡ **The rotor will rotate clockwise as the gears engage.**

2. Position the rotor slightly counter-clockwise of the matchmark made during removal.

3. Install the distributor.

4. Align the distributor housing and intake manifold matchmarks and check that the distributor housing matchmark and rotor are also aligned.

5. Install or connect the following:
- Distributor housing clamp and bolt. Tighten the bolt to 17 ft. lbs. (23 Nm).
- CMP sensor connector
- Distributor cap
- Air cleaner tube
- Negative battery cable

Timing Disturbed

2.5L ENGINE

1. Before servicing the vehicle, refer to the precautions in the beginning of this section.

2. Set the engine at Top Dead Center (TDC) of the No. 1 cylinder compression stroke.

3. Position the slot in the oil pump drive gear as shown.

4. Locate the alignment holes in the plastic ring and align the correct hole with the mating hole in the distributor housing as shown. Install a locking pin.

➡ **The distributor will rotate clockwise as the gears engage.**

5. Position the base mounting slot at the 1 o'clock position and install the distributor.

6. Check that the centerline of the mounting slot aligns with the centerline of the clamp bolt hole.

7. Install the clamp and bolt. Tighten the bolt to 17 ft. lbs. (23 Nm).

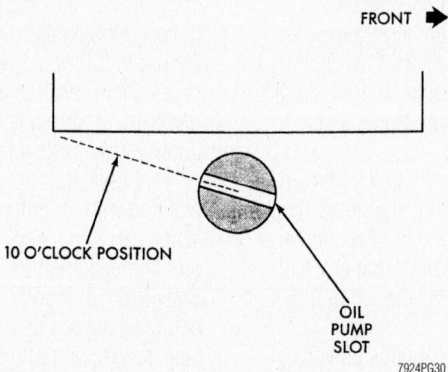

Slot in the oil pump gear at 10 o'clock position—2.5L engine

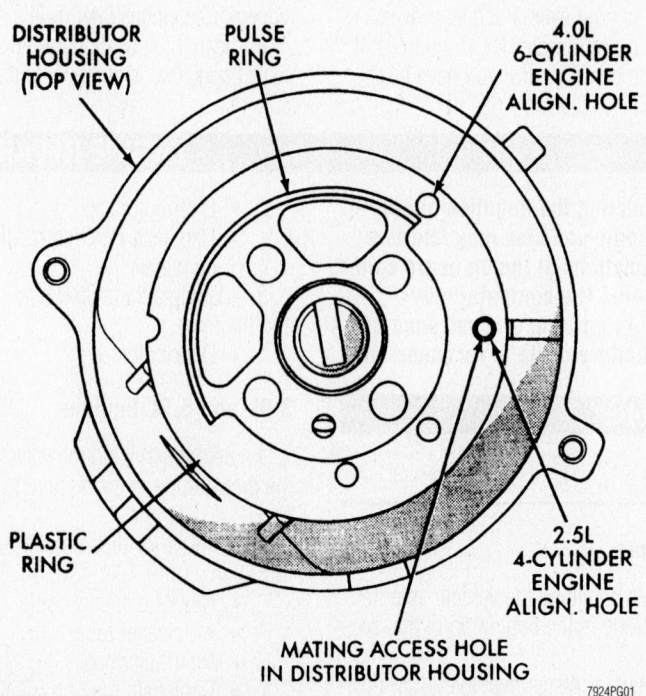

Distributor pin alignment holes—2.5L engine

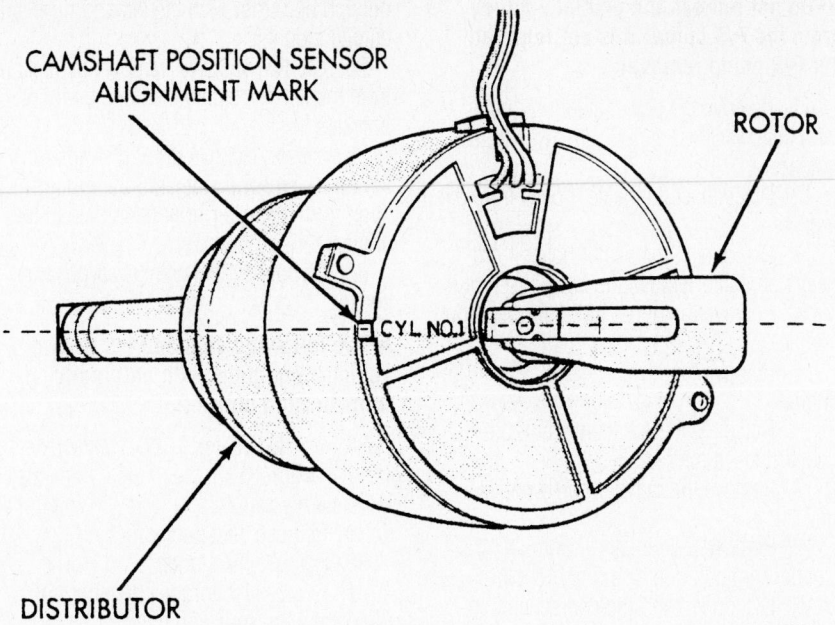

**CAMSHAFT POSITION SENSOR
ALIGNMENT MARK**

ROTOR

CYL NO.1

DISTRIBUTOR

9308PG01

Distributor rotor alignment—3.9L and 5.9L engines

8. Remove the locking pin.
9. Install or connect the following:
- CMP sensor connector
- Distributor cap
- Air cleaner tube
- Negative battery cable

3.9L AND 5.9L ENGINES

1. Before servicing the vehicle, refer to the precautions in the beginning of this section.
2. Set the engine at Top Dead Center (TDC) of the No. 1 cylinder compression stroke.
3. Install the distributor, clamp and bolt.
4. Rotate the distributor so that the rotor aligns with the No. 1 cylinder mark on the Camshaft Position (CMP) sensor. Tighten the clamp bolt to 17 ft. lbs. (23 Nm).
5. Install or connect the following:
- CMP sensor connector
- Distributor cap
- Air cleaner tube
- Negative battery cable
6. The final distributor position must be set with a scan tool. Follow the instructions supplied by the scan tool manufacturer.

Alternator

REMOVAL & INSTALLATION

2.5L, 3.9L and 5.9L Engines

1. Before servicing the vehicle, refer to the precautions in the beginning of this section.

2. Remove or disconnect the following:
- Negative battery cable
- Accessory drive belt
- Alternator harness connectors
- Alternator
3. Installation is the reverse of removal. Observe the following torques:
   a. 2.5L engine: 41 ft. lbs. (56 Nm).
   b. 3.9L, 5.9L engines: 30 ft. lbs. (41 Nm).

3.7L Engine

1. Before servicing the vehicle, refer to the precautions in the beginning of this section.
2. Remove or disconnect the following:
- Negative battery cable
- Accessory drive belt
- Alternator harness connectors
- Mounting bolts and alternator

➡There are 1 vertical and 2 horizontal bolts.

To install:
3. Before servicing the vehicle, refer to the precautions in the beginning of this section.
4. Install the alternator and tighten the bolts to the following specifications:
- Horizontal bolts to 42 ft. lbs. (57 Nm)
- Vertical bolt to 29 ft. lbs. (40 Nm)
5. Install or connect the following:
- Alternator harness connectors
- Accessory drive belt
- Negative battery cable

4.7L Engine

1. Before servicing the vehicle, refer to the precautions in the beginning of this section.
2. Remove or disconnect the following:
- Negative battery cable
- Accessory drive belt
- Alternator harness connectors
- Alternator
3. Installation is the reverse of removal. Observe the following torques:
- Vertical bolt and long horizontal bolt: 41 ft. lbs. (56 Nm)
- Short horizontal bolt: 55 ft. lbs. (74 Nm).

Ignition Timing

ADJUSTMENT

The ignition timing is controlled by the Powertrain Control Module (PCM) and is not adjustable.

Engine Assembly

REMOVAL & INSTALLATION

2.5L, 3.9L & 5.9L Engines

1. Before servicing the vehicle, refer to the precautions in the beginning of this section.
2. Drain the cooling system.
3. Drain the engine oil.
4. Relieve the fuel system pressure.
5. Remove or disconnect the following:
- Negative battery cable
- Hood
- Upper crossmember and top core support
- Radiator hoses
- Cooling fan and shroud
- Radiator
- Accelerator cable
- Cruise control cable, if equipped
- Transmission cable, if equipped
- Heater hoses
- Intake manifold vacuum lines
- Accessory drive belt
- Power steering pump
- A/C compressor, if equipped
- Engine control sensor harness connectors
- Engine block heater, if equipped
- Fuel line
- Exhaust front pipe
- Starter motor
- Torque converter, if equipped
- Transmission oil cooler lines, if equipped

- Engine mounts
- Transmission flange bolts. Support the transmission.
- Engine

✴✴ WARNING

Do not lift the engine by the intake manifold. Damage to the manifold may result.

To install:

6. Install or connect the following:
   - Engine. Tighten the engine mount bolts to 70 ft. lbs. (95 Nm).
   - Transmission flange bolts. Tighten the bolts to 40–45 ft. lbs. (54–61 Nm).
   - Transmission oil cooler lines, if equipped
   - Torque converter, if equipped. Tighten the bolts to 23 ft. lbs. (31 Nm).
   - Starter motor
   - Exhaust front pipe
   - Fuel line
   - Engine block heater, if equipped
   - Engine control sensor harness connectors
   - A/C compressor, if equipped
   - Power steering pump
   - Accessory drive belt
   - Intake manifold vacuum lines
   - Heater hoses
   - Transmission cable, if equipped
   - Cruise control cable, if equipped
   - Accelerator cable
   - Radiator
   - Cooling fan and shroud
   - Radiator hoses
   - Upper crossmember and top core support
   - Hood
   - Negative battery cable
7. Fill the crankcase to the correct level.
8. Fill the cooling system.
9. Start the engine and check for leaks.

3.7L Engine

1. Discharge the A/C system.
2. Drain the cooling system.
3. Release the fuel rail pressure.
4. Remove the air cleaner assembly.
5. Disconnect the battery.
6. Remove the upper fan shroud.
7. Remove the accessory drive belt.
8. Remove the viscous fan.
9. Remove the A/C compressor and position out of the way.
10. Remove the generator and secure away from engine.

➡ Do not remove the phenolic pulley from the P/S pump. It is not required for P/S pump removal.

11. Remove the power steering pump with lines attached and secure away from engine.
12. Disconnect the heater hoses from the engine.
13. Disconnect the heater hoses from heater core and remove the hose assembly.
14. Remove the upper radiator hose from engine.
15. Remove the lower radiator hose from engine.
16. Disconnect the transmission oil cooler lines at the radiator.
17. Remove the radiator core support bracket.
18. Remove the radiator assembly, A/C Condenser and transmission oil cooler.
19. Disconnect throttle and speed control cables.
20. Disconnect the engine to body ground straps at the left side of cowl.
21. Disconnect the engine wiring harness at the following points:
   - Intake air temperature (IAT) sensor
   - Fuel Injectors
   - Throttle Position (TPS) Switch
   - Idle Air Control (IAC) Motor
   - Engine Oil Pressure Switch
   - Engine Coolant Temperature (ECT) Sensor
   - Manifold Absolute Pressure MAP) Sensor
   - Camshaft Position (CMP) Sensor
   - Coil Over Plugs
   - Crankshaft Position Sensor
22. Remove the coil over plugs.
23. Remove fuel rail and secure away from engine.

➡ It is not necessary to release the quick connect fitting from the fuel supply line for engine removal.

24. Remove the PCV hose.
25. Remove the breather hoses.
26. Remove the vacuum hose for the power brake booster.
27. Disconnect the knock sensors.
28. Remove the engine oil dipstick tube.
29. Remove the intake manifold.
30. Install the engine lifting fixture, special tool 8247, using original fasteners from the removed intake manifold, and fuel rail.
31. Raise the vehicle on hoist.
32. Remove the exhaust crossover pipe from exhaust manifolds.
33. On 4wd vehicles, disconnect the axle vent tube from the left side engine mount.
34. Remove the through bolt retaining

nut and bolt from both the left and right side engine mounts.
35. On 4wd vehicles, remove the locknut from the left and right side engine mount brackets.
36. Disconnect two ground straps from the lower left hand side and one ground strap from the lower right hand side of the engine.
37. Disconnect the crankshaft position sensor.

➡ The following step applies to 4wd vehicles equipped with automatic transmission only.

38. On 4wd vehicles, remove the axle isolator bracket from the engine, transmission and the axle.
39. Remove the structural cover.
40. Remove the starter.
41. Remove the torque converter bolts (automatic transmission only).
42. Remove transmission to engine mounting bolts.
43. Disconnect the engine block heater power cable from the block heater, if equipped.
44. Lower the vehicle.
45. Remove throttle body resonator assembly and air inlet hose.
46. Disconnect the throttle and speed control cables.
47. Disconnect the tube from both the left and right side crankcase breathers. Remove the breathers.
48. Remove the generator.
49. Disconnect the two heater hoses from the timing chain cover and heater core.
50. Unclip and remove the heater hoses and tubes from the intake manifold.
51. Disconnect the engine harness at the following points:
   - Intake air temperature (IAT) sensor
   - Fuel Injectors
   - Throttle Position (TPS) Switch
   - Idle Air Control (IAC) Motor
   - Engine Oil Pressure Switch
   - Engine Coolant Temperature (ECT) Sensor
   - Manifold absolute pressure (MAP) Sensor
   - Camshaft Position (CMP) Sensor
   - Coil Over Plugs
52. Disconnect the vacuum lines at the throttle body and intake manifold.
53. Remove the power steering pump and position out of the way.
54. Disconnect the body ground strap at the right side cowl.
55. Disconnect the body ground strap at the left side cowl.

→It will be necessary to support the transmission in order to remove the engine.

56. Position a suitable jack under the transmission.

57. Remove the engine from the vehicle.

To install:

58. Position the engine in the vehicle. Position both the left and right side engine mount brackets and install the through bolts and nuts. On 2wd vehicles, tighten nuts to 95 Nm (70 ft. lbs.); on 4wd vehicles, to 102 Nm (75 ft. lbs.).

59. On 4wd vehicles, install the locknuts onto the engine mount brackets. Tighten the locknuts to 41 Nm (30 ft. lbs.).

60. Remove the jack from under the transmission.

61. Remove Engine Lifting Fixture Tool 8347.

62. Remove Special Tools 8400 Lifting Studs.

63. Position the generator wiring behind the oil dipstick tube, then install the oil dipstick tube upper mounting bolt.

64. Connect both left and right side body ground straps.

65. Install the power steering pump.

66. Connect the fuel supply line quick connect fitting.

67. Connect the vacuum lines at the throttle body and intake manifold.

68. Connect the engine harness at the following points:

- Intake Air Temperature (IAT) Sensor
- Idle Air Control (IAC) Motor
- Fuel Injectors
- Throttle Position (TPS) Switch
- Engine Oil Pressure Switch
- Engine Coolant Temperature (ECT) Sensor
- Manifold Absolute Pressure (MAP) Sensor
- Camshaft Position (CMP) Sensor
- Coil Over Plugs

69. Position and install the heater hoses and tubes onto intake manifold.

70. Install the heater hoses onto the heater core and the engine front cover.

71. Install the generator.

72. Install the A/C condenser, radiator and transmission oil cooler.

73. Connect the radiator upper and lower hoses.

74. Connect the transmission oil cooler lines to the radiator.

75. Install the accessory drive belt, fan assembly and shroud.

76. Install A/C compressor.

77. Install both breathers. Connect the tube to both crankcase breathers.

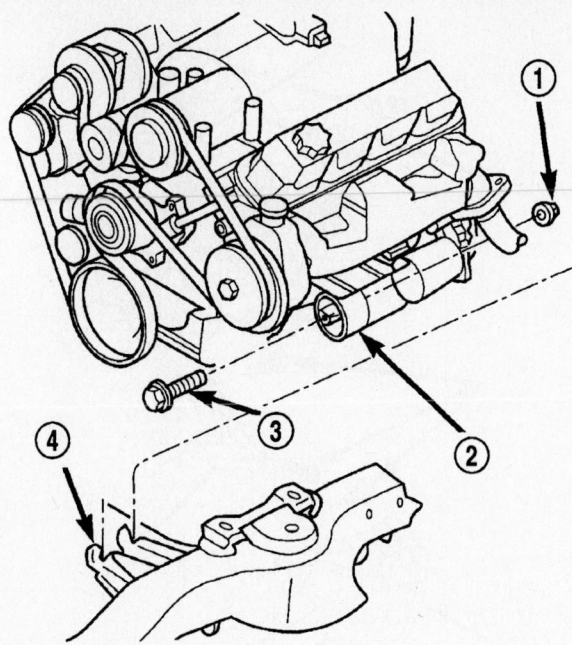

1 - LOCKNUT AND WASHER
2 - ENGINE MOUNT/INSULATOR
3 - THROUGH BOLT
4 - FRAME

67189-DAKO-G49

Engine mount through-bolt removal—3.7L engine w/2wd

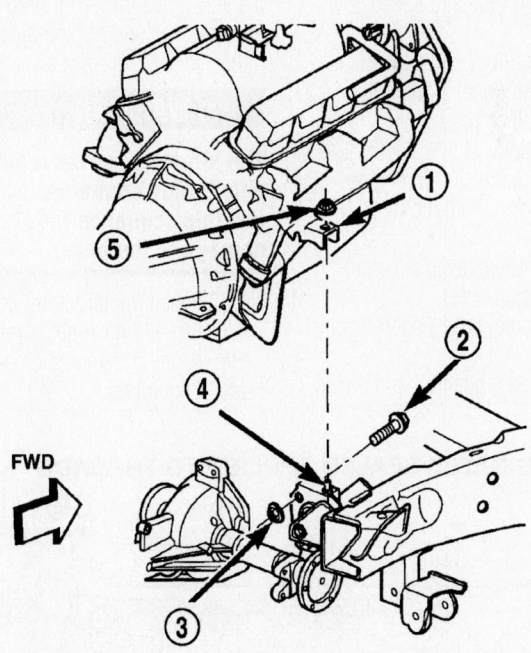

1 - ENGINE MOUNT BRACKET (2)
2 - THROUGH BOLT (2)
3 - LOCKNUT AND WASHER (2)
4 - ENGINE ISOLATOR TO ENGINE MOUNT BRACKET STUD (2)
5 - LOCKNUT (2)

67189-DAKO-G47

Engine mount through-bolt removal—3.7L engine w/4wd

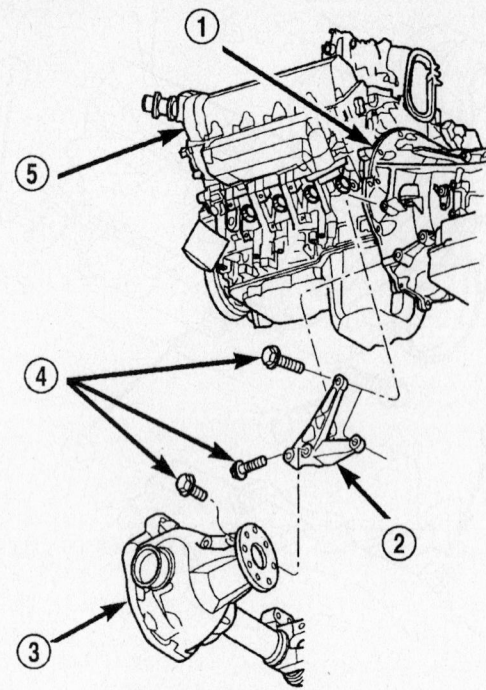

1 - TRANSMISSION
2 - AXLE ISOLATOR BRACKET
3 - FRONT AXLE 4X4 VEHICLES
4 - BOLTS
5 - ENGINE

67189-DAKO-G48

Axle isolator bracket removal—3.7L engine w/4wd

78. Connect the throttle and speed control cables.

79. Install the throttle body resonator assembly and air inlet hose. Tighten the clamps to 4 Nm (35 inch lbs.).

80. Raise the vehicle.

81. Install the transmission to engine mounting bolts. Tighten the bolts to 41 Nm (30 ft. lbs.).

82. Install the torque converter bolts (automatic transmission only).

83. Connect the crankshaft position sensor.

84. On 4wd vehicles, position and install the axle isolator bracket onto the axle, transmission and engine block.

85. Install the starter.

✳✳ WARNING

The structural cover requires a specific torque sequence. Failure to follow this sequence may cause severe damage to the cover.

86. Install the structural cover.

87. Install the exhaust crossover pipe.

88. Install the engine block heater power cable, if equipped.

89. On 4wd vehicles, connect the axle vent tube to the left side engine mount.

90. Lower the vehicle.

91. Check and fill engine oil.

92. Recharge the A/C system.

93. Refill the engine cooling system.

94. Install the battery tray and battery.

95. Connect the battery positive and negative cables.

96. Start the engine and check for leaks.

4.7L Engine

1. Disconnect the battery negative and positive cables.

2. Remove the battery and the battery tray.

3. Raise the vehicle on hoist.

4. Remove exhaust crossover pipe from the exhaust manifolds.

5. On 4wd vehicles, disconnect the axle vent tube from left side engine mount.

6. Remove the through bolt retaining nut and bolt from both the left and right side engine mounts.

7. On 4wd vehicles, remove the locknut from left and right side engine mount brackets.

8. Disconnect two ground straps from the lower left hand side and one ground strap from the lower right hand side of the engine.

9. Disconnect the crankshaft position sensor.

➡**The following step applies to 4X4 vehicles equipped with automatic transmission only.**

10. On 4wd vehicles, remove the axle isolator bracket from the engine, transmission and the axle.

11. Remove the structural cover.

12. Remove the starter.

13. Drain the cooling system.

14. Remove the torque converter bolts (automatic transmission only).

15. Remove the transmission to engine mounting bolts.

16. Disconnect the engine block heater power cable from the block heater, if equipped.

17. Lower the vehicle.

18. Remove the throttle body resonator assembly and air inlet hose.

19. Disconnect the throttle and speed control cables.

20. Disconnect the tube from both the left and right side crankcase breathers. Remove breathers.

21. Discharge the A/C system.

22. Remove the A/C compressor.

♦ **INDICATES SEALER APPLIED TO THREADS**

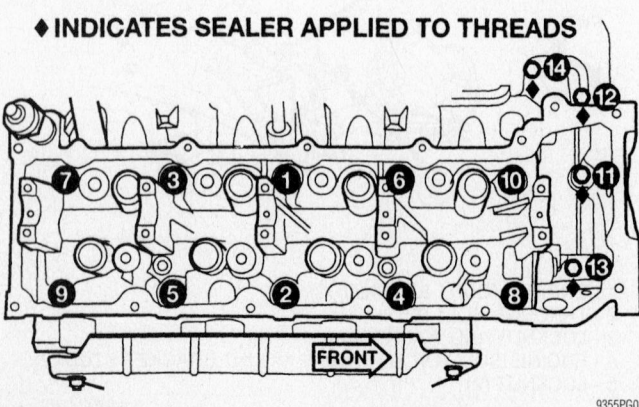

9355PG01

Tighten the structural cover bolts in this order—3.7L Engine

23. Remove the shroud, fan assembly and accessory drive belt.

24. Disconnect the transmission oil cooler lines at the radiator.

25. Disconnect the radiator upper and lower hoses.

26. Remove the radiator, A/C condenser and transmission oil cooler.

27. Remove the generator.

28. Disconnect the two heater hoses from the timing chain cover and heater core.

29. Unclip and remove the heater hoses and tubes from the intake manifold.

30. Disconnect the engine harness at the following points :

- Intake air temperature (IAT) sensor
- Fuel Injectors
- Throttle Position (TPS) Switch
- Idle Air Control (IAC) Motor
- Engine Oil Pressure Switch
- Engine Coolant Temperature (ECT) Sensor
- Manifold absolute pressure (MAP) Sensor
- Camshaft Position (CMP) Sensor
- Coil Over Plugs

31. Disconnect the vacuum lines at the throttle body and intake manifold.

32. Release the fuel rail pressure then disconnect the fuel supply quick connect fitting at the fuel rail.

33. Remove the power steering pump and position out of the way.

34. Install Special Tools 8400 Lifting Studs, into the cylinder heads.

35. Install Engine Lifting Fixture Special Tool 8347 following these steps:

- Holding the lifting fixture at a slight angle, slide the large bore in the front plate over the hex portion of the lifting stud.
- Position the two remaining fixture arms onto the two Special Tools 8400 Lifting Studs, in the cylinder heads.
- Pull forward and upward on the lifting fixture so that the lifting stud rest in the slotted area below the large bore.
- Secure the lifting fixture to the three studs using three 7/16—14 N/C locknuts.
- Make sure the lifting loop in the lifting fixture is in the last hole (closest to the throttle body) to minimize the angle of engine during removal.

36. Disconnect the body ground strap at the right side cowl.

37. Disconnect the body ground strap at the left side cowl.

➡ **It will be necessary to support the transmission in order to remove the engine.**

38. Position a suitable jack under the transmission.

39. Remove the engine from the vehicle.

To install:

40. Position the engine in the vehicle. Position both the left and right side engine mount

brackets and install the through bolts and nuts. On 2wd vehicles, tighten nuts to 95 Nm (70 ft. lbs.); On 4wd vehicles, tighten to 102 Nm (75 ft. lbs.).

41. On 4wd vehicles, install locknuts onto the engine mount brackets. Tighten the locknuts to 41 Nm (30 ft. lbs.).

42. Remove the jack from under the transmission.

43. Remove Engine Lifting Fixture Special Tool 8347.

44. Remove Special Tools 8400 Lifting Studs.

45. Position the generator wiring behind the oil dipstick tube, then install the oil dipstick tube upper mounting bolt.

46. Connect both left and right side body ground straps.

47. Install the power steering pump.

48. Connect the fuel supply line quick connect fitting.

49. Connect the vacuum lines at the throttle body and intake manifold.

50. Connect the engine harness at the following points:

- Intake Air Temperature (IAT) Sensor
- Idle Air Control (IAC) Motor
- Fuel Injectors
- Throttle Position (TPS) Switch
- Engine Oil Pressure Switch
- Engine Coolant Temperature (ECT) Sensor
- Manifold Absolute Pressure (MAP) Sensor
- Camshaft Position (CMP) Sensor
- Coil Over Plugs

51. Position and install the heater hoses and tubes onto intake manifold.

52. Install the heater hoses onto the heater core and the engine front cover.

53. Install the generator.

54. Install the A/C condenser, radiator and transmission oil cooler.

55. Connect the radiator upper and lower hoses.

56. Connect the transmission oil cooler lines to the radiator.

57. Install the accessory drive belt, fan assembly and shroud.

58. Install the A/C compressor.

59. Install both breathers. Connect the tube to both crankcase breathers.

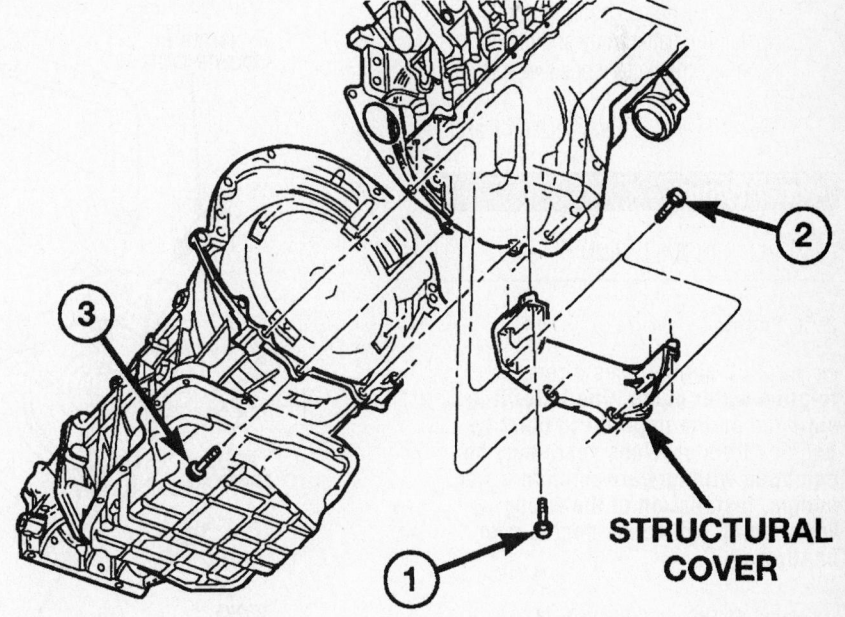

1 - BOLT

2 - BOLT

3 - BOLT

67189-DAKO-G46

Structural cover torque sequence—4.7L engine

60. Connect the throttle and speed control cables.

61. Install the throttle body resonator assembly and air inlet hose. Tighten clamps 4 Nm (35 inch lbs.).

62. Raise the vehicle.

63. Install transmission to engine mounting bolts. Tighten the bolts to 41 Nm (30 ft. lbs.).

64. Install the torque converter bolts (automatic transmission only).

65. Connect crankshaft position sensor.

66. On 4wd vehicles, position and install the axle isolator bracket onto the axle, transmission and engine block. Tighten the bolts to specification.

67. Install the starter.

✳✳ WARNING

The structural cover requires a specific torque sequence. Failure to follow this sequence may cause severe damage to the cover.

68. Install the structural cover.

69. Install the exhaust crossover pipe.

70. Install the engine block heater power cable, if equipped.

71. On 4wd vehicles, connect the axle vent tube to left side engine mount.

72. Lower the vehicle.

73. Check and fill the engine oil.

74. Recharge the A/C system.

75. Refill the engine cooling system.

76. Install the battery tray and battery.

77. Connect the battery positive and negative cables.

78. Start the engine and check for leaks.

Water Pump

REMOVAL & INSTALLATION

2.5L Engine

➡The 2.5L engine uses a reverse rotation water pump. The letter R is stamped on the impeller to identify. Engines from previous years may be equipped with forward rotation water pumps. Installation of the wrong water pump will cause engine over heating.

1. Before servicing the vehicle, refer to the precautions in the beginning of this section.

2. Drain the cooling system.

3. Remove or disconnect the following:
   - Negative battery cable
   - Accessory drive belt
   - Engine cooling fan and pulley

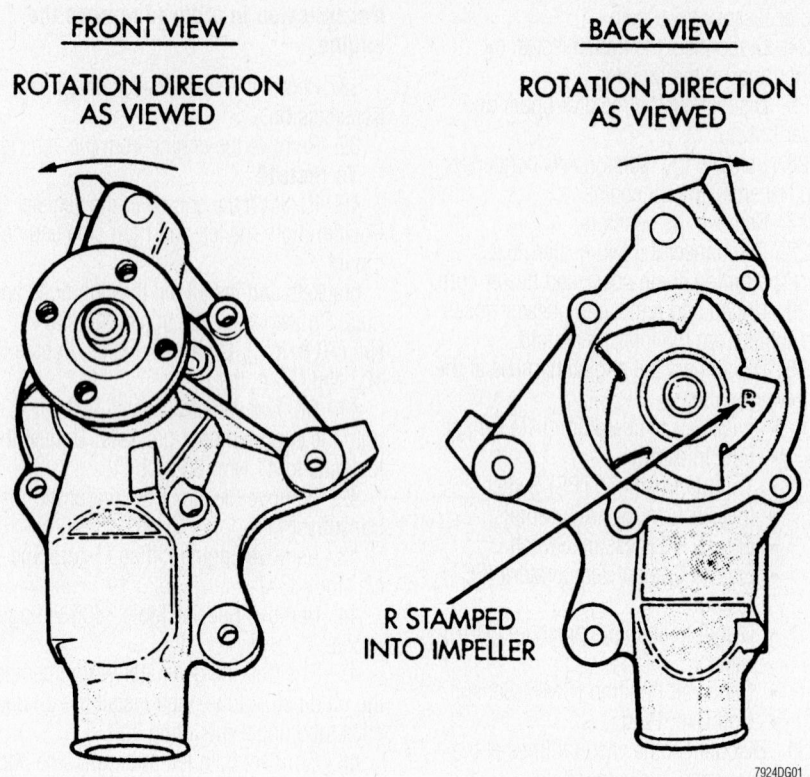

FRONT VIEW

ROTATION DIRECTION AS VIEWED

BACK VIEW

ROTATION DIRECTION AS VIEWED

R STAMPED INTO IMPELLER

7924DG01

Reverse rotation water pump—2.5L engine

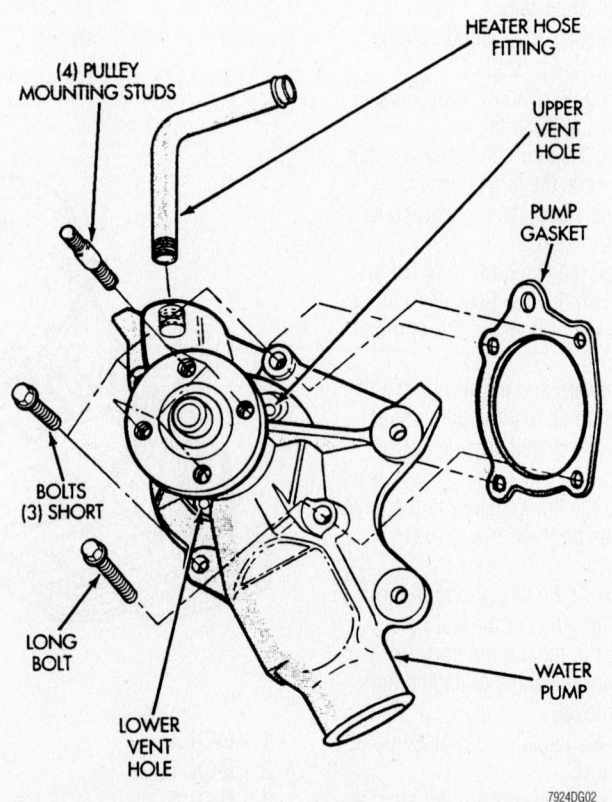

(4) PULLEY MOUNTING STUDS

HEATER HOSE FITTING

UPPER VENT HOLE

PUMP GASKET

BOLTS (3) SHORT

LONG BOLT

LOWER VENT HOLE

WATER PUMP

7924DG02

Water pump assembly—2.5L engine

➡Do not store the fan clutch assembly horizontally, silicone may leak into the bearing grease and cause contamination.

- Power steering pump
- Lower radiator hose
- Heater hose
- Water pump

➡One of the water pump bolts is longer than the others. Note the location for reassembly.

To install:

4. Install or connect the following:
   - Water pump using a new gasket. Tighten the bolts to 17 ft. lbs. (23 Nm).
   - Heater hose
   - Lower radiator hose
   - Power steering pump
   - Engine cooling fan and pulley
   - Accessory drive belt
   - Negative battery cable
5. Fill the cooling system.
6. Run the engine and check for leaks.

3.7L Engine

1. Before servicing the vehicle, refer to the precautions in the beginning of this section.
2. Drain the cooling system.
3. Remove or disconnect the following:

   - Negative battery cable
   - Fan and clutch assembly from the pump
   - Fan shroud and fan assembly. If you're reusing the fan clutch, keep it upright to avoid silicone fluid loss!
   - Lower hose
   - Water pump (8 bolts)
4. Installation is the reverse of removal. Tighten the bolts, in sequence, to 40 ft. lbs. (54 Nm).

4.7L Engine

1. Before servicing the vehicle, refer to the precautions in the beginning of this section.
2. Drain the cooling system.
3. Remove or disconnect the following:

   - Negative battery cable
   - Fan and fan drive assembly from the pump. Don't attempt to remove it from the vehicle, yet.

➡If a new pump is being installed, don't separate the fan from the drive.

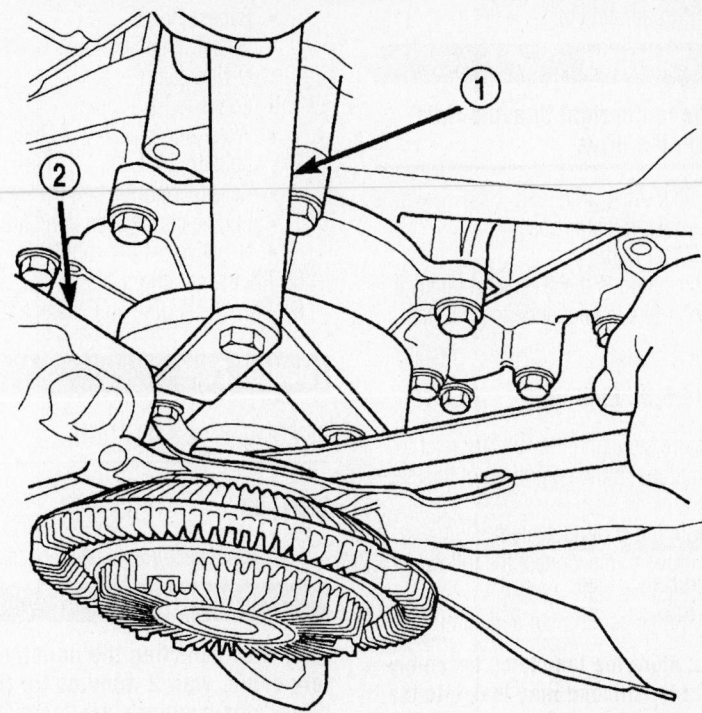

1 - SPECIAL TOOL 6958 SPANNER WRENCH WITH ADAPTER PINS 8346
2 - FAN

67189-DAKO-G99

Fan and Fan drive Removal—4.7L Engine

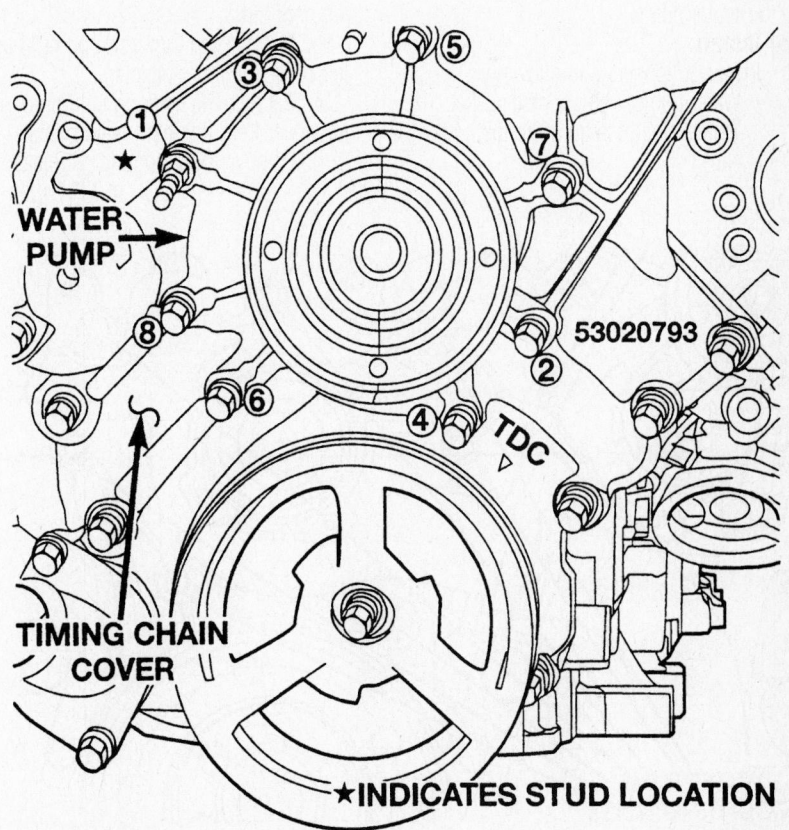

Water pump torque sequence—3.7L and 4.7L engines

9302PG06

- Shroud and fan

✻✻ WARNING

Keep the fan upright to avoid fluid loss from the drive.

- Accessory drive belt
- Lower radiator hose
- Water pump

4. Installation is the reverse of removal. tighten the bolts in sequence to 40 ft. lbs. (54 Nm).

3.9L and 5.9L Engines

1. Before servicing the vehicle, refer to the precautions in the beginning of this section.
2. Drain the cooling system.
3. Remove or disconnect the following:
   - Negative battery cable
   - Engine cooling fan and shroud

➡Do not store the fan clutch assembly horizontally, silicone may leak into the bearing grease and cause contamination.

- Accessory drive belt
- Water pump pulley
- Lower radiator hose
- Heater hose and tube
- Bypass hose
- Water pump

To install:

4. Install or connect the following:
   - Water pump, using a new gasket. Tighten the bolts to 30 ft. lbs. (40 Nm).

- Bypass hose
- Heater hose and tube. Use a new O-ring seal.
- Lower radiator hose
- Water pump pulley. Tighten the bolts to 20 ft. lbs. (27 Nm).
- Accessory drive belt
- Engine cooling fan and shroud
- Negative battery cable

5. Fill the cooling system.
6. Start the engine and check for leaks.

Heater Core

REMOVAL & INSTALLATION

2001–03

1. Disconnect the negative battery cable.

✻✻ CAUTION

After disconnecting the negative battery cable, wait 2 minutes for the driver's/passenger's air bag system capacitor to discharge before attempting to do any work around the steering column or instrument

2. Remove the instrument panel by performing the following procedure:

   a. Remove the trim from the right and left door sills.

   b. Remove the trim from the right and left cowl side inner panels.

   c. Remove the steering column opening cover from the instrument panel.

   d. Remove the 2 hood latch release handle-to-instrument panel lower reinforcement screws and lower the release handle to the floor.

   e. Disconnect the driver's side air bag module wire harness connector.

   f. If equipped, disconnect the overdrive lockout switch harness connector.

   g. Do not disassemble the steering column but remove the assembly from the vehicle.

   h. From under the driver's side of the instrument panel, disconnect or remove the following items:

   - The screw from the center of the headlight/dash-to-instrument panel bulkhead wire harness connector and disconnect the connector.
   - The 2 body wire harness connectors from the 2 instrument panel wire harness connector that secure the outboard side of the instrument panel bulkhead connector.
   - The 3 wire harness connectors from the 3 junction block connector receptacles located closest to the dash panel; 1 from the body wire harness and 2 from the headlight/dash wire harness.
   - The plastic park brake release linkage rod-to-rear parking brake release handle lever. Disengage the linkage rod from the handle lever.
   - The stoplight switch connector.
   - The vacuum harness connector from the left side of the heater/air conditioning housing assembly.

   i. Remove the instrument panel center support bracket.

   j. Remove the instrument panel wire harness ground screw located on the left side of the air bag control module (ACM) mount on the floor panel transmission tunnel.

   k. Disconnect the instrument panel wiring harness-to-ACM connector receptacle.

   l. Remove the glove box.

   m. Working through the glove box opening, disconnect or remove the following items:

   - The 2 halves of the radio antenna coaxial cable connector near the center of the lower instrument panel.
   - The antenna half of the radio antenna coaxial cable from the retainer clip near the outboard side of the lower instrument panel.
   - The blower motor wire harness connector located near the heater/air conditioning housing assembly support brace.

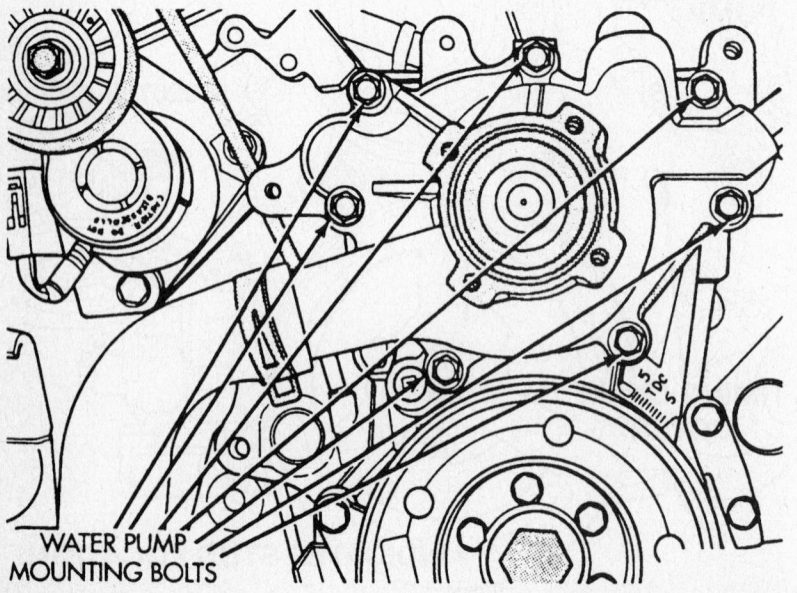

WATER PUMP MOUNTING BOLTS

7924DG03

Water pump mounting bolt locations—3.9L and 5.9L engines

n. From the passenger's side, disconnect or remove the following items:

- The 2 instrument panel wire harness connectors from the infinity speaker amplifier connector receptacles on the right cowl.
- The instrument panel wire harness radio ground eyelet-to-stud nut on the right cowl.

o. Loosen the right and left instrument panel cowl side roll-down bracket screws.

p. Remove the 5 top instrument panel-to-dash panel screws.

q. Pull the instrument panel rearward until the right and left cowl side roll-down bracket screws are in the roll-down slot position of both brackets.

r. Roll down the instrument panel and install a temporary hook in the center hole on top of the instrument panel; secure the other end to the top of the dash panel. The hook is to support the instrument panel in its rolled down position.

s. With the instrument panel in the rolled-down position, disconnect or remove the following items:

- The 2 instrument panel wire harness connector from the door jumper wire harness connectors located on the right bracket.
- The instrument panel wire harness

connector from the blower motor resistor connector receptacle.

- The temperature control cable flag retainer from the top of the heater/air conditioning housing assembly. Pull the cable core adjuster clip off the blend-air door lever.
- The demister duct flexible hose from the adapter on the top of the heater/air conditioning housing assembly.

t. With aid of an assistant, lift the instrument panel from the vehicle.

3. If equipped with air conditioning, perform the following procedure:

a. Discharge and recover the air conditioning system refrigerant.

b. Disconnect the refrigerant line from the evaporator inlet and outlet tubes. Plug the openings to prevent contamination.

4. Drain the cooling system into a clean container for reuse.

5. Disconnect the heater hoses from the heater core. Plug the openings.

6. Remove the 4 heater/air conditioning housing assembly-to-chassis nuts.

7. Remove the heater/air conditioning housing assembly-to-mounting brace nut; the nut is located on the passenger side of the vehicle.

8. Pull the heater/air conditioning

housing assembly rearward for the studs and drain tube to clear the dash panel hole.

9. Remove the heater/air conditioning housing assembly from the vehicle.

10. Remove the heater housing cover screws and the cover.

11. Remove the heater core from the heater/air conditioning housing assembly.

To install:

12. Install the heater core to the heater/air conditioning housing assembly.

13. Install the heater housing cover and the cover screws.

14. Install the heater/air conditioning housing assembly.

15. Install the heater/air conditioning housing assembly-to-mounting brace nut; the nut is located on the passenger side of the vehicle.

16. Install the 4 heater/air conditioning housing assembly-to-chassis nuts.

17. Connect the heater hoses to the heater core.

18. If equipped with air conditioning, perform the following procedure:

a. Using new gaskets, connect the refrigerant line to the evaporator inlet and outlet tubes.

b. Evacuate and charge the air conditioning system.

19. Install the instrument panel by performing the following procedures:

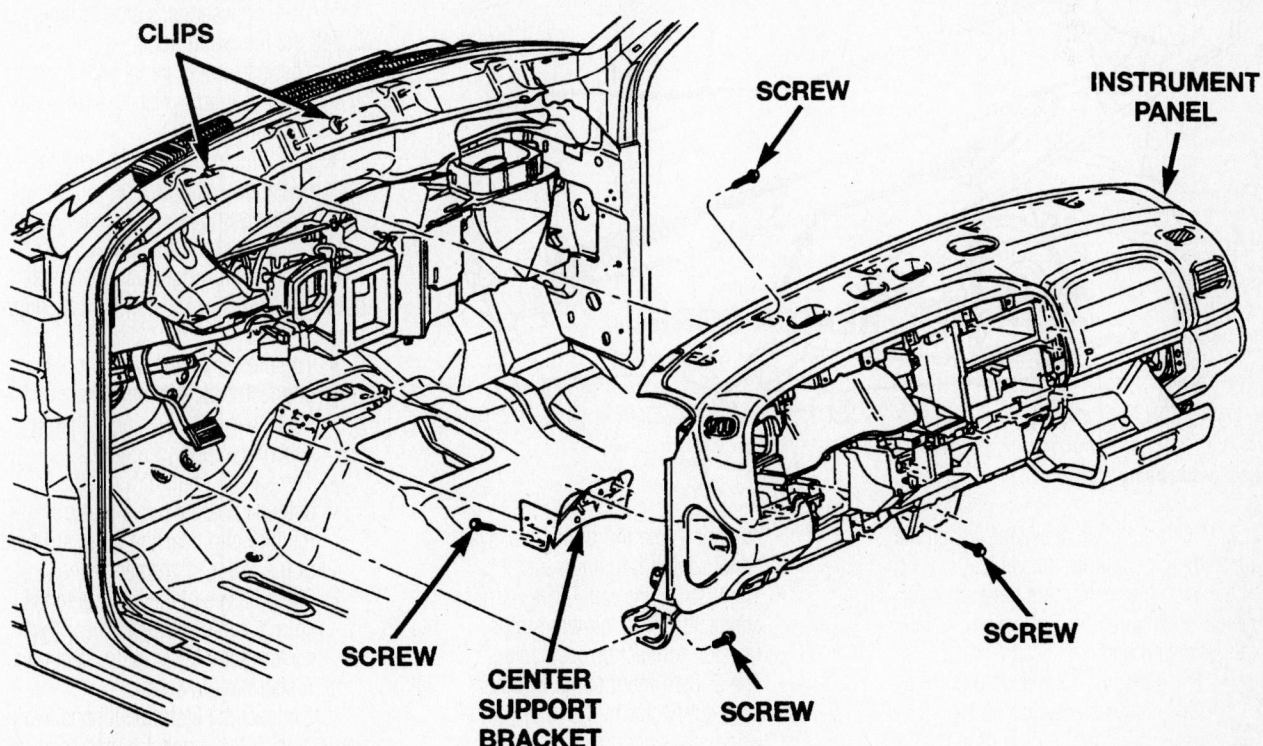

CLIPS **SCREW** **INSTRUMENT PANEL**

SCREW **CENTER SUPPORT BRACKET** **SCREW** **SCREW**

93113G92

View of the instrument panel assembly

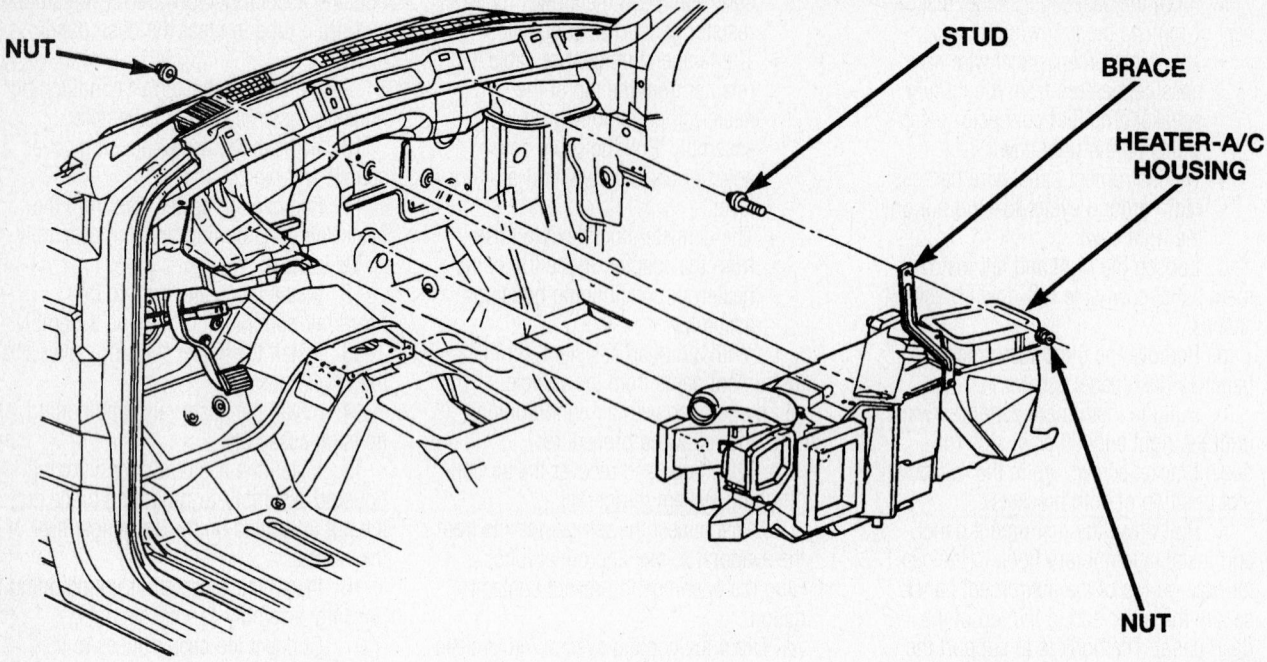

NUT

STUD

BRACE

HEATER-A/C HOUSING

NUT

93113G93

View of the heater/air conditioning assembly

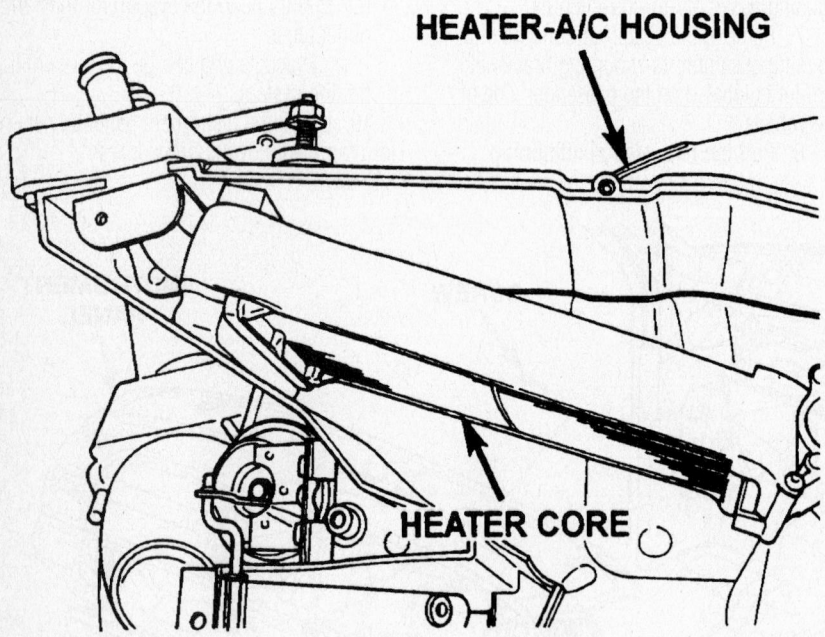

HEATER-A/C HOUSING

HEATER CORE

93113G94

View of the heater core

a. With aid of an assistant, install the instrument panel into the vehicle.

b. With the instrument panel in the rolled-down position, connect or install the following items:

- The demister duct flexible hose to the adapter on the top of the heater/air conditioning housing assembly.
- The temperature control cable flag retainer to the top of the heater/air conditioning housing assembly.

- The instrument panel wire harness connector to the blower motor resistor connector receptacle.
- The 2 instrument panel wire harness connector to the door jumper wire harness connectors located on the right bracket.

c. Move the instrument panel forward until the right and left cowl side roll-down bracket screws are in the roll-down slot position of both brackets.

d. Install the 5 top instrument panel-to-dash panel screws.

e. Install the right and left instrument panel cowl side roll-down bracket screws.

f. At the passenger's side, connect or install the following items:

- The instrument panel wire harness radio ground eyelet-to-stud nut on the right cowl.
- The 2 instrument panel wire harness connectors to the infinity speaker amplifier connector receptacles on the right cowl.

g. Working through the glove box opening, connect or install the following items:

- The blower motor wire harness connector located near the heater/air conditioning housing assembly support brace.
- The antenna half of the radio antenna coaxial cable to the retainer clip near the outboard side of the lower instrument panel.
- The 2 halves of the radio antenna coaxial cable connector near the center of the lower instrument panel.

h. Install the glove box.

i. Connect the instrument panel wiring harness-to-ACM connector receptacle.

j. Install the instrument panel wire harness ground screw located on the left side of the air bag control module (ACM) mount on the floor panel transmission tunnel.

k. Install the instrument panel center support bracket.

l. Under the driver's side of the instrument panel, connect or install the following items:

- The vacuum harness connector to the left side of the heater/air conditioning housing assembly.
- The stoplight switch connector.
- Engage the linkage rod to the handle lever. Install the plastic park brake release linkage rod-to-rear parking brake release handle lever.
- The 3 wire harness connectors to the 3 junction block connector receptacles located closest to the dash panel; 1 at the body wire harness and 2 at the headlight/dash wire harness.
- The 2 body wire harness connectors to the 2 instrument panel wire harness connector that secure the outboard side of the instrument panel bulkhead connector.
- Connect the connector. Install the screw at the center of the headlight/dash-to-instrument panel bulkhead wire harness connector.

m. Install the steering column assembly.

n. If equipped, connect the overdrive lockout switch harness connector.

o. Connect the driver's side air bag module wire harness connector.

p. Install the 2 hood latch release handle-to-instrument panel lower reinforcement screws and lower the release handle to the floor.

q. Install the steering column opening cover to the instrument panel.

r. Install the trim to the right and left cowl side inner panels.

s. Install the trim to the right and left door sills.

20. Refill the cooling system.

21. Connect the negative battery cable.

22. Run the engine to normal operating temperatures; then, check the climate control operation and check for leaks.

2004

✳✳ CAUTION

On vehicles equipped with airbags, disable the airbag system before attempting any steering wheel, steering column, or instrument panel component diagnosis or service. Disconnect and isolate the battery negative (ground) cable, then wait two minutes for the airbag system capacitor to discharge before performing further diagnosis or service. This is the only sure way to disable the airbag system. Failure to take the proper precautions could result in an accidental airbag deployment and possible personal injury.

1. Before servicing the vehicle, refer to the precautions in the beginning of this section.

2. Disconnect and isolate the negative battery cable.

3. Recover refrigerant from the refrigerant system.

4. Drain the engine cooling system.

5. Disconnect the liquid line from the evaporator inlet tube. Install plugs in, or tape over the opened liquid line fitting and the evaporator inlet tube.

6. Disconnect the accumulator from the evaporator outlet tube. Install plugs in, or tape over the opened refrigerant line fitting and the evaporator outlet tube.

7. Disconnect the heater hoses from the heater core tubes. Install plugs in, or tape over the opened heater core tubes.

8. Remove the four nuts that secure the HVAC housing to the dash panel in the engine compartment.

9. Remove the instrument panel from the vehicle.

10. Remove the two bolts that secure the HVAC housing to the dash panel in the passenger compartment.

11. Remove the HVAC housing from the vehicle.

12. Remove the foam seal from the heater core tubes.

13. Disassemble the HVAC housing to access the heater core.

14. Lift the heater core out of the HVAC housing.

To install:

15. Install the heater core into the HVAC housing.

16. Install the foam seal onto the heater core tubes.

17. Position the HVAC housing to the dash panel. Be certain that the evaporator condensate drain tube and the housing mounting studs are inserted into their correct mounting holes.

18. Install the two bolts that secure the HVAC housing to the dash panel in the passenger compartment. Tighten the bolts to 7 Nm (60 inch lbs.).

19. Install the instrument panel.

20. Install the four nuts that secure the HVAC housing to the dash panel in the engine compartment. Tighten the nuts to 7 Nm (60 inch lbs.).

21. Unplug or remove the tape from the heater core tubes. Connect the heater hoses to the heater core tubes and fill the engine cooling system.

22. Unplug or remove the tape from the accumulator refrigerant line fitting and the evaporator outlet tube. Install a new O-ring seal and connect the accumulator to the evaporator outlet tube.

23. Unplug or remove the tape from the liquid line fitting and the evaporator inlet tube. Install a new O-ring seal and connect the liquid line to the evaporator inlet tube.

24. Reconnect the negative battery cable.

25. Evacuate the refrigerant system.

26. Charge the refrigerant system.

27. Refill the cooling system.

Cylinder Head

REMOVAL & INSTALLATION

2.5L Engine

1. Before servicing the vehicle, refer to the precautions in the beginning of this section.

2. Drain the cooling system.

3. Remove or disconnect the following:

- Negative battery cable
- Crankcase Ventilation (CCV) hoses
- Air cleaner assembly
- Valve cover

➥**Keep valve train components in order for reassembly.**

- Rocker arms
- Pushrods
- Accessory drive belt
- A/C compressor and bracket, if equipped
- Power steering pump and bracket, if equipped
- Fuel line
- Combination manifold
- Thermostat housing coolant hoses
- Spark plugs
- Engine Coolant Temperature (ECT) sensor connector
- Cylinder head

To install:

✳✳ WARNING

Cylinder head bolts may only be reused one time. If reusing a cylinder head bolt, place a paint mark on the bolt after installation. If a cylinder head bolt has a paint mark, discard it and use a new bolt.

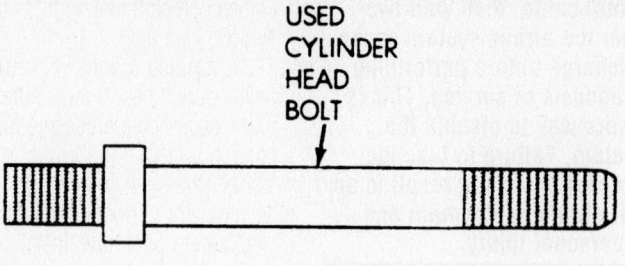

Fabricate 2 alignment dowels out of used cylinder head bolts—2.5L engine

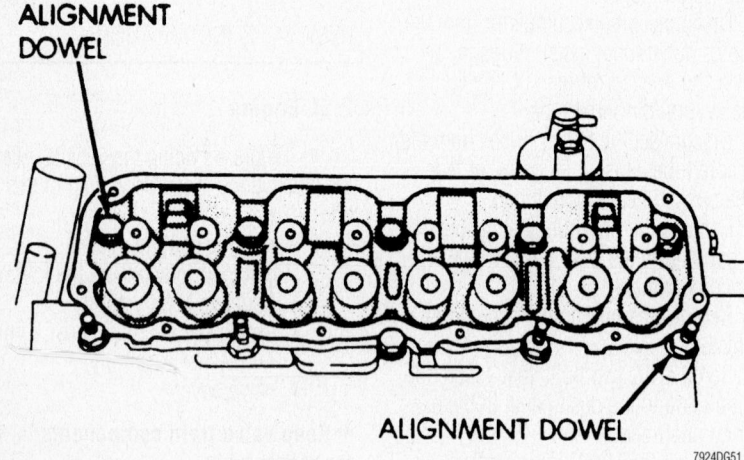

Alignment dowel locations—2.5L engine

4. Fabricate two alignment dowels from old cylinder head bolts. Cut the hex head off of the bolts, and cut a slot in each dowel to ease removal.

5. Install or connect the following:
- One dowel in bolt hole No. 8, and one dowel in bolt hole No. 10.
- Cylinder head and gasket.
- Cylinder head bolts except for No 8 and No 10. Coat the threads of bolt No. 7 with Loctite® 592 sealant.

6. Remove the alignment dowels and install the No. 8 and No. 10 head bolts.

✳✳ WARNING

During the final tightening sequence, bolt No. 7 will be tightened to a lower torque value than the rest of the bolts. Do not overtighten bolt No. 7.

7. Tighten the cylinder head bolts, in sequence, as follows:
- a. Step 1: 22 ft. lbs. (30 Nm).
- b. Step 2: 45 ft. lbs. (61 Nm).
- c. Step 3: 45 ft. lbs. (61 Nm).
- d. Step 4: Bolts 1–6 to 110 ft. lbs. (149 Nm).
- e. Step 5: Bolt 7 to 100 ft. lbs. (136 Nm).
- f. Step 6: Bolts 8–10 to 110 ft. lbs. (149 Nm).
- g. Step 7: Repeat steps 4, 5 and 6.

8. Install or connect the following:
- ECT sensor connector
- Spark plugs
- Thermostat housing coolant hoses
- Combination manifold
- Fuel line
- Power steering pump and bracket, if equipped
- A/C compressor and bracket, if equipped
- Accessory drive belt
- Pushrods and rocker arms in their original positions
- Valve cover
- Air cleaner assembly
- CCV hoses
- Negative battery cable

9. Fill the cooling system.
10. Start the engine and check for leaks.

3.7L Engine

LEFT SIDE

1. Before servicing the vehicle, refer to the precautions in the beginning of this section.

2. Drain the cooling system.

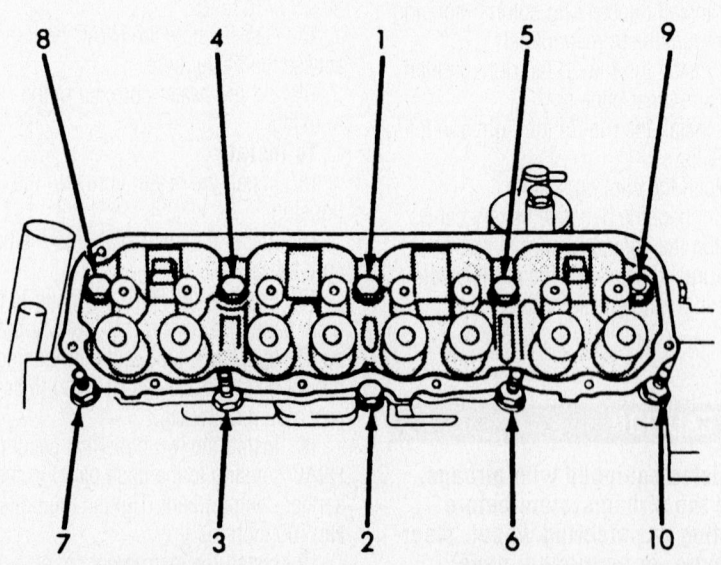

Cylinder head torque sequence—2.5L engine

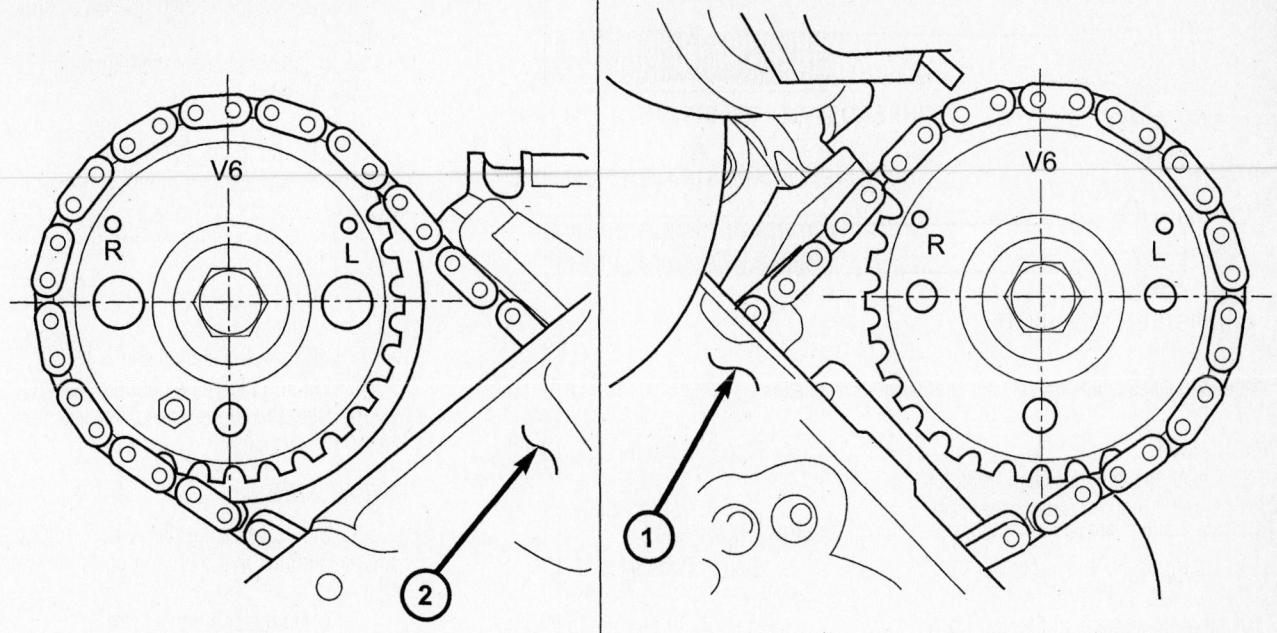

1 - LEFT CYLINDER HEAD
2 - RIGHT CYLINDER HEAD

9355PG04

Camshaft sprocket timing marks—3.7L

3. Properly relieve the fuel system pressure.

4. Remove or disconnect the following:
- Negative battery cable
- Exhaust Y-pipe
- Intake manifold
- Cylinder head cover
- Engine cooling fan and shroud
- Accessory drive belt
- Power steering pump

5. Rotate the crankshaft so that the crankshaft timing mark aligns with the Top Dead Center (TDC) mark on the front cover, and the **V6** marks on the camshaft sprockets are at 12 o'clock as shown.
- Crankshaft damper
- Front cover

6. Lock the secondary timing chain to the idler sprocket with Timing Chain Locking tool 8429.

7. Matchmark the secondary timing chain one link on each side of the V6 mark to the camshaft sprocket.
- Left secondary timing chain tensioner
- Cylinder head access plug
- Secondary timing chain guide
- Camshaft sprocket
- Cylinder head

➞The cylinder head is retained by twelve bolts. Four of the bolts are smaller and are at the front of the head.

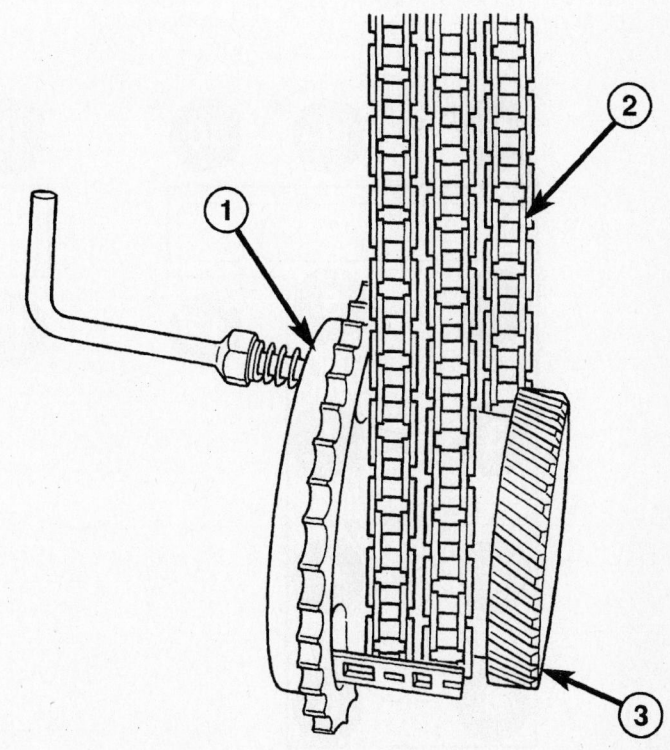

1 - SPECIAL TOOL 8429
2 - CAMSHAFT CHAIN
3 - CRANKSHAFT TIMING GEAR

9355PG05

Camshaft locking tool—3.7L

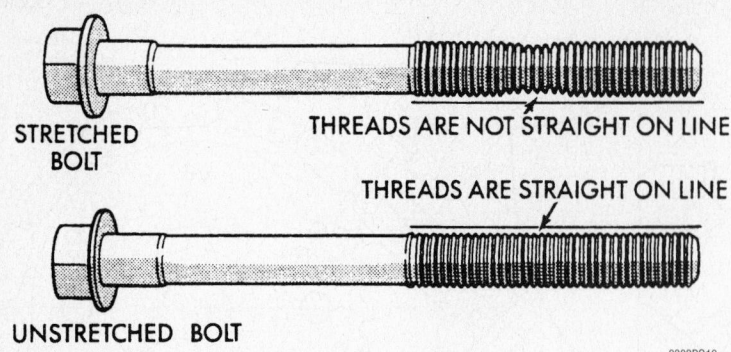

Examine the head bolts for signs of stretching—3.7L engine

To install:

8. Check the cylinder head bolts for signs of stretching and replace as necessary.

9. Lubricate the threads of the 11mm bolts with clean engine oil.

10. Coat the threads of the 8mm bolts with Mopar® Lock and Seal Adhesive.

11. Install the cylinder heads. Use new gaskets and tighten the bolts, in sequence, as follows:

a. Step 1: Bolts 1–8 to 20 ft. lbs. (27 Nm)

b. Step 2: Bolts 1–10 verify torque without loosening

c. Step 3: Bolts 9–12 to 10 ft. lbs. (14 Nm)

d. Step 4: Bolts 1–8 plus ¼ (90 degree) turn

e. Step 5: Bolts 9–12 to 19 ft. lbs. (26 Nm)

12. Install or connect the following:
- Camshaft sprocket. Align the secondary chain matchmarks and tighten the bolt to 90 ft. lbs. (122 Nm).
- Secondary timing chain guide
- Cylinder head access plug
- Secondary timing chain tensioner. Refer to the timing chain procedure in this section.

13. Remove the Timing Chain Locking tool.

14. Install or connect the following:
- Front cover
- Crankshaft damper. Torque the bolt to 130 ft. lbs. (175 Nm).
- Power steering pump
- Accessory drive belt
- Engine cooling fan and shroud
- Cover
- Intake manifold
- Exhaust Y-pipe
- Negative battery cable

15. Fill and bleed the cooling system.

16. Start the engine, check for leaks and repair if necessary.

RIGHT SIDE

1. Before servicing the vehicle, refer to the precautions in the beginning of this section.

2. Drain the cooling system.

3. Properly relieve the fuel system pressure.

4. Remove or disconnect the following:
- Negative battery cable
- Exhaust Y-pipe
- Intake manifold
- Valve cover
- Engine cooling fan and shroud

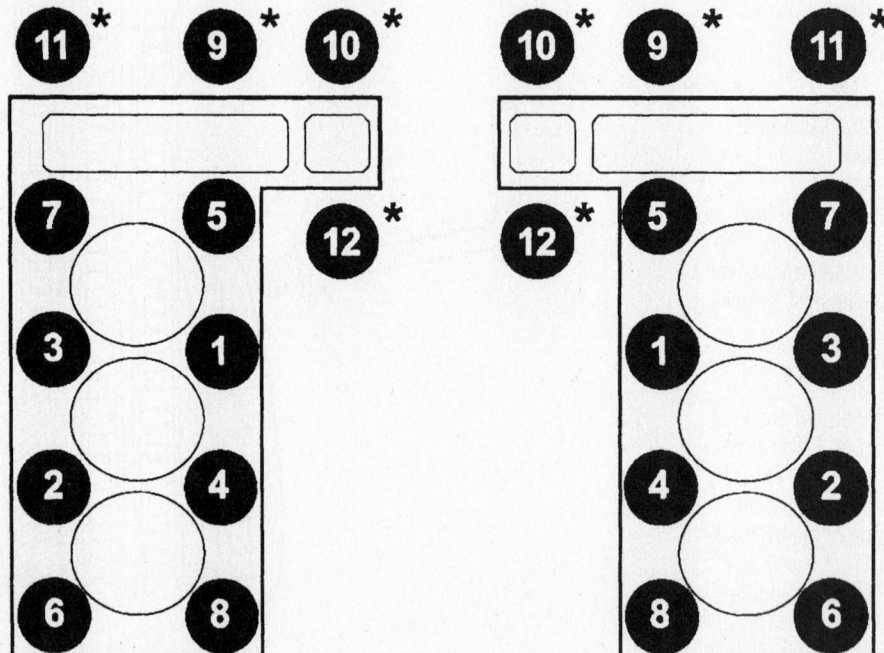

LEFT BANK   RIGHT BANK

* - INDICATES SEALANT ON THREADS

Cylinder head bolt torque sequence—3.7L

- Accessory drive belt
- Oil fill housing
- Power steering pump

5. Rotate the crankshaft so that the crankshaft timing mark aligns with the Top Dead Center (TDC) mark on the front cover, and the **V6** marks on the camshaft sprockets are at 12 o'clock as shown.

6. Remove or disconnect the following:
- Crankshaft damper
- Front cover

7. Lock the secondary timing chains to the idler sprocket with Timing Chain Locking tool 8429.

8. Matchmark the secondary timing chains to the camshaft sprockets.

9. Remove or disconnect the following:
- Secondary timing chain tensioners
- Cylinder head access plugs
- Secondary timing chain guides
- Camshaft sprockets
- Cylinder heads

➡**Each cylinder head is retained by eight 11mm bolts and four 8mm bolts.**

To install:

10. Check the cylinder head bolts for signs of stretching and replace as necessary.

11. Lubricate the threads of the 11mm bolts with clean engine oil.

12. Coat the threads of the 8mm bolts with Mopar® Lock and Seal Adhesive.

13. Install the cylinder heads. Use new gaskets and tighten the bolts, in sequence, as follows:

    a. Step 1: Bolts 1–8 to 20 ft. lbs. (27 Nm)

    b. Step 2: Bolts 1–10 verify torque without loosening

    c. Step 3: Bolts 9–12 to 10 ft. lbs. (14 Nm)

    d. Step 4: Bolts 1–8 plus ¼ (90 degree) turn

    e. Step 5: Bolts 9–12 to 19 ft. lbs. (26 Nm)

14. Install or connect the following:
- Camshaft sprockets. Align the secondary chain matchmarks and tighten the bolts to 90 ft. lbs. (122 Nm).

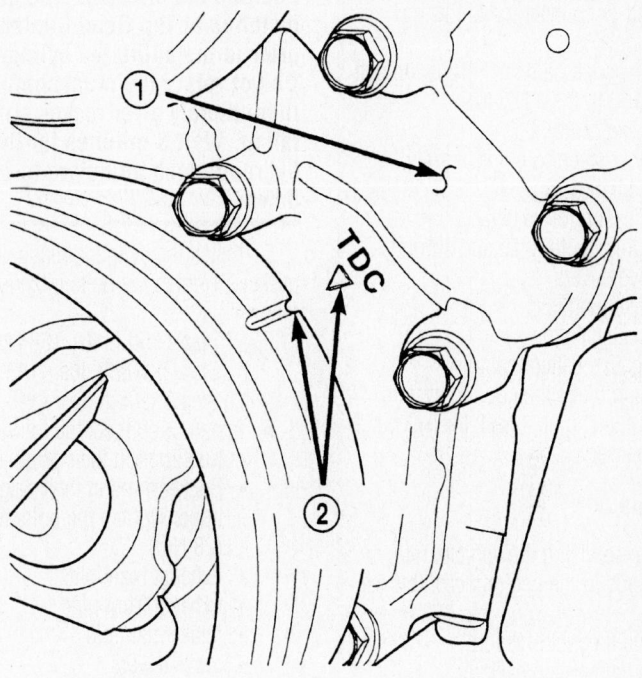

1 – TIMING CHAIN COVER
2 – CRANKSHAFT TIMING MARKS

9308PG04

Crankshaft timing marks—3.7L engine

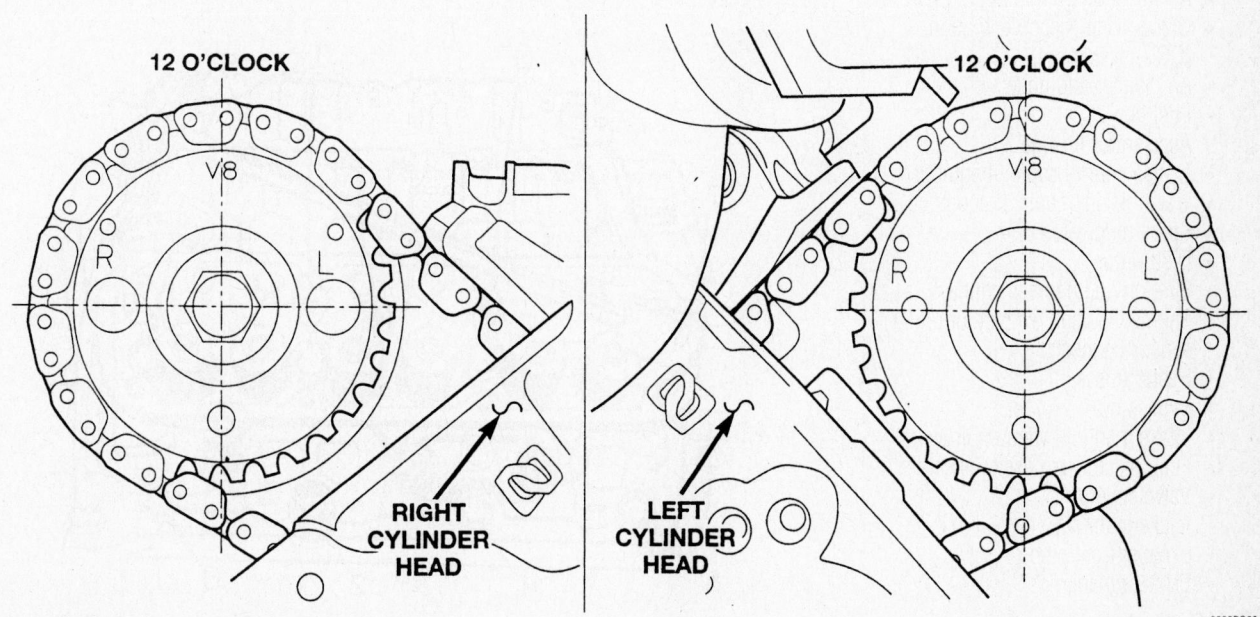

Camshaft positioning—4.7L engine

9302PG08

- Secondary timing chain guides
- Cylinder head access plugs
- Secondary timing chain tensioners. Refer to the timing chain procedure in this section.

15. Remove the Timing Chain Locking tool.

16. Install or connect the following:
- Front cover
- Crankshaft damper. Torque the bolt to 130 ft. lbs. (175 Nm).
- Rocker arms
- Power steering pump
- Oil fill housing
- Accessory drive belt
- Engine cooling fan and shroud
- Valve covers
- Intake manifold
- Exhaust Y-pipe
- Negative battery cable

17. Fill and bleed the cooling system.

18. Start the engine, check for leaks and repair if necessary.

3.9L Engine

1. Before servicing the vehicle, refer to the precautions in the beginning of this section.

2. Relieve the fuel pressure.

3. Drain the cooling system.

4. Remove or disconnect the following:
- Negative battery cable
- Accessory drive belt
- Alternator
- A/C compressor, if equipped
- Alternator and A/C compressor bracket
- Air injection pump, if equipped
- Closed Crankcase Ventilation (CCV) system
- Air cleaner and hose
- Fuel line
- Accelerator linkage
- Cruise control cable, if equipped
- Transmission cable, if equipped
- Spark plug wires
- Distributor
- Ignition coil harness connectors
- Engine Coolant Temperature (ECT) sensor connector
- Heater hoses
- Bypass hose
- Intake manifold vacuum lines
- Fuel injector harness connectors
- Valve covers
- Intake manifold
- Exhaust front pipe
- Exhaust manifolds

➡Keep all valvetrain components in order for assembly.

- Rocker arms
- Pushrods
- Cylinder heads

To install:

✳✳ WARNING

Position the crankshaft so that no piston is at Top Dead Center (TDC) prior to installing the cylinder heads. Do not rotate the crankshaft during or immediately after rocker arm installation. Wait 5 minutes for the hydraulic lash adjusters to bleed down.

5. Install the cylinder heads with new gaskets. Tighten the bolts in sequence as follows:
   a. Step 1: 50 ft. lbs. (68 Nm).
   b. Step 2: 105 ft. lbs. (143 Nm).
   c. Step 3: 105 ft. lbs. (143 Nm).

6. Install or connect the following:
- Pushrods in their original locations
- Rocker arms in their original locations. Tighten the bolts to 21 ft. lbs. (28 Nm).
- Exhaust manifolds
- Exhaust front pipe
- Intake manifold
- Valve covers
- Fuel injector harness connectors
- Intake manifold vacuum lines
- Bypass hose
- Heater hoses
- ECT sensor connector
- Ignition coil harness connectors
- Distributor
- Spark plug wires

- Transmission cable, if equipped
- Cruise control cable, if equipped
- Accelerator linkage
- Fuel line
- Air cleaner and hose
- CCV system
- Air injection pump, if equipped
- Alternator and A/C compressor bracket
- Alternator
- A/C compressor
- Accessory drive belt
- Negative battery cable

7. Fill the cooling system.

8. Start the engine and check for leaks.

4.7L Engine

LEFT SIDE

1. Before servicing the vehicle, refer to the precautions in the beginning of this section.

2. Drain the cooling system.

3. Remove or disconnect the following:
- Negative battery cable
- Exhaust pipe
- Intake manifold
- Cylinder head cover
- Fan shroud and fan
- Accessory drive belt
- Power steering pump

4. Rotate the crankshaft until the damper mark is aligned with the TDC mark. Verify that the V8 mark on the camshaft sprocket is at the 12 o'clock position.

5. Remove or disconnect the following:
- Vibration damper
- Timing chain cover

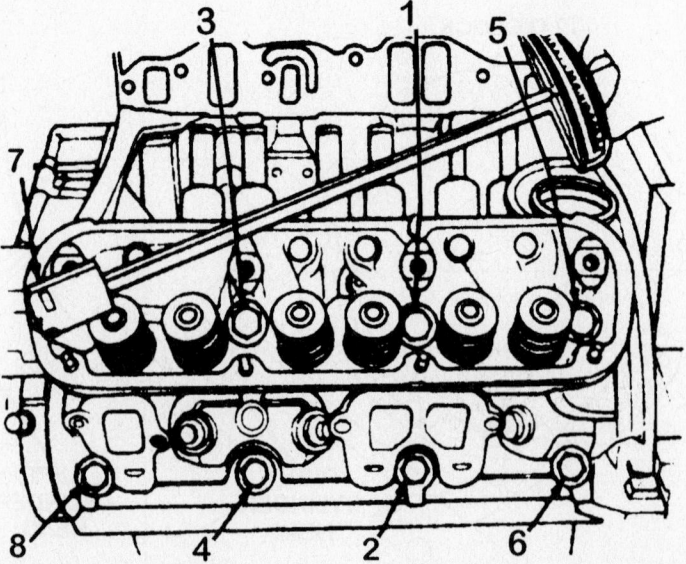

Cylinder head torque sequence—3.9L engine

7924DG06

6. Lock the secondary timing chains to the idler sprocket with tool 8515, or equivalent.

7. Mark the secondary timing chain, on link on either side of the V8 mark on the cam sprocket.

8. Remove the left side secondary chain tensioner.

9. Remove the cylinder head access plug.

10. Remove the chain guide.

11. Remove the camshaft sprocket.

➡ **There are 4 smaller bolts at the front of the head. Don't overlook these.**

12. Remove the head bolts and head.

※ WARNING

Don't lay the head on its sealing surface. Due to the design of the head gasket, any distortion to the head sealing surface will result in leaks.

13. Installation is the reverse of removal. Observe the following:
- Check the head bolts. If any necking is observed, replace the bolt.
- The 4 small bolts must be coated with sealer.
- The head bolts are tightened in the following sequence:

Step 1: Bolts 1-10 to 15 ft. lbs. (20 Nm)
Step 2: Bolts 1-10 to 35 ft. lbs. (47 Nm)
Step 3: Bolts 11-14 to 18 ft. lbs. (25 Nm)
Step 4: Bolts 1-10 90 degrees
Step 5: Bolts 11-14 to 22 ft. lbs.

RIGHT SIDE

1. Before servicing the vehicle, refer to the precautions in the beginning of this section.

2. Drain the cooling system.

3. Remove or disconnect the following:
- Negative battery cable
- Exhaust pipe
- Intake manifold
- Cylinder head cover
- Fan shroud and fan
- Oil filler housing
- Accessory drive belt

4. Rotate the crankshaft until the damper mark is aligned with the TDC mark. Verify that the V8 mark on the camshaft sprocket is at the 12 o'clock position.

5. Remove or disconnect the following:
- Vibration damper
- Timing chain cover

6. Lock the secondary timing chains to the idler sprocket with tool 8515, or equivalent.

7. Mark the secondary timing chain, on link on either side of the V8 mark on the cam sprocket.

8. Remove the left side secondary chain tensioner.

9. Remove the cylinder head access plug.

10. Remove the chain guide.

11. Remove the camshaft sprocket.

※ WARNING

Do not pry on the target wheel for any reason!

➡ **There are 4 smaller bolts at the front of the head. Don't overlook these.**

12. Remove the head bolts and head.

※ WARNING

Don't lay the head on its sealing surface. Due to the design of the head gasket, any distortion to the head sealing surface will result in leaks.

13. Installation is the reverse of removal. Observe the following:
- Check the head bolts. If any necking is observed, replace the bolt.
- The 4 small bolts must be coated with sealer.
- The head bolts are tightened in the following sequence:

Step 1: Bolts 1-10 to 15 ft. lbs. (20 Nm)
Step 2: Bolts 1-10 to 35 ft. lbs. (47 Nm)
Step 3: Bolts 11-14 to 18 ft. lbs. (25 Nm)
Step 4: Bolts 1-10 90 degrees
Step 5: Bolts 11-14 to 22 ft. lbs.

5.9L Engines

1. Before servicing the vehicle, refer to the precautions in the beginning of this section.

2. Drain the cooling system.

3. Remove or disconnect the following:
- Negative battery cable
- Accessory drive belt
- Alternator
- A/C compressor, if equipped
- Air injection pump, if equipped
- Air cleaner assembly
- Closed Crankcase Ventilation (CCV) system
- Evaporative emissions control system
- Fuel line
- Accelerator linkage
- Cruise control cable
- Transmission cable
- Distributor cap and wires
- Ignition coil wiring
- Engine Coolant Temperature (ECT) sensor connector
- Heater hoses
- Bypass hose
- Upper radiator hose
- Intake manifold
- Valve covers

➡ **Keep valvetrain components in order for reassembly.**

- Rocker arms
- Pushrods
- Exhaust manifolds
- Spark plugs
- Cylinder heads

To install:

※ WARNING

Position the crankshaft so that no piston is at Top Dead Center (TDC) prior to installing the cylinder heads. Do not rotate the crankshaft during or immediately after rocker arm installation. Wait 5 minutes for the hydraulic lash adjusters to bleed down.

♦ INDICATES SEALER APPLIED TO THREADS

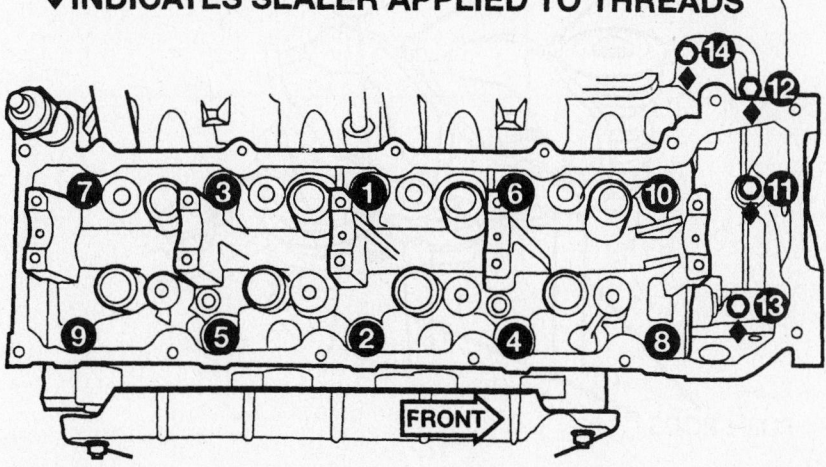

9355PG01

Cylinder head tightening sequence—4.7L

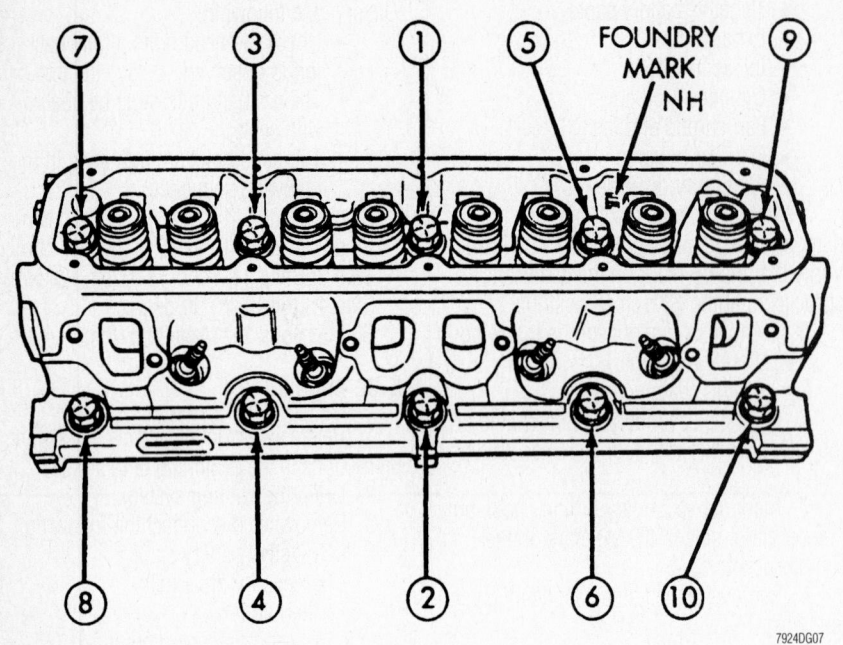

Cylinder head torque sequence—5.9L engines

7924DG07

4. Install the cylinder heads. Use new gaskets and tighten the bolts in sequence as follows:
   a. Step 1: 50 ft. lbs. (68 Nm).
   b. Step 2: 105 ft. lbs. (143 Nm).
   c. Step 3: 105 ft. lbs. (143 Nm).
5. Install or connect the following:
   • Spark plugs
   • Exhaust manifolds
   • Pushrods and rocker arms in their original positions
   • Valve covers
   • Intake manifold
   • Upper radiator hose
   • Bypass hose
   • Heater hoses
   • ECT sensor connector
   • Ignition coil wiring
   • Distributor cap and wires
   • Transmission cable
   • Cruise control cable
   • Accelerator linkage
   • Fuel line
   • Evaporative emissions control system
   • CCV system
   • Air cleaner assembly
   • A/C compressor, if equipped
   • Air injection pump, if equipped
   • Alternator
   • Accessory drive belt
   • Negative battery cable
6. Fill the cooling system.
7. Start the engine and check for leaks.

Rocker Arms/Shafts

REMOVAL & INSTALLATION

2.5L Engine

1. Before servicing the vehicle, refer to the precautions in the beginning of this section.
2. Remove or disconnect the following:
   • Negative battery cable
   • Accelerator cable
   • Transmission cable, if equipped
   • Cruise control cable, if equipped
   • Valve cover
   • Rocker arm bolts, loosen them evenly to avoid damaging the alignment bridges
   • Rocker arms

➡**Keep valvetrain components in order for reassembly.**

To install:
3. Install or connect the following:
   • Rocker arms, pivots and bridges in their original positions. Tighten the

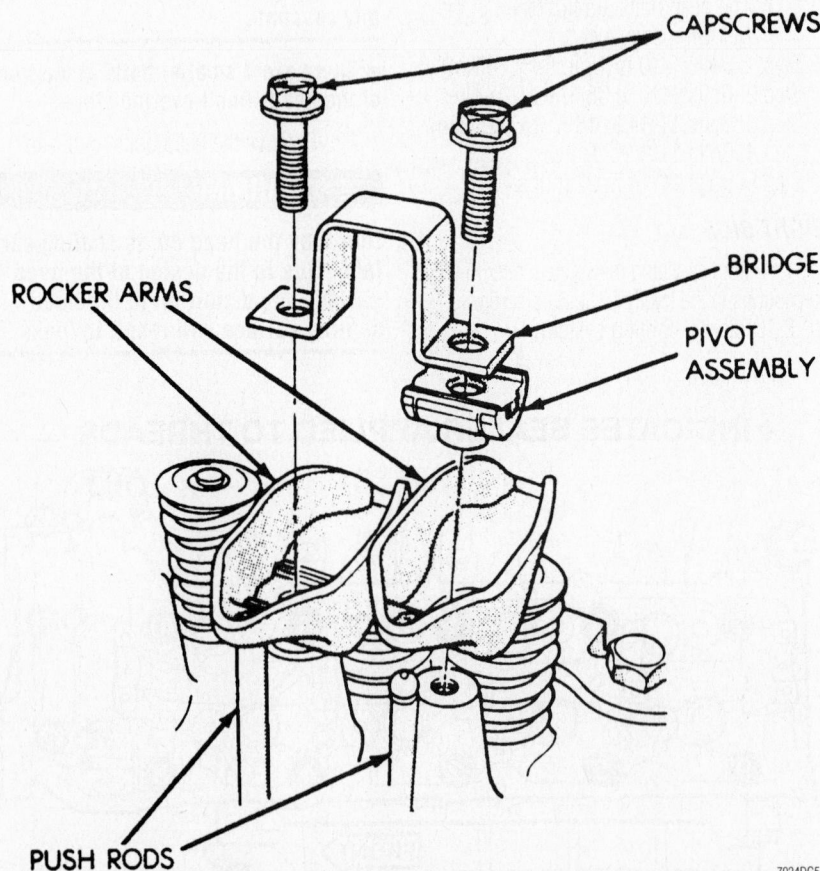

Exploded view of the rocker arm mounting—2.5L engine

7924DG54

bolts for each bridged pair one turn at a time to 21 ft. lbs. (28 Nm).
- Valve cover
- Cruise control cable, if equipped
- Transmission cable, if equipped
- Accelerator cable
- Negative battery cable

3.9L Engine

1. Before servicing the vehicle, refer to the precautions in the beginning of this section.
2. Remove or disconnect the following:
   - Negative battery cable
   - Valve covers
   - Rocker arms

➡**Keep all valvetrain components in order for assembly.**

To install:

3. Rotate the crankshaft so that the **V6** mark on the crankshaft damper aligns with the timing mark on the front cover. The **V6** mark is located 147 degrees **AFTER** Top Dead Center (TDC).
4. Install the rocker arms in their original positions and tighten the bolts to 21 ft. lbs. (28 Nm).

✳✳ CAUTION

Do not rotate the crankshaft during or immediately after rocker arm installation. Wait 5 minutes for the hydraulic lash adjusters to bleed down.

5. Install or connect the following:
   - Valve covers
   - Negative battery cable

3.7L and 4.7L Engines

1. Before servicing the vehicle, refer to the precautions in the beginning of this section.
2. Remove or disconnect the following:
   - Negative battery cable
   - Valve covers
3. Rotate the crankshaft so that the piston of the cylinder to be serviced is at Bottom Dead Center (BDC) and both valves are closed.
4. Use special tool 8516 to depress the valve and remove the rocker arm.
5. Repeat for each rocker arm to be serviced.

➡**Keep valvetrain components in order for reassembly.**

To install:

6. Rotate the crankshaft so that the piston of the cylinder to be serviced is at BDC.

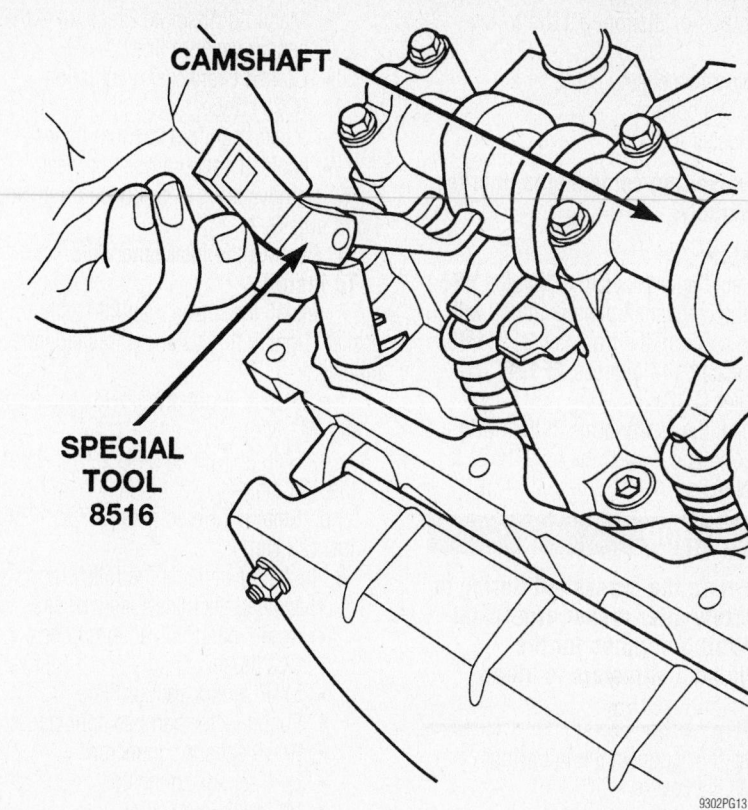

Rocker arm service—4.7L engine

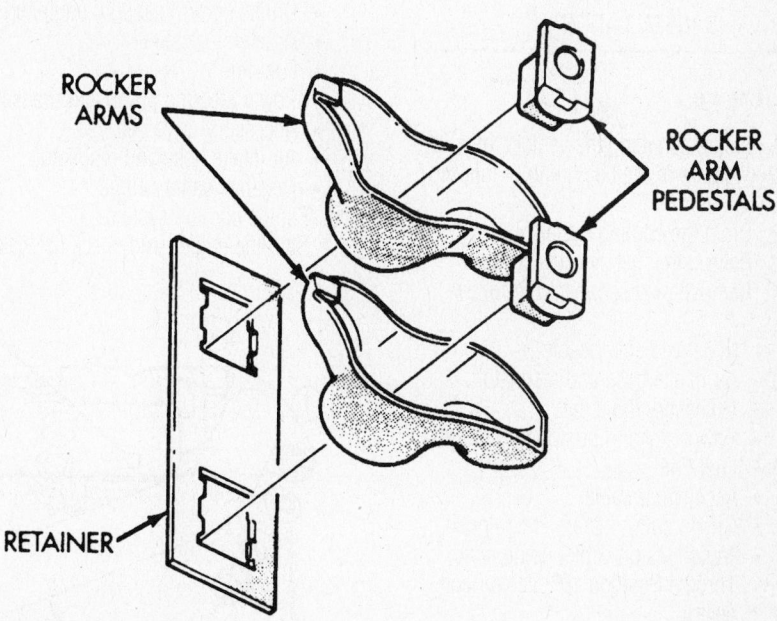

Exploded view of the rocker arm assembly—5.9L engine

7. Compress the valve spring and install each rocker arm in its original position.
8. Repeat for each rocker arm to be installed.
9. Install or connect the following:
   - Cylinder head cover
   - Negative battery cable

5.9L Engines

1. Before servicing the vehicle, refer to the precautions in the beginning of this section.

2. Remove or disconnect the following:
- Negative battery cable
- Valve covers
- Rocker arms

➡ **Keep valvetrain components in order for reassembly.**

To install:

3. Rotate the crankshaft so that the **V8** mark on the crankshaft damper aligns with the timing mark on the front cover. The **V8** mark is located 147 degrees **AFTER** Top Dead Center (TDC).

4. Install the rocker arms in their original positions and tighten the bolts to 21 ft. lbs. (28 Nm).

✳✳ CAUTION

Do not rotate the crankshaft during or immediately after rocker arm installation. Wait 5 minutes for the hydraulic lash adjusters to bleed down.

5. Install or connect the following:
- Valve covers
- Negative battery cable

Intake Manifold

REMOVAL & INSTALLATION

2.5L Engine

1. Before servicing the vehicle, refer to the precautions in the beginning of this section.
2. Drain the cooling system.
3. Relieve the fuel system pressure.
4. Remove or disconnect the following:
- Negative battery cable
- Air intake hose and resonator
- Accessory drive belt
- Power steering pump and brackets
- Fuel line
- Accelerator cable
- Cruise control cable, if equipped
- Transmission cable, if equipped
- Throttle Position (TP) sensor connector
- Intake Air Temperature (IAT) sensor connector
- Idle Air Control (IAC) valve connector
- Engine Coolant Temperature (ECT) sensor connector
- Heated Oxygen (HO2S) sensor connector
- Fuel injector harness connectors

- Manifold Absolute Pressure (MAP) sensor vacuum line
- Closed Crankcase Ventilation (CCV) hose
- Intake manifold vacuum hoses
- Molded vacuum hose harness

5. Remove bolts 2–5. Loosen bolt No. 1 and nuts 6–7.

6. Remove the intake manifold.

To install:

7. Install the intake manifold with a new gasket. Tighten the fasteners in sequence as follows:
   a. Step 1: Tighten bolt No. 1 to 30 ft. lbs. (41 Nm)
   b. Step 2: Tighten bolts 2–5 to 23 ft. lbs. (31 Nm)
   c. Step 3: Tighten nuts 6–7 to 17 ft. lbs. (23 Nm)

8. Install or connect the following:
- Molded vacuum hose harness
- Intake manifold vacuum hoses
- CCV hose
- MAP sensor vacuum line
- Fuel injector harness connectors
- HO2S sensor connector
- ECT sensor connector
- IAC valve connector
- IAT sensor connector
- TP sensor connector
- Transmission cable, if equipped
- Cruise control cable, if equipped
- Accelerator cable
- Fuel line
- Power steering pump and brackets
- Accessory drive belt
- Air intake hose and resonator
- Negative battery cable

9. Fill the cooling system.
10. Start the engine and check for leaks.

3.7L Engine

1. Before servicing the vehicle, refer to the precautions in the beginning of this section.
2. Drain the cooling system.
3. Remove or disconnect the following:
- Negative battery cable
- Air cleaner assembly
- Accelerator cable
- Cruise control cable
- Manifold Absolute Pressure (MAP) sensor connector
- Intake Air Temperature (IAT) sensor connector
- Throttle Position (TP) sensor connector
- Idle Air Control (IAC) valve connector
- Engine Coolant Temperature (ECT) sensor
- Positive Crankcase Ventilation (PCV) valve and hose
- Canister purge vacuum line
- Brake booster vacuum line
- Cruise control servo hose
- Accessory drive belt
- Alternator
- A/C compressor
- Engine ground straps
- Ignition coil towers
- Oil dipstick tube
- Fuel line
- Fuel supply manifold
- Throttle body and mounting bracket
- Cowl seal
- Right engine lifting stud
- Intake manifold. Remove the fasteners in reverse of the tightening sequence.

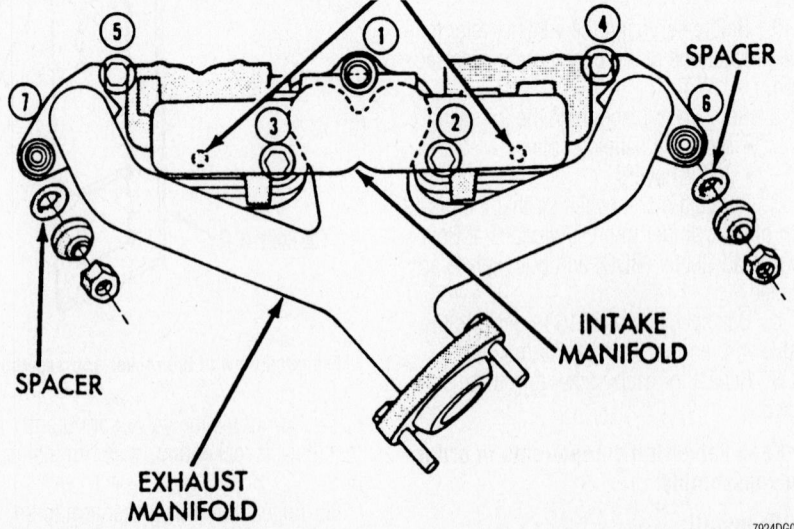

Manifold torque sequence—2.5L engine

7924DG09

To install:

4. Install or connect the following:
- Intake manifold using new gaskets. Tighten the bolts, in sequence, to 105 inch lbs. (12 Nm).
- Right engine lifting stud
- Cowl seal
- Throttle body and mounting bracket
- Fuel supply manifold
- Fuel line
- Oil dipstick tube
- Ignition coil towers
- Engine ground straps
- A/C compressor
- Alternator
- Accessory drive belt
- Cruise control servo hose
- Brake booster vacuum line
- Canister purge vacuum line
- PCV valve and hose
- ECT sensor
- IAC valve connector
- TP sensor connector
- IAT sensor connector
- MAP sensor connector
- Cruise control cable
- Accelerator cable
- Air cleaner assembly
- Negative battery cable
5. Fill the cooling system.
6. Start the engine and check for leaks.

3.9L Engine

1. Before servicing the vehicle, refer to the precautions in the beginning of this section.
2. Drain the cooling system.
3. Remove or disconnect the following:
- Negative battery cable
- Accessory drive belt
- Alternator
- A/C compressor
- Alternator and A/C compressor bracket
- Air cleaner assembly
- Fuel line
- Fuel supply manifold
- Accelerator cable
- Transmission cable
- Cruise control cable
- Distributor cap and wires
- Ignition coil wiring
- Engine Coolant Temperature (ECT) sensor connector
- Heater hose
- Upper radiator hose
- Bypass hose
- Closed Crankcase Ventilation (CCV) system
- Evaporative emissions system
- Intake manifold

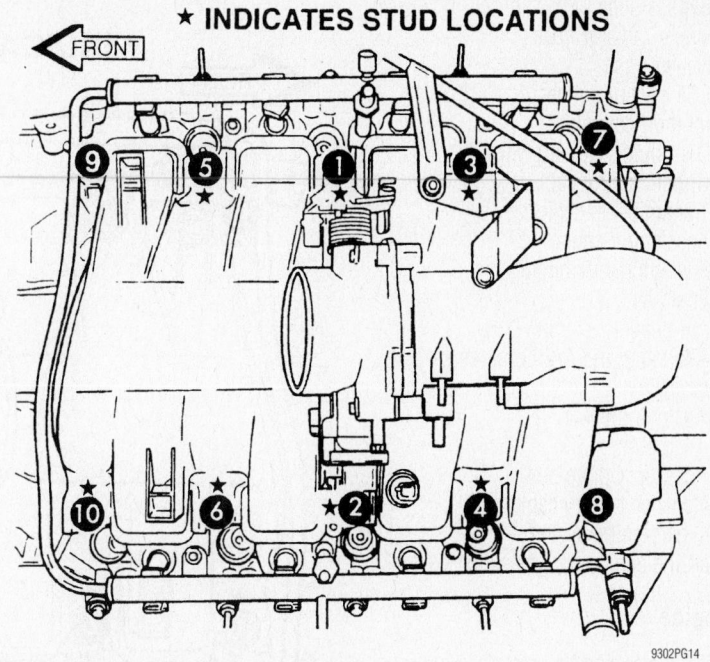

★ **INDICATES STUD LOCATIONS**

Intake manifold torque sequence—3.7L engine

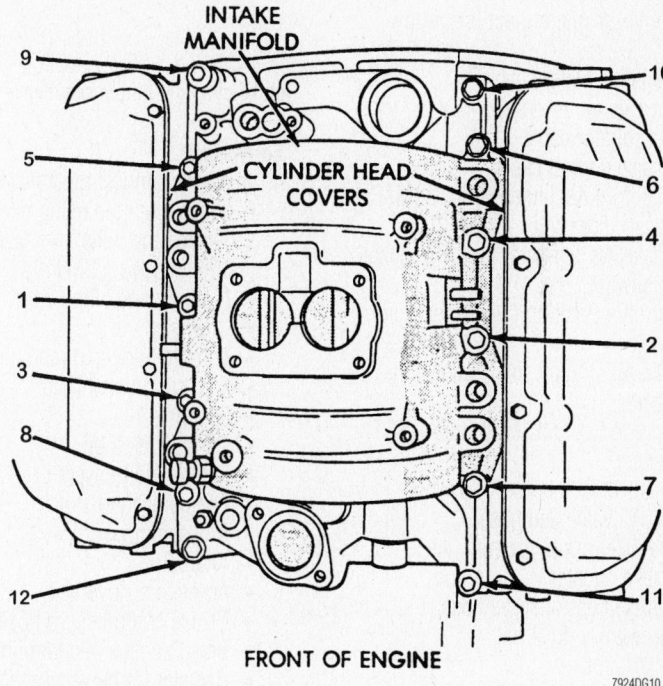

Intake manifold torque sequence—3.9L engine

To install:

4. Install the intake manifold. Use a new gasket and tighten the bolts in sequence as follows:

a. Step 1: Bolts 1–2 to 72 inch lbs. (8 Nm) using 12 inch lb. (1.4 Nm) increments.

b. Step 2: Bolts 3–12 to 72 inch lbs. (8 Nm).

c. Step 3: Bolts 1–12 to 72 inch lbs. (8 Nm).

d. Step 4: Bolts 1–12 to 12 ft. lbs. (16 Nm).

e. Step 5: Bolts 1–12 to 12 ft. lbs. (16 Nm).

5. Install or connect the following:
- Evaporative emissions system
- CCV system

- Bypass hose
- Upper radiator hose
- Heater hose
- ECT sensor connector
- Ignition coil wiring
- Distributor cap and wires
- Cruise control cable
- Transmission cable
- Accelerator cable
- Fuel supply manifold
- Fuel line
- Air cleaner assembly
- Alternator and A/C compressor bracket
- A/C compressor
- Alternator
- Accessory drive belt
- Negative battery cable
6. Fill the cooling system.
7. Start the engine and check for leaks.

4.7L Engine

1. Before servicing the vehicle, refer to the precautions in the beginning of this section.
2. Drain the cooling system.
3. Remove or disconnect the following:

- Negative battery cable
- Air cleaner assembly
- Accelerator cable
- Cruise control cable
- Manifold Absolute Pressure (MAP) sensor connector
- Intake Air Temperature (IAT) sensor connector
- Throttle Position (TP) sensor connector
- Idle Air Control (IAC) valve connector
- Engine Coolant Temperature (ECT) sensor
- Positive Crankcase Ventilation (PCV) valve and hose
- Canister purge vacuum line
- Brake booster vacuum line
- Cruise control servo hose
- Accessory drive belt
- Alternator
- A/C compressor
- Engine ground straps
- Ignition coil towers
- Oil dipstick tube
- Fuel line
- Fuel supply manifold
- Throttle body and mounting bracket
- Cowl seal
- Right engine lifting stud
- Intake manifold. Remove the fasteners in reverse of the tightening sequence.

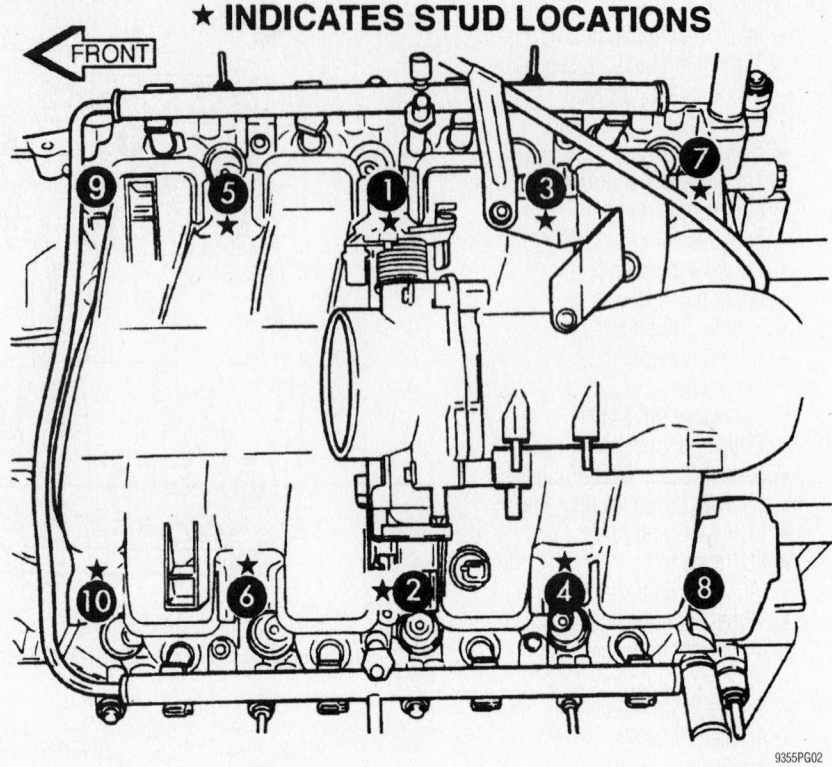

★ **INDICATES STUD LOCATIONS**

Intake manifold torque sequence—4.7L

To install:
4. Install or connect the following:
- Intake manifold using new gaskets. Tighten the bolts, in sequence, to 105 inch lbs. (12 Nm).
- Right engine lifting stud
- Cowl seal
- Throttle body and mounting bracket
- Fuel supply manifold
- Fuel line
- Oil dipstick tube
- Ignition coil towers
- Engine ground straps
- A/C compressor
- Alternator
- Accessory drive belt
- Cruise control servo hose
- Brake booster vacuum line
- Canister purge vacuum line
- PCV valve and hose
- ECT sensor
- IAC valve connector
- TP sensor connector
- IAT sensor connector
- MAP sensor connector
- Cruise control cable
- Accelerator cable
- Air cleaner assembly
- Negative battery cable
5. Fill the cooling system.
6. Start the engine and check for leaks.

5.9L Engines

1. Before servicing the vehicle, refer to the precautions in the beginning of this section.
2. Drain the cooling system.
3. Remove or disconnect the following:

- Negative battery cable
- Accessory drive belt
- Alternator
- A/C compressor
- Alternator and A/C compressor bracket
- Air cleaner assembly
- Fuel line
- Fuel supply manifold
- Accelerator cable
- Transmission cable
- Cruise control cable
- Distributor cap and wires
- Ignition coil wiring
- Engine Coolant Temperature (ECT) sensor connector
- Heater hose
- Upper radiator hose
- Bypass hose
- Closed Crankcase Ventilation (CCV) system
- Evaporative emissions system
- Intake manifold

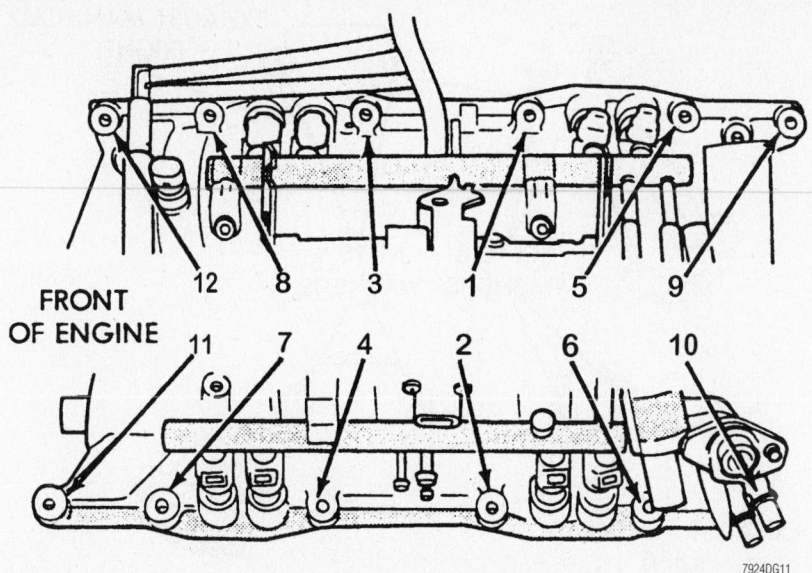

Intake manifold torque sequence—5.9L engines

To install:

4. Install the intake manifold. Use a new gasket and tighten the bolts in sequence as follows:

    a. Step 1: Bolts 1–4 to 72 inch lbs. (8 Nm) using 12 inch lb. (1.4 Nm) increments.

    b. Step 2: Bolts 5–12 to 72 inch lbs. (8 Nm).

    c. Step 3: Bolts 1–12 to 72 inch lbs. (8 Nm).

    d. Step 4: Bolts 1–12 to 12 ft. lbs. (16 Nm).

    e. Step 5: Bolts 1–12 to 12 ft. lbs. (16 Nm).

5. Install or connect the following:
- Evaporative emissions system
- CCV system
- Bypass hose
- Upper radiator hose
- Heater hose
- ECT sensor connector
- Ignition coil wiring
- Distributor cap and wires
- Cruise control cable
- Transmission cable
- Accelerator cable
- Fuel supply manifold
- Fuel line
- Air cleaner assembly
- Alternator and A/C compressor bracket
- A/C compressor
- Alternator
- Accessory drive belt
- Negative battery cable

6. Fill the cooling system.

7. Start the engine and check for leaks.

Exhaust Manifold

REMOVAL & INSTALLATION

2.5L Engine

1. Before servicing the vehicle, refer to the precautions in the beginning of this section.

2. Remove or disconnect the following:

- Negative battery cable
- Exhaust front pipe
- Intake manifold
- Exhaust manifold

To install:

3. Install or connect the following:
- Exhaust manifold
- Intake manifold
- Exhaust front pipe
- Negative battery cable

4. Start the engine and check for leaks.

3.7L Engines

1. Before servicing the vehicle, refer to the precautions in the beginning of this section.

2. Remove or disconnect the following:

- Negative battery cable
- Exhaust manifold heat shields
- Exhaust Gas Recirculation (EGR) tube
- Exhaust Y-pipe
- Exhaust manifolds

To install:

➡If the exhaust manifold studs came out with the nuts when removing the exhaust manifolds, replace them with new studs.

3. Install or connect the following:
- Exhaust manifolds. Torque the fasteners to 20 ft. lbs. (27 Nm), starting with the center nuts and work out to the ends.
- Exhaust Y-pipe
- EGR tube
- Exhaust manifold heat shields
- Negative battery cable

4. Start the engine, check for leaks and repair if necessary.

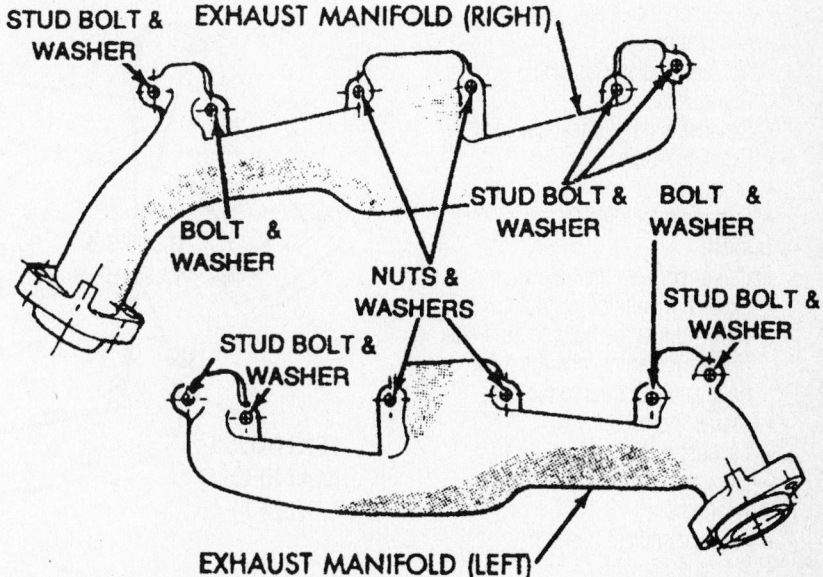

Exhaust manifold fastener locations—3.7L engines

3.9L Engine

1. Before servicing the vehicle, refer to the precautions in the beginning of this section.
2. Remove or disconnect the following:
   - Negative battery cable
   - Heated Oxygen (HO$_2$S) sensor connectors
   - Exhaust manifold heat shields
   - Exhaust front pipe
   - Exhaust manifolds

To install:

➡ **If the exhaust manifold studs came out with the nuts when removing the exhaust manifolds, replace them with new studs.**

3. Install or connect the following:
   - Exhaust manifolds. Tighten the fasteners to 25 ft. lbs. (34 Nm), starting with the center fasteners and working out to the ends.
   - Exhaust front pipe
   - Exhaust manifold heat shields
   - HO$_2$S sensor connectors
   - Negative battery cable
4. Start the engine and check for leaks.

4.7L Engine

1. Before servicing the vehicle, refer to the precautions in the beginning of this section.
2. Drain the cooling system.
3. Remove or disconnect the following:
   - Battery
   - Power distribution center
   - Battery tray
   - Windshield washer fluid bottle
   - Air cleaner assembly
   - Accessory drive belt
   - A/C compressor
   - A/C accumulator bracket
   - Heater hoses
   - Exhaust manifold heat shields
   - Exhaust Y-pipe
   - Starter motor
   - Exhaust manifolds

To install:

4. Install or connect the following:
   - Exhaust manifolds, using new gaskets. Tighten the bolts to 18 ft. lbs. (25 Nm), starting with the inner bolts and work out to the ends.
   - Starter motor
   - Exhaust Y-pipe
   - Exhaust manifold heat shields
   - Heater hoses
   - A/C accumulator bracket
   - A/C compressor
   - Accessory drive belt
   - Air cleaner assembly

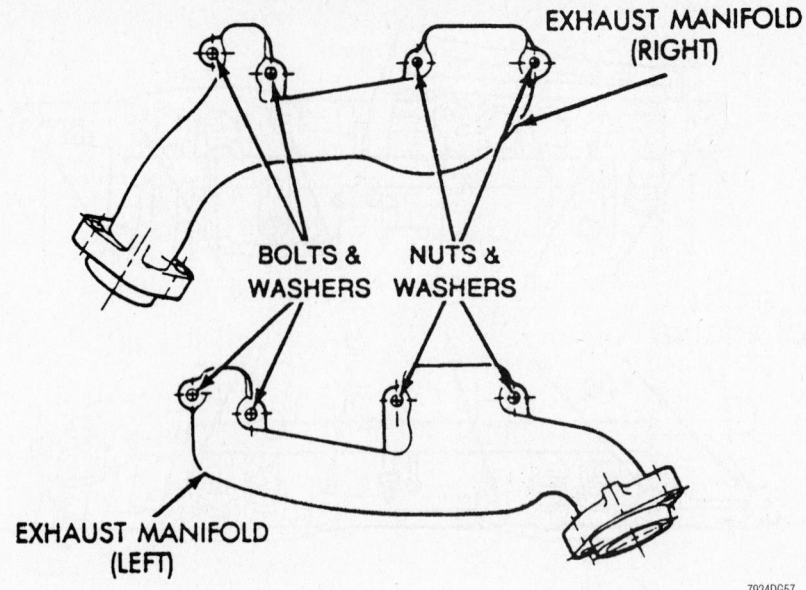

Stud and bolt locations—3.9L engine

   - Windshield washer fluid bottle
   - Battery tray
   - Power distribution center
   - Battery
5. Fill the cooling system.
6. Start the engine and check for leaks.

5.9L Engines

1. Before servicing the vehicle, refer to the precautions in the beginning of this section.
2. Remove or disconnect the following:
   - Negative battery cable
   - Exhaust manifold heat shields
   - Exhaust Gas Recirculation (EGR) tube

   - Exhaust Y-pipe
   - Exhaust manifolds

To install:

➡ **If the exhaust manifold studs came out with the nuts when removing the exhaust manifolds, replace them with new studs.**

3. Install or connect the following:
   - Exhaust manifolds. Tighten the fasteners to 20 ft. lbs. (27 Nm), starting with the center nuts and work out to the ends.
   - Exhaust Y-pipe
   - EGR tube
   - Exhaust manifold heat shields

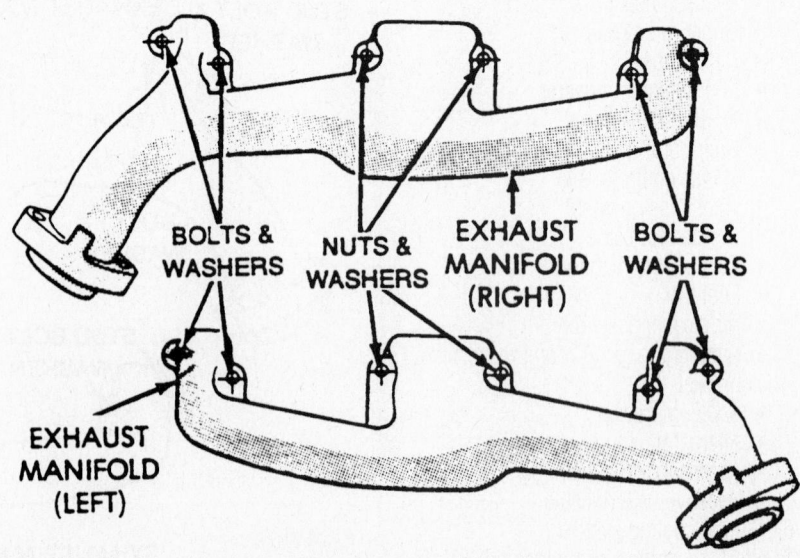

Exhaust manifold fastener locations—5.9L engines

• Negative battery cable
4. Fill the cooling system.
5. Start the engine and check for leaks.

Camshaft and Valve Lifters

REMOVAL & INSTALLATION

2.5L Engine

1. Before servicing the vehicle, refer to the precautions in the beginning of this section.
2. Drain the cooling system.
3. Recover the A/C refrigerant, if equipped with air conditioning.
4. Remove or disconnect the following:

- Negative battery cable
- Grille, if necessary
- Radiator
- A/C condenser, if equipped
- Distributor
- Valve cover

➡**Keep all valvetrain components in order for assembly.**

- Rocker arms and pushrods
- Hydraulic valve tappets
- Accessory drive belt
- Crankshaft damper
- Front cover
- Timing chain and gears
- Camshaft

To install:

➡**If the camshaft sprocket appears to have been rubbing against the cover, check the oil pressure relief holes in the rear cam journal for debris.**

5. Lubricate the camshaft with clean engine oil.
6. Install or connect the following:

- Camshaft
- Timing chain and gears
- Front cover
- Crankshaft damper
- Accessory drive belt
- Hydraulic valve tappets
- Rocker arms and pushrods
- Valve cover
- Distributor
- A/C condenser, if equipped
- Radiator
- Grille, if removed
- Negative battery cable

7. Fill the cooling system.
8. Recharge the A/C system, if equipped.
9. Start the engine and check for leaks.

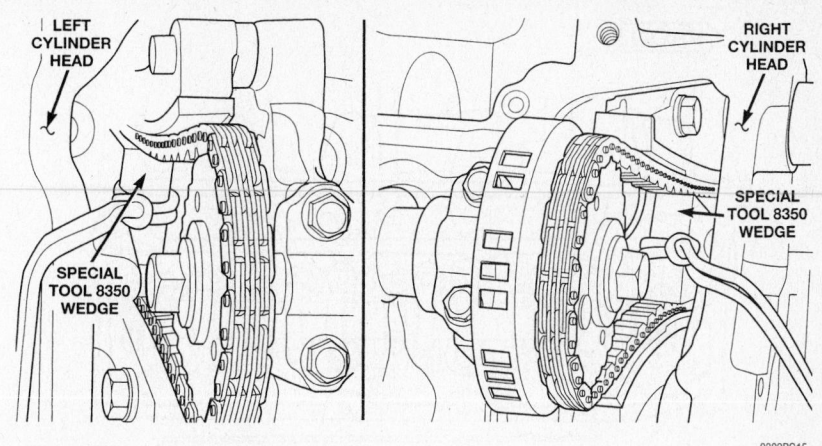

Chain Tensioner Retaining Wedges—3.7L engine

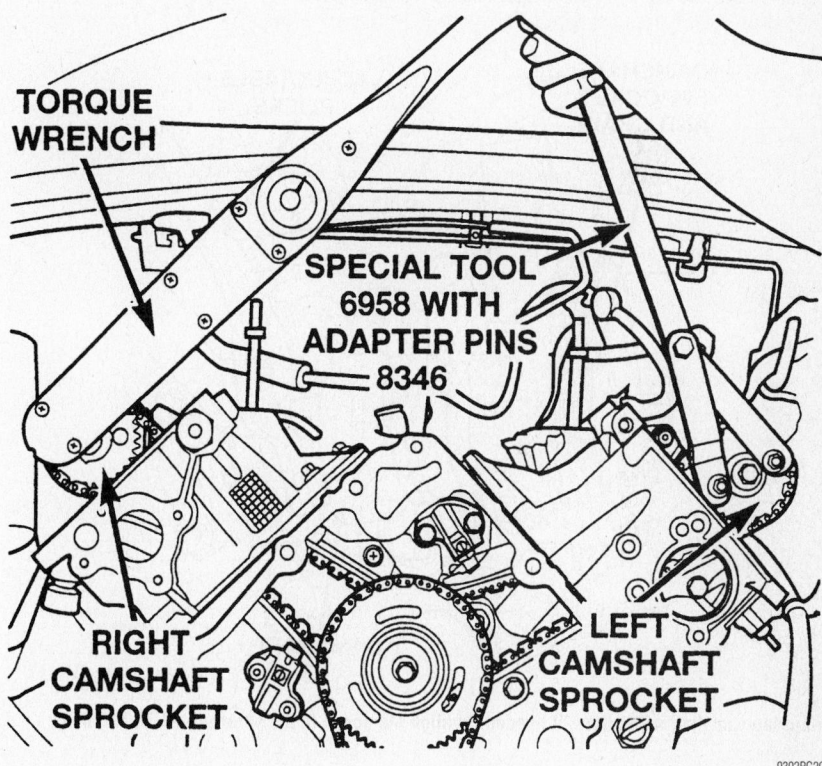

Hold the left camshaft sprocket with a spanner wrench while removing or installing the camshaft sprocket bolts—3.7L engine

3.7L Engines

1. Before servicing the vehicle, refer to the precautions in the beginning of this section.
2. Remove or disconnect the following:

- Negative battery cable
- Cylinder head covers
- Rocker arms
- Hydraulic lash adjusters

➡**Keep all valvetrain components in order for assembly.**

3. Set the engine at Top Dead Center (TDC) of the compression stroke for the No. 1 cylinder.
4. Install Timing Chain Wedge 8350 to retain the chain tensioners.
5. Matchmark the timing chains to the camshaft sprockets.
6. Install Camshaft Holding Tool 6958 and Adapter Pins 8346 to the left camshaft sprocket.
7. Remove or disconnect the following:

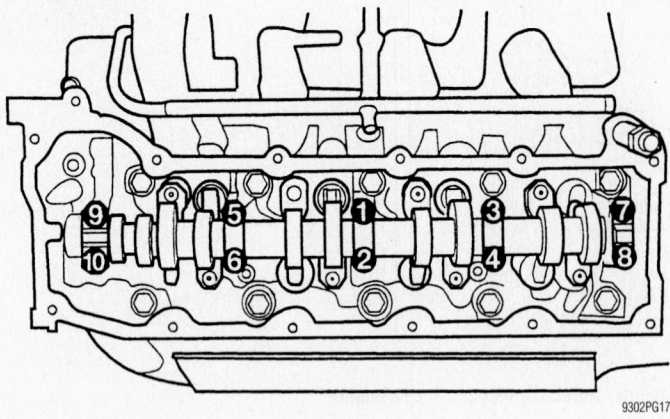

Camshaft bearing cap bolt tightening sequence—3.7L engine

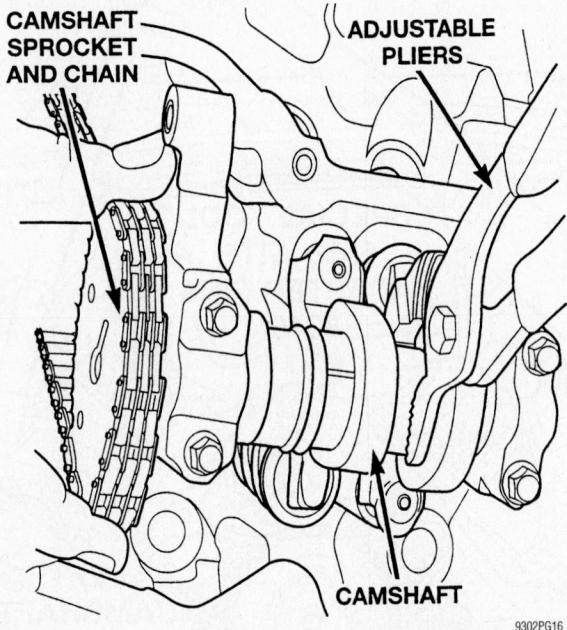

Turn the camshaft with pliers, if needed, to align the dowel in the sprocket—3.7L engine

- Right camshaft timing sprocket and target wheel
- Left camshaft sprocket
- Camshaft bearing caps, by reversing the tightening sequence
- Camshafts

To install:

8. Install or connect the following:
- Camshafts. Tighten the bearing cap bolts in ½ turn increments, in sequence, to 100 inch lbs. (11 Nm).
- Target wheel to the right camshaft
- Camshaft timing sprockets and chains, by aligning the matchmarks

9. Remove the tensioner wedges and tighten the camshaft sprocket bolts to 90 ft. lbs. (122 Nm).

10. Install or connect the following:
- Hydraulic lash adjusters in their original locations
- Rocker arms in their original locations
- Cylinder head covers
- Negative battery cable

4.7L Engine

1. Before servicing the vehicle, refer to the precautions in the beginning of this section.

2. Remove or disconnect the following:
- Negative battery cable
- Cylinder head covers
- Rocker arms
- Hydraulic lash adjusters

➥**Keep all valvetrain components in order for assembly.**

3. Set the engine at Top Dead Center (TDC) of the compression stroke for the No. 1 cylinder.

4. Install Timing Chain Wedge 8350 to retain the chain tensioners.

5. Matchmark the timing chains to the camshaft sprockets.

6. Install Camshaft Holding Tool 6958 and Adapter Pins 8346 to the left camshaft sprocket.

7. Remove or disconnect the following:
- Right camshaft timing sprocket and target wheel
- Left camshaft sprocket
- Camshaft bearing caps, by reversing the tightening sequence
- Camshafts

To install:

8. Install or connect the following:
- Camshafts. Tighten the bearing cap bolts in ½ turn increments, in sequence, to 100 inch lbs. (11 Nm).

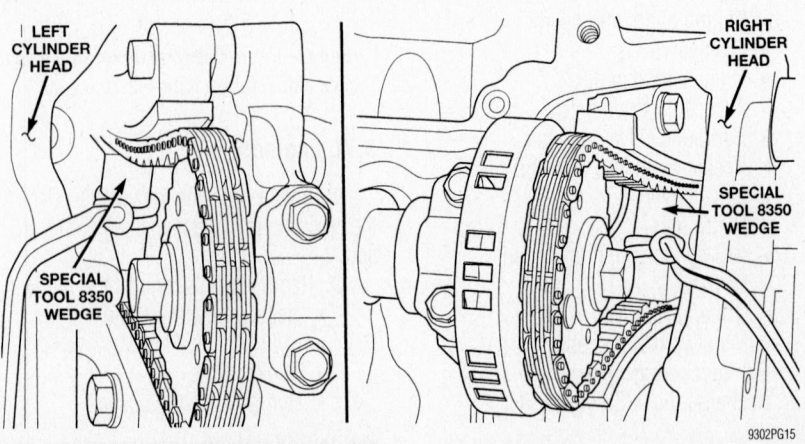

Chain Tensioner Retaining Wedges—4.7L engine

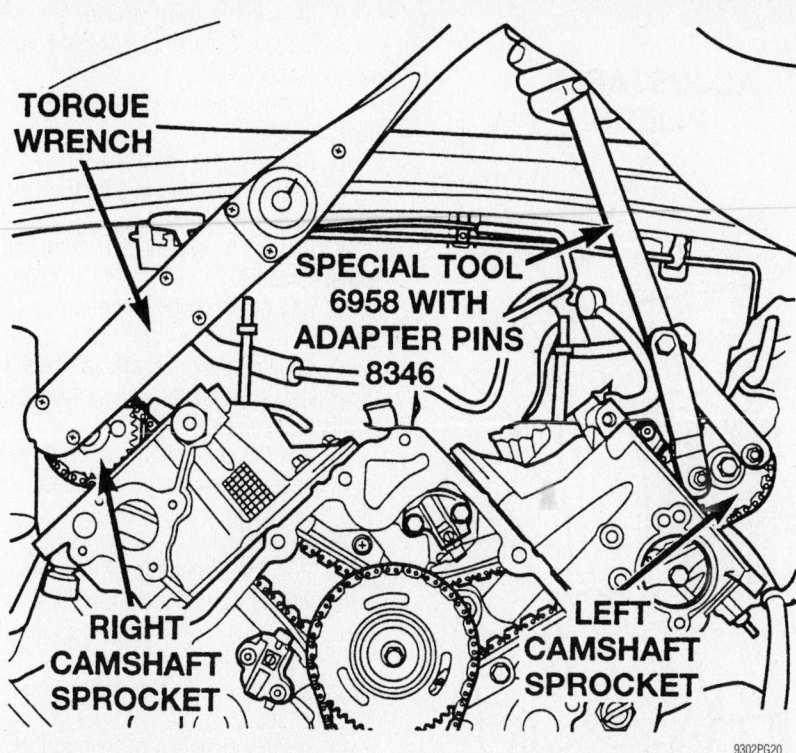

Hold the left camshaft sprocket with a spanner wrench while removing or installing the camshaft sprocket bolts—4.7L engine

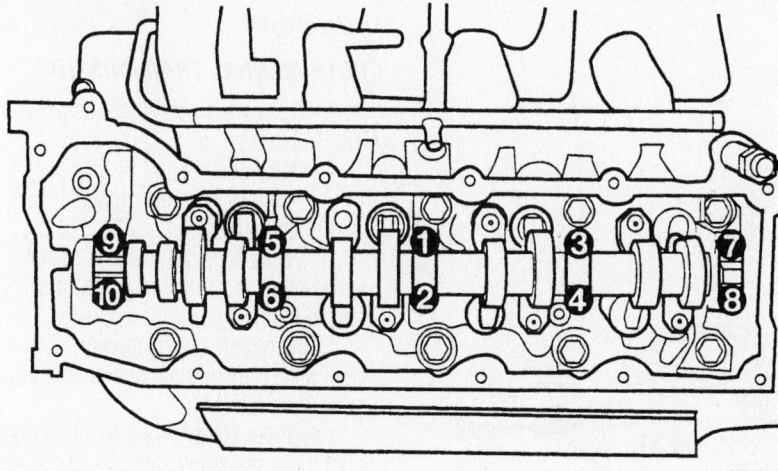

FRONT

Camshaft bearing cap bolt tightening sequence—4.7L engine

- Target wheel to the right camshaft
- Camshaft timing sprockets and chains, by aligning the matchmarks

9. Remove the tensioner wedges and tighten the camshaft sprocket bolts to 90 ft. lbs. (122 Nm).

10. Install or connect the following:
- Hydraulic lash adjusters in their original locations

- Rocker arms in their original locations
- Cylinder head covers
- Negative battery cable

3.9L and 5.9L Engines

1. Before servicing the vehicle, refer to the precautions in the beginning of this section.

2. Drain the cooling system.

3. Recover the A/C refrigerant, if equipped with air conditioning.

4. Set the crankshaft to Top Dead Center (TDC) of the compression stroke for the No. 1 cylinder.

5. Remove or disconnect the following:
- Negative battery cable
- Accessory drive belt
- Power steering pump
- Water pump
- Radiator
- A/C condenser
- Grille
- Crankshaft damper
- Front cover
- Valve covers
- Distributor
- Intake manifold

➡Keep all valvetrain components in order for assembly.

- Rocker arms and pushrods
- Hydraulic lifters
- Timing chain and sprockets
- Camshaft thrust plate and chain oil tab
- Camshaft

To install:

6. Install or connect the following:
- Camshaft
- Camshaft Holding Tool C-3509
- Camshaft thrust plate and chain oil tab. Tighten the bolts to 18 ft. lbs. (24 Nm).
- Timing chain and sprockets. Tighten the camshaft sprocket bolt to 50 ft. lbs. (68 Nm).

7. Remove the camshaft holding tool.

8. Install or connect the following:
- Hydraulic lifters in their original positions
- Rocker arms and pushrods in their original positions
- Intake manifold
- Distributor
- Valve covers
- Front cover
- Crankshaft damper
- Grille
- A/C condenser
- Radiator
- Water pump
- Power steering pump
- Accessory drive belt
- Negative battery cable

9. Fill the cooling system.

10. Recharge the A/C system, if equipped.

11. Start the engine and check for leaks.

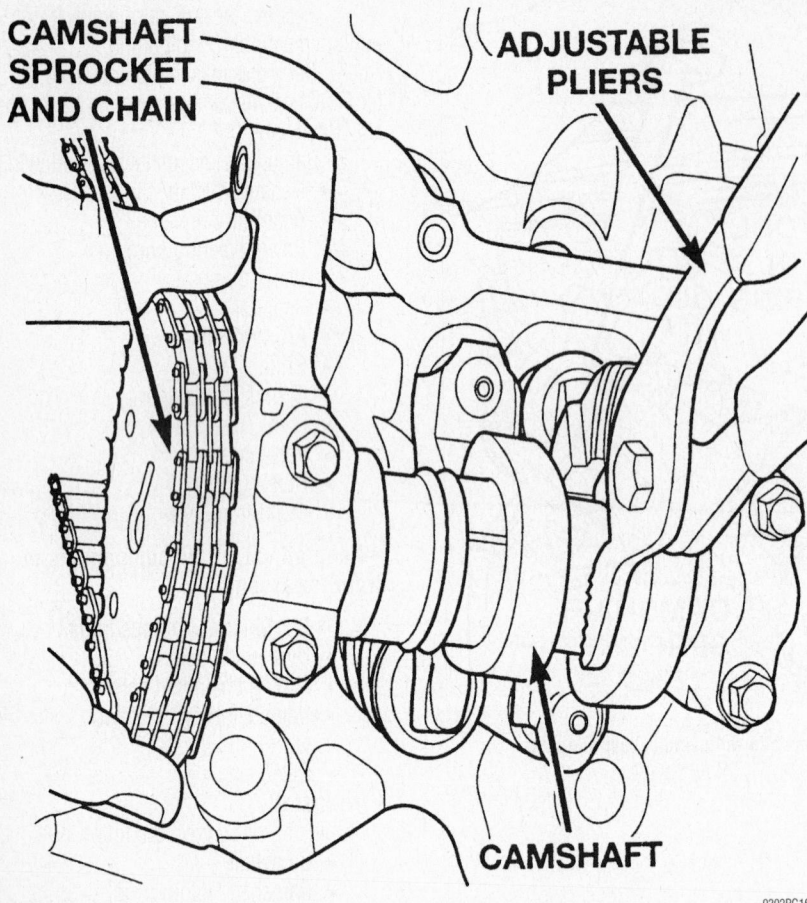

CAMSHAFT SPROCKET AND CHAIN

ADJUSTABLE PLIERS

CAMSHAFT

9302PG16

Turn the camshaft with pliers, if needed, to align the dowel in the sprocket—4.7L engine

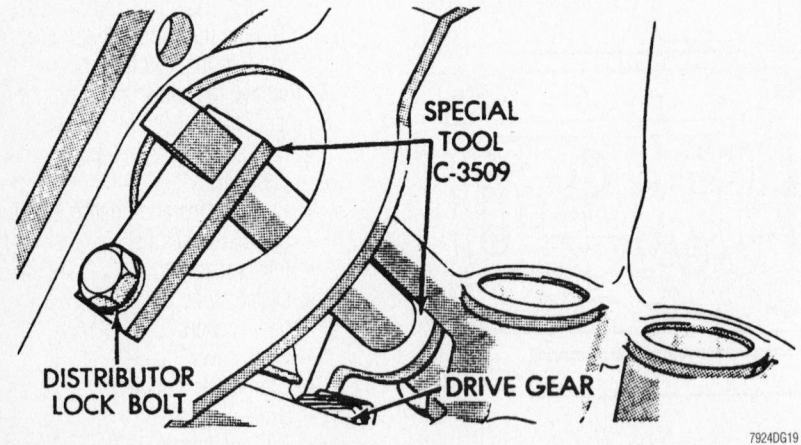

SPECIAL TOOL C-3509

DISTRIBUTOR LOCK BOLT

DRIVE GEAR

7924DG19

Camshaft holding tool C-3509—3.9L and 5.9L engines

Valve Lash

ADJUSTMENT

All gasoline engines covered in this section use hydraulic lifters. No maintenance or periodic adjustment is required.

Starter Motor

REMOVAL & INSTALLATION

3.7L Engine

1. Disconnect and isolate negative battery cable.

2. Raise and support vehicle.

3. Remove 2 starter heat shield bolts at side of starter.

4. Remove starter heat shield nut at front of starter.

5. Remove starter heat shield.

6. Remove solenoid wire from solenoid terminal.

7. Remove battery cable from stud on starter solenoid.

8. Remove 2 starter mounting bolts.

9. Position front of starter to face rear of vehicle. Rotate starter until solenoid position is located below starter.

10. Remove starter from vehicle by passing it between exhaust pipe and transmission bellhousing.

To install:

11. Position starter into bellhousing and install 2 bolts. Torque to 68 Nm (50 ft. lbs.).

12. Install battery cable and nut to stud on starter solenoid. Tighten nut to 13.6 Nm (120 inch lbs.).

13. Install solenoid wire connector to solenoid terminal.

14. Position starter heat shield and install nut at front of starter.

15. Install 2 starter heat shield bolts at side of starter.

16. Lower vehicle.

17. Connect negative battery cable.

4.7L Engine

WITH MANUAL TRANSMISSION

1. Disconnect and isolate negative battery cable.

2. Raise and support vehicle.

3. Remove nut securing starter motor to stud on transmission housing.

4. While supporting starter motor, remove bolt securing starter motor to transmission housing.

5. If equipped with automatic transmission, slide transmission cooler tube bracket forward on tubes far enough for starter motor to be removed from lower mounting stud.

6. Lower starter motor from front transmission housing far enough to access and remove nut securing battery cable eyelet to starter solenoid stud.

Always support starter motor during this process. Do not let starter motor hang from wire harness.

7. Remove solenoid wire solenoid terminal stud.

8. Disconnect battery cable solenoid wire from receptacle on starter solenoid.

9. Remove starter motor from transmission housing.

To install:

10. Position starter motor to transmission housing.

11. Connect battery cable solenoid terminal wire harness connector to connector receptacle on starter solenoid. Always support the starter motor during this process. Do not let the starter motor hang from the wire harness.

12. Install battery cable eyelet terminal onto solenoid B(+) terminal stud.

13. Install and tighten nut securing battery cable eyelet terminal to starter solenoid B (+) terminal stud. Tighten nut to 13.6 Nm (120 inch lbs.).

14. Position starter motor over stud on transmission housing.

15. If equipped with automatic transmission, slide automatic transmission cooler tube bracket rearward on tubes and into position over starter motor flange.

16. Loosely install the washers, bolt, and nut to starter. Tighten bolt and nut to 67.8 Nm (50 ft. lbs.).

17. Lower vehicle.

18. Connect negative battery cable.

WITH AUTOMATIC TRANSMISSION

1. Disconnect and isolate negative battery cable.

2. Raise and support vehicle.

3. Remove bolt and washer (rearward facing) securing starter motor to the transmission housing.

4. While supporting starter motor, remove bolt and washer (rearward facing) securing starter motor to the transmission housing.

5. Lower starter motor from front of transmission housing far enough to access and remove nut securing battery positive cable eyelet terminal to the starter solenoid B (+) terminal stud. Always support starter motor during this process. Do not let starter motor hang from wire harness.

6. Remove battery cable eyelet terminal from solenoid B (+) terminal stud.

7. Disconnect battery cable solenoid terminal wire harness connector from receptacle on starter solenoid.

8. Remove starter motor from transmission housing.

To install:

9. Position starter motor to transmission housing.

10. Connect battery cable solenoid terminal wire harness connector to connector receptacle on starter solenoid. Always support the starter motor during this process. Do not let the starter motor hang from the wire harness.

11. Install battery cable eyelet terminal onto solenoid B (+) terminal stud.

12. Install and tighten nut securing battery cable eyelet terminal to starter solenoid B (+) terminal stud. Tighten nut to 13.6 Nm (120 inch lbs.).

13. Position starter motor to transmission housing and loosely install two bolts/washers.

14. Tighten bolts to 67.8 Nm (50 ft. lbs.).

15. Lower vehicle.

16. Connect negative battery cable.

2.5L, 3.9L, and 5.9L Engines

1. Before servicing the vehicle, refer to the precautions in the beginning of this section.

2. Remove or disconnect the following:

- Negative battery cable
- Starter mounting bolts
- Starter solenoid harness connections
- Starter

To install:

3. Connect the starter solenoid wiring connectors.

4. Install the starter and tighten the bolts to the following specifications:

    a. 2.5L engine: 33 ft. lbs. (45 Nm).

    b. 4.7L engine: 40 ft. lbs. (54 Nm).

    c. 3.9L, 5.9L engines: 50 ft. lbs. (68 Nm).

5. Install the negative battery cable and check for proper operation.

Oil Pan

REMOVAL & INSTALLATION

2.5L Engine

1. Before servicing the vehicle, refer to the precautions in the beginning of this section.

2. Drain the engine oil.

3. Remove or disconnect the following:

- Negative battery cable
- Exhaust front pipe
- Starter motor
- Bell housing access cover
- Oil level sensor connector, if equipped
- Left and right motor mounts

4. Place a jack under the crankshaft

damper and raise the engine for clearance.

5. Remove the oil pan.

To install:

6. Fabricate 4 alignment dowels from 1½ inch x ¼ inch bolts. Cut the heads off the bolts and cut a slot into the top of the dowel to allow installation/removal with a screwdriver.

7. Install or connect the following:

- Dowels
- Oil pan, using a new gasket. Tighten the ¼ inch bolts to 85 inch lbs. (9.5 Nm) and the ⁵⁄₁₆ inch bolts to 11 ft. lbs. (15 Nm).

8. Replace the alignment dowels with ¼ inch bolts and tighten them to 85 inch lbs. (9.5 Nm).

9. Install or connect the following:

- Left and right motor mounts
- Oil level sensor connector, if equipped
- Bell housing access cover
- Starter motor
- Exhaust front pipe
- Negative battery cable

10. Fill the crankcase.

11. Start the engine and check for leaks.

3.7L Engine

1. Disconnect the negative battery cable.

2. Install an engine support fixture. Do not raise engine at this time.

3. Loosen both left and right side engine mount through bolts. Do not remove bolts.

4. Remove the structural dust cover, if equipped.

5. Drain engine oil.

6. Remove the front crossmember.

➡**Raise the engine just enough to p̶vide clearance for oil pan removal. Check for proper clearance at fan shroud to fan and cowl to intake m̶fold.**

7. Raise engine to provide clea̶ remove oil pan.

➡**Do not pry on oil pan or oil ̶ket. Gasket is integral to engi̶ windage tray and does not c̶ with oil pan.**

8. Remove the oil pan mo̶ bolts and oil pan.

9. Unbolt oil pump pick̶be and remove tube.

10. Inspect the integral ̶dage tray and gasket and replace as nee̶

To install:

11. Clean the oil pan gasket mating surface of the bedplate and oil pan.

12. Inspect integrated oil pan gasket, and replace as necessary.

13. Position the integrated oil pan gasket/windage tray assembly.

14. Install the oil pickup tube

15. If removed, install stud at position No. 9.

16. Install the mounting bolt and nuts. Tighten nuts to 28 Nm (20 ft. lbs.).

17. Position the oil pan and install the mounting bolts. Tighten the mounting bolts to 15 Nm (11 ft. lbs.) in the sequence shown.

18. Lower the engine into mounts.

19. Install both the left and right side engine mount through bolts. Tighten the nuts to 68 Nm (50 ft. lbs.).

20. Remove the lifting device.

21. Install structural dust cover, if equipped.

22. Install the front crossmember.

23. Fill engine oil.

24. Reconnect the negative battery cable.

25. Start engine and check for leaks.

3.9L Engine

2-WHEEL DRIVE MODELS

1. Before servicing the vehicle, refer to the precautions in the beginning of this sec-

Drain the engine oil.

Remove or disconnect the follow-

Negative battery cable

Distributor cap

Oil dipstick

Exhaust front pipe

Flywheel access panel, if equipped

Left and right motor mount through bolts

pan. Raise the engine as necessary for clearance.

1½ ate 4 alignment dowels from inch bolts. Cut the heads off the bo cut a slot into the top of the dowel installation/removal with a screwdri

5. Ins connect the following:
- A ent dowels
- Oil Replace the dowels with bolt tighten all bolts to 17 ft. lbs. Nm).
- Left a ight motor mount through bolts
- Flywhe ccess panel, if equipped
- Exhaust nt pipe

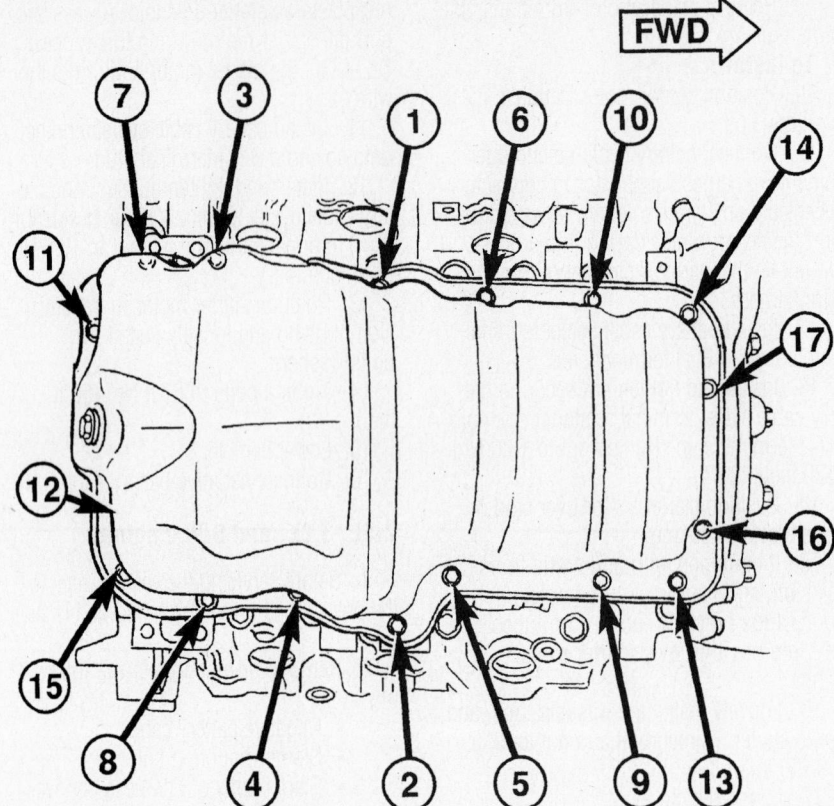

Oil pan bolt torque sequence—3.7L Engine

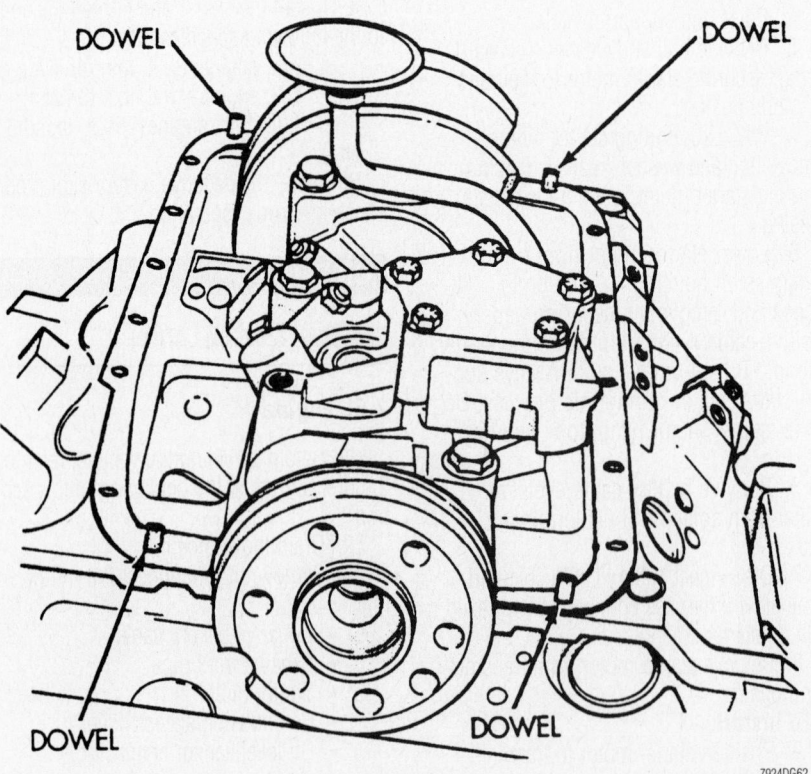

Oil pan alignment dowel placement—3.9L engine shown

- Oil dipstick
- Distributor cap
- Negative battery cable

6. Fill the crankcase to the correct level.

7. Start the engine and check for leaks.

4-WHEEL DRIVE MODELS

1. Before servicing the vehicle, refer to the precautions in the beginning of this section.

2. Drain the engine oil.

3. Remove or disconnect the following:

- Negative battery cable
- Engine oil dipstick
- Front axle. Support the engine
- Exhaust front pipe
- Flywheel access panel, if equipped
- Oil pan

To install:

4. Fabricate 4 alignment dowels from 1½ inch x ¼ inch bolts. Cut the heads off the bolts and cut a slot into the top of the dowel to allow installation/removal with a screwdriver.

5. Install or connect the following:

- Alignment dowels
- Oil pan. Replace the dowels with bolts and tighten all bolts to 17 ft. lbs. (23 Nm).
- Flywheel access panel, if equipped
- Exhaust front pipe
- Front axle
- Engine oil dipstick
- Negative battery cable

6. Fill the crankcase to the correct level.

7. Start the engine and check for leaks.

4.7L Engine

2-WHEEL DRIVE

1. Drain the cooling system.

2. Remove the upper fan shroud.

3. Remove the throttle body resonator and air inlet hose.

4. Remove the intake manifold.

5. Raise vehicle on hoist.

6. Disconnect exhaust pipe at exhaust manifolds.

7. Remove the structural dust cover using sequence shown.

8. Drain engine oil and remove oil filter.

9. Position suitable jack under engine.

10. Remove both left and right side engine mount through bolts.

11. Raise engine to provide clearance to remove oil pan.

12. Place blocks of wood between

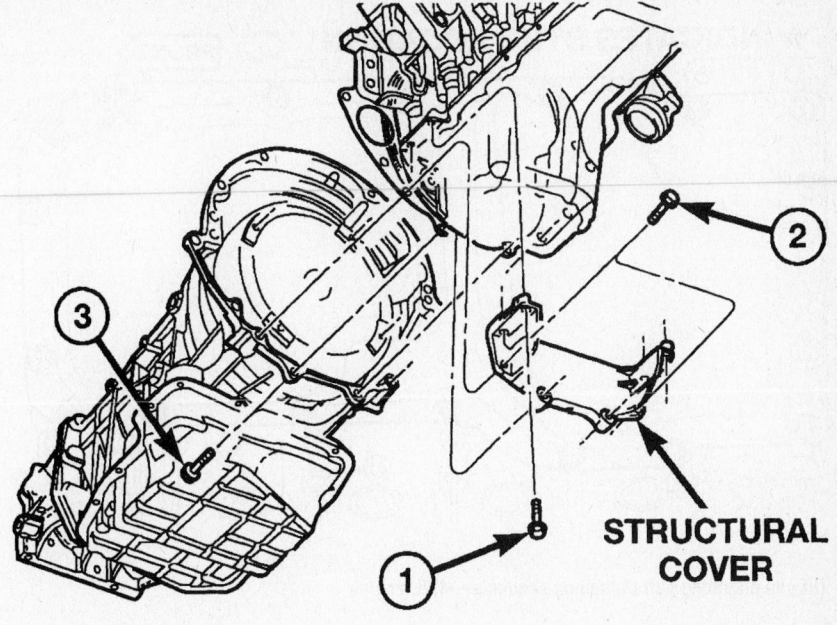

1 - BOLT
2 - BOLT
3 - BOLT

67189-DAKO-G45

Structural cover torque sequence—3.7L and 4.7L engines

engine brackets and lower mounts to provide stability to engine.

➡**Do not pry on oil pan or oil pan gasket. Gasket is mounted to engine and does not come out with oil pan.**

13. Remove the oil pan mounting bolts and oil pan.

14. Unbolt oil pump pickup tube and remove tube and oil pan gasket from engine.

To install:

15. Clean the oil pan gasket mating surface of the bedplate and oil pan.

16. Position the oil pan gasket and pickup tube with new o-ring. Install the mounting bolt and nuts.
Tighten bolt and nuts to 28 Nm (20 ft. lbs.).

17. Position the oil pan and install the mounting bolts. Tighten the mounting bolts to 15 Nm (11 ft. lbs.) in the sequence shown.

18. Raise the engine and remove the blocks of wood.

19. Lower engine and install both the left and right side engine mount through bolts. Tighten the nuts to 68 Nm (50 ft. lbs.).

20. Remove jack and install oil filter.

21. Install structural dust cover.

22. Install exhaust pipe onto exhaust manifolds.

23. Lower vehicle.

24. Install intake manifold.

25. Install throttle body resonator and air inlet hose.

26. Install upper fan shroud.

27. Fill cooling system.

28. Fill engine oil.

29. Start engine and check for leaks.

4-WHEEL DRIVE

➡**On 4wd vehicles, the front axle must be removed before the oil pan can be removed.**

1. Remove the front axle from vehicle.

2. Remove the structural dust cover using sequence shown.

3. Drain the engine oil and remove oil filter.

4. Remove the oil pan mounting bolts and oil pan.

5. Unbolt oil pump pickup tube and remove tube and oil pan gasket from engine.

To install:

6. Clean the oil pan gasket mating surface of the bedplate and oil pan.

7. Position the oil pan gasket and

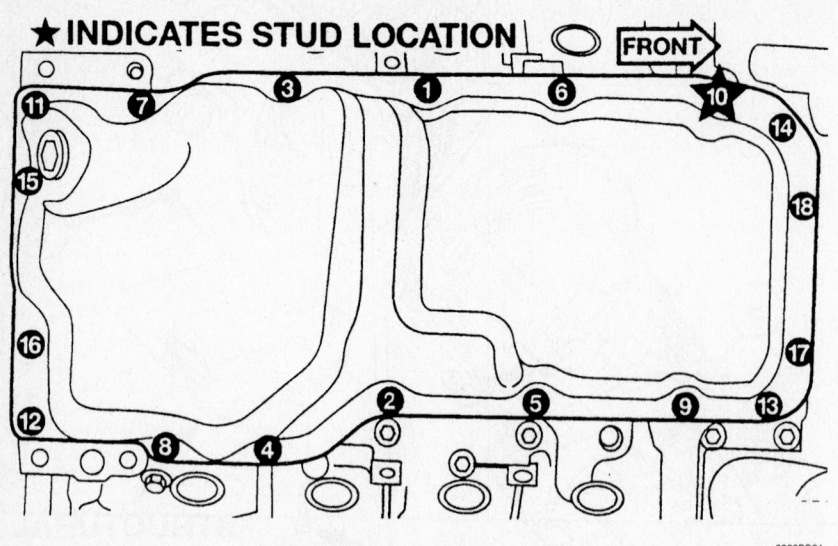

★ INDICATES STUD LOCATION

Oil pan mounting bolt tightening sequence—4.7L engine

pickup tube with new o-ring. Install the mounting bolt and nuts.

Tighten bolt and nuts to 28 Nm (20 ft. lbs.).

8. Position the oil pan and install the mounting bolts. Tighten the mounting bolts to 15 Nm (11 ft.

lbs.) in the sequence shown.

9. Install structural dust cover.
10. Install oil filter.
11. Install front axle.
12. Lower vehicle.
13. Fill engine oil.
14. Start engine check for leaks.

5.9L Engines

1. Before servicing the vehicle, refer to the precautions in the beginning of this section.
2. Drain the engine oil.
3. Remove or disconnect the following:
   - Oil filter
   - Starter motor
   - Cooler lines
   - Oil level sensor connector
   - Heated Oxygen (HO$_2$S) sensor connector
   - Exhaust Y-pipe
   - Oil pan

To install:

4. Install or connect the following:
   - Oil pan, using a new gasket. Tighten the bolts to 18 ft. lbs. (24 Nm).
   - Exhaust Y-pipe
   - Heated Oxygen (HO$_2$S) sensor connector
   - Oil level sensor connector
   - Cooler lines
   - Starter motor
   - Oil filter
5. Fill the crankcase.
6. Start the engine and check for leaks.

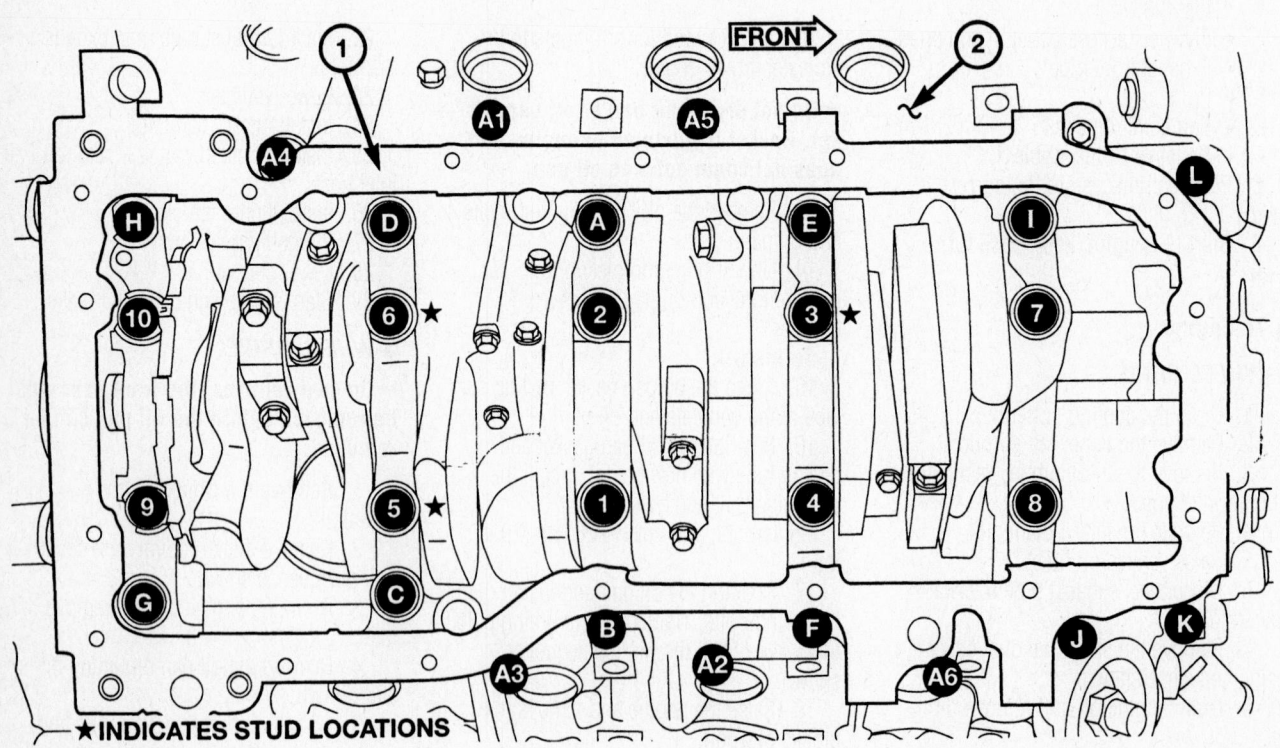

★ INDICATES STUD LOCATIONS

1 - BEDPLATE
2 - CYLINDER BLOCK

Bedplate bolt tightening sequence—4.7L engine

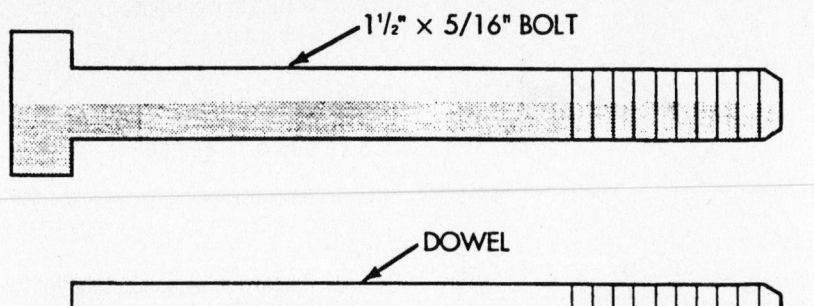

Oil pan alignment dowels—3.9L, 5.9L engines

Oil Pump

REMOVAL & INSTALLATION

2.5L Engine

1. Before servicing the vehicle, refer to the precautions in the beginning of this section.
2. Drain the engine oil.
3. Remove or disconnect the following:

  • Negative battery cable

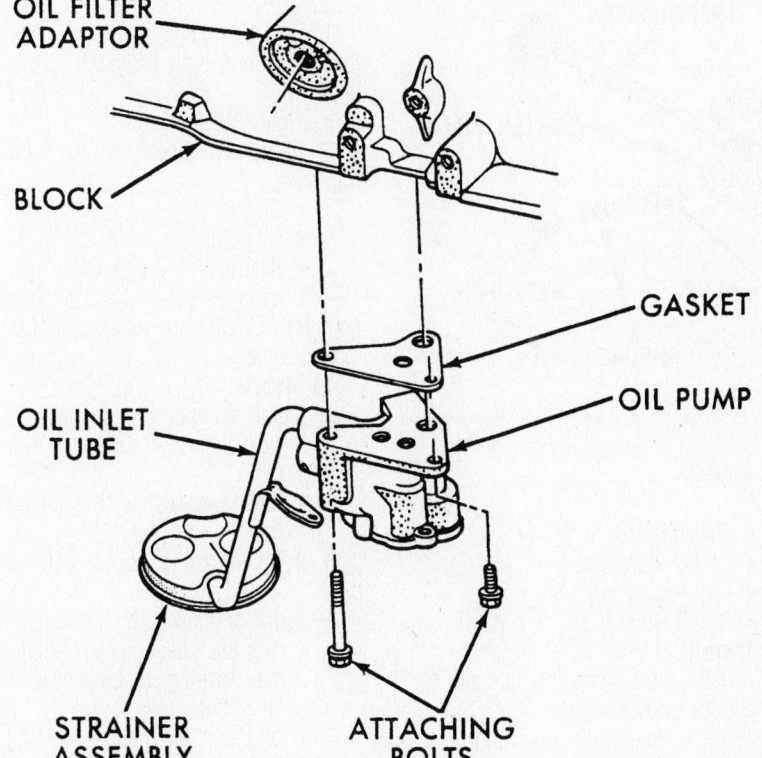

Exploded view of the oil pump assembly—2.5L engine

  • Oil pan
  • Oil pump and pickup tube

➡If the oil pump is not to be serviced, do not disturb the position of the oil inlet tube and strainer assembly in the pump body. If the tube is moved within the pump body, a replacement tube and strainer assembly must be installed to assure an airtight seal.

To install:
4. Install or connect the following:
  • Oil pump. Tighten the mounting bolts to 17 ft. lbs. (23 Nm).

  • Oil pan
  • Negative battery cable
5. Fill the crankcase to the correct level.
6. Start the engine and check for leaks.

3.7L Engine

1. Before servicing the vehicle, refer to the precautions in the beginning of this section.
2. Remove or disconnect the following:
  • Oil Pan
  • Timing chain cover
  • Timing chains and tensioners
  • Oil pump
3. Installation is the reverse of removal. Torque the pump bolts, in sequence, to 21 ft. lbs. (28 Nm),

4.7L Engine

1. Before servicing the vehicle, refer to the precautions in the beginning of this section.
2. Drain the engine oil.
3. Remove or disconnect the following:

  • Negative battery cable
  • Oil pan
  • Oil pump pick-up tube
  • Timing chains and tensioners
  • Oil pump
To install:
4. Install or connect the following:
  • Oil pump. Tighten the bolts to 21 ft. lbs. (28 Nm).
  • Timing chains and tensioners
  • Oil pump pick-up tube
  • Oil pan
  • Negative battery cable
5. Fill the crankcase to the correct level.
6. Start the engine and check for leaks.

3.9L and 5.9L Engines

1. Before servicing the vehicle, refer to the precautions in the beginning of this section.
2. Drain the engine oil.
3. Remove or disconnect the following:
  • Negative battery cable
  • Oil pan
  • Oil pump pick-up tube
  • Oil pump
To install:
4. Install or connect the following:
  • Oil pump. Tighten the bolts to 30 ft. lbs. (41 Nm).
  • Oil pump pick-up tube
  • Oil pan
  • Negative battery cable
5. Fill the crankcase to the correct level.
6. Start the engine and check for leaks.

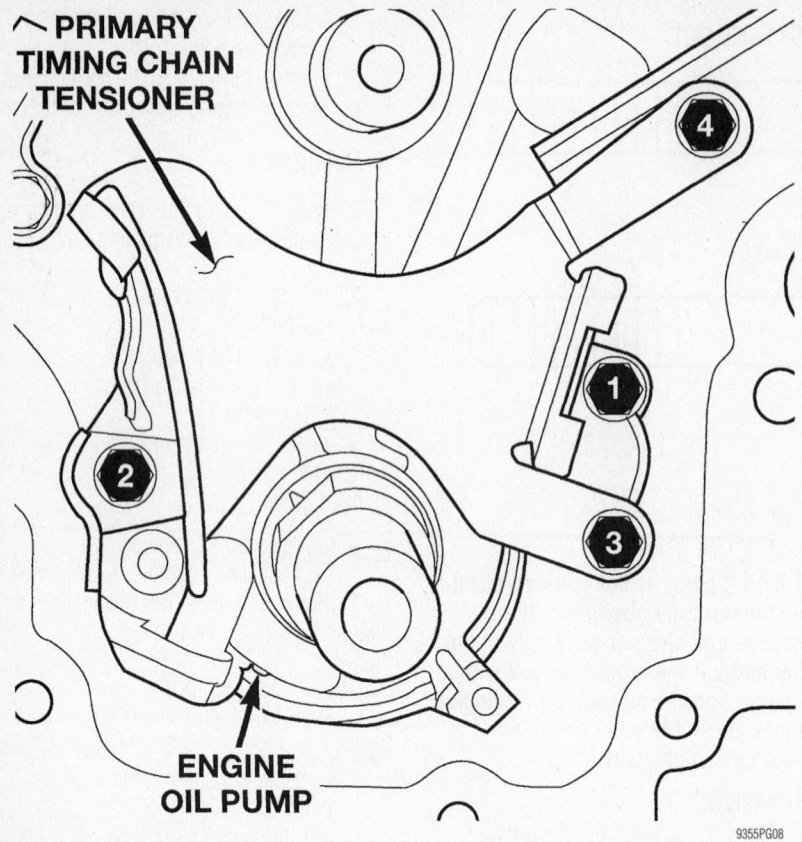

Oil pump bolt torque sequence—3.7L

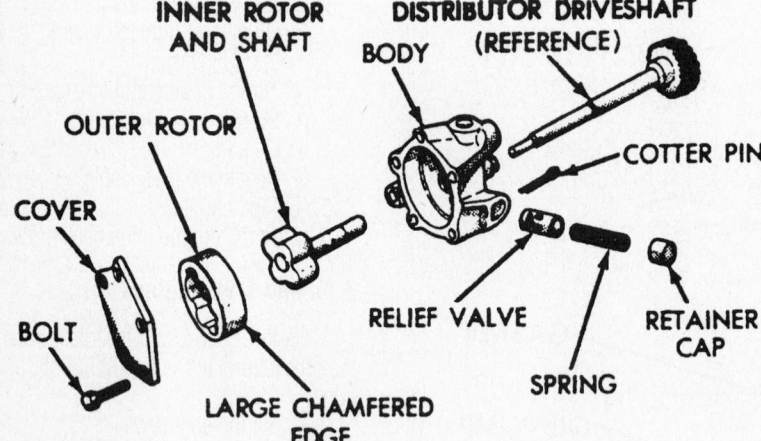

Exploded view of the oil pump assembly—3.9L and 5.9L engines

Rear Main Seal

REMOVAL & INSTALLATION

2.5L Engine

1. Before servicing the vehicle, refer to the precautions in the beginning of this section.
2. Remove or disconnect the following:

- Transmission
- Clutch, if equipped
- Flywheel
- Rear main seal

To install:

3. Install the rear main seal so that it is flush with the cylinder block.
4. Install or connect the following:

- Flywheel. Use new bolts and tighten to 50 ft. lbs. (68 Nm) plus 60 degrees

- Clutch, if equipped
- Transmission

5. Start the engine and check for leaks.

3.7L and 4.7L Engine

1. Before servicing the vehicle, refer to the precautions in the beginning of this section.
2. Remove or disconnect the following:

- Transmission
- Flexplate

3. Thread Oil Seal Remover 8506 into the rear main seal as far as possible and remove the rear main seal.

To install:

4. Install or connect the following:

- Seal Guide 8349-2 onto the crankshaft
- Rear main seal on the seal guide
- Rear main seal, using the Crankshaft Rear Oil Seal Installer 8349 and Driver Handle C-4171; tap it into place until the installer is flush with the cylinder block
- Flexplate. Tighten the bolts to 45 ft. lbs. (60 Nm).
- Transmission

5. Start the engine and check for leaks.

3.9L and 5.9L Engines

1. Before servicing the vehicle, refer to the precautions in the beginning of this section.
2. Drain the engine oil.
3. Remove or disconnect the following:

- Oil pan
- Oil pump
- Rear main bearing cap

4. Loosen the other main bearing cap bolts for clearance and remove the rear main seal halfs.

To install:

5. Install or connect the following:

- New upper seal half to the cylinder block
- New lower seal half to the bearing cap

6. Apply sealant to the rear main bearing cap.
7. Install or connect the following:

- Rear main bearing cap. Tighten **all** main bearing cap bolts to 85 ft. lbs. (115 Nm).
- Oil pump and oil pan

8. Fill the engine.
9. Start the engine and check for leaks.

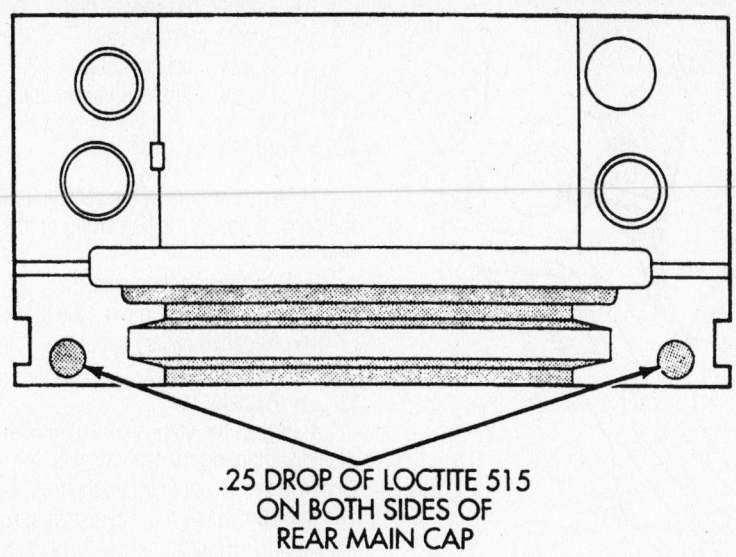

.25 DROP OF LOCTITE 515
ON BOTH SIDES OF
REAR MAIN CAP

7924DG24

Sealant application locations —3.9L and 5.9L engines

Timing Chain, Sprockets, Front Cover and Seal

REMOVAL & INSTALLATION

2.5L Engine

1. Before servicing the vehicle, refer to the precautions in the beginning of this section.
2. Remove or disconnect the following:
   - Negative battery cable
   - Accessory drive belt
   - Cooling fan and shroud
   - Crankshaft damper
   - Front crankshaft seal
   - Accessory brackets
   - Front cover
   - Oil slinger
3. Rotate the crankshaft so that the timing marks are aligned.
4. Remove the timing chain and sprockets.

To install:

5. Turn the timing chain tensioner lever to the unlock (down) position. Pull the tensioner block toward the tensioner lever to compress the spring. Hold the block and turn the tensioner lever to the lock (up) position.
6. Install the timing chain and sprockets with the timing marks aligned. Tighten the camshaft sprocket bolt to 80 ft. lbs. (108 Nm) for 2.5L engines or to 50 ft. lbs. (68 Nm) for 4.0L engines.
7. Release the timing chain tensioner.
8. Install or connect the following:
   - Oil slinger

- New front crankshaft seal to the front cover
- Front cover, using a new gasket
- Timing Case Cover Alignment and Seal Installation Tool 6139 in the crankshaft opening to center the front cover

9. Tighten the front cover bolts as follows:
   a. Step 1: Cover-to-block ¼ inch bolts to 60 inch lbs. (7 Nm).
   b. Step 2: Cover-to-block ⁵⁄₁₆ inch bolts to 16 ft. lbs. (22 Nm).
   c. Step 3: Oil pan-to-cover ¼ inch bolts to 85 inch lbs. (9.5 Nm).
   d. Step 4: Oil pan-to-cover ⁵⁄₁₆ inch bolts to 11 ft. lbs. (15 Nm).
10. Install or connect the following:
    - Accessory brackets
    - Crankshaft damper. Tighten the bolt to 80 ft. lbs. (108 Nm).

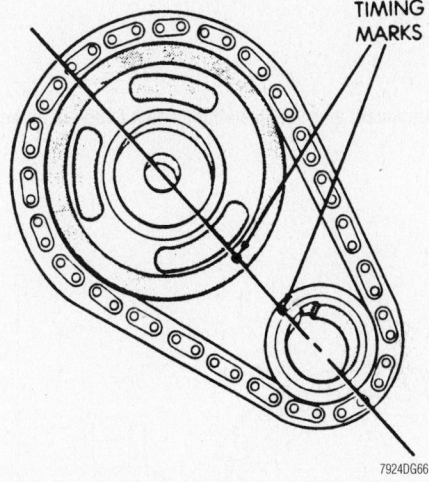

TIMING
MARKS

7924DG66

Timing mark alignment—2.5L engine

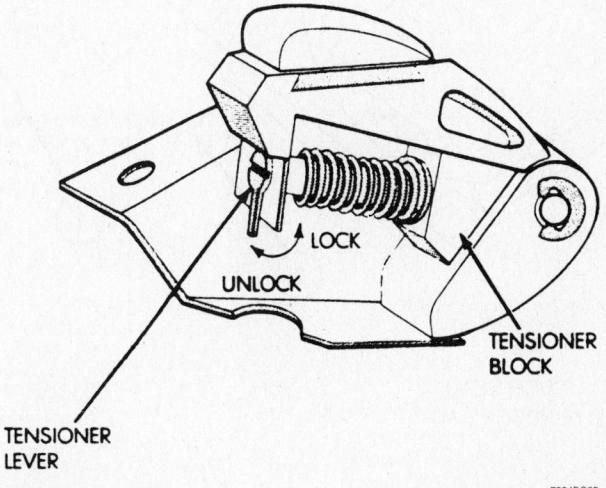

LOCK

UNLOCK

TENSIONER
BLOCK

TENSIONER
LEVER

7924DG65

Timing chain tensioner—2.5L engines

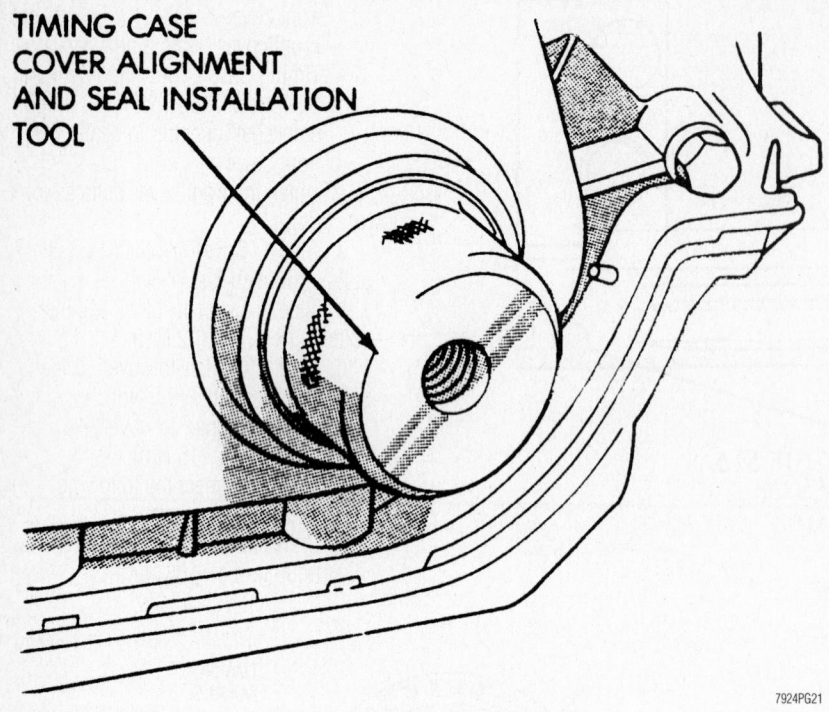

TIMING CASE COVER ALIGNMENT AND SEAL INSTALLATION TOOL

7924PG21

Timing Case Cover Alignment and Seal Installation Tool 6139—2.5L engine

- Cooling fan and shroud
- Accessory drive belt
- Negative battery cable

11. Start the engine and check for leaks.

3.7L Engines

1. Before servicing the vehicle, refer to the precautions in the beginning of this section.

2. Drain the cooling system.

3. Remove or disconnect the following:
- Negative battery cable
- Valve covers
- Radiator fan

4. Rotate the crankshaft so that the crankshaft timing mark aligns with the Top Dead Center (TDC) mark on the front cover, and the **V6** marks on the camshaft sprockets are at 12 o'clock.
- Power steering pump
- Access plugs from the cylinder heads
- Oil fill housing
- Crankshaft damper

5. Compress the primary timing chain tensioner and install a lockpin.

6. Remove the secondary timing chain tensioners.

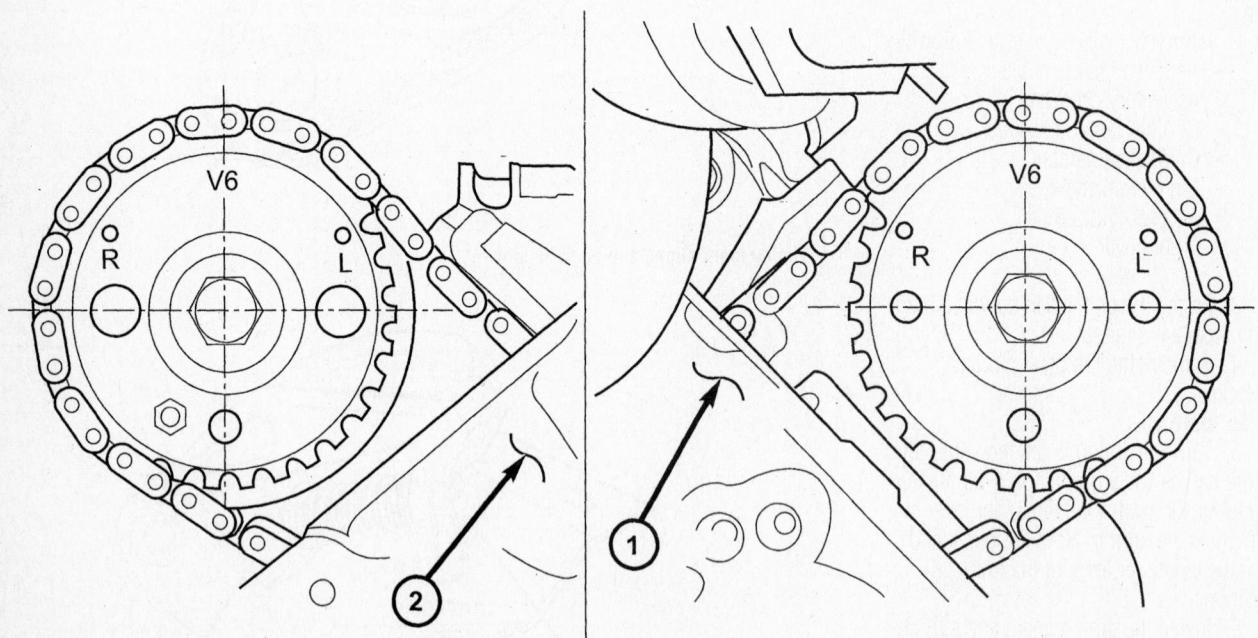

1 - LEFT CYLINDER HEAD
2 - RIGHT CYLINDER HEAD

9355PG09

Camshaft sprocket timing marks—3.7L

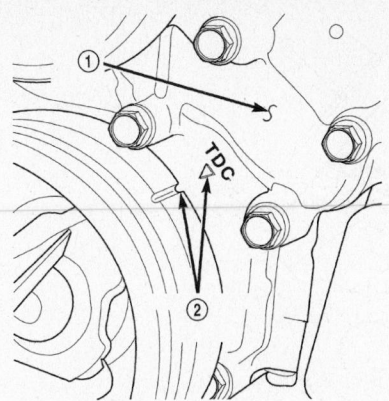

1 - TIMING CHAIN COVER
2 - CRANKSHAFT TIMING MARKS

9355PG10

Crankshaft timing marks—3.7L

7. Hold the left camshaft with adjustable pliers and remove the sprocket and chain. Rotate the **left** camshaft 15 degrees **clockwise** to the neutral position.

8. Hold the right camshaft with adjustable pliers and remove the camshaft sprocket. Rotate the **right** camshaft 45 degrees **counterclockwise** to the neutral position.

9. Remove the primary timing chain and sprockets.

To install:

10. Use a small prytool to hold the ratchet pawl and compress the secondary timing chain tensioners in a vise and install locking pins.

➥**The black bolts fasten the guide to the engine block and the silver bolts fasten the guide to the cylinder head.**

11. Install or connect the following:
   • Secondary timing chain guides. Tighten the bolts to 21 ft. lbs. (28 Nm).
   • Secondary timing chains to the idler sprocket so that the double plated links on each chain are visible through the slots in the primary idler sprocket

12. Lock the secondary timing chains to the idler sprocket with Timing Chain Locking tool as shown.

13. Align the primary chain double plated links with the idler sprocket timing mark and the single plated link with the crankshaft sprocket timing mark.

14. Install the primary chain and sprockets. Tighten the idler sprocket bolt to 25 ft. lbs. (34 Nm).

15. Align the secondary chain single plated links with the timing marks on the secondary sprockets. Align the dot at the **L** mark on the left sprocket with the plated link

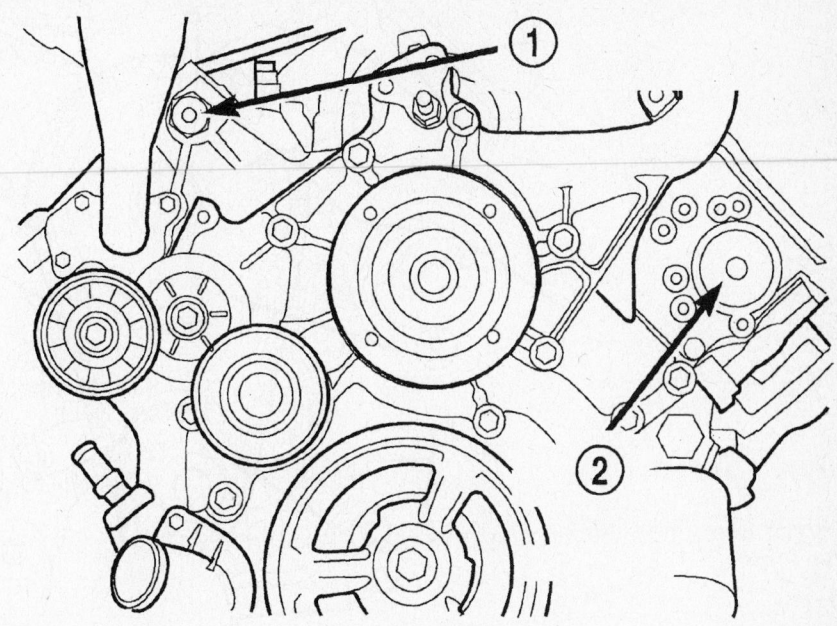

1 - RIGHT CYLINDER HEAD ACCESS PLUG
2 - LEFT CYLINDER HEAD ACCESS PLUG

9355PG11

Cylinder head access plugs—3.7L

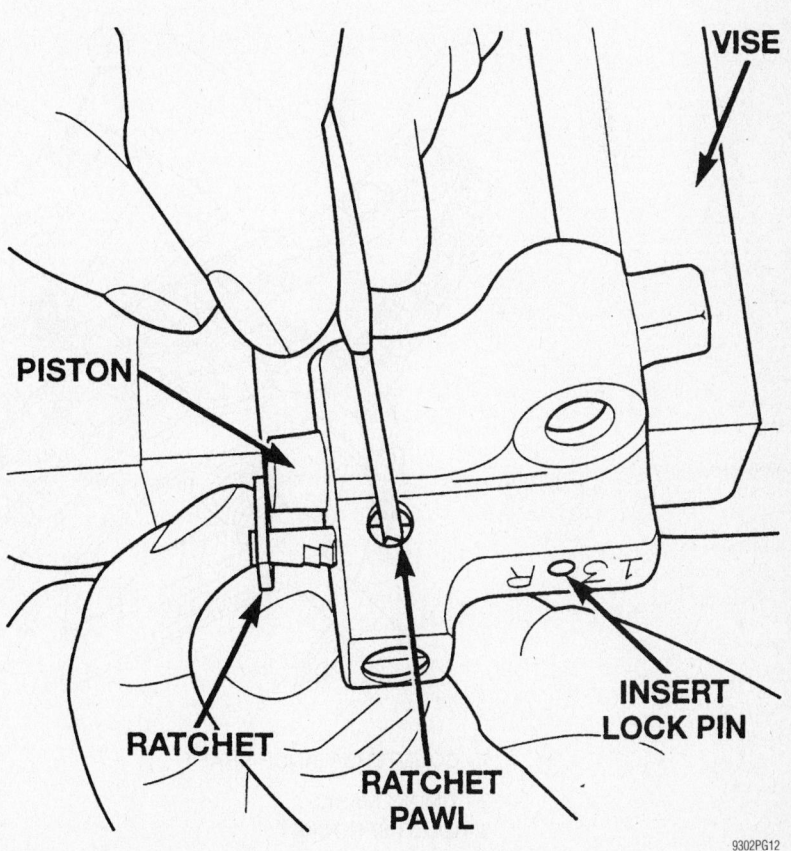

Secondary timing chain tensioner preparation—3.7L engine

9302PG12

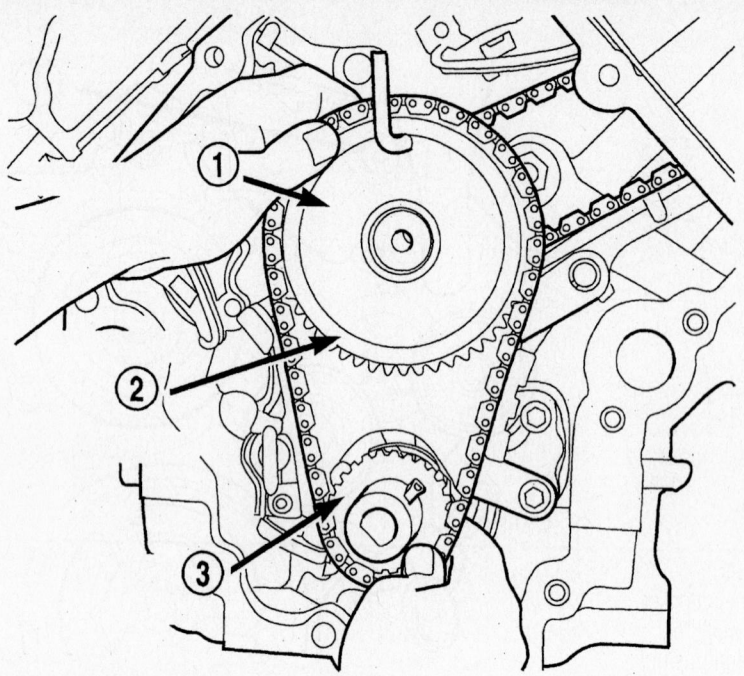

1 - SPECIAL TOOL 8429
2 - PRIMARY CHAIN IDLER SPROCKET
3 - CRANKSHAFT SPROCKET

9355PG12

Installing the idler gear and timing chains—3.7L

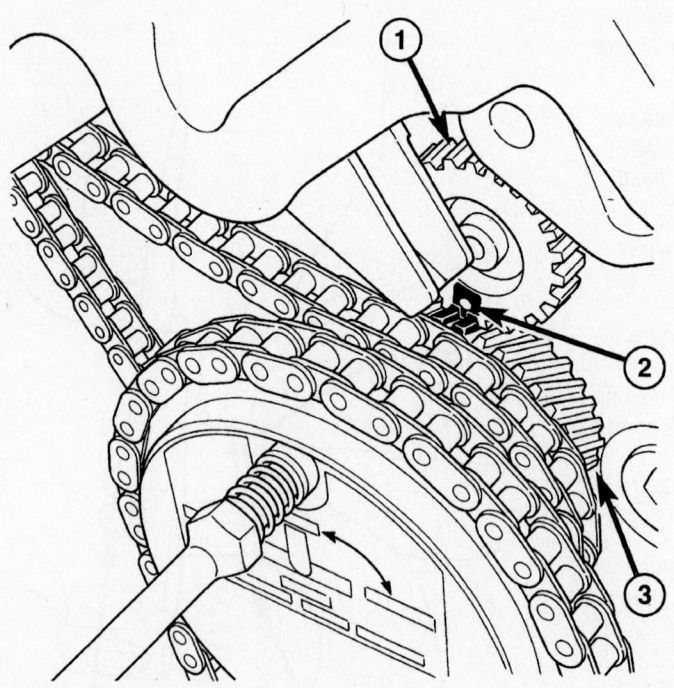

1 - COUNTERBALANCE SHAFT
2 - TIMING MARKS
3 - IDLER SPROCKET

9355PG13

Counterbalance shaft timing marks—3.7L

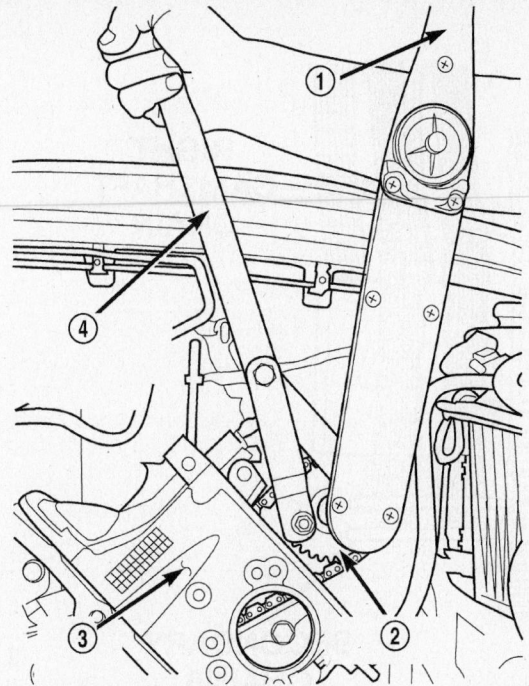

1 - TORQUE WRENCH
2 - CAMSHAFT SPROCKET
3 - LEFT CYLINDER HEAD
4 - SPECIAL TOOL 6958 SPANNER WITH ADAPTER PINS 8346

9355PG14

Tightening the left side camshaft sprocket—3.7L

1 - TORQUE WRENCH
2 - SPECIAL TOOL 6958 WITH ADAPTER PINS 8346
3 - LEFT CAMSHAFT SPROCKET
4 - RIGHT CAMSHAFT SPROCKET

9355PG15

Tightening the right side camshaft sprocket—3.7L

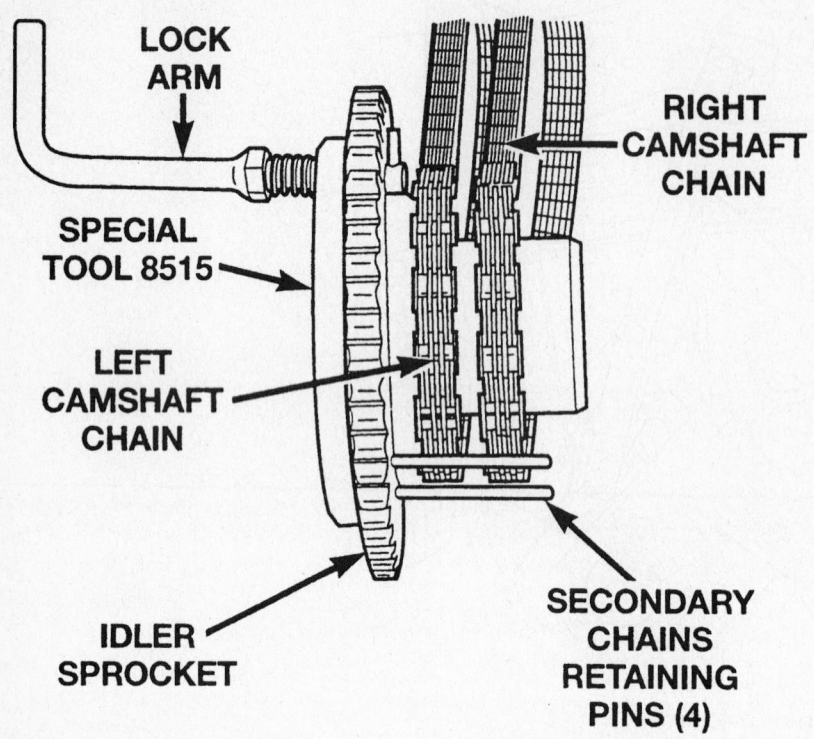

Use the Timing Chain Locking tool to lock the timing chains on the idler gear—3.7L engine

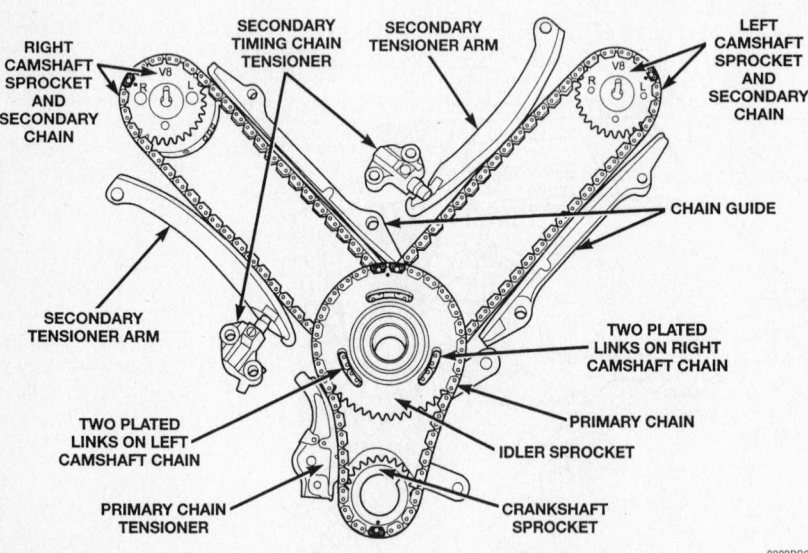

Timing chain system and alignment marks—4.7L engine shown, 3.7L similar

on the left chain and the dot at the **R** mark on the right sprocket with the plated link on the right chain.

16. Rotate the camshafts back from the neutral position and install the camshaft sprockets.

17. Remove the secondary chain locking tool.

18. Remove the primary and secondary timing chain tensioner locking pins.

19. Hold the camshaft sprockets with a spanner wrench and tighten the retaining bolts to 90 ft. lbs. (122 Nm).

20. Install or connect the following:
- Front cover. Tighten the bolts, in sequence, to 40 ft. lbs. (54 Nm).
- Front crankshaft seal
- Cylinder head access plugs
- A/C compressor
- Alternator
- Accessory drive belt tensioner. Tighten the bolt to 40 ft. lbs. (54 Nm).
- Oil fill housing
- Crankshaft damper. Tighten the bolt to 130 ft. lbs. (175 Nm).
- Power steering pump
- Lower radiator hose
- Heater hoses
- Accessory drive belt
- Engine cooling fan and shroud
- Camshaft Position (CMP) sensor
- Valve covers
- Negative battery cable

21. Fill and bleed the cooling system.

22. Start the engine, check for leaks and repair if necessary.

3.9L Engine

1. Before servicing the vehicle, refer to the precautions in the beginning of this section.

2. Drain the cooling system.

3. Remove or disconnect the following:
- Negative battery cable
- Accessory drive belt
- Radiator
- Cooling fan
- Water pump
- Crankshaft pulley
- Front crankshaft seal
- Front cover
- Timing chain and gears

To install:

4. Install or connect the following:
- Timing chain and gears. Align the timing marks and tighten the camshaft sprocket bolt to 35 ft. lbs. (47 Nm).
- Front cover. Tighten the bolts to 30 ft. lbs. (41 Nm).
- Front crankshaft seal
- Crankshaft pulley. Tighten the bolt to 135 ft. lbs. (183 Nm).
- Water pump
- Cooling fan
- Radiator
- Accessory drive belt
- Negative battery cable

5. Fill the cooling system.

6. Start the engine and check for leaks.

4.7L Engine

1. Before servicing the vehicle, refer to the precautions in the beginning of this section.

2. Drain the cooling system.

⭐ **INDICATES STUD LOCATIONS**

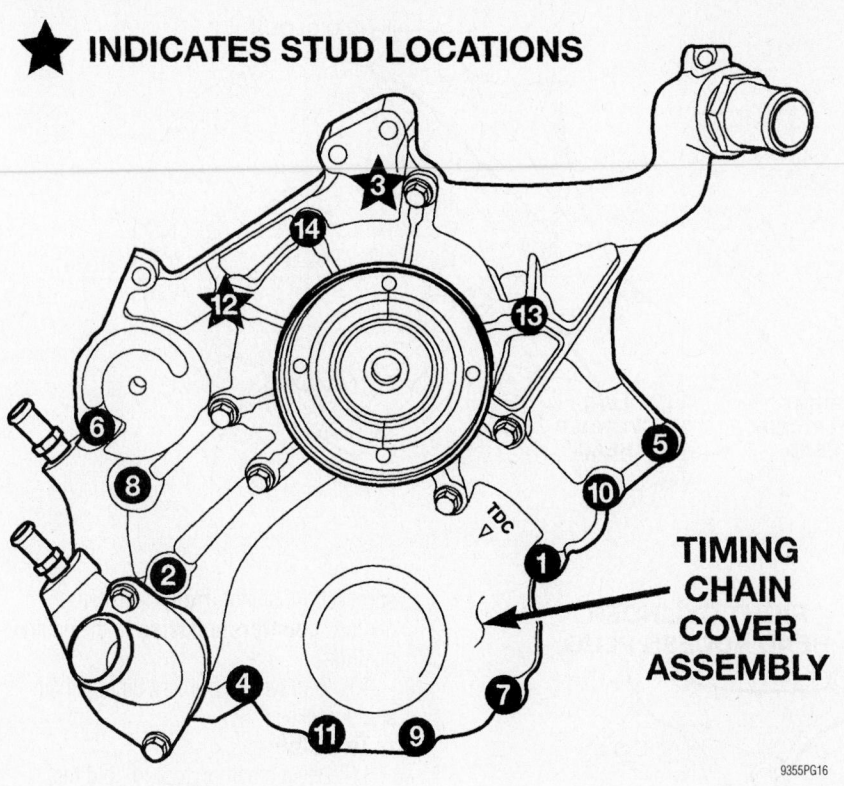

**TIMING
CHAIN
COVER
ASSEMBLY**

9355PG16

Timing cover bolt torque sequence—3.7L

- Heater hoses
- Lower radiator hose
- Power steering pump

4. Rotate the crankshaft so that the crankshaft timing mark aligns with the Top Dead Center (TDC) mark on the front cover, and the **V8** marks on the camshaft sprockets are at 12 o'clock.

5. Remove or disconnect the following:

- Crankshaft damper
- Oil fill housing
- Accessory drive belt tensioner
- Alternator
- A/C compressor
- Front cover
- Front crankshaft seal
- Cylinder head access plugs
- Secondary timing chain guides

6. Compress the primary timing chain tensioner and install a lockpin.

7. Remove the secondary timing chain tensioners.

8. Hold the left camshaft with adjustable pliers and remove the sprocket and chain. Rotate the **left** camshaft 15 degrees **clockwise** to the neutral position.

9. Hold the right camshaft with adjustable pliers and remove the camshaft

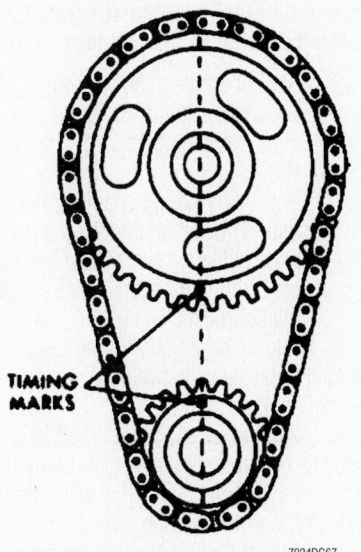

7924DG67

**Timing chain alignment marks—3.9L,
5.9L engines**

3. Remove or disconnect the following:
- Negative battery cable
- Valve covers
- Camshaft Position (CMP) sensor
- Engine cooling fan and shroud
- Accessory drive belt

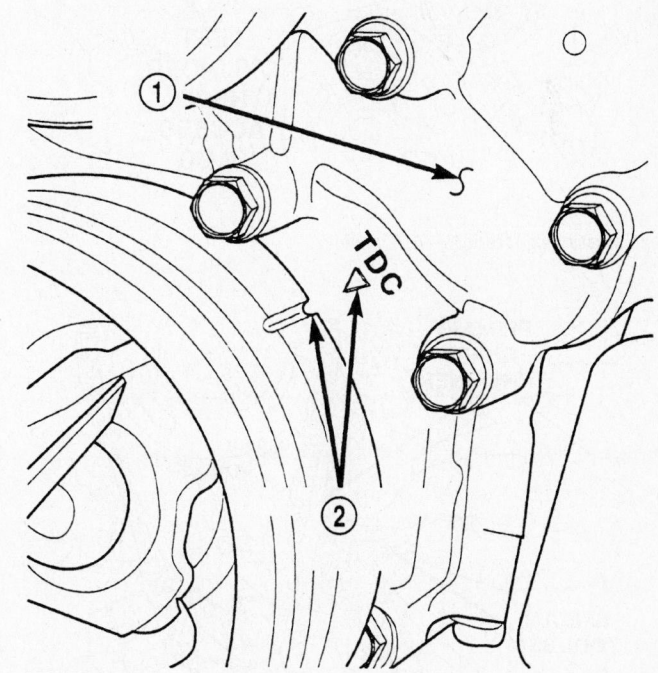

1 – TIMING CHAIN COVER
2 – CRANKSHAFT TIMING MARKS

9308PG04

Crankshaft timing marks—4.7L engine

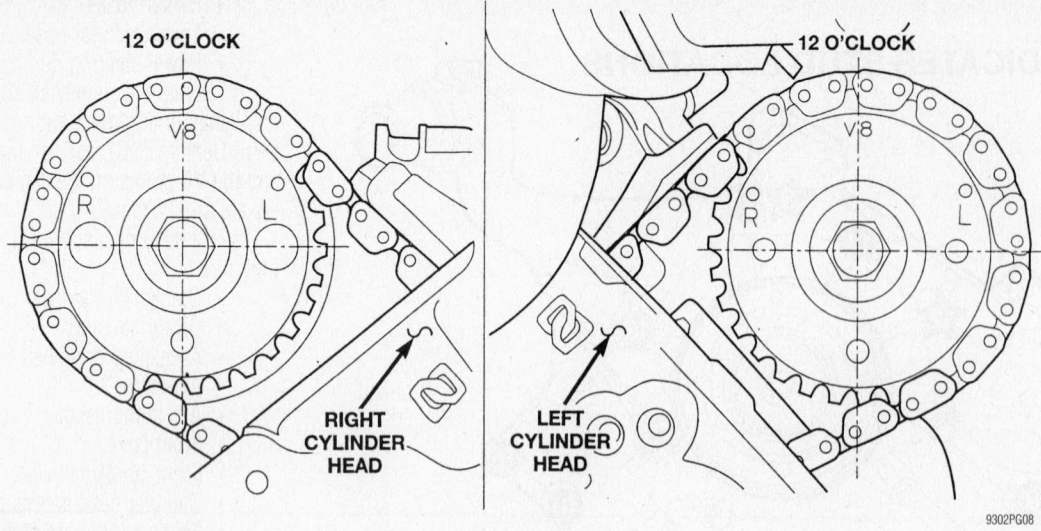

Camshaft positioning—4.7L engine

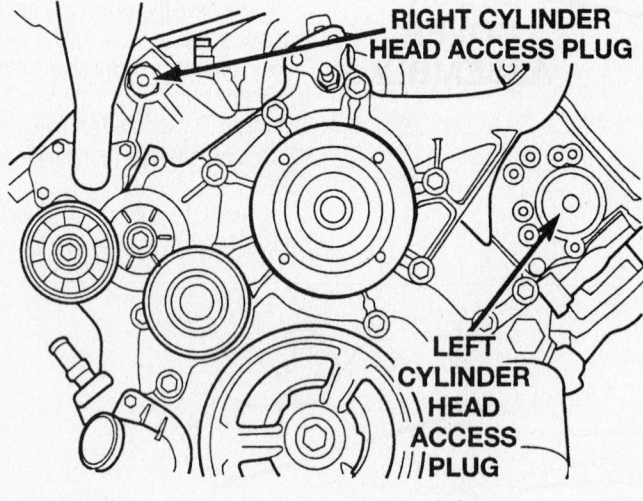

Cylinder head access plug locations—4.7L engine

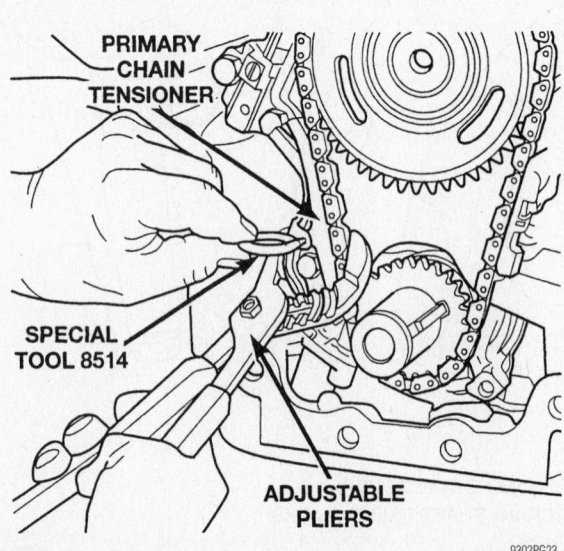

Compress and lock the primary chain tensioner—4.7L engine

sprocket. Rotate the **right** camshaft 45 degrees **counterclockwise** to the neutral position.

10. Remove the primary timing chain and sprockets.

To install:

11. Use a small prytool to hold the ratchet pawl and compress the secondary timing chain tensioners in a vise and install locking pins.

➡ **The black bolts fasten the guide to the engine block and the silver bolts fasten the guide to the cylinder head.**

12. Install or connect the following:
- Secondary timing chain guides. Tighten the bolts to 21 ft. lbs. (28 Nm).
- Secondary timing chains to the idler sprocket so that the double plated links on each chain are visible through the slots in the primary idler sprocket

13. Lock the secondary timing chains to the idler sprocket with Timing Chain Locking tool 8515 as shown.

14. Align the primary chain double plated links with the idler sprocket timing mark and the single plated link with the crankshaft sprocket timing mark.

15. Install the primary chain and sprockets. Tighten the idler sprocket bolt to 25 ft. lbs. (34 Nm).

16. Align the secondary chain single plated links with the timing marks on the secondary sprockets. Align the dot at the **L** mark on the left sprocket with the plated link on the left chain and the dot at the **R** mark on the right sprocket with the plated link on the right chain.

17. Rotate the camshafts back from the

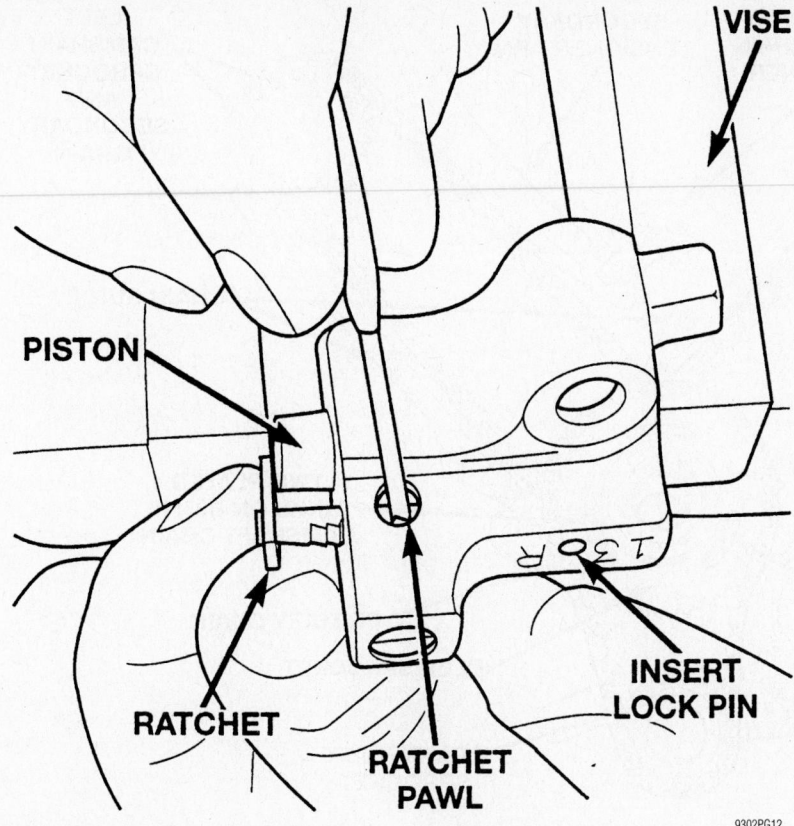

Secondary timing chain tensioner preparation—4.7L engine

9302PG12

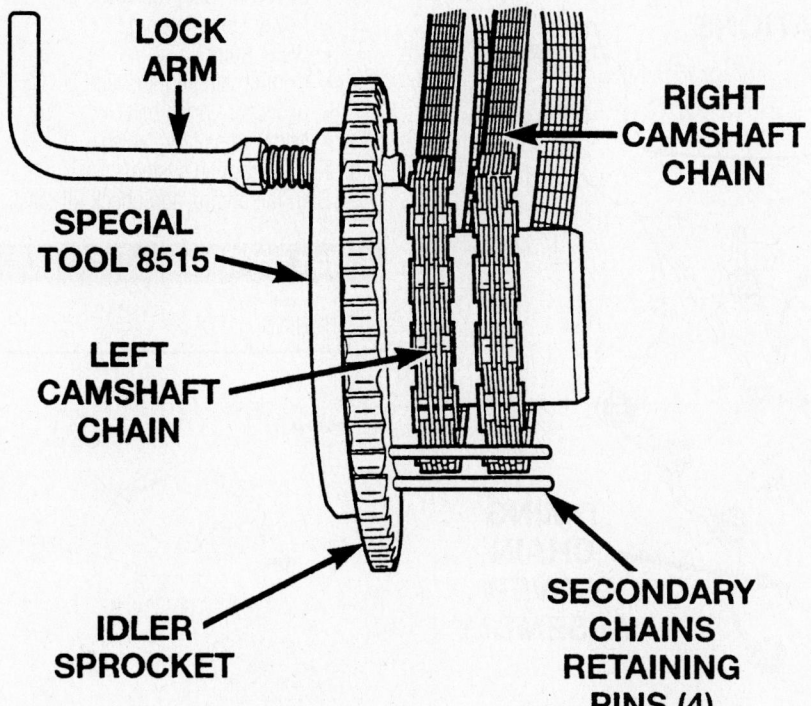

Use the Timing Chain Locking tool to lock the timing chains on the idler gear—4.7L engine

9302PG07

neutral position and install the camshaft sprockets.

18. Remove the secondary chain locking tool.

19. Remove the primary and secondary timing chain tensioner locking pins.

20. Hold the camshaft sprockets with a spanner wrench and tighten the retaining bolts to 90 ft. lbs. (122 Nm).

21. Install or connect the following:
- Front cover. Tighten the bolts, in sequence, to 40 ft. lbs. (54 Nm).
- Front crankshaft seal
- Cylinder head access plugs
- A/C compressor
- Alternator
- Accessory drive belt tensioner. Tighten the bolt to 40 ft. lbs. (54 Nm).
- Oil fill housing
- Crankshaft damper. Tighten the bolt to 130 ft. lbs. (175 Nm).
- Power steering pump
- Lower radiator hose
- Heater hoses
- Accessory drive belt
- Engine cooling fan and shroud
- Camshaft Position (CMP) sensor
- Valve covers
- Negative battery cable

22. Fill the cooling system.
23. Start the engine and check for leaks.

5.9L Engines

1. Before servicing the vehicle, refer to the precautions in the beginning of this section.

2. Drain the cooling system.

3. Remove or disconnect the following:
- Negative battery cable
- Accessory drive belt
- Cooling fan and shroud
- Water pump
- Power steering pump
- Crankshaft damper
- Front crankshaft seal
- Front cover

4. Rotate the crankshaft so that the camshaft sprocket and crankshaft sprocket timing marks are aligned.

5. Remove the timing chain and sprockets.

To install:

6. Install the timing chain and sprockets with the timing marks aligned. Tighten the camshaft sprocket bolt to 50 ft. lbs. (68 Nm).

7. Install or connect the following:
- Front cover. Tighten the cover bolts to 30 ft. lbs. (41 Nm) and the oil pan bolts to 18 ft. lbs. (24 Nm).

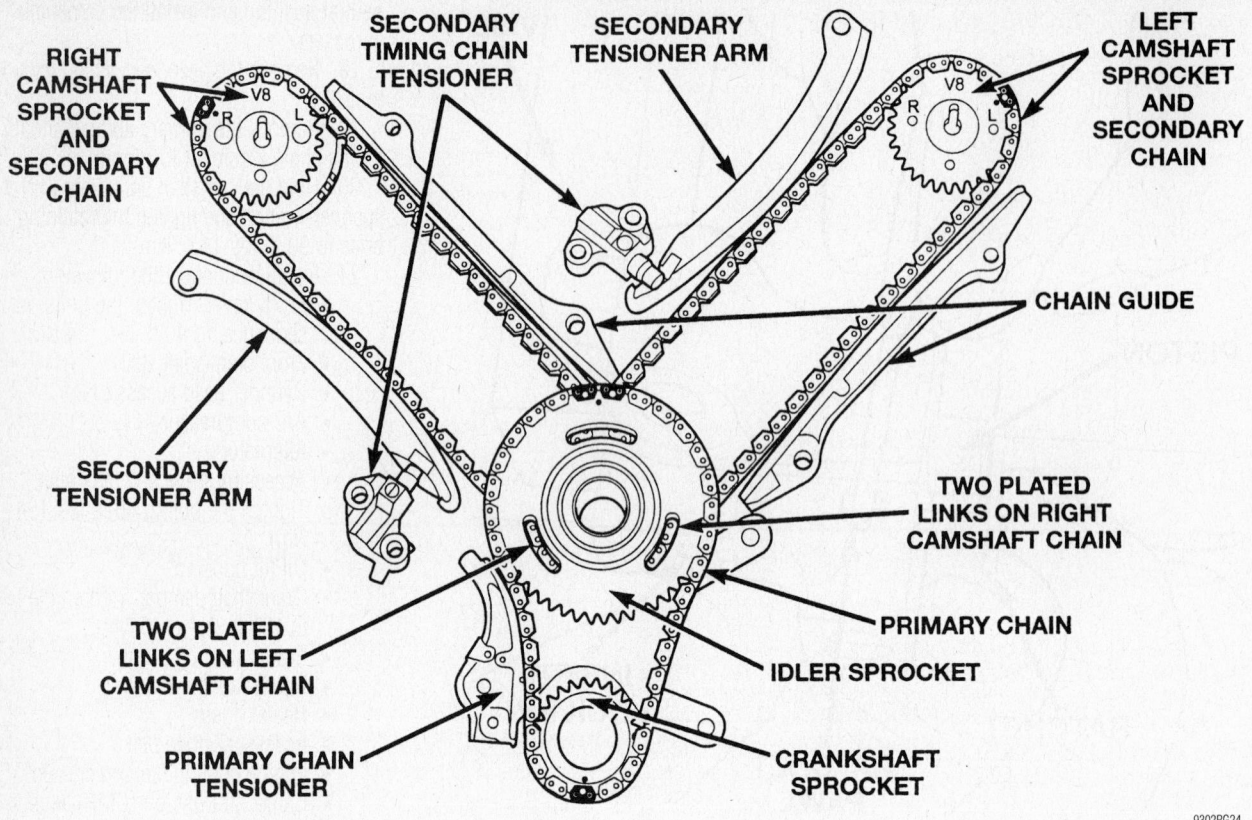

RIGHT CAMSHAFT SPROCKET AND SECONDARY CHAIN

SECONDARY TIMING CHAIN TENSIONER

SECONDARY TENSIONER ARM

LEFT CAMSHAFT SPROCKET AND SECONDARY CHAIN

CHAIN GUIDE

SECONDARY TENSIONER ARM

TWO PLATED LINKS ON RIGHT CAMSHAFT CHAIN

TWO PLATED LINKS ON LEFT CAMSHAFT CHAIN

PRIMARY CHAIN

IDLER SPROCKET

PRIMARY CHAIN TENSIONER

CRANKSHAFT SPROCKET

9302PG24

Timing chain system and alignment marks—4.7L engine

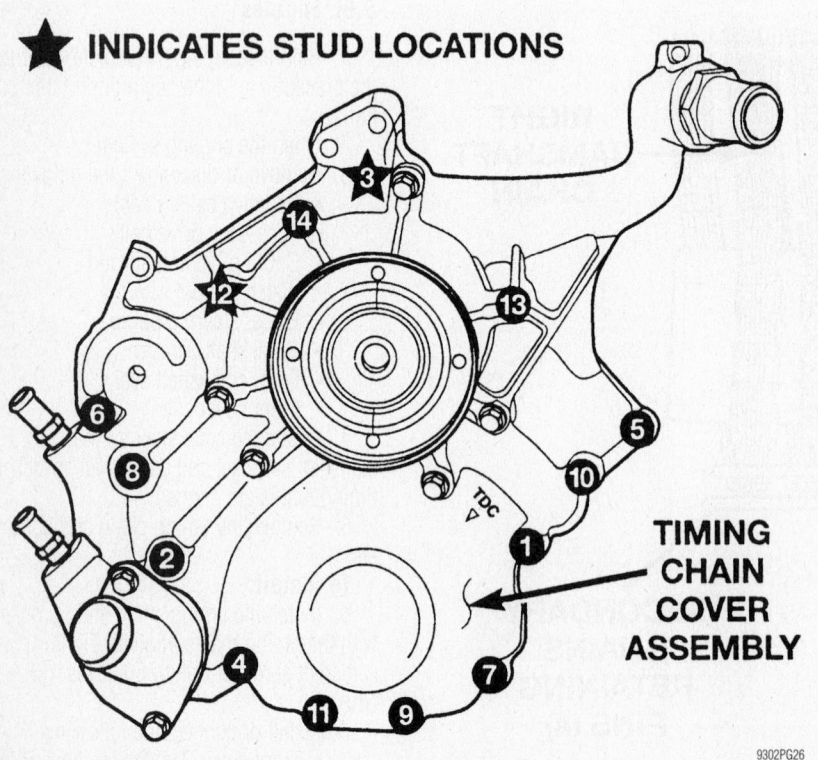

⭐ INDICATES STUD LOCATIONS

TIMING CHAIN COVER ASSEMBLY

9302PG26

Timing chain cover bolt torque sequence—4.7L engine

- Front crankshaft seal
- Crankshaft damper. Tighten the bolt to 135 ft. lbs. (183 Nm).
- Power steering pump
- Water pump
- Cooling fan and shroud
- Accessory drive belt
- Negative battery cable
8. Fill the cooling system.
9. Start the engine and check for leaks.

Piston and Ring

POSITIONING

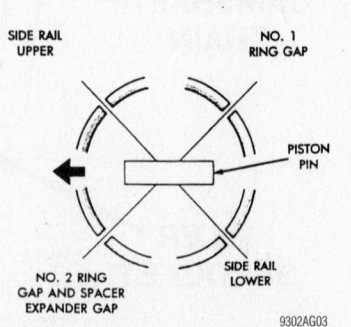

SIDE RAIL UPPER

NO. 1 RING GAP

PISTON PIN

NO. 2 RING GAP AND SPACER EXPANDER GAP

SIDE RAIL LOWER

9302AG03

Piston ring end-gap spacing. Position raised "F" on piston towards front of engine—3.7L and 4.7L engines

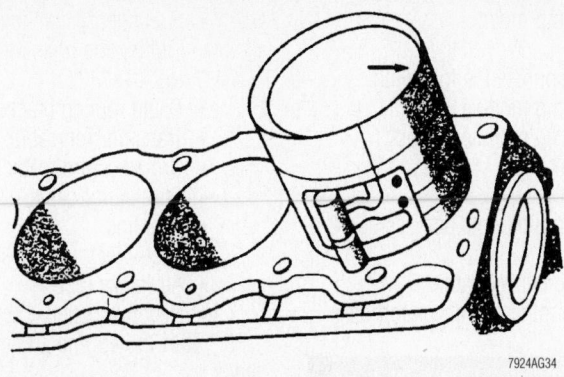

Piston to engine positioning—2.5L, 3.9L, 5.9L engines

7924AG34

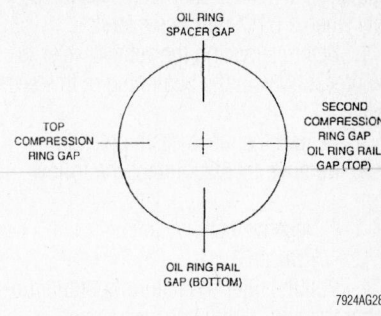

Piston ring end-gap spacing—2.5L, 3.9L, 5.9L engines

7924AG28

GASOLINE FUEL SYSTEM

Fuel System Service Precautions

Safety is the most important factor when performing not only fuel system maintenance but any type of maintenance. Failure to conduct maintenance and repairs in a safe manner may result in serious personal injury or death. Maintenance and testing of the vehicle's fuel system components can be accomplished safely and effectively by adhering to the following rules and guidelines.

• To avoid the possibility of fire and personal injury, always disconnect the negative battery cable unless the repair or test procedure requires that battery voltage be applied.

• Always relieve the fuel system pressure prior to detaching any fuel system component (injector, fuel rail, pressure regulator, etc.), fitting or fuel line connection. Exercise extreme caution whenever relieving fuel system pressure to avoid exposing skin, face and eyes to fuel spray. Please be advised that fuel under pressure may penetrate the skin or any part of the body that it contacts.

• Always place a shop towel or cloth around the fitting or connection prior to loosening to absorb any excess fuel due to spillage. Ensure that all fuel spillage (should it occur) is quickly removed from engine surfaces. Ensure that all fuel soaked cloths or towels are deposited into a suitable waste container.

• Always keep a dry chemical (Class B) fire extinguisher near the work area.

• Do not allow fuel spray or fuel vapors to come into contact with a spark or open flame.

• Always use a back-up wrench when loosening and tightening fuel line connection fittings. This will prevent unnecessary stress and torsion to fuel line piping.

• Always replace worn fuel fitting O-rings with new. Do not substitute fuel hose or equivalent, where fuel pipe is installed.

Before servicing the vehicle, also make sure to refer to the precautions in the beginning of this section as well.

Fuel System Pressure

RELIEVING

2001–2002

1. Before servicing the vehicle, refer to the precautions in the beginning of this section.
2. Remove the fuel filler cap.
3. Remove the fuel pump relay from the power distribution center.
4. Run the engine until it stalls.
5. Turn the key to **OFF**.

2003–04

1. Before servicing the vehicle, refer to the precautions in the beginning of this section.
2. Disconnect the negative battery cable.
3. Remove the fuel tank filler cap to release any fuel tank pressure.
4. Remove the fuel pump relay from the PDC.

5. Start and run the engine until it stops.
6. Unplug the connector from any injector and connect a jumper wire from either injector terminal to the positive battery terminal. Connect another jumper wire to the other terminal and momentarily touch the other end to the negative battery terminal.

✳✳ WARNING

Just touch the jumper to the battery. Powering the injector for more than a few seconds will permanently damage it.

7. Place a rag below the quick-disconnect coupling at the fuel rail and disconnect it.

Fuel Filter

REMOVAL & INSTALLATION

These engines are equipped with a fuel filter/pressure regulator. This unit is not a

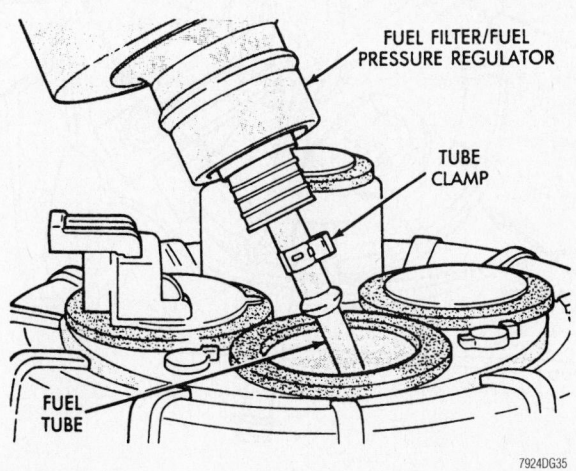

FUEL FILTER/FUEL PRESSURE REGULATOR

TUBE CLAMP

FUEL TUBE

7924DG35

Pull and twist the filter/regulator to remove it from the top of the fuel pump module

regularly maintained unit and is serviced only when a DTC indicates a fault.

1. Before servicing the vehicle, refer to the precautions in the beginning of this section.
2. Relieve the fuel system pressure.
3. Remove or disconnect the following:
   - Negative battery cable
   - Fuel tank
4. Pull the filter/regulator out of the rubber grommet. Cut the hose clamp and remove the fuel line.

To install:
5. Install the filter/regulator with a new clamp and push it into the rubber grommet.
6. Install or connect the following:
   - Fuel tank
   - Negative battery cable
7. Start the engine and check for leaks.

Fuel Pump

REMOVAL & INSTALLATION

1. Before servicing the vehicle, refer to the precautions in the beginning of this section.
2. Relieve the fuel system pressure.
3. Remove or disconnect the following:
   - Negative battery cable
   - Fuel pump module harness connector
   - Fuel line
   - Fuel tank
   - Fuel pump module locknut

- Fuel pump module

To install:
4. Install or connect the following:
   - Fuel pump module
   - Fuel pump module locknut
   - Fuel tank
   - Fuel line
   - Fuel pump module harness connector
   - Negative battery cable
5. Start the engine and check for leaks.

Fuel Injector

REMOVAL & INSTALLATION

2.5L Engine

1. Before servicing the vehicle, refer to the precautions in the beginning of this section.
2. Relieve fuel system pressure.
3. Remove or disconnect the following:
   - Negative battery cable
   - Air intake tube
   - Fuel injector connectors
   - Fuel line
   - Accelerator cable
   - Cruise control cable, if equipped
   - Transmission cable, if equipped
   - Cable routing bracket
   - Fuel supply manifold with injectors attached
   - Fuel injectors

To install:
4. Install or connect the following:
   - Fuel injectors, using new O-rings

- Fuel supply manifold with injectors. Tighten the bolts to 75–125 inch lbs. (8–14 Nm).
- Cable routing bracket
- Transmission cable, if equipped
- Cruise control cable, if equipped
- Accelerator cable
- Fuel line
- Fuel injector connectors
- Air intake tube
- Negative battery cable

5. Start the engine and check for leaks.

3.7L Engine

> **※※ CAUTION**
>
> **THE FUEL SYSTEM IS UNDER CONSTANT PRESSURE EVEN WITH ENGINE OFF.**

> **BEFORE SERVICING FUEL RAIL, FUEL SYSTEM PRESSURE MUST BE RELEASED.**

> **※※ CAUTION**
>
> **The left and right fuel rails are replaced as an assembly. Do not attempt to separate rail halves at connector tube. Due to design of tube, it does not use any clamps. Never attempt to install a clamping device of any kind to tube. When removing fuel rail assembly for any reason, be careful not to bend or kink tube.**

1. Remove fuel tank filler tube cap.
2. Perform Fuel System Pressure Release Procedure.
3. Remove negative battery cable at battery.
4. Remove air duct at throttle body air box.
5. Remove air box at throttle body.
6. Disconnect fuel line latch clip and fuel line at fuel rail. A special tool will be necessary for fuel line disconnection.
7. Remove necessary vacuum lines at throttle body.
8. Disconnect electrical connectors at all 6 fuel injectors. To remove connector, push red colored slider away from injector. While pushing slider, depress tab and remove connector from injector. The factory fuel injection wiring harness is numerically tagged (INJ 1, INJ 2, etc.) for injector position identification. If harness is not tagged, note wiring location before removal.

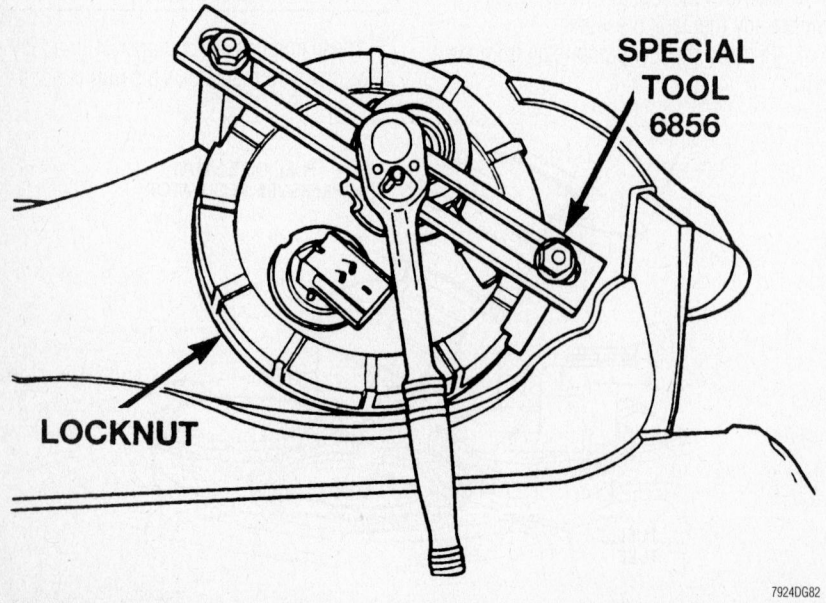

LOCKNUT

SPECIAL TOOL 6856

7924DG82

Fuel pump module locknut removal

9. Disconnect electrical connectors at throttle body sensors.

10. Remove 6 ignition coils.

11. Remove 4 fuel rail mounting bolts.

12. Gently rock and pull left side of fuel rail until fuel injectors just start to clear machined holes in cylinder head. Gently rock and pull right side of rail until injectors just start to clear cylinder head holes.

Repeat this procedure (left/right) until all injectors have cleared cylinder head holes.

13. Remove fuel rail (with injectors attached) from engine.

14. Disconnect clip(s) that retain fuel injector(s) to fuel rail.

To install:

15. Apply a small amount of clean engine oil to each fuel injector o-ring. This will help in fuel rail installation.

16. Install injector(s) and injector clip(s) to fuel rail.

17. Apply a small amount of engine oil to each fuel injector o-ring. This will help in fuel rail installation.

18. Position fuel rail/fuel injector assembly to machined injector openings in cylinder head.

19. Guide each injector into cylinder head. Be careful not to tear injector o-rings.

20. Push right side of fuel rail down until fuel injectors have bottomed on cylinder head shoulder. Push left fuel rail down until injectors have bottomed on cylinder head shoulder.

21. Install 4 fuel rail mounting bolts and tighten to 27 Nm (20 ft. lbs.).

22. Install ignition coils.

23. Connect electrical connectors to throttle body.

24. Connect electrical connectors to MAP and IAT sensors.

25. Connect electrical connectors at all fuel injectors. To install connector, push connector onto injector and then push and lock red colored slider. Verify connector is locked to injector by lightly tugging on connector.

26. Connect vacuum lines to throttle body.

27. Connect fuel line latch clip and fuel line to fuel rail.

28. Connect wiring to rear of generator.

29. Install air box to throttle body.

30. Install air duct to air box.

31. Connect battery cable to battery.

32. Start engine and check for leaks.

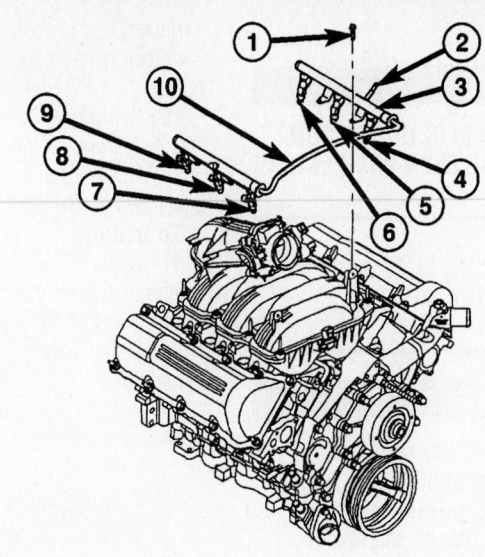

1 - MOUNTING BOLTS (4)
2 - QUICK-CONNECT FITTING
3 - FUEL RAIL
4 - INJ. #1
5 - INJ. #3
6 - INJ. #5
7 - INJ. #2
8 - INJ. #4
9 - INJ. #6
10 - CONNECTOR TUBE

67189-DAKO-G42

Fuel rail components—3.7L engine

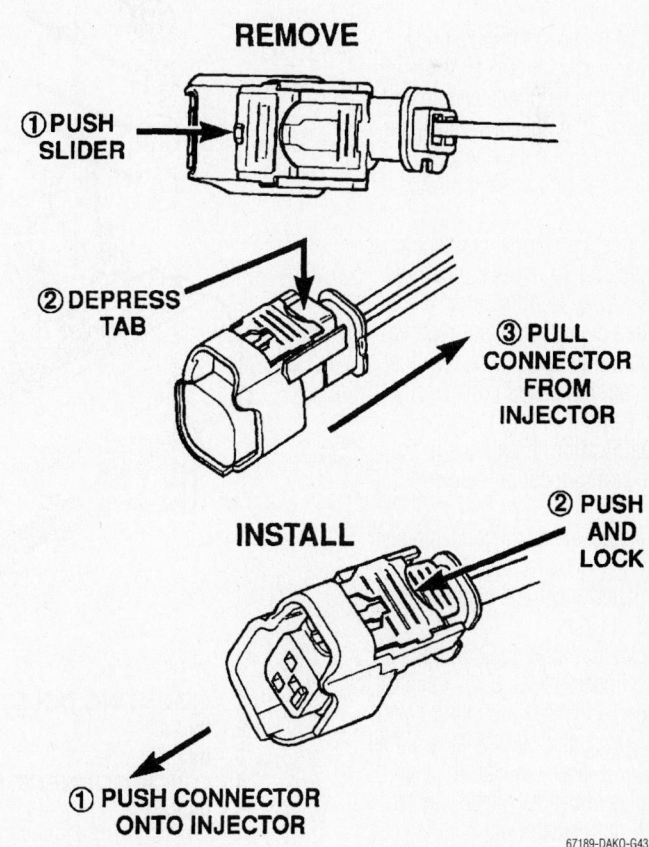

REMOVE

① PUSH SLIDER

② DEPRESS TAB

③ PULL CONNECTOR FROM INJECTOR

INSTALL

② PUSH AND LOCK

① PUSH CONNECTOR ONTO INJECTOR

67189-DAKO-G43

Fuel injector connector removal

4.7L Engine

> ※※ **CAUTION**
>
> THE FUEL SYSTEM IS UNDER CONSTANT PRESSURE EVEN WITH ENGINE OFF.
>
> BEFORE SERVICING FUEL RAIL, FUEL SYSTEM PRESSURE MUST BE RELEASED.

> ※※ **WARNING**
>
> The left and right fuel rails are replaced as an assembly. Do not attempt to separate rail halves at connector tube. Due to design of tube, it does not use any clamps. Never attempt to install a clamping device of any kind to tube. When removing fuel rail assembly for any reason, be careful not to bend or kink tube.

1. Remove fuel tank filler tube cap.
2. Perform Fuel System Pressure Release Procedure.
3. Remove negative battery cable at battery.
4. Remove air duct at throttle body air box.
5. Remove air box at throttle body.
6. Remove wiring at rear of generator.
7. Disconnect fuel line latch clip and fuel line at fuel rail. A special tool will be necessary for fuel line disconnection.
8. Remove vacuum lines at throttle body.
9. Disconnect electrical connectors at all 8 fuel injectors. To remove, push red colored slider away from injector. While pushing slider, depress tab and remove connector from injector. The factory fuel injection wiring harness is numerically tagged (INJ 1, INJ 2, etc.) for injector position identification. If harness is not tagged, note wiring location before removal.
10. Disconnect electrical connectors at throttle body.
11. Disconnect electrical connectors at MAP and IAT sensors.
12. Remove first three ignition coils on each bank (cylinders #1, 3, 5, 2, 4 and 6).
13. Remove 4 fuel rail mounting bolts.
14. Gently rock and pull left side of fuel rail until fuel injectors just start to clear machined holes in cylinder head. Gently rock and pull right side of rail until injectors just start to clear cylinder head holes.

Repeat this procedure (left/right) until all injectors have cleared cylinder head holes.
15. Remove fuel rail (with injectors attached) from engine.
16. Disconnect clip(s) that retain fuel injector(s) to fuel rail.

To install:

17. Apply a small amount of clean engine oil to each fuel injector O-ring. This will help in fuel rail installation.
18. Install injector(s) and injector clip(s) to fuel rail.
19. Apply a small amount of engine oil to each fuel injector o-ring. This will help in fuel rail installation.
20. Position fuel rail/fuel injector assembly to machined injector openings in cylinder head.
21. Guide each injector into cylinder head. Be careful not to tear injector o-rings.
22. Push right side of fuel rail down until fuel injectors have bottomed on cylinder head shoulder. Push left fuel rail down until injectors have bottomed on cylinder head shoulder.
23. Install 4 fuel rail mounting bolts and tighten to 27 Nm (20 ft. lbs.).
24. Install ignition coils.
25. Connect electrical connectors to throttle body.
26. Connect electrical connectors to MAP and IAT sensors.
27. Connect electrical connectors at all

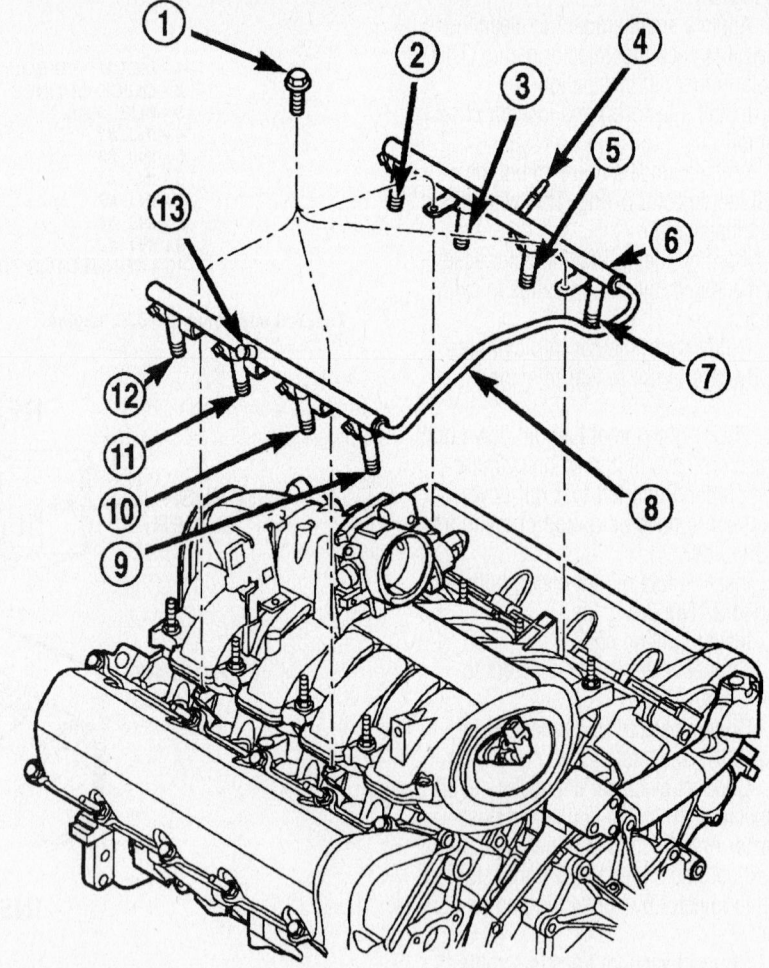

1 - MOUNTING BOLTS (4)
2 - INJ.#7
3 - INJ.#5
4 - QUICK-CONNECT FITTING
5 - INJ.#3

67189-DAKO-G44

Fuel rail components—4.7L engine

fuel injectors. To install connector, push connector onto injector and then push and lock red colored slider. Verify connector is locked to injector by lightly tugging on connector.

28. Connect vacuum lines to throttle body.

29. Connect fuel line latch clip and fuel line to fuel rail.

30. Connect wiring to rear of generator.

31. Install air box to throttle body.

32. Install air duct to air box.

33. Connect battery cable to battery.

34. Start engine and check for leaks.

3.9L and 5.9L Engines

1. Before servicing the vehicle, refer to the precautions in the beginning of this section.

2. Relieve fuel system pressure.

3. Remove or disconnect the following:
- Negative battery cable
- Air intake tube
- Throttle body
- A/C compressor bracket
- Fuel injector connectors
- Fuel line
- Fuel supply manifold with injectors
- Fuel injectors

To install:

4. Install or connect the following:
- Fuel injectors with new O-ring seals
- Fuel supply manifold with injectors. Tighten the bolts to 17 ft. lbs. (23 Nm).
- Fuel line
- Fuel injector connectors
- A/C compressor bracket
- Throttle body
- Air intake tube
- Negative battery cable

5. Start the engine and check for leaks.

DRIVE TRAIN

Transmission Assembly

REMOVAL & INSTALLATION

Manual

2001–03

1. Before servicing the vehicle, refer to the precautions in the beginning of this section.

2. Remove or disconnect the following:
- Negative battery cable
- Shift lever and tower assembly
- Crankshaft Position (CKP) sensor
- Skidplate, if equipped
- Rear driveshaft
- Front driveshaft, if equipped
- Transfer case shift linkage, if equipped
- Transmission mount and crossmember. Support the transmission.
- Exhaust front pipe
- Clutch slave cylinder
- Starter motor
- Vehicle Speed (VSS) sensor connector
- Reverse light switch connector
- Transmission flange bolts
- Transmission

To install:

3. Install or connect the following:
- Transmission. Tighten the flange bolts to 40–45 ft. lbs. (54–61 Nm).
- Reverse light switch connector
- Vehicle Speed (VSS) sensor connector
- Starter motor
- Clutch slave cylinder
- Exhaust front pipe
- Transmission mount and crossmember. Tighten the fasteners to 50 ft. lbs. (68 Nm).
- Transfer case shift linkage, if equipped

- Front driveshaft, if equipped
- Rear driveshaft
- Skidplate, if equipped
- CKP sensor
- Shift lever and tower assembly
- Negative battery cable

2004—2-WHEEL DRIVE

1. Disconnect battery negative cable.

2. Shift transmission into Neutral.

3. Remove floor console.

4. Remove shift lever boot.

5. Remove the shift lever extension from the shift tower and lever assembly.

6. Raise vehicle.

7. Remove skid plate, if equipped (4WD).

8. If transmission will be disassembled for repair, remove drain plug and drain lubricant from transmission.

9. Mark propeller shafts and companion flange for assembly reference.

10. Disconnect and remove propeller shafts.

11. Disconnect and remove exhaust system Y-pipe. Then disconnect and lower remaining exhaust pipes for clearance as necessary.

12. Disconnect backup light switch wires.

13. Remove bolts/nuts attaching transmission to rear mount.

14. Support transmission with a transmission jack. Secure transmission to jack with safety chains.

15. Remove rear crossmember.

16. Remove bolts attaching clutch slave cylinder to clutch housing. Then move cylinder aside for working clearance.

17. Remove starter.

18. Remove transmission dust shield.

19. Remove transmission harness wires from clips on transmission shift cover.

20. Lower transmission slightly.

21. Remove the bolts attaching the shift tower and lever assembly to the transmission housing. Then remove the shift tower and lever assembly.

22. Remove bolts attaching transmission to engine.

23. Slide transmission and jack rearward until input shaft clears clutch disc.

24. Lower transmission jack and remove transmission from under vehicle.

To install:

➡**If a new transmission is being installed, be sure to use all components supplied with the new transmission. For example, if a new shift tower is supplied with the new transmission, do not re-use the original shift tower.**

25. Apply light coat of Mopar® high temperature bearing grease to contact surfaces of following components:
- input shaft splines.
- release bearing slide surface of front retainer.
- release bearing bore.
- release fork.
- release fork ball stud.
- propeller shaft slip yoke.

26. Apply sealer to threads of drain plug, then install plug in case.

27. Mount transmission on jack and position transmission under vehicle. Secure transmission to jack with safety chains.

28. Raise transmission until input shaft is centered in clutch disc hub.

29. Move transmission forward and start input shaft in clutch disc.

30. Work transmission forward until seated against engine. Do not allow transmission to remain unsupported after input shaft has entered clutch disc.

31. Install and tighten transmission to engine bolts to 108 Nm (80 ft. lbs.).

32. Position transmission harness wires in clips on shift cover.

33. Install slave cylinder and shield, if equipped.

34. Install transmission mount on transmission or rear crossmember.

2004—4-WHEEL DRIVE

1. Disconnect battery negative cable.
2. Shift transmission into Neutral.
3. Remove floor console.
4. Remove shift lever boot.
5. Remove the shift lever extension from the shift tower and lever assembly.
6. Raise vehicle.
7. Remove skid plate, if equipped (4WD).
8. If transmission will be disassembled for repair, remove drain plug and drain lubricant from transmission.
9. Mark propeller shafts and companion flange for assembly reference.
10. Disconnect and remove propeller shafts.
11. Disconnect and remove exhaust system Y-pipe. Then disconnect and lower remaining exhaust pipes for clearance as necessary.
12. Disconnect backup light switch wires.
13. Support engine with adjustable safety stand.
14. Disconnect transfer case shift linkage at transfer case range lever.
15. Remove transfer case shift lever from transmission.
16. Remove bolts/nuts attaching transmission to rear support.
17. Remove crossmember bolts/nuts and remove crossmember.
18. Support transfer case with transmission jack. Secure transfer case to jack with safety chains.
19. Remove transfer case attaching nuts.
20. Move transfer case rearward until input gear clears transmission output shaft.
21. Lower transfer case assembly and move it from under vehicle.
22. Support transmission with transmission jack. Secure transmission to jack with safety chains.
23. Remove transmission harness from retaining clips on transmission shift cover.
24. Remove clutch slave cylinder splash shield, if equipped.
25. Remove clutch slave cylinder attaching nuts. Move cylinder aside for working clearance.
26. Remove starter.
27. Remove transmission splash shield.
28. Lower transmission slightly
29. Remove bolts attaching shift tower

and lever assembly to rear case. Then remove shift tower and lever as an assembly.

30. Remove bolts attaching transmission to engine.

31. Move transmission rearward until input shaft clears clutch disc.

32. Lower transmission and remove it from under vehicle.

To install:

➡If a new transmission is being installed, be sure to use all components supplied with the new transmission. For example, if a new shift tower is supplied with the new transmission, do not re-use the original shift tower.

33. Apply light coat of Mopar® high temperature bearing grease to contact surfaces of following components:
  • input shaft splines.
  • release bearing slide surface of front retainer.
  • release bearing bore.
  • release fork.
  • release fork ball stud.
  • propeller shaft slip yoke.

34. Apply sealer to threads of drain plug, then install plug in case.

35. Mount transmission on jack and position transmission under vehicle. Secure transmission to jack with safety chains.

36. Raise transmission until input shaft is centered in clutch disc hub.

37. Move transmission forward and start input shaft in clutch disc.

38. Work transmission forward until seated against engine. Do not allow transmission to remain unsupported after input shaft has entered clutch disc.

39. Install and tighten transmission to engine bolts to 108 Nm (80 ft. lbs.).

40. Position transmission harness wires in clips on shift cover.

41. Install slave cylinder and shield, if equipped.

42. Install transmission mount on transmission or rear crossmember.

43. Install transfer case shift lever on transmission.

44. Install rear crossmember.

➡Ensure wiring harness is clear before installing crossmember.

45. Remove transmission jack and engine support fixture.

46. Install transfer case on transmission jack. Secure transfer case to jack with safety chains.

47. Raise jack and align transfer case input gear with transmission output shaft.

48. Move transfer case forward and seat it on transmission.

49. Install and tighten transfer case attaching nuts. Tighten nuts to 41–47 Nm (30–35 ft. lbs.) if case has ⅜ studs, or 30–41 Nm (22–30 ft. lbs.) if case has 5/16 studs.

50. Connect backup light switch wires.

51. Install transmission dust cover.

52. Install starter.

53. Install transfer case shift lever to side of transfer case.

54. Connect transfer case shift lever to range lever on transfer case.

55. Align and connect propeller shafts.

56. Fill transmission with required lubricant. Check lubricant level in transfer case and add lubricant if necessary.

57. Install transfer case skid plate, if equipped, and crossmember. Tighten attaching bolts/nuts to 41 Nm (30 ft. lbs.).

58. Install exhaust system components.

59. Lower vehicle.

60. Install shift tower and lever assembly. Tighten shift tower bolts to 7–10 Nm (5–7 ft. lbs.).

61. Install shift lever boot.

62. Install floor console.

63. Connect battery negative cable.

Automatic

2001–03

1. Before servicing the vehicle, refer to the precautions in the beginning of this section.

2. Remove or disconnect the following:
  • Negative battery cable
  • Rear driveshaft
  • Crankshaft Position (CKP) sensor
  • Exhaust front pipe
  • Transmission braces, if equipped
  • Starter motor
  • Transmission oil cooler lines
  • Torque converter access cover
  • Torque converter
  • Transmission oil dipstick tube
  • Vehicle Speed (VSS) sensor connector
  • Park/Neutral switch connector
  • Shift cable
  • Throttle valve cable
  • Transmission mount and crossmember. Support the transmission.
  • Front driveshaft and transfer case, if equipped
  • Transmission flange bolts
  • Transmission

To install:

3. Install or connect the following:
  • Transmission. Tighten the flange bolts to 65 ft. lbs. (87 Nm).

- Front driveshaft and transfer case, if equipped
- Transmission mount and crossmember
- Throttle valve cable
- Shift cable
- Park/Neutral switch connector
- VSS sensor connector
- Transmission oil dipstick tube
- Torque converter. Tighten the bolts to 23 ft. lbs. (31 Nm) for 10.75 inch converters and to 35 ft. lbs. (47 Nm) for 12.2 inch converters.
- Torque converter access cover
- Transmission oil cooler lines
- Starter motor
- Transmission braces, if equipped. Tighten the bolts to 30 ft. lbs. (41 Nm).
- Exhaust front pipe
- CKP sensor
- Rear driveshaft
- Negative battery cable

2004 545RFE

1. Disconnect the negative battery cable.
2. Raise and support the vehicle
3. Remove any necessary skid plates.
4. Mark propeller shaft and axle companion flanges for assembly alignment.
5. Remove the rear propeller shaft.
6. Remove the front propeller shaft, if necessary.
7. Remove the engine to transmission collar.
8. Remove the exhaust support bracket from the rear of the transmission.
9. Disconnect and lower or remove any necessary exhaust components.
10. Remove the starter motor.
11. Rotate crankshaft in clockwise direction until converter bolts are accessible. Then remove bolts one at a time. Rotate crankshaft with socket wrench on dampener bolt.
12. Disengage the output speed sensor connector from the output speed sensor.
13. Disengage the input speed sensor connector from the input speed sensor.
14. Disengage the transmission solenoid/TRS assembly connector from the transmission solenoid/TRS assembly.
15. Disengage the line pressure sensor connector from the line pressure sensor.
16. Disconnect gearshift cable from transmission manual valve lever.
17. Disconnect transmission fluid cooler lines at transmission fittings and clips.
18. Disconnect the transmission vent hose from the transmission.
19. Support rear of engine with safety stand or jack.

20. Raise transmission slightly with service jack to relieve load on crossmember and supports.
21. Remove bolts securing rear support and cushion to transmission and crossmember.
22. Remove bolts attaching crossmember to frame and remove crossmember.
23. Remove transfer case.
24. Remove all remaining converter housing bolts.
25. Carefully work transmission and torque converter assembly rearward off engine block dowels.
26. Hold torque converter in place during transmission removal.
27. Lower transmission and remove assembly from under the vehicle.
28. To remove torque converter, carefully slide torque converter out of the transmission.

To install:
29. Check torque converter hub and hub drive flats for sharp edges burrs, scratches, or nicks. Polish the hub and flats with 320/400 grit paper and crocus cloth if necessary. Verify that the converter hub O-ring is properly installed and is free of any debris. The hub must be smooth to avoid damaging pump seal at installation.
30. If a replacement transmission is being installed, transfer any components necessary, such as the manual shift lever and shift cable bracket, from the original transmission onto the replacement transmission.
31. Lubricate oil pump seal lip with transmission fluid.
32. Align converter and oil pump.
33. Carefully insert converter in oil pump. Then rotate converter back and forth until fully seated in pump gears.
34. Check converter seating with steel scale and straightedge. Surface of converter lugs should be at least 13mm (½ in.) to rear of straightedge when converter is fully seated.
35. Temporarily secure converter with C-clamp.
36. Position transmission on jack and secure it with chains.
37. Check condition of converter driveplate. Replace the plate if cracked, distorted or damaged. Also be sure transmission dowel pins are seated in engine block and protrude far enough to hold transmission in alignment.
38. Apply a light coating of high temperature grease to the torque converter hub pocket in the rear pocket of the engine's crankshaft.

39. Raise transmission and align the torque converter with the drive plate and transmission converter housing with the engine block.
40. Move transmission forward. Then raise, lower or tilt transmission to align the converter housing with engine block dowels.
41. Carefully work transmission forward and over engine block dowels until converter hub is seated in crankshaft. Verify that no wires, or the transmission vent hose, have become trapped between the engine block and the transmission.
42. Install two bolts to attach the transmission to the engine.
43. Install remaining torque converter housing to engine bolts. Tighten to 68 Nm (50 ft. lbs.).
44. Install transfer case, if equipped. Tighten transfer case nuts to 35 Nm (26 ft. lbs.).
45. Install rear transmission crossmember. Tighten crossmember to frame bolts to 68 Nm (50 ft. lbs.).
46. Install rear support to transmission. Tighten bolts to 47 Nm (35 ft. lbs.).
47. Lower transmission onto crossmember and install bolts attaching transmission mount to crossmember. Tighten clevis bracket to crossmember bolts to 47 Nm (35 ft. lbs.). Tighten the clevis bracket to rear support bolt to 68 Nm (50 ft. lbs.).
48. Remove engine support fixture.
49. Connect gearshift cable to transmission.
50. Connect wires to solenoid and pressure switch assembly connector, input and output speed sensors, and line pressure sensor. Be sure transmission harnesses are properly routed.
51. Install torque converter-to-driveplate bolts. Tighten bolts to 31 Nm (270 inch lbs.).
52. Install starter motor and cooler line bracket.
53. Connect cooler lines to transmission.
54. Install transmission fill tube.
55. Install exhaust components.
56. Install the engine collar onto the transmission and the engine. Tighten the bolts to 54 Nm (40 ft. lbs.).
57. Align and connect propeller shaft(s).
58. Adjust gearshift cable if necessary.
59. Install any skid plates removed previously.
60. Lower vehicle.
61. Fill transmission with Mopar® ATF +4, Automatic Transmission Fluid.

2004 42RLE

1. Disconnect the negative battery cable.

2. Raise and support the vehicle

3. Remove any necessary skid plates.

4. Mark the propeller shaft and axle companion flanges for assembly alignment.

5. Remove the rear propeller shaft.

6. Remove the front propeller shaft, if necessary.

7. Disconnect the wires from the input and output speed sensors.

8. Disconnect the wires from the transmission range sensor and the solenoid/pressure switch assembly.

9. Remove the bolts holding the exhaust crossover pipe to the pre-catalytic converter pipe flanges.

10. Remove the bolts holding the exhaust crossover pipe to the catalytic converter flange.

11. Disconnect gearshift cable from transmission manual valve lever.

12. Disengage the shift cable from the cable support bracket.

13. Remove the starter motor.

14. Remove the engine to transmission collar.

15. Rotate the crankshaft in clockwise direction until the converter bolts are accessible. Then remove the bolts one at a time. Rotate the crankshaft with a socket wrench on dampener bolt.

16. Disconnect the transmission vent hose from the transmission.

17. Remove the transfer case.

18. Support the rear of engine with a safety stand or jack.

19. Raise the transmission slightly with a service jack to relieve the load on the crossmember and supports.

20. Remove the bolts securing the rear support and cushion to the transmission and crossmember.

21. Remove the bolts attaching the crossmember to frame and remove crossmember.

22. Disconnect the transmission fluid cooler lines at the transmission fittings and clips.

23. Remove all remaining converter housing bolts.

24. Carefully work the transmission and torque converter assembly rearward off the engine block dowels.

25. Hold the torque converter in place during transmission removal.

26. Lower the transmission and remove the assembly from under the vehicle.

27. To remove the torque converter, carefully slide the torque converter out of the transmission.

To install:

28. Check the torque converter hub and hub drive flats for sharp edges burrs, scratches, or nicks. Polish the hub and flats with 320/400 grit paper and crocus cloth if necessary. Verify that the converter hub O-ring is properly installed and is free of any debris. The hub must be smooth to avoid damaging pump seal at installation.

29. If a replacement transmission is being installed, transfer any components necessary, such as the manual shift lever and shift cable bracket, from the original transmission onto the replacement transmission.

30. Lubricate the oil pump seal lip with transmission fluid.

31. Align the converter and oil pump.

32. Carefully insert the converter in the oil pump. Then rotate the converter back and forth until fully seated in the pump gears.

33. Check the converter seating with a steel scale and straightedge. The surface of the converter lugs should be at least 13mm (½ in.) to the rear of straightedge when the converter is fully seated.

34. Temporarily secure the converter with a C-clamp.

35. Position the transmission on the jack and secure it with chains.

36. Check the condition of the converter driveplate. Replace the plate if cracked, distorted or damaged. Also, be sure the transmission dowel pins are seated in the engine block and protrude far enough to hold the transmission in alignment.

37. Apply a light coating of high temperature grease to the torque converter hub pocket in the rear pocket of the engine's crankshaft.

38. Raise the transmission and align the torque converter with the drive plate and transmission converter housing with the engine block.

39. Move the transmission forward. Then raise, lower or tilt transmission to align the converter housing with engine block dowels.

40. Carefully work the transmission forward and over the engine block dowels until the converter hub is seated in the crankshaft. Verify that no wires, or the transmission vent hose, have become trapped between the engine block and the transmission.

41. Install the two bolts to attach the transmission to the engine.

42. Install the remaining torque converter housing to engine bolts. Tighten to 68 Nm (50 ft. lbs.).

43. Install the transfer case, if equipped. Tighten the transfer case nuts to 35 Nm (26 ft. lbs.).

44. Install the rear transmission crossmember. Tighten the crossmember to frame bolts to 68 Nm (50 ft. lbs.).

45. Install the rear support to the transmission. Tighten the bolts to 47 Nm (35 ft. lbs.).

46. Lower the transmission onto the crossmember and install the bolts attaching the transmission mount to the crossmember. Tighten clevis bracket to crossmember bolts to 47 Nm (35 ft. lbs.). Tighten the clevis bracket to rear support bolt to 68 Nm (50 ft. lbs.).

47. Remove the engine support fixture.

48. Connect the gearshift cable to the support bracket and transmission manual lever.

49. Connect the input and output speed sensor wires.

50. Connect the wires to the transmission range sensor and the solenoid/pressure switch assembly.

51. Install the torque converter-to-driveplate bolts. Tighten the bolts to 88 Nm (65 inch lbs.).

52. Install the starter motor and cooler line bracket.

53. Connect the cooler lines to transmission.

54. Install the transmission fill tube.

55. Install the exhaust components.

56. Align and connect the propeller shaft(s).

57. Adjust the gearshift cable if necessary.

58. Install any skid plates removed previously.

59. Lower the vehicle.

60. Fill the transmission with the required amount of Mopar® ATF +4, Automatic Transmission Fluid.

Clutch

REMOVAL & INSTALLATION

2001–03

1. Before servicing the vehicle, refer to the precautions in the beginning of this section.

2. Remove or disconnect the following:
   - Negative battery cable
   - Transfer case, if equipped
   - Transmission
   - Pressure plate. Loosen the bolts evenly in ½ turn steps.
   - Clutch disc

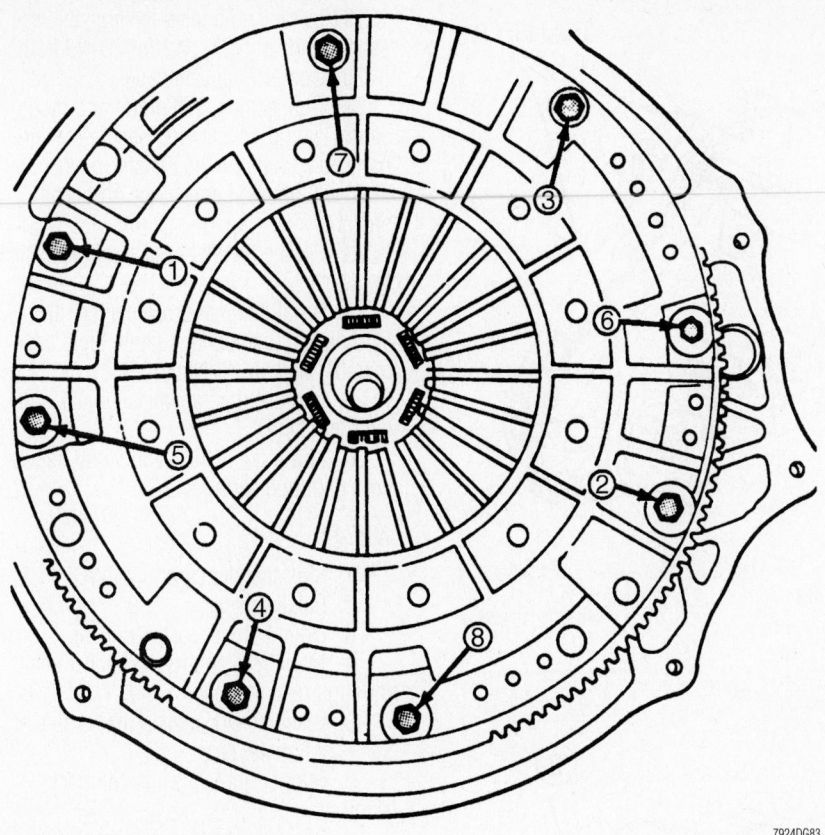

7924DG83

Pressure plate torque sequence

To install:

3. Install or connect the following:
- Clutch disc and pressure plate. Tighten the pressure plate bolts evenly in ½ turns to 21 ft. lbs. (28 Nm).
- Transmission
- Transfer case, if equipped
- Negative battery cable

2004

1. Raise vehicle.
2. Remove transmission and clutch housing as assembly.
3. If pressure plate is being removed for access to another component, mark position of pressure plate cover on flywheel with small punch marks.
4. Loosen pressure plate cover bolts evenly and in rotation to relieve spring tension. Loosen bolts a few threads at a time to avoid warping cover.
5. Remove cover bolts, pressure plate and clutch disc.

To install:

➡Clean flywheel surface with solvent. Scuff sand surface with 120/180 grit emery cloth to remove minor scratches and glazing.

6. Check new clutch disc for runout and free operation on input shaft splines.
7. Lubricate crankshaft pilot bearing with a NLGI—2 rated grease.
8. Position clutch disc on the flywheel.
9. Insert alignment tool or spare input shaft through clutch disc and into pilot bearing.
10. Verify that disc hub is positioned correctly. The raised portion of the hub faces away from the flywheel.
11. Position pressure plate cover over disc and on flywheel.
12. Install cover bolts finger tight.
13. Tighten cover bolts evenly (and in rotation) a few threads at a time. Cover bolts must be tightened evenly and to specified torque to avoid distorting cover.
14. Tighten cover bolts to:
- ⁵⁄₁₆ in. bolts to 23 Nm (17 ft. lbs.).
- ³⁄₈ in. bolts to 41 Nm (30 ft. lbs.).
15. Apply light coat of high temperature bearing grease to splines of transmission input shaft
and to release bearing slide surface of front bearing retainer.

➡Do not over-lubricate shaft splines. This could result in grease contamination of disc.

16. Install transmission as assembly.

Hydraulic Clutch System

BLEEDING

The system is self-bleeding. Press the clutch pedal repeatedly to release air from the fluid. The air will be vented from the reservoir.

Transfer Case Assembly

REMOVAL & INSTALLATION

2001–03

1. Before servicing the vehicle, refer to the precautions in the beginning of this section.
2. Shift the transfer case into **N**.
3. Remove or disconnect the following:
- Front and rear driveshafts
- Transmission mount and crossmember. Support the transmission.
- Vehicle Speed (VSS) sensor connector
- Shift linkage
- Vent hose
- Vacuum hose
- Indicator switch connector
- Transfer case attaching nuts
- Transfer case

To install:

4. Install or connect the following:
- Transfer case. Tighten the nuts to 26 ft. lbs. (35 Nm).
- Indicator switch connector
- Vacuum hose
- Vent hose
- Shift linkage
- VSS sensor connector
- Transmission mount and crossmember
- Front and rear driveshafts

2004

NV233

1. Shift the transfer case into 2WD.
2. Raise the vehicle.
3. Drain the transfer case lubricant.
4. Mark the front and rear propeller shafts for alignment reference.
5. Support the transmission with jack stand.
6. Remove the rear crossmember and skid plate, if equipped.
7. Disconnect the front and rear propeller shafts at transfer case.
8. Disconnect the transfer case shift motor and mode sensor wire connectors.
9. Disconnect the transfer case vent hose.

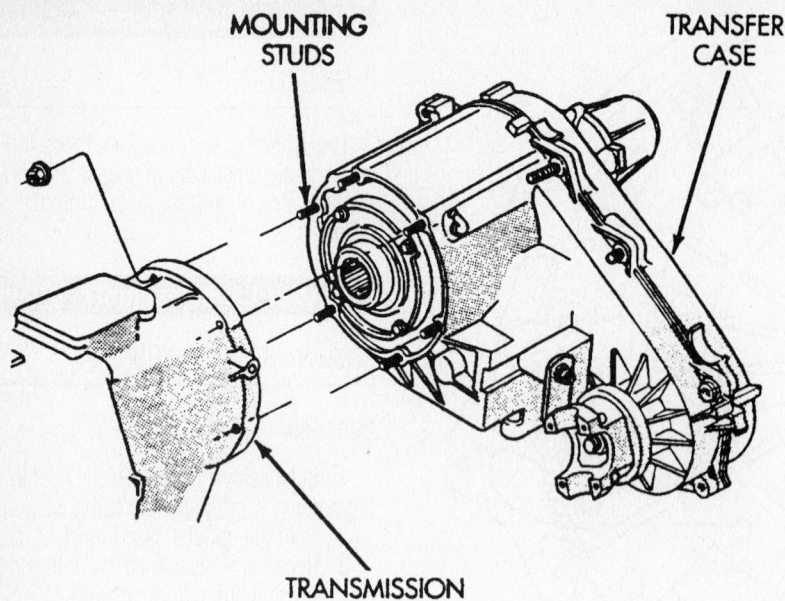

MOUNTING STUDS

TRANSFER CASE

TRANSMISSION

7924DG84

Typical transfer case mounting

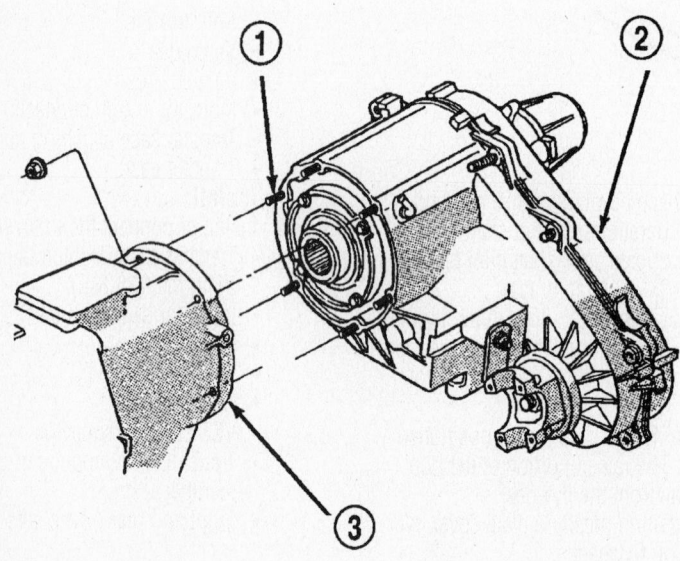

1 - MOUNTING STUDS
2 - TRANSFER CASE
3 - TRANSMISSION

67189-DAKO-G41

Typical transfer case mounting

10. Support the transfer case with a transmission jack.

11. Secure the transfer case to the jack with chains.

12. Remove the nuts attaching transfer case to the transmission.

13. Pull the transfer case and jack rearward to disengage the transfer case.

14. Remove the transfer case from under the vehicle.

To install:

15. Mount the transfer case on a transmission jack.

16. Secure the transfer case to the jack with chains.

17. Position transfer case under vehicle.

18. Align the transfer case and transmission shafts and install the transfer case onto the transmission.

19. Install and tighten the transfer case attaching nuts to 27–34 Nm (20–25 ft. lbs.).

20. Connect the vent hose.

21. Connect the shift motor and mode sensor wiring connectors. Secure the wire harness to clips on the transfer case.

22. Align and connect the propeller shafts.

23. Fill the transfer case with the correct fluid.

24. Install the rear crossmember and skid plate, if equipped. Tighten the crossmember bolts to 41 Nm (30 ft. lbs.).

25. Remove the transmission jack and support stand.

26. Lower the vehicle and verify transfer case shift operation.

NV244

1. Shift transfer case into AWD.

2. Raise vehicle.

3. Drain transfer case lubricant.

4. Mark front and rear propeller shafts for alignment reference.

5. Disconnect front and rear propeller shafts at transfer case.

6. Support transmission with jack stand.

7. Remove rear crossmember and skid plate, if equipped.

8. Disconnect transfer case shift motor and mode sensor wire connectors.

9. Disconnect transfer case vent hose.

10. Support transfer case with transmission jack.

11. Secure transfer case to jack with chains.

12. Remove nuts attaching transfer case to transmission.

13. Pull transfer case and jack rearward to disengage transfer case from the transmission adapte housing and output shaft.

14. Remove transfer case from under vehicle.

To install:

15. Mount transfer case on a transmission jack.

16. Secure transfer case to jack with chains.

17. Position transfer case under vehicle.

18. Align transfer case and transmission shafts and install transfer case onto the transmission.

19. Install and tighten transfer case attaching nuts to 27–34 Nm (20–25 ft. lbs.). Connect the vent hose.

20. Connect the shift motor and mode sensor wiring connectors. Secure wire harness to clips on transfer case.

21. Align and connect the propeller shafts.

22. Fill transfer case with correct fluid.

23. Install rear crossmember and skid plate, if equipped. Tighten crossmember bolts to 41 Nm (30 ft. lbs.).

24. Remove transmission jack and support stand.

25. Lower vehicle and verify transfer case shift operation.

Halfshaft

REMOVAL & INSTALLATION

2001–03

1. Before servicing the vehicle, refer to the precautions in the beginning of this section.

2. Remove or disconnect the following:
- Skid plate, if equipped
- Front wheel
- Split pin
- Nut lock
- Spring washer
- Hub nut
- Brake caliper and rotor
- Wheel speed sensor, if equipped
- Wheel bearing and hub assembly

3. Pry the inner tripod joint out of the differential and remove the axle halfshaft.

To install:

4. Install the axle halfshaft so that the snapring is felt to seat in the joint housing groove.

5. Install or connect the following:
- Wheel bearing and hub assembly
- Wheel speed sensor, if equipped
- Brake caliper and rotor
- Hub nut. Tighten the nut to 180 ft. lbs. (244 Nm).
- Spring washer
- Nut lock
- Split pin
- Front wheel
- Skid plate, if equipped

2004

1. Raise the vehicle.

2. Remove the wheel and tire assembly.

3. Remove the skid plate, if equipped.

4. Remove the cotter pin, nut lock, and spring washer from the stub shaft.

5. Remove the hub nut and washer from the stub shaft.

6. Remove the brake caliper and rotor.

7. Remove the ABS wheel speed sensor if equipped.

8. Remove the hub bearing bolts and hub bearing from the knuckle.

9. Support the half shaft at the CV joint housings.

10. Position two pry bars behind the

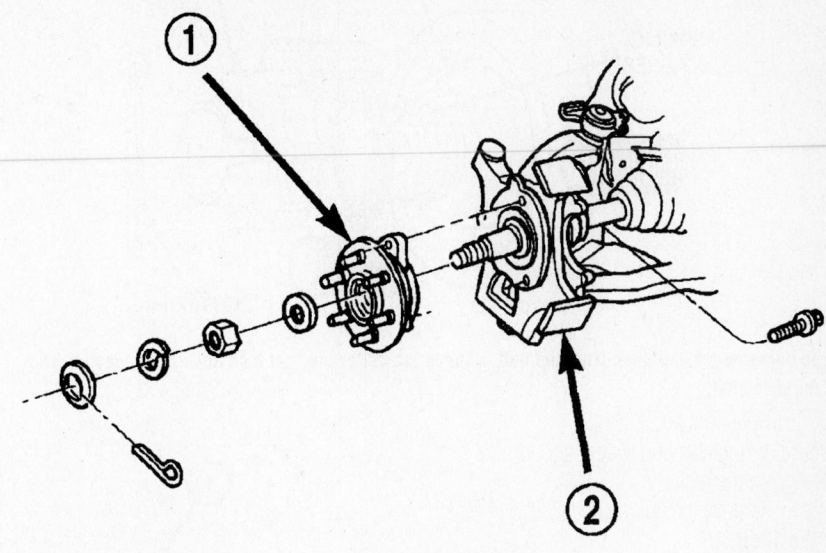

1 - HUB BEARING
2 - STEERING KNUCKLE

67189-DAKO-G39

Hub/bearing assembly

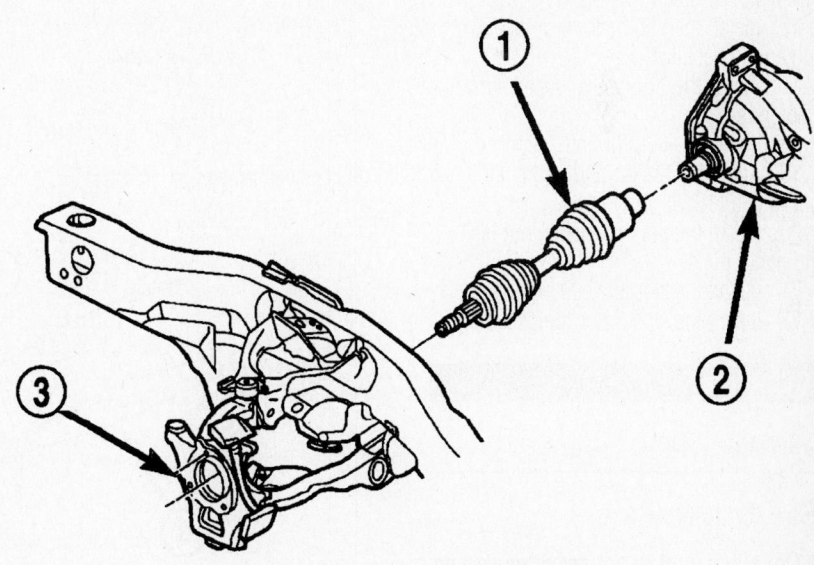

1 - HALF SHAFT
2 - FRONT AXLE
3 - STEERING KNUCKLE

67189-DAKO-G40

Halfshaft

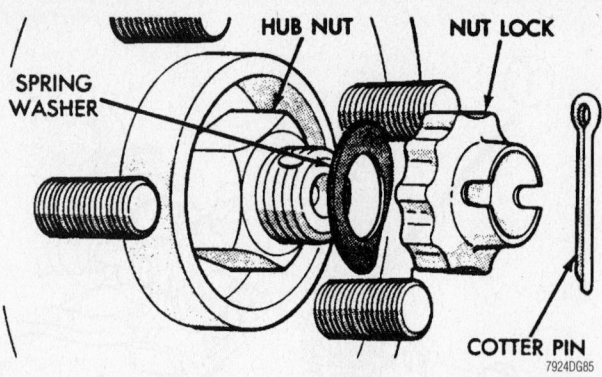

To separate the halfshaft from the hub, remove the cotter pin, nut lock and spring washer from the axle shaft

inner CV housing and disengage the CV joint from the axle.

11. Remove the half shaft from the vehicle.

To install:

12. Apply a light coating of wheel bearing grease on the axle splines.

13. Insert the half shaft stub through the steering knuckle and onto the axle. Verify the shaft snapring engages with the groove on the inside of the joint housing.

14. Clean the hub bearing bore and hub bearing mating surface of all foreign materials. Apply a light coating of grease to all mating surfaces.

15. Install the hub bearing onto the axle half shaft and steering knuckle.

16. Install the hub bearing bolts and tighten to specifications.

17. Install the ABS wheel speed sensor, if equipped.

18. Install brake rotor and caliper.

19. Apply the brakes and tighten hub nut to 244 Nm (180 ft. lbs.).

20. Install the spring washer, nut lock and cotter pin.

21. Install the skid plate, if equipped.

22. Install the wheel and tire assembly.

CV-Joints

OVERHAUL

Outer CV-Joint

➡The outer joint is not serviceable and must be replaced as a unit.

1. Before servicing the vehicle, refer to the precautions in the beginning of this section.

2. Remove or disconnect the following:

- Axle halfshaft from the vehicle
- CV-joint boot and clamps
- Snapring
- CV-joint

To install:

3. Install or connect the following:
- CV-joint
- Snapring
- CV-joint boot and clamps

4. Fill the joint housing and boot with grease and tighten the boot clamps.

5. Install the axle halfshaft.

Inner Tripod Joint

2001–03

1. Before servicing the vehicle, refer to the precautions in the beginning of this section.

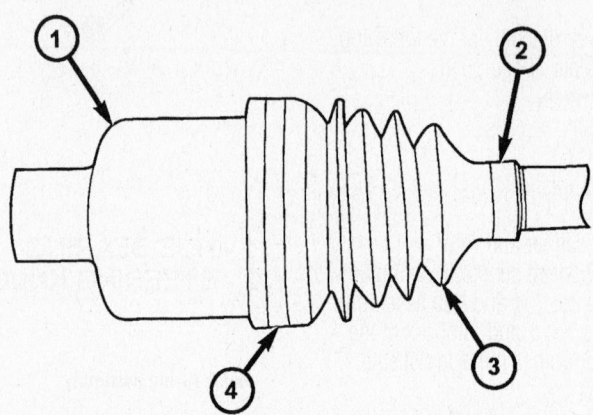

1 - C/V HOUSING
2 - CLAMP
3 - BOOT
4 - CLAMP

67189-DAKO-G33

Boot clamp location

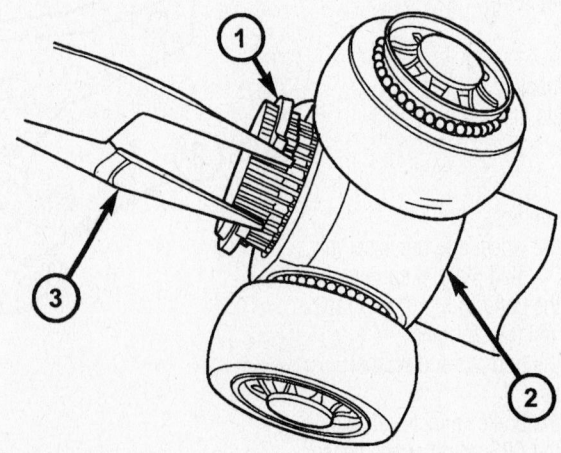

1 - SNAP RING
2 - TRIPOD
3 - PLIERS

67189-DAKO-G34

Removing the snapring

2. Remove or disconnect the following:
- Axle halfshaft from the vehicle
- Inner tripod joint boot clamps
- Tripod joint housing
- Snapring
- Circlip
- Tripod joint

To install:

➡**Use new snaprings, clips, and boot clamps for assembly.**

3. Install or connect the following:
- Tripod joint
- Circlip
- Snapring
- Tripod joint housing

4. Fill the tripod joint housing and boot with grease and tighten the boot clamps.

5. Install the axle halfshaft.

2004

1. Clamp the shaft in a vise with soft jaws and support the CV joint.

2. Remove the clamps with a cut-off wheel or grinder.

✳✳ WARNING

Do not damage the CV housing or half shaft with the cut-off wheel or grinder.

3. Remove the housing from the half shaft and slide the boot down shaft.

4. Remove the housing bushing from the housing.

5. Remove the tripod snapring.

6. Remove the tripod and boot from the halfshaft.

7. Clean and inspect the CV components for excessive wear and damage. Replace the tripod as a unit only if necessary.

To install:

8. Slide a new boot down the halfshaft.

9. Install the tripod and tripod snapring on the halfshaft.

10. Pack the grease supplied with the joint/boot into the housing and boot.

11. Coat the tripod with the supplied grease.

12. Install new bushing onto the housing.

13. Insert the tripod and shaft in the housing.

14. Position the boot on the joint in its original position.

➡**Verify the boot is not twisted and remove any excess air.**

15. Secure both boot clamps with Clamp Installer C-4975A, or equivalent. Place the tool on the clamp bridge and tighten the tool until the jaws of the tool are closed.

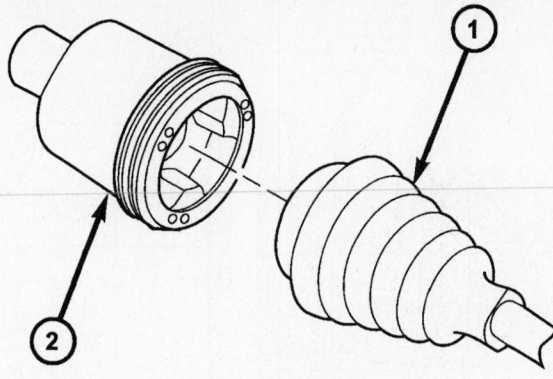

1 - BOOT
2 - HOUSING

67189-DAKO-G35

CV joint housing

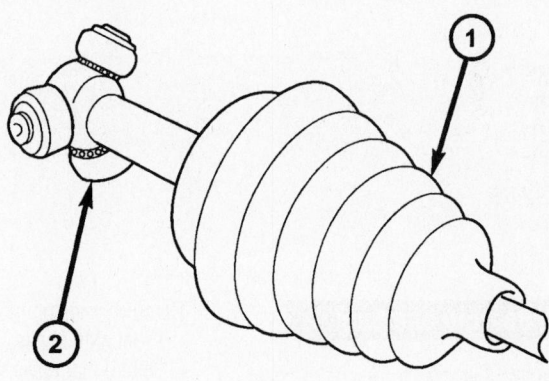

1 - BOOT
2 - TRIPOD

67189-DAKO-G36

Tripod joint

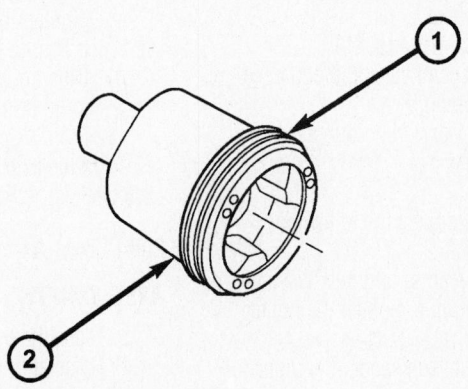

1 - HOUSING
2 - BUSHING

67189-DAKO-G37

Housing bushing

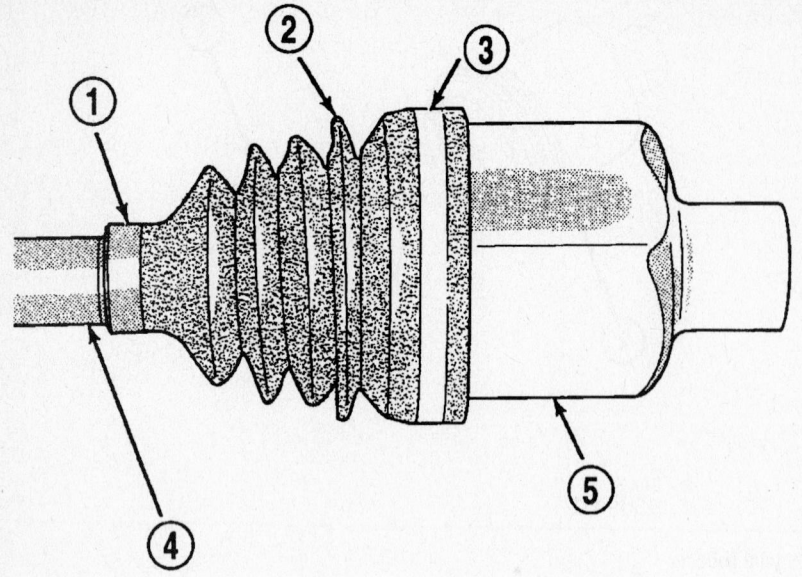

1 - CLAMP
2 - BOOT
3 - CLAMP
4 - SHAFT
5 - HOUSING

Inner CV joint boot

67189-DAKO-G38

Axle Shaft, Bearing and Seal

REMOVAL & INSTALLATION

2001–03 Front Axle

AXLE SHAFT

1. Before servicing the vehicle, refer to the precautions in the beginning of this section.
2. Remove or disconnect the following:
   - Front wheel
   - Brake caliper and rotor
   - Wheel speed sensor, if equipped
   - Axle hub nut
   - Wheel bearing and hub assembly
   - Axle shaft

To install:
3. Install or connect the following:
   - Axle shaft
   - Wheel bearing and hub assembly
   - Axle hub nut. Tighten the nut to 175 ft. lbs. (237 Nm).
   - Wheel speed sensor, if equipped
   - Brake caliper and rotor
   - Front wheel

SEAL

1. Before servicing the vehicle, refer to the precautions in the beginning of this section.

2. Remove or disconnect the following:
   - Front axle shafts
   - Differential cover
   - Differential and ring gear assembly
   - Axle seals

To install:
3. Press the axle seals into the differential housing with Turnbuckle 6797 and Disc set 8110.
4. Install or connect the following:
   - Differential and ring gear assembly. Tighten the bearing cap bolts to 45 ft. lbs. (61 Nm).
   - Differential cover. Tighten the bolts to 30 ft. lbs. (41 Nm).
   - Front axle shafts
5. Fill the axle assembly with gear oil and check for leaks.

2004 Front Axle

AXLE SHAFTS

1. Place the transmission in neutral.
2. Raise and support the vehicle.
3. Remove the half shaft from vehicle.
4. Remove skid plate, if equipped.
5. Clean the axle seal area.
6. Remove the snapring from the axle shaft.
7. Remove the axle with remover 8420A and slide hammer C-3752, or equivalent.

To install:

➡ **Use care to prevent shaft splines from damaging axle shaft seal lip.**

8. Lubricate the bearing bore and seal lip with gear lubricant.
9. Insert the axle shaft through seal, bearing, and engage it into the side gear splines. Push firmly on the axle shaft to engage the snapring.
10. Check the differential fluid level and add fluid if necessary.
11. Install the skid plate, if necessary.
12. Install the half shaft.

SEALS

1. Remove the half shaft and axle shaft.
2. Remove the axle shaft seal with a small pry bar.

To install:
3. Wipe the axle shaft tube bore clean.
4. Install a new axle shaft seal with installer 8402 and handle C-4171, or equivalent.
5. Install the axle shaft and half shaft.

BEARINGS

1. Remove the half shaft from vehicle.
2. Remove the skid plate, if equipped.
3. Clean the axle seal area.
4. Remove axle shaft O-ring.
5. Remove the axle shaft.
6. Remove the axle shaft seal.
7. Install the axle shaft bearing remover C-4660-A, or equivalent, in the bearing. Then tighten the nut to spread the remover in the bearing.
8. Install the bearing remove cup, bearing and nut. Then tighten the nut to draw the bearing out.
9. Inspect the axle shaft tube bore for roughness and burrs. Remove as necessary.

To install:
10. Wipe the axle shaft tube bore clean.
11. Install the axle shaft bearing with installer 5063 and handle C-4171, or equivalent.
12. Install a new axle shaft seal with installer 8402 and handle C-4171, or equivalent.
13. Install the axle shaft and half shaft.

2001–03 Rear Axle

C-CLIP TYPE

1. Before servicing the vehicle, refer to the precautions in the beginning of this section.
2. Remove or disconnect the following:
   - Rear wheel
   - Brake drum

- Differential cover
- Differential gear shaft retainer
- Differential gear shaft
- C-clip
- Axle shaft
- Axle seal
- Axle bearing

To install:

3. Install or connect the following:
- Axle bearing
- Axle seal
- Axle shaft
- C-clip
- Differential gear shaft. Use Loctite® and tighten the retainer to 14 ft. lbs. (19 Nm).
- Differential cover. Tighten the bolts to 30 ft. lbs. (41 Nm).
- Brake drum
- Rear wheel

4. Fill the axle assembly with gear oil and check for leaks.

NON C-CLIP TYPE

1. Before servicing the vehicle, refer to the precautions in the beginning of this section.

2. Remove or disconnect the following:
- Rear wheel
- Brake caliper and rotor, if equipped
- Brake drum, if equipped
- Axle retainer nuts
- Axle shaft, seal and bearing assembly

3. Split the bearing retainer with a chisel and remove the retainer ring.

4. Press the bearing off the axle shaft.

5. Remove the axle seal and retaining plate.

To install:

6. Install the retaining plate and axle seal onto the axle shaft.

7. Pack the wheel bearing with axle grease and press the bearing on to the axle shaft.

8. Press the retaining ring onto the axle shaft.

9. Install or connect the following:
- Axle shaft, seal and bearing assembly. Tighten the nuts to 45 ft. lbs. (61 Nm).
- Brake caliper and rotor, if equipped
- Brake drum, if equipped
- Rear wheel

10. Fill the axle assembly with gear oil and check for leaks.

2004 8¼ inch Rear Axle

AXLE SHAFTS

1. Place the transmission in neutral and raise and support the vehicle.

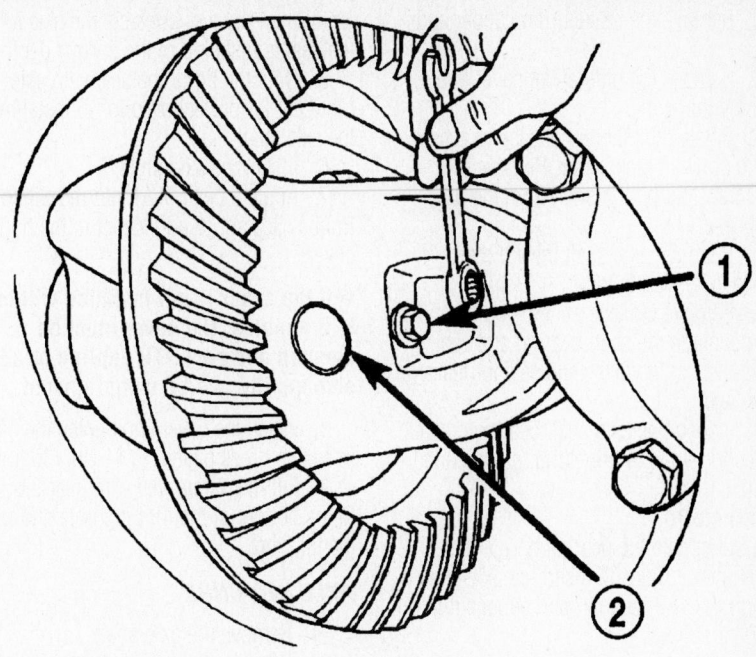

1 - LOCK SCREW
2 - PINION MATE SHAFT

67189-DAKO-G29

Pinion mate shaft lock screw

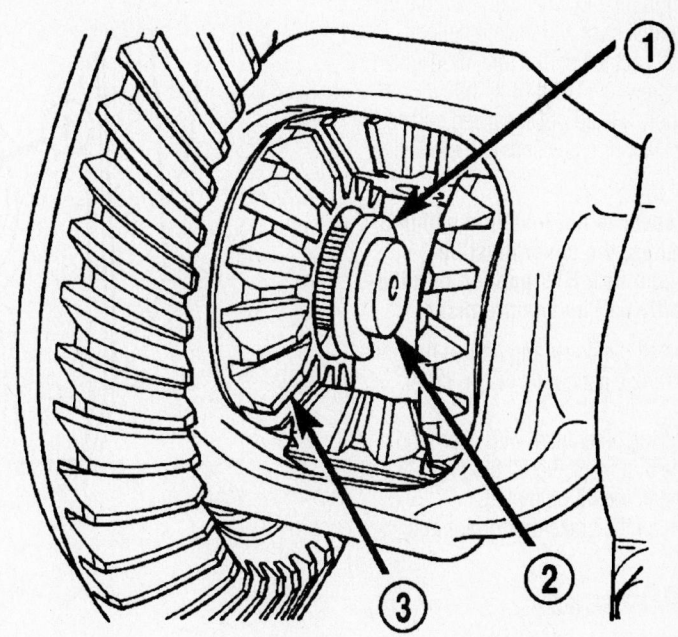

1 - C-LOCK
2 - AXLE SHAFT
3 - SIDE GEAR

67189-DAKO-G30

Axle shaft C-lock

2. Remove the brake drum/caliper and rotor.

3. Remove the differential cover and drain the lubricant.

4. Rotate the differential case to access the pinion shaft lock screw. Remove the lock screw and pinion shaft from the differential case.

5. Push the axle shaft inward then remove axle shaft C-lock.

6. Remove the axle shaft being careful not to damage the shaft bearing and seal.

7. Inspect the axle shaft seal for leakage or damage.

8. Inspect the axle shaft bearing contact surface for signs of brinelling, galling and pitting.

To install:

9. Lubricate the bearing bore and seal lip with gear lubricant. Insert the axle shaft through the seal, bearing and engage it into side gear splines.

➡**Use care to prevent shaft splines from damaging axle shaft seal lip.**

10. Insert the C-lock in end of axle shaft. Push the axle shaft outward to seat the C-lock in side gear.

11. Insert the pinion shaft into differential case and through thrust washers and differential pinions.

12. Align hole in shaft with the hole in the differential case and install the lock screw with Loctite® on the threads. Tighten the lock screw to 11 Nm (8 ft. lbs.).

13. Apply a bead of Mopar red silicone rubber sealant or equivalent to the housing cover.

➡**If the cover is not installed within 3 to 5 minutes, the cover must be cleaned and new RTV applied or adhesion quality will be compromised.**

14. Install the cover and tighten the bolts in a criss-cross pattern to 41 Nm (30 ft. lbs.).

15. Fill the differential with gear lubricant to the bottom of the fill plug hole.

16. Install the fill hole plug.

17. Install the brake drum/rotor and caliper.

AXLE SHAFT SEALS

1. Remove the axle shaft.

2. Remove the axle shaft seal from the end of the axle tube with a small pry bar.

To install:

3. Wipe the axle tube bore clean. Remove any old sealer or burrs from the tube.

4. Install a new axle seal with Installer C-4076-B and Handle C-4735-1, or equiva-

lent. When the tool contacts the axle tube, the seal is installed to the correct depth.

5. Coat the lip of the seal with axle lubricant for protection prior to installing the axle shaft.

6. Install the axle shaft.

7. Apply a bead of Mopar red Silicone Rubber Sealant or equivalent to the housing cover.

➡**If the cover is not installed within 3 to 5 minutes, the cover must be cleaned and new RTV applied or adhesion quality will be compromised.**

8. Install the cover and tighten the bolts in a criss-cross pattern to 41 Nm (30 ft. lbs.).

9. Fill the differential with gear lubricant to the bottom of the fill plug hole and install the fill plug.

AXLE BEARINGS

1. Remove the axle shaft.

2. Remove the axle shaft seal from the axle tube with a small pry bar.

➡**The seal and bearing can be removed at the same time with the bearing removal tool.**

3. Remove the axle shaft bearing with Bearing Removal Tool Set 6310 and Adapter Foot 6310-9, or equivalent.

To install:

4. Wipe the axle tube bore clean. Remove any old sealer or burrs from the tube.

5. Install the axle shaft bearing with Installer C-4198 and Handle C-4171, or equivalent.

➡**Install the bearing with part number against the installer.**

6. Install a new axle seal with Installer C-4076-B and Handle C-4735-1, or equivalent. When the tool contacts the axle tube, the seal is installed to the correct depth.

7. Coat the lip of the seal with axle lubricant and install the axle shaft.

8. Apply a bead of Mopar red Silicone Rubber Sealant or equivalent to the housing cover.

➡**If the cover is not installed within 3 to 5 minutes, the cover must be cleaned and new RTV applied or adhesion quality will be compromised.**

9. Install the cover and tighten the bolts in a criss-cross pattern to 41 Nm (30 ft. lbs.).

10. Fill the differential with gear lubricant to bottom of the fill plug hole and install fill plug.

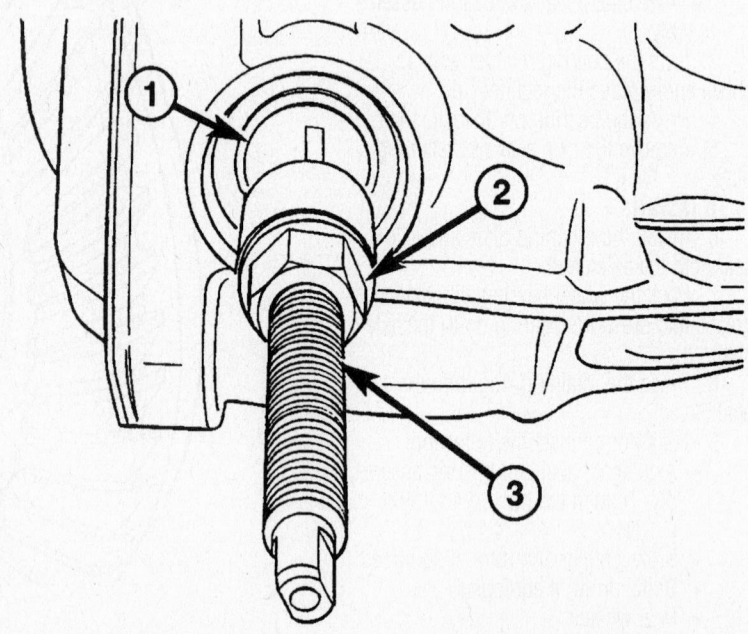

1 - AXLE BEARING
2 - NUT
3 - REMOVER

67189-DAKO-G31

Bearing remover

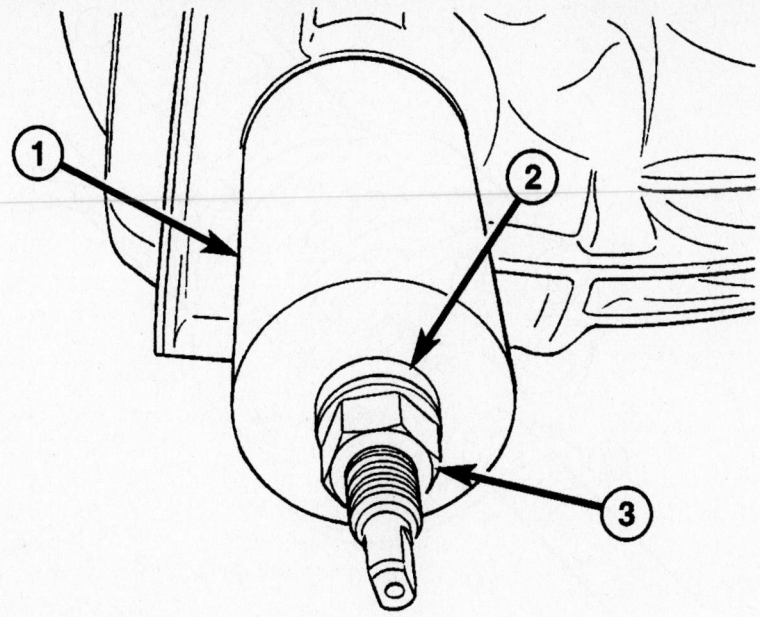

1 - REMOVER CUP
2 - BEARING
3 - NUT

Bearing cup remover

67189-DAKO-G32

2004 9¼ inch Rear Axle

AXLE SHAFTS

1. Place the transmission in neutral.
2. Remove the brake caliper, adapter and rotor.
3. Remove the differential housing cover and drain lubricant.
4. Rotate the differential case so the pinion mate shaft lock screw is accessible. Remove the lock screw and pinion mate shaft from the differential case.
5. Push the axle shaft inward and remove the axle shaft C-lock from the axle shaft.
6. Remove the axle shaft. Use care to prevent damage to the axle shaft bearing and seal in the axle tube.
   To install:
7. Lubricate the bearing bore and seal lip with gear lubricant. Insert the axle shaft through seal, bearing, and engage it into the side gear splines.

➡**Use care to prevent the shaft splines from damaging the axle shaft seal.**

8. Insert the C-lock in end of the axle shaft then push the axle shaft outward to seat the C-lock in the side gear.
9. Insert the pinion shaft into the differential case and through the thrust washers and differential pinions.

10. Align the hole in shaft with the hole in the differential case and install the lock screw with Loctite® on the threads. Tighten the lock screw to 11 Nm (8 ft. lbs.).
11. Install the differential cover and fill with gear lubricant.
12. Install the brake rotor, caliper adapter and caliper.

AXLE BEARINGS

1. Remove the axle shaft.
2. Remove the axle shaft seal from the end of the axle tube with a small pry bar.

➡**The seal and bearing can be removed at the same time with the bearing removal tool.**

3. Remove the axle shaft bearing with Bearing Remover 6310 and Foot 6310-9, or equivalent.
   To install:
4. Wipe the axle tube bore clean. Remove any old sealer or burrs from the tube.
5. Install the axle shaft bearing with Installer C-4198 and Handle C-4171, or equivalent. Drive the bearing in until tool contacts the axle tube.

➡**Bearing is installed with the bearing part number against the installer.**

6. Coat the lip of the new axle seal with axle lubricant and install with Installer C-4076-B and Handle C-4735-1, or equivalent.
7. Install the axle shaft.

AXLE SHAFT SEALS

1. Remove the axle shaft.
2. Remove the axle shaft seal from the end of the axle tube with a small pry bar.
   To install:
3. Wipe the axle tube bore clean. Remove any old sealer or burrs from the tube.
4. Coat the lip of the new seal with axle lubricant and install a seal with Installer C-4076-B and Handle C-4735-1, or equivalent. When the tool contacts the axle tube, the seal is installed to the correct depth.
5. Install the axle shaft.

Pinion Seal

REMOVAL & INSTALLATION

2001–03

C-CLIP TYPE

1. Before servicing the vehicle, refer to the precautions in the beginning of this section.
2. Remove or disconnect the following:
   - Wheels
   - Brake drums
   - Driveshaft
3. Check the bearing preload with an inch lb. torque wrench.
4. Remove the pinion flange and seal.
   To install:

➡**Use a new pinion nut for assembly.**

5. Install the new pinion seal and flange. Tighten the nut to 210 ft. lbs. (285 Nm).
6. Check the bearing preload. The bearing preload should be equal to the reading taken earlier, plus 5 inch lbs.
7. If the preload torque is low, tighten the pinion nut in 5 inch lb. increments until the torque value is reached. Do not exceed 350 ft. lbs. (474 Nm) pinion nut torque.
8. If the pinion bearing preload torque cannot be attained at maximum pinion nut torque, replace the collapsible spacer.
9. Install or connect the following:
   - Driveshaft
   - Brake drums
   - Wheels
10. Fill the axle assembly with gear oil and check for leaks.

NON C-CLIP TYPE—FRONT

1. Before servicing the vehicle, refer to the precautions in the beginning of this section.

2. Remove or disconnect the following:
   • Wheels
   • Brake rotors
   • Driveshaft

3. Check the bearing preload with an inch lb. torque wrench.

4. Remove the pinion flange and seal.

To install:

➤**Use a new pinion nut for assembly.**

5. Install the new pinion seal and flange. Tighten the nut to 160 ft. lbs. (217 Nm).

6. Check the bearing preload. The bearing preload should be equal to the reading taken earlier, plus 5 inch lbs.

7. If the preload torque is low, tighten the pinion nut in 5 inch lb. increments until the torque value is reached. Do not exceed 260 ft. lbs. (353 Nm) pinion nut torque.

8. If the pinion bearing preload torque can not be attained at maximum pinion nut torque, replace the collapsible spacer.

9. Install or connect the following:
   • Driveshaft
   • Brake rotors
   • Wheels

10. Fill the axle assembly with gear oil and check for leaks.

NON C-CLIP TYPE—REAR

1. Before servicing the vehicle, refer to the precautions in the beginning of this section.

2. Remove or disconnect the following:
   • Wheels
   • Brake rotors or drums
   • Driveshaft

3. Check the bearing preload with an inch lb. torque wrench.

4. Remove the pinion flange and seal.

To install:

➤**Use a new pinion nut for assembly.**

5. Install the new pinion seal and flange. Tighten the nut to 160 ft. lbs. (217 Nm).

6. Check the bearing preload. The bearing preload should be equal to the reading taken earlier, plus 5 inch lbs.

7. If the preload torque is low, tighten the pinion nut in 5 inch lb. increments until the torque value is reached. Do not exceed 260 ft. lbs. (353 Nm) pinion nut torque.

8. If the pinion bearing preload torque can not be attained at maximum pinion nut torque, remove one or more pinion preload shims.

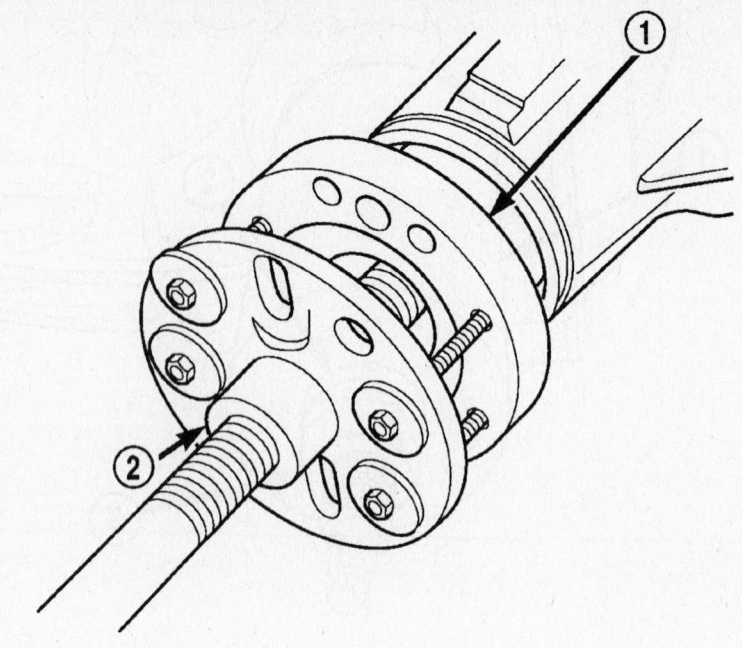

1 - COMPANION FLANGE
2 - PULLER

67189-DAKO-G28

Removing the companion flange

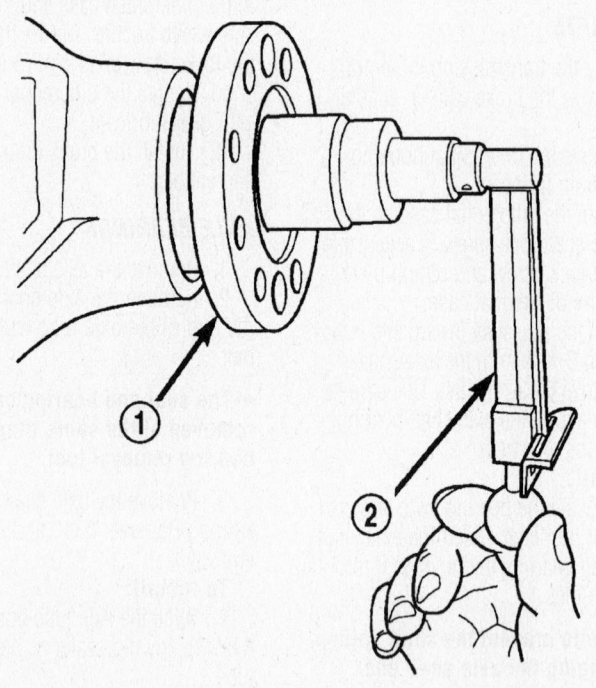

1 - COMPANION FLANGE
2 - TORQUE WRENCH

67189-DAKO-G25

Measuring pinion rotating torque

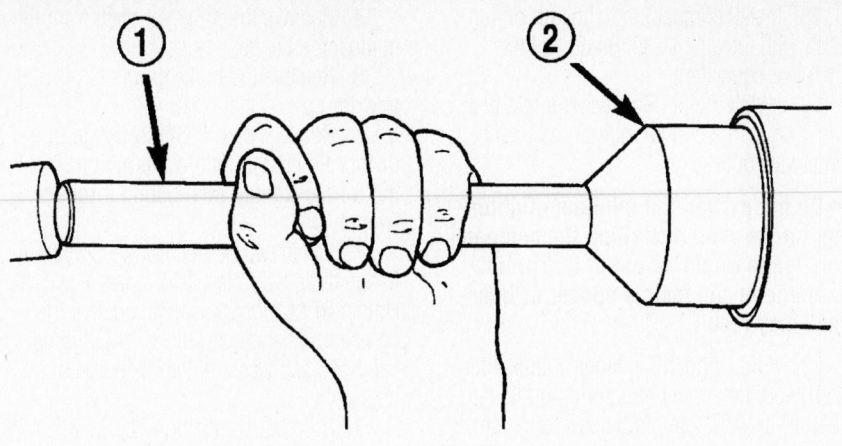

1 - HANDLE
2 - INSTALLER

67189-DAKO-G27

Pinion seal installer

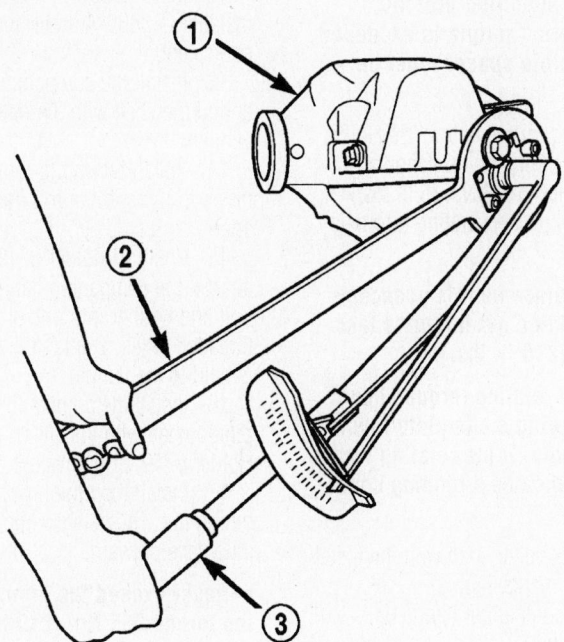

1 - DIFFERENTIAL HOUSING
2 - HOLDER
3 - TORQUE WRENCH

67189-DAKO-G26

Tightening the pinion nut

9. Install or connect the following:
- Driveshaft
- Brake rotors or drums
- Wheels

10. Fill the axle assembly with gear oil and check for leaks.

2004

FRONT AXLE

1. Raise and support the vehicle.
2. Remove skid plate, if equipped.
3. Remove both half shafts.

4. Mark the propeller shaft and pinion companion flange for installation reference.

5. Remove the front propeller shaft.

6. Rotate the pinion gear three or four times and verify pinion rotates smoothly.

7. Record pinion rotating torque with an inch pound torque wrench, for installation reference.

8. Position Holder 6719, or equivalent, against the companion flange and install a four bolts and washers into the threaded holes and tighten the bolts.

9. Remove the pinion nut.

10. Remove the companion flange with Remover C-452, or equivalent.

11. Remove pinion seal with a pry tool or a slide hammer mounted screw.

To install:

12. Apply a light coating of gear lubricant on the lip of pinion seal.

13. Install seal with Installer C-3972-A and Handle C-4171, or equivalent,

14. Install the companion flange onto the pinion with Installer C-3718 and Holder 6719A, or equivalent.

15. Position holder against the companion flange and install four bolts and washers into the threaded holes. Tighten the bolt and washer so that the holder is held to the flange.

16. Install a new pinion nut onto the pinion shaft and tighten the pinion nut until there is zero bearing end-play.

➡ **Do not exceed the minimum tightening torque when installing the companion flange at this point. Damage to the collapsible spacer or bearings may result.**

17. Tighten the nut to 271 Nm (200 ft. lbs.).

➡ **Never loosen pinion nut to decrease pinion bearing rotating torque and never exceed specified preload torque. If preload torque or rotating torque is exceeded a new collapsible spacer must be installed.**

18. Record the pinion rotating torque using a torque wrench. The rotating torque should be equal to the reading recorded during removal plus an additional 0.56 Nm (5 inch lbs.).

19. If the rotating torque is low, tighten the pinion nut in 6.8 Nm (5 ft. lbs.) increments until the proper rotating torque is achieved.

➡ **If the maximum tightening torque is reached prior to reaching the required rotating torque, the collapsible spacer may have been damaged. Replace the collapsible spacer.**

20. Install propeller shaft with reference marks aligned.

21. Add gear lubricant to differential housing if necessary.

22. Install half shafts.

8¼ INCH REAR AXLE

1. Raise and support the vehicle.

2. Mark the universal joint, companion flange and pinion shaft for installation reference.

3. Remove companion flange bolts and secure the shaft in an upright position to prevent damage to the rear universal joint.

4. Remove the wheel and tire assemblies.

5. Remove brake drums to prevent any drag.

6. Rotate companion flange three or four times and verify flange rotates smoothly.

7. Measure rotating torque of the pinion with an inch pound torque wrench and record the reading for installation reference.

8. Install bolts into two of the threaded holes in the companion flange 180° apart.

9. Position Holder 6719, or equivalent, against the companion flange and install a bolt and washer into one of the remaining threaded holes. Tighten the bolts so the Holder 6719, or equivalent, is held to the flange.

10. Remove the pinion nut and washer.

11. Remove companion flange with Remover C-452, or equivalent.

12. Remove pinion seal with a pry tool or slidehammer mounted screw.

To install:

➡ **The outer perimeter of the seal is pre-coated with a special sealant.**

13. Apply a light coating of gear lubricant on the lip of pinion seal.

14. Install new pinion seal with Installer C-4076-B and Handle C-4735-1, or equivalent.

15. Install companion flange on the end of the shaft with the reference marks aligned.

16. Install bolts into two of the threaded holes in the companion flange 180° apart.

17. Position Holder 6719, or equivalent, against the companion flange and install a bolt and washer into one of the remaining threaded holes. Tighten the bolts so Holder 6719 is held to the flange.

18. Install companion flange on pinion shaft with Installer C-3718 and Holder 6719, or equivalent.

19. Install the pinion washer and a new pinion nut. The convex side of the washer must face outward.

➡ **Do not exceed the minimum tightening torque when installing the companion flange retaining nut at this point. Damage to collapsible spacer or bearings may result.**

20. Hold companion flange with Holder 6719 and tighten the pinion nut to 285 Nm (210 ft. lbs.). Rotate pinion several revolutions to ensure the bearing rollers are seated.

21. Rotate pinion with an inch pound torque wrench. Rotating torque should be equal to the reading recorded during removal plus an additional 0.56 Nm (5 inch lbs.).

➡ **Never loosen pinion nut to decrease pinion bearing rotating torque and never exceed specified preload torque. If rotating torque is exceeded, a new collapsible spacer must be installed.**

22. If rotating torque is low use Holder 6719 to hold the companion flange and tighten pinion nut in 6.8 Nm (5 ft. lbs.) increments until proper rotating torque is achieved.

➡ **The seal replacement is unacceptable if final pinion nut torque is less than 285 Nm (210 ft. lbs.).**

➡ **The bearing rotating torque should be constant during a complete revolution of the pinion. If the rotating torque varies, this indicates a binding condition.**

23. Install propeller shaft with the installation reference marks aligned.

24. Tighten companion flange bolts to 108 Nm (80 ft. lbs.).

25. Install the brake drums.

26. Check the differential housing lubricant level.

27. Install wheel and tire assemblies and lower the vehicle.

9¼ INCH REAR AXLE

REMOVAL

1. Raise and support the vehicle.

2. Remove the wheel and tire assemblies.

3. Mark the universal joint, companion flange and pinion shaft for installation reference.

4. Remove the propeller shaft from the companion flange.

5. Remove the brake drums to prevent any drag.

6. Rotate the companion flange three or four times and record the pinion rotating torque with an inch pound torque wrench.

7. Install two bolts into the companion flange threaded holes, 180° apart. Position Holder 6719A, or equivalent, against the companion flange and install and tighten two bolts and washers into the remaining holes.

8. Hold the companion flange with Holder 6719A, or equivalent, and remove pinion nut and washer.

9. Remove the companion flange with Remover C-452, or equivalent.

10. Remove the pinion seal with pry tool or slidehammer mounted screw.

To install:

11. Apply a light coating of gear lubricant on the lip of pinion seal.

12. Install a new pinion seal with Installer C-3860-A and Handle C-4171, or equivalent.

13. Install the companion flange on the end of the shaft with the reference marks aligned.

14. Install two bolts into the threaded holes in the companion flange, 180° apart.

15. Position Holder 6719, or equivalent, against the companion flange and install a bolt and washer into one of the remaining threaded holes. Tighten the bolts so holder is held to the flange.

16. Install the companion flange on the pinion shaft with Installer C-3718 and Holder 6719, or equivalent.

17. Install the pinion washer and a new pinion nut. The convex side of the washer must face outward.

➡ **Never exceed the minimum tightening torque 285 Nm (210 ft. lbs.) when installing the companion flange retaining nut at this point. Damage to the collapsible spacer or bearings may result.**

18. Hold the companion flange with Holder 6719 and tighten the pinion nut with a torque set to 285 Nm (210 ft. lbs.). Rotate the pinion several revolutions to ensure the bearing rollers are seated.

19. Rotate the pinion with an inch pound torque wrench. Rotating torque should be equal to the reading recorded during removal plus an additional 0.56 Nm (5 inch lbs.).

➡**Never loosen the pinion nut to decrease pinion bearing rotating torque and never exceed the specified preload torque. If the rotating torque is exceeded, a new collapsible spacer must be installed.**

20. If the rotating torque is low, use Holder 6719 to hold the companion flange and tighten the pinion nut in 6.8 Nm (5 ft. lbs.) increments until the proper rotating torque is achieved.

➡**The bearing rotating torque should be constant during a complete revolution of the pinion. If the rotating torque varies, this indicates a binding condition.**

➡**The seal replacement is unacceptable if the final pinion nut torque is less than 285 Nm (210 ft. lbs.).**

21. Install the propeller shaft with the installation reference marks aligned.
22. Tighten the companion flange bolts to 108 Nm (80 ft. lbs.).
23. Install the brake drums, wheel and tire assemblies and lower the vehicle.
24. Check the differential lubricant level.

STEERING AND SUSPENSION

Air Bag

✳✳ CAUTION

Some vehicles are equipped with an air bag system. The system must be disarmed before performing service on, or around, system components, the steering column, instrument panel components, wiring and sensors. Failure to follow the safety precautions and the disarming procedure could result in accidental air bag deployment, possible injury and unnecessary system repairs.

PRECAUTIONS

Several precautions must be observed when handling the inflator module to avoid accidental deployment and possible personal injury.

• Never carry the inflator module by the wires or connector on the underside of the module.

• When carrying a live inflator module, hold securely with both hands, and ensure that the bag and trim cover are pointed away.

• Place the inflator module on a bench or other surface with the bag and trim cover facing up.

• With the inflator module on the bench, never place anything on or close to the module which may be thrown in the event of an accidental deployment.

Before servicing the vehicle, also make sure to refer to the precautions in the beginning of this section as well.

DISARMING

1. Disconnect and isolate the negative battery cable. Wait 2 minutes for the system capacitor to discharge before performing any service.
2. When repairs are completed, connect the negative battery cable.

Recirculating Ball Power Steering Gear

REMOVAL & INSTALLATION

1. Before servicing the vehicle, refer to the precautions in the beginning of this section.
2. Remove or disconnect the following:
   • Negative battery cable
   • Power steering pressure and return lines
   • Intermediate shaft
   • Pitman arm
   • Steering gear

To install:

3. Install or connect the following:
   • Steering gear. Tighten the bolts to 100 ft. lbs. (136 Nm).
   • Pitman arm. Tighten the nut to 175 ft. lbs. (237 Nm).
   • Intermediate shaft. Tighten the pinch bolt to 36 ft. lbs. (49 Nm).

   • Power steering pressure and return lines
   • Negative battery cable
4. Fill the power steering fluid reservoir.
5. Start the engine and check for leaks.

Rack and Pinion Steering Gear

REMOVAL & INSTALLATION

Except 2004 Models

1. Before servicing the vehicle, refer to the precautions in the beginning of this section.
2. Remove or disconnect the following:
   • Front wheels
   • Outer tie rod ends
   • Steering shaft coupler
   • Power steering hoses
   • Steering gear

To install:

3. Install or connect the following:
   • Steering gear. Tighten the bolts to 190 ft. lbs. (258 Nm).

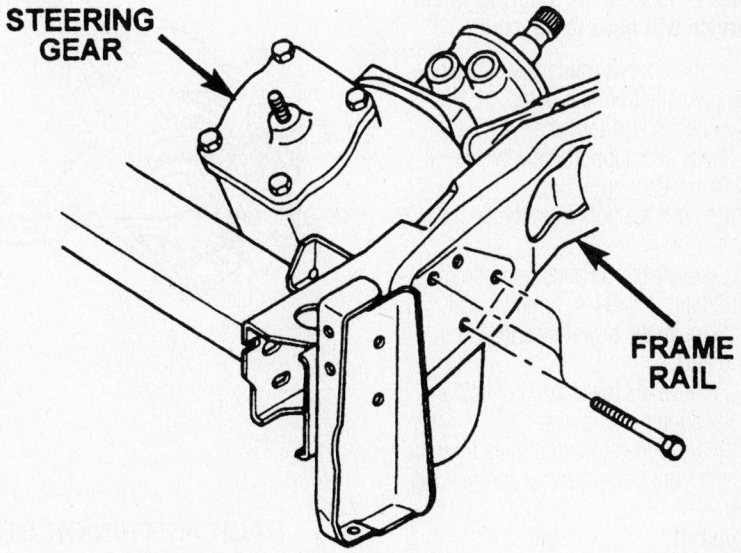

STEERING GEAR

FRAME RAIL

7924DG41

Typical recirculating ball power steering gear mounting

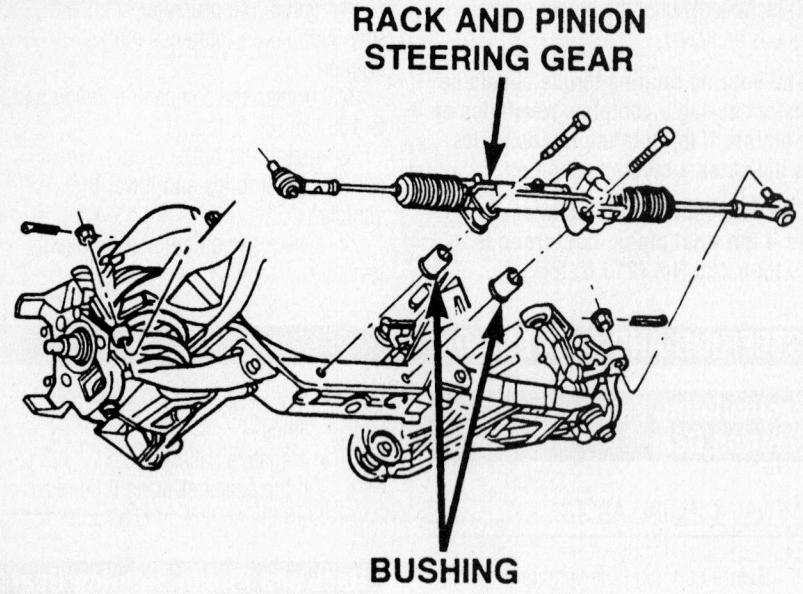

RACK AND PINION
STEERING GEAR

BUSHING

7924DG42

Rack and pinion steering gear mounting used on the 2002–03 2WD Dakota models

- Power steering hoses. Tighten the fittings to 25 ft. lbs. (35 Nm).
- Steering shaft coupler. Tighten the bolt to 36 ft. lbs. (49 Nm).
- Outer tie rod ends. Tighten the nuts to 65 ft. lbs. (88 Nm).
- Front wheels

2004 Models

2-WHEEL DRIVE

➡The steering column on vehicles with an automatic transmission may not be equipped with an internal locking shaft that allows the ignition key cylinder to be locked with the key. Alternative methods of locking the steering wheel for service will have to be used.

1. Siphon out as much power steering fluid as possible from the pump.
2. Lock the steering wheel.
3. Raise and support the vehicle.
4. Remove the front tires.
5. Remove the nuts from the tie rod ends.
6. Separate tie rod ends from the knuckles with Puller C-3894-A, or equivalent.
7. Remove the power steering lines from the gear.
8. Remove the lower coupler bolt and slide the coupler off the gear.
9. Remove the mounting bolts from the gear to the front crossmember and remove the gear.

   To install:

➡Before installing gear inspect bushings and replace if worn or damaged.

10. Install gear on front crossmember and tighten mounting bolts to 271 Nm (200 ft. lbs.).
11. Slide shaft coupler onto gear. Install new bolt and tighten to 49 Nm (36 ft. lbs.).
12. Clean tie rod end studs and knuckle tapers.
13. Install tie rod ends into the steering knuckles and tighten the nuts to 81 Nm (60 ft. lbs.).
14. Install power steering lines to steering gear and tighten the hose to 31 Nm (23 ft. lbs.).
15. Install the front tires.
16. Remove support and lower vehicle.
17. Unlock the steering wheel.
18. Fill system with fluid.
19. Adjust the toe position.

4-WHEEL DRIVE

➡The steering column on vehicles with an automatic transmission may not be equipped with an internal locking shaft that allows the ignition key cylinder to be locked with the key. Alternative methods of locking the steering wheel for service will have to be used.

1. Siphon out as much power steering fluid as possible from the pump.
2. Lock the steering wheel.
3. Raise and support the vehicle.
4. Remove the front tires.
5. Remove the nuts from the tie rod ends.
6. Separate tie rod ends from the knuckles with Puller C-3894-A, or equivalent.
7. Remove the splash shield from under the front end to gain access to the gear.

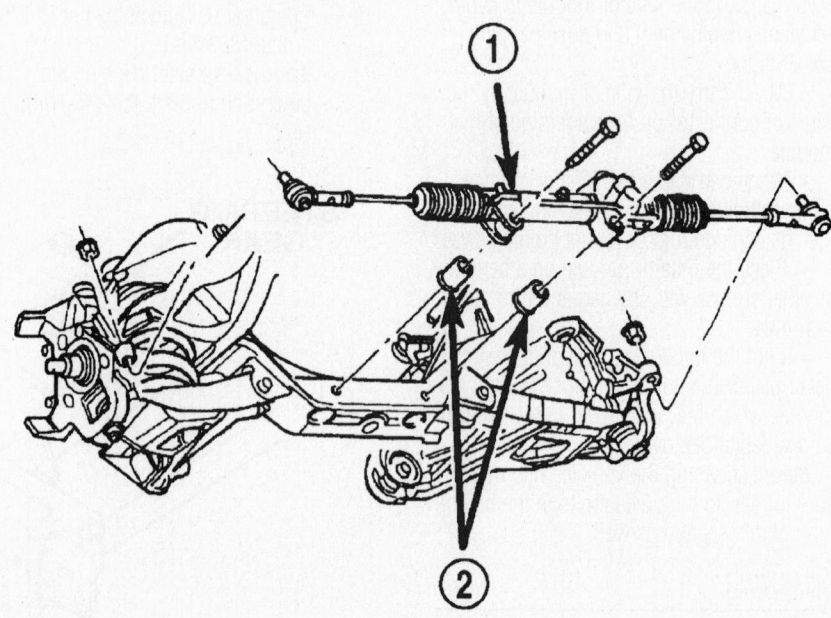

1 - RACK AND PINION STEERING GEAR
2 - BUSHING

67189-DAKO-G23

Steering gear mounting—2004 2 wheel drive

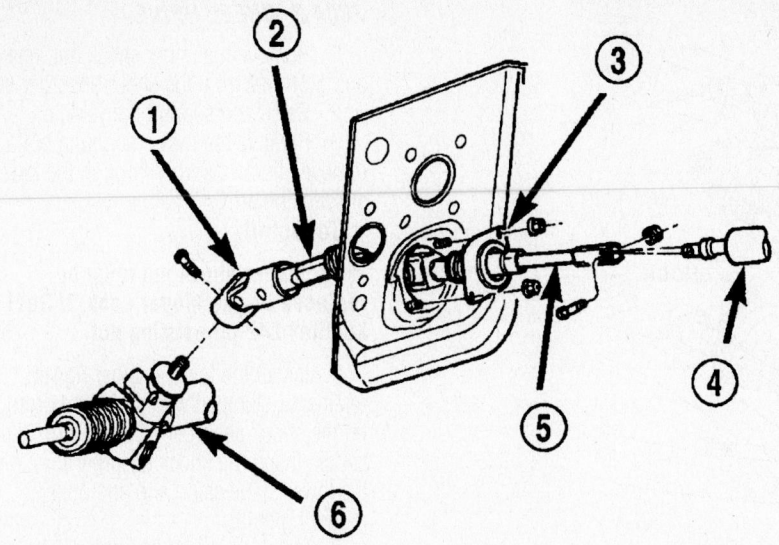

1 - COUPLER
2 - LOWER SHAFT
3 - TOE PLATE
4 - STEERING COLUMN
5 - UPPER SHAFT
6 - RACK AND PINION STEERING GEAR

67189-DAKO-G24

Steering gear coupler

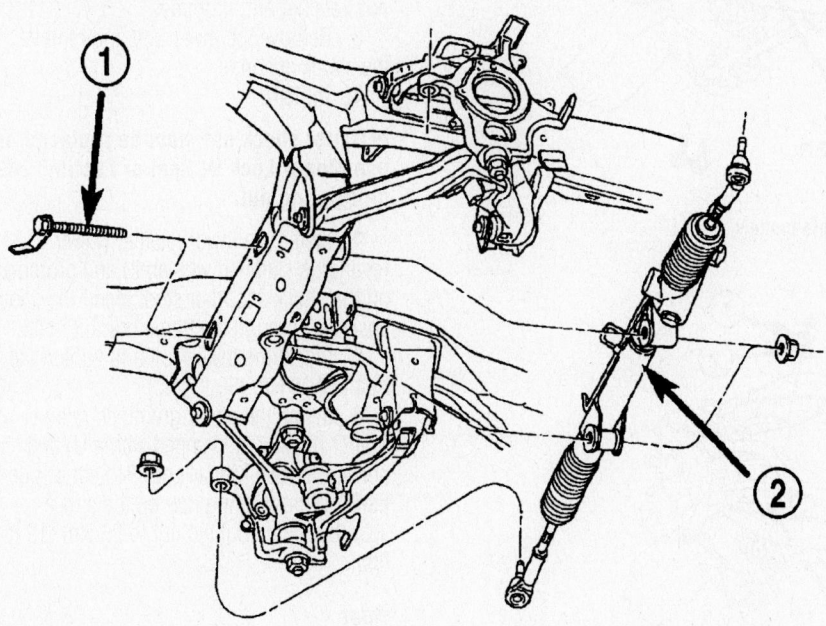

1 - MOUNTING BOLT
2 - RACK AND PINION STEERING GEAR

67189-DAKO-G22

Steering gear mounting—2004 4 wheel drive

8. Remove the skid plate.
9. Remove the power steering lines from the gear.
10. Remove the lower coupler bolt and slide the coupler off the gear.
11. Remove the mounting bolts from the gear to the front crossmember. Slide the gear to the right side of the vehicle. Then tilt the left end of the gear down and remove the gear.
12. Adjust the toe position.

To install:

➡**Before installing the gear inspect bushings and replace if worn or damaged.**

13. Install the gear on the front crossmember and tighten the mounting bolts to 271 Nm (200 ft. lbs.).
14. Slide the shaft coupler onto gear. Install a new bolt and tighten to 49 Nm (36 ft. lbs.).
15. Clean the tie rod end studs and knuckle tapers.
16. Install the tie rod ends into the steering knuckles and tighten the nuts to 81 Nm (60 ft. lbs.).
17. Install the power steering lines to steering gear and tighten the hose to 31 Nm (23 ft. lbs.).
18. Install the splash shield.
19. Install the skid plate.
20. Install the front tires.
21. Remove the support and lower the vehicle.
22. Unlock the steering wheel.
23. Fill the system with fluid.

Shock Absorber

REMOVAL & INSTALLATION

Front

2001–03

1. Before servicing the vehicle, refer to the precautions in the beginning of this section.
2. Remove or disconnect the following:
3. Install or connect the following:
   • Front wheel
   • Upper mount nut
   • Lower mount bolt
   • Shock absorber

To install:

4. Install or connect the following:
   • Shock absorber. Tighten the lower bolt to 100 ft. lbs. (136 Nm) and the upper nut to 30 ft. lbs. (41 Nm).
   • Front wheel

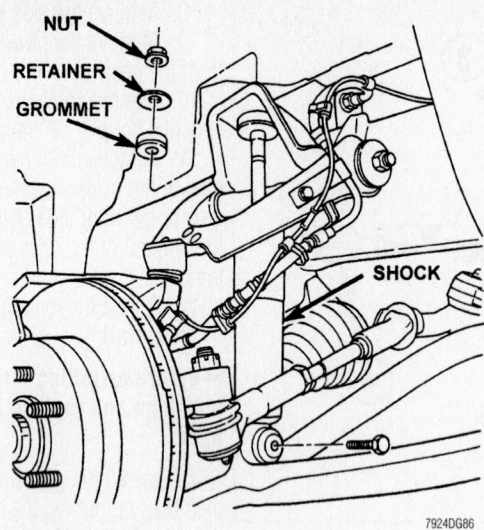

Front shock absorber mounting—2001–03 4WD Dakota models

7924DG86

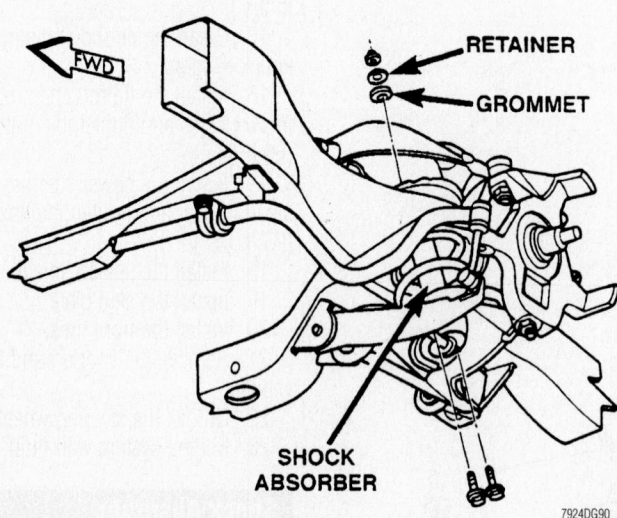

Front shock absorber mounting—2001–03 2WD Dakota models

7924DG90

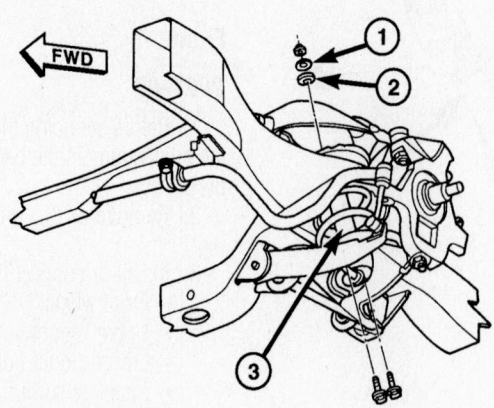

1 - RETAINER
2 - GROMMET
3 - SHOCK ABSORBER

67189-DAKO-G21

Front shock absorber mounting—2004 2 wheel drive

2004 2-WHEEL DRIVE

1. Remove the upper shock nut, retainer and grommet from the shock absorber stud.
2. Raise and support the vehicle.
3. Remove the lower mounting bolts and remove shock absorber through the lower suspension arm.

To install:

➡**The upper shock nut must be replaced or use Mopar Lock 'N Seal or Loctite® 242 on existing nut.**

4. Install the lower retainer (lower retainer is stamped with an L) and grommet on the shock absorber stud and extend the shock. Insert the shock absorber through the lower suspension arm and upper mounting hole.
5. Install the lower mounting bolts and tighten to 28 Nm (21 ft. lbs.).
6. Remove the support and lower the vehicle.
7. Install the upper grommet and retainer (upper retainer is stamped with a U) on the shock absorber stud. Install a new nut or use Mopar Lock 'N Seal or Loctite® 242 on the existing nut and tighten to 26 Nm (19 ft. lbs.).

2004 4-WHEEL DRIVE

1. Raise and support vehicle.
2. Remove the upper shock absorber nut, retainer and grommet.
3. Remove the lower bolt and remove the shock absorber.

To install:

➡**Upper shock nut must be replaced or use Mopar Lock 'N Seal or Loctite® 242 on existing nut.**

4. Install the lower retainer (lower retainer is stamped with an L) and grommet on the shock absorber stud. Insert the shock absorber through the frame bracket hole.
5. Install the lower bolt and tighten the bolt to 108 Nm (80 ft. lbs.).
6. Install the upper grommet, retainer (upper retainer is stamped with a U) and new nut or use Mopar Lock 'N Seal or Loctite® 242 on existing nut, on the shock absorber stud. Tighten nut to 26 Nm (19 ft. lbs.).

Rear

2001–03

1. Before servicing the vehicle, refer to the precautions in the beginning of this section.
2. Support the axle.
3. Remove or disconnect the following:
   • Upper bolt

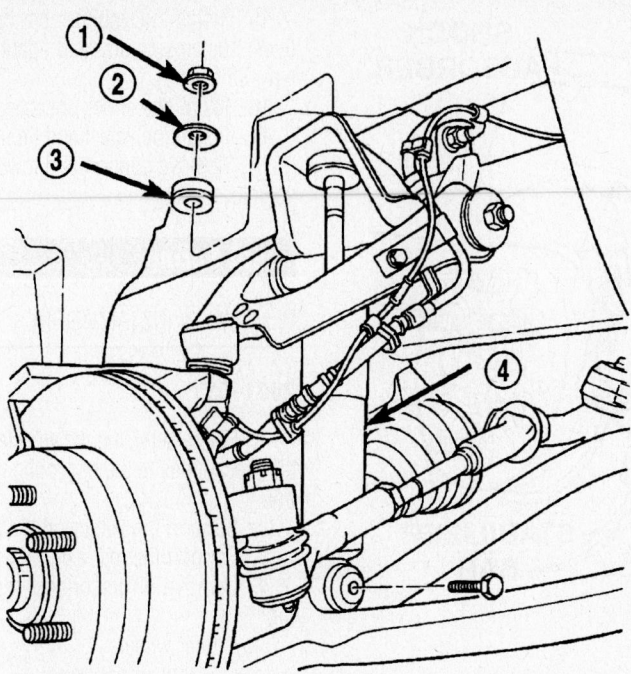

1 - NUT
2 - RETAINER
3 - GROMMET
4 - SHOCK

67189-DAKO-G20

Front shock absorber mounting—2004 4 wheel drive

- Lower bolt
- Shock absorber

To install:

4. Install the bolts through the brackets and shock and tighten them as follows:
- Tighten the lower bolt and nut to 60 ft. lbs. (81 Nm) and the upper bracket nuts to 20 ft. lbs. (27 Nm)

2004

1. Raise the vehicle and support rear axle.
2. Remove the shock absorber lower nut and bolt from the axle bracket.
3. Remove the shock absorber upper nut and bolt from the frame bracket and remove the shock absorber.

To install:

4. Install the shock absorber and upper mounting bolt and nut. Tighten the nut to 95 Nm (70 ft. lbs.).
5. Install the shock absorber into the axle bracket. Install the bolt and nut and tighten the nut to 95 Nm (70 ft. lbs.).
6. Remove the axle support and lower the vehicle.

Coil Spring

REMOVAL & INSTALLATION

2001–03

1. Before servicing the vehicle, refer to the precautions in the beginning of this section.
2. Support the lower control arm on a floor jack.
3. Remove or disconnect the following:
- Front wheel
- Shock absorber
- Stabilizer bar link
- Lower ball joint.
4. Lower the jack and remove the coil spring.

To install:

5. Install the coil spring and raise the control arm into position.
6. Install or connect the following:
- Lower ball joint. Tighten the nut to 135 ft. lbs. (183 Nm).
- Stabilizer bar link
- Shock absorber
- Front wheel

2004

1. Raise and support the vehicle.
2. Remove wheel and tire assembly.
3. Remove the stabilizer bar link from the lower suspension arm.
4. Remove the shock absorber.

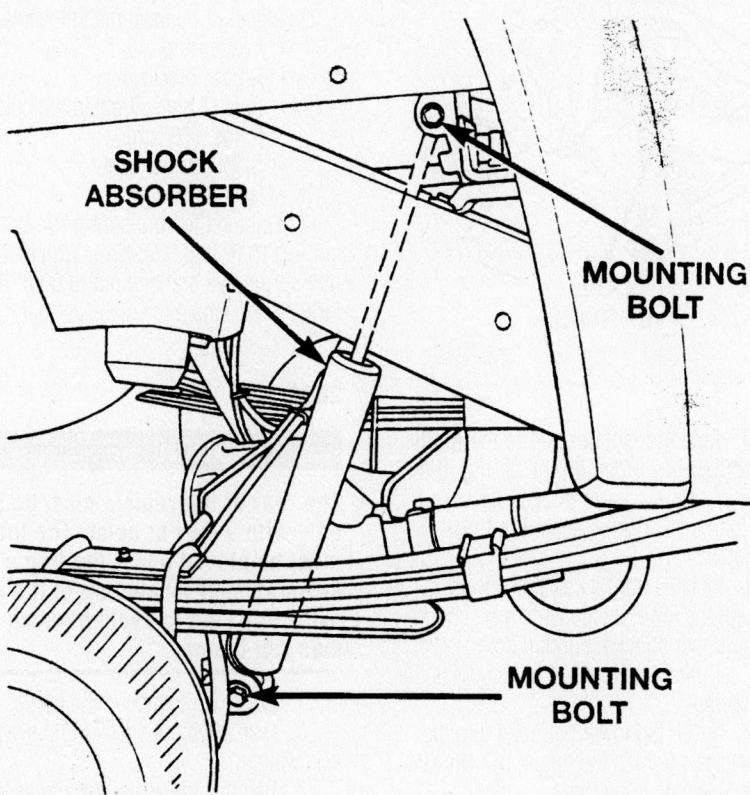

SHOCK ABSORBER

MOUNTING BOLT

MOUNTING BOLT

7924DG91

Rear shock absorber mounting—2001–03

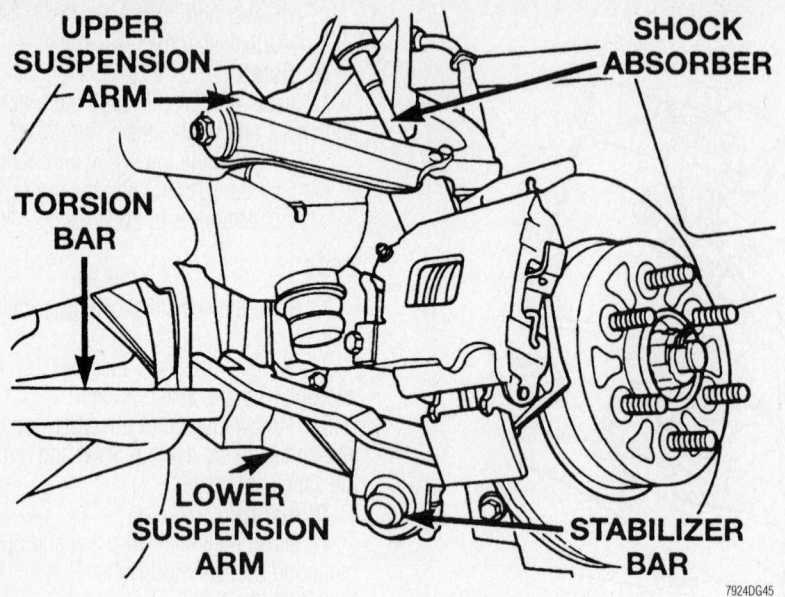

Front suspension components—2001–03 4WD models

7924DG45

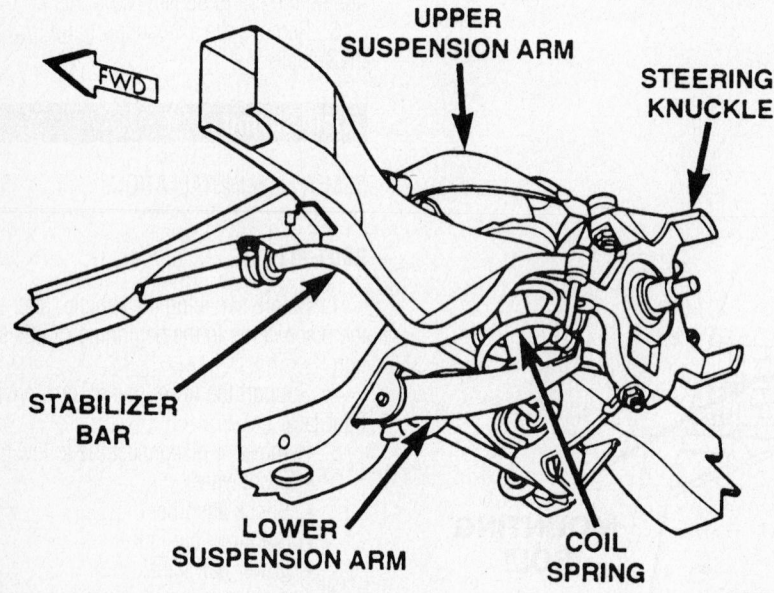

Front suspension components—2001–03 2WD models

7924DG46

5. Install a spring compressor up through the lower suspension arm, coil spring and upper shock mounting hole.

6. Tighten the tool lower nut to compress the coil spring.

7. Remove the lower ball joint nut and separate the ball joint from the knuckle.

8. Loosen the spring compressor lower nut to relieve spring tension.

9. Remove the tool and pull down on the lower suspension arm to remove the spring.

To install:

➡ **The ramped or open end of the coil spring is the bottom of the spring.**

10. Tape the isolator pad to the top of the coil spring. Position the spring in the lower suspension arm pocket. Be sure that the coil spring is seated in the pocket.

11. Install a spring compressor up through the lower suspension arm, coil spring upper shock mounting hole.

12. Tighten the tool nut to compress the coil spring.

13. Install the lower ball joint into the knuckle and tighten the nut to 127 Nm (94 ft. lbs.). Install cotter pin.

14. Remove the spring compressor tool.

15. Install the stabilizer bar link to the lower suspension arm and tighten nut to 47 Nm (35 ft. lbs.).

16. Install the shock absorber.

17. Install the wheel and tire assembly.

18. Remove support and lower the vehicle.

Leaf Spring

REMOVAL & INSTALLATION

2001–03

1. Before servicing the vehicle, refer to the precautions in the beginning of this section.

2. Support the vehicle at the frame rails.

3. Support the rear axle with a jack.

4. Remove or disconnect the following:
   - Rear wheel
   - Stabilizer bar link
   - Axle U-bolts
   - Spring bracket
   - Leaf spring

To install:

➡ **The weight of the vehicle must be supported by the springs when the spring eye and stabilizer bar fasteners are tightened.**

5. Install or connect the following:
   - Leaf spring
   - Spring bracket
   - Axle U-bolts. Tighten the nuts to 52 ft. lbs. (70 Nm).
   - Stabilizer bar link
   - Rear wheel

6. Tighten the front spring eye bolt and nut to 115 ft. lbs. (156 Nm). Tighten the rear spring eye bolt and nut to 80 ft. lbs. (108 Nm). Tighten the stabilizer bar nuts 55 ft. lbs. (74 Nm).

2004

✳✳ CAUTION

The rear of the vehicle must be lifted only with a jack or hoist. The lift must be placed under the frame rail crossmember located aft of the rear axle. Use care to avoid bending the side rail flange.

1. Raise the vehicle at the frame.

2. Use a hydraulic jack to relieve the axle weight.

3. Remove the wheel and tire assemblies.

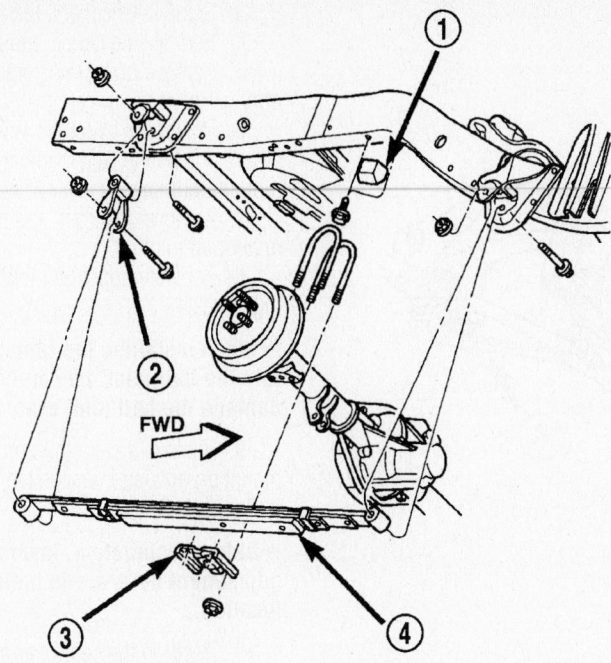

1 - JOUNCE BUMPER
2 - SHACKLE
3 - PLATE
4 - LEAF SPRING

67189-DAKO-G19

Leaf spring—2 wheel drive

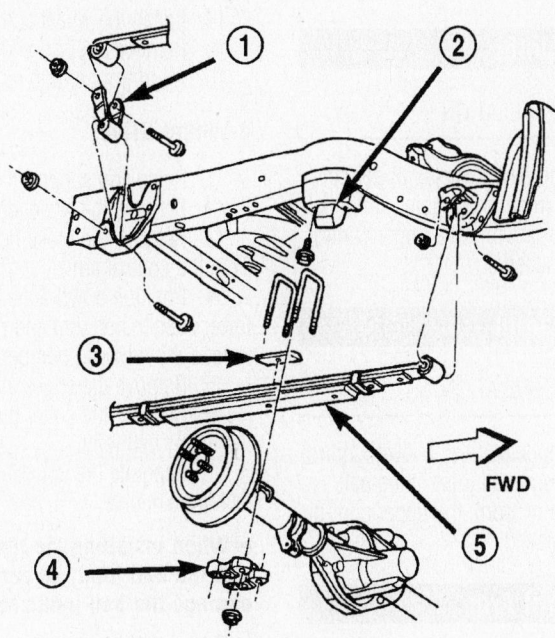

1 - SHACKLE
2 - JOUNCE BUMPER
3 - SEAT
4 - PLATE
5 - LEAF SPRING

67189-DAKO-G18

Leaf spring—4 wheel drive

4. Remove the nuts, the U-bolts and spring plate from the axle and.

5. Remove the nut and bolt from the spring front eye.

6. Remove the nut and bolt that attaches the spring shackle to the rear frame bracket.

7. Remove the spring from the vehicle.

8. Remove the shackle from the spring.

To install:

9. Install the spring shackle on the spring finger tight.

10. Position the spring on the rear axle pad. Make sure the spring center bolt is inserted in the pad locating hole.

11. Align front spring eye with the bolt hole in the front frame bracket. Install the spring eye bolt and nut and tighten the spring eye nut finger-tight.

12. Align spring shackle eye with the bolt hole in the rear frame bracket. Install the bolt and nut and tighten the spring shackle eye nut finger-tight.

13. Install the spring seat (4x4 only), U-bolts, spring plate and nuts.

14. Tighten the U-bolt nuts to 149 Nm (110 ft. lbs.).

15. Install the wheel and tire assemblies.

16. Remove the support stands from under the frame rails. Lower the vehicle until the springs are supporting the weight of the vehicle.

17. Tighten the spring eye pivot bolt nut and all shackle nuts to 163 Nm (120 ft. lbs.).

Torsion bar

REMOVAL & INSTALLATION

2001–03

1. Before servicing the vehicle, refer to the precautions in the beginning of this section.

2. Loosen the adjustment bolt to remove spring load. Note the number of turns for installation.

3. Remove or disconnect the following:

  • Adjustment bolt, swivel and bearing
  • Torsion bar and anchor

4. Separate the torsion bar and anchor.

To install:

5. Assemble the torsion bar and anchor.

6. Install or connect the following:

  • Torsion bar and anchor
  • Adjustment bolt, swivel and bearing

7. Tighten the adjustment bolt the recorded number of turns.

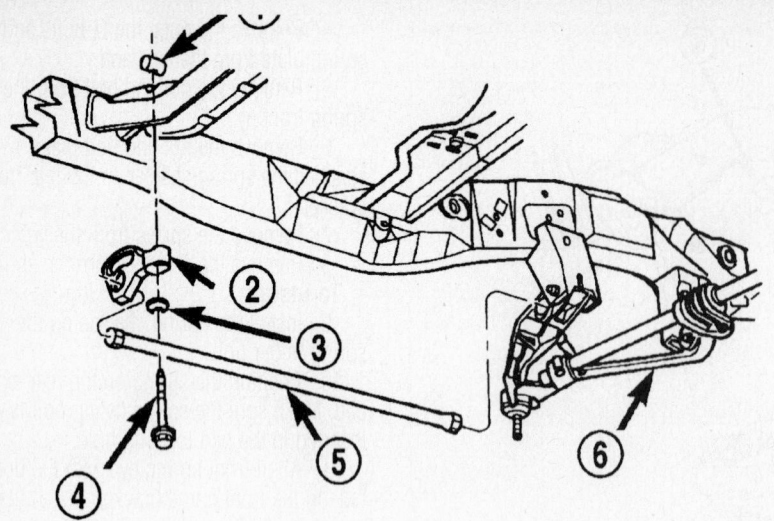

1 - SWIVEL
2 - ANCHOR
3 - BEARING
4 - ADJUSTMENT BOLT
5 - TORSION BAR
6 - LOWER SUSPENSION ARM

67189-DAKO-G17

Torsion bar

2004

→The left and right side torsion bars are NOT interchangeable. The bars are identified and stamped R or L, for right or left. The bars do not have a front or rear end and can be installed with either end facing forward.

1. Raise and support the vehicle with the front suspension hanging.

2. Turn the adjustment bolt counter-clockwise to release spring load.

→Count and record the number of turns for installation reference.

3. Remove the adjustment bolt from swivel.

4. Remove torsion bar and anchor. Remove anchor from torsion bar.

5. Remove all foreign material from torsion bar mounting in anchor and suspension arm.

6. Inspect adjustment bolt, bearing and swivel for damage.

To install:

7. Insert torsion bar ends into anchor and suspension arm.

8. Position anchor and bearing in frame crossmember. Install adjustment bolt through bearing, anchor and into the swivel.

9. Turn adjustment bolt clockwise the recorded amount of turns.

10. Lower vehicle and adjust the front suspension height.

Upper Ball Joint

REMOVAL & INSTALLATION

The Dakota models utilize an upper control arm with an integral ball joint. If the ball joint is damaged or worn, the upper control arm must be replaced.

Lower Ball Joint

REMOVAL & INSTALLATION

The Dakota models utilize a lower control arm with an integral ball joint. If the ball joint is damaged or worn, the upper control arm must be replaced.

Upper Control Arm

REMOVAL & INSTALLATION

2-Wheel Drive

1. Before servicing the vehicle, refer to the precautions in the beginning of this section.

2. Raise and support the vehicle.

3. Remove wheel and tire assembly.

4. Remove the caliper mounting bolts.

5. Remove brake hose bracket from the arm.

6. Position a hydraulic jack under the arm and raise the jack to unload the rebound bumper.

7. Remove cotter pin and nut from upper ball joint.

8. Separate upper ball joint from steering knuckle.

→When installing the remover to separate the ball joint, be careful not to damage the ball joint seal.

9. Remove the control arm pivot bar mounting nuts and remove the control arm.

To install:

→Before installation, insure pivot bar adjustment bolts are in their original location.

10. Position the control arm pivot bar on adjustment bolts. Install nuts and tighten to 210 Nm (155 ft. lbs.).

11. Position steering knuckle on upper ball joint. Tighten the upper ball joint nut to 81 Nm (60 ft. lbs.) and install a new cotter pin.

12. Install the brake hose bracket to the arm.

13. Install the caliper mounting bolts. Tighten the bolts to 30 Nm (22 ft. lbs.).

14. Install the wheel and tire assembly.

15. Remove support and lower vehicle.

16. Align front end to specifications.

4-Wheel Drive

1. Raise and support the vehicle.

2. Remove the wheel and tire assembly.

3. Remove the brake hose brackets from the control arm.

4. Position a hydraulic jack under the lower suspension arm and raise the jack to unload the rebound bumper.

5. Remove the shock absorber.

6. Remove the cotter pin and nut from the upper ball joint.

7. Separate the upper ball joint from the steering knuckle.

→When installing the remover to separate the ball joint, be careful not to damage the ball joint seal.

To install:

8. Position the control arm pivot bar on the mounting bracket. Install the bolts and tighten (temporarily) to 136 Nm (100 ft. lbs.).

9. Insert the ball joint in the steering knuckle and tighten the ball joint nut to 81 Nm (60 ft. lbs.) then install a new cotter pin.

10. Install the shock absorber.

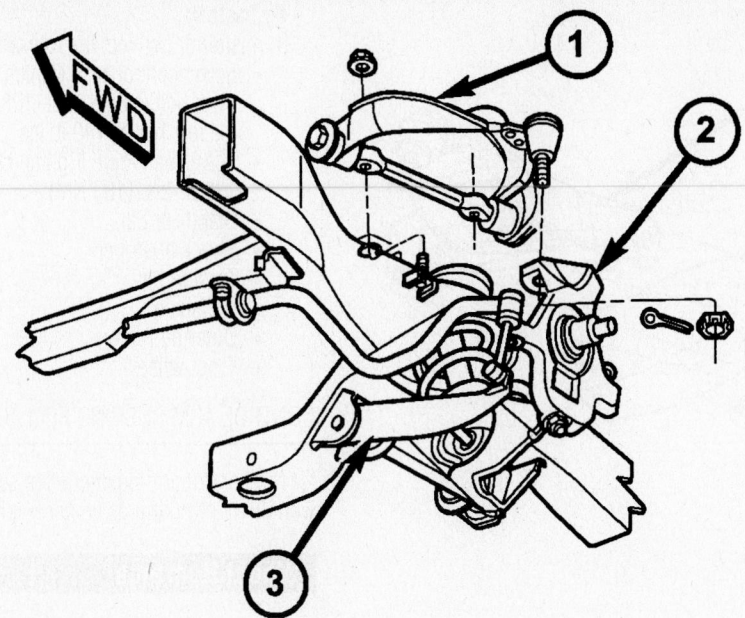

1 - UPPER CONTROL ARM
2 - STEERING KNUCKLE
3 - LOWER CONTROL ARM

67189-DAKO-G16

Upper control arm—2 wheel drive

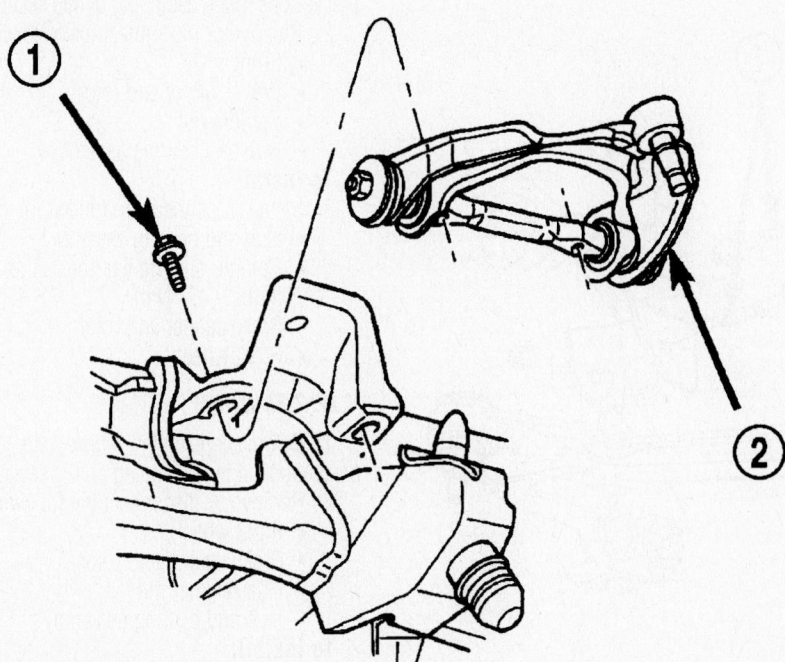

1 - PIVOT BAR BOLT
2 - UPPER SUSPENSION ARM

67189-DAKO-G14

Upper control arm—4 wheel drive

➡The upper shock nut must be replaced or use Mopar Lock 'N Seal or Loctite® 242 on the existing nut.

11. Remove the hydraulic jack.

12. Attach the brake hose brackets to the control arm.

13. Tighten the upper control arm pivot bolts to 204 Nm (150 ft. lbs.).

14. Install the wheel and tire assembly.

15. Remove the support and lower the vehicle.

16. Align the front suspension.

CONTROL ARM BUSHING REPLACEMENT

The control arm bushings are serviced with the control arm as an assembly.

Lower Control Arm

REMOVAL & INSTALLATION

2 WHEEL DRIVE

1. Before servicing the vehicle, refer to the precautions in the beginning of this section.

2. Remove or disconnect the following:
   - Front wheel
   - Shock absorber
   - Brake caliper and rotor
   - Stabilizer bar link
   - Coil spring
   - Inner mounting bolts
   - Lower control arm

To install:

3. Install or connect the following:
   - Lower control arm. Tighten the front bolt to 130 ft. lbs. (175 Nm) and the rear bolt to 80 ft. lbs. (108 Nm).
   - Coil spring. Tighten the lower ball joint nut to 94 ft. lbs. (127 Nm).
   - Stabilizer bar link
   - Brake caliper and rotor
   - Shock absorber
   - Front wheel

4 WHEEL DRIVE

1. Before servicing the vehicle, refer to the precautions in the beginning of this section.

2. Remove or disconnect the following:
   - Front wheel
   - Outer tie rod end
   - Halfshaft
   - Torsion bar
   - Shock absorber
   - Stabilizer bar
   - Lower ball joint
   - Pivot bolts
   - Lower control arm

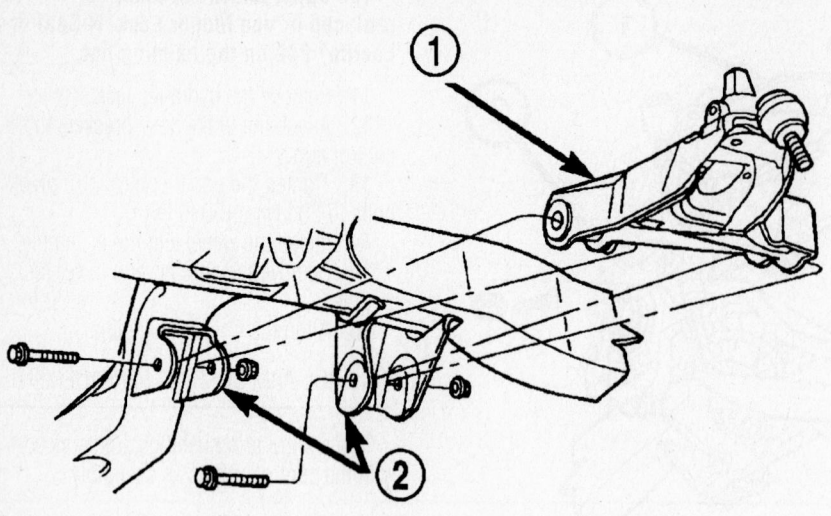

1 - LOWER CONTROL ARM
2 - FRAME MOUNTS

67189-DAKO-G15

Lower control arm—2 wheel drive

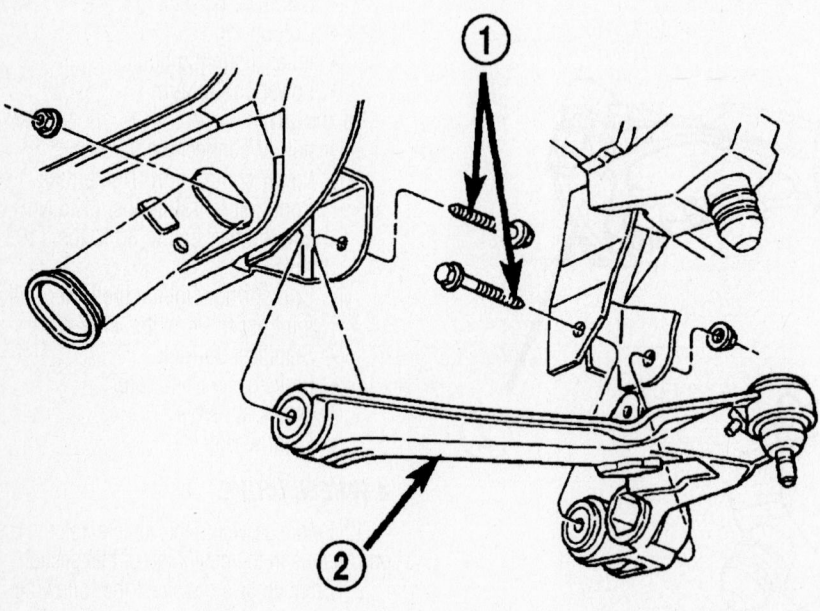

1 - PIVOT BOLTS
2 - LOWER CONTROL ARM

67189-DAKO-G13

Lower control arm—4 wheel drive

To install:
3. Install or connect the following:
- Lower control arm. Tighten the front pivot bolt to 80 ft. lbs. (108 Nm) and the rear bolt to 140 ft. lbs. (190 Nm).
- Lower ball joint. Tighten the nut to 135 ft. lbs. (183 Nm).
- Stabilizer bar
- Shock absorber
- Torsion bar
- Halfshaft
- Outer tie rod end
- Front wheel

CONTROL ARM BUSHING REPLACEMENT

The control arm bushings are serviced with the control arm as an assembly.

Wheel Bearings

ADJUSTMENT

The Dakota models utilize a hub/bearing assembly which is not adjustable.

REMOVAL & INSTALLATION

2 WHEEL DRIVE

1. Before servicing the vehicle, refer to the precautions in the beginning of this section.
2. Remove or disconnect the following:
- Front wheel
- Brake caliper and rotor
- Spindle nut
- Hub and bearing assembly

To install:
3. Install or connect the following:
- Hub and bearing assembly
- Spindle nut. Tighten the nut to 185 ft. lbs. (251 Nm).
- Brake caliper and rotor
- Front wheel

4 WHEEL DRIVE

1. Before servicing the vehicle, refer to the precautions in the beginning of this section.
2. Remove or disconnect the following:
- Front wheel
- Brake caliper and rotor
- Hub retainer nut
- Hub and bearing assembly

To install:
3. Install or connect the following:
- Hub and bearing assembly. Tighten the bolts to 123 ft. lbs. (166 Nm).
- Hub retainer nut. Tighten the nut to 173 ft. lbs. (235 Nm).
- Brake caliper and rotor
- Front wheel

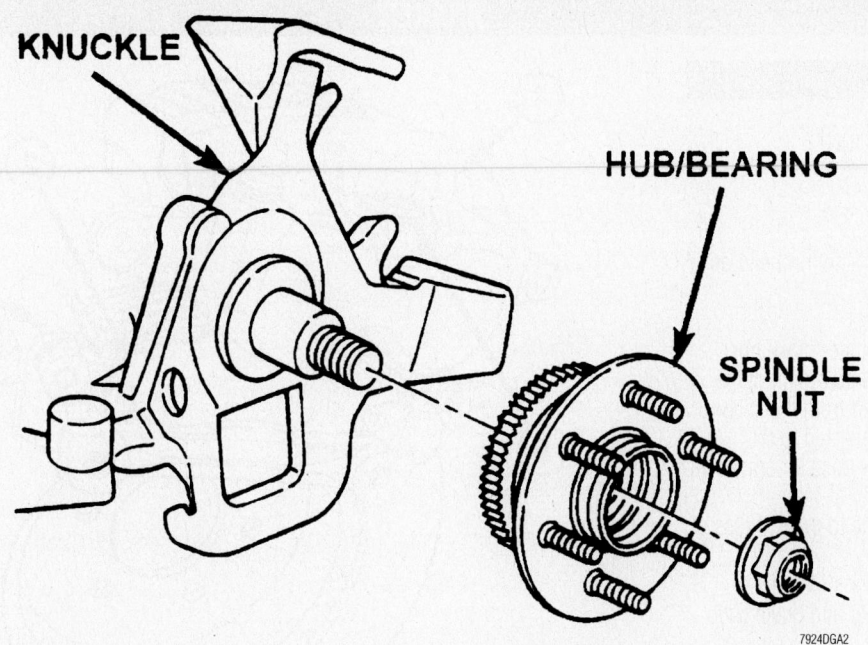

KNUCKLE

HUB/BEARING

SPINDLE
NUT

7924DGA2

Hub/bearing assembly—2 wheel drive

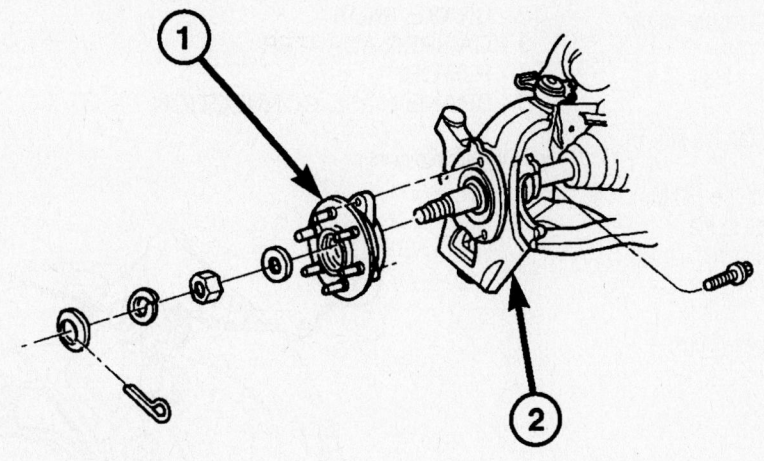

1 - HUB BEARING
2 - STEERING KNUCKLE

67189-DAKO-G12

Hub/bearing assembly—4 wheel drive

Brake Caliper

REMOVAL & INSTALLATION

Front

1. Raise and support the front end on jackstands.
2. Remove the wheels.
3. Disconnect the rubber brake hose from the tubing at the frame mount. If the pistons are to be removed from the caliper, leave the brake hose connected to the caliper. Check the rubber hose for cracks or chafed spots.
4. Plug the brake line to prevent loss of fluid.
5. Remove the caliper slide pins.
6. Remove the caliper and brake pads from the rotor adapter.

To install:

7. Position the outboard shoe in the caliper. The shoe should not rattle in the caliper. If it does, or if any movement is obvious, bend the shoe tabs over the caliper to tighten the fit.
8. Slide the caliper into position on the adapter and over the rotor.
9. Align the caliper and start the pins in by hand.
10. Tighten the pins to 22 ft. lbs. (30 Nm).
11. Connect the brake hose to the caliper. Use new washers to attach the hose fitting if the original washers are scored, worn or damaged.
12. Fill and bleed the brake system.
13. Install the wheels.
14. Lower the vehicle.

Rear

1. Install prop rod on the brake pedal to keep pressure on the brake system.
2. Raise and support vehicle.
3. Remove the wheel and tire assembly.
4. Drain small amount of fluid from master cylinder brake reservoir with suction gun.
5. Remove the brake hose banjo bolt if replacing caliper.
6. Remove the caliper mounting slide pin bolts.
7. Remove the caliper from vehicle.

To install:

8. Install the brake pads if removed.
9. Lubricate anti-rattle clips for the disc brake pads.
10. Install the caliper to the brake caliper adapter.

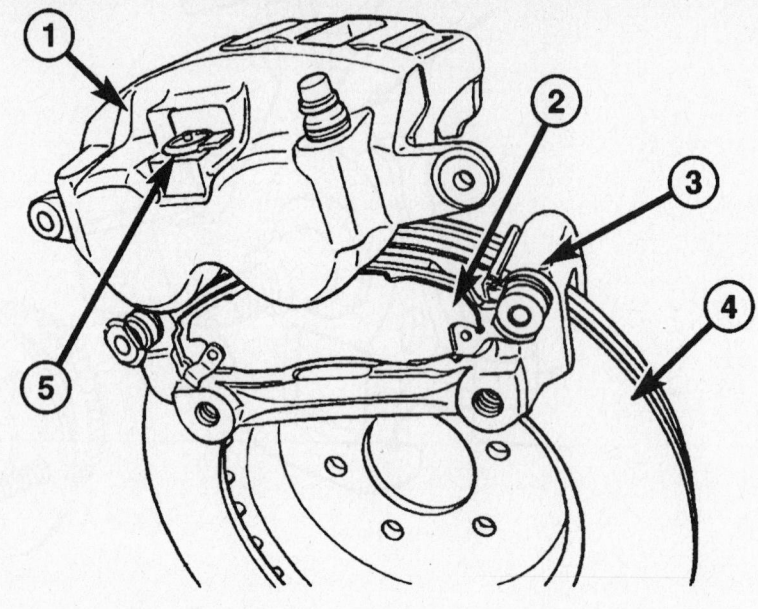

1 - CALIPER
2 - BRAKE PADS
3 - CALIPER ADAPTER
4 - ROTOR
5 - BRAKE HOSE CONNECTION

67189-DAKO-G11

Front caliper removed

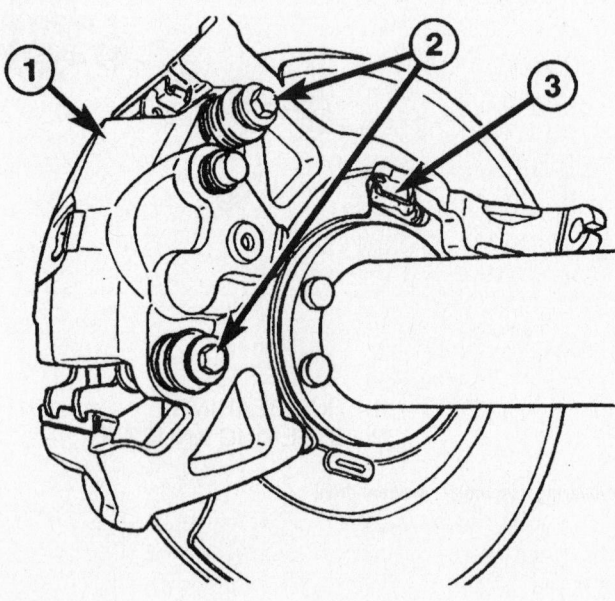

1 - CALIPER
2 - CALIPER SLIDE BOLTS
3 - PARK BRAKE SHOE LEVER

67189-DAKO-G10

Rear caliper installed

11. Coat the caliper mounting slide pin bolts with silicone grease. Then install and tighten the bolts to 25 Nm (18 ft. lbs.).

12. Install the brake hose banjo bolt if removed.

13. Install the brake hose to the caliper with new seal washers and tighten fitting bolt to 31 Nm (23 ft. lbs.).

➡**Verify brake hose is not twisted or kinked before tightening fitting bolt.**

14. Remove the prop rod from the vehicle.

15. Bleed the base brake system.

16. Install the wheel and tire assemblies.

Disc Brake Pads

REMOVAL & INSTALLATION

Front

1. Raise and support the front end on jackstands.

2. Remove the wheels.

3. Press the caliper piston back into the bore with a suitable prytool. Use a large C-clamp to drive the piston into the bore of additional force is required.

4. Remove the caliper mounting bolts.

5. Rotate the caliper rearward off the rotor and out from its mount.

6. Set the caliper on a crate or sturdy box, then remove the inboard and outboard brake pads. The inboard pad has a spring clip that holds it in the caliper. Tilt this pad out at the top to unseat the clip. The outboard pad has a retaining spring that secures it in the caliper. Unseat 1 spring end and rotate the pad out of the caliper.

7. Secure the caliper to a chassis or suspension component with a sturdy wire. Do not let it hang from the hose.

To install:

8. Clean the caliper and steering knuckle sliding surfaces with a wire brush. Then, apply a coat of Mopar® multi-mileage grease or equivalent.

9. Clean the caliper slide pins with brake cleaner or brake fluid. Then apply a light coating of silicone grease to the pins.

➡**If there is minor rust or corrosion on the pins, first polish them with a crocus cloth. If they are severely rusted, replace them.**

10. Install the inboard pad and its spring clip.

11. Install the outboard brake pad.

12. Install the caliper over the rotor and seat it in its original position until flush.

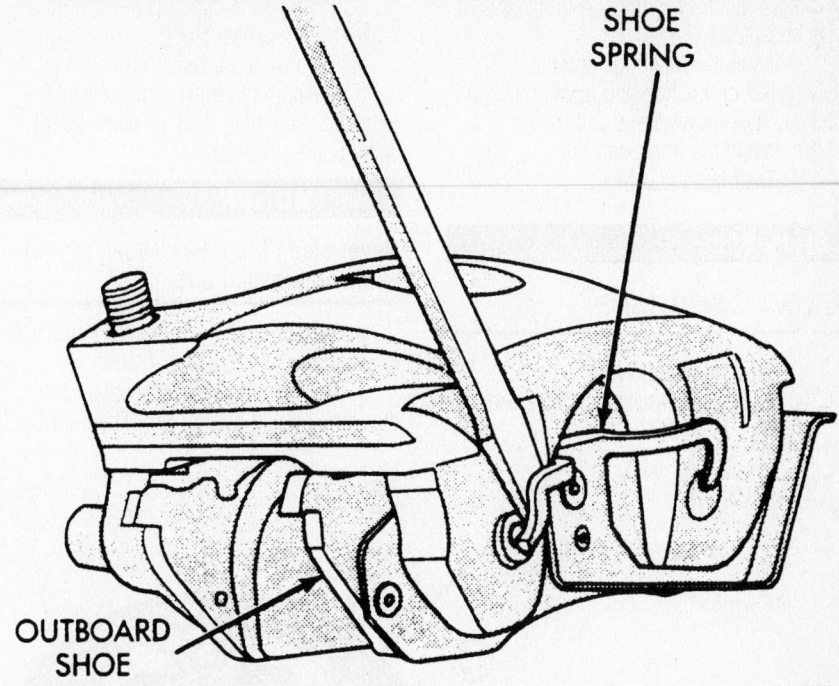

Prying the disc brake from the 4WD front brake caliper assembly

SHOE SPRING

OUTBOARD SHOE

93026G08

13. Final tighten the caliper slide pins to 22 ft. lbs. (30 Nm).

14. Install the wheels.

15. Lower the vehicle.

16. Pump the brakes several times to seat the pads.

Rear

1. Raise and support vehicle.

2. Remove the wheel and tire assemblies.

3. Compress the caliper.

4. Remove the caliper.

5. Remove the caliper by tilting the top up and off the caliper adapter.

➡**Do not allow brake hose to support caliper assembly.**

6. Support and hang the caliper.

7. Remove the inboard brake pad from the caliper adapter.

8. Remove the outboard brake pad from the caliper adapter.

➡**To install:**

9. Bottom the piston in the caliper bore with C-clamp. Place an old brake shoe between a C-clamp and caliper piston.

10. Clean the caliper mounting surface.

11. Install anti-rattle clips.

➡**When servicing the rear brake pads, replace the anti-rattle clips on the brake adapter. Anti-rattle clips are provided with the shoe & lining kit.**

12. Install inboard brake pad in adapter.

13. Install outboard brake pad in adapter.

14. Tilt the top of the caliper over rotor and secure to the mounting holes on the brake caliper adapter.

15. Install the caliper.

16. Install wheel and tire assemblies and lower vehicle.

17. Apply brakes several times to seat caliper pistons and brake shoes and obtain firm pedal.

18. Top off master cylinder fluid level.

Brake Drums

REMOVAL & INSTALLATION

1. Remove the axle shaft nuts, washers and cones. If the cones do not readily release, rap the axle shaft sharply in the center.

2. Remove the axle shaft.

3. Remove the outer hub nut.

4. Straighten the lockwasher tab and remove it along with the inner nut and bearing.

5. Carefully remove the drum.

To install:

6. Position the drum on the axle housing.

7. Install the bearing and inner nut. While rotating the wheel and tire, tighten the adjusting nut until a slight drag is felt.

8. Back off the adjusting nut 1/6 turn so

that the wheel rotates freely without excessive end-play.

9. Install the lockrings and nut. Place a new gasket on the hub and install the axle shaft, cones, lockwashers and nuts.

10. Install the wheel and tire.

11. Road-test the vehicle.

Brake Shoes

REMOVAL & INSTALLATION

1. Raise and support the vehicle.
2. Remove the wheel and tire assembly.

3. Remove the clip nuts securing the brake drum to wheel studs.

4. Remove the drum.

5. Vacuum the brake components to remove brake lining dust, or use a liquid spray brake cleaner.

❊❊ CAUTION

Never clean the brake shoes or parts with compressed air!

6. Remove the shoe return springs with brake spring pliers tool.

7. Remove the adjuster cable. Slide the cable eye off anchor pin. Then unhook

and remove the cable from the adjuster lever.

8. Remove the cable guide from the secondary shoe and anchor plate from anchor pin.

9. Remove the adjuster lever. Disengage the lever from spring by sliding the lever forward to clear the pivot and work the lever out from under the spring.

10. Remove the adjuster lever spring from the pivot.

11. Disengage and remove the shoe spring from the brake shoes.

12. Disengage and remove the adjuster screw assembly from the brake shoes.

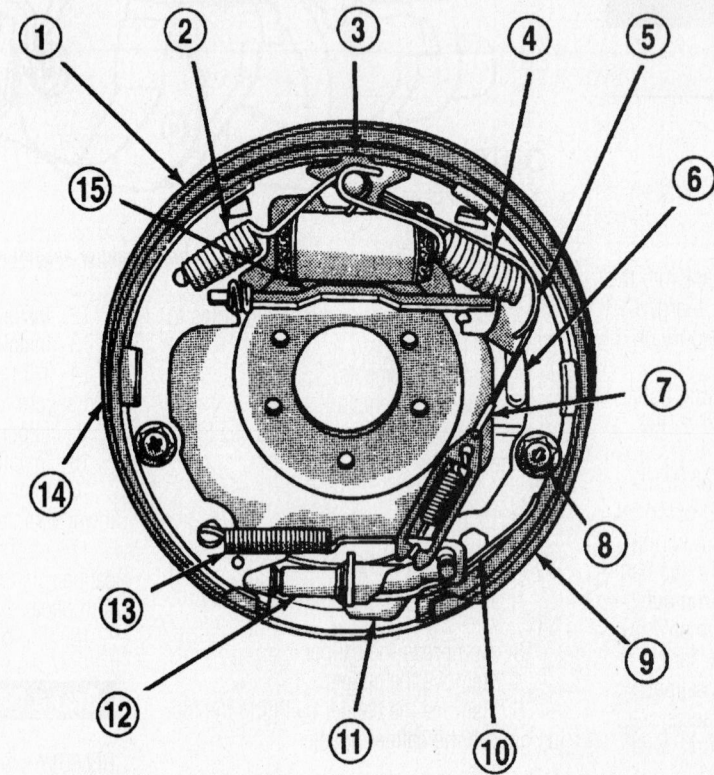

1 - SUPPORT PLATE
2 - RETURN SPRING
3 - ANCHOR PLATE
4 - RETURN SPRING
5 - CABLE GUIDE
6 - PARKING BRAKE LEVER
7 - ADJUSTER CABLE AND SPRING
8 - SHOE RETAINER, SPRING AND PIN
9 - SECONDARY SHOE AND LINING
10 - LEVER SPRING
11 - ADJUSTER LEVER
12 - ADJUSTER SCREW ASSEMBLY
13 - SHOE SPRING
14 - PRIMARY SHOE AND LINING
15 - PARKING BRAKE STRUT AND SPRING

67189-DAKO-G09

Brake shoes assembled

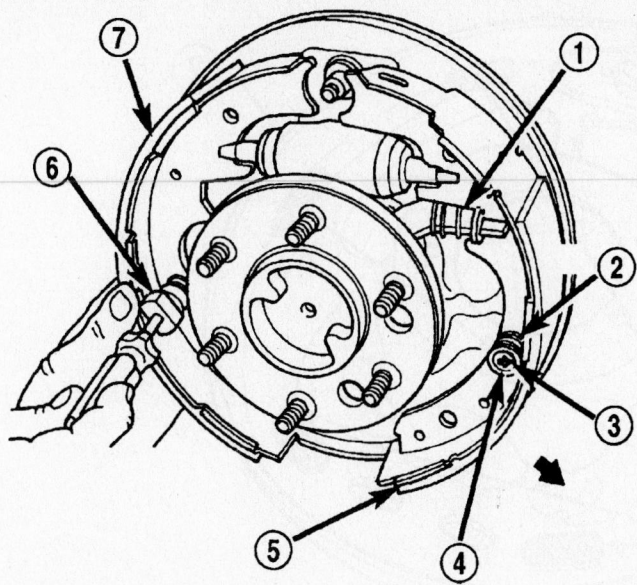

1 - STRUT AND SPRING
2 - SPRING
3 - PIN
4 - RETAINER
5 - PRIMARY SHOE AND LINING
6 - TOOL C-4070
7 - SECONDARY SHOE AND LINING

67189-DAKO-G08

Shoe retainers, springs and pins

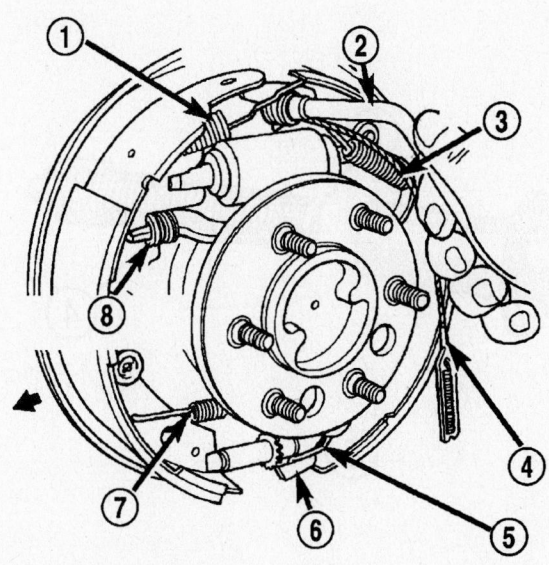

1 - SHOE RETURN SPRING
2 - SPECIAL TOOL
(REMOVING AND INSTALLING)
3 - SHOE RETURN SPRING
4 - ADJUSTER CABLE
5 - LEVER SPRING
6 - ADJUSTER LEVER
7 - SHOE TO SHOE SPRING
8 - ANTI-RATTLE SPRING

67189-DAKO-G07

Shoe return springs

13. Remove the brake shoe retainers, springs.

14. Remove the secondary brake shoe from the support plate.

15. Remove the strut and spring.

16. Remove the parking brake lever retaining clip from the secondary shoe and remove the lever.

17. Remove the primary shoe from the support plate.

18. Disengage the parking brake lever from the parking brake cable.

To install:

19. Clean and inspect individual brake components.

20. Lubricate anchor pin and brake shoe contact pads on support plate with high temperature grease.

21. Lubricate adjuster screw socket, nut, button and screw thread surfaces with grease.

22. Install the parking brake cable to the parking brake lever.

23. Install parking brake lever to the secondary shoe and install retaining clip.

24. Install primary shoe on support plate. Secure shoe with new spring retainers and pin.

25. Install spring on parking brake strut and engage strut in primary.

26. Install secondary shoe on support plate. Insert strut in shoe and guide shoe onto anchor pin. Temporarily secure shoe with retaining pin.

27. Install anchor plate and adjuster cable eyelet on support plate anchor pin.

28. Install cable guide in secondary shoe and position cable in guide.

29. Assemble adjuster screw. Then install and adjuster screw between the brake shoes.

➡Be sure the adjuster screws are installed on the correct brake unit. The adjuster screws are marked L (left) and R (right) for identification.

30. Install adjuster lever and spring and connect adjuster cable to lever.

31. Install secondary shoe retainers and spring.

32. Install shoe spring. Connect spring to secondary shoe first. Then to primary shoe.

33. Verify adjuster operation. Pull adjuster cable upward, cable should lift lever and rotate star wheel. Be sure adjuster lever properly engages star wheel teeth.

34. Adjust brake shoes to drum with brake gauge.

35. Install wheel and tire assembly.

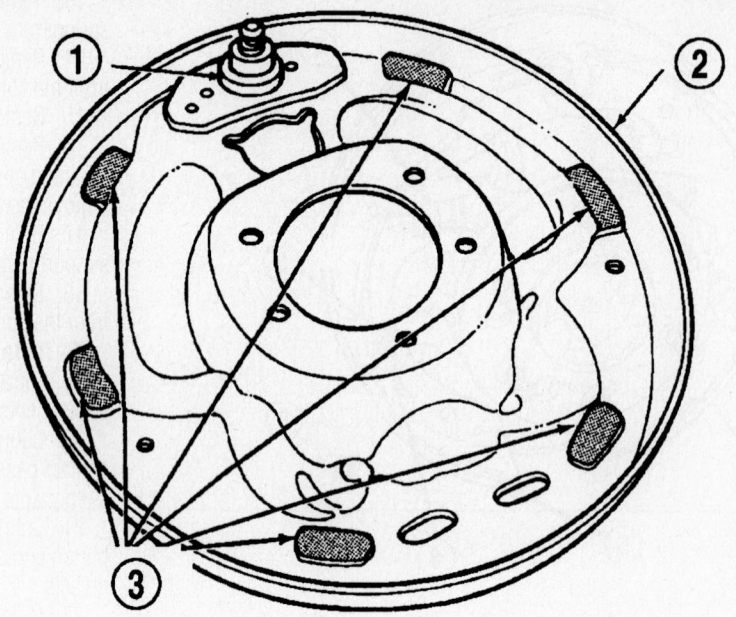

1 - ANCHOR PIN
2 - SUPPORT PLATE
3 - SHOE CONTACT SURFACES

67189-DAKO-G05

Shoe contact surfaces

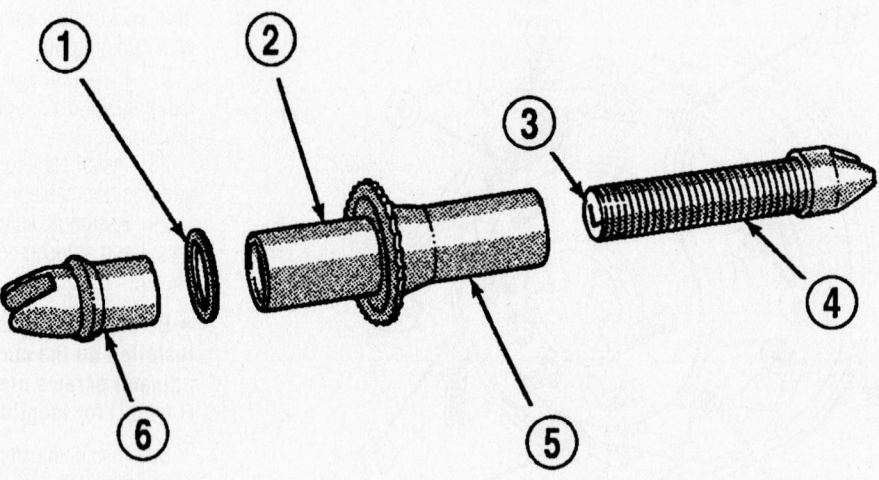

1 - WASHER
2 - SOCKET
3 - STAMPED LETTER
L-LEFT BRAKE
R-RIGHT BRAKE
4 - SCREW THREADS
5 - NUT
6 - BUTTON

67189-DAKO-G06

Adjuster screw

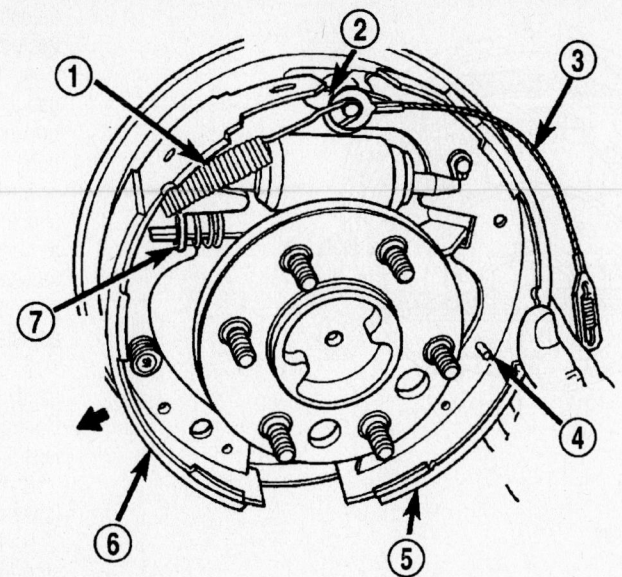

1 - SHOE RETURN SPRING
2 - ANCHOR PLATE
3 - ADJUSTER CABLE
4 - SHOE RETAINING PIN
5 - SECONDARY SHOE AND LINING
6 - PRIMARY SHOE AND LINING
7 - STRUT AND SPRING

67189-DAKO-G04

Brake shoe installation

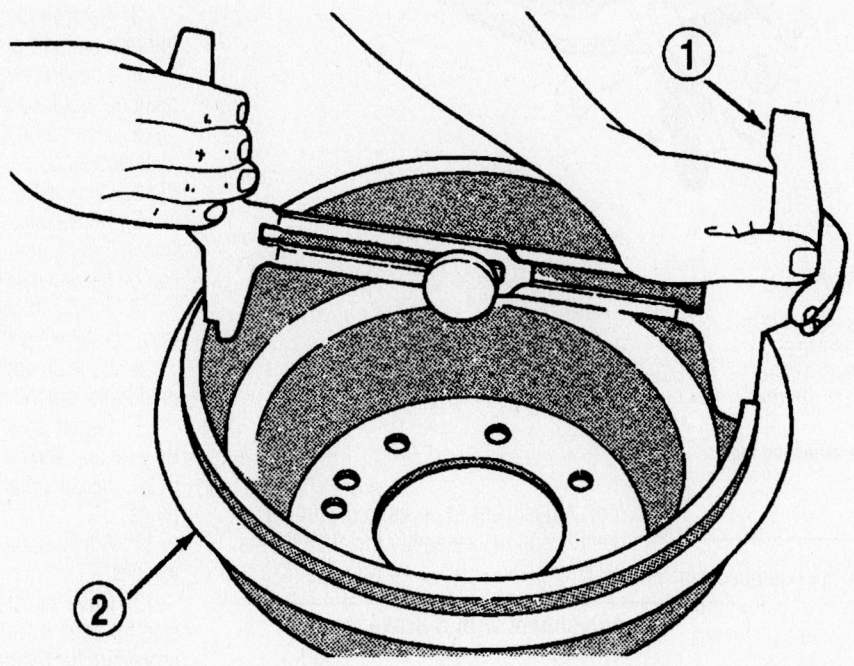

1 - BRAKE GAUGE
2 - BRAKE DRUM

67189-DAKO-G03

Adjusting gauge on the drum

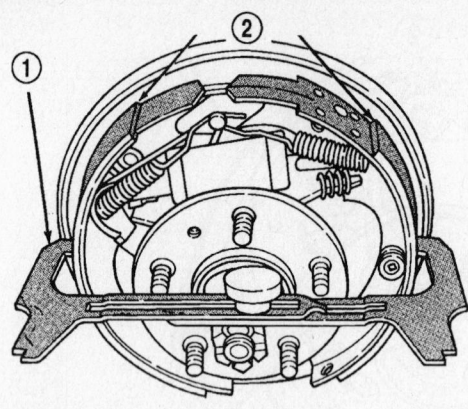

1 - BRAKE GAUGE
2 - BRAKE SHOES

67189-DAKO-G01

Adjustment with a brake adjusting gauge

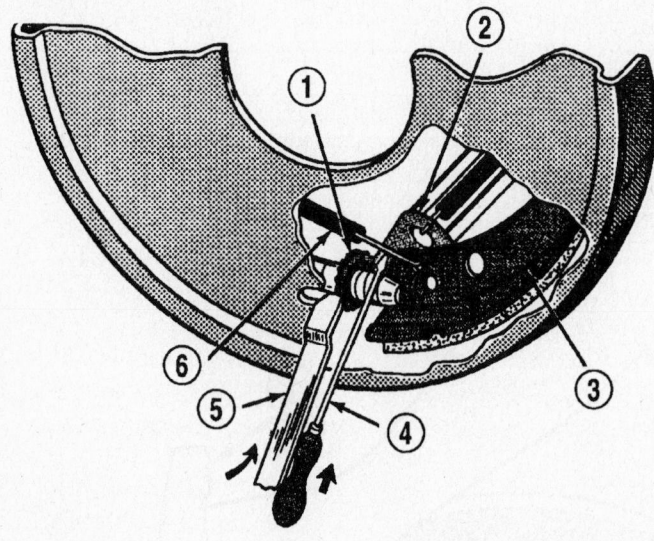

1 - STAR WHEEL
2 - LEVER
3 - BRAKE SHOE WEB
4 - SCREWDRIVER
5 - ADJUSTING TOOL
6 - ADJUSTER SPRING

67189-DAKO-G02

Adjustment with a brake adjusting tool

ADJUSTMENTS

The rear drum brakes are equipped with a self-adjusting mechanism. Under normal circumstances, the only time adjustment is required is when the shoes are replaced, removed for access to other parts, or when one or both drums are replaced. Adjustment can be made with a standard brake gauge or with adjusting tool. Adjustment is performed with the complete brake assembly installed on the backing plate.

Adjustment with a Brake Gauge

1. Be sure parking brakes are fully released.
2. Raise rear of vehicle and remove wheels and brake drums.
3. Verify that left and right automatic adjuster levers and cables are properly connected.
4. Insert brake gauge in drum. Expand gauge until gauge inner legs contact drum braking surface. Then lock gauge in position.
5. Reverse gauge and install it on brake shoes. Position gauge legs at shoe centers as shown. If gauge does not fit (too loose/too tight), adjust shoes.
6. Pull shoe adjuster lever away from adjuster screw star wheel.
7. Turn adjuster screw star wheel (by hand) to expand or retract brake shoes. Continue adjustment until gauge outside legs are light drag-fit on shoes.
8. Install brake drums and wheels and lower vehicle.
9. Drive vehicle and make one forward stop followed by one reverse stop. Repeat procedure 8-10 times to operate automatic adjusters and equalize adjustment.

➡ **Bring vehicle to complete standstill at each stop. Incomplete, rolling stops will not activate automatic adjusters.**

Adjustment with an Adjusting Tool

1. Be sure parking brake lever is fully released.
2. Raise vehicle so rear wheels can be rotated freely.
3. Remove plug from each access hole in brake support plates.
4. Loosen parking brake cable adjustment nut until there is slack in front cable.
5. Insert adjusting tool through support plate access hole and engage tool in teeth of adjusting screw star wheel.
6. Rotate adjuster screw star wheel (move tool handle upward) until slight drag can be felt when wheel is rotated.
7. Push and hold adjuster lever away from star wheel with thin screwdriver.
8. Back off adjuster screw star wheel until brake drag is eliminated.
9. Repeat adjustment at opposite wheel. Be sure adjustment is equal at both wheels.
10. Install support plate access hole plugs.
11. Adjust parking brake cable and lower vehicle.
12. Drive vehicle and make one forward stop followed by one reverse stop. Repeat procedure 8-10 times to operate automatic adjusters and equalize adjustment.

➡ **Bring vehicle to complete standstill at each stop. Incomplete, rolling stops will not activate automatic adjusters.**

SPECIFICATION CHARTS

ENGINE AND VEHICLE IDENTIFICATION

	Engine						Model Year	
Code ①	Liters (cc)	Cu. In.	Cyl.	Fuel Sys.	Engine Type	Eng. Mfg.	Code ②	Year
N	4.7 (4701)	287	8	SMFI	SOHC	Chrysler	1	2001
K	3.7 (3701)	226	6	MFI	SOHC	Chrysler	2	2002
D	5.7 (5653)	345	8	SMPI	OHV	Chrysler	3	2003
Z	5.9 (5899)	360	8	SMFI	OHV	Chrysler	4	2004
							5	2005

SOHC: Single overhead camshaft

OHV: Overhead Valve

SMFI: Sequential Multi-port Fuel Injection

① 8th position of VIN

② 10th position of VIN

67189-DURA-C01

GENERAL ENGINE SPECIFICATIONS

Year	Model	Engine Displ. Liters	Engine VIN	Net Horsepower @ rpm	Net Torque @ rpm (ft. lbs.)	Bore x Stroke (in.)	Com-pression Ratio	Oil Pressure @ rpm
2001	Durango	4.7	N	235@4800	295@3200	3.66x3.40	9.3:1	25-110@3000
		5.9	Z	230@4000	330@3250	4.00x3.58	9.1:1	30-80@3000
2002	Durango	4.7	N	235@4800	295@3200	3.66x3.40	9.3:1	25-110@3000
		5.9	Z	230@4000	330@3250	4.00x3.58	9.1:1	30-80@3000
2003	Durango	4.7	N	235@4800	295@3200	3.66x3.40	9.0:1	25-110@3000
		5.9	Z	230@4000	330@3250	4.00x3.58	9.1:1	30-80@3000
2004	Durango	3.7	K	210@5200	225@4200	3.66x3.40	9.1:1	25-110@3000
		4.7	N	235@4800	295@3200	3.66x3.40	9.0:1	25-110@3000
		5.7	D	345@5400	375@4200	3.91x3.58	9.6:1	25-110@3000

SMFI: Sequential Multi-port Fuel Injection

67189-DURA-C02

GASOLINE ENGINE TUNE-UP SPECIFICATIONS

Year	Engine Displ. Liters	Engine VIN	Spark Plug Gap (in.)	Ignition Timing (deg.)	Fuel Pump (psi)	Idle Speed (rpm)	Valve Clearance	
							Intake	Exhaust
2001	4.7	N	0.040	①	47-51	②	HYD	HYD
	5.9	Z	0.040	①	44.2-54.2	②	HYD	HYD
2002	4.7	N	0.040	①	47-51	②	HYD	HYD
	5.9	Z	0.040	①	44.2-54.2	②	HYD	HYD
2003	4.7	N	0.040	①	47-51	②	HYD	HYD
	5.9	Z	0.040	①	44.2-54.2	②	HYD	HYD
2004	3.7	K	0.042	①	44-54	②	HYD	HYD
	4.7	N	0.040	①	47-51	②	HYD	HYD
	5.7	D	0.045	①	49.0-49.4	②	HYD	HYD

NOTE: The Vehicle Emission Control Information (VECI) label often reflects specification changes made during production.

The label figures must be used if they differ from those in this chart.

HYD: Hydraulic

① Ignition timing is controlled by the PCM and is not adjustable.

② Idle speed is controlled by the PCM and is not adjustable

67189-DURA-C03

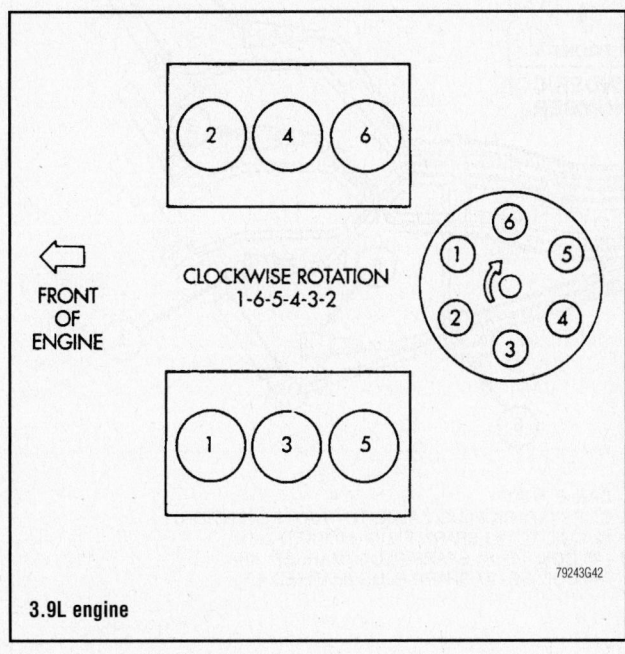

FRONT OF ENGINE

CLOCKWISE ROTATION
1-6-5-4-3-2

3.9L engine

79243G42

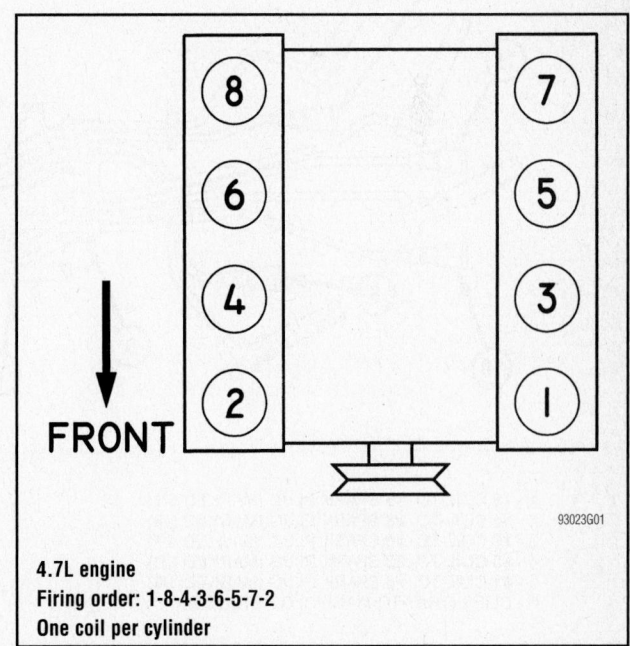

FRONT

93023G01

4.7L engine
Firing order: 1-8-4-3-6-5-7-2
One coil per cylinder

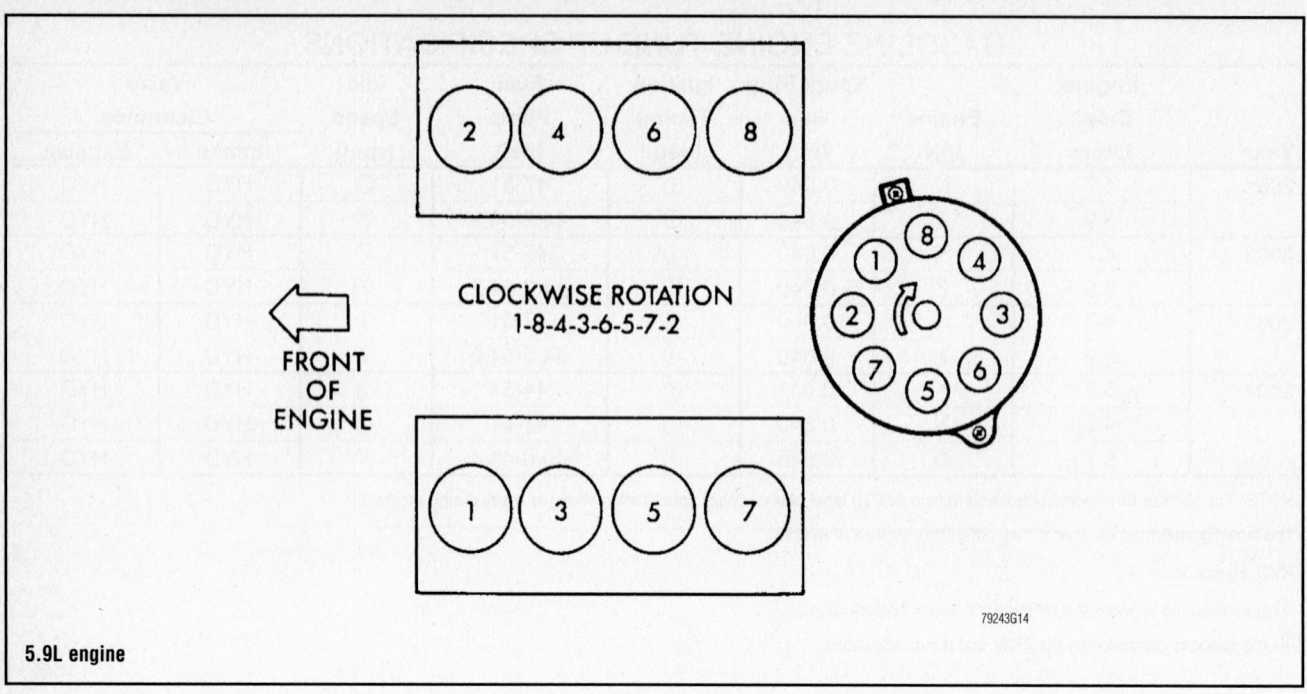

CLOCKWISE ROTATION
1-8-4-3-6-5-7-2

FRONT
OF
ENGINE

79243G14

5.9L engine

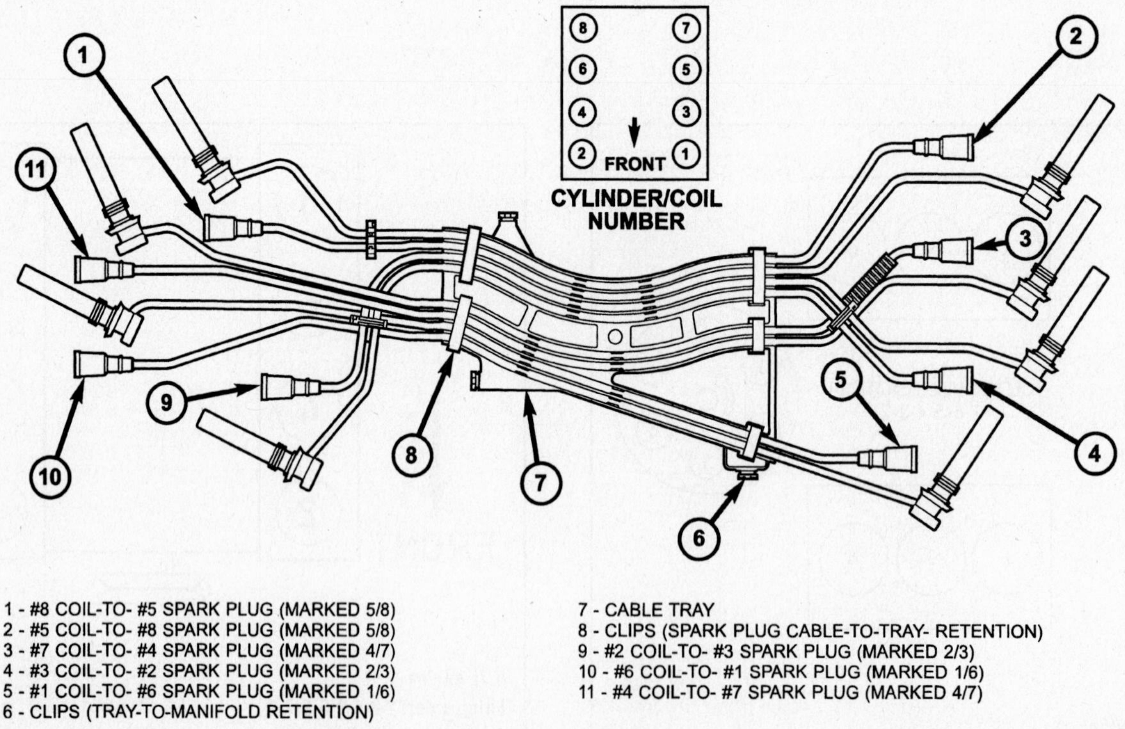

CYLINDER/COIL
NUMBER

FRONT

1 - #8 COIL-TO- #5 SPARK PLUG (MARKED 5/8)
2 - #5 COIL-TO- #8 SPARK PLUG (MARKED 5/8)
3 - #7 COIL-TO- #4 SPARK PLUG (MARKED 4/7)
4 - #3 COIL-TO- #2 SPARK PLUG (MARKED 2/3)
5 - #1 COIL-TO- #6 SPARK PLUG (MARKED 1/6)
6 - CLIPS (TRAY-TO-MANIFOLD RETENTION)

7 - CABLE TRAY
8 - CLIPS (SPARK PLUG CABLE-TO-TRAY- RETENTION)
9 - #2 COIL-TO- #3 SPARK PLUG (MARKED 2/3)
10 - #6 COIL-TO- #1 SPARK PLUG (MARKED 1/6)
11 - #4 COIL-TO- #7 SPARK PLUG (MARKED 4/7)

67189-DURA-G05

5.7L engine firing order

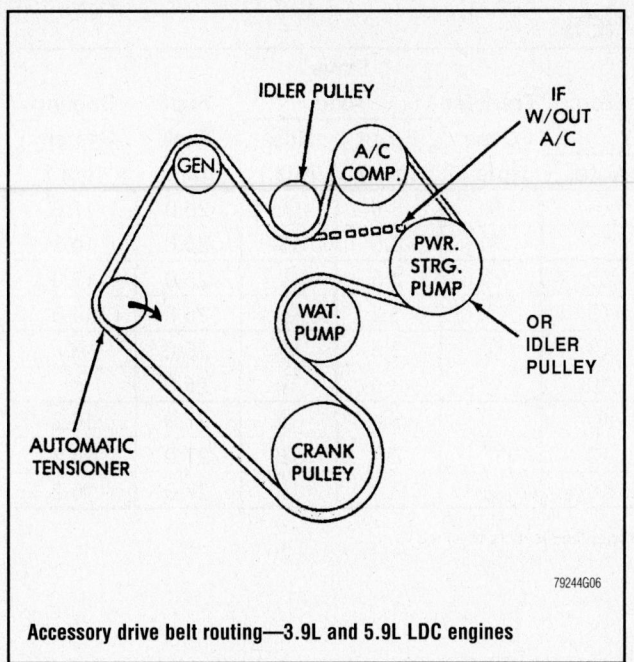

Accessory drive belt routing—3.9L and 5.9L LDC engines

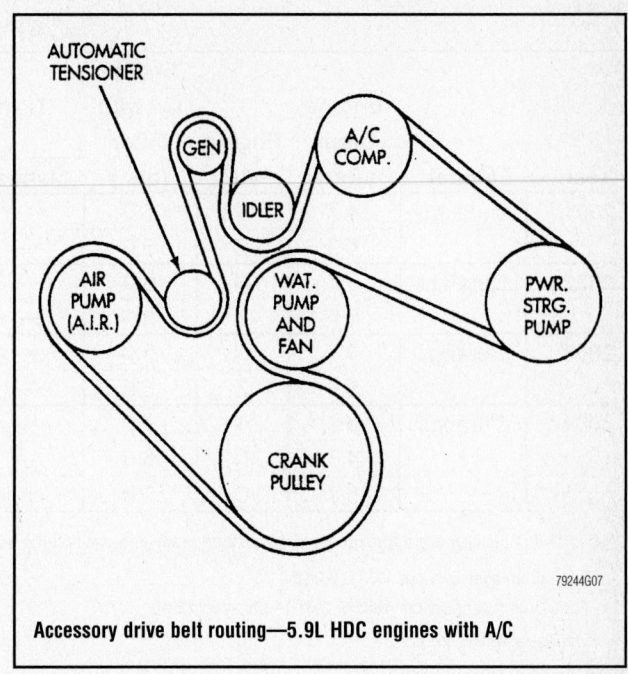

Accessory drive belt routing—5.9L HDC engines with A/C

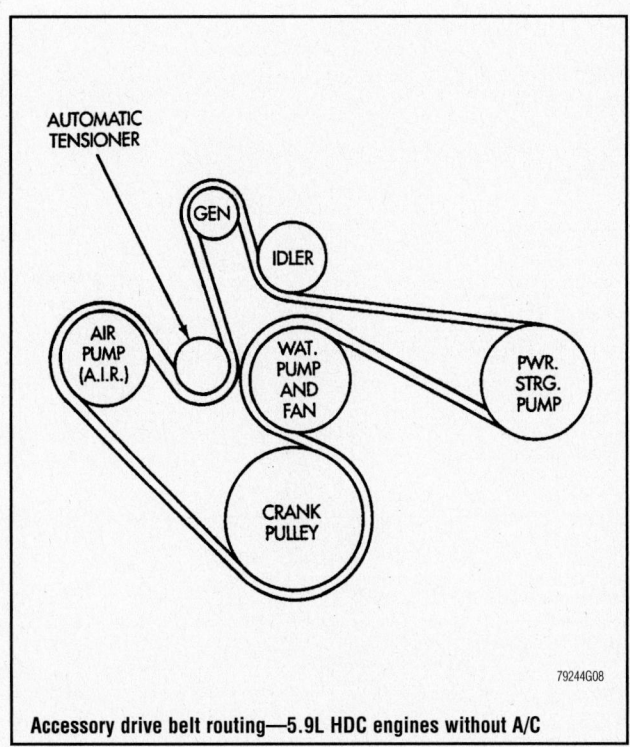

Accessory drive belt routing—5.9L HDC engines without A/C

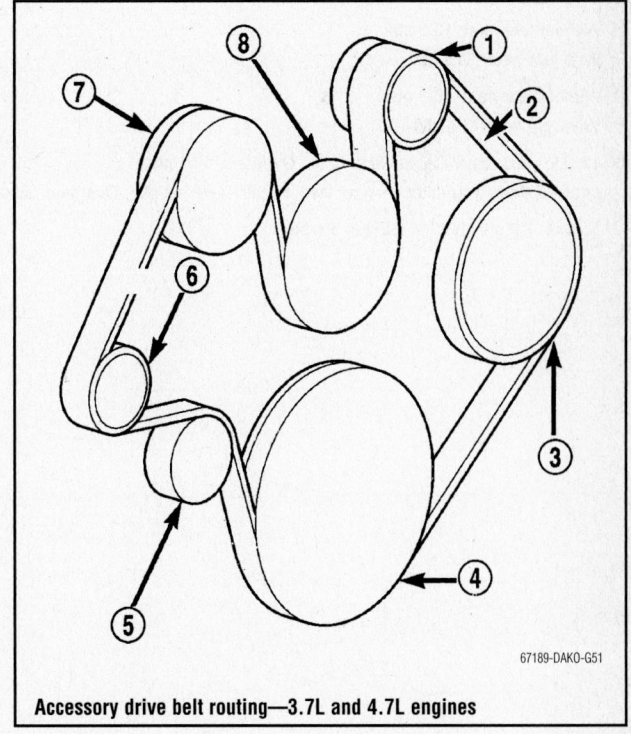

Accessory drive belt routing—3.7L and 4.7L engines

CAPACITIES

| Year | Model | Engine Displ. Liters | Engine VIN | Engine Oil with Filter (qts.) | Transmission (pts.) | | Transfer Case (pts.) | Drive Axle | | Fuel Tank (gal.) | Cooling System (qts.) |
					Manual	Auto.		Front (pts.)	Rear (pts.)		
2001	Durango	4.7	N	6.0	—	①	②	3.5	③	25.0	17.0
		5.9	Z	5.0	—	①	②	3.5	③	25.0	14.3
2002	Durango	4.7	N	6.0	—	①	②	3.5	③	25.0	17.0
		5.9	Z	5.0	—	①	②	3.5	③	25.0	14.3
2003	Durango	4.7	N	6.0	—	①	②	3.5	③	25.0	④
		5.9	Z	5.0	—	①	②	3.5	③	25.0	⑤
2004	Durango	3.7	K	5.0	—	⑥	⑦	3.4	③	27.0	16.2
		4.7	N	6.0	—	⑥	⑦	3.4	③	27.0	16.2
		5.7	D	7.0	—	⑥	⑦	3.4	③	27.0	16.2

NOTE: All capacities are approximate. Add fluid gradually and check to be sure a proper fluid level is obtained.

① Fluid drain/filter service: 46RE 8.0 pts.

  Fluid drain/filter service: 45RFE 2wd 11 pts; 4wd 13 pts.

  Overhaul, 46RE: 20 pts.

  Overhaul, 45RFE: 28 pts.

② NV133 and NV233: 2.5 pts.

  NV244: 2.85 pts.

③ The following values include 0.25 pt. of friction
  modifier for LSD axles.

  8.25 axle: 4.4 pts.

  9.25 axle: 4.9 pts.

④ Without rear heat: 13.3 qts.
  With rear heat: 14.1 qts.

⑤ Without rear heat: 16.3 qts.
  With rear heat: 17.2 qts.

⑥ 42RLE Fluid drain/filter service: 8pts.; Overhaul: 17.6 pts.

  545RFE Fluid drain/filter service, 2wd: 11 pts.; 4wd: 13 pts.; Overhaul: 28 pts.

⑦ NV144: 1.8 pts.; NV244 GEN II: 3.4 pts.

VALVE SPECIFICATIONS

Year	Engine Displ. Liters	Engine VIN	Seat Angle (deg.)	Face Angle (deg.)	Spring Test Pressure (lbs. @ in.)	Spring Installed Height (in.)	Stem-to-Guide Clearance (in.)		Stem Diameter (in.)	
							Intake	Exhaust	Intake	Exhaust
2001	4.7	N	44.5-45	45-45.5	176.7-193.3 @1.1670	1.601	0.0008-0.0028	0.0019-0.0039	0.2729-0.2739	0.2717-0.2728
	5.9	Z	44.25-44.75	43.25-43.75	200@1.212	1.640	0.0010-0.0030	0.0020-0.0040	0.3720-0.3730	0.3710-0.3720
2002	4.7	N	44.5-45	45-45.5	176.7-193.3 @1.1670	1.601	0.0008-0.0028	0.0019-0.0039	0.2729-0.2739	0.2717-0.2728
	5.9	Z	44.25-44.75	43.25-43.75	200@1.212	1.640	0.0010-0.0030	0.0020-0.0040	0.3720-0.3730	0.3710-0.3720
2003	4.7	N	44.5-45	45-45.5	176.7-193.3 @1.1670	1.601	0.0008-0.0028	0.0019-0.0039	0.2729-0.2739	0.2717-0.2728
	5.9	Z	44.25-44.75	43.25-43.75	200@1.212	1.640	0.0010-0.0030	0.0020-0.0040	0.3720-0.3730	0.3710-0.3720
2004	3.7	K	44.5-45	45-45.5	221-242@ 1.107	1.619	0.0008-0.0028	0.0019-0.0039	0.2729-0.2739	0.2717-0.2728
	4.7	N	44.5-45	45-45.5	176.7-193.3 @1.1670	1.601	0.0008-0.0028	0.0019-0.0039	0.2729-0.2739	0.2717-0.2728
	5.7	D	44.5-45	45-45.5	95@1.81	1.81	0.0008-0.0025	0.0019-0.0037	0.3120-0.3130	0.3110-0.3120

67189-DURA-C05

CRANKSHAFT AND CONNECTING ROD SPECIFICATIONS

All measurements are given in inches.

Year	Engine Displ. Liters	Engine VIN	Crankshaft				Connecting Rod		
			Main Brg. Journal Dia.	Main Brg. Oil Clearance	Shaft End-play	Thrust on No.	Journal Diameter	Oil Clearance	Side Clearance
2001	4.7	N	2.4996-2.5005	0.0008-0.0021	0.0021-0.0112	2	2.0076-2.0082	0.0006-0.0022	0.0040-0.0138
	5.9	Z	2.8095-2.8105	①	0.0020-0.0070	2	2.1240-2.1250	0.0005-0.0022	0.0060-0.0140
2002	4.7	N	2.4996-2.5005	0.0008-0.0021	0.0021-0.0112	2	2.0076-2.0082	0.0006-0.0022	0.0040-0.0138
	5.9	Z	2.8095-2.8105	①	0.0020-0.0070	2	2.1240-2.1250	0.0005-0.0022	0.0060-0.0140
2003	4.7	N	2.4996-2.5005	0.0008-0.0021	0.0021-0.0112	2	2.0076-2.0082	0.0006-0.0022	0.0040-0.0138
	5.9	Z	2.8095-2.8105	①	0.0020-0.0070	2	2.1240-2.1250	0.0005-0.0022	0.0060-0.0140
2004	3.7	K	2.4996-2.5005	0.0020-0.0034	0.0021-0.0112	2	2.2794-2.2797	0.0004-0.0019	0.0040-0.0138
	4.7	N	2.4996-2.5005	0.0008-0.0021	0.0021-0.0112	2	2.0076-2.0082	0.0006-0.0022	0.0040-0.0138
	5.7	D	2.5585-2.5595	0.0009-0.0020	0.0020-0.0110	3	2.1250-2.1260	0.0007-0.0023	0.0030-0.0137

① No. 1: 0.0005-0.0015
  Nos. 2-5: 0.0005-0.0020

67189-DURA-C06

PISTON AND RING SPECIFICATIONS
All measurements are given in inches.

Year	Engine Displ. Liters	Engine VIN	Piston Clearance	Ring Gap			Ring Side Clearance		
				Top Comp.	Bottom Comp.	Oil Control	Top Comp.	Bottom Comp.	Oil Control
2001	4.7	N	0.0014	0.0146-0.0249	0.0146-0.0249	0.0099-0.0300	0.0020-0.0037	0.0016-0.0031	0.0175-0.0185
	5.9	Z	0.0005-0.0015	0.0120-0.0220	0.0220-0.0310	0.0150-0.0550	0.0016-0.0033	0.0016-0.0033	0.0020-0.0080
2002	4.7	N	0.0014	0.0146-0.0249	0.0146-0.0249	0.0099-0.0300	0.0020-0.0037	0.0016-0.0031	0.0175-0.0185
	5.9	Z	0.0005-0.0015	0.0120-0.0220	0.0220-0.0310	0.0150-0.0550	0.0016-0.0033	0.0016-0.0033	0.0020-0.0080
2003	4.7	N	0.0014	0.0146-0.0249	0.0146-0.0249	0.0099-0.0300	0.0020-0.0037	0.0016-0.0031	0.0175-0.0185
	5.9	Z	0.0005-0.0015	0.0120-0.0220	0.0220-0.0310	0.0150-0.0550	0.0016-0.0033	0.0016-0.0033	0.0020-0.0080
2004	3.7	K	0.0014	0.0146-0.0249	0.0146-0.0249	0.0100-0.0300	0.0020-0.0037	0.0016-0.0031	0.0007-0.0091
	4.7	N	0.0014	0.0146-0.0249	0.0146-0.0249	0.0099-0.0300	0.0020-0.0037	0.0016-0.0031	0.0175-0.0185
	5.7	D	0.0008-0.0019	0.0090-0.0149	0.0137-0.0236	0.0059-0.0259	0.0007-0.0026	0.0007-0.0022	NA

67189-DURA-C07

TORQUE SPECIFICATIONS
All readings in ft. lbs.

Year	Engine Displ. Liters	Engine VIN	Cylinder Head Bolts	Main Bearing Bolts	Rod Bearing Bolts	Crankshaft Damper Bolts	Flexplate Bolts	Manifold Intake	Manifold Exhaust	Spark Plugs	Oil Pan Drain Plug
2001	4.7	N	①	②	③	130	45	④	18	20	25
	5.9	Z	⑤	85	45	18	55	⑥	25	30	25
2002	4.7	N	①	②	③	130	45	④	18	20	25
	5.9	Z	⑤	85	45	18	55	⑥	25	30	25
2003	4.7	N	⑦	②	③	130	45	④	18	20	25
	5.9	Z	⑤	85	45	18	55	⑥	25	30	25
2004	3.7	K	①	①	③	130	45	9	18	27	25
	4.7	N	⑦	②	③	130	45	④	18	20	25
	5.7	D	⑧	⑨	⑩	90	70	①	18	18	25

① See text

② Bed plate bolt sequence. Refer to illustration in text section
  Step 1: Bolts A-L to 40 ft. lbs.
  Step 2: Bolts 1-10 25 inch lbs.
  Step 3: Bolts 1-10 plus 90 degrees
  Step 4: Bolts A1-A6 20 ft. lbs.

③ 20 ft. lbs. plus 90 degrees

④ 105 inch lbs.

⑤ Step 1: 50 ft. lbs.
  Step 2: 105 ft. lbs.

⑥ See illustration in text section
  Step 1: 1-4 to 72 inch lbs. in 12 inch lb. Increments
  Step 2: bolts 5-12: 72 inch lbs.
  Step 3: Check that all bolts are at 72 inch lbs.
  Step 4: All bolts, in sequence, to 12 ft. lbs.
  Step 5: Check that all bolts are at 12 ft. lbs.

⑦ M8 bolts: 19 ft. lbs.
  M11 bolts: 60 ft. lbs.

⑧ M8

67189-DURA-C08

WHEEL ALIGNMENT

Year	Model	Caster Range (+/-Deg.)	Caster Preferred Setting (Deg.)	Camber Range (+/-Deg.)	Camber Preferred Setting (Deg.)	Toe-in (in.)
2001	2WD	0.50	+3.10	0.50	-0.25	0.10+/-0.06
	4WD	0.50	+3.30	0.50	-0.25	0.10+/-0.06
2002	2WD	0.50	+3.10	0.50	-0.25	0.10+/-0.06
	4WD	0.50	+3.30	0.50	-0.25	0.10+/-0.06
2003	2WD	0.50	+3.10	0.50	-0.25	0.10+/-0.06
	4WD	0.50	+3.30	0.50	0	0.10+/-0.06
2004	Front	0.50	+3.50	0.50	0	0.05+/-0.05
	Rear	—	—	0.35	-0.10	0.30+/-0.35

67189-DURA-C09

TIRE, WHEEL AND BALL JOINT SPECIFICATIONS

Year	Model	OEM Tires Standard	OEM Tires Optional	Tire Pressures (psi) Front	Tire Pressures (psi) Rear	Wheel Size	Ball Joint Inspection	Lug Nut Torque ft. lbs.
2001	Durango	P235/75R15XL	P255/65R16 P265/80R16 P275/60R17	NA	NA	NA	0.060 in. ①	100
2002	Durango	P235/75R15XL	P255/65R16 P265/80R16 P275/60R17	NA	NA	NA	0.060 in. ①	100
2003	Durango	P235/75R15XL	P255/65R16 P265/80R16 P275/60R17	NA	NA	NA	0.060 in. ①	100
2004	Durango	P245/70R17	P265/65R17	NA	NA	std: 7 opt: 8	0.020 in. ①	100

NA: Information not available

OEM: Original Equipment Manufacturer

PSI: Pounds Per Square Inch

STD: Standard

OPT: Optional

① Both upper and lower

BRAKE SPECIFICATIONS

All measurements in inches unless noted

Year	Model	Brake Disc Original Thickness	Brake Disc Minimum Thickness	Brake Disc Maximum Run-out	Brake Drum Original Inside Diameter	Brake Drum Max. Wear Limit	Brake Drum Maximum Machine Diameter	Minimum Lining Thickness Front	Minimum Lining Thickness Rear	Brake Caliper Bracket Bolts (ft. lbs.)	Brake Caliper Mounting Bolts (ft. lbs.)
2001	Durango	0.900	0.890	0.004	11.00	①	①	②	③	④	22
2002	Durango	0.900	0.890	0.004	11.00	①	①	②	③	④	22
2003	Durango	F 1.100 R 0.866	F 1.047 R 0.811	F 0.0009 R 0.0009	—	—	—	NA	NA	④	⑤
2004	Durango	F 1.100 R 0.866	F 1.039 R 0.811	F 0.0009 R 0.0009	—	—	—	NA	NA	⑥	F 24 R 11

NA: Not Available

① Maximum allowable drum diameter, either from wear or machining, is stamped on the drum.

② Riveted brake pads: 0.0625 in.
Bonded brake pads: 0.1875 in.

③ Riveted brake shoes: 0.031 in.
Bonded brake shoes: 0.0625 in.

④ Support plate bolts/nuts: 47 ft. lbs.
Adapter plate bolts: 130 ft. lbs.

⑤ Front and rear: 23.5 ft. lbs.

⑥ Adapter plate: front 130; rear 100
Support plate: 47

SCHEDULED MAINTENANCE INTERVALS
2001-03 DODGE DURANGO

TO BE SERVICED	TYPE OF SERVICE	VEHICLE MILEAGE INTERVAL (x1000)													
		7.5	15	22.5	30	37.5	45	52.5	60	67.5	75	82.5	90	97.5	100
Engine oil & filter	R	✓	✓	✓	✓	✓	✓	✓	✓	✓	✓	✓	✓	✓	✓
Front wheel bearings	S/I			✓			✓			✓			✓		
Brake linings	S/I			✓			✓			✓			✓		
Air cleaner element	R				✓				✓				✓		
Spark plugs	R				✓				✓				✓		
Transfer case fluid	R					✓					✓				
Engine coolant ①	R						✓						✓		
Spark plug wires (5.9L)	R								✓						
PCV valve	R								✓						
Drive belt tensioner ②	S/I								✓						
Automatic transmission fluid	R														✓
Automatic transmission bands	Adj.														✓

R: Replace   S/I: Service or Inspect

① Replace every 36 months, regardless of mileage.

② Replace if necessary

FREQUENT OPERATION MAINTENANCE (SEVERE SERVICE)

If a vehicle is operated under any of the following conditions it is considered severe service:

- Extremely dusty areas.

- 50% or more of the vehicle operation is in 32°C (90°F) or higher temperatures, or constant operation in temperatures below 0°C (32°F).

- Prolonged idling (vehicle operation in stop and go traffic.

- Frequent short running periods (engine does not warm to normal operating temperatures).

- Police, taxi, delivery usage or trailer towing usage.

Oil & oil filter change: change every 3000 miles.

Air filter: change every 24,000 miles.

Engine coolant level, hoses & clamps: check every 6,000 miles.

Exhaust system: check every 6000 miles.

Drive belts: check every 18,000 miles; replace every 24,000 miles.

Oxygen sensor: replace every 82,500 miles.

Automatic transmission fluid, filter & bands: change & adjust every 12,000 miles.

Rear axle fluid: change every 12,000 miles.

67189-DURA-C12

SCHEDULED MAINTENANCE INTERVALS
2004 DODGE DURANGO

TO BE SERVICED	TYPE OF SERVICE	VEHICLE MILEAGE INTERVAL (x1000)																
		6	12	18	24	30	36	42	48	54	60	66	72	78	84	90	96	102
Engine oil & filter	R	✓	✓	✓	✓	✓	✓	✓	✓	✓	✓	✓	✓	✓	✓			
Brake linings	S/I			✓			✓			✓			✓					
Air cleaner element	R					✓					✓					✓		
Spark plugs	R					✓					✓					✓		
Transfer case fluid	I					✓					✓							
Engine coolant	R	Replace every 60 months regardless of mileage																
Spark plug wires (5.7L)	R								✓									
PCV valve	S/I										✓							
Drive belt tensioner	S/I																✓	
Automatic transmission fluid ①	R																	✓
Transfer case fluid	R	Drain the fluid every 120,000 miles																

R: Replace   S/I: Service or Inspect

① On 4.7L and 5.7L, change the filter, if equipped

FREQUENT OPERATION MAINTENANCE (SEVERE SERVICE)

If a vehicle is operated under any of the following conditions it is considered severe service:

- Extremely dusty areas.
- Day or night time temperatures below 0°C (32°F).
- Prolonged idling (vehicle operation in stop and go traffic.)
- Frequent short running periods (engine does not warm to normal operating temperatures).
- Police, taxi, delivery usage or trailer towing usage.
- Off road or desert operation
- 50% or more of you driving is done in temperatures above 90 degrees F (32 deg. C)

Oil & oil filter change: change every 3000 miles.

Air filter: change every 15,000 miles and change as necessary.

Drive belts: check and replace as necessary every 60,000 miles.

Automatic transmission fluid, filter every 30,000 miles 4.7L and 5.7L.

Automatic transmission fluid, filter every 60,000 miles 3.7L.

Front and rear axle fluid: change every 15,000 miles.

Inspect PCV valve every 30,000 miles

67189-DURA-C13

PRECAUTIONS

Before servicing any vehicle, please be sure to read all of the following precautions, which deal with personal safety, prevention of component damage, and important points to take into consideration when servicing a motor vehicle:

• Never open, service or drain the radiator or cooling system when the engine is hot; serious burns can occur from the steam and hot coolant.

• Observe all applicable safety precautions when working around fuel. Whenever servicing the fuel system, always work in a well-ventilated area. Do not allow fuel spray or vapors to come in contact with a spark, open flame, or excessive heat (a hot drop light, for example). Keep a dry chemical fire extinguisher near the work area. Always keep fuel in a container specifically designed for fuel storage; also, always properly seal fuel containers to avoid the possibility of fire or explosion. Refer to the additional fuel system precautions later in this section.

• Fuel injection systems often remain pressurized, even after the engine has been turned **OFF**. The fuel system pressure must be relieved before disconnecting any fuel lines. Failure to do so may result in fire and/or personal injury.

• Brake fluid often contains polyglycol ethers and polyglycols. Avoid contact with the eyes and wash your hands thoroughly after handling brake fluid. If you do get brake fluid in your eyes, flush your eyes with clean, running water for 15 minutes. If eye irritation persists, or if you have taken brake fluid internally, IMMEDIATELY seek medical assistance.

• The EPA warns that prolonged contact with used engine oil may cause a number of skin disorders, including cancer! You should make every effort to minimize your exposure to used engine oil. Protective gloves should be worn when changing oil. Wash your hands and any other exposed skin areas as soon as possible after exposure to used engine oil. Soap and water, or waterless hand cleaner should be used.

• All new vehicles are now equipped with an air bag system, often referred to as a Supplemental Restraint System (SRS) or Supplemental Inflatable Restraint (SIR) system. The system must be disabled before performing service on or around system components, steering column, instrument panel components, wiring and sensors. Failure to follow safety and disabling procedures could result in accidental air bag deployment, possible personal injury and unnecessary system repairs.

• Always wear safety goggles when working with, or around, the air bag system. When carrying a non-deployed air bag, be sure the bag and trim cover are pointed away from your body. When placing a non-deployed air bag on a work surface, always face the bag and trim cover upward, away from the surface. This will reduce the motion of the module if it is accidentally deployed. Refer to the additional air bag system precautions later in this section.

• Clean, high quality brake fluid from a sealed container is essential to the safe and proper operation of the brake system. You should always buy the correct type of brake fluid for your vehicle. If the brake fluid becomes contaminated, completely flush the system with new fluid. Never reuse any brake fluid. Any brake fluid that is removed from the system should be discarded. Also, do not allow any brake fluid to come in contact with a painted surface; it will damage the paint.

• Never operate the engine without the proper amount and type of engine oil; doing so WILL result in severe engine damage.

• Timing belt maintenance is extremely important! Many models utilize an interference-type, non-freewheeling engine. If the timing belt breaks, the valves in the cylinder head may strike the pistons, causing potentially serious (also time-consuming and expensive) engine damage. Refer to the maintenance interval charts in the front of this manual for the recommended replacement interval for the timing belt, and to the timing belt section for belt replacement and inspection.

• Disconnecting the negative battery cable on some vehicles may interfere with the functions of the on-board computer system(s) and may require the computer to undergo a relearning process once the negative battery cable is reconnected.

• When servicing drum brakes, only disassemble and assemble one side at a time, leaving the remaining side intact for reference.

GASOLINE ENGINE REPAIR

➡**Disconnecting the negative battery cable on some vehicles may interfere with the functions of the on board computer system. The computer may undergo a relearning process once the negative battery cable is reconnected.**

Distributor

REMOVAL

5.9L Engines

1. Before servicing the vehicle, refer to the precautions in the beginning of this section.
2. Remove or disconnect the following:
   • Negative battery cable
   • Air cleaner tube

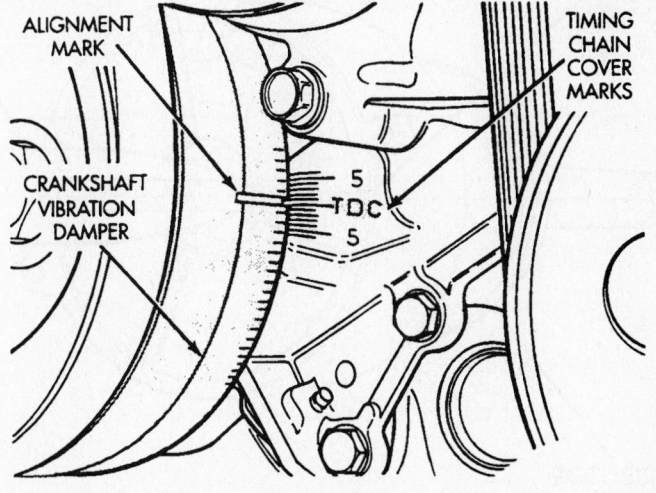

Crankshaft pulley and timing chain cover marks aligned at Top Dead Center (TDC)

7924DG47

- Distributor cap
- Camshaft Position (CMP) sensor connector

3. Matchmark the distributor housing and the rotor.

4. Matchmark the distributor housing and the intake manifold.

5. Remove the distributor.

INSTALLATION

Timing Not Disturbed

5.9L ENGINES

1. Before servicing the vehicle, refer to the precautions in the beginning of this section.

➡**The rotor will rotate clockwise as the gears engage.**

2. Position the rotor slightly counter-clockwise of the matchmark made during removal.

3. Install the distributor.

4. Align the distributor housing and intake manifold matchmarks and check that the distributor housing matchmark and rotor are also aligned.

5. Install or connect the following:
- Distributor housing clamp and bolt. Tighten the bolt to 17 ft. lbs. (23 Nm).
- CMP sensor connector
- Distributor cap
- Air cleaner tube
- Negative battery cable

Timing Disturbed

5.9L ENGINES

1. Before servicing the vehicle, refer to the precautions in the beginning of this section.

2. Set the engine at Top Dead Center (TDC) of the No. 1 cylinder compression stroke.

3. Install the distributor, clamp and bolt.

4. Rotate the distributor so that the rotor aligns with the No. 1 cylinder mark on the Camshaft Position (CMP) sensor. Tighten the clamp bolt to 17 ft. lbs. (23 Nm).

5. Install or connect the following:
- CMP sensor connector
- Distributor cap
- Air cleaner tube
- Negative battery cable

6. The final distributor position must be set with a scan tool. Follow the instructions supplied by the scan tool manufacturer.

Alternator

REMOVAL & INSTALLATION

3.7L and 4.7L Engines

1. Before servicing the vehicle, refer to the precautions in the beginning of this section.

2. Remove or disconnect the following:
- Negative battery cable
- Accessory drive belt
- Alternator harness connectors

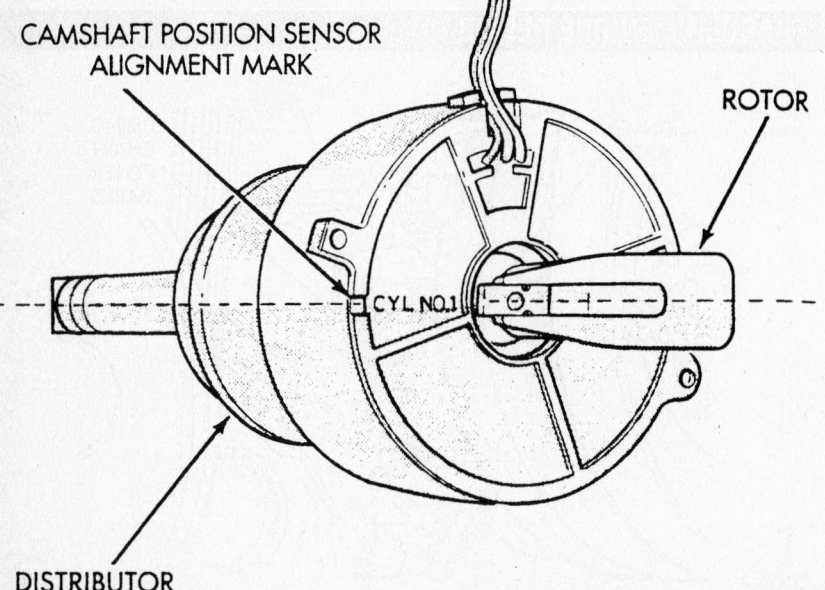

CAMSHAFT POSITION SENSOR ALIGNMENT MARK

ROTOR

CYL. NO. 1

DISTRIBUTOR

9308PG01

Distributor rotor alignment—5.9L engines

- Mounting bolts and alternator

➡**There are 1 vertical and 2 horizontal bolts.**

To install:

3. Before servicing the vehicle, refer to the precautions in the beginning of this section.

4. Install the alternator and tighten the bolts to the following specifications:
- Short horizontal bolts to 55 ft. lbs. (74 Nm)
- Vertical bolt and long horizontal bolt to 40 ft. lbs. (55 Nm)

5. Install or connect the following:
- Alternator harness connectors
- Accessory drive belt
- Negative battery cable

5.7L Engines

1. Before servicing the vehicle, refer to the precautions in the beginning of this section.

2. Remove or disconnect the following:
- Negative battery cable
- Accessory drive belt
- Alternator harness connectors
- Support bracket nuts and bolt
- Mounting bolts and alternator

3. Installation is the reverse of removal. Torque the bolts to 30 ft. lbs. (41Nm).

5.9L Engines

1. Before servicing the vehicle, refer to the precautions in the beginning of this section.

2. Remove or disconnect the following:
- Negative battery cable
- Accessory drive belt
- Alternator harness connectors
- Alternator

3. Installation is the reverse of removal. Tighten the bolts to 30 ft. lbs. (41 Nm).

Ignition Timing

ADJUSTMENT

The ignition timing is controlled by the Powertrain Control Module (PCM) and is not adjustable.

Engine Assembly

REMOVAL & INSTALLATION

3.7L Engine

1. Discharge the A/C system.
2. Drain the cooling system.

3. Release the fuel system pressure.

4. Remove the air cleaner assembly.

5. Disconnect the battery.

6. Remove the accessory drive belt.

7. Remove the viscous fan.

8. Remove the A/C compressor.

9. Remove the generator and secure away from engine.

➡**Do not remove the phenolic pulley from the P/S pump. It is not required for P/S pump removal.**

10. Remove the power steering pump with lines attached and secure away from engine.

11. Disconnect the heater hoses from the engine.

12. Disconnect the heater hoses from heater core and remove hose assembly.

13. Remove the upper radiator hose from engine.

14. Remove the lower radiator hose from engine.

15. Disconnect the transmission oil cooler lines at the radiator.

16. Disconnect the power steering cooler lines.

17. Remove the radiator assembly.

18. Disconnect the throttle and speed control cables.

19. Disconnect the engine to body ground straps at the left side of cowl.

20. Disconnect the engine wiring harness at the following points:

- Intake air temperature (IAT) sensor
- Fuel Injectors
- Throttle Position (TPS) Switch
- Idle Air Control (IAC) Motor
- Engine Oil Pressure Switch
- Engine Coolant Temperature (ECT) Sensor
- Manifold Absolute Pressure MAP) Sensor
- Camshaft Position (CMP) Sensor
- Coil Over Plugs
- Crankshaft Position Sensor

21. Remove the coil over plugs.

22. Remove fuel rail and secure away from engine.

➡**It is not necessary to release the quick connect fitting from the fuel supply line for engine removal.**

23. Remove the PCV hose.

24. Remove the breather hoses.

25. Remove the vacuum hose for the power brake booster.

26. Disconnect the knock sensors.

27. Remove the engine oil dipstick tube.

28. Remove intake manifold.

29. Install an engine lifting fixture, Tool

8247, using the original fasteners from the removed intake manifold, and fuel rail.

30. Raise the vehicle on hoist.

31. Disconnect the exhaust pipes from exhaust manifolds.

32. On 4wd vehicles disconnect axle vent tube from left side engine mount.

33. Remove the through bolt retaining nut and bolt (3) from both the left and right side engine mounts.

34. On 4wd vehicles Remove locknut from left and right side engine mount brackets.

35. Disconnect two ground straps from the lower left hand side and one ground strap from the lower right hand side of the engine.

36. Disconnect the crankshaft position sensor.

➡**The following step applies to 4wd vehicles equipped with automatic transmission only.**

37. On 4wd vehicles, remove the axle isolator bracket (2) from the engine, transmission and the axle.

38. Remove the structural cover.

39. Remove the starter.

40. Remove the torque converter bolts (automatic transmission only).

41. Remove the transmission to engine mounting bolts.

42. Disconnect the engine block heater power cable from the block heater, if equipped.

43. Lower the vehicle.

44. Remove throttle body resonator assembly and air inlet hose.

45. Disconnect throttle and speed control cables.

46. Disconnect the tube from both the left and right side crankcase breathers. Remove the breathers.

47. Remove the generator.

48. Disconnect the two heater hoses from the timing chain cover and heater core.

49. Unclip and remove the heater hoses and tubes from the intake manifold.

50. Disconnect engine harness at the following points:

- Intake air temperature (IAT) sensor
- Fuel Injectors
- Throttle Position (TPS) Switch
- Idle Air Control (IAC) Motor
- Engine Oil Pressure Switch
- Engine Coolant Temperature (ECT) Sensor
- Manifold absolute pressure (MAP) Sensor
- Camshaft Position (CMP) Sensor
- Coil Over Plugs

51. Disconnect the vacuum lines at the throttle body and intake manifold.

52. Remove power steering pump and position out of the way.

53. Disconnect body ground strap at the right side cowl (3).

54. Disconnect body ground strap at the left side cowl (2).

➡**It will be necessary to support the transmission in order to remove the engine.**

55. Position a suitable jack under the transmission.

56. Remove engine from the vehicle.

To install:

57. Position engine in the vehicle.

58. Position both the left and right side engine mount brackets and install the through bolts (3) and nuts. (1). Tighten nuts (1) to 95 Nm (70 ft. lbs.) on 2wd vehicles; 102 Nm (75 ft. lbs.) on 4wd vehicles.

59. On 4wd vehicles, install locknuts onto the engine mount brackets. Tighten locknuts to 41 Nm (30 ft. lbs.)

60. Remove jack from under the transmission.

61. Remove Engine Lifting Fixture Special Tool.

62. Remove Special Tools 8400 Lifting Studs.

63. Position generator wiring behind the oil dipstick tube, then install the oil dipstick tube upper mounting bolt.

64. Connect both left and right side body ground straps.

65. Install power steering pump.

66. Connect fuel supply line quick connect fitting.

67. Connect the vacuum lines at the throttle body and intake manifold.

68. Connect engine harness at the following points:

- Intake Air Temperature (IAT) Sensor
- Idle Air Control (IAC) Motor
- Fuel Injectors
- Throttle Position (TPS) Switch
- Engine Oil Pressure Switch
- Engine Coolant Temperature (ECT) Sensor
- Manifold Absolute Pressure (MAP) Sensor
- Camshaft Position (CMP) Sensor
- Coil Over Plugs

69. Position and install heater hoses and tubes (1) onto intake manifold.

70. Install the heater hoses onto the heater core and the engine front cover.

71. Install generator.

72. Install radiator assembly.

73. Connect radiator upper and lower hoses.

74. Connect power steering cooler lines.
75. Connect the transmission oil cooler lines to the radiator.
76. Install accessory drive belt.
77. Install A/C compressor.
78. Install both breathers. Connect tube to both crankcase breathers.
79. Connect throttle and speed control cables.
80. Install throttle body resonator assembly and air inlet hose. Tighten clamps 4 Nm (35 in. lbs.).
81. Raise vehicle.
82. Install transmission to engine mounting bolts.
   Tighten the bolts to 41 Nm (30 ft. lbs.).
83. Install torque converter bolts (Automatic Transmission Only).
84. Connect crankshaft position sensor.
85. On 4wd vehicles, position and install the axle isolator bracket onto the axle, transmission and engine block. Tighten bolts to specification.
86. Install starter.

✳✳ CAUTION

The structural cover requires a specific torque sequence. Failure to follow this sequence may cause severe damage to the cover.

87. Install structural cover.
88. Install exhaust crossover pipe.
89. Install engine block heater power cable, if equipped.
90. On 4wd vehicles, connect axle vent tube to left side engine mount.
91. Lower vehicle.
92. Check and fill engine oil.
93. Recharge the A/C system.
94. Refill the engine cooling system.

95. Install the battery tray and battery.
96. Connect the battery positive and negative cables.
97. Start the engine and check for leaks.

4.7L Engine

1. Before servicing the vehicle, refer to the precautions in the beginning of this section.
2. Drain the cooling system and engine oil.
3. Remove or disconnect the following:
   - Negative battery cable
   - Battery and tray
   - Exhaust crossover pipe
   - On 4wd, the axle vent tube
   - Left and right engine mount through bolts
   - On 4wd, the left and right engine mount bracket locknuts
   - Ground straps
   - CKP sensor
   - On 4wd, the axle isolator bracket
   - Structural cover
   - Starter
   - Torque converter bolts
   - Transmission-to-engine bolts
   - Engine block heater
   - Resonator and air inlet
   - Throttle and speed control cables
   - Crankcase breathers
   - A/C compressor
   - Shroud and fan assemblies
   - Transmission cooler lines
   - Radiator hoses
   - Radiator
   - Alternator
   - Heater hoses
   - Engine harness
   - Vacuum lines
   - Fuel system pressure

- Fuel line at the rail
- Power steering pump

4. Install lifting eyes and take up the weight of the engine with a crane.
5. Support the transmission with a jack.
6. Remove the engine.
7. Installation is the reverse of removal. Observe the following:
   - Left and right engine mount through bolts: 2wd 70 ft. lbs. (95 Nm); 4wd 75 ft. lbs. (102 Nm)
   - On 4wd, the bracket locknuts: 30 ft. lbs. (41 Nm)
   - Transmission-to-engine bolts: 30 ft. lbs. (41 Nm)

➡**The structural cover has a specific torque sequence.**

5.7L Engine

1. Before servicing the vehicle, refer to the precautions in the beginning of this section.
2. Drain the cooling system.
3. Drain the engine oil.
4. Relieve the fuel system pressure.
5. Remove or disconnect the following:
   - Negative battery cable
   - Hood
   - Air cleaner and resonator
   - Accessory drive belt
   - Engine fan
   - Radiator
   - Upper crossmember and top core support
   - A/C compressor, if equipped
   - Alternator
   - Intake manifold and IAFM as an assembly
   - Heater hoses
   - Power steering pump
   - Fuel line
   - Engine front mount thru-bolt nuts
   - Transmission oil cooler lines, if equipped
   - Exhaust pipes at the manifolds
   - Starter motor
   - Structural dust cover and transmission inspection cover
   - Torque converter-to-flexplate bolts
   - Transmission flange bolts. Support the transmission.
   - Engine

To install:

6. Install or connect the following:
   - Engine. Tighten the engine mount thru-bolt finger tight.
   - Transmission flange bolts. Tighten the bolts to 40–45 ft. lbs. (54–61 Nm). Then, tighten the mount bolt nuts to 70 ft. lbs. (95 Nm)
   - Transmission oil cooler lines, if equipped

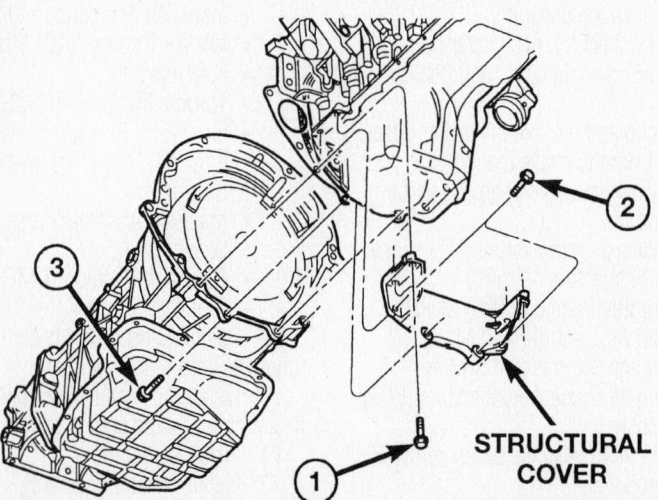

STRUCTURAL COVER

67189-DURA-G99

Tighten the structural cover bolts in this order—3.7L Engine

- Torque converter, if equipped. Tighten the bolts to 23 ft. lbs. (31 Nm).
- Structural dust cover and transmission inspection cover
- Starter motor
- Fuel line
- Power steering pump
- Heater hoses
- Intake manifold and IAFM as an assembly
- Alternator
- A/C compressor, if equipped
- Upper crossmember and top core support
- Radiator
- Engine fan
- Accessory drive belt
- Air cleaner and resonator
- Hood
- Negative battery cable

7. Fill the crankcase to the correct level.
8. Fill the cooling system.
9. Start the engine and check for leaks.

5.9L Engines

1. Before servicing the vehicle, refer to the precautions in the beginning of this section.
2. Drain the cooling system.
3. Drain the engine oil.
4. Relieve the fuel system pressure.
5. Remove or disconnect the following:
   - Negative battery cable
   - Hood
   - Upper crossmember and top core support
   - Radiator hoses
   - Cooling fan and shroud
   - Radiator
   - Accelerator cable
   - Cruise control cable, if equipped
   - Transmission cable, if equipped
   - Heater hoses
   - Intake manifold vacuum lines
   - Accessory drive belt
   - Power steering pump
   - A/C compressor, if equipped
   - Engine control sensor harness connectors
   - Engine block heater, if equipped
   - Fuel line
   - Exhaust front pipe
   - Starter motor
   - Torque converter, if equipped
   - Transmission oil cooler lines, if equipped
   - Engine mounts
   - Transmission flange bolts. Support the transmission.
   - Engine

❊❊ WARNING

Do not lift the engine by the intake manifold. Damage to the manifold may result.

To install:

6. Install or connect the following:
   - Engine. Tighten the engine mount bolts to 70 ft. lbs. (95 Nm).
   - Transmission flange bolts. Tighten the bolts to 40–45 ft. lbs. (54–61 Nm).
   - Transmission oil cooler lines, if equipped
   - Torque converter, if equipped. Tighten the bolts to 23 ft. lbs. (31 Nm).
   - Starter motor
   - Exhaust front pipe
   - Fuel line
   - Engine block heater, if equipped
   - Engine control sensor harness connectors
   - A/C compressor, if equipped
   - Power steering pump
   - Accessory drive belt
   - Intake manifold vacuum lines
   - Heater hoses
   - Transmission cable, if equipped
   - Cruise control cable, if equipped
   - Accelerator cable
   - Radiator
   - Cooling fan and shroud
   - Radiator hoses
   - Upper crossmember and top core support
   - Hood
   - Negative battery cable

7. Fill the crankcase to the correct level.
8. Fill the cooling system.
9. Start the engine and check for leaks.

Water Pump

REMOVAL & INSTALLATION

3.7L Engine

1. Before servicing the vehicle, refer to the precautions in the beginning of this section.
2. Drain the cooling system.
3. Remove or disconnect the following:
   - Negative battery cable
   - Fan and clutch assembly from the pump
   - Fan shroud and fan assembly. If you're reusing the fan clutch, keep it upright to avoid silicone fluid loss!
   - Lower hose

- Water pump (8 bolts)
4. Installation is the reverse of removal. Tighten the bolts, in sequence, to 40 ft. lbs. (54 Nm).

4.7L Engine

1. Before servicing the vehicle, refer to the precautions in the beginning of this section.
2. Drain the cooling system.
3. Remove or disconnect the following:
   - Negative battery cable
   - Fan and fan drive assembly from the pump. Don't attempt to remove it from the vehicle, yet.

➡**If a new pump is being installed, don't separate the fan from the drive.**

   - Shroud and fan

❊❊ WARNING

Keep the fan upright to avoid fluid loss from the drive.

   - Accessory drive belt
   - Lower radiator hose
   - Water pump
4. Installation is the reverse of removal. tighten the bolts in sequence to 40 ft. lbs. (54 Nm).

5.7L Engines

1. Before servicing the vehicle, refer to the precautions in the beginning of this section.
2. Drain the cooling system.
3. Remove or disconnect the following:
   - Negative battery cable
   - Accessory drive belt
   - Engine cooling fan
   - Coolant recovery bottle
   - Washer bottle
   - Fan shroud
   - A/C compressor and alternator brace
   - Idler pulleys
   - Belt tensioner
   - Radiator hoses
   - Heater hoses
   - Water pump

To install:

4. Install or connect the following:
   - Water pump. Tighten the bolts to 18 ft. lbs. (24 Nm).
   - Heater hoses
   - Radiator hoses
   - Idler pulleys
   - A/C compressor and alternator brace
   - Fan shroud
   - Washer bottle
   - Coolant recovery bottle

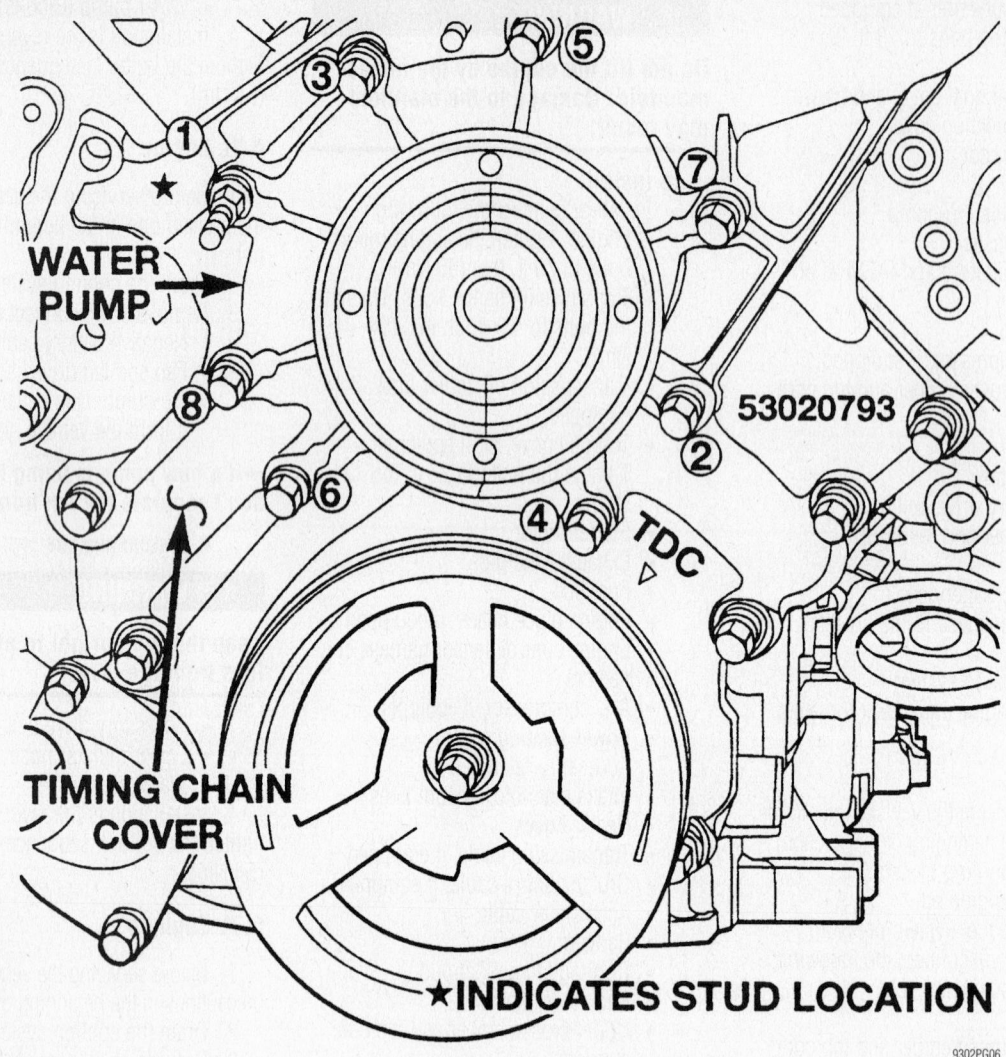

WATER PUMP

TIMING CHAIN COVER

TDC

53020793

★INDICATES STUD LOCATION

9302PG06

Water pump torque sequence—3.7L and 4.7L engine

- Accessory drive belt
- Negative battery cable
- Negative battery cable
5. Fill the cooling system.
6. Start the engine and check for leaks.

5.9L Engines

1. Before servicing the vehicle, refer to the precautions in the beginning of this section.
2. Drain the cooling system.
3. Remove or disconnect the following:
   - Negative battery cable
   - Engine cooling fan and shroud

➡Do not store the fan clutch assembly horizontally, silicone may leak into the bearing grease and cause contamination.

- Accessory drive belt
- Water pump pulley
- Lower radiator hose
- Heater hose and tube

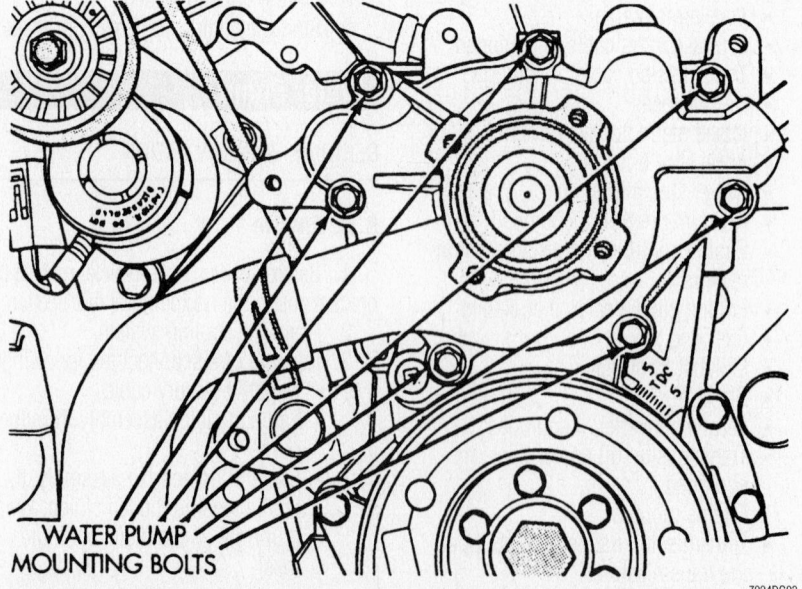

WATER PUMP MOUNTING BOLTS

7924DG03

Water pump mounting bolt locations—5.9L engines

- Bypass hose
- Water pump

To install:

4. Install or connect the following:
   - Water pump, using a new gasket. Tighten the bolts to 30 ft. lbs. (40 Nm).
   - Bypass hose
   - Heater hose and tube. Use a new O-ring seal.
   - Lower radiator hose
   - Water pump pulley. Tighten the bolts to 20 ft. lbs. (27 Nm).
   - Accessory drive belt
   - Engine cooling fan and shroud
   - Negative battery cable
5. Fill the cooling system.
6. Start the engine and check for leaks.

Heater Core

REMOVAL & INSTALLATION

2001–03

1. Disconnect the negative battery cable.

❋❋ CAUTION

After disconnecting the negative battery cable, wait 2 minutes for the driver's/passenger's air bag system capacitor to discharge before attempting to do any work around the steering column or instrument

2. Remove the instrument panel by performing the following procedure:

   a. Remove the trim from the right and left door sills.

   b. Remove the trim from the right and left cowl side inner panels.

   c. Remove the steering column opening cover from the instrument panel.

   d. Remove the 2 hood latch release handle-to-instrument panel lower reinforcement screws and lower the release handle to the floor.

   e. Disconnect the driver's side air bag module wire harness connector.

   f. If equipped, disconnect the overdrive lockout switch harness connector.

   g. Do not disassemble the steering column but remove the assembly from the vehicle.

   h. From under the driver's side of the instrument panel, disconnect or remove the following items:

   - The screw from the center of the headlight/dash-to-instrument panel bulkhead wire harness connector and disconnect the connector.
   - The 2 body wire harness connectors from the 2 instrument panel wire harness connector that secure the outboard side of the instrument panel bulkhead connector.
   - The 3 wire harness connectors from the 3 junction block connector receptacles located closest to the

dash panel; 1 from the body wire harness and 2 from the headlight/dash wire harness.

   - The plastic park brake release linkage rod-to-rear parking brake release handle lever. Disengage the linkage rod from the handle lever.
   - The stoplight switch connector.
   - The vacuum harness connector from the left side of the heater/air conditioning housing assembly.

   i. Remove the instrument panel center support bracket.

   j. Remove the instrument panel wire harness ground screw located on the left side of the air bag control module (ACM) mount on the floor panel transmission tunnel.

   k. Disconnect the instrument panel wiring harness-to-ACM connector receptacle.

   l. Remove the glove box.

   m. Working through the glove box opening, disconnect or remove the following items:

   - The 2 halves of the radio antenna coaxial cable connector near the center of the lower instrument panel.
   - The antenna half of the radio antenna coaxial cable from the retainer clip near the outboard side of the lower instrument panel.
   - The blower motor wire harness

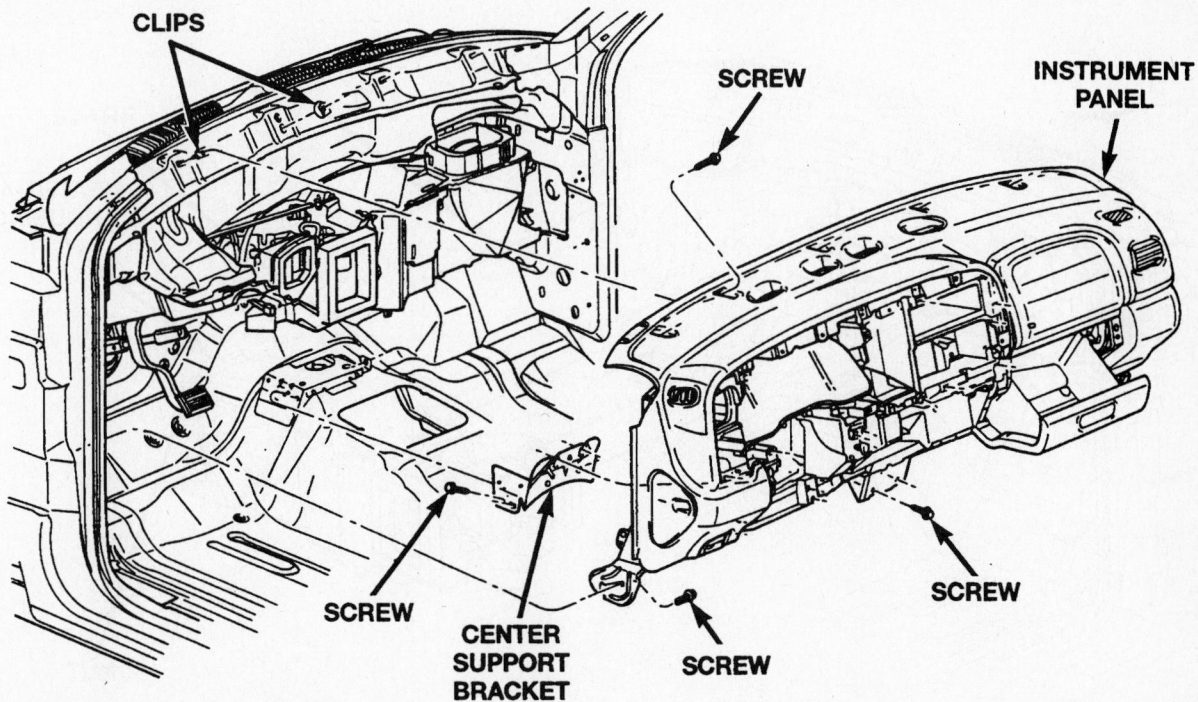

View of the instrument panel assembly

93113G92

connector located near the heater/air conditioning housing assembly support brace.

n. From the passenger's side, disconnect or remove the following items:

- The 2 instrument panel wire harness connectors from the infinity speaker amplifier connector receptacles on the right cowl.
- The instrument panel wire harness radio ground eyelet-to-stud nut on the right cowl.

o. Loosen the right and left instrument panel cowl side roll-down bracket screws.

p. Remove the 5 top instrument panel-to-dash panel screws.

q. Pull the instrument panel rearward until the right and left cowl side roll-down bracket screws are in the roll-down slot position of both brackets.

r. Roll down the instrument panel and install a temporary hook in the center hole on top of the instrument panel; secure the other end to the top of the dash panel. The hook is to support the instrument panel in its rolled down position.

s. With the instrument panel in the rolled-down position, disconnect or remove the following items:

- The 2 instrument panel wire harness connector from the door jumper wire harness connectors located on the right bracket.

- The instrument panel wire harness connector from the blower motor resistor connector receptacle.
- The temperature control cable flag retainer from the top of the heater/air conditioning housing assembly. Pull the cable core adjuster clip off the blend-air door lever.
- The demister duct flexible hose from the adapter on the top of the heater/air conditioning housing assembly.

t. With aid of an assistant, lift the instrument panel from the vehicle.

3. If equipped with air conditioning, perform the following procedure:

a. Discharge and recover the air conditioning system refrigerant.

b. Disconnect the refrigerant line from the evaporator inlet and outlet tubes. Plug the openings to prevent contamination.

4. Drain the cooling system into a clean container for reuse.

5. Disconnect the heater hoses from the heater core. Plug the openings.

6. Remove the 4 heater/air conditioning housing assembly-to-chassis nuts.

7. Remove the heater/air conditioning housing assembly-to-mounting brace nut; the nut is located on the passenger side of the vehicle.

8. Pull the heater/air conditioning housing assembly rearward for the studs

and drain tube to clear the dash panel hole.

9. Remove the heater/air conditioning housing assembly from the vehicle.

10. Remove the heater housing cover screws and the cover.

11. Remove the heater core from the heater/air conditioning housing assembly.

To install:

12. Install the heater core to the heater/air conditioning housing assembly.

13. Install the heater housing cover and the cover screws.

14. Install the heater/air conditioning housing assembly.

15. Install the heater/air conditioning housing assembly-to-mounting brace nut; the nut is located on the passenger side of the vehicle.

16. Install the 4 heater/air conditioning housing assembly-to-chassis nuts.

17. Connect the heater hoses to the heater core.

18. If equipped with air conditioning, perform the following procedure:

a. Using new gaskets, connect the refrigerant line to the evaporator inlet and outlet tubes.

b. Evacuate and charge the air conditioning system.

19. Install the instrument panel by performing the following procedures:

a. With aid of an assistant, install the instrument panel into the vehicle.

b. With the instrument panel in the

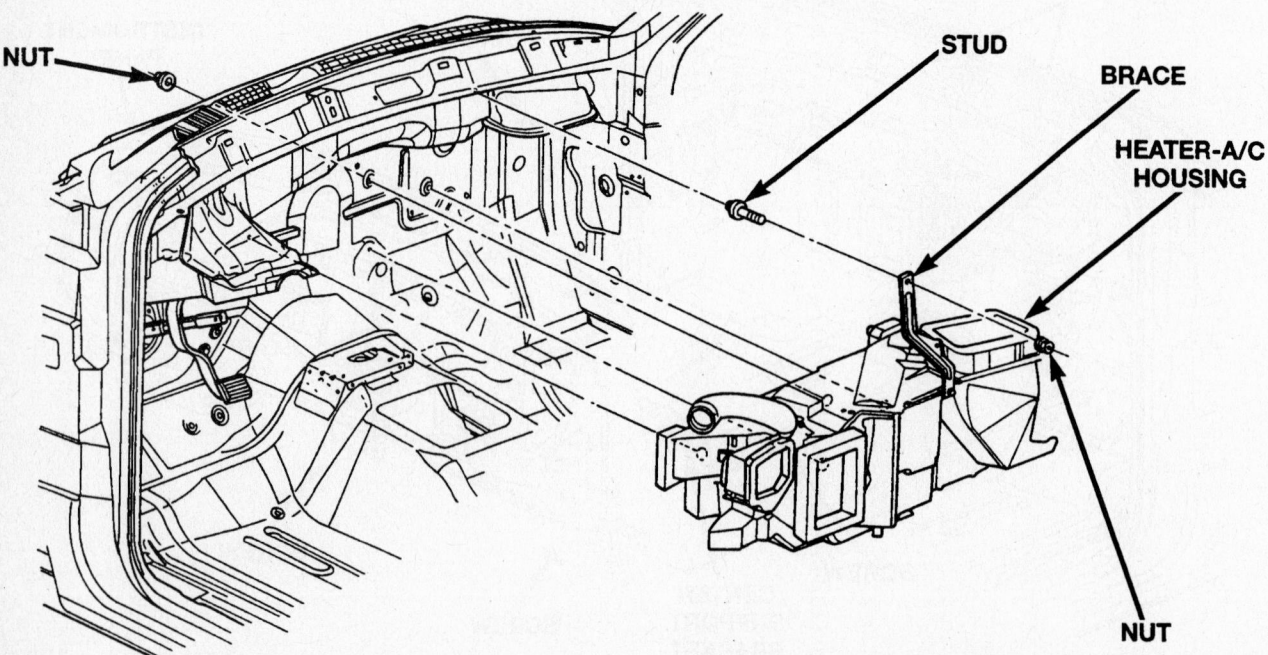

NUT · · · STUD · BRACE · HEATER-A/C HOUSING · NUT

View of the heater/air conditioning assembly

93113G93

HEATER-A/C HOUSING

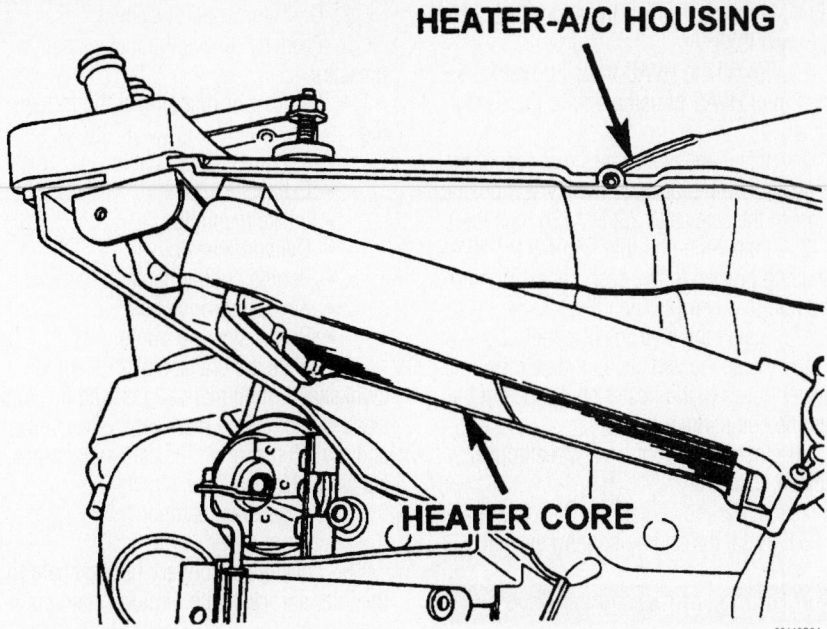

HEATER CORE

93113G94

View of the heater core

rolled-down position, connect or install the following items:

- The demister duct flexible hose to the adapter on the top of the heater/air conditioning housing assembly.
- The temperature control cable flag retainer to the top of the heater/air conditioning housing assembly.
- The instrument panel wire harness connector to the blower motor resistor connector receptacle.
- The 2 instrument panel wire harness connector to the door jumper wire harness connectors located on the right bracket.

  c. Move the instrument panel forward until the right and left cowl side roll-down bracket screws are in the roll-down slot position of both brackets.

  d. Install the 5 top instrument panel-to-dash panel screws.

  e. Install the right and left instrument panel cowl side roll-down bracket screws.

  f. At the passenger's side, connect or install the following items:

- The instrument panel wire harness radio ground eyelet-to-stud nut on the right cowl.
- The 2 instrument panel wire harness connectors to the infinity speaker amplifier connector receptacles on the right cowl.

  g. Working through the glove box opening, connect or install the following items:

- The blower motor wire harness connector located near the heater/air conditioning housing assembly support brace.
- The antenna half of the radio antenna coaxial cable to the retainer clip near the outboard side of the lower instrument panel.
- The 2 halves of the radio antenna coaxial cable connector near the center of the lower instrument panel.

  h. Install the glove box.

  i. Connect the instrument panel wiring harness-to-ACM connector receptacle.

  j. Install the instrument panel wire harness ground screw located on the left side of the air bag control module (ACM) mount on the floor panel transmission tunnel.

  k. Install the instrument panel center support bracket.

  l. Under the driver's side of the instrument panel, connect or install the following items:

- The vacuum harness connector to the left side of the heater/air conditioning housing assembly.
- The stoplight switch connector.
- Engage the linkage rod to the handle lever. Install the plastic park brake release linkage rod-to-rear parking brake release handle lever.
- The 3 wire harness connectors to the 3 junction block connector receptacles located closest to the

dash panel; 1 at the body wire harness and 2 at the headlight/dash wire harness.

- The 2 body wire harness connectors to the 2 instrument panel wire harness connector that secure the outboard side of the instrument panel bulkhead connector.
- Connect the connector. Install the screw at the center of the headlight/dash-to-instrument panel bulkhead wire harness connector.

  m. Install the steering column assembly.

  n. If equipped, connect the overdrive lockout switch harness connector.

  o. Connect the driver's side air bag module wire harness connector.

  p. Install the 2 hood latch release handle-to-instrument panel lower reinforcement screws and lower the release handle to the floor.

  q. Install the steering column opening cover to the instrument panel.

  r. Install the trim to the right and left cowl side inner panels.

  s. Install the trim to the right and left door sills.

20. Refill the cooling system.
21. Connect the negative battery cable.
22. Run the engine to normal operating temperatures; then, check the climate control operation and check for leaks.

2004

1. Drain the engine cooling system.
2. Raise and support the vehicle.
3. Remove the right front wheelhouse splash shield.
4. Remove the heater hoses from the heater core tubes in the engine compartment.
5. Lower the vehicle and remove the instrument panel.
6. Remove the bolt that secures the HVAC housing bracket to the dash panel.
7. Remove the two screws that secure the HVAC housing bracket to the top of the HVAC housing.
8. Remove the HVAC housing bracket from the vehicle.
9. Remove the screw that secures the heater core tube retaining bracket to the top of the HVAC housing.
10. Remove the heater core tube retaining bracket from the HVAC housing.
11. Remove the screw that secures the heater core tubes to the heater core.
12. Remove the heater core tubes from the heater core and the dash panel. Remove the O-ring seals from the heater core tube fittings and discard.

13. Remove the two screws that secure the heater core retaining bracket to the top of the HVAC housing.

14. Remove the heater core retaining bracket from the top of the HVAC housing.

15. Carefully lift the heater core out of the HVAC housing.

To install:

16. Carefully install the heater core and the heater core retaining bracket to the top of the HVAC housing. Make sure that the heater core insulator is properly positioned.

17. Install the two screws that secure the heater core and retaining bracket to the HVAC housing. Tighten the screws to 2.2 Nm (20 inch lbs.).

18. Lubricate new rubber O-ring seals with clean engine coolant and install them onto the heater core tube fittings. Use only the specified O-ring as it is made of a special material for the engine cooling system.

19. Install the heater core tubes through the dash panel and to the heater core.

20. Install the screw that secures the heater core tubes to the heater core. Tighten the screw securely.

21. Install the heater core tube retaining bracket to the top of the HVAC housing.

22. Install the screw that secures the heater core tube retaining bracket to the

HVAC housing. Tighten the screw to 2.2 Nm (20 inch lbs.).

23. Install the HVAC housing bracket to the top of HVAC housing and to the dash panel.

24. Install the two screws that secure the HVAC housing bracket to the HVAC housing. Tighten the screws to 2.2 Nm (20 inch lbs.).

25. Install the bolt that secures the HVAC housing bracket to the dash panel. Tighten the bolt to 3 Nm (26 inch lbs.).

26. Install the instrument panel.

27. Raise the vehicle and install the heater hoses to the heater core tubes in the engine compartment.

28. Install the right front wheelhouse splash shield.

29. Lower the vehicle.

30. Refill the engine cooling system.

Cylinder Head

REMOVAL & INSTALLATION

3.7L Engine

LEFT SIDE

1. Before servicing the vehicle, refer to the precautions in the beginning of this section.

2. Drain the cooling system.

3. Properly relieve the fuel system pressure.

4. Remove or disconnect the following:
- Negative battery cable
- Exhaust Y-pipe
- Intake manifold
- Cylinder head cover
- Engine cooling fan and shroud
- Accessory drive belt
- Power steering pump

5. Rotate the crankshaft so that the crankshaft timing mark aligns with the Top Dead Center (TDC) mark on the front cover, and the **V6** marks on the camshaft sprockets are at 12 o'clock as shown.
- Crankshaft damper
- Front cover

6. Lock the secondary timing chain to the idler sprocket with Timing Chain Locking tool 8429.

7. Matchmark the secondary timing chain one link on each side of the V6 mark to the camshaft sprocket.
- Left secondary timing chain tensioner
- Cylinder head access plug
- Secondary timing chain guide
- Camshaft sprocket
- Cylinder head

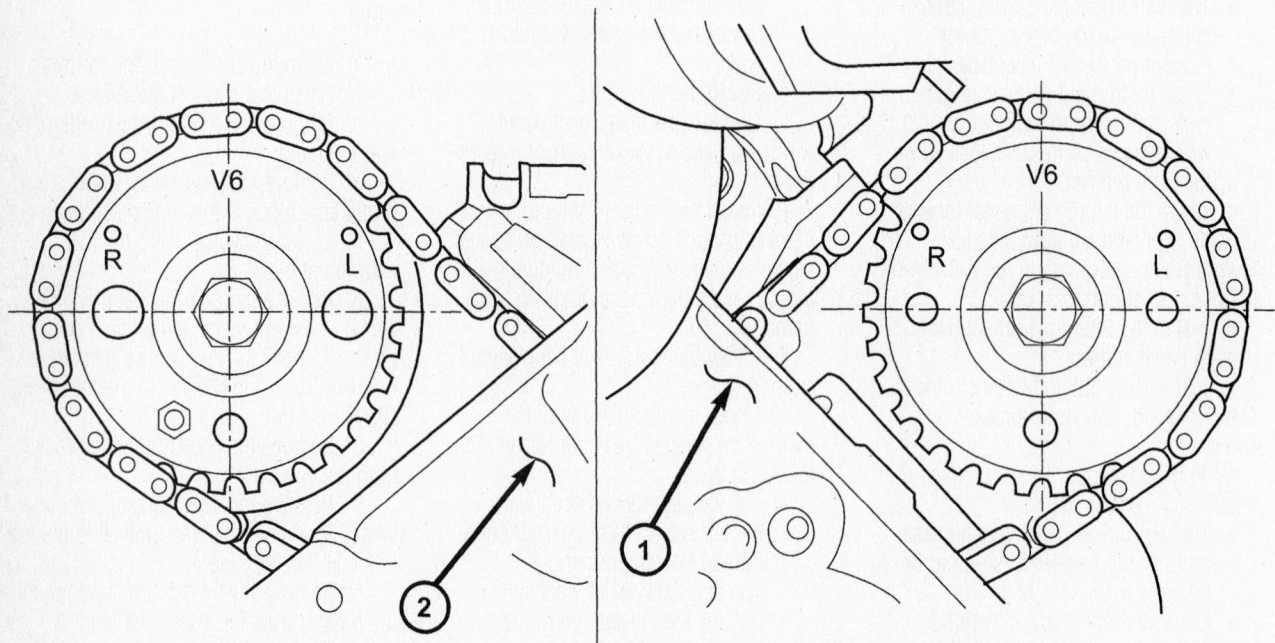

1 - LEFT CYLINDER HEAD
2 - RIGHT CYLINDER HEAD

Camshaft sprocket timing marks—3.7L

9355PG04

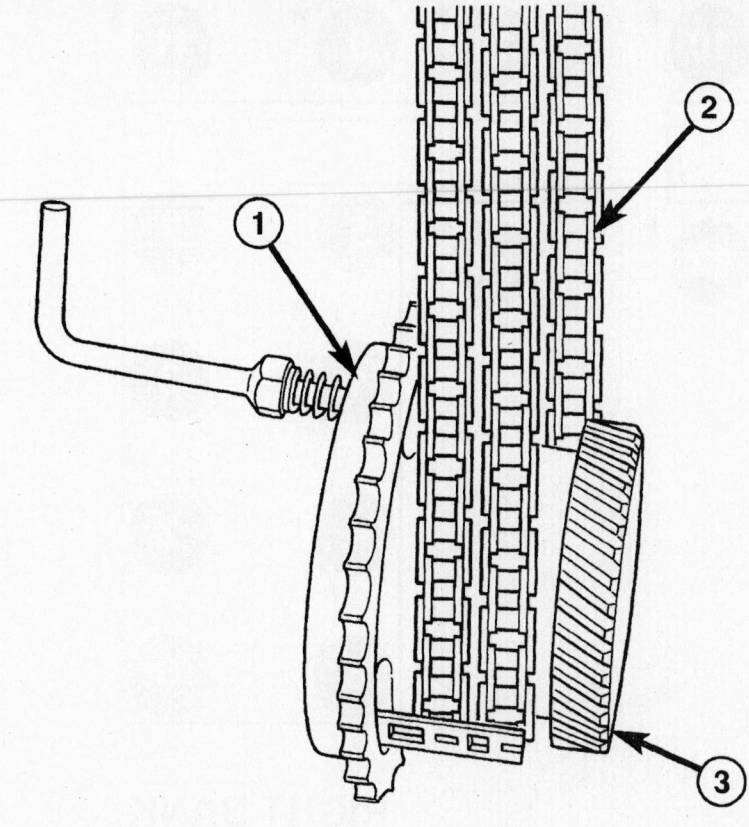

1 - SPECIAL TOOL 8429

2 - CAMSHAFT CHAIN

3 - CRANKSHAFT TIMING GEAR

9355PG05

Camshaft locking tool—3.7L

➡ **The cylinder head is retained by twelve bolts. Four of the bolts are smaller and are at the front of the head.**

To install:

8. Check the cylinder head bolts for signs of stretching and replace as necessary.

9. Lubricate the threads of the 11mm bolts with clean engine oil.

10. Coat the threads of the 8mm bolts with Mopar® Lock and Seal Adhesive, or equivalent.

11. Install the cylinder heads. Use new gaskets and tighten the bolts, in sequence, as follows:

a. Step 1: Bolts 1–8 to 20 ft. lbs. (27 Nm)

b. Step 2: Bolts 1–10 verify torque without loosening

c. Step 3: Bolts 9–12 to 10 ft. lbs. (14 Nm)

d. Step 4: Bolts 1–8 plus ¼ (90 degree) turn

e. Step 5: Bolts 9–12 to 19 ft. lbs. (26 Nm)

12. Install or connect the following:
   - Camshaft sprocket. Align the secondary chain matchmarks and tighten the bolt to 90 ft. lbs. (122 Nm).
   - Secondary timing chain guide
   - Cylinder head access plug
   - Secondary timing chain tensioner. Refer to the timing chain procedure in this section.

13. Remove the Timing Chain Locking tool.

14. Install or connect the following:
   - Front cover
   - Crankshaft damper. Torque the bolt to 130 ft. lbs. (175 Nm).
   - Power steering pump
   - Accessory drive belt
   - Engine cooling fan and shroud
   - Cover
   - Intake manifold
   - Exhaust Y-pipe
   - Negative battery cable

15. Fill and bleed the cooling system.

16. Start the engine, check for leaks and repair if necessary.

RIGHT SIDE

1. Before servicing the vehicle, refer to the precautions in the beginning of this section.

2. Drain the cooling system.

3. Properly relieve the fuel system pressure.

4. Remove or disconnect the following:
   - Negative battery cable
   - Exhaust Y-pipe
   - Intake manifold
   - Valve cover
   - Engine cooling fan and shroud
   - Accessory drive belt
   - Oil fill housing
   - Power steering pump

5. Rotate the crankshaft so that the crankshaft timing mark aligns with the Top Dead Center (TDC) mark on the front cover, and the **V6** marks on the camshaft sprockets are at 12 o'clock as shown.

6. Remove or disconnect the following:
   - Crankshaft damper
   - Front cover

7. Lock the secondary timing chains to

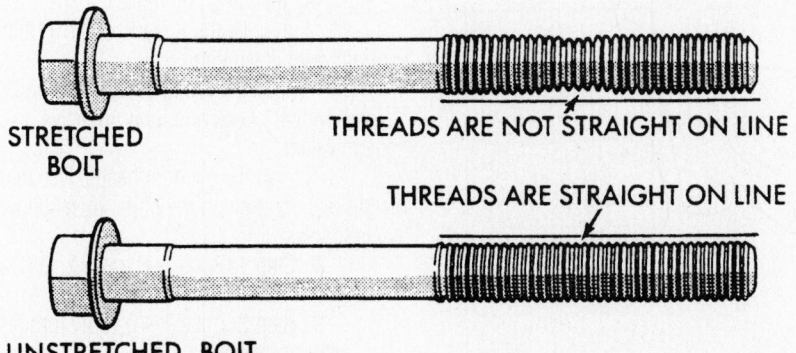

STRETCHED BOLT

THREADS ARE NOT STRAIGHT ON LINE

THREADS ARE STRAIGHT ON LINE

UNSTRETCHED BOLT

9302PG10

Examine the head bolts for signs of stretching—3.7L engine

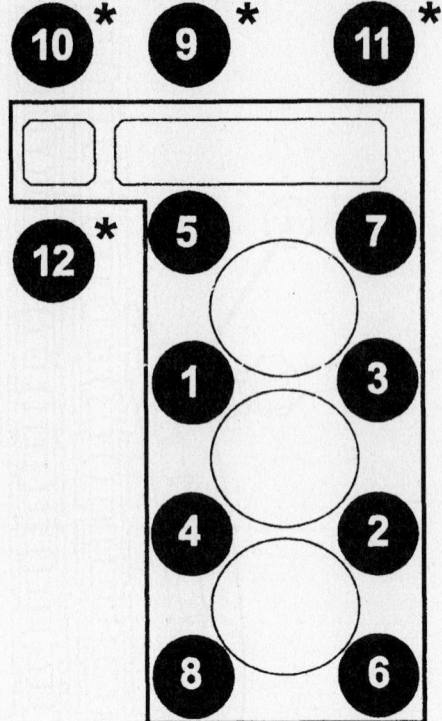

LEFT BANK

RIGHT BANK

67189-DURA-G98

Cylinder head bolt torque sequence—3.7L; bolts with asterisk require sealer

the idler sprocket with Timing Chain Locking tool 8429.

8. Matchmark the secondary timing chains to the camshaft sprockets.

9. Remove or disconnect the following:
- Secondary timing chain tensioners
- Cylinder head access plugs
- Secondary timing chain guides
- Camshaft sprockets
- Cylinder heads

➡**Each cylinder head is retained by 8 11mm bolts and four 8mm bolts.**

To install:

10. Check the cylinder head bolts for signs of stretching and replace as necessary.

11. Lubricate the threads of the 11mm bolts with clean engine oil.

12. Coat the threads of the 8mm bolts with Mopar® Lock and Seal Adhesive, or equivalent.

13. Install the cylinder heads. Use new gaskets and tighten the bolts, in sequence, as follows:

a. Step 1: Bolts 1–8 to 20 ft. lbs. (27 Nm)

b. Step 2: Bolts 1–10 verify torque without loosening

c. Step 3: Bolts 9–12 to 10 ft. lbs. (14 Nm)

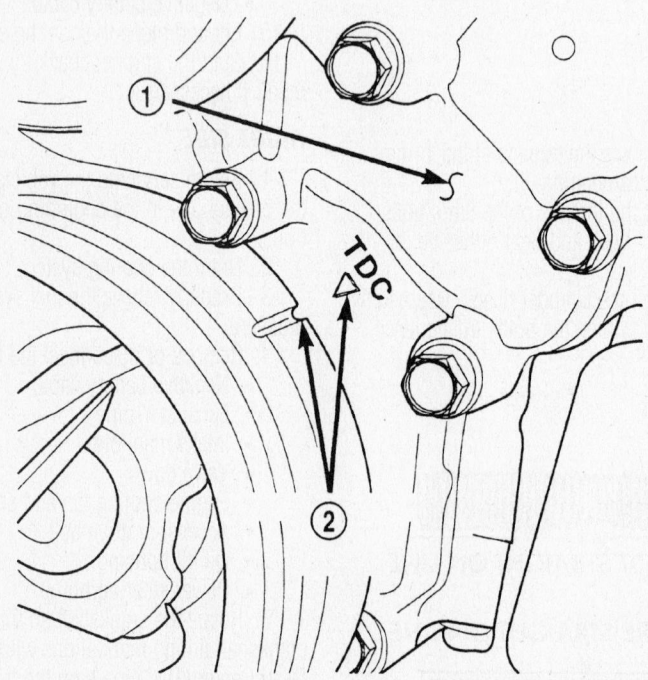

1 – TIMING CHAIN COVER
2 – CRANKSHAFT TIMING MARKS

9308PG04

Crankshaft timing marks—3.7L engine

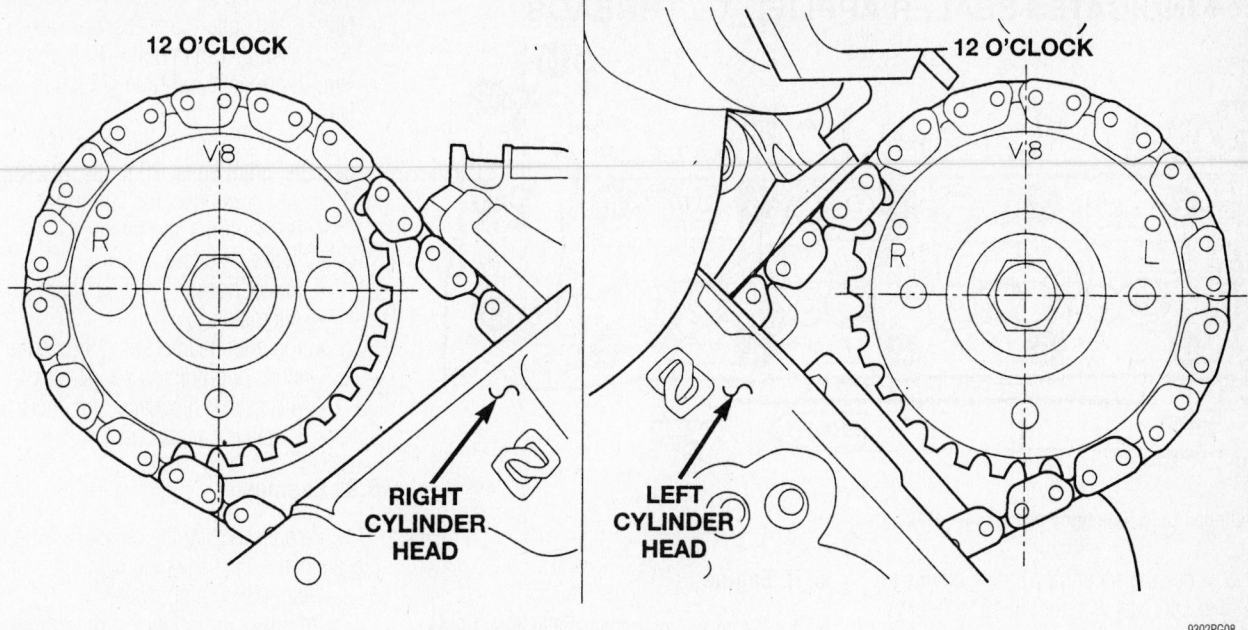

Camshaft positioning—4.7L engine shown; 3.7L engine similar

d. Step 4: Bolts 1–8 plus ¼ (90 degree) turn

e. Step 5: Bolts 9–12 to 19 ft. lbs. (26 Nm)

14. Install or connect the following:
- Camshaft sprockets. Align the secondary chain matchmarks and tighten the bolts to 90 ft. lbs. (122 Nm).
- Secondary timing chain guides
- Cylinder head access plugs
- Secondary timing chain tensioners. Refer to the timing chain procedure in this section.

15. Remove the Timing Chain Locking tool.

16. Install or connect the following:
- Front cover
- Crankshaft damper. Torque the bolt to 130 ft. lbs. (175 Nm).
- Rocker arms
- Power steering pump
- Oil fill housing
- Accessory drive belt
- Engine cooling fan and shroud
- Valve covers
- Intake manifold
- Exhaust Y-pipe
- Negative battery cable

17. Fill and bleed the cooling system.

18. Start the engine, check for leaks and repair if necessary.

4.7L Engine

LEFT SIDE

1. Before servicing the vehicle, refer to the precautions in the beginning of this section.

2. Drain the cooling system.

3. Remove or disconnect the following:
- Negative battery cable
- Exhaust pipe
- Intake manifold
- Cylinder head cover
- Fan shroud and fan
- Accessory drive belt
- Power steering pump

4. Rotate the crankshaft until the damper mark is aligned with the TDC mark. Verify that the V8 mark on the camshaft sprocket is at the 12 o'clock position.

5. Remove or disconnect the following:
- Vibration damper
- Timing chain cover

6. Lock the secondary timing chains to the idler sprocket with tool 8515, or equivalent.

7. Mark the secondary timing chain, on link on either side of the V8 mark on the cam sprocket.

8. Remove the left side secondary chain tensioner.

9. Remove the cylinder head access plug.

10. Remove the chain guide.

11. Remove the camshaft sprocket.

➡There are 4 smaller bolts at the front of the head. Don't overlook these.

12. Remove the head bolts and head.

✳✳ WARNING

Don't lay the head on its sealing surface. Due to the design of the head gasket, any distortion to the head sealing surface will result in leaks.

13. Installation is the reverse of removal. Observe the following:
- Check the head bolts. If any necking is observed, replace the bolt.
- The 4 small bolts must be coated with sealer.
- The head bolts are tightened in the following sequence:

Step 1: Bolts 1-10 to 15 ft. lbs. (20 Nm)
Step 2: Bolts 1-10 to 35 ft. lbs. (47 Nm)
Step 3: Bolts 11-14 to 18 ft. lbs. (25 Nm)
Step 4: Bolts 1-10 90 degrees
Step 5: Bolts 11-14 to 22 ft. lbs.

RIGHT SIDE

1. Before servicing the vehicle, refer to the precautions in the beginning of this section.

2. Drain the cooling system.

3. Remove or disconnect the following:
- Negative battery cable
- Exhaust pipe
- Intake manifold
- Cylinder head cover
- Fan shroud and fan
- Oil filler housing
- Accessory drive belt

4. Rotate the crankshaft until the damper mark is aligned with the TDC mark. Verify that the V8 mark on the camshaft sprocket is at the 12 o'clock position.

5. Remove or disconnect the following:
- Vibration damper
- Timing chain cover

♦ INDICATES SEALER APPLIED TO THREADS

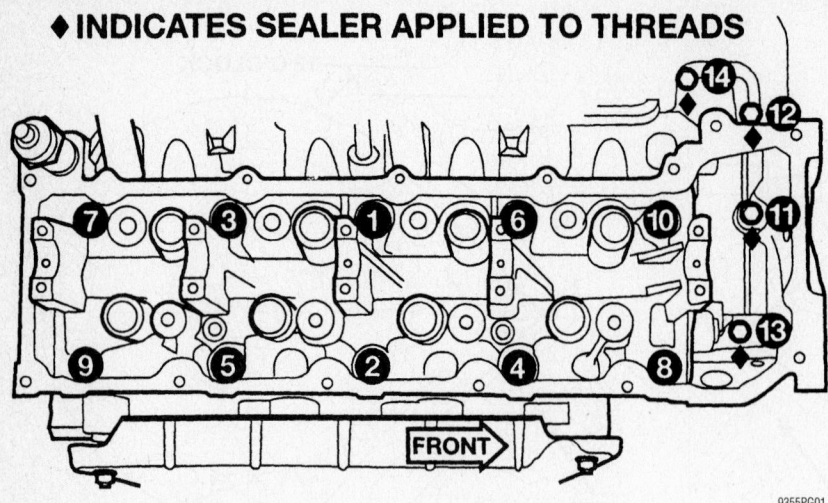

Cylinder head tightening sequence—4.7L

6. Lock the secondary timing chains to the idler sprocket with tool 8515, or equivalent.

7. Mark the secondary timing chain, on link on either side of the V8 mark on the cam sprocket.

8. Remove the left side secondary chain tensioner.

9. Remove the cylinder head access plug.

10. Remove the chain guide.

11. Remove the camshaft sprocket.

❋❋ WARNING

Do not pry on the target wheel for any reason!

➡There are 4 smaller bolts at the front of the head. Don't overlook these.

12. Remove the head bolts and head.

❋❋ WARNING

Don't lay the head on its sealing surface. Due to the design of the head gasket, any distortion to the head sealing surface will result in leaks.

13. Installation is the reverse of removal. Observe the following:

• Check the head bolts. If any necking is observed, replace the bolt.
• The 4 small bolts must be coated with sealer.
• The head bolts are tightened in the following sequence:
Step 1: Bolts 1-10 to 15 ft. lbs. (20 Nm)
Step 2: Bolts 1-10 to 35 ft. lbs. (47 Nm)
Step 3: Bolts 11-14 to 18 ft. lbs. (25 Nm)
Step 4: Bolts 1-10 90 degrees
Step 5: Bolts 11-14 to 22 ft. lbs.

5.7L Engine

1. Before servicing the vehicle, refer to the precautions in the beginning of this section.
2. Drain the cooling system.
3. Properly relieve the fuel system pressure.
4. Remove or disconnect the following:

• Negative battery cable
• Air cleaner resonator and ducts
• Alternator
• Closed crankcase ventilation system
• EVAP control system
• Heater hoses
• Cylinder head covers
• Intake manifold
• Rocker arms and pushrods
• Cylinder heads

To install:

➡The head gaskets are not interchangeable. They are marked "L" and "R".

5. Install the cylinder heads. Use new gaskets and tighten the bolts, in sequence, as follows:

a. Step 1: 12 mm bolts–25 ft. lbs. (34 Nm); 8mm bolts–15 ft. lbs. (20 Nm)
b. Step 2: 12mm bolts–40 ft. lbs. (54 Nm); 8mm bolts retorque–15 ft. lbs. (20 Nm)
c. Step 3: 12mm bolts–plus 90 degrees; 8mm bolts–25 ft. lbs. (34 Nm)

6. Install or connect the following:

• Rocker arms and pushrods
• Intake manifold
• Heater hoses
• Alternator
• Cylinder head covers. Torque the studs and bolts to 70 inch lbs.
• Air cleaner resonator and ducts
• Negative battery cable

5.9L Engines

1. Before servicing the vehicle, refer to the precautions in the beginning of this section.
2. Drain the cooling system.
3. Remove or disconnect the following:

• Negative battery cable
• Accessory drive belt
• Alternator
• A/C compressor, if equipped
• Air injection pump, if equipped
• Air cleaner assembly
• Closed Crankcase Ventilation (CCV) system
• Evaporative emissions control system
• Fuel line
• Accelerator linkage
• Cruise control cable
• Transmission cable
• Distributor cap and wires
• Ignition coil wiring
• Engine Coolant Temperature (ECT) sensor connector
• Heater hoses
• Bypass hose
• Upper radiator hose
• Intake manifold
• Valve covers

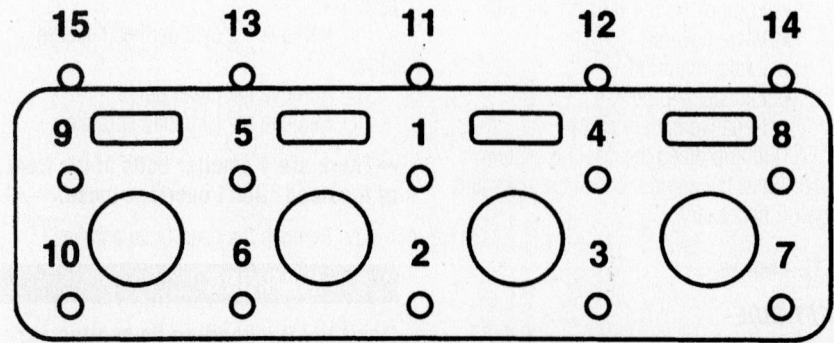

Cylinder head torque sequence—5.7L engine

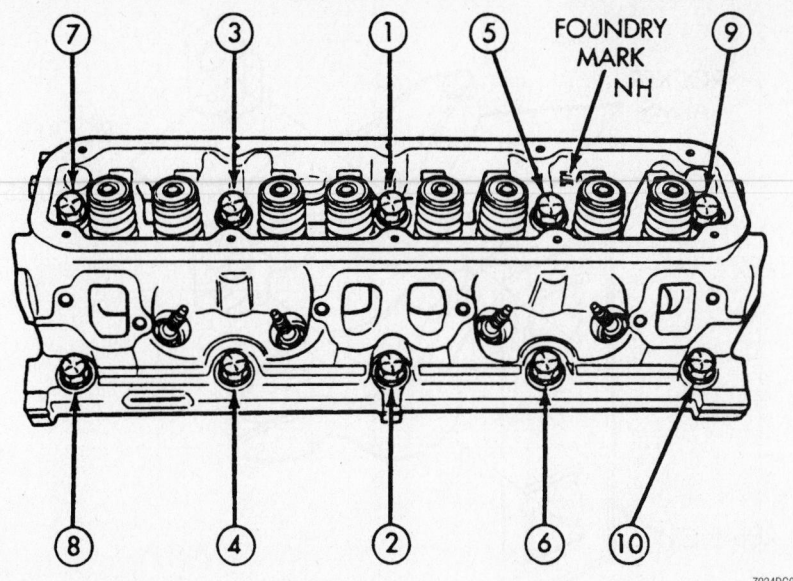

FOUNDRY MARK NH

Cylinder head torque sequence—5.9L engines

➡ **Keep valvetrain components in order for reassembly.**

- Rocker arms
- Pushrods
- Exhaust manifolds
- Spark plugs
- Cylinder heads

To install:

✴✴ WARNING

Position the crankshaft so that no piston is at Top Dead Center (TDC) prior to installing the cylinder heads. Do not rotate the crankshaft during or immediately after rocker arm installation. Wait 5 minutes for the hydraulic lash adjusters to bleed down.

4. Install the cylinder heads. Use new gaskets and tighten the bolts in sequence as follows:
   a. Step 1: 50 ft. lbs. (68 Nm).
   b. Step 2: 105 ft. lbs. (143 Nm).
   c. Step 3: 105 ft. lbs. (143 Nm).
5. Install or connect the following:
   - Spark plugs
   - Exhaust manifolds
   - Pushrods and rocker arms in their original positions
   - Valve covers
   - Intake manifold
   - Upper radiator hose
   - Bypass hose
   - Heater hoses
   - ECT sensor connector
   - Ignition coil wiring
   - Distributor cap and wires
   - Transmission cable
   - Cruise control cable
   - Accelerator linkage
   - Fuel line
   - Evaporative emissions control system
   - CCV system
   - Air cleaner assembly
- A/C compressor, if equipped
- Air injection pump, if equipped
- Alternator
- Accessory drive belt
- Negative battery cable
6. Fill the cooling system.
7. Start the engine and check for leaks.

Rocker Arms/Shafts

REMOVAL & INSTALLATION

3.7L and 4.7L Engine

1. Before servicing the vehicle, refer to the precautions in the beginning of this section.
2. Remove or disconnect the following:
   - Negative battery cable
   - Valve covers
3. Rotate the crankshaft so that the piston of the cylinder to be serviced is at Bottom Dead Center (BDC) and both valves are closed.
4. Use special tool 8516 to depress the valve and remove the rocker arm.
5. Repeat for each rocker arm to be serviced.

➡ **Keep valvetrain components in order for reassembly.**

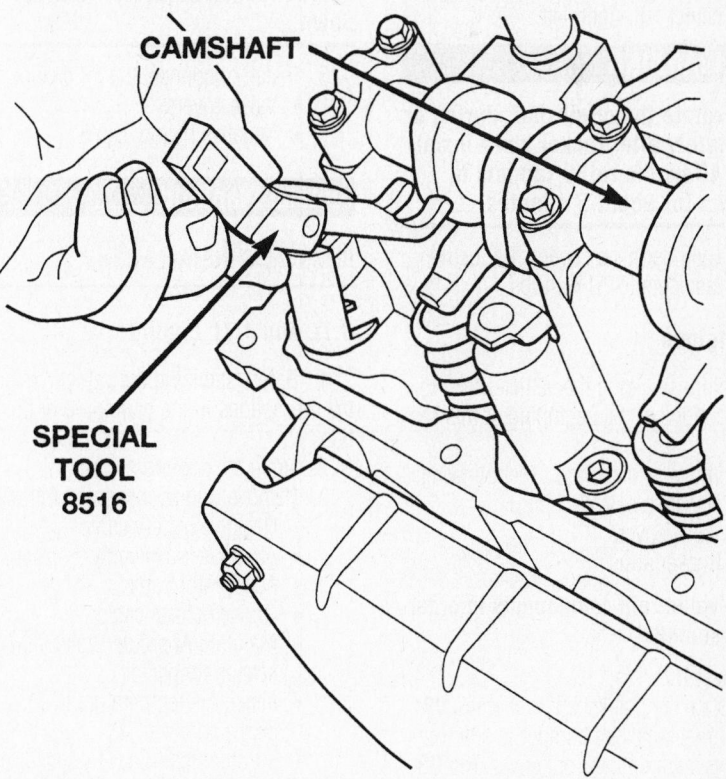

CAMSHAFT

SPECIAL TOOL 8516

Rocker arm service—3.7L and 4.7L engine

To install:

6. Rotate the crankshaft so that the piston of the cylinder to be serviced is at BDC.

7. Compress the valve spring and install each rocker arm in its original position.

8. Repeat for each rocker arm to be installed.

9. Install or connect the following:
- Cylinder head cover
- Negative battery cable

5.7L Engines

1. Before servicing the vehicle, refer to the precautions in the beginning of this section.

2. Remove or disconnect the following:
- Negative battery cable
- Cylinder head cover

3. Loosen the rocker shaft bolts as follows: center, center left, center right, left, right.

➡ **The rocker shaft assemblies are not interchangeable. The intake side is marked "I". The shorter pushrods are used on the intake side.**

To install:

4. Install or connect the following:
- Pushrods
- Rocker shaft assemblies. Torque the shafts to 16 ft. lbs. (22 Nm) in this sequence: center, center right, center left, right, left.

✳ WARNING

Do not rotate the crankshaft during or immediately after rocker shaft installation. Allow the roller tappets to leakdown for about 5 minutes.

- Cylinder head covers. Torque the fasteners to 70 inch lbs.

5.9L Engines

1. Before servicing the vehicle, refer to the precautions in the beginning of this section.

2. Remove or disconnect the following:
- Negative battery cable
- Valve covers
- Rocker arms

➡ **Keep valvetrain components in order for reassembly.**

To install:

3. Rotate the crankshaft so that the **V8** mark on the crankshaft damper aligns with the timing mark on the front cover. The **V8** mark is located 147 degrees **AFTER** Top Dead Center (TDC).

4. Install the rocker arms in their origi-

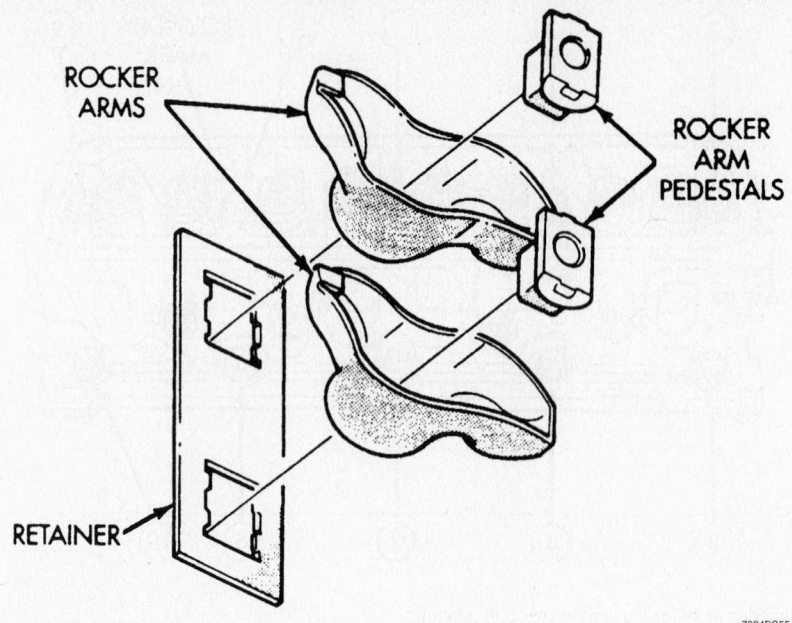

Exploded view of the rocker arm assembly—5.9L engine

7924DG55

nal positions and tighten the bolts to 21 ft. lbs. (28 Nm).

✳✳ CAUTION

Do not rotate the crankshaft during or immediately after rocker arm installation. Wait 5 minutes for the hydraulic lash adjusters to bleed down.

5. Install or connect the following:
- Valve covers
- Negative battery cable

Intake Manifold

REMOVAL & INSTALLATION

3.7L and 4.7L Engine

1. Before servicing the vehicle, refer to the precautions in the beginning of this section.

2. Drain the cooling system.

3. Remove or disconnect the following:
- Negative battery cable
- Air cleaner assembly
- Accelerator cable
- Cruise control cable
- Manifold Absolute Pressure (MAP) sensor connector
- Intake Air Temperature (IAT) sensor connector
- Throttle Position (TP) sensor connector
- Idle Air Control (IAC) valve connector
- Engine Coolant Temperature (ECT) sensor
- Positive Crankcase Ventilation (PCV) valve and hose
- Canister purge vacuum line
- Brake booster vacuum line
- Cruise control servo hose
- Accessory drive belt
- Alternator
- A/C compressor
- Engine ground straps
- Ignition coil towers
- Oil dipstick tube
- Fuel line
- Fuel supply manifold
- Throttle body and mounting bracket
- Cowl seal
- Right engine lifting stud
- Intake manifold. Remove the fasteners in reverse of the tightening sequence.

To install:

4. Install or connect the following:
- Intake manifold using new gaskets. Tighten the bolts, in sequence, to 105 inch lbs. (12 Nm).
- Right engine lifting stud
- Cowl seal
- Throttle body and mounting bracket
- Fuel supply manifold
- Fuel line
- Oil dipstick tube
- Ignition coil towers
- Engine ground straps
- A/C compressor
- Alternator
- Accessory drive belt

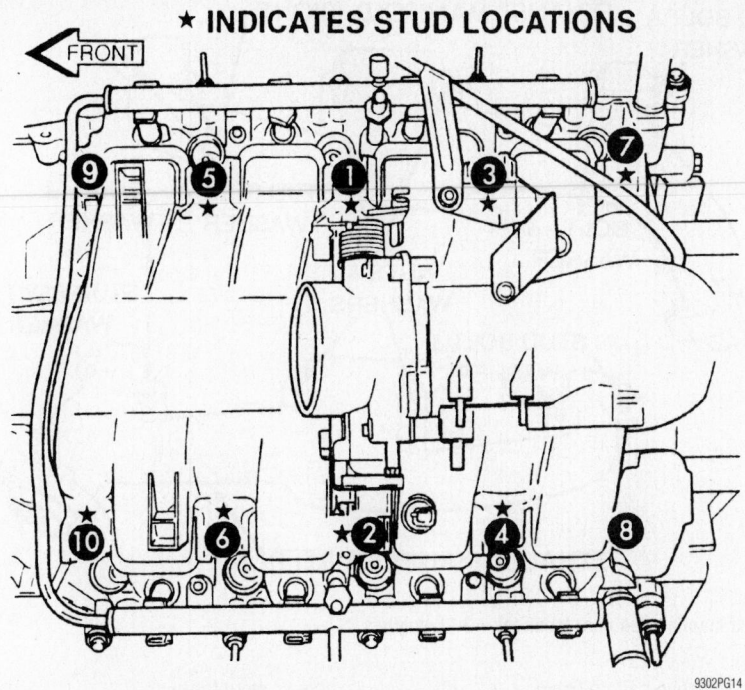

★ **INDICATES STUD LOCATIONS**

◀FRONT

Intake manifold torque sequence—3.7L and 4.7L engine

9302PG14

- Cruise control servo hose
- Brake booster vacuum line
- Canister purge vacuum line
- PCV valve and hose
- ECT sensor
- IAC valve connector
- TP sensor connector
- IAT sensor connector
- MAP sensor connector
- Cruise control cable
- Accelerator cable
- Air cleaner assembly
- Negative battery cable

5. Fill the cooling system.
6. Start the engine and check for leaks.

5.7L Engines

1. Before servicing the vehicle, refer to the precautions in the beginning of this section.
2. Drain the cooling system.
3. Relieve the fuel system pressure.
4. Remove or disconnect the following:
- Negative battery cable
- Air cleaner assembly
- Accessory drive belt
- MAP connector
- IAT connector
- TPS connector
- CTS connector
- Brake booster hose
- PCV hose
- Alternator
- A/C compressor
- Intake manifold bolts, in a criss-

cross pattern, from the outside to the center
- Intake manifold/IAFM

To install:

5. Position new intake manifold seals.
6. Install the intake manifold. Tighten the bolts in sequence from the center outwards, to 105 inch lbs. (12 Nm).
7. Install or connect the following:
- Electrical connectors
- Alternator
- A/C compressor
- Brake booster hose
- PCV hose

- Accessory drive belt
- Negative battery cable
- Air cleaner assembly

5.9L Engines

1. Before servicing the vehicle, refer to the precautions in the beginning of this section.
2. Drain the cooling system.
3. Remove or disconnect the following:
- Negative battery cable
- Accessory drive belt
- Alternator
- A/C compressor
- Alternator and A/C compressor bracket
- Air cleaner assembly
- Fuel line
- Fuel supply manifold
- Accelerator cable
- Transmission cable
- Cruise control cable
- Distributor cap and wires
- Ignition coil wiring
- Engine Coolant Temperature (ECT) sensor connector
- Heater hose
- Upper radiator hose
- Bypass hose
- Closed Crankcase Ventilation (CCV) system
- Evaporative emissions system
- Intake manifold

To install:

4. Install the intake manifold. Use a new gasket and tighten the bolts in sequence as follows:
   a. Step 1: Bolts 1–4 to 72 inch lbs. (8 Nm) using 12 inch lb. (1.4 Nm) increments.

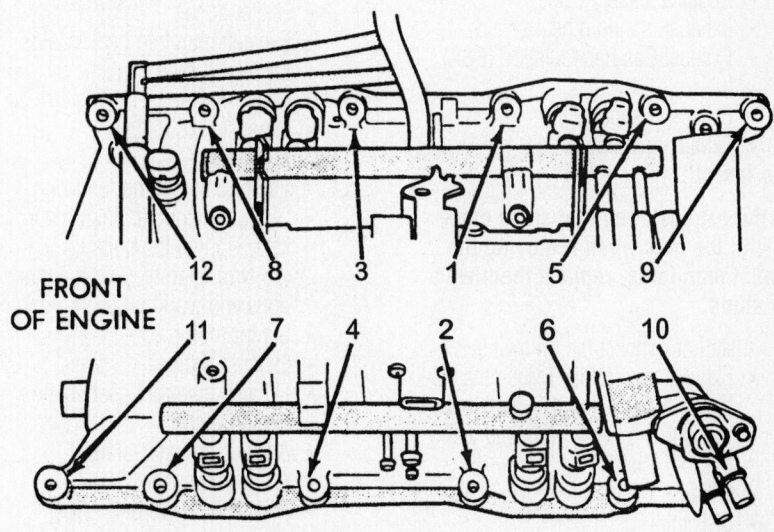

FRONT
OF ENGINE

Intake manifold torque sequence—5.9L engines

7924DG11

b. Step 2: Bolts 5–12 to 72 inch lbs. (8 Nm).

c. Step 3: Bolts 1–12 to 72 inch lbs. (8 Nm).

d. Step 4: Bolts 1–12 to 12 ft. lbs. (16 Nm).

e. Step 5: Bolts 1–12 to 12 ft. lbs. (16 Nm).

5. Install or connect the following:
- Evaporative emissions system
- CCV system
- Bypass hose
- Upper radiator hose
- Heater hose
- ECT sensor connector
- Ignition coil wiring
- Distributor cap and wires
- Cruise control cable
- Transmission cable
- Accelerator cable
- Fuel supply manifold
- Fuel line
- Air cleaner assembly
- Alternator and A/C compressor bracket
- A/C compressor
- Alternator
- Accessory drive belt
- Negative battery cable

6. Fill the cooling system.
7. Start the engine and check for leaks.

Exhaust Manifold

REMOVAL & INSTALLATION

3.7L Engines

1. Before servicing the vehicle, refer to the precautions in the beginning of this section.
2. Remove or disconnect the following:
- Negative battery cable
- Exhaust manifold heat shields
- Exhaust Gas Recirculation (EGR) tube
- Exhaust Y-pipe
- Exhaust manifolds

To install:

➡**If the exhaust manifold studs came out with the nuts when removing the exhaust manifolds, replace them with new studs.**

3. Install or connect the following:
- Exhaust manifolds. Torque the fasteners to 20 ft. lbs. (27 Nm), starting with the center nuts and work out to the ends.
- Exhaust Y-pipe
- EGR tube
- Exhaust manifold heat shields

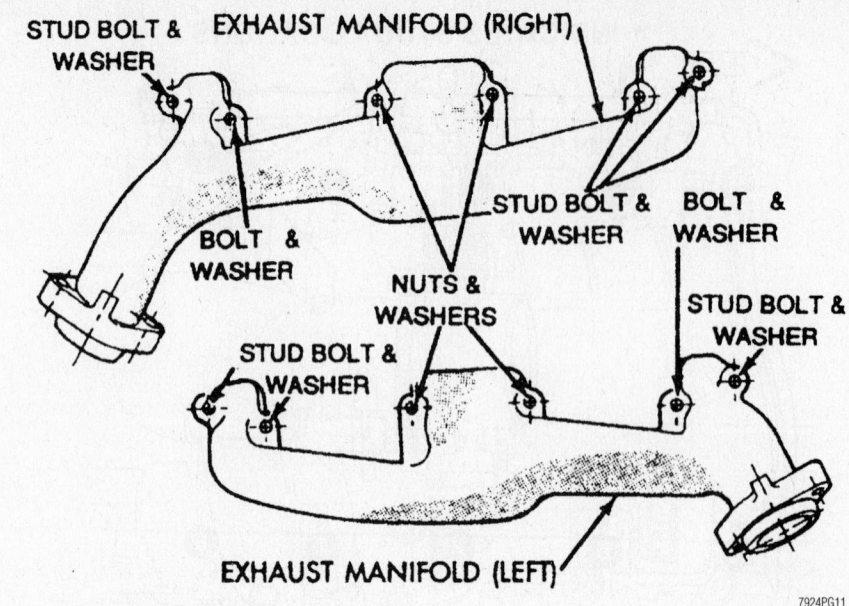

STUD BOLT & WASHER — EXHAUST MANIFOLD (RIGHT) — STUD BOLT & WASHER — BOLT & WASHER — STUD BOLT & WASHER — BOLT & WASHER — NUTS & WASHERS — STUD BOLT & WASHER — EXHAUST MANIFOLD (LEFT)

7924PG11

Exhaust manifold fastener locations—3.7L engines

- Negative battery cable

4. Start the engine, check for leaks and repair if necessary.

4.7L Engine

1. Before servicing the vehicle, refer to the precautions in the beginning of this section.
2. Drain the cooling system.
3. Remove or disconnect the following:
- Battery
- Power distribution center
- Battery tray
- Windshield washer fluid bottle
- Air cleaner assembly
- Accessory drive belt
- A/C compressor
- A/C accumulator bracket
- Heater hoses
- Exhaust manifold heat shields
- Exhaust Y-pipe
- Starter motor
- Exhaust manifolds

To install:

4. Install or connect the following:
- Exhaust manifolds, using new gaskets. Tighten the bolts to 18 ft. lbs. (25 Nm), starting with the inner bolts and work out to the ends.
- Starter motor
- Exhaust Y-pipe
- Exhaust manifold heat shields
- Heater hoses
- A/C accumulator bracket
- A/C compressor
- Accessory drive belt
- Air cleaner assembly
- Windshield washer fluid bottle

- Battery tray
- Power distribution center
- Battery

5. Fill the cooling system.
6. Start the engine and check for leaks.

5.7L Engines

1. Before servicing the vehicle, refer to the precautions in the beginning of this section.
2. Drain the cooling system.
3. Remove or disconnect the following:
- Battery
- Exhaust pipe-to-manifold bolts

4. Install an engine crane. Remove the right and left mount through bolts. Raise the engine just enough to provide clearance for manifold removal.

5. Remove or disconnect the following:
- Exhaust manifold heat shields
- Exhaust manifolds

To install:

6. Install or connect the following:
- Exhaust manifolds, using new gaskets. Tighten the bolts to 18 ft. lbs. (25 Nm), starting with the inner bolts and work out to the ends.
- Exhaust manifold heat shields

7. Install the right and left mount through bolts. Remove the crane.

8. Install or connect the following:
- Exhaust pipe-to-manifold bolts
- Battery

5.9L Engines

1. Before servicing the vehicle, refer to the precautions in the beginning of this section.

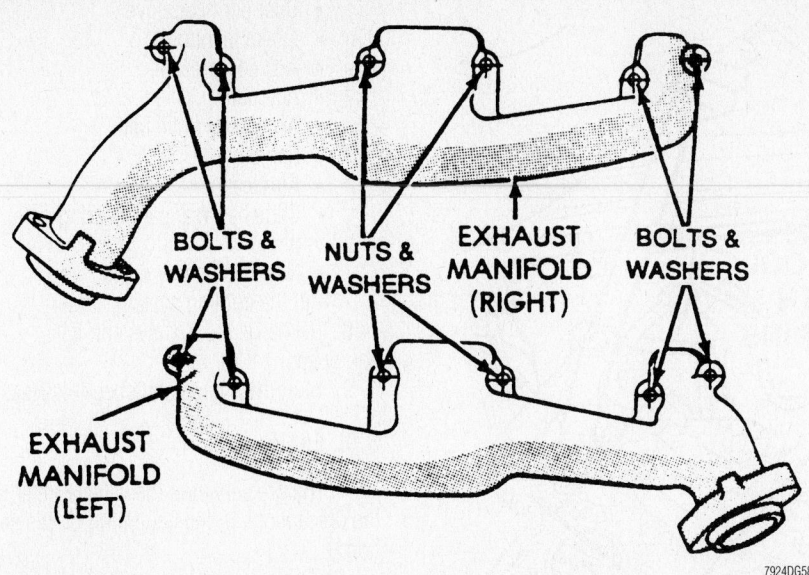

Exhaust manifold fastener locations—5.9L engines

2. Remove or disconnect the following:
- Negative battery cable
- Exhaust manifold heat shields
- Exhaust Gas Recirculation (EGR) tube
- Exhaust Y-pipe
- Exhaust manifolds

To install:

➡ **If the exhaust manifold studs came out with the nuts when removing the exhaust manifolds, replace them with new studs.**

3. Install or connect the following:
- Exhaust manifolds. Tighten the fasteners to 20 ft. lbs. (27 Nm), starting with the center nuts and work out to the ends.
- Exhaust Y-pipe
- EGR tube
- Exhaust manifold heat shields
- Negative battery cable
4. Fill the cooling system.
5. Start the engine and check for leaks.

Camshaft and Valve Lifters

REMOVAL & INSTALLATION

3.7L and 4.7L Engines

1. Before servicing the vehicle, refer to the precautions in the beginning of this section.
2. Remove or disconnect the following:
- Negative battery cable
- Cylinder head covers
- Rocker arms
- Hydraulic lash adjusters

➡ **Keep all valvetrain components in order for assembly.**

3. Set the engine at Top Dead Center (TDC) of the compression stroke for the No. 1 cylinder.
4. Install Timing Chain Wedge 8350 to retain the chain tensioners.
5. Matchmark the timing chains to the camshaft sprockets.
6. Install Camshaft Holding Tool 6958 and Adapter Pins 8346 to the left camshaft sprocket.
7. Remove or disconnect the following:
- Right camshaft timing sprocket and target wheel
- Left camshaft sprocket
- Camshaft bearing caps, by reversing the tightening sequence
- Camshafts

To install:
8. Install or connect the following:
- Camshafts. Tighten the bearing cap

bolts in ½ turn increments, in sequence, to 100 inch lbs. (11 Nm).
- Target wheel to the right camshaft
- Camshaft timing sprockets and chains, by aligning the matchmarks
9. Remove the tensioner wedges and tighten the camshaft sprocket bolts to 90 ft. lbs. (122 Nm).
10. Install or connect the following:
- Hydraulic lash adjusters in their original locations
- Rocker arms in their original locations
- Cylinder head covers
- Negative battery cable

5.7L Engines

1. Before servicing the vehicle, refer to the precautions in the beginning of this section.
2. Drain the cooling system.
3. Recover the A/C refrigerant, if equipped with air conditioning.
4. Set the crankshaft to Top Dead Center (TDC) of the compression stroke for the No. 1 cylinder.
5. Remove or disconnect the following:
- Negative battery cable
- Camshaft rear cam bearing core plug
- Air cleaner
- Accessory drive belt
- Alternator
- A/C compressor
- Radiator
- Intake manifold
- Cylinder head covers
- Cylinder heads
- Oil pan
- Front cover
- Oil pickup tube
- Oil pump
- Timing chain and sprockets

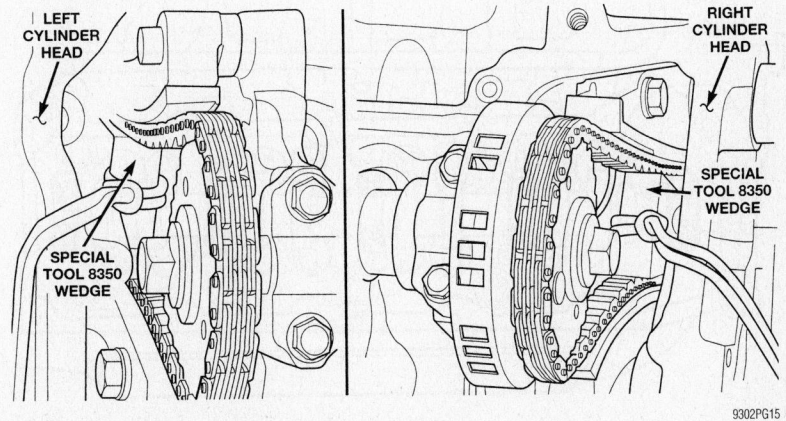

Chain Tensioner Retaining Wedges—4.7L engine

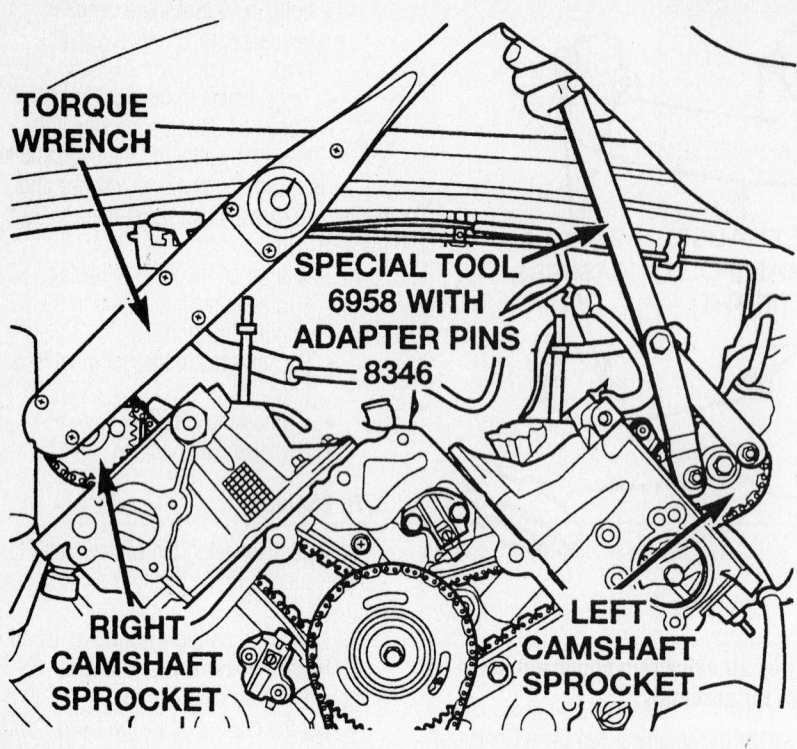

Hold the left camshaft sprocket with a spanner wrench while removing or installing the camshaft sprocket bolts—3.7L and 4.7L engine

- Camshaft thrust plate
- Hydraulic lifters
- Camshaft

To install:
6. Install or connect the following:
- Camshaft
- Camshaft thrust plate. Tighten the bolts to 21 ft. lbs. (28 Nm).
- Timing chain and sprockets
- Oil pump
- Oil pickup tube

➡**Lifters must be replaced in their original positions.**

- Hydraulic lifters
- Cylinder heads
- Pushrods
- Rocker arms
- Front cover
- Oil pan

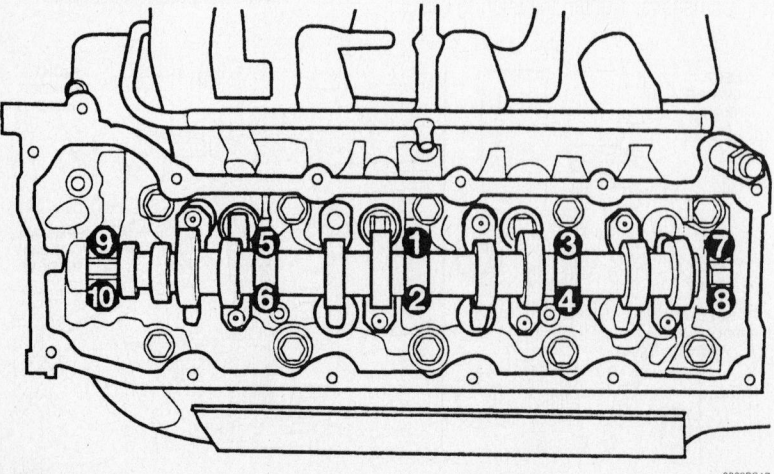

Camshaft bearing cap bolt tightening sequence—3.7L and 4.7L engine

- Cylinder head covers
- Intake manifold
- A/C compressor
- Alternator
- Accessory drive belt
- Radiator
- Air cleaner
- Camshaft rear cam bearing core plug
- Negative battery cable
7. Fill the cooling system.
8. Recharge the A/C system, if equipped.
9. Start the engine and check for leaks.

5.9L Engines

1. Before servicing the vehicle, refer to the precautions in the beginning of this section.
2. Drain the cooling system.
3. Recover the A/C refrigerant, if equipped with air conditioning.
4. Set the crankshaft to Top Dead Center (TDC) of the compression stroke for the No. 1 cylinder.
5. Remove or disconnect the following:
- Negative battery cable
- Accessory drive belt
- Power steering pump
- Water pump
- Radiator
- A/C condenser
- Grille
- Crankshaft damper
- Front cover
- Valve covers
- Distributor
- Intake manifold

➡**Keep all valvetrain components in order for assembly.**

- Rocker arms and pushrods
- Hydraulic lifters
- Timing chain and sprockets
- Camshaft thrust plate and chain oil tab
- Camshaft

To install:
6. Install or connect the following:
- Camshaft
- Camshaft Holding Tool C-3509
- Camshaft thrust plate and chain oil tab. Tighten the bolts to 18 ft. lbs. (24 Nm).
- Timing chain and sprockets. Tighten the camshaft sprocket bolt to 50 ft. lbs. (68 Nm).
7. Remove the camshaft holding tool.
8. Install or connect the following:
- Hydraulic lifters in their original positions

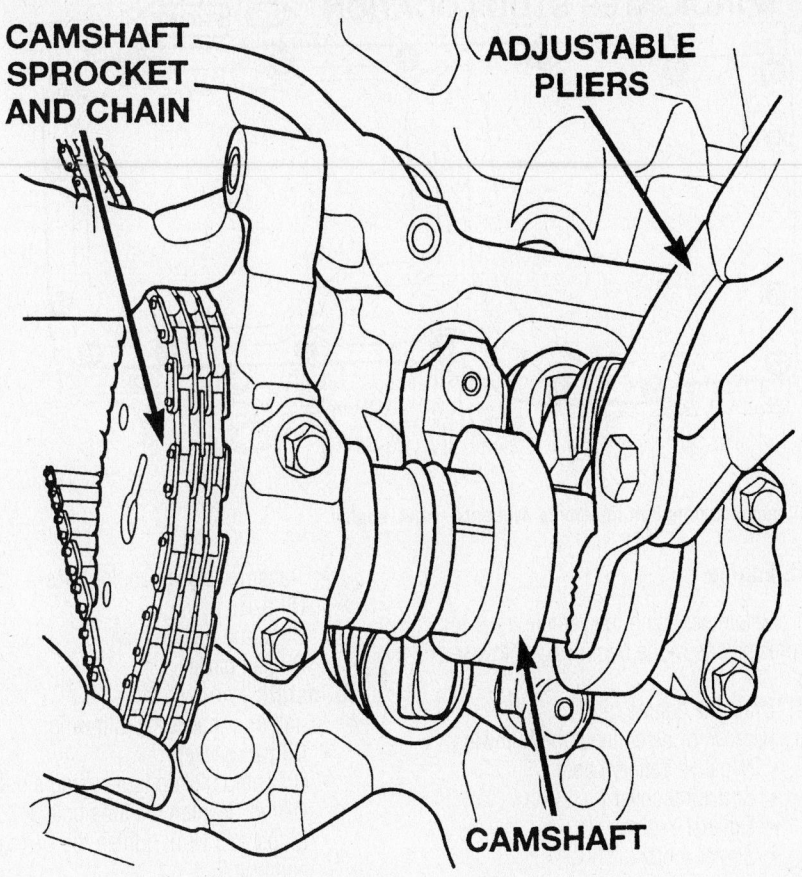

Turn the camshaft with pliers, if needed, to align the dowel in the sprocket—4.7L engine

- Rocker arms and pushrods in their original positions
- Intake manifold
- Distributor
- Valve covers
- Front cover
- Crankshaft damper
- Grille
- A/C condenser
- Radiator
- Water pump
- Power steering pump
- Accessory drive belt
- Negative battery cable
9. Fill the cooling system.
10. Recharge the A/C system, if equipped.
11. Start the engine and check for leaks.

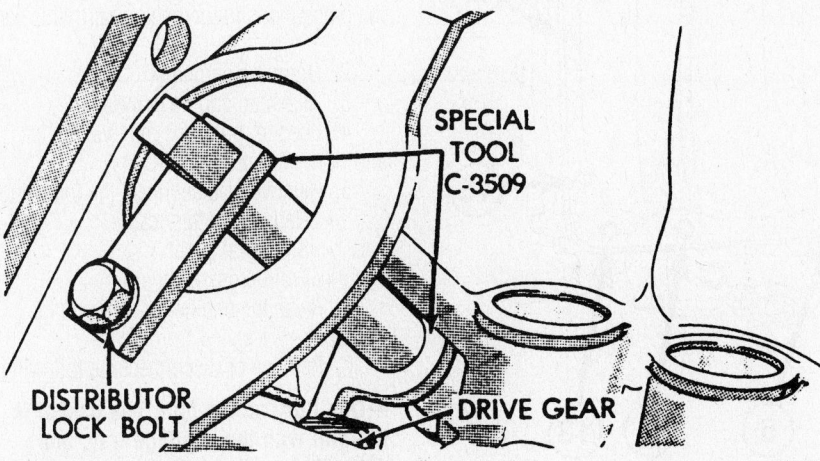

Camshaft holding tool C-3509—5.9L engines

Valve Lash

ADJUSTMENT

All gasoline engines covered in this section use hydraulic lifters. No maintenance or periodic adjustment is required.

Starter Motor

REMOVAL & INSTALLATION

3.7L Engine

1. Before servicing the vehicle, refer to the precautions in the beginning of this section.
2. Remove or disconnect the following:
   - Negative battery cable
   - Starter mounting bolts

➡**The left side exhaust pipe and front driveshaft must be disconnected.**

   - Starter solenoid harness connections
   - Starter

To install:
3. Connect the starter solenoid wiring connectors.
4. Install the starter and torque the bolts to 40 ft. lbs. (54 Nm).
5. Install the negative battery cable and check for proper operation.

4.7L and 5.9L Engines

1. Before servicing the vehicle, refer to the precautions in the beginning of this section.
2. Remove or disconnect the following:
   - Negative battery cable
   - Starter mounting bolts
   - Starter solenoid harness connections
   - Starter

To install:
3. Connect the starter solenoid wiring connectors.
4. Install the starter and tighten the bolts to the following specifications:
   a. 4.7L engine: 40 ft. lbs. (54 Nm).
   b. 5.9L engines: 50 ft. lbs. (68 Nm).
5. Install the negative battery cable and check for proper operation.

5.7L Engines

1. Before servicing the vehicle, refer to the precautions in the beginning of this section.
2. Remove or disconnect the following:
   - Negative battery cable

➡**Depending on drivetrain configuration, a support bracket may be used.**

- Starter mounting bolts
- Starter solenoid harness connections
- Starter

To install:

3. Connect the starter solenoid wiring connectors.

4. Install the starter and torque the bolts to 50 ft. lbs. (68 Nm).

5. Install the negative battery cable and check for proper operation.

Oil Pan

REMOVAL & INSTALLATION

3.7L Engine

1. Before servicing the vehicle, refer to the precautions in the beginning of this section.

2. Remove or disconnect the following:
   - Engine from the vehicle
   - Oil pan
   - Oil pump pickup tube
   - Oil pan gasket

3. Installation is the reverse of removal. Torque the bolts, in sequence, to 11 ft. lbs. (15 Nm).

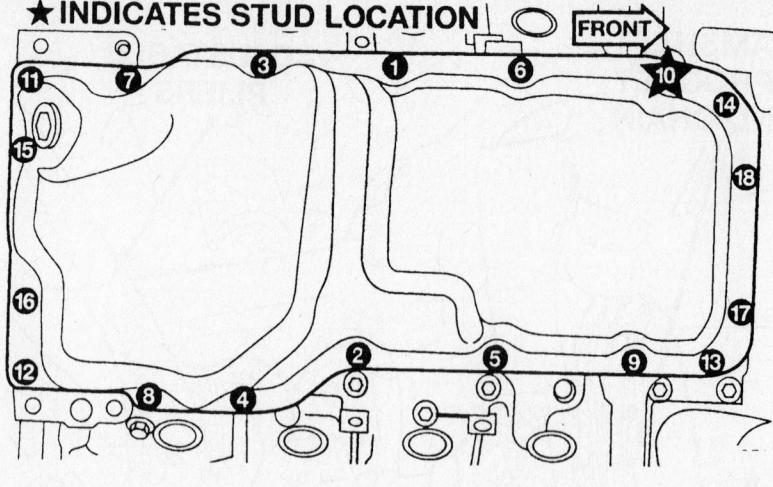

Oil pan mounting bolt tightening sequence—4.7L engine

4.7L Engine

1. Before servicing the vehicle, refer to the precautions in the beginning of this section.

2. Drain the engine oil.

3. Remove or disconnect the following:
   - Negative battery cable
   - Structural cover
   - Exhaust Y-pipe
   - Starter motor
   - Transmission oil cooler lines
   - Oil pan
   - Oil pump pickup tube
   - Oil pan gasket

To install:

4. Install or connect the following:
   - Oil pan gasket
   - Oil pump pickup tube, using a new O-ring. Tighten the tube bolts to 20 ft. lbs. (28 Nm); tighten the O-ring end bolt first.
   - Oil pan. Tighten the bolts, in sequence, to 11 ft. lbs. (15 Nm).
   - Transmission oil cooler lines
   - Starter motor
   - Exhaust Y-pipe
   - Structural cover
   - Negative battery cable

5. Fill the crankcase to the proper level with engine oil.

6. Start the engine and check for leaks.

5.7L Engine

1. Before servicing the vehicle, refer to the precautions in the beginning of this section.

2. Drain the engine oil.

3. Attach an engine crane.

4. Loosen, but don't remove, the left and right mount through-bolts.

5. Remove or disconnect the following:
   - Negative battery cable
   - Structural cover
   - Front crossmember

6. Raise the engine just enough for clearance.

7. Remove or disconnect the following:

➡**Don't pry on the pan. The gasket is integral with the windage tray, and doesn't come out with the pan.**

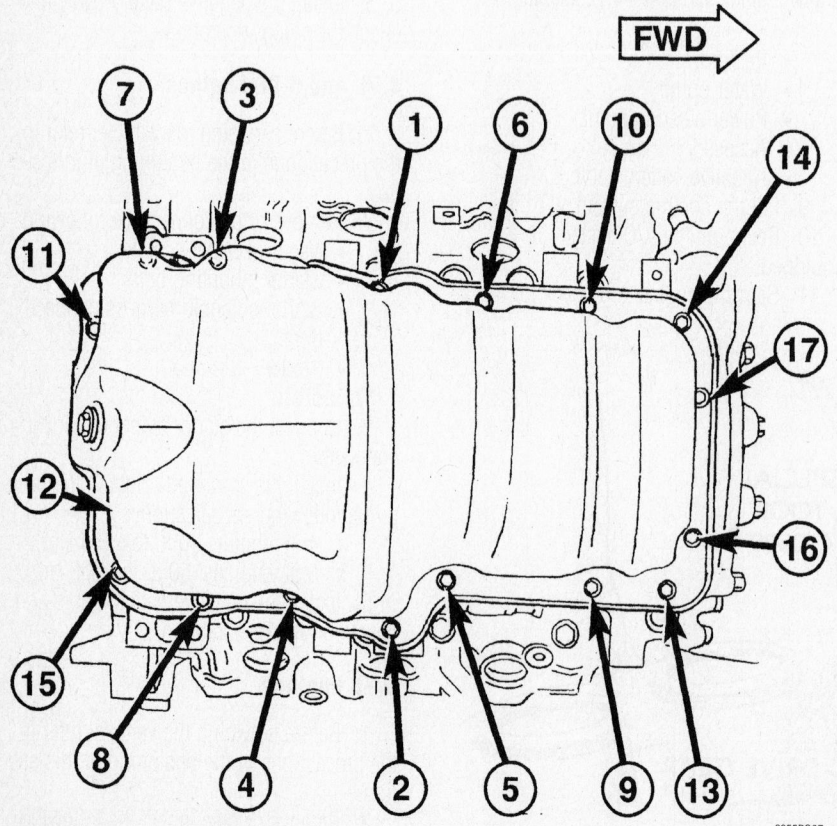

Oil pan bolt torque sequence—3.7L Engine

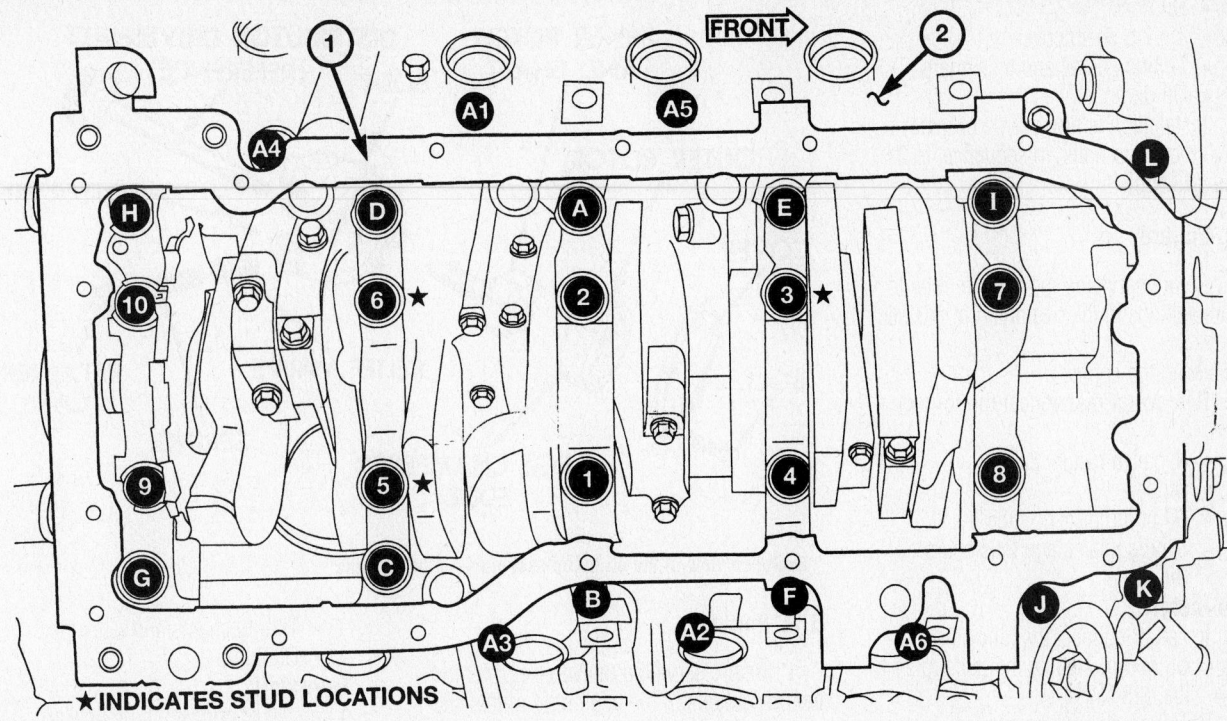

★ **INDICATES STUD LOCATIONS**

1 - BEDPLATE
2 - CYLINDER BLOCK

9355PG03

Bedplate bolt tightening sequence—4.7L engine

- Pan bolts and studs
- Pan

➡ **The double ended studs must be installed in their original locations.**

To install:

8. Install or connect the following:
   - Oil pan gasket
   - Oil pan. Tighten the bolts, in sequence, to 105 inch lbs. (12 Nm).
   - Front crossmember
   - Structural cover
   - Negative battery cable

9. Fill the crankcase to the proper level with engine oil.
10. Start the engine and check for leaks.

5.9L Engines

1. Before servicing the vehicle, refer to the precautions in the beginning of this section.
2. Drain the engine oil.
3. Remove or disconnect the following:
   - Oil filter
   - Starter motor
   - Cooler lines

- Oil level sensor connector
- Heated Oxygen (HO2S) sensor connector
- Exhaust Y-pipe
- Oil pan

To install:

4. Install or connect the following:
   - Oil pan, using a new gasket. Tighten the bolts to 18 ft. lbs. (24 Nm).
   - Exhaust Y-pipe
   - Heated Oxygen (HO2S) sensor connector
   - Oil level sensor connector
   - Cooler lines
   - Starter motor
   - Oil filter
5. Fill the crankcase.
6. Start the engine and check for leaks.

Oil Pump

REMOVAL & INSTALLATION

3.7L Engine

1. Before servicing the vehicle, refer to the precautions in the beginning of this section.
2. Remove or disconnect the following:

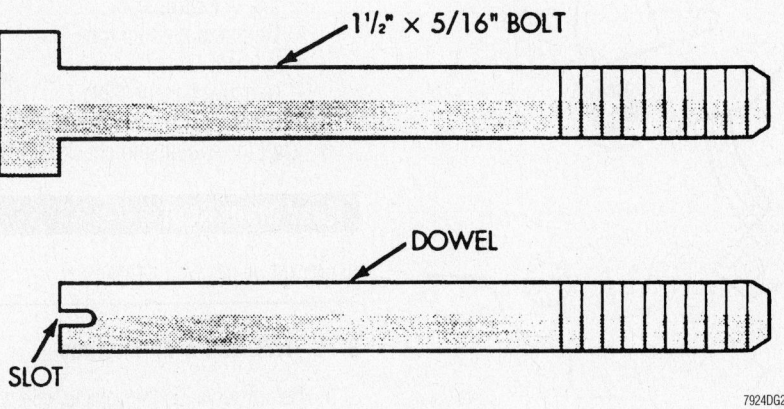

1½" × 5/16" BOLT

DOWEL

SLOT

7924DG20

Oil pan alignment dowels—5.9L engines

- Oil Pan
- Timing chain cover
- Timing chains and tensioners
- Oil pump

3. Installation is the reverse of removal. Torque the pump bolts, in sequence, to 21 ft. lbs. (28 Nm),

4.7L Engine

1. Before servicing the vehicle, refer to the precautions in the beginning of this section.
2. Drain the engine oil.
3. Remove or disconnect the following:

- Negative battery cable
- Oil pan
- Oil pump pick-up tube
- Timing chains and tensioners
- Oil pump

To install:

4. Install or connect the following:

- Oil pump. Tighten the bolts to 21 ft. lbs. (28 Nm).
- Timing chains and tensioners
- Oil pump pick-up tube
- Oil pan
- Negative battery cable

5. Fill the crankcase to the correct level.
6. Start the engine and check for leaks.

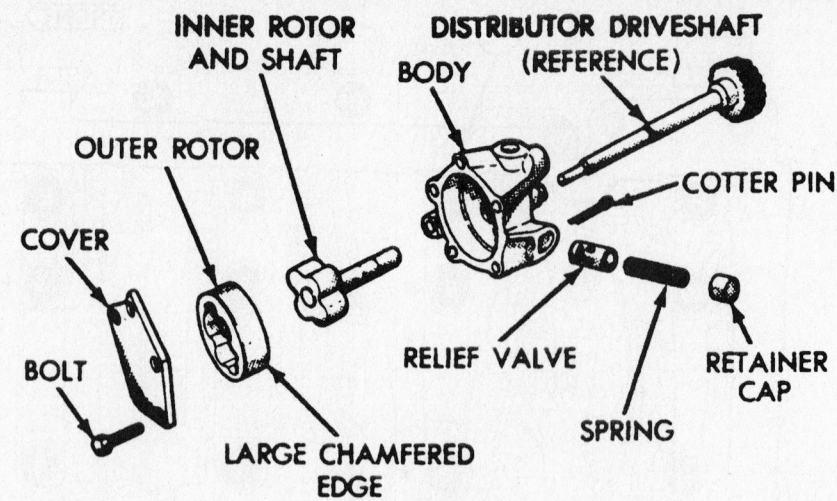

Exploded view of the oil pump assembly—5.9L engines

5.7L Engine

1. Before servicing the vehicle, refer to the precautions in the beginning of this section.
2. Drain the engine oil.
3. Remove or disconnect the following:

- Negative battery cable
- Oil pan
- Oil pump pick-up tube

- Timing chains and tensioners
- Oil pump

To install:

4. Install or connect the following:

- Oil pump. Tighten the bolts to 21 ft. lbs. (28 Nm).
- Timing chains and tensioners
- Oil pump pick-up tube
- Oil pan
- Negative battery cable

5. Fill the crankcase to the correct level.
6. Start the engine and check for leaks.

5.9L Engines

1. Before servicing the vehicle, refer to the precautions in the beginning of this section.
2. Drain the engine oil.
3. Remove or disconnect the following:

- Negative battery cable
- Oil pan
- Oil pump pick-up tube
- Oil pump

To install:

4. Install or connect the following:

- Oil pump. Tighten the bolts to 30 ft. lbs. (41 Nm).
- Oil pump pick-up tube
- Oil pan
- Negative battery cable

5. Fill the crankcase to the correct level.
6. Start the engine and check for leaks.

Rear Main Seal

REMOVAL & INSTALLATION

3.7L, 4.7L and 5.7L Engines

1. Before servicing the vehicle, refer to the precautions in the beginning of this section.

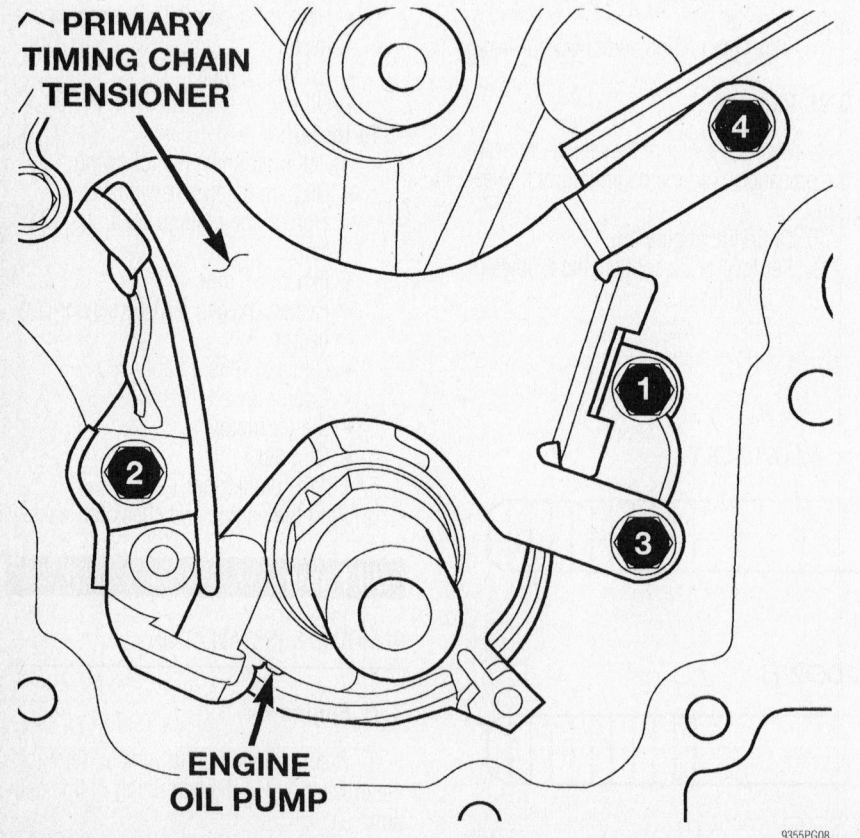

Oil pump bolt torque sequence—3.7L

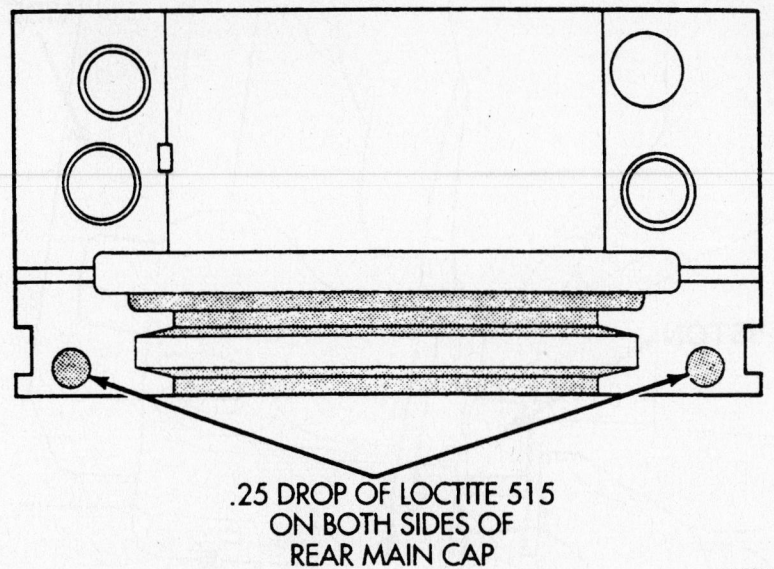

.25 DROP OF LOCTITE 515
ON BOTH SIDES OF
REAR MAIN CAP

7924DG24

Sealant application locations —5.9L engines

2. Remove or disconnect the following:
- Transmission
- Flexplate

3. Thread Oil Seal Remover 8506 into the rear main seal as far as possible and remove the rear main seal.

To install:

4. Install or connect the following:
- Seal Guide 8349-2 onto the crankshaft
- Rear main seal on the seal guide
- Rear main seal, using the Crankshaft Rear Oil Seal Installer 8349 and Driver Handle C-4171; tap it

into place until the installer is flush with the cylinder block
- Flexplate. Tighten the bolts to 45 ft. lbs. (60 Nm).
- Transmission

5. Start the engine and check for leaks.

5.9L Engines

1. Before servicing the vehicle, refer to the precautions in the beginning of this section.

2. Drain the engine oil.

3. Remove or disconnect the following:

- Oil pan
- Oil pump
- Rear main bearing cap

4. Loosen the other main bearing cap bolts for clearance and remove the rear main seal halfs.

To install:

5. Install or connect the following:
- New upper seal half to the cylinder block
- New lower seal half to the bearing cap

6. Apply sealant to the rear main bearing cap.

7. Install or connect the following:
- Rear main bearing cap. Tighten **all** main bearing cap bolts to 85 ft. lbs. (115 Nm).
- Oil pump and oil pan

8. Fill the engine.

9. Start the engine and check for leaks.

Timing Chain, Sprockets, Front Cover and Seal

REMOVAL & INSTALLATION

3.7L Engines

1. Before servicing the vehicle, refer to the precautions in the beginning of this section.

2. Drain the cooling system.

3. Remove or disconnect the following:
- Negative battery cable
- Valve covers
- Radiator fan

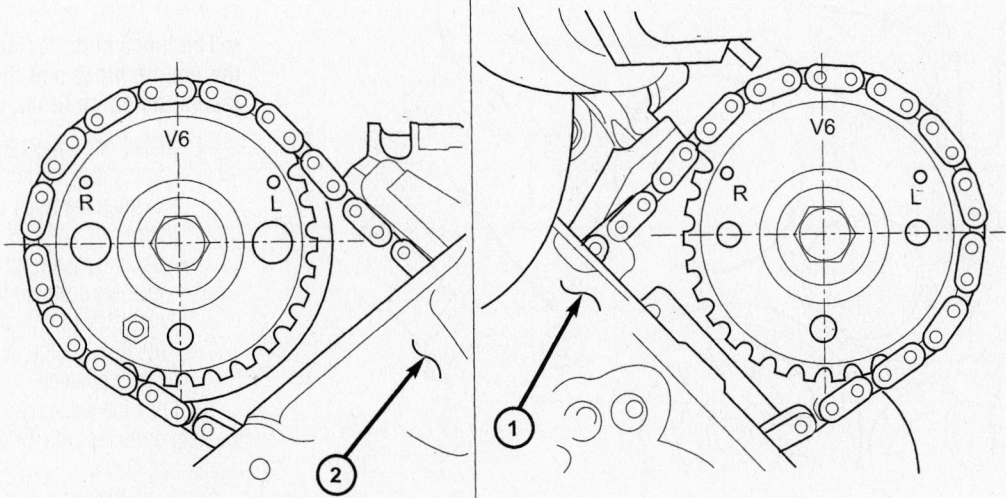

1 - LEFT CYLINDER HEAD
2 - RIGHT CYLINDER HEAD

9355PG09

Camshaft sprocket timing marks—3.7L

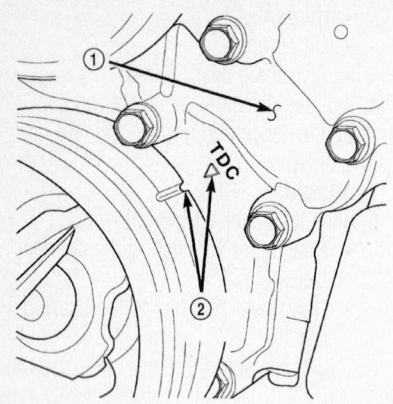

1 - TIMING CHAIN COVER
2 - CRANKSHAFT TIMING MARKS

9355PG10

Crankshaft timing marks—3.7L

4. Rotate the crankshaft so that the crankshaft timing mark aligns with the Top Dead Center (TDC) mark on the front cover, and the **V6** marks on the camshaft sprockets are at 12 o'clock.
  • Power steering pump
  • Access plugs from the cylinder heads
  • Oil fill housing
  • Crankshaft damper

5. Compress the primary timing chain tensioner and install a lockpin.

6. Remove the secondary timing chain tensioners.

7. Hold the left camshaft with adjustable pliers and remove the sprocket and chain. Rotate the **left** camshaft 15

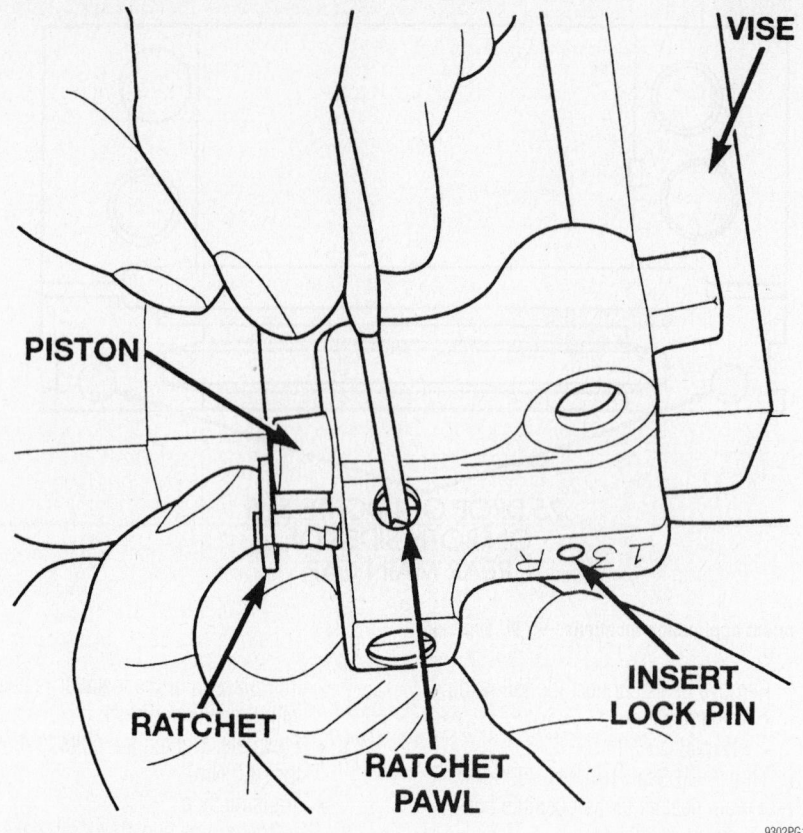

9302PG12

Secondary timing chain tensioner preparation—3.7L engine

degrees **clockwise** to the neutral position.

8. Hold the right camshaft with adjustable pliers and remove the camshaft sprocket. Rotate the **right** camshaft 45

degrees **counterclockwise** to the neutral position.

9. Remove the primary timing chain and sprockets.

To install:

10. Use a small prytool to hold the ratchet pawl and compress the secondary timing chain tensioners in a vise and install locking pins.

➡ **The black bolts fasten the guide to the engine block and the silver bolts fasten the guide to the cylinder head.**

11. Install or connect the following:
  • Secondary timing chain guides. Tighten the bolts to 21 ft. lbs. (28 Nm).
  • Secondary timing chains to the idler sprocket so that the double plated links on each chain are visible through the slots in the primary idler sprocket

12. Lock the secondary timing chains to the idler sprocket with Timing Chain Locking tool as shown.

13. Align the primary chain double plated links with the idler sprocket timing mark and the single plated link with the crankshaft sprocket timing mark.

14. Install the primary chain and sprock-

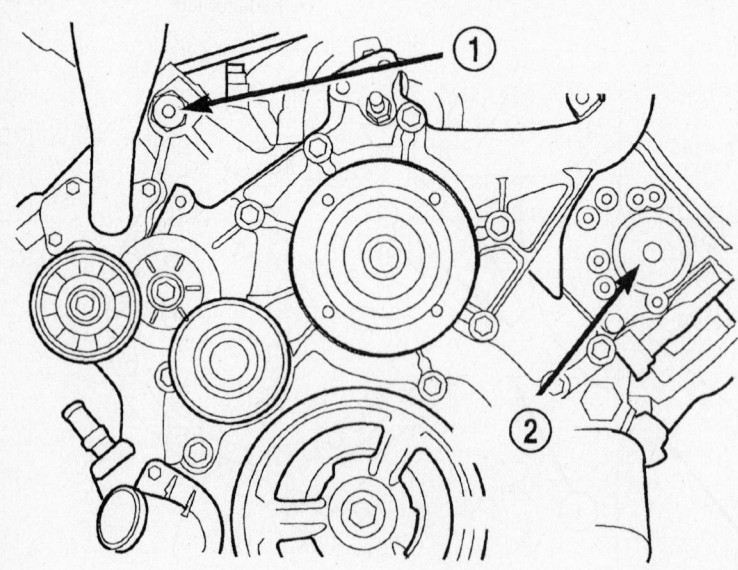

1 - RIGHT CYLINDER HEAD ACCESS PLUG
2 - LEFT CYLINDER HEAD ACCESS PLUG

9355PG11

Cylinder head access plugs—3.7L

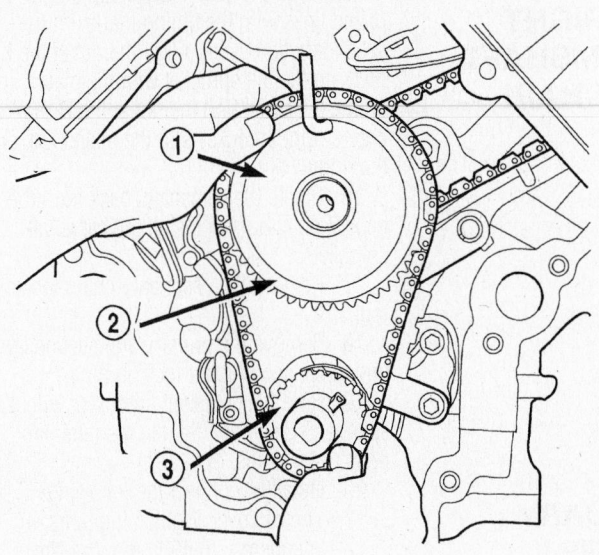

1 - SPECIAL TOOL 8429
2 - PRIMARY CHAIN IDLER SPROCKET
3 - CRANKSHAFT SPROCKET

9355PG12

Installing the idler gear and timing chains—3.7L

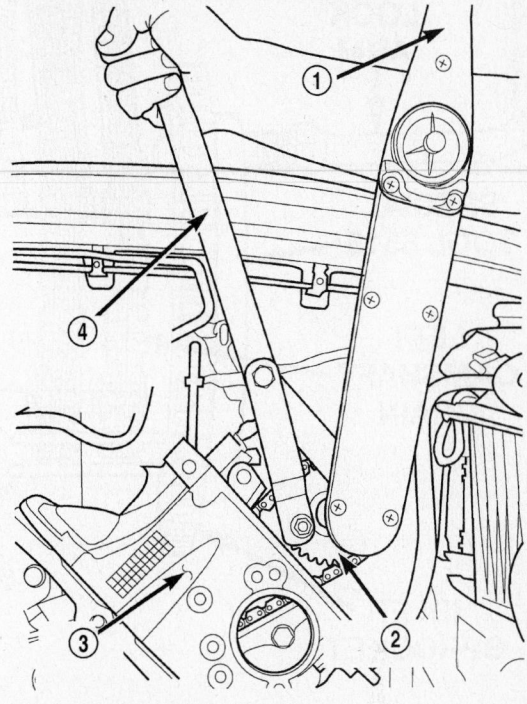

1 - TORQUE WRENCH
2 - CAMSHAFT SPROCKET
3 - LEFT CYLINDER HEAD
4 - SPECIAL TOOL 6958 SPANNER WITH ADAPTER PINS 8346

9355PG14

Tightening the left side camshaft sprocket—3.7L

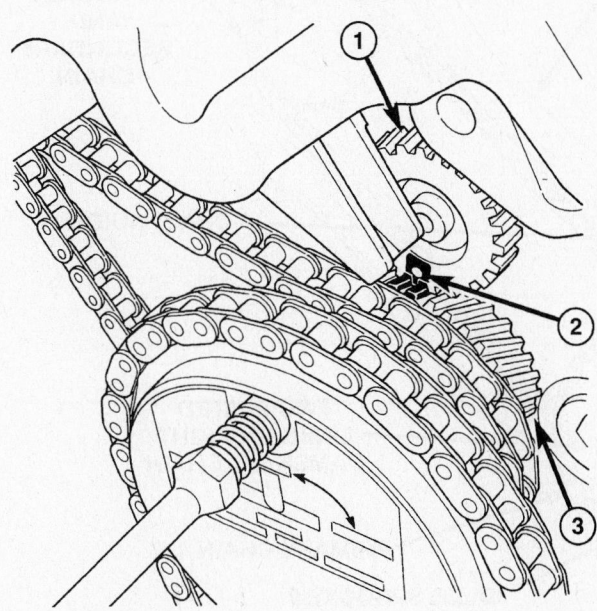

1 - COUNTERBALANCE SHAFT
2 - TIMING MARKS
3 - IDLER SPROCKET

9355PG13

Counterbalance shaft timing marks—3.7L

1 - TORQUE WRENCH
2 - SPECIAL TOOL 6958 WITH ADAPTER PINS 8346
3 - LEFT CAMSHAFT SPROCKET
4 - RIGHT CAMSHAFT SPROCKET

9355PG15

Tightening the right side camshaft sprocket—3.7L

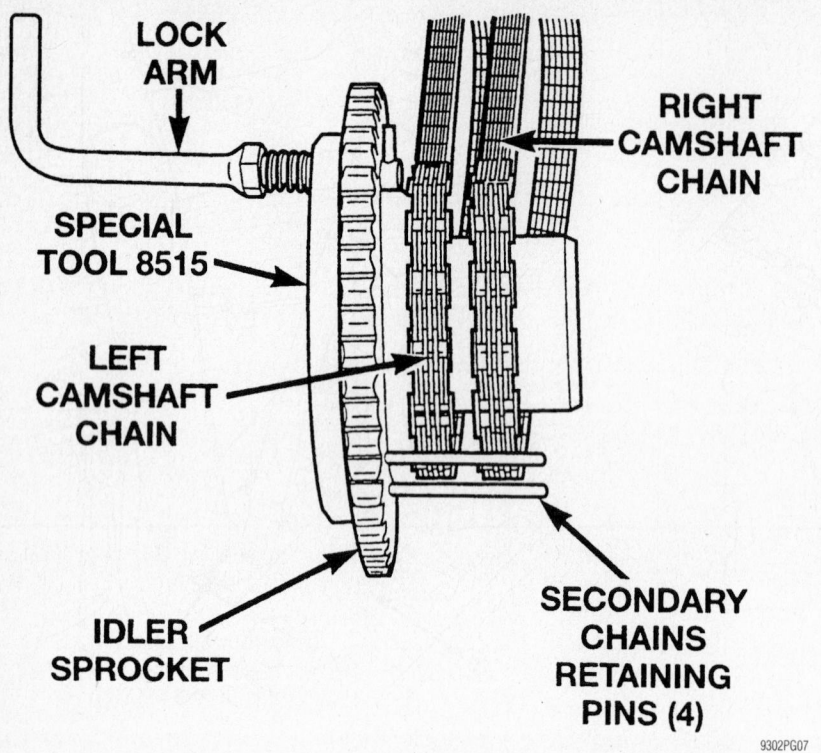

Use the Timing Chain Locking tool to lock the timing chains on the idler gear—3.7L engine

ets. Tighten the idler sprocket bolt to 25 ft. lbs. (34 Nm).

15. Align the secondary chain single plated links with the timing marks on the secondary sprockets. Align the dot at the **L** mark on the left sprocket with the plated link on the left chain and the dot at the **R** mark on the right sprocket with the plated link on the right chain.

16. Rotate the camshafts back from the neutral position and install the camshaft sprockets.

17. Remove the secondary chain locking tool.

18. Remove the primary and secondary timing chain tensioner locking pins.

19. Hold the camshaft sprockets with a spanner wrench and tighten the retaining bolts to 90 ft. lbs. (122 Nm).

20. Install or connect the following:
- Front cover. Tighten the bolts, in sequence, to 40 ft. lbs. (54 Nm).
- Front crankshaft seal
- Cylinder head access plugs
- A/C compressor
- Alternator
- Accessory drive belt tensioner.

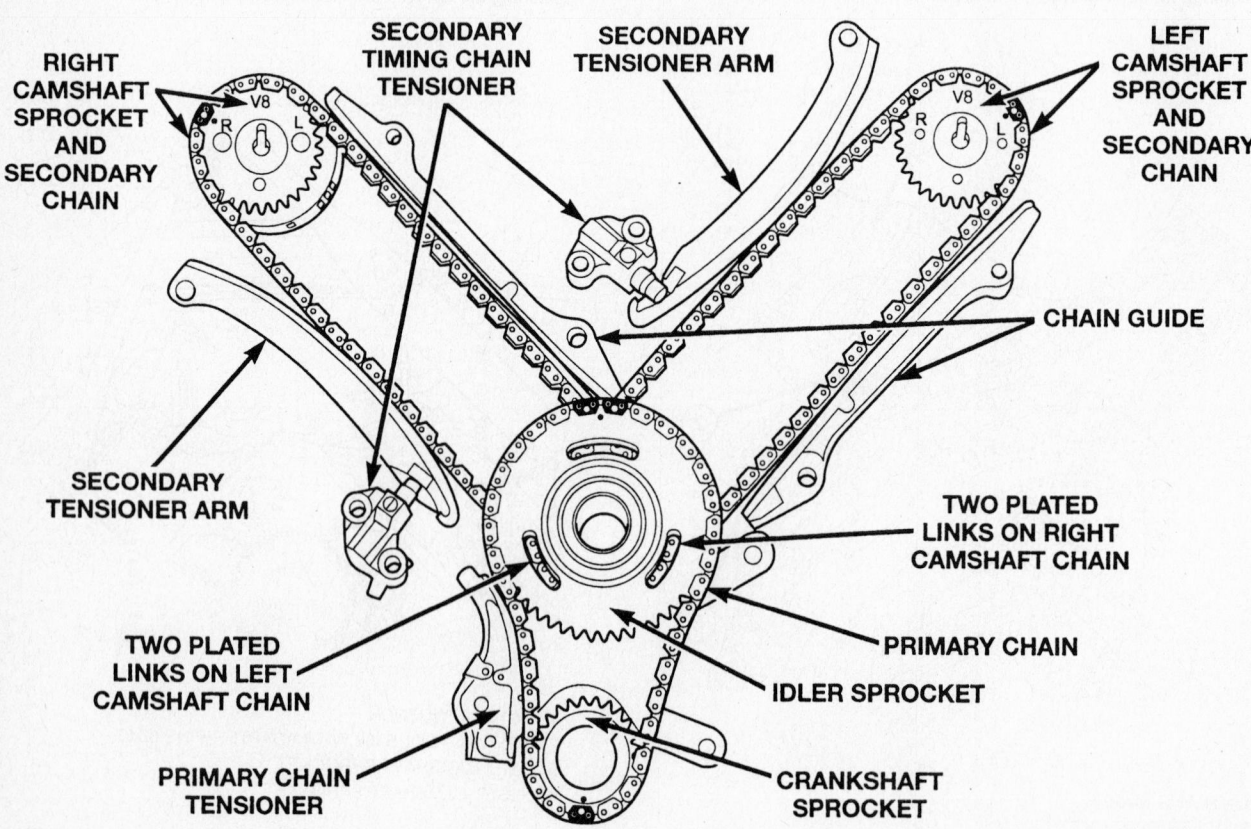

Timing chain system and alignment marks—4.7L engine shown; 3.7L engine similar

INDICATES STUD LOCATIONS

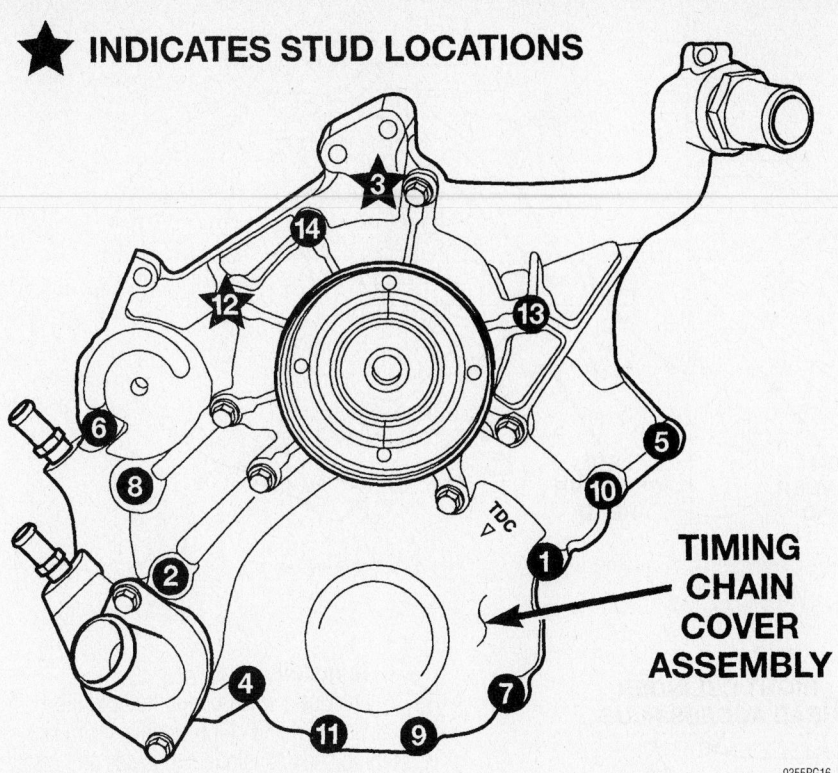

TIMING CHAIN COVER ASSEMBLY

9355PG16

Timing cover bolt torque sequence—3.7L

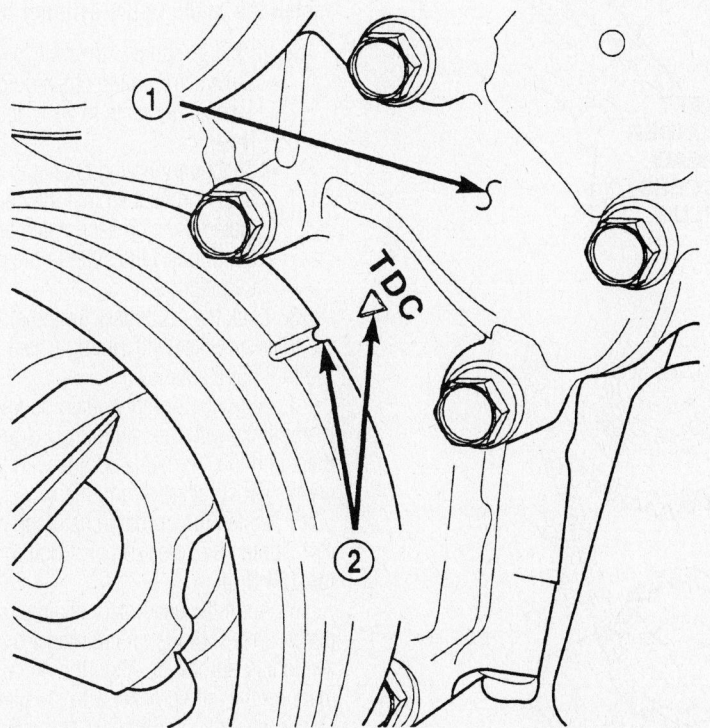

1 – TIMING CHAIN COVER
2 – CRANKSHAFT TIMING MARKS

9308PG04

Crankshaft timing marks—4.7L engine

Tighten the bolt to 40 ft. lbs. (54 Nm).
- Oil fill housing
- Crankshaft damper. Tighten the bolt to 130 ft. lbs. (175 Nm).
- Power steering pump
- Lower radiator hose
- Heater hoses
- Accessory drive belt
- Engine cooling fan and shroud
- Camshaft Position (CMP) sensor
- Valve covers
- Negative battery cable

21. Fill and bleed the cooling system.

22. Start the engine, check for leaks and repair if necessary.

4.7L Engine

1. Before servicing the vehicle, refer to the precautions in the beginning of this section.

2. Drain the cooling system.

3. Remove or disconnect the following:
- Negative battery cable
- Valve covers
- Camshaft Position (CMP) sensor
- Engine cooling fan and shroud
- Accessory drive belt
- Heater hoses
- Lower radiator hose
- Power steering pump

4. Rotate the crankshaft so that the crankshaft timing mark aligns with the Top Dead Center (TDC) mark on the front cover, and the **V8** marks on the camshaft sprockets are at 12 o'clock.

5. Remove or disconnect the following:
- Crankshaft damper
- Oil fill housing
- Accessory drive belt tensioner
- Alternator
- A/C compressor
- Front cover
- Front crankshaft seal
- Cylinder head access plugs
- Secondary timing chain guides

6. Compress the primary timing chain tensioner and install a lockpin.

7. Remove the secondary timing chain tensioners.

8. Hold the left camshaft with adjustable pliers and remove the sprocket and chain. Rotate the **left** camshaft 15 degrees **clockwise** to the neutral position.

9. Hold the right camshaft with adjustable pliers and remove the camshaft sprocket. Rotate the **right** camshaft 45 degrees **counterclockwise** to the neutral position.

10. Remove the primary timing chain and sprockets.

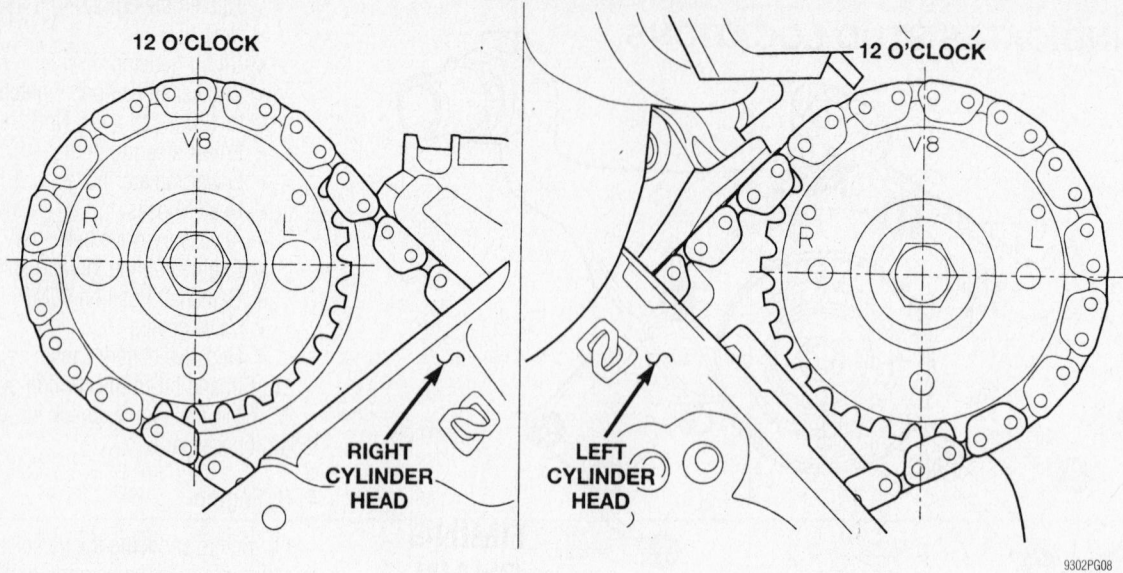

Camshaft positioning—4.7L engine

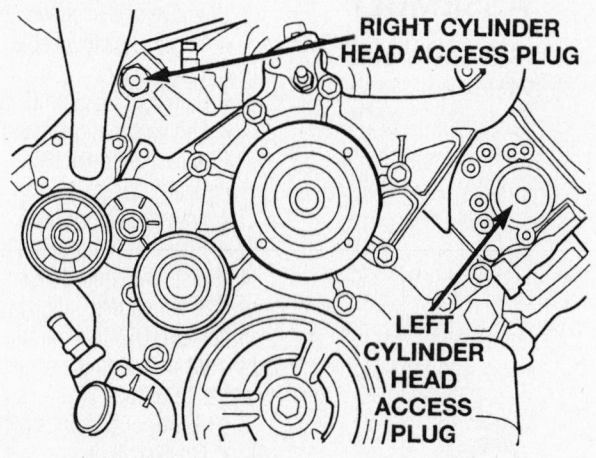

Cylinder head access plug locations—4.7L engine

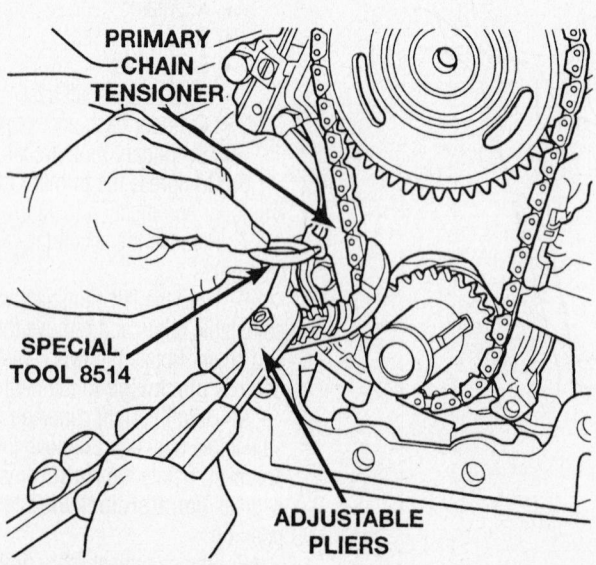

Compress and lock the primary chain tensioner—4.7L engine

To install:

11. Use a small prytool to hold the ratchet pawl and compress the secondary timing chain tensioners in a vise and install locking pins.

➡**The black bolts fasten the guide to the engine block and the silver bolts fasten the guide to the cylinder head.**

12. Install or connect the following:
- Secondary timing chain guides. Tighten the bolts to 21 ft. lbs. (28 Nm).
- Secondary timing chains to the idler sprocket so that the double plated links on each chain are visible through the slots in the primary idler sprocket

13. Lock the secondary timing chains to the idler sprocket with Timing Chain Locking tool 8515 as shown.

14. Align the primary chain double plated links with the idler sprocket timing mark and the single plated link with the crankshaft sprocket timing mark.

15. Install the primary chain and sprockets. Tighten the idler sprocket bolt to 25 ft. lbs. (34 Nm).

16. Align the secondary chain single plated links with the timing marks on the secondary sprockets. Align the dot at the **L** mark on the left sprocket with the plated link on the left chain and the dot at the **R** mark on the right sprocket with the plated link on the right chain.

17. Rotate the camshafts back from the neutral position and install the camshaft sprockets.

18. Remove the secondary chain locking tool.

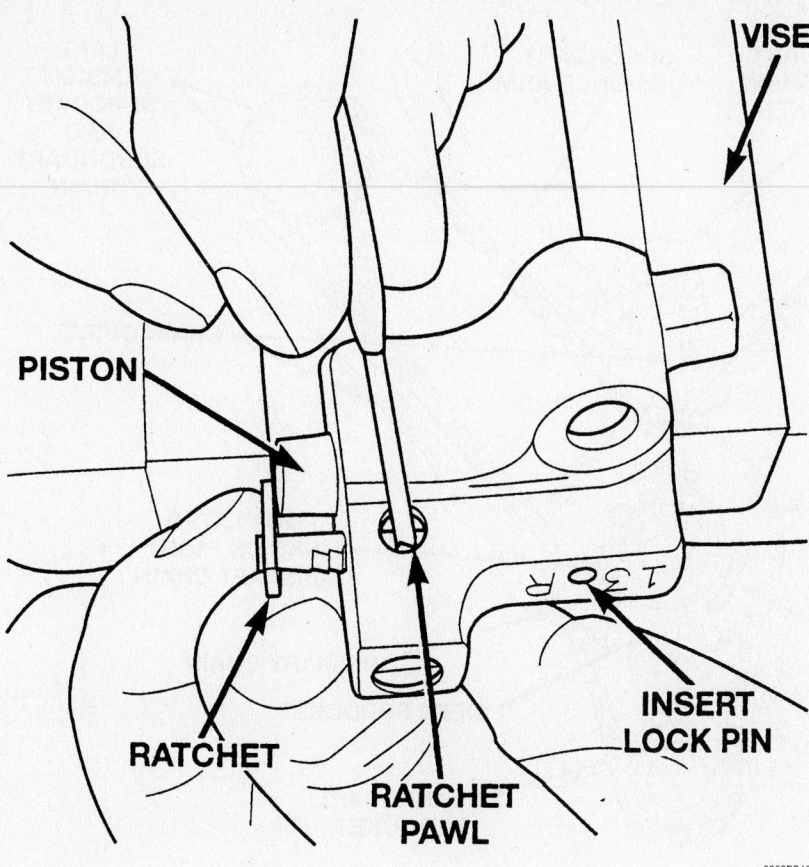

Secondary timing chain tensioner preparation—4.7L engine

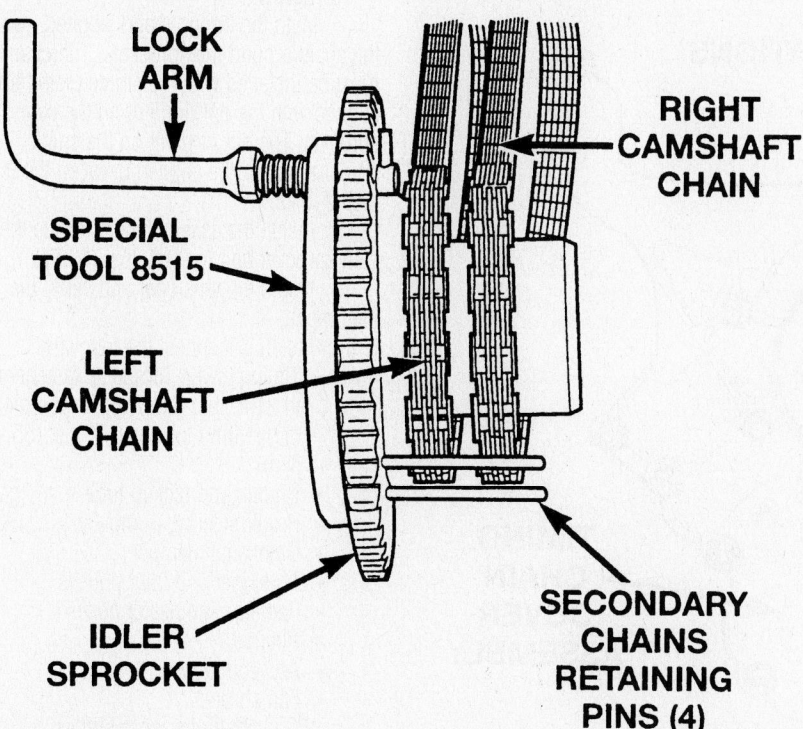

Use the Timing Chain Locking tool to lock the timing chains on the idler gear—4.7L engine

19. Remove the primary and secondary timing chain tensioner locking pins.

20. Hold the camshaft sprockets with a spanner wrench and tighten the retaining bolts to 90 ft. lbs. (122 Nm).

21. Install or connect the following:
- Front cover. Tighten the bolts, in sequence, to 40 ft. lbs. (54 Nm).
- Front crankshaft seal
- Cylinder head access plugs
- A/C compressor
- Alternator
- Accessory drive belt tensioner. Tighten the bolt to 40 ft. lbs. (54 Nm).
- Oil fill housing
- Crankshaft damper. Tighten the bolt to 130 ft. lbs. (175 Nm).
- Power steering pump
- Lower radiator hose
- Heater hoses
- Accessory drive belt
- Engine cooling fan and shroud
- Camshaft Position (CMP) sensor
- Valve covers
- Negative battery cable

22. Fill the cooling system.

23. Start the engine and check for leaks.

5.7L Engines

1. Before servicing the vehicle, refer to the precautions in the beginning of this section.

2. Drain the cooling system.

3. Remove or disconnect the following:
- Negative battery cable
- Drive belt
- Radiator fan
- Coolant and washer bottles
- Fan shroud
- A/C compressor
- Alternator
- Radiator and heater hoses
- Tensioner and idler pulleys
- Crankshaft damper
- Power steering pump
- Oil pan and pickup tube
- Timing cover
- Re-install the damper

4. Rotate the crankshaft so that the camshaft sprocket and crankshaft sprocket timing marks are aligned.

➡The camshaft pin and slot in the cam sprocket must be a 12 o'clock, the crankshaft keyway must be at 2 o'clock, and the dots or paint on the crank sprocket must be at 6 o'clock.

5. Pin back the tensioner shoe.

6. Remove the timing chain and sprockets.

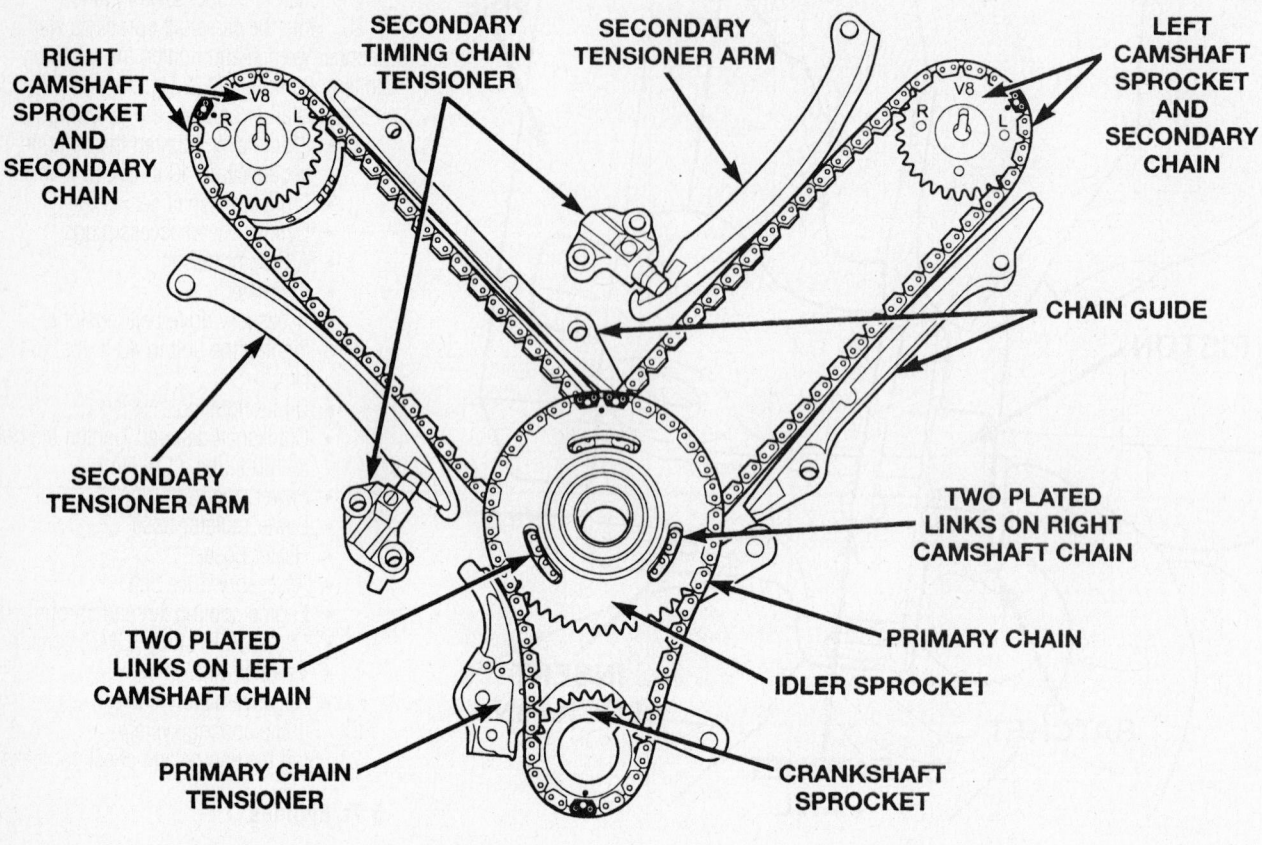

RIGHT CAMSHAFT SPROCKET AND SECONDARY CHAIN

SECONDARY TIMING CHAIN TENSIONER

SECONDARY TENSIONER ARM

LEFT CAMSHAFT SPROCKET AND SECONDARY CHAIN

CHAIN GUIDE

SECONDARY TENSIONER ARM

TWO PLATED LINKS ON RIGHT CAMSHAFT CHAIN

TWO PLATED LINKS ON LEFT CAMSHAFT CHAIN

PRIMARY CHAIN

IDLER SPROCKET

PRIMARY CHAIN TENSIONER

CRANKSHAFT SPROCKET

9302PG24

Timing chain system and alignment marks—4.7L engine

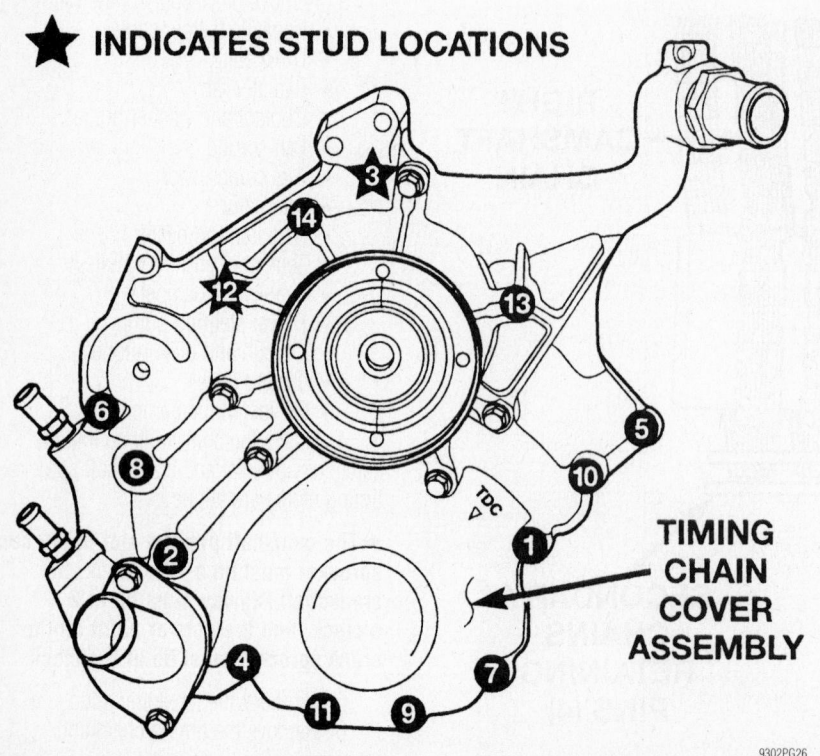

★ INDICATES STUD LOCATIONS

TIMING CHAIN COVER ASSEMBLY

9302PG26

Timing chain cover bolt torque sequence—4.7L engine

To install:

7. With the timing marks aligned, wrap the chain around the sprockets. The chain must be installed with the single plated link aligned with the dot or paint on the cam sprocket. The dot or paint on the crank sprocket should be aligned between the 2 plated links.

8. Install the assembly and torque the cam sprocket bolt to 90 ft. lbs. (122 Nm).

9. Unpin the tensioner and verify the alignment.

10. Install or connect the following:
- Timing cover. Torque all fasteners to 21 ft. lbs. (28 Nm). Torque the large lifting lug to 40 ft. lbs. (55 Nm)
- Oil pan and pickup tube
- Power steering pump
- Crankshaft damper
- Tensioner and idler pulleys
- Radiator and heater hoses
- Alternator
- A/C compressor
- Fan shroud
- Coolant and washer bottles
- Radiator fan

POSITIONING

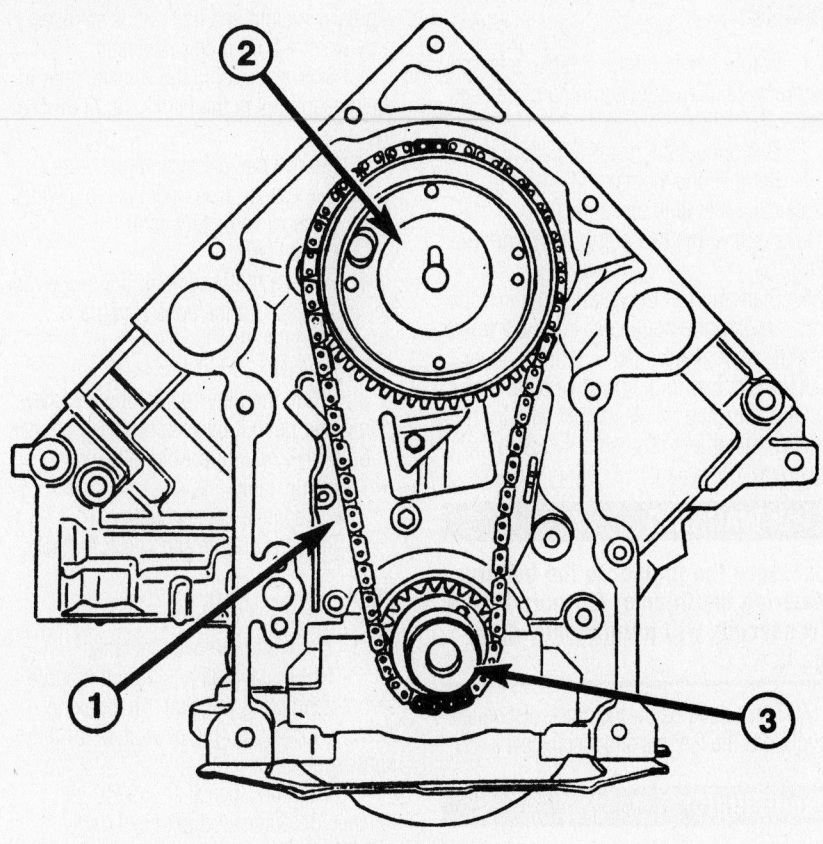

1 - Chain Tensioner
2 - Camshaft Sprocket
3 - Crankshaft Sprocket

2399PG01

Timing mark alignment—5.7L engine

- Drive belt
- Negative battery cable

5.9L Engines

1. Before servicing the vehicle, refer to the precautions in the beginning of this section.
2. Drain the cooling system.
3. Remove or disconnect the following:
   - Negative battery cable
   - Accessory drive belt
   - Cooling fan and shroud
   - Water pump
   - Power steering pump
   - Crankshaft damper
   - Front crankshaft seal
   - Front cover
4. Rotate the crankshaft so that the camshaft sprocket and crankshaft sprocket timing marks are aligned.

5. Remove the timing chain and sprockets.

To install:

6. Install the timing chain and sprockets with the timing marks aligned. Tighten the camshaft sprocket bolt to 50 ft. lbs. (68 Nm).
7. Install or connect the following:
   - Front cover. Tighten the cover bolts to 30 ft. lbs. (41 Nm) and the oil pan bolts to 18 ft. lbs. (24 Nm).
   - Front crankshaft seal
   - Crankshaft damper. Tighten the bolt to 135 ft. lbs. (183 Nm).
   - Power steering pump
   - Water pump
   - Cooling fan and shroud
   - Accessory drive belt
   - Negative battery cable
8. Fill the cooling system.
9. Start the engine and check for leaks.

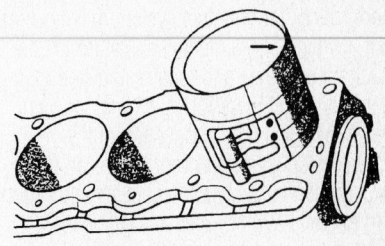

Piston to engine positioning—5.9L engines

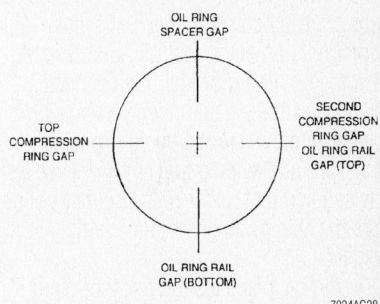

Piston ring end-gap spacing—5.9L engines

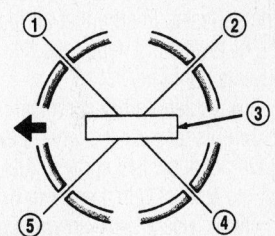

1 - SIDE RAIL UPPER
2 - NO. 1 RING GAP
3 - PISTON PIN
4 - SIDE RAIL LOWER
5 - NO. 2 RING GAP AND SPACER EXPANDER GAP

2399PG02

Piston ring end-gap spacing—5.7L engines

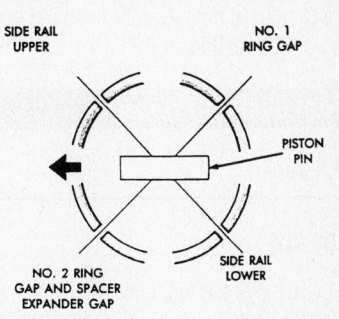

Piston ring end-gap spacing. Position raised "F" on piston towards front of engine—3.7L and 4.7L engine

FUEL SYSTEM

Fuel System Service Precautions

Safety is the most important factor when performing not only fuel system maintenance but any type of maintenance. Failure to conduct maintenance and repairs in a safe manner may result in serious personal injury or death. Maintenance and testing of the vehicle's fuel system components can be accomplished safely and effectively by adhering to the following rules and guidelines.

• To avoid the possibility of fire and personal injury, always disconnect the negative battery cable unless the repair or test procedure requires that battery voltage be applied.

• Always relieve the fuel system pressure prior to detaching any fuel system component (injector, fuel rail, pressure regulator, etc.), fitting or fuel line connection. Exercise extreme caution whenever relieving fuel system pressure to avoid exposing skin, face and eyes to fuel spray. Please be advised that fuel under pressure may penetrate the skin or any part of the body that it contacts.

• Always place a shop towel or cloth around the fitting or connection prior to loosening to absorb any excess fuel due to spillage. Ensure that all fuel spillage (should it occur) is quickly removed from engine surfaces. Ensure that all fuel soaked cloths or towels are deposited into a suitable waste container.

• Always keep a dry chemical (Class B) fire extinguisher near the work area.

• Do not allow fuel spray or fuel vapors to come into contact with a spark or open flame.

• Always use a back-up wrench when loosening and tightening fuel line connection fittings. This will prevent unnecessary stress and torsion to fuel line piping.

• Always replace worn fuel fitting O-rings with new. Do not substitute fuel hose or equivalent, where fuel pipe is installed.

Before servicing the vehicle, also make sure to refer to the precautions in the beginning of this section as well.

Fuel System Pressure

RELIEVING

2001–02

1. Remove the fuel filler cap.
2. Remove the fuel pump relay from the power distribution center.
3. Run the engine until it stalls.
4. Turn the key to **OFF**.

2003–04

1. Before servicing the vehicle, refer to the precautions in the beginning of this section.
2. Disconnect the negative battery cable.
3. Remove the fuel tank filler cap to release any fuel tank pressure.
4. Remove the fuel pump relay from the PDC.
5. Start and run the engine until it stops.
6. Unplug the connector from any injector and connect a jumper wire from either injector terminal to the positive battery terminal. Connect another jumper wire to the other terminal and momentarily touch the other end to the negative battery terminal.

❋❋ WARNING

Just touch the jumper to the battery. Powering the injector for more than a few seconds will permanently damage it.

7. Place a rag below the quick-disconnect coupling at the fuel rail and disconnect it.

Fuel Filter

REMOVAL & INSTALLATION

2001–03

These engines are equipped with a fuel filter/pressure regulator. This unit is not a regularly maintained unit and is serviced only when a DTC indicates a fault.

1. Before servicing the vehicle, refer to the precautions in the beginning of this section.
2. Relieve the fuel system pressure.
3. Remove or disconnect the following:
   • Negative battery cable
   • Fuel tank
4. Pull the filter/regulator out of the rubber grommet. Cut the hose clamp and remove the fuel line.
   To install:
5. Install the filter/regulator with a new clamp and push it into the rubber grommet.
6. Install or connect the following:
   • Fuel tank
   • Negative battery cable
7. Start the engine and check for leaks.

2004

1. Drain and remove fuel tank.
2. Note rotational position of module before attempting removal. An indexing arrow is located on top of module for this purpose.
3. Position Special Tool 9340 into notches on outside edge of lockring.
4. Install ½ inch drive breaker bar to tool 9340.
5. Rotate breaker bar counter-clockwise to remove lockring.
6. Remove lockring. The module will spring up slightly when lockring is removed.

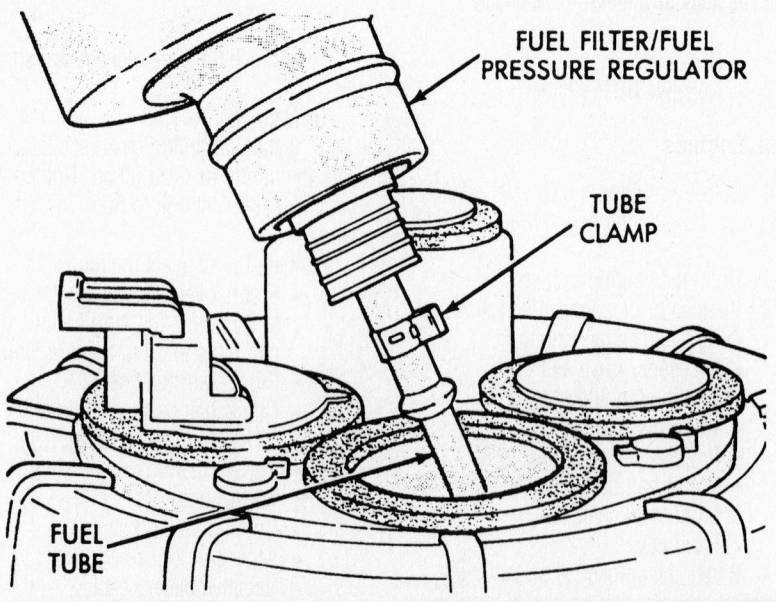

FUEL FILTER/FUEL PRESSURE REGULATOR

TUBE CLAMP

FUEL TUBE

Pull and twist the filter/regulator to remove it from the top of the fuel pump module

7924DG35

7. Remove module from fuel tank. Be careful not to bend float arm while removing.

To install:

⁂ CAUTION

Whenever the fuel pump module is serviced, the rubber seal (gasket) must be replaced.

8. Using a new seal (gasket), position fuel pump module into opening in fuel tank.

9. Position lockring over top of fuel pump module.

10. Rotate module until embossed alignment arrow points to center alignment mark. This step must be performed to prevent float from contacting side of fuel tank. Also be sure fuel fitting on top of pump module is pointed to driver's side of vehicle.

11. Install Special Tool 9340 to lockring.

12. Install ½ inch drive breaker (1) into Special Tool 9340.

13. Tighten lockring (clockwise) until all seven notches have engaged.

14. Install fuel tank.

Fuel Pump

REMOVAL & INSTALLATION

2001–03

1. Before servicing the vehicle, refer to the precautions in the beginning of this section.

2. Relieve the fuel system pressure.

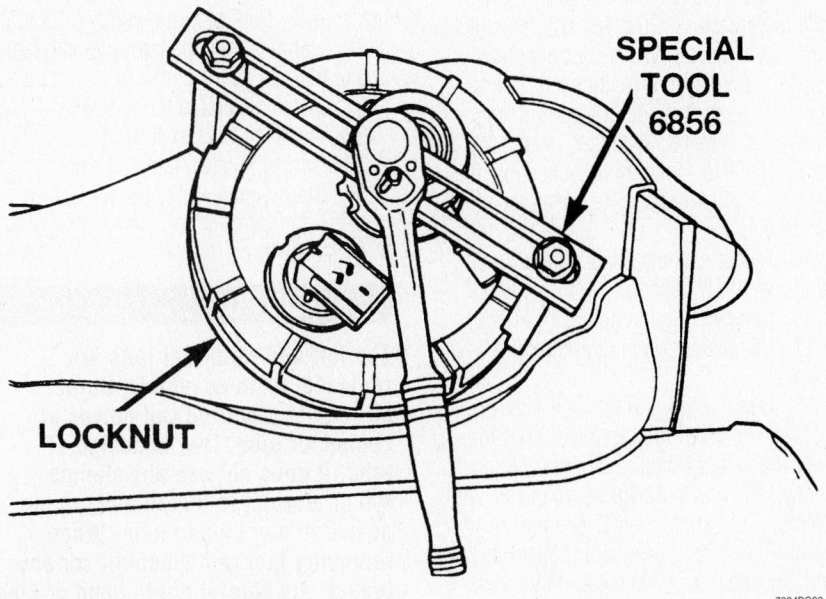

LOCKNUT

SPECIAL TOOL 6856

Fuel pump module locknut removal

3. Remove or disconnect the following:
- Negative battery cable
- Fuel pump module harness connector
- Fuel line
- Fuel tank
- Fuel pump module locknut
- Fuel pump module

To install:

4. Install or connect the following:
- Fuel pump module
- Fuel pump module locknut
- Fuel tank
- Fuel line
- Fuel pump module harness connector
- Negative battery cable

5. Start the engine and check for leaks.

2004

1. Drain and remove fuel tank.

2. Note rotational position of module before attempting removal. An indexing arrow is located on top of module for this purpose.

3. Position Special Tool 9340 into notches on outside edge of lockring.

4. Install ½ inch drive breaker bar to tool 9340.

5. Rotate breaker bar counter-clockwise to remove lockring.

6. Remove lockring. The module will spring up slightly when lockring is removed.

7. Remove module from fuel tank. Be careful not to bend float arm while removing.

To install:

⁂ CAUTION

Whenever the fuel pump module is serviced, the rubber seal (gasket) must be replaced.

8. Using a new seal (gasket), position fuel pump module into opening in fuel tank.

9. Position lockring over top of fuel pump module.

10. Rotate module until embossed alignment arrow points to center alignment mark. This step must be performed to prevent float from contacting side of fuel tank. Also be sure fuel fitting on top of pump module is pointed to driver's side of vehicle.

11. Install Special Tool 9340 to lockring.

12. Install ½ inch drive breaker (1) into Special Tool 9340.

13. Tighten lockring (clockwise) until all seven notches have engaged.

14. Install fuel tank.

Fuel Injector

REMOVAL & INSTALLATION

3.7L Engine

⁂ CAUTION

The left and right fuel rails are replaced as an assembly. Do not attempt to separate rail halves at connector tubes. Due to design of tubes, it does not use any clamps. Never attempt to install a clamping device of any kind to tubes. When removing fuel rail assembly for any reason, be careful not to bend or kink tubes.

1. Remove fuel tank filler tube cap.

2. Perform Fuel System Pressure Release Procedure.

3. Remove negative battery cable at battery.

4. Remove air duct at throttle body air box.

5. Remove air box at throttle body.

6. Remove air resonator mounting bracket at front of throttle body (2 bolts).

7. Disconnect fuel line latch clip and fuel line at fuel rail. A special tool will be necessary for fuel line disconnection.

8. Remove necessary vacuum lines at throttle body.

9. Disconnect electrical connectors at

all 6 fuel injectors. Push red colored slider away from injector. While pushing slider, depress tab and remove connector from injector. The factory fuel injection wiring harness is numerically tagged (INJ 1, INJ 2, etc.) for injector position identification. If harness is not tagged, note wiring location before removal.

10. Disconnect electrical connectors at all throttle body sensors.

11. Remove 6 ignition coils.

12. Remove four fuel rail mounting bolts.

13. Gently rock and pull left side of fuel rail until fuel injectors just start to clear machined holes in cylinder head. Gently rock and pull right side of rail until injectors just start to clear cylinder head

holes. Repeat this procedure (left/right) until all injectors have cleared cylinder head holes.

14. Remove fuel rail (with injectors attached) from engine.

15. Disconnect clip(s) that retain fuel injector(s) to fuel rail.

To install:

16. Install fuel injector(s) into fuel rail assembly and install retaining clip(s).

17. If same injector(s) is being reinstalled, install new o-ring(s).

18. Apply a small amount of clean engine oil to each injector o-ring. This will aid in installation.

19. Clean out fuel injector machined bores in intake manifold.

20. Apply a small amount of engine oil to each fuel injector o-ring. This will help in fuel rail installation.

21. Position fuel rail/fuel injector assembly to machined injector openings in cylinder head.

22. Guide each injector into cylinder head. Be careful not to tear injector o-rings.

23. Push right side of fuel rail down until fuel injectors have bottomed on cylinder head shoulder. Push left fuel rail down until injectors have bottomed on cylinder head shoulder.

24. Install 4 fuel rail mounting bolts and tighten.

25. Install 6 ignition coils.

26. Connect electrical connectors to throttle body.

27. Connect electrical connectors at all fuel injectors. Refer to graphic. Push connector onto injector and then push and lock red colored slider. Verify connector is locked to injector by lightly tugging on connector.

28. Connect necessary vacuum lines to throttle body.

29. Install air resonator mounting bracket near front of throttle body (2 bolts).

30. Connect fuel line latch clip and fuel line to fuel rail.

31. Install air box to throttle body.

32. Install air duct to air box.

33. Connect battery cable to battery.

34. Start engine and check for leaks.

4.7L Engine

✳✳ CAUTION

The left and right fuel rails are replaced as an assembly. Do not attempt to separate rail halves at connector tubes. Due to design of tubes, it does not use any clamps. Never attempt to install a clamping device of any kind to tubes. When removing fuel rail assembly for any reason, be careful not to bend or kink tubes.

1. Remove fuel tank filler tube cap.

2. Perform Fuel System Pressure Release Procedure.

3. Remove negative battery cable at battery.

4. Remove air duct at throttle body air box.

5. Remove air box at throttle body.

6. Remove air resonator mounting bracket at front of throttle body (2 bolts).

7. Disconnect fuel line latch clip and fuel line at fuel rail. A special tool will be necessary for fuel line disconnection.

8. Remove necessary vacuum lines at throttle body.

9. Disconnect electrical connectors at all 8 fuel injectors. Push red colored slider away from injector. While pushing slider, depress tab and remove connector from injector. The factory fuel injection wiring harness is numerically tagged (INJ 1, INJ 2, etc.) for injector position identification. If harness is not tagged, note wiring location before removal.

10. Disconnect electrical connectors at all throttle body sensors.

11. Remove 8 ignition coils.

12. Remove four fuel rail mounting bolts.

13. Gently rock and pull left side of fuel rail until fuel injectors just start to clear machined holes in cylinder head. Gently rock and pull right side of rail until injectors just start to clear cylinder head holes. Repeat this procedure (left/right) until all injectors have cleared cylinder head holes.

14. Remove fuel rail (with injectors attached) from engine.

15. Disconnect clip(s) that retain fuel injector(s) to fuel rail.

To install:

16. Install fuel injector(s) into fuel rail assembly and install retaining clip(s).

17. If same injector(s) is being reinstalled, install new o-ring(s).

18. Apply a small amount of clean engine oil to each injector o-ring. This will aid in installation.

19. Clean out fuel injector machined bores in intake manifold.

20. Apply a small amount of engine oil to each fuel injector o-ring. This will help in fuel rail installation.

21. Position fuel rail/fuel injector assembly to machined injector openings in cylinder head.

22. Guide each injector into cylinder head. Be careful not to tear injector o-rings.

23. Push right side of fuel rail down until fuel injectors have bottomed on cylinder head shoulder. Push left fuel rail down until injectors have bottomed on cylinder head shoulder.

24. Install 4 fuel rail mounting bolts and tighten.

25. Install 8 ignition coils.

26. Connect electrical connectors to throttle body.

27. Connect electrical connectors at all fuel injectors. Refer to graphic. Push connector onto injector and then push and lock red colored slider. Verify connector is locked to injector by lightly tugging on connector.

28. Connect necessary vacuum lines to throttle body.

29. Install air resonator mounting bracket near front of throttle body (2 bolts).

30. Connect fuel line latch clip and fuel line to fuel rail.

31. Install air box to throttle body.

32. Install air duct to air box.

33. Connect battery cable to battery.

34. Start engine and check for leaks.

5.7L Engine

✳✳ CAUTION

The left and right fuel rails are replaced as an assembly. Do not attempt to separate rail halves at connector tube. Due to design of tube, it does not use any clamps. Never attempt to install a clamping device of any kind to tube. When removing fuel rail assembly for any reason, be careful not to bend or kink tube.

1. Remove fuel tank filler tube cap.
2. Perform Fuel System Pressure Release Procedure.
3. Remove negative battery cable at battery.
4. Remove flex tube (air cleaner housing to engine).
5. Remove air resonator box at throttle body.
6. Disconnect all spark plug cables from all spark plugs and ignition coils. Do not remove cables from cable routing tray. Note original cable positions while removing.
7. Remove spark plug cable tray from engine by releasing 4 retaining clips. Remove tray and cables from engine as an assembly.
8. Disconnect electrical connectors at all 8 ignition coils.
9. Disconnect fuel line latch clip and fuel line at fuel rail. A special tool will be necessary for fuel line disconnection.
10. Disconnect electrical connectors at all 8 fuel injectors. Refer to graphic. Push red colored slider away from injector. While pushing slider, depress tab and remove connector from injector. The factory fuel injection wiring harness is numerically tagged (INJ 1, INJ 2, etc.) for injector position identification. If harness is not tagged, note wiring location before removal.
11. Disconnect electrical connectors at all throttle body sensors.
12. Remove four fuel rail mounting bolts (2) and hold-down clamps.
13. Gently rock and pull left side of fuel rail until fuel injectors just start to clear machined holes in intake manifold. Gently rock and pull right side of rail until injectors just start to clear intake manifold head

holes. Repeat this procedure (left/right) until all injectors have cleared machined holes.
14. Remove fuel rail (with injectors attached) from engine.
15. Disconnect clip(s) that retain fuel injector(s) to fuel rail (2).

To install:

16. Install fuel injector(s) into fuel rail assembly and install retaining clip(s).
17. If same injector(s) is being reinstalled, install new o-ring(s).
18. Clean out fuel injector machined bores in intake manifold.
19. Apply a small amount of engine oil to each fuel injector o-ring. This will help in fuel rail installation.
20. Position fuel rail/fuel injector assembly to machined injector openings in intake manifold.
21. Guide each injector into intake manifold. Be careful not to tear injector o-rings.
22. Push right side of fuel rail down until fuel injectors have bottomed on shoulders. Push left fuel rail down until injectors have bottomed on shoulders.
23. Install 4 fuel rail hold-down clamps and 4 mounting bolts.
24. Position spark plug cable tray and cable assembly to intake manifold. Snap 4 cable tray retaining clips into intake manifold.
25. Install all cables to spark plugs and ignition coils.
26. Connect electrical connector to throttle body.
27. Install electrical connectors to all 8 ignition coils.
28. Connect electrical connector to throttle body.
29. Connect electrical connectors at all

fuel injectors. Refer to graphic. Push connector onto injector and then push and lock red colored slider. Verify connector is locked to injector by lightly tugging on connector.
30. Connect fuel line latch clip and fuel line to fuel rail.
31. Install air resonator to throttle body (2 bolts).
32. Install flexible air duct to air box.
33. Connect battery cable to battery.
34. Start engine and check for leaks.

5.9L Engines

1. Before servicing the vehicle, refer to the precautions in the beginning of this section.
2. Relieve fuel system pressure.
3. Remove or disconnect the following:
   - Negative battery cable
   - Air intake tube
   - Throttle body
   - A/C compressor bracket
   - Fuel injector connectors
   - Fuel line
   - Fuel supply manifold with injectors
   - Fuel injectors

To install:

4. Install or connect the following:
   - Fuel injectors with new O-ring seals
   - Fuel supply manifold with injectors. Tighten the bolts to 17 ft. lbs. (23 Nm).
   - Fuel line
   - Fuel injector connectors
   - A/C compressor bracket
   - Throttle body
   - Air intake tube
   - Negative battery cable
5. Start the engine and check for leaks.

DRIVE TRAIN

Transmission Assembly

REMOVAL & INSTALLATION

2001–03

1. Before servicing the vehicle, refer to the precautions in the beginning of this section.

2. Remove or disconnect the following:
   - Negative battery cable
   - Rear driveshaft
   - Crankshaft Position (CKP) sensor
   - Exhaust front pipe
   - Transmission braces, if equipped
   - Starter motor
   - Transmission oil cooler lines
   - Torque converter access cover
   - Torque converter
   - Transmission oil dipstick tube
   - Vehicle Speed (VSS) sensor connector
   - Park/Neutral switch connector
   - Shift cable
   - Throttle valve cable
   - Transmission mount and crossmember. Support the transmission.
   - Front driveshaft and transfer case, if equipped
   - Transmission flange bolts
   - Transmission

To install:

3. Install or connect the following:
   - Transmission. Tighten the flange bolts to 65 ft. lbs. (87 Nm).
   - Front driveshaft and transfer case, if equipped
   - Transmission mount and crossmember
   - Throttle valve cable
   - Shift cable
   - Park/Neutral switch connector
   - VSS sensor connector
   - Transmission oil dipstick tube
   - Torque converter. Tighten the bolts to 23 ft. lbs. (31 Nm) for 10.75 inch converters and to 35 ft. lbs. (47 Nm) for 12.2 inch converters.
   - Torque converter access cover
   - Transmission oil cooler lines
   - Starter motor
   - Transmission braces, if equipped. Tighten the bolts to 30 ft. lbs. (41 Nm).
   - Exhaust front pipe
   - CKP sensor
   - Rear driveshaft
   - Negative battery cable

2004

42RLE

1. Disconnect the negative battery cable.

2. Raise and support the vehicle

3. Remove any necessary skid plates.

4. Mark propeller shaft and axle companion flanges for assembly alignment.

5. Remove the rear propeller shaft.

6. Remove the front propeller shaft, if necessary.

7. Disconnect wires from the input and output speed sensors.

8. Disconnect wires from the transmission range sensor.

9. Disconnect wires from the solenoid/pressure switch assembly.

10. Remove the bolts holding the exhaust crossover pipe to the pre-catalytic converter pipe flanges.

11. Remove the bolts holding the exhaust crossover pipe to the catalytic converter flange.

12. Disconnect gearshift cable from transmission manual valve lever.

13. Disengage the shift cable from the cable support bracket.

14. Remove the starter motor.

15. Remove the engine to transmission collar.

16. Rotate crankshaft in clockwise direction until converter bolts are accessible. Then remove bolts one at a time. Rotate crankshaft with socket wrench on dampener bolt.

17. Disconnect the transmission vent hose from the transmission.

18. Remove transfer case.

19. Support rear of engine with safety stand or jack.

20. Raise transmission slightly with service jack to relieve load on crossmember and supports.

21. Remove bolts securing rear support and cushion to transmission and crossmember.

22. Remove bolts attaching crossmember to frame and remove crossmember.

23. Disconnect transmission fluid cooler lines at transmission fittings and clips.

24. Remove all remaining converter housing bolts.

25. Carefully work transmission and torque converter assembly rearward off engine block dowels.

26. Hold torque converter in place during transmission removal.

27. Lower transmission and remove assembly from under the vehicle.

28. To remove torque converter, carefully slide torque converter out of the transmission.

To install:

➡ **Check torque converter hub and hub drive flats for sharp edges burrs, scratches, or nicks. Polish the hub and flats with 320/400 grit paper and crocus cloth if necessary. The hub must be smooth to avoid damaging pump seal at installation.**

29. If a replacement transmission is being installed, transfer any components necessary, such as the manual shift lever and shift cable bracket, from the original transmission onto the replacement transmission.

30. Lubricate oil pump seal lip with transmission fluid.

31. Align converter and oil pump.

32. Carefully insert converter in oil pump. Then rotate converter back and forth until fully seated in pump gears.

33. Check converter seating with steel scale and straightedge. Surface of converter lugs should be at least 13mm to rear of straightedge when converter is fully seated.

34. Temporarily secure converter with C-clamp.

35. Position transmission on jack and secure it with chains.

36. Check condition of converter driveplate. Replace the plate if cracked, distorted or damaged. Also be sure transmission dowel pins are seated in engine block and protrude far enough to hold transmission in alignment.

37. Apply a light coating of high temperature grease to the torque converter hub pocket in the rear pocket of the engine's crankshaft.

38. Raise transmission and align the torque converter with the drive plate and transmission converter housing with the engine block.

39. Move transmission forward. Then raise, lower or tilt transmission to align the converter housing with engine block dowels.

40. Carefully work transmission forward and over engine block dowels until converter hub is seated in crankshaft. Verify that no wires, or the transmission vent hose, have become trapped between the engine block and the transmission.

41. Install two bolts to attach the transmission to the engine.

42. Install remaining torque converter housing to engine bolts. Tighten to 68 Nm (50 ft. lbs.).

43. Install transfer case, if equipped. Tighten transfer case nuts to 35 Nm (26 ft. lbs.).

44. Install rear transmission crossmember. Tighten crossmember to frame bolts to 68 Nm (50 ft. lbs.).

45. Install rear support to transmission. Tighten bolts to 47 Nm (35 ft. lbs.).

46. Lower transmission onto crossmember and install bolts attaching transmission mount to crossmember. Tighten clevis bracket to crossmember bolts to 47 Nm (35 ft. lbs.). Tighten the clevis bracket to rear support bolt to 68 Nm (50 ft. lbs.).

47. Connect gearshift cable to support bracket and transmission manual lever.

48. Connect input and output speed sensor wires and the transmission range sensor.

49. Connect wires to the solenoid/pressure switch assembly.

❄❄ CAUTION

It is essential that correct length bolts be used to attach the converter to the driveplate. Bolts that are too long will damage the clutch surface inside the converter.

50. Install torque converter-to-driveplate bolts. Tighten bolts to 88 Nm (65 in. lbs.).

51. Install starter motor and cooler line bracket.

52. Connect cooler lines to transmission.

53. Install transmission fill tube.

54. Install exhaust components.

55. Align and connect propeller shaft(s).

56. Adjust gearshift cable if necessary.

57. Install any skid plates removed previously.

58. Lower vehicle.

59. Fill transmission with Mopar® ATF +4, Automatic Transmission Fluid or equivalent.

545RFE

1. Disconnect the negative battery cable.

2. Raise and support the vehicle

3. Remove any necessary skid plates.

4. Mark propeller shaft and axle companion flanges for assembly alignment.

5. Remove the rear propeller shaft.

6. Remove the front propeller shaft, if necessary.

7. Remove the engine to transmission collar.

8. Remove the exhaust support bracket from the rear of the transmission.

9. Disconnect and lower or remove any necessary exhaust components.

10. Remove the starter motor.

11. Rotate crankshaft in clockwise direction until converter bolts are accessible. Then remove bolts one at a time. Rotate crankshaft with socket wrench on dampener bolt.

12. Disengage the output speed sensor connector from the output speed sensor.

13. Disengage the input speed sensor connector from the input speed sensor.

14. Disengage the transmission solenoid/TRS assembly connector from the transmission solenoid/TRS assembly.

15. Disengage the line pressure sensor connector from the line pressure sensor.

16. Disconnect gearshift cable from transmission manual valve lever.

17. Remove the gearshift cable from the shift cable support bracket.

18. Disconnect transmission fluid cooler lines at transmission fittings and clips.

19. Support rear of engine with safety stand or jack.

20. Raise transmission slightly with service jack to relieve load on crossmember and supports.

21. Remove bolts securing rear support and cushion to transmission and crossmember.

22. Remove bolts attaching crossmember to frame and remove crossmember.

23. Remove transfer case.

24. Remove all remaining converter housing bolts.

25. Carefully work transmission and torque converter assembly rearward off engine block dowels.

26. Hold torque converter in place during transmission removal.

27. Lower transmission and remove assembly from under the vehicle.

28. To remove torque converter, carefully slide torque converter out of the transmission.

To install:

➡**Check torque converter hub and hub drive flats for sharp edges burrs, scratches, or nicks. Polish the hub and flats with 320/400 grit paper and crocus cloth if necessary. Verify that the converter hub o-ring is properly installed and is free of any debris. The hub must be smooth to avoid damaging pump seal at installation.**

29. If a replacement transmission is being installed, transfer any components necessary, such as the manual shift lever

and shift cable bracket, from the original transmission onto the replacement transmission.

30. Lubricate oil pump seal lip with transmission fluid.

31. Align converter and oil pump.

32. Carefully insert converter in oil pump. Then rotate converter back and forth until fully seated in pump gears.

33. Check converter seating with steel scale and straightedge. Surface of converter lugs should be at least 13mm to rear of straightedge when converter is fully seated.

34. Temporarily secure converter with C-clamp.

35. Position transmission on jack and secure it with chains.

36. Check condition of converter driveplate. Replace the plate if cracked, distorted or damaged. Also be sure transmission dowel pins are seated in engine block and protrude far enough to hold transmission in alignment.

37. Apply a light coating of high temperature grease to the torque converter hub pocket in the rear pocket of the engine's crankshaft.

38. Raise transmission and align the torque converter with the drive plate and transmission converter housing with the engine block.

39. Move transmission forward. Then raise, lower or tilt transmission to align the converter housing with engine block dowels.

40. Carefully work transmission forward and over engine block dowels until converter hub is seated in crankshaft. Verify that no wires, or the transmission vent hose, have become trapped between the engine block and the transmission.

41. Install two bolts to attach the transmission to the engine.

42. Install remaining torque converter housing to engine bolts. Tighten to 68 Nm (50 ft. lbs.).

43. Install transfer case, if equipped. Tighten transfer case nuts to 35 Nm (26 ft. lbs.).

44. Install rear transmission crossmember. Tighten crossmember to frame bolts to 68 Nm (50 ft. lbs.).

45. Install rear support to transmission. Tighten bolts to 47 Nm (35 ft. lbs.).

46. Lower transmission onto crossmember and install bolts attaching transmission mount to crossmember. Tighten clevis bracket to crossmember bolts to 47 Nm (35 ft. lbs.). Tighten the clevis bracket to rear support bolt to 68 Nm (50 ft. lbs.).

47. Remove engine support fixture.

48. Connect gearshift cable to transmission.

49. Connect the wiring harness connector to the solenoid and pressure switch assembly connector. Be sure transmission harnesses are properly routed.

50. Connect the wiring harness connector to the input speed sensor.

51. Connect the wiring harness connector to the output speed sensor.

52. Connect the wiring harness connector to the line pressure sensor.

✳✳ CAUTION

It is essential that correct length bolts be used to attach the converter to the driveplate. Bolts that are too long will damage the clutch surface inside the converter.

53. Install torque converter-to-driveplate bolts. Tighten bolts to 31 Nm (270 inch lbs.).

54. Install starter motor and cooler line bracket.

55. Connect cooler lines to transmission.

56. Install transmission fill tube.

57. Install exhaust components.

58. Install the structural dust cover onto the transmission and the engine.

59. Align and install the front propeller shaft, if necessary.

60. Align and install the rear propeller shaft.

61. Adjust gearshift cable if necessary.

62. Install any skid plates removed previously.

63. Lower vehicle.

64. Fill transmission with Mopar® ATF +4, Automatic Transmission Fluid.

Transfer Case Assembly

REMOVAL & INSTALLATION

2001–03

1. Before servicing the vehicle, refer to the precautions in the beginning of this section.

2. Shift the transfer case into **N**.

3. Remove or disconnect the following:
- Front and rear driveshafts
- Transmission mount and cross-member. Support the transmission.
- Vehicle Speed (VSS) sensor connector
- Shift linkage
- Vent hose
- Vacuum hose

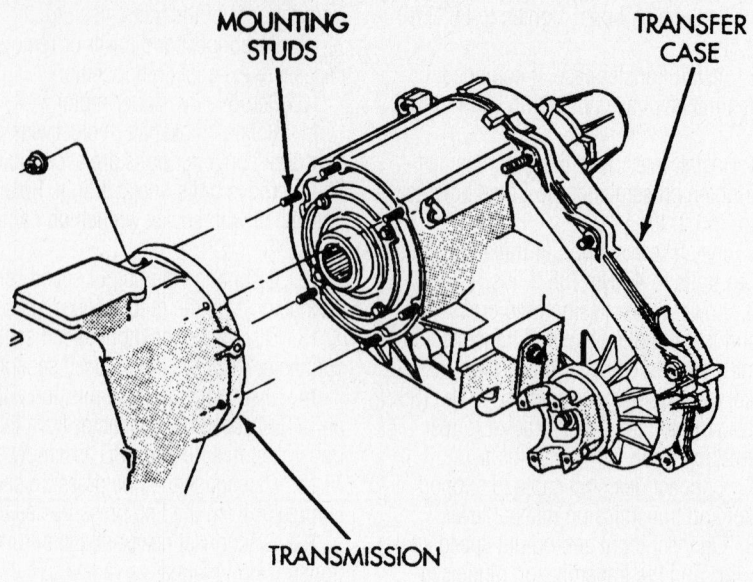

Typical transfer case mounting

- Indicator switch connector
- Transfer case attaching nuts
- Transfer case

To install:

4. Install or connect the following:
- Transfer case. Tighten the nuts to 26 ft. lbs. (35 Nm).
- Indicator switch connector
- Vacuum hose
- Vent hose
- Shift linkage
- VSS sensor connector
- Transmission mount and cross-member
- Front and rear driveshafts

2004

NV144

1. Shift transfer case into AWD.
2. Raise vehicle.
3. Drain transfer case lubricant.
4. Mark front and rear propeller shafts for alignment reference.
5. Support transmission with jack stand.
6. Remove the transfer case skid plate, if equipped.
7. Disconnect front and rear propeller shafts at transfer case.
8. Disconnect transfer case shift motor and mode sensor wire connectors.
9. Disconnect transfer case vent hose.
10. Support transfer case with transmission jack.
11. Secure transfer case to jack with chains.
12. Remove nuts attaching transfer case to transmission.

13. Pull transfer case and jack rearward to disengage transfer case.

14. Remove transfer case from under vehicle.

To install:

15. Mount transfer case on a transmission jack.

16. Secure transfer case to jack with chains.

17. Position transfer case under vehicle.

18. Align transfer case and transmission shafts and install transfer case onto the transmission.

19. Install and tighten transfer case attaching nuts to 27–34 Nm (20–25 ft. lbs.) torque.

20. Connect the vent hose.

21. Connect the shift motor and mode sensor assembly wiring connector. Secure wire harness to clips on transfer case.

22. Align and connect the propeller shafts.

23. Fill transfer case with correct fluid.

24. Install skid plate, if equipped.

25. Remove transmission jack and support stand.

26. Lower vehicle and verify transfer case shift operation.

Halfshaft

REMOVAL & INSTALLATION

2001–03

1. Before servicing the vehicle, refer to the precautions in the beginning of this section.

To separate the halfshaft from the hub, remove the cotter pin, nut lock and spring washer from the axle shaft

2. Remove or disconnect the following:
- Skid plate, if equipped
- Front wheel
- Split pin
- Nut lock
- Spring washer
- Hub nut
- Brake caliper and rotor
- Wheel speed sensor, if equipped
- Wheel bearing and hub assembly

3. Pry the inner tripod joint out of the differential and remove the axle halfshaft.

To install:

4. Install the axle halfshaft so that the snapring is felt to seat in the joint housing groove.

5. Install or connect the following:
- Wheel bearing and hub assembly
- Wheel speed sensor, if equipped
- Brake caliper and rotor
- Hub nut. Tighten the nut to 180 ft. lbs. (244 Nm).
- Spring washer
- Nut lock
- Split pin
- Front wheel
- Skid plate, if equipped

2004

1. With vehicle in neutral, position vehicle on hoist.

2. Remove skid plate, if equipped.

3. Remove hub nut from the halfshaft.

4. Remove brake caliper and rotor.

5. Remove wheel speed sensor if equipped.

6. Remove hub bearing bolts from the knuckle.

7. Remove hub bearing and brake shield from knuckle.

8. Support halfshaft at the C/V joint housings.

9. Position two pry bars behind the inner C/V housing and disengage the C/V joint from the axle.

10. Remove halfshaft through the knuckle.

To install:

11. Apply a light coating of wheel bearing grease on the axle splines.

12. Insert halfshaft through the steering knuckle and onto the axle. Verify shaft snapring engages with the groove on the inside of the joint housing.

13. Clean hub bearing bore and hub bearing mating surface. Lightly coat mating surfaces with grease.

14. Install hub bearing onto the axle halfshaft and into steering knuckle. Tighten hub bearing bolts to 163 Nm (120 ft. lbs.).

15. Install wheel speed sensor, if equipped.

16. Install brake rotor and caliper adapter with caliper.

17. Install halfshaft nut. Apply brakes and tighten shaft nut to 251 Nm (185 ft. lbs.).

18. Install skid plate, if equipped.

CV-Joints

OVERHAUL

Outer CV-Joint

1. Before servicing the vehicle, refer to the precautions in the beginning of this section.

2. Remove or disconnect the following:
- Axle halfshaft from the vehicle
- CV-joint boot and clamps
- Snapring
- CV-joint

To install:

3. Install or connect the following:
- CV-joint
- Snapring
- CV-joint boot and clamps

4. Fill the joint housing and boot with grease and tighten the boot clamps.

5. Install the axle halfshaft.

Inner Tripod Joint

1. Before servicing the vehicle, refer to the precautions in the beginning of this section.

2. Remove or disconnect the following:
- Axle halfshaft from the vehicle
- Inner tripod joint boot clamps
- Tripod joint housing
- Snapring
- Circlip
- Tripod joint

To install:

➡**Use new snaprings, clips, and boot clamps for assembly.**

3. Install or connect the following:
- Tripod joint
- Circlip
- Snapring
- Tripod joint housing

4. Fill the tripod joint housing and boot with grease and tighten the boot clamps.

5. Install the axle halfshaft.

Axle Shaft, Bearing and Seal

REMOVAL & INSTALLATION

Front—2001–03

AXLE SHAFT

1. Before servicing the vehicle, refer to the precautions in the beginning of this section.

2. Remove or disconnect the following:
- Front wheel
- Brake caliper and rotor
- Wheel speed sensor, if equipped
- Axle hub nut
- Wheel bearing and hub assembly
- Axle shaft

To install:

3. Install or connect the following:
- Axle shaft
- Wheel bearing and hub assembly
- Axle hub nut. Tighten the nut to 175 ft. lbs. (237 Nm).
- Wheel speed sensor, if equipped
- Brake caliper and rotor
- Front wheel

SEAL

1. Before servicing the vehicle, refer to the precautions in the beginning of this section.
2. Remove or disconnect the following:
   - Front axle shafts
   - Differential cover
   - Differential and ring gear assembly
   - Axle seals

To install:

3. Press the axle seals into the differential housing with Turnbuckle 6797 and Disc set 8110.
4. Install or connect the following:
   - Differential and ring gear assembly. Tighten the bearing cap bolts to 45 ft. lbs. (61 Nm).
   - Differential cover. Tighten the bolts to 30 ft. lbs. (41 Nm).
   - Front axle shafts
5. Fill the axle assembly with gear oil and check for leaks.

Front—2004

AXLE SHAFT

1. Remove half shaft from vehicle.
2. Clean axle seal area.
3. Remove snap ring from the axle shaft.
4. Remove axle shaft with Remover 8420A and slide hammer.

To install:

5. Lubricate bearing bore and seal lip with gear lubricant.
6. Install axle shaft and engage shaft into side gear. Push firmly on axle shaft to engage snap-ring.
7. Check the differential fluid level and add fluid if necessary.
8. Install half shaft.
9. Install skid plate, if necessary.

SEAL

1. Remove half shaft and axle shaft.
2. Remove axle shaft seal with a small pry bar.

To install:

3. Wipe axle shaft tube bore clean.
4. Install new axle shaft seal with Installer 8402 and Handle C-4171.
5. Install axle shaft and half shaft.

Rear—2001–03

C-CLIP TYPE

1. Before servicing the vehicle, refer to the precautions in the beginning of this section.
2. Remove or disconnect the following:
   - Rear wheel
   - Brake drum
   - Differential cover

- Differential gear shaft retainer
- Differential gear shaft
- C-clip
- Axle shaft
- Axle seal
- Axle bearing

To install:

3. Install or connect the following:
   - Axle bearing
   - Axle seal
   - Axle shaft
   - C-clip
   - Differential gear shaft. Use Loctite® and tighten the retainer to 14 ft. lbs. (19 Nm).
   - Differential cover. Tighten the bolts to 30 ft. lbs. (41 Nm).
   - Brake drum
   - Rear wheel
4. Fill the axle assembly with gear oil and check for leaks.

NON C-CLIP TYPE

1. Before servicing the vehicle, refer to the precautions in the beginning of this section.
2. Remove or disconnect the following:
   - Rear wheel
   - Brake caliper and rotor, if equipped
   - Brake drum, if equipped
   - Axle retainer nuts
   - Axle shaft, seal and bearing assembly

3. Split the bearing retainer with a chisel and remove the retainer ring.
4. Press the bearing off the axle shaft.
5. Remove the axle seal and retaining plate.

To install:

6. Install the retaining plate and axle seal onto the axle shaft.
7. Pack the wheel bearing with axle grease and press the bearing on to the axle shaft.
8. Press the retaining ring onto the axle shaft.
9. Install or connect the following:
   - Axle shaft, seal and bearing assembly. Tighten the nuts to 45 ft. lbs. (61 Nm).
   - Brake caliper and rotor, if equipped
   - Brake drum, if equipped
   - Rear wheel
10. Fill the axle assembly with gear oil and check for leaks.

Rear—2004 8¼ inch axle

AXLE SHAFT

1. With vehicle in neutral, position it on a hoist.

2. Remove brake caliper adapter with caliper and remove rotor.
3. Remove differential housing cover and drain lubricant.
4. Rotate differential case, to access pinion mate shaft lock screw. Remove screw and pinion mate shaft from differential case.
5. Push axle shaft inward and remove axle shaft C-lock.
6. Remove axle shaft.

To install:

7. Lubricate bearing bore and seal lip with gear lubricant.
8. Insert axle shaft through seal and engage into side gear splines.
9. Insert C-lock in end of axle shaft then push axle shaft outward to seat C-lock in side gear.
10. Insert pinion shaft into differential case and through thrust washers and differential pinions.
11. Align hole in shaft with hole in differential case and install lock screw with Loctite® on the threads. Tighten lock screw to 25 Nm (220 inch lbs.).
12. Install differential cover and fill with gear lubricant.
13. Install brake rotor and caliper adapter with caliper.

BEARING AND SEAL

1. Remove axle shaft.
2. Remove axle seal with pry bar.
3. Position bearing Receiver 9338 on axle tube.
4. Insert bearing Remover 6310 with Foot 6310-9 through receiver and bearing.
5. Tighten Remove 6310 nut to pull bearing into the receiver.

To install:

6. Remove any old sealer/burrs from axle tube.
7. Install axle shaft bearing with Installer 9337 and Handle. Drive bearing in until tool contacts the axle tube.

➡ Bearing is installed with the bearing part number against the installer.

8. Coat new axle seal lip with axle lubricant and install with Installer 9337 and Handle C-4171.
9. Install axle shaft.

Rear— 2004 9¼ inch axle

AXLE SHAFT

1. With vehicle in neutral, position it on a hoist.
2. Remove brake caliper adapter with caliper and remove rotor.
3. Remove differential housing cover and drain lubricant.

4. Rotate differential case, to access pinion mate shaft lock screw. Remove screw and pinion mate shaft from differential case.

5. Push axle shaft inward and remove axle shaft C-lock.

6. Remove axle shaft.

To install:

7. Lubricate bearing bore and seal lip with gear lubricant.

8. Insert axle shaft (1) through seal (2) and engage into side gear splines.

9. Insert C-lock in end of axle shaft then push axle shaft outward to seat C-lock in side gear.

10. Insert pinion shaft into differential case and through thrust washers and differential pinions.

11. Align hole in shaft with hole in differential case and install lock screw (4) with Loctite® on the threads. Tighten lock screw to 25 Nm (220 inch lbs.).

12. Install differential cover and fill with gear lubricant.

13. Install brake rotor and caliper adapter with caliper.

SEAL

1. Remove axle shaft.

2. Remove axle shaft seal from end of the axle tube with a pry bar.

To install:

3. Remove any old sealer/burrs from axle tube.

4. Coat new seal lip with axle lubricant and install seal with Installer 9337 and Handle C-4171.

5. Install axle shaft.

BEARING

1. Remove axle shaft.

2. Remove axle seal with pry bar.

3. Position bearing Receiver 9338 on axle tube.

4. Insert bearing Remover 6310 with Foot 6310-9 through receiver and bearing.

5. Tighten Remove 6310 nut to pull bearing into the receiver.

To install:

6. Remove any old sealer/burrs from axle tube.

7. Install axle shaft bearing with Installer 9337 and Handle C-417

8. Drive bearing in until tool contacts the axle tube.

➡**Bearing is installed with the bearing part number against the installer.**

9. Coat new axle seal lip with axle lubricant and install with Installer 9337 and Handle C-4171.

10. Install axle shaft.

Pinion Seal

REMOVAL & INSTALLATION

C-Clip Type—2001–03

1. Before servicing the vehicle, refer to the precautions in the beginning of this section.

2. Remove or disconnect the following:
   - Wheels
   - Brake drums
   - Driveshaft

3. Check the bearing preload with an inch lb. torque wrench.

4. Remove the pinion flange and seal.

To install:

➡**Use a new pinion nut for assembly.**

5. Install the new pinion seal and flange. Tighten the nut to 210 ft. lbs. (285 Nm).

6. Check the bearing preload. The bearing preload should be equal to the reading taken earlier, plus 5 inch lbs.

7. If the preload torque is low, tighten the pinion nut in 5 inch lb. increments until the torque value is reached. Do not exceed 350 ft. lbs. (474 Nm) pinion nut torque.

8. If the pinion bearing preload torque cannot be attained at maximum pinion nut torque, replace the collapsible spacer.

9. Install or connect the following:
   - Driveshaft
   - Brake drums
   - Wheels

10. Fill the axle assembly with gear oil and check for leaks.

Non C-Clip Type—2001–03

FRONT

1. Before servicing the vehicle, refer to the precautions in the beginning of this section.

2. Remove or disconnect the following:
   - Wheels
   - Brake rotors
   - Driveshaft

3. Check the bearing preload with an inch lb. torque wrench.

4. Remove the pinion flange and seal.

To install:

➡**Use a new pinion nut for assembly.**

5. Install the new pinion seal and flange. Tighten the nut to 160 ft. lbs. (217 Nm).

6. Check the bearing preload. The bearing preload should be equal to the reading taken earlier, plus 5 inch lbs.

7. If the preload torque is low, tighten the pinion nut in 5 inch lb. increments until the torque value is reached. Do not exceed 260 ft. lbs. (353 Nm) pinion nut torque.

8. If the pinion bearing preload torque can not be attained at maximum pinion nut torque, replace the collapsible spacer.

9. Install or connect the following:
   - Driveshaft
   - Brake rotors
   - Wheels

10. Fill the axle assembly with gear oil and check for leaks.

REAR

1. Before servicing the vehicle, refer to the precautions in the beginning of this section.

2. Remove or disconnect the following:
   - Wheels
   - Brake rotors or drums
   - Driveshaft

3. Check the bearing preload with an inch lb. torque wrench.

4. Remove the pinion flange and seal.

To install:

➡**Use a new pinion nut for assembly.**

5. Install the new pinion seal and flange. Tighten the nut to 160 ft. lbs. (217 Nm).

6. Check the bearing preload. The bearing preload should be equal to the reading taken earlier, plus 5 inch lbs.

7. If the preload torque is low, tighten the pinion nut in 5 inch lb. increments until the torque value is reached. Do not exceed 260 ft. lbs. (353 Nm) pinion nut torque.

8. If the pinion bearing preload torque can not be attained at maximum pinion nut torque, remove one or more pinion preload shims.

9. Install or connect the following:
   - Driveshaft
   - Brake rotors or drums
   - Wheels

10. Fill the axle assembly with gear oil and check for leaks.

2004 Front Axle

1. Remove both half shafts.

2. Mark propeller shaft and pinion flange for installation reference.

3. Remove front propeller shaft.

4. Rotate pinion gear three to four times, to verify pinion rotates smoothly.

5. Record pinion flange rotating torque with an inch pound torque wrench for installation reference.

6. Hold flange with Holder 6719 and four bolts and washers.

7. Remove pinion nut.

8. Remove flange with Remover C-452.

9. Remove pinion seal with a pry tool.

To install:

10. Apply a light coating of gear lubricant on the lip of pinion seal.

11. Install seal with Installer C-3972-A and Handle C-4171.

12. Install pinion flange onto the pinion with Installer C-3718 and holder.

13. Hold pinion flange with Holder 6719A.

14. Install new pinion nut and tighten nut until there is zero bearing end-play.

✸✸ CAUTION

Do not exceed the minimum tightening torque when installing the companion flange at this point. Damage to the collapsible spacer or bearings may result.

15. Tighten pinion nut to 271 Nm (200 ft. lbs.).

✸✸ CAUTION

Never loosen pinion nut to decrease pinion bearing rotating torque and never exceed specified preload torque. If preload torque or rotating torque is exceeded a new collapsible spacer must be installed.

16. Record pinion flange rotating torque, with a torque wrench. Rotating torque should be equal to the reading recorded during removal plus an additional 0.56 Nm (5 inch lbs.).

17. If rotating torque is low, tighten pinion nut in 6.8 Nm (5 ft. lbs.) increments until rotating torque is achieved.

✸✸ CAUTION

If maximum tightening torque is reached prior to reaching the required rotating torque, the collapsible spacer may have been damaged. Replace the collapsible spacer.

18. Install propeller shaft with reference marks aligned.

19. Install half shafts.

2004 Rear Axle

8¼ INCH AXLE

1. With vehicle in neutral, position vehicle on hoist.

2. Mark a reference line across the axle flange and propeller shaft flange.

3. Remove propeller shaft.

4. Remove brake calipers and rotors to prevent any drag.

5. Rotate flange three or four times and verify flange rotates smoothly.

6. Measure torque to rotating pinion flange with an inch pound torque wrench. Record reading for installation reference.

7. Install bolts into two of the threaded holes in the flange 180° apart.

8. Position Holder 6719 against the flange and install a bolt and washer into one of the remaining threaded holes. Tighten the bolts so the Holder 6719 is held to the flange.

9. Remove pinion nut and washer.

10. Remove flange with Remover C-452.

11. Remove pinion seal with a pry tool or slide-hammer mounted screw.

To install:

12. Apply a light coating of gear lubricant on the lip of pinion seal.

13. Install new pinion seal with Installer C-4076-B and Handle C-4735.

14. Install flange on the end of the shaft with the reference marks aligned.

15. Install bolts into two of the threaded holes in the flange 180° apart.

16. Position Holder 6719 against flange. Install a bolt and washer into one of the remaining threaded holes. Tighten bolts so Holder 6719 is held to the flange.

17. Install flange on pinion shaft with Installer C-3718 and Holder 6719.

18. Install pinion washer and a new pinion nut. The convex side of the washer must face outward.

✸✸ CAUTION

Do not exceed the minimum tightening torque when installing the companion flange retaining nut at this point. Failure to follow these instructions can damage the collapsible spacer or bearings.

19. Hold flange with Holder 6719 and tighten pinion nut to 285 Nm (210 ft. lbs.). Rotate pinion several revolutions to ensure bearing rollers are seated.

20. Rotate pinion flange with an inch pound torque wrench. Rotating torque should be equal to the reading recorded during removal plus an additional 0.56 Nm (5 inch lbs.).

✸✸ CAUTION

Never loosen pinion nut to decrease pinion bearing rotating torque and

never exceed specified preload torque. If rotating torque is exceeded, a new collapsible spacer must be installed. Failure to follow these instructions can damage the collapsible spacer or bearings.

21. If rotating torque is low use Holder 6719, to hold flange and tighten pinion nut in 6.8 Nm (5 ft. lbs.) increments until proper rotating torque is achieved.

➡ **The seal replacement is unacceptable if final pinion nut torque is less than 285 Nm (210 ft. lbs.).**

➡ **The bearing rotating torque should be constant during a complete revolution of the pinion. If the rotating torque varies, this indicates a binding condition.**

22. Install propeller shaft.

23. Install rear brake rotors components.

9¼ INCH AXLE

1. With vehicle in neutral, position it on a hoist.

2. Mark an installation reference line across the pinion flange and driveshaft flange.

3. Remove driveshaft.

4. Remove brake calipers and rotors to prevent any drag.

5. Rotate pinion flange three or four times and verify flange rotates smoothly.

6. Measure rotating torque of the pinion with an inch pound torque wrench and record reading.

7. Install two bolts into the pinion flange threaded holes, 180° apart. Position Holder 6719A, or equivalent, against the flange and install and tighten two bolts and washers into the remaining holes.

8. Hold the flange with Holder 6719A, or equivalent, and remove pinion nut and washer.

9. Remove companion flange with Remover C-452, or equivalent.

10. Remove pinion seal with pry tool or slide-hammer mounted screw.

To install:

11. Apply a light coating of gear lubricant on the lip of pinion seal.

12. Install new pinion seal with Installer C-4076-B and Handle C-4735, or equivalent.

13. Install pinion flange on pinion shaft with Installer C-3718, or equivalent.

14. Install two bolts into the threaded holes in the flange, 180° apart.

15. Position Holder 6719 against the flange and install a bolt and washer into one of the remaining threaded holes. Tighten the bolts so holder is held to the flange.

16. Install pinion washer and a new pinion nut. The convex side of the washer must face outward.

✵✵ CAUTION

Never exceed the minimum tightening torque 285 Nm (210 ft. lbs.) when installing the companion flange retaining nut at this point. Failure to follow these instructions will result in damage to collapsible spacer or bearings.

17. Hold companion flange with Holder 6719, or equivalent, and tighten pinion nut with a torque set to 285 Nm (210 ft. lbs.).

Rotate pinion several revolutions to ensure the bearing rollers are seated.

18. Rotate pinion with an inch pound torque wrench. Rotating torque should be equal to the reading recorded during removal plus an additional 0.56 Nm (5 inch lbs.).

✵✵ CAUTION

Never loosen pinion nut to decrease pinion bearing rotating torque and never exceed specified preload torque. If rotating torque is exceeded, a new collapsible spacer must be installed. Failure to follow these instructions will result in damage to the collapsible spacer

19. If rotating torque is low, use Holder 6719, or equivalent, to hold the companion flange and tighten pinion nut in 6.8 Nm (5 ft. lbs.) increments until proper rotating torque is achieved.

➡ **The bearing rotating torque should be constant during a complete revolution of the pinion. If the rotating torque varies, this indicates a binding condition.**

➡ **The seal replacement is unacceptable if the final pinion nut torque is less than 285 Nm (210 ft. lbs.).**

20. Install driveshaft.

STEERING AND SUSPENSION

Air Bag

✵✵ CAUTION

Some vehicles are equipped with an air bag system. The system must be disarmed before performing service on, or around, system components, the steering column, instrument panel components, wiring and sensors. Failure to follow the safety precautions and the disarming procedure could result in accidental air bag deployment, possible injury and unnecessary system repairs.

PRECAUTIONS

Several precautions must be observed when handling the inflator module to avoid accidental deployment and possible personal injury.

• Never carry the inflator module by the wires or connector on the underside of the module.

• When carrying a live inflator module, hold securely with both hands, and ensure that the bag and trim cover are pointed away.

• Place the inflator module on a bench or other surface with the bag and trim cover facing up.

• With the inflator module on the bench, never place anything on or close to the module which may be thrown in the event of an accidental deployment.

Before servicing the vehicle, also make sure to refer to the precautions in the beginning of this section as well.

DISARMING

1. Disconnect and isolate the negative battery cable. Wait 2 minutes for the system capacitor to discharge before performing any service.

2. When repairs are completed, connect the negative battery cable.

Recirculating Ball Power Steering Gear

REMOVAL & INSTALLATION

1. Before servicing the vehicle, refer to the precautions in the beginning of this section.

2. Remove or disconnect the following:
   • Negative battery cable
   • Power steering pressure and return lines
   • Intermediate shaft
   • Pitman arm
   • Steering gear

To install:

3. Install or connect the following:
   • Steering gear. Tighten the bolts to 100 ft. lbs. (136 Nm).

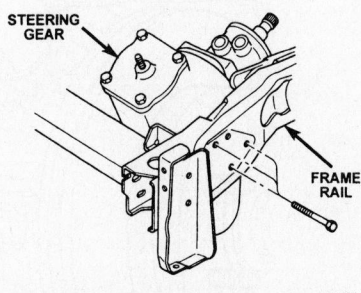

Typical recirculating ball power steering gear mounting

• Pitman arm. Tighten the nut to 175 ft. lbs. (237 Nm).
• Intermediate shaft. Tighten the pinch bolt to 36 ft. lbs. (49 Nm).
• Power steering pressure and return lines
• Negative battery cable

4. Fill the power steering fluid reservoir.

5. Start the engine and check for leaks.

Rack and Pinion Steering Gear

REMOVAL & INSTALLATION

2001–03

1. Before servicing the vehicle, refer to the precautions in the beginning of this section.

2. Remove or disconnect the following:
   • Front wheels
   • Outer tie rod ends
   • Steering shaft coupler
   • Power steering hoses
   • Steering gear

To install:

3. Install or connect the following:
   • Steering gear. Tighten the bolts to 190 ft. lbs. (258 Nm).
   • Power steering hoses. Tighten the fittings to 25 ft. lbs. (35 Nm).
   • Steering shaft coupler. Tighten the bolt to 36 ft. lbs. (49 Nm).
   • Outer tie rod ends. Tighten the nuts to 65 ft. lbs. (88 Nm).
   • Front wheels

2004

1. Siphon out as much power steering fluid as possible from the pump.

2. Lock the steering wheel.

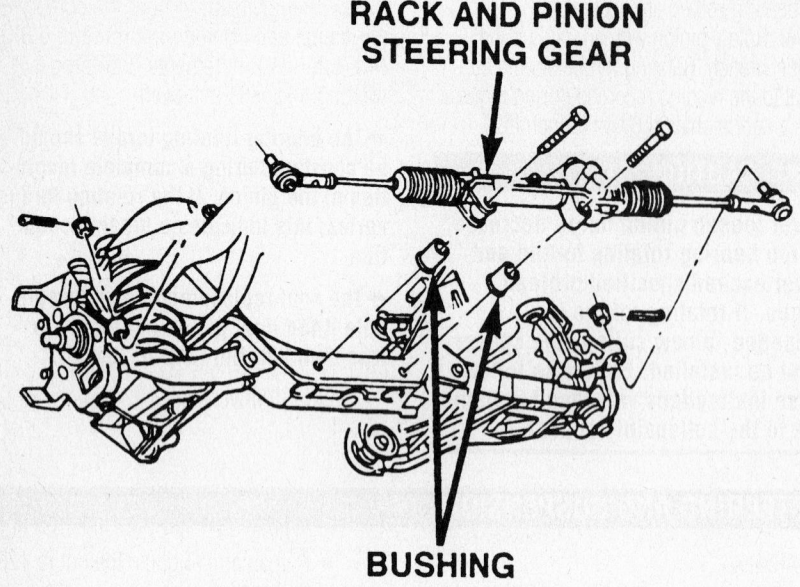

RACK AND PINION STEERING GEAR

BUSHING

7924DG42

Rack and pinion steering gear mounting used on the 2WD models

3. Raise and support the vehicle.
4. Remove the front tires.
5. Remove the nuts from the tie rod ends.
6. Separate tie rod ends from the knuckles.
7. Remove the steering gear pinch bolt.
8. Remove the lower steering coupling from the steering gear.
9. Turn the steering gear to the full right position.
10. Remove the exhaust Y-pipe (3.7L & 4.7L engines only).
11. Remove the power steering lines from the gear.
12. Remove the front crossmember.
13. Remove the steering gear mounting bolts, washers and nuts.
14. Tip the gear forward to allow clearance and move to the right then tip the gear downward on the left side to remove from the vehicle.

To install:

➡**Before installing gear inspect the bushings and replace if worn or damaged, also use new gear mounting bolts and nuts.**

15. Install gear to the vehicle and tighten mounting nuts and bolts to 258 Nm (190 ft. lbs.).
16. Install power steering lines to steering gear and tighten the pressure hose to 38 Nm (28 ft. lbs.) and tighten the return hose to 65 Nm (48 ft. lbs.).
17. Slide the shaft coupler onto gear. Install new bolt and tighten to 49 Nm (36 ft. lbs.).

18. Clean tie rod end studs and knuckle tapers.
19. Install tie rod ends into the steering knuckles and tighten the nuts to 75 Nm (55 ft. lbs.).
20. Install the Y-pipe (3.7L & 4.7L engines only).
21. Install the front crossmember.
22. Install the front tires.
23. Remove the support and lower the vehicle.

24. Unlock the steering wheel.
25. Fill system with fluid.
26. Adjust the toe position.

Shock Absorber

REMOVAL & INSTALLATION

2001–03
FRONT

1. Before servicing the vehicle, refer to the precautions in the beginning of this section.
2. Remove or disconnect the following:
3. Install or connect the following:
   • Front wheel
   • Upper mount nut
   • Lower mount bolt
   • Shock absorber

To install:

4. Install or connect the following:
   • Shock absorber. Tighten the lower bolt to 100 ft. lbs. (136 Nm) and the upper nut to 30 ft. lbs. (41 Nm).
   • Front wheel

REAR

1. Before servicing the vehicle, refer to the precautions in the beginning of this section.
2. Support the axle.
3. Remove or disconnect the following:
   • Upper bolt

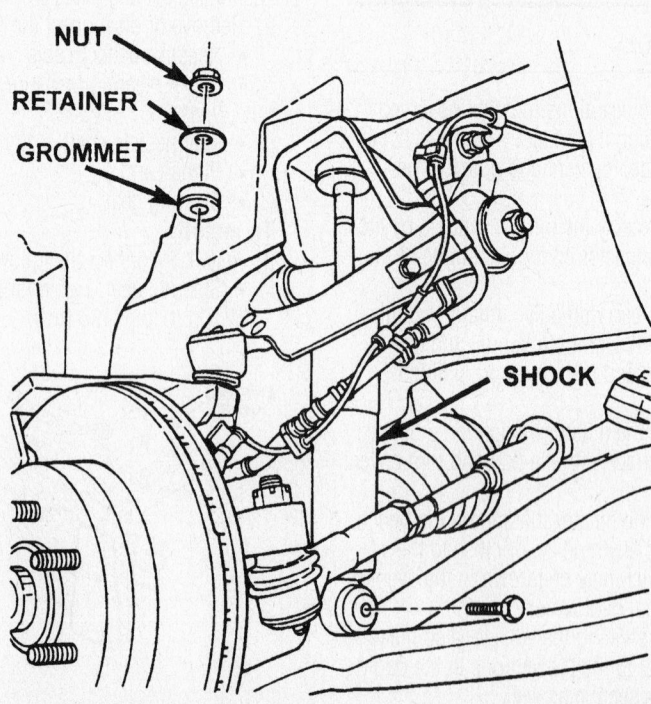

NUT
RETAINER
GROMMET
SHOCK

7924DG86

Front shock absorber mounting—4WD models

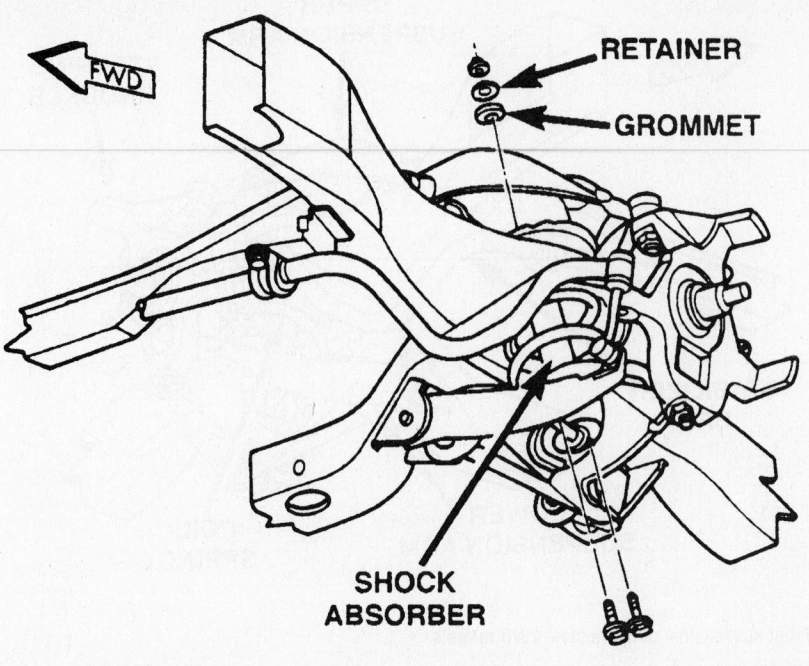

Front shock absorber mounting—2WD models

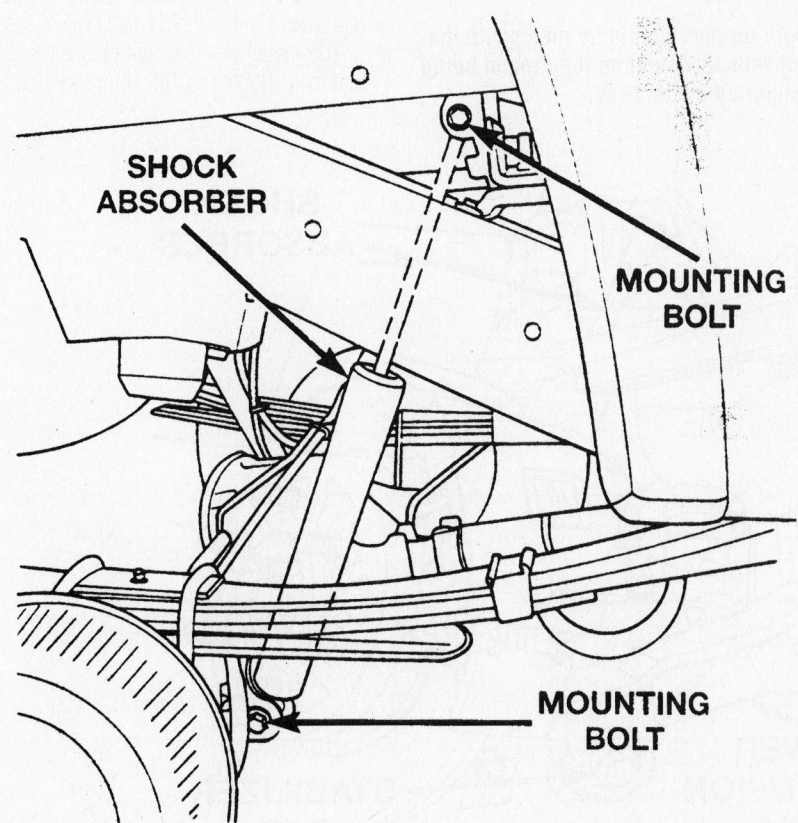

Rear shock absorber mounting

- Lower bolt
- Shock absorber

To install:

4. Install the bolts through the brackets and shock and tighten them as follows: Tighten the lower bolt and nut to 60 ft. lbs. (81 Nm) and the upper bracket nuts to 20 ft. lbs. (27 Nm)

2004

FRONT

1. Raise and support the vehicle.
2. Remove the tire and wheel assembly.
3. Support the lower control arm outboard end.
4. Remove the upper shock bolt and nut.
5. Remove the stabilizer link lower nut and then separate the stabilizer link from the lower control arm to gain access to the lower shock bolt.
6. Remove the lower shock bolt.
7. Remove the shock.

To install:

8. Install the upper part of the shock into the frame bracket.
9. Install the nut and bolt. Tighten to 102 Nm (75 ft. lbs.).
10. Install the lower part of the shock into the lower control arm and Tighten the bolt to 81 Nm (60 ft. lbs.).
11. Install the stabilizer link lower nut to the lower control arm.
12. Remove the support from the lower control arm outboard end.
13. Install the tire and wheel assembly.
14. Remove the support and lower the vehicle.

REAR

1. Raise vehicle and support the axle.
2. Lower the spare tire.

➡**This step must be done if replacing the left side shock.**

3. Remove the upper shock bolt and flag nut.
4. Remove the lower shock bolt and nut.
5. Remove the rear shock absorber from the vehicle.

To install:

6. Position the shock absorber in the brackets.
7. Install the bolts through the brackets and the shock. Install the flag nut on the top bolt and nut on lower bolt.
8. Tighten the upper and lower bolt/nuts to 102 Nm (75 ft. lbs.)
9. Raise the spare tire back in place if lowered for the left shock.

10. Remove the support and lower the vehicle.

Coil Spring

REMOVAL & INSTALLATION

Front

1. Before servicing the vehicle, refer to the precautions in the beginning of this section.
2. Support the lower control arm on a floor jack.
3. Remove or disconnect the following:
   - Front wheel
   - Shock absorber
   - Stabilizer bar link
   - Lower ball joint.
4. Lower the jack and remove the coil spring.

To install:

5. Install the coil spring and raise the control arm into position.
6. Install or connect the following:
   - Lower ball joint. Tighten the nut to 135 ft. lbs. (183 Nm).
   - Stabilizer bar link
   - Shock absorber
   - Front wheel

Rear

1. Raise and support the vehicle.
2. Support the axle with a suitable holding fixture.

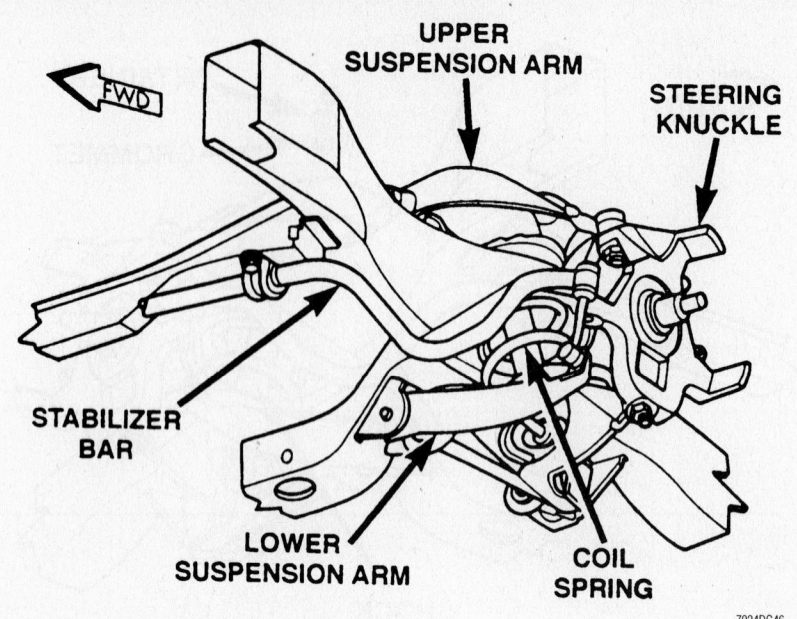

Front suspension components—2WD models

3. Remove the lower shock bolt.
4. Remove the bell crank bolt (4) from the rear axle.
5. Lower the jack to remove the spring and isolator from the vehicle.

To install:

➡ **All torques should be made with the full vehicle weight on the ground being supported by the tires.**

6. Position spring (3) to the vehicle on top of the isolator (2).
7. Align the springs (3) to the spring pockets.
8. Raise the rear axle into place.
9. Install the bell crank bolt (4) to the rear axle. Tighten to 251 Nm (185 ft. lbs.).
10. Install the lower shock bolts to the rear axle. Tighten to 102 Nm (75 ft. lbs.).

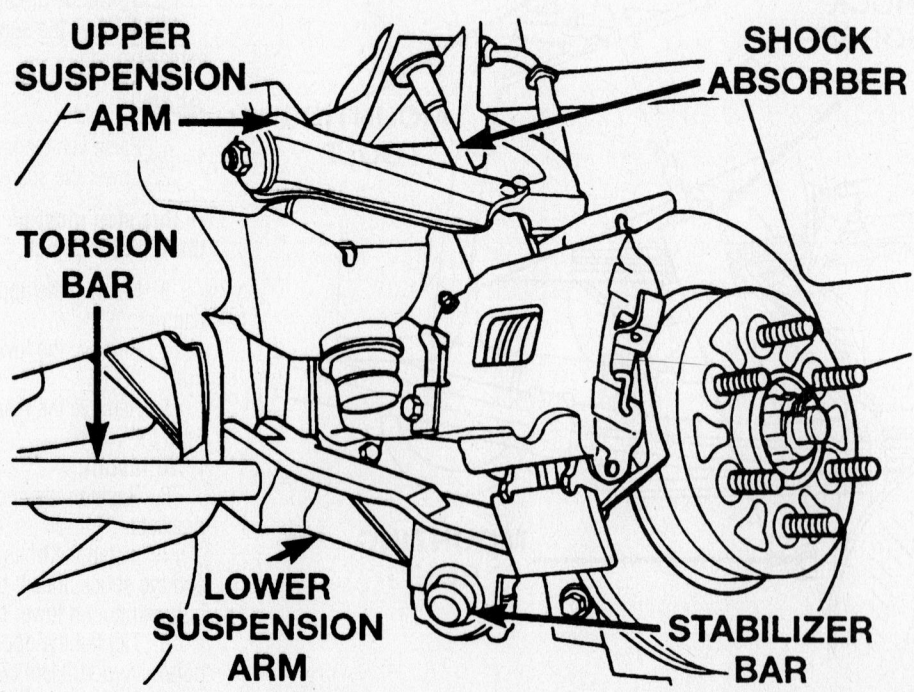

Front suspension components—4WD models

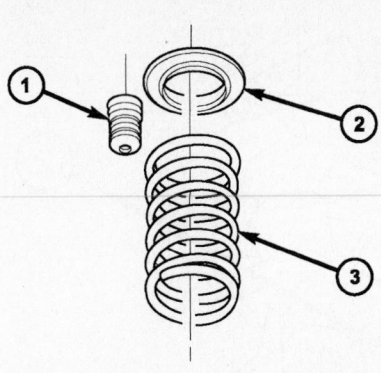

67189-DURA-G04

2004 rear coil spring removal

11. Remove the holding fixture for the rear axle.

Leaf Spring

REMOVAL & INSTALLATION

1. Before servicing the vehicle, refer to the precautions in the beginning of this section.
2. Support the vehicle at the frame rails.
3. Support the rear axle with a jack.
4. Remove or disconnect the following:

- Rear wheel
- Stabilizer bar link
- Axle U-bolts

- Spring bracket
- Leaf spring

To install:

➡**The weight of the vehicle must be supported by the springs when the spring eye and stabilizer bar fasteners are tightened.**

5. Install or connect the following:
- Leaf spring
- Spring bracket
- Axle U-bolts. Tighten the nuts to 52 ft. lbs. (70 Nm).
- Stabilizer bar link
- Rear wheel

6. Tighten the front spring eye bolt and nut to 115 ft. lbs. (156 Nm). Tighten the rear spring eye bolt and nut to 80 ft. lbs. (108 Nm). Tighten the stabilizer bar nuts 55 ft. lbs. (74 Nm).

Torsion Bar

REMOVAL & INSTALLATION

Except 2004

1. Before servicing the vehicle, refer to the precautions in the beginning of this section.
2. Loosen the adjustment bolt to remove spring load. Note the number of turns for installation.

3. Remove or disconnect the following:
- Adjustment bolt, swivel and bearing
- Torsion bar and anchor

4. Separate the torsion bar and anchor.

To install:

5. Assemble the torsion bar and anchor.
6. Install or connect the following:
- Torsion bar and anchor
- Adjustment bolt, swivel and bearing

7. Tighten the adjustment bolt the recorded number of turns.

2004

➡**The left and right side torsion bars are NOT interchangeable. The bars are identified and stamped R or L, for right or left. The bars do not have a front or rear end and can be installed with either end facing forward.**

1. Raise and support the vehicle with the front suspension hanging.
2. Remove the transfer case skid plate.

➡**Count and record the number of turns for installation reference.**

3. Mark the adjustment bolt setting.
4. Install Special Tool 8686 (1) to the anchor arm (2) and the cross member (3).
5. Increase the tension on the anchor arm tool 8686 (1) until the load is removed from the adjustment bolt and the adjuster nut.

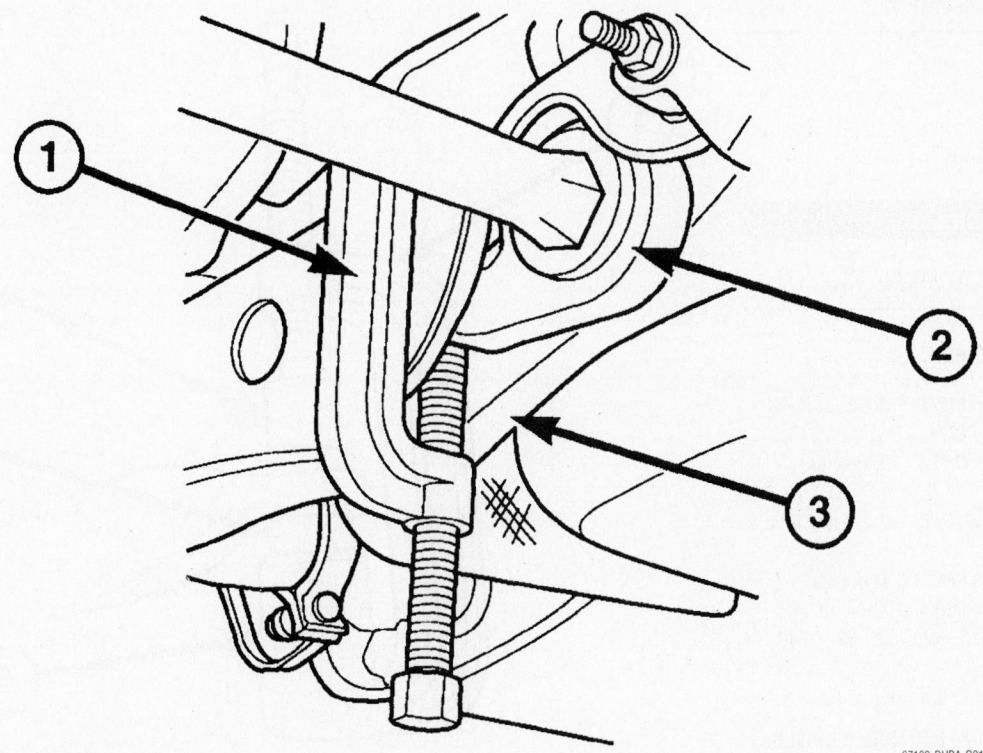

67189-DURA-G01

Unloading the torsion bar

6. Turn the adjustment bolt counter-clockwise to remove the bolt and the adjuster nut.

7. Remove the Special Tool 8686 (1), allowing the torsion bar to unload.

To install:

➡The left and right side torsion bars are NOT interchangeable. The bars are identified and stamped R or L, for right or left. The bars do not have a front or rear end and can be installed with either end facing forward.

8. Insert torsion bar ends into anchor and suspension arm.

9. Position the anchor (2) in the cross-member frame (3).

10. Install Special Tool 8686 (1) to the anchor (2) and the crossmember (3).

11. Increase the tension on the anchor in order to load the torsion bar.

12. Install the adjustment bolt and the adjuster nut.

13. Turn adjustment bolt clockwise the recorded amount of turns.

14. Remove tool 8686 (1) from the torsion bar crossmember (3).

15. Install the transfer case skid plate.

16. Lower vehicle and adjust the front suspension height.

17. Perform a wheel alignment.

Upper Ball Joint

REMOVAL & INSTALLATION

Durango models utilize an upper control arm with an integral ball joint. If the ball joint is damaged or worn, the upper control arm must be replaced.

Lower Ball Joint

REMOVAL & INSTALLATION

Replaceable Types

1. Remove the tire and wheel assembly.
2. Remove the brake caliper and rotor.
3. Disconnect the tie rod from the steering knuckle.
4. Separate the lower ball joint from the steering knuckle.
5. Remove the steering knuckle.
6. Move the halfshaft to the side and support the halfshaft out of the way (4wd only).
7. Chisel out the ball joint stakes.

➡Extreme pressure lubrication must be used on the threaded portions of the

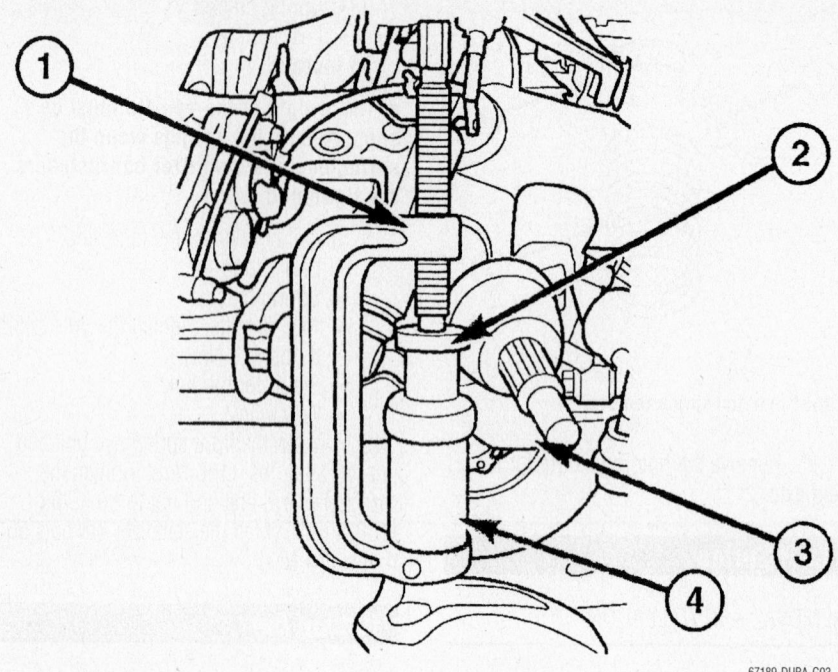

Removing the lower ball joint

tool. This will increase the longevity of the tool and insure proper operation during the removal and installation process.

8. Press the ball joint from the lower control arm (3) using special tools C-4212-F (press) (1), 9331-1 (driver) (2) and 9331-2 (receiver) (4).

To install:

➡Extreme pressure lubrication must be used on the threaded portions of the tool. This will increase the longevity of the tool and insure proper operation during the removal and installation process.

67189-DURA-G03

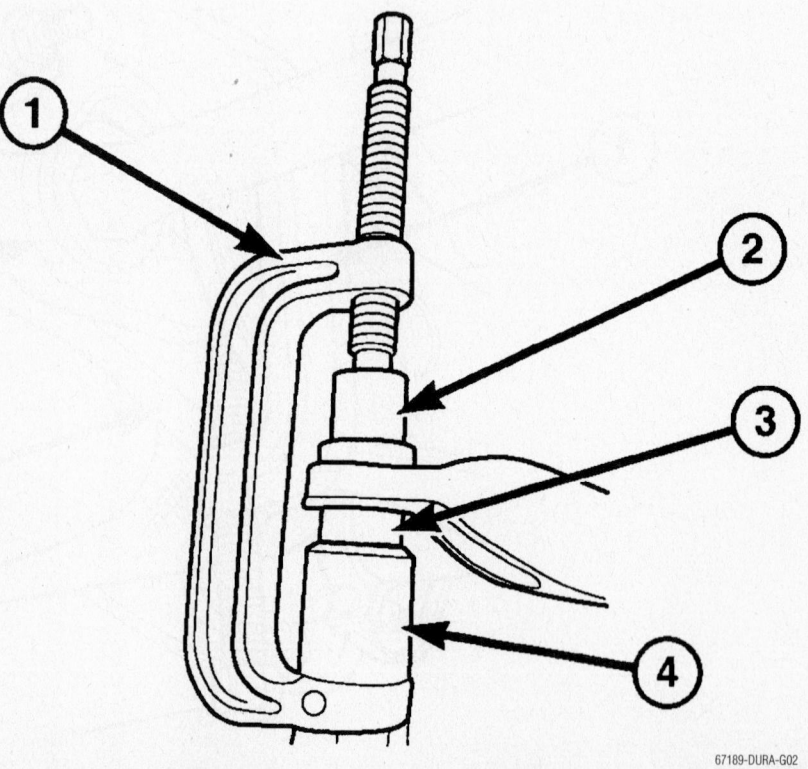

67189-DURA-G02

Installing the lower ball joint

9. Install the ball joint (3) into the control arm and press in using special tools C-4212-F (press) (1), 9331-1 (receiver) (2) and 9331-3 (driver) (4).

10. Stake the ball joint flange in four evenly spaced places around the ball joint flange, using a chisel and hammer.

11. Remove the support for the halfshaft and install into position (4wd only).

12. Install the steering knuckle.

13. Install the tie rod end into the steering knuckle.

14. Install and tighten the halfshaft nut to 251 Nm (185 ft. lbs.).

15. Install the brake caliper and rotor.

16. Install the tire and wheel assembly.

17. Check the vehicle ride height.

18. Perform a wheel alignment.

Upper Control Arm

REMOVAL & INSTALLATION

2001–03

1. Before servicing the vehicle, refer to the precautions in the beginning of this section.

2. Support the lower control arm.

3. Remove or disconnect the following:

- Front wheel
- Brake hose brackets
- Upper ball joint
- Pivot mounting nuts
- Upper control arm

To install:

4. Install or connect the following:

- Upper control arm. Tighten the pivot nuts to 155 ft. lbs. (210 Nm).
- Upper ball joint. Tighten the nut to 60 ft. lbs. (81 Nm).
- Brake hose brackets
- Front wheel

5. Check the wheel alignment and adjust as necessary.

2004

1. Raise and support vehicle.

2. Remove wheel and tire assembly.

3. Remove the nut from upper ball joint.

4. Separate upper ball joint from the steering knuckle.

➡**When installing the tool to separate the ball joint, be careful not to damage the ball joint seal.**

5. Remove the control arm pivot bolts and remove control arm.

To install:

6. Position the control arm into the frame brackets. Install bolts and tighten to 102 Nm (75 ft. lbs.).

7. Insert ball joint in steering knuckle and tighten ball joint nut to 75 Nm (55 ft. lbs.).

8. Install the wheel and tire assembly.

9. Remove the support and lower vehicle.

10. Perform a wheel alignment.

CONTROL ARM BUSHING REPLACEMENT

The control arm bushings are serviced with the control arm as an assembly.

Lower Control Arm

REMOVAL & INSTALLATION

2001–03

2-WHEEL DRIVE

1. Before servicing the vehicle, refer to the precautions in the beginning of this section.

2. Remove or disconnect the following:

- Front wheel
- Shock absorber
- Brake caliper and rotor
- Stabilizer bar link
- Coil spring
- Inner mounting bolts
- Lower control arm

To install:

3. Install or connect the following:

- Lower control arm. Tighten the front bolt to 130 ft. lbs. (175 Nm) and the rear bolt to 80 ft. lbs. (108 Nm).
- Coil spring. Tighten the lower ball joint nut to 94 ft. lbs. (127 Nm).
- Stabilizer bar link
- Brake caliper and rotor
- Shock absorber
- Front wheel

4-WHEEL DRIVE

1. Before servicing the vehicle, refer to the precautions in the beginning of this section.

2. Remove or disconnect the following:

- Front wheel
- Front driveshaft
- Torsion bar
- Shock absorber
- Stabilizer bar
- Lower ball joint
- Pivot bolts
- Lower control arm

To install:

3. Install or connect the following:

- Lower control arm. Tighten the front pivot bolt to 80 ft. lbs. (108 Nm) and the rear bolt to 140 ft. lbs. (190 Nm).
- Lower ball joint. Tighten the nut to 135 ft. lbs. (183 Nm).
- Stabilizer bar
- Shock absorber
- Torsion bar
- Front driveshaft
- Front wheel

2004

1. Raise and support the vehicle.

2. Remove the wheel and tire assembly.

3. Remove the disc brake caliper assembly.

4. Remove the disc brake rotor.

5. Disconnect the wheel speed sensor at the wheel well.

6. Disconnect the tie rod from the knuckle.

7. Remove the front halfshaft nut (4wd).

8. Remove the upper ball joint nut. Separate the upper ball joint from the steering knuckle.

9. Remove the lower ball joint nut. Separate the lower ball joint from the steering knuckle.

10. Remove the steering knuckle.

11. Unload the torsion bar using special tool 8686 (1).

12. Remove the torsion bar from the vehicle.

13. Remove the stabilizer bar link.

14. Remove the shock absorber lower bolt.

15. Remove the front and rear pivot bolts.

16. Remove the lower control arm from the vehicle.

To install:

17. Position the lower control arm at the frame rail brackets. Install the pivot bolts and nuts. Tighten the nuts finger-tight.

➡**The ball joint stud taper must be CLEAN and DRY before installing the knuckle.**

Clean the stud taper with mineral spirits to remove dirt and grease.

18. Install the steering knuckle.

19. Insert the lower ball joint into the steering knuckle. Install and tighten the retaining nut to 81 Nm (60 ft. lbs.).

20. Install the torsion bar.

21. Install shock absorber lower bolt and tighten to 81 Nm (60 ft. lbs.).

22. Install the front halfshaft nut (4wd).

23. Insert the upper ball joint into the steering knuckle. Install and tighten the retaining nut to 75 Nm (55 ft. lbs.).

24. Install the stabilizer bar link and tighten to 169 Nm (125 ft. lbs.).

25. Tighten the lower control arm pivot nut and bolts to 244 Nm (180 ft. lbs.).

26. Insert the outer tie rod end into the steering knuckle (1). Install and tighten the retaining nut to 75 Nm (55 ft. lbs.).

27. Install the disc brake rotor.

28. Install the disc brake caliper and adaptor assembly and tighten to 135 Nm (100 ft. lbs.).

29. Install the wheel and tire assembly.

30. Remove the support and lower the vehicle.

31. Adjust the front suspension height and perform a wheel alignment.

CONTROL ARM BUSHING REPLACEMENT

2001–03

The control arm bushings are serviced with the control arm as an assembly.

2004

1. Remove lower control.
2. Secure the control arm in a vise.
3. Press the bushings from the control arm.
4. Installation is the reverse of removal.

Wheel Bearings

ADJUSTMENT

Durango models utilize a hub/bearing assembly which is not adjustable.

REMOVAL & INSTALLATION

2001–03

2-WHEEL DRIVE

1. Before servicing the vehicle, refer to the precautions in the beginning of this section.

2. Remove or disconnect the following:
   • Front wheel
   • Brake caliper and rotor
   • Spindle nut
   • Hub and bearing assembly

To install:

3. Install or connect the following:
   • Hub and bearing assembly
   • Spindle nut. Tighten the nut to 185 ft. lbs. (251 Nm).
   • Brake caliper and rotor
   • Front wheel

4-WHEEL DRIVE

1. Before servicing the vehicle, refer to the precautions in the beginning of this section.

2. Remove or disconnect the following:
   • Front wheel
   • Brake caliper and rotor
   • Hub retainer nut
   • Hub and bearing assembly

To install:

3. Install or connect the following:
   • Hub and bearing assembly. Tighten the bolts to 123 ft. lbs. (166 Nm).
   • Hub retainer nut. Tighten the nut to 173 ft. lbs. (235 Nm).
   • Brake caliper and rotor
   • Front wheel

2004

1. Raise and support the vehicle.
2. Remove the wheel and tire assembly.
3. Remove the brake caliper and rotor.
4. Remove the ABS wheel speed sensor if equipped,
5. Remove the halfshaft nut, (4wd only).

➡**Do not strike the knuckle with a hammer to remove the tie rod end or the ball joint. Damage to the steering knuckle will occur.**

6. Remove the tie rod end nut and separate the tie rod from the knuckle.

7. Remove the upper ball joint nut and separate the upper ball joint from the knuckle.

8. Pull down on the steering knuckle to separate the halfshaft from the hub/bearing, (4wd)

9. Remove the three hub/bearing mounting bolts (1) from the steering knuckle.

10. Slide the hub/bearing out of the steering knuckle.

11. Remove the brake dust shield.

To install:

12. Install the brake dust shield.

13. Install the hub/bearing into the steering knuckle and tighten the bolts to 163 Nm (120 ft. lbs.).

14. Install the brake rotor and caliper.

15. Install the ABS wheel speed sensor if equipped.

16. Install the upper ball joint nut to the steering knuckle and tighten to 75 Nm (55 ft. lbs.).

17. Install the tie rod end nut to the steering knuckle and tighten to 81 Nm (60 ft. lbs.).

18. Install the halfshaft nut and tighten to 251 Nm (185 ft. lbs.) (4wd only).

19. Install the wheel and tire assembly.

20. Remove the support and lower vehicle.

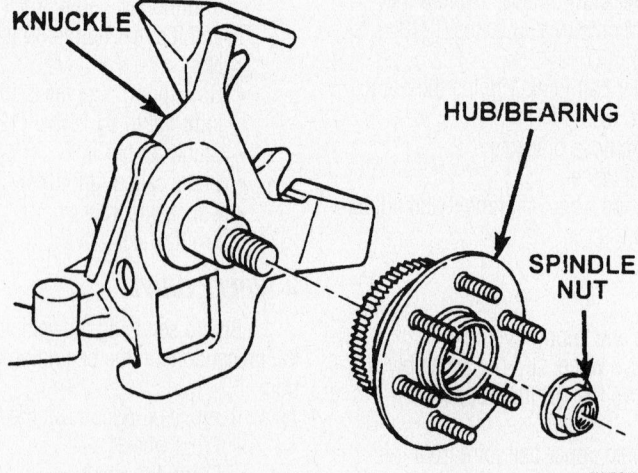

Hub/bearing assembly—2 wheel drive

BRAKES

Brake Caliper

REMOVAL & INSTALLATION

Front

2001–03

1. Remove the wheels.
2. Disconnect the rubber brake hose from the tubing at the frame mount. If the pistons are to be removed from the caliper, leave the brake hose connected to the caliper. Check the rubber hose for cracks or chafed spots.
3. Plug the brake line to prevent loss of fluid.
4. Remove the caliper slide pins.
5. Remove the caliper and brake pads from the rotor adapter.

To install:

6. Position the outboard shoe in the caliper. The shoe should not rattle in the caliper. If it does, or if any movement is obvious, bend the shoe tabs over the caliper to tighten the fit.
7. Slide the caliper into position on the adapter and over the rotor.
8. Align the caliper and start the pins in by hand.
9. Tighten the pins to 22 ft. lbs. (30 Nm) on 2001–02; 23.5 ft. lbs. on 2003 models.
10. Connect the brake hose to the caliper. Use new washers to attach the hose fitting if the original washers are scored, worn or damaged. Torque the bolt to 21 ft. lbs. (28 Nm).
11. Fill and bleed the brake system.
12. Install the wheels.
13. Lower the vehicle.

2004

1. Install prop rod on the brake pedal to keep pressure on the brake system.
2. Raise and support the vehicle.
3. Remove the tire and wheel assembly.
4. Compress the disc brake caliper.
5. Remove the banjo bolt and discard the copper washers.
6. Remove the caliper slide pin bolts.
7. Remove the disc brake caliper from the caliper adapter.

To install:

➡**Install a new copper washers on the banjo bolt when installing**

8. Install the disc brake caliper on the brake caliper adapter.

✳✳ CAUTION

Verify brake hose is not twisted or kinked before tightening fitting bolt.

9. Install the banjo bolt with new copper washers on the caliper. Tighten to 28 Nm (250 inch lbs.)
10. Install the caliper slide pin bolts. Tighten to 32 Nm (24 ft. lbs.)
11. Remove the prop rod.
12. Bleed the base brake system.
13. Install the tire and wheel assembly.

Rear

2001–03

1. Remove the wheels.
2. Disconnect the rubber brake hose from the tubing at the frame mount. If the pistons are to be removed from the caliper, leave the brake hose connected to the caliper. Check the rubber hose for cracks or chafed spots.
3. Plug the brake line to prevent loss of fluid.
4. Remove the caliper slide pins.
5. Remove the caliper and brake pads from the rotor adapter.
6. Remove the anti-rattle springs.

To install:

7. Position the outboard shoe in the caliper. The shoe should not rattle in the caliper. If it does, or if any movement is obvious, bend the shoe tabs over the caliper to tighten the fit.
8. Slide the caliper into position on the adapter and over the rotor.
9. Align the caliper and start the pins in by hand.
10. Tighten the pins to 22 ft. lbs. (30 Nm) on 2001–02; 23.5 ft. lbs. on 2003 models.
11. Connect the brake hose to the caliper. Use new washers to attach the hose fitting if the original washers are scored, worn or damaged.
12. Fill and bleed the brake system.
13. Install the wheels.
14. Lower the vehicle.

2004

1. Install prop rod on the brake pedal to keep pressure on the brake system.
2. Raise and support vehicle.
3. Remove the wheel and tire assembly.
4. Drain small amount of fluid from master cylinder brake reservoir with suction gun.
5. Remove the brake hose banjo bolt if replacing caliper.
6. Remove the caliper mounting slide pin bolts.
7. Remove the caliper from vehicle.

To install:

8. Install caliper (4) to the caliper adapter.
9. Coat the caliper mounting slide pin bolts with silicone grease. Then install and tighten the bolts to 15 Nm (11 ft. lbs.).
10. Install the brake hose banjo bolt and new copper seal washers if caliper was removed.
11. Install the brake hose to the caliper with and tighten fitting bolt to 28 Nm (250 inch lbs.).

✳✳ CAUTION

Verify brake hose is not twisted or kinked before tightening fitting bolt.

12. Remove the prop rod from the vehicle.
13. Bleed the base brake system.
14. Install the wheel and tire assemblies.
15. Remove the supports and lower the vehicle.
16. Verify a firm pedal before moving the vehicle.

Disc Brake Pads

REMOVAL & INSTALLATION

2001–03

FRONT

1. Remove the wheels.
2. Press the caliper piston back into the bore with a suitable prytool. Use a large C-clamp to drive the piston into the bore of additional force is required.
3. Remove the caliper mounting bolts with a ⅜ in. hex wrench or socket.
4. Rotate the caliper rearward off the rotor and out from its mount.
5. Set the caliper on a crate or sturdy box, then remove the inboard and outboard brake pads. The inboard pad has a spring clip that holds it in the caliper. Tilt this pad out at the top to unseat the clip. The outboard pad has a retaining spring that secures it in the caliper. Unseat 1 spring end and rotate the pad out of the caliper.
6. Secure the caliper to a chassis or suspension component with a sturdy wire. Do not let it hang from the hose.

To install:

7. Clean the caliper and steering knuckle sliding surfaces with a wire brush. Then, apply a coat of Mopar® multi-mileage grease or equivalent.

8. Clean the caliper slide pins with brake cleaner or brake fluid. Then apply a light coating of silicone grease to the pins.

➡**If there is minor rust or corrosion on the pins, first polish them with a crocus cloth. If they are severely rusted, replace them.**

9. Install the inboard pad and its spring clip.

10. Install the outboard brake pad.

11. Install the caliper over the rotor and seat it in its original position until flush.

12. Final tighten the caliper slide pins to 22 ft. lbs. (30 Nm).

13. Install the wheels.

14. Lower the vehicle.

15. Pump the brakes several times to seat the pads.

REAR

1. Remove the wheels.

2. Press the caliper piston back into the bore with a suitable prytool. Use a large C-clamp to drive the piston into the bore of additional force is required.

3. Remove the caliper mounting bolts with a ⅜ in. hex wrench or socket.

4. Rotate the caliper rearward off the rotor and out from its mount.

5. Set the caliper on a crate or sturdy box, then remove the inboard and outboard brake pads. The inboard pad has a spring clip that holds it in the caliper. Tilt this pad out at the top to unseat the clip. The outboard pad has a retaining spring that secures it in the caliper. Unseat 1 spring end and rotate the pad out of the caliper.

6. Secure the caliper to a chassis or suspension component with a sturdy wire. Do not let it hang from the hose.

To install:

7. Clean the caliper, anti-rattle springs and steering knuckle sliding surfaces with a wire brush. Then, apply a coat of Mopar® multi-mileage grease or equivalent.

8. Clean the caliper slide pins with brake cleaner or brake fluid. Then apply a light coating of silicone grease to the pins.

➡**If there is minor rust or corrosion on the pins, first polish them with a crocus cloth. If they are severely rusted, replace them.**

9. Install the inboard pad and its spring clip.

10. Install the outboard brake pad.

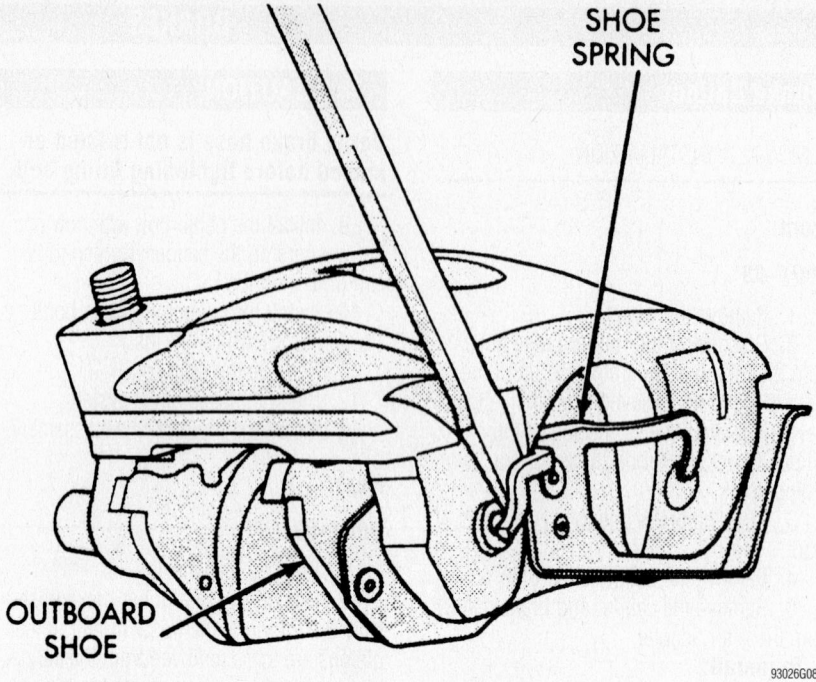

Prying the disc brake from the 4WD front brake caliper assembly

11. Install the caliper over the rotor and seat it in its original position until flush.

12. Final tighten the caliper slide pins to 22 ft. lbs. (30 Nm).

13. Install the wheels.

14. Lower the vehicle.

15. Pump the brakes several times to seat the pads.

2004

FRONT OR REAR

1. Raise and support vehicle.

2. Remove the wheel and tire assemblies.

3. Compress the caliper.

4. Remove the caliper.

5. Remove the caliper by tilting the top up and off the caliper adapter.

➡**Do not allow brake hose to support caliper assembly.**

6. Support and hang the caliper.

7. Remove the inboard brake shoe from the caliper adapter.

8. Remove the outboard brake shoe from the caliper adapter.

➡**Anti-rattle springs are not inter-changeable.**

9. Remove the top anti-rattle springs from the caliper adapter.

10. Remove the bottom anti-rattle springs from the caliper adapter.

To install:

11. Bottom pistons in caliper bore with

C-clamp. Place an old brake shoe between a C-clamp and caliper piston.

12. Clean caliper mounting adapter and anti-rattle springs.

13. Lubricate anti-rattle springs with brake grease.

➡**Anti-rattle springs are not inter-changeable.**

14. Install the bottom anti-rattle springs.

15. Install the top anti-rattle springs.

16. Install inboard brake shoe in adapter.

17. Install outboard brake shoe in adapter.

18. Tilt the top of the caliper over rotor and under adapter. Then push the bottom of the caliper down onto the adapter.

19. Install caliper.

20. Install wheel and tire assemblies and lower vehicle.

21. Apply brakes several times to seat caliper pistons and brake shoes and obtain firm pedal.

22. Top off master cylinder fluid level.

Brake Drums

REMOVAL & INSTALLATION

Chrysler Servo Type With Single Anchor

1. Raise and safely support the truck.

2. Remove the plug from the brake adjustment access hole.

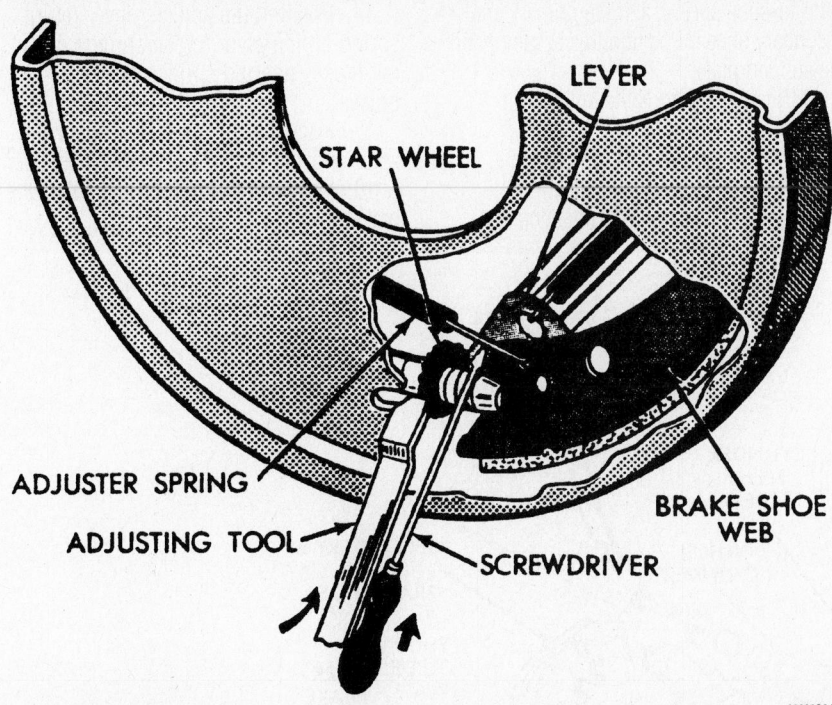

STAR WHEEL

LEVER

ADJUSTER SPRING

ADJUSTING TOOL

SCREWDRIVER

BRAKE SHOE WEB

93026G04

Use a lever releasing tool to depress the adjuster lever while turning the starwheel with a pry-tool—Bendix brakes

3. Insert a thin bladed screwdriver through the adjusting hole and hold the adjusting lever away from the starwheel.

4. Release the brake by prying down against the starwheel with a brake spoon.

5. Remove the rear wheel and clips from the wheel studs. Remove the brake drum.

6. Installation is the reverse of removal. Adjust the brakes.

Bendix Duo-Servo Type

1. Raise and safely support the vehicle.
2. Remove the rear wheel and tire.
3. Remove the axle shaft nuts, washers and cones. If the cones do not readily release, rap the axle shaft sharply in the center.
4. Remove the axle shaft.
5. Remove the outer hub nut.
6. Straighten the lockwasher tab and remove it along with the inner nut and bearing.
7. Carefully remove the drum.
To install:
8. Position the drum on the axle housing.
9. Install the bearing and inner nut. While rotating the wheel and tire, tighten the adjusting nut until a slight drag is felt.
10. Back off the adjusting nut 1/6 turn so that the wheel rotates freely without excessive end-play.

11. Install the lockrings and nut. Place a new gasket on the hub and install the axle shaft, cones, lockwashers and nuts.
12. Install the wheel and tire.
13. Road-test the vehicle.

Brake Shoes

REMOVAL & INSTALLATION

Servo Type With Single Anchor

1. Raise and support the vehicle.
2. Remove the rear wheel, drum retaining clips and the brake drum.
3. Remove the brake shoe return springs, noting how the secondary spring overlaps the primary spring.
4. Remove the brake shoe retainer, springs and nails.
5. Disconnect the automatic adjuster cable from the anchor and unhook it from the lever. Remove the cable, cable guide, and anchor plate.
6. Remove the spring and lever from the shoe web.
7. Spread the anchor ends of the primary and secondary shoes and remove the parking brake spring and strut.
8. Disconnect the parking brake cable and remove the brake assembly.
9. Remove the primary and secondary brake shoe assemblies and the star adjuster

as an assembly. Block the wheel cylinders to retain the pistons.
To install:
10. Measure the drum as described in this section.
11. Apply a thin coat of lubricant to the support platforms.
12. Attach the parking brake lever to the rear of the secondary shoe.
13. Place the primary and secondary shoes in their relative positions on a work-bench.
14. Lubricate the adjuster screw threads. Install it between the primary and secondary shoes with the star wheel next to the secondary shoe. The star wheels are stamped with an **L** (left) and **R** (right).
15. Overlap the ends of the primary and second brake shoes and install the adjusting spring and lever at the anchor end.
16. Hold the shoes in position and install the parking brake cable into the lever.
17. Install the parking brake strut and spring between the parking brake lever and primary shoe.
18. Place the brake shoes on the support and install the retainer nails and springs.
19. Install the anchor pin plate.
20. Install the eye of the adjusting cable over the anchor pin and install the return spring between the anchor pin and primary shoe.
21. Install the cable guide in the secondary shoe and install the secondary return spring. Be sure that the primary spring overlaps the secondary spring.
22. Position the adjusting cable in the groove of the cable guide and engage the hook of the cable in the adjusting lever.
23. Install the brake drum and retaining clips. Install the wheel and tire.
24. Adjust the brakes and road-test the truck.

Bendix Duo-Servo Type

1. Unhook and remove the adjusting lever return spring.
2. Remove the lever from the lever pivot pin.
3. Unhook the adjuster lever from the adjuster cable.
4. Unhook the upper shoe-to-shoe spring.
5. Unhook and remove the shoe hold-down springs.
6. Disconnect the parking brake cable from the parking brake lever.
7. Remove the shoes with the lower shoe-to-shoe spring and star wheel as an assembly.

To install:

8. The pivot screw and adjusting nut on the left side have left-hand threads and right-hand threads on the right side.

9. Lubricate and assemble the star wheel assembly. Lubricate the guide pads on the support plates.

10. Assemble the star wheel, lower shoe-to-shoe spring, and the primary and secondary shoes. Position this assembly on the support plate.

11. Install and hook the hold-down springs.

12. Install the upper shoe-to-shoe spring.

13. Install the cable and retaining clip.

14. Position the adjuster lever return spring on the pivot (green springs on left brakes and red springs on right brakes).

15. Install the adjuster lever. Route the adjuster cable and connect it to the adjuster.

16. Install the brake drum and adjust the brakes.

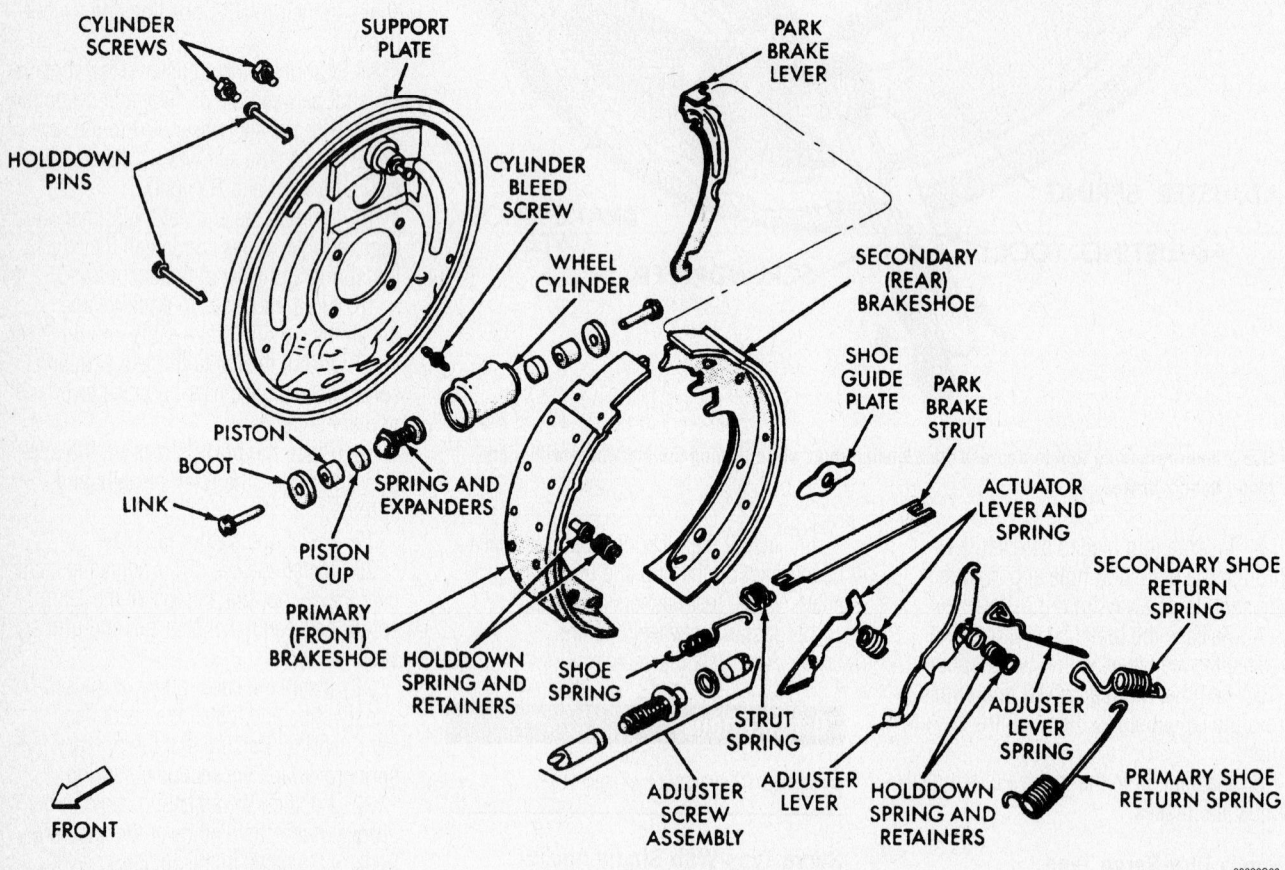

FRONT

Exploded view of the rear brake components

93026G09

DODGE AND PLYMOUTH

8

Neon

SPECIFICATION CHARTS

ENGINE AND VEHICLE IDENTIFICATION

Code ①	Liters	Cu. In. (cc)	Cyl.	Fuel Sys.	Engine Type	Eng. Mfg.
C	2.0	122 (1996)	4	MFI	SOHC	Chrysler
F	2.0	122 (1996)	4	MFI	SOHC	Chrysler
S	2.4	148 (2421)	4	MFI-Turbo	DOHC	Chrysler

Code ②	Year
1	2001
2	2002
3	2003
4	2004
5	2005

MFI: Multi-point Fuel Injection

SOHC: Single Overhead Camshaft

① 8th position of VIN

② 10th position of VIN

67189-NEON-C01

GENERAL ENGINE SPECIFICATIONS
All measurements are given in inches.

Year	Model	Engine Displacement Liters	Engine Series (ID/VIN)	Net Horsepower @ rpm	Net Torque @ rpm (ft. lbs.)	Bore x Stroke (in.)	Compression Ratio	Oil Pressure @ rpm
2001	Neon	2.0	C	132@6000	129@5000	3.44x3.26	9.8:1	25-80@3000
	Neon	2.0	F	150@4400	133@5500	3.44x3.26	9.8:1	25-80@3000
2002	Neon	2.0	C	132@6000	129@5000	3.44x3.26	9.8:1	25-80@3000
	Neon	2.0	F	150@4400	133@5500	3.44x3.26	9.8:1	25-80@3000
2003	Neon	2.0	C	132@6000	129@5000	3.44x3.26	9.8:1	25-80@3000
	Neon	2.0	F	150@4400	133@5500	3.44x3.26	9.8:1	25-80@3000
	Neon	2.4	S	230@5300	250@4400	3.44x3.97	8.1:1	25-80@3000
2004	Neon	2.0	C	132@6000	129@5000	3.44x3.26	9.8:1	25-80@3000
	Neon	2.0	F	150@4400	133@5500	3.44x3.26	9.8:1	25-80@3000
	Neon	2.4	S	230@5300	250@4400	3.44x3.97	8.1:1	25-80@3000

NA: Not available

67189-NEON-C02

ENGINE TUNE-UP SPECIFICATIONS

Year	Engine Displacement Liters	Engine ID/VIN	Spark Plugs Gap (in.)	Ignition Timing (deg.) MT	Ignition Timing (deg.) AT	Fuel Pump (psi)	Idle Speed (rpm) MT	Idle Speed (rpm) AT	Valve Clearance In.	Valve Clearance Ex.
2001	2.0	C	0.035	①	①	48	②	②	HYD	HYD
	2.0	F	0.035	①	①	48	②	②	HYD	HYD
2002	2.0	C	0.035	①	①	48	②	②	HYD	HYD
	2.0	F	0.035	①	①	48	②	②	HYD	HYD
2003	2.0	C	0.035	①	①	48	②	②	HYD	HYD
	2.0	F	0.035	①	①	48	②	②	HYD	HYD
	2.4	S	0.035	—	①	48	②	②	HYD	HYD
2004	2.0	C	0.035	①	①	48	②	②	HYD	HYD
	2.0	F	0.035	①	①	48	②	②	HYD	HYD
	2.4	S	0.035	—	①	48	②	②	HYD	HYD

NOTE: The Vehicle Emission Control Information label often reflects specification changes made during production. The label figures must be used if they differ from those in this chart.

HYD : Hydraulic

① Refer to the Vehicle Emission Control Information label for correct timing specifications with a range of +/- 2 degrees.

Ignition timing cannot be adjusted. Base engine timing is set at TDC during assembly.

② Refer to Vehicle Emissions Control Information label for proper specification

67189-NEON-C03

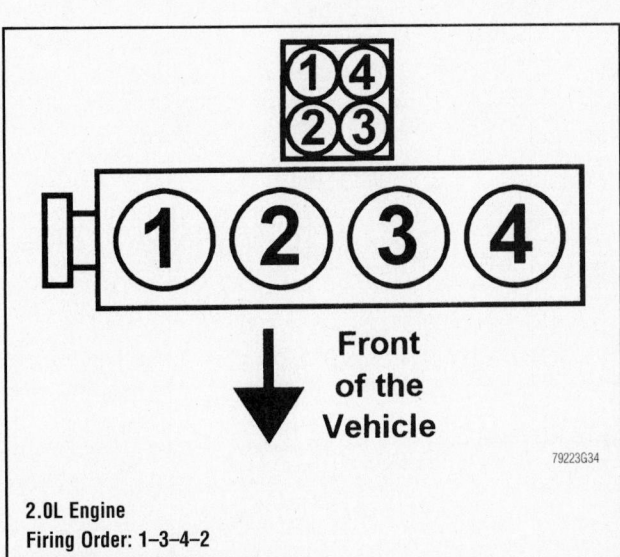

2.0L Engine
Firing Order: 1–3–4–2
Distributorless ignition system

Front of the Vehicle

79223G34

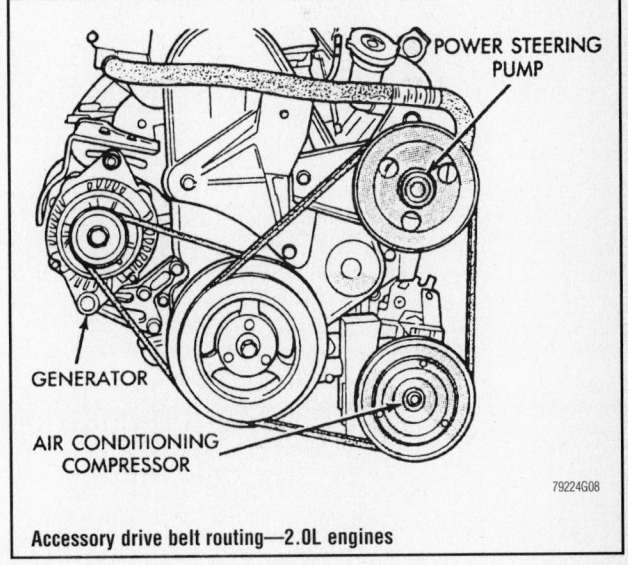

POWER STEERING PUMP

GENERATOR

AIR CONDITIONING COMPRESSOR

Accessory drive belt routing—2.0L engines

79224G08

CAPACITIES

Year	Model	Engine Displacement Liters	Engine ID/VIN	Engine Oil with Filter (qts.)	Transmission (pts.)		Fuel Tank (gal.)	Cooling System (qts.)
					5-Spd	Auto.		
2001	Neon	2.0	C	4.5	5.0-5.6	8.0	12.5	6.5
	Neon	2.0	F	4.5	5.0-5.6	8.0	12.5	6.5
2002	Neon	2.0	C	4.5	5.0-5.6	8.0	12.5	6.5
	Neon	2.0	F	4.5	5.0-5.6	8.0	12.5	6.5
2003	Neon	2.0	C	4.5	5.0-5.6	8.0	12.5	6.5
	Neon	2.0	F	4.5	5.0-5.6	8.0	12.5	6.5
	Neon	2.4	S	5.0	4.8-5.2 ①	—	12.5	6.5
2004	Neon	2.0	C	4.5	5.0-5.6	8.0	12.5	6.5
	Neon	2.0	F	4.5	5.0-5.6	8.0	12.5	6.5
	Neon	2.4	S	5.0	4.8-5.2 ①	—	12.5	6.5

NOTE: All capacities are approximate. Add fluid gradually and ensure a proper fluid level is obtained.

① Includes 4 oz. Of Mopar Limited Slip Additive (P/N 04318060AB)

67189-NEON-C04

VALVE SPECIFICATIONS

Year	Engine Displacement Liters	Engine ID/VIN	Seat Angle (deg.)	Face Angle (deg.)	Spring Test Pressure (lbs. @ in.)	Spring Installed Height (in.)	Stem-to-Guide Clearance (in.)		Stem Diameter (in.)	
							Intake	Exhaust	Intake	Exhaust
2001	2.0	C	45	45-45.5	70@1.57	1.58	0.0018-0.0025	0.0029-0.0037	0.234	0.233
	2.0	F	45	45-45.5	72@1.57	1.58	0.0018-0.0025	0.0029-0.0037	0.234	0.233
2002	2.0	C	45	45-45.5	70@1.57	1.58	0.0018-0.0025	0.0029-0.0037	0.234	0.233
	2.0	F	45	45-45.5	72@1.57	1.58	0.0018-0.0025	0.0029-0.0037	0.234	0.233
2003	2.0	C	45	45-45.5	70@1.57	1.58	0.0018-0.0025	0.0029-0.0037	0.234	0.233
	2.0	F	45	45-45.5	72@1.57	1.58	0.0018-0.0025	0.0029-0.0037	0.234	0.233
	2.4	S	45	45-45.5	75@1.49	1.49	0.0018-0.0025	0.0029-0.0037	0.233-0.234	0.232-0.233
2004	2.0	C	45	45-45.5	70@1.57	1.58	0.0018-0.0025	0.0029-0.0037	0.234	0.233
	2.0	F	45	45-45.5	72@1.57	1.58	0.0018-0.0025	0.0029-0.0037	0.234	0.233
	2.4	S	45	45-45.5	75@1.49	1.49	0.0018-0.0025	0.0029-0.0037	0.233-0.234	0.232-0.233

67189-NEON-C05

CRANKSHAFT AND CONNECTING ROD SPECIFICATIONS

All measurements are given in inches.

Year	Engine Displacement Liters	Engine ID/VIN	Crankshaft				Connecting Rod		
			Main Brg. Journal Dia.	Main Brg. Oil Clearance	Shaft End-play	Thrust on No.	Journal Diameter	Oil Clearance	Side Clearance
2001	2.0	C	2.0469-2.0475	0.0008-0.0024	0.0035-0.0094	3	1.8894-1.8900	0.0010-0.0023	0.0050-0.0150
	2.0	F	2.0469-2.0475	0.0008-0.0024	0.0035-0.0094	3	1.8894-1.8900	0.0010-0.0023	0.005-0.0015
2002	2.0	C	2.0469-2.0475	0.0008-0.0024	0.0035-0.0094	3	1.8894-1.8900	0.0010-0.0023	0.0050-0.0150
	2.0	F	2.0469-2.0475	0.0008-0.0024	0.0035-0.0094	3	1.8894-1.8900	0.0010-0.0023	0.005-0.0015
2003	2.0	C	2.0469-2.0475	0.0008-0.0024	0.0035-0.0094	3	1.8894-1.8900	0.0010-0.0023	0.0050-0.0150
	2.0	F	2.0469-2.0475	0.0008-0.0024	0.0035-0.0094	3	1.8894-1.8900	0.0010-0.0023	0.005-0.0015
	2.4	S	2.3620-2.3625	0.0007-0.0024	0.0035-0.0094	3	1.9680-1.9865	0.0009-0.0027	0.005-0.0015
2004	2.0	C	2.0469-2.0475	0.0008-0.0024	0.0035-0.0094	3	1.8894-1.8900	0.0010-0.0023	0.0050-0.0150
	2.0	F	2.0469-2.0475	0.0008-0.0024	0.0035-0.0094	3	1.8894-1.8900	0.0010-0.0023	0.005-0.0015
	2.4	S	2.3620-2.3625	0.0007-0.0024	0.0035-0.0094	3	1.9680-1.9865	0.0009-0.0027	0.005-0.0015

67189-NEON-C06

TORQUE SPECIFICATIONS
All readings in ft. lbs.

Year	Engine Displacement Liters	Engine ID/VIN	Cylinder Head Bolts	Main Bearing Bolts	Rod Bearing Bolts	Crankshaft Damper Bolts	Flywheel Bolts	Manifold		Spark Plugs	Oil Pan Drain Plug
								Intake	Exhaust		
2001	2.0	C	①	②	③	100	70	8.5	17	20	20
	2.0	F	①	②	③	100	70	8.5	17	20	20
2002	2.0	C	①	④	③	100	70	8.5	17	20	20
	2.0	F	①	④	③	100	70	8.5	17	20	20
2003	2.0	C	①	④	③	100	70	8.5	17	20	20
	2.0	F	①	④	③	100	70	8.5	17	20	20
	2.4	S	①	⑤	③	100	70	17	21	11-15	20
2004	2.0	C	①	④	③	100	70	8.5	17	20	20
	2.0	F	①	④	③	100	70	8.5	17	20	20
	2.4	S	①	⑤	③	100	70	17	21	11-15	20

① Step 1: 25 ft. lbs.
Step 2: 50 ft. lbs.
Step 3: 50 ft. lbs.
Step 4: plus 90 degrees

② M8: 25 ft. lbs.
M11: 60 ft. lbs.

③ Step 1: 20 ft. lbs.
Step 2: plus 90 degrees

④ Step 1:
M10 bolts: 15 ft. lbs.
M8 bolts: 11 ft. lbs.
Step 2:
M10 bolts: 30 ft. lbs.
Step 3:
M10 bolts: 30 ft. lbs.
Step 4:
M10 bolts: plus 90 degrees
M8 bolts: 22 ft. lbs.

⑤ M8 bolts: 21 ft. lbs.
M11 bolts: 55 ft. lbs.

67189-NEON-C07

PISTON AND RING SPECIFICATIONS
All measurements are given in inches.

Year	Engine Displacement Liters	Engine ID/VIN	Piston Clearance	Ring Gap Top Compression	Ring Gap Bottom Compression	Ring Gap Oil Control	Ring Side Clearance Top Compression	Ring Side Clearance Bottom Compression	Ring Side Clearance Oil Control
2001	2.0	C	0.0008-0.0020	0.0090-0.0200	0.0190-0.0310	0.009-0.026	0.0010-0.0026	0.0010-0.0026	0.0002-0.0070
	2.0	F	0.0008-0.0020	0.0090-0.0200	0.0190-0.0310	0.009-0.0260	0.0010-0.0026	0.0010-0.0026	0.0002-0.0070
2002	2.0	C	0.0008-0.0020	0.0090-0.0200	0.0190-0.0310	0.009-0.026	0.0010-0.0026	0.0010-0.0026	0.0002-0.0070
	2.0	F	0.0008-0.0020	0.0090-0.0200	0.0190-0.0310	0.009-0.0260	0.0010-0.0026	0.0010-0.0026	0.0002-0.0070
2003	2.0	C	0.0008-0.0020	0.0090-0.0200	0.0190-0.0310	0.009-0.026	0.0010-0.0026	0.0010-0.0026	0.0002-0.0070
	2.0	F	0.0008-0.0020	0.0090-0.0200	0.0190-0.0310	0.009-0.0260	0.0010-0.0026	0.0010-0.0026	0.0002-0.0070
	2.4	S	0.0009-0.0022	0.0078-0.0157	0.0070-0.0015	0.0050-0.0250	0.0010-0.0030	0.0010-0.0030	0.0010-0.0060
2004	2.0	C	0.0008-0.0020	0.0090-0.0200	0.0190-0.0310	0.009-0.026	0.0010-0.0026	0.0010-0.0026	0.0002-0.0070
	2.0	F	0.0008-0.0020	0.0090-0.0200	0.0190-0.0310	0.009-0.0260	0.0010-0.0026	0.0010-0.0026	0.0002-0.0070
	2.4	S	0.0009-0.0022	0.0078-0.0157	0.0070-0.0015	0.0050-0.0250	0.0010-0.0030	0.0010-0.0030	0.0010-0.0060

67189-NEON-C08

WHEEL ALIGNMENT

Year	Model		Caster Range (+/-Deg.)	Caster Preferred Setting (Deg.)	Camber Range (+/-Deg.)	Camber Preferred Setting (Deg.)	Toe-in (in.)
2001	Neon	F	+1.00	+2.60	+0.40	0	0.05 +/- 0.10
		R	—	—	+0.40	-0.25	0.15 +/- 0.10
2002	Neon	F	+1.00	+2.60	+0.40	0	0.05 +/- 0.10
		R	—	—	+0.40	-0.25	0.15 +/- 0.10
2003	Neon	F	+1.00	+2.60	+0.40	0	0.20 +/-0.20
		R	—	—	+0.40	-0.25	0.30 +/- 0.20
2004	Neon	F	+1.00	+2.60	+0.40	0	0.20 +/-0.20
		R	—	—	+0.40	-0.25	0.30 +/- 0.20

67189-NEON-C09

TIRE, WHEEL AND BALL JOINT SPECIFICATIONS

Year	Model	OEM Tires Standard	OEM Tires Optional	Tire Pressures (psi) Front	Tire Pressures (psi) Rear	Wheel Size	Ball Joint Inspection	Lug Nut (ft. lbs.)
2001	Neon, base	P185/65R14	None	30	30	5.5-JJ	①	100
	Neon SE	P185/60R15	None	32	32	5.5-JJ	①	100
	Neon ES	P185/60R15	None	32	32	6-JJ	①	100
	Neon ACR	P185/60R15	None	32	32	6-JJ	①	100
	Neon RT	P195/50R16	None	32	32	6-JJ	①	100
2002	Neon, base	P185/65R14	None	30	30	5.5-JJ	①	100
	Neon SE	P185/60R15	None	32	32	5.5-JJ	①	100
	Neon ES	P185/60R15	None	32	32	6-JJ	①	100
	Neon ACR	P185/60R15	None	32	32	6-JJ	①	100
	Neon RT	P195/50R16	None	32	32	6-JJ	①	100
2003	Neon, base	P185/65R14	None	30	30	5.5-JJ	①	100
	Neon SE	P185/60R15	None	32	32	5.5-JJ	①	100
	Neon ES	P185/60R15	None	32	32	6-JJ	①	100
	Neon ACR	P185/60R15	None	32	32	6-JJ	①	100
	Neon RT	P195/50R16	None	32	32	6-JJ	①	100
	Neon SRT-4	P205/50R17	None	32	32	6-JJ	①	100
2004	Neon, base	P185/65R14	None	30	30	5.5-JJ	①	100
	Neon SE	P185/60R15	None	32	32	5.5-JJ	①	100
	Neon ES	P185/60R15	None	32	32	6-JJ	①	100
	Neon ACR	P185/60R15	None	32	32	6-JJ	①	100
	Neon RT	P195/50R16	None	32	32	6-JJ	①	100
	Neon SRT-4	P205/50R17	None	32	32	6-JJ	①	100

OEM: Original Equipment Manufacturer

PSI: Pounds Per Square Inch

① Replace if any measurable movement is found.

67189-NEON-C10

BRAKE SPECIFICATIONS
All measurements in inches unless noted

Year	Model		Brake Disc Original Thickness	Brake Disc Minimum Thickness	Brake Disc Maximum Runout	Brake Drum Diameter Original Inside Diameter	Brake Drum Diameter Max. Wear Limit	Brake Drum Diameter Maximum Machine Diameter	Minimum Lining Thickness	Brake Caliper Bracket Bolts (ft. lbs.)	Brake Caliper Mounting Bolts (ft. lbs.)
2001	Neon	F	0.866	0.803	0.003	—	—	—	0.300	55	16
		R	0.354	0.285	0.003	7.88	NA	①	②	55	16
2002	Neon	F	0.866	0.803	0.005	—	—	—	0.300	55	16
		R	0.354	0.285	0.005	7.88	NA	①	②	55	16
2003	Neon	F	0.866	0.803	0.005	—	—	—	0.300	55	16
		R	0.354	0.285	0.005	7.88	NA	①	②	55	16
2004	Neon	F	0.866	0.803	0.005	—	—	—	0.300	55	16
		R	0.354	0.285	0.005	7.88	NA	①	②	55	16

NA: Not Available

F: Front

R: Rear

① Stamped on the outer edge of drum

② Disc brake pad total thickness: 0.281

Brake shoe lining thickness: 0.0625

67189-NEON-C11

SCHEDULED MAINTENANCE INTERVALS

2001-02 Dodge—Neon

TO BE SERVICED	TYPE OF SERVICE	VEHICLE MILEAGE INTERVAL (x1000)												
		7.5	15	22.5	30	37.5	45	52.5	60	67.5	75	82.5	90	97.5
Engine oil & filter	R	✓	✓	✓	✓	✓	✓	✓	✓	✓	✓	✓	✓	✓
Brake hoses	S/I	✓	✓	✓	✓	✓	✓	✓	✓	✓	✓	✓	✓	✓
Coolant level, hoses & clamps	S/I	✓	✓	✓	✓	✓	✓	✓	✓	✓	✓	✓	✓	✓
CV joints & front suspension components	S/I	✓	✓	✓	✓	✓	✓	✓	✓	✓	✓	✓	✓	✓
Exhaust system	S/I	✓	✓	✓	✓	✓	✓	✓	✓	✓	✓	✓	✓	✓
Manual transaxle oil	S/I	✓	✓	✓	✓	✓	✓	✓	✓	✓	✓	✓	✓	✓
Rotate tires	S/I	✓	✓	✓	✓	✓	✓	✓	✓	✓	✓	✓	✓	✓
Accessory drive belts	S/I		✓		✓		✓		✓		✓		✓	
Brake linings	S/I			✓			✓			✓			✓	
Air filter element	R				✓				✓				✓	
Spark plugs	R				✓				✓				✓	
Lubricate ball joints	S/I				✓				✓				✓	
Engine coolant	R						✓				✓			
PCV valve	S/I								✓				✓	
Ignition cables	R								✓					
Camshaft timing belt ①	R													

R: Replace   S/I: Service or Inspect

① Camshaft timing belt: replace at 105,000 miles

FREQUENT OPERATION MAINTENANCE (SEVERE SERVICE)

 If a vehicle is operated under any of the following conditions it is considered severe service:

- Extremely dusty areas

- 50% or more of the vehicle operation is in 32°C (90°F) or higher temperatures, or constant operation in temperatures below 0°C (32°F)

Prolonged idling (vehicle operation in stop and go traffic)

- Frequent short running periods (engine does not warm to normal operating temperatures)

- Police, taxi, delivery usage or trailer towing usage

Oil & oil filter change: change every 3000 miles

Rotate tires every 6000 miles

Brake linings: inspect every 12,000 miles

Air filter element: service or inspect every 15,000 miles

Automatic transaxle: change fluid & adjust bands every 15,000 miles

Manual transaxle fluid: replace every 15,000 miles

Engine coolant: replace at 36,000 miles and every 30,000 miles thereafter

67189-NEON-C12

SCHEDULED MAINTENANCE INTERVALS

2003-04 Dodge—Neon

TO BE SERVICED	TYPE OF SERVICE	VEHICLE MILEAGE INTERVAL (x1000)												
		6	12	18	24	30	36	42	48	54	60	66	72	78
Engine oil & filter	R	✓	✓	✓	✓	✓	✓	✓	✓	✓	✓	✓	✓	✓
Brake hoses	S/I	✓	✓	✓	✓	✓	✓	✓	✓	✓	✓	✓	✓	✓
Coolant level, hoses & clamps	S/I	✓	✓	✓	✓	✓	✓	✓	✓	✓	✓	✓	✓	✓
CV joints & front suspension components	S/I	✓	✓	✓	✓	✓	✓	✓	✓	✓	✓	✓	✓	✓
Exhaust system	S/I	✓	✓	✓	✓	✓	✓	✓	✓	✓	✓	✓	✓	✓
Manual transaxle oil	S/I	✓	✓	✓	✓	✓	✓	✓	✓	✓	✓	✓	✓	✓
Rotate tires	S/I	✓	✓	✓	✓	✓	✓	✓	✓	✓	✓	✓	✓	✓
Accessory drive belts	S/I		✓			✓			✓		✓		✓	
Brake linings	S/I			✓			✓			✓			✓	
Air filter element	R				✓				✓				✓	
Spark plugs	R				✓				✓				✓	
Lubricate ball joints	S/I				✓				✓				✓	
Engine coolant	R						✓				✓			
PCV valve	S/I								✓				✓	
Ignition cables	R								✓					
Camshaft timing belt ①	R													

R: Replace   S/I: Service or Inspect

① Camshaft timing belt: replace at 105,000 miles

FREQUENT OPERATION MAINTENANCE (SEVERE SERVICE)

If a vehicle is operated under any of the following conditions it is considered severe service:

- Extremely dusty areas
- 50% or more of the vehicle operation is in 32°C (90°F) or higher temperatures, or constant operation in temperatures below 0°C (32°F)

Prolonged idling (vehicle operation in stop and go traffic)

- Frequent short running periods (engine does not warm to normal operating temperatures)
- Police, taxi, delivery usage or trailer towing usage

Oil & oil filter change: change every 3000 miles

Rotate tires every 6000 miles

Brake linings: inspect every 12,000 miles

Air filter element: service or inspect every 15,000 miles

Automatic transaxle: change fluid & adjust bands every 15,000 miles

Manual transaxle fluid: replace every 15,000 miles

Engine coolant: replace at 36,000 miles and every 30,000 miles thereafter

67189-NEON-C13

PRECAUTIONS

Before servicing any vehicle, please be sure to read all of the following precautions, which deal with personal safety, prevention of component damage, and important points to take into consideration when servicing a motor vehicle:

• Never open, service or drain the radiator or cooling system when the engine is hot; serious burns can occur from the steam and hot coolant.

• Observe all applicable safety precautions when working around fuel. Whenever servicing the fuel system, always work in a well-ventilated area. Do not allow fuel spray or vapors to come in contact with a spark, open flame, or excessive heat (a hot drop light, for example). Keep a dry chemical fire extinguisher near the work area. Always keep fuel in a container specifically designed for fuel storage; also, always properly seal fuel containers to avoid the possibility of fire or explosion. Refer to the additional fuel system precautions later in this section.

• Fuel injection systems often remain pressurized, even after the engine has been turned **OFF**. The fuel system pressure must be relieved before disconnecting any fuel lines. Failure to do so may result in fire and/or personal injury.

• Brake fluid often contains polyglycol ethers and polyglycols. Avoid contact with the eyes and wash your hands thoroughly after handling brake fluid. If you do get brake fluid in your eyes, flush your eyes with clean, running water for 15 minutes. If eye irritation persists, or if you have taken brake fluid internally, IMMEDIATELY seek medical assistance.

• The EPA warns that prolonged contact with used engine oil may cause a number of skin disorders, including cancer! You should make every effort to minimize your exposure to used engine oil. Protective gloves should be worn when changing oil. Wash your hands and any other exposed skin areas as soon as possible after exposure to used engine oil. Soap and water, or waterless hand cleaner should be used.

• All new vehicles are now equipped with an air bag system, often referred to as a Supplemental Restraint System (SRS) or Supplemental Inflatable Restraint (SIR) system. The system must be disabled before performing service on or around system components, steering column, instrument panel components, wiring and sensors. Failure to follow safety and disabling procedures could result in accidental air bag deployment, possible personal injury and unnecessary system repairs.

• Always wear safety goggles when working with, or around, the air bag system. When carrying a non-deployed air bag, be sure the bag and trim cover are pointed away from your body. When placing a non-deployed air bag on a work surface, always face the bag and trim cover upward, away from the surface. This will reduce the motion of the module if it is accidentally deployed. Refer to the additional air bag system precautions later in this section.

• Clean, high quality brake fluid from a sealed container is essential to the safe and proper operation of the brake system. You should always buy the correct type of brake

fluid for your vehicle. If the brake fluid becomes contaminated, completely flush the system with new fluid. Never reuse any brake fluid. Any brake fluid that is removed from the system should be discarded. Also, do not allow any brake fluid to come in contact with a painted surface; it will damage the paint.

• Never operate the engine without the proper amount and type of engine oil; doing so WILL result in severe engine damage.

• Timing belt maintenance is extremely important! Many models utilize an interference-type, non-freewheeling engine. If the timing belt breaks, the valves in the cylinder head may strike the pistons, causing potentially serious (also time-consuming and expensive) engine damage. Refer to the maintenance interval charts in the front of this manual for the recommended replacement interval for the timing belt, and to the timing belt section for belt replacement and inspection.

• Disconnecting the negative battery cable on some vehicles may interfere with the functions of the on-board computer system(s) and may require the computer to undergo a relearning process once the negative battery cable is reconnected.

• When servicing drum brakes, only disassemble and assemble one side at a time, leaving the remaining side intact for reference.

• Only an MVAC-trained, EPA-certified automotive technician should service the air conditioning system or its components.

ENGINE REPAIR

Alternator

REMOVAL & INSTALLATION

2.0L Engines

1. Before servicing the vehicle, refer to the precautions in the beginning of this section.
2. Remove or disconnect the following:
   • Negative battery cable
   • Alternator jam nut, loosen only
   • Alternator adjustment nut, loosen only
   • Accessory drive splash shield
   • Alternator lower bolt, loosen only
   • Alternator drive belt

• Alternator field circuit wiring connector, push the **RED** locking tab to release
   • Battery positive terminal
   • Upper and lower bolts, move alternator off pivot bracket
   • Pivot bracket
   • Alternator

To install:
3. Install or connect the following:
   • Alternator
   • Pivot bracket and torque the bolts to 40 ft. lbs. (54 Nm)
   • Alternator, move it onto pivot bracket
   • Alternator field circuit wiring connector, push **RED** locking tab in until it snaps in place

• Battery positive terminal
   • Alternator drive belt
   • Tension the drive belt to 100 lbs. (used) or 135 lbs. (new)
   • Alternator adjustment bolt
   • Alternator jam nut and torque nut to 40 ft. lbs. (54 Nm)
   • Alternator mount bolts and torque the bolts to 40 ft. lbs. (54 Nm)
   • Accessory drive splash shield
   • Negative battery cable

2.4L Engines

➡The alternator is located above the oil filter and axle shaft.

1. Before servicing the vehicle, refer to the precautions in the beginning of this section.

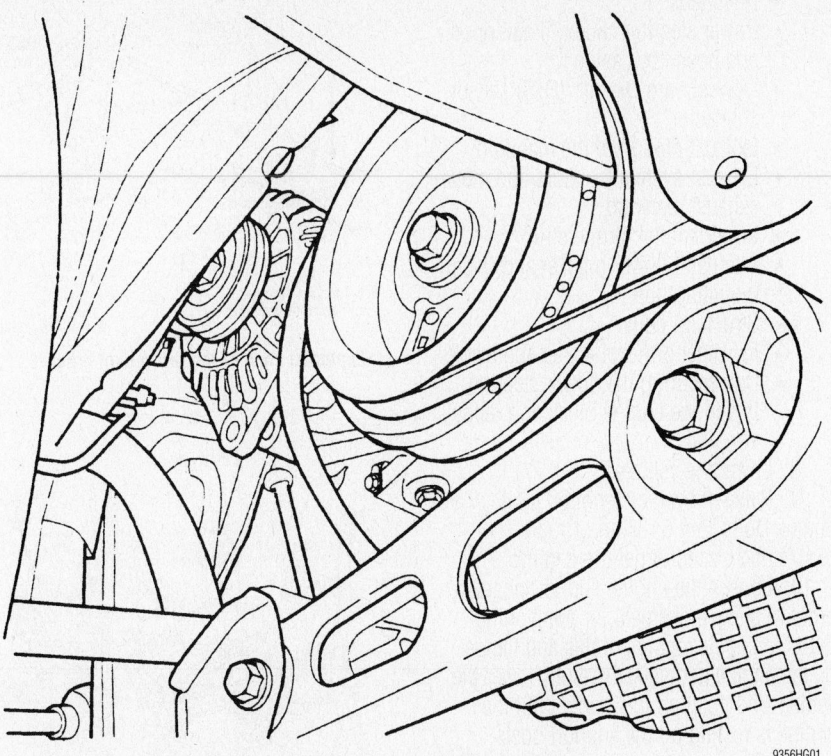

View of the alternator after removing the splash shield—2.0L engines

9356HG01

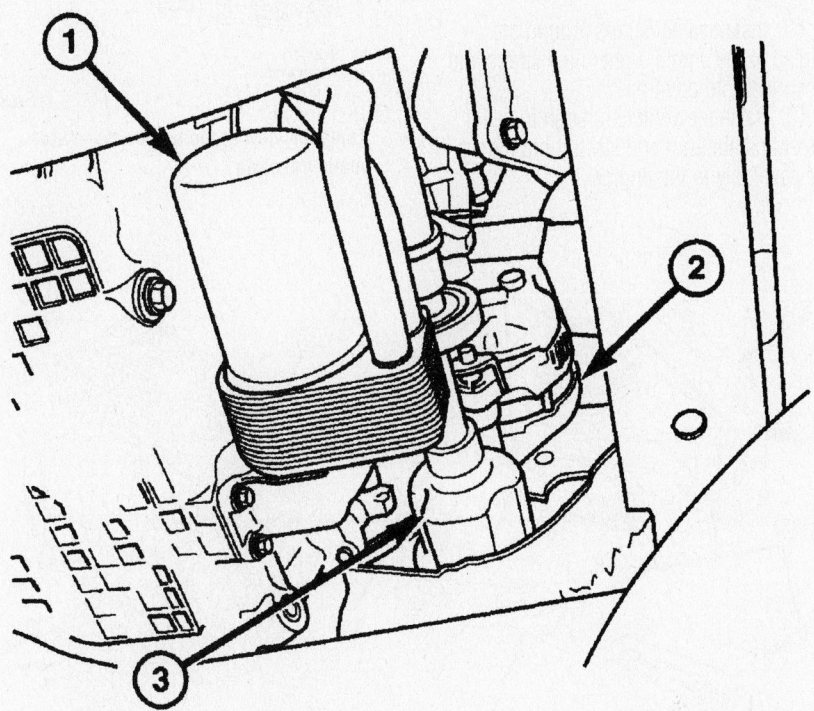

1 - Oil Filter
2 - Generator
3 - Axle Shaft

67189-NEON-G01

Alternator mounting—2.4L engine shown

2. Remove or disconnect the following:
   - Negative battery cable
   - 2 bolts from the top of the heat shield on the alternator
   - Nut from the upper T-bolt adjustment bracket
   - Right front wheel
   - Accessory drive splash shield
   - Axle retaining nut
   - Lower control arm from the steering knuckle
   - 2 bolts for the axle shaft bearing support
   - Axle shaft assembly. Put a pan under the transmission to catch the transmission fluid.
   - Lower heat shield bolt
   - Alternator heat shield
   - Field circuit from the alternator
   - B+ terminal nut and wire
   - Loosen the accessory drive belt T-bolt
   - Pencil strut
   - Lower alternator pivot bolt
   - Alternator belt
   - Alternator from the lower mounting bracket and position aside
   - Alternator lower mounting bracket
   - Alternator through the axle shaft hole

To install:
3. Install or connect the following:
   - Alternator into the vehicle through the axle shaft hole in the wheel well
   - Alternator on upper T-bolt and loosely install the nut
   - Lower mounting bracket to the block and tighten the bolts to 40 ft. lbs. (54 Nm)
   - Axle shaft into the steering knuckle
   - Lower control arm to the knuckle and secure with the bolt. Torque to 70 ft. lbs. (90 Nm)
   - Axle retaining nut and tighten to 180 ft. lbs. (240 Nm)
   - Alternator belt and adjust tension
   - Accessory belt T-bolt and tighten the nut to 40 ft. lbs. (54 Nm).
   - Lower generator pivot bolt and tighten to 40 ft. lbs. (54 Nm)
   - Pencil strut
   - Alternator heat shield
   - Lower heat shield bolt and tighten to 40 ft. lbs. (54 Nm)
   - Accessory drive splash shield
   - Right front wheel
   - Nut to the upper adjustment bracket and tighten to 18 ft. lbs. (25 Nm0
   - 2 heat shield bolts and tighten to 40 inch lbs. (4.5 Nm)
   - Negative battery cable

Ignition Timing

ADJUSTMENT

Ignition timing is controlled by the Powertrain Control Module (PCM). No adjustment is necessary or possible.

Engine Assembly

REMOVAL & INSTALLATION

2.0L Engines

1. Before servicing the vehicle, refer to the precautions in the beginning of this section.

➡After all components are installed on the engine, a DRB scan tool is necessary to perform the camshaft and crankshaft timing relearn procedure.

2. Properly recover the air conditioning system refrigerant.

3. Properly relieve the fuel system pressure.

4. Drain the engine oil.

5. Drain the cooling system.

6. Remove or disconnect the following:
- Battery and battery tray
- Air intake duct from the intake manifold
- Throttle cables
- Electrical connectors
- Air cleaner assembly
- Upper radiator hose
- Fan module assembly
- Lower radiator hose
- Cooler lines if equipped with an automatic transaxle
- Clutch cable or hydraulic line as necessary, if equipped with a manual transaxle
- Shift linkage from the transaxle
- Transaxle electrical connections
- Engine wiring harness
- Powertrain Control Module (PCM)
- Positive cable from the Power Distribution Center (PDC)
- Ground wires
- Heater hoses
- Brake booster vacuum hose
- Coolant recovery hose
- Accessory drive belts
- Power steering pump and reservoir, move them aside
- Air conditioning compressor, if equipped

7. Raise the vehicle.
- Front wheels
- Right inner splash shield

- Halfshafts
- Power steering cooler, if equipped, and position it aside
- Downstream Oxygen (O_2S) sensor connector
- Exhaust pipe from the manifold
- Exhaust system isolators and move exhaust rearward
- Lower engine torque strut
- Alternator, lower bracket and upper mounting bolt
- Structural collar
- Flywheel or flexplate, as applicable
- Bolt securing the power steering line to the engine block and reposition line.

8. Lower the vehicle.

9. Raise the vehicle enough to allow an Engine Dolly Tool 6135 and Cradle Tool 6710 to be placed under the engine.

10. Loosen the engine support posts in order to allow movement for positioning onto the engine locating holes and flange on the engine bedplate. Carefully lower the vehicle and position the cradle until the engine is resting on the support posts. Tighten the mounts to the cradle frame. This will keep the support posts from moving when removing or installing the engine and transaxle.

11. Install safety straps around the engine to the cradle; tighten the straps and lock them into position.

12. Raise the vehicle enough to see if the straps are tight enough to hold the cradle assembly to the engine.

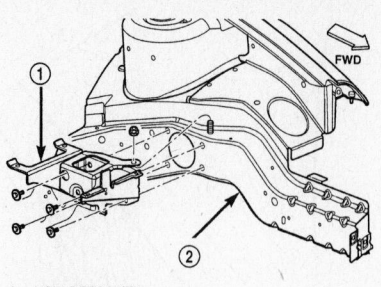

1 — MOUNT BRACKET
2 — BODY FRAME RAIL

9306EG14

Exploded view of the left mount bracket

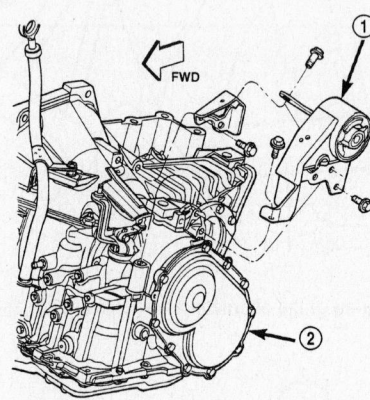

1 — MOUNT
2 — TRANSAXLE

9306EG15

Exploded view of the left mount—Automatic transaxle

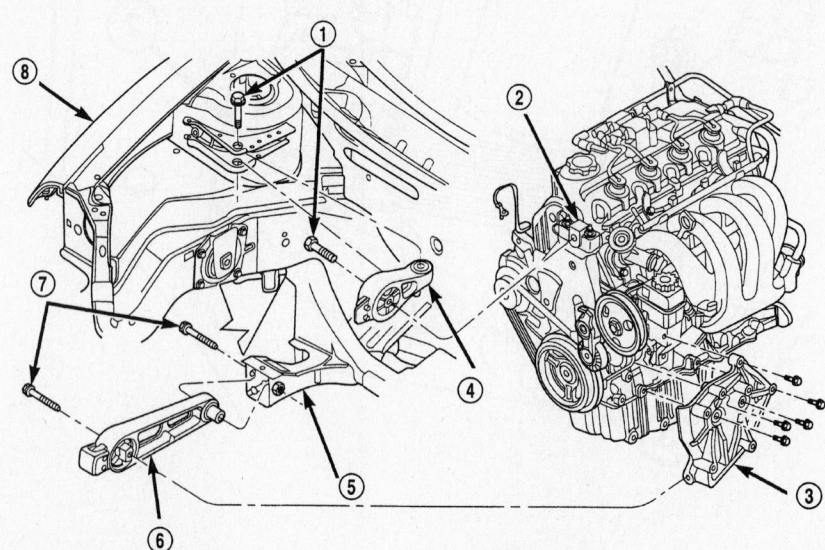

1 – BOLTS	5 – CROSSMEMBER
2 – ENGINE MOUNT BRACKET	6 – LOWER TORQUE STRUT
3 – TORQUE STRUT BRACKET	7 – BOLTS
4 – UPPER TORQUE STRUT	8 – RIGHT FENDER

9306EG13

Exploded view of the torque struts and bracket

13. Lower the vehicle so the weight of the engine and transaxle ONLY is on the cradle.

14. Remove the upper engine torque strut.

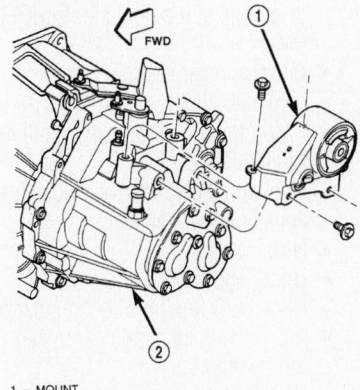

1 – MOUNT
2 – TRANSAXLE

9306EG16

Exploded view of the left mount—Manual transaxle

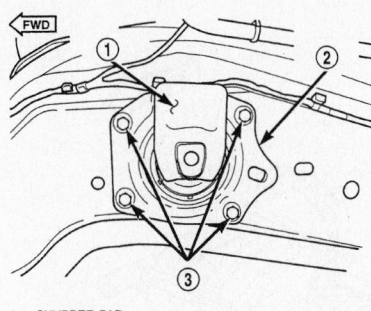

1 – SNUBBER PAD
2 – RIGHT ENGINE MOUNT
3 – BOLTS

9306EG17

Exploded view of the right mount

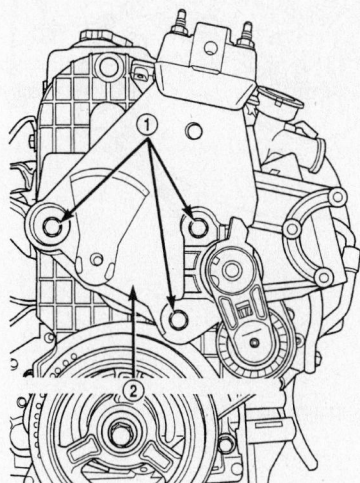

1 – BOLTS
2 – ENGINE MOUNT BRACKET ASSEMBLY

9306EG19

Exploded view of the right mount bracket

15. Remove the right and left engine/transaxle mount through-bolts.

16. Raise the vehicle slowly, until it is about 6 in. (15cm) above normal engine mount locations.

17. Remove the alternator.

18. Lower bracket and upper mounting bolt.

19. Continue raising the vehicle until the engine/transaxle assembly clears the compartment.

➡**If may be necessary to move the engine/transaxle assembly with the cradle to clear the body flanges.**

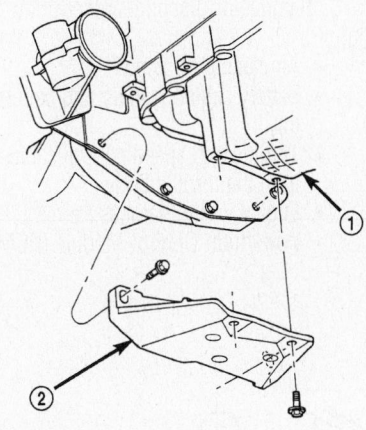

1 – OIL PAN
2 – STRUCTURAL COLLAR

9306EG18

Exploded view of the structural collar

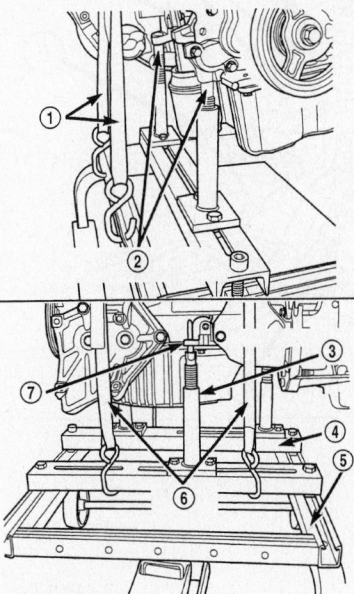

1 – SAFETY STRAPS
2 – PLACE REAR POSTS INTO LOCATING HOLES
3 – SPECIAL TOOL 6848
4 – SPECIAL TOOL 6710
5 – SPECIAL TOOL 6135
6 – SAFETY STRAPS
7 – PLACE FRONT POST UNDER BLOCK FLANGE

9306EG20

View of the engine cradle and dolly

To install:

20. Install or connect the following:
- Engine/transmission assembly to the vehicle and torque the engine mount bolts to 87 ft. lbs. (118 Nm)
- Upper engine torque strut and remove the engine safety straps

21. Slowly raise the vehicle enough to remove the engine dolly and cradle.
- Alternator, lower bracket and upper mounting bolt
- Halfshafts
- Flywheel or flexplate
- Bolt securing the power steering line to the engine block
- Power steering cooler screws, if equipped
- Dust shield
- Lateral bending brace

22. Install the structural collar and torque the bolts, in 3 steps, using the following procedure:
  a. Step 1: Collar-to-oil pan bolts to 30 inch lbs. (3 Nm).
  b. Step 2: Collar-to-transaxle bolts to 80 ft. lbs. (108 Nm).
  c. Step 3: Collar-to-oil pan bolts to 40 ft. lbs. (54 Nm).

23. Install or connect the following:
- Lower engine torque strut and torque the through-bolts to 87 ft. lbs. (118 Nm)
- Exhaust system isolators
- Exhaust pipe to the manifold. Tighten the exhaust fasteners to 21 ft. lbs. (28 Nm)
- Downstream O_2S sensor connector
- A/C compressor, if equipped
- Drive belts
- Right inner splash shield
- Front wheels, then lower vehicle
- Power steering pump and reservoir

➡**If equipped with a hydraulic clutch, it will not be necessary to bleed the system. The quick-connect fittings used on this system close immediately after disconnection and don't allow any fluid to leak out.**

- Clutch cable/hydraulic line if equipped with a manual transaxle
- Automatic transaxle cooler lines, if equipped
- Shift linkage to the transaxle
- Transaxle electrical connectors
- Fuel lines
- Heater hoses
- Ground wires
- PCM
- Engine wire harness

- Lower radiator hose
- Fan module assembly
- Upper radiator hose
- Air cleaner assembly
- Electrical connectors
- Throttle cables
- Air intake duct to the intake manifold
- Battery and battery tray

24. Fill the engine with clean oil.

25. Fill the cooling system.

26. After all components are installed, perform the camshaft and crankshaft timing relearn procedure as follows:

a. Connect a DRB scan tool to the DLC (located under the instrument panel, near the steering column).

b. Turn the ignition switch **ON**, and access the "miscellaneous" screen.

c. Select "re-learn cam/crank" option and follow the directions on the scan tool screen.

27. If equipped with air conditioning, recharge the system.

2.4L Engine

1. Before servicing the vehicle, refer to the precautions in the beginning of this section.

➡**After all components are installed on the engine, a DRB scan tool is necessary to perform the camshaft and crankshaft timing relearn procedure.**

2. Properly recover the air conditioning system refrigerant.

3. Properly relieve the fuel system pressure.

4. Drain the engine oil.

5. Drain the cooling system.

6. Remove or disconnect the following:

- Air cleaner housing
- Battery cables, battery and battery tray
- Throttle and speed control cables from the throttle body
- Engine wiring harness from the Powertrain Control Module (PCM)
- Positive cable from the Power Distribution Center (PDC) and ground wire from vehicle body
- Bolts attaching PDC and set aside
- Wiring connectors at lower battery tray support
- Ground wire from the vehicle body-to-engine at the right side strut tower
- Brake booster vacuum hose from intake manifold
- Proportional purge hose from the throttle body
- Coolant reserve/recovery hose from coolant outlet connector
- Heater hoses
- Upper and lower radiator hoses
- Upper A/C line from A/C condenser
- A/C lines at junction near upper torque strut
- Clutch hydraulic line, using Special Tool 6638A
- Transmission shift linkage
- Transmission electrical connectors
- Power steering hoses from radiator

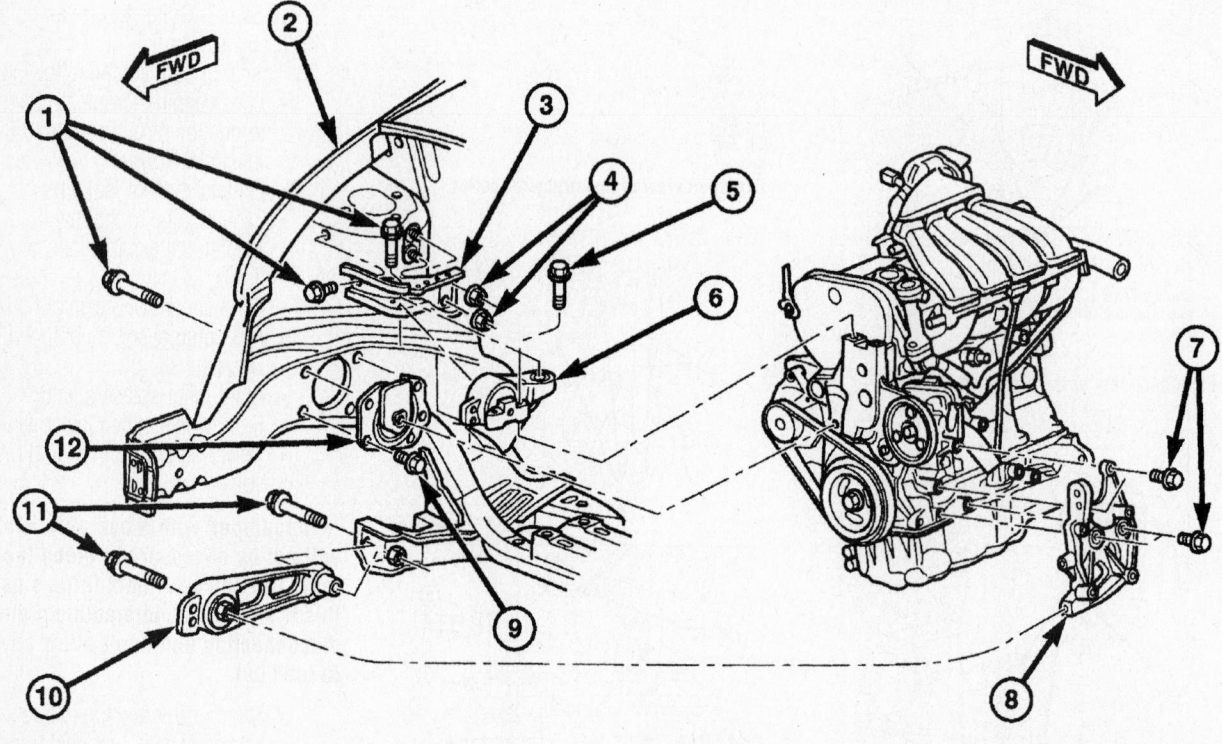

1 - BOLT
2 - RIGHT FENDER
3 - UPPER TORQUE STRUT BRACKET
4 - NUTS
5 - BOLT
6 - UPPER TORQUE STRUT

7 - BOLT
8 - LOWER TORQUE STRUT BRACKET
9 - BOLT
10 - LOWER TORQUE STRUT
11 - BOLT
12 - RIGHT ENGINE MOUNT

Exploded view of the torque struts—2.4L engine

67189-NEON-G07

- Radiator fan electrical connector and cooling module assembly
- Front wheels
- Right inner splash shield
- Halfshafts
- Accessory drive belts
- Alternator and support brackets
- Charge air cooler hoses
- Downstream Oxygen Sensor (O_2S) connector
- Exhaust system from manifold
- Both power steering hoses from steering gear
- Upper and lower heat shields, elbow support bracket, turbocharger support bracket, and elbow
- Structural collar

➡**Matchmark the drive plate-to-clutch module bolts before removing them.**

- Drive plate to clutch module bolts
- Lower engine torque strut
- A/C compressor
- Fluid lines from power steering pump and the pump

7. Raise vehicle enough to allow engine dolly, cradle, and posts, (Special Tools 6135, 6710, and 6848) to be installed under vehicle.

8. Loosen the engine support posts to allow movement for positioning onto engine

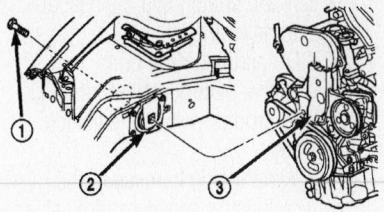

1 - BOLT
2 - RIGHT ENGINE MOUNT
3 - ENGINE MOUNT BRACKET

67189-NEON-G02

View of the right mount—2.4L engine

locating holes and flange on the engine bedplate. Lower the vehicle and position the cradle until the engine is resting on the support posts. Tighten the mounts to cradle frame. This will keep support posts from moving when removing or installing the engine and transmission assembly.

✳✳ CAUTION

You MUST use safety straps.

9. Install safety straps around the engine to cradle. Tighten straps and lock them into position.

10. Raise vehicle enough to determine if the straps are secure enough to hold the cradle assembly to the engine.

11. Lower the vehicle so weight of the

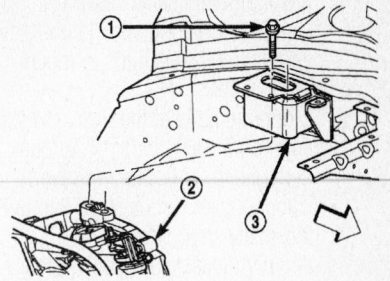

1 - BOLT
2 - TRANSAXLE
3 - LEFT MOUNT

67189-NEON-G03

View of the left mount—2.4L engine

engine and transmission ONLY is on the cradle assembly.

12. Remove or disconnect the following:
- Upper engine torque strut
- Right mount through bolt and left mount attaching bolts

13. Raise the vehicle slowly until the engine/transaxle assembly clears the engine compartment. You may need to move the assembly with the cradle to allow for removal around body flanges.

To install:

14. Install or connect the following:
- Position the engine and transmission assembly under the vehicle and slowly lower the vehicle over

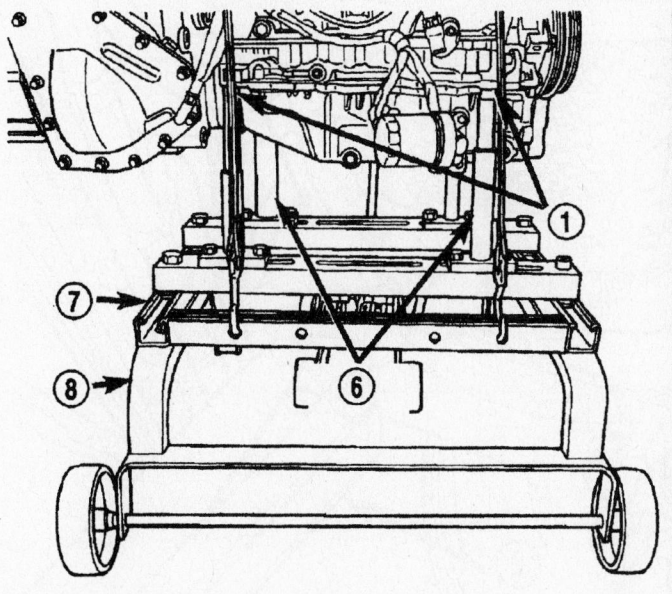

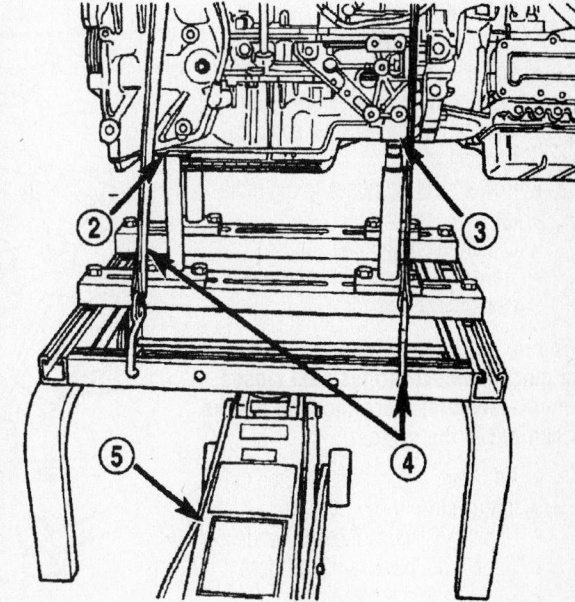

1 - POST LOCATING HOLES IN BLOCK
2 - POST POSITIONED UNDER BRACKET
3 - POST LOCATING HOLE IN STRUT
4 - SAFETY STRAPS

5 - FLOOR JACK
6 - SPECIAL TOOL 6848
7 - SPECIAL TOOL 6135
8 - SPECIAL TOOL 6710

67189-NEON-G04

Proper positioning of the engine cradle—2.4L engine

the engine/transaxle assembly. Continue lowering vehicle until the engine/transaxle aligns to mounting locations.

- Mounting bolts at the right and left engine/transaxle mounts. Torque the bolts to 87 ft. lbs. (118 Nm)
- Upper engine torque strut. Remove the safety straps from engine/transaxle assembly. Slowly raise the vehicle enough to remove the engine dolly and cradle.
- Power steering pump
- Fluid lines to power steering pump
- A/C compressor
- Lower engine torque strut
- Drive plate to clutch module bolts
- Structural collar

❄❄ WARNING

To prevent damaging the lower shield, you must install the upper shield first.

- Elbow, turbocharger support bracket, elbow support bracket, and upper and lower heat shields
- Power steering hoses to the steering gear
- Exhaust system to the manifold
- Downstream O$_2$S
- Charge air cooler hoses
- Alternator and mounting brackets
- Accessory drive belts
- Halfshafts
- Right inner splash shield
- Wheels
- Cooling module assembly, and connect the radiator fan electrical connector
- Power steering hoses to the radiator
- Clutch hydraulic line
- Transmission shift linkage and electrical connectors

➡**You do not have to bleed the clutch. The quick-connect fittings seal closed immediately after disconnection and no air can get in the system**

- A/C lines at junction near the upper torque strut
- Upper A/C line to the A/C condenser
- Upper and lower radiator hoses
- Fuel line and heater hoses
- Coolant reserve/recovery hose to the coolant outlet connector
- Brake booster vacuum hose to the intake manifold
- Proportional purge hose to the throttle body

- Ground straps, and connect all engine wiring
- PDC and retaining bolts
- Positive battery cable to the PDC and ground wire to the vehicle body
- Engine wiring harness to the PCM
- Throttle and speed control cables
- Battery tray, battery and cables
- Air cleaner housing assembly and clean air hose
- Oil filter. Fill engine with proper type and amount of oil

15. Fill power steering and engine cooling systems.

16. Evacuate and recharge A/C system.

17. Start the engine and run until operating temperature is reached. Check for leaks.

18. Perform torque strut adjustment procedure, as follows:

a. Remove the accessory drive belt splash shield.

b. Remove the pencil strut.

c. Loosen the upper and lower torque strut attaching bolt at the suspension crossmember and shock tower bracket.

d. Adjust the engine position by placing a floor jack on the forward edge of the transmission bell housing

e. With the engine supported, remove the upper and lower torque strut attach-

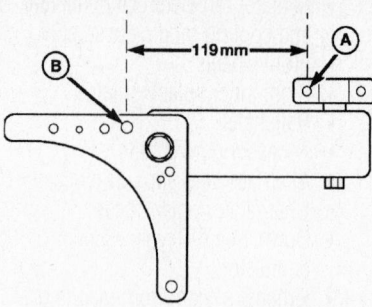

67189-NEON-G06

Measure the distance between the center of the rearmost attaching bolt on the engine mount bracket (point A) and the center of the hole on the shock tower bracket (point B)—2.4L engine

ment bolt(s) at shock tower bracket and suspension crossmember. Make sure the torque struts are free to move within the shock tower bracket and crossmember. Reinstall the torque strut bolt(s), but do not tighten yet.

f. Apply upward force, allowing the upper engine to rotate rearward until the distance between the center of the rearmost attaching bolt on the engine mount bracket (point A) and the center of the hole on the shock tower bracket (point B) is 4.70 in. (119mm)

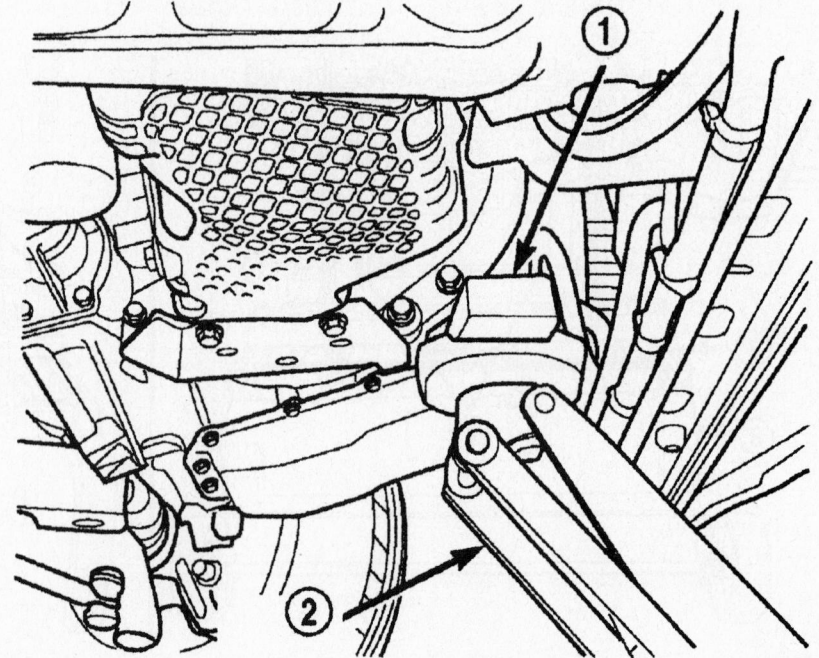

1 - WOOD BLOCK
2 - FLOOR JACK

67189-NEON-G05

Position the floor jack properly to prevent upward lifting of the engine—2.4L engine

※※ **WARNING**

Hold the engine in position with the jack until the upper and lower torque strut bolts are tightened.

g. With the engine held at the proper position, tighten both the upper and lower torque strut bolts to 85 ft. lbs. (115 Nm)

h. Remove the floor jack.

i. Install the pencil strut and tighten nuts to 43 ft. lbs. (58 Nm).

19. Install the accessory drive belt splash shield.

20. Adjust transmission linkage, if necessary.

Water Pump

REMOVAL & INSTALLATION

1. Before servicing the vehicle, refer to the precautions in the beginning of this section.

➡ **After all components are installed on the engine, a DRB scan tool is necessary to perform the camshaft and crankshaft timing relearn procedure.**

2. Drain the cooling system.

3. Remove or disconnect the following:
- Negative battery cable
- Right inner splash shield
- Accessory drive belts
- Power steering pump and support the engine from the bottom with a jack
- Upper and lower torque isolator struts
- Right engine mount
- Power steering pump bracket bolts,

move the pump/bracket assembly aside

➡ **It is not necessary to disconnect the power steering lines.**

- Right engine mount bracket
- Timing belt and timing belt tensioner
- Camshaft sprocket(s)
- Inner (rear) timing belt cover
- Water pump and discard the O-ring

4. Clean all mating surfaces of any residual gasket material.

To install:

5. Install a new O-ring in the water pump groove.

➡ **Hold the O-ring in place with a few small dabs of silicone sealant.**

※※ **WARNING**

Before proceeding, be sure the O-ring gasket is properly seated in the water pump groove before tightening the screws. An improperly installed O-ring could cause a coolant leak.

6. Install or connect the following:
- Water pump, torque the bolt to 105 inch lbs. (12 Nm). Use a pressure tester to pressurize the cooling system to 15 psi (103 kPa) and check the water pump shaft seal and O-ring for leaks.

➡ **Rotate the pump by hand to check for freedom of movement.**

- Inner (rear) timing belt cover
- Camshaft sprocket
- Timing belt tensioner and timing belt
- Right engine mount bracket and engine mount

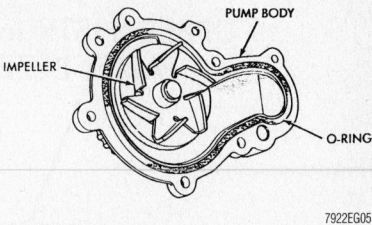

7922EG05

Be sure the O-ring is seated in the groove before installing the new pump

- Power steering pump
- Upper and lower torque isolator struts
- Accessory drive belts
- Negative battery cable

7. Refill the cooling system.

8. Use a DRB scan tool to perform the camshaft and crankshaft timing relearn procedure, as follows:

a. Connect the scan tool to the DLC, located under the instrument panel near the steering column.

b. Turn the ignition switch **ON**, and access the "miscellaneous" screen.

c. Select the "re-learn cam/crank" option, then follow the instructions on the scan tool screen.

Heater Core

REMOVAL & INSTALLATION

1. Before servicing the vehicle, refer to the precautions in the beginning of this section.

2. Disconnect the negative battery cable.

3. Discharge and recover the air conditioning system refrigerant.

4. Remove the instrument panel from the vehicle by removing or disconnecting the following:
- Push the seats back all the way
- Pry out the left and right A-pillar trim moldings and remove (using a trim stick tool)
- Upper instrument panel cover
- Pull up on the cluster bezel (carefully), and remove it from the vehicle
- Pull the instrument panel cover rearward (carefully), and remove it from the vehicle

※※ **WARNING**

Lock the steering wheel in the straight-ahead position; this will prevent damage to the clockspring

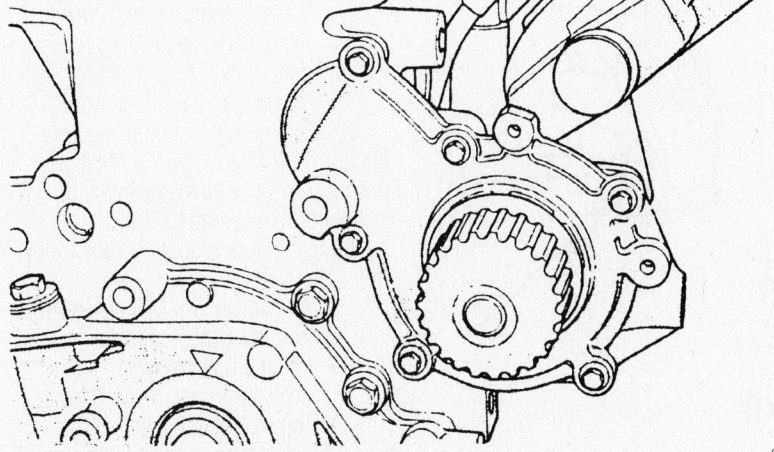

9356HG02

Water pump mounting—2002 Neon shown, other years similar

UPPER INSTRUMENT PANEL BEZEL

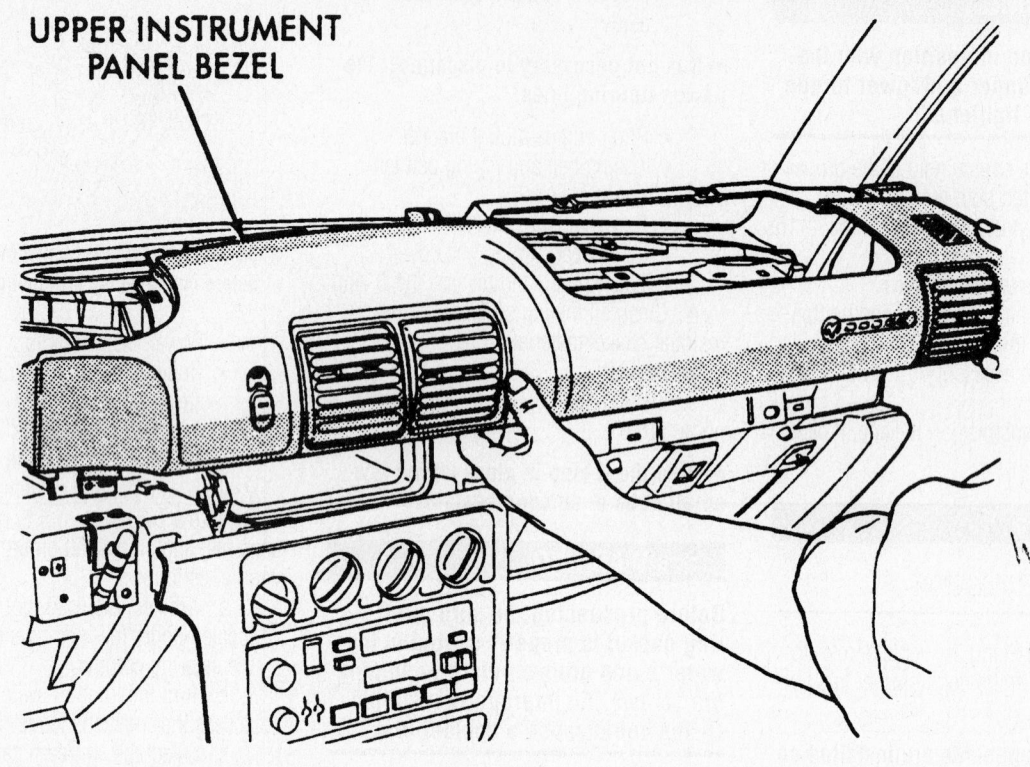

89716G11

Remove the right side upper instrument panel bezel

- Steering column as an assembly
- Left and right instrument panel end caps
- Center console
- Depress the Data Link Connector (DLC) sides and remove the DLC from the instrument panel reinforcement

- 4 bulkhead instrument panel screws
- 2 brake pedal support bracket bolts
- 2 center support mounting bolts
- Left and right A-pillar bolts; there are 2 on each side
- Antenna connector from the right side

- Left and right A-pillar door harness connectors
- 2 HVAC wiring harness connectors from the top right of the instrument panel
- Left side wiring harness connector from the top left of the instrument panel for the vanity and rear view mirrors
- Pull off the HVAC control head knobs
- 2 top front center bezel screws
- Using a trim stick, carefully pry out the instrument panel center bezel and remove it
- 2 HVAC control head screws
- Instrument panel wiring harness connector
- Vacuum harness connector

5. Pull the HVAC control head out of the instrument panel, twist it 90 degrees and push it back through the opening; do not disconnect the control cables.

6. Remove or disconnect the following:

- Air bag Control Module (ACM) from the center console
- Parking Brake Warning Lamp Switch from the center console
- Transmission Range Indicator Lamp from the center console

7. Using an assistant, pull the instru-

CENTER VENT DUCT

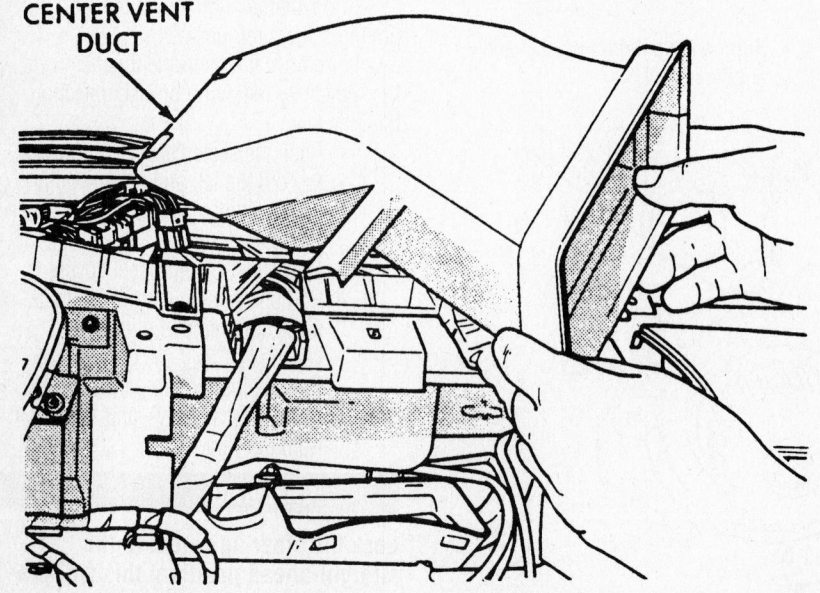

89716G12

You must remove the instrument panel center vent duct

ment panel rearward and remove it from the vehicle.

8. Drain the cooling system and remove the heater hoses at the dash panel. Place plugs in the heater core outlets to prevent coolant spillage during the unit housing removal.

9. Remove the suction line at the expansion valve. Place a piece of tape over the open refrigerant line to prevent moisture and/or dirt from entering the line.

10. Remove the expansion valve from the evaporator fitting to prevent moisture and/or dirt from entering the evaporator.

11. Remove or disconnect the following:
- Rubber drain tube extension from the condensation drain tube
- Vacuum harness from the power brake booster (if equipped)
- Defroster duct
- 3 retaining nuts located in the engine compartment, on the dash panel
- Right side retaining screw
- Remaining nut located on the dash panel stud
- Blue 5-way wiring harness connector from the plenum
- Heater/air conditioning assembly from the vehicle
- Separate the air distribution outlet-to-parting case line foam seals
- Evaporator lines and heater core tubes foam seals
- Clips and screws that hold the unit to the housing

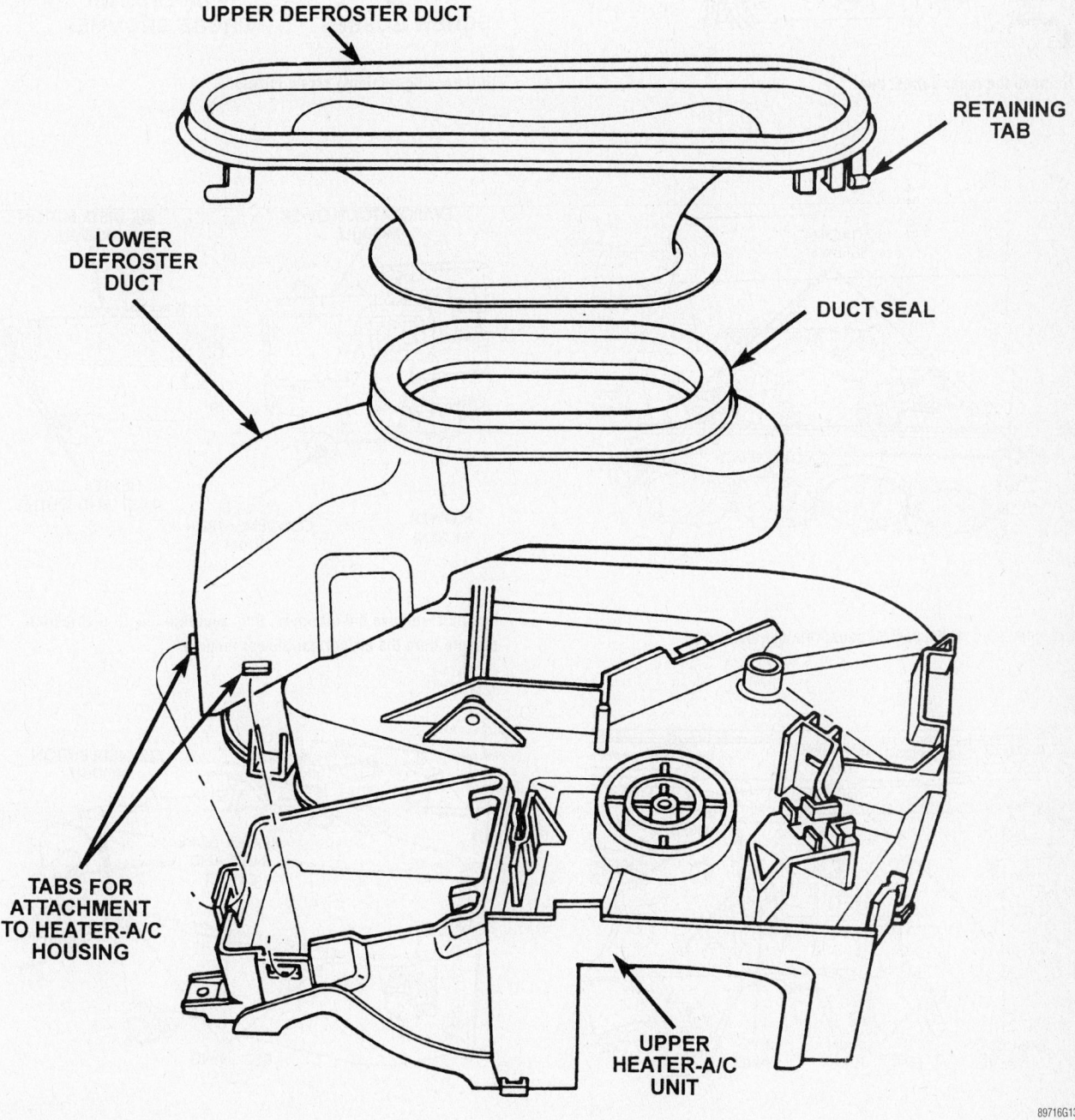

The upper defrost duct is secured with retaining tabs

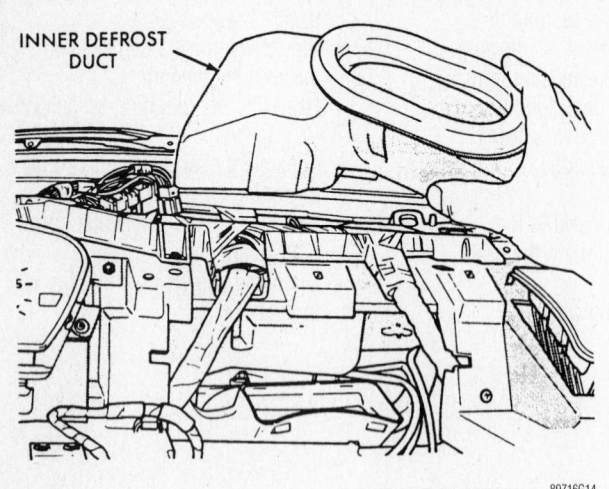

Remove the inner defrost duct

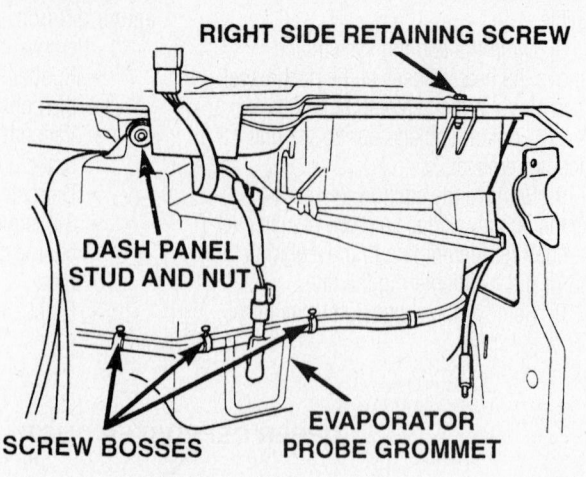

Unit housing retaining screw location

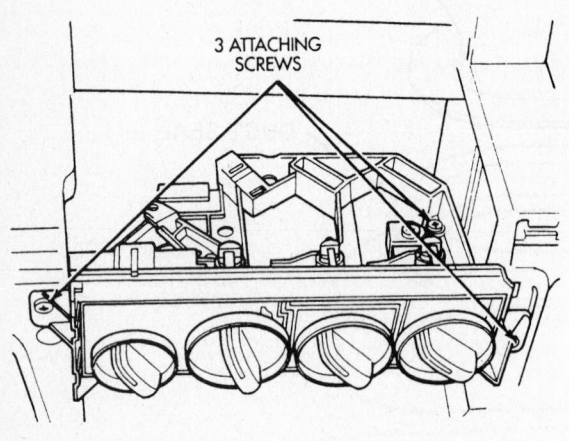

Exploded view of the control panel retainer locations

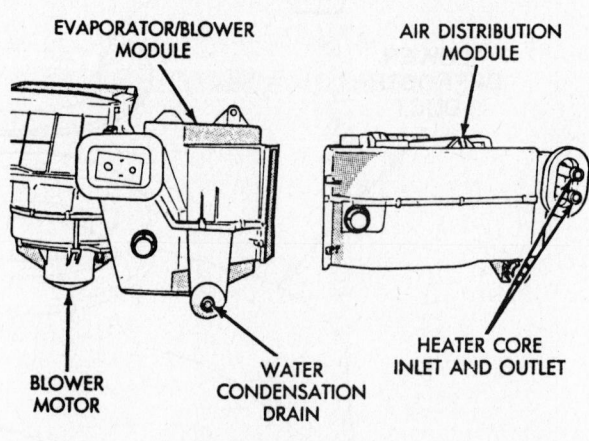

You must remove the retainers, then separate the air distribution module from the evaporator/blower module

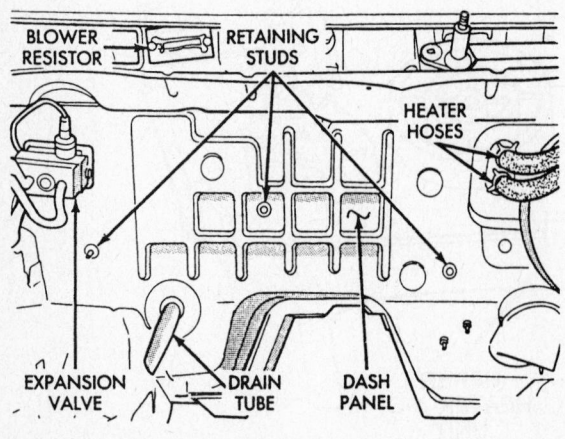

Location of the 3 dash panel retaining studs

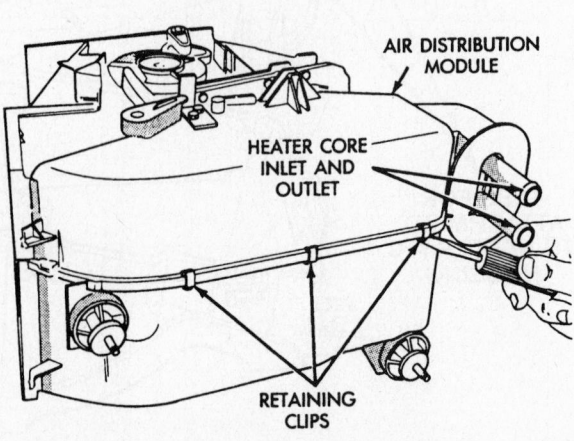

Remove the upper-to-lower housing retaining clips and screws

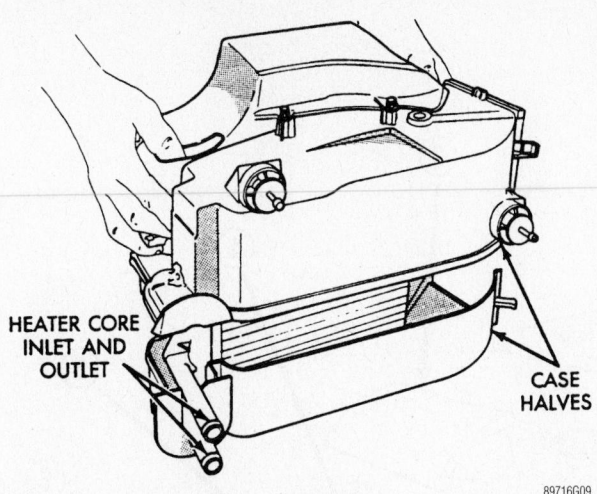

Then separate the 2 halves of the module

HEATER CORE INLET AND OUTLET

CASE HALVES

89716G09

The blower motor assembly is located under the passenger's side

89716P56

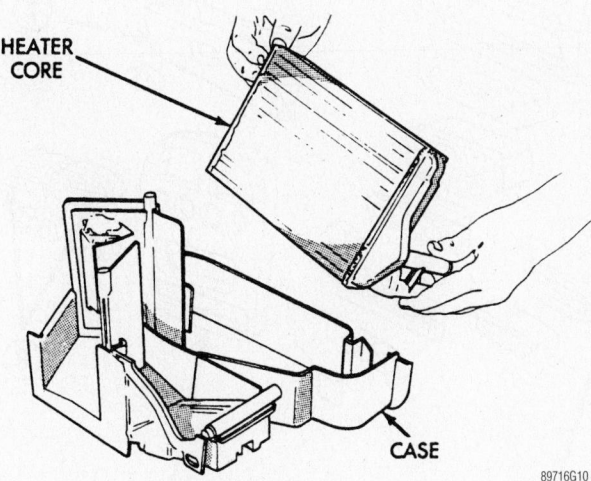

HEATER CORE

CASE

89716G10

Remove the heater core from the case

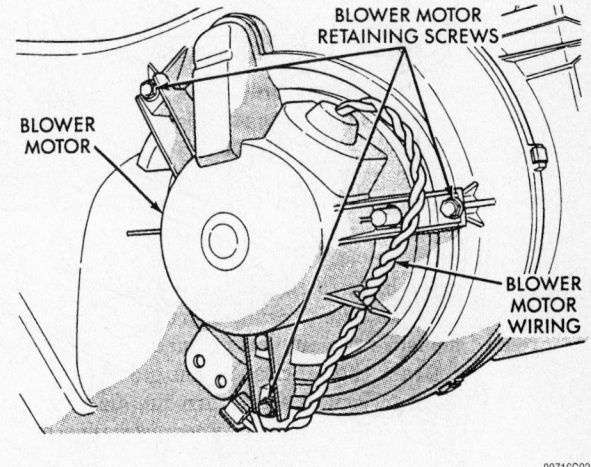

BLOWER MOTOR RETAINING SCREWS

BLOWER MOTOR

BLOWER MOTOR WIRING

89716G03

Blower motor mounting and location of the retaining screws and wiring vehicles with air conditioning

- 4 retaining screws from the inlet air duct on the module
- Air inlet and recirc door assembly from the module
- Sensing switch from its harness
- Upper-to-lower case retaining clips and screws
- Separate the case halves, and remove the evaporator
- Heater core

To install:

12. Install or connect the following:
- Heater core into the heater/air conditioning unit.
- Evaporator and the upper-to-lower case retaining clips and screws
- Sensing switch to its harness
- Recirc door assembly to the module and the air inlet

- Clips and screws that hold the unit to the housing
- Evaporator lines and heater core tubes foam seals.
- Air distribution outlet-to-parting case line foam seals
- Assembly into the vehicle
- Blue 5-way wiring harness connector to the plenum
- Nut on the dash panel stud
- Right side retaining screw
- 3 retaining nuts located in the engine compartment, on the dash panel
- Defroster duct
- If equipped, the vacuum harness to the power brake booster
- Rubber drain tube extension to the condensation drain tube

- Expansion valve to the evaporator fitting
- Suction line at the expansion valve

13. Evacuate, charge and leak-test the air conditioning system, if equipped.
14. Install the heater hoses.
15. Refill the cooling system.
16. Install the instrument panel to the vehicle by performing the following procedure:

- Air bag Control Module (ACM)
- Parking Brake Warning Lamp Switch
- Transmission Range Indicator Lamp
- HVAC control head into the instrument panel
- Vacuum harness connector

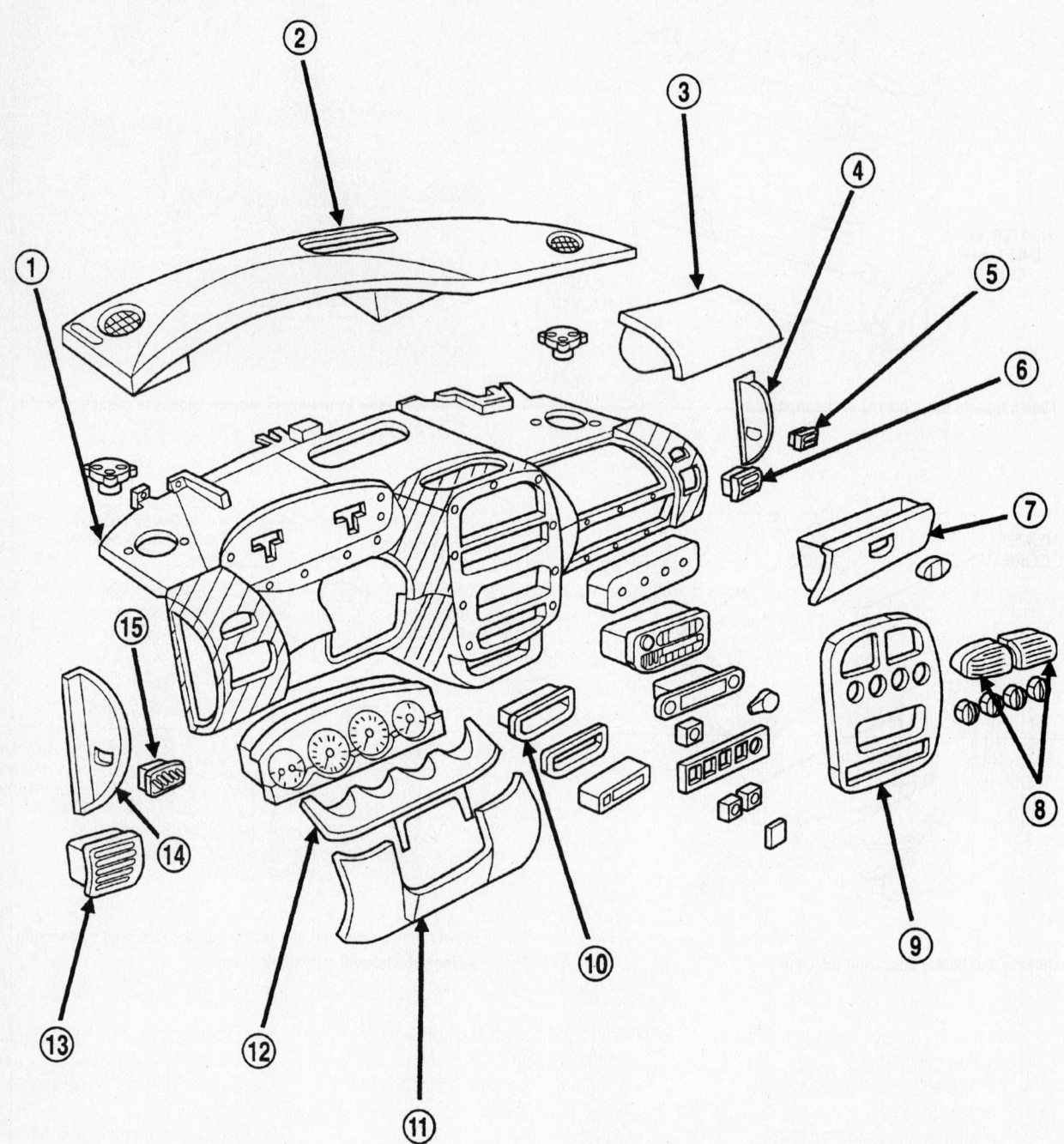

1 – INSTRUMENT PANEL ASSEMBLY
2 – UPPER COVER INSTRUMENT PANEL
3 – MODULE, PASSENGER SIDE AIRBAG
4 – END CAP, RIGHT
5 – DEMISTER GRILLE, RIGHT
6 – LOUVER, AIR OUTLET, RIGHT
7 – DOOR, GLOVE BOX
8 – LOUVER, AIR OUTLET, CENTER

9 – BEZEL INSTRUMENT PANEL, CENTER
10 – BIN, LOWER STORAGE
11 – COVER, LOWER INSTRUMENT PANEL
12 – CLUSTER BEZEL
13 – LOUVER, AIR OUTLET, LEFT
14 – END CAP, LEFT
15 – DEMISTER GRILLE, LEFT

93111G86

Exploded view of the instrument panel

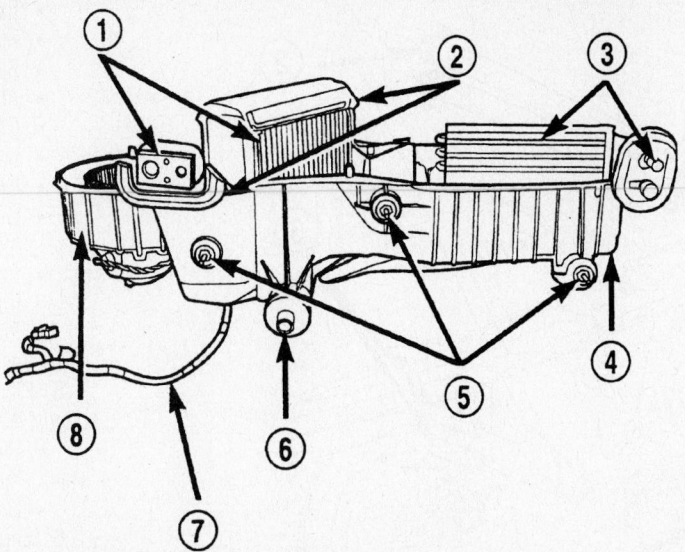

1 – EVAPORATOR AND CONNECTION
2 – FOAM SEALS
3 – HEATER CORE AND TUBES
4 – HVAC HOUSING LOWER CASE
5 – HOUSING MOUNTING STUDS
6 – HOUSING DRAIN
7 – WIRING
8 – BLOWER MOTOR AND WHEEL

93111G87

Sectional view of the heater/air conditioning assembly

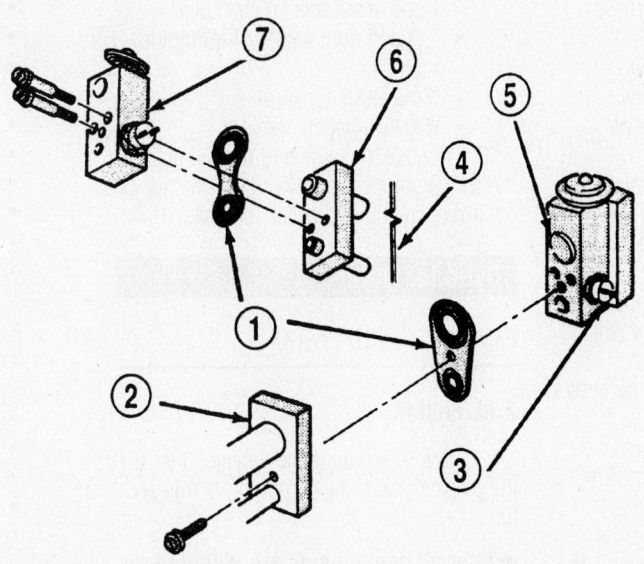

1 – ALUMINUM N-GASKET
2 – PLUMBING SEALING PLATE
3 – LOW/DIFFERENTIAL PRESSURE CUT-OFF SWITCH
4 – DASH PANEL
5 – H-VALVE
6 – EVAPORATOR SEALING PLATE
7 – H-VALVE

93111G88

Exploded view of the expansion valve

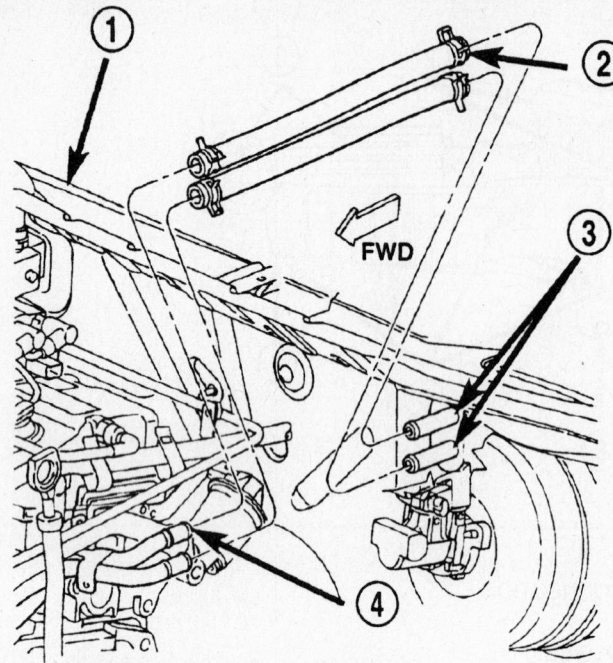

1 – COWL PANEL
2 – HEATER HOSE AND CLAMPS
3 – HEATER CORE TUBES
4 – HEATER HOSE SUPPLY AND RETURN TUBES

93111G89

View of the heater hoses and clamps

- Instrument panel wiring harness connector
- 2 HVAC control head screws
- Instrument panel center bezel
- 2 top front center bezel screws
- HVAC control head knobs
- Left side wiring harness connector to the top left of the instrument panel for the vanity and rear view mirrors
- 2 HVAC wiring harness connectors to the top right of the instrument panel
- Left and right A-pillar door harness connectors
- Antenna connector to the right side
- Left and right A-pillar bolts; there are 2 on each side
- 2 center support mounting bolts
- 2 brake pedal support bracket bolts
- 4 bulkhead instrument panel screws
- Data Link Connector (DLC) to the instrument panel reinforcement
- Center console
- Left and right instrument panel end caps
- Steering column as an assembly
- Instrument panel cover
- Cluster bezel
- Upper instrument panel cover
- Left and right A-pillar trim moldings
- Move seats forward
- Negative battery cable

17. Operate the engine to normal operating temperatures; then, check the climate control operation and check for leaks.

Cylinder Head

REMOVAL & INSTALLATION

2.0L Engine

1. Before servicing the vehicle, refer to the precautions in the beginning of this section.

➡**After all components are installed on the engine, a DRB scan tool is necessary to perform the camshaft and crankshaft timing relearn procedure.**

2. Properly relieve the fuel system pressure.
3. Drain the cooling system.
4. Remove or disconnect the following:
- Negative battery cable
- Power steering/air conditioning drive belt
- Exhaust pipe from the manifold
- Right front wheel
- Right side splash shield
- Alternator drive belt
- Crankshaft damper
- Lower torque strut
- Upper torque strut
- Ground strap from the engine mount bracket
- Power steering hose support clip from the engine mount bracket
- Power steering pump, move it aside

5. Place a jack under the engine and support it.
- Right side engine mount-to-bracket through-bolt
- Lower engine mount bracket bolt

6. Slightly raise the engine.
- Upper engine mount bracket bolts
- Engine mount bracket

➡**It may be necessary to raise and lower the engine until the bracket clears the engine components.**

- Front timing belt cover

7. Rotate the crankshaft and align the timing marks.
- Timing belt and tensioner
- Camshaft sprocket
- Rear timing belt cover
- Fuel line from the fuel rail
- Coolant recovery container
- Ground wire and from the cylinder head
- Upper radiator hose
- Intake manifold
- Ignition coil electrical connector
- Coil pack
- Spark plug wires
- Crankcase Closed Ventilation (CCV) hose from the valve cover
- Cam sensor electrical connector
- Coolant Temperature (CT) sensor electrical connector
- Heater tube from the cylinder head
- Heater hose from the thermostat housing connector
- Cylinder head cover
- Cylinder head bolts, working from the center out
- Cylinder head and discard gasket

➡**The cylinder head bolts must be inspected before they can be reused. If the threads of bolts are stretched, they must be replaced. Check for thread stretching by holding a scale or other straightedge against the threads. If all the threads do not contact the scale, the bolts must be replaced.**

> ⁕⁕ **WARNING**

Use only a plastic scraper to clean the mating surfaces. NEVER use metal, as this may gouge the surfaces and cause leaks!

8. Cover the combustion chambers, then use a plastic scraper to thoroughly and carefully clean the engine block and cylinder head mating surfaces.

To install:

9. Install the cylinder head with a new gasket after applying Mopar Gasket Sealant to both sides of the gasket.

10. Lubricate the cylinder head bolt threads. The 4 short bolts (164mm) are installed in positions 7, 8, 9 and 10.

11. Torque the cylinder head bolts, in the sequence, to:

  a. Step 1: 25 ft. lbs. (34 Nm).

  b. Step 2: 50 ft. lbs. (68 Nm).

  c. Step 3: 50 ft. lbs. (68 Nm).

  d. Step 4: An additional ¼ turn using a torque angle wrench.

12. Install or connect the following:

- Cylinder head cover
- Heater hose to the thermostat housing connector
- Heater tube to the cylinder head
- CT sensor electrical connector
- Cam sensor electrical connector
- CCV hose to the valve cover
- Coil pack
- Spark plug wires
- Ignition coil electrical connector
- Intake manifold
- Upper radiator hose
- Ground wire to the cylinder head
- Coolant recovery container
- Fuel line to the fuel rail
- Rear timing belt cover
- Camshaft sprocket

- Timing belt and tensioner
- Front timing belt cover
- Engine mount bracket
- Upper engine mount bracket bolts
- Lower engine mount bracket bolt
- Right side engine mount-to-bracket through-bolt and torque it to 87 ft. lbs. (118 Nm)

13. Remove the jack from under the engine.

- Power steering pump
- Power steering hose support clip to the engine mount bracket
- Ground strap to the engine mount bracket
- Upper torque strut
- Lower torque strut
- Crankshaft damper and torque the bolt to 100 ft. lbs. (136 Nm)
- Alternator drive belt
- Right side splash shield
- Right front wheel
- Exhaust pipe to the manifold
- Power steering/air conditioning drive belt
- Negative battery cable

14. Refill the cooling system.

15. Use a DRB scan tool to perform the camshaft and crankshaft timing relearn procedure, as follows:

  a. Connect the scan tool to the DLC, located under the instrument panel near the steering column.

  b. Turn the ignition switch **ON**, and access the "miscellaneous" screen.

  c. Select the "re-learn cam/crank" option, then follow the instructions on the scan tool screen.

2.4L Engine

1. Before servicing the vehicle, refer to the precautions in the beginning of this section.

2. Properly relieve the fuel system pressure.

3. Drain the cooling system.

4. Remove or disconnect the following:

- Negative battery cable
- Air intake hose and air cleaner housing
- Fuel supply line quick-connect from the fuel rail assembly
- Heater tube support bracket from the cylinder head
- Upper radiator hose
- Heater hoses from the thermostat housing
- Engine Coolant Temperature (ECT) sensor connector
- Accessory drive belts
- Exhaust pipe from the manifold
- Turbocharger heat shields
- Elbow support bracket
- Turbocharger support bracket
- Oil return tube and oil supply line
- Coolant supply and return lines
- Ignition coil wiring connector
- Ignition coil and plug wires
- Camshaft Position (CMP) sensor connector
- Timing belt and camshaft sprockets
- Timing belt idler pulley and rear timing belt cover
- Cylinder head (valve) cover
- Camshafts and rocker arms
- Cylinder head bolts, in the reverse of the tightening sequence
- Cylinder head and discard gasket

➡**The cylinder head bolts must be inspected before they can be reused. If the threads of bolts are stretched, they must be replaced. Check for thread stretching by holding a scale or other straightedge against the threads. If all the threads do not contact the scale, the bolts must be replaced.**

> ⁕⁕ **WARNING**

Use only a plastic scraper to clean the mating surfaces. NEVER use metal, as this may gouge the surfaces and cause leaks!

5. Cover the combustion chambers, then use a plastic scraper to thoroughly and carefully clean the engine block and cylinder head mating surfaces.

To install:

6. Install the cylinder head with a new gasket after applying Mopar Gasket Sealant to both sides of the gasket.

7. Lubricate the cylinder head bolt

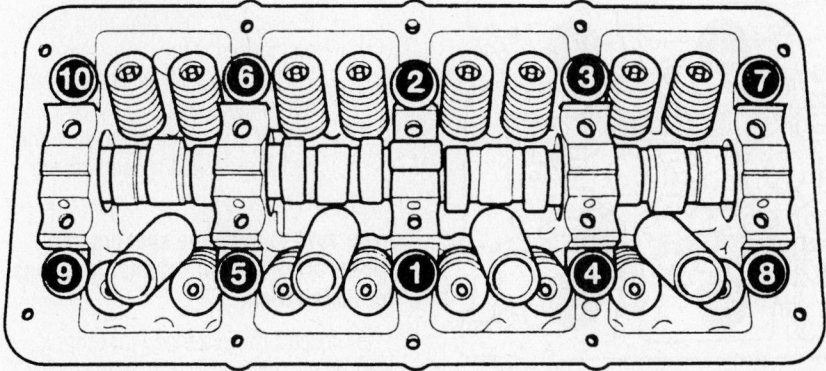

9356HG03

Cylinder head bolt torque sequence—2.0L engines

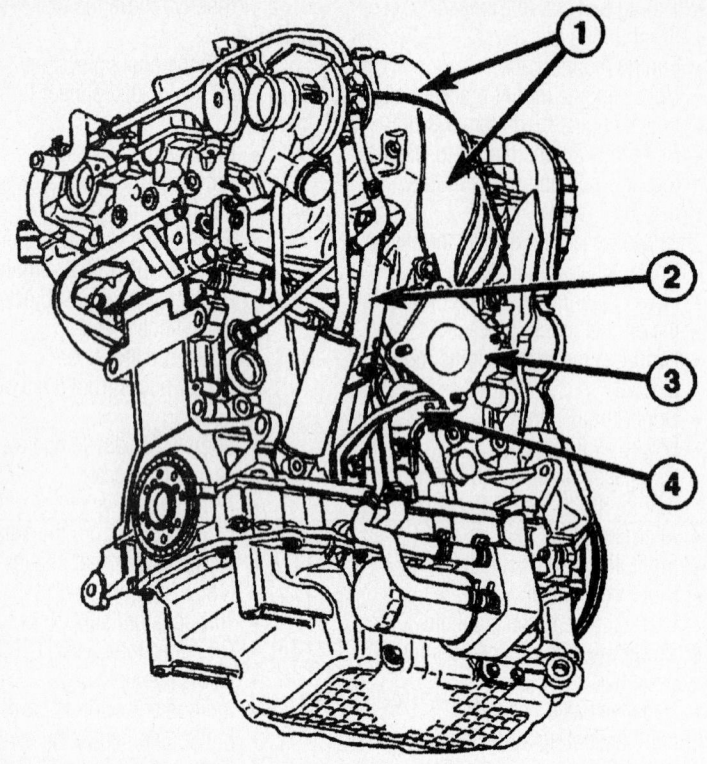

1 - UPPER/LOWER HEAT SHIELDS
2 - TURBOCHARGER SUPPORT BRACKET
3 - ELBOW
4 - ELBOW SUPPORT BRACKET

67189-NEON-G08

Turbocharger heat shield mounting—2.4L engine

threads. The 4 short bolts (164mm) are installed in positions 7, 8, 9 and 10.

8. Torque the cylinder head bolts, in the sequence, to:
   a. Step 1: 25 ft. lbs. (34 Nm).
   b. Step 2: 50 ft. lbs. (68 Nm).
   c. Step 3: 50 ft. lbs. (68 Nm).

   d. Step 4: An additional ¼ turn using a torque angle wrench.
9. Install or connect the following:
   • Rocker arms and camshafts
   • Cylinder head (valve) cover
   • Rear timing belt cover and timing belt idler pulley

• Camshaft sprockets and timing belt
• CMP sensor connector
• Ignition coil and plug wire
• Ignition coil wiring connector
• Coolant return and supply lines
• Oil supply line and return tube
• Exhaust elbow
• Turbocharger support bracket
• Elbow support bracket
• Turbocharger heat shields
• Exhaust pipe to the manifold. Tighten to 20 ft. lbs. (28 Nm)
• Accessory drive belts
• ECT sensor connector
• Upper radiator hose
• Heater hoses to the thermostat housing
• Heater tube support bracket to the cylinder head
• Lower intake manifold support bracket
• Quick-connect fuel lines to the fuel rail assembly
• Air intake hose and air cleaner housing
• Negative battery cable
10. Refill the cooling system.

Rocker Arms/Shafts

REMOVAL & INSTALLATION

   This procedure applies to Single Overhead Camshaft (SOHC) engines only. On Dual Overhead Camshaft (DOHC) engines, the valves are actuated directly by the camshafts and no rocker arms are used.
1. Before servicing the vehicle, refer to the precautions in the beginning of this section.
2. Remove or disconnect the following:
   • Negative battery cable
   • Cylinder head cover

➥Be sure to note the installed positions of the rocker arm shaft assemblies before removal.

   • Rocker arm shaft attaching fasteners, loosen them
   • Rocker arm shaft assembly from the cylinder head
   • Rocker arm assemblies, slide the rocker arms and spacers off the shaft

➥Be sure to keep the spacers and rocker arms in their original locations for installation.

3. Inspect the rocker arm for scoring, wear on the roller or damage to the rocker arm; replace any components showing dam-

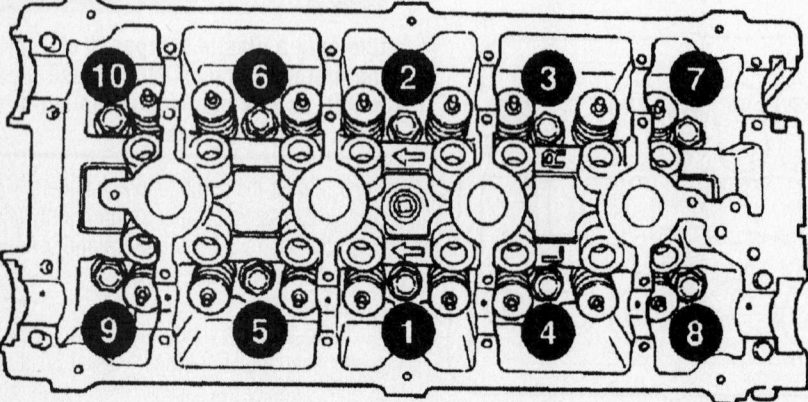

67189-NEON-G09

Cylinder head bolt torque sequence—2.4L engine

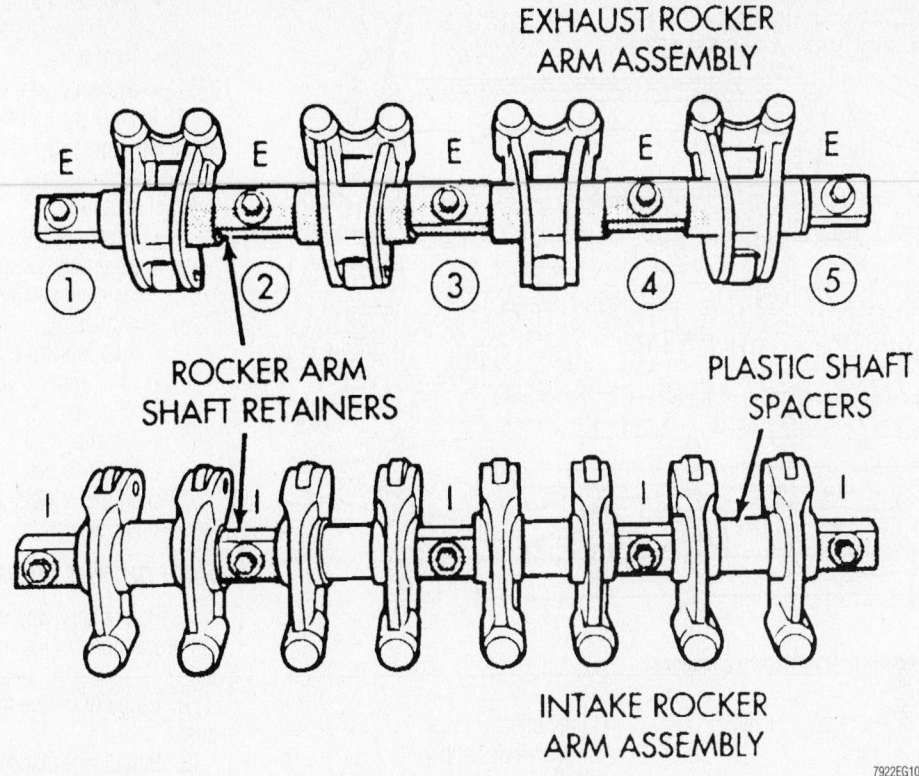

EXHAUST ROCKER
ARM ASSEMBLY

ROCKER ARM
SHAFT RETAINERS

PLASTIC SHAFT
SPACERS

INTAKE ROCKER
ARM ASSEMBLY

7922EG10

Be sure to note the rocker arm shaft component positions before disassembling the rocker arm shafts

age. Check the location where the rocker arms mount to the shafts for wear or damage. Replace if damaged or worn. The rocker arm shaft is hollow and is used as a lubrication oil duct. Check the oil holes for clogs with a small piece of wire, and clean as required. Lubricate the rocker arms and spacers. Be sure to install in their original locations.

To install:

⁑ **WARNING**

Set the crankshaft to 3 notches before Top Dead Center (TDC) before installing the rocker arm shafts, otherwise valvetrain damage may occur when the mounting bolts are tightened.

4. Set the crankshaft sprocket to TDC by aligning the mark on the sprocket with the arrow on the oil pump housing, then back off to 3 notches before TDC.

5. Install the rocker arm/hydraulic lash adjuster assembly making sure that the adjusters are at least partially full of oil. This is indicated by little or no plunger travel when the lash adjuster is depressed. If there is excessive plunger travel, submerge the rocker arm assembly into clean engine oil and pump the plunger until the

lash adjuster travel is taken up. If travel is not reduced, replace the assembly. The hydraulic lash adjuster and rocker arm are serviced as an assembly.

6. Install or connect the following:
   - Rocker arm and shaft assemblies. Position the rocker arm shafts with the **NOTCH FACING UP** and toward the timing belt side of the engine.
   - Retainers in their original positions on the exhaust and intake shafts.

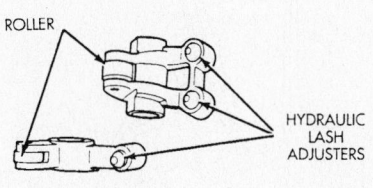

EXHAUST ROCKER ARM

ROLLER

HYDRAULIC
LASH
ADJUSTERS

INTAKE ROCKER ARM

7922EG11

Disassembled view of an intake and exhaust rocker arm

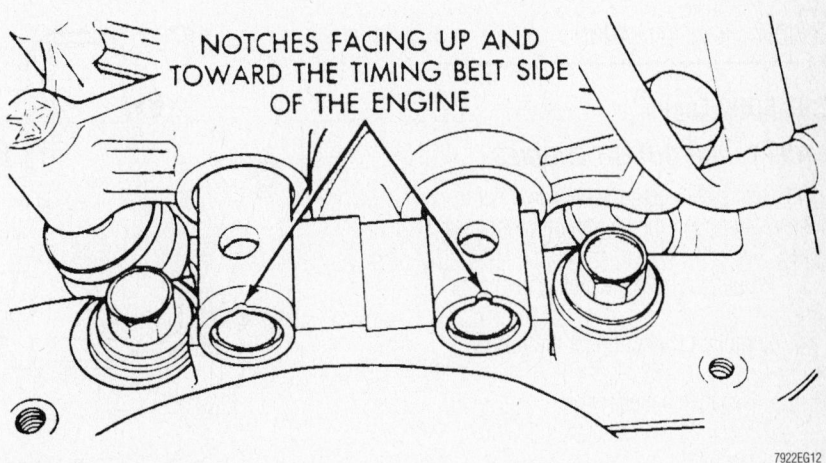

NOTCHES FACING UP AND
TOWARD THE TIMING BELT SIDE
OF THE ENGINE

7922EG12

Be sure the notches in the rocker arm shafts point up and toward the timing belt side of the engine

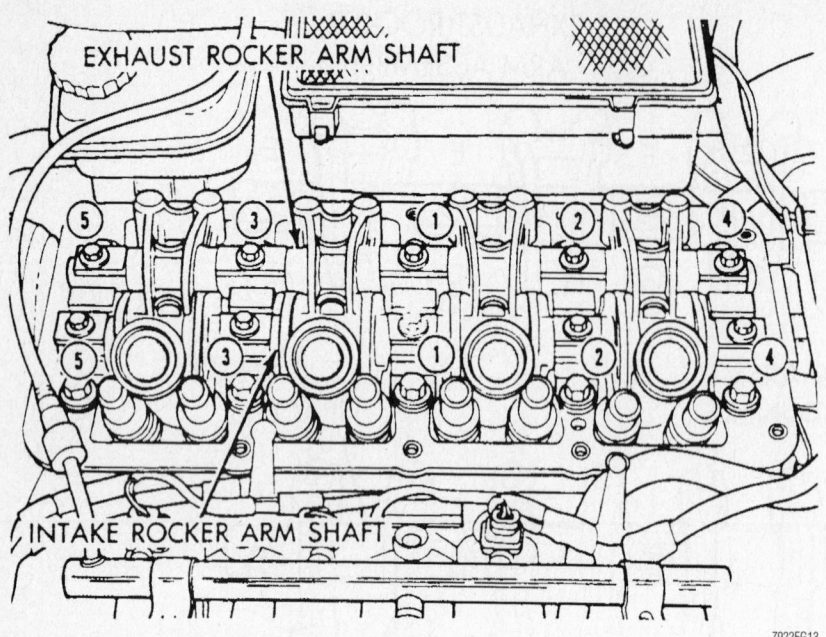

Rocker arm shaft assembly bolt tightening sequence

Tighten the bolts, in the sequence, to 21 ft. lbs. (28 Nm).

✴✴ WARNING

When installing the intake rocker arm shaft assembly, be sure the plastic spacers do not interfere with the spark plug tubes. If the spacers do interfere, rotate them until they are at the proper angle. To avoid damaging the spark plug tubes, do not try to rotate the spacers by forcing the shaft down.

7. Install the rocker arm (valve) cover.
8. Connect the negative battery cable.

Intake Manifold

REMOVAL & INSTALLATION

2.0L SOHC Engine

EXCEPT HIGH OUTPUT ENGINES

1. Before servicing the vehicle, refer to the precautions in the beginning of this section.
2. Properly relieve the fuel system pressure.
3. Remove or disconnect the following:

- Negative battery cable
- Fresh air inlet duct from the air cleaner
- Fuel supply line from the fuel tube assembly
- Fuel injector cover
- Fuel rail assembly
- Brake booster vacuum hose
- Positive Crankcase Ventilation (PCV) hose
- Knock Sensor (KS) electrical connector
- Starter wiring connectors
- Intake manifold and discard the gasket

4. Clean all mating surfaces of any residual gasket material.

To install:

5. Install or connect the following:
- New gaskets and seals
- Intake manifold and torque the new retainers, in sequence, to 95 inch lbs. (11 Nm)

- Fuel rail and torque the screws to 17 ft. lbs. (23 Nm)
- PCV hose
- Brake booster hose

6. Inspect the fuel line quick-connect fittings for damage and replace if necessary. Apply a small amount of clean engine oil to the fuel inlet tube.

- Fuel supply hose to the fuel rail assembly. Ensure the connection is fastened securely by pulling on the connector.
- Fuel injector cover
- MAP sensor wiring
- KS sensor connector
- Air inlet duct
- Negative battery cable

7. Start the vehicle, check for leaks and repair if necessary.

HIGH OUTPUT ENGINE

1. Before servicing the vehicle, refer to the precautions in the beginning of this section.
2. Properly relieve the fuel system pressure.
3. Remove or disconnect the following:
- Negative battery cable

✴✴ CAUTION

Wrap towels around the fitting to catch any spilled fuel.

- Fuel supply line quick-connect from the fuel tube assembly
- Fuel rail cover from the engine
- Fuel injector connector and harness from the injectors and fuel rail
- Positive Crankcase Ventilation (PCV) hose
- Throttle body-to-intake manifold air duct

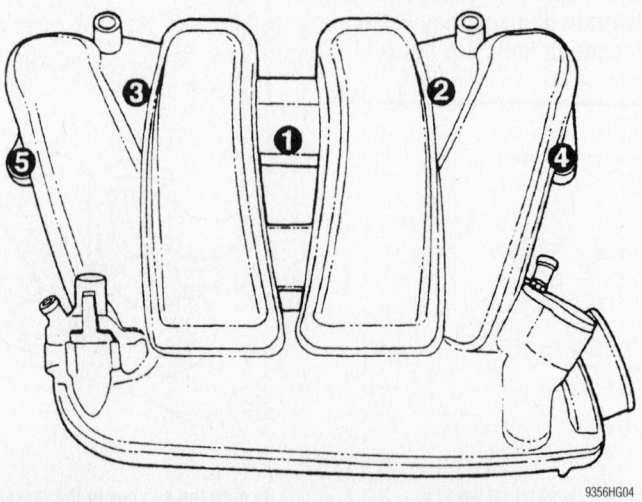

Intake manifold bolt torque sequence—2.0L (VIN C) SOHC engine

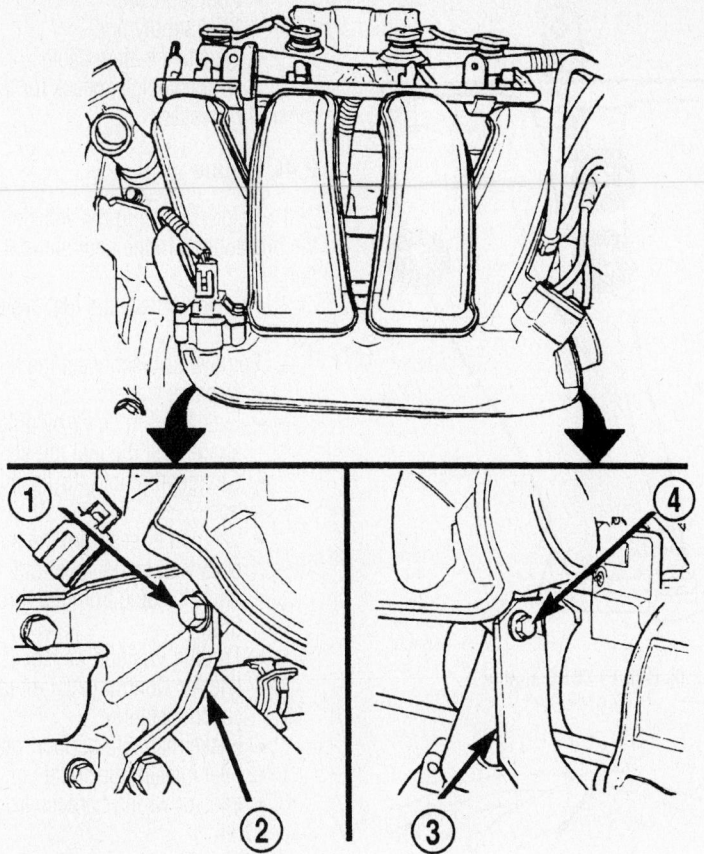

1 – BOLT
2 – BRACKET
3 – BRACKET
4 – BOLT

9306EG21

View of the intake manifold lower supports—SOHC engines

- Lower intake manifold support bracket bolt
- Oil dipstick
- Manifold Absolute Pressure (MAP) sensor wiring
- Manifold Tuning Valve (MTV) actuator wiring
- Intake manifold-to-cylinder head

bolts and move the manifold forward
- Knock (KS) Sensor wiring
- Brake booster hose
- Intake manifold and discard the gasket

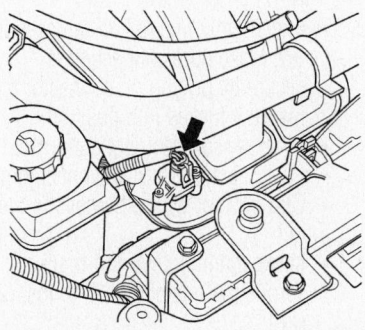

9306EG22

View of the Manifold Absolute Pressure (MAP) sensor—SOHC engines

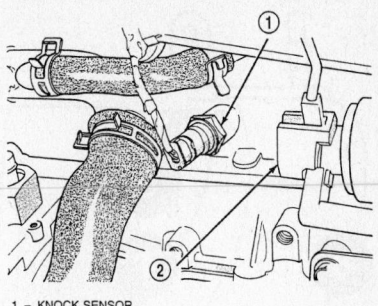

1 – KNOCK SENSOR
2 – STARTER MOTOR

9306EG23

View of the Knock Sensor (KS)—SOHC engines

4. If necessary, separate the upper manifold from the lower manifold.

To install:

5. Clean the mating surfaces of any residual gasket material.

6. Install or connect the following:
- Lower manifold to the upper manifold with a new gasket and torque the bolts to 105 inch lbs. (12 Nm)
- Intake manifold on the cylinder head with a new gasket
- Engine wiring harness between the middle intake runners
- KS wiring connector
- Brake booster hose and torque the intake manifold bolts, in sequence, to 105 inch lbs. (12 Nm)
- Lower manifold support bracket bolt and torque the bolt to 21 ft. lbs. (28 Nm)
- MAP sensor connector
- MTV actuator connector
- Oil dipstick
- Inlet air duct to the manifold and the throttle body
- PCV hose
- Fuel injector connectors and harness to the fuel rail

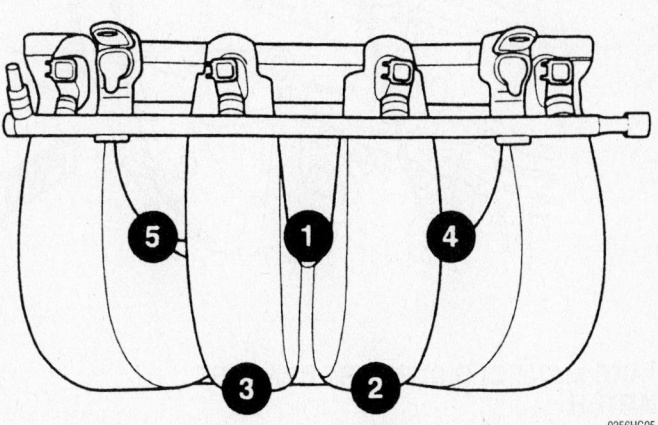

9356HG05

Lower intake manifold bolt torque sequence—High Output 2.0L (VIN F) SOHC engine

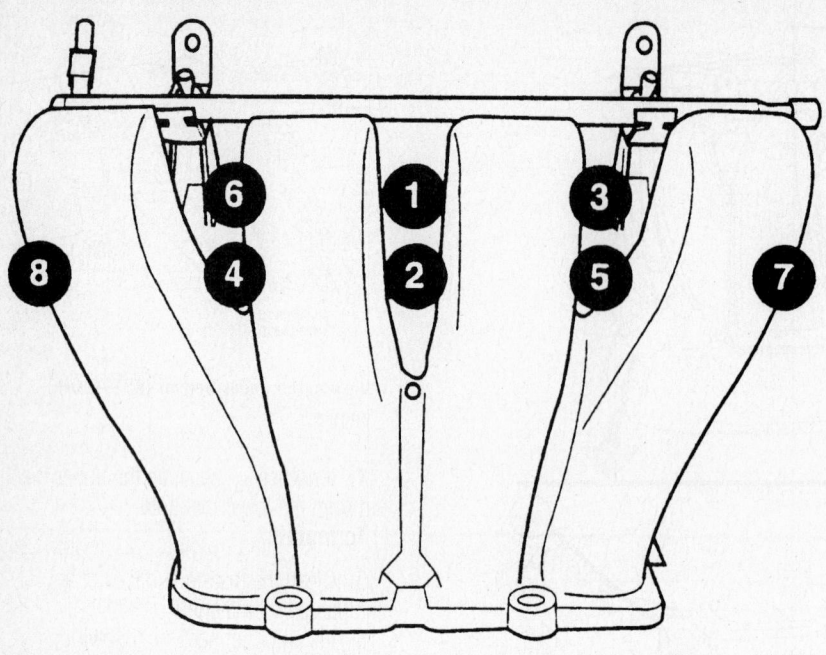

9356HG06

Upper intake manifold bolt torque sequence—High Output 2.0L (VIN F) SOHC engine

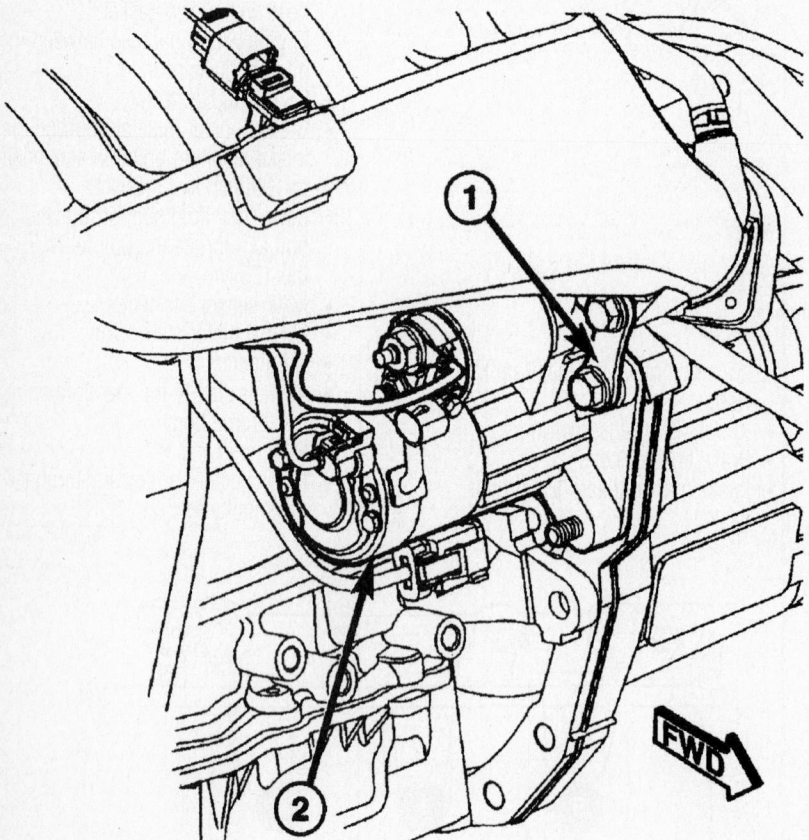

1 - INTAKE MANIFOLD SUPPORT BRACKET
2 - STARTER

67189-NEON-G10

Intake manifold support bracket—2.4L engine

- Fuel rail cover
- Fuel supply line
- Negative battery cable

7. Start the vehicle, check for leaks and repair if necessary.

2.4L Engine

1. Before servicing the vehicle, refer to the precautions in the beginning of this section.

2. Properly relieve the fuel system pressure.

3. Remove or disconnect the following:
- Negative battery cable
- Fuel rail trim cover by pulling straight off the fuel rail studs
- Charge air cooler-to-throttle body hose
- Vacuum hoses from the throttle body and intake manifold
- Throttle cable from the throttle body
- Throttle Position Sensor (TPS) and Idle Air Control (IAC) motor electrical connectors
- Intake manifold support bracket
- Fuel injector electrical connectors
- Injector wiring harness from fuel rail
- Manifold Absolute Pressure (MAP) sensor electrical connector
- Fuel supply line quick connect from the fuel rail assembly
- Intake manifold fasteners
- Intake manifold.
- Intake manifold and discard the gasket

4. Clean all mating surfaces of any residual gasket material.

To install:

5. Install or connect the following:
- New gaskets and seals
- Intake manifold and torque the new retainers to 17 ft. lbs. (23 Nm)
- Fuel rail assembly to the intake manifold, if removed. Tighten the retainers to screws to 17 ft. lbs. (23 Nm)
- Fuel supply line to fuel rail assembly. Make sure connection is secure by pulling on connector to insure it locked into position
- MAP sensor electrical connector
- Fuel injector electrical connectors
- Clip injector wiring harness to fuel rail
- Intake manifold support bracket. Torque the retainer to 17 ft. lbs. (23 Nm).
- TPS and IAC motor electrical connectors

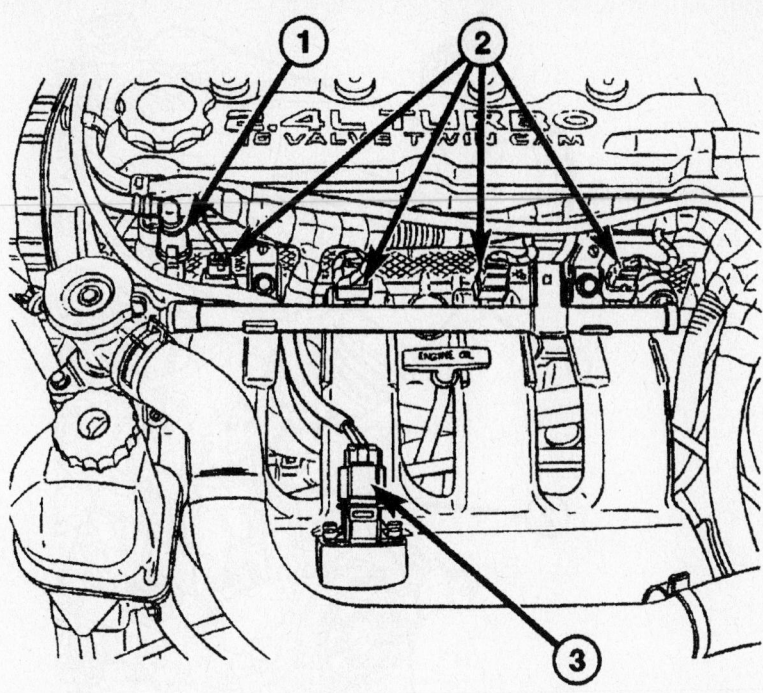

1 - FUEL SUPPLY LINE CONNECTION
2 - FUEL INJECTOR CONNECTORS
3 - MAP SENSOR

67189-NEON-G11

Location of some electrical connectors that need to be disconnected for intake manifold removal—2.4L engine

- Throttle cable to the throttle body
- Vacuum hoses to the throttle body and intake manifold
- Charge air cooler to the throttle body hose
- Fuel rail trim cover by pushing it over the fuel rail studs
- Negative battery cable

Exhaust Manifold

REMOVAL & INSTALLATION

2.0L Engines

1. Before servicing the vehicle, refer to the precautions in the beginning of this section.
2. Remove or disconnect the following:
   - Negative battery cable
   - Wiring harness heat shield-to-exhaust manifold support bracket bolt, if equipped
   - Wiring harness heat shield-to-exhaust manifold, if equipped
   - Exhaust manifold support bracket bolt and bracket, if equipped
   - Exhaust flex joint-to-manifold flange

fasteners. Move the exhaust system rearward to clear the flange studs
   - Make up air hose
   - Speed control vacuum reservoir, if equipped
   - Oxygen (O_2S) sensor connector and harness clip
   - Upper heat shield
   - Exhaust manifold bolts
   - Exhaust manifold and discard gasket

To install:

3. Install or connect the following:
   - Exhaust manifold with a new gasket and torque the bolts, in sequence, to 16 ft. lbs. (23 Nm)
   - Upper heat shield, if ULEV equipped and torque bolts to 16 ft. lbs. (23 Nm)
   - Upper and lower heat shields and torque the bolts to 95 inch lbs. (11 Nm)
   - O_2S sensor connector and harness clip
   - Cylinder head cover, if ULEV equipped
   - Speed control vacuum reservoir, if equipped

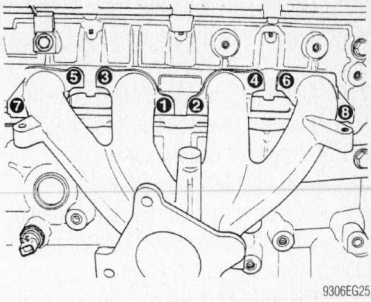

9306EG25

Exhaust manifold tightening sequence

- Make up air hose
- Exhaust manifold support bracket bolt, if LEV equipped. Torque the M10 bolt to 40 ft. lbs. (54 Nm), M12 bolt to 70 ft. lbs. (95 Nm) and the nut to 21 ft. lbs. (28 Nm).
- Cylinder head-to-exhaust manifold support bracket bolt, if equipped. Torque the bolts to 40 ft. lbs. (54 Nm).
- Wiring harness heat shield-to-exhaust manifold, if equipped
- Wiring harness heat shield-to-exhaust manifold support bracket bolt, if equipped
- Negative battery cable

2.4L Engines

On 2.4L turbocharged engines, the exhaust manifold is serviced as an assembly with the turbocharger. Refer to Turbocharger the removal and installation procedure.

Turbocharger Assembly

REMOVAL & INSTALLATION

2.4L Engine

✸✸ WARNING

If the turbocharger is replaced due to a bearing failure, you must also replace the oil pressure feed line. The oil return tube should also be cleaned.

➡The turbocharger and exhaust manifold are serviced as an assembly. Do NOT try to separate the turbocharger from the exhaust manifold. This will cause exhaust leaks. Chrysler recommends that the turbocharger elbow be replaced along with the turbocharger/exhaust manifold assembly.

1. Before servicing the vehicle, refer to the precautions in the beginning of this section.

2. Drain the engine cooling system.

3. Remove or disconnect the following:
- Negative battery cable
- Air cleaner housing and lid
- Clean air hose from turbocharger
- Throttle and speed control cables from the throttle body

4. Disconnect the electrical connectors from the following components:
- Inlet Air Temperature (IAT) Sensor
- Manifold Absolute Pressure (MAP) Sensor
- Idle Air Control (IAC) Motor
- Throttle Position Sensor (TPS)
- Ignition Coil Capacitor
- Upstream Oxygen Sensor (O$_2$S)
- Air inlet hose from the throttle body
- Vacuum hoses from the throttle body and upper intake manifold
- Upper intake manifold support bracket and upper intake manifold

➡**Cover the openings in the manifold to prevent debris from entering.**

- Turbocharger upper heat shield
- Oil supply line from the turbocharger
- Coolant return line
- Vacuum hoses from the turbocharger
- Muffler ground strap and muffler
- Downstream O$_2$S
- Fasteners securing the catalytic converter to the exhaust manifold
- Catalytic converter and intermediate pipe as an assembly
- Turbocharger to charge air cooler hose assembly
- Turbocharger elbow support bracket
- Turbocharger support bracket
- Oil return tube
- Turbocharger coolant supply line
- Turbocharger lower heat shield
- Turbocharger elbow
- Lower exhaust manifold fasteners that are accessible while vehicle raised
- Upper exhaust manifold fasteners
- Turbocharger/exhaust manifold assembly from above/between the engine and cowl panel. Discard the gasket

5. Inspect the turbocharger assembly for cracks and other damage and replace if necessary.

To install:

6. Install or connect the following:
- New gasket. The stainless steel

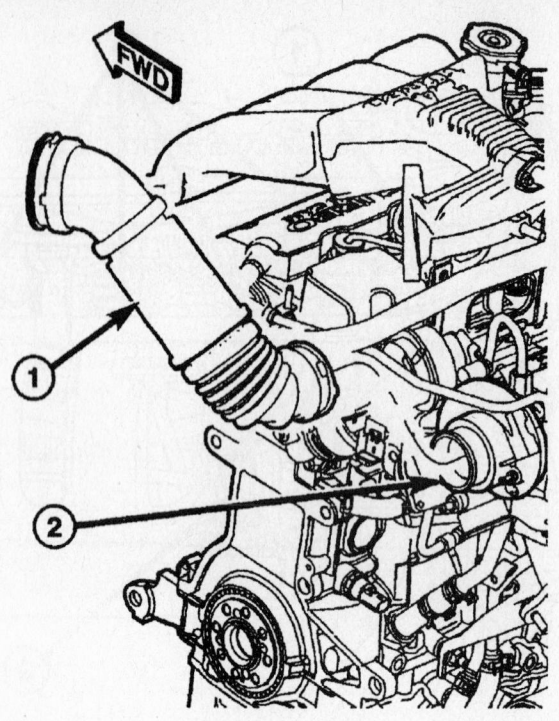

1 - CLEAN AIR HOSE
2 - TURBOCHARGER

67189-NEON-G12

Clean air hose—2.4L engine

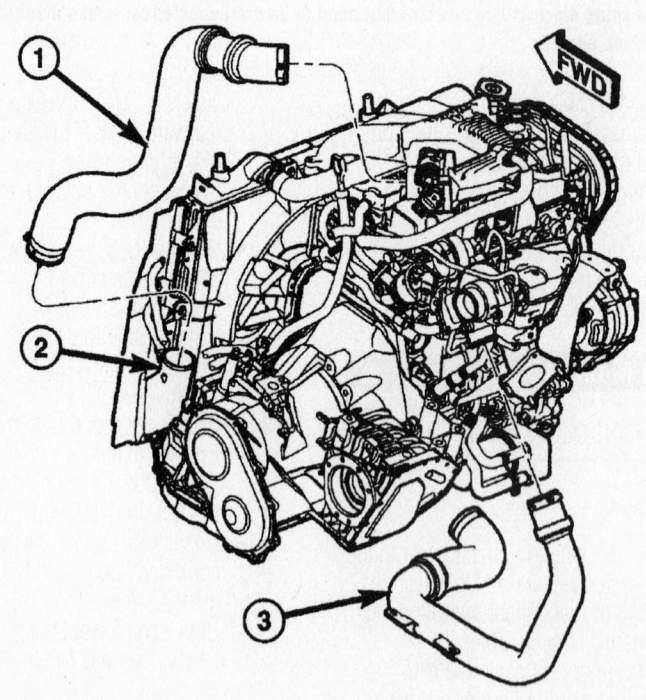

1 - HOSE - CHARGE AIR COOLER TO THROTTLE BODY
2 - CHARGE AIR COOLER
3 - HOSE - TURBOCHARGER TO CHARGE AIR COOLER

67189-NEON-G13

Charge air cooler hoses—2.4L engine

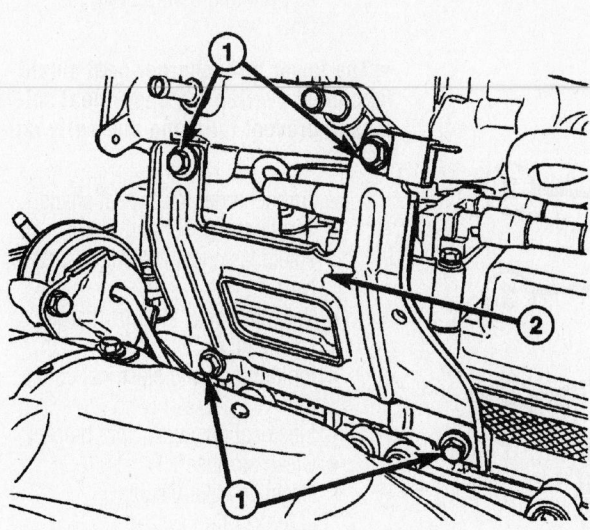

1 - FASTENERS
2 - UPPER INTAKE MANIFOLD SUPPORT BRACKET

67189-NEON-G14

Support bracket mounting—2.4L engine

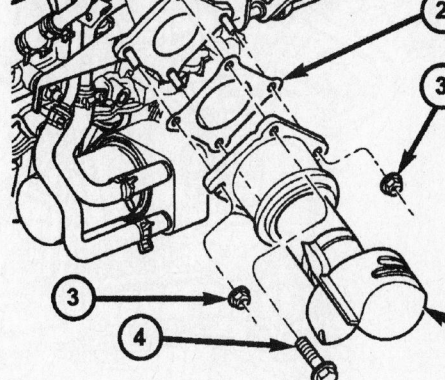

1 - FLAG NUT
2 - GASKET
3 - NUT
4 - BOLT
5 - CATALYTIC CONVERTER

67189-NEON-G16

Converter-to-exhaust manifold attachment—2.4L engine

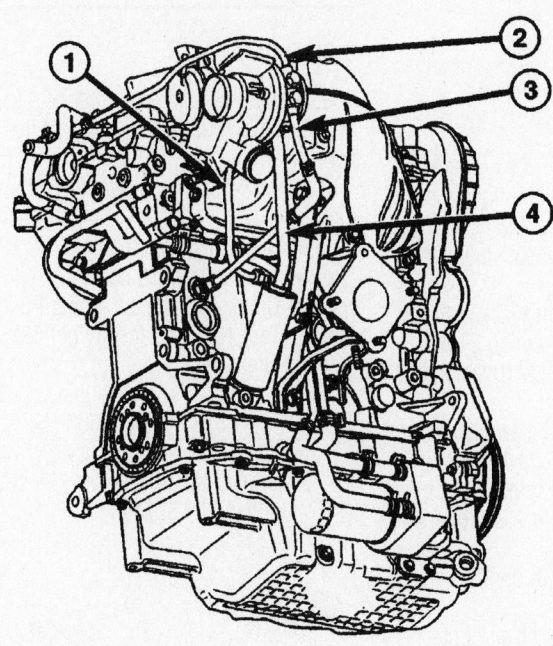

1 - OIL SUPPLY LINE
2 - COOLANT RETURN LINE
3 - COOLANT SUPPLY LINE
4 - OIL RETURN TUBE

67189-NEON-G15

Turbocharger lines and hoses—2.4L engine

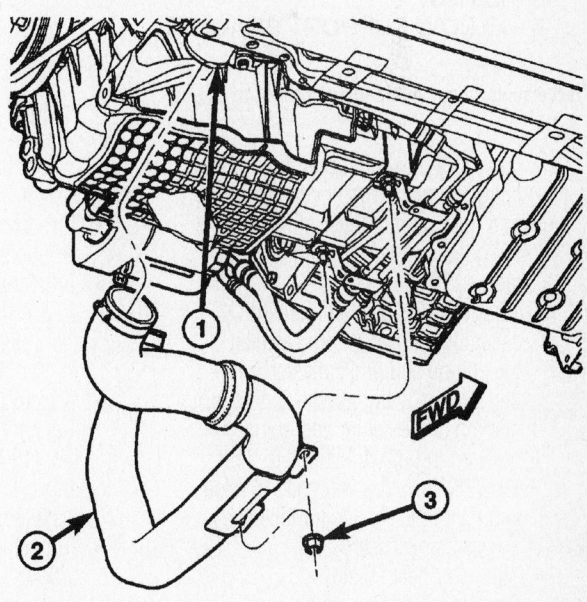

1 - CHARGE AIR COOLER
2 - HOSE - TURBOCHARGER TO CHARGE AIR COOLER
3 - NUT

67189-NEON-G17

Charge air cooler hose—2.4L engine

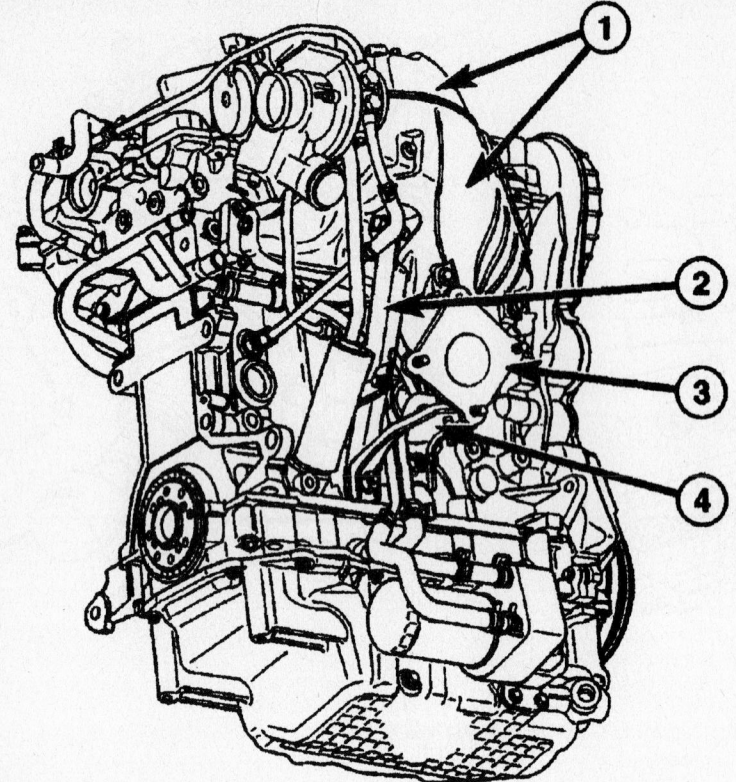

1 - UPPER/LOWER HEAT SHIELDS
2 - TURBOCHARGER SUPPORT BRACKET
3 - ELBOW
4 - ELBOW SUPPORT BRACKET

67189-NEON-G08

Turbocharger heat shield mounting—2.4L engine

layer of the exhaust manifold gasket goes against the cylinder head, and the graphite layer of the gasket goes against the manifold surface.

- Turbocharger/exhaust manifold assembly in place, from between the engine and brake master cylinder. Gradually tighten the fasteners, starting at the center and going outward in both directions to 21 ft. lbs. (28 Nm). You will need to raise and lower the vehicle for fastener access as necessary.
- Alternator heat shield
- Coolant return line, using new washers. Tighten the banjo fitting bolt to 27 ft. lbs. (37 Nm).
- Turbocharger coolant supply line, using new washers. Tighten the banjo fitting bolt to 27 ft. lbs. (37 Nm). Torque the flared fitting to 23 ft. lbs. (31 Nm).

- Turbocharger elbow and tighten the fasteners to 21 ft. lbs. (28 Nm)
- Turbocharger upper heat shield and fasteners, loosely.
- Turbocharger support bracket. Torque the M8 fasteners to 21 ft. lbs. (28 Nm) and the M10 fasteners to 40 ft. lbs. (54 Nm).
- Oil return tube with a new gasket. Tighten the fasteners to 105 inch lbs. (12 Nm). Make sure heat shield for oil return line is properly installed.
- Turbocharger elbow support bracket
- Turbocharger-to-charge air cooler hose assembly
- Catalytic converter with new gasket. Tighten the converter-to-manifold fasteners to 21 ft. lbs. (28 Nm).
- Downstream O2S
- Exhaust extension pipe to catalytic converter

- Power steering cooler fasteners
- Oil supply line to the turbocharger with new washers. Tighten the banjo fitting bolt to 27 ft. lbs. (37 Nm).

➡ **The lower turbocharger heat shield tabs must overlap the upper heat shield tabs to prevent fatiguing and early failure.**

- Turbocharger lower heat shield. Torque the upper and lower heat shield fasteners to 21 ft. lbs. (28 Nm)
- Vacuum hoses to turbocharger
- Upstream O2S
- Ignition coil and electrical connector
- Electrical connector into bracket
- Clean air hose
- Negative battery cable

7. Fill the cooling system
8. Change the oil and filter.
9. Start the engine and check for exhaust leaks. Repair leaks as necessary.
10. Inspect the exhaust system for contact with the body panels. Make the necessary adjustments, if needed.

Front Crankshaft Seal

REMOVAL & INSTALLATION

2.0L Engines

1. Before servicing the vehicle, refer to the precautions in the beginning of this section.
2. Remove or disconnect the following:

- Negative battery cable
- Accessory drive belts
- Crankshaft damper, using Puller Tool 1026 and Insert Tool 6827-A
- Timing belt
- Crankshaft sprocket using special Tools 6793 and C-4685-C2
- Crankshaft sprocket key
- Crankshaft oil seal, using a seal puller Tool 6771

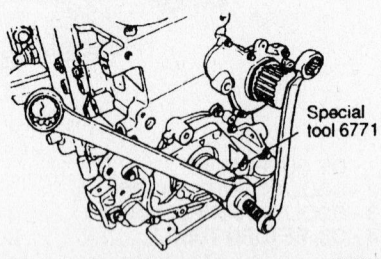

7922EG17

Removing the front crankshaft oil seal

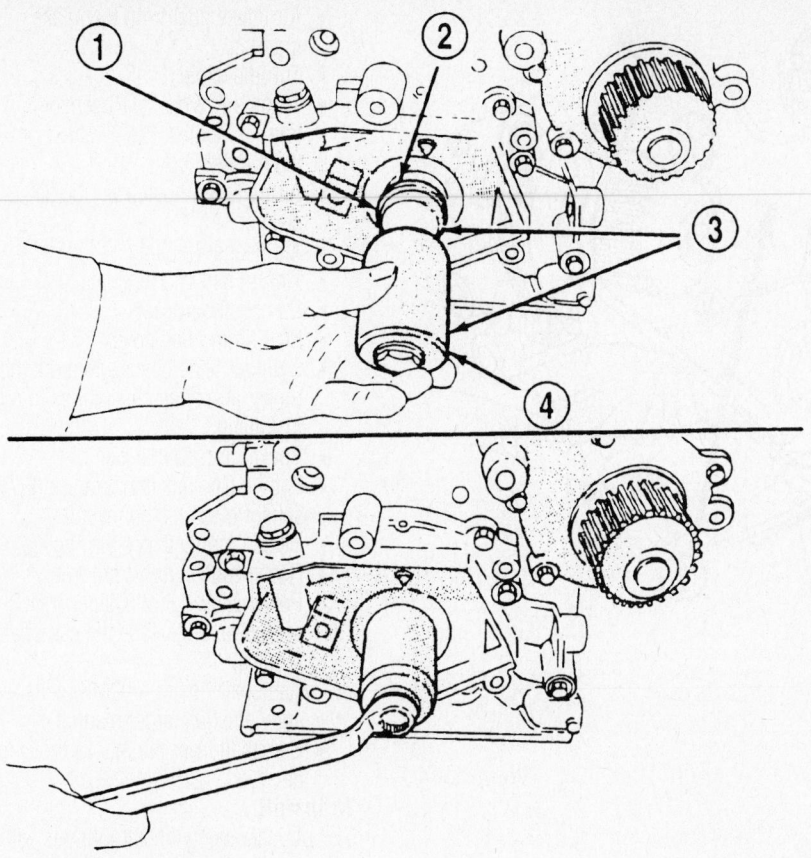

1 - PROTECTOR
2 - SEAL
3 - SPECIAL TOOL 6780-1
4 - INSTALLER

9356HG07

Installing a new seal, using seal installer 6780-1; proceed with caution if using substitute tools

To install:
3. Install or connect the following:
   • New front crankshaft oil seal, using Crankshaft Installer Tool 6780–1
   • Crankshaft key
   • Crankshaft sprocket using Installer Tool 6792

➡**Make sure the word FRONT on the crankshaft sprocket is facing outward.**

   • Timing belt
   • Crankshaft damper using thrust bearing washer and bolt from Installer Tool 6792. Torque the damper bolt to 100 ft. lbs. (136 Nm).
   • Accessory drive belts
   • Negative battery cable

2.4L Engine

1. Before servicing the vehicle, refer to the precautions in the beginning of this section.
2. Remove or disconnect the following:
   • Negative battery cable
   • Crankshaft vibration damper with special Tools 1026 and 6287-A
   • Timing belt
   • Crankshaft sprocket with special Tools 6793 and insert C-4685-C2

✳✳ WARNING

Be careful not to scratch or nick the seal surface or seal bore.

   • Crankshaft oil seal with special Tool 6771

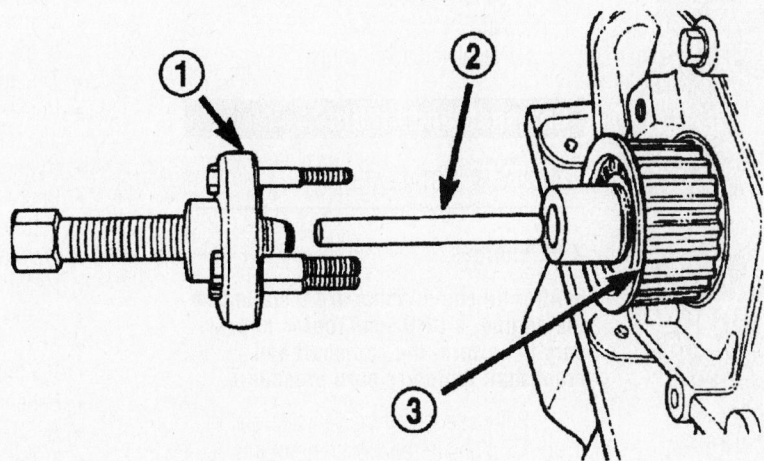

1 - SPECIAL TOOL 6793
2 - SPECIAL TOOL C-4685-C2
3 - CRANKSHAFT SPROCKET

67189-NEON-G18

Removing the crankshaft sprocket—2.4L engine

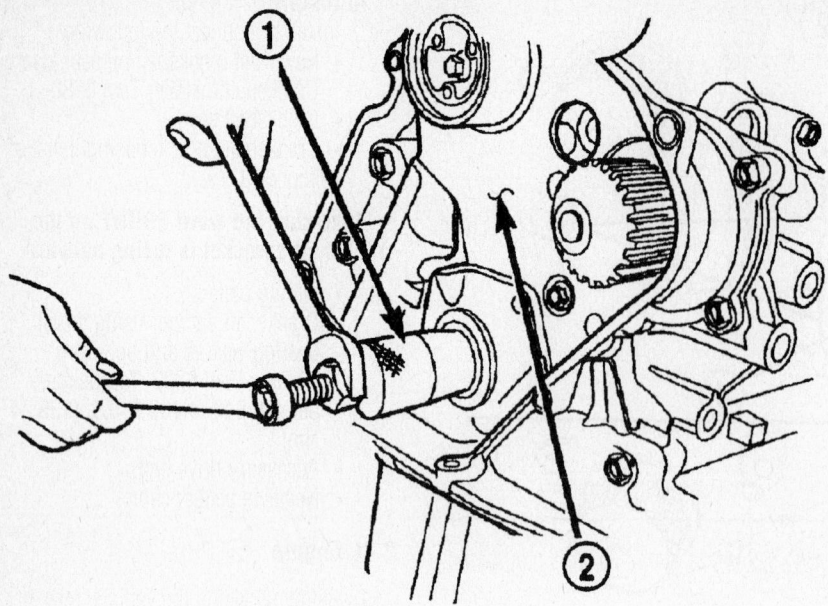

1 - SPECIAL TOOL 6771
2 - REAR TIMING BELT COVER

67189-NEON-G19

Front crankshaft oil seal removal—2.4L engine

To install:
3. Install or connect the following:
- Front crankshaft oil seal with special Tool 6780-1 until it is flush with the cover

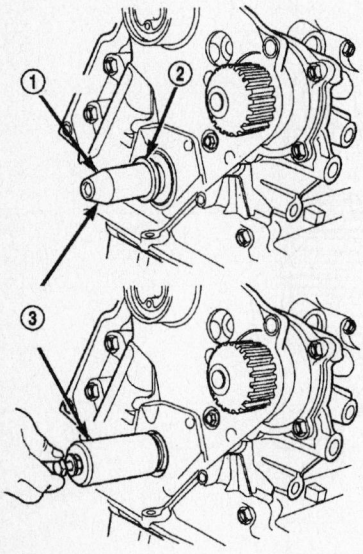

1 - PROTECTOR
2 - SEAL
3 - SPECIAL TOOL 6780

67189-NEON-G20

Installation of the front crankshaft oil seal—2.4L engine

- Crankshaft sprocket with special Tool 6792

➡**Make sure the word FRONT on the crankshaft sprocket is facing outward.**

- Timing belt
- Crankshaft vibration damper using thrust bearing washer and bolt from Installer Tool 6792 and torque the damper bolt to 100 ft. lbs. (135 Nm)
- Negative battery cable

Camshaft and Lifters

REMOVAL & INSTALLATION

2.0L Engines

➡**After all components are installed on the engine, a DRB scan tool is necessary to perform the camshaft and crankshaft timing relearn procedure.**

1. Before servicing the vehicle, refer to the precautions in the beginning of this section.
2. Relieve the fuel system pressure.
3. Drain the cooling system.
4. Remove or disconnect the following:
- Battery and tray
- Inlet Air Temperature (IAT) sensor electrical connector

- Air intake duct from the intake manifold
- Throttle cables
- Throttle Position (TP) sensor
- Idle Air Control (IAC) valve from the throttle body
- Air make-up hose from the air cleaner
- Air cleaner housing
- Timing belt and tensioner
- Camshaft sprocket
- Rear timing belt cover
- Cylinder head cover and mark the rocker arm shaft assemblies to ease installation
- Rocker arm shaft assemblies
- Engine Coolant Temperature (ECT) sensor electrical connector
- Heater supply and return hoses
- Heater tube support bracket
- Power Distribution Center (PDC) screws aside to ease the camshaft removal
- Camshaft Position (CMP) sensor and camshaft target magnet
- Camshaft from the rear of the cylinder head

To install:
5. Lubricate the camshaft journals with clean engine oil.
6. Install or connect the following:
- Camshaft and reposition the PDC and heater tubes
- Camshaft target magnet and torque the screw to 30 inch lbs. (3 Nm)
- CMP sensor
- Front camshaft seal, if removed
- Timing belt rear cover
- Camshaft sprocket and torque to 85 ft. lbs. (115 Nm)
- Timing belt and tensioner
- Rocker arms in the proper order and torque to 21 ft. lbs. (28 Nm)
- Cylinder head cover
- Ignition coil and spark plug cables and torque the fasteners to 105 inch lbs. (12 Nm)
- Heater tube bracket bolt
- Heater tubes
- ECT sensor electrical connector
- PDC attaching screws
- Battery and tray
- Air cleaner housing
- Throttle body connections
- IAT sensor

7. Fill the cooling system to the proper level.

8. Use a DRB scan tool to perform the camshaft and crankshaft timing relearn procedure, as follows:
    a. Connect the scan tool to the DLC

(located under the instrument panel, near the steering column).

    b. Turn the ignition switch **ON**, and access the "miscellaneous" screen.

    c. Select the "re-learn cam/crank" option, then follow the instructions on the scan tool screen.

9. Start the engine and check for leaks. Run the engine with the radiator cap off so as the engine warms and the thermostat opens, coolant can be added to the radiator. Test drive vehicle to check for proper operation.

2.4L Engine

1. Before servicing the vehicle, refer to the precautions in the beginning of this section.

2. Remove or disconnect the following:
- Cylinder head (valve) cover
- Camshaft Position (CMP) sensor and camshaft target magnet
- Timing belt
- Camshaft sprockets and timing belt rear cover
- Camshaft bearing caps, loosen the bearing caps in sequence, one camshaft at a time.

➡The bearing caps are identified for location. Remove the outside bearing caps first. If the bearing caps are difficult to remove, use a plastic hammer to gently tap the rear part of the camshaft.

- Intake and exhaust camshafts

To install:

3. Before installation, clean the cylinder head and cover mating surfaces. Make certain that the rails are flat.

4. Lubricate the bearing journals, rocker arms and cam lobes with clean engine oil.

✳✳ WARNING

Make sure that NONE of the pistons are at Top Dead Center (TDC) when installing the camshafts.

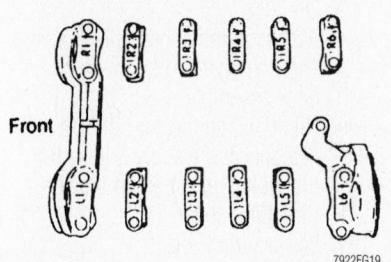

7922EG19

Identifying the camshaft bearing caps—2.4L engine

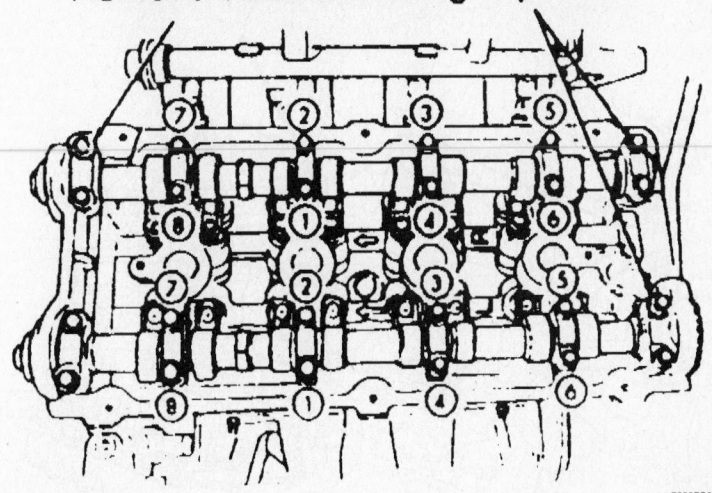

Remove outside bearing caps first

7922EG20

Bearing cap removal sequence—2.4L engine

5. Install or connect the following:
- Camshafts on cylinder head bearing journals.
- Right and left bearing caps No. 2–5 and No. 6. Torque the M6 fasteners, in sequence, to 105 inch lbs. (12 Nm)
- Apply Mopar Gasket Maker to the No. 1 and No. 6 bearing caps. Install the caps and tighten the M8 fasteners to 21 ft. lbs. (28 Nm).

➡The bearing end caps must be installed before the seals can be installed.

- Camshaft oil seals
- Camshaft sprockets
- Timing belt
- Camshaft target magnet and CMP sensor
- Cylinder head cover
- Negative battery cable

6. Start the engine and check for proper operation and leaks.

Valve Lash

ADJUSTMENT

The engines in these vehicles do not require periodic valve lash adjustment.

Starter Motor

REMOVAL & INSTALLATION

2.0L Engines

1. Before servicing the vehicle, refer to the precautions in the beginning of this section.

2. Remove or disconnect the following:
- Negative battery cable
- Inlet hose from the intake manifold and reposition the air cleaner box, on high output engines only
- Starter electrical connectors
- Starter-to-engine bolts
- Starter

To install:

3. Install or connect the following:
- Starter and torque bolts to 40 ft. lbs. (54 Nm)
- Starter electrical connectors
- Intake manifold inlet hose on high output engines
- Negative battery cable

2.4L Engine

1. Before servicing the vehicle, refer to the precautions in the beginning of this section.

2. Remove or disconnect the following:
- Negative battery cable
- Air cleaner assembly
- Throttle body
- Upper mounting bolt and ground wire
- Positive battery cable from the starter
- Solenoid connector from the starter
- Middle starter mounting bolt
- Upper bolt for the intake manifold support (loosen only) and swing the bracket out of the way
- Starter lower mounting bolt and starter

To install:

3. Install or connect the following:
- Starter and lower mounting bolt
- Loosen the upper bolt for the intake

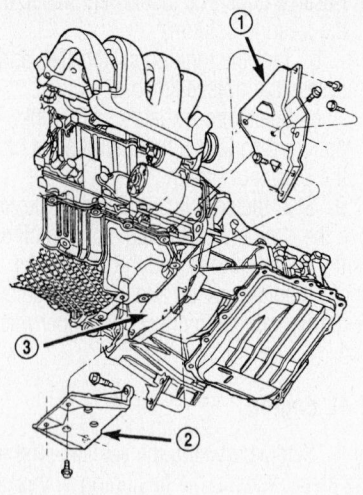

1 – LATERAL BENDING BRACE
2 – STRUCTURAL COLLAR
3 – DUST COVER

9306EG26

Exploded view of the bending brace, structural collar and dust cover

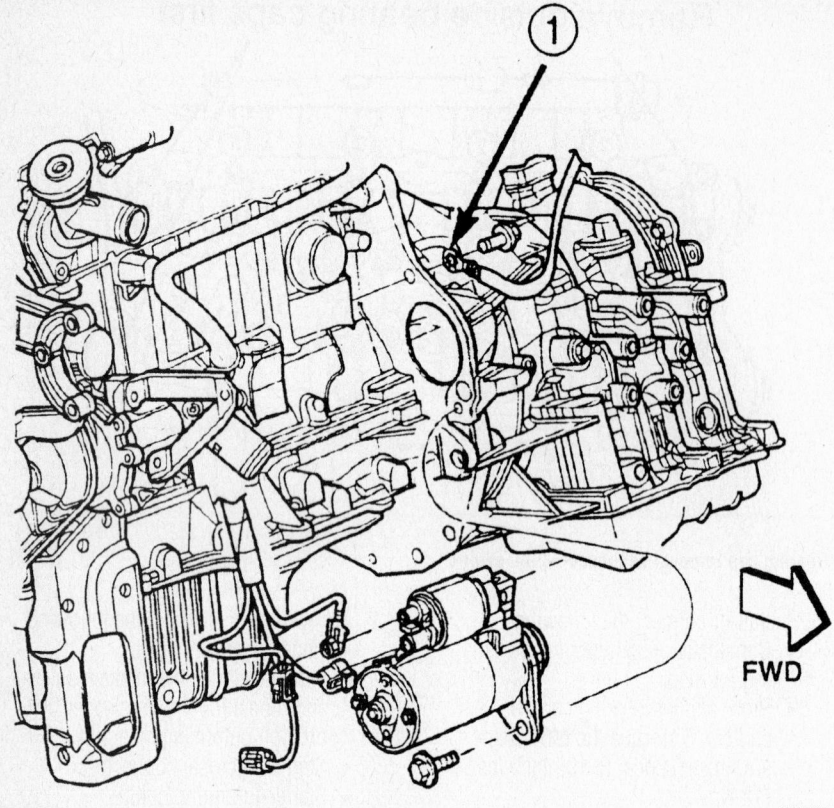

FWD

9356HG08

Starter mounting—2002 Neon shown, others similar

manifold support bracket and swing bracket back into place
- Middle starter mounting bolt
- Tighten the intake manifold support bracket bolt
- Tighten the starter mounting bolts to 40 ft. lbs. (54 Nm)
- Solenoid connector to the starter
- Positive battery cable to starter. Torque the captive nut to 89 inch lbs. (10 Nm)

- Upper mounting bolt and ground wire, then tighten the starter mounting bolt to 40 ft. lbs. (54 Nm)
- Throttle body
- Air cleaner assembly
- Negative battery cable

Oil Pan

REMOVAL & INSTALLATION

2.0L Engines

1. Before servicing the vehicle, refer to the precautions in the beginning of this section.
2. Raise and safely support the vehicle.
3. Drain the engine oil.
4. Remove or disconnect the following:

- Negative battery cable
- Oil filter
- Oil filter adapter
- Structural collar
- Lateral bending brace
- Transaxle lower dust cover
- Oil pan

5. Thoroughly, clean the gasket mating surfaces.

To install:

6. Apply silicone sealer to the oil pump-to-engine block parting line.
7. Install or connect the following:
- New oil pan gasket
- Oil pan and torque the screws to 105 inch lbs. (12 Nm)
- Transaxle lower dust cover
- Lateral bending brace
- Structural collar
- Oil filter adapter and torque the bolts to 60 ft. lbs. (80 Nm)
- Oil filter

8. Fill the engine with clean oil.
9. Start the engine, check for leaks and repair if necessary.

2.4L Engine

1. Before servicing the vehicle, refer to the precautions in the beginning of this section.
2. Raise and safely support the vehicle.
3. Drain the engine oil.
4. Remove or disconnect the following:

- Negative battery cable
- Oil filter
- Accessory drive belt splash shield
- Turbocharger-to-charge air cooler hose assembly
- Oil cooler connector bolt. Do NOT disconnect the coolant lines from oil cooler, just position it aside.
- Structural collar
- Lower torque strut
- Oil filter adapter and gasket
- Oil pan and gasket

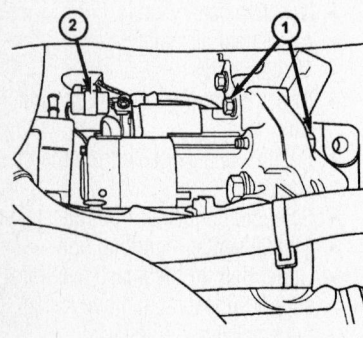

1 - Mounting Bolts
2 - Electrical Connections

67189-NEON-G21

Starter mounting and connector location—2.4L engine

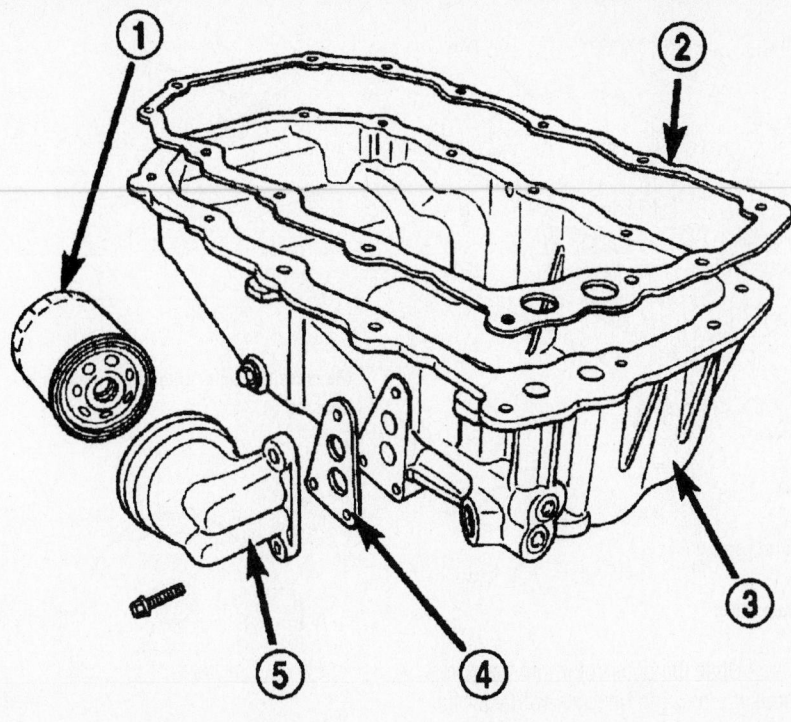

1 - OIL FILTER
2 - OIL PAN GASKET
3 - OIL PAN
4 - ADAPTER GASKET
5 - OIL FILTER ADAPTER

67189-NEON-G22

Exploded view of the oil pan—2.4L engine

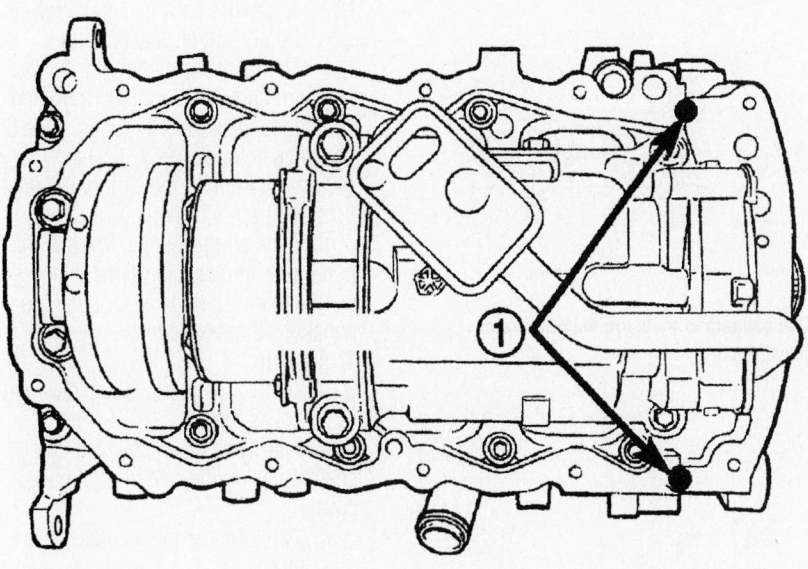

1 - SEALER LOCATIONS

67189-NEON-G23

Oil pan sealant locations—2.4L engine

5. Thoroughly clean the gasket mating surfaces.

To install:

6. Apply Mopar Engine RTV GEN II, or equivalent silicone sealer to the oil pump-to-engine block parting line.

7. Install or connect the following:

- New oil pan gasket
- Oil pan and torque the screws to 105 inch lbs. (12 Nm)
- Oil filter adapter and gasket. Torque the screws to 105 inch lbs. (12 Nm)
- Oil cooler seal. Lubricate seal and position oil cooler to oil filter adapter, aligning notch to tab.
- Oil cooler connector bolt and tighten to 41 ft. lbs. (55 Nm).
- Oil drain plug and oil filter
- Structural collar
- Lower torque strut
- Turbocharger-to-charge air cooler hose assembly

8. Fill the engine with clean oil

9. Start the engine, check for leaks and repair if necessary.

Oil Pump

REMOVAL & INSTALLATION

1. Before servicing the vehicle, refer to the precautions in the beginning of this section.

2. Drain the engine oil.

3. Remove or disconnect the following:

- Negative battery cable
- Crankshaft damper
- Timing belt
- Timing belt tensioner
- Camshaft sprocket and rear timing belt cover
- Oil pan
- Crankshaft sprocket, using Tool 6795 and Insert Tool C-4685-C2
- Oil pickup tube
- Oil pump
- Front crankshaft seal
- Oil pump cover screws and lift the cover off
- Oil pump rotors

4. Wash all parts in a solvent; then, inspect carefully for damage or wear, as follows:

a. Inspect the mating surface of the oil pump should be smooth. Replace the pump cover, if scratched or grooved.

b. Lay a straightedge across the pump cover surface. If a 0.003 in. (0.076mm) feeler gauge can be inserted

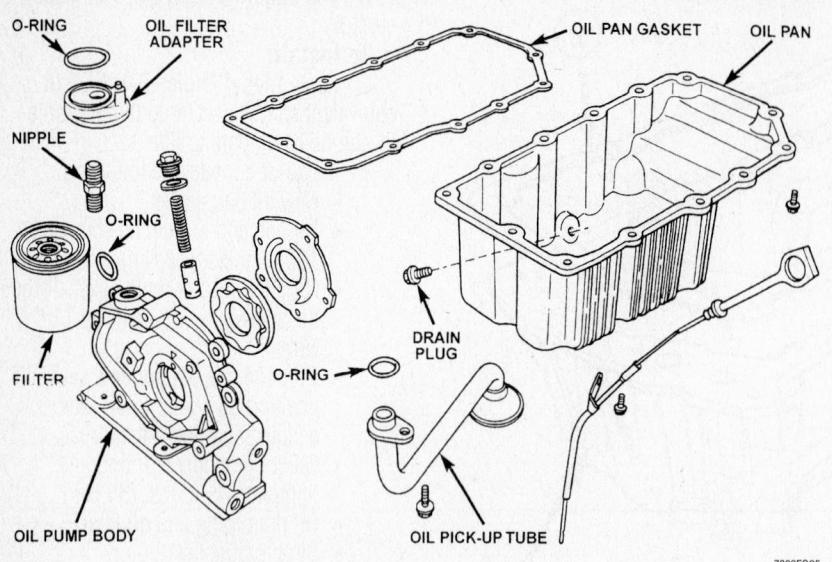

Exploded view of the oil pump and related component mounting

between the cover and the straightedge, the cover should be replaced.

c. Measure the thickness and diameter of the outer rotor. If the outer rotor thickness measures 0.301 in. (7.64mm) or less, or if the diameter is 3.148 in. (79.95mm) or less, replace the outer rotor.

d. If the inner rotor measures 0.301 in. (7.64mm) or less, replace the inner rotor.

e. Slide the outer rotor into the pump housing, press to one side with your fingers and measure the clearance between the rotor and the housing. If the measurement is 0.015 in. (0.39mm) or more, replace the housing only if the outer rotor is within specification.

f. Install the inner rotor into the pump housing, If the clearance between the inner and outer rotors is 0.008 in. (0.203mm) or more, replace both rotors.

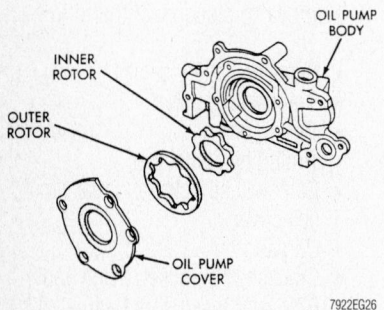

Exploded view of the oil pump assembly

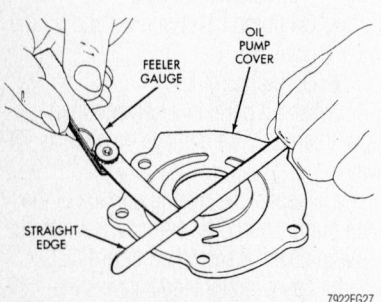

Use a feeler gauge and straightedge to check the oil pump cover for warpage

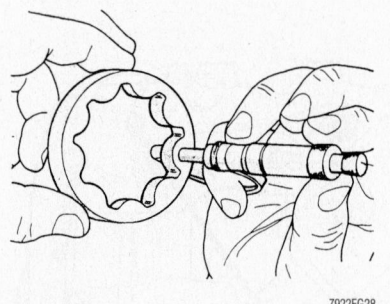

Use calipers to measure the outer rotor thickness

and the inner rotor thickness

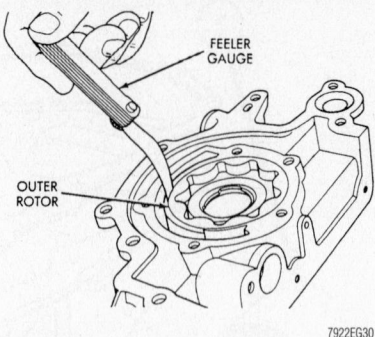

Measure the outer rotor clearance in the housing

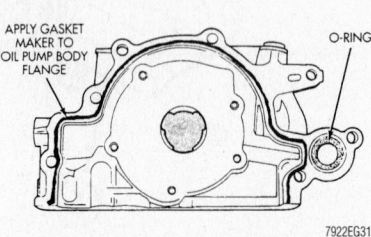

Apply a small amount of gasket maker to the pump body cover mounting surface

g. Place a straightedge across the face of the pump housing, between the bolt holes. If a feeler gauge of 0.004 in. (0.102mm) or more can be inserted between the rotors and the straightedge, replace the pump assembly.

h. Inspect the oil pressure relief valve plunger for scoring and free operation in its bore. Small marks may be removed with 400 grit wet or dry sandpaper.

i. The relief valve spring has a free length of about 2.39 in. (60.7mm) and should test between 18–19 lbs. (8.1–8.6 kg) when compressed to 1.60 in. (40.6mm). Replace the spring, if it falls outside of specifications.

j. If the oil pressure is low and the pump is within specifications, inspect for worn engine bearings or for other reasons for oil pressure loss.

To install:

5. Assemble the pump, using new parts as required, as follows:

a. Install the inner rotor with the chamfer facing the cast iron oil pump cover.

b. Apply Mopar gasket maker to the oil pump.

c. Install the oil ring into the oil pump body discharge passage.

6. Prime the oil pump before installation by filling the rotor cavity with engine oil.

7. Install or connect the following:

• Oil pump, align the rotor flats

with the crankshaft flats. Torque the pump bolts to 21 ft. lbs. (28 Nm).

※※ WARNING

The front crankshaft seal MUST be out of the pump to align or damage may result.

- New front crankshaft seal, using Seal Driver Tool 6780
- Crankshaft sprocket, using a Crankshaft Sprocket Installer Tool 6792
- Oil pump pickup tube
- Oil pan
- Rear timing belt cover and camshaft sprocket
- Timing belt tensioner
- Timing belt and front cover
- Crankshaft damper
- Accessory drive belts
- Negative battery cable

8. Fill the engine with clean oil.

9. Start the engine and check for leaks; then, recheck the fluid level and add as necessary.

Rear Main Seal

REMOVAL & INSTALLATION

1. Before servicing the vehicle, refer to the precautions in the beginning of this section.

2. Remove or disconnect the following:
- Transaxle
- Flexplate/flywheel
- Rear main seal. Insert a seal remover between the dust lip and the metal case of the crankshaft seal.

Angle the tool through the dust lip against the metal case of the seal. Pry out the seal.

※※ WARNING

DO NOT let the prytool contact the crankshaft seal surface. Contact of the tool blade against the crankshaft edge (chamfer) is permitted.

To install:

※※ WARNING

If the crankshaft edge (chamfer) has any burrs or scratches on the, clean it up with 400 grit sand paper to prevent seal damage during installation of the new seal.

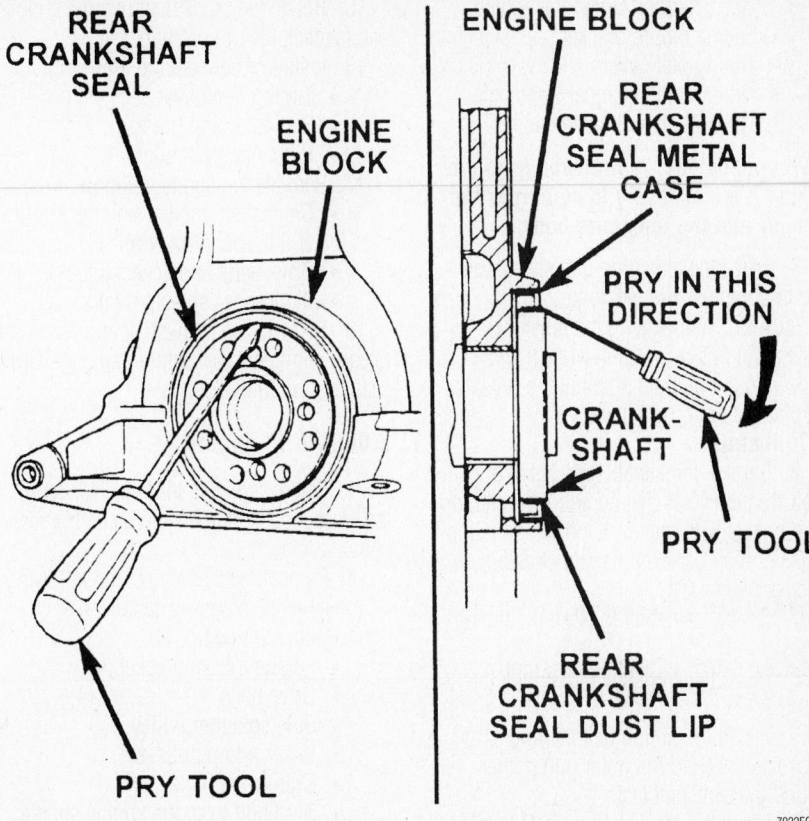

When prying the seal out, be sure to use the prytool at the proper angle

➡**No lubrication is necessary when installing the seal.**

3. Place Crankcase Seal Pilot Tool 6926–1 on the crankshaft; this is a pilot tool with a magnetic base.

4. Position the seal over the Pilot Tool; be sure the words THIS SIDE OUT on the seal can be read.

➡**The pilot tool should stay on the crankshaft during installation of the seal. Be sure the seal lip faces the crankcase during installation.**

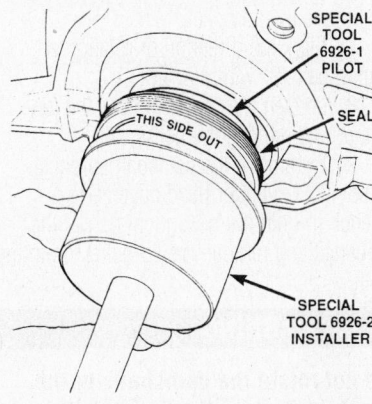

Place a proper size pilot tool with a magnetic base on the crankshaft

※※ WARNING

If the seal is driven in the block past flush, this may cause an oil leak.

5. Drive the seal into the block, using Crankshaft Seal Tool 6926-2 and Handle C-4171, until the tool bottoms out against the block.

6. Install the flexplate/flywheel. Apply Lock & Seal Adhesive to the bolt treads. Torque the bolts in a star pattern, to 70 ft. lbs. (95 Nm).

7. Install the transaxle.

Timing Belt

REMOVAL & INSTALLATION

2.0L (VIN C) Engine

1. Before servicing the vehicle, refer to the precautions in the beginning of this section.

2. Remove or disconnect the following:
- Negative battery cable
- Drive belts and accessories
- Right inner splash-shield
- Crankshaft damper
- Right engine mount

3. Place a support under the engine.
- Engine mount bracket
- Timing belt cover
- Timing belt tensioner bolts
- Timing belt and the tensioner

➡ **When tensioner is removed from the engine it is necessary to compress the plunger into the tensioner body.**

4. Place the tensioner in a soft-jawed vise to compress the tensioner.

5. After compressing the tensioner place a pin (a 5/64 in. Allen wrench will work) into the plunger side hole to retain the plunger until installation.

To install:

6. Set the crankshaft sprocket to Top Dead Center (TDC) by aligning the notch on the sprocket with the arrow on the oil pump housing, then back off the sprocket three notches before TDC.

7. Set the camshaft to align the timing marks.

8. Move the crankshaft to ½ notch before TDC.

9. Install the timing belt starting at the crankshaft, around the water pump, then around the camshaft last.

10. Move the crankshaft to TDC to take up the belt slack.

11. Reinstall the tensioner to the block but do not tighten it.

12. Using a torque wrench apply 250 inch lbs. (28 Nm) of torque to the tensioner pulley.

13. With torque being applied to the tensioner pulley, move the tensioner up against the tensioner bracket and tighten the fasteners to 275 inch lbs. (31 Nm).

14. Remove the tensioner plunger pin, the tension is correct when the plunger pin can be removed and replaced easily.

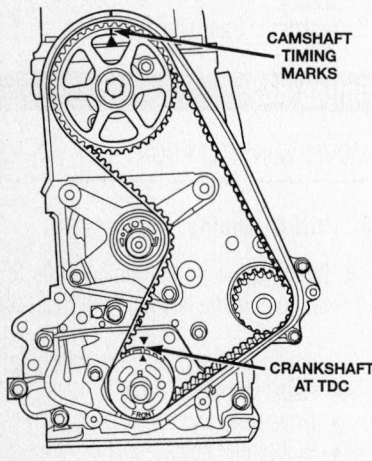

TDC alignment for timing belt installation—2.0L (VIN C) engine

15. Rotate the crankshaft two revolutions and recheck the timing marks.

16. Install or connect the following:
- Timing belt cover
- Engine mount bracket
- Right engine mount

17. Remove the engine support.
- Crankshaft damper and tighten to 105 ft. lbs. (142 Nm)
- Drive belts and accessories
- Right inner splash-shield

18. Perform the crankshaft and camshaft relearn alignment procedure using the DRB scan tool or equivalent.

2.0L (VIN F) Engine

1. Before servicing the vehicle, refer to the precautions in the beginning of this section.

2. Remove or disconnect the following:
- Negative battery cable
- Engine undercover
- Engine mount bracket
- Drive belts
- Belt tensioner pulley
- Water pump pulleys
- Crankshaft pulley.
- Stud bolt from the engine support bracket, then remove the timing belt covers

3. Rotate the crankshaft clockwise to align the camshaft timing marks. Always turn the crankshaft in the forward direction only.

4. Loosen the tension pulley center bolt.

➡ **If the timing belt is to be reused, mark the direction of rotation on the flat side of the belt with an arrow.**

5. Move the tension pulley towards the water pump and remove the timing belt.

6. Remove the crankshaft sprocket center bolt using special tool MB990767 to hold the crankshaft sprocket while removing the center bolt. Then, use MB998778 puller to remove the sprocket.

7. Mark the direction of rotation on the timing belt "B" with an arrow.

8. Loosen the center bolt on the tensioner and remove the belt.

9. To remove the camshaft sprocket, remove the cylinder head cover. Use a wrench to hold the hexagonal part of the camshaft and remove the sprocket mounting bolt.

✳✳ WARNING

Do not rotate the camshafts or the crankshaft while the timing belt is removed.

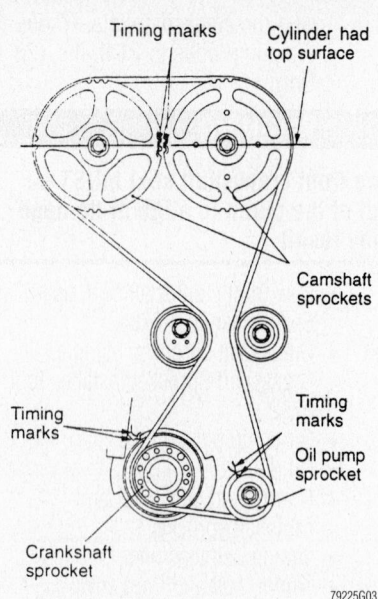

Notice the timing mark on the oil pump drive sprocket—2.0L (VIN F) engines

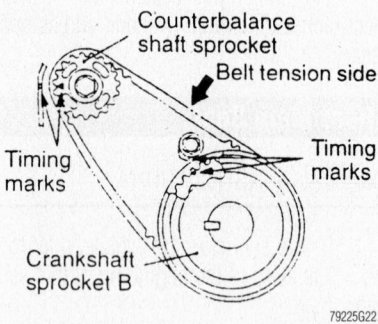

Timing belt "B" timing marks locations—2.0L (VIN F) engines

To install:

10. Use a wrench to hold the camshaft, then install the sprocket and mounting bolt. Tighten the bolt(s) to 65 ft. lbs. (88 Nm).

11. Install the cylinder head cover.

12. Place the crankshaft sprocket on the crankshaft. Use tool MB990767 to hold the crankshaft sprocket while tightening the center bolt. Tighten the center bolt to 80–94 ft. lbs. (108–127 Nm).

13. Align the timing marks on the crankshaft sprocket "B" and the balance shaft.

14. Install timing belt "B" on the sprockets. Position the center of the tensioner pulley to the left and above the center of the mounting bolt.

15. Push the pulley clockwise toward the crankshaft to apply tension to the belt and tighten the mounting bolt to 14 ft. lbs. (19 Nm). Do not let the pulley turn when tightening the bolt because it will cause excessive tension on the belt. The belt should

deflect 0.20–0.28 in. (5–7mm) when finger pressure is applied between the pulleys.

16. Install the crankshaft sensing blade and the crankshaft sprocket. Apply engine oil to the mounting bolt and tighten the bolt to 80–94 ft. lbs. (108–127 Nm).

17. Use a press or vise to compress the auto tensioner pushrod. Insert a set pin when the holes are aligned.

※※ WARNING

Do not compress the pushrod too quickly, damage to the pushrod can occur.

18. Install the auto tensioner on the engine.

19. Align the timing marks on the camshaft sprocket, crankshaft sprocket and the oil pump sprocket.

20. After aligning the mark on the oil pump sprocket, remove the cylinder block plug and insert a Phillips screwdriver in the hole to check the position of the counter balance shaft. The screwdriver should go in at least 2.36 in. or more. if not, rotate the oil pump sprocket once and realign the timing mark so the screwdriver goes in. Do not remove the screwdriver until the timing belt is installed.

21. Install or connect the following:
- Timing belt on the intake camshaft and secure it with a clip
- Timing belt on the exhaust camshaft. Align the timing marks with the cylinder head top surface using two wrenches. Secure the belt with another clip.
- Belt around the idler pulley, oil pump sprocket, crankshaft sprocket and the tensioner pulley

22. Turn the tension pulley so the pin-holes are at the bottom. Press the pulley lightly against the timing belt.

23. Screw the special tool into the left engine support bracket until it contacts the tensioner arm, then screw the tool in a little more and remove the pushrod pin from the auto tensioner. Remove the special tool and tighten the center bolt to 35 ft. lbs. (48 Nm).

24. Turn the crankshaft ¼ turn counter-clockwise, then clockwise until the timing marks are aligned.

25. Loosen the center bolt. Install special tool MD998767 on the tension pulley. Turn the tension pulley counterclockwise with a torque of 2.6 ft. lbs. (3.5 Nm) and tighten the center bolt to 35 ft. lbs. (48 Nm). Do not let the tension pulley turn with the bolt.

26. Turn the crankshaft two revolutions

to the right and align the timing marks. After 15 minutes, measure the protrusion of the pushrod on the auto tensioner. The standard measurement is 0.150–0.177 in. (3.8–4.5mm). If the protrusion is out of specification, loosen the tension pulley, apply the proper torque to the belt and retighten the center bolt.

27. Install or connect the following:
- Crankshaft pulley. Tighten the mounting bolts to 18 ft. lbs. (25 Nm).
- Water pump. Tighten the mounting bolts to 6.5 ft. lbs. (8.8 Nm).
- Drive belts, then adjust
- Engine mount bracket
- Engine undercover
- Negative battery cable

2.4L Engine

1. Before servicing the vehicle, refer to the precautions in the beginning of this section.

2. Remove or disconnect the following:
- Negative battery cable
- Upper torque strut attaching bolts and strut

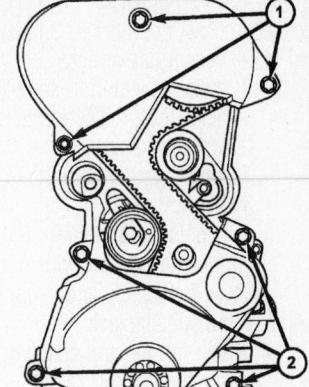

1 - UPPER COVER FASTENERS
2 - LOWER COVER FASTENERS

67189-NEON-G25

Upper and lower timing belt cover mounting—2.4L engine

- Torque strut bracket from strut tower
- Torque strut bracket from engine
- Upper timing belt cover fasteners and cover

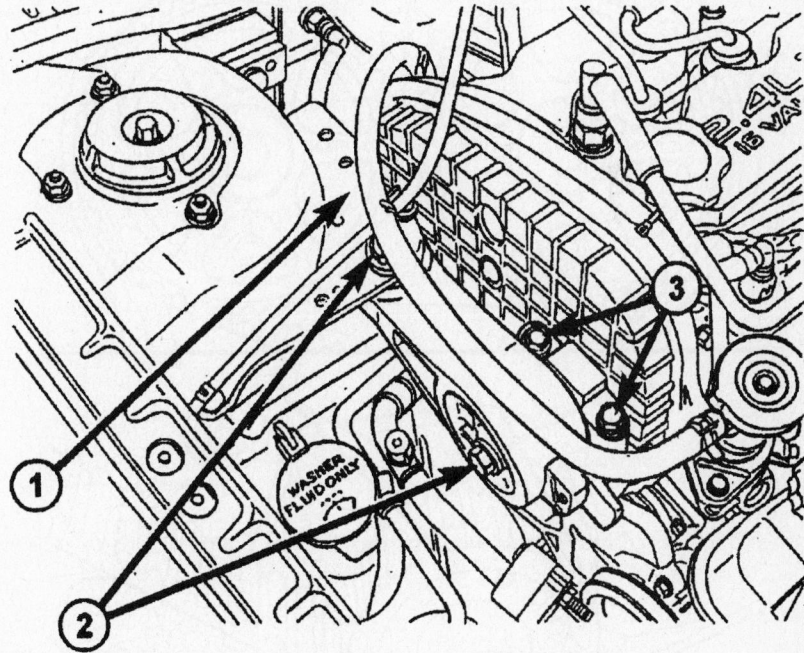

1 - TORQUE STRUT BRACKET
2 - TORQUE STRUT MOUNTING BOLTS
3 - ENGINE BRACKET MOUNTING BOLTS

67189-NEON-G24

Upper torque strut and bracket mounting—2.4L engine

- Right front wheel
- Accessory drive belt splash shield
- Accessory drive belts
- Crankshaft vibration damper with special Tools 1026 and 6287-A
- Lower torque strut
- Exhaust system from the exhaust manifold
- A/C pressure switch from the rear of the compressor housing

3. Lower the vehicle and support the engine with a suitable jack.

- Upper torque strut attaching bolts and remove strut
- Power steering pump and bracket. Set the pump aside. Do not disconnect the fluid lines from pump.
- Make sure the engine is properly supported, then remove the right engine mount through bolt
- Raise the engine with jack until you can access the engine support bracket bolts
- Engine support bracket

- Lower timing belt cover fasteners and cover

4. Rotate the crankshaft until the TDC mark on oil pump housing aligns with the TDC mark on crankshaft sprocket (trailing edge of sprocket tooth)

5. Loosen the timing belt tensioner lock bolt.

6. Insert a 6mm Allen wrench into the hexagon opening located on the top plate of the belt tensioner pulley. Rotate the top plate clockwise until there is enough slack in timing belt to allow for removal.

7. Remove or disconnect the following:
- Timing belt and camshaft sprockets.

✳✳ WARNING

If the timing belt was damaged due to incorrect alignment, you must replace the belt tensioner pulley and bracket as an assembly.

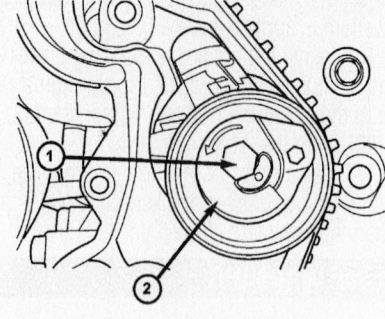

1 - LOCK BOLT
2 - TOP PLATE

67189-NEON-G27

Timing belt tensioner mounting—2.4L engine

- Crankshaft sprocket using Special Tool 6793 and insert C-4685-C2

To install:

➡ **The crankshaft sprocket is set to a pre-determined depth from the factory for correct timing belt tracking. If removed, use of Special Tool 6792 is required to set the sprocket to original installation depth. An incorrectly installed sprocket will result in timing belt and engine damage.**

8. Install or connect the following:
- Crankshaft sprocket using Special Tool 6792

✳✳ WARNING

Do not use an impact wrench to tighten camshaft sprocket bolts. Damage to the camshaft-to-sprocket locating dowel pin may occur.

- Camshaft sprockets. Hold the sprockets with Special Tool 6847 while tightening center bolt to 85 ft. lbs. (115 Nm)

9. Set the crankshaft sprocket to TDC by aligning the sprocket with the arrow on the oil pump housing. Set the camshafts timing marks so that the exhaust camshaft sprocket is a 1/2 notch below the intake camshaft sprocket. Make sure the arrows on the camshaft sprockets are facing up.

- Timing belt, starting at the crankshaft, go around the water pump sprocket, idler pulley, camshaft sprockets and then around the tensioner
- Move the exhaust camshaft sprocket counterclockwise to align the marks and take up any timing belt slack

10. Insert a 6mm Allen wrench into the

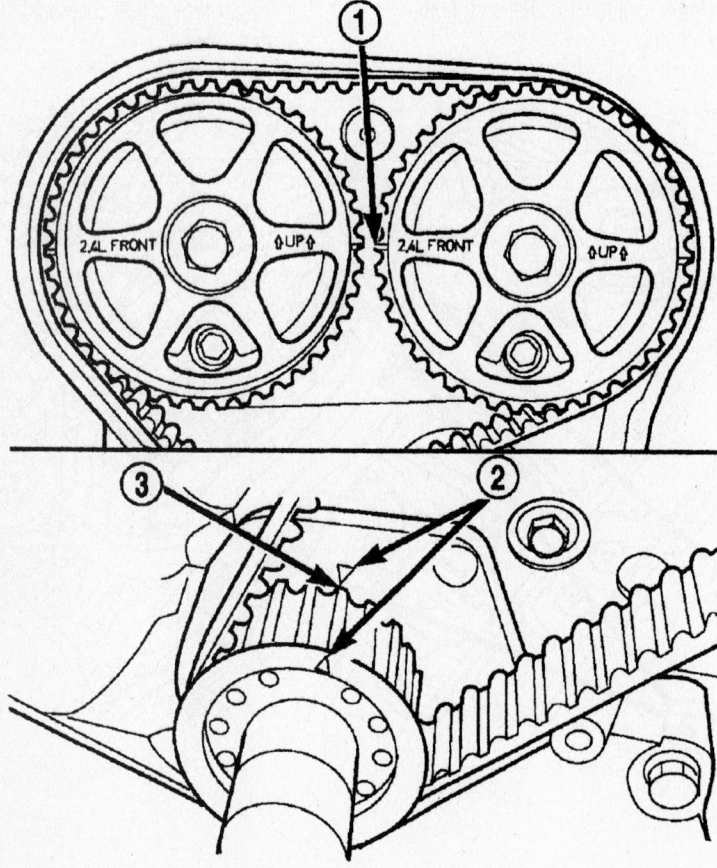

1 - CAMSHAFT TIMING MARKS
2 - CRANKSHAFT TDC MARKS
3 - TRAILING EDGE OF SPROCKET TOOTH

67189-NEON-G26

View of the proper timing alignment marks for timing belt removal—2.4L engine

hexagon opening located on the top plate of the belt tensioner pulley. Rotate the top plate counterclockwise. The tensioner pulley will move against the belt and the tensioner setting notch will eventually start to move clockwise. Watching the movement of the setting notch, continue rotating the top plate counterclockwise until the setting notch is aligned with the spring tang. Using the Allen wrench to prevent the top plate from moving, tighten the tensioner lock bolt to 18 ft. lbs. (25 Nm). The setting notch and spring tang should remain aligned after lock bolt is torqued.

11. Remove the Allen wrench and torque wrench.

✳✳ WARNING

Repositioning the crankshaft to the TDC position must be done only during the clockwise rotation movement. If TDC is missed, rotate a further two revolutions until TDC is achieved. Do NOT rotate crankshaft counterclockwise as this will make verification of proper tensioner setting impossible.

12. Rotate the crankshaft clockwise 2 complete revolutions manually for seating of the timing belt, until the crankshaft is

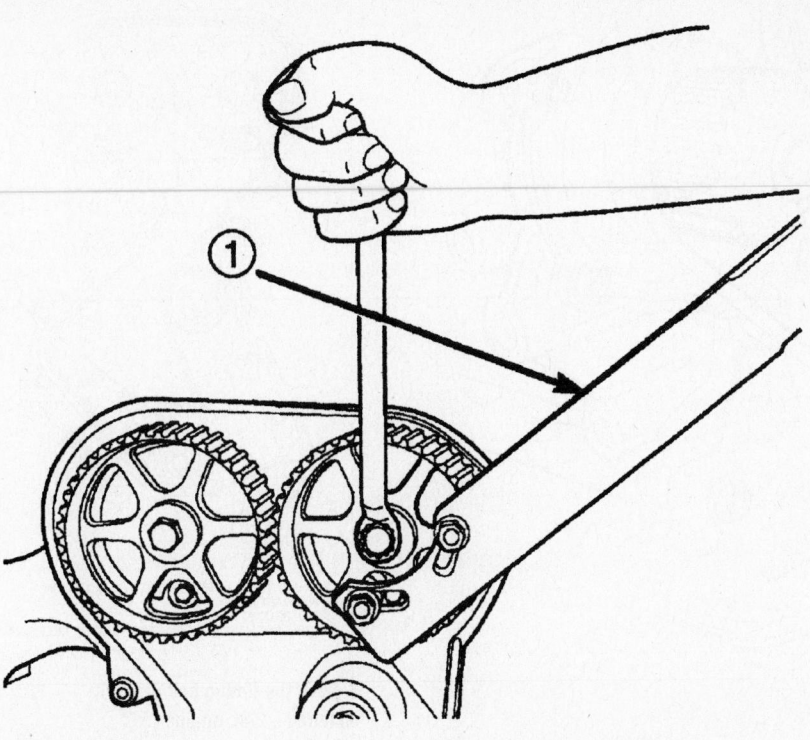

1 - SPECIAL TOOL 6847

67189-NEON-G28

Removal and installation of the camshaft sprockets—2.4L engine

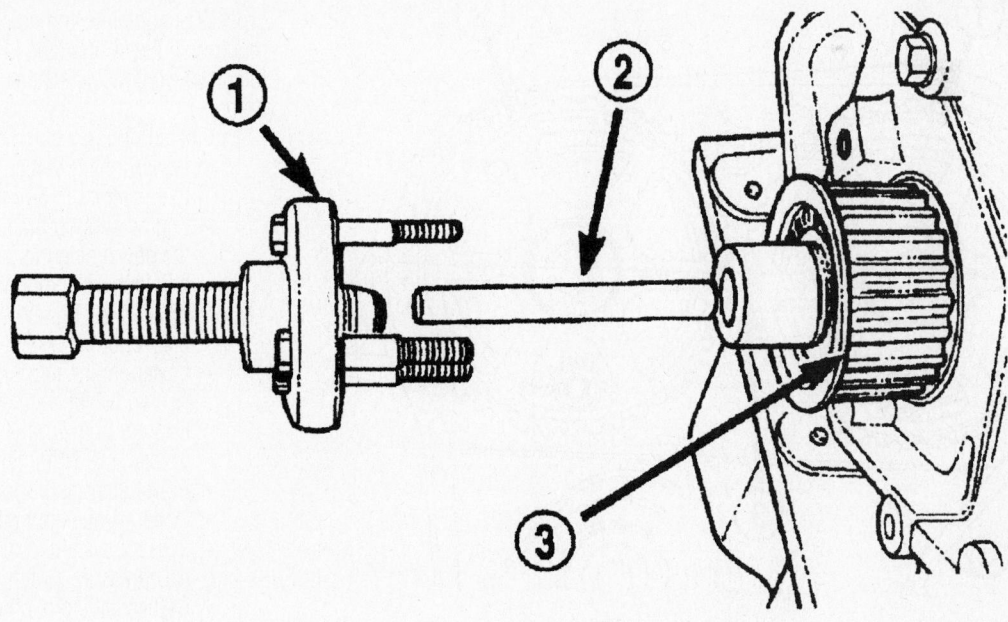

1 - SPECIAL TOOL 6793
2 - SPECIAL TOOL C-4685–C2
3 - CRANKSHAFT SPROCKET

67189-NEON-G29

Removal of the crankshaft sprocket—2.4L engine

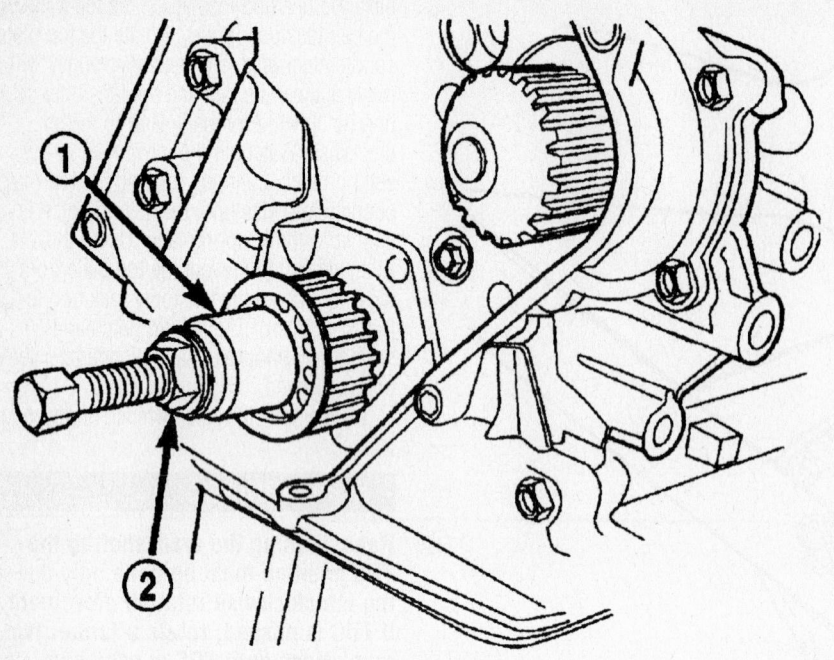

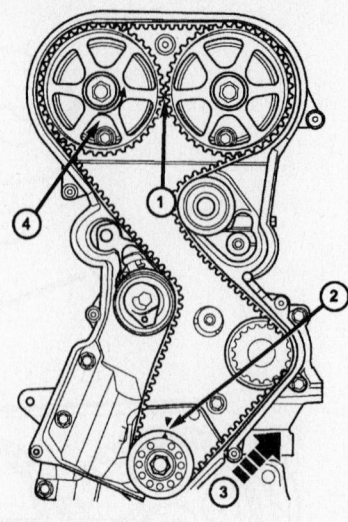

1 - CAMSHAFT TIMING MARKS 1/2 NOTCH LOCATION
2 - CRANKSHAFT AT TDC
3 - INSTALL BELT IN THIS DIRECTION
4 - ROTATE CAMSHAFT SPROCKET TO TAKE UP BELT SLACK

67189-NEON-G32

View of the timing belt properly installed—2.4L engine

1 - SPECIAL TOOL 6792
2 - TIGHTEN NUT TO INSTALL

67189-NEON-G30

Installation of the crankshaft sprocket—2.4L engine

repositioned at the TDC position. Make sure that the camshaft and crankshaft timing marks are in proper position.

13. Check if the spring tang is within the tolerance window. If the spring tang is within the tolerance window, the installation process is complete and nothing further is required. If the spring tang is not within the tolerance window, repeat the previous 3 Steps.

14. Install or connect the following
- Lower timing belt cover and tighten the retainers to 50 inch lbs. (6 Nm)
- Engine support bracket. Make sure the power steering pump is properly located in mounting location on bracket. Tighten the mount bracket bolts to 45 ft. lbs. (61 Nm).
- Lower engine into mounting position and right engine mount through bolt. Torque the bolt to 87 ft. lbs. (118 Nm)
- Power steering pump and bracket
- Upper torque strut attaching bolts
- Exhaust system to manifold
- A/C pressure switch connector
- Crankshaft vibration damper using thrust bearing washer and bolt from Installer Tool 6792 and torque the damper bolt to 100 ft. lbs. (135 Nm)
- Accessory drive belts
- Lower torque strut
- Accessory drive belt splash shield
- Right front wheel

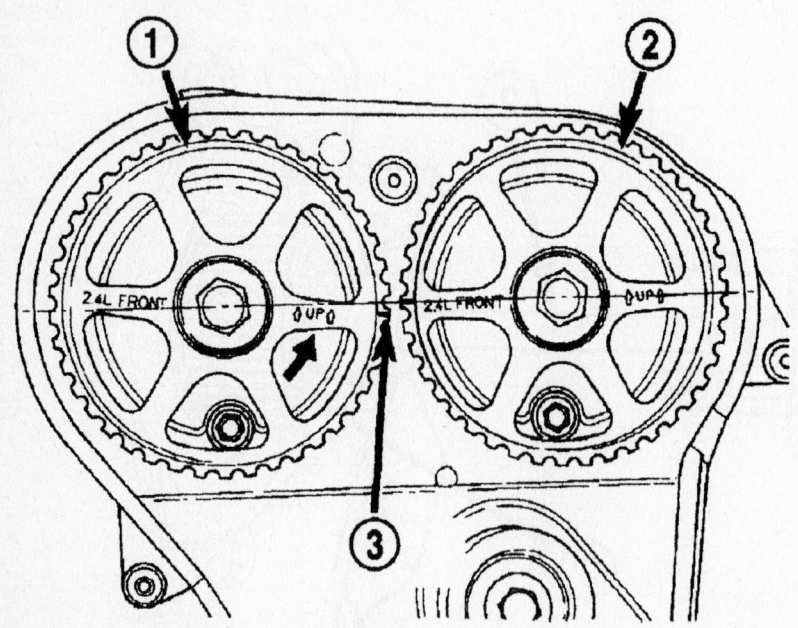

1 - CAMSHAFT SPROCKET-EXHAUST
2 - CAMSHAFT SPROCKET-INTAKE
3 - 1/2 NOTCH LOCATION

67189-NEON-G31

Make sure the marks are aligned before installing the timing belt—2.4L engine

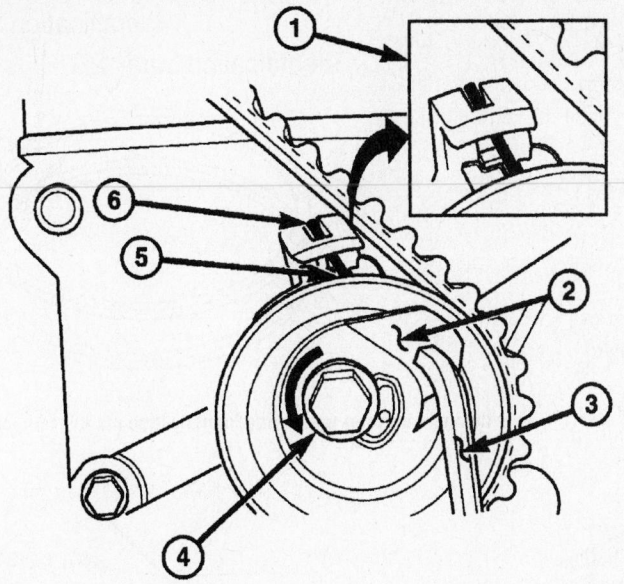

1 - ALIGN SETTING NOTCH WITH SPRING TANG
2 - TOP PLATE
3 - 6mm ALLEN WRENCH
4 - LOCK BOLT
5 - SETTING NOTCH
6 - SPRING TANG

67189-NEON-G33

Timing belt tensioner adjustment—2.4L engine

- Upper timing belt cover and tighten fasteners to 50 inch lbs. (6 Nm)
- Torque strut bracket to engine
- Torque strut bracket to the strut tower
- Upper torque attaching bolts

15. Perform torque strut adjustment procedure, as follows:

a. Remove the accessory drive belt splash shield.

b. Remove the pencil strut.

c. Loosen the upper and lower torque strut attaching bolt at the suspension crossmember and shock tower bracket.

d. Adjust the engine position by placing a floor jack on the forward edge of the transmission bell housing

e. With the engine supported, remove the upper and lower torque strut attachment bolt(s) at shock tower bracket and suspension crossmember. Make sure the torque struts are free to move within the shock tower bracket and crossmember. Reinstall the torque strut bolt(s), but do not tighten yet.

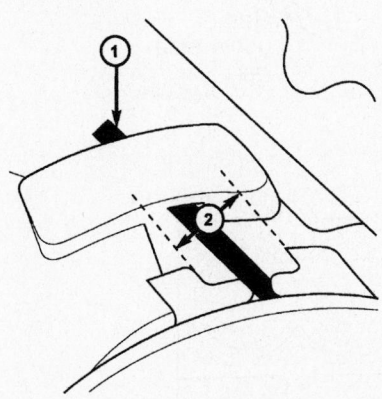

1 - SPRING TANG
2 - TOLERANCE WINDOW

67189-NEON-G34

Make sure the spring tang is within the tolerance window—2.4L engine

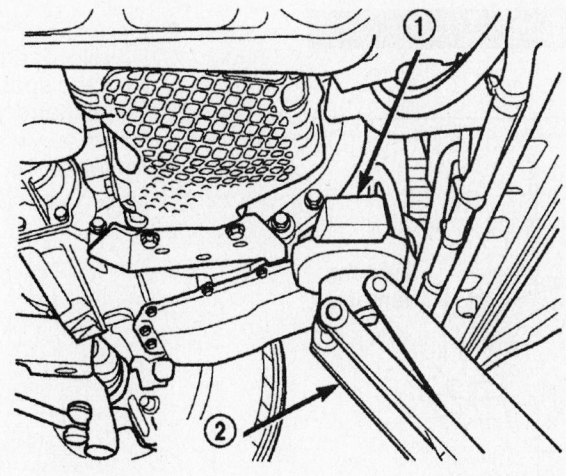

1 - WOOD BLOCK
2 - FLOOR JACK

67189-NEON-G05

Position the floor jack properly to prevent upward lifting of the engine—2.4L engine

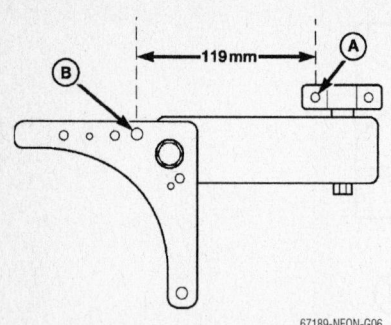

67189-NEON-G06

Measure the distance between the center of the rearmost attaching bolt on the engine mount bracket (point A) and the center of the hole on the shock tower bracket (point B)—2.4L engine

f. Apply upward force, allowing the upper engine to rotate rearward until the distance between the center of the rear-most attaching bolt on the engine mount bracket (point A) and the center of the hole on the shock tower bracket (point B) is 4.70 in. (119mm)

❋❋ WARNING

Hold the engine in position with the jack until the upper and lower torque strut bolts are tightened.

g. With the engine held at the proper position, tighten both the upper and lower torque strut bolts to 85 ft. lbs. (115 Nm)
h. Remove the floor jack.
i. Install the pencil strut and tighten nuts to 43 ft. lbs. (58 Nm).
16. Install the accessory drive belt splash shield.

Piston and Ring

POSITIONING

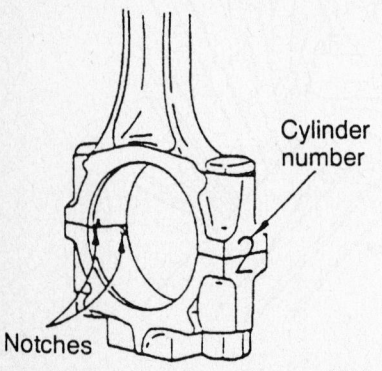

7922AG01

Chrysler engine connecting rod and cap installation—ensure to matchmark the cap and rod prior to disassembly

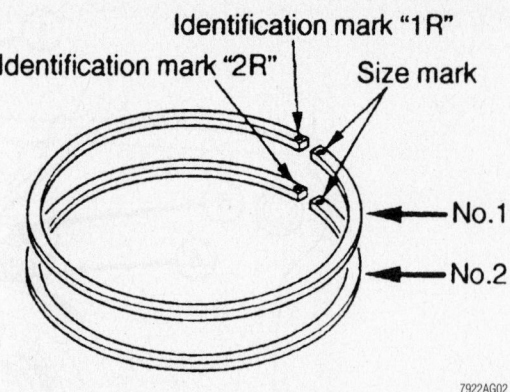

7922AG02

Common Chrysler piston ring identification mark locations

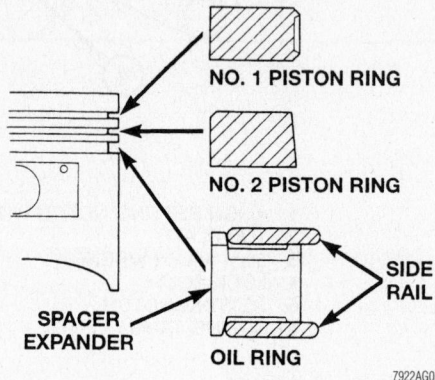

7922AG03

Piston ring orientation

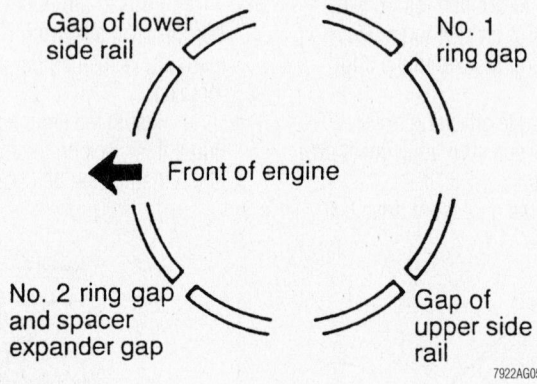

7922AG05

Piston ring end-gap spacing

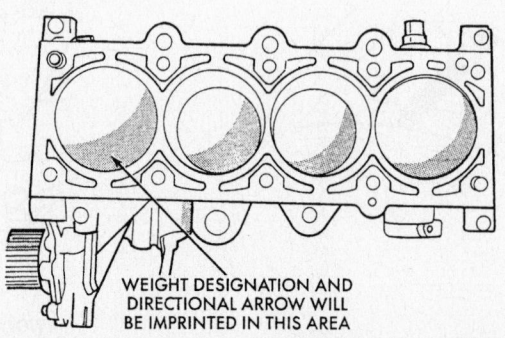

7922AG56

Piston positioning. The arrow or weight marking (L or H) must face the timing belt side of the engine—2.0L engines

FUEL SYSTEM

Fuel System Service Precautions

Safety is an important factor when servicing the fuel system. Failure to conduct maintenance and repairs in a safe manner may result in serious personal injury. Maintenance and testing of the vehicle's fuel system components can be accomplished safely and effectively by adhering to the following rules and guidelines:

• To avoid the possibility of fire and personal injury, always disconnect the negative battery cable unless the repair or test procedure requires that battery voltage be applied.

• Always relieve the fuel system pressure prior to disconnecting any fuel system component (injector, fuel rail, pressure regulator, etc.), fitting or fuel line connection. Exercise extreme caution whenever relieving fuel system pressure, to avoid exposing skin, face and eyes to fuel spray. Please be advised that fuel under pressure may penetrate the skin or any part of the body that it contacts.

• Always place a shop towel or cloth around the fitting or connection prior to loosening to absorb any excess fuel due to spillage. Ensure that all fuel spillage is quickly removed from engine surfaces. Ensure that all fuel soaked cloths or towels are deposited into a suitable waste container.

• Always keep a dry chemical (Class B) fire extinguisher near the work area.

• Do not allow fuel spray or fuel vapors to come into contact with a spark or open flame.

• Always use a back-up wrench when loosening and tightening fuel line connection fittings. This will prevent unnecessary stress and torsion to fuel line piping.

• Always replace worn fuel fitting O-rings. Do not substitute fuel hose where fuel pipe is installed.

Fuel System Pressure

RELIEVING

✳✳ CAUTION

Relieve the fuel system pressure before servicing any components of the fuel system. Service vehicles in well ventilated areas and avoid ignition sources. NEVER smoke while servicing the vehicle!

1. Before servicing the vehicle, refer to the precautions in the beginning of this section.

2. Remove the fuel pump relay from the Power Distribution Center (PDC). Location of the relay can be found on a label on the underside of the PDC cover.

➡**Removal of the fuel pump relay may cause Diagnostic Trouble Code(s) (DTC's) to set. You must use a DRB scan tool to clear codes if necessary.**

3. Start the engine and let run until it stalls.

4. Try to start the engine again until it will no longer run.

5. Turn the ignition key **OFF**.

6. Place a shop towel below the fuel line quick-connect fitting at the fuel rail.

7. Put the fuel pump relay back in the PDC.

8. The vehicle is now safe for servicing.

Fuel Filter

A combination fuel filter/pressure regulator assembly is used, which is located on the top of the fuel pump module.

REMOVAL & INSTALLATION

✳✳ CAUTION

Do not allow fuel spray or fuel vapors to come in contact with a spark or open flame. Keep a dry chemical fire extinguisher nearby. Never store fuel in an open container due to risk of fire or explosion.

1. Before servicing the vehicle, refer to the precautions in the beginning of this section.

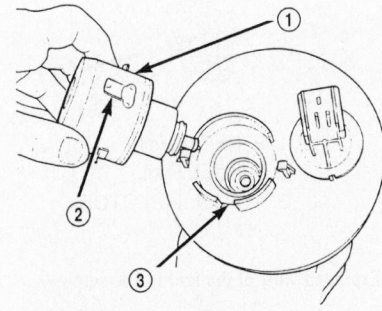

1 - FUEL FILTER/PRESSURE REGULATOR
2 - SPRING TAB
3 - LOCATING SLOT

9356HG09

Depress the spring tab, then rotate and pull the fuel filter assembly out

2. Properly relieve the fuel system pressure.

3. Disconnect the negative battery cable.

4. Disconnect the quick-connect fuel supply line from the filter/regulator nipple.

5. Depress the locking spring tab, located on the side of the fuel filter/regulator, then rotate 90 degrees and pull out. Be sure the upper and lower O-rings are still on the filter assembly.

To install:

6. Lightly coat the filter O-rings with clean engine oil. Insert the filter into the opening in the fuel pump module, then align the 2 hold-down tabs with the flange.

7. While applying downward pressure, rotate the filter clockwise until the spring tab catches in the locating slot.

8. Attach the fuel line to the filter/regulator assembly.

9. Connect the negative battery cable.

Fuel Pump

The fuel pump is integral with the pump module, which also contains the fuel reservoir, level sensor, inlet strainer and fuel pressure regulator. The inlet strainer, fuel pressure regulator and level sensor are the only serviceable items. If the fuel pump requires service, replace the entire fuel pump module.

REMOVAL & INSTALLATION

1. Before servicing the vehicle, refer to the precautions in the beginning of this section.

2. Properly relieve the fuel system pressure.

3. Drain the fuel tank.

4. Remove or disconnect the following:

• Negative battery cable
• Vapor line from the Evaporative Emissions (EVAP) canister tube
• EVAP canister

✳✳ WARNING

The fuel reservoir of the fuel pump module does not empty out when the tank is drained. The fuel in the reservoir will spill out when the module is removed.

• Fuel line from the fuel pump module by depressing the quick-con-

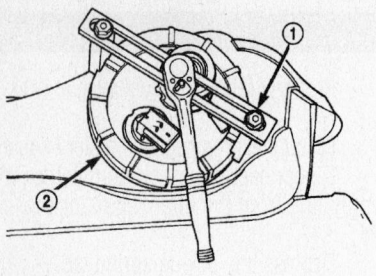

1 - SPECIAL TOOL 6856
2 - LOCKNUT

67189-NEON-G35

Fuel pump removal using the special tool—2004 vehicle shown

nect retainers with your thumb and forefinger
- Fuel pump module electrical connector by pushing down on the connector retainer and pulling the connector off of the module
- Fuel filler tube and filler vent tube from the fuel tank's filler hose

5. Support the fuel tank with a transmission jack.
- Fuel tank strap bolts; then, lower the tank slightly for access to the module
- Fuel pump module locknut using a ratchet and spanner wrench or special tool 6856, as necessary
- Fuel pump and discard the O-ring seal

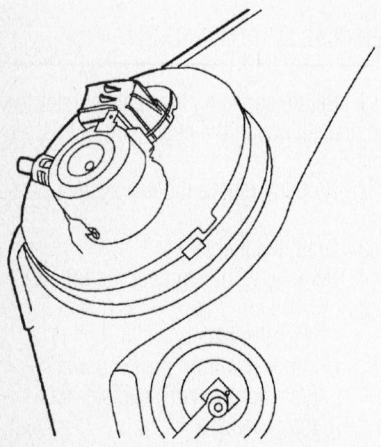

67189-NEON-G36

Be sure the alignment tab on the underside of the fuel pump module flange sits in the notch on the fuel tank

To install:

✳✳ WARNING

Take care to avoid wiping dirt or debris into the fuel tank opening. Such contaminants may clog or damage the new fuel pump.

6. Wipe the fuel pump mounting flange and surrounding area of the tank clean.
7. Install or connect the following:
- New O-ring seal in the tank opening
- Fuel pump

➡Be sure the alignment tab on the underside of the fuel pump module flange sits in the notch on the fuel tank.

- Fuel pump module locknut. Using the ratchet and spanner wrench, torque the locknut to 41 ft. lbs. (55 Nm).
- Fuel tank
- Fuel filler tube and filler vent tube to the fuel tank's filler hose
- Fuel pump module electrical connector
- Fuel line, to the fuel pump module
- EVAP canister
- Vapor line to the EVAP canister tube
- Negative battery cable

8. Start the vehicle, check for leaks and repair if necessary.

Fuel Injector

REMOVAL & INSTALLATION

1. Before servicing the vehicle, refer to the precautions in the beginning of this section.
2. Release the fuel system pressure.
3. Remove or disconnect the following:

- Negative battery cable
- Fuel line(s) from fuel rail
- Fuel injector electrical connectors
- Fuel rail mounting screws
- Fuel rail with the fuel injectors
- Fuel injector retainers
- Fuel injector(s) and discard the O-rings

To install:

4. Install or connect the following:
- Fuel injector(s) using new O-rings

➡Lubricate the O-rings with clean engine oil.

- Fuel injector retainers
- Fuel rail with the fuel injectors and torque the bolts to 15–19 ft. lbs. (20–25 Nm)
- Fuel injector electrical connectors
- Fuel line(s) to fuel rail
- Negative battery cable

5. Start the vehicle, check for leaks and repair if necessary.

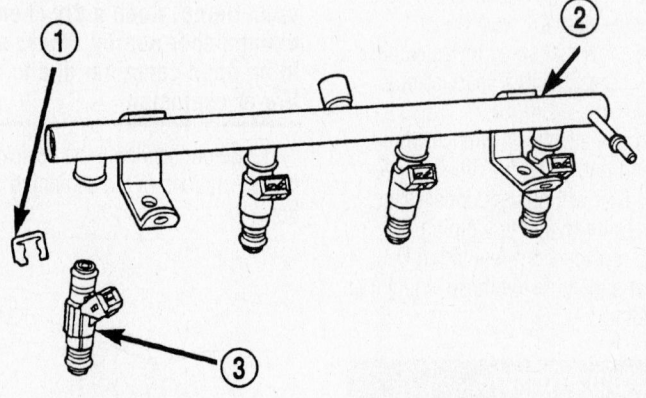

1 – RETAINER
2 – FUEL RAIL
3 – FUEL INJECTOR

9306EG01

Exploded view of the fuel rail assembly

DRIVE TRAIN

Transaxle Assembly

REMOVAL & INSTALLATION

Manual Transaxle

T-350 TRANSAXLE

1. Before servicing the vehicle, refer to the precautions in the beginning of this section.
2. Drain the transaxle fluid.
3. Remove or disconnect the following:
   - Battery
   - Air cleaner/throttle body assembly
   - Proportional Purge Solenoid (PSS)
   - Crankcase vent hose
   - Throttle Position Sensor (TPS) and Idle Air Control (IAC) connectors
   - Throttle body air duct from the intake manifold
   - Accelerator cable
   - Cruise control cable, if equipped
   - Battery tray from the bracket
   - Ground cable from the battery tray bracket
   - Back-up light switch connector
   - Hydraulic line quick-connect fitting
   - Shift cable-to-bracket clips
   - Shift selector and crossover cables from the levers and move aside
   - Halfshafts
   - Structural collar
   - Left engine-to-transaxle lateral bending brace
   - Bell housing dust cover
   - Right engine-to-transaxle lateral bending brace
   - Starter

- Driveplate-to-clutch module bolts
4. Place a jack and block of wood under the engine's oil pan and support the engine. Remove the transaxle upper mount through-bolt.

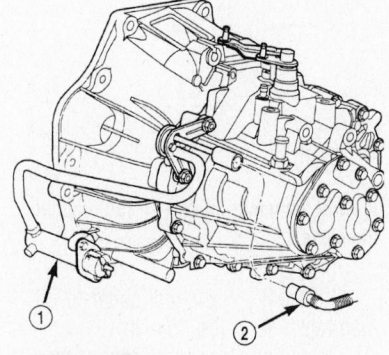

1 - SLAVE CYLINDER
2 - HYDRAULIC TUBE

9356HG10

View of the clutch slave cylinder connection —T-350 manual transaxle

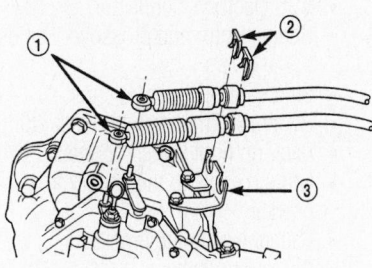

1 – SHIFT CABLES
2 – CLIPS
3 – BRACKET

9306EG29

View of the shift cables—T-350 manual transaxle

5. Lower the engine/transaxle assembly to provide clearance.
   - Transaxle-to-engine bolts, using an assistant to support the transaxle
   - Transaxle

To install:

6. If installing a new or replacement transaxle, transfer the upper mount to the new transaxle; then, torque the upper mount-to-transaxle bolts to 50 ft. lbs. (68 Nm).

7. Install or connect the following:
   - Transaxle
   - Transaxle-to-engine bolts, using an assistant to support the transaxle. Torque the transaxle-to-engine bolts to 70 ft. lbs. (95 Nm).

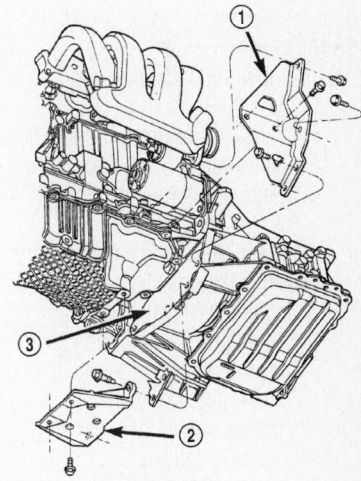

1 – LATERAL BENDING BRACE
2 – STRUCTURAL COLLAR
3 – DUST COVER

9306EG31

View of the left lateral bending brace and structural collar—T-350 manual transaxle

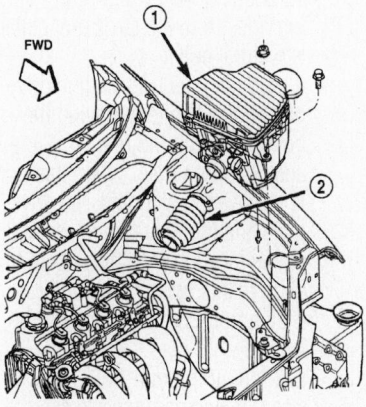

1 – AIR CLEANER ASSY.
2 – THROTTLE BODY DUCT

9306EG27

View of the air cleaner assembly and throttle body duct

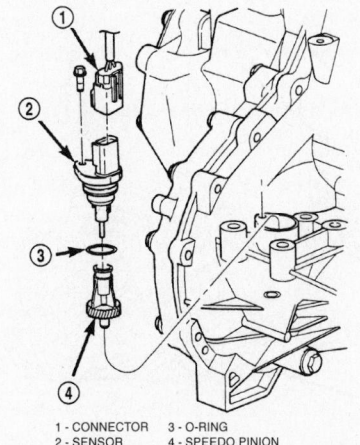

1 - CONNECTOR   3 - O-RING
2 - SENSOR   4 - SPEEDO PINION

9306EG30

View of the Vehicle Speed Sensor (VSS)—T-350 manual transaxle

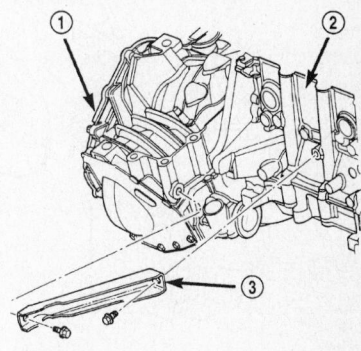

1 – TRANSAXLE
2 – ENGINE
3 – LATERAL BENDING BRACE

9306EG32

View of the right lateral bending brace—T-350 manual transaxle

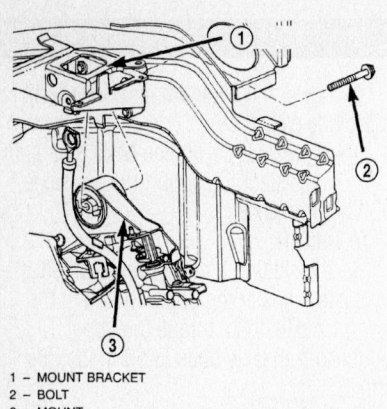

1 – MOUNT BRACKET
2 – BOLT
3 – MOUNT

9306EG33

**View of the upper mount through-bolt—
T-350 manual transaxle**

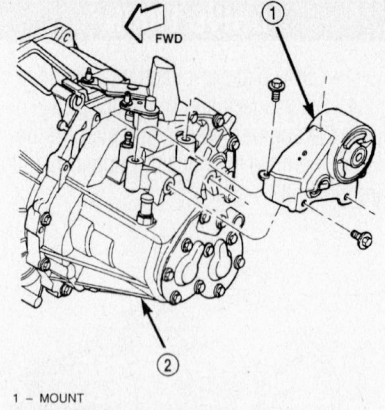

1 – MOUNT
2 – TRANSAXLE

9306EG35

**View of the upper mount—T-350 manual
transaxle**

8. Raise the engine/transaxle assembly until the upper mount aligns with the mount bracket hole. Torque the upper mount-to-mount bracket through-bolt to 80 ft. lbs. (108 Nm).

9. Remove the jack and wooden block.

- Driveplate-to-clutch module bolts and torque to 65 ft. lbs. (88 Nm)
- Starter and torque the bolts to 40 ft. lbs. (54 Nm)
- Starter electrical connectors and torque the positive cable bolt to 89 inch lbs. (10 Nm)
- Left engine-to-transaxle lateral bending brace and torque the bolts to 60 ft. lbs. (81 Nm)

10. Install the structural collar and torque the bolts as follows:

   a. Step 1—Structural collar-to-oil pan bolts: 30 inch lbs. (3 Nm).

   b. Step 2—Structural collar-to-transaxle bolts: 80 ft. lbs. (108 Nm).

   c. Step 3—Structural collar-to-oil pan bolts: 40 ft. lbs. (54 Nm).

11. Install or connect the following:

- Right engine-to-transaxle lateral bending brace and torque the bolts to 60 ft. lbs. (81 Nm)
- Halfshafts
- Bell housing dust cover
- VSS electrical connector
- Shift selector and crossover cables to the levers
- Shift cable-to-bracket clips
- Hydraulic line quick-connect fitting
- Back-up light switch connector
- Ground cable to the battery tray bracket
- Battery tray to the bracket
- Accelerator cable
- Cruise control cable, if equipped
- Air cleaner assembly and torque the nuts and bolts to 10 ft. lbs. (14 Nm)

- Throttle body air duct to the intake manifold
- TPS and IAC connectors
- Crankcase vent hose to the throttle body
- PSS to the throttle body
- Battery

12. Fill the transaxle to the proper level.

13. Road test the vehicle and inspect for leaks.

T-850 TRANSAXLE

1. Before servicing the vehicle, refer to the precautions in the beginning of this section.

2. Drain the transaxle fluid.

3. Remove or disconnect the following:

- Turbocharger solenoid pack from the air cleaner assembly and position it aside
- AAT sensor and remove air cleaner assembly
- Throttle body inlet hose from the inter-cooler
- Idle Air Control (IAC) and Throttle Position Sensor (TPS) connectors
- Accelerator cable from throttle body linkage
- Throttle body adapter
- Battery and battery tray
- Shift cables from the transaxle shift levers and bracket
- Both front wheel/tire assemblies
- Both Anti-lock Brake System (ABS) front wheel speed sensors
- Halfshafts and intermediate shaft
- Clutch hydraulic circuit quick-connect fitting, using Tool 6638A, or equivalent

4. Back-up lamp switch connector

- Starter motor-to-transaxle case bolts
- Front lateral bending brace
- Turbocharger pipe and power steering lines from the structural collar
- Structural collar
- 4 modular clutch-to-engine driveplate bolts. While removing the bolts, you will find one tight-tolerance (slotted) hole. When this bolt is removed, matchmark the driveplate and modular clutch assembly at this location, and make sure to align marks upon reassembly.

5. Install and secure a suitable transmission jack.

- Left mount through-bolt

6. Lower the engine/transaxle assembly enough to gain access to left mount and bellhousing bolts.

- Left mount-to-transaxle bolts and mount

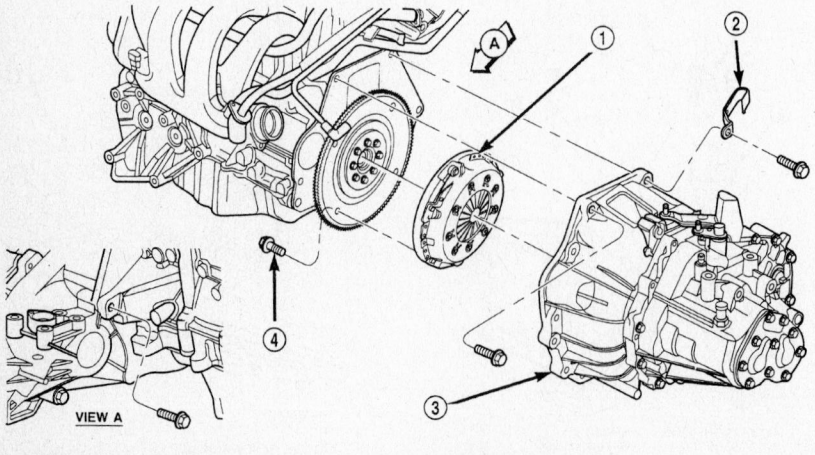

1 – MODULAR CLUTCH ASSEMBLY
2 – CLIP

VIEW A

3 – TRANSAXLE
4 – CLUTCH MODULE BOLT (4)

9306EG34

Exploded view of the transaxle and related components—T-350 manual transaxle

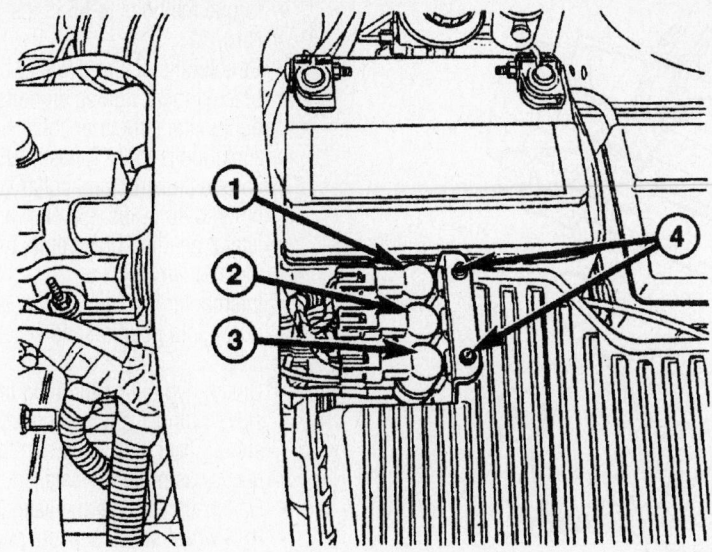

1 - TIP SOLENOID
2 - SURGE VALVE ACTUATOR SOLENOID
3 - WASTEGATE ACTUATOR SOLENOID
4 - BRACKET MOUNTING SCREWS

67189-NEON-G37

View of the turbocharger solenoids and mounting bracket—T-850 manual transaxle

7. Install a screw jack and wood block to support the engine after transaxle removal.

- Transaxle bellhousing-to-block bolts, then remove the transaxle and modular clutch assembly.

8. If replacing the transmission, transfer the impact blocker, back-up lamp switch, vehicle speed sensor and gearshift cable bracket.

To install:

9. Install or connect the following:

- Modular clutch to the input shaft assembly
- Transaxle assembly to a suitable transmission jack, making sure it is secure
- Transaxle into position. Torque the transaxle-to-block bolts to 80 ft. lbs. (108 Nm)
- Upper mount/bracket to transaxle and tighten the bolts to 50 ft. lbs. (68 Nm)

10. Raise the engine/transaxle assembly into position. Install and torque the through-bolt to 90 ft. lbs. (122 Nm).

- 4 modular clutch-to-driveplate bolts. Align the driveplate and modular clutch marks made during removal. Start with the tight-tolerance (slotted) hole, install and

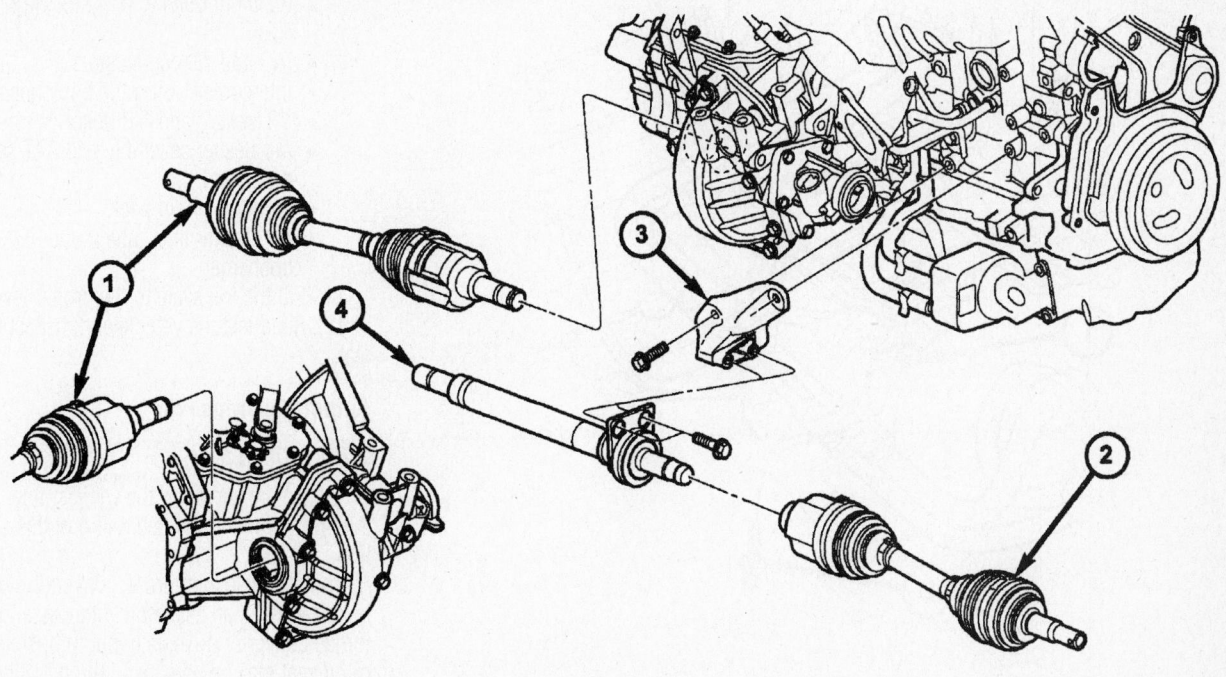

1 - HALFSHAFT - LEFT
2 - HALFSHAFT - RIGHT

3 - SUPPORT BRACKET - INTERMEDIATE SHAFT
4 - INTERMEDIATE SHAFT/BEARING ASSEMBLY

67189-NEON-G38

Exploded view of the halfshaft assemblies—2.4L turbo models

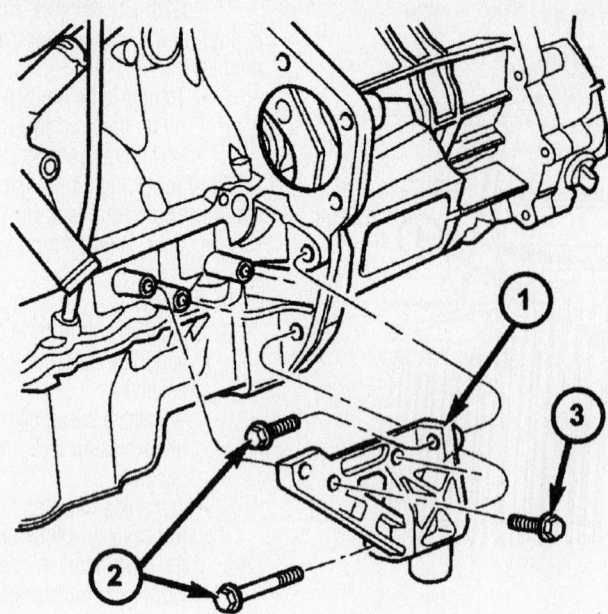

1 - LATERAL BENDING BRACE
2 - BOLT (1 LONG/1 SHORT)
3 - BOLT (2)

67189-NEON-G39

Lateral bending brace mounting—2.4L turbo models

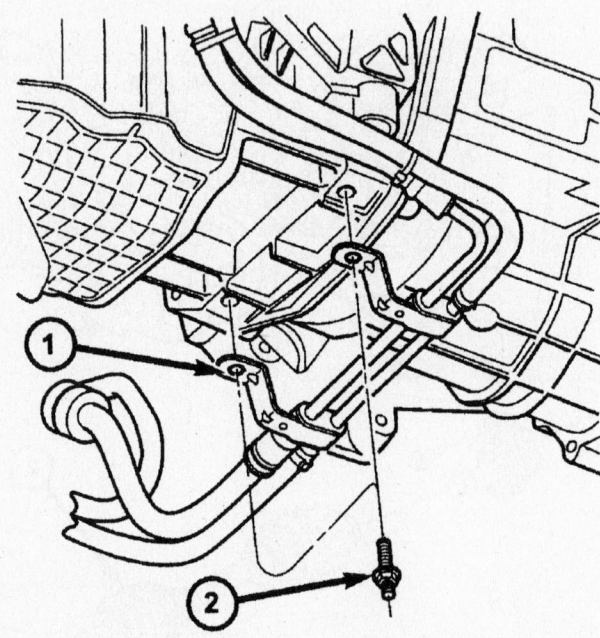

1 - POWER STEERING HOSE ASSY
2 - BOLT/STUD (2)

67189-NEON-G40

Structural collar/power steering hose retainers—2.4L turbo models

torque the bolts to 65 ft. lbs. (88 Nm).
* Structural collar and lateral bending brace. Finger-tighten all bolts, then tighten the structural collar horizontal bolts to 80 ft. lbs. (108 N). Torque the structural collar vertical bolts to 40 ft. lbs. (54 Nm) and the lateral bending brace bolts to 45 ft. lbs. (61 Nm).
* Intercooler pipe and power steering hoses into position and secure with nuts
* Clutch hydraulic plumbing to the slave cylinder quick-connect. You should hear a "click" when the connection is properly installed.
* Halfshafts and intermediate shaft
* ABS wheel speed sensors and torque the bolts to 105 inch lbs. (12 Nm)
* Gearshift cables to the transaxle bracket and secure with new retaining clips
* Battery tray
* Battery (with thermal wrap) and hold-down clamp/bolt. Connect battery temperature sensor.
* Throttle body adapter housing
* Accelerator cable while placing the throttle body into position. Torque the 2 throttle body-to-manifold bolts to 21 ft. lbs. (28 Nm).
* IAC and TPS connectors
* Intercooler-to-throttle body hose
* IAT sensor and TIP hose
* Air cleaner assembly and AAT sensor connector
* Turbo solenoid pack
* Air cleaner inlet tube to turbocharger

11. Fill the transaxle to the proper level.
12. Road test the vehicle and inspect for leaks.

Automatic Transaxle

2001 MODELS

1. Before servicing the vehicle, refer to the precautions in the beginning of this section.

The transaxle and torque converter must be removed as an assembly; otherwise the torque converter driveplate, pump bushing or oil seal may be damaged. The driveplate will not support a load; therefore, none of the weight of the transaxle should be allowed to rest on the plate during removal.

2. Drain the transaxle fluid.
3. Remove or disconnect the following:
* Battery

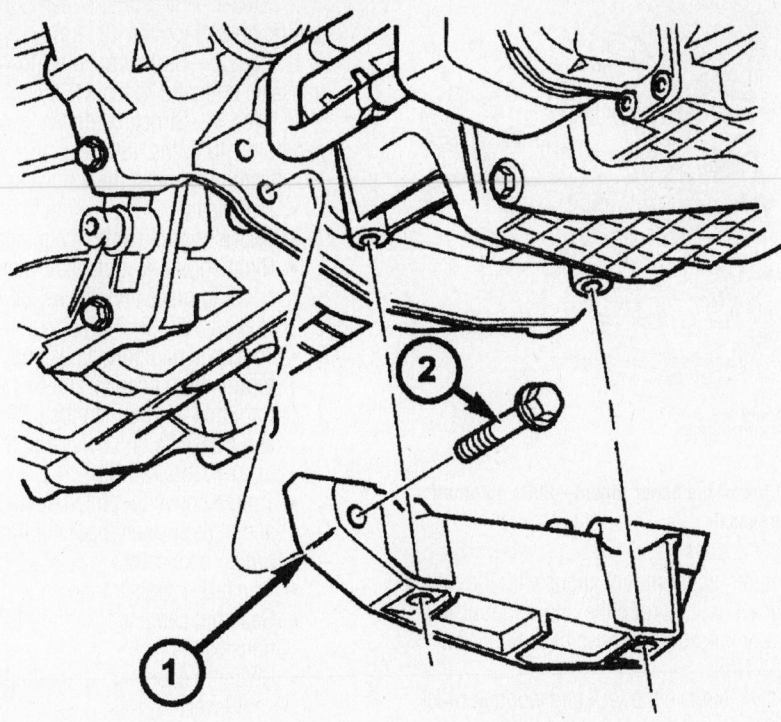

- Torque converter clutch solenoid connector
- Neutral safely/back-up light switch connector
- Transaxle oil cooler lines and plug the lines to prevent contamination
- Shift cable and bracket from the transaxle
- Kickdown cable and bracket from the transaxle
- Transaxle oil pan
- Halfshafts
- Structural collar
- Left engine-to-transaxle lateral bending brace
- Bell housing duct cover
- Starter
- Torque converter-to-driveplate bolts

6. Using a jack and a block of wood, support the engine/transaxle assembly.

7. Remove the transaxle upper mount through-bolt.

➡**The upper mount through-bolt may be accessed through the left side wheel house.**

**1 - STRUCTURAL COLLAR
2 - BOLT**

67189-NEON-G41

Structural collar mounting—2.4L turbo models

4. Remove the air cleaner/throttle body assembly by removing or disconnect the following:

a. Proportional Purge Solenoid (PSS)
b. Crankcase vent hose
c. Throttle Position Sensor (TPS) and Idle Air Control (IAC) connectors
d. Throttle body air duct from the intake manifold

e. Accelerator cable
f. Transaxle kickdown cable
g. Cruise control cable, if equipped
h. Air cleaner assembly

5. Remove or disconnect the following:
- Battery tray from the bracket

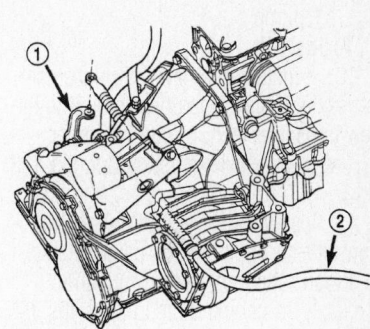

1 – SHIFT LEVER
2 – SHIFT CABLE

9306EG36

View of the gear shift cable—2001 automatic transaxle

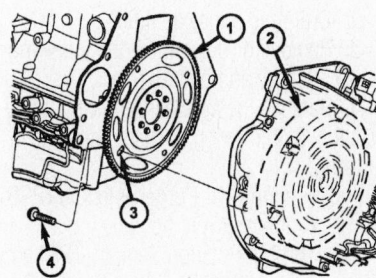

1 - DRIVEPLATE
2 - MODULAR CLUTCH ASSEMBLY
3 - TIGHT TOLERANCE HOLE
4 - BOLT (4)

67189-NEON-G42

Modular clutch-to-driveplate mounting—T-850 manual transaxle

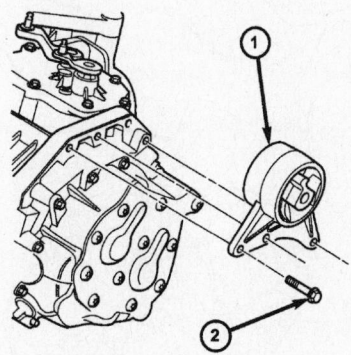

1 - MOUNT/BRACKET
2 - BOLT (3)

67189-NEON-G43

View of the transaxle upper mount and fasteners—T-850 manual transaxle

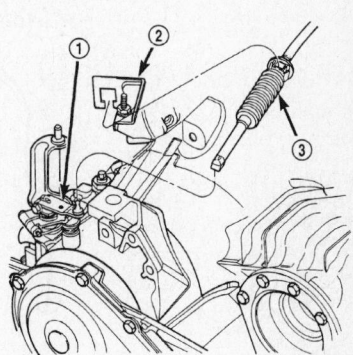

1 – LEVER
2 – BRACKET
3 – KICKDOWN CABLE

9306EG37

View of the kickdown cable—2001 automatic transaxle

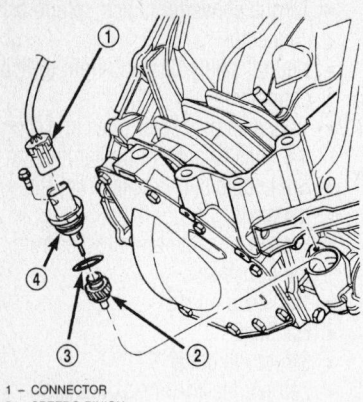

1 - CONNECTOR
2 - SPEEDO PINION
3 - O-RING
4 - SENSOR

9306EG38

View of the Vehicle Speed Sensor (VSS)—2001 automatic transaxle

8. Lower the engine/transaxle assembly for clearance.

9. Remove or disconnect the following:
- Transaxle-to-engine bolts, using an assistant to support the transaxle
- Transaxle
- Torque converter from the front pump

To install:

10. If installing a new or replacement transaxle, transfer the upper mount to the new transaxle; then, torque the upper mount-to-transaxle bolts to 50 ft. lbs. (68 Nm).

11. Install transaxle and tighten the transaxle-to-engine bolts, using an assistant to support the transaxle. Torque the transaxle-to-engine bolts to 70 ft. lbs. (95 Nm).

12. Raise the engine/transaxle assembly

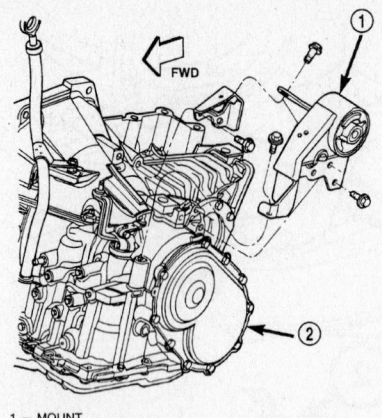

1 - MOUNT
2 - TRANSAXLE

9306EG40

View of the upper mount—2001 automatic transaxle

until the upper mount aligns with the mount bracket hole. Torque the upper mount-to-mount bracket through-bolt to 80 ft. lbs. (108 Nm).

13. Remove the jack and wooden block.

14. Install or connect the following:
- Torque converter-to-driveplate bolts and torque the bolts to 65 ft. lbs. (88 Nm)
- Starter and torque the bolts to 40 ft. lbs. (54 Nm)
- Starter electrical connectors and torque the positive cable bolt to 90 inch lbs. (10 Nm)
- Bell housing dust cover
- Left engine-to-transaxle bending brace and torque the bolts to 60 ft. lbs. (81 Nm)

15. Install the structural collar and torque the bolts as follows:

a. Step 1—Structural collar-to-oil pan bolts: 30 inch lbs. (3 Nm).

b. Step 2—Structural collar-to-transaxle bolts: 80 ft. lbs. (108 Nm).

c. Step 3—Structural collar-to-oil pan bolts: 40 ft. lbs. (54 Nm).

16. Install or connect the following:
- Halfshafts
- Speed control cable, if equipped
- Right engine-to-transaxle lateral bending brace and torque the bolts to 60 ft. lbs. (81 Nm)
- Transaxle oil cooler lines and torque the oil cooler line-to-radiator fitting to 108 inch lbs. (12 Nm) and transaxle oil cooler line fitting to 21 ft. lbs. (28 Nm)
- Torque converter clutch solenoid and neutral safely/back-up light switch connectors
- Transaxle dipstick tube
- Gearshift cable bracket bolt to the transaxle
- Gearshift cable end to the transaxle shift lever
- Transaxle kickdown cable to the lever and bracket
- Battery tray
- Battery
- Accelerator cable
- Transaxle kickdown cable
- Cruise control cable, if equipped
- Air cleaner assembly and torque the nuts and bolts to 10 ft. lbs. (14 Nm)
- TPS and IAC connectors
- Crankcase vent hose to the throttle body
- PSS to the throttle body

17. Fill the transaxle and check for leaks.

18. Be sure the vehicle's back-up lights and speedometer are working properly.

2002–04 MODELS

1. Before servicing the vehicle, refer to the precautions in the beginning of this section.

2. Drain the transaxle fluid.

3. Remove or disconnect the following:
- Battery and battery tray

4. Remove the air cleaner/throttle body assembly by removing or disconnect the following:

a. Proportional Purge Solenoid (PSS)

b. Crankcase vent hose

c. Throttle Position Sensor (TPS) and Idle Air Control (IAC) connectors

d. Throttle body air duct from the intake manifold

e. Accelerator cable

f. Transaxle kickdown cable

g. Cruise control cable, if equipped

h. Air cleaner assembly

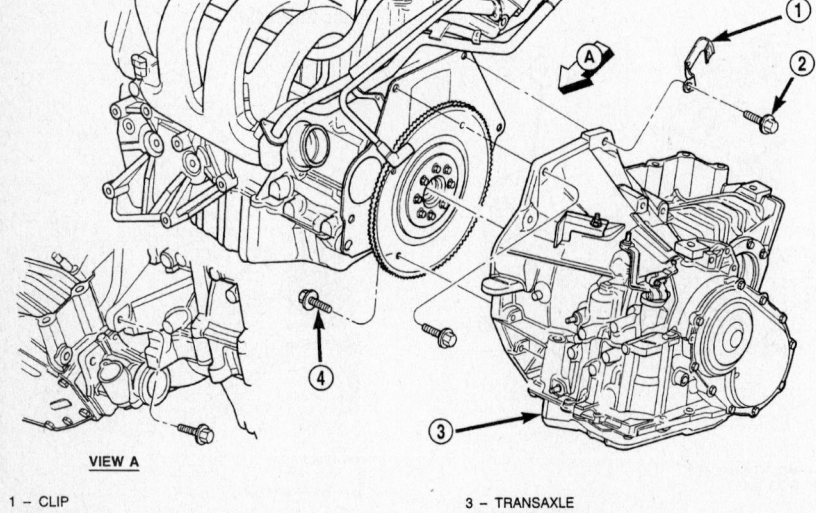

VIEW A

1 - CLIP
2 - BOLT (3)
3 - TRANSAXLE
4 - CONVERTER BOLT (4)

9306EG39

Exploded view of the transaxle and related components—2001 automatic transaxle

5. Remove or disconnect the following:
- Transaxle oil cooler lines. Use a cutter to cut the hoses off flush with the end of the fittings. You will need to install a service splice kit during installation.
- Input and output speed connectors
- Solenoid/pressure switch assembly connector
- Gearshift cable from manual valve
- Gearshift cable from upper mount bracket and position aside
- Upper starter mounting bolt
- Halfshafts
- Structural collar
- Left engine-to-transaxle lateral bending brace
- Starter
- Power steering cooler-to-cross-member fasteners and position the cooler aside
- Right lateral bending brace-to-transaxle bolt
- Torque converter dust shield
- Torque converter-to-driveplate bolts
- Both lower transaxle bellhousing-to-engine bolts

6. Using a jack and a block of wood, support the engine/transaxle assembly.

7. Remove the transaxle upper mount through-bolt.

➡**The upper mount through-bolt may be accessed through the left side wheel house.**

8. Lower the engine/transaxle assembly for clearance.

9. Remove or disconnect the following:
- Transaxle-to-engine bolts, using an assistant to support the transaxle
- Transaxle

To install:

10. Install transaxle and tighten the transaxle-to-engine bolts, using an assistant to support the transaxle. Torque the transaxle-to-engine bolts to 80 ft. lbs. (108 Nm).

11. Raise the engine/transaxle assembly until the upper mount aligns with the mount bracket hole. Torque the upper mount-to-mount bracket through-bolt to 87 ft. lbs. (118 Nm).

12. Remove the jack and wooden block.

13. Install or connect the following:
- Torque converter-to-driveplate bolts and torque the bolts to 65 ft. lbs. (88 Nm)
- Converter dust shield
- Starter and starter electrical connections
- Left lateral bending brace. Torque the brace-to-transaxle bolts to 80

ft. lbs. (108 Nm), the brace-to-engine bedplate bolt to 40 ft. lbs. (54 Nm) and the brace-to-intake manifold bolt to 21 ft. lbs. (28 Nm).
- Structural collar
- Power steering oil cooler
- Halfshafts
- Gearshift cable to mount. Secure the cable end to the manual valve lever
- Solenoid/pressure switch assembly connector
- Input and output speed connectors
- Cooler lines with service splice kit
- Air cleaner and throttle body assembly
- Battery tray and battery

14. Fill the transaxle and check for leaks.

15. Be sure the vehicle's back-up lights and speedometer are working properly.

Clutch

ADJUSTMENT

The clutches on these vehicles are self-adjusting. No manual adjustment is necessary or possible.

REMOVAL & INSTALLATION

1. Before servicing the vehicle, refer to the precautions in the beginning of this section.

2. Remove or disconnect the following:
- Negative battery cable
- Modular clutch-to-driveplate bolts and discard the bolts
- Transaxle assembly
- Modular clutch assembly from the transaxle input shaft

To install:

3. Install or connect the following:
- Modular clutch assembly onto the transaxle input shaft
- Transaxle assembly and torque the transaxle-to-engine bolts to 70 ft. lbs. (95 Nm)
- New modular clutch-to-driveplate bolts and torque the bolts to 65 ft. lbs. (88 Nm)
- Negative battery cable

Halfshafts

REMOVAL & INSTALLATION

Vehicles with 2.0L Engines

1. Before servicing the vehicle, refer to the precautions in the beginning of this section.

2. With the vehicle on the ground and brakes applied, loosen, but do not remove the stub axle-to-hub and bearing retaining nut.

➡**The front hub and driveshaft are splined together and retained by the hub nut.**

3. If equipped with Anti-lock Brake System (ABS), disconnect the front wheel speed sensor.

4. Remove or disconnect the following:
- Front wheel
- Front caliper-to-steering knuckle bolts
- Caliper from the steering knuckle

5. Support the caliper out of the way by suspending it with a piece of wire from the strut. DO NOT allow the caliper to hang by the brake hose.

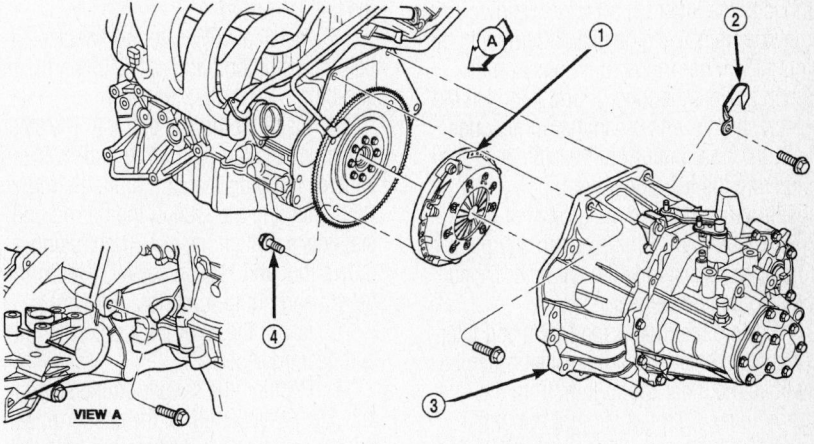

VIEW A

1 – MODULAR CLUTCH ASSEMBLY
2 – CLIP
3 – TRANSAXLE
4 – CLUTCH MODULE BOLT (4)

9306EG41

Exploded view of the manual transaxle, clutch module and related components

- Rotor from the hub
- Outer tie rod end-to-steering knuckle nut by holding the tie rod end stud with a 11/32 in. socket while loosening the nut
- Tie rod end stud from the steering knuckle
- Ball joint stud-to-steering knuckle nut and bolt

✷✷ WARNING

Be careful when separating the ball joint stud from the steering knuckle, so the ball joint seal does not get damaged.

- Ball joint stud from the steering knuckle by prying down on the lower control arm

✷✷ WARNING

Be careful when separating the inner CV-joint during this operation. Do not let the driveshaft hang by the inner CV-joint, the driveshaft must be supported.

6. Pull the steering knuckle assembly out and away from the outer CV-joint of the driveshaft assembly. Support the outer end of the driveshaft assembly.

7. Insert a prybar between the inner tripod joint and the transaxle case. Pry against the inner tripod joint until the joint retaining snapring is disengaged from the transaxle side gear.

➡**Inner tripod joint removal is easier by applying outward pressure on the joint while hitting the punch with a hammer.**

8. Remove the inner tripod joints from the transaxle side gears using a punch to dislodge the inner tripod joint retaining ring from the transaxle side gear. If removing the right side inner tripod joint, position the punch against the inner tripod joint. Hit the punch sharply with a hammer to dislodge the right inner joint from the side gear. If removing the left side inner tripod joint, position the punch in the groove of the inner tripod joint. Hit the punch sharply with a hammer to dislodge the left inner tripod joint from the side gear.

9. Hold the inner tripod joint and interconnecting shaft of the driveshaft assembly. Remove the inner tripod joint from the transaxle by pulling it straight out of the transaxle side gear and transaxle oil seal. When removing the tripod joint, do not let the spline or snapring drag across the seal-

ing lip of the transaxle-to-tripod joint oil seal.

✷✷ WARNING

The driveshaft, when installed, acts as a bolt which secures the front hub and bearing assembly. If the vehicle is to be supported or moved on its wheels with a driveshaft removed, install a proper-sized bolt and nut through the front hub. Tighten the bolt and nut to 135 ft. lbs. (183 Nm). This will ensure that the hub bearing cannot loosen.

To install:

10. Thoroughly clean the spline and oil seal sealing surface on the tripod joint. Lightly lubricate the oil seal sealing surface on the tripod joint with fresh, clean transmission fluid.

11. Holding the driveshaft assembly by the tripod joint and interconnecting shaft, install the tripod joint into the transaxle side gear as far as possible by hand.

12. Carefully align the tripod joint with the transaxle side gears. Then, grasp the driveshaft interconnecting shaft and push the tripod joint into the transaxle side gear until fully seated. Be sure the snapring is fully engaged with the side gear by trying to remove the tripod joint from the transaxle by hand. If the snapring is fully seated with the side gear, the tripod joint will not be removable by hand.

13. Clean all debris and moisture out of the steering knuckle.

14. Be sure that the outer CV-joint, which fits into the steering knuckle, has no debris or moisture on it before installing into the steering knuckle.

15. Slide the driveshaft back into the front hub. Install the steering knuckle into the ball joint stud.

16. Install a NEW steering knuckle-to-ball joint stud bolt and nut. Tighten the nut and bolt to 70 ft. lbs. (95 Nm).

17. Insert the tie rod end into the steering knuckle. Start the tie rod end-to-steering knuckle nut onto the stud of the tie rod end. While holding the stud of the tie rod end stationary, tighten the nut. Then, using a crow's foot and 11/32 in socket, tighten the tie rod end nut to 45 ft. lbs. (61 Nm).

18. Install the rotor back onto the hub and bearing assembly.

19. Position the caliper on the steering knuckle. Slide the top of the caliper under the top abutment on the steering knuckle, then install the bottom of the caliper against the bottom abutment of the steering knuckle.

20. Install the caliper-to-knuckle bolts and tighten to 23 ft. lbs. (31 Nm).

21. On vehicles equipped with ABS, connect the front wheel speed sensor.

22. Clean all foreign matter from the threads of the outer CV-joint stub axle. Install hub nut and washer onto the threads of the stub axle and tighten the nut.

23. With the vehicle's brakes applied to prevent the axle shaft from turning, tighten the hub nut to 180 ft. lbs. (244 Nm).

24. Install the front wheel and tire assembly. Install the lug nuts and tighten to 100 ft. lbs. (135 Nm).

25. Check the transaxle fluid level, lowering the vehicle as necessary.

Vehicles with 2.4L Engines

1. Before servicing the vehicle, refer to the precautions in the beginning of this section.

2. Disconnect the negative battery cable.

3. With the vehicle on the ground and brakes applied, loosen, but do not remove the stub axle-to-hub and bearing retaining nut.

➡**The front hub and driveshaft are splined together and retained by the hub nut.**

4. If equipped with Anti-lock Brake System (ABS), disconnect the front wheel speed sensor.

5. Remove or disconnect the following:
- Front wheel
- Cotter pin, nut lock, and spring washer, and hub nut from the end of the outer CV-joint stub axle
- Front wheel speed sensor and secure harness out of the way, if equipped with Anti-lock Brake System (ABS)
- Nut and bolt securing the ball joint stud into steering knuckle

✷✷ WARNING

Be careful when separating the ball joint stud from the steering knuckle to prevent damaging the components.

- Ball joint stud from steering knuckle by prying down on the lower control arm

✷✷ WARNING

Be careful not to separate the inner CV-joint during this operation. Do not let the halfshaft hang by inner CV-joint, it MUST be supported.

- Halfshaft from the steering knuckle by pulling outward on knuckle while pressing in on the halfshaft. Support the outer end of halfshaft assembly.

➡️**If you are encountering difficulty separating the halfshaft from the hub, use puller 1026, do NOT hit the shaft with a hammer.**

6. Support the outer end of the halfshaft assembly.

➡️**When the left halfshaft is removed from transaxle, some fluid may leak out.**

➡️**It is easier to remove the inner tripod joint if you apply outward pressure on the joint as you strike the punch with a hammer. Do not pull on the interconnecting shaft to remove, as the inner joint will become separated.**

7. Remove or disconnect the following:
- Left halfshaft by applying outward pressure on joint by hand, dislodge the inner tripod joint from the differential side gear by striking outward with a punch. When removing the tripod joint and halfshaft, do

not let the spline or snapring drag across axle seal.
- Right halfshaft by sliding the inner tripod joint off the intermediate shaft. If necessary, use a punch to dislodge the joint from the intermediate shaft.
- Intermediate shaft, if necessary, by removing the 2 intermediate shaft bearing-to-bracket bolts

To install:

✳️✳️ WARNING

The driveshaft, when installed, acts as a bolt which secures the front hub and bearing assembly. If the vehicle is to be supported or moved on its wheels with a driveshaft removed, install a proper-sized bolt and nut through the front hub. Tighten the bolt and nut to 135 ft. lbs. (183 Nm). This will ensure that the hub bearing cannot loosen.

8. Install or connect the following:
- Intermediate shaft/bearing, if removed. Torque the bearing-to-bracket bolts to 40 ft. lbs. (54 Nm).
9. On the left halfshaft, thoroughly

clean the spline and oil seal sealing surface on the left tripod joint. Lightly lubricate the oil seal sealing surface on the tripod joint with fresh, clean transmission fluid.

10. Holding the driveshaft assembly by the tripod joint and interconnecting shaft, install the tripod joint into the transaxle side gear as far as possible by hand.

11. Carefully align the tripod joint with the transaxle side gears. Then, grasp the driveshaft interconnecting shaft and push the tripod joint into the transaxle side gear until fully seated. Be sure the snapring is fully engaged with the side gear by trying to remove the tripod joint from the transaxle by hand. If the snapring is fully seated with the side gear, the tripod joint will not be removable by hand.

12. On the right halfshaft, thoroughly clean the right halfshaft tripod joint spline, and the intermediate shaft spline. While holding the halfshaft assembly by the tripod joint and interconnecting shaft, install tripod joint onto the intermediate shaft as far as possible by hand.

13. Slide the halfshaft back into front hub. Install the steering knuckle onto the ball joint stud

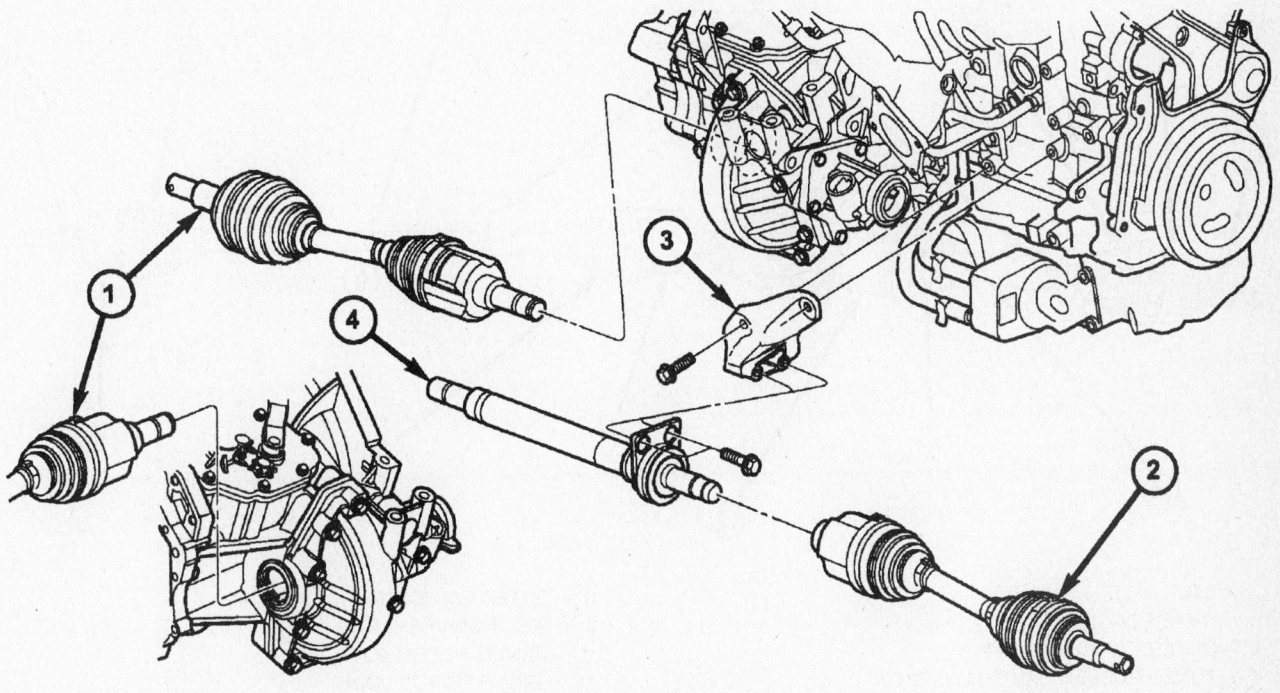

1 - HALFSHAFT - LEFT
2 - HALFSHAFT - RIGHT

3 - SUPPORT BRACKET - INTERMEDIATE SHAFT
4 - INTERMEDIATE SHAFT/BEARING ASSEMBLY

67189-NEON-G38

Exploded view of the halfshaft assemblies—2.4L engines

➡At this point, the outer joint will not seat completely into the front hub. The outer joint will be pulled into hub and seated when the hub nut is installed and torqued.

14. Install or connect the following:
- New steering knuckle-to-ball joint stud bolt and nut. Tighten to 70 ft. lbs. (95 Nm).

15. On vehicles equipped with ABS, connect the front wheel speed sensor.

16. Clean all foreign matter from the threads of the outer CV-joint stub axle. Install hub nut and washer onto the threads of the stub axle and tighten the nut.

17. With the vehicle's brakes applied to prevent the axle shaft from turning, tighten the hub nut to 180 ft. lbs. (244 Nm).
- Spring washer, nut lock, and cotter pin

18. Install the front wheel and tire assembly. Install the lug nuts and tighten to 100 ft. lbs. (135 Nm).

19. Check the transaxle fluid level, lowering the vehicle as necessary.

CV-Joints

OVERHAUL

Tri-Pod (Inner) Joint

1. Remove or disassemble the following:
- Halfshaft and place it in a soft-jawed vise
- Tri-pod joint boot clamps and the boot down the shaft

✸✸ WARNING

When removing the spider joint, hold the rollers in place on the trunnions to keep the rollers and needle bearings in place.

- Spider/shaft assembly from the tri-pod housing
- Snapring from the shaft
- Spider assembly

➡If necessary, tap the spider assembly from the shaft with a brass drift.

✸✸ WARNING

When removing the spider assembly, do not hit the outer bearings.

- Boot by sliding it off the shaft

✸✸ WARNING

If any parts show excessive wear, replace the halfshaft assembly; the component parts are not serviceable.

To assemble:

✸✸ WARNING

The Tri-pod sealing boots are made of 2 different types of material; silicon rubber (high temperature) which is soft and pliable or hytrel plastic (standard temperature) which is stiff and rigid. Be sure to replace the boot made of the correct material.

2. Install a new small boot clamp and slide it on the shaft.

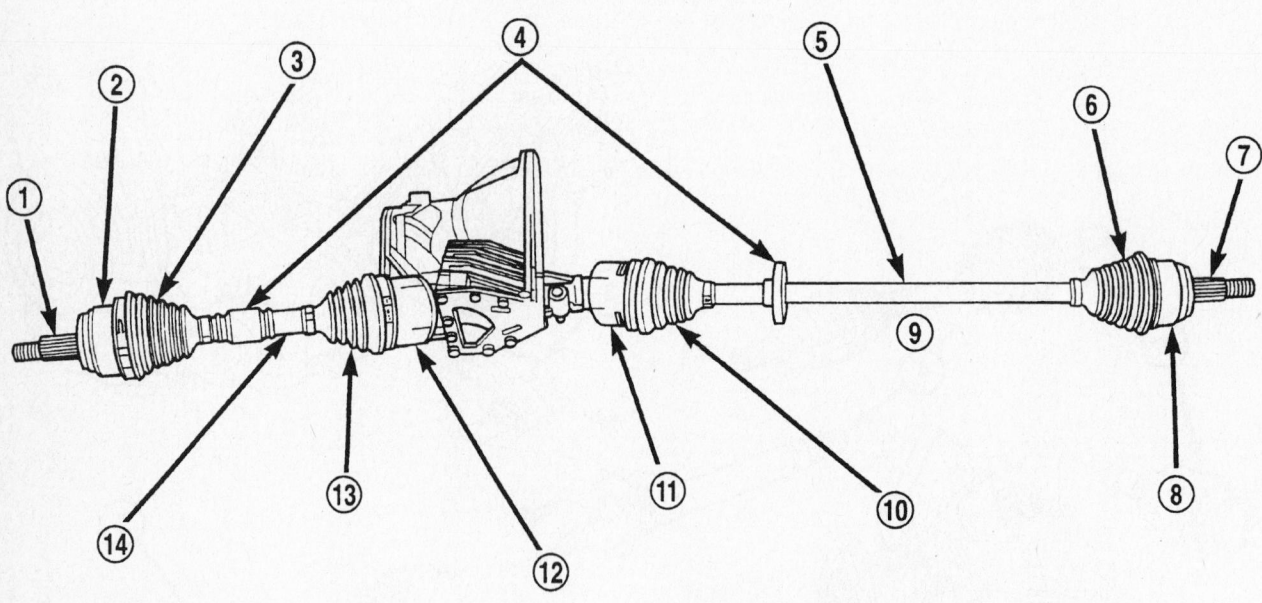

1 – STUB AXLE
2 – OUTER C/V JOINT
3 – OUTER C/V JOINT BOOT
4 – TUNED RUBBER DAMPER WEIGHT
5 – INTERCONNECTING SHAFT
6 – OUTER C/V JOINT BOOT
7 – STUB AXLE
8 – OUTER C/V JOINT
9 – RIGHT DRIVESHAFT
10 – INNER TRIPOD JOINT BOOT
11 – INNER TRIPOD JOINT
12 – INNER TRIPOD JOINT
13 – INNER TRIPOD JOINT BOOT
14 – INTERCONNECTING SHAFT LEFT DRIVESHAFT

9306EG02

View of the halfshaft assemblies—Typical

3. Position the boot so that the raised bead on the inside the boot seal is in the shaft groove.

4. If installing a new boot, distribute ½ of the grease in the service package inside the tri-pod housing and the other ½ inside the boot.

5. Install or connect the following:
- Spider assembly, face the chamfered side toward the shaft

❊❊ WARNING

If necessary, tap the spider assembly onto the shaft using a brass drift; be careful not to hit the outer bearings.

- Snapring making sure it is fully seated in the groove

- Spider/shaft assembly into the tripod housing
- New inner boot clamp and position it evenly on the sealing boot

6. Using a trim stick, adjust the boot length to 107mm (hytrel plastic) or 115mm (silicone rubber).

7. If installing a high profile boot clamp, perform the following procedure:

a. Using the Crimper Tool C-4975-A, place the tool over the clamp bridge, tighten the tool nut until the jaws are completely closed (face-to-face).

❊❊ WARNING

The seal must not be dimpled, stretched or out of shape. If necessary, equalize the seal pressure and shape it by hand.

b. Position the boot onto the tri-pod housing retaining groove and install the retaining clamp evenly on the boot.

c. Using the Crimper Tool C-4975-A, place the tool over the clamp bridge, tighten the tool nut until the jaws are completely closed (face-to-face).

8. If installing a low profile latching type boot clamp, position Snap-On® Clamp Locking Tool YA3050 prongs in the clamp

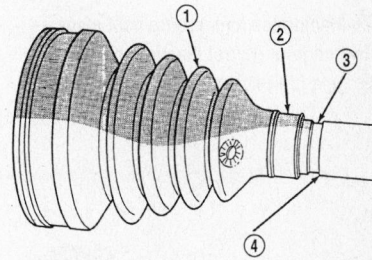

1 – SEALING BOOT
2 – RAISED BEAD IN THIS AREA OF SEALING BOOT
3 – GROOVE
4 – INTERCONNECTING SHAFT

9306EG04

Positioning the boot on the shaft—Tri-pod and Rzeppa joints

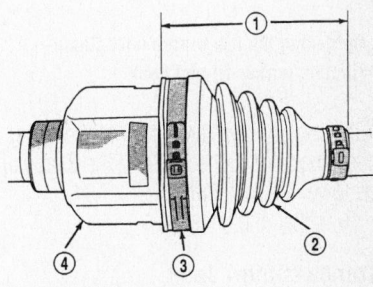

1 – 107 MILLIMETERS
2 – HYTREL SEALING BOOT
3 – SEALING BOOT CLAMP
4 – INNER TRIPOD JOINT

9306EG05

Adjusting the Tri-pod boot length—Hytrel plastic boot

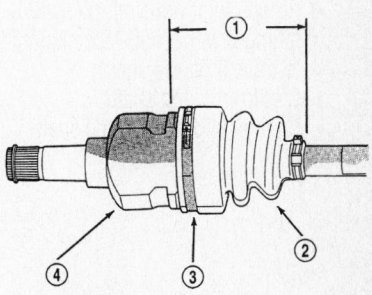

1 – 115 MILLIMETERS
2 – SILICONE SEALING BOOT
3 – CLAMP
4 – INNER TRIPOD JOINT

9306EG06

Adjusting the Tri-pod boot length—Silicone rubber boot

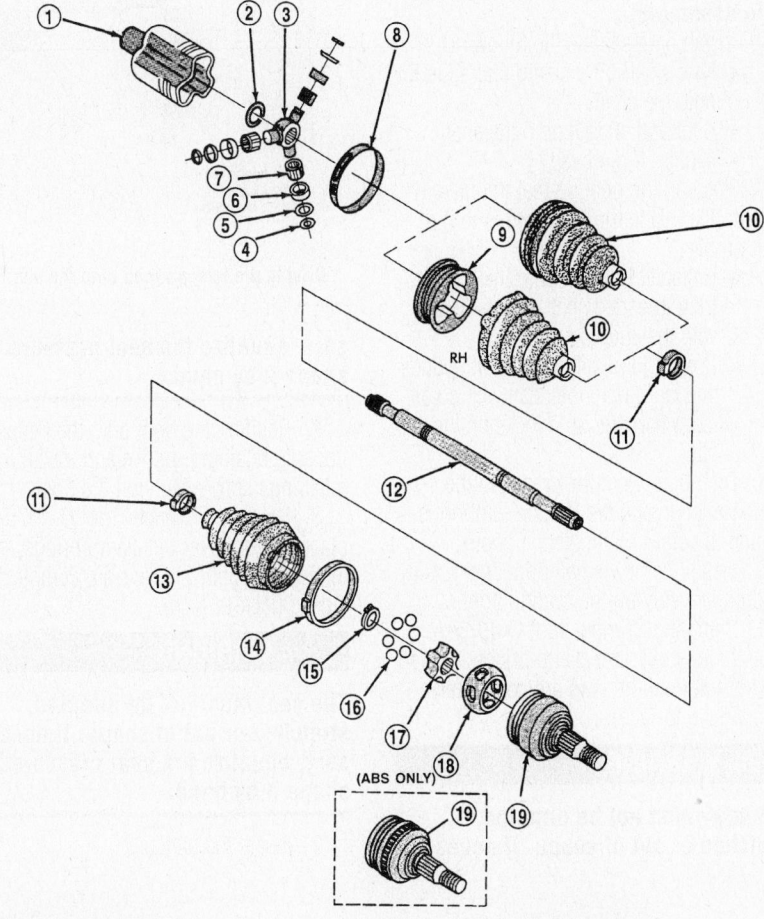

1 – HOUSING ASM, RETAINER &
2 – RING, SPACER
3 – SPIDER, TRIPOTD JOINT
4 – RING, RETAINING
5 – RETAINER, BALL & ROLLER
6 – BALL, TRIPOD JOINT
7 – ROLLER, NEEDLE
8 – CLAMP, SEAL RETAINING
9 – BUSHING, TRILOBAL TRIPOD
10 – SEAL, DRIVE AXLE INBOARD

11 – CLAMP, SEAL RETAINING
12 – SHAFT, AXLE (RH SHOWN, LH SIMILAR)
13 – SEAL, DRIVE AXLE OUTBOARD
14 – CLAMP, SEAL RETAINING
15 – RING, RACE RETAINING
16 – BALL, CHROME ALLOY
17 – RACE, C/V JOINT INNER
18 – CAGE, C/V JOINT
19 – RACE, C/V JOINT OUTER

9306EG03

Exploded view of the halfshaft assembly—Typical

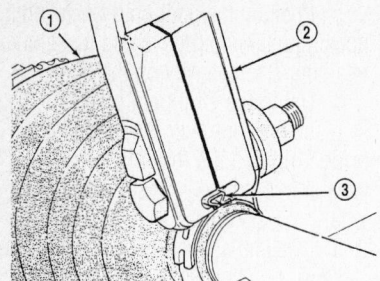

1 – SEALING BOOT
2 – SPECIAL TOOL C-4975
3 – CLAMP BRIDGE

9306BG13

Tightening the high profile boot clamp—Tri-pod joint (Hytrel plastic) boot and Rzeppa joint boots

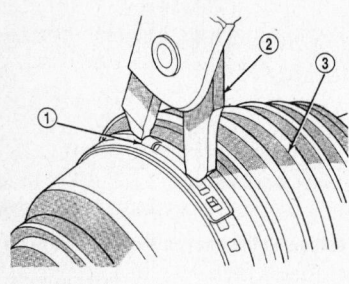

1 – CLAMP
2 – SPECIAL TOOL YA3050
3 – SEALING BOOT

9306EG07

Tightening the low profile boot clamp—Silicone rubber Tri-pod boot

holes and squeeze the tool until the upper clamp band is latched behind the 2 tabs on the lower clamp band.

9. Install the halfshaft.

Rzeppa (Outer) Joint

1. Remove or disassemble the following:

- Halfshaft and place it in a soft-jawed vise
- Rzeppa joint boot clamps and slide the boot down the shaft
- Rzeppa joint housing, by sharply hitting it with a soft-faced hammer to drive it off the shaft
- Circlip from the shaft
- Boot by sliding it off the shaft

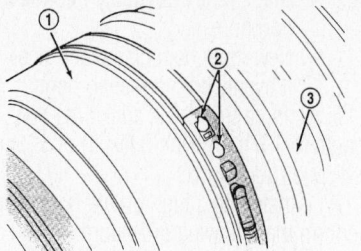

1 – INNER TRIPOD JOINT HOUSING
2 – TOP BANK OF CLAMP MUST BE RETAINED BY TABS AS SHOWN HERE TO CORRECTLY LATCH BOOT CLAMP
3 – SEALING BOOT

9306EG08

Secured the low profile boot clamp—Silicone rubber Tri-pod boot

> ✹✹ WARNING
>
> **If any parts show excessive wear, replace the halfshaft assembly; the component parts are not serviceable.**

To assemble:

2. Install or assemble the following:
- New small boot clamp and slide it onto the shaft
- Boot and slide it onto the shaft
- Circlip, if removed

3. Position the boot so that the raised bead on the inside the boot seal is in the shaft groove.

- Halfshaft hub nut onto the Rzeppa joint threaded shaft so it is flush with the end
- Rzeppa joint, align the shaft splines and tap it onto the shaft with a soft-faced hammer so it locks on the circlip

4. Distribute ½ of the grease in the service package inside the Rzeppa joint housing and the other ½ inside the boot.

5. Install the new small boot clamp and position it evenly on the sealing boot

6. Using the Crimper Tool C-4975-A, place the tool over the clamp bridge, tighten the tool nut until the jaws are completely closed (face-to-face).

> ✹✹ WARNING
>
> **The seal must not be dimpled, stretched or out of shape. If neces-**

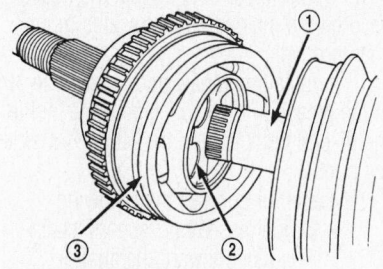

1 – INTERCONNECTING SHAFT
2 – CROSS
3 – OUTER C/V JOINT ASSEMBLY

9306EG09

Aligning the cross splines with the shaft splines—Rzeppa joint

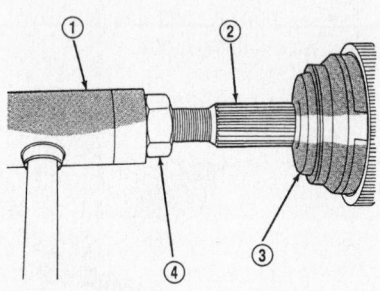

1 – SOFT FACED HAMMER
2 – STUB AXLE
3 – OUTER C/V JOINT
4 – NUT

9306EG10

Driving the Rzeppa joint onto the shaft

sary, equalize the seal pressure and shape it by hand.

7. Position the boot onto the Rzeppa housing retaining groove and install the retaining clamp evenly on the boot.

8. Using the Crimper Tool C-4975-A, place the tool over the clamp bridge, tighten the tool nut until the jaws are completely closed (face-to-face).

> ✹✹ WARNING
>
> **The seal must not be dimpled, stretched or out of shape. If necessary, equalize the seal pressure and shape it by hand.**

9. Install the halfshaft.

STEERING AND SUSPENSION

Air Bag

✳✳ CAUTION

Some vehicles are equipped with an air bag system. The system MUST BE disabled before performing service on or around system components, steering column, instrument panel components, wiring and sensors. Failure to follow safety and disabling procedures could result in accidental air bag deployment, possible personal injury and unnecessary system repairs.

PRECAUTIONS

Several precautions must be observed when handling the inflator module to avoid accidental deployment and possible personal injury:

- Never carry the inflator module by the wires or connector on the underside of the module.
- When carrying a live inflator module, hold securely with both hands, and ensure that the bag and trim cover are pointed away.
- Place the inflator module on a bench or other surface with the bag and trim cover facing up.
- With the inflator module on the bench, never place anything on or close to the module which may be thrown in the event of an accidental deployment.

DISARMING

Proper Supplemental Restraint System (SRS) disarming can be obtained by disconnecting and isolating the negative battery cable. Allow the air bag system capacitor at least 2 minutes to discharge before removing any air bag system components.

Rack and Pinion Steering Gear

These vehicles are designed and assembled using NET BUILD front suspension alignment settings. This means that the alignment settings are determined as the vehicle is designed by the location of the front suspension components in relation to the body. This is carried out when building the vehicle, by precisely locating the front crossmember to meter gauge holes located in the underbody of the vehicle. With this method of designing and building a vehicle,

it is no longer possible to adjust a vehicle's front suspension alignment settings to the required specifications. As a result, whenever the crossmember is removed from a vehicle, it MUST be replaced in the same location on the body of the vehicle it was removed from. The front suspension toe settings can still be adjusted by the outer tie rod ends.

REMOVAL & INSTALLATION

1. Before servicing the vehicle, refer to the precautions in the beginning of this section.
2. Place the steering wheel in the straight-ahead position. Lock the steering wheel in place, using a steering wheel holder.

➡**Locking the steering wheel keeps the clockspring in alignment position.**

3. Working inside the passenger compartment, remove the steering column coupling retainer pin, pinch bolt nut/bolt and separate the couplings.
4. Remove or disconnect the following:
   - Both front wheels
   - Outer tie rods-to-steering knuckle nuts

➡**Remove the tie rod nut by holding the stud securely.**

   - Outer tie rod ends from the steering knuckle using Remover Tool MB991113
   - Tie rod heat shield
   - Power steering fluid pressure switch wiring connector by releasing the locking tab
   - Power steering fluid pressure hose from the steering gear
   - Power steering fluid return hose from the steering gear, if not equipped with a power steering fluid cooler
   - Power steering fluid cooler hose from the steering gear, if equipped with a power steering fluid cooler
   - Power steering fluid return hose from the routing clip C-clamps, if not equipped with a power steering fluid cooler
   - Power steering fluid pressure hose from the steering gear's routing clips
   - Power steering cooler hose from the steering gear's right routing clip, if equipped

   - Both power steering cooler screws from the front suspension crossmember, if equipped, and move the cooler aside

➡**The screws are located behind the cooler and can be accessed from above.**

   - Power steering cooler, if equipped, and move it aside
   - Engine torque strut-to-front suspension crossmember bolt from the right forward corner of the crossmember

5. Matchmark the front suspension crossmember-to-chassis location.

✳✳ WARNING

If the front suspension crossmember-to-chassis location is not matchmarked, the front wheel alignment setting will be lost.

6. Place a transmission jack under the front crossmember and support it.
7. Remove both front suspension crossmember-to-frame rail bolts, one located at each side.
8. Loosen both rear suspension crossmember-to-frame rail bolts, one located at each side, until they release from the threaded tapping plates in the bolt.

✳✳ WARNING

Do not completely remove the rear bolts for they are designed to disengage from the body threads and will stay within the lower control arm rear isolator bushing.

➡**The threaded tapping plates allow the lower control arm to stay in place on the crossmember.**

9. Using the transmission jack, lower the front suspension crossmember enough to allow the power steering gear to be removed form the rear of the crossmember. Use the jack to support the crossmember's weight.
10. Remove or disconnect the following:
   - Lower steering column coupling-to-power steering gear pinion shaft's roll pin, using a roll pin punch
   - Lower steering column coupling from the power steering column pinion shaft
   - Pinion shaft dash cover seal from

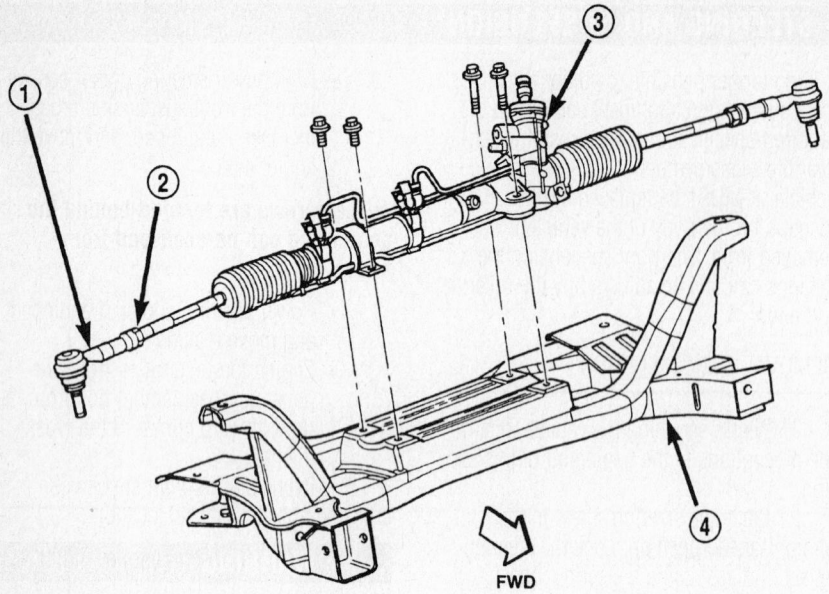

1 – OUTER TIE ROD
2 – JAM NUT

3 – STEERING GEAR
4 – FRONT SUSPENSION CROSSMEMBER

9306EG42

View of the power steering gear and crossmember

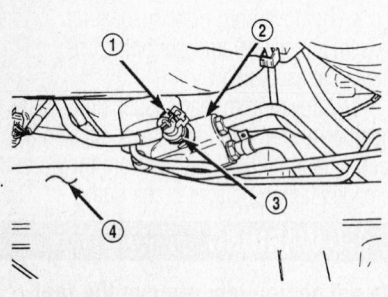

1 – WIRING HARNESS CONNECTOR
2 – POWER STEERING GEAR
3 – POWER STEERING FLUID PRESSURE SWITCH
4 – REAR OF FRONT SUSPENSION CROSSMEMBER

9306EG43

View of the power steering gear fluid pressure switch

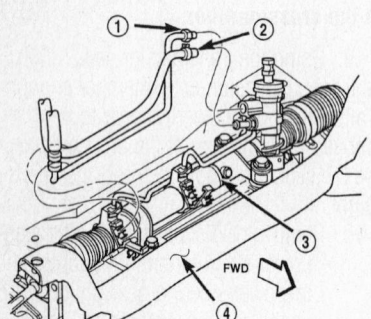

1 – PRESSURE HOSE TUBE NUT
2 – RETURN HOSE
3 – POWER STEERING GEAR
4 – FRONT SUSPENSION CROSSMEMBER

9306EG44

View of the power steering gear hoses and routing clips

the tabs cast into the power steering gear housing
• Power steering gear from the front suspension crossmember

To install:

11. Install or connect the following:
   • Power steering gear onto the front suspension crossmember and torque the bolts to 45 ft. lbs. (61 Nm)

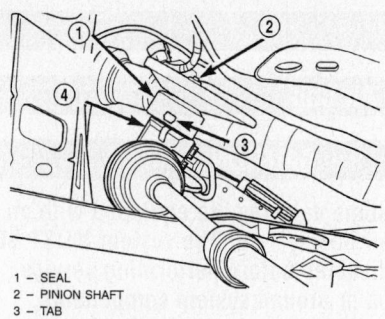

1 – SEAL
2 – PINION SHAFT
3 – TAB
4 – POWER STEERING GEAR

9306EG45

View of the pinion shaft dash cover seal

• Pinion shaft dash cover seal over the shaft and onto the power steering gear housing

➡**Align the seal holes with the tabs cast into the power steering gear housing.**

• Lower steering column coupling by aligning the coupling and steering gear pinion shaft flats
• Lower steering column coupling-to-pinion shaft's roll pin until it is centered

12. Center the power steering gear rack's travel.

• Front suspension crossmember/power steering gear assembly by raising it with the jack until is aligns with its matchmarks
• Lower steering column coupling, guide it through the dash panel hole as it is raised

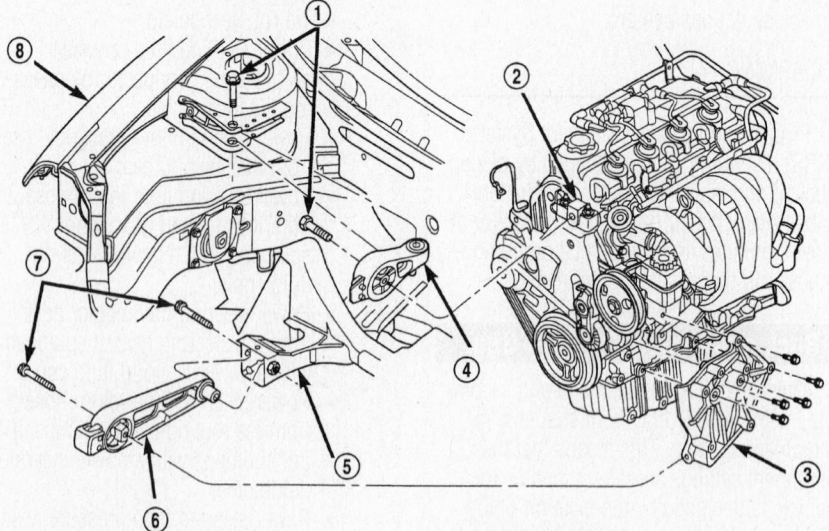

1 – BOLTS
2 – ENGINE MOUNT BRACKET
3 – TORQUE STRUT BRACKET
4 – UPPER TORQUE STRUT

5 – CROSSMEMBER
6 – LOWER TORQUE STRUT
7 – BOLTS
8 – RIGHT FENDER

9306EG46

View of the engine torque struts and related components

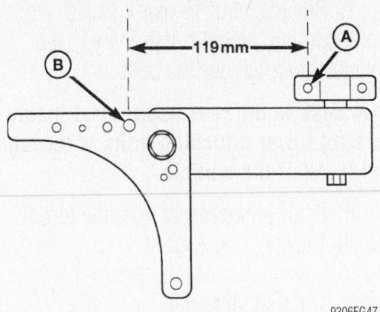

Measuring the engine torque bracket

- Both rear crossmember-to-tapping plate bolts
- Both front crossmember-to-frame rail bolts and torque them to 20 inch lbs. (2 Nm)

❊❊ WARNING

Be sure to align the front suspension crossmember-to-chassis match-marks; otherwise, the front wheel alignment setting will be lost.

- Once aligned, torque both rear crossmember-to-rear lower control arm bolts to 150 ft. lbs. (203 Nm) and both front crossmember bolts to 105 ft. lbs. (142 Nm)
- Engine torque strut to the right forward corner of the front suspension crossmember

13. Adjust the engine torque strut by performing the following procedure:

a. Loosen the upper torque strut at the shock tower bracket.

b. Position a floor jack on the forward edge of the bell housing to prevent the least amount of upward lifting of the engine.

c. Slowly, lift the assembly, allowing the engine to rotate rearward so the distance between center of the engine mount bracket's rearmost attaching stud (point A) and the center of the shock tower bracket's washer hose clip hole (point B) is 4.70 in. (119mm).

d. Torque the upper and lower torque strut bolts to 87 ft. lbs. (118 Nm).

e. Remove the floor jack.

14. Install or connect the following:

- Power steering hose-to-power steering gear using a new O-ring lubricated with power steering oil, if not equipped with a power steering cooler
- Power steering fluid cooler line-to-power steering gear, if equipped with a power steering cooler
- Power steering fluid return hose to the routing clip C-clamps, if not equipped with a power steering fluid cooler
- Power steering fluid pressure hose to the steering gear's routing clips
- Power steering cooler hose to the steering gear's right routing clip, if equipped
- Torque the power steering pressure hose-to-power steering gear nut to 25 ft. lbs. (34 Nm)
- Both power steering cooler screws to the front suspension crossmember, if equipped
- Power steering fluid pressure switch wiring connector be sure the locking tab is secure latched
- Tie rod heat shield, facing outboard
- Outer tie rod ends to the steering knuckle. Torque the nut, using a crowsfoot wrench, to 40 ft. lbs. (55 Nm), while holding the tie rod stationary
- Both front wheels and torque the lug nuts, in a crisscross pattern, to 100 ft. lbs. (135 Nm)
- Dash-to-lower coupling seal over the lower coupling's plastic collar

➡**Verify that the seal's lip shows grease at the coupling's plastic collar contact.**

- Steering column lower coupling-to-steering column upper coupling pinch bolt and torque the nut to 21 ft. lbs. (28 Nm)
- Pinch bolt retainer pin

15. Remove the steering wheel holder.

16. Fill and bleed the power steering system.

17. Check for leaks.

18. Check and/or adjust the front toe setting.

Strut

REMOVAL & INSTALLATION

Front

1. Before servicing the vehicle, refer to the precautions in the beginning of this section.

2. Remove the front wheels.

3. Mark each one right or left, as applicable, if both struts are being removed.

4. Remove the screw securing the ground strap to the rear of the strut, if equipped.

5. Remove the screw securing the ABS wheel speed sensor to the back of the strut, if equipped.

❊❊ WARNING

The steering knuckle-to-strut assembly attaching bolts are serrated and must not be turned during removal.

6. Remove or disconnect the following:

- Steering knuckle nuts while holding the bolts stationary

➡**If necessary, partially lower the vehicle for access to the upper mounting nuts.**

- 3 upper strut mount-to-strut tower nuts
- Strut assembly

To install:

7. Install or connect the following:

- Strut assembly into the strut tower by aligning the 3 upper strut mount studs with the shock tower holes. Torque the 3 upper strut mount nut/washer assemblies to 25 ft. lbs. (34 Nm).

❊❊ WARNING

The steering knuckle-to-strut assembly attaching bolts are serrated and must not be turned during installation.

- Lower end of the strut, in line with the upper end of the steering knuckle and align the mounting holes.
- Both strut-to-steering knuckle bolts and nuts. Torque both nuts, while holding the bolts stationary, to 40 ft. lbs. (53 Nm), plus an additional ¼ turn after the specified torque is met

➡**The bolts should be installed with the nuts facing the front of the vehicle.**

- ABS wheel speed sensor to the rearward ear of the strut, and secure with its mounting screw. Tighten the screw to 10 ft. lbs. (13 Nm)
- Hydraulic brake hose routing bracket and screw onto the strut damper bracket and torque the bracket bolts to 10 ft. lbs. (13 Nm), if necessary. If equipped with ABS, the hydraulic hose routing bracket is combined with the speed sensor cable routing bracket
- Screw securing the ground strap to the rear of the strut, if equipped.

- Front wheels and torque the lug nuts, in a criss-cross pattern, to 100 ft. lbs. (135 Nm)

Rear

1. Before servicing the vehicle, refer to the precautions in the beginning of this section.
2. Remove the rear wheel.
3. If equipped with drum brakes, remove the screw securing the brake hose bracket to the strut.
4. If equipped with Anti-lock Brake System (ABS), remove the screw securing the ABS wheel speed sensor bracket to the rear of the strut assembly.
5. Remove the nut from the end of the rear stabilizer bar link bolt. Pull the bolt out through the top of the link and remove the link.
6. If equipped with rear disc brakes, perform the following steps:
    a. Remove the 2 guide pins securing the caliper, then remove the caliper
    b. Use a piece of wire to suspend the caliper aside. Do NOT let the caliper hang by the brake hose.
    c. Remove the rotor.

❋❋ WARNING

The knuckle-to-strut attaching bolts are serrated and must not be turned during removal.

7. Remove strut-to-rear knuckle nuts while holding the bolts stationary in the knuckle, then remove the bolts.

➡**Access to the rear upper strut mount-to-strut tower attaching bolts is through the trunk of the vehicle.**

8. Remove the carpet from the top of the strut tower, if necessary.
9. Loosen, but do not remove the 3 upper strut mounting nuts.
10. Remove the 3 strut-to-chassis mount nuts, while supporting the strut.
11. Remove the strut from the knuckle by sliding it away from the knuckle, lowering it between the 2 lateral arms, then tipping the top outward and removing it through the wheel well opening.

To install:

12. Install or connect the following:
    - Strut and torque the 3 strut mount-to-body nuts to 25 ft. lbs. (34 Nm)
    - Carpeting on top of the strut tower, then close the trunk lid
13. Align the holes in the strut clevis bracket on the lower end of the strut with the mounting holes in the knuckle.
    - 2 bolts attaching the strut to the rear knuckle and torque to 65 ft. lbs. (88 Nm)
14. If equipped with rear disc brakes, perform the following:
    a. Install the brake rotor and caliper.
    b. Install the caliper guide bolts and tighten to 16 ft. lbs. (22 Nm).
15. Install the stabilizer bar link, as follows:
    a. Place the link center sleeve and bushings between the eye in the end of the stabilizer bar and the link mounting bracket on the strut.
    b. Start the stabilizer bar link bolt with bushing from the top, down through the stabilizer bar, inner link bushings and sleeve, and strut link mounting bracket.
    c. Install a lower bushing, then the nut, but don't tighten the nut yet.
16. If equipped with ABS, install the screw holding the wheel speed sensor bracket to the rear of the strut and tighten to 10 ft. lbs. (13 Nm).
17. If equipped with rear drum brakes, install the screw securing the brake hose bracket to the rear of the strut assembly and tighten to 23 ft. lbs. (31 Nm).
18. Rear wheel and torque the lug nuts evenly, in sequence, to 100 ft. lbs. (135 Nm)
19. Lower the vehicle, then tighten the stabilizer bar link nut to 17 ft. lbs. (23 Nm).
20. Check the alignment and adjust, if necessary.

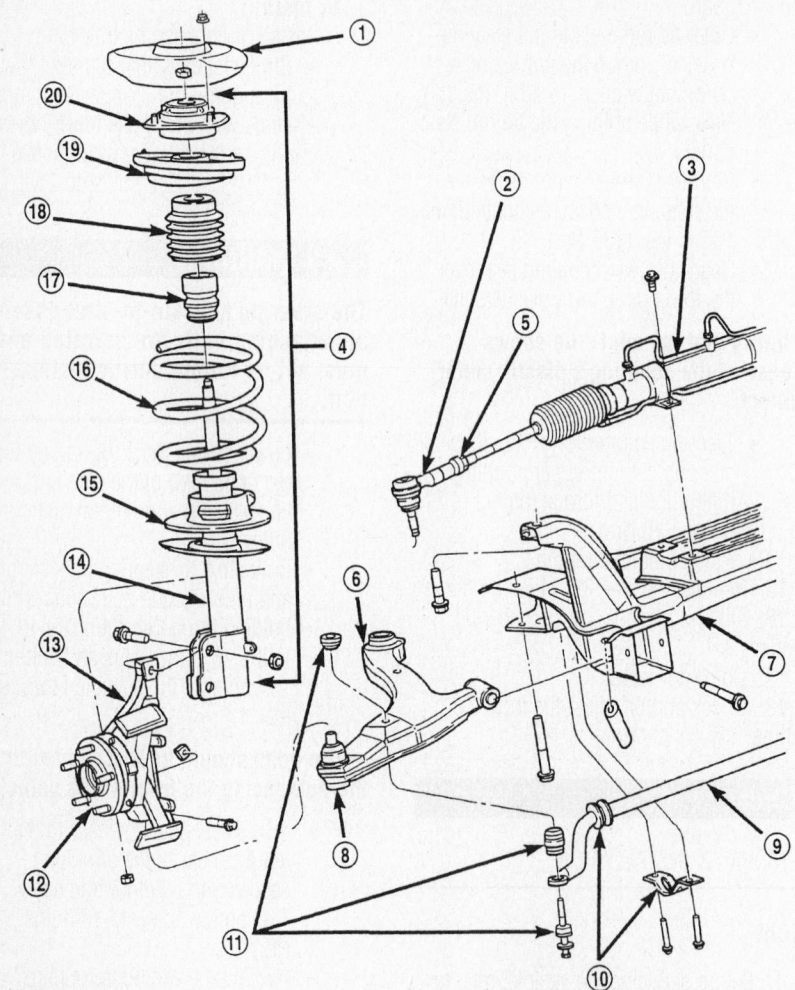

1 – VEHICLE STRUT TOWER
2 – OUTER TIE ROD
3 – STEERING GEAR
4 – STRUT ASSEMBLY
5 – JAM NUT
6 – LOWER CONTROL ARM
7 – CROSSMEMBER
8 – BALL JOINT
9 – STABILIZER BAR
10 – STABILIZER BAR CUSHION AND RETAINER
11 – STABILIZER BAR LINK
12 – HUB
13 – KNUCKLE
14 – STRUT
15 – LOWER SPRING ISOLATOR
16 – COIL SPRING
17 – JOUNCE BUMPER
18 – DUST SHIELD
19 – SPRING SEAT AND BEARING
20 – UPPER MOUNT

9306EG11

Exploded view of the front suspension

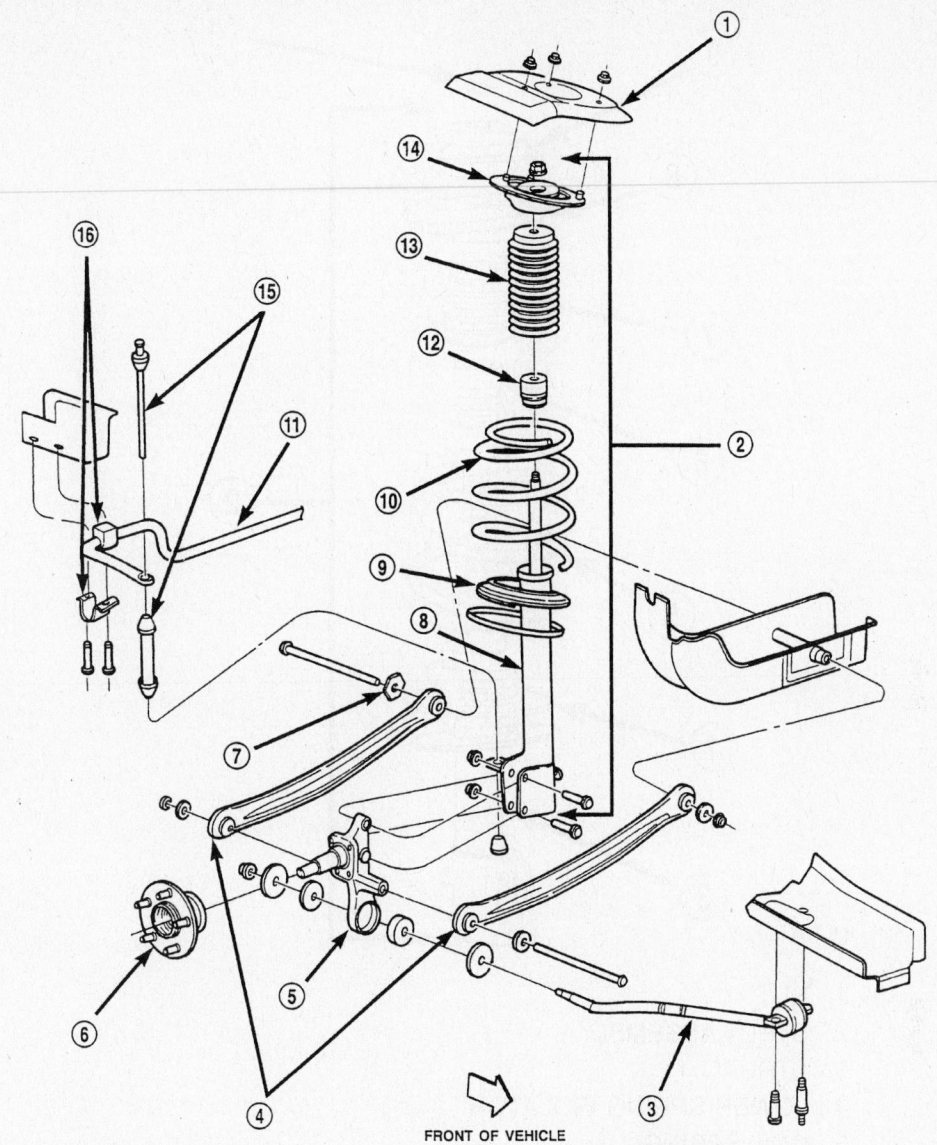

FRONT OF VEHICLE

1 – VEHICLE STRUT TOWER
2 – STRUT ASSEMBLY
3 – TENSION STRUT
4 – LATERAL ARMS
5 – KNUCKLE
6 – HUB AND BEARING
7 – WHEEL ALIGNMENT ADJUSTMENT CAM
8 – STRUT

9 – LOWER SPRING ISOLATOR
10 – COIL SPRING
11 – STABILIZER BAR
12 – JOUNCE BUMPER
13 – DUST SHIELD
14 – UPPER MOUNT
15 – STABILIZER BAR LINK
16 – STABILIZER BAR CUSHION AND RETAINER

9306EG12

Exploded view of the rear suspension

Coil Spring

REMOVAL AND INSTALLATION

1. Before servicing the vehicle, refer to the precautions in the beginning of this section.

2. Remove the strut assembly.

3. Install the strut assembly in a spring compressor.

4. Set the lower hooks first then the upper hooks.

5. Position the strut clevis bracket straight outward away from the compressor

6. Place a clamp on the lower end of the coil spring so the strut is held in place.

7. Compress the spring until all tension is removed from the upper mount.

8. Install a strut nut socket, Special Tool 6864, on the shaft retaining nut.

9. Install socket on the hex end of the strut shaft.

10. Hold the shaft from turning and remove the nut from the shaft.

11. Remove or disconnect the following:

- Upper mount from the strut shaft
- Upper spring seat, bearing and upper spring isolator as an assembly
- Dust shield
- Jounce bumper
- Clamp from the bottom of the coil spring
- Strut assembly through the bottom of the coil spring
- Lower spring isolator

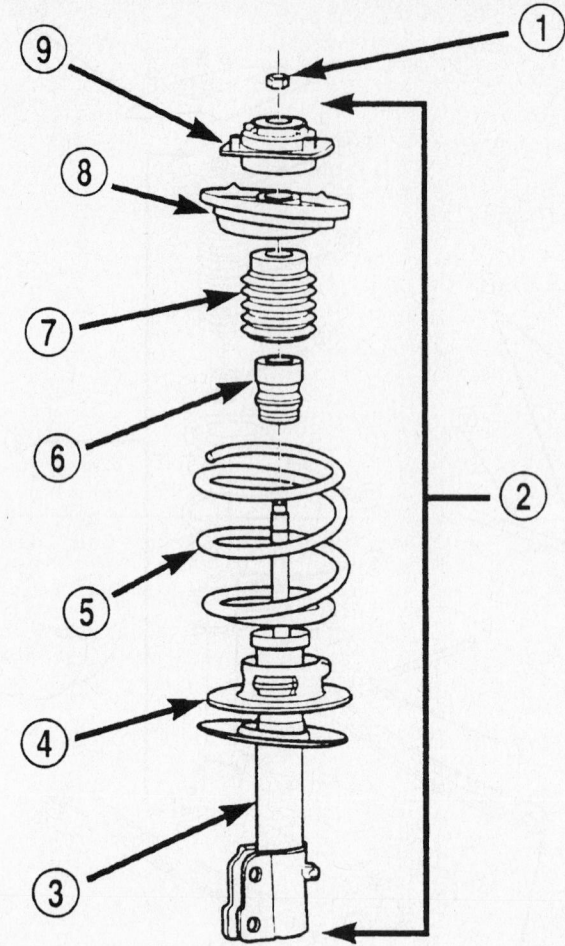

1 - NUT
2 - STRUT ASSEMBLY
3 - STRUT
4 - LOWER SPRING ISOLATOR
5 - COIL SPRING
6 - JOUNCE BUMPER
7 - DUST SHIELD
8 - SPRING SEAT AND BEARING (WITH SPRING ISOLATOR)
9 - UPPER MOUNT

9356HG11

Exploded view of the strut assembly

- Coil spring by releasing the tension on the compressor drive

To install:

12. Place the coil spring in a compressor and rotate the spring so that the end of the top coil is directly in back.

13. Slowly compress the coil spring until enough room is available for reassembly.

14. Install or connect the following:
- Lower spring isolator on the lower spring seat
- Strut through the bottom of the coil spring until the spring seat contacts the lower end of the coil spring

15. Rotate the strut, if necessary, until the clevis bracket is positioned straight outward and away from the compressor.
- Clamp on the lower end of the coil spring and strut
- Jounce bumper on the strut shaft with the small end facing down
- Dust shield on the strut shaft
- Upper spring isolator on the upper spring seat and bearing
- Upper spring seat and bearing on the top of the coil spring

- Strut upper mount over the strut shaft and on top of the upper spring seat and bearing
- Retaining nut on the strut shaft, loosely
- Strut nut socket Tool 6864 on the retaining nut
- Socket on the hex end of the shaft and torque the nut to 55 ft. lbs. (75 Nm)

16. Slowly release the tension from the from the coil spring and make certain that the coil spring is properly aligned.

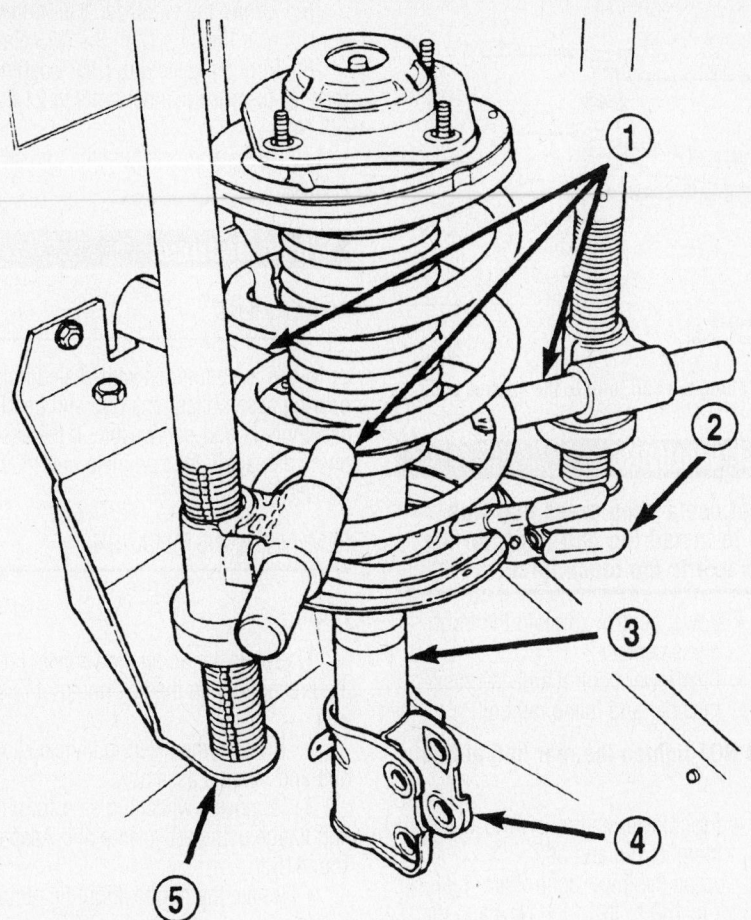

1 - LOWER HOOKS
2 - CLAMP
3 - STRUT ASSEMBLY
4 - CLEVIS BRACKET
5 - SPRING COMPRESSOR

9356HG12

Assemble the coil spring as shown

17. Remove the clamp from the lower end of the coil spring and strut.

18. Remove the upper and lower hooks from the compressor tool and remove the tool.

19. Install the strut assembly.

Tension Strut

REMOVAL & INSTALLATION

1. Remove or disconnect the following:
   • Rear wheels
   • Tension strut-to-knuckle nut by holding the strut with a wrench
   • Forward tension strut retainer
   • Rear tension strut bayonet bushing
   • Parking brake cable from tension strut bolt

   • Tension strut from the chassis
To install:
2. Install or connect the following:
   • Tension strut to the chassis and torque the bolts to 70 ft. lbs. (95 Nm)
   • Parking brake cable to tension strut nut and torque the bolts to 21 ft. lbs. (28 Nm)

➡**The mounting bolt with the stud on the head is installed on the inboard side.**

   • Rear tension strut bayonet bushing making sure that the stepped area faces toward the knuckle
   • Tension strut retainer
   • Tension strut-to-knuckle nut by holding the strut with a wrench and torque the nut to 70 ft. lbs. (95 Nm)

   • Rear wheels and torque the lug nuts to 100 ft. lbs. (135 Nm)
3. Check and/or align the rear wheels.

Lower Ball Joint

REMOVAL & INSTALLATION

The front suspension ball joints operate with no free-play. The ball joints are replaceable ONLY as an assembly. Do not attempt any type of repair on the ball joint assembly. The ball joint is a press fit into the lower control arm with the joint stud retained in the steering knuckle by the clamp bolt. To check the ball joint, with the weight of the vehicle resting on the road wheels, grasp the grease fitting and without using any tools, attempt to move the grease fitting. If the ball joint is worn the grease fitting will move easily. If movement is noted, replacement of the ball joint is recommended.

1. Before servicing the vehicle, refer to the precautions in the beginning of this section.

2. Remove or disconnect the following:
   • Wheel
   • Steering knuckle-to-ball joint stud's pinch bolt and nut
   • Stabilizer bar-to-lower control arm links

3. Loosen, but do not remove the bolts holding the stabilizer bar retainers to the crossmember. Then, rotate the stabilizer bar and attaching links away from the lower control arms.

✷✷ WARNING

Pulling the steering knuckle outward after releasing the ball joint can separate the inner CV-joint.

4. Remove or disconnect the following:
   • Ball joint from the steering knuckle using a prybar

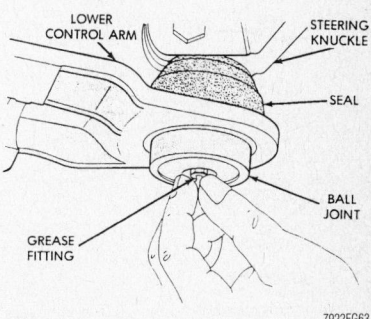

7922EG63

Wiggle the grease fitting with your fingers—if it moves, the ball joint should be replaced

✲✲ WARNING

Be careful when separating the ball joint stud from the knuckle, so the seal does not become damaged.

- Front lower control arm bushing-to-crossmember nut and bolt
- Rear lower control arm-to-crossmember bolt
- Lower control arm
- Ball joint using a prytool

5. Using a hydraulic press, press the ball joint from the lower control arm using tools:

- Receiver Tool 6908-2
- Adapter Tool 6804

To install:

6. Reinstall the ball joint into the lower control arm with the notch in the ball joint stud facing the front lower control arm bushing.

7. Using a hydraulic press, press the ball joint into the lower control arm using tools:

   a. Receiver Tool 6758 and Adapter Tool 6804.

8. Install or connect the following:

- Ball joint boot seal using a driver tool such as a large socket or suitable sized piece of pipe

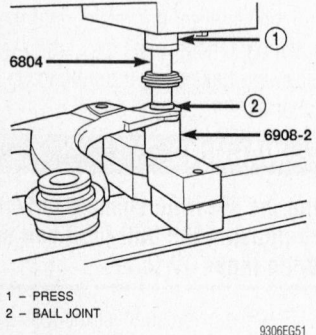

1 – PRESS
2 – BALL JOINT

9306EG51

Removing the ball joint from the control arms

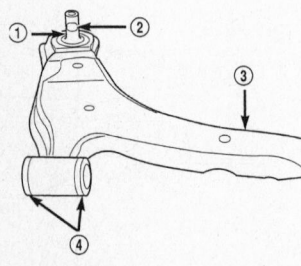

1 – BALL JOINT STUD
2 – NOTCH
3 – LOWER CONTROL ARM
4 – FRONT ISOLATOR BUSHING

9306EG52

Aligning the ball joint stud notch to the control arms

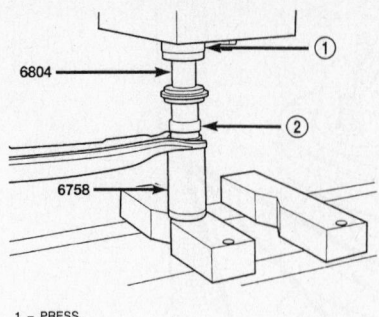

1 – PRESS
2 – BALL JOINT

9306EG53

Installing the ball joint to the control arms

✲✲ WARNING

Do not use a shop press that was used to install the ball joint, for the press exerts too much force.

- Lower control arm into the front crossmember
- Rear lower control arm-to-crossmember and frame rail bolt

➡ **DO NOT tighten the rear bolt at this time.**

- Front lower control arm-to-crossmember nut and bolt

9. Torque the lower control arm rear pivot bolt to 150 ft. lbs. (203 Nm) and the front pivot bolt to 120 ft. lbs. (163 Nm).

10. Install the ball joint stud into the steering knuckle. Torque the steering knuckle-to-ball joint stud pinch bolt and nut to 70 ft. lbs. (95 Nm).

11. Assemble the stabilizer bar-to-lower control arm link assemblies and bushings.

12. Rotate the stabilizer bar into position, installing the stabilizer bar links into the lower control arms. Install the top stabilizer bar link bushings and nuts. DO NOT tighten the link yet.

13. Install the wheel.

14. Lower the vehicle so the suspension is supporting the total weight of the vehicle.

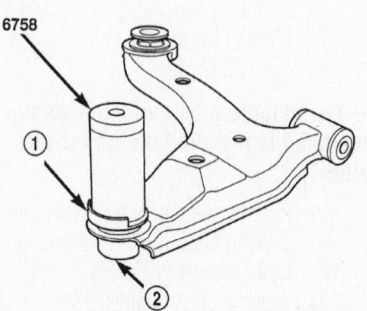

1 – SEAL BOOT UPWARD LIP
2 – BALL JOINT

9306EG54

Installing the ball joint boot seal

15. Torque the stabilizer bar-to-lower control arm links to 17 ft. lbs. (23 Nm).

16. Torque the stabilizer bar bushing retainer-to-crossmember bolts to 21 ft. lbs. (28 Nm).

17. Check and/or adjust the toe, as necessary.

Wheel Bearings

ADJUSTMENT

Neons are equipped with sealed hub and bearing assemblies. The hub and bearing assembly is non-serviceable. If the assembly is damaged, the complete unit must be replaced.

REMOVAL & INSTALLATION

Front

1. Before servicing the vehicle, refer to the precautions in the beginning of this section.

2. Remove the steering knuckle and hub and bearing assembly.

3. Remove a wheel lug stud from the hub flange using a C-clamp and Adapter Tool 4150A.

4. Rotate the hub to align the removed

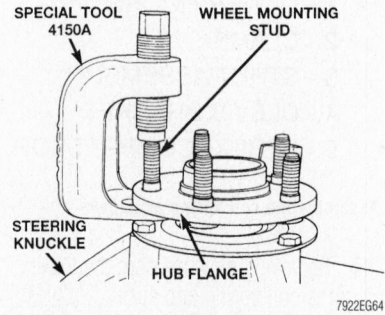

7922EG64

Use a proper C-clamp and adapter tool to press out one of the lug studs

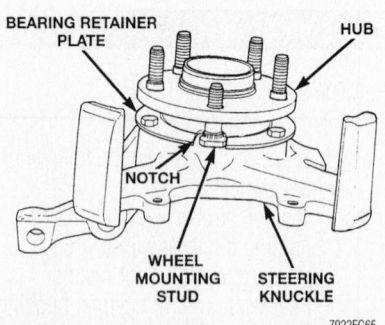

7922EG65

Rotate the hub in order to remove the lug stud

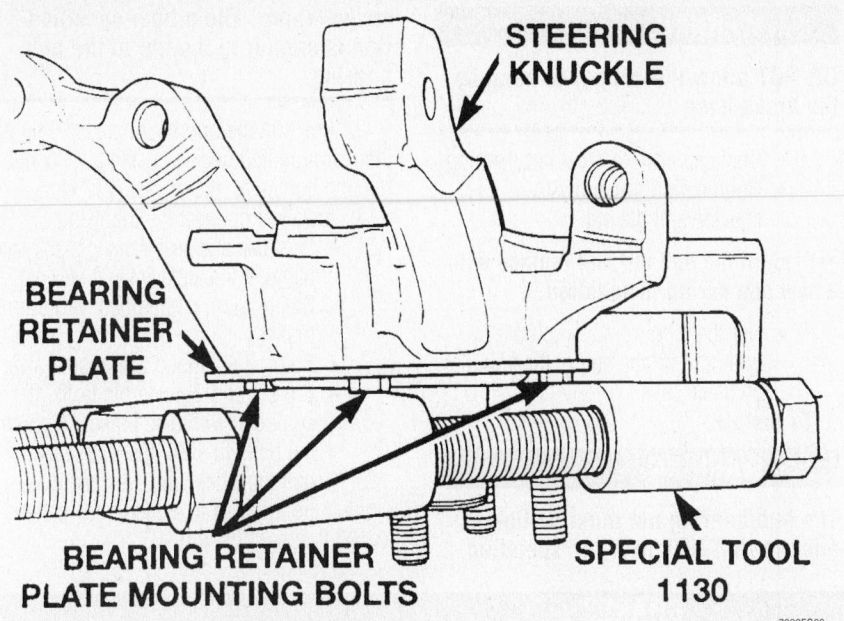

Proper installation of the bearing splitter

lug stud with the notch in the bearing retainer plate.

5. Rotate the hub so the stud hole is facing away from the brake caliper's lower rail on the steering knuckle.

6. Install ½ of a Bearing Splitter Tool 1130, between the hub and the bearing retainer plate. The threaded hole in this ½ is to be aligned with the caliper rail on the steering knuckle.

7. Install the remaining pieces of the bearing splitter on the steering knuckle.

Hand-tighten the nuts to hold the splitter in place on the knuckle.

8. When the bearing splitter is installed, be sure the 3 bolts attaching the bearing retainer plate to the knuckle are contacting the bearing splitter. The bearing retainer plate should not support the knuckle or contact the splitter.

9. Place the steering knuckle in a hydraulic press, supported by the bearing splitter.

10. Position a driver on the small end of

the hub. Using the press, remove the hub from the wheel bearing. The outer bearing race will come out of the wheel bearing when the hub is pressed out of the bearing.

11. Remove the bearing splitter tool from the knuckle.

12. Place the knuckle in a press supported by the press block. The blocks must not obstruct the bore in the steering knuckle so the wheel bearing can be pressed out of the knuckle. Place a driver on the outer race of the wheel bearing, then press the bearing out of the knuckle.

13. Install the bearing splitter on the hub. The splitter is to be installed on the hub so it is between the flange of the hub and the bearing race on the hub. Place the hub, bearing race and splitter in a press. Use a driver to press the hub out of the bearing race.

To install:

14. Use clean, dry cloth to wipe and grease or dirt from the bore of the steering knuckle.

15. Clean the rust preventative from the replacement wheel bearing using a clean, dry towel.

16. Place the new wheel bearing into the bore of the steering knuckle. Be sure the bearing is placed squarely into the bore. Place the knuckle in a press with a receiver tool, C-4698–2 supporting the steering knuckle. Place a driver tool on the outer race of the wheel bearing. Press the wheel bearing into the steering knuckle until it is fully bottomed in the bore of the steering knuckle.

➡**Only the original or original equipment replacement bolts should be used to mounting the bearing retainer to the knuckle. If a bolt requires replacement when installing the bearing retainer plate, be sure to get the proper type of replacement.**

17. Install the bearing retainer plate on the steering knuckle. Install the 3 bearing retainer mounting bolts. Tighten the bolts to 21 ft. lbs. (28 Nm).

18. Install the removed wheel lug stud into the hub flange.

19. Place the hub with the lug stud installed, in a press supported by Adapter Tool C-4698–1. Press the wheel lug stud into the hub flange until it is fully seated against the back side on the hub flange.

20. Place the steering knuckle with the wheel bearing installed, in a press with special Receiver Tool MB-990799 supporting the inner race of the wheel bearing. Place the hub in the wheel bearing, making sure it is square with the bearing. Press the hub

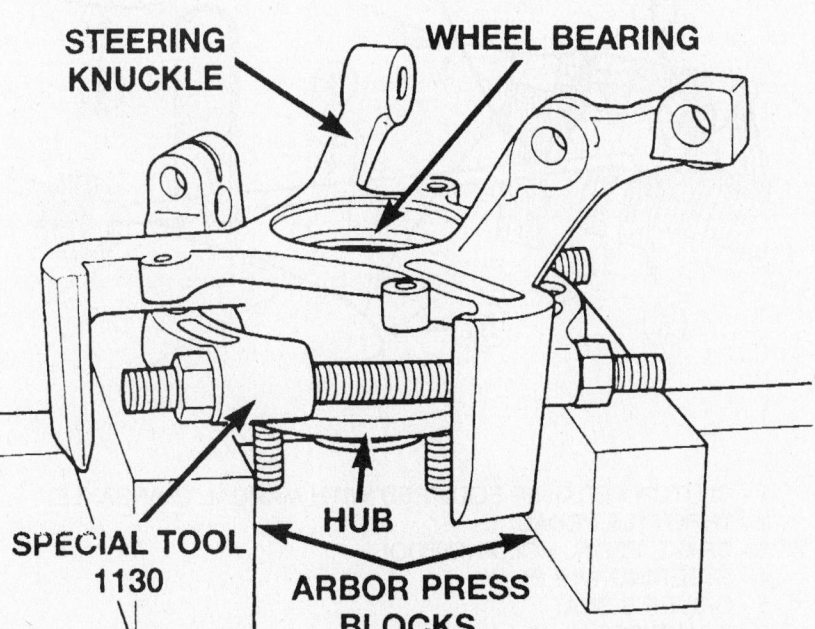

Properly support the steering knuckle for hub and bearing removal

into the wheel bearing until it is fully bottomed in the wheel bearing.

21. Install the steering knuckle and the wheel.

22. Check and/or adjust the front alignment.

Rear

1. Before servicing the vehicle, refer to the precautions in the beginning of this section.

2. Remove or disconnect the following:
- Wheel
- Rear brake drum, if equipped
- Caliper (suspend on a wire) and the rotor, if equipped with rear disc brakes

✳✳ WARNING

DO NOT allow the caliper to hang by the brake hose.

- Dust cap from the rear hub/bearing
- Hub/bearing assembly-to-knuckle/spindle nut

➡**Discard the hub nut and replace with a new one during installation.**

- Hub/bearing from the spindle by pulling it off the end of the spindle by hand

To install:

✳✳ WARNING

The hub/bearing nut must be tightened to, but NOT over, its specified

torque value. The proper specification is crucial to the life of the hub bearing.

3. Position the hub/bearing assembly on the rear spindle/knuckle. Install a NEW hub nut and tighten to 160 ft. lbs. (217 Nm).

4. Install or connect the following:
- Dust cap and seat it using a soft face hammer to carefully tap it into place
- Brake drum, if equipped with drum brakes
- Rotor, if equipped with disc brakes
- Caliper and 2 guide pin bolts, if equipped with disc brakes. Tighten the bolts to 16 ft. lbs. (22 Nm).

5. Install the wheel and tire assembly. Tighten the lug nuts in a crisscross pattern, to 100 ft. lbs. (135 Nm).

BRAKES

Brake Caliper

REMOVAL & INSTALLATION

Front

EXCEPT SRT-4

1. Before servicing the vehicle, refer to the precautions in the beginning of this section.

2. Isolate the master cylinder as follows:

    a. Use a brake pedal holding tool and depress the brake pedal past its first one inch of travel and hold it in this position. (This will keep brake fluid from draining from the master cylinder).

3. Remove or disconnect the following:
- Front wheels
- 2 caliper guide pin bolts
- Brake hose from the caliper
- Caliper from the steering spindle

To install:

4. Install or connect the following:
- Brake hose to the caliper and torque to 35 ft. lbs. (48 Nm)
- Caliper to the steering spindle
- Caliper guide pin bolts and torque to 16 ft. lbs. (22 Nm)

5. Remove the brake pedal holding tool.

6. Properly bleed the brake system.

7. Reinstall the front wheels and lower the vehicle.

8. Pump the brake pedal to seat the front brake pads before moving the vehicle.

9. Road test the vehicle and check for proper operation.

SRT-4

1. Before servicing the vehicle, refer to the precautions in the beginning of this section.

2. Isolate the master cylinder as follows:

    a. Use a brake pedal holding tool and depress the brake pedal past its first one

inch of travel and hold it in this position. (This will keep brake fluid from draining from the master cylinder).

3. Remove or disconnect the following:
- Front wheels
- Banjo bolt connecting the brake hose to the brake caliper. There are two washers (one on each side of

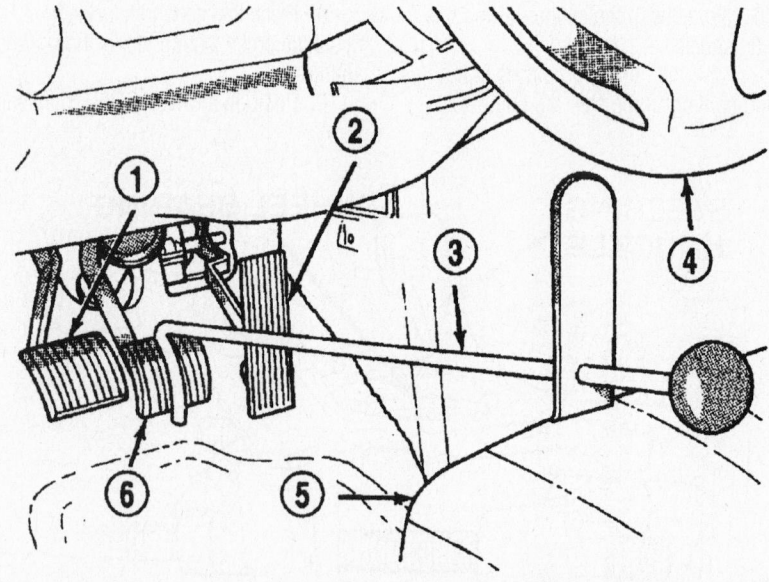

1 - CLUTCH PEDAL (IF EQUIPPED WITH MANUAL TRANSAXLE)
2 - THROTTLE PEDAL
3 - BRAKE PEDAL HOLDING TOOL
4 - STEERING WHEEL
5 - DRIVER'S SEAT
6 - BRAKE PEDAL

View of the brake pedal holding tool—all models

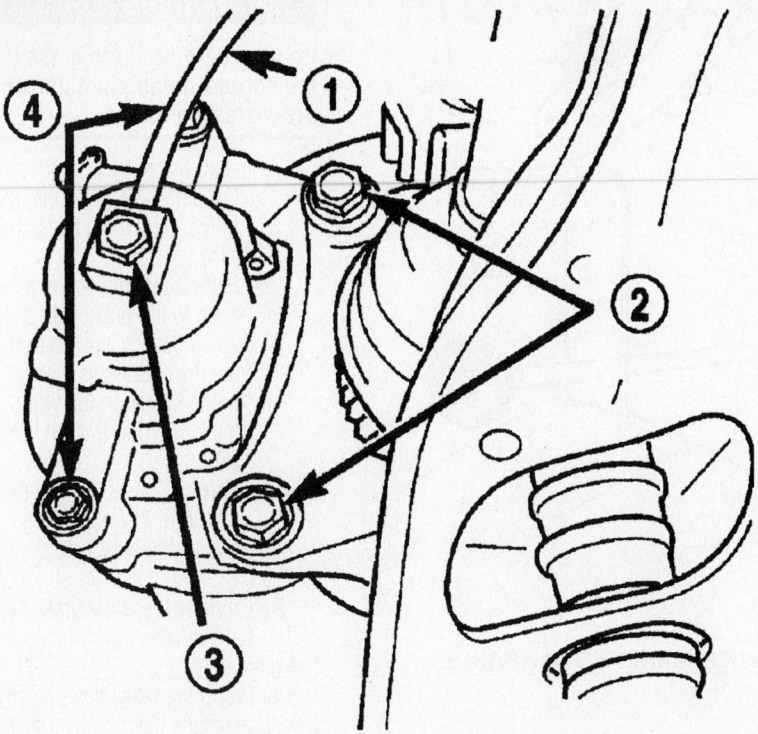

1 - BRAKE HOSE
2 - ADAPTER MOUNTING BOLTS
3 - BANJO BOLT
4 - CALIPER GUIDE PIN BOLTS

67189-NEON-G44

Front brake caliper mounting—SRT-4 models

the flex hose fitting) that will come off with the banjo bolt. Discard the washers.
- 2 caliper guide pin bolts
- Caliper from the disc brake adapter.

To install:

➡**When installing new brake components, be sure to use correct parts. Parts designed for the**

BR3 Performance Brake System must not be mixed with other brake systems.

4. If new pads were installed, completely retract the caliper piston back into the bore of the caliper. Use a C-clamp to retract the piston. Place a wood block over the piston before installing the C-clamp to avoid damaging the piston.

❊❊ WARNING

Be careful when installing the caliper onto the disc brake adapter to avoid damaging the boots on the caliper guide pins.

5. Install or connect the following:
- Caliper over the brake shoes on the brake caliper adapter. Make sure the springs on the shoes do not get caught in the hole formed into the center of the caliper housing.
- Align the caliper guide pin bolt holes with the guide pins.
- Caliper guide pin bolts and torque to 26 ft. lbs. (of 35 Nm)
- Banjo bolt connecting the brake hose to the brake caliper, using new washers on each side of the hose fitting as the banjo bolt is guided through the fitting. Torque the banjo bolt to 18 ft. lbs. (24 Nm).
- Tire and wheel. Torque the lug nuts to 100 ft. lbs. (135 Nm).

6. Remove the brake pedal holding tool.
7. Properly bleed the brake system.
8. Reinstall the front wheels and lower the vehicle.

9. Pump the brake pedal to seat the front brake pads before moving the vehicle.
10. Road test the vehicle and check for proper operation.

Rear

1. Before servicing the vehicle, refer to the precautions in the beginning of this section.
2. Isolate the master cylinder as follows:
   a. Use a brake pedal holding tool and depress the brake pedal past its first one inch of travel and hold it in this position. (This will keep brake fluid from draining from the master cylinder).
3. Remove or disconnect the following:

- Rear wheels
- Banjo bolt connecting the brake hose to the brake caliper. There are two washers (one on each side of the flex hose fitting) that will come off with the banjo bolt. Discard the washers.
- 2 caliper guide pin bolts
- Caliper assembly from the brake adapter by first rotating the top of the caliper away from the rotor, and then lifting the caliper assembly off the machined abutment on the adapter

To install:

➡**When installing new brake components, be sure to use correct parts. Parts designed for the**

BR3 Performance Brake System must not be mixed with other brake systems.

4. If new pad were installed, completely retract the caliper piston back into piston bore of the caliper.
5. Lubricate both adapter caliper slide abutments with a liberal amount of Mopar Multipurpose Lubricant, or an equivalent.

❊❊ WARNING

Be careful when installing the caliper assembly onto adapter so the guide pin bushings and sleeves do not get damaged by the mounting bosses on adapter.

6. Starting with the lower end, carefully lower the caliper and brake shoes over the brake rotor and catch the caliper's bottom edge behind the caliper slide abutment. Rotate the top of the caliper into mounting position on the adapter.

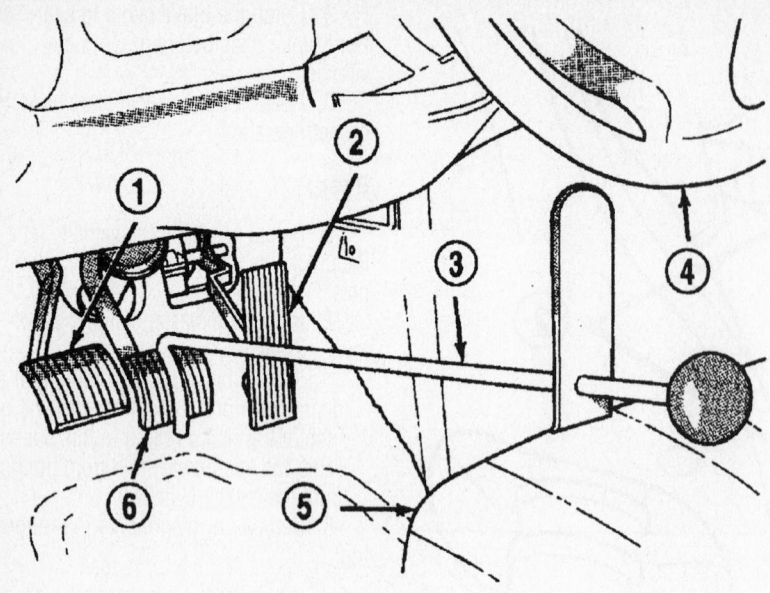

1 - CLUTCH PEDAL (IF EQUIPPED WITH MANUAL TRANSAXLE)
2 - THROTTLE PEDAL
3 - BRAKE PEDAL HOLDING TOOL
4 - STEERING WHEEL
5 - DRIVER'S SEAT
6 - BRAKE PEDAL

67189-NEON-G45

View of the brake pedal holding tool—all models

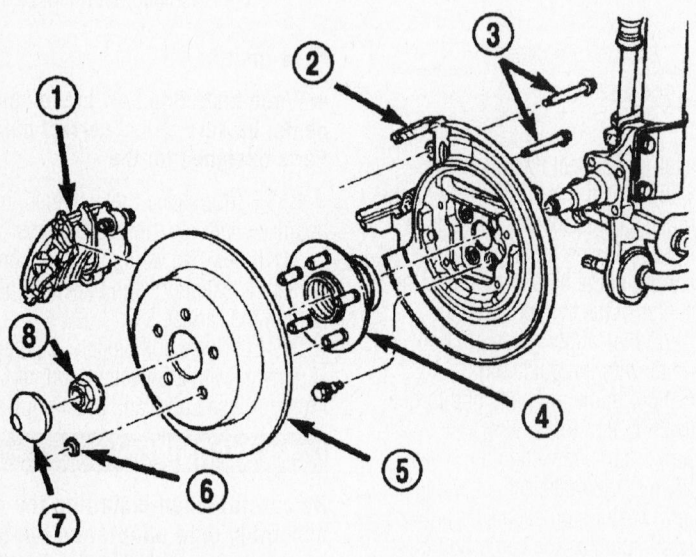

1 - DISC BRAKE CALIPER
2 - DISC BRAKE ADAPTER
3 - GUIDE PIN BOLTS
4 - HUB AND BEARING
5 - BRAKE ROTOR
6 - RETAINER CLIP
7 - DUST CAP
8 - NUT

67189-NEON-G46

Exploded view of the rear brake caliper mounting

✳✳ WARNING

Be very careful not to cross thread the caliper guide pin bolts when you are installing them!

7. Install or connect the following:
 • Caliper guide pin bolts, then carefully torque them to 16 ft. lbs. (22 Nm)
 • Banjo bolt connecting the brake hose to the brake caliper. Install new washers, one on each side of the hose fitting as the banjo bolt is guided through the fitting. Torque the banjo bolt to 18 ft. lbs. (24 Nm).
 • Tire and wheel. Torque the lug nuts to 100 ft. lbs. (135 Nm).
8. Remove the brake pedal holding tool.
9. Properly bleed the brake system.
10. Reinstall the front wheels and lower the vehicle.
11. Pump the brake pedal to seat the front brake pads before moving the vehicle.
12. Road test the vehicle and check for proper operation.

Disc Brake Pads

REMOVAL & INSTALLATION

Front

EXCEPT SRT-4

1. Before servicing the vehicle, refer to the precautions in the beginning of this section.
2. Remove the front wheels.
3. Remove the 2 caliper to steering knuckle guide pin bolts.
4. Lift the caliper away from the steering knuckle by first rotating the free end of the caliper away from the steering knuckle. Then, slide the opposite end of the caliper out from under the machined end of the steering knuckle.
5. Support the caliper from the upper control arm to prevent the weight of the caliper from being supported by the brake flex hose that will damage the hose.
6. Remove the brake pads from the caliper. Remove the outboard brake pad by prying the pad retaining clip over the raised area on the caliper. Then, slide the pad down and off the caliper. Pull the inboard brake pad away from the piston until the retaining clip is free from the cavity in the piston.
7. If required, the rotor can be removed

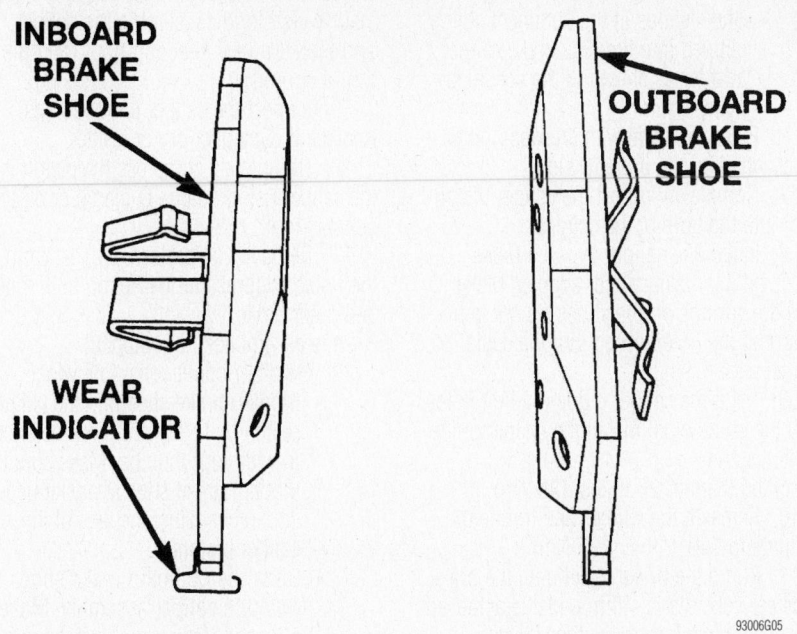

Disc brake pad identification

INBOARD
BRAKE
SHOE

OUTBOARD
BRAKE
SHOE

WEAR
INDICATOR

93006G05

by pulling it straight off the wheel mounting studs.

To install:

8. Clean all parts well. Inspect the caliper for piston seal leaks (brake fluid in and around the boot area and inboard lining) and for any ruptures of the piston dust boot. If the boot is damaged or fluid leak is visible, disassemble the caliper and install a new seal and boot (and piston, if scored).

9. Inspect the caliper pin bushings. Replace if damaged, dry or brittle.

10. Completely compress the piston into the caliper using a large C-clamp or other suitable tool.

11. Lubricate the area on the steering knuckle where the caliper slides with high temperature grease.

12. Reinstall the rotor if removed.

13. Reinstall the brake pads into the caliper. Note that the inboard and outboard pads are different. Make sure the inboard brake shoe assembly is positioned squarely against the face of the caliper piston.

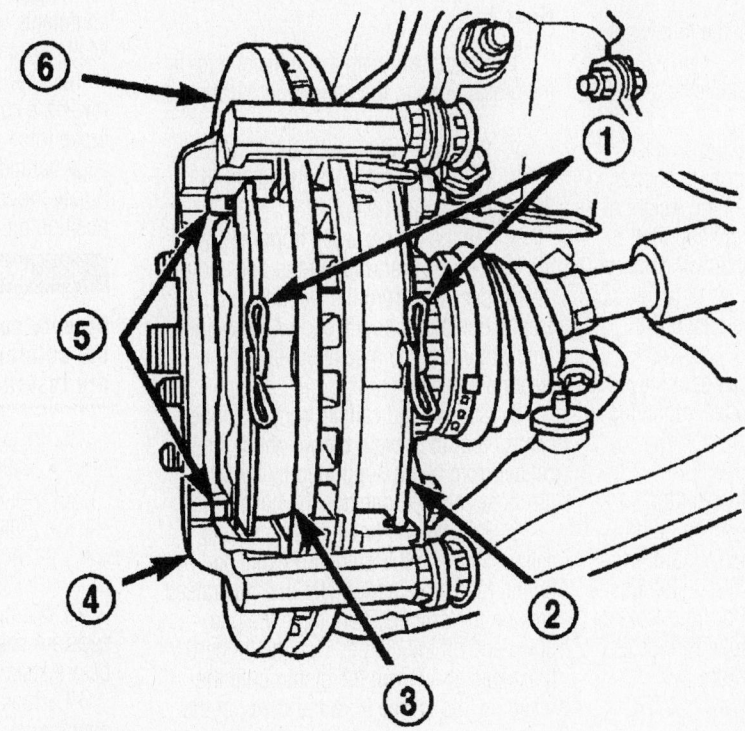

1 - SPRINGS
2 - INBOARD SHOE
3 - OUTBOARD SHOE
4 - DISC BRAKE ADAPTER
5 - ABUTMENT SHIMS
6 - BRAKE ROTOR

67189-NEON-G47

Install the brake shoes in the abutment shims clipped into the disc brake caliper adapter—SRT-4

➡ **Be sure to remove the noise suppression gasket paper cover if the pads come so equipped.**

14. Carefully position the caliper and brake shoe assemblies over the rotor by hooking the lower end of the caliper over the steering knuckle. Then, rotate the caliper into position at the top of the steering knuckle. Make sure the caliper guide pin bolts, bushings and sleeves are clear of the steering knuckle bosses.

15. Reinstall the caliper guide pin bolts and torque to 16 ft. lbs. (22 Nm).

16. Reinstall the wheels and torque the lug nuts to 100 ft. lbs. (135 Nm).

17. Pump the brake pedal until the brake pads are seated and a firm pedal is achieved before attempting to move the vehicle.

18. Road test the vehicle for proper operation.

SRT-4

1. Before servicing the vehicle, refer to the precautions in the beginning of this section.

2. Remove or disconnect the following:
- Front wheels
- 2 caliper to steering knuckle guide pin bolts
- Caliper from the steering knuckle. Support the caliper from the upper control arm to prevent the weight of the caliper from being supported by the brake flex hose that will damage the hose.
- Brake shoes from the disc brake caliper adapter

3. If required, the rotor can be removed by pulling it straight off the wheel mounting studs.

To install:

4. Clean all parts well. Inspect the caliper for piston seal leaks (brake fluid in and around the boot area and inboard lining) and for any ruptures of the piston dust boot. If the boot is damaged or fluid leak is visible, disassemble the caliper and install a new seal and boot (and piston, if scored).

➡ **The inboard brake shoes are not identical side-to-side. This is due to placement of the audible wear indicator on the end of each inboard shoe. Make sure that the audible wear indicators are placed toward the top when the inboard shoes are installed on each side of the vehicle.**

5. Install or connect the following:
- Brake shoes in the abutment shims clipped into the disc brake caliper adapter as shown in the accompanying figure

6. Place the shoe with the wear indicator attached on the inboard side

7. Completely retract the caliper piston back into the bore of the caliper.

8. Install the caliper over the brake shoes on the brake caliper adapter. Make sure the springs on the shoes do not get caught in the hole formed into the center of the caliper housing.

9. Align the caliper guide pin bolt holes with the guide pins. Install the caliper guide pin bolts and
tighten them to 26 ft. lbs. (35 Nm)

10. Reinstall the wheels and torque the lug nuts to 100 ft. lbs. (135 Nm).

11. Pump the brake pedal until the brake pads are seated and a firm pedal is achieved before attempting to move the vehicle.

12. Road test the vehicle for proper operation.

Rear

1. Before servicing the vehicle, refer to the precautions in the beginning of this section.

2. Remove the rear wheels.

3. Remove the 2 caliper to steering knuckle guide pin bolts.

4. Lift the caliper away from the steering knuckle by first rotating the free end of the caliper away from the steering knuckle. Then, slide the opposite end of the caliper out from under the machined end of the steering knuckle.

5. Support the caliper from the upper control arm to prevent the weight of the caliper from being supported by the brake flex hose that will damage the hose.

6. Remove the brake pads from the caliper. Remove the outboard brake pad by prying the pad retaining clip over the raised area on the caliper. Then, slide the pad down and off the caliper. Pull the inboard brake pad away from the piston until the retaining clip is free from the cavity in the piston.

7. If required, the rotor can be removed by pulling it straight off the wheel mounting studs.

To install:

8. Clean all parts well. Inspect the caliper for piston seal leaks (brake fluid in and around the boot area and inboard lining) and for any ruptures of the piston

dust boot. If the boot is damaged or fluid leak is visible, disassemble the caliper and install a new seal and boot (and piston, if scored).

9. Inspect the caliper pin bushings. Replace if damaged, dry or brittle.

10. Completely compress the piston into the caliper using a large C-clamp or other suitable tool.

11. Remove any protective paper from the noise suppression gasket on both inner and outer brake shoe assemblies (if equipped).

12. Install or connect the following:
- Inboard brake shoe into the caliper piston by firmly pressing the shoe in with your thumbs. Make sure the inboard brake shoe is positioned squarely against the face of the caliper piston.
- Slide the outboard brake shoe onto the caliper assembly. Make sure the retaining clip is squarely seated in the depressed areas on the caliper.

13. Lubricate both adapter caliper slide abutments with a liberal amount of Mopar Multipurpose Lubricant, or an equivalent.

14. Starting with the lower end, carefully lower the caliper and brake shoes over the brake rotor and catch the caliper's bottom edge behind the caliper slide abutment. Rotate the top of the caliper into mounting position on the adapter.

✷✷ WARNING

Be very careful not to cross thread the caliper guide pin bolts when they are installed.

15. Install or connect the following:
- Caliper guide pin bolts and tighten to 16 ft. lbs. (22 Nm)
- Rear tire and wheel assemblies. Torque the lug nuts to 100 ft. lbs. (135 Nm).

16. Pump the brake pedal until the brake pads are seated and a firm pedal is achieved before attempting to move the vehicle.

17. Road test the vehicle for proper operation.

Brake Drums

REMOVAL & INSTALLATION

1. Before servicing the vehicle, refer to the precautions in the beginning of this section.

2. Remove the rear wheels.

   a. Locate and remove the rubber plug from the top of the brake support plate (backing plate).

   b. Insert a small prying tool through the adjuster access hole and engage the teeth on the adjuster wheel.

   c. Rotate the adjuster wheel so it is moved toward the front of the vehicle. This will back off the adjustment of the rear brake shoes.

   d. Continue moving the adjuster wheel toward the front of the vehicle until it stops moving.

3. Remove the rear brake drum from the hub assembly.

To install:

4. Inspect the brake drums for cracks or signs of overheating. Measure the drum runout and diameter. If not to specification, resurface the drum. Runout should not exceed 0.006 inch (0.152mm). The diameter variation (oval shape) of the drum braking surface must not exceed either 0.0025 inch (0.0635mm) in 30° or 0.0035 inch (0.089mm) in 360°. All brake drums are marked with the maximum allowable brake drum diameter on the face of the drum.

5. Install the rear brake drum onto the hub assembly.

6. Reinstall the wheels and properly adjust the brakes.

7. Road test vehicle to check brake operation.

Brake Shoes

REMOVAL & INSTALLATION

1. Before servicing the vehicle, refer to the precautions in the beginning of this section.

2. Remove the rear wheels.

3. Remove the drums.

4. Remove the automatic adjuster spring and lever.

5. Remove the hold-down clips and pins.

6. Rotate the automatic adjuster star-wheel enough so both shoes move out far enough to be free of the wheel cylinder boots.

7. Disconnect the parking brake cable from the actuating lever.

8. Remove the lower shoe to shoe spring.

9. With the shoes held together by the upper shoe to shoe spring, remove them from the backing plate.

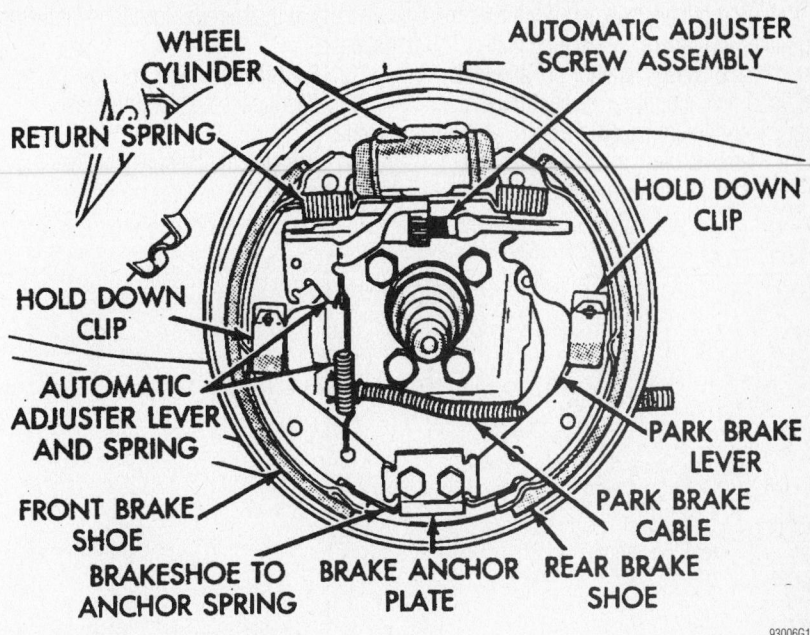

Kelsey Hayes rear brake assembly (left side shown)

93006G13

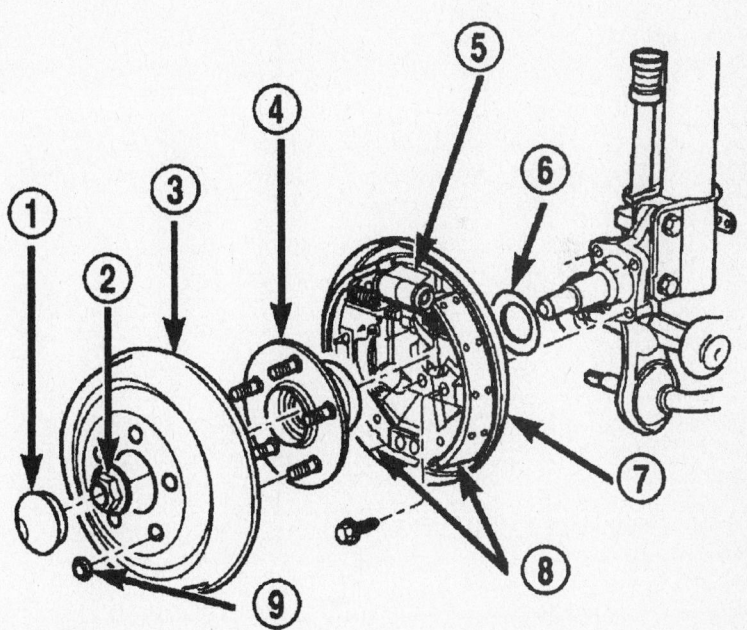

1 - DUST CAP
2 - NUT
3 - DRUM
4 - HUB AND BEARING
5 - WHEEL CYLINDER
6 - SEAL
7 - SUPPORT PLATE
8 - BRAKE SHOES
9 - RETAINER CLIP

67189-NEON-G48

Exploded view of the drum brake assembly—2004 model shown

To install:

10. Thoroughly clean and dry the backing plate. To prepare the backing plate, lubricate the bosses, anchor pin and parking brake actuating lever pivot surface lightly with lithium based grease.

11. Remove, clean and dry all brake components. Lubricate the starwheel shaft threads with anti-seize lubricant and transfer all parts to their proper locations on the new shoes.

12. Reinstall the lower spring.

13. Reconnect the parking brake cable.

14. Reinstall the automatic adjuster lever and spring.

15. Adjust the starwheel.

16. Remove any grease from the linings and install the drum.

17. Reinstall the wheels and properly adjust the brakes.

18. Check for proper brake system operation.

CHRYSLER

Pacifica

9

SPECIFICATION CHARTS

ENGINE AND VEHICLE IDENTIFICATION

Engine								Model Year	
Code ①	Liters (cc)	Cu. In.	Cyl.	Fuel Sys.	Engine Type	Eng. Mfg.		Code ②	Year
4	3.5 (3518)	215	V6	MFI	SOHC	Chrysler		1	2001

MFI: Multi-point Fuel Injection

SOHC: Single Overhead Camshaft

① 8th position of the Vehicle Identification Number (VIN)

② 10th position of VIN

Code ②	Year
2	2002
3	2003
4	2004
5	2005

67189-PACI-C01

GENERAL ENGINE SPECIFICATIONS
All measurements are given in inches.

Year	Model	Engine Displacement Liters (VIN)	Net Horsepower @ rpm	Net Torque @ rpm (ft. lbs.)	Bore x Stroke (in.)	Compression Ratio	Oil Pressure @ rpm
2003	Pacifica	3.5 (4)	250@6400	255@3950	3.78x3.18	10:01	45-105@3000
2004-05	Pacifica	3.5 (4)	250@6400	255@3950	3.78x3.18	10:01	45-105@3000

MFI: Multi-point Fuel Injection

67189-PACI-C02

ENGINE TUNE-UP SPECIFICATIONS

Year	Engine Displacement Liters (VIN)	Spark Plug Gap (in.)	Ignition Timing (deg.)	Fuel Pump (psi)	Idle Speed (rpm)	Valve Clearance	
						Intake	Exhaust
2003	3.5 (4)	0.048-0.053	①	58	①	HYD	HYD
2004-05	3.5 (4)	0.048-0.053	①	58	①	HYD	HYD

NOTE: The Vehicle Emission Control Information label often reflects specification changes made during production. The label figures must be used if they differ from those in this chart.

HYD: Hydraulic

① Controlled by the Powertrain Control Module (PCM)

67189-PACI-C03

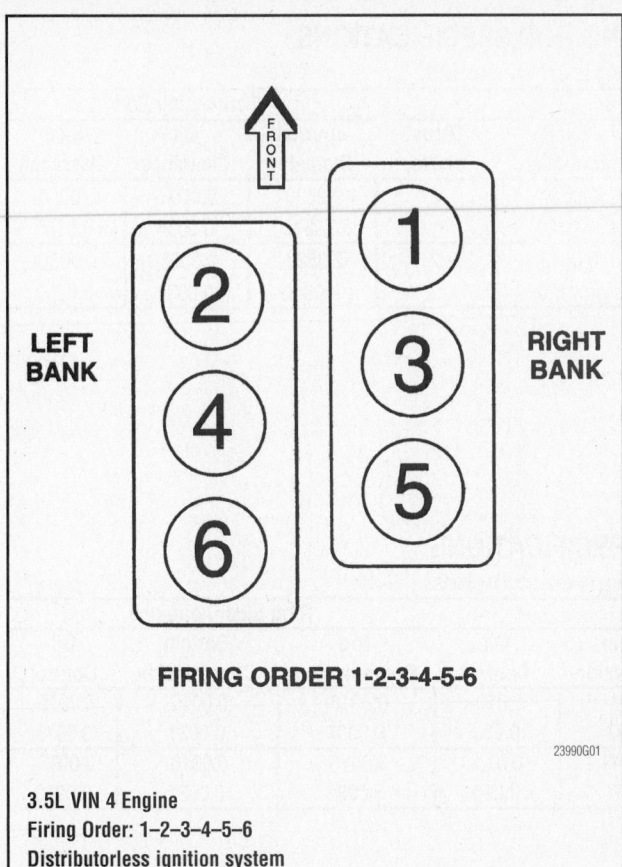

LEFT BANK

RIGHT BANK

FIRING ORDER 1-2-3-4-5-6

23990G01

3.5L VIN 4 Engine
Firing Order: 1–2–3–4–5–6
Distributorless ignition system

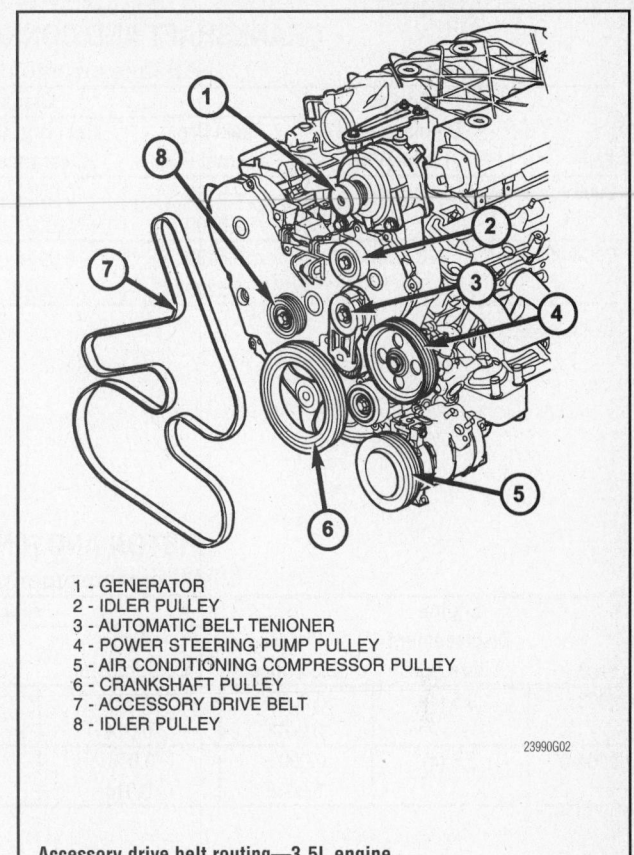

1 - GENERATOR
2 - IDLER PULLEY
3 - AUTOMATIC BELT TENIONER
4 - POWER STEERING PUMP PULLEY
5 - AIR CONDITIONING COMPRESSOR PULLEY
6 - CRANKSHAFT PULLEY
7 - ACCESSORY DRIVE BELT
8 - IDLER PULLEY

23990G02

Accessory drive belt routing—3.5L engine

CAPACITIES

Year	Model	Engine Displacement Liters (VIN)	Engine Oil with Filter (qts.)	Auto. Transmission (pts.) ①	Front Drive Axle (pts.)	Fuel Tank (gal.)	Cooling System (qts.)
2003	Pacifica	3.5 (4)	5.5	19.8	2.0	23.0	10.5
2004-05	Pacifica	3.5 (4)	5.5	19.8	2.0	23.0	10.5

NOTE: All capacities are approximate. Add fluid gradually and ensure a proper fluid level is obtained.

① Overhaul fill capacity with torque converter empty
  Estimated service fill: 8 pts.

67189-PACI-C04

VALVE SPECIFICATIONS

Year	Engine Displacement Liters (VIN)	Seat Angle (deg.)	Face Angle (deg.)	Spring Test Pressure (lbs. @ in.)	Spring Installed Height (in.)	Stem-to-Guide Clearance (in.)		Stem Diameter (in.)	
						Intake	Exhaust	Intake	Exhaust
2003	3.5 (4)	45-45.5	44.5-45	①	1.496	0.0009-0.0026	0.0020-0.0037	0.2730-0.2737	0.2719-0.2726
2004-05	3.5 (4)	45-45.5	44.5-45	①	1.496	0.0009-0.0026	0.0020-0.0037	0.2730-0.2737	0.2719-0.2726

① Intake: 69.5-80.5 lbs. @ 1.496 in. valve closed
  Intake: 188.0-204.0 lbs. @ 1.1594 in. valve opened
  Exhaust: 71-79 lbs. @ 1.4961 in. valve closed
  Exhaust: 130-144 lbs. @ 1.239 in. valve opened

67189-PACI-C05

CRANKSHAFT AND CONNECTING ROD SPECIFICATIONS
All measurements are given in inches.

| Year | Engine Displacement Liters (VIN) | Crankshaft | | | | Connecting Rod | | |
		Main Brg. Journal Dia.	Main Brg. Oil Clearance	Shaft End-play	Thrust on No.	Journal Diameter	Oil Clearance	Side Clearance
2003	3.5 (4)	2.5190 2.5200	0.0004- 0.0022	0.0040- 0.0120	2	2.2830- 2.2840	0.0008- 0.0034	0.0050- 0.0150
2004-05	3.5 (4)	2.5190 2.5200	0.0004- 0.0022	0.0040- 0.0120	2	2.2830- 2.2840	0.0008- 0.0034	0.0050- 0.0150

Max: Maximum

67189-PACI-C06

PISTON AND RING SPECIFICATIONS
All measurements are given in inches.

| Year | Engine Displacement Liters (VIN) | Piston Clearance | Ring Gap | | | Ring Side Clearance | | |
			Top Compression	Bottom Compression	Oil Control	Top Compression	Bottom Compression	Oil Control
2003	3.5 (4)	0.0003- 0.0018	0.008- 0.014	0.0091- 0.0197	0.010- 0.030	0.0016- 0.0031	0.0016- 0.0031	0.0015- 0.0073
2004-05	3.5 (4)	0.0003- 0.0018	0.008- 0.014	0.0091- 0.0197	0.010- 0.030	0.0016- 0.0031	0.0016- 0.0031	0.0015- 0.0073

67189-PACI-C07

TORQUE SPECIFICATIONS
All readings in ft. lbs.

| Year | Engine Displacement Liters (VIN) | Cylinder Head Bolts | Main Bearing Bolts | Rod Bearing Bolts | Crankshaft Damper Bolts | Flywheel Bolts | Manifold | | Spark Plugs | Oil Pan Drain Plug |
							Intake	Exhaust		
2003	3.5 (4)	①	②	③	70	70	④	⑤	20	20
2004-05	3.5 (4)	①	②	③	70	70	④	⑤	20	20

① Step 1: 45 ft. lbs.
　Step 2: 65 ft. lbs.
　Step 3: 65 ft. lbs.
　Step 4: Plus 1/4 turn
　Final torque should be over 90 ft. lbs.

② Main cap inside bolts: 15 ft. lbs. plus 1/4 turn
　Main cap outside bolts: 20 ft. lbs. plus 1/4 turn
　Main cap tie bolts: 250 inch lbs.

③ Step 1: 20 ft. lbs.
　Step 2: Plus 1/4 turn

④ Upper 105 inch lbs.
　Lower 250 inch lbs.

⑤ 200 inch lbs.

67189-PACI-C08

WHEEL ALIGNMENT

Year	Model		Caster Range (+/-Deg.)	Caster Preferred Setting (Deg.)	Camber Range (+/-Deg.)	Camber Preferred Setting (Deg.)	Toe-in (in.)
2003	Pacifica	F	+4.00 to + 5.00	+ 4.50	-0.60 to +0.20	-0.20	0.10 +/- 0.20
		R	—	—	-0.90 to +0.10	-0.40	0.10 +/- 0.30
2004-05	Pacifica	F	+4.00 to + 5.00	+ 4.50	-0.60 to +0.20	-0.20	0.10 +/- 0.20
		R	—	—	-0.90 to +0.10	-0.40	0.10 +/- 0.30

67189-PACI-C09

TIRE, WHEEL AND BALL JOINT SPECIFICATIONS

Year	Model	OEM Tires Standard	OEM Tires Optional	Tire Pressures (psi) Front	Tire Pressures (psi) Rear	Wheel Size	Ball Joint Inspection	Lug Nuts
2003	Pacifica	P235/65/R17	None	NA	NA	7-JJ	①	100
2004-05	Pacifica	P235/65/R17	None	NA	NA	7-JJ	①	100

OEM: Original Equipment Manufacturer

PSI: Pounds Per Square Inch

NA: Not Available

① Do not lift car. Grasp the grease fitting and attempt to move or rotate. Replace if any movement is found.

67189-PACI-C10

BRAKE SPECIFICATIONS
All measurements in inches unless noted

Year	Model		Brake Disc Original Thickness	Brake Disc Minimum Thickness	Brake Disc Maximum Runout	Brake Drum Diameter Original Inside Diameter	Brake Drum Diameter Max. Wear Limit	Brake Drum Diameter Maximum Machine Diameter	Minimum Lining Thickness	Brake Caliper Mounting Bolts (ft. lbs.)
2003	Pacifica	F	1.107	1.040	0.0014	—	—	—	0.250	32
		R	0.556	0.492	0.0055	—	—	—	0.280	17
2004-05	Pacifica	F	1.107	1.040	0.0014	—	—	—	0.250	32
		R	0.556	0.492	0.0055	—	—	—	0.280	17

F: Front

R: Rear

67189-PACI-C11

SCHEDULED MAINTENANCE INTERVALS
Chrysler—Pacifica

TO BE SERVICED	TYPE OF SERVICE	VEHICLE MILEAGE INTERVAL (x1000)												
		7.5	15	22.5	30	37.5	45	52.5	60	67.5	75	82.5	90	97.5
Engine oil & filter	R	✓	✓	✓	✓	✓	✓	✓	✓	✓	✓	✓	✓	✓
Exhaust system	S/I	✓	✓	✓	✓	✓	✓	✓	✓	✓	✓	✓	✓	✓
Brake hoses	S/I	✓	✓	✓	✓	✓	✓	✓	✓	✓	✓	✓	✓	✓
CV joints & front suspension components	S/I	✓	✓	✓	✓	✓	✓	✓	✓	✓	✓	✓	✓	✓
Rotate tires	S/I	✓	✓	✓	✓	✓	✓	✓	✓	✓	✓	✓	✓	✓
Coolant level, hoses & clamps	S/I	✓	✓	✓	✓	✓	✓	✓	✓	✓	✓	✓	✓	✓
Accessory drive belts	S/I		✓		✓		✓		✓		✓		✓	
Brake linings	S/I		✓	✓				✓	✓				✓	
Spark plugs	R				✓				✓				✓	
Air filter element	R				✓				✓				✓	
Lubricate steering linkage & tie rod ends	S/I				✓				✓				✓	
Engine coolant	R						✓				✓			
PCV valve	S/I								✓				✓	
Ignition cables	R								✓					
Camshaft timing belt	R								✓					

R: Replace   S/I: Service or Inspect

FREQUENT OPERATION MAINTENANCE (SEVERE SERVICE)

If a vehicle is operated under any of the following conditions it is considered severe service:

- Extremely dusty areas

- 50% or more of the vehicle operation is in 32°C (90°F) or higher temperatures, or constant operation in temperatures below 0°C (32°F)

Prolonged idling (vehicle operation in stop and go traffic)

- Frequent short running periods (engine does not warm to normal operating temperatures)

- Police, taxi, delivery usage or trailer towing usage

CV joints & front suspension components: check every 3000 miles

Oil & oil filter change: change every 3000 miles

Rotate tires: every 3000 miles

Brake linings: check every 9000 miles

Air filter element: change every 15,000 miles

Automatic transaxle fluid: change every 15,000 miles

Differential fluid: change every 15,000 miles

Tie rod ends & steering linkage: lubricate every 15,000 miles

PCV valve: check every 30,000 miles

67189-PACI-C12

PRECAUTIONS

Before servicing any vehicle, please be sure to read all of the following precautions, which deal with personal safety, prevention of component damage, and important points to take into consideration when servicing a motor vehicle:

• Never open, service or drain the radiator or cooling system when the engine is hot; serious burns can occur from the steam and hot coolant.

• Observe all applicable safety precautions when working around fuel. Whenever servicing the fuel system, always work in a well-ventilated area. Do not allow fuel spray or vapors to come in contact with a spark, open flame or excessive heat (a hot drop light, for example). Keep a dry chemical fire extinguisher near the work area. Always keep fuel in a container specifically designed for fuel storage; also, always properly seal fuel containers to avoid the possibility of fire or explosion. Refer to the additional fuel system precautions later in this section.

• Fuel injection systems often remain pressurized, even after the engine has been turned **OFF**. The fuel system pressure must be relieved before disconnecting any fuel lines. Failure to do so may result in fire and/or personal injury.

• Brake fluid often contains polyglycol ethers and polyglycols. Avoid contact with the eyes and wash your hands thoroughly after handling brake fluid. If you do get brake fluid in your eyes, flush your eyes with clean, running water for 15 minutes. If eye irritation persists, or if you have taken brake fluid internally, IMMEDIATELY seek medical assistance.

• The EPA warns that prolonged contact with used engine oil may cause a number of skin disorders, including cancer! You should make every effort to minimize your exposure to used engine oil. Protective gloves should be worn when changing oil. Wash your hands and any other exposed skin areas as soon as possible after exposure to used engine oil. Soap and water, or waterless hand cleaner should be used.

• All new vehicles are now equipped with an air bag system. The system must be disabled before performing service on or around system components, steering column, instrument panel components, wiring and sensors. Failure to follow safety and disabling procedures could result in accidental air bag deployment, possible personal injury and unnecessary system repairs.

• Always wear safety goggles when working with, or around, the air bag system. When carrying a non-deployed air bag, be sure the bag and trim cover are pointed away from your body. When placing a non-deployed air bag on a work surface, always face the bag and trim cover upward, away from the surface. This will reduce the motion of the module if it is accidentally deployed. Refer to the additional air bag system precautions later in this section.

• Clean, high quality brake fluid from a sealed container is essential to the safe and proper operation of the brake system. You should always buy the correct type of brake fluid for your vehicle. If the brake fluid becomes contaminated, completely flush the system with new fluid. Never reuse any brake fluid. Any brake fluid that is removed from the system should be discarded. Also, do not allow any brake fluid to come in contact with a painted surface; it will damage the paint.

• Never operate the engine without the proper amount and type of engine oil; doing so WILL result in severe engine damage.

• Timing belt maintenance is extremely important! Many models utilize an interference-type, non-freewheeling engine. If the timing belt breaks, the valves in the cylinder head may strike the pistons, causing potentially serious (also time-consuming and expensive) engine damage.

• Disconnecting the negative battery cable on some vehicles may interfere with the functions of the on-board computer system(s) and may require the computer to undergo a relearning process once the negative battery cable is reconnected.

• When servicing drum brakes, only disassemble and assemble one side at a time, leaving the remaining side intact for reference.

• Only an MVAC-trained, EPA-certified automotive technician should service the air conditioning system or its components.

ENGINE REPAIR

Alternator

REMOVAL & INSTALLATION

1. Remove or disconnect the following:
   • Negative battery cable
   • Engine cover
   • Drive belt
   • Alternator B+ terminal nut and wire
   • Alternator field circuit wiring by pushing the red locking tab
   • Upper mounting bracket
   • Lower mounting bolts and the alternator

To install:
2. Install or connect the following:
   • Alternator and lower mounting bolts, tighten to 40 ft. lbs. (54 Nm)
   • Upper mounting bracket and tighten the bolts 20 ft. lbs. (28 Nm)

• Alternator field circuit wiring making sure to engage the red locking tab
• Alternator B+ terminal nut and wire, tighten the nut to 110 inch lbs. (12 Nm)
• Drive belt
• Engine cover
• Negative battery cable

Ignition Timing

ADJUSTMENT

This model utilize a Distributorless Ignition System (DIS). It is a fixed ignition timing system, which means that basic ignition timing cannot be adjusted. All spark advance is permanently set by the Powertrain Control Module (PCM).

Engine Assembly

REMOVAL & INSTALLATION

1. Before servicing the vehicle, refer to the precautions in the beginning of this section.
2. Drain the engine oil.
3. Drain the engine coolant.
4. Properly relieve the fuel system pressure.
5. Remove or disconnect the following:
   • Negative battery cable
   • Hood
   • Cruise control servo
   • Radiator closure panel
   • Radiator core support
   • Upper radiator hose
   • Cooling fan
   • Air cleaner assembly

6. Recover the A/C system refrigerant using approved recycling equipment.

- Throttle and cruise control cable from the throttle body
- Cruise control and power brake booster vacuum hoses from the engine
- Transmission wiring harness from the solenoid pack, input and output sensors and the range sensor connectors
- Engine wiring harness grounds from the inner frame rail
- Transmission shift cable
- Transmission cooler lines
- Engine block heater connector, if equipped
- Coolant reservoir houses from the thermostat housing
- Heater hoses from the heater core
- Upper radiator hoses from the thermostat housing
- Brake lines from the Hydraulic Control Unit (HCU)
- A/C suction discharge hoses from the compressor and plug the openings
- Lower radiator hose from the engine outlet
- A/C clutch connection
- Oil pressure sender and alternator connections
- Fuel supply line from the fuel rail
- Ignition harness connection at the intake manifold
- Front wheel
- Left inner fender well
- Engine harness from the Powertrain Control Module (PCM)
- Left wheel speed sensor and retainer
- Left sway bar link from the strut
- Left drive axle nut
- Both left front steering knuckle pinch bolts
- Left steering knuckle from the strut and support the knuckle
- Negative battery cable from the transmission
- Left fascia screws
- Right inner fender well
- Right fascia screws
- Right wheel speed sensor and retainer
- Right sway bar link from the strut
- Right drive axle nut
- Both right front steering knuckle pinch bolts
- Right steering knuckle from the strut and support the knuckle
- Fascia driving lamps, if equipped
- Front fascia

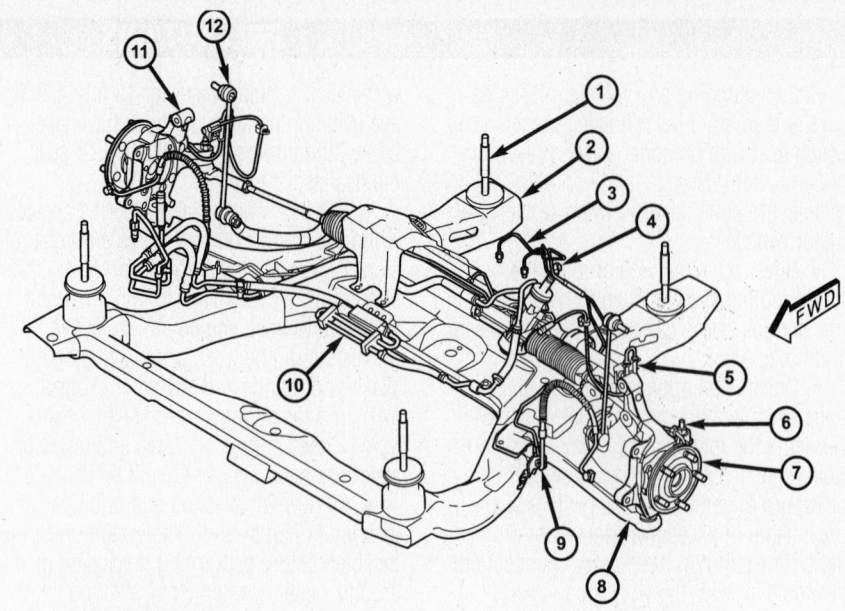

1 - FRONT CRADLE BOLT
2 - FRONT CRADLE ASSEMBLY
3 - FRONT BRAKE LINES
4 - STEERING GEAR
5 - WHEEL SPEED SENSOR HARNESS
.6 - TIE ROD END
7 - WHEEL HUB
8 - LOWER CONTROL ARM
9 - FRONT BRAKE HOSE
10 - POWER STEERING COOLER
11 - STEERING KNUCKLE
12 - STABILIZER LINK

23990G03

Exploded view of the front cradle assembly

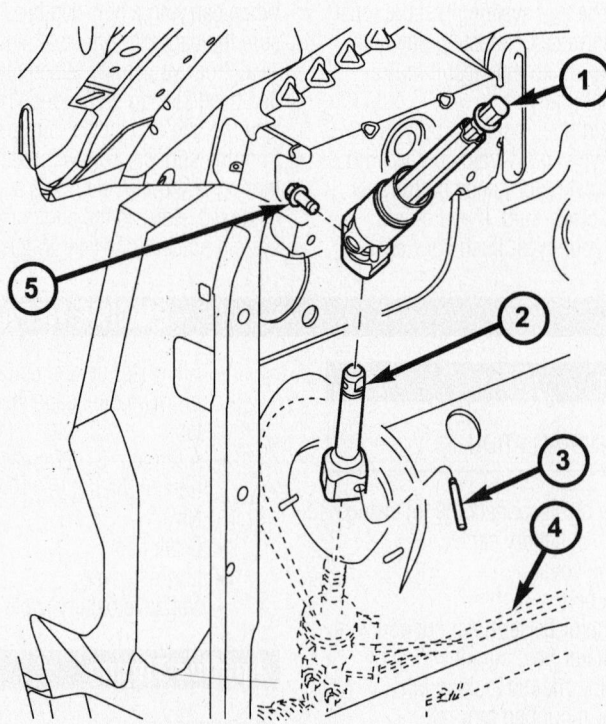

1 - LOWER STEERING COLUMN COUPLING ASSEMBLY
2 - INTERMEDIATE SHAFT TO STEERING GEAR
3 - PIN
4 - STEERING GEAR
5 - EXTENSION TO SHAFT BOLT

23990G04

Lower steering column coupling assembly

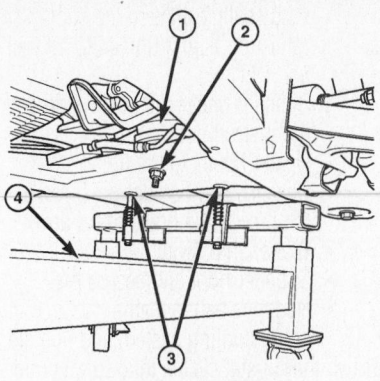

1 - RIGHT LOWER ENGINE MOUNT
2 - RIGHT LOWER ENGINE MOUNT NUT
3 - ALIGNMENT DOWEL ACCESS HOLES
4 - CRADLE SUPPORT FIXTURE

23990G05

Rear cradle support fixture mounting

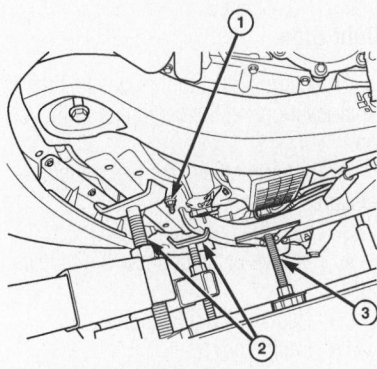

1 - LEFT LOWER ENGINE MOUNT NUT
2 - CRADLE SUPPORT FIXTURE ARMS
3 - CRADLE SUPPORT FIXTURE ENGINE STABILIZER

23990G06

Adjust the cradle support fixture to fit flush with the cradle

- Starter connectors and harness retainers
- Oxygen (O2S) sensor electrical wiring at the exhaust manifold and catalytic converter
- Exhaust from the manifold
- Driveshaft after marking the position at the front and rear of the shaft. This will help with installation and retain driveshaft balance.
- Engine-to-transmission plate
- Transmission inspection plate
- Flex plate-to-torque converter bolt

✳✳ WARNING

Secure the steering wheel to prevent it rotating which may cause damage to the steering column clock spring.

7. Remove or disconnect the following:
- Lower steering column coupling pin using tool 6831-A and separate the union
- Both lower engine mount bolts

8. Matchmark using paint the front cradle-to-body location and position the engine cradle support under the vehicle.

9. Remove or disconnect the following:

10. Lower the vehicle until it is just above the cradle

11. Align the cradle support dowels with cradle access holes and adjust the engine support fixture to fit flush with the oil pan and adjust the cradle support fixture support to fit flush with the cradle.

12. Carefully lower the vehicle onto the cradle.

13. Remove the upper engine mount

14. Very **carefully** remove the front and rear cradle mounting bolts and raise the vehicle to separate the engine/transmission and cradle from the vehicle.

15. Connect lifting brackets to the engine, separate the Power Transfer Unit (PTU), if equipped.

16. Separate the engine from transmission using the lift brackets.

To install:

17. Installation is the reverse of removal, please note the following important specifications.

18. Remove or disconnect the following:
- Cradle supporting bolts: 120 ft. lbs. (163 Nm)
- Upper engine mount bolt to timing cover to 40 ft. lbs. (54 Nm) and the bolt to right rail to 50 ft. lbs. (68 Nm)
- Drive axle-to-steering knuckle nut to 180 ft. lbs. (244 Nm)
- Steering knuckle-to-strut bolts to 180 ft. lbs. (244 Nm)
- Front speed sensor support bracket bolts to 105 inch lbs. (12 Nm)
- Brake line support bracket bolts to 105 inch lbs. (12 Nm)

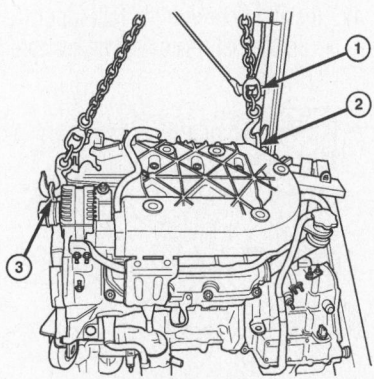

1 - SUITABLE LIFTING CHAIN
2 - 8537-7 PART OF KIT 8537-12
3 - 8537-15 PART OF KIT 8537-12

23990G07

Use suitable lift brackets and a hoist assembly to separate the engine from the transaxle

- Flex plate-to-torque converter bolts to 55 ft. lbs. (75 Nm)
- Engine-to-transmission support collar bolts to 40 ft. lbs. (54 Nm)

19. Fill the cooling system to the proper level.

20. Fill the engine with clean oil.

21. Bleed the brakes.

22. Recharge the A/C system

23. Check wheel alignment.

24. Start the vehicle, check for leaks and repair if necessary.

Water Pump

REMOVAL & INSTALLATION

The water pump has a die cast aluminum body and a stamped steel impeller. It bolts directly to the chain case cover using an O-ring for sealing. It is driven by the back side of the serpentine belt.

It is normal for a small amount of coolant to drip from the weep hole located on the water pump body (small black spot). If this condition exists, DO NOT replace the water pump. Only replace the water pump if a heavy deposit or steady flow of brown/green coolant is visible on the water pump body from the weep hole, which would indicate shaft seal failure. Before replacing the water pump, be sure to perform a thorough inspection. A defective pump will not be able to circulate heated coolant through the long heater hose.

1. Before servicing the vehicle, refer to the precautions in the beginning of this section.

2. Drain the cooling system.

✳✳ WARNING

Do not use pliers to open the plastic drain.

3. Remove or disconnect the following:
- Negative battery cable
- Coolant recovery cap and open the thermostat bleed valve
- Timing belt

➡**It is good practice to turn the crankshaft until the No. 1 cylinder is at Top Dead Center (TDC) of its compression stroke (firing position).**

- Water pump mounting bolts and pump. Discard the O-ring seal.

4. Clean the gasket sealing surfaces, being careful not to scratch the aluminum surfaces.

To install:

5. Install or connect the following:
- New O-ring and coat with dielectric grease prior to installation

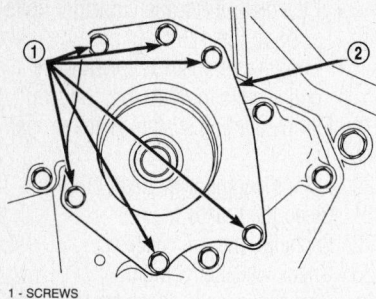

1 - SCREWS
2 - WATER PUMP BODY

23990G62

Location of the water pump mounting bolts

- Water pump with a new O-ring. Torque the water pump-to-engine bolts to 105 inch lbs. (12 Nm).

✳✳ WARNING

Rotate the pump and check for freedom of movement.

- Timing belt
6. Fill the cooling system.
7. Reconnect the negative battery cable. Start the engine and allow it to reach normal operating temperatures.
8. Check the cooling system for leaks and correct coolant level.

Heater Core

REMOAVL & INSTALLATION

1. Before servicing the vehicle, refer to the precautions in the beginning of this section.
2. Disable the airbag system, by disconnecting the negative battery cable. Tape the end of the cable to avoid it grounding and wait 2 minutes for the system capacitor to discharge before performing any service.

➡**If the vehicle is adjustable pedals, these must be removed prior to heater core removal.**

3. Drain the cooling system.

✳✳ WARNING

Do not use pliers to open the plastic drain.

4. Remove or disconnect the following:
- Negative battery cable if not already done to disable the air bag system
- Silencer boot fasteners located around the base of the lower steering shaft from the dash panel so it can be moved aside
- Brake lamp switch from the bracket

- Power brake booster push rod from the pin on the brake pedal arm
5. Place plastic below the heater core inside the vehicle to protect against spills.
6. Remove or disconnect the following:
- Screw that secures the heater core to tube sealing plate to the supply and return ports
- Heater core tubes by pushing simultaneously towards the dash panel and disengaging the fittings and cap the tube fittings and heater core ports
- Two screws that retain the heater core mounting plate to the distribution housing
7. Push the brake pedal downward and at the same time pull the accelerator pedal up to create enough clearance to remove the heater core.

To install:

8. Push the brake pedal downward and at the same time pull the accelerator pedal up to create enough clearance to install the heater core.
- Two screws that retain the heater core mounting plate to the distribution housing and tighten to 17 inch lbs. (2 Nm)
9. Remove the plugs from the heater core tubes and if using the same heater core the core as well. Position the heater core tubes and sealer plate as a unit beneath the instrument panel.
10. Position both heater core tubes and the sealing plate at the same time to the heater core supply and return ports.
11. The tubes have a slot that must be indexed to a location tab within each of the core ports. Adjust the tube position so that the sealing plate fits flush against the supply and return ports which will ensure the tubes are correctly indexed.
12. Install or connect the following:
- Screw that retains the heater core

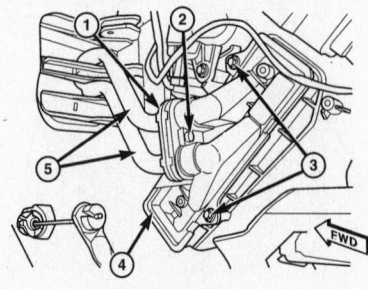

1 - SEALING PLATE
2 - SCREW
3 - SCREWS (2)
4 - HEATER CORE
5 - HEATER CORE TUBES

23990G08

Heater core mounting and related components

tube sealing plate to the supply and return ports and tighten to 27 inch lbs. (3 Nm).
- Silencer under the driver side end if the instrument panel
- Heater hoses to the tubes
- Power brake booster push rod to the pin on the brake pedal arm
- Brake lamp switch
- Silencer boot and secure the
- Negative battery cable
13. Fill the cooling system and operate for two thermostat cycles to assure elimination of air in the cooling system.

Cylinder Head

REMOVAL & INSTALLATION

Right Side

1. Before servicing the vehicle, refer to the precautions in the beginning of this section.
2. Properly relieve the fuel system pressure.
3. Drain the cooling system.
4. Remove or disconnect the following:
- Negative battery cable
- Engine cover
- Air cleaner assembly
- Fuel line from the rail
- Upper intake manifold
- Lower intake manifold
- Right front wheel
- Right inner splash shield
- Drive belt
- Crankshaft damper using the proper puller
- Lower drive belt idler pulley
- Power steering bolts and set the pump aside
- Lower timing belt cover bolts
- Catalytic converter nuts
- Oxygen (O_2S) sensor electrical wiring at the exhaust manifold and catalytic converter
- Muffler-to-tail pipe union
- Catalytic converter
- Exhaust cross over pipe lower bolts
- Right exhaust manifold
- Upper drive belt idler pulley
- Belt tensioner
5. Support the engine with jackstands and wood blocks.
- Upper engine mount
- Power steering reservoir bolts and set aside
- All remaining outer timing belt cover bolts

6. Rotate the engine until it is at Top Dead center (TDC).

➡ **Mark the timing belt running direction for installation. Align the camshaft sprockets with the marks on the rear covers.**

- Timing belt tensioner and reset the tensioner. Refer to the timing belt removal and installation procedure in this section.

7. Pre-load the timing belt tensioner as follows:

a. Place tensioner in a vise the same way it is mounted on the engine.

b. Slowly compress the plunger into the tensioner body.

c. Once the plunger is compressed, install a pin through the body and plunger to retain it in place until the tensioner is installed.

- Timing belt. Refer to the timing belt removal and installation procedure in this section.
- Right vale cover
- Exhaust Gas Recirculation (EGR) and tube assembly
- Right cylinder head cover
- Right rocker arm assembly
- Right rear camshaft thrust plate
- Holding the camshaft gear, unfasten the right cam gear retaining bolt

8. Push the camshaft about 3.5 inches out of the back of the cylinder head and remove the cam gear.

- Inner timing cover-to-cylinder head bolts
- Cylinder head bolts in the reverse of tightening sequence. Refer to the cylinder head bolt tightening sequence illustration.
- Cylinder head and gasket.

To install:

9. Thoroughly clean and dry the mating surfaces of the head and block.

10. Check the cylinder head for cracks, damage or engine coolant leakage. Check the head for flatness. End-to-end, the head should be within 0.002 in. (0.051mm) normally with 0.008 in. (0.203mm) the maximum allowed out of true. The resurface limit is 0.008 in. (0.203mm) maximum, the com-

bined total dimension of stock removal from the cylinder head, if any, and block top surface.

11. Place a new head gasket on the cylinder block locating dowels, being sure the gasket is on the correct side.

12. Inspect the cylinder head bolts for necking (stretching) by holding a straightedge against the threads of each bolt. If all of the threads are not contacting the scale, the bolt should be replaced.

13. Install the cylinder head into position on the engine block and over the dowels. Install the cylinder head bolts, lubricating the threads with clean engine oil prior to installation.

14. Torque the cylinder head bolts using the proper sequence as follows:

a. Step 1: torque in sequence to 45 ft. lbs. (61 Nm).

b. Step 2: torque in sequence to 65 ft. lbs. (88 Nm).

c. Step 3: torque in sequence to 65 ft. lbs. (88 Nm).

d. Step 4: torque in sequence an additional ¼ turn by using a torque angle meter.

➡ **Inspect the bolt torque after tightening. The torque should be over 90 ft. lbs. (122 Nm). If not, replace the cylinder head bolt.**

15. Install or connect the following:

- Inner timing cover-to-cylinder head bolts and tighten to 40 ft. lbs. (54 Nm)
- Cam gear, holding the camshaft gear, fasten the right cam gear retaining bolt to 75 ft. lbs. (102 Nm) plus an additional ¼ turn
- Rear camshaft thrust plate

16. Rotate the camshaft gear until it is properly aligned and check the left camshaft gear and crankshaft timing alignment marks.

- Timing belt. Refer to the timing belt removal and installation procedure in this section.
- Timing belt outer cover
- Power steering reservoir
- Crankshaft damper
- Upper engine mount
- Belt tensioner
- Upper drive belt idler pulley
- Right exhaust manifold
- Exhaust cross over pipe
- Catalytic converter
- Muffler-to-tail pipe union
- O_2S sensor electrical wiring at the exhaust manifold and catalytic converter
- Right rocker arm assembly
- Right cylinder head cover
- EGR and tube assembly
- Lower drive belt idler pulley
- Right vale cover
- Lower intake manifold and fuel rail
- Upper intake manifold
- Fuel line to the rail
- Power steering pump
- Drive belt
- Air cleaner assembly
- Engine cover
- Right inner splash shield
- Right front wheel
- Negative battery cable

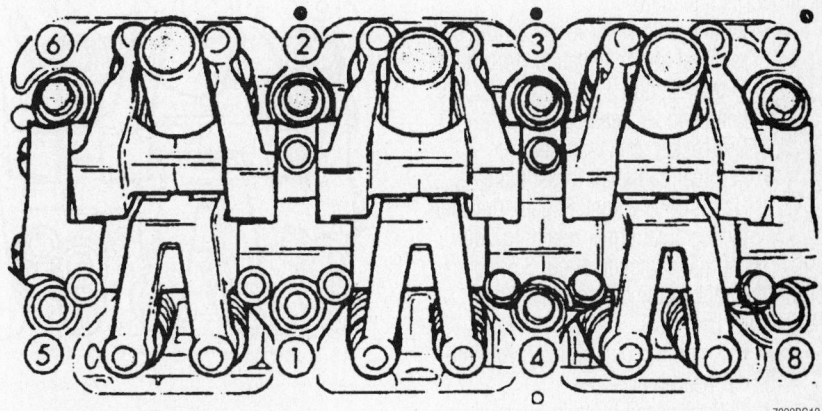

Cylinder head bolt tightening sequence

Left Side

1. Before servicing the vehicle, refer to the precautions in the beginning of this section.

2. Properly relieve the fuel system pressure.

3. Drain the cooling system.

4. Remove or disconnect the following:

- Negative battery cable
- Engine cover
- Air cleaner assembly
- Radiator close out panel
- Radiator core support and fan assembly
- Fuel line from the rail
- Upper intake manifold
- Lower intake manifold
- Crankshaft damper
- Upper engine mount
- Upper drive belt idler pulley
- Belt tensioner
- Left exhaust manifold
- Exhaust cross over pipe
- Outer timing belt cover

5. Rotate the engine until it is at Top Dead center (TDC).

➡**Mark the timing belt running direction for installation. Align the camshaft sprockets with the marks on the rear covers.**

- Timing belt tensioner and reset the tensioner. Refer to the timing belt removal and installation procedure in this section.

6. Pre-load the timing belt tensioner as follows:

a. Place tensioner in a vise the same way it is mounted on the engine.

b. Slowly compress the plunger into the tensioner body.

c. Once the plunger is compressed, install a pin through the body and plunger to retain it in place until the tensioner is installed.

- Timing belt. Refer to the timing belt removal and installation procedure in this section.
- Left cylinder head cover
- Left rocker arm assembly
- Left rear camshaft thrust plate
- Holding the camshaft gear, unfasten the right cam gear retaining bolt

7. Push the camshaft about 3.5 inches out of the back of the cylinder head and remove the cam gear.

- Front timing belt housing-to-cylinder head bolts
- Cylinder head bolts in the reverse of tightening sequence. Refer to the cylinder head bolt tightening sequence illustration.
- Cylinder head and gasket.

To install:

8. Thoroughly clean and dry the mating surfaces of the head and block.

✳✳ WARNING

When cleaning the cylinder head and block mating surfaces, do not use a metal scraper because the soft aluminum surfaces could be cut or damaged. Instead, use a scraper made of wood or plastic.

9. Check the cylinder head for cracks, damage or engine coolant leakage. Check the head for flatness. End-to-end, the head should be within 0.002 in. (0.051mm) normally with 0.008 in. (0.203mm) the maximum allowed out of true. The resurface limit is 0.008 in. (0.203mm) maximum, the combined total dimension of stock removal from the cylinder head, if any, and block top surface.

10. Place a new head gasket on the cylinder block locating dowels, being sure the gasket is on the correct side.

11. Inspect the cylinder head bolts for necking (stretching) by holding a straightedge against the threads of each bolt. If all of the threads are not contacting the scale, the bolt should be replaced.

✳✳ WARNING

Due to the cylinder head bolt torque method used, it is imperative that the threads of the bolts be inspected for necking prior to installation. If the threads are necked down, the bolt should be replaced. Failure to do so may result in parts failure or damage. New bolts are always recommended.

12. Install the cylinder head into position on the engine block and over the dowels. Install the cylinder head bolts, lubricating the threads with clean engine oil prior to installation.

13. Torque the cylinder head bolts using the proper sequence as follows:

a. Step 1: torque in sequence to 45 ft. lbs. (61 Nm).

b. Step 2: torque in sequence to 65 ft. lbs. (88 Nm).

c. Step 3: torque in sequence to 65 ft. lbs. (88 Nm).

d. Step 4: torque in sequence an additional ¼ turn by using a torque angle meter.

➡**Inspect the bolt torque after tightening. The torque should be over 90 ft. lbs. (122 Nm). If not, replace the cylinder head bolt.**

14. Install or connect the following:

- Inner timing cover-to-cylinder head bolts and tighten to 40 ft. lbs. (54 Nm)
- Cam gear, holding the camshaft gear, fasten the right cam gear retaining bolt to 75 ft. lbs. (102 Nm) plus an additional ¼ turn
- Rear camshaft thrust plate

15. Rotate the camshaft gear until it is properly aligned and check the left camshaft gear and crankshaft timing alignment marks.

- Timing belt. Refer to the timing belt removal and installation procedure in this section.
- Timing belt front cover

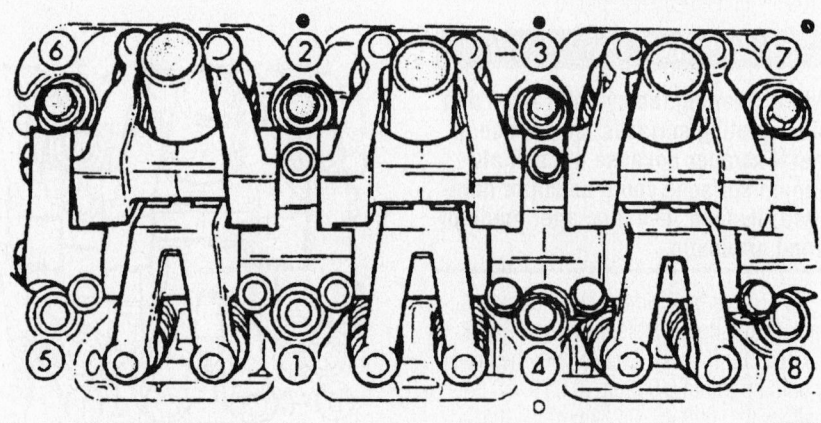

7922BG12

Cylinder head bolt tightening sequence

- Crankshaft damper
- Upper engine mount
- Belt tensioner
- Upper drive belt idler pulley
- Left exhaust manifold
- Exhaust cross over pipe
- Left rocker arm assembly
- Left cylinder head cover
- Lower drive belt idler pulley
- Lower intake manifold and fuel rail
- Upper intake manifold
- Fuel line to the rail
- Radiator core support and fan assembly
- Radiator close out panel
- Drive belt
- Air cleaner assembly
- Engine cover
- Negative battery cable

Rocker Arm/Shafts

REMOVAL & INSTALLATION

1. Before servicing the vehicle, refer to the precautions in the beginning of this section.
2. Relieve the fuel system pressure.
3. Remove or disconnect the following:
   - Negative battery cable
   - Air cleaner assembly and the intake manifold plenum

➡**Cover the lower intake manifold during service.**

   - Cylinder head covers
   - Rocker arm assembly
4. Inspect the rocker arms for wear or damage. Inspect the roller for scuffing or wear. Replace assembly as necessary.

❋❋ WARNING

Do not remove the lash adjusters from the rocker arm assembly. The rocker arm and the adjuster are serviced as an assembly.

5. Identify the rocker arm assemblies and rocker arms and disassemble the shaft as follows:
   a. Thread a nut, washer and spacer onto a 4mm screw.
   b. Insert and tighten the 4mm screw into the dowel pin on the shaft.
   c. Loosen the nut on the screw. This will pull the dowel pin from the shaft support.
   d. Remove the rocker arms and pedestals, keeping them in order.
   e. Check the oil holes for restrictions with a small wire and clean as required.

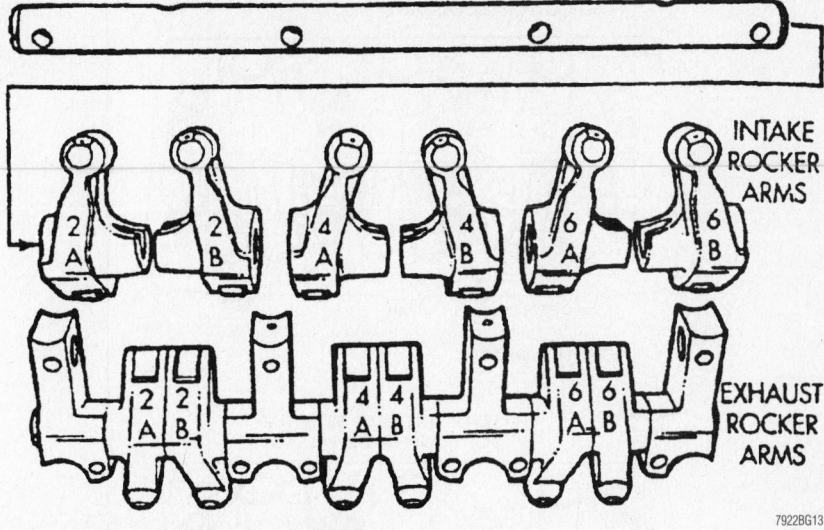

Left bank rocker arm and shaft identification

To install:

6. Assemble the rocker shaft as follows:
   a. Install the rocker arms and pedestals onto the shaft, keeping them in the original order.
   b. Press the dowel pins into the pedestals until they bottom out in the pedestals.
7. Position the camshaft so that the timing mark on the right camshaft timing belt sprocket aligns with the timing mark on the rear timing belt cover and the timing mark on the left sprocket is 45 degrees from the mark on the rear timing belt cover. There will be no load on the shaft during installation. Install the rocker shafts so the identification marks are facing toward the front of the engine.
8. Install the oil feed bolt in the correct location on the rocker shaft retainer. Torque the bolts in proper sequence to 23 ft. lbs. (31 Nm).
9. Install or connect the following:
   - Valve covers and torque the bolts to 105 inch lbs. (12 Nm)
   - Intake manifold plenum
   - Air cleaner assembly
   - Negative battery cable

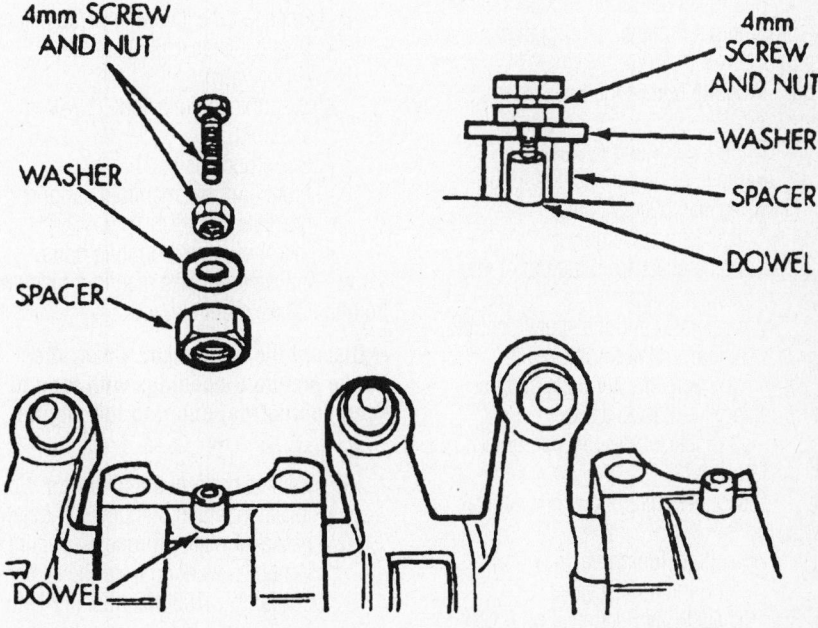

Remove the dowel pin using a 4mm screw, nut, spacer and washer installed into the pin

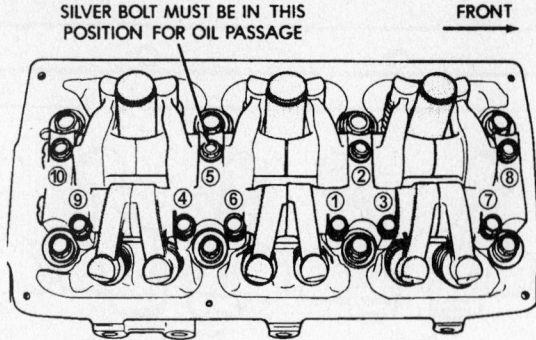

RIGHT SIDE SHOWN

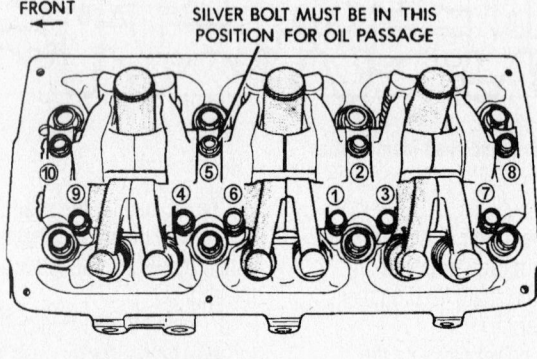

LEFT SIDE SHOWN

7922BG15

Proper torque sequence for the rocker arm and shaft assemblies

Intake Manifold

REMOVAL & INSTALLATION

Upper

1. Before servicing the vehicle, refer to the precautions in the beginning of this section.

2. Properly relieve the fuel system pressure.

3. Drain the cooling system.

4. Remove or disconnect the following:
   • Negative battery cable
   • Air cleaner assembly
   • Engine cover from the top of the intake manifold
   • Accelerator and the speed control cable from the throttle lever
   • Two cable bracket bolts and position the bracket aside
   • Oil dipstick tube-to-upper intake bolt

5. Disconnect the following electrical connections:
   • Manifold Tuning Valve (MTV)
   • Sort Runner (SR) valve
   • Exhaust Gas Recirculation (EGR) valve
   • Throttle Position (TP) sensor

   • Idle Air Control (IAC) motor
   • Intake Air Temperature (IAT) sensor
   • Manifold Absolute Pressure (MAP) sensor

6. Remove or disconnect the following:
   • Alternator support bracket
   • EGR tube

7. Disconnect the following vacuum lines:
   • Positive Crankcase Ventilation (PCV) valve
   • Evaporative Emissions (EVAP) solenoid
   • Brake booster
   • Front and rear manifold support brackets
   • Intake plenum mounting bolts

8. Remove the intake manifold plenum from the intake manifold.

➡ Discard the old gasket. Cover the intake manifold openings with tape to keep debris from entering the engine.

To install:

9. Install or connect the following:
   • Intake manifold plenum with a new gasket in place. Torque the mounting bolts, working from the center outward, to 105 inch lbs. (12 Nm)

➡ Do not overtighten bolts when working with light alloys.

   • Oil dipstick tube-to-upper intake bolt
   • EGR tube

10. Connect the following vacuum lines:
    • PCV valve
    • EVAP solenoid
    • Brake booster

11. Install or connect the following:
    • Alternator support bracket

12. Connect the following electrical connections:
    • MTV valve
    • Sort Runner (SR) valve
    • EGR valve
    • TP sensor
    • IAC motor
    • IAT sensor
    • MAP sensor

13. Install or connect the following:
    • Cable bracket and bolts
    • Accelerator and the speed control cable to the throttle lever
    • Air cleaner assembly
    • Engine cover
    • Negative battery cable

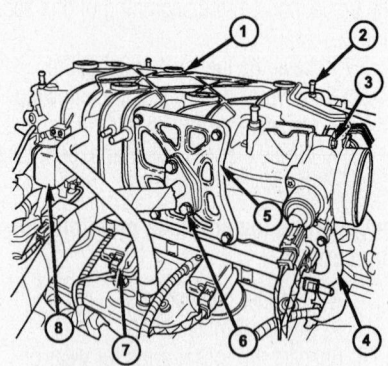

1 - UPPER INTAKE MANIFOLD
2 - ENGINE COVER MOUNTING STUD
3 - THROTTLE BODY
4 - FUEL RAIL
5 - UPPER INTAKE MANIFOLD RETAINING BRACKET
6 - EGR TUBE
7 - IGNITION COIL ASSEMBLY
8 - MANIFOLD TUNER VALVE ACTUATOR

23990G09

Upper intake manifold and related components

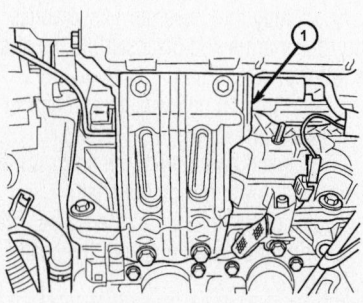

1 - INTAKE MANIFOLD SUPPORT

23990G10

Upper intake manifold support bracket—left shown, right similar

14. Fill and bleed the cooling system.
15. Change the engine oil and filter.
16. Test run the engine, check for fuel and coolant leaks and verify correct engine operation.

Lower

1. Before servicing the vehicle, refer to the precautions in the beginning of this section.
2. Remove or disconnect the following:
   - Upper radiator hose from the thermostat housing
   - Upper intake manifold
   - Power steering reservoir and position aside
   - Fuel injector and Coolant temperature (CTS) sensor electrical connections
   - Exhaust crossover bolt
   - Heater hose from the intake manifold
   - Coolant container hose at the thermostat housing
   - Fuel supply hose from the rail
   - Fuel rail bolts and rail with the injectors as an assembly
   - Lower intake manifold bolts and manifold

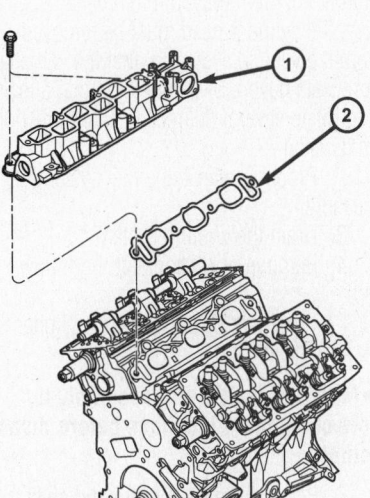

1 - LOWER INTAKE MANIFOLD
2 - GASKET

23990G11

Lower intake manifold and gasket

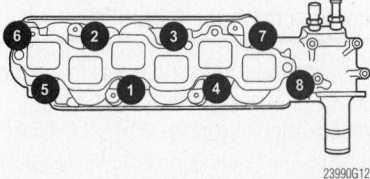

23990G12

Lower intake manifold tightening sequence

➡**Clean all gasket mating surfaces and inspect for distortion with a good straightedge.**

To install:

➡**Verify that all intake manifold and cylinder head sealing surfaces are clean.**

3. Install or connect the following:
   - Intake manifold gasket, then the lower manifold. Torque the bolts in the proper sequence to 21 ft. lbs. (28 Nm).
   - Fuel rail bolts and rail with the injectors as an assembly
   - Fuel supply hose to the rail
   - Heater hose to the intake manifold
   - Coolant container hose to the thermostat housing
   - Exhaust crossover bolt
   - Fuel injector and CTS sensor electrical connections
   - Power steering reservoir
   - Upper intake manifold
   - Upper radiator hose from the thermostat housing
4. Fill the cooling system and check for leaks.

Exhaust Manifold

REMOVAL & INSTALLATION

Left Side

1. Before servicing the vehicle, refer to the precautions in the beginning of this section.
2. Remove or disconnect the following:
   - Negative battery cable
   - Radiator close out panel
   - Radiator core support and cooling fan assembly
   - Oil dipstick tube bolt and move the tube aside
   - Exhaust manifold crossover pipe bolts
   - Exhaust manifold bolts and manifold
3. Inspect the manifold for damage or cracks. Check for distortion against a straight-edge or thickness gauge. Replace manifold if required.
4. Remove all traces of the old manifold gasket and clean both gasket mating surfaces.

To install:
   - Exhaust manifold and gasket, tighten the bolts to 200 inch lbs. (23 Nm)
   - Exhaust manifold crossover pipe

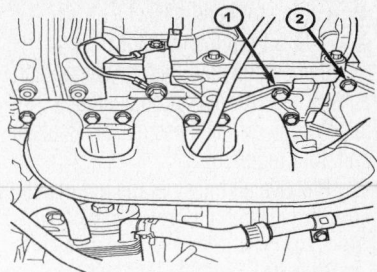

1 - OIL LEVEL INDICATOR TUBE RETAINING BOLT
2 - EXHAUST CROSS OVER PIPE UNION

23990G13

The oil dipstick tube and exhaust crossover pipe bolts must be removed prior to removing the manifold

and tighten the bolts to 275 inch lbs. (31 Nm)
   - Oil dipstick tube and secure the bolt
   - Radiator cooling fan assembly and core support
   - Radiator close out panel
   - Negative battery cable

Right Side

1. Before servicing the vehicle, refer to the precautions in the beginning of this section.
2. Remove or disconnect the following:
   - Negative battery cable
   - Exhaust manifold crossover pipe bolts
   - Downstream Oxygen (O_2S) sensor electrical wiring
   - Exhaust manifold flange bolts and the hangers from the pipe assembly
   - Exhaust system from the vehicle
   - Exhaust Gas Recirculation (EGR) tube support bracket bolt
   - Lower manifold bolts

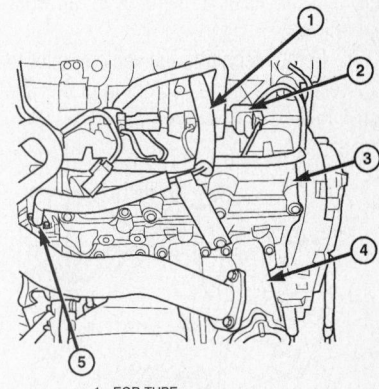

1 - EGR TUBE
2 - IGNITION COIL
3 - CYLINDER HEAD COVER
4 - RIGHT EXHAUST MANIFOLD
5 - EGR VALVE

23990G14

Exhaust Gas Recirculation (EGR) valve and tube assembly

- Manifold and gasket
- O_2S from the manifold

3. Inspect the manifold for damage or cracks. Check for distortion against a straight-edge or thickness gauge. Replace manifold if required.

4. Remove all traces of the old manifold gasket and clean both gasket mating surfaces.

To install:

5. Install or connect the following:
- Exhaust manifold and gasket, tighten the bolts to 200 inch lbs. (23 Nm)
- Exhaust system to the vehicle
- Exhaust manifold flange bolts and the hangers to the pipe assembly. tighten the flange bolts to 22 ft. lbs. (30 Nm).
- Downstream O_2S sensor electrical wiring
- EGR tube support bracket bolt
- Exhaust manifold crossover pipe bolts and tighten to 275 inch lbs. (31 Nm)
- O_2S to the manifold
- Negative battery cable

Front Crankshaft Seal

REMOVAL & INSTALLATION

Note that the timing belt must be removed from the vehicle to perform this service. Use care to be sure all valve timing marks are carefully aligned both before removing the belt and after belt installation and all service has been completed. It may be good practice to set the engine to Top Dead Center (TDC) No. 1 cylinder compression stroke (firing position) and aligning all timing marks before removing the timing belt. This serves as a reference for all work that follows.

1. Before servicing the vehicle, refer to the precautions at the beginning of this section.

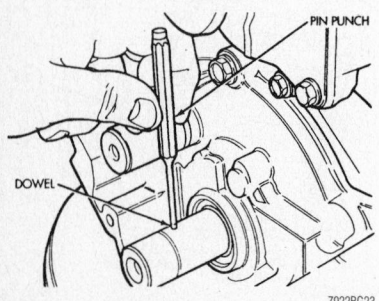

Removing the timing belt sprocket dowel pin from the crankshaft

2. Properly relieve the fuel system pressure.

3. Drain the cooling system.

4. Remove or disconnect the following:
- Negative battery cable
- Radiator and cooling fan module assembly
- Accessory drive belts
- Crankshaft damper bolt
- Timing belt front cover

➡The sealer on the timing belt front cover may be reusable and should not be removed. Use silicone rubber adhesive sealant to replace any missing sealer.

- Timing belt and tensioner. Refer to the appropriate procedure for removal and installation.
- Crankshaft timing belt sprocket

5. Locate the small dowel pin in the crankshaft. With a small punch, carefully tap out the dowel from the end of the crankshaft.

6. Remove the crankshaft seal using tool 6341A, taking care not to nick the shaft seal surface or seal bore during removal.

To install:

7. Inspect the crankshaft seal lip surface for varnish and dirt. Polish the area using 400 grit sandpaper to remove varnish as necessary.

8. Install or connect the following:
- Crankshaft seal using seal installer tool 6342
- Rear lower timing belt cover
- Dowel into the crankshaft so that it protrudes 0.047 in. (1.2mm)
- Timing belt sprocket at the crankshaft using tool C-4685C1, thrust bearing, washer and 12mm bolt or an equivalent setup to pull the sprocket onto crankshaft. Do not hammer on the sprocket.

9. Verify that all valve timing marks are aligned.
- Timing belt and tensioner using the recommended procedure

10. Rotate the crankshaft 2 complete turns and recheck the timing marks on the camshafts and crankshaft. The marks must align with their respective locations. If the marks do not align, repeat the timing belt installation procedure. When correct valve timing has been verified, install the timing belt covers.
- Crankshaft damper. Hold the crankshaft damper, using tool L-3281, and torque the bolt to 85 ft. lbs. (115 Nm).
- Accessory drive belts and adjust to the proper tension

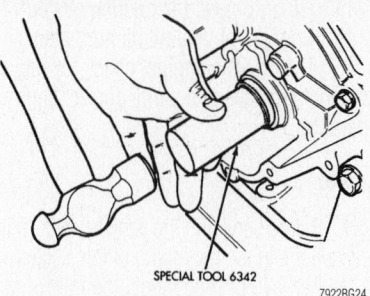

Installing the crankshaft oil seal

- Radiator and cooling fan assemblies
- Negative battery cable

11. Fill and bleed the cooling system.
12. Start the vehicle, check for leaks and repair if necessary.

Camshaft and Valve Lifters

REMOVAL & INSTALLATION

1. Before servicing the vehicle, refer to the precautions in the beginning of this section.

Camshafts are serviced from the rear of the cylinder head. Although the engine does not need to be removed for camshaft service, the cylinder head must be removed from the vehicle. Note too, that the camshaft sprockets have a D-shaped hole that allows it to rotate several degrees in each direction on its shaft.

2. Properly relieve the fuel system pressure.

3. Drain the cooling system.

4. Remove or disconnect the following:
- Cylinder head bolts and cylinder head

➡Mark the rocker arm assembly to note component locations before disassembly.

5. Remove the rocker arm and shaft assemblies.

6. Remove the rear camshaft cover and O-ring.

✳✳ WARNING

Carefully, remove the camshaft from the rear of the head taking care not to nick or scratch the journals.

7. Inspect camshaft journals for wear or damage. If wear is present, inspect the cylinder head for damage. Inspect the head oil holes for clogging. Replace the camshaft as required.

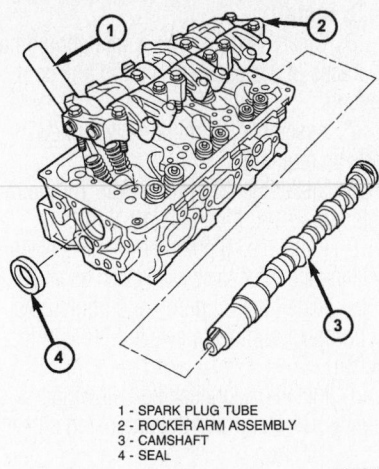

1 - SPARK PLUG TUBE
2 - ROCKER ARM ASSEMBLY
3 - CAMSHAFT
4 - SEAL

23990G15

Exploded view of the camshaft assembly

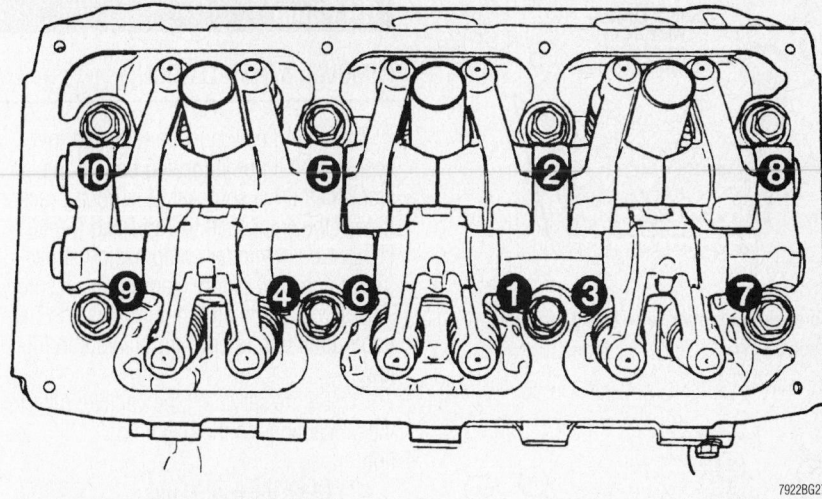

7922BG27

Rocker arm/shaft tightening sequence

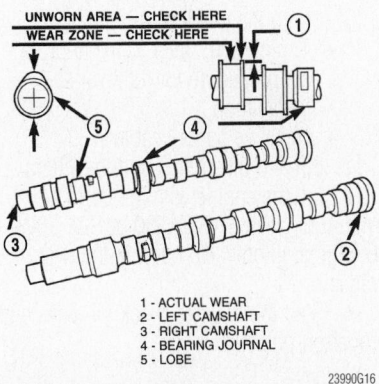

1 - ACTUAL WEAR
2 - LEFT CAMSHAFT
3 - RIGHT CAMSHAFT
4 - BEARING JOURNAL
5 - LOBE

23990G16

Measure the camshaft to ensure it meets specifications, if the camshaft is worn beyond the specifications it must be replaced

8. Measure the height of the cam using a micrometer. Measure in 2 places: the unworn area and in the wear zone. Subtract the figures to get cam wear. The standard specification is 0.001 in. (0.0254mm) with the wear limit being 0.010 in. (0.254mm). Replace the camshaft if it is worn beyond this specification.

To install:

9. Lubricate the camshaft journals and lobes with clean engine oil.

10. Install or connect the following:
   - Camshaft
   - Camshaft cover and O-ring and torque the bolts to 21 ft. lbs. (28 Nm)
   - Rocker arm assemblies
   - Cylinder head assembly
   - Negative battery cable

11. Fill and bleed the cooling system. An oil and filter change is recommended.

12. Start the vehicle, check for leaks and repair if necessary.

Valve Lash

ADJUSTMENT

These engines use hydraulic roller lifters to take up the free-play in the valve train system, therefore no lash adjustments are necessary.

Starter Motor

REMOVAL & INSTALLATION

1. Remove or disconnect the following:
   - Negative battery cable
   - Upper radiator crossmember
   - Radiator fan module
   - Wiring harness clip from the front mount
   - Coolant line clamp bolt
   - Upper nut from the front mount
   - Upper starter mounting bolt
   - Middle starter bolt
   - Lower front mount nut
   - Lower bracket-to-transmission bolt
   - Lower bracket-to-engine block bolt

2. Move the front mount and bracket out of the way.
   - Positive battery cable from the starter
   - Solenoid connector from the starter
   - Lower starter bolt and the starter.

To install:

3. Install or connect the following:
   - Starter and lower bolt. Tighten to 35 ft. lbs. (47 Nm).
   - Solenoid connector to the starter
   - Positive battery cable to the starter

4. Move the front mount and bracket into position.

- Middle starter bolt. Tighten to 35 ft. lbs. (47 Nm).
- Lower bracket-to-engine block bolt. Tighten to 50 ft. lbs. (67 Nm).
- Lower bracket-to-transmission bolt. Tighten to 50 ft. lbs. (67 Nm).
- Lower front mount nut. Tighten to 75 ft. lbs. (101 Nm).
- Upper starter mounting bolt. Tighten to 35 ft. lbs. (47 Nm).
- Upper nut to the front mount
- Coolant line clamp bolt
- Wiring harness clip from the front mount
- Radiator fan module
- Upper radiator crossmember
- Negative battery cable

Oil Pan

REMOVAL & INSTALLATION

1. Before servicing the vehicle, refer to the precautions in the beginning of this section.

2. Drain the engine oil and remove the oil filter.

3. Remove or disconnect the following:
   - Negative battery cable
   - Radiator close out panel
   - Radiator core support
   - Radiator fans
   - Top and bottom A/C compressor bolts. It is not necessary to evacuate the system. reposition the compressor.
   - A/C compressor bracket
   - Structural collar from the rear of the oil pan and transmission housing
   - Inspection shield from between transmission and oil pan

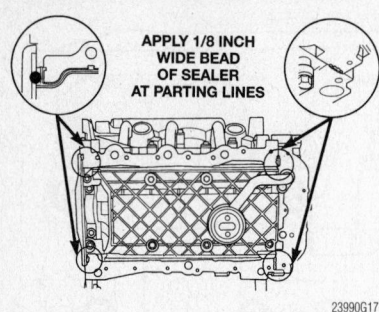

To ensure a proper seal, apply sealer as shown

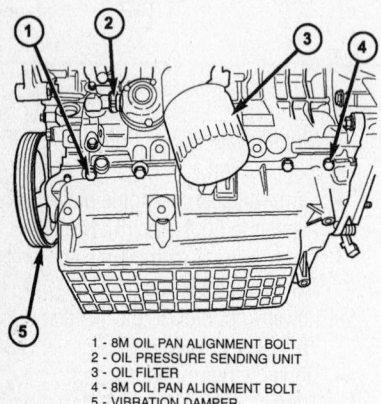

1 - 8M OIL PAN ALIGNMENT BOLT
2 - OIL PRESSURE SENDING UNIT
3 - OIL FILTER
4 - 8M OIL PAN ALIGNMENT BOLT
5 - VIBRATION DAMPER

Location of the oil pan alignment bolts

- Dipstick and housing
- Oil pan mounting bolts, oil pan and gasket

4. Clean the oil pan and all gasket surfaces.

To install:

5. Apply a ⅛ in. (3mm) bead of sealer at the parting line of the oil pump body and the rear seal retainer.

6. Install or connect the following:
- Oil pan and torque the M8 nuts/bolts to 21 ft. lbs. (28 Nm) and the M6 nuts/bolts to 105 inch lbs. (12 Nm)
- Dipstick and housing
- Inspection shield from between transmission and oil pan
- Structural collar to the rear of the oil pan and transmission housing
- A/C compressor bracket and tighten the bolts to 40 ft. lbs. (54 Nm)
- Top and bottom A/C compressor bolts to 250 inch lbs. (28 Nm)
- Radiator fans
- Radiator core support
- Radiator close out panel
- Negative battery cable

Oil Pump

REMOVAL & INSTALLATION

The timing belt must be removed to access the oil pump located behind the crankshaft drive sprocket. It is good practice to turn the crankshaft to Top Dead Center (TDC) No. 1 cylinder compression stroke (firing position) before starting disassembly. This should align all timing marks and be a good point of reference for all work to follow.

1. Before servicing the vehicle, refer to the precautions in the beginning of this section.

2. Drain the engine oil.

3. Drain the cooling system.

4. Remove or disconnect the following:

- Negative battery cable
- Timing belt. Refer to the appropriate procedure in this section.
- Crankshaft sprocket using tool L-4407-A
- Oil pan
- Oil pump pickup tube
- Oil pump-to-engine screws and pump
- Oil pump rotors

5. Wash all parts in solvent and inspect carefully for damage or wear.

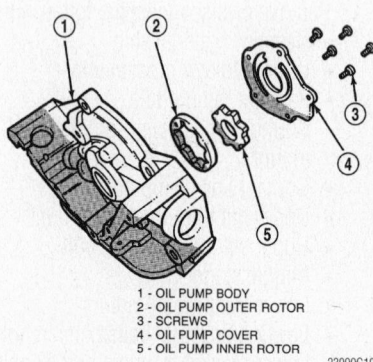

1 - OIL PUMP BODY
2 - OIL PUMP OUTER ROTOR
3 - SCREWS
4 - OIL PUMP COVER
5 - OIL PUMP INNER ROTOR

Exploded view of the oil pump assembly

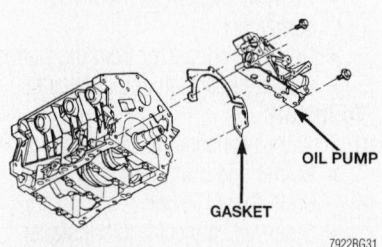

OIL PUMP

GASKET

Prime the oil pump before installation, because a dry pump will wear prematurely and cause low oil pressure

To install:

6. Clean all parts well. There should be no traces of old gasket/sealer on any components.

7. Assemble the oil pump with new parts as required.

8. Install the oil pump cover. Torque the fasteners to 108 inch lbs. (12 Nm).

9. Prime the oil pump prior to installation by filling the rotor cavity with clean engine oil.

10. Install the oil pump and tighten the oil pump-to-engine screws to 250 inch lbs. 28 Nm)

11. Install or connect the following:
- Oil pump pickup tube using a new O-ring
- Oil pan
- Crankshaft sprocket using tool C-4685C1, thrust bearing, washer and 12mm bolt to draw the sprocket onto the crankshaft
- Timing belt. Refer to the appropriate procedure for removal and installation.
- Negative battery cable

12. Fill and bleed the cooling system.

13. Fill the engine with the correct amount of clean SAE 5W-30 or SAE 10W-30 engine oil only. Do not mix the two grades of oil.

14. Start the engine, check for leaks and proper oil pressure.

Rear Main Seal

REMOVAL & INSTALLATION

1. Before servicing the vehicle, refer to the precautions in the beginning of this section.

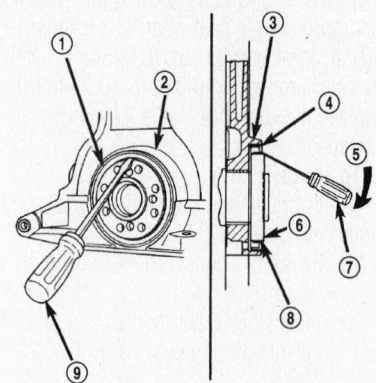

1 - REAR CRANKSHAFT SEAL
2 - ENGINE BLOCK
3 - ENGINE BLOCK
4 - REAR CRANKSHAFT SEAL METAL CASE
5 - PRY IN THIS DIRECTION
6 - CRANKSHAFT
7 - SCREWDRIVER
8 - REAR CRANKSHAFT SEAL DUST LIP
9 - SCREWDRIVER

Use a suitable prytool to remove the rear oil seal as shown

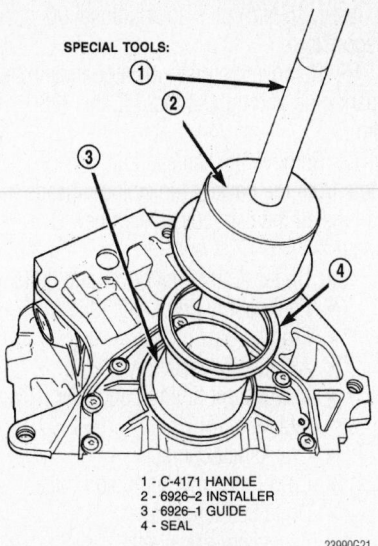

SPECIAL TOOLS:

1 - C-4171 HANDLE
2 - 6926-2 INSTALLER
3 - 6926-1 GUIDE
4 - SEAL

23990G21

Installing the rear main oil seal

2. Drain the transaxle fluid.

3. Remove the negative battery cable.

4. Remove the transaxle, inspection cover and flywheel/flexplate.

5. Using a small prytool, carefully pry out the rear oil seal. Be careful not to nick or damage the crankshaft flange seal surface or the retainer bore.

To install:

6. Place the Seal Pilot tool 6926-1 on the crankshaft.

7. Apply a light coating of engine oil to the entire circumference of the oil seal lip.

8. Place the seal over tool 6926-1 and use installer 6926-2 and handle C-4171 to drive the seal into place until it is flush with the housing surface.

9. Install the flexplate/flywheel and transaxle.

10. Connect the negative battery cable.

11. Fill the transaxle with the proper fluid.

12. Start the vehicle, check for leaks and repair if necessary.

Timing Belt

REMOVAL & INSTALLATION

Use care when servicing a timing belt. Valve timing is absolutely critical to engine performance. If the valve timing marks on all drive sprockets are not properly aligned, engine damage will result. If only the belt and tensioner are being serviced, do not loosen the camshaft drive sprockets unless they are to be replaced. The sprockets have oversized openings and can be rotated several degrees in each direction on their shafts. This means the sprockets must be re-timed, requiring some special tools.

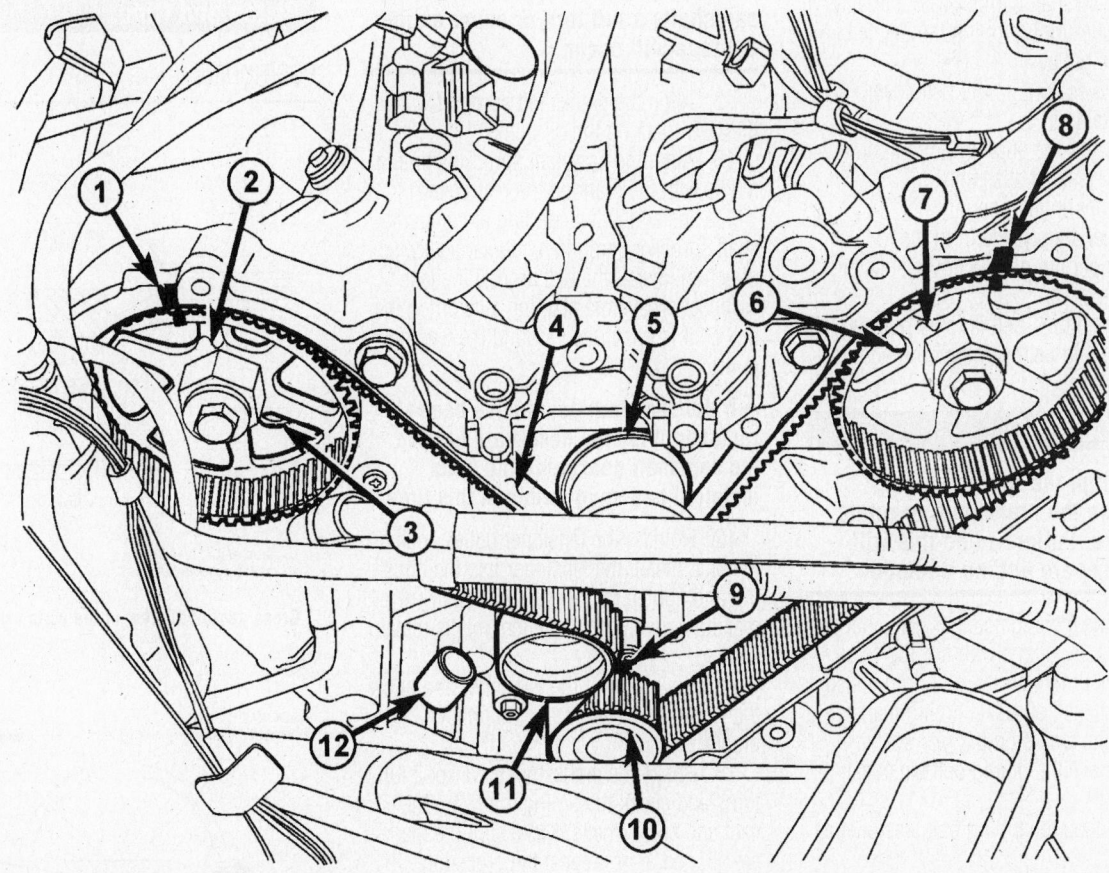

1 - RIGHT CAMSHAFT GEAR ALIGNMENT MARK
2 - RIGHT CAMSHAFT GEAR
3 - CYLINDER HEAD TO INNER TIMING BELT COVER BOLTS - RIGHT
4 - TIMING BELT
5 - WATER PUMP PULLEY
6 - CYLINDER HEAD TO INNER TIMING BELT COVER BOLTS - LEFT

7 - LEFT CAMSHAFT GEAR
8 - LEFT CAMSHAFT GEAR ALIGNMENT MARK
9 - CRANKSHAFT GEAR ALIGNMENT MARK

10 - CRANKSHAFT GEAR
11 - TIMING BELT TENSIONER PULLEY
12 - TIMING BELT TENSIONER

23990G22

View of the timing belt alignment marks

✲✲ CAUTION

Fuel injection systems remain under pressure, even after the engine has been turned off. The fuel system pressure must be relieved before disconnecting any fuel lines. Failure to do so may result in fire and/or personal injury.

1. Loosen the valve train rocker assemblies before servicing the timing components.

2. Release the fuel system pressure using the recommended procedure.

3. Remove or disconnect the following:
- Negative battery cable
- Cylinder head covers
- Accessory drive belts
- Drive belt tensioner
- Power steering pump bolts and position the pump aside
- Crankshaft damper with a quality puller tool gripping the inside of the pulley
- Lower front timing belt cover retainers

4. Support the engine with a floor jack.
- Air cleaner assembly
- Front engine mount
- Fuel supply line from the rail
- Upper timing belt cover bolts and the cover

5. If the timing belt is to be reused, mark the timing belt with the running direction for installation.

✲✲ WARNING

Always align the timing marks always use the crankshaft to rotate the engine. Failure to do this will result in severe engine damage.

6. Rotate the engine clockwise until the crankshaft mark aligns with the Top Dead Center (TDC) mark on the oil pump housing and the camshaft sprocket timing marks align with the marks on the rear cover.

7. Remove the timing belt tensioner and the belt.

8. Pre-load the timing belt tensioner as follows:

    a. Place tensioner in a vise the same way it is mounted on the engine.

    b. Slowly compress the plunger into the tensioner body.

    c. Once the plunger is compressed, install a pin through the body and plunger to retain it in place until the tensioner is installed.

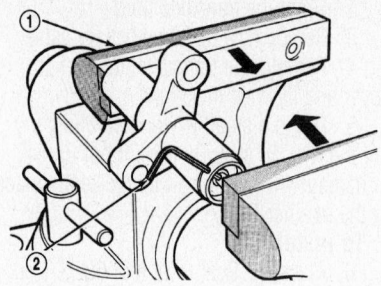

1 - VISE
2 - LOCKING PIN

23990G23

Pre-load the timing belt tensioner

To install:

✲✲ WARNING

If the camshafts have moved from the timing marks always rotate them towards the direction nearest to the the timing marks. Never turn the camshafts a full turn or sever engine damage will occur.

9. Align the crankshaft sprocket with the TDC mark on the oil pump.

10. Align the camshaft sprockets timing marks with the mark on the rear cover.

11. Install the belt starting at the crankshaft sprocket going in a counterclockwise direction. Install the belt around the last sprocket always maintaining tension on the belt as it is positioned around the tensioner pulley.

➡ **If the camshaft gears have been removed it is only necessary to have the camshaft gear retaining bolts installed to a snug torque at this time.**

12. Holding the tensioner pulley against the belt, install the tensioner into the housing and tighten to 250 inch lbs. (28 Nm) making sure all the alignment marks remain aligned.

Pull the pin from the tensioner and allow the tensioner to extend to the pulley bracket.

13. Rotate the crankshaft sprocket 2 full turns and check the timing marks on the cam and crank shafts. Make sure the marks are aligned, if not repeat the procedure.

14. if the camshafts were removed using a dial indicator, position the number 1 piston at TDC.

15. Hold the camshaft sprocket hex with a 36mm wrench and tighten the right camshaft sprocket bolt to 75 ft. lbs. (102 Nm) plus an additional 90 degree turn. Tighten the left camshaft sprocket bolt to 85

ft. lbs. (115 Nm) plus an additional 90 degree turn.

16. Remove the dial indicator. Install the spark plug and tighten to 20 ft. lbs. (28 Nm).

17. Remove the camshaft alignment tools from the back of the cylinder heads and install the cam covers with new O-rings.
- Upper timing belt cover and bolts
- Fuel supply line to the rail
- Front engine mount
- Air cleaner assembly
- Lower front timing belt cover retainers
- Crankshaft damper
- Power steering pump and bolts
- Drive belt tensioner
- Accessory drive belts
- Cylinder head covers
- Negative battery cable

Piston and Rings

POSITIONING

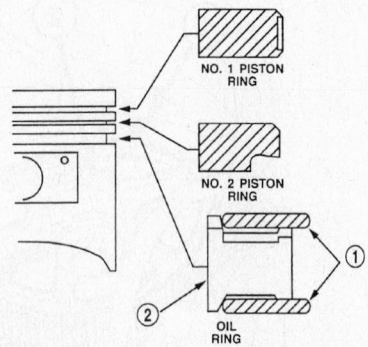

1 - SIDE RAIL
2 - SPACER EXPANDER

9306BG04

Cross-sectional view of the piston rings

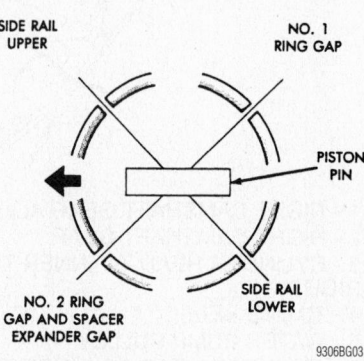

9306BG03

Piston ring gap positions

FUEL SYSTEM

Fuel System Service Precautions

Safety is the most important factor when performing not only fuel system maintenance but any type of maintenance. Failure to conduct maintenance and repairs in a safe manner may result in serious personal injury or death. Maintenance and testing of the vehicle's fuel system components can be accomplished safely and effectively by adhering to the following rules and guidelines:

• To avoid the possibility of fire and personal injury, always disconnect the negative battery cable unless the repair or test procedure requires that battery voltage be applied.

• Always relieve the fuel system pressure prior to disconnecting any fuel system component (injector, fuel rail, pressure regulator, etc.), fitting or fuel line connection. Exercise extreme caution whenever relieving fuel system pressure, to avoid exposing skin, face and eyes to fuel spray. Please be advised that fuel under pressure may penetrate the skin or any part of the body that it contacts.

• Always place a shop towel or cloth around the fitting or connection prior to loosening to absorb any excess fuel due to spillage. Ensure that all fuel spillage (should it occur) is quickly removed from engine surfaces. Ensure that all fuel soaked cloths or towels are deposited into a suitable waste container.

• Always keep a dry chemical (Class B) fire extinguisher near the work area.

• Do not allow fuel spray or fuel vapors to come into contact with a spark or open flame.

• Always use a back-up wrench when loosening and tightening fuel line connection fittings. This will prevent unnecessary stress and torsion to fuel line piping.

• Always replace worn fuel fitting O-rings with new. Do not substitute fuel hose, where fuel pipe is installed.

Before servicing the vehicle, also make sure to refer to the precautions in the beginning of this section as well.

Fuel System Pressure

RELIEVING

1. Before servicing the vehicle, refer to the precautions in the beginning of this section.
2. At the Power Distribution Center (PDC), remove the Fuel Pump Relay.

3. Operate the engine until the engine stalls; then, continue restarting the engine until it will no longer run.
4. Turn the ignition switch **OFF**.

✳✳ CAUTION

The previous steps must be performed to relieve the high pressure fuel from the fuel rail. The following steps must be performed to remove excess fuel from the fuel rail. Do not use the following steps to relieve high pressure for excessive fuel will be forced into a cylinder chamber.

5. Disconnect the electrical connector from any injector.
6. Connect 1 end of a jumper wire (with an alligator clip) to either injector terminal and the other end to the positive side of the battery.
7. Connect 1 end of a second jumper wire to the other injector terminal.

✳✳ WARNING

Applying power to an injector for more than a few seconds will damage the injector.

8. Momentarily, touch the other end of the jumper wire to a ground for no more than a few seconds.
9. At the Power Distribution Center (PDC), install the fuel pump relay.

➡ When the fuel pump relay is removed, 1 or more Diagnostic Trouble Codes (DTC's) may be stored in the Powertrain Control Module (PCM) memory. A DRB scan tool must by used to clear the DTC's.

Fuel Pump

REMOVAL & INSTALLATION

The in-tank fuel pump module contains the fuel pump and pressure regulator which adjusts fuel system pressure. Fuel pump voltage is supplied through the fuel pump relay.

The fuel pump is serviced as part of the fuel pump module. The fuel pump module is installed in the top of the fuel tank and contains the electric fuel pump, fuel pump reservoir, inlet strainer fuel gauge sending unit, fuel supply and return line connections and the pressure regulator. The inlet

strainer, fuel pressure regulator and level sensor are the only serviceable items. If the fuel pump requires service, replace the fuel pump module.

1. Before servicing the vehicle, refer to the precautions in the beginning of this section.
2. Properly relieve the fuel system pressure.
3. Remove or disconnect the following:
   • Negative battery cable
4. Fold the drivers side rear seat and tilt it forward.
   • Two bolts for the rear seat
   • Seat electrical connection
   • Seat
   • Rear console scuff plate
   • Rear console electrical connector
   • Fourteen screws retaining the rear console
   • Screws from the cup holder pockets
   • Screw from the console pocket under the rubber mat
   • Left rear door sill plate
5. Cut the carpet along the mounting bracket for the center console. The cut does not need to exceed a foot in length. Pull the carpet back to access the pump cover.

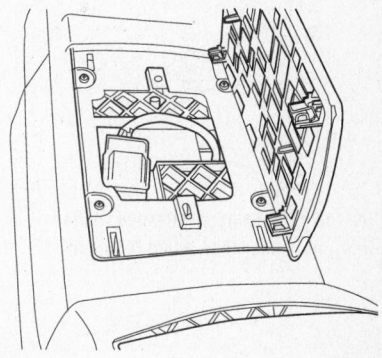

23990G24

Remove the rear console scuff plate

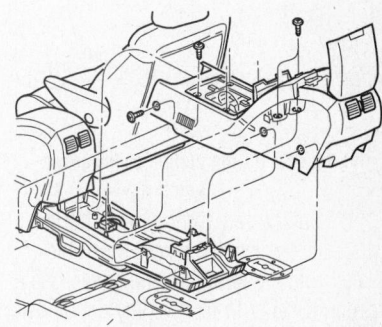

23990G25

Remove the rear console assembly

23990G26

Remove the screws from the cup holder pockets

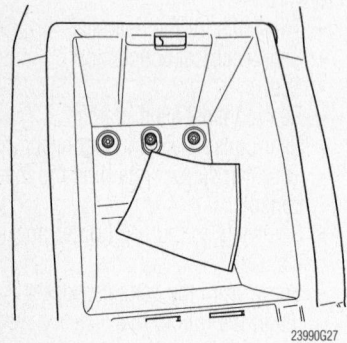

23990G27

Remove the screws from under the rubber mat in the console pocket

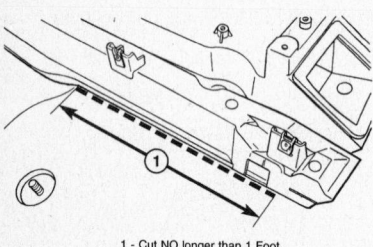

1 - Cut NO longer than 1 Foot

23990G28

Cut the carpet where indicated (dotted line), no longer than a foot in length

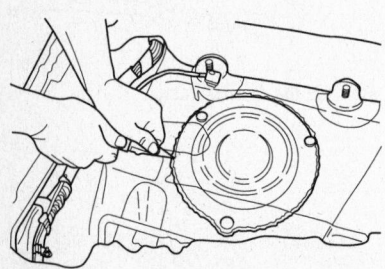

23990G29

Cut the seal for the pump access cover

6. Cut the seal for the pump access panel and remove the cover. Clean the sealer from the floor pan and cover. Use a vacuum cleaner to remove any debris before removing the pump lock ring.

7. Remove or disconnect the following:

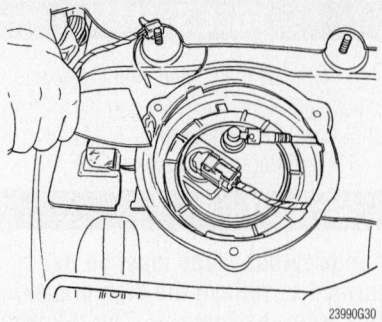

23990G30

Clean the sealer from the cover contact surfaces

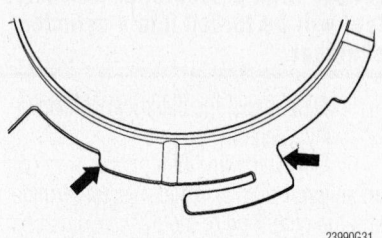

23990G31

Pump lock ring contact points, place the brass punch at these locations when removing/installing the ring

23990G32

Apply sealer to the outer edge of the access plate and the push pin holes

- Vapor line and electrical connection
- Pump lock ring using a brass punch
- Pump top

8. Drain the fuel tank.
- Electrical connections from the bottom of the pump top and the passenger side level sensor assembly
- Tabs attaching the return and supply lines
- Return line from the pump using a small prytool to pry the tab back and tip the hose to one side, then pry the tab on the other side to release the hose
- Fuel supply line from the top of the pump in the same manner as the return line
- Fuel pump

To install:

9. Install or connect the following:
- Fuel pump
- Fuel supply line and return lines
- Electrical connections to the bottom of the pump
- Pump top, gasket and lock ring
- Fuel line and electrical connection
- Negative battery cable

10. Fill the fuel tank.

11. Use a DRB III to pressurize the fuel system and check for leaks. If no leaks detected, disconnect the negative battery cable.

12. Clean the access cover and floorpan.

13. Apply sealer to the outer edge of access plate and the push pin holes.

14. Install or connect the following:
- Access cover and the push pins into the holes
- Carpet and left rear door sill plate
- Screws to the console pocket under the rubber mat
- Screws to the cup holder pockets
- Fourteen screws retaining the rear console
- Rear console electrical connector
- Rear console scuff plate
- Seat
- Seat electrical connection
- Two bolts for the rear seat
- Negative battery cable

Fuel Injector

REMOVAL & INSTALLATION

1. Before servicing the vehicle, refer to the precautions in the beginning of this section.

2. Relieve the fuel system pressure.

3. Remove or disconnect the following:
- Negative battery cable
- Intake manifold plenum and cover opening with a clean cloth

4. Tag the injectors to match them with their correct cylinder during installation.

5. Place a shop rag under the fuel rail's quick-connect fitting; then, squeeze the quick-connect fitting's retainer tabs together and pull the fitting assembly off of the fuel tube nipple.
- Fuel injector electrical connectors
- Fuel rail-to-engine bolts and fuel rail
- Fuel injector-to-fuel rail retainer clips
- Fuel injectors

To install:

6. Lubricate the injector O-rings with clean engine oil.

7. Install or connect the following:

- Fuel injector and secure with retaining clips
- Fuel rail onto cylinder head and press rail into place. Make sure that the injectors are fully seated.

- Fuel rail-to-cylinder head bolts and torque bolts to 250 inch lbs. (28 Nm)
- Fuel injector electrical connectors
- Intake plenum

8. Lubricate the quick-connect fitting's O-rings with clean engine oil; then, push the connector together until the retainer seats and a click is heard.

9. Connect the negative battery cable.

DRIVE TRAIN

Transaxle Assembly

REMOVAL & INSTALLATION

1. Before servicing the vehicle, refer to the precautions in the beginning of this section.

2. Remove or disconnect the following:
- Battery cables
- Battery and tray
- Gearshift cable from the manual valve lever
- Oil cooler off of hose fitting connection.

3. Use a suitable cutting tool, cut the oil cooler hoses off flush at the fittings. A service splice kit will be required during installation. Cap the hoses.
- Ground cable from the transaxle case
- Input and output speed sensor connections
- Crankshaft Position Sensor (CKP)
- Transaxle range sensor and solenoid/pressure switch connections
- Transaxle harness from retainers and position aside
- Coolant bypass tube-to-engine and transaxle retainers
- Front brake lines from the Hydraulic Control Unit (HCU) and position aside
- Intermediate shaft extension from the steering gear using tool 6831A to remove the roll pin and then slide the shaft extension off the gear

4. Install a powertrain support fixture such as 8534B and adapter kit 8534-12.

5. Oil dipstick tube-to-cylinder head bolt

6. Install lift support bracket 8534-8 and use the dipstick tube bolt to secure it.

7. Remove the Coolant Temperature Sensor (CTS) at the thermostat housing. Remove the engine harness-to-cylinder head bolt and position the harness aside.

8. Install lift support bracket 8534-7 and bolt.

9. Assemble mounting bracket/sleeve assemblies 8534-2 to support tube 8534-1 and install on the vehicle allowing the brackets to rest on the inner fender ledges.

10. Assemble cross bar 8534-3, clamp 8534-5 and support leg 8534-4 to support tube allowing the support leg to rest on the radiator upper support.

11. Tighten the cross bar-to-support tube clamp 8534-5 as well as mounting bracket/sleeve 8534-3.

12. Install lift bracket assemblies as shown in the accompanying illustration.

13. Raise the vehicle and remove the front wheels.

14. Remove or disconnect the following:
- ABS sensor connector and sensor brackets from the struts
- Sway bar links from struts

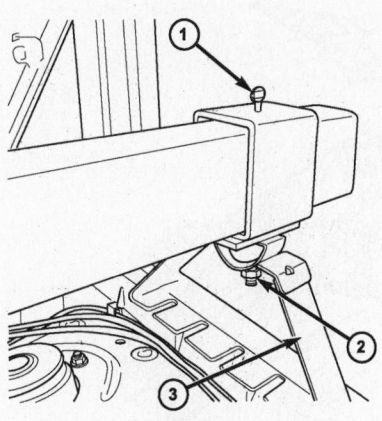

1 - THUMB SCREW
2 - PIVOT HEX NUT
3 - BRACKET/SLEEVE 8534-2

23990G33

Assemble the torque thumb screw and pivot nut bracket assembly as shown

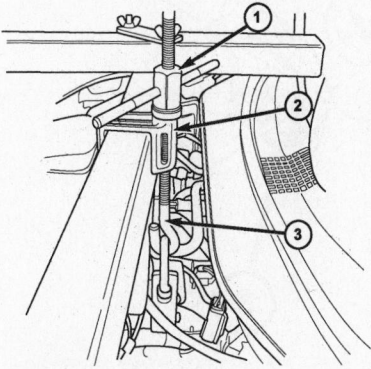

1 - T-HANDLE 8534-14
2 - LIFTING BRACKET 8534-13
3 - HOOK ASSEMBLY 8534-11

23990G34

Assemble the T-handle, bracket and hook assembly at the rear lift bracket as illustrated

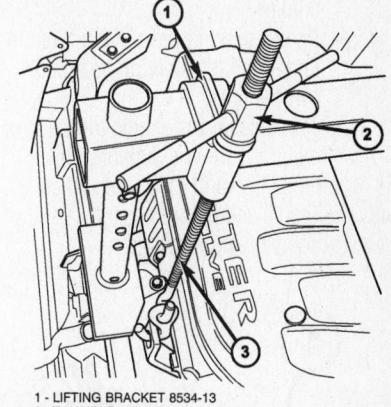

1 - LIFTING BRACKET 8534-13
2 - T-HANDLE 8534-14
3 - HOOK ASSEMBLY 8534-11

23990G35

Assemble the T-handle, bracket and hook assembly at the front lift bracket as shown

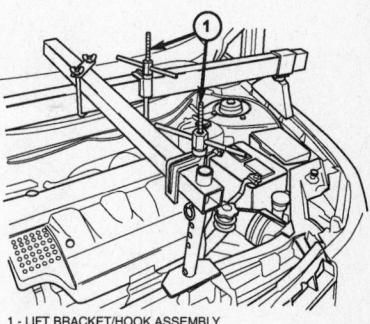

1 - LIFT BRACKET/HOOK ASSEMBLY

23990G36

The engine support system slack is taken up by using the T-handles

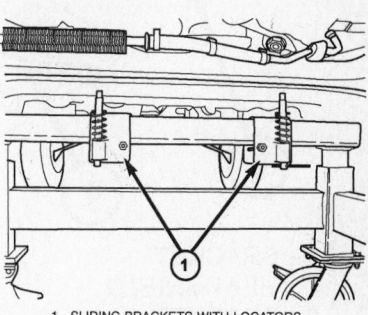

1 - SLIDING BRACKETS WITH LOCATORS

23990G37

Position drive line support table 8874 into position and lower the vehicle until the cradle and fixture engage

- Halfshafts
- Engine front and rear mount-to-cradle nuts
- Brake lines and brackets from the frame rails
- Power steering hydraulic line from the bracket at the cradle located on the passenger side
- Power steering oil cooler from the cradle
- Power steering pressure and return lines from the steering gear and cap the lines
- Transmission-to-cradle torque strut

15. Position drive line support table 8874 into position and lower the vehicle until the cradle and fixture engage as shown in the accompanying illustration.

16. Scribe alignment marks to reference cradle-to-body alignment to help during installation, then remove the four cradle-to-body bolts.

17. Slowly raise the vehicle and separate the cradle from the vehicle. Check the overhead fixture is secure on the fenders and radiator support.

18. Have an assistant guide the brake and power steering lines through as they remain attached.

19. Remove or disconnect the following:

- Engine front mount/bracket
- Starter motor

20. On All Wheel Drive (AWD) models remove or disconnect the following:

- Propeller shaft

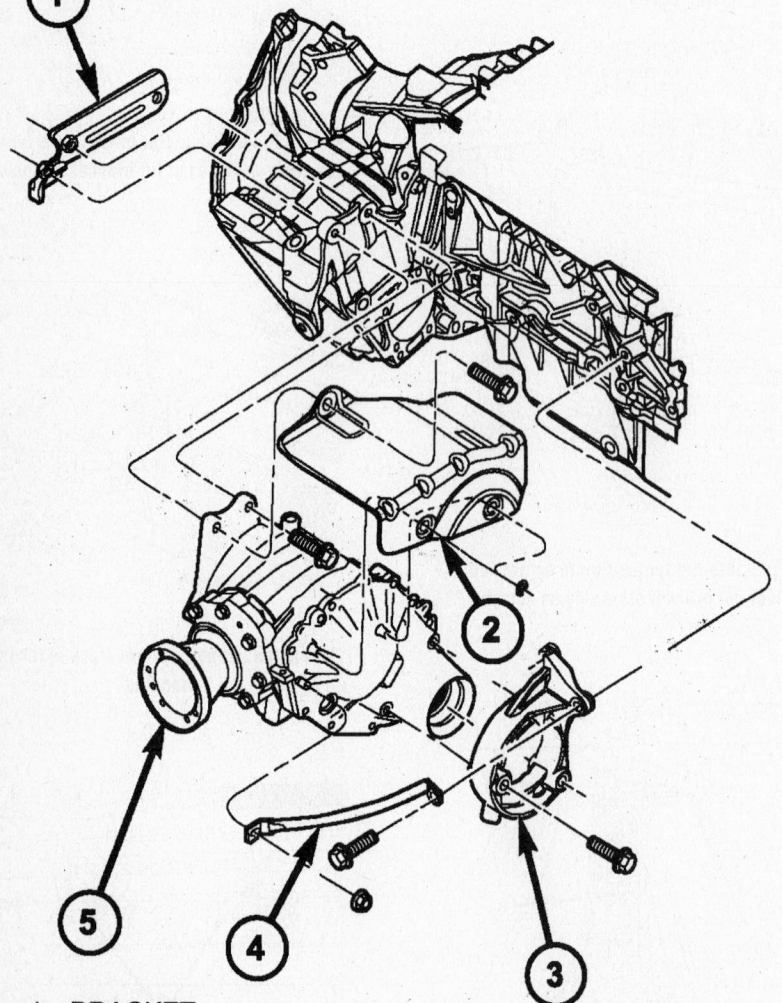

1 - BRACKET
2 - HEAT SHIELD
3 - BRACKET
4 - BRACE
5 - POWER TRANSFER UNIT

23990G38

Exploded view of the Power Transfer Unit mounting and related components

- Power Transfer Unit (PTU) rear mount bracket
- Oil pan-to-transaxle collar
- Heat shield
- PTU-to-transaxle upper bolts
- PTU-to-bracket lower bolts
- PTU

21. Remove or disconnect the following:

- Torque converter dust shield
- Four torque converter-to-flexplate bolts. Upon removing the bolts a tight tolerance (slotted) bolt will be found. Mark this location on the converter and flexplate to use as a reference during assembly.
- Four transaxle upper bellhousing bolts

22. Use the overhead fixture to lower the engine/transaxle assembly. and secure the transmission to a suitable jack.

23. Remove the two transaxle-to-engine lower bolts and remove the transaxle.

To install:

24. Install or connect the following:

- Transaxle
- Two transaxle-to-engine lower bolts and tighten to 70 ft. lbs. (95 Nm)

25. Remove the transmission jack.

- Four transaxle upper bellhousing bolts and tighten to 70 ft. lbs. (95 Nm)
- Four torque converter-to-flexplate bolts and tighten to 65 ft. lbs. (88 Nm)
- Torque converter dust shield

26. Loosely install the engine-to-transaxle collar.

27. On AWD models install or connect the following:

- PTU lower bracket and PTU
- PTU-to-bracket lower bolts and tighten to 21 ft. lbs. (28 Nm)
- PTU-to-transaxle upper bolts and tighten to 40 ft. lbs. (54 Nm)
- PTU rear mount bracket and tighten the bolts to 40 ft. lbs. (54 Nm)
- Heat shield
- Propeller shaft

28. Install or connect the following:

- Starter motor
- Engine front mount/bracket and tighten the bolts to 50 ft. lbs. (67 Nm)

29. Align the scribe alignment marks made to reference cradle-to-body alignment during removal, then install the four cradle-to-body bolts and tighten to 120 ft. lbs. (162 Nm).

30. Route the brake and power steering lines.

31. Remove the overhead support fixture.

32. Install or connect the following:

- Engine front and rear mount-to-cradle nuts and tighten to 40 ft. lbs. (54 Nm)
- Steering shaft-to-gear coupler using tool 6831A
- Power steering pressure and return lines to the steering gear and tighten the fittings to 23 ft. lbs. (31 Nm)
- Power steering oil cooler and hose routing clip to the cradle
- Transmission-to-cradle torque strut and tighten to 40 ft. lbs. (54 Nm)
- Halfshafts
- Sway bar links to the struts and tighten the nuts to 65 ft. lbs. (88 Nm)
- ABS sensor brackets and electrical connectors
- Both wheels
- Front brake lines to the HCU
- Coolant bypass tube fasteners
- Solenoid/pressure switch and tighten the screw to 35 inch lbs. (4 Nm)
- Transaxle range sensor connection
- Input and output speed sensor connections
- Transaxle harness using the retainers
- CKP sensor
- Transaxle oil cooler splice kit using the kit instructions

- Gearshift cable to the manual valve lever
- Battery tray, battery and cables
33. Bleed the brakes.
34. Add power steering fluid.
35. Check and fill the transmission fluid to the proper level.
36. If equipped with AWD, remove the PTU fill plug, check the fluid level. The PTU should have 2.1 pints of 75W-90 gear and axle lubricant. After checking the fluid level, install the fill plug and tighten to 26 ft. lbs. (35 Nm).
37. Check and adjust the wheel alignment.

Halfshaft

REMOVAL & INSTALLATION

Front

✳✳ WARNING

Allowing the CV-joint assemblies to dangle unsupported, or pulling or pushing the ends, can damage boots or CV-joints. Always support both ends of the halfshaft to prevent damage or disengagement of the Tri-pot joint.

1. Before servicing the vehicle, refer to the precautions in the beginning of this section.
2. Remove or disconnect the following:
- Negative battery cable
- Hub and bearing-to-stub axle cotter pin and retainer nut
- Front wheels
- Front caliper assembly from steering knuckle
- Front brake rotor from the hub
- Steering knuckle-to-strut bolts
- Steering knuckle from the strut bracket
- Knuckle down and away from the outer CV-joint while pulling the joint out of the bearing assembly
3. Dislodge the inner Tri-pot joint from the stub shaft retaining snapring on the transaxle. To do this, insert a prybar between the transaxle case and the inner Tri-pot joint and pry on Tri-pot joint.
To install:
4. Install the Tri-pot joint side of the halfshaft by performing the following procedure:
   a. Replace the inner Tri-pot joint retaining circlip and O-ring seal on the transaxle stub shaft. These components are not reusable and must be replaced whenever the halfshaft is removed.
   b. Apply an even coat of grease on the splines of the inner Tri-pot joint,

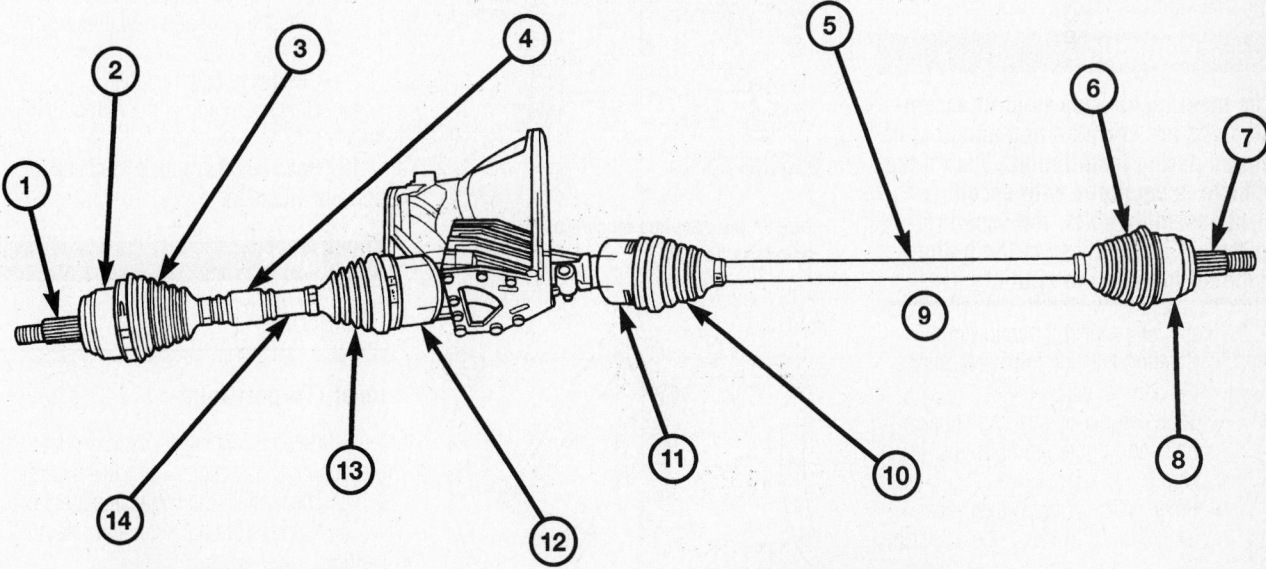

1 - STUB AXLE
2 - OUTER C/V JOINT
3 - OUTER C/V JOINT BOOT
4 - TUNED RUBBER DAMPER WEIGHT
5 - INTERCONNECTING SHAFT
6 - OUTER C/V JOINT BOOT
7 - STUB AXLE
8 - OUTER C/V JOINT
9 - RIGHT HALFSHAFT
10 - INNER TRIPOD JOINT BOOT
11 - INNER TRIPOD JOINT
12 - INNER TRIPOD JOINT
13 - INNER TRIPOD JOINT BOOT
14 - INTERCONNECTING SHAFT LEFT HALFSHAFT

23990G39

Exploded view of the front halfshaft and related components

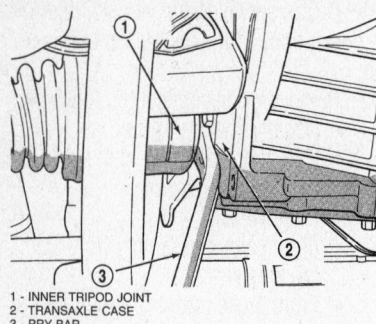

1 - INNER TRIPOD JOINT
2 - TRANSAXLE CASE
3 - PRY BAR

23990G40

Use a prybar to disengage the halfshaft from the transaxle

where the O-ring seats against the Tri-pot joint.

c. Grasp the inner joint in 1 hand and interconnecting shaft in the other. Align the inner Tri-pot joint spline with the stub shaft spline on the transaxle. Use a rocking motion with the inner Tri-pot joint to get it past the circlip on the transaxle stub shaft.

d. Continue pushing the Tri-pot joint onto transaxle stub shaft until it stops moving. The O-ring on the stub shaft should not be visible when the inner Tri-pot joint is fully installed. Check that the inner Tri-pot joint is locked in position by grasping the inner joint and pulling. If locked in position, the joint will not move on the stub shaft.

5. Install the outer CV-joint into the hub and bearing assembly.

※※ WARNING

The steering knuckle-to-strut assembly bolts are serrated and must not be turned during installation.. Also if the vehicle is equipped with eccentric strut assembly bolts, the eccentric bolt must be installed in the bottom (slotted) hole on the strut bract.

6. Install or connect the following:
- Steering knuckle in the bracket of the damper assembly
- Strut damper-to-steering knuckle bolts and tighten to 65 ft. lbs. (88 Nm)
- Rotor, caliper. Tighten the caliper adapter bolts to 125 ft. lbs. (169 Nm)
- Washer and bearing-to-stub axle bolt and hand-tighten.
- Wheels

7. Apply the brakes and tighten the bearing-to-stub axle bolt to 180 ft. lbs. (244 Nm) and install the cotter pin.

8. Road test the vehicle to check for noise or vibration.

Rear

※※ WARNING

The rear suspension and drivetrain design requires this procedure to be performed on a drive on hoist as the front and rear suspension must be compressed to the vehicle ride height.

1. Before servicing the vehicle, refer to the precautions in the beginning of this section.

2. Remove or disconnect the following:
- Negative battery cable
- Wheel cover cap
- Cotter pin, locknut and washer
- Exhaust system center hanger at the propeller shaft center bearing
- Exhaust system at the rear hanger and lower at least 10 inches (254mm) before securing with wire

3. Mark the propeller shaft and rear driveline module flanges for reference during installation.
- 3 propeller shaft-to-driveline module bolts. Do not disconnect the shaft

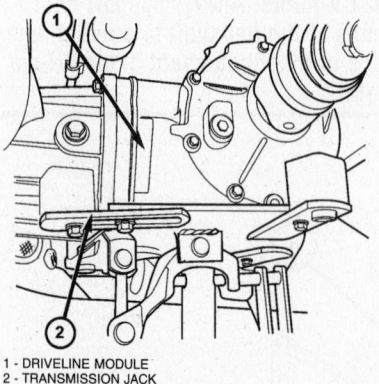

1 - DRIVELINE MODULE
2 - TRANSMISSION JACK

23990G41

support the driveline module using a transmission jack

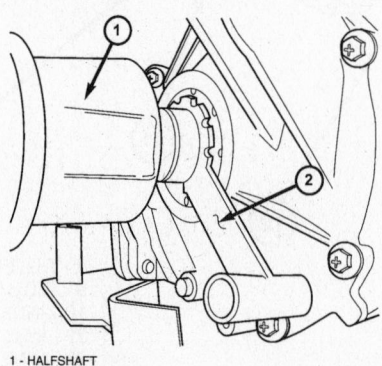

1 - HALFSHAFT
2 - SEAL PROTECTOR

23990G42

Use seal protector tool 9099 to protect against seal damage

from the module as it will be disconnected upon lowering the module.

4. Support the driveline module with a transmission jack.

5. Using a suitable prytool, partially dislodge the halfshaft from the differential.

6. Install seal protector tool 9099 to protect the seal during disassembly.

7. Remove the 3 driveline-to-crossmember bolts.

8. Lower the driveline module enough to remove the shaft from the differential, making sure tool 9099 engages the seal.

9. Disconnect the propeller shaft from the differential and secure it to the exhaust system.

10. Remove the halfshaft from the bearing assembly.

To install:

11. Remove the halfshaft from the bearing assembly. Install the hub nut and washer and hand tighten.

12. Attach the shaft to the differential, using tool 9099.

13. Connect the propeller shaft to the module flange. Install but do not tighten the 3 bolts

14. Position the module and tighten the module-to-cradle bolts to 75 ft. lbs. (102 Nm).

15. Tighten the propeller shaft-to-module bolts to 40 ft. lbs. (54 Nm)

16. Tighten the halfshaft nut to 180 ft. lbs. (244 Nm).

17. Install or connect the following:
- Washer, nut and new cotter pin
- Wheel cover cap
- Negative battery cable

18. Check and adjust the differential fluid.

19. Road test the vehicle to check for noise or vibration.

CV-Joints

OVERHAUL

Inner (Tri-pot) Joint

1. Before servicing the vehicle, refer to the precautions in the beginning of this section.

2. Disconnect the negative battery cable.

3. Remove the halfshaft and retaining clamps.

4. Slide the boot down the shaft away from the tri-pot housing.

➡**When separating the spider joint from the tri-pot joint housing, hold the rollers in place on the trunnions to prevent the rollers and needle bearings from falling away.**

5. Carefully, slide the shaft/spider assembly from the tri-pot housing.

6. Remove the spider assembly-to-shaft snapring; then, slide the spider assembly off the shaft.

✳✳ WARNING

If necessary, tap the spider assembly off the shaft using a brass drift; be careful not to hit the outer bearings.

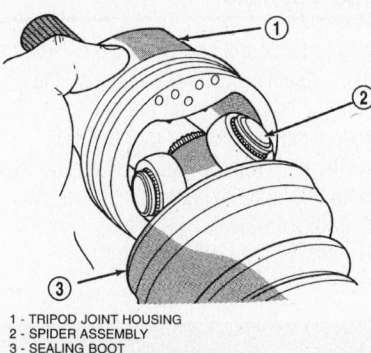

1 - TRIPOD JOINT HOUSING
2 - SPIDER ASSEMBLY
3 - SEALING BOOT

23990G43

Slide the boot down the shaft away from the tri-pot housing

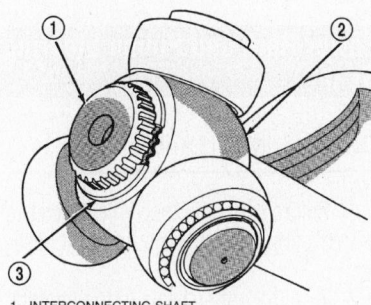

1 - INTERCONNECTING SHAFT
2 - SPIDER ASSEMBLY
3 - RETAINING SNAP-RING

23990G44

Remove the spider assembly-to-shaft snapring; then, slide the spider assembly off the shaft

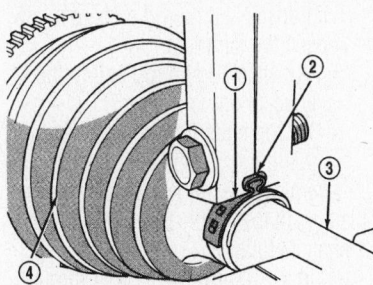

1 - CLAMP
2 - JAWS OF SPECIAL TOOL C-4975-A MUST BE CLOSED COMPLETELY TOGETHER HERE
3 - INTERCONNECTING SHAFT
4 - SEALING BOOT

23990G45

Securing the halfshaft boot clamp

7. Slide the boot off the shaft.

8. Thoroughly, inspect all parts for signs of excessive wear; if necessary, replace the halfshaft.

➡ **Component parts are not serviceable and must be replaced as an assembly.**

To install:

9. Slide the inner tri-pot boot clamp and boot onto the shaft; then, position the boot so that only the thinnest (sight) groove is visible on the shaft.

10. Install the spider assembly onto the shaft just far enough so that the snapring can be installed.

✳✳ WARNING

If necessary, tap the spider assembly onto the shaft using a brass drift; be careful not to hit the outer bearings.

11. Install the snapring onto the shaft; make sure that the snapring is fully seated in the groove.

12. If installing a new boot, distribute ½ of the grease in the service package inside the tri-pot housing and the other ½ inside the boot.

13. Carefully, slide the spider assembly and shaft into the tri-pot housing.

14. Position the inner boot clamp evenly on the sealing boot.

15. Using the Crimper tool C-4975,

place the tool over the clamp bridge, tighten the tool nut until the jaws are completely closed (face-to-face).

✳✳ WARNING

The seal must not be dimpled, stretched or out of shape. If necessary, equalize the seal pressure and shape it by hand.

16. Position the boot onto the tri-pot hoize the seal pressure and shape it by hand.

17. Position the boot onto the CV-joint housing retaining groove and install the retaining clamp evenly on the boot.

18. Using the Crimper tool C-4975, place the tool over the clamp bridge, tighten the tool nut until the jaws are completely closed (face-to-face).

19. Install the halfshaft into the vehicle.

Outer CV-Joint

1. Before servicing the vehicle, refer to the precautions in the beginning of this section.

2. Disconnect the negative battery cable.

3. Remove the halfshaft and the retaining clamps.

4. Slide the boot down the shaft away from the CV-joint housing.

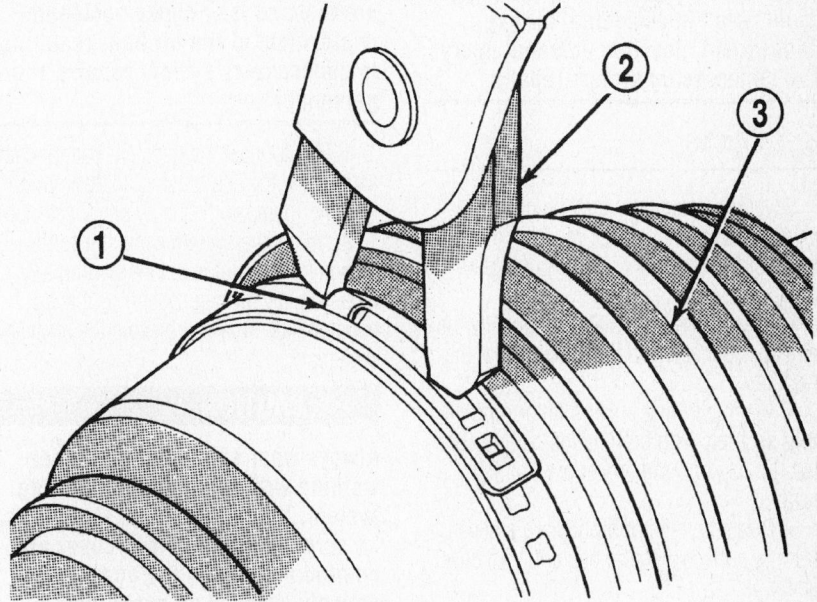

1 - CLAMP
2 - TOOL YA3050, OR EQUIVALENT
3 - SEALING BOOT

23995G46

The latch type boot clamp is tighten as illustrated....

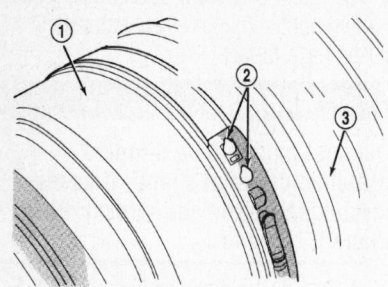

1 - INNER TRIPOD JOINT HOUSING
2 - TOP BAND OF CLAMP MUST BE RETAINED BY TABS AS SHOWN HERE TO CORRECTLY LATCH BOOT CLAMP
3 - SEALING BOOT

23995G47

.... and should look like this when properly installed

5. Remove the grease to expose the CV-joint-to-shaft retaining ring.

6. Spread the snapring ears apart and slide the CV-joint assembly off of the shaft.

7. Slide the boot off the shaft.

8. Thoroughly, clean and inspect all parts for signs of excessive wear; if necessary, replace the halfshaft.

➡**Component parts are not serviceable and must be replaced as an assembly.**

To install:

9. Slide the outer CV-joint boot clamp and boot onto the shaft; then, position the boot so that only the thinnest (sight) groove is visible on the shaft.

10. Slide the outer CV-joint assembly on the shaft, spread the snapring ears, position the CV-joint and verify that the snapring is fully seated in the shaft groove.

11. If installing a new boot, distribute ½ of the grease in the service package into the CV-joint housing and the other ½ inside the boot.

12. Position the outer boot clamp evenly on the sealing boot.

13. Using the Crimper tool C-4975, place the tool over the clamp bridge, tighten the tool nut until the jaws are completely closed (face-to-face).

✳✳ WARNING

The seal must not be dimpled, stretched or out of shape. If necessary, equalize the seal pressure and shape it by hand.

14. Position the boot onto the CV-joint housing retaining groove and install the retaining clamp evenly on the boot.

15. Using the Crimper tool C-4975, place the tool over the clamp bridge, tighten the tool nut until the jaws are completely closed (face-to-face).

16. Install the halfshaft into the vehicle.

STEERING AND SUSPENSION

Air Bag

✳✳ CAUTION

Some vehicles are equipped with an air bag system. The system must be disabled before performing service on or around system components, steering column, instrument panel components, wiring and sensors. Failure to follow safety and disabling procedures could result in accidental air bag deployment, possible personal injury and unnecessary system repairs.

PRECAUTIONS

Several precautions must be observed when handling the inflator module to avoid accidental deployment and possible personal injury.

• Never carry the inflator module by the wires or connector on the underside of the module.

• When carrying a live inflator module, hold securely with both hands, and ensure that the bag and trim cover are pointed away.

• Place the inflator module on a bench or other surface with the bag and trim cover facing up.

• With the inflator module on the bench, never place anything on or close to the module which may be thrown in the event of an accidental deployment.

Before servicing the vehicle, also make sure to refer to the precautions in the beginning of this section as well.

DISARMING

✳✳ CAUTION

The Air Bag system must be disarmed before repair and/or removal of any component in its immediate area including the air bag itself. Failure to do so may cause accidental deployment of the air bag, resulting in unnecessary system repairs and/or personal injury.

1. Disconnect the negative battery cable and isolate the cable using an appropriate insulator (wrap with quality electrical tape).

2. Allow the system capacitor to discharge for 2 minutes before starting any repair on any air bag system or related components. This will disable the air bag system.

✳✳ CAUTION

Always wear safety goggles when working with or around the air bag system. When carrying a live air bag, be sure the bag and trim cover are pointed away from the body. In the unlikely event of an accidental deployment, the bag will, then deploy with minimal chance of injury. When placing a live air bag on a bench or other surface, always face the bag and trim cover up, away from the surface. This will reduce the motion of the module if it is accidentally deployed.

Power Rack and Pinion Steering Gear

REMOVAL & INSTALLATION

1. Before servicing the vehicle, refer to the precautions in the beginning of this section.

2. Turn the front wheels to the straight-ahead position.

3. Remove or disconnect the following:
• Negative battery cable
• Cap from the power steering fluid reservoir and siphon out as much fluid as possible
• Front wheels

4. If transferring tie rod ends to a new gear, loosen the jam nuts.
• Nut attaching tie rod end to the knuckle on both sides
• Tie rod ends from the knuckle using tool C-3894-A

5. Remove the roll pin fastening intermediate shaft extension-to-power steering gear shaft as follows:
   a. Insert removal tool 631A through the roll pin attaching the intermediate shaft extension to the steering gear.
   b. Thread the knurled nut all the way onto the remover.

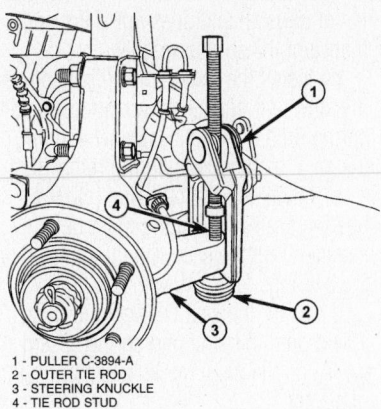

1 - PULLER C-3894-A
2 - OUTER TIE ROD
3 - STEERING KNUCKLE
4 - TIE ROD STUD

23990G48

Use tool C-3894-A to separate the tie rod ends from the knuckle

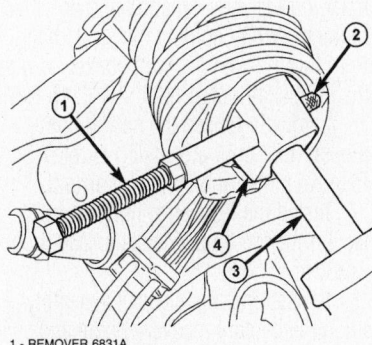

1 - REMOVER 6831A
2 - KNURLED NUT
3 - STEERING GEAR SHAFT
4 - INTERMEDIATE SHAFT EXTENSION

23990G49

Insert removal tool 631A through the roll pin attaching the intermediate shaft extension to the steering gear

c. Hold the remove head steady and turn the hex nut to pull the pin from the shafts and remove the tool.

6. Remove or disconnect the following:
- Intermediate shaft extension off the gear shaft
- Tube nut attaching return hose to the gear
- Return hose from the gear port
- Two gear mounting bolts and the gear.

7. If necessary, remove the tie rod ends counting how many turns it takes as this will aid during installation.

To install:

8. If removed, install the tie rod ends.
9. Install or connect the following:
- Gear through the left wheel well being careful of the tie rod end shields

➡**The drivers side gear mounting bolt is longer than the passenger side bolt.**

- Gear bolts after centering the gear over the mounting bosses over the holes in the cradle and tighten to 120 ft. lbs. (163 Nm)

10. Wipe the power steering hoses on the gear, the connections with a lint free cloth and replace the O-rings coated with clean power steering fluid.
- Pressure and return hoses to the gear and tighten the tube nut to 275 inch lbs. (31 Nm)

11. Find the gear's center of travel and match the splines inside the intermediate shaft extension with those on the gear shaft, then slide the extension onto the shaft.

12. Install the roll pin as follows:

a. Position installer 6381-A onto the extension and gear shaft.

b. Slide the roll pin onto the remover shaft.

c. Thread the knurled nut all the way onto the end of the installer.

d. While holding the install head, turn the hex nut to pull the pin into the shafts. Once the pin is centered in the intermediate shaft extension, remove the tool.

13. Install or connect the following:
- Shaft cover over the top of the gear
- Outer tie rod ends to the knuckle using new nuts and tighten to 35 ft. lbs. (47 Nm), then an additional 180 degree (½ turn)
- Snug the jam nuts
- Wheels
- Negative battery cable

14. Fill and bleed the power steering system.

15. Perform a wheel alignment and tighten the tie rod jam nuts to 55 ft. lbs. (75 Nm).

Strut

REMOVAL & INSTALLATION

Front

1. Before servicing the vehicle, refer to the precautions in the beginning of this section.

➡**Service of the coil spring requires the use of a coil spring compressor tool. It is required that 5 coils be captured within the jaws of the compressor tool.**

➡**Do not support the vehicle by placing supports under the suspension arms. The suspension arms must hang freely.**

2. Remove or disconnect the following:
- Negative battery cable
- Front wheel(s)
- Speed sensor wiring harness mounting bracket from the strut, if equipped with Antilock Brake System (ABS)

➡**When removing the nut from the stud of the stabilizer bar link, do not allow the stud to rotate in the socket by holding with an open ended wrench.**

- Strut-to-steering knuckle bolts and nuts
- Stabilizer bar attaching link at the strut assembly

✳✳ WARNING

The steering knuckle-to-strut assembly bolts are serrated and must not be turned during installation.

3. If removing the left strut assembly, unfasten the nuts attaching the coolant recovery bottle and position the bottle and hoses aside.

4. Remove the 3 strut assembly upper mount-to-shock tower mounting nuts and washers. Remove the strut from the vehicle.

5. Disassemble the strut by performing the following procedure:

a. Securely mount the strut assembly into a vise. Using paint, mark the strut unit, lower spring isolator, spring and upper strut mount for indexing of the parts at assembly.

b. Position the spring compressor tool onto the strut. Compress the coil spring until all load is off the upper strut mount assembly.

c. Install Strut Rod Socket tool 6864 on the strut shaft nut and a 10mm socket on the end of the strut shaft to prevent it from turning. Remove the strut shaft nut.

d. Remove the upper mount assembly, jounce bumper and seat bearing and dust shield as an assembly.

e. Remove the coil spring and compressor as an assembly from the strut. Remove the lower spring isolator from the strut assembly lower spring seat.

f. Inspect all components for abnormal wear, oil leakage or failure. Replace parts as required.

To install:

6. Assemble the strut by performing the following procedure:

a. Inspect the strut assembly for signs of leakage. Actual leakage will be a stream of fluid running down the side and dripping off the lower end of the strut. A slight amount of seepage

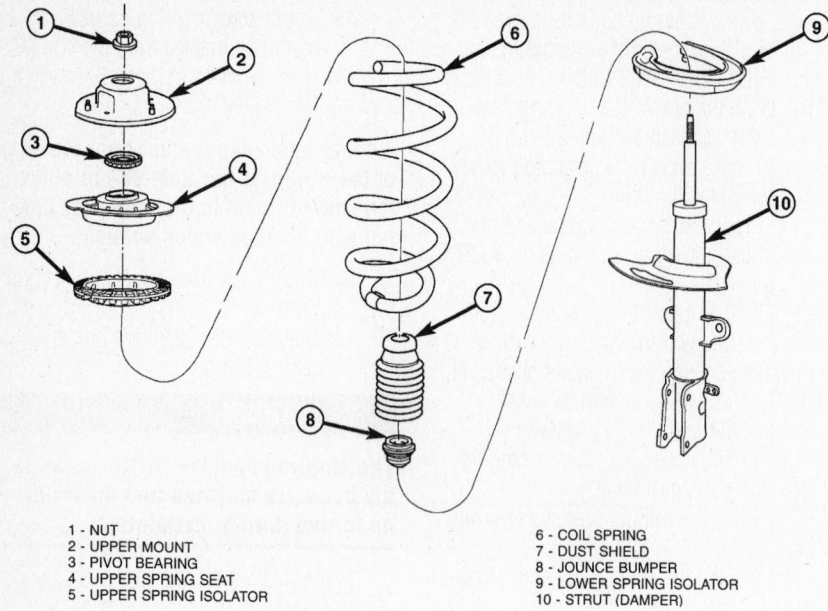

1 - NUT
2 - UPPER MOUNT
3 - PIVOT BEARING
4 - UPPER SPRING SEAT
5 - UPPER SPRING ISOLATOR

6 - COIL SPRING
7 - DUST SHIELD
8 - JOUNCE BUMPER
9 - LOWER SPRING ISOLATOR
10 - STRUT (DAMPER)

23990G50

Exploded view of the front strut assembly

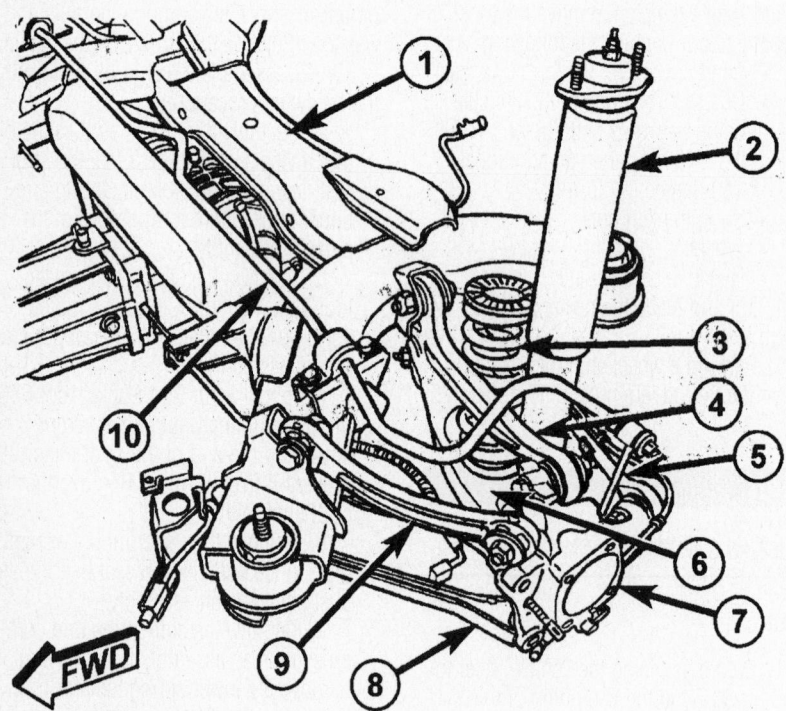

1 - CROSSMEMBER
2 - SHOCK ABSORBER (LOAD-LEVELING)
3 - COIL SPRING
4 - CAMBER LINK
5 - TOE LINK
6 - SPRING LINK
7 - KNUCKLE
8 - COMPRESSION LINK
9 - TENSION LINK
10 - STABILIZER BAR

23990G51

Exploded view of the rear suspension components

between the strut rod and strut shaft seal is not unusual and does not affect performance of the strut assembly.

b. Install the lower spring isolator on the strut unit. Install the compressed coil spring onto the strut assembly aligning the paint marks made during removal.

c. Install the strut bearing into the bearing seat. The bearing must be installed into the seat with the notches on the bearings facing down.

d. Lower the seat bearing and dust shield onto the strut and spring assembly. Align the paint marks made during removal.

e. Install the jounce bumper and upper mount on the strut shaft, aligning the paint marks.

f. Install the strut mount-to-shaft retainer nut. Inspect all alignment marks made during removal and align as required. While holding the strut shaft from turning with a 10mm socket, torque the strut shaft nut to 85 ft. lbs. (115 Nm).

g. Equally loosen the spring compressor tool until all tension is released. Remove the spring compressor tool.

7. Install the front strut into the strut tower. Torque the 3 upper nuts to 250 inch lbs. (28 Nm).

8. If installing the left strut assembly, position the coolant recovery bottle and hoses. Fasten the bottle retainers.

✳✳ WARNING

The steering knuckle-to-strut assembly bolts are serrated and must not be turned during installation.. Also if the vehicle is equipped with eccentric strut assembly bolts, the eccentric bolt must be installed in the bottom (slotted) hole on the strut bract.

➡The strut-to-steering are installed differently on each side. The left hand side bolts are installed from the vehicle rear to front and the right side bolts are installed from the vehicle front to rear.

9. Position the steering knuckle neck into the strut assembly. Install the strut assembly-to-steering knuckle bolts. Install the nuts onto the attaching bolts and torque to 60 ft. lbs. (81 Nm), then tighten an additional ¼ or 90 degrees turn.

➡When installing the nut to the stud of the stabilizer bar link, do not allow the stud to rotate in the socket by holding with an open ended wrench.

10. Install or connect the following:
- Stabilizer bar attaching link to the strut assembly, hand thread the nut then tighten to 65 ft. lbs. (88 Nm) using an open ended wrench on the machine surfaced to prevent from turning.
- Speed sensor wiring harness mounting bracket to the strut and tighten the retainers to 115 inch lbs. (13 Nm)
- Front wheel(s)
- Negative battery cable

11. Align the vehicle.

Rear

1. Before servicing the vehicle, refer to the precautions in the beginning of this section.

2. Remove or disconnect the following:
- Rear wheel

3. Position jack stands under the forward end of the engine cradle to support and stabilizer the vehicle.

4. Place a transmission jack under the center of the rear suspension crossmember on models not equipped with All Wheel Drive (AWD) or rear driveline module, if equipped with AWD.

5. Remove or disconnect the following:
- Two shock absorber upper bolts
- Lower shock absorber bolts
- Front and rear crossmember bolts

6. Lower the jack slowly enough so to allow the top of the shock absorber to clear the body flange and remove the shock absorber by tipping it outwards and lifting the lower end out of the pocket in the spring link.

7. Remove the coil spring and isolator lower end first, if necessary. Never lower the jack any more than necessary for removal and installation.

To install:

8. Install or connect the following:
- Coil spring isolator on top of the sprint, then the spring top end first. Match the lower end coil against the abutment in the spring link.
- Shock absorber by setting the lower end into the pocket in the spring link, then tipping the top inwards. Hand start the upper mounting bolts

9. Raise the jack slowly guiding the spring and lower end of the shock absorber into position. Once the bolt holes are aligned, stop jacking. Install and hand tighten the bolts.

10. Raise the jack until the two cross-

member bolts can be installed and tighten to 120 ft. lbs. (163 Nm).

11. Tighten the upper shock absorber bolts to 45 ft. lbs. (61 Nm) and remove the jack.

12. Install the wheels, position the vehicle on an alignment rack and raise the vehicle enough to access the lower shock absorber bolts. Tighten the bolts to 75 ft. lbs. (102 Nm).

13. Have the rear wheel toe set to specifications.

Coil Springs

REMOVAL & INSTALLATION

Front

Refer to the front strut removal and installation procedure for coil spring service information.

Rear

Refer to the rear strut removal and installation procedure for coil spring service information.

Lower Ball Joint

➡The lower ball joints on these vehicles are not serviced separately. The lower ball joints operate with no freeplay. If defective, the entire lower control arm must be replaced.

Lower Control Arm

REMOVAL & INSTALLATION

1. Before servicing the vehicle, refer to the precautions in the beginning of this section.

2. Remove or disconnect the following:
- Negative battery cable
- Front wheels
- Hub and bearing-to-stub axle cotter pin and retainer nut
- Front caliper assembly from steering knuckle
- Front brake rotor from the hub
- Wheel Speed Sensor (WSS) harness connector and unclip the connector and retaining clip from the frame rail
- Screw attaching the WSS routing bracket to the strut, open the routing clip at the knuckle and remove the cable

3. Push in the outer end of the halfshaft to disengage the splines from the hub splines.

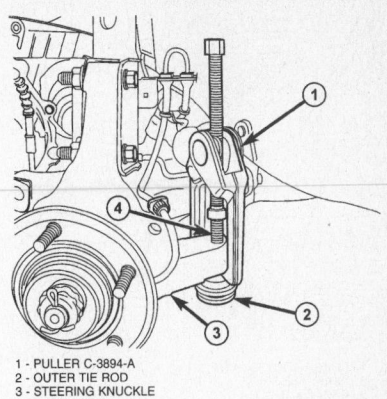

1 - PULLER C-3894-A
2 - OUTER TIE ROD
3 - STEERING KNUCKLE
4 - TIE ROD STUD

23990G48

Use tool C-3894-A to separate the tie rod ends from the knuckle

- Tie rod end nut from the steering knuckle by holding the tie rod while loosening the nut, then use tool C-3894-A to separate the tie rod from the knuckle
- Steering knuckle-to-strut bolts
- Steering knuckle from the strut bracket

4. Tip the knuckle outward at the top and remove the halfshaft from the hum and suspend the halfshaft using wire.
- Ball joint nut using an impact gun, then reinstall the nut until the top of the nut is even with the stud. this will prevent the stud from distorting while performing the next step.

5. Place removal tool C-4150A over the ball joint and stud and tighten the tool to

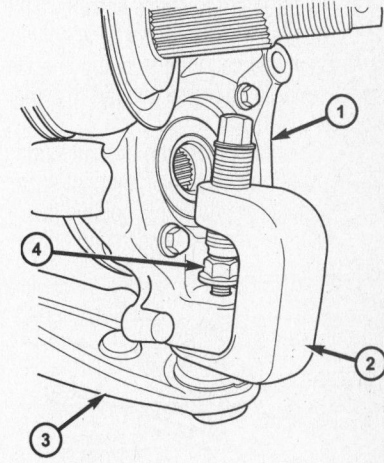

1 - ALUMINUM KNUCKLE
2 - SPECIAL TOOL C-4150A
3 - LOWER CONTROL ARM
4 - NUT INSTALLED ON BALL JOINT STEM

23990G52

Place removal tool C-4150A over the ball joint and stud and tighten the tool to release the stud. It may help to rotate the knuckle around so the inside of the knuckle faces outwards

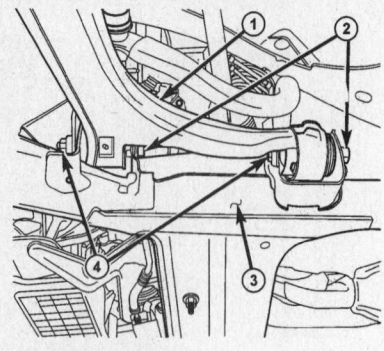

1 - LOWER CONTROL ARM
2 - MOUNTING BOLTS
3 - ENGINE CRADLE
4 - FLAG NUTS

23990G53

Exploded view of the lower control arm
mounting, note the orientation of the bolts
for installation purposes

release the stud. It may help to rotate the
knuckle around so the inside of the knuckle
faces outwards, then remove the tool.

- Knuckle
- Bolts attaching the lower control
  arm to the engine cradle and the
  control arm

To install:

➡**When installing the lower control
arm-to-cradle rear mounting bolts,
make sure the flag on the left control
arm nut are positioned upwards (above
the control arm and the right control
arm nut are positioned downwards
(below the control arm).**

6. Install or connect the following:
- New lower control arm and the
  control arm to cradle bolts from the
  rear as illustrated and hand tighten
7. Clean the ball joint stud and knuckle
contact surfaces of grease and dirt to avoid
damage.
- Steering knuckle on the ball stud
- New knuckle to ball joint nut, then
  tighten the nut while holding the
  stud with a hex wrench and tighten

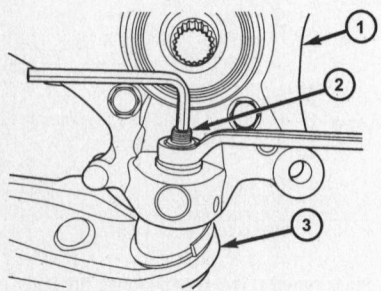

1 - KNUCKLE
2 - BALL JOINT STUD
3 - LOWER CONTROL ARM

23990G54

Tightening the ball joint nut

to 60 ft. lbs. (81 Nm) plus an additional 90 degrees (¼) turn
- Halfshaft into the hub assembly

✳✳ WARNING

**The steering knuckle-to-strut assem-
bly bolts are serrated and must not
be turned during installation.. Also if
the vehicle is equipped with eccentric strut assembly bolts, the eccentric bolt must be installed in the
bottom (slotted) hole on the strut
bract.**

➡**The strut-to-steering are installed
differently on each side. The left hand
side bolts are installed from the
vehicle rear to front and the right side
bolts are installed from the vehicle
front to rear.**

8. Position the steering knuckle neck
into the strut assembly. Install the strut
assembly-to-steering knuckle bolts. Install
the nuts onto the attaching bolts and torque
to 60 ft. lbs. (81 Nm), then tighten an additional ¼ or 90 degrees turn.
9. Clean the tie rod end and knuckle
contact areas to prevent damage.
10. Install or connect the following:
- Tie rod end to the knuckle and
  hand tighten the nut. While using a
  socket to hold the stud, tighten the
  nut to 35 ft. lbs. (47 Nm) plus an
  additional 180 degree (½ turn).
- WSS sensor clip, cable, routing
  clip and attach the connection
- Rotor and caliper
11. Have an assistant apply the brakes,
install the washer, and hub nut. Tighten the
nut to 180 ft. lbs. (244 Nm).
- Spring washer , new hub nut and
  cotter pin
- Wheels
12. Check and adjust the brake fluid and
brakes.
13. Align the vehicle.

CONTROL ARM BUSHING REPLACEMENT

The control arm bushings are an integral part of the assembly. If worn or damaged, the control arm assembly must be
replaced.

Wheel Bearings

ADJUSTMENT

These front wheel drive vehicles are
equipped with permanently sealed front and
rear wheel bearings. There is no periodic

lubrication or maintenance recommended
for these units.

REMOVAL & INSTALLATION

Front

1. Before servicing the vehicle, refer to
the precautions in the beginning of this section.
2. Remove or disconnect the following:
- Negative battery cable
- Front wheels
- Hub and bearing-to-stub axle cotter
  pin and retainer nut
- Front caliper assembly from steering knuckle
- Front brake rotor from the hub
- Wheel Speed Sensor (WSS) harness connector and unclip the connector and retaining clip from the
  frame rail
- Screw attaching the WSS routing
  bracket to the strut, open the routing clip at the knuckle and remove
  the cable
3. Push in the outer end of the halfshaft
to disengage the splines from the hub
splines.
- Four hub/bearing bolts from the
  rear of the knuckle
- Hub/bearing with the WSS
To install:
4. Clean all mating and mounting surfaces prior to installation to prevent damage.
5. Install or connect the following:
- Hub/bearing onto the halfshaft stub
  shaft and into the knuckle until it is
  squarely seated on the face of the
  knuckle.
- Four hub/bearing bolts and using a
  crisscross pattern tighten progres-
  sively and equally to 45 ft. lbs. (65
  Nm)
- WSS sensor clip, cable, routing
  clip and attach the connection
- Rotor and caliper

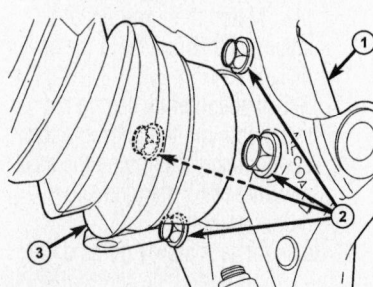

1 - KNUCKLE
2 - HUB AND BEARING MOUNTING BOLTS
3 - HALFSHAFT

23990G58

**Location of the front hub/bearing mounting
bolts**

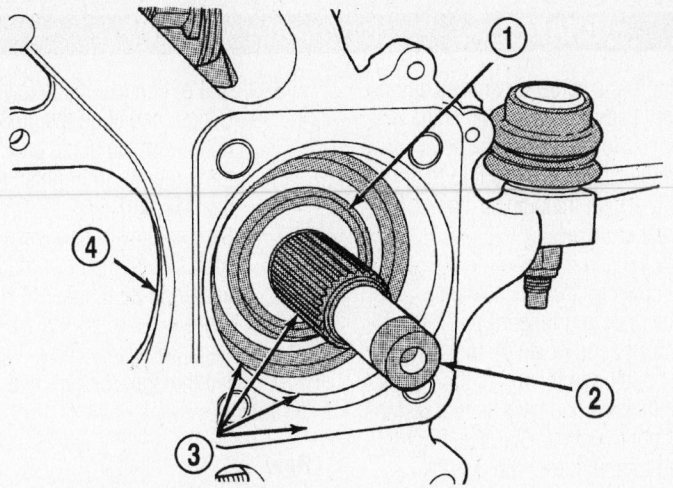

1 - HALFSHAFT OUTER C/V JOINT
2 - STUB SHAFT
3 - THESE SURFACES MUST BE CLEAN AND FREE OF NICKS
BEFORE INSTALLING BEARING ASSEMBLY
4 - STEERING KNUCKLE

23990G59

Exploded view of the front hub/bearing assembly

6. Have an assistant apply the brakes, install the washer, and hub nut. Tighten the nut to 180 ft. lbs. (244 Nm).
- Spring washer , new hub nut and cotter pin
- Wheels

7. Check and adjust the brake fluid and brakes.

8. Align the vehicle.

Rear

1. Before servicing the vehicle, refer to the precautions in the beginning of this section.

2. Remove or disconnect the following:
- Rear wheel

3. If equipped with All Wheel Drive (AWD), remove the cotter pin, nut and spring washer. Have someone apply the brake pedal and remove the hub nut.
- Brake caliper and rotor
- Wheel Speed Sensor (WSS) connector from the spare tire mounting support
- Two sensor cable routing clips along the toe link
- Sensor cable from the bracket on the brake support

- Loosen but do not remove the hub/bearing retaining bolts. Once loosened, push the bolts against the rear of the hub to keep the brake support in place when the hub is removed.
- Hub and thread the WSS cable through the hole in the brake support plate as you remove the hub

To install:

4. Install or connect the following:
- Hub/bearing bolts through the rear of the knuckle and parking brake support just enough to hold the support in place
- Hub/bearing first feeding the WSS cable through the hole in the brake support. At the same time, slide the hub/bearing onto the halfshaft, if equipped with AWD. Place the hub/bearing through the brake support onto the knuckle and align the bolt mounting holes with bolts and place the WSS head at the bottom.
- Four hub/bearing bolts and tighten to 60 ft. lbs. (81 Nm).
- Sensor cable onto the bracket on the brake support and route the cable

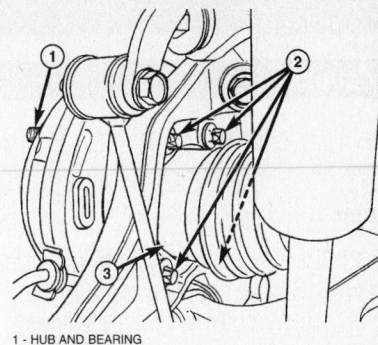

1 - HUB AND BEARING
2 - MOUNTING BOLTS
3 - HALF SHAFT (IF EQUIPPED)

23990G60

Location of the rear hub/bearing mounting bolts

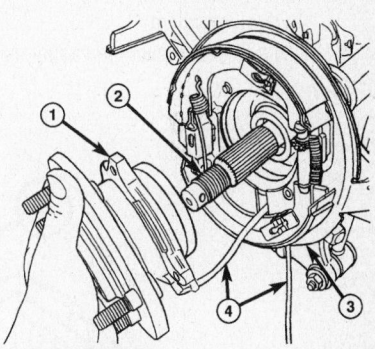

1 - HUB AND BEARING
2 - HALF SHAFT (IF EQUIPPED)
3 - BRAKE SUPPORT PLATE
4 - WHEEL SPEED SENSOR CABLE

23990G61

When removing/installing the rear hub/bearing assembly, thread the WSS cable through the hole in the brake support plate

- Two sensor cable routing clips along the toe link
- WSS connector to the spare tire mounting support
- Brake caliper and rotor

5. If equipped with AWD, install the hub nut, have someone apply the brake and tighten the nut to 180 ft. lbs. (244 Nm). Install the spring washer, nut lock and cotter pin.
- Wheels

6. Pump the brake pedal several times to ensure proper operation and road test the vehicle.

BRAKES

Caliper

REMOVAL & INSTALLATION

Front

1. Before servicing the vehicle, refer to the precautions in the beginning of this section.
2. Remove or disconnect the following:
   - Negative battery cable
   - Front wheels
   - Banjo bolt retaining the brake hose to the caliper. Be sure to plug the end of the brake hose or cover it with a plastic bag to prevent contamination from entering the hydraulic system.
3. Place a C-clamp over the caliper as illustrated, place a screw drive head against the outboard pad and hook against the rear of the caliper. Slowly tighten the screw drive and retract the caliper pistons into their bores and breaking the outboard brake pad from the caliper fingers.
4. Once the pad is free, slide the caliper in on the guides to provide clearance between the rotor and inboard pad.
5. Place an appropriate prytool through the center opening in the top of the caliper behind the inboard pad between the pistons using care not to contact the piston boots. Pry the pad to free it from the pistons.
6. Remove the two caliper guide pin bolts and caliper.

To install:

7. Completely compress the caliper pistons

8. Install or connect the following:
   - Caliper and align the guide pin bolt holes with the guide pins. Install the pin bolts and tighten to 32 ft. lbs. (43 Nm).
   - Banjo bolt with new washers on each side of the hose fitting and tighten to 35 ft. lbs. (47 Nm)
   - Wheels
   - Negative battery cable
9. Bleed the brake system and road test the vehicle.

Rear

1. Before servicing the vehicle, refer to the precautions in the beginning of this section.
2. Remove or disconnect the following:

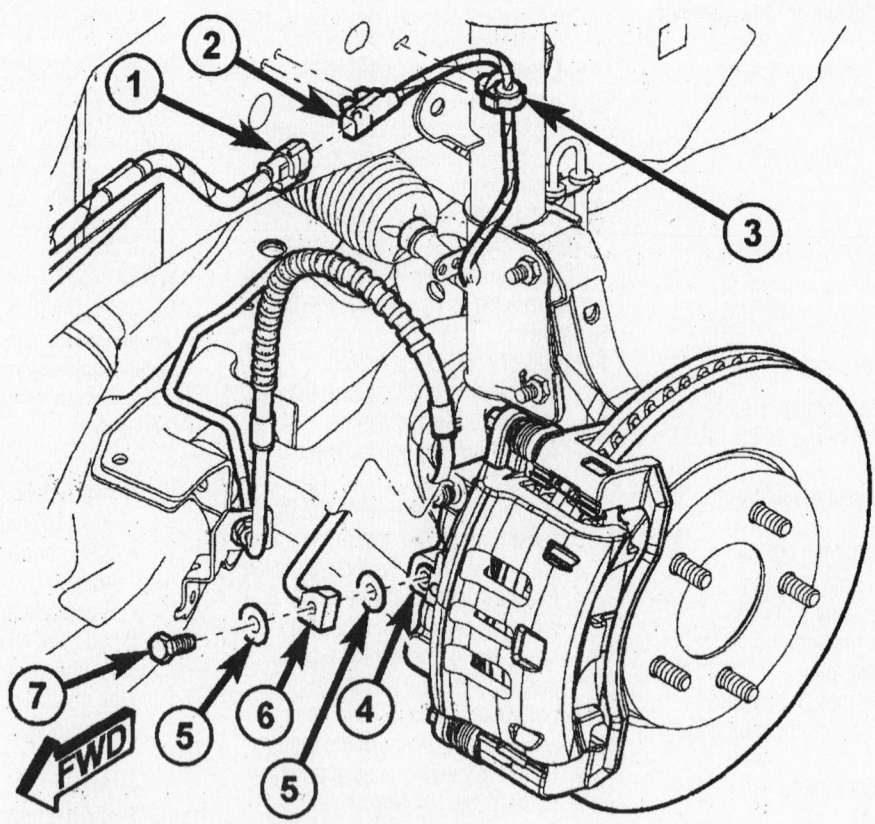

1 - WIRING HARNESS CONNECTOR
2 - WHEEL SPEED SENSOR CONNECTOR
3 - ROUTING CLIP
4 - PORT ON BRAKE CALIPER
5 - COPPER SEALING WASHER
6 - BRAKE HOSE BANJO FITTING
7 - BANJO BOLT

Exploded view of the banjo bolt and related components

23990G57

- Negative battery cable
- Rear wheels
- Banjo bolt retaining the brake hose to the caliper. Be sure to plug the end of the brake hose or cover it with a plastic bag to prevent contamination from entering the hydraulic system.
- Caliper guide pin bolts
- Caliper assembly from the brake adapter by rotating the bottom of the caliper away from the rotor, then lift the caliper with the pads away from the adapter abutment
- Brake pads by pushing (outboard) or pulling (inboard) from the caliper fingers and piston

To install:

3. Completely compress the caliper piston

4. Install or connect the following:
- Inboard pad clip against the piston

cavity and press the pad until the clip is seated making sure the pad backing plate is flush against the piston

➡**The outboard pads are side oriented, make sure the spring clip is installed so it is positioned downwards when the caliper is installed.**

- Outboard pad making sure the locating pins are positioned against the ramps. Slide the pad onto the caliper and ensure the locating pins are squarely seated into the holes on the caliper and the pad is flush against the caliper fingers.

5. Make sure the abutment shims are in place on both slide abutments.

6. Retract the caliper guide pins to clear the caliper adapter bosses.
- Brake caliper. Staring with the upper end, position the caliper and shoes

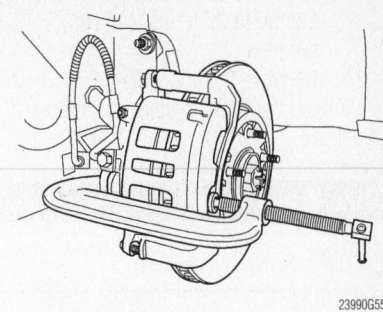

23990G55

Place a C-clamp over the caliper as illustrated

over the rotor and align the outboard pad upper edge with the caliper slide abutment. Rotate the lower end of the caliper into position
- Caliper guide pin bolts and tighten to 200 inch lbs. (23 Nm)
- Banjo bolt with new washers on

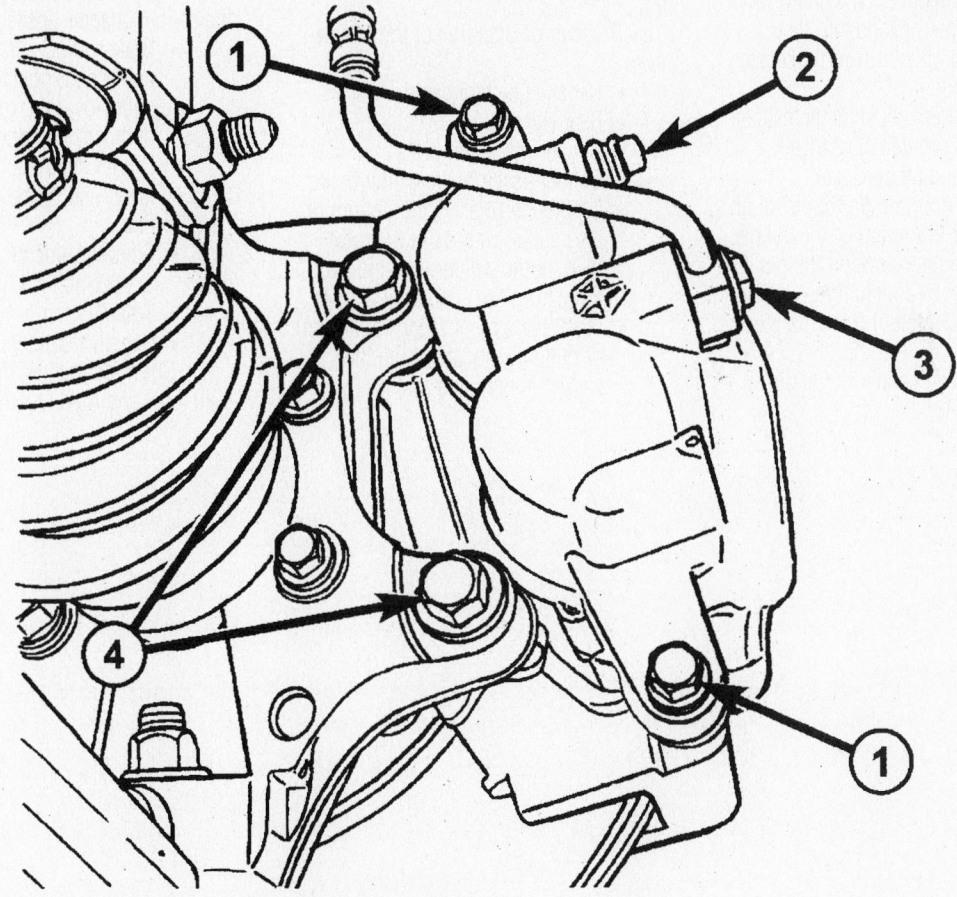

1 - CALIPER GUIDE PIN BOLTS
2 - BLEEDER SCREW
3 - BRAKE HOSE BANJO BOLT
4 - CALIPER ADAPTER MOUNTING BOLTS

23990G56

Exploded view of the caliper and adapter mounting

each side of the hose fitting and tighten to 35 ft. lbs. (47 Nm)
- Wheels
- Negative battery cable

7. Bleed the brake system and road test the vehicle.

Disc Brake Pads

REMOVAL & INSTALLATION

Front

1. Before servicing the vehicle, refer to the precautions in the beginning of this section.

2. Remove or disconnect the following:
- Negative battery cable
- Front wheels

3. Place a C-clamp over the caliper as illustrated, place a screw drive head against the outboard pad and hook against the rear of the caliper. Slowly tighten the screw drive and retract the caliper pistons into their bores and breaking the outboard brake pad from the caliper fingers.

4. Once the pad is free, slide the caliper in on the guides to provide clearance between the rotor and inboard pad.

5. Place an appropriate prytool through the center opening in the top of the caliper behind the inboard pad between the pistons using care not to contact the piston boots. Pry the pad to free it from the pistons.

6. Remove the two caliper guide pin

bolts and position the caliper aside. Remove the brake pads.

To install:

7. Completely compress the caliper pistons

8. Install or connect the following:
- Brake pads with anti-rattle clips onto the adapter
- Caliper and align the guide pin bolt holes with the guide pins. Install the pin bolts and tighten to 32 ft. lbs. (43 Nm).
- Banjo bolt with new washers on each side of the hose fitting and tighten to 35 ft. lbs. (47 Nm)
- Wheels
- Negative battery cable

9. Check and adjust the brake system fluid level and road test the vehicle.

Rear

1. Before servicing the vehicle, refer to the precautions in the beginning of this section.

2. Remove or disconnect the following:
- Negative battery cable
- Rear wheels
- Caliper guide pin bolts
- Caliper assembly from the brake adapter by rotating the bottom of the caliper away from the rotor, then lift the caliper with the pads away from the adapter abutment
- Brake pads by pushing (outboard) or pulling (inboard) from the caliper fingers and piston

To install:

3. Completely compress the caliper piston

4. Install or connect the following:
- Inboard pad clip against the piston cavity and press the pad until the clip is seated making sure the pad backing plate is flush against the piston

➡ **The outboard pads are side oriented. Make sure the spring clip is installed so it is positioned downwards when the caliper is installed.**

- Outboard pad making sure the locating pins are positioned against the ramps. Slide the pad onto the caliper and ensure the locating pins are squarely seated into the holes on the caliper and the pad is flush against the caliper fingers.

5. Make sure the abutment shims are in place on both slide abutments.

6. Retract the caliper guide pins to clear the caliper adapter bosses.
- Brake caliper. Staring with the upper end, position the caliper and shoes over the rotor and align the outboard pad upper edge with the caliper slide abutment. Rotate the lower end of the caliper into position
- Caliper guide pin bolts and tighten to 200 inch lbs. (23 Nm)
- Wheels
- Negative battery cable

7. Check and adjust the brake system fluid level and road test the vehicle.

CHRYSLER

PT Cruiser

10

SPECIFICATION CHARTS

ENGINE AND VEHICLE IDENTIFICATION

		Engine					Code ②	Model Year
Code ①	Liters (cc)	Cu. In.	Cyl.	Fuel Sys.	Engine Type	Eng. Mfg.		Year
B	2.4 (2429)	148	L4	SMFI	DOHC	Chrysler	1	2001
DOHC: Double Overhead Camshaft							2	2002
① 8th position of VIN							3	2003
② 10th position of VIN							4	2004

67189-PTCR-C01

GENERAL ENGINE SPECIFICATIONS

Year	Model	Engine Displacement Liters (VIN)	Net Horsepower @ rpm	Net Torque @ rpm (ft. lbs.)	Bore x Stroke (in.)	Compression Ratio	Oil Pressure @ rpm
2001	PT Cruiser	2.4 (B)	150@5200	164@4000	3.44x3.98	9.4:1	25-80@3000
2002	PT Cruiser	2.4 (B)	150@5200	164@4000	3.44x3.98	9.4:1	25-80@3000
2003	PT Cruiser	2.4 (B)	150@5200	164@4000	3.44x3.98	9.4:1	25-80@3000
2004-05	PT Cruiser	2.4 (B)	150@5200	164@4000	3.44x3.98	9.4:1	25-80@3000

SMFI: Sequential Multi-port Fuel Injection

67189-PTCR-C02

ENGINE TUNE-UP SPECIFICATIONS

Year	Engine Displacement Liters (VIN)	Spark Plug Gap (in.)	Ignition Timing (deg.)	Fuel Pump (psi)	Idle Speed (rpm)	Valve Clearance In.	Valve Clearance Ex.
2001	2.4 (B)	0.048-0.053	①	49	②	HYD	HYD
2002	2.4 (B)	0.048-0.053	①	49	②	HYD	HYD
2003	2.4 (B)	0.048-0.053	①	49	②	HYD	HYD
2004-05	2.4 (B)	0.048-0.053	①	49	②	HYD	HYD

NOTE: The Vehicle Emission Control Information label often reflects specification changes made during production. The label figures must be used if they differ from those in this chart.

HYD: Hydraulic

① Ignition timing is regulated by the Powertrain Control Module (PCM), and cannot be adjusted.

② Idle speed is controled by the Powertrain Control Module (PCM), and cannot be adjusted.

67189-PTCR-C03

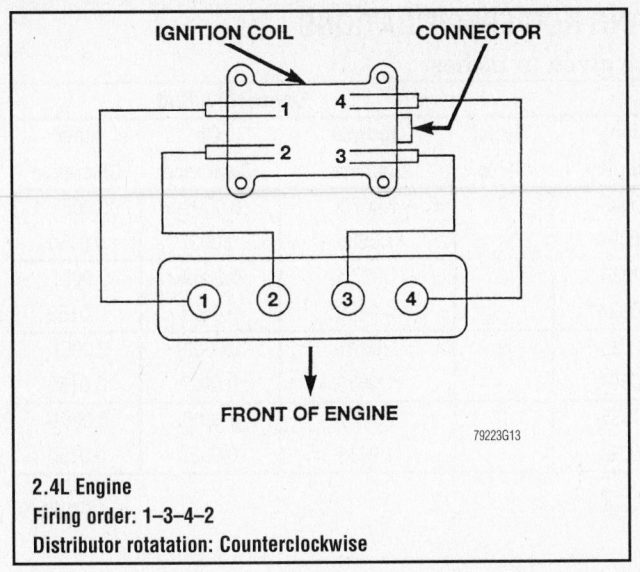

IGNITION COIL **CONNECTOR**

FRONT OF ENGINE

79223G13

2.4L Engine
Firing order: 1–3–4–2
Distributor rotatation: Counterclockwise

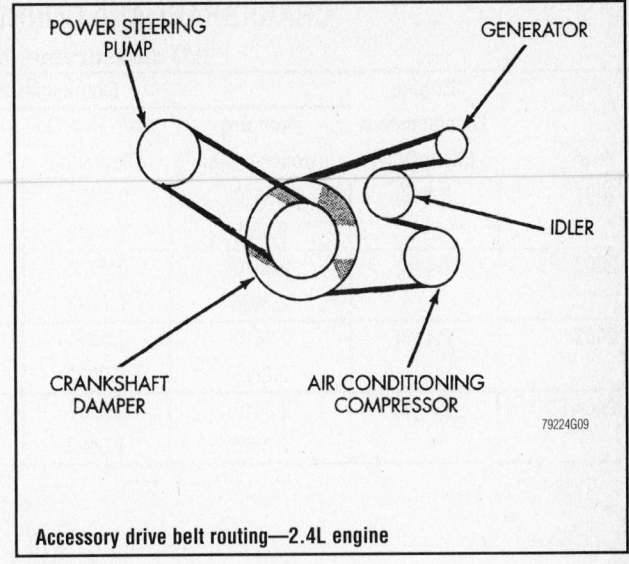

POWER STEERING PUMP — GENERATOR — IDLER — CRANKSHAFT DAMPER — AIR CONDITIONING COMPRESSOR

79224G09

Accessory drive belt routing—2.4L engine

CAPACITIES

Year	Model	Engine Displacement Liters (VIN)	Engine Oil with Filter (qts.)	Automatic Transaxle (qts.)	Power Transfer Unit (qts.)	Rear Drive Axle (pts.)	Fuel Tank (gal.)	Cooling System (qts.)
2001	PT Cruiser	2.4 (B)	4.5	①	1.22	②	20.0	9.5
2002	PT Cruiser	2.4 (B)	4.5	①	1.22	②	20.0	9.5
2003	PT Cruiser	2.4 (B)	4.5	①	1.22	②	20.0	9.5
2004-05	PT Cruiser	2.4 (B)	4.5	①	1.22	②	20.0	9.5

NOTE: All capacities are approximate. Add fluid gradually and check to be sure a proper fluid level is obtained.

① 31TH overhaul fill capacity with torque converter empty: 8.5 qts.

  41TE overhaul fill capacity with torque converter empty: 9.1 qts.

② Overrunning clutch: 0.75 pts.

67189-PTCR-C04

VALVE SPECIFICATIONS

Year	Engine Displacement Liters (VIN)	Seat Angle (deg.)	Face Angle (deg.)	Spring Test Pressure (lbs. @ in.)	Spring Installed Height (in.)	Stem-to-Guide Clearance (in.) Intake	Stem-to-Guide Clearance (in.) Exhaust	Stem Diameter (in.) Intake	Stem Diameter (in.) Exhaust
2001	2.4 (B)	45	44.5-45.0	129-143@ 1.17	1.50	0.0018-0.0025	0.0029-0.0037	0.2340	0.2330
2002	2.4 (B)	45	44.5-45.0	129-143@ 1.17	1.50	0.0018-0.0025	0.0029-0.0037	0.2340	0.2330
2003	2.4 (B)	45	44.5-45.0	129-143@ 1.17	1.50	0.0018-0.0025	0.0029-0.0037	0.2340	0.2330
2004-05	2.4 (B)	45	44.5-45.0	129-143@ 1.17	1.50	0.0018-0.0025	0.0029-0.0037	0.2340	0.2330

① Intake valve: 44.5 degrees

  Exhaust valve: 45 degrees

67189-PTCR-C05

CRANKSHAFT AND CONNECTING ROD SPECIFICATIONS

All measurements are given in inches.

| Year | Engine Displacement Liters (VIN) | Crankshaft | | | | Connecting Rod | | |
		Main Brg. Journal Dia.	Main Brg. Oil Clearance	Shaft End-play	Thrust on No.	Journal Diameter	Oil Clearance	Side Clearance
2001	2.4 (B)	2.3610-2.3625	0.0007-0.0023	0.0035-0.0094	2	1.9670-1.9685	0.0009-0.0027	0.0051-0.0150
2002	2.4 (B)	2.3610-2.3625	0.0007-0.0023	0.0035-0.0094	2	1.9670-1.9685	0.0009-0.0027	0.0051-0.0150
2003	2.4 (B)	2.3610-2.3625	0.0007-0.0023	0.0035-0.0094	2	1.9670-1.9685	0.0009-0.0027	0.0051-0.0150
2004-05	2.4 (B)	2.3610-2.3625	0.0007-0.0023	0.0035-0.0094	2	1.9670-1.9685	0.0009-0.0027	0.0051-0.0150

67189-PTCR-C06

PISTON AND RING SPECIFICATIONS

All measurements are given in inches.

| Year | Engine Displacement Liters (VIN) | Piston Clearance | Ring Gap | | | Ring Side Clearance | | |
			Top Compression	Bottom Compression	Oil Control	Top Compression	Bottom Compression	Oil Control
2001	2.4 (B)	0.0009-0.0022	0.0098-0.0200	0.0090-0.0180	0.0098-0.0250	0.0011-0.0031	0.0011-0.0031	0.0004-0.0070
2002	2.4 (B)	0.0009-0.0022	0.0098-0.0200	0.0090-0.0180	0.0098-0.0250	0.0011-0.0031	0.0011-0.0031	0.0004-0.0070
2003	2.4 (B)	0.0009-0.0022	0.0098-0.0200	0.0090-0.0180	0.0098-0.0250	0.0011-0.0031	0.0011-0.0031	0.0004-0.0070
2004-05	2.4 (B)	0.0009-0.0022	0.0098-0.0200	0.0090-0.0180	0.0098-0.0250	0.0011-0.0031	0.0011-0.0031	0.0004-0.0070

① Oil control ring side rails must be free to rotate after assembly

67189-PTCR-C07

TORQUE SPECIFICATIONS

All readings in ft. lbs.

| Year | Engine Displacement Liters (cc) | Engine ID/VIN | Cylinder Head Bolts | Main Bearing Bolts | Rod Bearing Bolts | Crankshaft Damper Bolts | Flywheel Bolts | Manifold | | Spark Plugs |
								Intake	Exhaust	
2001	2.4 (2429)	B	①	②	③	100	70	20	17	20
2002	2.4 (2429)	B	①	②	③	100	70	20	17	20
2003	2.4 (2429)	B	①	②	③	100	70	④	17	20
2004-05	2.4 (2429)	B	①	②	③	100	70	④	17	20

① Step 1: 25 ft. lbs.
  Step 2: 50 ft. lbs.
  Step 3: 50 ft. lbs.
  Step 4: Plus 1/4 turn

② M8 bolts: 20 ft. lbs.
  M11 bolts: 30 ft. lbs. plus 1/4 turn

③ Step 1: 20 ft. lbs.
  Step 2: Plus 1/4 turn

④ Lower manifold: non turbo models 105 inch lbs.
  Lower manifold: turbo models 250 inch lbs.
  Upper manifold: non turbo models 105 inch lbs.
  Upper manifold: turbo models 250 inch lbs.

67189-PTCR-C08

WHEEL ALIGNMENT SPECIFICATIONS

Year	Model		Caster Range (+/-Deg.)	Caster Preferred Setting (Deg.)	Camber Range (+/-Deg.)	Camber Preferred Setting (Deg.)	Toe-in (in.)
2001	PT Cruiser	F	+1.00	+2.45	+0.40	0.00	0.00 + 0.40
		R	—	—	+0.40	+0.20	+0.00 + 0.40
2002	PT Cruiser	F	+1.00	+2.45	+0.40	0.00	0.00 + 0.40
		R	—	—	+0.40	+0.20	+0.00 + 0.40
2003	PT Cruiser	F	+1.00	+2.45	+0.40	0.00	0.00 + 0.40
		R	—	—	+0.40	+0.20	+0.00 + 0.40
2004-05	PT Cruiser	F	+1.00	+2.45	+0.40	0.00	0.00 + 0.40
		R	—	—	+0.40	+0.20	+0.00 + 0.40

NA: Not Available

67189-PTCR-C09

TIRE, WHEEL AND BALL JOINT SPECIFICATIONS

Year	Model	OEM Tires Standard	OEM Tires Optional	Tire Pressures (psi) Front	Tire Pressures (psi) Rear	Wheel Size	Ball Joint Inspection	Lug Nuts
2001	PT Cruiser	P205/55R16	None	32	32	5.5-J	①	100
2002	PT Cruiser	P205/55R16	None	32	32	5.5-J	①	100
2003	PT Cruiser	P205/55R16	None	32	32	5.5-J	①	100
2004-05	PT Cruiser	P205/55R16	None	32	32	5.5-J	①	100

OEM: Original Equipment Manufacturer

PSI: Pounds Per Square Inch

① Replace if any measurable movement is found

67189-PTCR-C10

BRAKE SPECIFICATIONS
All measurements in inches unless noted

Year	Model		Brake Disc Original Thickness	Brake Disc Minimum Thickness	Brake Disc Maximum Run-out	Brake Drum Diameter Original Inside Diameter	Max. Wear Limit	Maximum Machine Diameter	Min. Lining Thickness	Caliper Guide Pin Bolts (ft. lbs.)
2001	PT Cruiser	F	0.902-0.909	0.803	0.005	—	—	—	0.313	26
		R	0.344-0.364	0.285	0.005	8.63-8.65	NA	NA	①	16
2002	PT Cruiser	F	0.902-0.909	0.803	0.005	—	—	—	0.313	26
		R	0.344-0.364	0.285	0.005	8.63-8.65	NA	NA	①	16
2003	PT Cruiser	F	0.902-0.909	0.803	0.005	—	—	—	0.313	26
		R	0.344-0.364	0.285	0.005	8.63-8.65	NA	NA	①	16
2004-05	PT Cruiser	F	0.902-0.909	0.803	0.005	—	—	—	0.313	26
		R	0.344-0.364	0.285	0.005	8.63-8.65	NA	NA	①	16

NA: Not Available

F: Front

R: Rear

① Rear Disc: 0.350 in.

  Rear Bonded Shoes: 0.062 in.

  Rear Riveted Shoes: 0.031 in.

67189-PTCR-C11

SCHEDULED MAINTENANCE INTERVALS
Chrysler—PT Cruiser

TO BE SERVICED	TYPE OF SERVICE	VEHICLE MILEAGE INTERVAL (x1000)												
		7.5	15	22.5	30	37.5	45	52.5	60	67.5	75	82.5	90	97.5
Engine oil & filter	R	✓	✓	✓	✓	✓	✓	✓	✓	✓	✓	✓	✓	✓
Brake hoses	S/I	✓	✓	✓	✓	✓	✓	✓	✓	✓	✓	✓	✓	✓
Coolant level, hoses & clamps	S/I	✓	✓	✓	✓	✓	✓	✓	✓	✓	✓	✓	✓	✓
CV joints & front suspension components	S/I	✓	✓	✓	✓	✓	✓	✓	✓	✓	✓	✓	✓	✓
Exhaust system	S/I	✓	✓	✓	✓	✓	✓	✓	✓	✓	✓	✓	✓	✓
Manual transaxle oil	S/I	✓	✓	✓	✓	✓	✓	✓	✓	✓	✓	✓	✓	✓
Rotate tires	S/I	✓	✓	✓	✓	✓	✓	✓	✓	✓	✓	✓	✓	✓
Accessory drive belts	S/I		✓		✓		✓		✓		✓		✓		
Brake linings	S/I			✓			✓			✓			✓		
Air filter element	R				✓				✓				✓		
Spark plugs	R				✓				✓				✓		
Lubricate ball joints	S/I				✓				✓				✓		
Engine coolant ①	R														
PCV valve	S/I								✓				✓		
Ignition cables	R								✓						
Camshaft timing belt ②	R														

R: Replace   S/I: Service or Inspect

① Engine coolant: flush and replace at 100,000 miles.

② Camshaft timing belt: replace at 120,000 miles.

FREQUENT OPERATION MAINTENANCE (SEVERE SERVICE)

If a vehicle is operated under any of the following conditions it is considered severe service:

- Extremely dusty areas.

- 50% or more of the vehicle operation is in 32°C (90°F) or higher temperatures, or constant operation in temperatures below 0°C (32°F).

- Prolonged idling (vehicle operation in stop and go traffic).

- Frequent short running periods (engine does not warm to normal operating temperatures).

- Police, taxi, delivery usage or trailer towing usage.

Oil & oil filter change: change every 3000 miles.

Rotate tires every 6000 miles.

Brake linings: inspect every 12,000 miles.

Air filter element: service or inspect every 15,000 miles.

Automatic transaxle: change fluid & adjust bands every 15,000 miles.

Manual transaxle fluid: replace every 15,000 miles.

Engine coolant: replace at 36,000 miles and every 30,000 miles thereafter.

67189-PTCR-C12

PRECAUTIONS

Before servicing any vehicle, please be sure to read all of the following precautions, which deal with personal safety, prevention of component damage, and important points to take into consideration when servicing a motor vehicle:

• Never open, service or drain the radiator or cooling system when the engine is hot; serious burns can occur from the steam and hot coolant.

• Observe all applicable safety precautions when working around fuel. Whenever servicing the fuel system, always work in a well-ventilated area. Do not allow fuel spray or vapors to come in contact with a spark, open flame, or excessive heat (a hot drop light, for example). Keep a dry chemical fire extinguisher near the work area. Always keep fuel in a container specifically designed for fuel storage; also, always properly seal fuel containers to avoid the possibility of fire or explosion. Refer to the additional fuel system precautions later in this section.

• Fuel injection systems often remain pressurized, even after the engine has been turned **OFF**. The fuel system pressure must be relieved before disconnecting any fuel lines. Failure to do so may result in fire and/or personal injury.

• Brake fluid often contains polyglycol ethers and polyglycols. Avoid contact with the eyes and wash your hands thoroughly after handling brake fluid. If you do get brake fluid in your eyes, flush your eyes with clean, running water for 15 minutes. If

eye irritation persists, or if you have taken brake fluid internally, IMMEDIATELY seek medical assistance.

• The EPA warns that prolonged contact with used engine oil may cause a number of skin disorders, including cancer! You should make every effort to minimize your exposure to used engine oil. Protective gloves should be worn when changing oil. Wash your hands and any other exposed skin areas as soon as possible after exposure to used engine oil. Soap and water, or waterless hand cleaner should be used.

• All new vehicles are now equipped with an air bag system, often referred to as a Supplemental Restraint System (SRS) or Supplemental Inflatable Restraint (SIR) system. The system must be disabled before performing service on or around system components, steering column, instrument panel components, wiring and sensors. Failure to follow safety and disabling procedures could result in accidental air bag deployment, possible personal injury and unnecessary system repairs.

• Always wear safety goggles when working with, or around, the air bag system. When carrying a non-deployed air bag, be sure the bag and trim cover are pointed away from your body. When placing a non-deployed air bag on a work surface, always face the bag and trim cover upward, away from the surface. This will reduce the motion of the module if it is accidentally deployed. Refer to the additional air bag system precautions later in this section.

• Clean, high quality brake fluid from a sealed container is essential to the safe and proper operation of the brake system. You should always buy the correct type of brake fluid for your vehicle. If the brake fluid becomes contaminated, completely flush the system with new fluid. Never reuse any brake fluid. Any brake fluid that is removed from the system should be discarded. Also, do not allow any brake fluid to come in contact with a painted surface; it will damage the paint.

• Never operate the engine without the proper amount and type of engine oil; doing so WILL result in severe engine damage.

• Timing belt maintenance is extremely important! Many models utilize an interference-type, non-freewheeling engine. If the timing belt breaks, the valves in the cylinder head may strike the pistons, causing potentially serious (also time-consuming and expensive) engine damage.

• Disconnecting the negative battery cable on some vehicles may interfere with the functions of the on-board computer system(s) and may require the computer to undergo a relearning process once the negative battery cable is reconnected.

• When servicing drum brakes, only disassemble and assemble one side at a time, leaving the remaining side intact for reference.

• Only an MVAC-trained, EPA-certified automotive technician should service the air conditioning system or its components.

ENGINE REPAIR

Distributor

The PT Cruiser uses a Direct Ignition System (DIS) and basic ignition timing is not adjustable.

Alternator

REMOVAL

Remove or disconnect the following:
• Negative battery cable
• Air cleaner lid
• Inlet Air Temperature (IAT) sensor and make-up hose
• Upper alternator adjustment lock nut, loosen only
• Right front wheel

• Splash shield
• Support bracket
• Lower pivot bolt, loosen only
• Drive belt T-bolt, loosen only
• Alternator wiring connectors
• Alternator belt
• Axle retaining nut
• Lower control arm from the steering knuckle
• Axle shaft
• Lower mounting bolt from the upper adjustment bracket
• Alternator

INSTALLATION

Install or connect the following:
• Alternator
• Alternator drive belt

• Axle shaft
• Lower control arm to the steering knuckle
• Lower ball joint nut and torque the nut to 70 ft. lbs. (95 Nm)
• Axle retaining nut and torque the nut to 120 ft. lbs. (163 Nm)
• Support bracket
• Alternator electrical connectors and torque the B+ terminal nut to 100 inch lbs. (11 Nm)
• Splash shield and right front wheel
• Right front wheel
• Upper pivot nut and torque to 40 ft. lbs. (54 Nm)
• Negative battery cable
• Air cleaner lid, IAT sensor and make up hose

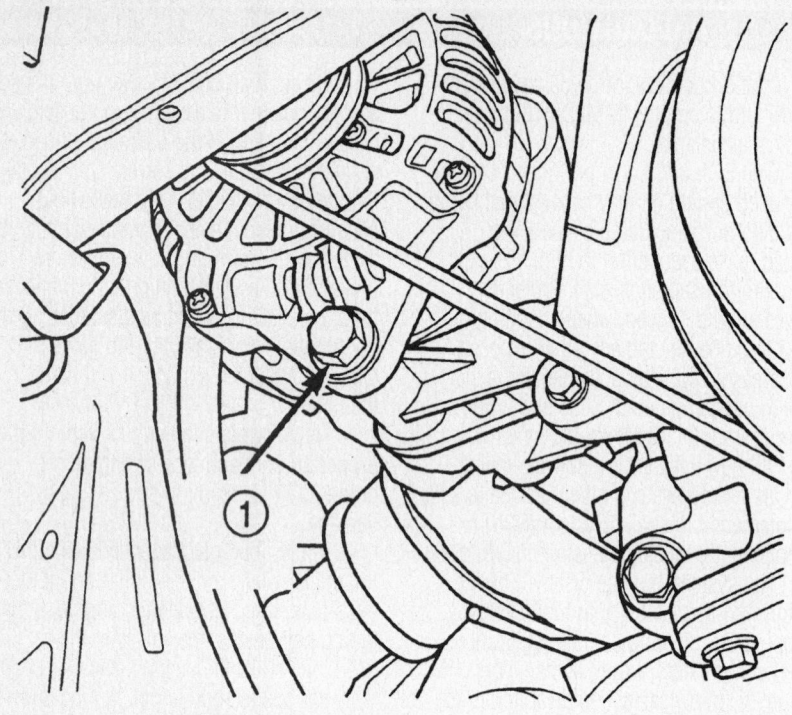

1 – LOWER PIVOT BOLT

Remove the lower pivot bolt from the alternator

9346JG01

Ignition Timing

ADJUSTMENT

Ignition timing is controlled by the Powertrain Control Module (PCM). No adjustment is necessary or possible.

Engine Assembly

REMOVAL & INSTALLATION

➡ **After all components are installed on the engine, a DRB scan tool is necessary to perform the camshaft and crankshaft timing relearn procedure.**

1. Before servicing the vehicle, refer to the precautions in the beginning of this section.

2. Properly recover the air conditioning system refrigerant.

3. Properly relieve the fuel system pressure.

4. Drain the cooling system.

5. Drain the engine oil.

6. Remove or disconnect the following:
- Battery and battery tray
- Throttle and cruise control cables
- Powertrain Control Module (PCM) wiring harness
- Positive cable from the Power Distribution Center (PDC) and ground wire
- Ground wire from the body to the engine
- Brake booster vacuum hose
- Proportional purge hoses from the intake manifold
- Coolant recovery hose
- Heater hoses
- Upper radiator support crossmember
- Upper and lower radiator hoses
- A/C line from the condenser
- A/C lines from the junction at the upper torque strut
- Cooler lines, if equipped
- Fan cooling module assembly
- Transmission shift linkage and electrical connectors
- Clutch hydraulic lines, if equipped
- Power steering hoses from the radiator on turbo models
- Radiator fan, radiator and condenser on 2003–05 models
- Both front wheels and the right inner splash shield
- Halfshafts
- Accessory drive belts
- Alternator and support brackets
- Charge air cooler hoses on turbo models
- Downstream Oxygen (O$_2$S) Sensor
- Exhaust system from the manifold
- Power steering pressure hose on non-turbo models
- Power steering hoses from the steering gear on turbo models
- Lower engine torque strut, if equipped
- Upper and lower heat shields, elbow support bracket, turbocharger support bract and elbow if equipped with a turbocharger
- Structural collar
- Torque converter bolts
- A/C compressor
- Power steering return line
- Power steering pump

7. Raise the vehicle enough to allow an engine dolly and cradle to be placed under the engine.

8. Loosen the engine support posts in order to allow movement for positioning onto the engine locating holes and flange on the engine bedplate. Lower the vehicle and position the cradle until the engine is resting on the support posts. Tighten the mounts to the cradle frame. This will keep the support posts from moving when removing or installing the engine and transaxle.

9. Install safety straps around the engine to the cradle; tighten the straps and lock them into position.

10. Raise the vehicle enough to see if the straps are tight enough to hold the cradle assembly to the engine.

11. Lower the vehicle so the weight of the engine and transaxle ONLY is on the cradle.

12. Remove the engine and transaxle mount through-bolts.

13. Raise the vehicle slowly, it might be necessary to move the engine/transaxle assembly with the cradle to allow removal around the body flanges.

To install:

14. Install or connect the following:
- Engine/transaxle assembly by lowering the vehicle over the assembly
- Engine and transaxle mounts and torque the bolts to 87 ft. lbs. (118 Nm)
- Upper torque strut

15. Remove the support fixtures.
- Alternator and support brackets
- Halfshafts

16. Install the structural collar and torque the bolts, in 3 steps, using the following procedure:

a. Step 1: Collar-to-oil pan bolts to 30 inch lbs. (3 Nm).

b. Step 2: Collar-to-transmission bolts to 80 ft. lbs. (108 Nm).

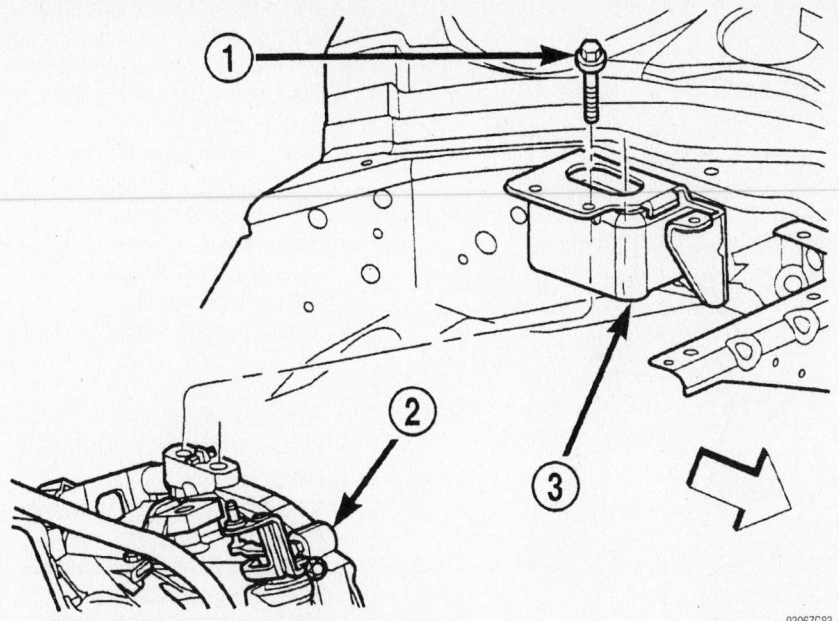

Front engine mount location and bolt identification

9306ZG83

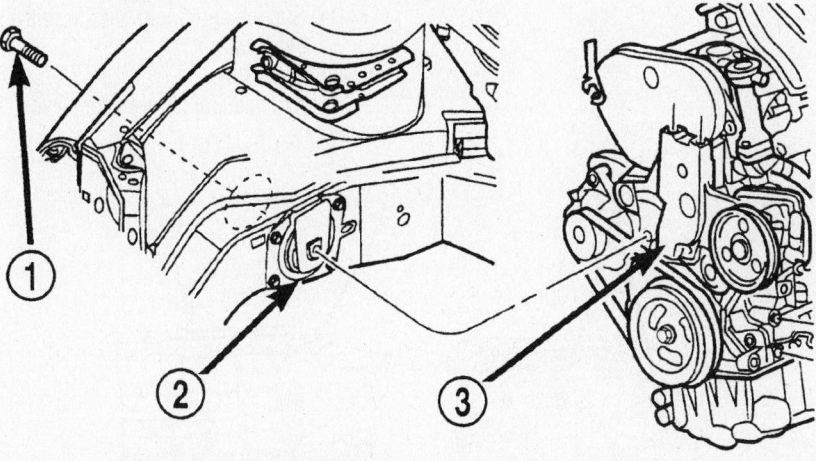

Exploded view of the left mount through bolt

9306ZG84

c. Step 3: Collar-to-oil pan bolts to 40 ft. lbs. (54 Nm).

17. Install or connect the following:
- Upper and lower heat shields, elbow support bracket, turbocharger support bract and elbow if equipped with a turbocharger
- Lower engine torque strut and torque the through bolts to 87 ft. lbs. (118 Nm)
- Exhaust pipe to the manifold
- Downstream O$_2$S sensor connector
- Charge air cooler hoses on turbocharged models
- Power steering pressure hose to the steering gear
- A/C compressor

- Power steering pump
- Drive belts
- Right inner splash shield and both front wheels
- Power steering return hose
- Clutch hydraulic line, electrical connectors and shift linkage, if equipped
- Shift linkage and cooler lines, if equipped
- Fuel lines
- Heater hoses
- Engine wiring harness and ground strap
- Lower and upper radiator hoses
- Fan module assembly
- Throttle and speed control cables

- PDC, positive battery cable and ground strap
- PCM
- Battery and battery tray
- Battery cables

18. Perform the camshaft and crankshaft synchronization procedure.

19. Fill the engine with clean oil.

20. Fill the cooling system.

21. Recharge the A/C system.

22. After all components are installed, perform the camshaft and crankshaft timing relearn procedure as follows:

a. Connect a DRB or equivalent, scan tool to the DLC (located under the instrument panel, near the steering column).

b. Turn the ignition switch **ON**, and access the "miscellaneous" screen.

c. Select "re-learn cam/crank" option and follow the directions on the scan tool screen.

23. Start the vehicle and check for leaks, repair if necessary.

Water Pump

REMOVAL & INSTALLATION

1. Before servicing the vehicle, refer to the precautions in the beginning of this section.

2. Drain the cooling system.

3. Remove or disconnect the following:

- Negative battery cable
- Timing belt
- Camshaft sprockets
- Rear timing belt cover
- Water pump and discard the O-ring

To install:

4. Install or connect the following:
- New O-ring in the water pump groove

❊❊ WARNING

Before proceeding, be sure the O-ring gasket is properly seated in the water pump groove before tightening the screws. An improperly installed O-ring could cause a coolant leak.

- Water pump and torque the bolts to 105 inch lbs. (12 Nm)

➡**Rotate the pump by hand to check for freedom of movement.**

- Rear timing belt cover
- Camshaft sprockets
- Timing belt
- Negative battery cable

5. Fill the cooling system.

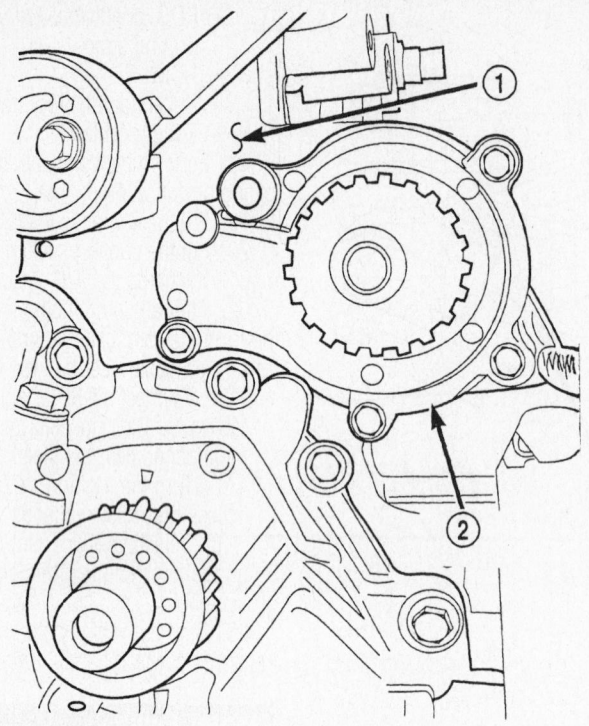

1 – CYLINDER BLOCK
2 – WATER PUMP

9306ZG85

Exploded view of the water pump

6. Start the vehicle and check for leaks, repair if necessary.

Heater Core

REMOVAL & INSTALLATION

2001 Vehicles

1. Before servicing the vehicle, refer to the precautions in the beginning of this section.

2. Remove the A/C refrigerant by using approved equipment.

3. Drain the cooling system.

4. Disconnect the negative battery cable.

5. Remove the instrument panel by removing or disconnecting the following:

- Left and right **A** pillar trim moldings using a trim stick
- Front power window switch
- Center bezel retaining screw
- Heating, Ventilation and Air Conditioning (HVAC) control knobs
- Center bezel using a trim stick
- Top cover retaining screws and pull it rearward to remove
- HVAC control unit from the instrument panel

- Retaining screws from the upper and lower steering column shrouds
- Left lower instrument panel bezel
- Steering column wire connectors
- Steering column
- Brake pedal support bracket
- Left side instrument panel end cap from the wiring connectors
- Rear power window switch from the console and pull the parking handle all the way up
- Auxiliary power outlet wire connector
- Center console
- Glove box assembly
- Right side instrument panel end cap
- 5 right side instrument panel wire connectors
- Instrument panel from the vehicle, with the help of an assistant

6. Remove or disconnect the following:

- Heater hoses from the heater core
- Reposition the vehicle speed control servo, if equipped
- Suction and liquid lines from the evaporator
- Drain tube
- A/C vacuum harness connector

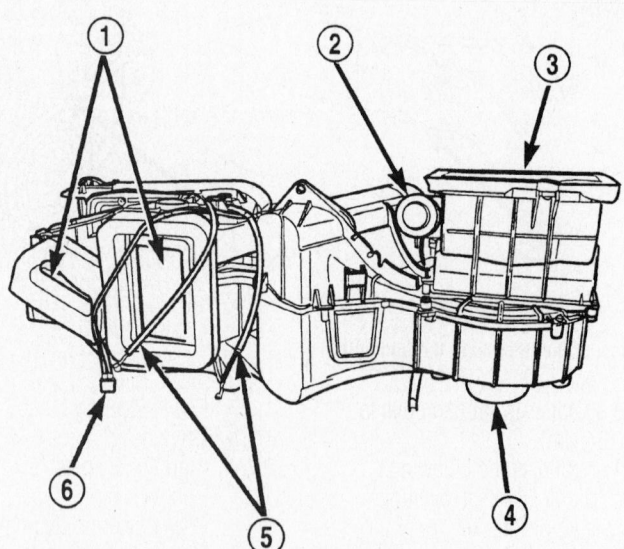

1 – AIR DISTRIBUTION
2 – RECIRCULATION DOOR VACUUM ACTUATOR
3 – AIR INLET
4 – BLOWER MOTOR
5 – CONTROL CABLES
6 – VACUUM HARNESS

9346BG01

Heater-A/C unit housing foam seals

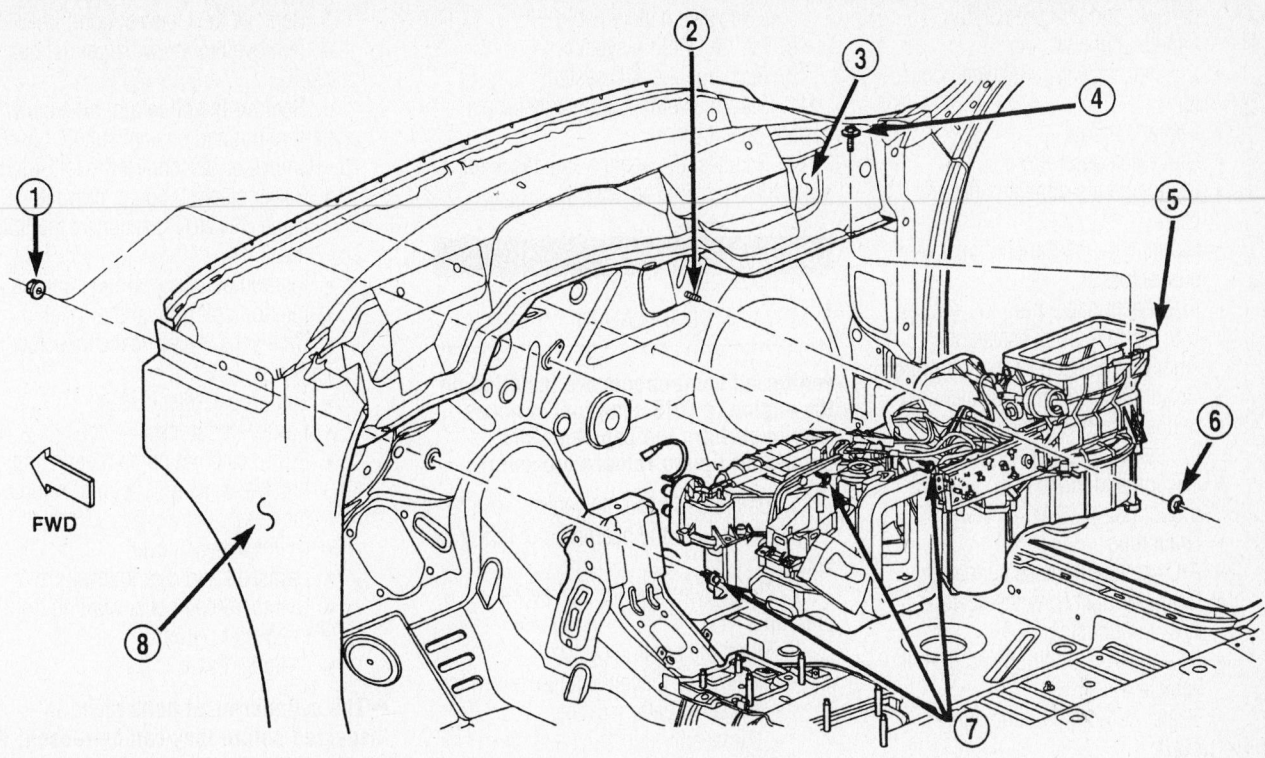

1 – NUT
2 – STUD
3 – PLENUM PANEL
4 – SCREW

5 – HEATER-A/C UNIT HOUSING
6 – NUT
7 – STUDS
8 – DASH PANEL

9346BG02

Heater-A/C unit housing removal

- Defroster duct from the heater—A/C unit housing
- Heater—A/C unit housing from the vehicle
- Heater core from the housing

To install:

7. Install or connect the following:
- Heater core to the housing
- Heater—A/C unit housing to the vehicle
- Defroster duct to the housing
- A/C vacuum harness
- Drain tube
- Suction and liquid lines to the evaporator
- Reposition the vehicle speed control servo, if equipped
- Heater hoses to the heater core

8. Install the instrument panel by installing or connecting the following:
- Instrument panel
- Right side instrument panel wire connectors
- Right side instrument panel end cap
- Glove box assembly
- Center console
- Auxiliary power outlet wire connector
- Rear power window switch
- Left side instrument panel end cap
- Left side wire connectors
- Brake pedal support bracket
- Steering column with a new pinch bolt
- Steering column wire connectors
- Left lower instrument panel bezel
- HVAC control unit
- Center bezel
- Front power window switch
- Left and right **A** pillar trim moldings
- Negative battery cable

9. Fill the cooling system.
10. Recharge the A/C system.
11. Start the vehicle and verify proper system operations.
12. Roadtest the vehicle and check for any rattles, repair if necessary.

2002–05 Vehicles

1. Before servicing the vehicle, refer to the precautions in the beginning of this section.

2. Remove the A/C refrigerant by using approved equipment.
3. Drain the cooling system.
4. Disconnect the negative battery cable.
5. Remove the instrument panel by removing or disconnecting the following:
- Left and right **A** pillar trim moldings using a trim stick
- Front power window switch
- Center bezel retaining screw
- Heating, Ventilation and Air Conditioning (HVAC) control knobs
- Center bezel using a trim stick
- Top cover retaining screws and pull it rearward to remove
- HVAC control unit from the instrument panel
- Retaining screws from the upper and lower steering column shrouds
- Left lower instrument panel bezel
- Steering column wire connectors
- Steering column
- Brake pedal support bracket
- Left side instrument panel end cap from the wiring connectors
- Rear power window switch from the

console and pull the parking handle all the way up
- Auxiliary power outlet wire connector
- Center console
- Glove box assembly
- Right side instrument panel end cap
- 5 right side instrument panel wire connectors
- Instrument panel from the vehicle, with the help of an assistant

6. Remove or disconnect the following:
- Heater hoses from the heater core
- Reposition the vehicle speed control servo, if equipped
- Suction and liquid lines from the evaporator
- Drain tube
- A/C vacuum harness connector
- Defroster duct from the heater—A/C unit housing
- Heater—A/C unit housing from the vehicle
- Heater core from the housing

To install:

7. Install or connect the following:
- Heater core to the housing
- Heater—A/C unit housing to the vehicle
- Defroster duct to the housing
- A/C vacuum harness
- Drain tube
- Suction and liquid lines to the evaporator
- Reposition the vehicle speed control servo, if equipped
- Heater hoses to the heater core

8. Install the instrument panel by installing or connecting the following:
- Instrument panel
- Right side instrument panel wire connectors
- Right side instrument panel end cap
- Glove box assembly
- Center console
- Auxiliary power outlet wire connector
- Rear power window switch
- Left side instrument panel end cap
- Left side wire connectors
- Brake pedal support bracket
- Steering column with a new pinch bolt
- Steering column wire connectors
- Left lower instrument panel bezel
- HVAC control unit
- Center bezel
- Front power window switch
- Left and right **A** pillar trim moldings

- Negative battery cable
9. Fill the cooling system.
10. Recharge the A/C system.
11. Start the vehicle and verify proper system operations.
12. Roadtest the vehicle and check for any rattles, repair if necessary.

Cylinder Head

REMOVAL & INSTALLATION

➡**After all components are installed on the engine, a DRB scan tool is necessary to perform the camshaft and crankshaft timing relearn procedure.**

1. Before servicing the vehicle, refer to the precautions in the beginning of this section.
2. Properly relieve the fuel system pressure.
3. Drain the cooling system.
4. Remove or disconnect the following:
- Negative battery cable
- Air cleaner inlet duct and air cleaner
- Inlet Air Temperature (IAT) sensor and make-up air hose
- Upper intake manifold
- Dipstick tube fastener
- Lower intake manifold support bracket on turbo models
- Fuel supply line from the fuel rail
- Heater tube support bracket
- Upper radiator hose
- Heater supply hoses
- Accessory drive belt
- Exhaust pipe from the manifold

5. On turbo models perform the following:

a. Remove the turbocharger heat shields.

b. Remove the elbow support bracket.
c. Remove the turbocharger support bracket.
d. Remove the oil return, oil supply, coolant return and coolant supply hoses.

6. Remove or disconnect the following:
- Power steering pump, move it aside. DO NOT disconnect the fluid lines.
- Ignition coil pack wiring connector
- Ignition coil pack and bracket
- Cam sensor electrical connector
- Timing belt
- Timing belt idler pulley
- Camshaft sprocket
- Power steering pump reservoir and bracket, if necessary on non-turbo models
- Cylinder head cover
- Camshaft and cam followers
- Cylinder head bolts, working from the center outward
- Cylinder head

➡**The cylinder head bolts must be inspected before they can be reused. If the threads of bolts are stretched, they must be replaced. Check for thread stretching by holding a scale or other straightedge against the threads. If all the threads do not contact the scale, the bolts must be replaced.**

✳✳ WARNING

Use only a plastic scraper to clean the mating surfaces. NEVER use metal, as this may gouge the surfaces and cause leaks.

7. Cover the combustion chambers, then use a plastic scraper to thoroughly and carefully clean the engine block and cylinder head mating surfaces.

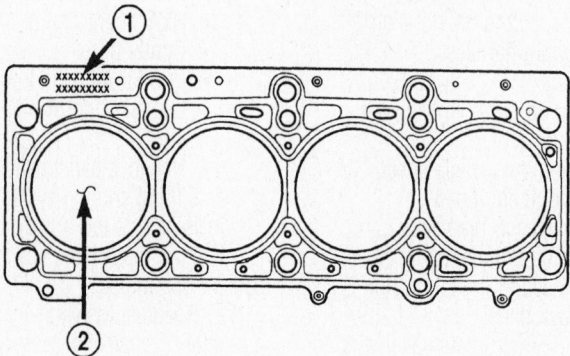

1 – PART NUMBER FACES UP
2 – NO. 1 CYLINDER

9346JG04

Install the new gasket with the part number facing upwards

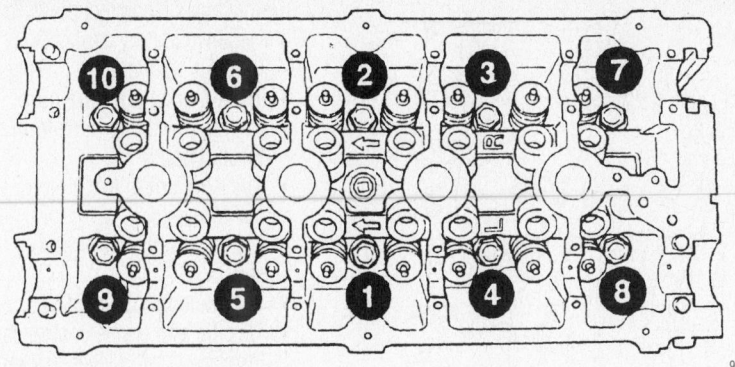

Cylinder head bolt torque sequence

To install:

8. Install the cylinder head with a new gasket. Make certain that the part number on the new gasket is facing up.

9. Lubricate the cylinder head bolt threads with clean engine oil.

10. Torque the cylinder head bolts, in the following sequence:

   a. Step 1: 25 ft. lbs. (34 Nm).

   b. Step 2: 50 ft. lbs. (68 Nm).

   c. Step 3: 50 ft. lbs. (68 Nm).

   d. Step 4: An additional ¼ turn.

11. Install or connect the following:
- Camshaft and cam follower assemblies
- Cylinder head cover
- Rear timing belt cover and pulley
- Camshaft sprockets
- Timing belt
- Cam sensor wiring connector
- Ignition coil and spark plug wires
- Power steering pump reservoir and bracket, if removed

12. On turbo models perform the following:

   a. Install the oil return, oil supply, coolant return and coolant supply hoses.

   b. Install the elbow support bracket.

   c. Install the turbocharger heat shields.

   d. Install the turbocharger support bracket.

13. Install or connect the following:
- Exhaust pipe to the manifold
- Accessory drive belts
- Lower intake manifold
- Upper radiator and heater supply hose
- Heater support bracket
- Lower intake manifold support bracket on turbo models
- Dipstick tube fastener
- Power brake vacuum hose to the intake manifold
- Fuel supply line to the fuel rail
- Vacuum lines and electrical wiring

- Upper intake manifold
- Air cleaner inlet duct and air cleaner
- IAT sensor and make up hose
- Negative battery cable

14. Fill the cooling system.

15. Turn the ignition switch **ON**, and access the "miscellaneous" screen.

16. Select the "re-learn cam/crank" option, then follow the instructions on the scan tool screen.

Intake Manifold

REMOVAL & INSTALLATION

Upper

NON-TURBO MODELS

1. Before servicing the vehicle, refer to the precautions in the beginning of this section.

2. Properly relieve the fuel system pressure.

3. Remove or disconnect the following:
- Negative battery cable
- Inlet Air Temperature (IAT) sensor and air hose
- Engine cover
- Throttle and cruise control cables from the throttle lever bracket
- Manifold Absolute Pressure (MAP) sensor
- Idle Air Control (IAC) motor electrical connector
- Throttle Position (TPS) sensor wiring connector

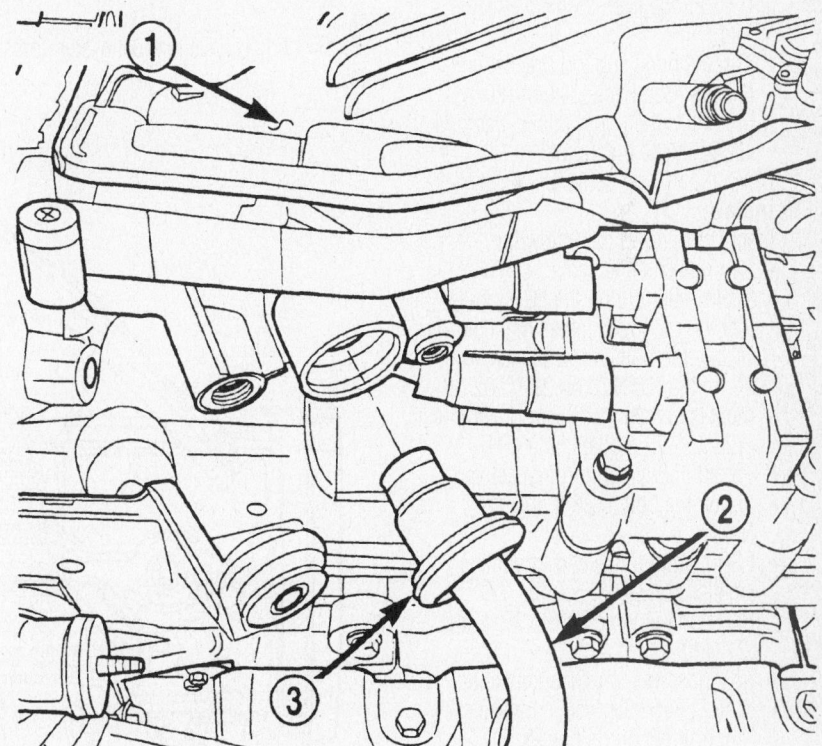

1 – INTAKE MANIFOLD

2 – EGR TUBE

3 – SEAL

Remove the EGR tube from the upper intake manifold

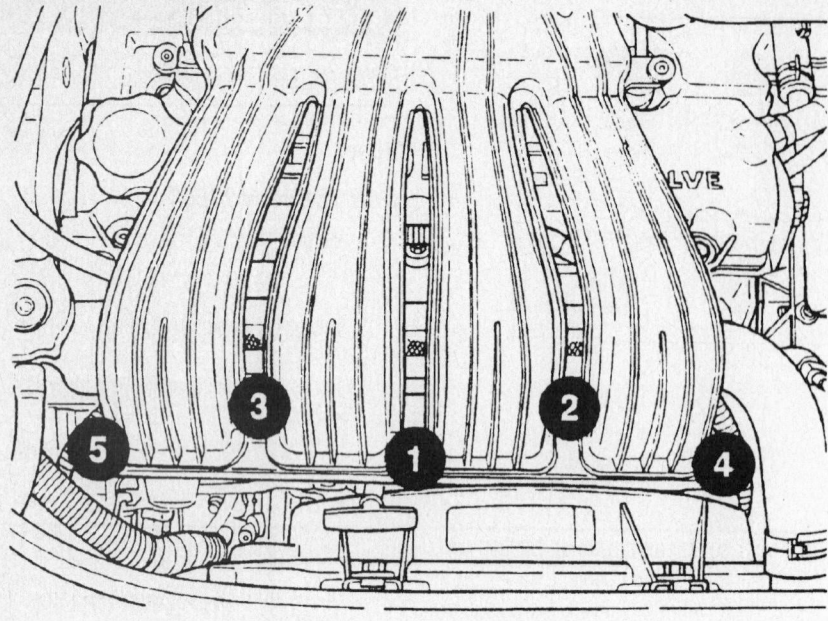

93062G87

Upper intake manifold bolt torque sequence—non-turbo models

- Proportional purge hoses
- Brake booster vacuum hose
- Positive Crankcase Ventilation (PCV) hose from the intake manifold
- Throttle body support bracket bolt
- Exhaust Gas Recirculation (EGR) tube from the upper intake manifold
- Upper intake manifold

4. Clean the mating surfaces.

To install:

5. Install or connect the following:
- New gaskets and seals
- Intake manifold on the EGR tube then on the lower intake manifold. Torque the bolts, in sequence, to 20 ft. lbs. (28 Nm) on 2001–02 models or 105 inch lbs. (12 Nm) on 2003–05 models.
- Throttle body support bracket and torque the bolt to 28 ft. lbs. (20 Nm)
- EGR retainer plate and torque the smaller bolt to 95 inch lbs. (11 Nm) and the large bolt to 28 ft. lbs. (21 Nm)
- PCV hose to the intake manifold
- MAP sensor electrical connector
- Proportional purge hoses
- Brake booster hose
- IAC motor and TPS connectors
- Throttle and speed control cables
- Air cleaner assembly
- Engine cover
- IAT sensor
- Negative battery cable

TURBO MODELS

1. Before servicing the vehicle, refer to the precautions in the beginning of this section.

2. Properly relieve the fuel system pressure.

3. Remove or disconnect the following:
- Negative battery cable
- Inlet Air Temperature (IAT) sensor
- Throttle inlet pressure hose
- Charge air cooler hose from the throttle body
- Idle Air Control (IAC) motor electrical connector
- Manifold Absolute Pressure (MAP) sensor
- Throttle control shield
- Throttle and cruise control cables from the throttle lever bracket
- Throttle cable bracket
- Brake booster vacuum hose
- Positive Crankcase Ventilation (PCV) hose from the intake manifold
- Purge solenoid hose from the throttle body
- Upper intake manifold support bracket and manifold

➡Cover the lower intake manifold to avoid dirt and other objects from entering.

4. Clean the mating surfaces.
To install:

5. Remove the cover from the lower intake manifold.

6. Install or connect the following:

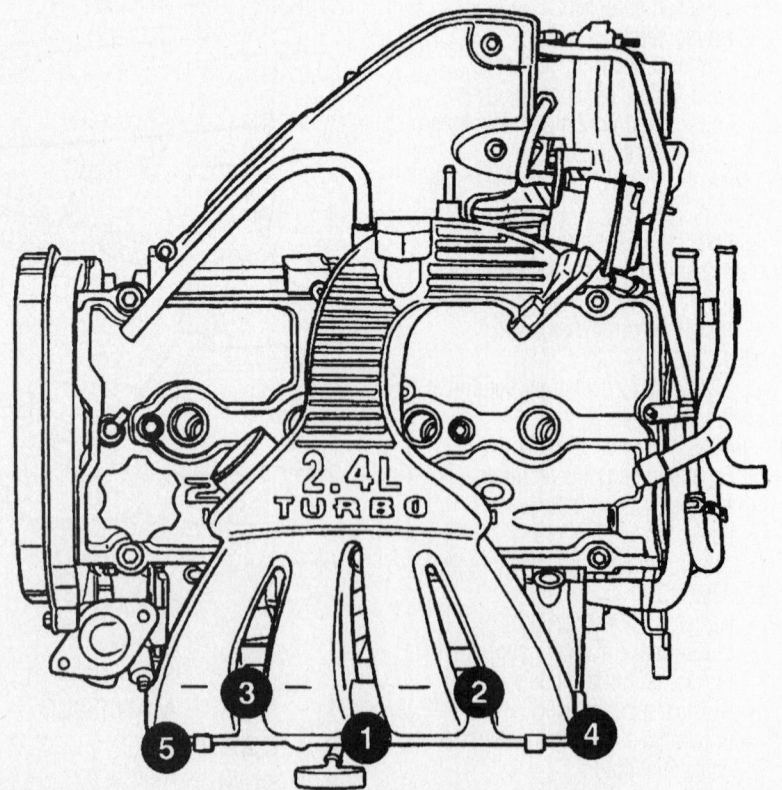

67189-PTCR-G05

Upper intake manifold bolt torque sequence—turbo models

- New gasket
- Intake manifold on the lower intake manifold. Torque the bolts, in sequence, to 250 inch lbs. (28 Nm).
- Upper intake manifold support bracket Torque the retainers to 250 inch lbs. (28 Nm).
- Purge solenoid hose to the throttle body
- Brake booster vacuum hose
- PCV hose
- Throttle cable bracket and tighten the screws to 105 inch lbs. (12 Nm)
- Throttle and cruise control cables to the throttle lever bracket
- Throttle control shield
- IAC motor electrical connector
- MAP sensor
- Charge air cooler hose to the throttle body
- IAT sensor
- Throttle inlet pressure hose
- Negative battery cable

Lower

1. Before servicing the vehicle, refer to the precautions in the beginning of this section.
2. Properly relieve the fuel system pressure.
3. Drain the coolant system.
4. Remove or disconnect the following:
   - Negative battery cable
   - Inlet Air Temperature (IAT) sensor and make-up hose

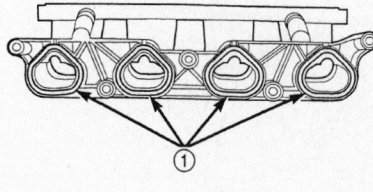

1 – SEALS

9346JG06

Install new seals on the lower intake manifold–non-turbo models

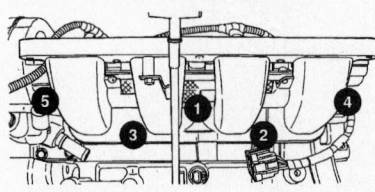

67189-PTCR-G06

Lower intake manifold bolt torque sequence

- Air cleaner
- Upper intake manifold
- Upper radiator hose and coolant outlet connector
- Fuel supply line quick-connect from the fuel rail
- Fuel injector wiring harness
- Oil dipstick tube from the lower intake manifold
- Intake manifold and discard the gaskets and seals

5. Thoroughly clean the gasket mating surfaces.

To install:

6. Install or connect the following:
   - New seals on non-turbo models
   - Intake manifold and torque the bolts in sequence to 105 inch lbs. (12 Nm) on non-turbo models
   - New gaskets on turbo models
   - Intake manifold and torque the bolts in sequence to 250 inch lbs. (28 Nm) on turbo models
   - Lower intake manifold support bracket retainers on turbocharged models to 40 ft. lbs. (54 Nm)
   - Fuel injector wiring harness
   - Fuel supply line quick-connect to the fuel tube assembly
   - Oil dipstick tube to the lower intake manifold
   - Upper radiator hose
   - Upper intake manifold
   - IAT sensor
   - Air cleaner assembly
   - Negative battery cable
7. Fill the coolant system.
8. Pressurize the fuel system.
9. Start the vehicle and check for leaks, repair if necessary.

Turbocharger

REMOVAL & INSTALLATION

1. Before servicing the vehicle, refer to the precautions in the beginning of this section.

➡**The exhaust manifold on turbocharged models is removed as an assembly with the turbocharger. Do not remove the turbocharger from the manifold always replace as a complete assembly.**

✳✳ CAUTION

If the turbocharger is being replaced due to bearing failure, the oil pressure feed line has to be replaced and the return tube should be cleaned.

2. Properly relieve the fuel system pressure.
3. Remove or disconnect the following:
   - Negative battery cable
   - Air cleaner housing
   - Clean air hose from the turbocharger
   - Throttle and cruise control cables from the throttle lever bracket
   - Inlet Air Temperature (IAT) sensor
   - Manifold Absolute Pressure (MAP) sensor
   - Idle Air Control (IAC) motor electrical connector
   - Throttle Position (TPS) sensor wiring connector
   - Ignition coil capacitor
   - Upstream Oxygen (O2S) sensor connector
   - Air inlet hose from the throttle body
   - Vacuum hoses from the throttle body and upper intake manifold
   - Upper intake manifold support bracket and manifold

➡**Cover the lower intake manifold to avoid dirt and other objects from entering.**

   - Turbocharger lower heat shield
   - Oil supply line from the turbocharger
   - Coolant return line
   - Vacuum hoses
   - Muffler ground strap
   - Downstream O2S sensor connector
   - Catalytic converter and intermediate pipe as an assembly
   - Turbocharger to charge air cooler hose assembly
   - Turbocharger, elbow support and support brackets
   - Oil return tube
   - Turbocharger coolant supply line and upper heat shield
   - Turbocharger elbow
   - Lower exhaust manifold fasteners from below, then lower the vehicle and remove the upper manifold fasteners
   - Turbocharger/manifold assembly from between the engine and cowl panel.
   - Discard the gasket
4. Clean the mating surfaces.
5. Install or connect the following:
   - New gasket. Use no sealer when installing the gasket.
   - Turbocharger/manifold assembly between the engine and cowl panel. Tighten the fasteners working from the center out in progressing in

both directions to 259 inch lbs. (28 Nm).

- Turbocharger elbow. Tighten the fasteners to 259 inch lbs. (28 Nm).
- Turbocharger upper heat shield. Tighten the fasteners to 259 inch lbs. (28 Nm).
- Coolant supply line using new washers. Tighten the banjo bolt to 22 ft. lbs. (30 Nm) and the flared fitting to 23 ft. lbs. (31 Nm).
- New oil return tube gasket, return tub and tighten to 105 inch lbs. (12 Nm). make sure the heat shield for the oil return line is installed properly.
- Turbocharger support bracket. Tighten the M8 fasteners to 259 inch lbs. (28 Nm) and the M10 fasteners to 40 ft. lbs. (54 Nm).
- Turbocharger elbow support bracket
- Turbocharger to charge air cooler hose assembly
- Catalytic converter and intermediate pipe as an assembly
- Muffler ground strap
- Downstream O_2S sensor connector
- Vacuum hose
- Coolant return line using new washers. Tighten the banjo bolt to 22 ft. lbs. (30 Nm)
- Oil supply line and tighten the flared fitting to 23 ft. lbs. (31 Nm)

➡**The lower heat shield tabs must overlap the upper heat shield to prevent fatigue and premature union failure.**

- Turbocharger lower heat shield. Tighten the fasteners to 259 inch lbs. (28 Nm).

6. Remove the cover placed on the lower intake manifold.

- Upper intake manifold and support bracket
- Vacuum hoses to the throttle body and upper intake manifold
- Air inlet hose to the throttle body
- IAT sensor connector
- MAP sensor
- TPS sensor wiring connector
- IAC motor electrical connector
- Ignition coil capacitor
- Upstream O_2S sensor connector
- Throttle and cruise control cables to the throttle lever bracket
- Clean air hose to the turbocharger
- Air cleaner housing
- Negative battery cable

7. Fill the cooling system.
8. Chain the oil and filter.

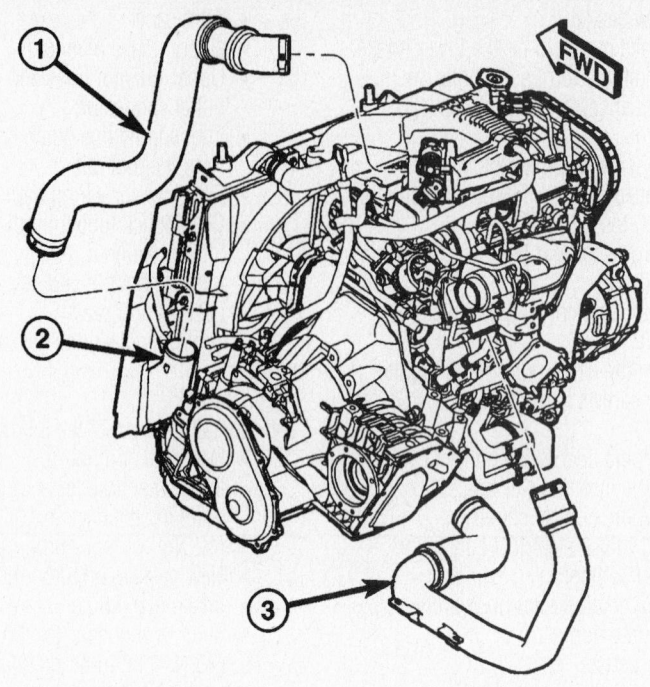

1 - HOSE - CHARGE AIR COOLER TO THROTTLE BODY
2 - CHARGE AIR COOLER
3 - HOSE - TURBOCHARGER TO CHARGE AIR COOLER

67189-PTCR-G01

Charge air cooler hoses—Turbo models

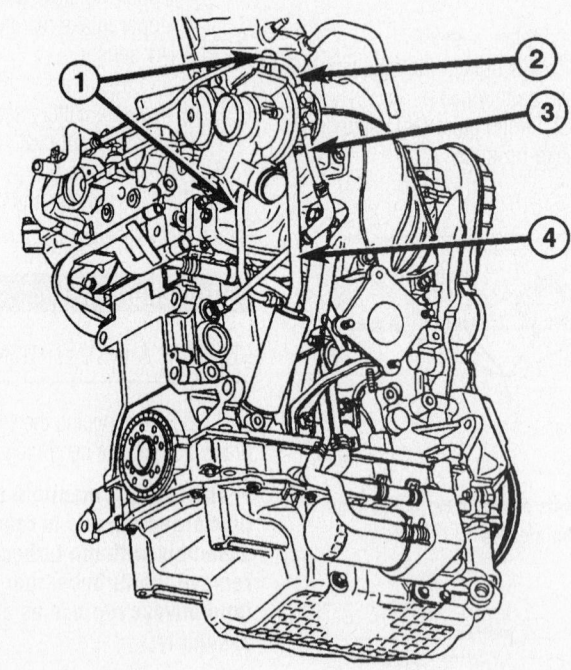

1 - OIL SUPPLY LINE
2 - COOLANT RETURN LINE
3 - COOLANT SUPPLY LINE
4 - OIL RETURN TUBE

67189-PTCR-G02

Turbocharger line and hose locations—Turbo models

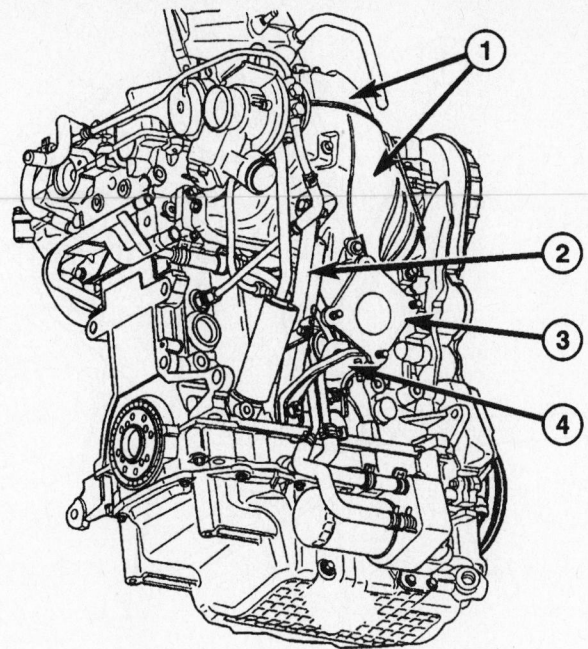

1 - UPPER/LOWER HEAT SHIELDS
2 - TURBOCHARGER SUPPORT BRACKET
3 - ELBOW
4 - ELBOW SUPPORT BRACKET

67189-PTCR-G03

Turbocharger bracket and heat shield locations—Turbo models

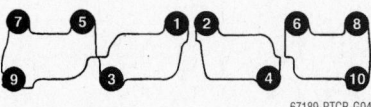

67189-PTCR-G04

Exhaust manifold torque sequence—Turbo models

9. Start the vehicle and check for exhaust system leaks and also check the system is not contacting any body panels. Adjust the system as necessary to avoid panel contact.

Exhaust Manifold

REMOVAL & INSTALLATION

Non Turbo Models

1. Before servicing the vehicle, refer to the precautions in the beginning of this section.
2. Remove or disconnect the following:
   - Negative battery cable
   - Air cleaner assembly and bracket
   - Throttle and speed control cables
   - Manifold Absolute Pressure (MAP) sensor electrical connector
   - Power steering reservoir, move it

aside. DO NOT disconnect the fluid lines.
   - Coolant recovery bottle
   - Exhaust manifold upper heat shield
   - Exhaust pipe from the manifold
   - Engine wiring heat shield
   - Manifold support bracket
   - Lower exhaust manifold heat shield
   - Upstream Heated Oxygen (HO2S) sensor connector
   - Exhaust manifold and discard the gasket

3. Thoroughly clean the mating surfaces.

To install:

4. Install or connect the following:
   - New gasket
   - Exhaust manifold and torque the bolts, in sequence, to 17 ft. lbs. (23 Nm)
   - Alternator bracket bolt, if loosened
   - Exhaust manifold heat shields and torque the bolts to 105 inch lbs. (12 Nm)
   - Exhaust manifold support bracket
   - Engine wiring heat shield
   - Upstream HO2 sensor wiring connector
   - Exhaust pipe to the manifold and

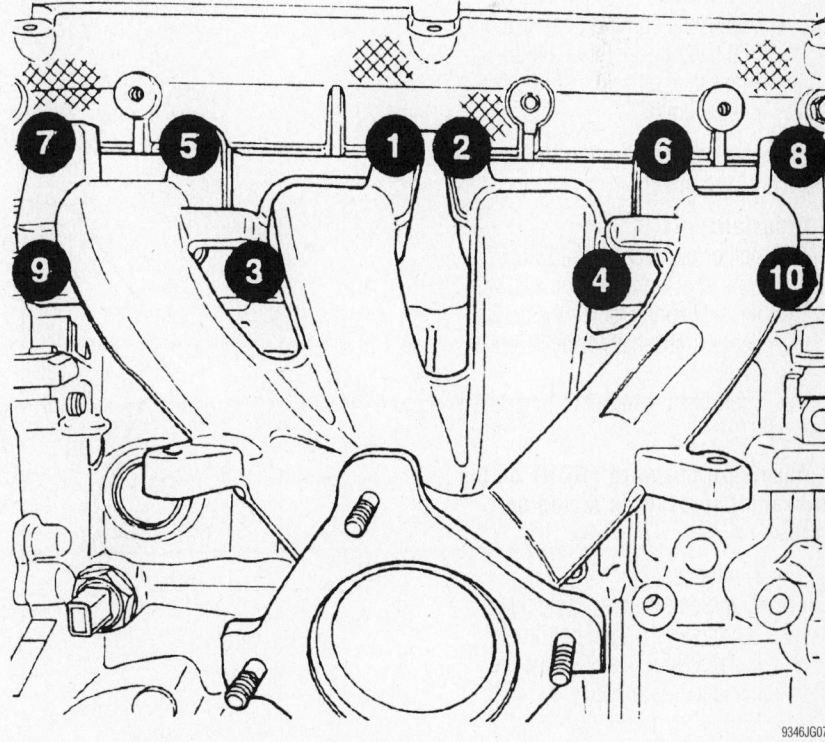

9346JG07

Torque the exhaust manifold bolts in sequence

torque the bolts to 21 ft. lbs. (28 Nm)

- Coolant recovery bottle
- Power steering pump reservoir
- MAP sensor connector
- Throttle and speed control cables
- Air cleaner bracket and assembly
- Negative battery cable

5. Start the vehicle and check for leaks, repair if necessary.

Turbo Models

The exhaust manifold on turbocharged models is removed as an assembly with the turbocharger. Refer to the turbocharger removal and installation procedure in this manual for manifold removal.

Front Crankshaft Seal

REMOVAL & INSTALLATION

1. Before servicing the vehicle, refer to the precautions in the beginning of this section.

2. Remove or disconnect the following:

- Negative battery cable
- Crankshaft damper bolt
- Crankshaft damper, using Puller Tool 1026 and Insert Tool 6827-A
- Timing belt
- Crankshaft sprocket, using Tool 6793 and Insert Tool C-4685-C2
- Crankshaft oil seal, using a seal puller Tool 6771

➡Be careful not to damage the seal surface of the cover.

To install:

3. Install or connect the following:
- New front crankshaft oil seal with the seal spring facing the engine, using Crankshaft Installer Tool 6780
- Crankshaft sprocket with special T 6792

➡Make sure the word FRONT on the crankshaft sprocket is facing outward.

- Timing belt
- Crankshaft damper using thrust bearing washer and bolt from Installer Tool 6792. Torque the damper bolt to 105 ft. lbs. (142 Nm).
- Negative battery cable

4. Start the vehicle and check for leaks, repair if necessary.

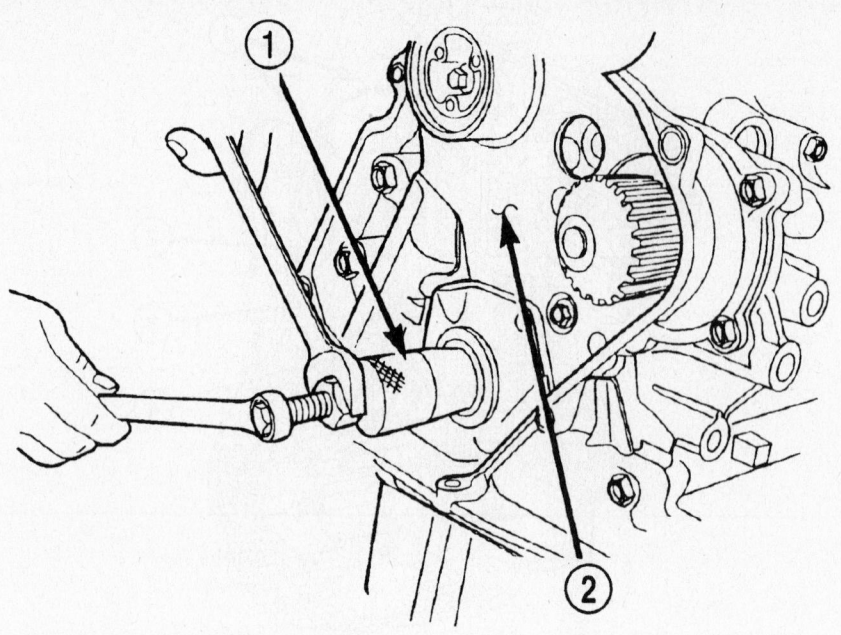

1 – SPECIAL TOOL 6771
2 – REAR TIMING BELT COVER

9346JG08

Removing the front crankshaft oil seal

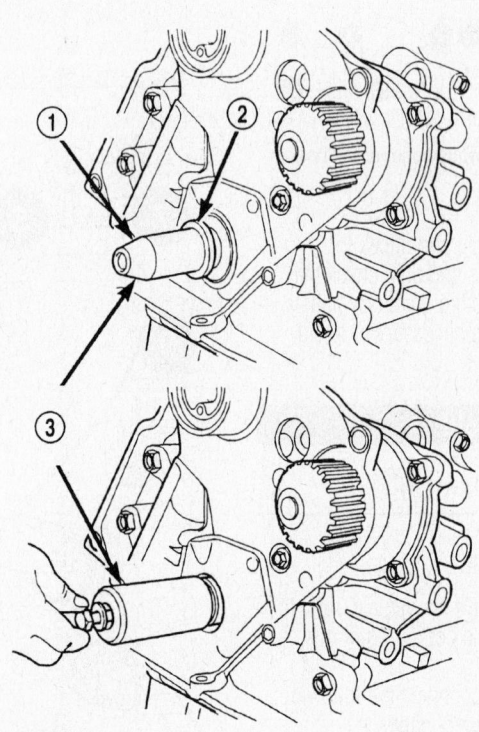

1 – PROTECTOR
2 – SEAL
3 – SPECIAL TOOL 6780

9346JG09

Installing a new seal using seal installer 6780-1; proceed with caution if using substitute tools

Camshaft and Lifters

REMOVAL & INSTALLATION

➡**After all components are installed on the engine, a DRB scan tool is necessary to perform the camshaft and crankshaft timing relearn procedure.**

1. Before servicing the vehicle, refer to the precautions in the beginning of this section.

2. Relieve the fuel system pressure.

3. Remove or disconnect the following:
- Negative battery cable
- Cylinder head cover
- Camshaft Position (CMP) sensor and target magnet
- Timing belt
- Camshaft sprocket
- Rear timing belt cover
- Loosen the camshaft bearing caps in sequence from the rear of the cylinder head
- Camshafts

To install:

Make certain that the pistons are **NOT** at Top Dead Center (TDC) before installing the camshafts.

4. Install the camshafts and cam follow-ers, lubricate the bearing journals thoroughly.

5. Install the left and right camshaft bearing caps, No's 2–5 and right side No. 6. Torque these fasteners to 105 inch lbs. (12 Nm). Apply Mopar® gasket maker to No.1 and left side No. 6. Torque these fasteners to 18 ft. lbs. (24 Nm) on 2001–02 models or 250 ft. lbs. (28 Nm) on 2003–05 models.

6. Install or connect the following:
- Camshaft seals
- Rear timing belt cover
- Camshaft sprockets and torque the bolt to 85 ft. lbs. (115 Nm)
- Timing belt
- Target magnet
- CMP sensor and torque the screws to 85 inch lbs. (9.6 Nm)
- Cylinder head cover
- Negative battery cable

➡**An oil and filter change are recommended.**

7. Use a DRB scan tool to perform the camshaft and crankshaft timing relearn procedure, as follows:

a. Connect the scan tool to the DLC (located under the instrument panel, near the steering column).

b. Turn the ignition switch **ON** and access the "miscellaneous" screen.

c. Select the "re-learn cam/crank" option, then follow the instructions on the scan tool screen.

8. Start the engine and check for leaks. Run the engine with the radiator cap off so as the engine warms and the thermostat opens, coolant can be added to the radiator. Test drive vehicle to check for proper operation.

Valve Lash

ADJUSTMENT

The engines in these vehicles do not require periodic valve lash adjustment.

Starter Motor

REMOVAL & INSTALLATION

Non Turbo Models

1. Before servicing the vehicle, refer to the precautions in the beginning of this section.

2. Remove or disconnect the following:
- Negative battery cable
- Air cleaner box cover
- Engine structural collar
- Starter electrical connectors
- Starter

To install:

3. Install the starter and torque the bolts to 40 ft. lbs. (54 Nm). Attach the starter electrical connectors.

4. Install the engine structural collar on models with a automatic transaxle as follows, refer to the illustration for bolt locations:

a. Place the collar in position and hand tighten the collar to transaxle bolt.

b. Position the power steering hose support bracket and install the collar to oil pan bolt.

c. Place the bending strut in place and hand start the bolt.

d. Install the bolt through the strut and collar and hand tighten.

e. Place the power steering hose support bracket in position and install and hand tighten the remaining bolts.

f. Tighten the collar-to-transmission bolts to 75 ft. lbs. (101 Nm).

g. Install the bolts through the strut and into the block.

h. Tighten the remaining bolts to 45 ft. lbs. (61 Nm).

5. Install the engine structural collar on

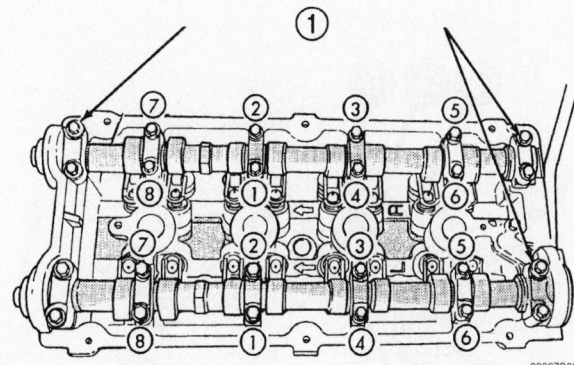

Remove the camshaft bearing caps in sequence

9306ZG89

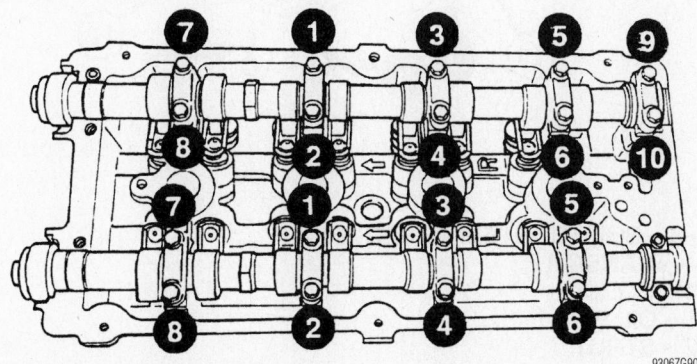

Camshaft bearing cap tightening sequence

9306ZG90

- Studs retaining the power steering lines
- Position the power steering lines aside
- Engine structural collar
- Starter electrical connectors
- Starter

To install:

3. Install the starter and torque the bolts to 40 ft. lbs. (54 Nm). Attach the starter electrical connectors.

4. Install the engine structural collar on models with a automatic transaxle as follows, refer to the illustration for bolt locations:

    a. Place the collar in position and hand tighten the collar to transaxle bolt.

    b. Position the power steering hose support bracket and install the collar to oil pan bolt.

    c. Place the bending strut in place and hand start the bolt.

    d. Install the bolt through the strut and collar and hand tighten.

    e. Place the power steering hose support bracket in position and install and hand tighten the remaining bolts.

    f. Tighten the collar-to-transmission bolts to 75 ft. lbs. (101 Nm).

    g. Install the bolts through the strut and into the block.

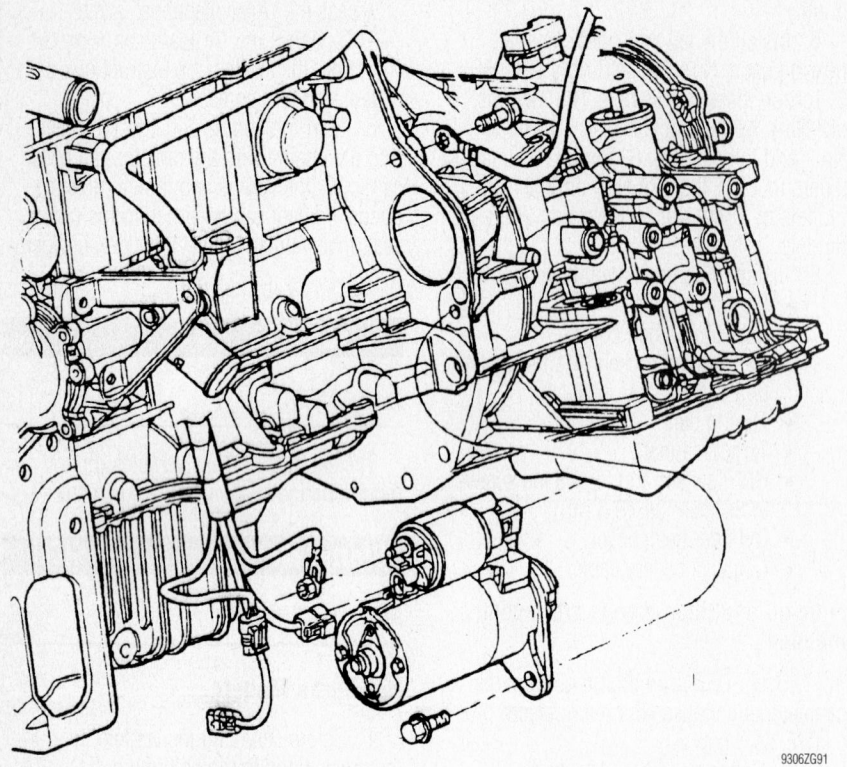

Removal of the starter motor mounting

models with a manual transaxle as follows, refer to the illustration for bolt locations:

    a. Place the collar in position and hand tighten the collar to transaxle bolt.

    b. Position the power steering hose support bracket and install the collar to oil pan bolt.

    c. Position the clutch slave cylinder into position and hand start the bolts.

    d. Tighten bolts (1) to 75 ft. lbs. (101 Nm).

    e. Tighten bolts (2 and 5) to 45 ft. lbs. (61 Nm).

    f. Tighten bolts (3 and 4) to 20 ft. lbs. (28 Nm).

6. Install the air cleaner box cover and connect the negative battery cable.

Turbo Models

1. Before servicing the vehicle, refer to the precautions in the beginning of this section.

2. Remove or disconnect the following:
- Negative battery cable
- Air cleaner box cover
- Upper starter bolt by pushing the inner cooler up and out of the way
- Inner cooler lower hose from the inner cooler
- Nuts retaining the inner cooler tube

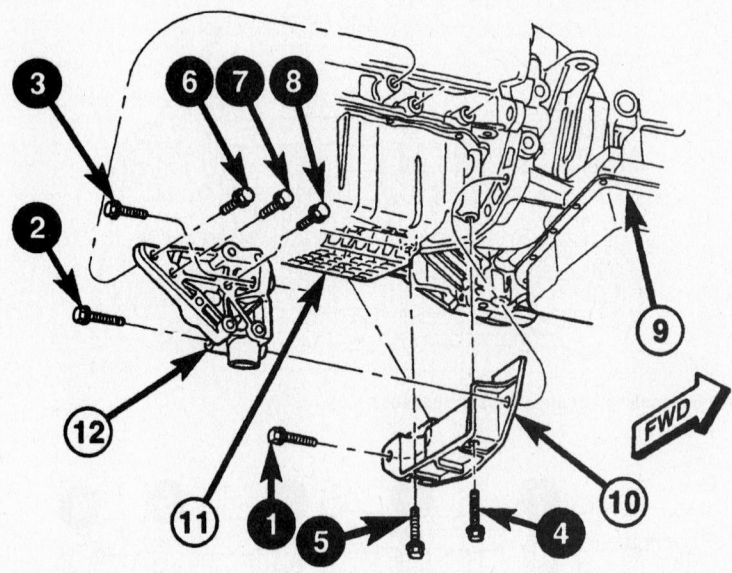

1–8 – BOLT TIGHTENING SEQUENCE
9 – TRANSAXLE
10 – COLLAR
11 – OIL PAN
12 – STRUT

Structural collar assembly—automatic transaxle

67189-PTCR-G07

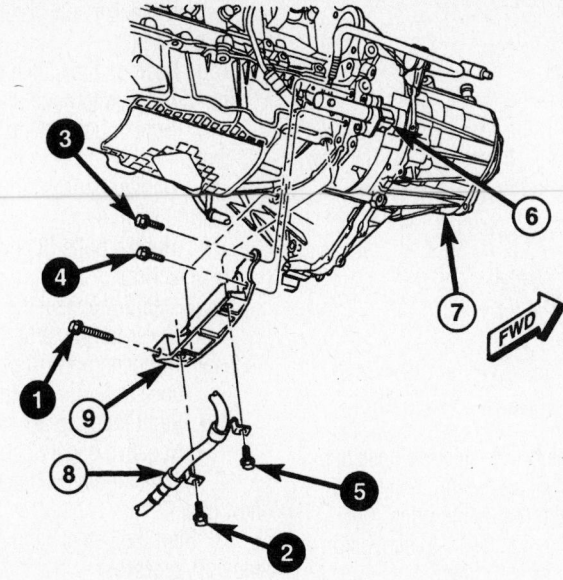

1–5 – BOLT TIGHTENING SEQUENCE
6 – HYDRAULIC CLUTCH SLAVE CYLINDER
7 – TRANSAXLE
8 – POWER STEERING HOSE
9 – COLLAR

67189-PTCR-G08

Structural collar assembly—manual transaxle

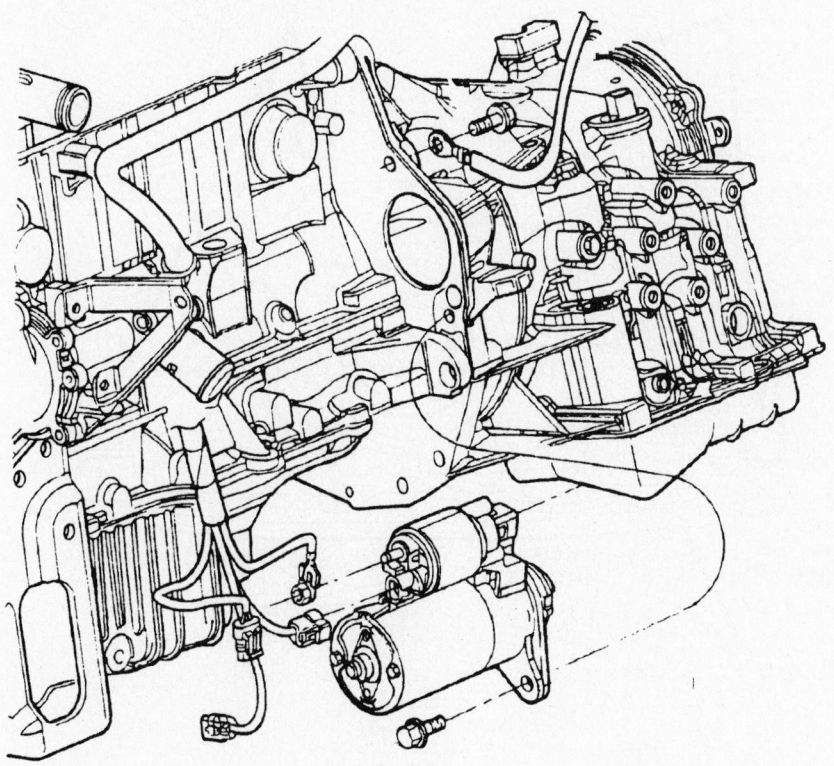

9306ZG91

Removal of the starter motor mounting

h. Tighten the remaining bolts to 45 ft. lbs. (61 Nm).

5. Install the engine structural collar on models with a manual transaxle as follows, refer to the illustration for bolt locations:

a. Place the collar in position and hand tighten the collar to transaxle bolt.

b. Position the power steering hose support bracket and install the collar to oil pan bolt.

c. Position the clutch slave cylinder into position and hand start the bolts.

d. Tighten bolts (1) to 75 ft. lbs. (101 Nm).

e. Tighten bolts (2 and 5) to 45 ft. lbs. (61 Nm).

f. Tighten bolts (3 and 4) to 20 ft. lbs. (28 Nm).

6. Install or connect the following:
- Power steering lines aside
- Studs retaining the power steering lines and tighten to 45 ft. lbs. (61 Nm)
- Nuts retaining the inner cooler tube
- Inner cooler lower hose to the inner cooler
- Upper starter bolt by pushing the inner cooler up and out of the way. Tighten to 40 ft. lbs. (54 Nm).
- Air cleaner box cover
- Negative battery cable

Oil Pan

REMOVAL & INSTALLATION

1. Before servicing the vehicle, refer to the precautions in the beginning of this section.

2. Drain the engine oil and remove the oil filter.

3. Support the powertrain assembly.

4. Remove or disconnect the following:
- Negative battery cable
- Right inner splash shield
- Turbocharger-to-charge air cooler hose assembly, if equipped
- Oil cooler connector bolt, if

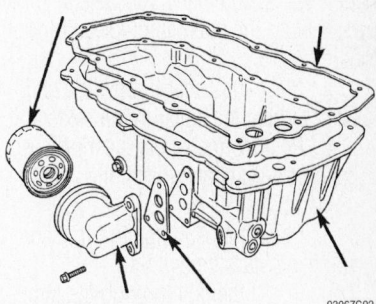

9306ZG92

Remove the oil filter adapter

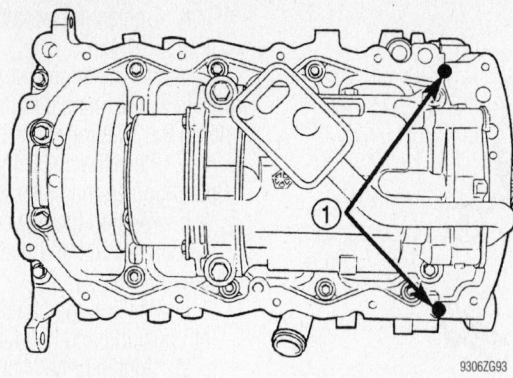

Silicone sealer application locations

equipped with a turbocharger. Do not disconnect the coolant lines from the oil cooler and reposition the cooler.

- Engine structural collar
- Lower torque strut
- Oil filter adapter
- Oil pan and gasket

5. Thoroughly clean all gasket mating surfaces.

To install:

6. Apply silicone sealer to the oil pump-to-engine block parting line.

7. Install or connect the following:

- New gasket on the oil pan
- Oil pan and torque the bolts to 105 inch lbs. (12 Nm)
- Oil filter and adapter and torque the screws to 105 inch lbs. (12 Nm)

8. If equipped with a turbocharger, replace the oil cooler seal. Lubricate the seal with and place the oil cooler-to-oil filter adapter in position making sure to align the notch on the tab. Install the oil cooler connector bolt and tighten to 41 ft. lbs. (55 Nm).

✳✳ WARNING

Follow the proper tightening sequence for the structural collar or damage to the collar or oil pan may occur.

9. Install the engine structural collar on models with a automatic transaxle as follows, refer to the illustration for bolt locations:

a. Place the collar in position and hand tighten the collar to transaxle bolt.

b. Position the power steering hose support bracket and install the collar to oil pan bolt.

c. Place the bending strut in place and hand start the bolt.

d. Install the bolt through the strut and collar and hand tighten.

e. Place the power steering hose support bracket in position and install and hand tighten the remaining bolts.

f. Tighten the collar-to-transmission bolts to 75 ft. lbs. (101 Nm).

g. Install the bolts through the strut and into the block.

h. Tighten the remaining bolts to 45 ft. lbs. (61 Nm).

10. Install the engine structural collar on models with a manual transaxle as follows, refer to the illustration for bolt locations:

a. Place the collar in position and hand tighten the collar to transaxle bolt.

b. Position the power steering hose support bracket and install the collar to oil pan bolt.

c. Position the clutch slave cylinder into position and hand start the bolts.

d. Tighten bolts (1) to 75 ft. lbs. (101 Nm).

e. Tighten bolts (2 and 5) to 45 ft. lbs. (61 Nm).

f. Tighten bolts (3 and 4) to 20 ft. lbs. (28 Nm).

11. Install or connect the following:

- Lower torque strut
- Turbocharger-to-charge air cooler hose assembly, if equipped
- Right inner splash shield
- Negative battery cable

12. Fill the engine with clean oil and a new filter.

13. Start the vehicle and check for leaks, repair if necessary.

Oil Pump

REMOVAL & INSTALLATION

1. Before servicing the vehicle, refer to the precautions in the beginning of this section.

2. Drain the engine oil.

3. Remove or disconnect the following:

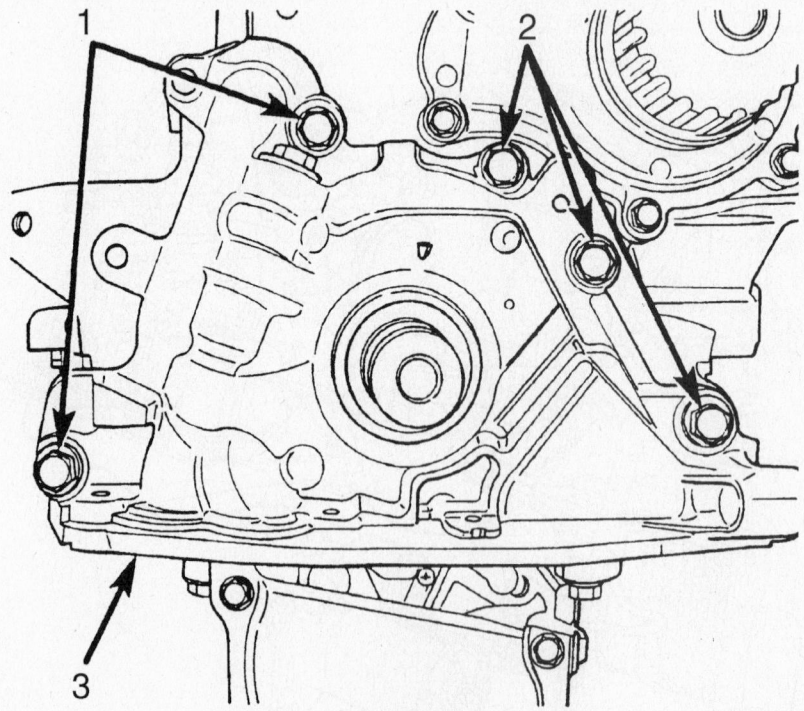

1 – BOLTS
2 – BOLTS
3 – OIL PUMP

Exploded view of the oil pump mounting bolts

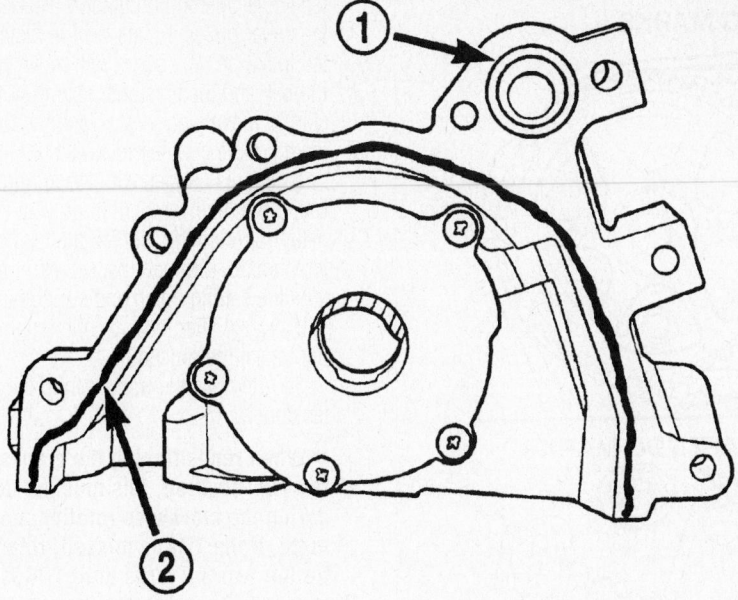

1 – O-RING
2 – SEALER LOCATION

67189-PTCR-G09

Apply a small amount of gasket maker to the pump body cover mounting surface—2001 models

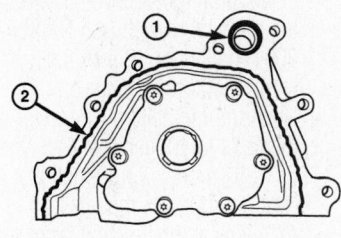

1 - O-RING
2 - SEALER LOCATION

67189-PTCR-G10

Apply a small amount of gasket maker to the pump body cover mounting surface— 2002–05 models

- Negative battery cable
- Timing belt and rear cover
- Oil pan
- Crankshaft sprocket, using Tool 6793 and Insert Tool C-4685-C2
- Crankshaft key
- Oil pickup tube
- Oil pump

To install:

4. Wash all parts in a solvent; then, inspect carefully for damage or wear, as follows:

   a. Inspect the mating surface of the oil pump should be smooth. Replace the pump cover, if scratched or grooved.

   b. Apply Mopar® gasket maker to the oil pump.

   c. Install the O-ring into the oil pump body discharge passage.

5. Prime the oil pump before installation by filling the rotor cavity with engine oil.

6. Install or connect the following:

   - Oil pump, align the rotor flats with the crankshaft flats and torque the bolts to 21 ft. lbs. (28 Nm).

❋❋ WARNING

The front crankshaft seal MUST be out of the pump to align or damage may result.

- New front crankshaft seal, using Seal Driver Tool 6780
- Crankshaft key
- Crankshaft sprocket, using a Crankshaft Sprocket Installer Tool 6792
- Oil pump pickup tube
- Oil pan
- Rear timing belt cover
- Timing belt
- Negative battery cable

7. Fill the engine with clean oil.

8. Start the engine and check for leaks; repair if necessary.

Timing Belt

REMOVAL & INSTALLATION

1. Before servicing the vehicle, refer to the precautions in the beginning of this section.

2. Remove the A/C refrigerant using approved equipment.

3. Remove or disconnect the following to remove the upper timing cover:

   - Negative battery cable
   - Upper torque strut bolts and set the strut aside
   - A/C lines at the junction block near the upper timing belt cover, if equipped with a turbocharger
   - Upper cover bolts and the cover

4. Remove or disconnect the following to remove the lower timing cover:

   - Right front wheel and inner splash shield
   - Accessory drive belts
   - Crankshaft damper
   - Lower torque strut
   - Exhaust system from the manifold
   - A/C compressor switch
   - Upper torque strut and bracket
   - Upper radiator support crossmember
   - Power steering pump and bracket without disconnecting the lines
   - Right engine mount through bolt after supporting the engine
   - Engine support bracket
   - Lower cover bolts and the cover

5. Rotate the crankshaft until the Top Dead Center (TDC) mark on the oil pump housing aligns with the TDC mark on the crankshaft sprocket.

6. Loosen the belt tensioner lock bolt.

7. Install an 6mm Allen wrench into the tensioner. Rotate the tensioner counterclockwise while pushing on the wrench until it slides into the locking hole.

8. Remove the timing belt.

To install:

9. Set the crankshaft sprocket at TDC by aligning the sprocket with the arrow on the oil pump housing.

10. Set the camshafts timing marks so that the exhaust camshaft sprocket is a ½notch below the intake camshaft sprocket.

11. Install the timing belt by starting at the crankshaft. Go around the water pump sprocket, idler pulley, camshaft sprockets and the tensioner.

12. Move the exhaust camshaft sprocket counterclockwise to align the marks and to remove any slack.

13. Insert a 6mm Allen wrench into the

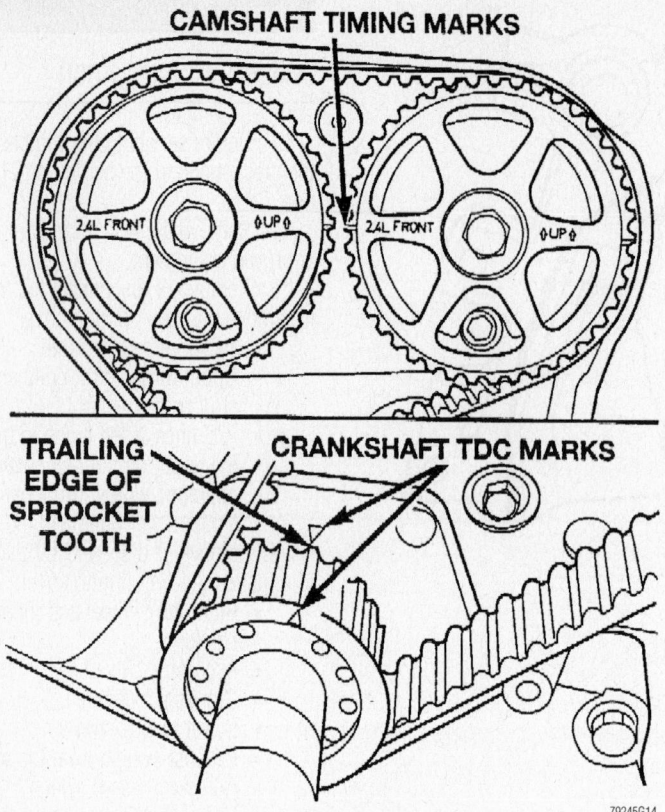

CAMSHAFT TIMING MARKS

TRAILING EDGE OF SPROCKET TOOTH

CRANKSHAFT TDC MARKS

79245G14

Camshaft and crankshaft alignment marks—2.4L engine

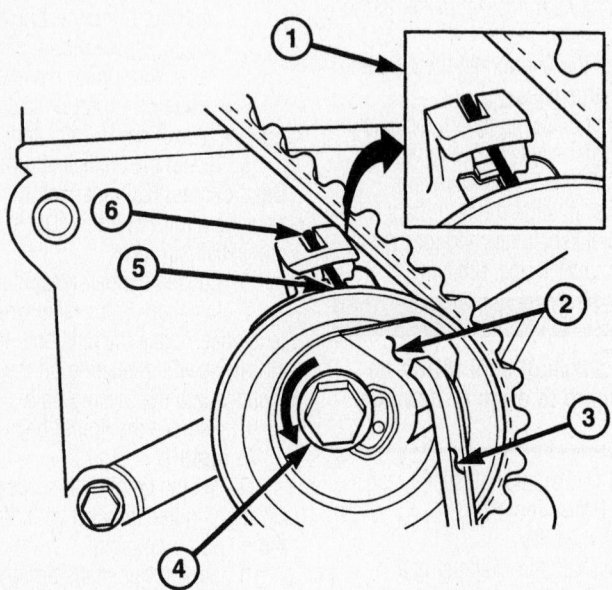

1 - ALIGN SETTING NOTCH WITH SPRING TANG
2 - TOP PLATE
3 - 6mm ALLEN WRENCH
4 - LOCK BOLT
5 - SETTING NOTCH
6 - SPRING TANG

67189-PTCR-G11

Install the belt tensioner

tensioner opening on the top plate of the tensioner pulley. Rotate the top plate counterclockwise. The pulley will move against the belt and the tensioner setting notch will start to move clockwise. Continue to move the top plate counterclockwise until the setting notch is aligned with the spring tang. Using the Allen wrench, to prevent the top plate from moving, tighten the tensioner lock bolt to 220 inch lbs. (25 Nm). Make sure the setting notch and spring tang are still aligned after the lock nut is tighten. If not repeat the procedure.

14. Remove the wrench from the belt tensioner.

➡**When repositioning the crankshaft to the TDC position, this must be done during the clockwise rotation movement. If the TDC is missed, rotate a further two full turns until TDC is reached. Do not rotate the crankshaft counterclockwise as this will result in improper tensioner settings.**

15. Rotate the crankshaft two full revolutions and verify that the TDC marks are properly aligned.

16. Check the spring tang is within the tolerance window, if not repeat the previous two steps.

17. Install or connect the following:
  - Lower timing belt cover and torque the bolts to 40 inch (4.5 Nm) on 2001 models or 50 inch lbs. (6 Nm) on 2002–05 models.
  - Upper timing belt cover and torque the bolts to 40 inch lbs. (4.5 Nm) on 2001 models or 50 inch lbs. (6 Nm) on 2002–05 models.
  - A/C lines at the junction block near the upper timing belt cover, if equipped with a turbocharger
  - Right engine support bracket and reposition the power steering pump. Torque the bracket bolts to 45 ft. lbs. (61 Nm).
  - Right engine mount through bolt and torque it to 87 ft. lbs. (118 Nm)
  - Upper radiator support crossmember
  - Torque strut bracket
  - Upper torque strut
  - A/C lines and pressure switch
  - Exhaust system to the manifold
  - Crankshaft damper
  - Accessory drive belts
  - Lower torque strut
  - Right splash shield and wheel
  - Negative battery cable

18. Recharge the A/C system.

19. Perform the camshaft/crankshaft synchronization procedure.

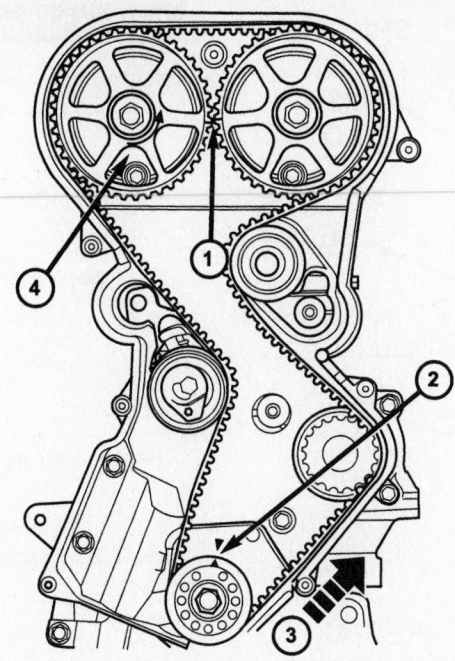

1 - CAMSHAFT TIMING MARKS 1/2 NOTCH LOCATION
2 - CRANKSHAFT AT TDC
3 - INSTALL BELT IN THIS DIRECTION
4 - ROTATE CAMSHAFT SPROCKET TO TAKE UP BELT SLACK
67189-PTCR-G12

Proper timing belt routing

Rear Main Seal

REMOVAL & INSTALLATION

1. Before servicing the vehicle, refer to the precautions in the beginning of this section.
2. Remove or disconnect the following:
   - Transmission
   - Flexplate/flywheel
   - Rear main seal
3. Insert a seal remover between the dust lip and the metal case of the crankshaft seal. Angle the tool through the dust lip against the metal case of the seal. Pry out the seal.

❊❊ WARNING

DO NOT let the prytool contact the crankshaft seal surface. Contact of the tool blade against the crankshaft edge (chamfer) is permitted.

To install:

❊❊ WARNING

If the crankshaft edge (chamfer) has any burrs or scratches on the, clean it up with 400 grit sand paper to prevent seal damage during installation of the new seal.

➡ No lubrication is necessary when installing the seal.

4. Place Crankcase Seal Pilot Tool 6926-1 on the crankshaft; this is a pilot tool with a magnetic base
5. Position the seal over the pilot Tool; be sure the words THIS SIDE OUT on the seal can be read.

➡ The pilot tool should stay on the crankshaft during installation of the seal. Be sure the seal lip faces the crankcase during installation.

❊❊ WARNING

If the seal is driven in the block past flush, this may cause an oil leak.

6. Drive the seal into the block, using Crankshaft Seal Tool 6926-2 and handle C-4171, until the tool bottoms out against the block.
7. Install or connect the following:
   - Flexplate/flywheel. Apply Lock & Seal Adhesive to the bolt treads and torque the bolts in a star pattern, to 70 ft. lbs. (95 Nm).
   - Transmission
   - Negative battery cable
8. Start the vehicle and check for leaks, repair if necessary.

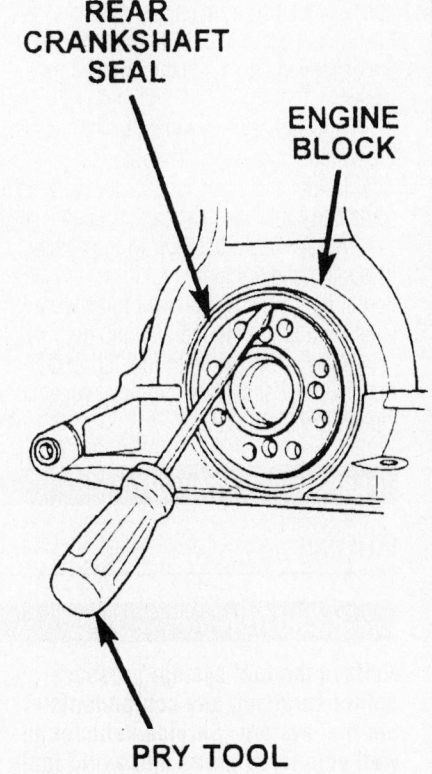

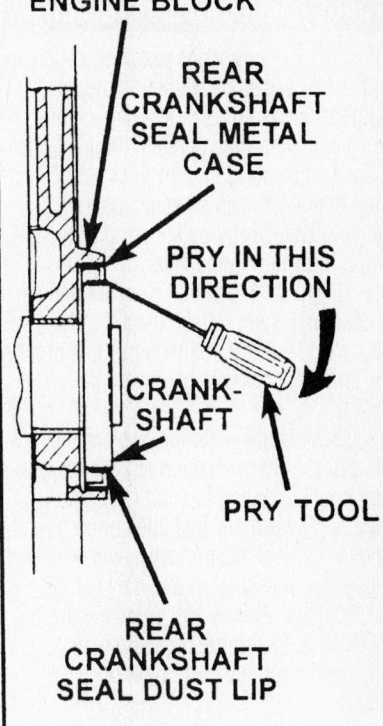

7922EG32

When prying the seal out, be sure to use the prytool at the proper angle

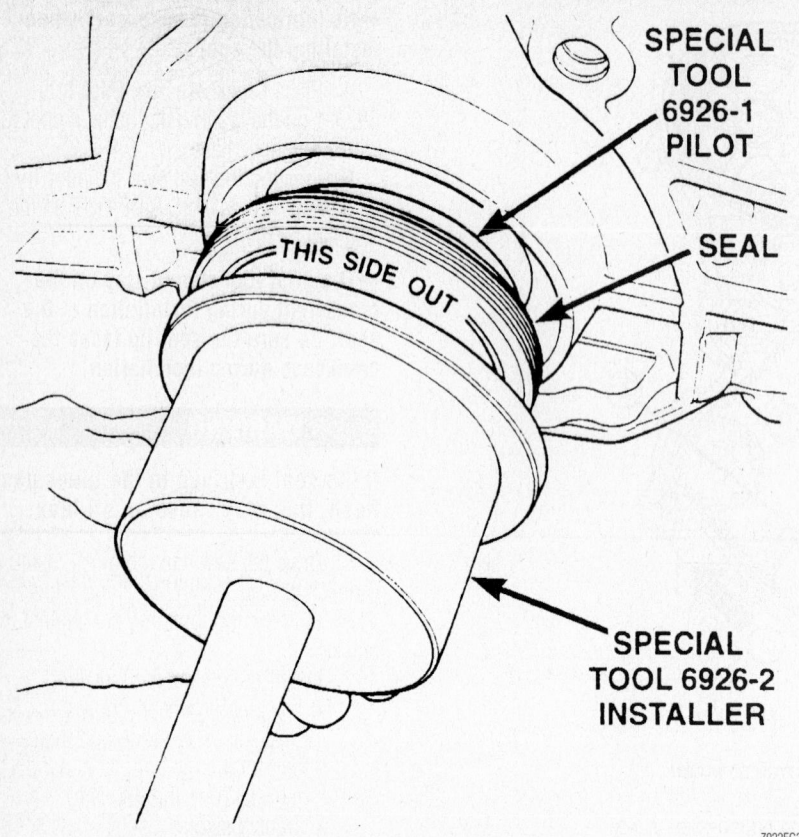

Place a proper size pilot tool with a magnetic base on the crankshaft

Piston and Ring

POSITIONING

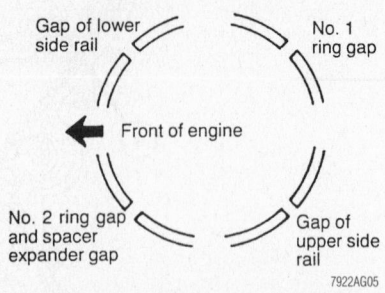

Piston ring end-gap spacing—2.4L engine

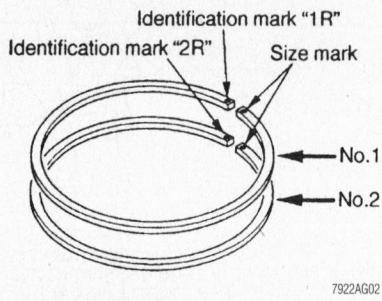

Common Chrysler piston ring identification mark locations

FUEL SYSTEM

Fuel System Service Precautions

Safety is an important factor when servicing the fuel system. Failure to conduct maintenance and repairs in a safe manner may result in serious personal injury. Maintenance and testing of the vehicle's fuel system components can be accomplished safely and effectively by adhering to the following rules and guidelines:

• To avoid the possibility of fire and personal injury, always disconnect the negative battery cable unless the repair or test procedure requires that battery voltage be applied.

• Always relieve the fuel system pressure prior to disconnecting any fuel system component (injector, fuel rail, pressure regulator, etc.), fitting or fuel line connection. Exercise extreme caution whenever relieving fuel system pressure, to avoid exposing skin, face and eyes to fuel spray. Please be advised that fuel under pressure may penetrate the skin or any part of the body that it contacts.

• Always place a shop towel or cloth around the fitting or connection prior to loosening to absorb any excess fuel due to spillage. Ensure that all fuel spillage is

quickly removed from engine surfaces. Ensure that all fuel soaked cloths or towels are deposited into a suitable waste container.

• Always keep a dry chemical (Class B) fire extinguisher near the work area.

• Do not allow fuel spray or fuel vapors to come into contact with a spark or open flame.

• Always use a back-up wrench when loosening and tightening fuel line connection fittings. This will prevent unnecessary stress and torsion to fuel line piping.

• Always replace worn fuel fitting O-rings. Do not substitute fuel hose where fuel pipe is installed.

Fuel System Pressure

RELIEVING

❊❊ CAUTION

Relieve the fuel system pressure before servicing any components of the fuel system. Service vehicles in well ventilated areas and avoid ignition sources. NEVER smoke while servicing the vehicle!

1. Before servicing the vehicle, refer to the precautions in the beginning of this section.
2. Remove the negative battery cable.
3. Remove the fuel pump relay from the Power Distribution Center (PDC).
4. Start and run the engine until it stalls.
5. Turn the ignition key to the OFF position.

Fuel Filter

The fuel filter is part of the fuel pump module located in the fuel tank. Refer to the fuel pump module procedure for the fuel filter.

Fuel Pump

The fuel pump is integral with the pump module, which also contains the fuel reservoir, level sensor, inlet strainer and fuel pressure regulator. The inlet strainer, fuel pressure regulator and level sensor are the only serviceable items. If the fuel pump requires service, replace the entire fuel pump module.

REMOVAL & INSTALLATION

1. Before servicing the vehicle, refer to the precautions in the beginning of this section.

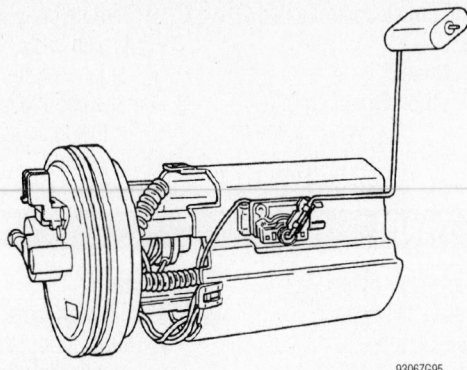

Exploded view of the fuel pump module

2. Properly relieve the fuel system pressure.

3. Remove or disconnect the following:

- Negative battery cable
- Air cleaner lid
- Inlet Air Temperature (IAT) sensor and make up air hose
- Fuel tank
- Fuel pump module and seal from the tank
- Locknut to release the fuel pump module, using a Ring Spanner Tool No. 6856
- Fuel filter from the fuel pump module
- Fuel pump and seal from the fuel tank

To install:

4. Install or connect the following:

- Fuel filter to the fuel pump module
- Fuel filter lines to the fuel pump module
- Fuel pump module in the tank
- Locknut while holding the fuel pump in position. Using Tool 6856, torque the nut to 56 ft. lbs. (75 Nm).
- Fuel tank
- IAT sensor and make-up hose
- Air cleaner lid
- Negative battery cable

5. Fill the fuel tank.

6. Start the vehicle and check for leaks, repair if necessary.

Fuel Injector

REMOVAL & INSTALLATION

Except Turbocharged Models

1. Before servicing the vehicle, refer to the precautions in the beginning of this section.

2. Release the fuel system pressure.

3. Remove or disconnect the following:

- Negative battery cable
- Remove the air cleaner lid
- Inlet Air Temperature (IAT) sensor and the make-up hose
- Engine cover or throttle control shield, if equipped
- Fuel supply tube from fuel rail
- Intake manifold
- Fuel injector electrical connectors
- Fuel rail with the fuel injectors
- Fuel injector(s) and discard the O-rings

To install:

➡**Lubricate the O-rings with clean engine oil.**

4. Install or connect the following:

- Fuel injector(s) to the fuel rail using new O-rings
- Fuel injector nozzles into the intake manifold and torque the fuel rail bolts to 8 ft. lbs. (12 Nm)
- Fuel injector electrical connectors
- Fuel supply tube to fuel rail
- Intake manifold
- Engine cover or throttle control shield, if equipped
- IAT sensor and make-up hose
- Air cleaner lid
- Negative battery cable

5. Start the vehicle and check for leaks, repair if necessary.

Turbocharged Models

1. Before servicing the vehicle, refer to the precautions in the beginning of this section.

2. Release the fuel system pressure.

3. Drain the cooling system.

4. Remove or disconnect the following:

- Negative battery cable
- Throttle body inlet hose from the throttle body
- Purge hose from the throttle body

- Electrical connections from the throttle body
- Throttle control shield
- Accelerator and cruise control cable from the throttle body
- Manifold Absolute Pressure (MAP) sensor connector
- Vacuum lines from the rear of the intake manifold
- 5 bolts from the front and 2 bolts from the rear of the intake manifold
- Intake manifold and cover the lower manifold to avoid contamination
- Upper radiator hose up and out of the way
- 2 small hoses from the thermostat housing
- 2 bolts from the thermostat housing and rotate the assembly up and out of the way
- Fuel lines from the rail
- Fuel injector electrical connectors and wire from the rail
- Fuel rail with the fuel injectors
- Fuel injector(s) and discard the O-rings

To install:

➡**Lubricate the O-rings with clean engine oil.**

5. Install or connect the following:

- Fuel injector(s) to the fuel rail using new O-rings
- Fuel injector nozzles into the intake manifold and torque the fuel rail bolts to 170 inch lbs. (19 Nm)
- Fuel injector electrical connectors and wiring to the fuel rail
- Fuel line to fuel rail
- Rotate the thermostat assembly into position and tighten the 2 bolts
- 2 small hoses to the thermostat housing
- Upper radiator hose

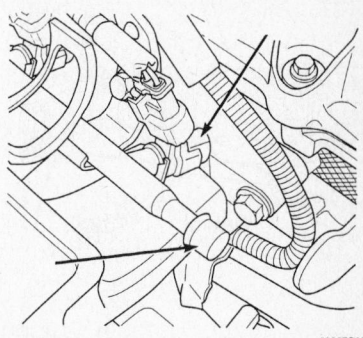

Remove the fuel rail and injectors as an assembly

- Intake manifold
- 5 bolts front and 2 rear bolts on the intake manifold
- Vacuum lines to the rear of the intake manifold
- MAP sensor connector

- Accelerator and cruise control cable to the throttle body
- Throttle control shield
- Electrical connections to the throttle body
- Purge hose to the throttle body

- Throttle body inlet hose to the throttle body
- Negative battery cable

6. Fill the cooling system.
7. Start the vehicle and check for leaks, repair if necessary.

DRIVE TRAIN

Transaxle Assembly

REMOVAL & INSTALLATION

Manual

T350 TRANSAXLE

1. Before servicing the vehicle, refer to the precautions in the beginning of this section.
2. Drain the transaxle fluid.
3. Remove or disconnect the following:

- Battery cables
- Battery and tray
- Air cleaner
- Back-up lamp switch wiring from the transaxle
- Shift selector and crossover cable and move them out of the way
- Vehicle Speed Sensor (VSS) wire

- Clutch master cylinder hydraulic tube from the slave cylinder
- Both halfshafts
- Bell housing dust cover
- Power steering hose from the structural collar
- Left side engine-to-transaxle lateral bending brace and structural collar
- Right side engine-to-transaxle lateral bending brace
- Starter
- Driveplate-to-clutch module bolts and support the engine at the oil pan
- Transaxle upper mount bolts and lower the powertrain assembly onto the support
- Transaxle-to-engine mounting bolts
- Transaxle
- Clutch module from the input shaft
- Slave cylinder
- Upper transaxle mount

To install:

4. Install or connect the following:

- Clutch module to the transaxle input shaft and place the transaxle into position
- Transaxle-to-engine mounting bolts and torque them to 80 ft. lbs. (108 Nm)
- Upper transaxle mount and torque the bolts to 45 ft. lbs. (62 Nm)
- Driveplate-to-clutch module bolts and torque to 65 ft. lbs. (88 Nm)
- Starter
- Starter electrical connectors
- Bell housing dust cover
- Left side engine-to-transaxle bending brace and structural brace. Torque the bolts to 60 ft. lbs. (81 Nm).
- Right lateral bending brace and torque the bolts to 60 ft. lbs. (81 Nm)

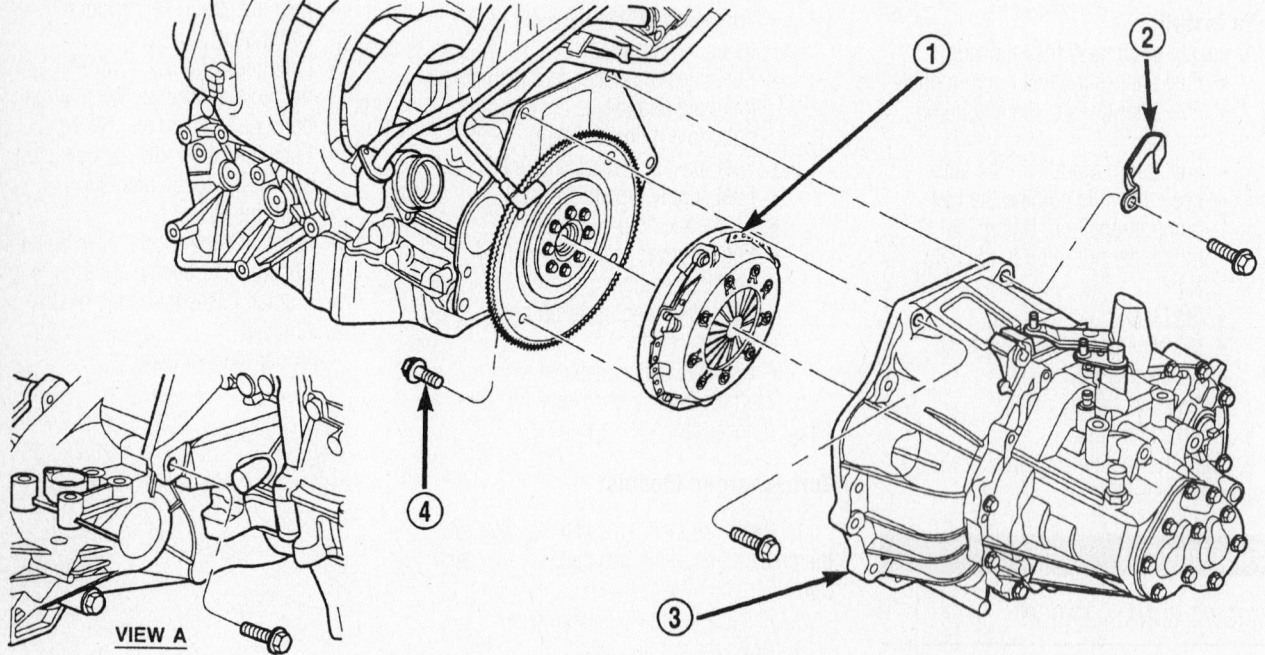

VIEW A

1 – MODULAR CLUTCH ASSEMBLY
2 – CLIP
3 – TRANSAXLE
4 – CLUTCH MODULE BOLT (4)

9306ZG0E

Remove the clutch module from the transaxle assembly

- Power steering hose to the structural collar
- Both halfshafts
- Clutch master cylinder tube to the slave cylinder
- VSS electrical connector
- Shift crossover and selector cables to the shift lever
- Cables to the bracket and install a new retainer clip
- Back-up lamp switch connector
- Battery tray
- Battery and cables
- Air cleaner assembly

5. Fill the transaxle fluid to the proper level.

6. Be sure the vehicle's back-up lights and speedometer are functioning properly.

7. Start the vehicle and check for leaks, repair if necessary.

G288 TRANSAXLE

1. Before servicing the vehicle, refer to the precautions in the beginning of this section.

2. Drain the transaxle fluid.

3. Remove or disconnect the following:
- Negative battery cable
- Air cleaner
- Power Distribution Center (PDC) from the bracket
- Air cleaner/PDC mounting bracket
- Gearshift cable from the shift mechanism
- Gearshift cable bracket
- Back-up lamp switch wiring from the transaxle
- Vehicle Speed Sensor (VSS) wire
- Clutch master cylinder hydraulic tube from the slave cylinder
- Upper transaxle bell housing bolts
- Transaxle drain plug, drain the fluid and reinstall the plug
- Both halfshafts
- Intermediate shaft/bearing bolts and the shaft assembly
- 2 intercooler connector pipe-to-oil pan bolts and position aside
- Oil pan-to-bell housing bolts

4. Place a screw jack with a piece of wood on top onto the oil pan.
- 2 transaxle upper mount-to-bracket bolts and lower the engine onto the wood and screw jack
- Transaxle upper mount bracket
- Attach a transaxle jack to the transaxle
- Starter lower bolt and ground cable
- 4 modular clutch drive-to-drive plate bolts. While removing the bolts one tolerance (slotted) drive plate hole will be seen. When this

bolt is removed, mark the drive plate and clutch assembly alignment to aid during reinstallation for proper alignment
- Remaining transaxle-to-engine bolts
- Transaxle

To install:
5. Install or connect the following:
- Transaxle into position
- Accessible transaxle-to-engine mounting bolts and torque them to 80 ft. lbs. (108 Nm)
- 4 modular clutch-to-drive plate bolts, align the drive plate and clutch assembly marks made during removal, start with the slotted hole first and tighten the bolts to 65 ft. lbs. (88 Nm)
- Upper mount bracket
- raise the transaxle using the jack until the mount bracket is aligned and install the bolts
- Starter motor bolt and ground cable
- Lower bellhousing bolt
- 2 intercooler connector pipe-to-oil pan bolts
- Remove the drain plug and fill the transaxle with the correct type and amount of fluid. Install the plug and tighten to 35 ft. lbs. (47 Nm).
- Intermediate/bearing assembly and tighten the bolts
- Both halfshafts
- Clutch master cylinder tube to the slave cylinder
- VSS electrical connector
- Gearshift cable bracket and tighten the bolts
- Gearshift crossover and selector cables and secure using new clips
- Back-up lamp switch connector
- PDC/air cleaner bracket
- PDC
- Air cleaner assembly
- negative battery cable

6. Be sure the vehicle's back-up lights and speedometer are functioning properly.

7. Start the vehicle and check for leaks, repair if necessary.

Automatic

1. Before servicing the vehicle, refer to the precautions in the beginning of this section.

2. Drain the transaxle fluid.

3. Remove or disconnect the following:
- Battery cables
- Air cleaner assembly
- Battery and tray
- Upper starter-to-transaxle bell housing bolt

- Transaxle dipstick and tube
- Gearshift cable end from the transaxle shift lever
- Gearshift cable bracket bolt from the transaxle
- Transaxle oil cooler lines. Plug the lines to prevent contamination.
- Input and output speed sensor electrical connectors
- Transaxle range sensor connector
- Solenoid/pressure switch assembly connector
- Both front wheels
- Left front splash shield
- Both halfshafts
- Power steering hose from the structural collar
- Left and right side engine-to-transaxle lateral brace and structural collar
- Starter motor electrical connectors
- Starter
- Gearshift cable bracket
- Driveplate-to-torque converter bolts and support the powertrain assembly
- Transaxle upper mount to bracket bolts and lower the powertrain assembly
- Transaxle

To install:
4. Install or connect the following:
- Transaxle and torque the bolts to 80 ft. lbs. (105 Nm) while supporting the transaxle with a jack
- Mount-to-transaxle bracket. Torque the bolts to 50 ft. lbs. (68 Nm) and remove the jack.
- Driveplate-to-torque converter bolts and torque the bolts to 65 ft. lbs. (88 Nm)
- Starter and hand tighten the bolts
- Dipstick tube and secure the bracket to the transaxle. Torque the upper starter bolt to 40 ft. lbs. (54 Nm).
- Cable bracket to the bell housing and torque the bolt to 45 ft. lbs. (61 Nm)
- Starter lower bolt and torque the bolt to 40 ft. lbs. (54 Nm)
- Starter electrical connections
- Bell housing dust cover
- Lower dust shield screws and torque the bell housing cover bolts to 108 inch lbs. (12 Nm)
- Right side lateral bending brace and torque the bolts to 60 ft. lbs. (81 Nm)
- Both halfshafts
- Left side splash shield and both wheels

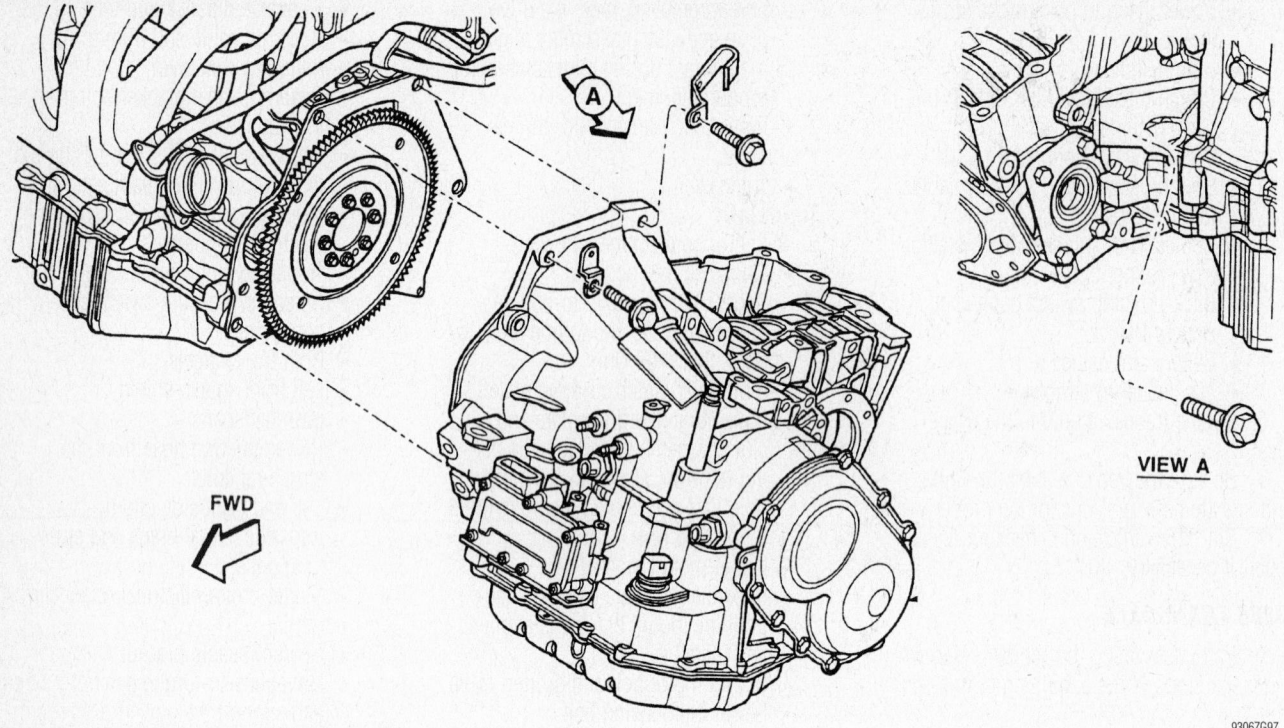

9306ZG97

Exploded view of automatic transaxle removal

- Cooler lines to the transaxle and secure with constant tension clamps
- Solenoid/pressure switch assembly
- Transmission range sensor connector
- Input/output sensor connectors
- Gearshift cable to the bracket and connect it to the manual valve lever
- Battery tray and battery
- Air cleaner assembly
- VSS wiring
- Battery cables

5. Fill the transaxle fluid to the proper level.

6. Be sure the vehicle's back-up lights and speedometer are working properly.

Clutch

ADJUSTMENT

This vehicle utilizes a modular clutch assembly located between the engine and the transaxle. The modular clutch is serviced as an assembly and is self-adjusting. The self-adjusting feature of the clutch relies on a sensor ring and adjuster ring.

REMOVAL & INSTALLATION

1. Before servicing the vehicle, refer to the precautions in the beginning of this section.

2. Remove or disconnect the following:

- Negative battery cable
- Air cleaner assembly
- Battery and tray
- Back-up lamp electrical connector
- Shift cable-to-bracket clips
- Shift lever and crossover cable from the levers. Move the cables out of the way.
- Vehicle Speed Sensor (VSS) electrical connector
- Clutch master cylinder tube from the slave cylinder using a clutch hydraulic quick connect Tool 6638A
- Both halfshafts
- Power steering hose from the structural collar
- Left side lateral bending brace and structural collar
- Bell housing dust cover
- Right side lateral bending brace
- Starter
- Driveplate-to-clutch module bolts and support the engine at the oil pan
- Transaxle upper mount bolts and lower the engine/transaxle assembly
- Modular clutch from the input shaft

To install:

3. Install or connect the following:

- Clutch module to the input shaft
- Transaxle-to-engine mount bolts and torque them to 80 ft. lbs. (108 Nm)

- Upper mount bolts and torque them to 65 ft. lbs. (88 Nm)
- Driveplate-to-clutch module bolts and torque them to 65 ft. lbs. (88 Nm)
- Starter and torque the bolts to 40 ft. lbs. (54 Nm). Make certain that the ground cable is fastened to the upper bolt.

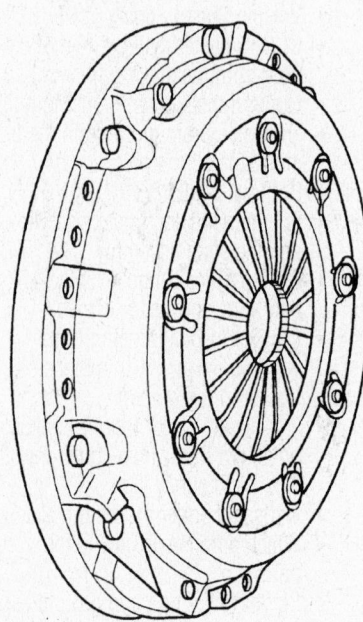

9306ZG98

Exploded view of the modular clutch assembly

- Starter electrical connectors
- Bell housing dust cover
- Left side lateral bending brace and structural collar
- Power steering hose to the structural collar
- Right side lateral bending brace and torque the bolts to 60 ft. lbs. (81 Nm)
- Both halfshafts
- Clutch master cylinder tube to the slave cylinder
- VSS electrical connector
- Back-up lamp switch electrical connector
- Air cleaner assembly
- Battery and tray
- Battery cables

4. Fill the transmission to the proper level.

5. Road test the vehicle and check for proper clutch operation.

6. Check the fluid level and adjust if needed.

Halfshafts

REMOVAL & INSTALLATION

1. Before servicing the vehicle, refer to the precautions in the beginning of this section.

2. Place the transaxle in the **P** position, for automatic and neutral for manual.

3. Remove or disconnect the following:
- Negative battery cable
- Front wheel
- Cotter pin, locknut and spring washer from the end of the outer Constant Velocity (CV) joint stub axle
- Driveshaft to hub and bearing nut
- Front wheel speed sensor, if equipped
- Steering knuckle from the ball joint
- Driveshaft from the steering knuckle and support the outer end of the driveshaft

✵✵ WARNING

Be careful when separating the ball joint stud from the steering knuckle, so the ball joint seal does not get damaged.

✵✵ WARNING

Be careful when separating the inner CV-joint during this operation. Do not let the driveshaft hang by the inner CV-joint, the driveshaft must be supported.

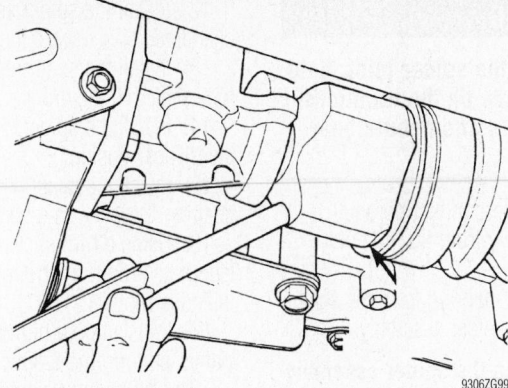

9306ZG99

Remove the Tri-Pot joint from the transaxle

➡**Inner Tri-Pot joint removal is easier by applying outward pressure on the joint while hitting the punch with a hammer.**

4. Inner Tri-Pot joints from the transmission side gears using a punch to dislodge the inner Tri-Pot joint retaining ring from the transmission side gear. If removing the right side inner Tri-Pot joint, position the punch against the inner Tri-Pot joint. Hit the punch sharply with a hammer to dislodge the right inner joint from the side gear. If removing the left side inner Tri-Pot joint, position the punch in the groove of the inner Tri-Pot joint. Hit the punch sharply with a hammer to dislodge the left inner Tri-Pot joint from the side gear.

5. Hold the inner Tri-Pot joint and interconnecting shaft of the driveshaft assembly. Remove the inner Tri-Pot joint from the transaxle by pulling it straight out of the transaxle side gear and transmission oil seal. When removing the Tri-Pot joint, do not let the spline or snapring drag across the sealing lip of the transaxle-to-Tri-Pot joint oil seal.

✵✵ WARNING

The driveshaft, when installed, acts as a bolt which secures the front hub and bearing assembly. If the vehicle is to be supported or moved on its wheels with a driveshaft removed, install a proper-sized bolt and nut through the front hub. Tighten the bolt and nut to 135 ft. lbs. (183 Nm). This will ensure that the hub bearing cannot loosen.

To install:

6. Thoroughly clean the spline and oil seal sealing surface on the Tri-Pot joint. Lightly lubricate the oil seal sealing surface on the Tri-Pot joint with fresh, clean transmission fluid.

7. Holding the driveshaft assembly by the Tri-Pot joint and interconnecting shaft, install the Tri-Pot joint into the transaxle side gear as far as possible by hand

8. Align the Tri-Pot joint with the transmission side gears, grasp the driveshaft interconnecting shaft and push the Tri-Pot joint into the transaxle side gear until fully seated. Be sure the snapring is fully engaged with the side gear by trying to remove the Tri-Pot joint from the transaxle by hand. If the snapring is fully seated with the side gear, the Tri-Pot joint will not be removable by hand.

9. Install or connect the following:
- Driveshaft back into the front hub
- Steering knuckle into the ball joint stud
- New steering knuckle-to-ball joint stud bolt and nut. Torque the nut and bolt to 70 ft. lbs. (95 Nm).
- Washer and hub nut to the stub axle and torque the nut to 180 ft. lbs. (244 Nm)
- Spring washer, locknut and cotter pin
- VSS, if equipped
- Front wheel
- Negative battery cable

10. Check the transaxle fluid and adjust if needed.

CV-Joints

OVERHAUL

Tri-Pot (Inner) Joint

1. Before servicing the vehicle, refer to the precautions in the beginning of this section.

2. Remove the negative battery cable.

3. Remove the halfshaft.

4. Remove the tri-pot joint boot clamps and slide the boot down the shaft.

✳ WARNING

When removing the spider joint, hold the rollers in place on the trunnions to keep the rollers and needle bearings in place.

5. Slide the interconnecting shaft and spider assembly from the tri-pot housing.

6. Remove the snapring from the shaft.

7. Remove the spider assembly.

➡ **If necessary, tap the spider assembly from the shaft with a brass drift.**

✳ WARNING

When removing the spider assembly, do not hit the outer bearings.

8. Remove the boot by sliding it off the shaft.

✳ WARNING

If any parts show excessive wear, replace the halfshaft assembly; the component parts are not serviceable.

To install:

✳ WARNING

The Tri-pot sealing boots are made of 2 different types of material, silicon rubber (high temperature) which is soft and pliable or hytrel plastic (standard temperature) which is stiff and rigid. Be sure to replace the boot made of the correct material.

9. Install a new small boot clamp and slide it on the shaft.

10. Install the boot and slide it on the shaft.

11. Position the boot so that the raised bead on the inside the boot seal is in the shaft groove.

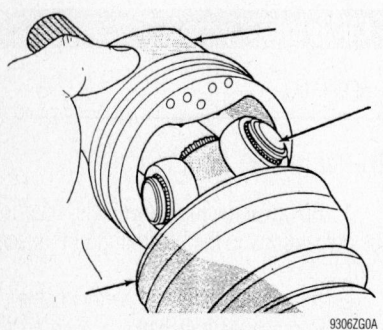

Install the Tri-Pot housing on to the spider assembly

12. Install the spider assembly, face the chamfered side toward the shaft.

13. Install the snapring making sure it is fully seated in the groove.

14. Install the spider/shaft assembly into the tri-pot housing.

15. Install a new inner boot clamp and position it evenly on the sealing boot.

16. Using a trim stick, adjust the boot length to 115mm (hytrel plastic) or 115mm (silicone rubber).

17. If installing a high profile boot clamp, perform the following procedure:

a. Using the Crimper Tool C-4975-A, place the tool over the clamp bridge, tighten the tool nut until the jaws are completely closed (face-to-face).

✳ WARNING

The seal must not be dimpled, stretched or out of shape. If necessary, equalize the seal pressure and shape it by hand.

b. Position the boot onto the tri-pot housing retaining groove and install the retaining clamp evenly on the boot.

c. Using the Crimper Tool C-4975-A, place the tool over the clamp bridge, tighten the tool nut until the jaws are completely closed (face-to-face).

18. If installing a low profile latching type boot clamp, position Snap-On® Clamp Locking Tool YA3050 prongs in the clamp holes and squeeze the tool until the upper clamp band is latched behind the 2 tabs on the lower clamp band.

19. Install the halfshaft.

20. Connect the negative battery cable.

Outer Joint

1. Before servicing the vehicle, refer to the precautions in the beginning of this section.

2. Remove the halfshaft.

3. Remove the clamps from the CV-joint boot and discard.

4. Remove the boot from the CV-joint housing and slide it down the interconnecting shaft.

5. Remove the outer CV-joint from the interconnecting shaft by sharply hitting it with a soft-faced hammer to drive it off the shaft.

6. Remove the circlip from the shaft.

7. Remove the CV-joint by sliding it off the shaft.

✳ WARNING

If any parts show excessive wear, replace the halfshaft assembly; the component parts are not serviceable.

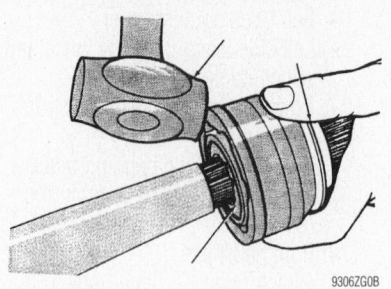

Remove the outer C/V joint from the interconnecting shaft

To install:

8. Install a new small boot clamp and slide it onto the shaft.

9. Install the boot and slide it onto the shaft.

10. Install the circlip.

11. Position the boot so that the raised bead on the inside the boot seal is in the shaft groove.

12. Install the halfshaft hub nut onto the joint threaded shaft so it is flush with the end.

13. Align the shaft splines and tap it onto the shaft with a soft-faced hammer so it locks on the circlip.

14. Distribute ½ of the grease in the service package inside the joint housing and the other ½ inside the boot.

15. Install a new small boot clamp and position it evenly on the sealing boot.

✳ CAUTION

Clamp the boot to the shaft using the Crimper Tool C-4975-A, place the tool over the clamp bridge, tighten the tool nut until the jaws are completely closed (face-to-face).

✳ WARNING

The seal must not be dimpled, stretched or out of shape. If necessary, equalize the seal pressure and shape it by hand.

16. Position the boot onto the retaining groove and install the retaining clamp evenly on the boot.

17. Clamp the boot to the outer CV-joint housing using the Crimper Tool C-4975-A, place the tool over the clamp bridge, tighten the tool nut until the jaws are completely closed (face-to-face).

18. Install the halfshaft

STEERING AND SUSPENSION

Air Bag

✳✳ CAUTION

Some vehicles are equipped with an air bag system. The system MUST BE disabled before performing service on or around system components, steering column, instrument panel components, wiring and sensors. Failure to follow safety and disabling procedures could result in accidental air bag deployment, possible personal injury and unnecessary system repairs.

PRECAUTIONS

Several precautions must be observed when handling the inflator module to avoid accidental deployment and possible personal injury:

1. Never carry the inflator module by the wires or connector on the underside of the module.

2. When carrying a live inflator module, hold securely with both hands, and ensure that the bag and trim cover are pointed away.

3. Place the inflator module on a bench or other surface with the bag and trim cover facing up.

4. With the inflator module on the bench, never place anything on or close to the module which may be thrown in the event of an accidental deployment.

DISARMING

Proper Supplemental Restraint System (SRS) disarming can be obtained by disconnecting and isolating the negative battery cable. Allow the air bag system capacitor at least 2 minutes to discharge before removing any air bag system components.

Rack and Pinion Steering Gear

REMOVAL & INSTALLATION

1. Before servicing the vehicle, refer to the precautions in the beginning of this section.

2. Place the steering wheel in the straight-ahead position. Lock the steering wheel in place, using a steering wheel holder.

➡**Locking the steering wheel keeps the clockspring in alignment position.**

3. Remove or disconnect the following:
- Silencer pad from below the knee blocker panel
- Knee blocker
- Steering column coupling retainer pin, pinch bolt nut/bolt and separate the couplings
- Both front wheels
- Outer tie rod-to-steering knuckle nuts
- Outer tie rod ends from the steering knuckle using remover Tool MB991113
- Tie rod heat shield
- Power steering fluid pressure switch wiring connector by releasing the locking tab
- Power steering fluid pressure hose from the steering gear
- Power steering fluid return hose from the steering gear, if not equipped with a power steering fluid cooler
- Power steering fluid cooler hose from the steering gear, if equipped with a power steering fluid cooler
- Power steering fluid return hose from the routing clip C-clamps, if not equipped with a power steering fluid cooler
- Power steering fluid pressure hose from the steering gear's routing clips
- Power steering cooler hose from the steering gear's right routing clip, if equipped
- Both power steering cooler screws from the front suspension crossmember, if equipped, and move the cooler aside
- Drive belt splash shield
- Engine torque strut-to-front suspension crossmember bolt from the right forward corner of the crossmember

4. Matchmark the front suspension crossmember-to-chassis location.

✳✳ WARNING

If the front suspension crossmember-to-chassis location is not matchmarked, the front wheel alignment setting will be lost.

5. Place a transmission jack under the front crossmember and support it.

6. Remove the front suspension cross-member-to-frame rail bolts, one located at each side.

7. Loosen both rear suspension crossmember-to-frame rail bolts, one located at each side, until they release from the threaded tapping plates in the bolt.

✳✳ WARNING

Do not completely remove the rear bolts for they are designed to disengage from the body threads and will stay within the lower control arm rear isolator bushing.

➡**The threaded tapping plates allow the lower control arm to stay in place on the crossmember.**

8. Using the transmission jack, lower the front suspension crossmember enough to allow the power steering gear to be removed form the rear of the crossmember. Use the jack to support the crossmember's weight.

9. Remove or disconnect the following:
- Lower steering column coupling-to-power steering gear pinion shaft's roll pin, using a roll pin punch
- Lower steering column coupling from the power steering column pinion shaft
- Pinion shaft dash cover seal from the tabs cast into the power steering gear housing
- Power steering gear from the front suspension crossmember

To install:

10. Install or connect the following:
- Power steering gear onto the front suspension crossmember and torque the bolts to 45 ft. lbs. (61 Nm)
- Pinion shaft dash cover seal over the shaft and onto the power steering gear housing. Align the seal holes with the tabs cast into the power steering gear housing.
- Lower steering column coupling by aligning the coupling and steering gear pinion shaft flats
- Lower steering column coupling-to-pinion shaft's roll pin until it is centered. Center the power steering gear rack's travel.
- Front suspension crossmember/power steering gear assembly by raising it with the jack until is aligns with its matchmarks
- Lower steering column coupling,

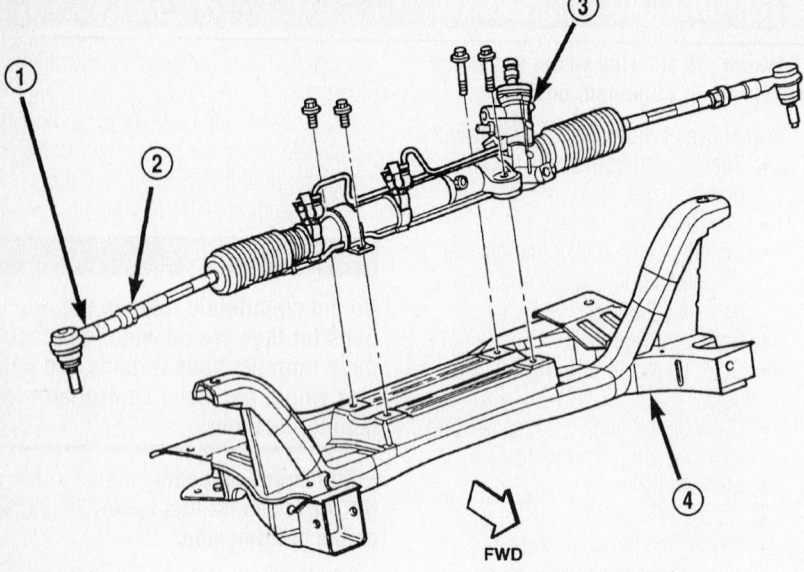

1 – OUTER TIE ROD
2 – JAM NUT
3 – STEERING GEAR
4 – FRONT SUSPENSION CROSSMEMBER

9306EG42

View of the power steering gear and crossmember

guide it through the dash panel hole as it is raised
- Both rear crossmember-to-tapping plate bolts
- Both front crossmember-to-frame rail bolts and torque the 4 bolts to 20 inch lbs. (2 Nm)

※※ WARNING

Be sure to align the front suspension crossmember-to-chassis matchmarks; otherwise, the front wheel alignment setting will be lost.

11. Once aligned, torque both rear crossmember-to-rear lower control arm bolts to 185 ft. lbs. (250 Nm) and both front crossmember bolts to 113 ft. lbs. (153 Nm).

12. Install the engine torque strut to the right forward corner of the front suspension crossmember.

13. Adjust the engine torque strut by performing the following procedure:

   a. Loosen the upper torque strut at the shock tower bracket.

   b. Position a floor jack on the forward edge of the bell housing to prevent the least amount of upward lifting of the engine.

   c. Slowly, lift the assembly, allowing the engine to rotate rearward so the distance between center of the engine mount bracket's rearmost attaching the and the center of the shock tower bracket's washer hose clip hole is 4.70 in. (119mm).

   d. Torque the upper and lower torque strut bolts to 87 ft. lbs. (118 Nm).

   e. Remove the floor jack.

14. Install or connect the following:
- Drive belt splash shield
- Power steering hose-to-power steering gear using a new O-ring lubricated with power steering oil, if not equipped with a power steering cooler
- Power steering fluid cooler line-to-power steering gear, if equipped with a power steering cooler
- Power steering fluid return hose to the routing clip C-clamps
- Tie rod ends to the steering knuckle. Torque the nut, using a crowsfoot wrench, to 40 ft. lbs. (55 Nm), while holding the tie rod stationary.
- Both front wheels
- Dash-to-lower coupling seal over the lower coupling's plastic collar

➡**Verify that the seal's lip shows grease at the coupling's plastic collar contact.**

- Steering column lower coupling-to-steering column upper coupling pinch bolt and torque the nut to 21 ft. lbs. (28 Nm)
- Pinch bolt retainer pin
- Knee blocker and silencer pad

15. Remove the steering wheel holder.

16. Fill and bleed the power steering system.

17. Start the vehicle and check for leaks, repair if necessary.

18. Check and/or adjust the front toe setting.

Strut

REMOVAL & INSTALLATION

Front

1. Before servicing the vehicle, refer to the precautions in the beginning of this section.

2. Install or connect the following:
- Negative battery cable
- Front wheels
- Mark each one right or left, as applicable, if both struts are being removed
- Ground strap from the rear of the strut
- Anti-lock Brake System (ABS) wheel speed sensor from the strut, if equipped

※※ WARNING

The steering knuckle-to-strut assembly attaching bolts are serrated and must not be turned during removal.

- Steering knuckle nuts while holding the bolts stationary
- 3 upper strut mount-to-strut tower nuts
- Strut assembly

To install:

3. Install or connect the following:
- Strut assembly into the strut tower by aligning the 3 upper strut mount studs with the shock tower holes. Torque the 3 upper strut mount nut/washer assemblies to 25 ft. lbs. (34 Nm).

※※ WARNING

The steering knuckle-to-strut assembly attaching bolts are serrated and must not be turned during installation.

- Steering knuckle nuts while holding the bolts stationary
- Steering knuckle arm and position it into the strut assembly by aligning the strut assembly-to-steering knuckle holes
- Both strut-to-steering knuckle bolts. Torque both bolts to 40 ft. lbs. (53 Nm), plus an additional 90 degrees after the specified torque is met.

➡**The bolts should be installed with the nuts facing the front of the vehicle.**

- ABS wheel sensor to the rear of the strut and torque the screw to 120 inch lbs. (13 Nm)

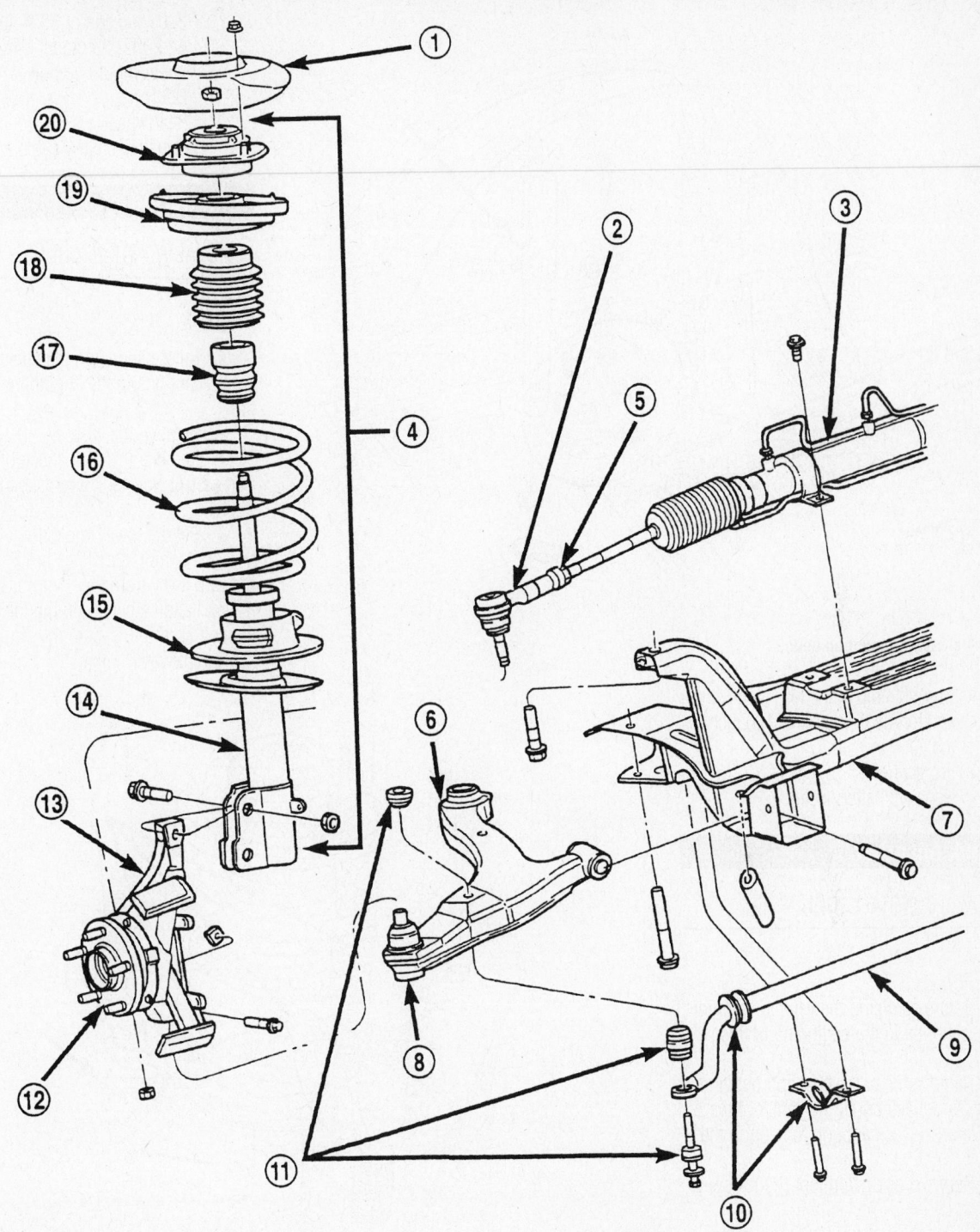

1 – VEHICLE STRUT TOWER
2 – OUTER TIE ROD
3 – STEERING GEAR
4 – STRUT ASSEMBLY
5 – JAM NUT
6 – LOWER CONTROL ARM
7 – CROSSMEMBER
8 – BALL JOINT
9 – STABILIZER BAR
10 – STABILIZER BAR CUSHION AND RETAINER

11 – STABILIZER BAR LINK
12 – HUB
13 – KNUCKLE
14 – STRUT
15 – LOWER SPRING ISOLATOR
16 – COIL SPRING
17 – JOUNCE BUMPER
18 – DUST SHIELD
19 – SPRING SEAT AND BEARING
20 – UPPER MOUNT

9306EG11

Exploded view of the front suspension

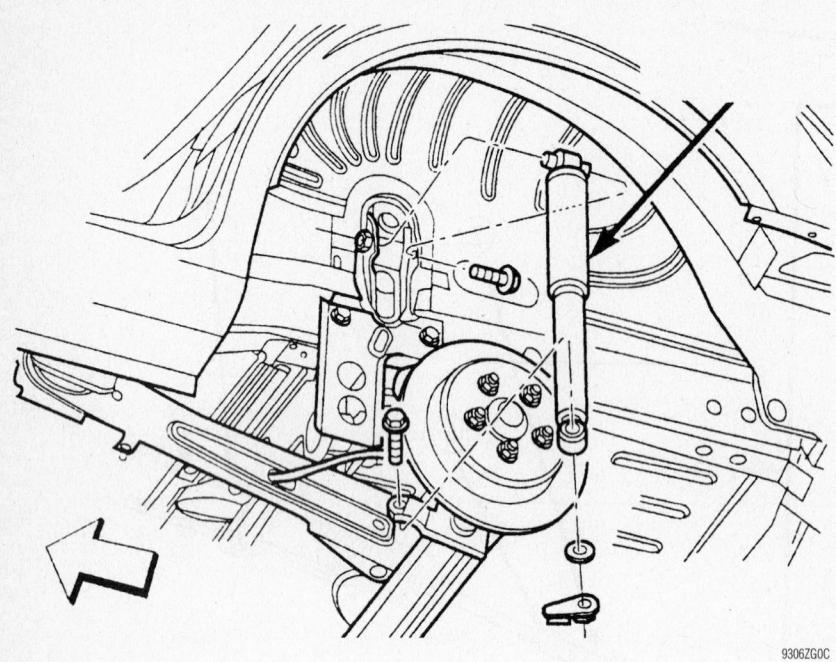

Shock absorber mounting bolts

- Ground strap to the rear of the strut and torque the screw to 120 inch lbs. (13 Nm)
- Front wheel
- Negative battery cable

Shock Absorber

REMOVAL & INSTALLATION

Rear

1. Before servicing the vehicle, refer to the precautions in the beginning of this section.

2. Remove the rear wheel and position a transmission jack under the center of the axle. Raise the jack enough to support the axle.

3. Remove or disconnect the following:

- Shock absorber lower mounting bolt from the axle
- Upper mounting bolt
- Shock absorber

To install:

4. Install or connect the following:

- Shock absorber eye to the body bracket. Hand tighten the upper mounting bolt.
- Lower the jack and install the lower mounting bolt through the axle flange and shock absorber. Torque the bolt to 50 ft. lbs. (68 Nm) on

2001 models or 65 ft. lbs. (88 Nm) on 2002–05 models. Torque the upper mounting bolt to 73 ft. lbs. (99 Nm).

- Rear wheel
- Negative battery cable

Coil Spring

REMOVAL & INSTALLATION

Front

1. Before servicing the vehicle, refer to the precautions in the beginning of this section.

2. Remove the wheel.

3. Remove the strut assembly and place it in a Strut Spring Compressor Tool W-7200.

4. Set the lower hooks then the upper hooks. Position the clevis bracket straight outward away from the compressor tool

5. Install a clamp on the lower end of the coil spring to secure the strut when the nut is removed.

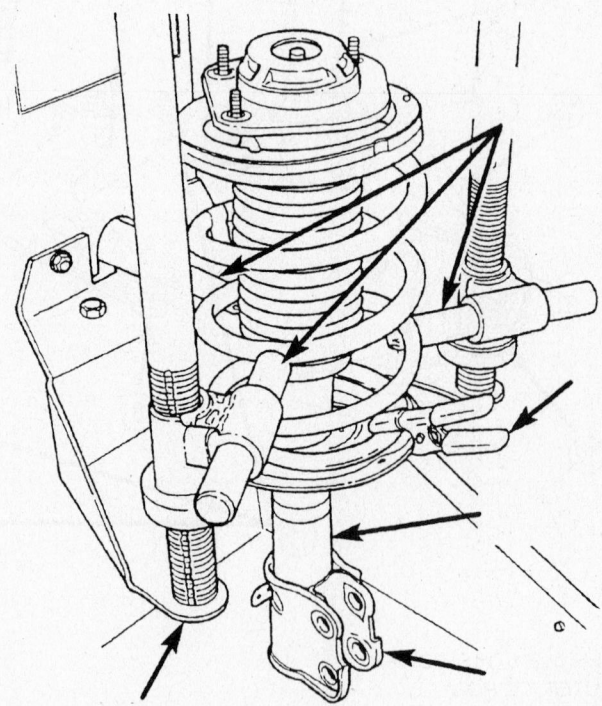

1 – LOWER HOOKS
2 – CLAMP
3 – STRUT ASSEMBLY
4 – CLEVIS BRACKET
5 – SPRING COMPRESSOR

9306ZG0D

Coil spring mounted in the coil spring compressor tool

FWD

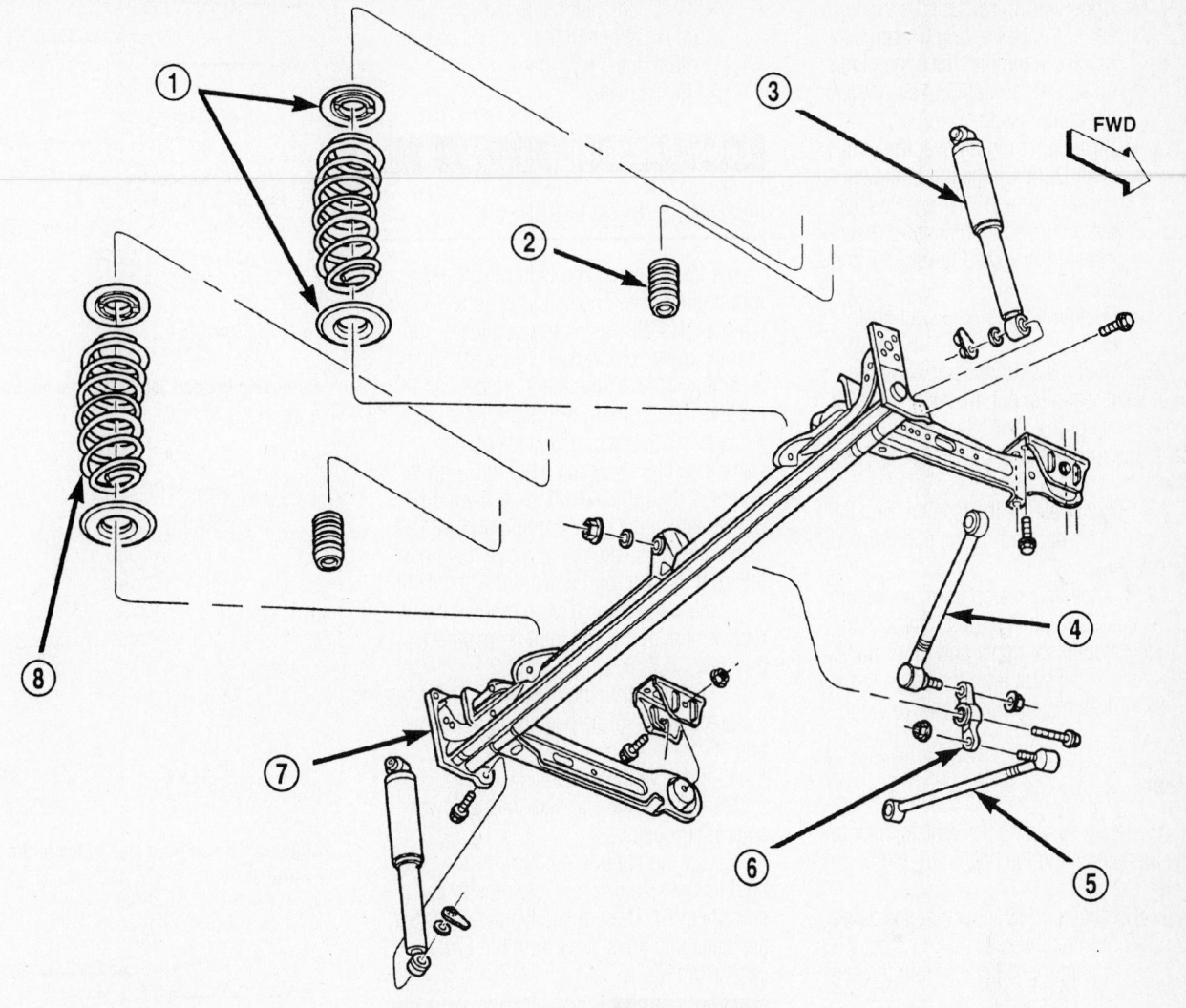

1 – ISOLATORS
2 – JOUNCE BUMPER
3 – SHOCK ABSORBER
4 – WATTS LINK (UPPER)

5 – WATTS LINK (LOWER)
6 – BELL CRANK
7 – AXLE
8 – COIL SPRING

9346JG12

Exploded view of the rear suspension

➡**Do not remove the strut shaft nut until the coil spring is compressed. The coil spring is under pressure and must be compressed before the shaft nut is removed.**

6. Compress the coil spring until all tension is removed from the upper mount.

7. Install a Strut Nut Socket Tool 6864 once the spring is compressed.

8. Install a socket on the hex end of the strut shaft and remove the nut.

9. Remove or disconnect the following:
- Upper mount from the strut shaft
- Upper spring seat, bearing and upper isolator as an assembly
- Dust shield and jounce bumper

- Clamp from the bottom of the coil spring
- Strut through the bottom of the coil
- Release the tension from the coil spring by backing off the compressor drive completely
- Coil spring

10. Inspect the coil spring for any signs of damage

To install:

11. Install or connect the following:
- Coil spring in the compressor. Rotate the spring so that the end of the top coil is directly in the front.
- Slowly compress the coil until enough room is available to install the strut

- Lower spring isolator on the lower spring seat
- Strut through the bottom of the of coil spring until the lower spring seat contacts the lower end of the coil spring. Rotate the strut until the clevis bracket is positioned straight outward away from the compressor.
- Clamp on the lower end of the coil spring and strut
- Jounce bumper on the strut shaft with the smaller end pointing downward
- Dust shield until the bottom of the shield snaps on to the retainer
- Upper spring isolator

- Upper spring seat and bearing on top of the coil spring. Position the notch formed into the edge of the upper seat straight out away from the compressor.
- Strut upper mount over the strut shaft and onto the top of the upper spring seat and bearing. Position the mount so that the third mounting stud is inward toward the compressor.
- Retaining nut on the strut shaft, loosely

12. Install the strut nut socket on the strut shaft retaining nut. Install a socket on the hex end of the shaft. Secure the strut shaft and torque the nut to 55 ft. lbs. (75 Nm).

13. Slowly release the tension from the coil spring by backing off the compressor completely.

14. Remove the clamp from the bottom of the coil spring and strut. Push back the spring compressor upper and lower hooks and remove the strut from the compressor.

15. Install the strut assembly to the vehicle.

Rear

1. Before servicing the vehicle, refer to the precautions in the beginning of this section.

2. Remove or disconnect the following:
- Both rear wheels
- Watts link bell crank from the center of the axle
- Sway bar cushion retainers
- Sway bar from the rear axle and place a jack under the rear axle
- Shock absorber and lower the jack
- Coil springs and rubber isolators

To install:

3. Install or connect the following:
- Rubber isolator on each end of the coil spring and wrap the finger around the coil
- Coil springs on top of the axle spring perches and make certain that the upper coils end near the outboard sides of the vehicle and not at 180 degrees of that location
- Coil springs into the spring mounting brackets
- Washer and nut on the lower mounting bolts and torque the bolts to 50 ft. lbs. (68 Nm)
- Lower end of the sway bar retainers in the slots at the back of the axle
- Mounting bolt through the cushion retainer and torque to 40 ft. lbs. (54 Nm)

- Watts link bell crank to the center of the axle and torque the bolts to 90 ft. lbs. (122 Nm)
- Rear wheels

Lower Ball Joint

REMOVAL & INSTALLATION

The front suspension ball joints operate with no free-play. The ball joints are replaceable ONLY as an assembly. Do not attempt any type of repair on the ball joint assembly. The ball joint is a press fit into the lower control arm with the joint stud retained in the steering knuckle by the clamp bolt. To check the ball joint, with the weight of the vehicle resting on the road wheels, grasp the grease fitting and without using any tools, attempt to move the grease fitting. If the ball joint is worn the grease fitting will move easily. If movement is noted, replacement of the ball joint is recommended.

1. Before servicing the vehicle, refer to the precautions in the beginning of this section.

2. Remove the wheel.

3. Remove the stabilizer bar-to-lower control arm links.

4. Loosen, but do not remove the bolts holding the stabilizer bar retainers to the crossmember. Then, rotate the stabilizer bar and attaching links away from the lower control arms.

✳✳ WARNING

Pulling the steering knuckle outward after releasing the ball joint can separate the inner CV-joint.

5. Remove the steering knuckle-to-ball joint stud's pinch bolt and nut.

6. Remove the ball joint from the steering knuckle using a prybar.

✳✳ WARNING

Be careful when separating the ball joint stud from the knuckle, so the seal does not become damaged.

7. If removing the right lower control arm, perform the following steps:
   a. Remove the drive belt splash shield.
   b. Remove the pencil strut from the right front corner of the crossmember
   c. Remove the engine torque strut.

8. Remove the pivot bolts attaching the lower control arm to the front crossmember.

9. Remove the lower control arm.

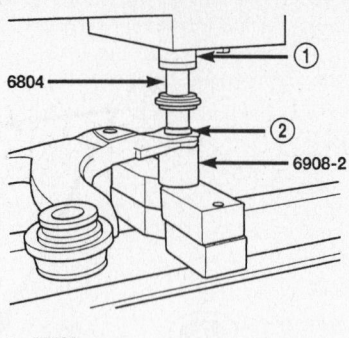

1 – PRESS
2 – BALL JOINT

9306EG51

Removing the ball joint from the control arm

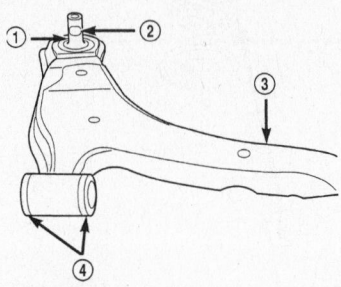

1 – BALL JOINT STUD
2 – NOTCH
3 – LOWER CONTROL ARM
4 – FRONT ISOLATOR BUSHING

9306EG52

Aligning the ball joint stud notch to the control arm

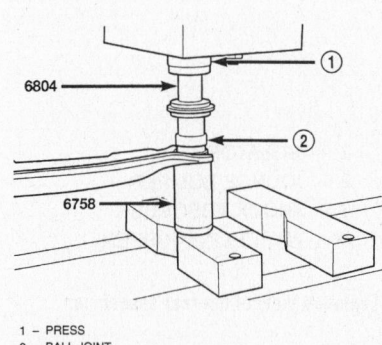

1 – PRESS
2 – BALL JOINT

9306EG53

Installing the ball joint to the control arm

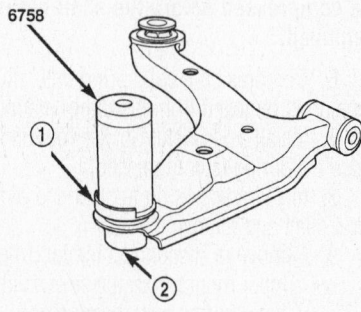

1 – SEAL BOOT UPWARD LIP
2 – BALL JOINT

9306EG54

Installing the ball joint boot seal

10. Remove the ball joint using a pry tool.

11. Using a hydraulic press, press the ball joint from the lower control arm using Receiver tool 6908-2 and Adapter Tool 6804.

To install:

12. Reinstall the ball joint into the lower control arm with the notch in the ball joint stud facing the front lower control arm bushing.

13. Using a hydraulic press, press the ball joint into the lower control arm using Receiver Tool 6758 and Adapter tool 6804.

14. Install the ball joint boot seal using a driver tool such as a large socket or suitable sized piece of pipe.

✳✳ WARNING

Do not use a shop press that was used to install the ball joint, for the press exerts too much force.

15. Install the lower control arm into the front crossmember.

16. Install the rear lower control arm-to-crossmember and frame rail bolt.

➡DO NOT tighten the rear bolt at this time.

17. Install the front lower control arm-to-crossmember nut and bolt.

18. Torque the lower control arm to rear pivot bolt to 185 ft. lbs. (250 Nm) and the front pivot bolt 120 ft. lbs. (163 Nm).

19. Install the ball joint stud into the steering knuckle. Torque the steering knuckle-to-ball joint stud pinch bolt/nut to 70 ft. lbs. (95 Nm).

20. If the right side lower control arm has been service, install the following:

21. Install the engine torque strut.

22. Install the pencil strut to the right front corner of the crossmember and torque the nuts to 43 ft. lbs. (58 Nm).

23. Install the drive belt splash shield.

24. Install the front fascia-to-reinforcement screws.

25. Install the stabilizer bar-to-lower control arm link assemblies and bushings.

26. Rotate the stabilizer bar into position, installing the stabilizer bar links into the lower control arms.

27. Install the top stabilizer bar link bushings and nuts. DO NOT tighten the link yet.

28. Install the wheel.

29. Lower the vehicle so the suspension is supporting the total weight of the vehicle.

30. Torque the stabilizer bar-to-lower control arm links to 21 ft. lbs. (28 Nm).

31. Torque the stabilizer bar bushing retainer-to-crossmember bolts to 21 ft. lbs. (28 Nm).

32. Check and/or adjust the toe, as necessary.

Wheel Bearings

ADJUSTMENT

The PT Cruiser is equipped with a sealed hub and bearing assemblies. The hub and bearing assembly is non-serviceable. If the assembly is damaged, the complete unit must be replaced.

REMOVAL & INSTALLATION

Front

1. Before servicing the vehicle, refer to the precautions in the beginning of this section.

2. Remove the steering knuckle and hub and bearing assembly.

3. Remove the wheel lug stud from the hub flange, using a C-clamp and Adapter Tool 4150A.

4. Rotate the hub to align the removed lug stud with the notch in the bearing retainer plate.

5. Rotate the hub so the stud hole is facing away from the brake caliper's lower rail on the steering knuckle.

6. Install ½ of a Bearing Splitter Tool 1130, between the hub and the bearing retainer plate. The threaded hole in this ½ is to be aligned with the caliper rail on the steering knuckle.

7. Install the remaining pieces of the bearing splitter on the steering knuckle. Hand-tighten the nuts to hold the splitter in place on the knuckle.

8. When the bearing splitter is installed, be sure the 3 bolts attaching the bearing retainer plate to the knuckle are contacting the bearing splitter. The bearing retainer plate should not support the knuckle or contact the splitter.

9. Place the steering knuckle in a hydraulic press, supported by the bearing splitter.

10. Position a driver on the small end of the hub. Using the press, remove the hub from the wheel bearing. The outer bearing race will come out of the wheel bearing when the hub is pressed out of the bearing.

11. Remove the bearing splitter tool from the knuckle.

12. Remove the 3 bolts mounting the bearing retainer plate to the steering knuckle.

13. Place the knuckle in a press supported by the press block. The blocks must not obstruct the bore in the steering knuckle so the wheel bearing can be pressed out of the knuckle. Place a driver on the outer race of the wheel bearing, then press the bearing out of the knuckle.

14. Install the bearing splitter on the hub. The splitter is to be installed on the hub so it is between the flange of the hub and the bearing race on the hub. Place the hub, bearing race and splitter in a press. Use a driver to press the hub out of the bearing race.

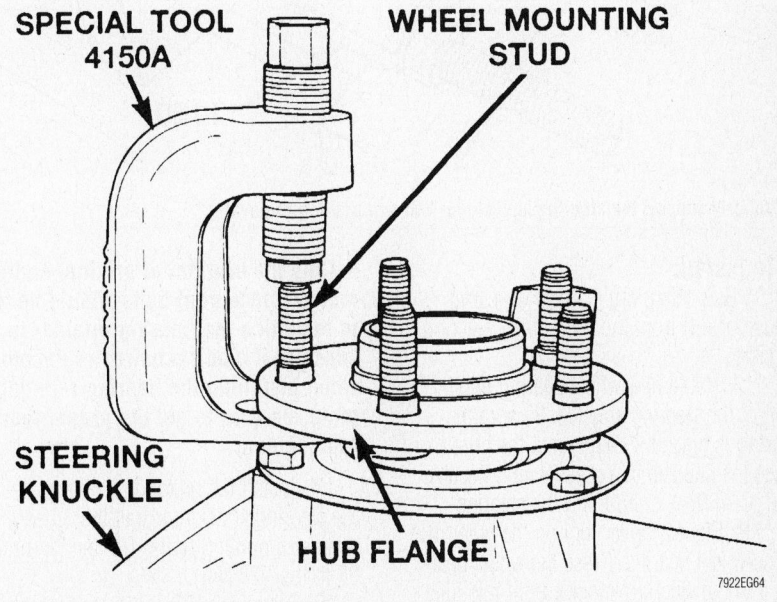

SPECIAL TOOL 4150A

WHEEL MOUNTING STUD

STEERING KNUCKLE

HUB FLANGE

7922EG64

Use a proper C-clamp and adapter tool to press out one of the lug studs

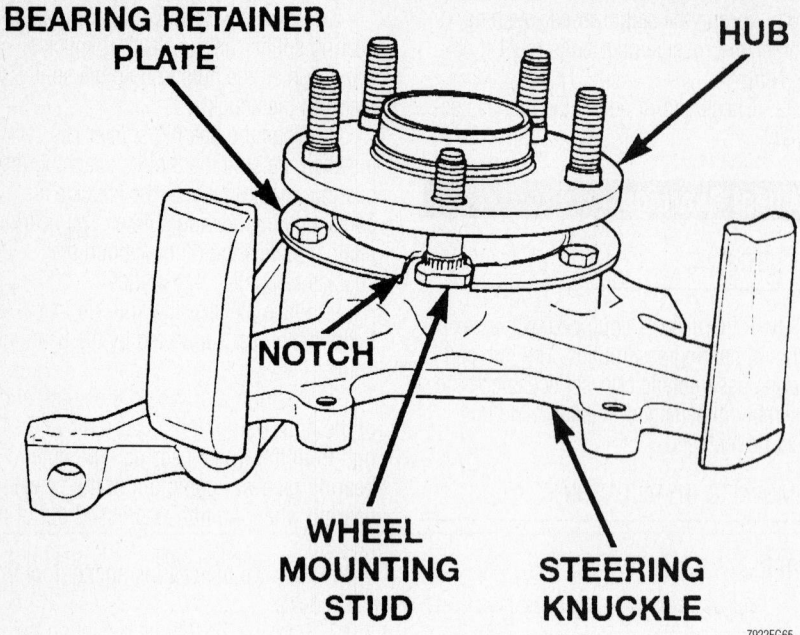

BEARING RETAINER PLATE **HUB** **NOTCH** **WHEEL MOUNTING STUD** **STEERING KNUCKLE**

7922EG65

Rotate the hub in order to remove the lug stud

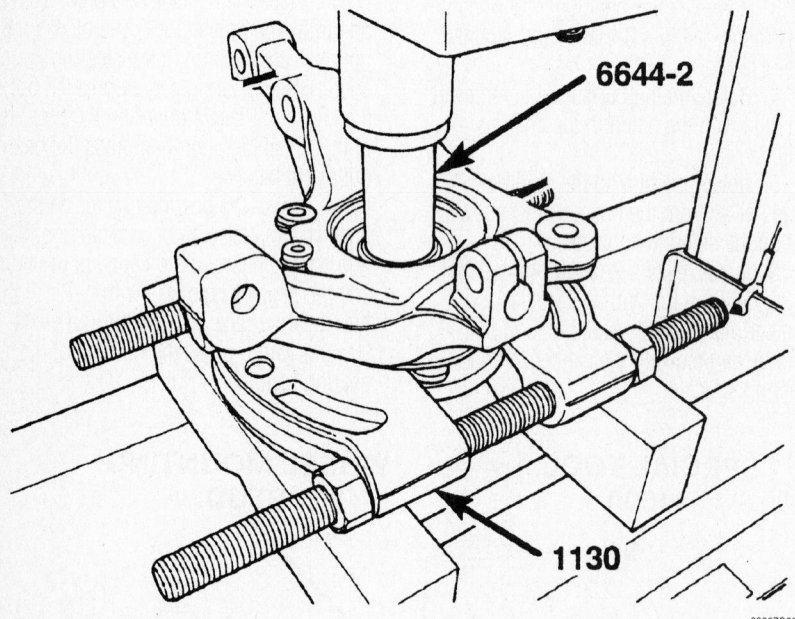

6644-2 1130

9306ZG82

Properly support the steering knuckle for hub and bearing removal

To install:

15. Use clean, dry cloth to wipe and grease or dirt from the bore of the steering knuckle.

16. Install a new wheel bearing into the bore of the steering knuckle. Be sure the bearing is placed squarely into the bore. Place the knuckle in a press with a Receiver Tool, C-4698-2 supporting the steering knuckle. Place a driver tool on the outer race of the wheel bearing. Press the wheel bearing into the steering knuckle until it is fully bottomed in the bore of the steering knuckle.

➡Only the original or original equipment replacement bolts should be used to mounting the bearing retainer to the knuckle. If a bolt requires replacement when installing the bearing retainer plate, be sure to get the proper type of replacement.

17. Install the bearing retainer plate on the steering knuckle. Install the 3 bearing retainer mounting bolts. Tighten the bolts to 21 ft. lbs. (28 Nm).

18. Install the wheel lug stud into the hub flange.

19. Place the hub with the lug stud installed, in a press supported by an Adapter Tool C-4698-1 and press the wheel lug stud into the hub flange until it is fully seated against the back side on the hub flange.

20. Place the steering knuckle with the wheel bearing installed, in a press with special Receiver Tool MB-990799 supporting the inner race of the wheel bearing. Place the hub in the wheel bearing, making sure it is square with the bearing. Press the hub into the wheel bearing until it is fully bottomed in the wheel bearing.

21. Install the steering knuckle and the wheel.

22. Check and/or adjust the front alignment.

Rear

1. Before servicing the vehicle, refer to the precautions in the beginning of this section.

2. Remove or disconnect the following:
   - Wheel
   - Rear brake drum, if equipped
   - Caliper (suspend on a wire) and the rotor, if equipped with rear disc brakes

✳✳ WARNING

DO NOT allow the caliper to hang by the brake hose.

   - Dust cap from the rear hub/bearing
   - Hub/bearing assembly-to-knuckle/spindle nut

➡Discard the hub nut and replace with a new one during installation.

   - Hub/bearing from the spindle by pulling it off the end of the spindle by hand

To install:

✳✳ WARNING

The hub/bearing nut must be tightened to, but NOT over, its specified torque value. The proper specification is crucial to the life of the hub bearing.

3. Position the hub/bearing assembly on the rear spindle/knuckle. Install a NEW hub nut and tighten to 160 ft. lbs. (217 Nm)

4. Install or connect the following:
   - Dust cap and seat it using a soft face hammer to carefully tap it into place
   - Brake drum, if equipped
   - Rotor, if equipped
   - Caliper and 2 guide pin bolts, if equipped with disc brakes. Torque the bolts to 16 ft. lbs. (22 Nm).

BRAKES

Caliper

REMOVAL & INSTALLATION

Front

1. Before servicing the vehicle, refer to the precautions in the beginning of this section.

2. Isolate the master cylinder as follows:
   a. Use a brake pedal holding tool and depress the brake pedal past its first one inch of travel and hold it in this position. (This will keep brake fluid from draining from the master cylinder).

3. Remove or disconnect the following:
   - Front wheels
   - 2 caliper guide pin bolts
   - Brake hose from the caliper
   - Caliper from the steering spindle

To install:

4. Install or connect the following:
   - Brake hose to the caliper and torque to 21 ft. lbs. (28 Nm)
   - Caliper to the steering spindle
   - Caliper guide pin bolts and torque to 26 ft. lbs. (35 Nm)

5. Properly bleed the brake system.
   - Front wheels and lower the vehicle.

6. Pump the brake pedal to seat the front brake pads before moving the vehicle.

7. Road test the vehicle and check for proper operation.

Rear

1. Before servicing the vehicle, refer to the precautions in the beginning of this section.

2. Isolate the master cylinder as follows:
   a. Use a brake pedal holding tool and depress the brake pedal past its first one inch of travel and hold it in this position. (This will keep brake fluid from draining from the master cylinder).

3. Remove or disconnect the following:
   - Front wheels
   - 2 caliper guide pin bolts
   - Brake hose from the caliper
   - Caliper from the steering spindle

To install:

4. Install or connect the following:
   - Brake hose to the caliper and torque to 210 inch lbs. (24 Nm)
   - Caliper to the steering spindle
   - Caliper guide pin bolts and torque to 192 inch lbs. (22 Nm)

5. Properly bleed the brake system.
   - Front wheels and lower the vehicle.

6. Pump the brake pedal to seat the front brake pads before moving the vehicle.

7. Road test the vehicle and check for proper operation.

Disc Brake Pads

REMOVAL & INSTALLATION

1. Before servicing the vehicle, refer to the precautions in the beginning of this section.

2. Remove or disconnect the following:
   - Front wheels
   - 2 caliper to steering knuckle guide pin bolts
   - Caliper away from the steering knuckle by first rotating the free end of the caliper away from the steering knuckle. Then, slide the opposite end of the caliper out from under the machined end of the steering knuckle.

3. Support the caliper from the upper control arm to prevent the weight of the caliper from being supported by the brake flex hose that will damage the hose.
   - Brake pads from the caliper
   - Outboard brake pad by prying the pad retaining clip over the raised area on the caliper. Then, slide the pad down and off the caliper
   - Inboard brake pad by pulling it away from the piston until the retaining clip is free from the cavity in the piston.

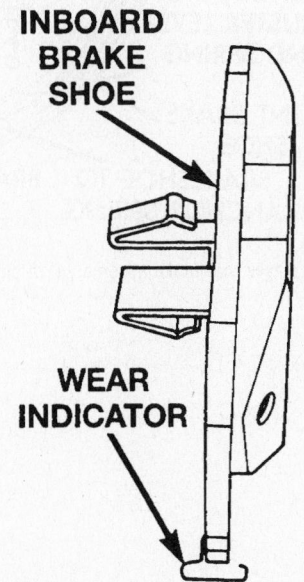

INBOARD BRAKE SHOE

WEAR INDICATOR

Disc brake pad identification

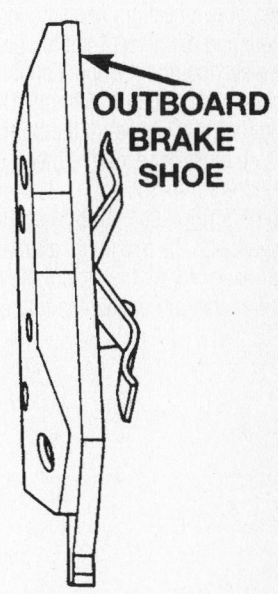

OUTBOARD BRAKE SHOE

- Rotor, if necessary, by pulling it straight off the wheel mounting studs

To install:

4. Clean all parts well. Inspect the caliper for piston seal leaks (brake fluid in and around the boot area and inboard lining) and for any ruptures of the piston dust boot. If the boot is damaged or fluid leak is visible, disassemble the caliper and install a new seal and boot (and piston, if scored).

5. Inspect the caliper pin bushings. Replace if damaged, dry or brittle.

6. Completely compress the piston into the caliper using a large C-clamp or other suitable tool.

7. Lubricate the area on the steering knuckle where the caliper slides with high temperature grease.

8. Install or connect the following:
   - Rotor if removed
   - Brake pads into the caliper. Note that the inboard and outboard pads are different. Make sure the inboard brake shoe assembly is positioned squarely against the face of the caliper piston.

➡ **Be sure to remove the noise suppression gasket paper cover if the pads come so equipped.**

   - Caliper and brake shoe assemblies over the rotor by hooking the lower end of the caliper over the steering knuckle. Then, rotate the caliper into position at the top of the steer-

93006G05

ing knuckle. Make sure the caliper guide pin bolts, bushings and sleeves are clear of the steering knuckle bosses.

- Caliper guide pin bolts and torque to 26 ft. lbs. (35 Nm) on the front caliper or 192 inch lbs. (22 Nm) on the rear caliper
- Wheels and torque the mounting bolts to 100 ft. lbs. (135 Nm)

9. Pump the brake pedal until the brake pads are seated and a firm pedal is achieved before attempting to move the vehicle.

10. Road test the vehicle for proper operation.

Brake Drums

REMOVAL & INSTALLATION

1. Before servicing the vehicle, refer to the precautions in the beginning of this section.

2. Remove or disconnect the following:
- Rear wheels
- Rubber plug from the top of the brake support plate (backing plate)

3. Insert a small prying tool through the adjuster access hole and engage the teeth on the adjuster wheel.

4. Rotate the adjuster wheel so it is moved toward the front of the vehicle. This will back off the adjustment of the rear brake shoes. Continue moving the adjuster wheel toward the front of the vehicle until it stops moving.
- Rear brake drum from the hub assembly

To install:

5. Inspect the brake drums for cracks or signs of overheating. Measure the drum runout and diameter. If not to specification, resurface the drum. Runout should not exceed 0.006 inch (0.152mm). The diameter variation (oval shape) of the drum braking surface must not exceed either 0.0025 inch (0.0635mm) in 30° or 0.0035 inch (0.089mm) in 360°. All brake drums are marked with the maximum allowable brake drum diameter on the face of the drum.

6. Install or connect the following:

- Rear brake drum onto the hub assembly
- Wheels and properly adjust the brakes

7. Road test vehicle to check brake operation.

Brake Shoes

REMOVAL & INSTALLATION

1. Before servicing the vehicle, refer to the precautions in the beginning of this section.

2. Remove or disconnect the following:
- Rear wheels
- Drums
- Automatic adjuster spring and lever
- Hold-down clips and pins

3. Rotate the automatic adjuster starwheel enough so both shoes move out far enough to be free of the wheel cylinder boots.
- Parking brake cable from the actuating lever

- Lower shoe-to-shoe spring
- Shoes from the backing plate, holding them together by the upper shoe-to-shoe spring

To install:

4. Thoroughly clean and dry the backing plate. To prepare the backing plate, lubricate the bosses, anchor pin and parking brake actuating lever pivot surface lightly with lithium based grease.

5. Remove, clean and dry all brake components. Lubricate the starwheel shaft threads with anti-seize lubricant and transfer all parts to their proper locations on the new shoes.

6. Install or connect the following:
- Lower spring
- Parking brake cable
- Automatic adjuster lever and spring

7. Adjust the starwheel.

8. Remove any grease from the linings and install the drum.
- Wheels and properly adjust the brakes

9. Check for proper brake system operation.

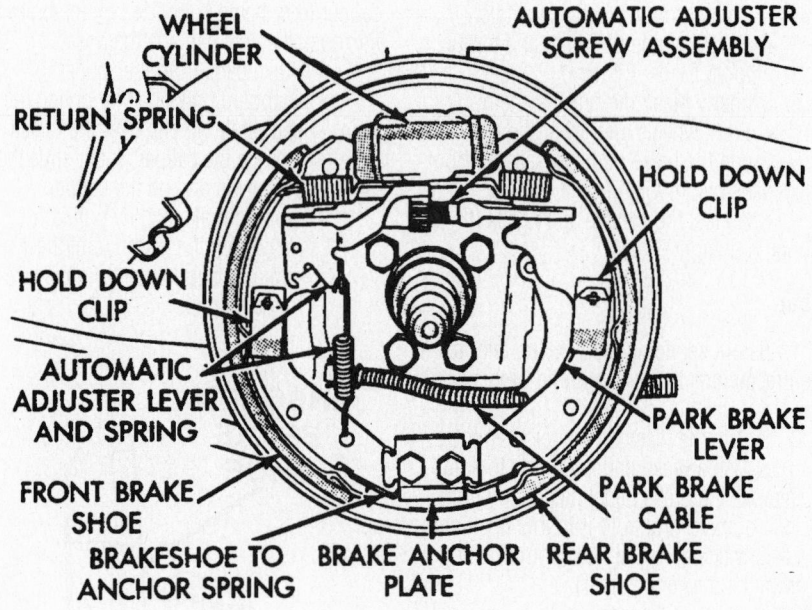

Kelsey Hayes rear brake assembly (left side shown)

DODGE

11

RAM Trucks • RAM Vans

SPECIFICATION CHARTS

ENGINE AND VEHICLE IDENTIFICATION

Engine								Model Year	
Code ①	Liters (cc)	Cu. In.	Cyl.	Fuel Sys.	Engine Type	Eng. Mfg.		Code ②	Year
5	5.9 (5899)	360	8	SMFI	OHV	Chrysler		1	2001
6	5.9 (5882)	359	6	DSL-24V Turbo	OHV	Cummins		2	2002
7	5.9 (5882)	359	6	DSL-24V Turbo	OHV	Cummins		3	2003
C	5.9 (5882)	359	6	DSL-24V Turbo	OHV	Cummins		4	2004
D	5.7 (5653)	345	8	SMPI	OHV	Chrysler		5	2005
K	3.7 (3701)	226	6	MPI	OHV	Chrysler			
N	4.7 (4701)	287	8	MPI	OHV	Chrysler			
W	8.0 (7994)	488	10	SMFI	OHV	Chrysler			
X	3.9 (3916)	238	6	SMFI	OHV	Chrysler			
Y	5.2 (5208)	318	8	SMFI	OHV	Chrysler			
Z	5.9 (5899)	360	8	SMFI	OHV	Chrysler			

OHV: Overhead Valve

DSL-24V: Diesel with 24-valve cylinder head

SMFI: Sequential Multi-port Fuel Injection

① 8th position of VIN

② 10th position of VIN

67189-RAMT-C01

GENERAL ENGINE SPECIFICATIONS

Year	Model	Engine Displacement Liters	Engine Series (ID/VIN)	Net Horsepower @ rpm	Net Torque @ rpm (ft. lbs.)	Bore x Stroke (in.)	Compression Ratio	Oil Pressure @ rpm
2001	Ram Truck 1500	3.9	X	175@4800	220@3200	3.91x3.31	9.1:1	30-80@3000
		5.2	Y	220@4400	300@3200	3.91x3.31	9.1:1	30-80@3000
		5.9	Z	230@4000	330@3250	4.00x3.58	9.1:1	30-80@3000
	Ram Truck 2500	5.9	6	①	②	4.02x4.72	16.5	30@2500
		5.9	7	245@2700	505@1600	4.02x4.72	17.0:1	30@2500
		5.9	Z	230@4000	330@3250	4.00x3.58	9.1:1	30-80@3000
		8.0	W	300@4000	450@2400	4.00x3.58	8.4:1	50-60@3000
	Ram Truck 3500	5.9	6	①	②	4.02x4.72	16.5	30@2500
		5.9	7	245@2700	505@1600	4.02x4.72	17.0:1	30@2500
		5.9	5	230@4000	330@2800	4.00x3.58	8.9:1	30-80@3000
		8.0	W	300@4000	450@2400	4.00x3.58	8.4:1	50-60@3000
	Ram Van 1500	3.9	X	175@4800	220@3200	3.91x3.31	9.1:1	30-80@3000
		5.2	Y	220@4400	300@3200	3.91x3.31	9.1:1	30-80@3000
	Ram Van 2500	3.9	X	175@4800	220@3200	3.91x3.31	9.1:1	30-80@3000
		5.2	Y	220@4400	300@3200	3.91x3.31	9.1:1	30-80@3000
		5.9	Z	230@4000	330@3250	4.00x3.58	9.1:1	30-80@3000
	Ram Van 3500	5.2	Y	220@4400	300@3200	3.91x3.31	9.1:1	30-80@3000
		5.9	Z	230@4000	330@3250	4.00x3.58	9.1:1	30-80@3000
2002	Ram Truck 1500	3.9	X	175@4800	220@3200	3.91x3.31	9.1:1	30-80@3000
		5.2	Y	220@4400	300@3200	3.91x3.31	9.1:1	30-80@3000
		5.9	Z	230@4000	330@3250	4.00x3.58	9.1:1	30-80@3000
	Ram Truck 2500	5.9	6	①	②	4.02x4.72	16.5	30@2500
		5.9	7	245@2700	505@1600	4.02x4.72	17.0:1	30@2500
		5.9	Z	230@4000	330@3250	4.00x3.58	9.1:1	30-80@3000
		8.0	W	300@4000	450@2400	4.00x3.58	8.4:1	50-60@3000
	Ram Truck 3500	5.9	6	①	②	4.02x4.72	16.5	30@2500
		5.9	7	245@2700	505@1600	4.02x4.72	17.0:1	30@2500
		5.9	5	230@4000	330@2800	4.00x3.58	8.9:1	30-80@3000
		8.0	W	300@4000	450@2400	4.00x3.58	8.4:1	50-60@3000
	Ram Van 1500	3.9	X	175@4800	220@3200	3.91x3.31	9.1:1	30-80@3000
		5.2	Y	220@4400	300@3200	3.91x3.31	9.1:1	30-80@3000
	Ram Van 2500	3.9	X	175@4800	220@3200	3.91x3.31	9.1:1	30-80@3000
		5.2	Y	220@4400	300@3200	3.91x3.31	9.1:1	30-80@3000
		5.9	Z	230@4000	330@3250	4.00x3.58	9.1:1	30-80@3000
	Ram Van 3500	5.2	Y	220@4400	300@3200	3.91x3.31	9.1:1	30-80@3000
		5.9	Z	230@4000	330@3250	4.00x3.58	9.1:1	30-80@3000
2003	Ram Truck 1500	3.7	K	210@5200	225@4200	3.66x3.40	9.1:1	25-110@3000
		4.7	N	235@4800	295@3200	3.66x3.40	9.3:1	25-110@3000
		5.7	D	345@5400	375@4200	3.91x3.58	9.6:1	25-110@3000
		5.9	Z	230@4000	330@3250	4.00x3.58	9.1:1	30-80@3000
	Ram Truck 2500	5.9	6	①	②	4.02x4.72	16.5	30@2500
		5.9	C	305@2900	555@1400	4.02x4.72	17.2:1	30@2500
		5.7	D	345@5400	375@4200	3.91x3.58	9.6:1	25-110@3000
		8.0	W	300@4000	450@2400	4.00x3.58	8.4:1	50-60@3000
	Ram Truck 3500	5.9	6	①	②	4.02x4.72	16.5	30@2500
		5.9	C	305@2900	555@1400	4.02x4.72	17.2:1	30@2500
		5.7	D	345@5400	375@4200	3.91x3.58	9.6:1	25-110@3000
		8.0	W	300@4000	450@2400	4.00x3.58	8.4:1	50-60@3000

GENERAL ENGINE SPECIFICATIONS

Year	Model	Engine Displacement Liters	Engine Series (ID/VIN)	Net Horsepower @ rpm	Net Torque @ rpm (ft. lbs.)	Bore x Stroke (in.)	Com-pression Ratio	Oil Pressure @ rpm
2003	Ram Van 1500	3.9	X	175@4800	220@3200	3.91x3.31	9.1:1	30-80@3000
		5.2	Y	220@4400	300@3200	3.91x3.31	9.1:1	30-80@3000
	Ram Van 2500	3.9	X	175@4800	220@3200	3.91x3.31	9.1:1	30-80@3000
		5.2	Y	220@4400	300@3200	3.91x3.31	9.1:1	30-80@3000
		5.9	Z	230@4000	330@3250	4.00x3.58	9.1:1	30-80@3000
	Ram Van 3500	5.2	Y	220@4400	300@3200	3.91x3.31	9.1:1	30-80@3000
		5.9	Z	230@4000	330@3250	4.00x3.58	9.1:1	30-80@3000
2004	Ram Truck 1500	3.7	K	210@5200	225@4200	3.66x3.40	9.1:1	25-110@3000
		4.7	N	235@4800	295@3200	3.66x3.40	9.3:1	25-110@3000
		5.7	D	345@5400	375@4200	3.91x3.58	9.6:1	25-110@3000
	Ram Truck 2500	5.9	6	①	②	4.02x4.72	16.5	30@2500
		5.9	C	305@2900	555@1400	4.02x4.72	17.2:1	30@2500
		5.7	D	345@5400	375@4200	3.91x3.58	9.6:1	25-110@3000
	Ram Truck 3500	5.9	6	①	②	4.02x4.72	16.5	30@2500
		5.9	C	305@2900	555@1400	4.02x4.72	17.2:1	30@2500
		5.7	D	345@5400	375@4200	3.91x3.58	9.6:1	25-110@3000

① AT: 215@2700rpm
  MT: 235@2700rpm

② AT: 420@1600rpm
  MT: 460@1600rpm

67189-RAMT-C03

GASOLINE ENGINE TUNE-UP SPECIFICATIONS

Year	Engine Displacement Liters	Engine ID/VIN	Spark Plug Gap (in.)	Ignition Timing (deg.)	Fuel Pump (psi)	Idle Speed (rpm)	Valve Clearance Intake	Exhaust
2001	3.9	X	0.040	①	44.2-54.2	②	HYD	HYD
	5.2	Y	0.040	①	44.2-54.2	②	HYD	HYD
	5.9	5	0.040	①	44.2-54.2	②	HYD	HYD
	5.9	Z	0.040	①	44.2-54.2	②	HYD	HYD
	8.0	W	0.045	①	44.2-54.2	②	HYD	HYD
2002	3.9	X	0.040	①	44.2-54.2	②	HYD	HYD
	5.2	Y	0.040	①	44.2-54.2	②	HYD	HYD
	5.9	5	0.040	①	44.2-54.2	②	HYD	HYD
	5.9	Z	0.040	①	44.2-54.2	②	HYD	HYD
	8.0	W	0.045	①	44.2-54.2	②	HYD	HYD
2003	3.7	K	0.042	①	44-54	②	HYD	HYD
	3.9	X	0.040	①	44.2-54.2	②	HYD	HYD
	4.7	N	0.040	①	47-51	②	HYD	HYD
	5.2	Y	0.040	①	44.2-54.2	②	HYD	HYD
	5.7	D	0.045	①	49.0-49.4	②	HYD	HYD
	5.9	Z	0.040	①	44.2-54.2	②	HYD	HYD
	8.0	W	0.045	①	44.2-54.2	②	HYD	HYD
2004	3.7	K	0.042	①	44-54	②	HYD	HYD
	4.7	N	0.040	①	47-51	②	HYD	HYD
	5.7	D	0.045	①	49.0-49.4	②	HYD	HYD

NOTE: The Vehicle Emission Control Information (VECI) label often reflects specification changes made during production.
The label figures must be used if they differ from those in this chart.

HYD: Hydraulic

① Ignition timing is controlled by the PCM and is not adjustable.

② Idle speed is controlled by the PCM and is not adjustable

67189-RAMT-C04

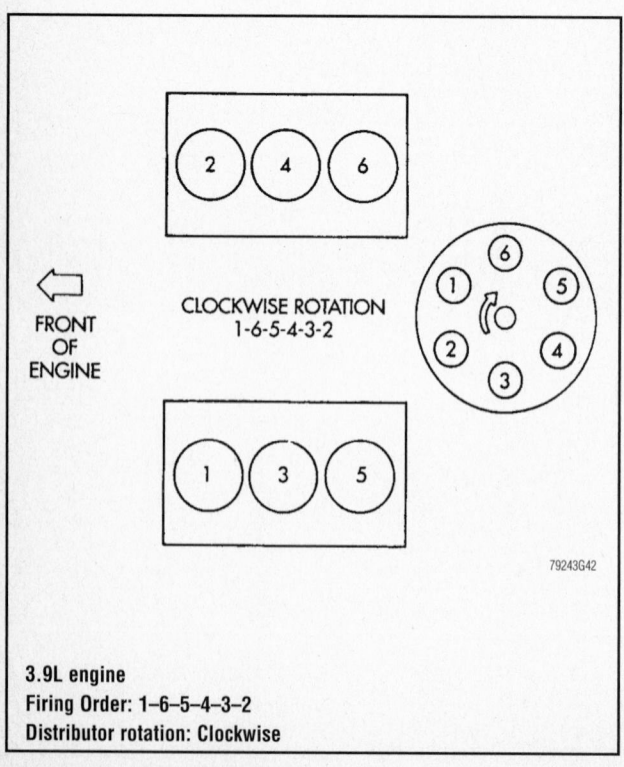

FRONT OF ENGINE

CLOCKWISE ROTATION
1-6-5-4-3-2

79243G42

3.9L engine
Firing Order: 1–6–5–4–3–2
Distributor rotation: Clockwise

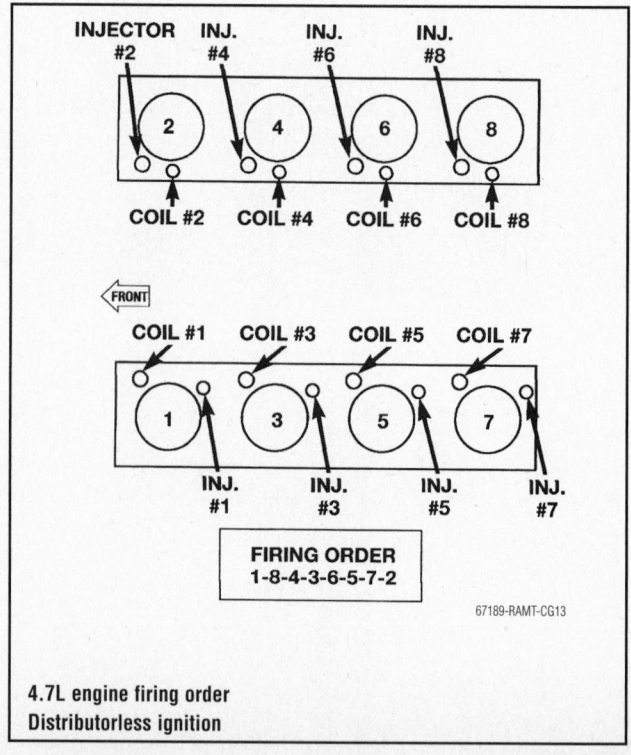

INJECTOR #2   INJ. #4   INJ. #6   INJ. #8

COIL #2   COIL #4   COIL #6   COIL #8

FRONT

COIL #1   COIL #3   COIL #5   COIL #7

INJ. #1   INJ. #3   INJ. #5   INJ. #7

FIRING ORDER
1-8-4-3-6-5-7-2

67189-RAMT-CG13

4.7L engine firing order
Distributorless ignition

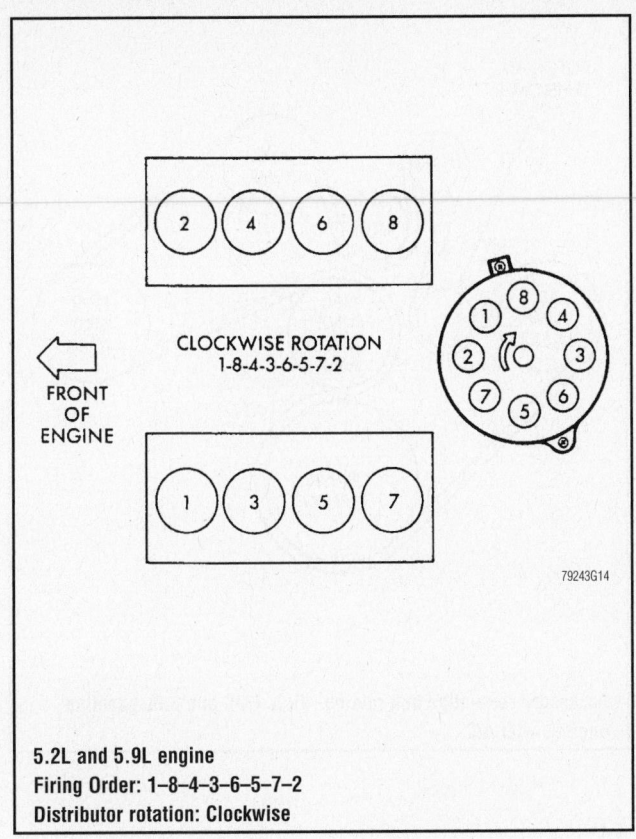

5.2L and 5.9L engine
Firing Order: 1–8–4–3–6–5–7–2
Distributor rotation: Clockwise

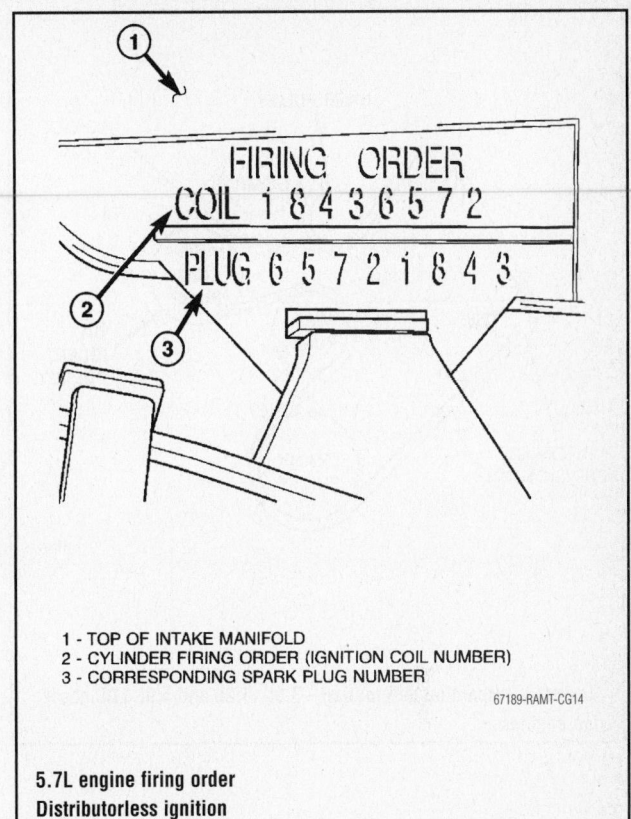

1 - TOP OF INTAKE MANIFOLD
2 - CYLINDER FIRING ORDER (IGNITION COIL NUMBER)
3 - CORRESPONDING SPARK PLUG NUMBER

67189-RAMT-CG14

5.7L engine firing order
Distributorless ignition

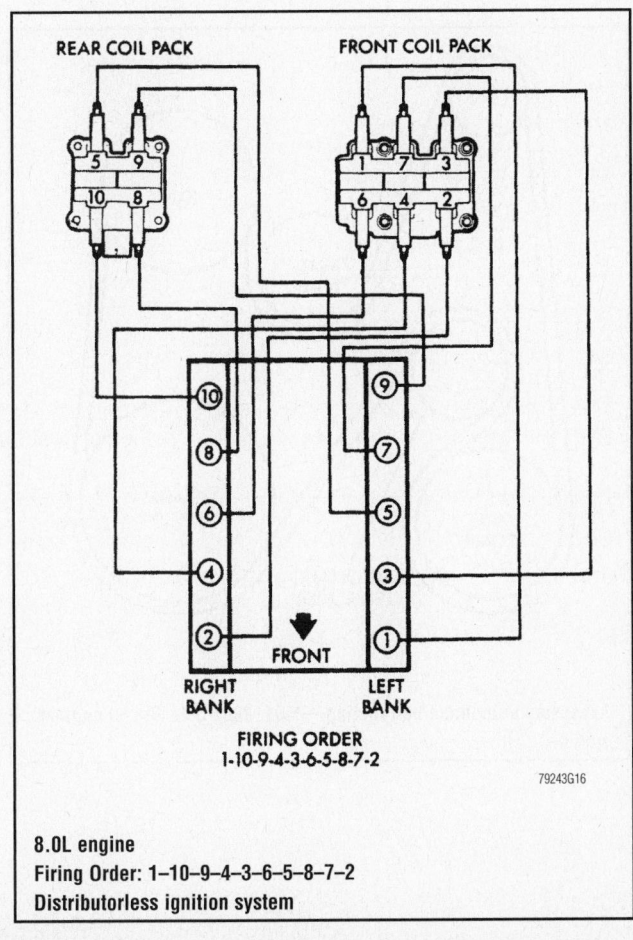

8.0L engine
Firing Order: 1–10–9–4–3–6–5–8–7–2
Distributorless ignition system

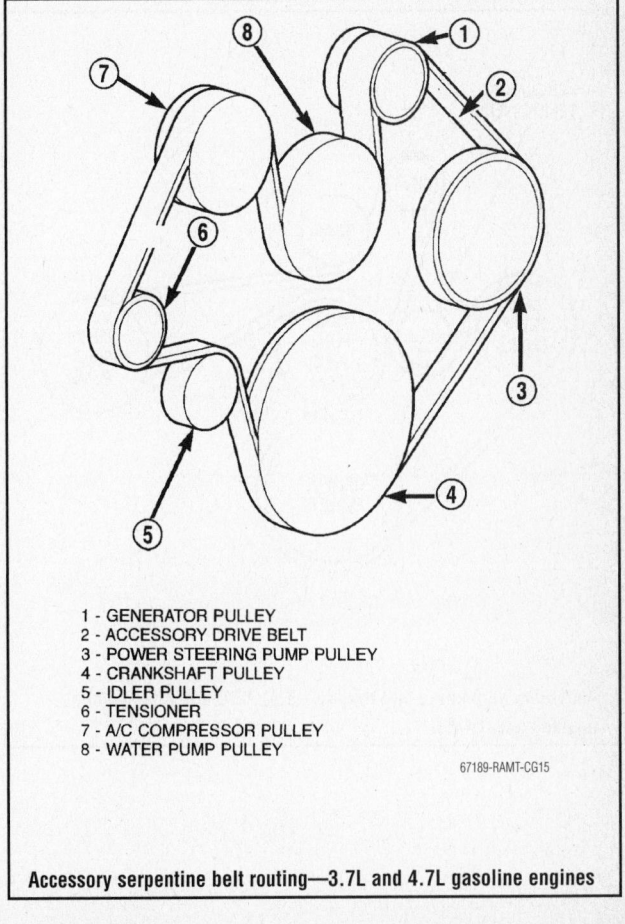

1 - GENERATOR PULLEY
2 - ACCESSORY DRIVE BELT
3 - POWER STEERING PUMP PULLEY
4 - CRANKSHAFT PULLEY
5 - IDLER PULLEY
6 - TENSIONER
7 - A/C COMPRESSOR PULLEY
8 - WATER PUMP PULLEY

67189-RAMT-CG15

Accessory serpentine belt routing—3.7L and 4.7L gasoline engines

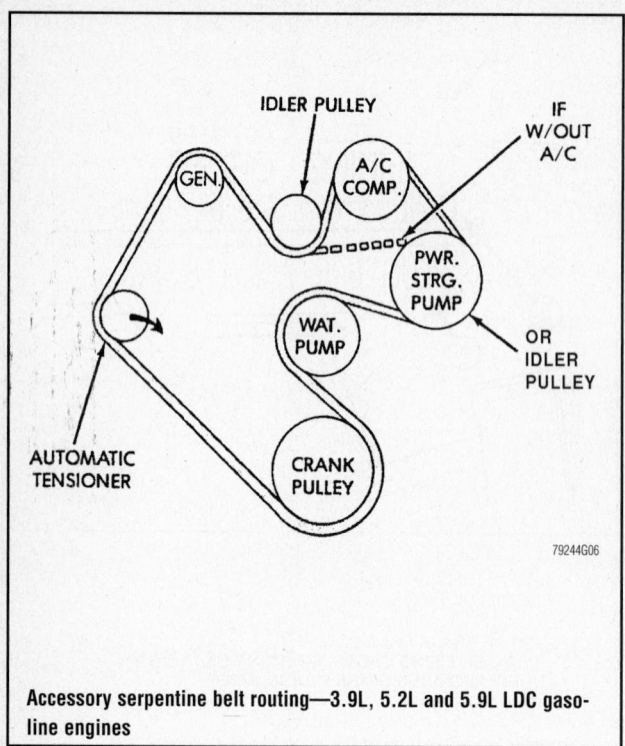

Accessory serpentine belt routing—3.9L, 5.2L and 5.9L LDC gasoline engines

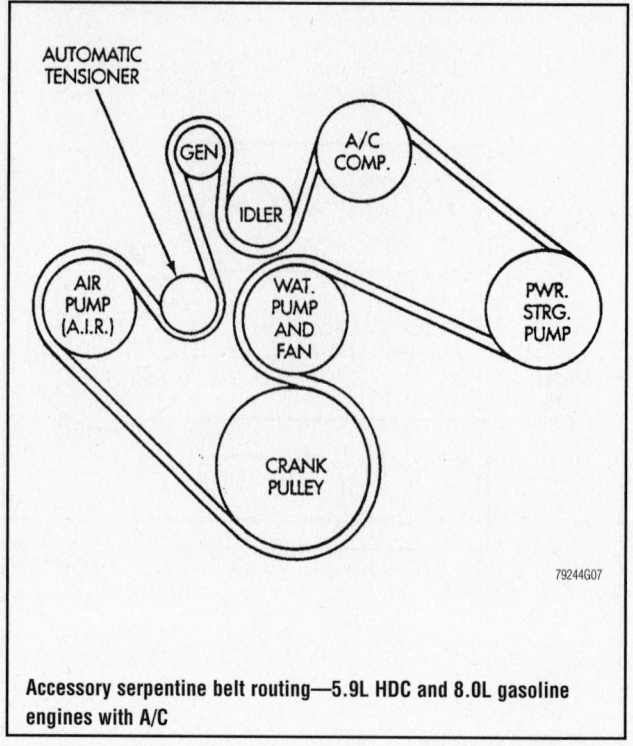

Accessory serpentine belt routing—5.9L HDC and 8.0L gasoline engines with A/C

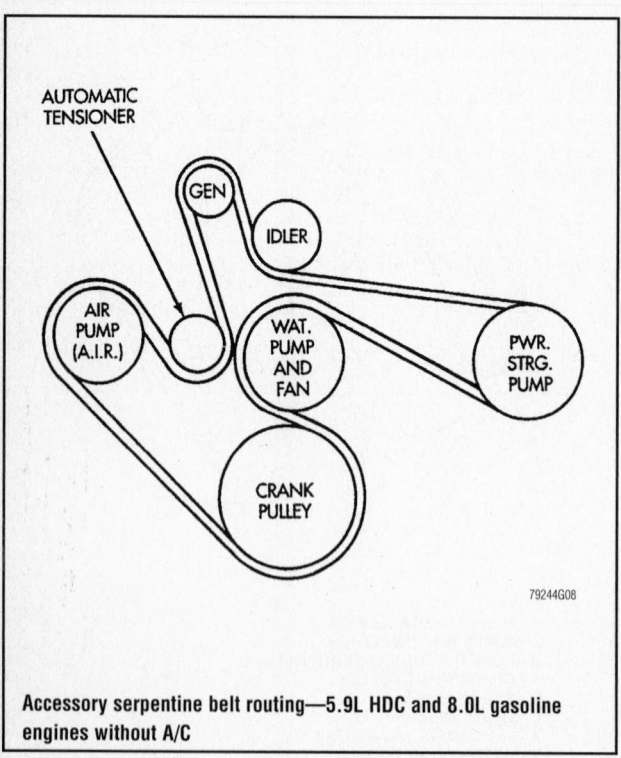

Accessory serpentine belt routing—5.9L HDC and 8.0L gasoline engines without A/C

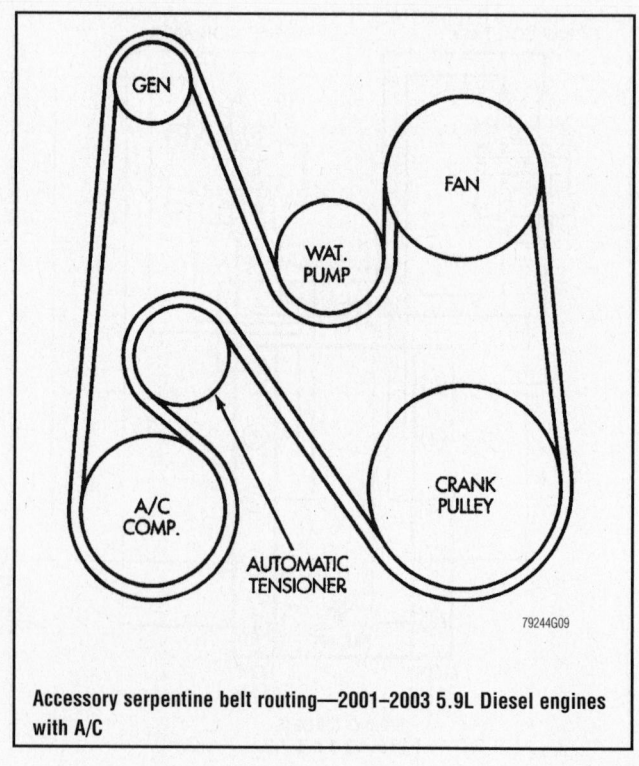

Accessory serpentine belt routing—2001–2003 5.9L Diesel engines with A/C

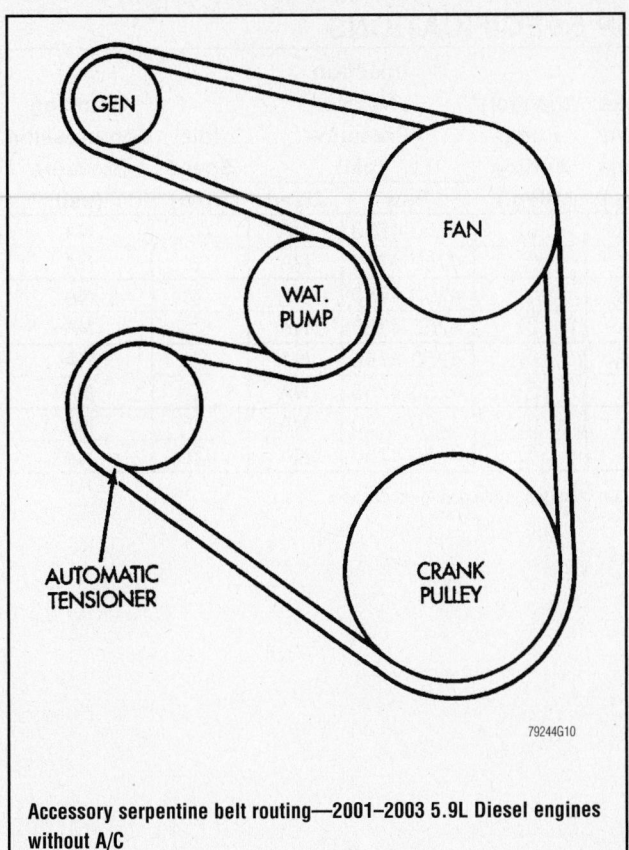

GEN
FAN
WAT.
PUMP
AUTOMATIC
TENSIONER
CRANK
PULLEY
79244G10

Accessory serpentine belt routing—2001–2003 5.9L Diesel engines without A/C

1 - GENERATOR PULLEY
2 - WATER PUMP PULLEY
3 - IDLER PULLEY
4 - POWER STEERING PUMP PULLEY
5 - RADIATOR FAN PULLEY
6 - CRANKSHAFT PULLEY
7 - AUTOMATIC TENSIONER
8 - A/C COMPRESSOR PUMP PULLEY
67189-RAMT-CG16

Accessory serpentine belt routing—2004 5.9L Diesel engines with A/C

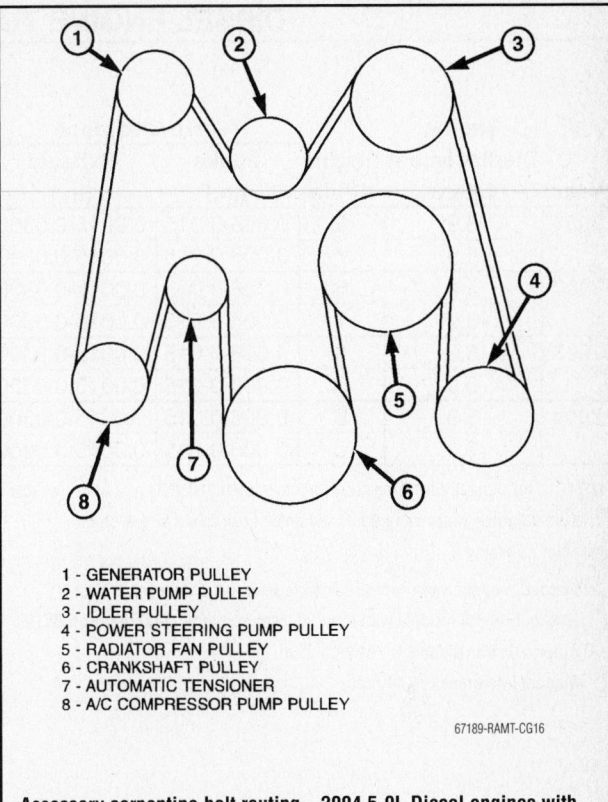

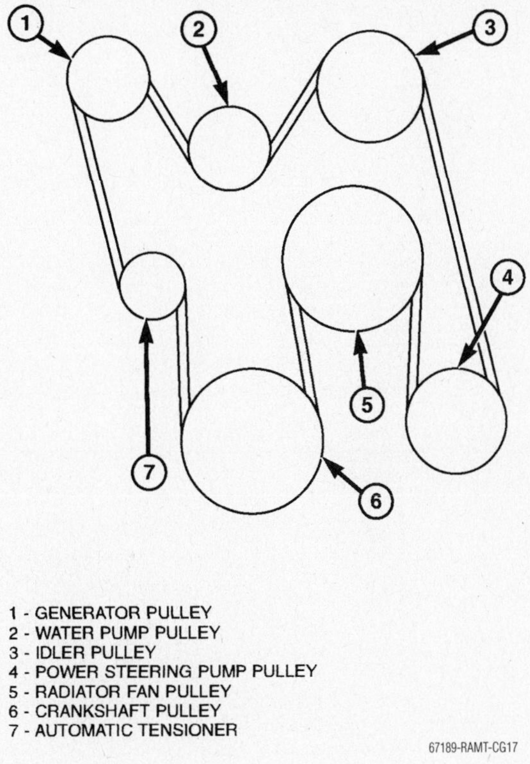

1 - GENERATOR PULLEY
2 - WATER PUMP PULLEY
3 - IDLER PULLEY
4 - POWER STEERING PUMP PULLEY
5 - RADIATOR FAN PULLEY
6 - CRANKSHAFT PULLEY
7 - AUTOMATIC TENSIONER
67189-RAMT-CG17

Accessory serpentine belt routing—2004 5.9L Diesel engines without A/C

DIESEL ENGINE TUNE-UP SPECIFICATIONS

Year	Engine Displacement Liters	Engine ID/VIN	Valve Clearance		Intake Valve Opens (deg.)	Injection Pump Setting (deg.)	Injection Nozzle Pressure (psi)		Idle Speed (rpm)	Cranking Compression Pressure (psi)
			Intake (in.)	Exhaust (in.)			New	Used		
2001	5.9	6	0.006-0.015	0.0015-0.0300	NA	①	4250-4750	NA	②	NA
	5.9	7	0.006-0.015	0.0015-0.0300	NA	①	4250-4750	NA	②	NA
2002	5.9	6	0.006-0.015	0.0015-0.0300	NA	①	4250-4750	NA	②	NA
	5.9	7	0.006-0.015	0.0015-0.0300	NA	①	4250-4750	NA	②	NA
2003	5.9	6	0.006-0.015	0.0015-0.0300	NA	①	4250-4750	NA	②	NA
	5.9	C	0.006-0.015	0.0015-0.0300	NA	①	4250-4750	NA	②	NA
2004	5.9	6	0.006-0.015	0.0015-0.0300	NA	①	4250-4750	NA	②	NA
	5.9	C	0.006-0.015	0.0015-0.0300	NA	①	4250-4750	NA	②	NA

NOTE: The Vehicle Emission Control Information (VECI) label often reflects specification changes made during production.

The label figures must be used if they differ from those in this chart

NA: Not Available

① Federal models with manual transmissions: 13.5 degrees BTDC

Except Federal models with manual transmissions: 14.0 degrees BTDC

② Automatic transmission: 750-800 rpm

Manual transmission: 780 rpm

CAPACITIES

Year	Model	Engine Displacement Liters	Engine ID/VIN	Oil with Filter (qts.)	Engine Transmission (pts.) Manual	Engine Transmission (pts.) Auto. ①	Transfer Case (pts.)	Drive Axle Front (pts.)	Drive Axle Rear (pts.)	Fuel Tank (gal.)	Cooling System (qts.)
2001	Ram Truck 1500	3.9	X	4.5	②	8.0	③	④	⑤	⑥	20.0
		5.2	Y	5.0	②	8.0	③	④	⑤	⑤	20.0
		5.9	Z	5.0	—	8.0	③	④	⑤	⑤	20.0
	Ram Truck 2500	5.9	6	11.0	②	8.0	③	④	⑤	⑥	24.0
		5.9	7	11.0	②	—	③	④	⑤	⑤	24.0
		5.9	Z	5.0	②	8.0	③	④	⑤	⑤	20.0
		8.0	W	7.0	②	8.0	③	④	⑤	⑥	26.0
	Ram Truck 3500	5.9	6	11.0	②	8.0	③	④	⑤	⑤	24.0
		5.9	7	11.0	②	—	③	④	⑤	⑥	24.0
		5.9	5	5.0	②	8.0	③	④	⑤	⑤	20.0
		8.0	W	7.0	②	8.0	③	④	⑤	⑤	26.0
	Ram Van	3.9	X	4.0	—	8.0	—	—	⑦	⑧	14.5
		5.2	Y	5.0	—	8.0	—	—	⑦	⑧	16.5
		5.9	Z	5.0	—	8.0	—	—	⑦	⑧	⑨
2002	Ram Truck 1500	3.9	X	4.5	②	8.0	③	④	⑤	⑥	20.0
		5.2	Y	5.0	②	8.0	③	④	⑤	⑤	20.0
		5.9	Z	5.0	—	8.0	③	④	⑤	⑤	20.0
	Ram Truck 2500	5.9	6	11.0	②	8.0	③	④	⑤	⑥	24.0
		5.9	7	11.0	②	—	③	④	⑤	⑤	24.0
		5.9	Z	5.0	②	8.0	③	④	⑤	⑤	20.0
		8.0	W	7.0	②	8.0	③	④	⑤	⑥	26.0
	Ram Truck 3500	5.9	6	11.0	②	8.0	③	④	⑤	⑤	24.0
		5.9	7	11.0	②	—	③	④	⑤	⑥	24.0
		5.9	5	5.0	②	8.0	③	④	⑤	⑤	20.0
		8.0	W	7.0	②	8.0	③	④	⑤	⑤	26.0
	Ram Van	3.9	X	4.0	—	8.0	—	—	⑦	⑧	14.5
		5.2	Y	5.0	—	8.0	—	—	⑦	⑧	16.5
		5.9	Z	5.0	—	8.0	—	—	⑦	⑧	⑨
2003	Ram Truck 1500	3.7	K	5.0	⑩	⑪	⑫	⑬	⑭	⑮	16.2
		4.7	N	6.0	⑩	⑪	⑫	⑬	⑭	⑮	16.2
		5.7	D	7.0	⑩	⑪	⑫	⑬	⑭	⑮	16.2
		5.9	Z	5.0	⑩	⑪	⑫	⑬	⑭	⑮	16.3
	Ram Truck 2500	5.9	6	11.0	⑩	⑪	⑫	⑬	⑭	⑮	29.5
		5.9	C	11.0	⑩	⑪	⑫	⑬	⑭	⑮	29.5
		5.7	D	7.0	⑩	⑪	⑫	⑬	⑭	⑮	16.2
		8.0	W	7.0	⑩	⑪	⑫	⑬	⑭	⑮	24.3
	Ram Truck 3500	5.9	6	11.0	⑩	⑪	⑫	⑬	⑭	⑮	29.5
		5.9	C	11.0	⑩	⑪	⑫	⑬	⑭	⑮	29.5
		5.7	D	7.0	⑩	⑪	⑫	⑬	⑭	⑮	16.2
		8.0	W	7.0	⑩	⑪	⑫	⑬	⑭	⑮	24.3
	Ram Van	3.9	X	4.0	—	8.0	—	—	⑦	⑧	14.5
		5.2	Y	5.0	—	8.0	—	—	⑦	⑧	16.5
		5.9	Z	5.0	—	8.0	—	—	⑦	⑧	⑨

67189-RAMT-C06

CAPACITIES

Year	Model	Engine Displacement Liters	Engine ID/VIN	Oil with Filter (qts.)	Engine Transmission (pts.) Manual	Engine Transmission (pts.) Auto. ①	Transfer Case (pts.)	Drive Axle Front (pts.)	Drive Axle Rear (pts.)	Fuel Tank (gal.)	Cooling System (qts.)
2004	Ram Truck 1500	3.7	K	5.0	⑩	⑪	⑫	⑬	⑭	⑮	16.2
		4.7	N	6.0	⑩	⑪	⑫	⑬	⑭	⑮	16.2
		5.7	D	7.0	⑩	⑪	⑫	⑬	⑭	⑮	16.2
	Ram Truck 2500	5.9	6	12.0	⑩	⑪	⑫	⑬	⑭	⑮	29.5
		5.9	C	12.0	⑩	⑪	⑫	⑬	⑭	⑮	29.5
		5.7	D	7.0	⑩	⑪	⑫	⑬	⑭	⑮	16.2
	Ram Truck 3500	5.9	6	12.0	⑩	⑪	⑫	⑬	⑭	⑮	29.5
		5.9	C	12.0	⑩	⑪	⑫	⑬	⑭	⑮	29.5
		5.7	D	7.0	⑩	⑪	⑫	⑬	⑭	⑮	16.2

NOTE: All capacities are approximate. Add fluid gradually and check to be sure a proper fluid level is obtained.

① For fluid drain and filter replacement only.

② NV3500: 4.2 pts.
NV4500: 8.0 pts.
NV4500 HD: 8.0 pts.
NV5600: 9.5 pts.

③ NV231HD: 2.5 pts.
NV241: 4.6 pts.
NV241HD: 6.5 pts.

④ 216FBI: 4.8 pts.
248FBI: 8.5 pts.

⑤ 9.25 in.: 4.5 pts.
248RBI (2wd) 6.3 pts.
248RBI (4wd): 7.0 pts.
267RBI (2wd): 7.0 pts.
267RBI (4wd): 7.5 pts.
286RBI (2wd): 6.8 pts.
286RBI (4wd): 10.1 pts.

⑥ 1500 series w/6.5 ft. box: 26.0
2500 series w/6.5 ft. box: 34.0
All others: 35.0

⑦ 8.25 in. 4.4 pts.
9.25 in.: 4.5 pts.
248RBI: 6.25 pts.

⑧ 109 in. wheelbase: 31.0
132 in. wheelbase: 35.0

⑨ With rear heater: 16.0

⑩ NV3500 2wd: 4.8 pts.
NV3500 4wd: 4.2 pts.
NV4500: 8.0 pts.
NV5600: 9.5 pts.

⑪ 46RE: 8.0 pts.
45RFE and 545RFE
2wd: 11.0 pts.
4wd: 13.0 pts.

⑫ NV241 GEN II: 3.4 pts.
NV243: 3.4 pts.
NV 271: 4.0 pts.
NV 273: 4.0 pts.

⑬ C205F: 3.5 pts.
9.25AA: 4.75 pts.

⑭ 9.25: 4.9 pts. W/Trac-Lok + 4 oz. Additive
10.5AA: 4.75 pts.
11.5AA: 7.65 pts.

⑮ Short bed: 26.0
Long bed: 35.0

67189-RAMT-C07

VALVE SPECIFICATIONS

Year	Engine Displacement Liters	Engine ID/VIN	Seat Angle (deg.)	Face Angle (deg.)	Spring Test Pressure (lbs. @ in.)	Spring Installed Height (in.)	Stem-to-Guide Clearance (in.)		Stem Diameter (in.)	
							Intake	Exhaust	Intake	Exhaust
2001	3.9	X	44.25-44.75	43.25-43.75	200@1.21	1.64	0.001-0.003	0.001-0.003	0.311-0.312	0.311-0.312
	5.2	Y	44.25-44.75	43.25-43.75	200@1.21	1.64	0.001-0.003	0.001-0.003	0.311-0.312	0.311-0.312
	5.9	Z	44.25-44.75	43.25-43.75	200@1.21	1.64	0.001-0.003	0.002-0.004	0.372-0.373	0.371-0.372
	5.9	6	①	①	76.4@1.39	1.39	0.002	0.002	0.274-0.276	0.274-0.276
	5.9	7	①	①	76.4@1.39	1.39	0.002	0.002	0.274-0.276	0.274-0.276
	8.0	W	44.5	45.0	200@1.212	1.64	0.001-0.003	0.001-0.003	0.311-0.312	0.311-0.312
	5.9	5	44.25-44.75	43.25-43.75	200@1.21	1.64	0.001-0.003	0.002-0.004	0.372-0.373	0.371-0.372
2002	3.9	X	44.25-44.75	43.25-43.75	200@1.21	1.64	0.001-0.003	0.001-0.003	0.311-0.312	0.311-0.312
	5.2	Y	44.25-44.75	43.25-43.75	200@1.21	1.64	0.001-0.003	0.001-0.003	0.311-0.312	0.311-0.312
	5.9	Z	44.25-44.75	43.25-43.75	200@1.21	1.64	0.001-0.003	0.002-0.004	0.372-0.373	0.371-0.372
	5.9	6	①	①	76.4@1.39	1.39	0.002	0.002	0.274-0.276	0.274-0.276
	5.9	7	①	①	76.4@1.39	1.39	0.002	0.002	0.274-0.276	0.274-0.276
	8.0	W	44.5	45.0	200@1.212	1.64	0.001-0.003	0.001-0.003	0.311-0.312	0.311-0.312
	5.9	5	44.25-44.75	43.25-43.75	200@1.21	1.64	0.001-0.003	0.002-0.004	0.372-0.373	0.371-0.372
2003	3.7	K	44.5-45	45-45.5	221-242@1.107	1.619	0.0008-0.0028	0.0019-0.0039	0.2729-0.2739	0.2717-0.2728
	3.9	X	44.25-44.75	43.25-43.75	200@1.21	1.64	0.001-0.003	0.001-0.003	0.311-0.312	0.311-0.312
	4.7	N	44.5-45	45-45.5	176.2-192.4@1.1532	1.602	0.0011-0.0017	0.0029	0.2728-0.2739	0.2717-0.2728
	5.2	Y	44.25-44.75	43.25-43.75	200@1.21	1.64	0.001-0.003	0.001-0.003	0.311-0.312	0.311-0.312
	5.7	D	44.5-45	45-45.5	95@1.81	1.81	0.0008-0.0025	0.0019-0.0037	0.3120-0.3130	0.3110-0.3120
	5.9	Z	44.25-44.75	43.25-43.75	200@1.21	1.64	0.001-0.003	0.002-0.004	0.372-0.373	0.371-0.372
	5.9	6	①	①	76.4@1.39	1.39	0.002	0.002	0.274-0.276	0.274-0.276
	5.9	C	①	①	76.4@1.39	1.39	0.002	0.002	0.274-0.276	0.274-0.276
	8.0	W	44.5	45.0	200@1.212	1.64	0.001-0.003	0.001-0.003	0.311-0.312	0.311-0.312

67189-RAMT-C08

VALVE SPECIFICATIONS

Year	Engine Displacement Liters	Engine ID/VIN	Seat Angle (deg.)	Face Angle (deg.)	Spring Test Pressure (lbs. @ in.)	Spring Installed Height (in.)	Stem-to-Guide Clearance (in.)		Stem Diameter (in.)	
							Intake	Exhaust	Intake	Exhaust
2004	3.7	K	44.5-45	45-45.5	221-234@ 1.107	1.579	0.0008- 0.0028	0.0019- 0.0039	0.2729- 0.2739	0.2717- 0.2728
	4.7	N	44.5- 45	45- 45.5	176.2-192.4 @1.1532	1.579	0.0008- 0.0028	0.0019- 0.0039	0.2728- 0.2739	0.2717- 0.2728
	5.7	D	44.5-45	45-45.5	95@1.81	1.81	0.0008- 0.0025	0.0019- 0.0037	0.3120- 0.3130	0.3110- 0.3120
	5.9	6	①	①	76.4@1.39	1.39	0.002	0.002	0.274- 0.276	0.274- 0.276
	5.9	C	①	①	76.4@1.39	1.39	0.002	0.002	0.274- 0.276	0.274- 0.276

① Intake: 30 degrees
  Exhaust: 45 degrees

67189-RAMT-C09

CRANKSHAFT AND CONNECTING ROD SPECIFICATIONS

All measurements are given in inches.

Year	Engine Displacement Liters	Engine ID/VIN	Crankshaft Main Brg. Journal Dia.	Crankshaft Main Brg. Oil Clearance	Crankshaft Shaft End-play	Thrust on No.	Connecting Rod Journal Diameter	Connecting Rod Oil Clearance	Connecting Rod Side Clearance
2001	3.9	X	2.4995-2.5005	①	0.0020-0.0070	2	2.1240-2.1250	0.0005-0.0022	0.0060-0.0140
	5.2	Y	2.4995-2.5005	①	0.0020-0.0070	2	2.1240-2.1250	0.0005-0.0022	0.0060-0.0140
	5.9	6	3.2662-3.2682	0.0037-0.0057	0.0040-0.0017	2	2.7150-2.7170	0.0035	0.0040-0.0120
	5.9	7	3.2662-3.2682	0.0037-0.0057	0.0040-0.0017	2	2.7150-2.7170	0.0035	0.0040-0.0120
	5.9	5	2.8095-2.8105	①	0.0020-0.0070	2	2.1240-2.1250	0.0005-0.0022	0.0060-0.0140
	5.9	Z	2.8095-2.8105	①	0.0020-0.0070	2	2.1240-2.1250	0.0005-0.0022	0.0060-0.0140
	8.0	W	2.9995-3.0005	0.0002-0.0023	0.0030-0.0120	2	2.1240-2.1250	0.0002-0.0029	0.0100-0.0180
2002	3.9	X	2.4995-2.5005	①	0.0020-0.0070	2	2.1240-2.1250	0.0005-0.0022	0.0060-0.0140
	5.2	Y	2.4995-2.5005	①	0.0020-0.0070	2	2.1240-2.1250	0.0005-0.0022	0.0060-0.0140
	5.9	6	3.2662-3.2682	0.0037-0.0057	0.0040-0.0017	2	2.7150-2.7170	0.0035	0.0040-0.0120
	5.9	7	3.2662-3.2682	0.0037-0.0057	0.0040-0.0017	2	2.7150-2.7170	0.0035	0.0040-0.0120
	5.9	5	2.8095-2.8105	①	0.0020-0.0070	2	2.1240-2.1250	0.0005-0.0022	0.0060-0.0140
	5.9	Z	2.8095-2.8105	①	0.0020-0.0070	2	2.1240-2.1250	0.0005-0.0022	0.0060-0.0140
	8.0	W	2.9995-3.0005	0.0002-0.0023	0.0030-0.0120	2	2.1240-2.1250	0.0002-0.0029	0.0100-0.0180
2003	3.7	K	2.4996-2.5005	0.0020-0.0034	0.0021-0.0112	2	2.2794-2.2797	0.0004-0.0019	0.0040-0.0138
	3.9	X	2.4995-2.5005	①	0.0020-0.0070	2	2.1240-2.1250	0.0005-0.0022	0.0060-0.0140
	4.7	N	2.4996-2.5005	0.0002-0.0013	0.0021-0.0112	2	2.0076-2.0082	0.0004-0.0019	0.0040-0.0138
	5.2	Y	2.4995-2.5005	①	0.0020-0.0070	2	2.1240-2.1250	0.0005-0.0022	0.0060-0.0140
	5.7	D	2.5585-2.5595	0.0009-0.0020	0.0020-0.0110	3	2.1250-2.1260	0.0007-0.0023	0.0030-0.0137
	5.9	6	3.2662-3.2682	0.0037-0.0057	0.0040-0.0017	2	2.7150-2.7170	0.0035	0.0040-0.0120
	5.9	C	3.2662-3.2682	0.0037-0.0057	0.0040-0.0017	2	2.7150-2.7170	0.0035	0.0040-0.0120
	5.9	Z	2.8095-2.8105	①	0.0020-0.0070	2	2.1240-2.1250	0.0005-0.0022	0.0060-0.0140
	8.0	W	2.9995-3.0005	0.0002-0.0023	0.0030-0.0120	2	2.1240-2.1250	0.0002-0.0029	0.0100-0.0180

67189-RAMT-C10

CRANKSHAFT AND CONNECTING ROD SPECIFICATIONS

All measurements are given in inches.

Year	Engine Displacement Liters	Engine ID/VIN	Crankshaft				Connecting Rod		
			Main Brg. Journal Dia.	Main Brg. Oil Clearance	Shaft End-play	Thrust on No.	Journal Diameter	Oil Clearance	Side Clearance
2004	3.7	K	2.4996-2.5005	0.0020-0.0034	0.0021-0.0112	2	2.2794-2.2797	0.0004-0.0022	0.0040-0.0138
	4.7	N	2.4996-2.5005	0.0008-0.0021	0.0021-0.0112	2	2.0076-2.0082	0.0006-0.0022	0.0040-0.0138
	5.7	D	2.5585-2.5595	0.0009-0.0020	0.0020-0.0110	3	2.1250-2.1260	0.0007-0.0023	0.0030-0.0137
	5.9	6	3.2662-3.2682	0.0037-0.0057	0.0040-0.0017	2	2.7150-2.7170	0.0035	0.0040-0.0120
	5.9	C	3.2662-3.2682	0.0037-0.0057	0.0040-0.0017	2	2.7150-2.7170	0.0035	0.0040-0.0120

① No.1: 0.0005-0.0015
All others: 0.0005-0.0020

67189-RAMT-C11

PISTON AND RING SPECIFICATIONS

All measurements are given in inches.

Year	Engine Displacement Liters	Engine ID/VIN	Piston Clearance	Ring Gap			Ring Side Clearance		
				Top Compression	Bottom Compression	Oil Control	Top Compression	Bottom Compression	Oil Control
2001	3.9	X	0.0005-0.0015	0.0100-0.0200	0.0100-0.0200	0.0100-0.0500	0.0015-0.0030	0.0015-0.0030	0.0020-0.0080
	5.2	Y	0.0005-0.0015	0.0100-0.0200	0.0100-0.0200	0.0100-0.0500	0.0015-0.0030	0.0015-0.0030	0.0020-0.0080
	5.9	6	NA	0.0138-0.0178	0.0335-0.0453	0.0098-0.0217	0.0037	0.0037	0.0033
	5.9	7	NA	0.0138-0.0178	0.0335-0.0453	0.0098-0.0217	0.0037	0.0037	0.0033
	5.9	5	0.0005-0.0015	0.0120-0.0220	0.0220-0.0310	0.0150-0.0550	0.0016-0.0033	0.0016-0.0033	0.0020-0.0080
	5.9	Z	0.0005-0.0015	0.0120-0.0220	0.0220-0.0310	0.0150-0.0550	0.0016-0.0033	0.0016-0.0033	0.0020-0.0080
	8.0	W	0.0005-0.0015	0.0100-0.0200	0.0100-0.0200	0.0150-0.0550	0.0029-0.0038	0.0029-0.0038	0.1020-0.1080
2002	3.9	X	0.0005-0.0015	0.0100-0.0200	0.0100-0.0200	0.0100-0.0500	0.0015-0.0030	0.0015-0.0030	0.0020-0.0080
	5.2	Y	0.0005-0.0015	0.0100-0.0200	0.0100-0.0200	0.0100-0.0500	0.0015-0.0030	0.0015-0.0030	0.0020-0.0080
	5.9	6	NA	0.0138-0.0178	0.0335-0.0453	0.0098-0.0217	0.0037	0.0037	0.0033
	5.9	7	NA	0.0138-0.0178	0.0335-0.0453	0.0098-0.0217	0.0037	0.0037	0.0033
	5.9	5	0.0005-0.0015	0.0120-0.0220	0.0220-0.0310	0.0150-0.0550	0.0016-0.0033	0.0016-0.0033	0.0020-0.0080
	5.9	Z	0.0005-0.0015	0.0120-0.0220	0.0220-0.0310	0.0150-0.0550	0.0016-0.0033	0.0016-0.0033	0.0020-0.0080
	8.0	W	0.0005-0.0015	0.0100-0.0200	0.0100-0.0200	0.0150-0.0550	0.0029-0.0038	0.0029-0.0038	0.1020-0.1080
2003	3.7	K	0.0014	0.0146-0.0249	0.0146-0.0249	0.0100-0.0300	0.0020-0.0037	0.0016-0.0031	0.0007-0.0091
	3.9	X	0.0005-0.0015	0.0100-0.0200	0.0100-0.0200	0.0100-0.0500	0.0015-0.0030	0.0015-0.0030	0.0020-0.0080
	4.7	N	0.0008-0.0020	0.0146-0.0249	0.0146-0.0249	0.0100-0.0500	0.0020-0.0041	0.0016-0.0032	0.0007-0.0091
	5.2	Y	0.0005-0.0015	0.0100-0.0200	0.0100-0.0200	0.0100-0.0500	0.0015-0.0030	0.0015-0.0030	0.0020-0.0080
	5.7	D	0.0008-0.0019	0.0090-0.0149	0.0137-0.0236	0.0059-0.0259	0.0007-0.0026	0.0007-0.0022	0.0007-0.0091
	5.9	6	NA	0.0138-0.0178	0.0335-0.0453	0.0098-0.0217	0.0037	0.0037	0.0033
	5.9	C	NA	0.0138-0.0178	0.0335-0.0453	0.0098-0.0217	0.0037	0.0037	0.0033
	5.9	Z	0.0005-0.0015	0.0120-0.0220	0.0220-0.0310	0.0150-0.0550	0.0016-0.0033	0.0016-0.0033	0.0020-0.0080
	8.0	W	0.0005-0.0015	0.0100-0.0200	0.0100-0.0200	0.0150-0.0550	0.0029-0.0038	0.0029-0.0038	0.1020-0.1080

67189-RAMT-C12

PISTON AND RING SPECIFICATIONS

All measurements are given in inches.

Year	Engine Displacement Liters	Engine ID/VIN	Piston Clearance	Ring Gap			Ring Side Clearance		
				Top Compression	Bottom Compression	Oil Control	Top Compression	Bottom Compression	Oil Control
2004	3.7	K	0.0014	0.0079-0.0142	0.0146-0.0249	0.0100-0.0300	0.0020-0.0037	0.0016-0.0031	0.0007-0.0091
	4.7	N	0.0008-0.0020	0.0079-0.0142	0.0146-0.0249	0.0100-0.0300	0.0020-0.0037	0.0016-0.0032	0.0007-0.0091
	5.7	D	0.0008-0.0019	0.0090-0.0149	0.0137-0.0236	0.0059-0.0259	0.0007-0.0026	0.0007-0.0022	0.0007-0.0091
	5.9	6	NA	0.0138-0.0178	0.0335-0.0453	0.0098-0.0217	0.0037	0.0037	0.0033
	5.9	C	NA	0.0138-0.0178	0.0335-0.0453	0.0098-0.0217	0.0037	0.0037	0.0033

67189-RAMT-C13

TORQUE SPECIFICATIONS
All readings in ft. lbs.

Year	Engine Displacement Liters	Engine ID/VIN	Cylinder Head Bolts	Main Bearing Bolts	Rod Bearing Bolts	Crankshaft Damper Bolts	Flywheel Bolts	Manifold Intake	Manifold Exhaust	Spark Plugs	Oil Pan Drain Plug
2001	3.9	X	①	85	45	18	55	②	25	30	③
	5.2	Y	①	85	45	18	55	②	25	30	③
	5.9	6	④	⑤	⑥	92	⑦	18	32	—	③
	5.9	7	④	⑤	⑥	92	⑦	18	32	—	③
	5.9	5	①	85	45	18	55	②	25	30	③
	5.9	Z	①	85	45	18	55	②	25	30	③
	8.0	W	⑧	⑨	45	230	55	⑩	16	30	③
2002	3.9	X	①	85	45	18	55	②	25	30	③
	5.2	Y	①	85	45	18	55	②	25	30	③
	5.9	6	④	⑤	⑥	92	⑦	18	32	—	③
	5.9	7	④	⑤	⑥	92	⑦	18	32	—	③
	5.9	5	①	85	45	18	55	②	25	30	③
	5.9	Z	①	85	45	18	55	②	25	30	③
	8.0	W	⑧	⑨	45	230	55	⑩	16	30	③
2003	3.7	K	⑪	⑫	⑬	130	45	9	18	27	③
	3.9	X	①	85	45	18	55	②	25	30	③
	4.7	N	⑭	⑮	15⑯	130	45	9	18	27	③
	5.2	Y	①	85	45	18	55	②	25	30	③
	5.7	D	⑰	⑱	⑲	90	70	⑪	18	18	③
	5.9	C	④	⑤	⑥	92	⑦	18	32	—	③
	5.9	6	④	⑤	⑥	92	⑦	18	32	—	③
	5.9	Z	①	85	45	18	55	②	25	30	③
	8.0	W	⑧	⑨	45	230	55	⑩	16	30	③
2004	3.7	K	⑪	⑫	⑬	130	45	9	18	27	③
	4.7	N	⑭	⑮	⑬	130	45	9	18	27	③
	5.7	D	⑰	⑱	⑲	130	55	⑪	18	18	③
	5.9	C	④	⑤	⑥	92	⑦	18	32	—	③
	5.9	6	④	⑤	⑥	92	⑦	18	32	—	③

① Step 1: 50 ft. lbs.
Step 2: 105 ft. lbs.

② Step 1: bolts 1-4 72 inch lbs.
Step 2: bolts 5-12 72 inch lbs.
Step 3: Recheck all bolts at 72 inch lbs.
Step 4: all bots to 12 ft. lbs.
Step 5: Recheck all bolts at 12 ft. lbs.

③ Gas engines:25 ft. lbs.
Diesel engine: 37 ft. lbs.

④ Step 1: 59 ft. lbs.
Step 2: 77 ft. lbs.
Step 3: Recheck at 77 ft. lbs.
Step 4: All bolts an additional 1/4 turn (90 degrees)

⑤ Step 1: 45 ft. lbs.
Step 2: 60 ft. lbs.
Step 3: additional 1/4 turn (90 degrees)

⑥ Step 1: 26 ft. lbs.
Step 2: 51 ft. lbs.
Step 3: 73 ft. lbs.

⑦ Manual transmission: 101 ft. lbs.
Automatic transmission: 32 ft. lbs.

⑧ Step 1: 43 ft. lbs.
Step 2: 105 ft. lbs.

⑨ Step 1: 20 ft. lbs.
Step 2: 85 ft. lbs.

⑩ Upper: 16 ft. lbs.
Lower: 40 ft. lbs.

⑪ See procedure

⑫ Bed plate: see procedure

⑬ 20 ft. lbs. plus 90 degrees

⑭ M11 bolts: 60 ft. lbs.
M8 bolts: 250 inch lbs.

⑮ Step 1: Bolts 1-10 to 25 inch lbs.
Step 2: Bolts 1-10 plus 90 degrees
Step 3: Bolts A-K to 40 ft. lbs.
Step 4: Bolts A1-A5 to 20 ft. lbs.

⑯ Plus 110 degrees

⑰ M8
Step 1: 15 ft. lbs.
Step 2: 25 ft. lbs.
M12
Step 1: 25 ft. lbs.
Step 2: 40 ft. lbs.
Step 3: plus 90 degree turn

⑱ M12 bolts: 12 ft. lbs. plus 90 degrees
M8 bolts: 13 ft. lbs.

⑲ 15 ft. lbs. Plus 90 degrees

BRAKE SPECIFICATIONS
All measurements in inches unless noted

Year	Model		Brake Disc			Brake Drum			Minimum Lining Thickness		Brake Caliper	
			Original Thickness	Minimum Thickness	Maximum Run-out	Original Inside Diameter	Max. Wear Limit	Max. Machine Diameter	Front	Rear	Bracket Bolts (ft. lbs.)	Mounting Bolts (ft. lbs.)
2001	B1500 Van		1.26	1.181	0.004	11.03	①	①	②	③	110	15
	B2500 Van		1.26	1.181	0.004	11.03	①	①	②	③	110	15
	B3500 Van		1.26	1.181	0.004	12.125	①	①	②	③	110	15
	Ram 1500 Pick-up		1.18	1.117	0.005	11.00	11.09	11.06	②	③	④	24
	Ram 2500 Pick-up	F	1.50	1.334	0.005	—	—	—	②	—	④	24
		R	1.18	1.117	0.005	—	—	—	—	NA	—	24
	Ram 3500 Pick-up	F	1.50	1.334	0.005	—	—	—	②	—	④	24
		R	1.18	1.117	0.005	—	—	—	—	NA	—	24
2002	B1500 Van		1.26	1.181	0.004	11.03	①	①	②	③	110	15
	B2500 Van		1.26	1.181	0.004	11.03	①	①	②	③	110	15
	B3500 Van		1.26	1.181	0.004	12.125	①	①	②	③	110	15
	Ram 1500 Pick-up		1.18	1.117	0.005	11.00	11.09	11.06	②	③	④	24
	Ram 2500 Pick-up	F	1.50	1.334	0.005	—	—	—	②	—	④	24
		R	1.18	1.117	0.005	—	—	—	—	NA	—	24
	Ram 3500 Pick-up	F	1.50	1.334	0.005	—	—	—	②	—	④	24
		R	1.18	1.117	0.005	—	—	—	—	NA	—	24
2003		NA	NA	NA	NA	NA	NA	NA	NA	NA	⑤	⑤
2004	Ram Pick-up LD	F	1.10	1.039	0.005	—	—	—	NA	—	④	24
		R	0.86	1.117	0.005	—	—	—	—	NA	—	11
	Ram Pick-up HD	F	1.39	1.334	0.005	—	—	—	NA	—	④	24
		R	1.18	1.117	0.005	—	—	—	—	NA	—	11

NA: Not Available

① Maximum allowable drum diameter, either from wear or machining is stamped on the drum

② Riveted brake pads: 0.0625 in.
　Bonded brake pads: 0.1875 in.

③ Riveted brake shoes: 0.031 in.
　Bonded brake shoes: 0.0625 in.

④ LD adapter: 130 ft. lbs.
　HD adapter: 210 ft. lbs.

⑤ See procedures in the text

67189-RAMT-C15

TIRE, WHEEL AND BALL JOINT SPECIFICATIONS

Year	Model	OEM Tires		Tire Pressures (psi)		Wheel Size	Wheel Lug Nut Torque
		Standard	Optional	Front	Rear		
2001	1500 PU 2wd	P225/75R16	P245/75R16C	35	35	7-J	①
			P275/60R17			9-J	
	2500 PU 2wd, w/6400 GVW	P225/75R16	P245/75R16C	35	35	7-J	①
			P275/60R17			9-J	
	2500 PU 2wd, w/8800 GVW	LT245/75R16E	None	40	40	6.5-J	①
				w/V10: 45	40		
				w/Diesel: 50	65		
				w/Club Cab: 45	80		
	1500 PU 4wd	P245/75R16	P265/75R16	35	35	7-J	①
	2500 PU 4wd, w/6400 GVW	P245/75R16	P265/75R16	35	35	7-J	①
	2500 PU 4wd w/8800 GVW	P245/75R16E	None	40	40	6.5-J	①
				w/V10: 45	40		
				w/Diesel: 50	65		
				w/Club Cab: 45	80		
	1500 Van	P235/75R15	None	35	40	6.5-J	①
	2500 Van	LT225/75R16	None	50	65	6.5-J	①
	3500 Van	LT225/75R16E	None	55	80	6.5-J	①
2002	1500 PU	P265/70R17	P275/55R20	②	②	Std: 8-J	①
						Opt: 9-J	
	2500 ST PU	LT245/75R16E	LT265/75R16E	②	②	Std: 7.5-J	①
						Opt: 8J	
	2500 SLT PU	LT265/75R16E	None	②	②	8-J	①
	3500 PU	LT235/85R16E	None	②	②	6-J	①
	1500 Cargo Van	P235/75R15	None	②	②	6.5-J	①
	2500 Cargo Van	LT225/75R16D	None	②	②	6.5-J	①
	3500 Cargo Van	LT225/75R16E	None	②	②	6.5-J	①
	1500 Ram Wagon	P235/75R15	None	②	②	6.5-J	①
	2500 Ram Wagon	LT225/75R16D	None	②	②	6.5-J	①
	3500 Ram Wagon	LT245/75R16E	None	②	②	6.5-J	①
2003	1500 PU	P265/70R17	P275/55R20	②	②	Std: 8-J	①
						Opt: 9-J	
	2500 ST PU	LT245/75R16E	LT265/75R16E	②	②	Std: 7.5-J	①
						Opt: 8J	
	2500 SLT PU	LT265/75R16E	None	②	②	8-J	①
	3500 PU	LT235/85R16E	None	②	②	6-J	①
	1500 Cargo Van	P235/75R15	None	②	②	6.5-J	①
	2500 Cargo Van	LT225/75R16D	None	②	②	6.5-J	①
	3500 Cargo Van	LT225/75R16E	None	②	②	6.5-J	①
	1500 Ram Wagon	P235/75R15	None	②	②	6.5-J	①
	2500 Ram Wagon	LT225/75R16D	None	②	②	6.5-J	①
	3500 Ram Wagon	LT245/75R16E	None	②	②	6.5-J	①
2004	1500 PU	P265/70R17	P275/55R20	②	②	Std: 8-J	①
						Opt: 9-J	
	2500 ST PU	LT245/75R16E	LT265/75R16E	②	②	Std: 7.5-J	①
						Opt: 8J	
	2500 SLT PU	LT265/75R16E	None	②	②	8-J	①
	3500 PU	LT235/85R16E	None	②	②	6-J	①

OEM: Original Equipment Manufacturer   ① 5 stud wheel: 95 ft. lbs.   ② See sticker on door
PSI: Pounds Per Square Inch   8 stud wheel: 135 ft. lbs.
STD: Standard   8 stud dual wheel: 145 ft. lbs.
OPT: Optional

WHEEL ALIGNMENT

Year	Model	GVW	Wheel Base (in.)	Caster Range (+/-Deg.)	Caster Preferred Setting (Deg.)	Camber Range (+/-Deg.)	Camber Preferred Setting (Deg.)	Toe-in (in.)
2001	2WD Pickup	6,400	118.7	1.00	+3.66	0.50	+0.50	0.10+/-0.10
	2WD Pickup	6,400	134.7	1.00	+3.89	0.50	+0.50	0.10+/-0.10
	2WD Pickup	6,400	138.7	1.00	+3.99	0.50	+0.50	0.10+/-0.10
	2WD Pickup	6,400	154.7	1.00	+4.17	0.50	+0.50	0.10+/-0.10
	2WD Pickup	8,800	134.7	1.00	+3.53	0.50	+0.50	0.10+/-0.10
	2WD Pickup	8,800	138.7	1.00	+3.59	0.50	+0.50	0.10+/-0.10
	2WD Pickup	8,800	154.7	1.00	+3.78	0.50	+0.50	0.10+/-0.10
	2WD Pickup	10,500	134.7	1.00	+3.33	0.50	+0.50	0.10+/-0.10
	2WD Pickup	10,500	154.7	1.00	+3.58	0.50	+0.50	0.10+/-0.10
	4WD Pickup	6,400	118.7	1.00	+2.86	0.50	NA	0.10+/-0.10
	4WD Pickup	6,400	134.7	1.00	+3.04	0.50	NA	0.10+/-0.10
	4WD Pickup	6,400	138.7	1.00	+3.19	0.50	NA	0.10+/-0.10
	4WD Pickup	6,400	154.7	1.00	+3.37	0.50	NA	0.10+/-0.10
	4WD Pickup	8,800	134.7	1.00	+2.68	0.50	NA	0.10+/-0.10
	4WD Pickup	8,800	138.7	1.00	+2.74	0.50	NA	0.10+/-0.10
	4WD Pickup	8,800	154.7	1.00	+2.88	0.50	NA	0.10+/-0.10
	4WD Pickup	10,500	134.7	1.00	+2.48	0.50	NA	0.10+/-0.10
	4WD Pickup	10,500	154.7	1.00	+2.63	0.50	NA	0.10+/-0.10
	Van	All	All	1.50	+2.75	0.30	0	0.25+/-0.25
2002	2WD Pickup	6,400	118.7	1.00	+3.66	0.50	+0.50	0.10+/-0.10
	2WD Pickup	6,400	134.7	1.00	+3.89	0.50	+0.50	0.10+/-0.10
	2WD Pickup	6,400	138.7	1.00	+3.99	0.50	+0.50	0.10+/-0.10
	2WD Pickup	6,400	154.7	1.00	+4.17	0.50	+0.50	0.10+/-0.10
	2WD Pickup	8,800	134.7	1.00	+3.53	0.50	+0.50	0.10+/-0.10
	2WD Pickup	8,800	138.7	1.00	+3.59	0.50	+0.50	0.10+/-0.10
	2WD Pickup	8,800	154.7	1.00	+3.78	0.50	+0.50	0.10+/-0.10
	2WD Pickup	10,500	134.7	1.00	+3.33	0.50	+0.50	0.10+/-0.10
	2WD Pickup	10,500	154.7	1.00	+3.58	0.50	+0.50	0.10+/-0.10
	4WD Pickup	6,400	118.7	1.00	+2.86	0.50	NA	0.10+/-0.10
	4WD Pickup	6,400	134.7	1.00	+3.04	0.50	NA	0.10+/-0.10
	4WD Pickup	6,400	138.7	1.00	+3.19	0.50	NA	0.10+/-0.10
	4WD Pickup	6,400	154.7	1.00	+3.37	0.50	NA	0.10+/-0.10
	4WD Pickup	8,800	134.7	1.00	+2.68	0.50	NA	0.10+/-0.10
	4WD Pickup	8,800	138.7	1.00	+2.74	0.50	NA	0.10+/-0.10
	4WD Pickup	8,800	154.7	1.00	+2.88	0.50	NA	0.10+/-0.10
	4WD Pickup	10,500	134.7	1.00	+2.48	0.50	NA	0.10+/-0.10
	4WD Pickup	10,500	154.7	1.00	+2.63	0.50	NA	0.10+/-0.10
	Van	All	All	1.50	+2.75	0.30	0	0.25+/-0.25
2003	1500 2wd	—	120.5	0.75	+4.0	0.50	0	0.10+/-0.10
	1500 2wd	—	140.5	0.75	+4.2	0.50	0	0.10+/-0.10
	1500 2wd	—	160.5	0.75	+4.4	0.50	0	0.10+/-0.10
	1500 4wd	—	120.5	0.75	+4.2	0.50	0	0.10+/-0.10
	1500 4wd	—	140.5	0.75	+4.4	0.50	0	0.10+/-0.10
	1500 4wd	—	160.5	0.75	+4.6	0.50	0	0.10+/-0.10
	2500 & 3500 2wd	—	140	0.75	+4.0	0.50	0	0.10+/-0.05
	2500 & 3500 2wd	—	160	0.75	+4.3	0.50	0	0.10+/-0.05
	2500 & 3500 4wd	—	140	0.75	+4.5	0.25	+0.25	0.10+/-0.05
	2500 & 3500 4wd	—	160	0.75	+4.7	0.25	0.25	0.10+/-0.05
	Van	All	All	1.00	+3.5	0.30	0	0.25+/-0.25

WHEEL ALIGNMENT

Year	Model	GVW	Wheel Base (in.)	Caster Range (+/-Deg.)	Caster Preferred Setting (Deg.)	Camber Range (+/-Deg.)	Camber Preferred Setting (Deg.)	Toe-in (in.)
2004	1500 2wd	—	120.5	0.75	+4.0	0.50	0	0.10+/-0.10
	1500 2wd	—	140.5	0.75	+4.2	0.50	0	0.10+/-0.10
	1500 2wd	—	160.5	0.75	+4.4	0.50	0	0.10+/-0.10
	1500 4wd	—	120.5	0.75	+4.2	0.50	0	0.10+/-0.10
	1500 4wd	—	140.5	0.75	+4.4	0.50	0	0.10+/-0.10
	1500 4wd	—	160.5	0.75	+4.6	0.50	0	0.10+/-0.10
	2500 & 3500 2wd	—	140	0.75	+4.0	0.50	0	0.10+/-0.05
	2500 & 3500 2wd	—	160	0.75	+4.3	0.50	0	0.10+/-0.05
	2500 & 3500 4wd	—	140	0.75	+4.5	0.25	+0.25	0.10+/-0.05
	2500 & 3500 4wd	—	160	0.75	+4.7	0.25	0.25	0.10+/-0.05

NA: Not adjustable

67189-RAMT-C18

SCHEDULED MAINTENANCE INTERVALS
2001-03 LIGHT DUTY RAM PICKUP (Exc. 8.0L)

TO BE SERVICED	TYPE OF SERVICE	VEHICLE MILEAGE INTERVAL (x1000)														
		7.5	15	22.5	30	37.5	45	52.5	60	67.5	75	82.5	90	97.5	100	105
Engine oil & filter	R	✓	✓	✓	✓	✓	✓	✓	✓	✓	✓	✓	✓	✓		✓
Engine coolant & hoses	I	✓	✓	✓	✓	✓	✓	✓	✓	✓	✓	✓	✓	✓		✓
Brake hoses	I	✓	✓	✓	✓	✓	✓	✓	✓	✓	✓	✓	✓	✓		✓
Steering linkage	L	✓	✓	✓	✓	✓	✓	✓	✓	✓	✓	✓	✓	✓		✓
Manual trans. fluid level	I	✓	✓	✓	✓	✓	✓	✓	✓	✓	✓	✓	✓	✓		✓
Exhaust system	I	✓	✓	✓	✓	✓	✓	✓	✓	✓	✓	✓	✓	✓		✓
Brake linings	I			✓				✓		✓				✓		
Front wheel bearings (2wd)	S/I			✓			✓			✓			✓			
Ball joints	L			✓			✓			✓			✓			
Air cleaner element	R				✓				✓				✓			
Spark plugs	R				✓				✓				✓			
Transfer case fluid	R					✓					✓					
Rear drum brakes	Adj	✓	✓	✓	✓	✓	✓	✓	✓	✓	✓	✓	✓	✓		✓
Tires	Rotate	✓	✓	✓	✓	✓	✓	✓	✓	✓	✓	✓	✓	✓		✓
Engine coolant ①	R							✓								
Spark plug wires	R								✓							
PCV valve	S/I								✓							
Drive belt tensioner	S/I										✓					
Automatic trans. fluid	R														✓	
Automatic trans. bands	Adj														✓	

R: Replace   S/I: Service or Inspect   Adj: Adjust   L: Lubricate

① Replace every 36 months, regardless of mileage.

FREQUENT OPERATION MAINTENANCE (SEVERE SERVICE)

If a vehicle is operated under any of the following conditions it is considered severe service:

- Extremely dusty areas.
- 50% or more of the vehicle operation is in 32°C (90°F) or higher temperatures, or constant operation in temperatures below 0°C (32°F).
- Prolonged idling (vehicle operation in stop and go traffic.
- Frequent short running periods (engine does not warm to normal operating temperatures).
- Police, taxi, delivery usage or trailer towing usage.

Oil & oil filter change: change every 3000 miles.

Air filter/air pump air filter: change every 24,000 miles.

Engine coolant level, hoses & clamps: check every 6,000 miles.

Exhaust system: check every 6000 miles.

Drive belts: check every 18,000 miles; replace every 24,000 miles.

Crankcase inlet air filter (6 & 8 cyl.): clean every 24,000 miles.

Oxygen sensor: replace every 82,500 miles.

Automatic transmission fluid, filter & bands: change & adjust every 12,000 miles.

Steering linkage: lubricate every 6000 miles.

Rear axle fluid: change every 12,000 miles.

67189-RAMT-C19

SCHEDULED MAINTENANCE INTERVALS
2001-03 Ram Van

TO BE SERVICED	TYPE OF SERVICE	VEHICLE MILEAGE INTERVAL (x1000)														
		7.5	15	22.5	30	37.5	45	52.5	60	67.5	75	82.5	90	97.5	100	105
Engine oil & filter	R	✓	✓	✓	✓	✓	✓	✓	✓	✓	✓	✓	✓	✓		✓
Engine coolant & hoses	I	✓	✓	✓	✓	✓	✓	✓	✓	✓	✓	✓	✓	✓		✓
Brake hoses	I	✓	✓	✓	✓	✓	✓	✓	✓	✓	✓	✓	✓	✓		✓
Steering linkage	L		✓		✓		✓		✓		✓		✓		✓	
Exhaust system	I	✓	✓	✓	✓	✓	✓	✓	✓	✓	✓	✓	✓	✓		✓
Brake linings	I			✓			✓			✓			✓			✓
Front wheel bearings	S/I			✓			✓			✓			✓			✓
Ball joints	L			✓			✓			✓			✓			✓
Air cleaner element	R				✓				✓				✓			
Spark plugs	R				✓				✓				✓			
Tires	Rotate	✓	✓	✓	✓	✓	✓	✓	✓	✓	✓	✓	✓	✓		✓
Engine coolant ①	R							✓					✓			
Spark plug wires	R							✓								
PCV valve	S/I							✓								
Drive belt tensioner	S/I							✓								
Automatic trans. fluid	R														✓	
Automatic trans. bands	Adj														✓	

R: Replace   S/I: Service or Inspect   Adj: Adjust   L: Lubricate

① Replace every 36 months, regardless of mileage.

FREQUENT OPERATION MAINTENANCE (SEVERE SERVICE)

If a vehicle is operated under any of the following conditions it is considered severe service:

- Extremely dusty areas.
- 50% or more of the vehicle operation is in 32°C (90°F) or higher temperatures, or constant operation in temperatures below 0°C (32°F).
- Prolonged idling (vehicle operation in stop and go traffic.
- Frequent short running periods (engine does not warm to normal operating temperatures).
- Police, taxi, delivery usage or trailer towing usage.

Oil & oil filter change: change every 3000 miles.

Air filter/air pump air filter: change every 24,000 miles.

Engine coolant level, hoses & clamps: check every 6,000 miles.

Exhaust system: check every 6000 miles.

Drive belts: check every 18,000 miles; replace every 24,000 miles.

Crankcase inlet air filter (6 & 8 cyl.): clean every 24,000 miles.

Oxygen sensor: replace every 82,500 miles.

Automatic transmission fluid, filter & bands: change & adjust every 12,000 miles.

Steering linkage: lubricate every 6000 miles.

Rear axle fluid: change every 12,000 miles.

67189-RAMT-C20

SCHEDULED MAINTENANCE INTERVALS
2001-03 Medium Duty (California 8.0L 2500 & 3500 only)

TO BE SERVICED	TYPE OF SERVICE	VEHICLE MILEAGE INTERVAL (x1000)														
		6	12	18	24	30	36	42	48	54	60	66	72	78	84	90
Engine oil and filter	R	✓	✓	✓	✓	✓	✓	✓	✓	✓	✓	✓	✓	✓	✓	✓
Brake linings	I			✓			✓			✓			✓			✓
Front wheel bearings (2wd)	S/I			✓			✓			✓			✓			✓
Auto trans fluid and filter	R				✓				✓				✓			
Auto trans bands	Adj				✓				✓				✓			
Air cleaner element	R					✓					✓					✓
Spark plugs	R					✓					✓					✓
Exhaust system	I	✓	✓	✓	✓	✓	✓	✓	✓	✓	✓	✓	✓	✓	✓	✓
Brake hoses	I	✓	✓	✓	✓	✓	✓	✓	✓	✓	✓	✓	✓	✓	✓	✓
Rear drum brakes	Adj	✓	✓	✓	✓	✓	✓	✓	✓	✓	✓	✓	✓	✓	✓	✓
Tires	Rotate	✓	✓	✓	✓	✓	✓	✓	✓	✓	✓	✓	✓	✓	✓	✓
Steering linkage	L	✓	✓	✓	✓	✓	✓	✓	✓	✓	✓	✓	✓	✓	✓	✓
Transfer case fluid	R						✓						✓			
Engine coolant ①	R						✓						✓			
Spark plug wires	R										✓					

R: Replace   S/I: Service or Inspect   Adj: Adjust   L: Lubricate

① Replace every 36 months, regardless of mileage.

FREQUENT OPERATION MAINTENANCE (SEVERE SERVICE)

If a vehicle is operated under any of the following conditions it is considered severe service:

- Extremely dusty areas.
- 50% or more of the vehicle operation is in 32°C (90°F) or higher temperatures, or constant operation in temperatures below 0°C (32°F).
- Prolonged idling (vehicle operation in stop and go traffic.
- Frequent short running periods (engine does not warm to normal operating temperatures).
- Police, taxi, delivery usage or trailer towing usage.

Oil & oil filter change: change every 3000 miles.

Air filter/air pump air filter: change every 24,000 miles.

Engine coolant level, hoses & clamps: check every 6,000 miles.

Exhaust system: check every 6000 miles.

Drive belts: check every 18,000 miles; replace every 24,000 miles.

Crankcase inlet air filter (6 & 8 cyl.): clean every 24,000 miles.

Oxygen sensor: replace every 82,500 miles.

Automatic transmission fluid, filter & bands: change & adjust every 12,000 miles.

Steering linkage: lubricate every 6000 miles.

Rear axle fluid: change every 12,000 miles.

67189-RAMT-C21

SCHEDULED MAINTENANCE INTERVALS
2001-03 Heavy Duty (Federal 2500 8.0L and 3500 5.9L & 8.0L)

TO BE SERVICED	TYPE OF SERVICE	VEHICLE MILEAGE INTERVAL (x1000)														
		6	12	18	24	30	36	42	48	54	60	66	72	78	82.5	84
Engine oil and filter	R	✓	✓	✓	✓	✓	✓	✓	✓	✓	✓	✓	✓	✓		✓
Air cleaner element (8.0L)	R		✓				✓				✓					✓
Brake linings	I			✓			✓			✓			✓			
Front wheel bearings (4wd)	S/I			✓			✓			✓			✓			
Air cleaner element (5.9L)	R				✓				✓				✓			
Air pump filter	R				✓				✓				✓			
Crankcase inlet air filter (5.9L)	C/L				✓				✓				✓			
Auto trans fluid & filter	R				✓				✓				✓			
Auto trans bands	Adj				✓				✓				✓			
Front wheel bearings (2wd)	S/I				✓				✓				✓			
Spark plugs	R					✓					✓					
Transfer case fluid	R						✓						✓			
Exhaust system	I	✓	✓	✓	✓	✓	✓	✓	✓	✓	✓	✓	✓	✓		✓
Brake hoses	I	✓	✓	✓	✓	✓	✓	✓	✓	✓	✓	✓	✓	✓		✓
Tires	Rotate	✓	✓	✓	✓	✓	✓	✓	✓	✓	✓	✓	✓	✓		✓
Steering linkage	L	✓	✓	✓	✓	✓	✓	✓	✓	✓	✓	✓	✓	✓		✓
Engine coolant ①	R						✓						✓			
Spark plug wires	R										✓					
PCV valve (5.9L)	R										✓					
Distributor cap & rotor (5.9L)	R										✓					
Oxygen sensor (5.9L)	R														✓	

R: Replace   S/I: Service or Inspect   Adj: Adjust   C/L: Clean & lubricate

① Replace every 36 months, regardless of mileage.

FREQUENT OPERATION MAINTENANCE (SEVERE SERVICE)

If a vehicle is operated under any of the following conditions it is considered severe service:

- Extremely dusty areas.
- 50% or more of the vehicle operation is in 32°C (90°F) or higher temperatures, or constant operation in temperatures below 0°C (32°F).
- Prolonged idling (vehicle operation in stop and go traffic.
- Frequent short running periods (engine does not warm to normal operating temperatures).
- Police, taxi, delivery usage or trailer towing usage.

Oil & oil filter change: change every 3000 miles.

Air filter/air pump air filter: change every 24,000 miles.

Engine coolant level, hoses & clamps: check every 6,000 miles.

Exhaust system: check every 6000 miles.

Drive belts: check every 18,000 miles; replace every 24,000 miles.

Crankcase inlet air filter (6 & 8 cyl.): clean every 24,000 miles.

Oxygen sensor: replace every 82,500 miles.

Automatic transmission fluid, filter & bands: change & adjust every 12,000 miles.

Steering linkage: lubricate every 6000 miles.

Rear axle fluid: change every 12,000 miles.

SCHEDULED MAINTENANCE INTERVALS
2004 RAM PICKUP

TO BE SERVICED	TYPE OF SERVICE	VEHICLE MILEAGE INTERVAL (x1000)																
		6	12	18	24	30	36	42	48	54	60	66	72	78	84	90	96	102
Engine oil & filter	R	✓	✓	✓	✓	✓	✓	✓	✓	✓	✓	✓	✓	✓	✓			
Brake linings	S/I			✓			✓			✓			✓					
Air cleaner element	R					✓					✓					✓		
Spark plugs	R					✓					✓					✓		
Transfer case fluid	I					✓					✓							
Engine coolant	R	Replace every 60 months regardless of mileage																
Spark plug wires (5.7L)	R										✓							
PCV valve	S/I										✓							
Accessory drive belt	S/I																✓	
Auto. Trans. fluid ①	R																	✓
Transfer case fluid	R	Drain the fluid every 120,000 miles																

R: Replace   S/I: Service or Inspect

① On 4.7L and 5.7L, change the fluid return filter, if equipped

FREQUENT OPERATION MAINTENANCE (SEVERE SERVICE)

If a vehicle is operated under any of the following conditions it is considered severe service:

- Extremely dusty areas.
- Day or night time temperatures below 0°C (32°F).
- Prolonged idling (vehicle operation in stop and go traffic.)
- Frequent short running periods (engine does not warm to normal operating temperatures).
- Police, taxi, delivery usage or trailer towing usage.
- Off road or desert operation
- 50% or more of you driving is done in temperatures above 90 degrees F (32 deg. C)

Oil & oil filter change: change every 3000 miles.

Air filter: change every 15,000 miles and change as necessary.

Drive belts: check and replace as necessary every 60,000 miles.

Automatic transmission fluid, filter every 30,000 miles 4.7L and 5.7L.

Automatic transmission fluid, filter every 60,000 miles 3.7L.

Front and rear axle fluid: change every 15,000 miles.

Inspect PCV valve every 30,000 miles

67189-RAMT-C23

SCHEDULED MAINTENANCE INTERVALS
2001-03 Turbo Diesel

TO BE SERVICED	TYPE OF SERVICE	VEHICLE MILEAGE INTERVAL (x1000)												
		7.5	15	22.5	30	37.5	45	52.5	60	67.5	75	82.5	90	97.5
Engine oil & filter	R	✓	✓	✓	✓	✓	✓	✓	✓	✓	✓	✓	✓	✓
Crankcase breather canister	D	✓	✓	✓	✓	✓	✓	✓	✓	✓	✓	✓	✓	✓
Water pump weep hole	I		✓		✓		✓		✓		✓		✓		
Fuel filter	R		✓		✓		✓		✓		✓		✓		
Fuel sensor	C		✓		✓		✓		✓		✓		✓		
Drive belts	I			✓			✓			✓			✓		
Brake linings	I			✓			✓			✓			✓		
Fan hub	I				✓				✓				✓		
Damper	I				✓				✓				✓		
Transfer case fluid	R						✓						✓		
Auto trans & filter	R				✓				✓				✓		
Auto trans bands	Adj				✓				✓				✓		
Front wheel bearings (2WD)	S/I				✓				✓				✓		
Engine coolant ①	R				✓				✓				✓		
Exhaust system	I	✓	✓	✓	✓	✓	✓	✓	✓	✓	✓	✓	✓	✓
Steering linkage	L	✓	✓	✓	✓	✓	✓	✓	✓	✓	✓	✓	✓	✓
Rear drum brakes	Adj	✓	✓	✓	✓	✓	✓	✓	✓	✓	✓	✓	✓	✓
Tires	Rotate	✓	✓	✓	✓	✓	✓	✓	✓	✓	✓	✓	✓	✓

R: Replace   S/I: Service or Inspect   D: Drain   C: Clean   Adj: Adjust

① Replace every 36 months, regardless of mileage

FREQUENT OPERATION MAINTENANCE (SEVERE SERVICE)

If a vehicle is operated under any of the following conditions it is considered severe service:

- Extremely dusty areas.

- 50% or more of the vehicle operation is in 32°C (90°F) or higher temperatures, or constant operation in temperatures below 0°C (32°F).

- Prolonged idling (vehicle operation in stop and go traffic.

- Frequent short running periods (engine does not warm to normal operating temperatures).

- Police, taxi, delivery usage or trailer towing usage.

Oil & oil filter change: change every 3000 miles.

Air filter/air pump air filter: change every 24,000 miles.

Engine coolant level, hoses & clamps: check every 6,000 miles.

Exhaust system: check every 6000 miles.

Drive belts: check every 18,000 miles; replace every 24,000 miles.

Crankcase inlet air filter (6 & 8 cyl.): clean every 24,000 miles.

Oxygen sensor: replace every 82,500 miles.

Automatic transmission fluid, filter & bands: change & adjust every 12,000 miles.

Steering linkage: lubricate every 6000 miles.

Rear axle fluid: change every 12,000 miles.

Valve lash: adjust every 150,000 miles.

67189-RAMT-C24

SCHEDULED MAINTENANCE INTERVALS
2004 Turbo Diesel

TO BE SERVICED	TYPE OF SERVICE	VEHICLE MILEAGE INTERVAL (x1000)												
		7.5	15	22.5	30	37.5	45	52.5	60	67.5	75	82.5	90	97.5
Engine oil & filter	R	①	①②	①	①②	①	①②	①	①②	①	①②	①	①②	①
Outer tie rod ends 2500/3500 4x4 only	L	✓	✓	✓	✓	✓	✓	✓	✓	✓	✓	✓	✓	✓
Water pump weep hole	I		✓		✓		✓		✓		✓		✓	
Fuel filter	R		✓		✓		✓		✓		✓		✓	
Fuel sensor	C		✓		✓		✓		✓		✓		✓	
Drive belts	I				✓		✓			✓			✓	
Brake linings	I				✓		✓			✓			✓	
Parking brake	I/A				✓		✓			✓			✓	
Fan hub	I					✓			✓				✓	
Damper	I					✓			✓				✓	
Transfer case fluid	I					✓			✓				✓	
Front wheel bearings (2WD)	S/I					✓			✓				✓	
Engine coolant	R	Flush and replace every 60 months regardless of mileage												
Transfer case fluid	R	Every 120,000 miles												
Auto trans & filter	R	Every 100,000 miles												
Auto trans bands	Adj	Every 100,000 miles												
Valve lash	Adj	Every 150,000 miles												

R: Replace   S/I: Service or Inspect   D: Drain   C: Clean   Adj: Adjust   L: Lubricate

① California LEV 235 hp engines only

② Tier 1 EPA 250 and 305 hp. Engines

FREQUENT OPERATION MAINTENANCE (SEVERE SERVICE)

 If a vehicle is operated under any of the following conditions it is considered severe service:

- Extremely dusty areas.

- 50% or more of the vehicle operation is in 32°C (90°F) or higher temperatures, or constant operation in temperatures below 0°C (32°F).

- Prolonged idling (vehicle operation in stop and go traffic.

- Frequent short running periods (engine does not warm to normal operating temperatures).

- Police, taxi, delivery usage or trailer towing usage.

Oil & oil filter change: California LEV 235 hp engines every 3750 miles; EPA Tier 1 250 hp and 305 hp engines every 7500 miles

Rear axle fluid: change every 15,000 miles.

Front axle fluid (4x4): change every 15,000 miles.

Brake linings: inspect every 15,000 miles

Parking brake: inspect and adjust every 15,000 miles

Automatic transmission fluid, filter & bands: change & adjust every 30,000 miles.

Transfer case fluid: drain and refill every 60,000 miles

Engine air filter canister: clean every 135,000 miles

PRECAUTIONS

Before servicing any vehicle, please be sure to read all of the following precautions, which deal with personal safety, prevention of component damage, and important points to take into consideration when servicing a motor vehicle:

• Never open, service or drain the radiator or cooling system when the engine is hot; serious burns can occur from the steam and hot coolant.

• Observe all applicable safety precautions when working around fuel. Whenever servicing the fuel system, always work in a well-ventilated area. Do not allow fuel spray or vapors to come in contact with a spark, open flame, or excessive heat (a hot drop light, for example). Keep a dry chemical fire extinguisher near the work area. Always keep fuel in a container specifically designed for fuel storage; also, always properly seal fuel containers to avoid the possibility of fire or explosion. Refer to the additional fuel system precautions later in this section.

• Fuel injection systems often remain pressurized, even after the engine has been turned **OFF**. The fuel system pressure must be relieved before disconnecting any fuel lines. Failure to do so may result in fire and/or personal injury.

• Brake fluid often contains polyglycol ethers and polyglycols. Avoid contact with the eyes and wash your hands thoroughly after handling brake fluid. If you do get brake fluid in your eyes, flush your eyes with clean, running water for 15 minutes. If eye irritation persists, or if you have taken brake fluid internally, IMMEDIATELY seek medical assistance.

• The EPA warns that prolonged contact with used engine oil may cause a number of skin disorders, including cancer! You should make every effort to minimize your exposure to used engine oil. Protective gloves should be worn when changing oil. Wash your hands and any other exposed skin areas as soon as possible after exposure to used engine oil. Soap and water, or waterless hand cleaner should be used.

• All new vehicles are now equipped with an air bag system, often referred to as a Supplemental Restraint System (SRS) or Supplemental Inflatable Restraint (SIR) system. The system must be disabled before performing service on or around system components, steering column, instrument panel components, wiring and sensors. Failure to follow safety and disabling procedures could result in accidental air bag deployment, possible personal injury and unnecessary system repairs.

• Always wear safety goggles when working with, or around, the air bag system. When carrying a non-deployed air bag, be sure the bag and trim cover are pointed away from your body. When placing a non-deployed air bag on a work surface, always face the bag and trim cover upward, away from the surface. This will reduce the motion of the module if it is accidentally deployed. Refer to the additional air bag system precautions later in this section.

• Clean, high quality brake fluid from a sealed container is essential to the safe and proper operation of the brake system. You

should always buy the correct type of brake fluid for your vehicle. If the brake fluid becomes contaminated, completely flush the system with new fluid. Never reuse any brake fluid. Any brake fluid that is removed from the system should be discarded. Also, do not allow any brake fluid to come in contact with a painted surface; it will damage the paint.

• Never operate the engine without the proper amount and type of engine oil; doing so WILL result in severe engine damage.

• Timing belt maintenance is extremely important! Many models utilize an interference-type, non-freewheeling engine. If the timing belt breaks, the valves in the cylinder head may strike the pistons, causing potentially serious (also time-consuming and expensive) engine damage. Refer to the maintenance interval charts in the front of this manual for the recommended replacement interval for the timing belt, and to the timing belt section for belt replacement and inspection.

• Disconnecting the negative battery cable on some vehicles may interfere with the functions of the on-board computer system(s) and may require the computer to undergo a relearning process once the negative battery cable is reconnected.

• When servicing drum brakes, only disassemble and assemble one side at a time, leaving the remaining side intact for reference.

• Only an MVAC-trained, EPA-certified automotive technician should service the air conditioning system or its components.

GASOLINE ENGINE REPAIR

➡**Disconnecting the negative battery cable on some vehicles may interfere with the functions of the on board computer system. The computer may undergo a relearning process once the negative battery cable is reconnected.**

Distributor

➡**The 3.7L, 4.7L and 5.7L engines do not use a distributor.**

REMOVAL

3.9L, 5.2L and 5.9L Engines

1. Before servicing the vehicle, refer to the precautions in the beginning of this section.
2. Remove or disconnect the following:

• Negative battery cable
• Air cleaner tube
• Distributor cap
• Camshaft Position (CMP) sensor connector
3. Matchmark the distributor housing and the rotor.
4. Matchmark the distributor housing and the intake manifold.
5. Remove the distributor.

INSTALLATION

Timing Not Disturbed

3.9L, 5.2L AND 5.9L ENGINES

1. Before servicing the vehicle, refer to the precautions in the beginning of this section.

➡**The rotor will rotate clockwise as the gears engage.**

2. Position the rotor slightly counterclockwise of the matchmark made during removal.
3. Install the distributor.
4. Align the distributor housing and intake manifold matchmarks and check that the distributor housing matchmark and rotor are also aligned.
5. Install or connect the following:

• Distributor housing clamp and bolt. Tighten the bolt to 17 ft. lbs. (23 Nm).
• CMP sensor connector
• Distributor cap
• Air cleaner tube
• Negative battery cable

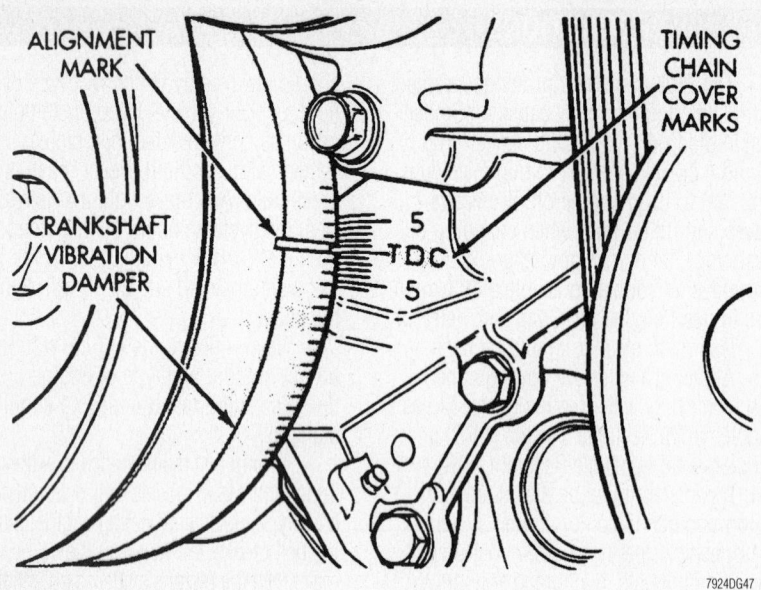

ALIGNMENT MARK

TIMING CHAIN COVER MARKS

CRANKSHAFT VIBRATION DAMPER

5
TDC
5

7924DG47

Crankshaft pulley and timing chain cover marks aligned at Top Dead Center (TDC)

Timing Disturbed

3.9L, 5.2L AND 5.9L ENGINES

1. Before servicing the vehicle, refer to the precautions in the beginning of this section.

2. Set the engine at Top Dead Center (TDC) of the No. 1 cylinder compression stroke.

3. Install the distributor, clamp and bolt.

4. Rotate the distributor so that the rotor aligns with the No. 1 cylinder mark on the Camshaft Position (CMP) sensor. Tighten the clamp bolt to 17 ft. lbs. (23 Nm).

5. Install or connect the following:
- CMP sensor connector
- Distributor cap
- Air cleaner tube
- Negative battery cable

6. The final distributor position must be set with a scan tool. Follow the instructions supplied by the scan tool manufacturer.

Alternator

REMOVAL

3.9L, 5.2L, 5.9L and 8.0L engines

1. Before servicing the vehicle, refer to the precautions in the beginning of this section.

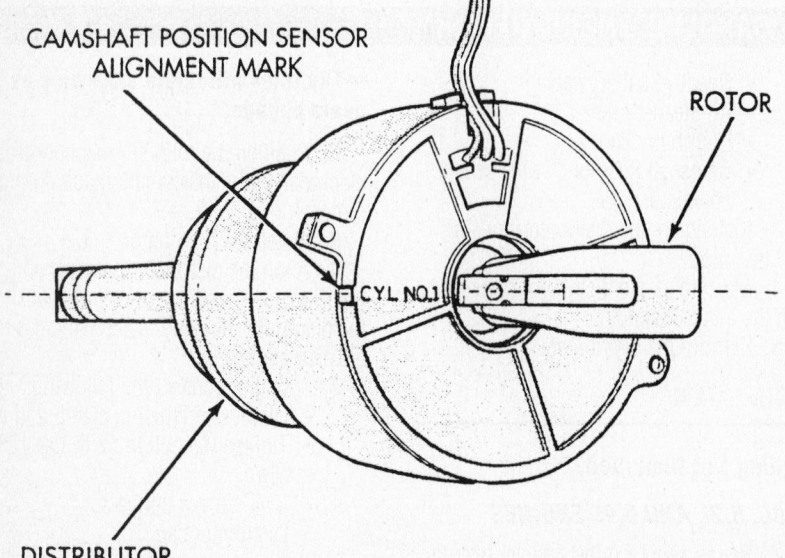

CAMSHAFT POSITION SENSOR ALIGNMENT MARK

ROTOR

CYL NO.1

DISTRIBUTOR

9308PG01

Distributor rotor alignment—3.9L, 5.2L and 5.9L engines

2. Remove or disconnect the following:
- Negative battery cable
- Accessory drive belt
- Alternator harness connectors
- Alternator

3.7L, 4.7L and 5.7L Engines

1. Before servicing the vehicle, refer to the precautions in the beginning of this section.

2. Remove or disconnect the following:
- Negative battery cable
- Accessory drive belt
- Alternator harness connectors
- Support bracket nuts and bolt
- Mounting bolts and alternator

➡ The 3.7L engine has 1 vertical and 2 horizontal bolts.

INSTALLATION

3.9L, 5.2L, 5.9L and 8.0L engines

1. Before servicing the vehicle, refer to the precautions in the beginning of this section.

2. Install the alternator and tighten the bolts to 30 ft. lbs. (41 Nm).

3. Install or connect the following:
- Alternator harness connectors
- Accessory drive belt
- Negative battery cable

3.7L and 4.7L engines

1. Before servicing the vehicle, refer to the precautions in the beginning of this section.

2. Install the alternator and tighten the bolts to the following specifications:
- 3.7L engine: Tighten the horizontal bolts to 42 ft. lbs. (57 Nm), then the vertical bolt to 29 ft. lbs. (40 Nm)
- 4.7L engine: Vertical bolt and long horizontal bolt to 41 ft. lbs. (56 Nm), short horizontal bolt to 55 ft. lbs. (74 Nm)

3. Install or connect the following:
- Alternator harness connectors
- Accessory drive belt
- Negative battery cable

5.7L engines

1. Before servicing the vehicle, refer to the precautions in the beginning of this section.

2. Install or connect the following:
- Alternator and tighten the bolts to 30 ft. lbs. (41 Nm)

- Support bracket and tighten the nuts and bolt to 30 ft. lbs. (41 Nm)
- Alternator harness connectors
- Accessory drive belt
- Negative battery cable

Ignition Timing

ADJUSTMENT

The ignition timing is controlled by the Powertrain Control Module (PCM) and is not adjustable.

Engine Assembly

REMOVAL & INSTALLATION

Ram Van

1. Before servicing the vehicle, refer to the precautions in the beginning of this section.
2. Drain the cooling system.
3. Recover the A/C refrigerant, if equipped.
4. Drain the engine oil.
5. Relieve the fuel system pressure.
6. Remove or disconnect the following:
- Negative battery cable
- Hood
- Grille
- Radiator support brace
- Inside engine cover
- Air cleaner assembly
- Transmission oil cooler, if equipped
- Radiator hoses
- Heater hoses
- Radiator and fan shroud
- A/C condenser
- Accessory drive belt
- Power steering pump
- Air injection pump
- Intake manifold vacuum lines
- Washer solvent bottle
- A/C compressor, if equipped
- Throttle linkage
- Engine control sensor harness connectors
- Alternator
- Cooling fan
- Distributor cap and spark plug wires
- Fuel line
- Throttle body
- Fuel supply manifold
- Intake manifold
- Exhaust front pipe
- Starter motor
- Transmission

- Engine mounts
- Engine

To install:

7. Install or connect the following:
- Engine. Tighten the mount bolts to 30 ft. lbs. (41 Nm) and the nuts to 75 ft. lbs. (101 Nm).
- Transmission
- Starter motor
- Exhaust front pipe
- Intake manifold
- Fuel supply manifold
- Throttle body
- Fuel line
- Distributor cap and spark plug wires
- Cooling fan
- Alternator
- Engine control sensor harness connectors
- Throttle linkage
- A/C compressor, if equipped
- Washer solvent bottle
- Intake manifold vacuum lines
- Air injection pump
- Power steering pump
- Accessory drive belt
- A/C condenser
- Radiator and fan shroud
- Heater hoses
- Radiator hoses
- Transmission oil cooler, if equipped
- Air cleaner assembly
- Inside engine cover
- Radiator support brace
- Grille
- Hood
- Negative battery cable

8. Fill the crankcase to the correct level.
9. Fill the cooling system.
10. Start the engine and check for leaks.

Ram Truck

3.9L, 5.2L, 5.9L AND 8.0L ENGINES

1. Before servicing the vehicle, refer to the precautions in the beginning of this section.
2. Drain the cooling system.
3. Drain the engine oil.
4. Relieve the fuel system pressure.
5. Remove or disconnect the following:
- Negative battery cable
- Hood
- Upper crossmember and top core support
- Radiator hoses
- Cooling fan and shroud
- Radiator
- Accelerator cable
- Cruise control cable, if equipped

- Transmission cable, if equipped
- Heater hoses
- Intake manifold vacuum lines
- Accessory drive belt
- Power steering pump
- A/C compressor, if equipped
- Engine control sensor harness connectors
- Engine block heater, if equipped
- Fuel line
- Exhaust front pipe
- Starter motor
- Torque converter, if equipped
- Transmission oil cooler lines, if equipped
- Engine mounts
- Transmission flange bolts. Support the transmission.
- Engine

✳✳ WARNING

Do not lift the engine by the intake manifold. Damage to the manifold may result.

To install:

6. Install or connect the following:
- Engine. Tighten the engine mount bolts to 70 ft. lbs. (95 Nm).
- Transmission flange bolts. Tighten the bolts to 40–45 ft. lbs. (54–61 Nm).
- Transmission oil cooler lines, if equipped
- Torque converter, if equipped. Tighten the bolts to 23 ft. lbs. (31 Nm).
- Starter motor
- Exhaust front pipe
- Fuel line
- Engine block heater, if equipped
- Engine control sensor harness connectors
- A/C compressor, if equipped
- Power steering pump
- Accessory drive belt
- Intake manifold vacuum lines
- Heater hoses
- Transmission cable, if equipped
- Cruise control cable, if equipped
- Accelerator cable
- Radiator
- Cooling fan and shroud
- Radiator hoses
- Upper crossmember and top core support
- Hood
- Negative battery cable

7. Fill the crankcase to the correct level.
8. Fill the cooling system.
9. Start the engine and check for leaks.

3.7L ENGINE

1. Before servicing the vehicle, refer to the precautions in the beginning of this section.

2. Properly relieve the fuel system pressure.

3. Drain the cooling system.

4. Drain the engine oil.

5. Remove or disconnect the following:
- Negative battery cable
- Hood
- Air cleaner assembly
- Radiator
- Electric and mechanical fan assemblies
- A/C compressor, if equipped, and secure it out of the way with the lines attached. DO NOT DISCHARGE!
- Power steering pump, with the lines attached
- Alternator
- Coolant bottle
- Heater hoses
- Accelerator and speed control cables
- Lower and upper radiator hoses
- Engine ground straps
- Intake Air Temperature (IAT) sensor
- Fuel injection wiring connectors
- Throttle Position (TP) sensor
- Idle Air Control (IAC) motor
- Oil pressure sender connector
- Engine Coolant Temperature (ECT) sensor
- Manifold Absolute Pressure (MAP) sensor
- Camshaft position sensor
- Ignition coil wiring connector
- Crankshaft Position (CKP) sensor
- Coil pack
- Fuel rail
- PCV hose
- Vacuum hoses from the intake manifold
- Knock sensor connectors
- Oil dipstick tube
- Intake manifold
- Heated Oxygen (HO2S) sensor connector
- Block heater connector
- Front driveshaft at the differential
- Starter
- Structural cover
- With a manual transmission, remove the transmission
- Torque converter bolts and match-mark the converter
- Automatic transmission-to-engine bolts
- Exhaust front pipes

- Left and right engine mounts

6. Place a support stand under the transmission.

7. Install an engine lift plate

8. Lift the engine out of the vehicle.

To install:

9. If equipped with a manual transmission, install the transmission

10. Lower the engine and install the mounts. Don't tighten the bolts yet.

11. If equipped with an automatic transmission, perform the following steps:

 a. Align the torque converter housing to the engine.

 b. Torque the bolts to 30 ft. lbs. (41 Nm).

12. Install or connect the following:
- Install the torque converter to flex-plate bolts. Torque the bolts to 50 ft. lbs. (68 Nm)
- Torque the through bolts to 45 ft. lbs. (61 Nm)
- Engine ground strap
- Starter motor
- CKP sensor
- Block heater cable
- Structural cover

✳✳ WARNING

The structural cover must be held tightly against the engine and bell-housing during tightening. The torque for all bolts is 40 ft. lbs. (54 Nm); the bolts must be tightened in the order shown.

- Exhaust pipes. New flange clamps MUST be used!

- HO2S sensor connectors
- KS sensors
- Intake Manifold
- Dipstick tube
- Vacuum hoses to the intake manifold
- PCV and breather hoses
- Fuel rail
- Ignition coil
- IAT sensor
- Fuel injector connectors
- TP sensor
- IAC motor
- Oil pressure sender
- ECT sensor electrical connector
- MAP sensor
- CMP sensor
- Radiator hoses
- Cruise control cable, if equipped
- Throttle cable
- Heater hoses
- Coolant bottle
- Power steering pump
- Alternator
- A/C compressor
- Radiator
- Fan assemblies
- Air cleaner assembly
- Negative battery cable

13. Fill and bleed the power steering system.

14. Fill the engine with clean oil.

15. Start the engine and check for leaks, repair if necessary.

4.7L ENGINE

1. Before servicing the vehicle, refer to the precautions in the beginning of this section.

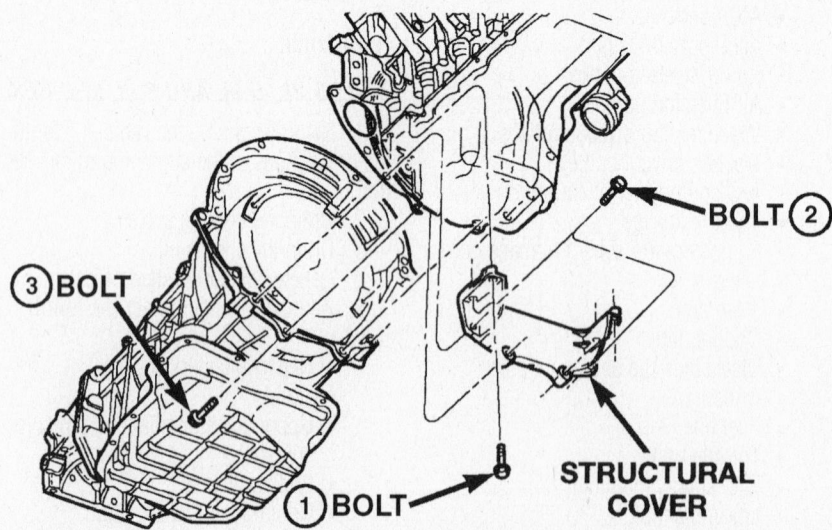

Tighten the structural cover bolts in this order—3.7L, 4.7L and 5.7L engines

67189-RAMT-CG01

2. Drain the cooling system and engine oil.
   3. Remove or disconnect the following:
   - Negative battery cable
   - Battery and tray
   - Exhaust crossover pipe
   - On 4wd, the axle vent tube
   - Left and right engine mount through bolts
   - On 4wd, the left and right engine mount bracket locknuts
   - Ground straps
   - CKP sensor
   - On 4wd, the axle isolator bracket
   - Structural cover
   - Starter
   - Torque converter bolts
   - Transmission-to-engine bolts
   - Engine block heater
   - Resonator and air inlet
   - Throttle and speed control cables
   - Crankcase breathers
   - A/C compressor
   - Shroud and fan assemblies
   - Transmission cooler lines
   - Radiator hoses
   - Radiator
   - Alternator
   - Heater hoses
   - Engine harness
   - Vacuum lines
   - Fuel system pressure
   - Fuel line at the rail
   - Power steering pump
   4. Install lifting eyes and take up the weight of the engine with a crane.
   5. Support the transmission with a jack.
   6. Remove the engine.
   7. To install
   8. Installation is the reverse of removal. Observe the following:
   - Left and right engine mount through bolts: 2wd 70 ft. lbs. (95 Nm); 4wd 75 ft. lbs. (102 Nm)
   - On 4wd, the bracket locknuts: 30 ft. lbs. (41 Nm)
   - Transmission-to-engine bolts: 30 ft. lbs. (41 Nm)

❄❄ WARNING

The structural cover must be held tightly against the engine and bellhousing during tightening. The torque for all bolts is 40 ft. lbs. (54 Nm); the bolts must be tightened in the order shown.

5.7L ENGINE

1. Before servicing the vehicle, refer to the precautions in the beginning of this section.

2. Drain the cooling system.
3. Drain the engine oil.
4. Relieve the fuel system pressure.
5. Remove or disconnect the following:
   - Negative battery cable
   - Hood
   - Air cleaner and resonator
   - Accessory drive belt
   - Engine fan
   - Radiator
   - Upper crossmember and top core support
   - A/C compressor, if equipped
   - Alternator
   - Intake manifold and IAFM as an assembly
   - Heater hoses
   - Power steering pump
   - Fuel line
   - Engine front mount thru-bolt nuts
   - Transmission oil cooler lines, if equipped
   - Exhaust pipes at the manifolds
   - Starter motor
   - Structural dust cover and transmission inspection cover
   - Torque converter-to-flexplate bolts
   - Transmission flange bolts. Support the transmission.
   - Engine

To install:

❄❄ WARNING

The structural cover must be held tightly against the engine and bellhousing during tightening. The torque for all bolts is 40 ft. lbs. (54 Nm); the bolts must be tightened in the order shown.

6. Install or connect the following:
   - Engine. Tighten the engine mount thru-bolt finger tight.
   - Transmission flange bolts. Tighten the bolts to 40–45 ft. lbs. (54–61 Nm). Then, tighten the mount bolt nuts to 70 ft. lbs. (95 Nm)
   - Transmission oil cooler lines, if equipped
   - Torque converter, if equipped. Tighten the bolts to 23 ft. lbs. (31 Nm).
   - Structural dust cover and transmission inspection cover
   - Starter motor
   - Fuel line
   - Power steering pump
   - Heater hoses
   - Intake manifold and IAFM as an assembly
   - Alternator

- A/C compressor, if equipped
   - Upper crossmember and top core support
   - Radiator
   - Engine fan
   - Accessory drive belt
   - Air cleaner and resonator
   - Hood
   - Negative battery cable
7. Fill the crankcase to the correct level.
8. Fill the cooling system.
9. Start the engine and check for leaks.

Water Pump

REMOVAL & INSTALLATION

3.7L and 4.7L Engines

1. Before servicing the vehicle, refer to the precautions in the beginning of this section.
2. Drain the cooling system.
3. Remove or disconnect the following:
   - Negative battery cable
   - Fan and clutch assembly from the pump
   - Fan shroud and fan assembly. If you're reusing the fan clutch, keep it upright to avoid silicone fluid loss!
   - Lower hose
   - 8 water pump bolts
4. Installation is the reverse of removal. Tighten the bolts, in sequence, to 40 ft. lbs. (54 Nm).

5.7L Engines

1. Before servicing the vehicle, refer to the precautions in the beginning of this section.
2. Drain the cooling system.
3. Remove or disconnect the following:
   - Negative battery cable
   - Accessory drive belt
   - Engine cooling fan
   - Coolant recovery bottle
   - Washer bottle
   - Fan shroud
   - A/C compressor and alternator brace
   - Idler pulleys
   - Belt tensioner
   - Radiator hoses
   - Heater hoses
   - Water pump
To install:
4. Install or connect the following:
   - Water pump and tighten the bolts to 18 ft. lbs. (24 Nm)
   - Heater hoses

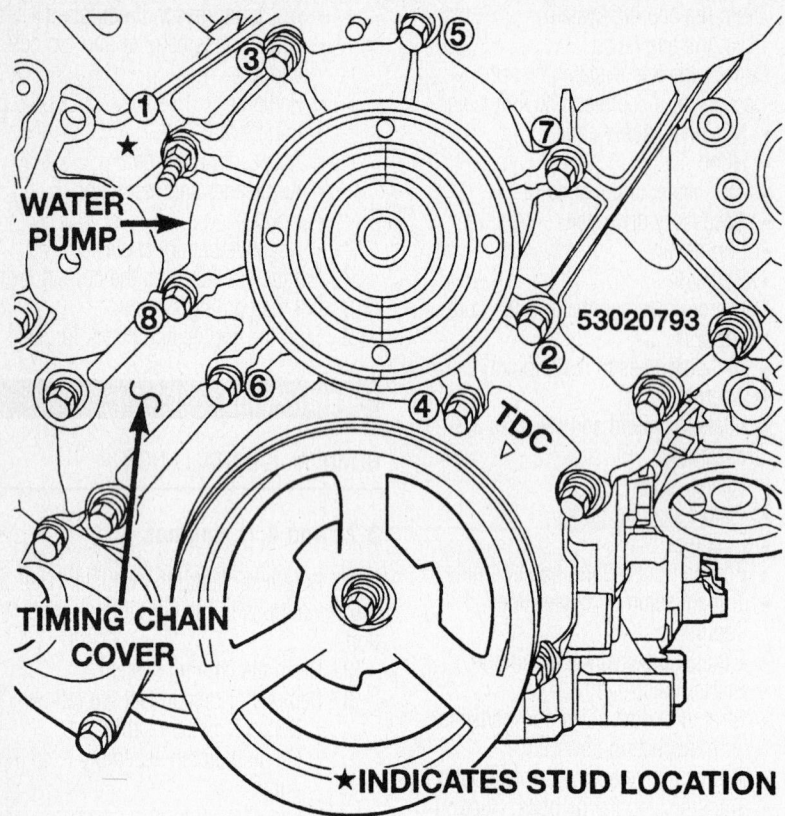

Water pump tightening sequence—3.7L and 4.7L engines

- Radiator hoses
- Idler pulleys
- A/C compressor and alternator brace
- Fan shroud
- Washer bottle
- Coolant recovery bottle
- Accessory drive belt
- Negative battery cable

5. Fill the cooling system.
6. Start the engine and check for leaks.

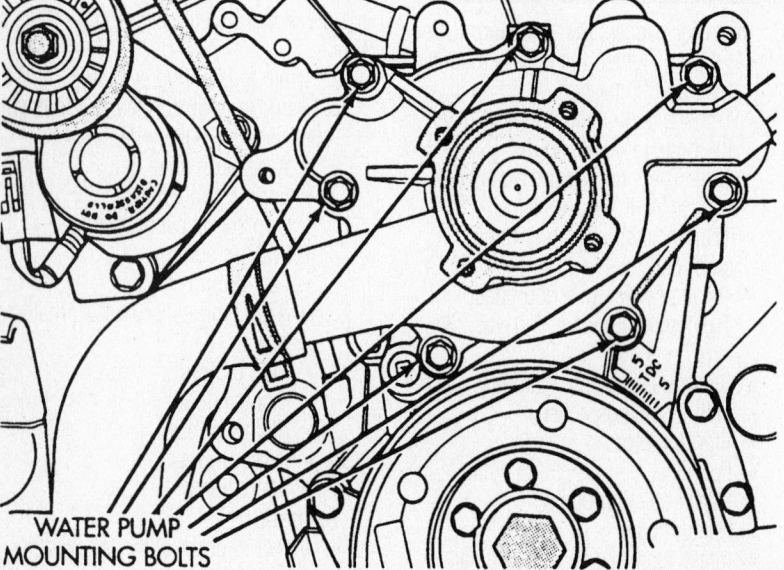

Water pump mounting bolt locations—3.9L, 5.2L and 5.9L engines, 8.0L engine is similar

3.9L, 5.2L and 5.9L Engines

1. Before servicing the vehicle, refer to the precautions in the beginning of this section.
2. Drain the cooling system.
3. Remove or disconnect the following:
   - Negative battery cable
   - Engine cooling fan and shroud

➡ **Do not store the fan clutch assembly horizontally, silicone may leak into the bearing grease and cause contamination.**

   - Accessory drive belt
   - Water pump pulley
   - Lower radiator hose
   - Heater hose and tube
   - Bypass hose
   - Water pump

To install:
4. Install or connect the following:
   - Water pump, using a new gasket. Tighten the bolts to 30 ft. lbs. (40 Nm).
   - Bypass hose
   - Heater hose and tube. Use a new O-ring seal.
   - Lower radiator hose
   - Water pump pulley. Tighten the bolts to 20 ft. lbs. (27 Nm).
   - Accessory drive belt
   - Engine cooling fan and shroud
   - Negative battery cable
5. Fill the cooling system.
6. Start the engine and check for leaks.

8.0L Engine

1. Before servicing the vehicle, refer to the precautions in the beginning of this section.
2. Drain the cooling system.
3. Remove or disconnect the following:
   - Negative battery cable
   - Washer solvent bottle
   - Upper radiator hose

➡ **The 8.0L engine is equipped with a fan clutch that threads directly onto the water pump shaft. This fan clutch is equipped with right-hand threads.**

➡ **Do not store the fan clutch assembly horizontally, silicone may leak into the bearing grease and cause contamination.**

   - Engine cooling fan and shroud
   - Accessory drive belt
   - Water pump pulley
   - Lower radiator hose
   - Heater hose
   - Bypass hose
   - Water pump

To install:

4. Install or connect the following:
   - Water pump. Use a new O-ring seal and tighten the bolts to 30 ft. lbs. (40 Nm).
   - Bypass hose
   - Heater hose
   - Lower radiator hose
   - Water pump pulley. Tighten the bolts to 16 ft. lbs. (22 Nm).
   - Accessory drive belt
   - Engine cooling fan and shroud
   - Upper radiator hose
   - Washer solvent bottle
   - Negative battery cable
5. Fill the cooling system.
6. Start the engine and check for leaks.

Heater Core

REMOVAL & INSTALLATION

Dodge Ram Pick-Up

1. Disconnect the negative battery cable.

✳✳ CAUTION

After disconnecting the negative battery cable, wait 2 minutes for the driver's/passenger's air bag system capacitor to discharge before attempting to do any work around the steering column or instrument

2. On diesel models, remove the passenger side battery and battery tray.

3. Remove the instrument panel by performing the following procedures:

   a. Remove the air bag control module (ACM) and bracket from the floor panel tunnel.

   b. Remove the trim from both cowl side inner panels.

   c. Remove the steering column-to-instrument panel opening cover.

   d. Remove the 2 hood latch release handle-to-instrument panel lower reinforcement screws and lower the handle to the floor.

   e. Disconnect the driver's side air bag module wiring harness connector from the lower instrument panel reinforcement.

   f. Place the wheels in the straight-ahead position and lock the steering wheel. Remove the steering column without disassembling it.

   g. From under the driver's side of the instrument, disconnect or remove the following items:
   - Parking brake release handle link-

age rod from the parking brake mechanism located on the left cowl side inner panel.
   - Instrument panel wiring harness connector from the parking brake switch located on the parking brake mechanism.
   - Three wiring harness connectors (body wiring harness, headlight, dash) from the 3 junction block connector receptacles located closest to the dash panel.
   - Head light/dash-to-instrument panel bulkhead wiring harness connector screw and disconnect the connector.
   - Instrument panel-to-door wiring harness connector located directly below the bulkhead wiring harness connector.
   - Infinity sound system wiring harness connector (if equipped), located at the outboard side of the

instrument panel bulkhead connector.
   - Stop light switch electrical connector.
   - Vacuum harness connector located near the left side of the heater/air conditioning housing.

   h. Under the passenger's side of the instrument panel, disconnect the radio antenna coaxial cable connector.

   i. Loosen both sides of the instrument panel cowl side roll-down bracket screws about ½ inch (13mm).

   j. Remove the 5 upper instrument panel-to-upper dash panel screws; remove the center screw last.

   k. Roll down the instrument panel and install a temporary hook in the center hole on top of the panel. Attach the other end to the center hole in the top of the dash panel. The opening should be approximately 18 inches (46cm).

   l. Disconnect the instrument panel-

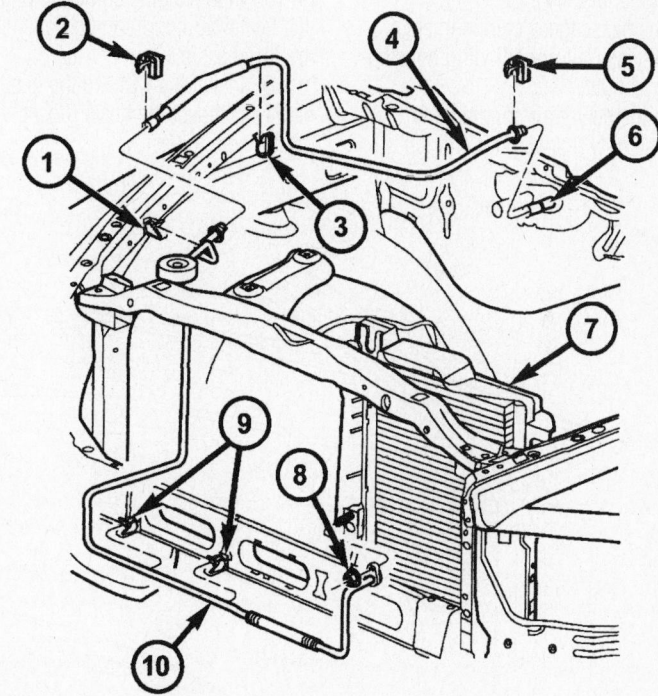

1 - BODY RETAINING CLIP
2 - SECONDARY RETAINING CLIP
3 - BODY RETAINING CLIP
4 - LIQUID LINE (REAR SECTION)
5 - SECONDARY RETAINING CLIP
6 - EVAPORATOR INLET TUBE
7 - A/C CONDENSER
8 - NUT
9 - BODY RETAINING CLIP
10 - LIQUID LINE (FRONT SECTION)

67189-RAMT-CG02

Exploded view of refrigerant line—Dodge Ram Pick-Up

to-heater/air conditioning housing assembly wiring harness connectors.

m. Using an assistant, remove the instrument panel from the vehicle.

4. If equipped with air conditioning, perform the following procedure:

a. Discharge and recover the air conditioning system refrigerant.

b. Disconnect the refrigerant lines from the evaporator. Plug the openings to prevent contamination.

c. Disconnect the refrigerant lines from the accumulator. Plug the openings to prevent contamination.

d. Remove the accumulator.

5. Drain the cooling system into a clean container for reuse.

6. Disconnect the heater hoses from the heater core tubes.

7. Remove the PCM from the dash panel and move it aside. Do not disconnect the PCM harness connector.

8. In the engine compartment, remove the heater/air conditioning housing assembly-to-chassis nuts.

9. In the passenger's compartment, remove the heater/air conditioning housing-to-dash panel nuts.

10. Pull the heater/air conditioning housing assembly rearward far enough to clear the studs and air conditioning drain tube holes.

11. Remove the heater/air conditioning housing assembly from the vehicle.

12. Remove the upper-to-lower heater/air conditioning housing screws and remove the upper housing.

13. Remove the heater core from the lower housing.

To install:

14. Install the heater core in the lower housing.

15. Install the upper housing and the upper-to-lower heater/air conditioning housing screws.

16. Install the heater/air conditioning housing assembly to the vehicle.

17. Push the heater/air conditioning housing assembly forward far enough to engage the studs and air conditioning drain tube holes.

18. In the passenger's compartment, install the heater/air conditioning housing-to-dash panel nuts.

19. In the engine compartment, install the heater/air conditioning housing assembly-to-chassis nuts.

20. Install the PCM to the dash panel.

21. Connect the heater hoses to the heater core tubes.

22. Refill the cooling system.

23. If equipped with air conditioning, perform the following procedure:

a. Install the accumulator.

b. Connect the refrigerant lines to the accumulator.

c. Connect the refrigerant lines from the evaporator.

d. Evacuate and charge the air conditioning system refrigerant.

24. Install the instrument panel by performing the following procedures:

a. Using an assistant, install the instrument panel to the vehicle.

b. Connect the instrument panel-to-heater/air conditioning housing assembly wiring harness connectors.

c. Roll-up the instrument panel and install a temporary hook in the center hole on top of the panel and the top of the dash panel. The opening should be approximately 18 inches (46cm).

d. Install the 5 upper instrument panel-to-upper dash panel screws; install the center screw first.

e. Install both sides of the instrument panel cowl side roll-down bracket screws to about ½ inch (13mm).

f. Under the passenger's side of the

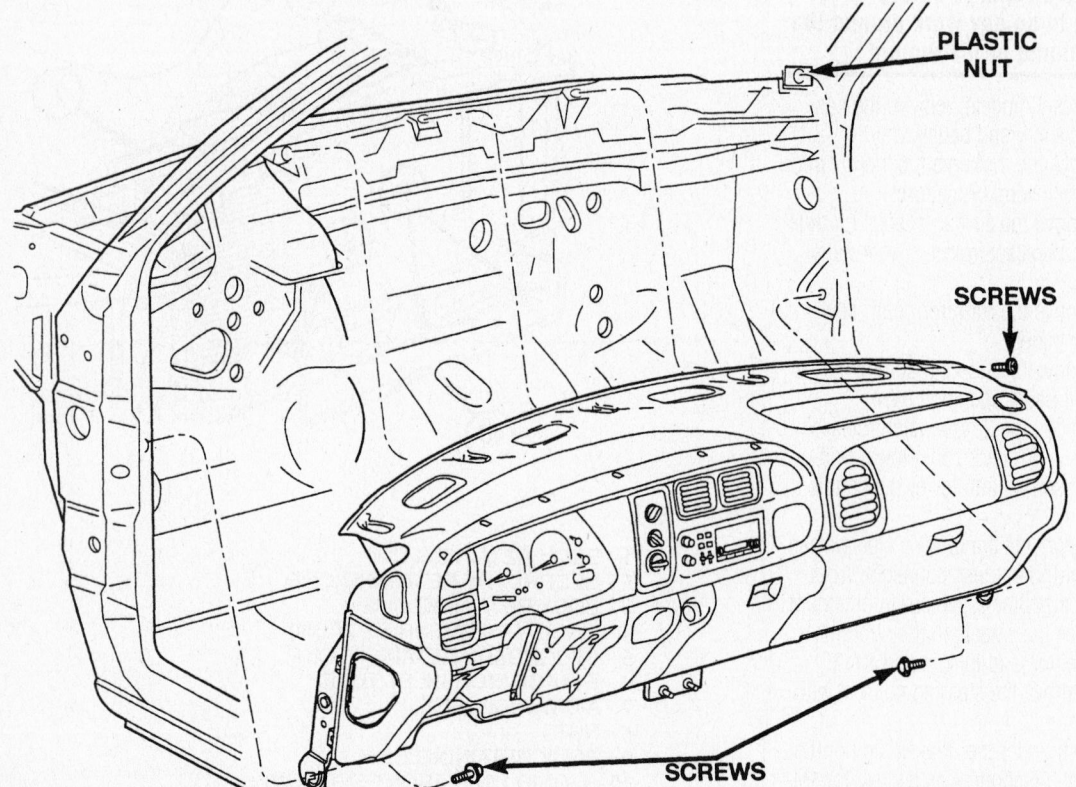

View of the instrument panel—Dodge Ram Pick-Up

93113GB5

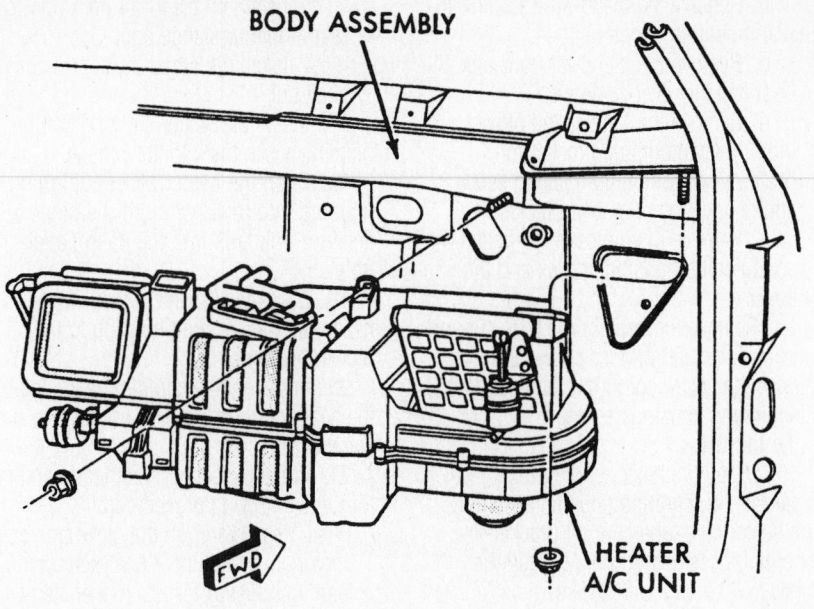

BODY ASSEMBLY

HEATER
A/C UNIT

FWD

93113GB6

View of the heater/air conditioning housing assembly—Dodge Ram Pick-Up

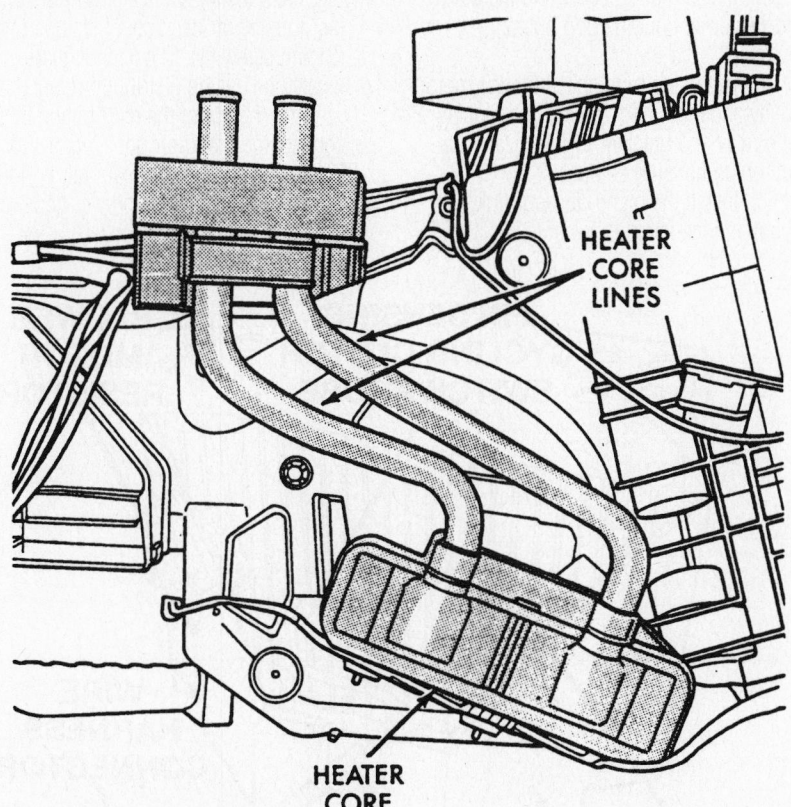

HEATER
CORE
LINES

HEATER
CORE

93113GB7

View of the heater core—Dodge Ram Pick-Up

instrument panel, connect the radio antenna coaxial cable connector.

g. Under the driver's side of the instrument, connect or install the following items:

- Vacuum harness connector located

near the left side of the heater/air conditioning housing.
- Stoplight switch electrical connector.
- Infinity sound system wiring harness connector (if equipped),

located at the outboard side of the instrument panel bulkhead connector.

- Instrument panel-to-door wiring harness connector located directly below the bulkhead wiring harness connector.
- Headlight/dash-to-instrument panel bulkhead wiring harness connector and install the screw.
- Three wiring harness connectors (body wiring harness, headlight, dash) from the 3 junction block connector receptacles located closest to the dash panel.
- Instrument panel wiring harness connector to the parking brake switch located on the parking brake mechanism.
- Parking brake release handle linkage rod to the parking brake mechanism located on the left cowl side inner panel.

h. Install the steering column.

i. Connect the driver's side air bag module wiring harness connector to the lower instrument panel reinforcement.

j. Install the 2 hood latch release handle and the handle-to-instrument panel lower reinforcement screws.

k. Install the steering column-to-instrument panel opening cover.

l. Install the trim to both cowl side inner panels.

m. Install the air bag control module (ACM) and bracket to the floor panel tunnel.

25. On diesel models, install the battery tray and battery.

26. Connect the negative battery cable.

27. Run the engine to normal operating temperatures; then, check the climate control operation and check for leaks.

Dodge Ram Van/Wagon

FRONT HEATER

1. Disconnect the negative battery cable.

✳✳ CAUTION

After disconnecting the negative battery cable, wait 2 minutes for the driver's/passenger's air bag system capacitor to discharge before attempting to do any work around the steering column or instrument

2. If equipped with air conditioning, perform the following procedure:

a. Discharge and recover the air conditioning system refrigerant.

b. Disconnect the suction line jumper from the evaporator

c. Remove the filter/drier and bracket-to-heater/air conditioning housing and cowl panel screws; then, move the filter/drier, bracket and refrigerant lines toward the center of the vehicle.

d. Disconnect the wire harness connector from the fin sensing cycling clutch switch.

e. Disconnect the wire harness connectors from the blower motor resistor and the high speed blower motor relay.

3. Cover the alternator to protect it from coolant spillage.

4. Drain the cooling system into a clean container for reuse.

5. Disconnect the heater hoses from the heater core tubes.

6. Remove the right headlight assembly, the grille panel, the right cowl grille support panel and the right radiator core support assembly.

7. Remove both lower heater/air conditioning housing flange-to-blower housing screws.

8. In the passenger's compartment, perform the following procedures:

a. From the top of the distribution duct, disconnect the blend-air door motor link.

b. Reach through the glove box opening and remove the heater/air conditioning housing-to-dash panel stamped nuts (2) and screw (1). The fasteners are located below the distribution duct.

c. Reach through the glove box opening; then, remove the 2 heater/air conditioning housing-to-dash panel stamped nuts. The nuts are located above the distribution duct.

9. In the engine compartment, remove the heater/air conditioning housing-to-dash stamped nut.

10. Carefully, separate the lower heater/air conditioning housing from the blower housing.

11. Pull the heater/air conditioning housing from the dash until the blend-air door link is clear of the dash panel hole.

12. Remove the heater/air conditioning housing assembly from the vehicle and place it on a bench with the top cover facing upward.

13. Disassemble the heater/air conditioning housing assembly by performing the following procedure:

a. Remove the blend-air door pivot shaft nut.

b. Remove the blend-air door link-to-rear mounting flange boot.

c. Remove the blend-air door link, the boot and the lever as a unit.

d. Remove the high speed blower motor relay mounting bracket-to-heater/air conditioning housing screw and remove the relay and bracket.

e. Remove the top cover-to-heater/air conditioning housing screws and the cover.

14. Remove the heater core tube support bracket-to-mounting boss screw and remove the heater core from the heater/air conditioning housing assembly.

To install:

15. Install the heater core to the heater/air conditioning housing assembly and heater core tube support bracket-to-mounting boss screw, then, tighten the screw to 20 inch lbs. (2.2 Nm).

16. Assemble the heater/air conditioning housing assembly by performing the following procedure:

a. Install the top cover and the cover-to-heater/air conditioning housing screws.

b. Install the high speed blower motor relay and mounting bracket, then install the relay and bracket -to-heater/air conditioning housing screw.

c. Install the blend-air door link, the boot and the lever as a unit.

d. Install the blend-air door link-to-rear mounting flange boot.

e. Install the blend-air door pivot shaft nut.

17. Install the heater/air conditioning housing assembly into the vehicle.

18. Push the heater/air conditioning housing into the dash until the blend-air door link falls into the dash panel hole.

19. Carefully, assemble the lower heater/air conditioning housing to the blower housing.

20. In the engine compartment, install the heater/air conditioning housing-to-dash stamped nut.

21. In the passenger's compartment, perform the following procedures:

a. Reach through the glove box opening; then, install the 2 heater/air conditioning housing-to-dash panel stamped nuts. The nuts are located above the distribution duct.

b. Reach through the glove box opening; then, install the heater/air conditioning housing-to-dash panel stamped nuts (2) and screw (1). The fasteners are located below the distribution duct.

c. At the top of the distribution duct, connect the blend-air door motor link.

22. Install both lower heater/air conditioning housing flange-to-blower housing screws.

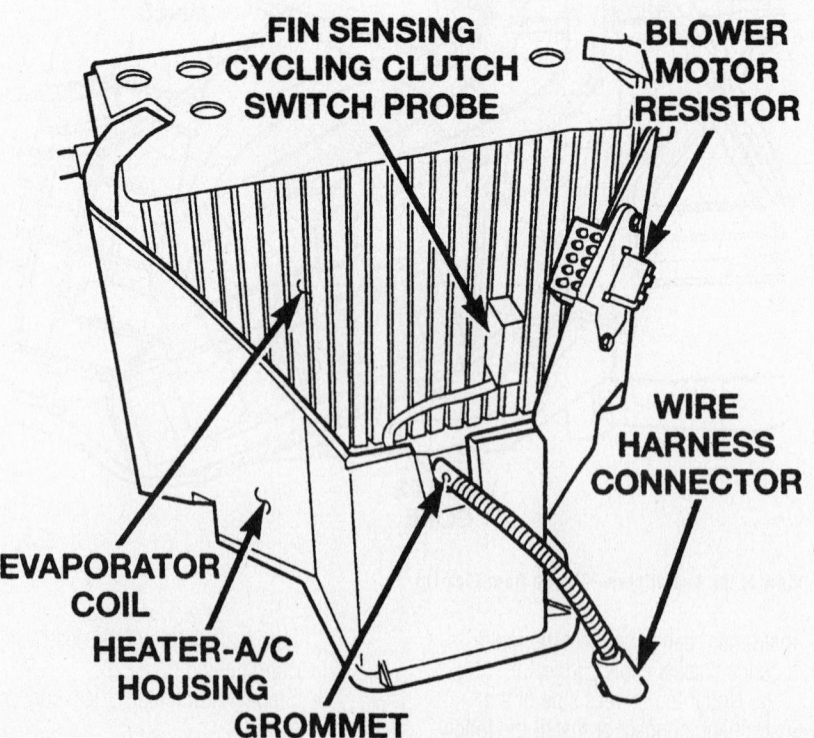

FIN SENSING CYCLING CLUTCH SWITCH PROBE

BLOWER MOTOR RESISTOR

WIRE HARNESS CONNECTOR

EVAPORATOR COIL

HEATER-A/C HOUSING

GROMMET

93113GA9

View of the air conditioning fin sensing cycling clutch switch—Dodge Ram Van/Wagon

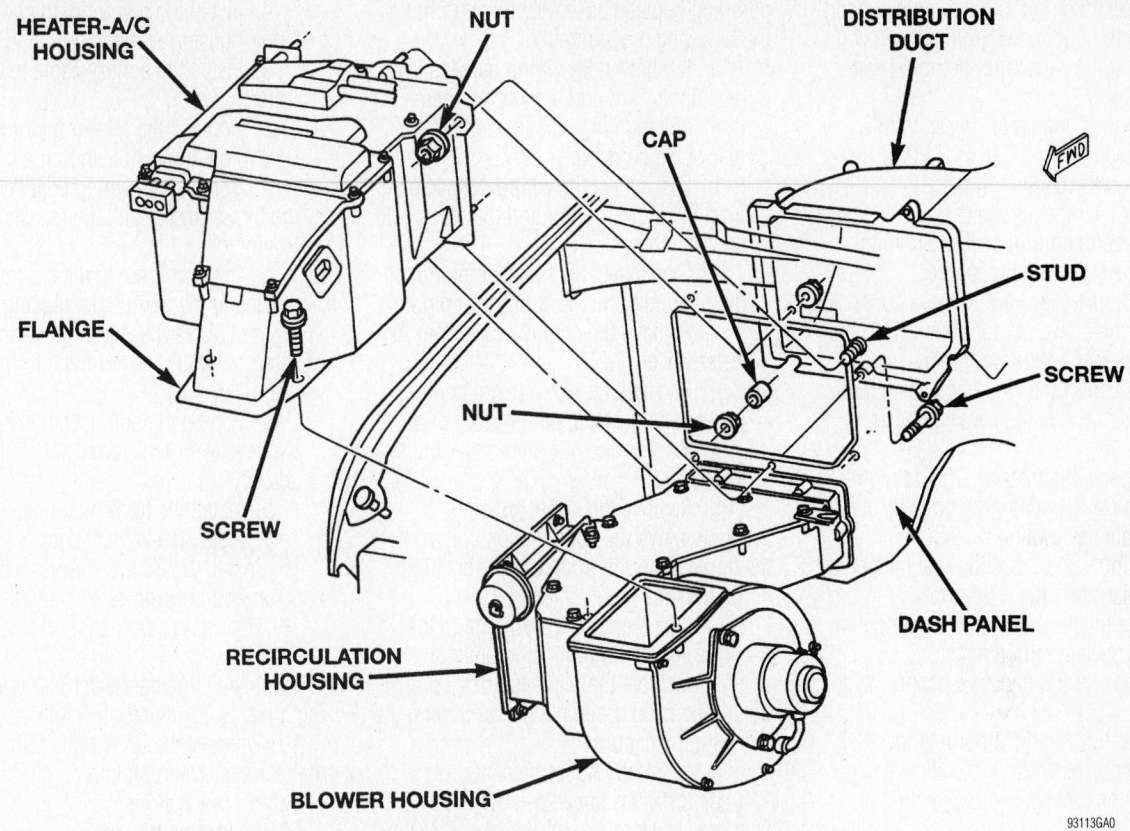

HEATER-A/C HOUSING

NUT

DISTRIBUTION DUCT

CAP

STUD

SCREW

FLANGE

NUT

SCREW

DASH PANEL

SCREW

RECIRCULATION HOUSING

BLOWER HOUSING

93113GA0

Exploded view of the heater/air conditioning housing and related components—Dodge Ram Van/Wagon

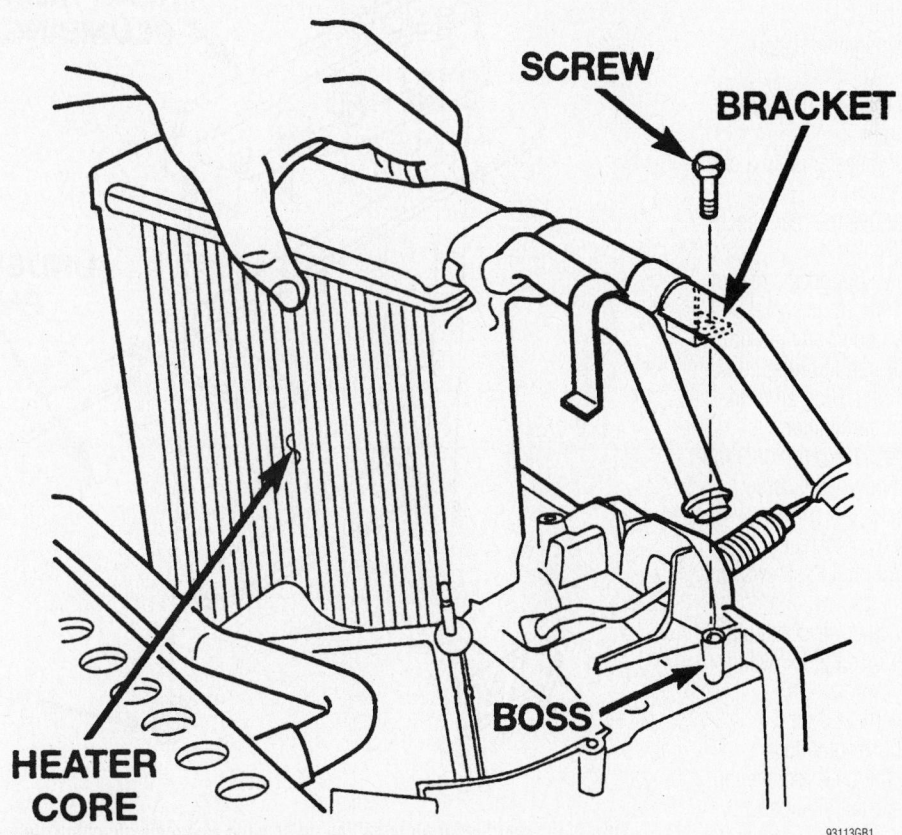

SCREW

BRACKET

HEATER CORE

BOSS

93113GB1

View of the heater core—Dodge Ram Van/Wagon

23. Install the right radiator core support assembly, the right cowl grille support panel, the grille panel and the right headlight assembly.

24. Connect the heater hoses to the heater core tubes.

25. Refill the cooling system.

26. Uncover the alternator.

27. If equipped with air conditioning, perform the following procedure:

a. Connect the wire harness connectors to the blower motor resistor and the high-speed blower motor relay.

b. Connect the wire harness connector to the fin sensing cycling clutch switch.

c. Install the filter/drier and bracket-to-heater/air conditioning housing and cowl panel screws.

d. Connect the suction line jumper to the evaporator

e. Evacuate and charge the air conditioning system refrigerant.

28. Connect the negative battery cable.

29. Run the engine to normal operating temperatures; then, check the climate control operation and check for leaks.

REAR AUXILIARY HEATER

The combination coil is used only with the optional rear heater/air conditioning housing assembly.

1. Disconnect the negative battery cable.

2. Discharge and recover the air conditioning system refrigerant.

3. Drain the cooling system into a clean container for reuse.

4. If equipped, remove the rear bench seat.

5. Raise and safely support the rear of the vehicle.

6. Disconnect the underbody plumbing from the rear heater/air conditioning housing plumbing connections. Plug all of the openings to prevent contamination.

7. Remove the rear heater/air conditioning housing-to-underbody panel screw.

8. Lower the vehicle.

9. Remove the 3 cover-to-rear heater/air conditioning housing screws and the cover.

10. From the rear heater/air conditioning housing assembly, perform the following procedures:

a. Remove the vertical duct.

b. Remove the horizontal duct.

c. From behind the unit, remove the ground wire eyelet-to-left side panel strainer screw.

11. Disassemble the rear heater/air con-

ditioning housing assembly by performing the following procedure:

a. Lift both relay wiring harness connectors upward and disconnect them from the mounting tabs located at the rear of the housing.

b. Disconnect the wiring connectors from the blower motor and the rear mode control motor.

c. Disconnect the blower motor cooling tube from the lower housing nipple.

d. Remove the control cable from the water valve.

e. Remove the rear inboard corner lower housing-to-upper housing clip.

f. Remove the upper-to-lower housing screws.

g. Remove the upper housing.

12. Remove the combination coil from the lower heater/air conditioning housing.

To install:

13. Install the combination coil to the lower heater/air conditioning housing.

14. Assemble the rear heater/air conditioning housing assembly by performing the following procedure:

a. Install the upper housing.

b. Install the upper-to-lower housing screws and torque to 20 inch lbs. (2.2 Nm).

c. Install the rear inboard corner lower housing-to-upper housing clip.

d. Install the control cable to the water valve.

e. Connect the blower motor cooling tube to the lower housing nipple.

f. Connect the wiring connectors to the blower motor and the rear mode control motor.

g. Connect them to the mounting tabs located at the rear of the housing.

15. At the rear heater/air conditioning housing assembly, perform the following procedures:

a. Behind the unit, install the ground wire eyelet-to-left side panel strainer screw.

b. Install the horizontal duct.

c. Install the vertical duct.

16. Install the 3 cover-to-rear heater/air conditioning housing screws and the cover.

17. Raise and safely support the rear of the vehicle.

18. Install the rear heater/air conditioning housing-to-underbody panel screw.

19. Connect the underbody plumbing to the rear heater/air conditioning housing plumbing connections.

20. Lower the vehicle.

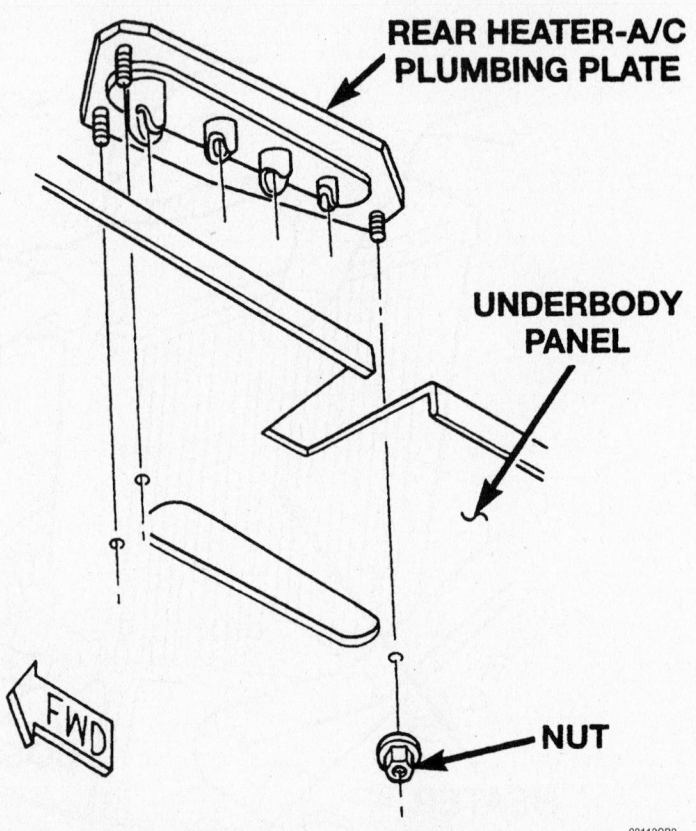

REAR HEATER-A/C PLUMBING PLATE

UNDERBODY PANEL

NUT

FWD

93113GB2

View of the rear heater/air conditioning housing assembly plumbing plate—Dodge Ram Van/Wagon

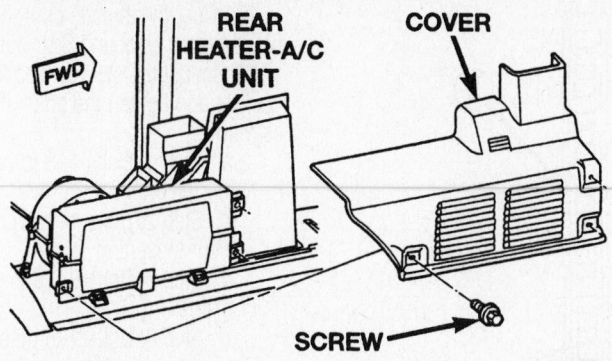

View of the rear heater/air conditioning housing assembly—Dodge Ram Van/Wagon

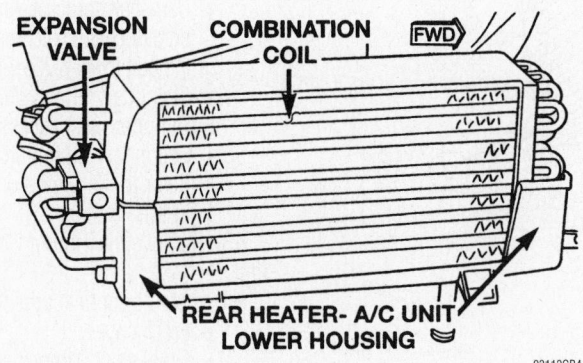

View of the combination coil—Dodge Ram Van/Wagon

21. If equipped, install the rear bench seat.
22. Refill the cooling system.
23. Connect the negative battery cable.
24. Evacuate and charge the air conditioning system refrigerant.
25. Run the engine to normal operating temperatures; then, check the climate control operation and check for leaks.

Cylinder Head

REMOVAL & INSTALLATION

3.7L Engine

LEFT SIDE

1. Before servicing the vehicle, refer to the precautions in the beginning of this section.
2. Drain the cooling system.
3. Properly relieve the fuel system pressure.
4. Remove or disconnect the following:
   - Negative battery cable
   - Exhaust Y-pipe
   - Intake manifold
   - Brake booster and master cylinder
   - Cylinder head cover
   - Engine cooling fan and shroud
   - Accessory drive belt
   - Power steering pump

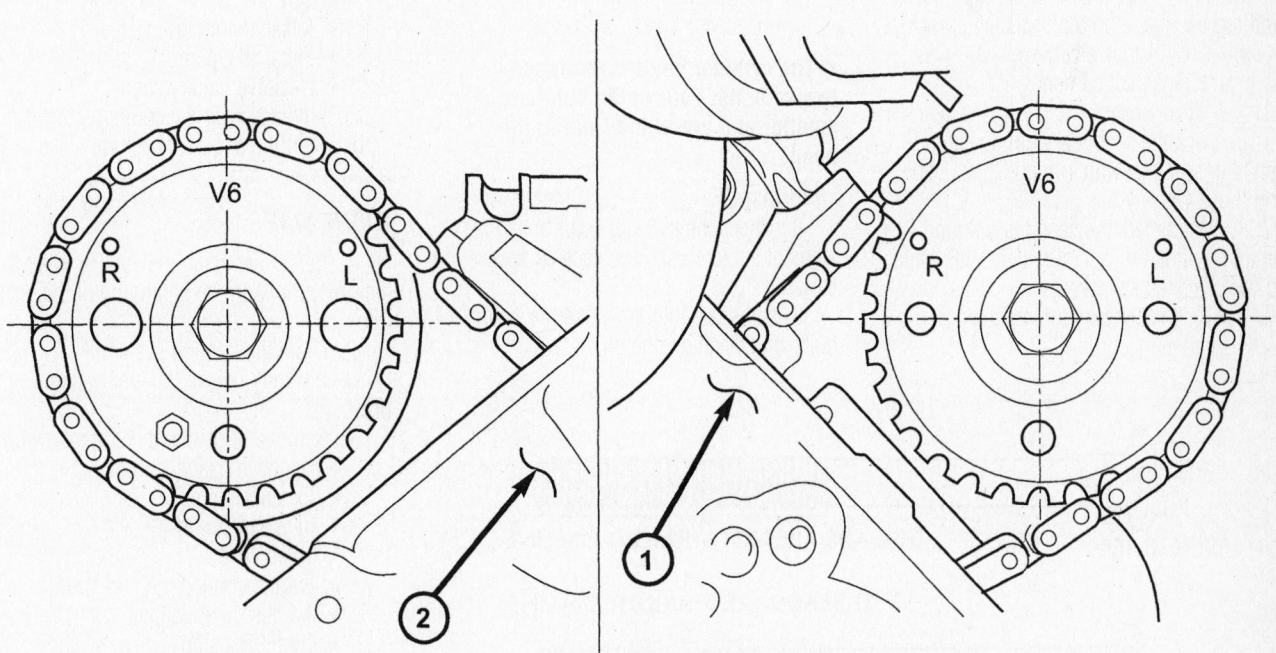

1 - LEFT CYLINDER HEAD
2 - RIGHT CYLINDER HEAD

Camshaft sprocket timing marks—3.7L

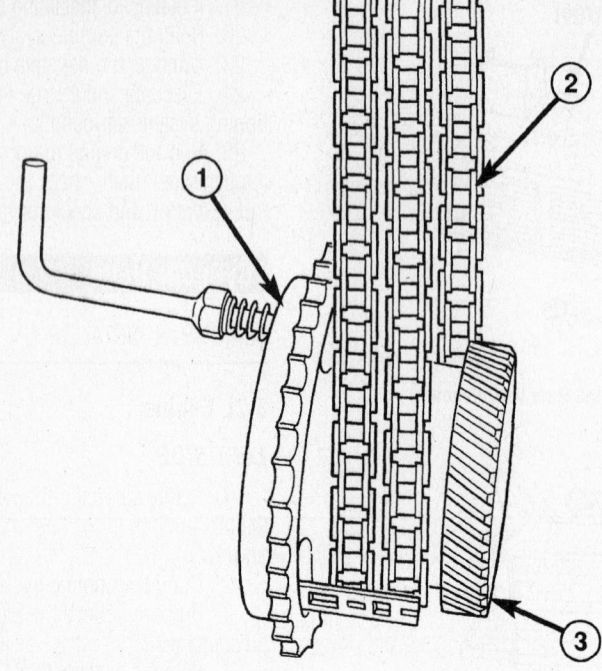

1 - SPECIAL TOOL 8429

2 - CAMSHAFT CHAIN

3 - CRANKSHAFT TIMING GEAR

9355PG05

Camshaft locking tool—3.7L

5. Rotate the crankshaft so that the crankshaft timing mark aligns with the Top Dead Center (TDC) mark on the front cover, and the **V6** marks on the camshaft sprockets are at 12 o'clock as shown.
- Crankshaft damper
- Front cover

6. Lock the secondary timing chain to the idler sprocket with Timing Chain Locking tool 8429.

7. Matchmark the secondary timing chain one link on each side of the V6 mark to the camshaft sprocket.
- Left secondary timing chain tensioner

- Cylinder head access plug
- Secondary timing chain guide
- Camshaft sprocket
- Cylinder head

➡**The cylinder head is retained by twelve bolts. Four of the bolts are smaller and are at the front of the head.**

To install:

8. Check the cylinder head bolts for signs of stretching and replace as necessary.

9. Lubricate the threads of the 11mm bolts with clean engine oil.

10. Coat the threads of the 8mm bolts with Mopar® Lock and Seal Adhesive.

11. Install the cylinder heads. Use new gaskets and tighten the bolts, in sequence, as follows:
   a. Step 1: Bolts 1–8 to 20 ft. lbs. (27 Nm)
   b. Step 2: Bolts 1–10 verify torque without loosening
   c. Step 3: Bolts 9–12 to 10 ft. lbs. (14 Nm)
   d. Step 4: Bolts 1–8 plus ¼ (90 degree) turn
   e. Step 5: Bolts 9–12 to 19 ft. lbs. (26 Nm)

12. Install or connect the following:
- Camshaft sprocket. Align the secondary chain matchmarks and tighten the bolt to 90 ft. lbs. (122 Nm).
- Secondary timing chain guide
- Cylinder head access plug
- Secondary timing chain tensioner. Refer to the timing chain procedure in this section.

13. Remove the Timing Chain Locking tool.

14. Install or connect the following:
- Front cover
- Crankshaft damper. Torque the bolt to 130 ft. lbs. (175 Nm).
- Power steering pump
- Accessory drive belt
- Engine cooling fan and shroud
- Cover
- Intake manifold
- Exhaust Y-pipe
- Negative battery cable

15. Fill and bleed the cooling system.

16. Start the engine, check for leaks and repair if necessary.

RIGHT SIDE

1. Before servicing the vehicle, refer to the precautions in the beginning of this section.

2. Drain the cooling system.

3. Properly relieve the fuel system pressure.

4. Remove or disconnect the following:
- Negative battery cable
- Exhaust Y-pipe
- Intake manifold
- Valve cover
- Engine cooling fan and shroud
- Accessory drive belt
- Oil fill housing
- Power steering pump

5. Rotate the crankshaft so that the crankshaft timing mark aligns with the Top Dead Center (TDC) mark on the front cover, and the **V6** marks on the camshaft sprockets are at 12 o'clock as shown.

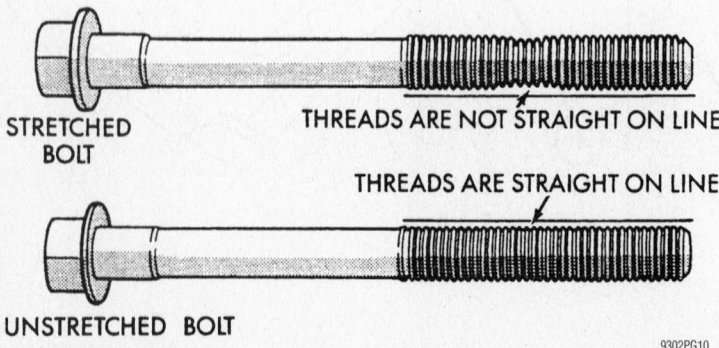

STRETCHED BOLT

THREADS ARE NOT STRAIGHT ON LINE

THREADS ARE STRAIGHT ON LINE

UNSTRETCHED BOLT

9302PG10

Examine the head bolts for signs of stretching—3.7L engine

LEFT BANK

RIGHT BANK

9355PG03

Cylinder head bolt torque sequence—3.7L

6. Remove or disconnect the following:
- Crankshaft damper
- Front cover

7. Lock the secondary timing chains to the idler sprocket with Timing Chain Locking tool 8429.

8. Matchmark the secondary timing chains to the camshaft sprockets.

9. Remove or disconnect the following:
- Secondary timing chain tensioners
- Cylinder head access plugs
- Secondary timing chain guides
- Camshaft sprockets
- Cylinder heads

➡ **Each cylinder head is retained by 8 11mm bolts and four 8mm bolts.**

To install:

10. Check the cylinder head bolts for signs of stretching and replace as necessary.

11. Lubricate the threads of the 11mm bolts with clean engine oil.

12. Coat the threads of the 8mm bolts with Mopar® Lock and Seal Adhesive.

13. Install the cylinder heads. Use new gaskets and tighten the bolts, in sequence, as follows:

a. Step 1: Bolts 1–8 to 20 ft. lbs. (27 Nm)

b. Step 2: Bolts 1–10 verify torque without loosening

c. Step 3: Bolts 9–12 to 10 ft. lbs. (14 Nm)

d. Step 4: Bolts 1–8 plus ¼ (90 degree) turn

e. Step 5: Bolts 9–12 to 19 ft. lbs. (26 Nm)

14. Install or connect the following:
- Camshaft sprockets. Align the secondary chain matchmarks and tighten the bolts to 90 ft. lbs. (122 Nm).
- Secondary timing chain guides
- Cylinder head access plugs
- Secondary timing chain tensioners. Refer to the timing chain procedure in this section.

15. Remove the Timing Chain Locking tool.

16. Install or connect the following:
- Front cover
- Crankshaft damper. Torque the bolt to 130 ft. lbs. (175 Nm).
- Rocker arms
- Power steering pump
- Oil fill housing
- Accessory drive belt
- Engine cooling fan and shroud
- Valve covers
- Intake manifold

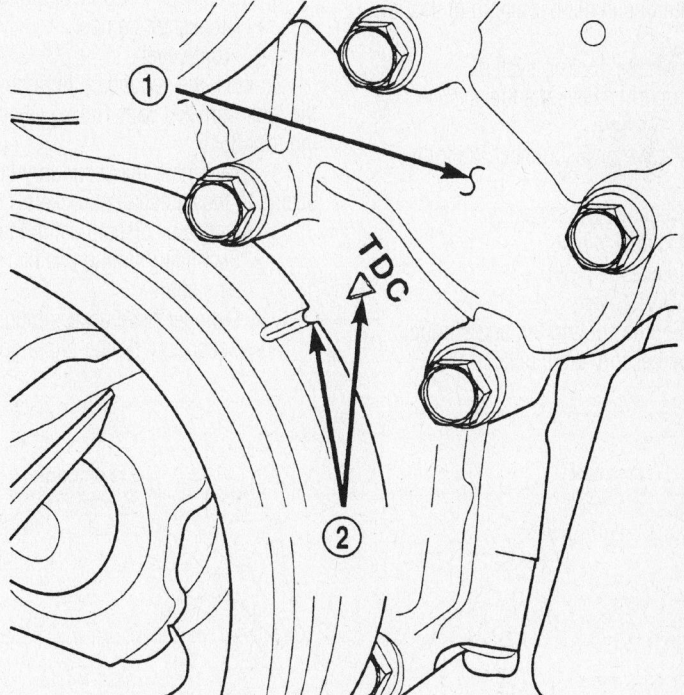

1 — TIMING CHAIN COVER
2 — CRANKSHAFT TIMING MARKS

9308PG04

Crankshaft timing marks—3.7L engine

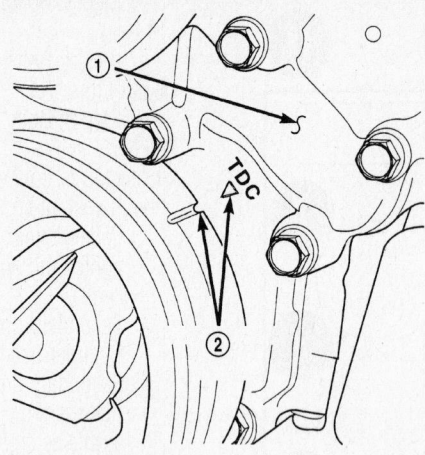

1 – TIMING CHAIN COVER
2 – CRANKSHAFT TIMING MARKS

9308PG04

Crankshaft timing marks—4.7L engine

- Exhaust Y-pipe
- Negative battery cable
17. Fill and bleed the cooling system.
18. Start the engine, check for leaks and repair if necessary.

4.7L Engine

1. Before servicing the vehicle, refer to the precautions in the beginning of this section.
2. Drain the cooling system.
3. Properly relieve the fuel system pressure.
4. Remove or disconnect the following:

- Negative battery cable
- Exhaust Y-pipe
- Intake manifold
- Valve covers
- Engine cooling fan and shroud
- Accessory drive belt

- Oil fill housing
- Power steering pump
- Rocker arms

5. Rotate the crankshaft so that the crankshaft timing mark aligns with the Top Dead Center (TDC) mark on the front cover, and the **V8** marks on the camshaft sprockets are at 12 o'clock as shown.
6. Remove or disconnect the following:

- Crankshaft damper
- Front cover

7. Lock the secondary timing chains to the idler sprocket with Timing Chain Locking tool 8515.
8. Matchmark the secondary timing chains to the camshaft sprockets.
9. Remove or disconnect the following:

- Secondary timing chain tensioners
- Cylinder head access plugs
- Secondary timing chain guides

- Camshaft sprockets
- Cylinder heads

➥ **Each cylinder head is retained by ten 11mm bolts and four 8mm bolts.**

To install:

10. Check the cylinder head bolts for signs of stretching and replace as necessary.
11. Lubricate the threads of the 11mm bolts with clean engine oil.
12. Coat the threads of the 8mm bolts with Mopar® Lock and Seal Adhesive.
13. Install the cylinder heads. Use new gaskets and tighten the bolts, in sequence, as follows:

    a. Step 1: Bolts 1–10 to 15 ft. lbs. (20 Nm)
    b. Step 2: Bolts 1–10 to 35 ft. lbs. (47 Nm)
    c. Step 3: Bolts 11–14 to 18 ft. lbs. (25 Nm)
    d. Step 4: Bolts 1–10 plus ¼ (90 degree) turn
    e. Step 5: Bolts 11–14 to 19 ft. lbs. (26 Nm)

14. Install or connect the following:

- Camshaft sprockets. Align the secondary chain matchmarks and tighten the bolts to 90 ft. lbs. (122 Nm).
- Secondary timing chain guides
- Cylinder head access plugs
- Secondary timing chain tensioners. Refer to the timing chain procedure in this section.

15. Remove the Timing Chain Locking tool 8515.
16. Install or connect the following:

- Front cover
- Crankshaft damper. Torque the bolt to 130 ft. lbs. (175 Nm).
- Rocker arms
- Power steering pump
- Oil fill housing
- Accessory drive belt
- Engine cooling fan and shroud
- Valve covers
- Intake manifold
- Exhaust Y-pipe
- Negative battery cable

17. Fill and bleed the cooling system.
18. Start the engine, check for leaks and repair if necessary.

5.7L Engine

1. Before servicing the vehicle, refer to the precautions in the beginning of this section.
2. Drain the cooling system.
3. Properly relieve the fuel system pressure.

12 O'CLOCK

V8

R   L

RIGHT CYLINDER HEAD

12 O'CLOCK

V8

R   L

LEFT CYLINDER HEAD

9302PG08

Camshaft positioning—4.7L engine

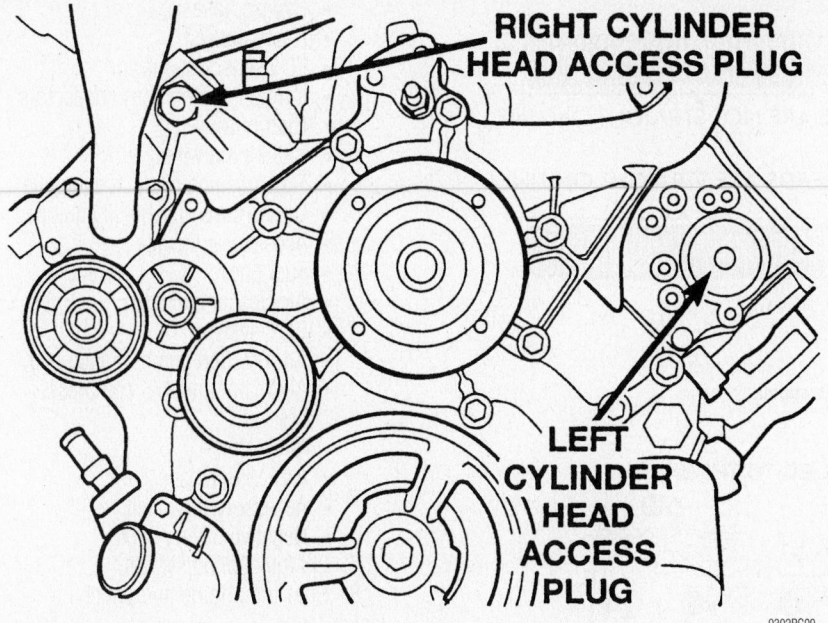

Cylinder head access plug locations—4.7L engine

9302PG09

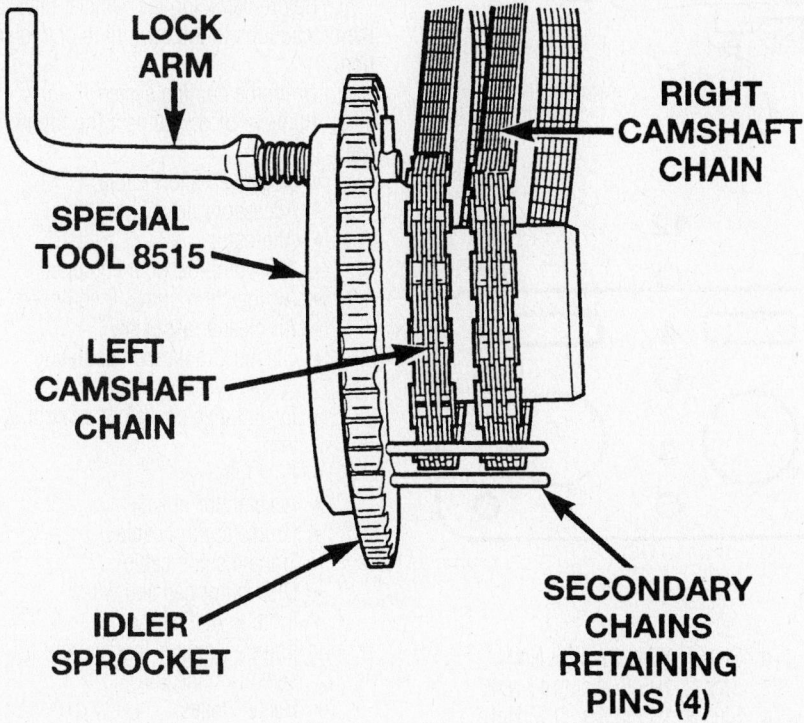

Use the special tool to lock the timing chains on the idler gear—4.7L engine

9302PG07

4. Remove or disconnect the following:
- Negative battery cable
- Air cleaner resonator and ducts
- Alternator
- Closed crankcase ventilation system
- EVAP control system
- Heater hoses

- Cylinder head covers
- Intake manifold
- Rocker arms and pushrods
- Cylinder heads

To install:

➡The head gaskets are not interchangeable. They are marked "L" and "R".

5. Install the cylinder heads. Use new gaskets and tighten the bolts, in sequence, as follows:

a. Step 1: 12 mm bolts–25 ft. lbs. (34 Nm); 8mm bolts–15 ft. lbs. (20 Nm)

b. Step 2: 12mm bolts–40 ft. lbs. (54 Nm); 8mm bolts retorque–15 ft. lbs. (20 Nm)

c. Step 3: 12mm bolts–plus 90 degrees; 8mm bolts–25 ft. lbs. (34 Nm)

6. Install or connect the following:
- Rocker arms and pushrods
- Intake manifold
- Heater hoses
- Alternator
- Cylinder head covers. Torque the studs and bolts to 70 inch lbs.
- Air cleaner resonator and ducts
- Negative battery cable

3.9L Engine

1. Before servicing the vehicle, refer to the precautions in the beginning of this section.
2. Relieve the fuel pressure.
3. Drain the cooling system.
4. Remove or disconnect the following:
- Negative battery cable
- Accessory drive belt
- Alternator
- A/C compressor, if equipped
- Alternator and A/C compressor bracket
- Air injection pump, if equipped
- Closed Crankcase Ventilation (CCV) system
- Air cleaner and hose
- Fuel line
- Accelerator linkage
- Cruise control cable, if equipped
- Transmission cable, if equipped
- Spark plug wires
- Distributor
- Ignition coil harness connectors
- Engine Coolant Temperature (ECT) sensor connector
- Heater hoses
- Bypass hose
- Intake manifold vacuum lines
- Fuel injector harness connectors
- Valve covers
- Intake manifold
- Exhaust front pipe
- Exhaust manifolds

➡Keep all valvetrain components in order for assembly.

- Rocker arms
- Pushrods
- Cylinder heads

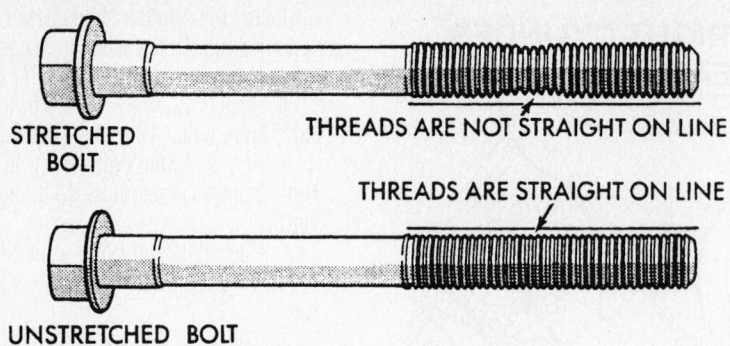

STRETCHED BOLT

THREADS ARE NOT STRAIGHT ON LINE

THREADS ARE STRAIGHT ON LINE

UNSTRETCHED BOLT

9302PG10

Examine the head bolts for signs of stretching—4.7L engine

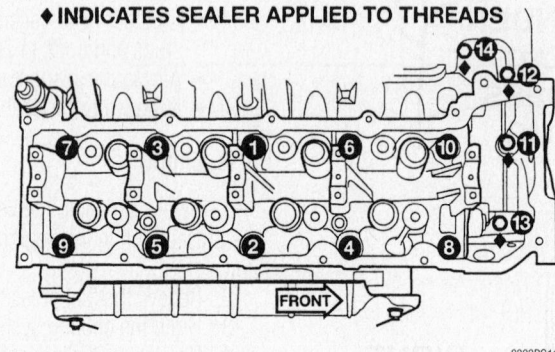

♦ INDICATES SEALER APPLIED TO THREADS

FRONT

9302PG11

Cylinder head torque sequence—4.7L engine

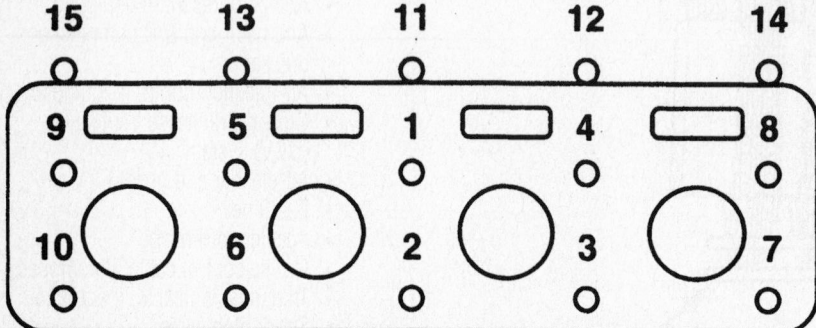

2399PG03

Cylinder head torque sequence—5.7L engine

To install:

✷✷ WARNING

Position the crankshaft so that no piston is at Top Dead Center (TDC) prior to installing the cylinder heads. Do not rotate the crankshaft during or immediately after rocker arm installation. Wait 5 minutes for the hydraulic lash adjusters to bleed down.

5. Install the cylinder heads with new gaskets. Tighten the bolts in sequence as follows:

a. Step 1: 50 ft. lbs. (68 Nm).
b. Step 2: 105 ft. lbs. (143 Nm).
c. Step 3: 105 ft. lbs. (143 Nm).

6. Install or connect the following:
- Pushrods in their original locations
- Rocker arms in their original locations. Tighten the bolts to 21 ft. lbs. (28 Nm).
- Exhaust manifolds
- Exhaust front pipe
- Intake manifold
- Valve covers
- Fuel injector harness connectors
- Intake manifold vacuum lines
- Bypass hose
- Heater hoses
- ECT sensor connector
- Ignition coil harness connectors
- Distributor
- Spark plug wires
- Transmission cable, if equipped
- Cruise control cable, if equipped
- Accelerator linkage
- Fuel line
- Air cleaner and hose
- CCV system
- Air injection pump, if equipped
- Alternator and A/C compressor bracket
- Alternator
- A/C compressor
- Accessory drive belt
- Negative battery cable

7. Fill the cooling system.
8. Start the engine and check for leaks.

5.2L and 5.9L Engines

1. Before servicing the vehicle, refer to the precautions in the beginning of this section.
2. Drain the cooling system.
3. Remove or disconnect the following:

- Negative battery cable
- Accessory drive belt
- Alternator
- A/C compressor, if equipped
- Air injection pump, if equipped
- Air cleaner assembly
- Closed Crankcase Ventilation (CCV) system
- Evaporative emissions control system
- Fuel line
- Accelerator linkage
- Cruise control cable
- Transmission cable
- Distributor cap and wires
- Ignition coil wiring
- Engine Coolant Temperature (ECT) sensor connector
- Heater hoses
- Bypass hose
- Upper radiator hose
- Intake manifold
- Valve covers

➡Keep valvetrain components in order for reassembly.

- Rocker arms
- Pushrods
- Exhaust manifolds
- Spark plugs
- Cylinder heads

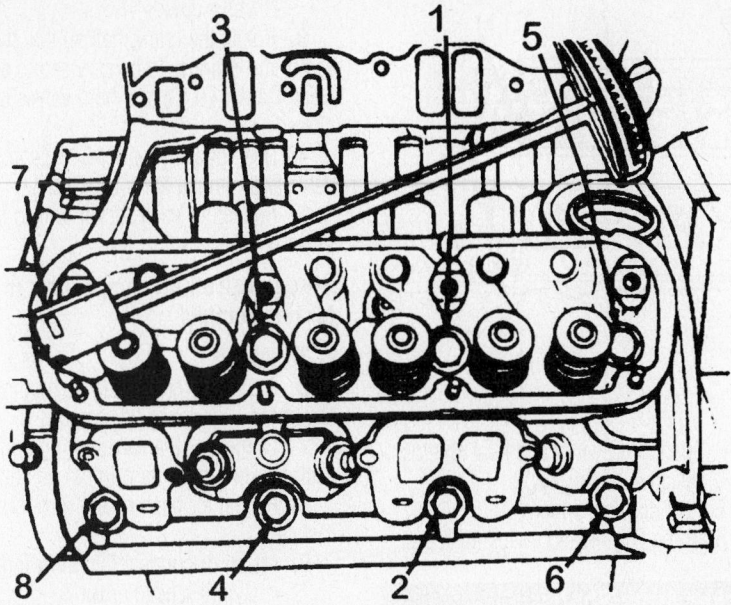

Cylinder head torque sequence—3.9L engine

To install:

> ※※ **WARNING**
>
> **Position the crankshaft so that no piston is at Top Dead Center (TDC) prior to installing the cylinder heads. Do not rotate the crankshaft during or immediately after rocker arm installation. Wait 5 minutes for the hydraulic lash adjusters to bleed down.**

4. Install the cylinder heads. Use new gaskets and tighten the bolts in sequence as follows:

   a. Step 1: 50 ft. lbs. (68 Nm).
   b. Step 2: 105 ft. lbs. (143 Nm).
   c. Step 3: 105 ft. lbs. (143 Nm).

5. Install or connect the following:
- Spark plugs
- Exhaust manifolds
- Pushrods and rocker arms in their original positions
- Valve covers
- Intake manifold
- Upper radiator hose
- Bypass hose
- Heater hoses
- ECT sensor connector
- Ignition coil wiring
- Distributor cap and wires
- Transmission cable
- Cruise control cable
- Accelerator linkage
- Fuel line
- Evaporative emissions control system

- CCV system
- Air cleaner assembly
- A/C compressor, if equipped
- Air injection pump, if equipped
- Alternator
- Accessory drive belt
- Negative battery cable

6. Fill the cooling system.
7. Start the engine and check for leaks.

8.0L Engine

1. Before servicing the vehicle, refer to the precautions in the beginning of this section.

2. Drain the cooling system.
3. Relieve the fuel system pressure.
4. Remove or disconnect the following:
- Negative battery cable
- Spark plug wires and heat shields
- Accessory drive belt
- Alternator
- A/C compressor, if equipped
- Alternator and A/C compressor bracket
- Air injection pump, if equipped
- Closed Crankcase Ventilation (CCV) system
- Evaporative Emissions (EVAP) canister vacuum lines
- Air cleaner and hose
- Fuel line
- Accelerator linkage
- Cruise control cable, if equipped
- Transmission cable, if equipped
- Ignition coil pack and bracket
- Temperature gauge sender connector
- Heater hoses
- Bypass hose
- Upper intake manifold
- Valve covers
- Exhaust Gas Recirculation (EGR) tube
- Lower intake manifold
- Exhaust front pipe
- Exhaust manifolds

➡**Keep all valvetrain components in order for assembly.**

- Rocker arms
- Pushrods
- Cylinder heads

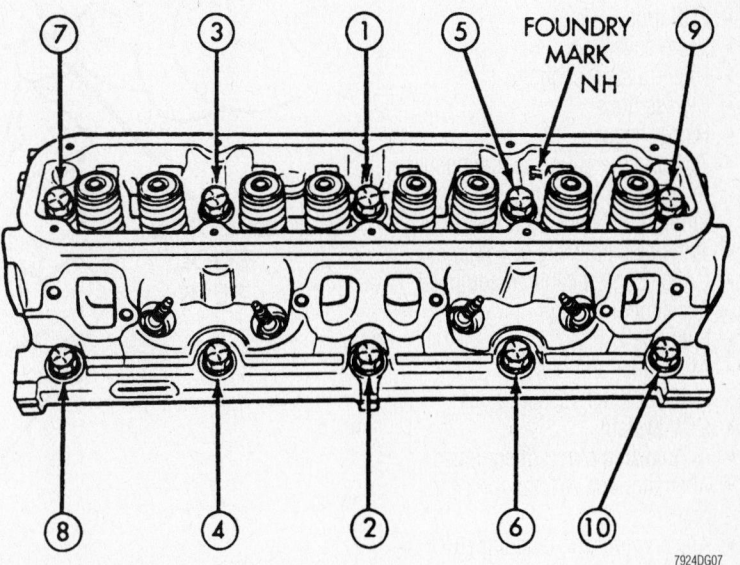

Cylinder head torque sequence—5.2L and 5.9L engines

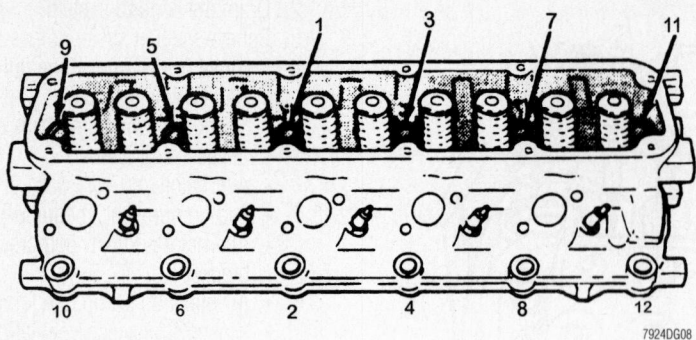

7924DG08

Cylinder head torque sequence—8.0L engine

To install:

❋❋ WARNING

Position the crankshaft so that no piston is at Top Dead Center (TDC) prior to installing the cylinder heads. Do not rotate the crankshaft during or immediately after rocker arm installation. Wait 5 minutes for the hydraulic lash adjusters to bleed down.

5. Install the cylinder heads with new gaskets. Tighten the bolts in sequence as follows:
   a. Step 1: 43 ft. lbs. (58 Nm).
   b. Step 2: 105 ft. lbs. (143 Nm).
6. Install or connect the following:
   • Pushrods in their original locations
   • Rocker arms in their original locations. Tighten the bolts to 21 ft. lbs. (28 Nm).
   • Exhaust manifolds
   • Exhaust front pipe
   • Lower intake manifold
   • EGR tube
   • Valve covers
   • Upper intake manifold
   • Bypass hose
   • Heater hoses
   • Temperature gauge sender connector
   • Ignition coil pack and bracket
   • Transmission cable, if equipped
   • Cruise control cable, if equipped
   • Accelerator linkage
   • Fuel line
   • Air cleaner and hose
   • EVAP canister vacuum lines
   • CCV system
   • Air injection pump, if equipped
   • Alternator and A/C compressor bracket
   • A/C compressor, if equipped
   • Alternator
   • Accessory drive belt

• Spark plug wires and heat shields
• Negative battery cable
7. Fill the cooling system.
8. Start the engine and check for leaks.

Rocker Arms/Shafts

REMOVAL & INSTALLATION

3.7L and 4.7L Engines

1. Before servicing the vehicle, refer to the precautions in the beginning of this section.
2. Remove or disconnect the following:
   • Negative battery cable

• Valve covers
3. Rotate the crankshaft so that the piston of the cylinder to be serviced is at Top Dead Center (TDC) and both valves are closed.
4. Use special tool 8516 to depress the valve and remove the rocker arm.
5. Repeat for each rocker arm to be serviced.

➡**Keep valvetrain components in order for reassembly.**

To install:

6. Rotate the crankshaft so that the piston of the cylinder to be serviced is at BDC.
7. Compress the valve spring and install each rocker arm in its original position.
8. Repeat for each rocker arm to be installed.
9. Install or connect the following:
   • Cylinder head cover
   • Negative battery cable

3.9L Engine

1. Before servicing the vehicle, refer to the precautions in the beginning of this section.
2. Remove or disconnect the following:
   • Negative battery cable
   • Valve covers
   • Rocker arms

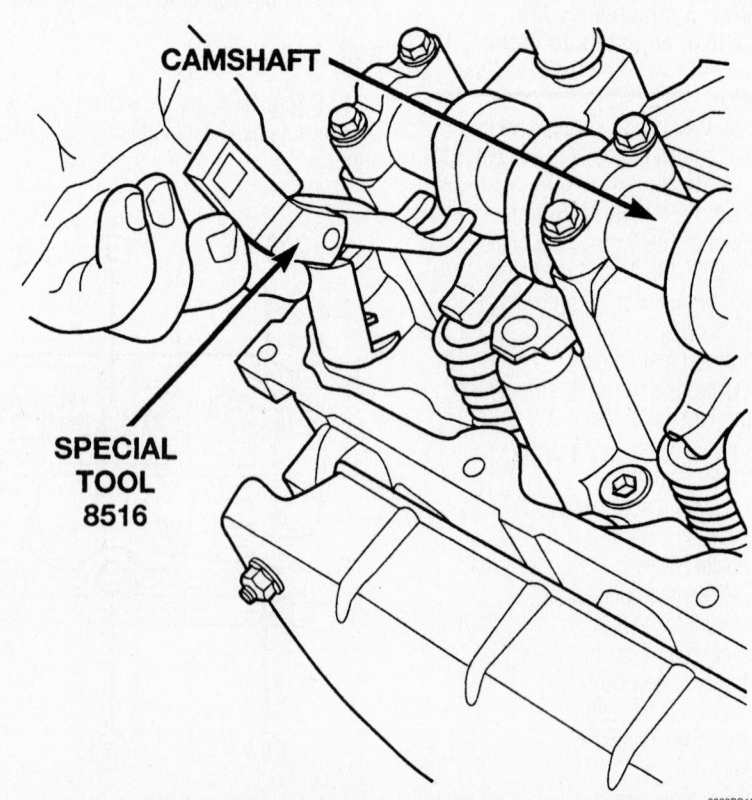

9302PG13

Rocker arm service—3.7L and 4.7L engines

➡Keep all valvetrain components in order for assembly.

To install:

3. Rotate the crankshaft so that the **V6** mark on the crankshaft damper aligns with the timing mark on the front cover. The **V6** mark is located 147 degrees **AFTER** Top Dead Center (TDC).

4. Install the rocker arms in their original positions and tighten the bolts to 21 ft. lbs. (28 Nm).

✳✳ CAUTION

Do not rotate the crankshaft during or immediately after rocker arm installation. Wait 5 minutes for the hydraulic lash adjusters to bleed down.

5. Install or connect the following:
   • Valve covers
   • Negative battery cable

5.7L Engines

1. Before servicing the vehicle, refer to the precautions in the beginning of this section.

2. Remove or disconnect the following:
   • Negative battery cable
   • Cylinder head cover

3. Loosen the rocker shaft bolts as follows: center, center left, center right, left, right.

➡**The rocker shaft assemblies are not interchangeable. The intake side is marked "I". The shorter pushrods are used on the intake side.**

To install:

4. Install or connect the following:
   • Pushrods
   • Rocker shaft assemblies. Torque the shafts to 16 ft. lbs. (22 Nm) in this sequence: center, center right, center left, right, left.

✳✳ WARNING

Do not rotate the crankshaft during or immediately after rocker shaft installation. Allow the roller tappets to leakdown for about 5 minutes.

   • Cylinder head covers. Torque the fasteners to 70 inch lbs.

5.2L and 5.9L Engines

1. Before servicing the vehicle, refer to the precautions in the beginning of this section.

---

2. Remove or disconnect the following:
   • Negative battery cable
   • Valve covers
   • Rocker arms

➡Keep valvetrain components in order for reassembly.

To install:

3. Rotate the crankshaft so that the **V8** mark on the crankshaft damper aligns with the timing mark on the front cover. The **V8** mark is located 147 degrees **AFTER** Top Dead Center (TDC).

4. Install the rocker arms in their original positions and tighten the bolts to 21 ft. lbs. (28 Nm).

✳✳ CAUTION

Do not rotate the crankshaft during or immediately after rocker arm installation. Wait 5 minutes for the hydraulic lash adjusters to bleed down.

5. Install or connect the following:
   • Valve covers
   • Negative battery cable

8.0L Engine

1. Before servicing the vehicle, refer to the precautions in the beginning of this section.

2. Remove or disconnect the following:
   • Negative battery cable
   • Valve covers
   • Rocker arms

➡Keep valvetrain components in order for reassembly.

To install:

✳✳ CAUTION

When installing the rocker arms, ensure that the piston is not at Top Dead Center. Tighten the rocker arm bolts slowly. Do not rotate the crankshaft during or immediately after rocker arm installation. Wait 5 minutes for the hydraulic lash adjusters to bleed down.

3. Install or connect the following:
   • Rocker arms in their original positions. Tighten the bolts to 21 ft. lbs. (28 Nm).
   • Valve covers
   • Negative battery cable

Intake Manifold

REMOVAL & INSTALLATION

3.7L and 4.7L Engines

1. Before servicing the vehicle, refer to the precautions in the beginning of this section.

2. Drain the cooling system.

3. Properly relieve the fuel system pressure.

4. Remove or disconnect the following:
   • Negative battery cable
   • Air cleaner assembly

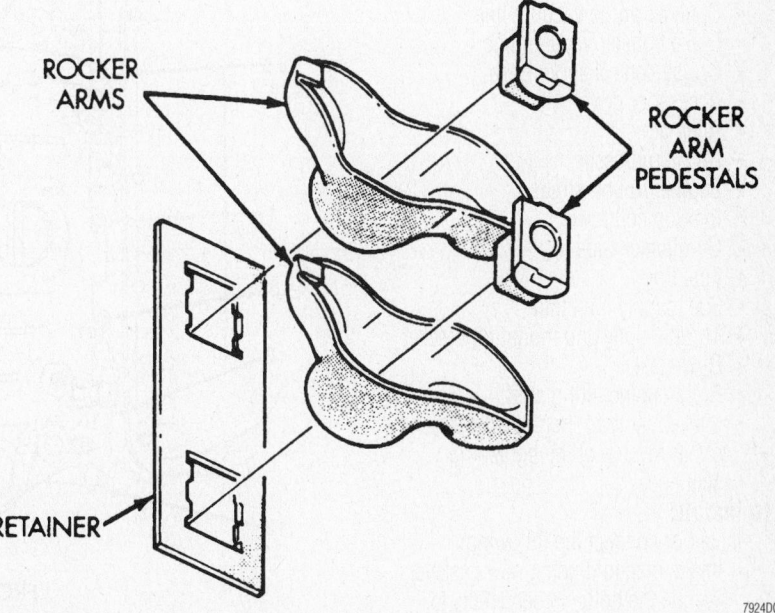

ROCKER ARMS

ROCKER ARM PEDESTALS

RETAINER

Exploded view of the rocker arm assembly—8.0L engine

7924DG55

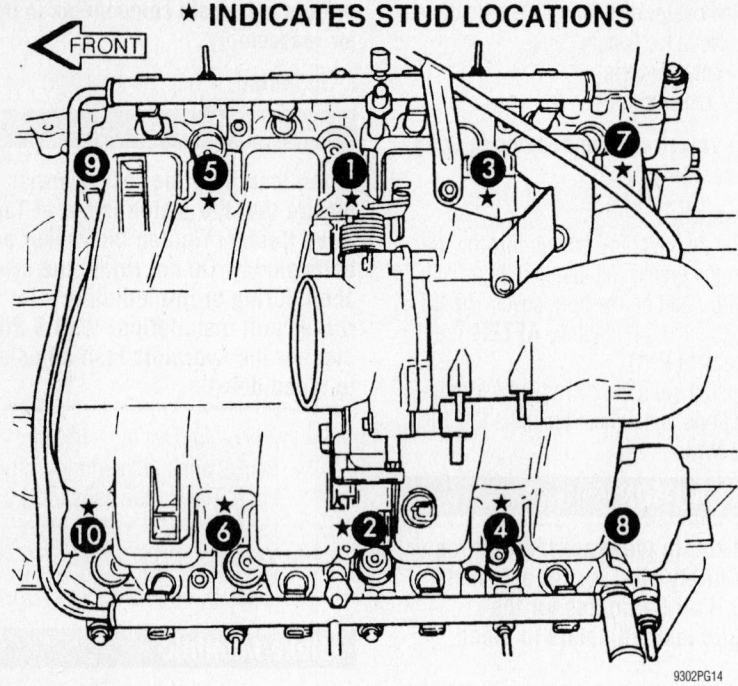

★ INDICATES STUD LOCATIONS

Intake manifold torque sequence—3.7 & 4.7L engines

- Accelerator cable
- Cruise control cable
- Manifold Absolute Pressure (MAP) sensor connector
- Intake Air Temperature (IAT) sensor connector
- Throttle Position (TP) sensor connector
- Idle Air Control (IAC) valve connector
- Engine Coolant Temperature (ECT) sensor
- Positive Crankcase Ventilation (PCV) valve and hose
- Canister purge vacuum line
- Brake booster vacuum line
- Cruise control servo hose
- Accessory drive belt
- Alternator
- A/C compressor
- Engine ground straps
- Ignition coil towers
- Oil dipstick tube
- Fuel line
- Fuel supply manifold
- Throttle body and mounting bracket
- Cowl seal
- Right engine lifting stud
- Intake manifold. Remove the fasteners in reverse of the tightening sequence.

To install:

5. Install or connect the following:
   - Intake manifold using new gaskets. Torque the bolts, in sequence, to 105 inch lbs. (12 Nm).
   - Right engine lifting stud

- Cowl seal
- Throttle body and mounting bracket
- Fuel supply manifold
- Fuel line
- Oil dipstick tube
- Ignition coil towers
- Engine ground straps

- A/C compressor
- Alternator
- Accessory drive belt
- Cruise control servo hose
- Brake booster vacuum line
- Canister purge vacuum line
- PCV valve and hose
- ECT sensor
- IAC valve connector
- TP sensor connector
- IAT sensor connector
- MAP sensor connector
- Cruise control cable
- Accelerator cable
- Air cleaner assembly
- Negative battery cable

6. Fill and bleed the cooling system.
7. Start the engine, check for leaks and repair if necessary.

3.9L Engine

1. Before servicing the vehicle, refer to the precautions in the beginning of this section.
2. Drain the cooling system.
3. Remove or disconnect the following:
   - Negative battery cable
   - Accessory drive belt
   - Alternator
   - A/C compressor
   - Alternator and A/C compressor bracket
   - Air cleaner assembly

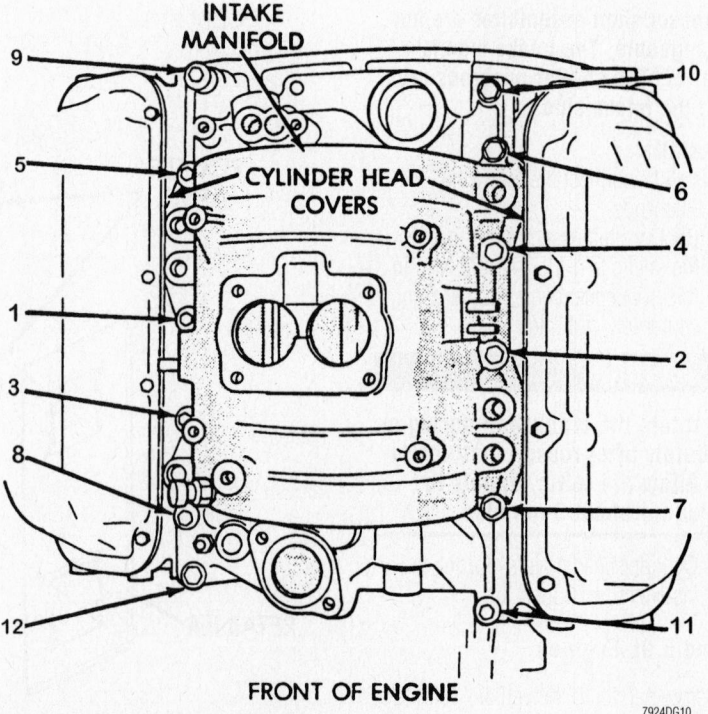

Intake manifold torque sequence—3.9L engine

- Fuel line
- Fuel supply manifold
- Accelerator cable
- Transmission cable
- Cruise control cable
- Distributor cap and wires
- Ignition coil wiring
- Engine Coolant Temperature (ECT) sensor connector
- Heater hose
- Upper radiator hose
- Bypass hose
- Closed Crankcase Ventilation (CCV) system
- Evaporative emissions system
- Intake manifold

To install:

4. Install the intake manifold. Use a new gasket and tighten the bolts in sequence as follows:

   a. Step 1: 48 inch lbs.
   b. Step 2: 84 inch lbs.
   c. Step 3: Recheck 84 inch lbs.

5. Install or connect the following:

- Evaporative emissions system
- CCV system
- Bypass hose
- Upper radiator hose
- Heater hose
- ECT sensor connector
- Ignition coil wiring
- Distributor cap and wires
- Cruise control cable
- Transmission cable
- Accelerator cable
- Fuel supply manifold
- Fuel line
- Air cleaner assembly
- Alternator and A/C compressor bracket
- A/C compressor
- Alternator

- Accessory drive belt
- Negative battery cable

6. Fill the cooling system.
7. Start the engine and check for leaks.

5.2L and 5.9L Engines

1. Before servicing the vehicle, refer to the precautions in the beginning of this section.

2. Drain the cooling system.
3. Remove or disconnect the following:

- Negative battery cable
- Accessory drive belt
- Alternator
- A/C compressor
- Alternator and A/C compressor bracket
- Air cleaner assembly
- Fuel line
- Fuel supply manifold
- Accelerator cable
- Transmission cable
- Cruise control cable
- Distributor cap and wires
- Ignition coil wiring
- Engine Coolant Temperature (ECT) sensor connector
- Heater hose
- Upper radiator hose
- Bypass hose
- Closed Crankcase Ventilation (CCV) system
- Evaporative emissions system
- Intake manifold

To install:

4. Install the intake manifold. Use a new gasket and tighten the bolts in sequence as follows:

   a. Step 1: Bolts 1–4 , 72 inch lbs.
   b. Step 2: Bolts 5–12 , 72 inch lbs.
   c. Step 3: Recheck all bolts at 72 inch lbs.

   d. Step 4: All bolts to 12 ft. lbs.
   e. Recheck all bolts at 12 ft. lbs.

5. Install or connect the following:

- Evaporative emissions system
- CCV system
- Bypass hose
- Upper radiator hose
- Heater hose
- ECT sensor connector
- Ignition coil wiring
- Distributor cap and wires
- Cruise control cable
- Transmission cable
- Accelerator cable
- Fuel supply manifold
- Fuel line
- Air cleaner assembly
- Alternator and A/C compressor bracket
- A/C compressor
- Alternator
- Accessory drive belt
- Negative battery cable

6. Fill the cooling system.
7. Start the engine and check for leaks.

5.7L Engines

1. Before servicing the vehicle, refer to the precautions in the beginning of this section.

2. Drain the cooling system.
3. Relieve the fuel system pressure.
4. Remove or disconnect the following:

- Negative battery cable
- Air cleaner assembly
- Accessory drive belt
- MAP connector
- IAT connector
- TPS connector
- CTS connector
- Brake booster hose
- PCV hose
- Alternator
- A/C compressor
- Intake manifold bolts, in a criss-cross pattern, from the outside to the center
- Intake manifold/IAFM

To install:

5. Position new intake manifold seals.
6. Install the intake manifold. Tighten the bolts in sequence from the center outwards, to 105 inch lbs. (12 Nm).

7. Install or connect the following:

- Electrical connectors
- Alternator
- A/C compressor
- Brake booster hose
- PCV hose
- Accessory drive belt

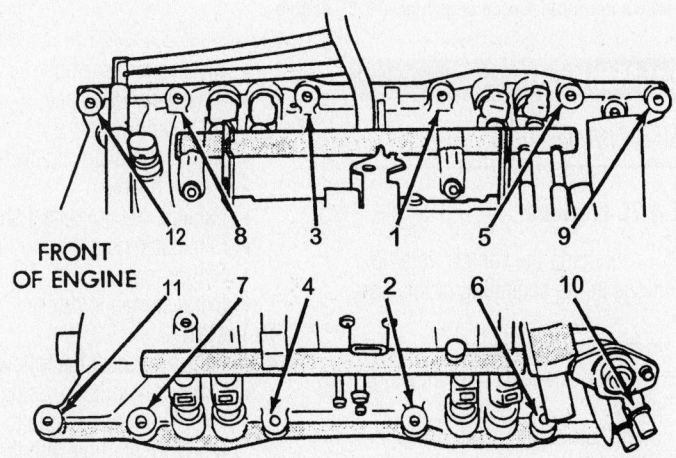

FRONT OF ENGINE

Intake manifold torque sequence—5.2L and 5.9L engines

- Negative battery cable
- Air cleaner assembly

8.0L Engines

1. Before servicing the vehicle, refer to the precautions in the beginning of this section.
2. Drain the cooling system.
3. Relieve the fuel system pressure.
- Negative battery cable
- Accessory drive belt
- Alternator and brace
- A/C compressor and brace
- Air cleaner housing
- Fuel line
- Accelerator linkage
- Cruise control cable, if equipped
- Transmission cable, if equipped
- Ignition coil pack and spark plug wires
- Intake manifold vacuum lines
- Engine control sensor harness connectors
- Heater hoses
- Bypass hose
- Evaporative Emissions (EVAP) system
- Closed Crankcase Ventilation (CCV) system
- Throttle body
- Upper intake manifold
- Lower intake manifold

To install:

4. Install or connect the following:
- Lower intake manifold. Tighten the bolts in sequence to 40 ft. lbs. (54 Nm).
- Upper intake manifold. Tighten the bolts in sequence to 16 ft. lbs. (22 Nm).
- Throttle body. Tighten the bolts to 17 ft. lbs. (23 Nm).
- CCV system
- EVAP system
- Bypass hose
- Heater hoses
- Engine control sensor harness connectors
- Intake manifold vacuum lines
- Ignition coil pack and spark plug wires
- Transmission cable, if equipped
- Cruise control cable, if equipped
- Accelerator linkage
- Fuel line
- Air cleaner housing
- A/C compressor and brace
- Alternator and brace
- Accessory drive belt
- Negative battery cable

5. Fill the cooling system.
6. Start the engine and check for leaks.

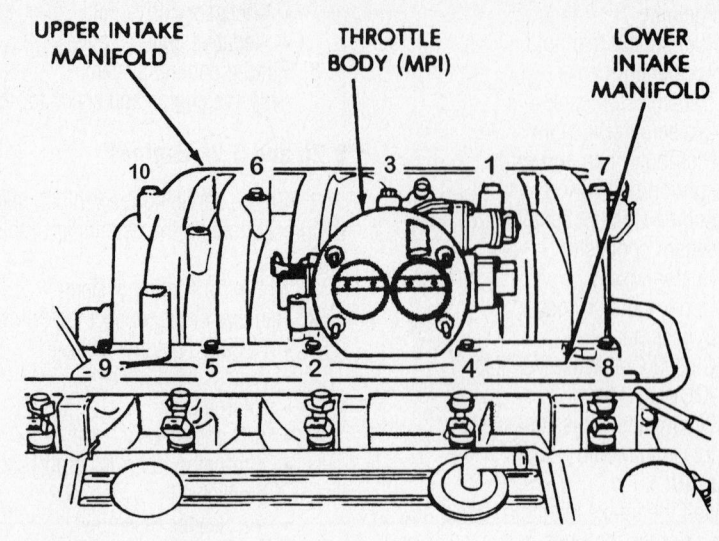

Upper intake manifold torque sequence—8.0L engine

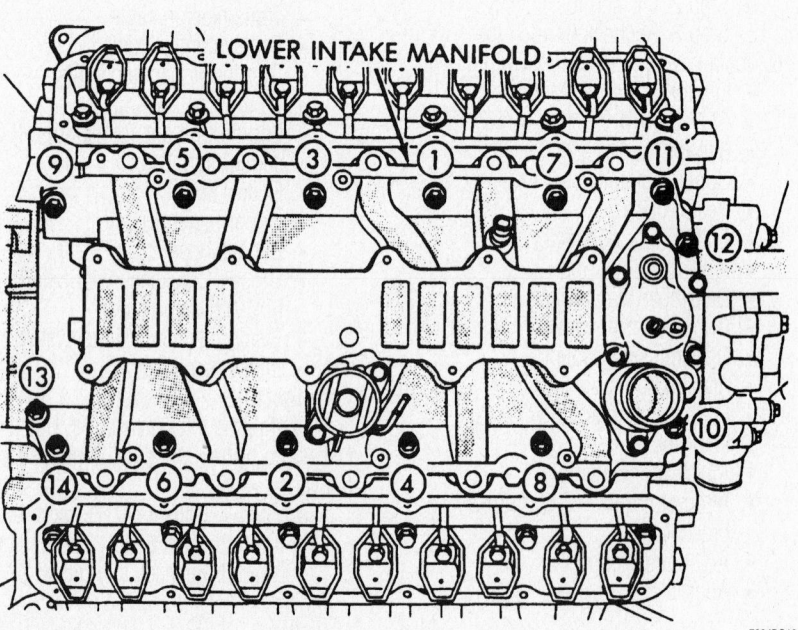

Lower intake manifold torque sequence—8.0L engine

Exhaust Manifold

REMOVAL & INSTALLATION

3.7 and 4.7L Engines

1. Before servicing the vehicle, refer to the precautions in the beginning of this section.
2. Drain the cooling system.
3. Remove or disconnect the following:
- Battery
- Power distribution center
- Battery tray
- Windshield washer fluid bottle
- Air cleaner assembly
- Accessory drive belt
- A/C compressor
- A/C accumulator bracket
- Heater hoses
- Exhaust manifold heat shields
- Exhaust Y-pipe
- Starter motor
- Exhaust manifolds

To install:

4. Install or connect the following:
- Exhaust manifolds, using new gaskets. Tighten the bolts to 18 ft. lbs. (25 Nm), starting with the inner bolts and work out to the ends.
- Starter motor

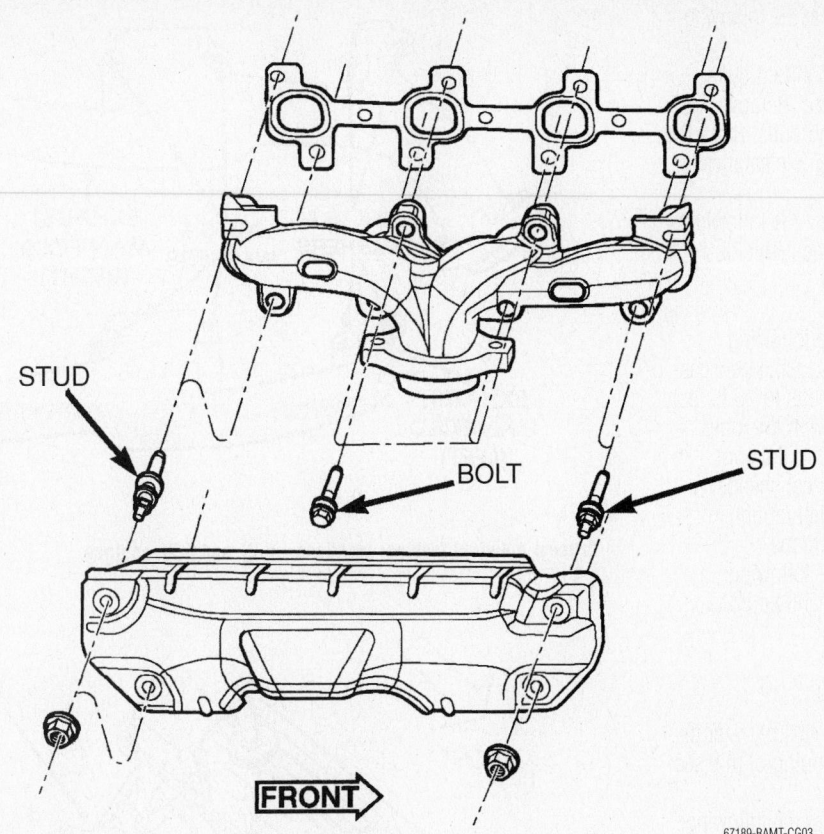

67189-RAMT-CG03

Right side exhaust manifold—4.7L engine, left side and 3.7L engine similar

- Exhaust Y-pipe
- Exhaust manifold heat shields
- Heater hoses
- A/C accumulator bracket
- A/C compressor
- Accessory drive belt
- Air cleaner assembly
- Windshield washer fluid bottle
- Battery tray
- Power distribution center
- Battery

5. Fill the cooling system.
6. Start the engine and check for leaks.

3.9L Engine

1. Before servicing the vehicle, refer to the precautions in the beginning of this section.
2. Remove or disconnect the following:
   - Negative battery cable
   - Heated Oxygen (HO$_2$S) sensor connectors
   - Exhaust manifold heat shields
   - Exhaust front pipe
   - Exhaust manifolds

To install:

➡**If the exhaust manifold studs came out with the nuts when removing the exhaust manifolds, replace them with new studs.**

3. Install or connect the following:
   - Exhaust manifolds. Tighten the fasteners to 25 ft. lbs. (34 Nm), starting with the center fasteners and working out to the ends.
   - Exhaust front pipe
   - Exhaust manifold heat shields
   - HO$_2$S sensor connectors
   - Negative battery cable
4. Start the engine and check for leaks.

5.7L Engines

1. Before servicing the vehicle, refer to the precautions in the beginning of this section.
2. Drain the cooling system.

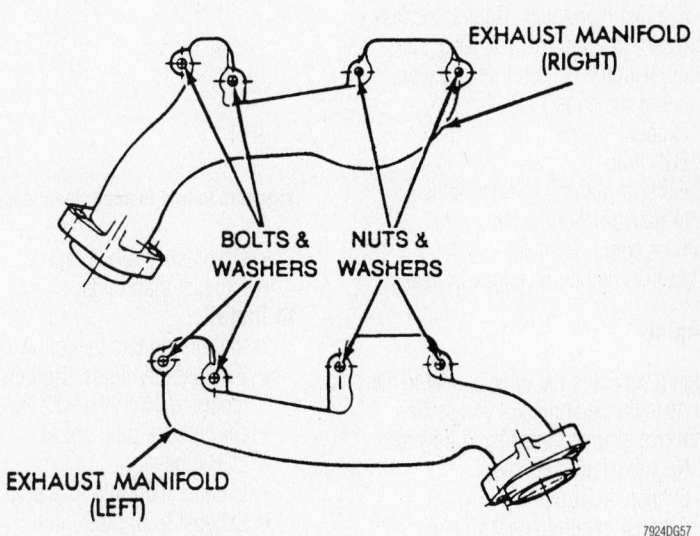

7924DG57

Stud and bolt locations—3.9L engine

3. Remove or disconnect the following:
- Battery
- Exhaust pipe-to-manifold bolts

4. Install an engine crane. Remove the right and left mount through bolts. Raise the engine just enough to provide clearance for manifold removal.

5. Remove or disconnect the following:
- Exhaust manifold heat shields
- Exhaust manifolds

To install:

6. Install or connect the following:
- Exhaust manifolds, using new gaskets. Tighten the bolts to 18 ft. lbs. (25 Nm), starting with the inner bolts and work out to the ends.
- Exhaust manifold heat shields

7. Install the right and left mount through bolts. Remove the crane.

8. Install or connect the following:
- Exhaust pipe-to-manifold bolts
- Battery

5.2L and 5.9L Engines

1. Before servicing the vehicle, refer to the precautions in the beginning of this section.

2. Remove or disconnect the following:
- Negative battery cable
- Exhaust manifold heat shields
- Exhaust Gas Recirculation (EGR) tube
- Exhaust Y-pipe
- Exhaust manifolds

To install:

➡️**If the exhaust manifold studs came out with the nuts when removing the exhaust manifolds, replace them with new studs.**

3. Install or connect the following:
- Exhaust manifolds. Tighten the fasteners to 25 ft. lbs. (27 Nm), starting with the center nuts and work out to the ends.
- Exhaust Y-pipe
- EGR tube
- Exhaust manifold heat shields
- Negative battery cable

4. Fill the cooling system.
5. Start the engine and check for leaks.

8.0L Engine

1. Before servicing the vehicle, refer to the precautions in the beginning of this section.

2. Remove or disconnect the following:
- Negative battery cable
- Exhaust front pipe
- Exhaust manifold heat shields
- Exhaust Gas Recirculation (EGR) tube

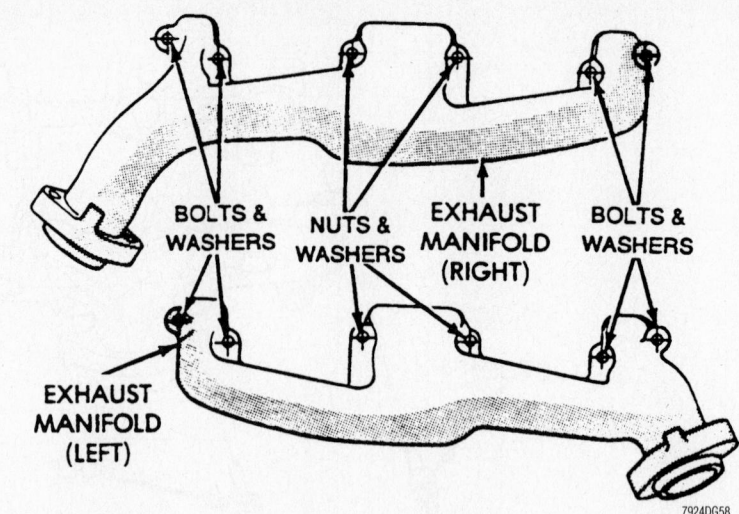

Exhaust manifold fastener locations—5.2L and 5.9L engines

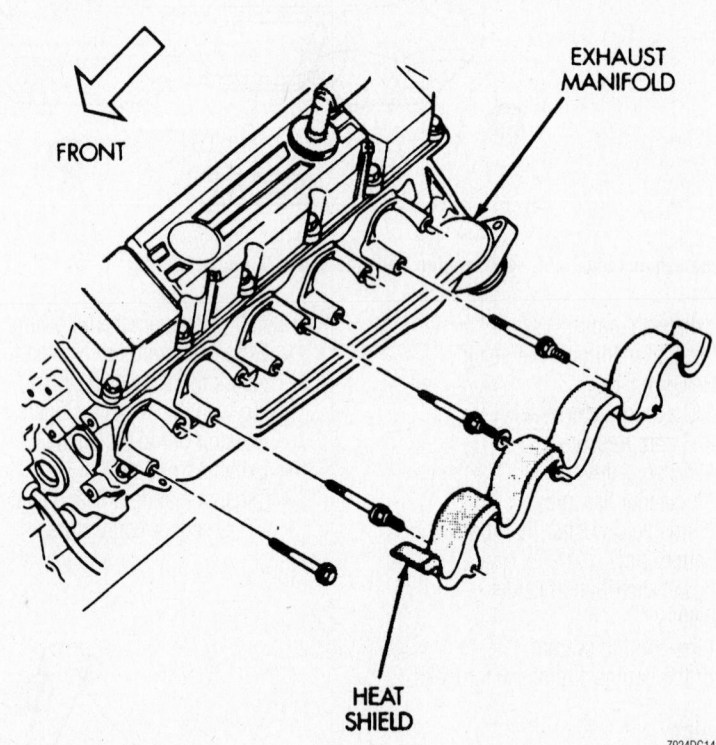

Exhaust manifold fastener locations—8.0L engine

- Oil dipstick tube bracket
- Exhaust manifolds

To install:

3. Install or connect the following:
- Exhaust manifolds. Tighten the fasteners to 16 ft. lbs. (22 Nm).
- Oil dipstick tube bracket
- EGR tube
- Exhaust manifold heat shields
- Exhaust front pipe
- Negative battery cable

4. Start the engine and check for leaks.

Camshaft and Valve Lifters

REMOVAL & INSTALLATION

3.7L and 4.7L Engines

1. Before servicing the vehicle, refer to the precautions in the beginning of this section.

2. Remove or disconnect the following:
- Negative battery cable

- Cylinder head covers
- Rocker arms
- Hydraulic lash adjusters

➡ **Keep all valvetrain components in order for assembly.**

3. Set the engine at Top Dead Center (TDC) of the compression stroke for the No. 1 cylinder.

4. Install Timing Chain Wedge (8350 4.7L; 8379 3.7L) to retain the chain tensioners.

5. Matchmark the timing chains to the camshaft sprockets.

6. Install Camshaft Holding Tool (6958 and Adapter Pins 8346 4.7L; 8428 3.7L) to the left camshaft sprocket.

7. On 3.7L engines, use tool 8428 and rotate the camshaft 5 degrees clockwise to eliminate valve load.

8. On 4.7L engines, use adjustable pliers to rotate the left camshaft 15 degrees clockwise to eliminate valve load. Use adjustable pliers to rotate the right camshaft 45 degrees counterclockwise to eliminate valve load.

9. Remove or disconnect the following:
- Right camshaft timing sprocket and target wheel
- Left camshaft sprocket
- Camshaft bearing caps, by reversing the tightening sequence
- Camshafts

To install:

10. Install or connect the following:
- Camshafts. Torque the bearing cap bolts in ½ turn increments, in sequence, to 100 inch lbs. (11 Nm).
- Target wheel to the right camshaft
- Camshaft timing sprockets and chains, by aligning the matchmarks

11. Remove the tensioner wedges and tighten the camshaft sprocket bolts to 90 ft. lbs. (122 Nm).

12. Install or connect the following:
- Hydraulic lash adjusters in their original locations
- Rocker arms in their original locations
- Cylinder head covers
- Negative battery cable

3.9L, 5.2L and 5.9L Engines

1. Before servicing the vehicle, refer to the precautions in the beginning of this section.

2. Drain the cooling system.

3. Recover the A/C refrigerant, if equipped with air conditioning.

4. Set the crankshaft to Top Dead Cen-

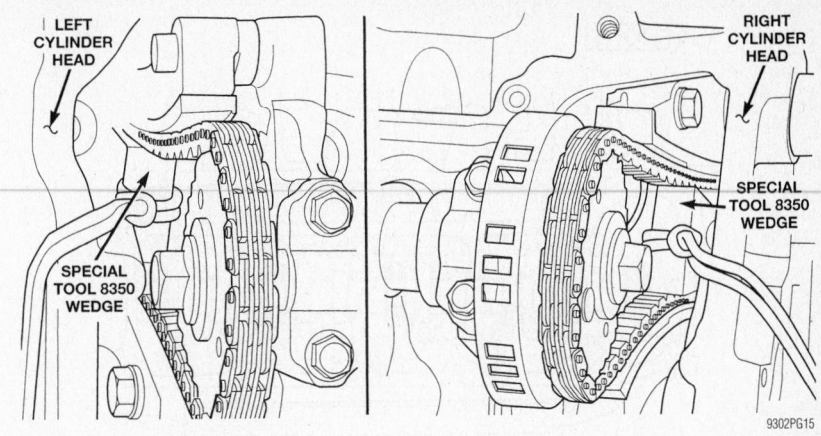

Chain Tensioner Retaining Wedges—3.7L and 4.7L engines

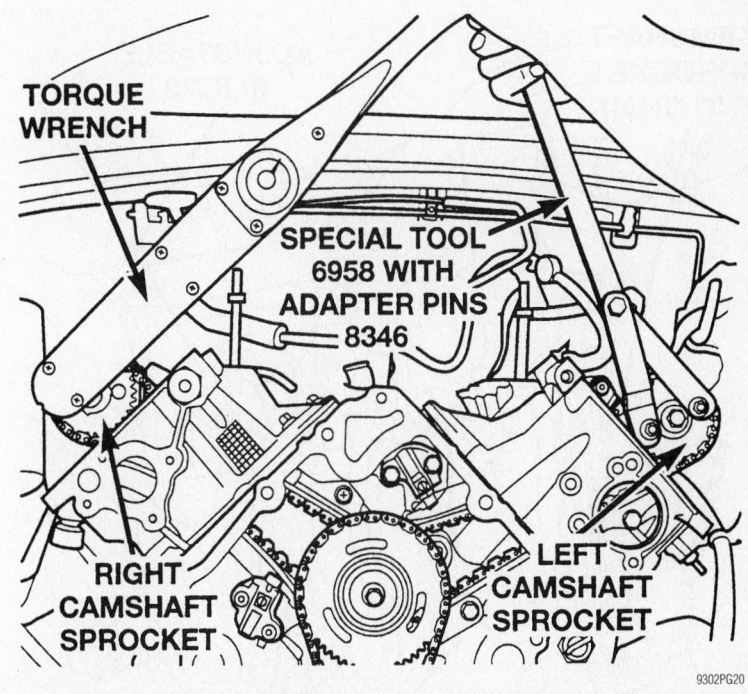

Hold the left camshaft sprocket with a spanner wrench while removing or installing the camshaft sprocket bolts—4.7L engine

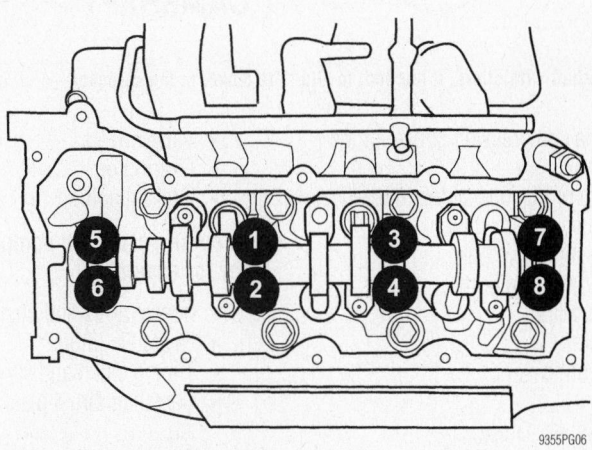

Camshaft bearing cap bolt tightening sequence—3.7L

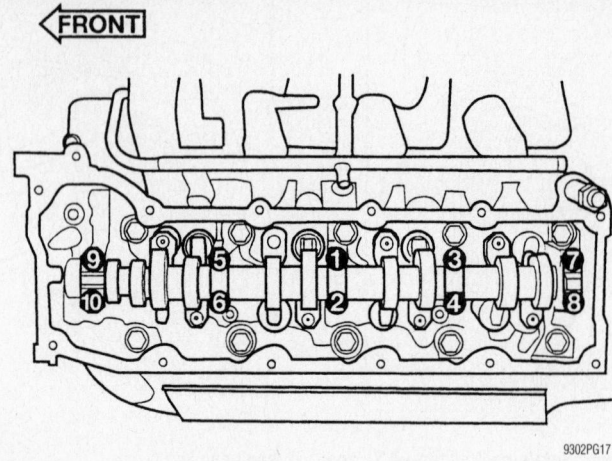

Camshaft bearing cap bolt tightening sequence—4.7L engine

9302PG17

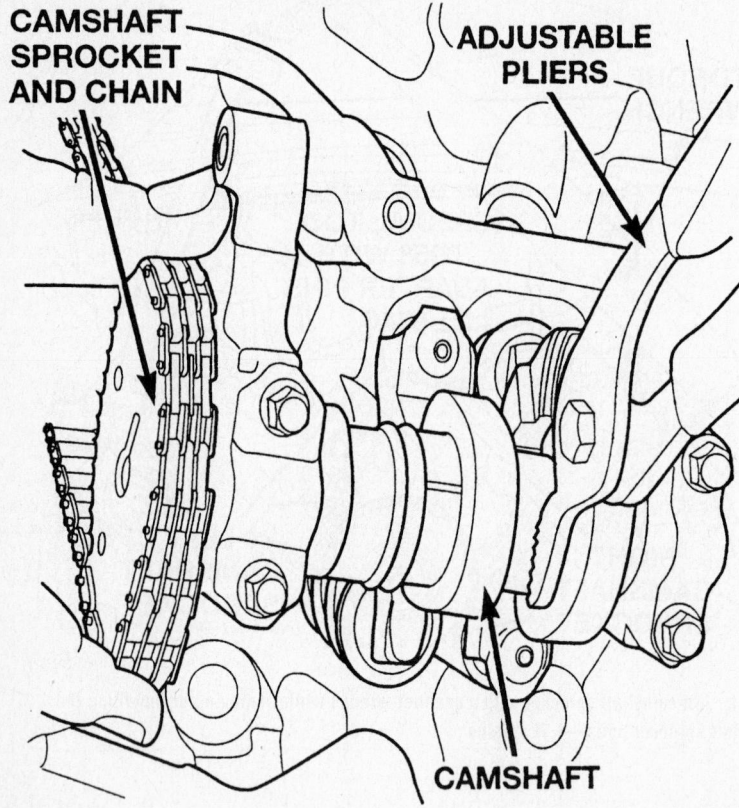

CAMSHAFT SPROCKET AND CHAIN

ADJUSTABLE PLIERS

CAMSHAFT

9302PG16

Turn the camshaft with pliers, if needed, to align the dowel in the sprocket—4.7L engine

ter (TDC) of the compression stroke for the No. 1 cylinder.

5. Remove or disconnect the following:
- Negative battery cable
- Accessory drive belt
- Power steering pump
- Water pump
- Radiator
- A/C condenser
- Grille
- Crankshaft damper
- Front cover
- Valve covers
- Distributor
- Intake manifold

➡ **Keep all valvetrain components in order for assembly.**

- Rocker arms and pushrods
- Hydraulic lifters
- Timing chain and sprockets
- Camshaft thrust plate and chain oil tab
- Camshaft

To install:

6. Install or connect the following:
- Camshaft
- Camshaft Holding Tool C-3509
- Camshaft thrust plate and chain oil tab. Tighten the bolts to 18 ft. lbs. (24 Nm).
- Timing chain and sprockets. Tighten the camshaft sprocket bolt to 50 ft. lbs. (68 Nm).
7. Remove the camshaft holding tool.
8. Install or connect the following:
- Hydraulic lifters in their original positions
- Rocker arms and pushrods in their original positions
- Intake manifold
- Distributor
- Valve covers
- Front cover
- Crankshaft damper
- Grille
- A/C condenser
- Radiator
- Water pump
- Power steering pump
- Accessory drive belt
- Negative battery cable
9. Fill the cooling system.
10. Recharge the A/C system, if equipped.
11. Start the engine and check for leaks.

5.7L Engines

1. Before servicing the vehicle, refer to the precautions in the beginning of this section.
2. Drain the cooling system.
3. Recover the A/C refrigerant, if equipped with air conditioning.
4. Set the crankshaft to Top Dead Center (TDC) of the compression stroke for the No. 1 cylinder.
5. Remove or disconnect the following:
- Negative battery cable
- Camshaft rear cam bearing core plug
- Air cleaner
- Accessory drive belt
- Alternator
- A/C compressor
- Radiator
- Intake manifold
- Cylinder head covers
- Cylinder heads
- Oil pan
- Front cover
- Oil pickup tube
- Oil pump
- Timing chain and sprockets
- Camshaft thrust plate

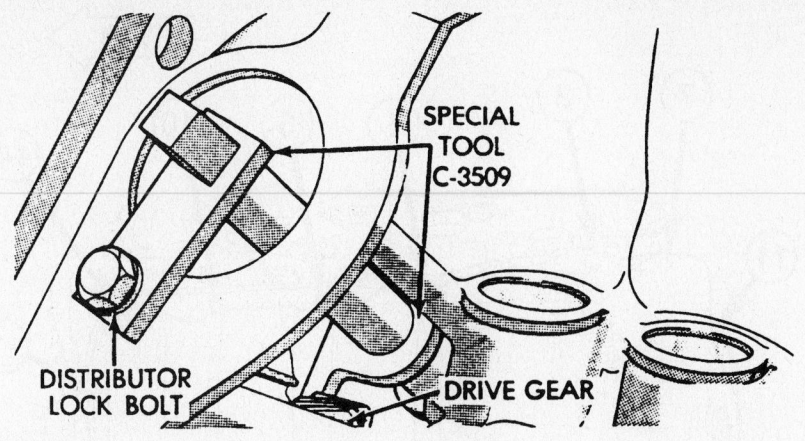

Camshaft holding tool C-3509—3.9L, 5.2L and 5.9L engines

- Hydraulic lifters
- Camshaft

To install:

6. Install or connect the following:
- Camshaft
- Camshaft thrust plate. Tighten the bolts to 21 ft. lbs. (28 Nm).
- Timing chain and sprockets
- Oil pump
- Oil pickup tube

➡**Lifters must be replaced in their original positions.**

- Hydraulic lifters
- Cylinder heads
- Pushrods
- Rocker arms
- Front cover
- Oil pan
- Cylinder head covers
- Intake manifold
- A/C compressor
- Alternator
- Accessory drive belt
- Radiator
- Air cleaner
- Camshaft rear cam bearing core plug
- Negative battery cable

7. Fill the cooling system.
8. Recharge the A/C system, if equipped.
9. Start the engine and check for leaks.

8.0L Engine

1. Before servicing the vehicle, refer to the precautions in the beginning of this section.
2. Recover the A/C refrigerant, if equipped with air conditioning.
3. Drain the cooling system.
4. Relieve the fuel system pressure.
5. Remove or disconnect the following:

- Negative battery cable
- Valve covers
- Upper and lower intake manifolds

➡**Keep all valvetrain components in order for assembly.**

- Rocker arms
- Pushrods
- Cylinder heads
- Valve lifters
- Crankshaft pulley
- Front cover
- Timing chain and sprockets
- Camshaft thrust plate
- Camshaft

To install:

6. Install or connect the following:
- Camshaft
- Camshaft thrust plate
- Timing chain and sprockets. Tighten the bolt to 55 ft. lbs. (75 Nm).
- Front cover
- Crankshaft pulley
- Valve lifters
- Cylinder heads
- Pushrods
- Rocker arms
- Upper and lower intake manifolds
- Valve covers
- Negative battery cable

7. Fill the cooling system.
8. Recharge the A/C system, if equipped.
9. Start the engine and check for leaks.

Valve Lash

ADJUSTMENT

All gasoline engines covered in this section use hydraulic lifters. No maintenance or periodic adjustment is required.

Starter Motor

REMOVAL & INSTALLATION

3.7L and 4.7L Engines

1. Before servicing the vehicle, refer to the precautions in the beginning of this section.
2. Remove or disconnect the following:
- Negative battery cable
- Starter mounting bolts

➡**On the 3.7L engine, the left side exhaust pipe and front driveshaft must be disconnected.**

- Starter solenoid harness connections
- Starter

To install:

3. Connect the starter solenoid wiring connectors.
4. Install the starter and torque the bolts to 40 ft. lbs. (54 Nm).
5. On the 3.7L engine, the left side exhaust pipe and front driveshaft
6. Install the negative battery cable and check for proper operation.

5.7L Engines

1. Before servicing the vehicle, refer to the precautions in the beginning of this section.
2. Remove or disconnect the following:
- Negative battery cable

➡**Depending on drivetrain configuration, a support bracket may be used.**

- Starter mounting bolts
- Starter solenoid harness connections
- Starter

To install:

3. Connect the starter solenoid wiring connectors.
4. Install the starter and torque the bolts to 50 ft. lbs. (68 Nm).
5. Install the negative battery cable and check for proper operation.

3.9L, 5.2L, 5.9L and 8.0L Engines

1. Before servicing the vehicle, refer to the precautions in the beginning of this section.
2. Remove or disconnect the following:
- Negative battery cable
- Starter mounting bolts
- Starter solenoid harness connections
- Starter

To install:

3. Connect the starter solenoid wiring connectors.

4. Install the starter and tighten the bolts to 50 ft. lbs. (68 Nm).

5. Install the negative battery cable and check for proper operation.

Oil Pan

REMOVAL & INSTALLATION

3.7L, 4.7L and 5.7L Engines

1. Before servicing the vehicle, refer to the precautions in the beginning of this section.

2. Drain the engine oil.

3. Attach an engine crane.

4. Loosen, but don't remove, the left and right mount through-bolts.

5. Remove or disconnect the following:
- Negative battery cable
- Structural cover
- Front crossmember

6. Raise the engine just enough for clearance.

7. Remove or disconnect the following:

➡ **Don't pry on the pan. The gasket is integral with the windage tray, and doesn't come out with the pan.**

- Pan bolts and studs
- Pan

➡ **The double ended studs must be installed in their original locations.**

To install:

8. Install or connect the following:
- Oil pan gasket
- Oil pan. Tighten the bolts, in sequence, to 105 inch lbs. (12 Nm).
- Front crossmember
- Structural cover
- Negative battery cable

9. Fill the crankcase to the proper level with engine oil.

10. Start the engine and check for leaks.

3.9L Engine

2-WHEEL DRIVE MODELS

1. Before servicing the vehicle, refer to the precautions in the beginning of this section.

2. Drain the engine oil.

3. Remove or disconnect the following:
- Negative battery cable
- Distributor cap
- Oil dipstick
- Exhaust front pipe

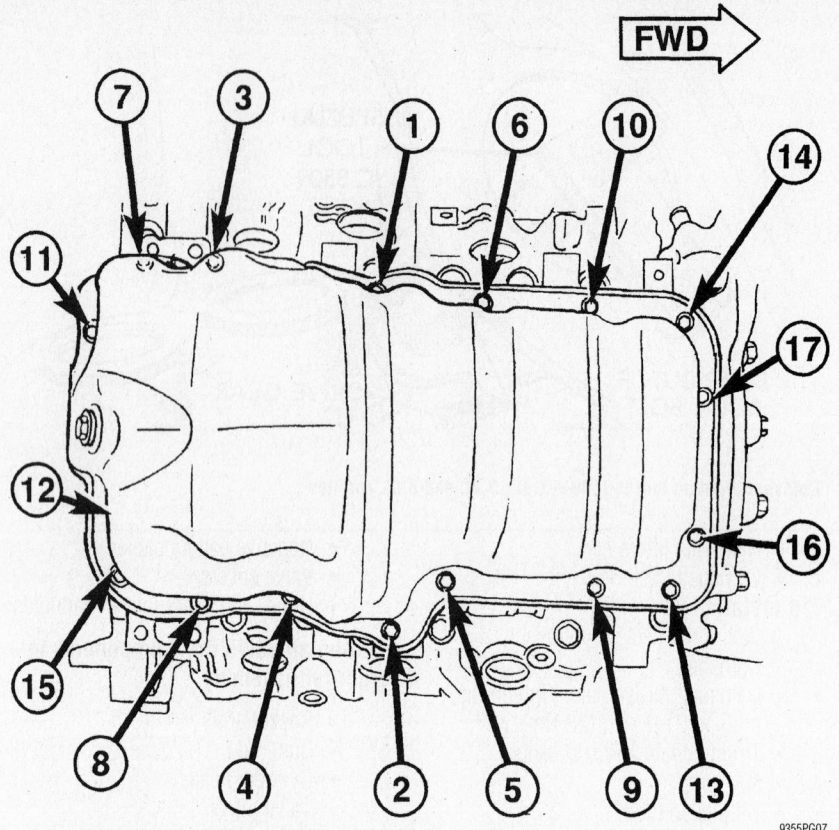

Oil pan bolt torque sequence—3.7L

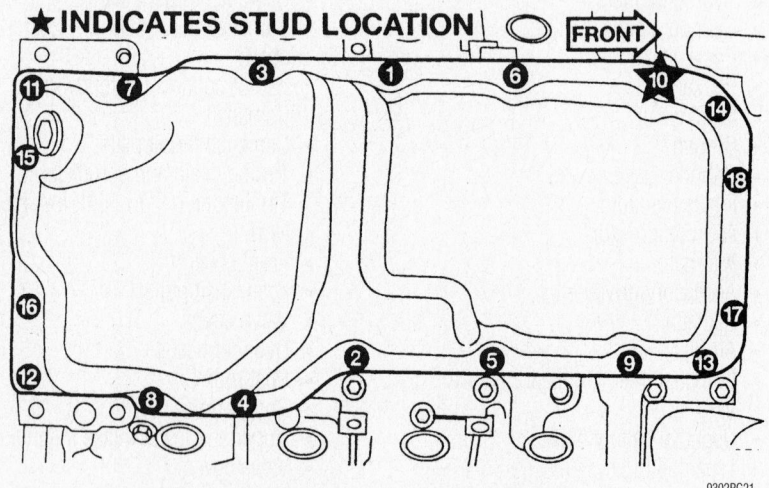

Oil pan mounting bolt tightening sequence—4.7L and 5.7L engine

- Flywheel access panel, if equipped
- Left and right motor mount through bolts
- Oil pan. Raise the engine as necessary for clearance.

To install:

4. Fabricate 4 alignment dowels from 1½ inch x ¼ inch bolts. Cut the heads off the bolts and cut a slot into the top of the dowel to allow installation/removal with a screwdriver.

5. Install or connect the following:
- Alignment dowels
- Oil pan. Replace the dowels with bolts and tighten all bolts to 17 ft. lbs. (23 Nm).
- Left and right motor mount through bolts
- Flywheel access panel, if equipped
- Exhaust front pipe
- Oil dipstick
- Distributor cap

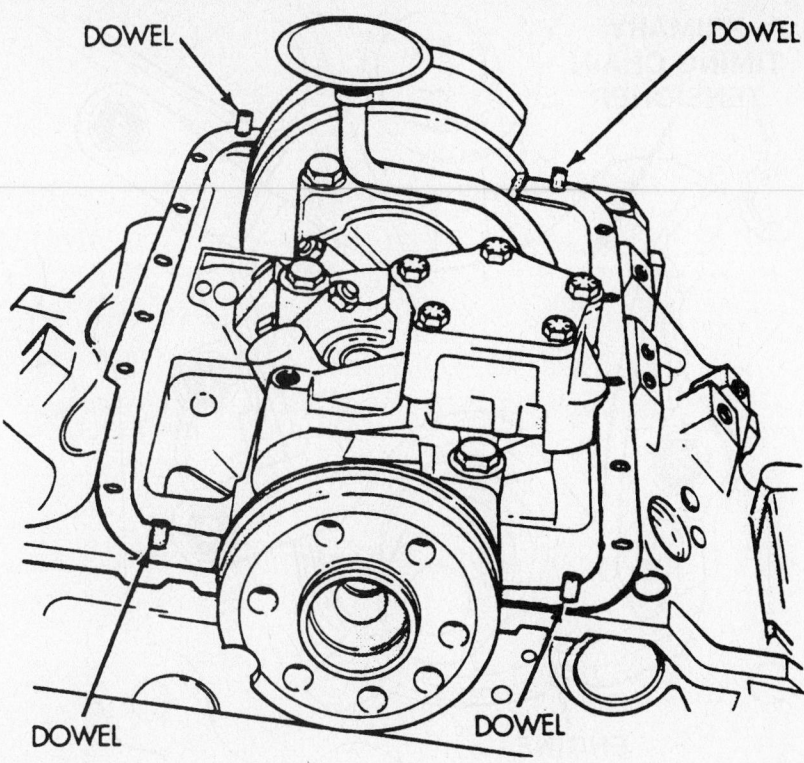

Oil pan alignment dowel placement—3.9L engine shown

- Negative battery cable
6. Fill the crankcase to the correct level.
7. Start the engine and check for leaks.

4-WHEEL DRIVE MODELS

1. Before servicing the vehicle, refer to the precautions in the beginning of this section.
2. Drain the engine oil.
3. Remove or disconnect the following:
   - Negative battery cable
   - Engine oil dipstick
   - Front axle. Support the engine
   - Exhaust front pipe
   - Flywheel access panel, if equipped
   - Oil pan

To install:

4. Fabricate 4 alignment dowels from 1½ inch x ¼ inch bolts. Cut the heads off the bolts and cut a slot into the top of the dowel to allow installation/removal with a screwdriver.
5. Install or connect the following:
   - Alignment dowels
   - Oil pan. Replace the dowels with bolts and tighten all bolts to 17 ft. lbs. (23 Nm).
   - Flywheel access panel, if equipped
   - Exhaust front pipe
   - Front axle
   - Engine oil dipstick
   - Negative battery cable

6. Fill the crankcase to the correct level.
7. Start the engine and check for leaks.

5.2L and 5.9L Engines

1. Before servicing the vehicle, refer to the precautions in the beginning of this section.
2. Drain the engine oil.
3. Remove or disconnect the following:
   - Oil filter
   - Starter motor
   - Cooler lines
   - Oil level sensor connector
   - Heated Oxygen (HO$_2$S) sensor connector
   - Exhaust Y-pipe
   - Oil pan

To install:

4. Install or connect the following:
   - Oil pan, using a new gasket. Tighten the bolts to 18 ft. lbs. (24 Nm).
   - Exhaust Y-pipe
   - Heated Oxygen (HO$_2$S) sensor connector
   - Oil level sensor connector
   - Cooler lines
   - Starter motor
   - Oil filter
5. Fill the crankcase.
6. Start the engine and check for leaks.

8.0L Engine

1. Before servicing the vehicle, refer to the precautions in the beginning of this section.
2. Drain the engine oil.
3. Remove or disconnect the following:
   - Negative battery cable
   - Engine oil dipstick
   - Left transmission brace
   - Oil pan

To install:

4. Install or connect the following:
   - Oil pan. Tighten the ⁵⁄₁₆ bolts to 12 ft. lbs. (16 Nm) and the ¼ bolts to 96 inch lbs. (11 Nm).
   - Left transmission brace
   - Engine oil dipstick
   - Negative battery cable
5. Fill the crankcase.
6. Start the engine and check for leaks.

Oil Pump

REMOVAL & INSTALLATION

3.7L Engine

1. Before servicing the vehicle, refer to the precautions in the beginning of this section.

1½" × 5/16" BOLT

DOWEL

SLOT

Oil pan alignment dowels—3.9L, 5.2L, 5.9L and 8.0L engines

2. Remove or disconnect the following:
- Oil Pan
- Timing chain cover
- Timing chains and tensioners
- Oil pump

3. Installation is the reverse of removal. Torque the pump bolts, in sequence, to 21 ft. lbs. (28 Nm).

4.7L Engine

1. Before servicing the vehicle, refer to the precautions in the beginning of this section.
2. Drain the engine oil.
3. Remove or disconnect the following:

- Negative battery cable
- Oil pan
- Oil pump pick-up tube
- Timing chains and tensioners
- Oil pump

To install:

4. Install or connect the following:
- Oil pump. Tighten the bolts to 21 ft. lbs. (28 Nm).
- Timing chains and tensioners
- Oil pump pick-up tube
- Oil pan
- Negative battery cable

5. Fill the crankcase to the correct level.
6. Start the engine and check for leaks.

5.7L Engine

1. Before servicing the vehicle, refer to the precautions in the beginning of this section.
2. Drain the engine oil.
3. Remove or disconnect the following:
- Negative battery cable
- Oil pan
- Oil pump pick-up tube
- Timing chains and tensioners
- Oil pump

To install:

4. Install or connect the following:
- Oil pump. Tighten the bolts to 21 ft. lbs. (28 Nm).
- Timing chains and tensioners
- Oil pump pick-up tube
- Oil pan
- Negative battery cable

5. Fill the crankcase to the correct level.
6. Start the engine and check for leaks.

3.9L, 5.2L and 5.9L Engines

1. Before servicing the vehicle, refer to the precautions in the beginning of this section.
2. Drain the engine oil.
3. Remove or disconnect the following:

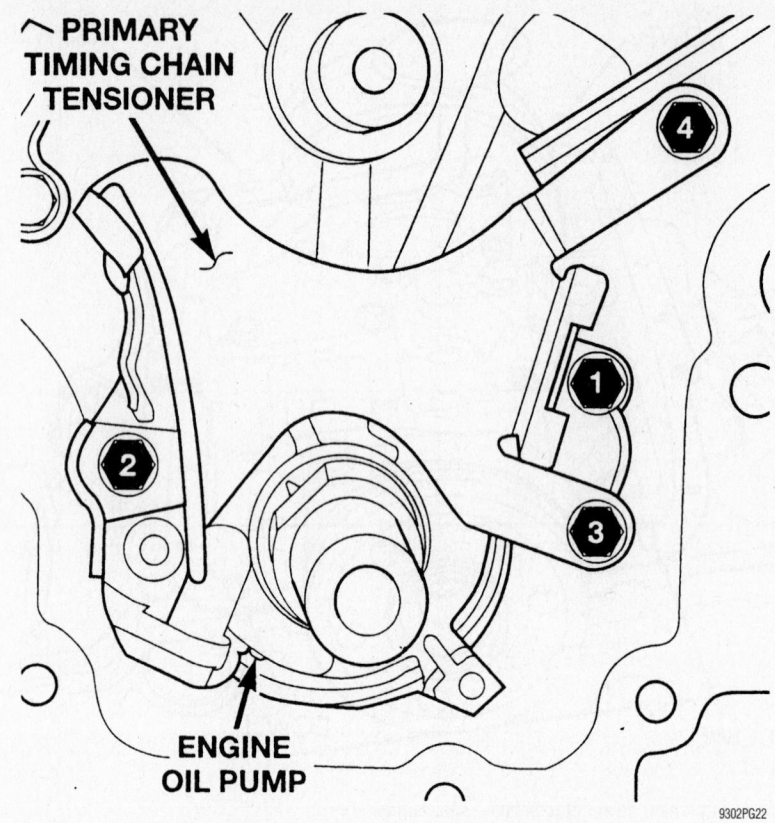

Oil pump and chain tensioner torque sequence—3.7L and 4.7L engines

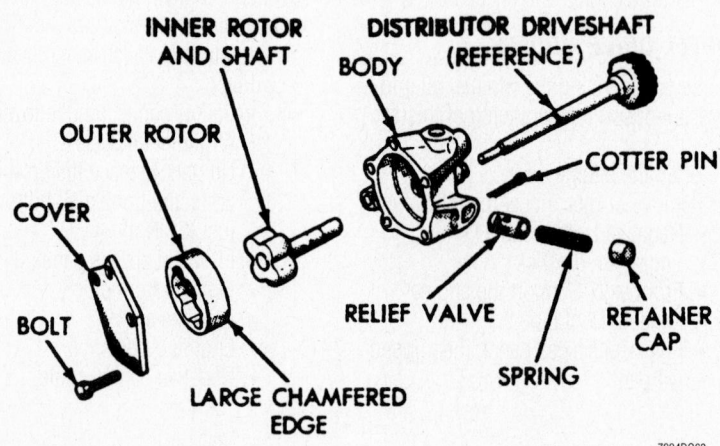

Exploded view of the oil pump assembly—3.9L, 5.2L and 5.9L engines

- Negative battery cable
- Oil pan
- Oil pump pick-up tube
- Oil pump

To install:

4. Install or connect the following:
- Oil pump. Tighten the bolts to 30 ft. lbs. (41 Nm).
- Oil pump pick-up tube
- Oil pan
- Negative battery cable

5. Fill the crankcase to the correct level.
6. Start the engine and check for leaks.

8.0L Engine

1. Before servicing the vehicle, refer to the precautions in the beginning of this section.
2. Drain the cooling system.
3. Remove or disconnect the following:
- Negative battery cable
- Accessory drive belt

- Cooling fan and shroud
- A/C compressor, if equipped
- Alternator
- Air injection pump
- Bracket assembly
- Water pump
- Crankshaft pulley
- Front oil pan bolts
- Front cover
- Oil pump pressure relief valve
- Oil pump cover
- Oil pump rotors

To install:

4. Install or connect the following:
   - Oil pump rotors
   - Oil pump cover. Tighten the bolts to 10 ft. lbs. (14 Nm).
   - Oil pump pressure relief valve. Tighten the plug to 15 ft. lbs. (20 Nm).
   - Front cover. Tighten the bolts to 35 ft. lbs. (47 Nm).
   - Front oil pan bolts. Tighten the bolts to 12 ft. lbs. (16 Nm).
   - Crankshaft pulley. Tighten the bolt to 135 ft. lbs. (183 Nm).
   - Water pump
   - Bracket assembly
   - Air injection pump
   - Alternator
   - A/C compressor, if equipped
   - Cooling fan and shroud
   - Accessory drive belt
   - Negative battery cable
5. Fill the cooling system.
6. Start the engine and check for leaks.

Rear Main Seal

REMOVAL & INSTALLATION

3.7L, 4.7L and 5.7L Engines

1. Before servicing the vehicle, refer to the precautions in the beginning of this section.
2. Remove or disconnect the following:
   - Transmission
   - Flexplate
3. Thread Oil Seal Remover 8506 into the rear main seal as far as possible and remove the rear main seal.

To install:

4. Install or connect the following:
   - Seal Guide 8349-2 onto the crankshaft
   - Rear main seal on the seal guide
   - Rear main seal, using the Crankshaft Rear Oil Seal Installer 8349 and Driver Handle C-4171; tap it into place until the installer is flush with the cylinder block

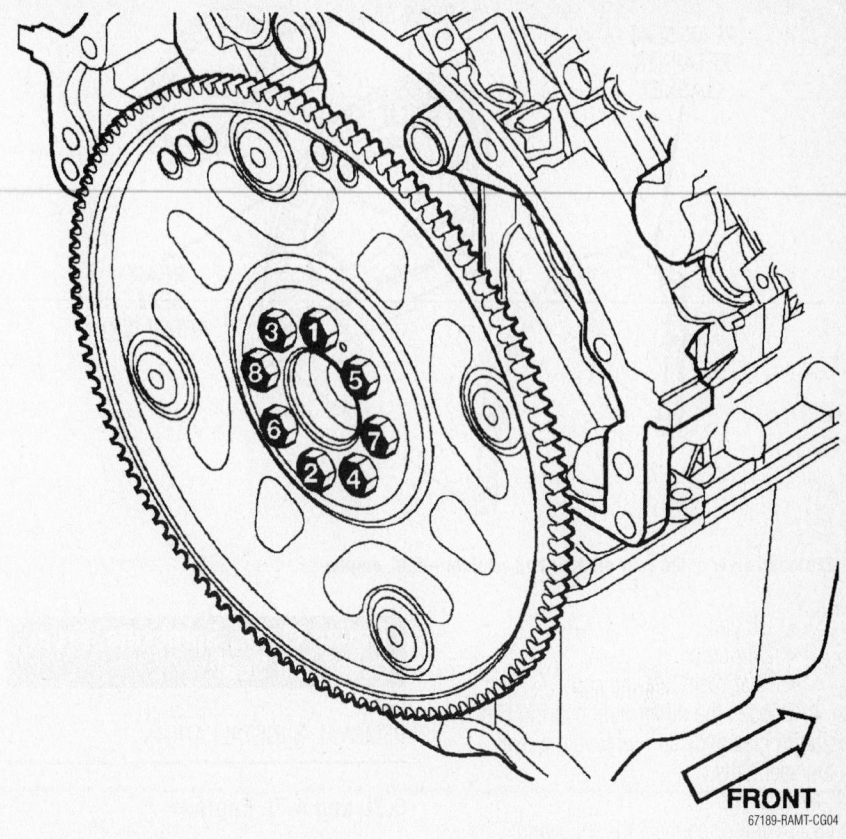

Flexplate bolt tightening sequence—3.7L, 4.7L and 5.7L engines

67189-RAMT-CG04

- Flexplate. Torque the bolts in the sequence shown to 45 ft. lbs. (60 Nm) on auto trans., or 70 ft. lbs. (95 Nm) for man. trans.
- Transmission
5. Start the engine, check for leaks and repair if necessary.

3.9L, 5.2L and 5.9L Engines

1. Before servicing the vehicle, refer to the precautions in the beginning of this section.
2. Drain the engine oil.
3. Remove or disconnect the following:

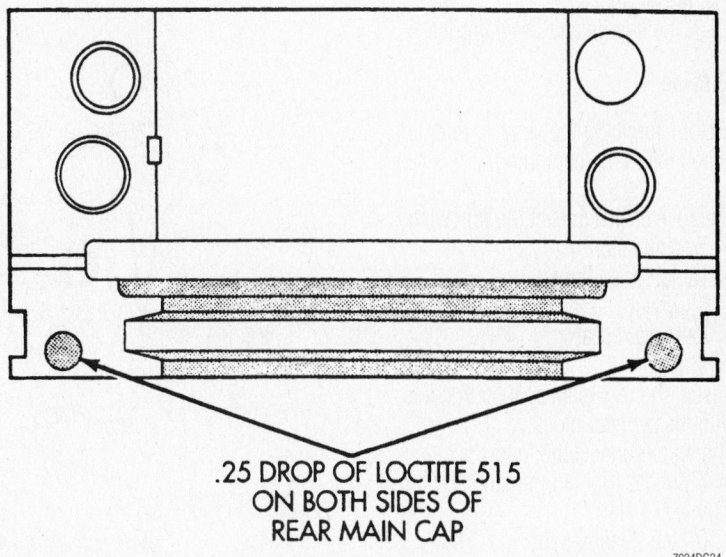

.25 DROP OF LOCTITE 515
ON BOTH SIDES OF
REAR MAIN CAP

Sealant application locations —3.9L, 5.2L and 5.9L engines

7924DG24

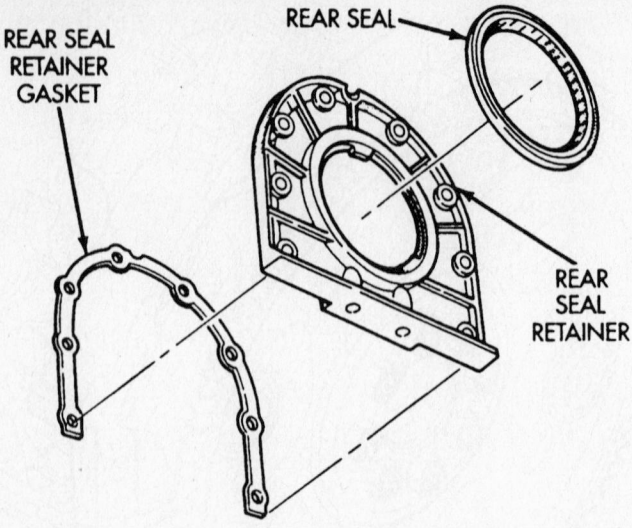

Exploded view of the rear oil seal and retainer—8.0L engine

- Oil pan
- Oil pump
- Rear main bearing cap

4. Loosen the other main bearing cap bolts for clearance and remove the rear main seal halfs.

To install:

5. Install or connect the following:
- New upper seal half to the cylinder block
- New lower seal half to the bearing cap

6. Apply sealant to the rear main bearing cap.

7. Install or connect the following:
- Rear main bearing cap. Tighten **all** main bearing cap bolts to 85 ft. lbs. (115 Nm).
- Oil pump and oil pan

8. Fill the engine.

9. Start the engine and check for leaks.

8.0L Engine

1. Before servicing the vehicle, refer to the precautions in the beginning of this section.

2. Remove or disconnect the following:
- Transmission
- Clutch, if equipped
- Flywheel
- Rear main seal

To install:

3. Install the rear main seal so that it is flush with the cylinder block.

4. Install or connect the following:
- Flywheel. Tighten the bolts to 55 ft. lbs. (75 Nm).
- Clutch, if equipped
- Transmission

5. Start the engine and check for leaks.

Timing Chain, Sprockets, Front Cover and Seal

REMOVAL & INSTALLATION

3.7L and 4.7L Engines

1. Before servicing the vehicle, refer to the precautions in the beginning of this section.

2. Drain the cooling system.

3. Remove or disconnect the following:
- Negative battery cable
- Valve covers
- Radiator fan
- Heater hoses
- Alternator
- Air conditioning compressor

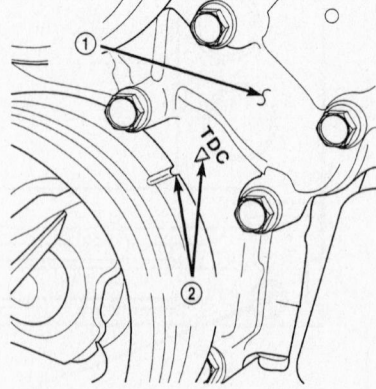

1 – TIMING CHAIN COVER
2 – CRANKSHAFT TIMING MARKS

Crankshaft timing marks—3.7L and 4.7L engines

- Power steering pump
- Front cover

4. Rotate the crankshaft so that the crankshaft timing mark aligns with the Top Dead Center (TDC) mark on the front cover, and the **V8 or V6** marks on the camshaft sprockets are at 12 o'clock.
- Access plugs from the cylinder heads
- Oil fill housing
- Crankshaft damper

5. Compress the primary timing chain tensioner and install a lockpin.

6. Remove the secondary timing chain tensioners.

7. Hold the left camshaft with adjustable pliers and remove the sprocket and chain. Rotate the **left** camshaft 15 degrees **clockwise** to the neutral position.

8. Hold the right camshaft with adjustable pliers and remove the camshaft sprocket. Rotate the **right** camshaft 45 degrees **counterclockwise** to the neutral position.

9. Remove the primary timing chain and sprockets.

To install:

10. Use a small prytool to hold the ratchet pawl and compress the secondary timing chain tensioners in a vise and install locking pins.

➡️**The black bolts fasten the guide to the engine block and the silver bolts fasten the guide to the cylinder head.**

11. Install or connect the following:
- Secondary timing chain guides. Tighten the bolts to 21 ft. lbs. (28 Nm).
- Secondary timing chains to the idler sprocket so that the double plated links on each chain are visible through the slots in the primary idler sprocket

12. Lock the secondary timing chains to the idler sprocket with Timing Chain Locking tool as shown.

13. Align the primary chain double plated links with the idler sprocket timing mark and the single plated link with the crankshaft sprocket timing mark.

14. Install the primary chain and sprockets. Tighten the idler sprocket bolt to 25 ft. lbs. (34 Nm).

15. Align the secondary chain single plated links with the timing marks on the secondary sprockets. Align the dot at the **L** mark on the left sprocket with the plated link on the left chain and the dot at the **R** mark on the right sprocket with the plated link on the right chain.

16. Rotate the camshafts back from the

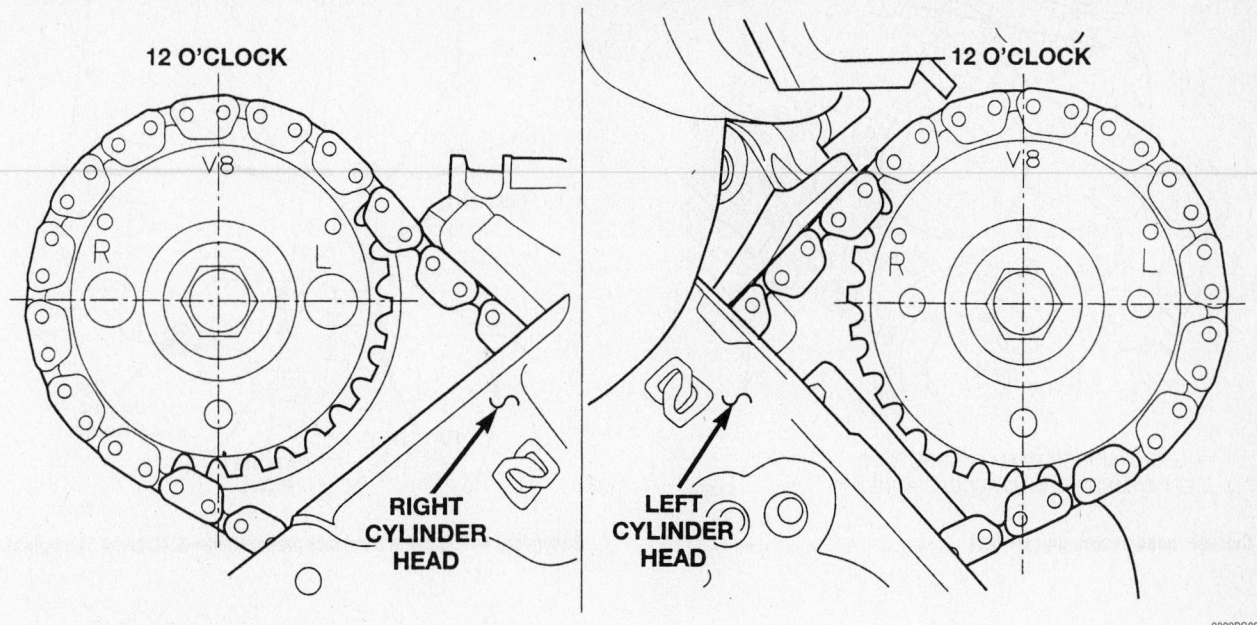

12 O'CLOCK

12 O'CLOCK

V8

V8

R L

R L

RIGHT CYLINDER HEAD

LEFT CYLINDER HEAD

9302PG08

Camshaft positioning—4.7L engine

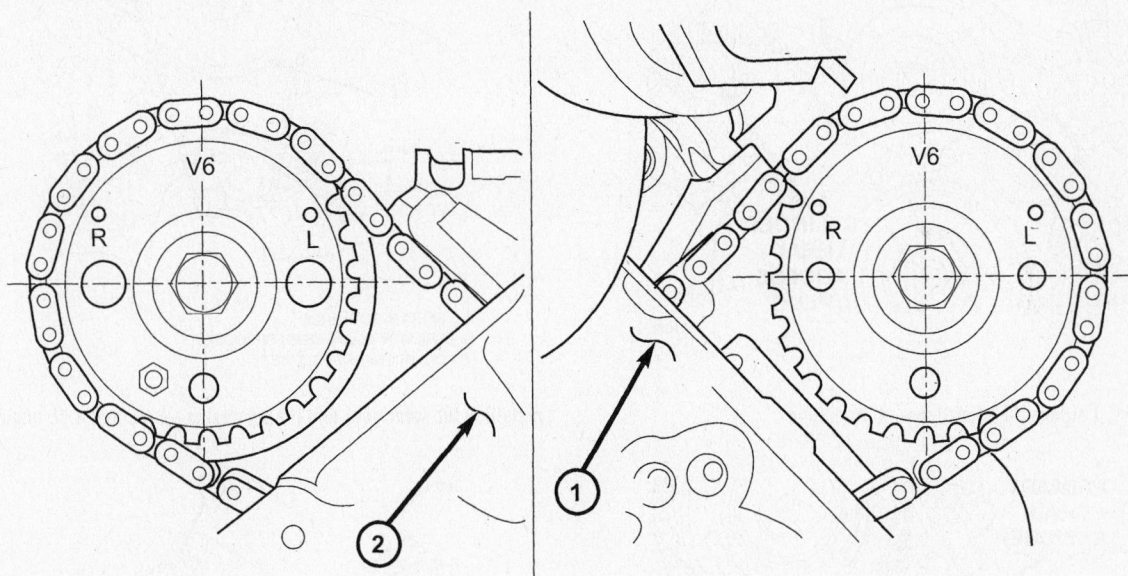

V6

V6

R L

R L

①

②

1 - LEFT CYLINDER HEAD
2 - RIGHT CYLINDER HEAD

9355PG09

Camshaft sprocket timing marks—3.7L

neutral position and install the camshaft sprockets.

17. Remove the secondary chain locking tool.

18. Remove the primary and secondary timing chain tensioner locking pins.

19. Hold the camshaft sprockets with a spanner wrench and tighten the retaining bolts to 90 ft. lbs. (122 Nm).

20. Install or connect the following:
- Front cover. Tighten the bolts, in sequence, to 40 ft. lbs. (54 Nm).
- Front crankshaft seal
- Cylinder head access plugs
- A/C compressor
- Alternator
- Accessory drive belt tensioner.

Tighten the bolt to 40 ft. lbs. (54 Nm).
- Oil fill housing
- Crankshaft damper. Tighten the bolt to 130 ft. lbs. (175 Nm).
- Power steering pump
- Lower radiator hose
- Heater hoses
- Accessory drive belt

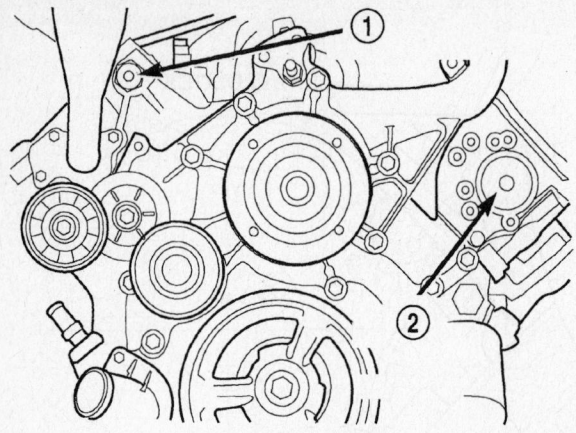

1 - RIGHT CYLINDER HEAD ACCESS PLUG
2 - LEFT CYLINDER HEAD ACCESS PLUG

9355PG11

Cylinder head access plugs—3.7L

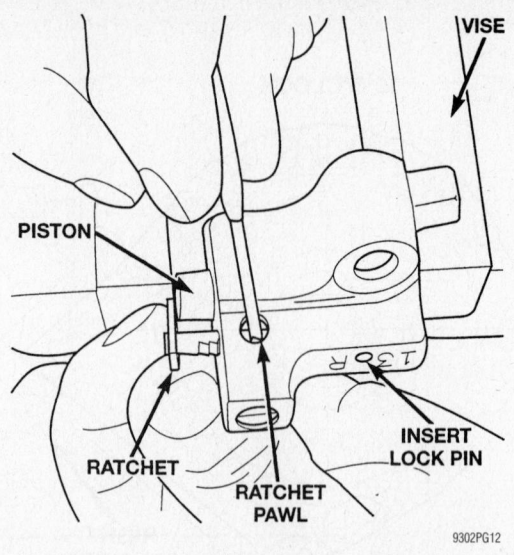

Secondary timing chain tensioner preparation—3.7L and 4.7L engines

9302PG12

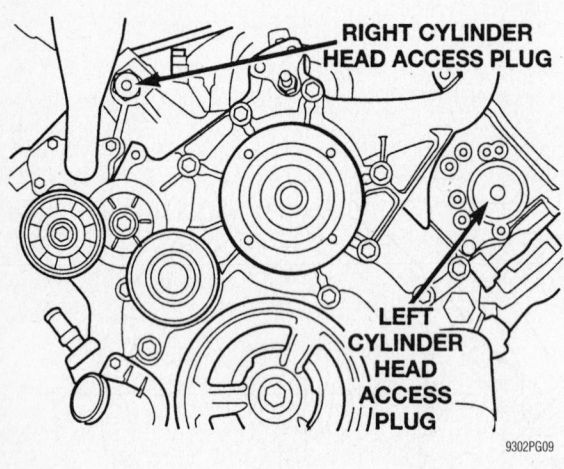

9302PG09

Cylinder head access plug locations—4.7L engine

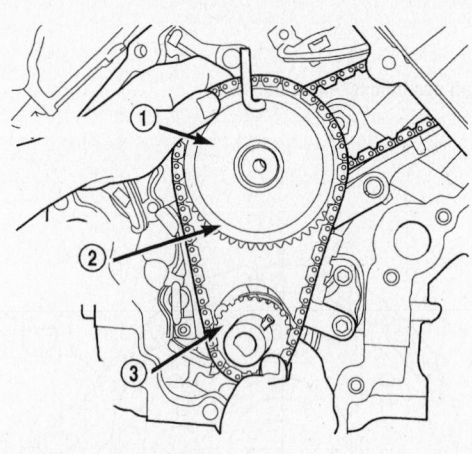

1 - SPECIAL TOOL 8429
2 - PRIMARY CHAIN IDLER SPROCKET
3 - CRANKSHAFT SPROCKET

9355PG12

Installing the idler gear and timing chains—3.7L and 4.7L engines

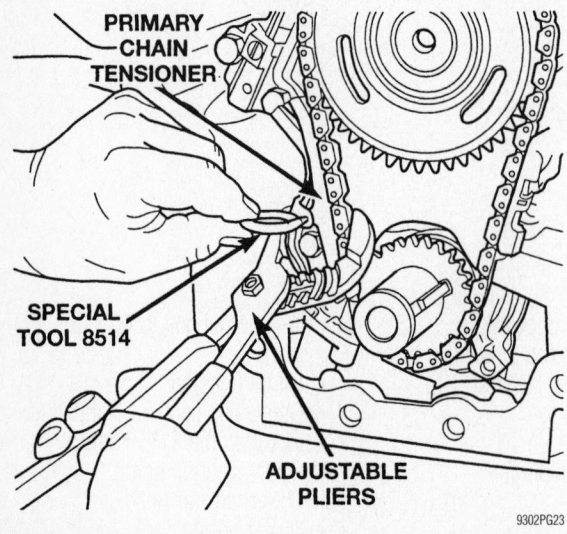

9302PG23

Compress and lock the primary chain tensioner—4.7L engine

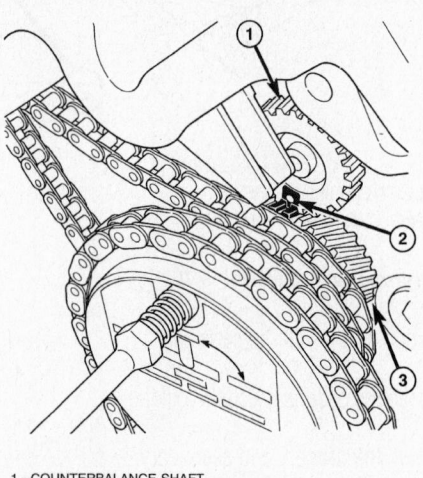

1 - COUNTERBALANCE SHAFT
2 - TIMING MARKS
3 - IDLER SPROCKET

9355PG13

Counterbalance shaft timing marks—3.7L

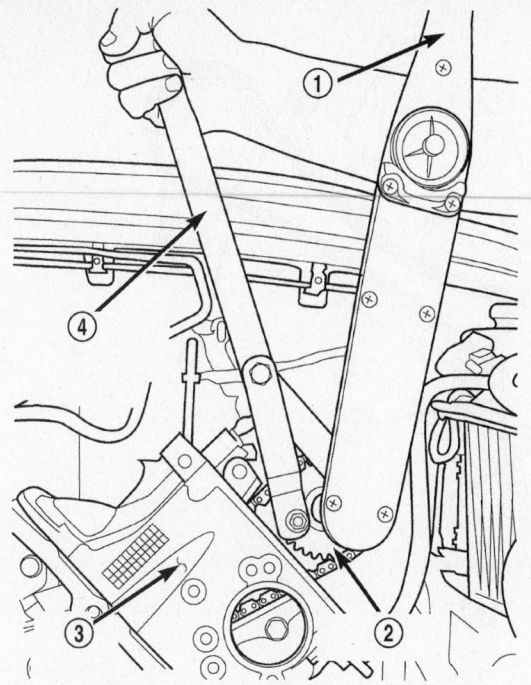

1 - TORQUE WRENCH
2 - CAMSHAFT SPROCKET
3 - LEFT CYLINDER HEAD
4 - SPECIAL TOOL 6958 SPANNER WITH ADAPTER PINS 8346

9355PG14

Tightening the left side camshaft sprocket—3.7L and 4.7L engines

1 - TORQUE WRENCH
2 - SPECIAL TOOL 6958 WITH ADAPTER PINS 8346
3 - LEFT CAMSHAFT SPROCKET
4 - RIGHT CAMSHAFT SPROCKET

9355PG15

Tightening the right side camshaft sprocket—3.7L and 4.7L engines

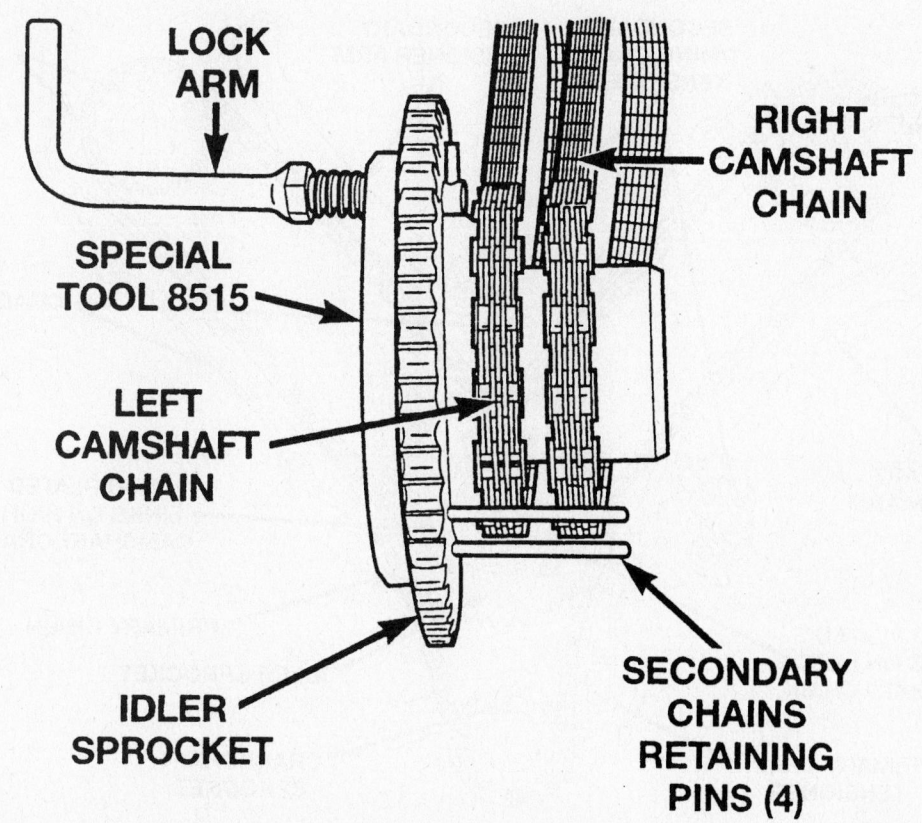

LOCK ARM

SPECIAL TOOL 8515

LEFT CAMSHAFT CHAIN

IDLER SPROCKET

RIGHT CAMSHAFT CHAIN

SECONDARY CHAINS RETAINING PINS (4)

9302PG07

Use the Timing Chain Locking tool to lock the timing chains on the idler gear—4.7L engine

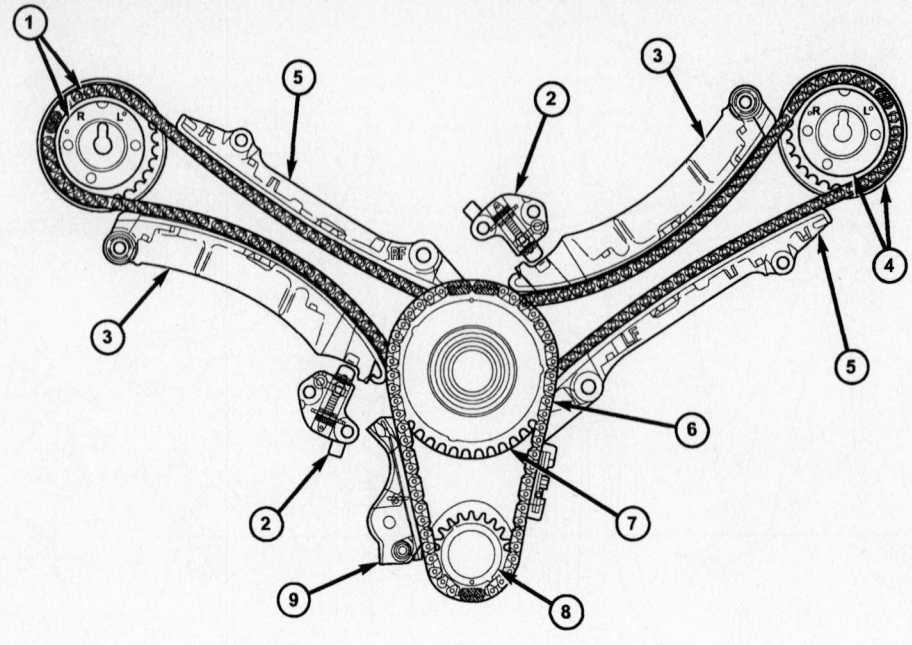

1 - RIGHT CAMSHAFT SPROCKET AND SECONDARY CHAIN
2 - SECONDARY TIMING CHAIN TENSIONER (LEFT AND RIGHT SIDE NOT INTERCHANGEABLE)
3 - SECONDARY TENSIONER ARM
4 - LEFT CAMSHAFT SPROCKET AND SECONDARY CHAIN
5 - CHAIN GUIDE (LEFT AND RIGHT SIDE ARE NOT INTERCHANGEABLE)

6 - PRIMARY CHAIN
7 - IDLER SPROCKET
8 - CRANKSHAFT SPROCKET
9 - PRIMARY CHAIN TENSIONER

67189-RAMT-CG05

Timing chain system and alignment marks—3.7L engine

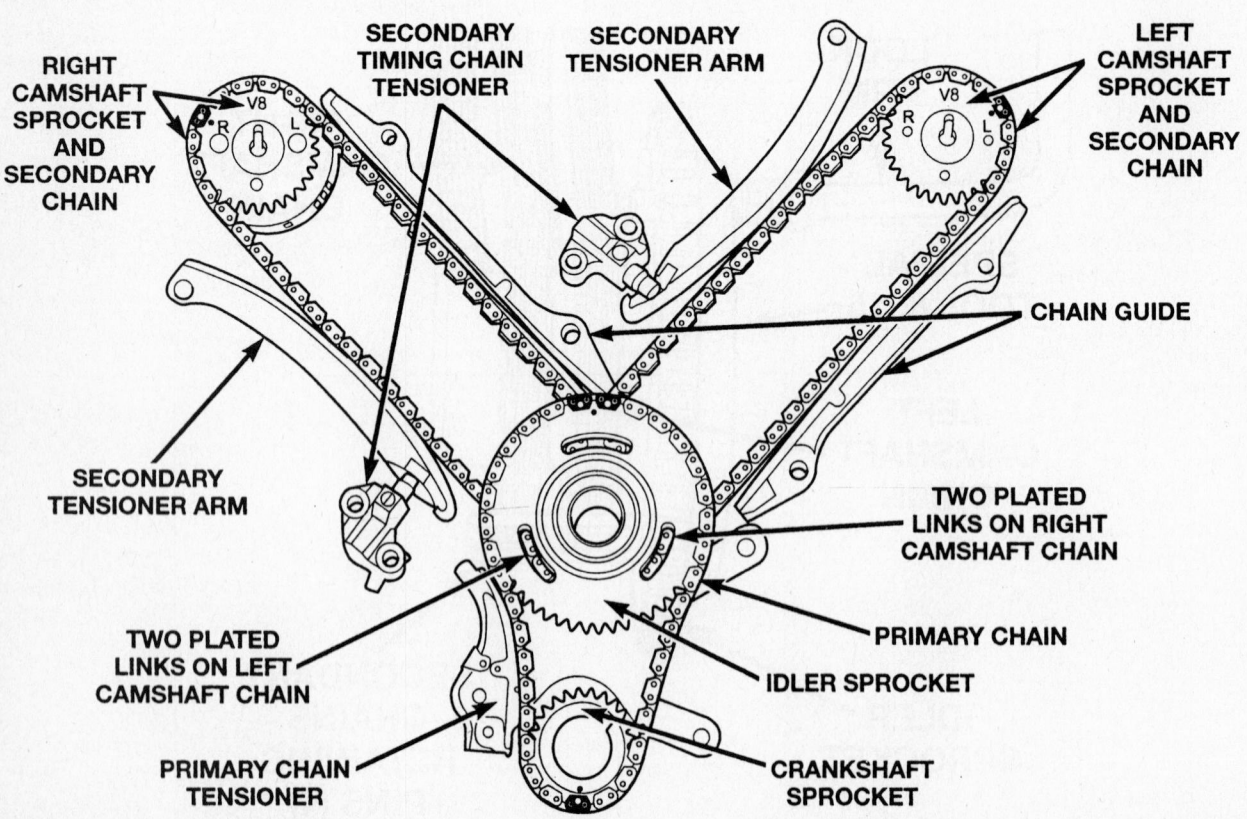

Timing chain system and alignment marks—4.7L engine

9302PG24

⭐ **INDICATES STUD LOCATIONS**

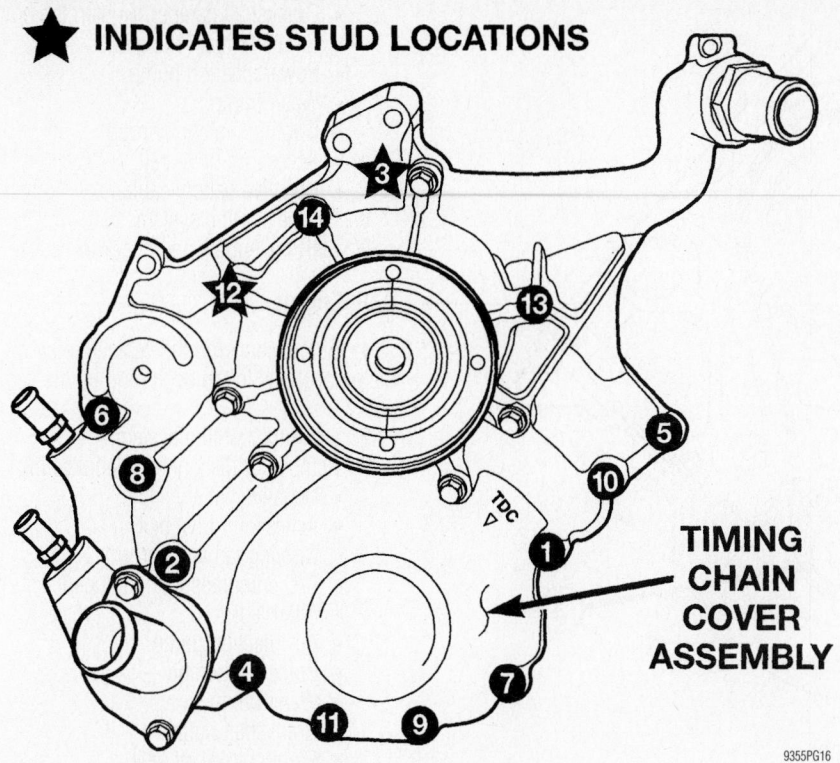

TIMING
CHAIN
COVER
ASSEMBLY

9355PG16

Timing cover bolt torque sequence—3.7L and 4.7L engines

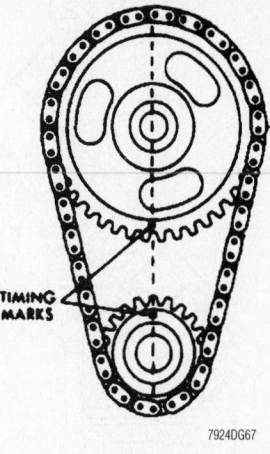

7924DG67

Timing chain alignment marks—3.9L, 5.2L, 5.9L and 8.0L engines

- Engine cooling fan and shroud
- Camshaft Position (CMP) sensor
- Valve covers
- Negative battery cable

21. Fill and bleed the cooling system.

22. Start the engine, check for leaks and repair if necessary.

3.9L Engine

1. Before servicing the vehicle, refer to the precautions in the beginning of this section.

2. Drain the cooling system.

3. Remove or disconnect the following:
- Negative battery cable
- Accessory drive belt
- Radiator
- Cooling fan
- Water pump
- Crankshaft pulley
- Front crankshaft seal
- Front cover
- Timing chain and gears

To install:

4. Install or connect the following:
- Timing chain and gears. Align the timing marks and tighten the camshaft sprocket bolt to 35 ft. lbs. (47 Nm).
- Front cover. Tighten the bolts to 30 ft. lbs. (41 Nm).
- Front crankshaft seal

- Crankshaft pulley. Tighten the bolt to 135 ft. lbs. (183 Nm).
- Water pump
- Cooling fan
- Radiator
- Accessory drive belt
- Negative battery cable

5. Fill the cooling system.

6. Start the engine and check for leaks.

5.7L Engines

1. Before servicing the vehicle, refer to the precautions in the beginning of this section.

2. Drain the cooling system.

3. Remove or disconnect the following:
- Negative battery cable
- Drive belt
- Radiator and cooling fan
- Coolant and washer bottles
- Fan shroud
- A/C compressor
- Alternator
- Radiator and heater hoses
- Tensioner and idler pulleys
- Crankshaft damper
- Power steering pump
- Oil pan and pickup tube
- Timing cover
- Re-install the damper

4. Rotate the crankshaft so that the

camshaft sprocket and crankshaft sprocket timing marks are aligned.

➡The camshaft pin and slot in the cam sprocket must be a 12 o'clock, the crankshaft keyway must be at 2 o'clock, and the dots or paint on the crank sprocket must be at 6 o'clock.

5. Pin back the tensioner shoe.

6. Remove the timing chain and sprockets.

To install:

7. With the timing marks aligned, wrap the chain around the sprockets. The chain must be installed with the single plated link aligned with the dot or paint on the cam sprocket. The dot or paint on the crank sprocket should be aligned between the 2 plated links.

8. Install the assembly and torque the cam sprocket bolt to 90 ft. lbs. (122 Nm).

9. Unpin the tensioner and verify the alignment.

10. Install or connect the following:
- Timing cover. Torque all fasteners to 21 ft. lbs. (28 Nm). Torque the large lifting lug to 40 ft. lbs. (55 Nm).
- Oil pan and pickup tube
- Power steering pump
- Crankshaft damper
- Tensioner and idler pulleys
- Radiator and heater hoses
- Alternator
- A/C compressor
- Fan shroud
- Coolant and washer bottles
- Radiator fan
- Drive belt
- Negative battery cable

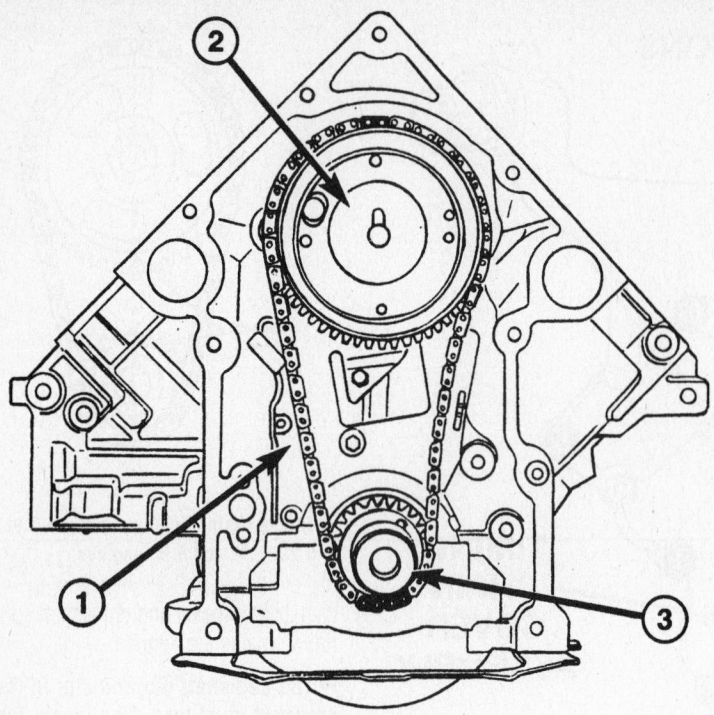

1 - Chain Tensioner
2 - Camshaft Sprocket
3 - Crankshaft Sprocket

2399PG01

Timing mark alignment—5.7L engine

5.2L and 5.9L Engines

1. Before servicing the vehicle, refer to the precautions in the beginning of this section.

2. Drain the cooling system.

3. Remove or disconnect the following:

- Negative battery cable
- Accessory drive belt
- Cooling fan and shroud
- Water pump
- Power steering pump
- Crankshaft damper
- Front crankshaft seal
- Front cover

4. Rotate the crankshaft so that the camshaft sprocket and crankshaft sprocket timing marks are aligned.

5. Remove the timing chain and sprockets.

To install:

6. Install the timing chain and sprockets with the timing marks aligned. Tighten the camshaft sprocket bolt to 50 ft. lbs. (68 Nm).

7. Install or connect the following:

- Front cover. Tighten the cover bolts to 30 ft. lbs. (41 Nm) and the oil pan bolts to 18 ft. lbs. (24 Nm).
- Front crankshaft seal

- Crankshaft damper. Tighten the bolt to 135 ft. lbs. (183 Nm).
- Power steering pump
- Water pump
- Cooling fan and shroud
- Accessory drive belt
- Negative battery cable

8. Fill the cooling system.

9. Start the engine and check for leaks.

8.0L Engine

1. Before servicing the vehicle, refer to the precautions in the beginning of this section.

2. Drain the cooling system.

3. Remove or disconnect the following:

- Negative battery cable
- Accessory drive belt
- Cooling fan and shroud
- A/C compressor, if equipped
- Alternator
- Air injection pump
- Bracket assembly
- Water pump
- Crankshaft pulley
- Front crankshaft seal
- Front oil pan bolts
- Front cover
- Camshaft sprocket and timing chain

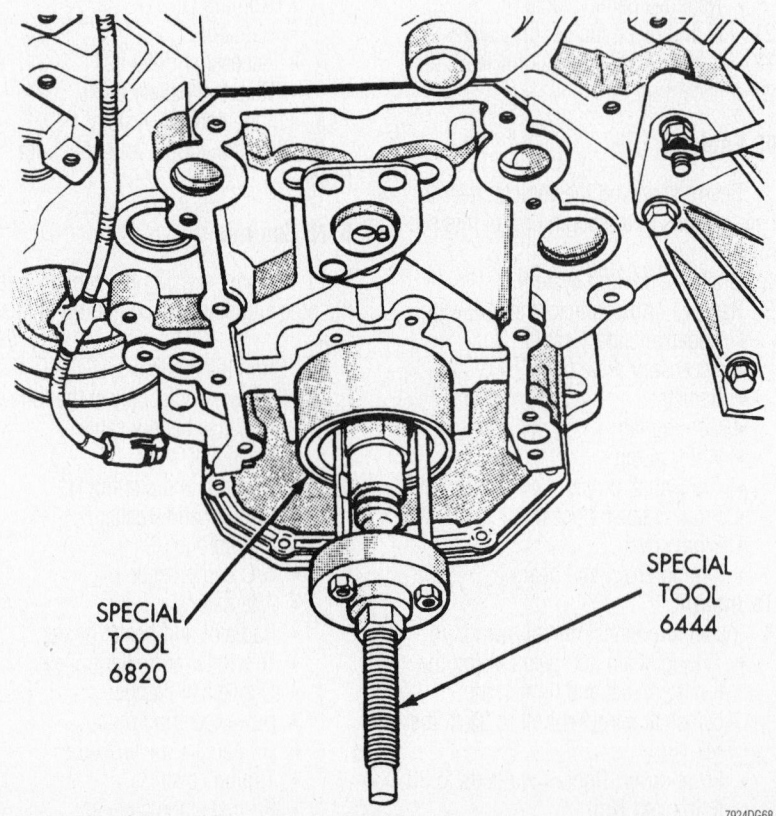

SPECIAL TOOL 6820

SPECIAL TOOL 6444

Crankshaft sprocket removal tools—8.0L engine

7924DG68

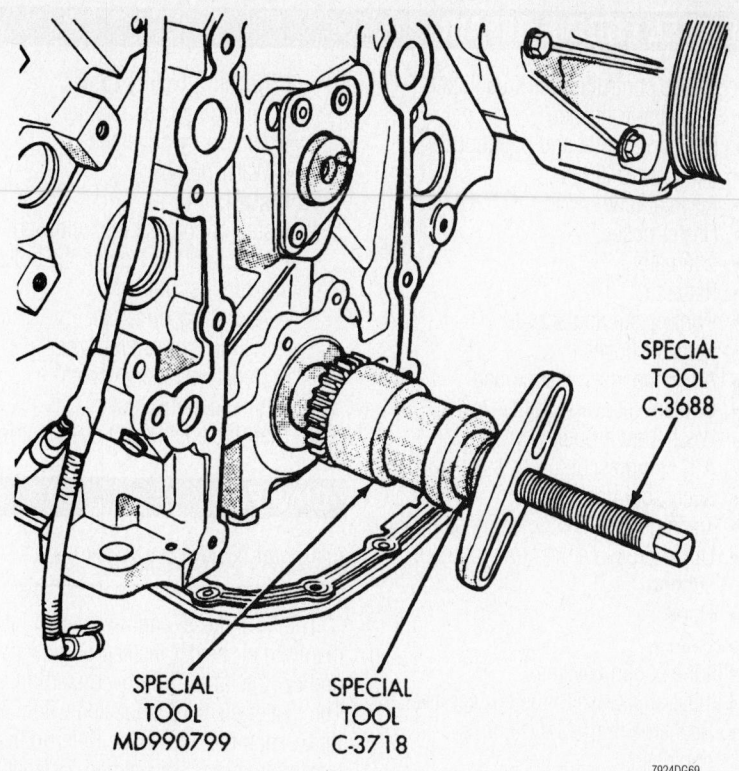

Crankshaft installation tools—8.0L engine

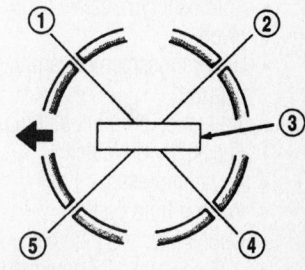

Piston ring end-gap spacing. Position raised "F" on piston toward front of engine—3.7L and 4.7L engines

4. Remove the crankshaft sprocket with Special tools 6444 and 6820.

To install:

5. Install the crankshaft sprocket with Special tools C-3688, C-3718 and MD990799.

6. Install or connect the following:
- Timing chain and camshaft sprocket with the timing marks aligned. Tighten the bolt to 45 ft. lbs. (61 Nm).
- Front cover. Tighten the bolts to 35 ft. lbs. (47 Nm).
- Front oil pan bolts. Tighten the bolts to 12 ft. lbs. (16 Nm).
- Front crankshaft seal
- Crankshaft pulley. Tighten the bolt to 135 ft. lbs. (183 Nm).
- Water pump
- Bracket assembly
- Air injection pump
- Alternator
- A/C compressor, if equipped
- Cooling fan and shroud
- Accessory drive belt
- Negative battery cable

7. Fill the cooling system.
8. Start the engine and check for leaks.

Piston and Ring

POSITIONING

1 - SIDE RAIL UPPER
2 - NO. 1 RING GAP
3 - PISTON PIN
4 - SIDE RAIL LOWER
5 - NO. 2 RING GAP AND SPACER EXPANDER GAP

Piston ring end-gap spacing—5.7L

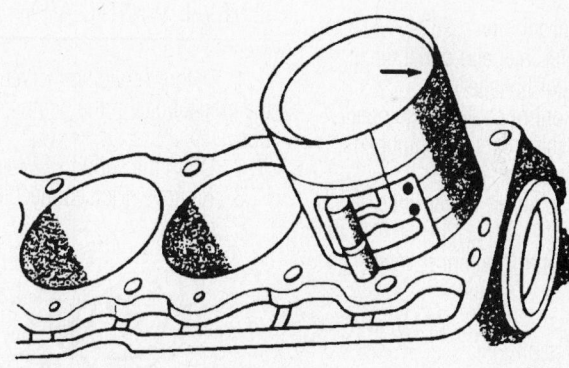

Piston to engine positioning—3.9L, 5.2L, 5.9L and 8.0L engines

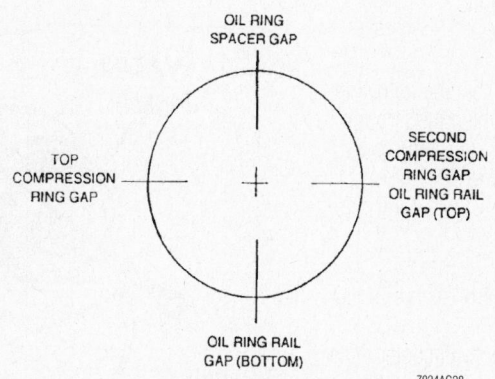

Piston ring end-gap spacing—3.9L, 5.2L, 5.9L and 8.0L engines

DIESEL ENGINE REPAIR

Engine Assembly

REMOVAL & INSTALLATION

1. Before servicing the vehicle, refer to the precautions in the beginning of this section.
2. Drain the cooling system.
3. Drain the engine oil.
4. Recover the A/C refrigerant, if equipped.
5. Remove or disconnect the following:
   - Both battery negative cables
   - Hood
   - Upper crossmember and top core support
   - Transmission oil cooler, if equipped
   - Accessory drive belt
   - A/C compressor
   - Washer fluid bottle
   - Coolant recovery bottle
   - A/C condenser, if equipped
   - Radiator hoses
   - Cooling fan and shroud
   - Radiator
   - Alternator
   - Heater hoses
   - Air inlet tube
   - Exhaust front pipe
   - Intercooler inlet and outlet ducts
   - Accelerator linkage
   - Cruise control cable, if equipped
   - Transmission cable, if equipped
   - Power steering hoses
   - Transmission oil cooler lines, if equipped
   - Engine control sensor harness connectors
   - Fuel lines
   - Transmission
   - Oil pan
   - Starter motor
   - Install engine hoist
   - Engine mounts
   - Engine

To install:

6. Install or connect the following:
   - Engine. Tighten the through bolts to 57 ft. lbs. (77 Nm).
   - Starter motor
   - Oil pan
   - Transmission
   - Fuel lines
   - Engine control sensor harness connectors
   - Transmission oil cooler lines, if equipped
   - Power steering hoses
   - Transmission cable, if equipped
   - Cruise control cable, if equipped
   - Accelerator linkage
   - Intercooler inlet and outlet ducts
   - Exhaust front pipe
   - Air inlet tube
   - Heater hoses
   - Alternator
   - Radiator
   - Cooling fan and shroud
   - Radiator hoses
   - A/C condenser, if equipped
   - Coolant recovery bottle
   - Washer fluid bottle
   - A/C compressor
   - Accessory drive belt
   - Transmission oil cooler, if equipped
   - Upper crossmember and top core support
   - Hood
   - Battery

7. Fill the cooling system.
8. Fill the crankcase to the correct level.
   - Recharge the A/C system, if equipped.
9. Start the engine and check for leaks.

Water Pump

REMOVAL & INSTALLATION

1. Before servicing the vehicle, refer to the precautions in the beginning of this section.
2. Drain the cooling system.
3. Remove or disconnect the following:
   - Negative battery cables
   - Wiring harness retainer
   - Accessory drive belt
   - Water pump

To install:

4. Install or connect the following:
   - Water pump. Tighten the bolts to 18 ft. lbs. (24 Nm).
   - Accessory drive belt
   - Wiring harness retainer
   - Negative battery cables
5. Fill the cooling system.
6. Start the engine and check for leaks.

Glow Plugs

REMOVAL & INSTALLATION

The 5.9L diesel engine uses an intake manifold air heater instead of glow plugs to preheat the air for improved starting ability. The heater element is located within the intake manifold top cover. Refer to the intake manifold removal and installation procedure to service the intake manifold air heater.

Cylinder Head

REMOVAL & INSTALLATION

1. Before servicing the vehicle, refer to the precautions in the beginning of this section.
2. Drain the cooling system.

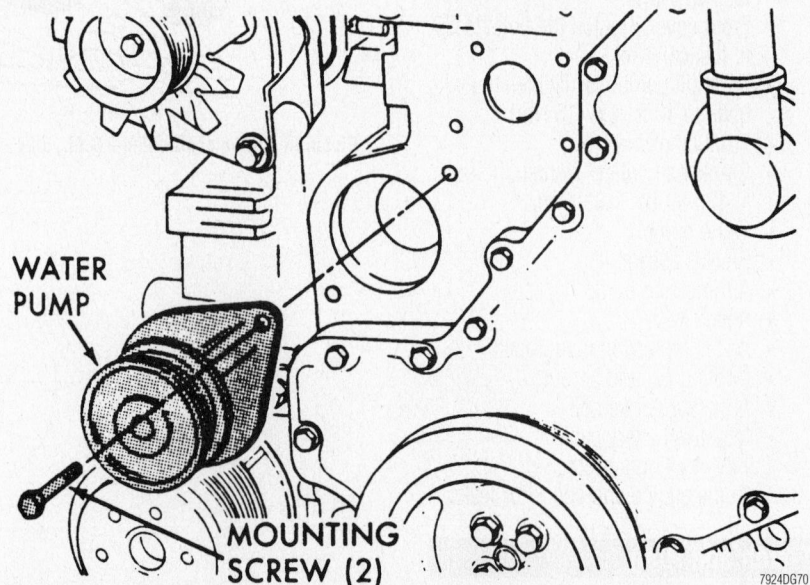

WATER PUMP

MOUNTING SCREW (2)

7924DG70

Exploded view of the water pump mounting—5.9L diesel engine

3. Drain the engine oil.
4. Remove or disconnect the following:

- Negative battery cables
- Radiator hoses
- Heater hoses
- Drive belt
- Turbocharger
- Exhaust Gas Recirculation (EGR) tube
- Exhaust manifold
- Accelerator pedal position sensor
- Fuel lines and injector nozzles
- Valve cover

➡**Keep all valvetrain components in order for assembly.**

- Rocker levers and pedestal assemblies
- Pushrods
- Fuel filter and water separator assembly

➡**If the cylinder head is hot, gradually loosen the cylinder head bolts using the TIGHTENING sequence. If the engine is cold, then the loosening sequence for the head bolts is not important.**

➡**The cylinder head bolts are different sizes. Note their locations for assembly.**

- Cylinder head bolts
- Cylinder head

To install:

➡**Check the cylinder head bolt length. If length exceeds 5.2 inches (132.1 mm), the bolt must be replaced.**

5. Install or connect the following:
- Cylinder head
- Pushrods
- Rocker levers and pedestal assemblies

6. Install the cylinder head bolts and tighten them in sequence as follows:
   a. Step 1: 59 ft. lbs. (80 Nm).
   b. Step 2: 77 ft. lbs. (104 Nm).
   c. Step 3: Retighten to 77 ft. lbs. (80 Nm).
   d. Step 4: Tighten an additional 90 degrees

7. Install or connect the following:
- Fuel filter and water separator assembly
- Valve cover
- Fuel lines and injector nozzles
- Accelerator pedal position sensor
- Exhaust manifold
- EGR tube
- Turbocharger

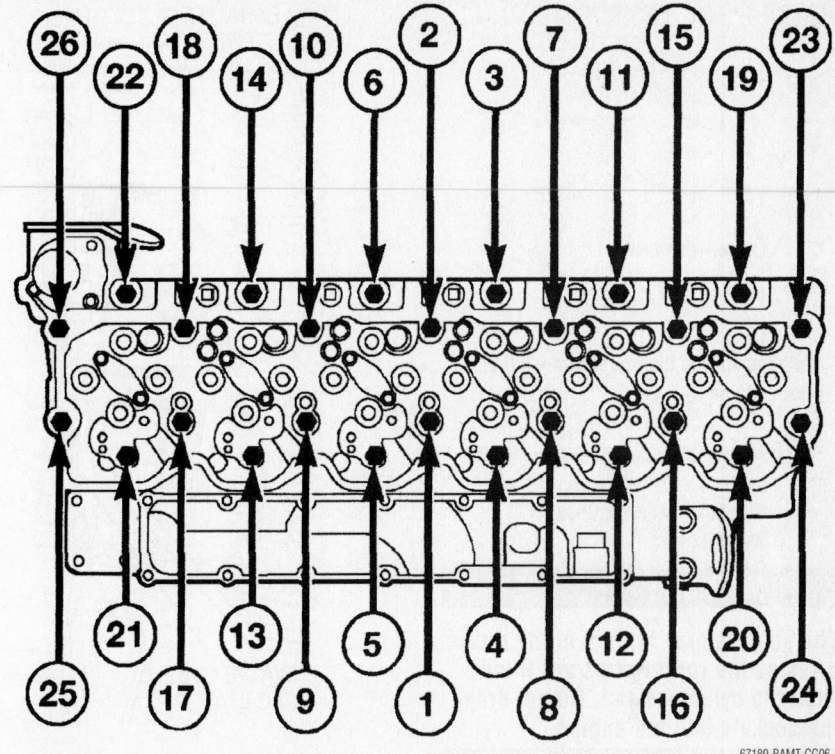

Cylinder head torque sequence—5.9L diesel engine

- Drive belt
- Heater hoses
- Radiator hoses
- Negative battery cables

8. Fill the crankcase to the correct level.
9. Fill the cooling system.
10. Start the engine and check for leaks.

Rocker Arms/Shafts

REMOVAL & INSTALLATION

1. Before servicing the vehicle, refer to the precautions in the beginning of this section.

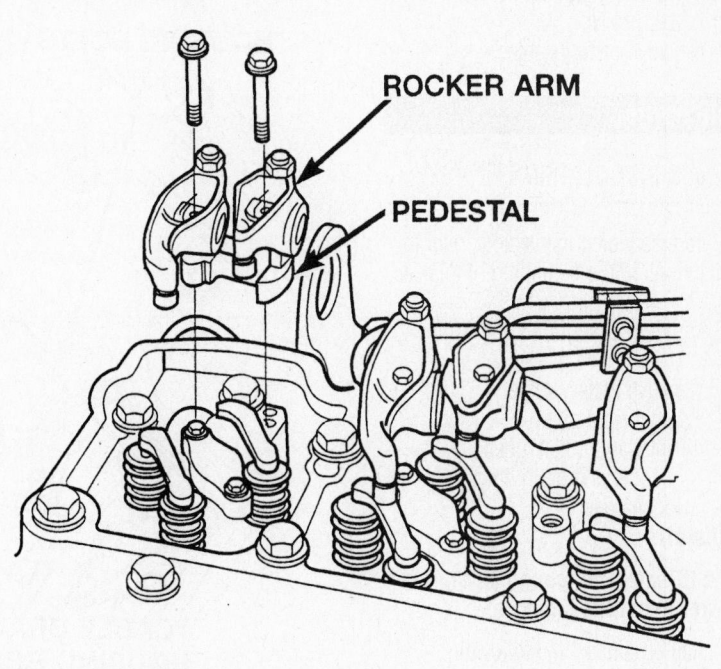

ROCKER ARM

PEDESTAL

Exploded view of the rocker arm mounting—5.9L diesel engine

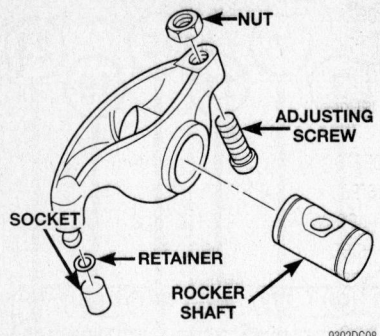

Exploded view of the rocker arm—5.9 L diesel engine

2. Remove or disconnect the following:

- Negative battery cables
- Valve cover

❄❄ WARNING

The sockets may fall out of the rocker arms as the rocker arms are lifted from the cylinder head. Do not drop the sockets into the engine.

➡**Keep all valvetrain components in order for assembly.**

- Rocker arms and pedestal assemblies

To install:

3. Install or connect the following:

- Rocker arm and pedestal assemblies. Tighten the bolts to 27 ft. lbs. (36 Nm).
- Valve cover. Tighten the bolts to 18 ft. lbs. (24 Nm).
- Negative battery cables

Turbocharger

REMOVAL & INSTALLATION

1. Before servicing the vehicle, refer to the precautions in the beginning of this section.

2. Remove or disconnect the following:

- Negative battery cable
- Exhaust front pipe
- Air inlet and outlet tubes
- Oil supply and drain lines
- Turbocharger

To install:

➡**Use anti-seize compound on the turbocharger mounting studs.**

3. Install or connect the following:

- Turbocharger. Tighten the nuts to 24 ft. lbs. (32 Nm).

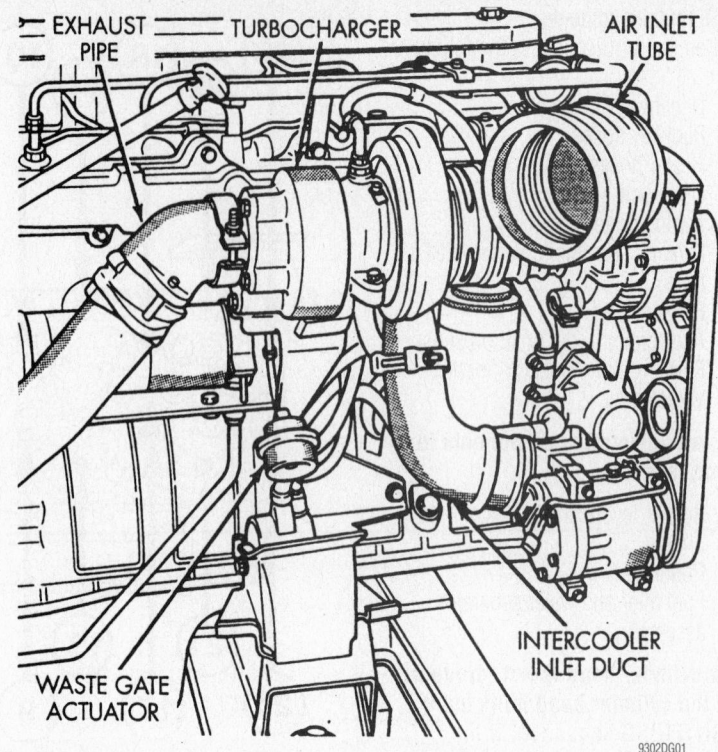

Turbocharger and related components—5.9L diesel engine

- Oil supply and drain lines
- Air inlet and outlet tubes
- Exhaust front pipe
- Negative battery cable

4. Start the engine and check for leaks.

Intake Manifold

REMOVAL & INSTALLATION

1. Before servicing the vehicle, refer to the precautions in the beginning of this section.

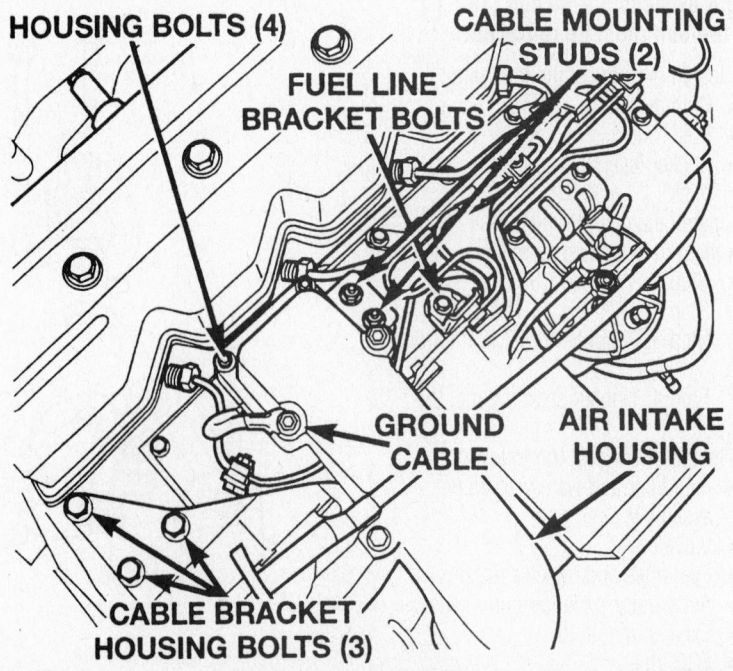

Intake air heater electrical connections—5.9L diesel engine

2. Remove or disconnect the following:
  - Negative battery cables
  - Intercooler outlet duct
  - Dipstick tube
  - Engine appearance cover
  - Air inlet housing
  - Exhaust Gas Recirculation (EGR) tube
  - Fuel line assembly
  - Air intake heater harness connectors
  - Charge air temperature sensor connector
  - Accelerator pedal position sensor bracket
  - Intake manifold cover

To install:

➡**Use liquid Teflon® sealer on the intake manifold cover bolts.**

3. Install or connect the following:
  - Intake manifold cover. Tighten the bolts to 18 ft. lbs. (24 Nm).
  - Accelerator pedal position sensor bracket. Tighten the bracket to 32 ft. lbs. (43 Nm).
  - Charge air temperature sensor connector
  - Air intake heater harness connectors. Tighten the nuts to 10 ft. lbs. (14 Nm).
  - Fuel line assembly
  - Exhaust Gas Recirculation (EGR) tube
  - Air inlet housing. Tighten the bolts to 18 ft. lbs. (24 Nm).
  - Engine appearance cover
  - Dipstick tube
  - Intercooler outlet duct
  - Negative battery cables
4. Start the engine and check for leaks.

Exhaust Manifold

REMOVAL & INSTALLATION

1. Before servicing the vehicle, refer to the precautions in the beginning of this section.
2. Remove or disconnect the following:
  - Negative battery cables
  - Turbocharger
  - Exhaust Gas Recirculation (EGR) tube
  - Cab heater supply and return lines
  - Exhaust manifold

To install:

➡**Use anti-seize compound on the exhaust manifold bolts.**

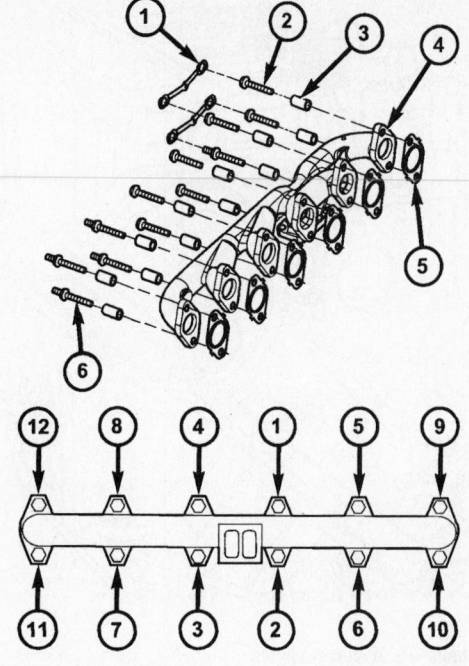

1 - RETAINING STRAP
2 - BOLT (7)
3 - SPACER
4 - MANIFOLD, EXHAUST
5 - GASKET
6 - BOLT (5)

67189-RAMT-CG07

Exhaust manifold torque sequence—5.9L diesel engine

3. Install or connect the following:
  - Exhaust manifold. Tighten the bolts in sequence to 32 ft. lbs. (43 Nm).
  - Cab heater supply and return lines
  - EGR tube. Tighten the bolts to 18 ft. lbs. (24 Nm).
  - Turbocharger
  - Negative battery cables
4. Start the engine and check for leaks.

Camshaft and Valve Lifters

REMOVAL & INSTALLATION

1. Before servicing the vehicle, refer to the precautions in the beginning of this section.
2. Recover the A/C refrigerant, if equipped.
3. Drain the cooling system.
4. Remove or disconnect the following:
  - Negative battery cables
  - Accessory drive belt and tensioner
  - Cooling fan and shroud
  - Radiator
  - A/C condenser, if equipped
  - Intercooler
  - Auxiliary transmission cooler
  - Upper radiator support

  - Exhaust Gas Recirculation (EGR) tube
  - Engine appearance cover
  - Valve cover

➡**Keep all valvetrain components in order for assembly.**

  - Rocker arms
  - Pushrods
  - Crankshaft pulley
  - Front cover
  - Lift pump
5. Insert dowel tools into the tappets. Raise the dowels and secure them with rubber bands.

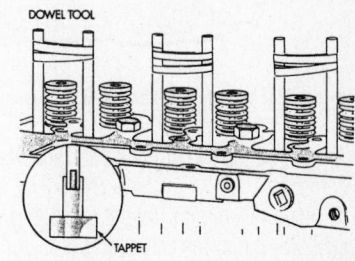

7924DG74

Use dowel tools and rubber bands to hold the lifters up in the bore while removing the camshaft—5.9L diesel engine

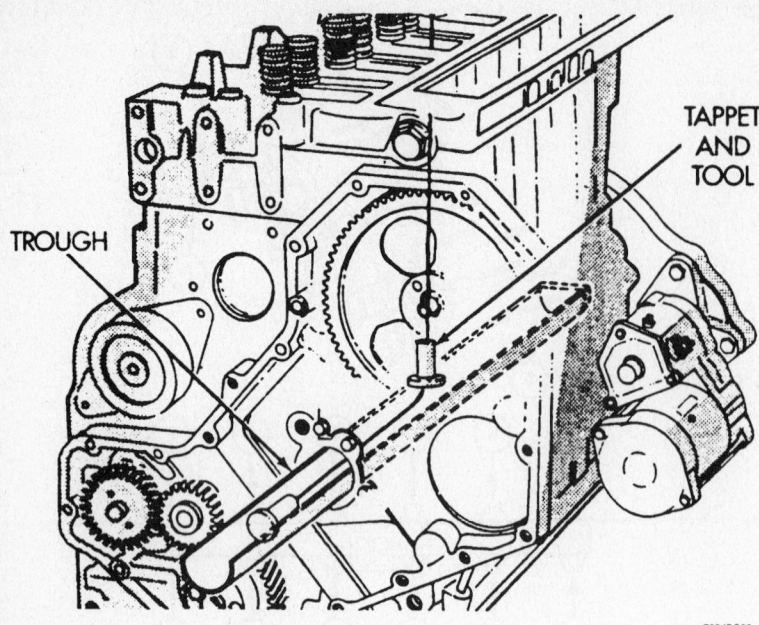

Tappet installation tools—5.9L diesel engine

7924DG29

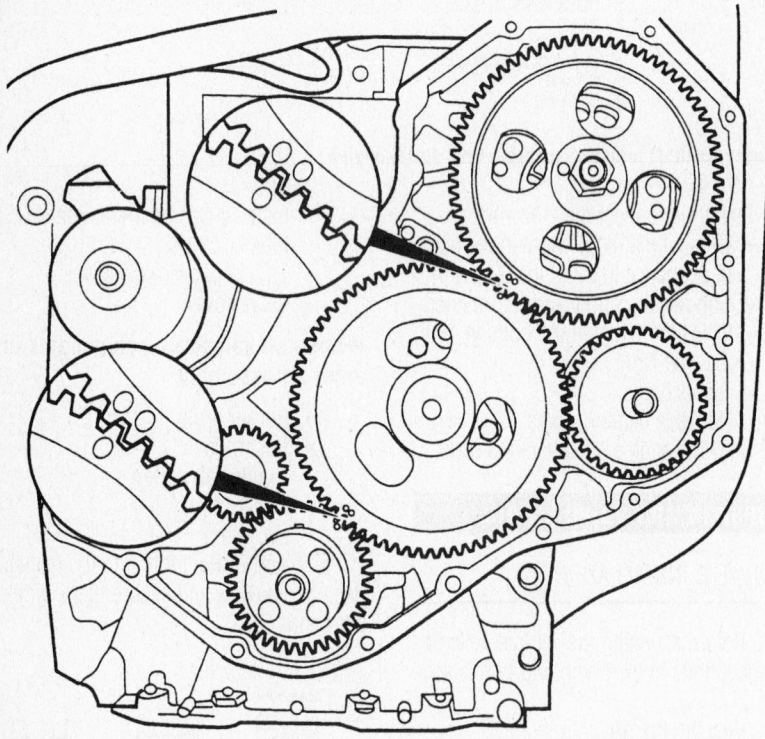

Timing gear alignment marks—5.9L diesel engine

9302DG06

6. Unbolt the thrust plate and remove the camshaft.

7. Install Cummins Tappet Changing Tool 3822513 into the camshaft bore.

8. Remove the dowel tools and remove the tappets. If the tappets are to be reused, note their locations for assembly.

To install:

9. Insert the tappet installation tool into the pushrod bore. Use the tappet trough to pull the installation tool to the front of the engine block.

10. Attach a tappet to the installation tool and pull the tappet through the camshaft bore and into position in the tappet bore.

11. Rotate the trough so that the round side holds the tappet up in the tappet bore.

12. Remove the tappet installation tool and install a dowel tool to hold the tappet in place.

13. Repeat for each tappet to be installed.

14. Install or connect the following:
- Camshaft and thrust plate. Align the timing marks and tighten the bolts to 18 ft. lbs. (24 Nm).
- Lift pump
- Front cover
- Crankshaft pulley. Tighten the bolt to 92 ft. lbs. (125 Nm).
- Pushrods
- Rocker arms
- Valve cover
- Engine appearance cover
- EGR tube
- Upper radiator support
- Auxiliary transmission cooler
- Intercooler
- A/C condenser, if equipped
- Radiator
- Cooling fan and shroud
- Accessory drive belt tensioner
- Accessory drive belt
- Negative battery cables

15. Fill the cooling system.

16. Start the engine and check for leaks.

Valve Lash

ADJUSTMENT

1. Before servicing the vehicle, refer to the precautions in the beginning of this section.

2. Remove or disconnect the following:
- Negative battery cables
- Valve cover
- Fuel pump gear access cover

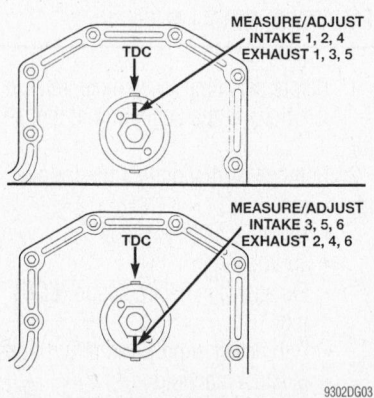

9302DG03

Adjust the specified valves when the mark on the pump gear is in either of the 2 positions—diesel engines

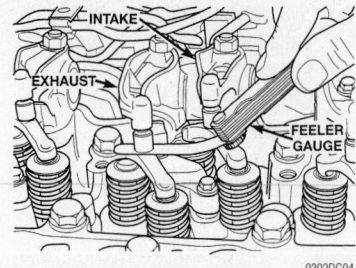

Use a feeler gauge to measure the valve lash—diesel engine

3. Position the gear as shown and measure the clearance of the indicated valves. No adjustment is necessary if the lash falls within the following specifications:

    a. Intake—0.006–0.015 inch (0.152–0.381mm).

    b. Exhaust—0.015–0.030 inch (0.381–0.762mm).

4. Install or connect the following:
- Fuel pump access cover
- Valve cover
- Negative battery cables

Oil Pan

REMOVAL & INSTALLATION

1. Before servicing the vehicle, refer to the precautions in the beginning of this section.

2. Drain the engine oil.

3. Remove or disconnect the following:

- Negative battery cables
- Starter motor
- Transmission
- Flywheel and transmission adapter plate.
- Oil pan bolts
- Oil pump suction tube
- Oil pan

To install:

4. Install or connect the following:
- Oil pan
- Oil pump suction tube. Tighten the bolts to 18 ft. lbs. (24 Nm).
- Oil pan bolts. Tighten the bolts to 18 ft. lbs. (24 Nm).

- Flywheel and transmission adapter plate.
- Transmission
- Starter motor

5. Fill the crankcase to the correct level.

Oil Pump

REMOVAL & INSTALLATION

1. Before servicing the vehicle, refer to the precautions in the beginning of this section.

2. Drain the cooling system.

3. Remove or disconnect the following:

- Negative battery cables
- Accessory drive belt
- Cooling fan and shroud
- Radiator
- Oil fill tube and adapter
- Crankshaft pulley
- Front cover
- Oil pump

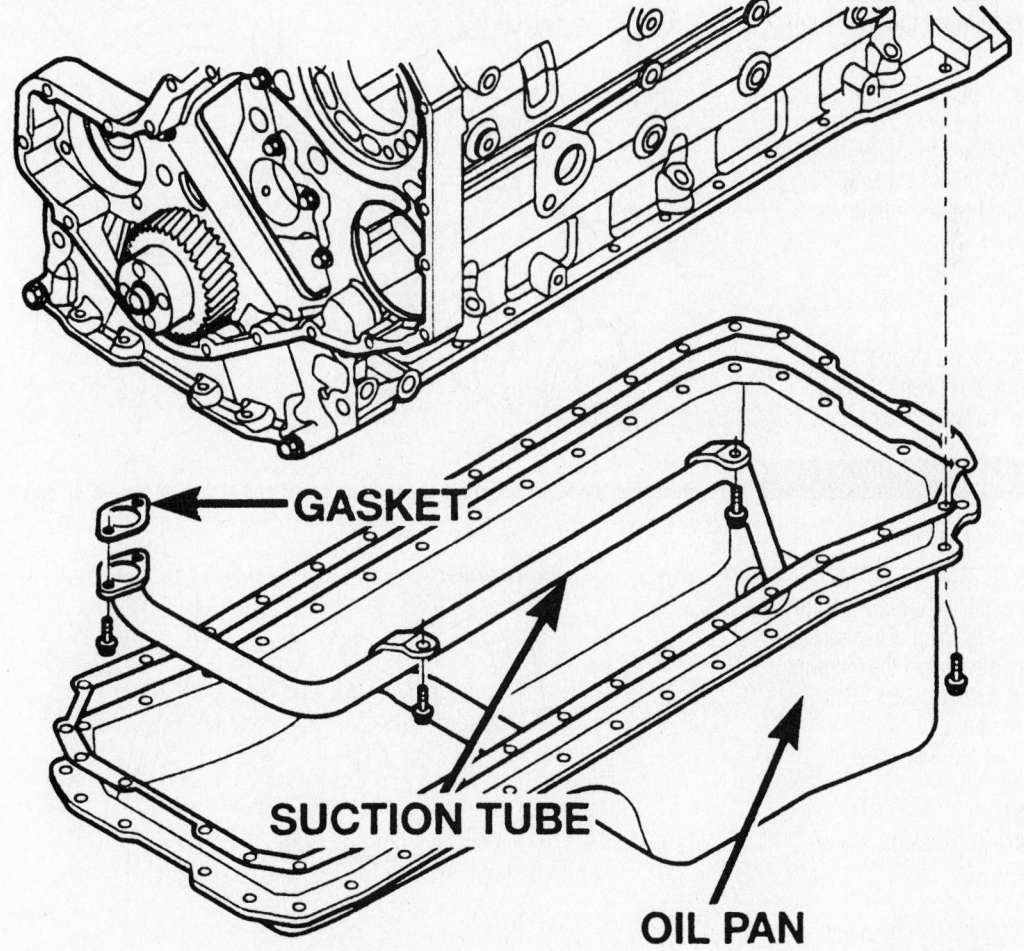

GASKET

SUCTION TUBE

OIL PAN

9302DG05

Exploded view of the oil pan mounting—diesel engines

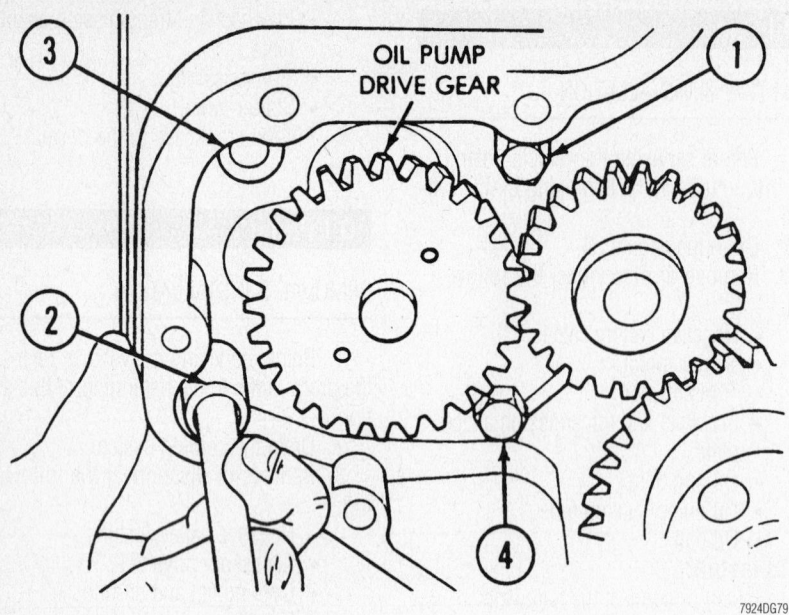

Oil pump torque sequence—5.9L diesel engine

To install:

➡**When the pump is correctly installed, the flange on the pump does not touch the block; the back plate on the pump seats against the bottom of the bore.**

4. Install the oil pump. Tighten the bolts in sequence as follows:
   a. Step 1: 44 inch lbs. (5 Nm).
   b. Step 2: 18 ft. lbs. (24 Nm).
5. Install or connect the following:
   - Front cover
   - Crankshaft pulley
   - Oil fill tube and adapter
   - Radiator
   - Cooling fan and shroud
   - Accessory drive belt
   - Negative battery cables

Rear Main Seal

REMOVAL & INSTALLATION

1. Before servicing the vehicle, refer to the precautions in the beginning of this section.
2. Remove or disconnect the following:
   - Negative battery cables
   - Transmission
   - Clutch and pressure plate, if equipped
   - Flywheel
   - Seal retainer housing
   - Rear main seal

To install:

3. Install or connect the following:
   - Rear main seal. Use the alignment tool supplied with the seal kit.

- Seal retainer housing. Tighten the bolts to 84 inch lbs. (9 Nm).
- Flywheel
- Clutch and pressure plate, if equipped
- Transmission
- Negative battery cables
4. Start the engine and check for leaks.

Timing Gears, Front Cover and Seal

REMOVAL & INSTALLATION

Front Cover and Seal

1. Before servicing the vehicle, refer to the precautions in the beginning of this section.
2. Remove or disconnect the following:
   - Negative battery cables
   - Accessory drive belt
   - Cooling fan and shroud
   - Accessory drive belt tensioner

Removing the rear seal with a sheet metal screw and slide hammer—5.9L diesel engine

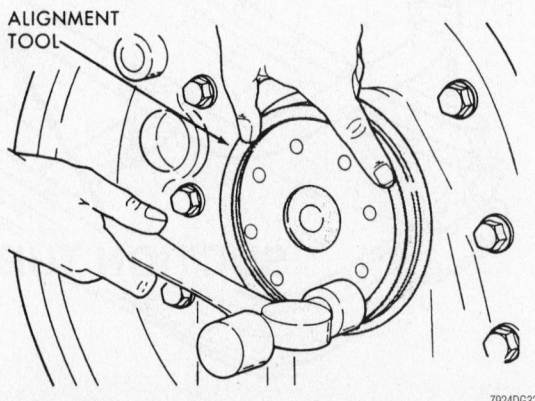

Place the alignment tool on the seal and tap the seal into place—5.9L diesel engine

- Oil fill tube and adapter
- Crankshaft pulley
- Front cover
- Front crankshaft seal

To install:

3. Install or connect the following:
- Front crankshaft seal
- Front cover. Tighten the bolts to 18 ft. lbs. (24 Nm).
- Crankshaft pulley
- Oil fill tube and adapter. Tighten the bolts to 32 ft. lbs. (43 Nm).
- Accessory drive belt tensioner. Tighten the bolts to 32 ft. lbs. (43 Nm).
- Cooling fan and shroud
- Accessory drive belt. Tighten the crankshaft pulley bolts to 92 ft. lbs. (125 Nm).
- Negative battery cables

4. Start the engine and check for leaks.

Timing Gears

1. Before servicing the vehicle, refer to the precautions in the beginning of this section.
2. Remove or disconnect the following:
- Negative battery cables
- Accessory drive belt
- Cooling fan and shroud
- Belt tensioner
- Oil fill tube and adapter
- Crankshaft pulley
- Front cover
- Camshaft

3. Press the camshaft out of the timing gear.

To install:

4. Install the camshaft key.
5. Heat the timing gear in an oven to 350°F (177°C) for 45 minutes.

➡**The camshaft gear will be permanently distorted if overheated. Do not exceed 350°F (177°C).**

6. Install the timing gear to the camshaft with the timing marks facing out and the gear seated on the camshaft shoulder.
7. Install or connect the following:
- Camshaft with the timing marks aligned
- Front cover. Tighten the bolts to 18 ft. lbs. (24 Nm).
- Crankshaft pulley. Tighten the bolt to 92 ft. lbs. (125 Nm).
- Oil fill tube and adapter. Tighten the mounting bolts to 32 ft. lbs. (43 Nm).
- Belt tensioner. Tighten the mounting bolts to 32 ft. lbs. (43 Nm).
- Cooling fan and shroud
- Accessory drive belt
- Negative battery cables

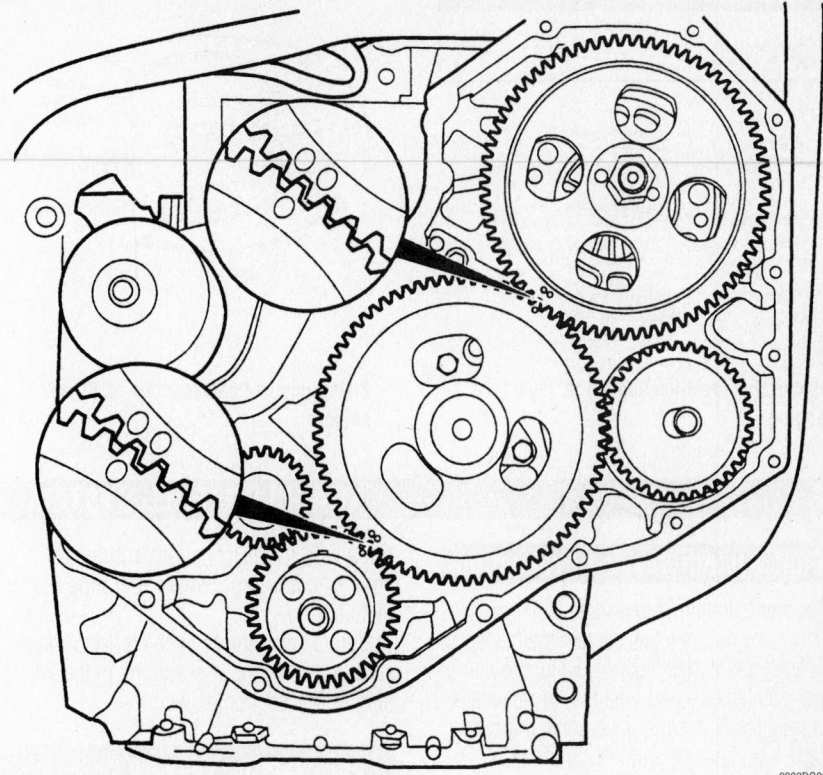

9302DG06

Timing gear alignment marks—5.9L diesel engine

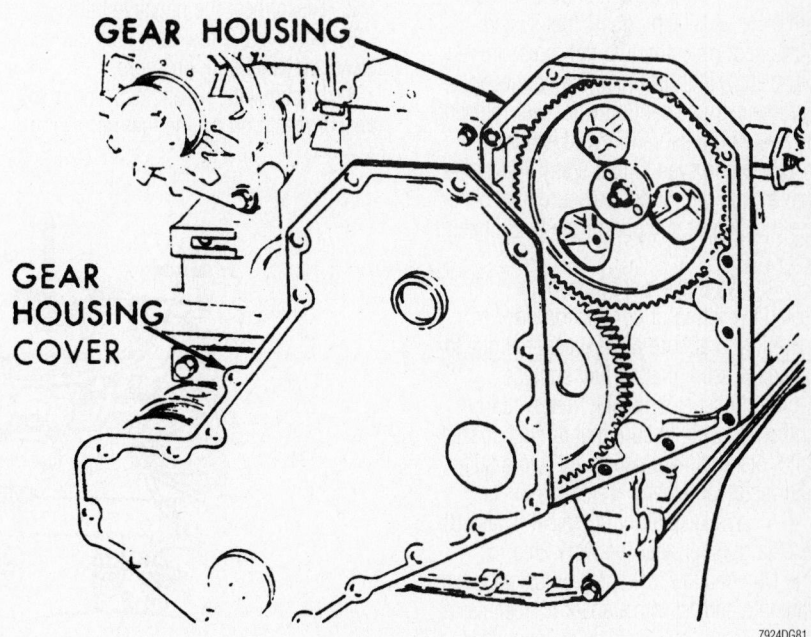

GEAR HOUSING

GEAR HOUSING COVER

7924DG81

Remove the gear cover to replace the timing gears—5.9L diesel engine

Piston and Ring

POSITIONING

TOP RING   TOP

INTERMEDIATE
RING

OIL CONTROL
RING

7924AG30

Piston ring identification—5.9L diesel engine

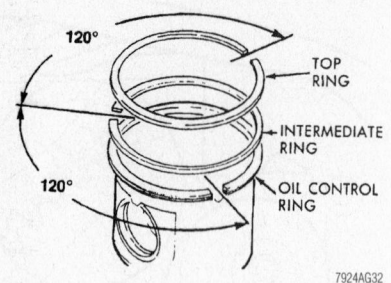

Piston ring end gap spacing—5.9L diesel engine

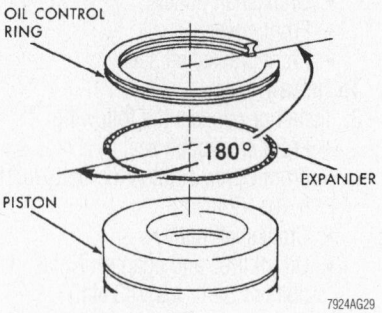

Oil control ring-to-spacer end gap spacing—5.9L diesel engine

GASOLINE FUEL SYSTEM

Fuel System Service Precautions

Safety is the most important factor when performing not only fuel system maintenance but any type of maintenance. Failure to conduct maintenance and repairs in a safe manner may result in serious personal injury or death. Maintenance and testing of the vehicle's fuel system components can be accomplished safely and effectively by adhering to the following rules and guidelines.

• To avoid the possibility of fire and personal injury, always disconnect the negative battery cable unless the repair or test procedure requires that battery voltage be applied.

• Always relieve the fuel system pressure prior to detaching any fuel system component (injector, fuel rail, pressure regulator, etc.), fitting or fuel line connection. Exercise extreme caution whenever relieving fuel system pressure to avoid exposing skin, face and eyes to fuel spray. Please be advised that fuel under pressure may penetrate the skin or any part of the body that it contacts.

• Always place a shop towel or cloth around the fitting or connection prior to loosening to absorb any excess fuel due to spillage. Ensure that all fuel spillage (should it occur) is quickly removed from engine surfaces. Ensure that all fuel soaked cloths or towels are deposited into a suitable waste container.

• Always keep a dry chemical (Class B) fire extinguisher near the work area.

• Do not allow fuel spray or fuel vapors to come into contact with a spark or open flame.

• Always use a back-up wrench when loosening and tightening fuel line connection fittings. This will prevent unnecessary stress and torsion to fuel line piping.

• Always replace worn fuel fitting O-rings with new. Do not substitute fuel hose or equivalent, where fuel pipe is installed.

Before servicing the vehicle, also make sure to refer to the precautions in the beginning of this section as well.

Fuel System Pressure

RELIEVING

2001–02

1. Before servicing the vehicle, refer to the precautions in the beginning of this section.
2. Disconnect the negative battery cable.
3. Remove the fuel tank filler cap to release any fuel tank pressure.
4. Unscrew the plastic cap from the pressure test port on the fuel rail. On the 8.0L engine, the test port is found at the front of the engine.
5. Obtain a fuel pressure gauge/hose from a fuel pressure gauge tool set No. 5069, or equivalent. Remove the gauge, then place the gauge end of the hose into a suitable gasoline container.
6. Place a shop towel under the test port.
7. Screw the other end of the hose onto the fuel pressure port to relieve the pressure.
8. When the pressure has been relieved, remove the hose and cap the port.

2003–04

1. Before servicing the vehicle, refer to the precautions in the beginning of this section.
2. Disconnect the negative battery cable.
3. Remove the fuel tank filler cap to release any fuel tank pressure.

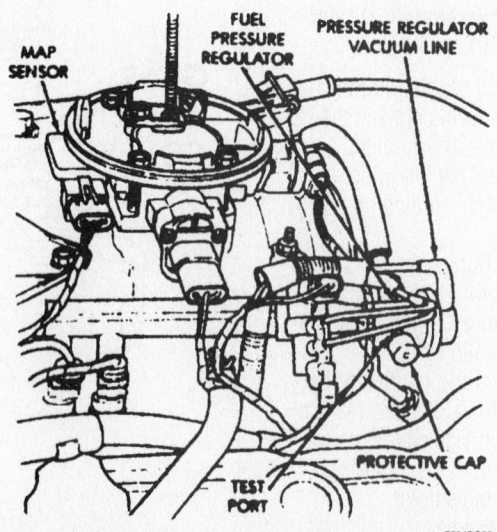

Fuel pressure test port—3.9L engine

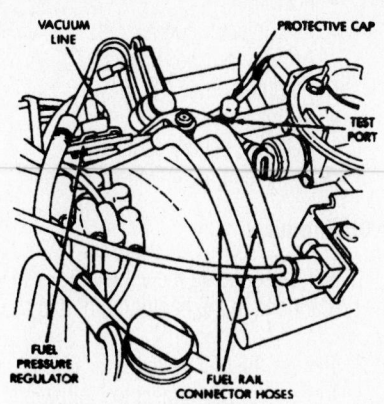

Fuel pressure test port—5.2L engine, 5.9L engine is similar

4. Remove the fuel pump relay from the PDC.

5. Start and run the engine until it stops.

6. Unplug the connector from any injector and connect a jumper wire from either injector terminal to the positive battery terminal. Connect another jumper wire to the other terminal and momentarily touch the other end to the negative battery terminal.

✳✳ WARNING

Just touch the jumper to the battery. Powering the injector for more than a few seconds will permanently damage it.

7. Place a rag below the quick-disconnect coupling at the fuel rail and disconnect it.

Fuel Filter

REMOVAL & INSTALLATION

1. Before servicing the vehicle, refer to the precautions in the beginning of this section.

2. Relieve the fuel system pressure.

3. Remove or disconnect the following:

- Negative battery cable
- Fuel tank

4. Pull the filter/regulator out of the rubber grommet. Cut the hose clamp and remove the fuel line.

To install:

5. Install the filter/regulator with a new clamp and push it into the rubber grommet.

6. Install or connect the following:

- Fuel tank
- Negative battery cable

7. Start the engine and check for leaks.

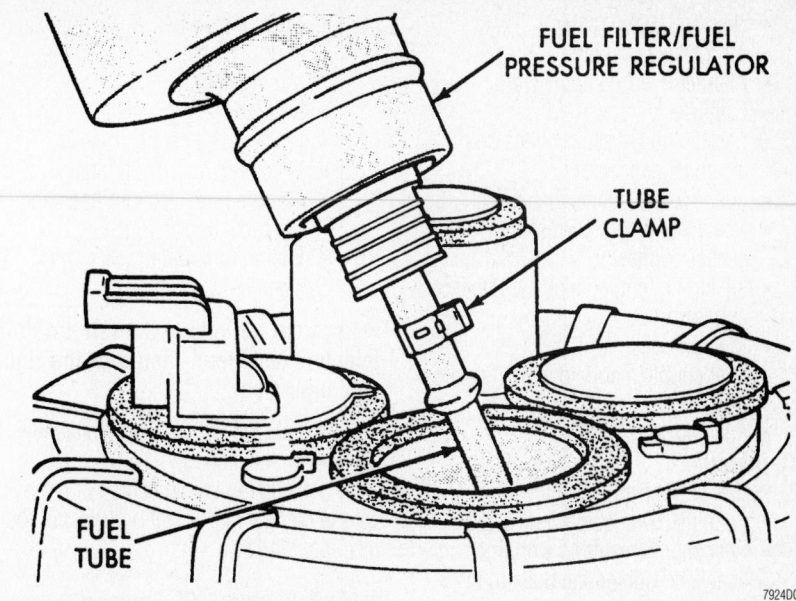

Pull and twist the filter/regulator to remove it from the top of the fuel pump module

Fuel Pump

REMOVAL & INSTALLATION

1. Before servicing the vehicle, refer to the precautions in the beginning of this section.

2. Relieve the fuel system pressure.

3. Remove or disconnect the following:

- Negative battery cable
- Fuel pump module harness connector
- Fuel line
- Fuel tank
- Fuel pump module locknut
- Fuel pump module

To install:

4. Install or connect the following:

- Fuel pump module
- Fuel pump module locknut
- Fuel tank
- Fuel line
- Fuel pump module harness connector
- Negative battery cable

5. Start the engine and check for leaks.

Fuel Injector

REMOVAL & INSTALLATION

3.7L and 4.7L Engines

1. Before servicing the vehicle, refer to the precautions in the beginning of this section.

2. Relieve fuel system pressure.

3. Remove or disconnect the following:

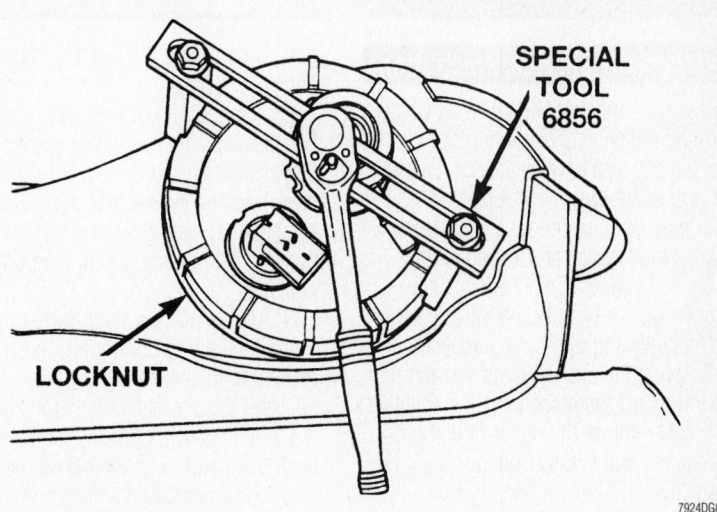

Fuel pump module locknut removal

- Negative battery cable
- Air intake assembly
- Alternator wiring connectors
- Fuel line
- Throttle body vacuum lines and electrical connectors
- Fuel injector connectors
- Manifold Absolute Pressure (MAP) sensor connector
- Intake Air Temperature (IAT) sensor connector
- Ignition coils
- Fuel supply manifold with injectors attached
- Fuel injectors

To install:

4. Install or connect the following:
- Fuel injectors, using new O-rings
- Fuel supply manifold with injectors attached. Tighten the bolts to 20 ft. lbs. (27 Nm).
- Ignition coils
- IAT sensor connector
- MAP sensor connector
- Fuel injector connectors
- Throttle body vacuum lines and electrical connectors
- Fuel line
- Alternator wiring connectors
- Air intake assembly
- Negative battery cable

5. Start the engine and check for leaks.

5.7L Engine

1. Before servicing the vehicle, refer to the precautions in the beginning of this section.
2. Relieve the fuel system pressure.
3. Remove or disconnect the following:
- Negative battery cable
- Air intake assembly

- Spark plug wires at the plugs and coils. . .but leave them in their routing clips.
- Routing tray
- Ignition coil connectors
- Injector wires. . .note positions
- Throttle body vacuum lines and electrical connectors
- Fuel rail mounting volts and clamps

➡ **Rock the left end of the rail until the injectors just clear, then rock the right side until clear.**

- Fuel rail with injectors attached
- Fuel injectors

4. Installation is the reverse of removal. Torque the fuel rail mounting bolts to 100 inch lbs. (11 Nm).

3.9L, 5.2L and 5.9L Engines

1. Before servicing the vehicle, refer to the precautions in the beginning of this section.
2. Relieve the fuel system pressure.
3. Remove or disconnect the following:
- Negative battery cable
- Air intake tube
- Throttle body
- A/C compressor bracket
- Fuel injector connectors
- Fuel line
- Fuel supply manifold with injectors
- Fuel injectors

To install:

4. Install or connect the following:
- Fuel injectors with new O-ring seals
- Fuel supply manifold with injectors. Tighten the bolts to 17 ft. lbs. (23 Nm).

- Fuel line
- Fuel injector connectors
- A/C compressor bracket
- Throttle body
- Air intake tube
- Negative battery cable

5. Start the engine and check for leaks.

8.0L Engine

1. Before servicing the vehicle, refer to the precautions in the beginning of this section.
2. Relieve the fuel system pressure.
3. Remove or disconnect the following:
- Negative battery cable
- Air cleaner housing and tube
- Throttle body
- Ignition coil pack and bracket
- Upper intake manifold
- Fuel injector harness connectors
- Fuel line
- Fuel supply manifold with injectors attached
- Fuel injector retainer clips
- Fuel injectors

To install:

4. Install or connect the following:
- Fuel injectors with new O-ring seals
- Fuel injector retainer clips
- Fuel supply manifold. Tighten the bolts to 11 ft. lbs. (15 Nm).
- Fuel line
- Fuel injector harness connectors
- Upper intake manifold
- Ignition coil pack and bracket
- Throttle body
- Air cleaner housing and tube
- Negative battery cable

5. Start the engine and check for leaks.

DIESEL FUEL SYSTEM

Fuel System Service Precautions

Safety is the most important factor when performing not only fuel system maintenance, but any type of maintenance. Failure to conduct maintenance and repairs in a safe manner may result in serious personal injury or death. Maintenance and testing of the vehicle's fuel system components can be accomplished safely and effectively by adhering to the following rules and guidelines.

- To avoid the possibility of fire and personal injury, always disconnect the negative battery cable unless the repair or test procedure requires that battery voltage be applied.
- Always relieve the fuel system pres-

sure prior to disengaging any fuel system component (injector, fuel rail, pressure regulator, etc.), fitting or fuel line connection. Exercise extreme caution whenever relieving fuel system pressure, to avoid exposing skin, face and eyes to fuel spray. Please be advised that fuel under pressure may penetrate the skin or any part of the body that it contacts.

- Always place a shop towel or cloth around the fitting or connection prior to loosening to absorb any excess fuel due to spillage. Ensure that all fuel spillage (should it occur) is quickly removed from engine surfaces. Ensure that all fuel soaked cloths or towels are deposited into a suitable waste container.

- Always keep a dry chemical (Class B) fire extinguisher near the work area.
- Do not allow fuel spray or fuel vapors to come into contact with a spark or open flame.
- Always use a back-up wrench when loosening and tightening fuel line connection fittings. This will prevent unnecessary stress and torsion to fuel line piping.
- Always replace worn fuel fitting O-rings with new. Do not substitute fuel hose or equivalent, where fuel pipe is installed.

Before servicing the vehicle, also make sure to refer to the precautions in the beginning of this section as well.

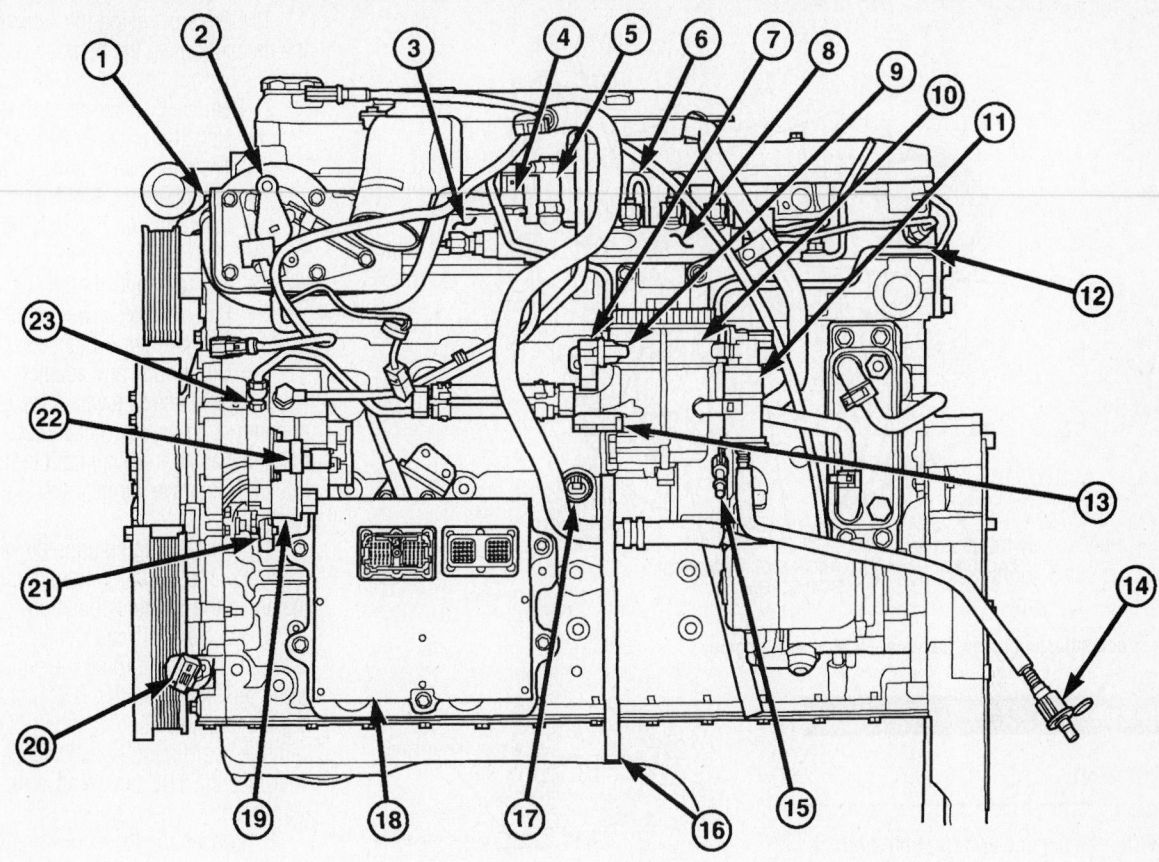

1 - ENGINE COOLANT TEMPERATURE (ECT) SENSOR
2 - THROTTLE LEVER BELLCRANK AND APPS (ACCELERATOR PEDAL POSITION SENSOR)
3 - INTAKE MANIFOLD AIR HEATER/ELEMENTS
4 - FUEL PRESSURE SENSOR
5 - FUEL PRESSURE LIMITING VALVE
6 - HIGH-PRESSURE FUEL LINES
7 - FUEL HEATER
8 - HIGH-PRESSURE FUEL RAIL
9 - FUEL HEATER TEMPERATURE SENSOR (THERMOSTAT)
10 - FUEL FILTER/WATER SEPARATOR
11 - FUEL TRANSFER (LIFT) PUMP
12 - FUEL DRAIN MANIFOLD (CYLINDER HEAD FUEL RETURN LINE)
13 - DRAIN VALVE

14 - FUEL SUPPLY LINE (LOW-PRESSURE, TO ENGINE)
15 - FUEL RETURN LINE CONNECTION (TO FUEL TANK)
16 - FUEL DRAIN TUBE
17 - OIL PRESSURE SWITCH
18 - ENGINE CONTROL MODULE (ECM)
19 - FUEL INJECTION PUMP
20 - CRANKSHAFT POSITION (ENGINE SPEED) SENSOR
21 - CAMSHAFT POSITION SENSOR (CMP)
22 - FUEL CONTROL ACTUATOR (FCA)
23 - CASCADE OVERFLOW VALVE

67189-RAMT-CG08

5.9L Diesel fuel system components—2004 shown, other years similar

Fuel System

BLEEDING AIR

1. Loosen the low pressure bleed bolt.

2. Operate the rubber push-button primer on the fuel transfer pump. Do this until the fuel exiting the bleed screw is free of air. If the primer button feels as if it is not pumping, rotate (crank) the engine approximately 90°, then continue pumping as described.

3. Tighten the low pressure bleed screw to 72 inch lbs. (8 Nm).

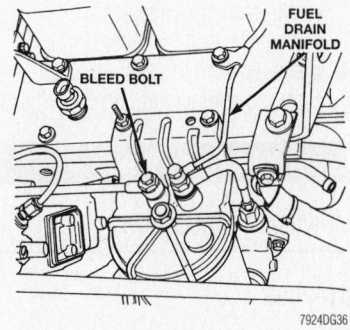

Location of the low pressure bleed bolt—5.9L diesel engine

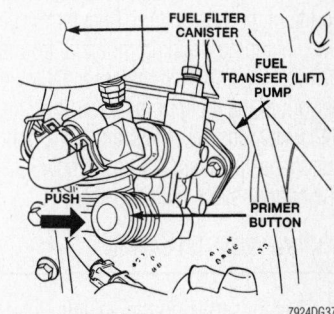

Operate the push-button primer on the fuel transfer pump until the escaping fuel is free of air

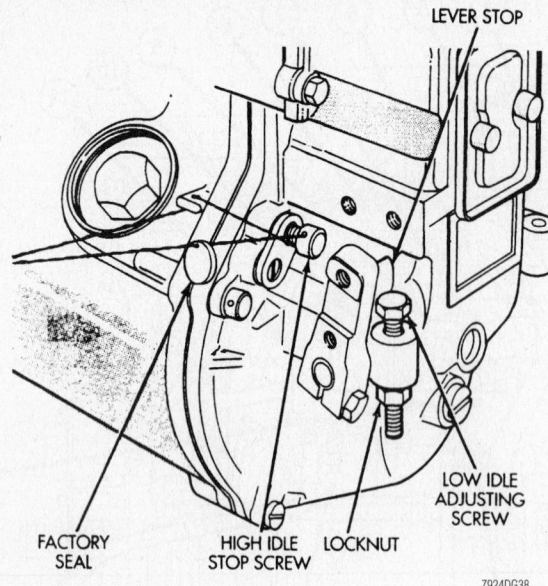

Idle speed adjusting screw location—5.9L diesel engine

Idle Speed

ADJUSTMENT

1. Start the engine and run until normal operating temperature is reached.
2. An optical tachometer must be used to read engine speed.
3. If equipped, turn the air conditioning **ON**.
4. Turn the idle speed screw until the desired idle speed is obtained. The specification for a vehicle equipped with automatic transmission is 700 rpm. The specification for a vehicle equipped with manual transmission is 750 rpm.

Fuel Water Separator/Filter

DRAINING WATER

Filtration and separation of water from the fuel is important for trouble-free operation and long life of the fuel system. Regular maintenance, including draining moisture from the fuel/water separator filter is essential to keep water out of the fuel pump. To remove the collected water, unscrew the drain at the bottom of the Water-In-Filter (WIF) assembly located at the bottom of the filter separator.

REMOVAL & INSTALLATION

1. Before servicing the vehicle, refer to the precautions in the beginning of this section.
2. Remove or disconnect the following:

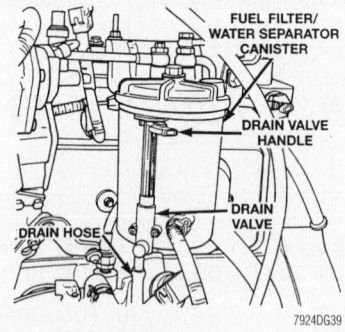

Fuel filter/water separator assembly and related components

- Negative battery cables
- Water In Filter (WIF) sensor connector
- Separator filter assembly

To install:

3. Install or connect the following:
- Separator filter assembly with a new O-ring seal
- WIF sensor connector
- Negative battery cables
4. Bleed air from the system.
5. Start the engine and check for leaks.

Diesel Injection Pump

REMOVAL & INSTALLATION

2001–2003

➡**The Bosch VE lever is indexed to the shaft during pump calibration. Do not remove it from the pump during removal.**

1. Before servicing the vehicle, refer to the precautions in the beginning of this section.
2. Remove or disconnect the following:
- Negative battery cables
- Throttle linkage and bracket
- Fuel drain manifold
- Injection pump supply line
- High pressure lines
- Fuel air control tube
- Fuel shut off valve connector
- Pump support bracket
- Oil fill tube and adapter

3. Place a shop towel in the gear cover opening in a position that will prevent the nut and washer from falling into the gear housing. Remove the gear retaining nut and washer.
4. Turn the engine until the keyway on the fuel pump shaft is pointing approximately in the 6 o'clock position.
5. Locate Top Dead Center (TDC) for cylinder No. 1 by turning the engine slowly while pushing in on the TDC pin. Stop turning the engine as soon as the pin engages with the gear timing hole. Disengage the pin after locating TDC and remove the turning equipment.
6. Loosen the lockscrew, remove the special washer from the injection pump and wire it to the line above it so it will not get misplaced. Retighten the lockscrew to 22 ft. lbs. (30 Nm) to lock the driveshaft.
7. Press the pump drive gear from the driveshaft.

➡**Be careful not to drop the drive gear key into the front cover when removing or installing the pump. If it does drop in, it must be removed before proceeding.**

8. Remove the 3 mounting nuts and remove the injection pump from the vehicle.
9. Remove the gasket and clean the mounting surface.

To install:

➡**The shaft of a new or reconditioned pump is locked so the key aligns with the drive gear keyway with cylinder No. 1 at TDC.**

10. Install the pump and finger-tighten the mounting nuts; the pump must be free to move in the slots.
11. Install the pump drive gear, washer and nut to the driveshaft. The pump will rotate slightly because of gear helix and clearance. This is acceptable providing the pump is free to move on the flange slots and the crankshaft does not move. Tighten the nut to 11–15 ft. lbs. (15–20 Nm). Do not overtighten.

12. If installing the original pump, rotate the pump to align the original timing marks and tighten the mounting nuts to 18 ft. lbs. (24 Nm).

13. If installing a replacement pump, take up gear lash by rotating the pump counterclockwise toward the cylinder head, and tighten the mounting nuts to 18 ft. lbs. (24 Nm). Permanently mark the new injection pump flange to match the mark on the gear housing.

14. Loosen the lockscrew and install the special washer under the lockscrew; tighten to 13 ft. lbs. (18 Nm). Disengage the TDC pin.

15. Install or connect the following:
- Pump support bracket. Tighten the pump drive gear nut to 48 ft. lbs. (65 Nm).
- Oil fill tube and adapter
- Fuel shut off valve connector
- Fuel air control tube
- High pressure lines
- Injection pump supply line
- Fuel drain manifold
- Throttle linkage and bracket
- Negative battery cables

16. Bleed air from the system.

17. Start the engine and check for leaks.

2004

1. Before servicing the vehicle, refer to the precautions in the beginning of this section.

2. Remove or disconnect the following:
- Negative battery cables
- Intake air tube
- Drive belt
- Pump wiring harness
- Injection pump supply line
- Fuel lines
- Drive gear access cover
- Drive gear mounting nut

3. Using a gear puller, remove the drive gear and leave it hanging within the timing gear cover.

4. Remove 3 pump mounting nuts and remove the pump.

To install:

5. Install a new O-ring coated with clean engine oil into the machined groove at the pump mounting area.

6. Install the pump to the mounting flange on the gear housing while aligning the pump shaft through the back of the pump gear.

7. Install the 3 pump mounting nuts and finger tighten.

8. Install the shaft washer and nut and tighten by hand.

9. Tighten the pump mounting nuts to 18 ft. lbs. (25 Nm).

10. Tighten the pump shaft nut to 77 ft. lbs. (105 Nm).

11. Install the drive gear cover and tighten to 71 INCH lbs. (8 Nm).

12. Install or connect the following:
- Fuel lines
- Injection pump supply line
- Pump wiring harness
- Drive belt
- Intake air tube
- Negative battery cables

13. Bleed air from the system.

14. Start the engine and check for leaks.

Fuel Injectors

REMOVAL & INSTALLATION

2001–2003

1. Before servicing the vehicle, refer to the precautions in the beginning of this section.

2. Remove or disconnect the following:
- Negative battery cables
- High pressure fuel lines
- Fuel drain manifold
- Fuel injector. Hold the injector with

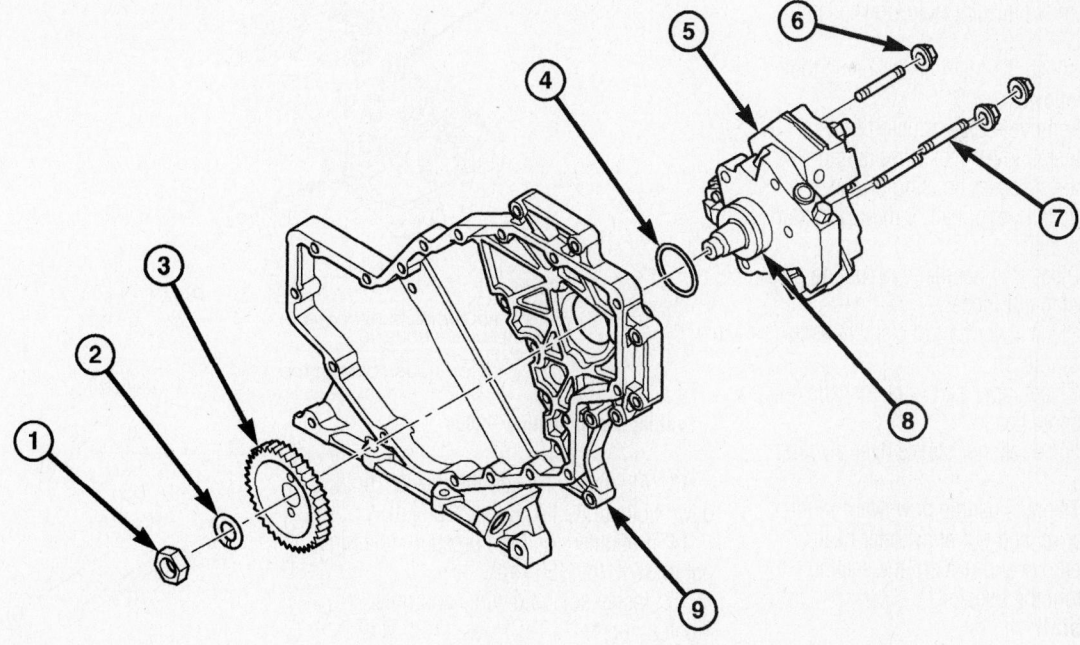

1 - PUMP DRIVE GEAR NUT
2 - WASHER
3 - PUMP DRIVE GEAR
4 - RUBBER O-RING
5 - FUEL INJECTION PUMP

6 - PUMP MOUNTING NUTS (3)
7 - PUMP MOUNTING STUDS (3)
8 - O-RING MACHINED GROOVE
9 - FRONT TIMING GEAR HOUSING

67189-RAMT-CG09

Exploded view of diesel injection pump—2004 models

a backing wrench while loosening the mounting nut.

To install:

3. Use a new copper washer and coat the backing nut threads with anti-seize compound.

4. Install or connect the following:
- Fuel injector. Tighten the mounting nut to 44 ft. lbs. (60 Nm).
- Fuel drain manifold. Use new copper gaskets and tighten the fitting screws to 84 inch lbs. (9 Nm).
- High pressure fuel lines
- Negative battery cables

5. Bleed the fuel system.

6. Start the engine and check for leaks.

2004

1. Before servicing the vehicle, refer to the precautions in the beginning of this section.

2. Remove or disconnect the following:
- Negative battery cables
- Valve cover
- Fuel line connector
- Connector retainer nut

3. Use high pressure tool 9015 and remove high pressure connectors.

4. Remove the necessary exhaust rocker arms.

5. Disconnect the injector solenoid wire nuts from top of injectors.

6. Remove injector hold-down clamp bolts.

7. Remove the fuel injector using tool 9010 as follows:

a. Remove rocker housing bolt.

b. Install lower half of mounting stud to center of rocker housing bridge.

c. Install upper half of mounting stud to lower half.

d. Place tool handle to mounting stud and install nut loosely.

e. Place lower part of clamshells to sides of injector.

f. Place upper part of clamshells onto tool handle head.

g. Slide retainer sleeve over pivoting handle.

h. Depress handle downward to force injector up and out of cylinder head.

8. Remove and discard the sealing washer from the injector.

To install:

9. Use a new sealing washer and O-ring on the injector.

10. Install the injector with the male connector port facing intake manifold.

11. Fuel injector clamp. Tighten the bolt to 44 INCH lbs. (5 Nm), then loosen the bolts.

12. Install the high pressure connector and tighten the nut to 11 ft. lbs. (15 Nm).

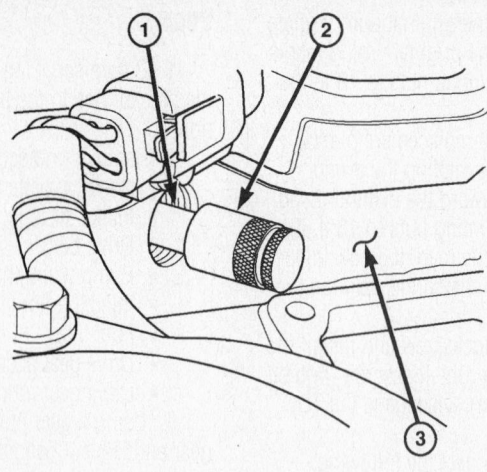

1 - CONNECTOR TUBE
2 - TOOL #9015
3 - CYLINDER HEAD (LEFT SIDE)

67189-RAMT-CG10

Using tool 9015 to remove high pressure connectors—2004

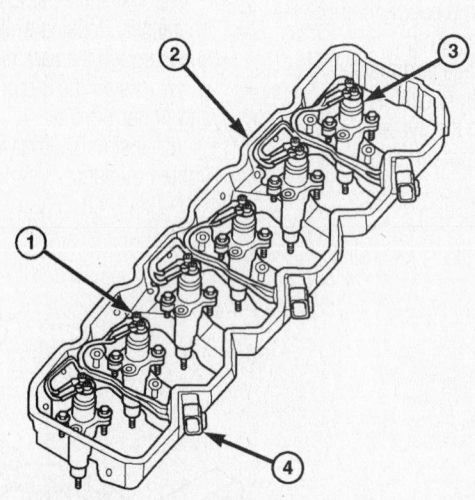

1 - SOLENOID CONNECTIONS
2 - ROCKER HOUSING
3 - FUEL INJECTOR
4 - PASSTHROUGH CONNECTOR

67189-RAMT-CG11

Fuel injector mounting—2004

13. Alternately tighten the injector hold down bolts to 89 INCH lbs. (10 Nm).

14. Retighten the high pressure retaining nut to 37 ft. lbs. (50 Nm).

15. Install solenoid wires and nuts. Tighten nuts to 11 INCH lbs. (1.25 Nm).

16. Install or connect the following:
- Exhaust rocker arms
- Check valve lash
- Fuel line connector
- Valve cover
- Negative battery cables

17. Bleed the fuel system.

18. Start the engine and check for leaks.

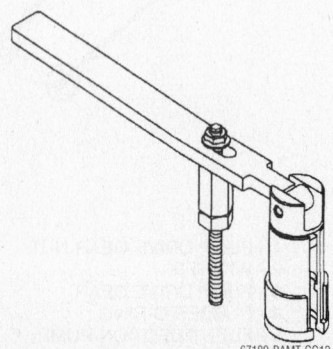

67189-RAMT-CG12

Fuel injector removal tool 9010—2004

DRIVE TRAIN

Transmission Assembly

REMOVAL & INSTALLATION

Automatic

1. Before servicing the vehicle, refer to the precautions in the beginning of this section.

2. Remove or disconnect the following:
   - Negative battery cable
   - Rear driveshaft
   - Crankshaft Position (CKP) sensor
   - Exhaust front pipe
   - Transmission braces, if equipped
   - Starter motor
   - Transmission oil cooler lines
   - Torque converter access cover
   - Torque converter
   - Transmission oil dipstick tube
   - Vehicle Speed (VSS) sensor connector
   - Park/Neutral switch connector
   - Shift cable
   - Throttle valve cable
   - Transmission mount and crossmember. Support the transmission.
   - Front driveshaft and transfer case, if equipped
   - Transmission flange bolts
   - Transmission

To install:

3. Install or connect the following:
   - Transmission. Tighten the flange bolts to 65 ft. lbs. (87 Nm).
   - Front driveshaft and transfer case, if equipped
   - Transmission mount and crossmember
   - Throttle valve cable
   - Shift cable
   - Park/Neutral switch connector
   - VSS sensor connector
   - Transmission oil dipstick tube
   - Torque converter. Tighten the bolts to 23 ft. lbs. (31 Nm) for 10.75 inch converters and to 35 ft. lbs. (47 Nm) for 12.2 inch converters.
   - Torque converter access cover
   - Transmission oil cooler lines
   - Starter motor
   - Transmission braces, if equipped. Tighten the bolts to 30 ft. lbs. (41 Nm).
   - Exhaust front pipe
   - CKP sensor
   - Rear driveshaft
   - Negative battery cable

Manual

1. Before servicing the vehicle, refer to the precautions in the beginning of this section.

2. Remove or disconnect the following:
   - Negative battery cable
   - Shift lever and tower assembly
   - Crankshaft Position (CKP) sensor
   - Skidplate, if equipped
   - Rear driveshaft
   - Front driveshaft, if equipped
   - Transfer case shift linkage, if equipped
   - Transmission mount and crossmember. Support the transmission.
   - Exhaust front pipe
   - Clutch slave cylinder
   - Starter motor
   - Vehicle Speed (VSS) sensor connector
   - Reverse light switch connector
   - Transmission flange bolts
   - Transmission

To install:

3. Install or connect the following:
   - Transmission. Tighten the flange bolts to 40–45 ft. lbs. (54–61 Nm).
   - Reverse light switch connector
   - Vehicle Speed (VSS) sensor connector
   - Starter motor
   - Clutch slave cylinder
   - Exhaust front pipe
   - Transmission mount and crossmember. Tighten the fasteners to 50 ft. lbs. (68 Nm).
   - Transfer case shift linkage, if equipped
   - Front driveshaft, if equipped
   - Rear driveshaft
   - Skidplate, if equipped
   - CKP sensor
   - Shift lever and tower assembly
   - Negative battery cable

Clutch

REMOVAL & INSTALLATION

1. Before servicing the vehicle, refer to the precautions in the beginning of this section.

2. Remove or disconnect the following:
   - Negative battery cable

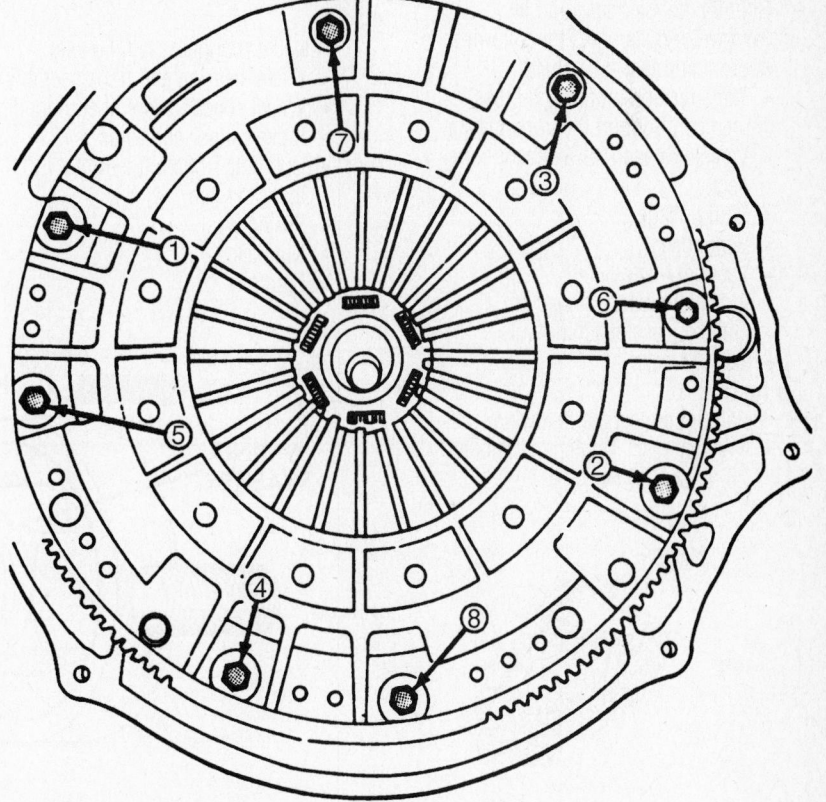

Pressure plate torque sequence

7924DG83

- Transfer case, if equipped
- Transmission
- Pressure plate. Loosen the bolts evenly in ½ turn steps.
- Clutch disc

To install:

3. Install or connect the following:
- Clutch disc and pressure plate. Tighten the pressure plate bolts evenly in ½ turns to 21 ft. lbs. (28 Nm) for 2001; 37 ft. lbs. (50 Nm) for 2002–04 V6 and V8, or 23 ft. lbs. (30 Nm) for V10.
- Transmission
- Transfer case, if equipped
- Negative battery cable

Hydraulic Clutch System

BLEEDING

The system is self-bleeding. Press the clutch pedal repeatedly to release air from the fluid. The air will be vented from the reservoir.

Transfer Case Assembly

REMOVAL & INSTALLATION

1. Before servicing the vehicle, refer to the precautions in the beginning of this section.
2. Shift the transfer case into **N**.
3. Remove or disconnect the following:
- Front and rear driveshafts
- Transmission mount and cross-member. Support the transmission.
- Vehicle Speed (VSS) sensor connector
- Shift linkage
- Vent hose
- Vacuum hose
- Indicator switch connector
- Transfer case attaching nuts
- Transfer case

To install:

4. Install or connect the following:
- Transfer case. Tighten the nuts to 26 ft. lbs. (35 Nm).

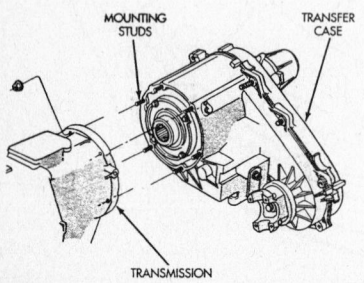

MOUNTING STUDS · TRANSFER CASE

TRANSMISSION

7924DG84

Typical transfer case mounting

- Indicator switch connector
- Vacuum hose
- Vent hose
- Shift linkage
- VSS sensor connector
- Transmission mount and cross-member
- Front and rear driveshafts

Halfshaft

REMOVAL & INSTALLATION

2001–02

1. Before servicing the vehicle, refer to the precautions in the beginning of this section.
2. Remove or disconnect the following:
- Skid plate, if equipped
- Front wheel
- Split pin
- Nut lock
- Spring washer
- Hub nut
- Brake caliper and rotor
- Wheel speed sensor, if equipped
- Wheel bearing and hub assembly
3. Pry the inner tripod joint out of the differential and remove the axle halfshaft.

To install:

4. Install the axle halfshaft so that the snapring is felt to seat in the joint housing groove.
5. Install or connect the following:
- Wheel bearing and hub assembly
- Wheel speed sensor, if equipped
- Brake caliper and rotor
- Hub nut. Tighten the nut to 180 ft. lbs. (244 Nm).
- Spring washer
- Nut lock
- Split pin

- Front wheel
- Skid plate, if equipped

2003–2004

1. Before servicing the vehicle, refer to the precautions in the beginning of this section.
2. Remove or disconnect the following:
- Skid plate, if equipped
- Front wheel
- Hub nut
- Brake caliper and rotor
3. Unload the suspension with a jack.
4. Remove or disconnect the following:
5. Lower shock absorber bolt
6. Upper ball joint
7. Pry the inner joint out of the differential and remove the axle halfshaft.

To install:

8. Install the axle halfshaft so that the snapring is felt to seat in the joint housing groove.
9. Install or connect the following:
- Ball joint
- Shock absorber
- Brake caliper and rotor
- Hub nut. Tighten the nut to 185 ft. lbs. (251 Nm).
- Front wheel
- Skid plate, if equipped

CV-Joints

OVERHAUL

Outer CV-Joint

1. Before servicing the vehicle, refer to the precautions in the beginning of this section.
2. Remove or disconnect the following:
- Axle halfshaft from the vehicle

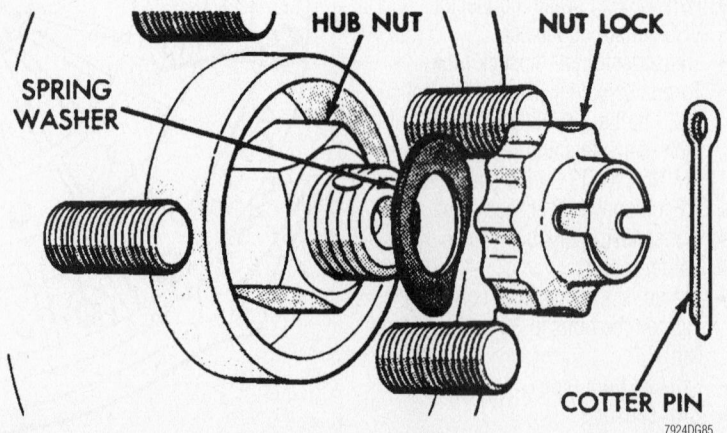

SPRING WASHER · HUB NUT · NUT LOCK · COTTER PIN

7924DG85

To separate the halfshaft from the hub, remove the cotter pin, nut lock and spring washer from the axle shaft

- CV-joint boot and clamps
- Snapring
- CV-joint

To install:
3. Install or connect the following:
- CV-joint
- Snapring
- CV-joint boot and clamps

4. Fill the joint housing and boot with grease and tighten the boot clamps.
5. Install the axle halfshaft.

Inner Tripod Joint

1. Before servicing the vehicle, refer to the precautions in the beginning of this section.
2. Remove or disconnect the following:
- Axle halfshaft from the vehicle
- Inner tripod joint boot clamps
- Tripod joint housing
- Snapring
- Circlip
- Tripod joint

To install:

➡**Use new snaprings, clips, and boot clamps for assembly.**

3. Install or connect the following:
- Tripod joint
- Circlip
- Snapring
- Tripod joint housing

4. Fill the tripod joint housing and boot with grease and tighten the boot clamps.
5. Install the axle halfshaft.

Axle Shaft, Bearing and Seal

REMOVAL & INSTALLATION

C-Clip Type

1. Before servicing the vehicle, refer to the precautions in the beginning of this section.
2. Remove or disconnect the following:
- Rear wheel
- Brake drum
- Differential cover
- Differential gear shaft retainer
- Differential gear shaft
- C-clip
- Axle shaft
- Axle seal
- Axle bearing

To install:
3. Install or connect the following:
- Axle bearing
- Axle seal

- Axle shaft
- C-clip
- Differential gear shaft. Use Loctite® and tighten the retainer to 14 ft. lbs. (19 Nm).
- Differential cover. Tighten the bolts to 30 ft. lbs. (41 Nm).
- Brake drum
- Rear wheel

4. Fill the axle assembly with gear oil and check for leaks.

Non C-Clip Type

1. Before servicing the vehicle, refer to the precautions in the beginning of this section.
2. Remove or disconnect the following:
- Rear wheel
- Brake caliper and rotor, if equipped
- Brake drum, if equipped
- Axle retainer nuts
- Axle shaft, seal and bearing assembly

3. Split the bearing retainer with a chisel and remove the retainer ring.
4. Press the bearing off the axle shaft.
5. Remove the axle seal and retaining plate.

To install:
6. Install the retaining plate and axle seal onto the axle shaft.
7. Pack the wheel bearing with axle grease and press the bearing on to the axle shaft.
8. Press the retaining ring onto the axle shaft.
9. Install or connect the following:
- Axle shaft, seal and bearing assembly. Tighten the nuts to 45 ft. lbs. (61 Nm).
- Brake caliper and rotor, if equipped
- Brake drum, if equipped
- Rear wheel

10. Fill the axle assembly with gear oil and check for leaks.

Pinion Seal

REMOVAL & INSTALLATION

C-Clip Type

1. Before servicing the vehicle, refer to the precautions in the beginning of this section.
2. Remove or disconnect the following:
- Wheels
- Brake drums
- Driveshaft

3. Check the bearing preload with an inch lb. torque wrench.
4. Remove the pinion flange and seal.

To install:

➡**Use a new pinion nut for assembly.**

5. Install the new pinion seal and flange. Tighten the nut to 210 ft. lbs. (285 Nm).
6. Check the bearing preload. The bearing preload should be equal to the reading taken earlier, plus 5 inch lbs.
7. If the preload torque is low, tighten the pinion nut in 5 inch lb. increments until the torque value is reached. Do not exceed 350 ft. lbs. (474 Nm) pinion nut torque.
8. If the pinion bearing preload torque cannot be attained at maximum pinion nut torque, replace the collapsible spacer.
9. Install or connect the following:
- Driveshaft
- Brake drums
- Wheels

10. Fill the axle assembly with gear oil and check for leaks.

Non C-Clip Type

1. Before servicing the vehicle, refer to the precautions in the beginning of this section.
2. Remove or disconnect the following:
- Wheels
- Brake rotors or drums
- Driveshaft

3. Check the bearing preload with an inch lb. torque wrench.
4. Remove the pinion flange and seal.

To install:

➡**Use a new pinion nut for assembly.**

5. Install the new pinion seal and flange. Tighten the nut to 160 ft. lbs. (217 Nm).
6. Check the bearing preload. The bearing preload should be equal to the reading taken earlier, plus 5 inch lbs.
7. If the preload torque is low, tighten the pinion nut in 5 inch lb. increments until the torque value is reached. Do not exceed 260 ft. lbs. (353 Nm) pinion nut torque.
8. If the pinion bearing preload torque can not be attained at maximum pinion nut torque, remove one or more pinion preload shims.
9. Install or connect the following:
- Driveshaft
- Brake rotors or drums
- Wheels

10. Fill the axle assembly with gear oil and check for leaks.

STEERING AND SUSPENSION

Air Bag

✷✷ CAUTION

Some vehicles are equipped with an air bag system. The system must be disarmed before performing service on, or around, system components, the steering column, instrument panel components, wiring and sensors. Failure to follow the safety precautions and the disarming procedure could result in accidental air bag deployment, possible injury and unnecessary system repairs.

PRECAUTIONS

Several precautions must be observed when handling the inflator module to avoid accidental deployment and possible personal injury.

• Never carry the inflator module by the wires or connector on the underside of the module.

• When carrying a live inflator module, hold securely with both hands, and ensure that the bag and trim cover are pointed away.

• Place the inflator module on a bench or other surface with the bag and trim cover facing up.

• With the inflator module on the bench, never place anything on or close to the module which may be thrown in the event of an accidental deployment.

Before servicing the vehicle, also make sure to refer to the precautions in the beginning of this section as well.

DISARMING

1. Disconnect and isolate the negative battery cable. Wait 2 minutes for the system capacitor to discharge before performing any service.

2. When repairs are completed, connect the negative battery cable.

Recirculating Ball Power Steering Gear

REMOVAL & INSTALLATION

1. Before servicing the vehicle, refer to the precautions in the beginning of this section.

2. Remove or disconnect the following:
   • Negative battery cable
   • Power steering pressure and return lines

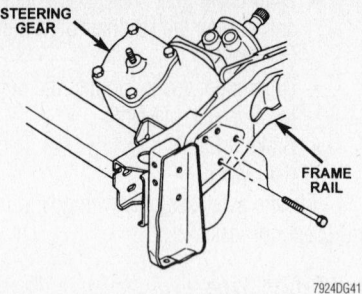

7924DG41

Typical recirculating ball power steering gear mounting

• Intermediate shaft
• Pitman arm
• Steering gear

To install:

3. Install or connect the following:
   • Steering gear. Tighten the bolts to 100 ft. lbs. (136 Nm).
   • Pitman arm. Tighten the nut to 175 ft. lbs. (237 Nm).
   • Intermediate shaft. Tighten the pinch bolt to 36 ft. lbs. (49 Nm).
   • Power steering pressure and return lines
   • Negative battery cable

4. Fill the power steering fluid reservoir.

5. Start the engine and check for leaks.

Rack and Pinion Steering Gear

REMOVAL & INSTALLATION

1. Before servicing the vehicle, refer to the precautions in the beginning of this section.

2. Remove or disconnect the following:
   • Front wheels
   • Skid plate, if equipped
   • Outer tie rod ends
   • Steering shaft coupler. Discard the pinch bolt.
   • Power steering hoses
   • Steering gear

To install:

3. Install or connect the following:
   • Steering gear. Tighten the bolts to 235 ft. lbs. (319 Nm).
   • Power steering hoses. Tighten the pressure line fitting to 23 ft. lbs. (32 Nm); return line to 52 ft. lbs. (71 Nm).
   • Steering shaft coupler. Tighten the bolt to 36 ft. lbs. (49 Nm).
   • Outer tie rod ends. Tighten the nuts to 45 ft. lbs. (61 Nm).
   • Front wheels

Shock Absorber

REMOVAL & INSTALLATION

Front

COIL SPRING SUSPENSION

1. Before servicing the vehicle, refer to the precautions in the beginning of this section.

2. Remove or disconnect the following:
   • Front wheel
   • Upper mounting bolt
   • Lower mounting bolts
   • Shock absorber

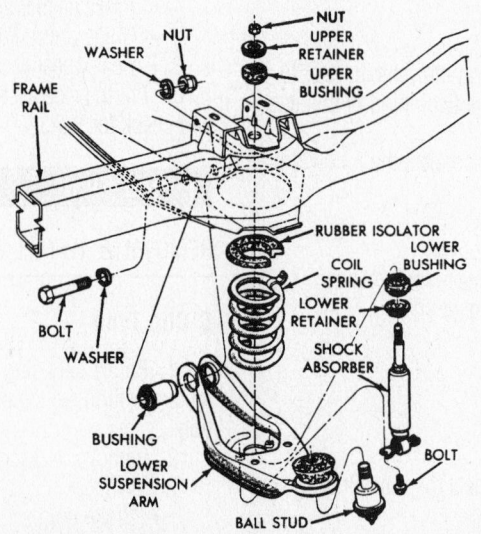

7924DG87

Front shock absorber mounting—Ram Van models

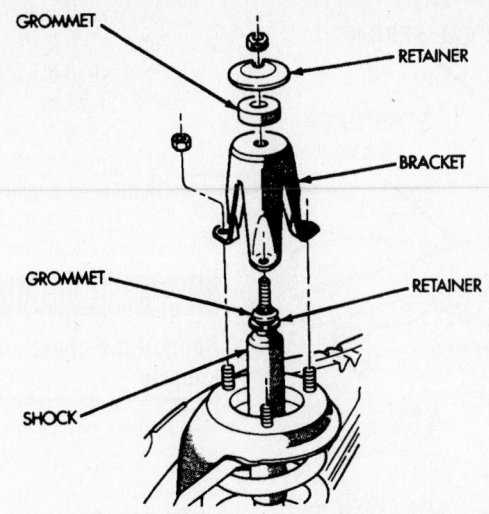

Upper shock absorber mounting—2001–02 Ram Truck models

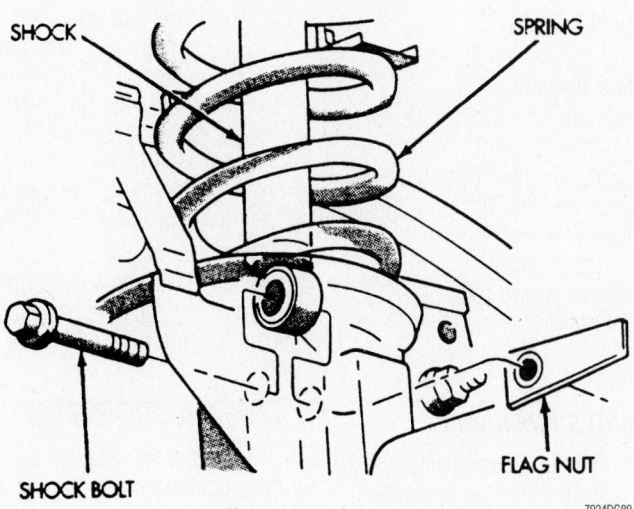

Lower shock absorber mounting—2001–02 Ram Truck models

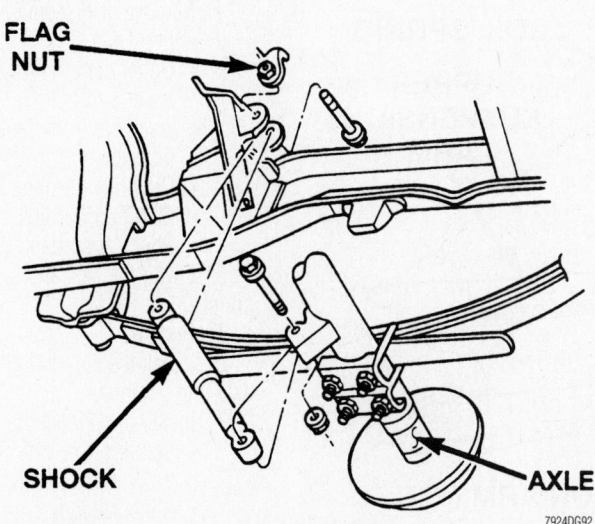

Rear shock absorber mounting—Ram Truck and Van models

To install:

➡ **On 2003–04 2wd pickups, the upper nut must be replaced, or if-reused, installed using Loctite 242.**

3. Install or connect the following:
   - Shock absorber. Tighten the upper bolt or nut to 25 ft. lbs. (34 Nm), except 2003–04 4wd, which is tightened to 40 ft. lbs. (54 Nm); and the lower bolt(s) to 15 ft. lbs. (20 Nm) on 2001 models; lower nuts on 2wd 2003 models to 25 ft. lbs. (34 Nm) and 100 ft. lbs. (135 Nm) on 2003–04 4wd.
   - Front wheel

LINK/COIL FRONT SUSPENSION

1. Before servicing the vehicle, refer to the precautions in the beginning of this section.
2. Remove or disconnect the following:
   - Upper mounting nut
   - 3 upper mounting nuts from the upper shock bracket
   - Lower mounting bolt
   - Shock absorber

To install:

3. Install or connect the following:
   - Shock absorber. Tighten the lower bolt to 100 ft. lbs. (135 Nm); upper bracket nuts to 55 ft. lbs. (75 Nm); Upper shock nut to 55 ft. lbs. (47 Nm)

Rear

1. Before servicing the vehicle, refer to the precautions in the beginning of this section.
2. Support the axle.
3. Remove or disconnect the following:
   - Upper bolt
   - Lower bolt
   - Shock absorber

To install:

4. Install the bolts through the brackets and shock and tighten them as follows:
   - Tighten the upper bolt to 70 ft. lbs. (95 Nm) and the lower bolt to 100 ft. lbs. (136 Nm) on 2001–02 models; 100 ft. lbs. (135 Nm) for upper and lower on 2003–04 models.

Coil Spring

REMOVAL & INSTALLATION

2001–2002

1. Before servicing the vehicle, refer to the precautions in the beginning of this section.

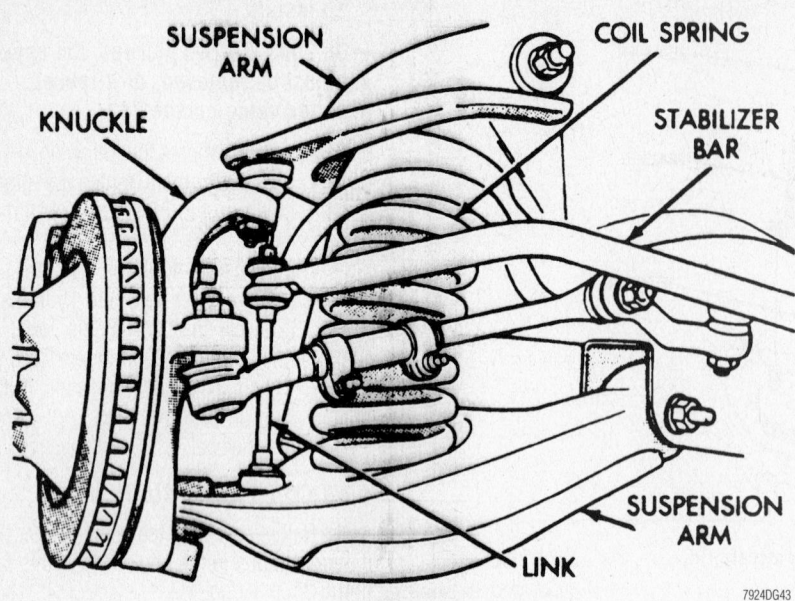

Independent front suspension components—2WD Ram Truck and Van models

2. Support the lower control arm on a floor jack.

3. Remove or disconnect the following:
- Front wheel
- Brake caliper and rotor
- Outer tie rod end
- Stabilizer bar link
- Lower ball joint
- Shock absorber

4. Lower the jack and remove the coil spring.

To install:

5. Install the coil spring and raise the control arm into position.

6. Install or connect the following:

- Shock absorber
- Lower ball joint. Tighten the nut to 95 ft. lbs. (129 Nm).
- Stabilizer bar link. Tighten the nut to 27 ft. lbs. (37 Nm).
- Outer tie rod end. Tighten the nut to 65 ft. lbs. (88 Nm).
- Brake caliper and rotor
- Front wheel

2003–2004

COIL SPRING SUSPENSION

1. Before servicing the vehicle, refer to the precautions in the beginning of this section.

Link and coil front suspension components—4WD Ram Truck and Van models

2. Support the lower control arm on a floor jack.

3. Remove or disconnect the following:
- Front wheel
- Shock absorber

4. Compress the spring.

5. Remove or disconnect the following:
- Stabilizer bar link
- Lower ball joint

Support the upper control arm and knuckle.

6. Lower the jack and tighten the compressor to allow coil spring removal. Catch the isolator pad.

To install:

7. Install the coil spring and raise the control arm into position.

8. Install or connect the following:
- Lower ball joint. Tighten the nut to 38 ft. lbs. (52 Nm), plus a 90 degree turn (1500 Series), or, 100 ft. lbs. (135 Nm) (HD Series)
- Shock absorber
- Stabilizer bar link. Tighten the nut to 27 ft. lbs. (37 Nm).
- Front wheel

9. Lower the vehicle and allow the suspension to take the weight. Tighten the front and rear control arm pivot bolts to 150 ft. lbs. (204 Nm) (LD); 210 ft. lbs. (285 Nm) HD

LINK/COIL SUSPENSION

1. Before servicing the vehicle, refer to the precautions in the beginning of this section.

2. Support the axle on a floor jack.

3. Place alignment marks on the lower arm adjuster and axle bracket.

4. Remove or disconnect the following:
- Front wheel
- Upper control arm and loosen the lower arm bolts
- Track bar from the frame rail bracket
- Drag link from the Pitman arm
- Stabilizer bar link
- Shock absorber

5. Lower the jack and remove the coil spring.

To install:

6. Install the coil spring and raise the axle into position.

7. Install or connect the following:
- Stabilizer bar link. Tighten the nut to 45 ft. lbs. (61 Nm).
- Shock absorber
- Track bar

- Drag link
- On 4wd, the front driveshaft
- Upper control arm and lower arm bolts. Upper arm nuts to 110 ft. lbs. (149 Nm); lower arm nuts to 160 ft. lbs. (217 Nm)

Torsion Bar

REMOVAL & INSTALLATION

1. Before servicing the vehicle, refer to the precautions in the beginning of this section.
2. Loosen the adjustment bolt to remove spring load. Note the number of turns for installation.
3. Remove or disconnect the following:
   - Adjustment bolt, swivel and bearing
   - Torsion bar and anchor
4. Separate the torsion bar and anchor.

To install:

5. Assemble the torsion bar and anchor.
6. Install or connect the following:
   - Torsion bar and anchor
   - Adjustment bolt, swivel and bearing
7. Tighten the adjustment bolt the recorded number of turns.

Leaf Spring

REMOVAL & INSTALLATION

1. Before servicing the vehicle, refer to the precautions in the beginning of this section.
2. Support the vehicle at the frame rails.
3. Support the rear axle with a jack.
4. Remove or disconnect the following:
   - Rear wheel
   - Stabilizer bar link
   - Axle U-bolts
   - Spring bracket
   - Leaf spring

To install:

➡The weight of the vehicle must be supported by the springs when the spring eye and stabilizer bar fasteners are tightened.

5. Install or connect the following:
   - Leaf spring
   - Spring bracket
   - Axle U-bolts. Tighten the nuts to 52 ft. lbs. (70 Nm) on 2001–02 models; 110 ft. lbs. (149 Nm) on 2003–04 models.
   - Stabilizer bar link
   - Rear wheel
6. On 2001–02 models, tighten the front spring eye bolt and nut to 115 ft. lbs. (156

Nm). Tighten the rear spring eye bolt and nut to 80 ft. lbs. (108 Nm). Tighten the stabilizer bar nuts 55 ft. lbs. (74 Nm). On 2003–04 models, torque the front and rear pivot bolt nuts to 120 ft. lbs. (163 Nm).

Upper Ball Joint

REMOVAL & INSTALLATION

➡The upper ball joint on 2003–04 coil spring suspensions, is not replaceable. If the ball joint is defective, the upper arm must be replaced.

2001–2002

2-WHEEL DRIVE

1. Before servicing the vehicle, refer to the precautions in the beginning of this section.
2. Support the lower control arm.
3. Remove the front wheel.
4. Remove the stud nut and separate the ball joint from the steering knuckle.
5. Unscrew the ball joint from the control arm with tool C-3561.

To install:

6. Thread the ball joint into the upper control arm and tighten it to 125 ft. lbs. (169 Nm). Tighten the upper ball stud nut to 135 ft. lbs. (183 Nm).
7. Install the front wheel.

4-WHEEL DRIVE

1. Before servicing the vehicle, refer to the precautions in the beginning of this section.
2. Remove or disconnect the following:
   - Front wheel
   - Brake caliper and rotor
   - Hub retainer
   - Axle shaft
   - Outer tie rod ends
   - Upper and lower ball joint stud nuts
3. If equipped with a Model 44 front axle, use a brass drift and hammer to separate the steering knuckle from the axle tube yoke. Use tool C-4169 to remove the sleeve from the upper yoke arm.
4. Remove the snapring from the ball joint. Install the knuckle in a vise and use tools D-150-1, D-150-3 and C-4212-L to remove the ball joint from the knuckle.
5. If equipped with a Model 60 front axle, remove the bolts from the knuckle lower cap. Dislodge the cap from the steering knuckle and axle tube yoke. Remove the steering knuckle. Use tool D-192 to remove the upper socket pin from the axle tube upper arm bore. Remove the seal.

To install:

6. If equipped with a Model 44 front axle, use tools C-4212-L and C-4288 to force the upper ball joint into the steering

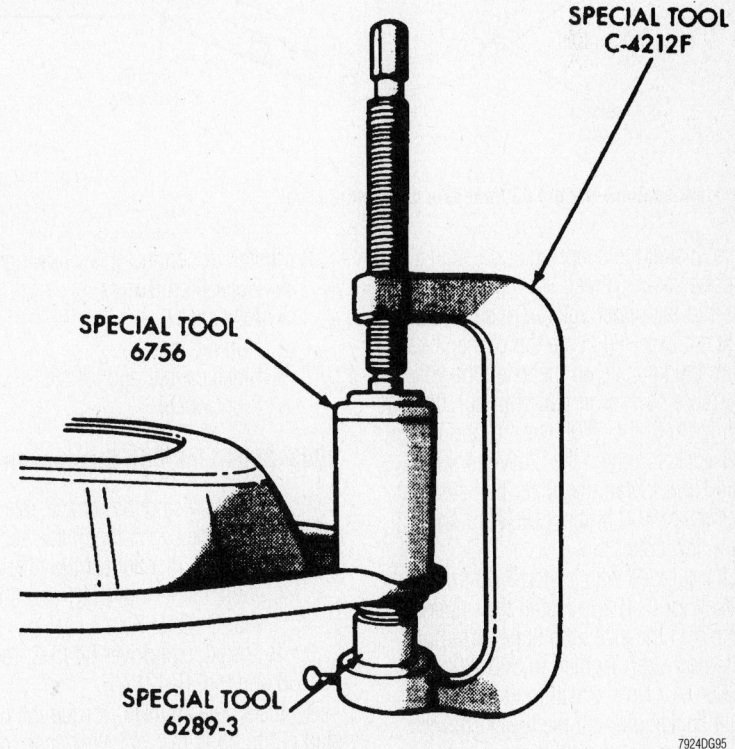

SPECIAL TOOL
C-4212F

SPECIAL TOOL
6756

SPECIAL TOOL
6289-3

7924DG95

Upper ball joint removal— Ram Truck and Ram Van

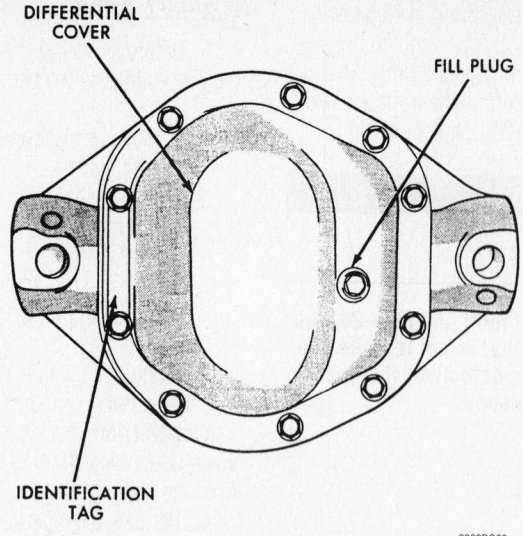

Axle identification—Model 44 front axle differential cover

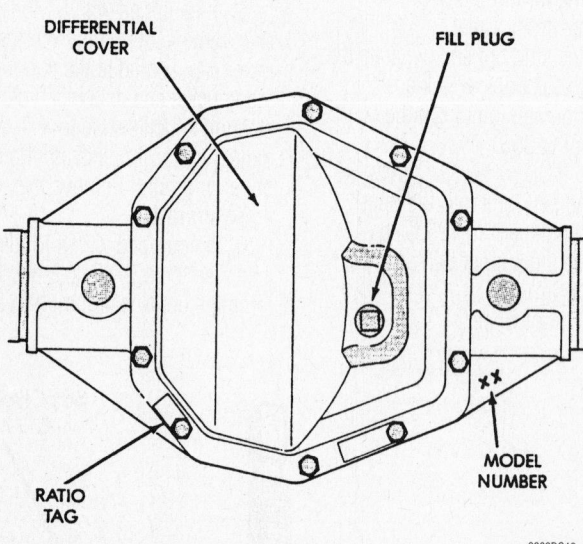

Axle identification—Model 60 front axle differential cover

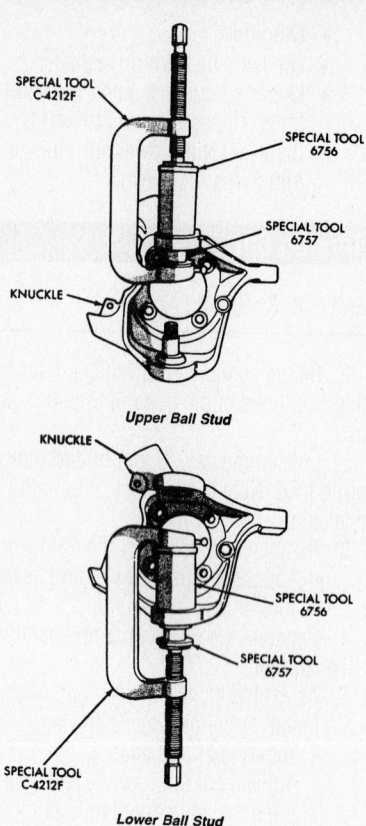

Typical ball joint removal using the special tools—248 FBI axle shown

Lower Ball Joint

REMOVAL & INSTALLATION

2001–02

2WD MODELS

- Front wheel
- Shock absorber
- Stabilizer bar link
- Coil spring

1. Press the ball joint out of the control arm.

To install:

2. Use the remover tool to press the ball joint into the arm.

3. Install or connect the following:
  - Coil spring. Tighten 11/16 lower ball joint nuts to 135 ft. lbs. (183 Nm). Tighten 3/4 nuts to 175 ft. lbs. (237 Nm).
  - Stabilizer bar link
  - Shock absorber
  - Front wheel

4WD MODELS

1. Before servicing the vehicle, refer to the precautions in the beginning of this section.

---

knuckle. Install the snapring and install a new rubber boot. Thread the replacement sleeve into the upper yoke bore so that 2 threads are exposed at the top of the yoke. Position the knuckle on the axle tube yoke and install a new lower ball stud nut, then tighten to 80 ft. lbs. (108 Nm). Using the special socket, tighten the sleeve to 40 ft. lbs. (54 Nm). Install the upper ball stud nut and tighten to 100 ft. lbs. (136 Nm) and install a new cotter pin.

7. If equipped with a Model 60 front axle, use tool D-192 to install the upper socket pin in the axle tube upper arm bore. Install a new seal. Tighten to 500–600 ft. lbs. (668–813 Nm). Position the knuckle over the socket pin. Fill the lower socket cavity with grease. Install the lower cap and tighten the bolts to 80 ft. lbs. (108 Nm).

8. Install or connect the following:
  - Outer tie rod ends
  - Axle shaft
  - Hub retainer
  - Brake caliper and rotor
  - Front wheel

2003–2004 Link/Coil Suspension

1. Before servicing the vehicle, refer to the precautions in the beginning of this section.

2. Remove or disconnect the following:
  - Joint, using tool 6761 and 8445-3 with C-4212-F

3. To install, use driver 8445-2, receiver 8975-5 and tool C-4212-F.

4. Install the knuckle. Torque the ball stud nut to 35 ft. lbs. (47 Nm), then torque it to 70 ft. lbs. (94 Nm) and install the pin.

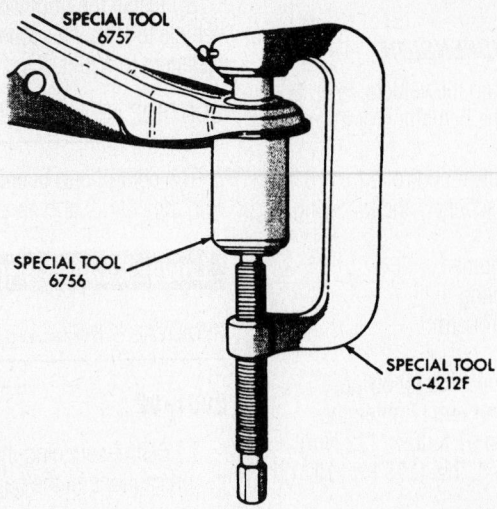

Lower ball joint removal—2WD Ram Truck and Van models

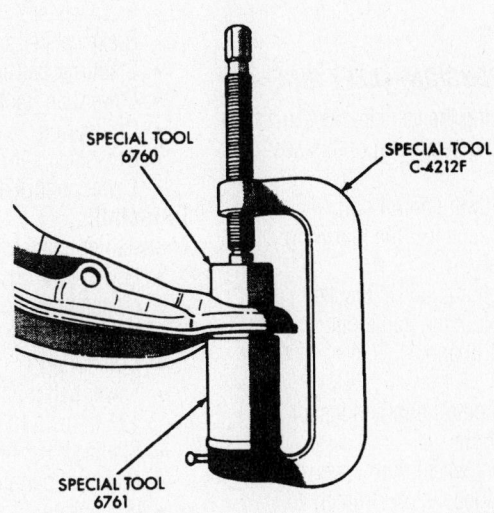

Lower ball joint installation—2WD Ram Truck and Van models

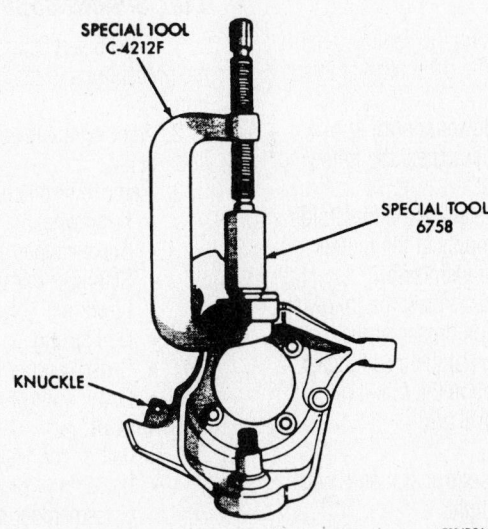

Lower ball joint installation—4WD Ram Truck and Van models

2. Remove or disconnect the following:
- Front wheel
- Brake caliper and rotor
- Hub retainer
- Axle shaft
- Outer tie rod ends
- Upper and lower ball joint stud nuts

3. If equipped with a Model 44 front axle, use a brass drift and hammer to separate the steering knuckle from the axle tube yoke.

4. Remove the snapring from the ball joint. Install the knuckle in a vise and use tools D-150-1, D-150–3 and C-4212-L to remove the ball joint from the knuckle.

5. If equipped with a Dana 60 front axle, use tools C-4212-L, C-4366–1 and C-4366-2 (or equivalents) to remove the lower ball joint.

To install:

6. If equipped with a Model 44 front axle, use tools C-4212-L and C-4288 to force the lower ball joint into the steering knuckle. Install the snapring and install a new rubber boot. Position the knuckle on the axle tube yoke and install a new lower ball stud nut. Tighten to 80 ft. lbs. (108 Nm). Install the upper ball stud nut and tighten to 100 ft. lbs. (136 Nm).

7. If equipped with a Model 60 front axle, use tools C-4212-L, C-4366-3 and C-4366-4 to install the seal and lower bearing cup into the axle tube yoke lower bore. Reposition the tools and install the lower bearing and seal into the bore. Position the knuckle over the socket pin. Fill the lower socket cavity with grease. Install the lower cap and tighten the bolts to 80 ft. lbs. (108 Nm).

8. Install or connect the following:
- Outer tie rod ends
- Axle shaft
- Hub retainer
- Brake caliper and rotor
- Front wheel

2003–2004

COIL SPRING SUSPENSION

1. Before servicing the vehicle, refer to the precautions in the beginning of this section.

2. Remove or disconnect the following:
- Front wheel
- Brake caliper and rotor
- Outer tie rod ends
- Steering knuckle
- Snapring from the ball joint (HD 2wd)

3. Use tools C-4212-F, 8698-2 and 8698-3 (or equivalents) to remove the lower ball joint.

➡**Use EP grease on the tool threads.**

To install:

4. Use tools C-4212-F, 8698-1 and 8698-3 (or equivalents) to install the lower ball joint.

5. On 2wd HD models, install a new snapring; on all others, stake the ball joint flange in 4 evenly spaced places.

6. Install or connect the following:
- Knuckle. Torque the ball stud nut to 38 ft. lbs. (52 Nm) plus 90 degrees on all except 2wd HD; on 2wd HD, torque the nut to 100 ft. lbs. (135 Nm).
- Outer tie rod ends
- Axle shaft
- Hub retainer
- Brake caliper and rotor
- Front wheel

LINK/COIL SUSPENSION

1. Before servicing the vehicle, refer to the precautions in the beginning of this section.

2. Remove or disconnect the following:
- Joint, using tool 6761 and 8445-3 with C-4212-F

3. To install, use driver 8445-2, receiver 8975-5 and tool C-4212-F.

4. Install the knuckle. Torque the ball stud nut to 35 ft. lbs. (47 Nm), then torque it to 70 ft. lbs. (94 Nm) and install the pin.

Upper Control Arm

REMOVAL & INSTALLATION

2001–02

1. Before servicing the vehicle, refer to the precautions in the beginning of this section.

2. Support the lower control arm.

3. Remove or disconnect the following:
- Front wheel
- Brake hose brackets
- Upper ball joint
- Pivot mounting nuts
- Upper control arm

To install:

4. Install or connect the following:
- Upper control arm. Tighten the pivot nuts to 155 ft. lbs. (210 Nm).
- Upper ball joint. Tighten the nut to 60 ft. lbs. (81 Nm).
- Brake hose brackets
- Front wheel

5. Check the wheel alignment and adjust as necessary.

2003–2004

COIL SPRING SUSPENSION

1. Before servicing the vehicle, refer to the precautions in the beginning of this section.

2. Support the lower control arm.

3. Remove or disconnect the following:
- Front wheel
- Upper ball joint
- Pivot mounting nuts
- Upper control arm

To install:

4. Install or connect the following:
- Upper control arm. Tighten the pivot nuts to 97 ft. lbs. (132 Nm) for LD; 125 ft. lbs. (170 Nm) HD 2wd
- Upper ball joint. Tighten the nut to 40 ft. lbs. (54 Nm). On 1500 series, turn the nut an additional 90 degrees.
- Front wheel

LINK/COIL SUSPENSION—LEFT SIDE

1. Before servicing the vehicle, refer to the precautions in the beginning of this section.

2. Support the lower control arm.

3. Remove or disconnect the following:
- Front wheel
- Nut and bolt at the axle bracket
- Nut and bolt at the frame rail
- Upper control arm

To install:

4. Position the control arm and install all fasteners finger-tight.

5. Install the front wheel. And lower the vehicle to load the suspension. Tighten all fasteners to 120 ft. lbs. (163 Nm).

LINK/COIL SUSPENSION—RIGHT SIDE

1. Before servicing the vehicle, refer to the precautions in the beginning of this section.

2. Support the lower control arm.

3. Remove or disconnect the following:
- Front wheel
- Exhaust system at the manifolds
- Exhaust mounts at the muffler

4. Support the transmission

5. Remove or disconnect the following:
- Transmission crossmember
- Nut and bolt at the axle bracket
- Nut and bolt at the frame rail
- Upper control arm

To install:

6. Position the control arm and install all fasteners finger-tight.

7. Install the exhaust system and crossmember.

8. Install the front wheel. And lower the vehicle to load the suspension. Tighten all fasteners to 120 ft. lbs. (163 Nm).

CONTROL ARM BUSHING REPLACEMENT

The control arm bushings are serviced with the control arm as an assembly.

Lower Control Arm

REMOVAL & INSTALLATION

2001–02

1. Before servicing the vehicle, refer to the precautions in the beginning of this section.

2. Support the lower control arm with a floor jack.

3. Remove or disconnect the following:
- Front wheel
- Brake caliper and rotor
- Stabilizer bar link
- Lower ball joint
- Coil spring
- Crossmember nuts
- Lower control arm

To install:

4. Install or connect the following:
- Lower control arm. Tighten the crossmember nuts to 145 ft. lbs. (196 Nm).
- Coil spring
- Lower ball joint. Tighten the nut to 135 ft. lbs. (183 Nm).
- Stabilizer bar link
- Brake caliper and rotor
- Front wheel

2003–2004

COIL SPRING SUSPENSION—2WD

1. Before servicing the vehicle, refer to the precautions in the beginning of this section.

2. Support the lower control arm with a floor jack.

3. Remove or disconnect the following:
- Front wheel
- Brake caliper and rotor
- Stabilizer bar link
- Lower ball joint
- Coil spring
- Crossmember nuts
- Lower control arm

To install:

4. Install or connect the following:
- Lower control arm. Tighten the crossmember nuts to 145 ft. lbs. (196 Nm).

- Coil spring
- Lower ball joint. Tighten the nut to 135 ft. lbs. (183 Nm).
- Stabilizer bar link
- Brake caliper and rotor
- Front wheel

COIL SPRING SUSPENSION—4WD

1. Before servicing the vehicle, refer to the precautions in the beginning of this section.
2. Remove or disconnect the following:
- Front wheel
- Upper ball joint from the knuckle
- Halfshaft
- Torsion bar
- Shock absorber lower bolt
- Stabilizer bar link
- Lower ball joint from the knuckle
- Lower control arm

To install:
3. Install or connect the following:
- Lower control arm. Tighten bolts finger-tight.
- Torsion bar

➡**The ball stud taper must be clean and dry before installation.**

- Lower ball joint. Tighten the nut to 38 ft. lbs. (52 Nm) (on 1500 series, plus 90 degrees).
- Shock absorber lower bolt. Torque it to 100 ft. lbs. (135 Nm).
- Halfshaft
- Upper ball joint. Tighten the nut to 40 ft. lbs. (54 Nm) (on 1500 series, plus 90 degrees).
- Stabilizer bar link
4. Tighten the pivot bolts to 150 ft. lbs. (204 Nm).
- Front wheel

COIL/LINK SUSPENSION

1. Before servicing the vehicle, refer to the precautions in the beginning of this section.
2. Paint or scribe matchmarks on the cam adjusters and suspension arm.
3. Remove or disconnect the following:
- Lower arm nut, cam and cam bolt
- Nut and bolt from the frame rail
- Lower arm

To install:
4. Install or connect the following:
- Lower control arm. Tighten bolts finger-tight. Align the reference marks.
5. Lower the truck to load the suspension
6. Tighten the nuts to 160 ft. lbs. (217 Nm).

CONTROL ARM BUSHING REPLACEMENT

The control arm bushings are serviced with the control arm as an assembly.

Front Wheel Bearings

ADJUSTMENT

2001–02 Pickups; 2001–04 Vans

1. Before servicing the vehicle, refer to the precautions in the beginning of this section.
2. Tighten the wheel bearing nut to 30–40 ft. lbs. (41–54 Nm) while turning the rotor.
3. Loosen the wheel bearing adjusting nut completely.
4. Tighten the nut finger-tight.
5. Check the wheel bearing end-play. The specification is 0.001–0.003 inch (0.025–0.076mm).
6. Install the nut lock and cotter pin.

2003–2004 Pickups

These are unitized hub/bearing assemblies. No adjustment is possible.

REMOVAL & INSTALLATION

2 Wheel Drive

2001–02

1. Before servicing the vehicle, refer to the precautions in the beginning of this section.
2. Remove or disconnect the following:
- Front wheel
- Brake caliper
- Grease cap
- Cotter pin
- Adjusting nut
- Washer
- Outer bearing
- Brake rotor and hub assembly
- Grease seal
- Inner bearing

To install:
3. Install or connect the following:
- Inner bearing
- Grease seal
- Brake rotor and hub assembly
- Outer bearing
- Washer
- Adjusting nut. Adjust the bearings.
- Cotter pin
- Grease cap
- Brake caliper
- Front wheel

4 WHEEL DRIVE

1. Before servicing the vehicle, refer to the precautions in the beginning of this section.
2. Remove or disconnect the following:
- Front wheel
- Brake caliper and rotor
- Hub retainer nut
- Hub and bearing assembly

To install:
3. Install or connect the following:
- Hub and bearing assembly. Tighten the bolts to 125 ft. lbs. (170 Nm).
- Hub retainer nut. Tighten the nut to 175 ft. lbs. (237 Nm).
- Brake caliper and rotor
- Front wheel

2003–2004 Pickups

COIL SPRING SUSPENSION—2 WHEEL DRIVE

1. Before servicing the vehicle, refer to the precautions in the beginning of this section.
2. Remove or disconnect the following:
- Front wheel
- Brake caliper and rotor
- Wheel speed sensor
- Hub/bearing mounting bolts
- Hub/bearing assembly

To install:
3. Install or connect the following:
- Hub and bearing assembly. Tighten the bolts to 120 ft. lbs. (163 Nm).
- Wheel speed sensor
- Brake caliper and rotor
- Front wheel

COIL SPRING SUSPENSION—4 WHEEL DRIVE

1. Before servicing the vehicle, refer to the precautions in the beginning of this section.
2. Remove or disconnect the following:
- Front wheel
- Brake caliper and rotor
- Wheel speed sensor
- Halfshaft nut
- Tie rod end from the knuckle
- Upper ball joint from the knuckle
- Knuckle from the halfshaft
- Hub/bearing mounting bolts
- Hub/bearing assembly

To install:
3. Install or connect the following:
- Hub and bearing assembly. Tighten the bolts to 120 ft. lbs. (163 Nm).
- Knuckle onto the halfshaft
- Upper ball joint on the knuckle. Torque the nut to 40 ft. lbs. (54

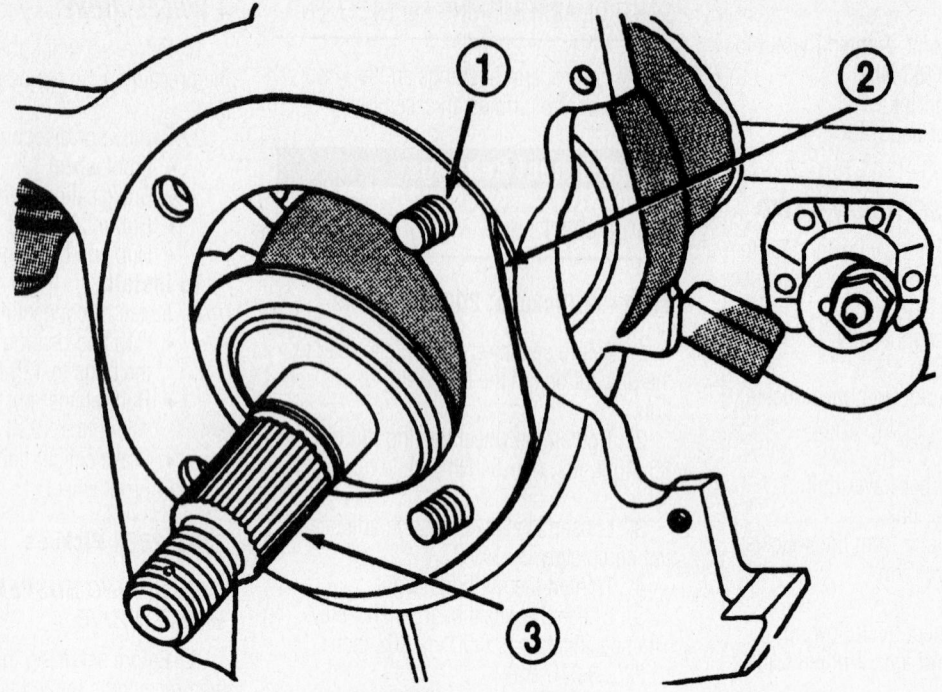

1 - ROTOR HUB BOLTS
2 - HUB SPACER (POSITION FLAT TO REAR)
3 - APPLY ANTI-SEIZE COMPOUND TO SPLINES

2399PG04

Hub bolt positioning

Nm) (on 1500 series, plus 90 degrees).
- Tie rod end to the knuckle. Tighten the nut to 45 ft. lbs. (61 Nm) plus 90 degrees.
- Halfshaft nut. Torque to 185 ft. lbs. (251 Nm).
- Wheel speed sensor
- Brake caliper and rotor
- Front wheel

COIL/LINK SUSPENSION

1. Before servicing the vehicle, refer to the precautions in the beginning of this section.

2. Remove or disconnect the following:
- Front wheel
- Hub extension from the rotor, if equipped
- Brake caliper and rotor
- Hub nut
- Wheel speed sensor wiring
- Hub/bearing mounting bolts ¼ each. Then, tap the bolts with a mallet to loosen the hub/bearing assembly.
- Hub/bearing assembly, wheel studs/extension studs, rotor, shield and spacer
- Wheel speed sensor

To install:

3. Install or connect the following:
- Wheel speed sensor
- Rotor on the hub/bearing
- Studs

4. Apply a liberal amount of anti-seize compound on the splines of the front shaft.

5. Insert the 2 rearmost, top and bottom rotor hub bolts in the knuckle, so they extend from the front face as shown.

6. Install or connect the following:
- Spacer and shield
- Hub assembly onto the shaft

7. Align the bolt holes in the flange with the installed bolts. Thread the bolts onto the flange far enough to hold the unit in place.

8. Install the remaining bolts. Tighten all hub/bearing bolts to 149 ft. lbs. (202 Nm).

9. Install the washer and shaft nut. Tighten the nut to 132 ft. lbs. (179 Nm). Rotate the axle 5 to 10 times to seat the bearings. Tighten the nut to 263 ft. lbs. (356 Nm). Install a new cotter pin, advancing the nut for alignment.

10. The remainder of installation is the reverse of removal.

BRAKES

Brake Caliper

REMOVAL & INSTALLATION

B-Series Van

FRONT

1. Raise and support the front end on jackstands.
2. Remove the wheels.
3. Press the caliper piston back into the bore with a suitable prytool. Use a large C-clamp to drive the piston into the bore of additional force is required.
4. Remove the caliper mounting bolts with a ⅜ in. hex wrench or socket.
5. Loosen the bolt that secures the front brake hose fitting bolt in the caliper.
6. Rotate the caliper rearward off the rotor and out from its mount.
7. Remove the front brake hose fitting bolt completely, then remove the caliper with the pads installed as an assembly. Take care not to drip fluid onto the pad surfaces.
8. Cover the open end of the front brake hose fitting to prevent dirt entry.

   To install:
9. Clean the caliper and steering knuckle sliding surfaces with a wire brush. Then, apply a coat of Mopar® multi-mileage grease or equivalent.
10. Lubricate the caliper mounting bolts, collars, bushings and bores with Dow 111® or GE 661® silicone grease or equivalent.
11. Install the caliper over the rotor and seat it in its original position until flush.
12. Install the mounting bolts by hand, then tighten them to 38 ft. lbs. (51 Nm) for 2001–02 models; 24 ft. lbs. (33 Nm) for 2003–04 models.
13. Install the wheels.
14. Lower the vehicle.
15. Pump the brakes several times to seat the pads.

REAR

1. Before servicing the vehicle, refer to the precautions in the beginning of this section.
2. Compress the piston.
3. Remove or disconnect the following:
   • Caliper pin bolts
   • Banjo bolt
   • Caliper
4. Installation is the reverse of removal. Use new copper washers. On 2001–02 models, torque the caliper pin bolts to 25 ft. lbs. (33Nm) and the banjo bolt to 28 ft. lbs.

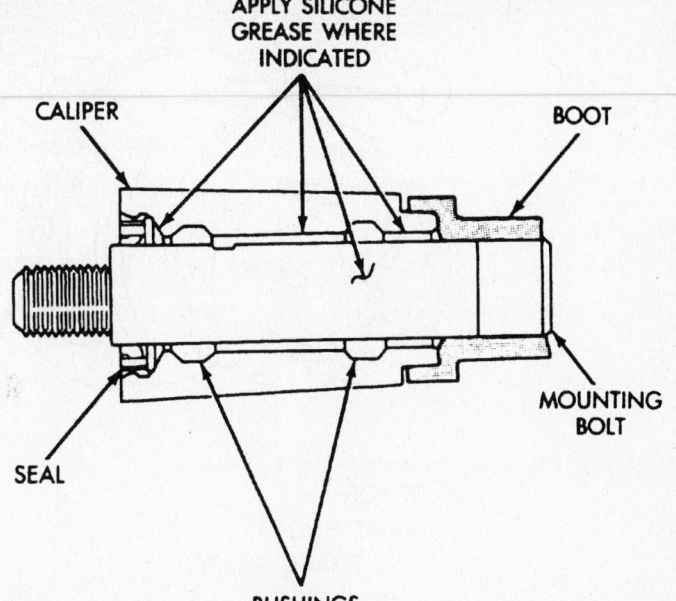

Mounting bolt lubrication points—Ram Pick-Up 75mm caliper

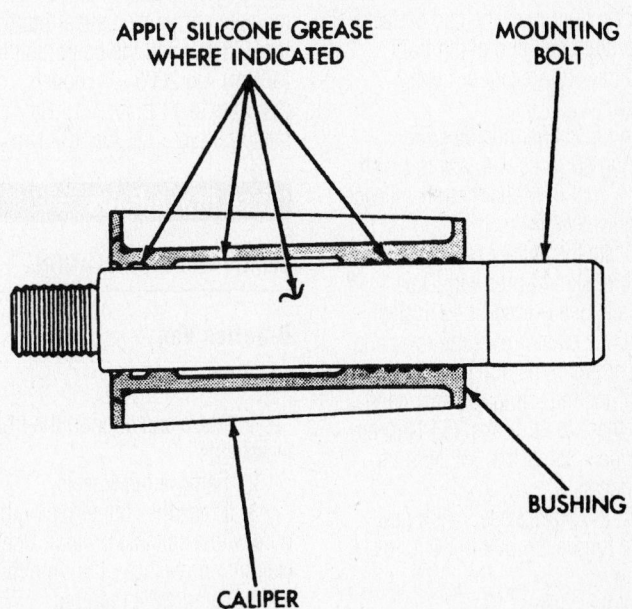

Mounting bolt lubrication points—Ram Pick-Up 80 or 86mm caliper

(38Nm). On 2003–04 models, torque the pin bolts to 24 ft. lbs. (32 Nm) and the banjo bolt to 21 ft. lbs. (28 Nm).

Ram Pick-Up

FRONT

1. Raise and support the front end on jackstands.
2. Remove the wheels.
3. Press the caliper piston back into the bore with a suitable prytool. Use a large C-clamp to drive the piston into the bore of additional force is required.
4. Remove the caliper mounting bolts with a ⅜ in. hex wrench or socket.
5. Loosen the bolt that secures the front brake hose fitting bolt in the caliper.
6. Rotate the caliper rearward off the rotor and out from its mount.
7. Remove the front brake hose fitting bolt completely, then remove the caliper

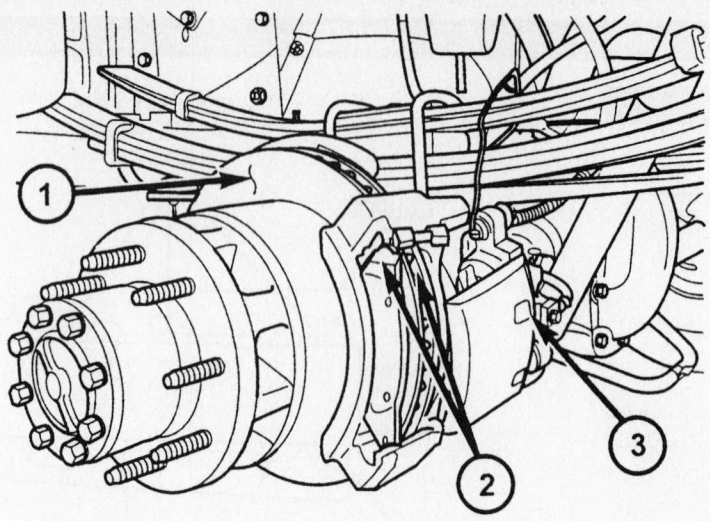

1 - ROTOR
2 - BRAKE SHOES
3 - DISC BRAKE CALIPER

9348DG01

Rear brake caliper

with the pads installed as an assembly. Take care not to drip fluid onto the pad surfaces.

8. Cover the open end of the front brake hose fitting to prevent dirt entry.

To install:

9. Clean the caliper and steering knuckle sliding surfaces with a wire brush. Then, apply a coat of Mopar® multi-mileage grease or equivalent.

10. Lubricate the caliper mounting bolts, collars, bushings and bores with Dow 111® or GE 661® silicone grease or equivalent.

11. Install the caliper over the rotor and seat it in its original position until flush.

12. Install the mounting bolts by hand, then tighten them to 38 ft. lbs. (51 Nm) on 2001–02 models; 24 ft. lbs. (32 Nm) on 2003–04 models.

13. Attach the brake hose, using new washers, and torque the bolt to 20 ft. lbs. (27 Nm).

14. Install the wheels.

15. Lower the vehicle.

16. Pump the brakes several times to seat the pads.

REAR

1. Before servicing the vehicle, refer to the precautions in the beginning of this section.

2. Compress the piston.

3. Remove or disconnect the following:
   - Caliper pin bolts
   - Banjo bolt
   - Caliper

4. Installation is the reverse of removal.

Use new copper washers. On 2001–02 models, torque the caliper pin bolts to 25 ft. lbs. (33Nm) and the banjo bolt to 28 ft. lbs. (38Nm). On 2003–04 models, torque the pin bolts to 11 ft. lbs. (15 Nm) and the banjo bolt to 23 ft. lbs. (32 Nm).

Disc Brake Pads

REMOVAL & INSTALLATION

B-Series Van

FRONT

1. Raise and support the front end on jackstands.

2. Remove the wheels.

3. Press the caliper piston back into the bore with a suitable prytool. Use a large C-clamp to drive the piston into the bore of additional force is required.

4. Remove the caliper mounting bolts with a ⅜ in. hex wrench or socket.

5. Rotate the caliper rearward off the rotor and out from its mount.

6. Set the caliper on a crate or sturdy box, then remove the inboard and outboard brake pads. The inboard pad has a spring clip that holds it in the caliper. Tilt this pad out at the top to unseat the clip. The outboard pad has a retaining spring that secures it in the caliper. Unseat 1 spring end and rotate the pad out of the caliper.

7. Secure the caliper to a chassis or

suspension component with a sturdy wire. Do not let it hang from the hose.

To install:

8. Clean the caliper and steering knuckle sliding surfaces with a wire brush. Then, apply a coat of Mopar® multi-mileage grease or equivalent.

➡**If there is minor rust or corrosion on the pins, first polish them with a crocus cloth. If they are severely rusted, replace them.**

9. Lubricate the caliper mounting bolts, collars, bushings and bores with Dow 111® or GE 661® silicone grease or equivalent.

10. Install the inboard pad and its spring clip.

11. Install the outboard brake pad.

12. Install the caliper over the rotor and seat it until flush in its original position.

13. Install the mounting bolts by hand, then tighten them to 38 ft. lbs. (51 Nm) for 2001–02 models; 24 ft. lbs. (33 Nm) for 2003–04 models.

14. Install the wheels.

15. Lower the vehicle.

16. Pump the brakes several times to seat the pads.

REAR

1. Before servicing the vehicle, refer to the precautions in the beginning of this section.

2. Compress the piston.

3. Remove or disconnect the following:
   - Caliper pin bolts

- Caliper
- Inboard and outboard pads
- Anti-rattle springs

➡**The springs are not interchangeable.**

4. Installation is the reverse of removal. Use brake grease on the springs.

Ram Pick-Up

FRONT

1. Raise and support the front end on jackstands.
2. Remove the wheels.
3. Press the caliper piston back into the bore with a suitable prytool. Use a large C-clamp to drive the piston into the bore of additional force is required.
4. Remove the caliper mounting bolts with a ⅜ in. hex wrench or socket.
5. Rotate the caliper rearward off the rotor and out from its mount.
6. Set the caliper on a crate or sturdy box, then remove the inboard and outboard brake pads. The inboard pad has a spring clip that holds it in the caliper. Tilt this pad out at the top to unseat the clip. The outboard pad has a retaining spring that secures it in the caliper. Unseat 1 spring end and rotate the pad out of the caliper.
7. Secure the caliper to a chassis or suspension component with a sturdy wire. Do not let it hang from the hose.

To install:

8. Clean the caliper and steering knuckle sliding surfaces with a wire brush.

Then, apply a coat of Mopar® multi-mileage grease or equivalent.

➡**If there is minor rust or corrosion on the pins, first polish them with a crocus cloth. If they are severely rusted, replace them.**

9. Lubricate the caliper mounting bolts, collars, bushings and bores with Dow 111® or GE 661® silicone grease or equivalent.
10. Install the inboard pad and its spring clip.
11. Install the outboard brake pad.
12. Install the caliper over the rotor and seat it until flush in its original position.
13. Install the mounting bolts by hand, then tighten them to 38 ft. lbs. (51 Nm) for 2001–02 models; 24 ft. lbs. (33 Nm) for 2003–04 models.
14. Install the wheels.
15. Lower the vehicle.
16. Pump the brakes several times to seat the pads.

REAR

1. Before servicing the vehicle, refer to the precautions in the beginning of this section.
2. Compress the piston.
3. Remove or disconnect the following:
   - Caliper pin bolts
   - Caliper
   - Inboard and outboard pads
   - Anti-rattle springs

➡**The springs are not interchangeable.**

4. Installation is the reverse of removal. Use brake grease on the springs.

Rear Disc Rotor

REMOVAL & INSTALLATION

2002

1. Before servicing the vehicle, refer to the precautions in the beginning of this section.
2. Compress the piston.
3. Remove or disconnect the following:
   - Caliper pin bolts
   - Caliper
   - With dual wheels, the axle shaft
   - 2500, the rotor
   - 3500, the hub/rotor assembly
4. Installation is the reverse of removal. Torque the rotor-to-hub bolts to 95 ft. lbs. (128Nm).

2003–2004

1. Before servicing the vehicle, refer to the precautions in the beginning of this section.
2. Remove or disconnect the following:
   - Wheel
   - Caliper
   - Caliper adapter
   - Rotor
3. Installation is the reverse of removal. Torque the caliper adapter bolts to 100 ft. lbs. (135 Nm).

Brake Drums

REMOVAL & INSTALLATION

B-Series Van

CHRYSLER SERVO TYPE WITH SINGLE ANCHOR

1. Raise and safely support the truck.
2. Remove the plug from the brake adjustment access hole.
3. Insert a thin bladed screwdriver through the adjusting hole and hold the adjusting lever away from the starwheel.
4. Release the brake by prying down against the starwheel with a brake spoon.
5. Remove the rear wheel and clips from the wheel studs. Remove the brake drum.
6. Installation is the reverse of removal. Adjust the brakes.

BENDIX DUO-SERVO TYPE

1. Raise and safely support the vehicle.
2. Remove the rear wheel and tire.

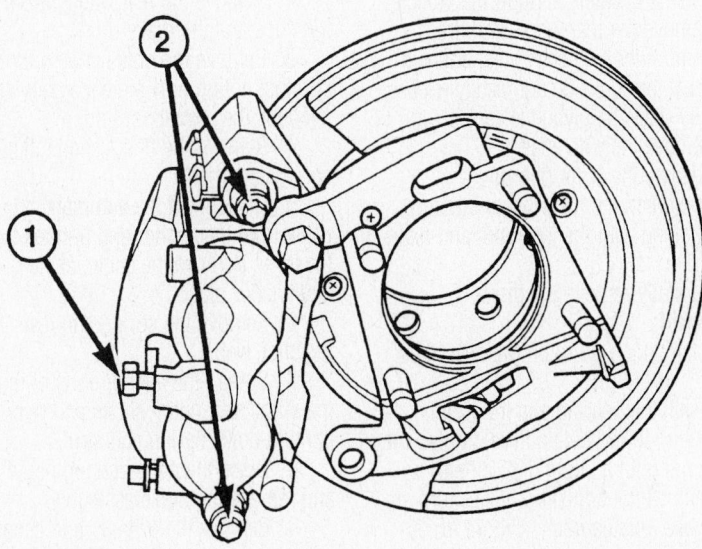

1 - Banjo Bolt
2 - Caliper Pin Bolts

9348DG02

Rear brake caliper assembly in position

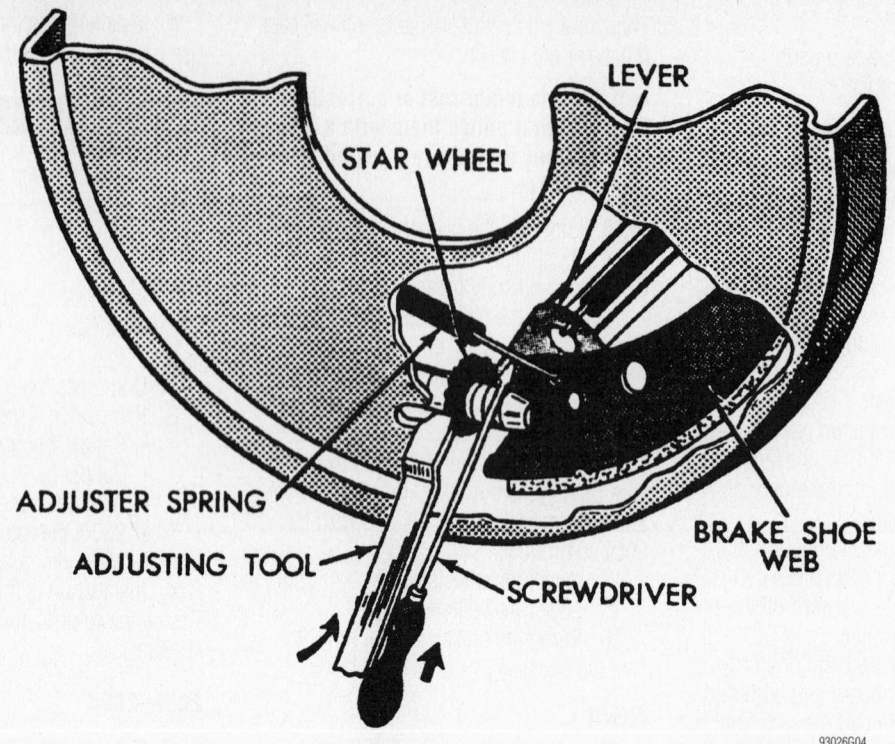

Use a lever releasing tool to depress the adjuster lever while turning the starwheel with a prytool—Bendix brakes

3. Remove the axle shaft nuts, washers and cones. If the cones do not readily release, rap the axle shaft sharply in the center.

4. Remove the axle shaft.

5. Remove the outer hub nut.

6. Straighten the lockwasher tab and remove it along with the inner nut and bearing.

7. Carefully remove the drum.

To install:

8. Position the drum on the axle housing.

9. Install the bearing and inner nut. While rotating the wheel and tire, tighten the adjusting nut until a slight drag is felt.

10. Back off the adjusting nut ⅙ turn so that the wheel rotates freely without excessive end-play.

11. Install the lockrings and nut. Place a new gasket on the hub and install the axle shaft, cones, lockwashers and nuts.

12. Install the wheel and tire.

13. Road-test the vehicle.

Ram Pick-Up

CHRYSLER SERVO TYPE WITH SINGLE ANCHOR

1. Raise and safely support the truck.

2. Remove the plug from the brake adjustment access hole.

3. Insert a thin bladed screwdriver through the adjusting hole and hold the adjusting lever away from the starwheel.

4. Release the brake by prying down against the starwheel with a brake spoon.

5. Remove the rear wheel and clips from the wheel studs. Remove the brake drum.

6. Installation is the reverse of removal. Adjust the brakes.

BENDIX DUO-SERVO TYPE

1. Raise and safely support the vehicle.

2. Remove the rear wheel and tire.

3. Remove the axle shaft nuts, washers and cones. If the cones do not readily release, rap the axle shaft sharply in the center.

4. Remove the axle shaft.

5. Remove the outer hub nut.

6. Straighten the lockwasher tab and remove it along with the inner nut and bearing.

7. Carefully remove the drum.

To install:

8. Position the drum on the axle housing.

9. Install the bearing and inner nut. While rotating the wheel and tire, tighten the adjusting nut until a slight drag is felt.

10. Back off the adjusting nut ⅙ turn so that the wheel rotates freely without excessive end-play.

11. Install the lockrings and nut. Place a new gasket on the hub and install the axle shaft, cones, lockwashers and nuts.

12. Install the wheel and tire.

13. Road-test the vehicle.

Brake Shoes

REMOVAL & INSTALLATION

B-Series Van

SERVO TYPE WITH SINGLE ANCHOR

1. Raise and support the vehicle.

2. Remove the rear wheel, drum retaining clips and the brake drum.

3. Remove the brake shoe return springs, noting how the secondary spring overlaps the primary spring.

4. Remove the brake shoe retainer, springs and nails.

5. Disconnect the automatic adjuster cable from the anchor and unhook it from the lever. Remove the cable, cable guide, and anchor plate.

6. Remove the spring and lever from the shoe web.

7. Spread the anchor ends of the primary and secondary shoes and remove the parking brake spring and strut.

8. Disconnect the parking brake cable and remove the brake assembly.

9. Remove the primary and secondary brake shoe assemblies and the star adjuster as an assembly. Block the wheel cylinders to retain the pistons.

To install:

10. Measure the drum as described in this section.

11. Apply a thin coat of lubricant to the support platforms.

12. Attach the parking brake lever to the rear of the secondary shoe.

13. Place the primary and secondary shoes in their relative positions on a workbench.

14. Lubricate the adjuster screw threads. Install it between the primary and secondary shoes with the star wheel next to the secondary shoe. The star wheels are stamped with an **L** (left) and **R** (right).

15. Overlap the ends of the primary and second brake shoes and install the adjusting spring and lever at the anchor end.

16. Hold the shoes in position and install the parking brake cable into the lever.

17. Install the parking brake strut and spring between the parking brake lever and primary shoe.

18. Place the brake shoes on the support and install the retainer nails and springs.

19. Install the anchor pin plate.

20. Install the eye of the adjusting cable over the anchor pin and install the return spring between the anchor pin and primary shoe.

21. Install the cable guide in the secondary shoe and install the secondary return spring. Be sure that the primary spring overlaps the secondary spring.

22. Position the adjusting cable in the groove of the cable guide and engage the hook of the cable in the adjusting lever.

23. Install the brake drum and retaining clips. Install the wheel and tire.

24. Adjust the brakes and road-test the truck.

BENDIX DUO-SERVO TYPE

1. Unhook and remove the adjusting lever return spring.

2. Remove the lever from the lever pivot pin.

3. Unhook the adjuster lever from the adjuster cable.

4. Unhook the upper shoe-to-shoe spring.

5. Unhook and remove the shoe hold-down springs.

6. Disconnect the parking brake cable from the parking brake lever.

7. Remove the shoes with the lower shoe-to-shoe spring and star wheel as an assembly.

To install:

8. The pivot screw and adjusting nut on the left side have left-hand threads and right-hand threads on the right side.

9. Lubricate and assemble the star wheel assembly. Lubricate the guide pads on the support plates.

10. Assemble the star wheel, lower shoe-to-shoe spring, and the primary and secondary shoes. Position this assembly on the support plate.

11. Install and hook the hold-down springs.

12. Install the upper shoe-to-shoe spring.

13. Install the cable and retaining clip.

14. Position the adjuster lever return spring on the pivot (green springs on left brakes and red springs on right brakes).

15. Install the adjuster lever. Route the adjuster cable and connect it to the adjuster.

16. Install the brake drum and adjust the brakes.

Ram Pick-Up

SERVO TYPE WITH SINGLE ANCHOR

1. Raise and support the vehicle.

2. Remove the rear wheel, drum retaining clips and the brake drum.

3. Remove the brake shoe return springs, noting how the secondary spring overlaps the primary spring.

4. Remove the brake shoe retainer, springs and nails.

5. Disconnect the automatic adjuster cable from the anchor and unhook it from the lever. Remove the cable, cable guide, and anchor plate.

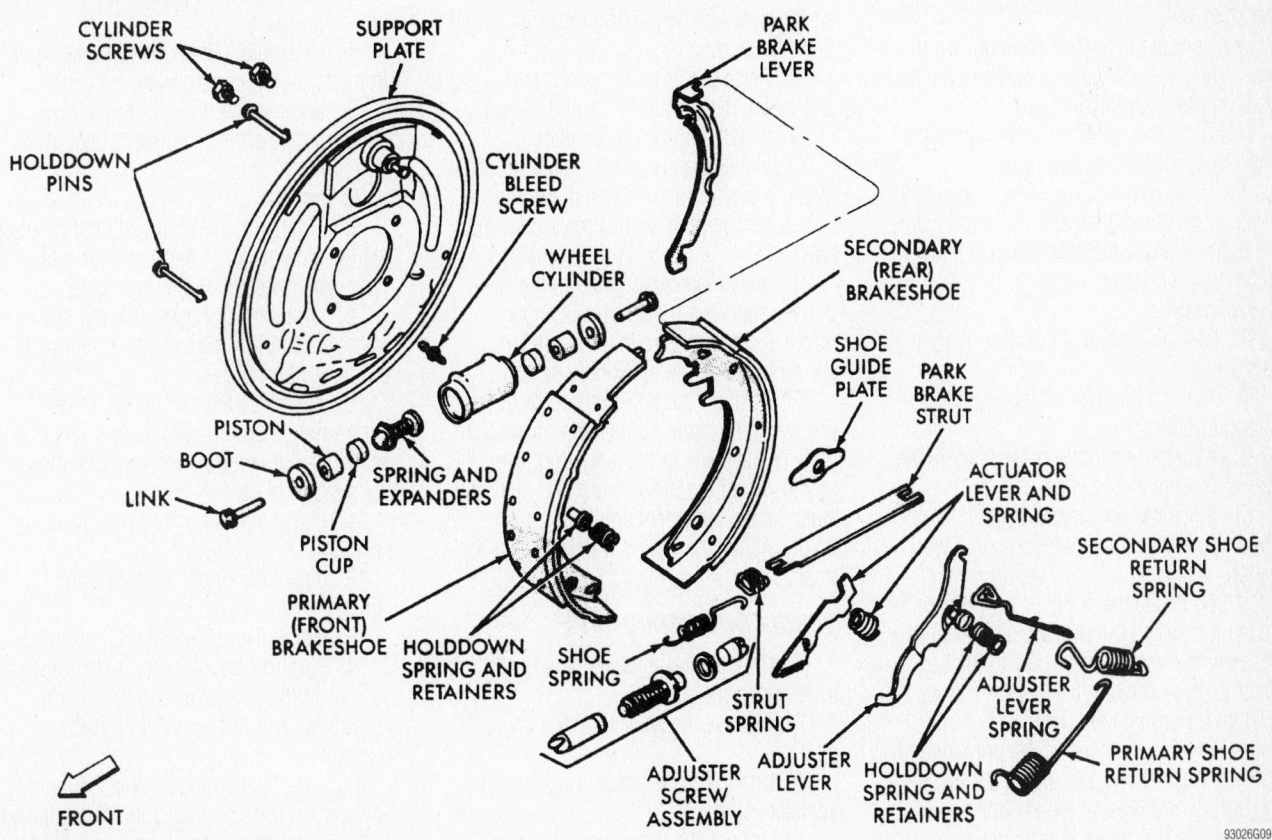

Exploded view of the rear brake components—B-Series Van

93026G09

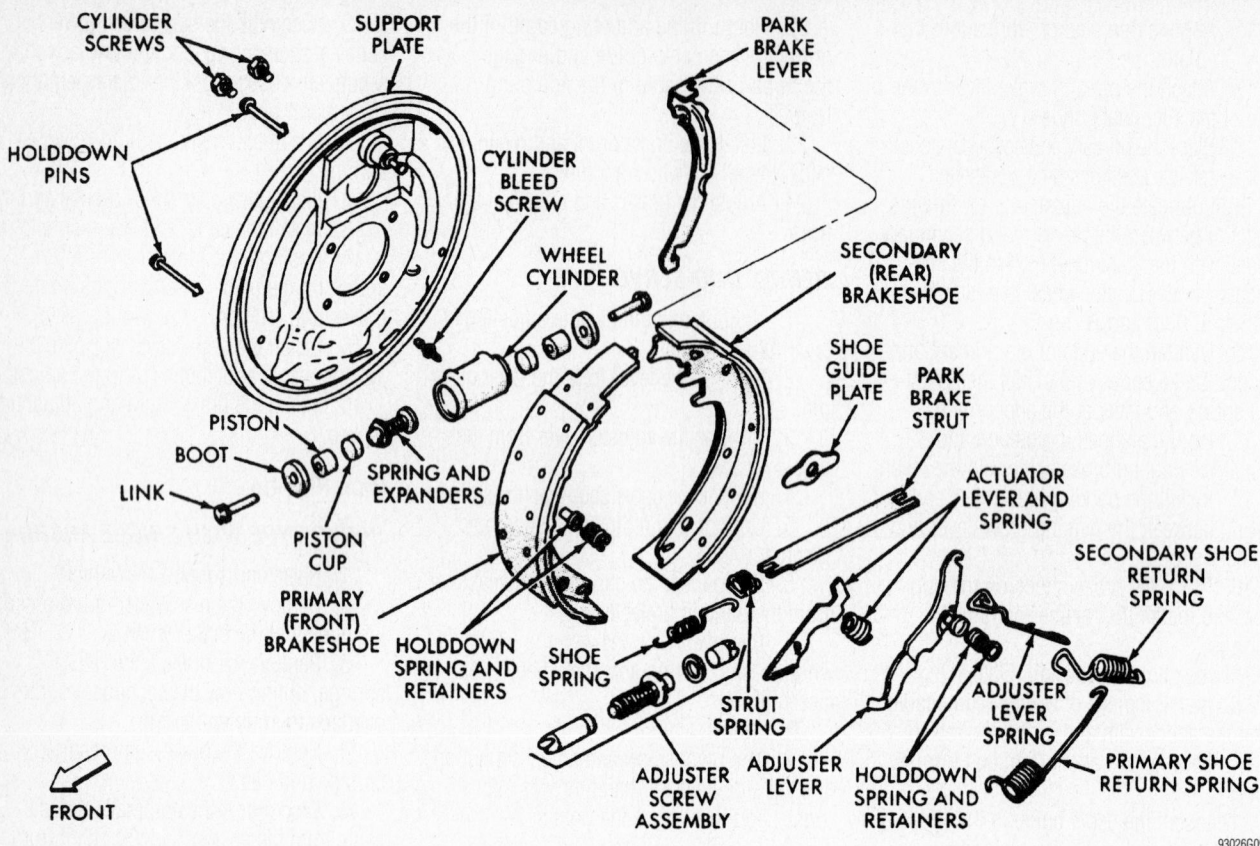

Exploded view of the rear brake components—Ram Pick-Up

6. Remove the spring and lever from the shoe web.

7. Spread the anchor ends of the primary and secondary shoes and remove the parking brake spring and strut.

8. Disconnect the parking brake cable and remove the brake assembly.

9. Remove the primary and secondary brake shoe assemblies and the star adjuster as an assembly. Block the wheel cylinders to retain the pistons.

To install:

10. Measure the drum as described in this section.

11. Apply a thin coat of lubricant to the support platforms.

12. Attach the parking brake lever to the rear of the secondary shoe.

13. Place the primary and secondary shoes in their relative positions on a workbench.

14. Lubricate the adjuster screw threads. Install it between the primary and secondary shoes with the star wheel next to the secondary shoe. The star wheels are stamped with an **L** (left) and **R** (right).

15. Overlap the ends of the primary and second brake shoes and install the adjusting spring and lever at the anchor end.

16. Hold the shoes in position and install the parking brake cable into the lever.

17. Install the parking brake strut and spring between the parking brake lever and primary shoe.

18. Place the brake shoes on the support and install the retainer nails and springs.

19. Install the anchor pin plate.

20. Install the eye of the adjusting cable over the anchor pin and install the return spring between the anchor pin and primary shoe.

21. Install the cable guide in the secondary shoe and install the secondary return spring. Be sure that the primary spring overlaps the secondary spring.

22. Position the adjusting cable in the groove of the cable guide and engage the hook of the cable in the adjusting lever.

23. Install the brake drum and retaining clips. Install the wheel and tire.

24. Adjust the brakes and road-test the truck.

BENDIX DUO-SERVO TYPE

1. Unhook and remove the adjusting lever return spring.

2. Remove the lever from the lever pivot pin.

3. Unhook the adjuster lever from the adjuster cable.

4. Unhook the upper shoe-to-shoe spring.

5. Unhook and remove the shoe hold-down springs.

6. Disconnect the parking brake cable from the parking brake lever.

7. Remove the shoes with the lower shoe-to-shoe spring and star wheel as an assembly.

To install:

8. The pivot screw and adjusting nut on the left side have left-hand threads and right-hand threads on the right side.

9. Lubricate and assemble the star wheel assembly. Lubricate the guide pads on the support plates.

10. Assemble the star wheel, lower shoe-to-shoe spring, and the primary and secondary shoes. Position this assembly on the support plate.

11. Install and hook the hold-down springs.

12. Install the upper shoe-to-shoe spring.

13. Install the cable and retaining clip.

14. Position the adjuster lever return spring on the pivot (green springs on left brakes and red springs on right brakes).

15. Install the adjuster lever. Route the adjuster cable and connect it to the adjuster.

16. Install the brake drum and adjust the brakes.

CHRYSLER AND DODGE

12

Sebring Coupe • Stratus Coupe

SPECIFICATION CHARTS

ENGINE AND VEHICLE IDENTIFICATION

Engine							Model Year	
Code ①	Liters (cc)	Cu. In.	Cyl.	Fuel Sys.	Engine Type	Eng. Mfg.	Code ②	Year
G	2.4 (2350)	143	I4	MFI	SOHC	Chrysler	1	2001
H	3.0 (2972)	181	V6	MFI	DOHC	Chrysler	2	2002
							3	2003
							4	2004
							5	2005

MFI: Multi-port Fuel Injection

DOHC: Double Overhead Camshaft

SOHC: Single Overhead Camshaft

① 8th position of the Vehicle Identification Number (VIN).

② 10th position of VIN.

67189-SEBR-C01

GENERAL ENGINE SPECIFICATIONS
All measurements are given in inches.

Year	Model	Engine Displacement Liters	Engine Series (ID/VIN)	Net Horsepower @ rpm	Net Torque @ rpm (ft. lbs.)	Bore x Stroke (in.)	Com- pression Ratio	Oil Pressure @ rpm
2001	Sebring Coupe	2.4	G	140@5500	158@4000	3.40 x 3.94	9.5:1	43-100@3500
		3.0	H	200@5500	205@4400	3.59 x 2.99	9.0:1	43-100@3500
	Stratus Coupe	2.4	G	140@5500	158@4000	3.40 x 3.94	9.5:1	43-100@3500
		3.0	H	200@5500	205@4400	3.59 x 2.99	9.0:1	43-100@3500
2002	Sebring Coupe	2.4	G	140@5500	158@4000	3.40 x 3.94	9.5:1	43-100@3500
		3.0	H	200@5500	205@4000	3.59 x 2.99	9.0:1	43-100@3500
	Stratus Coupe	2.4	G	140@5500	158@4000	3.40 x 3.94	9.5:1	43-100@3500
		3.0	H	200@5500	205@4000	3.59 x 2.99	9.0:1	43-100@3500
2003	Sebring Coupe	2.4	G	140@5500	158@4000	3.40 x 3.94	9.5:1	43-100@3500
		3.0	H	200@5500	205@4000	3.59 x 2.99	9.0:1	43-100@3500
	Stratus Coupe	2.4	G	140@5500	158@4000	3.40 x 3.94	9.5:1	43-100@3500
		3.0	H	200@5500	205@4000	3.59 x 2.99	9.0:1	43-100@3500
2004	Sebring Coupe	2.4	G	140@5500	158@4000	3.40 x 3.94	9.5:1	43-100@3500
		3.0	H	200@5500	205@4000	3.59 x 2.99	9.0:1	43-100@3500
	Stratus Coupe	2.4	G	140@5500	158@4000	3.40 x 3.94	9.5:1	43-100@3500
		3.0	H	200@5500	205@4000	3.59 x 2.99	9.0:1	43-100@3500

① California - 9.0:1

67189-SEBR-C02

ENGINE TUNE-UP SPECIFICATIONS

Year	Engine Displacement Liters	Engine ID/VIN	Spark Plug Gap (in.)	Ignition Timing (deg.)	Fuel Pump (psi)	Idle Speed (rpm)		Valve Clearance	
						MT	AT	In.	Ex.
2001	2.4	G	0.039-0.043	①	47-50	—	700	HYD	HYD
	3.0	H	0.039-0.043	①	47-50	700	700	HYD	HYD
2002	2.4	G	0.039-0.043	①	47-50	700	700	HYD	HYD
	3.0	H	0.039-0.043	①	47-50	700	700	HYD	HYD
2003	2.4	G	0.039-0.043	①	47-50	700	700	HYD	HYD
	3.0	H	0.039-0.043	①	47-50	700	700	HYD	HYD
2004	2.4	G	0.039-0.043	①	47-50	700	700	HYD	HYD
	3.0	H	0.039-0.043	①	47-50	700	700	HYD	HYD

NOTE: The Vehicle Emission Control Information label often reflects specification changes made during production. The label figures must be used if they differ from those in this chart.

HYD: Hydraulic

① Basic ignition timing is not adjustable.

67189-SEBR-C03

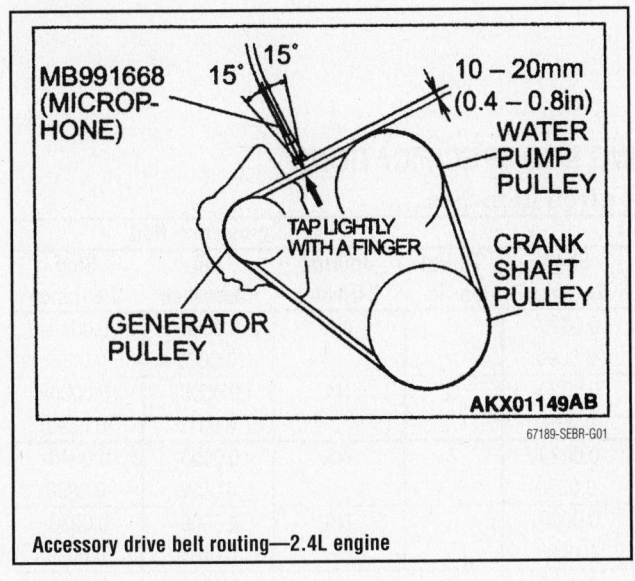

Accessory drive belt routing—2.4L engine

67189-SEBR-G01

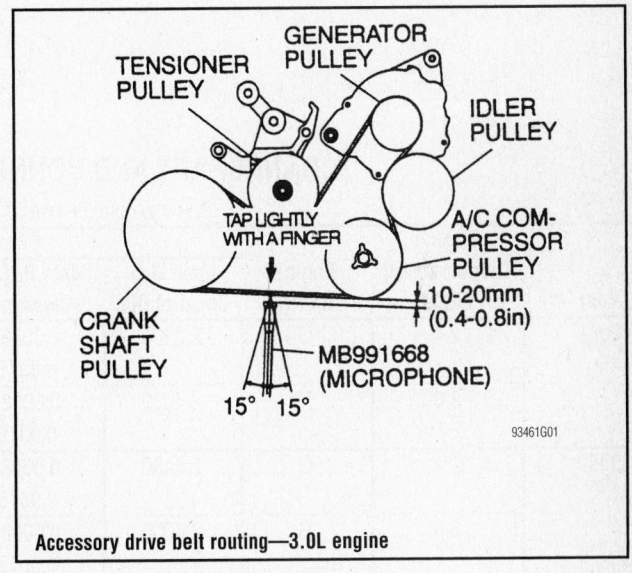

Accessory drive belt routing—3.0L engine

93461G01

CAPACITIES

Year	Model	Engine Displacement Liters	Engine ID/VIN	Engine Oil with Filter (qts.)	Transmission (pts.) Manual	Transmission (pts.) Auto.①	Fuel Tank (gal.)	Cooling System (qts.)
2001	Sebring Coupe	2.4	G	4.5	4.6	16.2	16.4	7.4
		3.0	H	4.5	6.0	17.0	16.4	8.5
	Stratus Coupe	2.4	G	4.5	4.6	16.2	16.4	7.4
		3.0	H	4.5	6.0	17.0	16.4	8.5
2002	Sebring Coupe	2.4	G	4.5	4.6	16.2	16.4	7.4
		3.0	H	4.5	6.0	17.0	16.4	8.5
	Stratus Coupe	2.4	G	4.5	4.6	16.2	16.4	7.4
		3.0	H	4.5	6.0	17.0	16.4	8.5
2003	Sebring Coupe	2.4	G	4.5	4.6	16.2	16.4	7.4
		3.0	H	4.5	6.0	17.0	16.4	8.5
	Stratus Coupe	2.4	G	4.5	4.6	16.2	16.4	7.4
		3.0	H	4.5	6.0	17.0	16.4	8.5
2004	Sebring Coupe	2.4	G	4.5	4.6	16.2	16.4	7.4
		3.0	H	4.5	6.0	17.0	16.4	8.5
	Stratus Coupe	2.4	G	4.5	4.6	16.2	16.4	7.4
		3.0	H	4.5	6.0	17.0	16.4	8.5

NOTE: All capacities are approximate. Add fluid gradually and ensure a proper fluid level is obtained.

① Overhaul fill capacity with torque converter empty

67189-SEBR-C04

CRANKSHAFT AND CONNECTING ROD SPECIFICATIONS

All measurements are given in inches.

Year	Engine Displacement Liters	Engine ID/VIN	Crankshaft Main Brg. Journal Dia.	Crankshaft Main Brg. Oil Clearance	Crankshaft Shaft End-play	Crankshaft Thrust on No.	Connecting Rod Journal Diameter	Connecting Rod Oil Clearance	Connecting Rod Side Clearance
2001	2.4	G	2.2400	0.0008-0.0015	0.0020-0.0090	3	NA	0.0008-0.0020	0.0040-0.0090
	3.0	H	2.4000	0.0008-0.0015	0.0020-0.0090	3	NA	0.0008-0.0019	0.0030-0.0090
2002	2.4	G	2.2400	0.0008-0.0015	0.0020-0.0090	3	NA	0.0008-0.0020	0.0040-0.0090
	3.0	H	2.4000	0.0008-0.0015	0.0020-0.0090	3	NA	0.0008-0.0019	0.0030-0.0090
2003	2.4	G	2.2400	0.0008-0.0015	0.0020-0.0090	3	NA	0.0008-0.0020	0.0040-0.0090
	3.0	H	2.4000	0.0008-0.0015	0.0020-0.0090	3	NA	0.0008-0.0019	0.0030-0.0090
2004	2.4	G	2.2400	0.0008-0.0015	0.0020-0.0090	3	NA	0.0008-0.0020	0.0040-0.0090
	3.0	H	2.4000	0.0008-0.0015	0.0020-0.0090	3	NA	0.0008-0.0019	0.0030-0.0090

67189-SEBR-C05

VALVE SPECIFICATIONS

Year	Engine Displacement Liters	Engine ID/VIN	Seat Angle (deg.)	Face Angle (deg.)	Spring Test Pressure (lbs. @ in.)	Spring Installed Height (in.)	Stem-to-Guide Clearance (in.)		Stem Diameter (in.)	
							Intake	Exhaust	Intake	Exhaust
2001	2.4	G	44-44.5	45-45.5	60@1.740	1.740	0.0008-0.0019	0.0012-0.0027	0.2400	0.2400
	3.0	H	44-44.5	45-45.5	60@1.740	1.740	0.0008-0.0019	0.0016-0.0027	0.2400	0.2400
2002	2.4	G	44-44.5	45-45.5	60@1.740	1.740	0.0008-0.0019	0.0012-0.0027	0.2400	0.2400
	3.0	H	44-44.5	45-45.5	60@1.740	1.740	0.0008-0.0019	0.0016-0.0027	0.2400	0.2400
2003	2.4	G	44-44.5	45-45.5	60@1.740	1.740	0.0008-0.0019	0.0012-0.0027	0.2400	0.2400
	3.0	H	44-44.5	45-45.5	60@1.740	1.740	0.0008-0.0019	0.0016-0.0027	0.2400	0.2400
2004	2.4	G	44-44.5	45-45.5	60@1.740	1.740	0.0008-0.0019	0.0012-0.0027	0.2400	0.2400
	3.0	H	44-44.5	45-45.5	60@1.740	1.740	0.0008-0.0019	0.0016-0.0027	0.2400	0.2400

67189-SEBR-C06

PISTON AND RING SPECIFICATIONS

All measurements are given in inches.

Year	Engine Displacement Liters	Engine ID/VIN	Piston Clearance	Ring Gap			Ring Side Clearance		
				Top Compression	Bottom Compression	Oil Control	Top Compression	Bottom Compression	Oil Control
2001	2.4	G	0.0008-0.0015	0.010-0.013	0.016-0.021	0.004-0.015	0.0008-0.0023	0.0012-0.0023	NA
	3.0	H	0.0008-0.0015	0.012-0.017	0.018-0.023	0.008-0.023	0.0012-0.0027	0.0008-0.0023	NA
2002	2.4	G	0.0008-0.0015	0.010-0.013	0.016-0.021	0.004-0.015	0.0008-0.0023	0.0012-0.0023	NA
	3.0	H	0.0008-0.0015	0.012-0.017	0.018-0.023	0.008-0.023	0.0012-0.0027	0.0008-0.0023	NA
2003	2.4	G	0.0008-0.0015	0.010-0.013	0.016-0.021	0.004-0.015	0.0008-0.0023	0.0012-0.0023	NA
	3.0	H	0.0008-0.0015	0.012-0.017	0.018-0.023	0.008-0.023	0.0012-0.0027	0.0008-0.0023	NA
2004	2.4	G	0.0008-0.0015	0.010-0.013	0.016-0.021	0.004-0.015	0.0008-0.0023	0.0012-0.0023	NA
	3.0	H	0.0008-0.0015	0.012-0.017	0.018-0.023	0.008-0.023	0.0012-0.0027	0.0008-0.0023	NA

NA: Not Available

67189-SEBR-C07

TORQUE SPECIFICATIONS
All readings in ft. lbs.

Year	Engine Displacement Liters	Engine ID/VIN	Cylinder Head Bolts	Main Bearing Bolts	Rod Bearing Bolts	Crankshaft Damper Bolts	Flywheel Bolts	Manifold Intake	Manifold Exhaust	Spark Plugs	Oil Pan Drain Plug
2001	2.4	G	①	18 ②	14 ②	87	98	14	③	18	20
	3.0	H	④	69	38	134	55	⑤	33	18	20
2002	2.4	G	①	18 ②	14 ②	87	98	14	④	18	20
	3.0	H	④	69	38	134	55	⑤	33	18	20
2003	2.4	G	①	18 ②	14 ②	87	98	14	④	18	20
	3.0	H	④	69	38	134	55	⑤	33	18	20
2004	2.4	G	①	18 ②	14 ②	87	98	14	④	18	20
	3.0	H	④	69	38	134	55	⑤	33	18	20

① Step 1: 55-61 ft. lbs.
  Step 2: Loosen all bolts
  Step 3: 14-16 ft. lbs.
  Step 4: 1/4 turn
  Step 5: 1/4 turn
② Plus 1/4 turn
③ M8: 22 ft. lbs.
  M10: 36 ft. lbs.

④ Step 1: 80 ft. lbs.
  Step 2: Loosen all bolts
  Step 3: 80 ft. lbs.
⑤ Step 1: Right bank nuts to 56 inch lbs.
  Step 2: Left bank nuts to 16 ft. lbs.
  Step 3: Right bank nuts to 16 ft. lbs.
  Step 4: Left bank nuts to 16 ft. lbs.
  Step 5: Right bank nuts to 16 ft. lbs.

67189-SEBR-C08

WHEEL ALIGNMENT

Year	Model		Caster Range (+/-Deg.)	Caster Preferred Setting (Deg.)	Camber Range (+/-Deg.)	Camber Preferred Setting (Deg.)	Toe-in (in.)
2001	ALL	F	+1.50	+4.31	+0.50	0	0 +/- 0.13
		R	—	—	+0.50	-1.34	0.13 +/- 0.13
2002	ALL	F	+1.50	+4.31	+0.50	0	0 +/- 0.12
		R	—	—	+0.50	-1.34	0.13 +/- 0.13
2003	ALL	F	+1.50	+4.31	+0.50	0	0 +/- 0.13
		R	—	—	+0.50	-1.34	0.13 +/- 0.13
2004	ALL	F	+1.50	+4.31	+0.50	0	0 +/- 0.13
		R	—	—	+0.50	-1.34	0.13 +/- 0.13

67189-SEBR-C09

TIRE, WHEEL AND BALL JOINT SPECIFICATIONS

Year	Model	OEM Tires Standard	OEM Tires Optional	Tire Pressures (psi) Front	Tire Pressures (psi) Rear	Wheel Size	Ball Joint Inspection	Lug Nuts (ft. lbs.)
2001	Sebring LX	P205/60HR16	None	32	29	6JJ	①	73
	Sebring Lxi	P215/50HR17	None	32	29	6.5JJ	①	73
	Stratus SE	P205/60HR16	None	32	29	6JJ	①	73
	Stratus R/T	P215/50HR17	None	32	29	6.5JJ	①	73
2002	Sebring LX	P205/60HR16	None	32	29	6JJ	①	73
	Sebring Lxi	P215/50HR17	None	32	29	6.5JJ	①	73
	Stratus SE	P205/60HR16	None	32	29	6JJ	①	73
	Stratus R/T	P215/50HR17	None	32	29	6.5JJ	①	73
2003	Sebring LX	P205/60HR16	None	32	29	6JJ	①	73
	Sebring Lxi	P215/50HR17	None	32	29	6.5JJ	①	73
	Stratus SE	P205/60HR16	None	32	29	6JJ	①	73
	Stratus R/T	P215/50HR17	None	32	29	6.5JJ	①	73
2004	Sebring LX	P205/60HR16	None	32	29	6JJ	①	73
	Sebring Lxi	P215/50HR17	None	32	29	6.5JJ	①	73
	Stratus SE	P205/60HR16	None	32	29	6JJ	①	73
	Stratus R/T	P215/50HR17	None	32	29	6.5JJ	①	73

OEM: Original Equipment Manufacturer

PSI: Pounds Per Square Inch

STD: Standard

OPT: Optional

L: Lower

U: Upper

① Replace if any measurable movement is found

67189-SEBR-C10

BRAKE SPECIFICATIONS
All measurements in inches unless noted

Year	Model		Brake Disc Original Thickness	Brake Disc Minimum Thickness	Brake Disc Maximum Run-out	Brake Drum Diameter Original Inside Diameter	Brake Drum Diameter Max. Wear Limit	Brake Drum Diameter Maximum Machine Diameter	Min. Lining Thickness	Brake Caliper Bracket Bolts (ft. lbs.)	Brake Caliper Mounting Bolts (ft. lbs.)
2001	Sebring	F	0.900	0.880	0.002	—	—	—	0.080	74	28
	Coupe	R	0.400	0.330	0.003	9.00	—	—	①	②	32
	Stratus	F	0.900	0.880	0.002	—	—	—	0.080	74	28
	Coupe	R	0.400	0.330	0.003	9.00	—	—	①	②	32
2002	Sebring	F	0.900	0.880	0.002	—	—	—	0.080	74	28
	Coupe	R	0.400	0.330	0.003	9.00	—	—	①	②	32
	Stratus	F	0.900	0.880	0.002	—	—	—	0.080	74	28
	Coupe	R	0.400	0.330	0.003	9.00	—	—	①	②	32
2003	Sebring	F	0.900	0.880	0.002	—	—	—	0.080	74	28
	Coupe	R	0.400	0.330	0.003	9.00	—	—	①	②	32
	Stratus	F	0.900	0.880	0.002	—	—	—	0.080	74	28
	Coupe	R	0.400	0.330	0.003	9.00	—	—	①	②	32
2004	Sebring	F	0.900	0.880	0.002	—	—	—	0.080	74	28
	Coupe	R	0.400	0.330	0.003	9.00	—	—	①	②	32
	Stratus	F	0.900	0.880	0.002	—	—	—	0.080	74	28
	Coupe	R	0.400	0.330	0.003	9.00	—	—	①	②	32

F: Front

R: Rear

① With rear disc: 0.080 in.
   With rear drum: 0.040

② Flange bolt: 44 ft. lbs.
   Bolt and washer: 41 ft. lbs.

67189-SEBR-C11

SCHEDULED MAINTENANCE INTERVALS
2001-02 Chrysler—Sebring Coupe, Dodge—Stratus Coupe

TO BE SERVICED	TYPE OF SERVICE	VEHICLE MILEAGE INTERVAL (x1000)												
		7.5	15	22.5	30	37.5	45	52.5	60	67.5	75	82.5	90	97.5
Engine oil & filter	R	✓	✓	✓	✓	✓	✓	✓	✓	✓	✓	✓	✓	✓
Coolant level, hoses & clamps	S/I	✓	✓	✓	✓	✓	✓	✓	✓	✓	✓	✓	✓	✓
Rotate tires	S/I	✓	✓	✓	✓	✓	✓	✓	✓	✓	✓	✓	✓	✓
Automatic transaxle fluid level	S/I		✓		✓		✓		✓		✓		✓	
Brake hoses & disc brake pads	S/I		✓		✓		✓		✓		✓		✓	
Driveshaft boots & front suspension components	S/I		✓		✓		✓		✓		✓		✓	
Air filter element	R				✓				✓				✓	
Engine coolant	R				✓				✓				✓	
Spark plugs (DOHC)	R				✓				✓				✓	
Spark plugs (SOHC) ①	R				✓				✓				✓	
Accessory drive belts	S/I				✓				✓				✓	
Ball joints & steering linkage seals	S/I				✓				✓				✓	
Exhaust system	S/I				✓				✓				✓	
Fuel hoses	S/I				✓				✓				✓	
Manual transaxle oil	S/I				✓				✓				✓	
PCV valve	S/I				✓				✓				✓	
Rear drum brake lining & rear wheel cylinders	S/I				✓				✓				✓	
Camshaft timing belt ②	R								✓					
Ignition cables	R								✓					
Distributor cap & rotor	S/I								✓					
EVAP system	S/I								✓					
Fuel system	S/I								✓					

R: Replace   S/I: Service or Inspect

① Spark plugs: replace every 100,000 miles.

② If equipped with the 3.0L engine, replace at 100.000 miles

FREQUENT OPERATION MAINTENANCE (SEVERE SERVICE)

If a vehicle is operated under any of the following conditions it is considered severe service:

- Extremely dusty areas.
- 50% or more of the vehicle operation is in 32°C (90°F) or higher temperatures, or constant operation in temperatures below 0°C (32°F).
- Prolonged idling (vehicle operation in stop and go traffic).
- Frequent short running periods (engine does not warm to normal operating temperatures).
- Police, taxi, delivery usage or trailer towing usage.

Oil & oil filter change: change every 3000 miles.

Disc brake pads: check every 6000 miles.

Air filter element: change every 15,000 miles.

Automatic transaxle fluid: change every 15,000 miles.

Rear drum brake linings & rear wheel cylinders: check every 15,000 miles.

Spark plugs: change every 15,000 miles.

SCHEDULED MAINTENANCE INTERVALS
2003-04 Chrysler—Sebring Coupe, Dodge—Stratus Coupe

TO BE SERVICED	TYPE OF SERVICE	VEHICLE MILEAGE INTERVAL (x1000)												
		6	12	18	24	30	36	42	48	54	60	66	72	78
Engine oil & filter	R	✓	✓	✓	✓	✓	✓	✓	✓	✓	✓	✓	✓	✓
Coolant level, hoses & clamps	S/I	✓	✓	✓	✓	✓	✓	✓	✓	✓	✓	✓	✓	✓
Rotate tires	S/I	✓	✓	✓	✓	✓	✓	✓	✓	✓	✓	✓	✓	✓
Automatic transaxle fluid level	S/I		✓		✓		✓		✓		✓		✓	
Brake hoses & disc brake pads	S/I		✓		✓		✓		✓		✓		✓	
Driveshaft boots & front suspension components	S/I		✓		✓		✓		✓		✓		✓	
Air filter element	R				✓				✓				✓	
Engine coolant	R				✓				✓				✓	
Spark plugs (DOHC)	R				✓				✓				✓	
Spark plugs (SOHC) ①	R				✓				✓				✓	
Accessory drive belts	S/I				✓				✓				✓	
Ball joints & steering linkage seals	S/I				✓				✓				✓	
Exhaust system	S/I				✓				✓				✓	
Fuel hoses	S/I				✓				✓				✓	
Manual transaxle oil	S/I				✓				✓				✓	
PCV valve	S/I				✓				✓				✓	
Rear drum brake lining & rear wheel cylinders	S/I				✓				✓				✓	
Camshaft timing belt ②	R								✓					
Ignition cables	R								✓					
Distributor cap & rotor	S/I								✓					
EVAP system	S/I								✓					
Fuel system	S/I								✓					

R: Replace   S/I: Service or Inspect

① Spark plugs: replace every 100,000 miles.

② If equipped with the 3.0L engine, replace at 100.000 miles

FREQUENT OPERATION MAINTENANCE (SEVERE SERVICE)

If a vehicle is operated under any of the following conditions it is considered severe service:

- Extremely dusty areas.

- 50% or more of the vehicle operation is in 32°C (90°F) or higher temperatures, or constant operation in temperatures below 0°C (32°F).

- Prolonged idling (vehicle operation in stop and go traffic).

- Frequent short running periods (engine does not warm to normal operating temperatures).

- Police, taxi, delivery usage or trailer towing usage.

Oil & oil filter change: change every 3000 miles.

Disc brake pads: check every 6000 miles.

Air filter element: change every 15,000 miles.

Automatic transaxle fluid: change every 15,000 miles.

Rear drum brake linings & rear wheel cylinders: check every 15,000 miles.

Spark plugs: change every 15,000 miles.

67189-SEBR-C13

PRECAUTIONS

Before servicing any vehicle, please be sure to read all of the following precautions, which deal with personal safety, prevention of component damage, and important points to take into consideration when servicing a motor vehicle:

• Never open, service or drain the radiator or cooling system when the engine is hot; serious burns can occur from the steam and hot coolant.

• Observe all applicable safety precautions when working around fuel. Whenever servicing the fuel system, always work in a well-ventilated area. Do not allow fuel spray or vapors to come in contact with a spark, open flame or excessive heat (a hot drop light, for example). Keep a dry chemical fire extinguisher near the work area. Always keep fuel in a container specifically designed for fuel storage; also, always properly seal fuel containers to avoid the possibility of fire or explosion. Refer to the additional fuel system precautions later in this section.

• Fuel injection systems often remain pressurized, even after the engine has been turned **OFF**. The fuel system pressure must be relieved before disconnecting any fuel lines. Failure to do so may result in fire and/or personal injury.

• Brake fluid often contains polyglycol ethers and polyglycols. Avoid contact with the eyes and wash your hands thoroughly after handling brake fluid. If you do get brake fluid in your eyes, flush your eyes with clean, running water for 15 minutes. If

eye irritation persists, or if you have taken brake fluid internally, IMMEDIATELY seek medical assistance.

• The EPA warns that prolonged contact with used engine oil may cause a number of skin disorders, including cancer! You should make every effort to minimize your exposure to used engine oil. Protective gloves should be worn when changing oil. Wash your hands and any other exposed skin areas as soon as possible after exposure to used engine oil. Soap and water, or waterless hand cleaner should be used.

• All new vehicles are now equipped with an air bag system, often referred to as a Supplemental Restraint System (SRS) or Supplemental Inflatable Restraint (SIR) system. The system must be disabled before performing service on or around system components, steering column, instrument panel components, wiring and sensors. Failure to follow safety and disabling procedures could result in accidental air bag deployment, possible personal injury and unnecessary system repairs.

• Always wear safety goggles when working with, or around, the air bag system. When carrying a non-deployed air bag, be sure the bag and trim cover are pointed away from your body. When placing a non-deployed air bag on a work surface, always face the bag and trim cover upward, away from the surface. This will reduce the motion of the module if it is accidentally deployed. Refer to the additional air bag system precautions later in this section.

• Clean, high quality brake fluid from a sealed container is essential to the safe and proper operation of the brake system. You should always buy the correct type of brake fluid for your vehicle. If the brake fluid becomes contaminated, completely flush the system with new fluid. Never reuse any brake fluid. Any brake fluid that is removed from the system should be discarded. Also, do not allow any brake fluid to come in contact with a painted surface; it will damage the paint.

• Never operate the engine without the proper amount and type of engine oil; doing so WILL result in severe engine damage.

• Timing belt maintenance is extremely important! Many models utilize an interference-type, non-freewheeling engine. If the timing belt breaks, the valves in the cylinder head may strike the pistons, causing potentially serious (also time-consuming and expensive) engine damage.

• Disconnecting the negative battery cable on some vehicles may interfere with the functions of the on-board computer system(s) and may require the computer to undergo a relearning process once the negative battery cable is reconnected.

• When servicing drum brakes, only disassemble and assemble one side at a time, leaving the remaining side intact for reference.

• Only an MVAC-trained, EPA-certified automotive technician should service the air conditioning system or its components.

ENGINE REPAIR

Alternator

REMOVAL

2.4L Engine

1. Before servicing the vehicle, refer to the precautions in the beginning of this section.
2. Remove or disconnect the following:
   • Negative battery cable
   • Engine under cover
   • Engine mount bracket
   • Oil pressure hose and tube clamp bolts
   • Oil return tube clamp bolt
   • Power steering and A/C drive belt
   • Alternator drive belt
   • Water pump pulley

   • Alternator connector
   • Alternator brace
   • Alternator bolt and nut
   • Alternator

3.0L Engine

1. Before servicing the vehicle, refer to the precautions in the beginning of this section.
2. Remove or disconnect the following:
   • Negative battery cable
   • Engine under cover
   • Engine mount bracket
   • Drive belt
   • Alternator belt
   • Oil level gauge unit
   • Alternator electrical connectors
   • Alternator brace
   • Alternator

INSTALLATION

2.4L Engine

1. Install or connect the following:
   • Alternator
   • Alternator bolt and nut. Tighten to 26–40 ft. lbs. (34–54 Nm).
   • Alternator brace
   • Alternator connector
   • Water pump pulley
   • Alternator drive belt
   • Power steering and A/C drive belt
   • Oil return tube clamp bolt
   • Oil pressure hose and tube clamp bolts
   • Engine mount bracket
2. Adjust the belt tension.
   • Negative battery cable
   • Under cover

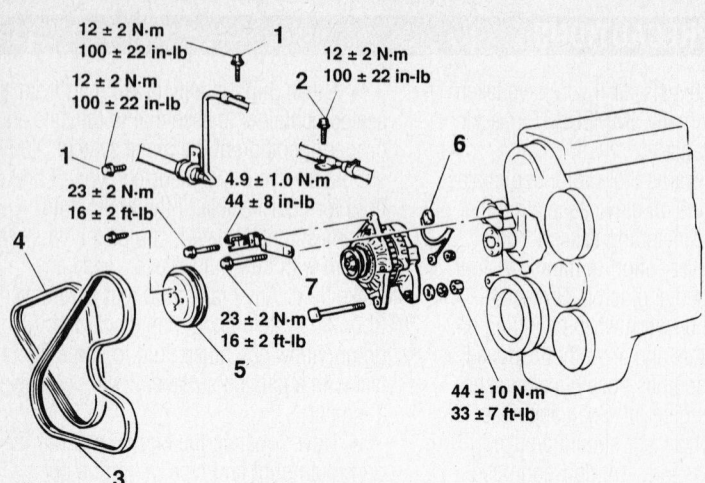

1. OIL PRESSURE HOSE AND TUBE ASSEMBLY CLAMP BOLTS
2. OIL RETURN TUBE ASSEMBLY CLAMP BOLT
3. DRIVE BELT (POWER STEERING, A/C)
4. DRIVE BELT (GENERATOR)
5. WATER PUMP PULLEY
6. GENERATOR CONNECTOR
7. GENERATOR BRACE

9356CG01

Exploded view of the alternator and related components—2.4L engine

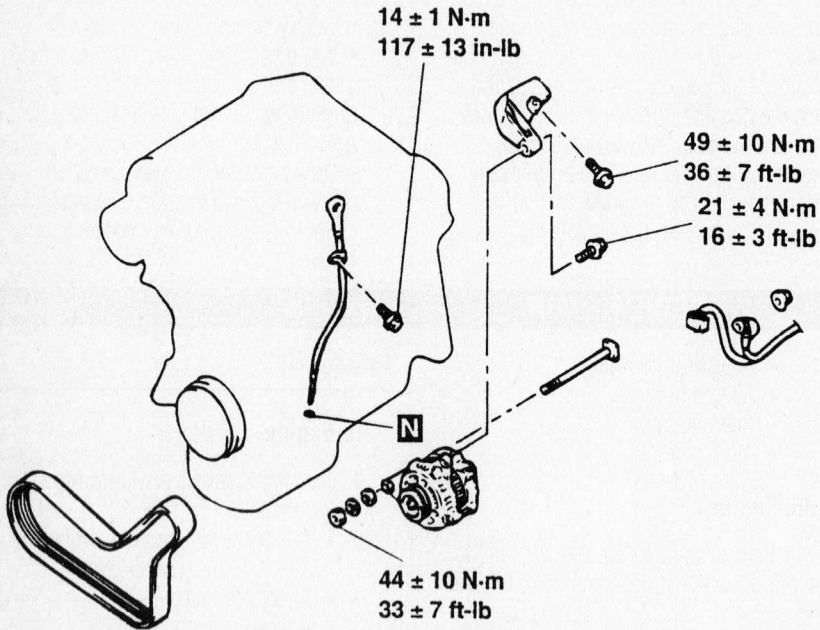

9356CG02

Exploded view of the alternator and related components—3.0L engine

3.0L Engine

1. Install or connect the following:
   - Alternator and torque the bolts to 33 ft. lbs. (44 Nm)
   - Alternator brace and torque the bolts to 36 ft. lbs. (49 Nm)
   - Oil level gauge unit and torque the bolt to 10 ft. lbs. (15 Nm)
   - Alternator belt
   - Drive belt
   - Engine mount bracket
2. Adjust the belt tension.

- Negative battery cable
- Under cover

Ignition Timing

Ignition timing is controlled by the Powertrain Control Module (PCM) and is not adjustable.

Engine Assembly

REMOVAL & INSTALLATION

2.4L Engine

1. Before servicing the vehicle, refer to the precautions in the beginning of this section.
2. Properly relieve the fuel system pressure.
3. Drain the engine oil.
4. Drain the cooling system.
5. Recover the A/C refrigerant.
6. Remove or disconnect the following:
   - Hood
   - Negative battery cable
   - Strut tower bar
   - Air cleaner assembly
   - Radiator and recovery tank
   - Front exhaust pipe
   - Accelerator cable connection
   - Purge hose connection
   - Brake booster vacuum hose connection
   - Ignition coil connector
   - Injector connectors
   - Manifold differential pressure sensor connector
   - Throttle Position Sensor (TP) connection
   - Heated Oxygen Sensor (HO2S) connection
   - Capacitor connection
   - Engine Coolant Temperature (ECT) connector
   - Camshaft Position (CMP) sensor connector
   - Knock Sensor (KS) connector
   - Engine coolant temperature gauge unit connector
   - Idle Air Control (IAC) motor connector
   - Evaporative emission purge solenoid valve
   - Exhaust Gas Recirculation (EGR) solenoid valve connector
   - High pressure and fuel return fuel hoses
   - Pressure hose connection
   - Oil dipstick and tube
   - Pressure hose connection

- Heater hose
- Alternator connector
- Oil pressure switch connector
- Drive belts
- Crankshaft Position (CKP) sensor connector
- Power steering pressure switch connector

7. Unbolt the power steering pump and bracket, and place aside, as an assembly with the hose attached.

✳✳ WARNING

Do NOT disconnect the A/C compressor refrigerant lines!

- A/C compressor connector. Unbolt and place the A/C compressor aside with the lines attached.

8. Remove the transaxle. Do NOT remove the flywheel mounting bolt shown in the accompanying figure. Removal of this bolt will throw the flywheel out of balance and damage it.

9. Remove the engine mount bracket, as follows:

   a. Support the engine with a suitable jack.

   b. Remove the special tools that were attached during transaxle removal

   c. Hold the engine assembly with a chain block or similar tool.

   d. Place a jack against the oil pan

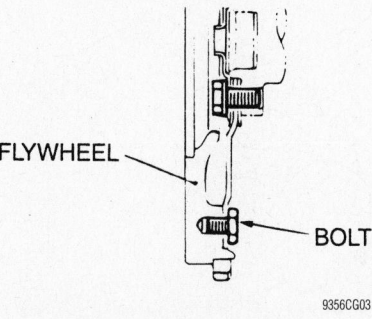

9356CG03

Do not remove the flywheel bolt shown, or you will damage the flywheel

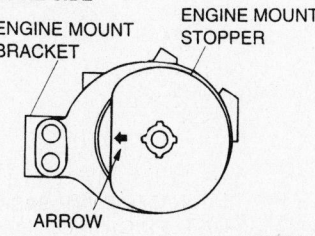

9356CG04

Correct installation of the engine mount stopper—2.4L engine

(with a block of wood in between), the jack up the engine so that the weight of the engine is no longer on the engine mount, then remove the engine mount bracket.

10. Remove the engine mount stopper.

11. Make sure that all cables, hoses and electrical connections are disconnected, slowly raise the engine up and out of the engine compartment.

To install:

12. Install or connect the following:

- Engine assembly
- Engine mount stopper and torque the bolts to 60 ft. lbs. (81 Nm). Clamp the mount stopper so that the arrow points in the direction shown in the accompanying figure.
- Engine mount bracket and torque the bolts to 64 ft. lbs. (86 Nm). Remove the engine lifting device
- Transaxle
- A/C compressor and connector
- Power steering pump and bracket
- Power steering pressure switch connector
- CKP sensor connector
- Drive belts
- Oil pressure switch connector
- Alternator connector
- Heater hose
- Pressure hose
- Oil dipstick tube and dipstick
- Pressure hose connector
- Fuel return hose
- High pressure fuel hose. Apply a small quantity of oil to the hose O-ring and then installing using a twisting motion. Make sure that the O-ring does not get twisted or clamped during installation.
- EGR solenoid valve connector
- Evap purge solenoid valve
- IAC motor connector
- ECT gauge unit connector
- KS connector
- CMP sensor connector
- ECT sensor connector
- Capacitor connector
- HO2S connector
- TP sensor connector
- Manifold differential pressure sensor connector
- Injector connector
- Ignition coil connector
- Vacuum hose connection
- Brake booster vacuum hose
- Purge hose
- Accelerator cable
- Front exhaust pipe
- Radiator reservoir tank and radiator
- Air cleaner

- Strut tower bar
- Negative battery cable
- Hood and adjust as needed

13. Fill the engine with clean oil and replace the filter.

14. Fill the cooling system to the proper level.

15. Recharge the A/C system with the proper refrigerant.

16. Start the vehicle, check for leaks and repair if necessary.

3.0L Engine

1. Before servicing the vehicle, refer to the precautions in the beginning of this section.

2. Properly relieve the fuel system pressure.

3. Drain the engine oil.

4. Drain the cooling system.

5. Recover the A/C refrigerant.

6. Remove or disconnect the following:

- Hood
- Negative battery cable
- Strut tower bar
- Air cleaner
- Radiator reservoir tank
- Front exhaust pipe
- Throttle cable
- Manifold differential pressure sensor connector
- Control wiring harness and power steering wiring harness combination connector
- Exhaust Gas Recirculation (EGR) valve connector
- Evaporative (EVAP) Emission purge solenoid valve connector
- Knock Sensor (KS) connector
- Crankshaft Position (CKP) sensor connector
- Right bank Heated Oxygen Sensor (HO2S)
- Fuel injector connector
- Distributor connector
- Control wiring/injector wiring harness combination connector
- Throttle Position Sensor (TPS) connector
- Idle Air Control (IAC) motor connector
- Ground wire
- Engine Coolant Temperature (ECT) gauge unit
- ECT sensor connector
- Left bank HO2S
- Brake booster vacuum hose
- Fuel supply and return hoses
- Heater hoses
- Starter electrical connectors
- Oil Pressure switch

- Alternator wiring
- Drive belts
- A/C compressor
- Power steering pressure switch
- Power steering oil pump. Do not disconnect the hose.
- Engine mount stay and install an engine Lifting Tool, such as MZ203827
- Engine mount bracket and stopper
- Engine assembly

To install:

7. Install or connect the following:
- Engine assembly
- Engine mount bracket and stopper and torque the bolts to 60 ft. lbs. (81 Nm)
- Engine mount stay and torque the bolts to 26 ft. lbs. (35 Nm). Remove the engine lifting device.
- Power steering oil pump and torque the bolts to 31 ft. lbs. (42 Nm)
- Power steering pressure switch
- A/C compressor
- Drive belts
- Alternator wiring
- Oil pressure switch
- Starter electrical connectors
- Heater hoses
- Fuel supply and return lines
- Brake booster vacuum hose
- Left bank HO_2S
- ECT sensor connector
- ECT gauge unit
- Ground wire
- IAC motor electrical connector
- TPS connector
- Control wiring/injector wiring harness combination connector
- Distributor connector
- Fuel injector connector
- Right bank HO_2S
- CKP sensor connector
- KS connector
- EVAP solenoid valve connector
- EGR valve connector
- Control wiring harness and power steering harness combination connector
- Manifold differential Pressure sensor connector
- Throttle cable
- Front exhaust pipe
- Radiator reservoir tank
- Air cleaner
- Strut tower bar
- Negative battery cable
- Hood and adjust as needed

8. Fill the engine with clean oil and replace the filter.

9. Fill the cooling system to the proper level.

10. Recharge the A/C system with the proper refrigerant.

11. Start the vehicle, check for leaks and repair if necessary.

Water Pump

REMOVAL & INSTALLATION

2.4L Engine

1. Before servicing the vehicle, refer to the precautions in the beginning of this section.

2. Drain the cooling system.

3. Remove or disconnect the following:
- Negative battery cable
- Timing belt tensioner pulley
- Alternator brace
- Water pump and discard the gasket and O-ring

To install:

4. Clean all mating surfaces of any residual gasket material

5. Install or connect the following:
- New O-ring gasket
- Water pump and torque the bolts to 117 inch lbs. (14 Nm)
- Alternator brace and torque the bolts to 17 ft. lbs. (23 Nm)

3.0L Engine

1. Before servicing the vehicle, refer to the precautions in the beginning of this section.

2. Drain the cooling system.

3. Remove or disconnect the following:
- Negative battery cable
- Timing belt
- Thermostat and bracket
- Water pump and discard the gasket and O-ring

To install:

4. Install or connect the following:
- Water pump with a new O-ring and gasket and torque the bolts to 17 ft. lbs. (23 Nm)
- Bracket and torque the bolts to 17 ft. lbs. (23 Nm)
- Thermostat
- Timing belt
- Negative battery cable

5. Fill the cooling system to the proper level.

6. Start the vehicle, check for leaks and repair if necessary.

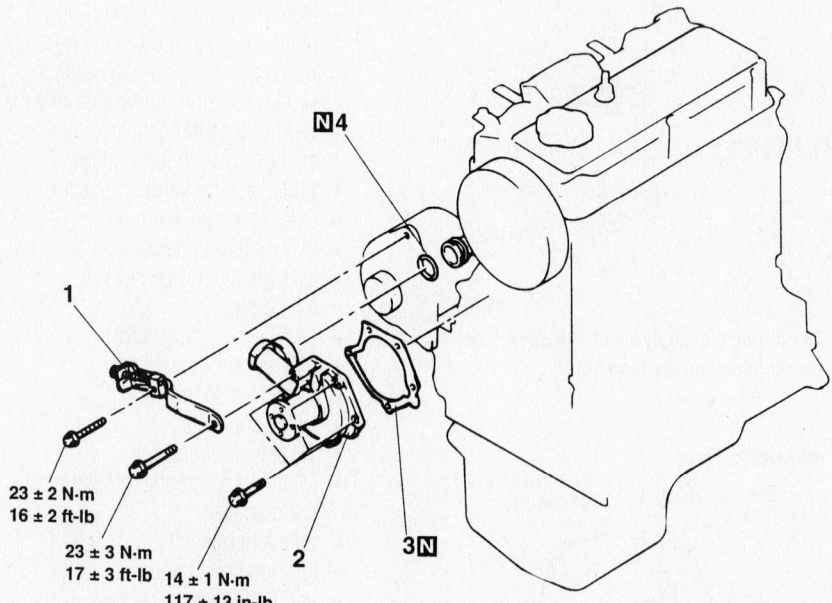

23 ± 2 N·m
16 ± 2 ft-lb

23 ± 3 N·m
17 ± 3 ft-lb

14 ± 1 N·m
117 ± 13 in-lb

1.	GENERATOR BRACE	3.	WATER PUMP GASKET
2.	WATER PUMP ASSEMBLY	4.	O-RING

9356CG05

Exploded view of the water pump—2.4L engine

Heater Core

REMOVAL & INSTALLATION

1. Before servicing the vehicle, refer to the precautions in the beginning of this section.

2. Disconnect the negative battery cable. Properly drain the cooling system.

3. Disconnect the heater hoses from the heater core.

4. Remove the floor console by removing or disconnecting the following:
   - Center console panel
   - Shift knob
   - Accessory box or ashtray
   - Floor console panel assembly
   - Shift lever cover assembly
   - Floor console assembly

5. From under the steering wheel, remove the hood lock release handle, the driver's side under cover and the lap cooler grille.

6. Remove the steering wheel by removing or disconnecting the following:
   - Air bag module-to-steering wheel screws (rear of the steering wheel) and the air bag module
   - Steering wheel-to-column nut and press the steering wheel from the column
   - Lower steering column cover screws, the lower cover, the upper cover and the column pad

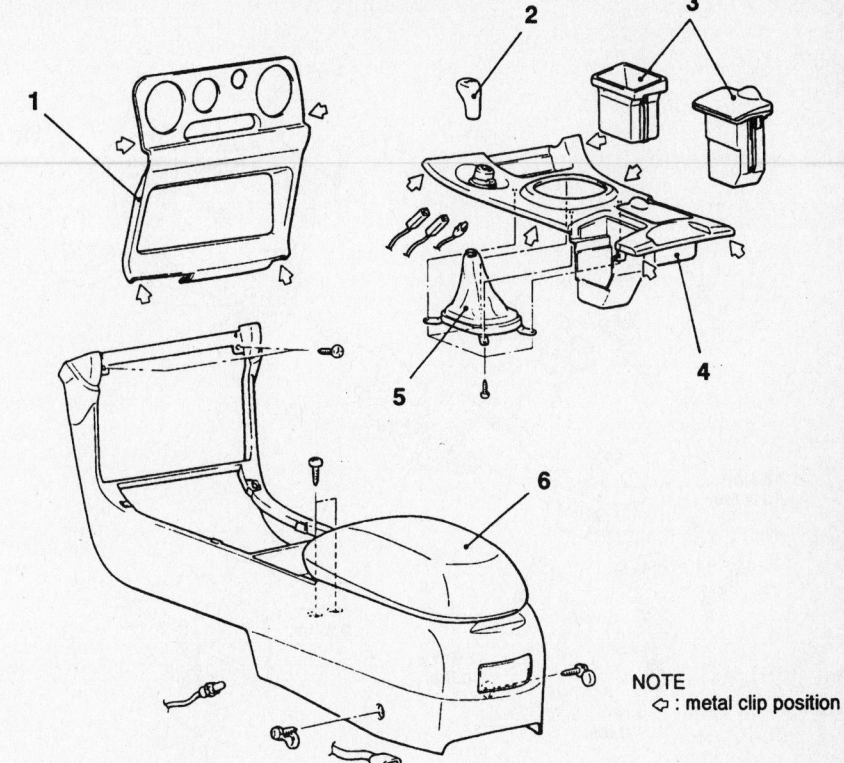

NOTE
◁ : metal clip position

1. Center console panel
2. Shift knob
3. Accessory box or ashtray
4. Floor console panel assembly
5. Shift lever cover assembly
6. Floor console assembly

93111G52

Exploded view of the center console

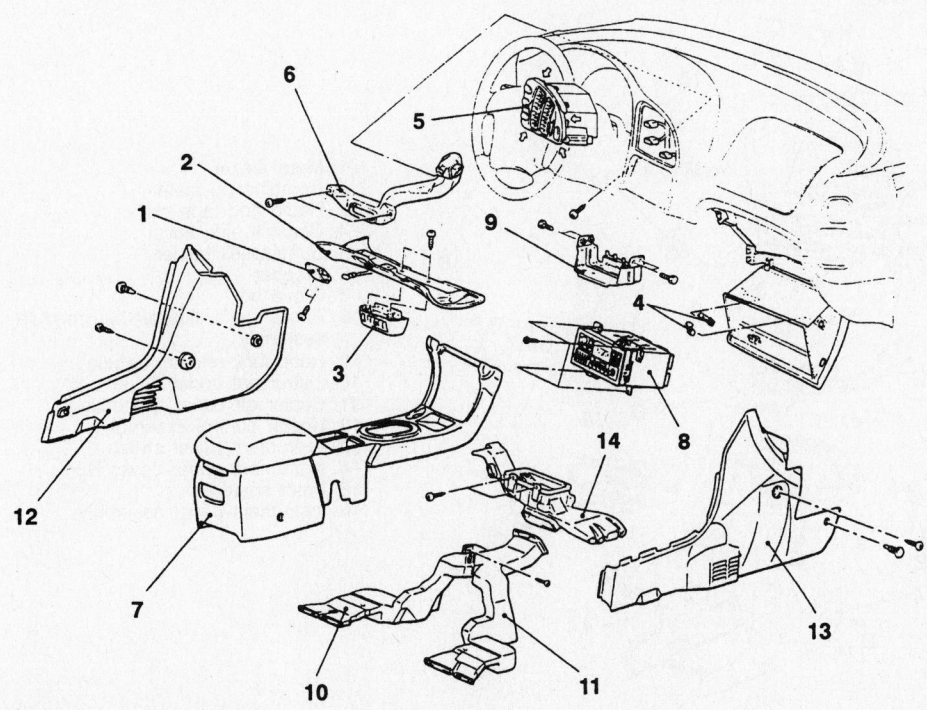

1. Hood lock release handle
2. Driver's side under cover
3. Lap cooler grille
4. Stopper
5. Center air outlet
6. Lap cooler duct
7. Floor console
8. Radio and tape player
9. Relay bracket
10. Rear heater duct (L.H.)
11. Rear heater duct (R.H.)
12. Console side cover (L.H.)
13. Console side cover (R.H.)
14. Foot distribution duct

93111G53

Exploded view of the ventilators and related components

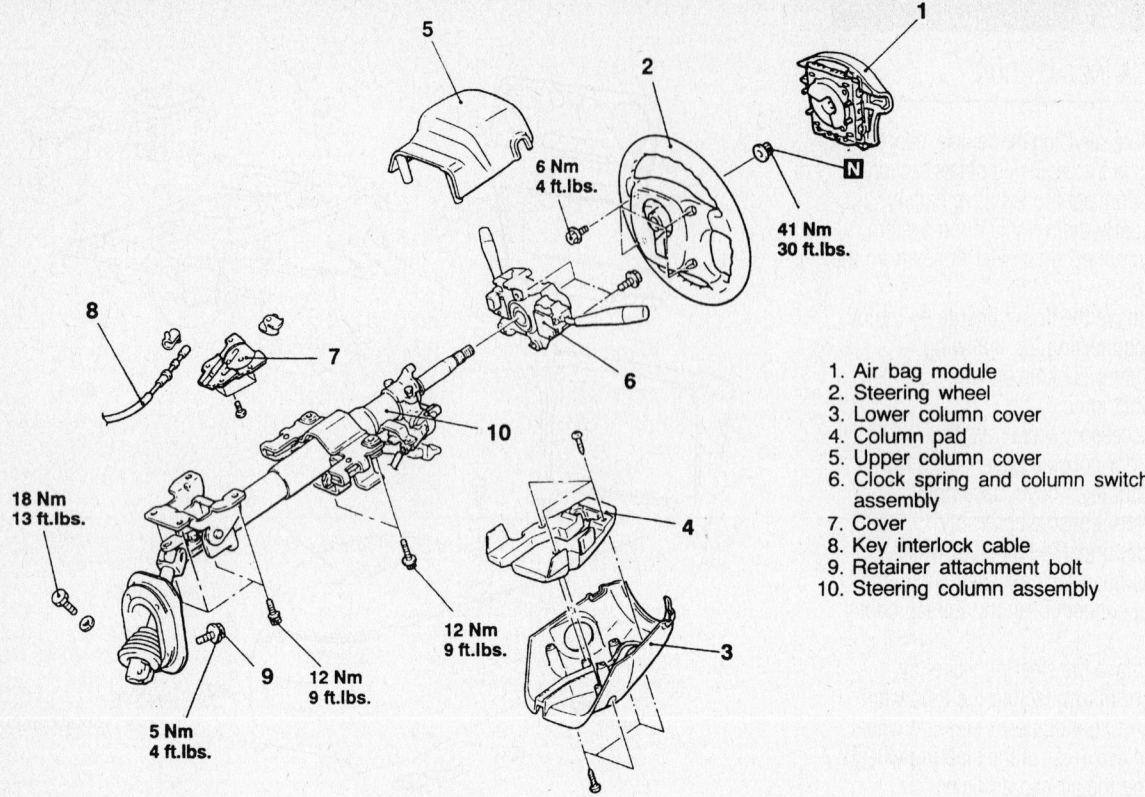

1. Air bag module
2. Steering wheel
3. Lower column cover
4. Column pad
5. Upper column cover
6. Clock spring and column switch assembly
7. Cover
8. Key interlock cable
9. Retainer attachment bolt
10. Steering column assembly

6 Nm
4 ft.lbs.

41 Nm
30 ft.lbs.

18 Nm
13 ft.lbs.

12 Nm
9 ft.lbs.

5 Nm
4 ft.lbs.

12 Nm
9 ft.lbs.

93111G54

Exploded view of the steering column and related components

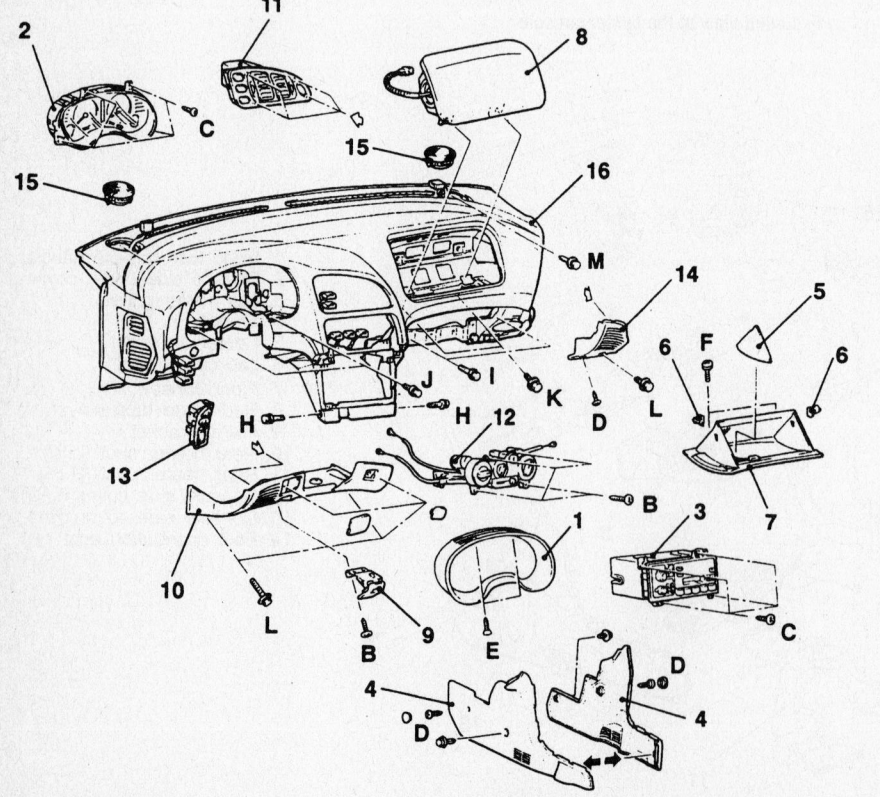

1. Meter bezel
2. Combination meter
3. Radio and tape player
4. Console side cover
5. Sunglasses holder
6. Stopper
7. Glove box
8. Passenger's side air bag module assembly
9. Hood lock release handle
10. Instrument under cover L.H.
11. Center air outlet assembly
12. Heater control assembly
13. Instrument panel switch
14. Instrument under cover R.H.
15. Front speaker
16. Instrument panel assembly

93111G55

Exploded view of the instrument panel and related components

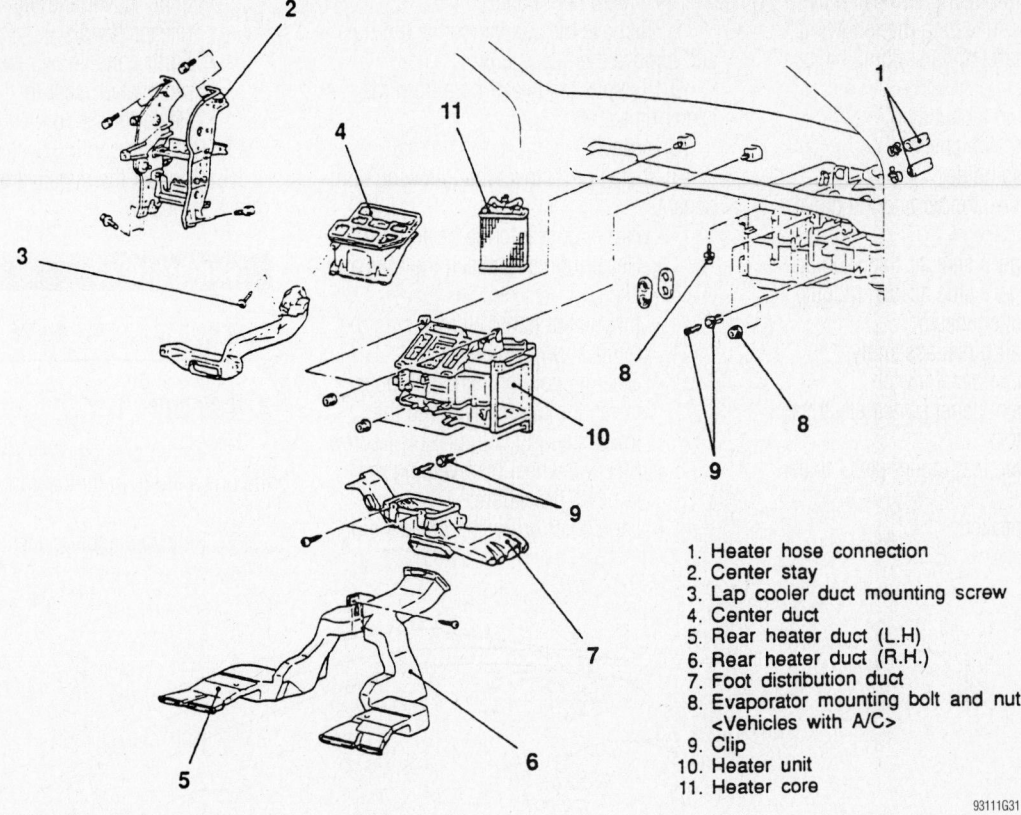

1. Heater hose connection
2. Center stay
3. Lap cooler duct mounting screw
4. Center duct
5. Rear heater duct (L.H)
6. Rear heater duct (R.H)
7. Foot distribution duct
8. Evaporator mounting bolt and nut
   <Vehicles with A/C>
9. Clip
10. Heater unit
11. Heater core

93111G31

Exploded view of the heater core housing and related components

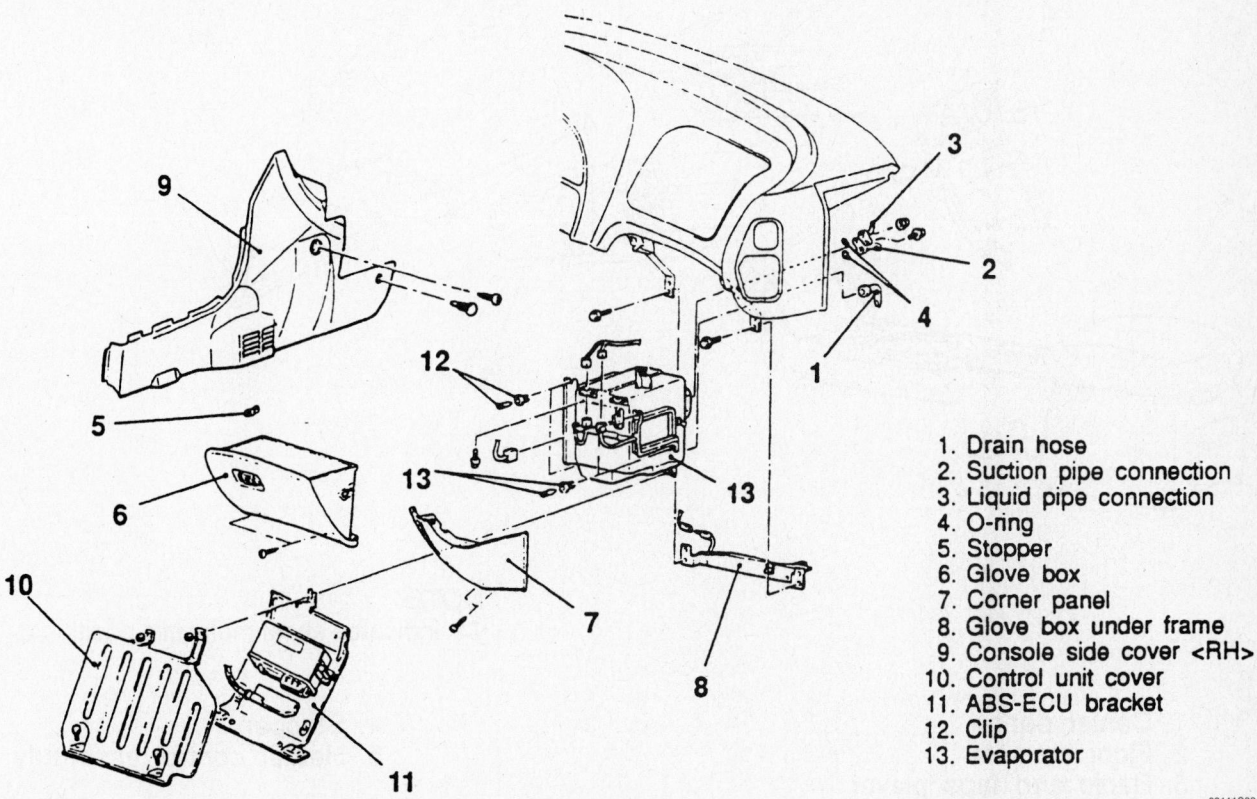

1. Drain hose
2. Suction pipe connection
3. Liquid pipe connection
4. O-ring
5. Stopper
6. Glove box
7. Corner panel
8. Glove box under frame
9. Console side cover <RH>
10. Control unit cover
11. ABS-ECU bracket
12. Clip
13. Evaporator

93111G32

Exploded view of the evaporator core housing and related components

7. Remove the instrument panel by removing or disconnecting the following:
- Meter bezel and the combination meter
- Radio and tape player
- Console side cover
- Sunglass holder
- Glove box stopper and the glove box
- Passenger's side air bag module
- Passenger's side air bag module electrical connector
- Center air outlet assembly
- Instrument panel switch
- Instrument panel switch electrical connectors
- Right side instrument panel under cover
- Front speaker

- Instrument panel

8. Remove the attaching clip and nuts and remove the heater unit.

9. Remove the heater core from the heater unit.

To install:

10. Install or connect the following components:
- Heater core into the heater unit
- Heater unit and attach the nuts and clip
- Instrument panel by reversing the removal procedures
- Steering column pad, the upper cover, the lower cover and the lower steering column cover screws
- Steering wheel by reversing the removal procedure
- Lap cooler grille, the driver's side

under cover and the hood lock release handle
- Floor console by reversing the removal procedure
- Heater hoses to the heater core

11. Refill the cooling system.

12. Connect the negative battery cable. Check for leaks.

Cylinder Head

REMOVAL & INSTALLATION

2.4L Engine

1. Before servicing the vehicle, refer to the precautions in the beginning of this section.

2. Relieve the fuel system pressure.

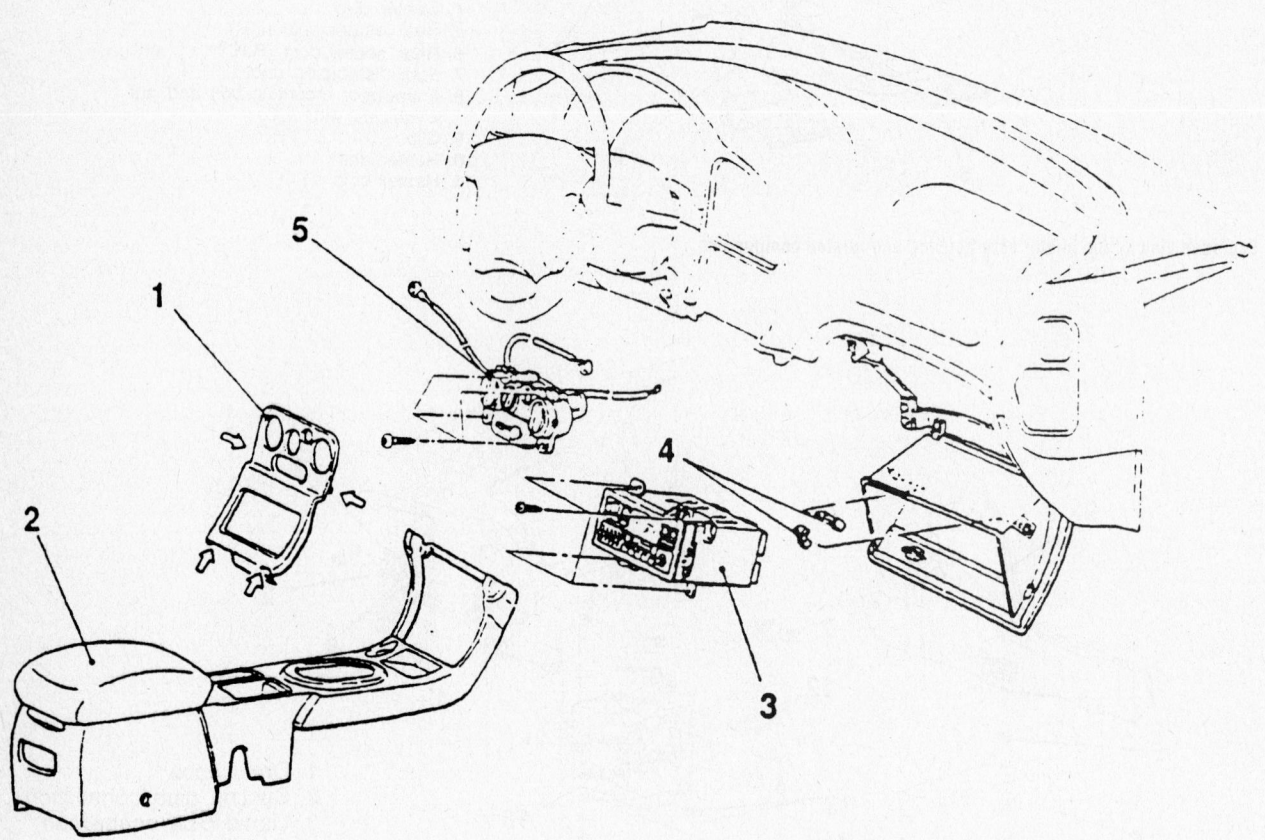

NOTE
⇐ indicates sheet metal clip positions.

1. Center panel
2. Floor console
3. Radio and tape player

4. Stopper
5. Heater control assembly

93111G33

Exploded view of the manual control head and related components

3. Drain the cooling system.
4. Drain the engine oil.
5. Remove or disconnect the following:
- Negative battery cable
- Strut tower bar
- Air cleaner assembly
- Thermostat case
- Front exhaust pipe
- Accelerator cable
- Purge hose
- Brake booster vacuum hose
- Ignition coil
- Fuel injector electrical connector
- Ignition failure sensor connector
- Manifold differential pressure sensor connector
- Throttle Position (TP) sensor connector
- Heated Oxygen Sensor (HO2S) connector
- Capacitor connector
- Engine Coolant Temperature (ECT) sensor
- Camshaft Position (CMP) sensor
- Idle Air Control (IAC) motor connector
- Evaporative (EVAP) emission purge solenoid valve
- Exhaust Gas Recirculation (EGR) valve
- Fuel supply and return hoses
- Pressure hose
- Oil dipstick and guide
- Ignition coil and spark plug cables
- Upper radiator hose
- Positive Crankcase Ventilation (PCV) hose
- Breather hose
- Rocker arm cover
- Spark plug guide oil seal
- Water hose
- Timing belt
- Power steering pressure switch
- Power steering pump and bracket
- Exhaust manifold bracket

➡**Loosen the cylinder head bolts, in 2 or 3 passes, in the proper sequence.**

- Cylinder head and discard the gasket

To install:
6. Clean all mating surfaces of any residual gasket material.
7. Lubricate the threads of the cylinder head bolts with clean engine oil.
8. Install or connect the following:
- Cylinder head with a new gasket
9. Torque the bolts, at a 90 degree angle with special Tool MB991654, using the following procedure:
a. Step 1: 58 ft. lbs. (79 Nm).
b. Step 2: Fully loosen all the bolts.

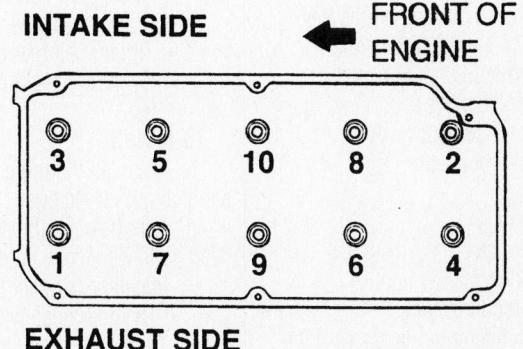

INTAKE SIDE ← FRONT OF ENGINE

| 3 | 5 | 10 | 8 | 2 |
| 1 | 7 | 9 | 6 | 4 |

EXHAUST SIDE

9356CG06

Cylinder head bolt loosening sequence—2.4L engine

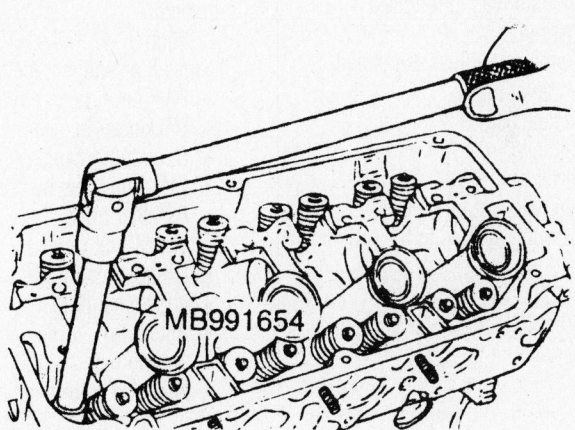

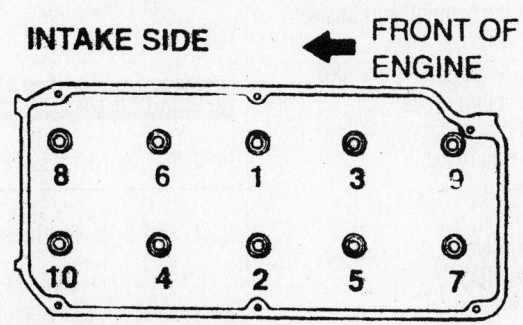

INTAKE SIDE ← FRONT OF ENGINE

| 8 | 6 | 1 | 3 | 9 |
| 10 | 4 | 2 | 5 | 7 |

EXHAUST SIDE

9346IG02

Cylinder head bolt torque sequence—2.4L engine

c. Step 3: 15 ft. lbs. (20 Nm).
d. Step 4: An additional 90 degrees.
e. Step 5: An additional 90 degrees.
- Exhaust manifold bracket and torque the bolts to 26 ft. lbs. (35 Nm)
- Power steering pump and bracket
- Power steering pressure switch
- Timing belt
- Water hose
- Spark plug guide oil seal
- Rocker arm cover
- Breather and PCV hoses
- Upper radiator hose
- Ignition coil and spark plug cables
- Pressure hose
- Oil dipstick and tube. Torque the bolt to 177 inch lbs. (14 Nm).
- Fuel return and supply hoses
- EGR solenoid valve
- EVAP purge solenoid valve
- IAC motor connector
- CMP sensor
- ECT sensor
- Capacitor
- HO_2S connector
- TP sensor connector
- Manifold differential pressure sensor connector
- Ignition failure sensor connector
- Ignition coil connector
- Brake booster vacuum hose
- Purge hose
- Accelerator and adjust as needed
- Front exhaust pipe
- Thermostat case
- Air cleaner
- Strut tower bar
- Negative battery cable

10. Fill the cooling system to the proper level.
11. Fill the engine with clean oil.
12. Start the vehicle, check for leaks and repair if necessary.

3.0L Engine

1. Before servicing the vehicle, refer to the precautions in the beginning of this section.
2. Relieve the fuel system pressure.
3. Drain the cooling system.
4. Remove or disconnect the following:
- Negative battery cable
- Timing belt
- Alternator
- Intake manifold
- Exhaust manifold
- Water inlet pipe
- Breather and blow by hoses
- Positive Crankcase Ventilation (PCV) hose

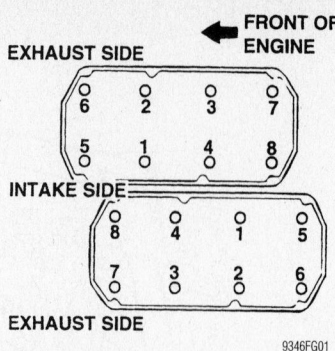

← FRONT OF ENGINE

EXHAUST SIDE

6 2 3 7
5 1 4 8

INTAKE SIDE

8 4 1 5
7 3 2 6

EXHAUST SIDE

9346FG01

Cylinder head bolt torque sequence—3.0L engine

- Spark plug cables
- Rocker arm cover
- Timing belt rear cover
- Cylinder head and discard the gasket

To install:

5. Clean all mating surfaces of any residual gasket material.
6. Lubricate the threads of the cylinder head bolts with clean engine oil.
7. Install or connect the following:
- Cylinder head with a new gasket
8. Torque the bolts, in sequence, as follows:
  a. Step 1: 80 ft. lbs. (108 Nm).
  b. Step 2: Loosen all bolts in sequence.
  c. Step 3: 80 ft. lbs. (108 Nm).
9. Install or connect the following:
- Rear timing belt cover
- Rocker arm cover
- Spark plug cables
- PCV and blow by hoses
- Breather hose
- Water inlet pipe
- Exhaust manifold
- Intake manifold
- Alternator
- Timing belt
- Negative battery cable

10. Fill the cooling system to the proper level.
11. Start the vehicle, check for leaks and repair if necessary.

Rocker Arm/Shafts

REMOVAL & INSTALLATION

2.4L and 3.0L Engines

1. Before servicing the vehicle, refer to the precautions in the beginning of this section.
2. Relieve the fuel system pressure using the recommended procedure.

3. Remove or disconnect the following:
- Negative battery cable
- Breather hose
- Positive Crankcase Ventilation (PCV) hose and valve
- Rocker arm cover
- Rocker shaft cap
- Rocker arms and shafts

To install:

4. Install or connect the following:
- Rocker arms and shafts. When properly aligned, torque the rocker shaft cap to 23 ft. lbs. (31 Nm).
- Oil seals
- Rocker arm cover with a new gasket
- PCV valve with a new gasket
- PCV hose
- Breather hose
- Negative battery cable

Intake Manifold

REMOVAL & INSTALLATION

2.4L Engine

The intake manifold is a long branch design made of cast aluminum. It is attached to the cylinder head with 8 fasteners.

1. Before servicing the vehicle, refer to the precautions in the beginning of this section.
2. Drain the cooling system.
3. Properly relieve the fuel system pressure.
4. Remove or disconnect the following:
- Negative battery cable
- Air cleaner assembly
- Throttle body assembly
- Thermostat case assembly
- Purge hose connection
- Brake booster vacuum hose
- Ignition coil connector
- Fuel injector electrical connectors
- Manifold differential pressure sensor connector
- Evaporative Emission (EVAP) purge solenoid valve connector
- Exhaust Gas Recirculation (EGR) solenoid valve connector
- High pressure fuel line and fuel return hose
- Oil dipstick and tube
- Positive Crankcase Ventilation (PCV) hose
- Fuel return pipe
- Fuel hose
- Fuel rail assembly, including the injectors and pressure regulator
- Insulator and vacuum pipe

- EGR valve
- Intake manifold retainers, manifold and gasket

5. If replacing the manifold, remove the following and transfer to the new manifold:
- Intake manifold stay
- Manifold differential pressure sensor
- Vacuum pipe
- Evaporative Emission (EVAP) purge solenoid valve
- EGR solenoid valve and vacuum control valve
- Accelerator cable clamp

To install:
6. Install or connect the following:
- Intake manifold with a new gasket

and torque the bolts, in sequence, to 14 ft. lbs. (19 Nm)
- EGR valve
- Vacuum pipe and insulator
- Fuel rail assembly
- Fuel hose and return pipe
- PCV hose
- Oil dipstick tube and dipstick
- Fuel return hose
- High pressure fuel hose. Apply a small quantity of oil to the hose O-ring and then installing using a twisting motion. Make sure that the O-ring does not get twisted or clamped during installation.
- EGR solenoid valve connector

- EVAP purge solenoid valve connector
- Manifold differential pressure sensor connector
- Injector connector
- Ignition coil connector
- Brake booster vacuum hose connection
- Purge hose connection
- Thermostat case assembly
- Throttle body assembly
- Air cleaner assembly
- Negative battery cable

7. Fill the cooling system to the proper level.

8. Start the vehicle, check for leaks and repair if necessary.

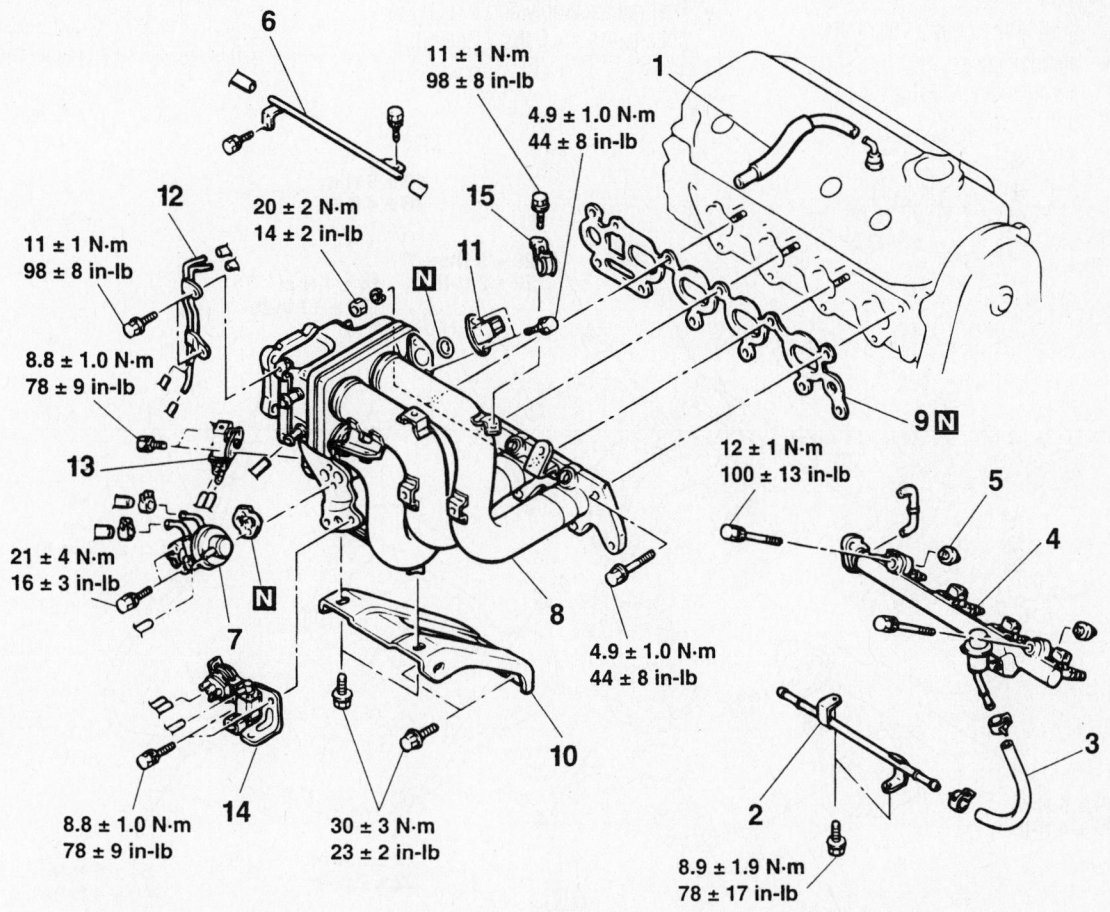

1. PCV HOSE
2. FUEL RETURN PIPE
3. FUEL HOSE
4. FUEL RAIL, INJECTOR AND FUEL PRESSURE REGULATOR
5. INSULATOR
6. VACUUM PIPE
7. EGR VALVE
8. INTAKE MANIFOLD
9. INTAKE MANIFOLD GASKET
10. INTAKE MANIFOLD STAY
11. MANIFOLD DIFFERENTIAL PRESSURE SENSOR
12. VACUUM PIPE
13. EVAPORATIVE EMISSION PURGE SOLENOID VALVE
14. EGR SOLENOID VALVE AND VACUUM CONTROL VALVE
15. ACCELERATOR CABLE CLAMP

Exploded view of the intake manifold—2.4L engine

9356CG07

3.0L Engine

1. Before servicing the vehicle, refer to the precautions in the beginning of this section.

2. Properly relieve the fuel system pressure.

3. Remove or disconnect the following:
   - Negative battery cable
   - Intake manifold plenum
   - Fuel injector connector
   - Fuel supply and return hoses
   - Vacuum hose
   - Fuel rail, injectors and pressure regulator as an assembly
   - Insulators
   - Positive Crankcase Ventilation (PCV) hose
   - Right front upper timing belt cover and bracket
   - Intake manifold and discard the gasket

To install:

4. Clean all mating surfaces of any residual gasket material.

5. Install or connect the following:
   - Intake manifold with a new gasket. Torque the nuts as follows:
   a. Right bank to 62 inch lbs. (7 Nm).
   b. Left bank to 16 ft. lbs. (22 Nm).
   c. Right bank to 16 ft. lbs. (22 Nm).
   d. Left bank to 16 ft. lbs. (22 Nm).
   e. Right bank to 16 ft. lbs. (22 Nm).
   - Right front upper timing belt cover and bracket and torque the bolts to 10 ft. lbs. (15 Nm)
   - PCV hose
   - Fuel rail assembly and torque the bolts to 102 inch lbs. (12 Nm)
   - Vacuum hose
   - Fuel supply and return hoses
   - Fuel injector electrical connector
   - Negative battery cable

6. Start the vehicle, check for leaks and repair if necessary.

Exhaust Manifold

REMOVAL & INSTALLATION

2.4L Engine

1. Before servicing the vehicle, refer to the precautions in the beginning of this section.

2. Remove or disconnect the following:
   - Negative battery cable
   - Front Heated Oxygen Sensor (HO2S), using a special tool

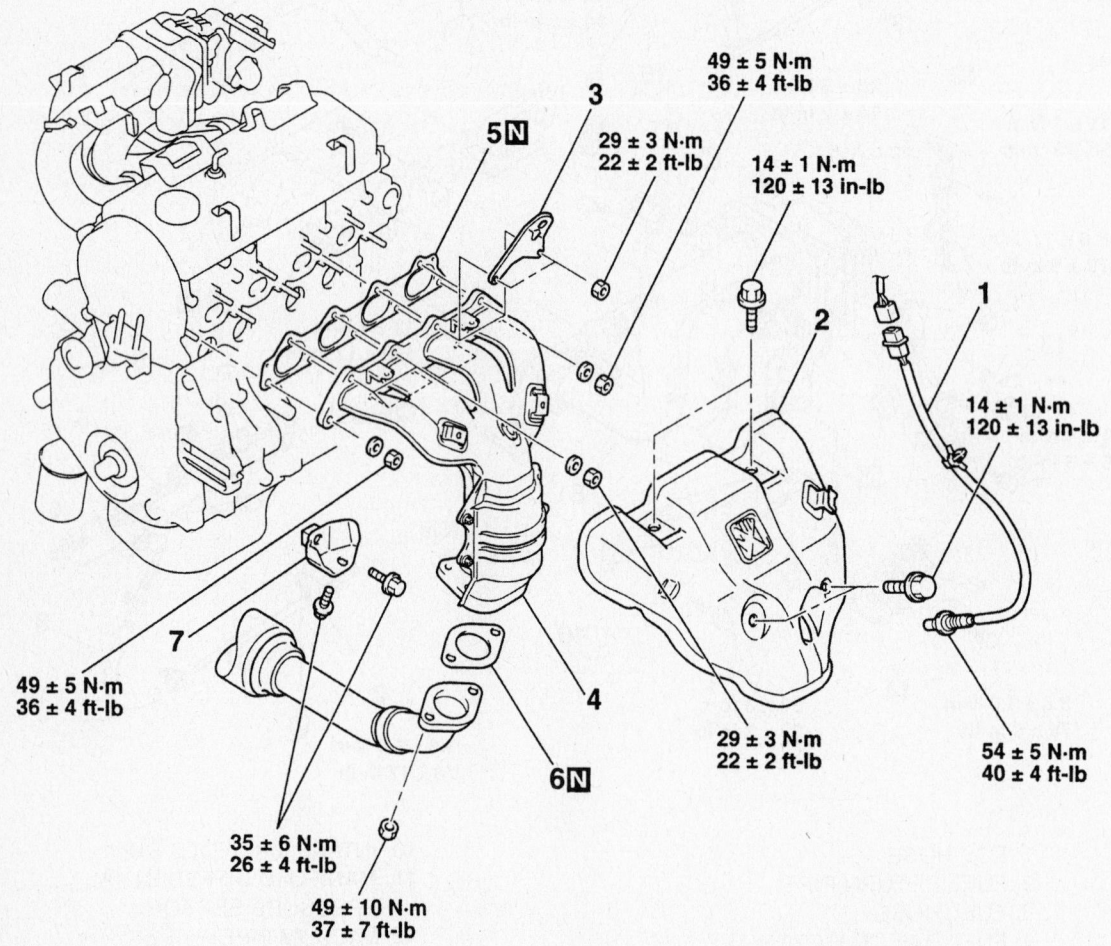

1. HEATED OXYGEN SENSOR (FRONT)
2. HEAT PROTECTOR
3. ENGINE HANGER
4. EXHAUST MANIFOLD
5. EXHAUST MANIFOLD GASKET
6. GASKET
7. EXHAUST MANIFOLD BRACKET

9356CG08

Exploded view of the exhaust manifold—2.4L engine

- Exhaust manifold heat shield
- Exhaust pipe from the exhaust manifold
- Exhaust manifold and discard the gaskets
- Exhaust manifold bracket, if necessary

To install:

3. Thoroughly clean all parts. Clean all sealing surfaces of the manifold and cylinder head. Check the manifold gasket surface for flatness with a straightedge and feeler gauge. The surface must be flat within 0.006 in. (0.15mm) per foot (30cm) of manifold length. Inspect the manifold for cracks or distortion. Replace if necessary.

4. Install or connect the following:
- Exhaust manifold bracket, if removed. Tighten the bolts to 26 ft. lbs. (35 Nm).
- New gaskets
- Exhaust manifold. Torque the M8 nuts to 22 ft. lbs.(29 Nm) and the M10 bolts to 36 ft. lbs.(49 Nm).
- Exhaust pipe to the exhaust manifold. Tighten the nuts to 37 ft. lbs. (49 Nm).
- Exhaust manifold heat shield
- HO$_2$S
- Negative battery cable

5. Start the engine and allow it to idle while inspecting the manifold for exhaust leaks.

3.0L Engine

1. Before servicing the vehicle, refer to the precautions in the beginning of this section.

2. Remove or disconnect the following:
- Battery and tray
- Front exhaust pipe
- Air cleaner assembly
- Oil dipstick guide
- Strut tower bar
- Upper and lower heat shields
- Exhaust Gas Recirculation (EGR) pipe, left side only
- Exhaust manifold and discard the gasket

To install:

3. Clean all mating surfaces of any residual gasket material.

4. Install or connect the following:
- Exhaust manifold with a new gasket and torque the bolts to 33 ft. lbs. (44 Nm)
- EGR pipe, if removed. Torque the bolt to 13 ft. lbs. (18 Nm).
- Upper and lower heat shields and

torque the bolts to 10 ft. lbs. (15 Nm)
- Strut tower bar
- Oil dipstick guide
- Air cleaner assembly
- Front exhaust pipe
- Battery and tray

5. Start the vehicle, check for leaks and repair if necessary.

Front Crankshaft Seal

REMOVAL & INSTALLATION

2.4L Engine

The timing belt must be removed for this procedure. Use care that all timing marks are aligned after installation or the engine will be damaged.

1. Drain the engine oil.

2. Before servicing the vehicle, refer to the precautions in the beginning of this section.

3. Disconnect the negative battery cable.

4. Remove or disconnect the following:
- Timing belt
- Crankshaft Position (CKP) sensor
- Crankshaft timing belt sprocket using a gear/sprocket puller
- Crankshaft sensing blade
- Timing belt "B"
- Crankshaft sprocket "B"

➡**Be careful not to nick the seal surface of the crankshaft or the seal bore.**

- Key
- Front crankshaft seal using a suitable spanner tool

➡**Be careful not to damage the seal contact area of the crankshaft.**

To install:

5. Lubricate the new oil seal lip with engine oil.

6. Install or connect the following:
- New crankshaft oil seal using Oil Seal Installer Tool MD998375, until it is flush with the front cover
- Crankshaft timing belt sprocket using Tool No. 6792
- Key

➡**To prevent the crankshaft bolt from loosening, clean the mating surfaces to the crankshaft, sprocket "B", crankshaft sensing blade and crankshaft at the positions shown in the accompanying figure.**

- Crankshaft sprocket "B"
- Timing belt "B"
- Crankshaft sensing blade, so it faces the direction shown by the figure
- Crankshaft sprocket

7. Apply a very small amount of engine oil to the seating surface and thread of the crankshaft bolt

8. Use the crankshaft pulley mounting bolt to secure the special spanner tool

9. Tighten the crankshaft sprocket bolt to 87 ft. lbs. (118 Nm).
- CKP sensor

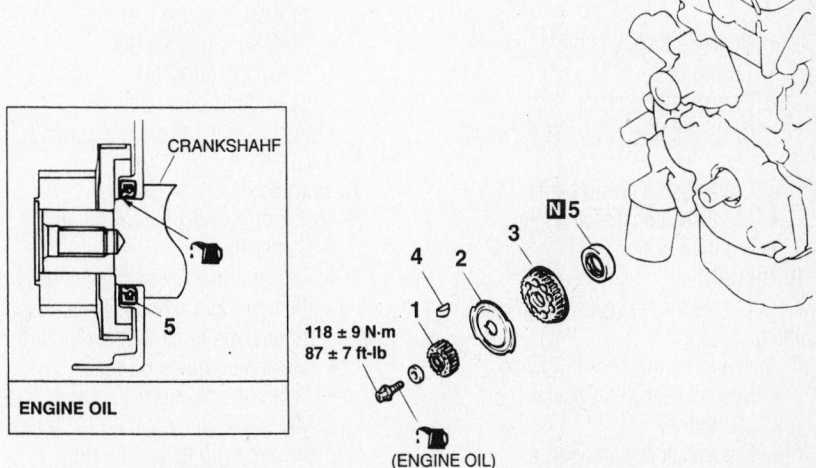

118 ± 9 N·m
87 ± 7 ft-lb

ENGINE OIL

CRANKSHAHF

(ENGINE OIL)

1. CRANKSHAFT SPROCKET
2. CRANKSHAFT SENSING BLADE
3. CRANKSHAFT SPROCKET B
4. KEY
5. CRANKSHAFT FRONT OIL SEAL

Exploded view of the front crankshaft oil seal—2.4L engine

9356CG09

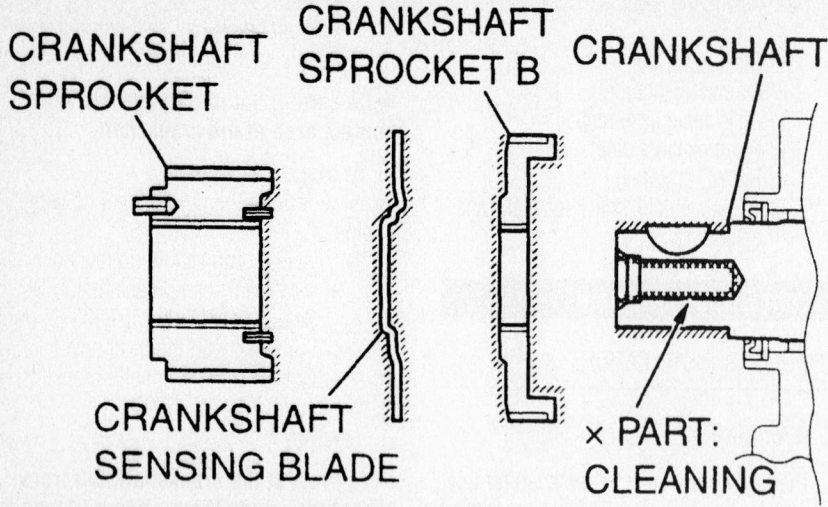

CRANKSHAFT SPROCKET

CRANKSHAFT SPROCKET B

CRANKSHAFT

CRANKSHAFT SENSING BLADE

× PART: CLEANING

SHADED PART: DEGREASE

9356CG10

View of the proper installation and mating order cleaning surfaces—2.4L engine

- Timing belt
- Negative battery cable

✳✳ WARNING

Operating the engine without the proper amount and type of engine oil will result in severe engine damage.

10. Fill the engine with clean oil.
11. Start the engine and check for leaks.

3.0L Engine

1. Before servicing the vehicle, refer to the precautions in the beginning of this section.
2. Remove or disconnect the following:

- Negative battery cable
- Timing belt
- Crankshaft sprocket
- Crankshaft Position (CKP) sensor
- Crankshaft sensing blade
- Crankshaft spacer and key
- Front oil seal

To install:
3. Lubricate the oil seal lip with clean engine oil.
4. Install or connect the following:

- New oil seal with Special Tool MD998717
- Crankshaft key and spacer
- Crankshaft sensing blade
- CKP sensor
- Crankshaft sprocket
- Timing belt
- Negative battery cable

Camshaft and Valve Lifters

REMOVAL & INSTALLATION

2.4L Engine

1. Before servicing the vehicle, refer to the precautions in the beginning of this section.
2. Remove or disconnect the following:

- Negative battery
- Air cleaner assembly
- Timing belt
- Ignition coils and spark plug cables
- Positive Crankcase Ventilation (PCV) hose
- Breather hose
- Rocker arm cover
- Camshaft Position (CMP) sensor support
- CMP sensing cylinder
- Camshaft sprocket
- Spark plug guide oil seal
- Rocker arm and shaft assembly
- Camshaft

To install:
3. Install or connect the following:

- Camshaft
- Rocker arm and shaft assemblies in there proper position. Torque the assemblies to 23 ft. lbs. (31 Nm).
- Spark plug guide oil seals
- Camshaft sprocket with special Tools MB990767 and MD998719. Torque the bolt to 65 ft. lbs. (88 Nm).
- CMP sensing cylinder and torque the bolt to 16 ft. lbs. (21 Nm)
- CMP support and torque the bolt to 10 ft. lbs. (15 Nm)
- Rocker arm cover

- Breather and PCV hoses
- Ignition coil and spark plug cables
- Timing belt
- Air cleaner assembly
- Negative battery cable

3.0L Engine

LEFT BANK

1. Before servicing the vehicle, refer to the precautions in the beginning of this section.
2. Remove or disconnect the following:

- Negative battery
- Timing belt
- Thermostat housing
- Blow-by hose
- Positive Crankcase Ventilation (PCV) hose
- Spark plug cables
- Rocker arm cover
- Rocker arm and shaft
- Camshaft sprocket
- Thrust case
- Camshaft

To install:
3. Install or connect the following:

- Camshaft
- Thrust case and torque the bolt to 109 inch lbs. (13 Nm)
- Camshaft sprocket and torque the bolt to 65 ft. lbs. (88 Nm)
- Rocker arm and shaft assembly and torque to 23 ft. lbs. (31 Nm)
- Rocker arm cover and torque the bolts to 35 inch lbs. (4 Nm)
- Spark plug cables
- PCV and blow-by hoses
- Thermostat housing
- Timing belt
- Negative battery cable

RIGHT BANK

1. Before servicing the vehicle, refer to the precautions in the beginning of this section.
2. Remove or disconnect the following:

- Negative battery
- Timing belt
- Intake manifold plenum
- Breather hose
- Blow-by hose
- Spark plug cables
- Rocker arm cover
- Rocker arm and shaft assembly with Tool MD998443
- Distributor
- Camshaft sprocket with Tool MB990767 and MD998715
- Camshaft

To install:
3. Install or connect the following:

- Camshaft and a new oil seal, if removed

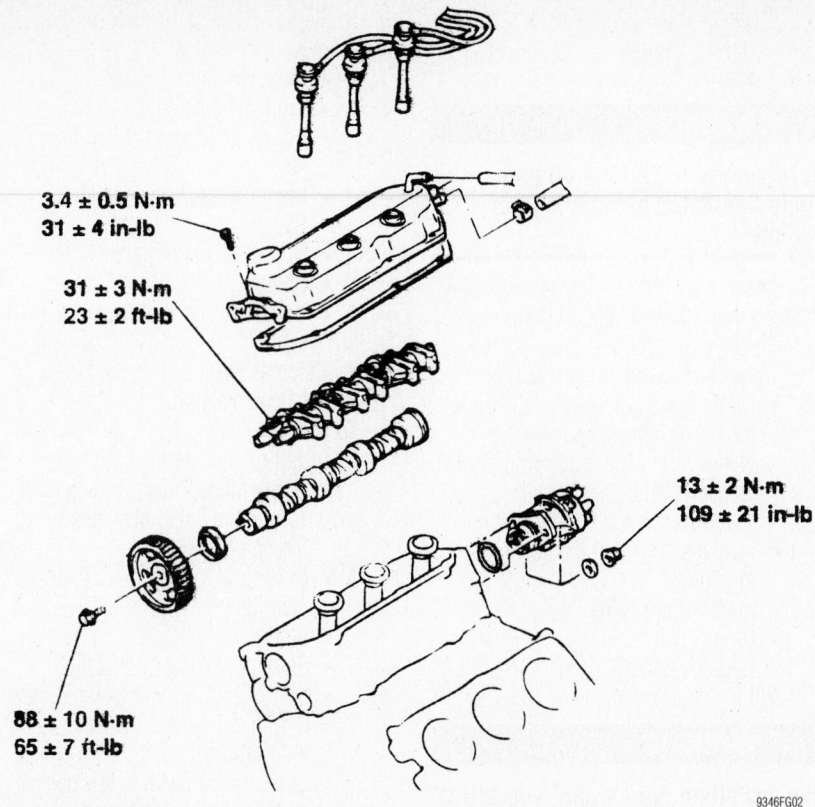

3.4 ± 0.5 N·m
31 ± 4 in-lb

31 ± 3 N·m
23 ± 2 ft-lb

13 ± 2 N·m
109 ± 21 in-lb

88 ± 10 N·m
65 ± 7 ft-lb

9346FG02

Exploded view of the rocker arm and camshaft assembly—3.0L engine

- Camshaft sprocket and torque the bolt to 65 ft. lbs. (88 Nm)
- Distributor

4. Rotate the camshaft until the dowel pin on the front end is properly positioned.

5. Install or connect the following:
- Rocker arm cover and torque the bolts to 35 inch lbs. (4 Nm)
- Spark plug cables
- Blow-by and breather hose connections
- Intake manifold plenum and torque the bolts to 13 ft. lbs. (18 Nm)
- Timing belt
- Negative battery cable

Valve Lash

ADJUSTMENT

The engines in these vehicles do not require periodic valve lash adjustment.

Starter Motor

REMOVAL & INSTALLATION

2.4L Engine

1. Before servicing the vehicle, refer to the precautions in the beginning of this section.

2. Remove or disconnect the following:
- Negative battery cable
- Air cleaner assembly
- Starter cover
- Battery cable from the starter
- Solenoid connector
- Starter-to-transaxle bolts
- Starter

To install:

3. Install or connect the following:
- Starter
- Starter-to-transaxle bolt(s) and torque the bolt to 23 ft. lbs. (30 Nm)
- Electrical connectors to the starter
- Starter cover and tighten the bolt to 44 inch lbs. (4.9 Nm)
- Air cleaner assembly
- Negative battery cable to the shock tower

3.0L Engine

1. Before servicing the vehicle, refer to the precautions in the beginning of this section.

2. Remove or disconnect the following:
- Negative battery cable from the shock tower
- Air cleaner resonator

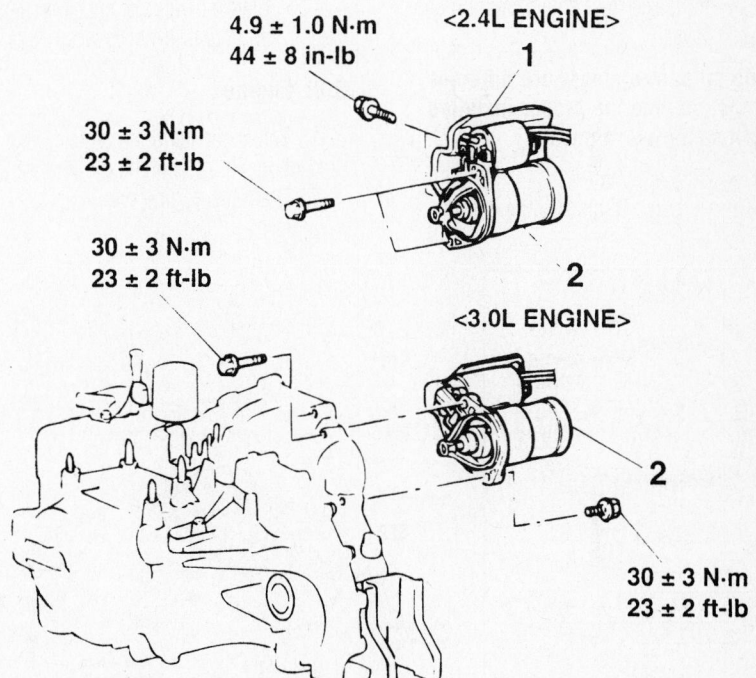

4.9 ± 1.0 N·m
44 ± 8 in-lb

<2.4L ENGINE>

30 ± 3 N·m
23 ± 2 ft-lb

30 ± 3 N·m
23 ± 2 ft-lb

<3.0L ENGINE>

30 ± 3 N·m
23 ± 2 ft-lb

1. STARTER COVER <2.4L ENGINE>
2. STARTER MOTOR

9356CG11

Exploded view of the starter and related components—2.4L engine

- Starter motor cover
3. Starter electrical connectors
- Starter assembly

To install:

4. Install or connect the following:
- Starter assembly. Torque the starter bolt to 25 ft. lbs. (33 Nm) and the rear bracket bolt to 8 inch lbs. (1 Nm).
- Electrical connectors to the starter
- Starter cover and torque the bolt to 15 inch lbs. (6 Nm)
- Air cleaner resonator
- Negative battery cable to the shock tower

Oil Pan

REMOVAL & INSTALLATION

2.4L Engine

1. Before servicing the vehicle, refer to the precautions in the beginning of this section.
2. Drain the engine oil. Install a new drain plug gasket.
3. Remove or disconnect the following:
- Negative battery cable
- Oil dipstick
- Front exhaust pipe
- Transaxle dust shield/bellhousing cover

➡**The oil pan retainers are different lengths, so note the proper installed locations during removal.**

- Oil pan retainers
- Oil pan and discard the gasket

4. Clean all mating surfaces of any residual gasket material.

To install:

✴✴ WARNING

Make sure to install the oil pan within 5 minutes of applying the sealant!

5. Using a suitable gasket sealant apply a 0.2 in. (4mm) bead at the oil pump-to-engine block parting line.
6. Install or connect the following:
- Oil pan with a new gasket. Torque the bolts to the specifications shown in the accompanying figure.
- Transaxle dust shield/bellhousing cover. Torque the bolts to the specifications shown in the accompanying figure.
- Front exhaust pipe
- Oil dipstick
- Negative battery cable
7. Fill the engine with clean oil.

✴✴ WARNING

After installing the oil pan, you MUST wait at least one hour before starting the engine.

- Start the vehicle, check for leaks and repair if necessary.

3.0L Engine

1. Before servicing the vehicle, refer to the precautions in the beginning of this section.
2. Drain the engine oil.

3. Remove the engine support module as follows:
- Negative battery cable
- Front exhaust pipe
- Lower oil pan and discard the gasket
- Starter electrical connectors
- Starter
- Oil dipstick and guide
- Cover
- Upper oil pan and discard the gasket

4. Clean all mating surfaces of any residual gasket material.

To install:

5. Install or connect the following:
- Upper oil pan with a new gasket. Torque the bolts, in sequence, to 53 inch lbs. (6 Nm).
- Cover and torque the bolts to 96 inch lbs. (11 Nm)
- Starter
- Starter electrical connectors
- Oil dipstick and guide with a new O-ring and torque the bolt to 36 ft. lbs. (48 Nm)
- Lower oil pan with a new gasket and torque the bolts to 96 inch lbs. (11 Nm)
- Front exhaust pipe
- Negative battery cable
6. Fill the engine with clean oil.
- Start the vehicle, check for leaks and repair if necessary.

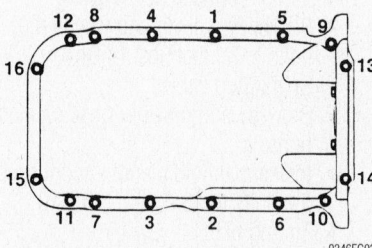

Tighten the upper oil pan bolts in sequence—3.0L engine

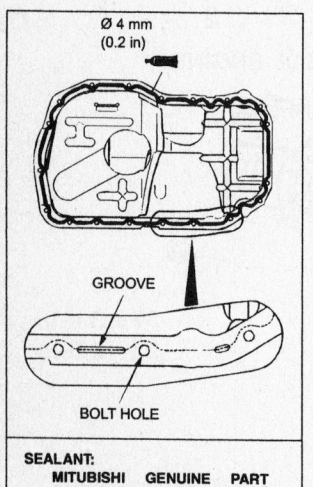

SEALANT:
MITUBISHI GENUINE PART
NO.MD970389 OR EQUIVALENT

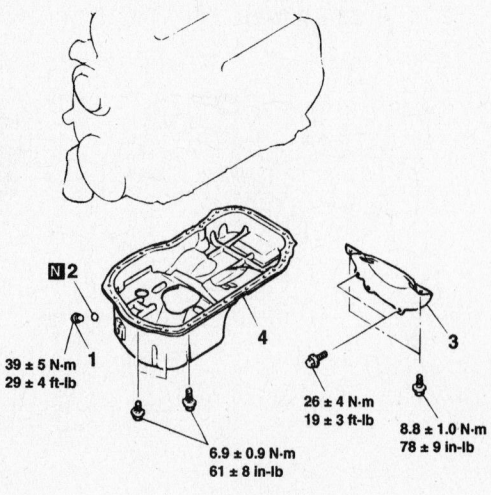

1. DRAIN PLUG
2. DRAIN PLUG GASKET
3. BELL HOUSING COVER
4. OIL PAN

67189-SEBR-G02

Exploded view of the oil pan and related components—2.4L engine

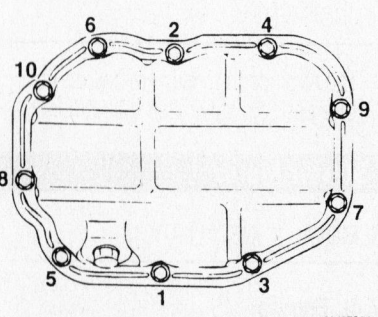

Tighten the lower oil pan bolts in sequence—3.0L engine

Oil Pump

REMOVAL & INSTALLATION

2.4L Engine

1. Before servicing the vehicle, refer to the precautions in the beginning of this section.

2. Drain the engine oil.

3. Remove the engine support module as follows:
   - Negative battery cable
   - Oil filter
   - Oil pressure switch
   - Oil pan and screen
   - Flange bolt
   - Relief plug, spring and plunger
   - Oil filter bracket
   - Oil pump case
   - Front case gasket
   - Oil pump cover
   - Oil pump driven gear
   - Oil pump drive gear
   - Oil pump

4. Clean all mating surfaces of any residual gasket material.

To install:

5. Prime the oil pump before installation.

6. Install or connect the following:
   - New oil seal into the front case
   - Oil pump
   - Oil pump drive gear
   - Oil pump driven gear
   - Oil pump cover
   - Front case gasket
   - Oil pump case
   - Oil filter bracket
   - Relief plunger, spring, gasket and plug
   - Flange bolt
   - Oil pan and screen
   - Oil pressure switch
   - New oil filter
   - Negative battery cable

7. Fill the engine with clean oil.

8. Start the vehicle, check for leaks and repair if necessary.

3.0L Engine

1. Before servicing the vehicle, refer to the precautions in the beginning of this section.

2. Drain the engine oil.

3. Remove the engine support module as follows:
   - Negative battery cable
   - Oil filter
   - Oil pressure switch
   - Oil filter bracket

- Lower and upper oil pan
- Baffle plate
- Oil screen and gasket
- Baffle plate
- Relief spring and plunger
- Front crankshaft oil seal
- Oil pump case
- Oil pump cover
- Oil pump outer and inner rotors
- Oil pump case

4. Clean all mating surfaces of any residual gasket material.

To install:

5. Prime the oil pump before installation.

6. Install or connect the following:
   - Oil pump case
   - Inner and outer rotors
   - Oil pump cover and torque the bolts to 89 inch lbs. (10 Nm)
   - Oil pump case assembly
   - New front crankshaft oil seal
   - Relief plunger and spring
   - Baffle plate
   - Oil screen with a new gasket
   - Baffle plate
   - Upper oil pan and cover
   - Lower oil pan
   - Oil filter bracket with a new gasket
   - Oil pressure switch
   - New oil filter
   - Negative battery cable

7. Fill the engine with clean oil.

8. Start the vehicle, check for leaks and repair if necessary.

Rear Main Seal

REMOVAL & INSTALLATION

2.4L Engine

1. Before servicing the vehicle, refer to the precautions in the beginning of this section.

2. Remove or disconnect the following:
   - Oil pan
   - Transaxle
   - Clutch cover and disc, if equipped
   - Adapter plate
   - Flexplate/flywheel
   - Adapter plate, if equipped with M/T
   - Crankshaft bushing
   - Crankshaft rear main seal

To install:

3. Apply a small amount of oil to the inside of the rear main seal lip.

4. Install or connect the following:
   - Rear main seal, using the proper installation tools
   - Crankshaft bushing
   - Adapter plate, if removed
   - Flywheel/flexplate
   - Adapter plate
   - Clutch cover and disc if equipped
   - Transaxle assembly
   - Oil pan

3.0L Engine

➡ Be sure to observe all cautions and warnings in the beginning of the section that may be related to this procedure.

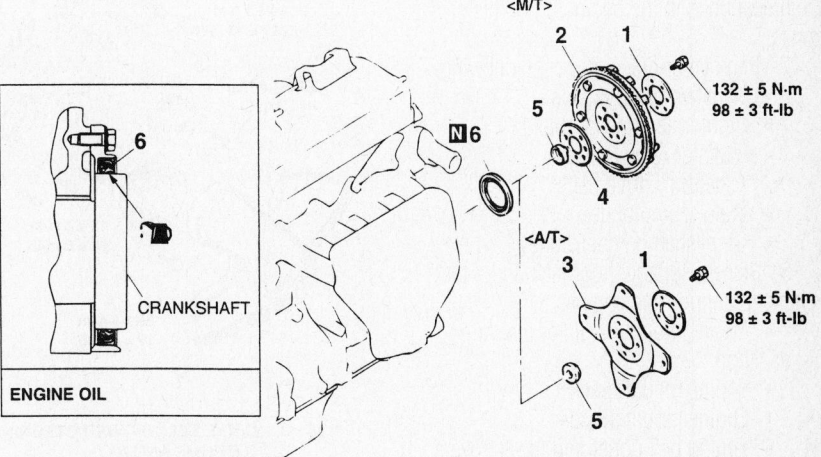

1. ADAPTER PLATE
2. FLYWHEEL <M/T>
3. DRIVE PLATE <A/T>
4. ADAPTER PLATE <M/T>
5. CRANKSHAFT BUSHING
6. CRANKSHAFT REAR OIL SEAL

9356CG12

Exploded view of the rear main seal—2.4L engine shown

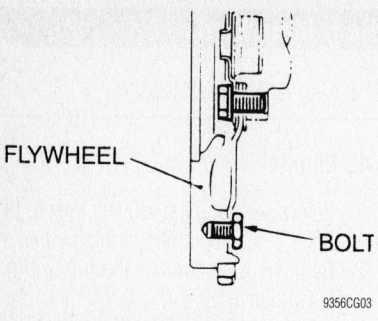

FLYWHEEL

BOLT

9356CG03

Do not remove the flywheel bolt shown, or you will damage the flywheel

1. Remove or disconnect the following:
   • Transaxle
   • Flexplate/flywheel
   • Rear crankshaft oil seal

➡ **Pry the oil seal from the housing using a suitable flat bladed prying tool.**

To install:

➡ **When installing the new seal there is no need to lubricate sealing surface.**

2. Install or connect the following:
   • Oil seal into housing using a suitable installation tool
   • Flexplate/flywheel and torque the bolts to 68 ft. lbs. (92 Nm)
   • Transaxle

Timing Belt

REMOVAL & INSTALLATION

2.4L Engine

1. Before servicing the vehicle, refer to the precautions in the beginning of this section.

2. Remove or disconnect the following:
   • Negative battery cable
   • Right inner splash-shield
   • Radiator reservoir tank
   • Accessory drive belts
   • Water pump pulley
   • Crankshaft damper/pulley

3. Place a suitable floor jack under the vehicle to support the engine.
   • Engine mount insulator mounting bolt
   • Engine mount bracket
   • Engine mount stopper
   • Timing belt upper and lower covers

➡ **Do not rotate the crankshaft or the camshafts after the timing belt has been removed. Damage to the valve components may occur. Before removing the timing belt, always align the timing marks.**

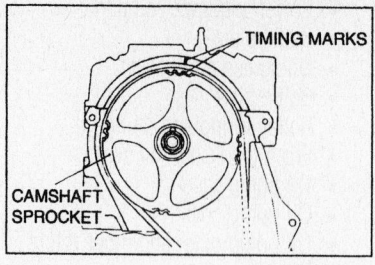

TIMING MARKS

CAMSHAFT SPROCKET

67189-SEBR-G03

Turn the crankshaft in the forward direction only to align the timing marks—2.4L engine

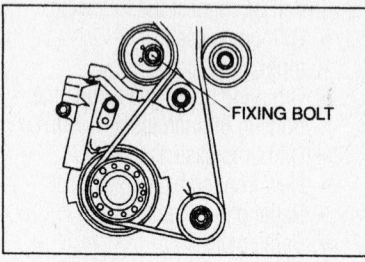

FIXING BOLT

67189-SEBR-G04

Location of the tension pulley fixing bolt—2.4L engine

4. Align the timing marks of the timing belt sprockets to the timing marks on the rear timing belt cover and oil pump cover.

➡ **If the timing belt is to be reused, draw an arrow indicating the direction of rotation on the back of the belt for reinstallation.**

5. Loosen the timing belt tension pulley fixing bolt, then move the tension pulley to the water pump side.

6. Remove or disconnect the following:
   • Timing belt
   • Tensioner pulley
   • Auto-tensioner
   • Camshaft timing belt sprockets
   • Crankshaft timing belt sprocket using special removal tool No. 6793

To install:

7. Install or connect the following:
   • Crankshaft timing belt sprocket onto the crankshaft, using special tool No. 6792
   • Camshaft sprockets onto the camshafts
   • Camshaft sprocket bolts, and torque to 75 ft. lbs. (101 Nm)

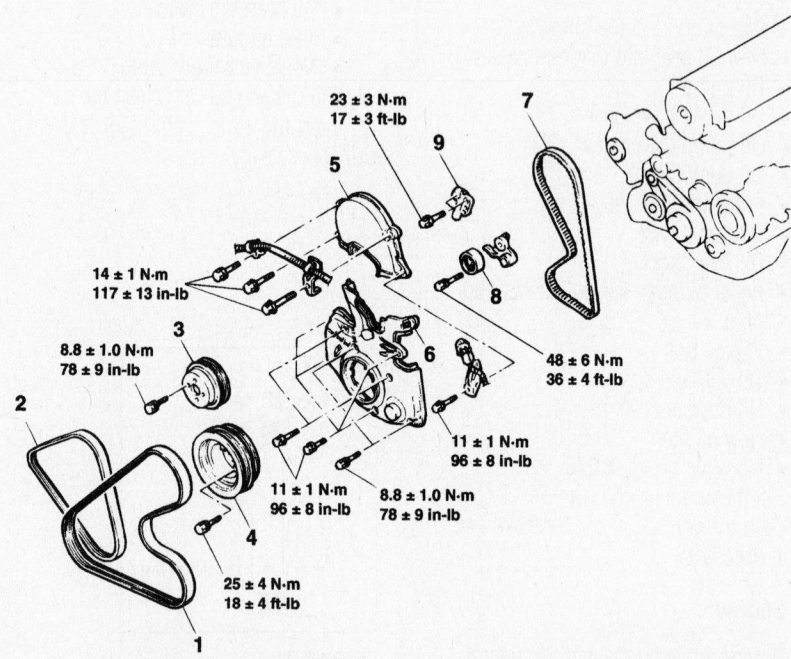

23 ± 3 N·m
17 ± 3 ft-lb

14 ± 1 N·m
117 ± 13 in-lb

8.8 ± 1.0 N·m
78 ± 9 in-lb

48 ± 6 N·m
36 ± 4 ft-lb

11 ± 1 N·m
96 ± 8 in-lb

11 ± 1 N·m
96 ± 8 in-lb

8.8 ± 1.0 N·m
78 ± 9 in-lb

25 ± 4 N·m
18 ± 4 ft-lb

1. DRIVE BELT (POWER STEERING OIL PUMP AND A/C COMPRESSOR)	6. TIMING BELT LOWER COVER ASSEMBLY
2. DRIVE BELT (GENERATOR)	• TIMING BELT TENSION ADJUSTMENT
3. WATER PUMP PULLEY	7. TIMING BELT
4. CRANKSHAFT PULLEY	8. TENSIONER PULLEY
5. TIMING BELT UPPER COVER ASSEMBLY	9. AUTO-TENSIONER

67189-SEBR-G07

Exploded view of the timing belt and related components—2.4L engine

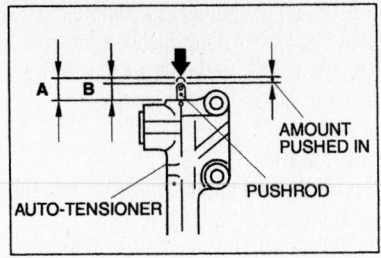

Measuring the auto tensioner—2.4L engine

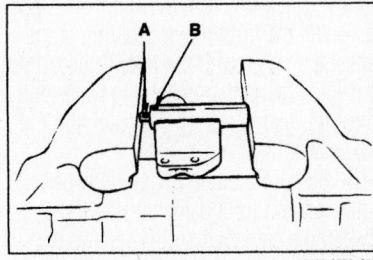

Compress the pushrod until holes A & B are aligned—2.4L engine

8. Apply 22–44 lbs. (98–196 N) of force to the pushrod of the auto tensioner by pushing it against a metal object (ex. engine block). Measure the movement of the pushrod, as shown in the accompanying figure. Standard value is within 0.04 in. (1mm). A is the length when it is free (not pressed). B is the length when it is pressed. A minus B equals movement. If the measurement is outside the standard, replace the tensioner.

9. Place the tensioner in a vise and slowly compress the pushrod until pin hole A of the pushrod and pinhole B of the tensioner cylinder are aligned. Take care not to damage the pushrod. When the holes are aligned, insert the set pin.

➡ **When replacing the tensioner with a new part, it will come with a pin.**

10. Align the timing marks on the camshaft sprocket, crankshaft sprocket and oil pump sprocket.

11. Remove the cylinder block plug and insert a Phillips head 0.3 inch (8mm) screwdriver. The screwdriver must go in 2.4 inches (60mm) or more. If the screwdriver only goes in 0.8–1.0 inch (20–25mm) hitting the counterbalance shaft, turn the sprocket once, realign the marks and insert the screwdriver again. Do not take the screwdriver out until the timing belt is installed.

12. Install or connect the following:

- Timing belt starting at the crankshaft, oil pump sprocket and camshaft sprocket, in that order, so there is no slack.

13. Set the tension pulley so that the pin holes are at the bottom, press the tension pulley lightly against the timing belt and tighten the fixing bolt.

14. Adjust belt tension as follows:

a. After turning the crankshaft ¼ revolution counterclockwise, turn it in the clockwise direction until the timing marks are aligned.

✳✳ WARNING

When tightening the fixing bolt, make sure the tension pulley does not turn with the bolt.

b. Loosen the tension pulley fixing bolt, and use special tool MD998767 or equivalent and a torque wrench to tighten the fixing bolt to 32–40 ft. lbs. (42–54 Nm) while applying tension of 31 inch lbs. (3.5 Nm) to the timing belt.

c. Remove the set pin from the tensioner.

d. Turn the crankshaft 2 revolutions clockwise so that the timing marks are aligned. After leaving it for 15 minutes, measure the amount of protrusion of the auto-tensioner. It should be 0.15–0.18 in. (3.8–5.5 Nm). If the measurement is outside the specifications, repeat steps A.-D.

e. Make sure the timing marks of each sprocket are still aligned.

- Front and rear timing belt covers
- Engine mount stopper
- Engine mount bracket
- Engine mount insulator mounting bolt and tighten to 60 ft. lbs. (81 Nm)

15. Remove the floor jack from under the vehicle.

- Crankshaft damper and tighten to 105 ft. lbs. (142 Nm)
- Water pump pulley
- Accessory drive belts
- Radiator reservoir tank
- Right inner splash-shield

16. Properly fill the cooling system.

17. Connect the negative battery cable.

18. Check for leaks and proper engine and cooling system operation.

3.0L Engine

1. Before servicing the vehicle, refer to the precautions in the beginning of this section.

2. Place a floor jack under the engine oil pan, with a block of wood in between,

and raise the engine so that the weight of the engine is no longer being applied to the engine support bracket.

- Upper engine mount. Spraying lubricant, slowly remove the reamer (alignment) bolt and remaining bolts and remove the engine support bracket.

➡ **The reamer bolt is sometimes heat-seized on the engine support bracket.**

3. Remove or disconnect the following:

- Negative battery cable
- Alternator
- Accessory drive belts
- Crankshaft bolt and pulley, using crankshaft holding tools MB990767 and MB998754/MD998715
- Heated oxygen sensor connection, if necessary
- Power steering pump with the hose attached and position it aside
- Power steering pump bracket
- Tensioner pulley assembly
- Upper right front timing belt cover
- Upper left front timing belt cover

➡ **If the timing belt is to be reused, draw an arrow indicating the direction of rotation on the back of the belt for reinstallation.**

4. Align the timing marks by turning the crankshaft with MD998769 crankshaft turning tool. Loosen the center bolt on the timing belt tensioner pulley and remove the belt.

✳✳ WARNING

Do not rotate the crankshaft or camshaft after removing the timing belt or valvetrain components may be damaged. Always align the timing marks before removing the timing belt.

5. Check the belt tensioner for leaks and check the pushrod for cracks.

6. If the timing belt tensioner is to be replaced, remove the retaining bolts and remove the timing belt tensioner. When the timing belt tensioner is removed from the engine, it is necessary to compress the plunger into the tensioner body.

7. Place the tensioner in a vise and slowly compress the plunger. Take care not to damage the pushrod.

➡ **Position the tensioner in the vise the same way it will be installed on the engine. This is to ensure proper pin orientation for when the tensioner is installed on the engine.**

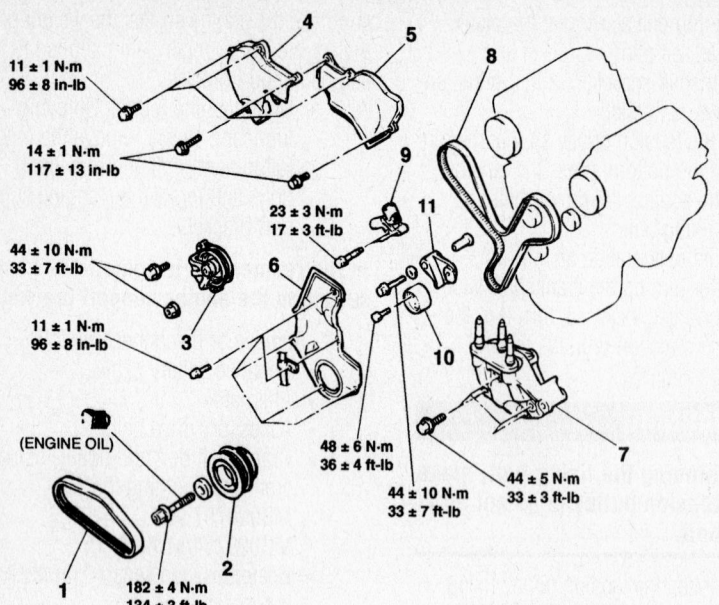

11 ± 1 N·m
96 ± 8 in-lb

14 ± 1 N·m
117 ± 13 in-lb

23 ± 3 N·m
17 ± 3 ft-lb

44 ± 10 N·m
33 ± 7 ft-lb

11 ± 1 N·m
96 ± 8 in-lb

(ENGINE OIL)

48 ± 6 N·m
36 ± 4 ft-lb

44 ± 10 N·m
33 ± 7 ft-lb

44 ± 5 N·m
33 ± 3 ft-lb

182 ± 4 N·m
134 ± 3 ft-lb

1. DRIVE BELT (POWER STEERING OIL PUMP)
2. CRANKSHAFT PULLEY
3. TENSIONER PULLEY ASSEMBLY (POWER STEERING OIL PUMP)
4. TIMING BELT FRONT UPPER COVER, RIGHT
5. TIMING BELT FRONT UPPER COVER, LEFT
6. TIMING BELT FRONT LOWER COVER
7. ENGINE SUPPORT BRACKET, RIGHT
8. TIMING BELT
9. AUTO-TENSIONER
10. TENSIONER PULLEY
11. TENSIONER ARM

67189-SEBR-G08

Exploded view of the timing belt and related components—3.0L engine

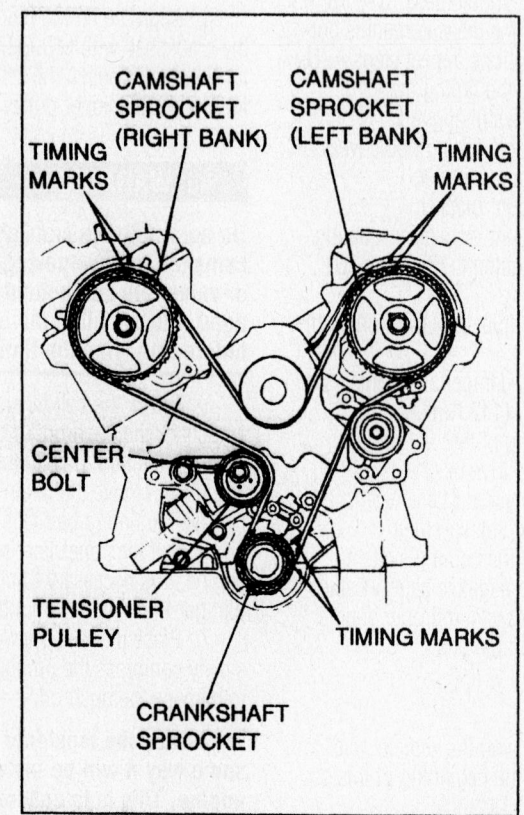

CAMSHAFT SPROCKET (RIGHT BANK)

CAMSHAFT SPROCKET (LEFT BANK)

TIMING MARKS

TIMING MARKS

CENTER BOLT

TENSIONER PULLEY

TIMING MARKS

CRANKSHAFT SPROCKET

67189-SEBR-G09

Alignment of the timing marks—3.0L engine

8. When the plunger is compressed into the tensioner body, install a pin through the body and plunger to hold the plunger in place until the tensioner is installed.

To install:

9. Install the timing belt tensioner and tighten the retaining bolts to 17 ft. lbs. (24 Nm), but do not remove the pin at this time.

10. Check that all timing marks are still aligned.

11. Use bulldog clips (large paper binder clips) or other suitable tool to secure the timing belt and to prevent it from slacking. Install the timing belt. Starting at the crankshaft, go around the idler pulley, then the front camshaft sprocket, the water pump pulley, the rear camshaft sprocket and the tensioner pulley.

12. Be sure the belt is tight between the crankshaft and front camshaft sprocket, between the camshaft sprockets and the water pump. Gently raise the tensioner pulley, so that the belt does not sag, and temporarily tighten the center bolt.

13. Move the crankshaft ¼ turn counterclockwise, then turn it clockwise to the position where the timing marks are aligned.

14. Loosen the center bolt of the tensioner pulley. Using MD998767 tensioner tool, and a torque wrench apply 3.3 ft. lbs. (4.4 Nm) tensional torque to the timing belt and tighten the center bolt to 35 ft. lbs. (48 Nm). When tightening the bolt, be sure that the tensioner pulley shaft does not rotate with the bolt.

15. Remove the tensioner plunger pin. Pretension is correct when the pin can be removed and installed easily. If the pin cannot be easily removed and installed it is still satisfactory as long as it is within its standard value.

16. Check that the tensioner pushrod is within the standard value. When the tensioner is engaged the pushrod should measure 0.149–0.177 in. (3.8–4.5mm).

17. Rotate the crankshaft two revolutions and check the timing marks. If the timing marks are not properly aligned remove the belt and repeat the installation steps.

18. Install or connect the following:
    • Timing belt covers
    • Engine mounting bracket

19. Lower the engine enough to install the engine mount onto bracket and remove the floor jack.
    • Power steering pump bracket and pump
    • Crankshaft pulley and tighten the retaining bolt to 13 ft. lbs. (18 Nm)

- Accessory drive belts
20. Properly fill the cooling system.
21. Connect the negative battery cable.
22. Check for leaks and proper engine and cooling system operation.

Piston and Ring

POSITIONING

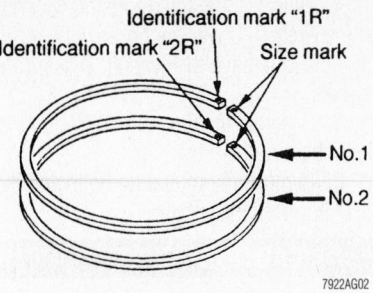

Piston ring identification mark locations

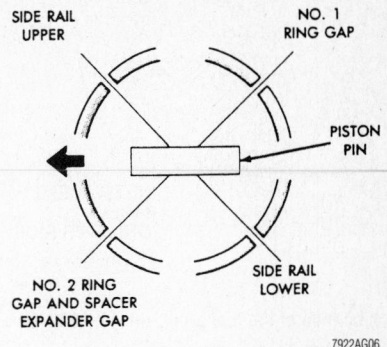

Piston ring end-gap spacing—3.0L engine

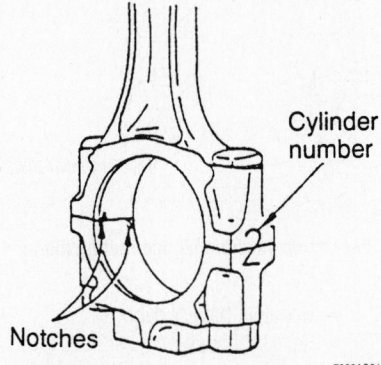

Connecting rod and cap installation— ensure to matchmark the cap and rod prior to disassembly

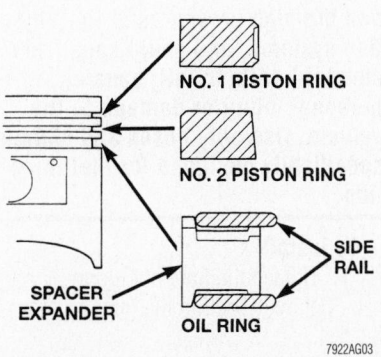

Piston ring orientation—2.4L engine

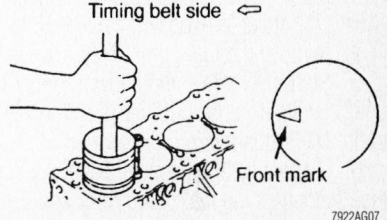

Piston positioning. The small arrows on the crown of the pistons must point toward the front of the engine

FUEL SYSTEM

Fuel System Service Precautions

Safety is the most important factor when performing not only fuel system maintenance but any type of maintenance. Failure to conduct maintenance and repairs in a safe manner may result in serious personal injury or death. Maintenance and testing of the vehicle's fuel system components can be accomplished safely and effectively by adhering to the following rules and guidelines.

- To avoid the possibility of fire and personal injury, always disconnect the negative battery cable unless the repair or test procedure requires that battery voltage be applied.
- Always relieve the fuel system pressure before disconnecting any fuel system component (injector, fuel rail, pressure regulator, etc.), fitting or fuel line connection. Exercise extreme caution whenever relieving fuel system pressure, to avoid exposing skin, face and eyes to fuel spray. Please be advised that fuel under pressure may penetrate the skin or any part of the body that it contacts.
- Always place a shop towel or cloth around the fitting or connection prior to loosening to absorb any excess fuel due to spillage. Ensure that all fuel spillage

(should it occur) is quickly removed from engine surfaces. Ensure that all fuel soaked cloths or towels are deposited into a suitable waste container.

- Always keep a dry chemical (Class B) fire extinguisher near the work area.
- Do not allow fuel spray or fuel vapors to come into contact with a spark or open flame.
- Always use a back-up wrench when loosening and tightening fuel line connection fittings. This will prevent unnecessary stress and torsion to fuel line piping.
- Always replace worn fuel fitting O-rings with new. Do not substitute fuel hose, where fuel pipe is installed.

Fuel System Pressure

RELIEVING

2001 Vehicles

1. Before servicing the vehicle, refer to the precautions in the beginning of this section.
2. Remove the fuel filler cap to release fuel tank pressure.
3. Remove the fuel pump relay from the junction block in the engine compartment.
4. Start the vehicle and allow it to run

until it stalls from lack of fuel. Turn the key to the **OFF** position.
5. Disconnect the negative battery cable, then reconnect the fuel pump relay.
6. Install the fuel filler cap.

✸✸ CAUTION

Always wrap shop towels around a fitting that is being disconnected to absorb residual fuel in the lines.

2002–04 Vehicles

1. Before servicing the vehicle, refer to the precautions in the beginning of this section.

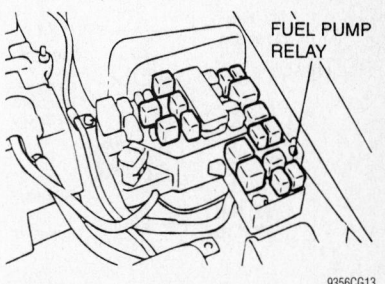

Location of the fuel pump relay—2001 vehicles

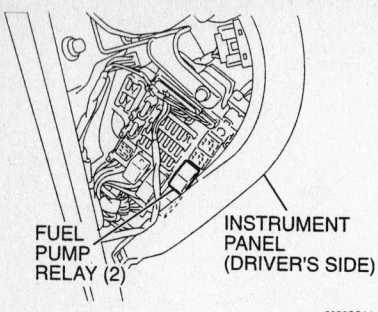

Location of the fuel pump relay—2002–04 vehicles

2. Remove the fuel filler cap to release fuel tank pressure.

3. Remove the driver's side instrument panel side cover to expose the fuel box.

4. Remove the fuel pump relay.

5. Start the vehicle and allow it to run until it stalls from lack of fuel. Turn the key to the **OFF** position.

6. Disconnect the negative battery cable, then reconnect the fuel pump relay and install the cover.

7. Install the fuel filler cap.

✳✳ CAUTION

Always wrap shop towels around a fitting that is being disconnected to absorb residual fuel in the lines.

Fuel Filter

REMOVAL & INSTALLATION

A replaceable fuel filter is located in the engine compartment, on the bulkhead, next to the brake booster.

1. Before servicing the vehicle, refer to the precautions in the beginning of this section.

2. Properly relieve the fuel system pressure.

3. Disconnect the negative battery cable and remove the air intake hose for access.

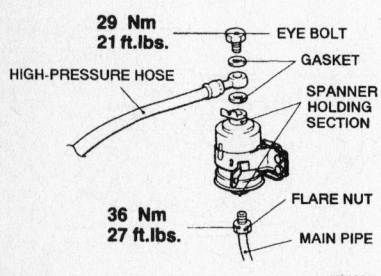

Exploded view of the fuel line-to-filter connection

4. Hold the fuel filter housing securely with a wrench. Cover the hoses with shop towels and remove the eyebolt. Discard the gaskets.

5. Separate the flare nut connection at the bottom of the filter.

6. Remove the mounting bolts and the fuel filter from the vehicle.

✳✳ CAUTION

Do not use conventional fuel filters, hoses or clamps when servicing fuel injection systems. They are not compatible with the injection system and the high pressures in fuel injection systems, and could cause substandard parts to fail, causing personal injury or damage to the vehicle. Use only hoses and clamps specifically designed for fuel injection.

To install:

7. Tighten the flare nut fitting by hand before mounting the filter on the bracket.

8. Install the filter on its bracket only finger-tight. Movement of the filter will ease attachment of the fuel lines.

9. Using new gaskets, connect the high pressure hose and eye bolt. While holding the fuel filter housing, torque the eye bolt to 22 ft. lbs. (29 Nm). Torque the flare nut to 27 ft. lbs. (36 Nm).

10. Tighten the filter mounting bolts fully.

11. Install the intake air hose, if removed.

12. Connect the negative battery cable, turn the key to the **ON** position to pressurize the fuel system and check for leaks.

13. Start the vehicle, check for leaks and repair if necessary.

Fuel Pump

REMOVAL & INSTALLATION

1. Before servicing the vehicle, refer to the precautions in the beginning of this section.

Do not use conventional fuel filters, hoses or clamps when servicing fuel injection systems. They are not compatible with the injection system and could fail, causing personal injury or damage to the vehicle. Use only hoses and clamps specifically designed for fuel injection.

2. Relieve the fuel system pressure.

3. Remove or disconnect the following:

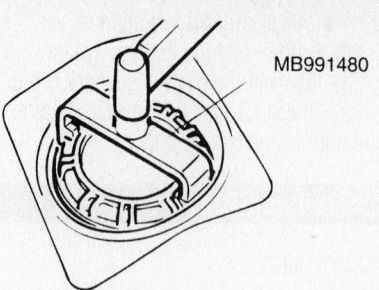

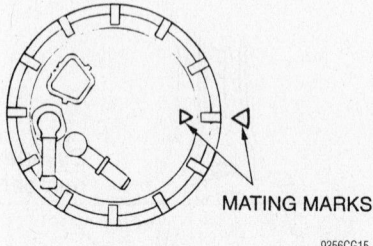

Fuel pump special tool and mating marks

- Negative battery cable
- Rear seat cushion

➡**Remove the seat cushion by pulling the stopper outward and lifting the lower cushion upward. There are 2 access covers underneath the seat. The panel on the far right side is for the fuel pump.**

- Access cover/protector
- Fuel pump wiring
- Return hose and high pressure fuel hose
- Fuel pump module, using special Tool No. MB991480, or equivalent

To install:

✳✳ WARNING

Install the packing to the fuel tank, then install the fuel pump module in the tank. Install the packing to the fuel pump module will damage the packing lip when installing the fuel pump module to the fuel tank and fuel leakage will occur.

4. Perform the following:

a. Check that the fuel tank is not damaged or deformed, then securely install the packing to the fuel tank. Replace the packing if it is damaged or deformed.

✳✳ WARNING

Do not tilt the fuel pump module during installation.

b. Apply soapy water to the inside of the packing, then install the fuel pump module.

❊❊ WARNING

When tightening, be careful not to let the fuel pump module turn together with the cap. If the mating marks are misaligned, the float may measure a remaining amount of fuel incorrectly, causing the low fuel warning light to malfunction.

    c. Use special Tool No. MB991480 to align the mating marks on the fuel tank and fuel pump module, then tighten the cap.

  5. Install or connect the following:
- High pressure hose, return hose and fuel pump wiring
- Negative battery cable

  6. Check the fuel pump for proper pressure and inspect the entire system for leaks.

  7. Apply sealant to the access cover and install the cover.

  8. Install the rear seat cushion.

  9. Pressurize the fuel system by turning the ignition key to the **ON** position.

  10. Start the engine to verify proper fuel pump performance.

Fuel Injector

REMOVAL & INSTALLATION

2.4L Engine

  1. Before servicing the vehicle, refer to the precautions in the beginning of this section.

  2. Relieve the fuel system pressure.

  3. Drain the engine coolant.

  4. Remove or disconnect the following:
- Air cleaner assembly
- Accelerator cable connection
- Throttle Position (TP) sensor connector
- Idle Air Control (IAC) motor connector
- Vacuum hose connector(s)
- Water hose
- Throttle body stay
- Throttle body and gasket
- Positive Crankcase Ventilation (PCV) hose connection
- Ignition coil connector
- Fuel injector connectors
- Manifold Differential Pressure (MDP) sensor
- Fuel lines
- Vacuum hose connections
- Fuel pressure regulator and O-ring
- Fuel rail with the injectors attached
- Insulators
- Injectors, O-rings and grommets

To install:

  5. Install or connect the following
- Grommets, O-rings and injectors to the fuel rail
- Insulators
- Fuel rail assembly
- Vacuum hose connections
- Fuel lines
- MDP sensor
- Injector connections
- Ignition coil connector
- PCV hose

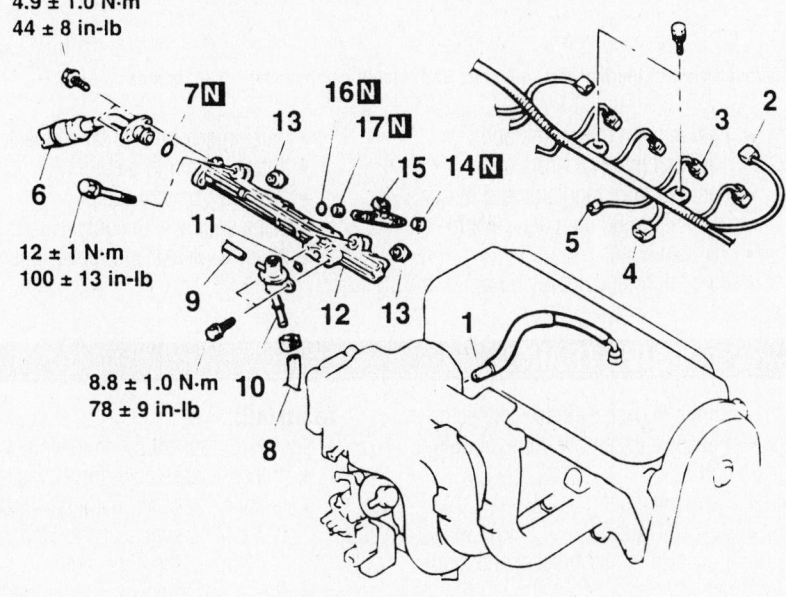

4.9 ± 1.0 N·m
44 ± 8 in-lb

12 ± 1 N·m
100 ± 13 in-lb

8.8 ± 1.0 N·m
78 ± 9 in-lb

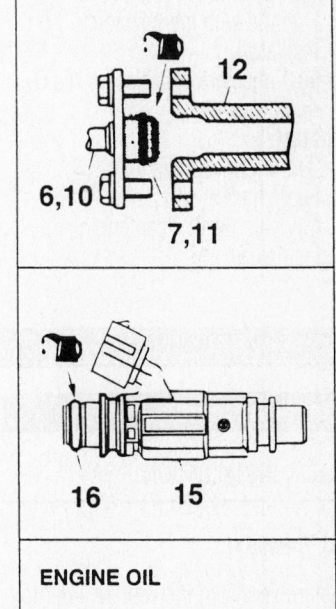

ENGINE OIL

1. PCV HOSE CONNECTION	8.	VACUUM HOSE CONNECTION
2. IGNITION COIL CONNECTOR	9.	FUEL PRESSURE REGULATOR
3. INJECTOR CONNECTOR	10.	O-RING
4. MANIFOLD DIFFERENTIAL PRESSURESENSOR CONNECTOR	11.	FUEL RAIL
5. HIGH-PRESSURE FUEL HOSE CONNECTION	12.	INSULATORS
6. O-RING	13.	INSULATORS
7. FUEL HOSE CONNECTION	14.	INJECTORS
	15.	O-RINGS
	16.	GROMMETS

9356CG16

Exploded view of the fuel rail, injectors and related components—2.4L engine

- Throttle body gasket and throttle body
- Throttle body stay
- Water hose
- Vacuum hose connector(s)
- IAC motor connector
- TP connector
- Accelerator cable connection
- Air cleaner assembly
- Negative battery cable.

6. Start the vehicle, check for leaks and repair if necessary.

3.0L Engine

1. Before servicing the vehicle, refer to the precautions in the beginning of this section.
2. Properly relieve the fuel system pressure.
3. Drain the engine coolant.
4. Remove or disconnect the following:
- Negative battery cable
- Intake manifold plenum
- Fuel injector electrical connector
- Fuel supply and return hoses
- Vacuum hose
- Fuel pressure regulator
- Fuel rail with the injectors
- Insulators
- Fuel injectors and discard the O-rings

To install:

5. Install or connect the following:
- Fuel injector with a new O-ring lubricated with clean engine oil
- Insulators

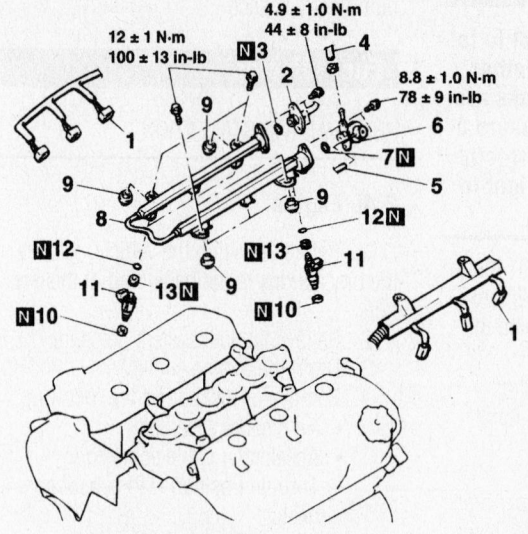

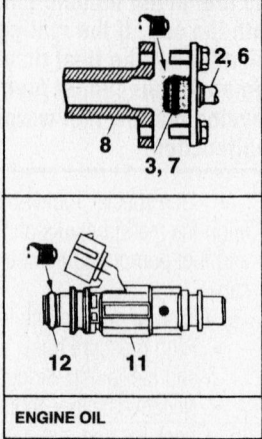

ENGINE OIL

1. INJECTOR CONNECTOR
2. FUEL HIGH-PRESSURE HOSE CONNECTION
3. O-RING
4. FUEL RETURN HOSE CONNECTION
5. VACUUM HOSE CONNECTION
6. FUEL PRESSURE REGULATOR
7. O-RING
8. FUEL RAIL
9. INSULATORS
10. INSULATORS
11. INJECTORS
12. O-RING
13. GROMMETS

67189-SEBR-G10

Exploded view of the fuel rail, injectors and related components—3.0L engine

- Fuel rail and torque the bolts to 100 inch lbs. (12 Nm)
- Fuel pressure regulator and torque the bolts to 80 inch lbs. (9 Nm)
- Vacuum hose
- Fuel return and supply hoses
- Fuel injector electrical connector
- Intake manifold plenum
- Negative battery cable

6. Fill the engine with coolant, start the vehicle, check for leaks and repair if necessary.

DRIVE TRAIN

Transaxle Assembly

REMOVAL & INSTALLATION

Manual Transaxle

1. Before servicing the vehicle, refer to the precautions in the beginning of this section.
2. Drain the transaxle fluid.
3. Remove or disconnect the following:
- Battery and tray
- Air cleaner assembly
- Front exhaust pipe, 3.0L engine
- Shift and selection cable connections
- Back-up light switch connection
- Vehicle Speed Sensor (VSS) connector
- Starter

- Clutch release cylinder connector
- Transaxle assembly upper coupling bolts
- Centermember
- Rear roll stopper
- Transaxle mount bracket and stopper
- Stabilizer link
- Wheel Speed Sensor (WSS), if equipped
- Brake hose clamp
- Tie rod end
- Lower arm
- Clutch release bearing engagement, 3.0L engine
- Driveshaft connection
- Upper oil pan bolt, 3.0L engine
- Bell housing cover
- Transaxle assembly lower coupling bolts
- Transaxle assembly

To install:

4. Install or connect the following:
- Transaxle assembly. Torque the lower coupling bolts to 36 ft. lbs. (48 Nm) on the 2.4L engine and to 52 ft. lbs. (71 Nm) for the 3.0L engine.
- Bell housing cover. Torque the cover—to—engine bolt to 80 inch lbs. (9 Nm) and the cover—to—transaxle bolt to 19 ft. lbs. (26 Nm).
- Upper oil pan bolt, 3.0L engine
- Driveshaft connection
- Clutch release bearing engagement
- Lower arm and torque the nut to 80 ft. lbs. (108 Nm)
- Tie rod end and torque the nut to 21 ft. lbs. (28 Nm)
- WSS, if equipped
- Stabilizer link and torque the nut to 33 ft. lbs. (45 Nm)

- Transaxle mount bracket and torque the nut to 42 ft. lbs. (57 Nm)
- Transaxle mount stopper and torque the nut to 60 ft. lbs. (81 Nm)
- Rear roll stopper and torque the nut to 33 ft. lbs. (45 Nm)
- Centermember
- Clutch release cylinder and torque to 13 ft. lbs. (18 Nm)
- Starter motor and electrical connectors
- VSS connector
- Back-up light switch connector
- Shift and selection cable connections
- Front exhaust pipe, 3.0L engine
- Air cleaner assembly
- Battery and tray

5. Fill the transaxle fluid to the proper level.

6. Start the vehicle, check for leaks and repair if necessary.

Automatic Transaxle

1. Before servicing the vehicle, refer to the precautions in the beginning of this section.
2. Drain the transaxle fluid.
3. Remove or disconnect the following:
- Battery and battery tray
- Battery brace
- Air cleaner and intake hoses
- Shifter lever-to-transaxle nut, cable retaining clip and cable from the transaxle
- Shifter cable mounting bracket
- Speedometer, solenoid, neutral safety switch (inhibitor switch), the pulse generator, kickdown servo switch and oil temperature sensor electrical connectors
- Transaxle oil cooler lines
- Transaxle dipstick and tube
- Starter and using Special Tool 7137 or C-4852, support the engine assembly
- Rear roll stopper mounting bracket
- Transaxle mount bracket
- Upper transaxle mounting bolts
- Front wheels
- Left side undercover
- Tie rod end from the steering knuckle
- Stabilizer bar link from the damper fork
- Damper fork from the lateral lower control arm
- Lateral lower arm and compression lower arm, lower ball joints from the steering knuckle
- Halfshafts from the transaxle and secure aside
- Bell housing cover

- Engine front roll stopper through-bolt
- Centermember
- Flexplate-to-torque converter bolts. Rotate the crankshaft to bring the bolts into a position for removal, one at a time.

➡ **To make installation easier, use chalk or paint to make matchmarks on the torque converter and flexplate. These marks will be used at assembly to realign the assembly, keeping these parts in balance.**

✳✳ WARNING

After removing the bolts, push the torque converter toward the transaxle. This will prevent the converter from remaining in contact with the engine, possibly damaging the converter.

4. Support the transaxle using a transmission jack (at the side of the case, NOT at the pan), and remove the transaxle lower coupling bolt.

➡ **The coupling bolt is inserted from the engine side into the transaxle and is located just above the halfshaft opening.**

5. Slide the transaxle rearward and carefully lower it from the vehicle.
To install:

6. After the torque converter has been mounted on the transaxle, install the transaxle assembly to the engine. Install the mounting bolts and tighten to 70 ft. lbs. (95 Nm).

7. Align the balance matchmarks and torque the torque converter-to-flexplate bolts to 55 ft. lbs. (75 Nm).

8. Install or connect the following:
- Bell housing cover and torque the bolts to 108 inch lbs. (12 Nm)
- Centermember and torque the front mounting bolts to 65 ft. lbs. (88 Nm) and the rear bolts to 51–58 ft. lbs. (69–78 Nm).
- Front engine roll stopper through-bolt and lightly tighten. Once the full weight of the engine is on the mounts, torque the bolt to 42 ft. lbs. (56 Nm).
- Halfshafts using new circlips

✳✳ WARNING

When installing the halfshaft, keep the inboard joint straight in relation to the axle to avoid damaging the oil seal lip of the transaxle with the splined part of the halfshaft.

- Tie rod and ball joints to the steering knuckle. Torque the ball joint self-locking nuts to 48 ft. lbs. (65 Nm), the tie rod end nut to 21 ft. lbs. (28 Nm) and secure with a new cotter pin.
- Damper fork to the lower control arm and torque the through-bolt to 65 ft. lbs. (88 Nm)
- Stabilizer link to the damper fork and torque the self-locking nut to 29 ft. lbs. (39 Nm)
- Left side undercover
- Wheels
- Transaxle mount bracket and torque the mounting nuts to 32 ft. lbs. (43 Nm)
- Rear roll stopper mounting bracket and remove the engine support
- Transaxle mount through-bolt to 51 ft. lbs. (69 Nm) and torque the front engine roll stopper bolt
- Upper transaxle mounting bolts and tighten to 35 ft. lbs. (48 Nm)
- Starter
- Dipstick tube and dipstick
- Shifter cable mounting bracket
- Shifter lever and torque the nut to 14 ft. lbs. (19 Nm)
- Oil cooler lines and secure with clamps
- Speedometer, solenoid, neutral safety switch (inhibitor switch), the pulse generator, kickdown servo switch and oil temperature sensor electrical connectors
- Air cleaner and air intake hose
- Battery tray and battery

9. Refill the transaxle with MOPAR® ATF PLUS transmission fluid. Start the engine and allow it to idle for 2 minutes. Apply the parking brake and move the selector through each gear position, ending in **N**. Recheck fluid level and add if necessary. Fluid level should be between the marks in the **HOT** range on the dipstick.

10. Check the transaxle for proper operation.

Clutch

ADJUSTMENT

Pedal Height and Free-Play

1. Measure the clutch pedal height from the face of the pedal pad to the floorboard. Compare the measured value with the desired distance of 6.88–7.00 in. (174.7–177.7mm).

2. Measure the clutch pedal clevis pin

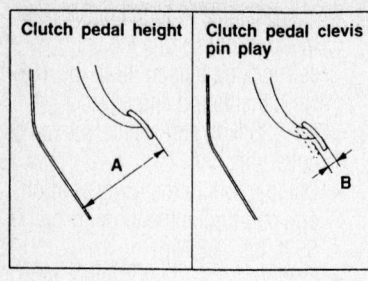

Clutch pedal height	Clutch pedal clevis pin play

9300CG01

Clutch pedal height and free-play adjustment measurements

play at the face of the pedal pad. Press the pedal lightly until resistance is met, and measure this distance. The clutch pedal clevis pin play should be within 0.040–0.120 in. (1–3mm).

3. If the clutch pedal height or clevis pin play, is not within the standard values, adjust as follows:

a. If not equipped with cruise control, turn and adjust the stop bolt so the pedal height is the standard value, then tighten the locknut.

b. If equipped with cruise control system, disconnect the clutch switch connector and turn the switch to obtain the standard clutch pedal height. Then, lock by tightening the locknut.

c. Turn the pushrod to adjust the clutch pedal clevis pin play to agree with the standard value and secure the pushrod with the locknut.

➡ **When adjusting the clutch pedal height or the clutch pedal clevis pin**

play, be careful not to push the pushrod toward the master cylinder.

d. Check that when the clutch pedal is depressed all the way, the interlock switch switches over from **ON** to **OFF**.

4. Move the clutch pedal until the resistance begins to increase; measure between this point and the pedal resting point to determine the clutch pedal free-play. The clutch pedal free-play measurement should be between 0.240–0.510 in. (6–13mm). With the pedal fully disengaged, check the distance between the bulkhead and the top of the pedal pad. The measurement should be 3.8 in. (95.4mm) for 2.4L engines or 3.5 in. (89.5mm) for 3.0L engines.

5. If the measurements are not within spec-

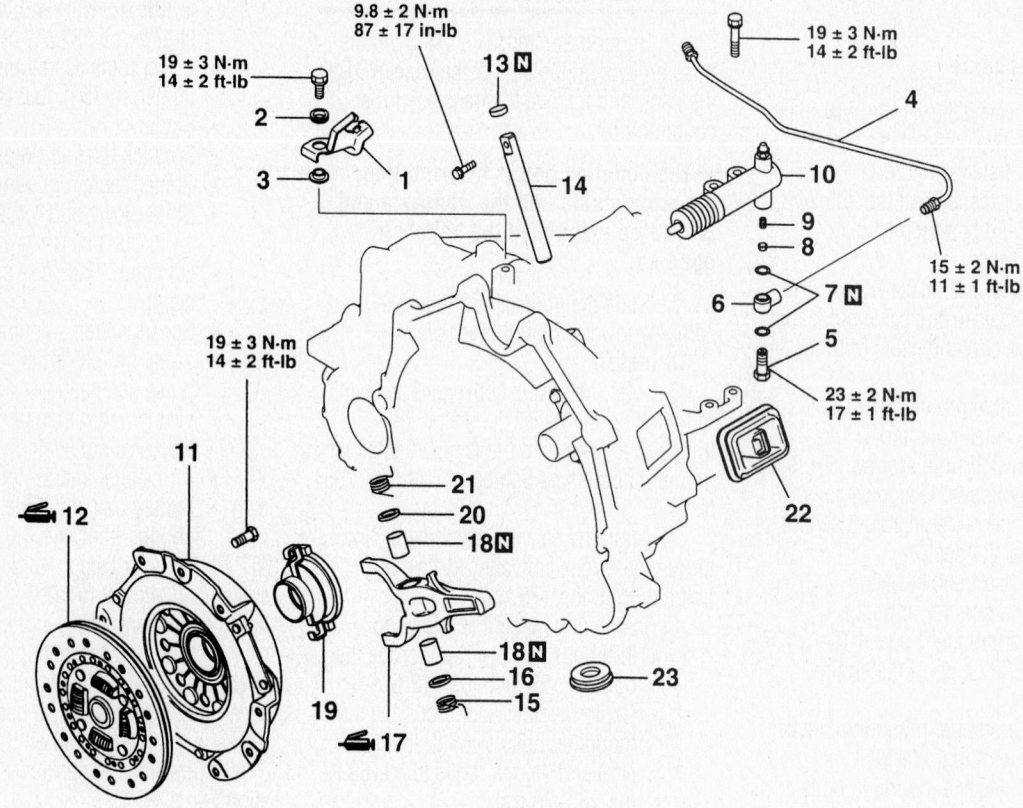

1.	CLUTCH FLUID LINE BRACKET	13.	SEALING CAP
2.	INSULATOR	14.	RELEASE FORK SHAFT
3.	WASHER	15.	SUPPORT SPRING (L)
4.	CLUTCH TUBE	16.	PACKING
5.	UNION BOLT	17.	RELEASE FORK
6.	UNION	18.	BUSHING
7.	GASKET	19.	CLUTCH RELEASE BEARING
8.	VALVE	20.	PACKING
9.	VALVE SPRING	21.	SUPPORT SPRING (R)
10.	CLUTCH RELEASE CYLINDER	22.	RELEASE FORK BOOT
11.	CLUTCH COVER	23.	MAINTENANCE HOLE COVER
12.	CLUTCH DISC		

9356CG17

Exploded view of the clutch assembly–3.0L engine shown, 2.4L similar

ification, bleed the clutch hydraulic system. If after bleeding the measurements are still not within specified range, there is a faulty component in the system, which must be replaced.

REMOVAL & INSTALLATION

1. Before servicing the vehicle, refer to the precautions in the beginning of this section.
2. Remove or disconnect the following:
   - Negative battery cable
   - Clutch fluid line bracket
   - Insulator, washer and clutch tube
   - Union, union bolt and gasket
   - Valve and spring, 3.0L engine
   - Clutch release cylinder
   - Clutch cover and disc
   - Release fork shaft, 3.0L engine
   - Left support spring, 3.0L engine
   - Release fork
   - Clutch release bearing
   - Release fork boot and fulcrum, 2.4L engine
   - Release fork boot and maintenance cover, 3.0L engine

To install:

3. Install or connect the following:
   - Release fork boot and maintenance cover, 3.0L engine
   - Release fork boot and fulcrum and torque the fulcrum to 26 ft. lbs. (35 Nm)
   - Clutch release bearing
   - Release fork
   - Left support spring and release fork shaft, 3.0L engine
   - Clutch disc and cover and torque the bolts to 14 ft. lbs. (19 Nm)
   - Clutch release cylinder
   - Valve and spring, 3.0L engine
   - Union, bolt and gasket and torque the bolt to 17 ft. lbs. (23 Nm)
   - Insulator, washer and clutch tube and torque the fastener to 11 ft. lbs. (15 Nm)
   - Fluid line bracket and torque the bolt to 14 ft. lbs. (19 Nm)
   - Negative battery cable

4. Check and top off the fluid level, if necessary.

Hydraulic Clutch System

BLEEDING

1. Before servicing the vehicle, refer to the precautions in the beginning of this section.

2. Fill the reservoir with clean DOT 3 or DOT 4 brake fluid.
3. Loosen the bleed screw, then have the clutch pedal pressed to the floor.
4. Tighten the bleed screw, then release the clutch pedal.
5. Repeat the procedure until the fluid is free of air bubbles.

➡ **It is suggested that a hose be attached to the bleed screw with the other end immersed in a container at least half full of brake fluid during the bleeding operation. Do not allow the reservoir to run out of fluid during bleeding.**

6. Refill the reservoir with clean brake fluid.
7. Check the clutch for proper operation.

Halfshaft

REMOVAL & INSTALLATION

1. Before servicing the vehicle, refer to the precautions in the beginning of this section.
2. Remove or disconnect the following:

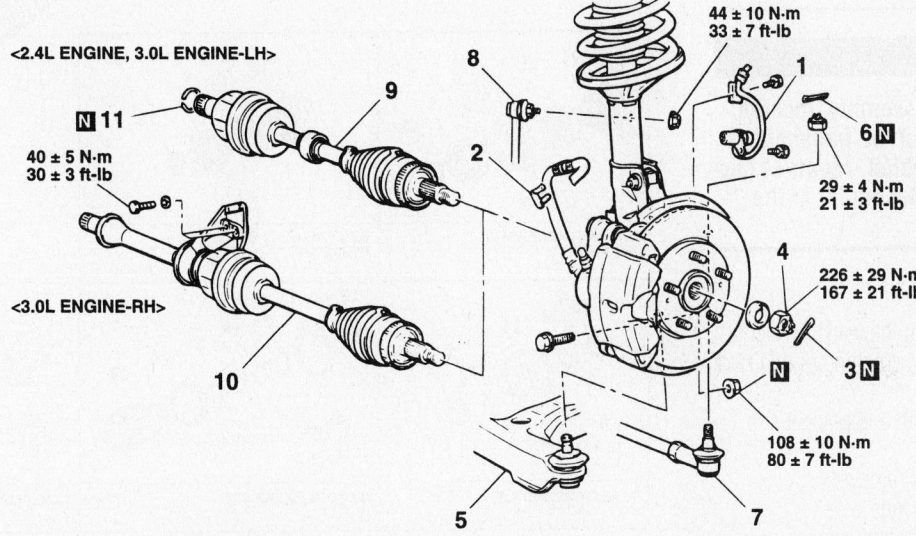

1. SPEED SENSOR CABLE CONNECTION <VEHICLES WITH ABS>
2. BRAKE HOSE CLIP
3. COTTER PIN
4. DRIVESHAFT NUT
5. LOWER ARM BALL JOINT CONNECTION
6. COTTER PIN
7. TIE ROD END CONNECTION
8. STABILIZER LINK CONNECTION
9. DRIVESHAFT
10. DRIVESHAFT AND INNER SHAFT
11. CIRCLIP

9356CG18

Exploded view of the left and right halfshaft assemblies

- Negative battery cable
- Front wheel
- Vehicle Speed Sensor (VSS) cable, if equipped
- Brake hose clip and cotter pin
- Driveshaft nut
- Lower arm ball joint connection
- Tie rod end
- Stabilizer link
- Driveshaft and circlip

To install:

3. Install or connect the following:
- Driveshaft
- Stabilizer link and torque the nut to 33 ft. lbs. (44 Nm)
- Tie rod end and torque the nut to 21 ft. lbs. (28 Nm)
- New cotter pin to the tie rod end
- Lower arm ball joint
- Driveshaft nut and torque it to a maximum of 167 ft. lbs. (226 Nm)
- New cotter pin to the driveshaft nut
- Brake hose clip
- VSS cable, if equipped
- Front wheel
- Negative battery cable

CV-Joints

OVERHAUL

✳✳ WARNING

The Birfield joint assembly, located on the wheel side of the halfshaft, is not to be disassembled; repair of this joint is only by replacement of the halfshaft.

Tri-Pot Joint

1. Before servicing the vehicle, refer to the precautions in the beginning of this section.

2. Remove halfshaft and place it in a soft jawed vise.

3. Disassemble or remove:
- Tri-pot boot bands
- Tri-pot case for left halfshaft or tri-pot case/inner shaft assembly for right halfshaft

➡ Wipe the grease from the tri-pot case.

- Halfshaft snapring
- Tri-pot spider assembly

✳✳ WARNING

Do not disassemble the spider assembly.

- Tri-pot boot

✳✳ WARNING

If the boot is to be reused, wrap plastic tape around the shaft splines to protect the boot from damage.

4. If working with the right halfshaft and it is necessary to replace the tri-pot case, press the case from the inner shaft assembly.

To install:

5. If installing a new tri-pot case onto the right halfshaft, perform the following procedure:

   a. Lubricate the inner shaft splines with Multi-Mileage Grease No. 2525035.

   b. Press the inner shaft assembly into tri-pot case.

   c. Secure the tri-pot case with special

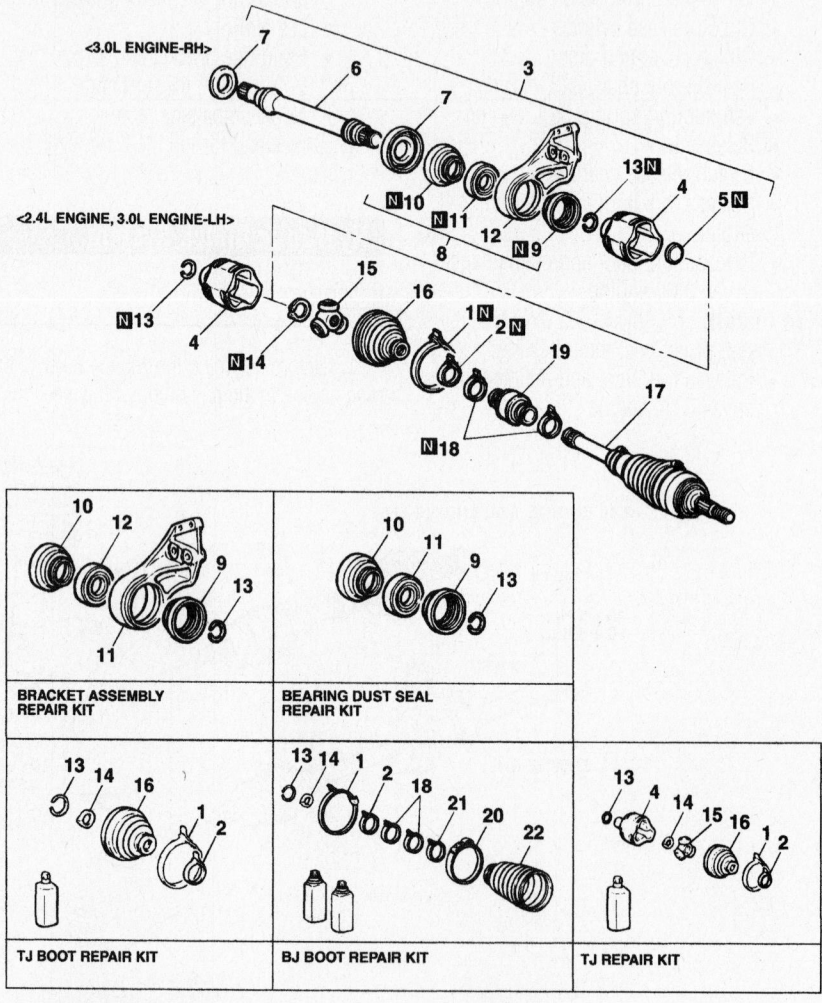

⚠ CAUTION
Never disassemble the BJ assembly except when replacing the BJ boot.

1. TJ BOOT BAND (LARGE)
2. TJ BOOT BAND (SMALL)
3. TJ CASE INNER SHAFT ASSEMBLY
4. TJ CASE
5. SEAL PLATE
6. INNER SHAFT
7. DUST COVER
8. BRACKET ASSEMBLY
9. DUST SEAL OUTER
10. DUST SEAL INNER
11. CENTER BEARING
12. CENTER BEARING BRACKET
13. CIRCLIP
14. SNAP RING
15. SPIDER ASSEMBLY
16. TJ BOOT
17. BJ ASSEMBLY
18. DAMPER BAND <2.4L ENGINE, 3.0L ENGINE-LH>
19. DYNAMIC DAMPER <2.4L ENGINE, 3.0L ENGINE-LH>
20. BJ BOOT BAND (LARGE)
21. BJ BOOT BAND (SMALL)
22. BJ BOOT

NOTE:
BJ: Birfield Joint
TJ: Tripod Joint

67189-SEBR-G11

Exploded view of the halfshaft assemblies

Tool MB991248 on a hydraulic press with the case facing upward.

    d. Position a new seal plate in the center of the tri-pot case.

    e. Place a 1.18 in. (30mm) pipe on the seal plate and press the seal into the tri-pot case.

✳✳ WARNING

Wrap plastic tape around the shaft splines to protect the boot from damage.

6. Install a new small boot clamp onto the halfshaft followed by the tri-pot boot.

7. Apply the specified repair kit grease between the spider axle and the roller.

8. Install the tri-pot spider assembly onto the shaft from the spline beveled section direction and secure with the snapring.

9. Distribute a portion of the 4.23 oz. (120 g) specified repair kit grease into the tri-pot case, insert the spider assembly and add the remaining grease.

10. Install the tri-pot boot by performing the following procedure:

    a. Position the boots large end on the tri-pot case.

    b. Position the large and small boot clamps onto the boot.

    c. Adjust the boot so that the bands are spaced at 3.03–3.27 in. (77–83mm).

    d. Tighten the band securely.

Center Bearing

1. Remove the right halfshaft and place it in a soft-jawed vise.

2. Remove the tri-pot spider assembly from the tri-pot case.

3. Remove the tri-pot case by performing the following procedure:

    a. Position the inner shaft/tri-pot case assembly on a hydraulic press supported by Tool MB991248 with the tri-pot case facing upward.

    b. Position a bar inside the tri-pot case, on the end of the inner shaft and press the case from the inner shaft assembly.

4. Place the inner shaft assembly in a soft-jawed vise.

5. Using a wheel puller, press the center bearing bracket from the inner shaft.

6. Using a hydraulic press, Installer Adapter Tool MB990932 and Snap-in Bar Tool MB990938, press the center bearing and inner dust seal from the center bearing bracket.

To install:

7. Using a hydraulic press, Installer Adapter Tool MB990932 and Snap-in Bar

Tool MB990938, press the center bearing into the center bearing bracket.

8. Using a hydraulic press, Installer Adapter Tool MB990933 and Snap-in Bar Tool MB990938, press the inner duct seal into the center bearing bracket.

9. Using a hydraulic press, Installer Adapter Tool MB990931 and Snap-in Bar Tool MB990938, press the outer duct seal into the center bearing bracket.

10. Using a hydraulic press and Adapter Tool MB991172, press the inner shaft into the center bearing bracket.

11. Install the tri-pot case onto the right halfshaft by performing the following procedure:

    a. Lubricate the inner shaft splines with Multi-Mileage Grease No. 2525035.

    b. Press the inner shaft assembly into tri-pot case.

    c. Secure the tri-pot case with special Tool MB991248 on a hydraulic press with the case facing upward.

    d. Position a new seal plate in the center of the tri-pot case.

    e. Place a 1.18 in. (30mm) pipe on the seal plate and press the seal into the tri-pot case.

12. Assemble the tri-pot joint assembly.

13. Install the right halfshaft.

Birfield Joint Boot

1. Remove halfshaft and place it in a soft jawed vise.

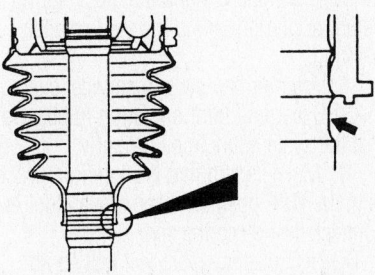

Positioning the birfield joint boot's small diameter

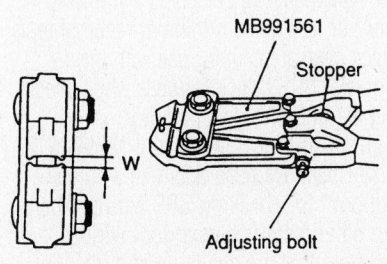

View of the band crimper tool

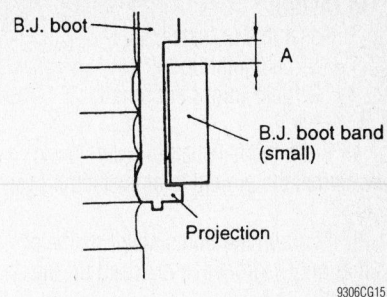

Positioning the birfield joint boot's small band

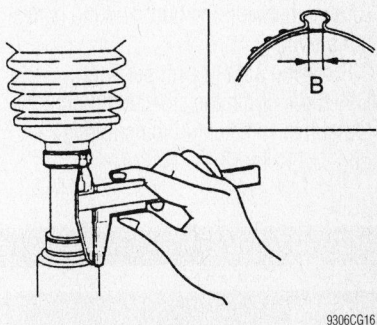

Measuring the small band's crimp

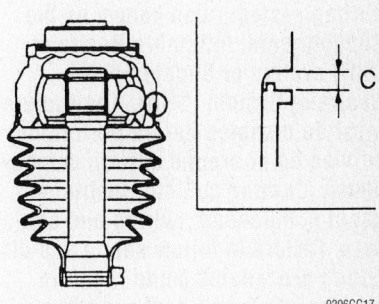

Positioning the birfield joint boot's large diameter

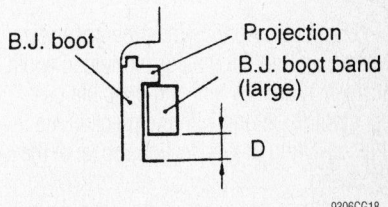

Positioning the birfield joint boot's large band

2. Disassemble or remove:
- Tri-pot joint
- Dynamic damper, if equipped
- Birfield joint boot bands
- Birfield joint boot

To install:

3. Assemble or install:
- Birfield joint boot
- Birfield joint boot small band

4. Place the halfshaft in a soft jawed vise so that the birfield joint is standing vertically.

5. Position the boots small diameter so that only 1 groove is exposed on the shaft.

6. Adjust the Band Crimper Tool MB991561 so that the jaw opening is 0.114 in. (2.9mm).

7. Position the small band so there is clearance between the boot and the bands mating edge.

8. Using a Band Crimper Tool MB991561, crimp the small band until the internal measurement of the crimp is 0.094–0.110 in. (2.4–2.8mm).

✳✳ WARNING

If the crimp dimension is not correct, remove the band install a new one.

9. Using 5.47 oz. (155g) specified amount of grease in the repair kit, pack the birfield boot.

10. Position the boot so 0.004–0.061 in. (0.1–1.55mm) of clearance exists between the boots large diameter end and the birfield joint housing shoulder.

11. Adjust the Band Crimper Tool MB991561 so that the jaw opening is 0.126 in. (3.2mm).

12. Install and position the large boot band so that it rests against the boot(s) projection and a gap exists between the clamp and the boot.

13. Using a Band Crimper Tool MB991561, crimp the small band until the internal measurement of the crimp is 0.094–0.110 in. (2.4–2.8mm).

✳✳ WARNING

If the crimp dimension is not correct, remove the band install a new one.

14. If equipped, install the dynamic damper by performing the following procedure:

a. Slide the damper onto the halfshaft with new bands.

b. Position the damper so the distance from the front of the birfield joint to the front edge of the damper assembly is 14.60–14.84 in. (371–377mm) for the right halfshaft or 7.52–7.76 in. (191–197mm) for the left halfshaft.

c. Tighten and secure the damper bands.

15. Install the tri-pot joint.

STEERING AND SUSPENSION

Air Bag

✳✳ CAUTION

Some vehicles are equipped with an air bag system, also known as the Supplemental Inflatable Restraint (SIR) system or Supplemental Restraint System (SRS). The system must be disabled before performing service on or around system components, steering column, instrument panel components, wiring and sensors. Failure to follow safety and disabling procedures could result in accidental air bag deployment, possible personal injury and unnecessary system repairs.

PRECAUTIONS

Several precautions must be observed when handling the inflator module to avoid accidental deployment and possible personal injury. Along with the precautions in the beginning of this section, observe the following:

1. Never carry the inflator module by the wires or connector on the underside of the module.

2. When carrying a live inflator module, hold securely with both hands, and ensure that the bag and trim cover are pointed away.

3. Place the inflator module on a bench or other surface with the bag and trim cover facing up.

4. With the inflator module on the bench, never place anything on or close to the module which may be thrown in the event of an accidental deployment.

5. Do not attempt to repair any of the air bag system wiring harness connectors. If any of the connectors or wires are faulty, replace that harness.

6. Air bag components should not be subjected to heat over 200°F (93°C). Remove the Supplemental Restraint System Air Bag Control Unit (SRS-ECU), the air bag modules themselves and the clock spring before drying or baking the vehicle after painting.

7. After air bag system service, check the SRS warning light operation to be sure that the system functions properly.

8. Make certain that the ignition switch is in the **OFF** position when a scan tool is connected or disconnected.

DO NOT use any electrical test equipment on or near any SRS components except those specified by Chrysler corporation:

9. Use a digital multi-meter for which the maximum test current is 2 milliamp (mA) or less at the minimum range of resistance measurement for use with the Chrysler SRS Check Harness when checking the SRS electrical circuitry.

10. Chrysler Special Tool MB991613 SRS Check Harness acts like a "breakout box" for checking SRS wiring. There are other factory special tool wiring adapters that are available and may be used.

11. DRB III scan tool for reading and erasing air bag diagnostic codes.

NEVER ATTEMPT TO REPAIR THE FOLLOWING COMPONENTS:

12. Air Bag Control Unit (SRS-ECU)

13. Air Bag Modules

14. If any of these components are diagnosed as faulty, they should only be replaced.

DISARMING

The system consists of 2 air bag modules, one located in the center of the steering wheel and another located above the glove box, which contains the folded air bag and an inflator unit. The air bag Electronic Control Unit (SRS-ECU) located under the floor console assembly monitors the system and which contains a safing G sensor and analog G sensor. An SRS warning light is located on the instrument panel which indicates the status of the air bag system. A clock spring interconnection is located within the steering column.

To deploy the air bags, the SRS-ECU must respond to the output signal from the analog G sensor and the safing G sensor must be ON. The SRS-ECU, then causes the air bag modules to ignite and deploy.

Service technicians should use care when working around any vehicle equipped with an air bag system, to avoid injury to the technician by inadvertent deployment of the air bag or to the driver by rendering the air bag system inoperative.

The SRS-ECU not only controls the air bag system, it can provide diagnostic information. The SRS-ECU monitors the air bag system and stores data concerning any detected faults in the system. When the ignition key is turned to the **ON** or **START** position, the SRS warning light should illuminate for about 7 seconds, then turn off. That indicates that the SRS system is in operating condition. If the SRS warning light does not illuminate as described or stays on for more than 7 seconds or if the SRS light illuminates while driving, immediate inspection is required. If the vehicle's SRS warning light is in any of these 3 conditions, the SRS system must be inspected, diagnosed and serviced.

To avoid injury from accidental deployment of the air bag during vehicle servicing, refer to all service precautions.

❊❊ CAUTION

The Air Bag system must be disarmed before removing many components. Failure to do so may cause accidental deployment of the air bag, resulting in unnecessary system repairs and/or personal injury.

1. Disarm the air bag system using the following procedure:

   a. Position the front wheels in the straight-ahead position and place the key in the **LOCK** position. Remove the key from the ignition lock cylinder.

   b. Disconnect the negative battery cable and insulate the cable end with high-quality electrical tape or similar non-conductive wrapping.

   c. Wait at least one minute before working on the vehicle. The air bag system is designed to retain enough voltage to deploy the air bag for a short period of time even after the battery has been disconnected.

Power Rack and Pinion Steering Gear

REMOVAL & INSTALLATION

1. Before servicing the vehicle, refer to the precautions in the beginning of this section.

2. Drain the power steering fluid.

3. Remove or disconnect the following:

   • Negative battery cable from the left shock tower

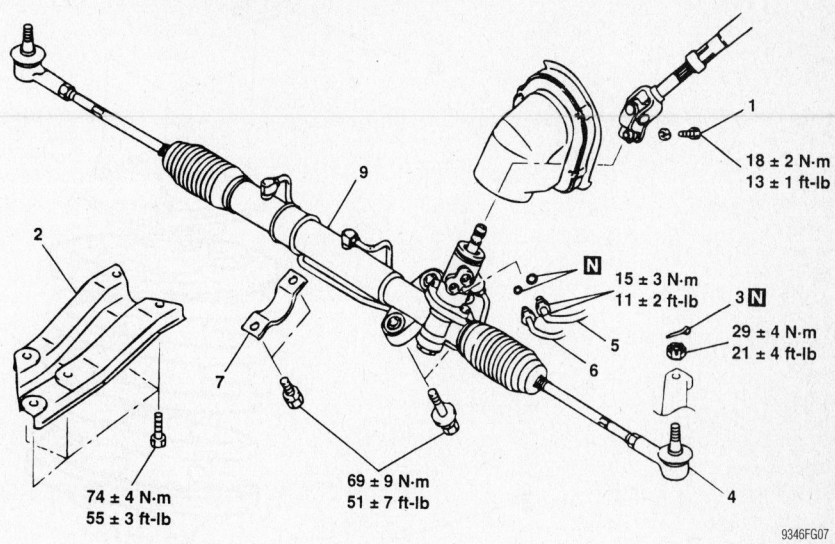

9346FG07

Exploded view of the steering gear assembly

• Front exhaust pipe
• Center member
• Sway bar, 2.4L engine only
• Steering shaft assembly
• Stay, 2.4L engine only
• Tie rod end from the steering knuckle
• Return hose and pressure tube connections
• Cylinder clamp
• Steering gear from the vehicle

To install:

4. Install or connect the following:
   • Steering gear to the vehicle
   • Cylinder clamp and torque the bolts to 51 ft. lbs. (69 Nm)
   • Pressure tube and return hoses and torque to 11 ft. lbs. (15 Nm)
   • Tie rod ends to the steering knuckle and torque the nut to 21 ft. lbs. (29 Nm)
   • Stay and torque the bolts to 55 ft. lbs. (74 Nm), 2.4L engine only
   • Steering shaft assembly and torque the bolt to 13 ft. lbs. (18 Nm)
   • Sway bar, 2.4L engine only
   • Center member
   • Front exhaust pipe, 2.4L engine only
   • Negative battery cable

5. Fill and bleed the power steering system.

6. Start the vehicle, check for leaks and repair if necessary.

7. Perform a front wheel alignment.

Strut

REMOVAL & INSTALLATION

Front

1. Before servicing the vehicle, refer to the precautions in the beginning of this section.

2. Remove or disconnect the following:
   • Negative battery cable
   • Front wheel
   • Brake hose clamp
   • Front speed sensor harness clamp, if equipped
   • Sway bar link
   • Lower control arm
   • Strut from the steering knuckle
   • Upper mounting bolts
   • Strut assembly

To install:

3. Install or connect the following:
   • Strut assembly and torque the upper bolts to 33 ft. lbs. (44 Nm)
   • Strut assembly to the lower control arm and steering knuckle. Torque the bolts to 221 ft. lbs. (300 Nm).
   • Sway bar link and torque the nut to 33 ft. lbs. (44 Nm)
   • Front speed sensor harness clamp, if equipped
   • Brake hose clamp
   • Front wheel
   • Negative battery cable

4. Check and adjust the front end alignment, if needed.

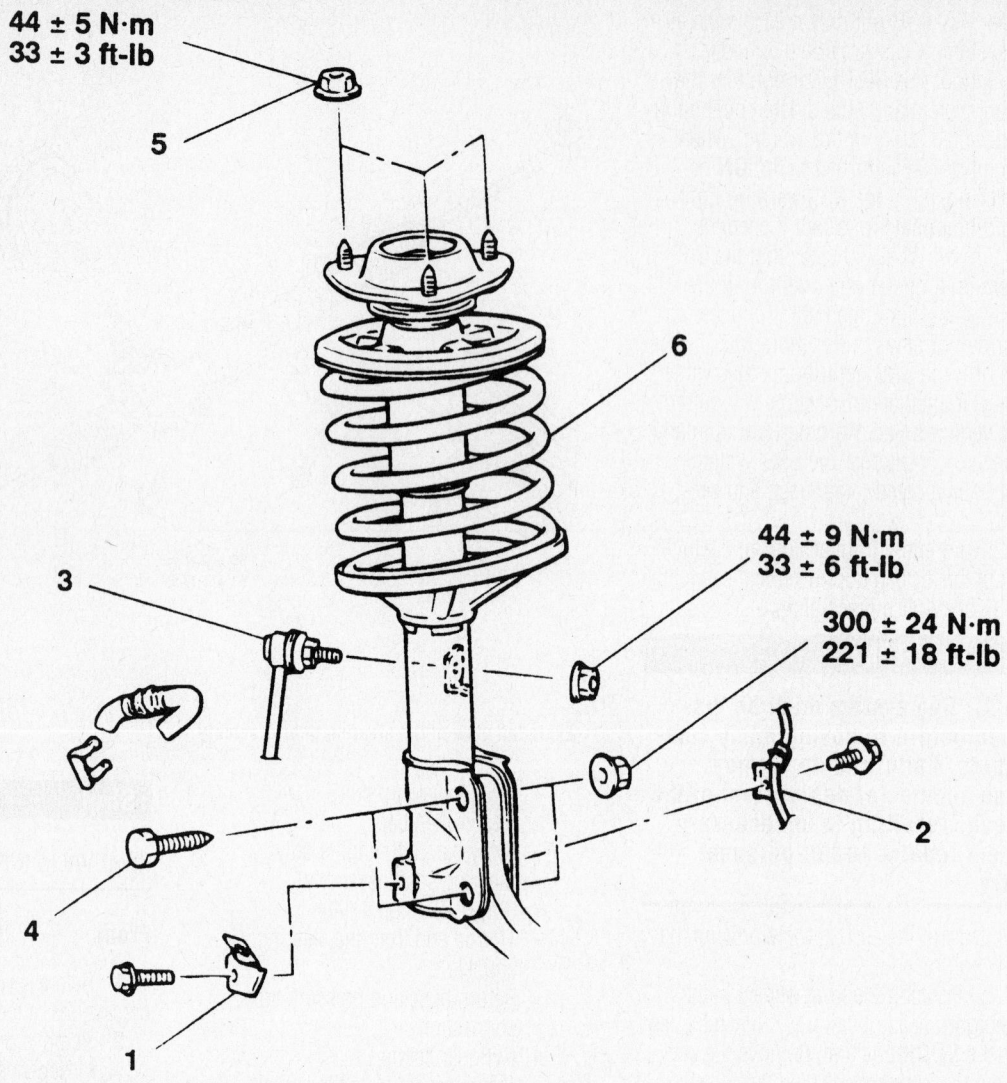

44 ± 5 N·m
33 ± 3 ft-lb

5

6

44 ± 9 N·m
33 ± 6 ft-lb

300 ± 24 N·m
221 ± 18 ft-lb

3

2

4

1

1. BRAKE HOSE CLAMP
2. FRONT SPEED SENSOR HARNESS
   CLAMP <VEHICLES WITH ABS>
3. STABILIZER LINK
4. BOLTS
5. NUT
6. STRUT ASSEMBLY

9346FG08

Exploded view of the front strut assembly

Shock Absorber

REMOVAL & INSTALLATION

Rear

1. Before servicing the vehicle, refer to the precautions in the beginning of this section.

2. Remove or disconnect the following:
   - Negative battery cable
   - Rear shelf trim
   - Cover from the upper shock tower
   - Upper mounting nuts
   - Shock from the rear steering knuckle
   - Shock assembly

To install:

3. Install or connect the following:
   - Shock absorber and torque the shock to steering knuckle bolt to 73 ft. lbs. (98 Nm)
   - Upper shock nuts and torque to 32 ft. lbs. (45 Nm)
   - Upper tower cover

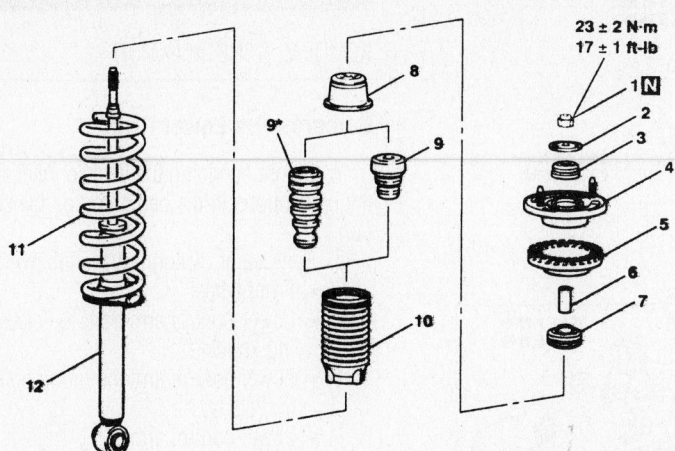

1. JAM NUT
2. WASHER
3. UPPER BUSHING A
4. UPPER BRACKET ASSEMBLY
5. UPPER SPRING PAD
6. COLLAR

7. UPPER BUSHING B
8. CUP ASSEMBLY
9. BUMP RUBBER
10. DUST COVER
11. COIL SPRING
12. SHOCK ABSORBER ASSEMBLY

9346FG09

Exploded view of the shock absorber

- Rear shelf trim
- Negative battery cable

Coil Spring

REMOVAL & INSTALLATION

Front

1. Before servicing the vehicle, refer to the precautions in the beginning of this section.
2. Remove the strut/shock assembly and place it in a suitable compressor tool.
3. Install Special Tool MB991176 to secure the strut and remove the dust cover and jam nut.
4. Remove the strut insulator.
5. Remove the upper spring seat and pad.
6. Remove the bump rubber and dust cover.
7. Remove the coil spring.
8. Remove the lower spring pad.

To assemble:

9. Install the lower spring pad and coil spring.
10. Install the dust cover and bump rubber.
11. Install the upper spring pad and seat.
12. Carefully compress the spring and install the insulator and jam nut.
13. When properly aligned, torque the jam nut to 47 ft. lbs. (64 Nm).

14. Install the dust cover and remove the compressor tools.
15. Install the strut/shock to the vehicle.

Rear

1. Before servicing the vehicle, refer to the precautions in the beginning of this section.
2. Disassemble as follows:
3. Remove the rear shock from the vehicle.
4. Install the assembly in a spring compressor tool.
5. Remove the jam nut, washer and upper bushing.
6. Remove the upper bracket assembly and spring pad.
7. Remove the collar and bushing.
8. Remove the cup, bump rubber and dust cover.
9. Remove the coil spring

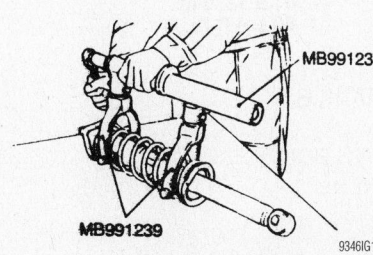

9346IG16

Compress the coil spring as shown

To assemble:

10. Install the coil spring to the shock absorber using Compressor Tools MB991237 and MB991239.
11. Install the dust cover and bump rubber.
12. Install the cup and bushing
13. Install the collar, supper spring pad and bracket assembly.
14. Install the upper bushing and washer.
15. When properly aligned, install the jam nut and torque to 17 ft. lbs. (23 Nm).
16. Remove the compressor tools and install the shock absorber.

Upper Ball Joint

REMOVAL & INSTALLATION

The upper ball joint is an integrated part of the upper control arm assembly, and cannot be serviced separately. A worn or damaged ball joint requires replacement of upper control arm assembly.

Lower Ball Joint

REMOVAL & INSTALLATION

On all vehicles, the ball joint cannot be serviced separately. If the ball joint is defective it will require replacement of the lower control arm.

Upper Control Arm

REMOVAL & INSTALLATION

1. Before servicing the vehicle, refer to the precautions in the beginning of this section.
2. Remove or disconnect the following:
   - Rear wheel
   - Upper control arm from the steering knuckle
   - Upper control arm assembly
   - Upper control arm bracket
   - Upper control arm

To install:

3. Install or connect the following:
   - Upper control arm and brackets and torque the bolts to 42 ft. lbs. (57 Nm)
   - Upper control arm assembly and torque the bolts to 29 ft. lbs. (39 Nm)
   - Upper control arm to the steering knuckle and torque the bolt 73 ft. lbs. (98 Nm)
   - Rear wheel

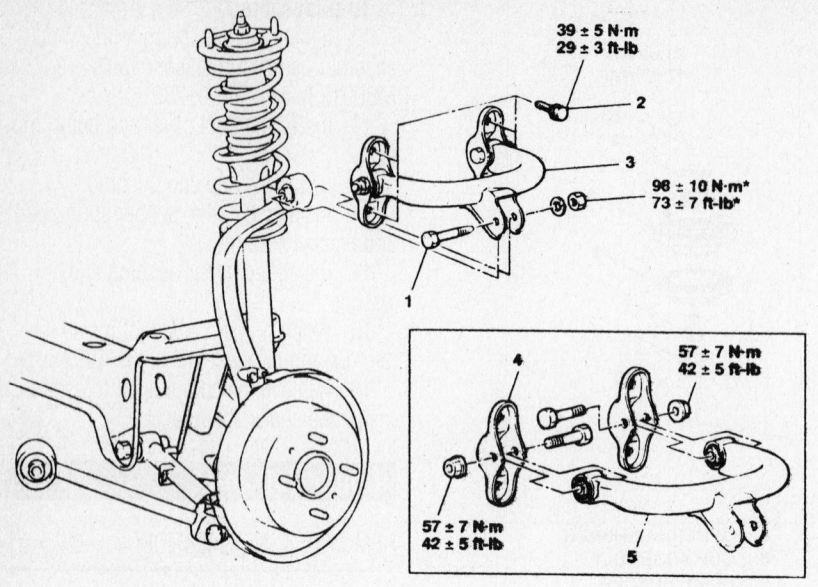

39 ± 5 N·m
29 ± 3 ft-lb

2

3

98 ± 10 N·m*
73 ± 7 ft-lb*

1

4

57 ± 7 N·m
42 ± 5 ft-lb

57 ± 7 N·m
42 ± 5 ft-lb

5

1. **UPPER ARM AND KNUCKLE CONNECTING BOLT**
2. **UPPER ARM ASSEMBLY MOUNTING BOLTS**

3. **UPPER ARM ASSEMBLY**
4. **UPPER ARM BRACKET**
5. **UPPER ARM**

9346FG10

Exploded view of the upper control arm assembly

Lower Control Arm

REMOVAL & INSTALLATION

Compression Lower Arm

1. Before servicing the vehicle, refer to the precautions in the beginning of this section.

2. Remove or disconnect the following:
   - Front wheel
   - Lower control arm from the steering knuckle
   - Lower control arm mounting bolt and clamp
   - Lower control arm

3. If replacing the bushing, perform the following:
   a. Apply soapy water between the shaft and old bushing and pry the old bushing out.

To install:

4. Apply soapy water to the shaft and new bushing and install the new bushing, if removed.

5. Install or connect the following:
   - Lower control arm

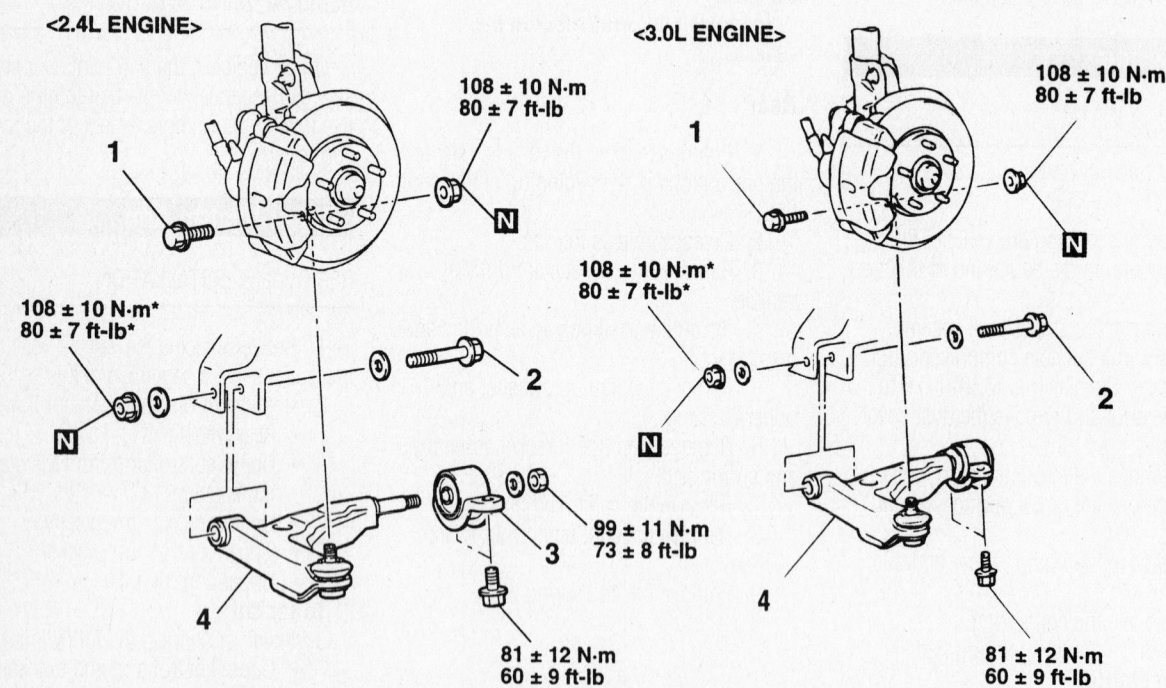

<2.4L ENGINE>

1

108 ± 10 N·m
80 ± 7 ft-lb

N

108 ± 10 N·m*
80 ± 7 ft-lb*

N

2

3

99 ± 11 N·m
73 ± 8 ft-lb

4

81 ± 12 N·m
60 ± 9 ft-lb

<3.0L ENGINE>

1

108 ± 10 N·m
80 ± 7 ft-lb

N

108 ± 10 N·m*
80 ± 7 ft-lb*

N

2

4

81 ± 12 N·m
60 ± 9 ft-lb

1. **LOWER ARM AND KNUCKLE CONNECTION**
2. **LOWER ARM MOUNTING BOLT**
3. **LOWER ARM CLAMP <2.4L ENGINE>**
4. **LOWER ARM**

9356CG19

Exploded view of the front lower control arm and related components

- Lower control arm clamp and torque the bolts to 60 ft. lbs. (81 Nm)
- Lower control arm to the steering knuckle and torque the bolts to 80 ft. lbs. (108 Nm)
- Front wheel

6. Check and adjust the front toe, if needed.

Rear

1. Before servicing the vehicle, refer to the precautions in the beginning of this section.
2. Remove or disconnect the following:
   - Rear wheel
   - Stabilizer link
   - Wheel Speed Sensor (WSS), if equipped
   - Lower control arm assembly to steering knuckle
   - Lower control arm

To install:

3. Install or connect the following:
   - Lower control arm and torque the bolt to 55 ft. lbs. (74 Nm)
   - Lower control arm to the steering knuckle and torque the bolt to 80 ft. lbs. (108 Nm)
   - WSS, if equipped
   - Stabilizer link and torque the bolt to 29 ft. lbs. (39 Nm)
   - Rear wheel

Wheel Bearings

ADJUSTMENT

Front

To check hub and bearing assembly end-play, remove the caliper and rotor. Position a dial indicator to bear against the hub flange near the center ridge. Wiggle the hub back and forth. If end-play exceeds 0.002 in. (0.05mm), replace the front hub and bearing assembly.

Rear

The rear hub and wheel bearing assembly is designed for the life of the vehicle and requires no type of adjustment or periodic maintenance. The bearing is a sealed unit with the wheel hub and can only be removed and/or replaced as one unit.

The following procedure may be used for evaluation of bearing condition:

1. Raise and safely support the vehicle.
2. Remove the rear wheels and brake drums.

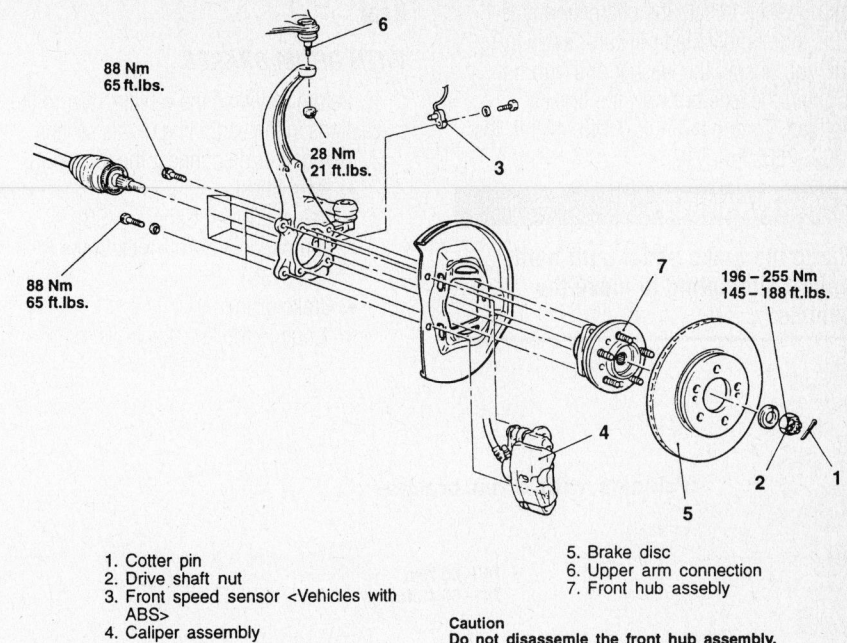

1. Cotter pin
2. Drive shaft nut
3. Front speed sensor <Vehicles with ABS>
4. Caliper assembly
5. Brake disc
6. Upper arm connection
7. Front hub assebly

Caution
Do not disassemle the front hub assembly.

7922CG37

Exploded view of the front hub assembly mounting and related components

3. Turn the hub flange carefully. Excessive roughness, lateral play or resistance to rotation may indicate dirt intrusion or bearing failure.
4. If the rear wheel bearings exhibit the conditions during inspection, the hub and bearing assembly should be replaced.
5. Damaged bearing seals and resulting excessive grease loss may also require bearing replacement. Moderate grease loss from the bearing is considered normal and should not require replacement of the hub and bearing assembly.

REMOVAL & INSTALLATION

Front

1. Before servicing the vehicle, refer to the precautions in the beginning of this section.
2. Remove the cotter pin, halfshaft nut and washer.
3. Remove or disconnect the following:

   - Front wheel
   - Vehicle Speed Sensor (VSS), if equipped
   - Caliper and brake pads; then, support the caliper out of the way using wire
   - Brake rotor from the hub assembly
   - Upper ball joint from the steering knuckle using a press type tool and pull the knuckle outward

✳✳ WARNING

Use of improper methods of joint separation can result in damage to joint, leading to possible failure. Never use wedge-type tools or the ball joint can be damaged.

   - 4 hub-to-steering knuckle bolts
   - Hub and bearing assembly from the knuckle

➡**The hub and wheel bearing assembly is not serviceable and should not be disassembled.**

To install:

4. Install or connect the following:
   - Hub to the steering knuckle and torque the bolts to 65 ft. lbs. (88 Nm)
   - Upper ball joint to the steering knuckle and torque the self-locking nut to 21 ft. lbs. (28 Nm)
5. Position the rotor on the hub. Install a couple of lug nuts and lightly tighten to hold the rotor on the hub.
   - Caliper holder and place the brake pads in the holder. Slide the caliper over the brake pads and install the guide pins. Once the caliper is secured, the lug nuts can be removed.
   - VSS, if equipped
   - Front wheel
   - New cotter pin and bend to secure
6. Examine the driveshaft (halfshaft)

hub washer. Locate the chamfered side. This side is installed outward, away from the hub. Install the washer and hub nut. Tighten the axle nut with the brakes applied. Torque the nut to 145–188 ft. lbs. (200–260 Nm).

✳✳ WARNING

Pump the brake pedal until hard, before attempting to move the vehicle.

Rear

WITH DRUM BRAKES

1. Before servicing the vehicle, refer to the precautions in the beginning of this section.
2. Remove or disconnect the following:
   - Rear wheel
   - Vehicle Speed Sensor (VSS), if equipped with Anti-lock Brake System (ABS)
   - Brake drum
   - 4 hub to the knuckle bolts
   - Hub and bearing assembly from the knuckle

➡**The hub assembly is not serviceable and should not be disassembled.**

3. If replacing the hub, use special Socket MB991248 and a press, to remove the wheel sensor rotor from the hub.

To install:

4. Press the wheel sensor rotor onto the hub.
5. Install or connect the following:

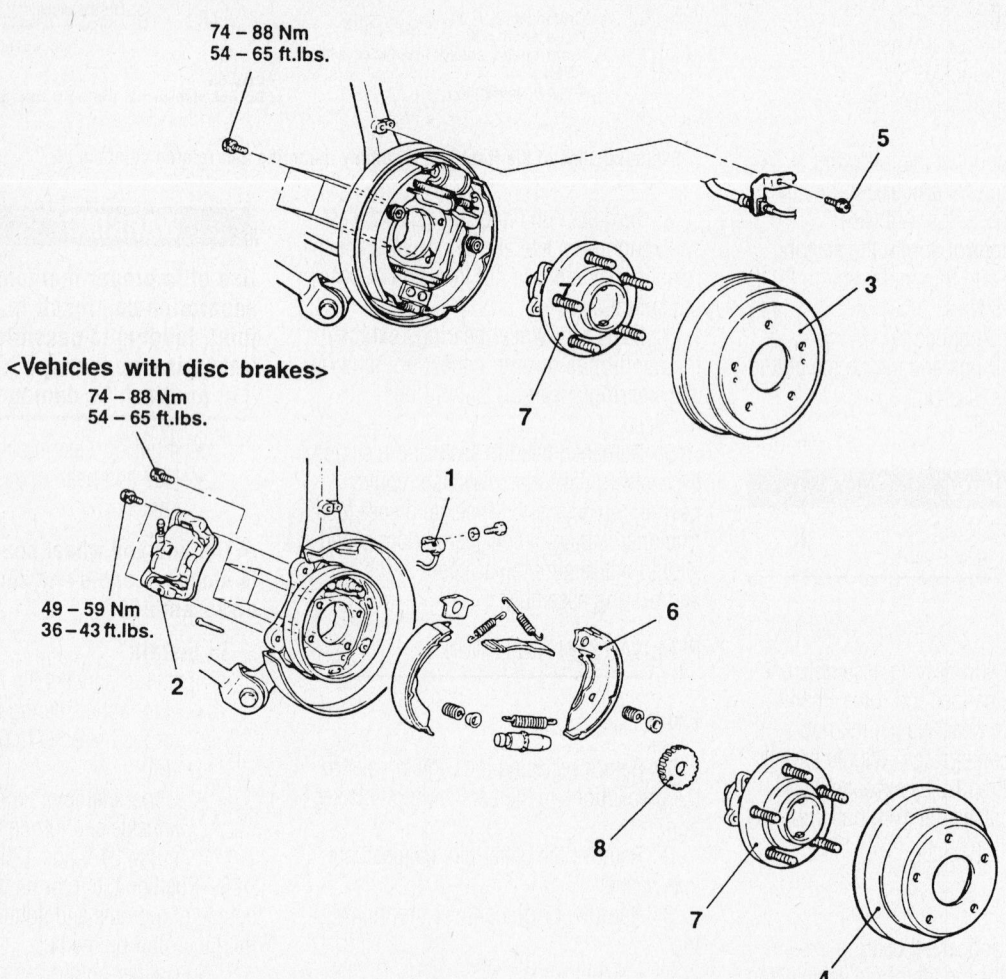

<Vehicles with drum brakes>

74 – 88 Nm
54 – 65 ft.lbs.

<Vehicles with disc brakes>

74 – 88 Nm
54 – 65 ft.lbs.

49 – 59 Nm
36 – 43 ft.lbs.

1. Rear speed sensor <Vehicles with ABS>
2. Caliper assembly
3. Brake drum
4. Brake disc
5. Clip mounting bolt
6. Shoe and lining assembly <Drum in disc brake>
7. Rear hub assembly
8. ABS-rotor <Vehicles with ABS>

Caution
Do not disassemble the rear hub assembly.

7922CG38

Exploded view of the rear hub assembly mounting

- Hub to the knuckle and torque the bolts to 54–65 ft. lbs. (74–88 Nm)
- Brake drum
- VSS, if equipped with ABS
- Rear wheel and lower the vehicle

WITH DISC BRAKES

1. Before servicing the vehicle, refer to the precautions in the beginning of this section.

2. Remove or disconnect the following:

- Rear wheel
- Vehicle Speed Sensor (VSS), if equipped with Anti-lock Brake System (ABS)
- Caliper and brake pads; then, support the caliper out of the way using wire
- Brake rotor

3. Remove the parking brake shoes as follows:

    a. Upper shoe-to-anchor springs.
    b. Lower shoe-to-shoe spring.
    c. Brake shoe hold-down springs.
    d. Parking brake cable from the actuating lever.

4. Remove the 4 hub-to-knuckle bolts.

5. Remove the hub and bearing assembly from the knuckle.

➡ **The hub assembly is not serviceable and should not be disassembled.**

6. If replacing the hub, use special Socket MB991248 and a press, to remove the wheel sensor rotor from the hub.

To install:

7. Press the wheel sensor rotor onto the hub.

8. Install the hub to the knuckle and torque the bolts to 54–65 ft. lbs. (74–88 Nm)

9. Install the parking brake shoes.

10. Position the rotor on the hub. Install a couple of lug nuts and lightly tighten to hold rotor on hub.

11. Install the caliper holder and place brake pads in holder. Slide the caliper over brake pads and install guide pins. Once caliper is secured, lug nuts can be removed.

12. Install the VSS, if equipped

13. Install the rear wheel.

BRAKES

Brake Caliper

REMOVAL & INSTALLATION

Front

1. Before servicing the vehicle, refer to the precautions in the beginning of this section.

2. Remove or disconnect the following:

- About half of the brake fluid from the master cylinder
- Wheel assembly

3. Position a C-clamp, or other suitable tool, over the caliper. Smoothly apply pressure, forcing the caliper piston into the caliper bore until it bottoms. Remove the C-clamp, if used.

- Brake hose attaching bolt and hose from the caliper, if the caliper is to be completely removed from the vehicle. Plug the hose to prevent fluid contamination or loss.
- Caliper mounting bolts and lift the caliper off of the support bracket
- Caliper from the vehicle. If the caliper is only removed for access to other components, support the caliper, with the brake hose attached, so that there is no strain on the brake hose.

To install:

4. Install or connect the following:

- Caliper to the steering knuckle. Lubricate and install the mounting bolts. Torque the bolts to 74 ft. lbs. (100 Nm).
- Brake line hose to the caliper, if removed, and torque the inlet fitting bolt to 22 ft. lbs. (29 Nm)
- Master cylinder with fresh brake fluid and, if the brake hose was removed, bleed the brake system.
- Wheel assembly

5. Depress the brake pedal 3–4 times to seat the brake linings and to restore pressure in the system.

❋❋ CAUTION

Do not move the vehicle until a firm pedal is obtained.

Rear

Unlike many rear disc brake designs, this system does not incorporate the parking brake system, into the rear brake caliper. The rear brake system is serviced the same as the front system.

1. Before servicing the vehicle, refer to the precautions in the beginning of this section.

2. Remove or disconnect the following:

- About half of the brake fluid from the master cylinder
- Wheel assembly

3. Position a C-clamp, or other suitable

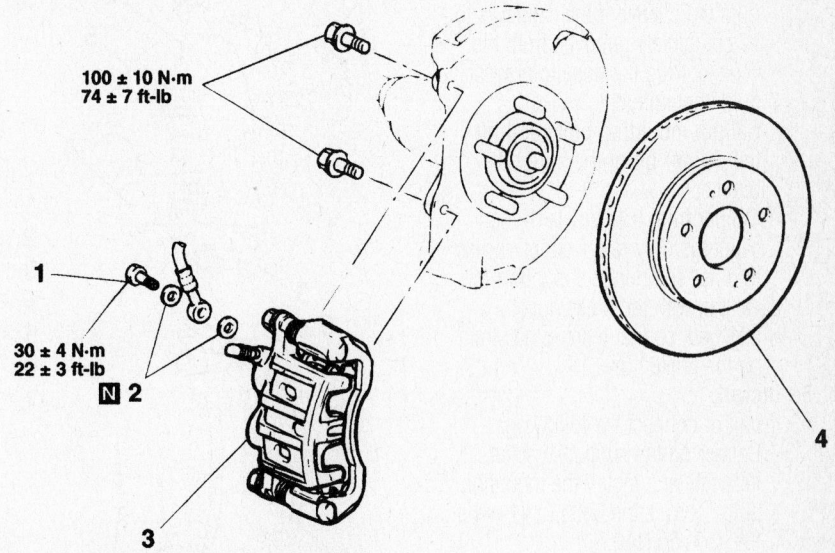

1. **BRAKE HOSE CONNECTOR BOLT**
2. **GASKET**
3. **FRONT BRAKE ASSEMBLY**
4. **BRAKE DISC**

Exploded view of the front disc brake assembly

67189-SEBR-G12

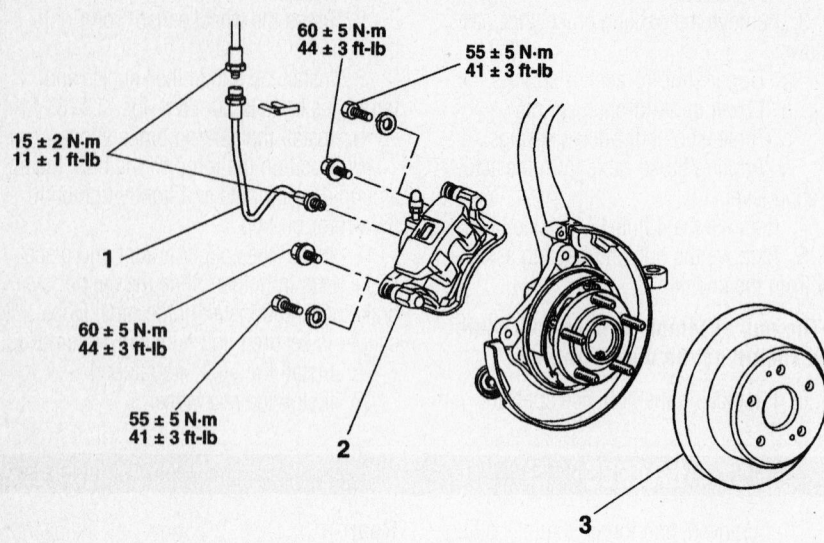

60 ± 5 N·m
44 ± 3 ft-lb

55 ± 5 N·m
41 ± 3 ft-lb

15 ± 2 N·m
11 ± 1 ft-lb

60 ± 5 N·m
44 ± 3 ft-lb

55 ± 5 N·m
41 ± 3 ft-lb

1. BRAKE HOSE
2. REAR BRAKE ASSEMBLY
3. BRAKE DISC

67189-SEBR-G13

Exploded view of the rear brake caliper mounting

tool, over the caliper. Smoothly apply pressure, forcing the caliper piston into the caliper bore until it bottoms. Remove the C-clamp, if used.

- Brake hose attaching bolt and hose from the caliper, if the caliper is to be completely removed from the vehicle. Plug the hose to prevent fluid contamination or loss.
- Caliper mounting bolts and lift the caliper off of the support bracket
- Caliper from the vehicle. If the caliper is only removed for access to other components, support the caliper, with the brake hose attached, so that there is no strain on the brake hose.

To install:

4. Install or connect the following:
- Caliper on the support bracket, lubricate and install the mounting bolts. Torque the bolts to 41–44 ft. lbs. (55–60 Nm).
- Brake line hose to the caliper, if removed, and torque the inlet fitting bolt to 22 ft. lbs. (29 Nm)
- Master cylinder with fresh brake fluid and, if the brake hose was removed, bleed the brake system

- Wheel assembly

5. Depress the brake pedal 3–4 times to seat the brake linings and to restore pressure in the system.

※ CAUTION

Do not move the vehicle until a firm pedal is obtained.

Disc Brake Pads

REMOVAL & INSTALLATION

Front

1. Before servicing the vehicle, refer to the precautions in the beginning of this section.

2. Remove or disconnect the following:

- Some of the brake fluid from the master cylinder reservoir. The reservoir should be no more than ½ full. When the pistons are depressed into the calipers, excess fluid will flow up into the reservoir.
- Tire and wheel
- Caliper guide and lock pins and lift

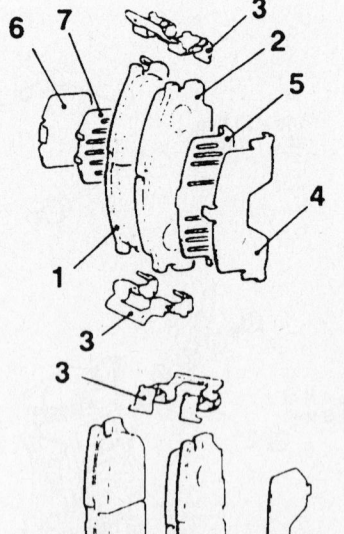

1. Pad & wear indicator assembly
2. Pad assembly
3. Clip
4. Outer shim (stainless)
5. Outer shim (coated with rubber)
6. Inner shim (stainless)
7. Inner shim (coated with rubber)

93006G02

Exploded view of the front brake pads and related components

the caliper assembly from the caliper support. Tie the caliper out of the way using wire. Do not allow the caliper to hang by the brake line.

➡**On some models the caliper can be flipped up by leaving the upper pin in place, using it as a pivot point.**

- Brake pads, spring clip and shims. Take note of positioning to aid installation.

3. Install the wheel lug nuts onto the studs and lightly tighten. This is done to hold the disc on the hub.

To install:

4. Use a large C-clamp to compress the piston(s) back into the caliper bore.

5. Install or connect the following:

- Brake pads, shims and spring clip onto the caliper support, after lubricating the slide points
- Caliper over the brake pads
- Lubricate caliper guide and lock pins in their original positions. Torque guide and locking pins to 74 ft. lbs. (100 Nm).
- Tire and wheel

✳✳ CAUTION

Pump brake pedal several times, until firm, before attempting to move the vehicle.

6. Road test the vehicle and check brakes for proper operation.

Rear

Unlike many rear disc brake designs, this system does not incorporate the parking brake system, into the rear brake caliper, therefore, the rear brake system is serviced the same as the front system.

1. Before servicing the vehicle, refer to the precautions in the beginning of this section.

2. Remove or disconnect the following:

- Some of the brake fluid from the master cylinder reservoir. The reservoir should be no more than ½ full. When the pistons are depressed into the calipers, excess fluid will flow up into the reservoir.
- Tire and wheel
- Caliper guide and lock pins and lift the caliper assembly from the caliper support. Tie the caliper out of the way using wire. Do not allow the caliper to hang by the brake hose.

➡**On some models, the caliper can be flipped up by leaving the upper pin in place, using it as a pivot point.**

- Brake pads, spring clip and shims. Take note of positioning to aid installation.

3. Install the wheel lug nuts onto the studs and lightly tighten. This is done to hold the brake disc on the hub.

To install:

4. Use a large C-clamp to compress the piston(s) back into the caliper bore.

5. Install or connect the following:

- Brake pads, shims and spring clip, after lubricating the slide points, onto the caliper support
- Caliper over the brake pads
- Lubricated caliper guide and lock pins in their original positions. Torque guide and locking pins to 41–44 ft. lbs. (55–60 Nm).
- Tire and wheel

✳✳ CAUTION

Pump brake pedal several times, until firm, before attempting to move the vehicle.

6. Road test the vehicle and check brakes for proper operation.

Brake Drums

REMOVAL & INSTALLATION

1. Before servicing the vehicle, refer to the precautions in the beginning of this section.

2. Remove or disconnect the following:

- Wheel assembly
- Brake drum detent (retaining) screw
- Drum from the axle

3. If difficulty is encountered in removing the drum, perform the following:

a. Verify the parking brake is released.

b. Loosen the parking brake cable.

c. Remove the access hole plug from the backing plate and move the parking brake lever until the lever stop rests on the brake shoe.

d. If necessary, turn the adjuster so that it draws in to allow more clearance between the shoe and drum.

To install:

4. Inspect all parts. Check the drum for cracks or excessive wear.

5. Turn the adjuster until it is drawn all the way in to the stop. Check that the

adjuster turns freely. The nut must NOT lock at the end of the adjuster. Check that the parking brake lever stops are against the edge of the shoe web.

6. Install or connect the following:

- Brake drum and detent screw
- Wheel assembly

7. Apply the foot brake at least 10 times until clicking of the adjustment actuator can no longer be heard. This procedure will automatically adjust the clearance between the shoe and drum.

Brake Shoes

REMOVAL & INSTALLATION

1. Before servicing the vehicle, refer to the precautions in the beginning of this section.

2. Remove or disconnect the following:

- Wheel assembly
- Brake drum detent (retaining) screw
- Drum from the axle

3. If difficulty is encountered in removing the drum:

a. Verify the parking brake is released.

b. Loosen the parking brake cable.

c. Remove the access hole plug from the backing plate and move the parking brake lever until the lever stop rests on the brake shoe.

d. If necessary, turn the adjuster so that it draws in to allow more clearance between the shoe and drum.

➡**Note the location of all springs and clips for proper reassembly.**

4. Remove or disconnect the following:

- Shoe-to-lever spring and the adjuster lever
- Auto adjuster assembly
- Retainer spring
- Hold-down springs, washers and pins
- Shoe-to-shoe spring
- Brake shoes from the backing plate

5. Using a flat-tipped tool, open up the parking brake lever retaining clip.

- Clip and washer from the pin on the shoe assembly and the shoe from the lever assembly.

To install:

6. Thoroughly clean and dry the backing plate. Lubricate the backing plate at the brake shoe contact points.

7. Lubricate backing plate bosses, anchor pin, and parking brake actuating mechanism with a lithium-based grease.

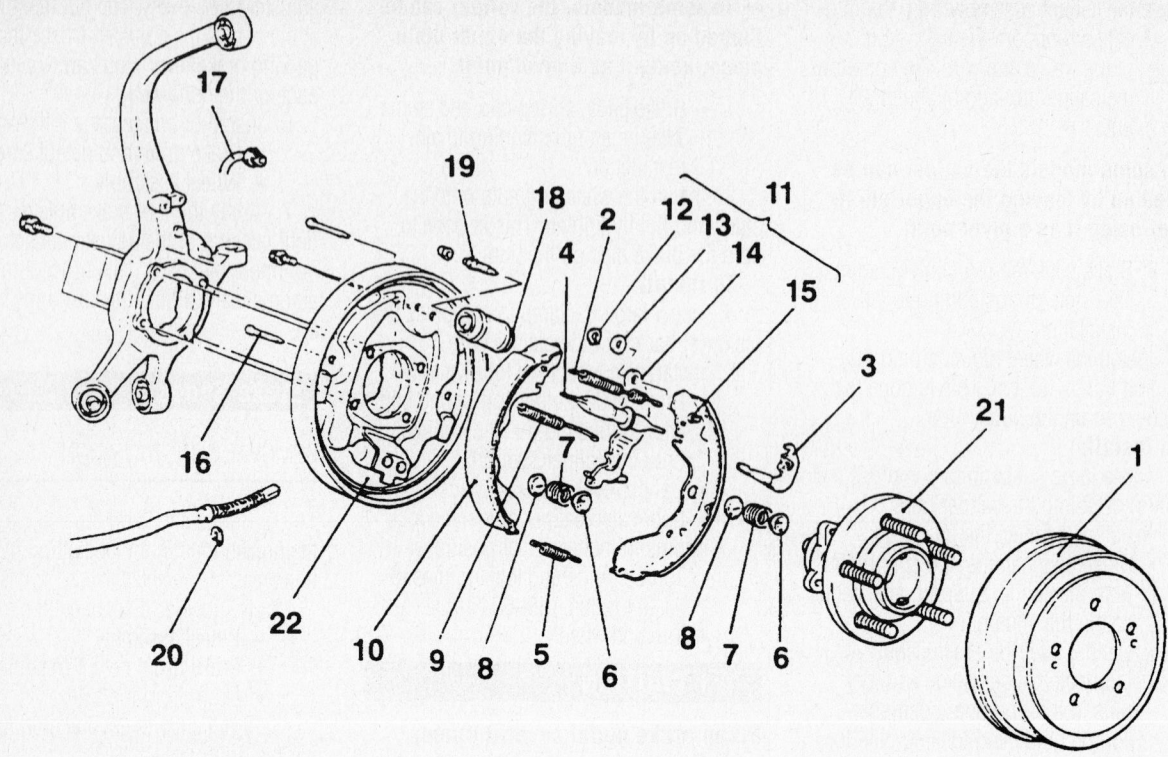

1. Brake drum
2. Shoe-to-lever spring
3. Adjuster lever
4. Auto adjuster assembly
5. Retainer spring
6. Shoe hold-down cup
7. Shoe hold-down spring
8. Shoe hold-down cup
9. Shoe-to-shoe spring
10. Shoe and lining assembly
11. Shoe and lever assembly
12. Retainer
13. Wave washer
14. Parking lever
15. Shoe and lining assembly
16. Shoe hold-down pin
17. Brake pipe connection
18. Wheel cylinder
19. Bleeder screw
20. Snap ring
21. Rear hub assembly
22. Backing plate

93006G10

Exploded view of the drum brake assembly

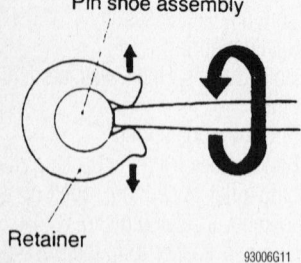

93006G11

Opening the retainer clip

8. Install or connect the following:
- Parking brake lever assembly on the lever pin
- Wave washer and a new retaining clip. Use pliers or the like to install

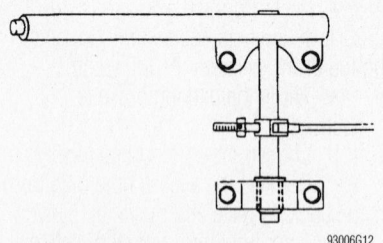

93006G12

Adjusting nut and nut holder

the retainer on the pin. If removed, connect the parking brake lever to the parking brake cable and verify that the cable is properly routed.

9. Clean and lubricate the adjuster assembly. Make sure the nut-adjuster is

drawn all the way to the stop, but the nut must NOT lock firmly at the end of the assembly.
- Brake shoes on the backing plate with the hold-down springs, washers and pins
- Shoe-to-shoe spring
- Retainer spring
- Auto adjuster assembly, the adjuster lever and the shoe-to-lever spring

10. Pre-adjust the shoes so the drum slides on with a light drag and install the brake drum.

11. Adjust the rear brake shoes and install the rear wheels.

12. Adjust the parking brake cable.

13. Check for proper brake operation.

ABS: Anti-lock braking system. An electro-mechanical braking system which is designed to minimize or prevent wheel lock-up during braking.

ABSOLUTE PRESSURE: Atmospheric (barometric) pressure plus the pressure gauge reading.

ACCELERATOR PUMP: A small pump located in the carburetor that feeds fuel into the air/fuel mixture during acceleration.

ACCUMULATOR: A device that controls shift quality by cushioning the shock of hydraulic oil pressure being applied to a clutch or band.

ACTUATING MECHANISM: The mechanical output devices of a hydraulic system, for example, clutch pistons and band servos.

ACTUATOR: The output component of a hydraulic or electronic system.

ADVANCE: Setting the ignition timing so that spark occurs earlier before the piston reaches top dead center (TDC).

ADAPTIVE MEMORY (ADAPTIVE STRATEGY): The learning ability of the TCM or PCM to redefine its decision-making process to provide optimum shift quality.

AFTER TOP DEAD CENTER (ATDC): The point after the piston reaches the top of its travel on the compression stroke.

AIR BAG: Device on the inside of the car designed to inflate on impact of crash, protecting the occupants of the car.

AIR CHARGE TEMPERATURE (ACT) SENSOR: The temperature of the airflow into the engine is measured by an ACT sensor, usually located in the lower intake manifold or air cleaner.

AIR CLEANER: An assembly consisting of a housing, filter and any connecting ductwork. The filter element is made up of a porous paper, sometimes with a wire mesh screening, and is designed to prevent airborne particles from entering the engine through the carburetor or throttle body.

AIR INJECTION: One method of reducing harmful exhaust emissions by injecting air into each of the exhaust ports of an engine. The fresh air entering the hot exhaust manifold causes any remaining fuel to be burned before it can exit the tailpipe.

AIR PUMP: An emission control device that supplies fresh air to the exhaust manifold to aid in more completely burning exhaust gases.

AIR/FUEL RATIO: The ratio of air-to-gasoline by weight in the fuel mixture drawn into the engine.

ALDL (assembly line diagnostic link): Electrical connector for scanning ECM/PCM/TCM input and output devices.

ALIGNMENT RACK: A special drive-on vehicle lift apparatus/measuring device used to adjust a vehicle's toe, caster and camber angles.

ALL WHEEL DRIVE: Term used to describe a full time four wheel drive system or any other vehicle drive system that continuously delivers power to all four wheels. This system is found primarily on station wagon vehicles and SUVs not utilized for significant off road use.

ALTERNATING CURRENT (AC): Electric current that flows first in one direction, then in the opposite direction, continually reversing flow.

ALTERNATOR: A device which produces AC (alternating current) which is converted to DC (direct current) to charge the car battery.

AMMETER: An instrument, calibrated in amperes, used to measure the flow of an electrical current in a circuit. Ammeters are always connected in series with the circuit being tested.

AMPERAGE: The total amount of current (amperes) flowing in a circuit.

AMPLIFIER: A device used in an electrical circuit to increase the voltage of an output signal.

AMP/HR. RATING (BATTERY): Measurement of the ability of a battery to deliver a stated amount of current for a stated period of time. The higher the amp/hr. rating, the better the battery.

AMPERE: The rate of flow of electrical current present when one volt of electrical pressure is applied against one ohm of electrical resistance.

ANALOG COMPUTER: Any microprocessor that uses similar (analogous) electrical signals to make its calculations.

ANODIZED: A special coating applied to the surface of aluminum valves for extended service life.

ANTIFREEZE: A substance (ethylene or propylene glycol) added to the coolant to prevent freezing in cold weather.

ANTI-FOAM AGENTS: Minimize fluid foaming from the whipping action encountered in the converter and planetary action.

ANTI-WEAR AGENTS: Zinc agents that control wear on the gears, bushings, and thrust washers.

ANTI-LOCK BRAKING SYSTEM: A supplementary system to the base hydraulic system that prevents sustained lock-up of the wheels during braking as well as automatically controlling wheel slip.

ANTI-ROLL BAR: See stabilizer bar.

ARC: A flow of electricity through the air between two electrodes or contact points that produces a spark.

ARMATURE: A laminated, soft iron core wrapped by a wire that converts electrical energy to mechanical energy as in a motor or relay. When rotated in a magnetic field, it changes mechanical energy into electrical energy as in a generator.

ATDC: After Top Dead Center.

ATF: Automatic transmission fluid.

ATMOSPHERIC PRESSURE: The pressure on the Earth's surface caused by the weight of the air in the atmosphere. At sea level, this pressure is 14.7 psi at 32°F (101 kPa at 0°C).

ATOMIZATION: The breaking down of a liquid into a fine mist that can be suspended in air.

AUXILIARY ADD-ON COOLER: A supplemental transmission fluid cooling device that is installed in series with the heat exchanger (cooler), located inside the radiator, to provide additional support to cool the hot fluid leaving the torque converter.

AUXILIARY PRESSURE: An added fluid pressure that is introduced into a regulator or balanced valve system to control valve movement. The auxiliary pressure itself can be either a fixed or a variable value. (See balanced valve; regulator valve.)

AWD: All wheel drive.

AXIAL FORCE: A side or end thrust force acting in or along the same plane as the power flow.

AXIAL PLAY: Movement parallel to a shaft or bearing bore.

AXLE CAPACITY: The maximum load-carrying capacity of the axle itself, as specified by the manufacturer. This is usually a higher number than the GAWR.

AXLE RATIO: This is a number (3.07:1, 4.56:1, for example) expressing the ratio between driveshaft revolutions and wheel revolutions. A low numerical ratio allows the engine to work easier because it doesn't have to turn as fast. A high numerical ratio means that the engine has to turn more rpm's to move the wheels through the same number of turns.

BACKFIRE: The sudden combustion of gases in the intake or exhaust system that results in a loud explosion.

BACKLASH: The clearance or play between two parts, such as meshed gears.

BACKPRESSURE: Restrictions in the exhaust system that slow the exit of exhaust gases from the combustion chamber.

BAKELITE®: A heat resistant, plastic insulator material commonly used in printed circuit boards and transistorized components.

BALANCED VALVE: A valve that is positioned by opposing auxiliary hydraulic pressures and/or spring force. Examples include mainline regulator, throttle, and governor valves. (See regulator valve.)

BAND: A flexible ring of steel with an inner lining of friction material. When tightened around the outside of a drum, a planetary member is held stationary to the transmission/transaxle case.

BALL BEARING: A bearing made up of hardened inner and outer races between which hardened steel balls roll.

BALL JOINT: A ball and matching socket connecting suspension components (steering knuckle to lower control arms). It permits rotating movement in any direction between the components that are joined.

BARO (BAROMETRIC PRESSURE SENSOR): Measures the change in the intake manifold pressure caused by changes in altitude.

BAROMETRIC MANIFOLD ABSOLUTE PRESSURE (BMAP) SENSOR: Operates similarly to a conventional MAP sensor; reads intake manifold pressure and is also responsible for determining altitude and barometric pressure prior to engine operation.

BAROMETRIC PRESSURE: (See atmospheric pressure.)

BALLAST RESISTOR: A resistor in the primary ignition circuit that lowers voltage after the engine is started to reduce wear on ignition components.

BATTERY: A direct current electrical storage unit, consisting of the basic active materials of lead and sulfuric acid, which converts chemical energy into electrical energy. Used to provide current for the operation of the starter as well as other equipment, such as the radio, lighting, etc.

BEAD: The portion of a tire that holds it on the rim.

BEARING: A friction reducing, supportive device usually located between a stationary part and a moving part.

BEFORE TOP DEAD CENTER (BTDC): The point just before the piston reaches the top of its travel on the compression stroke.

BELTED TIRE: Tire construction similar to bias-ply tires, but using two or more layers of reinforced belts between body plies and the tread.

BEZEL: Piece of metal surrounding radio, headlights, gauges or similar components; sometimes used to hold the glass face of a gauge in the dash.

BIAS-PLY TIRE: Tire construction, using body ply reinforcing cords which run at alternating angles to the center line of the tread.

BI-METAL TEMPERATURE SENSOR: Any sensor or switch made of two dissimilar types of metal that bend when heated or cooled due to the different expansion rates of the alloys. These types of sensors usually function as an on/off switch.

BLOCK: See Engine Block.

BLOW-BY: Combustion gases, composed of water vapor and unburned fuel, that leak past the piston rings into the crankcase during normal engine operation. These gases are removed by the PCV system to prevent the buildup of harmful acids in the crankcase.

BOOK TIME: See Labor Time.

BOOK VALUE: The average value of a car, widely used to determine trade-in and resale value.

BOOST VALVE: Used at the base of the regulator valve to increase mainline pressure.

BORE: Diameter of a cylinder.

BRAKE CALIPER: The housing that fits over the brake disc. The caliper holds the brake pads, which are pressed against the discs by the caliper pistons when the brake pedal is depressed.

BRAKE HORSEPOWER (BHP): The actual horsepower available at the engine flywheel as measured by a dynamometer.

BRAKE FADE: Loss of braking power, usually caused by excessive heat after repeated brake applications.

BRAKE HORSEPOWER: Usable horsepower of an engine measured at the crankshaft.

BRAKE PAD: A brake shoe and lining assembly used with disc brakes.

BRAKE PROPORTIONING VALVE: A valve on the master cylinder which restricts hydraulic brake pressure to the wheels to a specified amount, preventing wheel lock-up.

BREAKAWAY: Often used by Chrysler to identify first-gear operation in D and 2 ranges. In these ranges, first-gear operation depends on a one-way roller clutch that holds on acceleration and releases (breaks away) on deceleration, resulting in a freewheeling coast-down condition.

BRAKE SHOE: The backing for the brake lining. The term is, however, usually applied to the assembly of the brake backing and lining.

BREAKER POINTS: A set of points inside the distributor, operated by a cam, which make and break the ignition circuit.

BRINNELLING: A wear pattern identified by a series of indentations at regular intervals. This condition is caused by a lack of lube, overload situations, and/or vibrations.

BTDC: Before Top Dead Center.

BUMP: Sudden and forceful apply of a clutch or band.

BUSHING: A liner, usually removable, for a bearing; an anti-friction liner used in place of a bearing.

CALIFORNIA ENGINE: An engine certified by the EPA for use in California only; conforms to more stringent emission regulations than Federal engine.

CALIPER: A hydraulically activated device in a disc brake system, which is mounted straddling the brake rotor (disc). The caliper contains at least one piston and two brake pads. Hydraulic pressure on the piston(s) forces the pads against the rotor.

CAPACITY: The quantity of electricity that can be delivered from a unit, as from a battery in ampere-hours, or output, as from a generator.

CAMBER: One of the factors of wheel alignment. Viewed from the front of the car, it is the inward or outward tilt of the wheel. The top of the tire will lean outward (positive camber) or inward (negative camber).

CAMSHAFT: A shaft in the engine on which are the lobes (cams) which operate the valves. The camshaft is driven by the crankshaft, via a belt, chain or gears, at one half the crankshaft speed.

CAPACITOR: A device which stores an electrical charge.

CARBON MONOXIDE (CO): A colorless, odorless gas given off as a normal byproduct of combustion. It is poisonous and extremely dangerous in confined areas, building up slowly to toxic levels without warning if adequate ventilation is not available.

CARBURETOR: A device, usually mounted on the intake manifold of an engine, which mixes the air and fuel in the proper proportion to allow even combustion.

CASTER: The forward or rearward tilt of an imaginary line drawn through the upper ball joint and the center of the wheel. Viewed from the sides, positive caster (forward tilt) lends directional stability, while negative caster (rearward tilt) produces instability.

CATALYTIC CONVERTER: A device installed in the exhaust system, like a muffler, that converts harmful byproducts of combustion into carbon dioxide and water vapor by means of a heat-producing chemical reaction.

CENTRIFUGAL ADVANCE: A mechanical method of advancing the spark timing by using flyweights in the distributor that react to centrifugal force generated by the distributor shaft rotation.

CENTRIFUGAL FORCE: The outward pull of a revolving object, away from the center of revolution. Centrifugal force increases with the speed of rotation.

CETANE RATING: A measure of the ignition value of diesel fuel. The higher the cetane rating, the better the fuel. Diesel fuel cetane rating is roughly comparable to gasoline octane rating.

CHECK VALVE: Any one-way valve installed to permit the flow of air, fuel or vacuum in one direction only.

CHOKE: The valve/plate that restricts the amount of air entering an engine on the induction stroke, thereby enriching the air/fuel ratio.

CHUGGLE: Bucking or jerking condition that may be engine related and may be most noticeable when converter clutch is engaged; similar to the feel of towing a trailer.

CIRCLIP: A split steel snapring that fits into a groove to hold various parts in place.

CIRCUIT BREAKER: A switch which protects an electrical circuit from overload by opening the circuit when the current flow exceeds a pre-determined level. Some circuit breakers must be reset manually, while most reset automatically.

CIRCUIT: Any unbroken path through which an electrical current can flow. Also used to describe fuel flow in some instances.

CIRCUIT, BYPASS: Another circuit in parallel with the major circuit through which power is diverted.

CIRCUIT, CLOSED: An electrical circuit in which there is no interruption of current flow.

CIRCUIT, GROUND: The non-insulated portion of a complete circuit used as a common potential point. In automotive circuits, the ground is composed of metal parts, such as the engine, body sheet metal, and frame and is usually a negative potential.

CIRCUIT, HOT: That portion of a circuit not at ground potential. The hot circuit is usually insulated and is connected to the positive side of the battery.

CIRCUIT, OPEN: A break or lack of contact in an electrical circuit, either intentional (switch) or unintentional (bad connection or broken wire).

CIRCUIT, PARALLEL: A circuit having two or more paths for current flow with common positive and negative tie points. The same voltage is applied to each load device or parallel branch.

CIRCUIT, SERIES: An electrical system in which separate parts are connected end to end, using one wire, to form a single path for current to flow.

CIRCUIT, SHORT: A circuit that is accidentally completed in an electrical path for which it was not intended.

CLAMPING (ISOLATION) DIODES: Diodes positioned in a circuit to prevent self-induction from damaging electronic components.

CLEARCOAT: A transparent layer which, when sprayed over a vehicle's paint job, adds gloss and depth as well as an additional protective coating to the finish.

CLUTCH: Part of the power train used to connect/disconnect power to the rear wheels.

CLUTCH, FLUID: The same as a fluid coupling. A fluid clutch or coupling performs the same function as a friction clutch by utilizing fluid friction and inertia as opposed to solid friction used by a friction clutch. (See fluid coupling.)

CLUTCH, FRICTION: A coupling device that provides a means of smooth and positive engagement and disengagement of engine torque to the vehicle powertrain. Transmission of power through the clutch is accomplished by bringing one or more rotating drive members into contact with complementing driven members.

COAST: Vehicle deceleration caused by engine braking conditions.

COEFFICIENT OF FRICTION: The amount of surface tension between two contacting surfaces; identified by a scientifically calculated number.

COIL: Part of the ignition system that boosts the relatively low voltage supplied by the car's electrical system to the high voltage required to fire the spark plugs.

COMBINATION MANIFOLD: An assembly which includes both the intake and exhaust manifolds in one casting.

COMBINATION VALVE: A device used in some fuel systems that routes fuel vapors to a charcoal storage canister instead of venting them into the atmosphere. The valve relieves fuel tank pressure and allows fresh air into the tank as the fuel level drops to prevent a vapor lock situation.

COMBUSTION CHAMBER: The part of the engine in the cylinder head where combustion takes place.

COMPOUND GEAR: A gear consisting of two or more simple gears with a common shaft.

COMPOUND PLANETARY: A gearset that has more than the three elements found in a simple gearset and is constructed by combining members of two planetary gearsets to create additional gear ratio possibilities.

COMPRESSION CHECK: A test involving removing each spark plug and inserting a gauge. When the engine is cranked, the gauge will record a pressure reading in the individual cylinder. General operating condition can be determined from a compression check.

COMPRESSION RATIO: The ratio of the volume between the piston and cylinder head when the piston is at the bottom of its stroke (bottom dead center) and when the piston is at the top of its stroke (top dead center).

COMPUTER: An electronic control module that correlates input data according to prearranged engineered instructions; used for the management of an actuator system or systems.

CONDENSER: An electrical device which acts to store an electrical charge, preventing voltage surges.

2. A radiator-like device in the air conditioning system in which refrigerant gas condenses into a liquid, giving off heat.

CONDUCTOR: Any material through which an electrical current can be transmitted easily.

CONNECTING ROD: The connecting link between the crankshaft and piston.

CONSTANT VELOCITY JOINT: Type of universal joint in a halfshaft assembly in which the output shaft turns at a constant angular velocity without variation, provided that the speed of the input shaft is constant.

CONTINUITY: Continuous or complete circuit. Can be checked with an ohmmeter.

CONTROL ARM: The upper or lower suspension components which are mounted on the frame and support the ball joints and steering knuckles.

CONVENTIONAL IGNITION: Ignition system which uses breaker points.

CONVERTER: (See torque converter.)

CONVERTER LOCKUP: The switching from hydrodynamic to direct mechanical drive, usually through the application of a friction element called the converter clutch.

COOLANT: Mixture of water and anti-freeze circulated through the engine to carry off heat produced by the engine.

CORROSION INHIBITOR: An inhibitor in ATF that prevents corrosion of bushings, thrust washers, and oil cooler brazed joints.

COUNTERSHAFT: An intermediate shaft which is rotated by a mainshaft and transmits, in turn, that rotation to a working part.

COUPLING PHASE: Occurs when the torque converter is operating at its greatest hydraulic efficiency. The speed differential between the impeller and the turbine is at its minimum. At this point, the stator freewheels, and there is no torque multiplication.

CRANKCASE: The lower part of an engine in which the crankshaft and related parts operate.

CRANKSHAFT: Engine component (connected to pistons by connecting rods) which converts the reciprocating (up and down) motion of pistons to rotary motion used to turn the driveshaft.

CURB WEIGHT: The weight of a vehicle without passengers or payload, but including all fluids (oil, gas, coolant, etc.) and other equipment specified as standard.

CURRENT: The flow (or rate) of electrons moving through a circuit. Current is measured in amperes (amp).

CURRENT FLOW CONVENTIONAL: Current flows through a circuit from the positive terminal of the source to the negative terminal (plus to minus).

CURRENT FLOW, ELECTRON: Current or electrons flow from the negative terminal of the source, through the circuit, to the positive terminal (minus to plus).

CV-JOINT: Constant velocity joint.

CYCLIC VIBRATIONS: The off-center movement of a rotating object that is affected by its initial balance, speed of rotation, and working angles.

CYLINDER BLOCK: See engine block.

CYLINDER HEAD: The detachable portion of the engine, usually fastened to the top of the cylinder block and containing all or most of the combustion chambers. On overhead valve engines, it contains the valves and their operating parts. On overhead cam engines, it contains the camshaft as well.

CYLINDER: In an engine, the round hole in the engine block in which the piston(s) ride.

DATA LINK CONNECTOR (DLC): Current acronym/term applied to the federally mandated, diagnostic junction connector that is used to monitor ECM/PC/TCM inputs, processing strategies, and outputs including diagnostic trouble codes (DTCs).

DEAD CENTER: The extreme top or bottom of the piston stroke.

DECELERATION BUMP: When referring to a torque converter clutch in the applied position, a sudden release of the accelerator pedal causes a forceful reversal of power through the drivetrain (engine braking), just prior to the apply plate actually being released.

DELAYED (LATE OR EXTENDED): Condition where shift is expected but does not occur for a period of time, for example, where clutch or band engagement does not occur as quickly as expected during part throttle or wide open throttle apply of accelerator or when manually downshifting to a lower range.

DETENT: A spring-loaded plunger, pin, ball, or pawl used as a holding device on a ratchet wheel or shaft. In automatic transmissions, a detent mechanism is used for locking the manual valve in place.

DETENT DOWNSHIFT: (See kickdown.)

DETERGENT: An additive in engine oil to improve its operating characteristics.

DETONATION: An unwanted explosion of the air/fuel mixture in the combustion chamber caused by excess heat and compression, advanced timing, or an overly lean mixture. Also referred to as "ping".

DEXRON®: A brand of automatic transmission fluid.

DIAGNOSTIC TROUBLE CODES (DTCs): A digital display from the control module memory that identifies the input, processor, or output device circuit that is related to the powertrain emission/driveability malfunction detected. Diagnostic trouble codes can be read by the MIL to flash any codes or by using a handheld scanner.

DIAPHRAGM: A thin, flexible wall separating two cavities, such as in a vacuum advance unit.

DIESELING: The engine continues to run after the car is shut off; caused by fuel continuing to be burned in the combustion chamber.

DIFFERENTIAL: A geared assembly which allows the transmission of motion between drive axles, giving one axle the ability to rotate faster than the other, as in cornering.

DIFFERENTIAL AREAS: When opposing faces of a spool valve are acted upon by the same pressure but their areas differ in size, the face with the larger area produces the differential force and valve movement. (See spool valve.)

DIFFERENTIAL FORCE: (See differential areas)

DIGITAL READOUT: A display of numbers or a combination of numbers and letters.

DIGITAL VOLT OHMMETER: An electronic diagnostic tool used to measure voltage, ohms and amps as well as several other functions, with the readings displayed on a digital screen in tenths, hundredths and thousandths.

DIODE: An electrical device that will allow current to flow in one direction only.

DIRECT CURRENT (DC): Electrical current that flows in one direction only.

DIRECT DRIVE: The gear ratio is 1:1, with no change occurring in the torque and speed input/output relationship.

DISC BRAKE: A hydraulic braking assembly consisting of a brake disc, or rotor, mounted on an axle shaft, and a caliper assembly containing, usually two brake pads which are activated by hydraulic pressure. The pads are forced against the sides of the disc, creating friction which slows the vehicle.

DISPERSANTS: Suspend dirt and prevent sludge buildup in a liquid, such as engine oil.

DOUBLE BUMP (DOUBLE FEEL): Two sudden and forceful applies of a clutch or band.

DISPLACEMENT: The total volume of air that is displaced by all pistons as the engine turns through one complete revolution.

DISTRIBUTOR: A mechanically driven device on an engine which is responsible for electrically firing the spark plug at a pre-determined point of the piston stroke.

DOHC: Double overhead camshaft.

DOUBLE OVERHEAD CAMSHAFT: The engine utilizes two camshafts mounted in one cylinder head. One camshaft operates the exhaust valves, while the other operates the intake valves.

DOWEL PIN: A pin, inserted in mating holes in two different parts allowing those parts to maintain a fixed relationship.

DRIVELINE: The drive connection between the transmission and the drive wheels.

DRIVE TRAIN: The components that transmit the flow of power from the engine to the wheels. The components include the clutch, transmission, driveshafts (or axle shafts in front wheel drive), U-joints and differential.

DRUM BRAKE: A braking system which consists of two brake shoes and one or two wheel cylinders, mounted on a fixed backing plate, and a brake drum, mounted on an axle, which revolves around the assembly.

DRY CHARGED BATTERY: Battery to which electrolyte is added when the battery is placed in service.

DVOM: Digital volt ohmmeter

DWELL: The rate, measured in degrees of shaft rotation, at which an electrical circuit cycles on and off.

DYNAMIC: An application in which there is rotating or reciprocating motion between the parts.

EARLY: Condition where shift occurs before vehicle has reached proper speed, which tends to labor engine after upshift.

EBCM: See Electronic Control Unit (ECU).

ECM: See Electronic Control Unit (ECU).

ECU: Electronic control unit.

ELECTRODE: Conductor (positive or negative) of electric current.

ELECTROLYSIS: A surface etching or bonding of current conducting transmission/transaxle components that may occur when grounding straps are missing or in poor condition.

ELECTROLYTE: A solution of water and sulfuric acid used to activate the battery. Electrolyte is extremely corrosive.

ELECTROMAGNET: A coil that produces a magnetic field when current flows through its windings.

ELECTROMAGNETIC INDUCTION: A method to create (generate) current flow through the use of magnetism.

ELECTROMAGNETISM: The effects surrounding the relationship between electricity and magnetism.

ELECTROMOTIVE FORCE (EMF): The force or pressure (voltage) that causes current movement in an electrical circuit.

ELECTRONIC CONTROL UNIT: A digital computer that controls engine (and sometimes transmission, brake or other vehicle system) functions based on data received from various sensors. Examples used by some manufacturers include Electronic Brake Control Module (EBCM), Engine Control Module (ECM), Powertrain Control Module (PCM) or Vehicle Control Module (VCM).

ELECTRONIC IGNITION: A system in which the timing and firing of the spark plugs is controlled by an electronic control unit, usually called a module. These systems have no points or condenser.

ELECTRONIC PRESSURE CONTROL (EPC) SOLENOID: A specially designed solenoid containing a spool valve and spring assembly to control fluid mainline pressure. A variable current flow, controlled by the ECM/PCM, varies the internal force of the solenoid on the spool valve and resulting mainline pressure. (See variable force solenoid.)

ELECTRONICS: Miniaturized electrical circuits utilizing semiconductors, solid-state devices, and printed circuits. Electronic circuits utilize small amounts of power.

ELECTRONIFICATION: The application of electronic circuitry to a mechanical device. Regarding automatic transmissions, electrification is incorporated into converter clutch lockup, shift scheduling, and line pressure control systems.

ELECTROSTATIC DISCHARGE (ESD): An unwanted, high-voltage electrical current released by an individual who has taken on a static charge of electricity. Electronic components can be easily damaged by ESD.

ELEMENT: A device within a hydrodynamic drive unit designed with a set of blades to direct fluid flow.

ENAMEL: Type of paint that dries to a smooth, glossy finish.

END BUMP (END FEEL OR SLIP BUMP): Firmer feel at end of shift when compared with feel at start of shift.

END-PLAY: The clearance/gap between two components that allows for expansion of the parts as they warm up, to prevent binding and to allow space for lubrication.

ENERGY: The ability or capacity to do work.

ENGINE: The primary motor or power apparatus of a vehicle, which converts liquid or gas fuel into mechanical energy.

ENGINE BLOCK: The basic engine casting containing the cylinders, the crankshaft main bearings, as well as machined surfaces for the mounting of other components such as the cylinder head, oil pan, transmission, etc.

ENGINE BRAKING: Use of engine to slow vehicle by manually downshifting during zero-throttle coast down.

ENGINE CONTROL MODULE (ECM): Manages the engine and incorporates output control over the torque converter clutch solenoid. (Note: Current designation for the ECM in late model vehicles is PCM.)

ENGINE COOLANT TEMPERATURE (ECT) SENSOR: Prevents converter clutch engagement with a cold engine; also used for shift timing and shift quality.

EP LUBRICANT: EP (extreme pressure) lubricants are specially formulated for use with gears involving heavy loads (transmissions, differentials, etc.).

ETHYL: A substance added to gasoline to improve its resistance to knock, by slowing down the rate of combustion.

ETHYLENE GLYCOL: The base substance of antifreeze.

EXHAUST MANIFOLD: A set of cast passages or pipes which conduct exhaust gases from the engine.

FAIL-SAFE (BACKUP) CONTROL: A substitute value used by the PCM/TCM to replace a faulty signal from an input sensor. The temporary value allows the vehicle to continue to be operated.

FAST IDLE: The speed of the engine when the choke is on. Fast idle speeds engine warm-up.

FEDERAL ENGINE: An engine certified by the EPA for use in any of the 49 states (except California).

FEEDBACK: A circuit malfunction whereby current can find another path to feed load devices.

FEELER GAUGE: A blade, usually metal, of precisely predetermined thickness, used to measure the clearance between two parts.

FILAMENT: The part of a bulb that glows; the filament creates high resistance to current flow and actually glows from the resulting heat.

FINAL DRIVE: An essential part of the axle drive assembly where final gear reduction takes place in the powertrain. In RWD applications and north-south FWD applications, it must also change the power flow direction to the axle shaft by ninety degrees. (Also see axle ratio).

FIRING ORDER: The order in which combustion occurs in the cylinders of an engine. Also the order in which spark is distributed to the plugs by the distributor.

FIRM: A noticeable quick apply of a clutch or band that is considered normal with medium to heavy throttle shift; should not be confused with harsh or rough.

FLAME FRONT: The term used to describe certain aspects of the fuel explosion in the cylinders. The flame front should move in a controlled pattern across the cylinder, rather than simply exploding immediately.

FLARE (SLIPPING): A quick increase in engine rpm accompanied by momentary loss of torque; generally occurs during shift.

FLAT ENGINE: Engine design in which the pistons are horizontally opposed. Porsche, Subaru and some old VW are common examples of flat engines.

FLAT RATE: A dealership term referring to the amount of money paid to a technician for a repair or diagnostic service based on that particular service versus dealership's labor time (NOT based on the actual time the technician spent on the job).

FLAT SPOT: A point during acceleration when the engine seems to lose power for an instant.

FLOODING: The presence of too much fuel in the intake manifold and combustion chamber which prevents the air/fuel mixture from firing, thereby causing a no-start situation.

FLUID: A fluid can be either liquid or gas. In hydraulics, a liquid is used for transmitting force or motion.

FLUID COUPLING: The simplest form of hydrodynamic drive, the fluid coupling consists of two look-alike members with straight radial varies referred to as the impeller (pump) and the turbine. Input torque is always equal to the output torque.

FLUID DRIVE: Either a fluid coupling or a fluid torque converter. (See hydrodynamic drive units.)

FLUID TORQUE CONVERTER: A hydrodynamic drive that has the ability to act both as a torque multiplier and fluid coupling. (See hydrodynamic drive units; torque converter.)

FLUID VISCOSITY: The resistance of a liquid to flow. A cold fluid (oil) has greater viscosity and flows more slowly than a hot fluid (oil).

FLYWHEEL: A heavy disc of metal attached to the rear of the crankshaft. It smoothes the firing impulses of the engine and keeps the crankshaft turning during periods when no firing takes place. The starter also engages the flywheel to start the engine.

FOOT POUND (ft. lbs., lbs. ft. or sometimes, ft. lb.): The amount of energy or work needed to raise an item weighing one pound, a distance of one foot.

FREEZE PLUG: A plug in the engine block which will be pushed out if the coolant freezes. Sometimes called expansion plugs, they protect the block from cracking should the coolant freeze.

FRICTION: The resistance that occurs between contacting surfaces. This relationship is expressed by a ratio called the coefficient of friction (CL).

FRICTION, COEFFICIENT OF: The amount of surface tension between two contacting surfaces; expressed by a scientifically calculated number.

FRONT END ALIGNMENT: A service to set caster, camber and toe-in to the correct specifications. This will ensure that the car steers and handles properly and that the tires wear properly.

FRICTION MODIFIER: Changes the coefficient of friction of the fluid between the mating steel and composition clutch/band surfaces during the engagement process and allows for a certain amount of intentional slipping for a good "shift-feel".

FRONTAL AREA: The total frontal area of a vehicle exposed to air flow.

FUEL FILTER: A component of the fuel system containing a porous paper element used to prevent any impurities from entering the engine through the fuel system. It usually takes the form of a canister-like housing, mounted in-line with the fuel hose, located anywhere on a vehicle between the fuel tank and engine.

FUEL INJECTION: A system replacing the carburetor that sprays fuel into the cylinder through nozzles. The amount of fuel can be more precisely controlled with fuel injection.

FULL FLOATING AXLE: An axle in which the axle housing extends through the wheel giving bearing support on the outside of the housing. The front axle of a four-wheel drive vehicle is usually a full floating axle, as are the rear axles of many larger (1 ton and over) pick-ups and vans.

FULL-TIME FOUR-WHEEL DRIVE: A four-wheel drive system that continuously delivers power to all four wheels. A differential between the front and rear driveshafts permits variations in axle speeds to control gear wind-up without damage.

FULL THROTTLE DETENT DOWNSHIFT: A quick apply of accelerator pedal to its full travel, forcing a downshift.

FUSE: A protective device in a circuit which prevents circuit overload by breaking the circuit when a specific amperage is present. The device is constructed around a strip or wire of a lower amperage rating than the circuit it is designed to protect. When an amperage higher than that stamped on the fuse is present in the circuit, the strip or wire melts, opening the circuit.

FUSIBLE LINK: A piece of wire in a wiring harness that performs the same job as a fuse. If overloaded, the fusible link will melt and interrupt the circuit.

FWD: Front wheel drive.

GAWR: (Gross axle weight rating) the total maximum weight an axle is designed to carry.

GCW: (Gross combined weight) total combined weight of a tow vehicle and trailer.

GARAGE SHIFT: initial engagement feel of transmission, neutral to reverse or neutral to a forward drive.

GARAGE SHIFT FEEL: A quick check of the engagement quality and responsiveness of reverse and forward gears. This test is done with the vehicle stationary.

GEAR: A toothed mechanical device that acts as a rotating lever to transmit power or turning effort from one shaft to another. (See gear ratio.)

GEAR RATIO: A ratio expressing the number of turns a smaller gear will make to turn a larger gear through one revolution. The ratio is found by dividing the number of teeth on the smaller gear into the number of teeth on the larger gear.

GEARBOX: Transmission

GEAR REDUCTION: Torque is multiplied and speed decreased by the factor of the gear ratio. For example, a 3:1 gear ratio changes an input torque of 180 ft. lbs. and an input speed of 2700 rpm to 540 Ft. lbs. and 900 rpm, respectively. (No account is taken of frictional losses, which are always present.)

GEARTRAIN: A succession of intermeshing gears that form an assembly and provide for one or more torque changes as the power input is transmitted to the power output.

GEL COAT: A thin coat of plastic resin covering fiberglass body panels.

GENERATOR: A device which produces direct current (DC) necessary to charge the battery.

GOVERNOR: A device that senses vehicle speed and generates a hydraulic oil pressure. As vehicle speed increases, governor oil pressure rises.

GROUND CIRCUIT: (See circuit, ground.)

GROUND SIDE SWITCHING: The electrical/electronic circuit control switch is located after the circuit load.

GVWR: (Gross vehicle weight rating) total maximum weight a vehicle is designed to carry including the weight of the vehicle, passengers, equipment, gas, oil, etc.

HALOGEN: A special type of lamp known for its quality of brilliant white light. Originally used for fog lights and driving lights.

HARD CODES: DTCs that are present at the time of testing; also called continuous or current codes.

HARSH(ROUGH): An apply of a clutch or band that is more noticeable than a firm one; considered undesirable at any throttle position.

HEADER TANK: An expansion tank for the radiator coolant. It can be located remotely or built into the radiator.

HEAT RANGE: A term used to describe the ability of a spark plug to carry away heat. Plugs with longer nosed insulators take longer to carry heat off effectively.

HEAT RISER: A flapper in the exhaust manifold that is closed when the engine is cold, causing hot exhaust gases to heat the intake manifold providing better cold engine operation. A thermostatic spring opens the flapper when the engine warms up.

HEAVY THROTTLE: Approximately three-fourths of accelerator pedal travel.

HEMI: A name given an engine using hemispherical combustion chambers.

HERTZ (HZ): The international unit of frequency equal to one cycle per second (10,000 Hertz equals 10,000 cycles per second).

HIGH-IMPEDANCE DVOM (DIGITAL VOLT-OHMMETER): This styled device provides a built-in resistance value and is capable of limiting circuit current flow to safe milliamp levels.

HIGH RESISTANCE: Often refers to a circuit where there is an excessive amount of opposition to normal current flow.

HORSEPOWER: A measurement of the amount of work; one horsepower is the amount of work necessary to lift 33,000 lbs. one foot in one minute. Brake horsepower (bhp) is the horsepower delivered by an engine on a dynamometer. Net horsepower is the power remaining (measured at the flywheel of the engine) that can be used to turn the wheels after power is consumed through friction and running the engine accessories (water pump, alternator, air pump, fan etc.)

HOT CIRCUIT: (See circuit, hot; hot lead.)

HOT LEAD: A wire or conductor in the power side of the circuit. (See circuit, hot.)

HOT SIDE SWITCHING: The electrical/electronic circuit control switch is located before the circuit load.

HUB: The center part of a wheel or gear.

HUNTING (BUSYNESS): Repeating quick series of up-shifts and downshifts that causes noticeable change in engine rpm, for example, as in a 4-3-4 shift pattern.

HYDRAULICS: The use of liquid under pressure to transfer force of motion.

HYDROCARBON (HC): Any chemical compound made up of hydrogen and carbon. A major pollutant formed by the engine as a by-product of combustion.

HYDRODYNAMIC DRIVE UNITS: Devices that transmit power solely by the action of a kinetic fluid flow in a closed recirculating path. An impeller energizes the fluid and discharges the high-speed jet stream into the turbine for power output.

HYDROMETER: An instrument used to measure the specific gravity of a solution.

HYDROPLANING: A phenomenon of driving when water builds up under the tire tread, causing it to lose contact with the road. Slowing down will usually restore normal tire contact with the road.

HYPOID GEARSET: The drive pinion gear may be placed below or above the centerline of the driven gear; often used as a final drive gearset.

IDLE MIXTURE: The mixture of air and fuel (usually about 14:1) being fed to the cylinders. The idle mixture screw(s) are sometimes adjusted as part of a tune-up.

IDLER ARM: Component of the steering linkage which is a geometric duplicate of the steering gear arm. It supports the right side of the center steering link.

IMPELLER: Often called a pump, the impeller is the power input (drive) member of a hydrodynamic drive. As part of the torque converter cover, it acts as a centrifugal pump and puts the fluid in motion.

INCH POUND (inch lbs.; sometimes in. lb. or in. lbs.): One twelfth of a foot pound.

INDUCTANCE: The force that produces voltage when a conductor is passed through a magnetic field.

INDUCTION: A means of transferring electrical energy in the form of a magnetic field. Principle used in the ignition coil to increase voltage.

INITIAL FEEL: A distinct firmer feel at start of shift when compared with feel at finish of shift.

INJECTOR: A device which receives metered fuel under relatively low pressure and is activated to inject the fuel into the engine under relatively high pressure at a predetermined time.

INPUT: In an automatic transmission, the source of power from the engine is absorbed by the torque converter, which provides the power input into the transmission. The turbine drives the input(turbine)shaft.

INPUT SHAFT: The shaft to which torque is applied, usually carrying the driving gear or gears.

INTAKE MANIFOLD: A casting of passages or pipes used to conduct air or a fuel/air mixture to the cylinders.

INTERNAL GEAR: The ring-like outer gear of a planetary gearset with the gear teeth cut on the inside of the ring to provide a mesh with the planet pinions.

ISOLATION (CLAMPING) DIODES: Diodes positioned in a circuit to prevent self-induction from damaging electronic components.

IX ROTARY GEAR PUMP: Contains two rotating members, one shaped with internal gear teeth and the other with external gear teeth. As the gears separate, the fluid fills the gaps between gear teeth, is pulled across a crescent-shaped divider, and then is forced to flow through the outlet as the gears mesh.

IX ROTARY LOBE PUMP: Sometimes referred to as a gerotor type pump. Two rotating members, one shaped with internal lobes and the other with external lobes, separate and then mesh to cause fluid to flow.

JOURNAL: The bearing surface within which a shaft operates.

JUMPER CABLES: Two heavy duty wires with large alligator clips used to provide power from a charged battery to a discharged battery mounted in a vehicle.

JUMPSTART: Utilizing the sufficiently charged battery of one vehicle to start the engine of another vehicle with a discharged battery by the use of jumper cables.

KEY: A small block usually fitted in a notch between a shaft and a hub to prevent slippage of the two parts.

KICKDOWN: Detent downshift system; either linkage, cable, or electrically controlled.

KILO: A prefix used in the metric system to indicate one thousand.

KNOCK: Noise which results from the spontaneous ignition of a portion of the air-fuel mixture in the engine cylinder caused by overly advanced ignition timing or use of incorrectly low octane fuel for that engine.

KNOCK SENSOR: An input device that responds to spark knock, caused by over advanced ignition timing.

LABOR TIME: A specific amount of time required to perform a certain repair or diagnostic service as defined by a vehicle or after-market manufacturer.

LACQUER: A quick-drying automotive paint.

LATE: Shift that occurs when engine is at higher than normal rpm for given amount of throttle.

LIGHT-EMITTING DIODE (LED): A semiconductor diode that emits light as electrical current flows through it; used in some electronic display devices to emit a red or other color light.

LIGHT THROTTLE: Approximately one-fourth of accelerator pedal travel.

LIMITED SLIP: A type of differential which transfers driving force to the wheel with the best traction.

LIMP-IN MODE: Electrical shutdown of the transmission/ transaxle output solenoids, allowing only forward and reverse gears that are hydraulically energized by the manual valve. This permits the vehicle to be driven to a service facility for repair.

LIP SEAL: Molded synthetic rubber seal designed with an outer sealing edge (lip) that points into the fluid containing area to be sealed. This type of seal is used where rotational and axial forces are present.

LITHIUM-BASE GREASE: Chassis and wheel bearing grease using lithium as a base. Not compatible with sodium-base grease.

LOAD DEVICE: A circuit's resistance that converts the electrical energy into light, sound, heat, or mechanical movement.

LOAD RANGE: Indicates the number of plies at which a tire is rated. Load range B equals four-ply rating; C equals six-ply rating; and, D equals an eight-ply rating.

LOAD TORQUE: The amount of output torque needed from the transmission/transaxle to overcome the vehicle load.

LOCKING HUBS: Accessories used on part-time four-wheel drive systems that allow the front wheels to be disengaged from the drive train when four-wheel drive is not being used. When four-wheel drive is desired, the hubs are engaged, locking the wheels to the drive train.

LOCKUP CONVERTER: A torque converter that operates hydraulically and mechanically. When an internal apply plate (lockup plate) clamps to the torque converter cover, hydraulic slippage is eliminated.

LOCK RING: See Circlip or Snapring

MAGNET: Any body with the property of attracting iron or steel.

MAGNETIC FIELD: The area surrounding the poles of a magnet that is affected by its attraction or repulsion forces.

MAIN LINE PRESSURE: Often called control pressure or line pressure, it refers to the pressure of the oil leaving the pump and is controlled by the pressure regulator valve.

MALFUNCTION INDICATOR LAMP (MIL): Previously known as a check engine light, the dash-mounted MIL illuminates and signals the driver that an emission or driveability problem with the powertrain has been detected by the ECM/PCM. When this occurs, at least one diagnostic trouble code (DTC) has been stored into the control module memory.

MANIFOLD ABSOLUTE PRESSURE (MAP) SENSOR: Reads the amount of air pressure (vacuum) in the engine's intake manifold system; its signal is used to analyze engine load conditions.

MANIFOLD VACUUM: Low pressure in an engine intake manifold formed just below the throttle plates. Manifold vacuum is highest at idle and drops under acceleration.

MANIFOLD: A casting of passages or set of pipes which connect the cylinders to an inlet or outlet source.

MANUAL LEVER POSITION SWITCH (MLPS): A mechanical switching unit that is typically mounted externally to the transmission/transaxle to inform the PCM/ECM which gear range the driver has selected.

MANUAL VALVE: Located inside the transmission/transaxle, it is directly connected to the driver's shift lever. The position of the manual valve determines which hydraulic circuits will be charged with oil pressure and the operating mode of the transmission.

MANUAL VALVE LEVER POSITION SENSOR (MVLPS): The input from this device tells the TCM what gear range was selected.

MASS AIR FLOW (MAF) SENSOR: Measures the airflow into the engine.

MASTER CYLINDER: The primary fluid pressurizing device in a hydraulic system. In automotive use, it is found in brake and hydraulic clutch systems and is pedal activated, either directly or, in a power brake system, through the power booster.

MacPherson STRUT: A suspension component combining a shock absorber and spring in one unit.

MEDIUM THROTTLE: Approximately one-half of accelerator pedal travel.

MEGA: A metric prefix indicating one million.

MEMBER: An independent component of a hydrodynamic unit such as an impeller, a stator, or a turbine. It may have one or more elements.

MERCON: A fluid developed by Ford Motor Company in 1988. It contains a friction modifier and closely resembles operating characteristics of Dexron.

METAL SEALING RINGS: Made from cast iron or aluminum, their primary application is with dynamic components involving pressure sealing circuits of rotating members. These rings are designed with either butt or hook lock end joints.

METER (ANALOG): A linear-style meter representing data as lengths; a needle-style instrument interfacing with logical numerical increments. This style of electrical meter uses relatively low impedance internal resistance and cannot be used for testing electronic circuitry.

METER (DIGITAL): Uses numbers as a direct readout to show values. Most meters of this style use high impedance internal resistance and must be used for testing low current electronic circuitry.

MICRO: A metric prefix indicating one-millionth (0.000001).

MILLI: A metric prefix indicating one-thousandth (0.001).

MINIMUM THROTTLE: The least amount of throttle opening required for upshift; normally close to zero throttle.

MISFIRE: Condition occurring when the fuel mixture in a cylinder fails to ignite, causing the engine to run roughly.

MODULE: Electronic control unit, amplifier or igniter of solid state or integrated design which controls the current flow in the ignition primary circuit based on input from the pick-up coil. When the module opens the primary circuit, high secondary voltage is induced in the coil.

MODULATED: In an electronic-hydraulic converter clutch system (or shift valve system), the term modulated refers to the pulsing of a solenoid, at a variable rate. This action controls the buildup of oil pressure in the hydraulic circuit to allow a controlled amount of clutch slippage.

MODULATED CONVERTER CLUTCH CONTROL (MCCC): A pulse width duty cycle valve that controls the converter lockup apply pressure and maximizes smoother transitions between lock and unlock conditions.

MODULATOR PRESSURE (THROTTLE PRESSURE): A hydraulic signal oil pressure relating to the amount of engine load, based on either the amount of throttle plate opening or engine vacuum.

MODULATOR VALVE: A regulator valve that is controlled by engine vacuum, providing a hydraulic pressure that varies in relation to engine torque. The hydraulic torque signal functions to delay the shift pattern and provide a line pressure boost. (See throttle valve.)

MOTOR: An electromagnetic device used to convert electrical energy into mechanical energy.

MULTIPLE-DISC CLUTCH: A grouping of steel and friction lined plates that, when compressed together by hydraulic pressure acting upon a piston, lock or unlock a planetary member.

MULTI-WEIGHT: Type of oil that provides adequate lubrication at both high and low temperatures.

needed to move one amp through a resistance of one ohm.

MUSHY: Same as soft; slow and drawn out clutch apply with very little shift feel.

MUTUAL INDUCTION: The generation of current from one wire circuit to another by movement of the magnetic field surrounding a current-carrying circuit as its ampere flow increases or decreases.

NEEDLE BEARING: A bearing which consists of a number (usually a large number) of long, thin rollers.

NITROGEN OXIDE (NOx): One of the three basic pollutants found in the exhaust emission of an internal combustion engine. The amount of NOx usually varies in an inverse proportion to the amount of HC and CO.

NONPOSITIVE SEALING: A sealing method that allows some minor leakage, which normally assists in lubrication.

02 SENSOR: Located in the engine's exhaust system, it is an input device to the ECM/PCM for managing the fuel delivery and ignition system. A scanner can be used to observe the fluctuating voltage readings produced by an 02 sensor as the oxygen content of the exhaust is analyzed.

O-RING SEAL: Molded synthetic rubber seal designed with a circular cross-section. This type of seal is used primarily in static applications.

OBD II (ON-BOARD DIAGNOSTICS, SECOND GENERATION): Refers to the federal law mandating tighter control of 1996 and newer vehicle emissions, active monitoring of related devices, and standardization of terminology, data link connectors, and other technician concerns.

OCTANE RATING: A number, indicating the quality of gasoline based on its ability to resist knock. The higher the number, the better the quality. Higher compression engines require higher octane gas.

OEM: Original Equipment Manufactured. OEM equipment is that furnished standard by the manufacturer.

OFFSET: The distance between the vertical center of the wheel and the mounting surface at the lugs. Offset is positive if the center is outside the lug circle; negative offset puts the center line inside the lug circle.

OHM'S LAW: A law of electricity that states the relationship between voltage, current, and resistance. Volts = amperes x ohms

OHM: The unit used to measure the resistance of conductor-to-electrical flow. One ohm is the amount of resistance that limits current flow to one ampere in a circuit with one volt of pressure.

OHMMETER: An instrument used for measuring the resistance, in ohms, in an electrical circuit.

ONE-WAY CLUTCH: A mechanical clutch of roller or sprag design that resists torque or transmits power in one direction only. It is used to either hold or drive a planetary member.

ONE-WAY ROLLER CLUTCH: A mechanical device that transmits or holds torque in one direction only.

OPEN CIRCUIT: A break or lack of contact in an electrical circuit, either intentional (switch) or unintentional (bad connection or broken wire).

ORIFICE: Located in hydraulic oil circuits, it acts as a restriction. It slows down fluid flow to either create back pressure or delay pressure buildup downstream.

OSCILLOSCOPE: A piece of test equipment that shows electric impulses as a pattern on a screen. Engine performance can be analyzed by interpreting these patterns.

OUTPUT SHAFT: The shaft which transmits torque from a device, such as a transmission.

OUTPUT SPEED SENSOR (OSS): Identifies transmission/transaxle output shaft speed for shift timing and may be used to calculate TCC slip; often functions as the VSS (vehicle speed sensor).

OVERDRIVE: (1.) A device attached to or incorporated in a transmission/transaxle that allows the engine to turn less than one full revolution for every complete revolution of the wheels. The net effect is to reduce engine rpm, thereby using less fuel. A typical overdrive gear ratio would be .87:1, instead of the normal 1:1 in high gear. (2.) A gear assembly which produces more shaft revolutions than that transmitted to it.

OVERDRIVE PLANETARY GEARSET: A single planetary gearset designed to provide a direct drive and overdrive ratio. When coupled to a three-speed transmission/transaxle configuration, a four-speed/overdrive unit is present.

OVERHEAD CAMSHAFT (OHC): An engine configuration in which the camshaft is mounted on top of the cylinder head and operates the valve either directly or by means of rocker arms.

OVERHEAD VALVE (OHV): An engine configuration in which all of the valves are located in the cylinder head and the camshaft is located in the cylinder block. The camshaft operates the valves via lifters and pushrods.

OVERRUNCLUTCH: Another name for a one-way mechanical clutch. Applies to both roller and sprag designs.

OVERSTEER: The tendency of some vehicles, when steering into a turn, to over-respond or steer more than required, which could result in excessive slip of the rear wheels. Opposite of under-steer.

OXIDATION STABILIZERS: Absorb and dissipate heat. Automatic transmission fluid has high resistance to varnish and sludge buildup that occurs from excessive heat that is generated primarily in the torque converter. Local temperatures as high as 6000F (3150C) can occur at the clutch plates during engagement, and this heat must be absorbed and dissipated. If the fluid cannot withstand the heat, it burns or oxidizes, resulting in an almost immediate destruction of friction materials, clogged filter screen and hydraulic passages, and sticky valves.

OXIDES OF NITROGEN: See nitrogen oxide (NOx).

OXYGEN SENSOR: Used with a feedback system to sense the presence of oxygen in the exhaust gas and signal the computer which can use the voltage signal to determine engine operating efficiency and adjust the air/fuel ratio.

PARALLEL CIRCUIT: (See circuit, parallel.)

PARTS WASHER: A basin or tub, usually with a built-in pump mechanism and hose used for circulating chemical solvent for the purpose of cleaning greasy, oily and dirty components.

PART-TIME FOUR WHEEL DRIVE: A system that is normally in the two wheel drive mode and only runs in four-wheel drive when the system is manually engaged because more traction is desired. Two or four wheel drive is normally selected by a lever to engage the front axle, but if locking hubs are used, these must also be manually engaged in the Lock position. Otherwise, the front axle will not drive the front wheels.

PASSIVE RESTRAINT: Safety systems such as air bags or automatic seat belts which operate with no action required on the part of the driver or passenger. Mandated by Federal regulations on all vehicles sold in the U.S. after 1990.

PAYLOAD: The weight the vehicle is capable of carrying in addition to its own weight. Payload includes weight of the driver, passengers and cargo, but not coolant, fuel, lubricant, spare tire, etc.

PCM: Powertrain control module.

PCV VALVE: A valve usually located in the rocker cover that vents crankcase vapors back into the engine to be reburned.

PERCOLATION: A condition in which the fuel actually "boils," due to excessive heat. Percolation prevents proper atomization of the fuel causing rough running.

PICK-UP COIL: The coil in which voltage is induced in an electronic ignition.

PING: A metallic rattling sound produced by the engine during acceleration. It is usually due to incorrect ignition timing or a poor grade of gasoline.

PINION: The smaller of two gears. The rear axle pinion drives the ring gear which transmits motion to the axle shafts.

PINION GEAR: The smallest gear in a drive gear assembly.

PISTON: A disc or cup that fits in a cylinder bore and is free to move. In hydraulics, it provides the means of converting hydraulic pressure into a usable force. Examples of piston applications are found in servo, clutch, and accumulator units.

PISTON RING: An open-ended ring which fits into a groove on the outer diameter of the piston. Its chief function is to form a seal between the piston and cylinder wall. Most automotive pistons have three rings: two for compression sealing; one for oil sealing.

PITMAN ARM: A lever which transmits steering force from the steering gear to the steering linkage.

PLANET CARRIER: A basic member of a planetary gear assembly that carries the pinion gears.

PLANET PINIONS: Gears housed in a planet carrier that are in constant mesh with the sun gear and internal gear. Because they have their own independent rotating centers, the pinions are capable of rotating around the sun gear or the inside of the internal gear.

PLANETARY GEAR RATIO: The reduction or overdrive ratio developed by a planetary gearset.

PLANETARY GEARSET: In its simplest form, it is made up of a basic assembly group containing a sun gear, internal gear, and planet carrier. The gears are always in constant mesh and offer a wide range of gear ratio possibilities.

PLANETARY GEARSET (COMPOUND): Two planetary gearsets combined together.

PLANETARY GEARSET (SIMPLE): An assembly of gears in constant mesh consisting of a sun gear, several pinion gears mounted in a carrier, and a ring gear. It provides gear ratio and direction changes, in addition to a direct drive and a neutral.

PLY RATING: A. rating given a tire which indicates strength (but not necessarily actual plies). A two-ply/four-ply rating has only two plies, but the strength of a four-ply tire.

POLARITY: Indication (positive or negative) of the two poles of a battery.

PORT: An opening for fluid intake or exhaust.

POSITIVE SEALING: A sealing method that completely prevents leakage.

POTENTIAL: Electrical force measured in volts; sometimes used interchangeably with voltage.

POWER: The ability to do work per unit of time, as expressed in horsepower; one horsepower equals 33,000 ft. lbs. of work per minute, or 550 ft. lbs. of work per second.

POWER FLOW: The systematic flow or transmission of power through the gears, from the input shaft to the output shaft.

POWER-TO-WEIGHT RATIO: Ratio of horsepower to weight of car.

POWERTRAIN: See Drivetrain.

POWERTRAIN CONTROL MODULE (PCM): Current designation for the engine control module (ECM). In many cases, late model vehicle control units manage the engine as well as the transmission. In other settings, the PCM controls the engine and is interfaced with a TCM to control transmission functions.

Ppm: Parts per million; unit used to measure exhaust emissions.

PREIGNITION: Early ignition of fuel in the cylinder, sometimes due to glowing carbon deposits in the combustion chamber. Preignition can be damaging since combustion takes place prematurely.

PRELOAD: A predetermined load placed on a bearing during assembly or by adjustment.

PRESS FIT: The mating of two parts under pressure, due to the inner diameter of one being smaller than the outer diameter of the other, or vice versa; an interference fit.

PRESSURE: The amount of force exerted upon a surface area.

PRESSURE CONTROL SOLENOID (PCS): An output device that provides a boost oil pressure to the mainline regulator valve to control line pressure. Its operation is determined by the amount of current sent from the PCM.

PRESSURE GAUGE: An instrument used for measuring the fluid pressure in a hydraulic circuit.

PRESSURE REGULATOR VALVE: In automatic transmissions, its purpose is to regulate the pressure of the pump output and supply the basic fluid pressure necessary to operate the transmission. The regulated fluid pressure may be referred to as mainline pressure, line pressure, or control pressure.

PRESSURE SWITCH ASSEMBLY (PSA): Mounted inside the transmission, it is a grouping of oil pressure switches that inputs to the PCM when certain hydraulic passages are charged with oil pressure.

PRESSURE PLATE: A spring-loaded plate (part of the clutch) that transmits power to the driven (friction) plate when the clutch is engaged.

PRIMARY CIRCUIT: The low voltage side of the ignition system which consists of the ignition switch, ballast resistor or resistance wire, bypass, coil, electronic control unit and pick-up coil as well as the connecting wires and harnesses.

PROFILE: Term used for tire measurement (tire series), which is the ratio of tire height to tread width.

PROM (PROGRAMMABLE READ-ONLY MEMORY): The heart of the computer that compares input data and makes the engineered program or strategy decisions about when to trigger the appropriate output based on stored computer instructions.

PULSE GENERATOR: A two-wire pickup sensor used to produce a fluctuating electrical signal. This changing signal is read by the controller to determine the speed of the object and can be used to measure transmission/transaxle input speed, output speed, and vehicle speed.

PSI: Pounds per square inch; a measurement of pressure.

PULSE WIDTH DUTY CYCLE SOLENOID (PULSE WIDTH MODULATED SOLENOID): A computer-controlled solenoid that turns on and off at a variable rate producing a modulated oil pressure; often referred to as a pulse width modulated (PWM) solenoid. Employed in many electronic automatic transmissions and transaxles, these solenoids are used to manage shift control and converter clutch hydraulic circuits.

PUSHROD: A steel rod between the hydraulic valve lifter and the valve rocker arm in overhead valve (OHV) engines.

PUMP: A mechanical device designed to create fluid flow and pressure buildup in a hydraulic system.

QUARTER PANEL: General term used to refer to a rear fender. Quarter panel is the area from the rear door opening to the tail light area and from rear wheel well to the base of the trunk and roof-line.

RACE: The surface on the inner or outer ring of a bearing on which the balls, needles or rollers move.

RACK AND PINION: A type of automotive steering system using a pinion gear attached to the end of the steering shaft. The pinion meshes with a long rack attached to the steering linkage.

RADIAL TIRE: Tire design which uses body cords running at right angles to the center line of the tire. Two or more belts are used to give tread strength. Radials can be identified by their characteristic sidewall bulge.

RADIATOR: Part of the cooling system for a water-cooled engine, mounted in the front of the vehicle and connected to the engine with rubber hoses. Through the radiator, excess combustion heat is dissipated into the atmosphere through forced convection using a water and glycol based mixture that circulates through, and cools, the engine.

RANGE REFERENCE AND CLUTCH/BAND APPLY CHART: A guide that shows the application of clutches and bands for each gear, within the selector range positions. These charts are extremely useful for understanding how the unit operates and for diagnosing malfunctions.

RAVIGNEAUX GEARSET: A compound planetary gearset that features matched dual planetary pinions (sets of two) mounted in a single planet carrier. Two sun gears and one ring mesh with the carrier pinions.

REACTION MEMBER: The stationary planetary member, in a planetary gearset, that is grounded to the transmission/transaxle case through the use of friction and wedging devices known as bands, disc clutches, and one-way clutches.

REACTION PRESSURE: The fluid pressure that moves a spool valve against an opposing force or forces; the area on which the opposing force acts. The opposing force can be a spring or a combination of spring force and auxiliary hydraulic force.

REACTOR, TORQUE CONVERTER: The reaction member of a fluid torque converter, more commonly called a stator. (See stator.)

REAR MAIN OIL SEAL: A synthetic or rope-type seal that prevents oil from leaking out of the engine past the rear main crankshaft bearing.

RECIRCULATING BALL: Type of steering system in which recirculating steel balls occupy the area between the nut and worm wheel, causing a reduction in friction.

RECTIFIER: A device (used primarily in alternators) that permits electrical current to flow in one direction only.

REDUCTION: (See gear reduction.)

REGULATOR VALVE: A valve that changes the pressure of the oil in a hydraulic circuit as the oil passes through the valve by bleeding off (or exhausting) some of the volume of oil supplied to the valve.

REFRIGERANT 12 (R-12) or 134 (R-134): The generic name of the refrigerant used in automotive air conditioning systems.

REGULATOR: A device which maintains the amperage and/or voltage levels of a circuit at predetermined values.

RELAY: A switch which automatically opens and/or closes a circuit.

RELAY VALVE: A valve that directs flow and pressure. Relay valves simply connect or disconnect interrelated passages without restricting the fluid flow or changing the pressure.

RELIEF VALVE: A spring-loaded, pressure-operated valve that limits oil pressure buildup in a hydraulic circuit to a predetermined maximum value.

RELUCTOR: A wheel that rotates inside the distributor and triggers the release of voltage in an electronic ignition.

RESERVOIR: The storage area for fluid in a hydraulic system; often called a sump.

RESIN: A liquid plastic used in body work.

RESIDUAL MAGNETISM: The magnetic strength stored in a material after a magnetizing field has been removed.

RESISTANCE: The opposition to the flow of current through a circuit or electrical device, and is measured in ohms. Resistance is equal to the voltage divided by the amperage.

RESISTOR SPARK PLUG: A spark plug using a resistor to shorten the spark duration. This suppresses radio interference and lengthens plug life.

RESISTOR: A device, usually made of wire, which offers a preset amount of resistance in an electrical circuit.

RESULTANT FORCE: The single effective directional thrust of the fluid force on the turbine produced by the vortex and rotary forces acting in different planes.

RETARD: Set the ignition timing so that spark occurs later (fewer degrees before TDC).

RHEOSTAT: A device for regulating a current by means of a variable resistance.

RING GEAR: The name given to a ring-shaped gear attached to a differential case, or affixed to a flywheel or as part of a planetary gear set.

ROADLOAD: grade.

ROCKER ARM: A lever which rotates around a shaft pushing down (opening) the valve with an end when the other end is pushed up by the pushrod. Spring pressure will later close the valve.

ROCKER PANEL: The body panel below the doors between the wheel opening.

ROLLER BEARING: A bearing made up of hardened inner and outer races between which hardened steel rollers move.

ROLLER CLUTCH: A type of one-way clutch design using rollers and springs mounted within an inner and outer cam race assembly.

ROTARY FLOW: The path of the fluid trapped between the blades of the members as they revolve with the rotation of the torque converter cover (rotational inertia).

ROTOR: (1.) The disc-shaped part of a disc brake assembly, upon which the brake pads bear; also called, brake disc. (2.) The device mounted atop the distributor shaft, which passes current to the distributor cap tower contacts.

ROTARY ENGINE: See Wankel engine.

RPM: Revolutions per minute (usually indicates engine speed).

RTV: A gasket making compound that cures as it is exposed to the atmosphere. It is used between surfaces that are not perfectly machined to one another, leaving a slight gap that the RTV fills and in which it hardens. The letters RTV represent room temperature vulcanizing.

RUN-ON: Condition when the engine continues to run, even when the key is turned off. See dieseling.

SEALED BEAM: A automotive headlight. The lens, reflector and filament from a single unit.

SEATBELT INTERLOCK: A system whereby the car cannot be started unless the seatbelt is buckled.

SECONDARY CIRCUIT: The high voltage side of the ignition system, usually above 20,000 volts. The secondary includes the ignition coil, coil wire, distributor cap and rotor, spark plug wires and spark plugs.

SELF-INDUCTION: The generation of voltage in a current-carrying wire by changing the amount of current flowing within that wire.

SEMI-CONDUCTOR: A material (silicon or germanium) that is neither a good conductor nor an insulator; used in diodes and transistors.

SEMI-FLOATING AXLE: In this design, a wheel is attached to the axle shaft, which takes both drive and cornering loads. Almost all solid axle passenger cars and light trucks use this design.

SENDING UNIT: A mechanical, electrical, hydraulic or electromagnetic device which transmits information to a gauge.

SENSOR: Any device designed to measure engine operating conditions or ambient pressures and temperatures. Usually electronic in nature and designed to send a voltage signal to an on-board computer, some sensors may operate as a simple on/off switch or they may provide a variable voltage signal (like a potentiometer) as conditions or measured parameters change.

SERIES CIRCUIT: (See circuit, series.)

SERPENTINE BELT: An accessory drive belt, with small multiple v-ribs, routed around most or all of the engine-powered accessories such as the alternator and power steering pump. Usually both the front and the back side of the belt comes into contact with various pulleys.

SERVO: In an automatic transmission, it is a piston in a cylinder assembly that converts hydraulic pressure into mechanical force and movement; used for the application of the bands and clutches.

SHIFT BUSYNESS: When referring to a torque converter clutch, it is the frequent apply and release of the clutch plate due to uncommon driving conditions.

SHIFT VALVE: Classified as a relay valve, it triggers the automatic shift in response to a governor and a throttle signal by directing fluid to the appropriate band and clutch apply combination to cause the shift to occur.

SHIM: Spacers of precise, predetermined thickness used between parts to establish a proper working relationship.

SHIMMY: Vibration (sometimes violent) in the front end caused by misaligned front end, out of balance tires or worn suspension components.

SHORT CIRCUIT: An electrical malfunction where current takes the path of least resistance to ground (usually through damaged insulation). Current flow is excessive from low resistance resulting in a blown fuse.

SHUDDER: Repeated jerking or stick-slip sensation, similar to chuggle but more severe and rapid in nature, that may be most noticeable during certain ranges of vehicle speed; also used to define condition after converter clutch engagement.

SIMPSON GEARSET: A compound planetary gear train that integrates two simple planetary gearsets referred to as the front planetary and the rear planetary.

SINGLE OVERHEAD CAMSHAFT: See overhead camshaft.

SKIDPLATE: A metal plate attached to the underside of the body to protect the fuel tank, transfer case or other vulnerable parts from damage.

SLAVE CYLINDER: In automotive use, a device in the hydraulic clutch system which is activated by hydraulic force, disengaging the clutch.

SLIPPING: Noticeable increase in engine rpm without vehicle speed increase; usually occurs during or after initial clutch or band engagement.

SLUDGE: Thick, black deposits in engine formed from dirt, oil, water, etc. It is usually formed in engines when oil changes are neglected.

SNAP RING: A circular retaining clip used inside or outside a shaft or part to secure a shaft, such as a floating wrist pin.

SOFT: Slow, almost unnoticeable clutch apply with very little shift feel.

SOFTCODES: DTCs that have been set into the PCM memory but are not present at the time of testing; often referred to as history or intermittent codes.

SOHC: Single overhead camshaft.

SOLENOID: An electrically operated, magnetic switching device.

SPALLING: A wear pattern identified by metal chips flaking off the hardened surface. This condition is caused by foreign particles, overloading situations, and/or normal wear.

SPARK PLUG: A device screwed into the combustion chamber of a spark ignition engine. The basic construction is a conductive core inside of a ceramic insulator, mounted in an outer conductive base. An electrical charge from the spark plug wire travels along the conductive core and jumps a preset air gap to a grounding point or points at the end of the conductive base. The resultant spark ignites the fuel/air mixture in the combustion chamber.

SPECIFIC GRAVITY (BATTERY): The relative weight of liquid (battery electrolyte) as compared to the weight of an equal volume of water.

SPLINES: Ridges machined or cast onto the outer diameter of a shaft or inner diameter of a bore to enable parts to mate without rotation.

SPLIT TORQUE DRIVE: In a torque converter, it refers to parallel paths of torque transmission, one of which is mechanical and the other hydraulic.

SPONGY PEDAL: A soft or spongy feeling when the brake pedal is depressed. It is usually due to air in the brake lines.

SPOOLVALVE: A precision-machined, cylindrically shaped valve made up of lands and grooves. Depending on its position in the valve bore, various interconnecting hydraulic circuit passages are either opened or closed.

SPRAG CLUTCH: A type of one-way clutch design using cams or contoured-shaped sprags between inner and outer races. (See one-way clutch.)

SPRUNG WEIGHT: The weight of a car supported by the springs.

SQUARE-CUT SEAL: Molded synthetic rubber seal designed with a square- or rectangular-shaped cross-section. This type of seal is used for both dynamic and static applications.

SRS: Supplemental restraint system

STABILIZER (SWAY) BAR: A bar linking both sides of the suspension. It resists sway on turns by taking some of added load from one wheel and putting it on the other.

STAGE: The number of turbine sets separated by a stator. A turbine set may be made up of one or more turbine members. A three-element converter is classified as a single stage.

STALL: In fluid drive transmission/transaxle applications, stall refers to engine rpm with the transmission/transaxle engaged and the vehicle stationary; throttle valve can be in any position between closed and wide open.

STALL SPEED: In fluid drive transmission/transaxle applications, stall speed refers to the maximum engine rpm with the transmission/transaxle engaged and vehicle stationary, when the throttle valve is wide open. (See stall; stall test.)

STALL TEST: A procedure recommended by many manufacturers to help determine the integrity of an engine, the torque converter stator, and certain clutch and band combinations. With the shift lever in each of the forward and reverse positions and with the brakes firmly applied, the accelerator pedal is momentarily pressed to the wide open throttle (WOT) position. The engine rpm reading at full throttle can provide clues for diagnosing the condition of the items listed above.

STALL TORQUE: The maximum design or engineered torque ratio of a fluid torque converter, produced under stall speed conditions. (See stall speed.)

STARTER: A high-torque electric motor used for the purpose of starting the engine, typically through a high ratio geared drive connected to the flywheel ring gear.

STATIC: A sealing application in which the parts being sealed do not move in relation to each other.

STATOR (REACTOR): The reaction member of a fluid torque converter that changes the direction of the fluid as it leaves the turbine to enter the impeller vanes. During the torque multiplication phase, this action assists the impeller's rotary force and results in an increase in torque.

STEERING GEOMETRY: Combination of various angles of suspension components (caster, camber, toe-in); roughly equivalent to front end alignment.

STRAIGHT WEIGHT: Term designating motor oil as suitable for use within a narrow range of temperatures. Outside the narrow temperature range its flow characteristics will not adequately lubricate.

STROKE: The distance the piston travels from bottom dead center to top dead center.

SUBSTITUTION: Replacing one part suspected of a defect with a like part of known quality.

SUMP: The storage vessel or reservoir that provides a ready source of fluid to the pump. In an automatic transmission, the sump is the oil pan. All fluid eventually returns to the sump for recycling into the hydraulic system.

SUN GEAR: In a planetary gearset, it is the center gear that meshes with a cluster of planet pinions.

SUPERCHARGER: An air pump driven mechanically by the engine through belts, chains, shafts or gears from the crankshaft. Two general types of supercharger are the positive displacement and centrifugal type, which pump air in direct relationship to the speed of the engine.

SUPPLEMENTAL RESTRAINT SYSTEM: See air bag.

SURGE: Repeating engine-related feeling of acceleration and deceleration that is less intense than chuggle.

SWITCH: A device used to open, close, or redirect the current in an electrical circuit.

SYNCHROMESH: A manual transmission/transaxle that is equipped with devices (synchronizers) that match the gear speeds so that the transmission/transaxle can be downshifted without clashing gears.

SYNTHETIC OIL: Non-petroleum based oil.

TACHOMETER: A device used to measure the rotary speed of an engine, shaft, gear, etc., usually in rotations per minute.

TDC: Top dead center. The exact top of the piston's stroke.

TEFLON SEALING RINGS: Teflon is a soft, durable, plastic-like material that is resistant to heat and provides excellent sealing. These rings are designed with either scarf-cut joints or as one-piece rings. Teflon sealing rings have replaced many metal ring applications.

TERMINAL: A device attached to the end of a wire or cable to make an electrical connection.

TEST LIGHT, CIRCUIT-POWERED: Uses available circuit voltage to test circuit continuity.

TEST LIGHT, SELF-POWERED: Uses its own battery source to test circuit continuity.

THERMISTOR: A special resistor used to measure fluid temperature; it decreases its resistance with increases in temperature.

THERMOSTAT: A valve, located in the cooling system of an engine, which is closed when cold and opens gradually in response to engine heating, controlling the temperature of the coolant and rate of coolant flow.

THERMOSTATIC ELEMENT: A heat-sensitive, spring-type device that controls a drain port from the upper sump area to the lower sump. When the transaxle fluid reaches operating temperature, the port is closed and the upper sump fills, thus reducing the fluid level in the lower sump.

THROTTLE POSITION (TP) SENSOR: Reads the degree of throttle opening; its signal is used to analyze engine load conditions. The ECM/PCM decides to apply the TCC, or to disengage it for coast or load conditions that need a converter torque boost.

THROTTLE PRESSURE/MODULATOR PRESSURE: A hydraulic signal oil pressure relating to the amount of engine load, based on either the amount of throttle plate opening or engine vacuum.

THROTTLE VALVE: A regulating or balanced valve that is controlled mechanically by throttle linkage or engine vacuum. It sends a hydraulic signal to the shift valve body to control shift timing and shift quality. (See balanced valve; modulator valve.)

THROW-OUT BEARING: As the clutch pedal is depressed, the throwout bearing moves against the spring fingers of the pressure plate, forcing the pressure plate to disengage from the driven disc.

TIE ROD: A rod connecting the steering arms. Tie rods have threaded ends that are used to adjust toe-in.

TIE-UP: Condition where two opposing clutches are attempting to apply at same time, causing engine to labor with noticeable loss of engine rpm.

TIMING BELT: A square-toothed, reinforced rubber belt that is driven by the crankshaft and operates the camshaft.

TIMING CHAIN: A roller chain that is driven by the crankshaft and operates the camshaft.

TIRE ROTATION: Moving the tires from one position to another to make the tires wear evenly.

TOE-IN (OUT): A term comparing the extreme front and rear of the front tires. Closer together at the front is toe-in; farther apart at the front is toe-out.

TOP DEAD CENTER (TDC): The point at which the piston reaches the top of its travel on the compression stroke.

TORQUE: Measurement of turning or twisting force, expressed as foot-pounds or inch-pounds.

TORQUE CONVERTER: A turbine used to transmit power from a driving member to a driven member via hydraulic action, providing changes in drive ratio and torque. In automotive use, it links the driveplate at the rear of the engine to the automatic transmission.

TORQUE CONVERTER CLUTCH: The apply plate (lockup plate) assembly used for mechanical power flow through the converter.

TORQUE PHASE: Sometimes referred to as slip phase or stall phase, torque multiplication occurs when the turbine is turning at a slower speed than the impeller, and the stator is reactionary (stationary). This sequence generates a boost in output torque.

TORQUE RATING (STALL TORQUE): The maximum torque multiplication that occurs during stall conditions, with the engine at wide open throttle (WOT) and zero turbine speed.

TORQUE RATIO: An expression of the gear ratio factor on torque effect. A 3:1 gear ratio or 3:1 torque ratio increases the torque input by the ratio factor of 3. Input torque (100 ft. lbs.) x 3 = output torque (300 ft. lbs.)

TRACTION: The amount of usable tractive effort before the drive wheels slip on the road contact surface.

TORSION BAR SUSPENSION: Long rods of spring steel which take the place of springs. One end of the bar is anchored and the other arm (attached to the suspension) is free to twist. The bars' resistance to twisting causes springing action.

TRACK: Distance between the centers of the tires where they contact the ground.

TRACTION CONTROL: A control system that prevents the spinning of a vehicle's drive wheels when excess power is applied.

TRACTIVE EFFORT: The amount of force available to the drive wheels, to move the vehicle.

TRANSAXLE: A single housing containing the transmission and differential. Transaxles are usually found on front engine/front wheel drive or rear engine/rear wheel drive cars.

TRANSDUCER: A device that changes energy from one form to another. For example, a transducer in a microphone changes sound energy to electrical energy. In automotive air-conditioning controls used in automatic temperature systems, a transducer changes an electrical signal to a vacuum signal, which operates mechanical doors.

TRANSMISSION: A powertrain component designed to modify torque and speed developed by the engine; also provides direct drive, reverse, and neutral.

TRANSMISSION CONTROL MODULE (TCM): Manages transmission functions. These vary according to the manufacturer's product design but may include converter clutch operation, electronic shift scheduling, and mainline pressure.

TRANSMISSION FLUID TEMPERATURE (TFT) SENSOR: Originally called a transmission oil temperature (TOT) sensor, this input device to the ECM/PCM senses the fluid temperature and provides a resistance value. It operates on the thermistor principle.

TRANSMISSION INPUT SPEED (TIS) SENSOR: Measures turbine shaft (input shaft) rpm's and compares to engine rpm's to determine torque converter slip. When compared to the transmission output speed sensor or VSS, gear ratio and clutch engagement timing can be determined.

TRANSMISSION OIL TEMPERATURE (TOT) SENSOR: (See transmission fluid temperature (TFT) sensor.)

TRANSMISSION RANGE SELECTOR (TRS) SWITCH: Tells the module which gear shift position the driver has chosen.

TRANSFER CASE: A gearbox driven from the transmission that delivers power to both front and rear driveshafts in a four-wheel drive system. Transfer cases usually have a high and low range set of gears, used depending on how much pulling power is needed.

TRANSISTOR: A semi-conductor component which can be actuated by a small voltage to perform an electrical switching function.

TREAD WEAR INDICATOR: Bars molded into the tire at right angles to the tread that appear as horizontal bars when 1/16 in. of tread remains.

TREAD WEAR PATTERN: The pattern of wear on tires which can be "read" to diagnose problems in the front suspension.

TUNE-UP: A regular maintenance function, usually associated with the replacement and adjustment of parts and components in the electrical and fuel systems of a vehicle for the purpose of attaining optimum performance.

TURBINE: The output (driven) member of a fluid coupling or fluid torque converter. It is splined to the input (turbine) shaft of the transmission.

TURBOCHARGER: An exhaust driven pump which compresses intake air and forces it into the combustion chambers at higher than atmospheric pressures. The increased air pressure allows more fuel to be burned and results in increased horsepower being produced.

TURBULENCE: The interference of molecules of a fluid (or vapor) with each other in a fluid flow.

TYPE F: Transmission fluid developed and used by Ford Motor Company up to 1982. This fluid type provides a high coefficient of friction.

TYPE 7176: The preferred choice of transmission fluid for Chrysler automatic transmissions and transaxles. Developed in 1986, it closely resembles Dexron and Mercon. Type 7176 is the recommended service fill fluid for all Chrysler products utilizing a lockup torque converter dating back to 1978.

U-JOINT (UNIVERSAL JOINT): A flexible coupling in the drive train that allows the driveshafts or axle shafts to operate at different angles and still transmit rotary power.

UNDERSTEER: The tendency of a car to continue straight ahead while negotiating a turn.

UNIT BODY: Design in which the car body acts as the frame.

UNLEADED FUEL: Fuel which contains no lead (a common gasoline additive). The presence of lead in fuel will destroy the functioning elements of a catalytic converter, making it useless.

UNSPRUNG WEIGHT: The weight of car components not supported by the springs (wheels, tires, brakes, rear axle, control arms, etc.).

UPSHIFT: A shift that results in a decrease in torque ratio and an increase in speed.

VACUUM: A negative pressure; any pressure less than atmospheric pressure.

VACUUM ADVANCE: A device which advances the ignition timing in response to increased engine vacuum.

VACUUM GAUGE: An instrument used for measuring the existing vacuum in a vacuum circuit or chamber. The unit of measure is inches (of mercury in a barometer).

VACUUM MODULATOR: Generates a hydraulic oil pressure in response to the amount of engine vacuum.

VALVES: Devices that can open or close fluid passages in a hydraulic system and are used for directing fluid flow and controlling pressure.

VALVE BODY ASSEMBLY: The main hydraulic control assembly of the transmission/transaxle that contains numerous valves, check balls, and other components to control the distribution of pressurized oil throughout the transmission.

VALVE CLEARANCE: The measured gap between the end of the valve stem and the rocker arm, cam lobe or follower that activates the valve.

VALVE GUIDES: The guide through which the stem of the valve passes. The guide is designed to keep the valve in proper alignment.

VALVE LASH (clearance): The operating clearance in the valve train.

VALVE TRAIN: The system that operates intake and exhaust valves, consisting of camshaft, valves and springs, lifters, pushrods and rocker arms.

VAPOR LOCK: Boiling of the fuel in the fuel lines due to excess heat. This will interfere with the flow of fuel in the lines and can completely stop the flow. Vapor lock normally only occurs in hot weather.

VARIABLE DISPLACEMENT (VARIABLE CAPACITY) VANE PUMP: Slipper-type vanes, mounted in a revolving rotor and contained within the bore of a movable slide, capture and then force fluid to flow. Movement of the slide to various positions changes the size of the vane chambers and the amount of fluid flow. **Note:** GM refers to this pump design as variable displacement, and Ford terms it variable capacity.

VARIABLE FORCE SOLENOID (VFS): Commonly referred to as the electronic pressure control (EPC) solenoid, it replaces the cable/linkage style of TV system control and is integrated with a spool valve and spring assembly to control pressure. A variable computer-controlled current flow varies the internal force of the solenoid on the spool valve and resulting control pressure.

VARIABLE ORIFICE THERMAL VALVE: Temperature-sensitive hydraulic oil control device that adjusts the size of a circuit path opening. By altering the size of the opening, the oil flow rate is adapted for cold to hot oil viscosity changes.

VARNISH: Term applied to the residue formed when gasoline gets old and stale.

VCM: See Electronic Control Unit (ECU).

VEHICLE SPEED SENSOR (VSS): Provides an electrical signal to the computer module, measuring vehicle speed, and affects the torque converter clutch engagement and release.

VESPEL SEALING RINGS: Hard plastic material that produces excellent sealing in dynamic settings. These rings are found in late versions of the 4T60 and in all 4T60-E and 4T80-E transaxles.

VISCOSITY: The ability of a fluid to flow. The lower the viscosity rating, the easier the fluid will flow. 10 weight motor oil will flow much easier than 40 weight motor oil.

VISCOSITY INDEX IMPROVERS: Keeps the viscosity nearly constant with changes in temperature. This is especially important at low temperatures, when the oil needs to be thin to aid in shifting and for cold-weather starting. Yet it must not be so thin that at high temperatures it will cause excessive hydraulic leakage so that pumps are unable to maintain the proper pressures.

VISCOUS CLUTCH: A specially designed torque converter clutch apply plate that, through the use of a silicon fluid, clamps smoothly and absorbs torsional vibrations.

VOLT: Unit used to measure the force or pressure of electricity. It is defined as the pressure

VOLTAGE: The electrical pressure that causes current to flow. Voltage is measured in volts (V).

VOLTAGE, APPLIED: The actual voltage read at a given point in a circuit. It equals the available voltage of the power supply minus the losses in the circuit up to that point.

VOLTAGE DROP: The voltage lost or used in a circuit by normal loads such as a motor or lamp or by abnormal loads such as a poor (high-resistance) lead or terminal connection.

VOLTAGE REGULATOR: A device that controls the current output of the alternator or generator.

VOLTMETER: An instrument used for measuring electrical force in units called volts. Voltmeters are always connected parallel with the circuit being tested.

VORTEX FLOW: The crosswise or circulatory flow of oil between the blades of the members caused by the centrifugal pumping action of the impeller.

WANKEL ENGINE: An engine which uses no pistons. In place of pistons, triangular-shaped rotors revolve in specially shaped housings.

WATER PUMP: A belt driven component of the cooling system that mounts on the engine, circulating the coolant under pressure.

WATT: The unit for measuring electrical power. One watt is the product of one ampere and one volt (watts equals amps times volts). Wattage is the horsepower of electricity (746 watts equal one horsepower).

WHEEL ALIGNMENT: Inclusive term to describe the front end geometry (caster, camber, toe-in/out).

WHEEL CYLINDER: Found in the automotive drum brake assembly, it is a device, actuated by hydraulic pressure, which, through internal pistons, pushes the brake shoes outward against the drums.

WHEEL WEIGHT: Small weights attached to the wheel to balance the wheel and tire assembly. Out-of-balance tires quickly wear out and also give erratic handling when installed on the front.

WHEELBASE: Distance between the center of front wheels and the center of rear wheels.

WIDE OPEN THROTTLE (WOT): Full travel of accelerator pedal.

WORK: The force exerted to move a mass or object. Work involves motion; if a force is exerted and no motion takes place, no work is done. Work per unit of time is called power. Work = force x distance = ft. lbs. 33,000 ft. lbs. in one minute = 1 horsepower

ZERO-THROTTLE COAST DOWN: A full release of accelerator pedal while vehicle is in motion and in drive range.

Commonly Used Abbreviations

2

2WD	Two Wheel Drive

4

4WD	Four Wheel Drive

A

A/C	Air Conditioning
ABDC	After Bottom Dead Center
ABS	Anti-lock Brakes
AC	Alternating Current
ACL	Air cleaner
ACT	Air Charge Temperature
AIR	Secondary Air Injection
ALCL	Assembly Line Communications Link
ALDL	Assembly Line Diagnostic Link
AT	Automatic Transaxle/Transmission
ATDC	After Top Dead Center
ATF	Automatic Transmission Fluid
ATS	Air Temperature Sensor
AWD	All Wheel Drive

B

BAP	Barometric Absolute Pressure
BARO	Barometric Pressure
BBDC	Before Bottom Dead Center
BCM	Body Control Module
BDC	Bottom Dead Center
BPT	Backpressure Transducer
BTDC	Before Top Dead Center
BVSV	Bimetallic Vacuum Switching Valve

C

CAC	Charge Air Cooler
CARB	California Air Resources Board
CAT	Catalytic Converter
CCC	Computer Command Control
CCCC	Computer Controlled Catalytic Converter
CCCI	Computer Controlled Coil Ignition
CCD	Computer Controlled Dwell
CDI	Capacitor Discharge Ignition
CEC	Computerized Engine Control
CFI	Continuous Fuel Injection
CIS	Continuous Injection System
CIS-E	Continuous Injection System - Electronic
CKP	Crankshaft Position
CL	Closed Loop
CMP	Camshaft Position
CPP	Clutch Pedal Position
CTOX	Continuous Trap Oxidizer System
CTP	Closed Throttle Position
CVC	Constant Vacuum Control
CYL	Cylinder

D

DBC	Dual Bed Catalyst
DC	Direct Current
DFI	Direct Fuel Injection
DIS	Distributorless Ignition System
DLC	Data Link Connector
DMM	Digital Multimeter
DOHC	Double Overhead Camshaft
DRB	Diagnostic Readout Box
DTC	Diagnostic Trouble Code
DTM	Diagnostic Test Mode
DVOM	Digital Volt/Ohmmeter

E

EBCM	Electronic Brake Control Module
ECM	Engine Control Module
ECT	Engine Coolant Temperature
ECU	Engine Control Unit or Electronic Control Unit
EDIS	Electronic Distributorless Ignition System
EEC	Electronic Engine Control
EEPROM	Electrically Erasable Programmable Read Only Memory
EFE	Early Fuel Evaporation
EGR	Exhaust Gas Recirculation
EGRT	Exhaust Gas Recirculation Temperature
EGRVC	EGR Valve Control
EPROM	Erasable Programmable Read Only Memory
EVAP	Evaporative Emissions
EVP	EGR Valve Position

F

FBC	Feedback Carburetor
FEEPROM	Flash Electrically Erasable Programmable Read Only Memory
FF	Flexible Fuel
FI	Fuel Injection
FT	Fuel Trim
FWD	Front Wheel Drive

G

GND	Ground

H

HAC	High Altitude Compensation
HEGO	Heated Exhaust Gas Oxygen sensor
HEI	High Energy Ignition
HO2 Sensor	Heated Oxygen Sensor

I

IAC	Idle Air Control
IAT	Intake Air Temperature
ICM	Ignition Control Module
IFI	Indirect Fuel Injection
IFS	Inertia Fuel Shutoff
ISC	Idle Speed Control
IVSV	Idle Vacuum Switching Valve

Commonly Used Abbreviations

K

KOEO	Key On, Engine Off
KOER	Key ON, Engine Running
KS	Knock Sensor

M

MAF	Mass Air Flow
MAP	Manifold Absolute Pressure
MAT	Manifold Air Temperature
MC	Mixture Control
MDP	Manifold Differential Pressure
MFI	Multiport Fuel Injection
MIL	Malfunction Indicator Lamp or Maintenance
MST	Manifold Surface Temperature
MVZ	Manifold Vacuum Zone

N

NVRAM	Nonvolatile Random Access Memory

O

O2 Sensor	Oxygen Sensor
OBD	On-Board Diagnostic
OC	Oxidation Catalyst
OHC	Overhead Camshaft
OL	Open Loop

P

P/S	Power Steering
PAIR	Pulsed Secondary Air Injection
PCM	Powertrain Control Module
PCS	Purge Control Solenoid
PCV	Positive Crankcase Ventilation
PIP	Profile Ignition Pick-up
PNP	Park/Neutral Position
PROM	Programmable Read Only Memory
PSP	Power Steering Pressure
PTO	Power Take-Off
PTOX	Periodic Trap Oxidizer System

R

RABS	Rear Anti-lock Brake System
RAM	Random Access Memory
ROM	Read Only Memory
RPM	Revolutions Per Minute
RWAL	Rear Wheel Anti-lock Brakes
RWD	Rear Wheel Drive

S

SBC	Single Bed Converter
SBEC	Single Board Engine Controller
SC	Supercharger
SCB	Supercharger Bypass
SFI	Sequential Multiport Fuel Injection
SIR	Supplemental Inflatible Restraint
SOHC	Single Overhead Camshaft
SPL	Smoke Puff Limiter
SPOUT	Spark Output
SRI	Service Reminder Indicator
SRS	Supplemental Restraint System
SRT	System Readiness Test
SSI	Solid State Ignition
ST	Scan Tool
STO	Self-Test Output

T

TAC	Thermostatic Air Cleaner
TBI	Throttle Body Fuel Injection
TC	Turbocharger
TCC	Torque Converter Clutch
TCM	Transmission Control Module
TDC	Top Dead Center
TFI	Thick Film Ignition
TP	Throttle Position
TR Sensor	Transaxle/Transmission Range Sensor
TVV	Thermal Vacuum Valve
TWC	Three-way Catalytic Converter

V

VAF	Volume Air Flow, or Vane Air Flow
VAPS	Variable Assist Power Steering
VRV	Vacuum Regulator Valve
VSS	Vehicle Speed Sensor
VSV	Vacuum Switching Valve

W

WOT	Wide Open Throttle
WU-TWC	Warm Up Three-way Catalytic Converter

ENGLISH TO METRIC CONVERSION: TORQUE

To convert foot-pounds (ft. lbs.) to Newton-meters (Nm), multiply the number of ft. lbs. by 1.36
To convert Newton-meters (Nm) to foot-pounds (ft. lbs.), multiply the number of Nm by 0.7376

ft. lbs.	Nm	ft. lbs.	Nm	ft. lbs.	Nm	ft. lbs.	Nm
0.1	0.1	34	46.2	76	103.4	118	160.5
0.2	0.3	35	47.6	77	104.7	119	161.8
0.3	0.4	36	49.0	78	106.1	120	163.2
0.4	0.5	37	50.3	79	107.4	121	164.6
0.5	0.7	38	51.7	80	108.8	122	165.9
0.6	0.8	39	53.0	81	110.2	123	167.3
0.7	1.0	40	54.4	82	111.5	124	168.6
0.8	1.1	41	55.8	83	112.9	125	170.0
0.9	1.2	42	57.1	84	114.2	126	171.4
1	1.4	43	58.5	85	115.6	127	172.7
2	2.7	44	59.8	86	117.0	128	174.1
3	4.1	45	61.2	87	118.3	129	175.4
4	5.4	46	62.6	88	119.7	130	176.8
5	6.8	47	63.9	89	121.0	131	178.2
6	8.2	48	65.3	90	122.4	132	179.5
7	9.5	49	66.6	91	123.8	133	180.9
8	10.9	50	68.0	92	125.1	134	182.2
9	12.2	51	69.4	93	126.5	135	183.6
10	13.6	52	70.7	94	127.8	136	185.0
11	15.0	53	72.1	95	129.2	137	186.3
12	16.3	54	73.4	96	130.6	138	187.7
13	17.7	55	74.8	97	131.9	139	189.0
14	19.0	56	76.2	98	133.3	140	190.4
15	20.4	57	77.5	99	134.6	141	191.8
16	21.8	58	78.9	100	136.0	142	193.1
17	23.1	59	80.2	101	137.4	143	194.5
18	24.5	60	81.6	102	138.7	144	195.8
19	25.8	61	83.0	103	140.1	145	197.2
20	27.2	62	84.3	104	141.4	146	198.6
21	28.6	63	85.7	105	142.8	147	199.9
22	29.9	64	87.0	106	144.2	148	201.3
23	31.3	65	88.4	107	145.5	149	202.6
24	32.6	66	89.8	108	146.9	150	204.0
25	34.0	67	91.1	109	148.2	151	205.4
26	35.4	68	92.5	110	149.6	152	206.7
27	36.7	69	93.8	111	151.0	153	208.1
28	38.1	70	95.2	112	152.3	154	209.4
29	39.4	71	96.6	113	153.7	155	210.8
30	40.8	72	97.9	114	155.0	156	212.2
31	42.2	73	99.3	115	156.4	157	213.5
32	43.5	74	100.6	116	157.8	158	214.9
33	44.9	75	102.0	117	159.1	159	216.2

METRIC TO ENGLISH CONVERSION: TORQUE

To convert foot-pounds (ft. lbs.) to Newton-meters (Nm), multiply the number of ft. lbs. by 1.36
To convert Newton-meters (Nm) to foot-pounds (ft. lbs.), multiply the number of Nm by 0.7376

Nm	ft. lbs.	Nm	ft. lbs.	Nm	ft. lbs.	Nm	ft. lbs.	Nm	ft. lbs.
0.1	0.1	34	25.0	76	55.9	118	86.8	160	117.6
0.2	0.1	35	25.7	77	56.6	119	87.5	161	118.4
0.3	0.2	36	26.5	78	57.4	120	88.2	162	119.1
0.4	0.3	37	27.2	79	58.1	121	89.0	163	119.9
0.5	0.4	38	27.9	80	58.8	122	89.7	164	120.6
0.6	0.4	39	28.7	81	59.6	123	90.4	165	121.3
0.7	0.5	40	29.4	82	60.3	124	91.2	166	122.1
0.8	0.6	41	30.1	83	61.0	125	91.9	167	122.8
0.9	0.7	42	30.9	84	61.8	126	92.6	168	123.5
1	0.7	43	31.6	85	62.5	127	93.4	169	124.3
2	1.5	44	32.4	86	63.2	128	94.1	170	125.0
3	2.2	45	33.1	87	64.0	129	94.9	171	125.7
4	2.9	46	33.8	88	64.7	130	95.6	172	126.5
5	3.7	47	34.6	89	65.4	131	96.3	173	127.2
6	4.4	48	35.3	90	66.2	132	97.1	174	127.9
7	5.1	49	36.0	91	66.9	133	97.8	175	128.7
8	5.9	50	36.8	92	67.6	134	98.5	176	129.4
9	6.6	51	37.5	93	68.4	135	99.3	177	130.1
10	7.4	52	38.2	94	69.1	136	100.0	178	130.9
11	8.1	53	39.0	95	69.9	137	100.7	179	131.6
12	8.8	54	39.7	96	70.6	138	101.5	180	132.4
13	9.6	55	40.4	97	71.3	139	102.2	181	133.1
14	10.3	56	41.2	98	72.1	140	102.9	182	133.8
15	11.0	57	41.9	99	72.8	141	103.7	183	134.6
16	11.8	58	42.6	100	73.5	142	104.4	184	135.3
17	12.5	59	43.4	101	74.3	143	105.1	185	136.0
18	13.2	60	44.1	102	75.0	144	105.9	186	136.8
19	14.0	61	44.9	103	75.7	145	106.6	187	137.5
20	14.7	62	45.6	104	76.5	146	107.4	188	138.2
21	15.4	63	46.3	105	77.2	147	108.1	189	139.0
22	16.2	64	47.1	106	77.9	148	108.8	190	139.7
23	16.9	65	47.8	107	78.7	149	109.6	191	140.4
24	17.6	66	48.5	108	79.4	150	110.3	192	141.2
25	18.4	67	49.3	109	80.1	151	111.0	193	141.9
26	19.1	68	50.0	110	80.9	152	111.8	194	142.6
27	19.9	69	50.7	111	81.6	153	112.5	195	143.4
28	20.6	70	51.5	112	82.4	154	113.2	196	144.1
29	21.3	71	52.2	113	83.1	155	114.0	197	144.9
30	22.1	72	52.9	114	83.8	156	114.7	198	145.6
31	22.8	73	53.7	115	84.6	157	115.4	199	146.3
32	23.5	74	54.4	116	85.3	158	116.2	200	147.1
33	24.3	75	55.1	117	86.0	159	116.9	201	147.8

ENGLISH/METRIC CONVERSION: TEMPERATURE

To convert Fahrenheit (F°) to Celsius (C°), take F° temperature and subtract 32, multiply the result by 5 and divide the result by 9
To convert Celsius (C°) to Fahrenheit (F°), take C° temperature and multiply it by 9, divide the result by 5 and add 32

F°	C°	F°	C°	C°	F°	C°	F°
-40	-40.0	150	65.6	-38	-36.4	46	114.8
-35	-37.2	155	68.3	-36	-32.8	48	118.4
-30	-34.4	160	71.1	-34	-29.2	50	122
-25	-31.7	165	73.9	-32	-25.6	52	125.6
-20	-28.9	170	76.7	-30	-22	54	129.2
-15	-26.1	175	79.4	-28	-18.4	56	132.8
-10	-23.3	180	82.2	-26	-14.8	58	136.4
-5	-20.6	185	85.0	-24	-11.2	60	140
0	-17.8	190	87.8	-22	-7.6	62	143.6
1	-17.2	195	90.6	-20	-4	64	147.2
2	-16.7	200	93.3	-18	-0.4	66	150.8
3	-16.1	205	96.1	-16	3.2	68	154.4
4	-15.6	210	98.9	-14	6.8	70	158
5	-15.0	212	100.0	-12	10.4	72	161.6
10	-12.2	215	101.7	-10	14	74	165.2
15	-9.4	220	104.4	-8	17.6	76	168.8
20	-6.7	225	107.2	-6	21.2	78	172.4
25	-3.9	230	110.0	-4	24.8	80	176
30	-1.1	235	112.8	-2	28.4	82	179.6
35	1.7	240	115.6	0	32	84	183.2
40	4.4	245	118.3	2	35.6	86	186.8
45	7.2	250	121.1	4	39.2	88	190.4
50	10.0	255	123.9	6	42.8	90	194
55	12.8	260	126.7	8	46.4	92	197.6
60	15.6	265	129.4	10	50	94	201.2
65	18.3	270	132.2	12	53.6	96	204.8
70	21.1	275	135.0	14	57.2	98	208.4
75	23.9	280	137.8	16	60.8	100	212
80	26.7	285	140.6	18	64.4	102	215.6
85	29.4	290	143.3	20	68	104	219.2
90	32.2	295	146.1	22	71.6	106	222.8
95	35.0	300	148.9	24	75.2	108	226.4
100	37.8	305	151.7	26	78.8	110	230
105	40.6	310	154.4	28	82.4	112	233.6
110	43.3	315	157.2	30	86	114	237.2
115	46.1	320	160.0	32	89.6	116	240.8
120	48.9	325	162.8	34	93.2	118	244.4
125	51.7	330	165.6	36	96.8	120	248
130	54.4	335	168.3	38	100.4	122	251.6
135	57.2	340	171.1	40	104	124	255.2
140	60.0	345	173.9	42	107.6	126	258.8
145	62.8	350	176.7	44	111.2	128	262.4

LENGTH CONVERSION

To convert inches (in.) to millimeters (mm), multiply the number of inches by 25.4

To convert millimeters (mm) to inches (in.), multiply the number of millimeters by 0.04

Inches	Millimeters	Inches	Millimeters	Inches	Millimeters	Inches	Millimeters
0.0001	0.00254	0.005	0.1270	0.09	2.286	4	101.6
0.0002	0.00508	0.006	0.1524	0.1	2.54	5	127.0
0.0003	0.00762	0.007	0.1778	0.2	5.08	6	152.4
0.0004	0.01016	0.008	0.2032	0.3	7.62	7	177.8
0.0005	0.01270	0.009	0.2286	0.4	10.16	8	203.2
0.0006	0.01524	0.01	0.254	0.5	12.70	9	228.6
0.0007	0.01778	0.02	0.508	0.6	15.24	10	254.0
0.0008	0.02032	0.03	0.762	0.7	17.78	11	279.4
0.0009	0.02286	0.04	1.016	0.8	20.32	12	304.8
0.001	0.0254	0.05	1.270	0.9	22.86	13	330.2
0.002	0.0508	0.06	1.524	1	25.4	14	355.6
0.003	0.0762	0.07	1.778	2	50.8	15	381.0
0.004	0.1016	0.08	2.032	3	76.2	16	406.4

ENGLISH/METRIC CONVERSION: LENGTH

To convert inches (in.) to millimeters (mm), multiply the number of inches by 25.4

To convert millimeters (mm) to inches (in.), multiply the number of millimeters by 0.04

Inches		Millimeters	Inches		Millimeters	Inches		Millimeters
Fraction	Decimal	Decimal	Fraction	Decimal	Decimal	Fraction	Decimal	Decimal
1/64	0.016	0.397	11/32	0.344	8.731	11/16	0.688	17.463
1/32	0.031	0.794	23/64	0.359	9.128	45/64	0.703	17.859
3/64	0.047	1.191	3/8	0.375	9.525	23/32	0.719	18.256
1/16	0.063	1.588	25/64	0.391	9.922	47/64	0.734	18.653
5/64	0.078	1.984	13/32	0.406	10.319	3/4	0.750	19.050
3/32	0.094	2.381	27/64	0.422	10.716	49/64	0.766	19.447
7/64	0.109	2.778	7/16	0.438	11.113	25/32	0.781	19.844
1/8	0.125	3.175	29/64	0.453	11.509	51/64	0.797	20.241
9/64	0.141	3.572	15/32	0.469	11.906	13/16	0.813	20.638
5/32	0.156	3.969	31/64	0.484	12.303	53/64	0.828	21.034
11/64	0.172	4.366	1/2	0.500	12.700	27/32	0.844	21.431
3/16	0.188	4.763	33/64	0.516	13.097	55/64	0.859	21.828
13/64	0.203	5.159	17/32	0.531	13.494	7/8	0.875	22.225
7/32	0.219	5.556	35/64	0.547	13.891	57/64	0.891	22.622
15/64	0.234	5.953	9/16	0.563	14.288	29/32	0.906	23.019
1/4	0.250	6.350	37/64	0.578	14.684	59/64	0.922	23.416
17/64	0.266	6.747	19/32	0.594	15.081	15/16	0.938	23.813
9/32	0.281	7.144	39/64	0.609	15.478	61/64	0.953	24.209
19/64	0.297	7.541	5/8	0.625	15.875	31/32	0.969	24.606
5/16	0.313	7.938	41/64	0.641	16.272	63/64	0.984	25.003
21/64	0.328	8.334	21/32	0.656	16.669	1/1	1.000	25.400
			43/64	0.672	17.066			

THOMSON
DELMAR LEARNING

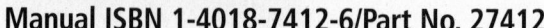

Manual ISBN 1-4018-7412-6/Part No. 27412

With the *Chilton® 2005 Labor Guide*, professional technicians gain access to labor times for vehicle brands and models that conform to current Automotive Aftermarket Industry Association standards. Thousands of labor times for 1981 through 2005 domestic and imported vehicles reflect technicians' use of aftermarket tools and training. Updates based on technical hotline input, Original Equipment Manufacturer (OEM) warranty times, and technical editor evaluation include more diagnostic labor times than ever before. Labor operations have been rewritten to conform to the most recent industry standards. Prior model coverage has been re-evaluated by experts to ensure accuracy. Chilton labor times are accepted by insurance and extended warranty companies.

Labor Guide Manual Benefits:

• 2,500 pages of Chilton labor times
• each OEM is arranged alphabetically by section for easy reference
• improved indexing means easier access to today's repair industry standards

Hardcover manual is 8 7/8" x 11", ©2005

Labor Guide CD-ROM Benefits:

• easy-to-use software to create and print professional-quality estimates and invoices
• three user-defined levels of labor rates correspond to different types of job scenarios, for "real-world" application
• functions as a database of aftermarket labor times for monitoring warranty and insurance claims
• software keeps track of customers and prior estimates for time-saving recall
• customizable application allows service writers to add labor operations and times, and parts companies to add labor times to existing parts ordering systems

CD-ROM ISBN 1-4018-7818-0/Part No. 27818

Previous Year Editions

Chilton 2004 Labor Guide Manual, ISBN 1-4018-4356-5/Part No. 24356

Chilton 2004 Labor Guide CD-ROM, ISBN 1-4018-4357-3/Part No. 24357

For the most up-to-date service and repair information anywhere, look no further than the newly updated *Chilton® 2005 Mechanical Service Manuals – Annual Editions*! Still the lowest-priced professional repair manuals on the market, this series of manufacturer-based books now features an easier-to-handle, two-volume Asian Manual set. Increased model coverage over the 2004 editions is supported by more illustrations in each section, making fast, accurate repairs and reassembly easier than ever before. With modernized content, it's no wonder that more professionals trust Chilton Professional Manuals for their mechanical service and repair needs.

Mechanical Service Manual Benefits:

- all books are grouped by manufacturer to make accessing information simple
- step-by-step procedures from drive train to chassis and related components help yield fast accurate results
- comprehensive, technically-detailed content is organized by model and system, and is supported by exploded-view illustrations, diagrams, and specification charts for added clarity
- most mechanical systems are included, such as engines, suspensions, steering components, and more
- special tools are described and clearly illustrated so that performing repairs is as easy and quick as possible

Chilton 2005 Ford Mechanical Service Manual
   ISBN 1-4018-6719-7/Part No. 26719
Chilton 2005 General Motors Mechanical Service Manual
   ISBN 1-4018-7146-1/Part No. 27146
Chilton 2005 Chrysler Mechanical Service Manual
   ISBN 1-4018-6718-9/Part No. 26718
Chilton 2005 Asian Mechanical Service Manual (Complete Set of 2 manuals)
   ISBN 1-4018-7180-1/Part No.
Chilton 2005 Asian Mechanical Service Manual, Acura - Mazda
   ISBN 1-4018-6716-2/Part No. 26716
Chilton 2005 Asian Mechanical Service Manual, Mitsubishi - Toyota
   ISBN 1-4018-6717-0/Part No. 26717
Chilton 2005 European Mechanical Service Manual
   ISBN 1-4018-6720-0/Part No. 26720

Manuals are 8 1/2" x 11", ©2005

The *Chilton® Perennial Editions* contain repair and maintenance information for popular mechanical systems that may not be available elsewhere. They offer a wide range of repair information on cars, trucks, vans, and SUVs dating back to the early 1960s, and as current as 2002. Information for 1993 and later model years includes scheduled maintenance interval charts.

Benefits:

- covers the most common vehicle models found in the repair aftermarket today
- gain quick understanding of systems using exploded-view illustrations, diagrams, and charts
- simplify tough jobs with easy-to-follow removal and installation instructions for heater core and other components
- obtain complete coverage of repair procedures from drive train to chassis and associated components

Auto Repair Manual, 1998-2002, 1,426 pages
ISBN 0-8019-9362-8/Part No. 9362
Auto Repair Manual, 1993-1997, 2,064 pages
ISBN 0-8019-7919-6/Part No. 7919
Auto Repair Manual, 1988-1992, 1,284 pages
ISBN 0-8019-7906-4/Part No. 7906
Auto Repair Manual, 1980-1987, 1,344 pages
ISBN 0-8019-7670-7/Part No. 7670

Import Car Repair Manual, 1998-2002, 1,792 pps
ISBN 0-8019-9363-6/Part No. 9363
Import Car Repair Manual, 1993-1997, 2,080 pps
ISBN 0-8019-7920-X/Part No. 7920
Import Car Repair Manual, 1988-1992, 1,632 pages
ISBN 0-8019-7907-2/Part No. 7907
Import Car Repair Manual, 1980-1987, 1,488 pages
ISBN 0-8019-7672-3/Part No. 7672

Truck & Van Repair Manual, 1998-2002, 1,408 pages
ISBN 0-8019-9364-4/Part No. 9364
Truck & Van Repair Manual, 1993-1997, 2,096 pages
ISBN 0-8019-7921-8/Part No. 7921
Truck & Van Repair Manual, 1991-1995, 1,664 pages
ISBN 0-8019-7911-0/Part No. 7911
Truck & Van Repair Manual, 1986-1990, 1,536 pages
ISBN 0-8019-7902-1/Part No. 7902
Truck & Van Repair Manual, 1979-1986, 1,440 pages
ISBN 0-8019-7655-3/Part No. 7655

SUV Repair Manual, 1998-2002, 1,292 pages
ISBN 0-8019-9365-2/Part No. 9365

Hardcover manuals are 8 1/2" x 11".

Chilton Collector's Editions - *Reference Manuals for Vintage Vehicles*
Auto Repair Manual, 1964-1971, ISBN 0-8019-5974-8/Part No. 5974,
Truck & Van Repair Manual, 1961-1971, ISBN 0-8019-6198-X/Part No. 6198
Truck & Van Repair Manual, 1971-1978, ISBN 0-8019-7012-1/Part No. 7012

ASE Test Preparation Series

Thomson Delmar Learning
ISBN 1-4018-5182-7
Part No. 25182

(Complete Set: A1-A8, L1, P2 X1, C1)
Thomson Delmar Learning has developed comprehensive ASE Test Preparation Manuals to help automotive technicians increase their success on these certification programs. The material covers the topics one might find during the test process. The booklets include many review questions and answers, as well as detailed descriptions of the repairs involved. Designed to look like the actual test, participants will feel more comfortable with practice, which will translate into greater success in taking the actual tests. The design of the Delmar Learning product also includes helpful test taking hints and student preparation ideas designed to enhance success.

BENEFITS
- The history of the ASE
- Test-taking strategies
- Tasks lists and overview
- Sample test questions
- ASE-style exams
- Explanations to the answers (right and wrong)
- Glossary of terms

(A1) Automotive Engine Repair, 2E
1-4018-2040-9
Part No. 22040

General Engine Diagnosis, Cylinder Head and Valve Train Diagnosis and Repair, Engine Block Diagnosis and Repair, Lubrication and Cooling Systems Diagnosis and Repair, and Fuel, Electrical, Ignition and Exhaust Systems Inspection and Service.

(A2) Automotive Transmissions and Transaxles, 2E
1-4018-2041-7
Part No. 22041

General Transmission/ Transaxle Diagnosis (Mechanical/Hydraulic Systems and Electronic Systems), Transmission/Transaxle Maintenance and Adjustment, In-Vehicle Transmission/Transaxle Repair, Off-Vehicle Transmission/Transaxle Repair.

(A3) Automotive Manual Drive Trains and Axles, 2E
1-4018-2042-5
Part No. 22042

Clutch Diagnosis and Repair, Transmission Diagnosis and Repair, Transaxle Diagnosis and Repair, Drive Shaft/Half Shaft and Universal Joint/Constant Velocity (CV) Joint Diagnosis and Repair (Front and Rear Wheel Drive), Rear Axle Diagnosis and Repair, Four Wheel Drive/All Wheel Drive Component Diagnosis and Repair.

(A4) Automotive Suspension and Steering, 2E
1-4018-2043-3
Part No. 22043

Steering Systems Diagnosis and Repair (Steering Columns and Manual Steering Gears, Power Assisted Steering Units, Steering Linkage), Suspension Systems Diagnosis and Repair (Front Suspensions, Rear Suspensions, Miscellaneous Services), Wheel Alignment Diagnosis, Adjustment and Repair, and Wheel and Tire Diagnosis and Repair.

(A5) Automotive Brakes, 2E
1-4018-2044-1
Part No. 22044

Hydraulic System Diagnosis and Repair, Drum Brake Diagnosis and Repair, Disc Brake Diagnosis and Repair, Power Assist Units Diagnosis and Repair, Miscellaneous Systems Diagnosis and Repair, Antilock Brake Systems (ABS) Diagnosis and Repair.

(A6) Automotive Electrical-Electronic Systems, 2E
1-4018-2045-X
Part No. 22045

General Electrical/Electronic Systems Diagnosis, Battery Diagnosis and Service, Starting Systems Diagnosis and Repair, Charging Systems Diagnosis and Repair, Lighting Systems Diagnosis and Repair, Gauges, Warning Devices and Driver Information Systems Diagnosis and Repair, Horn and Wiper/Washer Diagnosis and Repair.

(A7) Automotive Heating and Air Conditioning, 2E
1-4018-2046-8
Part No. 22046

The manual for A7 includes the following topics: A/C System Diagnosis and Repair, Refrigeration System Component Diagnosis and Repair, Heating and Engine Cooling Systems Diagnosis and Repair, Operating Systems and Related Controls Diagnosis and Repair, Refrigerant Recovery, Recycling, Handling and Retrofit.

(A8) Automotive Engine Performance, 2E
1-4018-2047-6
Part No. 22047

The manual for A8 includes the following topics: General Engine Diagnosis, Ignition System Diagnosis and Repair, Fuel, Air Induction, and Exhaust Systems Diagnosis and Repair, Emissions Control Systems Diagnosis and Repair (Including OBDII), Computerized Engine controls Diagnosis and Repair (Including OBDII), Engine Electrical Systems diagnosis and Repair.

(L1) Automotive Advance Engine Performance, 2E
1-4018-2049-2
Part No. 22049

The manual for L1 includes the following topics: General Powertrain Diagnosis, Computerized Powertrain Controls Diagnosis (Including OBDII), Ignition System Diagnosis, Fuel Systems and Air Induction Systems Diagnosis, Emission Control Systems Diagnosis, I/M Failure Diagnosis.

(P2) Automobile Parts Specialist, 2E
1-4018-2048-4
Part No. 22048

The manual for P2 includes the following topics: General Operations, Customer Relations and Sales Skills, Vehicle Systems Knowledge, Vehicle Identification, Cataloging Skills, Inventory Management, Merchandising.

(X1) Exhaust Systems
1-4018-2050-6
Part No. 22050

Exhaust Systems includes the following topics: Exhaust Systems Inspection and Repair, Emissions Systems Diagnosis, Exhaust System Fabrication, Exhaust System Installation, Exhaust System Repair Regulations.

(C1) Service Consultant
See next page for details

ASE Test Preparation Series in Español!

Thomson Delmar Learning
ISBN 1-4018-1530-8

(Complete Set: A1-A8, L1, P2, X1)
Now available in Español – the first of its kind for Spanish-speaking technicians! This comprehensive package of ASE test preparation booklets are intended for any Spanish-speaking automotive technician who is preparing to take an ASE examination. The series includes questions that relate to each competency required for certification by ASE. In addition to a multitude of questions, the reason why each answer is right or wrong is explained, along with task lists and overview, test-taking strategies, and more.

(A1) Reparación de Motores, 2A Edición
1-4018-1014-4/Part No. 21014

(A2) Transmision Automática/ Eje de Transmision Automática, 2A Edición
1-4018-1015-2/Part No. 21015

(A3) Tren de y Mando Ejes Manuales, 2A Edición
1-4018-1016-0/Part No. 21016

(A4) Suspensión y Dirección, 2A Edición
1-4018-1017-9/Part No. 21017

(A5) Frenos, 2A Edición
1-4018-1018-7/Part No. 21018

(A6) Sistemas Eléctricos/ Electrónicos, 2A Edición
1-4018-1019-5/Part No. 21019

(A7) Calefacción y Aire Acondicionado, 2A Edición
1-4018-1020-9/Part No. 21020

(A8) Funcionamiento de Motores, 2A Edición
1-4018-1021-7/Part No. 21021

(L1) Especialista en el Funciommiato Avansado de Motores, 2A Edición
1-4018-1022-5/Part No. 21022

(P2) Especialista en Partes de Automovil, 2A Edición
1-4018-1023-3/Part No. 21023

(X1) Sistemas de Escape, 2A Edición
1-4018-1024-1/Part No. 21024

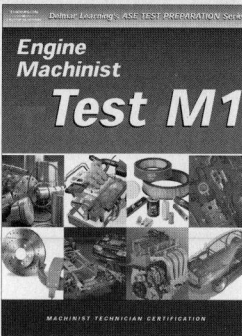

NEW!

ASE Test Preparation Manual - C1 Service Consultant
Thomson Delmar Learning
ISBN 1-4018-2029-8/
Part No.22029

Prepare to pass the new Service Consultant ASE Exam with help from this new test preparation booklet. The new C1 Exam is designed to measure systems knowledge and people skills of those who come in contact with the customer. It will contain questions on Communications, Product Knowledge, Sales Skills, and Shop Operations.

Service Consultant ASE Test Preparation Manual Benefits:

• the ASE task list is fully up-to-date, while current test prep questions reflect the most recent ASE task changes for the broadest knowledge possible
• hundreds of ASE-style exam questions adequately prepare readers to successfully pass the ASE exam
• readers are given multiple opportunities to check their understanding of critical concepts through sample problems, refresher materials, and competency-specific test questions
• overviews of each task provide a great reference point to help answer difficult ASE questions
• explanations for each answer help the user understand why the response is correct or incorrect

Softcover manual is
8 1/2" x 11", ©2004

ASE Test Preparation Manuals - Engine Machinist
Thomson Delmar Learning
ISBN 0-7668-6283-6/
Part No. 16283
(Complete Set: M1-M3)

With an abundance of up-to-date content, Thomson Delmar Learning's ASE Test Preparation Series contains the most current ASE test preparation material available. Each manual combines refresher materials with an abundance of sample test questions, as well as a wealth of information regarding test-taking strategies and the types of questions found in an ASE exam. In addition to the questions, thorough explanations are provided as to why each answer is correct or incorrect.

Benefits:

• The History section explains why the exams are important to the industry
• test-taking strategies help prepare technicians for the environment they will encounter during the actual exam experience testing first-hand

(M1) Cylinder Head Specialist
0-7668-6280-1/Part No. 16280

(M2) Cylinder Block Specialist
0-7668-6281-X/
Part No. 16281

(M3) Assembly Specialist
0-7668-6282-8/
Part No. 16282

Softcover manuals are
8 1/2" x 11", ©2002

ASE Test Preparation Manuals - Collison Repair
Thomson Delmar Learning
ISBN 1-4018-5120-7/Part No. 25120
(Complete Set: B2-B6)

This fully expanded third edition has been completely updated to provide the most current ASE test preparation material for collision repair and refinishing available anywhere. Each book in the series provides valuable preparation for automotive technicians seeking certification in one or more of the ASE collision repair areas. Readers are afforded scores of opportunities to ascertain their knowledge of critical concepts, through the extensive array of sample problems, ASE-style exams, and competency-specific test questions required for certification by ASE.

Benefits:

• all ASE task lists associated with collision repair and refinishing are fully up-to-date to help sufficiently prepare users for the ASE certification exam
• current, job-related ASE-style exam questions reflecting the most recent ASE task changes test the skills that technicians need to know on the job
• each book contains a general knowledge pretest, a sample test, and additional practice learning that add up to the most real-test practice time available

(B2) Painting and Refinishing, 2E
1-4018-3664-X/Part No. 23664
(B3) Non-Structural Analysis and Damage Repair, 2E
1-4018-3665-8/Part No. 23665
(B4) Structural Analysis and Damage Repair, 2E,
1-4018-3666-6/Part No. 23666
(B5) Mechanical and Electrical Components, 2E,
1-4018-3667-4/Part No. 23667
(B6) Damage Analysis and Estimation, 2E,
1-4018-3668-2/Part No. 23668

Softcover manuals are 8 1/2" x 11", ©2005

Automotive ASE Preparation Video Series

Thomson Delmar Learning

ISBN 0-7668-3168-X *(Complete Set of 12 Tapes)*
ISBN 0-7668-8042-7 *(Complete Set of 3 CD-ROMs)*

Thomson Delmar Learning's Automotive ASE Test Prep Videos present test takers with a review of the A1-A8, L1, and P2 tests prior to taking the exam. Each tape summarizes key topics and key task areas through live action and animation. Actual technicians, authentic automotive shops, and late-model vehicles are featured for an up-to-date look and feel. Safety is emphasized throughout each tape. An overview tape introduces test takers to the ASE testing style.

BENEFITS OF THE VIDEO SERIES

- lively, easy to follow videos emphasize safety throughout
- covers major task areas and topics for each of the ASE exams
- accompanying Instructor's Guide helps users comprehend and retain information presented

Complete Set of 12 Tapes (with Instructor's Guide), ©2001

Tape 1: Overview of ASE, 0-7668-2484-5
Tape 2: A1 Engine Repair, 0-7668-2485-3
Tape 3: A2 Automatic Transmission, 0-7668-2498-5
Tape 4: A3 Manual Transmission, 0-7668-2499-3
Tape 5: A4 Steering and Suspension, 0-7668-2500-0
Tape 6: A5 Automotive Brakes, 0-7668-2501-9
Tape 7: A6 Electricity/Electronics, 0-7668-2493-4
Tape 8: A7 Air Conditioning, 0-7668-2486-1
Tape 9: A8 Engine Performance, 0-7668-2494-2
Tape 10: P2 Parts Specialist, 0-7668-2487-X
Tape 11: L1 Advanced Engine Performance (Part 1), 0-7668-2491-8
Tape 12: L1 Advanced Engine Performance (Part 2), 0-7668-2492-6

BUNDLES

Bundle 1: Specialty Topics (Set of 4 Tapes) includes Overview of ASE, A1 Engine Repair, A7 Air Conditioning, and P2 Parts Specialist, 0-7668-2483-7
Bundle 2: Engine Performance/Electronics (Set of 4 Tapes) includes L1 Part 1, L1 Part 2, A6 Electricity/ Electronics, and A8 Engine Performance, 0-7668-2490-X
Bundle 3: Undercar (Set of 4 Tapes) includes A2 Automatic Transmissions, A3 Manual Transmissions, A4 Steering and Suspension, and A5 Automotive Brakes, 0-7668-2497-7

CD-ROM COURSEWARE

Based on the ASE Test Prep Series, the CD-ROMs offer the following in addition to the video content:

- Gradebook
- Pre-test/Post-test
- Ability to modify
- Video Glossary
- Variety of question types
- Remediation
- Video File Server compatible

CD-ROM 1: Specialty Topics CD-ROM includes Overview of ASE, A1 Engine Repair, A7 Air Conditioning, and P2 Parts Specialist, 0-7668-2489-6
CD-ROM 2: Engine Performance/Electronics CD-ROM includes L1 Part 1, L1 Part 2, A6 Electricity/ Electronics, and A8 Engine Performance, 0-7668-2496-9
CD-ROM 3: Undercar CD-ROM includes A2 Automatic Transmissions, A3 Manual Transmissions, A4 Steering and Suspension, and A5 Automotive Brakes, 0-7668-2503-5

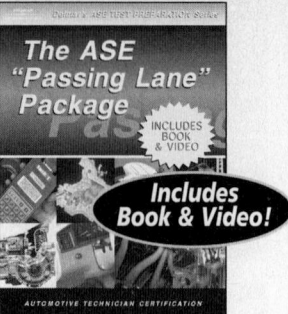

The ASE "Passing Lane" Package

Thomson Delmar Learning

ISBN 0-7668-4338-6

(Complete Set: A1-A8, L1, P2)

The most comprehensive test preparation for Automotive Tests A1-A8, L1, and P2. Combining the most thorough ASE Test Preparation books with the latest in ASE videos, this package provides a program of self-study for the automotive ASE Tests.

EACH BOOK IN THE SERIES BENEFITS:

- test-taking strategies
- tasks lists and overview
- sample test questions
- ASE-style exams
- explanations to the answers
- glossary of terms

EACH VIDEO IN THE SERIES BENEFITS:

- lively, easy to follow videos emphasize safety throughout
- covers major task areas and topics for each of the ASE exams
- accompanying Activity Sheets help comprehend and retain information

(A1) Automotive Engine Repair Book/Video, 0-7668-4181-2
(A2) Automotive Transmissions and Transaxles Book/Video, 0-7668-4182-0
(A3) Automotive Manual Drive Trains and Axles Book/Video, 0-7668-4183-9
(A4) Automotive Suspension and Steering Book/Video, 0-7668-4184-7
(A5) Automotive Brakes Book/Video, 0-7668-4185-5
(A6) Automotive Electrical-Electronics Systems Book/Video, 0-7668-4186-3
(A7) Automotive Heating and Air Conditioning Book/Video, 0-7668-4187-1
(A8) Automotive Engine Performance Book/Video, 0-7668-4188-X
(L1) Automotive Advanced Engine Performance Book/Video, 0-7668-4189-8
(P2) Automobile Parts Specialist Book/Video, 0-7668-4190-1

Automotive Technician Certification Test Preparation Manual, 2E

Don Knowles

**ISBN 0-7668-1948-5/
Part No. 11948**

The second edition of Certified ASE Master Technician Don Knowles' popular ASE test preparation book adds coverage of the L1 Advanced Engine Performance test to its coverage of automotive tests A1 through A8. All nine tests covered in this book reflect year 2000 task lists, including the updated composite vehicle in the L1 test. This revised edition contains at least one practice question for every ASE task in the tests. Also included is the updated and expanded coverage of electronic automatic transmissions, electronically controlled automatic transmissions, electronically controlled 4 wheel drive and steering, ABS systems, wiring diagrams, and repairing electronic components.

BENEFITS

- a new section has been added on computer-controlled automatic transmissions and transaxles including those used in OBD II vehicles
- new information has been included on electronically-controlled 4WD systems and ABS systems
- the chapter on Electrical/Electronic Systems has been expanded to include information on reading wiring diagrams and inspecting, testing, and repairing electronic components
- a complete chapter has been added to prepare technicians for the Advanced Engine Performance (L1) test

CONTENTS

Engine Repair Automatic Transmission/Transaxle. Manual Drive Train and Axles. Suspension and Steering. Brakes. Electrical/Electronic Systems. Heating, Ventilation, and Air Conditioning Systems. Engine Performance. Advanced Engine Performance.

788 pp, 8½" x 11", softcover, ©2001

This pioneering eight-book series offers automotive repair shop owners and those wanting to be shop owners the necessary business and customer service skills to run a successful automotive service facility.

The series covers three main topical areas: personnel management, business management, and sales and marketing. Each book provides a framework to help technicians make consistent, high-quality, and productive service a part of every day shop operations. According to the author, "Great performance coupled with increased customer loyalty, trust, and operational excellence will almost always result in increased profits."

Automotive Service Management Series Benefits:

- real-world approach reflects author's experience as a fourth generation technician, a repair & service company owner, and an automotive industry trainer
- all-inclusive coverage spans from designing an automotive repair facility floor plan through financial management techniques, customer/staff relations, and more
- length of each book makes it easy to incorporate this series into workshops, seminars, and training/education courses
- information is available "as is" or for customization

Total Customer Relationship Management
   ISBN 1-4018-2657-1/Part No. 22657
From Intent to Implementation
   ISBN 1-4018-2658-X/Part No. 22658
Operational Excellence
   ISBN 1-4018-2659-8/Part No. 22659
Building a Team
   ISBN 1-4018-2660-1/Part No. 22660
The High Performance Shop
   ISBN 1-4018-2661-X/Part No. 22661
Safety Communications
   ISBN 1-4018-2662-8/Part No. 22662
Managing Dollars with Sense
   ISBN 1-4018-2663-6/Part No. 22663
Operations Management
   ISBN 1-4018-2665-2/Part No. 22665
Entire Set of 8 Books
   ISBN 1-4018-2499-4/Part No. 2499

Softcover manuals are 8 1/2" x 11", ©2003

ABOUT THE AUTHOR

Mitch Schneider is a fourth generation mechanic/technician and is a frequent speaker at major conventions and meetings of automotive industry trade organizations. Schneider is also an award-winning journalist and is a regular contributor and senior contributing editor for *Motor Age* magazine. He provides commentary on the evolving relationship between service dealers, jobbers, warehouse directors and manufacturers.

Schneider has also appeared on the TNN cable show "Truckin' USA" where he hosted the "Tech Tips" segment. In addition to operating the award-winning Schneider's Automotive for 22 years in Simi Valley, CA, he is also the president and founder of Schneider's Future-Tech, a service company specializing in conducting management seminars for automotive service dealers, jobbers, warehouse distribution companies, and manufacturers.

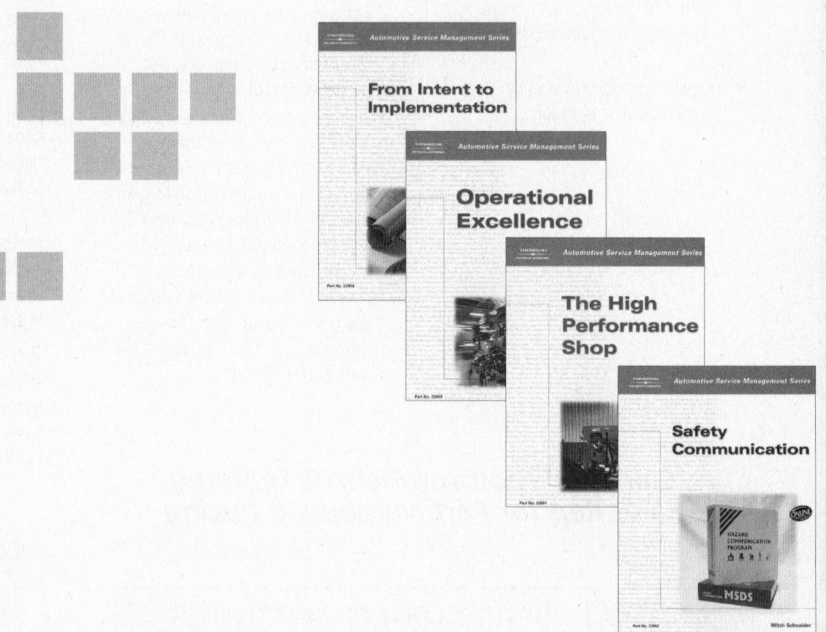

Delmar's Automotive Dictionary
David W. South &
Boyce Dwiggins
ISBN 0-8273-7405-4

This handy, ready-reference dictionary provides the automotive engineer, technician, mechanic, student, enthusiast or layperson with a single source for the most up-to-date definitions available of technical, professional and informal terminology used in today's automotive world. It is descriptive and covers the wide scope of terms pertinent to the automotive field. With multiple definitions and aids, and proper pronunciation of terms, this dictionary is a must for all!

BENEFITS

- over 3000 terms comprehensively covering more than 100 subject areas
- enhanced by a list of acronyms and abbreviations
- up-to-date definitions of today's automotive terminology
- aids for proper pronunciation
- each term has multiple definitions

281 pp, 6" x 9", softcover, ©1997

Practical Problems in Mathematics for Automotive Technicians, 5E
George Morre, Todd Sformo &
Larry Sformo
ISBN 0-8273-7944-7

By showing how to apply math solutions to everyday problems, this all-in-one math reference transforms the "remove it and replace it" mechanic into a complete automotive technician. The book builds from math basics to cover more complex topics--not to mention such workplace issues as invoices and scale reading of test meters. Each easy-to-read chapter features step-by-step instructions, diagrams, charts and examples to make the problem-solving process a snap.

256 pp, 7⅞" x 9¼", softcover, ©1998
Instructor's Manual   **0-8273-7945-5**

Math for the Automotive Trade, 3E
John C. Peterson &
William deKryger
ISBN 0-8273-6712-0

Math for Automotive Trades, 3E provides excellent examples and problems that reflect technological requirements of workers in automotive technology. The text has three parts: review of basic mathematics skills, math applications to specific automotive situations, and an examination of measurement aspects beginning with angle and linear measurements and ending with an extensive look at measurement tools used in the automotive trade.

345 pp, 8½" x 11", softcover, ©1995
Instructor's Manual   **0-8273-6713-9**